MAYNARD'S INDUSTRIAL ENGINEERING HANDBOOK

Other McGraw-Hill Books of Interest

MAYNARD'S INDUSTRIAL ENGINEERING HANDBOOK

William K. Hodson Editor in Chief

Fourth Edition

McGRAW-HILL, INC.

New York St. Louis San Francisco Auckland Bogotá
Caracas Lisbon London Madrid Mexico Milan
Montreal New Delhi Paris San Juan São Paulo
Singapore Sydney Tokyo Toronto

Library of Congress Cataloging-in-Publication Data

Maynard's industrial engineering handbook / William K. Hodson, editor
    in chief. — 4th ed.
          p.   cm.
      Rev. ed. of: Industrial engineering handbook / edited by H.B.
    Maynard, 1971
      ISBN 0-07-041086-0
      1. Industrial engineering—Handbooks, manuals, etc.   I. Maynard,
    Harold Bright (date).   II. Hodson, William K.   III. Industrial
    engineering handbook
    T56.M39   1992
    658.5—dc20   92-13512
                                                       CIP

Third edition published as Industrial Engineering Handbook, *edited by*
H. B. Maynard.

1 2 3 4 5 6 7 8 9 0   DOC/DOW   9 8 7 6 5 4 3 2

ISBN   0-07-041086-0

The sponsoring editor for this book was Gail F. Nalven, the editing
supervisor was Peggy Lamb, and the production supervisor was
Suzanne W. Babeuf. It was set in Times Roman. It was composed by
McGraw-Hill's Professional Book Group composition unit.

Printed and bound by R. R. Donnelley & Sons Company.

To all industrial engineers who, by their competent application of traditional industrial engineering procedures and their readiness to learn to use skillfully the newer techniques and procedures which are constantly expanding the effectiveness and fields of application of industrial engineering, are contributing so importantly to the ever-increasing usefulness of their profession.

CONTENTS

Section 4   Work Measurement Techniques

Section 5   Work Measurement Application and Control

Section 10   Planning and Control

Section 11   Quality Control

Section 12   Use of Computers

Section 15 Special Industry Applications

ABOUT THE EDITORS

William K. Hodson is an internationally recognized management consultant specializing in industrial engineering applications. In a career that spans more than 40 years, he has headed engineering consulting firms ranging from 100 to 5000 people. He was formerly president of the H. B. Maynard and Company, Inc., Planning Research Corporation (PRC), and the Institute of Industrial Engineers (IIE). He is the recipient of the Frank and Lillian Gilbreth Award of the IIE. He received his B.S. from Lehigh University and currently resides in Indian Wells, California.

The late **Harold B. Maynard** was the founder of the H. B. Maynard Company of Pittsburgh, Pennsylvania, and a major contributor to the advancement of the industrial engineering profession.

CONTRIBUTORS

John W. Adams *Associate Professor, Department of Industrial Engineering, Lehigh University, Bethlehem, Pennsylvania* (Sec. 14, Chap. 3)

Susan L. Albin *Department of Industrial Engineering, Rutgers University, Piscataway, New Jersey* (Sec. 11, Chap. 2)

Suraj M. Alexander *Professor, Department of Industrial Engineering, University of Louisville, Kentucky* (Sec. 11, Chap. 4)

Arvind Ballakur *Supervisor, Operations Analysis, AT&T Bell Laboratories, Holmdel, New Jersey* (Sec. 15, Chap. 2)

J. C. Banaag *Biomechanics Corporation of America, Melville, New York* (Sec. 8, Chap. 1)

Donald S. Bloswick *Assistant Professor, Department of Mechanical Engineering, University of Utah, Salt Lake City, Utah* (Sec. 8, Chap. 6)

Arthur P. Brief *Professor of Oganizational Behavior and Psychology, Tulane University, New Orleans, Louisiana* (Sec. 2, Chap. 2)

Chester L. Brisley *Director, Industrial Engineering Program, Marquette University, Milwaukee, Wisconsin* (Sec. 4, Chap. 3)

David W. Buker *Chairman, David W. Buker, Inc., and Associates Antioch, Illinois* (Sec. 10, Chap. 3)

Dr. Albert F. Celley *A. F. Celley & Associates, Toledo, Ohio* (Sec. 4, Chap. 1; Sec. 6, Chap. 2)

Tien-Chien Chang *Associate Professor, School of Industrial Engineering, Purdue University, West Lafayette, Indiana* (Sec. 12, Chap. 3)

Debra I. Danner *Attorney at Law, Matkov, Salzman, Madoff & Gunn, Chicago, Illinois* (Sec. 8, Chap. 4)

Herbert Davis *Herbert Davis and Company, Englewood Cliffs, New Jersey* (Sec. 13, Chaps. 6 & 7)

Ralph L. Disney *Professor, Industrial Engineering Department, Texas A&M University, College Station, Texas* (Sec. 14, Chap. 2)

Eric D. Dmytrow *Vice President, Juran Institute, Inc., Wilton, Connecticut* (Sec. 11, Chap. 1)

John F. Doran *Manufacturing Consultant, Raytheon Company, Lexington, Massachusetts* (Sec. 15, Chap. 3)

E. A. Elsayed *Department of Industrial Engineering, Rutgers University, Piscataway, New Jersey* (Sec. 11, Chap. 2)

Richard L. Engwall *Manager, Advanced Manufacturing Initiatives, Electronic Systems Group, Westinghouse Electric Corporation, Columbia, Maryland* (Sec. 5, Chap. 8)

Robert I. Felch *Professor Emeritus, College of Business and Economics, Radford University, Radford, Virginia* (Sec. 9, Chap. 5)

Michael D. Ferrell *Vice President Process Improvement, IBP Dakota City, Nebraska* (Sec. 1, Chap. 1; Sec. 6, Chap. 4)

William E. Fillmore *Senior Consultant, Richard Muther & Associates, Kansas City, Missouri* (Sec. 13, Chap. 1)

Robert Fitzpatrick *Consulting Psychologist, AMANSCO Incorporated, Mars, Pennsylvania* (Sec. 8, Chap. 3)

G. A. Fleischer *Professor, Department of Industrial and Systems Engineering, University of Southern California, Los Angeles, California* (Sec. 9, Chap. 1)

George Foo *Manufacturing and Engineering Director, American Telephone and Telegraph Co., Little Rock, Arkansas* (Sec. 10, Chap. 4)

Dr. Nikhil Gandhi *Vice President, XENERGY, Burlington, Massachusetts* (Sec. 13, Chap. 8)

Duane C. Geitgey *Principal, H. B. Maynard and Co., Inc., Pittsburgh, Pennsylvania* (Sec. 3, Chap. 2)

Michael Geurts *J. Darwin Gunnel Professor of Business Management, Marriott School of Management, Brigham Young University, Provo, Utah* (Sec. 14, Chap. 4)

R. S. Goonetilleke *Biomechanics Corporation of America, Melville, New York* (Sec. 8, Chap. 1)

Mikell P. Groover *Professor, Department of Industrial Engineering, Lehigh University, Bethlehem, Pennsylvania* (Sec. 7, Chap. 5)

C. M. Gross *Chief Executive Officer, Biomechanics Corporation of America, Melville, New York* (Sec. 8, Chaps. 1 & 2)

Yash P. Gupta *Frazier Family Professor, Department of Management, University of Louisville, Louisville, Kentucky* (Sec. 7, Chap. 11)

Richard A. Guzzo *Associate Professor of Psychology and Business Management, University of Maryland at College Park, College Park, Maryland* (Sec. 2, Chap. 2)

Hubert King Hardin *Manager, A. T. Kearney, Inc., Chicago, Illinois* (Sec. 13, Chap. 2)

R. Terry Hays *Vice President, Value Analysis Inc., Newport Beach, California* (Sec. 7, Chap. 2)

Harold C. Herriman *Consultant, H. B. Maynard and Co., Inc., Pittsburgh, Pennsylvania* (Sec. 5, Chap. 3)

Walter Hoberecht *Graduate Student, Department of Industrial and Management Systems Engineering, The Pennsylvania State University, University Park, Pennsylvania* (Sec. 12, Chap. 3)

William K. Hodson, P.E. *Consultant, Indian Wells, California* (Sec. 5, Chap. 5)

John R. Huffman, Ph.D., P.E. *President, John R. Huffman, P.E., Los Angeles, California* (Sec. 16, Chap. 1)

Thomas P. Huizenga *Vice President, Juran Institute, Inc., Wilton, Connecticut* (Sec. 11, Chap. 1)

Sanjay Joshi *Assistant Professor, Department of Industrial and Management Systems Engineering, The Pennsylvania State University, University Park, Pennsylvania* (Sec. 12, Chap. 3)

Swatantra K. Kachhal *Chairman, Department of Industrial and Systems Engineering, University of Michigan-Dearborn, Dearborn, Michigan* (Sec. 15, Chap. 1)

J. Patrick Kelly *K Mart Professor of Retailing, Wayne State University, Detroit, Michigan* (Sec. 14, Chap. 4)

W. J. Kennedy *Professor of Industrial Engineering, Clemson University, Clemson, South Carolina* (Sec. 1, Chap. 2)

Larry Kinney *Operations Director, American Telephone and Telegraph Co., North Andover, Massachuetts* (Sec. 10, Chap. 4)

Patrick W. Kocian *Attorney at Law, Matkov, Salzman, Madoff & Gunn, Chicago, Illinois* (Sec. 8, Chap. 5)

W. J. Kolarik *Industrial Engineering Department, Texas Tech University, Lubbock, Texas* (Sec. 11, Chap. 3)

Richard E. Kopelman *Professor of Management, The City University of New York, New York, New York* (Sec. 10, Chap. 5)

Paul W. Krueger *Plant Manager, Bush Hog Division of Allied Products Corporation, Selma, Alabama* (Sec. 2, Chap. 1)

Louis M. Kuh *Principal, Kuh & Associates, Stamford, Connecticut* (Sec. 6, Chap. 1)

Way Kuo *Chairman, Department of Industrial and Manufacturing Systems Engineering, Iowa State University, Ames, Iowa* (Sec. 11, Chap. 4)

Kenneth D. Lawrence *Department of Industrial Engineering and the Graduate School of Management, Rutgers University, Piscataway, New Jersey* (Sec. 14, Chaps. 4 & 7)

Peter S. Lucking *Contract Consultant, H. B. Maynard and Co., Inc., Pittsburgh, Pennsylvania* (Sec. 5, Chap. 1)

Eric M. Malstrom *Professor and Head, Department of Industrial Engineering, University of Arkansas, Fayetteville, Arkansas* (Sec. 10, Chap. 2)

James W. Mason, Jr. *Senior Vice President, H. B. Maynard and Co., Inc., Pittsburgh, Pennsylvania* (Sec. 4, Chap. 6)

George J. Matkov, Jr. *Attorney at Law, Matkov, Salzman, Madoff & Gunn, Chicago, Illinois* (Sec. 8, Chaps. 4 & 5)

William E. Mayo *Vice President, H. B. Maynard and Co., Inc., Pittsburgh, Pennsylvania* (Sec. 5, Chap. 5)

Edmund J. McCormick, Jr. *Chairman, McCormick & Company, Summit, New Jersey* (Sec. 9, Chaps. 2 & 3)

Mary T. McKinney *Director, Small Business Development Center, Duquesne University, Pittsburgh, Pennsylvania* (Sec.12, Chap. 1)

Charles W. McNichols *Dalton Professor of Business Administration, Radford University, Radford, Virginia* (Sec. 12, Chap. 2)

K. K. Menon *Biomechanics Corporation of America, Melville, New York* (Sec. 8, Chap. 1)

Joseph Metz *Vice President, Booz-Allen & Hamilton, Inc., Beloit, Wisconsin* (Sec. 7, Chap. 3)

Deborah Mitta *Assistant Professor, Department of Industrial Engineering, Texas A&M University, College Station, Texas* (Sec. 14, Chap. 1)

Marvin E. Mundel *Principal, M. E. Mundel & Associates, Silver Spring, Maryland* (Sec. 15, Chap. 5)

Richard Muther *President, Richard Muther & Associates, Inc., Kansas City , Missouri* (Sec. 13, Chaps. 3 & 5)

James L. Nevins *Arthur D. Little Inc., The Charles Stark Draper Laboratory, Inc., Cambridge, Massachusetts* (Sec. 7, Chap. 7)

J. E. Nicks *President, MiCAPP Inc., Big Rapids, Michigan* (Sec. 9, Chap. 4)

Nickolas G. Odrey *Professor, Industrial Engineering Department, Lehigh University, Bethlehem, Pennsylvania* (Sec. 7, Chap. 10)

O. Geoffrey Okogbaa *Professor, Department of Industrial Engineering and Management Systems, University of South Florida, Tampa, Florida* (Sec. 11, Chap. 4)

Joel Orr *President, Orr Associates, Inc., Virginia Beach, Virginia* (Sec. 7, Chap. 8)

Joseph Peake *Senior Consultant, Booz-Allen & Hamilton, Inc., Beloit, Wisconsin* (Sec. 7, Chap. 3)

Mark O. Presnell *Engineer Consultant, Electronic Data Systems Corporation, Warren, Michigan* (Sec. 12, Chap. 4)

Jayant Rajgopal *Department of Industrial Engineering, University of Pittsburgh, Pittsburgh, Pennsylvania* (Sec. 14, Chap. 6)

George F. Raymond *Consulting Supervisor, H. B. Maynard and Co., Inc., Pittsburgh, Pennsylvania* (Sec. 3, Chap. 1)

Lloyd B. Raymond *Principal, H. B. Maynard and Co., Inc., Pittsburgh, Pennsylvania* (Sec. 15, Chap. 4)

Joseph H. Redding *Senior Vice President, H. B. Maynard and Co., Inc., Pittsburgh, Pennsylvania* (Sec. 15, Chap. 4)

Gary R. Reeves *Deparment of Management Science, University of South Carolina, Columbia, South Carolina* (Sec. 14, Chap. 7)

Frank J. Riley *Senior Vice President, The Bodine Corporation, Bridgeport, Connecticut* (Sec. 7, Chap. 1)

Heikki Rinne *Professor, Scaggs Institute, Marriott School of Management, Brigham Young University, Provo, Utah* (Sec. 14, Chap. 4)

Alan J. Rowe *Professor, Department of Management and Oganization, University of Southern California, Los Angeles, California* (Sec. 12, Chap. 5)

Vinod K. Sahney *Corporate Vice President, Henry Ford Health Care Corp., Detroit, Michigan* (Sec. 15, Chap. 1)

Shigeyasu Sakamoto *President, Productivity Partner Inc., Nara, Japan* (Sec. 3, Chap. 3)

Blair H. Schlender *Manager-TQM & Production Engineeering, Martin Marietta Corporation, Orlando, Florida* (Sec. 15, Chap. 3)

Benjamin Schneider *Professor of Psychology and Business Management, University of Maryland at College Park, College Park, Maryland* (Sec. 2, Chap. 2)

Thomas Seidel *Grschaftsfuhrenender Director, ROI Management Consulting AG, Zurich, Switzerland* (Sec. 5, Chap. 7)

Clifford N. Sellie *Chairman/CEO, Standards International, Northbrook, Illinois* (Sec. 2, Chap. 1; Sec. 4, Chaps. 2 & 4)

Richard R. Shell *Professor of Industrial Engineering, Department of Mechanical, Industrial, and Nuclear Engineering, University of Cincinnati, Cincinnati, Ohio* (Sec. 5, Chaps. 2, 4, & 7)

E. Ralph Sims, P.E. *Chairman, The Sims Consulting Group, Inc., Lancaster, Pennsylvania* (Sec. 13, Chap. 4)

Cecil Smith, Jr., *President, Cecil Smith Inc., Baton Rouge, Louisanna* (Sec. 7, Chap. 9)

William K. Spence, Ph.D.   *Communications Consultant, Plano, Texas* (Sec. 2, Chap. 4)

Robert L. Staehle   *Chairman Emeritus, McCormick & Company, Summit, New Jersey* (Sec. 9, Chap. 2)

Katherine E. Stecke   *Graduate School of Business Administration, Ann Arbor, Michigan* (Sec. 7, Chap. 6)

Marlin U. Thomas   *Professor and Head, Department of Industrial Engineering, Lehigh University, Bethlehem, Pennsylvania* (Sec. 14, Chap. 8)

Curtis J. Tompkins   *University Professor and President, Michigan Technological University, Houghton, Michigan* (Sec. 1, Chap. 3)

Gregory L. Tonkay   *Assistant Professor, Department of Industrial Engineering, Lehigh Universaity, Bethlehem, Pennsylvania* (Sec. 7, Chap. 4)

Laurens van den Muyzenberg   *Managing Director, Muyzenberg Management Consultants, Windsor, United Kingdom* (Sec. 2, Chap. 3)

Hugh E. Warren   *Professor, School of Business and Economics, California State University Los Angeles, Los Angeles, California* (Sec. 14, Chap. 5)

Francis M. Webster, Jr.   *Management and Computer Consultant, Cullowhee, North Carolina; Editor-in-Chief, Project Management Institute, Drexel Hill, Pennsylvania* (Sec. 10, Chap. 7)

Roger M. Weiss   *Managing Partner, H. B. Maynard and Co., Inc, Hartford, Connecticut* (Sec. 6, Chap. 3)

Thomas A. Westerkamp   *Contract Consultant, H. B. Maynard and Co., Inc., Pittsburgh, Pennsylvania* (Sec. 5, Chap. 6; Sec. 10, Chap. 6)

George R. Wilson   *Professor, Department of Industrial Engineering, Lehigh University, Bethlehem, Pennsylvania* (Sec. 14, Chap. 8)

B. M. Worrall   *Professor, Industrial Engineering Department, Technical University of Nova Scotia, Halifax, Nova Scotia, Canada* (Sec. 4, Chap. 4)

Kjell B. Zandin   *Senior Vice President, H. B. Maynard and Co., Inc., Pittsburgh, Pennsylvania* (Sec. 4, Chap. 5)

PREFACE

The first edition of the *Industrial Engineering Handbook,* edited by H. B. Maynard and published by McGraw-Hill, appeared in 1956. The second edition followed in 1963 and the third in 1971. Mike Maynard died in 1975, delaying a fourth edition. In 1989 McGraw-Hill asked me to take on editing a fourth edition.

Mike Maynard had been a close friend since I joined his consulting firm in 1946. I wrote three chapters in the third edition and served on the editorial advisory committee for the handbook. We were coauthoring a book on standard data when Maynard died.

Maynard made very substantial contributions to the field of industrial engineering, including several books on operation analysis, methods engineering, and methods time measurement (MTM). He received the Frank and Lillian Gilbreth Award of the Institute of Industrial Engineers (IIE) for these contributions. I succeeded Maynard as President of H. B. Maynard and Company shortly after he retired. Spending the bulk of my working experience in the area of industrial engineering, I worked actively in IIE and became National President in 1964–65. I also received the Frank and Lillian Gilbreth Award for my work in the field. Because of my close relationship to Maynard and my strong interest in industrial engineering, I decided to take on editing this fourth edition. I can now say that it has been one of the most rewarding experiences of my career.

After I prepared an outline of this book, it was reviewed by the technical staff of the Institute of Industrial Engineers who made many valuable suggestions. I then distributed this outline to several past presidents of IIE and to friends who I felt were knowledgeable, for their comments and suggestions. I am particularly indebted to Professor Richard A. Dudek of Texas Tech University and Duane C. Geitgey of H. B. Maynard and Company, Inc., who made a number of substantial contributions to the outline.

Technical handbooks of this type are normally revised every seven or eight years. A revision normally consists of one third new chapters, one third revised chapters, and one third unrevised chapters. Since it has been more than 20 years since the third edition, this fourth edition practically constitutes a brand new book. More than 90 percent of the book consists of completely new chapters. Only seven chapters are revised from the third edition. They have been revised to bring them up to date. So, more than 90 percent of the material in the handbook is completely new. In effect, it *is* a new book.

Much of the material contained in the third edition is still useful, but could not be retained because space had to be made for all the new developments. Those who possess copies of the third edition are encouraged to keep them since they will be useful as a reference source for some time to come.

It has been interesting to look at the changes that have taken place in the field of industrial engineering in the past 20 years. As might be expected, not too many changes have taken place in the basic technologies such as charting techniques, motion study, work simplification, methods engineering, and work measurement. One significant exception has taken place in the field of quality control. While the basic principles and concepts of statistical quality control were well known 20 years ago, it took the Japanese to demonstrate the power of these tools. The entire concept of total quality control did not exist 20 years ago. Today, it is widely applied and has been widely promoted by the Department of Commerce under the banner of the Malcolm Baldrige National Quality Award.

In the last 20 years, computers have had a greater impact on industrial engineering than any other single development. The development of high-speed computers has led to many improvements in numerical controlled machine tools, process controls, and sensors. Computer integrated manufacturing (CIM) could not exist without the advent of high-speed computers and extended memory capacities. Computers also led to the development of computer aided design (CAD) and computer aided manufacturing (CAM). Many of the developments in MRP, JIT, inventory planning and control systems, and network planning and control could not have been made without the use of computers. The list goes on to include computer simulation, artificial intelligence, expert systems, plant layout, and product identification and tracking systems. Twenty years ago the third edition of the handbook had an entire chapter "Operation of the Slide Rule." Today few engineering students have ever seen a slide rule. The computer has not only made the slide rule obsolete but it has greatly expanded the field of quantitative methodologies by providing a means of implementing these sophisticated techniques. It has also made an impact on some of the more traditional industrial engineering functions such as work measurement. Computers are now being used to automatically establish work standards.

The first sentence of the original definition of industrial engineering stated, "Industrial engineering is concerned with the design, improvement, and installation of integrated systems of men, materials, and equipment." Obviously, when that definition was composed in 1955 we were not as conscious of sexual equality as we are today. So subsequently, "men" was quickly replaced by "people." While "men" was listed first in this definition, industrial engineers spent a great deal more time on materials and equipment than they did on men.

Another significant change has been the emphasis on people in the industrial engineering equation. This shows up in the emphasis on ergonomics, human factors, health, safety, and the emphasis on the importance of establishing a climate for productivity if productivity is to thrive.

Another significant trend is for industrial engineers to become much more aware of developments in other areas of management. One such development is in the area of concurrent engineering and design and cross functional teams. The functional form of organization has long been the standard in most organizations. But this form of organization naturally leads to a serial form of development in such things as bringing a product to market. The marketing department frequently comes up with its interpretation of what the customer wants. This is passed along to the design department who designs the new product and passes the design along to the manufacturing group. The manufacturing group then plans on how the product can be made but without any revisions to the basic design. This sequence of operations takes a great deal of time, does not always produce what the customer really wants, and results in a product that is expensive to manufacture.

This has led many companies to reorganize their organization to provide for the concurrent operations of engineering and design. This practice can greatly reduce the new product cycle time and produce a product that is more economical to manufacture and better meets the customers' needs. It permits the industrial engineer to apply the knowledge of work simplification and methods improvement while the product is still in the design stage.

Another development has been in the field of accounting, but is one where the industrial engineer has a great deal to offer. When labor costs constituted a major portion of the cost of manufacturing it made sense to use labor costs as the basis for allocating overhead costs. But in some highly automated factories labor costs only amount to 5 percent of the total cost. Because of the high cost of automation, overhead cost might be as much as 50 percent or more of the cost of goods sold. So it does not make much sense to use labor costs as the basis for allocating overhead. One solution to the problem is to analyze the activities that go to make up overhead cost and then allocate the cost to the products to the extent that they use the various activities involved. Industrial engineers are much better suited to allocating these costs than are accountants. This new concept is called activity based costing (ABC).

Another trend of the last 20 years has been the application of industrial engineering technology to the service industries. While many banks and insurance companies made good use of industrial engineers in the past, the service sector now accounts for a greater portion of the gross national product. The application of industrial engineering techniques to this sector has increased accordingly. In general, the experience has been that the concepts and principles of industrial engineering are very well applied to all types of service industries.

This handbook is the result of the cooperative efforts of a large number of people. The efforts of the staff of the IIE have already been acknowledged. But the primary effort in making this handbook possible was the contributions by more than 100 authors. They were all carefully selected for their how-to-do-it knowledge of the techniques and procedures used in their specialized fields of knowledge. I wish to thank each of them individually and collectively for their valuable contributions and for a job well done.

William K. Hodson
Editor-in-Chief

THE INDUSTRIAL ENGINEERING FUNCTION

CHAPTER 1
HISTORY, DEVELOPMENT, AND SCOPE OF INDUSTRIAL ENGINEERING

Michael D. Ferrell
Former President
H. B. Maynard and Company, Inc.
Pittsburgh, Pennsylvania

Industrial engineering, as the name is used in industry, commerce, and government throughout the world, may be the broadest of all the modern management functions. Time study people may consider themselves to be industrial engineers, as might process planners, manufacturing systems analysts, or rate setters. All of them, no doubt, are performing duties which fall within the broad range of activities generally considered to be part of the industrial engineering function. In fact, the range of industrial engineering activities is so broad that one prominent industrialist said that "industrial engineering consists of all of the engineering and management control activities that cannot be clearly designated as a part of other engineering or accounting functions."

The truth is that the field of industrial engineering is a large umbrella that includes a wide variety of tasks established for the purpose of designing, implementing, and maintaining management systems for effective operations. Mechanical engineers are generally perceived as individuals who design mechanical products or improvements to equipment, and their technical training is pointed largely toward that end. Similarly, electrical engineers design electrical systems or apparatus, and they are trained accordingly. Many people who call themselves industrial engineers, on the other hand, may never design anything. They may spend a whole career taking time studies, making methods studies, or conducting layout studies of factories or offices. Still, they are all involved in performing some aspect of industrial engineering work and may rightly feel justified in using the title industrial engineer in describing their jobs.

The ambiguity over what constitutes industrial engineering probably has its roots in the way that it developed as a profession. That, of course, dates back many decades before the name "industrial engineering" was ever coined in the years of the industrial revolution.

Much has been written about the early pioneers of management who emerged during and after the industrial revolution in England and the United States. Prior to the industrial revolution, goods were produced by individual craftsmen and the so-called cottage system. In those days, factory management was certainly not a problem. As new apparatus was developed and better power sources were discovered, however, it became practical to organize factories that could take advantage of the innovations. Perhaps the earliest pioneer in factory organization was Sir Richard Arkwright (1732–1792), who was the inventor of the spinning frame in England. He developed and implemented what was probably the first management control system to regulate production and the output of factory workers.

About the same time that Arkwright was installing his control system, another British inventor, James Watt, together with an associate, Matthew Boulton, was organizing a factory in Soho to produce steam engines. They instituted skills' training for craftsmen that far surpassed any such training that was available at that time. They also contributed much to the way that factories should be managed. Subsequently, their sons, James Watt, Jr., and Matthew Robinson Boulton, established the first complete machine manufacturing factory in the world. Following their fathers' example, they preplanned and built an integrated manufacturing facility that was years before its time. Among other things, they instituted a cost control system that was designed to decrease waste and improve productivity.

Another Englishman, Charles Babbage (1792–1891), made significant contributions to industrial engineering lore. He developed analytical systems for improving operations that were published in his book, *The Economy of Machinery and Manufacturers,* which was widely distributed in England, Europe, and the United States. The analytical methods that he originated were the state of the art for decades in the field of productivity improvement and bear some resemblance to the work of Frederick W. Taylor, which came much later.

The work of these British pioneers apparently was quite successful, particularly when applied to their own enterprises. Although there surely must have been exchanges of successful ideas among business leaders of the day (many of whom were related), there was no general movement by other entrepreneurs to adapt the successes of these individuals to their own businesses. Thus manufacturing in Great Britain, despite its being called "the workshop of the world," remained rather crude and unsophisticated. Toward the end of the nineteenth century the same primitive methods in general use in England were also in vogue in the United States.

The great impetus in changing the way that factory work was performed in America, and later in Europe, was started by Frederick W. Taylor. His successful experiments in improving manual methods of handling materials in steel mills led to startling gains in productivity. His papers on the subject presented to the American Society of Mechanical Engineers (ASME) gained wide attention.[1] His writings also gained a number of followers who built on his teachings. At the same time, Taylor gained a number of critics who felt that his philosophy on how work should be organized and managed was dehumanizing. Taylor became known as the "Father of Scientific Management" when he published his last book in 1911, *The Principles of Scientific Management.* He developed what he called a formula for maximum productions, which stated that "maximum production results when a worker is given a definite task to be performed in a definite time and in a definite manner."[2]

Although its form has changed somewhat, Taylor's formula is still an important part of industrial engineering. Taylor's formula emphasized that work must be well organized and the worker must be given a specific assignment and a specific method to be followed. Unfortunately, some practitioners following Taylor often achieved outstanding gains in labor productivity simply by establishing piecework and other wage incentive plans based on production standards. These schemes fell out of favor in some factories later because unscrupulous engineers and managers arbitrarily cut the production standards or piece rates in order to make the worker produce more output for the same or less money. The natural result was that workers would resist any and all efforts by management to change their production standards, even when there were legitimate reasons for the changes. Many worker attitudes that were created by "speed-up" tactics are unchanged to this day in many companies, particularly those with "featherbedding" or restrictive labor practices.

Although Taylor recognized and advocated the importance of shop methods, it was not until Frank and Lillian Gilbreth came along that the importance of motion study really began to be widely recognized. This was probably because no well-defined system was available for the study of motions. The Gilbreths isolated and identified the basic motions that make up all human activity and called them "therbligs" (Gilbreth spelled backward). They also specified that each of the 18 elemental motions, or "therbligs," should be accomplished within a definite range of time. This supported Taylor's belief that a handbook of universal time values (based on predetermined methods) could be established that would be applicable to any industry. This belief, of course, has never worked out universally but indeed has become a practical matter for some classes of work, such as maintenance operations, which have a fair degree of commonality from company to company and even from some industries to other industries. Gilbreth's "therbligs" later formed the basis for the research that ultimately led to the development of methods-time measurement (MTM), which is still widely used by industrial engineers.

Another industrial engineering pioneer was Harrington Emerson, who was an advocate of efficient operations and premium pay for increased production. His book, *The Twelve Principles of Efficiency*,[3] laid out the basis for effective operations. His 12 principles, which paralleled Taylor's teachings somewhat, were:

1. Clearly defined ideals
2. Common sense
3. Competent counsel
4. Discipline
5. Fair deal
6. Reliable, immediate, and adequate records
7. Dispatching
8. Standards and schedules
9. Standardized conditions
10. Standard operations
11. Written standard practice instruction
12. Efficiency reward

There is no question that the 12 principles espoused by Emerson in 1911 are just as valid today as they were then.

The spectacular increases in production that came from early incentive plans, and that later were sustained by unscrupulous rate cutting, led to two important aftereffects. First, because the increases were so easy to come by, very little attention was paid to the importance of good methods in production. The second aftereffect was the reaction of the workers and the public to the "speed-up" tactics and charges that were leveled. The workers pegged their output so that their earnings did not appear to be excessive, thereby avoiding the opportunities for management to try to cut their rates. Many public and government people also reacted to the so-called dehumanizing aspects of industrial engineering, and legislation was passed limiting the use of time standards in government operations.

These reactions led to an increased interest in the benefits of methods studies. The Gilbreth's efforts in the field of motion study had previously been considered to be rather theoretical and impractical. In the 1920s and 1930s, there was a renewed interest in their work and in the work of other industrial engineers. In 1927, H. B. Maynard, G. J. Stegemerten, and S. M. Lowry, wrote *Time and Motion Study,* which pointed out the importance of motion study and good methods. In 1932, A. H. Mogensen published *Common Sense Applied to Time and Motion Study,* in which he stressed his work simplification principles. R. M. Barnes published *Motion and Time Study,* which put particular emphasis on the motion study aspect of industrial engineering. During this period there was somewhat of a polarization between advocates of time study and others who felt that motion study was more important. Indeed, this polarization existed even in some of America's largest corporations, resulting in separate and largely equal time study departments and motion study departments. This separation continued in a number of companies and still exists today in some corporations. Predetermined time systems have diminished this polarization to some extent because a methods analysis must be made of an operation, with the work standard being a natural by-product of the methods analysis.

In 1934, H. B. Maynard and his associates coined the name "methods engineering," which they defined as follows:

> Methods engineering is the technique that subjects each operation of a given piece of work to close analysis in order to eliminate every unnecessary operation and in order to approach the quickest and best method of performing each standard method; when all this has been done, and not before, it determines by accurate measurement the number of standard hours in which an operator working with standard performance can do the job; finally, it usually, although not necessarily, devises a plan for compensating labor which encourages the operator to attain or surpass standard performance.[4]

This is a classic definition of methods engineering that is still valid today. Unfortunately in countless instances shortcuts have been taken and some valuable part of the definition disregarded.

During the 1930s depression years, many engineers were working on finding better ways to improve operations. Notable at the time was the 12-year study conducted by Western Electric Company at its Hawthorne Works.[5] In this study, the effects of multiple changes in methods and working conditions on the output of a group of workers were measured carefully so that the optimum output could be achieved. In most cases, a change in methods or working conditions resulted in increased output. One conclusion reached from the study was that workers generally respond favorably when they are involved and attention is paid to them.

Industrial engineering authorities during the 1930s were becoming very interested in improving the ability of industrial engineers to analyze and improve operations. Allan Mogensen developed his work simplification procedure, which concentrated on using the talents of shopworkers to improve methods.[6] His approach was to train key manufacturing people at his Lake Placid Work Simplification Conferences so that they could, in turn, conduct similar training in their own plants for managers and workers. The trainees would apply the techniques they were taught to actual shop operations, and countless improvements resulted.

During this same period, Maynard and Stegemerten wrote a book entitled *Operations Analysis* which detailed a procedure whereby an industrial engineer could systematically analyze all the conditions surrounding an operation to arrive at the best method (at the time) for doing a job. Along with improved methods and time study procedures, various job evaluation plans were developed that systematically and logically determined wage rates that were closely related to job content.

In 1943, the Work Standardization Committee of the Management Division of the American Society of Mechanical Engineers (ASME) drew up a chart depicting the functions of industrial engineering. A version of this chart is shown in Fig. 1.1.

The scope of the industrial engineering function began to expand rapidly in the years immediately following World War II and has continued to expand since then. A very significant industrial engineering development which gained prominence in the late 1940s and 1950s came with the publication of information on the use of predetermined motion time systems. Actually, the first of these systems, motion time analysis (MTA), was developed many years earlier by A. B. Segur. Segur, however, published little information on the use of MTA, preferring to use it only in his consulting practice, with his clients pledged to secrecy regarding details of the system. Consequently, MTA never gained wide public acceptance.

A new predetermined motion time system called work factor (WOFAC) was described

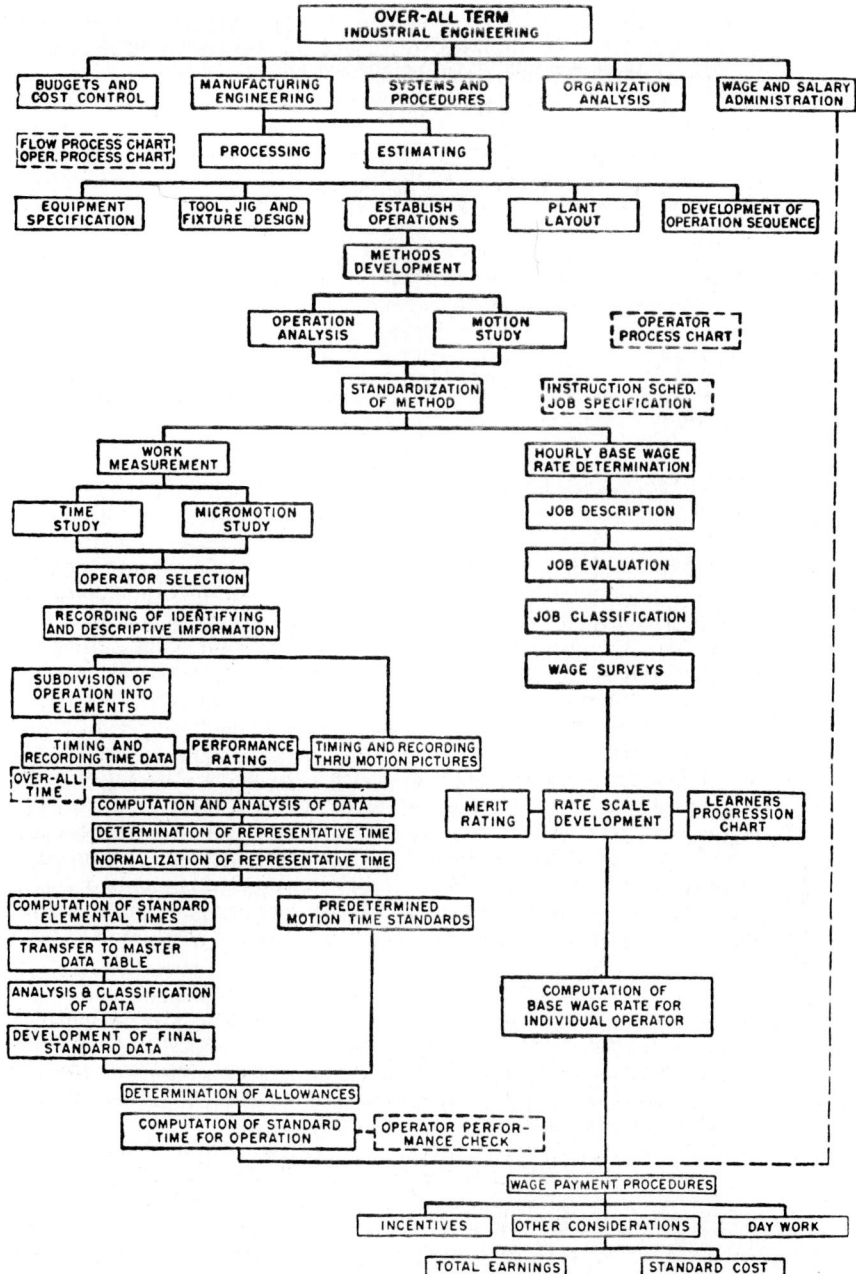

FIGURE 1.1 The field of industrial engineering. (*Adapted from ASME Work Standardization Committee, 1943.*)

in an article which appeared in *Factory Management and Maintenance* magazine in 1945.[7] Work factor was developed by J. H. Quick, W. J. Shea, and R. E. Koehler at the Radio Corporation of America (RCA) plant in Camden, N.J. Subsequently, the authors established the Work Factor Company (now called SMC Wofac, a division of Science Management Corporation) to promote the use of the work factor system.

The most prominent predetermined motion time system, methods-time measurement (MTM), was developed by H. B. Maynard, G. J. Stegemerten, and J. L. Schwab as the result of an intensive motion-time study sponsored by Westinghouse Electric Corporation. The developers authored the McGraw-Hill book entitled *Methods-Time Measurement* in 1948. The MTM system, promoted by Maynard's company, Methods Engineering Council (now H. B. Maynard and Company, Inc.), gained a great deal of attention around the world when *Fortune* magazine published an article on the new MTM system.[8] The MTM Association for Standards and Research was formed in America in 1951, to be followed by the formation of similar national associations in European countries.

Other variations of predetermined motion time systems with features similar to MTM followed. Basic Motion Times (BMT) was developed by Ralph Barnes in Canada. Engineers at General Motors, General Electric, and other companies came up with systems similar to MTM for use in their internal operations.

An important benefit that was stressed when predetermined motion time systems were first introduced was the practical motion study advantages of the system. An additional benefit was the ability to establish more accurate time standards than existing time study practices permitted, plus the elimination of the use of stopwatches for the measurement of most manual work. Another benefit which soon became apparent was the use of such systems to develop higher-level data systems to speed up the work study process. Thus such systems as general purpose data (GPD), MTM 2, and ready work factor were put into use. MOST®,* a newer MTM-based system, was developed by Maynard engineers in Europe in the late 1960s and early 1970s.

A more recent development has been the computerization of predetermined motion-time systems. Westinghouse Electric Corporation industrial engineers developed the 4M system based on MTM. Wofac engineers developed a computerized version of work factor. Computerized applications of MOST systems were completed by Maynard engineers in the United States and in Europe. CATS (computer assisted time standards), a computerized work study system, was developed for use within the U.S. Department of Defense. Other computerized work measurement systems, such as Autorate developed by IBM, and UniVation developed by Management Science, Inc., of Appleton, Wis., do not rely on predetermined motion-time systems.

The next logical step in the field of work measurement was the integration of computerized work study systems with automated process planning and other forms of computer-assisted design and manufacturing (CAD-CAM). This integration became a reality with the advent of systems such as AutoMOST (H. B. Maynard and Company, Inc.), which processes information from other manufacturing systems to set standards automatically. This is accomplished as a by-product of design and process planning activities and thereby frees industrial engineering time for other activities.

MODERN INDUSTRIAL ENGINEERING

During the time that the evolution of work measurement systems was taking place, many significant changes were occurring in other areas of industrial engineering. Great strides were made in the application of mathematical techniques and accounting solutions to manufacturing problems and costs. Computers improved the effectiveness of industrial engi-

---

*MOST is a registered trademark of H. B. Maynard and Company, Inc.

neers, resulting in improved productivity of the industrial engineering function. Along with computer technology, innovative management structures which incorporate team approaches and quality programs have also increased productivity through communication and cooperation by involving all levels of personnel in the improvement process. All of these new techniques are having a positive effect on the industrial engineering profession. The challenge today is to better integrate these tools and people resources into unified systems.

Development in recent years of concepts to improve efficiency and productivity has helped industrial engineers achieve their goals. Value analysis techniques have been developed to identify and apply industrial engineering approaches, and to eliminate unnecessary costs in all types of operations.

Prior to 1940, engineers were primarily concerned with the design and implementation of machines and processes and were not as concerned with the resources expended to produce the final products.[9] Success in today's production facilities depends on the mastery of basic principles of finance and accounting in order to justify factory improvement. These economic considerations have taken on added significance in the decision-making process of the industrial engineer. The concepts of engineering economy provide a tool to evaluate potential solutions to a production or manufacturing problem using accounting principles to see which solution is the most economically feasible. Engineering economy concepts cover topics such as return on investment, cash flow, working capital, and profitability.

Not only do engineers need to know the language and techniques of cost analysis in order to justify equipment and systems, they must master mathematical techniques and tools. Linear programming is a mathematical technique which deals with the efficient use of resources. Early commercial applications were made in petroleum refining and in the manufacture of livestock feeds. It has since spread to many other applications.

Queuing theory is another mathematical technique utilized in the industrial setting. Original work in waiting-line, or queuing, theory was used to determine the effect of fluctuating demand on equipment. Queuing theory explores the factors of lead time, setup costs, cost per unit, and demand in determining the proper inventory levels that must be maintained. Simulation, another helpful technique to industrial engineers, is the establishment of production system models. Simulation makes use of computers, queuing theory, and other mathematical techniques to study the effect of varying conditions on a production system. By simulating an environment, engineers are able to identify key elements which are problems within a system and the effect that varying these elements would have on the system. They may then use the information from the simulated environment to solve problems which may occur on an actual installation of the system or improve an existing system. The bottom line is to use these tools and techniques to improve efficiency, minimize throughput time, and reduce costs.

Automation has become more prevalent and viable in recent years because of the reduction of system costs and ever-relaxing attitudes of management toward the use of automation. Tasks that lend themselves best to automation are those that are highly repetitive or unpleasant to the worker. There are many advantages to automation. It can increase productivity by increasing the duty cycle, thereby yielding more machine hours per day. It can increase product quality by minimizing rework and waste. Limitations to automation, such as the high cost of automated machines and their vulnerability to downtime, must be considered, however. As these limitations are reduced, automated systems will be more widely utilized.[10] While automation is normally associated with high-volume production, the concept of flexible manufacturing systems (FMS) is used primarily for low-volume production. FMS is a multimachine system integrated via an automated material handling system, all under the control of one or more computers capable of producing a variety of parts with minimal setup.

Computer integrated manufacturing (CIM) provides a unified network of computer controls to support and/or monitor an organization. The use of the computer has been expanded to product coding and tracking, aided by the development of bar coding systems. These systems have enhanced the ability to monitor inventory, work-in-progress, and resource allocation. Bar coding can also monitor employee attendance and labor utilization, and then calculate payroll and provide a tighter control over inventory than traditional systems. Bar coding can also aid just-in-time manufacturing through the provision of real-time production data.

Robots, artificial intelligence, and expert systems are ways to improve manufacturing. Early robot models were used for simple material handling tasks, such as handling radioactive materials. Today's robots perform a variety of tasks including welding, machining, and painting. Work in the area of artificial intelligence (AI) is enabling the computer to solve problems in a humanlike manner. AI applications include problem solving, logical reasoning, learning, and expert systems. AI is used in developing expert systems, the most popular application of AI today.

When industrial engineers are called to lay out work areas and design tools for use by robots, they are employing the same types of skills and analyses used to improve human work.[11] The question of human work versus the use of robots is an issue industrial engineers will continue to face. Human workers' safety will also continue to be an issue for some time despite the use of robots to handle hazardous work. Industrial engineers must consider the workers' safety and comfort when designing a method or facility. One topic that is a major consideration in method and workplace design is ergonomics, or human factors engineering. Ergonomics is "the study of the...interaction between humans and the objects they use and the environments which they function in."[12] The goal of ergonomics is to design a method that will maximize the safety and comfort of the worker. Benefits from utilizing ergonomics include decreasing injuries and lost work time, lowering material and medical costs, and improving quality of work.[13]

Total quality management (TQM) is a concept that enables a company to achieve higher quality levels and elimination of waste. TQM is essentially a system whereby the qualities of products or services are produced economically to meet the requirements of the purchaser. TQM is possible only when all levels of personnel are integrated and communication is encouraged throughout the organization.

One way to integrate people and encourage communication is through the use of cross-functional teams. This approach is being implemented to solve problems and improve operations through better communication between all levels of an organization. Cross-functional teams range from machine operators to members of top management. The team can also consist of representatives from vendors or customers. The main objective is for all people involved in a project to meet, express ideas, and develop workable solutions to real problems.

The most widely accepted definition of industrial engineering was developed by the Institute of Industrial Engineers (IIE). It states:

> Industrial engineering is concerned with the design, improvement, and installation of integrated systems of men, materials and equipment. It draws upon specialized knowledge and skill in the mathematical, physical, and social sciences, together with the principles and methods of engineering analysis and design, to specify, predict, and evaluate the results to be obtained from such systems.[14]

Within that broad definition the importance of the industrial engineering function in business and industry has been growing constantly. In fact, a study conducted by the National Research Council indicates that while all of the engineering fields are growing, industrial engineering has sustained the highest growth rate since 1960.[15]

This significant growth has occurred during a period of dramatic shift in the nature of American business. At one time, practically all industrial engineering efforts were expended on manufacturing problems. As a result of international competition, the working population has moved rapidly from the factory to service areas. In the early 1960s, the ratio of blue-collar to white-collar workers reached 50-50. In 1990, the Bureau of Labor Statistics showed that 65 percent of the work force was in the white-collar or service area and estimated that by the year 2000, some 90 percent of the work force will be white-collar workers.[16]

This change in the working population will present industrial engineers with both great challenges and great opportunities. The techniques and procedures that have served industrial engineers so well in the past will continue to serve them in the future. There will undoubtedly be much greater emphasis on overall systems design, systems integration, and quality attainment on the impact of the workplace on worker safety and well-being and on the personal involvement of people in these design processes.

It will be an exciting period for industrial engineers.

REFERENCES

1. Taylor, Frederick Winslow, "The Present State of the Art of Industrial Management," *Transactions American Society of Mechanical Engineers,* vol. 34, 1912.

2. Taylor, Frederick Winslow, *Principles of Scientific Management,* Harper, New York, 1911.

3. Emerson, Harrington, *The Twelve Principles of Efficiency,* The Engineering Magazine, New York, 1911.

4. Maynard, Harold B., and G. J. Stegemerten, *Operation Analysis,* McGraw-Hill, New York, 1939, pp. 1–2.

5. Roethlisberger, Fritz J., and William J. Dickson, *Management and the Worker: An Account of a Research Program Conducted by the Western Electric Company, Hawthorne Works, Chicago,* Harvard University Press, Cambridge, 1939.

6. Mogensen, Allan H., *Common Sense Applied to Motion and Time Study,* McGraw-Hill, New York, 1932.

7. Quick, Joseph H., W. J. Shea, and R. E. Koehler, "Motion-Time Standards," *Factory Management and Maintenance,* May 1945.

8. "Timing a Fair's Day Work," *Fortune,* October 1949, pp. 129–139.

9. Riggs, James L., *Engineering Economics,* McGraw-Hill, New York, 1982, p. 2.

10. Considine, Douglas Maxwell, and Glenn Considine, eds., *Standard Handbook of Industrial Automation,* Chapman and Hall, New York, 1986, p. 4.

11. Ottinger, Lester V., "Robotics for the Industrial Engineer: Terminology Types of Robots," Edward L. Fisher, ed., *Robotics and Industrial Engineering: Selected Readings,* Institute of Industrial Engineers, Atlanta/Norcross, 1983, p. 1.

12. Alexander, David C., and Mustafa Babur Pulat, *Industrial Ergonomics: A Practitioners Guide,* Institute of Industrial Engineers, Atlanta/Norcross, 1985, p. 2.

13. Alexander, David C., and Mustafa Babur Pulat, *Industrial Ergonomics: Case Studies,* Institute of Industrial Engineers, Atlanta/Norcross, 1991, p. 4.

14. Institute of Industrial Engineers, 25 Technology Park, Atlanta/Norcross, Ga.

15. National Research Council, *Engineering Employment Characteristics,* National Academy Press, Washington D.C., 1985.

16. U.S. Department of Labor, *Outlook 2000: Projections of Occupational Employment, 1988–2000,* Bureau of Labor Statistics, April 1990.

BIBLIOGRAPHY

Alexander, David C., and Mustafa Babur Pulat, *Industrial Ergonomics: A Practitioner's Guide,* Institute of Industrial Engineers, Atlanta/Norcross, 1985.

Alexander, David C., and Mustafa Babur Pulat, *Industrial Ergonomics: Case Studies,* Institute of Industrial Engineers, Atlanta/Norcross, 1991.

American National Standards Committee, ANSI Z94.0-1982, *Industrial Engineering Terminology,* Institute of Industrial Engineers, Atlanta/Norcross, 1983.

Babbage, Charles, *On The Economy of Machinery and Manufactures,* C. Knight, London, 1835.

Barnes, Ralph M., *Motion and Time Study,* Wiley, New York, 1937.

Considine, Douglas Maxwell, and Glenn Considine, eds., *Standard Handbook of Industrial Automation,* Chapman and Hall, New York, 1986.

Emerson, Harrington, *The Twelve Principles of Efficiency,* The Engineering Magazine, New York, 1911.

Emerson, Howard P., and Douglass C. E. Naehring, *Origins of Industrial Engineering: The Early Years of a Profession,* Institute of Industrial Engineers, Atlanta/Norcross, 1988.

Fisher, Edward L., ed., *Robotics and Industrial Engineering: Selected Readings.* Institute of Industrial Engineers, Atlanta/Norcross, 1985.

Lowry, Stewart M., Harold B. Maynard, and G. J. Stegemerten, *Time and Motion Study and Formulas for Wage Incentives,* McGraw-Hill, New York, 1927.

Maynard, Harold B., G. J. Stegemerten, and John L. Schwab, *Methods-Time Measurement,* McGraw-Hill, New York, 1948.

Maynard, Harold B., and G. J. Stegemerten, *Operation Analysis,* McGraw-Hill, New York, 1939.

Mogensen, Allan H., *Common Sense Applied to Motion and Time Study,* McGraw-Hill, New York, 1932.

National Research Council, *Engineering Employment Characteristics,* National Academy Press, Washington, D.C., 1985.

Quick, Joseph H., W. J. Shea, and R. E. Koehler, "Motion-Time Standards," *Factory Management and Maintenance,* May 1945.

Riggs, James L., *Engineering Economics,* McGraw-Hill, New York, 1982.

Roethlisberger, Fritz J., and William J. Dickson, *Management and the Worker; An Account of a Research Program Conducted by the Western Electric Company, Hawthorne Works, Chicago,* Harvard University Press, Cambridge, 1939.

Taylor, Frederick Winslow, *Principles of Scientific Management,* Harper, New York, 1911.

Taylor, Frederick Winslow, "The Present State of the Art of Industrial Management," *Transactions of the American Society of Mechanical Engineers,* vol. 34, 1912.

"Timing a Fair Day's Work," *Fortune,* October 1949, pp. 129–139.

U.S. Department of Labor, *Outlook 2000: Projections of Occupational Employment, 1988–2000,* Bureau of Labor Statistics, April 1990.

Urwick, Lyndall F., "Development of Industrial Engineering," in Harold B. Maynard, ed., *Industrial Engineering Handbook,* 2d ed., McGraw-Hill, New York, 1963.

Western Electric Company, *Elemental Time Standards for Basic Manual Work,* New York, 1949.

CHAPTER 2
THE ROLE OF THE INDUSTRIAL ENGINEER

W. J. Kennedy, Ph.D., P.E.
Professor of Industrial Engineering
Clemson University
Clemson, South Carolina

As industry evolved, so did the role of the industrial engineer. Before World War II, the industrial engineer performed layout studies, did work measurement and time and motion analyses, and provided technical expertise in such manufacturing areas as statistical quality control, production control, and inventory control. Now, the broadened role of the industrial engineer includes systems analysis, the use of advanced statistics, and the development and use of simulation models. These subjects, promoted in industrial engineering curricula and made computationally feasible by advances in computer technology, have substantially enhanced the contributions industry can expect from a modern industrial engineer. The tasks performed by registered professional engineers in industry reflect this enhanced capability. The new qualifications and the new role of the industrial engineer can be illustrated in two ways: first by an examination of a modern IE curriculum in an accredited program, second by a review of a professional analysis of tasks performed by industrial engineers.

THE MODERN IE CURRICULUM

The objectives of a modern IE curriculum are to educate a person in the skills needed to contribute to modern society and to develop competence in those skills peculiar to industrial engineering. The courses developed to meet those needs depend upon the faculty of a particular school, subject to some constraints. The first constraint is the time it takes a student to graduate. Students (with the concurrence of their parents!) would like to finish in 4 years. The faculty would like to turn out students prepared for the technical challenges of the outside world, and the courses necessary to meet this goal generally require at least 5 years. What usually happens is that some number of credit hours is agreed upon by the faculty with the student able to complete these credit hours reasonably in 4 years. The student then takes this curriculum and, because of changes in major and various other disruptions, manages to finish in about 4 1/2 years.

ABET Guidelines—General.   A second constraint on courses is that the curriculum must meet ABET guidelines. ABET is the Accreditation Board for Engineering and Technology, and ABET accreditation demonstrates that the curriculum has met at least minimum standards imposed by a national accrediting body. In following ABET standards, a typical IE curriculum includes one or more courses in each of the categories shown in Fig. 2.1.

An ABET accreditation inspection includes an examination of all materials for each course, a study of the administration of the undergraduate program, discussion of the facilities, and an examination of faculty allocated to undergraduate teaching, and their qualifications. Meeting or exceeding ABET guidelines is a major goal of IE curriculum development. Once a department and curriculum has passed its first inspection, subsequent formal inspections are every 3 or 6 years.

The details of courses in a typical ABET accredited curriculum, and their rationale, are as follows:

English.   Most engineering curricula have at least three English courses—two in the freshman year common to all students in the school, with an additional course in technical writing. These courses train students in methods of organizing and in accepted standards of correct grammar and punctuation and are intended to develop the habits of clear thought and expression.

Mathematics.   A 2-year sequence of mathematics is required, starting from calculus and generally including differential equations. This sequence provides the basic vocabulary and concepts needed in most engineering courses. Problem-formulation and -solving skills are also enhanced in these courses.

Other mathematics courses required of industrial engineers include linear algebra—useful in model development, linear programming, and multivariate statistics—and advanced courses in probability and statistics.

Topics in the probability course (calculus-based) include discrete and continuous probability distributions such as the Poisson, binomial, normal, exponential, and others, multivariate distributions, functions of random variables, and the central limit theorem.

The statistics course requires the probability course as a prerequisite and includes hypothesis testing, interval estimation, analysis of variance, linear regression, and some nonparametric statistics.

Basic Sciences.   These include chemistry and physics and may include biology. A student leaves these courses with an enhanced vocabulary and with some knowledge of fundamental concepts of chemistry and physics. The physics courses have calculus as a prerequisite and include a brief introduction to electric circuits. The vocabulary and concepts

Subject	General Engineering Requirements	Additional IE Requirements
English	Two semesters	
Mathematics	Two years of calculus Probability and statistics	Linear algebra
Basic sciences	Chemistry Calculus-based physics	Biology (sometimes)
Humanities	(Each school sets its own policies for humanities)	
Miscellaneous		Engineering economy
Engineering sciences	Statics, dynamics, strength of materials, thermodynamics, fluid mechanics, electrical engineering basics	
Engineering design	Specific to the discipline	Facilities planning and design, production planning and control

FIGURE 2.1 Typical ABET approved IE curriculum.[1]

of these courses are useful. The courses are prerequisites to some of the basic engineering science courses. Chemistry and biology also provide a basis for understanding many present safety regulations and the chemical processes used in industry.

Humanities and Physical Education.   Humanities requirements change between colleges and between years in a given college, depending upon the perceptions of what is needed by students and by society. The same is true of physical education requirements.

General ABET Guidelines—Engineering Sciences.   To qualify as an engineering science, a course must provide a bridge from basic science or mathematics to application in engineering. A subset of these consists of courses applicable to more than one discipline of engineering. Such courses include

Introduction to computers and computing

Engineering economics

Statics

Dynamics

Strength of materials

Fluid mechanics

Thermodynamics

Basic electricity concepts

Materials of engineering

The courses include the following material:

Introduction to computing.   Required of all engineering students. Generally includes an introduction to BASIC or FORTRAN, some use of a spreadsheet, and some word processing.

Engineering economics.   Required of all IEs and most other engineering graduates. Includes interest, taxes, inflation, comparing the economics of different alternatives, and some cost accounting.

Statics.   Concepts of static equilibrium, decomposition of forces into components, free-body diagrams, analysis of structures.

Dynamics.   Force and acceleration, including rotation. Moments of inertia, Coriolis force.

Strength of materials.   How flexible bodies deform and break under various forces. Stresses and strains. Calculation of shear forces and turning moments. Analysis of beams, bars, and columns.

Fluid mechanics.   Fluid properties and fluid flow. Viscous flow. Turbulent flow. Lift and drag, and open-channel flow. Quantifying the fluid flow effects of friction in pipes.

Thermodynamics and heat transfer.   Thermodynamic properties of liquids and gases. Power and refrigeration cycles. The basics of combustion. Conduction, convection, and radiation. Quantifying the way that heat energy is stored in steam and stored energy in steam is changed into work. Insulation.

Basic electrical engineering.   AC and dc circuits, three-phase current, microprocessors, and microcomputers.

Materials of engineering.   A description of the microstructure of ferrous alloys and various ceramics, and how the properties of these alloys are influenced by heating and cooling.

The number of these basic science courses required in a curriculum depends upon the way the faculty at each school perceives their importance and upon ABET guidelines. All curricula include the basic computer course and engineering economics; most have statics, strength of materials, electricity, and fluid mechanics, thermodynamics, or materials of engineering.

The value of the basic science courses is in the vocabulary and concepts they teach, rather than in the mastery they provide. For example, designers of a material handling system need to know that weight and bending are an important consideration, even if they do not do the specific calculations associated with these. A person with these courses is also better equipped to understand new technology and regulations. As another example, one of the newest problems confronting IEs is handling and storing hazardous waste. Many of the problems associated with such waste are chemical, and a good knowledge of chemistry provides an IE with a key to understanding the regulations and any measures needed to mitigate the hazards. These courses also provide an industrial engineer with new ways of looking at complex systems.

ABET Guidelines for IE: Engineering Sciences.   In addition to the above topics, a number of engineering sciences are taught as a unique part of the industrial engineering discipline. These include

Manufacturing processes

Human factors engineering

Deterministic models

Probabilistic models

Systems modeling

Manufacturing Processes.   The manufacturing processes traditionally taught have been those involving the shaping and assembly of metal parts, and metal-removal processes. Recently, the emphasis has shifted to include complex combinations of machines such as those found in flexible manufacturing systems.

Human Factors Engineering.   This course is designed to enable students to realize the limits of human performance and take these limits into account when designing systems for people. Specific attention is given to information input and processing, controls, hand tools, workplace design, and the effects on people of noise, illumination, and other environmental factors.

Deterministic Models.   This course introduces students to modeling of systems using models which do not include probability. Modeling techniques introduced include linear programming, network analysis (CPM), and dynamic programming. All of these have proved useful in practice in formulating and solving systems problems.

Probabilistic Models.   The probabilistic models course develops ways of thinking about problems where the dominant mode of behavior is probabilistic. Such problems include queuing, some inventory problems, reliability, forecasting, and Markov analysis.

Systems Modeling.   Computer languages have been developed that enable the development of remarkably accurate models of the behavior of manufacturing systems. If a model of a system is good enough, it is possible to refine a possible new policy on the model before implementing the policy on the real system. The purpose of this course is to show students what goes into a systems model and to enable them to develop good models on their own, with all of the data collection, model validation, and model verification steps that are needed. Students also come out of the course with a good working knowledge of a specific simulation language.

ABET Guidelines for IE: Engineering Design.   This category includes courses that are open-ended, that is, have more than one right answer, that involve the development of evaluation criteria, and that demand significant judgment. Courses generally considered to have design content are:

Work methods and measurement

Quality assurance

Production planning and control

Facility planning and design

Work Methods and Measurements.   Topics in this basic IE course include flow process charts, methods engineering, micromotion study, predetermined data systems, operator payment and incentive systems, and setting time standards. Problems associated with collecting real data from systems with people are explained and illustrated.

Quality Assurance.   The pervasiveness of the quality criterion has caused this course to be broadened from an application of statistics into several courses which together embrace the entire manufacturing organization. The tools of statistical process control are taught, along with other control and management techniques needed to attain continuously improving quality.

Production Planning and Control.   A production system is one of the most complicated environments created by modern man. This course is aimed at giving students an understanding of such a system. The course includes such topics as production scheduling, line balancing, inventory control, personnel scheduling, forecasting, and the economic analysis of process improvement.

Facility Planning and Design.   Designing a facility includes determining how many machines are needed, where the machines are to be placed, where people will be in the system, what utilities are needed and where, how material is to be handled and moved, how operations are to be sequenced, and how information flows will be handled. These questions, and many others, are addressed in this course. Facility planning and design is one of the most demanding tasks in industrial engineering, and this course usually comes after a student has been exposed to all the modeling courses and the methods and quality assurance courses described above. The course generally includes a project where the students face the ambiguities and data demands of a real system.

Electives.   Electives are usually divided into those that are technical, nontechnical, and free, with specific guidelines for each category set by each college or department. The list of possible technical electives depends upon the size of the school, with large schools generally offering more. The list is usually controlled by the department, with students encouraged to take electives within the department. Nontechnical electives are filled by the humanities and social sciences and often are used to meet the college's requirements in those areas.

IE IN PRACTICE

What IEs Do.   The IE curriculum gives a good idea of what capabilities can be expected from a graduate of an accredited BSIE program. Many of the topics were not in the IE curriculum 20 years ago, and computer technology has changed those that were.

The IE curriculum has changed. So has IE practice. One of the better sources of information on IE practice was *Analysis of Professional Activities and Requirements of the Engineering Profession,* a study by the National Council of Examiners for Engineering and Surveying (NCEES), published in 1989.[2] The intent of this study was to compare the content of present professional engineering (PE) examinations with the activities of practicing engineers. The results of this study were then used to develop specifications for the questions on the PE examinations. The next few paragraphs describe the study and give a brief review of the test specifications for the IE professional engineering exam. (The author had the privilege of being a participant in each phase of this study as one of the engineers chosen by the Institute of Industrial Engineers and by NCEES.)

IE in Practice—the NCEE Study.   In 1987, NCEE (National Council of Engineering Examiners—the name was changed in 1989) commissioned a study to determine the dimensions of professional practice in each of the 14 branches of engineering represented by one of the professional engineering examinations prepared by NCEE. Interviews were con-

TABLE 2.1 Percentage of Time Spent on Professional Activity Categories

Professional activity	Percent who perform		Average percent time spent	
	PEs	Non-PEs	PEs	Non-PEs
Research and development	45.1	38.9	10.1	11.7
Conceptual design and planning	79.9	61.1	19.7	15.6
Preliminary and final design	61.8	55.6	12.2	6.7*
Construction engineering	23.0	5.6	4.5	0.6*
Facility engineering	57.4	55.6	12.5	10.0
Operations	79.4	94.4	33.2	53.3*
Applications and sales engineering	32.4	16.7	7.7	2.2*

*Difference significant at .01 level.
Source: Reprinted with permission from *Analysis of Professional Activities and Requirements of the Engineering Profession,* copyright 1989, National Council of Examiners for Engineering and Surveying.

ducted with groups of engineers from each of the 14 branches of engineering. These interviews identified professional activities, areas of practice, and professional requirements. Once categories had been identified for each of these dimensions, a survey was developed and administered to a sample of practicing engineers selected from a list obtained from the appropriate engineering professional society (for industrial engineers, the IIE). For industrial engineering, the original survey was sent to 417 PEs and to 156 non-PEs; of these, 208 (49.8%) usable responses were returned by PEs and 29 (18.6%) by non-PEs. The survey data were analyzed and were used to prepare the examination specifications.

Professional Activities.   The list of professional activities was common to all engineers and is given in Table 2.1, along with the mean time spent on each professional activity by the IE respondents. Since there were differences between the PE respondents and those without PEs, these are also shown.

Areas of Practice.   The original panel of IEs identified areas of specialization within IE. These areas were later refined and used in the survey. The results of the survey are shown in Table 2.2.

Professional Requirements.   As another task, the original committee identified a series of professional requirements—areas of knowledge, skills, and abilities used by engineers. These were included in the survey and are listed in Tables 2.3 and 2.4, with survey results.

Results.   The end result of the analysis described above was a set of specifications of the 20 questions to be asked on the PE examination. Each specification included the major

TABLE 2.2 Percentage of Time Spent on Professional Practice Area

Area of practice	Percent who practice in area		Average percent time spent	
	PEs	Non-PEs	PEs	Non-PEs
Facilities	67.6	55.6	17.9	10.0*
Management systems	69.1	66.7	16.1	13.3
Manufacturing systems	61.3	61.1	14.8	20.0
Production planning and control	51.0	72.2	9.5	10.6
Inventory planning and control	34.3	33.3	5.5	5.0
Quality assurance and safety	51.5	38.9	8.7	8.9
Work methods and measurement	53.9	83.3	9.2	17.8†
Human factors	40.7	33.3	5.9	3.9
Computer and information systems	52.9	44.4	12.4	10.6

*Difference significant at .05 level.
†Difference significant at .01 level.
Source: Reprinted with permission from *Analysis of Professional Activities and Requirements of the Engineering Profession,* copyright 1989, National Council of Examiners for Engineering and Surveying.

TABLE 2.3 Professional Requirements and Their Usage

Professional requirement	Percent who use	
	PEs	Non-PEs
Ethics—Canon of Ethics of Professional/Technical Society	86.9	64.7
Ethics—Rules of Professional Conduct of State Reg. Board	82.9	43.8
Engineering Economics	97.5	100.0
Communication—Oral Communications	100.0	100.0
Communication—Written Communications	100.0	100.0
Communications—Drawing and Graphics	98.5	100.0
Mathematics and Statistics—Calculus	69.7	50.0
Mathematics and Statistics—Differential Equations	46.5	27.8
Mathematics and Statistics—Probability and Statistics	94.0	94.1
Mathematics and Statistics—Vectors and Matrix Theory	58.3	44.4
Mathematics and Statistics—Transform Methods	31.5	22.2
Mathematics and Statistics—Numerical Techniques	71.2	61.1
Physical/Engineering Sciences—Chemistry	62.4	55.6
Physical/Engineering Sciences—Statics	72.1	55.6†
Physical/Engineering Sciences—Dynamics	62.4	50.0*
Physical/Engineering Sciences—Thermal Science	55.9	44.4
Physical/Engineering Sciences—Fluid Mechanics	53.8	22.2
Physical/Engineering Sciences—Optics	37.6	11.1
Physical/Engineering Sciences—Electrical Theory	73.7	44.4*
Computer Science—Hardware	87.9	83.3
Computer Science—Operating Systems	89.1	94.4
Computer Science—Software	95.0	100.0
Material Science—Physical Properties	69.0	38.9
Material Science—Chemical Properties	61.4	38.9
Material Science—Mechanical Properties	70.9	44.4
Material Science—Electrical Properties	63.8	35.3
Material Science—Phase Equilibria	23.1	16.7
Material Science—Diffusion	22.6	16.7
Measurement and Instrumentation	83.2	61.1
Codes and Standards	89.8	66.7
Management Principles	98.5	100.0
Ergonomics	82.0	72.2
Operations Research	84.6	88.9
Work Methods and Measurement Techniques	86.2	88.9
Manufacturing Processes	85.6	77.8
Systems Design/Analysis	93.5	88.9
Statistical Quality Control	87.0	88.9
Cost Analysis	97.5	94.4
Manufacturing Materials	70.9	55.6
Kinematics and Machine Design	53.8	38.9
Optimization Methods	85.1	77.8
Human Engineering	83.5	72.2
Reliability and Failure Analysis	72.7	66.7

*Difference significant at .05 level.
†Difference significant at .01 level.
Source: Reprinted with permission from *Analysis of Professional Activities and Requirements of the Engineering Profession,* copyright 1989, National Council of Examiners for Engineering and Surveying.

work behavior to be tested by that question, a description of the knowledge(s) required for that work, and the judgment level of each knowledge required. Helpful, though not required, knowledge was also specified. For example, for question E, the major work behavior to be tested is the conceptual design and planning of management systems. Neces-

TABLE 2.4   Professional Examination Topics for Industrial Engineering

Question	Topics
	Facilities
A	Conceptual design and planning
B	Preliminary design and planning
C	Facility engineering
D	Operations
	Management systems
E	Conceptual design and planning
F	Preliminary and final design
G	Operations
	Manufacturing systems
H	Conceptual design and planning
I	Preliminary and final design
J	Facility engineering
K	Operations
	Production planning and control
L	Conceptual design and planning
M	Operations
	Inventory planning and control
N	Operations
	Quality assurance and safety
O	Conceptual design and planning
P	Operations
	Work methods and measurement
Q	Operations
	Human factors
R	Preliminary and final design
	Computer and information systems
S	Conceptual design and planning
T	Operations

sary knowledges (the word used in the study), with their judgment levels, are codes and standards, management principles, systems design and analysis, and cost analysis. Helpful knowledges required were work methods and measurement techniques and human engineering. The judgment levels required were the ability to apply codes and standards and the two helpful knowledges, with a moderate amount of judgment required on application of the other three knowledges.

The full examination specification can be obtained from NCEES. In brief, the questions are as shown in Table 2.4.

CONCLUSION

Every practicing industrial engineer should know what colleges are teaching as industrial engineering. Every person teaching industrial engineering should know what practicing industrial engineers are doing. Industrial engineering is both the schooling and the practice. These have been described above.

REFERENCES

1. Minutes of the ABET Board Meeting, October 19, 1990, Denver, Colo.
2. National Council of Examiners for Engineering and Surveying, *Analysis of Professional Activities and Requirements of the Engineering Profession,* August 1989.

CHAPTER 3
EDUCATIONAL PROGRAMS FOR THE INDUSTRIAL ENGINEER

Curtis J. Tompkins, P.E., Ph.D.
University Professor and President
Michigan Technological University
Houghton, Michigan

Educational programs for the industrial engineer exist in many countries around the world; more industrial engineering educational programs are being created as developing countries seek to become competitive in the world economy and as more universities in developed countries respond to the needs of students and industry. This chapter concentrates primarily on industrial engineering programs in the United States, and unless explicitly stated otherwise, information presented in this chapter refers to the United States. Continuing education opportunities for the industrial engineer are increasing in quantity and accessibility, particularly with telecommunications linkages that span large distances.

For the majority of industrial engineers in the early 1990s, the bachelor of science degree was the only formal post-secondary educational credential achieved. In 1990, institutions awarded 4306 B.S., 2489 M.S., and 200 Ph.D. degrees in industrial engineering. Relative interest in industrial engineering as an undergraduate field of study has been growing steadily and virtually monotonically since it began in the early 1900s. Whereas in 1978, 4.5 percent of the B.S. degrees awarded in engineering were in industrial engineering, in 1990 that proportion had grown to 6.5 percent. Whereas in 1978 approximately 8.6 percent of the M.S. degrees and 3.8 percent of the Ph.D. degrees in engineering were in industrial engineering, in 1990 those proportions had changed slightly to 9.2 and 3.7, respectively. According to Turner, Mize, and Case,[1] before 1960 fewer than 100 total doctoral degrees in industrial engineering had been granted. By the mid-1970s, approximately 100 students received the doctoral degree each year; by 1990, that figure had grown to over 200 doctoral graduates per year. Tables 3.1 through 3.4 present detailed data on degrees awarded.

In 1990, the Accreditation Board for Engineering and Technology (ABET) listed 93 accredited undergraduate degree programs in industrial engineering and 12 accredited 4-year undergraduate degree programs in industrial engineering technology. Tables 3.5 and 3.6 list institutions with accredited programs awarding degrees in industrial engineering or industrial engineering technology in 1990.

TABLE 3.1 Bachelor's Degrees in Engineering, by Curriculum, 1978–1990

Curriculum	1978	1979	1980	1981	1982	1983	1984	1985	1986	1987	1988	1989	1990
Aerospace	977	1,145	1,376	1,587	1,731	2,207	2,364	2,663	2,747	2,845	2,949	3,065	2,971
Agricultural	559	619	779	666	711	704	655	598	638	456	362	330	317
Architectural	394	396	413	474	469	568	341	472	381	400	356	346	375
Biomedical	389	408	403	496	541	577	611	607	618	649	636	677	695
Ceramic	156	216	237	291	260	294	361	283	247	328	368	289	348
Chemical	4,621	5,837	6,555	6,863	7,039	7,499	7,685	7,244	6,148	5,129	4,082	3,711	3,622
Civil	9,168	10,030	10,191	10,547	10,330	10,484	9,877	9,468	8,798	8,388	7,714	7,688	7,587
Computer	1,546	1,510	1,816	2,356	2,666	2,643	3,499	4,248	4,999	5,012	4,275	4,398	4,355
Electrical and electronic	10,702	12,213	13,594	14,558	16,094	18,590	20,495	22,135	24,514	25,198	24,367	22,929	21,385
Engineering, general	1,744	2,065	2,029	2,169	2,360	1,923	2,037	1,847	1,385	1,315	1,085	1,058	1,239
Engineering, science	955	1,033	1,114	1,067	1,641	1,298	2,349	1,253	1,194	1,155	1,378	1,339	1,045
Environmental	269	284	232	248	254	292	301	224	182	124	192	138	137
Industrial and manufacturing	2,054	2,433	2,710	3,225	3,695	3,808	3,923	4,330	4,645	4,572	4,584	4,519	4,306
Marine, naval architecture and oceanography	703	792	794	854	698	699	739	736	602	537	549	468	475
Materials and metallurgical	706	829	1,045	1,081	914	1,085	1,011	988	1,011	885	877	842	857
Mechanical	8,786	10,076	11,916	13,462	14,178	16,484	17,214	17,152	16,702	16,056	15,610	15,369	14,969
Mining and mineral	721	868	1,001	1,054	1,078	1,019	978	926	769	628	404	274	168
Nuclear	576	534	529	444	426	420	434	429	400	324	306	303	264
Petroleum	617	812	915	1,031	1,256	1,420	1,587	1,550	1,381	1,064	612	436	286
Systems	199	199	211	235	346	210	222	442	594	458	474	457	362
Other	249	299	257	227	303	247	248	297	223	212	206	188	204
Total	46,091	52,598	58,117	62,935	66,990	72,471	76,931	77,892	78,178	75,735	71,386	68,824	65,967

Source: American Association of Engineering Societies/Engineering Manpower Commission, 1990.

TABLE 3.2 Master's and Engineer Professional Degrees in Engineering, by Curriculum, 1978–1990

Curriculum	1978	1979	1980	1981	1982	1983	1984	1985	1986	1987	1988	1989	1990
Aerospace	408	381	406	390	492	512	533	644	652	733	843	865	1,016
Agricultural	117	133	156	157	144	146	189	224	171	154	183	160	189
Architectural	21	20	52	53	42	30	35	73	48	33	34	29	33
Biomedical	184	205	211	184	199	178	211	272	233	270	257	303	310
Ceramic	43	48	34	54	55	55	87	75	79	102	96	100	80
Chemical	1,228	1,151	1,314	1,326	1,285	1,509	1,570	1,618	1,430	1,314	1,274	1,220	1,140
Civil	2,755	2,825	2,835	3,042	3,046	3,317	3,351	3,416	3,197	3,052	3,041	3,050	2,940
Computer	986	1,074	1,262	1,301	1,371	1,420	1,533	2,232	2,243	2,670	2,881	2,930	3,265
Electrical and electronic	3,475	3,335	3,736	3,845	4,281	4,730	5,393	5,592	5,926	6,780	7,335	7,520	7,691
Engineering, general	639	669	617	727	649	670	887	748	513	499	565	601	496
Engineering science	560	482	529	489	634	408	434	459	641	683	646	675	701
Environmental	521	445	518	487	412	456	424	382	351	337	329	427	471
Industrial and manufacturing	1,387	1,369	1,510	1,608	1,441	1,410	1,262	1,415	1,798	1,948	2,140	2,404	2,489
Marine, naval architecture and oceanography	160	173	180	174	202	163	186	163	197	132	120	121	146
Materials and metallurgical	479	482	469	453	507	563	583	622	658	661	623	672	671
Mechanical	1,910	2,026	2,181	2,495	2,573	3,001	3,160	3,315	3,462	3,511	3,767	3,855	3,994
Mining and mineral	188	183	195	151	207	280	242	256	256	296	282	237	192
Nuclear	482	408	368	311	317	311	282	268	284	272	221	245	236
Petroleum	112	125	134	161	147	227	260	240	249	199	220	197	162
Systems	404	415	465	437	490	430	469	380	499	523	636	662	692
Other	123	87	57	69	49	93	105	108	138	121	123	139	120
Total	16,182	16,036	17,229	17,914	18,543	19,909	21,226	22,502	23,025	24,290	25,616	26,412	27,034

Source: American Association of Engineering Societies/Engineering Manpower Commission, 1990.

TABLE 3.3 Doctorate Degrees in Engineering, by Curriculum, 1978–1990

Curriculum	1978	1979	1980	1981	1982	1983	1984	1985	1986	1987	1988	1989	1990
Aerospace	114	93	93	114	103	97	131	112	114	127	150	167	189
Agricultural	36	55	67	52	43	51	69	57	60	67	61	97	94
Architectural	0	0	0	1	0	0	0	3	0	0	0	0	0
Biomedical	59	49	54	54	43	50	49	55	46	56	70	86	103
Ceramic	17	20	17	18	12	18	18	14	14	30	24	25	38
Chemical	259	309	304	312	319	379	380	461	534	599	657	680	667
Civil	284	297	294	357	368	390	402	400	439	496	518	554	539
Computer	123	190	159	171	129	102	131	127	176	205	262	277	339
Electrical and electronic	524	545	523	503	549	608	693	714	779	811	1003	1139	1262
Engineering, general	86	112	98	123	114	76	55	51	57	63	66	81	58
Engineering science	212	238	211	187	256	170	151	156	182	217	191	238	239
Environmental	36	49	53	49	68	68	79	52	42	43	63	49	51
Industrial and manufacturing	97	116	109	109	98	108	124	117	120	158	149	204	200
Marine, naval architecture, and oceanography	21	24	18	22	16	18	21	29	21	21	30	26	21
Materials and metallurgical	213	181	227	206	196	228	241	272	272	337	310	326	392
Mechanical	304	335	307	339	341	399	476	540	565	691	738	795	900
Mining and mineral	30	21	35	45	32	54	33	45	59	55	56	74	79
Nuclear	94	102	107	112	127	114	115	98	112	98	102	90	115
Petroleum	19	22	21	13	23	14	11	14	18	19	38	29	54
Systems	40	42	52	48	44	56	49	64	61	69	63	56	69
Other	5	15	4	6	6	3	6	2	15	13	20	24	15
Total	2573	2815	2753	2841	2887	3023	3234	3383	3686	4175	4571	5017	5424

Source: American Association of Engineering Societies/Engineering Manpower Commission, 1990.

TABLE 3.4 Field Specialties of Women and Men B.S. Engineering Graduates, 1990

Specialty	Number of graduates			Specialization choices of 1990 engineering graduates, by sex (%)			Relative percentages of graduates, by specialties		
	Men	Women	Total	Men	Women	Total	Men	Women	Total
Aerospace	2,655	316	2,971	4.8	3.1	4.5	89.4	10.6	100.0
Agricultural	268	49	317	0.5	0.5	0.5	84.5	15.5	100.0
Architectural	306	69	375	0.5	0.7	0.6	81.6	18.4	100.0
Biomedical	488	207	695	0.9	2.0	1.1	70.2	29.8	100.0
Ceramic	273	75	348	0.5	0.7	0.5	78.4	21.6	100.0
Chemical	2,569	1,053	3,622	4.6	10.4	5.5	70.9	29.1	100.0
Civil	6,486	1,101	7,587	11.6	10.9	11.5	85.5	14.5	100.0
Computer	3,576	779	4,355	6.4	7.7	6.6	82.1	17.9	100.0
Electrical	18,785	2,600	21,385	33.6	25.7	32.4	87.8	12.2	100.0
Engineering science	844	201	1,045	1.5	2.0	1.6	80.8	19.2	100.0
Environmental	81	56	137	0.1	0.6	0.2	59.1	40.9	100.0
General	1,024	215	1,239	1.8	2.1	1.9	82.6	17.4	100.0
Industrial	3,099	1,207	4,306	5.6	11.9	6.5	72.0	28.0	100.0
Marine	448	27	475	0.8	0.3	0.7	94.3	5.7	100.0
Materials	643	214	857	1.2	2.1	1.3	75.0	25.0	100.0
Mechanical	13,237	1,732	14,969	23.7	17.1	22.7	88.4	11.6	100.0
Mining	141	27	168	0.3	0.3	0.3	83.9	16.1	100.0
Nuclear	230	34	264	0.4	0.3	0.4	87.1	12.9	100.0
Petroleum	266	20	286	0.5	0.2	0.4	93.0	7.0	100.0
Systems	262	100	362	0.5	1.0	0.5	72.4	27.6	100.0
Other	156	48	204	0.3	0.5	0.3	76.5	23.5	100.0
Total	55,837	10,130	65,967	100.1	100.1	100.0	84.6	15.4	100.0

Source: Engineering Manpower Commission 1990 degree survey. Percentages may not total exactly 100.0 due to rounding.

TABLE 3.5 Accredited Programs in Industrial Engineering

Engineering Management
Missouri-Rolla, University of
Southern Methodist University
United States Military Academy

Industrial and Management Engineering
Montana State University
Rensselaer Polytechnic Institute

Industrial and Operations Engineering
Michigan, University of; Ann Arbor

Industrial and Systems Engineering
Alabama in Huntsville, University of
Florida, University of
Michigan-Dearborn, University of
Ohio State University
Ohio University
San Jose State University
Southern California, University of

Industrial Engineering
Alabama, University of
Alfred University, New York State
Arizona State University
Arizona, University of
Arkansas University
Auburn University
Bradley University
California Polytechnic State University, San Luis
  Obispo
California State Polytechnic University, Pomona
California State University, Fresno
California, Berkeley; University of
Central Florida, University of
Cincinnati, University of
Clemson University
Cleveland State University
Columbia University
Fairleigh Dickinson University, Teaneck Campus
Florida International University

Industrial Engineering (Cont.)
Georgia Institute of Technology
GMI Engineering and Management Institute
Hofstra University
Houston, University of
Illinois at Chicago, University of
Illinois at Urbana-Champaign, University of
Iowa State University
Iowa, University of
Kansas State University
Lamar University
Lehigh University
Louisiana State University
Louisiana Tech University
Miami, University of
Mississippi State University
Missouri-Columbia, University of
Nebraska-Lincoln, University of
New Haven, University of
New Jersey Institute of Technology
New Mexico State University
New York at Buffalo, State University of
North Carolina Agricultural and Technical
  State University
North Carolina State University at Raleigh
North Dakota State University
Northeastern University
Northwestern University
Oklahoma, University of
Oregon State University (Basic Option)
Pennsylvania State University
Pittsburgh, University of
Polytechnic University
Puerto Rico, Mayaguez Campus; University of
Purdue University, West Lafayette
Rhode Island, University of
Rochester Institute of Technology
Rutgers-The State University of New Jersey
St. Mary's University
South Florida, University of
Stanford University

Industrial Engineering (Cont.)
Tennessee Technological University
Tennessee at Knoxville, University of
Texas A&M University
Texas Tech University
Texas at Arlington, University of
Texas at El Paso, University of
Toledo, University of
Utah, University of
Washington, University of
Wayne State University
West Virginia University
Western Michigan University, Kalamazoo
  Campus
Western New England College
Wichita State University
Wisconsin-Madison, University of
Wisconsin-Milwaukee, University of
Youngstown State University

Industrial Engineering and Management
Oklahoma State University

Industrial Engineering and Operations Research
Massachusetts at Amherst, University of
Virginia Polytechnic Institute and State
  University of

**Manufacturing Engineering Option in Industrial
Engineering**
Kansas State University
Oregon State University

Operations Research and Industrial Engineering
Cornell University

**Manufacturing Engineering Option in Industrial
Engineering**
Kansas State University
Oregon State University

TABLE 3.6 Accredited Programs in Industrial Engineering Technology

---

Industrial Engineering Technology
  Dayton, University of
  Georgia Southern College
  Indiana University—Purdue University at Fort Wayne
  Indiana University—Purdue University at Indianapolis
  New York, State University of, College of Technology
  New York, State University of; College of Technology, Utica Extension
    at Hudson Valley Community College
  Purdue University Calumet
  Southern College of Technology
  Southern Mississippi, University of
  Southern Mississippi, University of—Gulf Park Campus
  Trenton State College

Industrial Engineering Technology
Option in Engineering Technology
  Kansas State University

HISTORY OF EDUCATION IN INDUSTRIAL ENGINEERING

An interesting summary of the history of industrial engineering education is presented by Emerson and Naehring.[2] An entire issue of the *Journal of Industrial Engineering* in 1962 was devoted to describing all of the accredited programs in industrial engineering, including some historical aspects of most of those programs.[3]

The first industrial engineering course offered by a college or university was designed and taught by Prof. Hugo Diemer in the department of mechanical engineering at the University of Kansas in 1901. In 1904, Prof. Dexter Kimball introduced an elective course in works management for mechanical engineering seniors at Cornell University, and in 1907, Prof. Walter Rautenstrauch presented an engineering course entitled "Business Methods" at Columbia University.

Hugo Diemer moved from Kansas to the Pennsylvania State University as head of mechanical engineering in 1907; there he immediately led the design and introduction of the first baccalaureate program in industrial engineering. In 1909, Diemer became head of the new, separate 4-year program in industrial engineering at Penn State (Ref. 2, pp. 44–45). The description of the industrial engineering program in the 1909 to 1910 Penn State course catalog has some remarkable similarities to portions of descriptions found in many college catalogs more than 80 years later (as quoted in Ref. 2):

> This course is intended especially to prepare for positions that deal with the side of industrial organizations that has to do with business management, works management, superintendence, purchasing and sales. It deals largely with the application of the sciences and humanistic studies to industrial ends. It prepares for the competent handling of such subjects as the determining of costs, depreciation, statistics, proper distribution of expense, economic production, systems of remunerating labor and raising labor efficiency, the handling and records of stock and orders, sales, purchasing, corporation accounting and allied work.
>
> For the purpose indicated, well-trained engineers are required; hence the course includes all the fundamental engineering work in mathematics, drawing, physics, chemistry, machine design, heat engineering, mechanics, hydraulics, and structures that are common to all other engineering courses. *But more time should be and is given to such general studies as modern languages, English, economics, logic, and psychology and specialized work in accounting, factory management, shop time study, machine tools and methods and shop practice in general.* (Emphasis added.)

The first book using the words "industrial engineering" to describe the profession (Ref. 2, p. 47) was *Principles of Industrial Engineering* by Charles Going,[4] published in 1911 and based on his lectures at Columbia. That publication occurred a year after Hugo Diemer published *Factory Organization and Administration,* the first textbook in the field of industrial engineering.[5]

Table 3.7 (from Ref. 2) lists the dates when autonomous departments of industrial engineering were established at institutions in Canada and the United States. After 1937, when the Engineers' Council for Professional Development (ECPD), the forerunner of ABET, began accrediting programs, dates of initial accreditation are used in the absence of other information.

TABLE 3.7 Start-up Dates of Industrial Engineering Curricula (United States and Canada)

Date of first autonomous department	Institution
1908	Pennsylvania State College
1912	University of Kansas
1919	Columbia University
	New York University
1921	University of Pittsburgh
1922	University of Alabama
	Lafayette College
1924	University of Michigan
	Oklahoma A & M College
1925	GMI Engineering & Management Institute
	Lehigh University
	Montana State University
	Ohio State University
	Syracuse University
1929	Virginia Polytechnic Institute and State University
1931	Cornell University
1932	North Carolina State University (Raleigh)
1934	University of Florida
	Oregon State University
	Texas Tech University
1939	Northeastern University
1940	Illinois Institute of Technology
	Polytechnic Institute of New York (New York University)
1941	Texas A & M University
1942	University of Southern California
1945	Georgia Institute of Technology
	Stanford University
1946	Johns Hopkins University
	State University of New York at Buffalo
1947	University of Miami
	University of Rhode Island
	Rutgers, The State University of New Jersey
	University of Washington
1948	Bradley University
	University of Houston
	University of Massachusetts at Amherst
	University of Tennessee
	Wayne State University
	Wichita State University

TABLE 3.7 Start-up Dates of Industrial Engineering Curricula (United States and Canada) (Continued)

Date of first autonomous department	Institution
1949	University of Arkansas
	Lamar University
1950	Southern Methodist University
	University of Toledo
1952	North Dakota State University
1955	University of Puerto Rico, Mayaguez
	Purdue University
	University of Illinois, Urbana—Champaign
	West Virginia University
1956	Arizona State University
	University of California, Berkeley
	California Polytechnic State University—San Luis Obispo
	Fairleigh Dickinson University
	Iowa State University
1958	University of Missouri—Columbia
	Northwestern University
1959	University of New Haven
	Kansas State University
	University of Toronto
	Western Michigan University
1960	University of Arizona
	New Jersey Institute of Technology
	St. Mary's University
	San Jose State University
	University of Texas at Arlington
1962	University of Central Florida
	University of Iowa
	Mississippi State University
	Newark College of Engineering
	University of South Florida
	Tennessee Technological University
1963	University of Oklahoma
	Western New England College
1964	Auburn University
	University of Alabama in Huntsville
1965	Louisiana State University
	Technical University of Nova Scotia
1967	Louisiana Tech University
	Ohio University
1968	Rochester Institute of Technology
	University of Windsor
1969	University of Texas at El Paso
	New Mexico State University
1970	California State University—Fresno
	Cleveland State University
	University of Nebraska
	The University of Utah
	University of Wisconsin—Madison
1972	University of Illinois at Chicago
	Ecole Polytechnique de Montreal
1973	University of Michigan—Dearborn
1974	Southern Illinois University at Edwardsville

TABLE 3.7   Start-up Dates of Industrial Engineering Curricula (United States and Canada) (Continued)

Date of first autonomous department	Institution
1975	University of Cincinnati
1976	University of Wisconsin—Milwaukee
	California State Polytechnic University—Pomona
1977	University of Louisville
	North Carolina Agriculture and Technical State University
1978	Rensselaer Polytechnic Institute
	University of Bridgeport
1983	Clemson University
	St. Mary's University
1984	University of Regina

Source: Emerson and Naehring.[2]

Many of the founders and pioneers of industrial engineering methods and educational programs were active participants during the early 1900s in the Society for the Promotion of Engineering Education (SPEE), which was founded in 1893 and later was renamed the American Society for Engineering Education (ASEE). Two examples of such involvement appear to be particularly noteworthy.

In June 1912, Frank Gilbreth chaired a session at the twentieth annual meeting of SPEE cohosted by MIT, Harvard, and Wentworth in Boston. Speakers included Prof. Walter Rautenstrauch (of Columbia), who presented a paper entitled "Teaching the Principles of Scientific Management," and Hugo Diemer, Harrington Emerson, Henry Gantt, H. K. Hathaway, Robert Kent, William Kent (who started the IE program at Syracuse University), and H. J. Porter.[6] Many of the fundamental tenets on which industrial engineering programs were established and developed prior to World War II were discussed in that session.

The second instance displays Frank Gilbreth's passion for excellence in engineering education and his intolerance for the traditional educational practices of the early 1900s. Gilbreth spoke during the annual meeting of SPEE in December 1918 in Cambridge, Massachusetts; some of his remarks to the assembled educators were reported in the February 1919 issues of *Engineering Education*.[7] "I think you are a sorry lot," Gilbreth thundered. "I think you do not dare to face the facts; I think this is a mutual admiration society, where you come together with the idea of stealing half of the wrong ideas from each other." He threatened that unless they immediately improved engineering education (that is, found the "one best way") he would do it himself! Until his death in 1924 at the age of 56, Frank Gilbreth was an active member of ASEE (SPEE) and a popular speaker on college campuses.

Interest in aspects of what eventually became known as *engineering economy* began in the early 1900s. Rautenstrauch stressed break-even analysis in his courses at Columbia in 1912 and wrote a book on *The Economics of Business Enterprise*. John C. L. Fish, a civil engineer, wrote a book entitled *Engineering Economics*[8] in 1915.*

Eugene L. Grant published his textbook on *Principles of Engineering Economy*[9] in 1930 based on his course by that same title at Stanford University during the 1920s; later editions were coauthored with W. Grant Ireson. H. G. Thuesen published the first edition of his popular textbook on *Engineering Economy*[10] in 1950.

World War II brought intense attention to new statistical and mathematical methods and related course content which slowly emerged in industrial engineering degree programs in the 1950s and early 1960s. Many of these methods had roots in earlier course offerings and materials. For example, Thornton C. Fry's book on *Probability and Its En-*

---

*Grant dedicated his book, *Statistical Quality Control*,[15] to the memory of J. C. L. Fish.

gineering Uses[11] was published in 1928, and Rautenstrauch's article on a linear programming type of solution to "A Comparison of Lathe Headstock Characteristics"[12] was written in 1910. However, post-World War II transformation of industrial engineering education was characterized by the widespread development of courses in such subjects as computer programming, inventory theory, linear programming, probability theory, queuing theory, simulation, and statistical quality control. Textbooks in these and related subjects were written in the 1950s and early 1960s by industrial engineering faculty such as Apple,[13] Barnes,[14] and Grant[15] and by individuals from other disciplines such as Bellman,[16] Charnes,[17] Churchman,[18] Dantzig,[19] Feller,[20] Magee,[21] and Morse.[22]

Whereas prior to World War II, industrial engineering programs typically grew from mechanical engineering departments and were taught by faculty with degrees in mechanical engineering, the advent of operations research, probability and statistical methods, and computers brought different perspectives and backgrounds to bear on industrial engineering education. Furthermore, the industrial engineering function in industry generally did not keep pace with the changing academic programs in the 1950s and 1960s. Concurrently faculty members in other disciplines such as business, applied mathematics, statistics, and the emerging field known as computer science viewed some of the newer industrial engineering subjects as being separable from industrial engineering. Thus many academic disciplines became involved in and felt ownership in the areas of "management science" and "operations research." The Operations Research Society of America (ORSA) and The Institute of Management Science (TIMS) attracted industrial engineering faculty members interested in these emerging subjects.

Interest in applications of industrial engineering to nonmanufacturing activities grew during the 1960s and 1970s. Inspired by Lillian Gilbreth, Harold E. Smalley[23] and others developed courses and books in hospital industrial engineering; the Hospital Management Systems Society (HMSS) was formed in 1961 and AIIE created its hospital division in 1964. The fields of banking, nonprofit organizations,[24] printing and publishing, public administration,[25] retailing, and transportation and distribution were also fertile areas for industrial engineering; academic programs increasingly reflected those diverse interests during the late 1960s and into the early 1980s.

Statistical quality control became a required part of many industrial engineering academic programs during the 1950s and 1960s. Interestingly, during the 1970s and early 1980s that subject was not treated as one of the high priorities of most IE academic programs and in fact was no longer required at many colleges and universities.

Frank Gilbreth's book on *Motion Study*[26] indicated that there were three categories of variables involved in any job: the worker, the surroundings, and motion. While Gilbreth's main focus was on motion analysis, he and his wife Lillian (who earned her Ph.D. in psychology) also paid considerable attention to variables of the worker (psychology and physiology) and the worker's surroundings. The human factors, ergonomic, and industrial hygiene aspects of industrial engineering grew considerably after World War II, and academic programs added courses to reflect that growth. These subjects were often taught by individuals who had been trained in psychology, industrial hygiene, or physiology. Faculty members with backgrounds in one or more of those fields were added to departments of industrial engineering during the 1960s and 1970s. Most of the early textbooks in various aspects of this broad dimension of industrial engineering were not written by industrial engineers. Two examples are *Human Factors Engineering*[27] by Ernest J. McCormick, a professor of psychology at Purdue University, and *Human Performance in Industry*[28] by K. F. H. Murrell, a professor of physiology in Bristol, England.

Looking back at the pioneers, the innovators, the developers, the leaders, the teachers, and the researchers involved in industrial engineering education from 1900 to the 1980s, one recognizes mechanical engineers, psychologists, mathematicians, physiologists, statisticians, economists, and, yes, industrial engineers. Strangely, perhaps, there were very few *managers,* that is, individuals actually trained in schools of management. *Scientific management* was started by Frederick Taylor, a man without a college degree who went on to be president of the American Society of Mechanical Engineers. Frank Gilbreth, another man without a college degree, followed and expanded beyond Taylor, dealing with management issues. Since that beginning, in a broad sense, most if not all aspects of industrial engineering and industrial engineering education have dealt with management aspects of engineering or engineering aspects of management. But what about management per se, that is, courses in *management* taught in departments of industrial engineering for industrial engineering students? Indeed, many departments of industrial engineering have had what have been variously titled engineering management, management of technology, operations management, or, simply, management courses. Some of these have focused on *project management,* often built around the critical path method and PERT (Moder and Phillips[29]). Some have emphasized *organizational change* and group processes (Morris[30]). Others have covered *operations management* (Timms[31]), *operations planning and control* (Greene[32]), and *work systems design* (Nadler[33]). Professor Paul E. Torgersen has taught a popular course on management for industrial engineering students at Virginia Polytechnic Institute and State University for more than 25 years based largely on *The Functions of the Executive,*[34] written by Chester I. Barnard in 1938. Torgersen's textbook (with Ira T. Weinstock) on management[35] reflects much of the content of his long-standing course.

By the late 1980s, most departments of industrial engineering offered courses in the following subject categories:

1. "Classical" industrial engineering
2. Decision sciences (operations research)
3. Human factors (ergonomics)
4. Management
5. Manufacturing systems

While various schools may place somewhat different labels on categories of subject matter, the key point is that only one of the areas listed above is fairly clearly the primary province of industrial engineers in a strict, single-discipline focused sense. The other categories have always been and will always be multidisciplinary. Industrial engineering has evolved and developed into a very broad cross-disciplinary field; academic programs necessarily reflect that breadth and diversity.

INDUSTRIAL ENGINEERING PROGRAMS IN THE 1990S

Degree programs in industrial engineering involve faculty with backgrounds in industrial engineering, mechanical engineering, ergonomics, psychology, physiology, industrial hygiene, computer science, statistics, operations research, management, and other fields, thus reflecting the broad diversity of the curricula, research, and applications of industrial engineering. No other engineering discipline displays such a degree of diversity and breadth of coverage. The systems studied by industrial engineering students consist of a network of both human and physical elements, whereas in most engineering disciplines, the focus is on systems composed predominantly of physical elements. As is true of chemical engineers, the industrial engineer is usually concerned with the design and improvement of processes, in contrast to the design or improvement of products or equipment by mechanical engineers and electrical engineers or the design or improvement of structures, transportation, or environmental systems by civil engineers. This process orientation combined with a "systems integration" perspective form a substantial portion of good academic programs for industrial engineers.

One way to illustrate the nature of a modern curriculum is to revisit the first IE program, namely, the one started in 1908 by Hugo Diemer. The 1991–1992 undergraduate course catalog of The Pennsylvania State University describes the program in Industrial and Management Systems Engineering as follows:

> This major prepares students who intend to enter the technical area of industrial or commercial enterprises, or government services. The fundamentals of engineering are supplemented by sequences of courses essential to production and management positions where a scientific and engineering background is necessary. The following fields of study are included in the general program.
>
> 1. Management Systems: Management information systems, manufacturing and distribution systems utilizing mathematical models, programming a computer to do simulations, and relevant quantitative operations research and management science techniques.
> 2. Manufacturing Engineering: Automation, material removal, casting, forging, plastic working of metals, materials joining, fabrication of polymeric materials, tool engineering, robotics, computer-aided manufacturing, and manufacturing systems analysis.
> 3. Management Controls: Engineering economy, inventory control, quality control and reliability, production control and cost control.
> 4. Methods: Work simplification and measurement, factory planning and materials handling and data processing.
> 5. Ergonomics Engineering: Human factors, work physiology, biomechanics, industrial safety.

This description is typical of industrial engineering undergraduate programs in the early 1990s. Beyond those subjects taught in the department of industrial engineering, a student majoring in industrial engineering will typically take courses in mathematics, chemistry, physics, humanities and social sciences, mechanical engineering, electrical engineering, and English.

ACCREDITATION BOARD FOR ENGINEERING AND TECHNOLOGY

The Accreditation Board for Engineering and Technology (ABET) was founded in 1932 under the name Engineers' Council for Professional Development (ECPD). Its primary mission is to monitor, evaluate, and certify the quality of engineering and engineering-related education in colleges and universities in the United States. ABET accredits both engineering and engineering technology programs. The legal recognition of ABET comes from the U.S. Department of Education and the Council on Postsecondary Accreditation (COPA).

> The objectives established by ABET are:
>
> 1. To serve the public, industry, and the profession generally by stimulating the development of improved engineering education; including encouraging curricular improvement in existing programs and helping develop educational models for establishing new engineering programs.
> 2. To identify for prospective students, student counselors, parents, potential employers, public bodies, and officials, engineering and engineering technology programs which meet the minimum ABET criteria.

The accreditation process is initiated by a school requesting accreditation for a particular program(s) in engineering or engineering technology. Then a team of experienced engineering educators and practitioners is assigned to visit the school. The team must have at least one expert member for each engineering program being reviewed. Thus, when an industrial engineering program is being evaluated, an experienced industrial engineer from academia or from industry will be on the team.

After having the opportunity to review a detailed self-evaluation questionnaire prepared by the school beforehand, the team conducts an on-site visit for approximately 2 1/2 days. While on campus the team examines both the academic and professional credentials of the faculty, adequacy of computer facilities, equipment, library facilities, and service departments. In this evaluation process the team meets with administrators, faculty, and students. An analysis of both the qualitative and quantitative aspects of program content is conducted to ensure that the program satisfies the minimum ABET criteria.

The preliminary report of the ABET team is submitted to the school for its "due process" review and comment. Final action on each engineering program is taken at the annual meeting of the Engineering Accreditation Commission (EAC), and on each engineering technology program at the annual meeting of the Technology Accreditation Commission (TAC). The Institute of Industrial Engineers (IIE) has three representatives on EAC and two representatives on TAC.

Another commission, the Related Accreditation Commission (RAC), was established by ABET in 1983 to evaluate programs that are closely related to engineering or engineering technology but do not fit their precise specifications.

The three commissions, EAC, TAC, and RAC, report to the ABET board of directors. Two IIE representatives are on this board.

To be accredited, a particular program, for example, industrial engineering, must satisfy both the general criteria and the program criteria. A distinction is made in criteria for engineering and for engineering technology. Both 2-year and 4-year engineering technology programs can be accredited.

The curriculum should show a progression in the coursework and an application of the fundamental scientific and other training of the earlier years in the later engineering courses. The institution has the option to request program accreditation at either the basic (bachelor's degree) level or the advanced (master's degree) level. Relatively few institutions have requested accreditation at the advanced level.

Those programs for which accreditation is sought at the basic level should have the following curricular content.

1. At least 2 1/2 years of study in the area of mathematics, science, and engineering. The coursework should consist of approximately 1/2 year of mathematics beyond trigonometry; 1/2 year of basic sciences; 1 year of engineering sciences; and at least 1/2 year of engineering design (1/2 year of study is considered to be equivalent to 16 semester credit hours or 24 quarter hours).

2. At least 1/2 year in the area of humanities and social sciences.

For the basic level the program normally requires at least 128 semester hours (192 quarter hours). Advanced-level accreditation requires 1 additional year of study beyond the basic level program to include at least 1/3 year of advanced engineering design and at least 1/3 year of advanced mathematics, basic sciences, engineering sciences, or engineering design. Laboratory experience appropriate for the particular program must be included in the curriculum.

The engineering graduate is expected by ABET to have competency in oral and written communication skills in the English language. The program to be accredited must include at least an introduction to the ethical, social, and economic considerations in the practice of engineering. ABET also has stated criteria applicable to the faculty, student body, administration, institutional commitment, and facilities.

In addition to the general criteria which apply to all engineering disciplines, separate program criteria have been established for each of the respective disciplines. The program criteria for industrial engineering include a capstone engineering design experience, appropriate use of computers integrated throughout the curriculum, a minimum of three full-time equivalent faculty members, a reasonable faculty workload, and a department head qualified as an industrial engineer by education and/or experience.

Programs in engineering technology encompass technical education between engineering and industrial technology. Engineering technology has been defined as "that part of the technological field which requires the application of scientific and engineering knowledge and methods combined with technical s'   in support of engineering activities; it lies in the occupational spectrum between the craftsman and the engineer at the end of the spectrum closest to the engineer.''

Programs in engineering technology may be accredited at the associate degree level or at the baccalaureate level. Differential criteria in course requirements are specified for each level.

Associate degree (2-year program)

1. At least 60 semester hours (90 quarter hours)
2. 30 semester hour credits (40 quarter hours) in technical courses
3. 15 semester hour credits (22 quarter hours) in basic sciences and mathematics
4. 8 semester hour credits (12 quarter hours) in social sciences and humanities

Baccalaureate degree (4-year program)

1. At least 120 semester hours (180 quarter hours)
2. 45 semester hour credits (68 quarter hours) in technological courses
3. 23 semester hour credits (35 quarter hours) in basic sciences and mathematics
4. 21 semester hour credits (32 quarter hours) in social sciences/humanities

For engineering technology programs specific criteria also relate to faculty, student body, administration, satisfactory employment, financial support, and facilities. Program criteria have also been developed for industrial engineering technology programs.

The Accreditation Board for Engineering and Technology (ABET) recognizes the quality of the educational programs leading to degrees in engineering as accredited by the Canadian Accreditation Board (CAB), which is a standing committee of the Canadian Council of Professional Engineers. ABET considers the CAB accreditation decisions as acceptable for the educational preparation for the practice of engineering.

At the graduate level, ABET has accredited the master's degree program at the University of Louisville and the master's degree programs in engineering management and in systems engineering at the Air Force Institute of Technology. Otherwise, graduate programs in industrial engineering are not accredited. It appears that this long-standing American tradition of not seeking accreditation for graduate programs in engineering will continue for the foreseeable future.

The predominant areas of emphasis in graduate study and research at most institutions are manufacturing systems engineering, ergonomics engineering, and management systems and controls engineering, although different labels may be applied among the various schools. Most of the institutions listed in Table 3.5 provide graduate programs in industrial engineering.

CONTINUING EDUCATION

Recognizing that the major objectives of industrial engineering are greater productivity concomitant with improved quality, industrial engineering at all levels involves substantial use of computers in a variety of ways. Telecommunications capabilities at most universities have made it possible to share graduate education and continuing education courses among institutions. The largest effort of this type is that of the National Technological University, administered from Colorado State University and involving more than 30 universities and 30,000 students annually. Additionally, most of the more populous states have telecommunication networks to provide graduate and continuing education in engineering.

Lifelong learning is increasingly being recognized as important for personal and professional development, effectiveness, and competitiveness. A good source of information about available seminars, conferences, and courses is the Institute of Industrial Engineers.[36] Many management consulting firms also have substantial industrial engineering training programs which contribute to the continuing education of the industrial engineer.

The Industrial Engineering and Management Press of IIE also publishes many books and other materials which are helpful to industrial engineers who wish to expand their knowledge of the profession. A catalog is available from IIE.

Attendance at and participation in professional and technical society meetings provide additional opportunities for education. In addition to IIE, other relevant societies include the American Society for Quality Control (ASQC), the Human Factors Society, the Society of Manufacturing Engineering, The Institute of Management Society, and the Operations Research Society of America.

Several major corporations have developed intensive in-house training programs for their industrial engineers. Some of the continuing education courses taken by industrial engineers are of interest to many other corporate employees, especially in subjects such as total quality management, systems integration, concurrent engineering, and organizational change.

COUNCIL OF IE ACADEMIC DEPARTMENT HEADS

The Council of IE Academic Department Heads (CIEADH) is comprised of the heads of all of the academic departments which have accredited programs in industrial engineering. CIEADH is affiliated with and supported by IIE and has its major annual meeting concurrently with the spring conference of IIE. CIEADH shares information among its members on a wide variety of issues of mutual interest, including the continual improvement of IE education. Current information regarding CIEADH may be gained from IIE headquarters.[36]

IE DIVISION OF ASEE

The IE Division of ASEE is comprised of members of ASEE interested in industrial engineering education. The IE Division publishes an annual newsletter describing recent development in each of the accredited IE programs and holds its annual meeting concurrently with the ASEE conference in June of each year. To receive current information about the IE Division, inquiries should be made to ASEE headquarters.[37]

IE PROGRAMS IN OTHER COUNTRIES

There are numerous industrial engineering degree programs in many countries around the world. Current exceptions include the USSR, the Peoples Republic of China, and most other socialist countries. There are IIE chapters in 39 countries, all of which have academic programs in industrial engineering. These include Canada, Mexico, Singapore, Hong Kong, Australia, Saudi Arabia, Israel, Indonesia, Holland, and Denmark. A complete list of countries with IIE chapters can be requested from IIE.[36]

LARGEST PROGRAMS IN THE UNITED STATES

Tables 3.8 through 3.11 display listings of the largest industrial engineering programs in terms of number of degrees granted and amount of research support received in 1988–1989 and 1989–1990.

TABLE 3.8 The Twenty-One Largest Producers of Bachelor's Degrees in Industrial Engineering in 1989–1990

Institution	1989–1990	1988–1989
1. Georgia Tech (1)*	218	227
2. Purdue (2)	178	185
3. Penn State (7)	130	110
4. Cornell (10)	124	97
5. Texas A & M (8)	112	105
6. N. C. State (5)	103	116
7. Virginia Tech (6)	102	114
8. Iowa State (3)	101	127
9. Puerto Rico (4)	100	117
10. Michigan (11)	97	90
11. Wisconsin (9)	94	103
12. Ohio State (13)	81	80
13. Tennessee (12)	74	81
14. Illinois (nr)	71	54
15. GMI (15)	67	73
16. SUNY—Buffalo (20)	63	57
17. Florida (nr)	57	52
18. Cal Poly SLO (nr)†	56	28
18. Lehigh (16)†	56	71
20. Miss State (nr)†	53	50
20. Northwestern (14)†	53	79

*( ) 1988–1989 ranking: nr = not ranked.
†Indicates tie.
Source: American Association of Engineering Societies/Engineering Manpower Commission, 1990.

TABLE 3.9 The Twenty-Two Largest Producers of Master's Degrees in Industrial Engineering in 1989–1990

Institution	1989–1990	1988–1989
1. Stanford (1)*	68	139
2. Northeastern (4)	60	58
2. Purdue (11)	60	38
4. Texas A & M (12)	52	37
5. AL—Huntsville (20)	50	27
6. George Washington (6)	48	48
6. SUNY—Buffalo (15)	48	33
8. Arizona State (9)	41	45
8. Georgia Tech (2)	41	77
10. Virginia Tech (5)	39	52
11. Penn State (10)	39	44
12. Tennessee (3)	38	59
13. Cornell (20)	37	27
14. Michigan (8)	33	46
15. Columbia (13)	31	36
16. UC—Berkeley (nr)	30	22
17. Iowa (nr)†	29	12
17. New Mexico State (18)†	29	28
19. Texas (nr)	28	9
20. San Jose State (nr)†	26	12
20. Southern Cal (16)†	26	32
20. Wisconsin (nr)†	26	7

*( ) 1988–1989 ranking; nr = not ranked.
†Indicates tie.
Source: American Association of Engineering Societies/Engineering Manpower Commission, 1990.

TABLE 3.10 The Twenty-One Largest Producers of Doctoral Degrees in Industrial Engineering in 1989–1990

Institution	1989–1990	1988–1989
1. Virginia Tech (2)	16	15
2. Texas A & M (3)	12	12
3. Stanford (1)	11	18
4. Purdue (8)	10	10
5. UC—Berkeley (5)†	9	11
5. Georgia Tech (12)†	9	7
7. Cornell (3)	8	12
8. Michigan (5)†	7	11
8. Ohio State (13)†	7	5
10. Arizona State (13)†	6	5
10. Columbia (nr)†	6	1
10. George Washington (nr)†	6	1
10. Iowa State (11)†	6	8
10. Penn State (5)†	6	11
10. Pittsburgh (nr)†	6	4
16. Massachusetts (nr)	5	0
17. Nebraska (nr)†	4	3
17. Oklahoma State (nr)†	4	2
17. Texas Tech (13)†	4	5
17. West Virginia (nr)†	4	4
17. Wichita State (nr)†	4	0

*( ) 1988–1989 ranking; nr = not ranked.
†Indicates tie.
Source: American Association of Engineering Societies/Engineering Manpower Commission, 1990.

TABLE 3.11 The Twenty Largest Industrial Engineering Research Programs in 1989–1990

Institution	1989–1990	1988–1989
1. Virginia Tech (1)*	7512	5310
2. Texas A & M (2)	3740	4146
3. Georgia Tech (3)	3277	3278
4. Wisconsin (4)	3005	2609
5. Purdue (6)	2687	2153
6. Penn State (10)	2070	1534
7. Illinois (†)	1896	1744
8. Central Florida (13)	1650	1159
9. Michigan (7)	1640	1940
10. Ohio State (12)	1620	1355
11. UC—Berkeley (14)	1438	1023
12. Cornell (11)	1433	1493
13. Stanford (8)	1420	1837
14. Toledo (nr)	1286	196
15. N. C. State (15)	1006	913
16. Lehigh (9)	800	1542
17. Oklahoma State (nr)	777	626
18. West Virginia (19)	743	731
19. Texas Tech (20)	691	714
20. Massachusetts (nr)	657†	657

*( ) 1988–1989 ranking; nr = not ranked.
†Amounts in $000. 1989 data used by American Society for Engineering Education for 1990.
Source: American Society for Engineering Education, March 1991.

REFERENCES

1. Turner, Wayne C., Joe H. Mize, and Kenneth E. Case, *Introduction to Industrial and Systems Engineering,* Prentice-Hall, Englewood Cliffs, N.J., 1978, p. 22.

2. Emerson, Howard P., and Douglas C. E. Naehring, *Origins of Industrial Engineering: The Early Years of a Profession,* Institute of Industrial Engineers, Atlanta, Ga., 1988. (Chapter 5, "Education in Industrial Engineering," pp. 43–62.)

3. *Journal of Industrial Engineering,* American Institute of Industrial Engineers, New York, vol. 13, no. 5, 1962.

4. Going, Charles, *Principles of Industrial Engineering,* McGraw-Hill, New York, 1911.

5. Diemer, Hugo, *Factory Organization and Administration,* McGraw-Hill, New York, 1910.

6. *Proceedings of the Twentieth Annual Meeting, 1912,* Society for the Promotion of Engineering Education.

7. Gilbreth, Frank B., "The Training Required for Engineers, Discussion," *Engineering Education,* February 1919, pp. 238–239, Society for the Promotion of Engineering Education.

8. Fish, J. C. L., *Engineering Economics,* McGraw-Hill, New York, 1915.

9. Grant, Eugene L., *Principles of Engineering Economy,* The Ronald Press, New York, 1930.

10. Thuesen, H. G., *Engineering Economy,* Prentice-Hall, Englewood Cliffs, N.J., 1950.

11. Fry, Thornton C., *Probability and Its Engineering Uses,* Van Nostrand, 1928, 476 pp.

12. Rautenstrauch, Walter, "A Comparison of Lathe Head-Stock Characteristics," *American Society of Mechanical Engineers Journal,* vol. 32, 1910, American Society of Mechanical Engineers, New York.

13. Apple, James M., *Plant Layout and Materials Handling,* The Ronald Press, 2d ed, New York, 1963.

14. Barnes, Ralph M., *Work Sampling,* Wiley, New York, 1956.

15. Grant, Eugene L., *Statistical Quality Control,* McGraw-Hill, New York, 1946.

16. Bellman, Richard E., *Dynamic Programming,* Princeton University Press, Princeton, N.J., 1957, 342 pp.

17. Charnes, A., W. W. Cooper, and A. Henderson, *An Introduction to Linear Programming,* Wiley, New York, 1953, 74 pp.

18. Churchman, C. W., R. A. Ackoff, and E. L. Arnoff, *Introduction to Operations Research,* Wiley, New York, 1957, 645 pp.

19. Dantzig, George, *Linear Programming and Extensions,* Princeton University Press, Princeton, N.J., 1963, 632 pp.

20. Feller, William, *Introduction to Probability and Its Applications,* Wiley, 3d ed., vol. 1, New York, 1968.

21. Magee, John F., *Production Planning and Inventory Control,* McGraw-Hill, New York, 1958, 333 pp.

22. Morse, Philip M., and G. E. Kimball, *Methods of Operations Research,* Wiley, New York, 1951, 158 pp.

23. Smalley, Harold E., and J. R. Freeman, *Hospital Industrial Engineering,* Reinhold, New York, 1966, 460 pp.

24. Grayson, L. E., and C. J. Tompkins, *Management of Public Sector and Nonprofit Organizations,* Prentice-Hall, Englewood Clifs, N.J., 1984, 376 pp.

25. Byrd, Jack, Jr., *Operations Research Models for Public Administration,* D.C. Heath, Lexington, Mass., 1975, 277 pp.

26. Gilbreth, Frank, *Motion Study,* Van Nostrand, New York, 1911.

27. McCormick, Ernest J., *Human Factors Engineering,* McGraw-Hill, New York, 1957, 491 pp.

28. Murrell, K. F. Hywel, *Human Performance in Industry,* Reinhold, New York, 1965, 496 pp.

29. Moder, Joseph J., and C. R. Phillips, *Project Management with CPM and PERT*, Reinhold, New York, 1970.

30. Morris, William T., *Implementation Strategies for Industrial Engineers*, Grid Publishing, Columbus, Ohio, 1979, 252 pp.

31. Timms, Howard L., *Introduction to Operations Management*, Richard D. Irwin, Homewood, Ill., 1967, 159 pp.

32. Greene, James H., *Operations Planning and Control*, Richard D. Irwin, Homewood, Ill., 1967, 175 pp.

33. Nadler, Gerald, *Work Systems Design: The Ideals Concept*, Richard D. Irwin, Homewood, Ill., 1967, 183 pp.

34. Barnard, Chester I., *The Functions of the Executive*, Harvard University Press, Cambridge, Mass., 1938, 334 pp.

35. Torgersen, Paul E., and I. T. Weinstock, *Management: An Integrated Approach*, Prentice-Hall, Englewood Cliffs, N.J., 1972, 498 pp.

36. Institute of Industrial Engineers, 25 Technology Park, Norcross, Ga. 30092.

37. American Society for Engineering Education, Eleven DuPont Circle, Suite 200, Washington, D.C. 20036.

INDUSTRIAL ENGINEERING IN PRACTICE

CHAPTER 1
ORGANIZING FOR INDUSTRIAL ENGINEERING

Clifford N. Sellie
C.E.O., Standards International Inc.
Chicago, Illinois

Paul W. Krueger
Plant Manager, Bush Hog Division of
Allied Products Corporation
Selma, Alabama

Industrial engineering's role in business is expanding rapidly today. With the changes in favored motivational theories moving from classical to human resource to even more participative management styles, the skills of the industrial engineer have been in great demand. Increased competition, the broader specter of a world economy, changes in demographics, technological advances, and the higher educational levels of the workplace have all combined to create broad sweeping changes in organization strategies in the last decade. The industrial engineering focus has always been on the controlled management of change with the purpose of creating a continuously improving environment.

Organizing for industrial engineering should not violate the accepted principles of organizing any company function, but because of its open multivariate role in the organization much more care is required in the definition of its roles within the overall organizational goals and strategies. This chapter is primarily concerned with the concepts of organization for industrial engineering, but any reader interested in organization theory as such is referred to the many excellent texts on the subject in the Bibliography.

The organization of an industrial engineering department demands a minimum of the following four steps:

1. The industrial engineering department must be aligned with the overall company goals and strategies. The industrial engineering department is an integral *part* of the overall business strategy. It uniquely serves two roles in this process: implementing company business strategy and providing feedback for formulating the business strategy.

2. Because the industrial engineering department often overlaps the responsibilities of other functions and departments, a clear understanding of how these responsibilities will be integrated in the overall organization is required.

3. To implement the departmental goals and strategies effectively, the authority, responsibility, and accountability of this department must be clearly defined.

4. Personnel skills should be matched with the departmental organizational requirements

and strategies. Team attitude is usually more important than genius or the physical structure of an organizational chart. This may mean the choice of modifying the organizational structure to match personnel or forcing personnel to adapt to the requirements of the organization structure. Which is chosen may depend upon whether you are recruiting new personnel or utilizing team members.

With this introduction to the general problem, the subject can now be examined more closely. To that end, this chapter is divided into the following main sections:

1. Role of industrial engineering in business
2. Place of industrial engineering department in company organization structure
3. Internal organization of the industrial engineering department
4. Special forms of organization of the industrial engineering department
5. Administration of the industrial engineering department
6. Industrial engineering department personnel
7. Getting action on recommended programs
8. Conclusion

THE ROLE OF INDUSTRIAL ENGINEERING IN BUSINESS

—the business of change—

A glance at the table of contents of this handbook will show the variety of activities an industrial engineering department can perform. Rarely, however, does a specific department do more than a part of those listed. The scope of its activities in any company is a deciding factor in determining where to fit the industrial engineering department in the company plan or organization and how to organize it internally. Equally important considerations are the nature of the services provided and the organization plan of the company.

Traditional.   Traditionally the basic objective of an industrial engineering department is usually twofold: (1) to establish methods for controlling production costs, and (2) to develop programs for reducing those costs. Both methods and programs are carried out by line management.

In most cases, the industrial engineering department exists primarily to provide specialized services to the production division. The services may be few or many and normally include such functions as work standards, methods and processing studies, and the development and maintenance of measured daywork and wage incentive programs. The industrial engineering department is ordinarily held fully responsible for the applicability and accuracy of the programs developed and recommended.

Other divisions may use its services infrequently or extensively. Should their demands consistently be at a high level, it is often advisable to set up service departments in the other divisions so that industrial engineering can concentrate on its primary responsibility to the production division. If this step is taken, formal organizational relationships between the service departments are seldom necessary because they will have very little in common beyond a uniform approach to problem solving. When industrial engineering is occasionally asked to take on work in addition to its continuing services, it customarily acts only as an expert adviser. Responsibility for the actions taken rests jointly with the department requesting assistance and the industrial engineering department.

In some cases, the industrial engineering department is assigned to head projects that will probably never recur. In other cases, it may participate in a less important role. The reputation of the department will undoubtedly have an important influence on the scope of its responsibilities in these studies. Whoever is in charge, however, should define the role responsibility of every participant. Industrial engineering can then execute its task and assume its defined share of the total task.

Operation Research.   With the advent of the computer age came the ability to manipulate large amounts of data and numbers realistically. Mathematical modeling became an effective tool in forecasting and problem solving. For a short period there appeared to be two fields of industrial engineering, and many traditionalists feared an overobsession with quantification as a simplistic answer to business problems. Both approaches have tended to mold together in recent years, resulting in a more powerful arsenal of tools from which the industrial engineer can predict and predetermine the outcome of alternative methods of problem solving.

Modern Approach.   In the more competitive world market that exists today, there is a greater awareness of the need to integrate the functions of an organization toward the needs of the customer. To achieve this end, manufacturing, design, and distribution functions have become more integrated and driven by marketplace demands and business market strategies. More and more companies in mature industries seek to differentiate their products through superior customer service. This requires better products at a lower cost available in a time frame dictated by the consumer. New product introductions must be accomplished in shorter time frames and complement a simplified shorter-cycle manufacturing strategy.

Just-In-Time (JIT), focus factories, flexible manufacturing cells, MRP II, C.I.M., CAD, CAM, and SPC are just a few of today's manufacturing techniques that reduce the polarity between business functions. As a result, the role of the industrial engineer has broadened.

The basic objectives of the industrial engineering department of establishing the production methods and reducing costs remain the same. The strategies for accomplishing these goals have expanded. The manufacturing process is integrated at the product design stage. The overall investment factors are integrated into the material handling and inventory strategies in the manufacturing process. The marketplace demands cannot be changed, and therefore the manufacturing process must be altered to accommodate the needs of the customer. The services of industrial engineering as a result are more generally utilized by other business functions in addition to direct manufacturing.

The Industrial Engineering Approach.   In carrying out its functions, the industrial engineering department uses what is sometimes referred to as the scientific approach. In other words, industrial engineers gather and analyze facts, form tentative conclusions, compare and test the alternatives, and finally reach and present their findings, conclusions, and recommendations.

To ensure that the actions it recommends are beneficial to the company as a whole and have a good chance for adoption, the industrial engineering department must operate with as much objectivity as possible. The suggestions here illustrate how that objectivity may be put into practice:

1. In approaching a problem, the industrial engineering department must listen to and evaluate objectively the viewpoints of all the departments affected. In making recommendations, it should support its elected course of action by sound reasons proving that the proposal submitted offers the best possible solution.

2. The industrial engineering department must be prepared to meet prejudiced points of view and to treat them understandably but firmly. If the department has succeeded in gaining the confidence of line management, it should be able to explain why the prejudice is ill-founded. It should never, of course, ridicule or expose the prejudiced person.

3. Although the industrial engineering department must never arbitrarily disregard the opinion of line management, it should not lose sight of the fact that its first concern is to strengthen the overall operations of the company.

The Department Must Weigh Companywide Effects of Its Recommendations. Beyond the specific and obvious responsibility of the industrial engineering department to tie in its work in the production function with other business functions is its responsibility to recognize how the effects of its recommendations may reach beyond the area studied. The operating functions are concerned with regular planning and execution of agreed-upon company programs; each function plays a well-understood part. The industrial engineering department, however, is for the most part an advocate of change. When it recommends changes in the production function, it must anticipate the probable consequences of those changes for other business functions. This point is emphasized for two reasons:

1. The action recommended in a production department may open up improvement possibilities in other business functions. Sometimes, of course, these secondary improvement possibilities turn out to be unimportant. Often, however, they can lead to dramatic changes that not only benefit the nonproduction department but reinforce the action taken in the production department.

2. Occasionally the value of an apparent improvement in the production departments may be more than offset by the problems and additional costs it creates in another department. Sometimes the bad effects can be erased by modifying the recommendations; other times the recommendations should not be made.

PLACE OF INDUSTRIAL ENGINEERING DEPARTMENT IN COMPANY ORGANIZATION STRUCTURE

With an idea of the nature of the industrial engineering department, its responsibilities, and its working relationships within a company, where it belongs in the organization structure can now be considered. One guideline might be mentioned here, apparent though it may seem: the industrial engineering department should report to the executive that has line responsibility for the departments it regularly serves. Thus a vice president for operations who is responsible for coordinating research, engineering, production planning, sales, quality control, personnel, and so forth, should probably have industrial engineering, too. If, however, there is a works manager to whom production activities report, the industrial engineering department also should report to him (see Fig. 1.1).

This guideline admits many interpretations; for, practically speaking, there can be no hard-and-fast rule governing the place of industrial engineering in the organization structure of all companies and plants. To learn what forms have been effective, a number of

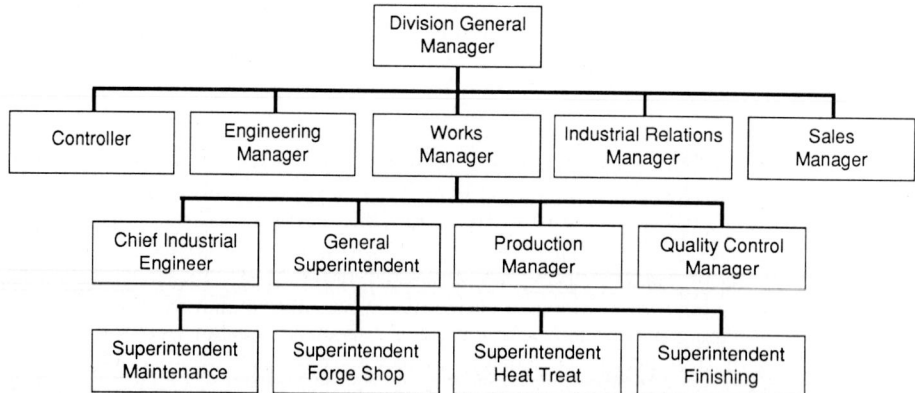

FIGURE 1.1 Organization chart illustrating a typical location for the chief industrial engineer in a division of approximately 1000 employees.

companies were reviewed on this question: "To what executive does the industrial engineering department report, and what other executives are on the same level as that executive?" At the same time, an explanation was sought of any line or functional direction the industrial engineering department received from any other executive, from a headquarters industrial engineering group, and the like.

Although the review showed no single, typical position in the organization structure for the industrial engineering department, it did point up the factors that influence its position. These are (1) the scope of industrial engineering activities, (2) the number of direct labor employees, and (3) the complexity of manufacturing operations.

Scope of Industrial Engineering Activities.   The activities of the industrial engineering department influence its place in the organization structure to this extent: it reports to the executive that is responsible for the departments in which it does the bulk of its work. In most cases, however, it increasingly provides services to other departments.

Number of Direct Labor Employees.   The formal recognition of industrial engineering as a full-time activity grows as the number of employees grows; at the same time, however, its organizational rank seems to become lower.

In small companies, it was indicated that the top manufacturing person did all the industrial engineering, in addition to other duties, and that no formal department existed.

In companies of 300 to 500 employees, the people performing industrial engineering activities reported variously to executive vice presidents and production or operations vice presidents. In these cases, the chief industrial engineer was on the same organizational level as the superintendent and chief product engineer (but not necessarily considered as important).

In companies with 600 and over hourly paid employees, industrial engineering was formally recognized as a department and occupied a lower place in the organization plan. In these companies, the chief manufacturing executive did not have authority over engineering, quality control, and other activities, which are often grouped under one executive in smaller companies, and the industrial engineering department reported to a factory manager, works manager, or superintendent.

Complexity of Manufacturing Operations.   The review showed that as manufacturing operations grew more complex, the size of the industrial engineering department increased, but the scope of its activities decreased, with more attention going to strictly production problems. The increased complexity of manufacturing operations and the corresponding growth of the department usually meant that industrial engineering reported at lower organizational echelons than it did in companies with less complex production operations. In companies electing the more modern style of participative management, the reporting level occurs at a higher echelon than in those companies utilizing the more traditional management style. This phenomenon appears to result from the expanded role of industrial engineering in the participative environment.

INTERNAL ORGANIZATION OF THE INDUSTRIAL ENGINEERING DEPARTMENT

Grouping of Related Functions.   When related functions are grouped, the industrial engineering department is divided into sections, each responsible for related functions. For example, work relating to work standards may be in one section, methods work in another, and so on. For the most part, the members of a section work regularly on all the activities in the section. Although they normally do not work on activities in another section, they may be temporarily transferred on special assignments.

A typical plan of organization for an industrial engineering department, internally organized by the grouping of related functions, is shown in Fig. 1.2.

As an organization device, the grouping of related functions offers these advantages:

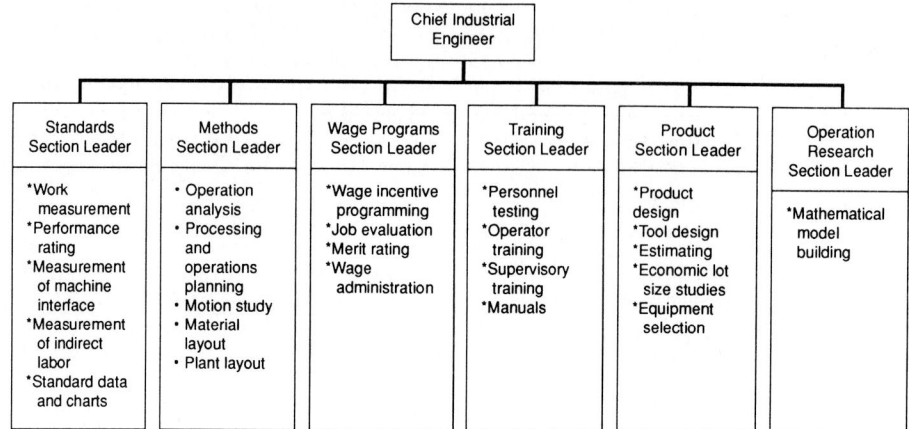

FIGURE 1.2   Industrial engineering department organization by grouping of related functions.

1. Because department engineers specialize within a narrow field, they rapidly develop high technical proficiency.
2. By limiting the range of its engineers' activities, the department can operate successfully with personnel of more limited experience.
3. Engineers can be easily assigned to tasks that best suit their talents and interests.
4. Line supervisors, assigned to the department for special training before advancement, can learn much about industrial engineering in a short time.

The disadvantages of such an organization are:

1. Because the engineers become specialists within a narrow field and work with all departments, they are less likely to develop a close working relationship with line supervisors in any single department.
2. Specialists are less likely to be able to relate the effect of their work to the overall operation of the department in which they are working.
3. It is more difficult to transfer a person with specialized experience from one section of the industrial engineering department to another or from the department into a line supervisory position.
4. If there are insufficient specialists in one section of the industrial engineering department, the solution of high-priority problems in some operating departments may be delayed until they become available.

Organization Paralleling That of Production Division.   The other common pattern for the internal organization of the industrial engineering department is to parallel the organization of the production division. In a plant that has, for example, a heavy machine department, a light machine department, and an assembly department, industrial engineering may be organized in sections to provide all or most all the industrial engineering services to each (see Fig. 1.3). The engineers in each section work with flexibility on all the problems of their allied production department. They are also transferred much more freely among sections than are their counterparts in departments organized by a grouping of related functions.

It is interesting to note that all companies with this type of organization usually had one ''floating'' section that was not tied to any specific production department. It was ordinarily concerned with such activities as supervisory training, job evaluation, and merit rat-

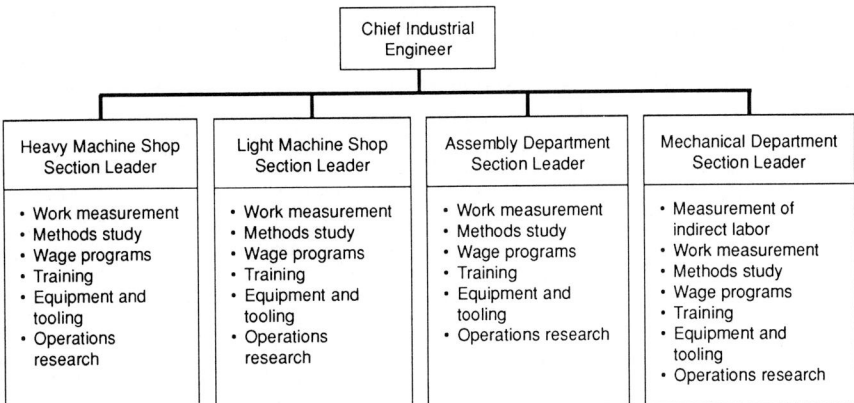

FIGURE 1.3   Industrial engineering organization paralleling that of production division.

ing. The engineers working on those activities did so in all production departments served by the industrial engineering department.

Several advantages are ascribed to an organization patterned along production division lines:

1. This arrangement ensures that the highest-priority problem in each production department gets earliest attention. Rarely are all engineers in a section engaged on a special project.
2. An easy and friendly working relationship usually develops earlier between production department supervisors and the industrial engineers.
3. The supervisors gain a good perspective of the production department's operations.
4. This form of organization provides excellent specialized training for individuals being groomed as line supervisors of the department served.
5. When people are transferred between sections of the industrial engineering department, they bring a fresh viewpoint that facilitates problem solving.
6. People working on all aspects of a department's problem receive very broad training and a good picture of how industrial engineering services compete and complement the production department. This enlarges their growth opportunities and promotes a better industrial engineering department.

Some of the disadvantages of this form of organization are:

1. The department must have available personnel with broader interest and greater talent than those required under a system which permits individuals to specialize in a single industrial engineering technique.
2. An individual's proficiency in any one field is not likely to be obtained so quickly or in so high a degree as that of a specialist who works with only one kind of problem.

Focus Factory/JIT Environments.   This currently more popular approach to manufacturing emphasizes family grouping of products and is therefore organized similar to paralleling of the production division, discussed previously. However, several important differences occur in this environment. The most important is that the scope of the process is expanded to include all of the production phases from the fabrication of raw materials to the final assembly process. A greater emphasis is placed on integrating manufacturing and product design, material scheduling and planning techniques, and an understanding of the marketing requirements. The industrial engineer operating in this environment must necessarily be more broad-based in skills than in the other two organization modes we have discussed. Second, there is less distinction between the disciplines of engineering support. The design engineering, manufacturing engineering, tool engineering, industrial engineering, material handling engineering, and quality engineering functions may all be a part of a support team reporting to the focus factory manager and may be generically referred to as simply engineering. See. Fig. 1.4.

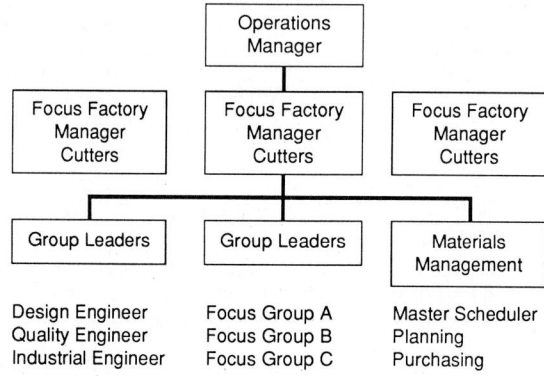

FIGURE 1.4 Organizational chart illustrating industrial engineering in a focus factory environment.

Some of the advantages for this type of organizational structure are:

1. This arrangement ensures the most immediate attention to problems within the focus factory.
2. A team environment between all functions of the business is developed.
3. All members of the team have a broader understanding of the business.
4. Communication lines are shorter and more concise.
5. Several viewpoints to a problem are available.
6. Exchanges of personnel between functions and promotions are less disruptive to the business segment.

Some of the disadvantages of this form of organization are:

1. The personnel required to function in this type of environment must have broad interests and greater skills.
2. It is difficult to create the culture to enhance this environment, especially where another culture currently exists.

Organization on a Project Basis.   Regardless of how the industrial engineering department is organized, all companies assemble special teams from time to time to conduct certain projects. Some companies regularly maintain a floating section for just that purpose. In most cases, however, the special project team includes engineers temporarily recruited from the various sections of the industrial engineering department. Project teams may be formed for an assignment in the production division—for example, to overhaul the wage incentive plan—or for special work in another division not ordinarily served by the department.

Factors Governing Choice of Organization Pattern.   Why a company chooses one form of industrial engineering department organization in preference to the other is due to some of these considerations:

1. *Type of industry.* Companies tend to organize based on patterns, established within their type of industry, that have been found to support their business objectives and process. Four basic categories of industry seem to prevail; fabrication assembly (machine shop and electromechanical assembly), process (steel, paper, and chemical), conversion (textile, papercoating, and printing), and service (banks, hospitals, and airlines). The underlying reason for similarity may often be the product of transferred style among companies exchanging employees via the recruitment process. Care needs to be taken to ensure that the style of organization complements and supports the business goals and strategies of the company.

2. *Size.* Small companies (50 to 300 direct labor) commonly operate with an industrial engineering department of one or two engineers. The industrial engineering activity is typically organized as shown in Fig. 1.5. The senior industrial engineer should have a wide variety of capabilities because the functions of industrial engineering that need to be performed in a small company are limited not in number but only in degree. Typically, the emphasis will be on work standards, methods, and cost reduction. Plant layout, estimating, processing, material handling, and the like will receive less emphasis. One alternative is the focus factory concept shown in Fig. 1.4.

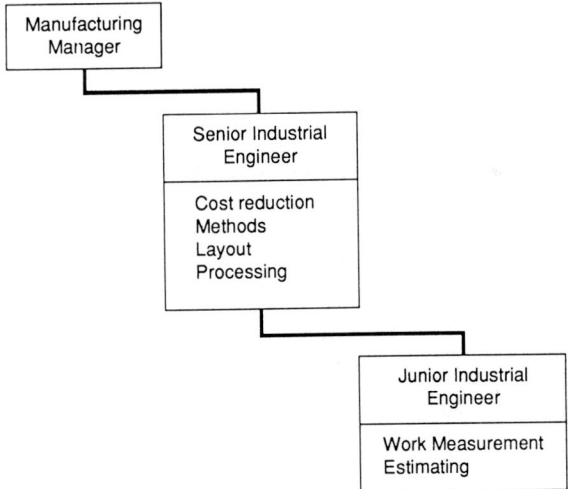

FIGURE 1.5   Typical organization for small companies.

Medium-sized companies (300 to 600 direct labor) generally take the form of individual sections for the various facets of industrial engineering. The need for sophistication in controlling flow through the more numerous and larger manufacturing departments, the need for greater control of labor hours, and the need for better methods, equipment, and training all tend to create a demand for the industrial engineer who is an expert in one of the functions of this field. Hence the organization takes on the appearance of Fig. 1.2. Functions are well defined, and the title "industrial engineer" generally used in the small company is of less prominence. Such titles as work measurement engineer, methods analysts, process engineer, and material handling engineer are used to indicate the division of responsibility for industrial engineering service.

An alternative is to break the organization into a product family orientation and utilize the focus factory concept illustrated in Fig. 1.4.

The internal organization of a large company (over 600 direct labor) often assumes the form of Fig. 1.2. In the large company, increased emphasis is placed on such functions as value engineering, operations research, training, and wage programs, none of which could feasibly be done in depth under the paralleling of production method. The necessity for increasing the depth, intellectually and in numbers, and yet retaining control of the various activities, is still another reason for using the method of grouping by related functions. The focus factory concept or reducing a continuous process to focus groups may be the most advantageous.

3. *Complexity*. Complexity of production operations may make it inadvisable to move an industrial engineer from one operation to another. In such a case, it is best to have an industrial engineering organization which parallels that of the production division so that an engineer need learn only one phase of the total process in detail.

Where the industrial engineering department is the major service group in the company and regularly works in other divisions besides production, it is usually organized by a grouping of related functions so that specialists will be available for assignment anywhere in the company.

Multiplant companies with a home office industrial engineering group organized on a functional basis (that is, by grouping of related activities) tend to organize the industrial engineering departments of the plants in a similar way. This is especially apparent where the plants produce different products, the reason being that it enables headquarters to coordinate the plants effectively despite the dissimilarities in production. In multiplant companies with highly autonomous plants, the pattern is the reverse. There industrial engineering is organized to parallel the department organization of production divisions.

If there are natural subdivisions of the business such as fabricating, machinery, and welding, the industrial engineering organization should follow those lines. If the business is highly integrated, however, with activities in one area having a substantial impact on activities in another, a centralized department will ensure that the business is treated as a single unit. Another variation of this centralization theme is the focus factory approach.

4. *Centralization vs. decentralization*. The choice as to whether the industrial engineering organization should be centralized or decentralized depends on several factors. A large company may find it necessary to decentralize its industrial engineering to promote a more detailed understanding of operations by the industrial engineer and to improve relations with operating personnel. However, when profits are low, a company may not be able to afford a large decentralized industrial engineering staff but must content itself with a small centralized staff which emphasizes profit improvement and cost reduction projects.

If the industrial engineering department is in its early growth stages, the training of large numbers of new personnel can often best be carried out with a centralized form of organization. However, when labor costs represent a large proportion of sales price, the profit improvement potential of industrial engineering is highest. A large decentralized staff with a centralized core of specialists will often assure that profit improvement opportunities in all manufacturing areas are given proper attention.

5. *The corporate level*. Perhaps the most important function of the corporate industrial engineering group is to coordinate the efforts of the industrial engineering groups in all the corporation's subsidiaries, divisions, or plants. The corporate group should also be responsible for initiating special projects and programs as required. It should be in a position to report objectively to corporate management on conditions and progress of industrial engineering activities in the suborganizations.

These functions can be accomplished by organizing the corporate group into three basic subgroups, with the corporate manager of industrial engineering reporting to the

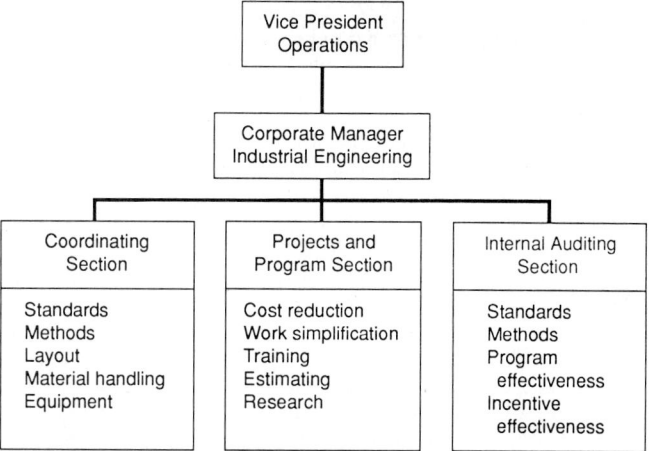

FIGURE 1.6 Organization at corporate level.

vice president of operations, as shown in Fig. 1.6. The three basic subgroups are designated as (1) the coordinating section, (2) the projects and programs sections, and (3) the internal auditing section. The size of the corporate group will depend upon the corporation size, the number of outlying plants, financial return on industrial engineering effort, and the like. The basic functions remain, however, no matter what the size.

With a corporation industrial engineering group in place (Fig. 1.7), the industrial engineering groups in the subsidiaries, divisions, or plants are often organized to parallel the production division organization, as shown in Fig. 1.3.

6. *Management style.* Perhaps one of the most critical but most often overlooked factors in selecting a proper industrial engineering department organization structure is making sure it matches the overall management style of the business. The focus factory approach illustrated in Fig. 1.4 is designed for a participative style of management and will not work well in an autocratic or functional environment. The structures in Fig. 1.2 and 1.3 are more suited for functional or autocratic management styles. Again it is important to emphasize the importance of selecting the organizational structure on the basis of how it serves the business objectives and strategies.

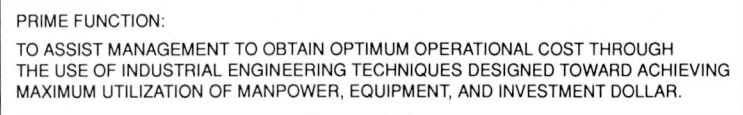

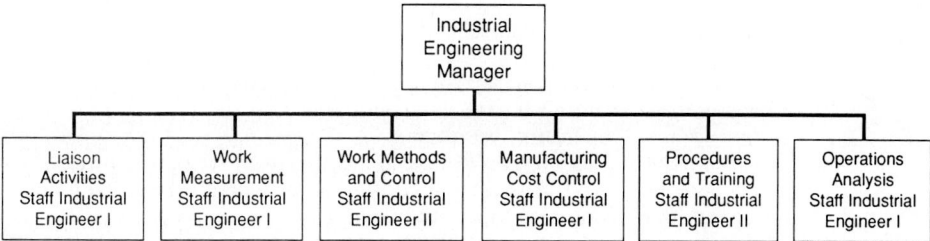

FIGURE 1.7 Industrial engineering organization—corporate level. (*Courtesy of Navistar Corp.*)

SPECIAL FORMS OF ORGANIZATION OF THE INDUSTRIAL ENGINEERING DEPARTMENT

The organization form chosen by any company will be governed partly by the philosophies of its management group, partly by the nature of the work it performs, and partly by the size of the company. An examination of some special forms of organization will show this clearly.

Banks. Figure 1.8 shows how a bank has organized its industrial engineering function into three divisions: technical planning, methods research, and operations analysis.

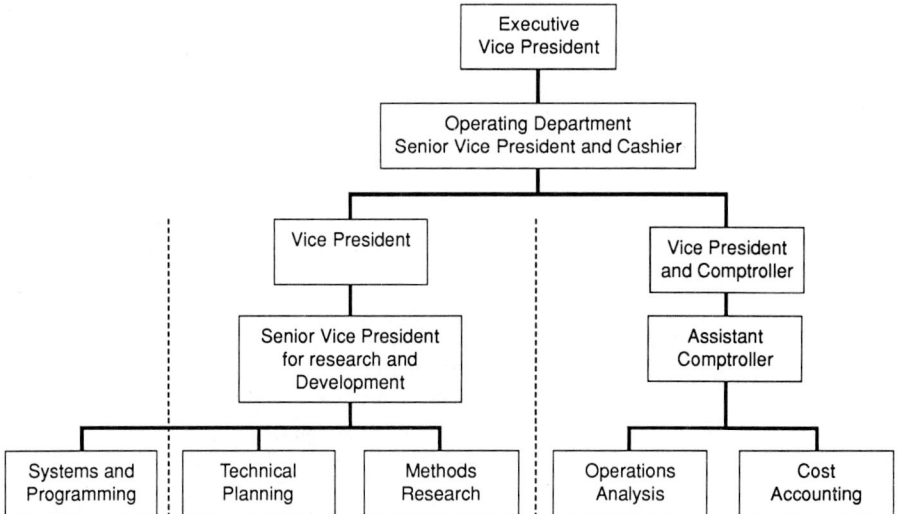

FIGURE 1.8 Organization of the industrial engineering function in a bank. (*Courtesy of Alfred E. Clem, Public Affairs Officer, Continental Illinois National Bank and Trust Company of Chicago.*)

All three divisions are in the operating department, which is headed by a senior vice president and cashier, who reports to an executive vice president. The operating department has the largest number of employees within the bank and is responsible for the organization and execution of the bank's many services to customers, employees, and the general public. All physical facilities as well as supporting services are within the operating department's responsibilities.

Two divisions—technical planning and methods research—report to a second vice president, who in turn reports to a vice president in the operating department.

The third group—operating analysis—is part of the accounting division, reporting through an assistant comptroller to the vice president and comptroller, and thus to the senior vice president and cashier who heads the operating department.

The technical planning division has two primary functions: quantitative studies and computer hardware evaluation. "Quantitative studies" is the bank's term for what many other firms call operations research.

The methods research division also has two main jobs: evaluation of equipment (other than computers) and management consulting, including the determination of costs associated with the operations under study.

As part of the accounting function, the operations analysis group examines the bank's manual systems, performs time study analyses, makes work simplification recommendations, conducts division-level operation analyses, and provides as a by-product time standards which are used to establish standard costs.

Corporate and Plant Combination.   The establishment of industrial engineering departments at both corporate and plant levels is quite common in large corporations. Figures 1.7 and 1.9 show how the industrial engineering organizations at the corporate and plant levels vary in design and primary functions in one large plant. At the corporate level, the industrial engineering organization is designed on a functional basis, while the plant organization is formulated according to specific production activities.

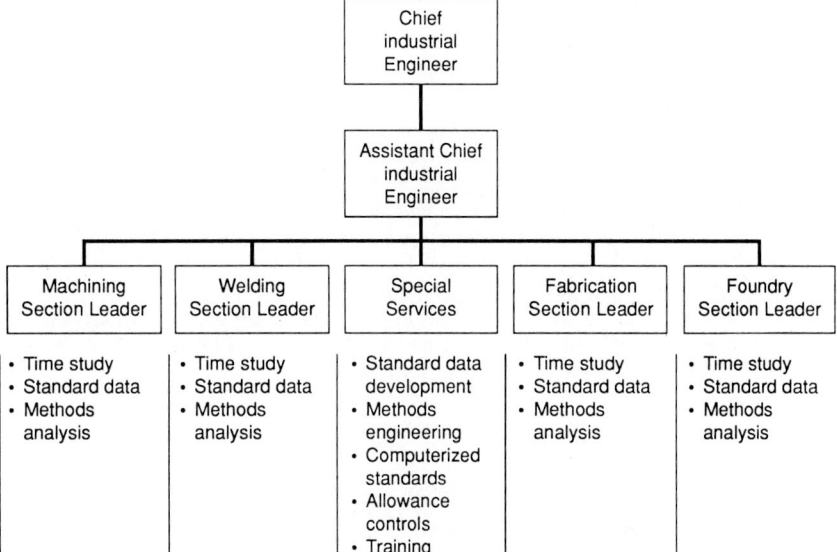

FIGURE 1.9   Industrial engineering organization—plant level. (*Courtesy of Navistar Corp.*)

The prime purpose of the corporate industrial engineering section is to provide consulting services for the divisions and plants on all phases of industrial engineering and manufacturing analysis. It is designed to promote and develop the use of new industrial engineering techniques in the manufacturing areas of industrial operations.

The corporate industrial engineering section provides corporate and plant management with feasibility studies designed to reveal potential cost reductions through the utilization of various management science techniques, including computer applications. It further provides training and guidance in such areas as work measurement, work methods, manufacturing costs, and operations analysis activities. In labor relations, it develops and recommends labor contract language and renders counsel on grievances. These functions are divided into six major areas of industrial engineering activities, with a staff industrial engineer responsible for each.

The major function of the industrial engineering department at the plant level is to establish measured daywork or incentive production standards by means of stopwatch time studies or the application of predetermined time standards.

Other areas of activity include the development of new standard data, methods analysis studies, manufacturing cost studies, and the administration and control of direct labor allowances.

Training courses are conducted periodically by designated instructors in work measurement, methods engineering, labor allowance control, and application of standard data. Performance rating sessions are also conducted on a regular basis. Performance rating sessions are essential for stopwatch time study people to assure that the time study standards are consistently fair. Performance rating sessions are helpful for all staff and line supervisors, as a basis for evaluating productivity, regardless of the standards setting techniques.

The section is organized by specific production activities within the manufacturing plant, such as machining, welding, fabrication, and foundry. Each production activity is supervised by a section head who is responsible for all industrial engineering activities within an assigned department.

ADMINISTRATION OF THE INDUSTRIAL ENGINEERING DEPARTMENT

Goals and Strategies in Perspective. Industrial engineering is the business of change and continual improvement. Successful administration of the industrial engineering department is perhaps more difficult to achieve than in the case of a line department. One reason is that it is less comfortable to measure some of the nonquantitative functions of the department and administer standards of performance and to measure attainment. That inherent difficulty may, however, be turned to the advantage of the department. Because department members receive less direction and control than operating personnel, they must be self-starters for work to get done. Their pride in doing a job well and quickly is a great asset to successful administration.

Foremost in the perspective of the department must be its understanding of the role it plays in ensuring the success of the company business objectives. The industrial engineering department exists to support, enhance, and in some cases help to formulate the company goals and objectives, not vice versa.

The Dilemma of Limited Resources. As with all functions of a business entity, the industrial engineering function must compete for and operate with limited resources. While all industrial engineering departments use their special techniques and expertise to measure the effective administration of other business functions, a successful industrial engineering department also utilizes these same industrial engineering techniques to measure itself.

The industrial engineering department cost is an investment that must have a satisfactory return to be justified. To obtain this return, care must be taken to ensure that the resources of the industrial engineering department are invested where they will achieve the greatest return on that investment.

As an example, if the industrial engineering department were charged with the responsibility of reducing cost in a given production department, it might approach that task by annualizing the added value of labor or material cost of each task or group of similar tasks within the department. These tasks could then be arranged from that task, resulting in the highest annualized cost to that task having the least. This arrangement is commonly referred to as a Pareto distribution. Almost universally the top 20 percent of the tasks listed will result in accounting for 80 percent or more of the department's annual cost. These tasks will then yield the highest probably cost reduction per given unit of the industrial engineering department's resource.

Similarly an evaluation can be made to determine the most cost-effective method of work measurement technique to use. Would you use a detailed predetermined time system to measure the process to build one unique unit of production which was never to be repeated? Priorities for the industrial engineering department should be made and recommended on the basis of the return on investment analysis and the requirements of the business strategy.

Credibility. The entire success of the industrial engineering department depends upon the credibility it has with the personnel in the business functions it serves. One of the most important actions the industrial engineering department can take to enhance its credibility is to involve and solicit comment from those who are affected by even the smallest of the department's activities or projects. The industrial engineer should begin even a work measurement study by explaining thoroughly to both the operator and the supervisor the purpose of and the procedure to be used in the study. They should both be invited to participate in the development of the standard method to be used, and at the conclusion of the study both should be allowed to review the results before their finalization.

Keeping projects on schedule and on budget are two additional aspects of credibility. Often there is a conflict between the required time frame of a project and the resource requirements for achieving a given level of quality or accuracy. The industrial engineer should carefully determine from the initiator of the project which is more important. If the time frame is of the greatest concern, alternative approaches to the project and their consequences can be discussed and understood at the front end of the project.

The measure of the industrial engineering department accomplishments should be conservative. Integrity and believability are the essential ingredients to the marketing of the industrial engineering department's credibility.

INDUSTRIAL ENGINEERING DEPARTMENT PERSONNEL

The same principles of personnel administration that apply to other line and functional departments apply likewise to industrial engineering. The reader is undoubtedly familiar with them but may not realize the special personnel considerations that set industrial engineering a little apart from the general run of departments. The major distinctions are discussed below.

Industrial Engineering as a Career.   Often the industrial engineering department is staffed largely by recent college graduates whose major qualification is educational background. They may lack the stature and maturity to influence action. Maturity and firsthand experience are beneficial in selling incentive or daywork standards, in gaining the confidence of line supervisors, and in persuading executives to act. It therefore appears desirable that the well-staffed department have enough thoroughly experienced senior engineers to balance the younger group.

Training.   An industrial engineering department can provide excellent training for people who will eventually assume line responsibility, but too often nothing is done to ensure that the training is complete and effective. This is as shortsighted as turning workers loose in a research laboratory and hoping that, with all the experiments going on, they will learn a lot. The experience may be interesting but not productive.

Training in industrial engineering should be formalized. The first obvious step is a good orientation program in which new workers learn what the department's mission is. If they are not familiar with operations in the departments which industrial engineering serves, they need that orientation too. Then they must get specific training in the following areas:

1. Methods and procedures used by the industrial engineering department
2. The formal and informal organization structures of the company, the production division, and the industrial engineering department
3. The technique of operating in a service atmosphere and dealing with workers, supervisors, management, and union officials
4. Pertinent technical phases of production operations
5. Company policies on work standards, wage administration, areas of participation by the union, craft observance, and the like

A new employee should also have the regular company indoctrination.

Conduct of Department Members Important to Their Success.   Industrial engineering can materially enhance its standing in the company by the conduct of its staff. Some hints on promoting good relationships with line personnel are:

Recognize the Service Role of the Department.   Members of the industrial engineering department must recognize that they hold a unique position in the company. As troubleshooters with special training, their work usually calls for upsetting the status quo. Because that may be an unpopular assignment, successful industrial engineers must build up confidence, take care not to disparage production efforts, and observe department protocol if they wish to be accepted by line management.

Steer Clear of Company Politics.   Industrial engineers are occasionally assigned problems loaded with political implications. In such a case, it is important for them to be completely objective both in attitude and in action. Any other course will jeopardize their future acceptability and effectiveness.

Respect Line Management's Prerogatives.   In their service capacity, industrial engineers must work through line supervision. Before undertaking a study in any department, they must see the supervisor, explain the purpose of the study, and solicit assistance. The department supervisor should be made a part of its staff.

Give Proper Credit for Contributions.   Industrial engineers should be quick to give others credit for contributing to the development of sound recommendations. Such recognition does not detract from their own reputation as idea people; on the contrary, it enhances their standing. It gives them a reputation as catalysts and will also make their work easier, for operating people will come to them more readily with good suggestions.

Provide Recognition.   Recognition of both the department's and the individual's contributions is important. First, it stimulates further good work. Second, it builds the reputation of the department and thus encourages operating departments to make wider use of its services.

To ensure that its contributions and accomplishments are properly recognized, the industrial engineering department might take these steps:

1. Prepare annual programs of work and anticipated attainments in dollar savings, just as a line department does; prepare periodic progress control reports listing accomplishments to date.

2. Promote internal recognition of accomplishment and outside-the-company publicity on work done, and where applicable, give the names of the department members making major contributions. Technical society papers and magazine articles are excellent outlets. Internal recognition can be gained by reports which point up individual and group accomplishments, assignment to greater responsibilities or more complex problems, and pay increases.

Use of Consultants.   Consultants can be a beneficial supplement to the industrial engineering department. They can supply expertise on a short-term basis and at a higher level of expertise than may be economically prudent to have on permanent staff. Some of the factors to consider when selecting a firm are:

1. Do you have a good definition of what you want the consulting firm to accomplish?
2. Can they readily supply references for similar work in a similar industry?
3. Will they commit to a time frame, cost, and minimum return on investment?
4. What are the qualifications of the consulting engineers to be assigned to your project? Insist on résumés.
5. Most important, will they work in conjunction with the engineers in your department, so that the work they develop will be maintainable by the existing industrial engineering staff?
6. Do they have a follow-up policy of review 1 year after project *completion?*

GETTING ACTION ON RECOMMENDED PROGRAMS

Because industrial engineering is a service department without line authority, it often has difficulty getting people to act on its recommended programs. This can be minimized by the manner in which the study is made, that is, whether or not the department sells as it goes along. It can also be minimized if the recommendation is in the form of an action program that makes it easy for management to act. Other steps the industrial engineering department can take to induce action include:

Emphasize Benefits to Entire Company.   The broader the range of significant benefits, the more likely that action will result. First, more members of management will be interested in achieving the benefits; second, the action cannot die through the inertia of one executive.

Top management is looking for more than "warm fuzzies." When recommending changes or solutions to a problem, the first question they will ask is: "What impact will your proposal have on the bottom line?" You had better be prepared with an answer. The *only* way to address this issue properly is with contribution segment and profitability analysis.

Involvement of the Participants (The Presell).   The more involvement by those whom the change is affecting in the development of the action plan, the more probability of success. Management hates surprises. The conference room is not the place to gather a group of executives to obtain approval of a project they have never been exposed to before. The conference room is where you confirm a consensus after you have reviewed it with each decision maker involved and understand their position.

Consideration of Alternatives.   There is almost always more than one acceptable way to solve a problem. The industrial engineer must always be open-minded to alternative approaches. Often a less perfect alternative may be more successful because the person responsible for its implementation has ownership in it. An analysis is not complete until the alternatives have been weighed for their pluses and minuses.

Point Up Compensating Benefits to Individuals.   If a recommended action program will relieve executives of problems, the desirability of action will be enhanced. If the recommendations reduce the number of workers reporting to executives or otherwise lessen their prestige or authority, the attendant benefits must be strongly emphasized to get them to act.

Spell Out the Action Program.   The action program should describe each step and designate each participant's responsibility and the date projected for completion of an assignment. Such a program will make it easy for management to take action.

Provide Progress Checkpoints in the Installation Plan for Use by Line Supervision.   The installation plan should have a procedure for controlling the progress of the installation similar to that recommended for projects within the industrial engineering department. Where practicable, a member of line operating management, rather than a member of the industrial engineering department, should follow up the installation program.

Don't Find a Fault;

Find a Remedy!

—*Henry Ford*

CONCLUSION

This chapter covers the broad subject of organizing for industrial engineering. After reading it, the reader will, it is hoped, agree with the conclusion that the effectiveness with which an industrial engineering department operates depends largely on how well the following factors are understood and weighed. The industrial engineering department must be considered in terms of its:

1. Role in the business
2. Working relationships with other departments
3. Place in the company plan of organization
4. Form of organization
5. Personnel
6. Administration
7. Getting action on recommended programs

Where these factors are properly related, a company can expect continuing improvements in control over production costs and steady reductions in actual production costs.

BIBLIOGRAPHY

Miner, John B., *Organizational Behavior,* Random House, New York, 1988.

Simon, H. A., *Administrative Behavior: A Study of Decision Making Processes in Administrative Organization,* Free Press, New York, 1976.

Ellis, R. B., "Dollarize your Priority List," *American Production and Inventory Control Society 26th Annual Conference Proceedings.* New Orleans, 1983, pp. 430–432.

Lambert, Douglas M., and Jay V. Sterling, "What Type of Profitability Reports Do Marketing Managers Receive?" *Industrial Marketing Management,* vol. 17, no. 1, 1988.

Monden, Yasuhiro, *Toyota Production System,* Institute of Industrial Engineers, Norcross, Ga., 1983.

Likert, Rensis, *New Patterns of Management,* McGraw-Hill, New York, 1961.

Galbraith, J. R., *Organization Design,* Addison-Wesley, Reading, Mass., 1977.

Goldratt, E. M., and J. Cox, *The Goal,* North River Press, Inc., 1984.

CHAPTER 2
CLIMATE FOR PRODUCTIVITY IMPROVEMENT

Benjamin Schneider
Professor of Psychology and Business Management
University of Maryland at College Park
College Park, Maryland

Richard A. Guzzo
Associate Professor of Psychology and Business Management
University of Maryland at College Park
College Park, Maryland

Arthur P. Brief
Professor of Organizational Behavior and Psychology
Tulane University
New Orleans, Louisiana

People who work in and/or observe organizations know that one organization "feels" like a dynamic and interesting place to work while another organization "feels" stodgy and unproductive. The labeling of an organization according to what it "feels" like is also an area of intense interest to academicians who study the phenomenon under the label *climate*. Climate refers to the "feel" of organizations. More specifically, climate refers to the perceptions of organizational members about how organizations function and/or what is important in their organizations. These global perceptions are based on the conditions people experience in their organizations—the events, practices, procedures, and rewarded, supported, and expected behaviors that characterize organizations.

Thus some organizations have a climate of energy and others a stultifying climate; some organizations have a climate for quality and others a climate for service (Schneider and Gunnarson, 1991). Climates are based on thousands of clues members of organizations pick up about their workplaces. The clues are associated with events, practices, and procedures that characterize an organization and the kinds of behaviors that get rewarded, supported, and expected in them. In what follows, we abbreviate events, practices, and procedures by the single word "routines" and the rewards, support, and expectations people experience by the single word "rewards."

The focus of this chapter is on establishing the climate for productivity improvement.

We recognize that the term "productivity" can mean different things in different organizations. Definitional problems of "productivity" and "improvement" notwithstanding, the terms are the focus of this chapter.

To focus on productivity improvement, however, we need to present some more information about what we mean by climate. In addition, we present an overview of a similar concept, organizational culture. This introduction to the climate and culture literature is followed by a review of what we know about behavioral science approaches to productivity improvement. The review addresses two different kinds of productivity improvement efforts, the macro approach and the micro approach. In the macro approach to productivity improvement, entire organizations undergo change wherein many of the routines and rewards of organizations are simultaneously, or relatively simultaneously, changed. In the micro approach, a selected aspect or a few aspects of organizational routines and rewards are changed. Finally, we draw all this material together in a concluding section on total quality management (TQM) that includes specification of the myriad factors of organizational functioning to which attention must be paid if productivity improvement through climate and culture change is to be attained by organizations.

In the chapter it is crucial that the reader keep in mind the idea that climate is in the heads of the members of organizations. That is, climate is subjective, though indeed it is a function of the organizational routines and rewards members experience.

ORGANIZATIONAL CLIMATE

The earliest published research on climate was conducted by Kurt Lewin and his associates in the late 1930s (Lewin, Lippitt, and White, 1939). Lewin, who left Germany as Hitler and Nazism began to take hold, was a social psychologist interested in the effects of environments on behavior. He was particularly interested in the influence of different leadership styles on people's behavior. His interests led him to conduct an interesting experiment in which he contrasted the effects of authoritarian, democratic, and laissez-faire leadership on productivity and the reactions of group members. He showed that, under democratic and authoritarian leadership styles, people were essentially equally productive but worked much more harmoniously and were more satisfied under the democratic leader.

Lewin and his colleagues called this an experiment in "social climate." By this title they meant that different leadership styles result in different social routines and rewards, that the style of the leader created conditions for different social arrangements and patterns of behavior.

Lewin's early work, done with ten-year-old boys, resulted in numerous applications of the climate concept to business and industry. For example, Chris Argyris (1957) showed that some banks create a climate such that only certain kinds of people fit in; he called this the creation of a "right type" climate. Douglas McGregor (1960) also used the concept of climate in his famous book *The Human Side of Enterprise*. He argued that managers create a "managerial climate" that reflects their beliefs about workers. To the degree that managers believe that workers need to be cajoled into working, work only for money, and are not to be trusted to make important decisions, McGregor said they create routines and rewards in keeping with their beliefs. This set of beliefs he called "Theory X." Theory Y, on the other hand, describes managers who believe in the inherent maturity and creativity of their employees and, in accord with these beliefs, create conditions for participation in decisions affecting work and the workplace.

These early ideas about climate resulted in many studies of different kinds of climate. Thus there are studies regarding the climate for safety, the climate for service, the climate for innovation, and so forth (Schneider and Gunnarson, 1990). In just about all these studies, the range of routines and rewards studied is very broad indeed. The issues usually mentioned in research on organization climate include the following (James, James, and Asch, 1990; Kopelman, Brief, and Guzzo, 1990):

The Nature of Interpersonal Relationships in the Organization.   Are interpersonal relationships characterized by mutual sharing and trust or conflict and mistrust; are relationships between functional units (for example, between production and sales) cooperative or competitive; is the organization characterized by effective and supportive orientation and socialization of newcomers or a sink-or-swim approach to socialization; do people feel that their personal welfare is important to those around them and to top management?

The Hierarchical Nature of the Organization.   Are decisions affecting work and the workplace made only by top management or are such decisions made through participation of those affected by the decision; is the organization characterized by a team approach to work or by an every-person-for-themselves approach; does top management have special prerequisites that separate them from their subordinates (such as special parking, food facilities) that keep the workers in their place?

The Nature of Work in the Organization.   Is the work people do described as challenging or boring; are jobs set up so that they are adaptable by the people in them or are they rigidly defined so that everyone must do them the same way; does the organization provide workers with the necessary resources (tools, information, support) to get the work done?

The Outcomes of Work on Which the Organization Targets Its Routines and Rewards.   Is innovation or routine supported and rewarded; is getting the work done or getting the work done *right* supported and rewarded; what do the human resources practices (for example, selection, training, performance appraisal) emphasize; are the goals of work and the standards of excellence widely known and shared?

It is obvious from this wide range of issues that have been studied that climate is determined by a diverse set of organizational routines and rewards. Because of this, climate is very difficult to create and change. Indeed, one might question the inclusion of such issues as the nature of interpersonal relationships or the nature of jobs if the focus of the present chapter is on productivity improvement. Why not just focus on productivity improvement?

The answer to this is that researchers have discovered that interpersonal relationships, the nature of the work done, and hierarchical structures all ultimately affect the effectiveness of an organization in achieving its outcomes. Thus it turns out to be true that regardless of the way an organization defines productivity improvement (for example, as improvements in quality, service, innovation; as decreases in accidents, absenteeism, or turnover) the probability of achieving the goal is enhanced when people feel their work is challenging, when they can participate in decisions determining how the goal will be achieved, and when their interpersonal relationships at work are characterized by mutual trust (Denison, 1990; Lawler, 1986).

ORGANIZATIONAL CULTURE

It is one thing to identify the facets of organizational climate and it is another thing to understand *why* an organization has the climate it has. Why do organizations differ in the routines and rewards that characterize them and the outcomes on which those routines and rewards are targeted? Answers to these kinds of questions are found in a literature that has some similarities to the organizational climate literature. This allied area is called organizational culture.

Because the *climate* literature comes out of a more psychological approach to understanding behavior at work. It has focused on routines and rewards more than the antecedents or causes of those routines and rewards. Conversely, the *culture* perspective on organizations has emerged from more anthropological approaches to understanding behavior. This perspective focuses more on the values and beliefs of the members of different cultures as the determinants of the "feel" of the culture. Indeed some research on organizational culture has emphasized such issues as dress, symbols, myths, and stories as explanations for why organizations look and feel as they do (Deal and Kennedy, 1982).

Other research has placed great emphasis on the values and beliefs that guided the founding of an organization. This work, associated most closely with Edgar Schein (1985), traces the antecedents of existing routines and rewards to the values of the organizational founder. "Strong" cultures, cultures in which there appears to be great sharing of the values by which behavior is governed, appear to have had strong founders. For example, the cultures of Hewlett-Packard, IBM, and J. C. Penney are often cited as examples of how a strong founder leaves a long legacy in terms of organizational behavior—a particular style if you will, characterized by particularistic routines and rewards all traceable to the founder's imprint.

This imprint of the founder is frequently difficult for current organizational members to identify. Thus, when asked why they do their jobs the way they do, members are likely to say things like "Well, that's the way we do things around here." This difficulty in identifying why people do things the way they do them has created problems for people interested in changing their organizational culture. For example, the topic of culture was popularized by Peters and Waterman (1982) in their best-selling book *In Search of Excellence,* but companies that have tried to become "excellent" have found it extraordinarily difficult. They have found it difficult because culture is not directly manipulable. Thus, if culture is rooted in the value systems of the founder one cannot go back and declare that the founder had a different value system!

A challenge for us in this chapter is to isolate ways that culture might be changed. Basically, we propose that culture can be changed through a focus on routines and rewards—by a focus on climate. Climate represents the tangible facets of culture, the kinds of things that happen to and around employees that they are able to describe. Our proposal is that changing the tangible facets of an organization can actually impact the kinds of belief systems that guide employee behavior.

We show how the various attempts at improving productivity, however productivity is defined, can be interpreted as attempts to impact the climate and thus the culture of organizations. In addition, we argue that it is only by changing the tangible routines and rewards of organizations that productivity enhancements can occur. By this we refer to the idea that change will *not* occur by having new mission statements, through speeches, through newsletters, and/or through a big party to kick off a new way of doing things in an organization. To communicate new ways of doing things and new values and beliefs will require changing the tangibles; deeds, not words, are tangible.

In the next two sections, we review some of the tangible ways that behavioral scientists have attempted to have an effect on productivity. First, we explore the more micro approaches to enhancing productivity, such as changes in selection, training, pay systems, and job design. For the most part, as noted earlier, these changes have been made one at a time. The benefit of one-at-a-time changes in organizations is that the effects, if any, are directly attributable to the change that is made. As the reader will see, there is good evidence that many different kinds of changes can have positive effects on productivity. At least of equal interest will be the idea that even micro changes can have important macro consequences.

The more macro changes described in the second section typically target whole organizations and the way they function. These changes include interventions designed to enhance relationships between functions and across levels, the implementation of new technology throughout an organization, and redirecting organizations from, for example, efficiency of production to quality of production (for example, total quality management, or TQM) as the important outcome. While it is frequently difficult to identify precisely what does and does not work in these macro changes, the idea is to try to have an *effect*. Some precision is thus lost in the macro change efforts in attributing what had the effect, but the belief is that change in the whole organization is more likely when the various subsystems of the organization are simultaneously changed.

MICRO CHANGES TO ENHANCE PRODUCTIVITY

We recognize that distinguishing between micro and macro organizational changes for productivity improvement is convenient but often imprecise. The distinction is convenient because it helps bring order to the many types of interventions and innovations that can be implemented to enhance productivity. It is imprecise, though, because there is considerable variation between organizations in how productivity-enhancing changes get made. For example, management may initially target employee involvement practices at just one component of an organization, only to see those practices be successful and quickly spread to other parts of the organization. This is what happened at Ford Motor Company (Banas, 1988). Quality circles, too, have been implemented in a "contained" way (that is, at only one level and in one function of an organization) but, once the genie is out of the bottle, some organizations begin to change more thoroughly (Shea, 1986).

These instances are not surprising if we think of organizations as systems made up of many subsystems. Some productivity improvement programs have as their target from the outset the total organizational system, or at least much of it. These are the macro approaches. Other attempts, the focus of this section, are the micro approaches, those which target only one subsystem in an organization. That subsystem might concern how certain employees are hired, paid, motivated, or directed, for example (see Kopelman, 1986, for an excellent review). Micro-oriented changes seek to change the routines and rewards in parts (subsystems) of the organization, though there may be reverberations and spillover into the broader organizational system. Several examples of micro-oriented productivity interventions and their implications for a productive organizational climate and culture are discussed next.

Earlier we identified four facets that appear to capture much of what constitutes the routines and rewards that characterize climate:

- The nature of interpersonal relationships in the organization
- The hierarchical nature of the organization
- The nature of work in the organization
- The outcomes of work on which the organization targets its routines and rewards

The relation of these facets to the climate-relevant consequences of various micro-oriented productivity interventions is discussed here.

Organizational Entry.   Some micro attempts to enhance productivity impact the entry of employees into the organization, such as programs for recruitment, selection, and socialization into a job. Selection techniques are typically quite specific to particular jobs or families of jobs (for reasons of practicalities as well as legalities). That is, the interviews, tests, and other methods of assessing the qualities of job applicants are geared toward finding out who among the applicants will be the most productive in a particular kind of job owing to their skills, aptitudes, and knowledge (Schneider and Schmitt, 1986). Although selection programs are locally oriented in organizations, they can affect organizational climate more broadly (Argyris, 1957).

Schneider (1990) presents an interesting example of the ways a selection system can have broad organizational systemic effects. He describes a telemarketing organization that implemented a new selection program for hiring telemarketers only to discover that this selection system yielded very broad organizational change. Hiring more competent and service-oriented telemarketers brought about the need to change the training program for those newly hired persons and, because the combination of selection and training produced superior workers, *supervisors* of those workers had to undergo new training and a new supervisory promotion system was also required. The new telemarketers were so successful on the job that a new career program was put in place, with a new compensation package, to keep them as telemarketers. Finally, the turnover of telemarketers was cut by 30 percent.

Programmatic selection programs often yield productivity improvements. The magnitude of those improvements is often difficult to estimate, however. Some of the best-documented figures come from the work of Schmidt, Hunter, Outerbridge, and Trattner (1986). They estimated that the productivity of white-collar workers in the federal government selected on the basis of cognitive ability tests exceeded the productivity of their fellow white-collar workers who were *not* selected on the basis of the tests by a value of $1700 to $16,000 per year, per person.

Cascio (1987) summarizes the ways the improvements in productivity from formal selection programs can be calculated. But the important point in the work of Schneider, Schmidt, et al., and Cascio is their emphasis on the concept of a *formal* selection system. Thus, they are speaking about selection systems based on extensive and intensive job analyses, carefully and appropriately chosen paper and pencil tests and work simulations (such as assessment centers), and carefully developed structured interviews. Ten-minute unstructured interviews do not improve productivity in organizations, though it is often difficult to convince people that their interview-based impressions are poor predictors of future job performance.

Our emphasis on selection should not obscure the potential gains to be had from recruitment and socialization interventions. For example, Wanous (1980) shows how turnover in organizations can be significantly reduced by providing people with realistic recruitment. By realistic recruitment Wanous means "telling it like it is" to potential employees *before* they get to the job. Wanous shows that the typical decrease in turnover rates from these realistic previews exceeds 10 percent.

With respect to socialization practices in organizations, Louis (1990) documents the ways by which people learn the culture of the organization—the difference between right and wrong there, what is valued and what is devalued, and so forth. Following up on Louis' work Laliberte and Schneider (1990) show that people who experience a more useful socialization experience, by both their own actions and the actions of others, are actually evaluated by their supervisor as being superior performers on the job.

Our belief is that when it comes to the issue of climate and culture for productivity many managers in organizations devalue the importance of these entry processes for employees. We have tried to show how absolutely critical these are for the future of an organization. After all, it is the people in an organization who will ultimately determine what that organization is and can become; who you let in the door is largely who you will be (Schneider, 1987). If we are what we eat, we are also who we hire!

Training.   Another route to improved productivity is through training and development. Training and development activities take many forms and have many objectives (Goldstein, 1989). Some training is quite specific to the requirements of a particular job (for example, learning to operate an automated inventory control system). Other training and development activities, in contrast, have rather broad aims. Training and development for managers often is of this sort.

Of 11 productivity-improvement practices reviewed by Guzzo, Jette, and Katzell (1985), training was found to have the largest average effect on productivity. The other practices with which training was compared included such interventions as feedback programs, goal setting, incentive pay plans, and flexible work scheduling. Training for managers in particular also was found to have significant productivity payoffs. A more specific analysis of training for managers has been done by Burke and Day (1986). Many of the instances of training they examined were characterized by rather broad training objectives, such as changing the "people skills" of managers or changing their styles of information sharing and decision making. Burke and Day concluded that indeed training aimed at managers, including quite general management training, increased not only knowledge but also actual performance.

Burke and Day's analysis raises several interesting points. One is that management training often pays off. Another is that managers' skills, knowledge, and attitudes acquired through training are often quite generalizable. That is, what managers acquire through training in their present job will often be transferable to future assignments and settings, and thus the effects of training become more pervasive in the organization (McCall, Lombardo, and Morrison, 1988). Over time, then, managerial training can change the climate and culture of an organization as more and more of the practices, routines, and rewards under management's control show the impact of prior management training. This can happen even though training programs are not explicitly aimed at changing organizational climate.

In this light it is interesting to consider some facets of management that may be less amenable to training. Here we refer to what has come to be called "vision," an attribute of leaders who are said to be "visionary" or "charismatic" (Bass, 1985; Bennis and Nanus, 1985; Conger and Kanungo, 1988). Apparently few if any people are charismatic in all settings and under all circumstances. More generally, the rule seems to be that some persons are able to capture the minds, hearts, and abilities of a group of followers *when a particular set of circumstances emerges and they are present to act on it*. The example of the charismatic usually given is Winston Churchill who, both before and after World War II, was unable to hold office but during the war was able to rally an entire nation, if not the western world, to victory.

Management training, compared with charismatic leadership training, seems to be a useful endeavor. Indeed, it may be true that training in management practices and procedures can eventually yield an appellation of visionary. Thus, it may be that teaching certain management skills prepares persons to be visionary when certain situations occur; management training might be the preparation for being able to capitalize on a particular confluence of conditions and events.

For present purposes, leadership and management training are potentially critical because of the central role leaders and managers are thought to play in the creation and maintenance of climate and culture (see McGregor, 1960; Schein, 1985). In a real sense, leaders and managers determine the direction organizations take and the means by which they proceed. By determining both goals and the means of goal attainment, leaders have great impact on routines and rewards and the climate and culture of organizations. Leadership and management training, then, is another source of rippling effects for change in organizations.

Goal Setting.   A well-known and often-practiced micro approach to productivity improvement is the setting of specific, difficult goals at work. Many studies of the impact of goal setting at work have been done and the results overwhelmingly show that important productivity gains can be realized through goal setting. A good summary of this work can be found in Locke and Latham (1990).

Goals that work are quite job- and individual-specific. The actual goals set might concern error rates, speed, quantity of output, or any other criterion of performance. The cycle for goal attainment tends to be rather short-term. For example, "number of phone calls per day" is an example of a work goal relevant to, say, a telemarketing job. However, there are examples of successful goal setting implemented with rather long-cycle jobs, such as the jobs of research and development scientists.

The productivity effects of goal setting programs are in large part due to the clarification of the ends to be achieved through work. That is, goals are statements about what is to be accomplished within a certain amount of time, with the consumption of a fixed amount of resources, and so on. Ideally, the goals explicitly set for a person or job are closely aligned with broader business unit goals and with the strategic objectives of the organization; we address this in greater detail later. Goals thus help make clear where organizations are headed, in terms of their business outputs, and the goals set throughout an organization can affect organizational climate because they help focus people's energies and competencies and thus pinpoint which accomplishments are likely to be rewarded.

Goal setting programs, however, not only clarify *what* is to be accomplished but also *how* to accomplish those objectives. Setting goals has the effect of sparking consideration of how to attain those goals. Such consideration may result in clarification of the optimal work process, increased efficiency, and innovations in the work process. In this way the routine aspects of organizational climate can change as a result of goal setting.

The greatest change in organizational climate will occur when goal setting programs are implemented in conjunction with other changes, such as new methods of performance appraisal or reward (Guzzo, 1988). Goal attainment can become a basis for periodic performance review and compensation adjustments, as has been attempted in management by objectives (MBO) programs in many organizations. Once again, this is an example of how a singular micro change, in this case the establishment of formal goal setting, can lead to other changes within the organizational system.

Work Redesign. The notion that routine, unchallenging jobs may be a source of alienation and diminished productivity has been in circulation for several decades. A response has been to enrich jobs, to make them more complex and challenging, typically by giving job incumbents more responsibility, feedback, autonomy, and variety in their work (Hackman and Oldham, 1980). Over the last two decades systematic research evidence has accumulated demonstrating that the tactic of enriching jobs can indeed pay off in terms of productivity gains, quite apart from increases in satisfaction at work. Often the productivity gains take the form of reduced turnover and absenteeism.

Redesigning work pushes responsibility, authority, information, and the like down the hierarchy, investing them in people at lower levels of the organization rather than just in those employees' supervisors. This has implications for the broader organizational system. Thus, while expanding the duties and demands of a single class of jobs is a micro approach to productivity improvement, taking such actions cannot help but be felt at least one level up the hierarchy. Again, a micro change in organizational practices—job enrichment—can send ripples that are felt well beyond the locus of the change.

The converse of the warning just identified needs to also be mentioned. That is, while there are systems implications for the implementation of an effective job enrichment program it is also true that job redesign will fail when implementation is attempted in a nonsupportive system (Hackman and Oldham, 1980). Like many other attempts at productivity improvement, job redesign requires careful implementation if the intervention is to have the intended effect. For example, supervisors who are not trained in ways to handle personally the fact that their subordinates are now making decisions previously restricted to themselves may sabotage the job redesign effort. Or, suppose that jobs are made extraordinarily more challenging than they were in the past—were they made *too* challenging for incumbents to handle with their own perhaps limited competencies?

Our point here is that micro interventions like those being described here can fail to have their intended effect because they are implemented poorly rather than because they are inherently ineffectual interventions.

Incentive Pay. The implementation of pay-for-performance is a classically popular approach to enhancing productivity. Most pay-for-performance plans are predicated on individual performance. Piece-rate pay and commission pay are traditional ways of tying pay to individual performance. Other methods include merit increases in salary to reflect year-end performance appraisals, though the pay-performance connection is often weaker here. In these ways incentive pay is often used as a micro approach to productivity improvement (Lawler, 1981).

Incentive pay, however, can also take a more macro approach. Organizationwide incentive pay, where payouts to employees are based on overall, system performance (as measured by annual profit, for example) rather than individual accomplishments, is of this sort. Of the many forms of organizationwide pay plans, the Scanlon plan is popular. A Scanlon plan uses money as a reward but also requires extensive employee participation in the design and execution of the plan. Participation also extends to planning and decision making that affects overall organizational performance (Tyler and Fisher, 1983). Thus significant changes in the routines of organizations can be brought about by the "simple" act of establishing an incentive pay plan of this sort.

There is lots of evidence on the positive effects of incentive pay on productivity. Guzzo (1988) reviews the evidence and finds that, while incentive pay frequently is associated with improved productivity, the actual impact of financial incentives on productivity is quite variable. In some organizations the implementation of incentive pay raises productivity while in others it has little or even opposite effects. What causes this variability? Guzzo cites many plausible factors that could be operating. Some of these are social in nature, such as group norms, shared attitudes, and trust of management. Other factors concern the adequacy of the criteria against which performance is assessed to determine monetary rewards. Still other factors relate to the nature of the individuals subject to an incentive pay plan.

Lawler (1990) asserts that how individuals are rewarded strongly influences an organization's culture. He argues that the impact often is greatest in the early days of an organization and that, over time, pay practices can influence such things as the culture's tolerance of risk taking. Pay practices become a part of the culture, Lawler points out, which also means that cultural change in organizations may be brought about by changing how people are rewarded.

Teams. Teams and work groups are very important to most organizations today. Modern management practice involves an increasing reliance on team-based work. Examples include the adoption of quality circles, multifunction (or "cross-function") teams, task forces, and autonomous (or "self-managed") work groups. Some forms of team-based work have a faddish flavor. Quality control circles, for example, sometimes seem to have been implemented as a management technique more because they were "in vogue" than because they were rationally determined to be a useful addition to an organization. Nonetheless, team-based organizations are here to stay.

Quality Circles. Quality circles (QCs) are teams of individuals from a common work area who meet regularly to identify and solve problems of quality, productivity, or other matters. Recommendations made by QCs may or may not be implemented by the organization. The early experiences of QCs are often quite positive. There are many reports of initial successes in solving critical problems and saving a company significant amounts of money. The return on investment of QC programs initially is very high, even though such formal programs involve extensive training of members, QC leaders, and facilitators. Barrick and Alexander (1987) suggest that the return is realized after 4 to 8 months of QC operation. However, QCs do not last long. Ledford, Lawler, and Mohrman (1988) review several firms' experiences with QCs and suggest that QCs rarely last more than 2.5 years and that the average age of existing QCs is about 1.5 years. The short-lived nature of QCs is apparent even in firms like Lockheed whose initial successes with QCs ignited management's interest in the device. Griffin's (1988) systematic evaluation of one company's QC experiences also shows that the positive consequences of QCs diminish over time.

Why are the effects of QCs short-lived? Ledford et al. (1988) describe QCs as structures that are created in parallel to the existing organizational structure. That is, QCs are appended to the existing organization rather than integrated into it. The fact that QCs rarely have the authority to implement their recommendations indicates the "parallel form" of QCs, as does the fact that QC performance is rarely tied to the formal appraisal and reward systems. Because they are parallel to rather than integral with the organization, QCs often may not be sustainable and may be expected to elicit little change in the organizational climate or culture, although there are exceptions (Shea, 1986). Teams, however, can be vehicles for sustained quality and productivity improvement as part of total quality management (TQM) practices. TQM is discussed in the context of macro organizational change efforts because it seeks quality improvements through means that are far more encompassing than QCs, though QC-like teams often are a part of TQM.

Autonomous Work Groups (AWGs). Unlike QCs, autonomous work groups (AWGs) are integrated into the formal organization. Whereas QCs are usually an add-on, something in which employees participate for a few hours a month when they are not performing their regular work, AWGs *are* their jobs. When autonomous work groups are implemented, members' jobs are defined by virtue of their place in the work group. Rather than spending a few hours a month in them, employees spend their entire workdays in AWGs (Hackman, 1986).

Because they are not a parallel structure the way QCs often are, the implementation of AWGs has significant ramifications for other parts of the organization. One thing that changes with the implementation of AWGs is supervision. As the name implies, AWGs take on the responsibility for making many of the decisions previously made only by superiors. One consequence of the implementation of AWGs is a reduction in the number and levels of management. Two examples of this are recorded by Dumaine (1990): AWGs at a General Mills plant eliminated the need for any managers to be present during certain shifts, and the implementation of such teams at Aetna Life and Casualty was associated with a change in the ratio of middle managers to workers from 1 to 7 to 1 to 30. Additionally, AWGs can change the nature of supervisory work, calling for supervisors to do more facilitating and consulting to the group compared with imposing their decisions on groups.

Other changes that occur with the adoption of AWGs include changes in compensation. More specifically, skill-based pay plans often are implemented with AWGs. Skill-based pay plans compensate individuals on the basis of their measurable mastery of specific skills or proficiencies. Skills are "priced," and thus the number of and types of skills mastered determines actual compensation. As Lawler (1990) discusses, skill-based pay is often well adapted for use with employees working in teams. AWGs are usually implemented with additional training, often of the sort that is done with QCs (for example, training in the use of statistical quality control and productivity data, training in group leadership skills). Selection systems change, too, with the adoption of AWGs. The Saturn Corporation, a subsidiary of General Motors, manufactures automobiles by relying on work teams. Consequently, its methods of selection and hiring are quite different from the traditional methods used in individually based automobile production plants. At Saturn, job applicants have been selected on the basis of their capacity to work in teams as judged by their performance in innovative lower-level employee selection systems like work simulations. In these simulations prospective employees are assessed by people like themselves with whom they may share membership in a team if they are hired.

Research on the productivity impact of AWGs is not very extensive. However, a review of research by Goodman, Devadas, and Hughson (1988) resulted in the conclusion that productivity does indeed increase through the use of such teams. Their review also suggests that such teams bring about changes in organizational climate and culture, although this conclusion must be regarded as speculative in the absence of directly confirming research evidence. However, considering the extent of change in supervision, pay, training, and selection that the implementation of team-based methods of productivity improvement can bring about, it is quite likely that changes in climate and culture for productivity occur. In fact, AWGs serve as a useful introduction to the macro changes to be discussed later.

Summary. There are many micro changes for productivity improvement. Such changes usually are not implemented for the purpose of changing organizational climate or culture. However, it is clear that what we call micro changes in organizational practices indeed can and often do affect climate and culture. Selection programs, training and development, incentive pay schemes, job redesign, team-based approaches, and to a lesser extent, goal setting programs have effects that carry into the larger organizational system. In fact, it may be that micro changes which have the most enduring productivity effects are those which influence and become part of an organization's climate and culture.

We turn now to a consideration of organizational interventions for productivity improvement that are more macro in nature. These interventions explicitly seek change in the whole organization or involve changes of multiple rather than single organizational subsystems.

MACRO CHANGES TO ENHANCE PRODUCTIVITY

Numerous macro or total organizational interventions have been espoused as means of improving productivity. Examples include quality of work life projects (Lawler and Ledford, 1982), organizational development and change (French and Bell, 1984), sociotechnical systems (Cummings, 1978), management by objectives (French and Hollmann, 1975), total organizational pay systems (Lawler, 1981), collateral organizations (Zand, 1974), and total quality management (Deming, 1986). The intent of this section is not to review the efficacy of all of these many available alternatives. Rather, our goal is to offer a perspective for thinking about total organizational changes aimed at enhancing productivity. We show how these macro organizational change efforts may be understood through the climate and culture lenses.

To begin, we further clarify what the phrase "macro change" implies. A macro change is one designed to affect multiple policies, practices, procedures, rewards, and other features of multiple functions, units, and levels of an organization. Borrowing from Beckhard's (1969, p.9) treatment of organizational development, a macro change will be defined as "an effort (1) *planned,* (2) *organization-wide,* and (3) *managed* from the *top,* to (4) increase *organization effectiveness* and *health* through (5) *planned interventions in* the organization's processes, using *behavioral-science* knowledge." Unlike most micro changes, macro changes are *intended* to ripple throughout the entire organization, altering routines and rewards at different levels, in different functions, and across all persons. In this way, macro changes may be seen to be efforts aimed at modifying an organization's climate and, eventually, its culture. It is this view of macro change we pursue below.

It is important to note that most approaches to macro change, like approaches to micro change, have not consciously operated from a climate or culture perspective. Thus the changes designed to enhance productivity discussed earlier, as well as those discussed here, have explicitly been designed to affect productivity without necessarily also being designed explicitly to change climate and culture. We think of most approaches to productivity change in organizations as *engineering* approaches because they assume a direct cause-and-effect relationship between changes made and (the hoped-for) changes in productivity. Understanding the changes in the psychology of the work force required before changes in productivity will also occur is not typically factored into these approaches.

In the climate and culture approach to productivity enhancement, the psychology of the work force is critical. We argue that unless great attention is paid to the psychology of the work force, hoped-for changes will remain hopes. This is especially true for changes that emphasize the media strategy of change—newsletters, speeches, video and audio tapes, and so forth. The climate and culture approach to change stresses the importance of implementing practices and procedures, of designing reward systems, of supporting and expecting certain changes by deed, not word.

In what follows, we describe in some detail three approaches to macro change that have received considerable attention by both researchers and managers. These approaches have made *actual* change in organization, but they have tended to not emphasize climate and culture issues. Nevertheless, all three have produced some effects on productivity (Nicholas, 1982). Our argument is that, had the approaches emphasized the issues of work force climate and culture, the effects might have been more impressive.

Macro change efforts reflect different philosophies of what an organization requires to be productive. Three different philosophies have guided these change efforts, the *human potential* philosophy, the *sociotechnical* philosophy, and the *total quality* philosophy.

Change Based on a Human Potential Philosophy.   The changes emanating from the *human potential philosophy* perspective grew out of the human relations and sensitivity training movements of the 1950s and 1960s (for example, Bennis, Benne, and Chin, 1961) and the self-actualization movements of the 1960s and 1970s (see Beer, 1980). These two streams of orientation to productivity from a behavioral science perspective have come to be called the organizational development (OD) perspective. This movement rests on a number of assumptions about effective organizational functioning (based on French and Bell, 1984):

1. People desire growth and development and can be creative when they have the opportunity.
2. Interpersonal interaction, both with peers and with superiors, is important to people, making the formal and informal nature of such relationships important.
3. Interpersonal trust, support, and cooperation are preconditions for effective functioning; the creation and maintenance of zero-sum, win-lose competition in organizations is not useful.

Assumptions like these (see French and Bell, 1984, for an expanded list) have produced an effective technology of change in organizations. Thus research suggests that organizations, when they learn to operate under assumptions like those just noted, can function more effectively in terms of employee attitudes and productivity (see Nicholas, 1982) as well as financial performance (Denison, 1990).

Readers familiar with the OD research literature might argue with our earlier claim that OD fails to emphasize the climate and culture of the organization. Thus recent approaches in OD, especially those connected to the quality of work life perspective (for example, Seashore, 1981), have emphasized employee attitudes. But these employee attitudes have focused primarily on such issues as job satisfaction, organizational commitment, and job involvement and have failed to identify the outcomes of work on which the organization is seen by its members to target routines and rewards. OD, then, has made the assumption that attention to the social and self-actualizing nature of people will produce enhanced productivity; what people see as the imperatives of their employing organization, its goals, and its direction has not been a focus of OD. As noted earlier, our definition of climate includes the *focus* of routines and rewards, not just the *nature* of routines and rewards. Focus, also as noted earlier, can be service, innovation, turnover, or the more general term "productivity"; our point is that there needs to be a productivity-related focus of macro change for the change to yield productivity improvements.

Change Based on the Sociotechnical Philosophy. The *sociotechnical philosophy* is based on the idea that paying attention to either social issues or technological issues alone in trying to improve productivity is not useful. The sociotechnical approach argues for the *integration* of the social and the technical for effective productivity change. This conceptualization of productivity change emerged out of post-World War II efforts in England to improve productivity in the workplace (for example, Rice, 1958; Trist and Bamforth, 1951) as part of the long and fruitful history of the Tavistock Institute. The Trist and Bamforth research provides a good flavor for the sociotechnical philosophy.

Trist and Bamforth studied the consequences of introducing a new method of mining coal in England, called the long-wall method. In the long-wall method coal is mined by an automated blade which slices coal off the *side* of the coal seam. The blade deposits the excised coal on a belt which takes the coal to the surface. In contrast, the traditional method of mining coal was for teams of men to work progressively *into* the seam of coal, extracting the coal by first placing it into trains. When the long-wall method was installed, the teams were dismantled and workers were placed at strategic places along the belt, alone, to ensure the smooth functioning of the belt.

The result of the intervention of the new automated process was worker distress and a fall-off in productivity. Trist and Bamforth attributed these results to the dismantling of the work team and the poor way in which the new procedures were introduced. That is, there was no worker participation in either the design or the implementation of the new technology. It is interesting to note that, while Trist and Bamforth conducted their research in the 1940s, Klein, Hall, and Laliberte (1989) report remarkably similar results for the introduction of computer-aided design and drafting in the 1980s.

The message emerging from the sociotechnical systems perspective on organizational change is that the technology and the social nature of work do not stand alone. To the degree that the two can be integrated, to that degree technology will actually yield improved productivity (see Trist, 1981, for a review of the sociotechnical systems perspective). Indeed, recent reviews of the literature show the sociotechnical systems perspective to consistently produce effective changes in productivity (Guzzo et al., 1985; Nicholas, 1982).

The sociotechnical approach to change, like OD, does not emphasize what is in the heads of employees in the organization except insofar as that is connected to how they feel about their work and their social relationships at work. Again, as with OD, the matrix of issues addressed have not included what employees believe the organization's goals to be, at what they believe they should be directing their energies and competencies, and for what kinds of behaviors they are being rewarded and supported.

Change Based on the Total Quality Management (TQM) Philosophy. The TQM approach to productivity, as articulated by Crosby (1984), Deming (1986), Juran (1987), and others, entails the modification of a great many facets of organizational functioning. Thus, whereas the human potential and the sociotechnical approaches to productivity enhancement focus on relatively narrow, albeit organizationwide issues, TQM focuses on the *customer* and requires *all* and *every* facet of the organization to be involved. With this *goal* and *focus*, TQM comes closest of the macro change efforts in the literature to providing a frame of reference for employees regarding the climate and culture of the organization. Also, because TQM is an organizationwide effort to meet the needs of the customer, it emphasizes the broadest range of issues and activities. These include such elements as (based on Hayes and Wheelwright, 1984):

1. Preplanning during the product design stage that involves engineering, production, quality assurance, and marketing
2. Training employees to be able to deliver high quality
3. Developing expectations of high quality in employees through selection, training, reward systems, and by supplying them with the necessary supplies, equipment, and information to deliver quality
4. Encouraging employees to surface, discuss, and resolve quality issues, not hide them
5. Working with suppliers to assure defect-free parts and accepting only perfect parts
6. Purging the concept of an acceptable quality level and, instead, adopting the concept of continuous improvement with the goal of zero defects

Indeed, TQM models go so far as to identify the principal barrier to the successful implementation of a TQM program, as suggested by the work of Gronroos (1990), "weak management." The characteristics of such a management are:

1. Management has structured an organization in which both operational as well as strategic decisions often must be pushed up through many layers to the top for resolution. Deviations from this protocol are punished. Thus decision making in the organization is sluggish and frequently ill-informed.
2. Management has not adequately updated operating systems and technology, resulting in the organization's being barely able to stay afloat even without adding on a TQM type of effort. The fiscal and human resources required to modernize at this point seem out of reach. Simply, management has allowed the infrastructure of the organization to decay so far that necessary repairs may be out of the question.
3. Management's dealings with lower-level employees can be described as less than completely honest and open, sometimes harsh, and typically disrespectful. Employee morale, understandably, is low. They therefore generally are unwilling to endorse virtually any initiative undertaken by management.

4. Management has a history of initiating change efforts which often do not come to fruition. This is so, at least in part, because management has failed to articulate its objectives; and, therefore, employees do not know why changes are initiated and do not know toward what goal they should focus their energies and competencies. Without such focus, the change process appears chaotic.

5. Management often has avoided making tough decisions and/or has been unwilling to stick to difficult courses of action. This has led to management's being seen as lacking determination, courage, and strength. Perhaps this negative perception of management constitutes management's greatest weakness.

In sum, the above represents real obstacles to implementing TQM or, in fact, any macro change. The list, while presented as barriers, identifies the elements of management necessary for TQM to actually have an effect.

While this description of management requirements and the previous list of issues constituting a typical TQM program in no way capture all alternative forms of TQM, it is representative of many such efforts. Perhaps more importantly for our purpose here it shows the very wide range of issues that need to be simultaneously addressed to bring about a total organizational change directed at quality as perceived by the customer. One can think of all these changes as an attempt to create a climate and culture for quality. Thinking about it this way frames the issue and indicates the absolute necessity of sending the message of quality to everyone through as many actions and changes as possible.

It becomes increasingly clear in discussing TQM that macro changes entail altering the many facets of organizational functioning rather than one or two dimensions of the organization. By example, we can return to our outline of the four major elements of climate presented earlier in the chapter and review the implications of TQM for the changes that might be required. Recall that the four dimensions presented were labeled as follows:

- The nature of interpersonal relationships in the organization
- The hierarchical nature of the organization
- The nature of work in the organization
- The outcomes of work on which the organization targets its routines and rewards

If one were putting into place TQM based on these climate dimensions, all of the following would be relevant: Goals would have to be set both within work units and across organizational levels *and* be supported by appropriate modifications in such other dimensions of climate as the outcomes of work on which routines and rewards would be focused—that is, routines and rewards targeted on customers. The change in what gets rewarded might also entail movement away from a noncontingent pay plan toward a group incentive pay plan employing a cash bonus contingent upon the attainment of the customer-defined goals. The focus on customers might also require change in the hierarchical nature of the organization since it has been shown that lower-level employees (like tellers in banks) may have the best information about customers (Schneider and Bowen, 1985). In a similar vein change might necessarily involve altering the nature of interpersonal relationships at work—if supervisors are to pay attention to subordinates because the subordinates have access to important information, then the nature of mutual sharing and trust that characterizes such relationships may need to be changed. Similarly, a way of improving such relationships might be through tangibly demonstrating support for lower-level workers through updating equipment and reformulating maintenance service schedules.

This sample of changes collectively would contribute to what could be called a "strong" climate for productivity vis-à-vis customers. Creating such a strong climate should yield an effect on productivity more powerful than any given micro change. But macro changes, in several ways, are more costly and therefore inherently more risky. Moreover, this risk, as will be shown, is compounded by the sheer complexity attached to implementing a macro change.

Of course, how much a change in climate TQM actually involves is a function of where an organization stands prior to the intervention. We suspect, however, that TQM represents a radical departure from the status quo for many, many organizations. Why then would an organization undertake such a massive program? The possible answers are numerous:

- To maintain a competitive edge—to survive
- To drive up revenues by selling quality
- To drive down cost by reducing waste
- To maintain the organization's integrity by delivering on its promises of quality

The considerable mental, physical, financial, and, perhaps, even emotional resources that TQM demands, however, by no means lead to a guaranteed payoff.

Suggestions for Implementing Macro Change.   In what follows we present some suggestions for ways that macro changes might be more successfully managed to create the kinds of climate and culture required for productivity enhancement. Goodman and Dean (1989), in addressing why productivity efforts fail, provide a number of suggestions for how to make a macro change self-sustaining. Some of these suggestions are summarized below.

1. Ensure that the organization is prepared to handle a major organizational change. As our previous discussion of barriers to TQM indicated, the following questions must be asked and answered in the affirmative: "Is employee morale high?" "Does management have a track record of successfully implementing major changes?" and "Is management known for confronting tough decisions and sticking to difficult courses of action?" If the answers to such questions are not affirmative, then moving ahead with a proposed macro change should be rethought and efforts directed at getting one's "organizational house" in order should be programmatically instituted. Macro change to enhance productivity in an environment that is unready for it will fail.

2. Be aware that proposed macro changes incongruent with existing organizational climate and culture require tremendous amounts of time and effort. Analyze the proposed change in terms of the four dimensions of climate previously listed. Try to understand, in advance, the magnitude of movement to be required in regard to the nature of interpersonal relationships in the organization, the hierarchical nature of the organization, the nature of work in the organization, and the outcomes of work on which the organization targets its routines and rewards. If, as in TQM, the customer is the ultimate focus of all levels and functions, ensure that all levels and functions understand their link to the end-user consumer.

3. Plan the change in as much detail as possible and communicate that plan widely, by deed and not by word; put your money where your mouth is. Begin by specifying, in writing, the goals of the change. Follow the goals up with written systems and procedures for implementation. Follow the written procedures with routines and rewards that implement the goals. Remember that implementing only technical or only social systems will produce incomplete results.

4. Pay particularly close attention to the organization's reward systems in order to ensure that employees are being motivated to implement and to sustain the change. Determine specifically what behaviors will need to be rewarded. Plan for how meaningful amounts of both extrinsic and intrinsic rewards (nonmonetary, like job attributes, supervisor-subordinate, and peer relationships) will be linked to the specified behaviors.

5. Recognize that, for the change to be sustained, resources must be allocated for maintenance as well as implementation. For instance, the costs of periodically updating operating systems and technologies need to be considered. Moreover, for example, the cost of periodically training personnel in order to reinforce the change should be treated as an essential budget item.

6. Design the mechanics to monitor periodically the effectiveness of the change. Recognize that the change will need to be adjusted over time. Do *not* assume that the change never will be abandoned.

7. Capitalize fully on the literature emanating from the micro changes enumerated earlier. Thus, in addition to incentive and compensation systems and training, it is critical to also implement recruiting, selection, and socialization systems, team systems, job design initiatives, and so forth. These micro changes *one at a time* have validated productivity enhancement outcomes; put in place *collectively* against a known goal or focus (for example, customer quality), they can be powerful indeed.

Summary. The literature reviewed here on macro changes reveals some potentially powerful techniques for enhancing productivity in organizations through climate and culture change using planned behavioral-science-based techniques. These techniques work because they send particular messages to employees, messages that convey the imperatives at which they should use their competencies and energies. The example we focused on, quality for customers, is but one example we could have used. Additional indexes of productivity effectiveness abound—accidents and safety, innovation, turnover, bottom line profitability both long- and short-term, ability to attract the best people, corporate social responsibility, customer account retention, and so forth. Some have argued that the ultimate index of effectiveness is long-term corporate survival.

Our message here is that for an organization to achieve the goals it sets it must create and maintain a climate and culture that fosters goal accomplishment. It may very well be true that different climates and cultures can be equally effective against different goals and in different markets. We would subscribe to this possibility. However, and this is our major point, it is what *people* in an organization *believe* are the goals of an organization that ultimately determines goal accomplishment. What top management says is irrelevant—the kind of climate top management creates through the routines and rewards it establishes will determine goal accomplishment.

CONCLUDING REMARKS

One common stereotype of engineers depicts them as skillful professionals who tenaciously pursue solutions to the problems they confront. These stereotypic engineers seem driven to isolate *the* variable they can manipulate to produce *the* outcome desired. They desperately want the certainty of knowing *why* that desired outcome is under their control. If such engineers do, in fact, exist and if they have read our chapter to this point, we guess they are none too happy. Their need for certainty has not been fulfilled.

Rather, altering climate and culture entails manipulating a bundle of variables to produce a set of outcomes, without the ability to isolate which manipulated variable was the cause of a particular outcome. This uncertainty is compounded by the fact that these changes in outcomes occur over time, with the most significant ones occurring more distally. This is so because the natural tendency of an organization is to maintain a steady course. This tendency constitutes what some call resistance to change. It is not so much resistance as desiring the steady state, an inclination that results in change proceeding very slowly before the organization really begins clearly moving in the intended direction.

The uncertainty we have described is inherent in dealing with climate and culture for one simple reason. Climate and culture are qualities of human systems. One is tempted to simplify the nature of these human systems qualities in order to construct a short list of straightforward principles for managing them. We hope we have not succumbed too much to this temptation and thus have portrayed adequately the wide variety of both human and technological issues required to establish a climate for productivity improvement. While we recognize the appeal of simple solutions, we believe firmly that in the current case they constitute a disservice to those bearing the responsibilities of organizational leadership.

We also have tried to portray that, even with these ideas in hand, success is far from assured. Additionally, one must be committed to sustaining, over an extended period of time, a seemingly remarkable high level of effort focused on producing the intended changes in climate. The necessity for such an effort becomes more comprehensible by once again considering how we have characterized climate: the nature of interpersonal relationships in the organization; the hierarchical nature of the organization; the nature of work in the organization; and the outcomes of work on which the organization targets its routines and rewards. Merely changing relationships, structure, the work itself, *or* routines and rewards represents a significant task. To do it all requires the type of commitment we have tried to describe here.

Is such a sustained focus of mental and physical energy worth it? If one's goal is to create *and* maintain productivity improvements, the answer is an unqualified yes. Indeed, the essence of our argument is that, given this goal, one has little choice. In fact, it is our position that, without periodically undertaking such efforts, organizations run the risk of stagnation, perhaps decline, or even death.

BIBLIOGRAPHY

Argyris, C., *Personality and Organization,* Harper, New York, 1957.

Banas, P. A., "Employee Involvement: A Sustained Labor/Management Initiative at Ford Motor Company," in J. P. Campbell and R. J. Campbell, eds., *Productivity in Organizations,* Jossey-Bass, San Francisco, 1988, pp. 388–416.

Barrick, M. R., and R. A. Alexander, "A Review of Quality Circle Efficacy and the Existence of Positive-Findings Bias," *Personnel Psychology,* vol. 40, pp. 579–592, 1987.

Bass, B. M., *Leadership and Performance beyond Expectations.* Free Press, New York, 1985.

Beckhard, R., *Organization Development: Strategies and Models,* Addison-Wesley, Reading, Mass., 1969.

Beer, M., *Organization Change and Development: A Systems View,* Goodyear, Santa Monica, Calif., 1980.

Bennis, W., and B. Nanus, *Leaders: The Strategies for Taking Charge,* Harper & Row, New York, 1985.

Bennis, W. G., K. D. Benne, and R. Chin, eds., *The Planning of Change: Readings in the Applied Behavioral Sciences.* Holt, Rinehart and Winston, New York, 1961.

Burke, M. J., and R. R. Day, "A Cumulative Study of the Effectiveness of Managerial Training," *Journal of Applied Psychology,* vol. 71, pp. 232–245, 1986.

Campbell, J. P., M. D. Dunnette, E. E. Lawler, and K. E. Weick, *Managerial Behavior, Performance, and Effectiveness,* McGraw-Hill, New York, 1970.

Campion, M. A., and P. W. Thayer, "Job Design: Approaches, Outcomes, and Trade-offs," *Organizational Dynamics,* vol. 15, pp. 66–79, 1987.

Cascio, W., *Applied Psychology in Personnel Management,* 3d ed., Prentice-Hall, Englewood Cliffs, N.J., 1987.

Cole, R. E., and D. S. Tachiki, "Forging Institutional Links: Making Quality Circles Work in the U.S.," *National Productivity Review,* vol. 3, pp. 417–429, 1984.

Conger, J. A., and R. N. Kanungo, eds., *Charismatic Leadership,* Jossey-Bass, San Francisco, 1988.

Crosby, P. B., *Quality without Tears,* New American Library, New York, 1984.

Cummings, T. E., "Self-Regulating Work Groups: A Socio-technical Synthesis," *Academy of Management Review,* vol. 3, pp. 625–634, 1978.

Deal, T. E., and A. A. Kennedy, *Corporate Culture: The Rites and Rituals of Organizational Life,* Addison-Wesley, Reading, Mass., 1982.

Deming, W. E., *Out of the Crisis,* Massachusetts Institute of Technology, Cambridge, Mass., 1986.

Denison, D. R., *Corporate Culture and Organizational Effectiveness,* Wiley, New York, 1990.

Dumaine, B., "Who Needs a Boss?" *Fortune,* May 7, pp. 52–55, 56, 58, 1990.

French, W. L., and C. H. Bell, *Organization Development: Behavioral Science Interventions for Organization Improvement,* 3d ed., Prentice-Hall, Englewood Cliffs, N.J., 1984.

French, W. L., and R. W. Hollmann, "Management by Objectives: The Team Approach." *California Management Review*, vol. 17, pp. 13–22, 1975.

Goldstein, I. L., ed., *Training and Development in Organizations*, Jossey-Bass, San Francisco, 1989.

Goodman, P. S., and J. W. Dean, "Why Productivity Efforts Fail," in W. L. French, C. H. Bell, and R. A. Zawacki, eds., *Organizational Development*, BPI/Irwin, Homewood, Ill., 1989.

Goodman, P. S., R. Devadas, and T. L. G. Hughson, "Groups and Productivity: Analyzing the Effectiveness of Self-managing Teams," in J. P. Campbell and R. J. Campbell, eds., *Productivity in Organizations*, Jossey-Bass, San Francisco, pp. 295–327, 1988.

Griffin, R. W., "Consequences of Quality Circles in an Industrial Setting: A Longitudinal Assessment," *Academy of Management Journal*, vol. 31, pp. 338–358, 1988.

Gronroos, C., *Service Management and Marketing*, Lexington Books, Lexington, Mass., 1990.

Guzzo, R. A., "Financial Incentives and Their Varying Effects on Productivity," in P. Whitney and R. B. Ochsman, eds., *Psychology and Productivity*, Plenum, New York, 1988.

Guzzo, R. A., R. D. Jette, and R. A. Katzell, "The Effect of Psychologically Based Intervention Programs on Worker Productivity: A Meta-analysis," *Personnel Psychology*, vol. 38, pp. 275–291, 1985.

Hackman, J. R., "The Psychology of Self-management in Organizations," in M. S. Pallak and R. Perloff, eds., *Psychology and Work: Productivity, Change, and Employment*. American Psychological Association, Washington, D.C., 1986.

Hackman, J. R., and G. R. Oldham, *Work Redesign*, Addison-Wesley, Reading, Mass., 1980.

Hays, R. H., and S. C. Wheelwright, *Restoring Our Competitive Edge: Competing through Manufacturing*, Wiley, New York, 1984.

James, L. R., L. A. James, and D. K. Ashe, "The Meaning of Organizations: The Role of Cognition and Values," in B. Schneider, ed., *Organizational Climate and Culture*, Jossey-Bass, San Francisco, 1990.

Juran, J. M., *On Quality Leadership*, Juran Institute, Wilton, Conn., 1987.

Kerr, S., "Some Characteristics and Consequences of Organizational Reward," in F. D. Schoorman and B. Schneider, eds., *Facilitating Work Effectiveness*, Lexington Books, Lexington, Mass., 1988.

Klein, K. J., R. J. Hall, and M. Laliberte, "Training and the Organizational Consequences of Technological Change: A Case Study of Computer Aided Design and Drafting," in U. Gattiker and L. Larwood, eds., *Technological Innovation and Human Resources: End-User Training*, Walter de Gruyter, New York, 1990.

Kopelman, R. E., A. P. Brief, and R. A. Guzzo, "The Role of Climate and Culture in Productivity," in B. Schneider, ed., *Organizational Climate and Culture*, Jossey-Bass, San Francisco, 1990.

Kopelman, R. E., *Managing Productivity in Organizations: A Practical, People-Oriented Perspective*, McGraw-Hill, New York, 1986.

Laliberte, M., and B. Schneider, *Comparing Newcomer Strategies for Gathering Information: An Individual Perspective on Newcomer Socialization to Work*, unpublished manuscript, University of Maryland, College Park, Md., 1991.

Lawler, E. E., III, *Strategic Pay*, Jossey-Bass, San Francisco, 1990.

Lawler, E. E., III, *High Involvement Management*, Jossey-Bass, San Francisco, 1986.

Lawler, E. E., III, *Pay and Organizational Development*, Addison-Wesley, Reading, Mass., 1981.

Lawler, E. E., III, and G. E. Ledford, "Productivity and the Quality of Work Life," *National Productivity Review*, vol. 1, pp. 23–36, 1982.

Ledford, G. E., E. E. Lawler, III, and S. A. Mohrman, "The Quality Circle and Its Variations," in J. P. Campbell and R. J. Campbell, eds., *Productivity in Organizations*, Jossey-Bass, San Francisco, 1988, pp. 255–294.

Lewin, K., R. Lippitt, and R. K. White, "Patterns of Aggressive Behavior in Experimentally Created Social Climates," *Journal of Social Psychology*, vol. 10, pp. 271–299, 1939.

Locke, E. A., and G. P. Latham, *A Theory of Goal Setting and Task Performance*, Prentice-Hall, Englewood Cliffs, N.J., 1990.

Louis, M. R., "Acculturation in the Workplace: Newcomers as Lay Ethnographers," in B. Schneider, ed., *Organizational Climate and Culture*, Jossey-Bass, San Francisco, 1990.

McCall, M. W., Jr., M. M. Lombardo, and A. M. Morrison, *The Lessons of Experience: How Successful Executives Develop on the Job*, Lexington Books, Lexington, Mass., 1988.

McGregor, D. M., *The Human Side of Enterprise,* McGraw-Hill, New York, 1960.

Nicholas, J. M., "The Comparative Impact of Organization Development Interventions on Hard Criteria Measures," *Academy of Management Review,* vol. 7, pp. 531–542, 1982.

Payne, R. L., and D. S. Pugh, "Organizational Structure and Climate," in M. D. Dunnette, ed., *Handbook of Industrial and Organizational Psychology,* Rand-McNally, Chicago, 1976.

Peters, T. J., and R. Waterman, *In Search of Excellence: Lessons from America's Best Run Companies,* Harper & Row, New York, 1982.

Rice, A. K., *Productivity and Social Organization: The Ahmedabad Experiment,* Tavistock Publications, London, 1958.

Schein, E. A., *Culture and Leadership,* Jossey-Bass, San Francisco, 1985.

Schmidt, F. L., J. E. Hunter, A. N. Outerbridge, and M. H. Trattner, "The Economic Impact of Job Selection Methods on Size, Productivity, and Payroll Costs of the Federal Work Force: An Empirically Based Demonstration," *Personnel Psychology,* vol. 39, pp. 1–29, 1986.

Schneider, B., "The Climate for Service: An Application of the Climate Construct," in B. Schneider, ed., *Organization Climate and Culture,* Jossey-Bass, San Francisco, 1990.

Schneider, B., "The People Make the Place," *Personnel Psychology,* vol. 40, pp. 437–453, 1987.

Schneider, B., and D. F. Bowen, "Employee and Customer Perceptions of Service in Banks: Replication and Extension," *Journal of Applied Psychology,* vol. 70, pp. 423–433, 1985.

Schneider, B., and S. Gunnarson, "Organizational Climate and Cultures: The Psychology of the Workplace," in J. W. Jones, B. D. Steffy, and D. Bray, eds., *Applying Psychology in Business: The Manager's Handbook.* Lexington Books, Lexington, Mass., 1990.

Schneider, B., and N. Schmitt, *Staffing Organizations,* 2d ed., Scott Foresman, Glenview, Ill., 1986.

Seashore, S. E., "Quality of Working Life Perspective," in A. H. Van de Ven and W. F. Joyce, eds., *Perspectives on Organization Design and Behavior,* Wiley, New York, 1981.

Shea, G. P., "Quality Circles: The Danger of Bottled Change," *Sloan Management Review,* spring, pp. 33–46, 1986.

Trist, E. L., "The Sociotechnical Perspective," in A. H. Van de Ven and W. F. Joyce, eds., *Perspectives on Organization Design and Behavior,* Wiley, New York, 1981.

Trist, E. L., and K. W. Bamforth, "Some Social and Psychological Consequences of the Longwall Method of Coal-Getting," *Human Relations,* vol. 4, pp. 3–38, 1951.

Tyler, L. S., and B. Fisher, "The Scanlon Concept: A Philosophy as Much as a System," *Personnel Administrator,* vol. 28, pp. 33–37, 1983.

Wanous, J. P., *Organizational Entry: Recruitment, Selection and Socialization of Newcomers,* Addison-Wesley, Reading, Mass., 1980.

Zand, D., "Collateral Organization: A New Change Strategy," *Journal of Applied Behavioral Science,* vol. 10, pp. 63–89, 1974.

CHAPTER 3
PROJECT MANAGEMENT

Laurens van den Muyzenberg
Managing Director
Muyzenberg Management Consultants, Ltd.
Windsor, United Kingdom

Project management is an important tool for industrial engineers. It facilitates the selection of projects to be undertaken, improves the management of the projects themselves, and ensures their successful completion. Project management is one element of the organization and management of industrial engineering covered in Sec. 2, Chap. 1.

An industrial engineering department carries out many different tasks. One category of work concerns the maintenance of the systems installed (for example, time standards); another category concerns the continuous improvement of work methods, and a third category concerns large projects such as a major change of an existing system or the installation of a new one. Projects of the third category typically require more than 20 person weeks of industrial engineering effort. Project management is of particular importance for this category.

DEFINITION OF PROJECT MANAGEMENT

Project management concerns the organizational and procedural aspects of:

• Selection of projects
• Scheduling of projects
• Implementation of the recommendations resulting from the project analysis
• Management of the individual projects and of the total portfolio of projects

In this chapter we first examine the benefits of project management and then describe how to establish project management, covering both the organizational and procedural aspects.

THE BENEFITS OF PROJECT MANAGEMENT

Project management:

- Maximizes the savings and other benefits produced by the industrial engineering department through the systematic selection of the most promising projects
- Reduces the calendar time elapsing between start of a project and completed implementation through careful planning and scheduling
- Increases the success rate through a careful progress review procedure and the involvement of the appropriate levels of management
- Raises the respect for the industrial engineering function because all, or at least most, projects are completed on time, according to expectations and delivered on time with a precise documentation of the results achieved

HOW TO ESTABLISH PROJECT MANAGEMENT

We cover the setting up of project management under two headings:

- Organizational aspects
- Procedural aspects

The organizational and procedural aspects of project management are interdependent; both have to be put in place. For example, project management requires the appointment of project managers, an organizational measure, as well as project scheduling procedures. Successful project management depends on the careful implementation of *all* aspects that will be dealt with in the following paragraphs.

ORGANIZATIONAL ASPECTS

The organizational aspects of project management include:

1. The project manager
2. The project team
3. The steering committee
4. Task forces
5. Responsibility for implementation

The Project Manager.   The core of the project management organization consists of the project manager, the project team, and the steering committee. Figure 3.1 shows an example of what a project organization can look like. The project manager is accountable for the timely completion of the assigned project or projects. If the project is very large, it may be necessary to appoint a full-time project manager for its duration. In smaller projects one project manager can manage several projects. For certain projects it can be feasible to appoint a section head or even the chief industrial engineer, who can handle this task in addition to the normal ones.

The responsibility and authority of the project manager who is also the hierarchical head of the team members assigned is straightforward. In many cases the team members are assigned only temporarily to the project; they continue to report to their former superiors on promotions, salary reviews, and disciplinary matters. When it comes to the professional direction of the activities of the team members, the ultimate responsibility generally also continues to reside with the "regular boss." Because of this complexity, it is important to think through and to specify the responsibility and authority of the project manager.

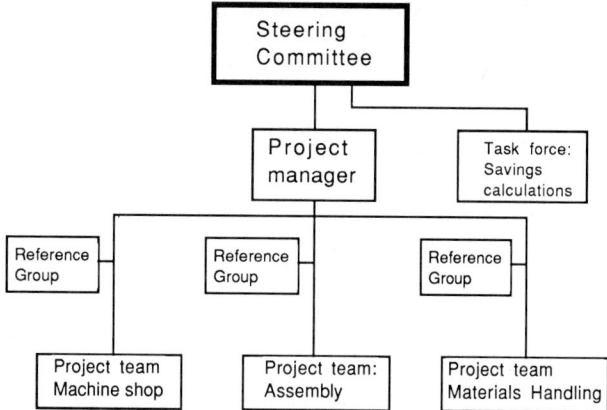

FIGURE 3.1 Project organization for large cost-reduction program covering several areas.

It is quite clear, however, that the project manager decides on what tasks and in what sequence the engineers assigned to the project will work. The task of the project manager described is delicate: full accountability for the results but with rather limited authority. This means that the selection of project managers with the right type of personality is crucial for success. Project managers must be driven by a strong ambition to meet deadlines, inspire confidence and loyalty in those assigned to them, and be persuasive and diplomatic in dealing with the "regular bosses" of the assigned team members.

Project managers are sometimes called "project coordinators" to avoid misunderstandings about their authority. The disadvantage of the name is that it weakens the image of an important job. Project management requires the development of leadership skills for complex situations, and it is therefore excellent training for engineers as part of their career development.

In large projects that concern the interests of several departments it is quite common that the project manager reports to a steering committee (discussed in a later section). For smaller projects the project manager generally reports in the capacity of project manager to the chief industrial engineer.

Project Teams.   The project teams contain the engineers that are assigned full time to the project. In large projects several teams can be assigned to different parts of the projects.

The Steering Committee.   A steering committee is frequently appointed for a large project. Membership of the committee depends on the nature of the project and the organization structure. In principle is should contain the heads of those departments or functions that will play a vital role in the success of the project. In a large cost-reduction program in manufacturing it could contain the chief industrial engineer, the head of production, the plant engineer, and the project manager. The membership of the steering committee should be chosen so that decisions can be made in the meetings concerning most issues raised. It is desirable that the chairman is the hierarchical head of most of the members of the steering committee.

The committee will meet approximately once a month. The tasks of the committee vary through the life cycle of the project. In the beginning they review and approve the plans for the project; in the intermediate period they review and follow up progress, and in the implementation phase they support the project manager in pushing through changes. In very large programs it may even be necessary to appoint subcommittees to the steering committee. For example, if the cost-reduction program covers the assembly operations, the machine shop, and the plate shop, it may be advisable to appoint subcommittees for each of the three shops: this type of subcommittee is sometimes called a reference group.

Designing the right committee structure is a balancing act. On the one hand, it is important to provide line managers and others concerned with a forum to participate in the project by reviewing plans and progress. On the other hand, if too many committees are created that hold many meetings, the project team will have to spend a disproportionate amount of time on preparing for and attending meetings. Designing the right committee structure is frequently an essential ingredient of success.

Task Forces.   A task force is different from a team. Members of a project team work fully or at least most of their time on a project. A task force consists of members who spend only a small portion of their time on an issue and most of it in task force meetings. Task forces are not an alternative to project teams but complementary. For example, it may be found that there are diverging views on how the labor savings of projects should be calculated. A good solution can be the appointment of a task force with representatives of the accounting department, personnel, and industrial engineering who hammer out a proposal that establishes guidelines for laborsaving calculations. The guidelines are presented for approval to the steering committee.

Responsibility for Implementation.   In many, if not most, instances the industrial engineering department does not have the authority to implement its conclusions. The "end product" is frequently a recommendation.

Recommendations that are not implemented are useless or even worse: they are a negative investment. It is therefore extremely important that the industrial engineering department carries out the projects such that their chances of implementation are high. A fairly common cause of lack of success is the belief that projects will be installed if the analysis clearly shows that there will be savings or other concrete benefits. The result of most industrial engineering projects is that employees have to change the way they work. This will be resisted unless the employees concerned are convinced that the new situation represents an improvement also from their point of view. The issue of how to overcome resistance to change is not the subject of this chapter, but project management can make a very important contribution in this area. The process of implementation should be built into the project from the very beginning. Some of the important methods of doing this are:

- Participation of line managers in the steering committee and reference groups
- Planning of information meetings in the course of the program with those who will be affected
- Holding appreciation courses in the case of the introduction of new techniques
- Involving the project manager and teams in the implementation of the recommendations
- Keeping the steering committee "alive" until most of the recommendations are installed

The project manager is in a unique position in having an overview of the entire project and all its facets; this allows foreseeing the human consequences and preparing for them. It is quite common to assign to the project manager responsibility not only for the development of the recommendations but also for their implementation.

PROCEDURAL ASPECTS

The procedural aspects of project management consist of the following items:

1. Project selection procedure
2. Project definition, project phases, and scheduling
3. Savings control

4. Role of the steering committee in the follow-up of results

5. Project portfolio management

Project Selection Procedure.   As stated in the beginning of this chapter, the workload of the industrial engineering department consists of the regular tasks of keeping different systems in good trim, making continuous improvements, and projects. There are two sources for projects: requests from different departments and initiatives taken by the industrial engineering department. There is considerable overlap between these two categories because the chief industrial engineer on many occasions will suggest that a department start a project. Based on such a suggestion the department will make the request.

Whether or not to consider a task as a project is a question of judgment. A task will be assigned project status depending on such factors as size and the attention required of senior management. There are three aspects to consider:

- Choosing the most promising projects
- Keeping the customers happy
- Considering subcontracting

These three aspects will be considered in turn.

Choosing the Most Promising Projects.   A well-performing industrial engineering department will generally be asked to undertake more work than it has the capacity to handle. The department should select those projects that are of the greatest value to the company. Many of the projects that are carried out will result in cost reductions. One factor that should be considered is the relation between the expected savings and the "investment" in the project in terms of industrial engineering time and investments in tooling and other hardware if applicable. This relationship can be expressed as a payout period or as a return on investment. In principle those projects should be chosen that give the best financial return. There are, however, other types of projects where no financial return can be calculated, for example, the making of a layout for a new factory unit, the improvement in the quality of products, or an increase in delivery reliability. When the choice is made, the tangible and intangible benefits should be carefully considered. Another factor to be considered is the likelihood of success; generally it is preferable to undertake those projects that can count on strong support of line management.

Keeping the Customer Happy.   As stated earlier, in most industrial departments there is a demand for more projects than the department can handle. This poses the problem of which projects to give priority. Every customer wishes to have the highest priority. There is of course the principle of first come first served. If establishing priorities becomes a big problem, it may be wise to involve the persons that are requesting the projects as a group in the setting of priorities.

Subcontracting.   Certain companies have a small industrial engineering staff and hire industrial engineers on a temporary basis depending on the workload. Even large industrial engineering departments do the same when confronted with peak loads. This is one of many different forms of subcontracting. A project can be subcontracted in its totality, in which case the subcontractor also takes on the responsibility for managing the project and for the professional quality; another approach is to hire temporary staff, as additional capacity, to expand team capacity.

Project Definition, Project Phases, and Scheduling.   The subjects of project definition, project phases, and scheduling will be described in the following paragraphs.

Project Definition.   Every project should be properly specified before it is launched. In most cases a small survey should be made to prepare the project definition. The project definition should include the following subjects:

1. Present situation

2. Objectives of the project

3. Scope

4. Analysis method and analysis steps

5. Project organization (for example, steering committee)

6. Expected industrial engineering resources required and other costs and investments if applicable

7. Expected savings and other benefits

8. Time schedule

In some companies a project will be started only when such a project definition has been made and formally signed by the person requesting the project and by the chief industrial engineer.

Large projects can be broken down into smaller projects. For example, a manufacturing cost-reduction project can be broken down into projects for the different areas like assembly and machining. These areas in turn can be broken down in areas such as workplace layout and cutting feeds and speeds. Unfortunately there is no clear terminology to distinguish "projects" at the different levels of aggregation. In some cases the highest level is called a program rather than a project. The extent to which large projects should be broken down into smaller projects is a matter of judgment. The guiding principle is that each project at the lowest level should be a logical unit that can be assigned to a team or a team member where they can take on the task relatively independently from the other projects at the same level. Each project goes through a number of phases over time which will be described in the next paragraph. It is again a matter of judgment at what levels of project aggregation different phases should be distinguished.

Project Phases. A project typically passes through the following phases:

1. Definition of project

2. Analysis of the present situation

3. Development of improvements

4. Preparation and presentation of the recommendations

5. Approval of the recommendations

6. Installation of the recommendations

7. Final report

These phases represent a more detailed planning of the project that has been broadly defined in connection with the project definition. The end of each phase is a milestone in the life cycle of the project and forms the basis for the scheduling of the project.

Scheduling. A schedule should be prepared for each project covering every phase. Such a schedule can be made using a Gantt chart or network planning. Figure 3.2 shows an example of a Gantt chart. Network planning is described in Sec. 10, Chap. 7. Network planning has the advantage of showing the interrelationships between the activities on the schedule as well as the critical path. Gantt charts have the advantage of providing an easy to understand overview. It is possible to indicate a few interdependencies in a Gantt chart. In Fig. 3.2 for example, the arrow from materials handling to assembly means that the new material handling methods must be installed before the new assembly methods can be installed. In the example shown, the total project is broken down into three subprojects, with a schedule of the phases for each of them. In many cases it would be practical to break the subprojects into sub-subprojects and make the schedule at that level. In addition to the schedule of the project a schedule should be prepared that shows the assignment of the industrial engineers to the different projects. The schedules are the basis for follow-up of the project. At the end of every month the tasks accomplished are compared with the plan. In Fig. 3.2 a follow-up, made at the end of May, is shown. It shows that the project is behind schedule in machining but ahead of schedule in materials handling. These follow-

Manufacturing Cost reduction Program													
Code	Project	Project Manager	Jan	Feb	Mar	May	Jun	Jul	Aug	Sep	Oct	Nov	
MS	**Machine Shop**	Green											
MS1	• Present situation			- - -									
MS2	• Development				- - - - - - -	- - - - -						
MS3	• Recommendations								- - -				
MS4	• Installation									- - - - - - - -		
	Assembly	Dixon											
AS1	• Present situation			- - -									
AS2	• Development				- - - - - - - -							
AS3	• Recommendations						- -						
AS4	•Installation							- - - - - - - - -				
MH	**Materials handling**	Clark											
MH1	• Present situation			- - -									
MH2	• Development				- - - - - -							
MH3	• Recommendations						- - -						
MH4	•Installation						- -						
	Steering Committee meetings		•	•	•	•	•	•	•	•	•	•	•
	- - - - Plan ——— Progress	▲ Review date					▲						

FIGURE 3.2   Schedule for a large industrial engineering project, showing the status at the end of May.

ups serve as the basis for taking corrective action. A similar follow-up should be made of the industrial engineering person days used as compared with the budget. Quite commonly a shortfall in results achieved is caused by an "underspend" in terms of industrial engineering resources.

Savings Control.   The achievement of savings or cost reduction is the objective of many projects. If a project consists of many subprojects and sub-subprojects, a formal procedure for savings control is essential. The project definition will generally specify the savings that are expected to be achieved. It is generally not acceptable to wait until the end of a project and determine if the projected savings actually were achieved. It is necessary to make a plan (a budget) for the achievement of savings. Such a plan is relatively easy to make if the project can be broken down into many small projects each "producing" savings. The purposes of savings control are as follows:

• To set targets for savings to be achieved in the course of the project
• To monitor the actual achievements as compared with the target

The process of achieving a saving goes in principle through the following steps:

1. Development of an improved method.
2. Calculation of savings that will be the result of the installation of the improved methods. The savings at this stage are called "recommended savings."
3. Presentation of the improved methods to a person or body, for example, the steering committee, with the authority to decide upon the installation of the proposed changes and the corresponding savings. After approval the savings are referred to as "approved savings."
4. The new methods are installed after approval. This installation can involve a change of

a layout, training of operators, and/or the installation of new tooling. When the employees are working in accordance with the new methods the approved savings are compared with the actual savings. The savings after installation are referred to as "installed savings."

There are many ways to calculate savings. An easy method for savings control purposes is to calculate the savings on an annual basis. The annual savings are the reduction in costs that will result from introducing the new method during one year. In savings control the savings are reported as installed savings from the point in time that the new methods are installed. It will obviously take 1 year from the start of the installation before the annual savings are realized.

One of the tools used in savings control is a savings control chart. An example of such a chart is shown in Fig. 3.3. The chart shows the budgeted savings compared with the results achieved. There is a time lag between the approved and the installed savings which corresponds to the time it takes after the approval until the new method is installed. When the program is started, a projection is made of the approved savings ("budget approved savings") and of the installed savings ("budget installed savings"). A review is made of all projects that are part of a program at the end of every month, and the savings of the

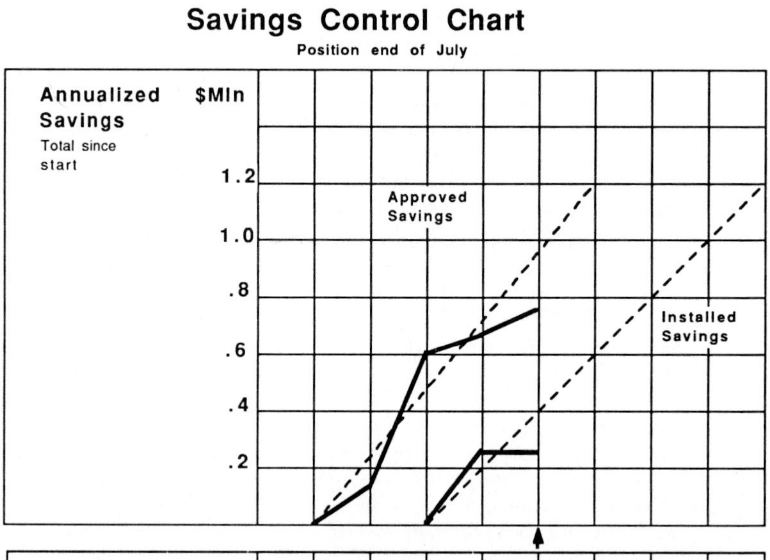

Savings Control Chart

Position end of July

End of Month		Mar	Apr	May	Jun	Jul	Aug	Sep	Oct	Nov
Approved Savings	Budget -----		.24	.48	.72	1.20				
	Actual		.15	.60	.68	.75				

Installed Savings		Mar	Apr	May	Jun	Jul	Aug	Sep	Oct	Nov
Installed Savings	Budget -----				.20	.40				
	Actual				.28	.28				

FIGURE 3.3 Savings control chart.

projects that have been approved or installed are added to the results of the previous month. This chart gives at a glance an overview of the status of the total program. It will highlight delays in getting savings approved as well as in installation. Top management is always interested in reviewing this chart. In Fig. 3.3 the situation is shown at the end of July. The approved savings and installed savings are behind budget.

The foundation of reviewing and monitoring results is the project definition and the division of projects into phases. The process is almost "automatic" if a steering committee is appointed.

Role of the Steering Committee in the Follow-up of Results.   In most cases the steering committee will meet once a month to review progress. A progress report should be presented at each meeting containing the following:

* Follow-up of the project time schedule
* Follow-up of the resources used as compared with budget
* Savings achieved as compared with budget (savings control)

Project Portfolio Management.   Projects in the context of the portfolio refers to the projects (or programs) and not to the subprojects. The chief industrial engineer is responsible for many different projects that are active at the same time. The portfolio content is determined by the project selection procedure. There are two points still to be made about managing the portfolio: the distribution and the annual review.

Distribution of Projects.   An industrial engineering department generally serves many departments. It is advisable to see to it that there are projects in all departments so as to have a wide distribution. It is not desirable to have projects concentrated in a few of the departments year after year.

Annual Report.   Many successful chief industrial engineers prepare an annual report. In such a report an overview is given of the results achieved with completed projects and the highlights of current projects. Such a report gives other departments an excellent insight into and an overview of the activities of the industrial engineering department and can be distributed to senior management in the company. It also serves as a marketing tool inside the company for the industrial engineering function.

CONCLUSION

Project management is one of the most important tools chief engineers have to manage the department. It allows them to direct and monitor the important activities in the department. Project management is also vital to organize and monitor relations with the customers of the industrial engineering department.

BIBLIOGRAPHY

Adams, J. R., and B. W. Campbell, *Roles and Responsibilities of the Project Manager,* Project Management Institute, Drexel Hill, Pa., 1982.

Baker and Wilemon, "A Summary of Major Research Findings Regarding the Human Element in Project Management," *Project Management Quarterly,* vol. 8, pp. 34–40, 1977.

Cleland, D. I., and W. R. King, *Project Management Handbook,* Van Nostrand Reinhold, New York, 1983.

Goldwerger and Paroush, "Capital Budgeting of Independent Projects," *Management Science,* vol. 23, pp. 1242–1246, July 1977.

Kernzner, H., *Project Management: A Systems Approach to Planning, Scheduling, and Controlling,* 2d ed., Van Nostrand Reinhold, New York, 1984.

Kimmons, Robert L., and James H. Loweree, *Project Management: A Reference for Professionals,* Dekker, New York, 1989.

Saaty, T. L., *The Analytical Hierarchy Process: Planning, Priority Setting, and Resource Allocation,* McGraw-Hill, New York, 1980.

Stuckenbruck, Linn C., *The Implementation of Project Management,* Addison-Wesley, Reading, Mass., 1981.

Thamhain and Wilemon, "Leadership Effectiveness in Program Management," *Project Management Quarterly,* vol. 8, pp. 25–31, 1977.

CHAPTER 4
EFFECTIVE COMMUNICATION

William K. Spence, Ph.D.
Communications Consultant
Plano, Texas

In the 1980s one of the most crowded highways in the nation was U.S. highway 75 in Dallas, Texas During rush hour tempers often flared and occasionally words would be exchanged between motorists. One hot summer afternoon in 1989, a motorist came close to hitting another while fighting traffic on U.S. 75. The motorist who was nearly hit managed to pull his car alongside the other and, honking his horn, raised his hand and gestured to express his displeasure. He then pulled away and was continuing his trip home when he heard a horn honk. Looking out the window to his left, he saw the driver he had just reprimanded. This time, however, the other driver's window was rolled down and the barrel of a gun extended out the window. The second motorist fired the gun, seriously injuring the first motorist, causing him to run off the road into an embankment. He eventually died of injuries suffered in the incident.

This story is true, and I relate it here to illustrate the difference between effective communication and the effective use of language. In this brief sequence of events, there was effective communication between the two drivers even though language was not used. The communication doesn't seem to be particularly civilized, but it was effective.

OBJECTIVES OF THIS CHAPTER

Objective One: The Evolutionary Nature of Language. In the story above, a distinction is made between communication and the use of language. In the section that follows, you'll find background information regarding the evolutionary nature of language. The expectation is that if we understand something about the unique characteristics of language, we can take advantage of the almost magical power it offers.

Objective Two: Listening and Rapport. Our second objective is to cover some of the major factors involved in developing rapport and being an effective listener. People are often surprised to find that there is very little difference between these two skills.

Objective Three: Effective Presentations—Making the Sale. The last part of this chapter is concerned with the output side of communication. We'll go through the steps for preparing and delivering a presentation that will optimize your chance to make the sale.

THE EVOLUTIONARY NATURE OF LANGUAGE

In the story of the two motorists, the consequence of the confrontation was tragic. The tragedy is that a rational discussion or a simple expression of concern might have ended the charged emotional state of the two drivers. However, the actions described in the story were not driven by a rational process but by a more fundamental physiological process that evolved to increase the probability that we will survive. We call this behavior the *fight or flight* syndrome.

In situations where one's life may be in danger, nature did not leave survival up to a rational decision process. Imminent danger requires action be taken quickly. Fortunately for us, the process of natural selection chose an elegant mechanism to increase our chances of survival.

Imagine for a moment that it's 10,000 B.C. and you're walking down a path through the woods carrying a stick with a sharpened rock tied to its end. You're wearing a loin cloth and little else. Suddenly, a bear crashes through the underbrush and charges toward you. Nature was not kind to those who entered into a rational decision-making process under these circumstances. If you take time to decide whether the bear is hungry or has sharp teeth or bad breath, it could be the last rational decision you have an opportunity to make.

For most of us, we make the decision to run in an instant, without the intervention of any logical process. We choose to fight only after our escape options are eliminated. Survival is paramount, and the process of natural selection allowed those who chose the most appropriate survival action to procreate and pass the survival behavior on to their offspring. That survival behavior is still with us. Today, however, our lives are seldom threatened by an outside force. If you see a bear, it's likely to be in a cage. Few of us have ever experienced a situation where our lives depended on our ability to run or fight. In a civilized society, fight or flight is most often invoked because someone interacts with us verbally. Although language is the distinguishing feature of humans, nature has not yet separated pure language from the ancestral representations it can be used to create. Even as adults, loud noises still frighten us. Insults still anger us. Threats still frighten or anger us.

Indeed, even the sight or sound of something that is life-threatening invokes part of our autonomic nervous system, and catabolic, or "energy using," processes are initiated. Epinephrine is released into the bloodstream, and in short order our heart is pounding. Epinephrine also causes a certain class of small muscles to contract, reducing the blood supply to less important parts of the body such as the skin and other noncritical extremities. This combined action causes more blood to be delivered to the large muscles of the body to prepare for fight or flight. Simultaneously, a glucocorticoid is released by the adrenal glands. This particular glucocorticoid aids in the conversion of glycogen to glucose. Glycogen is stored in the muscles but must be converted to glucose before it can be effectively used by the muscles. In less than a second several things happen that allow us to perform physical feats that we might normally consider impossible.

In animals, the process described above still plays an important role in survival. Humans, on the other hand, rarely require such a physical transformation to survive. Though seldom used for the purposes it evolved to serve, the "fight or flight" mechanism still exists in humans and can be initiated by factors other than those which threaten our survival; we are more likely to get angry over something someone says than we are over someone trying to harm us or our offspring; we are more likely to become fearful because of social forces (standing in front of an audience, for example) than we are because our life is in danger.

The physiological process that prepares us for fight or flight also impairs our ability to think rationally. Our ability to think is affected in the following way when we become frightened or angry.

First, the release of epinephrine causes the blood supply to our extremities to be reduced. Our neocortex (Greek for "new bark"), which has evolved over the older brain stem, is treated as an extremity and receives a reduced blood supply. Our neocortex is the thinking part of our brain. The reduced blood supply to our neocortex reduces our ability to think clearly. Therefore, in times of stress, whether caused by fear or anger, nature chose action over thinking.

Second, concurrent with the release of epinephrine, the adrenal glands release a glucocorticoid which aids in the conversion of glycogen to glucose. Glucose is used by our muscles as energy. The combined effect is that we can run faster or fight harder and be less aware of distractions such as pain. Normally, we would take action to avoid pain, but if we are fighting or running for our life, pain is a distractor.

It's easy to see that this process evolved for survival and not for civilized interaction. Unfortunately, we sometimes invoke this survival mechanism when it is inappropriate, as in the case of the two motorists. And, it isn't too difficult to imagine that some of these biological activities are invoked when we're under stress for a different reason, such as giving a presentation.

Since humans have the capacity for language, we also have the ability to invoke the same biological survival mechanisms through the use of language. If someone challenges a statement we make or makes a comment that appears to attack us or someone we care about, we may find ourselves getting angry. Or, if we're threatened, we may get frightened. In either case, we invoke primitive survival mechanisms and our ability to think clearly is diminished. Anger and fear will cause us to do things we would not do under ordinary circumstances. Murders, for example, are seldom planned. They are most often spontaneous acts committed during periods of anger or fear.

This survival process and the resulting inability to think clearly are described here to provide background information for some mental activities you will be encouraged to engage in when you are confronted with something that you disagree with, or when you are preparing to give a presentation.

Language: Universal and Species Specific.   There is general agreement among cognitive and developmental psychologists that language evolved to serve social needs. Although we use language for other purposes today, it is still deeply rooted in the business of facilitating interpersonal relationships.

Communication between and among higher forms of life is universal. Language, on the other hand, is both universal and species-specific. It is specific to human beings, and while efforts to teach chimpanzees and other primates to use a language have received considerable attention, none have produced language in a primate—or in any other animal. Some animals have been taught to vocalize a few words (a parrot, for example) and some have been taught to express a feeling or desire by signing (chimpanzees). However, no animals use syntax, grammar, or abstractions. Nor are they generative with the words they do learn. In short, they *appear* to develop a few behaviors that mimic language, but nothing more. This stark reality is in sharp contrast to the fact that any normally developed human will learn the language to which he or she is exposed. In fact, no other species has the neurological and physical equipment to produce language. Other species produce sounds to communicate, but sounds are to a language as paint is to a painting. The two should not be confused.

Language Production: A Complex, Distinguishing Feature.   The production and comprehension of language is an enormously complex activity involving many billions of neurons operating in concert to either produce or decode a thought. Indeed, much of the brain is involved in the production of speech, and some areas serve no other purpose, especially in parts of the left temporal lobe. Since Karl Wernicke (1874) discovered that people with damage to a particular area of the left posterior temporal lobe lost the ability to speak coherently, we have known that the brain has specialized areas which serve language. For some reason (probably a good one) nature selected the left hemisphere to be the primary server of language function. This piece of information has led to some general misinformation about the location and nature of language. There is, for example, a widely held belief that since approximately 94 percent of adults are right-handed and 95 percent of adults have verbal skills in the left hemisphere, left-handed people must have verbal skills in the right hemisphere. These facts can be misleading. More right-handed people have verbal skills in the right hemisphere than do left-handers. Most left-handers have verbal skills in the left hemisphere, and some people produce language bilaterally. Whatever the cause of left- or right-handedness, left-handers do have a higher incidence of left hemisphere dominance for language, but contrary to the popular belief that left-handers are *right-brained,* the incidence of right hemisphere language dominance in left-handers is only about 15 percent.

Perhaps we are still evolving, or perhaps nature is still trying to decide where language should reside. Whatever the case, language is the most obvious and observable feature distinguishing humans from all other animal species.

Language: Defining Features.   The characteristics that define language include its ability to displace things in time and space; to present abstractions, a continuous flow of events and ideas, or to pass knowledge from one person or generation to another. In contrast, other species communicate the "here and now" in concrete terms with knowledge passed to new generations through genetic encoding.

In cognitive psychology there exists a school of thought which argues that language is the basis for what we refer to as *thinking.* Even though many people prefer to believe in an ethereal force that produces thought as a separate entity (the philosophy of dualism), a great deal of empirical evidence supports the proposition that language and thought are inseparable. However, our purpose here is not to discuss the nature of language but to discuss how we can use language to communicate effectively in a civilized manner.

Language: Communicating Effectively.   There are two parts to communication: input and output. Since we're not very knowledgeable when we're born, the input part of communication is unquestionably most important when we're young. In fact, our ability to receive and process information verbally continues to be a distinguishing factor at almost every level of accomplishment.

One survey of over 200 leading executives who reached significant career milestones early in life (senior managers at an average age of 32 and chief executives at an average age of 41) attempted to determine what differentiated these 200 people from other would-be chief executives. One section of the study asked them to tell what they had to learn in order to progress as rapidly as they did. There were eight distinct categories of responses. Each category used four or five words to describe the activities. The five words used to describe the most important set of skills were "patience, understanding, listening, tact, and tolerance." An observation to be made here is that these five words represent skills that are not part of any formal instruction in most academic institutions. They are skills learned as we are processed by life. In this chapter we address these five skills, paying special attention to listening and the effect it has on establishing rapport.

On the output side, we are sometimes called on to share our knowledge or information. The ability to produce a presentation that sells, educates, or motivates is a skill that is universally admired. We quote those who have issued utterances that we find intellectually invigorating. The phrase, "Ask not what your country can do for you, ask what you can do for your country" still rings in the ears of those who heard John Kennedy when he gave his inaugural address in January of 1961. "Four score and seven years ago our fathers brought forth..." is the beginning of another speech that is universally admired. It is, of course, the opening sentence of Lincoln's Gettysburg Address. The question to be asked is, what makes these lines so memorable; why do some stand apart and others fade from memory?

In the first example, the rhetorical device is called a chiasmus. The "chi" is Greek for the letter "X," and the word literally means a crossing of the words. If you analyze the sentence you will find that it consists of two parallel phrases. The same words are used in both phrases, but they're reordered in the second phrase. The result gives the phrase a lyrical quality when done properly. A chiasmus can be very effective when used under the right circumstances.

In the second example, Lincoln's Gettysburg Address, the rhetorical device is called alliteration, sometimes referred to as head rhymes. If you look carefully at the construction of the sentences in this speech, you will find it filled with alliteration. Again, the construction of the sentence produces a lyrical, poetic quality, and while we may not be aware of the underlying cause, it has an emotion-generating quality and is very memorable.

If you're interested in adding some of these rhetorical devices to a presentation you plan to give, find a reference book on rhetoric and spend a few minutes reviewing the different techniques that can be used to enhance your presentation.

LISTENING AND RAPPORT

If you practice the techniques we cover in this section, you can change your life for the better by changing the way you relate to others. If you choose, you can change your relationships with your employees, manager, peers, spouse, children, or anyone. The people may be different, but the techniques are the same.

Why Should I Listen?   One reason to listen is to find out what the other person knows that you don't know. There are a lot of *other* smart people in the world. Perhaps at one point in our evolution we didn't need the intellectual leverage that multiple minds give us, but the complexity of the world today requires focus and specialization. To be successful in the intensely competitive and complex environment of today, we need the leverage of other minds.

Effective listening also demonstrates that you care about the other person. Our ability to develop rapport is a measure of our ability to be an effective listener.

A Case of Changed Listening Habits.   Imagine that you're back in school, sitting in a classroom on a bright, sunny, spring afternoon. You're looking out the window at a crystal clear, blue sky with a few puffy white clouds. You're daydreaming, not paying much attention to the activities in the classroom. This condition is referred to as ''passive'' listening. In this state, mental activity is subdued and energy consumption minimized. You're in a relaxed state and tuned to hear only those sounds that have significance for you. As you watch the activities outside, you suddenly hear something you recognize—your name! Suddenly, you've gone from *passive* to *active* listening. If you were measuring brain wave activity during the transition from passive to active listening, you would see a dramatic difference in the two states. In the active state, brain wave amplitude is higher and more frequent and you burn a lot more energy than you do in a passive state. Indeed, neural tissue is the body's most expensive in terms of energy consumption per unit weight.

As you snap back into the classroom, you realize that your professor just called your name. At this point, most of us are scrambling to recall what was said just before our name, and according to research that started with an elegant study by George Miller in 1958, that is entirely within the realm of possibility. The research demonstrates that we have a short-term, echoic buffer that can retain seven, plus or minus two, ''chunks'' of information for a few milliseconds.

Now, suppose you can't recall what was said just before your name; that is, your buffer is empty. Most of us have developed a stock question to get us out of this predicament. We ask something like, ''Would you repeat the question please?''

The situation described above actually happened to me when I was in high school, and my teacher responded with, ''I didn't ask a question,'' and launched a fully loaded eraser at some part of my body. The eraser ricocheted off my desk, leaving a cloud of chalk dust and a room full of snickering young men. I paid attention for most of the rest of the semester and was amazed at how much I learned. I didn't pay attention because I was afraid of failing, I paid attention because I was concerned about eraser missiles—or something worse. My high school was an all boys school taught by the Christian Brothers. They're wonderful people who had a well-deserved reputation for being somewhat forceful when necessary. I relate the incident here to illustrate the following two points.

First, we can classify listening into two different types: *active* and *passive*. In the case above, I was engaged in *passive* listening. Since we burn more energy when we are actively listening, nature gives us alternatives. We can listen passively and burn less energy, or we can listen actively. Passive listening is sometimes referred to as selective listening, but the term *passive* imparts the notion of energy conservation that is more appropriate. When we are passively listening, we appear to process only the information that we have determined to be important (we do hear all the sounds, we just don't process them all). The determination of importance is unlikely to be a conscious process. For example, the mother of a small baby can sleep very soundly through sounds that would seemingly wake the dead. However, if her baby cries in the next room, in the middle of the night, she probably will hear it—the baby is important to her. As the baby gets older, she may still hear it but allow Dad to make the determination of importance.

Selective listening also accounts for what is sometimes called the *cocktail party phenomenon*. This occurs when you are in a crowded, noisy room, engaged in conversation, and hear your name mentioned across the room. To everyone else, your name might have been additional noise, but to you it was important. We have an innate tendency to tune in to sounds or information that we think are important. If you plan to spend your day outside, for example, there's a good chance you are interested in the weather. If the radio happens to be on, you may find yourself listening only when the weather report comes on.

With equal ease we can tune out those sounds we deem unimportant. If you've been married for a long time, you may notice that your spouse seems to tune out when he or she hears your voice. Keep in mind that this behavior is not gender-specific—we all do it! When we do, we effectively tell the other person that what they have to say is not important. When we do listen actively, we send a message to the other person that what they have to say is important to us, and by extension, that *they* are important. By listening, we place value on what the other person has to say; they feel they have value when they're with us and we've taken a giant step toward building rapport.

Second, our stock response of "would you repeat the question please?" illustrates how sophisticated our patterns of habituation can become. We habituate to most of the routine activities we perform, including our verbal responses. We may catch ourselves using the same response to our spouse without regard to the comment or question. In the early comedy shows, we often heard an unconscious "yes dear" from a husband or wife as an indication that they weren't listening. Our children may hear the stock answer of "no" to so many of their requests that they stop making requests. Our habituated responses reveal that we are reacting to, but not processing, information from the other person. To the other person, this is an indication that they, and the things they say, are unimportant to us.

In the survey of the 200 executives mentioned earlier, you will recall that listening and the associated cognitive skills were among the most important things they felt they had to learn. Of course, these people were obviously intelligent and other factors were involved in their rapid rise to positions of influence and power, but the fact that they viewed these skills as crucial to their success should not be surprising. The reason is a very logical one. Once you leave the process of "doing," that is, executing a routine set of activities, the more you must have concern for human relations simply because you are no longer in the production, or "doing" end of the business, you are in the "people" business. Successful executives must be able to deal effectively with human relations problems because it's a major part of their job.

In 1979, I experienced the effect that good listening skills can have on your sense of self-worth. At the time, I had been employed by a major computer company for 16 years and had worked for 17 different managers. I was located in Washington, D.C., and had been there for 4 years. Although I liked Washington as a city, I decided it was time I moved on, and so I set about to locate another position in either Dallas or Atlanta, preferably in an educational organization. Over a period of a few months I put together a résumé and prepared a short presentation that covered five points I wanted to make each time I talked to a prospective new employer. The points related things like why I wanted to leave Washington, why I wanted to come to Dallas, what skills I thought I would bring to the job, and so on. When I got an opportunity to travel to Dallas a few months later, I tracked down a few managers who were putting together educational groups and made an appointment to talk to them while I was there. When I arrived I had a casual, noninterview conversation with several of them to get a feel for the managers and the opportunities available. Before I left Washington, a friend had asked me to look up a gentleman I'll call Al, and to pass on a warm ''hello'' for him. I agreed and before I left Dallas I dropped by Al's office. Al invited me in and asked how things were going in Washington. I passed on greetings from his friend and after spending a few minutes talking with him, I decided to give him my five-part presentation.

Two things should be noted here; one, I had never met Al before, and two, I didn't know if he had an opening on his staff.

During the course of our conversation Al did some things that I had never experienced in my previous 16 years with any of my 17 managers. Al listened intently to the reasons I gave for leaving Washington. After I made each point he would stop me and paraphrase the reason back to me. Each time I made one of the points he repeated the process. Al understood everything I said, but more importantly I was getting the impression that he was concerned that I was making the right decision for me; that he wasn't looking for a skill that I could bring to get him promoted, get him a raise, or the like. Something else happened that I had not experienced before. The phone rang on the credenza behind Al and he didn't bother to answer it. During my 16 years and 17 managers I had learned that whoever was on the other end of the phone was always more important than I was. I stopped, but Al didn't even blink. This had never happened to me before, and I wasn't sure how to proceed. Fortunately, the phone switched over to his secretary, allowing me to continue.

At the end of the conversation, Al pushed himself back from his desk, looked up and spent about two minutes summarizing his understanding of the five points that I had made. The effect of that short conversation was dramatic. Fifteen minutes earlier I had never met Al. After our short talk, I was going to be disappointed if I didn't go to work for him. The reason was simple. He acted as if he cared, he was concerned, and he wanted to make sure that I was doing the right thing for me—not just bringing some skill that he could use. As luck would have it, Al offered me a position on his staff. I accepted. I found out later that he only did those things during interviews.

The Interpersonal Gap.   We are all unique individuals. We grew up with different parents, went to different schools and churches, and had different friends and siblings. We have different thoughts, feelings, and experiences. Sometimes we have information we want to pass along. Since we are all individuals, we have our own unique way of encoding and decoding verbal information. The fact that everyone is so unique makes interacting with others exciting, but it can also create problems when we try to communicate. The problem occurs when the encoding scheme of person A does not match the decoding scheme of person B. Figure 4.1 depicts the transfer of a message from person A to person B. When we say something to someone else, we deliver our message using three basic communication modalities. Few people are surprised to find that a message consists of a verbal part, a tone, and body language, but many are surprised to discover the percentage of the message conveyed by each of these modalities.

If we study communication between people who know each other well, in relationships such as spousal, parent-child, or employee-manager, we find about 7 percent (5 to 8 percent) of the message to be verbal. Tone carries 38 percent (35 to 40 percent) and 55 percent of the message is body language. In the case of the two motorists, 100 percent of the communication was a combination of body and car language. Actions truly do speak louder than words.

If you've ever had the experience of coming home at 7:00 P.M. when you were expected around 6:00 P.M. because you and your spouse were to meet friends for dinner at 6:45 P.M., you know that tone can carry more than 38 percent of a message. If you didn't call and let your spouse know that you were going to be late, the greeting you receive might be less than friendly when you walk in the door. If you're greeted with something like, ''I hope you've had a good time,'' odds are that the verbal portion of the message means very little. In this case, the tone carries the message.

When the three modalities are congruent, that is, all three carry the same meaning, the message is believable and we interpret it as being sincere. However, when one modality is incongruent, or doesn't match the other two, we interpret the message as being sarcastic. Sarcasm is the term we use to describe a message where the verbal portion says one thing, but the tone and body language say something else.

In Fig. 4.1, the interpersonal gap is represented as the difference between the encoding process of person A and the decoding process of person B. The longer we know someone, the greater the probability that our encoding and decoding will be appropriate. But even with certainty of understanding, there is a need to provide feedback. Feedback lets the other person know that you care, that you're interested, and it tells them that they are important to you. In the story regarding Al and my move to Dallas, it was clear that he understood everything I said. It was his feedback that made me feel as if he was genuinely interested in me as a person.

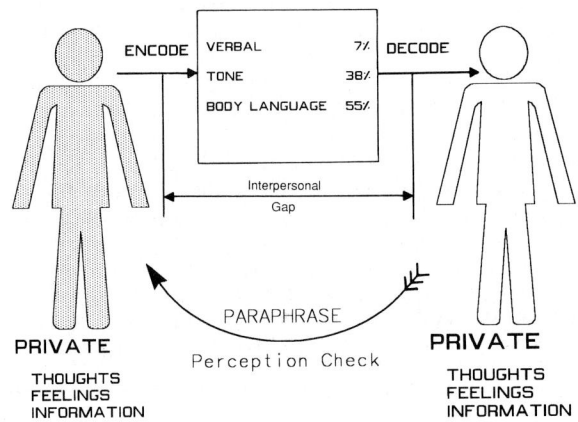

FIGURE 4.1 The interpersonal gap.

There are several techniques that we can consciously develop to make the communication between two people a closed-loop system. Paraphrasing is the most common and perhaps the most effective.

Paraphrasing occurs when one party restates, in his or her own words, a thought or concept expressed by the other person. Paraphrasing should not be confused with parroting. Parroting is simply repeating the words used by the other person. Paraphrasing forces us to understand the message. As we continue paraphrasing, we begin to think like the other person. If we do it properly, with sensitivity, we move toward a state of cognitive congruence with our discourse companion; that is, we are both processing information in the same modality.

Another technique we can use to close the communication loop is called *perception checking*. A perception check is an attempt to confirm your perception of what the other person wants to tell you. A perception check is used in a conversation when we get the impression (or have the perception) that the other person wants to tell us something but is hesitant to do so. If, for example, you are in a conversation with someone and sense that they don't like their current job, but they haven't said that to you directly, you can ask them a direct question, that is, "Are you trying to tell me that you don't like your job?" There are only two responses they can give. One is "yes," which means your perception was correct. The other is "no," which is usually followed by a statement of what they really want to talk to you about, that is, "what I'm really trying to say is...." In either case, there is no penalty for the perception check. In making a perception check, you have expressed interest in the other person and what they have to say. A perception check effectively opens the channel of communicati

An Exercise in Listening.   In the seminars on listening and rapport that I conduct, I frequently include an exercise to practice the paraphrasing and perception checking techniques mentioned above. As a topic for discussion, I relate a story about four people who get stranded on opposite sides of a river in South America and a sailor who happens to come down the river (Fig. 4.2). The audience listens to the story and then ranks the participants in the order that they would like to have them as a friend. It's an enjoyable exercise and the point it makes is better experienced than explained. I'll relate the story here. At the end of the story, you rank the participants, with number 1 being the person you would most like to have as a friend and number 5 being the person you would least like to have as a friend.

Once upon a time, four people lived on a river. On one side of the river lived Ann and Jack. On the other side of the river lived Ralph and Mike. None of the people could cross the river because they had no boat and there were no bridges or safe places to swim across. The river was infested with crocodiles and piranhas. It was possible to talk across the river, and over a period of time, Ann and Ralph fell in love with each other. They became engaged but had no way of getting together.

Ann and Jack were friends and Jack was Ann's confidant. On the other side of the river, Mike and Ralph were also friends.

One day a sailor named Sinbad came down the river on his boat. He was hailed by Ann, who asked him to take her across the river so that she could be with Ralph, her betrothed. Sinbad agreed to do this on one condition—that she sleep with him.

Ann was placed in a deep conflict by this offer and she sought help from Jack, her friend and confidant. Jack spent several hours talking with Ann. He was most sympathetic to her plight but essentially communicated that he was confident she would make the right decision.

Ann decided to take up Sinbad's offer. She spent the night with Sinbad on his boat and the next day Sinbad took her across the river and dropped her off on the opposite shore. Ralph was awaiting her landing and they embraced at once.

After a while, Ralph asked her how she had managed to convince Sinbad to bring her across the river. Ann told him the whole story.

When he heard the story, Ralph pushed Ann away, and said he would have nothing more to do with her.

Just at this time, Mike came by. He overheard what had happened and when he saw Ralph pushing Ann away, he moved in and beat up Ralph thoroughly.

Your task is to rank order the characters in this fable on the basis of how much you like each person. Your number 1 rank will be the person you like best; number 5 will be the person you like the least.

Rank order:
1. _____   is the person I would most like to have as a friend.
2. _____
3. _____
4. _____
5. _____   is the person I would least like to have as a friend.

Before continuing, rank the five people in this story yourself to see how your rankings stack up against the general population. You may or may not agree with the average rankings, but when you get large groups together the rankings are always similar and always interesting. Over the years I've asked over 20,000 people (in groups of 10 to 600) to rank the characters in this story. I'll give you the results at the end of this discussion.

In the workshop, discussion groups are formed in triads and each person in the group has an assignment. One person is the speaker and is charged with telling how they ranked the protagonists in the story. Another person is the listener, whose job is to practice paraphrasing and perception checking. The third person is an observer who watches the listener and takes notes on what they see.

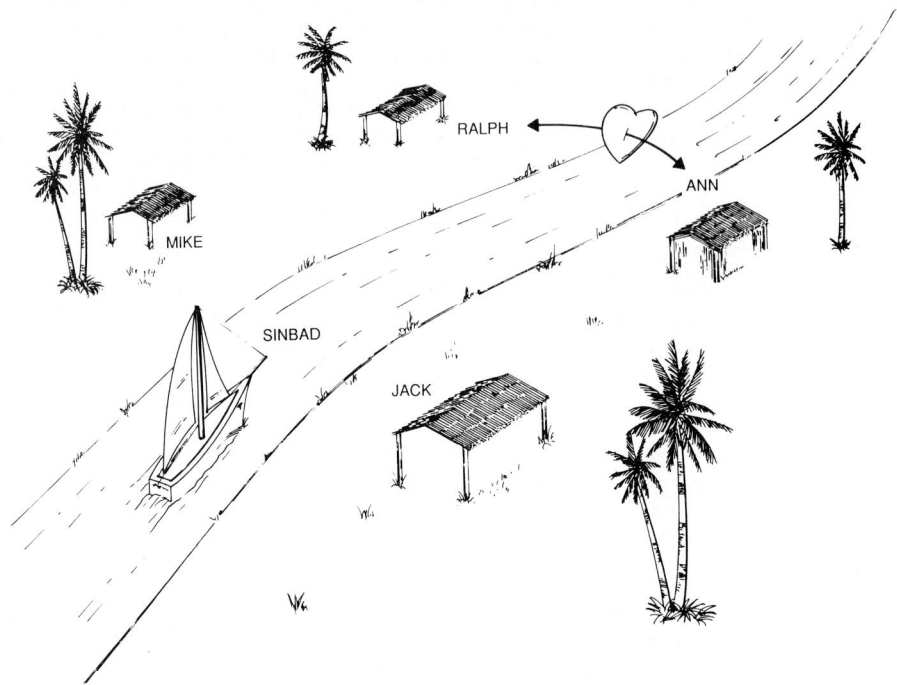

FIGURE 4.2   Illustration of a story on an exercise in listening.

The exercise starts and continues for about 5 minutes; then the conversation is stopped and they do a short debriefing. Roles are changed and they do it again. Almost everyone learns something about themselves during this exercise. First, they find that their natural inclination is to try to change the speaker's mind rather than listening to find out what caused the speaker to rank them in a particular order.

Second, the observer finds it very difficult to watch the listener. It is so difficult that few people watch the listener until they've been reminded several times.

Third, even though everyone has an intellectual understanding of paraphrasing and perception checking, they find it awkward at first. After a little practice, they develop a few paraphrasing stems (the first few words that lead into a paraphrase, such as ''Do you mean...'') and paraphrasing gets much easier. There are several stages we go through when we develop a new behavior. At first we feel awkward. Then it becomes intentional, that is, we know we're doing it, but it's not uncomfortable. After a while, it's integrated into our normal conversation. At some advanced stage, paraphrasing becomes a habituated behavior.

Fourth, the awkward feeling the listener has is not apparent to the speaker. Indeed, the speaker often gets the sense that the listener is genuinely interested in how he or she ranked the characters in the story, and by extension, that the listener is interested in the speaker.

After the exercises, I collect the rankings of the group. With very few exceptions, the order is always Jack, Ann, Mike, Ralph, and Sinbad. In general, about 50 percent of the people rank Jack as number 1. Ann gets about 25 percent of the votes for number 1, and Mike gets about 20 percent. It's rare that anyone has Ralph at the top of his or her list. Sinbad will frequently get number 1 votes even though he always gets the most number 5 votes.

The question I pose to the audience is ''Why do so many people want Jack to be their friend?'' The answer? Because Jack supported Ann and was nonjudgmental. Perhaps he offered his opinion, but in the end said he would support whatever decision she made.

Another question considered is "Why do so many people dislike Ralph?" The answer? Because he was judgmental. The discussions are always interesting and frequently humorous, but the point is made. We like people who will listen to our point of view. They don't have to agree, just listen.

If you want your family, peers, or management to know that you are interested in the subject, or care about them, take time to listen. Paraphrase even if you think you understand. Do a perception check occasionally. The feedback does much more than make sure you understand what the other person is saying. It tells them that you are interested, that you care about them as a person. Try it with your children. Listen to the full story before making a decision. Sometimes all a child wants is someone there to care.

Dealing with Emotion Generating Words and Phrases.   Someone has defined a snappy comeback as something we think of 24 hours after someone has verbally attacked us. Why can't we think of the comeback when we need it? Sometimes, it's because we're angry or flustered and can't think at all.

We all have convictions and biases and we all respond to emotional words and phrases. Occasionally, someone will say something that invokes the survival mechanism that was discussed earlier in this chapter. The comment may be something that intended to be an innocent observation, or it may be a comment intended to make us angry. In either case, the result is the same. When we get angry, we lose our ability to think clearly and quickly.

In a civilized, verbal society, we should be able to use and accept verbal comments without getting angry. However, a comment made at the wrong time or in the wrong tone or with the wrong body language may send us into an emotionally charged state. As you recall, when we are in a high-stress state, such as that caused by anger or fear, we lose our ability to think clearly. Our survival instincts have prepared us to act, not think. If we respond while we're in an emotionally charged state we occasionally say or do something that we regret later.

Politicians, as a group, have an amazing ability to handle comments that frequently border on personal attacks. If you observe the behavior of politicians in public, you will note that they can produce articulate answers to questions that would cause some of us to attack the questioner. Successful politicians have no choice. Richard Nixon once commented that "friends come and go, enemies accumulate."

Politicians can't afford to make additional enemies by becoming angry. The good ones have developed the ability to ignore critical comments. Some of us worry about what others think of us, but as someone once observed, "we wouldn't worry about what others think of us if we realized how seldom they did."

Fogging: Acknowledging the Element of Truth in a Critical Comment.   For many of us, our day can be ruined by a comment someone makes to us before we leave home in the morning. The comment may not be intended to be anything more than an observation, but when we accept it as personally directed criticism, we may get angry. When we do, we often respond by escalating, which results in a further escalation by the second party, which results in another escalation, and so forth. If you're married, you probably recognize the pattern. Let's say that you're getting dressed to go to work in the morning and your spouse makes some observation about the tie you've chosen, like, "That tie doesn't match your suit." In reality, if your spouse makes that observation, something probably is wrong with your choice of ties. However, we don't always respond to the truth in the statement; we respond to the way it's worded or the tone that's used. We could be clever and respond with something like "Your face doesn't match your body either," but that probably wouldn't end the conversation. Indeed, the odds are that your spouse will raise the ante, and before you know it, both of you are angry. By now, the *fight or flight* mechanism is in full operation and neither of you is thinking very clearly. You both say things that you really don't mean and you hear things that were better left unsaid. At some point, you find that you were the second choice for a marriage partner anyway. The next step is an attorney and a divorce.

Now think for a moment. Was there anything rational about the escalation process? We have invoked a powerful survival mechanism to resolve a disagreement over a necktie! For the price of an attorney you can buy an entire store full of ties. We consider ourselves to be rational yet we get a divorce over the color or pattern of a tie. Worse, when it's over we won't even remember that the tie started the argument. Worse still, the last comment you made will be remembered forever. It will be repeated when you have another conflict 20 years later...if you're still married to the same person.

Even if we don't get a divorce, we're still angry when we get to work and we're not happy being unhappy by ourselves. We spread it around!

One of the ways we can handle critical comments and other emotion-generating statements and phrases is to use a technique called *fogging*. Fogging is a technique whereby we acknowledge the element of truth in a comment and ignore the rest. We accomplish two things by fogging. One, our personal integrity remains intact because we only acknowledge the truth in the comment. Second, we acknowledge that the other party has expressed an opinion. We don't have to agree with their opinion; we just acknowledge that it was made. In the case above, we could have responded with "I'm sure there are ties that will match better," or something similar. In this case you're acknowledging that the other party is entitled to his or her opinion without agreeing or defending your selection of ties.

With a little practice you can become adept at fogging. After you become accustomed to searching for the element of truth in critical comment, you begin to realize that you are in control of your attitude. Since you own your attitude, controlling it is your responsibility. Unfortunately, we often let others dictate how we feel by letting an offhand, or worse, an intentional comment invoke the fight or flight portion of our autonomic nervous system.

The fogging technique consists of two steps. First, we look for an element of truth in the comment. Second, we acknowledge only the element of truth, accepting it as the other party's opinion. We then ignore the hyperbole surrounding the remark. We don't have to agree with it; we just acknowledge it.

Fogging allows you to continue to think rationally. If you choose to escalate intentionally, you may do so without getting emotionally charged. We can always think better in an unemotional state than we can when we're angry.

Practice fogging. In time you learn not to let a critical comment or other uncontrollable events determine how you feel.

PREPARING AN EFFECTIVE PRESENTATION

Delivering an effective presentation has less to do with the presenter than it does with the presenter's subject matter expertise, preparation, and rehearsal. In this section we cover each step required to prepare and deliver an effective presentation. The key to success is to prepare each step completely, in order, and not concern ourselves with the next step until the current step is complete.

Define Your Objectives.   Before you begin preparation, ask yourself these two questions:

1. What is my objective? If you don't have an objective, spend your time doing something else. If your objective is to sell your audience on a new technique or process, identify *precisely* what you want your audience to agree to when you finish.

2. How will this presentation change my audience? What will they be able to do after the presentation that they couldn't, or wouldn't, do before? If the answer is "nothing," don't give the presentation. Among the reasons to give a presentation are to inform, to entertain, to motivate, to sell, to build rapport, and the like. Whatever the reason, you should have that reason articulated and written down before you begin to prepare you material. Having a definite purpose focuses the mind for information that contributes to the accomplishment of that purpose. If you read your written objective occasionally while researching and preparing your presentation, you can continually ask yourself "does this information contribute to the accomplishment of my objective?'' It's a simple technique that can save hours or even days of writing and research.

Evaluate Your Audience.   What does your audience expect to get from this presentation? Will you be using terminology that your audience can understand? How can you get their attention; that is, how can you make sure they will be interested in what you have to say? Understanding your audience will help you choose the correct terminology and the proper stories and metaphors to make your presentation easy to follow. Although you may be enthusiastic, that may not be enough to sustain the interest of your audience.

A brief example of how important audience evaluation can be occurred in the spring of 1991 when my youngest son, Tippah, was asked to give a motivational speech. His audience was to be 65 young competitive swimmers at a spring swimming camp. The central theme of his talk dealt with goal setting and the importance of having a positive attitude.

At one point in the presentation he told several stories to illustrate how a goal keeps your mind focused. One of the stories related a desire he'd had since he was twelve years old—to own one of the muscle cars of the 1960s—a 1969 Hurst Olds 442. He related how he had written his goal on a 3 by 5 card (at the suggestion of his swim coach) and had kept it pinned to the bulletin board in his room. He told about how the car appeared in his garage, piece by piece, over the next 7 years. He described the 455 cubic inch engine and the Offenhauser manifold, the camshaft, the 800 cfm, four-barreled carburetor and the special lifters that he had saved up to buy. Then one day, after 7 years of accumulating and assembling parts, the 500-horsepower engine roared to life. He described how he would start it and SLOWLY back it out of his garage; how the police would surround him and he would SLOWLY pull it back into the garage.

His enthusiasm and depth of knowledge must have made an impression on several of the 14- and 15-year-old members of his audience. About 5 minutes after he changed topics one of the young men in the audience raised his hand and, unable to sit passively any longer, asked ''when are you going to tell us some more about your car?'' Several of the other young men enthusiastically nodded agreement.

Had he stopped to do an in-depth evaluation of his audience, he probably would have spent more time relating goal setting to issues that 14- and 15-year-old boys were really interested in, such as the Olds 442, and related the accomplishment of owning the Olds to how he had learned to set goals as a competitive swimmer.

Organizing the Presentation.   Effective presentations are divided into three parts: (1) establishing the topic (opening), (2) developing your case (the body), (3) asking for the sale (the close).

Establishing the Topic of the Presentation.   Effective communication requires that you and your audience be thinking about the same topic. Even if they are aware of the general topic of the talk, you should still make a conscious effort to help them get mentally prepared to assimilate the information you plan to give them.

Depending on your presentation medium, setting the stage for your talk can take several forms. If, for example, you are giving a presentation using visuals, such as flip charts or an overhead projector, it's common practice for the first visual after the cover sheet to be an outline of the talk. The more abstract or difficult the material, the greater the necessity to establish the proper framework.

As the presenter, *you* know what you will be talking about, but unless you're presenting to a group of mind readers, your audience will have only the clues generated by the title of your talk. Titles aren't always very informative. I once went to a presentation titled, ''After this, there must be dragons.'' The title was an attention-getter. However, I never discovered the central theme of the talk. The speaker had mastered the skill of titling, but he had work to do elsewhere.

Comprehension is dramatically affected by the existence of an organizing theme or principle. The following example, taken from Dooling and Lachman (1971), illustrates the difficulty we have comprehending a passage without an organizing principle.

> With hocked gems financing him, our hero bravely defined all scornful laughter that tried to prevent his scheme. "Your eyes deceive," he had said, "an egg, not a table, typifies this unexplored planet." Now three sturdy sisters sought proof, forging along through calm vastness, yet more often over turbulent peaks and valleys. Days became weeks as many doubters spread fearful rumors about the edge. At last from nowhere welcome winged creatures appeared signifying momentous success.

The absence of an organizing theme makes this piece of narrative difficult to understand. We are also forced to allocate much more attention, or cognitive capacity, than might be necessary if we knew the theme. Even then we still may not understand what is being discussed. Eventually, we lose interest.

However, once we know what the passage refers to, the story seems to leap into focus. If you read it with the knowledge that the "him" in the story is Christopher Columbus, it makes much more sense. It's the same with messages that we try to deliver. When we tell a story or make a presentation, *we* know the setting. If we share that information with our audience, they'll find it much easier to follow the flow of information.

Outlines, a statement of objectives, and the like serve the purpose of establishing a discourse domain. They give our audience a context in which to interpret our comments.

The Body: Developing Your Case for Change.   Now that you have a stated objective and an outline of your presentation, it's time to start adding meat. If, for example, you are trying to sell a change in the way of doing things, this is where you will carefully construct your case for change.

Creating the Basis for Comparison: Why a Change Is Required.   More often than not, presentations attempt to convince the audience that they should change something. You may have a better way to do something, a philosophy that you want them to adopt, a product that you think they should buy, and so on. There are exceptions, of course, but in general we prepare a presentation to convince someone to make a change in themselves or in their environment.

Since we want them to change the current environment, we should take time to describe it and the shortcomings that exist. In a religious setting, the spiritual leader describes the moral environment and the attendant problems to convince the congregation that a change is in order. Salespeople describe the problems that their product will solve.

Describing the current environment may seem unnecessary when the audience is involved in the environment on a daily basis, but a description of the current process, or product, and the associated problems does more than pass on information. By refreshing the current environment, you have created a basis for comparison. A vacuum cleaner salesperson takes time to show you how much dirt accumulates in your carpet; advertisements may imply problems without stating them directly, that is, "Quality is our most important product," implying that the product you own has quality-related problems; many advertisements start with "Are you tired of..." or a similar phrase to remind you of current conditions; military buildups and weapons systems are most often based on perceived or projected needs. An effective sales presentation does not assume that the audience or client is fully aware of all the problems in the current or projected environment. Often, we want to make changes before the problems occur.

If you're making a presentation in response to a request which has enumerated a set of problems, take the time to restate the problems even though they were stated to you. The restatement may be as simple as:

> I was asked to find out why the reject rate on assembly line 4 is 40 percent higher than the other three assembly lines. As you know, all four lines use identical machinery. The reject rate of lines 1, 2, and 3 is averaging 4 percent. However, the reject rate for line 4 has been running at 5.6 percent over the last 3 months, causing us to incur an additional $16,000 per month in rework. After investigating each phase of the assembly process, we've identified several causes for the variance.
>
> I'd like to spend the next 30 minutes reviewing our findings and our proposal for correcting the problems we've identified.

At this point you would open your presentation by revealing your first visual. Don't comment on the 4 percent reject rate of lines 1 through 3. Prudent salespeople do not point out problems that they do not plan to solve with their proposal. The reasons are simple.

First, if you can't solve the problem, no useful purpose is served by bringing it up during the presentation. Even if it's a minor problem it may become a major distractor. People who are in the problem solving business, such as engineers, may give an unsolved problem more attention than it deserves. However, you should be prepared to acknowledge the existence of an issue or concern if someone in the audience brings one up. Spend as little time as possible discussing problems you don't plan to solve.

Second, if you mention a problem that your proposal won't solve, your audience may start looking for a more complete solution. The concept of *completeness* is a very persuasive cognitive ploy. If, for example, you propose a solution that will reduce the reject rate of an assembly line from 4 to 2 percent, you have proposed a 2 percent reduction in defects. If you propose a solution reducing defects from 1 to 0 percent, you have only proposed a 1 percent reduction, but the 0 percent represents *completeness* and can be very persuasive. By pointing out the remaining 2 percent defect rate, we encourage our audience to search for a complete solution while ignoring a solution which will reduce the reject rate by 50 percent. To paraphrase a comment someone once made, "we sometimes let 'best' stand in the way of 'better'."

To summarize, don't point out problems you don't plan to solve.

Developing the Outline and Flow of Your Presentation.   Now that you've established the current conditions as a basis for comparison, you have an opportunity to present your solution to the problems you've just outlined. The first thing you should do is to prepare an outline. Just jot down a rough set of points you want to cover in outline form and mentally go through each point, subvocalizing the comments you plan to make about each point. By subvocalizing, you can tell if the presentation has an easy, coherent flow that you're comfortable with. If you're not comfortable with the flow, your audience won't be either. By the time you finish outlining your proposal, you will have a rough draft of the visuals you will want to use.

Construct the flow of your presentation so problems will be addressed in the same order they were raised when you described the current environment. Skipping around will force your audience to work harder to relate your solutions to the problems you described.

As you prepare each point, anticipate questions and answer them in the context of presenting your proposal. If each point naturally follows the preceding point, your audience will be on the verge of asking about the next point you plan to make. The presentation should proceed logically from one issue to the next, leading the audience from topic to topic with very little effort required on their part.

When making a presentation in a business environment, leave out irrelevant or superfluous comments. Stick to the main theme. Stories, analogies, and metaphors are excellent rhetorical devices that can help our audience reach the conclusion we want, but if they aren't relevant to the main theme, they may do more harm than good by disrupting the coherence of your presentation.

The Close: Asking for the Sale.   Earlier, when you wrote your objectives, you developed a written statement of what you expect your audience to do after they hear your presentation. When you finish, you should be prepared to articulate a clear, concise request for *action*. If possible, phrase your request so that your audience, or the decision maker, can make a decision on the spot. Make the options clear and head off any anticipated objections.

A closing statement might have the following format:

Thanks for your attention....I hope I've explained the current process and the concerns we have so that everyone here fully understands the reason we need to make some changes. The proposed changes will take care of *every* problem we described in our report. If we (1) get an O.K. from you to begin implementation today, we can have the changes in place by (date). I'll keep you informed on a weekly (daily, or the like) basis regarding our progress. If you have no other questions I'd like your permission to begin implementation immediately (Monday, for example).

Stop and wait for an answer. You've either done your job or you haven't. Don't volunteer any additional options. Force them to make a decision. If you've addressed all concerns, you should get a yes. If concerns still exist, you may get additional questions before you get permission to go ahead. In any event, if you've prepared your case thoroughly, you should relax and wait for an answer.

Those who aren't confident with their proposal often begin to offer compromises after a few seconds of silence. During the pause, assume that the decision maker is mentally reviewing the proposal you just made. If you begin to offer compromises without being asked to do so, you may appear unsure of your proposal. Remember, you know more about the topic than anyone else in the audience. After all, you did the study and prepared the presentation. You've done more thinking about the subject than anyone else in the room.

Preparation: Putting the Presentation Together. No amount of organization or visual cleverness can overcome a lack of subject matter expertise. You should be prepared to answer any reasonable subject matter related question from anyone in the audience.

Anticipating Questions and Comments. As you assemble your presentation, try to imagine the questions that could be asked. Write them down. Ask a peer to review your presentation. I've found that most people are very willing to review a presentation and make comments. They almost always have questions that I didn't anticipate and will often fail to understand a point that I thought I was making very clear. Most people feel honored by a request to review a presentation before it is given. If you don't care for their suggestions, ignore them. However, I suggest that you seriously consider any comment that is honestly given.

After you have your list of possible questions, prepare a crisp, articulate answer for each. Rehearse the answer aloud. This will do wonders for your self-confidence. It will also help you answer questions before they are asked. It takes only one weak, groping answer to destroy the credibility of your proposal.

On the other hand, a crisp, confident, factual answer to a difficult question can set aside any concerns about your credibility and your proposal. If you have proposed a process and are asked, for example, if the process has ever been used before, you will fare better by responding with something like, "There are no records of anyone ever having used this process and no patents have been issued. I also checked with companies A, B, C, and none of them has tried it..." or if the process has been used, "Yes, this process is similar to the one being used at companies A and B for the last 18 months. They are very pleased with the results." If you respond with "I don't know but I'll find out," you have just demonstrated a lack of homework and will probably have to do it anyway.

Take time to think about the questions you might get. Have answers prepared before giving the presentation.

Visuals. Preparing visuals should be the easiest part of the presentation, but inexperienced presenters often spend more time preparing visuals than they do preparing content. After you give several hundred presentations, preparing visuals becomes a routine task.

Although visuals may not be the most important part of your presentation, you should use them whenever possible. They can be a very powerful part of your presentation. The prosecution recognized the value of good visuals during the Contra-Gate hearings in the 1980s. At one point, the prosecutor condemned a slide presentation frequently given by Lt. Colonel Oliver North as being an illegal solicitation of funds. The presentation dealt with the situation in Central America and had some very poignant slides of the issues and problems faced by forces that the Reagan administration was trying to support. North's attorney jumped at the opportunity to have the slide show given at the hearing, knowing that it would be captured by television and broadcast nationally. The prosecutors apparently realized that this would be a serious mistake and grudgingly agreed to allow the slide presentation to be shown—without the slides. The room, they said, was too well lit for slides to show up on television. Therefore, North was forced to describe what they would see if the slides could have been used.

In this particular case, the visuals were actual slides of the destruction and death being administered by the Sandanista forces and had the potential of swinging more support toward Colonel North and making the prosecution look as if they were on a witch hunt. After all, people were donating millions of dollars to support the Contras after seeing the slide presentation.

In business presentations, visuals are rarely so dramatic. They do provide a constant reminder of the current topic and keep the presentation flowing according to a plan.

The visual medium should be suited to the audience and the facility. Each visual should have a heading and a *maximum* of seven lines. If you go beyond five or seven lines, you run the risk of making the visual too complex. Each line should be bulletized and should contain the essence of a point you want to make about the major subject of the visual.

The last line should lead you into a transition statement to the next visual. As you finish your comments about the last line on the visual, you make a verbal transition before revealing the next visual.

Assume you have covered the visual illustrated in Fig. 4.3. After explaining why the new process results in 90 percent fewer rejects, you would lead into the next page with a transition statement such as: ''You would expect these reductions in inventory, work in process, cycle time, and rejects to reduce cost...(slight pause while you remove the visual)...and they do!'' At this point you reveal the next visual, which might be titled ''Projected Savings'' followed by bullets noting the amount to be saved in each area.

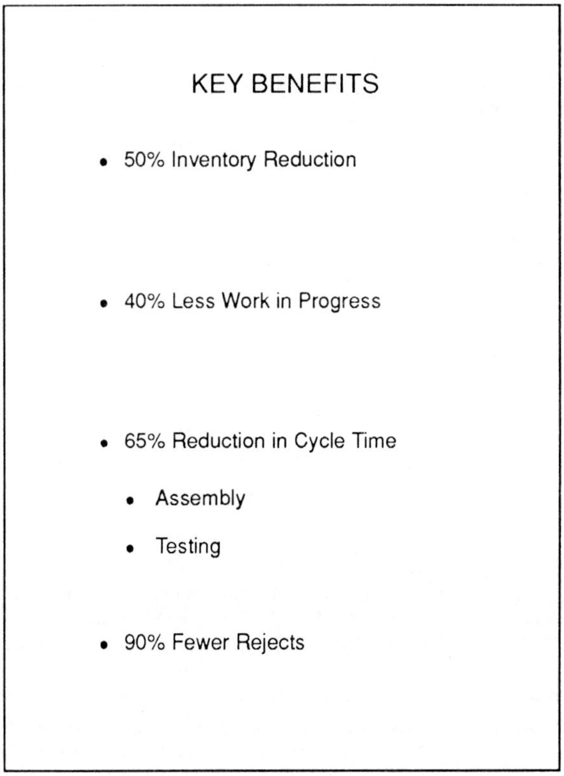

KEY BENEFITS

- 50% Inventory Reduction

- 40% Less Work in Progress

- 65% Reduction in Cycle Time
  - Assembly
  - Testing

- 90% Fewer Rejects

FIGURE 4.3 Sample visual.

Coloring your visuals can be an excellent way of providing visual organization. Pick three colors and use them in the same way throughout the presentation. Choose one for headings, one for major points, and one for clarifying points, and stick to the same scheme on each visual. For purposes of organization, the bullets should be the same color of the next higher level (see Fig. 4.4). Keep the number of colors to two or three. Color is used only as an aid to help establish the relationship of the points on the visual. Excessive or randomly placed colors will only confuse your audience.

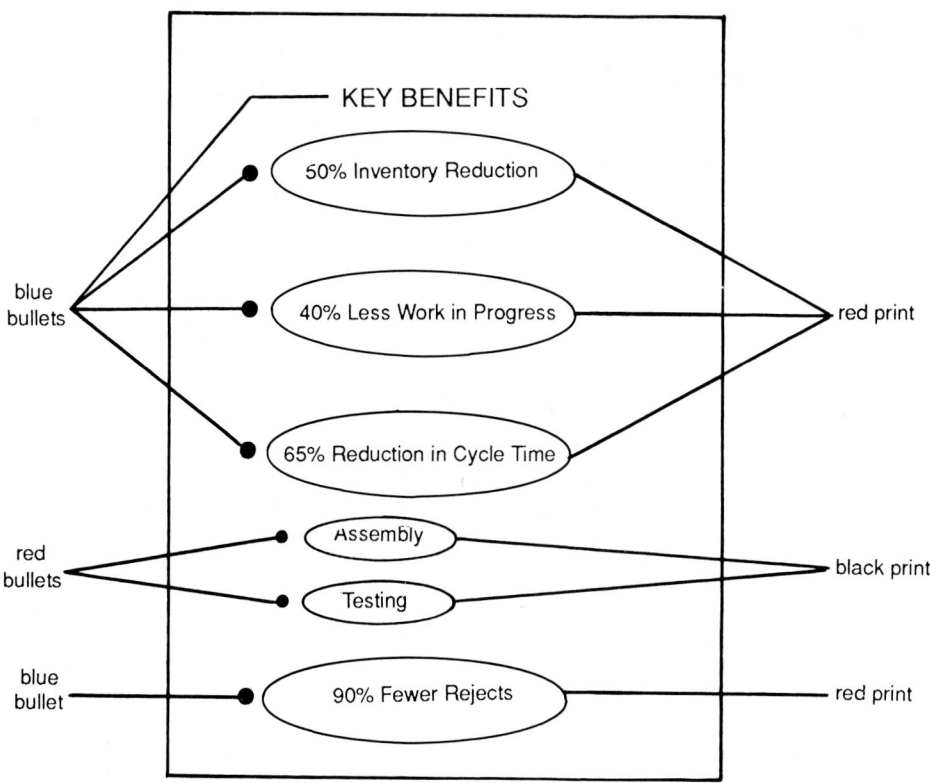

FIGURE 4.4 Typical color scheme for visuals.

Whenever possible, use font sizes to emphasize the importance of the line. The most important line would have the largest letters. Figure 4.4 KEY BENEFITS would use the largest font size on the visual. The major bullets would be one size smaller and the subtopics would be smaller still.

In small groups, *flips charts* may be appropriate. I personally wouldn't use flip charts if the audience was more than four or five rows deep or more than 20 feet from the easel. They are excellent for spontaneous work and can add color and character if used creatively.

If you do use flip charts, be sure to use dark colors such as red, blue, or black (avoid pastels) and keep the same color scheme throughout the presentation. When giving a presentation using flip charts, you should stand to the audience's left of the flip chart stand to avoid having to reach across the visual to point at the beginning of a line.

As a visual medium, *transparencies* have a lot to recommend them. They are physically small, can project a large image, are easily transported, are inexpensive, and can be prepared quickly. In addition, you can write on them with either permanent or water-soluble markers while they are being projected. Also, if you use overhead transparencies, your audience size is virtually unlimited.

You can get transparencies that make blue images on a clear background, black on clear, clear on black, yellow on blue, and a host of other combinations. All you need is an overhead projector and an infrared transparency maker.

Each transparency should be made the same way, that is, if they are hand-drawn, they should all be hand-drawn; if you are making them with an infrared transparency maker, make them all the same way. Find a type and style you like and stick with it until you find something you like better. Consistency and flow are more important than clever visuals.

If you choose to use transparencies and an overhead projector, be sure to clean the surface of the projector before your presentation. If you are right-handed, find a small table to put to the left of the projector (as you face the audience) on which to keep your visuals and notes. By having the table to the left you will not have to stand in front of the projector to change visuals. Also, if you choose to write on your visuals during your presentation, your markers and visuals will be located in front of you as you bend over to write on the transparency. If your notes and markers are on the opposite side of the projector, you will probably manage to look directly into the projector at some point in your presentation. Being blind for a minute or so can be disconcerting if you're in the middle of an important point. Besides, it's not good for your eyesight.

Slides should be reserved for very formal presentations or presentations that require photographic quality. Slides can be beautiful, but producing the original graphics is often an expensive, time-consuming process. It is not unusual for a quality slide presentation to cost $150 per usable slide (in 1991 dollars).

Other drawbacks include having to darken the room during the presentation and being locked into a flow that cannot be easily changed. If someone in the audience asks a question about slide 5 and you're on slide 9, it's awkward to go back to slide 5 and it disrupts the flow of the presentation. The effect is akin to building up to the punch line of a joke and someone asking you to clarify some earlier part—it destroys the flow. Another drawback is the time it takes to produce a quality graphic for a slide. Turnaround time for the final version of a glass-mounted slide is often 3 days or more.

If you do require photographic images, slides are the medium to use, and they can be inexpensive if special graphics aren't required. Sometimes they're the only appropriate medium, as in the Oliver North presentation mentioned earlier. Potential plant sites, photos of people, unusual formations and processes, anything that can be photographed may require slides.

Computer-generated graphics have become very popular in organizations that make extensive use of personal computers. Benefits include the ease of production and modification, the use of motion, and very vivid colors.

The array of software products available to produce graphics is vast and growing. One of the early PC software products to produce graphics was called "StoryBoard" and was produced by IBM (they now have other products that are far more powerful, including one called "HOLLYWOOD"). Once familiar with StoryBoard, it was possible to produce a complex color graphic in 15 minutes or less. The graphic could easily be incorporated into a series of graphics to produce a complete set of visuals for a presentation. The visuals could be arranged and projected as overheads directly from the computer with the aid of a projection unit.

Drawbacks include the problem of having to darken the room for viewing if the image will be projected onto a screen and the need for special equipment to project the computer-generated image. On balance, computer-generated visuals can be very effective and economical if the equipment to produce them is readily available.

When possible, provide a *handout* that elaborates on your presentation. The handout should follow the same flow as the presentation and should provide room for notes. I have found that audiences seldom use handouts after the presentation, but they always seem to want one. Provide them when you can.

Avoid giving presentations without visuals if possible. Visuals provide a second sensory input that reinforces the verbal points you make. They also provide an outline for the presenter to keep them on the subject. Comedians, politicians, preachers, and motivators don't use visuals (they don't deal with hard facts either) but there aren't many other groups of speakers who can operate effectively without some type of visual aid.

Rehearsal. Rehearsal should not be an option. After preparing your material you will serve your purpose well by going over the entire presentation aloud—from beginning to end. Your notes will serve as verbal stems to start each line on the visual. By speaking aloud, you will exercise the same neural circuitry that you will use when you give the presentation. Words or phrases that may be in your reading or writing vocabulary may be difficult or awkward to articulate when you say them for the first time—even though you may have read or heard them many times before.

By rehearsing aloud you may discover words or phrases that you need to practice or change before giving your talk. I have been surprised to find that I couldn't fluidly pronounce a word in the middle of a compound sentence that I could pronounce quite easily by itself. In one case, I was using the term "representativeness" in the middle of a compound sentence and discovered that the middle part of the word didn't come out right. I had to practice saying the word in the context of the sentence over the period of a day before I could say it properly. Sometimes the easiest solution is to rearrange the sentence. Representativeness is a long word, but it isn't a difficult word to pronounce. It just isn't used very often.

Notes. As you rehearse, you can create a set of bulletized notes to guide you through your presentation. The best way to create the notes is to articulate the introductory sentence for each point on the visual and write it down just as you said it. This sentence should reflect the language you use in normal conversation. Be yourself and you won't have to worry about memorizing a line. All you need to get you started is a verbal stem (the first few words of a sentence) for each bullet on the visual. If you use the same language you use in normal conversation, the rest of the sentence will flow automatically after you've rehearsed several times.

You should make notes regarding each bullet on the visual. They needn't be written word for word, but they should capture the essence of what you want to say about the point. As you rehearse you will probably change your articulation several times before it feels comfortable.

Your transition statements are particularly important to the flow of the presentation. Have the stem for the transition statement written so you can see it just before you get ready to change visuals. On flip charts you can write a transition statement in light pencil at the top of the chart, just inside the shadow of the curl in the pages that have been flipped over. For slides you will need a script. If you are using transparencies, the best approach is to use a sheet of paper to separate each transparency and write your transition statement at the bottom of the sheet of paper. As a personal convention, I always put a capital TR: at the bottom of the page followed by my transition statement. With transparencies, I use a dark transparency marker to write the stems on the separator page.

Imagine an open book for a moment. Organization of your notes should be such that you can see the notes for the current visual on the left-hand page (the even-numbered page). The transition to the next visual will be at the bottom of the page. The left-hand page has the notes for the transparency contained on the right-hand page. As you remove the transparency from the right side, you can glance at the note you made for the first point on the visual. When you remove the transparency from the projector to replace it in your set of visuals, you will be able to glance at the transition statement just before you place the visual on top of it. The transition statement can be articulated as you replace the visual on top of the left-hand side and flip the next separator page to gain access to the next visual. At this point you can remove the visual and place it on the projector while glancing at the note you've made for the first line on the visual. In case I haven't explained this properly, just imagine the notes and transition statement being only on the left-hand page. When the visual is on the projector, the right-hand page is blank.

In case you have trouble coming up with a transition statement, use the beginning of the first sentence you will use when you begin to explain the next visual. Whatever it is, it should provide a fluid bridge between the current visual and the next visual.

Taking Stage and Taking Charge.   Now that you've prepared and rehearsed your presentation, it's time for action. An hour or so before you are to give your presentation, review it from beginning to end. Think about the points you want to emphasize. Say the first few opening sentences and the difficult lines aloud. Subvocalize the transition statements. The objective is to be very confident of the first minute or so you will be in front of your audience. A little rehearsal can eliminate the nervousness (remember the fight or flight syndrome) that can cause us not to think very clearly. After a minute or so you should be into the flow of the presentation and your nervousness, if it ever existed, will be gone.

Just before you take your place in front of the audience, while you're still seated, glance at the first few sentences you will use. After you're introduced, take a few seconds to get everything in place. Inexperienced speakers will often start talking before they get to the front of the room to put their notes in place, sometimes they start talking as soon as they get out of their chair. The first few seconds of the presentation is the time to take stage, take charge, and develop rapport with the audience. They want you to be successful.

There may not be a best way to take stage, but there are some better ways. A proven technique is to calmly and purposefully move to your position in front of your audience, keep a smile on your face, and if possible, casually make light conversation with someone nearby while getting your notes and visuals in place. As you position your notes and visuals, glance at the stem for the opening sentence you plan to use. Pause, straighten up, and look around at your audience. Smile and make eye contact with as many people as possible. The slight pause, perhaps 5 seconds, the light comment to someone up front, combined with the deliberate scanning and evaluation of your audience, establishes you as confident, relaxed, and ready.

Open your presentation with the few lines you've rehearsed. Smile and be natural as you proceed through the rest of the presentation.

When you reach the end, ask for the order and keep quiet. You've done your part. It's time for them to make a decision. Good luck.

There are only two powers in the world, the sword and the pen;

and in the end the former is always conquered by the latter.

—Napoleon

METHODS ENGINEERING

CHAPTER 1
CHARTING PROCEDURES

George F. Raymond
Consulting Supervisor
H. B. Maynard and Company, Inc.
Pittsburgh, Pennsylvania

The term "charting procedures" refers to a family of charts, including operation process charts, flow process charts, multiple activity (operator and machine or work planning) charts, and workplace (right- and left-hand) charts.

OBJECTIVES OF CHARTING PROCEDURES

Process charts provide a systematic description of a process or work cycle, with sufficient detail for analysis to develop methods improvements. Each member of the process chart family is designed to help the analyst clearly visualize the present procedure. A standardized format provides a common language so that several people can visualize problems together. This stimulates an exchange or cross pollination of ideas. Most charts combine written, graphic, and pictorial visualization which promotes full participation by everyone concerned. Finally, the charts are excellent tools for presentation of proposals for improved methods to all levels of management.

PROCESS CHARTS AND ACTIVITIES DEFINED

Operation Process Charts.   An operation process chart is a graphic representation of the points at which materials are introduced into the process, and the sequence of inspections and all operations except those involved in material handling. It includes information desirable for analysis such as time required and location.

Flow Process Chart.   A flow process chart is a graphic representation of the sequence of all operations, transportation, inspections, delays, and storage occurring during a process or procedure; it includes information considered desirable for analysis such as time required and distance moved.

The material type presents the process in terms of the events which occur to the material.

The operator type presents the process in terms of the activities of the operator. For analytical purposes and to aid in detecting and eliminating inefficiencies, it is convenient to classify the actions which occur during a process into five classifications. These are known as operations, transportation, inspections, delays, and storage. The following definitions cover the meaning of these classifications under the majority of conditions which will be encountered in process charting work.

○
To change

Operation. An operation occurs when an object is changed in any of its physical or chemical characteristics, is assembled or disassembled from another object, or is arranged or prepared for another operation, transportation, inspection, or storage. An operation also occurs when information is given or received or when planning or calculating takes place.

▷
To move

Transportation. A transportation occurs when an object is moved from one place to another, except when such movements are a part of the operation or are caused by the operator at the workstation during an operation or inspection.

□
To verify

Inspection. An inspection occurs when an object is examined for identification or is verified for quality or quantity in any of its characteristics.

D
To wait

Delay. A delay occurs to an object when conditions except those which intentionally change the physical or chemical characteristics of the object, do not permit or require immediate performance of the next planned action.

▽
To protect

Storage. A storage occurs when an object is kept protected against unauthorized removal.

⬠
Combined Activity. When it is necessary to show activities performed either concurrently or by the same operator at the same work station, the symbols from those activities are combined as shown to represent a combined operation and inspection.

ROLE OF THE PROCESS CHART IN PROBLEM SOLVING

The six-step pattern approach to solving production problems is:

Step 1. Select and define the problem.
Step 2. Break down and visualize in detail.
Step 3. Question with an open mind.
Step 4. Develop an improvement proposal.
Step 5. Install the proposal.
Step 6. Follow up the installation.

The process chart is used to aid in carrying out step 2. Most modern flow process charts have preprinted symbols and include the question part of step 3. Some provide space for the idea part of step 4.

OPERATION PROCESS CHARTS

Operation process charts are used by engineers, chemists, cost accountants, plant managers, and others who want an overall view of the entire process. Because of the wide range of applications, no preprinted form has been devised for general use. A plain sheet of wide paper can be used. All steps should be listed in proper sequence for each component, working vertically from top to bottom. The major component or chassis is conventionally shown at the far right, and all other components are allotted space to the left of this component. The image is like a conveyor line with components fed into the chassis in proper sequence. See the diagram for material feeding into the process (Fig. 1.1).

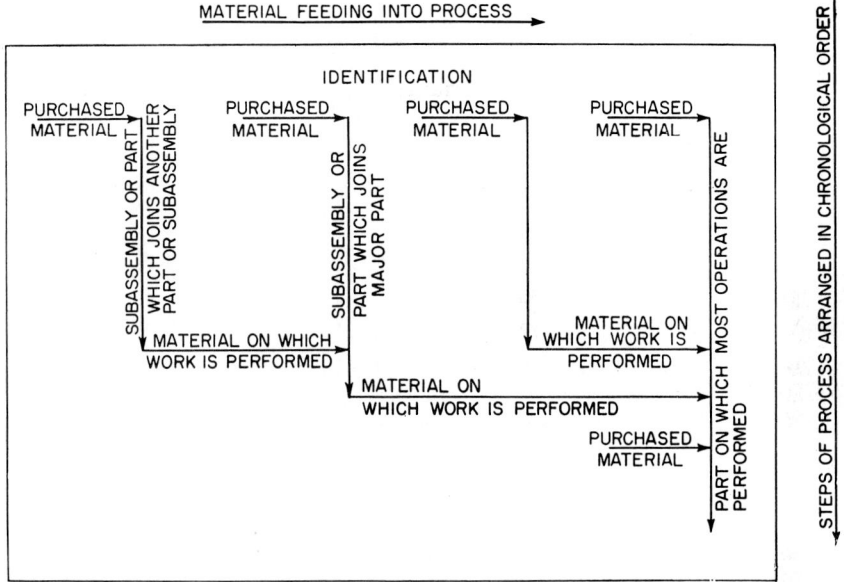

FIGURE 1.1   Graphic representation of principles of operation process chart construction.

To aid in a thorough analysis of the materials, all important information on alloy, finish, shape, and the like should be included. The descriptions of the operations or inspections should be brief. Use descriptive shop terminology (drill, bore, tap, and so on) with the name of the departmental location. For inspections, indicate whether for quantity or quality, sampling or 100 percent, and for what characteristics.

The only symbols used in this chart are for operations and inspections. The symbols are numbered in sequence, beginning with the first step on the major part or chassis, as indicated in Fig. 1.2 for a mechanical assembly. Note that the numbering starts with the first step on the chassis and continues up to the point where the first component is assembled. The numbering then shifts to the component, continues up to the point of assembly, and then shifts back to the chassis.

Time values are usually expressed in TMU [TMU (time measurement unit) = 0.00001 hour] for both operations and inspections. This helps evaluate the significance of each step in terms of potential savings.

Analyzing the Operation Process Chart.   With only four major considerations, materials, operations, inspections, and time, the subject of material is analyzed first. All alternative materials, finishes, and tolerances are evaluated as to function, reliability, service, and cost. Next, the operations are reviewed for possible alternative processing, fabrication, machining or assembly methods, and changed tooling and equipment. Can operations be eliminated, combined, changed, or simplified? Inspections are reviewed for quality level, for replacement with in-process sampling techniques, or by job enlargement or related operations. Time values are reviewed in terms of alternative methods, tooling, and of course, use of outside services for special-purpose equipment. For a more comprehensive discussion, see Sec. 3, Chap. 2, "Operation Analysis."

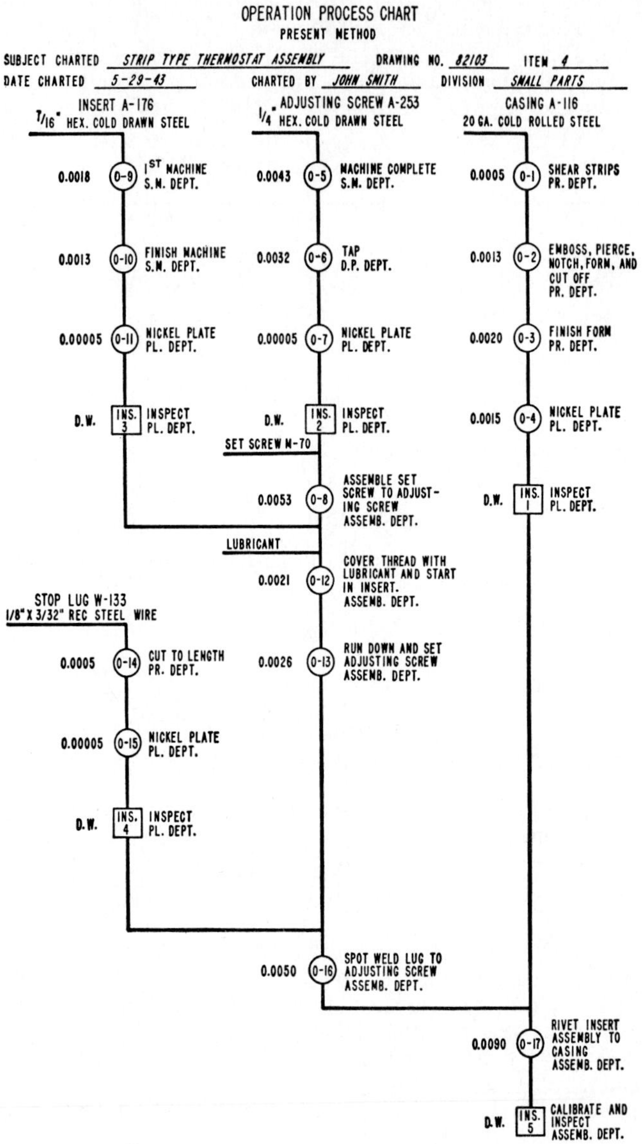

OPERATION PROCESS CHART
PRESENT METHOD

SUBJECT CHARTED *STRIP TYPE THERMOSTAT ASSEMBLY*   DRAWING NO. *82103*   ITEM *4*
DATE CHARTED *5-29-43*   CHARTED BY *JOHN SMITH*   DIVISION *SMALL PARTS*

INSERT A-176
7/16" HEX. COLD DRAWN STEEL

ADJUSTING SCREW A-253
1/4 HEX. COLD DRAWN STEEL

CASING A-116
20 GA. COLD ROLLED STEEL

Insert A-176	Adjusting Screw A-253	Casing A-116
0.0018 (O-9) 1ST MACHINE S.M. DEPT.	0.0043 (O-5) MACHINE COMPLETE S.M. DEPT.	0.0005 (O-1) SHEAR STRIPS PR. DEPT.
0.0013 (O-10) FINISH MACHINE S.M. DEPT.	0.0032 (O-6) TAP D.P. DEPT.	0.0013 (O-2) EMBOSS, PIERCE, NOTCH, FORM, AND CUT OFF PR. DEPT.
0.00005 (O-11) NICKEL PLATE PL. DEPT.	0.00005 (O-7) NICKEL PLATE PL. DEPT.	0.0020 (O-3) FINISH FORM PR. DEPT.
D.W. [INS. 3] INSPECT PL. DEPT.	D.W. [INS. 2] INSPECT PL. DEPT.	0.0015 (O-4) NICKEL PLATE PL. DEPT.

SET SCREW M-70

ASSEMBLE SET SCREW TO ADJUSTING SCREW 0.0053 (O-8) ASSEMB. DEPT.

D.W. [INS. 1] INSPECT PL. DEPT.

LUBRICANT

0.0021 (O-12) COVER THREAD WITH LUBRICANT AND START IN INSERT. ASSEMB. DEPT.

STOP LUG W-133
1/8" X 3/32" REC STEEL WIRE

0.0026 (O-13) RUN DOWN AND SET ADJUSTING SCREW ASSEMB. DEPT.

0.0005 (O-14) CUT TO LENGTH PR. DEPT.

0.00005 (O-15) NICKEL PLATE PL. DEPT.

D.W. [INS. 4] INSPECT PL. DEPT.

0.0050 (O-16) SPOT WELD LUG TO ADJUSTING SCREW ASSEMB. DEPT.

0.0090 (O-17) RIVET INSERT ASSEMBLY TO CASING ASSEMB. DEPT.

D.W. [INS. 5] CALIBRATE AND INSPECT ASSEMB. DEPT.

FIGURE 1.2 Typical operation process chart.

FLOW PROCESS CHARTS

The material type chart follows the steps performed on one component or material during the process or procedure. The operator type chart follows one person, indicating all the activities performed. The material type chart is most useful for a bird's-eye view of production operations, while the operator type chart is better for maintenance or service operations. These should be separate charts.

When using a preprinted form as shown in Fig. 1.3, the data required are obvious. Information should be gathered by actually following the object charted. No attempt should

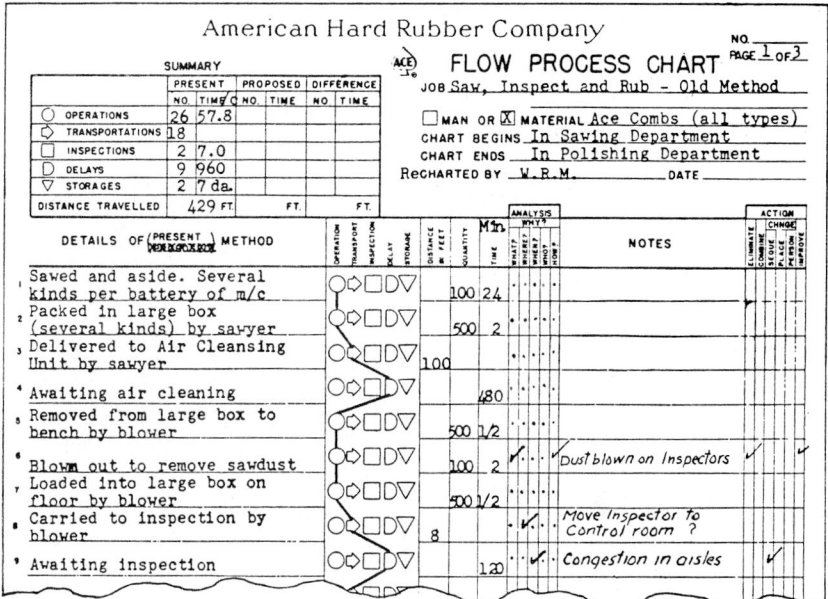

FIGURE 1.3 Flow process chart (material type) using dot and check technique and preprinted symbols.

be made to chart from memory. Descriptions should be brief. For the operator type chart, the active voice is used, as drills, taps, grinds, and so on. For the material type chart, the passive voice is used, as drilled, tapped, ground, and the like.

Time and distance may be shown for all important steps but may be omitted for minor steps. Everything that happens at a workstation during an operation or inspection should be shown on one line. Avoid breaking an operation into minor details such as ''place in basket'' or ''remove from basket.'' These details are best examined in a workplace chart. Delays should be listed when important, but omitted when trivial. The chart should not be cluttered with minor details. Greater detail is permitted on the operator type chart (Fig. 1.4). The notes column can be used to continue the description when it cannot be condensed into the details column. Otherwise, this column is available to record ideas developed during analysis.

Each chart should be marked to indicate whether it portrays the present or the proposed method. The symbols selected for each item should be connected. This will emphasize the relative value of each step and aid in totaling the data for the summary at the top of the chart. Operations have the greatest customer value, with decreasing value for each symbol moving to the right.

Analyzing Flow Process Charts.   To avoid resistance to change, it is desirable to use the six questions: why, where, what, when, who, and how. The questions, in proper sequence, and actions expected are as follows:

Question	Followed by	Action expected
1. What is the purpose?	Why?	1. Eliminate unnecessary activity.
2. Where should this be done?	Why?	2. Combine or change place.
3. When should this be done?	Why?	3. Combine or change time or sequence.
4. Who should do this?	Why?	4. Combine or change person.
5. How should this be done?	Why?	5. Simplify or improve method.

NO. 1
PAGE 1 OF 1

SUMMARY

	PRESENT		PROPOSED		DIFFERENCE	
	NO.	TIME	NO.	TIME	NO.	TIME
○ OPERATIONS	50	6.6				
⇒ TRANSPORTATIONS	43	21.3				
□ INSPECTIONS	17	21.9				
D DELAYS	1	5.5				
▽ STORAGES	-	-				
DISTANCE TRAVELED	1471	FT.		FT.		FT.

FLOW PROCESS CHART

JOB Receive air freight package and bring
to outgoing freight area
☒ MAN OR ☐ MATERIAL Baggage handler
CHART BEGINS At receiving dock
CHART ENDS Outgoing freight area
CHARTED BY A.S. DATE 9/26/—

DETAILS OF (PRESENT) METHOD	OPERATION	TRANSPORT	INSPECTION	DELAY	STORAGE	DISTANCE IN FEET	QUANTITY	MIN. TIME	ANALYSIS — WHAT? / WHERE? / WHEN? / WHO? / HOW? (WHY?)					NOTES	ELIMINATE	COMBINE	SEQUE.	PLACE	PERSON	IMPROVE
1 Other duties	○	⇒	□	D	▽															
2 Goes to equipment area for hand truck	○	⇒	□	D	▽	62		1.0	·	√				PLACE NEAR USE AREA				√		
3 Grasps hand truck and returns to receiving dock	○	⇒	□	D	▽	62		1.0	√					"		√				
4 Loads packages on H.T.	○	⇒	□	D	▽		4	.2	·	·	·	√		USE SEMI- LIVE SKID						√
5 Pushes H.T. to receiving dock scale	○	⇒	□	D	▽	21		.5	·	·	·	√		"						√
6 Tips packages off H.T. onto scale	○	⇒	□	D	▽		4	–	√					PAINT WEIGHT ON SKID		√				
7 Checks weight of each package	○	⇒	□	D	▽		4	.8	√					"		√				
8 Checks packages for cond.	○	⇒	□	D	▽		4	1.8	·	·	√			CHECK AS LOADED ON SKID		√				
9 Loads packages on H.T.	○	⇒	□	D	▽		4	.2	√							√				
10 Pushes to check-in area	○	⇒	□	D	▽	32		.3	·	·	·									
11 Tips packages off H.T.	○	⇒	□	D	▽		4	–	√					LEAVE ON SKID		√				
12 Returns with H.T. to receiving dock	○	⇒	□	D	▽	26		.3	·	·	·	√								√
Items 4-11 repeated 7 times Item 12 repeated 6 times	○	⇒	□	D	▽															
75 Rec. air bill from truck driver and checks no. of pkgs. with A.B.	○	⇒	□	D	▽		32	1.1	·	·	·									
76 Goes with driver to billing office	○	⇒	□	D	▽	48		.65	√					USE WIRE BASKET ON OVERHEAD CABLE		√				
77 Waits while bill is processed and lot labels prepared	○	⇒	□	D	▽			5.5	√							√				
78 Returns to pkgs. with processed copy of bill and lot labels	○	⇒	□	D	▽	48		.65	√							√				
79 Pastes lot labels to each pkg. and air bill to one	○	⇒	□	D	▽	32		1.8	·	·	·	√		USE STAMPING MACHINE						√
80 Loads packages on H.T.	○	⇒	□	D	▽		4	.2	√							√				
81 Pushes H.T. to outgoing freight area	○	⇒	□	D	▽	41		.6	·	·	·	√								√
82 Tips packages off H.T.	○	⇒	□	D	▽		4	–	√							√				
83 Returns with H.T. to check-in area	○	⇒	□	D	▽	41		.6	·	·	·	√								√
Items 80-82 repeated 7 times Item 83 repeated 6 times	○	⇒	□	D	▽															
111 Returns H.T. to equip. area	○	⇒	□	D	▽	30		.5	√							√				
	○	⇒	□	D	▽															

FIGURE 1.4 Flow process chart (operator type).

A simple but effective way to apply the six questions to each item on a flow process chart has been developed for supervisory and management use. This method is called the "dot and check technique." The analyst rests the pencil in each of the question columns, leaving a pencil dot as they think through the implications of the questions as applied to this particular item. If they get an idea from this study, they place a pencil check mark opposite the proper question. In the proper action column, they check "eliminate," "combine," or any of the indicated actions as shown in Fig. 1.3 and places supplementary details in the notes column. This approach is particularly effective when discussion groups review the chart, because it fixes attention on one item at a time. Everyone may participate in developing the proposed method. Guided by the action checked plus the supplementary notes, a "proposed method" flow process chart may be constructed.

The Flow Diagram. The flow diagram is a sketch of the layout of floors and buildings which shows the location of all activities on the flow process chart. The path of the material or operator that has been flow process charted is traced on the flow diagram by lines or string. Each activity is located and identified on the flow diagram by symbol and number corresponding to the flow process chart. The direction of movement is shown by placing arrows so that they point in the direction of flow.

If a movement backtracks over the same path or is repeated in the same direction, separate lines should be drawn for each movement to give emphasis to this backtracking. If string is used, it may be wound around pins and laid up in layers to show repetitive movement. Figure 1.5 is an example of a flow diagram.

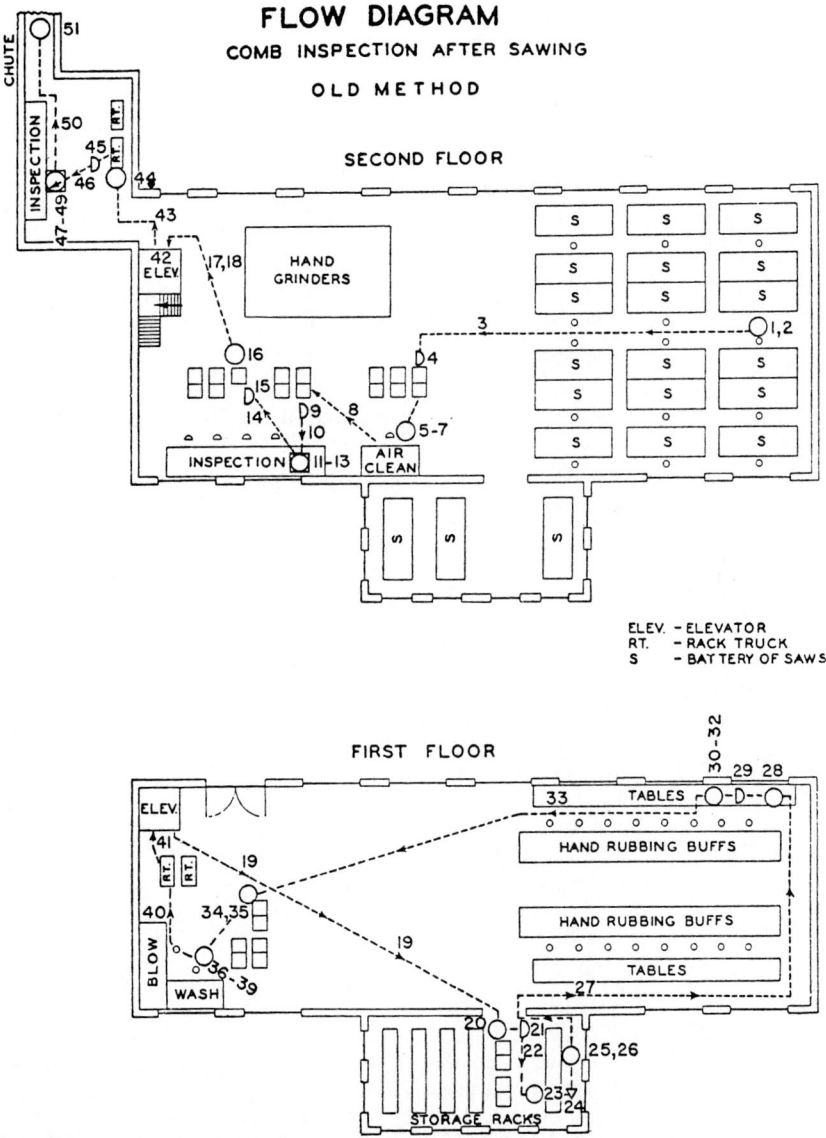

FIGURE 1.5 Flow diagram for flow process chart shown in Fig. 1.3.

When it is desirable to show the movement of more than one item or person on the same flow diagram, each may be identified with a different-colored line or string. If one item or person is being followed, one color may be used for the present method and another color for the proposed method.

The flow diagram becomes a necessary adjunct to the flow process chart wherever movement is an important factor. It shows up backtracking, excessive travel, and points of traffic congestion and acts as a guide for an improved layout.

When a relayout is contemplated, it is customary to use floor, building, or yard plans drawn to scale and templates of all machines and equipment made to the same scale. For the nonmechanical supervisor or executive, it is better to use three-dimensional models. These permit greater participation in the development of a new layout. This may produce a better layout and create better acceptance of it, because more of those affected by it were able to take part in its development.

Charting Paperwork. Single-copy forms can be flow process charted (FPC) like a material. Multicopy forms are first charted individually (FPC) and assembled into a multicolumn procedure chart (MPC). In preparation for the MPC, some modifications of symbols are necessary. The operation symbol will have three variations:

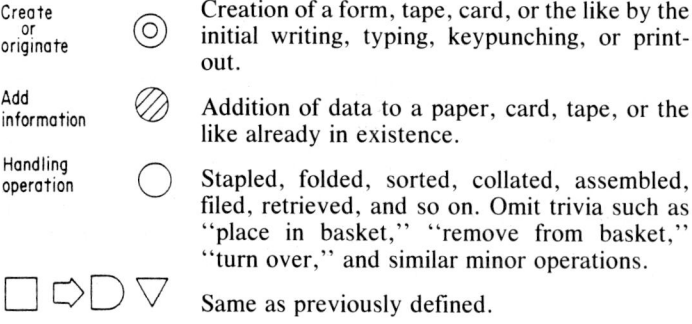

Create
or
originate
Creation of a form, tape, card, or the like by the initial writing, typing, keypunching, or printout.

Add
information
Addition of data to a paper, card, tape, or the like already in existence.

Handling
operation
Stapled, folded, sorted, collated, assembled, filed, retrieved, and so on. Omit trivia such as "place in basket," "remove from basket," "turn over," and similar minor operations.

Same as previously defined.

Delay should be shown only when significant. Listing minor delays may clutter up the chart so that the procedure cannot be clearly seen.

Remember that a storage of paperwork is always preceded by the operation "filed" and followed by the operation "retrieved" or "destroyed." When a paper is destroyed, it can be indicated thus:

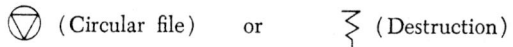

(Circular file)   or   (Destruction)

Flow Lines. FPC, in preparation for an MPC, will require modification of the flow line technique. For all steps that occur directly on the individual paper charted, the regular preprinted symbols are used. They will be modified by superimposing the double circle or crosshatching. For events that occur on affected papers, draw the symbol by hand as indicated in Fig. 1.6. Items 11, 12, and 13 occur on affected papers and are caused by the paper being charted. Instructions on assembling a procedure chart follow.

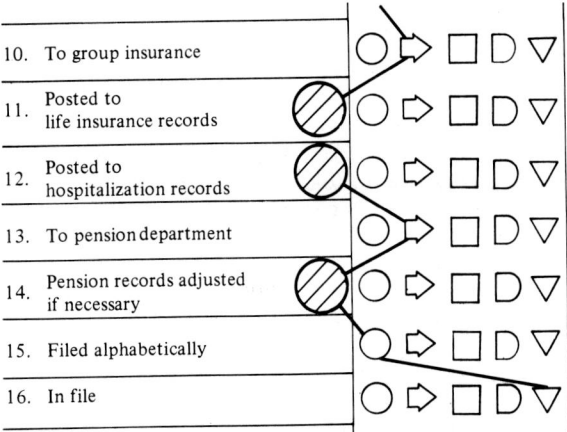

FIGURE 1.6 Convention for showing events that occur on affected papers.

MULTICOLUMN FLOW PROCESS CHARTS

The regular flow process chart is designed to visualize the action on one material or by one person during a process or a procedure. If there are several people working in a gang or several components in a product or procedure, there may be great value in viewing the overall relationship. The operation process chart does this but lacks the detailed information on transportation, delays, and storage shown on the flow process chart. One way to get the overall picture is to assemble a composite chart of all components, as shown in Fig. 1.7.

Because of the inconvenience of handling a complex assembly of flow process charts by a vertical-type chart, and to standardize on the construction, the multicolumn flow process chart was developed. This is a preprinted form with a horizontal line of symbols for each item charted. Mounted on the wall, the entire chart is at eye level. When explaining the chart or during analysis by a group, one can jump back and forth without losing perspective. Viewed on a desk, the horizontal chart is again easier to scan for an overall view. There are two types of multicolumn flow process charts: procedure type and gang type.

Procedure-Type Charts. The individual flow process charts for each paper in the procedure are constructed as outlined earlier under "Charting Paperwork." Because each detail will be brainstormed while in flowchart form, the multicolumn chart condenses the description of each step. Its function is to show relationships. See Fig. 1.8 for a seven-part procedure for receiving returned goods. For simplicity in visualizing the action and to avoid crossed lines where possible, arrange the individual charts on a desk so that those related to each other are positioned together. For example, if copy 3 causes things to happen to copy 5, they should be on adjacent lines or tracks.

Now the flow lines and the symbols can be transferred from the FPC to the MPC. Figuratively, the vertical line of symbols is picked up and placed horizontally on the MPC. The preprinted symbols will fall on the track reserved for the individual paper. Then hand-drawn symbols will fall between the tracks. In this manner, all steps that happen in a procedure are arranged in chronological sequence, reading from left to right, and there is a full view of the action on the multicopy set charted and a partial view of the action on all affected papers.

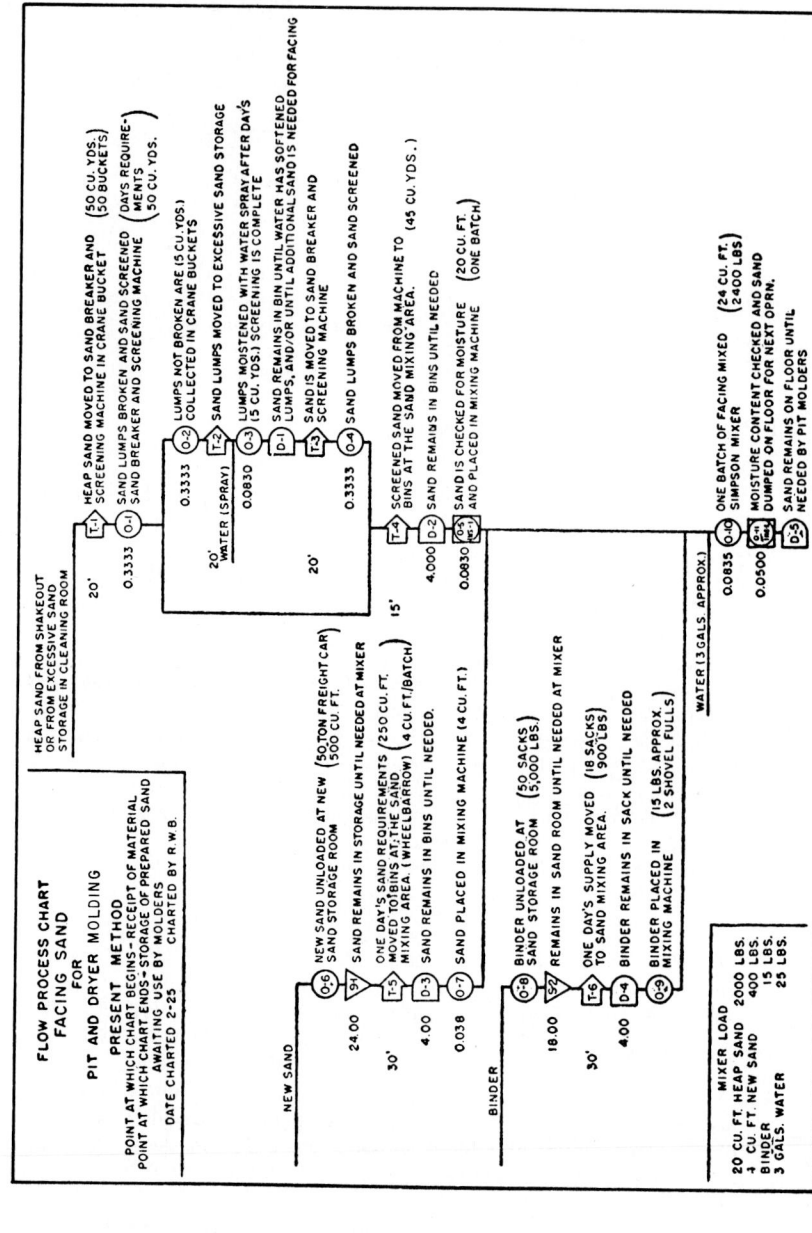

FIGURE 1.7 Flow process chart of the material type showing manner in which several components are processed and brought together.

Charting Techniques

Origin of a Set of Forms. This symbol is used each time a new record is created. If, during the procedure, a three-part form is written, it would be shown as follows. (Note that each part is indicated and each has its own flow line upon which are shown the symbols indicating what is happening to that part.)

ORIGIN

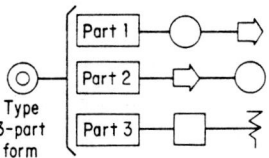

Adding Information. This symbol is used to show anything added to a record. (It includes posting, stamping, signing, and the like.) When one of the papers being charted is responsible for an "add to" operation on another record, it is shown as follows.

"ADD TO" OPERATIONS

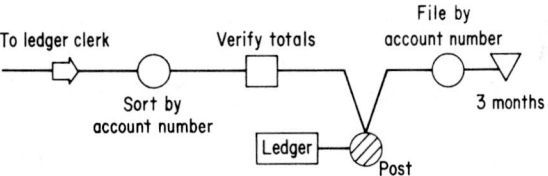

One Form Affecting Another. This line, going from one flow line to another (or from a flow line to a separate record as shown above), indicates that one paper in the system has an effect upon another. The V that the affect line creates may go up or down and may be as long as needed to reach the flow line of the paper affected. The following illustration shows a two-part form, with one part having an effect upon the other.

AFFECT LINE

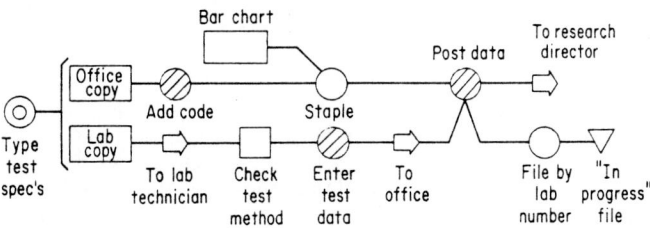

Simultaneous Action. When the same action is taken on two papers at the same time, this construction is used. In this case, it shows that the inspection was performed equally on both papers represented by the two flow lines.

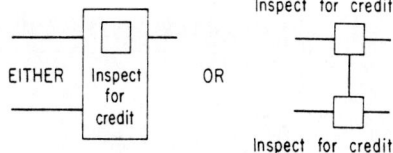

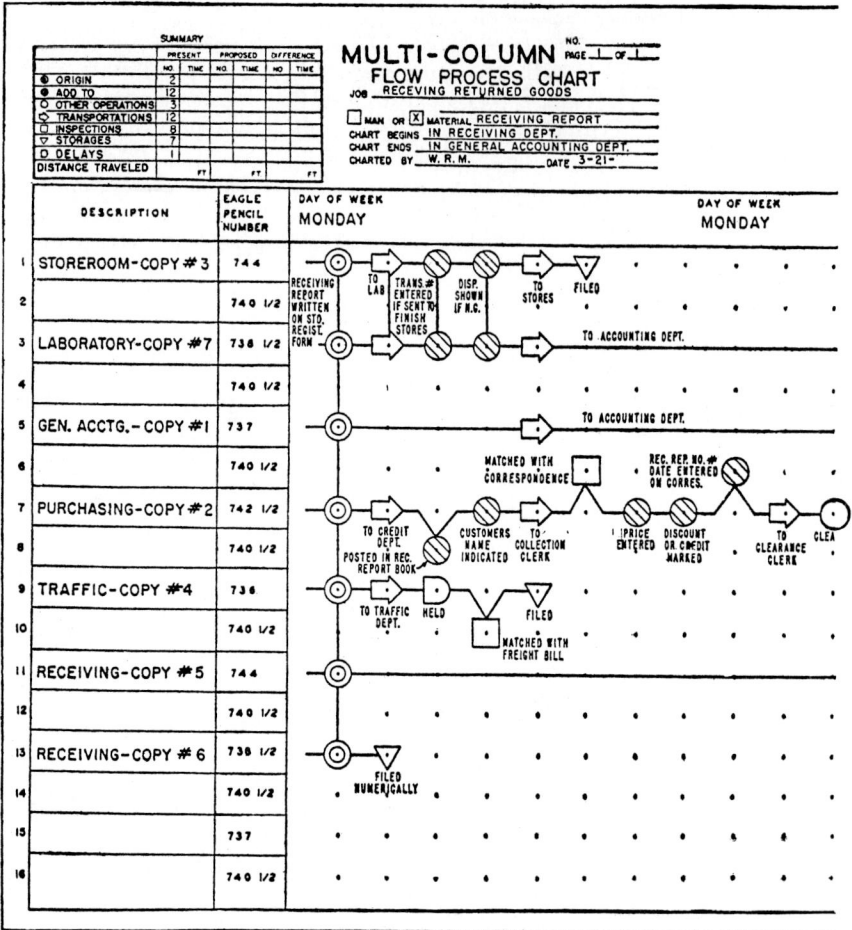

FIGURE 1.8   Multicolumn flow process

Alternate Action.   This construction is used when a piece of paper may be processed in different ways, depending on circumstances. The normal or most common way is shown on the regular flow line. The exception is shown on the alternate flow line.

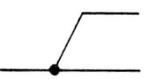

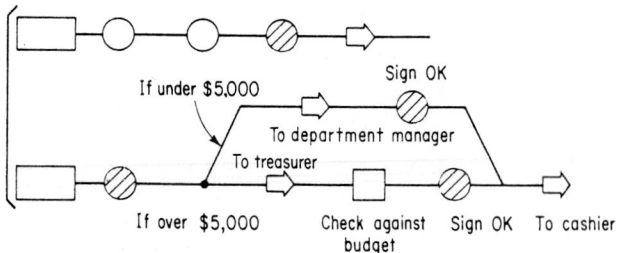

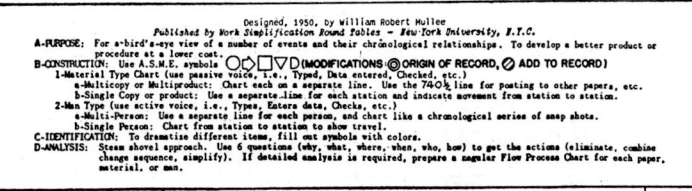

chart of the procedure type.

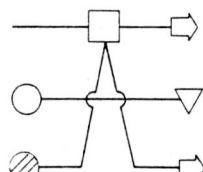

Skipping Over.   The skip is used, usually with an affect line, as a charting convenience. It may not always be practical to have forms which affect each other on adjacent flow lines. It may be necessary to skip over a flow line to get to the line of the form affected.

Physically Attached Set of Forms.   Often, a carbon-interleaved set of forms goes through a number of steps in a procedure as a set before being separated. Rather than draw many individual symbols or simultaneous action blocks, the flow lines can be brought together and the symbols shown on one flow line for the entire set. This technique may also be used when two or more forms are temporarily attached by staple or paper clip and go through several steps as a set.

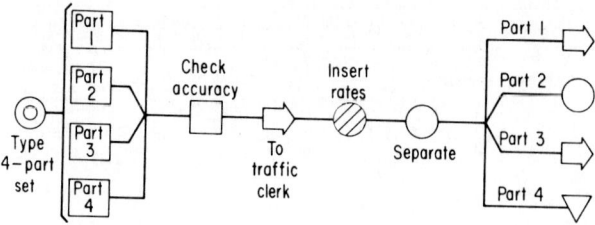

Chronological Time.   As far as possible, multicolumn charts should be drawn to show the relative time sequence of the various steps. If one part of a form is idle for some time, its flow line should be blank for some distance.

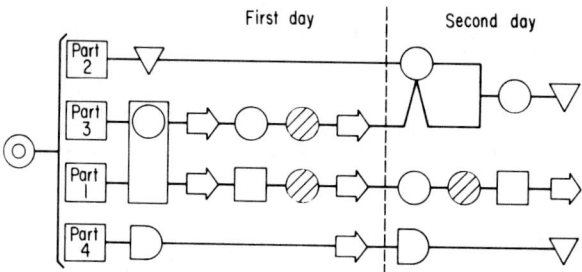

Gang Charts.   For groups of workers on warehousing, maintenance, rigging, or other material handling operations, the multicolumn form can be used as a gang chart as shown in Fig. 1.9. The regular ASME symbols are used, and a line on the chart is assigned to each member of the gang. Plotted horizontally, the symbols are arranged like parallel flow process charts, one above the other. Moving from left to right, a vertical column of symbols represents simultaneous activity of all members of the gang, like a snapshot of the gang's activity. These snapshots can be taken at frequent intervals so that all changes in activity are reflected by changes of symbols. Gang operations are usually not planned to the same degree as individual jobs, and the chart may show delay symbols. Substantial cost reduction may be had by rearrangement of symbols to cancel out delays. Eliminating delays permits transfer of excess personnel and shortens the production cycle. Combining operations or rearranging sequences saves make-ready and put-away. Study of individual operations or inspections permits many simplifications, followed by further rearrangement of the gang operation to reduce the overall cycle. Symbols are numbered to explain work details. The storage symbol is not used on gang charts.

PRINCIPLES AND PRACTICES FOR CONSTRUCTION OF MULTIPLE ACTIVITY CHARTS

A multiple activity chart, also called worker and machine chart or work planning chart, is a graphic representation of the coordinated working and waiting time of two or more workers or any combination of working and waiting time of two or more workers or any combination of workers and machines. The duration of the activities is represented by bars drawn to length against a time scale.

Activities Defined.   It is helpful to distinguish between the work of an operator when working on a machine or with another operator and when working independently of a machine or another operator. Similarly, it is helpful to distinguish between the operating time of a machine when operating independently of an operator and when being operated or serviced by an operator. It is also useful to distinguish between the time when a machine is waiting to be serviced and when it is being set up, loaded, or unloaded. To provide for this, the following classifications of working and waiting, with their graphic representations, are used.

Independent Work.   For the operator, this classification means working independently of the machine or other operator, such as when getting and preparing material, inspecting finished product, and performing other work not connected with the operation of the machine. For the machine, it includes the time it is actually doing its work without the services of the operator.

Combined Work.   For the operator, this classification includes working with a machine or other operator while setting up, loading, and operating a machine with hand feed or working in cooperation with other operators. For the machine, it includes the time it is operating and requiring the services of an operator and the time it is being set up, loaded, or unloaded.

It is convenient, when analyzing a cycle of worker and machine time, to differentiate between their times when working independently (independent work) and when one depends upon the other (combined work). The blocks of time representing independent work may be shifted around independently of each other, whereas the blocks representing combined work must not be shifted with reference to each other.

Waiting.   This classification includes waiting on the part of either an operator or a machine. It occurs when one is waiting for the other. Work of an operator which prevents a machine from running, but which might be rearranged to allow the machine to operate, should be classed as independent work, and the corresponding time of the machine should be classed as waiting. This classification and graphic code place emphasis on real waiting of the machine and focus attention on work of the operator which may be rearranged to occur during machine operating time and thereby reduce machine waiting.

Also solid red

An example of a multiple activity chart covering a worker and several machines is shown in Fig. 1.10. The details into which the activities are divided might follow the pattern of a time study, using a stopwatch for measuring time. However, where several workers or machines are being studied, it is usually more convenient to record the activities on a motion-picture film. This also has the advantage of recording and measuring time for smaller elements.

A simpler type of multiple activity chart does not give the actual time for each element on a time scale. Instead, it uses the symbols for working time and delays, balancing these as well as possible by estimating relative times.

Another form without time values, which uses the flow process chart symbols progressing in a horizontal direction, has already been described and is shown by Fig. 1.9 as the gang process chart.

Analyzing Multiple Activity Charts.   The same procedure is followed in analyzing multiple activity charts as is used in analyzing flow process charts, namely, the questioning technique and group participation described previously. However, before challenging each step of the work items, substantial savings may often be found by eliminating waiting time

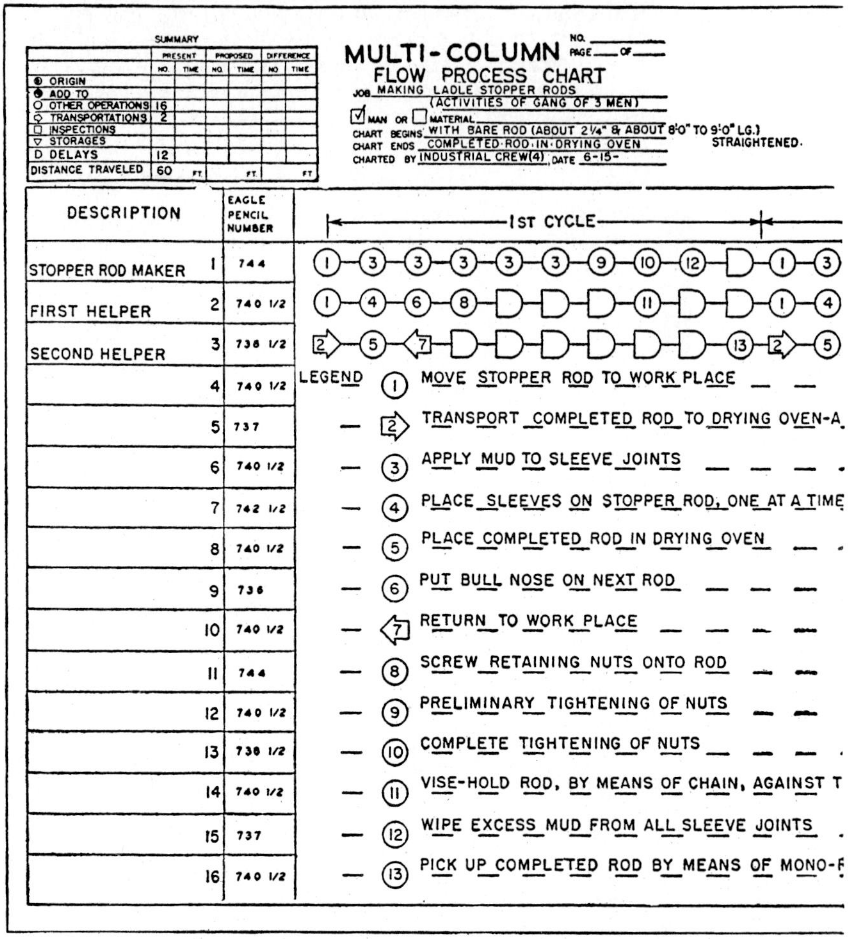

FIGURE 1.9 Multicolumn flow

for worker and machine. This may frequently be done by a simple rearrangement of the work cycle or by giving the worker other work to do.

PRINCIPLES AND PRACTICES FOR CONSTRUCTION OF WORKPLACE CHARTS

A workplace or right- and left-hand chart is a graphic representation of the coordinated activities of the right and left hands. Where the job is sufficiently repetitive to warrant a detailed study of the right and left hands, a workplace study may be made. Moves, operations, holds, or delays performed by each hand may be charted, using flow process chart symbols.

Figure 1.11 shows a workplace chart, also known as an RH-LH chart or an operator process chart. In parallel vertical columns, the elements are charted to show the simultaneous activity in each hand. A brief description is provided alongside the regular ASME symbols. Whenever a change occurs to either or both hands, it is shown on the next line of the chart, regardless of how short or long the time element.

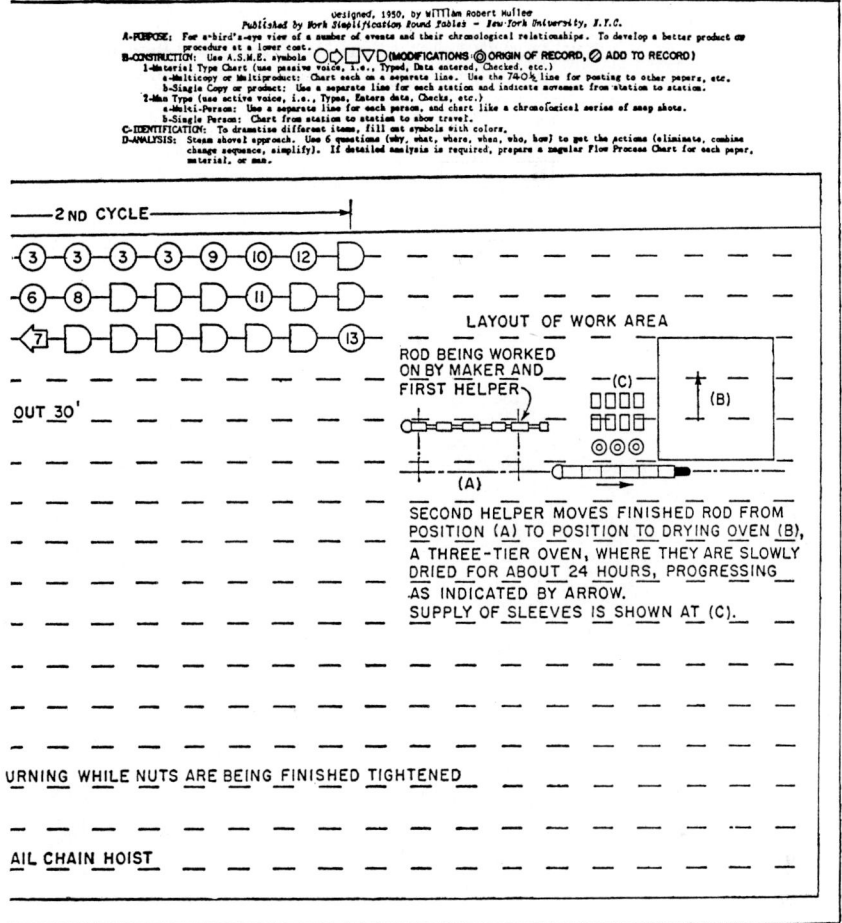

process chart of the gang type.

Counting the symbols and recording the totals in a summary at the top of the chart provides a comparison between the present and the proposed methods. Although the time for different elements will vary, the overall count of the symbols is a reasonable measure of the comparable time. The workplace chart may be considered as two detailed flow process charts, one for each hand. The symbols are later connected to form an activity pattern. This permits comparison of both hands as to similarity of work performed. A distance column on the form provides a record of the travel of each hand. Moving the work closer to the operator may improve the method and reduce the time.

The layout of the workplace is indicated by a grid of ¼-inch squares, against which the arrangement of bins, fixtures, and parts may be shown. The convenient work area is indicated by two semicircles described by the operator's forearms when seated at the workplace. Provision is made for parts sketches in the upper right-hand portion of the chart.

The position of each part on the layout is indicated by L_1, L_2, and so on, for items to the left of the operator, with L_1 nearest to center. The same convention is used for items to the right of the operator: R_1, R_2, and so on. Items directly in front are called C. If a second row or level is used, the designation is R_{1A}, L_{1A}, C_{1A}, and so on. The contents of each location are listed in the table at the upper left-hand side of the chart.

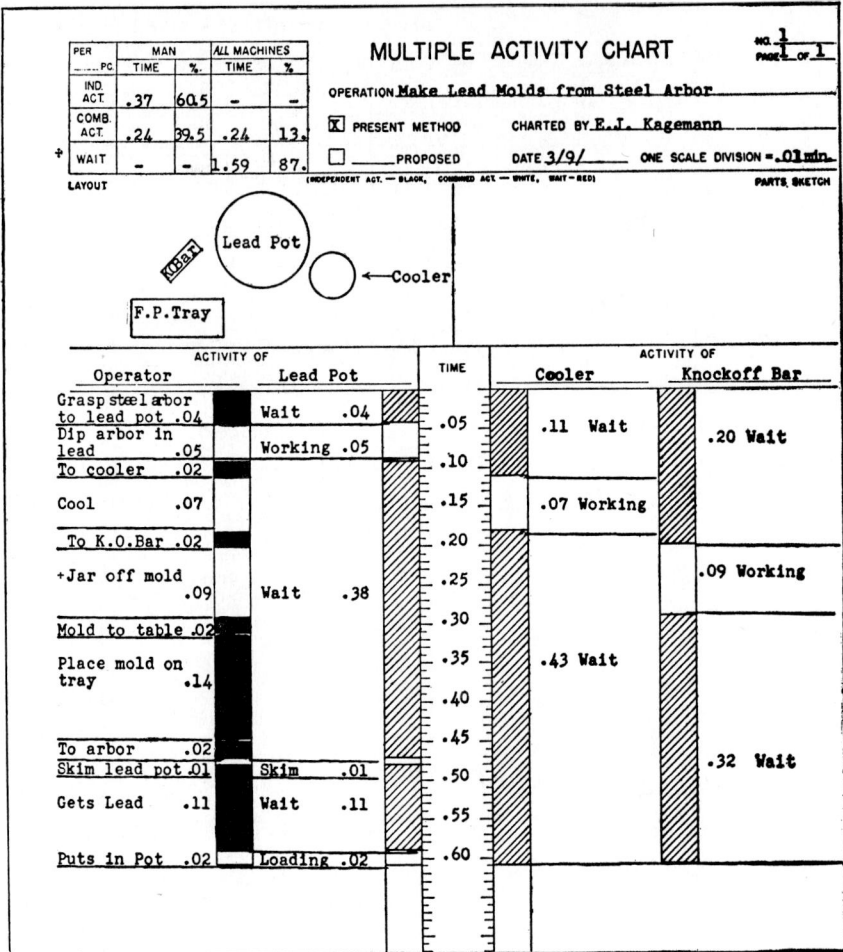

FIGURE 1.10   Multiple activity chart of a worker and several machines.

TWENTY PRINCIPLES OF MOTION ECONOMY

To help analyze the chart for improvements in method, the twenty principles of motion economy are shown in abbreviated form along the left-hand margin. This list of principles is a modification of the original Gilbreth list and has been found helpful in developing methods improvements for factory operations. The complete wording of these principles follows.

1. Begin each element simultaneously with both hands.
2. End each element simultaneously with both hands.
3. Use simultaneous arm motions, in opposite and symmetrical directions.
4. Use hand motions of lowest classification for satisfactory operations.
5. Keep motion path within normal working area.
6. Avoid sharp changes of direction. Plan a smoothly curved motion path.

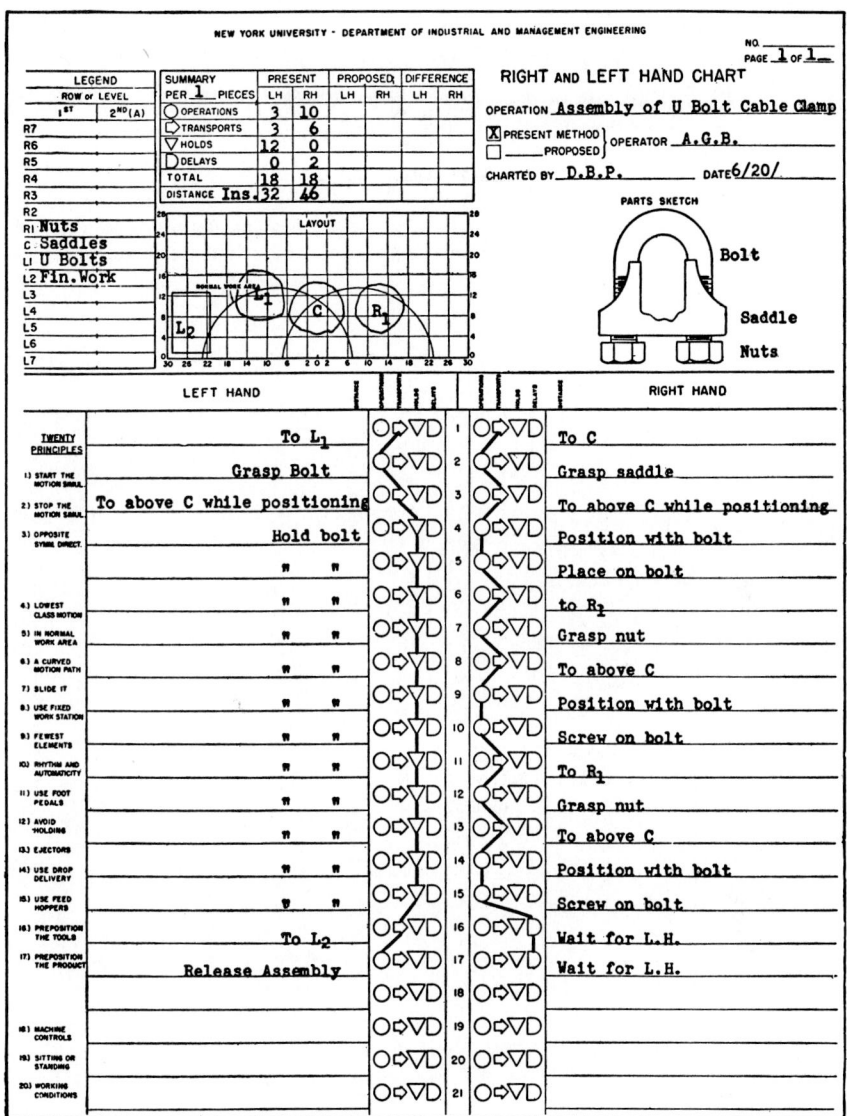

FIGURE 1.11 Workplace (RH-LH) chart.

7. Slide small objects. Avoid pickup and carry.
8. Locate tools and materials in proper sequence, at fixed workstations.
9. Use fewest elements to obtain shortest time.
10. Use rhythm and automaticity to increase output and lessen fatigue.
11. Relieve hands with foot pedals where possible.
12. Avoid holding. Use vise or fixture, freeing hands to move pieces.
13. Provide ejectors to remove finished pieces.
14. Use drop delivery where possible.

15. Shorten transports by keeping materials nearby in gravity-feed hoppers.
16. Preposition tools for quick grasp.
17. Preposition product for next operation.
18. Locate machine controls nearby for ease of operation.
19. Design workplace height for sitting-standing arrangement, and provide proper height chair with comfortable seat and backrest for good posture.
20. Provide pleasant working conditions, considering illumination, temperature, humidity, dust, fumes, ventilation, noise level, color scheme, orderliness, and the like.

CHAPTER 2
OPERATION ANALYSIS

Duane C. Geitgey
Principal
H. B. Maynard and Co., Inc.
Pittsburgh, Pennsylvania

The factors that surround the simplest process or operation are many and varied. Accordingly, small progress will be made toward methods improvement and automation if the job is studied as a whole. The first step in any study that will produce results is to resolve the job into its component parts or elements. Each part may then be considered separately, and the study of the process or operation becomes a series of studies of fairly simple problems.

This kind of analytical work is covered by the term "operation analysis." The study of each process or operation really consists of two analyses. The primary analysis breaks the job down into such factors as material, inspection requirements, and material handling. Each of these factors is then examined critically to discover the broad possibilities for methods improvements and automation. The secondary analysis is essentially a more detailed examination of some of the same factors, with the emphasis on the manual motions, or on the motions made by automatic equipment, required to do the job.

The operation analysis procedure is the basis for all the manufacturing research work that is being done in industry today.

OPERATION ANALYSIS DEFINED

Operation analysis may be defined as "a systematic procedure, employed to study all the factors which affect the method of performing an operation, to achieve maximum overall economy. Through this study, the best available method of performing each necessary part of an operation is found, and new manufacturing and maintenance developments are incorporated as they become available, in the continuing effort to move every job one step closer to continuous automatic accomplishment."

APPLICATIONS AND LIMITATIONS OF OPERATION ANALYSIS

The feeling which is often prevalent in the mind of managers who are acquainted only in a general way with methods engineering techniques is that, although operation analysis may be able to produce worthwhile accomplishments in some lines of work or in certain industries, their work is different and the techniques will be of little or no value to them.

Human nature is such that we all feel that "our work is different." The best-intentioned managers with a cost problem consequently feel that the technique will work better in every other situation than it will in their own. However, an industrial engineer will have sufficient experience with the technique to know of the beneficial results that are always obtained as a result of applying it.

The principles of operation analysis are fundamental. They can be applied to any kind or class of work. It makes no difference if the manager's cost problem is in the maintenance area or in a partially mechanized high-volume production line.

This broad application is possible because all work may be resolved in terms that are more or less basic. Working methods used on widely varying jobs present points of remarkable similarity when closely analyzed. The motion made by a maintenance mechanic in reaching for a power drill is the same as that made by a cutting room operator in reaching for a pair of scissors. Similarly, the problems involved in lifting a ladle of metal with a crane in a high-production foundry are the same as the problems encountered by the maintenance crew in lifting a ladle of molten metal that will become a new bearing surface. Every manager can cite countless similar examples that can be found in the various jobs that are being done in the area every day. A look at the steps of the operation analysis approach shown below emphasizes the fact that the technique may be applied to any job and that the principles of operation analysis are not limited in any way by the nature of the work being done.

Operation Analysis Approach to Improvement and Automation

1. Observe or visualize operation.
2. Ask questions.
3. Estimate degree of improvement or automation possible.
4. Investigate ten approaches to improvement and automation:
    a. Design of part or assembly
    b. Material specification
    c. Process of manufacture
    d. Purpose of operation
    e. Tolerances and inspection requirements
    f. Tools and speed, feed, and depth of cut
    g. Equipment analysis
    h. Workplace layout and motion analysis
    i. Material flow
    j. Plant layout
5. Compare old and new methods.

The repetitiveness of the operation is another factor that must be considered. If a large number of work-hours are being expended in a certain type of activity, a 1 percent saving may be important. On the other hand, a 10 percent saving on work that is infrequently performed may not offset the cost of making the analysis. It is obviously more profitable to study the work with the greatest activity; however, this does not mean that only mass production work can be studied. This is true because activity is measured on a type of work taken as a whole, rather than on individual jobs.

For example, the industrial engineer frequently finds operations that are repetitive from an operation analysis viewpoint even in maintenance work. In this instance, when a number of different jobs are reduced to their elements, the engineer finds that several elements are common to many jobs. If such common elements can be shortened through the selection of better materials, equipment, or manual methods, a saving is obtained each time the elements are performed, regardless of the larger task that the maintenance worker may be doing.

The rapid progress that is being made in all fields—materials, tools, and manufacturing processes—requires that every manager and every industrial engineer constantly search for continuing job improvement. They should never speak of the "best" method without using some qualifying clause which implies that some improvement is possible, even though economic reasons may make it impractical to make the improvement at the present time. This principle applies to all types of work. As a result, operation analysis is not limited to mass production work but may be applied to produce savings in any line of work in which a fair number of work-hours are expended. Conversely, it will probably not be profitable to study a varied line of work if only one worker is engaged on it only part time.

TYPES OF METHODS STUDIES

The manager who is searching for the solution to a cost problem has many industrial engineering procedures available. The number of effective industrial engineering procedures or tools has multiplied over the years. Some of the tools make it possible to achieve an accuracy that was impossible with older ones, with only a slight increase in cost. Others enable a manager to obtain needed information at a cost far less than previously possible, with only a slight reduction in accuracy. The industrial engineer, then, has a wide variety of tools to call on when undertaking a methods improvement project. In selecting the proper tools, it can generally be said that the savings that the study will produce must equal or exceed the cost of making the study. The manager and the industrial engineer must determine which tools to use, and to what extent each should be used, on the basis of their estimate of the degree of methods improvement that is possible in the particular job to be studied.

The large number of available industrial engineering tools may be combined in many ways. However, for practical purposes these may be combined to yield six classes of methods study that, with only minor modification, may be used to cover all types of activity. These six types are:

1. Written analysis using process charts and operation analysis charts. Methods analysis employing Maynard operation sequence technique (MOST®)*. Detailed analysis of all available automation devices, tools, and equipment. Operator methods training utilizing audiovisual devices.

2. Written analysis using operation analysis charts. Methods analysis using MOST. Analysis of all tools and equipment. Operator methods training by supervisors or trainers familiar with the MOST procedure. Providing MOST-type written methods instructions to operators.

3. Mental analysis using the points described on the operation analysis chart as a guide. Methods analysis based on MOST. Operator training given by supervisors or trainers familiar with the MOST procedure.

4. Written job analysis of class of work, using process charts and operation analysis charts for representative jobs. Methods analysis of representative jobs, using MOST to determine best methods. Operator training on specific jobs by supervisors or trainers familiar with MOST.

5. Mental analysis during general survey of class of work, using points on operation analysis chart as a guide. Methods analysis, using second generation predetermined time data. Operator training in use of standardized tools for the class of work by supervisors familiar with MOST.

6. Use of second generation predetermined time data as a guide to performance.

---

*MOST® is a registered trademark of H. B. Maynard and Company, Inc.

All the industrial engineering techniques mentioned in these six types of methods studies are described in detail in other chapters of this handbook.

FACTORS DETERMINING FIELD OF APPLICATION OF THE SIX TYPES OF METHODS STUDIES

The kind and amount of study that can be justified on any job or class of work are determined by three principal factors. These are the repetitiveness of the job or class of work, the amount of human attention required, and the expected life of the job or class of work. These factors must be considered together in selecting the type of methods study to be used, because no one of them is sufficient for the determination.

Repetitiveness. For the purpose of determining the field of application of the various types of methods studies, the repetitiveness of the job or class of work may be divided into four classes: high, medium, low, and jobbing. The following general descriptions can serve as a guide in fixing the repetitiveness of the job or class of work being considered.

High. A job or class of work may be considered highly repetitive if it occurs at least 2000 times a year and requires a total of not less than 1000 hours to perform.

Medium. A job or class of work may be said to be mediumly repetitive if it occurs at least 500 times per year and covers an elapsed time of 1 to 6 months.

Low. A job or class of work may be said to be of low repetitiveness if it occurs at least 50 times per year and covers an elapsed time of 2 weeks to 1 month.

Jobbing. A job or class of work may be said to be jobbing if it occurs less than 50 times per year, lasts less than 2 weeks, and is not expected to be repeated in the foreseeable future.

Judgment must be used in applying these descriptions because in many instances the job or class of work will not fit any of them exactly. However, the descriptions will serve as a guide.

Human Attention. The portion of the job or class of work that requires human attention has an important bearing on the type of study that should be made. The term "human attention" includes any part of the job or class of work that is manually performed by human labor, and it also includes the time when operators must be attentive to the equipment (watching or listening) to ensure its proper operation, even though they may have no specific manual motions to make.

The human attention required by a job or a class of work may be classified as high, medium, or low. The maximum condition, of course, is when all parts of the job or class of work are performed by the operator by hand or with simple, unpowered hand tools. The minimum condition is when the job is done entirely automatically by machinery, where the machine stops itself and signals the operator if there is a malfunction, so that the operator's attention does not need to be focused continuously on any one machine. The class into which any job falls with respect to human attention may be determined as follows:

High. Where human attention is required by the individual job or class of work more than 75 percent of the time.

Medium. Where human attention is required by the individual or class of work between 25 and 75 percent of the time.

Low. Where human attention is required less than 25 percent of the time.

Life of Job. The life of the job or the class of work is another factor which must be considered along with repetitiveness and human attention. The more detailed types of methods study are expensive, and the manager and industrial engineer must determine whether or not the estimated life of the job will justify the expenditure.

The length of life of the job or class of work can be divided into three classes: over 12 months; from 6 to 12 months; under 6 months.

TABULATION OF FACTORS

Table 2.1 has been compiled to aid in the selection of the type of methods study which is economically justified under any given conditions. Remember that the manager or industrial engineer who must make the choice must use judgment and be guided by the particular conditions surrounding the job or the class of work to be studied.

The most difficult factor to determine when using the table is the repetitiveness of the job. The degree of human attention and the expected life of the job or class of work may be quickly checked. Repetitiveness, in the sense in which it is used in connection with the table, is affected by the number of occurrences per year, the length of the particular cycle of the job or class of work being studied, and the total length of the job.

In determining the repetitiveness of a job or a class of work, the number of occurrences, the hours required to complete the work, and the time allowed per occurrence must be considered. By definition, a job or class of work is considered to be highly repetitive if it consists of not less than 2000 occurrences and requires not less than 1000 work-

TABLE 2.1   Tabulation of Factors Which Determine the Type of Methods Study to Be Employed

Repetitiveness of job or class of work	Human attention	Life of job, months	Type of study indicated
High	High	Over 12	1
		6 to 12	1 or 2
		Under 6	2 or 3
	Medium	Over 12	1 or 2
		6 to 12	2 or 3
		Under 6	3
	Low	Over 12	2
		6 to 12	2 or 3
		Under 6	3
Medium	High	Over 12	2
		6 to 12	2 or 3
		Under 6	3
	Medium	Over 12	2 or 3
		6 to 12	3
		Under 6	3 or 4
	Low	Over 12	3 or 5
		6 to 12	3, 5, or 6
		Under 6	6
Low	High	Over 12	3 or 4
		6 to 12	3, 4, or 5
		Under 6	3 or 5
	Medium	Over 12	3, 4, or 5
		6 to 12	3 or 5
		Under 6	3, 5, or 6
	Low	Over 12	3 or 5
		6 to 12	3, 5, or 6
		Under 6	6
Jobbing	High	Under 6	5
	Medium	Under 6	5 or 6
	Low	Under 6	6

hours to complete. However, because the time allowed per occurrence should be considered, the three factors can only be related algebraically. Therefore, if the following formula is satisfied, the job may be classed as highly repetitive:

$$\frac{N \times T}{1000} \geq 1$$

where N = number of pieces (not less than 2000)
 T = time allowed

By definition, a job is mediumly repetitive if it has not less than 500 pieces per year and lasts 1 to 6 months. To be considered mediumly repetitive, therefore, the following formula must be satisfied:

$$\frac{N_1 \times T}{167} \geq 1$$

where N_1 = number of pieces (not less than 500)

A job of low repetitiveness consists of not less than 50 pieces per year and lasts 2 weeks to 1 month. The formula which must be satisfied is

$$\frac{N_2 \times T}{80} \geq 1$$

where N_2 = number of pieces (not less than 50)

The use of the formulas and the table may be illustrated by the following example. In a production machine shop doing miscellaneous work, several representative jobs are selected to test the type of methods study that is economical to make. On the first job considered, the activity is estimated to be 5000 pieces per year. The first operation is a lathe operation that requires 0.392 hour to perform. Substituting these figures in the formula for highly repetitive jobs, a value of 1.96 is obtained. The expression is thus satisfied, and the operation is classed as highly repetitive.

There are several long cuts involved in the operation during which the machine is in complete control. The human attention required during the whole operation is estimated at 45 percent and is classed as medium. There is every indication that this operation will continue to be performed for several years in the future; thus its life is over 12 months. Referring to the table, it can be seen that a highly repetitive job, requiring medium human attention, lasting over 12 months, calls for a type 1 or a type 2 study. In this particular case, the manager or the industrial engineer would consider the fact that the human attention is comparatively low, would estimate from experience that the possibilities for improvement through detailed motion study or the application of additional available mechanical devices appear limited, and would specify a type 2 study.

As pointed out in this example, the manager or the industrial engineer must recognize that the judgment should be used when applying the formulas and using the table.

TO DO SUCCESSFUL ANALYSIS WORK, A SUITABLE MENTAL ATTITUDE MUST BE DEVELOPED

Human nature is such that the proper attitude toward analysis work does not develop naturally. Instead, people tend to get smug about their knowledge of a particular activity. They feel that they have reached the goal and do not need to strive any more. This attitude may be commendable as a means of securing peace of mind in everyday affairs, but it makes successful analysis impossible. An analyst who knows everything about a certain point and does not need to consider it further ensures that there will be no improvement on that point. To improve any process or operation, the analyst must approach it with a firm conviction that it can be improved.

As a result of many experiences with continuing job improvement, industrial engineers never speak of the "best" method. Rather, they refer to the "best available method" or the "best method yet devised." Carrying this thought to its logical conclusion, it might be stated: "Every time workers use their hands, there is a continuing opportunity for methods improvement. This opportunity exists until the operation is mechanized to the extent that human attention is completely eliminated and the mechanical devices used are of ultimate simplicity."

This statement makes it clear that simple, automatic operation is the ultimate goal of any methods improvement program. The best method of doing an operation from an economy point of view is reached only when the human attention required has been reduced to zero and all complicated production equipment has been eliminated or simplified. Until this point has been reached, further improvement is always possible.

This principle furnishes the foundation for a sound approach to universal operation analysis for methods improvement and automation. The analyst who appreciates its logic will have an open mind in accepting it and will not be bothered with such mental obstacles as "It won't work" and "We tried it before and it can't be done." Lack of success in improving or automating any job should not be interpreted to mean that the job cannot be improved. Such an occurrence is only an indication that the analyst is not aware of any developments that would improve the job or that available equipment is still too expensive to be economical. Acceptance of the continuous-opportunity-for-improvement principle will combat any tendency to feel content with things as they are, and it will inspire fresh attacks from new angles. It leads to progress.

An open mind paves the way for successful analytical work, but it is not sufficient in itself. One can be open-minded in the passive sense of being receptive to suggestions, but this type of open-mindedness will not lead to accomplishment. To get results, the analyst must take the initiative in originating suggestions.

In a world where it is often said that there is nothing new, the greatest amount of originality—or what passes for originality—comes from people who have an inquiring turn of mind. Someone who constantly asks questions and takes nothing for granted disturbs the complacent members of the organization but originates new and better ways of doing things. Progress begins with doubt. Improvement begins with analyzing what is being done and then inquiring into what new techniques are available so that it may be done better.

Once this point is understood, the industrial engineer should conscientiously develop what is known as the "questioning attitude." The questioning attitude is a state of mind that prevents anything being taken for granted in the investigation of a job. It questions everything and determines answers on the basis of facts. It guards against the influence of emotions, likes or dislikes, and prejudices.

Anyone who is successful in bringing about improvements has only one deep-seated conviction—that the method can be improved—and accepts nothing as being right just because it exists but asks questions and gathers answers. By evaluating the various possible answers in the light of knowledge and experience and questioning everything, the industrial engineer investigates all phases of the job to the extent that time permits and asks questions when the answers appear obvious, because the obvious things frequently hide valuable improvement opportunities.

The questions that the industrial engineer asks take the general form of what, why, how, who, where, and when. What is the operation? Why is it performed? How is it done? Who does it? Where is it done? When is it done in relation to other operations? These questions, in one form or another, should be asked about every factor connected with the job or class of work being analyzed.

When a job is examined in systematic detail and all factors related to it are questioned, possibilities for improvement are certain to be uncovered. The action that is taken on these possibilities will depend on whether the person who uncovers them has the authority to take action and approve expenditures. Someone with authority will undoubtedly go ahead and make the improvement without delay. Without that authority, ideas must be presented in the form of suggestions to the person or persons who do have that authority.

There are certain pitfalls to be avoided when making suggestions. In the first place, the real advantage of every suggested methods improvement should be carefully evaluated before it is offered. An individual who establishes a reputation for offering only meritorious suggestions will be assured of an attentive hearing. On the other hand, one who continually offers just *any* idea will find that those who receive suggestions will soon stop taking the time to separate the good from the impractical, and reject all offerings.

The best way to prove the merit of any suggestion is to make an estimate of both the cost of making the improvement and the total yearly savings that the improvement will produce. If an improvement will cost $1000 to adopt and will save $100 a year, it is not worth presenting—unless it would solve some pressing related problem such as operator safety. If, on the other hand, the expenditure will be returned in a reasonable length of time, the suggestion is worthy of careful consideration. Most companies have established criteria for determining what constitutes a reasonable length of time, and many have developed forms to standardize the manner in which the information on costs and savings is developed.

TEN PRIMARY POINTS OF ANALYSIS

When analyzing a job or an activity, there are so many questions which should be asked that, unless a systematic procedure is followed, it is quite possible that certain points may be forgotten. More than one analysis has proceeded to the point where elaborate suggestions for improvement have been presented only to have all the work discarded because some simple question like "Are all parts necessary to the function of the assembly?" was not previously asked, and the person to whom the suggestion was submitted recognized that the job should be eliminated rather than improved.

To avoid wasted effort and to make sure that all important points are considered, the analyst should keep clearly in mind the factors that should be examined in every operation. These factors should be considered in detail whether the analysis is mental or written. The ten main points or factors which should be considered in every operation—arranged in order of consideration—are as follows:

1. Purpose of operation
2. Design of part
3. Process analysis
4. Inspection requirements
5. Material
6. Material handling
7. Workplace layout, setup, and tool equipment
8. Common possibilities for job improvement
9. Working conditions
10. Method

When actually making an analysis, it is seldom possible to complete the analysis of one of these factors at a time and then leave it for good. Almost all the factors are interdependent, and a change in one will cause a change in one or more of the others. The list, however, indicates in a general way the course along which the analysis will best proceed.

THE ANALYSIS SHEET

To simplify the work of making an analysis, a form known as the operation analysis sheet has been designed. Wherever the form is regularly used, the number of suggestions for improvement increase. The form, of course, does not accomplish this through any mystic property of its own, but its use ensures that none of the factors which should be considered will be neglected. A typical form of the operation analysis sheet is shown in Fig. 2.1.

The form is equally useful whether analysis is to be mental or written. The mental analysis, made with the form as a guide, is quicker, but it is also less satisfactory than a written analysis. When a mental analysis is made, records are seldom kept, and if they are, they are usually not systematic or complete. This lack of records is a liability in the event that a change in the type of study required is later decided on, in which case, the analysis will have to be repeated. However, mental analyses made systematically with the form as a guide will produce many good results on jobs where low activity or low human attention makes it uneconomical to undertake a more elaborate study.

A written analysis using the operation analysis sheet has several obvious advantages. The written analysis is more likely to be carefully made. The fact that the answer to each question must be committed to writing will ensure that proper consideration is given to each factor. The data that are usually collected in the preparation of a written analysis will support the suggestions made for improvement of the job or class of work.

It should be unnecessary to stress the importance of identifying all the supporting paperwork connected with an analysis, but experience shows that, unless this point is emphasized and reemphasized, the identification of the supporting papers is seldom complete.

USE OF THE ANALYSIS FORM

The analysis form acts as a guide to systematic operation analysis. It directs the person making the analysis through the factors to be considered and ensures that none of them will be overlooked.

The analysis itself actually takes place in the mind of the analyst, who questions each point as it is raised, gathers all the known facts, and combines these facts with knowledge of alternatives, in this way arriving at suggestions for improvement. The nature and extent of these suggestions depend on the analyst's knowledge of what is taking place in the areas of new materials, tools, and manufacturing techniques. However, the systematic procedure outlined on the analysis form will help achieve maximum results. As the analysis is made all facts and ideas for improvement are recorded at the time they occur. The form should include enough detail to provide a record of the conditions that exist at the time of the analysis and to suggest any improvements that come to mind. All descriptions should be recorded clearly and concisely.

Space is provided in the heading of the operation analysis form for recording all information necessary to identify the job or class of work. In the following paragraphs, each of the ten factors to be considered will be discussed in detail. Specific questions to be asked by the analyst are shown on the operation analysis form in Fig. 2.1.

Purpose of Operation.   Although most operations are properly set up the first time a job is performed, changes in design or material specification may cause an operation to become incorrect or even unnecessary. In industry and business, as in other phases of life, nothing remains constant for any great length of time. As a result, slight changes in preceding or subsequent processes may affect the efficiency of or the necessity for a particular operation. In fact, the application of the operation analysis procedure has uncovered a surprising number of operations found to be unnecessary after further study. Unfortunately, those most familiar with the unnecessary operations often fail to recognize that they are

Date started _____ Department _____

Dwg. or spec. _____ Item or part no. _____ Material _____

Description of part _____

_____

Operation _____

_____

Yearly activity _____ Expected life _____ Yearly labor cost per .0001 hr. _____

DETERMINE AND DESCRIBE	DETAILS OF ANALYSIS	ACTION
1. PURPOSE OF OPERATION _____ _____ _____ _____ _____	Is the operation necessary? Does the operation accomplish the intended result? Can the operation be eliminated by doing a better job on preceding operations? Can the material supplier perform the operation more economically? Can the operation accomplish additional results to simplify succeeding operations?	
2. DESIGN OF PART (suggest improvements, make sketches where necessary)	Are all parts necessary? Could standard parts be substituted? Does design permit least costly processing and assembly? What design features do competitors use? Will design allow eventual automation?	

3. PROCESS ANALYSIS (complete list of all operations performed on part)
 	
 No.   Description   Work Sta.   Dept.
 	
 1. _____
 2. _____
 3. _____
 4. _____
 5. _____
 6. _____
 7. _____
 8. _____
 9. _____
 10. _____
	Can operation being analyzed
	    be eliminated?
	    be combined with another?
	    be performed during idle period of another?
	 Is sequence of operations best possible?
	 Should operation be done in another dept. to save cost or handling?	

FIGURE 2.1   Operation analysis form.

4. INSPECTION REQUIREMENTS

Tolerances and specifications _____

_____

_____

_____

Inspection procedures (suggest improvements) _____

_____

_____

_____

_____

Are tolerance, allowance, finish, and other requirements
necessary?
too costly?
suitable to purpose?
Should statistical quality control be used?
Is inspection procedure effective and efficient?

5. MATERIAL (suggest better material)

_____

_____

How can scrap costs be reduced? _____

_____

_____

Processing materials _____

_____

_____

Consider size, suitability, straightness, and condition.
Can cheaper material be substituted?
Will tool modifications permit use of lighter material or thinner sections?
Would a more expensive material lower machining and processing costs?
Is packaging suitable?

6. MATERIAL HANDLING (suggest improvements)

Brought by _____

_____

_____

_____

_____

Removed by _____

_____

_____

_____

Handled at work stations by _____

_____

_____

_____

_____

_____

Can incoming materials be delivered directly to the work station?
Can signals such as lights or bells be used to notify material handlers that material is ready to be moved?
Should crane, gravity conveyors, tote pans, or special trucks be used?
Consider layout with respect to distance moved.
Are containers correctly sized?

FIGURE 2.1   (*Continued*) Operation analysis form.

superfluous. The analyst, then, must be alert to the possibility that work that is being performed is no longer necessary. In some instances, the material supplier can perform the operation more economically, or the operation can be eliminated by doing a better job on either preceding or subsequent operations.

7. WORKPLACE LAYOUT, SETUP, AND TOOL EQUIPMENT
(suggest improvements, making sketches where necessary)

Arrangement of work area

Placement of tools, materials, supplies

How are dwgs. and tools secured?

Can setup be improved?

Trial pieces

Machine adjustments

TOOLS

Suitable?

Provided?
    Ratchet tools
    Power tools
    Special purpose tools
    Jigs, vises
    Special clamps
    Fixtures
    Multiple
    Duplicate

8. COMMON POSSIBILITIES FOR JOB IMPROVEMENT (consider the following)

RECOMMENDED ACTION

1. Install gravity delivery chutes.

2. Use drop delivery.

3. Compare methods if more than one operator is working on same job.

4. Provide correct chair for operator.

5. Improve jigs and fixtures by providing ejectors, quick-acting clamps, and the like.

6. Use foot-operated mechanisms.

7. Arrange for two-handed operation.

8. Arrange tools and parts within normal working area.

9. Change layout to eliminate backtracking and to permit coupling of machines.

10. Utilize all improvements developed for other jobs.

9. WORKING CONDITIONS (suggest improvements)

Light

Heat

Ventilation, fumes

Drinking fountains

Washrooms

Safety aspects

Design of part

Clerical work required (to fill out time cards and so on)

Probability of delays

Probable mfg. quantities

FIGURE 2.1 (*Continued*) Operation analysis form.

Design of the Part. Although the people making the analysis are seldom design engineers, it is important that they consider the design before proceeding to the other points of analysis. Often, the design engineer does not have time to reconsider the design after the decision to manufacture is made. Therefore, the person making the operation analysis must

10. **METHOD** (accompany with sketches or process charts if necessary)

   a. Before analysis and motion study _____

   _____

   _____

   _____

   b. After analysis and motion study _____

   _____

   _____

   _____

   _____

   _____

Are hand motions symmetrical?

Are parts transferred between hands?

Is a more detailed motion study needed?

Has safety been considered?

Working posture

Does method follow Laws of Motion Economy?

Are lowest classes of movements used?

RECOMMENDATIONS FOR FURTHER IMPROVEMENTS IF THE ACTIVITY INCREASES:

_____

_____

_____

_____

_____

_____

_____

RECORD OF ACTION TAKEN:

Proposal	Date	Referred To:	Action Taken

COMMENTS:_____

_____

_____

_____

_____

_____

_____

_____

_____

Date completed _____ Analyzed by _____

FIGURE 2.1 (*Continued*) Operation analysis form.

ensure that the design is correct and desirable. Consideration of this point can ensure that expensive details originally designed into the part are still necessary. Many unnecessary design features have been eliminated, with resultant large dollar savings, because of attention paid to this factor of the analysis.

To analyze this factor adequately, one must take enough time to understand the essential functions of the part and the assembly being studied to ensure that the design of the part and the assembly is a least-cost design.

Process Analysis. No single operation can be studied by itself. It must be considered as a part of the total process. The effect of any changes that are suggested must be considered in the light of the process. Only in this way can the analyst be sure that the suggested improvement will produce results. By carefully reviewing all the operations performed on a part, the analyst can determine whether the operation under study can be eliminated, combined with another, or performed during the idle time of another operation.

Because of the rapid development of new processes and techniques, analysts must keep abreast of the newest developments in the area of study. With this knowledge, they will be able to recommend changes which will simultaneously improve quality and reduce cost by improving or eliminating outmoded or unnecessary operations.

Inspection Requirements. Quality requirements established by the designer or originator of a process play an important part in the selection of operations and methods to be used. In fact, these quality requirements often force the selection of a specific process and method. On machined metal parts, for example, if a designer allows little or no variation from parallel for several holes through a part, the process engineer may be forced to specify the use of a precision boring mill rather than a drill press. Similarly, if the designer specifies too high a pressure test for welds on plant power piping, shop welding techniques may have to be substituted for simpler, less costly in-place welding.

The inspector, of course, also plays an important part in determining methods, because too literal an interpretation of the quality specifications can result in a more costly method.

Through application of the operation analysis procedure, the analyst will determine whether the quality requirements are consistent with the use to which the finished job will be put. After it is determined that the requirements are consistent with the use, the analyst can find whether the operation under study will produce a result that will meet the requirement economically. In this way, the company will not pay for unnecessary requirements, and properly established requirements will be uniformly enforced and met.

Material. Material costs are an important part of the total cost of any job or class of work. The kind of material from which parts are made is usually fixed by the nature of the part and the service conditions that it must withstand. However, materials that were originally specified by the designer or originator of a process may no longer be the most suitable. Unfortunately, design budgets seldom provide for periodic review of materials. Thus, the investigation of materials during the conduct of an operation analysis can sometimes result in significant savings. The analyst must be familiar with recent developments in new materials in order to recognize when a currently specified material is no longer the best material available for the job. During the course of the study, the analyst must consider the size, suitability, and condition of existing materials and the possibility of substitute materials. The use of supplies related to the operation should also be considered.

Material Handling. The flow of material through a plant or business is usually accomplished by a number of separate transportations. These transportations may be into and out of storage locations or to and away from workstations. The analyst, by carefully studying the need for transporting the material and the nature of the material handling activity, can often significantly reduce this major cost.

Many devices have been developed through research to expedite the flow of materials and eliminate the problems connected with material handling. The bulkier the part, the more advantageous it is to think in terms of orderly, continuous flow rather than batch handling. For example, storage conveyors are now used for almost every imaginable type of material to provide desired flow and to allow selection of items brought from storage areas without manual handling.

Workplace Layout, Setup, and Tool Equipment.   The workplace layout provided for an operator determines the motions used while performing a job. Almost every industrial engineer is familiar with the attention given to manually performed bench-type operations. They also recognize that most machine tool manufacturers now acknowledge the importance of locating controls in the most effective manner.

Despite the emphasis on workplace layout, there are still many examples of unplanned work areas and lack of standardization. Maintenance work, in particular, often suffers from poorly conceived workplace layout. Although some people contend that this class of work does not lend itself to reasonable workplace layout, many companies have provided maintenance crews with methods training and suitable tool carts and other equipment which minimize manual motions.

The statement, "With sufficient study, any method can be improved," certainly applies to workplace layout. In studying the workplace layout, the analyst must consider the placement and use of all materials and tools. Also important are such factors as the manner in which the job is assigned; how the operator receives job instruction; and how auxiliary equipment such as drawings, special tools, and measuring devices is obtained. During operator instruction and learning time on certain kinds of repetitive work, it may be desirable to provide an audiovisual training device. Although these devices take up valuable space, experience with them has proved their effectiveness. Whether the workplace layout involves a machine tool, a clerk's desk, or a bench, the application of operation analysis will result in an improved arrangement of the work area.

Common Possibilities for Job Improvement.   During the application of operation analysis, certain factors to be considered are particularly effective in improving almost any type of operation. These factors, which are based on the principles of motion economy, are considered as common possibilities for job improvement. They involve consideration of the use of such devices as delivery chutes, ejectors, quick-acting clamps, and foot-operated mechanisms. They also guide the analyst in the consideration of operator comfort and the motion pattern employed during the performance of the operation. Although these factors may be covered during the consideration of the other points of primary analysis, they have resulted in such significant improvements that they are listed separately as item 8 on the operation analysis form.

Working Conditions.   Although much attention is paid to the motions that a worker must perform and to the requirements for an effective process, the environment in which work is carried out also plays an important part in maintaining worker comfort and efficiency. Extremes of heat or light, poor ventilation, or safety hazards may cause unnecessary operator fatigue or concern. These factors have a direct bearing on output. To be most effective, an operator should have optimum environmental conditions. During the operation analysis procedure, the analyst must consider the effect of factors associated with operator comfort, safety, and well being.

Method.   Although it may seem unusual to consider method last during the application of operation analysis, each of the preceding points of primary analysis directly affects the final step, which is establishing the best method. When considering method, the analyst must first carefully examine the present method to find its weaknesses. Each of the ten points of primary analysis helps in the examination. After a thorough analysis of the present method, the analyst is prepared to develop the improved method. The operation analysis form provides space for description of both the original method and the improved method. It also provides space for recommendations for further improvements, if the activity of the operation increases. Periodic review of the operation analysis form and comparisons between levels of activity will help to pinpoint subsequent methods improvements.

The operation analysis form also includes a record of action taken. This record assists management in evaluating the disposition of methods improvements recognized during the operation analysis.

OFFICE OPERATION ANALYSIS SHEET

Date _____ Dept. _____

Operation _____ Operator _____

DETERMINE AND DESCRIBE	DETAILS OF ANALYSIS
	Why is the operation performed?
1. PURPOSE OF OPERATION	Can purpose be accomplished better otherwise?
2. COMPLETE LIST OF ALL OPERATIONS PERFORMED IN PROCEDURE	
No. Description Work station Dept.	Can operation being analyzed be eliminated? be combined with another?
1. _____	
2. _____	
3. _____	
4. _____	Is sequence of operations best possible?
5. _____	
6. _____	
7. _____	Should operation be done in another dept. to save time or handling?
8. _____	
9. _____	
10. _____	
3. QUALITY REQUIREMENTS	
a. Of previous operation	Are audit and other detail requirements necessary? too loose? too tight? suitable to purpose?
b. Of this operation	
c. Of next operation	
4. MATERIAL COSTS – FORMS AND SUPPLIES	Consider size, cost, condition, and advantage of standardization.
	Should form be prepared?
	Are forms best suited for insertion of data? ease in use?
5. TRANSMISSION OF INFORMATION	Consider means of transmitting data.
a. Brought by	Are means adequate? Is there excess backtracking? Is it done by "overqualified" personnel?
b. Removed by	Would conveyor system be practical?
6. WORKPLACE ARRANGEMENT (accompany description with sketches if necessary)	Can workplace arrangement be improved? Can distances be shortened?
A. Equipment	Equipment
Present	Should hand operation be mechanized? **computerized** Is equipment suitable? in good condition? used to best advantage?
Suggestions	Typewriters **word process** Other office machines File drawers and cabinets Reproducing equipment

FIGURE 2.2   Office operation analysis sheet.

Office Operation Analysis.   Although operation analysis has most often been applied to manufacturing operations, its principles can also be effectively applied to office operations. Although the points of primary analysis differ somewhat from the operation analysis

7. CONSIDER THE FOLLOWING POSSIBILITIES:	RECOMMENDED ACTION
1. Compare methods if more than one operator is working on same job. 2. Reassign duties a. To even work load? b. To handle peak loads? c. For better use of personal qualifications? 3. Use of snap-out forms. 4. Revision of existing form to serve additional purposes. 5. Arrange for two-handed operation. 6. Utilize all improvements developed for other jobs. 7. USE OF STAMPS, STAPLER, SIMPLE TOOLS.	
8. WORKING CONDITIONS a. Other conditions	Light Heat Ventilation Drinking fountains Washrooms Safety aspects Probability of delays
9. METHOD (accompany with sketches or process charts if necessary) a. Before analysis and motion study. b. After analysis and motion study.	Arrangement of work area Placement of: equipment materials supplies Working posture Does method follow Laws of Motion Economy? Are lowest classes of movements used? See supplementary report entitled Date

Observer _____ Approved by _____

FIGURE 2.2 (*Continued*) Office operation analysis sheet.

performed for manufacturing operations, the principles remain the same. The office operation analysis form is illustrated in Fig. 2.2. It will provide a systematic, written analysis of clerical operations where the level of activity justifies such a written analysis.

OPERATION ANALYSIS CHECK SHEET

The operation analysis sheet previously referred to is a guide to be used by the industrial engineer when analyzing any operation. The sheet is in abbreviated outline form; to make a more thorough analysis of any operation, the analyst must expand on the list of questions that appears on the form. With an abbreviated form, there is a danger that the analysis may be made too hurriedly and that proper consideration will not be given to each of the factors involved. To overcome this, an operation analysis check sheet may be developed.

The check sheet is an expansion of the shorter analysis form. On it are listed all the important questions which should be covered when considering the ten major points of analysis. Because a check sheet is quite detailed, a great deal of time is required to fill it in properly. Accordingly, it should be used only when it can be economically justified. Ordinarily, the check sheet would not be used unless a type 1 study, or occasionally a type 2 study, is indicated.

It must be remembered that the operation analysis form and the analysis check sheet are merely tools which can be employed when analyzing any operation. They are designed principally to guide the analyst and to display clearly the points to study and seek to improve. However, any improvements that are made will be the result of the ability and knowledge of the analyst rather than the tools used, for no form will take the place of sound reasoning, constructive thinking, and creative ability based on knowledge.

The opportunities for improving industrial operations are unlimited. The operation analysis technique is a powerful tool for accomplishing the methods improvement that every company constantly needs.

BIBLIOGRAPHY

Hammond, Ross W., "Industrial Engineering," sec. 10, chap. 2, in H. B. Maynard, ed., *Handbook of Modern Manufacturing Management,* McGraw-Hill, New York, 1970.

Maynard, H. B., ed., *Handbook of Business Administration,* sec. 7, McGraw-Hill, New York, 1967.

O'Donnell, Paul D., and John C. Martin, "Developing Improved Methods," sec. 3, chap. 8, in H. B. Maynard, ed., *Handbook of Modern Manufacturing Management,* McGraw-Hill, New York, 1970.

CHAPTER 3
DESIGN CONCEPT FOR METHODS INNOVATION (METHODS DESIGN CONCEPT: MDC)

Shigeyasu Sakamoto, P. E.
Director of Production Management
Research Institute of Management Innovation
Japan Management Association
Tokyo, Japan
President
Productivity Partner Incorporated
Nara, Japan

The methods design concept (MDC) for solving productivity problems is unique when compared with the traditional approach, generally called the scientific approach. The scientific approach is as follows:

Phase 1—Analysis of the Present Methods. There are many analysis and charting techniques in industrial engineering. They involve going to the shop floor first and collecting and/or analyzing the present operation methods as precisely and quantitatively as possible. This process normally takes a great deal of time, although a large part of such analysis is not helpful in providing any improvement ideas—it only identifies the present methods. It may, in fact, be no guarantee of being any use for improvement at all.

Phase 2—Find Defects or Weak Points in Present Methods through Analysis. The analyst looks for good operations, any inefficiencies in the daily work, and opportunities for cost reduction. These ideas normally result in the simplification of present methods but do not work as a guide for innovation in the present situation. Good ideas and effective improvement will help achieve simplification only of present methods.

Phase 3—Rearrange Methods. This is the synthetic stage of implementation of ideas as a particular new method. The resultant new method might be no fundamental change or little development of innovation, just a change of the previous method. Of course, ideas for improvement are unlimited. They depend on the knowledge and experiences of industrial engineers who use such a traditional approach. But, unfortunately, the results are always based on present methods, as illustrated by Fig. 3.1.

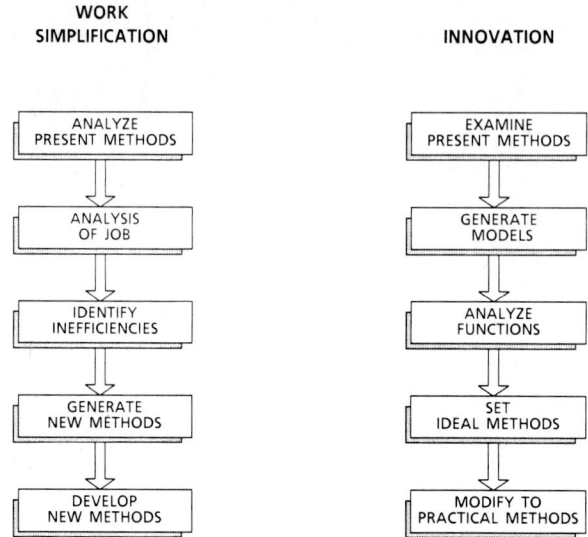

WORK
SIMPLIFICATION   INNOVATION

ANALYZE
PRESENT METHODS

ANALYSIS
OF JOB

IDENTIFY
INEFFICIENCIES

GENERATE
NEW METHODS

DEVELOP
NEW METHODS

EXAMINE
PRESENT METHODS

GENERATE
MODELS

ANALYZE
FUNCTIONS

SET
IDEAL METHODS

MODIFY TO
PRACTICAL METHODS

FIGURE 3.1   Two different approaches to develop new methods.
Work simplification and methods improvement versus innovation.

One way to avoid such a result, which is mainly simplification of present methods, is not to start from detailed analysis of present working methods. It is to start with basic design instead of simple improvement based on past experience. The general procedure for problem solving is as follows:

- Formulate the problem.
- Search for alternatives for a creative solution.
- Specify solution.

Let's look at a simple example of the difference between simplification based on present method and designing a new approach. No matter what the specific field of engineering, solutions for certain technological problems can be attained based on a worker's experience without any apparent theoretical knowledge. This experience falls into the frame of acquired skill. On the other hand, even without actually touching a machine and without a long history in a limited, narrow field, the same results can also be realized by application of technological fundamentals and principles drawn from experience. This is the design engineering approach. An important characteristic of engineering is the fact that the codified principles can be used to form a basis for reaching creative and innovative advances that would be difficult through the application of experience alone. The MDC approach is applying this design process to find innovative and/or fundamental improvement from present methods.

THE STEPS OF THE METHODS DESIGN CONCEPT (MDC)

The methods design concept consists of seven steps, as illustrated by Fig. 3.2.

Step 1—Develop Purpose Modules (Work Blocks).   Choosing the right improvement area for MDC is very important. There are two points in considering which modules to concentrate

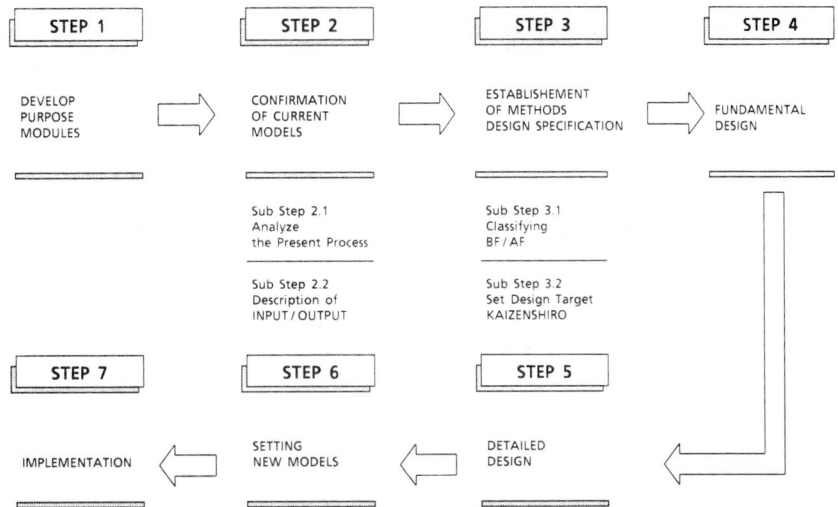

FIGURE 3.2 Steps of methods design concept (MDC).

on: first is the number of personnel in the module and second the job family. The number of workers in a module is important because the module must be easy to identify and define. It is difficult to visualize and understand the structure of a module if it is larger than 15 workers. The industrial engineer has to visualize the borders and image of the modules as the operation methods are designed.

If in doubt, however, it is better to have large modules rather than too small. Too small a group will limit the possible improvement potential. A group of three workers gives a maximum potential to work with of only three workers, while a bigger group creates a wider range of possible improvements, and hence the potential is greater.

Rather than retaining existing work blocks (modules), totally new modules must be established through new design. Ideally, modules should be established by considering the operation and its methods. Once a single MDC is developed, its results can be applied to multiple modules. By skillfully defining modules, points that appear to be different under the current organization or line structure can take on partial commonality or similarity (not necessarily identical, but similar). It is therefore desirable that modules that are as similar as possible be established.

Let's look at an example of module setting: Suppose there is a machine shop and beyond this shop there is another shop, the finishing shop of the machined processed part. They are on separate shop floors, but these two shops are in a very close functional relationship. This means the product flows from the first shop to the second. In this case, both of the shops should be seen as one module. If they are seen as separate modules, the input definition of the second module is just process machined materials and the output is parts that are ready to use in assembly work. But, if they are seen as one module, the input is raw materials and output is parts ready for the assembly shop. The necessity for finishing is not important for designing new working methods.

A similar example is circuit board automounting, inspection, and adjustment. Those are separate shops but should be considered as a single module for the MDC. Inspection and adjustment work themselves are not the purpose of the work. The ideal definitions of input and output are various kinds of supplied parts and mounted printed circuit boards ready for assembly.

Step 2—Confirmation of the Current Model.   What is a model and what does it mean? A model is a simpler representation of the real world. Moshe F. Rubinstein described this in

Patterns of Problem Solving (Prentice-Hall, Englewood Cliffs, N.J., 1975, pp. 192–193). The concept of a model is so fundamental to problem solving that it is present at all stages—from problem definition to solution. It is a concept characterized by uniqueness. The words and symbols we use, and the responses recorded by our senses, are all models.

A model is an abstract description of the real world. It is a simple representation of more complex forms, processes, and functions of physical phenomena or ideas.

For an existing object or product being treated by method design, it is necessary to generate totally new creative models within the existing design because the operation currently exists. The purpose of confirmation of the current model is as follows:

- To understand the details of current operating methods.
- To understand the amount of time and work required by current operating methods.
- To establish a benchmark in order to evaluate the effects of the newly designed methods.

A model should (although without details) be representative of the final shape and give the same impression.

Analyze the Present Process. The process to be analyzed and improved is described in operation sheet A illustrated in Fig. 3.3. (Note that the present process is to be described without any consideration of possible improvement potential already known at this stage.) Operation sheet A (Fig. 3.3) has the following principal columns:

Process: Give the designation of the whole process for which the module is made. If the process is large, it is advisable to subdivide it into operation steps and thereby describe the elements in the process. In this breakdown of work, no consideration is given to who performs the work. If there is any overlap between different elements, frequencies of elements, or the like, this is not noted.

Standard times must be collected for the elements which are shown in the W/U (work unit) column. The frequency of each operation is written in the W/C (work count) column. Choose a suitable time unit for standards in order to make it easy to talk about the process in further discussion. The current model rather than current condition is required in operation sheet A. The model is a standardization of current operation methods, or a confirmation of operation methods as a desired practice under current conditions. The details include the content of the operation, the amount of time per occurrence (work unit), and the required frequency per cycle (work count).

The steps of the MDC unfold along identical lines. Consequently, the existing model should not contain the many day-to-day variations that occur. Exceptions to the general rules and specific conditions need not be included in the model. Instead, the model should be a simple representation of the process, not an analysis of detailed conditions (Fig. 3.4).

Description of Input and Output. Assume the process to be developed is to go from city A to city B. When the process is expressed as "to go from A to B," it leaves scope for creativity. If, on the other hand, the process is called "how to reduce the time to go by train from A to B," the phrasing imposes severe restrictions and limits creativity.

A general definition of input and output is as follows:

Input: Power or energy put into a machine or system for storage or for conversion in kind, or conversion of characteristics with the intent of sizable recovery in the form of output.

Output: Power or energy delivered by a machine or system for storage, conversion in kind, or conversion of characteristics.

Defining the input and output means clearly defining the conditions coming into the model and those coming out. During this definition process, the tendency is to confuse the *input* or *output* of products and components with that of the methods. Remember that what we are designing here is the work methods. Therefore, the definition covers what conditions (*input*) are processed by the model to produce what conditions (*output*). In or-

FORMAT A

MODEL METHODS

WRITTEN BY Q.S.Z. A.A.   MODULE AT10-1   ICT   SEC. 15'   1/2

PROCESS			OPERATION			FUNCT		W/U	W/C		NUMBER OF	MEMO FOR ALTERNATE METHODS		
No	NAME	FUNCTION	No	NAME	FUNCTION	BF	AF	SEC.		O/A×D/A	WORKERS	I	II	III
1	attach hook button	connect	1	picking panel base	fixing place		√	1.609	1.	1				
			2	fixing panel box and attite	separate		√	2.015	1	2.015				
			3	attach hook button	connect	√		1.440	1.	1.440				
2	attach mike assy.	connect	1	attach mike assy.	connect	√		3.412	1	3.412				
3	attach packaging	connect	1	connect pkg.	connect	√		3.744	1	3.744				
			2	fasten screw	fixing		√	4.107	1	4.107				
4	wire connect to speaker	connect	1	join connector	connect	√		2.060	1	2.40				
			2	tidy wire	protect wire trouble	√		1.062	1.	1.062				

FIGURE 3.3 An example of operation sheet A.

3.45

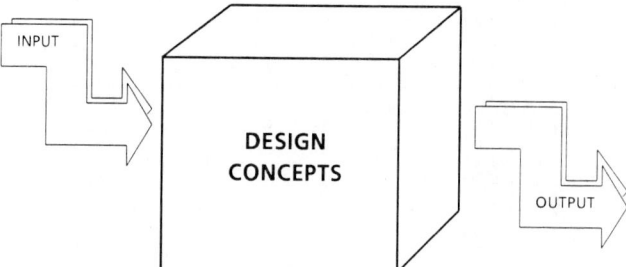

FIGURE 3.4   Engineering design means changing from input to output.

der to avoid confusion, we make it a rule to define products and components, as well as methods *input* and *output*. Both definitions may be written in operation sheet B, illustrated by Fig. 3.5.

Step 3—Establishment of the MDC Specifications.   The purpose of the MDC approach is *not* to reform or improve existing conditions. Rather, a creative approach is taken to design a *new* manufacturing process. This means that step 3 represents the first step of the actual MDC process. The MDC specifications are shown in operation sheet B.

Classifying Basic Function (BF) and Auxiliary Function (AF).   Once input and output are defined, function classification (that is, BF and AF) can be performed for each respective item, such as the processes and operations of the work performed in the model. These results are filled in on operation sheet B.

Basic functions are activities which directly contribute to the purpose of the process in order to change from "input" to "output," considering any restrictions if applicable. Basic functions must always be performed in order to obtain the desired change of situation, input and output in the model. If there are no basic functions, a purpose model has not been created.

Basic functions can be reduced by simplification, combinations, and the like. However, they can never be eliminated with the present technical solutions without changing the input and output in the purpose model. Basic functions can be eliminated only by new technical solutions, for example, by changing the product design. When developing an improved or entirely new process, you must start by considering the necessary basic functions.

Auxiliary functions are activities which are necessary to assist the basic functions when forming a complete process. While auxiliary functions do not contribute directly to the purpose of the process, they are not "waste" or unnecessary activities.

It is easy to explain these two functions in the simple assembly operations of a pen. They can be analyzed as:

1. Get body and cap.
2. Assemble body and cap.
3. Inspect product.
4. Place product on a table.

This process is illustrated in Fig. 3.6. In this analysis, "Assemble body and cap" is the only motion which is a basic function. Only this operation contributes to increasing the output of the product, with a direct connection between the input and output condition. When you take this MDC away from the work area layout, motion economy soon becomes a low level of improvement that is no longer of interest. When you create new operation methods which contain only the definitions of input and output and a few restrictions, it is like drawing a picture on a blank sheet of paper.

FIGURE 3.5 An example of operation sheet B.

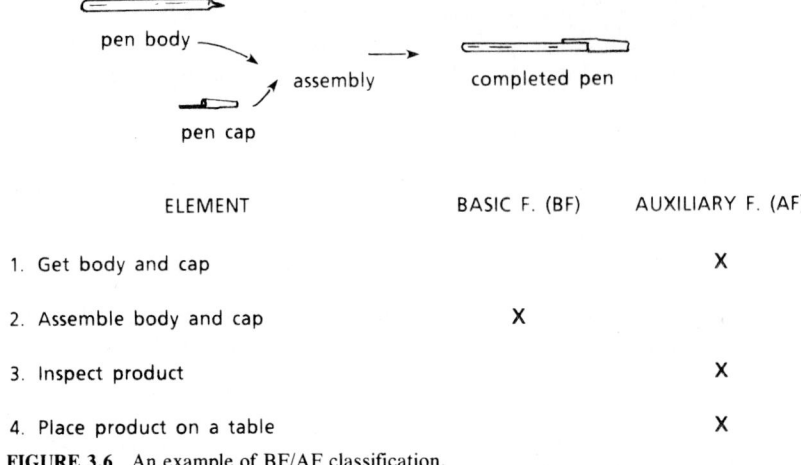

ELEMENT	BASIC F. (BF)	AUXILIARY F. (AF)
1. Get body and cap		X
2. Assemble body and cap	X	
3. Inspect product		X
4. Place product on a table		X

FIGURE 3.6 An example of BF/AF classification.

Returning to operation sheet B, various specifications may be established for production. Operation sheet B will be used to record not current conditions but rather the specifications which will be established as a reference to highlight the change from the current conditions. It is not necessary to consider all these items of specifications, since the desired results of method design will be established as design specification detail in the future. These will be the design requirements stated as objectives of the person who is in charge of the methods design. Operation sheet B also answers the question of how the design specification is to be determined. The most desirable case would be for a top-down approach, with management establishing a policy that a design target is extremely important and making the MDC different from the conventional work improvement approach. It is not concerned with addressing what to do, what to aim for, or what to achieve, or with attempting to improve on past accomplishments.

In operation sheet A, the labor power value is calculated by dividing (work unit × work count = labor power) by ICT, ideal cycle time. This conversion is performed for the labor power reduction purpose of the MDC application. The ICT as used here stands for production speed necessary to maintain the required work or output. If the ICT is ignored, it is possible that the production volume will drop along with any reduction in labor power.

The final aspect of the MDC specifications to be discussed is concerned with restrictions. These are not defined above, but they must be included in the design of the new methods. A representative example of this is shown at the bottom of operation sheet A. Through the new method, changes in cycle time, cost, and output conditions are achieved. This differs according to the level of mechanization and automation stated as a condition or design specification when designing a new method. Since the MDC is not merely aimed at the conversion of work contents into a mechanical process, the desirable method keeps capital investment to a minimum.

Set Design Target: Kaizenshiro. In a machined component, the quantity of scrap material is determined by subtracting the finished size from the stock material size. No matter what kind of material is used, the size of the finished product resulting from the machining process is the same. What changes is the amount of scrap, in accordance with the quantity of the original stock. More scrap results in higher cost, and vice versa. This point of explanation also applies for the MDC. It is the *Kaizenshiro*.

Let's take for an example of the *Kaizenshiro* the pen and cap assembly in Fig. 3.6. Assume each element requires 1 second. This means:

1. Get body and cap   1 second

2. Assemble body and cap   1 second

3. Inspect product   1 second

4. Place product on a table   1 second

Only the second element, "Assemble body and cap," is the basic function's element. *Kaizenshiro* = total present cycle time − BF element time. So *Kaizenshiro* in this example is 4 seconds − 1 second = 3 seconds. *Kaizenshiro* is the potential for improvement according to the basic function of the MDC approach.

The amount of scrap metal produced differs according to the original size of materials and specified finishing size. The specified finishing size is fixed and the supplied materials must be cut to the specifications regardless of their original size. This means that materials thicker than necessary will require more to be removed, while thinner materials require less to be removed. For the explanation of this concept, the size of original material is the current cycle time or personnel assigned to the task and the finished size is the basic function (BF). Within the MDC the basic function is not intended to be improved. This means that the amount of scrap and *Kaizenshiro* are not exactly the same, but since no other practical concept is available, the *Kaizenshiro* can be regarded as the scrap; that is, the *Kaizenshiro* equals the current work volume (cycle time times number of workers) minus the work volume of the basic function.

Let's look at another practical example where the MDC is applied for improvement of the labor force on operation sheets A and B (Figs. 3.3 and 3.5). There are five workers now and the total number of workers needed for the BF is 2.4. This means that 2.6 (5 − 2.4 = 2.6) is the *Kaizenshiro* in this example.

Accordingly, the most efficient work process is considered to be that which brings material to the finished size using only a basic function. In other words, based on the definition of the basic functions, all work processes are tied to increases in output. Of course, such conditions actually do not exist in real life, but methods design should be approached considering this as the logical design target.

Step 4—Fundamental Design.   The design steps of the methods design concept are divided into fundamental design and detailed design. Design is a process in which ideas are presented and design target achieved by incorporating them into the design.

The fundamental design is the stage at which the basic details are designed using only the basic functions. Since the auxiliary functions that accompany the basic functions also exist, the existence of auxiliary functions and the weight of their influence differ according to the basic function. Therefore, at this early stage, only the design specifications of the basic functions are established.

It can be said that an infinite number of ideas are possible for the new methods, precisely reflecting the fact that the potential for improvement is also unlimited. Limits do, however, actually exist in accordance with the MDC restrictions mentioned earlier. This means that even if the idea appears to be good, it may be too expensive to implement, and experiments, or some other confirmation, may be required. Moreover, the MDC is aimed not at repeatedly producing improvement ideas but rather at establishing local targets and producing ideas that are sure to be well within the realm of possibility. This roughly resembles a combination of work simplification and changing methods, but the difference lies in the fact that ideas are generated and developed while confirming their objective and the extent of the achievement of the objective.

In the beginning of the pursuit for ideas, the objective of the processes and operations of the basic functions within the model established by operation sheet A must be recognized. What? Why? Why is something performed? What would happen if it was discontinued? Application of the brainstorming process may be used here to search for new ideas. Can we eliminate the target? If we can, the work method itself disappears and is therefore the most effective of ideas.

Elimination and simplification principles of methods improvement should not be tried in order to find improvement ideas while brainstorming because, as described before, basic functions are functions which directly connect input conditions and output conditions. This means that basic function elimination is impossible. The principle of simplification is also not important in the fundamental design stage because there are no limited ideas. One example would be automated methods where such automated ideas may or may not be an improvement compared with the *Kaizenshiro* as a design target. This is why the simplification principle should not be picked up in the brainstorming of fundamental design.

The first step is to challenge the basic functions with the questioning technique and creativity. The reason for concentrating on the basic function is that we have noticed that method improvement itself is not sufficient, and methods development is necessary. Thus we have the possibility of starting on a new and higher platform.

The strategy behind methods development in the MDC is built around the "black box," the principle of the purpose model. Instead of improving the existing methods, take a step backward and begin with your purpose and a clean sheet of paper.

Step 5—Detailed Design.   When you have found or designed a new alternative basic process, it is time to take into account the necessary auxiliary functions. We can eliminate auxiliary functions from the old method that are no longer needed. This way of working can be compared with zero-base budgeting. You start at an absolute minimum, the core part, and then dress it up with whatever is needed, but no more. At this stage, we concentrate on elimination since auxiliary functions often can be eliminated. Using the total questioning technique and creative thinking, you can develop the new process including both basic and auxiliary functions.

Working with the questioning technique is analytical but exacting work. Openness and tenacity are needed, and you must not give it up until you have found the best possibilities for eliminating, combining, rearranging sequences, and simplifying. If your new solution does not lead to the target set by the design specification, you must again review the method with the questioning technique and creativity to develop an even better proposal. If, despite several attempts to make improvements, the method still cannot achieve the target, the present technical solutions must be reevaluated. We have to return to the process model in step 2 with input and output descriptions, in order to continue.

A significant point of detailed design is finding creative ideas to meet *Kaizenshiro,* such as a design target. This step of brainstorming should be continued until the total amount of created ideas meet the *Kaizenshiro.* This is very important in the creation of effective methods design. Brainstorming should be developed not only to find ideas but also to exactly meet *Kaizenshiro* such as a design target.

In practice, it might be difficult to meet the full standards of *Kaizenshiro* because of unexpected practical limitations, such as the cost of implementation, excessive implementing time, and engineering problems, but at least 80 percent of the accomplishment of the original amount of *Kaizenshiro* should be possible. It is normally possible to reach full or 80 percent of *Kaizenshiro* accomplishment, but it is important that the target-oriented ideas are created through brainstorming rather than traditional analytical approaches.

Step 6—Setting New Models.   As the design is realized, one new working method is developed. This is, however, a model that does not include any detailed working methods. You can find the description models in "Step 2—Confirmation of the Current Model."

Step 7—Implementation.   The last step is implementation of the new methods as a model. The small differences between the model and the actual method that is fully implemented and/or directed should be managed by shop floor supervisors. To implement the newly developed methods successfully, designing management systems is a key factor. Good supervision, instruction by supervisors, and fine tuning new methods are also significant to success. During the early weeks and months of the implementation of a new production model, worker performance will be low and will fluctuate widely. Considerable control will need to be implemented by both supervisors and industrial engineers to monitor and improve the system.

CHARACTERISTICS OF THE METHODS DESIGN CONCEPT (MDC)

Generally speaking, the following are the characteristics of the MDC.

1. To establish design goals that are logically calculated to produce the desired result. This is known as *Kaizenshiro* in the MDC process.
2. To pursue ideas not aimed at improving the current operation or methods, but rather aimed at the function (task objective).
3. Small investment compared with general experiences of productivity improvement can show a dramatic improvement of 40 to 50 percent.

Kaizenshiro (Potential for Improvement). It is the target of the MDC to improve *Kaizenshiro* and to reduce the cost of such improvement. It is a process whereby a logical model is determined within the confines of controlled input and output. Even if the same task objective is present, a change in the input or the output content or condition results in a change of basic function and improvement.

Kaizenshiro is a target to create ideas in the process of the MDC. The MDC is not interested in endless continuous small improvement but requires that industrial engineers be creative at the beginning of the MDC process.

Functions Prior to Manufacturing. Rather than spending time trying to figure out how to make a particular operation more efficient, it would probably be intrinsically more effective to consider what the objective of the operation is. In the automobile industry, for example, the basic operations for vehicle assembly will be the same in any factory. However, the time spent in producing a particular product and the breakdown of time spent on a single shift can differ according to the factory, company, drawings, production system, the supervisors' instruction, or the performance of the workers themselves. The contents of this fundamental operation can be considered as resembling the basic function of the MDC.

If we consider the problem of joining two items together, here are a few options: "glue," "nut and bolt," "welding," "hinging," and so on. Rather than taking an approach based upon preconceived notions about the process, greater improvements can be realized by questioning the process and asking why it must be nut and bolt. This is to say that an approach in which the objective is neglected while attention is paid only to function should be avoided. Improvements will be made only on present processes where this is the case (Fig. 3.7).

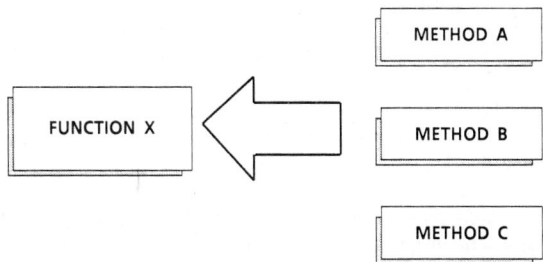

FIGURE 3.7 A function has many possible operation methods.

Small Investment. When new production processes and manufacturing methods are designed, the cost allowed by the MDC specifications should be determined as a design condition. The ultimate selling price and manufacturing cost must be determined beforehand, bearing in mind such factors as the product's application. Since such an approach is developed in the MDC, it can be shown that the MDC provides outstanding results at a very low cost. To follow the MDC specifications is an important part of design, and through this, industrial engineers can maintain improvements at relatively low cost.

AREAS OF DESIGN THROUGH THE METHODS DESIGN CONCEPT

There are three areas of the methods design concept:

1. Manufacturing methods
2. Manufacturing systems
3. Management systems

Manufacturing Methods.   First there are manufacturing methods which involve machines, facilities, tools, and so on. There are many types of machines, for example, and the problem for engineers is which machine to use for each particular function within the methods. There will be several machine tool combinations for any particular function of production methods.

A company in which the MDC was adopted developed a totally new shape of cutting chip for a machine tool. The design target of the cutting operation was to reduce the operation cycle time from 31 to 18 seconds, a 42 percent improvement. The industrial engineer tried to find an effective cutting tool that both was fast in removing metal and could be changed quickly. Because adequate cutting tools were not easily found on the market, the engineer asked a world-famous cutting chips producer if they could produce one. The producer replied that it would be very difficult, but they did succeed in developing such a chip.

This is a simple example where the MDC process did not just accept improvements attainable through easily available existing tooling. Instead, the original design targets were met by trying and succeeding in the development of their new tool. The tool also had another feature which was easy to set up because of the special shape of the design of the chip. Before improvement it took 10 minutes. Now it takes less than 2 minutes.

Manufacturing Systems.   Questions involved in the manufacturing system are: Why do you adopt a continuous flow line for particular products? Why do you install a certain number of robots in the production lines? What is the relation between workers and machines? Do you apply a straight or U-shaped line or layout? and so on. There is a lot of room to create new and effective systems. Manufacturing people prefer to install continuous, long, straight lines in their plant. But have they researched why their solution is more effective than another solution? The answer is usually no—even though everybody agrees that it is necessary. The question is how to do it and not what to do. One of the practical solutions is to change the approach from the experience-based approach to a design approach as in the MDC.

Management Systems.   Even if good manufacturing methods and systems are adopted, there is still one other point of design, the management system. How do you manage if unusual things happen? We have to be prepared for any uncertainties on the shop floor before the total system can work well. A good example is industrial engineers preparing a production line. They calculate the loss in balancing a line compared with the cycle time of each workstation. This kind of line balancing is called static line balancing (SLB). The actual practice of line balancing is quite different, however, depending on the workers' performance, interference, work mix, and so on. Hence industrial engineers must use more complete techniques such as dynamic line balancing (DLB). But how do you manage unskilled workers who are assigned to a particular workstation, unusual material conditions, machine conditions, and so on?

The MDC has been designed to leave a choice of alternatives of manufacturing tools, manufacturing systems, and management systems. If it is easy to find alternatives, all alternatives should be compared as thoroughly as possible. If it is difficult to find innovative ideas with the usual easy approach, then using the MDC application, mechanical engineers, electromechanical engineers, production engineers, and industrial engineers are almost always experienced enough to develop effective manufacturing systems and manufacturing methods. Their experience concerning management systems for the workers, materials, machines, and so on are, however, poor because there is no simple step-by-step approach to management systems.

THREE STAGES OF DEVELOPING EFFECTIVE METHODS

Three stages are involved in developing more effective operating methods. You can go to almost any shop floor and find some inefficiency. It may be improved immediately in many cases, and implementation is not difficult. Other types of improvement must be developed, however, to achieve the full potential for improvement. They may be classified into the following categories:

1. Work simplification
2. Methods engineering
3. Innovation

Figure 3.8 illustrates the typical relationship between improvement modes and time required to achieve them.

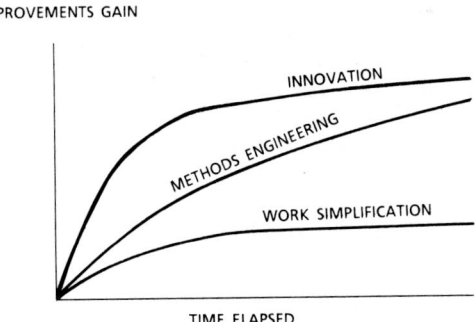

FIGURE 3.8 Relation between improvement gains and time elapsed among three stages of developing effective methods.

Work Simplification.   This kind of improvement, which is the lowest level of improvement, is normally focused on eliminating unnecessary work or minimizing any inefficiencies involved in an operation. Small group activities such as quality circles and the analysis of current methods or processes all belong in this category. While this area provides a great opportunity to achieve results through a large number of varied ideas, only a small percentage of the total improvement potential will be achieved through these techniques. Typically, these improvements do not require major changes in existing machines, facilities, or layouts and consequently do not involve a great deal of capital expenditure.

This approach requires time to be spent in analysis of the present situation, to identify problem areas, and to apply common sense to eliminate the waste and poor practices by developing new methods. Traditional industrial engineering training puts the emphasis on the above points, and most industrial engineering training courses and books are about improvement of the present work methods, relying upon small group projects. The advantage of "work simplification" is that the cost and time required to apply these methods are relatively small. The weakness of this approach is that it does not focus on the major opportunities for improving productivity.

Methods Engineering. Once methods have been improved through work simplification, it may be felt that the resulting working practices should be comparable with other companies or proved management techniques. However, this may not be the case, since varying levels of work simplification potential exist in different situations. In order to match or exceed the productivity of other systems which are being used as a benchmark for acceptable productivity levels, other improvement techniques must be used.

The methods engineering category will require minor changes to be made to production hardware, for example, improvement to workplace layout, fixtures, machines and tooling, and the like. This often requires a small amount of investment, which may be easily justified.

An effective improvement technique developed by Shigeo Shingo was named the SMED or single minute exchange die approach. This approach had a target setup time of less than 10 minutes. Many companies reduced their setup time at their press shop, for example, to less than 10 minutes. However, only the first company, who developed the die setting idea, had a real competitive advantage over the other companies, who were some years behind the original ideas. To make significant progress, therefore, we must develop a new way of thinking concerning methods improvement.

Innovation. Another category is innovation for methods change. It is not just work simplification and methods engineering ideas that are based on present working methods, machines, fixtures, and so on. It is easy to understand the difference between improvement and innovation of a product. Let's look at such an example.

There are two useful presentation devices. They are the 35-mm slide projector and the overhead projector. Using a slide projector is the traditional way of making presentations. Some modern features have been added such as remote control, forward, backward, focusing, and cartridges. These kinds of improvements have made it easy to use the slide projector for speakers at presentations, for example. However, there is still an inconvenience with the slide projector: the main switch. One can turn a slide projector on and off only on its shell and not at the speaker's position. It is very easy to control the master switch at the speaker's position if you install a switch. This kind of improvement is classified as work simplification and/or methods engineering.

On the other hand, there is the overhead projector (OHP). The OHP is easily turned off and on at the speaker's position, and all other points of inconvenience with slide projectors are overcome. The question is whether the OHP was developed as an improvement of the slide projector. The answer is no. The OHP was developed without improving the slide projector. The specification of OHP is defined and engineers developed the OHP instead of making an accumulation of continuous small improvements of slide projectors. This is an example of innovation of products and product design.

Another example of innovation is the air-powered nailing machine which is fed by a magazine or cartridge of nails. The conventional approach to improving the nailing of wooden structures is to develop improved methods by studying the present operation. The operation is then improved by simplifying the grasping and positioning of a nail and by experimenting with different lengths of hammer handles and hammer head shapes and weights until the optimum combination is obtained. This follows the procedure Taylor followed in his famous shovel experiments. It is the conventional methods engineering approach.

The innovation approach is to study the basic functions of the operation, to develop a completely new method of performing them without regard to the present method, and to accomplish the basic function in a minimum of time. This is the approach that would lead to the development of a nailing machine to replace a conventional hammer and nail operation.

With previously described types of methods improvement, you may have a large number of very small ideas which can only take you so far. The approach must be changed, or efficient and creative ideas will not be found to raise the potential for efficiency. This level is achieved through the design approach, such as the methods design concept.

An improvement target must be set as high as possible and by the improvement project itself rather than through the copying of other projects' attainments. The methods design concept may not be applied to current methods since innovative improvements do not work through analysis of present methods. In place of this approach, suitable methods for defined output should be created so that innovation results. Having gone through the methods design concept, innovation will reach the ideal level.

Within manufacturing, the operations and process are means of converting from an input status to an output status. At a lower level are operational elements and motions, some of which contribute directly to meeting output demands and others which support prior to upstream work contents. These two categories have been defined as basic function (BF) and auxiliary function (AF).

Three methods of change activities are summarized in Fig. 3.9. The most significant differences between the three activities are decreased nonworking time and/or auxiliary function work or increased basic function work. It looks as if they are similar, but real results achieved from each of them are totally different. The MDC increases basic function work for higher productivity improvement and innovative ideas.

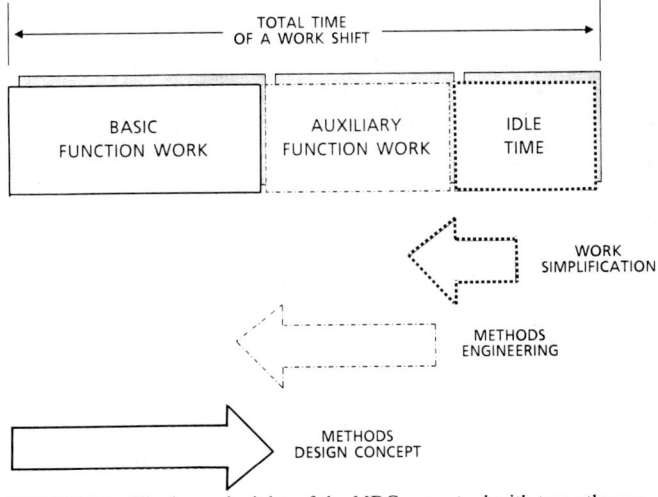

FIGURE 3.9 The key principles of the MDC compared with two other approaches.

When you use traditional methods improvement such as work simplification, you will be interested in the distance for getting and putting parts, utilizing both hands, and improvements without changing main parts of jigs and fixtures. If you follow the MDC, however, there is more opportunity to find out a totally different way of improving methods that is not just following the present basic operation methods. The point of this difference is whether you concentrate on the basic function or not. This is why the MDC will lead to a jump to a new higher platform as illustrated by Fig. 3.10.

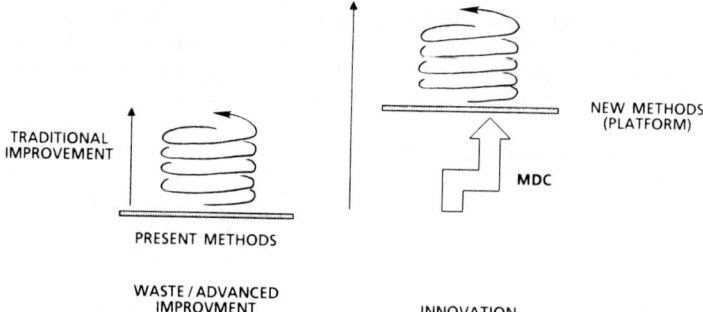

FIGURE 3.10   MDC aims to jump up to a higher platform instead of continuous small improvements.

MDC APPLICATION EXAMPLE

Objective of MDC Example.   Figure 3.11 describes the principal flow in a tube mill. The material, a steel tube blank, is heated up to 1200°F in a rotary furnace and transported on a conveyor to a steel tube mill. The company has an objective of reducing the batch sizes considerably and thereby going from stock order manufacturing to customer order manufacturing. Because of the present changeover times, the capacity in the tube mill is not enough to reduce the batch size to the size of customer orders. The target is therefore to reduce drastically the time for resetting or changing over from one size to another.

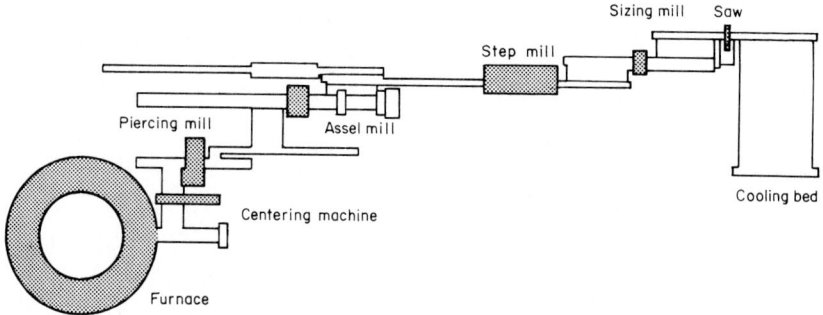

FIGURE 3.11   An example of the MDC application layout of tube mills.

Present Operation.   The conclusion of the investigation is that the step mill has the highest priority. The step mill comprises 12 stands. Every second stand has vertical as opposed to horizontal rolls. Within a limited area the rolls in a stand can be adjusted on site. For larger changes in dimensions the stand must be replaced with stands with other rolls. A separate workplace for changing rolls in the stands is located next to the step mill. When stands are changed, the old ones are pushed out with hydraulic pistons from the mill. The stands are disconnected from the piston and the piston is pulled into the mill. The hydraulic pistons are operated from a control panel. Prior to pushing out the stands, the cooling hoses have to be disconnected and the driving clutches for vertical rolls disengaged.

The stands are pushed out on a cart which is moved alongside the mill. When the cart with the old stands has been pushed aside, a cart with the new stands will be pushed into place. The hydraulic pistons are pushed out again, connected to the new stands, and the new stands are pulled hydraulically into the mill. The driving clutches must be engaged on the vertical stands prior to starting up the mill. The cooling hoses can be connected after the mill has been started. The cart is driven by wires. This results in poor positioning prior to pulling the stands into the mill, hence forcing the setters to adjust the final position of the cart manually by crowbars. The hydraulic pistons are connected to the stands by means of locking pins. The stands are locked in the mill by means of locking bolts. All operations of pistons for pushing out and pulling in stands, and for engaging and disengaging driving clutches is made hydraulically from the operating panel. Two or three setters are active simultaneously when resetting, and one of them is operating the control panel.

The space in front of the mill for the setter to disconnect and connect the hydraulic pistons with the locking pin on the stands is very limited. It is also dark and dirty. For safety reasons, only one piston and stand may be operated at a time when a setter is within the operation area. The input and output definition of the step mill is "set stands for next batch."

Calculation of **Kaizenshiro.**   The activities of resetting are shown in the accompanying table.

Activity	Work unit (minute)	Description
1	1.00	Stop mill (long stop due to large masses to be stopped)
2	0.30	Disengage driving clutch
3	0.25	Disengage cooling hoses
4	0.65	Push out stand (operated from control panel)
5	0.10	Disconnect and connect stand (from hydraulic piston manually)
6	0.60	Pull in piston (controlled from panel)
7	0.55	Cart aside, length mill (wire driven)
8	0.60	New cart in (across mill, operation from panel)
9	0.55	Push out piston
10	0.65	Pull in stand (operated from panel)
11	0.40	Connect cooling hoses
12	0.35	Engage driving clutch
13	0.10	Start mill

After critical path analysis of the resetting of the step mill, the time for changing three stands is 10.85 minutes. These activities directly contribute to the purpose of the input and output definition:

• Activity 10—pull in stand with hydraulic cylinder
• Activity 12—engage driving clutch
• Activity 13—start mill

Activity 11—connect cooling hoses—is not a basic function because the purpose model for the resetting restricts capacity. The cooling hoses can be connected after starting up the mill.

The sum of the required basic functions for the resetting in this example is:

• Activity 10 $3 \times 0.65 = 1.95$
• Activity 12 $1 \times 0.35 = 0.35$
• Activity 13 $1 \times 0.10 = 0.10$
               Total = 2.40 minutes

This result gives *Kaizenshiro* as $10.85 - 2.40 = 8.45$ minutes. *Kaizenshiro* indicates a potential reduction to approximately one-quarter of the original setting time.

Results of MDC Application. The target from a management point of view is as follows: Lately customer orders have been down to seven tubes per dimension. In order to manage these small batches, the changeover time must be reduced to a maximum of 5 minutes. The *Kaizenshiro* and its minimum time of the basic functions are consequently within the limit of the target for resetting time in the step mill.

The following ideas have been obtained:

- Assemble pneumatic disk brakes on the motors in order to stop the mill in 0.1 minute instead of the present 1.0 minute.
- Install new hydraulic pump with higher capacity, facilitating three stands to be pushed out and pulled in half the time of what presently can be done for one stand.
- Develop a mechanical device to connect and disconnect the stands from the hydraulic cylinder without the need of a setter in the mill.
- Exchange the wire-driven cart for hydraulic power with rack and pinion, allowing a much higher precision in positioning and eliminating all adjustments.
- Install new engaging and disengaging devices for the clutches (vertical stands only) so that the engagement can be increased.

The time for the new methods is 4.2 minutes. This is less than 5 minutes—the management target. Improved time value is 6.65 minutes ($10.85 - 4.2 = 6.65$ minutes) or 79 percent of *Kaizenshiro* ($6.65 \div 8.45 = 0.79$). This attained percent of *Kaizenshiro* is good enough. To discontinue searching for other ideas, 80 percent of *Kaizenshiro* is a practical benchmark. This is based on MDC experience. This is an example of the MDC application.

The productivity increase of the cycle time is 258 percent. It is not a small improvement. It still has not met the *Kaizenshiro* but is only 39 percent of the original resetting time. The key reason for getting such a good result is the dependence on the MDC, especially the target that is based on the basic functions.

A reduction of the changeover time from 10.85 to 4.2 minutes does not give any major economic improvement for the company. If the company utilizes the increased capacity by reducing the batches and manufacturing according to customer orders instead of stock orders, the improvement in changeover efficiency has resulted in considerable economic advantage in reduced inventories and capital investment and improved customer service.

APPLICATION OF THE MDC

The benefits of analyzing basic functions at the product design stage rather than at the production stage are far greater with MDC. According to an investigation, more than 70 percent of manufacturing costs are fixed before production methods are defined, that is, at the design stage. So how is it possible to use the MDC at the product design stage?

Product design should be developed and reviewed with basic functions in mind, leading to the evaluation of both producibility and productivity. Through the use of CAPP (computer-aided process planning), especially in assembly work, a number of production methods may be analyzed and compared, leading to a best method for a specific set of design specifications or targets.

Only through a close working relationship between design, manufacturing, and industrial engineering will significant innovation in production methods and systems be achieved. The basic functions defined as part of the MDC are a useful point of mutual contact and reference between these different engineers.

CONCLUSION

MDC is a productivity improvement technique which concentrates on achieving the basic function of a product or process through design rather than incremental improvement of existing methods. It is a technique that offers an effective combination of design engineering, value engineering, producibility, and methods improvement elements to solve productivity problems.

WORK MEASUREMENT TECHNIQUES

CHAPTER 1
IMPLEMENTING AND MAINTAINING A WORK MEASUREMENT PROGRAM

Albert F. Celley, Ph.D.
A. F. Celley and Associates
Toledo, Ohio

This chapter has as its essential purpose a discussion and prescription for implementing and maintaining a work measurement program. Work measurement of manually performed tasks consists of several major techniques: stopwatch time study, predetermined time systems (MTM, MOST, Work Factor, and the like), work sampling, and standard data. These tools are covered in detail in other sections of this handbook. They have the primary objective of determining a standard period of time to perform a defined quantity of work. A good work measurement program should consist of all the following functions: proper layout of the department and workplace, methods analysis and improvement, employee selection and training, ergonomic considerations, the correct equipment and tools, and so forth, along with the accurate application of whatever time measuring technique or techniques are used to establish the standard time allowed to perform the task.

Definitions. The implementation and maintenance of a successful work measurement program can be defined as bringing such a program into practical application and, having once accomplished this, keeping such a program performing its objective by providing a realistic measure of how much time should be required to perform a defined quantity of work.

Communication. Implementation consists basically of communicating the goal of the organization by management to its employees at all levels. Successful implementation commences early in the development of a work measurement program. Merely to complete all the technical components of such a program, then announce by letter, a notice, or a meeting, just prior to the installation of such a program, will not readily accomplish good implementation.

At all times in the history of human relations both management and workers have been suspicious of new ideas, approaches, and changes in the status quo. In a goods or services producing organization, where workloads and compensation may be affected, this suspicion and resistance to change may exhibit itself early and continuously. The literature of management is replete with discussions of this phenomenon, but because implementation of a work measurement program can be so important, a brief review of this problem will be undertaken here.

The first essential requirement of changing methods and determining task time standards consists, without reservation, of having top management's initial and permanent support. Such support must be maintained even though the organization may encounter many problems, human and technical, throughout the total process. Even among all members of top management and their descending levels of supervision, support for the program must be clearly identified and unanimity of purpose must be sustained. It is just as important for the industrial engineer to sell and resell management on the advantages of a work measurement program as it is to sell the workers who essentially will be directly impacted by such a program.

First, the advantages of such a program must be established and understood throughout the organization. This includes in addition to the hourly work force all office staff personnel as well as sales, marketing, research, and all the functional staff which offhand may appear to be unrelated or unaffected by the direct production process. The advantages of a work measurement program in general include the determination of:

1. Credible product cost
2. Labor requirements and unit labor costs
3. Product pricing
4. Tooling
5. Capital equipment investment
6. Quality attainment
7. Scheduling of both labor and material
8. Believable delivery promises
9. Effective organization size and structure
10. Compensation and incentive payment
11. Work design and human factors considerations
12. Planning, control, and budgeting
13. Production processes
14. All other internal human organizational interfaces

These advantages, while seemingly having the inherent ability to sell themselves in the form of a work measurement program because of logic and so-called common sense, become a problem in the implementation phase because of human intangibles. As with most organizational management problems, the technical problems we solve in general with little effort, the human problems we must solve over and over again.

Excellent communications throughout the organization is the beginning of successful implementation. Good human relations result from good communications. Successful implementation is based on good selling from the top manager down through the work methods and measurement analyst. Well-practiced persuasive skills are at a premium in the implementation process. While few individuals have natural selling skills, all managers can be given instruction and training in the basic skills of conveying the messages required to achieve all the advantages of a well-founded work measurement program. One of the most significant contributions a work measurement analyst can make is the ability to practice smooth persuasive skills—in short, good salesmanship.

Program Implementation Plan.   Before a credible program can be established, the following activities and/or tasks generally must take place or at least be considered to establish the proper environment for successful implementation.

1. The organization must have defined the purpose for having a work measurement program.

2. All members of the organization must understand this purpose.

3. All members must be afforded the opportunity to contribute to the establishment of such a program both at the beginning and during the process that is being developed and implemented.

4. Continuous appropriate communication and persuasion should be practiced to avoid or at best minimize resistance to the program.

5. Both the dollar and human cost should be recognized and kept in mind as the program develops.

6. All the various industrial engineering and work measurement analysis techniques such as methods engineering, process chart procedures, operations analysis, motion study, the appropriate type of measurement (stopwatch time study, predetermined time systems, standard data systems, and the like), and proper documentation should be accomplished by trained analysts.

7. Continuous follow-up meetings and explanations as appropriate should be maintained at all times throughout the implementation process. If the organization is represented in whole or in part by a union, then it is important to maintain good communication with the local officers and their international representatives. The key to avoiding resistance to change is enhanced by avoiding surprises, avoiding misunderstandings about the purpose and the techniques, and being sensitive at all times to good human relations.

8. The contribution of ideas, techniques, and better approaches should be readily welcomed and acknowledged at all times.

9. Training of all individuals when required undergirds acceptance of any change or new program at any level in the organization.

10. Any program can best demonstrate its vitality by selecting an area that can be easily and successfully installed early in the total implementation program.

11. It is important to build into the program a reporting system that is readily understood by all levels in the organization, and with the results made available to the workers as well as their supervisors.

12. A good reporting system requires careful and accurate reporting of units produced, who produced them, the labor standard applied, standard hours developed, actual hours charged, percent efficiency earned, and incentive earned in hours and dollars. What is actually contained in a reporting system of course will depend on the type of system used and the needs of the organization. The variations here can be numerous, but the significant issue is that adequate records be preserved to facilitate comparisons over subsequent time periods and various jobs run. It is essential that where a reporting and control system is to be computerized the computer program fit the process, not the reverse.

13. Reporting should be developed to facilitate a proper audit trail. For example, the total units reported as produced by single or multiple operators and/or equipment should be compared with the total units scheduled, shipped, or warehoused with losses reconciled into such a total.

14. Trial runs should be scheduled along with the training activities to serve as a learning process for the analysts, the workers, their supervisors, and whoever plays a role in the implementation.

15. Never, I repeat never, completely stop using an old system before all the bugs are eliminated or the necessary changes are made in the new program. Run a dual or parallel system until the new work measurement program is functioning satisfactorily. This debugging process is in itself a major training process at all levels.

Selling the Plan. Proper implementation is the achievement of professional work measurement techniques, and the secret of its success is the selling of the program to all those who will be directly and indirectly affected in the organization. Selling is the most important step. The implementation process is largely a selling process, the selling of the program throughout the organization from top to bottom—from top management down through all levels to the basic operators who produce the products and deliver the services.

Excellent salesmanship is a responsibility of the industrial engineer. Not every individual has the natural talent or skill to easily persuade others to accept a new idea or program—especially a program that can be perceived as threatening to many in the organization.

However, successful selling can be practiced and achieved, even by an industrial engineer who does not consider him or herself to be a salesperson in the usual meaning of the term. No apology need be made for the term ''selling.'' It is as much a tool for the industrial engineer or any member of management as is the most technical procedure used in analysis or direction of the work force. It is a fact of management life that many worthwhile projects have fallen by the wayside because they were not properly and persuasively sold to all the affected members of the organization.

The major steps in selling a work measurement program consist of the following:

1. Always avoid a threatening approach. Always emphasize the positive aspects of the program.

2. Understand each individual's and each group's (management, union, worker's, and the like) vested interest in the success or failure of the program. Insight into ''What's in it for me?'' is a basic step in selling a product, program, or idea.

3. The industrial engineer must demonstrate the desirability of the work measurement program to each identified individual or group. The most significant single aid to the selling function is the ability to communicate ideas clearly, concisely, and logically. This is much more easily stated than accomplished; therefore, careful preparation and practice is essential.

4. While many types of media are available—letters, posted notices, handouts, house organs, and the like, in general—a one-on-one personal presentation is the most successful when practical. This is especially significant where the human ''resistance to change'' is expected to be encountered. The one-on-one approach allows the seller to gain insight into the recipient and the surrounding circumstances so that unique tailoring can be pressed during this one-on-one process. The best approach is to present the program by taking advantage of any benefits to be gained by the delivery of the message in any combination of the media available. A one-on-one or a presentation to a small group avoids the problem of having dissident subgroups obstruct the presentation when such a presentation is made to a large group—especially where the exhibition of union or dissident politics may be a factor in the group. Selling is essentially an implementive process and can in no way be slighted when introducing any new program.

Training Process. To reiterate, the secret of success in the implementation of any new program, decision, procedure, or change is the proper communication of such activity to everyone in the organization. This important task generally requires more than just a letter or notice placed on a bulletin board.

1. All members of top management must understand the purpose and benefits of work measurement as the basis of many economic and operating decisions and processes in the organization. It is as important to have continuing top management knowledge and support of a work measurement program as it is to have a professionally designed and installed program. Top managers require meetings to explain the process of analysis, measurement, and application of time standards. They need the opportunity to ask questions and take part in discussions of the total work measurement procedure.

2. Middle managers such as division and department heads, along with first-line supervisors, most of all require special preparation for the introduction of methods analysis and work measurement programs. Because these managers will be directly impacted by such programs, well-designed and -organized training sessions should be established and professionally presented. The middle management group will be directly impacted, not only by the operating decisions that a work measurement program will engender, but also by the human and union problems, pressures, and unrest that labor standards and especially wage incentive systems can engender. Training sessions for this group should center around exercises that allow the group to analyze operations, perform basic work measurement activities, and be provided with answers to field the many questions that will be thrust upon them by the work force.

All too often this most important aspect of implementation is given a last-minute "lick and a promise." Then the organization cannot understand why this program that once appeared to have so much promise for improving management and productivity has now fallen on its back and lost the support of the middle management group so necessary to its success and payoff. Remember, do not leave out staff departments such as accounting, purchasing, and personnel, because their support is also vitally needed. Many excellent ideas and suggestions can come from this staff group concerning successful facilitation and implementation of a new or revised program.

3. The hourly and operating work force also requires special attention at the beginning of any methods and work measurement program. Resistance to change because of fear of how such a program will affect available work, crew sizes, long-term employment, and wages is a key factor to overcome before the program can commence. If a reduction in the work force is expected, they should be notified up front.

Under most circumstances the operating secrets, production know-how, and motivation vested in the hourly work force can make as great a contribution to improved productivity and profitability as that of management and the industrial engineering profession, if not more.

The significant aspects of training this group include the technical as well as the human activities of such an improvement program. Training sessions should be conducted that allow participants to take part in methods analysis examples and simplified work measurement exercises. Questions should be answered even if it again means acknowledging that over the long run the work force may be reduced if additional production volume is not forthcoming.

First impressions are difficult to overcome if they can ever be completely eliminated. Therefore, it is critical to communicate early and often—at all levels—at the commencement of and during the development of a methods improvement and work measurement program. This establishes the proper attitudes and understanding of such a program at the start. The benefits to all individuals in the organization should be constantly stressed through all media—meetings, letters to employees, notices, and the firm's house organ. New employees should find a part of their orientation program contains an explanation of such industrial engineering activities.

By application of a vigorous and continuing training and communication program at the very start of a work measurement program, many of the human problems of implementation can be significantly decreased or eliminated.

The industrial engineering staff will be responsible for the largest share of the communicating and training process, as well as the technical aspects of such a program. To repeat, no industrial engineer will be immediately successful unless such an individual is a good salesperson. Selling ideas and persuading others is a skill often born within many personalities, but even if such a skill does not come readily, many aspects of this process can be developed or improved through practice by the industrial engineer as well as management in general. Practice, both inside and outside the organization, will significantly improve every individual's skill in presenting new ideas and programs.

Documentation.   The documentation of work measurement systems is one of the most critical aspects of implementing such a program. The term ''documentation'' is defined here to include a complete and carefully written description of the program including all appropriate illustrations, statistics, forms, records, and examples of development and application. Such documentation preserves the integrity, understanding, and continuity of such programs over time, and for the turnover of personnel who may be involved. When changes in the system, its development and application—especially with wage incentive systems—are made, a basis for making the change can be compared and recorded against the original method. One only has to be involved in a work standards grievance or in contract negotiations to appreciate the value of a carefully written and permanently maintained documentation of the work standards program.

Manuals should be written in a complete but concise manner so as to forestall misinterpretation. Good documentation also has the value of enforcing the credibility both of the system and of those individuals who have been responsible for creating or maintaining the system.

Where feasible much developmental data can be preserved and pulled up from computer files when such data are in active use. However, for permanent reference, such data should be printed, bound, and kept as shelf copies to avoid their accidental or inadvertent loss in the computer system for any number of reasons.

MAINTENANCE OF WORK MEASUREMENT SYSTEMS

Proper maintenance of work measurement systems is as important to the success of an organization as is their proper implementation. Work measurement systems that are not kept in line with current methods and practices can soon lead to erroneous costs, incentive payments, cost estimates, scheduling, and a general loss of credibility. Maintaining accurate and representative labor standards is as significant a function and responsibility of good management as is the responsibility of maintaining the integrity of any other information system that supports effective operations decision making. Such standards are the basis for excellent communication of the results of decision making, and serve as basic guidance for further decision making as operations continue.

Significance of Good Maintenance.   In the third edition of Maynard's *Industrial Engineering Handbook* maintenance was defined generally as to keep up, to continue, to carry on, to keep in existence, and to preserve. Maintenance is an integral part of implementation, and good implementation in itself implies continued good maintenance. All too frequently management is guilty of expending significant sums of money, time, and organizational stress to develop and implement a work measurement program. Then it fails to pay the continuing costs of maintaining this important and often hard-won measurement, information, communication, and control systems. Work measurement systems require maintenance, and their maintenance should begin on the first day they are implemented. Maintenance should then become a continuous part of the management information and decision-making process.

Once the integrity of the work measurement system has been weakened because of a lack of good maintenance, it may be more costly and difficult to correct such a system than it was to originally develop and implement the program. Maintenance therefore requires exceptionally strong management. Again, maintenance of the work measurement and standards program should be a continuing process.

Records and Reports.   It goes without saying that to effectively maintain a work measurement program it is necessary to establish at the beginning proper ongoing records and reports. Such records should include an accurate definition of the method (workstation, setup, equipment used, inspection and quality requirements, manual procedures, working environment, and the like) upon which each labor standard was developed. The work measurement system is essentially maintained by conducting periodic audits of the application of these standards to the operations performed and reported. This is especially critical for the purposes of determining compensation, job estimates, shop schedules, budgets, and all other related management information.

Application Audits.   Determination of when or how often to conduct a formal audit is largely a function of how much impact the application of the standards has upon the organization and especially the compensation paid to the employees. In general, firms operating wage incentive programs audit the application of labor standards at no less than once per year. If there are ongoing changes to equipment or methods and materials used, the audit period may be more often than once a year. But primarily this generally is a judgment call, and the audit may be made only when incentive earnings or other standards-based decision information becomes out of line.

The actual audit in the case of incentive earnings should cover no less than one pay period, or longer if obtaining a representative period of activity is required. In an extensive incentive compensation shop, it may be appropriate to select jobs, departments, or audit periods by applying a simple random sampling approach. A key to effective auditing is always that the area or activity to be audited should never be known in advance to prevent any deviation in the normal practices of reporting production or hours charged to the work activities.

Audit findings should be kept as a permanent record so that they can be followed up and compared with future audits. As part of good audit procedures, recommendations should be made and documented to assure that questionable reporting of production, downtime, hours charged, and the like is brought to the attention of management and also to the attention of those employees involved.

Meaningful management reports of incentive earnings are necessary for the determination of how well the standards represent the work being performed, the work effort being expended, and the incentive paid. Whether such information is recorded and reported manually or in a computerized system, it must be in sufficient detail so that current and past activities can be resurrected for analysis and comparison.

Statistical Tools.   Unusually high or low incentive earnings can be readily identified through the application of a statistically based limit chart to establish what the limits are when large numbers of individual incentive earnings are calculated and reported over a given period of time. This is the same approach taken to establish upper and lower control limits in a statistical quality control program. Such a program is established by calculating the average percent efficiency or percent incentive earned along with the standard deviation to determine the upper and lower control limits within which these efficiencies or earnings should fall on a random basis.

Whenever the percents fall outside the control limitations more than once or twice on a random basis during a representative period, the standards, methods, or time charged should be investigated more closely.

Creeping Changes.   It is well understood that when significant changes are made in the methods, materials, speeds, specifications, or equipment used, the production standards should be changed. If this is not done, cost estimating, labor performance, labor and machine utilization, and incentive earnings, among other management decision information, will not be representative of the operations.

A well-known problem has always existed concerning when to revise the standards when so-called creeping or minor changes occur to raise efficiencies or earnings or, in some circumstances, to lower them. For example, a material or speed change of a marginal increment may not result in a significant change in performance efficiency or incentive earnings; however, six or seven such changes, that in each case serve to increase the efficiency of an operation, will result in a change that should warrant a revision in standards to maintain standards that reasonably reflect actual conditions in the operation. Often a small incremental change can occur in speeds, methods, and materials specifications simultaneously, thus compounding the spread between the applied standards results and actual operations. The periodic audit is especially important to stay in control of creeping changes.

The major problem with allowing creeping changes to exist over any unrealistic period of time is that once they become institutionalized, any attempt to correct such a situation leads to resentfulness and stress among employees. To mask the creeping changes employees often reduce their productive output, and such practice may even be abetted by their supervisors to avoid having to face up to another nagging problem.

Union Cooperation.   To effectively revise methods, speeds, materials, equipment, or other operating changes, and then revise the standards that represent these operations, it is a necessary practice to have the organized bargaining unit agree that this is an accepted part of management's responsibility. This action is facilitated by having first a standard management rights clause stated in the labor contract and, more specifically, a clause that states management shall have the right to implement, remove, change, and maintain all labor standards. Some negotiations have resulted in a contract provision that sets a limit of an average percent change, for example, 5.0 percent, before the standard can be revised. If possible, however, such contract statements should be limited to allowing management to make a revision for any type of change—except employee skill and experience development—without specifying a set percentage of performance efficiency or incentive earnings growth.

As with all management endeavors, the labor contract should never be a substitute for effective communications, employee participation, and plain good salesmanship. Extremes in incentive earnings because of poor standards maintenance can only lead to conflict between workers, the union, and management. Remember, the local union is basically what local management makes it.

It is common practice for organizations to provide employees with a handbook that states among other things the conditions of employment. The obligation of management to revise standards based on both implemented changes in materials, methods, equipment, and the like, as well as creeping changes, should be clearly and unequivocally stated in the handbook.

In summary, maintenance of a standards program is as important to the success of the program as is the original professional design and implementation of such a system. One of the key measures of a successful standards program over time is how well it is maintained. Only good maintenance will keep such a program permanently performing as it was designed.

SUMMARY

Effective implementation is as important to the organization as are the technical aspects of a work measurement program. The most professionally developed production standards program will not deliver useful information to management or to the organization as a whole until the program is accepted and smoothly installed. Underlying successful implementation is successful communication. No program, no matter how technically well developed, will sell itself. Successful selling is a human art and a managerial art that has no peer. This is especially true when it comes to introducing new ideas, measures, and procedures. It is essential that communication and training be conducted throughout the organization from top to bottom. The union should be included in this process and can be made one of management's staunchest allies if proper communication is practiced. Along with and as part of excellent and thorough communications is the complete documentation of the standards program. This will allow effective comparison among these changes as they occur.

The maintenance of a work measurement program is as significant and essential as its proper implementation. Changes in methods, equipment, processes, and materials will be constant—some major and many of a creeping nature. The measure of a good system is that it reflects and accurately measures the reality of the production process at all times. A well-designed program will provide for good maintenance through effective documentation, ongoing reports, and application audits. In large systems statistical sampling will provide practical coverage for effective auditing.

BIBLIOGRAPHY

Aft, Lawrence S., *Productivity Measurement and Improvement,* Reston, Reston, Va., 1983, 429 pp.

Barnes, Ralph M., *Motion and Time Study,* 7th ed., Wiley, New York, 1980, 689 pp.

Karger, Delmar W., and Franklin H. Bayha, *Engineered Work Measurement,* 2d ed., Industrial Press, New York, 1966, 722 pp.

Krick, Edward V., *Methods Engineering,* Wiley, New York, 1962, 530 pp.

Maynard, H. B., and Company, Inc., Editor-in-Chief, *Industrial Engineering Handbook,* 3d ed., McGraw-Hill, New York, 1971.

Niebel, Benjamin W., *Motion and Time Study,* 7th ed., Irwin, Homestead, Ill., 1982, 756 pp.

Zandin, Kjell B., *MOST Work Measurement Systems,* Dekker, New York, 1980, 204 pp.

CHAPTER 2
STOPWATCH TIME STUDY

Clifford N. Sellie
Chairman/CEO
Standards International Inc.
Northbrook, Illinois

Time study is used to determine time standards (targets) for planning, costing, scheduling, hiring, productivity evaluation, pay plans, and the like. Time standards may be determined by a number of different time study techniques: (1) They can be based upon historical records of time taken in the past to perform the task. These calculations of historical times can be based on straight arithmetic averages or sophisticated statistical analyses. (2) Another technique (sometimes called reasonable expectancies) is use of estimates by a knowledgeable individual of the time that it would take a qualified operator working at an acceptable performance level to do the job. (3) A third technique is predetermined times. Here the tasks are analyzed as to the work content and then "predetermined" times for the work segments are summed up to get the total time for the task. (4) The fourth and most often used technique is stopwatch time study.

Stopwatch time study is the most popular method of work measurement. It was first developed by Frederick W. Taylor before the turn of the twentieth century. It is now used worldwide to determine the time required to do work.

TIME STUDY

Time study may be defined as follows: Time study is a procedure used to measure the time required by a qualified operator working at the normal performance level to perform a given task in accordance with a specified method.[1] In practice, time study usually includes methods study. The industrial engineer (time study analyst) has to observe the methods while making a time study. The definition of time study states that the task measured is performed with a specified method. It is desirable, while the time study is being conducted, for the analyst to also look for opportunities for methods improvement. Many companies use the term "methods/standards analyst" to encourage the time study person to look for methods improvements while observing the methods for time study purposes. This chapter reviews the principles and techniques in using stopwatch time studies to establish allowed times. Section 3 of this handbook reviews principles and techniques of methods analyses.

Figure 2.1 presents a graphic analysis of the steps involved in establishing a stopwatch

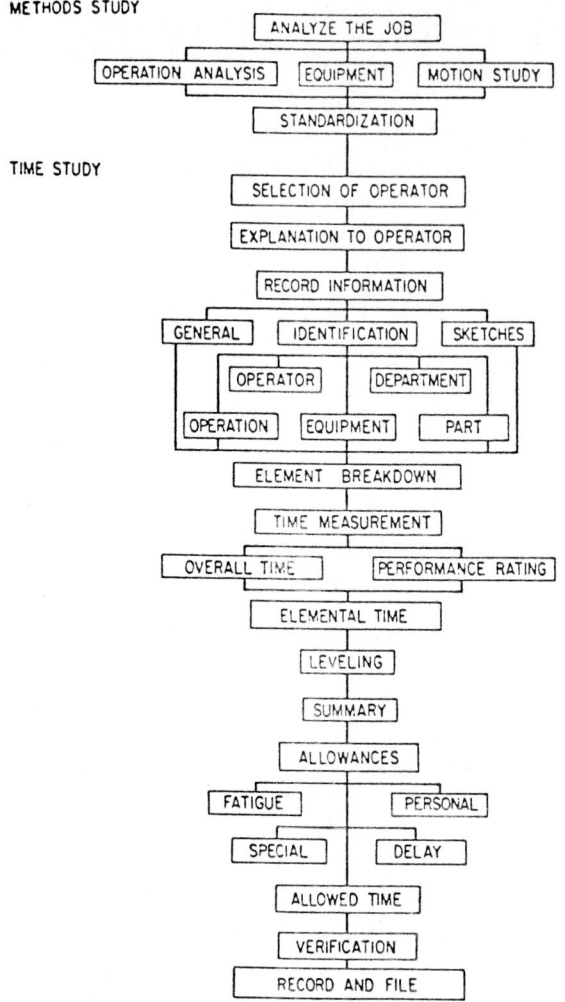

FIGURE 2.1 Graphic analysis of the steps involved in establishing a time standard.[2]

time standard. The first step is "methods study" and the second step "time study." Methods study is shown first to emphasize the fact that the method should always be studied, improved, and standardized before the time study is begun. Time study begins with the category "selection of operator."

STOPWATCH TIME STUDY TOOLS

The stopwatch equipment used to make a time study varies widely. It has often been said, and sometimes proved, that a good time study technician can make a usable time study with only the back of an envelope, a wristwatch, and a stubby pencil. This is a brag that has been responsible for many poor quality standards and the failure of many a time study analyst.

It is very desirable that the time study be accurate, understandable, and verifiable. The time study tools used can help or hinder the analyst in meeting these requirements. A few of the essential tools that are needed by the time study analyst to make a good time study include:

1. Time study watch—digital display (electronic) or sweep hand (mechanical).
2. Clipboard with bracket—to hold the time study forms.
3. Time study form(s)—repetitive and nonrepetitive, that provide for the written details to be covered in the time study.
4. Pencil.
5. Tape, ruler, or micrometer—depending on the distances involved and the precision with which they need to be measured.
6. Stroboscope—for measuring machine and equipment settings. Strobe lights (electronic stroboscopes) are the most accurate and usually the easiest to use.
7. Calculator or a programmed personal computer (PC)—to do the arithmetic calculations involved in time study.

Stopwatches, Basic Types.   The watch used when making a time study is the most important of these tools. The type of watch used should vary with the purposes of the time study. An ordinary wristwatch may be adequate for overall times and/or lengthy cycles. For most time studies, stopwatches are required. Sweep hand (mechanical) watches provide reasonable accuracy and readability (for cycles 0.03 minute and over). Most digital watches use quartz crystals that provide accuracy of plus and minus 0.00005. The digital display of numbers (electronic stopwatches) is much easier to read, since the numbers displayed can be frozen while the time study analyst reads and writes down the time. Also, the recorded time values tend to be more accurate when based on display of numbers.

The most common stopwatch (mechanical or electronic) is a decimal minute watch. Decimal hour watches and decimal second watches are also available. Decimal seconds are common in sports timing. The decimal hour watch is often used in conjunction with methods time measurement (MTM) studies, since MTM time values (TMUs) are in decimal hours.

However, decimal minutes are usually preferred for industrial time studies. It is easy for most of us to visualize a time interval in decimal minutes: a tenth of a minute, a half of a minute, or a minute (as contrasted with thousandths of an hour or 1.2 seconds).

Stopwatches are available in two modes:

- Snapback mode—the watch shows the time for each element, and automatically resets to zero for the start of each element.
- Cumulative mode (continuous mode)—the watch shows the total elapsed time from the start of the first element.

Some digital watches (offered by Meylan, New York, and Faehr Electronic Timers, Union, Ill.) are built into a clipboard that provides two displays—the time for each event (snapback mode) and the total time (cumulative mode) (see Fig. 2.2).

The most popular stopwatches continue to be single display, whether sweep hand or digital (see Figs. 2.3 and 2.4).

Comparison of Sweep Hand and Digital Watches.   There are advantages to the sweep hand or mechanical stopwatch and the digital or electronic watch. The sweep hand, the most commonly used, is manufactured in large quantities, which cuts manufacturing costs and sales prices. Electronic watches are produced in large volume in sports timing models, but in small quantities in good industrial timing models. Accordingly, the price of a quality sweep hand stopwatch for industrial use is about one-half the price of a comparable quality electronic watch.

Mechanical stopwatches also have a psychological advantage. More people are ac-

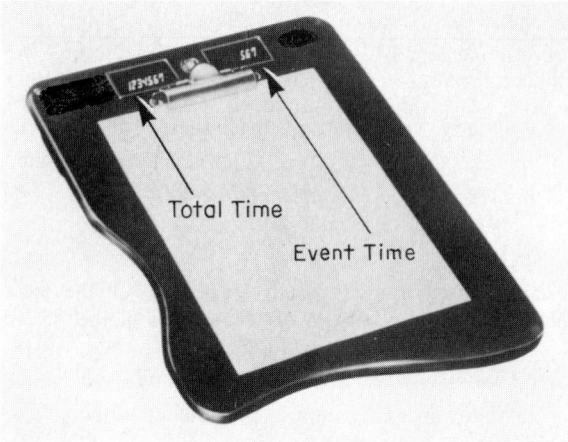

FIGURE 2.2 Dual display digital boardwatch. (*Courtesy of FAEHR Electronic Timers, Union, Ill.*)

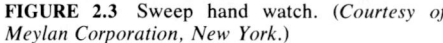

FIGURE 2.3 Sweep hand watch. (*Courtesy of Meylan Corporation, New York.*)

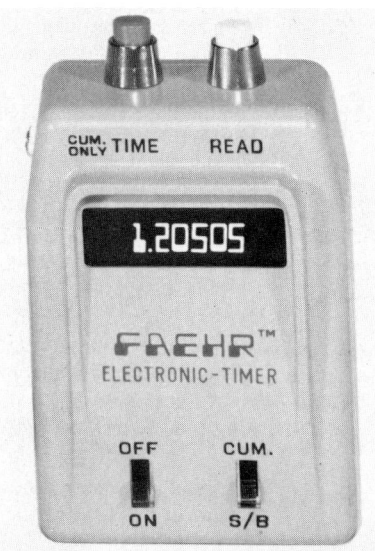

FIGURE 2.4 Single display digital watch. (*Courtesy of FAEHR Electronic Timers, Union, Ill.*)

quainted with the sweep hand type of stopwatch and most of us are more comfortable with tools with which we are familiar.

   This psychological advantage is gradually diminishing as the result of increasing exposure to electronic stopwatches. There are some technical advantages to the digital display of electronic stopwatches. Digital stopwatches provide frozen display of exact time in actual numbers. That is much easier to read then trying to read the exact time on a sweep hand. Digital displays tend to avoid reading errors and to reduce disputes about the readings.

The action of an electronic timer is practically instantaneous. The time required to snap a mechanical watch back to zero, while not large, is considerably greater than that required by the electronic timer. There is an inherent error in using mechanical watches for snapback time studies due to the time needed to snap the watch back to zero. Laboratory studies made with the aid of slow motion pictures show that an error of 3 to over 9 percent will occur on each element of 0.006 minute duration when using a mechanical stopwatch.[3]

The objection to lost time in snapback studies is minimal when using electronic timers. The lost time is less than 0.0003 minute per snapback when using an electronic stopwatch.[4]

Time Study Clipboard.   The observer generally stands and moves about when making a stopwatch time study. The observer must simultaneously: (1) watch the movements of the operator, (2) keep the stopwatch within the line of vision so the time can be observed at the end of each element, and (3) record the individual time readings on the time study form. A variety of time study clipboards are available from different suppliers. Dual display digital timers build the timer into the clipboard (see Fig. 2.2). Figure 2.5 shows a separate time study board (clipboard) available with a clamp. This ''body form'' shape is a little more comfortable for use then the customary straight rectangle shape.

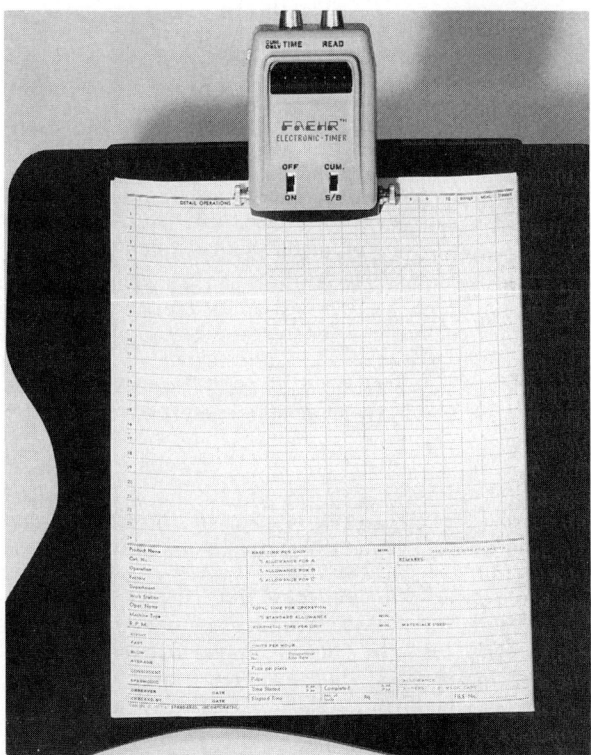

FIGURE 2.5   Separate time study board (clipboard) with clamp. (*Courtesy of FAEHR Electronic Timers, Union, Ill.*)

Computer-Assisted Electronic Stopwatch.   One of the latest developments in this field is the COMPU-RATE™ timer. It is shown in Fig. 2.6. The timer is so designed that once the time study observations have been entered into the timer, the data can be transmitted elec-

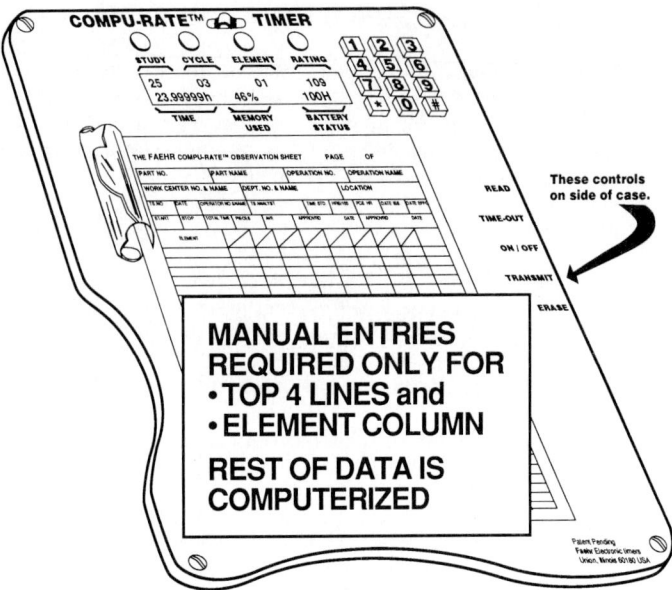

FIGURE 2.6   The FAEHR electronic timer. (*Courtesy of FAEHR Electronic Timers, Union, Ill. 60180.*)

tronically to an IBM compatible PC and the PC can then perform all of the calculations required to complete the time study.

The operation to be studied is divided into elements in the normal fashion. The element description is handwritten on the form. The element number is entered into the timer with the keyboard. The study is started by pressing the READ button on the right side of the board. The read button is then pressed at the end of each element. When the last element is entered, the CYCLE LCD window automatically changes to cycle 2 and the study is continued for as many cycles as needed. The performance rating, or leveling factor, can be applied to each element by using the keyboard.

The TIME counter LCD window is like a continuous reading stopwatch. The readings can be recorded manually on the form to record foreign elements, elements performed out of order, or any other abnormal occurrence in the normal cycle.

The time readings can be in hours to five decimal places (same as MTM-TMU times); in decimal minutes to three decimal places; or in hours, minutes, and seconds to two decimal places.

The windows on the top of the board are LCD displays which can be contrast adjusted for any lighting conditions. The MEMORY USED and BATTERY STATUS alert the analyst to take the necessary action to either dump the memory or recharge the battery.

At the end of the study the analyst turns off the timer. This clears the top four LCD displays, and the timer is ready for another study.

The timer can be connected to an IBM compatible PC that has the COMPU-RATE™ software system installed. The elemental data can then be sent from the timer memory to the PC. The PC can then:

1. Determine the individual element times plus mean and median element averages.

2. Adjust for the performance rating.

3. Summarize the results to determine the time standard per piece, and the pieces per hour.

The software can be adjusted to fit individual company needs.

Time Study Forms. The time study forms that are used in Figs. 2.7 and 2.8 are typical. Many different forms are available from suppliers. Select or design one to fit your needs. The form should provide space for certain descriptive information that must be recorded at the time the study is made, if the study is to be of value in the future. This information falls into two categories:

1. One category provides preliminary basics, such as the product, name of operator studied, process routing, machines used, tools used, workplace layout, date, and observer's name.

2. The second category describes the study, identifies the elements studied, lists the stopwatch readings, gives the performance rating and the standards calculations, and the like. Ideally, the element descriptions on the time study form provide a narrative description of everything the worker does to correctly accomplish the job.

The description of the method used on the job should include available methods improvements that can be installed for the current job. This should include all methods improvements: previously discovered methods ideas, reported and unreported; plus current ideas discovered by the methods standards analyst; plus ideas recommended by the operator, the supervisor, and others.

It is desirable to have a methods suggestion file or computer listing, where approved but not implemented methods improvements are noted. Approved but not implemented ideas are often forgotten unless there is a convenient, reliable reminder the next time a job is done. In a competitive world, neither poor methods nor poor standards can be afforded.

Examples of Filled-in Forms. Figures 2.7 and 2.8 show a time study that was made using a decimal hour watch. The continuous method of recording watch readings was followed. As a result, the study tells the complete story of the sequence in which all elements of the operation were performed. Every moment of time consumed during the period in which the study was made is accounted for. When the normal sequence of operations was interrupted, both the time consumed and a description of the nature of the interruption were recorded.

The completed study reads like the page of a book. It starts at the top left, reads across to the right, down one line, and so on. Foreign elements or interruptions appear as marginal notes.[7]

The usual number of readings on a repetitive time study form are a minimum of 8, up to and including 12. Ten observations are the most common, probably because the arithmetic is simpler using 10 observations. Figures 2.7 and 2.8 show copies, front and back, of a repetitive time study form.

Occasionally, it is desirable to take only a few complete readings. This is most often done on low-volume or nonrepetitive work. Figure 2.9 is a copy of one side of a nonrepetitive time study form. This form provides space for listing a large number of elements per operation. The back of the form is usually the same whether for repetitive or nonrepetitive time study.

Any time study form that meets the requirement of permitting the analyst to clearly tell the complete story is satisfactory. The work *clearly* is emphasized. It should be possible for anyone familiar with the operations to reconstruct from the time study description the methods used when the stopwatch study was made. This methods information can be useful to the operator, setup person, and line supervisor in setting up the job the next time so the methods used by the operator will be the same as the methods on which the standard is set. This methods information is essential in proving the standard right (or wrong), if the standard is questioned—especially if the standard goes to arbitration. It is also essential when methods and standards changes are to be made.

A good test of a time study write-up is to use it to reconstruct a job in accordance with the methods specified on the time standard. Provision is made on the front and back of the

FIGURE 2.7 Time study form—front.[5]

4.20

STUDY NO. __2__ DATE __7-20-__

OPERATION: PACK TERMINAL BLOCK

DEPARTMENT: M12

OPERATOR: ~~Man~~ Woman — NAME M. SMITH — No. 167

EQUIPMENT

MOULD: TB 9/3 — DIE
PATTERN — INS. SPEC. 22A 207
PART DESCRIPTION: TERMINAL BLOCK 463207

DWG. B 7194
STYLE 463207 — SUB. 9 — ITEM 1-4
L. SPEC.
MATERIAL R.S.M.

No.	ELEMENTS	SMALL TOOL NOS. FEED SPEED, DEPTH OF CUT, ETC	ELEMENTAL TIME ALLOWED (BOTTOM LINE OTHER SIDE) OR CYCLE	OCCURRENCES PER PIECE	TOTAL TIME ALLOWED
1	GET BOX D412. GET NUMBER STAMP & STAMP SERIAL # ON FLAP AT TOP OF BOX	INK PAD	.00164	1	.00164
2	CREASE & CLOSE BOTTOM OF BOX		.00173	1	.00173
3	GET, VISUAL INSPECT, & PUT BLOCK IN BOX		.00159	1	.00159
4	GET BAG #1714 & 2 SCREWS, PUT SCREWS IN BAG		.00117	1	.00117
5	FOLD OVER BAG		.00262	1	.00262
6	PUT BAG IN BOX		.00099	1	.00099
7	CREASE & CLOSE TOP OF BOX		.00174	1	.00174
8	SET BOX IN TOTE PAN		.00045	1	.00045

SPECIAL TOOLS, JIGS, FIXTURES, ETC.

MACHINE TOOL NO.

CONDITIONS

OBSERVER

APPROVED BY

TIME ALLOWED, SET UP

EACH PIECE .0119 HR — TOTAL .01193

REMARKS: MATERIAL HANDLER KEEPS BINS REPLENISHED, MOVES FULL TOTE PANS TO SHIPPING, & SUPPLIES EMPTY TOTE PANS.

CHECK WITH "METHODS" TO IMPROVE ELEMENT 5

SKETCH

BOX D 412 (FLAT)
BAG 1714
SCREWS
12"
16"
TERMINAL BLOCKS
SELF-INKING NUMBER STAMP
TOTE PAN
26" HIGH RACK UNDER
OPERATOR SEATED

OBSERVATION SHEET

FIGURE 2.8 Time study form—back.[6]

4.21

NON-REPETITIVE TIME STUDY

OPERATION ______ STUDY NO. ______ DATE ______ SHEET ___ OF ___

EQUIPMENT UTILIZED ______ OBSERVER ______ WORKER ______ CLOCK NO. ______

NO	ELEMENT DESCRIPTION	READING	ELEMENTAL TIME	TIME STUDY RATING	NORMAL TIME	CLASSIFICATION	NO.	ELEMENT DESCRIPTION	READING	ELEMENTAL TIME	TIME STUDY RATING	NORMAL TIME	CLASSIFICATION
1							27						
2							28						
3							29						
4							30						
5							31						
6							32						
7							33						
8							34						
9							35						
10							36						
11							37						
12							38						
13							39						
14							40						
15							41						
16							42						
17							43						
18							44						
19							45						
20							46						
21							47						
22							48						
23							49						
24							50						
25							51						
26							52						

FIGURE 2.9 Nonrepetitive time study form. (*Courtesy of Standards, International, Time Laboratory, Chicago, Ill.*[8])

time study form to record identifying and other pertinent information. These data should be recorded at the time the study is made. Records should be made to show complete identification of the operator; the part upon which the operation is being performed; the machines, tools, and equipment being used; the operation; and the department in which the operation is performed. Sketches, for which space is provided, are generally a desirable and necessary adjunct to verbal descriptions. These sketches preferably should show the workplace layout, or they can illustrate the part upon which the operation is performed. Some companies attach a photograph of the workplace or of the part to the time study form to illustrate the conditions at the time the study was made.

The recording of complete information is of the utmost importance and cannot be too highly stressed. Figure 2.8 shows typical information recorded on the time study form. Note that under "Remarks" the method of bringing material to the workplace and removing it from the workplace has been described. Such data, unless recorded at the time the study is made, become lost in limbo when later trying to reconstruct the conditions to which the standard applies.

Most of the headings of the spaces used to record information are self-explanatory. The "Ins. Spec." heading is used to record the number of the inspection specification, if such data are available in the plant. The "L. Spec." heading identifies the electrical specifications number where it is pertinent.

TIME STUDY PROCEDURE

When the method has been established, conditions standardized, and the operators trained to follow the approved method, the job is ready for a stopwatch study.

Selection of the Operator.   The operator studied is important. Time studying the wrong operator can (1) double the difficulty of making the study and (2) decrease the accuracy of the standard. The operator should be one who works with good skill and effort using the approved method. The time study analyst, by applying the performance rating procedure correctly, can arrive at the same final time standard within practical limits, regardless of whether the operator works fast or slowly. From every viewpoint, however, it is better if the stopwatch standard is based upon observation of an effective, cooperative worker performing at an acceptable performance level. As a rule of thumb, it is undesirable to try to rate an operator working a greater variance than 25 percent above or below 100 percent.

Show the Work Methods and Time Study Readings.   Time study should not be considered a secret document confined to the use of the analyst. It should be an accurate recording of informative data that cover the best and most efficient manner of doing the work under the conditions expected to be in effect when the job is run. It should be a set of instructions that can be used (1) by supervisors and setup personnel in arranging for the job to be run, and (2) also by the workers in performing their jobs. Time studies should be correct and subject to proof of correctness.

The operator, the supervisor, and the steward (in a union shop) should be shown the completed results. All details should be freely discussed with them; check studies should be made of any elements which may be questioned; and effort made to fully answer any questions concerning the study. An air of mystery or of hiding facts leads to distrust and "we-against-them" attitudes that prejudice both the study and the working conditions.

Explanation to the Operator and the Line Supervisor.   The manner in which the operator is approached at the beginning of the study is important. The analyst should be courteous and honest, show a recognition of and a respect for the operator's problems. The analyst must be frank in dealing with the operator on questions about the operations being studied and on the time studies. The analyst must be able to explain, in clear, nontechnical words, all steps in the actual stopwatch procedure.

When the analyst is not familiar with the operation, the analyst must not try to hide that unfamiliarity but should ask questions of the operator or supervisor or of anyone else in position to explain the operation. The analyst should be open and above board. Efforts to hide stopwatch studies can be self-defeating. They create distrust of the analyst. They cause "game playing" by the operator to trick the analyst in turn.

The analyst should stand alongside the worker, unless the latter's duties require moving sideways in the performance of the work. The analyst should not stand directly behind the operator while making the study. This practice causes the operator to become restless and suspicious as to what is going on behind the operator's back.

Operators who are not accustomed to stopwatch studies are apt to imagine all sorts of things. Some will believe that the analyst is continually operating the stopwatch and collecting information on all workers within the range of vision.

Whenever a worker is being timed, the stopwatch should be in plain sight. Some analysts prefer the use of a special time study board with an attached bracket that holds the stopwatch. The bracket holder holds the watch securely but allows ease of operation. Carrying the watch on this type of board allows it to remain in plain sight of the worker and in convenient position for the analyst's use.

The analyst is dealing with facts and has nothing to hide. There is nothing in correctly done stopwatch time study procedure which should be considered anything but fair by anyone who understands it. When time study work is properly handled, in cases where the correctness of a time study value is questioned, the analyst should willingly conduct a check time study in a sincere effort to settle the question with fairness to all concerned.

SELECTING THE ELEMENTS

For time study purposes, the work performed by the operator is divided into elements. An element is a distinct constituent part of a specific activity or task. It can be made up of one or more fundamental motions and/or machine or process activities selected for convenience of observation and timing.

Advantages of Time Studying Elements.   Dividing the work into elements makes it possible to:

- Performance rate more accurately.
- Determine changes in work elements or element sequences when checking standards in the future.
- Develop standard time values for frequently recurring elements. These elements can be checked against existing data, which helps maintain consistency.
- Identify nonproductive work.

Defining the Elements.   Elements should be clearly defined. Preferably the element description should indicate the beginning point, the specific work included, and the ending point. The basic rules for selecting elements are:

- Start and end each element with easily detected end points such as a definite sound or movement.
- The elements, depending on the need for detail, should be as small as is convenient to time.
- Time units of 0.03 to 0.04 minute are usually considered the minimum time for reading sweep hand stopwatches. Digital displays are somewhat easier to read, and minimum

times of 0.01 to 0.03 minute are realistic. The skill and reflexes of the analyst also need to be considered in selecting the minimum time to be used.

- The elements should be as coordinated as possible. Elements should consist of a pattern of motions performed in sequence on a single object.
- Preferably an element should cover motions only for one object. Restated, an element should seldom, preferably never, include motions for more than one object.
- Manual and machine times should be separate elements.
- Irregular and foreign elements should be separate from repetitive elements.

Several types of elements are usually found in a time study. Each time is handled differently in computing the standard and in developing standard time data. The types of elements are as follows:

- Regular—occur in every cycle.
- Irregular—do not occur in every cycle, but may occur at regular or irregular intervals, such as stock up or put away. Frequencies should be shown. As an example: stock up—"every 100 parts"; put away—"end of work order," and so on.
- Foreign—not a necessary part of the work. These cover items such as errors and idle.

The skilled stopwatch analyst plans ahead. Often, if the time studies are arranged in spreadsheet formats, additional rates can be set and/or validated by interpolating. It is desirable to classify the work elements as constant or variable.

- Constant—time for the element does not vary significantly between jobs in the same class of work.
- Variable—time varies with some characteristic of the product, equipment, or process (such as weight, size, or material) within the same class of work.

For example, the time to start up a computer for word processing is a constant regardless of the length of the letter. But the time to type the letter is a variable that depends on the format and length of the letter. Measurement of the length of the letter may be in general terms, such as the number of lines. But if the volume of typing is large, it may be desirable to measure the length of the letter more precisely, such as the number of characters (letters plus spaces and numbers plus spaces) and the format of the letter (the number of indents).

Minimum and Maximum Lengths for Elements. There are no hard-and-fast rules, just commonsense concepts, on how long or how short the time lengths for a stopwatch element should be. Selection of the work elements and the length of the elements should follow the criteria listed below:

1. *Easily detected and definite end points:* It helps if the observer can hear the end point as well as see it. This could be a noise caused by laying a part aside or starting or stopping a machine, by flashing a light, or by a readily observed difference in the operator's movements.

2. *As short as is convenient to time:* Rule of thumb is usually 0.02 to 0.03 minute. The observer is apt to have difficulty observing and recording times shorter than these, unless the observer's reaction times are quicker than average. The type of stopwatch used will also affect the minimum time that is practical. See "Comparison of Sweep Hand and Digital Watches," above.

3. *As unified as possible:* The element descriptions and watch readings are simplified when the work performed in an element is uniform in motions and sequence. Maximum length of time can be very long where the element consists of the same work performed

over and over again. The actual time spent reading the watch for a long element of uniform motions can be greatly reduced. The total time for the element can be calculated by correctly identifying the initial work content and number of times it is performed. Automatic drilling or doing computer calculations provide typical examples where all the work performed (between starting and stopping the machine) may be the same time duration. Where there are a number of variable work patterns in an element, it is best to subdivide the work element by uniform work patterns used within that element. The first rule of good time study is that the time values can be assigned to specific work sequences and methods descriptions.

Observation Principles. The following principles should be followed in making a time study.

Manual and Machine Times Should Be Recorded Separately. Where manual and machine times are performed simultaneously, manual times should be separate from machine times. Also, if two or more operators are working together, the elements should be divided according to the break between their simultaneous times and their nonsimultaneous times.

Regular and Irregular Elements Should Be Recorded Separately. There should be few, if any, variations in sequence of regular elements. By their nature, irregular elements often vary in sequence and occurrence.

Operators Should Follow Agreed-on Work Sequences. When the analyst, line supervisor, and operator have established the work sequence, the operator should follow it. Departure from the regular order interferes with the recording of the watch reading and complicates the study. Several alternatives are available. The observer can skip that cycle. This is why it is an advantage to use a form with more than 10 columns for recording cycles. Or the observer can continue to record the times per elements, coding the elements to show the sequence for that particular cycle. Or the observer and the line supervisor can decide whether the operator should be replaced by one who will follow the established sequence.

Irregular or foreign elements may be a necessary part of the work or may be completely unnecessary. For example, where the operator is required to get new material or supplies or orders, this can be necessary for the performance of the job. However, it may be difficult to determine if it is necessary or unnecessary. (For example, stopping work to talk to a fellow employee may be necessary or unnecessary. Don't guess; find out.)

Unnecessary irregular or foreign element occurrences should be duly recorded but should not be included in the standard for the job. Indicate the occurrence of an interruption on the time study by inserting a letter symbol opposite the time. Preferably these letter symbols have been preselected before the timing, so the observer will lose minimum time in jotting down the reason as well as the time observed.

OBSERVING THE OPERATIONS

Element Description. The first step in making observations is to subdivide the operation into a number of smaller operations which will be studied and timed separately. These subdivisions are known as elements or elemental operations. An element is a subdivision of an operation that is distinct and measurable; it contains a logical portion of the work.

An element, to be usable, must meet all the qualifications stated in the definition. For example, the element description "Move piece to vise" is neither distinct nor does it contain a logical portion of the work. In this example, the end point of the element is indefinite. At what point over the vise does the element end? This point can vary in the eyes of the individual observer, as well as being indefinite in the eyes of other observers. A more acceptable element would be "Move piece and place in vise." In this case, the end point of the element is definite. The element ends when the piece is in the vise.

Many companies develop a list of standard elements that are completely described. In such companies, the standard description of the element "Move piece and place in vise" might be: "The element begins as the operator grasps the part to be moved to the vise with one or both hands, depending on size, shape, and weight. It includes the total time required to move the part to the vise and insert it between the vise jaws, and ends as the part is located between the vise jaws."

With standard element descriptions of this type available to all the time study observers in the company, a consistency is obtained in all time studies that contain the standard elements. Data collected in such a manner are readily usable in developing standard data and making comparisons and in checking the variables.

When the element breakdown has been completed for a study, a short description of each element is recorded on the front of the time study form in the space provided. A more complete description of the elements is recorded on the back of the time study form in the "Elements" column.

Number of Observations.   Many businesses have, through either company policy or labor contracts, established the number of cycles to be studied or have specified a minimum elapsed time for the study. There are also mathematical methods for determining the number of observations which must be made to determine time at a desired confidence level. Nomographs and equations have been developed to aid in determining the number of observations to be made. Most time study books provide the equations and/or nomographs for the statistical calculations. However, statistical rules cannot take into account all the factors that affect the study. The most important factor is the degree of variability in the time values between the different cycle readings.

The best practical guide to the number of observations required is obtained by charting the element times. Out of 10 cycles, a reasonable number of the time values for an element should be grouped together. Plot the time values from minimum time to maximum time, grouping the observed times in 5 to 10 percent ranges. The time values should follow a normal distribution, with one or two at the low end and one or two at the top end and with 5 to 7 grouped together in the middle. This would be indicative of a normal distribution from observing 10 cycles for an element. If histograms for your elements follow normal distribution charts, it is reasonable to assume the number of observations is adequate.

However, suppose the time observed had a flat distribution, with one to three at a very low time and one to three at a very high time, and the balance of the studies scattered somewhat irregularly in between. This would indicate that the operator is following a number of different methods or is deliberately trying to confuse the analyst. Probably the best thing under these circumstances is to time study another operator.

Performance Rating.   Performance rating is invariably required when setting standards by stopwatch time study. The objective of a time standard is to show "the amount of time required to accomplish a unit of work, following a specified method, when working at the desired performance pace." Seldom in actual practice will these given conditions be universally observed.

- The operator being observed may work faster than average (naturally quicker, or showing off, or speeding up due to nervousness from being studied).
- The operator being observed may work more slowly than the average (just naturally slower, or slowing down on purpose to confuse the time study analyst, or excessive interruptions, and so on).

Some companies allow time studies to be made without performance rating or leveling, on the assumption that it is not necessary. Every audit we have made of standards at companies that do not use performance rating for stopwatch studies shows a tremendous inconsistency in standards. Also, most of their standards are considerably lower (read that "looser") than that company's desired performance concept.

Some performance leveling systems try to encompass a wide variety of factors in their performance leveling. These factors include skill, interruptions, operator speed, and working conditions. The customary yardstick for good performance leveling is a consistency of ±5 percent. While this may seem unusually liberal, in actual practice most time study people have trouble living up to that consistency goal. It is usually better to narrow down the number of variables considered in performance leveling and concentrate on rating operator performance.

Methods variables can be solved by asking the operator to follow the methods specified. "The methods specified should be that which the company expects employees to be able to follow to earn their pay scale." The working conditions can be best provided for by a job evaluation.

Performance Leveling Concepts. There are two different concepts as to performance level:

1. Dedicated effort
2. Normal effort

Most of the original performance leveling concepts were related to "dedicated effort." This is "the time required by an operator following the specified method, working with dedicated skill and dedicated effort to accomplish the task."

"Normal effort" or "average effort" is often less than "dedicated effort." Rule of thumb is that "normal effort" is 20 percent less than "dedicated effort." Restated, "average effort" takes about 25 percent more time than "dedicated effort."

This difference in average times supposedly reflects the stopwatch traditions that 25 percent more time will be taken by an average operator working under daywork conditions than the time taken by the same operator under incentive conditions.

H. B. Maynard was widely recognized for his technical contributions to stopwatch leveling, usually at "average" or "daywork" performance times. He very strongly influenced the predominant tendency today for using "normal effort" as the benchmark.

Currently, most performance rating concepts are based on average efforts. Most of the performance rating in the 1920s, 1930s, and 1940s was based on a concept of "dedicated effort." There are advantages to both concepts.

Leveling films are available from Tampa Manufacturing, Holmes Beach, Fla.; FAEHR Electronic Timers, Union, Ill.; Barnes Management Training Services, Ft. Lauderdale, Fla.; Marvin Mundel, Silver Spring, Md.

Which Performance Level Is Best? This decision depends on the management practices and concepts. The times are interchangeable.

- Adding 25 percent to required times equates to average times (or multiplying by 125 percent).
- Subtracting 20 percent from average time equates to required times (or multiplying by 80 percent).

Practicality of "Average" (Daywork) Time. It is easier to sell the standard, because it is easier to attain the standard. There is a psychological advantage in daywork operations in starting with motion-time values called "daywork" times. It sounds fair and reasonable, even though the resulting performance standards call for increased output over past averages.

Practicality of "Required" (Incentive) Time. On the other hand, there is also a psychological advantage in daywork operations in starting with required times and adding 25 percent. The 25 percent addition on the required time helps convince the supervisor and operator that the time standards are adequate, even generous.

Should Standards Be at Required Time, Even in Daywork Shops? There is a tendency among some United States companies, and often among Japanese and other Far Eastern companies, to set the performance standards at required time—even for daywork operations. The reasoning: it is possible for the average experienced person, working with dedicated skill and effort, to do the work at required time. Managements who prefer "required time" standards identify the difference between performance at required time and performance at daywork time as "waste." They believe that it is the responsibility of management to reduce that waste by better training, better scheduling, and better supervision.

USING THE STOPWATCH

Timing Methods.   There are two different methods of operating a stopwatch during a time study.

- Cumulative time or continuous (also known as split timing)
- Snapback timing

In cumulative timing, the watch accumulates the time. Each reading shows total elapsed time from start of the first event. The watch is started at the beginning of the first element and is not stopped until the entire study is completed. The watch is read at the end of each element, without resetting, and the time value is recorded on the study sheet. Thus, the study sheet reflects only progressively larger watch readings. After the observations have been completed, the individual element times are calculated by a series of subtractions (to "split the time" per cycle into the individual element times).

In snapback timing, the watch is started at the beginning of the first element of the first cycle. At the end of each element, the watch shows the time for that element and is "snapbacked" to zero. This procedure is followed for each element throughout the study. It is a good habit in snapback studies to record the time at the start and end of the study.

The total of the element times and other activities noted in the study should add up to the total time from start of the study to the end. In practice, it tends to be somewhat less, owing to incorrect readings and missed elements. If there is significant difference, that time study is suspect.

Electronic timers are available (see Fig. 2.2 on digital watches) that record cumulative time and snapback time. This combination watch provides the advantages of both types.

Comparison of Timing Methods

Cumulative	Snapback
Advantages:	Advantages:
Easy to teach	Good for irregular cycles
Gives accurate total performance time	Not hindered by delays
Employees more confident that all elements are included	Saves subtraction calculations
	Variations in element times readily apparent
Disadvantages:	Disadvantages:
Operator variations become confusing	More open to human error
Irregular elements become confusing	Operators and supervisors less confident that all elements are included
Delays become confusing	Operators and supervisors apt to be more accustomed to comparing cycle times than element times
More calculations, since subtraction is required to get the time for each element	
Variations in element times not readily apparent as the study is made	

Customary Stopwatch Calculations. There are many different time study forms. Most stock forms have space on the observation sheet for the following calculations.

The "R" and "T" Spaces. The "R" spaces are intended for posting cumulative watch "Readings," with the watch running continuously throughout the study. After the observations are completed, the elapsed time intervals "T" are obtained by subtracting each reading from the next. The "T" values preferably should be in ink, or a different color from the "R" readings to avoid subsequent confusion.

If the observer chooses to use snapback readings, the elapsed times may be posted directly into the respective "T" spaces, leaving the "R" spaces blank; or the "R" spaces may be used for individual performance "Ratings" or for other notations.

Abnormal Times. Among the observed times for the same element over several cycles, some individual values may be considered "abnormal" for some reason, and accordingly are temporarily or permanently discarded. A line may be drawn through such a discarded value, but leaving the value readable for future reference.

Quantity Deviations. A circled number within the observation grid may be used to indicate a "quantity" of output, number of occurrences, or the like which differs from the quantity implied in the element description. This special quantity will be considered in the subsequent calculations of elemental average.

Elemental Calculations. At the bottom of the sheet are lines for analyzing the elemental times. Across the bottom, the *first line* provides space for the "Totals T" of the nondiscarded observed element times over several different cycles.

The *second line* provides space for the "Number of Values Used" in the "Totals T."

The *third line* is for the arithmetic "Average T," using the data from the first and second lines. This "Average T" is generally recommended as being the most valid indicator of the typical observed time for the element. However, some observers avoid calculating the average and simply arbitrarily "select" one of the observed values as being typical for the element, in which case the value on the third line would be the "Selected T" for the element.

Additional lines, for Minimum "T" and Maximum "T," are sometimes provided for evaluating the timing and the leveling. Abnormally low minimum times or high maximum times need to be evaluated. Were they caused by mistakes in watch readings? Or were they due to undetected methods changes? Often very low or very high times are omitted from the calculations for Average "T."

Letter (Adjective) Ratings. The "Rating" line is for companies using a multivariable rating system. It provides space for posting the "letters" indicating the various degrees of skill, speed, conditions, and rhythm, for any particular element which the observer considers should deserve a special performance rating different from the operator's overall performance during the study. If no significant performance deviations occur for individual elements, this line ordinarily is left blank.

Leveling (Rating) Factors. This line, "Leveling Rating Factor," is the sum of 1.00, plus the algebraic sum of the plus-or-minus leveling guide values selected from Table 2.1. This factor provides the numerical means for adjusting observed times toward a common standard level of performance. As implied earlier, many experienced observers avoid the check-off guides and directly estimate a rating factor. This rating may be for the overall time study, for the individual cycles, or for certain individual elements as well. The more frequent rating during a study tends to give more accurate results.

Leveled Time per Element. The next line, "Element Base Time," is the time for the element average adjusted by the observer's estimate of the operator's level of performance. The result represents the observer's concept of a typical average time for the element, when performed as described and without interruption. This "element base time" ordinarily will be used for analysis and development of standard time data and/or sometimes for direct synthesis of the base time for the operation.

Elemental Times Distributed to Operation Cycles or Units of Output. The "Occurrences per Unit" space is used to indicate the number of times that the element is performed for each measured unit of output, which usually means for each completed cycle of the operation. For example, if a prescribed inspection occurs once on every fifth unit, the

TABLE 2.1   Performance Rating Table for Leveling

Skill			Effort		
+0.15 +0.13	A1 A2	Superskill	+0.13 +0.12	A1 A2	Excessive
+0.11 +0.08	B1 B2	Excellent	+0.10 +0.08	B1 B2	Excellent
+0.06 +0.03	C1 C2	Good	+0.05 +0.02	C1 C2	Good
0.00	D	Average	0.00	D	Average
−0.05 −0.10	E1 E2	Fair	−0.04 −0.08	E1 E2	Fair
−0.16 −0.22	F1 F2	Poor	−0.12 −0.17	F1 F2	Poor

Source: S. M. Lowry, H. B. Maynard, and G. J. Stegemerten, *Time and Motion Study and Formulas for Wage Incentive*, 3d ed., McGraw-Hill, New York, 1940, p. 233.

leveled time to perform one inspection is shown as ''Element Base Time''; the fraction ''1/5'' is posted as ''Occurrence per Unit,'' to indicate that the frequency of occurrence is ''one in five.'' The occurrences per unit should be shown as a common fraction rather than as a decimal fraction. For example, if an element occurs 3 times in 10 units (or cycles), show the common fraction ''3/10'' rather than the decimal ''0.3.'' Both fractions yield the same numerical result, but the decimal fraction does not clearly describe the actual observed frequency.

Occasionally identical elements will occur more than once in the cycle. Such elements customarily will be analyzed together to obtain a uniform (common) element base time. If such an element occurs twice per cycle, the ''Occurrence per Unit'' value would be 2. In most cases, however, the Occurrences per Unit will be 1.0.

Leveled Time per Unit.   The ''Base Time per Unit'' for each element is the element base time multiplied by the occurrences per unit.[9]

Stopwatches are available with built-in capability of being connected to PCs that do the tedious calculations and determine the standards. Sources include Royal J. Dossett, Excelsior, Minn.; and FAEHR Electronic Timers, Union, Ill.

MANAGERIAL ALLOWANCES ON LEVELED TIMES

Customary Allowances.   Managerial allowances added to leveled time per unit include

- Daywork allowances
- Incentive allowances
- Personal, fatigue, and miscellaneous delay

PF&D Allowances.   PF&D is the usual abbreviation for personal, fatigue and miscellaneous delay allowances. The same PF&D allowances are customarily used whether the performance rating concepts used are at required time or daywork time. Customary percentage allowances are as follows:

- Personal—3 to 5 percent

- Fatigue—3 to 5 percent
- Delay, miscellaneous—3 to 5 percent

This provides a total range of 9 to 15 percent.

There are no firm rules or guidelines on the percentages, only habits and traditions. Usually the percentages are negotiated, based on past experience of the negotiating parties.

Rules of Thumb for Calculating PF&D. The following rules should be followed in calculating PF&D.

Personal allowances can be readily calculated. Take the number of rest breaks to come up with the time to be allowed, and convert that into a percentage.

Miscellaneous delay can be done the same way. Usually miscellaneous delays of 6 minutes and over are recorded as time out or nonstandard time and not included in the production time. Accordingly, the only provision needed for miscellaneous delays in the standard time are the short interruptions of under 6 minutes. Theoretically, this is a small number per day; in practice it is usually quite a few minutes per day.

Managers who prefer "Required" time standards are also apt to consider miscellaneous delays as "waste" and consider the time lost as a managerial expense that can be eliminated by effective management practices.

Fatigue allowance is the only percentage that cannot be determined by observation. Numerous studies have been made without success in an effort to arrive at statistically sound conclusions as to fatigue allowances needed for average working conditions.

Special provision is needed for fatigue for excess discomfort due to extreme working conditions, or greater than average physical or visual effort. Provision for these conditions usually is made in the job evaluation plan or in work scheduling.

Another reason for an allowance for fatigue (even if not believed necessary) is that it provides additional minutes for miscellaneous delays. This is usually good, because no matter how carefully studies are made of miscellaneous delays, they tend to be underestimated.

General recommendation is that you use an allowance for all three as a combined percentage. In other words, if you decide to use 5-5-5, use a total of 15 percent. Do not break up the percentages into one amount for personal; one amount for fatigue; and one amount for delay. It is easier to verify that the total is adequate than to verify each percentage. Percentages 5 percent and under can be too readily disrupted by minor variations in the daily operations.

OVERCOMING GENERAL OBJECTIONS TO WORK MEASUREMENT

Everyone in the time study field is aware personally or through the experiences of others of the hostility toward work measurement by operators, by union leaders, by line supervisors, or by human relations personnel. Sociologists, "expert" consultants, union leadership, and company leadership have criticized the use of standards as damaging and deteriorating influences on human relations and employee productivity. Is this objection justified? Sad to say, many times it is.

Can the criticisms and the objections be overcome? Definitely. Overcoming the objections requires a combination of good standards, and thorough understanding by all who are involved in the standards program of the appropriate techniques for standards setting, for using standards, and for proper use of the standards.

The majority of the time, the first step toward better acceptance of work measurement is better standards. Do not expect good acceptance of your time standards if the time standards are not good. Consistency is an important criterion for good standards.

The next step for good acceptance of standards includes understanding of standards techniques and usage, plus proper use of standards.

1. It is undesirable to explain the standards techniques if the standards techniques are faulty.

2. It is irrelevant to understand how to use standards when good standards are not available.

The first step, therefore, is to verify or validate that the standards are correct. This should be done first by the time study analyst. Then when the analyst has verified that the standards are consistent, they need to be verified by the line supervisor and the operator.

Verification by the Time Study Analyst.   There is nothing wrong with making mistakes. The big problem is when the mistakes are not caught by the analyst before somebody else finds them. The best thing a time study analyst can do to maintain credibility and build rapport is to double check the time study results before the standards are issued. There are several easy steps to catch time study mistakes. The most common and the biggest mistakes in time study are in (1) faulty methods and (2) errors in performance rating.

Methods Verification.   Methods errors tend to cause the biggest errors in standards. Methods errors are the easiest to find. Review the method used in your time study one more time. Does it still agree with the method being used by the operator? If not, correct the methods specified in the time study, if you made an error.

The second methods verification step by the time study analyst duplicates the verification step for line supervisors and operators: Review the method used in the time study with (1) the line supervisor's opinion of how the job should run, and (2) the operator of how the job should run. Note: Be careful that the operator is not listening while the supervisor goes through the methods review, or the operator may merely mimic the line supervisor.

Performance Rating Verification.   Verifying performance rating can be either easy or tough depending on the techniques used by the time study analyst. If histograms are prepared, clues are readily apparent as to the correct leveling percentages. Where the histograms follow a normal distribution curve, they offer good clues to proper leveling percentages for that study.

Where histograms have not been prepared or do not follow a normal distribution curve, the correctness of the time value for an element can be assessed if (1) it is an element similar to others time studied, and (2) a reference system is available so that comparable elements from other studies can be retrieved. If there is good comparison, with acceptable time studies of the comparable elements, the analyst can have reasonable assurance that the current time study and the performance rating are correct.

If the elements have not been cataloged and filed so that this verification can be done readily, an alternative is to prepare a spreadsheet of time studies by families of parts and see if the spreadsheet comparisons show logical similarities or progressions.

Validating the Performance Rating.   The correctness of the performance rating time standard can also be checked by counting the pieces produced while watching the operator. Multiply the allowed time by the number of pieces completed. Divide this figure by the elapsed time. The resulting percentage will give the average performance of the operator, unless many or long foreign elements were included in the study. If this percentage parallels the performance rating that was used, it is probable that the performance rating was acceptable.

Verification by Line Supervisors and Operators.   The first step for verification with line supervisors and with operators is to help them check for the reasonableness of the time study standard by reviewing element by element. Where elements are comparable with other time studies, both line supervisor and operator can more quickly agree that the time study is correct for that element. It is usually easier to get agreement on individual elements than it is on overall times.

When verification of the elements shows they are reasonably correct, then is the time to validate the total time. This can be more difficult than for the element times. The opportunities for random variations in methods, parts, interference, performance, and the like can occur without being observed, unless there is very close attention through the entire cycle. This can be difficult to achieve, especially if the cycle is 10 to 15 minutes or longer. However, it is essential. The operator and the line supervisor may agree on the individual elements and still not be convinced without proof of the pudding. At that time, it is necessary that both the line supervisor and the time study analyst stay with the operator when the operator is validating the standard time by doing the work. In the final analysis, only proof of the pudding counts.

But do your preliminaries first so that your final check can be as successful as possible. Usually, actual performance working diligently, under close knowledgeable supervision, will come within ±5 percent of the leveled time study.

Validation by Following Up on Results.   If the operator on a job is failing to achieve the standard, note if there are:

1. Material difficulties?
2. Material handling problems?
3. Machine or equipment problems?
4. Quality problems?
5. Unfavorable attitude as evidenced by faulty performance on that job?
6. Unfavorable attitude or unqualified operator, as evidenced by below par performance on other jobs?
7. Diligence? Is the effort applied by the operator acceptable? If not, have the line supervisor provide someone who is willing to do the work with acceptable effort and correct methods.
8. Job aptitude? Does the job require better skills or coordination than the operator brings to the task?
9. Job knowledge? Does the operator have the required skill and talent to do the job, but is not following or has not been trained in the correct methods?

The time study standard is supposed to be correct for an operator working with good skill and effort, following the prescribed method, producing quality product from quality material. The standard should be validated in accordance with those specifications.

Use of Time Study Standards.   Once the reasonableness of the time study standard has been validated, the standards can become viable tools. First step to get good productivity is to use the standards positively. Employees want (1) personal recognition and (2) fair treatment consistent with the rest of their peer group.

They want to know what is a good day's performance. They want to be praised for good performance. They do not want to be criticized. Practicing this basic tenet of good management, "praise first," has contributed strongly to increased acceptance of good work measurement standards.

Overcoming Supervisors' Objections to Work Measurement.   Two important steps can improve the acceptance and use of work measurement from the supervisor's point of view:

1. Do measure the supervisor by the cost standards.
2. Do not measure the supervisor by the productivity standards.

One of the common burdens for the time study analyst is the hostility and opposition of line supervisors. This hostility and opposition by front-line managers is logical and to be expected in most operations. Why?

1. Usually installing standards calls for increased output for the supervisor's work area

and more work for the supervisor—in training operators, in having everything the operator needs on hand, in being sure everything is in good working order, and so on.

2. This is particularly annoying when the standards ask for more output in an inconsistent pattern compared with historical output. And where the supervisors cannot understand how the standards were set, there is a fear and suspicion that unfair standards are being installed.

3. Most front-line supervisors, where work measurement exists, are measured by their department's performance against time standards. Reducing the time standards coverage decreases the supervisor's workload. Increasing the time standards coverage increases the workload.

When operators are measured by time standards and line supervision and staff are measured by standard costs, there will be greater cooperation on good time standards along with mutual effort to reduce costs.

Management, aided by accounting and industrial engineering, must coordinate the standard time system and the standard cost system. Computer improvement in information gathering will permit appropriate cost information to be furnished quickly to management, line supervisors, and engineers.

Alert management will use bottom-line labor costs as a measurement of the effectiveness of both supervisor and engineer. When this is done, the work measurement engineer's job can be more enjoyable and more effective.

Recording and Filing.   The time standard, as determined by the time study, is recorded on the routing sheet, manufacturing information sheet, and other permanent papers related to the operation. The time study itself is placed on a permanent file and is available for reference at all times.

Record Maintenance.   Record maintenance preferably is simple and direct. There are a few practices that assist in good maintenance of time study standards. Finalize the time study—record the dates, job numbers, and the like. Be sure that all of the blanks on the time study form have been filled in. Be sure of the time study calculations and then recheck. It is usually best to let somebody else recheck the calculations. It is human tendency to repeat mathematical mistakes.

Have the time study and standard been approved according to accepted practice? Approval varies from company to company; typical are the chief industrial engineer, the line supervisor, and the department head.

Copies of approved methods and standards should be distributed to all parties concerned. Typical are line supervisor (for attaching to machine or workplace), department file, design engineering, production planning and scheduling, the accounting department, industrial engineering department, or central file as called for by company practice.

Where company practice calls for a review of a standard on a periodic basis (1 year, 2 years, 5 years) a follow-up file should be established for that particular month of review. It can be as easy but cumbersome as filing an extra set of the study in a follow-up calendar file or simply by entering the part number, time study number, and date of study into a computerized "update" list.

TIME STUDY MANUAL

Every time study department should have a time study manual which sets forth the policies, procedures, and rules for the use of time study at that company. Advantage should be taken of every opportunity to explain the contents of the manual to all concerned. Time study has nothing to hide. The better this is understood, the more harmonious will be the relationships among time study engineers, supervisors, and the work force. The time study manual can be as brief as copies of good studies, or can be a complete write-up of approved company practices and policies.

We have found it best to start out with an outline manual and add instructions and practices as time progresses so that the manual fits your particular operations. This can be done by support personnel using any type of word processing program. The data can be stored on a floppy disk and updated when necessary.

What should be in the manual? Minimum should be copies of the forms to be used, the performance leveling procedures to be used (required time or daywork time), general practices on personal, miscellaneous delay, fatigue allowances plus incentive allowances or measured daywork allowances.

It is important that the time standards practices are not "cast in stone." As time progresses, changes in company operations, industry practices, and time study techniques may warrant a change. Many a company, by issuing strict rules instead of general recommendations, has handicapped its ability to improve as conditions change.

An important caveat: Do *not* make your company time study manual part of a union contract. The company policy on work measurement can be adequately defined in the contract by stating (1) what the performance leveling concepts are, (2) that time standards will be set by generally accepted standards techniques, and (3) that the correctness of the standards can be verified by those techniques or other generally accepted techniques.

A union contract is best kept simple and focused on objectives. A time study manual dedicated to the objectives of the union contract but with provision for change as conditions change is most apt to keep your time study standards effective tools for setting consistent standards, increasing productivity, improving methods, helping to train operators, and reducing costs.

CONCLUSION

This chapter has reviewed the basic principles and practices for setting fair, consistent stopwatch standards. It is extremely important that this be done fairly and consistently. That is important to the company, important to the employee, important to you. Doing the methods work and performance evaluation that are required in time study provides practical skill and knowledge that will serve you well in any occupation. Do the job well, for both your immediate satisfaction and your personal growth.

Do read some of the reference books listed in the Bibliography. The management books that are listed will enhance your knowledge of how standards play a vital role in industry. The time study books that are listed will provide additional insight into effective stopwatch studies. The list includes both current publications and some of the older ones. Read some of the old ones, if you can find them, to give you an understanding of the differences and the consistencies in practices over 60 years, from the early 1920s through the 1980s. Professor Niebel is generally recognized as today's leading authority on time study. Professor Barnes in the 1960s and Lowry, Maynard, and Stegemerten in the 1940s were the recognized leaders. Lichtner's book was published in the 1920s. His practical advice on the human relations aspects of time study coincides with today's concepts, even though there have been tremendous changes in technology and techniques.

REFERENCES

1. Maynard, H. B., *Industrial Engineering Handbook,* 3d ed., McGraw-Hill, New York, 1971, p. 3–12.
2. Maynard, H. B., p. 3–13.
3. Maynard, H. B., p. 3–19.

4. James Gregie, vice president—engineering, Circuit Service, Inc., Des Plaines, Ill., unpublished paper.

5. Maynard, H. B., p. 3–15.

6. Maynard, H. B., p. 3–16.

7. Maynard, H. B., p. 3–14.

8. Copyrighted form, Standards, International Inc., Chicago, Ill.

9. Pass-out material used in time study seminars by H. B. Rogers, Northwestern University, Evanston, Ill., and C. N. Sellie, Standards, International Inc.

FURTHER READING

Managerial References

Industrial Management

Suzaki, Kiyoski, *The New Manufacturing Challenge: Techniques for Continuous Improvement,* Free Press, New York, 1987.

Line and Staff Supervisor

Caldwell, Charles M., *New Employee Orientation,* Crisp Publications, Los Altos, Calif., 1988.

Gagnon, Gene, *Supervising on the Line: A Self-Help Guide for First Line Supervisors,* Margo, Minnetonka, Minn., 1987.

Parsons, James A., *Practical Mathematical and Statistical Techniques for Production Managers,* Prentice-Hall, Englewood Cliffs, N.J., 1973.

Time Study References

Barnes, Ralph M., *Motion and Time Study: Design and Measurement of Work,* Wiley, New York, 1967.

Carroll, Phil, *Timestudy Fundamentals for Foremen,* McGraw-Hill, New York, 1951.

Holmes, Walter G., *Applied Time and Motion Study,* Ronald Press, New York, 1945.

Introduction to Work Study, International Labour Organisation, International Labour Office, Geneva, Switzerland, 1978.

Lichtner, William O., *Time Study and Job Analysis,* Ronald Press, New York, 1921.

Lowry, S. M., H. B. Maynard, and G. J. Stegemerten, *Time and Motion Study,* McGraw-Hill, New York, 1940.

Mundel, Marvin E., *Motion and Time Study,* Prentice-Hall, Englewood Cliffs, N.J., 1960.

Nadler, Gerald, *Motion and Time Study,* McGraw-Hill, New York, 1955.

Niebel, Benjamin W., *Motion and Time Study,* 8th ed., Irwin, Homewood, Ill., 1988.

Smith, George L., Jr., *Work Measurement: A Systems Approach,* Grid, Columbus, Ohio, 1978.

CHAPTER 3
WORK SAMPLING AND GROUP TIMING TECHNIQUE

Chester L. Brisley
Director
Industrial Engineering Program
Marquette University
Milwaukee, Wisconsin

Random work sampling and group timing technique (GTT), a fixed work sampling procedure, are both employed for work measurement and cost reduction analysis. No stopwatch is used, yet these techniques can often replace traditional stopwatch time study, providing equal or better data at a lower cost.

Work sampling is based on the law of probability. It works because a smaller number of chance occurrences tends to follow the same distribution pattern than a larger number produces. Work sampling has been developed under various names.

It was introduced in England by a statistician L. H. C. Tippett.[1] In the May 1953, issue of *Time and Motion Study* published in London, England,[2] he reviews his experience:

> Round About 1927 I was making surveys in weaving sheds to discover the causes and durations of loom stoppages with a view to estimating how much of the productive capacity was lost for various causes. At first I used the obvious method of timing looms with a stopwatch. This caused no difficulty from the operators because I was timing the looms and not the weavers and no one thought of my activities as having any connection with time study as conventionally understood. The work was tedious, and as it was practical to record only two, or three, or four looms at a time, I had to move about the shed and observe many looms in turn before a reasonably reliable average could be determined.
>
> One day a weaving manager remarked: "I can tell at a glance whether the weaving in the shed is good. If most of the weavers are bent over their looms mending warp breaks, weaving is bad; if the weavers are mostly watching running looms, weaving is good." In a moment "the penny dropped." It became clear that a snapshot of the state of the looms in a shed taken at any instant was in some way an indication of the rate of production in a short interval surrounding that instant due to various causes.
>
> After a little thought, I decided that the proportion (or percentage) of looms snapped as running was equal to the proportion (or percentage) of time the looms ran, on the average, during that short interval, and thus estimated directly the loom running efficiency. Likewise, the percentage of looms snapped as being stopped for any given cause estimated directly the percentage of the time looms were stopped for that cause, on the average, during a short time surrounding the instant of observation.
>
> Thus was started the snap-reading method. An observer progresses round a mill and as he comes to each loom he takes a snap-reading of its state, whether working or stopped, and if

stopped, the cause of the stop. In this way he collects several thousand snap-readings, classifies them, and estimates the various percentages.

Tippett employed basically the same approach that we use today, except today we are using electronic data collectors and computers. The method has been increasingly applied in many areas that were not formerly measured. Because his original paper is not available in most libraries and because many people interested in work sampling desire to study it, Ralph M. Barnes reproduced it with the permission of the author in his book on work sampling.[3,4] The technique was introduced in the United States by Robert Lee Morrow, professor at New York University, in 1941. He changed the name from "a snap-reading method" to "ratio delay study."[5] He concentrated on sampling various production delays.

The author, along with the editor of *Factory Management and Maintenance,* Harry L. Waddell,[6] coined the name "work sampling" in an article on the subject in 1952. This name has remained popular; however, researchers may find articles employing other titles such as activity sampling, chance observation, and snap-readings.

Professor Morrow, in the second edition of his book, *Motion Economy and Work Measurement,* 1957,[7] could not bring himself to change his choice of names completely; but he compromised and wrote Chapter 23, "The Ratio Delay Study," and Chapter 24, "Work Sampling." In the ratio delay chapter, he reviews the effort of Taylor, Merrick, Barth, and Tippett in the establishment of delay and variation allowances to be added to time studies. He proceeds to use the term "ratio delay study" to describe Tippett's snap-reading method as a technique for establishing allowances. In his next chapter, he reviews the new term:

"Work sampling is the term now generally applied to what heretofore was called the ratio delay study. The reason for the change in terminology is that the ratio delay technique is no longer confined to delays. Work sampling is a more descriptive term because it covers the more general application and varied uses of this technique." He then proceeded to describe its various uses in conjunction with work simplification, cost reduction, and elevator traffic studies.

Professor Richardson has had great success in employing work sampling as one of the first steps in a cost reduction program.[8] He has followed the pattern of using the supervisor in indirect labor areas as an observer.

He suggests that we make work sampling one of the first steps in a cost reduction program; and, consequently, management will have a much better chance of success. He also recommends if management will pay strict attention to measuring work units of output, they are able to produce a situation in which the most sensible course open to the supervisor, from any point of view, is to give as valid a study as is possible. This is true because the first work sampling study will be used as a benchmark from which to direct effort toward improvement and also to develop some broad "productive time per unit of output" measure. Therefore, these two factors tend to balance one another in that any distortion of the results can lead to conclusions which will work some hardship on the supervisor. The most successful approach, in the experience of Richardson and others, is being patient and truthful and not "cutting the throat" of the supervisor with the results of the work sampling study.

Work sampling is currently being used in every segment of society. Many universities and professional societies have conducted educational sessions on work sampling. More and more institutions and industries are using this broad-gauged tool of work measurement for cost reduction and labor power budgeting.

WHAT IS WORK SAMPLING AND GROUP TIMING TECHNIQUE (GTT)?

Work sampling consists of random observations to determine an estimate of the ratio of those observations of various delays and elements of work to the total number of observations in the process. The ratio or percentage of observations recorded in a given state tends to measure the average percentage of time it is in that state. The number of observations depends on how accurate the answers need to be. A larger number of observations provides a greater accuracy.

The group timing technique was developed by George Dew.* Group timing technique is a fixed interval work measurement procedure for multiple activities that enables one observer using a stopwatch to make a detailed elemental time study on from two to fifteen employees or machines at the same time. Continuous elemental observations are made at predetermined fixed intervals and are recorded as tallies on a form listing the elements of the job. Elements that will vary in time because of operator performance may be leveled. The techniques of work sampling and/or group timing technique have been used by private firms, hospitals, educational institutions, and government for improving individual and general efficiency. Among other areas, research has been focused upon chance occurrence and behavioral patterns of engineers, managers, nurses, clerical, and faculty employees.

HOW TO PREPARE FOR A WORK SAMPLING STUDY

Communication is important in initiating work sampling. In nearly every organization, if there is any failure regarding work sampling, it is in communication. Harold Smiddy, former vice president of General Electric, said, "A great deal could be accomplished if we would consider communication in the light of this simple TTTG four-word formula, "talk to the guys" or "talk to the girls."

First, Sell Work Sampling.   Although work sampling may seem simple enough, many people will not believe that accurate information can be collected by sampling methods. Management and employees have been oriented toward stopwatch studies on a continuous basis. So often, management takes the general attitude that the installation of work sampling is its prerogative and that it is not necessary to relate to the union, or to those to be observed, just what this technique is all about. Although much progress has developed with regard to refinements of work sampling, one of the greatest needs today is in this matter of gaining acceptance of the programs from the personnel involved. In short, it is necessary to "talk to the guys" more.

Two grievances that were received in one company on the subject of work sampling follow:

FIRST GRIEVANCE: The aggrieved and all our unit people are aroused and highly perturbed by the actions of the company in regard to the so-called survey now being conducted in the shops. The answer to the grievance in the first step was that only the facility was being checked. We say this is false. The checker in our shop on one occasion was running around going "nuts," not for the facility, but for the operator, when for quite a spell he was nowhere to be seen.

No matter what heading or title tags it—"industrial engineering," "work samples," "spying" "brainwashing," or whatever it is—it is a violation of our agreement. We request this survey be stopped at once.

SECOND GRIEVANCE: It is the contention of the aggrieved and of the union that the way the company is making this survey or work sample is not conducive to good employee-company relations.

To help clarify this, a member of the industrial engineering department walks along a prearranged route to this department and others and checks predetermined facilities (and operators) approximately every 6–10 minutes for the course of the entire day. We have been informed that this is to continue for three weeks.

The way this type of survey is being done, it is undemocratic, un-American in principle, costly, detrimental to the union people, and will not be condoned by the union.

---

*One of the originators of the group timing technique. See *Industrial Engineering Handbook,* 2d ed., sec. 3, chap. 7.

The union requests that this new method, survey, work sample, or whatever the company may wish to apply to it, cease immediately.

The union requests the company use the accepted methods of time studying to determine work loads, crane waits, etc., instead of these police state methods instituted by the industrial engineering department.

COMPANY ANSWER: In these grievances, the union is strenuously objecting to the "random work sampling" program instituted by management of various machine facilities. They contend that union employees are being subjected to undue pressure, harassment, intimidation, etc., by this type of observation.

As thoroughly explained during discussion, work sampling is a method of gathering data pertinent to manufacturing operations and is widely used by industries. Information gathered during this program will be used for the sole purpose of improving operating efficiencies and in no way can this program be construed as a means of harassment, intimidation, or coercing of our employees.

We do not expect employees to conduct themselves during the course of this "work sampling program" in any other manner than is normally expected of them. Accordingly, there should be no cause for alarm. We do not agree that the program places any undue pressure on our employees. Accordingly, the request that this study be discontinued is not granted, and any alleged violation of the agreement is denied.

The reaction of employees reflects a great need to do a better job with respect to the sociological and psychological aspects of the work of industrial engineering. Just what happened here? Having had some unfavorable experience in the past of applying work sampling, the chief industrial engineer attempted to do a good job in this case, but his attempt failed. The approach that he used was to talk to the general superintendents about the work sampling study. He then asked them to convey to the superintendents, the general supervisors, and the supervisors the approach that the observers expected to use in making the work sampling study. The chief industrial engineer asked them to relate to the employees through the line of command that "a work sampling was to be made for the purpose of evaluating both the equipment and the personnel to determine how they expended their time."

On all work sampling studies since this one, the chief industrial engineer now asks for an opportunity to explain this information to each level of the personnel below them. He "talks to the guys" directly, with the line organization members present.

There is no need to emphasize further how important it is to have the complete understanding and confidence of the people who are concerned with the results of a work sampling study. It is strongly recommended that no work sampling studies be taken without the knowledge of those being studied. The supervision directly over the operators should explain to them what the study is and what the purpose of it is. After the studies have been completed, the supervision and those studied should be enlisted in improving the conditions that are pinpointed by the study.

The best way to sell work sampling is in the explanation to "Why does work sampling work?" By all means explain the principles of probability such as that when a coin is tossed, the result is one of two possibilities, heads or tails. The law of chance says there should be 50 heads and 50 tails in 100 tosses of a coin. That is the ratio of the average possibility. It does not mean it will come out 50-50 on the button every 100 tosses. The score may be 60-40, 45-55, or some other ratio. It has been proved that the law of probability becomes increasingly accurate as the number of tosses increases and the percentage of possible error decreases.

Another example of how the law of probability works is dice throwing. It is a little more complex than coin tossing because one throw of two dice has 36 possible results, instead of two. It is an idea that will help sell work sampling to others by helping them understand it. Thirty-six throws will tend to produce one 2, two 3s, three 4s, four 5s, five 6s, six 7s, five 8s, and so on. Note that it *tends* to produce. Each series of 36 throws will not duplicate the pattern exactly. The results get closer to the pattern (probability curve) as the number of series of 36 throws is increased; the percentage of error decreases.

Define the Problem.   Determine exactly what information is required. It is usually well to make a preliminary survey—observe the operation for a day—to get a list of the operation elements. For instance, if the causes and amounts of downtime on a machine are sought, it will be necessary to define all the possible causes. It is much the same as the preparation for a time study.

Make an Observation Recording Form for the Job.   The form used for recording the observations made during the course of a work sampling study must be individually designed in each case. Its design will depend upon the number of workstations or people to be observed and the classification of the activities upon which it is desired to obtain data. Figure 3.1 shows a typical observation record form which was designed for a study of draw-bench activities.

STUDY	DATE		OBSERVER		
ITEM	BENCH NO. A-24	BENCH NO. A-23	BENCH NO. A-22	BENCH NO. A-21	TOTAL
CYCLE	ЖЖ ЖЖ IIII	ЖЖ ЖЖ ЖЖ I	ЖЖ ЖЖ ЖЖ ЖЖ	ЖЖ ЖЖ IIII	71
SET UP	I	III		I	5
NOT OPERATING	ЖЖ ЖЖ IIII	III		ЖЖ ЖЖ III	30
OPERATOR ABSENT			I		1
OPERATOR IDLE			II		2
POINT — HEAVY					
SMALL					
BENT		I	I		2
JAWS — NOT GRABBING		I			1
NOT RELEASING					
STOCK — TANGLED			I		1
BREAKING			I		1
HANDLING	II	I	II	II	7
WAITING — CRANE	I	I			2
STOCK	I	I		I	3
TAIL IN DIE					
THREADING MANUALLY			I		1
HOOK WON'T ENGAGE					
TUBE RELEASED TOO SOON					
GUIDE TUBE INTO DIE					
MAINTENANCE		I		I	2
ADJUSTING RODS				I	1
CHECKING PINS & DIES				I	1
CLEAN UP	I	I	I		3
REC. INSTRUCTIONS			II		2
					136

FIGURE 3.1   Observation record form designed for study of draw-bench activities.

Select the Frequency of Observation

Nature of the Operation.   If it is a short-cycle, repetitive operation in which all the desired elements occur frequently, the observations can be spread out over a period of time. If it is a nonrepetitive operation or one in which some elements occur infrequently, it is better to make more observations in a day. This improves the chance of getting all the details.

Physical Limits.   If there is just one observer and a long route is necessary to make one round of observations, the observer will be able to make relatively few observations in a day. For instance, a study of maintenance crews by one observer would probably require a long route.

Total Number of Observations Required and Time Limit.   If 1600 observations are needed for desired accuracy, and there are only 10 working days to make them, it is evident that an average of 160 observations a day will be needed.

Determine Time of Trips

On a Random Basis.   Making 20 random samplings which follow no set pattern is quite often difficult. The safest way is to use a table of random numbers, because the human mind has a tendency to follow a set pattern. Also, pocket or hand-held computers and microcomputers have programs to generate random numbers.

Sampling can be randomized by the day, within an hour, or within any other period of time—90 minutes, 2 hours, or the like. For example, 20 random samplings per hour could be observed (called stratified random sampling), or 160 random samplings per day could be made, with certain hours having more observations than others. Randomness of observation is stressed to reduce sampling errors.

Estimate the Number of Observations That Will Be Needed.   This information is needed to plan the frequency, number of observers, and length of the study. The number depends on how accurate the answers need to be. A larger number of observations provides greater accuracy. Experience with work sampling and a knowledge of the operation will enable the analyst to make a fair off-the-cuff estimate. As the study progresses, observers can check the results to see when they have enough observations. Later in this chapter, a simple chart that signals the end is explained.

There is a mathematical method of preestimating the number of observations needed to give the practical accuracy desired, which is described later in this chapter. The theory on which it is based is the same as that used in statistical quality control. Absolute accuracy is shown on the alignment chart for determining sample size (Fig. 3.2).

The key to the accuracy of the work sampling study is in the number of observations. A greater number of observations provides a higher degree of accuracy, provided the study is designed to reduce bias. But nearly all plant or business problems have a point beyond which greater accuracy of data is not worthwhile.

Evaluate Methods by Which Biased Readings May Be Reduced.   It should be pointed out that the inefficient motions or elements of an operation will not necessarily show up through a work sampling study. In some instances, the operator may be working during the downtime of the machine when this work could be done during the machine operating cycle. The work sampling observation would indicate that the operator was working, but a more refined motion study would show that this work could be performed during the machine cycle time. It should also be pointed out that a work sampling study will not show if the operators are limiting production or pacing themselves. If the observers are qualified to do so, some indication of effort level can be obtained by rating operators on those observations when they are recorded as performing the working elements of the job. It should be stressed to the observers that it is very important to make their observations at the same spot each time so that their readings will not be biased by distance from the designated spot.

Quite often, certain elements of work occur only once a year and that may be during

ALIGNMENT CHART FOR DETERMINING SAMPLE SIZE

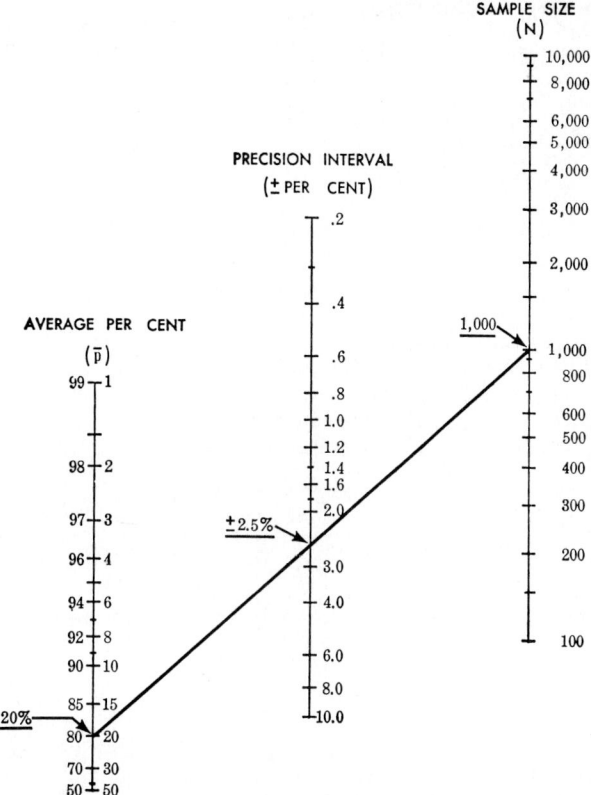

FIGURE 3.2 Alignment chart for determining number of observations required to obtain percent occurrences within given absolute limits of error and confidence level of 95 percent.

the period of the work sampling study, for example, machines down for a major overhaul. This abnormal maintenance might be weighted accordingly if it is known, for example, that the machine is usually down for abnormal maintenance because of overhaul for 5 days out of a year, and that this overhaul occurred during the 15-day study. Assuming that there are 252 working days to the year, the weighted percentage of downtime because of maintenance would be 2 percent instead of the recorded 33.3 percent. Obviously, it is necessary to rely on fairly accurate maintenance records to weight these factors with any degree of accuracy.

The question has often been raised with respect to the effect of an operator seeing an observer coming and then going to work. It is true that a study is thrown off at the beginning because of this situation. However, it has been found through experience that as observers pass the operators day in and day out over a sufficiently long period of time, this influence on the readings levels off.

Prior to the Start of a Study, Have a Session with the Observer or Observers.   Clearly define and discuss each element to be observed and recorded. This step is very important where two or more observers study the same operation. Without it, they may not be consistent in how they designate what they see.

EXAMPLE OF WORK SAMPLING

As an example of work sampling, assume it is desired to find out how much time a selected machine operator spends on operations, setup, maintenance, and delay. Using the work sampling technique, the machine is visited a predetermined number of times a day or hour. Assume that 10 random samplings that follow no set pattern are wanted during the day. The safest way to accomplish this is to use a table of random numbers such as Table 3.1, assigning times to these numbers. The numbers must be arranged in sequence.

Assuming that the period in which the study will be made is from 8:00 A.M. until 5:00 P.M., not including the lunch period, the 8 hours or 480 minutes may be divided into forty-eight 10-minute periods. Time intervals of 10 minutes each, beginning with 8:00 A.M. and going to 8:10, 8:20, and so on, will be numbered consecutively from 1 to 48. Analysts then make use of the random numbers by choosing as many as the number of observations they wish to make during the day. In Table 3.1, column 3, for example, ignoring numbers over 48, the first 10 numbers would be 43, 24, 17, 12, 07, 38, 40, 28, 17, and 18.

The numbers are then arranged in sequence and used to determine the observation times. The intervals between the random numbers determine the number of 10-minute periods between each observation as shown.

Sequence	Time
07	9:00
12	9:50
15	10:20
17	10:40
18	10:50
24	11:50
28	1:30
38	3:10
40	3:30
43	4:00

There are various ways of randomizing observations. Numbers from one to the number of periods into which the day has been divided can be written on slips of paper. By drawing them from a convenient container, random numbers will be selected. In the above example, these slips would be numbered from 1 to 48. Another method of assuring a random selection of observation times would be to utilize a tumbler and numbered disks so common to Bingo games. Likewise, a hand-held programmable calculator may be used to generate the random number.

The element which is occurring at the instant of each visit is recorded. At the end of 10 days, the record may read:

Element	Observations	Percent of total
Operation	60	60
Setup	18	18
Maintenance	10	10
Delay	12	12
	100	100

The percentage of distribution of the various elements, as they occurred during the example (100 observations), may or may not be enough, depending upon the accuracy required.

TABLE 3.1 A Table of Random Numbers

```
03 47 |43| 73 86   36 96 47 36 61   46 98 63 71 62   33 26 16 80 45   60 11 14 10 95
97 74 |24| 67 62   42 81 14 57 20   42 53 32 37 32   27 07 36 07 51   24 51 79 89 73
16 76 |62| 27 66   56 50 26 71 07   32 90 79 78 53   13 55 38 58 59   88 97 54 14 10
12 56 |85| 99 26   96 96 68 27 31   05 03 72 93 15   57 12 10 14 21   88 26 49 81 76
55 59 |56| 35 64   38 54 82 46 22   31 62 43 09 90   06 18 44 32 53   23 83 01 30 30

16 22 |77| 94 39   49 54 43 54 82   17 37 93 23 78   87 35 20 96 43   84 26 34 91 64
84 42 |17| 53 31   57 24 55 06 88   77 04 74 47 67   21 76 33 50 25   83 92 12 06 76
63 01 |63| 78 59   16 95 55 67 19   98 10 50 71 75   12 86 73 58 07   44 39 52 38 79
33 21 |12| 34 29   78 64 56 07 82   52 42 07 44 38   15 51 00 13 42   99 66 02 79 54
57 60 |86| 32 44   09 47 27 96 54   49 17 46 09 62   90 52 84 77 27   08 02 73 43 28

18 18 |07| 92 46   44 17 16 58 09   79 83 86 19 62   06 76 50 03 10   55 23 64 05 05
26 62 |38| 97 75   84 16 07 44 99   83 11 46 32 24   20 14 85 88 45   10 93 72 88 71
23 42 |40| 64 74   82 97 77 77 81   07 45 32 14 08   32 98 94 07 72   93 85 79 10 75
52 36 |28| 19 95   50 92 26 11 97   00 56 76 31 38   80 22 02 53 53   86 60 42 04 53
37 85 |94| 35 12   83 39 50 08 30   42 34 07 96 88   54 42 06 87 98   35 85 29 48 39

70 29 |17| 12 13   40 33 20 38 26   13 89 51 03 74   17 76 37 13 04   07 74 21 19 30
56 62 |18| 37 35   96 83 50 87 75   97 12 25 93 47   70 33 24 03 54   97 77 46 44 80
99 49 57 22 77      88 42 95 45 72   16 64 36 16 00   04 43 18 66 79   94 77 24 21 90
16 08 15 04 72      33 27 14 34 09   45 59 34 68 49   12 72 07 34 45   99 27 72 95 14
31 16 93 32 43      50 27 89 87 19   20 15 37 00 49   52 85 66 60 44   38 68 88 11 80

68 34 30 13 70      55 74 30 77 40   44 22 78 84 26   04 33 46 09 52   68 07 97 06 57
74 57 25 65 76      59 29 97 68 60   71 91 38 67 54   13 58 18 24 76   15 54 55 95 52
27 42 37 86 53      48 55 90 65 72   96 57 69 36 10   96 46 92 42 45   97 60 49 04 91
00 39 68 29 61      66 37 32 20 30   77 84 57 03 29   10 45 65 04 26   11 04 96 67 24
29 94 98 94 24      68 49 69 10 82   53 75 91 93 30   34 25 20 57 27   40 48 73 51 92

16 90 82 66 59      83 62 64 11 12   67 19 00 71 74   60 47 21 29 68   02 02 37 03 31
11 27 94 75 06      06 09 19 74 66   02 94 37 34 02   76 70 90 30 86   38 45 94 30 38
35 24 10 16 20      33 32 51 26 38   79 78 45 04 91   16 92 53 56 16   02 75 50 95 98
38 23 16 86 38      42 38 97 01 50   87 75 66 81 41   40 01 74 91 62   48 51 84 08 32
31 96 25 91 47      96 44 33 49 13   34 86 82 53 91   00 52 43 48 85   27 55 26 89 62

66 67 40 67 14      64 05 71 95 86   11 05 65 09 68   76 83 20 37 90   57 16 00 11 66
14 90 84 45 11      75 73 88 05 90   52 27 41 14 86   22 98 12 22 08   07 52 74 95 80
68 05 51 18 00      33 96 02 75 19   07 60 62 93 55   59 33 82 43 90   49 37 38 44 59
20 46 78 73 90      97 51 40 14 02   04 02 33 31 08   39 54 16 49 36   47 95 93 13 30
64 19 58 97 79      15 06 15 93 20   01 90 10 75 06   40 78 78 89 62   02 67 74 17 33

05 26 93 70 60      22 35 85 15 13   92 03 51 59 77   59 56 78 06 83   52 91 05 70 74
07 97 10 88 23      09 98 42 99 64   61 71 62 99 15   06 51 29 16 93   58 05 77 09 51
68 71 86 85 85      54 87 66 47 54   73 32 08 11 12   44 95 92 63 16   29 56 24 29 48
26 99 61 65 53      58 37 78 80 70   42 10 50 67 42   32 17 55 85 74   94 44 67 16 94
14 65 52 68 75      87 59 36 22 41   26 78 63 06 55   13 08 27 01 50   15 29 39 39 43

17 53 77 58 71      71 41 61 50 72   12 41 94 96 26   44 95 27 36 99   02 96 74 30 83
90 26 59 21 19      23 52 23 33 12   96 93 02 18 39   07 02 18 36 07   25 99 32 70 23
41 23 52 55 99      31 04 49 69 96   10 47 48 45 88   13 41 43 89 20   97 17 14 49 17
60 20 50 81 69      31 99 73 68 68   35 81 33 03 76   24 30 12 48 60   18 99 10 72 34
91 25 38 05 90      94 58 28 41 36   45 37 59 03 09   90 35 57 29 12   82 62 54 65 60
```

Over a sufficiently long study, the number of times an operator or machine is observed—idle, working, or in any other condition—tends to equal the percentage of time in that state. This is true whether the occurrences are very short or extremely long, regular or irregular, or many or few. It should be emphasized that the study can be as detailed as one cares to make it; but the more detailed it is, the greater are the number of observations necessary to obtain the degree of accuracy that might be desired for all the elements.

ACCURACY AND PRECISION OF WORK SAMPLING

Work sampling recognizes the variability inherent in work measurement. *Accuracy* is the measure of the degree of bias in measuring. *Bias* is the amount by which the long-run observed mean value of a set of measurements differs from the "true" value of the quantity. *Precision* is a measure of the reproducibility of the measured value of a given quantity without regard to the "true" value of that quantity.

Bias can be prevented only by the proper design and execution of the sampling process. Possible sources of bias are in:

1. The precise definition of the population to be sampled
2. The ambiguity of the definition of various states of activity
3. The latitude on the part of the observer in choosing the moment of observation
4. The method of selecting the observation times
5. The extent that the worker can anticipate the time of observation and is able to alter the state of activity that will be observed

During the design stage of the study, a period should be selected that will avoid some unusual circumstance. The period of study should be at least as long as the longest period of any cyclical behavior of the characteristic being studied. Likewise, the population upon which the estimate is based must be similar to and representative of the period to which the estimate is to be applied.

The amount that the observer contributes to bias can be investigated by having two, three, or more observers perform simultaneous studies on the same operations. Likewise, through multiple studies, individual workers can be evaluated. Some studies have been made using continuous time study along with the application of work sampling to determine the degree of bias.

The purpose of work sampling is to establish the value of p in the binomial distribution. The normal distribution describes the probabilities of the various values of p that might occur. It has a mean average of p. p is the percentage occurrence of any element selected, expressed as a decimal. Usually the most important element is selected; the total of all the elements equals one.

For example, in Fig. 3.1 the "cycle time" of the machine would be selected. Seventy-one observations divided by a total of 136 observations = 0.522 = $\hat{p}$.

$$1 - \hat{p} = 0.478$$

$$N = 136$$

$$\sigma = \sqrt{\frac{0.522\,(0.478)}{136}}$$

$$= 0.0428$$

$1.96\,\sigma = 0.084$ at 95 percent confidence level; $\hat{p} = 0.522 \pm 0.084$. The parameters of p and N (number of observations) are used in the binomial distribution, and p and σ (sigma) in the normal distribution. However, the binomial values p and N are used to measure the parameter σ. The relation between p, N, and σ is given as the equation

$$\sigma = \sqrt{\frac{p\,(1 - p)}{N}}$$

Sigma (σ) is the standard deviation. The statistical derivation which shows that 68 percent of the time an observation can be expected not to deviate from the mean in the normal distribution by more than ± sigma can also show the probability, associated with more than one sigma.

The percentage of the area under the normal curve *a*, or the selected level of confidence *a*, between a perpendicular erected at the arithmetic mean and a perpendicular erected at specific points to the left and right of the mean may be determined from Table 3.2. The distance between the arithmetic mean and the selected point is expressed in terms of standard deviations as *C*.

TABLE 3.2 Selected Level of Confidence

C (+ and −)	a
1.000	0.68
1.645	0.90
1.960 (usually rounded to 2)	0.95
2.567	0.99

The formula to be used to solve for *N*, the number of observations required, is as follows:

$$N = \frac{C^2 p\,(1 - p)}{\sigma^2}$$

Sigma is the standard error, arbitrarily determined. Therefore, as the standard error selected is increased, the number of observations required is decreased. Elinor S. Pape covers the more detailed mathematics of work sampling.[9]

ELECTRONIC DATA COLLECTORS/DESKTOP COMPUTERS[10]

Hand-held, battery-operated computers, specially made to collect data, are being used more frequently by industrial engineers. These units can generate a random or fixed interval of time for each tour. During a tour the label of the next study unit will appear on the display to prompt the observer as to when the next tour should begin. The work sampling software features automatic scheduling of samples, full element labeling, performance rating, summary of statistics, and transfer of data to micro-, mini-, or main-frame computers where printed reports are produced to the desired format. Some electronic data collectors will produce a printed report immediately in the proper format by simply connecting the collector to any printer with an RS 232, serial interface. Other features such as number of observations, date, time of day, uppercase, lowercase, numbers, and ranges of upper and lower control limits can be obtained for selected elements. Any data can be reviewed or edited for errors. The units are equipped with continuous memory that preserves the collected data. These data may be transferred to an audio cassette or disk file and stored for future use.

Microcomputers and Data Collectors.   Royal Dossett[*] has developed one of these data collectors. He makes the following observation: "The Datawriter model DW02[†] is about the size and weight of a TV remote control. It has only a few large keys. Taking studies is as simple as entering four-digit codes for the subject/element observed. The backspace key removes typo errors. The observer identification, date, and tour times are easily entered in

---

*Royal Dossett, Datawriter Corp., Exselsior, Minn., contributed much of the information in this section.
†Another electronic data collector is the Psion Organizer II.

the Datawriter.'' His organization has prepared a software package called CAS which offers five functions, each selected from a main menu by pressing a single key. These functions are: Datawriter simulator, Receive data, Edit data, Design tours, and Summary. Dossett has been involved with the Extensor Corporation that concentrates in management self-administered studies employing the Datawriter equipment. The president of Extensor Corporation,* Robert H. Scarlett, predicts that this will be the work sampling procedure of the future.

''Multidimensional work sampling'' is explained by John A. Robertsen, vice president.[11]

> One of the most commonly heard questions regarding the use of self-administered work measurement system is: ''What keeps the participants from cheating?'' The most powerful constraint is peer pressure, especially by those who know the participant. If the participant states that he or she completed 30 patient files and the technician records having processed 10, it is rather obvious that something is amiss. The consequences resulting from exposure to one's peers of such self-serving contradictions is generally perceived as a price too costly to pay.

Jeff S. MacMillan, corporate director of industrial engineering with the Standard Products Company, made an extensive work sampling study at Children's Hospital in Milwaukee, Wis. He worked with the administration in training observers in the use of the Radio Shack TRS-80 Model 102 lap-top computer to collect data.

The program which collected these data was written in TRS BASIC, which produced a sequential ASCII coded file. This was transmitted via the RS 232 serial port to an IBM-PC class desktop minicomputer for summary and statistical analysis. The study comprised collecting over 8000 observations in 7 departments, consisting of 15 nursing activities over a 6-week period. The nursing personnel who were observed wore colorful arm bands with numbers to aid the observers. This kind of dress was possible because it was a children's hospital where the administration thought that it would liven up the ward even though it might appear unprofessional.

The data collection program required a two-digit activity code. The codes were structured to facilitate easy recall by the observers.

The account codes were:

DIRECT CARE (with patient and family)

10 ''Direct Care''

11 ''Off Unit, Patient Related''

INDIRECT CARE (traceable to patient or family)

20 ''Patient Related Communications''

21 ''Planning and Documentation''

22 ''Preparation for Direct Care''

23 ''Off—Unit—Other''

24 ''Wait Time''

25 ''Looking for Equipment''

UNIT RELATED (general, productive tasks)

30 ''Administration and Supervision''

31 ''Unit Environmental Maintenance''

32 ''Clerical Tasks''

33 ''Unit Travel Time''

---

*The Extensor Corporation, located in Minnetonka, Minn., furnished the information in this section.

34 "Staff Education"

35 "Off Unit Meetings"

PERSONAL TIME (nonproductive activities)

40 "Personal Time"

The work sampling study at Children's Hospital in Milwaukee led to developing a package to conduct similar studies in other children's hospitals.

This program facilitates the uploading of data files from the lap-top computer to an MS DOS desktop computer for summary and analysis. The desktop computer must have at least one asynchronous serial port, one 360K disk drive, and a hard drive.

The program includes many features which were added to specifically address the unique needs of a hospital facility:

- Extensive data entry error checking with a beep prevents entry of invalid codes.
- Descriptions of the activities upon entry.
- Display of last four entries. This feature allows the observer to verify the codes and make any notes for later editing and correction (up to a 19-character message).
- Two observation counters, current and previous unit.
- Continuous time-of-day display.
- Ability to enter class of care giver.
- Time of day automatically written to file every 12 minutes.
- Census data for patients and staff by department.
- Ability to power-off in mid-study to conserve battery life.
- Continuous readout of available memory. The program automatically beeps and displays a warning message if available memory drops below 256 bytes.

The computer analyzes each of the 15 account codes, showing the percentage of time for each activity with a two-sigma deviation. The study determines ways by which patient care can be improved, and is a scheduling tool and a means of measuring the effectiveness of the registered nurses, assistant nurses, and interns.

BASIC Language.   BASIC is suitable for writing a work sampling design program. Once the calculations have been programmed and tested, the task of making the study design user-oriented must be undertaken. This design, showing all variables on the screen at once, allows the user to use the "what-if" tool, to change one at a time.

What if we can accept a 10 percent error instead of 5 percent?

What if our focal activity is 45 percent instead of 50 percent?

What if we sample 30 machines per tour instead of 20?

What if we allow 30 minutes per tour instead of 20?

Once the design looks satisfactory to the user, the program should print out all parameters along with the printout of tour-begin times. Some form of editing of mistakes is usually required to facilitate keyboard entry of manually taken data. The data are recorded on disk files for later processing. The data should be divided into time frames, say one shift, each tour by a recorded time of day each hour, or each day.

Overall Summary.   Summary reports can be formatted in any manner desired. A summary can be printed of the entry data collected, showing percentage observed for each activity for the entire population. These reports are frequently formatted in a matrix or spreadsheet layout with activities along one axis and operators and machines along the other axis. The intersection of a row and column then supplies the percentage of time a given code was recorded. Summary rows and columns can be added along with other relevant data. Deviation can be calculated for each activity.

Time frame analysis is very useful in uncovering inefficiencies that occur only, say, in the morning, or perhaps early after lunch, or on the second shift, or the day after a holiday, etc.

Work sampling on a desktop computer offers the industrial engineer an excellent tool for every conceivable type of analysis.

CONTROL CHART

The control chart, similar to those used in quality control, is advocated by many who apply work sampling to ascertain that the daily percent plots are within one-, two-, or three-sigma limits. This chart (Fig. 3.3) enables the work sampling observers to know that the data are in a state of statistical control and are homogeneous and consistent. The control chart is considered pertinent in determining equitable delay allowances. It has an additional and important advantage in that the effect of a change in operating conditions can be checked to determine whether it produces a significant change in the delay percentage. The observer should be alert to strive to bring about greater control as a result of the use of a control chart.

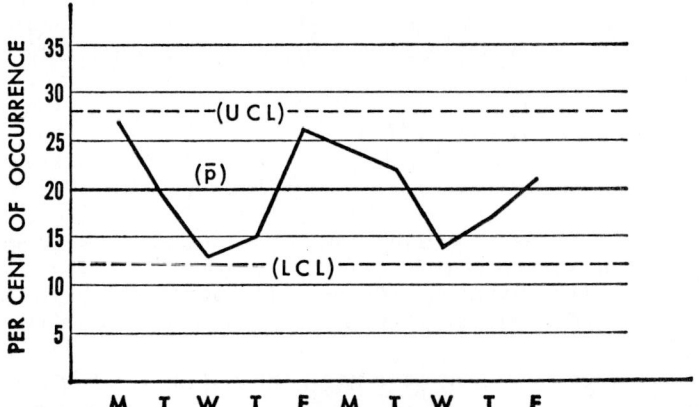

FIGURE 3.3 Control chart on a daily basis using constant upper and lower control limits.

Control charts may be determined on a daily basis, using constant upper and lower control limits for p. The first few random trips of observations produce percentage $p1$, $p2$, $p3,\ldots, pn$, or the average of p. This p is set up as the center line of the control chart. The upper and lower control limits are the $p \pm 1\sigma$, 2σ, 3σ, depending upon the confidence level desired. Also, these control charts are programmed into the electronic data collectors or computers.

Example.   If $p = 20$ percent and $N = 100$ observations per day, the upper control limits (UCL) and lower control limits (LCL) are:

Control Limits	
1 day—100 observations	10 days—1000 observations
$\sigma = \dfrac{p(1-p)}{N}$	$\sigma = \dfrac{p(1-p)}{N}$
$= \dfrac{0.20(0.80)}{100}$	$= \dfrac{0.20(0.80)}{1000}$
$= 0.04$	$= 0.0126$
$2\sigma = 0.08$	$2\sigma = 0.0252$
$\pm 2\sigma = 12$ to 28%	$\pm 2\sigma = 17.48$ to 22.52%

Another approach to emphasize that the percentage of error reduces as the number of observations increases is to compute 2σ deviations on cumulative number of daily observations (Fig. 3.4). Assuming 100 observations per day, 2σ is determined for 100, 200, 300, and so on, observations. This chart may also be used as an empirical method of determining the length of the study. When the variation from day to day is reached to the desired level for the element chosen to control the study, it can be assumed that enough observations have been gathered.

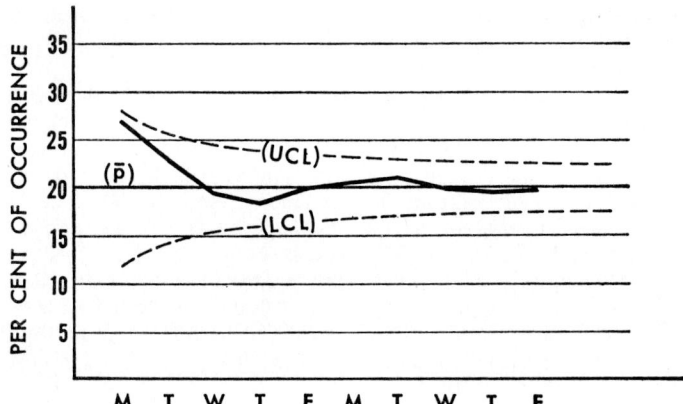

FIGURE 3.4   Control chart based on a cumulative number of daily observations.

GROUP TIMING TECHNIQUE (GTT)*

The author, as well as many others, have employed work sampling on a continuous basis whereby the same operation or operator may be observed every 1 minute, 3 minutes, or some other small fixed interval of time. This fixed interval sampling technique was called group timing technique by George Dew, who, along with others, developed it to be fundamentally the same process as measuring with a stopwatch where the intervals of sampling are shorter than the activity duration. Group timing technique (GTT) is an efficient and versatile measuring procedure easily applicable to work measurement and quantitative surveying tasks in industry.

---

*Most of the information relating to GTT in this chapter was developed by Rolf Tiefenthal.

GTT can often replace traditional stopwatch time study, providing equal or better data at a lower cost. We present the basic method of GTT, identify with examples its most important industrial applications, and describe application procedures. The mathematical theory of GTT is briefly presented, as well as some useful formulas for control of the statistical accuracy of time values derived from GTT application.

Basic Procedure.   Assume that a certain production operation is performed separately by four operators. It consists of three work categories, the duration of which is to be ascertained:

1. Preparation and finishing time
2. Process time
3. Allowances

After this element selection, each of the operators is observed at fixed time intervals of 2 minutes. At each observation and for each operator, a record is made of one of the three activities taking place.

The number of observations necessary to obtain a result that is acceptably close to the result that would be obtained through a continuous study can be computed. The statistical accuracy needed for the purpose of the study can be chosen. The length of the study is then determined by the number of observations needed and the observation interval, modified if necessary to allow for the inherent variability of the process itself.

After completion of the study, the tallies indicating the number of observations for each individual work category, activity, or element are summarized. The elapsed time, or duration of one activity as a percent of the total time of the study, is equal to the number of observations for the activity as a percent of total observations.

If the total number of observations is 1000 and activity B accounts for 600 of these, it may be concluded that process time accounts for 60 percent of the total time.

Absolute measurements can also be made. In the example, the total duration of activity B should be close to 600 registrations of approximately 2 minutes duration each, that is, 1200 minutes, or 20 hours. If the number of units processed during the study was 40 units, the average process time per unit will be close to 0.5 hour.

During the study, the observer can performance-rate operators and elements so that normalized time standards can be established. The total GTT procedure can be symbolized as shown in Fig. 3.5.

Where and When GTT Can Be Used.   GTT can be used for many different tasks and purposes. It has proved to be particularly useful for investigating or measuring easily observable activities taking place within limited localities where several operators, machines, or other activity centers are to be studied.

GTT can be used in production, maintenance, and offices for the following purposes:

1. Determination of leveled each-piece times for setting work standards, production norms, and incentives
2. Determination of allowances, including fatigue, along with a simultaneous check on standards previously set through other means called a control study
3. Determination of the load on groups of operators or machines
4. Surveying or fact-finding studies on organization and processes

Some typical examples of GTT application are described in the following paragraphs.

Use of GTT to Determine Each-Piece Time.   Figure 3.6 shows the front page of a study made of two men shaking out sand molds in a steel foundry. If desired, tallies could have

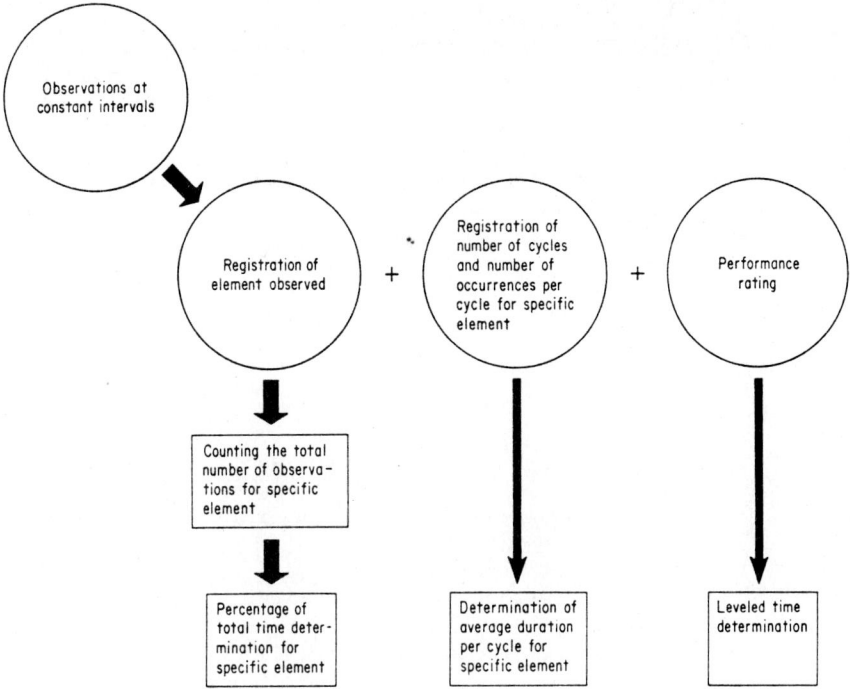

FIGURE 3.5   Chart summarizing the GTT procedure.

been recorded each 0.0050 hour, 18-second intervals instead of 0.0100 hour, 36-second intervals, and the tallies for each operator could have been recorded in separate columns. The 20 percent allowance had been established by earlier 8-hour GTT studies in the foundry. Most of the idle avoidable delay time occurred soon after 8:00 P.M. when these workers started their work shift. These operators were dayworkers. The job was completed at 9:45 P.M.; so the study was limited to 1.75 hours in length.

Figure 3.7 is the back of Fig. 3.6 and follows the design of a widely used time study form. Because 12 flasks were shaken out, the occurrences per flask on each element were 1/12, except on element 11, where the operator carried aside the 12 bottom boards from these flasks plus 8 bottom boards remaining from an earlier job, making a frequency of 1/20.

STEPS IN MAKING A GTT STUDY

GTT has many similarities to stopwatch time study. The steps employed in making a GTT study are as follows:

Decide on the Purpose of the Study.   It might be one of the following:

1. To provide quick information on idle time, proper crew size, waiting time, minor work performed, and similar data
2. To measure percent delays
3. To measure percent fatigue
4. To measure leveled time or allowed time

GROUP TIMING TECHNIQUE STUDY

Tallies recorded every .0100 elapsed hour
Operation: Shake Out with Jib Crane

Day	Wednesday
Date	February 22
GTT Study No.	S-64
Sheet No.	1 of 1 Sheets
Observer	George Dew

Operators	Skill	Effort	L.F.
G. Appleby	C1 +.06	C2 +.02	1.08
T. Zutic	D .00	E1 -.04	.96
		Ave. L.F.	1.02

Study Finished 9:45 PM
Study Started 8:00 PM
Over-all time 1.7500 hours
2 x 1.7500 = 3.500 man-hours

SUMMARY

175 Tallies x .0100 = 1.7500 Hours

Line	Elements	Total Tallies	Total Man-hrs.	Leveling Factor	Leveled Hours	% Allowance	Allowed Hours	Line
1	Remove clamps and wedges	8	.0800	1.02	.0816	20%	.0979	1
2	Wedges, gaggers, etc. aside	6	.0600	"	.0612	"	.0734	2
3	Move empty crane to flask	5	.0500	"	.0510	"	.0612	3
4	Hook flask with chain	12	.1200	"	.1224	"	.1469	4
5	Move flask by crane to shake out area	20	.2000	"	.2040	"	.2448	5
6	Mallet flask	33	.3300	"	.3366	"	.4039	6
7	Operate crane during shake out	4	.0400	"	.0408	"	.0490	7
8	Set down empty flask on 2nd flask	8	.0800	"	.0816	"	.0979	8
9	Stack empty flasks by crane	22	.2200	"	.2244	"	.2693	9
10	Unhook flask	7	.0700	"	.0714	"	.0857	10
11	Bottom board aside by hand	19	.1900	"	.1938	"	.2326	11
12	Move crane to casting	3	.0300	"	.0306	"	.0367	12
13	Hook casting with chain	14	.1400	"	.1428	"	.1714	13
14	Move casting aside by crane	2	.0200	"	.0204	"	.0245	14
15	Unhook casting	4	.0400	"	.0408	"	.0490	15
16	Set down moulds brought by cab crane	4	.0400	"	.0408	"	.0490	16
17	Wood flasks aside by hand	32	.3200	"	.3264	"	.3917	17
	Foreign Elements							
A	Get tools	2	.0200	—	—	—	—	A
B	Clean work area	2	.0200	—	—	—	—	B
C	Idle A.D.	109	1.0900	—	—	—	—	C
D	Idle U.D.	8	.0800	—	—	—	—	D
E	Personal delays	26	.2600	—	—	—	—	E
	TOTALS	350	3.5000					

FIGURE 3.6 Front of GTT study form used to establish an incentive time standard.

Study No.	S-64	Date February 22			

Operation: Shake out with Jib Crane

Department Steel Fdry.	Operator			
	Man	Name	Appleby Zutic	

Equipment: Jib crane with chain slings

3 fabricated steel flasks 24 x 36 x 20"
5 fabricated steel flasks 20 x 32 x 29"
4 fabricated steel flasks 18 x 24 x 16"

	Dwg. Style		Sub.
Mould		Die	
Pattern	Ins. Spec.	L. Spec.	Sub. Item
Part Description	Moulds made on J & J Roll Over Machine		

Material Manganese Steel Castings

No.	Elements	Elem. Time Allowed	Occurrences /pc	Total Time Allowed
1	Remove clamps and wedges	.0979	1/12	.0082
2	Wedges, gaggers, etc. aside	.0734	1/12	.0061
3	Move empty crane to flask	.0612	1/12	.0051
4	Hook flask with chain	.1469	1/12	.0122
5	Move flask by crane to shake out area	.2448	1/12	.0204
6	Mallet flask	.4039	1/12	.0337
7	Operate crane during shake out	.0490	1/12	.0041
8	Set down empty flask on 2nd flask	.0979	1/12	.0082
9	Stack empty flasks by crane	.2693	1/12	.0224
10	Unhook flask	.0857	1/12	.0071
11	Bottom board aside by hand	.2326	1/20	.0116
	TOTAL ALLOWED HOURS for Magnetic Castings			.1391
12	Move crane to casting	.0367	1/12	.0031
13	Hook casting with chain	.1714	1/12	.0143
14	Move casting aside by crane	.0245	1/12	.0020
15	Unhook casting	.0490	1/12	.0041
	TOTAL ALLOWED HOURS for Non-Magnetic Castings			.1626

Special tools, jigs, fixtures, etc. — Mach. Tool No. — Wood mallets

Conditions: Average

Observer: Dew — Sketch — Approved by

Time All., Set Up	Each Piece	Total

Remarks: Elements 16 and 17 are foreign to this study. Magnetic castings are lifted out of the sand by a magnet later on. Non-magnetic castings, such as some manganese steels are hooked with a chain by the shake-out men and set aside to be hooked with a chain by the cab crane later.

FIGURE 3.7 Back of GTT study form shown in Fig. 3.6.

Select the Operation, Group of Operators, and Time Periods to Be Studied

Decide Whether to Separate the Time for Each Operator.   If not, the GTT form will be easier to design and the observations can be recorded in less time, thus permitting the use of a shorter time interval between observations if desired.

Decide on the Recording Method.   One of the following procedures should be chosen: by observer(s), or by operator(s), or by memomotion filming.* Using observers is frequently the most suitable procedure. They register tallies on a study form, or employ an electronic data collector for subsequent data processing. Recordings can be made by operators if the purpose of the study is suitable and will not interfere too much with the work. In this case, the element breakdown must be held to a minimum. The format of the form should be designed in advance.

Select and Define Elements.   If unfamiliar with the operation, preliminary observations should be made to determine what elements to include. At this time, it should also be decided how detailed the study needs to be. The number of elements should be minimized without jeopardizing the purpose of the study. Some open space should be left on the study form for later addition of unexpected elements. The element breakdown should be logical and with natural limits easily definable and observable. For large studies, particularly when using several observers, it may be desirable to work out a brief written element definition. In many cases, element definitions can be standardized for repeated use within a company.

Design and Reproduce Study Form.   GTT is used in so many varying situations that no standard form suitable for all purposes exists (compare Figs. 3.6 and 3.7). A suitable form should be designed for each study. Listing all elements on one 8½ by 11 sheet is impossible. If not, two sheets may be used, but this will require turning the sheets back and forth while making the study. The forms should be reproduced in sufficient number to allow for spoilage and for summarizing the final results. In making the study, each observer will usually start a new form each hour.

Select Interval Size.   The proper interval size is primarily dependent on the number of operators or machines and their location. Observers should not be so pressured that they do not have time to observe thoughtfully what is going on. The following table suggests intervals that an observer can use comfortably during an 8-hour study. The interval, however, must be shorter than the smallest element to be measured. Also, any rigid cycle time inherent in the process should not coincide with multiples of the observation interval, to prevent possible overrepresentation of some activities.

Number of workers in group	Interval measured on stopwatch	
	Decimal hour watch	Decimal minute watch
1	Use time study or predetermined time system	
2	0.0050 hour	0.5 minute
3–6	0.0100 hour	1.0 minute
7–10	0.020 hour	2.0 minutes
	0.0300 hour	
11–15	0.0400 hour	3.0 minutes
	0.0500 hour	
Over 15	Use 2 or more observers	

---

*S. M. Lowry, H. B. Maynard, and G. J. Stegemerten, *Time and Motion Study and Formulas for Wage Incentives,* McGraw-Hill, New York, 1940, p. 259.

Determine Study Periods and Duration.   Several factors influence the length of the study. The minimum number of total observations or observations for the critical activity should be computed as described later under "Statistical Accuracy of GTT." The minimum duration of the study is the minimum number of total observations needed, multiplied by the observation interval.

To ensure that the data obtained are truly representative of normal conditions, the duration of the study must be sufficient to include and level out the natural, long-cycle variations that may exist in the process. Where such variations exist, the total number of observations can be divided into a suitable number of substudies and spread out to cover a longer period of time or a wider range of conditions.

Daily plotting of diagrams such as shown in Fig. 3.3 should be considered. This continuous check on the daily and accumulated values of one or more critical activities may be helpful in determining the stability of conditions and evaluating the necessary duration of the study.

Inform Users of Its Scope and Purpose and How It Is Employed

Consider a Trial Study.   A brief trial study will provide a good test of the soundness of the preparatory decisions such as element selection, element definition, and interval size.

Make the Study.   Record carefully the names or numbers of the operations, operators, interval size, and other identifying information. Start the stopwatch and allow it to run continuously. Record the starting time from a regular watch or clock to the closest ½ minute.

Make distinct observations and mark tallies at the predetermined interval. If five men are being timed each 0.01 hour, first record a tally in the overall tallies box when the decimal hour stopwatch reads 0.01. Then record a tally for each of the five men. Repeat this action in the same sequences each 0.01 hour for the duration of the study.

If the number of operators is large or if the observer has to move between different locations, it is possible to establish fixed time delays between groups of observations to allow for comfortable observation.

Change sheet at fixed intervals, such as every hour or second hour. Record the time for each switch on both the old and the new form. If desired, the performance rating for each operator is recorded once on each sheet.

On a ½-day or all-day GTT study, account for the full 4 hours or 8 hours of each employee by means of tallies. This may require elements such as start late or quit early, coffee break, or safety meeting. This will make the study more understandable, reduce errors, and make the study easier to summarize.

An exception may be made to this rule in the case of a "floating" worker who is sometimes a member of the group and at other times is outside the group or the observation area.

The element "work on foreign operation" is used for operators temporarily performing some task that is not part of their jobs.

When more than two or three employees are studied, an element such as "absent—reason unknown" is often required. Every effort should be made to hold the number of tallies recorded for this element to a minimum by discovering its reason.

On an each-piece study, the piece count and other pertinent information should be recorded.

At the completion of the study, the time finished and the skill and effort ratings for the last sheet or sheets should be recorded.

The study should be worked up the same as a time study.

USE OF GTT TO MEASURE ALLOWANCES AND FATIGUE

As has already been pointed out, the GTT procedure can be used to measure the allowances which should be made for fatigue and special, unavoidable, and personal delays.

Figure 3.8 (study 314) shows one sheet of an 8-hour study taken on eight sewing machine operators. The allowances can readily be calculated from this study. First, the results of the all-day study are summarized in the first table on page 4.62, and next, the allowances for special, unavoidable, and personal delays are calculated in the second table on page 4.62.

Finally, a reasonable allowance for fatigue is determined, using the method developed by Lowry, Maynard, and Stegemerten for use with all-day allowance time studies.[*]

$$\text{Percent fatigue} = OL/NS - 1 \times 100$$

where O = overall working element time
  L = leveling factor at the point of maximum performance during the day
  N = number of pieces produced during the day
  S = leveled time per piece

OL for study 314 is calculated in the third table on page 4.62.
NS is calculated as follows:
The 8 operators worked on 15 operations during the all-day study. Their piece counts N were obtained from the verified production reports submitted to the payroll department. The leveled time for each operation was taken from previously established time standards.

NS is obtained by multiplying the number of pieces produced on each operation by the leveled time for the operation. The total NS for the 8 operators is 60.24 hours.

$$60.24/8 = 7.5300 \ NS \text{ for one operator}$$

Substituting in the equation,

$$\% \text{ fatigue} = \left(\frac{OL}{NS} - 1\right)100 = \left(\frac{7.8650}{7.5300} - 1\right)100$$

$$\% \text{ fatigue} = 4.45\%$$

Total allowances for fatigue and special, unavoidable, and personal delays are therefore

	Percent
Special delays	4.51
Unavoidable delays	1.04
Personal delays	4.82
Fatigue	4.45
	14.82
Rounded	15

STATISTICAL ACCURACY OF GTT

Fixed versus random observation intervals—relates the two methods to entirely different mathematical models. The computations that relate the number of observations to the statistical error are different for the two models. It can be shown that, for the same level of accuracy and all other features being equal, GTT will normally call for a considerably smaller number of total observations than work sampling.

Being a sampling procedure, GTT produces time estimates containing some statistical error in relation to the true time values of a pattern of activity. These errors can be determined.

[*]Ibid.

Study on 8 Sewing Machine Operators   Sheet Started 10:30 A.M.   Finished 11:30 A.M.   GTT Study No. 314

Sheet 4 of 8 Sheets   Each Tally = .0200 Hour   Wednesday, March 26   Observer George Dew

Over-all Tallies: THL THL THL THL THL THL THL THL THL THL THL THL

No.		1 Grace	2 Sarah	3 Verna	4 Lizzie	5 Esther	6 Margaret	7 Bessie	8 Dornelda	No.
	Operator	Grace	Sarah	Verna	Lizzie	Esther	Margaret	Bessie	Dornelda	
	Machine	Flatlock	U. Special Sew Tubing	Button Machine	Flatlock	U. Special Sew Elastic	Flatlock	Flatlock	Flatlock	
	Skill-Effort L.F.	B1B2 1.19	C2C 1.065	C1C1 1.11	C2C 1.065	C1C1 1.11	E2C .935	B2C1 1.13	C1C1 1.11	
1	Sew	46	43	46	34	44	45	43	35	1
2	Handle bundle	1	4	1	2	1	1	1	1	2
3	Arrange work	1		1	1	3	1		1	3
4	Change thread									4
5	Change tubing or elastic									5
6	Thread breakage	2	1	2	13		4		2	6
7	Down-machine trouble									7
8	Talk to supervisor				1	1				8
9	Idle - Unavoidable									9
10	Start late									10
11	Quit early									11
12	Rest period									12
13	Work during rest period									13
14	Personal time		1			2				14
15	Idle - Avoidable							6	11	15
16	Clip parts apart									16
17	Clean up machine									17
18	Fill in PW Report									18
	Totals	50	50	50	50	50	50	50	50	

FIGURE 3.8 GTT study form used to measure percent allowances and fatigue.

Element no.	Description	Elapsed hours		Classification of time
		Total, 8 girls	Average, 1 girl	
1	Sew	51.94	6.4925	Working element
2	Handle bundle	2.22	0.2775	Working element
3	Arrange work	0.86	0.1075	Working element
4	Change thread	0.24	0.0300	Special delay
5	Change tubing or elastic	0.16	0.0200	Special delay
6	Thread breakage	1.16	0.1450	Special delay
7	Down—machine trouble	0.10	0.0125	Special delay
8	Talk to supervisor	0.14	0.0175	Unavoidable delay
9	Idle—unavoidable	0.46	0.0575	Unavoidable delay
10	Start late	0.02	0.0025	Avoidable delay
11	Quit early	0.20	0.0250	Avoidable delay
12	Rest period	1.68	0.2100	⎰Consider all as personal
13	Work during rest period	0.88	0.1100	⎱ delays
14	Additional personal time	0.22	0.0275	Personal delay
15	Idle—avoidable	0.16	0.0200	Avoidable delay
16	Clip parts apart	2.62	0.3275	Working element
17	Clean up machine	0.68	0.0850	Special delay
18	Fill in P. W. report	0.26	0.0325	Special delay
	Total...............	64.00	8.0000	

Working elements		Special delays		Unavoidable delays		Personal delays	
(1)	6.4925	(4)	0.0300	(8)	0.0175	(12)	0.2100
(2)	0.2775	(5)	0.0200	(9)	0.0575	(13)	0.1100
(3)	0.1075	(6)	0.1450			(14)	0.0275
(16)	0.3275	(7)	0.0125				
		(17)	0.0850				
		(18)	0.0325				
Total.......	7.2050		0.3250		0.0750		0.3475
% allowances.........		$\frac{0.3250}{7.2050} \times 100 = 4.51\%$		$\frac{0.0750}{7.2050} \times 100 = 1.04\%$		$\frac{0.3475}{7.2050} \times 100 = 4.82\%$	

Operator	Elapsed working element hours	Maximum leveling factor for day	Leveled hours at maximum
1	7.32	1.19	8.71
2	6.96	1.08	7.52
3	7.30	1.11	8.10
4	7.04	1.065	7.50
5	7.34	1.11	8.15
6	7.30	0.95	6.94
7	7.14	1.13	8.07
8	7.24	1.095	7.93
	Total for 8 operators....		62.92
	OL (average for 1 operator)....		7.8650

The formulas discussed below will provide an answer to questions such as:

What accuracy did we get by this study?
How many observations do we need in order to measure safely within a given accuracy level?

They will also give GTT practitioners an increased understanding of what they are doing, so that they can determine, for example, if some modification of the study procedure that they are considering for practical reasons is permissible within the basic procedure.

Explanations. The following symbols and abbreviations will be used:

T total length (time) of study
T_a total time for activity a
t time for one work cycle
t_a time for one occurrence of activity a
N total number of observations
N_a total number of observations for activity a
K_a total number of occurrences for activity a
C_a number of occurrences during one work cycle for activity a
i interval size
r_{T_a} relative error for T_a at the 95 percent confidence level
r_T relative error for T_a in percent of total time T at the 95 percent confidence level ($r_T = r_T \times T_a/T$)
s_{T_a} standard deviation for T_a
n_a number of observations for one occurrence of activity a
a_a time between the previous observations and the actual start of activity a
b_a time between the actual end of activity a and the following observation

Formulas. The following formulas (all valid at the 95 percent confidence level) are used in the practical planning and evaluation of GTT studies.

Determination of Total Time

$$T_a = N_a \times i \tag{1}$$

$$T = N \times i \tag{2}$$

Determination of Statistical Error

$$r_{T_a} = \pm \frac{80\sqrt{K_a}}{N_a}\% \tag{3}$$

$$r_T = \pm \frac{80\sqrt{K_a}}{N}\% \tag{4}$$

Determination of Number of Observations

$$N = \frac{6400}{r_{T^2}} \times \frac{i \times C_a}{t} \tag{5}$$

$$N = \frac{6400 \times i \times t \times C_a}{r_{T^2} \times t_a^2} \tag{6}$$

$$N_a = \frac{6400 \times i \times C_a}{r_{T_a}^2 \times t_a} \tag{7}$$

Theory.   Three different cases of observation exist for a specific occurrence j of activity a according to Fig. 3.9. For all three cases

$$t_{aj} = (n_{aj} + 1)i - (a_{aj} + b_{aj})$$

Because, according to definition,

$$T_a = \sum_{j=1}^{K_a} t_{a_j}$$

we get

$$T_a = \sum_{j=1}^{K_a} [(n_{a_j} + 1)i - (a_{a_j} + b_{aj})]$$

$$= N_a \times i + K_a \times i - \sum_{j=1}^{K_a} (a_{a_j} + b_{a_j})$$

a_a and b_a, however, are rectangularly distributed between the limit values 0 and i (equal probability for a_{a_j} of b_{a_j} to assume any value between 0 and i). Consequently, with increasing K_{a_j},

$$\sum_{j=1}^{K_a} a_{a_j} \quad \text{and} \quad \sum_{j=1}^{K_a} b_{a_j} \rightarrow 0, 5 \times i \times K_a$$

and

$$T_a = N_a \times i + (K_a \times i) - (K_a \times i) \tag{1}$$

$$= N_a \times i$$

Similarly,

$$T = N \times i \tag{2}$$

To establish the error of T_a in formula (1), we study the standard deviation s for the rectangular distribution between the limit values 0 and i. $s = i/\sqrt{2}$ *according to statistical reference literature.*

For t_{a_j}, the standard deviation $s_{t_{a_j}}$ is the square root of the square sum of two standard

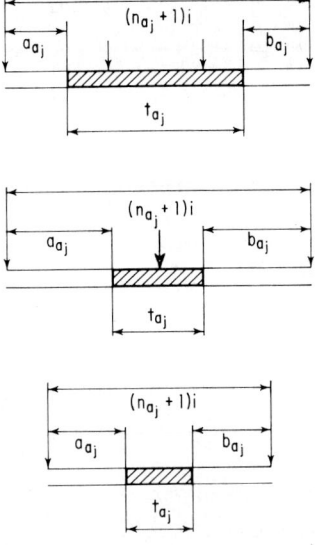

FIGURE 3.9 Three different cases of observation for a specific occurrence of activity a.

deviations corresponding to the rectangular distributions in the beginning and in the end of t_{a_j}:

$$s_{t_{a_j}} = \sqrt{\left(\frac{i}{\sqrt{12}}\right)^2 + \left(\frac{i}{\sqrt{12}}\right)^2} = \frac{i}{\sqrt{6}}$$

Further,

$$s_{T_a} = \sqrt{\sum_{j=1}^{K_a} s_{t_{a_j}}^2} = \sqrt{K_a \times \frac{t^2}{6}} = \frac{i \times \sqrt{K_a}}{\sqrt{6}}$$

At the 95 percent confidence level,

$$r_{T_a} = \frac{1.96 \times s_{T_a}}{T_a} \times 100 = \frac{1.96 \times \dfrac{i\sqrt{K_a}}{\sqrt{6}}}{N_a \times i} \times 100$$

$$= \frac{80\sqrt{K_a}}{N_a} \tag{3}$$

Similarly, according to definition,

$$r_T = \frac{1.96 \times s_{T_a}}{T} \times 100$$

$$= \frac{80\sqrt{K_a}}{N} \tag{4}$$

If, for cyclic work, the number of occurrences during one work cycle for activity a is C_a, the total number of occurrences

$$K_a = \frac{T \times C_a}{t} = \frac{N \times i \times C_a}{t}$$

This will turn formula (4) into

$$r_T = 80\sqrt{\frac{i \times C_a}{N \times t}}$$

$$N = \frac{6400 \times i \times C_a}{r_T^2 \times t} \tag{5}$$

which can be transformed into

$$N = \frac{6400 \times i \times t \times C_a}{r_{T_a}^2 \times t_a^2} \tag{6}$$

and

$$N_a = \frac{6400 \times i \times C_a}{r_{T_a}^2 \times t_a} \tag{7}$$

Application of Theory.   According to formula (3), the largest statistical error will occur on an activity showing a high value for K_a and a low one for N_a. The critical activity showing these properties is typically one with frequent occurrences, each occurrence being of short duration.

Formulas (3) to (7) are used for two practical purposes:

1. To estimate before the GTT study is begun the necessary number of observations needed to reach a certain accuracy level of measurement
2. To determine the statistical accuracy of measurement after the study has been made

For practical reasons, the control of accuracy will be based on the selection of either one or a few critical activities or elements or on the basis of an average activity or element.

The computations of statistical accuracy give control of only the methodical error in the GTT itself. The actual result may be influenced by a number of other and more intricate sources of error, such as an unrepresentative period of study, errors made by the observer in classification and performance rating, and the like. This reservation should call attention to these matters, but it in no way detracts from the obvious value of being able to forecast and control statistical deviation in GTT measurements in the simple manner illustrated by the following examples.

Case A. Forecasting the Accuracy of a Delay Study. Repair Work.   The study is planned to take 27 hours, with an observation interval of 2 minutes. The shortest delay time is estimated to be 6 minutes, and the average to be 9 minutes (according to available statistics or a limited pilot study). Total delay time is expected to be 15 percent of available time. Determine the relative error in the value for total delay time which will be determined by this study.

$$T = 27 \text{ hours} = 1620 \text{ minutes}$$

$$i = 2 \text{ minutes}$$

$$t_{\text{average}} = 9 \text{ minutes}$$

$$N_a = \frac{T_a}{i} = \frac{15 \times 1620}{100 \times 2} = 122 \qquad \text{formula (1)}$$

$$K_a = \frac{T_a}{t_{\text{average}}} = \frac{15 \times 1620}{100 \times 9} = 27$$

$$r_{T_a} = \frac{80 \times \sqrt{K_a}}{N_a} = \frac{80\sqrt{27}}{122} = \pm\, 3.4\% \qquad \text{formula (3)}$$

Case B. Planning a Study to Establish Piecework Standards.   Assembly work by two operators and one assistant. The smallest suboperation (element) occurring once per cycle was estimated to be 3 minutes, and the total cycle time to be 30 minutes. A relative error of no less than 5 percent is desired for the smallest element. Observation interval: 0.5 minute.

Determine the number of observations, the duration of the study, and the number of work cycles.

$$i = 0.5 \text{ minute}$$

$$t = 30 \text{ minutes}$$

$$r_{T_a} = \pm\, 5\%$$

$$t_a = 3 \text{ minutes}$$

$$C_a = 1$$

$$N = \frac{6400 \times i \times t \times C_a}{r_{T_a}^2 \times t_a^2} = \frac{6400 \times 0.5 \times 30 \times 1}{25 \times 9} = 427 \qquad \text{formula (6)}$$

$$T = N \times i = 427 \times 0.5 = 214 \text{ minutes}$$

$$K_a = \frac{T}{t} = \frac{214}{30} = 7.1, \text{ or } 8$$

CONCLUSION

Random and fixed interval work sampling are convenient methods of studying worker and machine processes in industry and elsewhere. The procedure is simple and easy to learn and can be partly automated if used extensively. In cases where systematic process improvement through detailed methods, tool, equipment, and workplace design is sought, alternative techniques which consider methods may be preferred.

REFERENCES

1. L. H. C. Tippett, "Use of the Binomial and Poisson Distribution: A Snap Reading Method of Making Time Studies of Machines and Operations in Factory Surveys," *Shirley Institute Memoirs,* vol. 13, pp. 35–93, November 1934.

2. L. H. C. Tippett, "The Ratio Delay Technique," *Time and Motion Study,* May 1953, pp. 10–19, Sawell Publications Ltd., 4 Ludgate Circus, London, E.C. 4.

3. Ralph M. Barnes, *Work Sampling,* 2d ed., Wiley, New York, 1957.

4. Ralph M. Barnes, *Motion and Time Study,* 7th ed., Wiley, New York, 1980.

5. R. L. Morrow, "Ratio Delay Study," *Mechanical Engineering,* vol. 63, no. 4, pp. 302–303, April 1941.

6. C. L. Brisley, "How You Can Put Work Sampling to Work," *Factory Management and Maintenance,* vol. 110, no. 7, pp. 84–89, July 1952. H. L. Waddell, "Work Sampling—A New Tool to Help Cut Costs, Boost Productivity, Make Decisions" (editorial), *Factory Management and Maintenance,* vol. 110, no. 7, p. 83, July 1952.

7. Robert Lee Morrow, *Motion Economy and Work Measurement,* 2d ed., Chaps. 23, 24.

8. Wallace J. Richardson, "Work Sampling and Indirect Cost Reduction," *Industrial Management Society, Industrial Engineering and Management Clinic,* Nov. 3–5, 1971, pp. 21–24.

9. E. S. Pape, "Work/Activity Sampling—Contemporary Design Analysis Methodology and Applications Part II—Work Sampling Calculations Revisited," *AIIE, 1979 Fall Industrial Engineering Conference Proceedings,* Norcross, Ga., 1979.

10. C. L. Brisley and R. Dossett, "Computer Use and Nondirect Labor Measurement Will Transform Profession in the Next Decade," *Industrial Engineering,* vol. 12, no. 8, pp. 34–43, August 1980.

11. John A. Robertsen, "Multidimensional Work Sampling: A New Tool for Pharmacy Management," *Topics in Hospital Management,* August 1982.

CHAPTER 4
PREDETERMINED MOTION TIME SYSTEMS

Clifford N. Sellie
Chief Executive Officer
Standards International Inc.
Chicago, Illinois

B. M. Worrall
Professor
Industrial Engineering Department
Technical University of Nova Scotia
Halifax, Nova Scotia, Canada

Many different methods are available for determining time standards. Traditional methods are

1. Stopwatch time study
2. Historical records
3. Reasonable expectancies
4. Work sampling
5. Standard data development
6. Predetermined times

Varying degrees of reliability and accuracy are claimed for all of them. Generally speaking, predetermined times are recognized as the most important from a viewpoint of methods specification and accuracy. Figure 4.1 provides a summary of the average degree of reliability in these different techniques. These are generally averages and are not "guaranteed" for any individual company or operations.

Predetermined time standards is the primary concern of this chapter. Well-known systems of predetermined times are: motion-time analysis (MTA); Work Factor (WF); basic motion time study (BMT); methods time measurement (MTM); universal analyzing system (UAS); Maynard's operations sequence techniques (MOST); micro motion analysis (MICRO); modular arrangement of PTS (MODAPTS); and macro motion analysis (MACRO). These systems are generally available to the public. Other systems of restricted distribution are Western Electric's elemental time standards (ETS); and the General Electric systems: Engstrom, motion time standards (MTS), and dimensional motion times (DMT).

Work Measurement Techniques:	When Set	Customary Trends
Historical records	± 30%	20% tight to 60% loose
Reasonable expectancies	± 20%	10% tight to 45% loose
Stopwatch studies	± 10%	5% tight to 35% loose
Predetermined times	± 5%	5% tight to 20% loose

Influences Affecting Standards Reliability:

- Standard data formats
- Methods specifications
- Work sampling
- Competitive needs

- Management experience
- Management controls
- Line and staff knowledge
- Union knowledge

The three modifying forces that have the strongest influence on the quality of a firm's work standards usually are:

- Competitive needs.
- Methods specifications.
- Management experience.

FIGURE 4.1   How are work standards set, and how well?

FAMILY TREE OF PREDETERMINED TIME SYSTEMS

Work elements can be quickly analyzed through the micro motion process used by nearly all predetermined time systems to arrive at time standards. The following analysis shows the typical steps followed in detailed predetermined time systems. Typical analysis is as follows:

- Work element and detailed motion-time analysis.

1. Get part from tote pan — Reach, Search, Select, Grasp

2. Place part in fixture — Move, Preposition, Assemble, Release

At one time approximately 50 different predetermined time systems were available. Figure 4.2 shows the chronological development of today's leading systems. While these individual systems were developed through independent research, most of them have been influenced by the independent systems of the early 1920s and 1930s. Significant sums of money are still spent on continuing research to explore new applications and develop new areas of use, but the major funds for initial research in the development of predetermined times appear to have been invested in the early 1920s and 1930s.

MAJOR DIFFERENCES IN PREDETERMINED TIME SYSTEMS

The currently available predetermined time systems can be classified in two categories:

1. Detailed. Provides precise subdivisions of body-member motions. MTA is one of the most detailed systems. MTM-1, MICRO, and Work Factor also fit in the detailed category.
2. Condensed tables. These are tables of combined motion times. They provide less de-

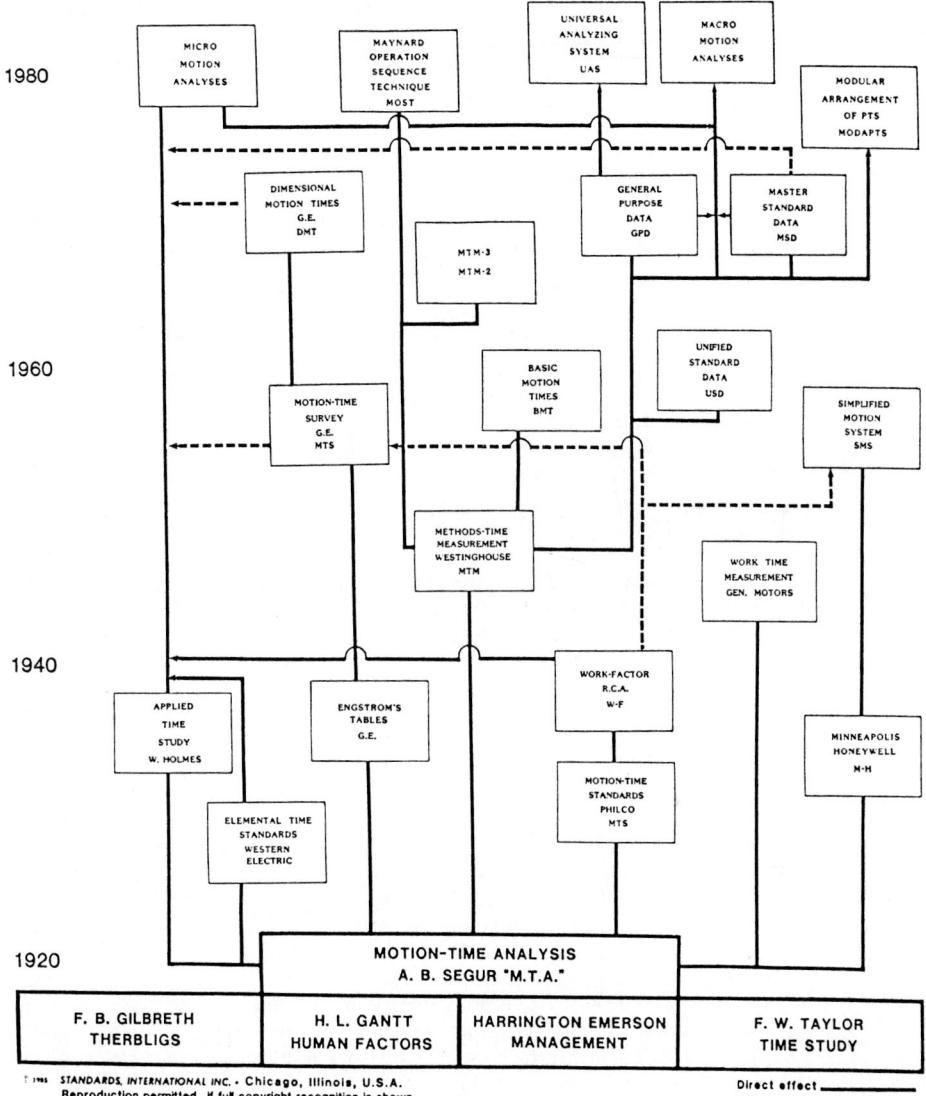

FIGURE 4.2 Family tree of predetermined times.

tailed methods analysis. Because they are average combinations, the time values' accuracy varies with the extent of averaging of the motion combinations. In many cases this is a satisfactory trade-off for use of less detailed tables and less detailed methods analysis.

OUTLINE DESCRIPTION OF SOME OF THE SYSTEMS

Methods Time Analysis (MTA). A. B. Segur of Oak Park, Ill., was one of the first to establish the relationship between the time element and the motion itself. His ambition to

Left hand	Motion symbol		Right hand
1. Move hand to tote pan	TE	TE	1. Move hand to tote pan
2. Grasp bracket	G	G	2. Grasp stud
3. Move bracket to work area	TL	TL	3. Move stud above hole in bracket
4. Hold bracket in work area	H	PP	4. Preposition stud
5. Hold bracket in work area	H	A	5. Turn stud in hole
6. Dispose assembly to tote pan	TL	BD	6. Balancing delay

FIGURE 4.3   Assemble stud to a bracket.

integrate time with motions led to the development of his MTA system. From his research, he discovered the "law of fundamental times," which finally made Taylor's dream of universally applicable standards in industry a working possibility. The law was stated as follows: "Within reasonable limits, the time required by experts to perform a fundamental motion is a constant."

This discovery enabled Segur to develop an analytical method which could be applied to a great variety of manual or manual and machine operations. He emphasized that the time required to accomplish an act depends on the method used by the operator. Segur insisted that the only realistic way to determine the time for an act was to know precisely how the act is performed.

Segur stated that the method must be well defined before an attempt is made to time analyze the motions involved. He developed a table of improvement principles involving many of his basic motions, such as hold, grasp, preposition, position, avoidable delay, and balance delay. Figure 4.3 shows the motion breakdown for the operation of starting a stud into the tapped hole of a single bracket.

The improvement principle involved here is in the elimination of the left hand as a holding device. Segur suggests the possibility of a simple mechanical fixture where two brackets could be assembled simultaneously, thus eliminating the hand hold.

The Work Factor System (WF).   During the years 1935 and 1936, a group of time study engineers were working on the development of a "second operation" punch press formula to be used in establishing rates for piercing, forming, and other types of punch press operations following the original blanking. Complete and detailed information was recorded for each work motion involved in the operations. These data consisted of such information as the distance moved; the body member used; the weight or resistance involved; and size and type of tools, jigs, and fixtures required. After several months of work on the second operation formula, it became evident that the type of data collected could be applied to many operations. This led to a broadening of the project into other types of factory operations. Hundreds of different types of work motions were studied and recorded.

Methods Time Measurement (MTM).   The methods time measurement procedure may be defined as follows: "Methods time measurement is a procedure which analyzes any manual operation or method into the basic motions required to perform it and assigns to each motion a predetermined time standard which is determined by the nature of the motion and the conditions under which it was made."

The primary objective of MTM is to improve methods of operation. Methods work is often correction of some previous method established by a worker, supervisor, or engineer. But MTM establishes methods accurately, before production starts, by determining correct times and motions for operations. Because most operators object to changes, it is of inestimable value to establish correct methods at the start.

Basic Motion Time Study (BMT).   Basic motion time study was developed and is taught by J. P. Woods and Gordon, Limited, Toronto, Canada. Like other predetermined motion time systems, all manual activity has been divided into basic motions.

A basic motion, according to Woods and Gordon, is defined as "Any motion which starts from rest, moves through space, and ends at rest." Hand activity is divided into three classifications:

(Type 1). *Reach:* The basic element employed when the predominant purpose is to move the hand or finger from one position to another.

(Type 2). *Move:* The basic element employed when the predominant purpose is to move an object from one position to another.

(Type 3). *Turn:* The motion employed to turn the hand, either empty or full, by a movement that rotates the hand, wrist, and forearm about the long axis of the forearm.

The body motion and symbols are very similar to the body motions employed by MTM. The only difference lies in the Side Step, where the distance measured is the distance the foot travels.

In using BMT, the basic Reaches, Moves, and Turns are classified and recorded, and additional allowances are added for any Precision, Simultaneous Allowance, or Force (weight) that is involved.

The variables which affect the time to perform a Reach or Move are:

1. Distance
2. Control
3. Precision
4. Force
5. Change Direction

Micro Motion Analyses (MICRO).   Standards, International Inc. of Chicago developed MICRO in the early 1980s to embody the best features of motion time analysis, work factor, elemental time standards, and dimensional motion times. Primary objectives were: (1) maintain their detailed accuracy, (2) simplify learning and applying the motion time tables, (3) speed up systems application. These objectives were accomplished by improving the motion time table formats and clarifying the application rules.

The MICRO table provides for

1. Body member used: finger-hand, arm, forearm, foot, leg, trunk, knee motion, eye motions, and mental processes. Special provision is made for highly repetitive arm-elbow motions
2. Motion combinations frequently used: grasps, plug-target assemblies, preposition (one hand and both hands), walking
3. Motion difficulties: care, deceleration, steer, and change direction

Master Standard Data (MSD).   Master standard data (MSD) is a "condensed" predetermined time system. It was developed by Crossan and Nance of the Serge A. Birn Company in the late 1950s and published in 1962. Master standard data was developed by statistically studying all motions.

MSD tables focus on the common motions that are most apt to be encountered and provide for those motions. The developers concluded that the majority of the motions that an operator does with hands and/or arms are

1. Get something.
2. Put it someplace else.

Their second conclusion was that there really are

1. Only two ways to pick things up.
2. Only three places to put them down.
   MSD Motion Categories. MSD is classified according to the degree of control required, and provides for three categories of control:

1. Little control required.

   - Reach to an object in a fixed location or an object in the other hand.
   - Simple pickup grasp made by closing of the fingers. The object is by itself and easily grasped.
   - Move an object against a stop.
   - Release or relinquish control of an object.

2. Some mental control required.

   - Reach to an object in a location that may vary from one reach to the next.
   - Move an object to an approximate location.

3. Highly controlled, either mentally or visually.

   - Reach to an object jumbled with other objects or to a very small object.
   - Grasp an object that cannot be controlled by a simple closing of the fingers.
   - Move an object to an exact location.
   - Align carefully one object with another object.

The MSD time values are in one-hundred-thousandths of an hour. This is the familiar TMU, or time measurement unit, found in MTM. Each unit = 0.00001 hour or 0.0006 minute or 0.036 second.

Master standard data became one of the first popular condensed data systems based on MTM-1. The GPD condensed elemental data developed by the MTM Association coincided with MSD concepts.

Maynard Operation Sequence Technique (MOST). Maynard Operation Sequence Technique was developed by the Swedish Division of H. B. Maynard and Company, Inc. in 1967–1972. It was introduced in the United States in 1974. The development of MOST was the result of an extensive review of MTM data. This review revealed that similarities existed in the sequence of MTM-defined motions whenever any object was handled. It was found that the same general sequence of motions required the same set of basic motions. Discovery of this phenomenon raised the questions as to whether this tendency for motions to fall into the same general sequence could be used to develop a new way to analyze methods and to measure operation times.

The next several years verified that the movement of objects tended to follow certain consistently repeating patterns, such as reach, grasp, move, and position. This tendency provided the basis for developing the sequence models used in MOST. These general patterns found in moving an object were identified and arranged as a sequence of events (or subactivities).

The MOST work measurement system is applicable for any cycle length and repetitiveness, *as long as there are variations in the motion pattern* from one cycle to another. The MOST system employs a small number of selected levels of fixed activity sequences which cover practically all aspects of manual activity. The differences between the levels are the multipliers. Identical index numbers are applied on all levels. The multipliers are as follows:

1. Basic sequence models (basic MOST) = multiplier 10.
2. Bridge cranes and wheeled trucks = multiplier 100.
3. Job preparation, and so on = multiplier 1000.

See Sec. 4, Chap. 5, for more information on MOST.

Universal Analyzing System (UAS).   MTM-UAS was developed between 1976 and 1978 by a consortium which included the German MTM Association, the Swiss MTM Association, and the Austrian MTM Group. The data were developed with the following objectives in mind:

- High analyzing speed
- Reproducibility of methods
- Sufficient accuracy
- Universal application

The MTM-UAS analyzing system consists of seven activity groups:

- Get and place
- Place
- Handle tool
- Operate
- Motion cycles
- Body motions
- Visual control

In addition, standard data are available covering the following areas:

- Fastening
- Assembling standard parts
- Transport
- Marking
- Adhering/bonding
- Cleaning
- Packing
- Applying agents
- Assembling cable and harness

The time elements comprise a motion sequence which is well defined, that is, get and place an object. All time elements also contain necessary "auxiliary motions," such as apply pressure, disengage, and regrasp.

MTM-UAS achieves ±5 percent accuracy with a 95 percent confidence level at an absolute balance time of 5.34 minutes. Manual application times for MTM-UAS ranges from approximately 15 to 30 times the cycle time depending on the extent of use of second-level data.

Modular Arrangement of Predetermined Time Standards (MODAPTS).   MODAPTS (modular arrangement of predetermined time standards) was developed by the Australian Association of Predetermined Time Standards and Research (AAPTSR). This association was formed in 1964 to research possible improvements upon the two major PMTSs employed in Australia at the time: MSD and MTM, plus MTM-2 when it was introduced by the International Directorate in 1965. Chris Heyde acted as general director of the Australian Association. The basic research in developing distance-time matrices and spatial arrangements from this study of the three systems led to the development of MODAPTS, a "condensed" system.

The basic unit in MODAPTS is a simple finger movement. All other activities are expressed in terms of this finger movement or module. There are only eight different values: 0, 1, 2, 3, 4, 5, 17, 30 MODS. These eight MOD values are applied to 21 types of activities derived from movements of fingers, limbs, body, and eyes.

MODAPTS elements are presented in three groups: movement elements, terminal elements, and supporting elements. There are elements for both small/light objects and for large/heavy objects.

In MODAPTS the first identification is the class of movement, and the second tag or label is that which happens at the end of the movement, or the "terminal activity." There are two categories of terminal activities: "obtaining control," and "things to destination." Each category has three different MOD values, which are selected based on the type of terminal activity.

The unit of time used by MODAPTS is called a MOD. (MODAPTS does contain a few fractional values, but not many.) The developers made every possible attempt to avoid fractional values to make the system simple to memorize and apply.

The difficult task was that of determining the correct MOD value for each motion. This was done by first assigning all reasonable values to each motion. Then, the one found to produce satisfactory results was selected.

Every motion in MODAPTS is identified by a two-part code. The first part is a letter of the alphabet which indicates the body part involved. The second part is a number which, when multiplied by 0.129 second, is the time awarded to the motion. Because of the simple coding system, and the fact that there are so few elements, MODAPTS can be completely memorized in 2 days.

Current research efforts in MODAPTS improvements are focused on the training manuals. This is being sponsored by the International MODAPTS Association located at Western Michigan University in Kalamazoo, Mich.

Macro Motion Analyses (MACRO).   MACRO is a condensed system, with motion-time values at "required" time. It was developed by Standards, International Inc. as a companion system to MICRO, which is a detailed system.

MACRO provides a quick analytical tool for setting standards on work with varying motion patterns. Also it is recommended where accuracy of ±10 percent is satisfactory. MICRO and MACRO are interchangeable. MICRO can be used on those elements where accuracy of ±5 percent is desired, MACRO on those elements where greater tolerance range is acceptable. The MACRO tables provide time values for Obtain and Place; Place Only; Assemble Plug and Target (Loose Fit and Normal Fit); Circular Motions; Walking; Visual and Mental Processes; and Tool Handling.

PERFORMANCE LEVELS

Some of the systems MTA, W-F, MICRO, MODAPTS, and MACRO time values are based on "required" (incentive) time. Most of the predetermined time systems are based on "average" (daywork) time. Which is the best for your company?

The tables are interchangeable:

Adding 25 percent to "required time" equates to "average time" (or multiplying by 125 percent).

Subtracting 20 percent from "average time" equates to "required time" (or multiplying by 80 percent).

Practicality of "Required (Incentive) Time."   In incentive shops, there is an advantage to adding a percentage for incentives. This clearly identifies the incentive percentage that the average operator can earn when working with good skill and dedicated effort.

There is also an advantage in starting with "required time" and adding 25 percent. The 25 percent addition helps convince the operator and the supervisor that the standards are adequate, even generous.

There is also an advantage in setting the performance standards at "required time"— even for daywork operations. It is possible for the average experienced person, working with good skill and dedicated effort, to do the work at "required time." Managers who prefer "required time" standards for daywork operations identify the difference between performance at required time and performance at "daywork time" as managerial waste. They identify performance below required time as waste that could be eliminated by better selection and assignment of employees, better scheduling, better training, and better supervision.

THE USES OF PREDETERMINED TIME STANDARDS

Performance standards have many uses. When set realistically, and tied to work methods and sequences, they are valuable tools for

- Work scheduling
- Cost estimating
- Operator evaluation
- Operator training
- Incentive and gain sharing programs
- "What if" evaluation of work methods
- Evaluating employee skills
- Evaluating labor requirements
- Determining production capacity
- Settling grievances
- Providing yardsticks for impartial supervision decisions

Predetermined time standards give management more confidence in using the company's performance standards for the above items. Predetermined time standards are more consistent than performance standards set by other methods. Therefore, supervision can use predetermined time performance standards with more confidence and assurance. Since predetermined times are methods-based, they also give supervision more facts for "what if" evaluations involved in the above uses of standards.

Developing Effective Methods in Advance of Beginning Production.   The problem of developing the best method in advance of performing an operation has been perplexing and time consuming. However, techniques in the field of predetermined time standards have given tremendous aid to engineers in deciding what method should be used to attain the least cost and the highest production per unit of time.

The time study groups of Taylor and the motion study groups of Gilbreth both emphasized the importance of using correct methods in the performance of a task. Job descriptions were given to operators to aid them in following the prescribed methods. But the one item lacking in these early job descriptions was that of time.

It was not easy to compare one method with another from a time standpoint until after the job was placed in production. The procedures of Taylor and Gilbreth considered method and time separately, but it is evident that the method used determines the time required to perform a task. With predetermined times established for the various motions necessary to perform a task, comparing one method with another becomes relatively easy, even before the job is placed in production.

Preproduction methods will not eliminate all changes in methods, because a certain amount of change is inevitable in industry. Also, it is often necessary to see the part in production before a change may suggest itself. These production changes will, in the long run, be beneficial. But a great many of these changes can be anticipated by proper methods study before production starts.

Before a study of a particular production method is begun, management services such as schedule control, tools and equipment, working conditions, and supervision should be considered, so that the worker will not be delayed even though the correct production method is installed. From this preliminary survey, it will become evident if distances between motions are longer than they should be, if mechanical devices can be used to replace manual motions, and if some motions can be eliminated.

When industrial engineers approach the problem of engineering effective methods in advance of beginning production, they are confronted immediately with a number of different ways of performing the same job. By visualizing the movements necessary to perform the operation, they can develop the best workplace layout, determine the type and position of tools, and prepare an instruction sheet for training the operator in the best method. This is true because the use of predetermined times for each required motion allows the industrial engineer to determine the cycle time for the best method. Where machine time is necessary, it is added to the motion time to get the cycle time per unit.

Establishing Time Standards. Establishing correct time standards for industrial operations is important to several necessary phases of successful manufacturing. Different phases in which time standards can be used to advantage include the following:

1. Basic management record of time to perform operations
2. Cost estimate to get business
3. Cost check on measured productive labor
4. Line balance of operations
5. Calculation of number of machines an operator can use effectively
6. Calculation of load in plant for scheduling purposes
7. Basis for incentive pay
8. Calculation of percent efficiency of labor operations
9. Determination of correct method
10. Time formula derivation

Thus, when one considers all the varied uses of time standards, one can see that they are essential to the day-to-day operation of a successful business. The technique of using predetermined times for establishing time standards can be summarized as follows:

1. Securing necessary information
2. Dividing operation into elements
3. Dividing elements into motions
4. Applying predetermined times to each motion
5. Determining allowances (personal, fatigue, and unavoidable delays)
6. Calculating the standard time

Recording the necessary information involves a description of the method of work, material characteristics, a sketch of the piece and workplace layout, and machine characteristics.

Elements should begin and end at well-defined points in the cycle and should not include more than twenty motions. Ten are preferable.

Manual handling should be separated from machine time. Where power feeds are used, cutting formulas can be applied as a check against stopwatch times for the machining part of the cycle. The separation of manual from machine time is essential for the construction of formulas.

Constants should be separated from variables. A constant is independent of the size and weight of a part, while a variable depends upon part characteristics.

After the elements have been divided properly, the motions for each element must be determined. These motions must be broken down into various reaches, grasps, moves, positions, releases, and other basic motions.

When the predetermined times are applied to each motion, the elemental times for the operation can be established. By adding the elemental times together, the normal time for the operation is determined. Then the allowances necessary for fatigue, personal, and unavoidable delays must be added to the normal time to get the standard time for the operation.

DEVELOPING STANDARD DATA

Standard data is work simplification applied to predetermined times. Predetermined time analyses identify a specific method and the times for that method. Standard data is designed to cover not one but many operations belonging to common groups and common work methods.

Standard data involves cataloging predetermined time analyses into work sheets or spreadsheets so that a time standard can be set for a job by (1) determining the work elements and time values for the job, (2) selecting the elements and time values from the standard data work sheet or spreadsheet, and (3) adding the controlling time values for the work elements to get the total time for the job.

STANDARD DATA DEVELOPMENT

Standard data catalogs are developed by:

- Studying the work methods
- Classifying the work methods significant variables, such as distances reached or moved; size of parts, weight of parts, and so on
- Arranging the methods and times in a work sheet or spreadsheet that is convenient to use

To avoid duplication of data, elemental descriptions should reflect methods, not items. "Obtain and aside pliers" specifies an item. "Obtain and aside tool, wrap grasp" reflects the method of grasping and could apply to many objects, including pliers.

Standard data developed from predetermined times can be built with the tolerance required: whether 2, 5, 10 percent, and so on. The tolerance percentages can be established in advance and modified as conditions change.

Standard data should be built by trained predetermined time analysts. The standard data should be built so that it can be used effectively by persons without formal training in predetermined times, if they are acquainted with the work operations, and if the standard data are expressed in customary terminology for the work area. See Sec. 4, Chap. 6, for further information.

Training Supervisors to Become Methods Conscious.   A common definition of methods study is the analysis of operations so that the maximum output of the desired quality can be produced in the shortest time.

The purposes of methods study are generally considered to include the following:

1. Elimination of unnecessary motions
2. Reduction of effort and fatigue
3. Improvement of working conditions
4. Training of supervisors and operators in correct methods
5. Reduction of time by better methods
6. Design of necessary controls so levers are in normal work areas
7. Reduction of long motions by keeping activity within easy reach

The efficient supervisor must be able to teach the correct method to the worker, help solve mechanical difficulties, and give and encourage cooperation. To teach correct methods, the supervisor should be able to break a job down into the motions required, arrange the workplace to correspond to these motions, select the best worker for the job, explain the operation to the worker in terms of motions, and demonstrate the method.

It has been stated by competent engineers that of the saving in time from methods analysis, 20 percent comes from the workplace layout and 80 percent comes from operator training in the use of the correct method. Proper training of the operator by the supervisor is essential, because each operator will otherwise perform differently. Some are well coordinated; others are not. Bad work habits should be avoided by setting up correct methods at the start. Correct methods usually result from a study of the motions used by the operator. Faculty research project No. 420 at the University of Michigan showed some of the reasons for the differences in time between fast and slow workers. It was observed, through film analysis of several operators, that the slow worker used:

1. Distinct and abrupt changes in direction
2. No eye anticipation
3. No merging of movements such as grasp and move
4. No circularity of movements
5. High arcs for reach and move above the working plane

The fast operator used the reverse in each of the five points. To teach correct methods, the supervisor must recognize and correct faulty adjustment to an operation, and then establish precise motion habits.

The supervisor can be trained to observe the practice of faulty methods through the use of predetermined elemental times. A supervisor who becomes familiar with the times for various motions will be able to reduce motions to a minimum and attain the best method for the job. Inefficient motions such as transfer grasps can often be eliminated by a change in workplace layout. Training in elemental times can be extended to tool designers, time study engineers, supervisors, and others interested in better methods and least cost.

FUTURE TRENDS IN LABOR STANDARDS

There are many contradictory opinions about the future: (1) as to the need for labor standards and (2) as to the standards techniques that will be used. Many people claim labor standards are unnecessary, a waste of time and money. However, there are so many different uses for labor standards that every company is apt to find at least one or two functions that do need current labor standards. For example:

- Marketing needs current labor cost standards to know how much a product or service costs if they are to price the product or service competitively yet profitably.
- Production control, supervision, and personnel need to know how many people will be needed to achieve a required production schedule or staff a service function.
- Productivity management needs current labor standards. "Creeping methods changes" are a universal phenomenon. As a result, performance that once reflected dedicated effort may now reflect poor or inadequate skill and effort. Updated labor standards and improved productivity are needed if a company wishes to survive in a competitive world.

Companies in countries with captive and thus less competitive markets frequently can continue to exist for a while without current standards and without good productivity. As international competition begins to take away customers, the inclination is to simply lower prices. But on which products? The company needs to know which are the highly profitable products on which they can lower prices and still make a profit. They need to know both parts and labor costs. Parts costs are easy to determine. Not so labor costs. Current labor standards are needed for good productivity and for good cost information.

- Manufacturing engineering needs current labor standards. How can they set up a production line if they do not know how long it takes to perform each component operation? How can they tell whether to make or buy? How can they balance the line for optimum productivity and minimum loss to balance? How can they set up production schedules without current information on labor requirements? How can they establish staffing schedules for service functions without knowing how long certain routine procedures require?
- Despite all the predictions that labor standards are passé, three things are certain:

1. Labor standards are an integral part of the future.
2. Computerized labor standards will be an increasingly important technique for setting labor standards (see "Computer-Aided Work Measurement" section of this chapter for information on this vital subject).
3. Predetermined time techniques will continue as an important technique in setting manual labor standards.

COMPUTER APPLICATION OF PMTS

The use of computerized labor standards simplifies enormously the cost involved in setting standards. But up-to-date labor standards are required as a good base for a good computerized standards program. Predetermined time standards are one of the most effective ways of achieving reliable, consistent computerized labor standards.

While predetermined motion time systems (PMTS) have existed since about 1920, the application of PMTS using computers started in the early 1970s. Many systems have been developed for use on the PC or microcomputers since the middle eighties. The objective of this presentation is to:

1. Review the advantages of using computerized PMTS
2. Explain what to look for in a system
3. Develop criteria for selection

It is assumed that the reader has a knowledge of some PMTS systems so that no explanation of any symbols used will be needed. Further, no one system will be reviewed in detail but emphasis will be placed on the good points.

Some salespeople emphasize that the computer version of a PMTS gives an engineered standard and is therefore correct and right. One salesperson, when pressed as to how one would know that one had a good method and being reminded of engineering failures, replied that it depended upon the analyst's industrial engineering knowledge. The analyst, therefore, must be methods conscious and have had training in the application of the PMTS prior to its use on a computer. If the analyst cannot produce a good method or analysis manually, the computer will not do so either, that is, garbage in, garbage out (GIGO). The PMTS analysis should be used to develop good methods rather than solely to determine a time for an operation.

The importance of methods is well illustrated in MIL-STD-1567A. The toughest requirement is not accuracy but rather the 80 percent touch-labor coverage. (Touch-labor is often identified as "direct labor.")

The requirement for an acceptable standard is that it be accurate within ±10 percent with a 90 percent confidence level. Stopwatch time studies, standard data, and PMTS, when correctly applied, meet or exceed this requirement. However, it is easier to verify that the PMTS has been correctly applied. These coverage requirements are seldom met in stopwatch studies, where the focus is apt to be more on reading the watch than on documentation of methods. These coverage requirements tend to be met automatically with PMTS. The methods must be documented for a PMTS standard to be set correctly.

A list of the PMTS evaluated and validated, together with a number of PC computerized implementations, can be found in USAF report AD-A108 205 titled "Manufacturing Work Measurement System Evaluation—Final Report and Reference Guide." Among the validated systems were MTS, the MTM family, the MOST family, MODAPTS Plus, MANPRO, CUE, MSD, Work Factor, UNIVEL, and AM Cost Estimator. Among the list of PC systems, some are no longer sold, such as ADAM, and the report does not include others now available, like FAST family.

The most common or popular computer systems are based upon the MTM family, MOST, and more recently MODAPTS. In this case, two or more different companies have developed computer PMTS applications on each of the above PMTS.

This brings up the next question. Should one build one's own computer system based upon a PMTS, which you know, or should a commercial system be purchased? While it is possible to build your own system, the time required to develop, debug, modify, and produce a manual for a minimal computer system would probably exceed 200 hours (spread over 8 to 9 months) and using a conservative estimate of $30 per hour, the cost would exceed $6000, which is in the range of the lower-priced commercial software. Also, in view of the rapid continuing advances in software and the emphasis on ease of data entry, screen displays, and the integration of information into other parts of the company operations, it would be better in most cases to investigate performance, price of commercial software, and also upgrading and/or maintenance costs.

Advantages of Computerized PMTS.   There are two types of computerized PMTS, those systems which are concerned only with the PMTS and those which are concerned with integrating the PMTS into a company's database. An example of the first type is Taskmaster, and examples of the second type are FAST, MOST, 4M, and EASE. Some of the latter type may require more than one module (an extra expense).

Although there are some distinct advantages in using a computerized PMTS over the manual version, the main advantage is when it can be integrated into a company's database. This is illustrated in Fig. 4.4.

We first concentrate upon the singular advantages of the computerized PMTS over the manual one. First, the computer PMTS will add up the times correctly and apply the al-

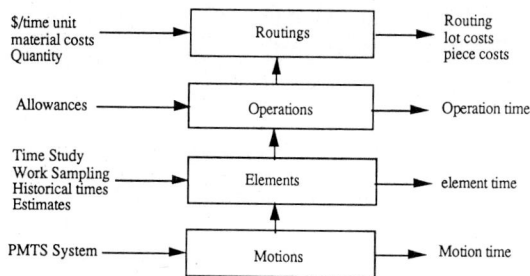

FIGURE 4.4 Integration of PMTS in company database.

lowances to determine the standard. Second, the computer will take a set of motions to form an element. These motions will be frequently performed sequences of motions in a large number of elements, such as get, move, and place screwdriver in slot, where the screwdriver could be either manual or power-driven. The time of this element is constant no matter what the operation is. Elements and motions are then brought together to form the operation. Next the computer will calculate the times for symmetrical operations and weight factor, for example:

In 4M, the weight factor for a move and place is given by

$$\text{Time} = 5 + 3.5w + 0.011wT$$

where w = weight per hand
T = time, no weight

and for cranks

$$\text{Time} = 52 + nt \qquad \text{continuous cranking}$$

$$= nt \qquad \text{intermittent cranking}$$

where n = number of revolutions
t = time per revolution; depends on diameter of crank

This is a big improvement over doing it manually.

Next, standards can be built faster, as editing will allow for the correction of mistakes and enable updates to be made faster. Standards are automatically updated at the time the editing is carried out or by a special update procedure for all operations.

There are three distinct types of approaches to the analysis:

1. MOST—uses a workplace layout noting the locations for parts, tools, and so on, distances between locations, and other information. The analysis is based upon the workplace and is based upon a sentence structure of nouns (objects), verbs (action, what is done), and prepositions. The time is calculated using the information for the workplace layout and the syntax of the sentence.

2. Most other methods use the motion symbols, such as P310-2 (4M), PC2 (MTM-2), and V3 (MODAPTS), and a description of the operation. If a workplace layout can be drawn, it is used as a guide and not as an integral part of the system calculation.

3. Some systems, such as EASE (MTM-2), use only the symbol and no descriptions are used. In this case one has used the routings and one has to have a knowledge of the process in order to know what the descriptions should be. In addition, the analyst has to make decisions on how simultaneous motions are to be calculated, that is, whether motions can be done simultaneously or whether some adjustment is needed.

MOST is the only software that has used a workplace as an integral part of the analysis, but a new system MODCAD appears to follow the same principle.

Sample Problem.   The left hand reaches 13 inches and grasps a 3/4- by 1/8-inch bolt jumbled with other bolts. Simultaneously the right hand travels 10 inches and picks up a nut. Both hands now travel 7 inches, and the nut is placed on the bolt.

The results for a 4M analysis are shown in Fig. 4.5. The first statement is comprised of two G42 motions with times of 245 MU and 230 MU. The net manual motion is 351 MU comprised of G42-10 and G42-F as the two grasps cannot be performed simultaneously. The second statement is comprised of a P01-7 and a P310-7 with times of 89 MU and 206 MU, a time of 206 MU. In this case both motions can be performed simultaneously.

There are some other features to be noted in Fig. 4.5. The GET, MOVE, and PLACE notations are automatic literals and do not have to be included in the written description, that is, for ''GET BOLT'' only ''G42-13'' and ''BOLT'' was entered into the computer. The process time in this analysis is 0, but where it does occur and the times for motions, which are performed at the same time as the process occurs, are not included in the total time.

The MAI, RMB, GRA, and POS are unique to 4M, and are classed as the Motion Assignment Index; Grasp, Release, and Apply; Position; and Reach, Move, and Body Motions. The percentage of time spent on each index is recorded at the bottom of the analysis. If the percentage for one of the indexes is high, it could call for a methods improvement. For example, if the GRA was high, it would indicate a large number of complex grasps and a methods improvement would be called for to reduce the number of complex grasps.

```
4M UNIVERSITY SYSTEM   NN715B   4M ELEMENT ANALYSIS        REQUESTED BY BMW  TIME 11.47.16  DATE 08/17/90   PAGE 1

ELEMENT ASBEOP01    JOIN BOLT TO NUT                          LEARNING LEVEL 100
                                                                     TOTAL MU - MANUAL         557
              PRACTICE OPPORTUNITY              UPDATED                       - PROCESS          0

                                                                 PROCESS                  NET
         LH MOTIONS                        RH OR BODY MOTIONS   FREQ.  TIME   LH   RH   MANUAL
010    GET    BOLT       G42-13  G42-10  GET    NUT           1.000         245  230    351
020    MOVE   BOLT       P01 -7  P310-7  PLACE  NUT OR BOLT   1.000          89  206    206

                                                         TOTAL MU         334  436    557

MAI  69 PERCENT          RMB  62 PERCENT  GRA  26 PERCENT  POS  12 PERCENT              PROC  0 PERCENT
```

FIGURE 4.5   4M element analysis.

The result for a MTM-2 analysis using FAST is shown in Fig. 4.6. In this case the first line is comprised of a GC18 and a GC12 motion. The motion is reduced to a GC18 (27 TMU) and a GC2. The in the column prior to the GC12 denotes that the time will be reduced and the ''o'' in the column after frequency denotes the overlap time is included. The second statement is comprised of a PC12 (30 TMU) and a PA12 (11 TMU). The in the column prior to the PA12 denotes that the time for this motion will be reduced to zero, as

```
ASBEOP02 JOIN NUT TO BOLT                                             0.043
GRASP NUT AND BOLT WITH SEPARATE HANDS AND JOIN TOGETHER
GET A NUT AND BOLT                              GC18  \GC12    41o    0.025   1
PLACE NUT ON BOLT                               \PA12  PC12    30     0.018   1
```

FIGURE 4.6   An MTM-2 analysis.

the motions can be performed simultaneously, giving a total time of 30 TMU. The next column gives the time in minutes, and the final column is the frequency.

An example using MODAPTS is a partial analysis using Taskmaster (Fig. 4.7). The V3, D3, and the like, denote the motions. Taskmaster and FAST (MODAPTS) versions are somewhat similar to MOST in that several motions can be put on the same line, such as M5G1, M3P0, M1G1, M2P2. These motions are supplemented by a written version describing the motion. The time for MODAPTS is measured in MODS. This is converted to seconds, minutes, or hours.

```
┌──────────────────────────────────────────────────────────────────────────────┐
│ PUT OUT THE CAT                                                                │
│ TaskMaster Case Study 1                                                        │
│      Station:  3-B                              Operator : Brian Worrall       │
├──────────────────────────────────────────────────────────────────────────────┤
│                                                                    Freq.       │
│                                                         Freq.      X MODS       │
│  Step      MODS Value                                                          │
│                                                                                │
│  1  V3                                                90.000       270.000      │
│     Vocalise about who puts out the cat                                        │
│                                                                                │
│  2  D3                                                10.000       30.000       │
│     DECIDE it must be my turn                                                  │
│                                                                                │
│  3  S30                                                0.500       15.00        │
│     SIT and STAND = get up from my soft chair, plush,                          │
│     extremely comfortable, and expensive easy chair                            │
│                                                                                │
│  4  W5                                                12.000       60.000       │
│     Walk to cat                                                                │
│                                                                                │
│ Total MODS:   375                                                              │
│                                                                                │
│ 1 Cycle time in:     Secs.        Mins.        Hrs.        Cycles/hr.          │
│ No Rest              48.375       .806         .0134       74.4186             │
│ W/ Rest              53.575       .893         .0149       67.1955             │
│ Rest Allowance: 10.75%                                                         │
└──────────────────────────────────────────────────────────────────────────────┘
```

FIGURE 4.7   MODAPTS analysis.

Design of System Software.   There are a number of desirable attributes to look for in the software. The first and foremost is the ease of handling data, like, how do you find your way around the software. This is one area of notable improvement in the more recent versions of software available. The second is that the analysis should be easy to edit, that is, correct mistakes, add or delete activities. There are two types of errors:

1. Nonlogic error is one in which the incorrect motion is entered, but is actually an allowed code.

2. Logic error—the activity does not exist.

The computer should pick up the logic errors. Some systems such as 4M highlight the error, and the error has to be corrected before the data entry can be made. FAST (MTM-2) operates in a similar way. Other systems wait until the analysis is completed before the error detect procedure is implemented.

A personal preference is to be able to see the whole analysis, that is, all lines, and to be able to edit the analysis in the same way as word processing. Only FAST and MOST have this capability. However, one difficulty with MOST is that if you have analysis data stored, you are unable to change the workplace without duplicating it and giving the duplicate a new name.

Another essential part is to have good and easy manuals to read, with clear instructions. Fortunately, they are getting better. One does not want to spend hours going through a manual to correct a problem which has arisen. In this connection one should ensure that one can get service and backup from the selling organization. This has not been a problem with the systems with which the author is familiar.

The next point is the directory. If one is building element and operation standards, it is important to be able to see the list of standards, subcomponents, or elements and which operations they are listed in. The directory listing should give both the file name and any other key word that is needed to access, for example, in 4M, the file name (code), and the operation number has to be known to call up the operation standard. Most software needs only the file name.

It is also important to be able to add times such as process times, times determined by time study, and times calculated from formulas. Most systems have the ability to insert process times, but not all are able to use formulas, that is, the time required to wind a wire on a core, which would depend upon the diameter, the length of core, the diameter of the wire, and the speed of rotation. If the size of the core varies, one would not want to have to calculate the time for every core size, and the calculations should be done by the computer.

Lastly, the PMTS system selected depends upon the volume of production, cycle time, the need for accuracy, and the urgency for the solution. The system should be able to handle more than one PMTS system. It would usually be from the same family. The following systems reflect an accuracy of ±5 percent at 95 percent confidence for cycle times of the following duration:*

MTM-1,4M   700 TMU
MTM-2   1473 TMU
MINI- MOST   500 TMU
BASIC MOST   3245 TMU

This is much tighter than the relatively easy ±10 percent at 90 percent confidence level allowed by MIL-STD-1567A.

At cycle times below these levels the accuracy decreases (range $\hbar \pm 5$ percent) and the confidence level decreases. Therefore, we would use MINI-MOST and 4M for high-volume, low-cycle time products, BASIC MOST and MTM-UAS (4M) would be used for low-volume, small-lot-size operations. It is usually much faster to do an analysis using a lower accuracy PMTS, and therefore one may decide to select speed over accuracy.

System Operation.   A typical system can be represented as shown in Fig. 4.8. The data card is not usually called up onto the screen. The screen viewed will depend upon the activity selected. Upon initiating the program, a menu will appear for the analyst to decide upon a course of action. The illustration shown in Fig. 4.9 is for 4M, Fig. 4.10 for FAST, and Fig. 4.11 for Taskmaster. Some systems will allow the analysis to be carried out be-

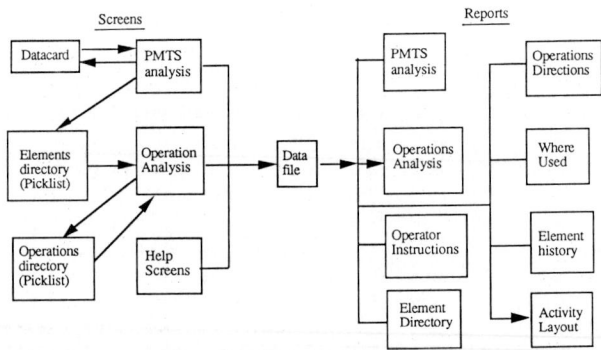

FIGURE 4.8 Diagrammatic representation of computer system.

*Source: Manufacturing Work System Measurement—System Evaluation USAF AD-A108 205, NTIS, U.S. Department of Commerce, Springfield, Va., 1987.

```
4M UNIVERSITY SYSTEM                                    09/20/90 14.44.35 (S000XA)
                              - - - 4M MASTER MENU - - -

11 - ELEMENT MASTER FILE MAINTENANCE          70 - 4M ANALYSIS REQUEST
21 - ANALYSIS DETAIL FILE MAINTENANCE         80 - OPERATION STANDARD REQUEST
31 - OPERATION MASTER FILE MAINTENANCE        90 - WHERE-USED REPORT REQUEST
41 - OPERATION ELEMENT FILE MAINTENANCE       95 - WHERE-USED SCREEN DISPLAY

12 - ELEMENT MASTER FILE DISPLAY              13 - ELEMENT MASTER FILE PRINT
32 - OPERATION MASTER FILE DISPLAY            33 - OPERATION MASTER FILE PRINT
42 - FORMULA ELEMENT DISPLAY                  43 - FORMULA ELEMENT FILE PRINT
52 - STUDY INDEX FILE DISPLAY                 53 - STUDY INDEX FILE PRINT
72 - SUB-ELEMENT FILE DISPLAY                 73 - SUB-ELEMENT FILE PRINT

66 - ELEMENT COPY                             88 - PRINT 4M REPORT FILES
77 - OPERATION STANDARD COPY                  94 - MASS/CHANGE REQUEST
00 - 4M/PC UTILITY                            99 - END 4M PROCESSING


PLEASE SELECT ACTIVITY

F1 - RESET SCREEN              F8 - SUBMIT DATA
```

FIGURE 4.9 Menu screen for 4M.

```
FAST w/MTM2   STUDENT VERSION  -  MAXIMUM 100 WORKSHEETS

Filename:   FASTDEMO
LastAdd:    09-06-1990
LastChg:    09-06-1990
Password  level 0


        ┌─────────────────┐
        │   2 PROCESSES   │
        └─────────────────┘
                 ▲
        ┌─────────────────┐
        │  84 ELEMENTS    │
        └─────────────────┘

New         :enter new element or process worksheet
Existing    :edit/print existing element or process worksheet
Directory   :list existing element and/or process worksheets
Whereused   :list/update whereused process worksheets
Update      :mass update all process worksheets
Quit        :return to DOS

                        Press F1 for help screens
```

FIGURE 4.10 FAST menu screen.

```
  FILE    EDIT    PRINT    OPTIONS        FILE: CATOUT.TSK

                PUT OUT THE CAT
                TaskMaster Case Study 1
          3-B                        Brian Worrall

Step Number: 3                      Frequency: .5
Description:   SIT and STAND =get up from my soft chair, plush, extremely,
               comfortable and expensive easy chair


MOD string: S30

Step Totals    MODS      SEC.    MINS.    HOURS       CYCLES/HR
No Rest        15        1.935   .032     .0005       1860.465
W/ Rest        16.612    2.143   .036     .0006       1679.888

MOVES, GETS, PUTS  OFFICE 1   OFFICE 2  SPECIAL 1  SPECIAL 2  RA%:10.750
        SMALLER OBJECTS          LARGER OBJECTS              IN
        MOVES  GETS  PUTS        MOVES GETS PUTS          GENERAL
No Rest    375        48.375     .806       .0134      74.4186
W/ Rest    415.312    53.575     .893       .0149      67.1955
```

FIGURE 4.11 Taskmaster screen.

fore the file is named. In the case of 4M the element code and name has to be entered first and a similar procedure is used for the operations standard.

4M has 24 activities. The various directories are covered by activities 12, 32, 42, 52, and 72. In order to carry out a 4M analysis, the code and name for the element is entered through activity 11 and analysis can then be entered by invoking activity 21. A similar sequence of events is carried out for developing an operation standard using activities 31 and 41. To develop a time standard activities for the element and the operation, activities 70 and 80 are used.

The FAST menu gives the number of processes (or operations) and the number of elements stored in the system. The next screen is invoked by entering N, E, D, W, U, or Q, which is the first letter of activity chosen. The analysis shown in Fig. 4.6 is exactly as it appears on the screen. The help screens can be called up at any time with no loss of information.

Taskmaster operates differently. A screen ready to start the analysis is presented. The activity shown in Fig. 4.11 represents the third statement in the analysis shown in Fig. 4.7. The new files are given the name of NEWTASK.TSK which is altered just prior to saving the analysis. The FILE, EDIT, PRINT, and OPTIONS represent pull-down menus. The help files are listed at the bottom of the screen, and one clicks on the appropriate help file.

After the course of action is decided, such as a PMTS analysis will be carried out, a new screen will be presented allowing the information on the analysis to be entered. Some systems calculate the time immediately (FAST, Taskmaster), while others (4M) calculate it later. After the analysis is complete, the information is saved. Both Taskmaster and 4M use a new screen for each line of code.

The reports and the ease of generating reports are an integral part of the system. The operations standard is prepared in a similar way to the elements. In the operation, one enters the elements rather than the individual notions, although sometimes individual motions can be entered as well. At this stage the allowances are usually added. The operation standards for 4M and FAST are shown in Figs. 4.12 and 4.13. The appearance of the 4M report has had to be slightly altered, but it contains all the relevant information. The operation standard for Taskmaster is shown in Fig. 4.7; that is, it is the same as the analysis. Some systems require 132-column printers, but most require only an 80-column printer.

```
4M UNIVERSITY SYSTEM  R5815B   OPERATION STANDARD REPORT

PART  NUTBOLT           NUT AND BOLT JOINED                ORIGINATED 08/17/90 BY BMW
OPERATION 0000001 JOIN NUT TO BOLT                         REVISION    / /   BY
DEPT. ASSEM   MACHINE       TOOLING MTM-4M UNIVERSITY DEMO VERSION
ALLOWANCE  - MANUAL  .100              STANDARD HOURS - SETUP  .00000  RUN  .00061
           - PROCESS  .100                        UNITS PER HOUR 1631.32

          ELEMENT                    PRACTICE                        EXTENDED
          CODE        DESCRIPTION    OPPORTUNITY  TIME   FREQUENCY     TIME

                       -- RUN --

   010    ASBWOP01  JOIN NUT TO BOLT      Y        557     1.0000      557

          TOTAL RUN MU  - MANUAL  557          WITH ALLOWANCE    613
                        - PROCESS    0         WITH ALLOWANCE      0
                                        TOTAL STANDARD RUN TIME   613 MU
                                               CYCLE QUANTITY    1.00
                                        UNIT STANDARD RUN TIME   613 MU
                                                               .00061 HOURS
                                                               .037 MINUTES
                                            UNITS PER HOUR    1631.32

MAI    69 PERCENT   RMB 61 PERCENT  GRA 26 PERCENT  POS 12 PERCENT
```

FIGURE 4.12   4M operation standard report.

```
ROUTING                                        Setup   Run    Wrkrs
12345678  ASSEMBLE NUT TO BOLT                 0.000   0.049C   1


--PROCESS WORKSHEET -----------------------------------------Setup----Run-----Wrkrs
SETUP$/HR:  15.00   RUN$/HR: 7.5   LOTSIZE: 1500                              1
--OPERATION 1----------------------------------- SWM-----Setup----Run-----Wrkrs
ASSEMBLE NUT TO BOLT                               0     0.000    0.049      1
ASBE0P02  JOIN NUT TO BOLT                                        0.043      1
```

FIGURE 4.13   FAST process standard and route sheet.

Element and Operation Codes or Names.   If one is going to take elements and operations and later build them into larger operations one has to be able to view the element and operation directories in order to select the appropriate one. This means that the element coding or name of the unit is important.

The number of characters allowed to make up an element code varies from 8 to 20 (one supplier allows up to 80). Much thought should be given to setting up the code so that it is meaningful and will enable the elements viewed to be selected in a meaningful way. A great deal will depend upon the system to be used. If each analysis is stored as a computer file, an 8-character code would be selected. One code could be selected as

UULAMD12

1, 2 Department, for example, machine shop

34 Type of machine, for example, lathe

56 Class of operation, for example, machine rod to correct diameter

78 Use of operation, for example, allows for variations in length and diameter of rod

Other variations could be used. One should be careful in setting up the codes to allow for expansions. If one uses a numeric system only, the above would give 10^8 possible codes. Using alphanumeric codes, and omitting 1 and 0, the choice would be 34^8. Although one would never use all the possible codes, one should give thought as to how they will be used. The alphanumeric codes can be easily deciphered by the mind without having to look at the coding system.

Reports. The two primary reports needed are the PMTS analysis report and the operations report. Other ancillary reports needed are "Where Used" to determine where elements and suboperations are used, and an "Activity Log" depicting the history of elements and operations. A useful additional report is an operator instructions report giving the details of the operation but not the times. Some systems also have the ability to produce graphics of the workplace. This may not be true graphics, but a use of regular keyboard characters to give pseudo-graphic layouts. One can expect that better graphical layouts will be developed in the future, as was done with MOST.

Miscellaneous Items. Some systems are looking at the development of "expert" systems, that is, EASE and Labour Standard Builder expert. In these systems the user is led through a set of questions and responses to determine the activity, activity class, distance, and the like. This is usually accomplished by up/down arrows and the enter key. However, if one is really familiar with the PMTS system being used, one might want to be able to dispense with the system and enter the code directly.

Most available systems go through the bottom 3 sections (motions, elements, operations) in Fig. 4.4. Some do have the ability to add process routings, although they may involve extra modules at extra cost. There is an emphasis in the latest developments to develop routings and process standards (EASE, FAST, and Labour Standard Builder) and to allow access from mainframe computers for MRP or costing systems. This is done by either printing the required information for manual entry into the mainframe routing system or sending them to disk file, which can be read when desired, on the mainframe system. The alternative procedure is to generate the routings, a combination of operations (see Fig. 4.14) and costing. Because it is linked, costings can also be obtained.

ROUTINGS 93		Setup	Run	Wrkrs.
87654321-010 BRIDGE/RIM ASSEMBLY	EJB	257.250	6.219~	1
87654321-020 SCREWPOST ASSY WITH BEARINGS, SPRINGS	J	0.000	5.559~	1
87654321-030 ASSEMBLE LEG TO BASE, INSTALL CASTORS	J	0.000	5.612~	1
87654321-040 ASSEMBLE SEAT TO BASE	J	0.000	6.999~	4
87654321-050 ASSEMBLE ARMS TO CHAIR	%JB	0.000	5.851~	1
87654321-060 PACK CHAIR	%JB	0.000	5.512~	1
	Cycle time (min.)		6.999	
	Cycles/hour		8.57	

printed 09-27-1990 07:53:02

FIGURE 4.14 Routing report using FAST.

Some systems, FAST and Labour Standard Builder, also provide data collectors to collect time study data which can be fed into the data manager for use with the PMTS system.

Line balancing is available with some systems. This enables elements and motions to be interchanged between workstations to balance the operators' workload. Some systems use graphics to illustrate the balance or lack of balance.

Selection Criteria. The following points should be considered in selecting a computerized PMTS system.

1. If you are familiar and satisfied with a given PMTS system, you will probably want to stay with that system.
2. If you are tied into another operation (plant) that is using a certain PMTS and standards are interchanged between plants, you may want to use that system.
3. Look at your operation, decide what is needed, and then look for a system which meets your requirements. A system that is suitable for one type of industry or firm may not be suitable for another.
4. Select a small set of problems to be analyzed using working demo disks. This will enable you to see how the system operates. Select a list of possible vendors and talk to them. If a working demo disk is not available, you will have to meet with sales representatives. Sales demonstrations look good, as the demonstrator knows the system intimately, but try to repeat what you have been shown! Make sure that you can use the system for your set of problems.

Cost of Systems. The cost is comprised of three parts:

1. Hardware cost
2. Software cost
3. Investigation and implementation cost

The hardware cost will be the smallest. Most software can be implemented on a PC type of computer. While most systems can be implemented upon a smaller and slower system, the minimum system usually purchased now is a PC-AT type of computer and 40-megabyte hard disk. The speed and disk storage is more than ample for many organizations. The software currently costs from around $1000 to about $40,000 and several are in the range of $3000 to $10,000. The personnel costs of investigating, selecting, and implementing a computerized PMTS will probably exceed the cost of the software. Although the cost of the personnel time is high, the time required should not be skimped because the decision will affect the future operations of the company.

CONCLUSIONS

It is impossible in this short discussion to discuss all the computerized PMTS. Accordingly important aspects of computerized PMTS have been illustrated using particular packages. All PMTS software packages have some advantages and disadvantages. It is up to the prospective user to decide upon which characteristics are important.

If one is using or contemplating using a PMTS, it is advisable to select a computerized system. First select the PMTS most appropriate to your operation and then select the computer software. In selecting software one should look at how it extends its uses beyond the PMTS system. Many attractive packages are available at reasonable cost. One should be able to select an appropriate system for the organization.

REGISTERED TRADEMARKS

MTS—General Electric Company

MOST—H. B. Maynard & Co. Ltd.

MODAPTS—Heyde Dynamics Pty. Ltd.

MANPRO—Methods Management

CUE—General Analysis Inc.

MICRO—Standards International, Inc.

MSD—Sirge A. Burn Company

WORK FACTOR—Science Management Corporation

UNIVEL—Management Science, Inc.

AM COST ESTIMATOR—Costcom, Inc.

ADAM—MTM Association

FAST—Royal J. Dossett Corporation

4M—MTM Association

Taskmaster—Artifacts Software

Labour Standard Builder—Applied Computer Services, Inc.

EASE—Ease Inc.

MODCAD—R. Wygant and R. Dawood

MTM—generally public domain or MTM association

MACRO—Standards International, Inc.

SUGGESTED READING

Computerized Work Measurement, Industrial Engineering and Management Press, Atlanta, 1984.

Crossan, R. M., and H. W. Nance, *Master Standard Data,* McGraw-Hill, New York, 1962.

Hamlin, J. L., ed., *Success Stories in Productivity Improvements,* Industrial Engineering and Management Press, Atlanta, 1985.

International MODAPTS Association, *MODAPTS Student Training Manual,* Western Michigan University, Kalamazoo, Mich., 1991.

Karger, D. W., *Engineered Work Measurement: The Principles, Techniques, and Data of Methods-Time Measurement, Modern Time and Motion Study, and Related Applications Engineering Data,* 3d ed., Industrial Press, New York, 1977.

Maynard, H. B., G. J. Stegemerten, and J. L. Schwab, *Methods Time Measurement,* McGraw-Hill, New York, 1948.

Presgrave, R., and G. B. Bailey, "Basic Motion Timestudy," in H. B. Maynard, ed., *Industrial Engineering Handbook,* 2d ed., McGraw-Hill, New York, 1963.

Quick, J. H., J. H. Duncan, and J. A. Malcolm, Jr., *Work-Factor Time Standards,* McGraw-Hill, New York, 1962.

Segur, A. B., *Motion Time Analysis Instruction Manual,* A. B. Segur Company, Oak Park, Ill., 1946.

Sellie, C. N., "Why Standards Programs Fail," *NSRP 1987 Ship Production Symposium Proceedings,* NSRP no. 0281, Paper no. 12, NSRP Publications Coordinator, Ann Arbor, Mich.

Shell, R. L., ed., *Work Measurement: Principles and Practices,* Industrial Engineering and Management Press, Atlanta, 1986.

Zandin, Kjell B., *MOST Work Measurement Systems,* Dekker, New York, 1980.

CHAPTER 5
MOST®WORK MEASUREMENT SYSTEMS*

Kjell B. Zandin

Senior Vice President—H. B. Maynard and Company, Inc.
Pittsburgh, Pennsylvania

Because industrial engineers are trained that with sufficient study any method can be improved, many efforts have been made to simplify the analyst's work measurement task. This has, for instance, led to a variety of work measurement systems now in use. These achievements also led us to examine the whole concept of work measurement to find a better way for analysts to accomplish their mission. This in turn led to a new approach later to be known as MOST—Maynard Operation Sequence Technique.

THE 'MOST' CONCEPT

Work to most of us means exerting energy, but we should add, to accomplish some task or to perform some useful activity. In the study of physics, we learn that work is defined as the product of force times distance ($W = f \times d$) or, more simply, work is the displacement of a mass or object. This definition applies quite well to the largest portion of the work accomplished every day, like pushing a pencil, lifting a heavy box, or moving the controls on a machine. Thought processes, or thinking time, is an exception to this concept, as no objects are being displaced. For the overwhelming majority of work, however, there is a common denominator from which work can be studied, the displacement of objects. All basic units of work are organized (or should be) for the purpose of accomplishing some useful result by simply moving objects. That is what work is. MOST is a system to measure work; therefore, MOST concentrates on the movement of objects.

Work, then, is the movement of objects—maybe we should add, following a tactical production outline. Efficient, smooth, productive work is performed when the basic motion patterns are tactically arranged and smoothly choreographed (methods engineering). It was noticed that the movement of objects follows certain consistently repeating patterns, such as reach, grasp, move, and position the object. These patterns were identified and arranged as a sequence of events (or subactivities) manifesting the movement of an

*Parts of this chapter have been extracted from the textbook, *MOST® Work Measurement Systems*, by Kjell B. Zandin with the permission and courtesy of the publisher, Marcel Dekker, Inc., 270 Madison Avenue, New York, New York.

object. A model of this sequence is made and acts as a standard guide in analyzing the movement of an object. It was also noted that the actual motion contents of the sub-activities in that sequence vary independently of one another.

This concept provides the basis for the MOST sequence models. The primary work units are no longer basic motions as in MTM (methods time measurement), but are fundamental activities (collections of basic motions) dealing with moving objects. These activities are described in terms of subactivities fixed in sequence. In other words, to move an object, a standard sequence of events occurs. Consequently, the basic pattern of an object's movement is described by a universal sequence model instead of an aggregate of detailed basic motions synthesized at random.

Objects can be moved in only one of two ways: either they are picked up and moved freely through space, or they are moved in contact with another surface. For example, a box can be picked up and carried from one end of a workbench to the other or it can be pushed across the top of the workbench. For each type of move, a different sequence of events occurs; therefore, a separate MOST activity sequence model applies. The use of tools is analyzed through a separate activity sequence model which allows the analyst the opportunity to follow the movement of a hand tool through a standard sequence of events, which, in fact, is a combination of the two basic sequence models.

Consequently, only three activity sequences are needed for describing manual work. The basic MOST work measurement technique therefore is comprised of the following sequence models:

- The General Move Sequence—for the spatial movement of an object freely through the air
- The Controlled Move Sequence—for the movement of an object when it remains in contact with a surface or is attached to another object during the movement
- The Tool Use Sequence—for the use of common hand tools
- A fourth sequence model—the Manual Crane Sequence for the measurement of moving heavy objects by using, for instance, a jib crane, is also part of the Basic MOST System, although used less frequently than the three first sequence models.

THE Basic MOST WORK MEASUREMENT TECHNIQUE

Sequence Models. General Move is defined as moving objects manually from one location to another freely through the air. To account for the various ways in which a General Move can occur, the activity sequence is made up of four subactivities:

A—Action distance (mainly horizontal)

B—Body motion (mainly vertical)

G—Gain control

P—Placement

These subactivities are arranged in a "sequence model" consisting of a series of parameters organized in a logical arrangement. The sequence model defines the events or actions that always take place in a preset order when an object is being moved from one location to another. The General Move sequence model which is the most commonly used of all available sequence models is defined as follows:

A Action distance	B Body motion	G Gain control	A Action distance	B Body motion	P Placement	A Action distance

These subactivities, or sequence model parameters, as they are called, are then assigned time-related index values based on the motion content of the subactivity. This approach provides complete analysis flexibility within the overall control of the sequence model. For each object moved, any combination of motions could occur, and, using MOST, any combination could be analyzed. For the General Move sequence, these index values are easily memorized from a brief data card (Fig. 5.1). A fully indexed General Move sequence, for example, might appear as follows:

$$A_6 \ B_6 \ G_1 \ A_1 \ B_0 \ P_3 \ A_0$$

where A_6 = walk three to four steps to object location
B_6 = bend and arise
G_1 = gain control of a light object
A_1 = move object a distance within reach
B_0 = no body motion
P_3 = place and adjust object
A_0 = no return

This example could, for instance, represent the following activity: "Walk three steps to pick up a bolt from floor level, arise, and place the bolt in a hole."

General Move is by far the most frequently used of the three sequence models. Roughly 50 percent of all manual work occurs as a General Move, with the percentage running higher for assembly and material handling work, and lower for machine shop operations.

The second type of move is described by the Controlled Move sequence. This sequence is used to cover such activities as operating a lever or crank, activating a button or switch, or simply sliding an object over a surface. In addition to the A, B, and G parameters from the General Move sequence, the sequence model for Controlled Move contains the following subactivities:

M—move controlled

X—process time

I—align

As many as one-third of the activities occurring in machine shop operations may involve Controlled Move sequences. A typical activity covered by the Controlled Move sequence is the engaging of the feed lever on a milling machine. The sequence model for this activity might be indexed as follows:

$$A_1 \ B_0 \ G_1 \ M_1 \ X_{10} \ I_0 \ A_0$$

where A_1 = reach to the lever a distance within reach
B_0 = no body motion
G_1 = get hold of lever
M_1 = move lever up to 12 in. (30 cm) to engage feed
X_{10} = process time of approximately 3.5 seconds
I_0 = no alignment
A_0 = no return

The third sequence model comprising the basic MOST technique is the Tool Use sequence model. This sequence model covers the use of hand tools for such activities as fasten or loosen, cutting, cleaning, gauging, and recording. Also, certain activities requiring the use of the brain for mental processes can be classified as Tool Use, such as reading and thinking. As indicated above, the Tool Use sequence model is a combination of General Move and Controlled Move activities. It was developed as a part of the basic MOST Systems, merely to simplify the analysis of activities related to the use of hand tools. It will

Basic MOST® System — GENERAL MOVE

A B G A B P A

INDEX × 10	A — ACTION DISTANCE: PARAMETER VARIANT	A: KEYWORD	B — BODY MOTION: PARAMETER VARIANT	B: KEYWORD	G — GAIN CONTROL: PARAMETER VARIANT	G: KEYWORD	P — PLACEMENT: PARAMETER VARIANT	P: KEYWORD	INDEX × 10
0	≤2 in ≤5 cm	CLOSE					Hold, Toss	THROW CARRY — TOSS PICKUP	0
1	Within reach				Light object, Light objects simo	GRASP (optional)	Lay aside, Loose fit	MOVE PUT	1
3	1 - 2 steps	1 STEP, 2 STEPS	Bend and arise 50% occ.	PBEND	Non Simo, Heavy / Bulky, Blind, Disengage — Obstructed, Interlocked, Collect	GET DISENGAGE FREE COLLECT	Adjustments, Light pressure, Double placement	PLACE REPLACE	3
6	3 - 4 steps	3 STEPS, 4 STEPS	Bend and arise	BEND			Care, Blind, Heavy pressure, Intermediate moves — Precision, Obstructed	POSITION REPOSITION	6
10	5 - 7 steps	5 STEPS, 6 STEPS, 7 STEPS	Sit or stand	SIT STAND					10
16	8 - 10 steps	8 STEPS, 9 STEPS, 10 STEPS	Through Door, Climb on or off, Stand and bend, Bend and sit	DOOR CLIMB/DESCEND STAND AND BEND BEND AND SIT					16

FIGURE 5.1 Basic MOST—general move data card.

later become obvious to the reader that any hand tool activity is made up of General and Controlled Moves.

The use of a wrench, for example, might be described by the following sequence:

$$A_1\ B_0\ G_1\ A_1\ B_0\ P_3\ F_{10}\ A_1\ B_0\ P_1\ A_0$$

where A_1 = reach to wrench
$\quad\quad B_0$ = no body motion
$\quad\quad G_1$ = get hold of wrench
$\quad\quad A_1$ = move wrench to fastener a distance within reach
$\quad\quad B_0$ = no body motion
$\quad\quad P_3$ = place wrench on fastener
$\quad\quad F_{10}$ = tighten fastener with wrench
$\quad\quad A_1$ = move wrench a distance within reach
$\quad\quad B_0$ = no body motion
$\quad\quad P_1$ = lay wrench aside
$\quad\quad A_0$ = no return

Time Units.   The time units used in MOST are identical to those used in the basic MTM (methods time measurement system) and are based on hours and parts of hours called TMU (time measurement unit). One TMU is equivalent to 0.00001 hour.

The time value in TMU for each sequence model is calculated by adding the index values and multiplying the sum by 10. In our previous General Move sequence example, the time would be

$$(6 + 6 + 1 + 1 + 0 + 3 + 0) \times 10 = 170\ \text{TMU}$$

corresponding to approximately 0.1 minute. The time values for the other two examples are computed in the same way. The Controlled Move totals up to

$$(1 + 0 + 1 + 1 + 10 + 0 + 0) \times 10 = 130\ \text{TMU}$$

and the Tool Use

$$(1 + 0 + 1 + 1 + 0 + 3 + 10 + 1 + 0 + 1 + 0) \times 10 = 180\ \text{TMU}$$

All time values established by MOST reflect the pace of an average skilled operator working at an average performance rate. This is often referred to as the 100 percent performance level that in time study is achieved by using "leveling factors" to adjust time to defined levels of skill and effort. Therefore, when using MOST, it is not necessary to adjust the time values unless they must conform with particular high or low task plans used by some companies. This also means that if a time standard for an operation is properly established by using either MOST, MTM, or stopwatch time study, the TMU values should be identical or almost identical for the three techniques.

The analysis of an operation will consist of a series of sequence models describing the movement of objects to perform the operation. See Fig. 5.2 for an example. Total time for the complete MOST analysis is arrived at by adding the computed sequence times. The operation time may be left in TMU or converted to minutes or hours. Again, this time would reflect pure work content (no allowances) at the 100 percent performance level. The final time standard will include the allowance factor consisting of P (personal time), R or F (rest or fatigue factor), and D for unavoidable delays (often determined by a work sampling study).

MOST-calculation

Code	
Date	7/29/87
Sign.	A.A.
Page	1/1

ELECTRONIC ASSEMBLY

Activity INSTALL CONNECTOR ON PC-BOARD

Conditions EDGE CONNECTORS ONLY

No.	Method	No.	Sequence Model	Fr	TMU
1	POSITION EDGE	1	$A_1 B_0 G_1 A_1 B_0 P_6 A_0$		90
		3	$A_1 B_0 G_1 A_1 B_0 P_3 A_0$	2	120
	CONNECTOR TO BOARD	4	$A_1 B_0 G_1 A_1 B_0 P_1 A_0$	4	160
		7	$A_1 B_0 G_1 A_1 B_0 P_3 A_0$		60
2	ALIGN CONNECTOR TO		A B G A B P A		
			A B G A B P A		
	ACCURATE LOCATION		A B G A B P A		
			A B G A B P A		
3	PLACE SCREW TO HOLE		A B G A B P A		
			A B G A B P A		
	IN CONNECTOR F2		A B G A B P A		
			A B G A B P A		
4	MOVE WASHER TO SCREW		A B G A B P A		
			A B G A B P A		
	ON BOARD F4		A B G A B P A		
			A B G A B P A		
5	FASTEN NUTS 2 SPINS		A B G A B P A		
		2	$A_0 B_0 G_0 M_3 X_0 I_{16} A_0$		190
	USING FINGERS F2		A B G M X I A		
			A B G M X I A		
6	FASTEN 2 SCREWS		A B G M X I A		
			A B G M X I A		
	5 SPINS USING		A B G M X I A		
			A B G M X I A		
	SCREWDRIVER		A B G M X I A		
7	PLACE BOARD TO RACK	5	$A_1 B_0 G_1 A_1 B_0 P_1 F_3 A_1 B_0 P_0 A_0$	2	140
		6	$A_1 B_0 G_1 A_1 B_0 L_3 A_1 F_{10} A_1 B_0 P_1 A_0$	(2)	330
			A B G A B P A B P A		
			A B G A B P A B P A		
			A B G A B P A B P A		
			A B G A B P A B P A		
			A B G A B P A B P A		
			A B G A B P A B P A		
			A B G A B P A B P A		
			A B G A B P A B P A		
			A B G A B P A B P A		
			A B G A B P A B P A		
			A B G A B P A B P A		

TIME = .65	millihours (mh.) / minutes (min.)	1090

FIGURE 5.2 Basic MOST analysis example (electronic assembly).

Application Speed.   MOST was designed to be considerably faster than other work measurement techniques. Because of its simpler construction, under ideal conditions Basic MOST requires only 10 applicator hours per measured hour.

Accuracy.   The accuracy principles that apply to MOST are the same as those used in statistical tolerance control. That is, the accuracy to which a part is manufactured depends on its role in the final assembly. Likewise, with MOST, time values are based on calculations that guarantee the overall accuracy of the final time standard. Based on these principles, MOST provides the means for covering a high volume of manual work with an accuracy that can be determined and controlled.

Method Sensitivity.   MOST is a method-sensitive technique; that is, it is sensitive to the variations in time required by different methods. This feature is very effective in evaluating alternative methods of performing operations with regard to time and cost. The MOST analysis will clearly indicate the more economical and less fatiguing method.

The fact that MOST Systems is method-sensitive greatly increases its worth as a work measurement tool. Not only does it indicate the time needed to perform various activities, it also provides the analyst with an instant clue that a method should be reviewed. The results are clear, concise, easily understood time calculations that indicate the opportunities for saving time, money, and energy.

Documentation.   One of the most burdensome problems in the standards development process is the volume of paperwork required by the most widely used predetermined work measurement systems. MOST has shown that where the more detailed systems require between 40 and 100 pages of documentation, MOST requires as few as 5. The substantially reduced amount of paperwork enables the analysts to complete studies faster and to update standards more easily. It is interesting to note that the reduction of paper generated by MOST does not lead to a lack of definition of the method used to perform the task. On the contrary, the method description found with MOST Systems is a clear, concise, plain-language description of the activity. These method descriptions can very well be used for operator training and instruction.

Applicability.   In what situations can MOST be used? Because manual work normally includes some variation from one cycle to the next, MOST, with its statistically established time ranges and time values, can produce times comparable with those of more detailed systems for the majority of manual operations. Therefore, MOST is appropriate for any manual work that contains variation from one cycle to another regardless of cycle length. Basic MOST should not be used in situations in which a short cycle (usually up to 10 seconds or 280 TMU long) is repeated identically over an extended period of time. In these situations, which, by the way, do not occur very often, the more detailed Mini MOST version should be chosen as the proper work measurement tool. In fact, Mini MOST was developed to cover highly repetitive, short-cycled work measurement tasks. At the other end of the spectrum, Maxi MOST was developed to measure long cycle (2 minutes or more), nonrepetitive operations such as heavy assembly, maintenance, and machine setups.

THE GENERAL MOVE SEQUENCE MODEL

The General Move sequence deals with the spatial displacement of one or more object(s). Under manual control, the object follows an unrestricted path through the air. If the object is in contact with, or restrained in any way by another object during the move, the General Move sequence is not applicable.

Characteristically, General Move follows a fixed sequence of subactivities identified by the following steps:

1. Reach with one or two hands a distance to the object(s), either directly or in conjunction with body motions.
2. Gain manual control of the object(s).
3. Move the object(s) a distance to the point of placement, either directly or in conjunction with body motions.
4. Place the object(s) in a temporary or final position.
5. Return to workplace.

These five subactivities form the basis for the activity sequence describing the manual displacement of the object(s) freely through space. This sequence describes the manual events that can occur when moving an object freely through the air and is therefore known as a "sequence model." The major function of the sequence model is to guide the attention of the analyst through an operation, thereby adding the dimension of having a preprinted and standardized analysis format. The existence of the sequence model provides for increased analyst consistency and reduced subactivity omission.

The Sequence Model.   The sequence model takes the form of a series of letters representing each of the various subactivities (called parameters) of the General Move activity sequence. With the exception of an additional parameter for body motions, the General Move sequence is the same as the above five-step pattern:

$$A\ B\ G\ A\ B\ P\ A$$

where A = Action distance
  B = Body motion
  G = Gain control
  P = Placement

Parameter Definitions

A   Action Distance.   This parameter covers all spatial movement or actions of the fingers, hands, and/or feet, either loaded or unloaded. Any control of these actions by the surroundings requires the use of other parameters.
  B   Body Motion.   This parameter refers to either vertical (up and down) motions of the body or the actions necessary to overcome an obstruction or impairment to body movement.
  G   Gain Control.   This parameter covers all manual motions (mainly finger, hand, and foot) employed to obtain complete manual control of an object(s) and to subsequently relinquish that control. The G parameter can include one or several short-move motions whose objective is to gain full control of the object(s) before it is to be moved to another location.
  P   Placement.   This parameter refers to actions at the final stage of an object's displacement to align, orient, and/or engage the object with another object(s) before control of the object is relinquished.

Parameter Indexing.   Index values for the above four parameters included in the General Move sequence model can be found in Fig. 5.1. Definitions of all available index values for the four General Move parameters can be found in the MOST textbook.[1] The definitions for *A*—Action Distance are included below as an example.

Action Distance (A).   Action Distance covers all spatial movement or actions of the fingers, hands, and/or feet, either loaded or unloaded. Any control of these actions by the surroundings requires the use of other parameters.

A_0 *<2 in. (5 cm).*   Any displacement of the fingers, hands, and/or feet a distance less than or equal to 2 in. (5 cm) will carry a zero index value. The time for performing these short distances is included within the Gain control and placement parameters. Example: Reaching between the number keys on a pocket calculator or placing nuts or washers on bolts located less than 2 in. (5 cm) apart.

A_1 *Within Reach.*   Actions are confined to an area described by the arc of the outstretched arm pivoted about the shoulder. With body assistance—a short bending or turning of the body from the waist—this "within reach" area is extended somewhat. However, taking a step for further extension of the area exceeds the limits of an A_1 and must be analyzed with an A_3 (one to two steps). Example: With the operator seated in front of a well-laid-out workbench, all parts and tools can be reached without displacing the body by taking a step.

The parameter value A_1 also applies to the actions of the leg or foot reaching to an object, lever, or pedal. If the trunk of the body is shifted, however, the action must be considered a step (A_3).

A_3 *One to Two Steps.*   The trunk of the body is shifted or displaced by walking, stepping to the side, or turning the body around using one or two steps. Steps refer to the total number of times each foot hits the floor.

Index values for longer-action distances involving walking on flat surfaces as well as up or down ladders can be found in Fig. 5.1 for up to 10 steps. This will satisfy the need for action distance values for most work areas in a manufacturing plant. Should, however, longer walking distance occur, the table can be extended. All index values for walking are based on an average step length of 2½ ft (0.75 m).

General Move Examples

1. A man walks four steps to a small suitcase, picks it up from the floor, and without moving further places it on a table located within reach.

$$A_6 \, B_6 \, G_1 \, A_1 \, B_0 \, P_1 \, A_0 \qquad 150 \text{ TMU}$$

2. An operator standing in front of a lathe walks six steps to a heavy part lying on the floor, picks up the part, walks six steps back to the machine, and places it in a three-jaw chuck with several adjusting actions. The part must be inserted 4 in. (10 m) into the chuck jaws.

$$A_{10} \, B_6 \, G_3 \, A_{10} \, B_0 \, P_3 \, A_1 \qquad 330 \text{ TMU}$$

3. From a stack located 10 ft (3 m) away, a heavy object must be picked up and moved 5 ft (2 m) and placed on top of a workbench with some adjustments. The height of this stack will vary from waist to floor level. Following the placement of the object on the workbench, the operator returns to the original location, which is 11 ft (3.5 m) away.

$$A_6 \, B_3 \, G_3 \, A_3 \, B_0 \, P_3 \, A_{10} \qquad 280 \text{ TMU}$$

THE CONTROLLED MOVE SEQUENCE MODEL

The Controlled Move sequence describes the manual displacement of an object over a controlled path. That is, movement of the object is restricted in at least one direction by contact with or an attachment to another object.

The Sequence Model.   The sequence model takes the form of a series of letters representing each of the various subactivities (called parameters) of the Controlled Move activity sequence.

$$A\ B\ G\ M\ X\ I\ A$$

where A = action distance
  B = body motion
  G = gain control
  M = move controlled
  X = process time
  I = align

Parameter Definitions.   Only three new parameters are introduced, as the A, B, and G parameters were discussed with the General Move sequence and remain unchanged.

  M *Move Controlled.*   This parameter covers all manually guided movements or actions of an object over a controlled path.

  X *Process Time.*   This parameter occurs as that portion of work controlled by processes or machines and not by manual actions.

  I *Align.*   This parameter refers to manual actions following the controlled move or at the conclusion of process time to achieve the alignment of objects.

  The index value definitions for the above parameters (M, X, and I) can be found in the textbook, *MOST Work Measurement Systems.*[1]

Controlled Move Examples

1. From a position in front of a lathe, the operator takes two steps to the side, turns the crank two revolutions, and sets the machining tool against a scale mark.

$$A_3\ B_0\ G_1\ M_6\ X_0\ I_6\ A_0 \qquad 160\ \text{TMU}$$

2. A milling cutter operator walks four steps to the quick-feeding cross level and engages the feed. The machine time following the 4-in. (10-cm) lever action is 2.5 seconds.

$$A_6\ B_0\ G_1\ M_1\ X_6\ I_0\ A_0 \qquad 140\ \text{TMU}$$

3. A material handler takes hold of a heavy carton with both hands and pushes it 18 in. (45 cm) across conveyor rollers.

$$A_1\ B_0\ G_3\ M_3\ X_0\ I_0\ A_0 \qquad 70\ \text{TMU}$$

4. Using the foot pedal to activate the machine, a sewing machine operator makes a stitch requiring 3.5 seconds process time. (The operator must reach the pedal with the foot.)

$$A_1\ B_0\ G_1\ M_1\ X_{10}\ I_0\ A_0 \qquad 130\ \text{TMU}$$

THE TOOL USE SEQUENCE MODEL

The Tool Use sequence is composed of subactivities from the General Move sequence, along with specially designed parameters describing the actions performed with hand tools or, in some cases, the use of certain mental processes. Tool use follows a fixed sequence of subactivities occurring in five main activity phases:

1. Get object or tool
2. Place object or tool in working position
3. Use tool

4. Put aside object or tool

5. Return to workplace

The Sequence Model. These five activity phases form the basis for the activity sequence describing the handling and use of hand tools. The sequence model takes the form of a series of letters representing each of the various subactivities of the Tool Use activity sequence:

A B G	A B P		A B P	A
Get object or tool	Place object or tool	Use tool	Aside object or tool	Return

where A = action distance
B = body motion
G = gain control
P = placement

The space in the sequence model "Use tool" is provided for the insertion of one of the following Tool Use parameters. These parameters refer to the specifications of using the tool and are:

$$F = \text{fasten}$$

$$L = \text{loosen}$$

$$C = \text{cut}$$

$$S = \text{surface treat}$$

$$M = \text{measure}$$

$$R = \text{record}$$

$$T = \text{think}$$

Tool Use Examples for "Fasten and Loosen"

1. Obtain a nut from a parts bin located within reach, place it on a bolt, and run it down with seven finger actions.

$$A_1 \ B_0 \ G_1 \ A_1 \ B_0 \ P_3 \ F_{10} \ A_0 \ B_0 \ P_0 \ A_0 \qquad 160 \text{ TMU}$$

2. Obtain a power wrench from within reach, run down four ⅜-in. (10-mm) bolts located 6 in. (15 cm) apart, and set aside wrench.

$$A_1 \ B_0 \ G_1 \ A_0 \ B_0 \ (P_3 \ A_1 \ F_6) \ A_1 \ B_0 \ P_1 \ A_0 \ (4) \qquad 440 \text{ TMU}$$

3. From a position in front of an engine lathe, obtain a large T-wrench located five steps away and loosen one bolt on a chuck on the engine lathe with both hands, using five arm actions. Set aside the T-wrench from the machine (but within reach).

$$A_{10} \ B_0 \ G_1 \ A_{10} \ B_0 \ P_3 \ L_{24} \ A_1 \ B_0 \ P_1 \ A_0 \qquad 500 \text{ TMU}$$

THE MOST SYSTEMS FAMILY

In addition to the Basic MOST System, several application oriented versions of MOST are now members of the MOST Systems Family: Mini MOST, Maxi MOST, and Clerical MOST. A new version, Mega MOST, is under development for future applications.

The Mini MOST System.   Basic MOST was not designed to measure short-cycled operations, although the original MOST version can be applied to nonidentical operations of 10 seconds or less and still meet the accuracy criteria.

Therefore, the Mini MOST version of MOST Work Measurement Systems was developed to satisfy higher accuracy requirements that apply to very short cycled, highly repetitive, identical operations. Such operations may be only from 2 to 10 seconds long and often are performed over long periods of time.

Mini MOST consists of two sequence models:

General Move—*A B G A B P A*

Controlled Move—*A B G M X I A*

These sequence models are identical to the two basic sequence models in the Basic MOST version. There is one major difference, however. The multiplier for the index value total is 1 for Mini MOST. Therefore, if the sum of the applied index values is 64, this is also the total TMU value for the sequence model. Another difference compared with Basic MOST is that distances in Mini MOST will be measured in inches.

The application speed of Mini MOST is about 25:1 under "ideal" conditions compared with about 10:1 for Basic MOST.

The definitions and descriptions of the parameters and elements in Mini MOST have been excluded because of space considerations. The second edition of *MOST Work Measurement Systems*[2] includes a complete review of Mini MOST Work Measurement Systems.

The Maxi MOST System.   In order to satisfy the need for a fast, less detailed but still accurate and consistent system for the measurement of long-cycled, nonrepetitive, nonidentical operations, Maxi MOST was developed.

Maxi MOST consists of five sequence models with a multiplier of 100. The sequence models are:

- Part handling
- Tool and equipment use
- Machine handling
- Transport with powered crane
- Transport with wheeled truck

For a complete listing of Maxi MOST sequence models and parameters, see Fig. 5.3.

An example of a typical Maxi MOST analysis has been included as Fig. 5.4. Maxi MOST has a measurement factor of 3 to 5 hours to 1 (analyst hours to measured hours) and is therefore a very cost-effective technique to use in a large number of cases where minute

ACTIVITY	SEQUENCE MODEL	SUB-ACTIVITY
PART HANDLING	A B P	A - ACTION WALKING DISTANCE B - BODY MOTION
TOOL USE	A B T	P - GET AND PLACE PARTS T - TOOL USE
MACHINE HANDLING	A B M	M - OPERATE MACHINE OR FIXED EQUIPMENT
POWERED CRANE TRANSPORT	A T K T P T A	A - ACTION WALKING DISTANCE T - TRANSPORT K - HOOKUP AND UNHOOK P - PLACE OBJECT
WHEELED TRUCK TRANSPORT	A S T L T L T A	A - ACTION WALKING DISTANCE S - START AND STOP T - TRANSPORT L - LOAD OR UNLOAD

FIGURE 5.3   Maxi MOST systems—sequence models.

MOST® SYSTEMS CALCULATION 100X		Code			
Area 735 ENGINE LINE		**Date** 4-2-85	**Sign** TWF	**Page** 1/1	
Operation POWER STEERING PUMP					

Title ASSEMBLE POWER STEERING COMPRESSOR TO ENGINE	**Time mh TMU (min)** 3.0

Conditions CUMMINS ENGINE	**Crew** 1/1	**Per** UNIT

No.	Method Description	Sequence	Fr	mh
	OPERATOR TO BEGIN: OP-1 BEGINS AT: BENCH			
1	MOVE COUPLING AND RETAINING RING AT BENCH	$A_0 B_0 P_1$		1
2	POSITION RETAINING RING TO COUPLING USING PLIERS	$A_0 B_0 T_3$		3
3	MOVE BOLT INTO COUPLING	$A_0 B_0 P_1$		1
4	FASTEN RETAINING RING TO COUPLING USING HAMMER, 5 STRIKES	$A_0 B_0 T_1$		1
5	MOVE COUPLING AND POWER STEERING PUMP TO ENGINE USING DOLLY - 6 STEPS - 1 BEND	$A_1 B_1 P_1$		3
6	LOOSEN 2 BOLTS FROM COVER PLATE USING IMPACT - SOCKET CHANGE - 8 STEPS - 3 BENDS	$A_1 B_3 T_3$		7
7	MOVE 2 BOLTS W/WASHERS, COVER PLATE AND GASKET FROM ENGINE TO BINS AND RETURN TO BENCH - 20 STEPS	$A_3 B_0 P_1$		4
8	MOVE 2 BOLTS W/WASHERS, GASKET AND TUBE OF LOCTITE TO DOLLY - 4 STEPS - 1 BEND	$A_1 B_1 P_1$		3
9	APPLY LOCTITE TO ENGINE IN TWO LOCATIONS	$A_0 B_0 T_3$		3
10	POSITION GASKET TO COMPRESSOR - POSITION COMPRESSOR TO ENGINE	$A_0 B_0 P_3$		3
11	INSTALL 2 BOLTS TO ENGINE - 1 BEND	$A_0 B_1 T_{10}$		11
12	TIGHTEN 2 BOLTS USING WRENCH - 15 TURNS - 10 STEPS - 1 BEND	$A_3 B_1 T_6$		10

PLACE PART A B P	TOOL/EQUIPMENT USE A B T	MACHINE HANDLING A B M	TRANSPORT WITH CRANE A T K T P T A	TRANSPORT WITH TRUCK A S T L T L T A	Total mh

© HBMCo T118-81

FIGURE 5.4 Maxi MOST example (truck assembly).

details are unnecessary or even detrimental to proper work instructions. The recommendation is to use Maxi MOST for nonidentical cycles that are 2 minutes or longer.

The definitions and descriptions of the parameters and elements in the Maxi MOST System have been excluded because of space considerations. The second edition of *MOST Work Measurement Systems*[2] includes a complete explanation of the Maxi MOST work measurement system.

Clerical MOST.   MOST clerical systems is based on three sequence models identical to those in basic MOST:

- General move
- Controlled move
- Tool and equipment use (two data cards)

The parameters for these sequence models are shown in Fig. 5.5.

ACTIVITY	SEQUENCE MODEL	SUB - ACTIVITIES
GENERAL MOVE	A B G A B P A	A - ACTION DISTANCE B - BODY MOTION G - GAIN CONTROL P - PLACEMENT
CONTROLLED MOVE	A B G M X I A	M - MOVE CONTROLLED X - PROCESS TIME I - ALIGNMENT
EQUIPMENT USE	A B G A B P A B P A	H - LETTER/PAPER HANDLING T - THINK R - RECORD K - CALCULATE W - TYPE
TOOL USE	A B G A B P A B P A	F - FASTEN L - LOOSEN C - CUT M - MEASURE

FIGURE 5.5   MOST clerical systems—sequence models.

Mega MOST.   The main purpose of adding a version of MOST on the 1000 multiplier level is to simplify and accelerate the standard setting for long (over 20 minutes) nonrepetitive operations in areas such as assembly and maintenance.

While Mini MOST is totally generic and basic MOST about 60 to 80 percent generic, Maxi MOST is primarily tool-oriented, and Mega MOST is part- and operations-oriented. Mega MOST will be adopted for automated calculation of standards by the computer.

PRINCIPLES AND PROCEDURES FOR DEVELOPING TIME STANDARDS BASED ON MOST SYSTEMS

A standard MOST calculation form should be used for all analysis work using Basic MOST. (Similar forms have been designed for use with Mini MOST and Maxi MOST.) As can be seen for the included example (Fig. 5.2), this form consists of four sections:

1. A header identifying the activity to be measured and the work center (area) in which it is being performed
2. A method description step by step (left half)

3. Preprinted sequence models in three groups—General Move, Controlled Move, and Tool Use

4. A field for the time value or time standard for the activity (bottom part).

Note. The activity time or standard does not include any allowances at this stage. Prior to applying this time standard, the time value on the form should be multiplied by the appropriate allowance factor, thereby constituting the standard time for the operation.

A frequency factor (*Fr*) for each sequence model can be specified in the column next to the TMU value column for the sequence model.

Normally, the space provided on one page of the MOST calculation form will allow for analyses up to approximately 1 minute.

MOST can either be applied for direct work measurement of defined operations or be used as a basis for standard data. In the case of short-cycled, "unique" operations, like subassemblies, the direct approach is preferred. On the other hand, if a great variety of the operations that are being performed at one work center (a lathe or drill press) occurs, the standard data approach is the most efficient and economical. A work sheet composed of standard data units, each one backed up by a MOST analysis, will provide a fast and simple way to calculate standards. Initially, the desired accuracy level for the resulting standards should be determined and the work sheet designed accordingly. This means that the tighter the accuracy requirements are the more elements and the more decisions have to be made in order to set a standard. A multipage detailed work sheet will take more time and cost more to use than a single-page work sheet with few elements designed for a lower accuracy level. Consequently, the economics of setting standards is a direct function of the required accuracy of the output.

For instance, if the required accuracy is ±5 percent with 95 percent confidence over an 8-hour period, the work sheet may consist of 75 different elements while a ±10 percent accuracy with 90 percent confidence over a 40-hour period may produce a work sheet with only 10 to 15 elements. The difference in application time will be substantial, and since standard setting normally is an ongoing activity, the cost saving potential is considerable.

In all situations where MOST is being used, the "top-down" approach should be followed. A two-step decision model can be put to use with the first generation being: Is it appropriate and practical to do direct measurements? If "yes," the work should be measured by using the MOST calculation form. If the answer is "no," a sample of the typical operations or activities for the work center should be broken down into logical suboperations. Each such suboperation will then be measured using MOST and placed on a work sheet for the calculation of time standards. In some instances, suboperations may have to be broken down still one more level and later combined into combined suboperations before assigning them onto the work sheet.

By following the "top-down" approach, the database with standard data (suboperation data) will remain compact and more manageable than if the conventional "bottom-up" procedure is applied.

MOST is an application-oriented or "user-friendly" system that will require some unlearning and rethinking by users experienced in conventional work measurement. It is a new concept not only regarding work measurement but also in the application areas.

MOST COMPUTER SYSTEMS

The logical sequence model approach lends itself very well to a computerized application. Therefore, in 1976 the first lines of code were written in an effort to develop a software program that would advance the state of the art of work measurement. While other computerized systems use element symbols or numerical data as input, MOST Computer Systems uses method descriptions expressed in plain English. In other words, MOST Computer Systems is a language-based system. Today the computerized MOST program reminds one of an "expert system," although this term was not generally known when the development started.

Computerized MOST Analysis. The input for a computer MOST analysis consists of (1) work area data and (2) a method description. Based on this information, the computer will produce a MOST analysis as output; that is, the computer actually completes the work measurement task "automatically." A simple but representative work area layout sketch is also part of the output. A typical example of a work area description is shown in Fig. 5.6 and the MOST analysis for an operation performed in that work area in Fig. 5.7.

In designing the program, the basic philosophy of establishing a time standard as a direct function of the work conditions was followed. The computer was therefore programmed to produce a time standard based on well-defined and complete user work conditions. The computer was also programmed not to allow the change of a time value without a change of the underlying work conditions. A change of, for instance, a distance or a "gain control" or a "placement" of an object or a body motion will result in a different standard. This discipline has proved to increase the uniformity and consistency of the method descriptions and analyses. Equally important is the fact that one does not have to read both the method description and the MOST index values to interpret an analysis. A review of the method is adequate: the index values and the time standard are a by-product and a direct function of the method.

How is it possible for the computer to generate a MOST analysis from the input of "only" work area data and a method description? How does the computer select the right sequence model and the correct index values?

As explained above, and as can be seen from the example in Fig. 5.6, all "action distances" and "body motions" are specified as part of the work area data. Therefore, the A and B parameters in the sequence models will be assigned an index value from the work area information.

Three additional variables remain to be determined: (1) sequence model selection, (2) index value for the G-parameter, and (3) index value for the P-parameter. This required information has been compounded into one word: *a keyword.* This keyword, always found in the beginning of each method step, has been chosen from a list of commonly used English activity words such as MOVE, PLACE, and POSITION. For instance, the keyword PLACE will mean the "General Move sequence model" and a combination of "G_1" and "P_3" to the computer. MOVE indicates the same sequence model with a $G_1 P_1$ combination and POSITION, a $G_1 P_6$ combination. A GET preceding MOVE, PLACE, and POSITION will render a $G_3 P_1$, $G_3 P_3$, and $G_3 P_6$, respectively.

Similar keywords are available for all sequence models in MOST Computer Systems. The knowledge of approximately 30 to 50 keywords for Basic MOST will provide the analyst with a sufficient "vocabulary" to be able to perform most of the analysis work.

Since both the work area data and the method description are entered within a well-structured format, it is possible to dictate this information using a hand-held tape recorder. A person can, in most situations, talk as fast as or faster than an operator can perform an assembly or a machining operation. Therefore, the data collection becomes much more efficient. The conventional handwriting of methods is usually cumbersome and inefficient. While the "dictation" of a method in principle will require the observation of just one cycle, the "writing" of the same method will require observation of several cycles. The information on the tape will then be transcribed by the analyst or a typist on a CRT terminal as input to the program. In the future, when a voice recognition system becomes available for practical applications, this "intermediate step" can be eliminated.

Data Management. The major advantage of a computerized application of MOST lies in the databases (suboperations and standards). These are accumulated as a result of the MOST analysis work and calculation of standards. The filing, searching, retrieving, and updating of the data becomes extremely efficient and fast compared with a manual system. Some functions requiring manipulations of data such as mass updating, simulations, and history of standards are very impractical or impossible to execute manually, while the computer can perform them routinely and quickly.

A complete database system for filing and retrieving suboperations and time standards

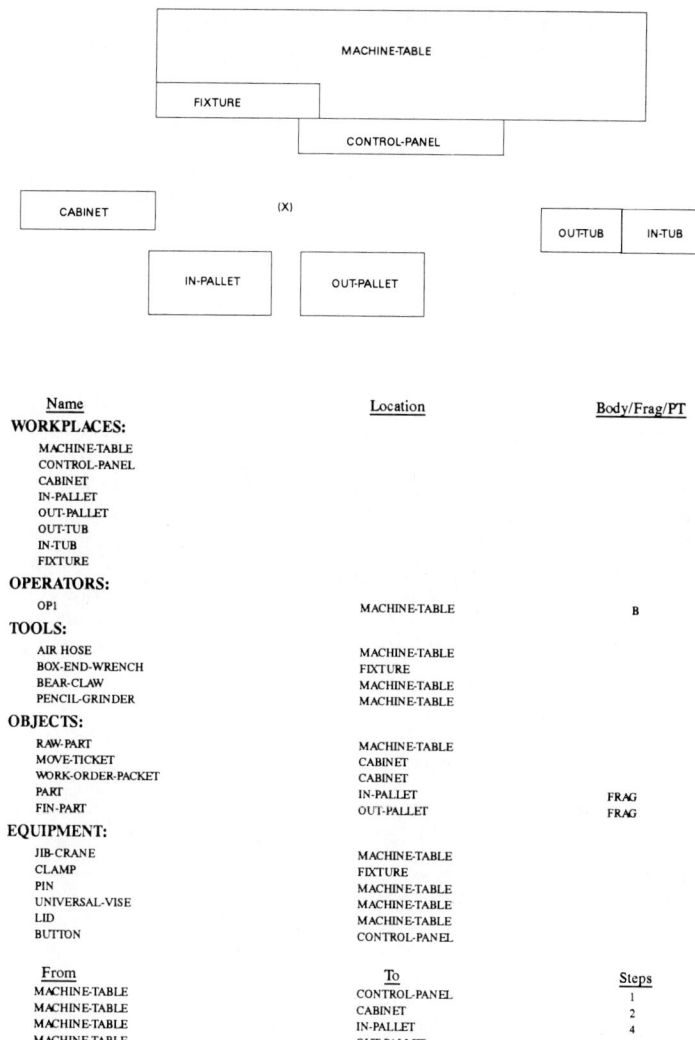

Name	Location	Body/Frag/PT
WORKPLACES:		
MACHINE-TABLE		
CONTROL-PANEL		
CABINET		
IN-PALLET		
OUT-PALLET		
OUT-TUB		
IN-TUB		
FIXTURE		
OPERATORS:		
OP1	MACHINE-TABLE	B
TOOLS:		
AIR HOSE	MACHINE-TABLE	
BOX-END-WRENCH	FIXTURE	
BEAR-CLAW	MACHINE-TABLE	
PENCIL-GRINDER	MACHINE-TABLE	
OBJECTS:		
RAW-PART	MACHINE-TABLE	
MOVE-TICKET	CABINET	
WORK-ORDER-PACKET	CABINET	
PART	IN-PALLET	FRAG
FIN-PART	OUT-PALLET	FRAG
EQUIPMENT:		
JIB-CRANE	MACHINE-TABLE	
CLAMP	FIXTURE	
PIN	MACHINE-TABLE	
UNIVERSAL-VISE	MACHINE-TABLE	
LID	MACHINE-TABLE	
BUTTON	CONTROL-PANEL	

From	To	Steps
MACHINE-TABLE	CONTROL-PANEL	1
MACHINE-TABLE	CABINET	2
MACHINE-TABLE	IN-PALLET	4
MACHINE-TABLE	OUT-PALLET	4
MACHINE-TABLE	OUT-TUB	1
MACHINE-TABLE	IN-TUB	1
MACHINE-TABLE	FIXTURE	0
CONTROL-PANEL	CABINET	4
CONTROL-PANEL	IN-PALLET	4
CONTROL-PANEL	OUT-PALLET	4
CONTROL-PANEL	OUT-TUB	3
CONTROL-PANEL	IN-TUB	2
CONTROL-PANEL	FIXTURE	1
CABINET	IN-PALLET	3
CABINET	OUT-PALLET	7
CABINET	OUT-TUB	9
CABINET	IN-TUB	10
IN-PALLET	OUT-PALLET	5
IN-PALLET	OUT-TUB	6
IN-PALLET	IN-TUB	8
IN-PALLET	FIXTURE	2
OUT-PALLET	OUT-TUB	3
OUT-PALLET	IN-TUB	4
OUT-TUB	IN-TUB	1
OUT-TUB	FIXTURE	4

FIGURE 5.6 Work area data example (machining).

```
LOAD PART IN FIXTURE WITH BOX END WRENCH AT MULTI SPINDLE VERTICAL DRILL 2000
PER PART                                        OFG: 2  13-Jul-90
   OP1 BEGINS AT MACHINE-TABLE

1 PLACE PART FROM IN-PALLET TO FIXTURE
                    A3  B0  G1  A3  B0  P3  A0        1.00     100.
2 PUSH SLIDE CLAMP AT FIXTURE
                    A1  B0  G1  M1  X0  I0  A0        1.00      30.
3 FASTEN 2 NUTS AT FIXTURE WITH 4 ARM-TURNS USING BOX-END-WRENCH AND ASIDE
    A1  B0  G1  A0  B0  (P3  A1  F10 )A1  B0  P1  A0  (2) 1.00   320.
4 FASTEN SCREW FASTENER AT FIXTURE 1 SPIN USING FINGERS
          A1  B0  G1  A1  B0  P1  F1  A0  B0  P0  A0    1.00      50.

                                        TOTAL TMU        500.
```

FIGURE 5.7   MOST analysis example (machining).

is the backbone of MOST Computer Systems. The database has to date been "pushed" to handle up to 1/2 million standards and suboperations. Even larger installations are planned.

The filing system for the database also uses the "word" concept. All suboperation data are filed and retrieved under well-defined words in five categories: activity, object/component, equipment/product, tool, and work area origin. The filing system for standards is in all cases being customized to fit the user's requirements and includes such conventional header items as part number, operation number, and work center number.

MOST Computer Systems—A Complete System.   MOST Computer Systems is a complete program for measuring work and calculating time standards as well as documenting and updating these standards. It consists of a basic program and a set of supplementary modules (see Fig. 5.8). The basic program includes the following features:

1. Work measurement (Basic MOST, Mini MOST, Maxi MOST)
2. Suboperation database
3. Time standards calculation

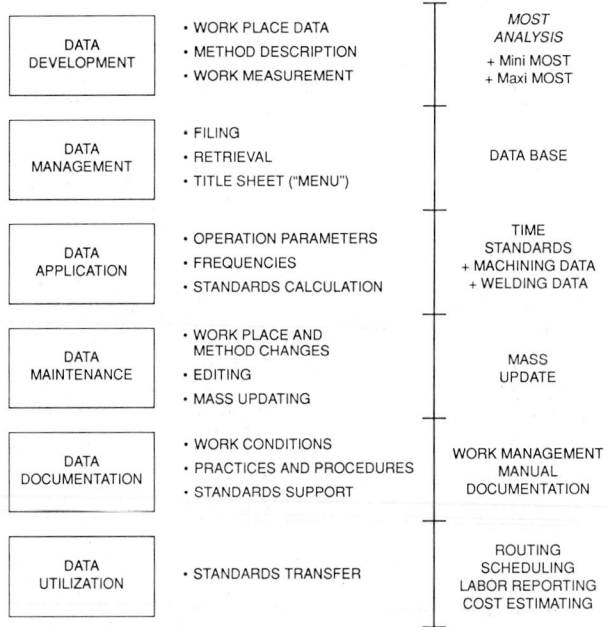

FIGURE 5.8   MOST Computer Systems overview.

4. Standards database

5. Mass update

6. Documentation of work conditions (Work Management Manual)

Supplementary modules are:

1. Machining data (feeds, speeds, and process times)

2. Welding data

3. Line balancing

4. Process planning (can be expanded to include variant and generative process planning)

5. Cost estimating

6. Performance reporting

The objective with MOST Computer Systems is to adapt the system to cover all possible aspects of establishing time standards in a wide range of situations. Another objective is to make the updating and maintenance of standards efficient and simple. Our intention is also to stimulate the industrial engineer in industry, services, and universities and colleges to adopt a positive attitude toward a fundamental and widely used discipline: the measurement of work.

SUMMARY AND FUTURE TRENDS

During the 1950s and 1960s, the work measurement field became inflated with "conventional" derivatives of the original MTM system (MTM-1). That trend has continued with one exception: MOST. In the mid-1960s, we felt that a new approach, a more practical and "user-friendly" method and more importantly a faster and simpler technique was necessary to maintain a reasonably high level of interest in work measurement. MOST seems to have been the answer. Over 18,000 persons representing more than 4000 organizations have become certified MOST users. MOST has been translated into at least 15 languages and is in use in more than 25 countries around the world. MOST satisfies all the criteria of simplicity, speed, accuracy, consistency, applicability, integrity, and universality that can be put on a modern work measurement technique and system. MOST Computer Systems represents the state of the art in the areas of work measurement and time standards. And the users enthusiastically endorse and support MOST.

A renaissance in work measurement has been noticeable during the past few years, perhaps because of a military standard (MIL-STD-1567A) issued by the Department of Defense in 1983. Since then, defense contractors are obligated to comply with this standard on major contracts. Compliance will include fully documented time standards (80 percent coverage) based on a recognized predetermined motion time system. MOST Systems has been used very successfully by a large number of defense contractors to satisfy the requirements of MIL-STD-1567A.

Also, service industries have shown an increased interest in work standards for staffing, labor power planning, budgeting, and the like. Despite the efforts by industry to increasingly automate manufacturing operations, the measurement of work done by people is here to stay for many more years. The advantages of knowing and being able to plan from realistic and consistent standards are just too great to dismiss. However, the work measurement and standard setting disciplines have to become simpler, faster, and more integrated with other functions to attract the attention they deserve.

MOST Systems and MOST Computer Systems have proven to meet those requirements to a great extent. But more can be done and more will be done. Today's computer technology has reached a level that cannot be ignored by work measurement specialists. If they take advantage of this technology, time standards could and should become a logical and integral part of any business system, as is the case in many companies already.

The general trend in industry is automation. Therefore, it is obvious we will see fully automated procedures for calculating and updating time standards based on data developed and maintained by industrial engineers. A direct link to a CAD system with the purpose of producing process plans and cost estimates based on these standards is likely to become a reality within the next few years.

REFERENCES

1. Zandin, Kjell B., *MOST Work Measurement Systems,* Dekker, New York, 1980.
2. Zandin, Kjell B., *MOST Work Measurement Systems, Basic MOST, Mini MOST, Maxi MOST,* 2d ed., Dekker, New York, 1990.
3. *Industrial Engineering Handbook,* 3d ed., McGraw-Hill, New York, 1971.
4. Cleland, David I., and Bopaya Bidanda, *The Automated Factory Handbook: Technology and Management,* TAB Books, Blue Ridge Summit, Pa., 1990.

CHAPTER 6
STANDARD DATA CONCEPTS AND DEVELOPMENT

James W. Mason, Jr.
Senior Vice President
H. B. Maynard and Company, Inc.
Pittsburgh, Pennsylvania

Standard data is defined as a structured collection of normal time values for work elements codified in tabular or graphic form. The data are used as a basis for determining time standards on work similar to that from which the data were collected without making additional time studies. Standard data is the organization of work elements into useful, well-defined building blocks. The size, content, and number of these building blocks depend on the accuracy desired, nature of the work, and the flexibility required.

The time required to establish standards by the detailed study of individual jobs is considerable. This fact has led industrial engineers to seek ways of reducing standards setting time. The goal has been to find a means of establishing standards quickly without unduly affecting the accuracy of the results. A number of different procedures have been developed for accomplishing this. The more successful and widely used of these techniques will be discussed in this chapter.

During the discussion, predetermined motion time systems, such as methods time measurement (MTM) or Maynard operation sequence technique (MOST®), will be referred to frequently. Those who are not acquainted with these procedures will find it helpful to study the other chapters in Section 4 of this handbook.

The development of standard data involves five primary phases that will be discussed under the subtopics of:

1. Benefits and limitations of standard data
2. Principles of standard data development
3. Procedures and techniques used to develop standard data
4. Input requirements to apply standard data
5. Standard data documentation

BENEFITS AND LIMITATIONS OF STANDARD DATA

The benefits of standard data are readily recognized by industrial engineers and production management. The benefits are frequently expressed in terms of:

- Standard setting time can be reduced by 40 to 80 percent depending upon the operations and prior standard time setting technique used.
- More consistent standard time values are produced.
- It is possible to attain a much higher percentage of coverage on work performed by direct and indirect workers.
- Maintenance of standard times to reflect changes in method and/or conditions is greatly simplified, especially when data application is computerized and the system has mass updating capability.

The benefits of standard data are substantial, but caution must be used to avoid improper use of the technique. Often, in an attempt to expand standard time coverage, limitations or restrictions of the data may be ignored. The most common mistakes are:

- "Stretching" the application to products or operations not intended to be included when the data were originally developed. Standard data will seldom permit application to 100 percent of the operations to a work center for which the data were developed.
- Developing standard data for applications not suited for this technique. Very high volume short-cycle operations may not be the best application of standard data.
- Failure of management to recognize the need to maintain the standard data to reflect changes in methods, equipment, tooling, or organization. The standard data approach is a "dynamic" tool that needs to be maintained to retain the benefits.

PRINCIPLES OF STANDARD DATA DEVELOPMENT

Three essentials are required for the successful development and application of standard data.

- Industrial engineering understanding of building block concepts of standard data development.
- Line management support and involvement during the development and implementation phases.
- Proper documentation of the data during the development and application steps.

Building Block Concepts.   The definition of standard data refers to work elements. A work element is defined as a subdivision of the work cycle, composed of a sequence of one or several fundamental motions and/or machine or process activities, which is distinct, described, and measured.

Figure 6.1 illustrates the building block concept showing the four basic levels. Level 1 is the basic motions and provides the smallest building block commonly used. This level is represented by MTM-1 and other basic motion systems. It requires the most time to apply and is used as a base for developing other larger building blocks. Figure 6.2 illustrates the same basic motions for MTM-1 and MOST.

Level 2 building blocks, called work elements, are data derived from time study stopwatch, MOST, and/or the combining of basic motions. Figure 6.3 illustrates a work element developed using MTM.

The MTM analysis (Fig. 6.3) shows a typical motion pattern for the element "sit down on chair or stool at workbench." This occurs when one stands in front of a desk or bench with a chair behind. To sit at the desk or bench, it is necessary to bend, reach down, grasp the chair, and lower the body to sit. This requires four basic motions: B; R__B; GIA; and Sit. The R__B is combined with the Bend. Before work can be done at the desk or bench, it is necessary to pull the chair forward. This is done by first moving both feet forward and

BUILDING BLOCK CONCEPT

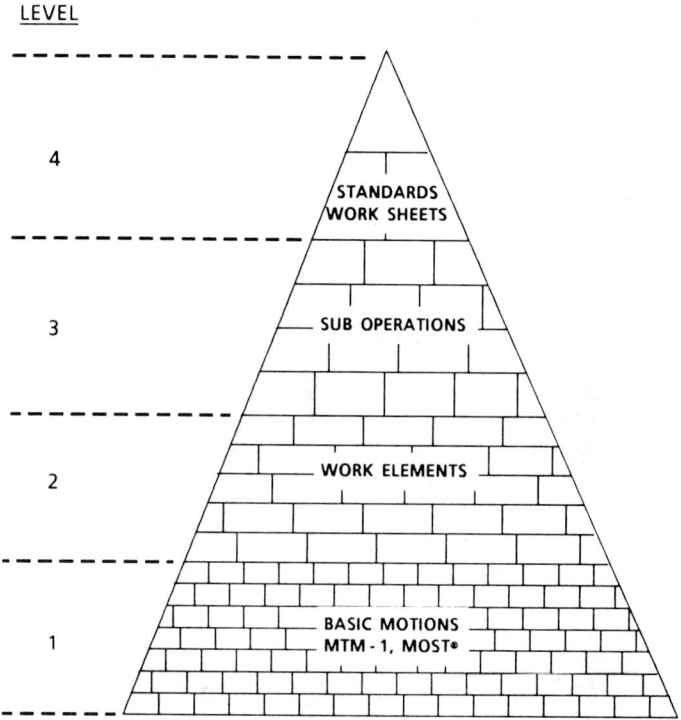

FIGURE 6.1 Building block concept.

then pulling the chair toward the desk. The arms are then placed in position on top of the desk as the feet shift to a comfortable position away from the chair. All this takes five additional motions after the body is seated.

A study of a number of jobs will show that this motion pattern, with slight variations, will occur again and again. Analyzing this pattern with basic MTM each time it occurs requires considerable time. A better approach is to record the motion pattern carefully once only. This then becomes a standard data element. The element can be described and coded and the value rounded off to the nearest TMU—112. It can then be used whenever "Sit" occurs. It is never necessary to go through the detailed MTM analysis of this particular motion pattern again.

Level 3 of Fig. 6.1 is identified as suboperations. This level recognized major work tasks that are repetitive within a defined work center, regardless of the part or product being produced. These suboperations are developed by combining selected level 2 work elements to create a suboperation. Figures 6.4 and 6.5 illustrate several suboperations that have been developed by combining work elements.

Level 4 of Fig. 6.1 is identified as "standards work sheets." This level is the final product of the development phase and represents the methodology and documentation by which a standard time is calculated from the data. Many formats can be used to display data for the calculation of standards. The format depends upon the operations involved, parts and product variations, product volumes, and desired retrieval documentation. The more common formats used to present standard data for the standards applications are:

MTM-1 BASIC MOTION TABLES

TABLE I – REACH – R

Distance Moved Inches	Time TMU				Hand In Motion		CASE AND DESCRIPTION
	A	B	C or D	E	A	B	
3/4 or less	2.0	2.0	2.0	2.0	1.6	1.6	A Reach to object in fixed location, or to object in other hand or on which other hand rests.
1	2.5	2.5	3.6	2.4	2.3	2.3	
2	4.0	4.0	5.9	3.8	3.5	2.7	
3	5.3	5.3	7.3	5.3	4.5	3.6	B Reach to single object in location which may vary slightly from cycle to cycle.
4	6.1	6.4	8.4	6.8	4.9	4.3	
5	6.5	7.8	9.4	7.4	5.3	5.0	
6	7.0	8.6	10.1	8.0	5.7	5.7	
7	7.4	9.3	10.8	8.7	6.1	6.5	C Reach to object jumbled with other objects in a group so that search and select occur.
8	7.9	10.1	11.5	9.3	6.5	7.2	
9	8.3	10.8	12.2	9.9	6.9	7.9	
10	8.7	11.5	12.9	10.5	7.3	8.6	
12	9.6	12.9	14.2	11.8	8.1	10.1	
14	10.5	14.4	15.6	13.0	8.9	11.1	

TABLE II – MOVE – M

Distance Moved Inches	Time TMU			Hand In Motion B	Wt. Allowance			CASE AND DESCRIPTION
	A	B	C		Wt. (lb.) Up to	Dynamic Factor	Static Constant TMU	
3/4 or less	2.0	2.0	2.0	1.7				
1	2.5	2.9	3.4	2.3	2.5	1.00	0	
2	3.6	4.6	5.2	2.9				A Move object to other hand or against stop.
3	4.9	5.7	6.7	3.6	7.5	1.06	2.2	
4	6.1	6.9	8.0	4.3				
5	7.3	8.0	9.2	5.0	12.5	1.11	3.9	
6	8.1	8.9	10.3	5.7				
7	8.9	9.7	11.1	6.5	17.5	1.17	5.6	
8	9.7	10.6	11.8	7.2				
9	10.5	11.5	12.7	7.9	22.5	1.22	7.4	B Move object to approximate or indefinite
10	11.3	12.2	13.5	8.6				
12	12.9	13.4	15.2	10.0	27.5	1.28	9.1	
14	14.4	14.6	16.9	11.4				

TABLE III A – TURN – T

Weight	Time TMU for Degrees Turned										
	30°	45°	60°	75°	90°	105°	120°	135°	150°	165°	180°
Small – 0 to 2 Pounds	2.8	3.5	4.1	4.8	5.4	6.1	6.8	7.4	8.1	8.7	9.4
Medium – 2.1 to 10 Pounds	4.4	5.5	6.5	7.5	8.5	9.6	10.6				
Large – 10.1 to 35 Pounds											

TABLE IV – GRASP – G

TYPE OF GRASP	Case	Time TMU	DESCRIPTION	
PICK-UP	1A	2.0	Any size object by itself, easily grasped	
	1B	3.5	Object very small or lying close against a flat surface	
	1C1	7.3	Diameter larger than 1/2''	Interference with Grasp on bottom and one side of nearly cylindrical object.
	1C2	8.7	Diameter 1/4'' to 1/2''	
	1C3	10.8	Diameter less than 1/4''	
REGRASP	2	5.6	Change grasp without relinquishing control	
TRANSFER	3	5.6	Control transferred from one hand to the other	
	4A	7.3	Larger than 1'' x 1'' x 1''	Object

TABLE V – POSITION* – P

	CLASS OF FIT	Symmetry	Easy To Handle	Difficult To Handle
1 – Loose	No pressure required	S	5.6	11.2
		SS	9.1	14.7
		NS	10.4	16.0
2 – Close	Light pressure required	S	16.2	21.8
		SS	19.7	25.3

FIGURE 6.2 Examples of basic motion data tables for MTM-1 and MOST.

MOST BASIC MOTION TABLES

GENERAL MOVE

ABGABPA									
Index ×10	**A** Action Distance		**B** Body Motion		**G** Gain Control		**P** Placement		Index ×10
	Parameter Variant	Keyword	Parameter Variant	Keyword	Parameter Variant	Keyword	Parameter Variant	Keyword	
0	≤2 in. ≤5 cm	CLOSE					Hold Toss	THROW TOSS CARRY PICKUP	0
1	Within reach				Light object Light objects simo	GRASP (optional)	Lay aside Loose fit	MOVE PUT	1
3	1-2 steps	1 STEP 2 STEPS	Bend and arise 50% occ	PBEND	Non Simo Obstructed Heavy Bulky Interlocked Bind Collect Disengage	GET DISENGAGE FREE COLLECT	Adjustments Light pressure Double placement	PLACE REPLACE	3
6	3-4 steps	3 STEPS	Bend and arise	BEND			Care Bind	Precise	6

CONTROLLED MOVE

ABGMXIA								
Index ×10	**M** Move Controlled			**X** Process Time			**I** Alignment		Index ×10
	Push Pull Pivot	Keyword	Crank (Revs.)	Seconds	Minutes	Hours	Object	Keyword	
1	≤12 inches (30 cm) Button Switch Knob	PUSH PULL ROTATE		5	01	0001	To 1 Point	ALIGN POINT	1
3	>12 inches (30 cm) Resistance Seat or Unseat High Control 2 Stages >12 in. (30 cm)	SLIDE SHUT SHIFT TURN SEAT PRESS OPEN UNSEAT PUSH - PULL (INCHES CM OR STAGES)	1	15	02	0004	To 2 Points ≤ 4 inches (10 cm)	ALIGN POINTS CLOSE	3
6	2 Stages >12 in. (30 cm) With 1-2 Steps	OPEN - SHUT OPERATE					To 2 Points	ALIGN POINTS	6

FIGURE 6.2   (*Continued*) Examples of basic motion data tables for MTM-1 and MOST.

1. Formula or equation—Fig. 6.6
2. Single or multiple pick-off charts—Fig. 6.7*a* and *b*
3. Work sheets—Fig. 6.8*a* and *b*

Figure 6.6 illustrates the equation method for standard time application. The use of computerized systems has increased the equation approach in recent years. The graphics capabilities of computers will increase the pick-off chart applications (Fig. 6.7*a* and *b*) as computers are introduced into the industrial engineering function of companies.

Figures 6.8*a* and *b* are an example of a work sheet that has been developed to set standard times for parts routed to 60- to 80-inch Niles boring mills. Side *A* provides the standard time for all manually controlled work tasks, and side *B* provides the process time for the cuts required. A third work sheet (not shown) would provide gauging time for the appropriate special instruments required. These work sheets include the annotation needed to support the standard time for setup and each piece and to trace the development to basic motion, if necessary.

Line Management Involvement.   The development of standard data will achieve the most successful results if line management becomes involved via:

- Participation in training programs designed to provide all levels of line management an understanding of standard data principles and the benefits they provide to the company and its workers.

DESCRIPTION – LEFT HAND	No.	LH	TMU	RH	No.	DESCRIPTION – RIGHT HAND
M1 Sit down on chair or stool at workbench						
			29.0	B		Bend over
		R-B	--	R-B		Reach to side of chair
		G1A	2.0	G1A		Grasp chair
			34.7	SIT		Lower body to chair
		LM12	14.3	LM12		Move legs forward
		M10B	12.2	M10B		Move chair forward
		RL1	2.0	RL1		Release chair
		R22E	18.0	R22E		Place arms on bench
		LM12	--	LM12		Move feet under bench
			112.2			

METHODS ANALYSIS CHART   REFERENCE No._____

PART_____ DATE _____ STUDY No. _____

OPERATION_____ ANALYST _____ SHEET No. __OF__ SHEETS

FIGURE 6.3   Typical MTM motion pattern for using a chair or stool.

- Standard operating practices committees that are established to identify and document the support activities and other manufacturing practices that are common to all products or parts produced in a defined area.
- Information meetings designed to review and verify facts developed by the industrial engineer relative to methods, workplace layouts, tooling, quality requirements, and safety procedures. These meetings should be conducted for the review of existing methods and/or proposed revisions to methods.
- Review of documented material related to the respective responsibilities of line management, staff groups, and workers to achieve the maximum results from the application of standard time data.

Documentation.   When standard data are used to set standard time values, good documentation procedures must be used to support the standard data development and to support the application of the data to generate standard times. A recommended procedure for documenting standard data development is provided later in this chapter. The procedure produces a formal report called a work management manual. This manual contains the data and support information which must be maintained to keep the standard data valid.

In addition to the formal report, a separate file folder should be established for each prepared report. This file should contain all of the backup papers or other data that would permit a total recall of conditions existing at the time the data were developed. This backup documentation is vital for proper standard data maintenance and as support to justify future data changes as a result of new methods or operating conditions.

The application of the standard data to establish a standard time is accomplished using the work sheets, equations, or pick-off sheets previously discussed. These represent the documentation methods to support the standard time established. Failure to utilize these types of approaches to documentation could make if difficult to trace the logic employed to establish a standard time value. These data application records should be carefully filed for future retrieval in case of standards grievances, or to assist proper standard time revisions.

| | | OPERATION SYNTHESIS | | | |
| | | | | CODE | 2222.01 |
SYM	REF	OPERATION OR ELEMENT DESCRIPTION	TMU	FREQ.	TOTAL
K32		Set stop — hex turret			.013
	KW2	Loosen and tighten lock screw			483.1
	SKM3	Get and return tool — toolbox			685.6
	Y1	Pick up wrench			35.0
	A2	Lay aside wrench			17.8
	Chart 6-11	Turn screw in or out, 6/12/8/2			123.7
					1345.2
K33		Set tool to cut — hex turret (additional cuts)			.002
	Chart 6-1	Start and stop spindle, 14/2			23.0
	Chart 6-2	Engage and disengage feed, 14/2 + AP	43.6	2	87.2
	Chart 6-2	Move to job, 14/4	38.5	2	77.0
					187.2
K34		Set tool to cut — hex turret, first cut			.003
	K33	Set tool to cut			187.2
	Chart 6-2	Index turret, 16/14	39.9	2	79.8
					267.0
K35		Set up sliding head — hex turret			.013
	SKL3	Get and return tool holder			664.6
	Y1	Pick up wrench	35.0	2	70.0
	A2	Lay aside wrench	17.8	2	35.6
	KW2	Loosen and tighten bolt (1)			483.1
	U11	Place holder in head			47.1
	X11	Remove holder			27.4
					1327.8
K36		Change tool — Jacobs chuck			.011
	SKM3	Get and return tool			756.6
	J4	Loosen chuck			125.2
	K4	Tighten chuck			133.8
	Y1	Pick up tool			35.0
	A2	Lay aside tool			17.8
	U11	Position key in chuck			33.0
					1101.4
K37		Place sleeve — hex turret			.007
	Chart 2-1	Place and remove sleeve, Class III/A			113.5
	SKL3	Get and return sleeve			593.6
					707.1

FIGURE 6.4 Typical operation synthesis.

PROCEDURES AND TECHNIQUES TO DEVELOP STANDARD DATA

Developing standard data involves an eight-step procedure which is valid regardless of the measurement techniques used or the task or operation for which the standard data are being developed. These eight step are:

1. Conduct activity analysis.
2. Establish elements.

765. INSPECT PART FOR DRILL PRESSES AT MULTIPLE SPINDLE DRILL
      PER PART OFG: 1 02-JUN-82
          CHECK 1:1 WITH CHECKING GAUGE
        * CHECK 1:1.
        * 687 - 16747T GAUGE.
      OP BEGINS AT MACH - 2

1   PICKUP PEN FROM HEAD - 2 TO OP SIMO SET UP

              <A1 B0 G1 A1 B0 P0 A0>   1.00   0.

2   MOVE GAUGE FROM OUT - P - 1 TO OUT - P - 2 SIMO SET UP

              <A10 B0 G1 A10 B0 P1 A0>   1.00   0.

3   POSITION PART FROM TABLE - 2 TO GAUGE AT OUT - P - 2

              A3 B0 G1 A3 B0 P6 A0   1.00   130.

4   INSPECT PART 3 POINTS

              A0 B0 G0 A0 B0 P0 T3 A0 B0 P0 A0   1.00   30.

5   HOLD + PULL PART FROM GAUGE AT OUT - P - 2 TO OUT - P - 2 ASIDE

              A0 B0 G0 M1 X0 I0 A1   1.00   20.

6   MARK ON PART AT OUT - P - 2 1 DIGIT = (CHECKMARK) USING PEN ASIDE

              A1 B0 G1 A1 B0 P1 R3 A1 B0 P1 A0   1.00   90.

7   GET + PLACE PART FROM OUT - P - 2 TO OUT - P - 2 (STACK) AND RETURN TO MACH - 2

              A1 B0 G3 A1 B0 P3 A3   1.00   110.

                                                  TOTAL TMU   380.

FIGURE 6.5 Example of suboperation sheet.

Standard Time = [.0016 + Table 1 value + .0021H + .0015T] 1.15
  per piece
  in hours
                    where   H = Number of holes
                              T = Number of threaded holes

TABLE 1
PART HANDLING

Weight of part in pounds	Hours/Piece
to 1	.0025
> 1 - 5	.0033
> 5 - 7½	.0048

(Values shown are only for illustration)

FIGURE 6.6 Example of time formula or equations.

3. Develop and/or collect elemental standard times.
4. Design work sheet.
5. Develop task times.
6. Test and refine data.
7. Develop allowances.
8. Prepare work management manual.

CHART 3-4
CLAMP PART IN POSITION FOR MACHINING

Number of clamps required	TYPE I CLAMPS		TYPE II CLAMPS		TYPE III CLAMPS		TYPE IV	ALL TYPES
	Nut and clamp remain on stud KN2	Nut and clamp removed from stud KR2	Nut and clamp remain on stud KP2	Nut and clamp removed from stud KS2	Nut and clamp remain on stud KQ2	Nut and clamp removed from stud KT2	Nut or bolt remains — clamp removed KU2	Relieve strain of clamp KV2
1	.006	.010	.006	.010	.007	.011	.006	.003
2	.012	.019	.011	.019	.014	.021	.011	.006
3	.017	.027	.016	.027	.020	.031	.016	.009
4	.022	.036	.021	.036	.026	.041	.020	.012
5	.028	.045	.026	.045	.032	.051	.025	.015
6	.033	.053	.031	.053	.039	.061	.030	.018
7	.038	.062	.036	.062	.045	.071	.035	.021
8	.043	.071	.041	.071	.051	.081	.040	.024
9	.049	.080	.046	.080	.057	.091	.045	.027
10	.054	.088	.051	.088	.064	.101	.050	.030
Condition	A	B	A	B	A	B	C	
Brief descriptive sketch	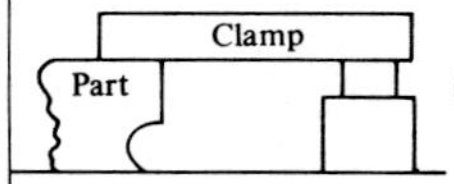		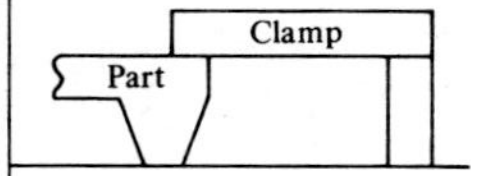		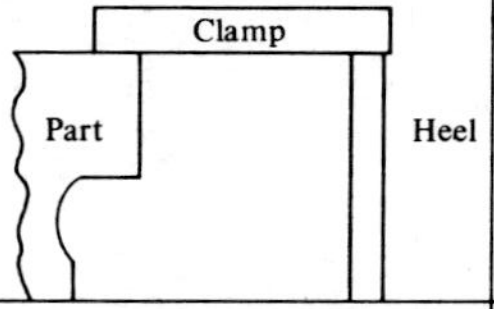		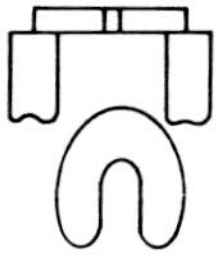	
Basic characteristic	Several blocks stacked to build up heel		Clamp and heel one unit		Clamp considerably above table level — heel will in most cases not stand by itself		Slotted washer	

The chart includes time to pick up wrench, loosen clamp, lay aside wrench, run nut off stud or provide more clearance, reverse clamp, replace clamp, run nut down or back on stud, pick up wrench, tighten nut, lay aside wrench.

FIGURE 6.7a Example of pick-off chart—elemental data covering time required to clamp parts into position prior to machining.

.03 Group D .05		.05 Group E .07		.07 Group F .12		.12 Group G .19	
	.04		.06 ·		.10		.15

24401 05 4/60 (.034)	**25120 05 4/60 (.053)**	**37988 05 4/60 (.091)**	**36090 05 3/60 (.133)**
SOCKET Assemble socket, post, washer, sleeve & PIN; assem screw nut & peen. PEEN THREAD; screw knobs onto "T" handle; SCREW into post (6)S (3)M (4) Thd. con.	LATERAL CASSETTE HOLDER UNWRAP tray & tube; DRILL & PIN tube to support; remove paper around screw holes & screw tray to support (2)S (1)M (1) Drill & pin (3) Screws Drill press	WEINBERGER HAND TRACTION UNWRAP angle & plate; RIVET angle to plate; bolt (5) rubber wedges to plate & PEEN ends (7)S (4) Rivets (5) Screws, washers, nuts (5) Peen	SANITIZER RACK Assemble (8) sets (shoulder pins, spacers, washers, wheels) and STAKE to rack (40)S (1)M (8) Stake
22669 05 5/59 (.039)	**12855 05 4/62 (.058)**	**B7201 05 6/60 (.099)**	**53207 05 8/63 (.151)**
WASTE VALVE STEM Assem DRILL, PIN handle to stem; assem screw nut, gland packing, cover, washer, disc; DRIVE PIN thru stem (5)S (3)M (1) Groove pin (1) Drill & pin	SWITCH ASSEMBLY DRILL kick fork; RIVET kick fork to shaft; assem roller & pin to kick fork & PEEN ends; DRIVE PIN in shaft; DRIVE weight lever thru shaft & PEEN; RIVET spring to frame w/clip (10)S (2)M (5) Rivets (3) Peen (1) Pin (1) Fit	ADJUSTABLE FLASK HOLDER UNWRAP flask handle & knob; SCREW thrust nut to handle; assem plunger, plunger screw, knob & guide screw to holder; DRILL & PIN knob to plunger screw; assem (2) rubber bumpers; TEST & WRAP (6)S (1)M (1) Screw (1) Drill & pin (2) Thd. con. Drill press	VACAMATIC SHAFT ASSEMBLE & ALIGN (6) cams to shaft w/set screws; DRILL & PIN (6) cams set to shaft (6)S (1)M (6) Set screws (6) Drill & pins Drill press Fixture
33191 05 12/62 (.044)			**53153 05 12/62 (.157)**
HANDLE & PINION ASSEMBLY Assem DRILL & PIN pinion to shaft; assem plate to shaft; assem DRILL, PIN "T" hand to shaft (3)S (1)M (2) Drill pins	OIL CHECK AEROFLUSH UNWRAP cap & tube; assem spring, gasket & cap to plunger; PRESS FIT sleeve to plunger; DRILL, TAPER, REAM & PIN connector to plunger; FILL TUBE with oil, insert plunger, screw up cap tight	**22036 05 12/59 (.110)** COMPER KNEE & FOOT REST UNWRAP foot rest & knee crutch; (2) DRILL & PIN foot rest to rod; REAM & assem foot rest, wing nut assem & stop screw to knee crutch; assem stud, set screw, post assem & wing nut to knee crutch; CLEAN & prepare for pack	HEAD REST Assem, DRILL & PIN (2) socket assemblies to weldment; assem coupling, washers, bolt & wing nut to weldment; DRILL & PIN stud to weldment; CLEAN and assem 3 rubber pads (11)S (1)M (3) Drill & pins (1) Speed nut & pin (1) Thd. con. Drill press
59123 05 11/61 (.050)	**18422 05 4/61 (.053)**		**30142 05 8/60 (.160)**
NEUROSURGICAL ATTACHMENT Screw support assembly in to vertical support; DRILL & PIN; CLEAN (2)M (1) Groove pin	(1) Drill taper pin no. 2 (1) Press fit Drill Arbor press	(3)S (4)M (2) Screws (2) Drill & pin (1) Thd. con. Drill press	UNIVERSAL SOCKET ASSEM "T" handle to socket & SCREW on (2) knob ends, insert pad in socket; ASSEM & PIN bolt to sleeve; screw swivel on bracket & screw in set screw; assem bolt assy. & washer to swivel & screw on "T" handle assy: assem stop pin & set screw; CLEAN UNIT (8)S (2)M (3) Screws (1) Roll pin (4) Thd. con. Fixture
		26186 05 3/60 (.096)	
		SUPPORT ASSEMBLY UNWRAP clamp & socket assembly, screw in stud, PIN knob to stud; assemble ball socket, set screw & wing nut (8)S (1)M (1) Screw (1) Drive pin (3) Thd. con.	

Dept. ASSEMBLY	Task Area SUBASSEMBLY	Activity ASSEMBLE, DRILL, PIN, PEEN, RIVET, PRESS FIT	Code No. 59400T10

FIGURE 6.7b (*Continued*) Example of pick-off chart—spreadsheet for short-cycle light assembly work.

Figure 6.9 is a graphic presentation in a chronological order of the steps to develop standard data. The eight-step approach to standard data development is the same whether the data are being developed to cover a family of parts, a job, or operations. These three recognized classifications are defined as follows:

- *Family:* Standard data designed to provide a standard time for all operations required to produce a variety of parts of similar design, for example, shafts, gear blanks, doorknobs.
- *Job:* (A subdivision of the family data.) Job standard data which have been designed to cover a specified operation performed upon a number of different but similar parts.
- *Operation:* Standard data designed to cover all work that can be performed to produce a variety of parts at one or more workstation having the same capability.

Before presenting the details of the eight-step procedure, we should list some basic concepts related to standard data development that have considerable influence on achieving a successful result. These are:

1. Elements of work are seldom performed in exactly the same manner when examined over time.

DATE: *10/5/—* APPLICATOR: *aℑ3* APPROVED:
VERTICAL BORING MILLS WORKSHEET "A" 60" and 80" NILES
BP NO. *52762-RH* OPER. NO. *10* CLASS *2*

LINE		ELEMENT DESCRIPTION	SETUP				EACH PIECE			
			Sym.	Unit hrs.	Occ	Total hours	Sym.	Unit hrs.	Occ	Total hours
1		Shop setup - 1st chucking	K216	.139	*1*	*.139*				
2		Shop setup - sub. chucking	K217	.116	*1*	*.116*				
	GET AND LAY ASIDE PART	**O.D. I.D. HEIGHT**								
3		To 36" -- To 18"					K5	.042		
4		24" - 36" 18" to 48"					K8	.082		
5		36" to 72" -- To 18"					K7	.061		
6		To 72" Over 36" 18" to 48"					K14	.082		
7		Over 72" -- To 18"					K10	.070	*2*	*.140*
8		Over 72" -- 18" to 48"					K16	.091		
9		Put aside fin. mach. part					K144	.031	*2*	*.062*
10	CLAMP AND TRUE	Sgl. stop - 1st chucking	K222	.328	*1*	*.328*	K218	.152	*1*	*.152*
11		Sgl. stop - sub. chucking	K222	.328			K220	.105	*1*	*.105*
12		Stop change	K223	.158						
13		Dbl. stop - (thin section)	K224	.486			K227	.140		
14		Forgings - 1st chucking								
15		sub. chucking								
16	TOOLING	Setup tool-rgh. or knife tool	K197	.028			K210	.019	*12*	*.228*
17		Setup tool-form or slice	K221	.038			K242	.029	*2*	*.058*
18		Set tool to 1st cut on surface					K201	.022	*9*	*.198*
19		Set tool to add. cut on surface					K202	.013	*13*	*.169*
20		Set tool to 1st cut-knife tool only					K198	.020	*6*	*.120*
21		Set tool to add. cut - knife tool only					K215	.005		
22		Trial cuts-tol. > ± 1/64					K225	.027		
23		Trial cuts-tol. < ± 1/64					K226	.055		
24		Set head to angle - normal	K234	.046			K234	.046		
25	GAGING	Trammel or rod	K98	.025	*4*	*.100*	K70	.015	*4*	*.060*
26		I.D. micrometer-to 36" dia.	K181	.052			K136	.034		
27		I.D. micrometer-over 36" dia.	K181	.052			K137	.051		
28		Template or square					K138	.012	*2*	*.024*
29		Calipers	K187	.010	*2*	*.020*	K134	.015	*3*	*.045*
30	MISC.	Change speed					K207	.005	*9*	*.045*
31		Change feed					K208	.003	*4*	*.012*
32		TOTAL LEVEL MANUAL TIME				*.703*				*1.418*
33		MACHINE TIME - WORKSHEET "B"				*—*				*2.99*
34		TOTAL LEVELED TIME				*.70*				*4.41*
35		% ALLOWANCE				*07*				*.44 / .60*
36		TOTAL ALLOWED TIME	SETUP			*.77*	EACH PIECE			*5.45*

FIGURE 6.8a Illustration of a worksheet—data for manual elements.

2. Time values need only represent a logical sequence of motions necessary to perform the work following a specified method.

3. Variables created by changes in the work or part should be viewed as different work tasks.

DATE:
APPLICATOR:
APPROVED:

VERTICAL BORING MILL MACHINING TIME
WORKSHEET "B"
MACHINE GROUP – V

VBM-1 – VI
B.P. NO. 5276-2-RH
OPER. NO. 10

LINE	TOOL MAT. (a)	No. (b)	CUT DESCRIPTION (c)	DIAMETER START (d)	DIAMETER FINISH (e)	STOCK (f)	DEPTH OF CUT (g)	SPEED SFM (h)	SPEED RPM (l)	FEED (m)	HOURS PER INCH (n)	LENGTH OF CUT (p)	HOURS PER CUT (q)	NO. OF CUTS (r)	TOTAL HOURS (s)	HEAD I (t)	HEAD II (u)	MAX. STOCK (v)
1	C	1	RF (OD)	72 7/8	62 1/4	3/8	1/4	180	9	.041	.045	5 1/4	.24	2	.48		(4)	7
2	"	2	RT To 3"BUTT	72 7/8	72 1/8	.	"	'	'	'	'	3 3/4	.17	2	.34	34		73 1/2
3	"	3	RB	62 1/4	62 3/4	1/4	"	'	'	'	'	6 3/4		1	.30	30		62 1/4
4	"	3	SFB	62 3/4	63	1/8	"	155	10	.053	.020	6 1/2		1	.13	13		–
5	HSS	4	KF	72 1/8	63	–	—	75	3.8	-	(	4 1/2		1	.09	09		(
6	"	5	KT BEGIN COLLAR		72 1/8	–		'	'	-	(	3		1	.06	06		(
7	"	6	KB		63	–		'	4.6	-	(	6 1/2		1	.09	09		(
8	C	7	RF (OD)	72 7/8	63	3/8	1/4	180	9	.041	.045	5	.23	2	.46	46		(20)
9	"	8	RT BUTT	72 7/8	72 1/8	"	"	'	'	'	'	3	.14	2	.28	28		
10	"	7-9	RS HUB	72 1/8	64 1/4	3/4	3/8	180	10	.031	.056	4	.22	7	1.78	110		(18)
11	HSS	10	"	64 1/4	66 1/4	–		'	'	"	(	3/4		1	.18	18		–
12	HSS	11	KF	64 1/4	63	–		75	4.6	(	(	3/4		1	.02	02		–
13	"	11	KF COLLAR	72 1/8	66 1/4	–		'	3.8	(	(	3		1	.06	06		–
14	"	12	KHUB	64 1/4	66 1/4	–		'	4.2	(	—	3/4		1 1/2	.08	08		–
15	C	8	FS 1/2 R(1) / FS WELD BEVEL		66 3/4 / 62 1/4										.05 / .03	.05 / .03		–

MATERIAL DETAILS		SPECIAL REQUIREMENTS	TOTAL HOURS	w
SPEC. 121-I	LINE		LIMITING MACHINE TIME	y
HEAT TREAT -	TOL.		LEVEL MACHINE TIME	z
BRINELL -	RMS			
GROUP 5	REMARKS			

TOTAL HOURS 2.99   2.99

FIGURE 6.8b (Continued) Illustration of a worksheet—cutting data and time allowed.

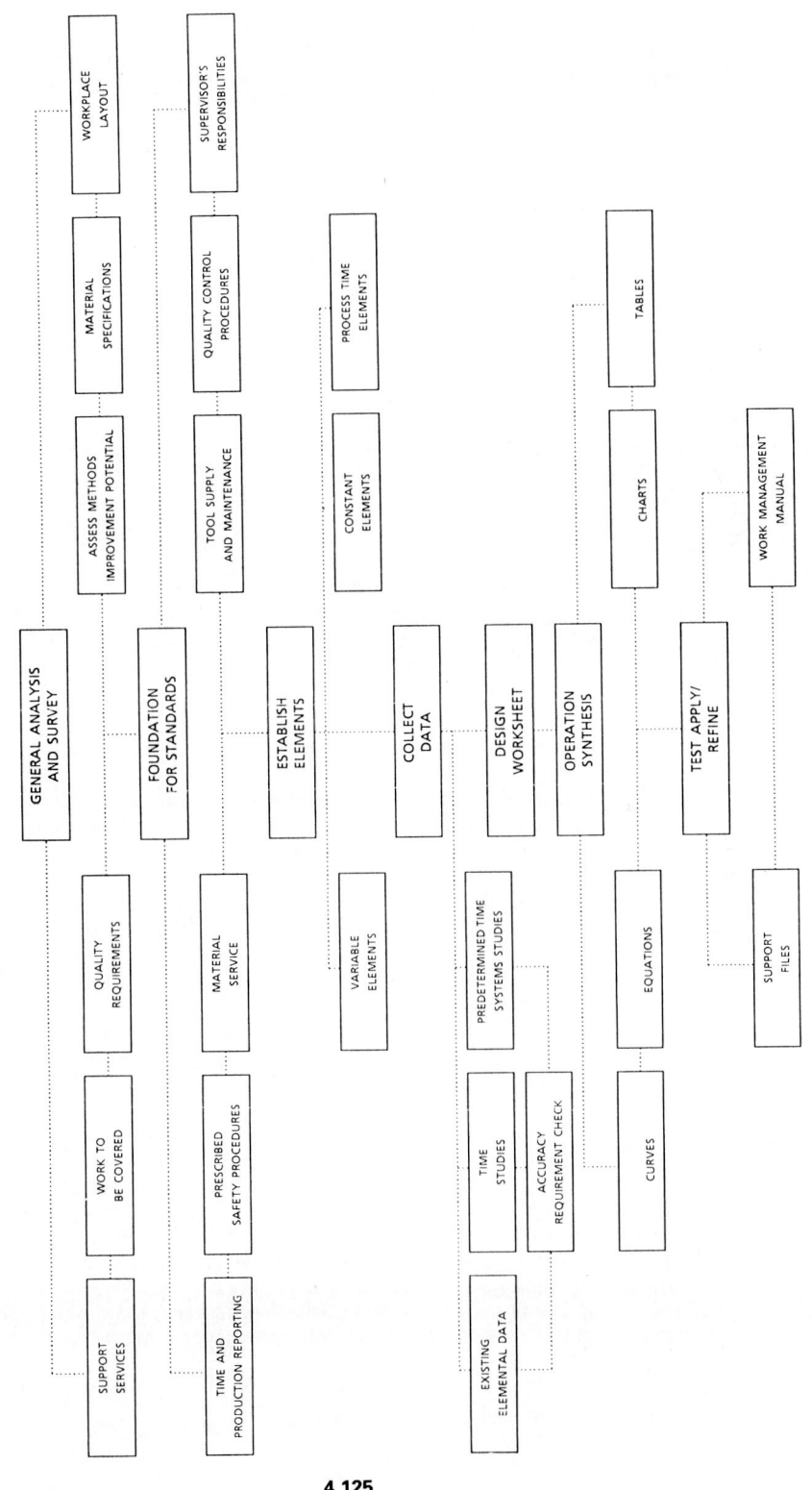

FIGURE 6.9 Steps of standard data development.

4. Averaging is required and necessary in standard data development. An exact standard time is impossible.

5. All standard data which are developed must be supported with sufficient documentation to permit duplication of the existing conditions at a future time.

Step 1—Activity Analysis. The objective of this step is to obtain a detailed understanding of the work that is to be included in the standard time being developed. The activity analysis entails obtaining information on the following items:

- Scope of product or parts and materials
- Primary manufacturing equipment used
- Operations performed
- Auxiliary equipment used
- Expendable materials required
- Workplace layout(s)
- Primary operator duties
- Secondary (auxiliary) operator duties
- Shop information package
- Timekeeping and unit count procedure
- Standard manufacturing practices
- General observations

Figures 6.10 and 6.11 present formats of forms which can be used to collect the activity analysis data. The illustrations are machine-shop-oriented, but the same concepts can be followed in preparing a similar format for other production areas.

Step 2—Establish Elements. This step involves the listing of all work element data required to set up the equipment to produce a product or parts, to perform an operation, and to perform those elements required to produce the part. Figure 6.12 illustrates a typical list of work elements for a simple "mill slot in shafts" operation.

The process of developing the work elemental data requires that the data be classified into either a constant element or a variable element. If an element is of such a nature that a constant or nearly constant time is required to perform it, the element is classified as a constant. If the time required to perform the element varies with one or more variable factors or parameters, the element is classified as a variable.

Constant Elements. A good example of a constant elemental time value is the element "Sit down on chair or stool at workbench." This is an example of an element that will require an almost constant time whenever it is performed—"almost" because the distances reached or the lengths of the leg motions can vary slightly. For practical purposes, however, the element can be considered as a constant. In many cases in actual practice an elemental time value will not be truly a constant time value.

Fortunately, the accuracy requirements of most time standards permit treating many "almost" constant elements as constant values. When an elemental time value is constant, it greatly simplifies the subsequent use of the element in developing standard data.

Variable Elements. There are some elements of work whose time values will vary substantially in relation to one or more parameters of variable factors. Many machine time or process time elements are of this nature. For example, the time required to drill a hole in metal will vary in proportion to a number of variable factors such as:

1. Size of the hole
2. Depth of the hole
3. Type of metal
4 Type of drill

STANDARD DATA DEVELOPMENT

ACTIVITY ANALYSIS

DEPARTMENT OR WORK CENTER NAME:_____NO._____

FOREMAN NAME(S):_____;_____;_____

SUPT. NAME:_____I.E._____

ITEM	INCLUDES	ANALYSIS FINDINGS & EXAMPLES	WMM SECTION
SCOPE	PRODUCT/PART NAME(S)		
	PRODUCT/PART SIZE RANGE	WT. CU. FT. OTHER	
	RAW MATERIAL CONFIGURATION (INCOMING MAT'L)	CASTINGS FORGINGS BAR STOCK BAR STOCK CUTOFFS SOFT GOODS PACKAGED ITEMS COILS STRIPS ETC.	
	RAW MATERIAL "CHEMICAL" SPECS		
	NORMAL SHOP ORDER VOLUMES	LOW VOLUME RANGE _____ HI VOLUME RANGE _____	
	MATERIALS REQUIRED AND BECOME PART OF THE PRODUCT AS A RESULT OF THE OPERATION	WELDING ROD AND SPECS WELDING WIRE AND SPECS THREAD ADHESIVE	
PRIMARY EQUIPMENT	LISTING OF: 1. EQUIPMENT BRAND NAME 2. COMPANY NO. 3. SIZE 4. H.P. 5. SPEED SELECTS 6. FEED SELECTS 7. RESTRICTIONS 8. CONDITION OPINION	RECORD HERE OR REFER TO SEPERATE EQUIPMENT SPEC SHEET(S) AS PER EXAMPLE	

FIGURE 6.10 Standard data development—activity analysis.

5. Speed of drill
6. Pressure on drill spindle
7. Sharpness of drill
8. Angle of drill tip
9. Relief angle of drill
10. Lubricant used

It is obvious that the determination of elemental machine time values can become quite complex because of the great number of variables that must be considered. A similar prob-

ITEM	INCLUDES	ANALYSIS FINDINGS & EXAMPLES	WMM SECTION
OPERATIONS PERFORMED	LIST OF ALL OPERATIONS NORMALLY PERFORMED	THE OPERATIONS INCLUDE THE NECESSARY TASKS REQUIRED TO SET UP AND RUN THE EQUIPMENT TO PRODUCE PIECES INVOLVING DRILL, TAP, ETC.	
AUXILIARY EQUIPMENT	ALL OTHER PIECES OF EQUIPMENT, HAND TOOLS, POWER TOOLS NEEED BY OPERATOR TO PERFORM THE OPERATIONS	OVERHEAD CRANES JIB CRANES HAND TRUCKS ETC. AIR HOSES	
EXPENDABLE MATERIALS	LIST MATERIALS THAT ARE USED BUT NOT PART OF THE PRODUCT	CUTTING OILS RAG WASTE LUBRICATING OILS	
WORKPLACE LAYOUT(S)	DETAIL SKETCH OF PROPER WORKPLACE LAYOUT - WITH TOOL LOCATIONS, JIB CRANE, ETC. INDICATED	USE FORM NO _____ FOR DOCUMENTATION OF THIS LAYOUT DIFFERENT LAYOUTS MAY BE REQUIERD TO ACCOMODATE VARIATIONS IN RAW MATERIAL CONFIGURATIONS AND/OR CONTAINERS	
PRIMARY OPERATOR DUTIES	LIST BASIC TASKS REQUIRED BY THE OPERATOR TO PRODUCE THE PRODUCT/PART (NOT THE DETAIL METHODS) LIST THOSE RELATED TO SETUP - AND EACH PIECE		
AUXILIARY OPERATOR DUTIES	LIST TASKS REQUIRED TO SUPPORT PRIMARY DUTIES THAT OCCUR INFREQUENTLY		
SHOP PACKAGE	LISTING OF INFORMATION PROVIDED THE OPERATOR TO PRODUCE THE PRODUCT/PART	EXAMPLES OF INFO PROVIDED AND/OR SOURCE OF INFO	
PRODUCTION TIME & COUNT RECORDING SYSTEM	REFERENCED TO COMPANY MANUAL OR BRIEF DESCRIPTION OF SYSTEM AND FORMS USED		
OBSERVATION COMMENTS	LISTING ANY OBSERVATIONS RESULTING FROM CONDUCTING THE ACTIVITY ANALYSIS	POOR HOUSEKEEPING - TOOL CUPBOARDS ARE VERY DISORGANIZED MATERIAL EVERYWHERE - NO ORGANIZATION MAJOR METHOD IDEAS	
STANDARD PRACTICES INITIATION	CREATION OF STANDARD PRACTICES COMMITTEE	STANDARD PRACTICES	

FIGURE 6.10   (*Continued*) Standard data development—activity analysis.

lem exists in developing elemental time values for process times. Handspray painting is a good example. Some of the variables that must be considered are the area to be covered, the shape of the part, whether the coat is a base coat or a finish coat, the finish quality requirements, the viscosity of the paint, and the pattern generated by the nozzle.

There are also manually controlled elements that are variable. The factor causing the variability may be the size of the part, the nature of the material, or the shape or weight of the part. From a work measurement standpoint, the problem is first to identify the variable and then to determine the relationship between the variable and the time required to perform the element.

ACTIVITY ANALYSIS

DATA COLLECTION FOR: _____

TYPE EQUIPMENT OR OPERATION_____

LOCATION(S) OF EQUIPMENT

PLANT(S)	WORK CENTERS WITHIN EACH PLANT

TYPE OPERATION PERFORMED

LIST TYPE OPERATION	(√) WERE PERFORMED BY WORK CENTER					

MISCELLANEOUS DATA RELATED TO MACHINE WORK CENTER,
DEPARTMENT OR PLANT

ITEM	YES	NO	EXPLAIN AS APPROPRIATE	APPLIC. SCOPE		
				W/C	AREA	PLT.
RAW MATERIAL DELIVERY						
FIN. PARTS REMOVED						
TOOL KITS PROVIDED						
TOOL KITS DELIVERY						
USED TOOL KITS REMOVED						
COOLANT USED/SPEC.						
COOLANT ADDED BY:						
TOOLS OBTAINED FROM TOOL CRIB						
TOOLS RETURNED TO TOOL CRIB						
TOOLS SHARPENED BY:						
QUALITY CONTROL SPECS. AVAILABLE						
AVAILABLE Q.C. SPEC. ACCEPTABLE FOR WORK MEASUREMENT						
PROCESS SHEETS AVAILABLE						
PROCESS SHEETS ACCEPTABLE						
DRAWINGS PROVIDED						
TIME RECORDING SYSTEM						
CHIP/SCRAP REMOVAL						
EQUIPMENT LUBRICATION						

FIGURE 6.11 Example of format used to collect activity analysis data.

MACHINE SPECIFICATION SHEET

MACHINE DATA NEW _____ REVISED _____ TYPE OF MACHINE _____	
MACHINE NUMBER _____ CONDITION CODE _____ COOLANT SYSTEM YES☐ NO☐	
HORSEPOWER OF MOTOR/MOTORS _____	
COST CENTER _____ NUMBER OF SPINDLES _____	
DO ALL SPINDLES HAVE SAME SPEED? YES☐ NO☐	
FEED? YES☐ NO☐	
MULTIPLE DRILLS ONLY	
NUMBER OF SPEEDS IN EACH RANGE _____ RAPID TRAVEL RATE _____	

		SPEED SETTINGS								
SPINDLE NUMBER	HORSEPOWER	SPEED SET AND ACTUAL								
		SET								
		ACT								
		SET								
		ACT								
		SET								

FIGURE 6.11   (*Continued*) Example of format used to collect activity analysis data.

In addition to the constant or variable classification, notations should also be made when developing the list as to the probable frequency of application and the factor(s) that will affect the frequency of application.

Determine if the element will occur once per setup, once per part, or less than once per part as in many gauging operations which occur according to established inspection frequencies. The brush vice element (*L*) as shown in Fig. 6.12 could have a less than one frequency, depending upon established operating practices for the mill slot operation.

Step 3—Develop and/or Collect Elemental Standard Times.   Step 2 involves the listing of the work elements included in the operations for which the standard data are to be developed. The objective of step 3 is to establish the standard times for elements identified and for any additional elements not identified during step 2. The time for each element can be developed by two basic approaches or a combination thereof:

- Select existing studies from which elemental time values can be extracted.
- Develop new elemental time values following the procedures of the work measurement technique used.

The application procedures for the available work measurement techniques are presented in other chapters of this handbook. The two approaches still used to develop standard times for work elements are time study and predetermined time systems.

Development of Standard Data for Work Elements Using Time Study.   The use of time study requires the taking of an appropriate number of complete time studies covering

PRODUCT/PARTS DESCRIPTIONS (BY COMPANY NAME)

RAW MATERIAL DESCRIPTION

MATERIAL SPECIFICATIONS

PART/PRODUCT SIZE RANGE (CIRCLE VARIABLE USED)
(FOR MAJORITY OF PRODUCT/PARTS)

W/C NO.	SIZE RANGE BY: WEIGHT; VOLUME; OR_____
	FROM: TO:

LOT SIZE RANGE (FOR MAJORITY OF ORDERS)

W/C NO	LOT SIZE RANGE
	FROM: TO:

MATERIAL CONTAINERS USED
LIST TYPE CONTAINERS AND, IF ASSIGNED, USAGE BY TYPE/SIZE PRODUCT OR WORK CENTER AS APPROPRIATE

TYPE CONTAINER	USED PRIMARILY FOR:

FIGURE 6.11 (*Continued*) Example of format used to collect activity analysis data.

the scope of the work to be included in the standard data. The number of studies necessary will vary with the scope of the data and can be determined by statistical analysis (Fig. 6.13). See Sec. 4, Chap. 2 for a typical time study taken for purposes of standard data development.

In many plants time studies that have been taken in the past may be suitable to use, and thereby reduce the time spent on this development step. As the time studies are taken, they are posted on a master table of detailed time studies—illustrated in Fig. 6.14. The

REF SYMBOL	ELEMENT DESCRIPTION	CONSTANT	VARIABLE	APPLIC. FREQ.
A	PICK UP PART FROM TABLE		X	I
B	PLACE IN VISE		X	I
C	TIGHTEN VISE	X		I
D	START MACHINE	X		I
E	RUN TABLE FORWARD - PER INCH	X		F
F	ENGAGE FEED	X		-
G	MILL SLOT		X	I
H	STOP MACHINE	X		F
I	RETURN TABLE - PER INCH	X		I
J	RELEASE VISE	X		I
K	LAY ASIDE PART IN TOTE PAN		V	F
L	BRUSH VISE	C		
	SET UP			
M	GET TIME CARD	X		
N	GET JOB AND DRAWING	X		
O	CLOCK TIME ON CARD	X		
P	GET TOOLS FROM TOOL ROOM	X		
Q	CHECK OPERATION WITH DRAWING	X		
R	CLEAN VICE AND TABLE	X		
S	PUT VICE ON TABLE	X		
T	GET TWO BOLTS AND SET IN TABLE SLOT	X		
U	TIGHTEN VICE TO TABLE (2 BOLTS)	X		
V	PUT ON CUTTER AND COLLARS	X		
W	LOOSEN VICE FROM TABLE	X		
X	REMOVE 2 BOLTS TO LOCKER	X		
Y	REMOVE VICE FROM TABLE	X		
Z	REMOVE CUTTER AND COLLARS	X		

FIGURE 6.12 Typical list of work elements.

DATE 2-1 —
STUDY NO. 1
SHEET NO. 1
OF 1
SHEETS

S-1

COLLECTING AND TABULATING DATA
MILL SLOT FORMULA

NUMBER	1		2		3		4		5		6		7		8		9		10		11		12	
ELEMENTS	PICK UP PART FROM TABLE		PLACE IN VISE		TIGHTEN VISE		START MACHINE		RUN TABLE FORWARD 3"		ENGAGE FEED		MILL SLOT		STOP MACHINE		RETURN TABLE 5.5"		RELEASE VISE		LAY ASIDE PART IN TOTE PAN		BRUSH VISE	
LINE	T	R	T	R	T	R	T	R	T	R	T	R	T	R	T	R	T	R	T	R	T	R	T	R
1	06	06	06	12	23	35	03	38	08	46	02	48	72	120	16	36	13	49	12	61	09	70	11	81
2	05	86	08	94	21	215	03	18	09	27	03	30	76	306	13	19	12	31	14	45	07	52	06	58
3	07	65	07	72	21	93	02	95	10	405	03	08	70	78	12	90	-	M	-	M	-	513	08	21
4	05	26	A 07	46	14	60	03	63	08	71	03	74	73	647	13	60	16	76	12	88	08	96	08	704
5	07	11	07	18	19	37	04	41	09	50	02	52	71	823	11	34	14	48	09	57	07	64	08	72
6	07	79	08	87	17	904	03	07	10	17	02	19	72	91	11	1002	13	15	11	26	09	35	—	
7	07	42	08	50	22	72	03	75	12	87	03	90	71	1161	16	77	14	91	09	1200	06	06	07	13
8	05	18	06	24	15	39	03	42	B 10	64	02	66	76	1342	12	54	16	70	10	80	06	86	06	92
9	06	98	08	1406	12	18	02	20	13	33	03	36	74	1510	14	24	15	39	10	49	06	55	06	61
10	06	67	08	75	14	89	03	92	12	1604	04	08	71	79	12	91	16	1707	11	18	07	25	08	33
11	05	38	06	44	16	60	02	62	09	71	03	74	75	1849	13	62	12	74	09	83	08	91	07	98
12	C 05	1918	08	26	14	45	03	48	12	60	04	64	71	2035	15	50	16	66	13	79	07	86	07	99/72
13	06	92/86	07	2106/99	14	20	04	24	13	37	02	39	75	2214	14	28	15	43	11	54	08	62	09	71
14	(15)	86	07	93	13	2306	03	09	12	21	04	25	71	96	13	2409	15	24	10	34	09	43	07	50
15	06	56	08	64	17	81	03	84	13	97	03	2500	D 72	2694	16	2710	16	26	10	36	08	44	08	52
16																								
17																								
18																								
19																								
20																								

SUMMARY

	1	2	3	4	5	6	7	8	9	10	11	12
	.0083	.0109	.0257	.0044	.0160	.0043	.1090	.0201	.0203	.0151	.0105	.0095
NO. OBSERVATIONS	14	15	15	15	15	15	15	15	14	14	14	13
AVERAGE "T"	.00059	.00073	.00171	.00029	.00107	.00029	.00727	.00134	.00145	.00108	.00075	.00073
MINIMUM "T"	.0005	.0006	.0012	.0002	.0008	.0002	.0070	.0011	.0012	.0009	.0006	.0006
MAXIMUM "T"	.0007	.0008	.0023	.0004	.0013	.0004	.0076	.0016	.0016	.0014	.0009	.0009
RATING (S.E.C. & CY.)	D C1											
LEVELING FACTOR	1.05											
L.F. x AVE. "T"	.00062	.00077	.00180	.00030	.00112	.00030	.00727	.00141	.00152	.00113	.00079	.00077
% ALLOWANCE	15	15	15	10	10	10	10	10	10	15	15	15
TIME ALLOWED	.0007	.0009	.0021	.0003	.0012	.0003	.0080	.0015	.0017	.0013	.0009	.0009

FOREIGN ELEMENTS

SYM	R	T	DESCRIPTION
A	539 / 526	13	PIECE IN VISE WRONG
B	1256 / 1244	12	MACHINE STOPPED
C	1913 / 1898	15	CLEAN HANDS
D	2656 / 2534	122	PIECE PULLED OUT OF VISE
E			
F			
G			
H			
I			
J			
K			
L			
M			
N			
O			

SKILL			EFFORT		
	A1, A2	SUPER		A1, A2	EXCESSIVE
	B1, B2	EXCELLENT		B1, B2	EXCELLENT
	C1, C2	GOOD	✓ (C1)	C1, C2	GOOD
✓	D	AVERAGE		D	AVERAGE
	E1, E2	FAIR		E1, E2	FAIR
	F1, F2	POOR		F1, F2	POOR

CONDITIONS		CONSISTENCY	
A	IDEAL	A	PERFECT
B	EXCELLENT	B	EXCELLENT
C	GOOD	C	GOOD
D	AVERAGE	D	AVERAGE
E	FAIR	E	FAIR
F	POOR	F	POOR

GENERAL RATING FOR STUDY	SKILL	EFFORT	COND	CONSY
	.00	+.05		

STUDY STARTED	STUDY FINISHED	OVERALL TIME
2:54 P.M.	3:10½ P.M.	.28 HRS.

FIGURE 6.13*a* Detailed time study of a simple milling machine operation—front.

MILL SLOT

DEPARTMENT	OPERATOR		
10	MAN WOMAN	NAME G ROSS	NO. 33

EQUIPMENT #3 LE BLOND HORIZONTAL

MILLING MACHINE

MACHINE TOOL NO. 3589

SPECIAL TOOLS, JIGS, FIXTURES, ETC. 6" DIA. SPL SIDE CUTTER

CONDITIONS SOME CASTINGS HAVE ROUGH SPOTS ON SIDES WHICH MAKE THEM HARD TO HOLD IN VISE

MATERIAL SUPPLY LIGHT, TEMPERATURE 8 VENTILATION ~ AVE

OBSERVER   APPROVED BY

MOULD

PATTERN 9341-R

PART DESCRIPTION CLAMP FOR REGULATOR - TYPE X - 4

DIE

DWG. 22299   STYLE ____   SUB. 1   ITEM 1

INS. SPEC.   L. SPEC.   SUB.

MATERIAL COMMON BRASS

NO.	ELEMENTS	SMALL TOOL NOS, FEED SPEED, DEPTH OF CUT, ETC.	ELEMENTAL TIME ALLOWED (BOTTOM LINE OTHER SIDE)	OCCURRENCES PIECE OR CYCLE	TOTAL TIME ALLOWED
1	PICK UP PART FROM TABLE				.0007
2	PLACE IN VISE				.0009
3	TIGHTEN VISE				.0021
4	START MACHINE				.0005
5	RUN TABLE FORWARD 3"				.0012
6	ENGAGE FEED				.0003
7	MILL SLOT	6" DIA. SPL SIDE CUTTER 190 R.P.M. 6"/MIN.			.0080
8	STOP MACHINE				.0015
9	RETURN TABLE 5.5"				.0017
10	RELEASE VISE				.0013
11	LAY ASIDE PART IN TOTE PAN				.0007
12	BRUSH VISE				.0009

TIME ALLOWED, SET UP

EACH PIECE   TOTAL .0198

REMARKS: OPERATOR REMOVES PARTS FROM TOTE PAN AND PLACES THEM ON TABLE WHILE MACHINE IS MAKING CUT. HE ALSO CLEANS CUTTINGS FROM TABLE AT THIS TIME. CUTTING SPEED FOR THIS LINE OF WORK IS HELD CONSTANT AT 190 R.P.M. FEED VARIES WITH WIDTH AND DEPTH OF CUT. ON THIS JOB, FEED IS 6" PER MINUTE

SKETCH

1/2"   1 3/8"   1/2"   7/8"   3 1/4"

OBSERVATION SHEET

FIGURE 6.13b (Continued) Detailed time study of a simple milling machine operation—back.

MASTER TABLE OF DETAIL TIME STUDIES

Job Characteristics

FORMULA Course C #1
DATE Oct. 14, 19 —
PART Brass Clamps for Type X Regulators
OPERATION Mill Slot
PERFORMED ON Horizontal Milling Machines
COMPILED BY M.E.C.

STUDY	S-1	S-2	S-3	S-4	S-5	S-6	S-7	S-8	S-9	S-10
	#1 6/1/	#2 6/1/	#1 6/4/	#2 6/4/	#3 6/4/	#1 6/5/	#2 6/5/	#1 6/6/	#2 6/6/	#1 6/7/
OPERATOR	GROSS	WILLIAMS	SMITH	GROSS	SMITH	WILLIAMS	WILLIAMS	SMITH	WILLIAMS	GROSS
SKILL EFFORT	D C1	D D	C C	D C1	C C	D D	D C	C C2	C1 D	C1 D
MACHINE	#3 Le Blond	#2 Cincinnati	#2 Milwaukee	#3 Le Blond	#2 Milwaukee	#2 Cincinnati	#2 Cincinnati	#2 Milwaukee	#2 Cincinnati	#3 Le Blond
DWG. NO.	22289	325907	89210	61918	99201	33213	82112	63800	92678	55210
ITEM	1	9	4	5	16	3	11	6	4	10
FEED	6"/Min.	4.75"/Min.	5.5"/Min.	5.5"/Min.	5.00"/Min.	5"/Min.				
SPEED	140 R.P.M.	140 R.P.M.	140 R.P.M.	140 R.P.M.	140 R.P.M.	140 R.P.M.				
CUTTER DIA.	6"	6"	5 1/2"	5 1/2"	5"	6"	6"	5 1/2"	6"	5"
MACHINE NO.	3589	863	248	3589	248	863	863	248	863	3589
CLAMP VOLUME	2.23	1.20	11.8	30.9	6.1	63.9				
SLOT DIMENSIONS	7/8" 1 3/8"	4" 1 1/2"	2 1/4" 1"	3" 1 1/8"	4 1/2" 7/8"	3 1/2" 1 1/4"				

Symbol	Operation Description	Time Allowed (Hours)	Reference	Operation Class	S-1	S-2	S-3	S-4	S-5	S-6	S-7	S-8	S-9	S-10
A	Pick up part from table	CURVE A		V	.0007	.0015	.0010	.0012	.0009	.0014				
B	Place in vise	CURVE B		V	.0009	.0013	.0009	.0010	.0009	.0011				
C	Tighten vise	.0024	S-3,5	C	.0021	.0028	.0024	.0026	.0024	.0023				
D	Start machine	.0003	S-1,2,3,4,6	C	.0003	.0003	.0003	.0003	.0004	.0003				
E	Run table forward - per inch	.0007	S-2,4,5	C	.0003	.0007	.0006	.0007	.0007	.0004				
F	Engage feed	.0003	S-1,2,4,5,6	C	.0003	.0003	.0004	.0003	.0003	.0003				
G	Mill slot	SEE MACHINING TABLE		V	.0080	.0178	.0108	.0129	.0082	.0158				
H	Stop machine	.0014	S-3,4,6	C	.0015	.0012	.0014	.0014	.0013	.0014				
I	Return table - per inch	.0005	S-2,3,6	C	.0003	.0005	.0005	.0006	.0004	.0005				
J	Release vise	.0009	S-3,6	C	.0013	.0008	.0009	.0008	.0010	.0009				
K	Lay aside part in tote pan	CURVE C		V	.0009	.0020	.0011	.0014	.0010	.0017				
L	Brush vise	.0010	S-3,6	C	.0009	.0012	.0010	.0011	.0009	.0010				
	SET UP													
M	Get time card	.0406	S-9								.0398	.0450	.0406	.0400
N	Get job and drawing	.0500	S-7								.0500	.0483	.0513	.0505
O	Clock time on card	.0178	S-8								.0136	.0178	.0201	.0179
P	Get tools from tool room	.1303	S-10								.1435	.1181	.1250	.1303
Q	Check operation with drawing	.0122	S-9								.0125	.0098	.0122	.0135
R	Clean vise and table	.0105	S-10								.0115	.0094	.0101	.0105
S	Put vise on table	.0084	S-9								.0075	.0091	.0084	.0087
T	Get 2 bolts and set in table slot	.0180	S-9								.0199	.0171	.0180	.0176
U	Tighten vise to table (2 bolts)	.0156	S-7,9								.0156	.0124	.0156	.0182
V	Put on cutter and collars	.0087	S-7,10								.0087	.0091	.0085	.0087
W	Loosen vise from table	.0078	S-8								.0091	.0078	.0072	.0074
X	Remove 2 bolts to locker	.0045	S-9								.0045	.0060	.0052	.0047
Y	Remove vise from table	.0117	S-10								.0118	.0105	.0125	.0117
Z	Remove cutter and collars	.0077	S-10								.0078	.0080	.0074	.0077

FIGURE 6.14 Time study spreadsheet.

master table becomes the analysis to verify and/or classify the work elements into constant values or variable values. The variable classification requires a determination on how to develop the allowed standard times in relationship to an identified variable's cause. Figure 6.15 illustrates two approaches used to plot the same data. A detailed analysis of these plots (Fig. 6.15) indicates that the variable of "location" would result in treating the work element as three constants, depending upon the plot location.

When an element is affected by two variable job characteristics and the method of treating one of them as a series of constants does not yield satisfactory accuracy, another way of treating the data must be found. For example, when standard data were being developed for the element "Fill core box with sand and peen," it was anticipated that filling and peening time should vary with the volume of sand handled. Accordingly, points were plotted of volume against time with the result shown by Fig. 6.16. Inspection showed that some other factor undoubtedly affected the data.

Further analysis indicated that the relationship between the height and the thickness of the core might also affect the time. It was reasoned that where the thickness of the core is great in comparison with the height, all sand may be put in the box and peened at one time. Where the thickness of the core is small compared with the height, sand must be put in a little at a time and peened frequently. Thus it seemed evident that, in addition to core volume, the ratio of height to thickness would affect filling and peening time.

An attempt was next made to classify the work according to this ratio, plotting one curve for ratios up to one, another for ratios between one and two, and so on. The results were unsatisfactory, however, and this approach was abandoned.

It was then decided to use two curves in conjunction with each other. A curve of time against ratio of height to thickness was plotted for cores having an approximately constant volume. Because volume was constant, the curve (Fig. 6.17) showed a true relationship unaffected by volume. Note that the curve intercepts the Y axis at 0.0020 hour. The time scale was changed by calling this point of interception "1"; the point at which the time doubled, or 0.0040 hour, "2"; and so on. This produced the factor X scale shown on the right of Fig. 6.17.

Next, to get a curve of volume against time unaffected by the height and thickness of the core, each time value taken from the original data was divided by a factor as determined from the curve by the ratio of height to thickness. These values plotted against the corresponding volumes yielded a time-volume curve unaffected by the height and thickness of the core. The points now lined up into two sets as shown by Fig. 6.18.

Additional analysis of the data showed that the higher values were all obtained from studies of cores which were difficult to make. A further classification of jobs as simple or complex permitted the plotting of the final curves shown by Fig. 6.19.

With the curves now available, to arrive at filling and peening time for any core, a factor X is found corresponding to the ratio of the height and the thickness of the core, and a base time for the volume is found from the time-volume curve. These two values multiplied together give the true time required for filling and peening. To illustrate, a simple core having a ratio of height to thickness of 4.2 and a volume of 130 cubic inches would have a standard time for filling and peening of $2.4 \times 0.0092 = 0.0021$ hour.

A more sophisticated solution can be achieved by the use of multivariate analysis. For example, the time to pick parts from a bin might be represented as:

$$Y = a_0 + a_1 X_1 + a_2 X_2 + a_3 X_3$$

where Y = normal time to obtain one part
$\quad X_1$ = number of parts grasped at one time
$\quad X_2$ = location of bin
$\quad X_3$ = weight of part

To calculate or estimate the coefficients (A_0, a_1, a_2), it is necessary to collect elemental data for a variety of conditions or variables. In this case, a minimum of four data sets are

PICK UP PART AND PLACE IN CHUCK

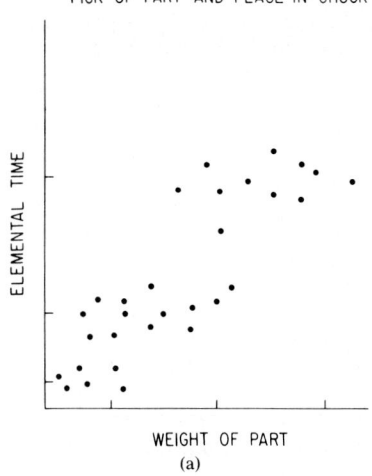

(a)

PICK UP PART AND PLACE IN CHUCK

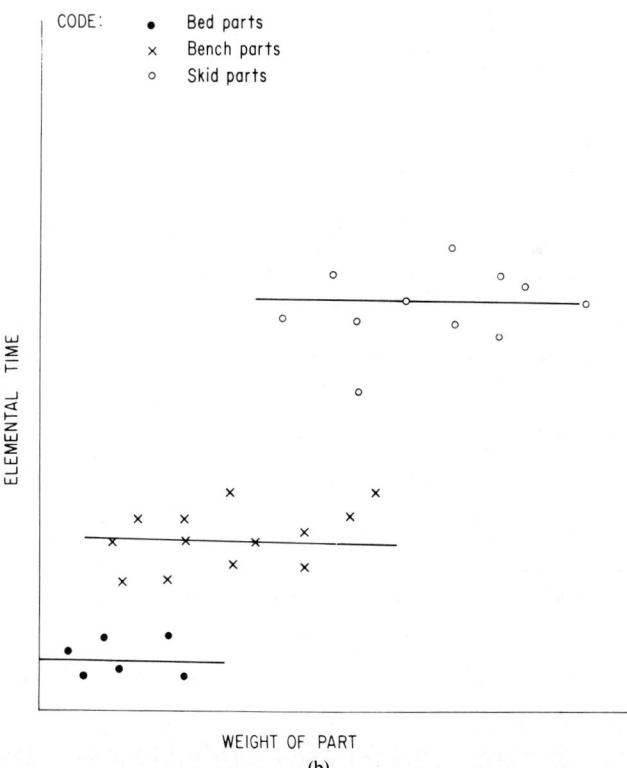

(b)

FIGURE 6.15 Plot of time value for "pick up part and place in chuck."

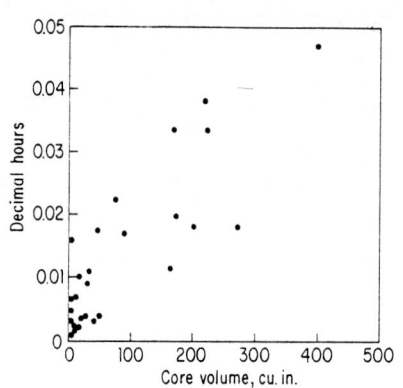

FIGURE 6.16   Points plotted for time study data of time required to fill core box with sand and peen.

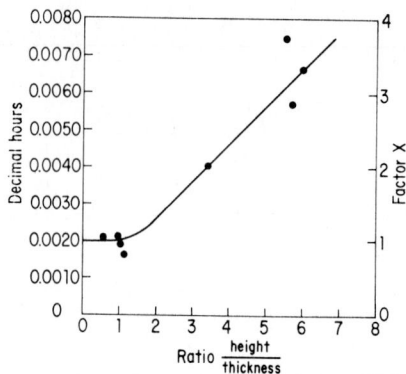

FIGURE 6.17   Curve of ratio of height to thickness versus decimal hours for cores of from 5 to 10 cubic inches volume for element "Fill core box with sand and peen."

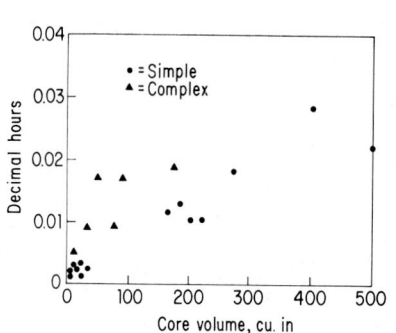

FIGURE 6.18   Points shown in Fig. 6.16 when corrected for effect of ratio of height to thickness of core for element "Fill core box with sand and peen."

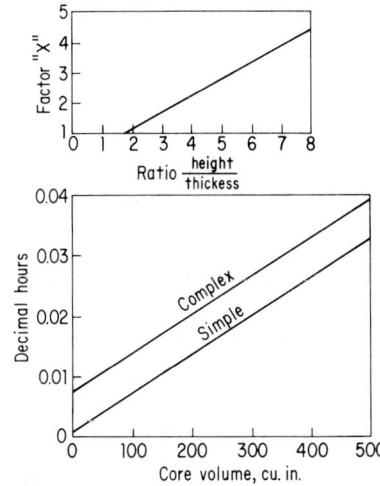

FIGURE 6.19   Final curve sheet for element "Fill core box with sand and peen."

needed to provide a solution. The greater the number of data sets collected, the higher the reliability of the estimates. Either manual or computer solutions may be used to solve the equation.

Although the above equation is for a linear relationship, nonlinear relationships may also be solved by the same model. For example, for the equation

$$Y = a_0 + a_1 X_1 + a_2 X_2 + a_3 X_1^2 + a_4 X_1 X_2 + a_5 X_2^2$$

one can set $X_1 = Z_1$; $X_2 = Z_2$; $X_1^2 = Z_3$; $X_1 X_2 = Z_4$; and in this way, proceed to estimate the linear regression equation described above.

Development of Standard Data for Work Elements Using Predetermined Time Systems.   When using a predetermined time system such as MTM, Work Factor, and MOST, it is generally not necessary to take a complete study of an operation to develop allowed time for work elements. The predetermined time systems enable industrial engineers to develop the motion patterns for the most logical method to accomplish the task related to the defined work element. The most logical method assumes a good workplace layout is or can be implemented, proper tools are or can be made available, and the sequence of motions to accomplish the task can be taught to an average operator.

The use of predetermined time systems will also permit the treating of variable work elements as a series of constants. Figure 6.20 illustrates how the normal variable of load, unload, secure, and release a variable weight part in a three- or four-jaw chuck is organized into a series of constants. The development of the work element time values involves a number of analytical decisions related to:

- Reasons for classifying work elements as variables or constants

Work Center: 1520   Part Name: SHAFT   Apparatus: PUMP   Date: 10-2-—

LATHES RAM TYPE   WORKSHEET "A"

Part No.: 390620   Oper. No.: 30   Material: CL. 2   Applicator: W. B

OPERATION DESCRIPTION								Occ.	Setup	Occ.	Each Piece
Shop setup:					K1	.108		/	.108		
LOAD AND UNLOAD CHUCK	Up to 5#	5.1 to 20#	20.1 to 60#	Over 60# Jib Hoist	Over 60# Chain Hoist						
A Load, unloaded, and 3-Jaw	K62 .006	K62 .009	K63 .012	K65 .042	K66 .069					/	.0120
B Secure in chuck 4-Jaw	K67 .010	K68 .013	K69 .016	K70 .046	K71 .073						
E Turn part end for end 3-Jaw	K72 .006	K73 .008	K74 .009	K75 .024	K76 .051						
and secure in chuck 4-Jaw	K77 .010	K78 .012	K79 .013	K80 .028	K81 .055						
Load and unload fixture	K2 .003	K4 .006	K6 .009	K9 .039	K11 .066						
C Tighten and loosen bolts K61	1 .005	2 .009	3 .012	4 .016	5 .020	6 .024	7 .027	8 .031			
Set up plug fixture				K44 .090							
Use plug fixture				K45 .005							
Advance bar stock and lay aside piece	K82 .005	K83 .007	K84 .010	-	-						
D Load bar stock	K85 .026	K86 .029	K87 .032	K88 .062	K89 .089						
Reset bar feed per 32" of stock used				K58 .006							
Change collet pads				K59 .028				/	.028		
Adjust chuck jaws		3-jaw K13 .027		4-jaw K19 .052							
Change jaws		3-jaw K16 .071		4-jaw K22 .060							
Reverse jaws		3-jaw K17 .055		4-jaw K23 .041							
Remove and replace chuck		3-jaw K15 .141		4-jaw K21 .149							
Cut soft jaws		3-jaw K18 .144		-							
Use copper shims		-		4-jaw K24 .019							
Relieve strain		3-jaw K55 .002		4-jaw K56 .004							
Set up tool			square turret K25 .028					3	.084		
Set up formed tool			square turret K26 .033					/	.033		
Set stop			square turret K27 .014					5	.070		
Index turret – 1st cut			square turret K28 .006							4	.0240
Each additional cut			square turret K29 .003							2	.0060
Set up tool holder			hexagon turret K30 .023								
Place tool in holder			hexagon turret K31 .017								
Set stop			hexagon turret K32 .013								
Set up sliding head			hexagon turret K35 .013								
Place sleeve on holder			hexagon turret K37 .007								
Change tool – Jacobs chuck			hexagon turret K36 .011								
Set up die head			hexagon turret K52 .204					/	.204		
Use die head or roller turner			hexagon turret K53 .003								
Place and remove drill			hexagon turret K47 .017								
Use tap in holder			hexagon turret K46 .001								
Set up roller turner			hexagon turret K91 .022								
Remove flanged tool holder or roller turner			hexagon turret K92 .039							/	.0030
Index turret – 1st cut			hexagon turret K34 .003								
Each additional cut			hexagon turret K33 .002							/	.0030
Change speed			K39 .0003							6	.0018
Change feed			K40 .0008							5	.0040
Set dial clip			K48 .001					5	.005		
Adjust coolant			K50 .0007								
Set up indicator			K42 .015								
Use indicator per spot			K43 .035								
Use rail indicator			K51 .005								
Align splits per split (surface gage)			K60 .051								
Polish, file, or deburr per 1" length × 1" diameter			K54 .0004							/	.0004
Index or change carbide insert			K90 .011								
GAGE PART – WORKSHEET C									.062		.0460
COMMENTS				Total net time					.594		.1002
				Allowances					15%		15%
				Allowed standard time					.683		.1152
				Machine time worksheet B							.0383
				Machine time allowed							.0440
				Total standard time					.683		.159
				Total standard time – per piece							.159

FIGURE 6.20 Worksheet A for setup and handling elements of turret lathe formula.

- Choosing times or developing representative motion patterns for constants
- Determining approaches for developing process times
- Determining work elements that can be best included in the final standard by a percentage allowance
- Determining approaches to develop percentage allowance and the base time to which the allowance is applied

The conclusions reached during the analysis must be completely recorded for future reference.

This step concludes with the preparation of a summary list of the work elements showing the reference symbol, element description, and the allowed normal time. Figure 6.21 illustrates a summary list of work elements.

ENGINE LATHE

REF SYM	WORK DESCRIPTION	TMU's	HOURS
424	TIGHTEN AND LOOSEN CHUCK JAWS	13648.0	.1365
425	CONSTANT TO USE OVERHEAD CRANE - SMALL ROLLS	12410.0	.1241
426	USE CRANE TO LIFT ACCESSORY	8450.0	.0845
427	PUT CAT-HEAD ON LARGE ROLL AND TRUE UP	32872.6	.3287
428	REMOVE CAT-HEAD FROM ROLL (EITHER SIZE)	11313.7	.1131
429	PLACE AND REMOVE LARGE STEADY REST (SET UP)	54571.8	.5457
430	USE LARGE STEADY REST - EACH OCCURRENCE	26091.2	.2609
431	MANUAL OPERATION TO CHASE THREADS	1014.1	.0101
432	COARSE ALIGNMENT OF STYLUS	4872.1	.0487
433	FINE ALIGNMENT OF STYLUS	1439.9	.0144

FIGURE 6.21 Illustrates a summary list of work elements.

Recent developments in computerized work measurement systems have simplified the process for developing the normal times for work elements. Systems such as MOST have considerably reduced the need to develop "work elemental" level data. The system is designed to initiate the standard data development process at the suboperation level as illustrated in Fig. 6.1, level 3. Figure 6.22 illustrates the relative difference between element level and suboperation level. The element level will in all probability be combined at a later step in the data development process to arrive at a suboperation level time value.

Using Statistical Significance to Determine Elemental or Suboperation Times. When developing standard data, we are often faced with an activity where the time required is not one precise standard but a range of times. The methods followed and the times required may vary depending upon part weight or dimensions, distances moved, tools used, or fasteners used. One can use considerable time developing detailed standard times for each identifiable method. Developing so many standards not only wastes time, but the application of different standards for such variations can be impractical.

We often see cases where many analyses are made and standard times developed in an effort to identify a "representative method" for an average standard. This is not necessary. We need to identify the range of times covering the probable extremes of the activity being measured and, then, determine statistically how many standards are needed to cover that range of times. We can minimize the number of standards needed and identify the maximum allowable time range covered by a standard and still preserve the required ac-

ELEMENT LEVEL			SUBOPERATION LEVEL		
REF. SYM	WORK DESCRIPTION	TIME	REF. SYM	WORK DESCRIPTION	TIME
A	PICK UP PART AND PLACE IN VICE		100	LOAD PART IN VICE, TIGHTEN VICE, LOOSEN VICE, REMOVE AND ASIDE PART	
B	PICK UP, POSITION VICE WRENCH AND TIGHTEN VICE				
C	PICK UP - POSITION VICE WRENCH AND LOOSEN				
D	REMOVE PART AND LAY ASIDE				

FIGURE 6.22 Comparison between elements and suboperation level.

curacy. We can then develop the standard time for a representative method which falls within the acceptable time range—not for all of the variations which might occur.

For example, we may need to develop standards for setting up a machine where the standard time can vary from 6 to 30 minutes. In order to minimize the number of standards to be applied, we need to know how far we dare go toward averaging setup times. We can classify these setups into time groupings, but how many classifications do we need? How large can the time increments between standards be and still preserve the needed accuracy?

In another case we may have a material handling activity where the storage location and distances traveled may vary depending upon the congestion, available rack space, or type of product. If the time per trip can vary from 1 to 10 minutes, how many separate delivery standards do we need to provide ±5 percent accuracy over a calculation period? Can we use a weighted average for all trips? Do we need to identify many different locations and apply a separate standard for each location? What is the minimum number of standards we can use for the total range and still preserve ±5 percent accuracy?

The accuracy required for most industrial applications is ±5 percent for the time period over which performance is calculated. This calculation period is usually an 8-hour period or, in some cases, a 40-hour period.

To calculate the accuracy required in an individual standard or "percentage allowed deviation," we can use the following formula:

$$r_t = r_T \sqrt{\frac{T}{n \times t}}$$

where r_t = percent allowed
r_T = required accuracy (usually ±5 percent)
T = time period required for the accuracy level to be reached
n = number of occurrences of the measured activity over the T period
t = standard time for the measured activity

If we have a measured activity which takes 0.2 standard hour and it occurs once per 8-hour period, we can calculate the allowed deviation for that one activity standard as follows:

$$r_t = 0.05 \sqrt{\frac{8}{1 \times 0.2}} = +0.32 \text{ or} +32\%$$

This means that we could apply that one standard (0.2 standard hour) to cover variations in work content of ± 32 percent and still maintain ± 5 percent accuracy in the result of an 8-hour time period calculation.

$$0.2 + 32\% = 0.264$$

$$0.2 - 32\% = 0.136$$

We can use a 0.2 standard hour standard to cover activities ranging in required time from 0.136 to 0.264 standard hour.

If we have a setup activity which occurs once per 8-hour period and has a measured standard time of 0.3 standard hour, the allowed deviation is

$$0.5 \sqrt{\frac{8}{1 \times 0.3}} = 0.26 \text{ or } 26\%$$

This means we can use that one standard to cover setups varying in time from that 0.3 standard hour by as much as 26 percent.

$$0.3 + 26\% = 0.378$$

$$0.3 - 26\% = 0.222$$

In this case, if we determine that the steps required in the various types of setup would vary the standard time from 0.222 standard hour for the easiest to 0.378 for the most complex, we can use 0.3 standard hour for all setups.

We may analyze setup activities and find that the most complex setup requires 0.5 hour and the simplest requires 0.2 hour. Assuming the expected frequency to be two times per day, we calculate the allowed deviations for the 0.5 hour as

$$r_t = 0.05 \sqrt{\frac{8}{2 \times 0.5}} = 0.141 \text{ or } 14.1\%$$

$$0.5 - (0.5 \times 0.141) = 0.43$$

And for the 0.2 hour as

$$r_t = 0.05 \sqrt{\frac{8}{2 \times 0.2}} = 0.224 \text{ or } 22.4\%$$

$$0.2 + (0.2 \times 0.224) = 0.245$$

Our two standards are not enough to cover the range from 0.2 to 0.5. The 0.2 standard would be limited to a maximum of 0.245 standard hour and the 0.5 standard could cover a minimum of 0.43 standard hour. This means we must identify some in-between levels of difficulty of setup in order to cover the total range.

To determine the number of standards needed to cover the variation from 0.20 to 0.5 hour with a frequency of two per 8-hour day, acceptable time groupings or ranges of time can be identified as follows:

- *Calculate the allowed deviation for 0.20 hour.*

$$r_t = 0.05\sqrt{\frac{8}{2 \times 0.20}} = +0.224 \text{ or} + 22.4\%$$

$$0.20 \times 0.224 = \pm 0.045 \text{ standard hour allowed deviation}$$

- *Calculate a time range.* If 0.2 hour is the minimum time needed, the time range should be a range where 0.2 hour is 0.045 standard hour below the average and the top of the range is 0.045 standard hour above the average for the range. The range in this case would be:

0.2 minimum time for the range
0.2 + 0.045 or 0.245 for the average of the range
0.245 + 0.045 or 0.290 for the top of the range

Calculation of standard time groups needed to cover times ranging from 0.2 to 0.5 standard hour—assuming two occurrences per calculation period of 8 hours—follows.

	Standard time range, standard hours	Allowed standard for the range, standard hours
$r_r = 0.05\sqrt{\dfrac{8}{2 \times 0.20}} = 0.224 \text{ or } + 22.4\%$	0.200–0.290	0.245
Allowed deviation $= 0.2 \times 0.224 = 0.045$		
Bottom of range $= 0.20$		
Middle of range $= 0.20 + \text{Dev. } 0.045$ $= 0.245$		
Top of range $= 0.245 + \text{Dev. } 0.045$ $= 0.290$		
$r_t = 0.05\sqrt{\dfrac{8}{2 \times 0.290}} = +0.186 \text{ or } + 18.6\%$	0.290–0.398	0.344
Allowed deviation $= 0.29 \times 0.186 = 0.054$		
Bottom of range $= 0.290$		
Middle of range $= 0.29 + 0.054 = 0.344$		
Top of range $= 0.344 + 0.054 + 0.398$		
$r_t = 0.5\sqrt{\dfrac{8}{2 \times 0.398}} = +0.159 \text{ or } + 15.9\%$	0.398–0.524	0.461
Allowed deviation $= 0.398 \times 0.159 = 0.063$		
Bottom of range $= 0.398$		
Middle of range $= 0.398 + 0.063 = 0.461$		
Top of range $= 0.461 + 0.063 = 0.524$		

Having identified the number of time groupings needed, one can develop standards for a representative method (or benchmark) illustrating the activities covered by the standard. After that, all identifiable variations in types of setup on that machine can be slotted into those three time groupings.

These calculations can be simplified by referring to the curves shown on Figs. 6.23 and 6.24, where you can read off the percentage allowed deviation closely enough for the required accuracy. To use Fig. 6.23, determine the percent of the total leveling period likely to be occupied by the activity being measured.

Standard time 0.5 hour × 2 occurrences per day = 1.0 standard hour per day

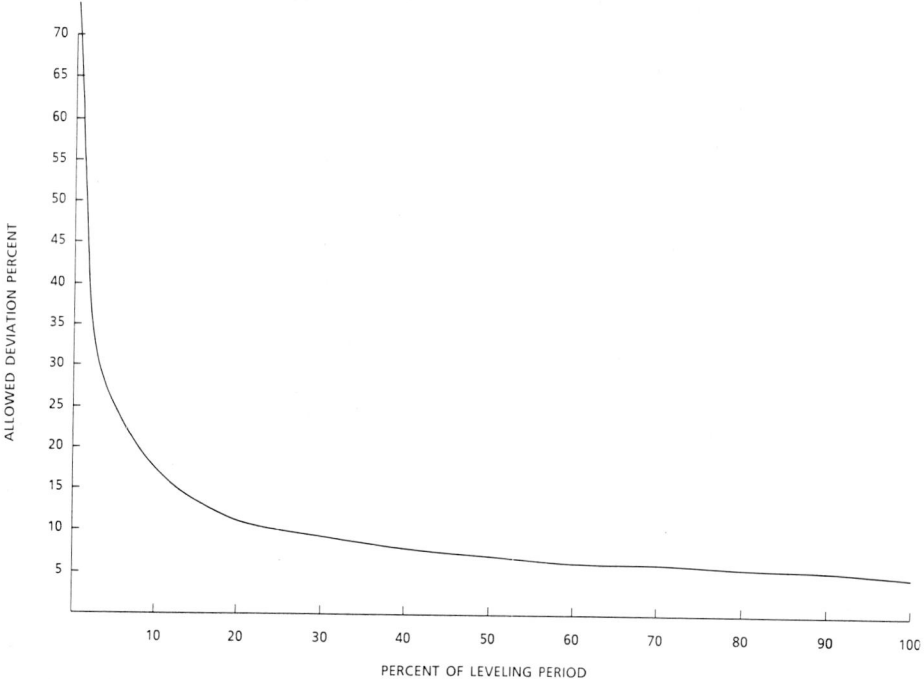

FIGURE 6.23 Allowed deviation based on leveling period.

Assuming the leveling period to be 8 hours, the percentage of leveling period is $2 \times 0.5/8$, or 12.5 percent. Find 12.5 on the scale across the bottom; then find where a vertical line above that point would intersect the curve. See where a horizontal line from that point would intersect that left side scale. This will give the plus and minus allowed percent deviation. As an example, for an element occupying 12.5 percent of the calculation period, the allowed deviation shown in Fig. 6.22 is +14 percent.

The curve shown in Fig. 6.24 is based on an 8-hour calculation period. To use it one doesn't need to calculate the "percent of leveling period." Instead, multiply the standard time per occurrence by the expected frequency per 8-hour period to get "time." Find this "time" figure on the scale at the bottom of the page. Find the point on the curve directly above the time on the bottom scale, and read off the percentage allowed deviation on the left scale directly to the left of that intersection on the curve.

Step 4—Work Sheet. A work sheet, also described as a pick-off sheet, is used to select the standard data values required to establish a standard time for a defined operation. When a manual documentation standard setting system is used, the work sheet becomes the record to support the established standard time. When a computerized standard setting system is used, the work sheet serves as a selection of locator numbers (addresses) that represent a task time value. The locator numbers are typed into the computer as a step in the standard setting process when a computerized standard system is used.

The objective of this fourth step is to design a basic format for the work sheet as previously illustrated in Fig. 6.8a. A general knowledge of the work sheet format and inclusion of work element descriptions will facilitate developing the standard times for each task time. The work sheets must also include a format for collecting and calculating the process time portion of a standard time. Figure 6.25 illustrates a process time development format for lathe operations.

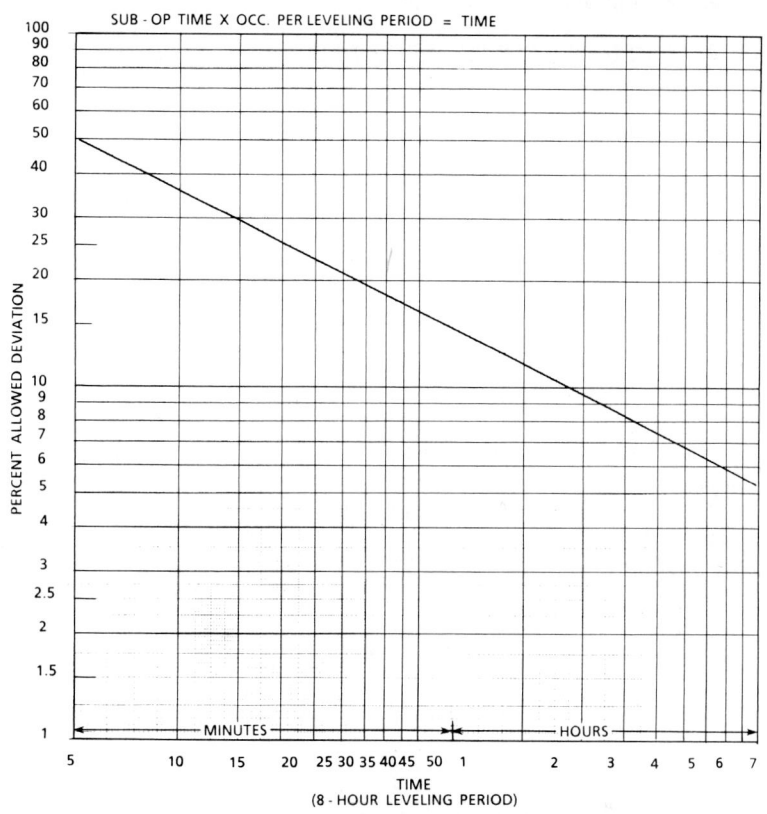

FIGURE 6.24 Percent allowed deviation based on 8-hour leveling period.

TURRET LATHES
WORKSHEET - MACHINE TIMES

Part No. _____
Applicator _____
Date _____

DESCRIPTION OF CUT SEQ. NO.	TOOL		DIAM. CUT	DEPTH CUT	SPEED		LENGTH OF CUT	TOOL LEAD	TOOL LENGTH	ALLOWED HRS. PER		NO. OF CUTS	TOTAL HOURS
	DESC.	MATL.			RPM	FEED				INCH	CUT		

FIGURE 6.25 Process time development work sheet for lathe operations.

The identification of other types of work sheets may also be considered as part of this step. Other types identified may include work sheets for specialized activities such as gauging, "family" work sheets that are designed to permit standard setting for that family only, charts that provide standard time for a total operation, or a portion of an operation. Figure 6.26 illustrates a block design of a standard time chart for single-point threading performed on lathes.

TURRET LATHES
CHART W
APPLICATION CHART - SINGLE POINT THREADING

DIAMETER	THREADS	THREADING TIME PER INCH BASED ON MATERIAL CLASS											HANDLING TIME
		1	2	3	4	5	6	7	8	9	10	11	
#8 (.164)	32	.0155	.0155	.0155	.0155	.0155	.0155	.0155	.0155	.0155	.0155	.0155	.0125
	36	.0170	.0170	.0170	.0170	.0170	.0170	.0170	.0170	.0170	.0170	.0170	.0125
#10 (.190)	24	.0189	.0161	.0161	.0161	.0161	.0161	.0161	.0161	.0161	.0161	.0161	.0175
	32	.0175	.0155	.0155	.0155	.0155	.0155	.0155	.0155	.0155	.0155	.0155	.0125
#12 (.216)	24	.0210	.0161	.0161	.0161	.0161	.0161	.0161	.0161	.0161	.0161	.0161	.0175
	28	.0175	.0135	.0135	.0135	.0135	.0135	.0135	.0135	.0135	.0135	.0135	.0125
1/4 (.250)	20	.0203	.0154	.0133	.0133	.0133	.0133	.0133	.0133	.0133			
	28	.0205	.0155	.0135	.0135	.0135	.0135	.0135					
5/16 (.313)	18	.0224	.0175	.0119	.0119	.0119							
	24	.0308	.0231	.0161									
3/8 (.375)	16	.0245											

FIGURE 6.26 Chart for determining single-point threading time.

Step 5—Develop Task Times (Synthesis). The objective of this step in standard data development is to provide the support documentation of:

- Time values as shown on the application work sheet
- Procedures used to develop and support process times

The end product is the final design of the work sheet as preconceived in step 4, including all time values with retrieval and support reference symbols shown. Figure 6.20 illustrates a final work sheet format for setup and handling work tasks for a ram turret lathe. A part of the time values shown in Fig. 6.20 was probably derived by combining work elements or suboperations to develop the values shown. One approach to this combination is the equation approach where the reference symbols for which the time values are to be added are presented as in algebraic equation:

$$K1 = A1 + B1 + C1 + D2 + F5$$
$$K1 = 110 + 50 + 75 + 80 + 95$$
$$K1 = 410 \text{ TMU}$$

A more popular approach in recent years, especially with the advent of the computer, is a listing of the synthesis as previously shown in Fig. 6.24. This approach is much better than the equation in that the elements are described to provide the reader a quick reference on what is included in the final time value for the work task.

Another phase of the synthesis step is to prepare charts and/or equations to develop process times. Charts may be feed and speed recommendations for developing process times as illustrated in Fig. 6.27. The source of charts as shown in Fig. 6.27 should be documented along with any application restraints or reference to standard practice manuals

RECOMMENDED CUTTING SPEEDS (SFM)
SINGLE-POINT CARBIDE AND HIGH-SPEED TOOLS

DEPTH OF CUT (in.)		0.040			0.150			0.300			0.625		
		HSS	CARBIDE		HSS	CARBIDE		HSS	CARBIDE		HSS	CARBIDE	
			COATED	UNCOATED		COATED	UNCOATED		COATED	UNCOATED		COATED	UNCOATED
MATERIAL	AVG. BHN	SPEED / FEED			SPEED / FEED			SPEED / FEED			SPEED / FEED		
LEADED STEEL 11L17	175	230 / .007	1450 / .007	975 / .007	170 / .015	950 / .015	750 / .020	135 / .020	775 / .020	590 / .030	105 / .030	--- / ---	460 / .040
LOW CARBON STEEL 1020	200	145 / .007	850 / .007	570 / .007	115 / .015	550 / .015	450 / .020	95 / .020	450 / .020	350 / .030	75 / .030	---	265 / .040
ALLOY STEEL 4140	225	130 / .007	700 / .007	530 / .007	100 / .015	550 / .015	430 / .020	80 / .020	425 / .020	275 / .030	60 / .030	---	250 / .040
HEAT TREATED ALLOY 8620	200	135 / .007	725 / .007	560 / .007	105 / .015	575 / .015	440 / .020	80 / .020	450 / .020	350 / .030	65 / .030	---	275 / .040
TOOL STEEL	250	50 / .005	325 / .007	280 / .007	40 / .010	300 / .015	225 / .015	30 / .015	225 / .020	175 / .020	25 / .020	---	135 / .030
TYPE 410 STAINLESS	200	145 / .007	850 / .007	570 / .007	115 / .015	550 / .015	450 / .015	90 / .020	450 / .020	350 / .030	70 / .030	---	265 / .040
CAST IRON MALLEABLE	180	130 / .007	775 / .010	600 / .010	100 / .015	600 / .015	450 / .020	80 / .020	475 / .020	375 / .030	65 / .030	---	300 / .040
CAST IRON ALLOY/DUCTILE	165	200 / .007	950 / .010	775 / .010	150 / .015	775 / .020	600 / .020	125 / .020	650 / .030	500 / .030	100 / .030	---	400 / .040
ALUMINUM FREE MACHINING	125	1000 / .007	---- / ----	MAX / .010	900 / .015	--- / ---	MAX / .020	800 / .030	---- / ----	MAX / .040	500 / .040	--- / ---	MAX / .080

FIGURE 6.27   Chart of recommended cutting speeds.

that provide machine process time development rules. An example of these rules is ''All finish cuts with a .003″ tolerance will use a feed of .007 inches per minute (ipm) and the carbide surface feet per minute (sfm) as specified in Figure 6.27.'' Reference is made in Fig. 6.27 to material class, which identifies the existence of a chart that presents the metal specifications grouped into the 11 material classes. The development of such a chart is included in this step of the development process.

This step is completed when the final work sheet with reference symbols and time values is designed, and all related data required to set a standard time are available.

Step 6—Test and Refine Data.   The objective of this step is to test the data application results. This is best done by establishing standard times for a variety of parts and operations for the range of product scope for which the standard data were developed. This step also includes the preparation of data application instructions that will result in a consistent application of the standard data.

The testing of the data consists of setting a standard time and verifying by floor observation that a qualified operator can produce the part in the standard time established from the standard data. The floor check should include verification of items such as illustrated in Fig. 6.28. This type of summary should be prepared for each test application.

The form shown in Fig. 6.28 is used as a summary record of the findings during an observation period. Any checks in the ''NOT OK'' column would be a definite indication that action is needed before an operator can be expected to meet the standard time. The action column is used to reference separate notes on action plans, responsibility, and dates. Incorrect workplace layouts are normally the responsibility of the supervisor. Proper equipment function is also a responsibility of the supervisor. These ''NOT OK'' items should not require data adjustments, but rather a more disciplined effort to adhere to accepted manufacturing practices and good methods.

The standard time test procedure also provides additional opportunity to get the supervisor involved in the overall standard data development phases. The ''standard time test record'' format (Fig. 6.28) provides space to record the results of review with supervisors and their conclusions.

The test step will identify the completeness of the data or the need for adjustments to task times, or that additional task times are required to attain suitable data. During the test

STANDARD TIME REVIEW RECORD

DATE:_____

DEPT. NO._____   MACH. NO._____   OPERATOR_____   CLOCK NO._____

PART NO._____   OPER NO._____   STD. TIME SETUP_____   EACH PIECE_____

	OK	NOT OK	ACTION NEEDED		OK	NOT OK	ACTION NEEDED
WORK PLACE LAYOUT				REVIEW WITH FOREMAN			
MATERIAL DELIVERY				WORKERS ATTITUDE			
MATERIAL REMOVAL				FOREMANS CONCLUSION			
TOOL DELIVERY				STANTARD TIME QUALITY CONTROL			
TOOL KIT QUALITY				WORK TASK ALLOWED			
Q. C. SERVICE				FREQUENCY APPLICATION			
INSTR. SHEET QUALITY				MATHEMATICS			
EQUIPMENT OPERATION				WORK TASK NOT COVERED			
TOOL/FIXTURE OPERATION				METHOD DETAIL			
SYSTEM DISCIPLINE				PROPERLY FOLLOWED			
TIME KEEPING				METHOD CHANGED			
UNIT COUNT				NON-STANDARD CONDITION			
"PUNCH-OUT" PROCEDURE				MATERIAL SPEC.			
NON-STANDARD- PROCEDURE				TOOL PERFORMANCE			
WORKER DISCIPLINES				EQUIPMENT PERFORMANCE			
PROPER METHOD USED				SETUP QUALITY			
WORK TIME PER AGREEMENT							
WORK ATTITUDE				PRODUCTION NO.			
OTHER (EXPLAIN)				PCS. PRODUCED IN_____HRS. OF OBSERVED TIME			

WORKER REVIEW OF FINDINGS	YES	NO	PERFORMANCE	$\frac{\text{TIME EARNED}}{\text{TIME OBSERVED}}$ = ____ % = ____ %
REVIEW SUMMARY:			REVIEW SUMMARY:	

FIGURE 6.28   Illustration of standard time test record.

SYMBOL	DESCRIPTION	APPLICATION INSTRUCTION
01	Get and load piece in fixture with hoist and secure. - Channels 151 lbs. and up.	The frequency for this element will always be one per occurrence when loading piece with hoist and weighing from 151 lbs. and greater. This element applies to large and heavy pieces such as channels.
02	Get and load piece in fixture by hand and secure.	The frequency for this element will always be one per occurrence when loading any piece by hand, weighing 15 lbs. to and including 60 lbs.
03	Get, load and secure piece in Universal Vise.	The frequency for this element will always be one per occurrence when loading any piece by hand, weighing 1 lb. to and including 15 lbs.
04	Get, load piece with fixture with hoist and secure. - 60 lbs. to 150 lbs.	The frequency for this element will always be one per occurrence when loading piece with hoist and weighing from 60 lbs. to 150 lbs. This element applies to parts which fall within the above weight range.

FIGURE 6.29   Instructions for calculating time standards—worksheet A.

application, emphasis should be placed on developing the application instructions for each task time (reference symbol) shown on the application work sheet. A typical format for recording these instructions is illustrated in Fig. 6.29. The application instructions should, to the maximum extent possible, eliminate individual judgment in the selection of work tasks or the frequency of application. Statements such as "Apply once per part as required" provide no application instruction.

The instructions should be precise. A typical instruction might read, "Apply once per part when the ratio on the length of the part to maximum diameter of the part exceeds 10." The adequacy of the application instructions should be tested by letting several other industrial engineers, not involved with data development, establish a standard time for the same part. Consistent results by different individuals applying the data indicate the application instructions are appropriate. Should the test result in inconsistent standard times, the instructions need to be clarified and improved until the desired consistency in standard time application is achieved.

Step 7—Develop Allowances.   The standard data developed during the previous steps have not included any allowances and represent the normal time to perform the work tasks. An allowance factor is generally added to the normal time to arrive at the standard time for an operation. The allowance factor is applied to compensate the workers for interruption to productive activities and/or in accordance with labor agreements. The most common allowance factors applied are for work interruptions caused by personal requirements, unavoidable delays, and rest periods. The combination of these three categories of delay is usually specified in union contracts. The most common allowance factor is 1.15, which represents 63 minutes in a 480-minute workday. The 1.15 factor as stated is not to imply that this is an industry norm; there are other factors for personal, unavoidable, and rest interruptions depending upon type of industry and labor agreements.

In addition to personal, delay, and rest allowances, there are other activities that are performed by workers that can be added to the normal time with an allowance factor. These are broadly classified as minor work elements. These elements occur at random and/or specified times throughout the workday. To charge these activities to part or products on the basis of when they occur would not provide a true labor cost for that part. For this reason, an allowance factor is developed that allocates the costs to all products produced in the most equitable manner. The activities that can be classified minor work elements for purposes of application by an allowance factor are:

• Adding coolant to machine reservoirs at random intervals
• Work area clean-up at the end of shifts
• Cutting tool maintenance
• Restacking auxiliary supplies that occurs daily and at random times
• Waiting for overhead cranes
• Waste removal from containers or the exchange of containers

Minor work elements can be included in the standard as "frequency" elements rather than being covered by an allowance. When an element occurs once for each 125 pieces produced, each piece standard includes 1/125 of the time to perform the work element. Experience indicates minor work elements are best handled on a frequency basis when practical, because they are usually more visible in this form than when they are included as an allowance, hence more likely to be revised when conditions change.

There is also a labor agreement allowance factor that is applied in some situations to machine process controlled operations. The allowance applies only to incentive pay plans and is referred to as an "incentive allowance" on process times. The objective of the allowance is to provide the machine operator with an incentive opportunity on the process controlled portion of the total standard times. This allowance also assumes that the machine operator has no opportunity to speed up the process time.

The presence of process time in a standard time may have an impact on the personal, rest, and unavoidable allowance factor applied. In some instances, the allowance portion for rest is not included when the process time provides a rest opportunity for the worker.

The application of the process time allowance will usually require that allowance factors be applied at two levels on the work sheet. One factor is applied to the total of the manually controlled elements and another factor to the total of process time. Caution must be exercised that the application of the process time allowance factor is not compounded by the application of any other allowance factor. By compounding we mean multiplying first by one factor and the result by another. The factors should be added together and the result used to increase the normalized elemental time value.

Work sampling or group timing technique (GTT) studies are normally used to develop the various allowance factors when they are not already specified. Time study can also be used, but this technique requires considerably more time than either work sampling or GTT. The detail approaches on the application of the three techniques are presented in other chapters of this handbook.

Step 8—Prepare Work Management Manual.   The final step in the standard data development process is to prepare a work management manual. The primary objective of this manual is a documentation of everything required to support the normal times shown on the work sheet. The manual is and has been called a time formula report, standard data report, or work measurement manual. The reference to the manual by the various tables does not alter the intended objective.

The work management manual table of contents (Fig. 6.30) is intended as a content description guide to use in writing standard data reports. It incorporates commonly accepted practices that you will find helpful in developing clear, understandable, well-organized standard data manuals. This outline is certainly not the only way to develop a good manual. But one of the most important features of a good standard data program is consistency in documentation.

The development process described in the previous eight steps primarily followed a logical approach to standard data development: input data collection, development of normal times, application output document design formats (work sheets), and allowances. The work management manual table of contents is organized to guide the reader through a logical sequence of events to achieve a full understanding of the developed standard data.

The contents of each section as outlined on the following pages are to be used as guidelines and are not intended to be all-inclusive. Referencing company policy manuals, process manuals, and operating procedure manuals may reduce the need to include such documentation in the work management manual. The manual contents are described below.

SCOPE (SECTION 1)

The purpose of this section is to describe the types of work covered in the manual, to identify the areas in which they are performed, and to describe the products and components which are affected. Basic guidelines for accomplishing the above are included in the subsections which follow.

Plant Area, Department, Work Center, Cost Center.   Describe where the work is performed using any or all of the above which are appropriate. Include all names and numbers of areas, departments, cost centers, and work centers which will help to establish where the work is performed. If any of these are not appropriate in your plant, use names, numbers, and terminology which are commonly used and are easily understood.

Products and Components.   Describe the parts or items—giving range of sizes and/or weights; design characteristics; and any other information, such as model number or families of parts, which will help to identify the products.

TABLE OF CONTENTS

SECTION 1	SCOPE	
	1.1	PLANT AREA, DEPARTMENT, WORK CENTER, COST CENTER
	1.2	PRODUCTS AND COMPONENTS
	1.3	OPERATIONS
	1.4	MANNING
SECTION 2	STANDARD PRACTICES AND POLICIES	
	2.1	CARE OF EQUIPMENT AND WORK AREA
	2.2	QUALITY CONTROL AND INSPECTION
	2.3	MATERIAL SERVICE
	2.4	SUPPLY AND MAINTENANCE OF TOOLS
	2.5	WORK ASSIGNMENTS
	2.6	TIME AND PRODUCTION REPORTING
	2.7	SET-UP AND TEAR DOWN
	2.8	SAFETY REGULATIONS
	2.9	SUPERVISORY RESPONSIBILITIES
SECTION 3	FACILITIES AND EQUIPMENT	
	3.1	PRODUCTION EQUIPMENT AND SPECIFICATIONS
	3.2	MATERIALS HANDLING EQUIPMENT
SECTION 4	WORK AREA LAYOUTS	
SECTION 5	PROCESS DATA	
	5.1	DERIVATION OF PROCESS TIMES
SECTION 6	MANUAL METHODS	
SECTION 7	STANDARD TIME CALCULATION	
	7.1	TITLESHEETS
	7.2	HOW TO CALCULATE TIME STANDARDS
SECTION 8	DATA SYNTHESIS AND BACK-UP	
	8.1	SUMMARY
	8.2	SYNTHESIS AND ANALYSIS
SECTION 9	ALLOWANCES	
SECTION 10	STANDARDS APPLICATION	
	10.1	INSTALLATION
	10.2	PROCEDURE FOR MAINTAINING LABOR STANDARDS
	10.3	REVISIONS

FIGURE 6.30 Work management manual table of contents.

Materials. List all direct materials and specification numbers which relate to the parts or product. Materials such as lubricants, cutting fluid, brushes, and solvents which are expendable will be listed in Section 3 and are not listed in this section.

Operations. List the operations that will be covered by the standard time data included in this manual. Operations which can be performed on the equipment or workstations but are not covered by the data should not be listed.

STANDARD PRACTICES AND POLICIES (SECTION 2)

This section is used to document standard practices and policies which affect or are applicable to the work covered in the manual. It should specify those cases where the operator is responsible for following established policy and practices. The manual should not be used as a document in which all company policy and practices can be found since these are best recorded in a separate policy and practices manual. The work management manual need only make reference to the policy manual as required. If all of the subsections included here are covered by the company policy or standard practices manual, a statement to that effect in Section 2 will be sufficient.

Care of Equipment and Work Area. Describe the operator's responsibility for cleaning the work area and equipment, and for the maintenance, lubrication, adjustment, and repair of machinery and equipment. Specify the frequency of those activities.

Describe how the time spent on the care of area and equipment will be recorded. It will usually be covered in one of three ways:

1. It can be covered by time standards or included as elements in standards.
2. It can be included in allowances (if so, include the same information in Section 9).
3. It can be performed off standard.

Quality Control and Inspection. Specify the operator's responsibilities for inspecting the parts or items processed. Include inspection frequencies required and the inspection criteria. When inspection is performed by someone else, describe how this is handled and how to account for the time required.

Material Service. Briefly describe how parts and materials are brought into, moved within, and carried away from the work center. Identify the operator's responsibility in this activity. If the operator has no responsibility for the material service, that fact should be noted. Specify materials handling requirements which can affect operator performance.

Supply and Maintenance of Tools. Identify the operator's responsibility for getting and returning tools, and designate the tool storage area(s) for each work area. Describe the operator's responsibility for cleaning, repairing, adjusting, and/or reconditioning tools; and specify the frequency with which this is done. "Tools" include manual and power hand tools, gauges, and tooling—such as jigs, fixtures, dies, and cutting tools for machining and drilling operations.

Identify personnel other than the operators who are responsible for the supply and maintenance of tooling, and describe those responsibilities in the same terms as stated above.

Work Assignments. Describe how work is assigned to the operator and specify who does it. The use of any forms or paperwork should be explained and copies or facsimiles of such should be included. If this is described in some other document, make reference to it so that, in effect, it becomes part of the work management manual.

Time and Production Reporting. Any procedures which are unique to the work center should be included here, along with any special forms used. Identify the operator's responsibilities for reporting run time, setup time, downtime, and production quantities. Procedures common to all work centers should be included in the plant's standard practice manual and only referenced here.

Set Up and Tear Down. Describe the responsibilities of the operator and other personnel who may be responsible for performing set-up and/or tear-down work. Define the limits of responsibility in each case and describe the procedures for getting and returning any tools, gauges, or fixtures used in this work. If such equipment is issued or supplied routinely to the operator, describe the procedure.

Safety Regulations. Identify the protective clothing and safety devices required for use by the operator to comply with company, state, and federal regulations. Cover such things as putting on and removing aprons, masks, shields, gloves, glasses, and helmets. Note any required inspections or adjustments to safety devices. Specify the frequency of occurrence of each task.

If the safety requirements in this manual are the same as those in common use in other areas or throughout the plant, a statement to that effect and reference to the appropriate sections of the company safety policy will be sufficient.

Supervisor's Responsibilities.   The supervisor is responsible for the following:

1. The assignment of work to the operator.
2. The accuracy of all information reported by the operator such as delays, piece counts, and start and stop times.
3. The quality of finished work and the rejection of defects (in conjunction with quality control personnel).
4. Adherence to safety regulations.
5. Adherence to the prescribed method.
6. Informing the industrial engineering department of any change in conditions.

If the supervisor's responsibilities are described adequately in a company policy manual or some other document, it will be sufficient to make reference to it here.

FACILITIES AND EQUIPMENT (SECTION 3)

The purpose of this section is to identify and locate the equipment and facilities needed to perform the work covered by this manual. You should identify and provide specifications for:

- Production equipment
- Auxiliary equipment
- Materials handling equipment

Production Equipment.   List all production machines and equipment and provide specifications such as, but not limited to, the following:

1. Manufacturer's name
2. Model and serial number
3. Size
4. Capacity
5. Limiting dimensions:
   - Distance between centers
   - Table size
   - Swing
6. Heat range
7. Speed range
8. Horsepower
9. Amperage and voltage range
10. Tonnage
11. Height

Auxiliary Equipment.   Auxiliary equipment is secondary or supporting equipment required to perform the work covered by the manual. Examples of such equipment are as follows:

1. *Tools:* Hand and power tools, soldering iron, bench tools, gauges.
2. *Special:* Heating oven, glue dispenser, refrigerator, coolant cleaners, fans.
3. *Expandables:* Solvents, lubricants, rags, brushes.

List all auxiliary equipment and provide the manufacturer's name, model, and serial number, machine or asset number, and key details.

Materials Handling Equipment.   List the materials handling equipment needed by the operator to perform the work covered in the manual.

This will include the equipment related to the scope of work covered by the work management manual. List the equipment by brand name or manufacturer and machine or asset number; and give key details and specifications.

LAYOUTS (SECTION 4)

The purpose of this section is to provide a detail operator workplace layout(s) for operation. The detail workplace layout should include the principal workplace of the operator, material locations or bins, tool cabinets related to the workplace, job crane locations, and any other items considered part of the workplace. The detail layout should also include detailed reference tables to nonworkplace locations that provide support to the workplace and to which the operator has to travel, such as tool cribs, inspection centers, and stockrooms.

Work Areas.   Hand-drawn or computer layouts for each work area are needed. A work area layout will contain the following information:

1. Name and location of each workplace
2. Distance in steps between workplaces
3. List of tools and where located
4. List of equipment and where located
5. Number of operators and where located if more than one is assigned to a workplace
6. List of parts and where located
7. Location and designation of parts containers

PROCESS DATA (SECTION 5)

The purpose of this section is to describe how process times were derived. Process time is that part of an operation which is considered to be beyond the control of the operator, even though it is understood that by changing settings or adjustments on machines the operator can, in fact, influence process times. A few common examples of process time operations are:

• Welding arc times
• Spray painting
• Heat treating
• Machining time, including feeds and speeds tables
• Electroplating
• Sewing (machine)

Derivation of Process Times.   All mathematical calculations used (such as least squares, regression analyses, and standard deviation) should be shown. Any supporting data such as lists and tables of observed times and the method used to compile them should also be included.

Technical Processes.   This section should also include a description of any special processes of a technical nature. Examples of these are electrochemical plating, heat treating, casting, and molding. Reference should be made to sources of information, such as manufacturer's manuals, technical bulletins, and reports so that these can be consulted if necessary.

Tool Life.   Provide information on tool life for the tools and materials covered in the manual. This will form the basis for the time allowed for tool changes.

MANUAL METHODS (SECTION 6)

The purpose of this section is to provide a general description and primary sequential steps that a worker will follow in performing the operation. This section provides in a summary format the work activities that the worker is expected to perform and for which a standard time will be allowed.

   This section is not intended to provide a detailed method description on how to produce the variety of parts for which standard time can be established from these standard data.

STANDARD TIME CALCULATION (SECTION 7)

This section describes how to set a standard. It will include the use of:

* Title sheets (listing of suboperations)
* Work sheets (standards application sheets, data application sheets)
* Direct-read tables and charts
* Spreadsheets and standard time groupings (where appropriate)

DATA SYNTHESIS AND BACKUP (SECTION 8)

This section will contain all the data and supporting information used in developing the standard time calculations in Section 7. This will include all elements, constants, suboperations, and combined suboperations which were developed using Manual MOST, MOST computer systems, MTM, time study, or other work measurement systems.

   The purpose in requiring this degree of detail is not only to show how standard times were developed but also to provide potential users of the manual with enough information to determine the extent to which the data will be transferable, as well as to provide a complete audit trail.

   When the suboperation or constant is made up of several activities or elements at any level above the basic motion pattern analysis (MOST, MTM, or basic time study elements), the synthesis is included. This represents the minimum acceptable level of methods documentation.

Element Summary.   This includes the MTM analyses, MOST calculation sheets, time studies, and any other detailed documents required to support the elemental time values used in the development of the standard data. This provides a summary listing of all elements contained in the data used to develop application sheets, direct-read tables, and slotted time sheets used in the synthesis. This summary includes the title of the element, its source reference code, and corresponding time value.

Constants, Suboperations, and Final Suboperations. This is a summary listing of all constants, suboperations, and final suboperations utilized in the development of values which appear on the data application sheets, direct-read tables, and/or slotted time sheets. Each constant and final suboperation is to be identified by title, source reference code, and time value.

Synthesis of Constants and Suboperation Times. This subsection documents the synthesis of all constants listed in the constants summary. It is to include the title of the suboperation and the component elements used to build or synthesize the corresponding time. All synthesis components must be identified by title, source reference code, time, and frequency.

ALLOWANCES (SECTION 9)

Allowances fall within two categories:

- *Regular:* These are intended to cover time for personal needs, time for rest to overcome the effects of fatigue or monotony, time lost due to unavoidable delays, and loss of incentive opportunity (as in process time).
- *Special:* These will cover conditions not normally encountered, such as extreme heat or cold, smoke, paint spray or fumes, and the use of restrictive clothing and equipment, to mention a few.

Specify the type of allowance, what it is intended to cover, and the percentage. State the authority by which the allowance is given, such as a contract or agreement, or if by company policy. If the allowances were developed by work sampling or some other type of study, include the supporting data in this section; or if a separate study, identify the source. One example covering the application of allowances should be included in this section. Other examples should be shown in Section 7.

STANDARDS APPLICATION (SECTION 10)

If it is essential to include all important information in a manual, it is equally important to maintain it in an up-to-date condition. Changes in equipment, methods, and working conditions occur frequently; they create the need for revising the material contained in the manual. It is therefore necessary to provide an effective means for ensuring that all revisions are made in a timely manner and that they are promptly disseminated to all users.

Responsibility for Maintaining Standards. It is the responsibility of the manager of the industrial engineering department to ensure that all copies of the work management manual are maintained in a complete and up-to-date condition.

Revisions to the manual and/or time standards will be considered if changes occur in any of the following:

1. Manual methods
2. Process time
3. Tooling
4. Work area layout
5. Quality requirements
6. Safety requirements

7. Material handling service
8. Material handling equipment
9. Routing of part
10. Allowances
11. Material specifications
12. Tolerances or dimensions
13. Machine assignment

Maintenance of the Manual and Time Standards. When any condition changes which might affect the contents of the manual or the calculation of time standards, a member of the industrial engineering department will be assigned to review the change to determine what effect, if any, it will have on the manual or on time standards. Where warranted, the necessary revisions will be made and submitted to the manager of the industrial engineering department for approval.

Procedure for Maintaining the Manual and Standards. The procedure for maintaining the work management manual and standards is as follows:

1. Make necessary revisions.
2. Enter date and approved signature on the specification sheet to be revised.
3. Post revision date and page number on title page in appropriate space.
4. Update process sheet to reflect standard revision when applicable.
5. Issue standard change notice showing old standard, revised standard, reason for change, and effective date of change to supervisor with copy to other departments as required.
6. Change routing file to reflect new standard.
7. Maintain copy of standard change sheet on file in the industrial engineering department.
8. Issue copies of all pages containing revisions to all people who have copies of the manual. Remove and destroy those pages which have been superseded.
9. Retain one copy of all obsolete or superseded material in a permanent file maintained in the industrial engineering department.

Distribution. This section is to provide a record of the distribution of the work management manual within the company or other divisions of the company, in accordance with the following control sheet format.

Copy no.	Division	Issued to (name and dept.)	Date
————	————	————————	————
————	————	————————	————
————	————	————————	————
————	————	————————	————
————	————	————————	————

Responsibilities. Each management function has a role in the standard data program and each must understand its role and participate in fulfilling its responsibilities.
   The standard hour program consists of three phases:

1. Development
2. Installation
3. Maintenance

Within each phase are program contents for which each management function may have responsibilities.

Revisions. The maintenance of the work management manual is important to assure the continuity and accuracy of the data. It is therefore necessary to keep a record of the revisions made, not only for maintenance of the data but also to satisfy a contractual agreement with the union related to standard changes. It is suggested that the following format be used to record the revisions made in sufficient detail to support standards adjustments.

Date	Change description	Approval
————	————————————	————
————	————————————	————
————	————————————	————
————	————————————	————
————	————————————	————

Section 10 as described may be part of the industrial engineering department policy manual and need not be repeated in a work management manual. However, reference should be made to the appropriate policy manual covering standard data maintenance procedure.

CONCLUSION

It may be seen from this discussion of standard data concepts that the possibilities of reducing standards setting costs and improving the accuracy and consistency of the standards are many and varied. A number of techniques have been developed and successfully applied which eliminate the necessity of using individual studies to develop time standards. The best technique to use in any given situation will be governed by the nature of the work, degree of repetition, length of cycle, and many other factors. There is opportunity for considerable ingenuity and even creativeness in designing the time measurement system which will best meet the requirements of a given standards setting application. For this reason, standard data development can be one of the most interesting and challenging tasks of the industrial engineer.

BIBLIOGRAPHY

Atlanta/Norcross, Institute of Industrial Engineers, Industrial Engineering Terminology, "Engineering Economy," 1983, pp. 41–50.

Barnes, Donald W., and Thomas Nowicki, "Standards by the Pound," *Industrial Engineering,* June 1974, p. 19.

Bostion, W. H., "Standard Data Development and Application," *Fall 1981 I.E. Conference Proceedings,* Dec. 6–9, p. 117.

Clark, Daniel O., "Standard Data and Its Maintenance Today," *1981 Spring Annual I. E. Conference Proceedings,* Detroit, p. 594.

Eady, Karl, *Standard Data Systems and Their Construction,* MTM Association for Standards and Research, 1977.

Gerharz, Michael R., "Simplification of Work Measurement through a Statistical Approach to Standard Data," *International I.E. Conference Proceedings,* 1987, pp. 676–681.

Hoey, R., and D. N. Rao, "Standard Data—A Great Time Saver for Industrial Engineering," *Fall 1976 I.E. Conference Proceedings,* Boston, Mass., Dec. 1–3, p. 172.

Karger, Delmar W., and Franklin H. Bayha, *Engineered Work Measurement,* Industrial Press, 1987, chap. 4, "Data Types and Levels," pp. 72–74.

Mayo, W., and R. Horne, "Considerations in Developing Long Cycle Assembly Standards," *Industrial Engineering,* March 1990, pp. 38–42.

Zandin, Kjell, *MOST Work Measurement Systems,* Dekker, New York, 1990.

WORK MEASUREMENT
APPLICATION AND
CONTROL

CHAPTER 1
ADMINISTRATIVE AND CONTROL PROCEDURES

Peter S. Lucking
Senior Consultant
H. B. Maynard and Company, Inc.
Pittsburgh, Pennsylvania

The purpose of this chapter is to provide information on the everyday procedures and routine required for the effective operation of an industrial engineering department. The details of administrative procedures and clerical routine will vary from plant to plant; the principles that guide these routines are addressed here. These principles and procedures are discussed from the standpoint of their application to the typical industrial engineering department.

The chapter begins with a discussion of the key component in administering an industrial engineering department, namely, what are the responsibilities of the department? Determining the responsibilities of the department sets the parameters for staff size, the qualifications required, work assignments, and paperwork organization. A project planning and control system is described in the second part of the chapter. This system can serve as the method for planning work, measuring progress to an estimate of the work, and presenting summaries so progress can be reviewed and priorities determined. The chapter continues with a discussion of time standards in the third part of the chapter. It concludes with a discussion of labor reporting and control procedures.

ADMINISTRATION OF THE INDUSTRIAL ENGINEERING DEPARTMENT

There are four parameters in the administration of the industrial engineering department: the size of the staff, their qualifications, their work assignments, and how documentation is organized. The key component for each of these parameters is defining the responsibilities of the department. The tasks the department will perform depend on the responsibilities of the department and the parameters for administering the department. The successful department will have tasks that are clearly defined, understood, agreed upon, and accepted.

Size of the Staff. The size of the staff depends on the responsibilities of the department, the size of the organization it serves, and the nature of that organization. A number of other factors influence the size of the staff, but these three are the most critical.

The information listed below is a ranking of 32 techniques used by industrial engineering departments who responded to a survey of industrial engineering practice.[1] A review of the list shows that there is a wide variety of activities that a department may perform. No department can handle all the activities with equal skill, so it is essential that the priorities of the department be established.

Below are the results of a survey of management techniques used, in order of frequency.[1]

- Method study
- Work measurement (direct)
- Incentive application
- Layout studies
- Design of forms
- Materials handling problems
- Development of information systems
- Cost-benefit analysis
- Work measurement (indirect)
- Choice of materials handling equipment
- Organization studies
- Job evaluation
- Choice of office equipment
- Management development
- Systems analysis
- Inventory and stock control analysis
- Computer programming
- Use of networks for project control
- Use of networks for planning
- Work measurement on the office
- Motion economy
- Management by objectives
- Value analysis
- Use of networks for resource allocation
- Ergonomics
- Group technology
- Hazard and operability studies
- Simulation
- Photographic and filming
- Linear programming
- Queuing
- Risk analysis

The activities that the department will be responsible for determine the expertise that will be required in the department. Expertise in the profession is interrelated; for example, engineers involved with incentive application will have an understanding of direct work measurement, method studies, layout studies, and a knowledge of information systems, since these areas are all related.

The activities of the department also depend on what is currently happening in the organization. An organization that is currently installing a measured daywork plan will undertake activities that are different from those of an organization that has a mature measured daywork plan. To determine the activities of the department, and thus the tasks they will be performing, it is essential to understand the organization and where it is headed. This understanding does not need to be exact or long-term (there is no reason to be intimidated by a five-year plan), but it is important to provide the organization with what it needs, so some understanding of the goals and objectives is essential.

Once the goals and objectives of the organization are understood, priorities can be determined. These priorities will determine the options available to the department. Since developing the responsibilities of the department is an interactive process, options are important. All organizations have dynamic needs, and the definition of the tasks that will be performed must meet the needs of the organization, even as those needs are changing. When the options for the department are identified, tasks can be determined that will fulfill the responsibilities of the department.

It is not enough to merely determine the tasks for the department; they must also be defined, understood, agreed upon, and accepted. The tasks of the department will completely define the responsibilities of the department. Organizations that can employ enough industrial engineers to have a department will also have some sort of goal setting or management by objectives procedure that will allow for the communication of the department tasks. The easiest way to communicate the department's responsibilities and gain acceptance for them is to communicate the tasks the department will perform, or the objectives that the department will meet for the planning horizon. Stating the tasks is the most effective way to deal with specifying the responsibilities of the department. It replaces a description of what the department would like to do with a concrete description of what the department will accomplish.

The list provided above can be used as a checkoff to determine what must be accomplished for the organization. Knowing the organization's current goals and how it is currently structured will allow for the development of a list of priority areas in which work is to be performed. These priorities can be revised and further specified until a list of concrete tasks for the department is available. Defining the responsibilities for the department is a part of the goal setting procedure, and though it often seems that management dictates what a department should do, the list of tasks that define the responsibilities of the department is created by the department.

Although it often seems that management dictates to a technical department (such as industrial engineering) what must be done, this is seldom the case. In actuality, the department determines 80 percent of what they want to do or will do. In general, management dictates the broad outline of what must happen for the organization but does not specify the specific tasks required to make things happen. For example, management may determine that a measured daywork plan is vital to the effective operation of the organization. If no plan exists, they will agree to develop one but are certain to leave most of the details of the development to the department. If a plan already exists, then certainly the details of maintaining the plan will be left to the department to develop.

The size of the organization (usually measured in terms of operating personnel) does not of itself dictate the size of the industrial engineering department. There are no cut and dried ratios that say that an organization of 500 people needs 6 engineers or that one of 5000 people needs 60 engineers; rather the size of the organization is a useful guide to the size and structure of the industrial engineering group but does not dictate the size of the group. It should also be obvious that larger organizations will have larger staffs of industrial engineers, but it is important to remember that the mere size of the organization served by the industrial engineering group does not, in and of itself, dictate the size of the industrial engineering staff.

The nature of the organization's operations (along with the responsibilities of the department) will have the greatest impact on determining the size of the industrial engineering staff. No matter what the organization is (a manufacturer or a service organization) the industrial engineering group will classify their activities in one of three groups: highly repetitive, fairly repetitive, or job lots. This description of an organization can give the most guidance in determining the size of the industrial engineering group, since experience has developed reasonable rules of thumb for staff size depending on these criteria. The following table shows the number of operating personnel handled by one industrial engineer for each of the above classifications.[2]

Classification	Number of operating personnel per industrial engineer
Highly repetitive	200–300
Fairly repetitive	100–200
Job lots	50–100

This approximation of the required staff size has been found to work in most manufacturing organizations. In a service organization, the same approach to developing staff size can be used. For example, in a hospital many unique activities take place, but some activities are very routine and highly repetitive, such as housekeeping or janitorial type of activities. Defining activities in the classification described above (highly repetitive, fairly repetitive, and job lots) makes it easy to see that a relatively small staff of engineers can handle its responsibilities in even a large hospital. In all organizations, the definition of the departments' responsibilities plays a key role in determining what will be classed as highly repetitive, fairly repetitive, or job lots.

The size of the industrial engineering staff is determined by the responsibilities of the department, the size of the organization being served, and the nature of the organization. Determining the responsibilities of the department in such a way that the tasks the department will perform are understood, agreed upon, and accepted is essential to the successful industrial engineering department. The mere size of the organization served by the industrial engineering department does not dictate the size of the engineering staff. The size of the staff is determined by what will be done (responsibilities of the department) and how much must be done (nature of the organization).

Qualifications of Personnel.   The qualifications required of the industrial engineering staff depend on the responsibilities of the department, the technical expertise of the existing staff, and their familiarity with the products or processes of the organization. These three factors are interrelated. The responsibilities of the department cannot exceed the expertise of the staff or their familiarity with the product. Developing a qualified staff is a continuous activity. All groups tend to be dynamic, so both personnel and priorities will change with time. Despite these changes, personnel with the proper qualifications must always be available. In addition, consideration must be given to the promotional path for the staff and the depth of expertise that should be kept available.

Shortages in skills can be relieved by training, in either technical areas or product or process familiarization, but such efforts will almost always be minimized. This creates a situation which is always changing. People with fewer skills must be trained by those with greater skills, who are, at the same time, being trained to develop additional skills of their own.

In the short term, the responsibilities that the department has identified provide the basis for the qualifications that the department requires. This means that responsibilities will interact with the qualifications of the existing staff, so that in the short term the responsibilities of the department will reflect the qualifications of the staff. In the long term, the qualifications of the staff must change since the responsibilities of the department will change.

Technically qualified personnel can be brought in from outside the organization, borrowed from other parts of the organization, or developed within the organization, or the expertise can be solicited on a short-term basis by retaining specialized consultants. The approach to be used depends on the responsibilities that the department defines for both the short and the long term. What is important to recognize is that the options for getting required skills are almost endless and depend only on the skills required and the length of time they will be used.

More difficult to obtain are personnel qualified in the products or processes of the organization that the department serves. Here the normal method to acquire expertise is through on-the-job training, a much longer process than the technical training that is readily available outside the organization (if necessary). This training is normally acquired by spending some time in the actual production of the product or the operation of the processes. The methods for acquiring personnel trained in the product or processes of an organization include obtaining personnel who are already trained (such as supervisors or personnel already trained in another organization), training engineering personnel by selection of their assignments or involvement, and developing and conducting training sessions for engineering personnel.

The effective department will have a satisfactory mix of skills available at all times. This requires a mix of personnel in the department, so that those with engineering expertise are available as well as those with product or process expertise. There are a variety of ways to obtain the required expertise, and an effective department will utilize all of them. The mix of skills will depend on the responsibilities of the department. Larger organizations with larger staffs will have the mixture of skills required to meet their responsibilities but will have a staff with more depth.

There must also exist a clear promotional path for the engineers on the staff, one that is clear enough and certain enough to keep experienced personnel interested in the department. This path will be in line with the responsibilities of the department. There can be nothing more frustrating for the experienced professional than to find that the challenging position that should exist is actually as high as that person will ever get in the organization. There is also nothing more frustrating for the person experienced in the product and process of an organization than to find that the position taken to expand knowledge of the organization is actually a niche of expertise with little or no future. The work assignments given to the department personnel become a key development tool in assuring people that the department has their development in mind.

In summary, the development of qualified personnel is an ongoing effort for the department. Since any organization is dynamic, with personnel, processes, and the organization always changing, the development of qualified personnel must also be a dynamic process. In larger organizations it is frequently handled with assistance from industrial relations, who help to identify and push ahead individuals that the organization feels are worth the effort. In any case, the department that does not develop its personnel will become a revolving door, with constantly changing faces and a low level of expertise in both the technical and product and process areas.

Work Assignments.   In general, there are two methods for assigning work. One will develop personnel who are expert in functional areas of the organization's product or process; the other will develop personnel with technical expertise that can be used in any area of the organization. The successful department will mix work assignments, with the most frequently used method depending on the responsibilities of the department and how the department is organized for routine work.

In the small organization, with a smaller staff, work assignments are hardly a problem; everyone is familiar with what everyone else is doing. The personnel in such a department are interchangeable, and work assignments are as frequently based on who has the available time as on who has the expertise. As the size of a department increases, a certain amount of specialization develops, and engineers tend to specialize both in areas of the organization and in technical specialties. In a very large organization a great deal of technical specialization occurs, but this is a function of depth and not of work assignments.

In general, having technical specialists is not the most effective use of laborpower and skills. One of the situations it creates is a limitation in flexibility; the engineers become specialized tools with limited uses. Another problem is that specialists are limited, but the problems encountered are not. Someone specialized in training may be useful but will need the help of the layout expert if developing training in plant layout, will need the help of work measurement specialists if developing training in standards, and so on. Replacing specialists is also a problem when technical specialists are relied upon too strongly.

The problems with technical specialization can be relieved by using people with general skills who can handle most industrial engineering tasks. If work assignments are made so that the engineer is assigned to one area or group of areas and then performs all the engineering work for that area, problems similar to technical specialization arise. The engineer loses flexibility, specializing in just a few products or processes, and the engineering work tends to be limited to what just one person (or one small group) can handle. Replacement is a problem, since the functional specialist will develop a rapport with the other people in the area and will be looked on as the one person to be trusted. An additional disadvantage is a tendency to lose the necessary questioning attitude, shutting out any chance for a fresh perspective in an area.

In addition to the problems in assignments discussed above, certain organization structures lend themselves to creating technical specialists or functional specialists. Figure 1.1 shows a centralized organization structure. A structure of this type lends itself to the creation of technical specialists. Figure 1.2 shows a decentralized organization structure. A structure of this type lends itself to the creation of functional specialists. Figure 1.3 shows how functional specialists can be developed even in a centralized organization, and Fig. 1.4 shows technical specialists in a centralized organization. Routine work can be handled in a variety of ways but generally is handled so that either technical experts or functional experts are developed. Either method tends to limit the flexibility of the department, creating problems when replacements need to be made and limiting the training and promotional opportunities for the engineers.

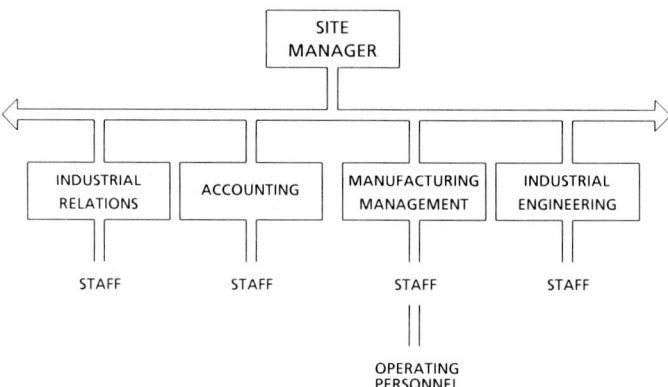

FIGURE 1.1 A centralized organization structure.

The next section of the chapter deals with the assignment of work as projects. This method makes each task unique, so attention can be focused on meeting the department responsibilities, reducing the chance of creating a department which is too inflexible to meet the dynamic demands of the organization. How the work is assigned is the heart of an industrial engineering department. Achieving a balance between technical expertise and product or process expertise is the challenge in managing a successful department.

Paperwork Organization.   There are three methods for organizing a department's paperwork. One is by part or print number, next is by department or area, and third is filing by project or report. The paperwork organization interacts with the other parameters in administering the department, but no matter how the department is administered, paperwork can still be organized in only these three ways.

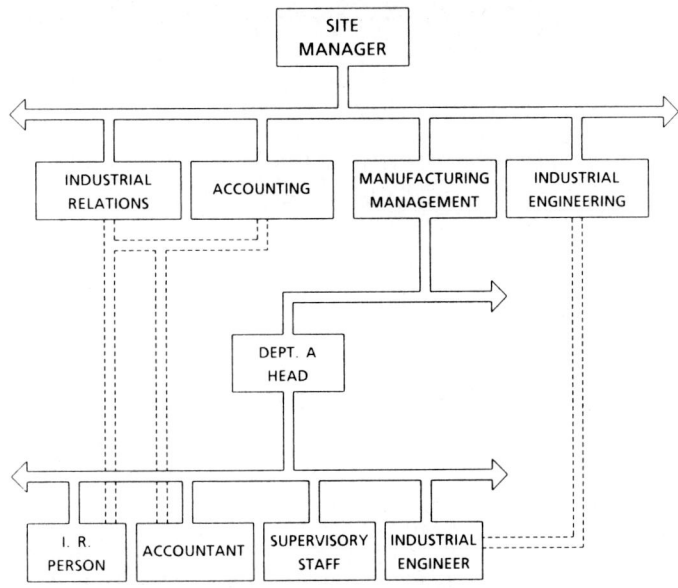

FIGURE 1.2 A decentralized organization structure.

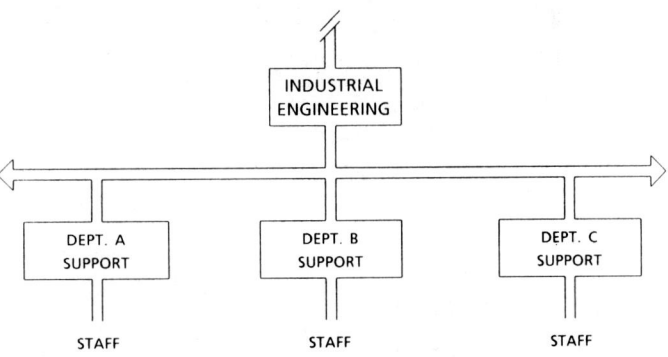

FIGURE 1.3 How functional specialists can be developed in a centralized organization.

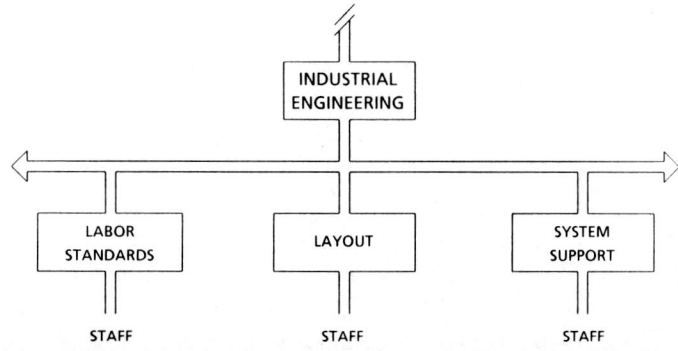

FIGURE 1.4 How technical specialists can be utilized in a centralized organization.

Most of the paperwork handled by a department, especially one involved in work measurement, is organized by part or print number. The bulk of the work can be filed in this manner. This includes standards, job instructions, route sheets, requests for studies, special allowance authorizations, and so on. The paperwork is organized by part or print number, but separate files may be maintained for each type of paperwork. Standards may be in one file, route sheets another, and so on.

Some data will be filed on a department or area basis. Such things as daily or weekly labor reports, cost reduction summaries, and method improvement summaries, lend themselves to this type of filing rather than by part number. In fact, data that relate to a department as a whole tend to be some type of summary, so that filing them will require both a different physical file and a different method. Other information is developed which fits into neither category. Such special reports and projects will demand special attention, and yet a third different file, in both physical type and method. Fortunately, this type of file is relatively small, so even if a search of the file becomes necessary, it is usually not very time-consuming.

With the advent of the computer the physical forms of filing systems and paperwork organization have changed a great deal. The largest and most difficult to manage file, information stored by part or print number, can be unified and easily managed with the computer. Data can be stored so that printed copies are produced on demand, greatly reducing the physical space required for storing information. All types of data (standards, job instructions, and the like) that can be referenced by part or print number can probably be stored in one computer system, or at least cross referenced in one system, which simplifies the data retrieval. (See Sec. 12, Chap. 3 for further information on database management.)

An effective filing system must be carefully planned and strictly adhered to. When the files are such that original records can be removed or revised, it may be necessary to limit access to records to control what happens to them. Computer systems lend themselves to maintaining secure data, especially when information is available for anyone to copy, but only a limited number can revise data. Limiting the categories of files will make retrieval easier and more certain, but retrieval can occur only if the data have been properly filed in the first place.

PROJECT PLANNING AND CONTROL

There are two components to controlling projects. The first is the project plan (based on a reasonable estimate of the work), and the second is the review of the project progress. Both components must be in place to manage and control projects. If all the work a department performs in the organization is treated as project work, the department can be managed with a minimum of record keeping. The system described here is used to manage all the work of a department using the same forms and formats for all the work that is reported. This allows for a complete review of the work on a weekly basis, which provides a frequent assessment of progress.[3]

Objective and Resources for the Project.   When the project is assigned, the objective for the project and the resources that are available to complete the objective are communicated to the responsible engineer or team who will perform the work. To ensure that the project is clearly communicated, the responsible personnel can develop an estimate of the time required to complete the project. The project estimate and control form shown in Fig. 1.5 can be used to develop the estimate, determine the resources required, and track the progress of the project. Instructions for the PROJECT ESTIMATE AND CONTROL FORM:

1. The personnel assigned to the project enter their names after "RESPONSIBILITY." Each step of the project is numbered, and an ESTIMATE of the person-days required

PROJECT ESTIMATE AND CONTROL

RESPONSIBILITY:_____SAMPLE COPY_____

STE	DESCRIPTION	ESTIMATE	ACTUAL 2/2	ACTUAL 2/9	ACTUAL 2/16	ACTUAL 2/23	ACTUAL 3/2	ACTUAL 3/9	ACTUAL	ACTUAL	ACTUAL	ACTUAL	ACTUAL
1	DEVELOP ESTIMATE	.5	.3	.3	.3	.3	.3	.3					
2	REVISE LAYOUT	2		1	3	3	3	3					
3	DEVELOP NEW DATA	5		2	4	4	5	5					
4	INSTALL IN NEW LOCATION	2					1	2					
5	TRAIN OPERATORS TO USE	3					1	4					
6	UPDATE ROUTE SHEETS	2			1	1	2	2					
7	REVISE & PRINT MANUAL	5						.2					
	APPROVAL	WHEN REVIEWED WEEKLY THE MANAGER MAY ENTER INITIALS, GIVES A RECORD OF COMPLETED REVIEWS											
	TOTAL COMPLETE EST./ACTUAL	15	.5 / .3	3.5 / 3.3	7.5 / 8.3	7.5 / 8.3	11.5 / 12.3	15 / 16.5					
	% COMPLETE EST./ACTUAL		3% / 2%	23% / 22%	50% / 55%	50% / 55%	77% / 82%	100% / 110%					

PROJECT NAME:_____AN EXAMPLE_____   PROJECT NUMBER:_____1_____

FIGURE 1.5 A sample project estimate and control form.

to complete each step is entered. The person-days to complete the project are totaled on the TOTAL COMPLETE line. The project is given a short descriptive PROJECT NAME.

2. The original of the estimate is reviewed with the responsible manager, who assigns a PROJECT NUMBER. This review allows the manager to determine if the project will meet the objective with the assigned resources.

3. Each week, enter the week ending date under ACTUAL, and record the accumulated person-days (to the nearest tenth) for each step to date.

4. Total the accumulated person-days on the TOTAL COMPLETE line. Enter both the EST.(imate) and ACTUAL person-days total. Estimated person-days complete are equal to the actual person-days complete, until the step is complete or the actual days, at that step, exceeds the estimate. Estimated person-days cannot exceed the estimate at a step, since the estimate is the standard for completing the project. Completion to the estimated person-days cannot exceed 100 percent; in fact it will not equal 100 percent until the project is complete.

5. Calculate the percent complete for EST.(imate) and ACTUAL. Divide the TOTAL COMPLETE by the total ESTIMATE person-days. The percent complete for the EST.(imate) cannot exceed 100 percent, although the percent complete for ACTUAL will, if the project takes longer than estimated.

6. Turn in the updated form to the responsible manager with the weekly time report. If no work was done on the project during the week, the form should be updated, and accumulated days brought forward.

7. All work in the department should have an estimate, because all hours will be charged to a project. If no project exists for a task, the time must be charged to another existing project.

Description of completed PROJECT ESTIMATE AND CONTROL form. Week ending (See Fig. 1.5):

2/2: Estimate is made, approved and receives a PROJECT NUMBER. In the first week, 0.3 person-day was spent on completing the project estimate. Since the step is complete, the project is ahead of schedule (3 percent complete to EST.(imate), with only 2 percent of the ACTUAL time used).

2/9: Work is done on steps 2 and 3, but neither step is completed. Progress to the EST.(imate) is: 0.5 for step 1, 1 for step 2, and 2 for step 3. Progress to ACTUAL is: 0.3 for step 1, 1 for step 2, and 2 for step 3.

2/16: Work is done on steps 2, 3, and 6. Five days were spent on the project this week. Progress to the EST.(imate) is 0.5 for step 1, 2 for step 2, 4 for step 3, and 1 for step 6. Progress to ACTUAL is the sum of the ACTUAL person-days: $0.3 + 3 + 4 + 1 = 8.3$. Note that work did not occur in the sequence originally estimated.

2/23: No work was done on the project this week, so the person-days are brought forward.

3/2: Work is done on steps 3, 4, 5, and 6. Four days were spent on the project. Progress to the EST.(imate) is 0.5 for step 1, 2 for step 2, 5 for step 3, 1 for step 4, 1 for step 5, and 2 for step 6. Progress to ACTUAL is the sum of the ACTUAL person-days: $0.3 + 3 + 5 + 1 + 1 + 2 = 12.3$.

3/9: Work is done on steps 4, 5, and 7. A little more than 4 days were spent on the project. It is 100 percent complete, since the progress to the EST.(imate) is reported as 100 percent. Progress to the EST.(imate) is the sum of the estimated person-days, 15 person-days. Progress to ACTUAL is the sum of the actual time spent on the project: $0.3 + 3 + 5 + 2 + 4 + 2 + 0.2 = 16.5$ person-days.

The project took more laborpower than anticipated, but this was apparent in the third week of the project (2/16). The decision to add or change assignments is not apparent here. What is important is that it was known when the project was only 50 percent complete. If changes were required, they could be made well in advance of the completion date. The project was originally estimated to take 3 person-weeks to complete. Clearly this did not mean 3 calendar weeks, since progress in the first 2 weeks showed that this would not be achieved, and nothing was done. The need for a flexible method of project control is demonstrated by the fact that no activity took place for the week of 2/23. In spite of spending 5 days on the project for the week of 2/16, no time was spent on it during the following week. Systems that tie estimated person-days to elapsed calendar time would show the project being well behind schedule, when this information was obviously not relevant.

Time Report.   A time record is useful when there are several multiperson assignments in a department. The assignments occur because of extra laborpower on projects to finish sooner or to train someone in a specific area or assignment. The time report allows detailed information to be passed to the person in charge of the project, with a minimum of effort. A time report like the one described in Fig. 1.6 is less useful when the scope of a project is fixed and repetitive or the department has one-person projects. The time report should be used only if it will speed up department communications. Instructions for IND. ENG. TIME REPORT:

1. Enter name after header IND. ENG. TIME REPORT FOR: Enter week ending date after W/E. Under the header for each day, enter the date for that day.
2. Enter the PROJECT NO. and DESCRIPTION for each project worked on during the week.
3. Report the time on a project to the nearest hour. Fractions are too small to be reported, but all time must be reported to some project.
4. Calculate and record all totals, and turn in the form to the person managing the project.

IND. ENG. TIME REPORT FOR:		SAMPLE COPY				W/E:	2/9		
PROJECT NO. **DESCRIPTION**	SUN.	MON.	TUES.	WED.	THUR.	FRI.	SAT.	**TOTAL**	
	2/3	2/4	2/5	2/6	2/7	2/8	2/9		
1 REVISE LAYOUT		4			4			8	
1 NEW DATA		4	6			6		16	
3 ESTIMATE			2	2				4	
2 STANDARD INVEST.				6	4			10	
3 DATA REVIEW						2		2	
TOTAL FOR WEEK		8	8	8	8	8		40	

FIGURE 1.6   A sample industrial engineering time report.

It must be remembered that the reason for using the time report is to simplify communications of progress. When using the form seems like too much effort, it probably is, and progress should be reported directly on the project estimate form.

Weekly Project Summary.   The project summary report provides the information necessary to control projects, since it provides the current status of all the projects. Knowledge of the current status of all the projects, and the resources committed to the projects, allows the manager to decide if there are problems with any projects. If no problems are apparent, or if the problems are within acceptable limits, no action is required. If there is a problem, and action must be taken, it is essential to know the impact of that action. Only when all the work is reported and progress for all of it is known, can the manager "juggle" resources to meet commitments. A sample of a WEEKLY PROJECT SUMMARY is shown as Fig. 1.7.

Description of WEEKLY PROJECT SUMMARY: The information on the report is described by header for each column. Under PROJ # is the project number from the PROJECT ESTIMATE AND CONTROL form. The number should be unique for each project. Under MNDYS (for the older term man-days) is the estimated person-days to complete the project. Under RESPONSIBILITY is the name of the engineer responsible for the project. The DESCRIPTION is a brief description of the project. Under WEEK is an identifier for the week reported. The week shown in Fig. 1.7 is the week of November 13. Under the column headed MNDY is the person-days worked on the project for that week. Under EST MNDY and ACTL MNDY are the estimated and actual person-days to date, from the project estimate form. The % EST and % ACTUAL give a percentage complete estimation for the project. This is done by dividing EST MNDY and ACTL MNDY by the estimate for the project under the column MNDYS. The AREA and the PRIORITY provide additional information about the projects on the summary form. When the weekly summary is done on a spreadsheet, it can be sorted by name, project number, area, or priority, depending on the information that is to be reviewed.

The summary is produced weekly, providing a frequent and timely review of the status of all projects. Since the environment in which most industrial engineering work is done is one of constant (and purposeful) change, detailed information on what is happening now is essential. Work will change in both scope and priority, so constant adjustments must be

WEEKLY PROJECT SUMMARY

PROJ. #	MNDYS	RESPONSIBILITY	DESCRIPTION	WEEK	MNDY	EST MNDY	ACTL MNDY	% EST	% ACTUAL	AREA	PRIORITY
6	24	B	PREP I - CHECK IN	N-13	1	18	18	75%	75%	1A	2
7	15	B	PREP I - FERRULES	N-13	3	14	17	93%	113%	1A	2
13	54	B	PANEL AREA - WIRE	N-13	1	42	42	78%	78%	1B	1
11	34	B2	PIN & COMB	N-13	5	23	23	68%	68%	1A	1
14	20	B3	BUNDLE	N-13	5	8	8	40%	40%	1A	1
4	115	H	WIRE HARNESS	N-13	8	68	68	9%	59%	1D	1
5	25	H2	BRADY TAPE	N-13	0	24	25.5	96%	102%	1A	2
9	27	H2	KINGSLEY	N-13	1	26	30.6	96%	113%	1A	2
20	22	H2	PANEL AREA - CLOSE UP	N-13	3	5	5	23%	23%	1B	1
19	15	H3	WIRE HARNESS - 2ND FERRL	N-13	2	9.3	10.3	62%	69%	1D	2
8	17	N	THERMAL FIT INSTALL	N-13	0	16	22	94%	129%	1A	2
12	16	N	WIRECUT	N-13	0	15	13	94%	81%	1A	2
18	23	N	PANEL AREA - COMPONENT	N-13	5	19	19	83%	83%	1B	1
15	35	P	SHEET METAL ASSEMBLY	N-13	5	18	18	51%	51%	1B	1
3	35	S	WATERPROOFING	N-13	0	32	32	91%	91%	1A	2
10	20	S	CLOSE UP (WIRE PREP)	N-13	0	17	17	85%	85%	1A	2
16	21	S	PREP II - SEQUENCE	N-13	3	5	5	24%	24%	1A	2
17	21	S	BRAID	N-13	2	5	5	24%	24%	1F	1

FIGURE 1.7   A sample weekly project summary report.

made. To adjust, it is essential to know how long a project will take (the project estimate) and what is the current status (the weekly summary). Since there is no way to know which project will change priority, it is essential to track all projects. In a dynamic environment it is essential to know the present status so that adjustments can be made quickly and effectively. (For additional information on project management and control, see Sec. 2, Chap. 3 and Sec. 10, Chap. 7.)

TIME STANDARDS AND TIMEKEEPING

The techniques used by the industrial engineering department are interrelated, so that applying or using one technique implies the ability to use several techniques. Engineers who develop and apply time standards tend to be familiar with method studies, direct work measurement, incentive application, material handling problems, information systems, indirect work measurement, motion economy, ergonomics, and filming operations. These are 9 of the 32 techniques listed previously. In addition, job evaluations and organization studies often are part of a complete work measurement package, so that time standards and timekeeping still tend to be a major area of industrial engineering effort and the administrative tasks associated with them deserve separate treatment.

Three elements of the tasks will be discussed here: setting standards, controlling standards, and timekeeping and payroll systems. The details of actually performing these tasks are fully discussed in other chapters of the handbook, so the discussion here will focus on how the tasks should be administered, not on how they are actually performed.

Setting Standards.   Time standards provide the basis for laborpower costs, standard costs, production control, machine utilization, staffing requirements, budgets, and wage incentives (if they are applied). When standards are initially set, they are accurate and consistent, but changes over time lead to inconsistent standards. Any department must be administered in such a way that standards accuracy and consistency is maintained.

This means that the department must be administered in such a way that change is handled and controlled, since change is inevitable. There are many ways to develop time standards that are accurate and consistent, but for purposes of administering the department, they break down into two types of standards, those that can be preapplied and those that must be postapplied. Preapplied standards exist when standard data have been developed, the task to be analyzed can be broken down into its standard data components, and an accurate standard can be developed prior to the part's being produced in normal production. Postapplied standards are those where standard data do not exist. The best that can be done is to develop an estimate for the labor involved and develop an accurate standard after normal production has started.

Speed and consistency in standard setting are enhanced with the use of standard data, but standard data can be costly to develop. When change is infrequent and relatively minor, estimates can be used (even in incentive applications) and standards can be developed when accuracy is ensured. Most departments operate with a mixture of both types of standards. Departments with substantial banks of standard data will still have areas where standard data have not been developed, so the department must be administered in a way that accommodates both types of standards.

In the administrative procedures for standards two classes of standards are of concern. Standards are either new (for new products, parts, or processes) or are already existing and must be revised when a change is identified. The general steps to deal with change are

1. Identify the change.
2. Review and approve the change.
3. Establish the standard.
4. Apply the standard.

The administrative procedures for each class of standard will be reviewed here.

New standards, whether preapplied or postapplied, are the easiest to control. The key step in setting the standard, identifying the change, is taken care of by the introduction of a new print, part, or physical process which does not already exist. This means new records must be generated to handle the new part or process, so there is no chance that anyone will question the need for the new standard. The department needs to be informed of the change as soon as possible, so it should be a part of the approval process in developing changes. The department can proceed to review the change, recommending any changes that are suitable and gathering information about the change at the same time.

Once the change is approved, the standard can be developed. In the case of preapplied standards, the standard data can be used to set the new standard, with relatively high confidence that the standard will not have to be revised. All that remains after this is to follow up on the application of the standard, to ensure that the standard data were correct and correctly applied. In the case of postapplied standards, the procedure is similar, but application of the standard is more complicated. In this case a temporary standard is set, and the permanent standard is developed during the initial application of the standard. Problems will occur if the new permanent standard differs greatly from the estimate, so procedures must be in place to ensure that the estimated standard is at least close to the permanent standard, or the difference between the two is well documented and explained.

In the case of existing standards, one of the biggest problems encountered is identifying the need for a change and what that change should be. Whether the standard comes from preapplied data or from actual studies in the shop, the identification of the need to change is the biggest problem encountered in administering consistent standards. Identifying changes or the need for change is discussed more fully below. Once the need for the change has been identified, the change needs to be introduced. If the change has been extensive, the procedures used in the organization to review and approve the changes may be used to make everyone aware of the change and to ensure that everyone approves it. Often the case is that the standard is now in error and the change that made the error is already accepted by everyone involved. In that case the new standard is established and applied.

Any procedure for introducing new or revised standards will naturally include the notification of all those affected by the change. Usually this procedure is part of the first three steps described above. In the case of new standards, it may include the fourth step described above. The biggest problem with time standards is in dealing with changes that are not new but reflect minor revisions that have been made and ignored, or changes that have taken place over time but have not been officially recognized.

In summary, whether the time standard setting technique involves standard data (which can be applied before normal production) or uniquely determined standards (which must be finalized after production starts) the general procedure for establishing standards is the same. The change must be identified, approved, and introduced. This means the department needs to establish administrative procedures to handle revisions from others, and to make others aware of revisions that the department is making. The exact procedures will vary from organization to organization but always will include some sort of approval process that industrial engineering must be a part of. Whatever the process is, the department should ensure that its role in it is always completed in a timely fashion. The department may find it worthwhile to develop internal procedures that push changes through only one or a few individuals. This will ensure that the organization's process for completing a change is always followed.

Time Standard Controls.   When a standard is authorized and put in place, it is assumed to be accurate for costs, production control, machine utilization, budgets, and the like, so standards must be accurate and consistent. To ensure that only the correct, accurate, consistent standard is applied, there must be controls on entering time standards into the systems where they are used. Control of standards is possible only if special procedures are put in place to authorize standards.

The environment in which standards are applied is dynamic. A standard may be changed by changes to method, equipment, design, machine settings, work area layouts, standard procedures, materials, material handling equipment or methods, tools, dies, fixtures; in fact, anything that will change in an area may impact the standard. Ensuring that standards stay accurate and consistent, despite this dynamic environment, is the challenge in maintaining standards. The primary method for controlling changes is continuous audits of existing standards, to ensure that they stay accurate and consistent.

Authorizing Standards.   Many methods can be used to authorize standards, and they depend on the types of systems that the organization uses. The purpose here is not to describe in detail all the various ways that standards can be authorized but rather to look at how the department needs to be administered to ensure that the proper standards are authorized.

We can consider a standard authorized when it is entered into the systems of the organization that use the standard. Standards can be authorized when they are entered on a cost card, but with computers they are more commonly authorized when entered into a restricted payroll or routing program. The controls necessary to ensure that the correct standard is entered need to occur prior to the standard's being entered on the card or in the computer program.

The introduction of standards must be controlled, so that accurate and consistent standards can be maintained. This means there must be a record of the old standard, the new standard, and the date the standard was introduced. This record should be kept in a way that allows progress or change reports to be developed. A good computerized standard setting system will keep such records as part of the history of the standard, but what is usually needed is a separate log of this information.

Any system of standards is continuously subject to pressures that tend to liberalize the standards. The pressure comes from those whose output is measured by the standard as well as those who are charged with managing the output. The pressure is acute when standard data are not used, and the potential for loosening all new standards in an area exists. When standard data are used, some of this pressure is relieved, since accepted and proved data will be less likely to be disputed or grieved.

In any system of standards, there is always a need for temporary standards, because of some temporary change in the conditions that surround a standard. All standard setting systems need to make provisions for the use of temporary standards. This helps to relieve some of the pressures on engineers to liberalize standards, since temporary problems are recognized. The best way for temporary standards to be applied is with some type of time limit, after which the permanent standard is put back in place.

The first element of control of time standards is to have only the industrial engineering department that sets the standards responsible for authorizing its use. This fixes the responsibility for the standard with one group, the same group that has a vested interest in the application of accurate and consistent standards. The next element in controlling standards is to ensure that a record of the change is kept. The best computerized standard setting systems maintain this record automatically, but the department may need a separate log for its own use, and to summarize the record of change. Changes made should be summarized and reported on a weekly or monthly basis. This will simplify a review of standards changes by the department managers. The final element is to ensure that the standard released into the organization records for using the standard is correct. This can be done only by double checking every standard that is entered into the records. This is a small job when done as part of the follow-up of a new or revised standard, but a major job when a typographical error calls into question all the standards set in an area.

Auditing Standards.   The objective of a standards audit is not to find errors or changes and revise standards but to assess the current status of the system, so that the best course of action can be taken. This of course does not mean that standards will not be changed as a result of errors uncovered in audits (of course they will be changed). It is a given that standards accuracy must be maintained, but prior to the change, a thorough review must be conducted and the nature and extent of the deviations must be known.

The purposes of an audit of standards are to detect change, evaluate the change, and provide decision data for those who will decide what to do about the change. The reasons for changes in standards (listed above) are too numerous to realistically assume that all changes will be brought to the attention of those responsible at the time that changes are made. It is essential that standards be reviewed as part of the routine in the department responsible for maintaining standards.

The frequency of audits is always a question. The following information on audit frequency has been published and provides a useful guide to when audits should take place. It is based on the number of hours that an operation is performed, so that those that represent the greatest labor investment are reviewed most frequently.[4]

Hours per year	Frequency of audit
0–10	3 years
10–50	2 years
50–600	1 year
Over 600	Twice per year

To detect change, it will be necessary to compare the documented standard with what is actually happening. The detailed methods for doing this depend on the extent of the documentation available and the availability of the operation that will be reviewed. It is not enough to determine how the operation should be done; it is essential to actually view it being done. This review of the operation being performed allows an evaluation of change to be made at the same time that the review to detect change is completed.

Collecting the information that will be used to provide the decision data that will form the basis for the recommendations on the audit should be done at the same time that the audit is performed. The best method is to develop some type of log sheet or document that will aid the engineer in gathering all the data needed when an audit is performed. The department should develop a documented standard practice that describes what will be done during an audit and provides a checklist of the information to be collected during an audit. This will help in conducting the audit, since those who may be affected by an audit (whether operating personnel or management personnel) must understand that an audit is routine and will be done regardless of what action they may or may not take.

The changes that will erode standards are never reported; they must be found out. The best method of determining the status of standards is to audit them, comparing them with existing operations. Good records are critical in conducting audits, since the information developed is used to determine the current health of the time standards program. Formal systems for controlling changes are vital to maintaining the accuracy and consistency of standards when dramatic changes occur, but the long-term effect of ignoring creeping change in operations will be standards that are no longer consistent. This will result in a time standards program that must be corrected at an expense that far exceeds that of proper maintenance. (Also see Sec. 6, Chap. 4.)

Timekeeping and Payroll Systems.   Key to an operations control system is determining what happened when operations were performed and the standard was applied. To manage and control an operation it is necessary to know what is happening. Integrated reporting systems allow management to determine what is happening with a minimum of expense and effort. To be truly useful, the reporting system needs to be fully integrated, so that one record, collected at the lowest level of the operation, provides the information needed for performance measurement, payroll and attendance records, and production control and scheduling information.

There are many methods of reporting, ranging from manual systems, through those run by timekeepers or dispatching centers, to highly automated computer systems. All systems collect the same basic information. The system used (computers, manual system, and the like) makes little difference, since it is a function of what other systems exist in the organization and the amount of money that should be spent to acquire the needed information. The basic information to be collected is always the same.

The information that must be collected is

- Department or area where the work is performed
- Part number (or print number)
- Operation number
- Start time, and stop time for the work
- The pieces produced
- A code to categorize the time spent
- Identification for the people doing the work

The department or area is collected for summary information; it attaches a summary identification to the record. The part number and operation number combine to uniquely identify a standard to be applied. The start and stop time are collected to determine the elapsed time that was worked. The pieces produced measure the output. A code for the time will separate productive time and nonproductive and will allow the time spent to be categorized. Identifying the people who worked the job details who did the work. These eight pieces of data are all that is needed to determine what is happening in the operation and to develop the reports management needs.

Any type of system that is used will require these inputs from operating personnel. Additional information is collected only if other systems are not integrated with the system that collects operating data. For example, if the part number and operation number identify a unique standard, but there is no way for the system to know that standard, then it should be reported at the same time that the part number and operation number are reported.

Some systems attempt to reduce the information required in the record by doing something like assigning operating personnel to a department or area, so that reporting of the operator's identification is sufficient to identify a department or area. On the surface, this appears to reduce the record, but the exceptions that will occur to utilize operating personnel (such as transfers or temporary work outside the area) will require stringent control over these activities, which will create more problems than the reduction in reporting can justify.

Different formats may be developed for reporting individual activities (involving one or a few operating personnel) and those activities that involve a large group of operating personnel (such as assembly line activities). Most organizations are faced with the problem of having to report both types of activities, and wind up with systems that either cloud the individual activities or make it difficult to readily interpret group activities. Experience has shown that it is easier to learn how to interpret reports for group activities than it is to clarify individual reports. It is important to recognize that no system has yet been developed that is perfect for both types of activities. The best that can be done is to reduce group activity reporting to one person, whose reported information is then spread out over the whole group.

Once the basic information is collected, management reports must be developed. The computer has greatly aided in the development of accurate and timely data, since it is able to quickly process a great deal of information. There are limits to this information, however, and these limits should be recognized and accounted for in the reporting system.

Without exception, the first report that operating personnel will see will have some errors in it that must be corrected. Unless the reporting system is an on-line real-time system, corrections to a completed report will have to be made. The reporting system used must be flexible enough to allow for easy corrections to data, even after a report has been developed and delivered. Sometimes errors are not noticed for days or weeks, so an integrated system must have effective correction capabilities.

The basic data that are collected are usually fed into a program or calculation procedure that summarizes and develops the information for reports to various management levels. Reports to management are detailed in the next section. Information generated from the raw data must provide an audit trail, so that information can be traced back to the source. This eliminates any problems with the system and simplifies responses to questions that may arise.

Integrated timekeeping systems are worth the efforts to develop them and tailor them to the organization's needs. Many systems already exist, or the organization can develop one of its own. The key information to be picked up from operating personnel are the eight bits of data described above. When the department has responsibility for collecting the information, personnel devoted to the task, who will become expert in it, must be available. If the department assists in maintaining the information that such systems require, it is best to ensure limited access to changing the data, and develop as many experts in the system as possible.

PERFORMANCE CONTROLS AND PROCEDURES

Summarizing the information collected from operations, and presenting it to management, is the prime concern of a department with work measurement responsibilities. Information relevant to each level of management must be presented to that level, and the same information summarized for higher levels of management. Dealing with the problems that are shown in the reports, or dealing with programs that will improve operations (whose progress is shown in the reports) are beyond the scope of this section. Here we present only the information required at various levels of management, and the procedures the department needs to have in place to administer this task. The reports that are issued, and the information summarized for all levels of management, are the prime focus for departments involved with work measurement. The department must ensure that the reports are adequate, accurate, timely, understood, and routinely produced.

Daily Reports.   The information supplied from timekeeping records provides the data to be used for developing the reports. The same format for the reports should be used for all levels of management. The only thing that should change is the information summarized

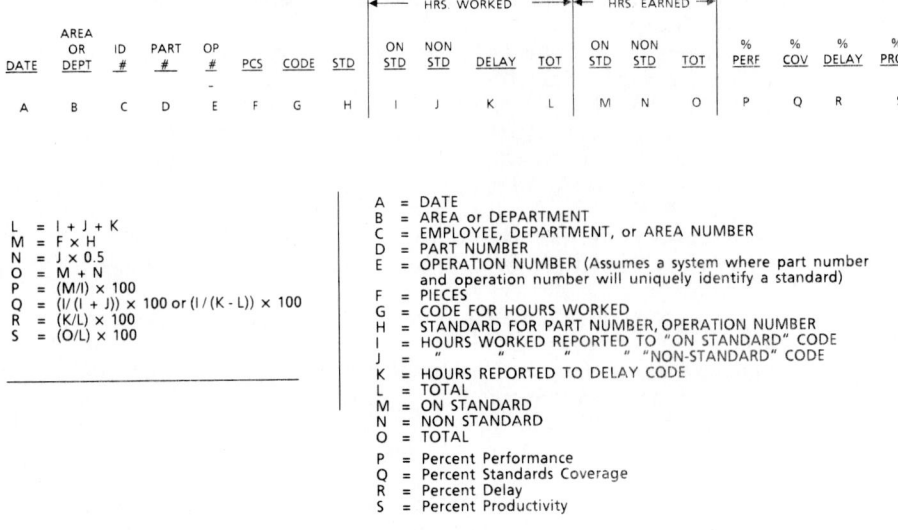

DATE	AREA OR DEPT	ID #	PART #	OP #	PCS	CODE	STD	HRS WORKED				HRS EARNED			% PERF	% COV	% DELAY	% PROD
								ON STD	NON STD	DELAY	TOT	ON STD	NON STD	TOT				
A	B	C	D	E	F	G	H	I	J	K	L	M	N	O	P	Q	R	S

L = I + J + K
M = F × H
N = J × 0.5
O = M + N
P = (M/I) × 100
Q = (I/(I + J)) × 100 or (I/(K - L)) × 100
R = (K/L) × 100
S = (O/L) × 100

A = DATE
B = AREA or DEPARTMENT
C = EMPLOYEE, DEPARTMENT, or AREA NUMBER
D = PART NUMBER
E = OPERATION NUMBER (Assumes a system where part number and operation number will uniquely identify a standard)
F = PIECES
G = CODE FOR HOURS WORKED
H = STANDARD FOR PART NUMBER, OPERATION NUMBER
I = HOURS WORKED REPORTED TO "ON STANDARD" CODE
J =   "   "   "   "   "NON-STANDARD" CODE
K = HOURS REPORTED TO DELAY CODE
L = TOTAL
M = ON STANDARD
N = NON STANDARD
O = TOTAL

P = Percent Performance
Q = Percent Standards Coverage
R = Percent Delay
S = Percent Productivity

FIGURE 1.8   Format for a labor performance report.

on the reports. This is so that everyone will be looking at the information presented in the same way, which makes it easier to train them in the use of the report and makes it easier for operating personnel, since the reports are all similar no matter what the area or level of the organization. A sample of a performance report is shown in Fig. 1.8. This report is for the basic level, in which each record is shown as a separate line.

The data in items A through G in Fig. 1.8 come from the record of the operation. The information for item B is from the original record or is made available from another file. Items I, J, and K are from the elapsed time records (stop time minus start time), categorized by the code in item G. Item L is the sum of I, J, and K. It is reported in this way to ensure that all the records are collected, since item L can be compared with attendance records. Item M is calculated by multiplying the number of pieces reported by the standard to arrive at the earned standard hours (or minutes). Item N is the time reported in item J, factored by a number (0.5 is used in the example) less than 1, making it an estimate for performance when no standard exists. Items P through S are calculated as shown in Fig. 1.8, and represent the minimum indicators required to manage labor effectively.

DEPARTMENT:_____   DATE:_____

ON STANDARD HOURS			NON-STANDARD HOURS			DELAY HOURS		
CODE	DESCRIPTION	HRS.	CODE	DESCRIPTION	HRS.	CODE	DESCRIPTION	HRS.
10	ON STANDARD		20	NON-STANDARD		30	LACK MATERIAL	
11	TEMPORARY STANDARD		21	MACHINE		31	MACHINE BREAKDOWN	
			22	TOOL OR FIXTURE		32	NO ASSIGNMENT	
			23	NON-STANDARD MATERIAL		33	NO MATERIAL HANDLING EQUIPMENT	
						34	INSPECTION	
						35	MEETING	
	TOTAL ___			TOTAL ___			TOTAL ___	

FIGURE 1.9   Format for a report on hours worked summaries by reporting codes.

IN THIS SECTION IS INFORMATION
TO IDENTIFY THE REPORT
SUMMARY SHOWN.

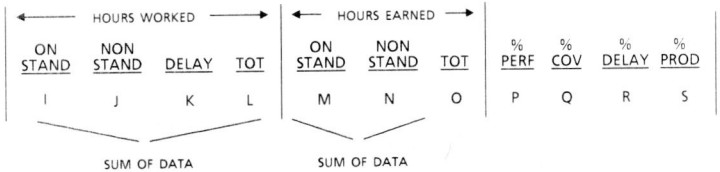

L, O, P, Q, R and S are calculated as shown in Figure 1.8

FIGURE 1.10   A summary report format for information shown in Fig. 1.8.

Figure 1.9 shows a report of the hours worked summarized by reporting code. This information should also be routinely provided, along with the report described above, to give management a clear picture of what is happening. It is mentioned here because this type of report has been found to be so useful that it will frequently be requested if it is not routinely produced. The codes to be used are entirely up to the organization. No one set of codes has been found to be ideal in all circumstances. Codes can be changed or added, but owing to their basic nature (they are part of the information in each record) and the need for wide understanding of the codes, the change procedure will be involved enough to ensure that only worthwhile codes are added or changed.

The summary by code can be an effective tool for identifying problems or opportunities for improvement, and for tracking progress over time. Once operating personnel have the labor indicators under control, with acceptable performance, delay, coverage, and productivity, the search for opportunities to improve will be guided by the summary of hours worked by code report. It becomes necessary because to effectively improve, management time must be prioritized, and time must be spent in the areas that will provide the greatest return. This means focusing on the areas (shown in the report by code) that are easiest to correct and show the greatest return. Since the time operating management can spend on correcting problems is very limited, they must know the areas to concentrate on, and this report will provide that information.

Summary information for the report shown in Fig. 1.8 is developed by adding the columns and developing new indicators. Items A through H would be changed (or left blank) to identify the summary information. Items I, J, and K would have new numbers based on adding the columns, as would items M and N. The calculations for L, O, P, Q, R, and S will be the same, but obviously the data will change. Figure 1.10 is an example of the summary that can be developed. Just the columns that are meaningful at the summary levels are shown here. The best way to communicate the information is to have the summary shown on the last line of the report, as all the information for one operator is summarized by day. The next level of the report would take this summary line and display it as one line of information on the new report, as one line for each operator in the same department and on the same shift. The last line would again be a summary, as for all of a department on the same shift, which would become a line of information on another summary report. These summaries would continue until one line shows all the information for the organization.

If the report is to be complete, all the information must be available for review and verification, but since much of the information will be correct as reported, the amount of data to be generated is a constant question. The summary information, beyond the detail of each reporting transaction, will take up little space, so the question becomes how much of the detailed transactions need to be shown to provide verification of the data reported. What is often done to shorten the detailed information that is reported is to suppress the detailed report, if certain summary criteria are met. This may mean, for instance, that only the operator's summary is shown if performance is between 100 and 150 percent, delay is less than 10 percent, and coverage is above 85 percent, or some such combination of acceptable criteria. Then only the exceptions will appear for verification, and the volume of reports reproduced for verification is reduced.

Summary reports also should be prepared of the information reported by code. Of course, this information begins at a higher level than the individual transaction or operator, but there should be added summaries beyond the basic information by code. Thus the plant manager can expect to see a report that would provide one line of information for each major reporting area, a one-line summary of the total, and a summary report by code.

The table below shows the summary report levels and who they should be directed to. This is, of course, just a guide or starting point for the summary information. In actuality, the data summarized will be developed for a specific organization's needs.

Level	Reports	Summary type
Supervisor	Daily performance report	Detailed transactions and area of responsibility
	Labor code report	Area of responsiblity
Superintendent	Daily performance report	By supervisors' area and area of responsibility
	Labor code report	By supervisors' area and area of responsibility
Managers	Daily performance report	By superintendents' area and area of responsibility
	Labor code report	By supervisors' area and area of responsibility
Plant manager	Daily performance report	By managers' area and total
	Labor code report	By managers' area and total

Weekly Reports.   In addition to the current data, it is important to review trends that are taking place in various areas. The most effective way to do this is to put the key indicators on graphs, plotting the points of every week. When plotting the trend, a common practice is to graph the key indicators for the last 13 weeks. The weekly information for hours worked and hours earned is summarized for the last 13 weeks, and indicators that cover the last quarter are calculated. This provides two lines, the current week and the trend over 13 weeks.

Historic Reports.   In addition to the routine reports, special-purpose reports also need to be developed. These will be developed as needed, and often for only one or a few special reports, such as a year-end summary or a status report developed just prior to updating annual costs, or as part of the budget process. The best way to deal with this type of report is to consistently use the same formulas, and train management in the use of the report. Few things are more irritating to managers than to have the same information reported in visually different formats. If data are reported in a histogram formatted as a bar chart one quarter and as a line graph the next quarter, managers will feel that they are not familiar with the data, even though you may have extensively trained them in the last quarter. The best way to handle special-purpose reports is to develop a system to present them as routine reports, even if they are presented only monthly. This will develop more opportunities to train management in the use of the reports and will ensure that they are familiar with the reports, even though the staff may frequently change.

To develop this type of report, it is essential that a substantial amount of detailed information is available if it is required. Detailed data covering the last 3 months should be readily available. After that they can be archived with little impact to routine information. Remember, it is not possible to predict when the history of performance to a specific standard is suddenly valuable. Having such information readily available can often hold off problems. Storing readily accessible data for the last 3 months will prove beneficial for producing reports that may be required on demand, such as performance for a specific operation or person.

A primary concern is always who will be responsible for developing these reports and delivering the information. The development of routine reports can often become a less than challenging, routine type of task, and few people look forward to doing it. Competent clerical personnel can be trained to produce the information, and where the volume of routine reports justifies it, they are ideal choices to handle routine reports and many special-purpose reports. We should not overlook the learning opportunity that these reports provide for new personnel. Actually compiling the reports or reviewing those developed by others can provide a valuable introduction to the facility or organization for the new engineer.

Providing reports that are adequate, accurate, timely, understood, and routinely produced is the key to effective control. The reports must be routinely produced, if possible, so that their use is understood. Unusual or special-purpose reports (such as a history of performance by operation) cannot be routinely produced, but the formats for such reports should be the same all the time. Paperwork should be reduced whenever possible. This can be done in a variety of ways that really depend on the abilities of the people receiving the information and the storage and retrieval systems available. Above all reports must be accurate. Errors that are overlooked on this week's report will call into question similar reports for other areas and previous weeks. If errors are encountered, the engineers should take the lead on informing everyone concerned and providing the necessary corrections. Finally, the reports must be timely. Effective control of operations requires that the status of the operation is frequently reviewed, so problems are uncovered and can be rectified before they become major.

SUMMARY

This chapter provides a review of the principles that guide the routines used in the day-to-day administration of the industrial engineering department. No two departments will ever run exactly the same, nor should they, but the general principles that guide the department will be the same. All departments must tailor their practices to fit the industry they are part of, the size of the organization they are part of, the size of the organization served, the available support resources, what has happened in the past (in both the department and the organization), and what future plans are.

The size of the industrial engineering staff is determined by the responsibilities of the department, the size of the organization being served, and the nature of the organization. The development of qualified personnel for the staff is an ongoing and dynamic activity, providing both the manager (or chief engineer) and the engineers with the challenge of developing themselves to meet the evolving needs of the organization. Managing the paperwork that a modern department generates is a challenge, since time spent handling papers is time away from improving the operation, and in a dynamic world, improvement is necessary for survival.

Several methods of tracking progress exist. The one detailed in this chapter uses an estimate of the work, developed by the engineer, as the standard for completing a project. Progress is measured to this estimate and reviewed weekly. Time standards, timekeeping, and controlling performance and productivity are still high-priority items for most industrial engineers, so this chapter concerned itself with the general principles of administering those specific functions.

REFERENCES

1. Harris, Neville, *Management Services in the United Kingdom,* The Institute of Management Services, Enfield, Middlesex, England, 1979.
2. Hodson, William K., "Administrative and Control Procedures," in H. B. Maynard, ed., *Industrial Engineering Handbook,* 3d ed., McGraw-Hill, New York, 1971.

3. Lucking, Peter, "Planning and Controlling the Industrial Engineering Function," *IIE Integrated Systems Conference Proceedings,* 1989.

4. Forberg, Richard A., "Effective Control of the Industrial Engineering Function," *Proceedings Twelfth Management Engineering Conference,* SAM-ASME, New York, April 1957, p. 214.

CHAPTER 2
MEASUREMENT OF LOW-QUANTITY WORK

Richard L. Shell
Professor of Industrial Engineering
Department of Mechanical, Industrial, and Nuclear Engineering
University of Cincinnati
Cincinnati, Ohio

Since Taylor's early work at the beginning of the 1900s, work measurement and methods analysis efforts have been directed toward higher-volume repetitive work which could be accurately measured and controlled. In this situation, the time (and cost) of measurement and control was easily cost justified by the economic and other operational benefits realized from having accurate production standards established for major work tasks, such as staffing, scheduling, and capacity analysis.

When conventional work measurement and methods analysis was initially applied to lower-quantity less repetitive work, the economic benefits rapidly diminished. Until recently, most engineers and managers would categorically agree that conventional work measurement and methods analysis was simply not economically justified for low-quantity work because of the excessive time required for formulation of the production standard. Many stories in job shops where formal work measurement programs existed tell how it took more time for the engineer to determine the production standard than for the machine operator to produce the entire lot of parts. Clearly, there must be a better way.

During recent years, three major changes have occurred that now make the measurement of low-quantity work cost-effective. These are

1. Computerization and CAD evolution, permitting database construction, standard data development, ease of production standard calculation, and the ability to design correct methods, tooling, and workplace rapidly

2. Work measurement technique development, namely, the formulation of high-level predetermined time systems

3. Increased automation levels in many low-quantity manufacturing and service environments, permitting the establishment of production standards with techniques other than conventional work measurement, that is, CAM database, built-in timing of cycle time, and other electronic timing or counting such as computer terminal output performance monitoring

Requirements for Work Measurement.   In general, work measurement may be defined as the application of techniques designed to estimate the time for a qualified worker to conduct a specified task at a defined level of performance to produce a minimum acceptable quality output. Properly practiced, the field of work measurement encompasses correct methods definition that specifies the human interface with all necessary tools and equipment. In addition to determining *how and with what to perform the task,* the engineer should make sure that the workplace meets *acceptable standards of ergonomic design and of occupational safety and health.* A final requirement of professional work measurement is to ensure worker *cooperation* in the measurement process and *involvement* in the creation of the total job environment. The importance of applying good *interpersonal skills* by the work measurement engineer or analyst cannot be overstated.

In low-quantity work situations, the requirements are usually the same as summarized in the preceding paragraph for the general case. The major difference is work measurement technique selection and application.

Work Measurement Technique Selection.   Several engineered work measurement techniques are enumerated in Sec. 4 of this handbook. Each of these will be discussed later in this chapter as related to low-quantity work, along with two other techniques not included in Sec. 4 (judgment estimating and the use of historical data). Conceptually, the process of work measurement cost optimization is very straightforward. One needs to develop relationships for the cost of establishing production standards and the cost for having inaccurate or no production standards for a given production volume. The addition of these two relationships yields the total cost as shown in Fig. 2.1.

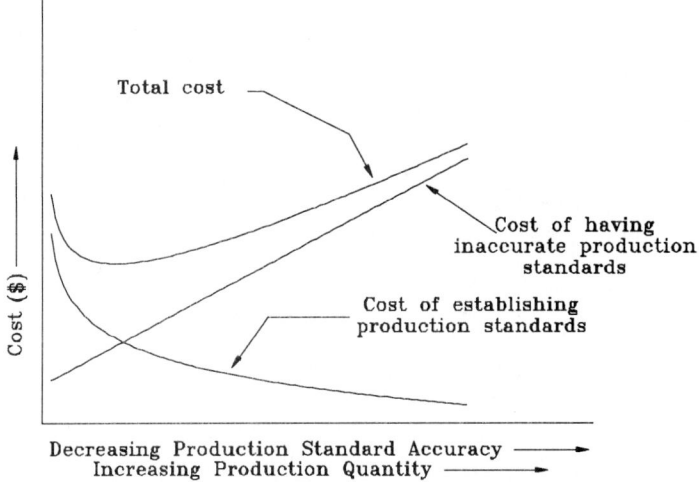

FIGURE 2.1   Work measurement cost optimization concept.

An important aspect of the work measurement cost optimization concept is the cost curve for establishing a production standard. Figure 2.2 depicts work measurement techniques as related to the cost or time to establish production standards versus decreasing production standard accuracy. Another aspect of the work measurement cost optimization concept is determining the cost of having inaccurate or no production standards. Figure 2.3 shows a family of cost relationships based on accuracy of the production standards. As the production quantity increases, the cost of having production standard inaccuracy increases. Each curve represents a production standard established with a given work measurement technique. This cost (time) difference has been well documented for most work tasks involving human labor. For example, individuals performing tasks with inaccurate or

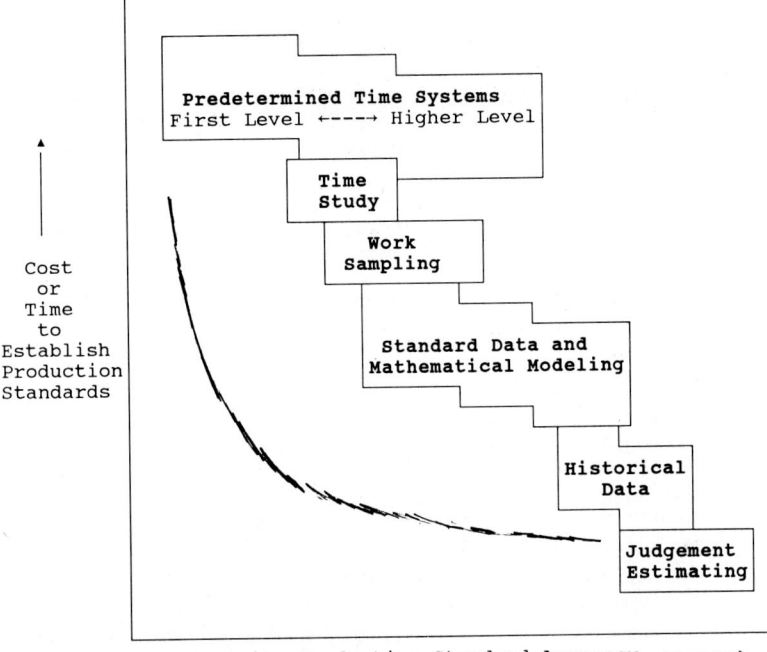

FIGURE 2.2   Work measurement techniques as related to cost, time, and accuracy.

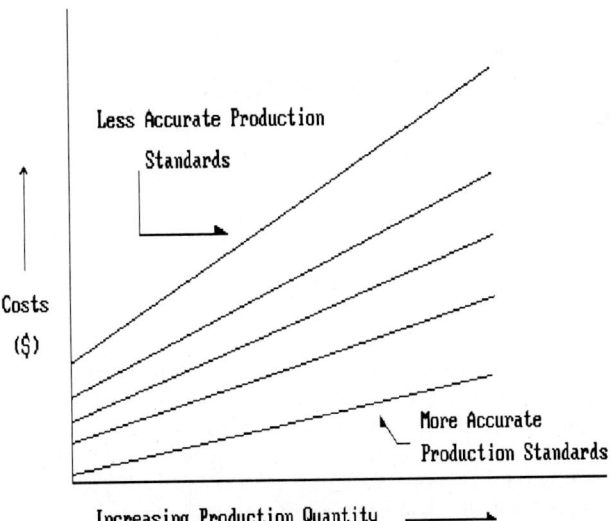

FIGURE 2.3   Family of cost relationships versus production quantity with varying accuracy of production standards.

no production standards will typically produce only 40 to 70 percent of the output as compared with individuals working with fair, accurate standards. Additional gains in output can be realized with the implementation of incentive systems as discussed in Sec. 6 of this handbook.

During the past several years, most of the published literature concerning production standards has stressed either the need for accuracy or the need for economy. The key for establishing production standards in low-quantity work situations is selection of the work measurement technique(s) that tend to optimize total costs, that is, cost of establishing production standards and cost of inaccurate production standards. While all of the techniques may be useful for specific aspects of a low-quantity work measurement program, the following will be most appropriate:

1. Higher-level predetermined motion time systems
2. Standard data and mathematical modeling
3. Historical data

Time study and work sampling may be used to develop standard data and allowances. Also, judgment estimating could be used to develop approximate estimates in situations where time is not available to use other techniques and accuracy requirements permit considerable error.

Accurate Production Standards and Cost Estimating.   An important consideration is the relationship between the establishment of accurate production standards and cost estimating for new business. Since the cost of the product, or service, may be based upon the production standard for the job, it is of paramount importance to define accurate standards. Setting too low a standard could result in a financial loss, as the actual cost to complete the job could exceed the estimated cost. Conversely, excessively high production standards result in a larger profit margin but could ultimately result in a financial loss, as market share will likely diminish. Therefore, the establishment of accurate standards allows the low-quantity work estimator to produce competitive bids while still retaining healthy profits. Another reason for accurate production standards is that they are often used as a basis for incentive pay.

THE LOW-QUANTITY WORK ENVIRONMENT

Low-quantity work is found throughout the U.S. economy including manufacturing, service organizations, and government. Low-quantity work is found in all sizes of organizations and businesses ranging from small job shops to very large corporations. The work is performed by direct as well as indirect or support personnel.

Manufacturing.   Low-quantity work in manufacturing is found in a variety of job shops. Some job shops make proprietary products of their own design while others design and manufacture products per customer requirements. Other job shops produce only components and/or assemblies for customers. Work measurement requirements vary in each of these situations. Despite the difficulty in classification, it is possible to identify basic common types of manufacturing job shops.

Classification Type I.   The first type of job shop is one that is described as having technology complex equipment. Products of these job shops are intricate units made up of hundreds or thousands of different parts and components, each having its own drawing number. Their customers usually order a small quantity of these units, sometimes only one. These companies' products are generally very complex and expensive. Example products of this type of job shop would be jet engines or machine tools. This type of job shop usually warrants higher-accuracy production standards and in most cases can cost justify standards based on higher-level predetermined time systems or well-defined standard data developed from fundamental predetermined time systems and/or time studies.

Classification Type II.   The second type of job shop is one that typically makes less complex end products and components. These companies usually produce larger quantities on any one order but are still considered job shops because of the infinite variations of size, shape, color, style, configuration, material, and function. Example products manufactured by this type of job shop include tires, shoes, and furniture. As with the first type of job shop, this type can usually cost justify higher-accuracy production standards based on predetermined time systems or well-defined standard data developed from fundamental predetermined time systems and/or time studies.

Classification Type III.   A third type of job shop is one that is characterized as making custom components and/or assemblies. This type of job shop commonly supplies the first two types of companies. They specialize in one or more specific processes and fill their shops with customer-designed parts containing many different configurations. This type of job shop is able to fill most orders in a relatively short period of time because of previous orders from the same customer. Example items manufactured include castings, stampings, and machined components. For items that are repeat ordered, this type of job shop can usually cost justify higher-accuracy production standards based on higher-level predetermined time systems or standard data. For items that are first-time orders, higher-accuracy production standards are probably not cost justifiable. In these cases standard data or historical data would usually be the most appropriate work measurement technique.

Classification Type IV.   The fourth type of job shop is a company whose services are subcontracted. The only major difference between this type and the previous customer components and/or assemblies shop is the size of the shop and scope of operations. This category of job shop is usually a smaller independent business that specializes in a particular operation or process. Typically, repeat order items represent a smaller part of total business volume. Example operations or processes provided by this type of shop include electroplating, welding, and tool and die. Usually, this type of shop cannot cost justify higher-accuracy production standards. Therefore, standard data, historical data, or in certain situations judgment estimating would be the most appropriate work measurement technique.

A major benefit of establishing production standards to all low-quantity work job shops, independent of the work measurement technique(s) utilized, is the requirement of defining the method (operation process sheet). This occurs *before* the production standard is established and provides improved operation efficiencies and effectiveness in scheduling, staffing, planning, and overall management control.

Service Industries.   Low-quantity work in service industries is found in a wide variety of businesses. During the past ten years, 90 percent of the new jobs created in the United States were in service organizations. Example industries with low-quantity work tasks include health care (Sec. 15, Chap. 1), retailing, hotel and food service, construction, and information processing. Much of the work force in service industries are clerical and white collar (professional). For work measurement and control for these workers, see Sec. 5, Chap. 7. Other workers in service industries would be classified as indirect or support. For work measurement and control for this group, see Sec. 5, Chap. 5.

Government.   Low-quantity work is found throughout the government sector: local, state, and federal. Like service industries, much of the work force is clerical and white collar (professional). Other government workers are classified as indirect or support. There are additional categories of workers in government, such as military, security, and technical. Other chapters in this handbook pertaining to workers in government include Sec. 5, Chap. 7; Sec. 5, Chap. 5; and Sec. 15, Chap. 5.

The Mental Workload Component.   A common characteristic of workers in all low-quantity work environments is that the majority of their tasks include considerable mental processing. This is not to say that workers performing tasks in higher-volume work situations do not think, but the nature of low-quantity work output usually requires more time for activities such as referring to instructions, information analysis, and communication. Longer cycle, irregular methods, less workplace structure, the use of less dedicated general-purpose tooling and equipment, and the frequency of job setup activities all tend to increase the amount of mental processing and human to human interaction.

Because of the mental thought processes and verbal communication found in low-quantity work, predetermined time systems, stopwatch time study, and conventionally established standard data are generally less useful techniques. Work sampling and historical data are typically more useful and provide a range of accuracy for production standard setting.

MEASUREMENT TECHNIQUES FOR LOW-QUANTITY WORK

The following five categories of work measurement techniques are discussed in the paragraphs below:

1. Predetermined time systems
2. Direct observation timing with performance rating (stopwatch time study)
3. Work sampling
4. Historical data (includes accounting records and self-logging)
5. Judgment estimating

The first three are considered engineered work measurement techniques. The last two, historical data and judgment estimating, are often used to *approximate* standard time values. However, these techniques have decreased accuracy and little underlying theory or standardized procedures and consequently are not considered engineered work measurement practice. An additional technique, standard data and mathematical modeling, is also useful in the establishment of low-quantity work standards. The engineer must be aware of the accuracy required for a given production standard or other work measurement application when using any of these techniques.

To varying degrees, all of the above techniques may be used for the measurement of low-quantity work, depending primarily on accuracy requirements but also considering availability of human resources, time to determine the production standard, and management objectives. It is important to know the strengths and limitations of each technique. This is useful in technique selection when evaluating the cost of establishing production standards versus the cost of having inaccurate or no production standards (Figs. 2.1, 2.2, and 2.3).

Predetermined Time Systems.   By definition, a predetermined time system is "an organized body of information, procedures, techniques, and motion times employed in the study and evaluation of manual work elements. The system is expressed in terms of the motions used, their general and specific nature, the conditions under which they occur, and their previously determined performance times" (ANSI Standard Z94.12). Section 4, Chaps. 4 and 5, contain general descriptions and more detailed information about predetermined time systems. The material here discusses their application to low-quantity work situations.

Predetermined time systems can be classified as generic, functional, or application specific. Generic systems are not restricted in any way and are widely used by many types of organizations. Functional systems are adapted for a specific activity type, such as machining, clerical, or microassembly. Application specific systems are designed for the operational needs of a particular industry or organization, like aircraft engines, banking, health care, or individual companies.

Numerous predetermined time systems have been developed over the years. Examples include methods-time measurement (MTM), work-factor, motion time study (MTS), modular arrangement of predetermined time standards (MODAPTS), master standard data (MSD), and robot time and motion (RTM). While each has certain features, most of the systems have similarities in their structure. Because of its wide usage, versions of the MTM system will be discussed here for application to low-quantity work.

Several predetermined time systems are based on a fundamental or detail structure. These are considered first-level generic systems. Higher-level systems are those developed with elements that include multimotions or combinations. For MTM, the fundamental level is MTM-1. Higher-level generic variations more useful for low-quantity work include:

MTM-2: Based on MTM-1 with about twice the speed of analysis but with less accuracy than MTM-1.

MTM-GPD (general purpose data): Based on MTM-1 to group common motion patterns. Less analysis time and lower accuracy than MTM-1.

MTM-3: Based on MTM-1 with about seven times the speed of analysis and less accuracy than MTM-1. Ideal for long-cycle, low-quantity work.

Examples of MTM functional variations include:

MTM-M: For subminiature assembly completed in a microscopic field.

MTM-C: Clerical activities data system.

MTM-V: Machine tool use data system.

Other MTM outgrowths include micro-matic methods and measurement (4M), and Maynard operational sequence technique (MOST®). A computerized work measurement system developed at Westinghouse, 4M retains MTM-1 accuracy and element description while providing a faster speed of analysis. The MOST system utilizes a larger number of motions in elements than MTM-2 and therefore requires less time for analysis, especially in its computerized version. Both 4M and MOST represent excellent examples of the positive effects of computerizing a predetermined time system. They are well suited for low-quantity work applications which typically have longer, nonrepetitive cycle times.

The key to using predetermined time systems for the measurement of low-quantity work is the degree of computerization and its interface or integration with computer-aided design and manufacturing systems. In manually developed production standards, low-quantity work situations demand the use of higher-level predetermined time systems. As computerization increases, it is possible to utilize more fundamental systems to maintain accuracy as well as minimize the time for analysis.

Stopwatch Time Study.   The Industrial Engineering Terminology Standard (ANSI Z94.12) defines time study as "a work measurement technique consisting of careful time measurement of the task with a time measuring instrument, adjusted for any observed variance from normal effort or pace and to allow adequate time for such items as foreign elements, unavoidable or machine delays, rest to overcome fatigue, and personal needs. Learning or process effects may also be considered. If the task is of sufficient length, it is normally broken down into short, relatively homogeneous work elements, each of which is treated separately as well as in combination with the rest." See Sec. 4, Chap. 2 for detail information about this important work measurement technique.

While stopwatch time study is usually not cost-effective for establishing production standards for low-quantity work, it does have certain applications. Included among these are the following:

• Development of standard data time elements
• Determination of times for repetitive activities including setup operations
• Production standards for short-cycle, "higher" production quantity operations

Work Sampling.   The Industrial Engineering Terminology Standard (ANSI Z94.12) defines work sampling as "an application of random sampling techniques to the study of work activities so that the proportions of time devoted to different elements of work can be estimated with a given degree of statistical validity." The technique can be applied to humans, machines, or any observable state or condition of an operation. The underlying assumption of work sampling is that the sampling percentage of any observed state of nature estimates the actual time spent in that condition. See Sec. 4, Chap. 3 for detail information about this important work measurement technique.

Work sampling with performance rating can be used in manufacturing, service organizations, and government for cost-effective low-quantity work measurement. The technique is particularly useful for the following:

- Determining allowances
- Determining machine and equipment and/or facility utilization
- Production standards for indirect and support personnel
- Production standards for long-cycle, "higher" production quantity operations
- Development of standard data

Generally, when work sampling is used for determining production standards for low-quantity work, lower accuracy and greater error limits are designed into the study. This reduces the number of observations (and time) required to complete the study. Common guidelines for the confidence interval and error limits are 95 percent and plus or minus 5 percent, respectively. However, in all work sampling applications the engineer must use judgment about the activity or state being observed and the overall accuracy requirements to determine the correct confidence interval and error limits.

Historical Data.   While not an engineered work measurement technique, historical data in most low-quantity work situations is very cost-effective for the development of production standards. Historical data are a special, less accurate form of standard data (see Sec. 4, Chap. 6). Times are developed for work activities from previous actual recorded times (job accounting records). Task times may also be developed by workers self-logging times for activities. These time data can be statistically averaged and evaluated.

Historical data based production standards can be used in any work environment: manufacturing, service organizations, or government. The time and cost to develop standards with this technique are generally minimal compared with other work measurement techniques (see Fig. 2.2). The obvious limitation is accuracy of the resulting production standards. See Sec. 9, Chap. 4 for more information on estimating techniques.

Judgment Estimating.   As with historical data, judgment estimating is not an engineered work measurement technique. It is the least accurate but requires the least amount of time to establish production standards. Many practitioners dismiss judgment estimating because of its accuracy limitations. Experience has shown that production standards established with judgment deviate from the true standard by at least 25 to 50 percent. There is, however, a methodology useful to improve judgment estimating.

The recommended five-step procedure for developing a judgment estimate is the following:

1. Subdivide the total job into major activities (tasks). This helps define what has to be done and enhances the estimating process.
2. Clearly understand the physical setting of each activity (task) to be performed, such as the workplace, tools, equipment to be used, and the job environment.
3. Think in terms of producing *one unit of output,* that is, a *production cycle* following any setup or get ready activities. As with all work measurement production standard setting, setup or get ready activities should be estimated independently.

4. Assume the task will be performed by an *average* or *normal* operator.

5. Complete the task time estimate. This can best be accomplished with a process commonly used in critical path scheduling models such as program evaluation and review technique (PERT). These models commonly use judgment estimating to determine the time values for each activity (task) identified within the network. In place of just estimating one value for the total task, three estimates are completed: an optimistic time, a pessimistic time, and a most likely time. The optimistic and pessimistic time estimates require the engineer to mentally establish minimum and maximum time boundaries considering the major factors that could cause variation in task completion. The task time estimate conforms to the beta distribution and can be calculated with the following equations:

$$T_e = \frac{t_o + 4t_m + t_p}{6}$$

where T_e = mean of the beta distribution, the mean time estimate for the task
   t_o = optimistic time estimate
   t_m = most likely time estimate
   t_p = pessimistic time estimate

and for the standard deviation,

$$\sigma = \left(\frac{t_p - t_o}{6}\right)^2$$

STANDARD DATA AND MATHEMATICAL MODELING

MIL-STD-1567A defines standard data as "A compilation of all the elements that are used for performing a given class of work with normal elemental time values for each element. The data are used as a basis for determining time standards on work similar to that from which the data were determined without making actual time studies." Standard data can have several levels of refinement much like predetermined time systems. The fundamental level provides times for specific *motions*. Higher-level standard data provide times for *elements* of work, and the highest level for complete *tasks*. In practice, standard data can be constructed from any of the five categories of work measurement previously discussed. Ranging from the lowest level (more refinement and more accuracy) to the highest level (less refinement and less accuracy): predetermined time systems, stopwatch time study, work sampling, historical data, and judgment estimating. Most commonly, standard data are developed from time study, particularly for higher-volume work. Standard data also include the development of normal times through the use of mathematical equations and tabulated element or task times. See Sec. 4, Chap. 6 for additional information about standard data.

Computerized standard data for low-quantity work is usually higher-level, that is, developed for common elements or complete tasks. Depending on accuracy requirements, standards for work elements are typically developed from time studies or predetermined time systems. Standards for work tasks are commonly developed from work sampling or historical data. Higher-level task standards can also be determined from judgment estimating for extremely low-quantity work situations. The principal advantages of computerized standard data systems are that they afford a fast and consistent methodology to estimate normal time values, and they can be tailored to meet the needs of an individual organization. The major disadvantage is that standard data systems have a fairly high development cost.

Developing standard data for low-quantity work requires the following steps:

1. Determine accuracy requirements for the production standards to be computed using the standard data and select the appropriate work measurement technique(s) to develop the standard data.

2. Review existing conditions to determine whether the methods, facilities, and total work environment are standardized or consistent. The standard data times will apply only for workplace conditions under which they were developed. If possible, the standard data should be established for types of operations independent of the product.

3. Determine and code all work elements or tasks in a database (see Sec. 12, Chap. 2). As standard data are developed, they should be indexed and filed. In this activity, it is necessary to separate constant elements from variable elements. For task-level standard data, it often is desirable to combine constants with variables, thus expressing the total normal time to perform the entire operation. Setup or get ready standard data can also frequently be combined at the task level.

4. Determine element and/or task times and build database.

5. Test for validity.

6. Perform ongoing audit to ensure that the standard data are based on defined methods and workplace configurations, and that the appropriate work measurement technique was correctly applied.

Application Procedure. For element based standard data, application requires subdividing the entire task or operation into elements (manual and machine) that are provided for by the system. Then the engineer uses the computerized standard data system to obtain normal time values for each element and all element values are summed to compute the total task or operation normal time. The computerized system should be designed to sum all task times to compute the total job normal time.

For task based standard data, application requires subdividing the total job into tasks or operation times that are provided for by the system. The engineer then uses the computerized standard data system to obtain normal time values for each task and all task values are summed to compute the total job normal time.

The computerized system can also add allowances to obtain the standard time. In addition, the computerized system should produce some level of operation sheets useful for low-quantity work processing. These provide documentation that minimizes processing errors and aids methods improvement. Computerization of the standard data also enables persons with limited experience and training to apply the data successfully in a low-quantity work environment.

Mathematical Modeling. A simplified aspect of standard data is mathematical modeling and formula construction. This aspect of standard data is particularly well suited for low-quantity work. Mathematical formulas can be used to establish time values for production standards throughout manufacturing, service, and government for both direct and indirect labor operations. Their advantages include having reasonable accuracy, being easy to computerize to obtain reduced time for standard setting, and having standards that can be determined by less experienced personnel. The main disadvantage is the fact that many elements or tasks have built-in constants and thus do not closely follow the mathematical relationship. Another limitation is that standard data formulas are at times applied to ranges beyond the original data used to establish the relationship.

The procedure used to develop the mathematical model requires identification of the dependent variable and the independent variable(s). The dependent variable is usually time and the independent variable(s) usually relate to some physical aspect of the item being produced, such as length, weight, or surface area. Independent variables can also relate to the process, equipment being used, tooling, and workplace. The following steps are recommended to develop computerized mathematical models for low-quantity work:

1. Define the job or group of jobs that time estimates will be determined by computerized mathematical models.

2. Collect the data to develop the mathematical relationship. The data can be obtained from time studies, predetermined standard times, historical data, or judgment estimates. Selection of the data and the work measurement technique used to develop the data depends on accuracy requirements for the completed model.

3. Determine constants and variables within the database, and define the dependent and independent variables.

4. Construct the mathematical relationship that best fits the data. Several examples of curve fitting are discussed later in this section.

5. Develop computer software for the models, or work into an existing CAD/CAM system.

6. Test and evaluate each relationship for accuracy and operational correctness.

Curve Fitting.   An important step in developing mathematical models for estimating is fitting the work measurement data to an appropriate curve. The objective is to understand how the data and independent variables influence the dependent variable. For certain curve fitting situations the classical graphic analysis can still be very useful.

Linear regression using the least-squares method is commonly used to fit a straight line through a group of data points. The principle of least squares requires that if a dependent variable y is a linear function of an independent variable x, the best location of the $y = a + b_x$ will exist when the sum of the squares of deviations of all data points from that location is minimized.

The values of a and b may be obtained by solving the normal equations:

$$a = \frac{\Sigma x^2 \Sigma y - \Sigma x \Sigma xy}{n\Sigma x^2 - (\Sigma x)^2}$$

and

$$b = \frac{n\Sigma xy - \Sigma x \Sigma y}{n\Sigma x^2 - (\Sigma x)^2}$$

For linear regression, the correlation coefficient r explains how well the variables are satisfied by a given relationship. The correlation coefficient may be computed as shown:

$$r = \frac{n\Sigma xy - \Sigma x \Sigma y}{[n\Sigma x^2 - (\Sigma x)^2][n\Sigma y^2 - (\Sigma y)^2]^{1/2}}$$

The value of r must be in the range $0 \leq |r| \leq 1$. The magnitude of the correlation coefficient indicates the strength of the relationship between variables. A word of caution: it should be noted that the correlation coefficient is a measure of relationship strength and is purely mathematical. It does *not* determine any cause or effect implications.

The method of least squares permits the computation of sampling errors and provides for the determination of the estimates' reliability. Confidence limits can be determined by computing the variance of y from the estimated value of y by the regression line with the following equations:

$$S_y^2 = \frac{\Sigma \epsilon_i^2}{v}$$

where S_y^2 = confidence interval
ϵ_i^2 = square of the deviation
   = $y_i - (a + bx_i)$

v = degrees of freedom

Additional confidence limits and significance tests may be determined for the regression equation depending on the estimating accuracy requirements.

For two or more independent variables, multiple linear regression may be used as expressed by the following equation:

$$y = a + b_1x_1 + b_2x_2 + \cdots + b_mx_m$$

where a = a constant

b_i = partial regression coefficients

A word of caution: single or multiple independent variable linear regression requires the following assumptions:

1. The regression of y on x_i is linear.
2. The work measurement data are representative of the job(s) time values.
3. There are no extraneous variables.
4. The deviations from y are mutually independent and have the same variance for any value of x_i.
5. The deviations are normally distributed.

Any error introduced by the above assumptions not being met will usually fall within the overall work measurement accuracy requirement for low-quantity work. Therefore, regression is a highly useful procedure for establishing mathematical models for determining low-quantity production standards.

Many situations in the development of mathematical models for low-quantity production standards require curve fitting for nonlinear data. As is the case for linear regression, nonlinear regression is best performed with computerized assistance. The following are examples of curvilinear relationships useful in developing time estimating models:

Exponential or logarithmic: $y = ax^b$

Semilogarthmic: $y = ae^{bx}$

Hyperbolic: $y = x/(a + bx)$

Polynomial: $y = a + b_1x + b_2x^2 + \cdots + b_mx^m$

A number of nonlinear relationships may be transformed to linear expressions. For example, nonlinear data points following the exponential form $y = ax^b$ on cartesian coordinates may be transformed to a linear relationship on logarithmic coordinates with the following equation:

$$\log y = \log a + b \log x$$

All these functions can be statistically evaluated for their fit to the data. Commonly used measures include the standard error of y, correlation coefficient, and confidence limits.

Slotting and Comparative Estimating.   Another form of standard data useful for establishing low-quantity work standards is comparative estimating. A single time value is used for the normal or standard time of all tasks that are determined to lie in a work group with a defined time range. This time range is termed a *time slot*, each containing a time. All of the time slots with their respective times for each work group make up the *time-slotting scale*.

Developing comparative estimating standards for low-quantity work requires the following steps:

1. Determine the overall range time for jobs to be estimated, for example, all jobs requiring 32 hours or less would be included.
2. Determine the number of time slots and the time range for each slot. Some comparative estimating systems, like universal indirect labor standards (UILS) and universal maintenance standards (UMS), have divided the time-slotting scale into 20 time slots.
3. Assign representative (benchmark) sample jobs to each time slot. Each benchmark job should have detailed work measurement based standards. The number of benchmark jobs for each time slot depends on accuracy requirements and the nature of work. Experience has shown that several hundred jobs with standards are required for 20 time slots.
4. Arrange the benchmark standards for individual jobs in increasing order based on time and assign to the time slots. Various statistical distribution assumptions are used to assign benchmark jobs into a work group time slot, such as normal, uniform, log-normal, and gamma. The commonly used assumption is the uniform distribution. Experience has shown that the normal distribution produces improved results.

Figure 2.4 depicts an example universal maintenance standard table that illustrates 20 time slots established with the uniform distribution. Benchmark jobs with standards would be assigned to each work group (A through T). This table could be used to estimate any job similar to a benchmark job, and has a range up to 30.0 hours. Comparative estimating has been widely used to estimate normal or standard time values for maintenance jobs. This type of standard data is also highly useful for most low-quantity work including manufacturing production, toolroom, service and indirect activities, and other support operations.

Work Groups with Benchmark Jobs	Standard (hrs.)	Time Range (hrs.)
A	0.1	0.00-0.15
B	0.2	0.15-0.25
C	0.4	0.25-0.50
D	0.7	0.5-0.9
E	1.2	0.9-1.5
F	2.0	1.5-2.5
G	3.0	2.5-3.5
H	4.0	3.5-4.5
I	5.0	4.5-5.5
J	6.0	5.5-6.5
K	7.3	5.6-8.0
L	9.0	8.0-10.0
M	11.0	10.0-12.0
N	13.0	12.0-14.0
O	15.0	14.0-16.0
P	17.0	16.0-18.0
Q	19.0	18.0-20.0
R	22.0	20.0-24.0
S	26.0	24.0-28.0
T	30.0	28.0-32.0

FIGURE 2.4 Example of universal maintenance standards for comparative estimating.

CASE EXAMPLES

***Standard Data Development for Manufacturing.**   LTV Aircraft Products Group in Dallas, Tex., manufactures aircraft subassemblies and components for military and commercial use and does extensive modernization and upgrading of existing military aircraft.

LTV is implementing MIL-STD-1567A, which requires type I standards of all touch labor operations (type I standards are standards which have an accuracy of plus or minus 10 percent with a confidence level of 90 percent). LTV will have to set approximately 500,000 standards within the next 4 years to complete this task. Extensive industrial engineering resources will be required to develop the standard data and apply these data to create times standards. LTV recognizes the importance of taking advantage of all the available techniques to simplify this task while still maintaining the required accuracy.

A five-axis three-spindle profiler located in LTV's machine shop will be used as an example. For several years LTV standards applicators used profiler work sheets which had 15 standard data elements just for the "per piece" work.

Per piece elements	Normal hours
Hoist load and unload	0.393
Hoist load and hand unload	0.234
Hand load and unload	0.090
Hand load and unload small parts	0.043
Change tools and set depth	0.028
Position tape and return to set point	0.053
Set for cut	0.006
Change fixture angle	0.026
Reload and secure	0.032
Reclamp parts	0.061
Position and secure clamps	0.010
Remove clamps	0.005
Position and secure bolts	0.007
Remove bolts	0.005
Reposition clamps	0.003
Machining process time	Variable

The standards applicators had to make many decisions just to set the run or "per piece" time standard. To develop a standard, the applicator studied blueprints, machining procedures, cutting tool requirements, materials, securing devices, and other factors which influence the making of a part. This total procedure could take as long as 4 hours, depending upon the parts complexity.

Obviously, the time needed to set each time standard could be reduced if the number of choices on the work sheet are minimal. The only real constraint would be to maintain the required accuracy of the time standards.

LTV is using MOST computer systems to develop the standard data and to apply them to create and maintain the time standards. The Maynard technique for developing standard data includes the use of a statistical formula for calculating the scope for application of each standard data element on a work sheet. This formula, called the "allowed deviation formula," is used to determine the minimum number of standard data elements needed to cover the range from the simplest case for each activity on a work sheet to the most difficult case.

Steps taken to develop the standard data for the LTV profilers were as follows:

*Source: J. E. May, and R. Jackson, June 1989. "Industrial Engineering Philosophies and Practices Can Increase Shop Floor Productivity . . . and Our Own," *Industrial Engineering*, pp. 39–43. Reprinted from *Industrial Engineering* magazine, June 1989. Copyright Institute of Industrial Engineers, 25 Technology Park/Atlanta, Norcross, Ga. 30092.

Step 1. Make a List of All the Operator's Activities at the Workplace. After observing the operator, a list was made of all the activities needed to make any part.

1. Load by hand (under 40 lb)   Small part
2. Load by hand (under 40 lb)   Medium part
3. Load by hoist (over 40 lb)   Small part
4. Load by hoist (over 40 lb)   Medium part
5. Load by hoist (over 40 lb)   Large part
6. Unload by hand   Small part
7. Unload by hand   Medium part
8. Unload by hoist   Small part
9. Unload by hoist   Medium part
10. Unload by hoist   Large part
11. Air-clean part
12. Air-clean fixture
13. Deburr part   Small size
14. Deburr part   Medium size
15. Deburr part   Large size
16. Wipe down part   Small size
17. Wipe down part   Medium size
18. Wipe down part   Large size
19. Install cutter and set depth
20. Install or remove clamps
21. Adjust clamps
22. Install or remove vacuum hose
23. Adjust jack screws
24. Install or remove bolts
25. Change fixture angle
26. Start tape on terminal
27. Identify part with pen
28. Install eye hook into part

Step 2. Document Worker's and Management's Responsibilities and Standard Practices. In order to set time standards, the engineer must know what the operator is and is not responsible for. Any required activities that were not observed were added to the list. Any activities observed in step 1 which were someone else's responsibility were omitted. Only those activities that are required to do the job were considered.

Step 3. Develop Standard Data. Next, the constant activities on our activity list were measured using the MOST technique. Constant activities are those which do not vary in the way they are performed. Items 19 through 28 were constant.

All constants which occur at the same frequency in all standards should be combined into one work sheet element. These were 27 and 28.

We were now ready to begin measuring the variable activities. These are called variable because the method for performing each one of these activities can vary from part to part. As mentioned before, valuable time and money could be saved by reducing the number of choices on the work sheet. This would also reduce the chance of applicator error, since there would be fewer choices to pick from on the standard setting work sheet.

It was known that the smallest parts required the smallest number of clamps to secure the part and the least cleaning. Also, the larger parts required more clamps and more cleaning. Once the relationship between the size of the part, the number of clamps, and the amount of cleaning was established, the loading and unloading of the parts with securing and cleaning was combined into a common standard data element.

The allowed deviation formula was used to determine the minimum number of standard data elements needed while still meeting the required accuracy demands. First the easiest and most difficult cases were identified for loading, securing, and cleaning a part. Once these two cases were measured, the extremes that the standard data must represent were known.

Easiest case:

1. Load or unload small part by hand (up to 40 lb).
2. Adjust 8 clamps.
3. Start NC tape.
4. Wipe down 4 sq ft part.
5. Identify part.
6. Air-clean 4 sq ft part.
   Total time = 9530 TMU (5.7 minutes)

Hardest case:

1. Load or unload large part by hoist (over 40 lb).
2. Adjust 14 clamps.
3. Install or remove 8 clamps.
4. Adjust 6 jack screws.
5. Start NC tape.
6. Wipe down 36 sq ft part.
7. Identify part.
8. Air-clean 36 sq ft part.
   Total time = 46,300 TMU (27.8 minutes)

At this point, the allowed deviation formula was applied. This determined the minimum number of standard data elements or "slots" needed to statistically represent all possible ways of loading or unloading, and securing a part.

The allowed deviation formula is defined as

$$r_1 = r\sqrt{\frac{T}{(n)t}}$$

where r_1 = percent allowed deviation

r = required accuracy for T. At LTV, an accuracy of plus or minus 10 percent is required for each standard.

T = time period required for accuracy level to be reached[*]

n = number of occurrences of measured activity during each operation

---

[*]T is normally the operator's pay period of 8, 40, or 80 hours. It is at the end of this period that plus or minus 5 percent accuracy is normally required. For defense contractors (as in LTV's case) plus or minus 10 percent accuracy, 90 percent confidence level is required at the operation level. This means that 90 percent of the time, standards must be accurate with plus or minus 10 percent. For the profilers example, the run or "per piece" time standard includes manual activities plus N/C machine or "tape" time. The average profiler operation had historically been 6.7 hours. From this, a 6.7-hour "balancing time," or T value, was used to form the structure of the standard data.

t = normal time for measured activity

In the example:

$r = 0.1$ (10 percent)
$T = 6.7$ standard hours (670,000 TMU)
$n = 1$
$t = 0.095$ standard hour (9533 TMU)

First, the time for the easiest case was substituted into the formula to calculate the allowed deviation:

$$r = 0.1 \sqrt{\frac{6.7}{0.0953}} = 0.840, \text{ or } 84.0 \text{ percent}$$

Allowed deviation = $0.840 \times 0.095 = 0.080$ hour

Next, the time range or slot was calculated for the work sheet:

Minimum time of slot = 0.095 hour (easiest case)
Average time = 0.095 + 0.080 = 0.175 hour
Maximum time = 0.175 + 0.080 = 0.255 hour

This means that a standard data element of 0.175 hour would statistically represent all loading or unloading, securing, and cleaning activities whose normal time falls within the range of 0.095 to 0.255 hour.

Next, the allowed deviation for the second slot was calculated for the work sheet:

$$r = 0.1 \sqrt{\frac{6.7}{0.255}} = 0.513, \text{ or } 51.3 \text{ percent}$$

Allowed deviation = $0.513 \times 0.255 = 0.131$ hour

Finally, the calculated time range for the second slot was calculated:

Minimum time = 0.255 hour
Average time = 0.255 + 0.131 = 0.386 hour
Maximum time = 0.386 + 0.131 = 0.517 hour

Since the second time range includes the hardest case (0.463 hour), only two standard data elements were required to represent all possibilities of the loading or unloading, cleaning, and securing of any part. The average of each slot was the target for writing the standard data analysis. An analysis was written to represent the first slot on the work sheet, while adjusting each variable element to make the standard time equal or close to the calculated average time.

The following elements were included:

1. Load or unload part by hand (up to 40 lb).
2. Adjust 8 clamps.
3. Install or remove 4 clamps.
4. Start NC tape.
5. Wipe down 16 sq ft part.
6. Identify part.
7. Air-clean 16 sq ft part.

Total time = 17,540 TMU (0.175 hour or 10.5 minutes)

An analysis for the second slot was written which contained the following elements:

1. Load or unload part by hoist (over 40 lb).
2. Adjust 10 clamps.
3. Install 6 clamps.
4. Adjust 2 jack screws.
5. Start NC tape.
6. Wipe down 30 sq ft part.
7. Identify part.
8. Air-clean 30 sq ft part.

Total time = 38,410 TMU (0.384 hour or 23.0 minutes)

Step 4. Prepare Standards Work Sheet.   Once the standard data were developed, they were organized on a standards work sheet. The work sheet included the two standard data elements for loading a small or large part, securing it, and unloading it after the machining is complete. It also included a variable for each cutter change and the machining process time. The instructions for the standards applicators were written to describe how to select the correct standard data element for loading, unloading, and securing a part. The choice is based on the weight of the part.

Per piece elements	Normal hours	Frequency	Total
1. Load or unload part (less than 40 lb)	0.175	_____	_____
(Apply a frequency of 1 for each part which is loaded or unloaded by hand. Securing or unsecuring part with clamps, identifying, and cleaning are included)			
2. Load or unload part (over 40 lb)	0.384	_____	_____
(Apply a frequency of 1 for each part which is loaded or unloaded by overhead hoist. Securing or unsecuring part with clamps, adjustment of any jack screws, identifying, and cleaning part are included)			
3. Change cutter	0.022	_____	_____
(Apply a frequency of 1 for each cutting tool changed at spindle)			
4. Machine process time			_____
	Total normal hours per piece		_____
	PF&D allowance		_____
	Standard hours per piece		_____

Eventually the standards applicators will set time standards directly on the computer. This will further reduce the time to set standards.

Productivity Increases Achieved.   A standards applicator can now set a time standard by selecting from just four choices versus the sixteen required with the old profilers work sheet. The choices have been reduced to only the weight of the part, the number of cutters used, and the machining "tape" time. This simplification of the standard setting work sheet will provide substantial savings and increased accuracy.

Standard Data Development for Toolroom.   This example provides information for the construction of computerized standard data for the fitting, assembly, and repair of tools and dies typical of low-quantity work. MTM was used to establish the normal time values

S.B.,F.P.,D.D.,S.O. Dies   Bench Work Order
Date _____   Work Order No. B_____
Die No. _____   Die Class (S)
Die Description _____
Regular Service ◯ Replacement ◯ Change ◯ Rebuild ◯

		Tool Room	Die Vault	Press Room	Clean
Complete	)	◯ .125	◯ .245	◯ .275	◯ .096
Top or Btm. Shoe	) From	◯ .076		◯ .226	◯ .036
Other Sub—assem.	)	◯ .008		◯ .060	◯ .012
Complete	)	◯ .125	◯ .245	◯ .275	
Top or Btm. Shoe	) To	◯ .076		◯ .226	
Other Sub—assem.	)	◯ .008		◯ .060	

TOP SHOE / BOTTOM SHOE

Cradle	x .066 =	Change: From	
Die Sections	x .061 =	To	
Bumper Blocks	x .042 =	Replace New Parts:	
Full Stripper	x .086 =	Piercing Punch	x .098 =
Stock Equalizer	x .078 =	P.O. Slot Punch	x .098 =
Other Sections	x .066 =	K.O. Punch	x .098 =
Punch Holders	x .104 =	(Ex. C.L.)	
Other Sections	x .066 =	Contour Punch	x .098 =

PUNCHES

Piercing.	x .039 =	Trimming Punch	x .270 =
Knock-out	x .039 =	Die Section	x .303 =
Stencil	x .049 =	Bushings	x .098 =
Trimming	x .068 =	Pilot Pin	x .098 =
Lancing	x .059 =	Round Insert	x .098 =
Dimpling	x .039 =	Dowel Pin	x .095 =
Extruding	x .039 =	Quill	x .098 =
Embossing	x .039 =		

		Restore Radius - Hard
		Clearance - Inches x .020 =

BUSHINGS, INSERTS & MISC.

Stencil Bumper	x .039 =	Draw or Form - Hand	
Forming Insert	x .074 =	Stock | x 3 =	
Cut-off Insert	x .069 =	Plus .020	
Embossing Insert	x .049 =	Inches | x | =	
Bushing	x .045 =	BROKEN BOLTS:	
Pilot	x .039 =	Large	Small
Spring Pad	x .104 =	Proj. .300	.250
Lifter Pin	x .056 =	Flush	
Push Back	x .056 =	Blind 1.000	.750
Quill	x .039 =	Flush	
Spring Stripper	x .152 =	Open .750	.500
Guard	x .043 =	BROKEN SET SCREWS:	

OTHER PARTS

Method of Mounting:		1.000	
Bolted Only	x .043 =	Get Parts Welded	◯ .250 =
Ditto w/springs		Use Die Flipper	◯ .747 =
or shims	x .068 =	Re-dowel each part ☐	x .366 =
Bolted & Dowelled	x .063 =	Ear Form (Octagon	◯ .458 =
Ditto w/springs		Adjust (Utility	◯ .308 =
or shims	x .078 =		
Press Fit	x .045 =		
Slide or Tap Fit	x .039 =		
		Job Constant	.120
		Total	
		Times 1.40 = Allowed Hrs.	

FIGURE 2.5   Standard data sheet for toolroom benchwork standards.

shown in Fig. 2.5. The time values, elements, and work tasks could be incorporated into a computerized estimating system to calculate the total assembly time.

A second example represents one of several standard data sheets necessary for establishing toolroom standards. The standard data time values must be used with a frequency determined by toolroom procedures and the die drawings. Figure 2.6 depicts example standard data information to compute the time for repair of punch, trim, blank, and draw dies. All time values shown are MTM derived. The necessary repair tasks and frequency would be determined for each repair job. The computerized program would then summarize all selected tasks and frequencies, and produce an output of necessary work tasks and the total estimated times.

Symbol _____ Order No. _____ Quantity _____ Date _____

A. Prepare Parts.	Time	Allow for each	Use	Lev. Mins.
1. Job preparation	34.50	Job		
2. Move large parts to work area.	5.42	Part over 50 lbs		
3. Check with B.M. for parts.	1.00	Item on B M		
4. Check completion of mach. parts.	1.00	Machined part		
5. Deburr parts. Small (under 50 lbs)	.74	Small part		
Large (over 50 lbs)	2.86	Large part		
6. Trial assm. of mating parts.	.68	Trial assm. crit. fit		
7. File parts to make a fit or match profile.	8.63	Part to be fit or profile		
PLUS	1.25	Inch to be filed		
8. Layout (a) Using comb. square	.86	Pair crosslines		
(b) Using angle plate	.11	Use of angle plate		
(c) Using height gage	.50	Line scribed/H.G.		
(d) Temp. or comp. part	3.81	Scribe around profile		
9. Move large part to layout.	.32	Large part to layout		
10. Apply bluing.	6.62	Area over 360 sq. in.		
B. Fitting and Assembly		TOTAL "A"		
1. Extra trips to tool crib.	8.72	Job		
2. Locate part approx. Hand	2.62	Parts to be located		
Hoist	6.97	for drill & assmb.		
3. Apply & remove "C" clamp or similar device.	1.40	Located part		
4. Move part to mach. & back.				
Hand	1.14	Part under 50 lbs		
Hoist	9.10	Part over 50 lbs		
5. Drill & tap fixing hole.	3.69	Machine set-up		
	2.22	Hole tapped		
6. Locate parts accurately.	7.33	Part accurately located		
7. Drill & ream dowel hole.	3.69	Machine set-up		
	2.53	Hole reamed		
8. Fit dowels, locating pins, taper pins, etc.	.40	Pin fitted		
9. Bolt all parts in position & tighten.	.29	Bolt used		
10. Hand operations (ave.)	2.84	Hand drill setup		
Hand drill	1.65	Hole drill. by hand		
Hand ream	.51	Reamer used		
	2.55	Hole hand reamed		
Hand tap hole	3.37	Hole tapped		
Clean out tapped hole	2.00	Tap. hole–Hard. pc.		
11. Fit bushing & pressed in parts.	1.47	Pressed–in part		
12. Ream bearings (1" diam. x 2" lg.)	1.02	Reamer set-up		
	5.10	Bearing to be reamed		
13. Check location of dowel parts.	2.62	Doweled part		
14. File scrape–seat to make square or flat.	10.57	10 sq." of surface		
15. Stencil symbol number, etc.	7.20	Unit up to 75 letters		
16. Final tryout of working parts, (Simple)	1.60	Simple assembly		
gaging of assembly, planning (Average)	10.00	Average assembly		
minor corrective action. (Complex)	40.00	Complex assembly		
		TOTAL "B"		

Total assembly time = Total A + Total B = _____ + _____ = _____ x 1.24 allowance

FIGURE 2.6   Standard data sheet for toolroom die repair standards.

BIBLIOGRAPHY

Barnes, R. M., *Motion and Time Study: Design and Measurement of Work,* Wiley, New York, 1980.

Martin, D. D., and R. L. Shell, *Management of Professionals,* Dekker, New York, 1988.

May, J. E., and R. Jackson, ''Industrial Engineering Philosophies and Practices Can Increase Shop Floor Productivity...and Our Own,'' *Industrial Engineering,* June 1989, pp. 39–43.

Mundel, M. E., *Motion and Time Study: Improving Productivity,* Prentice-Hall, Englewood Cliffs, N.J., 1985.

Niebel, B. W., *Motion and Time Study,* Irwin, Homewood, Ill., 1988.

Shell, R. L., ed., *Work Measurement: Principles and Practice,* Industrial Engineering and Management Press, Norcross/Atlanta, Ga., 1986.

CHAPTER 3
MEASUREMENT OF REPETITIVE WORK*

Harold C. Herriman, P.E.
Industrial Engineering Specialist

The measurement of repetitive work is fairly common throughout industry. The benefits which stem from work measurement are widely recognized, and the majority of managers look to work measurement, with or without the use of incentive wage payment, as a practical way of increasing labor productivity and reducing labor costs. Indeed, the benefits which have resulted from the measurement of repetitive work have been so apparent that over the years the trend has been to extend work measurement to areas where the work is not repetitive.

In spite of the wide acceptance of the value of measuring repetitive work and in spite of the long and varied experience which management has had in this area, too many applications of measurement have been something less than satisfactory. This chapter discusses some of the problems that exist in the measurement of repetitive work and points out some of the pitfalls that should be avoided.

DECEPTIVE SIMPLICITY OF MEASURING REPETITIVE WORK

Repetitive work is apparently very easy to measure. There are still companies who feel that it is so easy that formal work measurement is unnecessary, and they rely on the supervisor to establish production standards or piece rates on the basis of past performance records, some overall checks of floor-to-floor time, or perhaps just "experience." These methods sometimes seem quite satisfactory. The value of standards to any company is in cost control and planning information that can help a company compete. Under incentive programs, if the workers peg their production, their earnings on incentive can be made to appear consistent and within the range where management expects them to be. More refined work measurement, however, will usually show that productivity is well below what is readily achievable.

A somewhat more sophisticated approach is to use time study for work measurement purposes. This is certainly better than no objective measurement at all, but once again the measurement task is likely to appear deceptively simple. Management often feels that it

---

*This chapter draws heavily on Sec. 4, Chap. 1, by H. B. Maynard, in the Third Edition.

can be accomplished by clerks after a brief period of training in the techniques of stopwatch time study. Even industrial engineers themselves often feel that work measurement is a low-level task which should be delegated to technicians, while they concentrate on the more exotic aspects of industrial engineering.

It is true that the time study procedure itself can be mastered by one of average or better intelligence in a comparatively short time. But the proper application of the time study procedure requires an understanding in depth of many, many factors. Contrary to common belief, the more repetitive an operation is, the more difficult it is to measure satisfactorily.

ACCURACY OF MEASUREMENT

The typical pattern of a work measurement installation on repetitive work, particularly where incentive payment is used, is somewhat like the following. The operators who are doing the work are time studied, and time standards are established. When standards are first announced, there is usually a feeling shared by supervisors and workers alike that they are too tight. After a period of argument and discussion, sometimes accompanied by the adjustment of a few standards, the standards are accepted. The operators then go to work, and for a while everything seems quite satisfactory. Earnings climb to the anticipated level, the operators exert a good incentive effort, and production rises and costs decrease.

Presently, however, particularly if they feel that they will be supported by a strong union, the earnings of some workers will begin to climb. Some workers will earn more than others, perhaps with seemingly less effort, and dissatisfaction will begin to be expressed with the tighter standards. As earnings levels continue to rise, management will begin to feel that it is paying too much for the value of the work being done. If standards are guaranteed, as they usually are in a modern installation unless methods or conditions change, there will be a continuing effort on the part of management to prove that there have been changes in methods and a continuing countereffort on the part of the workers to prove that there have not. If the time studies do not show clearly the methods and conditions which were in effect when the study was made—and they seldom do—it is difficult to reach agreement on the propriety of changing the standards. The situation continues to deteriorate until remedial action can no longer be postponed. Incentive payment may be discontinued, management may seek a new location to make a fresh start, or management and the workers may agree to a program of complete restudy which will result in a return to more realistic standards and earnings.

In a daywork shop the pressure to perform is not as intense for operators as it is if they were working under an incentive program. Similar problems arise when standards are released in that operators feel the standards are too tight. There is a discussion, and possibly the standards are adjusted. At this point operators go to work and everything seems fine. After some time the operators will try different methods to beat the standard. Operators who find a way to beat the standard usually will peg their performance. Symptoms of this malady are extended breaks or personal time or early quits. If operators are challenged about slack time, they often will pace themselves so they are busy all day but performance is out of line with the effort expended. Many managers ignore this problem until other operators complain about inconsistencies in standard effort requirements. If left unattended, this situation often deteriorates to labor problems in the form of dissatisfied employees and to higher manufacturing costs than are necessary.

A number of factors cause the pattern just described to develop. In most cases, however, there is a tendency to blame the accuracy of the original work measurement. The solution, in part at least, appears to lie in the development of more accurate measurement procedures. Performance rating, for example, is a notorious area of uncertainty; so performance rating procedures are overhauled and training in using them is given to all time study personnel. The adequacy of the sample studied is often questioned; mathematical formulas are developed to show how many cycles must be studied to achieve a given degree of accuracy. Statistical tests are introduced to ascertain the limits and confidence level of a given study.

Certainly it is desirable to be as accurate as is reasonably practical when making time studies, but the fundamental problem goes much deeper than that. A time study measures what is being done at the time the study is made. There is no assurance that what is being done at that time will continue to be done 1 month, 6 months, or 1 year later. Indeed, experience shows that there will almost inevitably be change.

METHODS CHANGE

The fact that methods will change with the passing of time is increasingly recognized. In the early days of time study, this was not so. Although the procedure was called time and motion study almost from the start, many practitioners concentrated largely on the time and accepted whatever motions the worker happened to be using. The result was that when the standard was set the worker could usually improve the method rather easily and could thus increase earnings substantially without increasing effort. With the advent of the methods engineering concept which preceded the study of time by a study of methods, much of the potential for subsequent methods improvement was engineered out of the job in advance. This was certainly an important step in the right direction and helped reduce the magnitude of swing from satisfactory to unsatisfactory described above. It did not fully solve the problem, however, especially in the case of highly repetitive work. When the improved method was standardized and taught to the worker, it was still necessary to study what was being done at the time. Again there was no assurance that this is what would be done at any given time in the future.

THE LEARNING PROBLEM

Research has shown that methods seldom if ever remain the same over a period of time. Even if the method appears to the untrained observer to be the same from one time to another, refined motion study will usually show that changes have occurred. As an operation is performed over and over again, the operator, the material handlers, the setup people, the toolmakers, and others in the production operation, either consciously or unconsciously, constantly learn how to perform it a little better each time. The result is that performance time continues to decrease with repetition. There may be several plateaus where little or no learning occurs for a period of time. There may be periods of regression caused, for example, by experimentation, where time even increases for a while. In general, however, learning continues, probably for the life of the longest jobs.

The methods changes which ordinarily occur with practice may be classed for discussion purposes as:

1. Easily observable methods changes
2. Subtle methods changes

Continuing research, particularly into the second classification of methods change, is giving increased understanding of what happens in specific cases to make the learning curve behave the way it does. The following generalized discussion is by no means exhaustive, but it may help the industrial engineer to understand more clearly the problems he faces when he measures repetitive work.

Easily Observable Methods Changes. The term "easily observable methods changes" is something of a misnomer. This type of change is easily recognized by the trained observer who is looking for it but may not be recognized by the untrained observer or the trained observer who is not consciously looking closely for method changes. When the change is discovered and pointed out, however, it is easily observable to almost anyone.

This type of change occurs when an operator learns to omit one or more motions from a cycle. This may be possible because the operator finds the motions are unnecessary or may introduce changes which make the motions unnecessary. For example, an assembler may learn to seat two parts by grasping the parts a convenient way and eliminate several regrasps and positioning motions. Sometimes these methods changes are teachable to other operators and sometimes they can be used only by those with superior coordination.

In a plant making first-aid products, an industrial engineer changed the methods of applying sealing wax to the end of a cardboard tube containing an iodine swab from a one-handed method to a two-handed method. The operator had been picking up a tube with the right hand, dipping it in hot sealing wax, tapping it on the edge of the container to break the thread that trailed along from the sealing wax, turning the tube to an upright position, twirling it in the fingers to form a smooth sealing wax button, and setting it aside.

The industrial engineer recognized that this could be done by the right and left hands simultaneously and instructed the operator how to do this. Motion pictures were taken of the one- and two-handed methods so the improved method could be demonstrated. Over the next few months, the pictures were shown to others several times, and thus the industrial engineer became thoroughly familiar with the method the operator had used. One day the engineer happened to pass the operator while walking through the department. Because of familiarity with it, the engineer immediately recognized a method change. The operator was no longer tapping the tubes on the side of the container to break off the trailing sealing-wax thread. In fact there was no thread.

Investigation showed that the operator had discovered that, if the sealing wax was a few degrees hotter than before, no thread would form. Thus the operator introduced a methods change that was easily recognizable once it was spotted. The change also permitted another change which, while easily observable when one knew what to look for, was not quite so obvious. With the sealing wax at a high temperature, the operator could form a smooth button with two twirling motions instead of three.

Changes of this kind are teachable once they are known. If they had not been recognized, there would have been an apparent loose standard and unexplainably high earnings. Recognition of the change made it possible to adjust the standard and teach the method to others.

Sometimes, operators find it possible to change whole motion sequences. For example, the assembly of an electronic circuit board required placing components with 12 legs into sockets in the board. When the operation first started, operators began placing each leg of the component one at a time. Within a short time, the operators learned to position 2 legs on opposite sides of the diameter of the component. This method reduced the number of placements from 12 to 2, thus reducing operator effort and fatigue. The other assemblers copied the method, and soon most of them were able to use it.

Subtle Methods Changes.   What are here called subtle methods changes arise chiefly from four sources:

1. Skill development through practice
2. Muscular development
3. Unique physical characteristics
4. Changes in environmental conditions

Anyone who has attempted to master a repetitive manual operation—even a home handicraft operation like knitting—or who has watched someone else learning it will recognize how greatly skill can develop with practice. At first, each motion is performed separately with hesitations and obvious mental activity between motions. The motions are slow, awkward, and poorly controlled. After a period of practice, the hesitations begin to disappear and the motions begin to blend into one another. Speed is gained and awkwardness disappears. With further practice, additional improvement is observable. Not only are the motions blended into one seemingly continuous easy pattern, but they can be performed for the most part without visual guidance.

The development of predetermined motion time systems has helped gain an understanding of what happens. In the MOST system, for example, a placement with adjustments requires two to three short adjustments at the point of placement to orient or align the object. It is a motion requiring visual control. A lay aside or loose fit placement, on the other hand, is a placement with up to one adjustment at the point of placement which may or may not require visual control. The lay aside placement is faster than the placement with adjustments.

An operator who first begins to learn an operation will use placements with adjustments to place objects in fixed locations. Being unfamiliar with the workplace, the operator will have to look toward every object reached for. Shortly, however, the operator becomes oriented to the workplace and begins to make placements to objects in fixed locations without looking toward them. The operator has learned to replace the "placements with adjustments" with the easier "lay aside or loose fit placements" and thus has learned to perform work in a shorter time.

During the original MTM research, for example, the motions used by a highly skilled machine molder in a foundry were studied. The molder, after using the bag of parting sand, would lay it down on the yoke of the machine. This put it in a location which would normally be expected to vary slightly from cycle to cycle. Film analysis, however, showed that the molder was able to reach for the parting sand again without looking. The operator used kinesthetic sense to find the parting sand bag. With practice, the operator knew by feel where the bag was placed. After years of repetition, the molder had learned to place the bag of parting sand nearly in the same location each time it was laid aside and could reach for it again as though it actually were in a fixed location.

Operators often acquire the ability to eliminate the positioning motions in placement as they repeatedly bring the same parts together over a long period of time.

MTM recognizes the fact that three different types of motions are employed when making Reaches and Moves. In type 1 motion the hand is not in motion at either the beginning or end of the Reach or Move. A type 2 motion occurs when the hand is in motion at either the beginning or the end of the motion. In type 3 motions, the hand is moving at both the beginning and end of the motion. With practice, operators are often able to replace type 1 motions with type 2 motions which, of course, require less time to perform. Occasionally they are able to develop sufficient skill to use type 3 motions. These are subtle differences which are difficult to detect without concentrated study. The untrained observer, while recognizing the smooth blending of motions, would probably say that the method which used type 1 motions and the method which used type 2 and type 3 motions were the same. Actually they are not. They would be described differently by an MTM analysis, and the time required would be different.

A further area for improvement through subtle methods changes lies in the overlapping or combining of motions. Inexperienced operators tend to work first with one hand and then with the other. With practice, they begin to learn to do two things simultaneously with both hands. First motions learned are simultaneous motions that are for the same hand, as reaching for and picking up parts with the right and left hands simultaneously instead of first with one hand then with the other. With further practice, they become able to do different things with each hand at the same time. Improvement through overlapping or combining motions can continue for a long time, months or even years if the operation lasts that long.

Frank and Lillian Gilbreth in their original motion study research found that operators would use different motion patterns when working quickly than when working slowly.

More recent research, when developing predetermined motion times, has confirmed this finding. Therefore, in addition to practice opportunity, the pace at which the operators work will have an influence on method, usually in the area of the subtle changes just discussed.

This has an important bearing on the standards setting problem. When an operation is new, the working pace is quite likely to be slow while the operators are developing their skills. After practice, when the motions are coming more easily, the slow pace may continue for a while. Eventually, however, under the stimulus of incentive payment or when some change occurs which arouses an interest in better performance, the pace will pick up. When this occurs, production will increase not only because the operators are working faster, but because at the faster pace they will use a superior motion pattern, usually unconsciously. The result is a two-way gain, fewer and often better motions performed with a higher degree of effort. Improvements of this sort can be made more than once. Indeed, over a long period of time, they may occur several times as the operators move from one performance plateau to another.

Subtle methods improvements can result from muscular development which occurs as practice continues. Heavy jobs become easier to handle as operators develop new muscles and learn to apply their strength properly without waste or strain. Working postures which may have seemed tiring at first become comfortable after a period of time. Skin hardens, callouses form, and other physical changes take place which enable the operator to perform the work in an approved manner. For example, an operator who is giving the final twist to seat a screw being driven with a hand screwdriver will at first usually regrasp the screwdriver before applying the tightening pressure. This is done for a variety of reasons, one of which is to get the flesh of the hand and fingers distributed in a way which will avoid painful pinching as the final tightening pressure is applied. As the hand becomes toughened with constant repetition of this motion, the regrasp of the screwdriver handle becomes less necessary. Eventually it may be discontinued altogether, with an accompanying increase in productivity.

The third factor mentioned as leading to subtle methods changes was the physical characteristics of the operator. It is quite evident that a tall operator can reach to high locations more easily than a short operator. An operator with long arms can reach farther without body motion than an operator with short arms.

In any well-run manufacturing facility, efforts to improve are continuous. Changes come from improved tooling, better maintenance of machinery due to experience in supporting the operations, engineering changes, changes in quality requirements, and a host of other factors making up the work environment. Many of these are beyond the control of the operator but affect ability to produce a quality product in the standard time because they change the working conditions.

In a plant manufacturing roller chain, the chain assembly machines were jamming because of components that were out of size and hung up in the feeder tracks. This jamming added 60 percent to the time it took to assemble a foot of chain. Investigation to eliminate this problem found an incoming quality level of better than 99 percent. Feeders were reworked to capture the out of size components and the jamming effect was reduced to 40 percent of assembly time. After 6 months of improvement effort, standards were set with the allowance for time to clear jamups included in the standards. Several years later the same operation was found to be experiencing 20 percent or less time spent on jamups. The setup workers reworked the feeder tracks, the punch press department had improved tooling and improved quality, and operators had learned to clear jams faster. Individually these changes were not significant yet their combined effect contributed by several members of the organization significantly changed the conditions the operator was working under. In this case, management, operators, and support functions introduced changes that needed to be reflected in revised standards, or a significant inequity in standards and earning opportunity would be evident to other operators in the shop. These changes occurred gradually over a long time period. A continuing audit program supported by top management is needed to make the necessary changes to maintain fair and equitable standards.

FACTORS AFFECTING OUTPUT

From the foregoing discussion it may seem that many factors affect output and therefore complicate the problem of work measurement. In the first place, there is the original method established for doing the work. Industrial engineers must study this carefully and develop the best method available at the time, watching carefully thereafter for methods changes introduced by others if they wish to keep the standards in line with the method currently being used.

Even so, if engineers establish a standard based on what the operator is actually doing toward the start of a long-run job, they can expect to find that the standard will seem more and more loose as time goes on if they judge it on the basis of the performance efficiency of the operator. The reasons for this should now be evident. With practice opportunity, the operators and the supporting organization will constantly develop improved methods. Some of these are recognizable and teachable and present no particular difficulties if there is an established procedure for revising standards when methods change.

The more subtle improvements are harder to handle. They are often developed unconsciously by the operator. The operator will have difficulty recognizing these subtle improvements as methods change and usually will feel that they are a function of increasing skill which should be rewarded by higher earnings.

This problem is more acute in incentive payment plans than it is in measured daywork plans. On measured daywork, earnings do not vary with output. When operators have learned to produce the quota specified by the standard, they feel they have done what is required of them by management. The incentive to improve beyond that point, therefore, is not strong. Any subtle methods improvements which may be introduced will mean merely that the day's work can be done more easily. Improvements are likely to occur more slowly and for a shorter duration than in the case of a well-administered incentive plan. The administrative problem caused by inconsistent earnings due to methods changes does not occur. This is one of the reasons why some companies favor measured daywork. At the same time, productivity is not likely to increase so much, because of the lack of incentive to keep on improving.

When incentives are used, the desire of the operator to improve is strong. An operator may introduce ideas to get management to improve equipment or simplify the work. Individuals will try from time to time to increase the work pace, which, as has been seen, will result in additional subtle methods changes. Operators will make use of any exceptional physical capability they may have to increase output still further.

As a result of all this, the earnings of people on highly repetitive work are likely to become higher than management originally contemplated when the incentive plan was first established. In addition, they are likely to be less consistent from operator to operator because all operators are not endowed with the same skills, attitudes, and physical characteristics.

The longer a given job lasts, the greater the earnings and inconsistencies are likely to be. Management can perhaps accept the high earnings because it receives high output in return. The inconsistencies, however, cause greater problems. No operator likes to see another earning much more for equivalent work. It is the human problems caused by inconsistent earnings which cause management the greatest difficulty. These same problems cause difficulties for union leadership as well, for dissatisfactions which are not resolved by management are quickly referred to the union.

STANDARDS BASED ON ELEMENTAL TIMES

It has been shown how difficult it is to establish a standard by time study for a repetitive job which will be satisfactory over a period of time. The time study person must measure the methods in use at the time of the study. To try to adjust the data by performance rating and give a standard which will not quickly become loose requires such a low rating as to be unrealistic when compared with the early operator performance and hence unacceptable to supervision and workers alike. Even if the low performance rating were accepted because the reasons for it were understood, it would be at best an estimate unsupported by any factual data.

Standards based on predetermined elemental times can help solve some but not all of the problems which have been discussed. With a procedure like MTM or MOST, industrial engineers can establish a standard for any method they believe will be used at any time in the future. Using predetermined elemental times will provide a factual and defensible basis for standards which is not possible for future methods if time study is the only tool of measurement.

In predicting what the method will be at some future time, the industrial engineer must be careful to distinguish between teachable methods and methods which can be followed only by exceptionally endowed operators. The earnings of an operator who obviously has greater skills and capabilities because of unusual coordination, exceptional size or strength, or whatever, are seldom resented by those of lesser ability. Caution must be exercised in predicting future method improvements through a learning curve. Management must aggressively challenge the organization to make method improvements for real progress on the learning curve to occur. Changes in complexities of manufacturing and technological changes can accelerate or stagnate progress along the learning curve. Technological changes often are accompanied by large investments. Large financial outlays are needed to make changes in these technologies so organizations may stagnate rather than consider changes.

MAINTAINING CONSISTENT PERFORMANCE

A number of different approaches have been used to maintain a consistent earnings situation under incentive payment or a consistent effort under measured daywork. The objective in all cases has been largely to reduce the human problems which are always present when inconsistencies exist, rather than to seek cost reductions by tightening standards.

Typical approaches to maintaining consistency are as follows:

1. Revise standards as methods change.
2. Establish different methods and different standards for different amounts of repetition.
3. Establish different "normal" performance levels for different amounts of repetition.
4. Establish a single standard and a decreasing learning factor.
5. Use the equitable bonus plan or equivalent.

Revise Standards as Methods Change.   There is generally a provision in the administrative policies governing a standards installation that if the method is changed significantly on any job the standard will be changed. Sometimes this provision is accompanied by restrictions such as that the change in method must result in 5 percent or more change in the standard before the standard will be changed or that only the elements affected by the changed method will be revised.

The provision that standards may be changed makes it possible for management to keep standards consistent and in line. It requires continuing watchfulness and firm administration. Some companies have established the position of standards auditor. The auditor may be an employee of the company or an impartial outsider. In any case the auditor has no standards setting responsibility but has the task of auditing standards periodically to determine if the methods on which they were based are still in effect. Audits may be done on a continual basis at the rate of so many per day or per week, or standards may be sampled periodically in larger quantities every 3 to 6 months. A sound policy followed by some companies is that no standard can be in effect for more than a year without an audit. If, as the result of the audit, methods are found to have changed, the standard must be revised immediately. This is the only way of maintaining consistency.

Obvious as this is, many managers may hesitate to take action for what may appear at the time to be good reasons. Available industrial engineering time may be needed for developing new standards, and this may seem more important than changing existing standards. The supervisor may rationalize the situation by deciding that, because other standards are tight, it is only right to have a few loose ones. Management may justify inaction by pointing out that contract negotiations will begin shortly and that it would be unwise to "stir things up" by seeking to revise standards "at this time." With resistance to change almost certain on the part of the affected operators, it is all too easy to find good reasons for not carrying out the established administrative policy of changing standards when methods change.

Hence, the maintenance of consistent standards and earnings, where there is provision for revising standards as methods change, is largely a function of the skill and aggressiveness of management in carrying out its own policies. If a satisfactory installation is allowed to deteriorate, placing the blame on the industrial engineer or the union is seldom justified.

In the past few years the applications for computers in work measurement have expanded dramatically. Common applications include developing standard data elements, storage and retrieval of standard data, word processing for documentation of standard methods and procedures, standards generation from standard data, automatic updates of large families of standards affected by a common change, simulation for methods improvements, and computing assistance in standards setting. The computer has made maintaining standards easier and faster. The information system allows fundamental planning data to be easily integrated into budgets and scheduling programs. In measuring repetitive work the advantages of computer speed, accuracy, rigid discipline, and abilities to copy and edit files are essential tools for the modern industrial engineer. Methods updates can be made in an afternoon that under a manual system would take months to achieve, and then with uncertainty of their completeness.

Artificial intelligence and expert systems software has been developed to assist in routing parts. Some companies have utilized software to classify parts and to set standards automatically. Applications are expanding beginning with simpler operations to complex assembly and machining operations. These applications increase the speed of setting and maintaining standard data and reduce the clerical load of industrial engineers so they can spend more time on methods improvements.

When standards and methods are allowed to get out of line and when no corrective action is taken for a long period of time, it becomes increasingly difficult to start enforcing the policy of revising standards. The operators will point out forcefully that management accepted the standards as being satisfactory for months or perhaps years and will usually sincerely feel that the belated attempt to change them is nothing but a thinly disguised rate cut. Usually in a situation of this kind it is better to have a complete change through what is often called a "rationalization program" rather than to attempt to bring standards in line one at a time.

Establish Different Standards for Different Amounts of Repetition.   Some companies have approached the problem of maintaining consistent standards and earnings opportunity by establishing different standards based on different methods to be applied under different amounts of repetition. Higher-volume jobs usually are routed across the more efficient equipment and low-volume jobs run on equipment with simpler setups and less efficient operation. One way of administering this is to establish a standard and limit the number of pieces to which it is applicable. A set of standard data designed for a jobbing machine shop, for example, might carry a limiting note such as the following:

Application. The standard data for small engine lathes apply to engine lathe work on nonferrous parts as performed in department 0-4 on work orders of not more than 100 pieces.

If orders of more than 100 pieces are received, a new (and lower) standard is developed and applied.

Another approach is to establish and publish a list of standards which decrease as the number of pieces produced increases. A typical series of standards would appear as follows:

Piece no.	Standard
1–10	0.0100
11–100	0.0090
101–500	0.0081
501–2000	0.0073
2001–10,000	0.0065
10,000 up	0.0063

This approach is an attempt to establish standards based on alternate routings. It is technically feasible to do this, particularly if predetermined time standards are used. The approach has been acceptable to operators in situations where it has been used. Because of obvious administrative difficulties, the use of this approach has been limited.

Establish Different "Normal" Performance Levels for Different Amounts of Repetition. When measured daywork is used (see Chap. 1 of Sec. 6), a standard is established for each operation which represents the amount of production that is considered normal. The standard in effect indicates what management considers to be a "fair day's work."

The performance of the operator is determined by dividing the standard hours allowed by the hours worked. When the standard hours allowed are equal to the hours worked, the performance of the operator is said to be 100 percent. This is the goal which all operators but the learners are expected to attain under many measured daywork plans.

The 100 percent performance level recognized by a measured daywork plan is not necessarily the same as the average skill and effort level used for performance rating purposes during the time study. (See Chap. 2 of Sec. 4.) The latter average or normal performance level is a fixed level established by definition. It is not influenced by the present or anticipated future repetitiveness of the work.

When work is not very repetitive, the average or 100 percent performance level expected from the operator on measured daywork may be the same. As the work grows more repetitive, however, the output of the operator may be expected to increase for the reasons already discussed. Therefore, the amount of production which will be considered to represent a fair day's work changes. For a certain degree of repetition, 100 percent performance on the part of the measured dayworker—the normal expected quota—might be rated by a time study analyst as 105 percent in terms of skill and effort. In other words, greater production on highly repetitive work is the normal expectancy.

The concept is virtually the same as that discussed under "Establish Different Standards for Different Amounts of Repetition" except it is not tied to a prescribed method. It may not be quite as easy to grasp or explain to operators, however, because the several different normal performance levels are all considered to represent 100 percent performance and the time is not closely associated to a specified method.

Establish a Single Standard and a Decreasing Learning Factor. Another approach to coping with the realities of the learning problem is to establish a standard which will represent the time in which experienced operators may be expected to do it toward the end of the life of the job, and then apply a learning factor which is high at the start and decreases toward unity as time goes on. The hours earned at any time are determined from the formula:

$$N \times L \times S$$

where N is the number of pieces produced, L is the learning factor, and S is the standard time allowed.

If a standard which will be correct several months later is established at the start of the job, the learning factors will usually have to be rather high at first. An example of a table of learning factors might be as follows:

Period	Learning factor
1st week	2.50
2nd week	2.00
3rd week	1.60
4th week	1.35
5th week	1.20
6th week	1.10
7th week	1.06
8th week	1.03
9th week	1.00

It should be understood that these are illustrative figures only and that they are not to be applied to any specific job. The correct learning factors for any job can be determined only by a study of that job. (See Chap. 8 of Sec. 5 for more on learning factors.) Again, the objective of the approach is to approximate the realities of the learning curve as closely as possible. It is an approach which has been used frequently when a class of work is first changed from unmeasured daywork to incentive. It gives the operators a chance to make incentive earnings during the period when they are learning to come up to the incentive level of performance and avoids the discouragement which would exist if only their true performance against standard were reported.

This approach is equally applicable when standards are to be established for highly repetitive work. Some years ago, a time study person said, "I have found that no standard is any good if the operators can meet it in under three months." He recognized, without knowing fully the reasons for it, that output and hence earnings tend to increase continuously as time goes on. Because he felt responsible for maintaining earnings levels within a range acceptable to management, his solution was to establish standards which were impossible to meet at first.

This approach was acceptable in the 1920s. The operators were willing to keep plugging away at their minimum guaranteed rates, trying hard to meet the standard for a long period of time until they at last succeeded. It is doubtful if they would do so in a modern environment where immediate results are expected by almost everyone. If, after trying for a few days to meet the standard, the operators found that they could not come close to it, the normal reaction would be to enter a grievance against the standard and then work at a daywork pace until something was done about it. Where an attitude such as this is likely to prevail, the approach of establishing a single standard and providing a decreasing learning factor has much to recommend it.

Use the Equitable Bonus Plan.   Another approach to the problem of maintaining a consistent earnings picture has been through the design of the wage payment plan. Many years ago, the Rowan plan was developed to prevent runaway earnings. The bonus earned on any job was computed from the formula

$$\frac{\text{Time allowed} - \text{time taken}}{\text{time allowed}} \times 100 = \text{percent bonus}$$

If the time taken were reduced to zero, the bonus would become a maximum of 100 percent. Actually it was much lower even when standards were very loose.

A plan of this type puts a ceiling on earnings. It also puts a ceiling on production, for the operators quickly learn that it does not pay them to produce more than a certain amount. The same result can be accomplished under any incentive payment formula by putting an arbitrary ceiling on earnings.

To avoid the limitation of production which approaches of this sort produce, a plan known as the equitable bonus plan has been developed. It has several interesting features. First, management and the union take joint responsibility for seeing that an acceptable level of working effort is maintained.

The operators work in small groups whose performance is computed for a pay period, usually 1 week. Any group that fails to achieve 100 percent performance is paid at its guaranteed rate. The performance of all groups who make incentive—that is, exceed 100 percent performance—is averaged, and a single departmentwide performance figure is computed. All groups that make incentive are then paid on the basis of this average performance.

If any group exceeds 135 percent performance, this is taken as evidence that a methods change has occurred, the work is restudied, and new standards are established. However, the net savings resulting from the methods improvement for the next 6 months is paid to the operator or operators who developed the improved method. Thus there is strong incentive to do as well as possible at all times, for only good can come from it.

Any number of variations to this plan can be introduced to meet specific situations and conditions. The approach, however, offers yet another method of avoiding the difficult human problems which result when serious inconsistencies in earnings develop.

CONCLUSION

A few conclusions may be drawn from this discussion of the measurement of repetitive work.

First, if repetitive work is to be measured successfully, industrial engineers must understand thoroughly all the problems inherent in the situation. In particular, they must recognize the inevitability of continuing learning and the reasons which underlie it.

Next, they must have a thorough understanding of the work they are to measure. If they are to be able to develop standards which will be somewhat correct at any point of the learning curve, they must be able to recognize the subtle methods changes which may be expected to occur as time goes on. A knowledge of a predetermined elemental time system like MTM or MOST will be essential, but they will further need to know the elemental motions used for the specific class of work they are studying, work attitudes and motivations, general effort levels, and any other factors which may influence the learning process.

Armed with this knowledge, industrial engineers must be able to develop the application policies and procedures which best fit the requirements of the specific situations with which they are dealing. Here they run into management policies and attitudes, the educational level of the supervisors and the work force, union attitudes, and many other things.

It can be seen, therefore, that the measurement of highly repetitive work is not a simple task which can be handled by a poorly trained technician but rather that it is something which calls upon the skills and judgment to be found only in the competent industrial engineer. Any extra expense incurred in employing the more qualified engineer will quickly be recovered several times over by the superior work done.

BIBLIOGRAPHY

Gevarter, William B., *Artificial Intelligence, Expert Systems, Computer Vision, and Natural Language Processing,* Noyes Publications, Mill Road, Park Ridge, N.J., 1984.

Karger, Delmar W., and Franklin H. Bayha, *Engineered Work Measurement,* Industrial Press Inc., 200 Madison Ave., New York, N.Y. 10016, 1977.

Niebel, Benjamin W., *Motion and Time Study,* 4th ed., Irwin, Homewood, Ill., 1967.

Webber, Ross A., *Management,* Irwin, Homewood, Ill., 1975.

Zandin, Kjell B., *MOST Work Measurement Systems,* Dekker, New York, 1980.

CHAPTER 4
MEASUREMENT OF AUTOMATED PROCESSES

Richard L. Shell
Professor of Industrial Engineering
Department of Mechanical, Industrial, and Nuclear Engineering
University of Cincinnati
Cincinnati, Ohio

Automated processes today include a wide variety of computerized mechanization. Automation is found in varying degrees throughout manufacturing industries, the service sector, and government. The degree of mechanization may range from very little, for example, one piece of automated equipment in a discrete component manufacturing setting, to a very high level that would be found in a completely automatic controlled process plant. Today, automated processes are utilized by companies producing low and high volumes of output. Automation is found in large corporations as well as small companies producing many different kinds of goods and services. *The main objective of measuring automated processes is to enhance overall productivity.*

The principal components of mechanization include a diverse range of computer software and hardware, sensors, machine tools, robots, automated material handling and positioning equipment, and other mechanized machinery. Most automation designs require modules of intelligent control linked to machinery that performs work tasks. In larger-scale automation, these modules are interfaced to form a total system.

The Need to Measure Automated Processes. If it has been correctly determined that a process should be automated, then *a work measurement approach to determine the production output is clearly justified.* A false belief, still held true by some engineers and managers, is that the more automatic a production system is, the less important it is to measure it, or to expend resources or provide motivation or incentives for the operators involved. The simple rationale for this false belief is that the process is computer and machine controlled and paced, and therefore human operators have little or no influence on the output. This rationale will almost always lead to low equipment utilization (excessive downtime), poor-quality output, and high system maintenance cost. Often these inefficiencies combined will so adversely impact costs that the predicted economic justification for the automated system is not realized. Experience has shown that several automated systems after installation have had higher unit costs than the production processes that they replaced. Most of these situations can be avoided with the proper application of work measurement and human motivational techniques.

THE NATURE OF AUTOMATION

Systems that are automated make it possible to transfer physical and mental work from humans to machines. Properly designed *and* utilized, they can reduce the unit cost of production output. The automated system should also have positive effects on quality, throughput time, and the ability to respond to rapidly changing customer requirements. The level of automation is also influenced by material and energy utilization, operator safety and health considerations, availability of capital and qualified human resources, and the estimated life of the products or services being produced.

Several methods may be used to evaluate possible automation opportunities and to aid in making the final acceptance decision. Most of these methods include a cost-benefit analysis comparing the proposed automated system with the existing or previous production technology utilized. The cost part of the analysis should consider costs of labor, material, and overhead, and all capital and other implementation costs. These calculations can be assisted with an engineering economy approach (see Sec. 9, Chap. 1) and work measurement analysis.

In addition to the conventional cost analysis mentioned above, proposed automation system evaluation should also include intangibles or opportunity costs. For example, loss of unproduced units or profits, or loss of market share because of system downtime and failure to satisfy customer requirements. Remember, the major objective of automated processes measurement is to realize the maximum potential output of the total system.

Benefit and Concerns.   Many factors justify automation in any work situation. The more common factors include the need to increase production output (market expansion), reduce labor costs, and increase profits. Example benefits that usually have a positive cost impact include reduced throughput time, increased use of "standard" parts and tooling, and greater utilization of equipment and physical plant.

Major limitations and concerns associated with automation include large capital investment, rapidly changing technology, and personnel problems. The large capital investment has to be amortized into the product or service cost structure. Most automation modules or systems are continuously being improved and with each design become more cost-effective. Consequently, there is always a risk of obsolescence. Perhaps the most difficult aspect of automation limitations and concerns is personnel problems. The most common human issue associated with automation implementation include resistance to change, and training. Developing attitudes to accept automation and training is necessary for all levels of personnel: hourly, supervision, and management.

Common Characteristics.   The determination of how to measure automated processes requires an understanding of the system and its impact on specific business needs. The paragraphs below summarize common characteristics of automated modules and systems.

In addition to the factors that justify automation as mentioned above, reasons for mechanization include improved quality, greater production uniformity, and safety and health considerations. Facility construction costs place an increasing emphasis upon better space utilization by consolidation of individual machines or manual operations into an automated system. Human behavioral and safety considerations include automating out of the job those elements that are of a highly repetitive nature (prevention of cumulative trauma injury) and have a high output volume.

Capital investment is usually very high in comparison with manual or partially mechanized systems. In fact, costs can range from tens of thousands of dollars per worker to a multi-million-dollar installation which is monitored, managed, and serviced by a small work group.

Variability in production is inversely influenced by the number of stand-alone producing units. When many common individual units are producing, variability tends to be off-set from one unit to another. Thus a certain level of production can be closely predicted and relied upon. Conversely, a single but larger production system is on either run time or downtime on an on-off basis. When on, it may or may not be producing 100 percent good output, but when off, the system production is zero.

Probability of failure or downtime is much greater in a single highly complex system. For example, if a group of single machines have a 95 percent probability of being in run mode, then 95 percent of a number of those machines will likely be producing. If a single, complex system has several interdependent modules arranged in series, each with a 95 percent probability of running, the overall probability of run time is the product of the independent probabilities. Thus a four-module system would have a probability of running of only 0.815, or 81.5 percent of the shift time.

Investment justification usually requires a high degree of utilization. Typically, this is a two- or three-shift operation, sometimes for 7 days a week. Where overtime formerly could compensate for production breakdowns on one- or two-shift operation, the new system compresses all variability into one unit running around the clock, and time lost is much more difficult or impossible to replace. Fluctuating production schedules may compound the problem of lost production time and capacity, even resulting in lost customer sales. These facts often necessitate backup equipment, requiring even higher capital investment.

Labor content in highly mechanized processes is low, relative to depreciation or amortization costs, which continue even during a shutdown. Consequently, measurement must focus on system uptime, not just individual workers.

Maintenance support is critical for any automated process. The failure of a single component (even minor) can cause the shutdown of the module or system. Maintenance costs are more in automated systems than in less mechanized operations, because of higher technical skills needed, longer shutdowns during production runs, and the requirement of around-the-clock troubleshooting service. If the maintenance service reports to other than the production manager, this will be a source of functional and jurisdictional difficulties. The best management approach is to include maintenance personnel in the automated module or system work group.

The Human Component.   Knowledge and human know-how is vital to the successful operation of any automated module or system. Therefore, the work environment must be structured to ensure employee involvement and team building. Examples of work activities that strongly influence the overall productivity of the automated module or system include the following:

1. Monitoring and control of quality to preserve the level of quality engineered into the product or service.
2. Maintaining raw material supply, inventory, and waste disposal conditions to extend run time.
3. Exercising on-line preventive maintenance to avoid downtime through minor servicing and adjustments that do not require shutdowns.
4. Troubleshooting to minimize downtime as quickly as possible whenever breakdowns occur. This requires maximum knowledge and communication among operators and maintenance or other service people.
5. Changing products or machine setups as quickly as possible to avoid excessive downtime. This requires maximum communication and cooperation among members of a work group or among several groups.
6. Rendering effective assistance to coworkers or other specialized service staff during scheduled maintenance periods.

EMPLOYEE INVOLVEMENT

This concept is not a new idea. For example, in the 1930s Joseph W. Scanlon developed the Scanlon plan. Its major purpose was to measure overall productivity improvement and distribute a share of any gains to the employees. In addition, the plan included *labor-management committees* to effect improvements throughout the manufacturing operation. Beginning in the 1960s and continuing today, behavioral scientists have showed the benefits of democratic management and worker participation. These benefits have been further reinforced by the Japanese management practice, and European work groups, for example, Volvo in Sweden. The outgrowth of this has yielded self-managing (involvement) work teams throughout manufacturing and the service sector.

Types of Programs.   Involvement teams have emerged in the United States in many forms and are a vital part of successful automation. Several distinct forms range from small problem-solving groups of only hourly workers to well-organized self-managed work teams comprised of all skills necessary to run the automated system. The work team has complete responsibility and decision-making authority for their part of the automated system. Experience has shown that increased involvement leads to more effective decisions in most complex situations.

Quality Circles.   This involvement group usually consists of 8 to 10 members who do similar work and agree to meet on a regular basis to discuss work and quality problems. The group analyzes causes of problems and recommends solutions to management. Group members may also take action to assist implementation of recommended solutions. Quality circle activities usually enhance motivation and the overall quality of work life.

The quality circle leader is commonly a supervisor of all or several of the group members. In addition, each circle has a facilitator who is responsible for training the members and the leader. The facilitator forms a link between each circle and the rest of the organization, and may also take an active role in the measurement and evaluation of improvements.

Work Teams.   This involvement group usually consists of all the workers necessary to totally manage and operate an automated module or system. They are self-managed and are given considerable decision-making responsibility over their work area. Unlike quality circles the team is assigned full-time to a specific work area. Work measurement is essential to properly determine staffing levels by skill type for a self-managed group.

MEASUREMENT TECHNIQUES FOR AUTOMATION

Measuring automated processes is necessary to determine the correct human support staffing requirements, and to establish a basis to evaluate total output performance. The measurement system should provide performance data on the automated process for feedback to the self-managed work group and higher management. The system should be structured and used in such a way as to permit the application of incentives (see Sec. 6, Chaps. 2 and 3) and to enhance motivation. In addition, it should be a dynamic process of information generation whose primary goal is to establish direction for continuing improvement in output productivity and quality, lower costs, improved worker safety and health, and better products and services for timely customer delivery.

Measurement Technique Selection.   The work activities of humans in automated processes vary depending on the level of system control. Karwowski and Ward (1989) have defined four levels of control: manual, semiautomatic, supervisory, and cognitive as depicted in Fig. 4.1.

The manual system (level 1) shows the operator controlling and monitoring output of

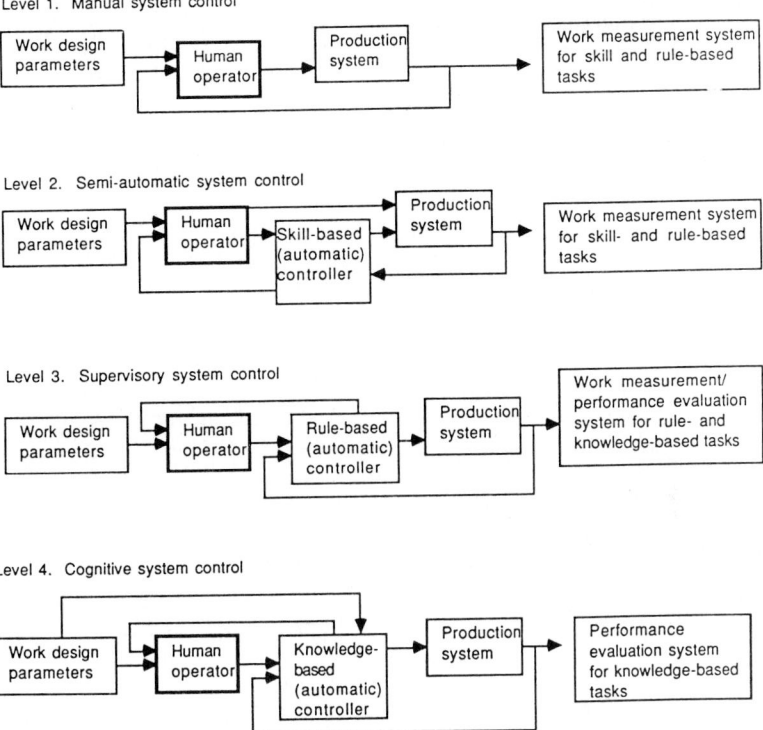

FIGURE 4.1 Work measurement for different levels of computer-aided automation. (*Reprinted from W. Karwowski and T. L. Ward, "Work Design and Measurement: Critical Issues for Advanced Manufacturing," Proceedings, International Industrial Engineering Conference, Toronto, 1989, p. 514. Copyright Dr. W. Karwowski, Department of Mechanical Engineering, Tampere University of Technology, Tampere, Finland.*)

the production system. In this nonautomated case the work measurement technique selection would be more conventional and would depend on quantity produced, length of production run, and type of job function.

Level 2 depicts system control commonly found in modular automation such as a single NC machine tool. In this case the operator(s) would be performing job functions such as load, unload, monitoring, setup, and maintenance. Therefore, work measurement selection should generally follow the guidelines for low-quantity work discussed in Chap. 2 of this section.

The supervisory and cognitive systems (levels 3 and 4) are typical of a flexible manufacturing cell (level 3) or a fully automated process plant (level 4) that require human interface only with the controller. The controller then instructs and receives feedback from the production system through computer control and various system state sensors. The primary activities of the work group include monitoring of the process and quality, maintaining supply, maintenance and troubleshooting, setup and changeover, and other support functions. While all work measurement techniques may have some utility, work sampling and predetermined time systems have the most applicability for automated processes.

Work Sampling.   As previously mentioned, the automated process (system) requires a variety of diversified human activities. Most of these are irregular and long-cycle, and require considerable mental effort. Consequently, work sampling is highly effective in determining the activity state of workers during the automatically controlled machine cycle, or during downtime periods. Common guidelines for the work sampling confidence interval and error limits are 95 percent and plus or minus 5 percent, respectively. Measurement of automated processes generally warrants at least this level of confidence and error, or more accuracy, that is, higher confidence and lower error percentages. For a detail discussion of work sampling, see Sec. 4, Chap. 3. Chapter 2 in this section has additional recommendations for work sampling applications.

Predetermined Time Systems. Several human work activities supporting the automated process (system) may be measured with predetermined time systems. The longer-cycle activities can be measured with higher-level predetermined systems. First-level systems may be used for shorter-cycle work activities. Some of the work activities for automated cycle times may be determined with functional or specialized predetermined systems. For example, MTM-C may be used for clerical support work, and Robot Time and Motion (RTM) for determining robot cycle times. For a detail discussion of predetermined time systems, see Sec. 4, Chaps. 4 and 5. Additional application information for predetermined time systems is included in Chap. 2 of this section.

USE OF MEASUREMENT DATA

Measurement of automated processes is useful for staffing determination and control, production output standards, work group building, and incentive pay determination. In addition, a measurement program is helpful for the following:

1. Evaluating personal, fatigue, and delay allowances
2. Debugging the new system for improved performance
3. Evaluating equipment changeover
4. Designing preventive maintenance procedures and schedules
5. Redesigning marginal components of the system
6. Developing job descriptions and training procedures
7. Enhancing cooperation among the work group, including operators, maintenance, and management
8. Determining interference patterns within the work group and/or between the human-equipment interface
9. Estimating labor and equipment capacity, and scheduling capabilities
10. Evaluating and measuring indirect and support personnel
11. Improving methods and procedures
12. Determining if the automated system should be provided with a manual or module insert replacement for backup during extended periods of breakdown
13. Providing data to compare the total cost effectiveness between various levels of operation or automation to produce a given output

THE FUTURE

As computer and sensor technology develop and combine to realize lower-cost intelligent systems, automation applications will continue to increase in the years ahead. Clearly the future holds higher levels of automation for manufacturing and the service sector. The trend in automated manufacturing is the ability to process different components, models, or families of assembled items on the same system, and in very small lot sizes as required by the customer. The service sector is moving toward completing information or service transactions with less (or no) human interaction. It is important to note, however, that even the most automated production process will still require the human for various support functions. Consequently, the need for work measurement in automated processes will still exist, but there will be changes. These changes are briefly summarized in the paragraphs below.

Stopwatch Time Study.   The requirement to conduct direct observation timing and the associated rating or leveling of human performance will decrease because of the increased machine and computer controlled processing cycles. Most cycle times will be automatically determined from internal clocks and calendars within the control device.

Allowances.   While the determination of personal, fatigue, and delay allowances will be accomplished as in the past (work sampling), there will be changes as automation levels increase. In general, physical fatigue will decrease while mental fatigue of workers will increase. Unavoidable delays in highly automated processes will be different from traditional manufacturing plants or service operations.

Work Sampling.   The use of this proved work measurement technique will increase as automation continues to evolve. The practice of work sampling will be simplified because of the ability to electronically monitor and record the activity state of many elements of the automated system.

Predetermined Time Systems.   This work measurement technique will still be used in the traditional form primarily for indirect and support personnel. Computerized synthetic time systems will be developed to improve application to the activities of knowledge workers. Specialized predetermined time systems, such as RTM, will become more useful for future automation applications.

Standard Data.   For estimating and planning purposes standard data will probably be developed for major segments of the automated system as contrasted to small elements of an individual processing cycle. With major segments of processing times included in a computerized database, cost comparison estimates could be easily made for alternate methods of production.

BIBLIOGRAPHY

Barnes, R. M., *Motion and Time Study: Design and Measurement of Work,* Wiley, New York, 1980.

Karwowski, W., and T. L. Ward, "Work Design and Measurement: Critical Issues for Advanced Manufacturing," *Proceedings, International Industrial Engineering Conference,* Toronto, 1989.

Martin, D. D., and R. L. Shell, *Management of Professionals,* Dekker, New York, 1988.

Mundel, M. E., *Motion and Time Study: Improving Productivity,* Prentice-Hall, Englewood Cliffs, N.J., 1985.

Niebel, B. W., *Motion and Time Study,* Irwin, Homewood, Ill., 1988.

Shell, R. L., ed., *Work Measurement: Principles and Practice,* Industrial Engineering and Management Press, Norcross/Atlanta, Ga., 1986.

CHAPTER 5
FUNDAMENTALS OF INDIRECT LABOR MEASUREMENT AND LABOR FORECASTING

William K. Hodson
Consultant
Indian Wells, California

William E. Mayo
Vice President Consulting
H. B. Maynard and Company, Inc.
Pittsburgh, Pennsylvania

Historically, work measurement and even labor forecasting has been concerned with the measurement of direct labor. One reason is that some direct labor measurement is required in most cases for product costing and pricing. Another major reason is that direct labor historically accounted for the bulk of the hourly labor in a plant and was the most significant cost to measure and control. From an industrial engineering viewpoint, the measurement of direct labor operations is a great deal easier and consequently less costly than measuring indirect labor. For these reasons, the emphasis in the past has been on the measurement of direct labor operations.

Both evolutionary and revolutionary changes have been taking place in the United States economy. These changes tend to place a much greater emphasis on measurement of indirect labor. Because of the impact of automation, the traditional relationships between direct and indirect labor are often reversed. It is no longer unusual to find a factory where indirect labor workers outnumber direct workers and even salaried personnel outnumber direct workers. This condition is further compounded as clerical and technical and administrative activities become a larger part of the economy. These activities predominate in the service industries, which are growing at a greater rate than the rest of the economy. Undoubtedly, the computer and advanced technological improvements such as numerical controlled machines are accelerating the rate of indirect labor growth. All these developments have created a greater need for indirect labor measurement.

DISTINCTION BETWEEN DIRECT AND INDIRECT LABOR

The manufacturing cost of a product is usually divided into three major components: labor, material, and overhead. Labor cost normally is composed of only the direct labor.

This is the cost of labor directly involved in the physical manufacturing of the product. Each direct labor operation has some effect on the part or product. As a consequence, the amount of labor required can be measured directly.

A major portion of overhead cost is the cost of indirect labor. A typical indirect labor cost is janitorial. Although the cost of janitorial labor can be easily measured, it is difficult to relate the cost of this labor to a specific part or product. As a result the cost is usually allocated to specific products on a percentage basis. A typical approach is to allocate indirect labor costs in proportion to their direct labor hours. Some types of indirect labor are material handling, shipping, receiving, warehousing, tool cribs, toolmaking, maintenance, janitorial service, clerical workers, inspection, laboratory workers, and drafters.

FACTORS INFLUENCING INDIRECT LABOR MEASUREMENT

Although the basic work measurement techniques are the same as normally used for direct labor, there is a significant difference in emphasis and in the method of application. Other, more serious differences exist in related areas such as methods, work counts, standard hour calculation, and group application of standards. An analysis of these factors will provide a better understanding of the differences between direct and indirect labor measurement.

Methods. Both direct and indirect labor measurements require that the method be standardized before the standard time is established. The application of this rule varies substantially between direct and indirect work. On very repetitive work, it is not at all unusual to spend from 50 to 100 analysis hours on methods work for every 1 hour on work measurement. Furthermore, greater emphasis is placed upon training the operator to follow the established method with a higher degree of precision. In some cases, each individual motion is analyzed and subjected to the operation analysis and motion study approach. The number of motions required to perform the work are reduced to a minimum, and the motions that remain are reduced to the simplest form of motion. Unless an operator is specifically trained to follow the method, motion by motion, it will be difficult to perform the operation in the standard time allowed.

The other extreme of methods analysis exists in most maintenance activities. Although methods study can contribute substantially to improving maintenance work, the emphasis is different. This principle also applies to any long-cycle work performed infrequently. The longer cycle and the job shop environment of low volume per production period decreases the possibility of repeating any specific method. This has led to the development of Occurrence Frequency Groups (OFG) used in the MOST (Maynard Operation Sequence Technique) predetermined time system. A brief explanation of the use of OFG tables follows. A table has been developed which is read by determining the cycle time of an operation and how often this operation will occur in a period of time. Once we know this information we can simply read the table and determine the accuracy of the method required for either plus or minus 5 or 10 percent accuracy. The specific use of OFG tables is covered in Chap. 5 of Sec. 4. It is a given that the opportunity to practice creates a higher methods level. Instead of concentrating on the methods to perform a specific task or operation, attention is focused on the overall methods used by the craftsperson. The overall methods are applicable to all work. For example, considerable time should be devoted to studying the various tasks used on the job and the methods of transporting and storing tools. Frequency studies can be used to determine which tools are used most often. It can then be determined which tools should be attached to the tool belt and which tools should be placed in a toolbox. It may be appropriate to create carts or other methods of storing tools. In addition an overall review of tool availability and the kind of tools to be used should be made. Some long-cycle direct labor tasks may fall into the same category as maintenance work. The method studies should be designed so as to have universal application and to minimize studies for specific operations. Some training may be required in the use of both new tools and the location system designed and installed.

Maintenance work is an example of nonrepetitive as contrasted to janitorial work. Not all janitorial work is repetitive but a large portion is. The approach to the methods study should be in accordance with the repetitiveness of the work. As a general rule one will discover that very little methods work has been performed for indirect labor and as a result will find any methods study will be profitable. If there is no other reason for indirect labor measurement it would be to provide for methods improvements.

Units of Work.   Another characteristic of indirect labor that makes it difficult to measure is the lack of a simple unit of work. The unit of work for most direct operations is the part count. A standard time can be expressed in minutes or hours per piece, and the number of pieces produced by the operator can be counted. The pieces then multiplied by the time per piece will give the number of standard hours generated by the worker.

Consider an order picker. When we set standards for this operation, we must ask if we are going to use line items, orders, or both for counts. If we consider both, we must use the OFG tables and slot standards using line items per order. Using this method, we can obtain accurate standards as well as minimize the standard setting time. The standard setting can then be defined so that it is merely a simple clerical function or could be part of an order entry program and be set without any cost. This approach can be used on almost any indirect operation for setting standards and monitoring efficiencies.

Units of work must be verifiable and countable. If we cannot determine if the reported work units were accomplished, we may as well have no standard. In the order picking example we can verify quantities and even line items per order through the invoicing system. This is one method of automatically reporting against the standards. With some ingenuity, the job of collecting the work counts and applying the standards can and should be simplified. If the system installed is very complex and difficult to apply, it will surely fail. Any system can and must be made simple if this can be accomplished without sacrificing accuracy.

Measurement of the Work.   The standard techniques for measuring indirect labor are the same as for direct labor. These are time study, predetermined time systems, standard data, and work sampling. In developing direct labor standards the industrial engineer will have ample opportunity to observe an operation. In the case of indirect labor some operations will occur infrequently. This will become very frustrating to the industrial engineer who normally establishes direct labor standards.

Because of the nonrepetitive nature of indirect labor operations, it is imperative that the work be carefully analyzed before any attempt is made to measure it. One of the primary purposes of this preplanning is to develop methods improvements and to standardize the methods used for performing the work. The standardization of work is important because it enables the predetermination and synthesis of what the operator should do to perform a given task. A secondary benefit of this analysis is a greater insight into the nature of the work and its characteristics, the major elements, and the sequence of these elements. All this information is essential in developing the time standards.

The variable sequence of elements and the variables that may exist make it essential to rely upon standard data and the building blocks of these data. Development of standard data is explained in detail in Chap. 6, Sec. 4.

Multiple Regression Analysis.   The procedures discussed in this chapter for measuring indirect labor emphasize the traditional approach to the problem. This approach involves studying each activity of the indirect labor operation and establishing time standards for each by the use of some recognized time standards approach. Although the return from indirect labor measurement programs is well worthwhile, the measurement procedure is often very tedious, time-consuming, and expensive. One procedure that has been used to simplify work measurement is multiple regression analysis combined with work sampling.

Two or more units of measure must be used to measure the output of indirect labor operations accurately. The multiple regression analysis technique provides a means of correlating several units of measure, called "predictors" in multiple regression terminology, with time required to do the job.

The multiple regression model is an equation that relates the predictors (units of measure or variables) to the response (standard time to perform the job). The standard format is

$$y = b_0 + b_1x_1 + b_2x_2 + b_3x_3 + \cdots$$

Assume that a shipping operation consists of filling orders by use of a fork truck. The truck operator picks up an empty pallet, proceeds to the various stock locations, and loads the necessary cases for the orders on the pallet until the orders are filled or the pallet is loaded. The operator then transports the pallet to an order assembly area. The objective is to determine a method for measuring the time required per pallet load. A preliminary analysis of the operation indicates that the predictors may be any of the following:

$$X_1 = \text{number of cases packed per pallet}$$

$$X_2 = \text{number of orders per pallet load}$$

$$X_3 = \text{weight of packed material per pallet load}$$

$$X_4 = \text{volume of cases per pallet load}$$

A work sampling study is then made of the operation. During the course of the study, observations are made of the work and nonwork elements for each of the operators. The data were readjusted by performance rating factors to provide a leveled observation percentage of the working time. Records are maintained of the actual hours worked, the number of cases packed, and the volume of the cases for each pallet loaded. All the data for each pallet load are then loaded into a computer program for multiple regression analysis. Because of the complexity of the regression analysis, a computer is the only economical way of using multiple regression for problems of this magnitude.

The computer solution provides values for the coefficients b_0, b_1, b_2,...,b_i. The program also provides correlation coefficients for each of the predictors.

By examining the various correlation coefficients and the correlation between the predictors, an individual firmly grounded in the mathematics of regression analysis can determine which of the predictors are significant and which are not. After such an analysis, it may be decided that the predictors of weight and space provide very little additional predictive capability to the predictors of number of cases and number of orders.

The final formula might look like this:

$$Y + 5.02 + 3.37X_1 + 0.45X_2$$

where Y = standard minutes per pallet loaded
$\quad X_1$ = number of cases packed per pallet
$\quad X_2$ = number of orders per pallet load

The values for the coefficients b_0, b_1, b_2 are respectively, 5.02, 3.37, and 0.45 minutes. The standard time for a pallet made up of six cases and two orders can be calculated by inserting the predictor quantities into the formula and solving for the standard time.

$$Y = 5.02 + 3.37(6) + 0.45(2) = 26.14 \text{ minutes}$$

Although regression analysis is a powerful tool and shows great promise in the field of indirect labor measurement, it has the potential of being grossly misapplied by industrial engineers who use the programs on a "cookbook" basis. It should not be used without the counsel of an individual who is completely familiar with the mathematics. Further information on this technique may be found in the Bibliography.

Group or Individual Application.   Once the units of work have been established and the time standards have been determined for each unit of work, the question arises of how to apply the standards. Should they be applied to individual operators or groups of operators? There are many points of view on this question, and the subject is treated more thoroughly in Sec. 6. There is a definite tendency to apply standards to an indirect operation on a group basis. One reason is the difficulty in establishing units of measure that are directly related to individual performance. This occurs whenever the work is performed by a team of people and not by individuals. John Donne's statement probably rings true: "No man is an island."

Another reason arises from the manner in which the standards are derived. Because of variations in methods and conditions of performing indirect jobs, it is frequently necessary to rely heavily on averages. If standards make extensive use of averaging conditions or methods, the group application is desirable. If standards are applied on an individual basis, the operators may be inclined to pick and choose from available jobs with variations on the favorable side of the average and let the more difficult jobs wait until later. If the performance of the group as a whole is measured, they will be less likely to do this. This is especially true when credit is given for completion of a group of tasks or quality levels are factored into the earned hours. The work of one operator can have significant influence on the efforts of an entire group. If this is the case, group incentives are essential.

An example of this arises in a toolroom. A part for a die is made by a machinist and milled to approximate size, heat-treated, and then ground to a close tolerance by a toolmaker. If the machinist leaves too much material on the block, the toolmaker will require excessive grinding time. Under a group application both the machinist and the toolmaker would suffer, whereas in an individual system the machinist's efficiency would be excessive and the toolmaker's efficiency would be substandard.

In some cases, however, individual application of standards is quite logical. In most janitorial installations, a specific schedule of work is prepared for each person for each day of the week. A worker who completes every item on time will be working at the desired level of performance. Even in this case, had the total plant been on a group incentive plan, this work could be reduced if the other workers were less careless and reduced the amount of cleanup required. The selection of the unit of work will have a bearing on the selection of group or individual system of measurement.

Incentive or Measured Daywork.   It is often asked if the design of the system for time standards needs to be altered in any way, depending upon whether the standards are to be used for incentives or measured daywork controls. Although there may be some specific exceptions, the general answer is "no." There should be no difference in the design of the time standards. It may be desirable to have some variation in the work count procedures if incentives are used, but the time standards should be of equal quality. If the standards for a measured daywork installation are less accurate, the weakness will become apparent to both the workers and their supervisors. The result will be an ineffective program.

Quality Control.   It is just as important to install some form of quality control over the work produced by indirect workers as it is to check quality of direct workers. Because many indirect labor operations are more in the nature of a service rather than the producing of a product, the job of establishing quality controls is more difficult. Define "clean." This one example is enough to make it easy to recognize the problem. Maybe it is not so difficult: quality is an attitude. A rating check sheet may be used to evaluate performance and quality of janitorial work. Figure 5.1 illustrates a form for quality and frequency of a janitorial task. The frequency is indicated by a D for daily, a W for weekly, and 2W for twice weekly, as shown.

The quality of other indirect operations can sometimes be measured by the end result. Will the die make a good part? It may be necessary to rely on the number and type of complaints or compliments received from internal and external customers. The customer

Formula: ___1090.01___

Date ___7-28-___

Sheet __1__ of __3__ Sheets

Building __Main office__ Floor __1st floor__ Room __Men's washroom__ Sq. ft. ___200___

REFERENCE	ELEMENT DESCRIPTION	NO. ITEMS	FREQ.	HOURS	TIME		
					DAILY	OTHER	STD. HRS.
20.0601	Clean drinking fountain			.0083			
20.0401	Clean dispenser	4	D 1.0	.0071	.0284		.0284
70.0101	Service dispenser	4	D 1.0	.0079	.0316		.0316
70.0201	Service eye glass cleaning station			.0098			
20.0416	Clean paper cup dispenser			.0098			
70.0104	Service paper cup dispenser			.0038			
70.0103	Service toilet tissue dispenser	2	2W .4	.0021		.0042	.0017
20.0412	Preflush urinal or commode	2	D 1.0	.0008	.0016		.0016
50.0103	Empty and wipe wall-hung ashtray			.0086			
50.0104	Empty and wipe wall-hung ashtray, water type			.011			
50.0106	Empty and refill sand urn			.0177			
50.0105	Remove butts from sand urn			.0089			
40.0003	Pick up heavy debris	1	D 1.0	.0006	.0006		.0006
50.0202	Empty wastebasket			.0045			
20.0502	Clean wastebasket			.0097			
50.0201	Empty oily rag receptacle			.0014			
50.0307	Empty waste receptacle liner - metal or cloth	1	D 1.0	.0092	.0092		.0092
20.0501	Clean exterior of waste receptacle	1	W .2	.0144		.0144	.0029

FIGURE 5.1 Evaluation report and standards calculation sheet for janitorial work.

is the final inspector. The quality of indirect work is as good as that of the products shipped to customers. No amount of inspection will make quality happen.

WAREHOUSE OPERATION

A typical application of indirect labor measurement may be illustrated by a warehousing operation. The warehouse is a large single-story building of approximately 400,000 square feet. It is serviced by an 800-foot railroad siding and 12 truck docks. Prior to any work measurement the operation employed 38 warehouse workers.

The products stored consist of a variety of large industrial equipment. The equipment was of a standardized shape and configuration but of different sizes. The various kinds of equipment were divided into seven general classes.

The equipment was manufactured in adjacent assembly buildings and transported to the warehouse by trailer trains. It was palletized after assembly and stacked and handled within the warehouse by the use of fork trucks.

As a first step in the design of the work measurement system, the warehouse operations were broken down into five basic activities:

1. Transport equipment from assembly to warehouse.
2. Receive and store incoming material.
3. Pick items to fill orders and accumulate in holding areas.
4. Prepare items for shipment by crating, labeling, banding, stenciling, etc.
5. Load trucks and railroad cars, including cribbing and bracing cars.

Methods Analysis. The first step in this installation was to carry out an intensive activity (methods) analysis program to standardize the work and develop improvements.

After the preliminary study, it quickly became apparent that there were substantial variations in the loads transported by identical fork trucks. One operator picked up and transported one unit at a time; another operator transported a stack of three units at a time. This problem was assigned to a committee composed of the warehouse supervisor, the safety engineer, the shipping clerk, group leaders, and an industrial engineer. They reviewed each commodity stored in the warehouse and then established a standard unit load in terms of the number of pieces that could safely be transported at one time. These unit loads were then used as a basis for standard development.

A significant improvement was made by combining the jobs of accumulating the equipment and the operators involved in the labeling of the equipment in the accumulating area. Previously, the labelers lost a great deal of time in locating the material. Further, it was difficult to balance the work loads so that some idle time did not exist. Combining the jobs solved both problems.

A study of the railroad car cribbing and blocking operation resulted in a simplified method that not only saved on labor but resulted in a significant savings in lumber blocking materials as well.

The location of material within the warehouse and the methods of stacking the equipment were changed. This improved the use of cubic capacity of the warehouse and eliminated a substantial amount of restacking as well. Changing item locations permitted the storage of fast-moving items closer to the assembly area, which reduced the transport times.

A simple light was installed, with a switch in the assembly area. The light signaled the trailer train operator when a load was ready to be moved. This reduced the amount of waiting time for the driver as well as reducing delays in the assembly area. Each improvement should be made to either standardize the method or simplify the work to be done. Normally standardization will result in simplified work. Quality and efficiencies are improved.

Work Measurement. Work measurement could not begin until standardization and methods improvements were completed. The work was not measured operation by operation. For example, studies were not made of unloading the trailer trains and placing the equipment in the storage area. Nor were studies made of moving equipment from storage and placing it in accumulating areas, or of moving equipment from material handling areas to trucks or railroad cars. Instead, sets of standard data were established for each of the basic operations or pieces of equipment.

The data developed include

1. Tow truck operating data (Fig. 5.2)
2. Fork truck operating data
3. Steel banding
4. Railroad car cribbing
5. Stenciling
6. Crating, etc.

By developing standard data first, a tremendous amount of flexibility was gained, and as a result standards could be developed to fit the units of work used to measure the output of the warehouse.

This approach also resulted in a library of standard data that simplified the development of material handling standards in the manufacturing departments. Later these data were used in developing the MOST powered truck sequences that are covered in Sec. 4, Chap. 5.

TOW TRUCK				CODE 3002.04	
Operation	Variable or type	Range	Sym.	.0001 hours	
Trip constant	Start and stop - empty or loaded		01	22	
	†Hook and unhook - tow truck to lead trailer	–	02	95	
Travel time	Open level floor or roadway *and little or no congestion	Ea. 10 ft.	11	3	
Open and close door	Free swinging and/or bumper type		21	82	
	†Sliding, overhead or equiv., and manually operated †Power contr. door - actuating switch at door	–	22	185	
	Power contr. door - act. sw. on approach path to door		23	59	
Miscellaneous	Mount and dism. tow truck	–	31	24	

†Includes mount and dismount of tow truck

FIGURE 5.2 Tow truck data.

Units of Work. The cost of having a clerk check each load shipped to determine the number of units of work was prohibitive. The idea of having the warehouse workers report the unit loads handled each day was also rejected. An independent source of information on the quantity of work completed was desired and necessary.

Two established reports seemed to cover most of the warehouse activities. One was a report on the number of trailer loads of equipment shipped from the final assembly area to the warehouse. The other was the shipping order prepared by the order entry group in the sales department and completed by the shipping department. Normally, one will find that, if there is enough information to manage the business, there is enough to report against as well as develop standards.

Using these two documents to measure units of work, standards were developed for each item on the shipping orders and the trailer train reports. Every operation performed in the warehouse was related in some way to these two documents. We discuss later the use of similar information for forecasting indirect labor requirements. Because the standard data were already available, the time standards were developed by taking frequency studies to determine what elements of work were required and how frequently they were performed for each item handled.

Performance Measurement. A procedure for measuring the units of work and time standards for each unit was now available. The time standards multiplied by the units of work gave the standard hours of work produced, which were then compared with the actual hours spent in carrying out the work. A ratio of the actual hours to the standard hours of work produced provides a performance index or a measure of the effectiveness of the operation.

These methods of measurement were applied to the previous months to establish a past performance level. This past performance level provided a benchmark by which future progress and savings could be determined. Next, a dry run period of 4 weeks was carried out to test completely the operation procedures and the adequacy of the standards. This period of time allows oversights to be discovered and corrected. In addition, the warehouse people had an opportunity to evaluate the system. This practice is essential for both the workers and the supervisors to develop an understanding of the use of the standard data and the fairness of the system. Figure 5.3 is a graphic representation of the before and after performances. The performance for the group improved from 70 percent to the 125 percent range.

Results.   As a result of this program, the number of warehouse workers required to perform the work was reduced from 38 to 22. The surplus workers were transferred to other departments. Since this was an incentive installation, the earnings of the group were increased by 25 percent and the overall cost of the operation was reduced by about one-third. Unexpected savings resulted from reducing fork trucks from 26 to 17. This savings alone paid the cost of the measurement program. The total savings was in excess of $150,000 annualized. The entire project required a total of 55 labor weeks to develop and install. Because of simplification of the work count procedures and the standards application procedure the system costs are less than 1 percent of the labor costs. Figure 5.3 illustrates performance before, during, and after application of the standards.

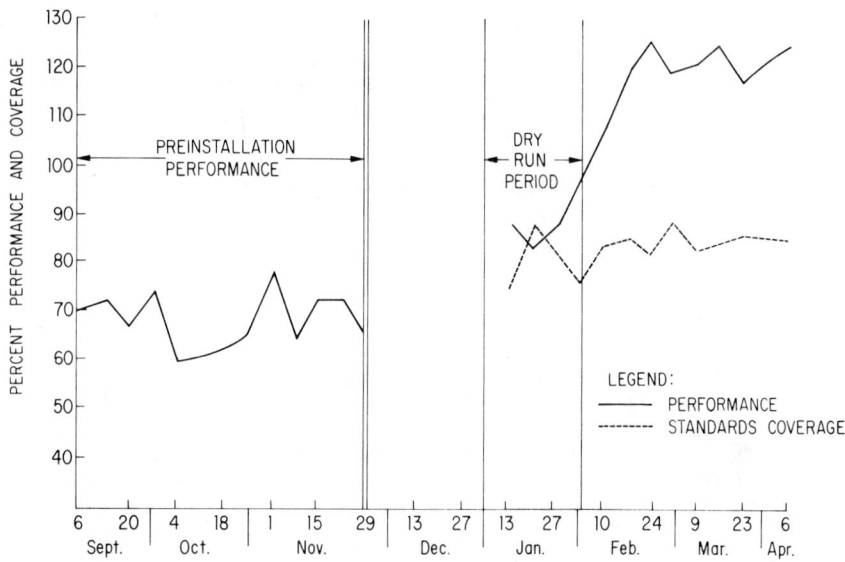

FIGURE 5.3   Record of increase in performance under wage incentive plan.

TOOLROOM

One of the characteristics of indirect labor is the wide variety of work performed. This necessitates flexibility in the design of the work measurement program which will be best suited to the type of work being measured. The fundamental concepts remain the same, but the application varies, and requires considerable ingenuity on the part of the industrial engineer to devise methods for measurement and reporting of the work performed. The discussion that follows is typical and is taken from a toolroom study.

Importance of Planning.   One feature common to almost all indirect labor work is that the work is performed with almost no planning. In a typical production machine shop, it is normal practice to develop a route sheet for each part manufactured, showing the sequence of operations to be performed. In a toolroom, however, it is common to do little or no planning. Typically one toolmaker who may be given the entire job of making a fixture or a die from beginning to end personally performs each operation on each part and does all the assembly and bench work as well. About the only thing that might be done by another operator is some specialized work such as heat treating or plating.

This approach does simplify the supervisor's work and provides the advantages of the job enlargement concept to the toolmaker. It also has some serious disadvantages. Contrary to popular belief, it does not produce the best-quality tool. The details on the drawings will constantly be revised to compensate for the toolmaker's mistakes. If a shaft is ground undersized, the bearing must be made to fit. This fitting and trying is time-consuming. Furthermore, the problems involved in replacing broken or worn parts are multiplied if the original parts were not made to dimension. Parts made separately by a group of operators, but made to print will produce a far better tool.

Requiring each operator to use every piece of equipment in the toolroom will make all operators competent on all machines without having expertise in any. The use of a fully qualified toolmaker to turn a simple shaft on a lathe is a waste of expensive talent.

Another myth is that it is more economical to have the toolmaker do the complete job because this does not require the overhead of planners. It is obvious that the planning must be done by someone, either in advance in a timely manner or prior to starting today's work. Many toolmakers have never learned shop arithmetic and geometry. They struggle through somehow, though it is a slow and error-prone process. The mistakes are usually put in the scrap bin out of sight. The average toolmaker is certainly no competition for a planner trained specifically to do planning 100 percent of the time. That planning is done by a toolmaker does not mean that the planning cost has been eliminated; quite the contrary, the cost has probably doubled or tripled or....

Operators assigned to specific machines or equipment soon become extremely proficient, thus outperforming the most skilled toolmaker. A program designed to standardize toolroom operations and improve methods frequently involves a major reorganization. A toolroom that uses detailed planning is far more efficient than one that uses the conventional method, which assigns one person to do the entire job. This organization is usually difficult to sell but is essential to the establishment of standard practices.

Machine Work. The bulk of work performed in most toolrooms can be classified under one or two categories: machine work and benchwork. The machine work can be broken down into classes or types of machine tools. Time standards, usually in the form of a work sheet, can be developed for each type of tool. Another type of work sheet or series can be developed for the various types of benchwork performed. A few examples will illustrate this point.

Figure 5.4 shows a generalized work sheet which can be used for a variety of machine tools. It is designed primarily to be used when calculating the machining time for an operation. The machining module of the MOST computer system will calculate machine time automatically. Supplementary sheets will be required to calculate setup, gauging, and part handling. All of the times are of course added to determine the total time for a piece. Tables of standard data are also needed to determine surface speeds, feeds, and rpm for a variety of materials and cutting tools. Typical charts of this type of data are shown in Fig. 5.5. A work sheet is prepared to develop time for each operation. The work sheets are completed by the planner, who must also lay out and plan each operation of each part of the tool to be manufactured. A route sheet is then prepared for each part to be manufactured in the shop.

Drawings. Many toolrooms attempt to work with a minimum of drawings. They rely on sketches and a toolmaker's experience to translate a sketch into a completed part, including all the proper tolerances and finishes. The competent toolmaker has the ability to do this but is in effect acting as a drafter. The idea that having a tookmaker make sketches saves drafting cost is another myth of the traditional toolroom. The use of a toolmaker as a drafter results in high-priced poor-quality drafting, which results in numerous errors that are converted directly into scrap. The end result is that no as-built drawings are completed and replacement parts must be made from a damaged part.

DRILL PRESS   WORKSHEET   MACHINE TIME

M.W.O. _____ S.O. _____ Job description _____

Tag no. _____ Mat. class _____ Dwg. no. _____ Item no. _____ Date _____

1	2	3	4	5	6	7	8 (6×7)	9	10 (8×9)
Operation description	Diameter of cut	SFM	RPM	Feed	Actual time/ inch	Length of cut	Time /cut	No. of cuts	Total time/ piece
						Total machine time			

Operation description code:

D = drill, R = ream, T = tap, CS = countersink, RD = redrill, RLF = relief, BT = blind tap, CB = counterbore

A. (Setup time ☐ + setup gaging time ☐) × 1.21 allow.   =   [____]

B. Total mach. time/pc. ☐ × 1.25 allow. × no. pcs. ○   =   [____]

C. Total const. time/pc. ☐

    ⊢ Total gaging time/pc. ☐ × 1.25 allow. × no. pcs. ○   =   [____]

                                    Total std. hours/job   =   [____]

FIGURE 5.4 Generalized work sheet for calculating machine tool time.

The answer is not in elaborate or overly detailed drawings. Simplified techniques can be used to a definite advantage. Drawings define the completed product. They are a universal language and as a result will give the product desired. The making of drawings also requires that the final product be simulated. The simulation is less expensive on paper than from tool steel. Figure 5.6(*a*) shows a die stripper part drawn in a conventional manner, while Fig. 5.6(*b*) shows the same part in a simplified manner. The use of coordinates in lieu of dimension lines greatly reduces the calculation time of the jig bore operator. All the coordinates are already calculated. The lack of clutter also reduces the time required for an operator to read and interpret the drawings. Simplified drawings have a double effect. They simplify both the drafting work and the shop work.

Benchwork.   Because of the many variations, benchwork in a toolroom is by far the most difficult to measure and control. The variety of benchwork can be simplified somewhat by separating the work into several different classifications and then developing work sheets for each classification. An obvious classification is the type of tools. Dies, jigs, fixtures, and special tools might be appropriate. The classifications should be tailored to the shop being measured. The purpose is to establish parameters which can be used to eliminate variations and tend to standardize and simplify the standards setting.

Figure 5.7 shows a work sheet designed to be used when developing standards on a very difficult type of toolroom benchwork, namely, the task of servicing and repairing dies. In making a new die, the work to be performed is clearly specified. In die repair work, however, the work to be performed is frequently unknown. If a die is not functioning properly in a press, it is pulled from the press and sent to the die repair section for repair. A brief description of the die trouble might be recorded on the work order, but that

CHART I
SF/M RECOMMENDATIONS

Material	Drill or redrill	Tap	Counterbore	Ream
Class I	300	70	200	200
Class II	200	70	130	130
Class III	90	50	60	60
Class IV	80	40	50	50
Class V	40	25	30	30
Transite*	160	40	100	

* Use feed equal to twice drill feed of soft metal:
Class I —aluminum
Class II —bronze and brass
Class III—cast iron
Class IV—soft steel
Class V—hard steel

NOTE: To counterbore—(*a*) flat bottom drill bit—use standard drill to depth desired minus length of drill point, then flat bottom drill at counterbore SFM and drill feed for distance equal to the length of the point for the diameter of the drill used in the redrilling; (*b*) "fly-cutting"—use counterbore SFM and .010 feed.

CHART II
FEEDS—DRILLING AND REAMING: DRILL POINT LENGTH

Drill dia.	Drill pt.	Drill feed/ rev.	Reamer feed	Redrill feed*
1/8	.04	.002	.004	
5/32	.05	.002	.004	
3/16	.06	.004	.008	
7/32	.06	.004	.008	
1/4	.08	.006	.012	
5/16	.10	.006	.012	
3/8	.11	.008	.016	
7/16	.13	.008	.016	
1/2	.15	.010	.020	.012
9/16	.17	.010	.020	.012
5/8	.19	.010	.020	.012
11/16	.21	.012	.024	.015
3/4	.23	.012	.024	.015
13/16	.24	.012	.024	.015
7/8	.26	.012	.024	.015
15/16	.28	.014	.028	.017
1	.30	.014	.028	.017
1 1/16	.32	.014	.028	.017
1 1/8	.34	.014	.028	.017
1 1/4	.38	.017	.034	.021
1 3/8	.41	.017	.034	.021
1 1/2	.45	.017	.034	.021
1 5/8	.49	.017	.034	.021
1 3/4	.53	.017	.034	.021
1 7/8	.57	.017	.034	.021
2	.60	.017	.034	.021
2 1/8	.64	.017	.034	.021
2 1/4	.68	.017	.034	.021
2 3/8	.71	.017	.034	.021
2 1/2	.75	.017	.034	.021
2 5/8	.79	.017	.034	.021
2 3/4	.83	.017	.034	.021
2 7/8	.86	.017	.034	.021
3	.90	.017	.034	.021

* 120 % of drill feed.

CHART III
DRILL PRESS—COUNTERSINK HOLE

Flat head screw size	Countersink diameter	Countersink depth	.010 feed RPM	Decimal hours
#4 (.112)	.25	.092	686	.0002
#12 (.216)	.50	.190	506	.0006
3/8	.78	.270	393	.0011
1/2	1.00	.333	280	.0020
5/8	1.25	.417	166	.0042
3/4	1.50	.500	140	.0060

FIGURE 5.5 Typical standard data for drilling operations.

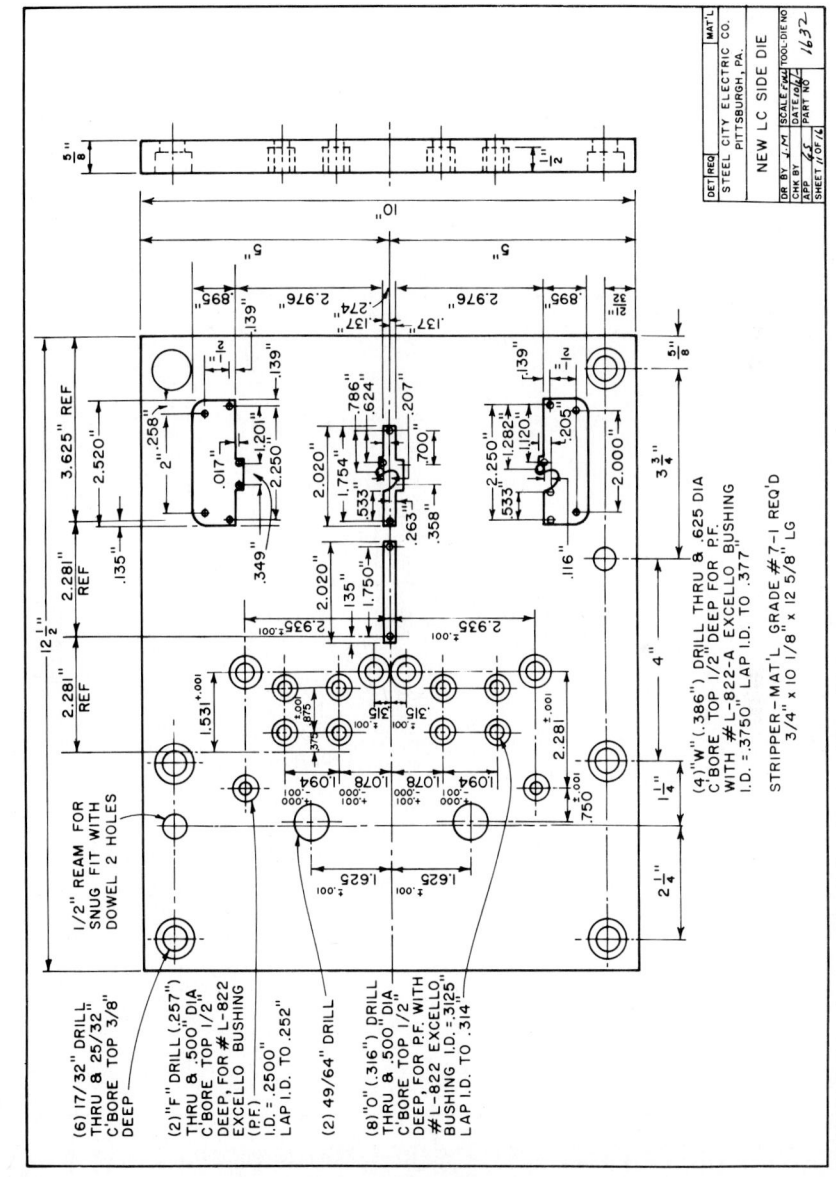

CONVENTIONAL

FIGURE 5.6(a) A conventional drawing for a die stripper part.

5.81

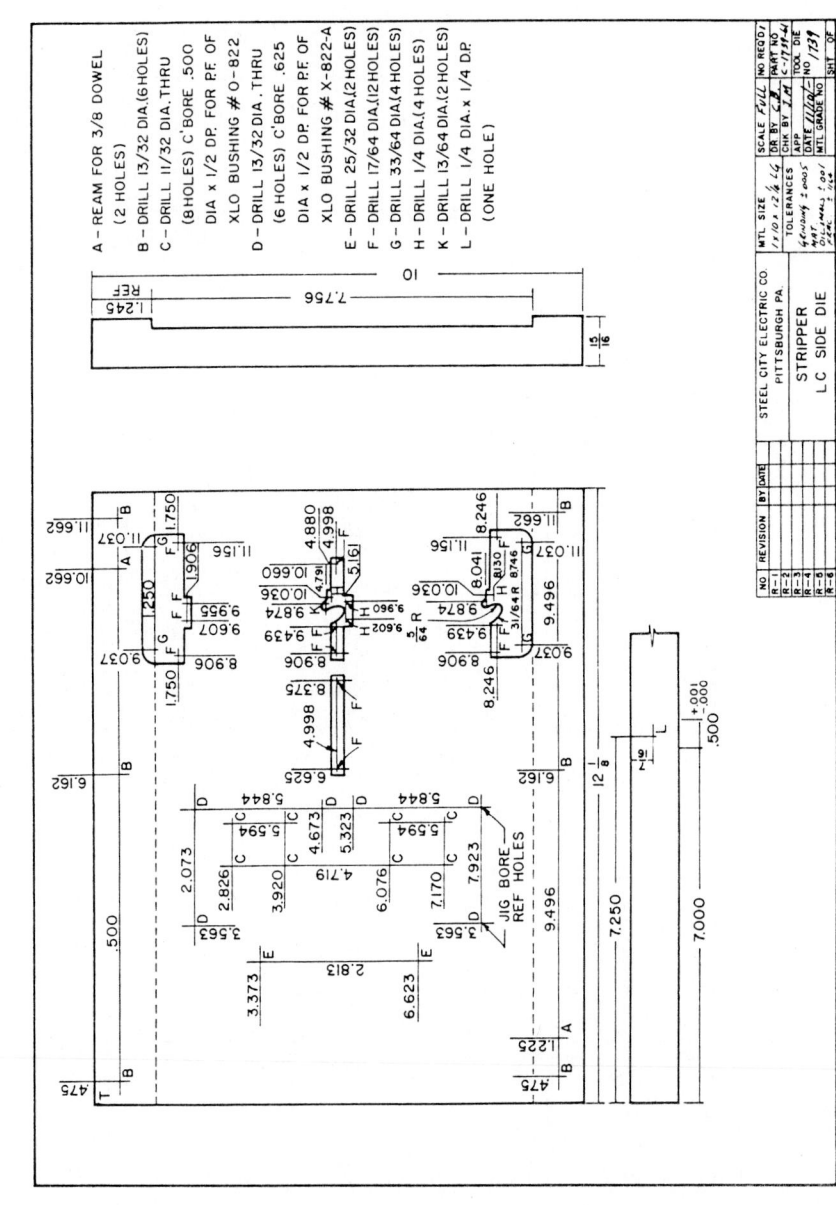

SIMPLIFIED

FIGURE 5.6(b)   A simplified version of Fig. 5.6a.

5.82

S.B., F.P., D.D., S.O. dies   Benchwork order-3-

Date _____   Work order No. B

Die no. _____   Die class S

Die description _____

Regular service ◯ replacement ◯ change ◯ rebuild ◯

	Toolroom	Die vault	Press room	Clean
Complete ⎫	◯ .125	◯ .245	◯ .275	◯ .096
Top or btm. shoe ⎬ from	◯ .076		◯ .226	◯ .036
Other subassem. ⎭	◯ .008		◯ .060	◯ .012
Complete ⎫	◯ .125	◯ .245	◯ .275	
Top or btm. shoe ⎬ to	◯ .076		◯ .226	
Other subassem. ⎭	◯ .008		◯ .060	

TOP SHOE / BOTTOM SHOE	Cradle	× .066 =	Change: from	
	Die sections	× .061 =	to	
	Bumper blocks	× .042 =	Replace new parts:	
	Full stripper	× .086 =	Piercing punch	× .098 =
	Stock equalizer	× .078 =	P.O. slot punch	× .098 =
	Other sections	× .066 =	K.O. punch (Ex. C.L.)	× .098 =
	Punch holders	× .104 =		
	Other sections	× .066 =	Contour punch	× .098 =
PUNCHES	Piercing	× .039 =	Trimming punch	× .270 =
	Knockout	× .039 =	Die section	× .303 =
	Stencil	× .049 =	Bushings	× .098 =
	Trimming	× .068 =	Pilot pin	× .098 =
	Lancing	× .059 =	Round insert	× .098 =
	Dimpling	× .039 =	Dowel pin	× .095 =
	Extruding	× .039 =	Quill	× .098 =
	Embossing	× .039 =		
			Restore radius—hard	
			Clearance—inches	× .020 =
BUSHINGS, INSERTS, & MISC.	Stencil bumper	× .039 =	Draw or form—hand	
	Forming insert	× .074 =	Stock × 3 =	
	Cutoff insert	× .069 =	plus .020	
	Embossing insert	× .049 =	inches × =	
	Bushing	× .045 =	Broken bolts:	
	Pilot	× .039 =		Large	Small
	Spring pad	× .104 =	Proj. .300	.250
	Lifter pin	× .056 =	Flush	
	Push back	× .056 =	blind 1.000	.750
	Quill	× .039 =	Flush	
	Spring stripper	× .152 =	open .750	.500
	Guard	× .043 =	Broken set screws:	
OTHER PARTS	Method of mounting:		|1.000|	
	Bolted only	× .043 =	Get parts welded	◯ .250 =
			Use die flipper	◯ .747 =
	Ditto w/springs or shims	× .068 =	Re-dowel each part	☐ × .366 =
	Bolted & dowelled	× .063 =	Ear form ⎰ octagon	◯ .458 =
	Ditto w/springs or shims	× .078 =	Adjust ⎱ utility	◯ .308 =
	Press fit	× .045 =		
	Slide or tap fit	× .039 =		
			Job constant	.120
			Total	
			Times 1.40 = allowed hrs.	

FIGURE 5.7   Work sheet for benchwork operations on repairing and servicing dies.

is all the information the repair person will have available. How are time standards set under these conditions? It is really not difficult. This is true of many indirect labor situations. After a first look at a problem, it may be thought that the work cannot be standardized. Further analysis will lead to a solution.

In this particular case, the various dies were broken down into a number of families. A form of group technology was developed for each family as shown on the work sheet in Fig. 5.7. This was done by observing the various types of repairs and faults that could occur in a die. These repairs were then grouped by the major sections of the die. The sections were the bottom shoe, top shoe, punches, bushings, inserts, pins, etc. Further, the work was generally limited to assembly, disassembly, and inspecting parts. Either all the broken parts were replaced from stock or an order was written to make a replacement part. By limiting the work in this manner it was not difficult to develop time values for disassembling the die, removing or replacing parts, and then reassembling them. These are the time values shown next to each part or item on the work sheet shown in Fig. 5.7.

A major problem in this case is to determine the frequencies of the work units. After some study, it was determined that the operator would record the frequency of work directly on the work sheet. The work sheet was extended by a clerk. Because this work was performed under a wage incentive program, it was feared that the operators might pad the frequencies so as to produce higher pay. To eliminate this temptation, a system of spot checks was established. These inspections checked both quantity and quality of the work units. It was quite simple for a qualified person to determine how many new parts had to be replaced. The results of inspections revealed that there was more underreporting than overreporting. This is because many minor items were overlooked when the recording was done.

MAINTENANCE AND REPAIR ACTIVITIES

From a work measurement and control viewpoint, one of the most difficult indirect labor activities is plant maintenance and repair. A unique approach to the measurement of maintenance work and similar nonrepetitive activities has been developed. This approach involves the use of benchmark jobs and a range of time concept. Instead of attempting to measure a job to an accuracy of a hundredth of a minute, jobs are slotted into standardized time ranges. The average of the time range is used as the standard time for the job. This concept is explained in some detail in Sec. 5, Chap. 6.

USES OF INDIRECT LABOR MEASUREMENT

It has been said that management is based on measurement plus control. The primary essential to controlling any process is measurement. On direct labor operations, the simplest method of measuring is to forecast and count the number of pieces. This measurement does provide a basic means of control. With the advent of the scientific management movement, Taylor and others developed more sophisticated means of measuring labor in relation to some established norms of performance. Thus, the use of piece counts as a relative measure of work gave way to the concept of standard hour of work as an absolute measure of performance. Because piece counts have little application to the measurement of indirect labor operations, the standard hour is the only way to measure indirect labor operations. This means that even salaried, clerical, and management personnel can be measured. Later in this chapter these kinds of forecasting and measurements will be discussed.

The primary use of indirect measurement is in the management of the performance of individuals or groups of individuals, but it is also used to measure the value of improved methods. It also provides a means of planning for the future when coupled with reliable forecasting indicators.

MEASURED DAYWORK AND INCENTIVES

Indirect labor measurement is used in two basic ways as a means of controlling the performance of the work force. One method is to use the standards as the basis for measured daywork controls and the other is for incentive payment. The many pros and cons as to the use of these two basic methods of control are discussed in Sec. 6.

An example illustrating the results to be expected from indirect labor measurement will show the use of both methods of control. This is a particularly interesting case. The management of a company decided to measure as many of the indirect operations as practical from a work measurement standpoint and to let each group of employees determine by a vote whether they wanted an incentive plan or not. If they decided not to go for incentives, management would use the work measurement results as a basis for a measured daywork program.

At the time the standards were developed, the industrial engineers had no way of knowing whether the group involved was going to decide for or against incentives. As a consequence, all the time standards and the work measurement procedures were identical for both measured daywork and incentive use. A comparison of the two applications could therefore be made with the assurance that there were no basic differences in the time standards or in the performance level concept. In some cases, the employees first elected not to go on incentive and later changed their minds and elected to accept incentives. In these cases, the same time standards were used without change or adjustment.

Results of Indirect Labor Measurement.   At the start of the indirect labor measurement program, the plant work force was made up of 1400 direct workers and 1050 indirect workers. The indirect work force was 75 percent of the direct workers. Company records showed that the percentage of indirect to direct had been increasing for the past 10 years. Management began to realize that, if something was not done to change this trend, the cost of indirect labor would soon exceed the cost of direct labor. Although all types of controls existed for direct workers, very little control, other than budgets, was applied to indirect labor. Despite an excellent system of budgetary controls, the cost of the indirect labor continued to grow each year. Management finally decided that a more effective control was needed to halt the steady increase in costs.

Eight years after management made a decision to undertake a program of indirect measurement, about 60 percent of all indirect workers had been measured. During this period the volume had increased to 1500 direct workers. Had the company maintained the 75 percent ratio they would have had approximately 1100 indirect workers. The number was actually 800. This represents a savings of 300 people, or about 27 percent. Most managers felt that they probably would have employed even more than the 1100 based on the past history and the established trend of increasing indirect to direct ratios.

Figure 5.8 shows the number of employees covered by incentives and measured daywork each year and the net savings accrued from the program each year. The cost of installing the program was deducted from the gross savings to arrive at the net savings.

By the end of the first year, the program was just breaking even. The cost of the installations equaled the gross savings. By the end of the second year, the program showed a net savings of about $500,000. By the end of 8 years, the program covered a total of 480 people and was producing savings at an annual rate of $1,700,000 per year. The cumulative savings for the 8-year period amounted to $6,000,000. These savings would have been almost twice that amount if all employees had elected the incentive system. Employee take-home pay was about 25 percent greater for the incentive personnel. The daywork plan people performed at about 80 percent while the incentive performance was 125 percent.

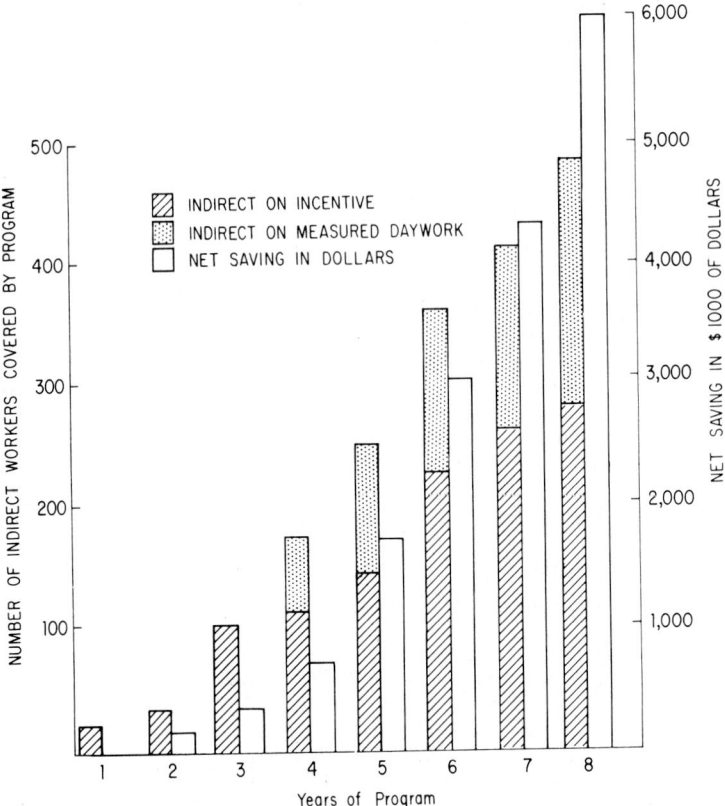

FIGURE 5.8 Results obtained in a typical indirect labor standards program.

IMPORTANCE OF MAINTAINING STANDARDS

The importance of maintaining standards to reflect changes in working methods or conditions cannot be overemphasized. If wage incentives are employed, management will be subjected to considerable pressure from time to time to relax the standards in one way or another to permit higher earnings for the same effort or equal earnings for less effort. Despite good intentions upon embarking on an incentive program, managers sometimes find it extremely difficult to resist the pressure.

A typical example of the pressure that can be exerted on management is illustrated by Fig. 5.9. This chart shows the performance of a group of repair people working under a group incentive plan. The work involved overhaul and repair of machinery requiring factory servicing and rebuilding. The group had been on incentive about a year before the beginning of the chart, and performance consistently averaged about 125 percent of standard.

Because of plant layout changes, the rebuild and service operations were moved to a new location. At the same time, a number of new methods, tools, and equipment were introduced to simplify work and reduce costs. The industrial engineers completely reworked the old standards to reflect the changes.

Referring to Fig. 5.9, prior to the move in month 9 the group had received credit for cleanup of miscellaneous work, which accounts for the higher than normal efficiency in

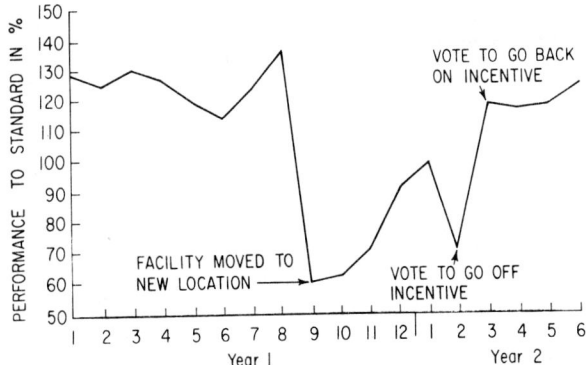

FIGURE 5.9 Effect of change in conditions on incentive earnings.

month 8. Immediately following the move, the new standards were applied to the work, and the employees promptly filed a grievance claiming the standards to be too tight. Production dropped off and performance fell to 60 percent.

Management then had industrial engineering review all the new standards to be certain that they were truly reflecting the working conditions. This took about 1 month, and a few changes were made. Management appealed to the employees to give the new standards a fair trial. This trial period took place over months 11, 12, and 1. During this trial period, performance gradually climbed toward 100 percent. In the meantime, the backlog of work built up and customers were complaining about deliveries. The employees were also complaining about loss of earnings. The industrial engineers firmly maintained their position that the standards were correct and that the group could achieve their previous earnings if they really went to work.

Management was clearly on the spot. What could be done? Should management instruct the industrial engineers to loosen the standards or should they stick to their guns? While management wrestled with this decision the union voted to go off the incentive plan.

Performance immediately dropped to 70 percent in month 2. After 2 weeks of being off incentive, one employee asked to be transferred to another department. The remaining employees then voted to go back on incentive. In the next 2 weeks performance went from 70 to 118 percent, and in a few months it climbed back to the 125 percent range.

This is typical of the problems that managers face in the administration of an incentive plan. It also illustrates the pressures that will be created by the workers to loosen standards. If management gives in to these pressures, degeneration of the plan will soon follow and costs will quickly get out of hand. An incentive plan is a powerful tool for controlling costs so long as it is administered intrepidly.

COST OF MAINTAINING AND OPERATING A STANDARDS PROGRAM

The cost of operating and maintaining a work measurement program on indirect labor operations is a factor to be considered. Although the savings which accrue can be substantial, a portion of the savings will be required for maintenance of the system. These costs will vary in accordance with the units of work used in applying the standards, the stabilization of the product line, and many other factors. An estimate of these costs for a typical indirect labor operation is

	Percent
Day-to-day application costs	3
Maintenance of the standard database	2
Timekeeping and payroll	1
Total	6

The total cost of operating and maintaining an incentive program on indirect labor operations will be about 6 percent of the indirect labor costs. About half the cost required is to develop the standards for the day-to-day system. About 30 percent is required for maintenance of the system. About 30 percent is required for administration of the system.

As in any new endeavor, a learning curve will apply. The costs will decrease as all those involved become more experienced in the use of the system. Figure 5.10 illustrates a typical trend over several years.

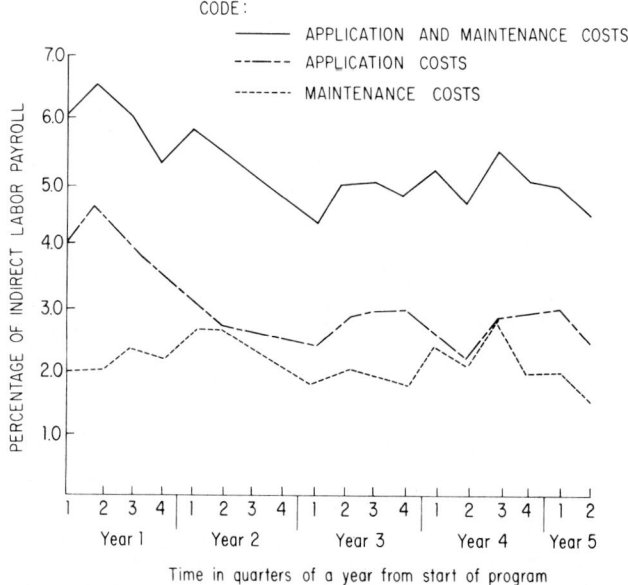

FIGURE 5.10 Effect of learning on cost of applying and maintaining time standards.

LONG-RANGE RESULTS

One of the major advantages of indirect labor controls based on sound work measurement practices is to provide the basis for long-term cost reduction and continuous control. Many managers have been tempted to use quick and easy approaches to controlling indirect labor costs. This approach frequently involves the establishment of approximate or estimated standards and concentrates on the scheduling of work in small batches with close supervision and follow-up. This method has produced amazing results in short periods of time. The results are usually short-lived because the standards are not soundly established or based on any specific set of conditions. What may appear to be an easy solution normally produces short-term cost reductions and creates problems when future work measurement programs are introduced.

For long-range results, there are three basic requirements: accurate method based standards; forecasting indicators for correct staffing; and integrity of management. If any of these are missing control problems will eventually occur. The next several pages will be devoted to the forecasting of indirect labor.

INTRODUCTION TO LABOR FORECASTING

For the most part we do a reasonably good job of forecasting direct labor. The reason for our ability to forecast direct labor is

We have direct labor standards by operation.

We have a production forecast by unit.

We know our performance level.

The equation then becomes

Total standard hours per unit $\times$ units $\times$ effectiveness

$$= \text{total hours of direct labor required}$$

We then can calculate the number of direct workers by dividing by hours per period covered by the unit forecast and adding an allowance for absentees. This process is simple because we have all the necessary information at hand. But, what about indirect labor? The common answer to this is that we do not have a forecast of work units. We often decide that we cannot forecast the kinds of units of work required to calculate indirect hours. And then you know that we certainly cannot predict when a bearing will fail in a specific machine or when a light bulb will go out or when the wind will blow off a section of the roof.

Indirect labor does present some unique challenges. It does require some different approaches to developing indicators (forecasts). Please notice that the words indicators and forecasts are used interchangeably. This is the key to forecasting indirect labor. Forecasts become indicators. The plural is used, indicating several forecasts. Each type of indirect labor will require different indicators. The units currently used as indicators, such as the number of units or dollars of production, probably will not give accurate forecasts. The standards will also be designed so that each matches the indicator. We will pursue this concept in detail.

After we have mastered the techniques involved in forecasting indirect labor, the natural progression is to clerical, technical, and even management. The same approaches that we develop for indirect labor will be used to make these forecasts.

A total labor forecast can be developed by period which will be the sum of direct, indirect, clerical, technical, and management personnel. All this sounds entirely too complex to actually do in a period of time as short as 1 month. In the next several pages a process is presented that will prove that this effort need not be complex and will actually improve the effectiveness of our direct labor by providing the necessary services to support the direct labor function.

THE CONCEPT OF FORECASTING INDIRECT LABOR

Personnel in any operation are necessary only to support a schedule. Without a schedule of units or activities we have no work to do and therefore require no personnel.

In the manufacturing situation, we have certain functions which do not relate to scheduled volume. These kinds of duties are facility related. Some of the many examples are security staff, fire protection inspections, OSHA inspections, and receptionist. All these functions are related to the facility required to perform the manufacturing operations.

These activities then become necessary to support a schedule. Also, these activities become unnecessary when the schedule disappears and thus the plant closes. The first principle of the concept is:

> "WORK IS REQUIRED ONLY TO SUPPORT A SCHEDULE"

The second principle is that the

> "PRODUCT OR SERVICE BEING GENERATED WILL DICTATE THE NECESSARY INDIRECT FUNCTIONS"

In order to determine what function, what work elements, what times should be applied, and what frequencies are required, some initial investment in a study of the operation is required.

Remember our direct labor formula was standard time multiplied by units to be produced in a period. Indirect labor forecasting also requires that we have standard times per element of work and the frequency with which each will occur. In order to develop these data so that they can be manipulated, we must standardize the information. This standardization will consist of indicators, schedules, and work elements.

Indicators. The third principle is to

> "USE INDICATORS FROM THE MATERIAL REQUIREMENTS PLANNING SYSTEM AS FORECASTS"

An indicator is something which will quantify a specific work element. If the work element is repair of a punch press, the indicator would be machine run hours. Likewise, if the work element is placing orders with vendors, the indicator is the number of line items to be purchased. Some information may not be *readily* available; however, it is available as shown:

Machine run hours by machine

Setup hours required by machine

Number of production releases by operation

Items to be purchased by product

Items purchased to be received and intervals

Parts to be stocked on the assembly line

Units by product to be produced

Units by product to be warehoused

Units by product to be shipped

Direct hours by machine, work center, department

Capacities by shift by work center

Fixed work elements per shift

Schedules. All of the above are needed to develop indirect labor forecasts. The source for most of these data is in the MRP system that schedules materials, production, and shipping. These data can be retrieved from the system by either programming and securing special reports or by manually creating the data using lists printed from the computer system. It, of course, would be better to program the reports. In any event, these data are obtainable without too much difficulty.

Work Elements.   The fourth principle is to

> "USE WORK ELEMENTS THAT MATCH FORECAST INDICATORS"

Work elements must be designed to complement the indicators available through the plant's material requirements planning (MRP) system. The work elements must be broad enough to cover the many small activities that are required by the indicators.

The work elements must be standardized so that the system can be computerized to reduce the amount of clerical effort required to use the forecasting system. Figure 5.11 is a typical departmental work element format. The heading definitions follow:

Department number	As required for accounting
Description	Example is machining
Classification	Worker skill code
Fixed hours	If the element is fixed, the hours required per period
Variable factor	Units in hours that will multiply by the indicator
Forecast	The indicator coded to for a specific forecast indicator
Required department	The number of the department requiring the indirect labor
Shift	Shift number, either 1, 2, 3
Idle fixed	Fixed time that is idle such as tending or guarding
Need code	How important is this function: M = must, N = need, W = want
Standard code	Evaluation of the standard accuracy: S = good standard, E = estimate, U = undetermined

WORK ELEMENT FORMAT

DEPT NO.	DESCRIPTION	GLASS	FIXED HOURS	VARIABLE FACTOR	FORECAST INDICATOR	REQ'D DEPT	SHIFT	IDLE FIXED	STD. CODE

FIGURE 5.11   Work element format.

Further definitions of need codes are necessary. "Must" is an indirect labor requirement that cannot be eliminated even in the short range such as required by law for health or safety. "Need" is required for continuing operation such as lubrication of a machine that may be discontinued for a short period of time. "Want" is work that if eliminated will not hurt the operation, such as a company driver.

The standard codes describe the accuracy of the standard. These are determined by the actual standard accuracy and the required accuracy of the standard. "S" for good standard data describes a standard that is normally plus or minus 5 percent accurate. Normally these standards are established using a predetermined time system such as MOST. "E" for estimated are time values that are established using at least a single time cycle taken by using a stopwatch. Usually little or no methods analysis has been done. "U" for undetermined is a time value normally by someone making an estimate. As the system matures these codes can be used to set priorities in standards review and updating.

SYSTEMS REQUIREMENTS

From the previous discussions, it can be concluded that the forecasting of indirect labor requires three major elements:

Indicators

Schedules

Work elements

Indicators are developed by using the product information contained in the standard process planning (routing) for any parts or several parts that make a complete product. An example could be an assembly made from two manufactured parts and two purchased parts. The manufactured parts require two different kinds of materials. It is planned to purchase material in monthly quantities. In addition the cycle time for manufacturing each part is 1.00 minute per piece and the parts are manufactured simultaneously, each on a separate machine. A pallet load of the raw material will make 1000 pieces. After the material is processed through the machine it is much bulkier and a pallet load will contain only 12 parts. After assembly, which requires a cycle of 0.90 minute per piece, the carton contains one completed assembly, and the skidded unit will contain 12 assemblies.

In addition to the two purchased parts, cartons, tape, banding, and two other miscellaneous items are required. Each of these items is purchased monthly on the average. The schedule is for 15,000 per month; the average material handling work element has been developed by industrial engineering and determined to be 0.10 hour per pallet load. The mechanical repair has historically been at the rate of 1 hour for each 40 hours of machine load hours.

This kind of information is obviously available in your material requirements planning (MRP) system. The kinds of information are as follows:

Purchases required are

2 raw material items

2 purchased items

1 carton

1 tape item

1 banding item

2 miscellaneous items

Total of nine items are to be purchased. If the average purchase order requires 30 minutes to prepare, we have just forecasted 4.5 hours of purchasing time for this product.

In addition, we can calculate the direct machine hours required by multiplying the 15,000 unit production forecast (schedule) by 1.00 minute per piece by 2 pieces required per assembly and dividing by 60 minutes per hour. This calculates to 500 hours at 100 percent effectiveness. If these machines operate at 90 percent, the 500 hours is divided by 0.90, which equates to 556 hours for the machine department.

This 500 hours can then be used to calculate the mechanical repair time required. This calculation is 500 hours multiplied by 0.025 hour per machine hour (1 hour repair per 40 machine hours), variable mechanical maintenance factor, equating to 12.5 hours for the mechanic. The variable factor was calculated using the data that determined that 1 hour of mechanical maintenance was required for each 40 hours of machine run time; that is, 1 divided by 40.

Material handling time required can be calculated by determining the number of pallet loads to be moved and multiplying by 0.1 hour per pallet. The following loads of material are required per assembly:

Raw material = 15,000 monthly production × 2 parts per assembly divided by 1000 pieces per pallet load equals 30 loads.

Fabricated parts = 15,000 times 2 per assembly divided by 12 per pallet equals 2500 loads.

Finished goods = 15,000 monthly production divided by 12 parts per pallet load equals 1250 loads.

Cartons = 15,000 production divided by (assume 100 per pallet) 100 cartons per pallet equals 150 pallet loads.

Miscellaneous = 5 percentage of total loads was determined to be the frequency of additional pallet loads.

Total loads are 30 + 2500 + 1250 + 150 + 0.05(3950) equals 4127 loads per month.

The material handling labor can then be forecasted by multiplying the 0.1 hour per load by 4127 loads indicated, equaling 413 hours. Notice that for 15,000 units of production 4127 pallet loads of material are required to be moved. This equals 0.275 pallet per unit of scheduled production. Since we have a standard of 0.1 hour per pallet, we can multiply by 0.1 hour by 0.275 pallet per unit of production.

At this point we should stop and consider what we already know about forecasting indirect labor:

Direct machine hours	556.0
Mechanical maintenance hours	13.9
Material handling hours	413.0
Purchasing hours	4.5

If we had not approached this forecasting problem by using the schedule and product information readily available, we would not have made these kinds of forecasts. Every function in business can be analyzed using this method.

SYSTEM DEVELOPMENT

The development of the system should follow an orderly sequence so that all necessary work is included and all unnecessary work is eliminated. Any reviewing of work will reveal better methods and unnecessary steps in the processes.

Development Process.   The steps listed should be followed:

List by department every job classification and function performed by that department.

Transfer these data to the work element format.

Determine what information is needed to complete this work element format.

Develop a plan for obtaining the missing information.

Insist that this information (especially the forecast indicators) be easy to determine.

Plan to use some form of computer for the actual calculations. Any spreadsheet will do.

You will by now have recognized that the above process is the same that one would use for setting direct labor standards and using those standards to develop labor forecasts. The only real differences is the size of the standard, and the indicator units are not usually pieces or whole goods finished.

DISADVANTAGES OF SHORT INTERVAL SCHEDULING FORECASTS

Short interval scheduling techniques can be used for developing frequencies and even estimated time values for each work element. Some companies have attempted to use short interval scheduling to forecast indirect labor. This method is okay if the schedule and effectiveness remain the same. If the schedule changes either up or down, we have the wrong number of indirect labor people. Short interval scheduling in an ever-changing product schedule environment and especially with a product mix change is a lot like driving by using the rear view mirror. If the curve is behind, we continue to curve and if the road behind is straight, we run off the road. Only on a constant scheduling condition will short interval scheduling be satisfactory.

We should use a short interval scheduling technique to determine frequencies to the forecast indicators that are most appropriate. Once we have defined the work elements and know how to obtain the forecast indicators, we are ready to develop standard times for each element.

We will be amazed at how many of these work elements can be measured using a system such as BASIC or Maxi-MOST. Some of these work elements will occur so infrequently that a good estimate will be appropriate. Others will require the simple taking of a cycle time, while others will require a MOST type standard time. We will discover that Pareto's law will apply and that 20 percent of our activities will cover 80 percent of our work.

METHODS IMPROVEMENT

Earlier it was mentioned that when listing our work elements, one of the requirements was to establish a "need code." The definition of the codes are as follows:

Want   This work element does not add to or take away from the business at hand.

Need   This work element adds to the business but in the short term can be delayed or eliminated.

Must   If this work is not accomplished we do harm to our business or facility.

When we begin to apply these simple definitions, several things happen. We first find some sacred tasks. We discover things we are doing and have no idea why we do them. The need for this work element has long been satisfied. We discover some elements that should be done more frequently and some that we should add. The net effect is a reduction in the work elements and a reduction in the frequency of work elements.

DIRECT LABOR EFFECTIVENESS

If we asked, "Why do we have indirect labor?" the response would be to support the direct function. We have maintenance to repair machines and equipment and we have warehousing people to store and retrieve materials and we would go on explaining the purpose of indirect labor. This kind of explanation is needed to develop our forecast indicators. In addition we soon discover that these reasons for indirect labor are why their functions should be timely.

Obviously the sooner a broken machine is repaired the more effective its operator will be. How many of us have been involved in an overkill to eliminate some downtime problem and have done so by adding extra maintenance personnel? This approach solved the downtime problem and created a maintenance cost problem when what was really needed was maintenance people available soon after a breakdown.

The question is how can we accomplish lower maintenance cost and still have high uptime? The answer is balance. We cannot hire or afford to hire enough people to eliminate all our downtime. A reasonable alternative would be to plan our work so that we do not have any emergency failures carry over and that our routine maintenance does not have a growing backlog or a schedule projecting too far into the distant future. If we can gather data that can be used to predict maintenance per machine run hour, we can use our machine load from the MRP system and calculate the hours required.

Using the work element data and the forecast indicators, we can predict how many of any classification we need by shift by department to support any schedule that is within the plant practical capacity. We are then in a position to support productive labor in a manner that will improve effectiveness without overspending. Forecasting of indirect labor by classification or skill will inform our training department or recruiting department specifically when and how many of a particular skill are needed. Many times we recognize that we needed a particular talent yesterday. This statement is not meant to be humorous. The truth is that the need for indirect labor is not readily apparent and as a result catches us by surprise. These surprises can be very expensive.

FORECASTING SUMMARY

Many businesses fail because of inability to hire or train skilled workers. Training requires time and expense. If we had the vision to recognize that our needs exceeded our ability to staff with workers, we would then know when to accept orders. We would know when the demand exceeded the supply of our product or services. We would know how to establish lead times and thus increase our credibility. Customers will not accept mistakes in delivery promises. If we produce an excellent quality product and back our product in the field, we do not have to worry about being the lowest price or the earliest delivery. It is far more important to deliver as promised and have the product perform properly. Forecasting indirect labor is one of several keys to success.

BIBLIOGRAPHY

Banks, Paul G., and Russell G. Heikes, *Handbook of Graphs and Tables for Industrial Engineers and Managers*, Reston Publishing, Reston, Va., 1984.

Bittel, Lester R., *Encyclopedia of Professional Management*, McGraw-Hill, New York, 1978.

Crockwell, D. C., "Intercorrelation and Multiple Regression," *Industrial Engineering*, January 1967.

Dhavale, Dipleep G., "Indirect Costs Take on Greater Importance; Require New Accounting Methods with CIM," *Industrial Engineering*, July 1988, pp. 41–43.

Hamlin, Jerry L., "Productivity Appraisal of Management Center," *Industrial Engineer,* September 1979, pp. 41–45.

Hoel, Paul G., and Raymond J. Jesson, *Basic Statistics for Business and Economics,* Wiley, New York, 1973.

Koop, John E., "Indirect Labor Incentives Pay Off," *Industrial Engineering,* February 1977, pp. 28–30.

Odiorne, George S., "Measuring the Unmeasurable: Setting Standards for Management Performance," *Business Horizons,* July–August 1987, pp. 69–75.

Salem, M. D., "Multiple Linear Regression Analysis for Work Measurement," *Journal of Industrial Engineering,* May 1967, p. 314.

Secor, H. W., "Regression Analysis—Standards Efficiency," *Journal of Industrial Engineering,* January 1966, p. 33.

Seifert, Deborah J., and Stan F. Settles, "Are Industrial Engineers Working on the Right Things in Manufacturing?" *Industrial Engineering,* February 1989, pp. 46–47.

Thelwel, D. R., "Linear Programming and Multiple Regression Analysis in Estimating Manpower Needs," *Journal of Industrial Engineering,* March 1967, p. 227.

Walton, Richard E., "Work Innovation in the United States," *Harvard Business Review,* July–August 1979, pp. 88–94.

Wilkinson, John J., "How to Manage Maintenance," *Harvard Business Review,* March–April 1968.

Williams, J. A., "Master Plan for Office Incentives," *Administrative Management,* April 1965.

CHAPTER 6
MEASUREMENT OF MAINTENANCE ACTIVITIES

Thomas A. Westerkamp
Contract Consultant
H. B. Maynard and Company, Inc.
Pittsburgh, Pennsylvania

This chapter describes the measurement of time to perform maintenance work. Measurement establishes the relationship between work content of a maintenance job and the time required to perform the job. Also described are methods of dividing a job into its components, applying incremental times, and totaling these times to calculate the complete task time. The unique aspects of measurement to pay maintenance workers according to output are discussed as well as auditing the measurement system. Planning and scheduling maintenance work is discussed in Sec. 10, Chap. 6 of this handbook.

DEFINITION OF MEASUREMENT

Measurement of maintenance activities is the process of establishing a standard method and time, or goal, which define the quantity and quality of work output expected from a given labor or machine time duration input. Two examples of work standards developed for maintenance activities are:

> EXAMPLE 1: Remove and replace rotating element, horizontal split case centrifugal pump, size 6 × 8, one millwright, 3.7 standard hours.
> EXAMPLE 2: Remove and replace single-pole single-throw electric light switch, one electrician, 0.5 standard hour.

Nine Ways Measurement Can Be Usefully Applied. Work measurement can be applied in any situation where maintenance planning, scheduling, and control are required. Control cannot be exercised without measurement of some kind, either a rough mental estimate or a carefully developed work standard determined by work measurement. Careful measurement provides the opportunity to achieve a very high degree of control over the maintenance activities.

The following are some of the major uses of work measurement:

1. Planning and scheduling maintenance activities

2. Measuring and evaluating supervisory and worker performance
3. Establishing average costs and estimating project time and cost
4. Preparing the annual operating budget
5. Comparison with other periods and other methods
6. Indicating trends such as performance, coverage, delays, and cost
7. Wage incentive plans that pay according to output
8. Measuring workload backlog by craft
9. Construction project planning and scheduling

Importance of Proper Methods. A method is a procedure followed in performing a task. A maintenance method is a specific combination of layout and working conditions; materials, equipment, and tools; and a standardized motion pattern followed to accomplish a maintenance task. Before measurement begins, proper standard methods and procedures are developed and employees are trained to follow them. This is an integral part of work standards development, and workers should be continually encouraged to follow the established method.

Quality and Quantity Limits of Acceptability. While quality and quantity are important aspects of measurement and control, application to maintenance may be somewhat subjective. Some examples are: How high can the grass grow before cutting? How much can the paint chip around exterior trim? Technical factors may be involved—How much tamping is required when backfilling an excavation? Management policy regarding these matters determines frequency and therefore has a significant influence on the annual time and cost to perform these maintenance activities.

MEASUREMENT TECHNIQUES AND THEIR USES

A variety of measurement techniques have been used with varying effectiveness and can be applied to measuring and gauging performance of maintenance labor commonly used in industrial, commercial, or government operations. The introduction of a new maintenance measurement system in an organization is a highly complex task and is often left to professionals. Training in-house managers and technical staff to expand, operate, and maintain the system is included as a part of the program. Measurement techniques determine the time it should take an average worker to do a given job under average conditions. A choice of which technique to employ is a matter of accuracy needed and how much you can afford to pay for this type of information versus the payback benefits. The higher the accuracy and the broader the coverage required, the higher the cost and the greater the return on investment will be. Often either accuracy or coverage alone is erroneously used as the sole basis for a work measurement program. Many times different methods are used in combination to arrive at the most cost effective program. The common measurement techniques include comparisons, historical records, estimates, time study, work sampling, predetermined time systems, standard data, and statistical techniques.

Comparisons. Comparisons are used to compare work in one area with the same work performed in another area, for example, a plant at another location. An estimate has already been established so that you can use that established time for your own operation.

Historical Records. Historical records are documentation of past experiences. The labor hours required to produce a given amount of work output are recorded in equipment history records and used later when the same repair is performed.

Estimates.   The use of estimates assumes that a qualified estimator can determine a reasonable approximation of the time it will actually take to do a job. Estimates may be combinations of in-house comparison and history or published estimating manuals such as those used frequently in construction projects.

All three of the above methods are based on actual times and tend to perpetuate delays previously experienced. These methods are most frequently used in maintenance because they are fast; however, they often produce inaccuracy and inconsistency. See Sec. 9, Chap. 4 for further details.

Time Study.   A time study is performed by observing and timing a large enough number of cycles of both manual and process time for an operation to calculate a correct average. The pace of the operator is judged and the average time is adjusted up or down depending on whether the operator's pace was faster or slower than normal. Time study has been applied to maintenance activity but has generally been too costly and is very difficult to maintain because of the volume of data required to cover all of the maintenance work. See Sec. 4, Chap. 2 for further details.

Work Sampling.   Work sampling is a measurement technique performed by collecting a series of instantaneous observations of each member in a group of maintenance workers. The observations are classified into activities such as direct work, getting instructions, transport material, and delays. The technique is used to establish delay allowances for maintenance work and as a quick method to determine the group's performance. This technique is used at temporary work sites such as during some refinery turnarounds. The job duration is too short to establish work standards. Measurement can begin as soon as the shutdown begins without having to develop standards first. The disadvantage is that observations must continue as long as measurement of performance is required. A number of observers is usually assigned full-time throughout the project as required according to the size of the facility and observation frequency desired. See Sec. 4, Chap. 3 for further details.

Predetermined Time Standards.   Work standards can be developed using predetermined time values that establish the normal time required to perform the basic manual motions and process cycles for all maintenance work. Techniques such as MOST®,* MTM, Work Factor, and MODAPTS, described in other chapters of this handbook, are used. Computer applications such as MOST Computer Systems and 4M are used to automate the process of establishing maintenance standards using a database of predetermined times. See Sec. 4, Chap. 4 for further details.

Statistical Approaches.   Statistical approaches rely on modern mathematical techniques such as slotting, Monte Carlo sampling, linear regression, or computer simulation. These approaches use a sample of known statistical accuracy to approximate the results achieved by direct measurement. Slotting is used in combination with predetermined times to produce a system that is fast and accurate when applied to long-cycle, low-volume work. Slotting is described in detail below.

DEVELOPMENT OF UNIVERSAL MAINTENANCE STANDARDS

Since the 1950s, the universal maintenance standard (UMS) technique has been gaining in popularity. Today it is the most frequently used and most effective method of establishing and maintaining engineering standards for maintenance work.

---

*MOST is a registered trademark of H. B. Maynard and Company, Inc.

Standard Data Approach.   The UMS technique is a predetermined standard data approach using the principles of work content comparison and range of time to apply work measurement to long-cycle, low-volume work.

Work Content Comparison.   If two jobs require nearly identical methods, the time to do one of them can be applied to both. For example, replacing a wall-mounted light switch requires nearly the same work content as replacing a two-plug wall receptacle. The time standard for the light switch can be used for the receptacle job and any other similar job. Thus all new work can be compared with a few benchmark jobs, representing various typical jobs sorted by craft. Using this principle, a few precise standards can be applied accurately and economically to a large variety and volume of work.

Range of Time.   The range of time principle recognizes that maintenance is much more variable than most production work. Therefore, rather than exact time standards, a range of time is used to specify how long a job should take. For example, valve replacement may involve removing rusted or painted-over fasteners. Depending on their condition, more or less time is required. Rather than saying the standard is 2 hours and 45 minutes, the UMS approach uses a range of time—say 2½ to 3½ hours. The mean of the range, or 3 hours, is used as the standard or planned time for the job. Over a period of time the pluses and minuses average out so that statistically, reasonable accuracy for planning, scheduling, performance measurement, and even incentive payment is achieved.

Standard Data Format.   The UMS approach described may be based on thc MOST system, developed in the 1960s by Kjell B. Zandin; the methods time measurement (MTM) predetermined time system, developed by Maynard, Stegemerten, and Schwab and published in the 1940s; or other predetermined time systems. A typical standard data arrangement may be viewed as a series of building blocks in the shape of a pyramid with MOST or MTM as its base (see Fig. 6.1). Illustrations of typical tables and backup used are

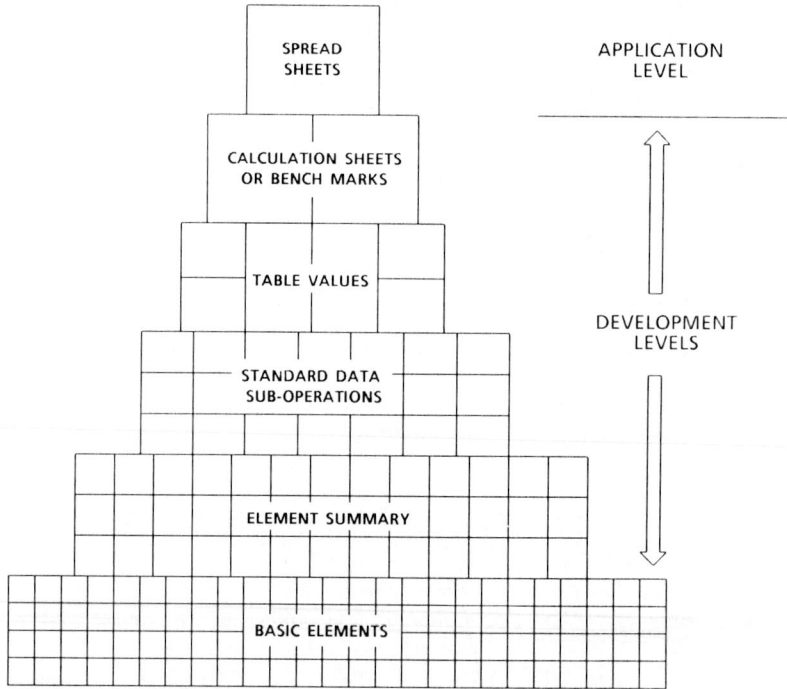

FIGURE 6.1   UMS building blocks for development of standard methods and times, and application to maintenance work orders.

Wrench Data				Code 0006.02	
TYPE OF FASTENER	**TOOL**	**OPERATION**	**SIZE**	**SYMBOL**	**HOURS**
BOLTS AND NUTS	ADJUSTABLE AND FIXED	ASSEMBLE OR REMOVE	Up to 3/8" diameter	(01)	.0061
			Over 3/8" to 5/8" diameter	02	.0085
			Over 5/8" to 1" diameter	03	.0121
			Over 1" to 1 1/2" diameter	04	.0232
			Over 1 1/2" diameter	05	.0293
		LOOSEN OR TIGHTEN SEVERAL THREADS	Up to 5/8" diameter	06	.0043
			Over 5/8" to 1" diameter	07	.0052
			Over 1" to 1 1/2" diameter	08	.0072
			Over 1 1/2" diameter	09	.0098
	RATCHET	ASSEMBLE OR REMOVE	Up to 3/8" diameter	10	.0050
			Over 3/8" to 5/8" diameter	11	.0072
			Over 5/8" to 1" diameter	12	.0103
			Over 1" diameter	13	.0185
* LAG BOLTS	ADJUSTABLE AND FIXED		Up to 2" length	14	.0192
			Over 2" length	15	.0313
ADJUSTING SCREW		TURN ONE REVOLUTION	Up to 1" diameter	16	.0017

Holes pre-drilled to a depth of about 3/4 to 7/8 of the bolt length

The above values cover the use of the wrench -- they do not include tool or part handling.

OPERATIONS SYNTHESIS				Code 0006.02		
SYM	REF	OPERATION OR ELEMENT DESCRIPTION		TMU	FREQ	TOTAL
(01)		Assemble or remove nut or bolt up to 3/8" diameter and 10 threads -- fixed or adjustable wrench				(.0061)
	A	Start nut or bolt with fingers -- finger motion				57.3
	C	Turn nut or bolt one revolution with fingers-- finger motion		18.8	8	150.4
	F	Turn nut or bolt one revolution with small wrench -- 5 to 6" long		(146.7)	2	293.4
	M	Tighten or loosen with small wrench				81.1
	R	Adjust opening of adjustable wrench (pro-rated)		75.8	1/3	25.3
						607.5

ELEMENT ANALYSIS CHART					Code 0006.02		
DESCRIPTION - LEFT HAND	NO	LH	TMU	RH	NO	DESCRIPTION - RIGHT HAND	
(F)Turn nut or bolt one revolution with small wrench about 5" to 6" long and a 120° turn (4" radius)			35.4	M8C	3	Relocate and position wrench	
			59.1	P2SSE	3		
			40.2	M8B5	3	Turn nut	
			12.0	D1E	3	Off nut	
			(146.7)				

FIGURE 6.2 Development of wrench data from basic MTM.

shown in Fig. 6.2. It illustrates how basic wrench data are developed from basic elements (MTM). A work sheet for calculated welding times is shown in Fig. 6.3. These data were developed with MTM and time study.

		OPERATION DESCRIPTION	SYM OR TABLE	HRS. PER OCCUR.	SET-UP FREQ.	SET-UP HRS.	EACH PIECE FREQ.	EACH PIECE HRS.
		SUB-TOTAL from page 1 Set-up & Each Pc. Time	-	-	-	-	-	-
15	Use of Positioner	Tilt Positioner per 45°	K-9	.0083				
		Rotate Positioner per 45°	K-10	.0180				
		Elevate Positioner (2-6')	K-11	.0289				
16		GET OR ASIDE ASSEMBLY	-	-	-	-	-	-
	Manual 0-50#	Within Work Area K32 .0017 Outside Work Area K33 .0062						
	Mechanical Over 50#	Jib-51-2000# K234 .0263 * Cab over 2000# K235 .0177						
		Floor 101-3000# K237 .0409 * Cab over 2000# K236 *.0354						
17		TURN ASSEMBLY OVER	-	-	-	-	-	-
	Manual	0-1--# K34- .0023 | 101-200# K207- .0059						
	Mechanical	Jib-101-2000# K238- .0221 Cab over 2000# K239-.0168						
		*Cab over 2000# K241- *.0336 Cab-Dble Flop K240- *.0672						
		FLOOR OPERATED BRIDGE CRANE 101-3000#	K-242	.0281				

COMMENTS:

Set-Up Time Sub-Total _____
Allowance _____ %
Total Set-Up Time Allowed

Each Piece Time _____
Allowance _____ %

Total Allowed Time Per Piece

FIGURE 6.3 Metal arc, automatic, and semiautomatic weld work sheet.

Standard Work Groupings Using Range of Time.   Because of the very nature of maintenance and construction work, it is impractical to strive for accurate individual standards. Workers do not do a given job with exactly the same motion patterns. The varying conditions of work and worker experience prevent this. It would be impractical to attempt to specify the exact method, motion by motion, as in repetitive work. Instead, a representative method must be used, allowing the worker some leeway in the actual method, which might vary slightly owing to working conditions.

The variables dictate the need for a practical way to measure nonrepetitive work. Instead of trying to set standards with pinpoint accuracy, they are set based on a range of time during which a given job will be done by a qualified worker. A job will be done in 1.5 to 2.5 hours, or 8.0 to 10.0 hours, with assurance that it will be correct 95 times out of 100. Jobs slotted into standard work groups in this manner are called benchmark jobs.

The set of standard work groups appears in Fig. 6.4. The columns from left to right show the slot or work group (A, B, C, etc.), the lower limit of the time range of the slot, the standard time to be applied for work falling in this slot, and the high limit of the slot time range. The time range is interpreted to read "up to and including" the top limit of the range.

Benchmarks.   Benchmarks or typical maintenance jobs are developed on the benchmark analysis sheet (see Fig. 6.5). They are developed using basic operations data, craft operations, and previously developed benchmarks. A sketch (see Fig. 6.6) is often necessary to clarify a benchmark. It is attached as the last page of the benchmark.

Developing Spreadsheets.   Spreadsheets as shown in Fig. 6.7 are used to classify benchmarks by type of work and range of time. Specific types of work are brakes, lighting, mo-

SLOTTING TIME RANGES FOR SPREAD SHEET

SLOT	HOURS		
	MIN.	STD.	MAX.
A	0.0	0.1	0.15
B	.15	0.2	0.25
C	.25	0.4	0.5
D	.50	0.7	0.9
E	.9	1.2	1.5
F	1.5	2.0	2.5
G	2.5	3.0	3.5
H	3.5	4.0	4.5
I	4.5	5.0	5.5
J	5.5	6.0	6.5
K	6.5	7.3	8.0
L	8.0	9.0	10.0
M	10.0	11.0	12.0
N	12.0	13.0	14.0
O	14.0	15.0	16.0
P	16.0	17.0	18.0
Q	18.0	19.0	20.0
R	20.0	22.0	24.0
S	24.0	26.0	28.0
T	28.0	30.0	32.0
U	32.0	34.0	36.0
V	36.0	38.0	40.0
W	40.0	42.0	44.0
X	44.0	46.0	48.0
Y	48.0	50.0	52.0
Z	52.0	54.0	56.0
AA	56.0	58.0	60.0
BB	60.0	63.0	66.0
CC	66.0	69.0	72.0
DD	72.0	75.0	78.0
EE	78.0	81.0	84.0
FF	84.0	87.0	90.0

FIGURE 6.4 Standard work groupings.

tors, and the like, or specific types of equipment such as sewing machines, scales, or cranes. The specific types of work or equipment are classified as the task area on the spreadsheet form.

The spreadsheet provides an easy-to-use format for setting standards. This is done by comparing the job content of a work order with the job content on the spreadsheet. When this comparison is made, the work order is slotted according to the range of time on the spreadsheet. Greater detail on setting standards is covered in the next section of this chapter.

Ideally there are three benchmarks for each standard work grouping in a task area. When the maximum exceeds five, over-development or the need for another task area may be indicated. Related benchmarks are arranged horizontally as illustrated in Fig. 6.7 so that those with expanding work content stand out and may be referenced quickly. Spreadsheets are bound together in separate craft binders and tab indexed by task area for quick retrieval.

Data Coding. The universal maintenance standards (UMS) coding system consists of eight digits broken down into four series of two digits each. A basic operations data code example is shown in Fig. 6.8.

If this same code number were written

$$0006.0103$$

it would mean

Assemble or remove a 3/8- to 1/2-inch machine screw using a regular or Phillips head screwdriver. The time value of 0.0067 hour is shown in the basic operations data manual.

Other basic operations data codes are listed in Fig. 6.9. An example of the coding system for craft operations data is shown in Fig. 6.10.

This is more clearly shown through examples. Let's take an actual craft code number and go through it. Code 1850.1009 in Fig. 6.10 shows the relationship between the code number and operation description.

APPLICATION OF UNIVERSAL MAINTENANCE STANDARDS

The application of universal maintenance standards is more involved than the act of setting a single standard. Approximately 80 percent of the maintenance work orders repre-

Description	Air cylinder, 18" diameter, 12 – 3/4" bolts:		Date:6/5/xx		B M#0290–101
	Replace gasket		Craft: Mechanical		
			Dwgs: 3421–02		
		No. of Men 1	Analyst: GTS	Sh. 1	of 3

Line	Men	Operation Description	Reference Symbol	Unit Time	Freq.	Total Time
1		Remove cover				(.1868)
2		Get and aside wrench	05.0003			25
3		Loosen and remove 12 – 3/4" nuts	06.0203	121	12	1452
4		Aside 12 nuts	04.0103	17	12	204
5		Remove 12 bolts	12.0001	4	12	48
6		Aside 12 bolts, 3 at a time	04.0103	17	4	68
7		Get and aside screwdriver	05.0003			25
8		Pry cover loose	06.1402			22
9		Remove cover	12.0004			7
10		Aside cover	04.0103			17
11		Remove gasket, clean cylinder and cover, replace gasket				(.1189)
12		Pry off gasket or portion thereof	06.1402	22	4	88
13		Aside gasket	04.0103			17
14		Get emery paper and place on block	10.0002			84
15		Rough clean approximately 60 square inches	10.0004	6	60	360
16		Wipe clean same area	09.0003			61
17		Wipe inside of cylinder, approximately 2 sq. ft.	09.0004			26
18			09.0005			12
19		Get gasket	04.0103			17
20		Place gasket in position	12.0003			10
21		Body displacement to cover on floor	03.0103			9
22		Clean cover gasket surface, lines 14, 15, and 16	---			505
23		Assemble cover				(.1874)
24		Move cover to cylinder	04.0103			17
25		Place cover on cylinder	12.0006	17	2	34
26		Get and assemble 12 bolts, lines 5 and 6				116
27		Obtain 12 nuts and place on top of cover	04.0103			17
28			04.0101	3	11	33
29		Get and aside wrench	05.0003			25
30		Get 12 nuts	04.0101	3	12	36

Notes		Bench Mark Time	.4931
See sketch		Standard Work Group	C

FIGURE 6.5 Benchmark analysis sheet—describes a typical maintenance job.

sent only about 20 percent of the total maintenance workload. Therefore, it is important that the bulk of the paperwork is quickly processed so that more time can be spent on the larger jobs. For planning purposes, a large job becomes a series of smaller jobs. More time is spent to analyze and plan the sequence of steps which leads to better communication. In applying UMS times, the planner must be thoroughly familiar with the maintenance work and maintenance procedures.

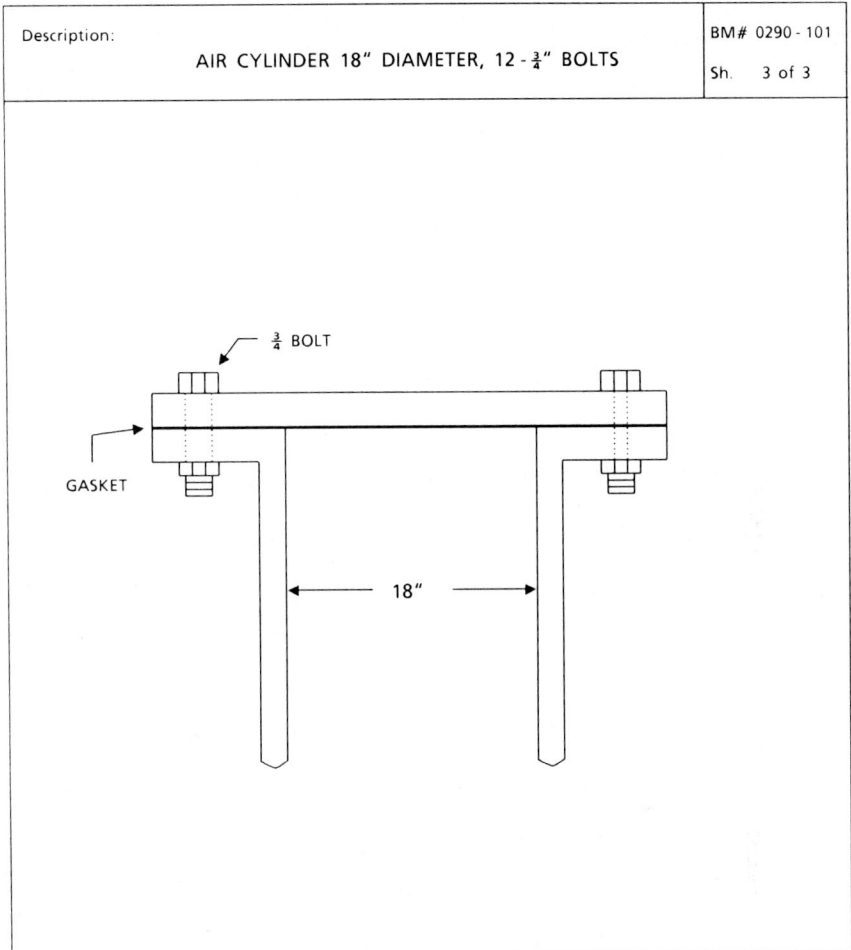

FIGURE 6.6 Benchmark sketch sheet—accompanies a benchmark to clarify it.

Three Ways to Determine Job Work Content.   The chief role of the planner is to determine work content and time. In order to do this, the planner determines the scope of work for each work order. This is accomplished in several ways:

1. The planner reviews the information from the requester and visualizes the full scope of the work required, assuming that the initial information provided by the requester is adequate.

2. The planner consults with production supervision, the maintenance supervisor, requester, engineering, or other planners when additional information is required.

3. Visits to the job site may be made to supplement existing information. These visits can occur prior to the start of a job, during the job, or both.

Determining the Operations.   After a clear picture of a job is achieved, the total requirements are divided into operations. These are logical blocks of work that can normally be started and completed without a break. On small jobs, one or two tasks will often handle the total work requirements. However, on large repair or project work, the overall job

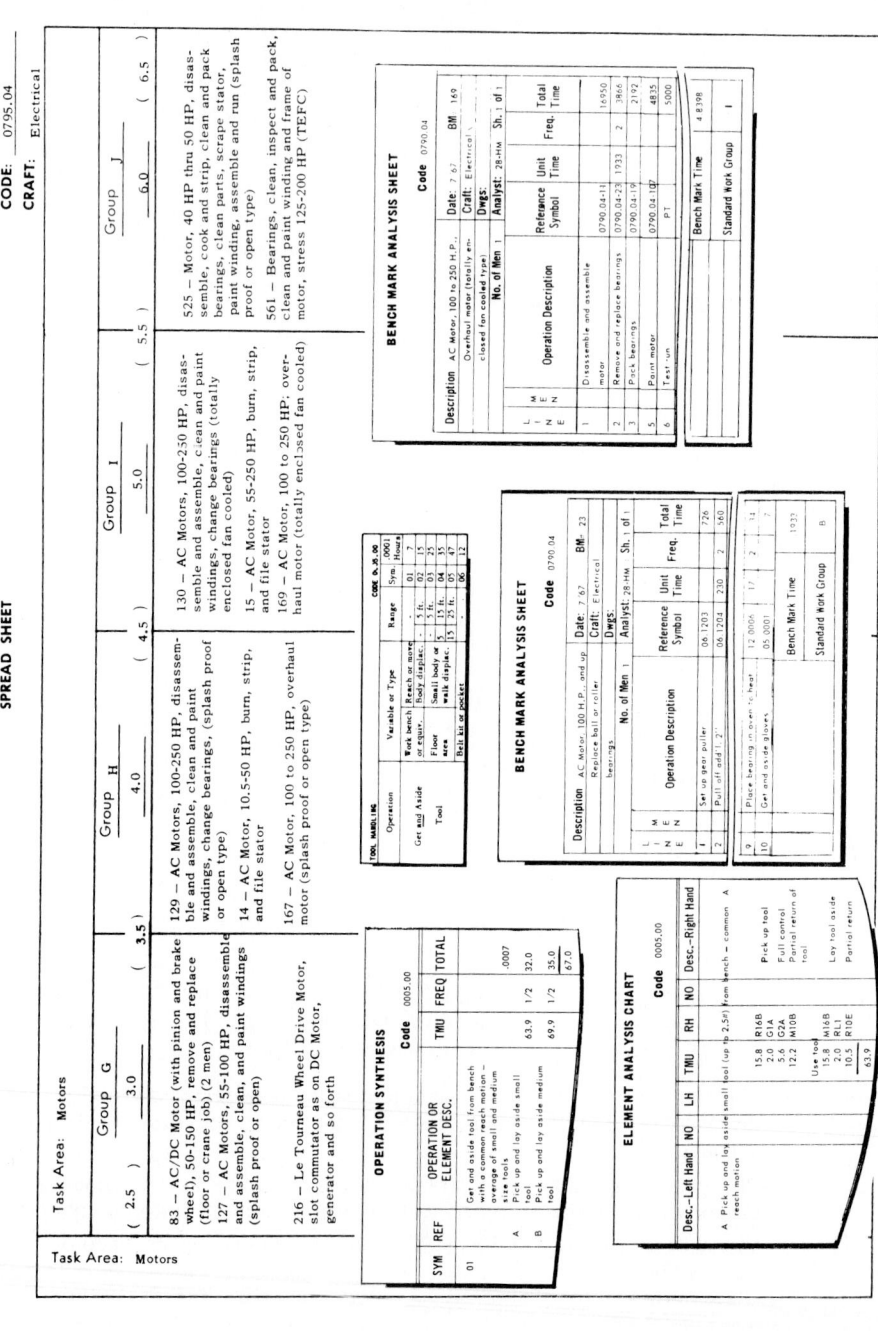

FIGURE 6.7 Spreadsheet for the motor task area.

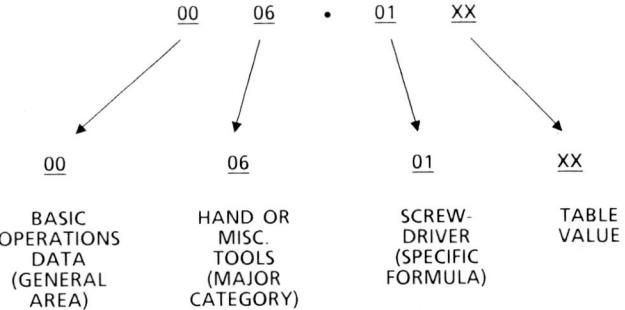

FIGURE 6.8 Basic operation data coding.

BASIC OPERATION	CODE
JOB PREPARATION	0001.00
AREA TRAVEL	0002.00
COMMON BODY DISPLACEMENTS	0003.01
LADDER DATA	0003.02
PART HANDLING	0004.01
OBTAIN PARTS--INTRA-STOREOOM	0004.02
TOOL HANDLING	0005.00
SCREWDRIVER DATA	0006.01
WRENCH DATA	0006.02
HAMMER DATA	0006.03
PLIER DATA	0006.04
PIPE WRENCH DATA	0006.05
STAR DRILL DATA (SEE 0006.03)	0006.06
TAP AND DIE DATA	0006.07
TIN SNIP DATA	0006.08
BOLT CUTTER DATA	0006.09
REAMER DATA	0006.10
HACKSAW DATA	0006.11
GEAR PULLER DATA	0006.12
FILE DATA	0006.13
PRY BAR DATA	0006.14
VISE AND CLAMP DATA	0006.15
HAND JACK DATA	0006.16
ELECTRIC DRILL	0007.01
ELECTRIC HAMMER	0007.02
IMPACT WRENCH	0007.03
MEASURING AND GAUGING	0008.01
CLEAN PARTS	0009.00
EMERY CLEANING AND POLISHING	0010.00
RIGGING--HAND LINE, BLOCK AND TACKLE, AND ELECTRIC AND CHAIN HOISTS	0011.01
ASSEMBLE AND DISASSEMBLE	0012.00
INSPECT AND ADJUST	0013.00
LUBRICATE	0014.00
COMPLETE FASTENER DATA	0015.00

FIGURE 6.9 Basic operations data and codes.

content is of such a magnitude that it is divided into a larger number of tasks. In effect, a sequence of suboperations is developed and each is handled like a separate job. The procedure eliminates extended work assignments, which tend to "drag out" and are more difficult to control.

Also, it is easier to establish job times when a large job is divided into smaller jobs. It is much easier to visualize the work content, making the large job easier to evaluate.

There will be occasions when it is impossible to determine the necessary operations. This can occur on some emergency or troubleshooting jobs. The overall work content may be known in a general way, but the specific operations cannot be determined until there

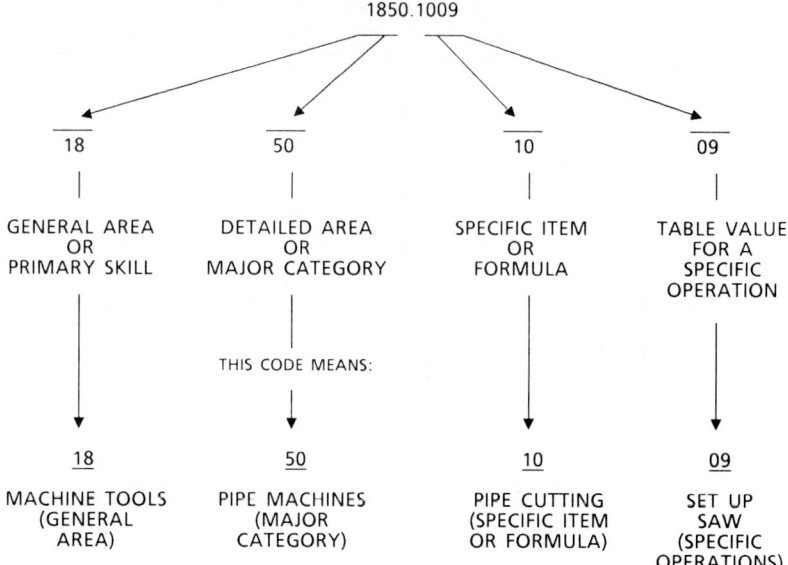

FIGURE 6.10   Example of an actual craft code for sawing in the machine tool data.

has been a partial dismantling and inspection of the equipment. When this occurs, it is feasible to plan the operations as the job progresses.

When a large job is segmented into tasks, use a benchmark analysis sheet. This represents a summary list of suboperations, and it can be backed up with other benchmark analysis sheets to show the work content of an operation. Then you have the complete job record which can be used as a guide to analyze other similar jobs.

Mechanics of Establishing Universal Maintenance Standard (UMS) Times.   The mechanics of establishing a UMS time can be divided into two steps. The first step involves the determination of the work group time to complete an operation or task. It is the leveled time to do a job at the job site. The second step involves the application of job preparation, area travel, and allowance. The sum of the two steps is the UMS (planned) time for the job.

Three Methods for Determining the Work Group.   The mechanics of determining the work group, or leveled time, for specific operations will be accomplished by using one or more of three methods.

1. Use the prepared spreadsheets and the work content comparison concept. Instead of attempting to set standards of pinpoint accuracy, standards are set that are based on a range of time in which a given job will probably be done by a qualified worker working at a normal pace. A given job is compared with the benchmark jobs on spreadsheets. Time is not compared. The work content of the job must be considered on an overall basis. This technique is known as "slotting" the job. An important point to remember is that it is not necessary to select a specific benchmark for comparison, but rather that the planner have an overall "grasp" of the work content of all the benchmarks in a standard work group. For example, a job is classified as a C job, not necessarily because of any specific benchmark in standard work group C, but because of the work content of the benchmarks in the two adjoining work groups, B and D. In effect, the job has been slotted C because of the work content of all the benchmarks in groups B, C, and D. If this principle is not adhered to, approximately five to ten times as many benchmarks will be required. On those installations where coverage progressed poorly, it was the result of either a breakdown in communications (poor job descriptions) or the fact that the planner was limiting slotting to a specific benchmark that, for all practical purposes, was almost identical to the job at hand. The planner's reaction to poor coverage is "I don't have enough benchmarks." Therefore, it is extremely important to continually apply the slotting technique.

This slotting approach is applied to all routing work and, to some degree, to project work. This is by far the most important method of establishing UMS times, and it will handle the bulk of work orders.

2. Use some of the craft formulas. For example, in painting or laying block or brick, these data are normally set up on a per square foot basis. In these and other comparable cases, it is easier to apply the data directly, once the job content has been determined. On the other hand, some work is not as simple to evaluate as painting or laying block. For example, the layout and installation of piping or wiring systems can become quite involved with different sets of data. In the early stages of an installation, these jobs are usually evaluated on a benchmark analysis sheet.

This is necessary because in preparing benchmarks of this type for an installation, it is usually difficult to obtain a sufficient number to cover the broad scope of work. This type of benchmark work continues and expands as a UMS program installation progresses. As more and more benchmarks of this type are developed, it can be expected that a limited number of them will eventually be set up on spreadsheets and used in whole or in part for slotting future work.

When the benchmark analysis sheet is used as a calculation sheet, as in any of the above examples, the time calculated for these jobs is slotted in the same manner as any benchmark development. That is, the time calculated is the "benchmark time," and this, in turn, is designated by its proper "work group."

This method will also be used in conjunction with the first method, the slotting procedure. When evaluating a job with craft operations data, it is permissible to include, as part of the analysis, another benchmark. When this situation occurs, use the calculated time value from the other benchmark, not the slot time.

3. Use an exact time to establish UMS for repetitive work. In spite of the general nonrepetitive nature of maintenance work, there will be service jobs and other maintenance jobs that can be classified as highly repetitive. For example, a repair worker in one instance spent approximately 80 percent of the job time repairing nut runners.

An analysis of this job reveals that it would be practical to establish a single time for each runner repaired, even though the work content of each nut runner varied. The analysis also reveals that work group C (0.4 hour) was too tight and work group D (0.7 hour) was too loose. The value established for this activity is 0.5 hour per gun.

Job Planning Using the Job Register.   Control of work orders by using the job planning register (Fig. 6.11) is described later in this section. The register also provides information to planners so that they can evaluate the improvement in their planning activity. On the right-hand side of the job planning register are columns labeled "UMS" and "Non-UMS." Under "UMS" the planner marks the data reference number used to plan each job. Under "Non-UMS" the planner marks a check under the reason for not planning each unplanned job. Analysis of this information will identify problems such as need for more benchmark data (no data) or too many benchmark data (too many exact benchmarks used); or if a large number of jobs are not planned because the work content is not known (no description), steps can be taken to improve the work content descriptions.

Using the Standard Hour Calculation Sheet.   In order to minimize time to calculate standards, a standards calculation sheet is used to apply four categories of time.

- Area travel time
- Job preparation (and cleanup time)
- Job site time
- Personal, rest, and delay time

JOB PLANNING REGISTER   AREA   DATE

WORK ORDER	PRE-PRINTED NUMBER	DESCRIPTION	BLDG.	CRAFT	UMS HOURS	UMS		NON-UMS			DATE COMP
						EXAC BENCHMARK	WORK CONT. COMPARISON	NO DATA	NO DESC.	NO TIME	

FIGURE 6.11 Job planning register.

During the data development, the planner team validates and/or develops standard data time and benchmarks for the first three categories—area travel, job preparation, and job site time—and management establishes policies for proper personal, rest, and delay allowance. This information is used to prepare the standard hour calculation sheet shown in Fig. 6.12. The following steps are taken to develop the total planned time for a maintenance work order:

1. Study the job description on the work order. If the work content is not clear, field check the job site, talk with supervision and/or engineering, and make sure agreement exists on the work involved.

2. Determine the job site time for this job by using the spreadsheets and the work content comparison technique.

3. Select the appropriate area travel and job preparation values for this job site location and work content.

4. From the standards calculation sheet, select the allowed hours for this combination of crew size, area travel, job preparation, job site time, and personal, rest, and delay allowance.

5. Mark the time in the "Planned Hours" block on the work order form and forward the job to the maintenance supervisor along with supporting drawings or sketches and the daily schedule.

How to Monitor Applicator Progress.   In order to measure the planners' improvement, each planner updates a learning curve weekly as shown in Fig. 6.13. The indicator used for this purpose is the ratio of hours spent applying UMS times to hours planned.

$$\text{Hours application time per UMS hour applied} = \frac{\text{hours application time}}{\text{UMS hours applied}}$$

Application time is recorded on the weekly activity report (Fig. 6.14). After sufficient experience is gained, a trained planner can apply, on the average, 20 to 30 UMS hours per hour of application time. In highly repetitive maintenance work, planners have achieved levels as high as 100 UMS hours planned for each hour spent planning.

Wage Incentives for Maintenance Work.   A virtually untapped area of potential productivity improvement and cost reduction, wage incentive plans for maintenance workers offer management an excellent motivational tool. When introduced with proper communication and training for all employees—higher management, supervision, and hourly workers—a wage incentive plan can provide better earnings, lower costs, and improved labor relations.

Principles.   The intent of an incentive plan is to provide maintenance crews with incentive earning opportunity based on work produced. The incentive plan should be designed to pay a guaranteed base rate for hours worked and a bonus for performance above standard.

In order to provide permanent benefit to both the worker and the company, the incentive plan must be based on sound principles. These principles are discussed in Sec. 6 of this handbook and are incorporated in all effective incentive plans.

Because of the difficulty in administering individual incentive plans for maintenance workers, the plans used in maintenance organizations are based on a small group incentive concept. Groups are structured along common work or craft lines, for example:

Electricians   Machine shop
Millwrights   Lube and preventive maintenance
Welders

STANDARD HOUR CALCULATION SHEET																								
TRAVEL & PREP. TIME			ALL BASIC SKILLS EXCEPT PAINTERS																				ADDITIONAL TIME REQUIRED PER MAN	
			A	B	C	D	E	F	G	H	I	J	K	L	M	N	O	P	Q	R	S	T		
ZONE	JOB PREP	MEN	0.1	0.2	0.4	0.7	1.2	2.0	3.0	4.0	5.0	6.0	7.3	9.0	11.0	13.0	15.0	17.0	19.0	22.0	26.0	30.0	TRAVEL	PREP.
IN SHOP	SIMPLE	1	.2	.3	.6	.9	1.5	2.5	3.7	4.9	6.1	7.3	8.9	10.1	13.3	15.7	18.1	20.5	22.9	26.5	31.3	36.1		.10
JOB NO	(.08 HRS.)	2	.3	.4	.7	1.0	1.6	2.6	3.8	5.0	6.2	7.4	9.0	11.0	13.4	15.8	18.2	20.6	23.0	26.6	31.4	36.2		
TRAVEL	AVERAGE	1	.3	.4	.7	1.0	1.6	2.6	3.8	5.0	6.2	7.4	9.0	11.0	13.4	15.8	18.2	20.6	23.0	26.6	31.4	36.2	NONE	.20
TIME	(.17 HRS.)	2	.5	.6	.9	1.2	1.9	2.8	4.0	5.2	6.4	7.6	9.2	11.2	13.6	16.0	18.4	20.8	23.2	26.8	31.6	36.4		
	COMPLEX	1	.5	.6	.9	1.2	1.8	2.8	4.0	5.2	6.4	7.6	9.2	11.2	13.6	16.0	18.4	20.8	23.2	26.8	31.6	36.4		
	(.33 HRS.)	2	.9	1.0	1.3	1.6	2.2	3.2	4.4	5.6	6.8	8.0	10.0	11.6	14.0	16.4	18.8	21.2	23.6	27.2	32.0	36.3		.40
1	SIMPLE	1	.3	.5	.7	1.1	1.7	2.6	3.8	5.0	6.2	7.4	9.0	11.0	13.4	15.8	18.2	20.6	23.0	26.6	31.0	36.2		.10
(.11 HRS.)	(.08 HRS.)	2	.6	.7	.9	1.3	1.9	2.9	4.0	5.3	6.5	7.7	9.2	11.3	13.7	16.1	18.5	20.9	23.3	26.9	31.7	36.5		
	AVERAGE	1	.5	.6	.8	1.2	1.8	2.7	3.9	5.1	6.3	7.5	9.1	11.1	13.5	15.9	18.3	20.7	23.1	26.7	31.5	36.3	.13 HRS.	.20
	(.17 HRS.)	2	.8	.9	1.1	1.5	2.1	3.1	4.3	5.5	6.7	7.9	9.4	11.5	13.9	16.3	18.7	21.1	23.5	27.1	31.9	36.7		
	COMPLEX	1	.6	.8	1.0	1.4	2.0	2.9	4.1	5.3	6.5	7.7	9.3	11.3	13.7	16.1	18.5	20.9	23.3	26.9	31.7	36.5		
	(.33 HRS.)	2	1.2	1.3	1.5	1.9	2.5	3.5	4.7	5.9	7.1	8.3	9.8	11.9	14.2	16.7	19.1	21.5	23.9	27.5	32.3	37.1		.40
2	SIMPLE	1	.5	.6	.9	1.2	1.8	2.8	4.0	5.2	6.4	7.6	9.1	11.2	13.6	16.0	18.4	20.8	23.2	26.8	31.6	36.4		.10
(.24 HRS.)	(.08 HRS.)	2	.9	1.0	1.2	1.6	2.2	3.2	4.4	5.6	6.8	8.0	9.5	11.6	14.0	16.4	18.8	21.2	23.6	27.2	32.0	36.9		
	AVERAGE	1	.6	.7	1.0	1.3	1.9	2.9	4.1	5.3	6.5	7.7	9.3	11.3	13.7	16.1	16.5	20.9	23.3	26.9	31.7	36.5	.25 HRS.	.20
	(.17 HRS.)	2	1.1	1.2	1.5	1.8	2.4	3.4	4.6	5.8	7.0	8.2	9.7	11.8	14.2	16.6	19.0	21.4	23.8	27.4	32.2	37.0		
	COMPLEX	1	.8	.9	1.2	1.5	2.1	3.1	4.3	5.5	6.7	7.9	9.4	11.5	13.9	16.3	18.7	21.1	23.5	27.1	31.9	36.7		
	(.33 HRS.)	2	1.5	1.6	1.8	2.2	2.8	3.7	5.0	6.1	7.3	8.5	10.1	12.2	14.5	16.9	19.3	21.7	24.1	27.7	32.5	37.3		.40
3	SIMPLE	1	.7	.8	1.0	1.4	2.0	3.0	4.1	5.3	6.5	7.7	9.3	11.3	13.7	16.1	18.5	21.0	23.3	27.0	31.7	36.5		.10
(.40 HRS.)	(.08 HRS.)	2	1.3	1.4	1.7	2.0	2.6	3.6	4.8	6.0	7.2	8.4	10.0	12.0	14.4	16.8	19.2	21.6	24.0	27.6	32.4	37.2		
	AVERAGE	1	.8	.9	1.2	1.5	2.1	3.1	4.3	5.5	6.7	7.9	9.5	11.5	13.9	16.3	18.7	21.1	23.5	27.1	31.9	36.7	.48 HRS.	.20
	(.17 HRS.)	2	1.5	1.6	1.9	2.2	2.8	3.8	5.0	6.2	7.4	8.6	10.2	12.2	14.6	17.0	19.4	21.8	24.2	27.8	32.6	37.4		
	COMPLEX	1	1.0	1.1	1.4	1.7	2.3	3.3	4.5	5.7	6.9	8.1	9.7	11.7	14.1	16.5	18.9	21.3	23.7	27.3	32.1	36.9		
	(.33 HRS.)	2	1.9	2.0	2.3	2.6	3.2	4.2	5.4	6.6	7.8	9.0	10.5	12.6	15.0	17.4	19.8	22.2	24.6	28.2	33.0	37.8		.40

a. Additional Travel & Preparation time includes personal, delay and rest time of 20%.
b. Reference to 0001.00 Basic Formula for Job Preparation.
c. Reference to 0002.00 Basic Formula for Area travel.
d. Table Values Are Determined: [Std. Work Group Man hours + (one man std. job preparation + one man area travel hours) × crew size] × 1.20 Personal, Rest and Delay Factor.
e. Additional Travel Time and/or Job Preparation as required for:
(1) Each crew member over two (2) Jobs spanning two or more work periods

FIGURE 6.12 Standard hour calculation sheet.

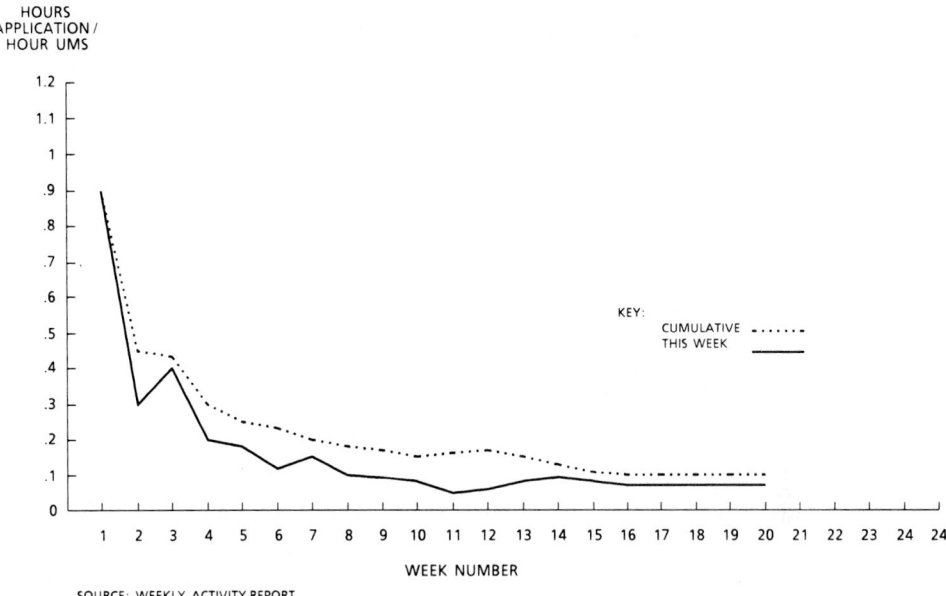

FIGURE 6.13   Planner learning curve.

Since it is generally better to keep the group size as small as practical to ena group members to better relate their own efforts to bonus earnings, craft groupings may be further broken down by shift.

Bonus calculations are made over a 1- or 2-week closeout period, and paid at regular pay intervals. This calculation period allows the statistical accuracy of maintenance standards to operate.

Coverage, the number of hours worked against a standard as a percent of total hours, should be no less than 85 percent, because drops in coverage result in a corresponding drop not only in output but also in incentive earnings opportunity. There will be pressure for average earnings payments if coverage goes too low. Average earnings payments are nothing more than payment of current incentive earnings for nonstandard work and will destroy an incentive plan, since payment is not tied to output.

There should be no "guarantees" of a specific coverage level; however, if the incentive plan is to operate and provide sufficient bonus opportunity for the workers, 85 percent should be regarded as a minimum acceptable level for each weekly time period. The planning goal should be 100 percent coverage on every job.

Changes in time reporting procedures are needed if there is no provision for delay or job time reporting. Individual maintenance workers will also be required to report any changes in job scope from that originally planned. They do this by describing additional work performed on their copy of the work order. The supervisor authorizes this additional work by initialing the work order, and the planner adjusts the original planned hours to reflect the additional work before the bonus calculation is made.

The maintenance standards used as a basis for determining bonus earnings are structured at the 100 percent performance level in a standard hour plan. That means that average, trained workers working with average skill and effort at a pace they can easily maintain all day can do a job in the time the standard allows. The standards also contain an allowance—set by management policy at 15 to 20 percent—to cover minor unavoidable delays, personal time, and rest. Using a 15 percent allowance, out of a 480-minute day, a total of 63 minutes [480 − (480/1.15)] is allowed for lunch, coffee breaks, minor delays, and just

WEEKLY ACTIVITY REPORT

NAME _____   WEEK ENDING _____

Section A
Program Project Team Distribution

PROJECT NUMBER	MON.	TUES.	WED.	THURS.	FRI.	SAT.	TOTAL FOR WEEK	ACCUM. TOTAL	
								PREV.	CURRENT
DEVELOP CRAFT DATA									
DEVELOP BENCH MARKS									
DEVELOP CHECKLIST MSR		2¼							
MEETINGS			1	1					
WRITE MSR'S	1		½	1½	1½				
POST JOB PLAN REGISTER	½	1							
SUMMARIZE BACKLOG	1								
PLAN MATERIAL AND EQUIPMENT	1	1½	1	2	2				
FIELD CHECK	1½	1½	½	1½	1		6		
PRE-APPLY UMS HRS.		½	4	1	2			} 8	
POST APPLY UMS HOURS	½				½		¾		
DAILY SCHEDULE	1½	1	1	1	1				
TOTAL	8	8	8	8	8		40		
UMS HRS. PLANNED	30.0	60.0	135	75.0	105		465		

Section B
Time Distribution for Non-Program Projects

WORK DESCRIPTION	MON.	TUES.	WED.	THURS.	FRI.	SAT.	TOTAL FOR WEEK	ACCUM. TOTAL	
								PREV.	CURRENT
TOTALS									

Summary of Time Distribution

TOTALS FROM SECTION "A" & "B"	TOTAL FOR WEEK	ACCUM. TOTAL	
		PREV.	CURRENT
SECTION "A"			
SECTION "B"			
GRAND TOTAL			

FIGURE 6.14 Planner's weekly activity report.

slowing down at the end of the day. It is not necessary, therefore, to "clock out" for these activities.

The incentive plan should not be installed until staffing has been adjusted to the current available workload.

COMPUTER-AIDED MAINTENANCE MEASUREMENT

Use of the computer has an important and fitting application in measuring maintenance activities. Because of the large volume of data which must be absorbed daily by the maintenance manager, the computer is put to work storing all the pieces of information in its memory and compiling and calculating summary information rapidly. Three applications described here are UMS standards, management control reports, and an expert systems approach to auditing the maintenance system.

Computer UMS for Maintenance Activities.   The entire process of maintaining a benchmark database and application of UMS has been programmed for rapid administration on the computer. The automated UMS system main menu is shown in Fig. 6.15.

F1 HELP	F2 SPREAD SHEETS		UNIVERSAL MAINTENANCE STANDARDS SYSTEM
F3 BENCH MARKS	F4 ESTIMATES	HELP	- Display help screen.
		SPREADSHEETS	- Retrieve and use a UMS spreadsheet. Set a standard.
F5 UTILITIES	F6	BENCHMARKS	- Maintain the benchmark library.
F7	F8	ESTIMATES	- Utilize the estimates file.
		UTILITIES	- UMS System Utilities.
F9	F10 EXIT	EXIT	- Exit the UMS System.

FIGURE 6.15   Automated UMS main menu.

As each new benchmark is developed, it is added to the craft library of benchmarks (Fig. 6.16) and spreadsheets according to craft and task area (Fig. 6.17). The planner adds

INSTALL UMS CRAFT LIBRARY	CRAFT CODES
	00 - Basic USD & Support Data
	01 - Special Operations
	02 - Millwright
	03 - Carpentry
	04 - Painters
Craft Code	05 - Masons
	06 - Pipefitters
	07 - Electricians
	08 - Metal Workers
	09 - Welders
	10 - Janitorial
	11 - Roads & grounds
Please Insert a Maynard UMS Craft Library Disk in the A: drive.	12 - Refrigeration & A.C.
	13 - Heating
	14 - Auto & Crane Repairs
	15 - Electrical Utilities
Enter Craft Code when ready to begin. Program will return to main menu when complete.	16 - Instrumentation
	17 - Gas & Water Utilities
	18 - Tool & Cutter Grinding
	19 - Preventive Maintenance

FIGURE 6.16   Craft library in automated UMS.

UMS SPREADSHEET RETRIEVAL	0290.XX - TASK CODES
	* 01 - Couplings
	* 02 - Pumps
	* 03 - Motors & Gear Reducers
ENTER CRAFT CODE 01 TASK CODE	* 04 - Fans and Blowers
	* 05 - Tanks and Chests
Enter the appropriate task code from the list at the right. Those tasks with an * next to them are currently available on your system.	* 06 - Agitators
	* 07 - Conveyors
	* 08 - Air Compressors
Hit Enter to Return!	* 09 - Bearing
	* 10 - Miscellaneous

FIGURE 6.17 Automated UMS task codes for the millwright craft.

preparation, travel, and allowances times to benchmark work group time (Fig. 6.18) to establish a standard time for a work order based on the method to be performed (Fig. 6.19). The computer's ability to retrieve data and calculate standard times rapidly, along with fast library updates, makes it easier to set up and maintain UMS data.

Craft: Millwright   Tasks: Pumps   Code: 0290.02

Group D (0.5) 0.7 (<0.9)	Group E (0.9) 1.2 (<1.5)	Group F (1.5) 2.0 (<2.5)
BM# 011 0.59 MEN: 1 PUMP, BEARINGS - HORIZONTAL SPLIT SMALL BOTH ENDS - REMOVE & REPLACE FROM SHAFT	BM# 001 0.94 MEN: 1 PUMP, HORIZONTAL SPLIT, SMALL REMOVE & REPLACE ROTATING ELEMENT THRU 3" DISCHARGE SIZE	BM# 006 1.91 MEN: 2 PUMP, VERTICAL SPLIT, MEDIUM, REMOVE & REPLACE SHAFT SLEEVE OVER 3" DISCHARGE SIZE
BM# 014 0.70 MEN: 1 SLEEVE, SHAFT, SMALL, HORIZONTAL SPLIT PUMP, BOTH ENDS, REMOVE & REPLACE FROM SHAFT	BM# 017 0.99 MEN: 2 PUMP, BEARINGS - HORIZONTAL SPLIT, MEDIUM BOTH ENDS - REMOVE & REPLACE FROM SHAFT	BM# 007 1.79 MEN: 2 PUMP, VERTICAL SPLIT, MEDIUM, REMOVE & REPLACE ROTATING ELEMENT OVER 3" DISCHARGE SIZE
BM # 015 0.58 MEN: 1 PUMP, BEARINGS - HORIZONTAL SPLIT, MEDIUM COUPLING END - REMOVE & REPLACE FROM SHAFT	BM# 036 1.12 MEN: 1 WATER PUMP, 25 HP, REPACK	BM# 019 1.80 MEN: 1 SLEEVE, SHAFT, MEDIUM, HORIZONTAL SPLIT PUMP, BOTH ENDS, REMOVE & REPLACE FROM SHAFT (INCL. BEARINGS & COUPLINGS)
	BM# 047 1.33 MEN: 1 WORTHINGTON 5 G A GEAR PUMP, REMOVE, OVERHAUL	

↔ TO MOVE AROUND SPREADSHEET   F2 -CREATE ESTIMATE   F10 -RETURN

FIGURE 6.18 Automated UMS benchmarks on a spreadsheet for pump repair.

WORK ORDER # 123-321-11	CODE 0290.02		BM # 017	
WORK TIME 1.20	YOU CAN ENTER A CLIENT SPECIFIC TABLE. SAMPLE TRAVEL TIME TABLE			
PREP TIME 0.08				
TRAVEL TIME 0.30		BLDG A	BLDG B	BLDG C	BLDG D
ALLOWANCE 0.24	BLDG A - 0.1 0.3 0.2			
	BLDG B 0.2 - 0.4 0.1			
TOTAL TIME 1.82	BLDG C 0.4 0.5 - 0.3			
	BLDG D 0.3 0.5 0.4 -			
SAVE ESTIMATE Y	WALKING - TOP RIGHT CORNER HAND TRUCK - LOWER LEFT CORNER			

FIGURE 6.19 Automated UMS standard times include work time, preparation, travel, and personal and delay allowances.

Computer Used for Control Reports.   Higher management is more limited in day-to-day associations and must depend more on reports to keep abreast of trends and conditions. These reports, along with information communicated to higher management by supervision, provide the facts used to decide future improvement action.

Definition of Control.   To control is to exercise a directing or restraining power over an activity. To control is to regulate. In order to control it is necessary to:

- Be aware of conditions, activities, and trends.
- Take corrective action.

A control report supplies information which, when combined with knowledge gained from other sources, indicates the corrective action required.

Information Needed for Control.   Maintenance control results are measured in terms of four factors:

- Performance, planned hours divided by actual hours
- Coverage, actual hours on planned work divided by actual hours on all work
- Delays (nonproductive activity that cannot be controlled by the worker)
- Cost per planned hour, actual average labor cost to produce one planned hour of work

High performances are largely the result of training, using good methods, correct crew size, and good worker skill and effort. High coverage is attained through early detection and preplanning of needed work. Minimum delay time (good utilization) is achieved through correct use of staff, proper planning, and effective scheduling.

The amount of emergency work should also be closely controlled through an effective preventive maintenance system. Good methods will result if proper training, procedures, and on-the-job supervision are carried out. These measures will further enhance performance and utilization.

Therefore, if the reports show the performance, the coverage, the delay time, and the cost, and give information to calculate correct labor requirements, management will be able to determine the trends in each area of responsibility. They will be able to spot weak areas that need improvement and, through further analysis and reflection, will be able to take corrective action.

The Weekly Maintenance Control Report.   The most important report in the maintenance management system is the weekly maintenance control report. This weekly report is prepared by the planning and scheduling function during the week following the activity being reported. All completed maintenance work orders are returned to the planning office to be included in the report for a given week. This report is used by the maintenance supervision to measure current results in their areas of responsibility and guide corrective action. It is also the basis for reports that keep higher management informed about how well supervision is using available laborpower.

An example of this report form is shown later in Fig. 6.20. Each line on the report will contain a week's set of figures for the person shown in the left-hand column. The last line at the bottom of the page is the summary of the indicators for all departments included in the program. Below the data section of the report is an explanation of the source or calculation used for figures in each column and block in the report. The information in the body of the report is divided into groups:

- *Actual:* The "actual" group contains hours on planned work, column (a), which are actual hours spent on completed jobs that had planned hours applied; hours on unplanned work (b) hours on completed jobs that did not have planned hours applied; delay hours by type and nonjob hours (c) hours spent on nonwork activity such as training, union business, and total attended hours (d) the sum of a, b, and c.
- *Credits:* The "credits" section of the report contains the planned hours credited or produced. Hours on planned work (e) is the total planned hours produced on completed

jobs covered by planned hours. Hours on unplanned work (f) is the estimated planned hours produced on completed jobs not covered by planned hours. (This figure decreases as coverage increases.) Total credit hours (g) is the sum of (e) and (f).

- *Control Indexes:* This section presents indicators which reflect results achieved in the previous period. Since a number of different conditions interact to complete the picture of current effectiveness, all these conditions are shown. Performance (h) is the planned hours produced divided by the actual hours taken (excluding delays); coverage (i) is the actual hours on planned work divided by the total actual hours on planned and unplanned work; percent delay (j) equals delay hours divided by total attended hours; percent productivity (k) is calculated by dividing total attended hours into total credit hours. It represents productivity of the group for a period reported.

- *Costs per Standard Hour (n):* This is the total planned hours produced divided into the total compensation including all payroll related items (fringe benefits):

$$\text{Cost per planned hour produced} = \frac{\text{total compensation}}{\text{total planned hours produced}}$$

- *Backlog:* The "backlog" section includes total standard (planned) hours (o) from the planners' backlog file, and weeks of backlog (p) is calculated by dividing (o) by (g), the total standard hours produced this week.

  Calculating Savings from Productivity Improvement. Savings are determined by comparing the cost per standard hour for the base period with this week's cost per standard hour. For example, assume that:

$$\text{Base period cost per standard hour} = \$10$$

$$\text{This week's cost per standard hour} = \$7$$

$$\text{Savings per standard hour} = \$10 - \$7$$

$$= \$3 \text{ saved for each planned hour produced}$$

SUPVR.	ACTUAL MAN-HOURS ON COMPLETED JOBS											UMS HOURS CREDIT			CONTROL INDICES					BACKLOG	
	MAN-HRS ON PLANNED WORK	MAN-HRS ON UN-PLANNED WORK	DELAY MAN-HRS BY TYPE CHARGED TO JOB						TOTAL DELAY MAN HRS	TOTAL MAN HRS	STD HRS COMPLETE PLANNED WORK	STD HRS CREDIT UN-PLANNED WORK	TOTAL STD HRS CREDIT	% PFM	% COV	% DIV	% PRD	COST PER STD HR	STD HRS	WKS
			1	2	3	4	5	6												
	A	B	1	2	3	4	5	6	C	D	E	F	G	H	I	J	K	N	O	P
AB	208	76	0	1	0	0	0	2	3	287	162	51	213	78	73	1	74	16.43	179	2.1
CD	578	303	0	0	0	0	0	0	0	881	472	219	691	82	66	0	78	15.55	584	2.1
GS-A	786	379	0	1	0	0	0	2	3	1168	634	270	904	81	67	0	77	15.76	763	2.1
EF	836	864	0	0	57	0	2	10	69	1769	626	575	1201	75	49	4	68	17.97	695	2.2
GH	111	518	0	1	31	7	2	0	50	1671	723	299	1022	65	68	2	61	19.95	705	2.1
IJ	11295	283	7	0	45	1	0	0	53	1631	900	177	1077	69	82	3	66	18.48	500	1.5
KL	1973	222	0	0	3	0	0	0	3	2198	1654	168	1822	84	90	0	83	14.72	856	1.4
GS-B	5217	1887	7	1	136	8	4	10	166	7270	3903	1219	5122	75	73	2	70	17.31	2756	1.7
MN	1190	296	10	14	47	2	14	15	102	1588	995	225	1220	84	80	6	77	15.88	4320	9.2
OP	1351	523	12	9	39	0	38	8	106	1980	1249	437	1686	92	72	5	85	14.33	2712	4.8
QR	1859	22	0	2	1	6	1	8	18	1899	1680	19	1699	90	99	0	89	13.64	2393	4.7
GS-C	4400	841	22	25	87	8	53	31	226	5467	3924	681	4605	89	84	4	84	14.48	9425	6.1
ST	2251	224	9	0	3	0	0	0	12	2487	1424	129	1553	63	91	0	62	19.54	1185	3.4
UV	2069	202	6	6	16	0	15	9	52	2323	1831	155	1986	88	91	2	86	14.26	1030	2.1
GS-D	4320	426	15	6	19	0	15	9	64	4810	3255	284	3539	75	91	1	74	16.58	2215	2.6
M. MGR.	14723	3533	44	33	242	16	72	52	459	18715	11716	2454	14170	80	81	2	76	16.11	15159	3.5

A B C D E F G) On work order COMPLETED THIS WEEK
A) Actual hours on work order that were REVIEWED BY PLANNER   B) Actual hours on work order NOT Reviewed by Planner
C) Total Delay hours = Sum of Delays by type
D) A + B + C   E) Std Hrs on work order that were REVIEWED BY PLANNER   F) B × H × .90
G) E + F   H) E ÷ A   I) A ÷ (A + B) J) C ÷ D   K) G ÷ D
N) Theoretical-Maint-Cost-per-Hour ($12.00) ÷ K
O) From Weekly Backlog Report   P) From Weekly Backlog Report

FIGURE 6.20 Computer-generated maintenance control report.

The total savings for the week is savings per standard hour multiplied by total standard hours (g) produced. For example, assume:

$$\text{Savings per standard hour} = \$3$$

$$\text{Standard hours produced} = 2000$$

$$\text{Savings this week} = \$3 \times 2000$$

$$= \$6000$$

Annual savings rate is savings this week ($6000) times weeks of operation (50), or $300,000 per year.

Control Charts for Evaluating Trends.   A control chart uses the computer's graphic presentation capability to show trends of control indexes from one reporting period to the next. Review of trends at regular and frequent intervals is essential to timely control action, improved maintenance service, and optimum costs.

Scheduling Control Chart Reviews.   The three most basic control indexes are performance, coverage, and delays. From these indexes all other trend indicators are developed.

Figure 6.21 shows a typical maintenance control chart preparation and review schedule. Each set of charts is posted where everyone with results responsibility can observe the trend. The supervisors post their set of charts weekly at the job assignment board so that workers can see their contribution to the team effort.

LEVEL OF MANAGEMENT	FREQUENCY	PREPARED BY
ASST. GENERAL MANAGER, OPERATIONS	EVERY 4 WEEKS	PLANNING SECTION
MANAGER, MAINT. DIV.	EVERY 4 WEEKS	PLANNING SECTION
SUPERINTENDENT	WEEKLY	PLANNING SECTION
MAINT. SUPERVISOR	WEEKLY	SUPERVISOR
MAINT. GEN. FOREMEN	WEEKLY	GENERAL FOREMEN
MAINTENANCE FOREMEN	WEEKLY	MAINT. FOREMEN

FIGURE 6.21 Maintenance control chart review frequency.

The performance percentage is the comparison between the planned hours to do the work and the actual time taken by the workers. Low performance will result from poor worker skill or effort, poor planning and scheduling, too many workers assigned to a given job, idle time that is controllable by the worker, and from using poor methods of performing the work. A reasonably good performance figure for a group of workers working on a measured daywork basis and having the benefit of good supervision and good planning and scheduling, and who have been trained to use good methods, is from 85 to 90 percent. In most maintenance areas without engineered standards the performance percentages will probably range much lower. The goal of every maintenance supervisor should be to raise a group's performance for the week to a minimum acceptable level of 85 percent through better planning and supervision. Experience has shown that the better supervisor is able to accomplish this in a matter of several months after the installation of engineered standards.

Figure 6.22 shows the performance percentage trend. All classes of work that are covered by planned hours are reflected in the performance figures. Work orders having no standards on them are not included in the calculation to determine the performance.

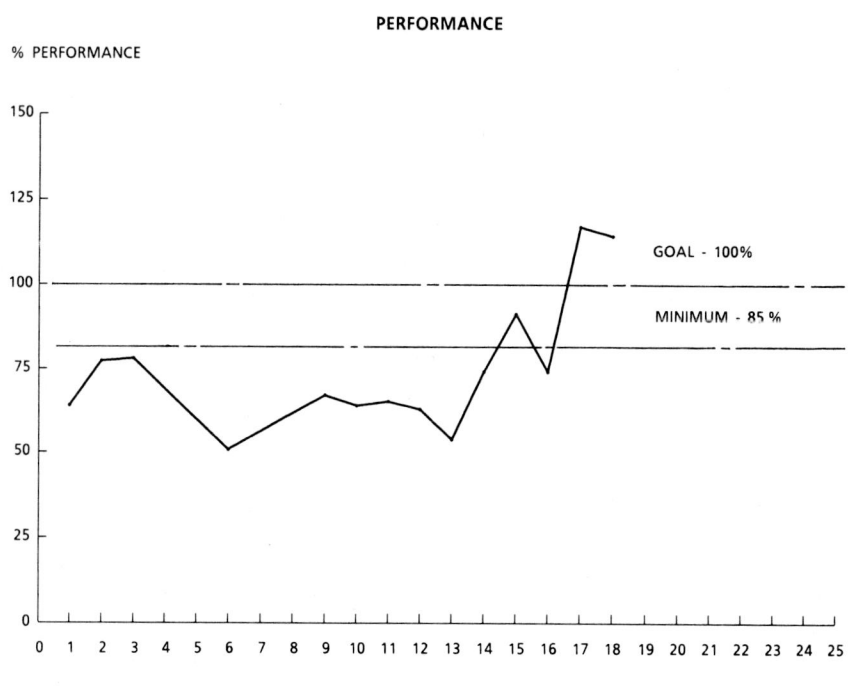

FIGURE 6.22   Performance trend chart.

The coverage percentage (see Fig. 6.23) shows the actual hours on planned work excluding delays compared with the total actual hours worked. As coverage increases, a larger portion of work benefits from the new planning techniques, and consequently, the cost of that work decreases and equipment reliability increases. Two factors come into play in attempting to obtain high coverage. They are: (1) definition of the work to be done, and (2) application of standard hours. Definition of the work is the responsibility of the supervisor in consultation with the requester. Application of standard hours is the responsibility of the planner. The ideal situation would be to have 100 percent of the work covered by standard hours because this preplanning would ensure greater effectiveness of the worker on the job. Some maintenance work, such as emergency work or troubleshooting, cannot be planned in advance. However, experience has shown that 85 percent of all maintenance (repair) work and 95 percent of all project (new or improvement) work can realistically be covered by planned hours. With good supervision, and close control over utilization of all the workers assigned to each supervisor, these coverage goals can be achieved in several months after the beginning of the installation. The greatest deterrent to high coverage is broad job assignments that do not contain specific work content. Where this condition exists, it is almost always accompanied by low performance and poor utilization of the work force.

The delay percentage (see Fig. 6.24) is the percentage of time that the maintenance

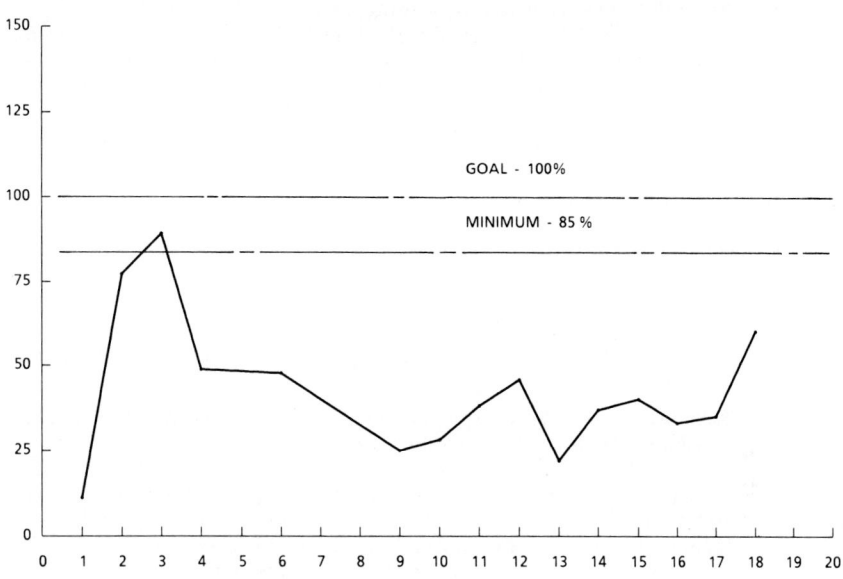

FIGURE 6.23 Coverage trend chart.

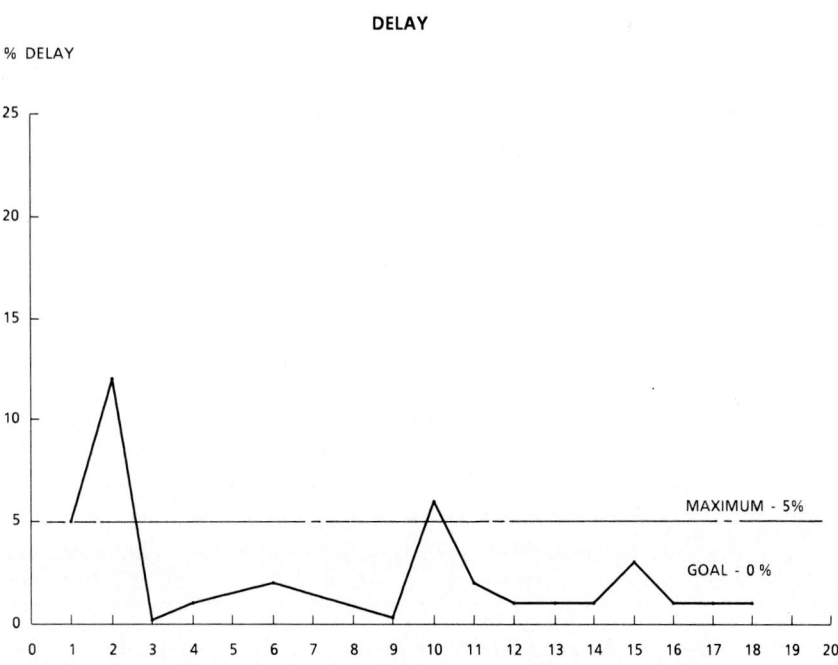

FIGURE 6.24 Delay trend chart.

force was idle for reasons beyond the control of the individual worker. Waiting for a crane or truck to deliver material, waiting for another worker, waiting for production personnel to release equipment, or waiting for an assignment are examples of delay time not controllable by the worker. Delays can be minimized through better planning and scheduling, through better supervision. Only excess delays are reported. The planned hours contain allowances for the expected "normal" amount of waiting time inherent in maintenance work. Lost time that is controllable by the worker is not included in "Delays" but must be charged against time worked. High avoidable delay time will lower performance and increase maintenance cost.

Auditing the Maintenance Program.   Like any complex machine, the maintenance program needs periodic tune-ups. The computer is used to assist in three ways by:

1. Recording the current status of the system components.
2. Comparing the status with an expert systems database of current maintenance practices.
3. Scoring the system status, highlighting areas for improvement, and calculating potential savings from corrective action.

One system uses eight audit areas (see Fig. 6.25), each containing a series of questions and multiple choice answers. Scores for each area are compiled and presented on a maintenance program profile graph (see Fig. 6.26). Low scores compared with the practical potential for a well-managed system highlight areas for improvement. The point values scored for each answer in an audit area show where a system is operating effectively and where improvement action can make a system even better. When specific information about the size of the department and wage costs are entered, productivity (from the profile) and savings potential are automatically calculated (see Fig. 6.27). Periodic reauditing and comparison with previous audit profiles shows the maintenance manager's progress over time.

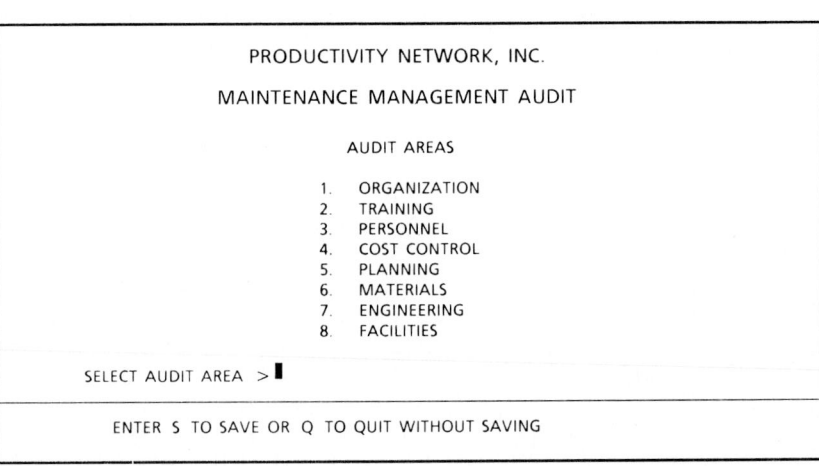

FIGURE 6.25   Computer main menu for audit of maintenance management system.

CONCLUSION

Maintenance costs billions of dollars annually. The application of work measurement to maintain activities is most certainly worthwhile when one considers that every dollar

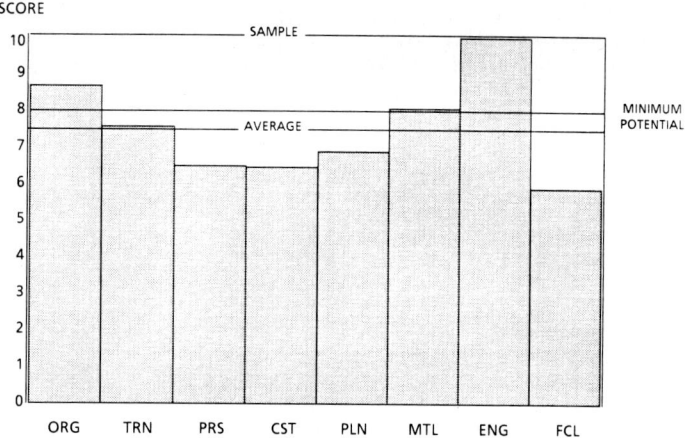

FIGURE 6.26 Computer screen showing the result of an audit in the form of a profile graph. Lower than minimum scores indicate areas for improvement.

```
                    PRODUCTIVITY NETWORK, INC.

                  MAINTENANCE MANAGEMENT AUDIT


    YOUR PRESENT PRODUCTIVITY IS 68 PERCENT
    OPTIMUM PRODUCTIVITY IS 80 PERCENT
    WHAT IS YOUR AVERAGE HOURLY RATE ?  10
    WHAT IS THE AVERAGE HOURS PER YEAR PER HOURLY MAINTENANCE EMPLOYEE ?  2000
    HOW MANY HOURLY EMPLOYEES DO YOU HAVE ?  35


       YOUR SAVINGS POTENTIAL IS $   105,000.00


    PRESS ANY KEY TO CONTINUE
```

FIGURE 6.27 Computer audit screen showing savings potential calculation.

saved goes right to the profit column or to improve service and reliability. Measurement is the only path to control, and control results in more service at lower costs.

Two predictions can logically be made about the maintenance manager's charter in coming years: (1) pressure will continue to come from higher management to reduce costs even while equipment and processes become more complex and environmental requirements more demanding; and (2) with a comprehensive plan and strategy, maintenance managers will continue to meet these challenges in innovative ways.

BIBLIOGRAPHY

Heintzelman, J. E., *The Complete Handbook of Maintenance Management*, Prentice-Hall, Englewood Cliffs, N.J., 1976.

Higgins, L. R., and L. C. Morrow, eds., *Maintenance Engineering Handbook,* 3d ed., McGraw-Hill, New York, 1977.

Westerkamp, T. A., *Maintenance Manager's Standard Manual and Guide,* Prentice-Hall, Englewood Cliffs, N.J. (in press).

Zandin, Kjell B., *MOST Work Measurement Systems,* 2d ed., Dekker, New York, 1990.

CHAPTER 7
DEVELOPING CLERICAL AND ADMINISTRATIVE STANDARDS

Thomas Seidel
Geschäftsführender Direktor
ROI Management Consulting AG
Zurich, Switzerland

Richard L. Shell
Professor of Industrial Engineering
Department of Mechanical, Industrial, and Nuclear Engineering
University of Cincinnati
Cincinnati, Ohio

During the early days of work measurement and following World War II, methods improvement and standards development were almost exclusively found in manufacturing plants and were applied only to direct hourly workers. Typically, managers in clerical and administrative areas did not understand the benefits to be derived from improved methods and/or production standards. Two major changes occurred in the job mix throughout manufacturing, government, and other service areas that created an interest in measuring the clerical and administrative work force. These changes were: First, as a percentage of work force, manufacturing jobs were declining and were accounting for less of the total factory cost to produce products. Second, clerical and administrative jobs were rapidly increasing with the service sector and government expansion, and the majority were not measured. A large number of manufacturing jobs had been successfully measured and substantial productivity gains had been realized. This along with job mix changes prompted the question, "If methods analysis and work measurement application are effective for manufacturing, will they work for clerical and administrative jobs?" The answer given by industrial engineers and managers in most progressive organizations was "yes."

Historical Perspective.   There were recognized differences and similarities between manufacturing and clerical or administrative work tasks. While all of the work measurement techniques were useful for selected aspects of nonmanufacturing work, there was no well-defined approach to developing standards for clerical and administrative personnel. Commonly, these tasks ranged from short-cycle highly repetitive to long-cycle nonrepetitive activities. While basic predetermined time systems could be used for many of the short-cycle office activities, they were not effective for the nonrepetitive tasks. Other measurement problems and limitations that existed included the following:

- Objectives concerning required coverage and accuracy levels were not realistically defined.
- The concept of measurement was not well accepted in areas with complex or creative work content.
- There were doubts about clerical and administrative standards being reproducible, owing to variable work methods.
- Before computerization and database management, standards maintenance was an almost impossible task.
- Daily, weekly, monthly, and annual capacity planning methods were little known and rarely used in clerical areas.
- Methods engineering and work measurement in many settings unfortunately acquired a low image within the last 20 years as compared with more attractive technologies and concepts, such as computer integrated design and manufacturing, information systems, artificial intelligence, and flexible automation.

Generally speaking most clerical or office measurement applications remained "one-shot approaches," whereby these installations deteriorated rapidly and were economically unsound.

Today, most highly repetitive office work activities are completed by either automation or low-paid clerical personnel. There has also been an increase in the number of mixed workplaces characterized by growing complexity and longer-cycle work activities. Consequently, two different kinds of clerical and administrative personnel have emerged.

In general, union representation so far has opposed standards in clerical and administrative areas as well as any kind of incentives. The increase in capital investment per workplace has been faster in offices than elsewhere in business. Yet equipment utilization continues to be rather low. Examples include insufficient terminal on-line activity, low utilization of printers and copiers, and generally poor return on office capital investment.

On the positive side there has been a growing use of standard software packages in office work activities. Therefore, jobs have been adjusted to meet software demands, and not vice versa. Consequently, more homogeneous work activities are developing, leading to basically more favorable conditions for setting office standards.

The Need to Measure Clerical and Administrative Personnel.   The number of workers and the relative importance of the administrative sector have been growing constantly for the past several years in the highly industrialized countries. All sorts of statistics support this well-known fact, be it by irony such as Mr. Parkinson's publications or through the very serious worldwide drain on factory union membership. Even though such a development was well predictable in advance, it does surprise greatly to realize that efficiency measurement methods, as well as administrative control procedures and especially administrative cost accounting, have reached to date only very primitive stages. This is partly due to complex office work tasks where it is difficult to find and rather unrealistic to define strictly repetitive operation flows. Such work flows prevail, however, in many computer-based systems. Thus banks, for example, have been natural leaders in administrative cost accounting and in the application of administrative control procedures. Banks as well as insurance companies also comply fully with the second main condition for such procedures, namely, a well-developed and -installed computer-based information system covering the whole organization. Both are prerequisites for success when introducing or adopting new administrative and control procedures, first, well-defined and strictly repetitive operation flows and second, a well-developed information system, providing data of all kinds on the given operations and activities through all stages of processing.

The most controversial discussions related to new procedures concern the measurement of creative mental work. H. B. Maynard proved some 40 years ago, through his pioneering studies in the research and development area, that only about 20 percent of high-caliber scientific tasks are on the whole required. It is furthermore a proven fact that even in this small segment of highly scientific work the truly creative part is almost negligible, since most considerations and conclusions are based on analogies and highly educated comparisons, which in the last analysis are measurable. It is by no coincidence that "Brainstorm," "Fulguration," and "Geistesblitz" are expressions used to describe truly innovative and creative activities. Since precision levels for most work measurement activities are set normally plus or minus 5 to 10 percent, it is quite obvious that the purely creative work content of clerical and administrative personnel can be neglected from a work measurement point of view. The psychological component, however, should be taken into account, and it is advisable to define and allocate a certain percentage of total time as a constant to cover once and for all those "elite" activities.

The need to measure clerical and administrative personnel is further underscored from the economic benefit. Experience has shown that productivity gains of 20 to 80 percent can be realized from a well-designed and -implemented work measurement program for clerical and administrative personnel.

Justification for Clerical and Administrative Standards.   While the basic justification for work measurement and higher productivity in the clerical area is economic benefits, it is also important to raise some philosophical arguments. Humanization of work, high degree of autonomy, quality circles, and self-control can work properly and be of real permanence only when there is a guarantee for high morale and constant discipline of the single employee.

A profound and accurate knowledge of work content and time spent is required at the very time when measurements and controls must be developed and established. Why is it unthinkable to require these control elements when working under otherwise accurately agreed-upon conditions during most of the active day? Do we *not* expect to count and measure every action during sports, off the job in our leisure time? Such a difference between expectations can be neither understood nor accepted.

A closer look at work content in the clerical field is appropriate at this point. Three types of workplaces should be considered, as shown in Fig. 7.1. Methods engineering approaches are obviously best suited for application to the sociotechnical group. It is imperative to overcome the so-called creativity myth in order to apply the principle of "complete coverage" to the sociotechnical group. Figure 7.2 shows the results of a study made by Siemens of West Germany on 100,000 white-collar employees. Of all activities in the overhead areas 75 percent are suited for standard setting. Only a small fraction of truly creative tasks can be located within the areas with predominantly creative thinking requirements. Therefore, work measurement can be applied to most sociotechnological systems. The following areas of standards application are listed in order of relative importance:

Workplace Type	Most Important Related Elements	Examples	
		Workplace Description	System Purpose
technical	office materials and tools	computer magnetic plate	to express information
social	human resources	head-offices, meetings, training	to consult, to coordinate, to ensure training
socio-technical	human resources and office materials	machine workplace, typewriter, copying machine telephone switchboard	to write reports, to make copies, to arrange communications

FIGURE 7.1 Workplace classification.

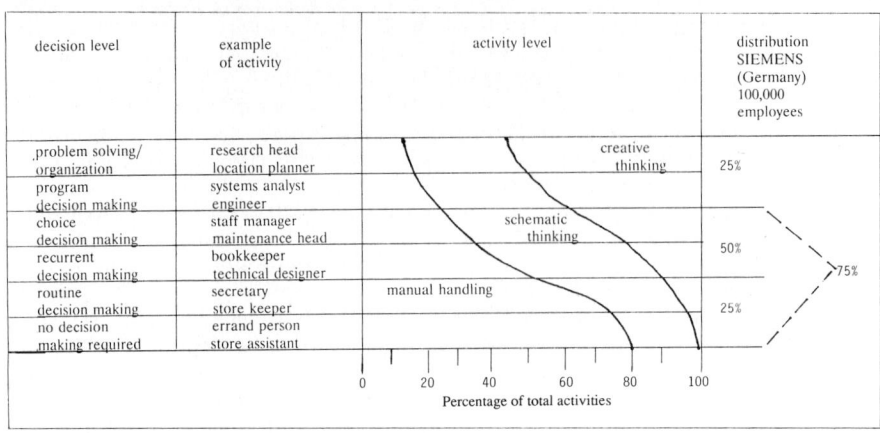

Source: Siemens Corp. (Germany)

FIGURE 7.2 Decision and work activity levels for white-collar employees. (*Siemens Corp., Germany.*)

1. Objective evaluation of employee performance
2. Work scheduling and time management
3. Methods improvement and workplace design
4. Personnel planning and budgeting
5. Investment planning
6. Productivity analysis
7. Cost accounting
8. Training and job documentation
9. Organization analysis

One practical yardstick for economics of application of any given work measurement system is the time necessary to establish a standard for 1 hour of work content. Using engineered standards, 2½ hours are typically necessary to analyze 1 hour of work content. Using a mix of measurement techniques, this can be reduced to a relation of about 1:1. Having understood this experienced based estimate, the importance of sound judgment in the establishment of clerical and administrative standards at this point has to be remembered. In many cases not so much the work output is the only means of measuring performance, but rather the quality. As mentioned, ideas and true creative work cannot be measured in time units. The value of clerical and administrative work is very much a combination of quality and quantity, and with this fact in mind, the economy of establishing and maintaining performance control systems becomes fundamental.

Favorable Preconditions for Successful Measurement. Performance control in white-collar areas should be considered as one of a set of management tools. An organizational philosophy based on management by objectives (MBO) will have advantages for establishing work measurement based standards. Well-developed control systems and sound statistical data are also important preconditions. Standards in the office, and in consequence data for cost accounting of administrative activities, are comparable with management tools such as job rating, merit rating, and other important control systems. In this area more advanced systems such as PBKS give a good and a continuous indication of the degree of organization, the organizational level of any area or department measured (PBKS is a performance control system created by ROI, Zurich). It is also useful to have a deep and widespread understanding of the principles of project management and project organization for a company embarking on clerical and administrative measurement. Resistance to these types of process controls will cause difficulty in the measurement program. This resistance can be counterbalanced by establishing the two key factors of any successful project, namely, the power promoter (a defined source of power) and the know-how promoter (the manager in charge of the technical details). This combination of management and expertise support is a must. There is no area more sensitive in the field of engineered work measurement than setting standards for office workplaces.

Preparing for a Standard Setting Project.   The project planning and objectives should include the following:

1. It is of practical importance when conducting work measurement studies first to choose the departments with homogeneous and relatively simple work content.
2. Choose the engineer or analyst with extensive knowledge of the areas to be analyzed. The know-how required is much more extensive than manufacturing work measurement, practically not comparable, since much of the work content is not readily detectable on the outside.
3. Engineers or analysts, with basic knowledge of the areas, have to be well trained in standard setting methods. Their task has to be considered as continuous. The need for experienced personnel for the clerical area is constantly rising. They are an asset to the future for any forward planning organization.
4. Departmental and/or group managers often use their own figures in order to control performance of their units. It is fundamental to convince these persons of the advantage of a more sophisticated and also more broadly applicable system. A change of attitude and an understanding of the principles involved will be the key to any permanent success.
5. Existing control systems not only will be the basis for development in this area but also will stand in constant close relation with it. Therefore, the personality of the controller, quality of data, capacity planning systems, and job and merit rating systems will also influence the overall project success.
6. The final responsibility for the standards system should be with the operating management, not industrial engineering or the personnel department.
7. Complete standards coverage of all area activities, and flexible continuous application, are needed.
8. The standards program should be operated economically.

In organizing for a successful standard setting project it is advisable, if possible, to build homogeneous work groups, thus helping the process of work measurement. Relocating work content and activities or other method improvements should also be an outcome of sound work measurement practice.

Work Areas for Application.   Clerical and administrative jobs are found in all sectors of the economy: manufacturing, service, and government. The following are example areas well suited for work measurement and standards development:

- Banking, investments, and pension funds
- Transportation, warehousing, and shipping
- Insurance
- Local, state, and federal government offices
- Communications (telephone, media, telex)
- Postal, packaging, and delivery services

- Health care
- Utilities
- Electronics
- Aerospace and defense
- Accounting
- Purchasing
- Maintenance and repair
- Engineering and research
- Education
- Data processing operations
- Real estate administration
- Tourist offices
- Laboratories

WORK MEASUREMENT TECHNIQUES

Overview.   Jobs in the clerical and administrative workplace include short- and long-cycle activities with varying degrees of complexity and quantity of output. In addition, the jobs range from all manual effort to higher amounts of mental processing. The level of automation also varies considerably in clerical and administrative work. Generally automation levels increase as the quantity of output becomes larger. Consequently, for establishing standards for clerical and administrative work, see Chaps. 2 and 4 in this section.

Two basic principles should be applied when using work measurement techniques in the clerical and administrative field:

1. Economy of development, application, and standards maintenance.
2. Transparency in order to facilitate worker acceptance. The work situation should become objective through the use of standards, rather than be blurred by them.

Most of the techniques mentioned below are well known. Therefore, comments and explanations refer to specific rules of application, areas of best use, advantages and disadvantages, and practical guidelines. These techniques include:

1. Work sampling and group timing technique
2. Time study
3. Comparisons and judgment estimating
4. Structured self-reporting and historical data
5. Predetermined time systems

All these techniques have obvious advantages and well-known accuracy limitations. Some are based on time measurement of existing situations; others develop optimized method and workplace standards. A rule that is reasonable to follow states: The higher the time share of a given activity, the more accuracy standards require.

In all practical respects, it is advisable to employ combinations of several techniques according to their relative advantages. Figure 7.3 shows a comparison of the various techniques useful for office work measurement.

All time measurements are based on the analysis of operation flows or sequences. This basic fact must be understood clearly and kept in mind when applying the different work measurement techniques. The logical steps are listed below.

Factor	Work Sampling	Time Study	Comparisons and Judgement Estimates	Self-reporting and Historical Data	Predetermined Time Systems
skill of analysts/engineers	low	average	high	low	high
time to install	low-average	high	average	low	average
administrative cost	average	average	low	average	average
savings potential	low	average	low	low	high
methods improvement potential	low	average	low	low	high
accuracy of standards	low	average	low	low	high
applicability to:					
high volume	high	high	low	high	high
low volume	average	average	average	low	average
employee acceptance	average	low-average	high	average	average
supervisory acceptance	average-high	low-average	average-high	average-high	average-high
flexibility (as changes occur)	low	average	low	low	high

FIGURE 7.3   Comparison of office work measurement techniques.

1. Definition of activity or task and its work content

2. Establishment of optimized methods

3. Time measurement

For any organization to become effective and efficient, operation sequences must be optimized at their core. Only when these sequences are well defined and controlled will consistent time values be realized. It is a proven fact that efficiency varies to a much greater extent in the office environment than on the production floor.

Work Sampling and Group Timing Technique.   When applied to develop office standards, work sampling and its allied group timing technique (GTT) provide extensive coverage in a relatively short time. Work sampling is characterized by a number of random or systematic observations of clerical work activities.

The engineer or analyst first determines and codes the different tasks performed by a group of office workers. Group size and work content breakdown must be well considered to avoid overloading the engineer or analyst and in order to achieve statistically valid results within reasonable time periods. The number of units completed by task or activity during the study is reported by the employees concurrently with the work sampling observations. Time values are arrived at through simple division of total time spent on each task or activity by the number of finished units. See Sec. 4, Chap. 3 for more detail information about work sampling and GTT. These work measurement techniques have certain limitations as summarized below.

• Results reflect current conditions.

• Main result is an indication of the overall laborpower utilization only by major activity.

• Indications are obtained concerning possible reduction of nonproductive times, yet none toward methods improvement.

• Statistical parameters, such as confidence levels, may be unsound and (similar to time studies) may not withstand mathematical proof.

GTT has proved to be useful in practical applications designed to set allowances, to search for gross organizational and work distribution unbalances, and to control service levels of waiting-line workplaces with heavy client affluence, such as bank tellers, counselors, and public offices.

Time Study.   This technique has not been widely used to measure office work. There appears to be no economical justification for the use of time study for measuring such highly variable work. Typical office work flows usually have several branching points. The engineer or analyst must therefore study more cycles of office work to cover the complete job than would be necessary for a direct labor job of comparable length. In the clerical and administrative areas it is rather problematic to arrive at performance ratings, because of extreme differences in efficiency. Above all, it is almost unbearable for the engineer or analyst to take time studies in the modern office environment. See Sec. 4, Chap. 2 for detail information about stopwatch time study.

Comparisons and Judgment Estimating.   This measurement technique is particularly well suited for tasks with low repetitiveness and when qualified analysts and engineers are available. The task to be analyzed is split up into clearly defined elements by defining very precisely starting and ending points. A group of work measurement experts estimate the various units. In some cases, according to the incidence of flow branching points, it is advisable to make three different estimates: optimistic, most likely average, and pessimistic (see Judgment Estimating in Chap. 2 of this section). Leveling out of errors compensation may be expected through the splitting of the total work content into smaller units, to which the estimating procedure is applied. The advantages of this technique are speed and a surprisingly high degree of acceptance, when professionally applied.

At this point it is important to rethink a previous statement, which referred to the combined use of different techniques. Judgment time estimates, as described above, are often an informal combination of process time elements (arrived at through time studies), allowances (work sampling or GTT observations), and predetermined time data.

Structured Self-Reporting and Historical Data.   This type of analysis is recommended in every instance when setting office standards. It covers all employees in the department under study and is undertaken during a 2- to 4-week period. It uses a structured and codified activity list. It gives the total time spent in the department for a given task during the observation period. In all possible cases, definite outputs are associated with each task. Since the number of completed tasks is counted and summed up for all employees, a time per task or activity may also be calculated.

It should be obvious that these results are not directly applicable for setting standards, since they were arrived at based on existing methods and organizational environments. When mainly applied in combination with predetermined times, use of the so-called homogenization factor is recommended. Depending on the factor chosen, an assumption is made concerning the methods improvement potential. For example, a factor of 0.6 would indicate that the self-reported time could be reduced by 40 percent based on a comparison with engineered predetermined time standards.

We now enter a field of very pragmatic approaches where skill and acceptance of the engineer or analyst is critical. However, historical based standards have to be taken very seriously since a complete coverage of any department can be achieved only by applying as well these less scientific methods.

There is a modified version of historical data. It may be defined as "engineered" historical data using in all cases structured interviews. The method applies merely to repetitive and clearly defined work content which for several reasons cannot be analyzed with other methods. Historical data, therefore, can be considered as complementary. In order to be systematic, some organized form of registration of all occurring cases should be developed. The time standards used should be derived as means of the data collected or, for example, the average between the total means and the last input. Standards of this origin obviously are not very dependable. Therefore, in many cases "engineered" historical data are recommended, whereby the engineer or analyst together with a skilled employee eliminate disturbances and unnecessary work content in order to obtain "purified' historical standards. For example, for a certain type of credit in a bank it is necessary to set an objective for the time used on giving advice to potential clients. Using more time on this type of credit with small profit margins would create a loss position and would be unacceptable.

Predetermined Time Systems.   The establishment of office cost controls based on predetermined time systems was generally hailed as a significant breakthrough. Yet such office standards had limitations and were not always successful. Predetermined time systems do have obvious advantages in the clerical field, including:

- The engineer or analyst does not need to observe the individual worker and estimate his or her performance level.
- High methods sensitivity is inherent.
- Data development may precede installations of new work flows and procedures.
- Standards may be used for evaluating alternative future processes, such as possible automation.
- Standard accuracy is acceptable and may be adjusted to meet various goals.
- The standards are reproducible.

Perhaps the most important advantage is based on the fact that predetermined time systems are definitely future-oriented and tend to exclude any discussions about existing situations.

The classical (lower-level) predetermined time systems have not proved to be as successful in the office areas as they have been in manufacturing. Accuracy and the respective cost of development of office standards have to be in sound economical relation. Therefore, higher-level predetermined time systems are very well suited for clerical and administrative applications. These systems have been specifically developed for operations or tasks with widely varying work content. An early example was the universal maintenance standards (UMS) developed by H. B. Maynard and Company.

In the early 1970s H. B. Maynard and Company developed the Maynard Operational Sequence Technique (MOST), a higher-level system based on MTM (thereby based on the principles of a well-accepted predetermined time system) which gained wide acceptance in the United States and also in large parts of Europe. ROI Management Consulting AG further improved MOST and arrived at ROM II (ROI Operational Sequence Technique in the clerical field). The system has been accepted well, especially in banks and insurance companies, and also in some of the leading industrial firms in Europe.

In 1984, H. B. Maynard and Company developed a clerical MOST system that was well suited for determining standards for many office activities. Figure 7.4 depicts the time values for the equipment use activity area. For more detail information concerning predetermined time systems, see Sec. 4, Chaps. 4 and 5.

DEVELOPMENT AND IMPLEMENTATION

The process of developing and implementing clerical and administrative standards includes a sequence of the following critical steps: objective setting and collection of basic information, structured self-reporting, selection of measurement technique(s), operation flow analysis, handling special conditions, and method development and improvement.

Objective Setting and Collection of Basic Information.   In order to decide upon the necessary depth, coverage, and accuracy of the project, it is obviously of first importance to define clear objectives. Different organizations may have quite different objectives. Examples of work measurement applications include the following:

- Staffing
- Performance control

EQUIPMENT USE

ABGABP ABPA

Index	Filing				Letter/Paper Handling Operation (H)				Handling		Index
	Select	O/C Select	File	O/C File	Operate	Jog	Tap	Staple	Meter	Stamp	
1						1	1	Electric Staple			1
3					Open Envelope	3	3	Punch Hand Staple Remove			3
6	1				Weigh Letter	6	6		1	1 Ink	6
10	3		1		Seal Envelope Change Postage Machine	10	10		6	2	10
16	6	2	4	1			15		13	3	16
24	9	6	7	5					23	5	24
32	12	9	10	8					33	7	32
42	17	12	15	11					45	9	42
54									59	11	54

FIGURE 7.4 Clerical MOST data card. (*MOST is a registered trademark of H. B. Maynard and Company, Inc.*)

EQUIPMENT USE

ABGABP ABPA

Index	Typewrite W			Calculate K	Index
	Correct	Set	Words	Key	
1				3	1
3	Key		1	7	3
6		Tab	2 Date	14	6
10	Tape	Margin	3	23	10
16	Liquid		6	36	16
24		Platen	8	52	24
32			11	69	32
42			14	89	42
54			18	113	54
67				138	67

EQUIPMENT USE

ABGABP ABPA

Index	Read T				Record R					Index
	Text of Words	Digits	Compare	Data	Mark Digits	Mark Words	Copy	Write or Print Digits	Words	
1	3	1	1			Check Mark		1		1
3	8	3	2	Gauge	1	Symbols Scribe Line	1	2		3
6	15	6	4	Scale Date/Time	2		3	4	1	6
10	24	12	8		3	1	5	6		10
16	38		13	Table	5		8	9 Date/Signature	2	16
24	54				7	2	10	13	3	24
24	72				10		14	18	4	32
42	94				13	3	18	23	5	42
54	119				16	4	22	29	7	54

FIGURE 7.4 (*Continued*) Clerical MOST data card. (*MOST is a registered trademark of H. B. Maynard and Company, Inc.*)

- Basis for merit rating
- Cost accounting
- Capacity planning
- Scheduling

Work measurement can also be used for a "one-shot evaluation" or establishing balance for a new continuous system. Two important applications are performance control and cost accounting, whereby the latter is consistently growing in relative importance.

In order to get acquainted with the area to be investigated, some preliminary information should be collected. Typically, this includes:

- Predominant type of work, for example, highly repetitive, irregular project-type, or customer service
- Forms used
- Amount of meetings, formal and informal
- Office layout
- Preliminary evaluation of work space
- Computer systems presently utilized and in the planning phase
- Uniformity and simultaneity of the workload
- Job descriptions (clearly defined work content)
- Existing flow sheets
- Reports, statistics, and other output volume indicators
- Overall organizational definition including number and size of smaller entities
- Number and design of typical workplaces

Structured Self-Reporting.   In order to get quick, reliable, and well-documented information about the distribution of all types of work and tasks, as well as actual count of the units (of key activities) produced, self-reporting is a necessity. Using existing job descriptions and catalogs of activities of other companies, together with key managers, a coded catalog of activities for the actual situation may be developed usually within 1 or 2 days (see Fig. 7.5).

Structured self-reporting is a method used to determine the share represented by individual activities of the whole job, through the employee's own information. The results may be evaluated from different perspectives. The main interest in self-reporting lies in the organizational units (sections, groups) and not in individual performances. Work volumes and times registered serve to define improvement goals (simplify, combine, eliminate) and to build approximate ratios (work content per unit occurrence) for cost-benefit considerations.

Self-reporting does not represent a performance control. Unproductive times should not be recorded; these data do not have the purpose of making statements on productivity. The smallest time unit should be 0.25 hour. The weekly reports must be filled in under the responsibility of each employee. The data should be entered immediately, for example, twice daily AM and PM. In the time column in Fig. 7.5, Monday through Friday, hours should be entered for each activity. The sum of all time entries should be 8 hours per day, respectively, 40 hours per week (rounded-off weekly working time without pauses). Frequency of occurrence should also be recorded along with quantities of work (see calculation of unit times in Fig. 7.5).

In order to get full coverage and good acceptance, it is recommended to have everybody in the group, including management and personal secretaries, participate in self-reporting. Self-reporting is usually done during a 2- or 4-week period. In many cases it may be necessary to cover work at month end or other cyclic time periods. Evaluations can be carried out with the help of simple computer-generated programs. Since in some cases the periods of self-reporting are not representative for the yearly workload, adjustments have to be made to cover the distribution over the entire year. The results of self-reporting data are summarized in the paragraphs that follow.

In most clerical and administrative areas there are commonly three types of work to be recorded:

1. Infrastructural activities, such as management tasks: personnel, planning, information; as well as unspecified administrative secretarial tasks and information necessary to maintain the state of the art of the business.

2. The typical activities of the department, usually highly repetitive and well-defined tasks, for instance, the purchasing process, the accounting process, preparing outgoing mail, writing vouchers, or checking documents

3. Project-type work, including participation in new developments, usually in teams, training sessions, or special task forces

Once this three-score classification is well understood, the results of self-reporting can be read like an x-ray. Irregularities can be readily recognized. The percentage of activities in item 2 above gives a picture of the degree of consistent organization, for example, a well-designed and closely followed work content susceptible to engineered work measurement technique application. On the other hand, if these activities are relatively small and the activities in item 1 above exceed 20 percent of the total, one can safely assume the existence of many small groups, unclear organization, and power structures. Obviously these interpretations lead to direct consequences regarding the choice of technique(s) and the accuracy needed in the work measurement process. In addition, a ranking of relative importance, based on percentage of total time, can be established.

By summing up the numbers reported respective to the activities it is possible to arrive at times per unit for quantity of output (see Fig. 7.5). Also, and finally very important, by making the analysis per person or workplace one can easily see the level of organization. Where there is no predominant activity for most of the workplaces, the work contents are badly distributed. Redistribution and/or reorganization is indicated.

Other methods used to arrive at this intermediate step are the use of estimates instead of structured self-reporting, the estimating done by the heads of groups and/or departments. This method is necessarily applied if self-reporting is not accepted by either the employees or the management. It is definitely considered inferior to self-reporting, since its basic data are not transparent and not reproducible. A less accurate approach is the use of an ABC analysis, using count of units and estimates of time consumption or estimates of work distribution throughout the organization.

Selection of Measurement Technique(s).   This is the third decisive step for getting good results. As mentioned before, economics, necessary accuracy, and the possibility of applying certain techniques are the key ingredients for successful project completion. The engineer or analyst must be in command of all possible methods in order to find the right mix. In many situations it is strongly recommended to adapt to the sensitivity of any investigated area. The technique suited best neither can nor should always be applied.

In deciding upon the measurement technique, it is also of high importance to understand the different frequencies of tasks done at the same workplace. Some activities might occur in high numbers, high repetitiveness meriting application of exact and more sophisticated techniques. Some activities will occur just several times, some only two or three times per year, and the respective level of method will be low and the accuracy of standard setting less. In order to avoid getting into many details, a method has been developed whereby only the so-called key activities are exactly measured, while supporting activities of rare appearance are measured by estimates.

				Mo		Tu		We		Th		Fr		Total	Share	Freq
		XYZ Co.				SELF-REPORTING		WEEKLY REPORT								

Department: Tools and Machinery   Group: Administration   Name: BH   Week: 23

No.	Code	Activity List	Qty.	Mo AM	Mo PM	Tu AM	Tu PM	We AM	We PM	Th AM	Th PM	Fr AM	Fr PM	Total Time	Share in %	Freq (see below)
	600,000	Tools and Machinery Administration														
1	601,000	Get Bids for Tools and Machinery		-	-	-	-	-	-	-	-	-	-	-	-	M
2	602,000	Write Requisition	6*	2	-	-	-	-	-	-	-	-	-	2	5%	W
3	603,000	Work on T/M Orders		-	-	-	-	-	-	-	-	-	-	-	-	M
4	604,000	Check Delivery Times	32⊕	-	-	1	-	-	1	-	-	1	-	3	7.5%	D
5	605,000	Prepare Repair and Maintenance Orders		-	-	-	-	-	-	-	-	-	-	-	-	Y
6	606,100	Add New Items to Installation List		-	-	-	2	-						2	5%	W
7	606,200	Delete Items from Installation List		2	-	-								2	5%	W
8	606,300	Compute Changes in Value		-	-	-	-	4						4	10%	W
9	607,100	Organize Purchasing of Small Tools		-	4	-	-	-	-	-	-	-	-	4	10%	W
10	607,200	Account Petty Cash for Small Tools		-	-	-	-	-	-	-	-	-	4	4	10%	W
11	608,000	Equipment Sale		-	-	-	--	-	-	-	-	-		-	-	M
12	609,000	Large-Equipment Rental Accounting		-	-	-	-	-	-	-	-	-		-	-	M
13	610,100	Check Equip-No. (File Maintenance)		-	-	2	-	-	-	2	-	2	-	6	15%	D
14	610,200	Write Delivery Slips		-	-	-	2	-	3	2	2	1	-	10	25%	D
15	610,300	Maintain Location File		-	-	1	-	-	-	-	-	-	-	1	2.5%	W
16	610,410	Determine Equip-No. (Lost Items)		-	-	-	-	-	-	-	1	-	-	1	2.5%	W
17	610,420	Determine Equipment Loss		-	-	-	-	-	-	-	1	-	-	1	2.5%	W
		Totals		4	4	4	4	4	4	4	4	4	4	40	100%	

Calculation of Unit Times
* 6 Requisitions in 2 Hours = 20 min/req.
⊕ 32 Checks in 3 Hours = 6 min/check

Frequency: D = Daily   W = Weekly   M = Monthly   Y = Yearly

FIGURE 7.5 A typical self-reporting form.

Operation Flow Analysis.   This being the fourth important step, now the skilled engineer or analyst should produce clear work flow schemes. These schemes in most cases should represent the so-called key activities and are the base for time measurement steps. It is recommended that PC software be used in preparing the necessary flow drawings.

Handling Special Conditions.   In this fifth step a sound knowledge of the type of work is the basis for success. The following special conditions have to be considered:

- Varying work volumes: they can appear daily, with changing hours, weekly, and in many cases monthly, quarterly, or yearly.

- Dependence on preceding groups or departments: in some cases it has been found that a group was waiting every morning for 1 hour to get papers which were produced or handled by another department the day before. Flexible working times or shifting the beginning times by 1 or 2 hours can sometimes be the solution.

- Level of training, experience, and know-how in a given group: some administrative jobs need a very long period of training; the differences of output are much higher in administrative areas compared with any field of manufacturing.

- Key persons: sometimes the heads of groups or departments tend to be bottlenecks. This has to be considered in reorganization and setting of time standards.

  Workplaces in the office or service area may have priorities other than performance.

For example, in banks all records have to be produced on the same day. In other fields of business, service reaction times have to meet certain requirements. In banks or other users of counters, a certain amount of "standby time," "readiness," or "availability" has to be considered in setting time standards.

Methods Development and Improvement.   In the clerical field detailed methods work is not as important as in manufacturing. However, as previously mentioned, there is a need for correct analysis of work flows. They are the basis for the following steps of methods analysis:

1. Before going to the mechanics of methods engineering, the usual critical questions of methods improvement should be asked:
   - Can this work (content) be eliminated?
   - Should it be transferred to another department or area?

2. Information is obtained from the department or group supervisor about work content and flow of operations to be analyzed.

3. The supervisor together with the engineer or analyst determine the workplace where methods work should be carried out; it is always recommended to choose one of the best employees working on a given work content.

4. Information about the employee to be analyzed, and the objectives of the analysis are stated.

5. The employee demonstrates the work in its usual routine and explains reasons for certain work steps and details.

6. When writing down the detailed work content on the analysis sheet, the engineer or analyst makes notes of possible improvements, special conditions, and percentages of branchings. He or she should also make notes of time-consuming work content.

7. The analysis is discussed with the employee after completion:
   - Have all necessary steps been recorded?
   - Have all possible special conditions been considered?
   - Can the employee recognize his or her own work by reading the analysis?

8. Necessary additions are made; the analysis of the work content is completed.

9. The engineer or analyst seeks methods improvements and records them.

10. Reductions in overall time originating from these possible methods improvements are calculated.

11. Improvements are presented to the group or department supervisor; cost savings and decisions as to the improved new flow of operations and its completeness are discussed and documented.

12. Time standards are set using predetermined time systems such as MTM, MOST, ROM II, or other appropriate techniques.

13. Changes in existing standards, job descriptions, and information to employees involved are carried through by the engineer or analyst and group leader.

It is good practice after having gone through this sequence of tasks and after having arrived at new standards to check on their plausibility (a test of judgment). In the clerical and administrative area, always "working according to rules" may lead to unreasonable, unrealistic time standards.

Some Practical Guidelines for Setting Standards.   In the process of setting time standards in manufacturing, allowances are often determined by negotiation between unions and management according to their impact on wages. It is recommended that allowances *not* be negotiated in the clerical and administrative area. The following guidelines for establishing allowances are recommended:

1. The establishment of personal allowances should follow the usual rules of measurement (work sampling or GTT).

2. General allowances may be required. In the administrative field a rather broad spectrum of occurrences such as uneven workload, interruptions of work through telephone and other means of communication, percentage of unfinishable work (due to missing information), degree of training (for instance, on new work content), and other delays may be experienced.

3. According to the method of setting standards, allowances should be applied to a varying degree. When analyzing with predetermined time systems, allowances should be handled generously. When predominantly less accurate techniques are applied, only small allowances should be included in the standard time.

4. Workplaces with a continuous workload and rather homogeneous work content will have smaller allowances. In the case of heterogeneous workplaces, or where interruptions are the rule, higher allowances will be necessary.

The application of differentiated allowances is one way to "homogenize" time standards derived from different techniques with different degrees of accuracy. Experience in comparing work content time standards derived by more accurate measurement techniques with historical data and other less accurate techniques have led to the method of homogenization. Since engineered standards do not contain interruptions, disturbances, lack of information and/or training deficiencies, organizational shortcomings, or uneven workload, the practitioner will, for 1 hour recorded by historical means, commonly find 0.4 to 0.6 hour by engineered work measurement. In the homogenization process, historical data (representing organizational situations as is) may be multiplied by 0.4 to 0.6 in order to make them "comparable" with engineered standards, thus evening out (equalizing) differences in methods.

The vast differences between results of different time setting systems prove the following: There is a broad field for organizational and method improvement in the clerical and administrative area. Also, there are remarkable performance differences between a highly professional and an average employee, more in the clerical field than in the production field, where machines or automated procedures often set the pace.

Standard data can be constructed for a variety of clerical and administrative tasks. These data can be derived from any work measurement technique or a combination of techniques. Figure 7.6 is an example of standard data for tasks commonly found in purchasing, engineering, or administrative offices. The standard time values were developed from ROM II, a higher-level predetermined time system.

USE AND BENEFITS OF STANDARDS

Besides their main use as a means of performance control, standards, because of their minute description of work content and methods, can be a valuable aid for training new employees. As previously mentioned, other applications include capacity planning, scheduling, methods improvement and workplace design, and a basis for administrative cost accounting.

Documentation and Computerization. Working with a wide range of sources of data, the importance of exact documentation is obvious. Any mix of data seems acceptable, as long as the source and method used are identified. Computer-based documentation is recom-

TIME RECORD	TASK	STD. TIME (min.)
TS 4501	Sort and Distribute Incoming Post (each mailing)	0.90
TS 450101	Sort and Open Incoming Post	0.39
TS 4502	Make Reservation on Screen	9.63
	Stick Unfolded Document in Internal Distribution	
	Envelope, Address, and File Document	
TS 4502	Clerk Prepares Order Confirmation	1.26
	First Entry of Order Confirmation	
	WID Entry of Buyer or Screen Mutation	
	Register Plant Supply Requisition	
TS 4513	Register Delivery Slips (Small Supply Orders)	
TS 4521	Order Printout	0.63
	Order Mailing	
	Stick Document in Internal Post Map and Send	
TS 4524	Require Prepayment	14.19
	Typewrite Letter (10 Lines) Obtain Two	
	Signatures Stick in Envelope and Send	4.39
	Control Prepayments/Bills	
TS 4526	Account Prepayment on Corresponding Bill	
TS 4532	Enter Fixed Data on Screen, Mutate	7.41
	EDP - Maintenance (Per Day/Hr.)	
	Open GP, Information Requests	
	Information Requests	
TS 4544	Files, Printer, Copier	
TS 4545	Management Tasks (Per Day/Hr.)	1.30
	Other Tasks	
TS 4550	Work on Order Confirmation	2.06
	Order Printout	
	Require Signature on Incoming Bill	
TS 4501	Circulation Control of Incoming Bill	8.16
TS 456101	Bring Working Papers Within 10 meter - Area	0.32
TS 456102	Transcribe 3 Words From One List Into Another	
	and Colour - Mark the Line	0.32
TS 456105	Typewrite a Table of 42 Lines on Two Pages	16.75

* Standard times developed from ROM II, ROI Management Consulting AG, Zurich Switzerland

FIGURE 7.6 Example standard data for purchasing, engineering, and administration.

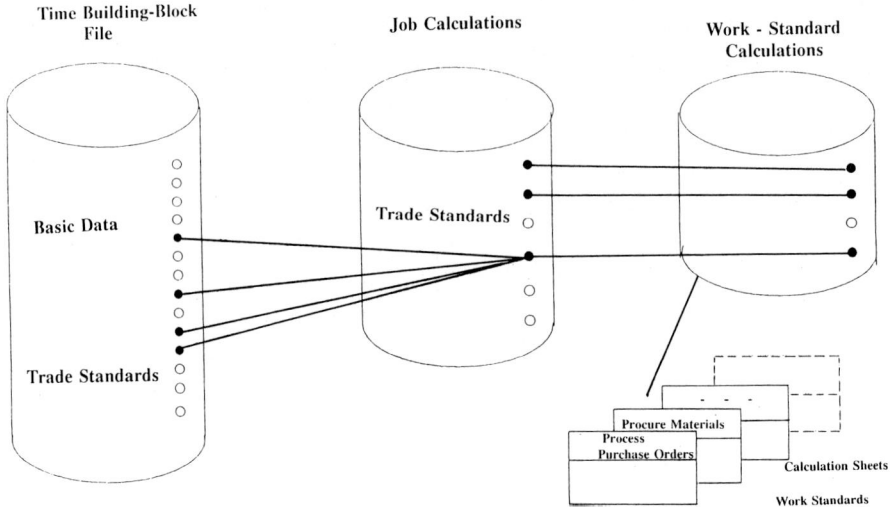

FIGURE 7.7   Computer-based documentation of office standards.

mended. Figure 7.7 depicts the structure for computer-based office standards. Computerization provides advantages in situations such as:

- Changing work content
- Changing layouts and/or workplace design
- New work content and applications
- New machinery or automation
- Improved methods of setting standards
- Change in counting units or key activities

In establishing clerical and administrative standards, it will sometimes be necessary (following the rules of sound economics) to cover some areas by using less accurate techniques. After 2 to 3 years following installation, it is recommended to go over these standards and in a second step improve the overall quality. This can be done simply by going through the documentation and identifying time standards generated with less accurate techniques or those that have not been well accepted. This process of standard improvement obviously is a continuous one and can be applied as necessary.

Control Procedures.   Summing up and partly repeating the basic principles and applications of clerical and administrative standards based on practical experience, the following can be said:

- The measurement system should provide total coverage.
- The choice of measurement technique(s) should be determined by overall economy and the type of work to be analyzed.
- Continuity of application is strongly recommended. Because of the relative accuracy of standards, the real test as to success or failure will depend on learning, experience, consistency in application, and flexible adaptation according to actual needs.
- Performance controls based on office work measurement are first-priority management tools. They are tools for establishing corrective measures on low or uneven performances of work groups rather than larger organizational units.

Benefits.   Measurement of clerical and administrative work will realize improvements throughout the organization. Particular benefits can be found in human resource utilization, methods, and overall performance.

Utilization.   Improvements include the following:

• Improved work scheduling
• Improved individual assignments
• More even workloads
• Redistribution of work content
• Necessary (desirable) changes in working times

Methods.   Improvements include the following:

• Detailed work descriptions
• Improved training conditions
• High methods sensitivity when applying predetermined time systems
• Improved office layouts, work flows, and communications
• Better basis for evaluating methods changes

Performance.   Improvements include the following:

• Objective tool of merit rating for individuals and groups
• Ready counterbalance of substandard performances through training or job rotation
• Basis for making organizational changes building more homogeneous and larger organizational units
• By using the system for cross comparisons, a more competitive attitude established within organizational units
• Giving the employee the feeling the company is basing competitiveness, decisions, and his or her own appreciation on modern objective management tools

Finally, some of the "by-products" of clerical and administrative work measurement and control procedures are identified below:

• Work counts are handled more accurately.
• Unmeasured times are reduced.
• The necessary lead time is gained in order to make personnel adjustments.
• Trends can be recognized.
• Actions are replaced by sound routines.

CHAPTER 8
LEARNING CURVES

Richard L. Engwall
Manager, Advanced Manufacturing Initiatives
Electronic Systems Group
Westinghouse Electric Corporation
Columbia, Maryland

The Industrial Engineering Terminology Definitions (ANSI Standard Z94.0-1982) defines learning curves as "a plot of productive output or unit work times of individual or group as a function of time or output per unit time; used to predict the learning rate in starting a new job or project. A learning curve is usually exponential and flattens out with time." Start-up curve is defined as "a learning curve applied to a job or process to adjust for work times longer than standard, or average, as a result of the introduction of a new job or new worker(s)." In the aerospace and defense industry the terms learning curve, improvement curve, and realization factor are sometimes used interchangeably. However, learning curve usually refers only to improvement due to operator learning while improvement curve is the result of cumulative gains in effectivity made by the organization, many of which are not impacted by the operator. Realization factor is the difference between estimated actual time versus standard time earned for either a particular of cumulative number of units produced. In the commercial world, the term start-up curve is more commonly used, since learning is one of the factors influencing start-up costs.

WHAT IS STANDARD TIME?

ANSI Standard Z94.0-1982 defines standard time as "a unit value time for the accomplishment of a work task as determined by the proper application of appropriate work measurement techniques by qualified personnel. Generally established by applying appropriate allowances to normal time." Normal time is "the time required by a qualified operator to perform a task at a normal pace to complete an element, cycle or operation using a prescribed method." Allowance is "a time value or percentage of time by which the normal time is increased, for the amount of nonproductive time applied, to compensate for justifiable causes or policy requirements which necessitate performance time not directly measured for each element or task." It usually includes irregular elements, incentive opportunity on machine control time, minor unavoidable delays, rest time to overcome fatigue, and times for personal needs.

Standard Time. Standard time assumes a fully trained operator is available. Operator learning is based on process, methods, and product familiarity.

Process Familiarity. The time required to train an operator in the basic skill requirements of the type of work being performed, this is sometimes included in manufacturing overhead cost.

Methods Familiarity. The number of repetitive cycles for an average operator to become sufficiently familiar with a specific method to be able to perform at standard; a certain amount of this learning may be transferred with an operator who starts to perform work which is similar to a previous job.

Product Familiarity. A number of repetitive cycles are required for a normal operator to apply the process skill acquired and methods familiarity to perform a standard operation on a specific item. The more complex the product, the greater the number of items, the longer the cycle time, the longer to learn.

Commercially accepted predetermined time standard data accuracy and reliability criteria are based on a 2-minute average cycle time which is mostly applicable to high-volume commercial industry. The complexity of the operation, length of job cycle time, redundancy of identical patterns of motions done repeatedly in the course of performing the operation, and process capability to meet design requirements all influence the ability or inability of an operator to achieve the normal standard methods level of learning proficiency.

Thus in layman's terms, standard is the time required for a fully trained operator to complete a defined element of work to known specifications following a prescribed method utilizing specific equipment, tools, material, and workplace layout, including average personal, fatigue, and delay time allowance. Standard time is theoretically achievable when rework and scrap are eliminated, all production rate tooling is available, engineering requirements and manufacturing processes have stabilized, machine downtime is eliminated, labor power is perfectly matched to schedule requirement, parts and material shortage problems have been eliminated, workers have become familiar with all aspects of the job, and work has been repeated in a stable environment many times, that is, hundred to thousands. All these conditions are seldom, if ever, achieved; yet standards are commonly met after sufficient learning on high-volume, short-cycle, mature and stable products operations.

IMPROVEMENT CURVE ELEMENTS

Actual time variance from standard time can be categorized as learning, logistics, technical, and miscellaneous variance. It should be noted, as shown in Figs. 8.1 and 8.2, that these categories of variance from standard all approach standard over the period of time that it takes to produce X units. Each of these categories of variance from standard comprises the real improvement opportunity, requiring specific management action to eliminate the cost drivers.

Learning. Three key factors affect learning—operator methods proficiency, product and process familiarity, and loss of learning.

Operator Methods Proficiency. Four levels of operator method proficiency are recognized in the development of MTM predetermined time standard data. See Fig. 8.3. Several current MTM-based systems as well as other method predetermined systems account for the various levels in standard time development.

1. *The normal standard methods level.* That standard time which is established based on performing simultaneous manipulations during the reach move.

2. *The low methods level.* That standard time which is 169 percent of normal standard time based on sequential (single-hand) moves, no manipulation action during the reach move.

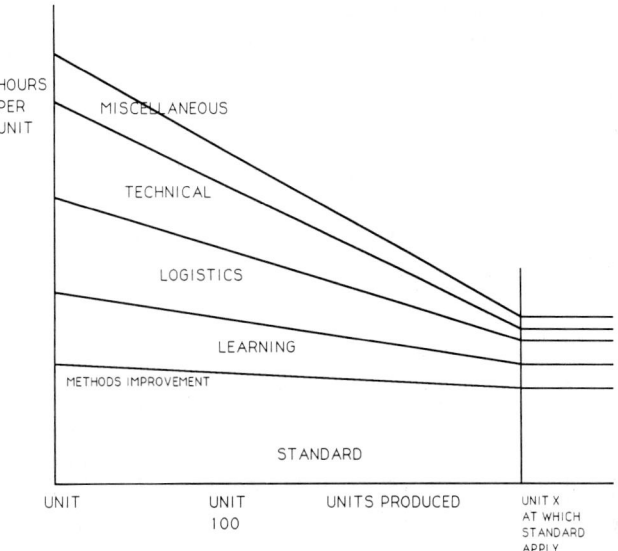

FIGURE 8.1   Typical improvement curve.

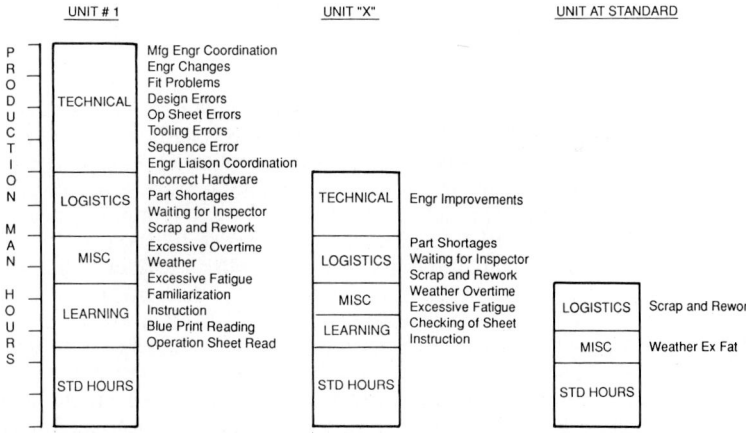

FIGURE 8.2   Improvement curve elements.

3. *The high methods level.*That standard time which is 138 percent of normal standard time that requires simultaneous two-hand moves, no manipulation actions during the reach move.

4. *The optimum methods level.*That standard time which is 94 percent of normal standard time, is based on higher-class actions using the eye, brain, and/or ear to be accomplished during the reach move.

As stated in MTM research literature, the improvement in performance (reduction in cycle time) for manual work results from a combination of classical and instrumental conditioning achieved by the practice of motor skills made significant by perceptual input within the information processing capability of the learner. Until the perceptual capacity of the manual learner has progressed, usually by a process of instruction followed by trial-

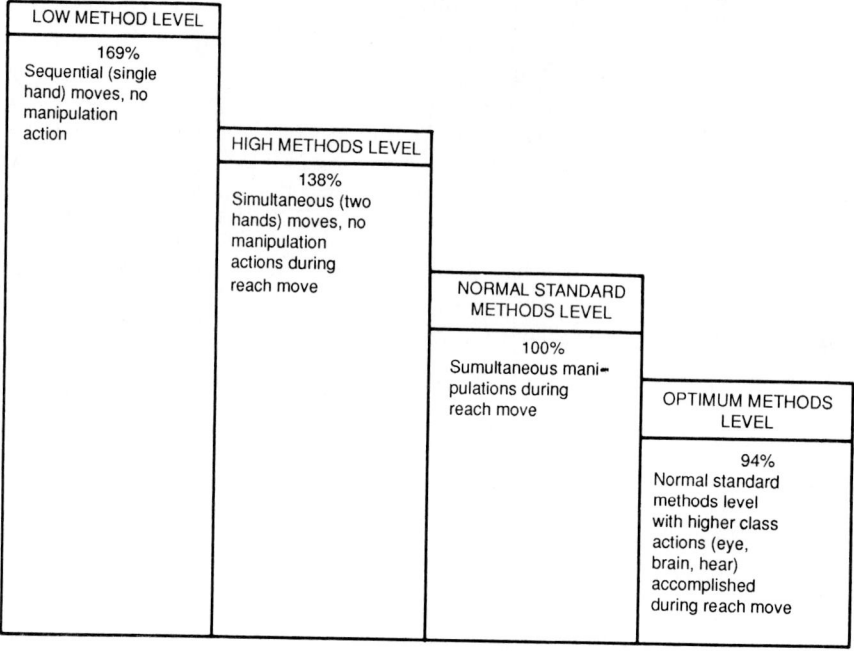

FIGURE 8.3 Learning proficiency. (*MTM-I*) (*Source: MTM Association*)

and-error practice, to the point at which the learner can perform a task unaided and without hesitation, the learner's performance will be in what is considered the threshold reinforcement region. Having crossed this barrier into the reinforcement region, the manual learner repeats enough task cycles to develop movements automatically. When the rate of cycle performance reaches that required by a fully trained operator with normal skill under normal conditions, the standard MTM time is achieved.

It is unreasonable to think that an operator can achieve normal standard methods level without repetitive production experience. In the aerospace and defense industry standard time is quoted as being achievable at the 1000th unit (Wright theory). In commercial industry application, with an average 2-minute cycle time, 1000 units represent 33.3 hours or 1 week elapsed time. Conversely for the aerospace and defense industry average assembly cycle time (90 minutes standard time) 1500 hours or a minimum of 9 months is required of repetitive production to achieve the status of a fully trained worker.

Product and Process Learning. Two of the key factors influencing the level of operation methods proficiency are familiarity with the product and process. Standard time can best be achieved when an operator is performing the same operation (process) each day or the same parts or assemblies (product). Each process and product can have unique technical performance requirements necessitating a different method to be performed.

1. *Product familiarity* is that number of repetitive cycles to become fully trained in the idiosyncrasies of specific product requirements different from other similar products performed by the same operator. This is particularly true of aerospace and defense industry products with their myriad of military specifications, many times unique to a specific product division of a specific service.

2. *Process familiarity* is that number of repetition cycles to become fully trained in the basic skill requirements of the type of work being performed for a specific skilled pro-

cess. Each major process area has similar methods, tooling, equipment, workplace layout, work instructions, and the like, regardless of the product being manufactured.

For both product and process familiarity, the complexity of the product, length of cycle time, and variability of similar products all significantly influence the number of repetitive cycles needed to achieve the level of a fully trained operator which the standard time is based on.

Loss of Learning.   Loss of learning due to break-in production continuity generally occurs more frequently in the aerospace and defense industry than in commercial industry. The commonly accepted Anderlohr theory (see Fig. 8.4) recognizes the changes from prior production that occur because of breaks in production that involve operator learning and supervisory learning as well as those changes in methods, tooling, and plant layout that adversely impact achieving prior cost performance. To apply the loss of learning factor one estimates the percentage of learning loss due to the following five factors.

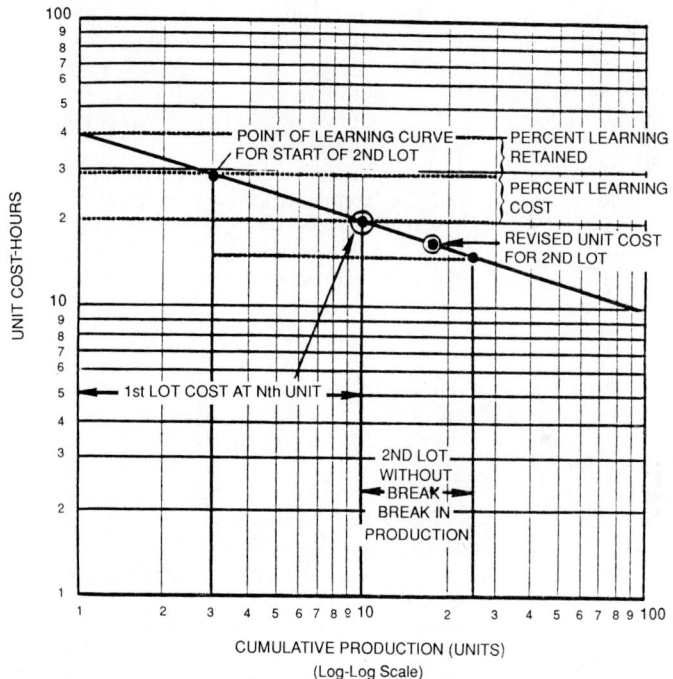

FIGURE 8.4   Loss of learning due to product breaks (George Anderlohr theory).

1. *Operator learning* covers actually forgetting work procedures, hiring untrained replacement personnel, and rehire of old personnel.

2. *Supervisory learning* loss is the loss resulting from transfer of supervisor, limited knowledge of new hires, and reduced guidance they furnish because of lost familiarity with the job.

3. *Continuity of production* relates to the physical establishment of production lines, the position adjustments for optimal working conditions, and work in progress buildup.

4. *Methods* concerns rerouting of operations due to in-plant changes since the last production lot.

5. *Special tooling* includes short-run versus long-run tooling, replacement of modified tools, and the effect of transition time.

Logistics.   Logistics factors contributing to failure to achieve standard time include parts availability, manufacturing complexity, production rates and schedules, and machine availability and facility layout.

Parts Availability.   This is a problem for new or recently developed processes or techniques which take time to settle down into a routine. As the engineering design matures, the impact of design changes and fixes reduces dramatically. During the production cycle, there are times when tool breakdowns or wear and tear can impact the continuous work flow. Inadequate time to procure material resources can influence the continuous work flow. Because of the low quantity requirements and more demanding production specifications for parts procured for the aerospace and defense industry relative to commercial industry, vendors place lower priority on needs.

Manufacturing Complexity.   Because of the complexity of items produced, the process yield has a high variability which is difficult to forecast. Company and customer inspection availability where required cannot always be coordinated perfectly with the production mechanic. Effective production control support is necessary to assure the availability of hardware, parts, power or hand tools, assembly tools, and assembly paperwork to achieve standard. The production operation sequence data package must have the necessary specification call-outs and operation work instructions which reflect the most efficient operation. This requires added time for reading the appropriate documentation, not always included in the standard time.

Production Rates and Schedules.   The coordination of unit cost reduction and production rate buildup must be in concert in order to minimize laborpower requirement fluctuations. Production cycle variations which are increasing and/or decreasing from a requirement standpoint cause inefficient laborpower utilization. The stops and restarts of terminations and gaps in the product life cycle cause a relearning process and a loss of progress in the improvement curve.

Machine Availability and Facility Layout.   Sometimes machines unexpectedly break down and/or are not available due to higher schedule priorities, which causes production to move to less efficient machines. Because of the uncertainties of future work, the most efficient plant layout must be developed and reevaluated for constant changes to avoid congestion, improve work flow, and best utilize space.

Technical.   Technical factors contributing to failure to achieve standard time include engineering, planning and tooling, and quality, all impacted by the state-of-the-art technology, product complexity, and producibility related issues.

Design Changes.   These result in tooling changes, process specification changes, product flow changes, worker unlearning and relearning, rework of units or parts built to early configuration, work stoppages pending revised paperwork and components, materials, and expendables not available and/or not to specification. These design changes are caused by unpredictable design, tolerance stack-ups, early start-up of production (that is, prior to completion of design verification), and lack of funding, time, and/or front end emphasis on producibility versus functionality of the design.

Tooling and Planning.   These problems create rework and repair caused by tools improperly designed or built, rework caused by process planning errors or wrong call-outs, and alternate or work around methods (that is, open setups versus jigs and fixture, conventional operations versus numerical control, single-spindle versus multiple-spindle, layout versus templates, labor-intensive operations versus automation and tool liaison interruptions). Tooling or planning changes are caused by schedule, funding, and business constraints:

1. *Schedule Constraints.* They include tools designed from preliminary engineering releases, lead time for tool fabrication, time availability for tool-proofing, and inflexibility of schedule changes.

2. *Funding Constraints.* They include producibility review of tool designs, expense tooling budgets, and automation capital availability.

3. *Business Constraints.*They include peaks and valleys in requirements for highly skilled workers in planning, tooling, and first article inspection.

Unclear and/or Unrealistic Quality Acceptance Criteria.   They result in usable hardware being rejected, excessive and/or redundant inspection buy-offs, and added rework and reinspection costs being incurred. These quality changes are caused by a low confidence factor on new product, unclear product requirement criteria, part or product complexity, process capability versus state-of-the-art designs and incomplete training—new workers, new products or processes.

Miscellaneous.   Even if we can control all learning, logistics, and technical factors, there still are times, particularly in the new product start-up phase, that excessive overtime is required to meet schedules. Similarly, excessive fatigue, created by excessive overtime and/or having to correct errors caused by others, adds time over and above standard. Also, if there is a snowstorm or some other act of God, people must be paid a minimum number of hours. These costs are sometimes covered in overhead but, if not, need to be factored into estimated total expected hours.

LEARNING CURVE EQUATION

Wright Theory.   The most common and useful learning curve model was developed in 1925 by Colonel L. McDill at McCook Field (Wright Patterson Air Force Base, Dayton, Ohio). T. P. Wright of Curtis Wright Corporation in Buffalo, New York, first applied what is now commonly called the Wright curve. He observed the cumulative unit cost to produce an airplane decreased at a predictable rate as the quantity produced increased. Wright noticed a predictable cost reduction rate in which the cumulative unit cost decreased by a constant percentage as the cumulative quantity increased by another constant percentage. Wright's original model has remained popular mainly because of the simplicity of its application. Other models became too complex to apply to typical production problems even for mathematicians utilizing programmable computers. To be effective, the models' mathematics applications must be understood by the typical manager, engineer, and factory worker. Also, the user must be able to apply the mathematics easily to a whole range of situations while experiencing minimal error.

Crawford Theory.   The other most common application is the unit cost curve (Crawford theory). The cumulative average curve (see Fig. 8.5) identifies the cumulative average cost for the nth unit and quantity in production, plotted cumulatively. The unit curve (see Fig. 8.6) identifies the unit cost for that particular nth unit of production, again plotted cumulatively. In both instances, the "improvement curve theory," simply stated, is that as the cumulative production quantity doubles, a specific cost improvement will occur. A 90 percent improvement curve assumes that as the cumulative production quantity doubles, the cost [either unit (Crawford) or cumulative average (Wright) cost] will be 90 percent of the former cost, which is a 10 percent cost reduction. Similarly, for an 80 percent improvement curve, a 20 percent cost reduction will occur as the cumulative production quantity is doubled.
    The expression of the Wright model is an algebraic power function and has a mathematical form:

$$\text{Cumulative unit cost } c = FX^{-n} \qquad (1)$$

where F = cost of first unit produced
    X = cumulative quantity of units produced
    n = learning rate

The value of n is usually related to learning curve percent by the equation

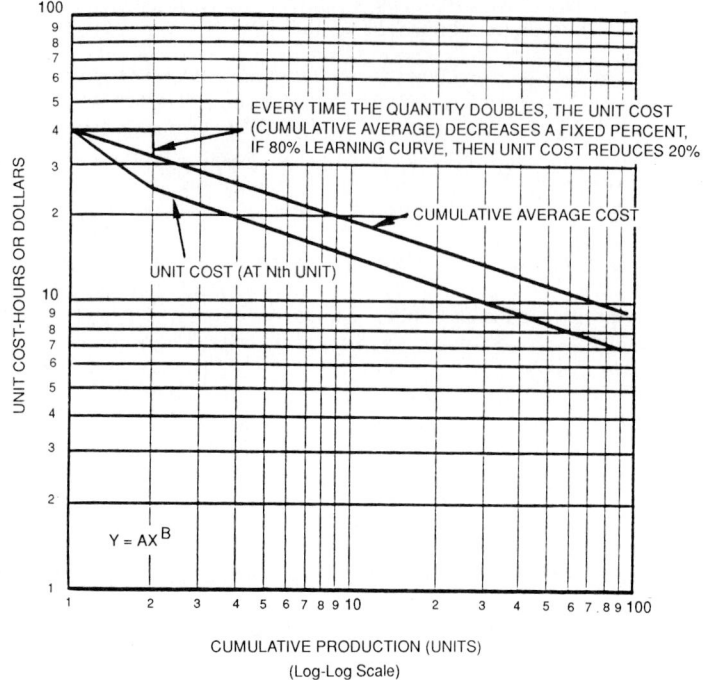

FIGURE 8.5 Cumulative average learning curve (Wright theory).

$$n = \frac{\log \text{ learning curve rate (or percentage)}}{2} \qquad (2)$$

expressed as a decimal, that is, 0.8 equals 80 percent learning curve.

Table 8.1 shows quantities and costs for one particular learning curve rate (80 percent) which is provided as an example. When the equation for either of the two curves ($c = FX^{-n}$) is plotted on regular (arithmetic) graph paper, a curve is obtained. (See Fig. 8.7.) However, if log-log graph paper is used to plot the equation, a straight line will result, with descending slope as the cumulative production increases on the X axis, the unit cost, hours or dollars, being plotted on the Y axis. When the factors based on the Wright (cumulative average) theory are used in developing the improvement curve, the cumulative average is linear, and the results at unit cost arrived are nonlinear for the first units. The inverse is true for the Crawford (unit) theory; the unit cost being linear and the cumulative average cost being nonlinear for the first units. See Figs. 8.5 and 8.6. In either theory, Crawford or Wright, the cumulative average and unit cost are the same for the first unit. However, for subsequent production the cumulative average is always greater than the unit cost for the nth cumulative unit of production. Nevertheless, as the production quantity increases, at the nth unit, the difference between the cumulative average and unit cost stabilizes.

ISSUES

Changing Standard Time Base.   The standard time base for a given product generally changes significantly during the early stages of production. The completeness of the initial product design release has a significant impact on estimating cost. For complex products, engineering

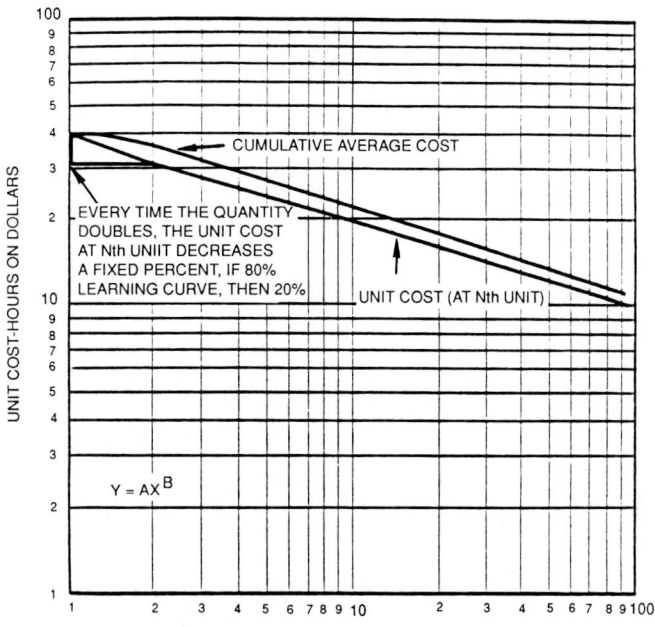

FIGURE 8.6   Unit cost learning curve (Crawford theory).

TABLE 8.1   Example of Cost Quantity Relationship for an 80 Percent Learning Curve

Cumulative quantity produced	Cumulative unit hours or cost	Learning curve rate (%)
1	40.0	
2	32.0	80
4	25.6	80
8	20.5	80
16	16.4	80
32	13.2	80
64	10.6	80

historically underestimates its requirements, causing the standard time to increase until materials and tooling changes can be instituted to reduce standard time. If the estimates proposed in the various stages of development can be compared with each other, a growth factor can most likely be developed to improve the estimating accuracy of initial production lots.

This changing standard time base can best be seen in Fig. 8.8. The current standard time base should always be used as the basis for all new estimates. However, prior cost variance was predicted on the then-effective standard time. If different allowances are included in standard time, or any inspection, rework, or similar activities are changed, the realization factor would be impacted but total actual time might not be changed. When this occurs, any prior percentage relationships used in estimating can be distorted. During the

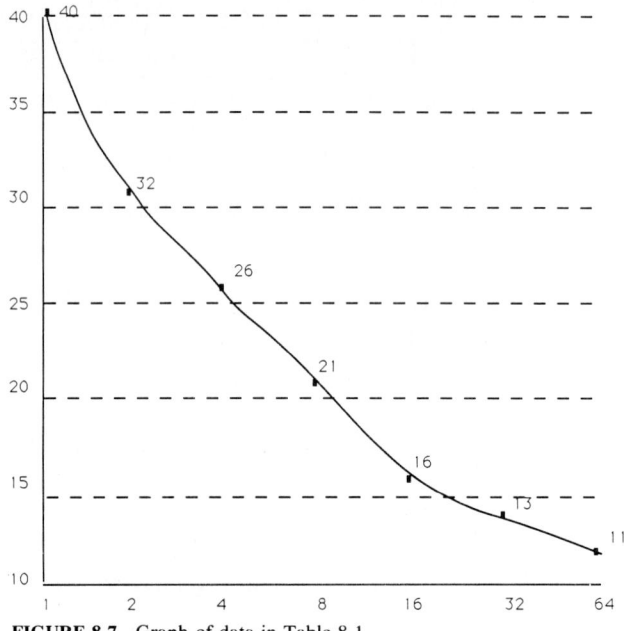

FIGURE 8.7   Graph of data in Table 8.1.

initial product design prove-out stage, different processes and methods are utilized to eliminate need for costly production tooling that might need to be changed or even voided by subsequent engineering changes. Thus the standard time and expected realization factor would most likely be different from that realized in subsequent production tooled processes.

Product producibility capability significantly impacts total labor hours, both standard time and realization factor, depending on the percentage of rework that is required to produce acceptable parts. This unpredictability of producibility problems severely impacts learning and logistics costs, in addition to technical costs themselves.

Automation.   Automation has an impact on learning curves. Typically the more automated the process the more controllable, particularly with respect to repeatability. Machine and tool prove-out are additional one-time setup costs, but once the process and method are proved out there is little learning, except for the person time component of cost, that is, materials and tool handling, perishable tool replacement, setup, and the like. Many of the technical and logistic cost drivers impacting the learning curve are "controllable" by the more automated process. Thus processes like machining, sheet metal, and robotic assembly approach a 100 percent learning curve, while final assembly and test of large complex items, with long cycle times, approaches a 70 percent learning curve.

Individual company labor relations policies may have a significant effect on allowable person-machine ratio. Manual time for setup, tool replacement, chip rework, and inspection can still be significant in automated processes. Fully automated material handling and tool setup, as well as perishable tool replacement, are being introduced to eliminate any person time.

Product Life Cycle Impact.   If one reviewed the learning curve relationships for the full product life cycle of a complex, high-technology new product, it probably would not follow a linear curve. Figure 8.9 attempts to graphically describe what most likely occurs. Usually the learning curve is steeper in the initial production stages because of the tech-

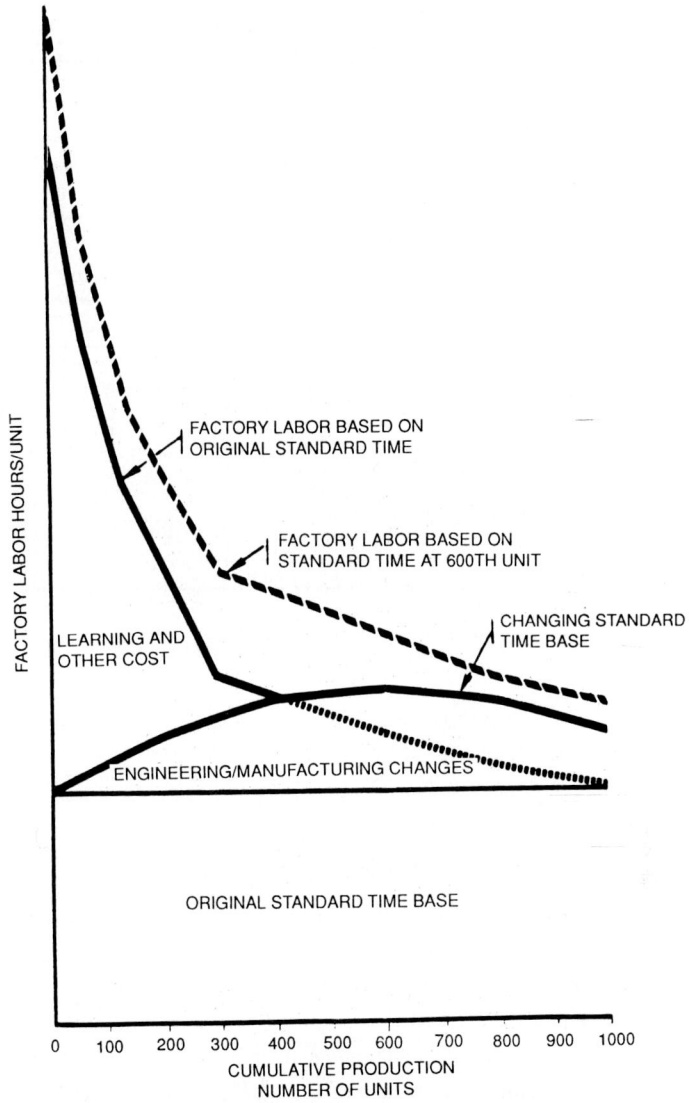

FIGURE 8.8   Changing standard time base.

nical and logistic cost driver activities previously defined. It then flattens out in three more stages to the point that only the learning cost driver activities remain. Finally, the improvement curve actually realizes a labor cost growth near the end of its product life as production rate and volume decrease, necessitating process and method changes. Machine utilization and tool repair costs don't warrant more automated processes anymore. This relationship is highly influenced by

• Production rate and volume changes
• Manufacturing methods, tooling and equipment utilized
• Product complexity

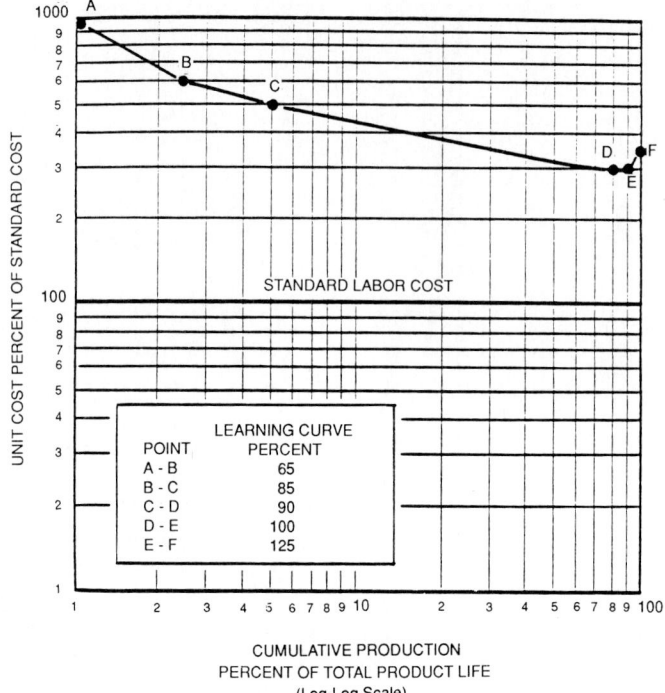

FIGURE 8.9   New product overall learning curve.

- Producibility
- Value engineering changes
- Make-versus-buy decisions

Estimating Accuracy.   The accuracy and reliability of estimating factory labor is normally highly dependent upon the previously defined factors:

- Well-defined and consistently applied methodology
- Accuracy and consistency of standard time estimate
- Improvement curve elements—technical, logistics, and learning
- Maintaining configuration control of a changing standard time base
- Automation
- Product life cycle impact

Findings and evaluations point out the need for refinement in all six areas identified above. There appears to be more of a tendency to understate rather than overstate the estimated cost if the stated methodology was used. Measuring the standard time to an accuracy of ±5 percent at a 95 percent confidence level is possible. This is especially true if an acceptable predetermined time system is used properly. However, the definition of standard time presupposes a fully trained operator, with normal skill, with normal working conditions, working to a prescribed method. And even more importantly one needs to "learn" the product and process and repeat the operation over and over again without a break in continuity in order to achieve standard.

The learning curve phenomenon comprises many variables, most of which contribute

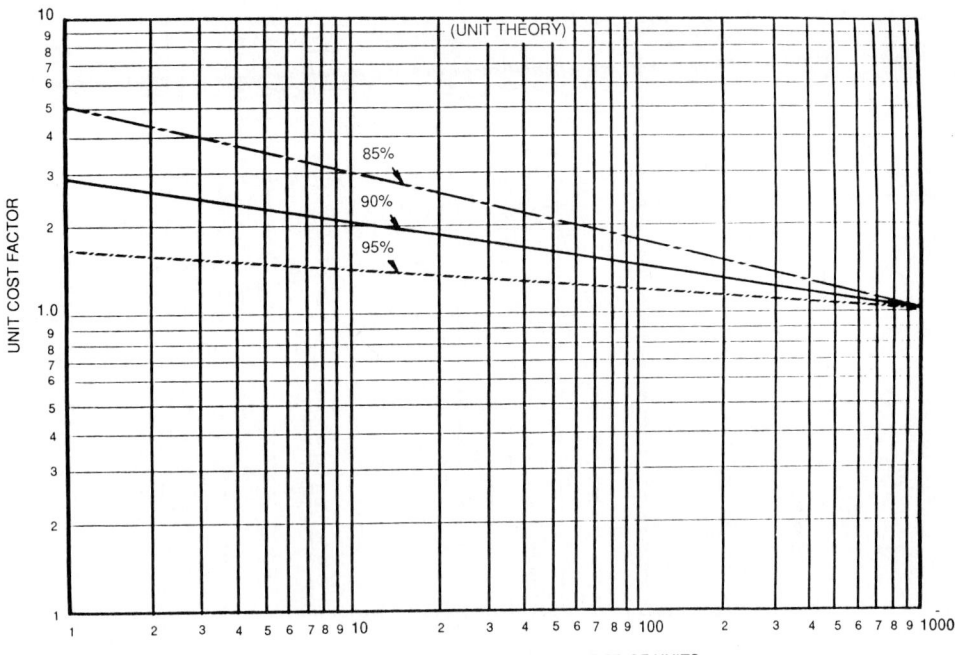

FIGURE 8.10   Variance learning curve (standard at unit 1000).

to taking longer to perform a standard operation than standard time. Thus it is considered more of an "art" than a "science" to predict the point on the learning curve at which the standard time applies. In addition, the percentage of improvement varies significantly in applying the learning curve theory.

Figures 8.10, 8.11, and 8.12 highlight the unit cost factor comparison for the following widely recognizing assumptions:

- Figure 8.10: based on various learning curves (85 to 95 percent) standard base at unit 1000
- Figure 8.11: based on standard being met at various nth unit quantities (100 to 1000), learning curve assumption 90 percent
- Figure 8.12: learning curves 85 to 95 percent point on learning curve applicable 100 to 1000

The variability of just the point at which the standard time applies, assuming a 90 percent learning curve relationship, varies from 46 percent at unit 10 to 40 percent at unit 100 to 28 percent at unit 1000. Similarly, the variability of the present learning curve relationship used, assuming the standard time applied to the 1000th unit, varies from 114 percent at unit 10 to 46 percent at unit 100 to no variance at unit 1000. The overall variability of these two parameters, when combined, varies from 150 percent at unit 10 to 70 percent at unit 100 to 66 percent at unit 1000. Table 8.2 summarizes the unit cost factor for unit 1, based on the same overall parameters defined above.

In addition, there is further potential variability in being able to consistently estimate the break-in-production continuity impact on loss of learning. The same is true for determining the impact of change in production rate and volume.

In summary, there is a significant degree of vulnerability in not being able to accurately estimate factory labor, when considering all the factors involved. However, fortunately,

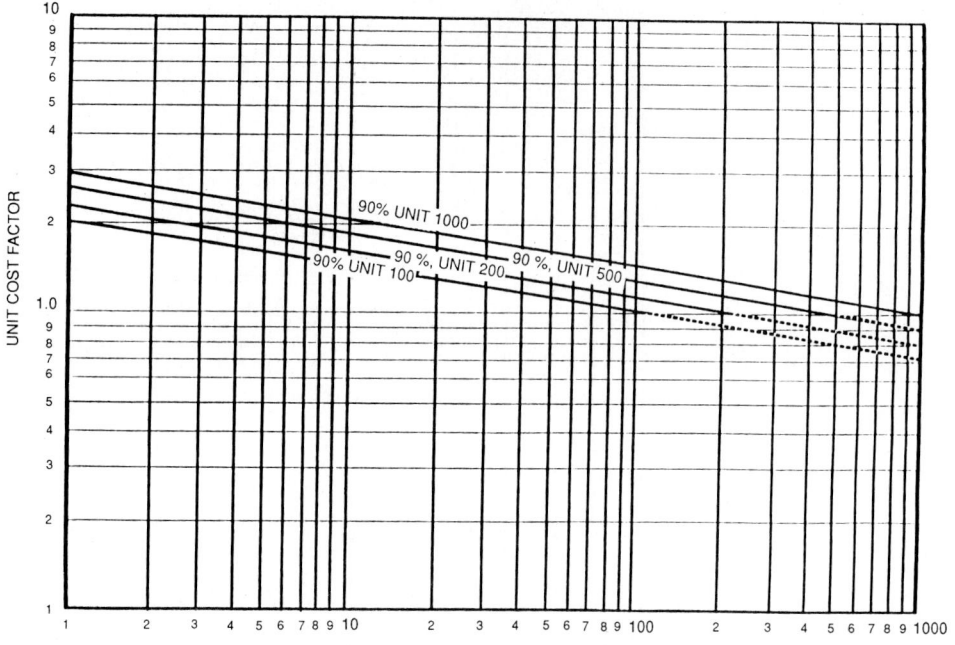

FIGURE 8.11   Standard at *n*th quantity level (90 percent learning curve).

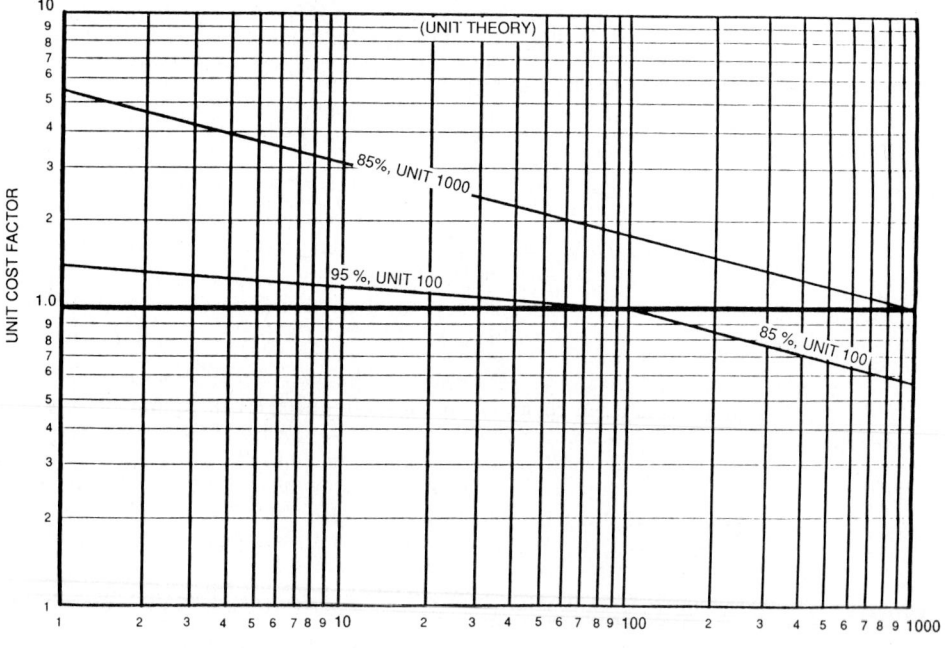

FIGURE 8.12   Learning curves 85 to 95 percent point on learning curve applicable 100 to 1000.

TABLE 8.2   Unit Cost Factor for Unit 1

Learning curve (%)	Cumulative production unit at which standard applies			
	100	200	500	1000
85	2.94	3.46	4.29	5.05
88	2.34	2.65	3.15	3.58
90	2.01	2.24	2.57	2.85
92	1.74	1.89	2.11	2.30
95	1.41	1.48	1.58	1.67
97	1.22	1.26	1.31	1.35

factory labor is a relatively small percentage of the inventory cost of sales. Total manufacturing "hands on" direct labor, including test and inspection, typically as factory labor, accounts for 5 to 15 percent of total direct cost of sales. Manufacturing overhead, again for factory, test, and inspection labor, typically accounts for 10 to 30 percent of direct cost of sales.

A more methodical utilization of learning curve theory will realize a significant improvement in estimating accuracy. Of particular interest should be the vast improvement in supportability of the factory labor estimate in negotiations, as well as having better internal visibility of the actual cost being incurred as compared with estimate. Thus percentage relationships can ultimately be empirically derived to further improve the accuracy of the estimate.

LEARNING CURVE APPLICATIONS

The most significant application of learning curves is in estimating future costs. T. P. Wright noticed a predictable cost reduction (improvement) rate in which the cumulative unit cost decreased by a constant percentage as the cumulative quantity increased by another constant percentage. Initially most people related this learning curve phenomenon to improved worker performance due to the repetition of performing the same task over and over again. However, more modern theory recognizes cost improvement is mostly due to contributions from management, engineering, and other support departments to attack the factors other than operator, product, and process learning, namely, technical and logistics factors. Figure 8.2 highlights the activities that need to be eliminated or at least significantly reduced if one wants to ultimately achieve standard time. In today's total quality management process action team environment, identification of all major non-value-added activities is necessary. Tracking of improvement curves performance in terms of unit hours or cost versus cumulative quantities of production is essential to continuous process improvement.

Learning curve is really a mathematical, graphical, or tabular representation of how resources are reduced as production of a product is repeated. Thus the learning or improvement curve is used mostly to predict production costs from known historical costs of producing products. The unit of measure is typically unit hours of worker labor required. However, materials cost, facilities floor space, energy, and the like can be similarly estimated when quantity of repetition improves performance.

In manufacturing, improvement curve representatives are used to plan laborpower needs, estimate expected actual labor performance (hour per unit), establish sale prices and make or buy decisions, judge wage incentive payments if applicable, evaluate organizational efficiency, evaluate capital equipment proposals, create more accurate production delivery schedules, and estimate new product start-up costs to be amortized in standard cost of product. In the aerospace and defense industry, improvement curves are used extensively to estimate future unit hours. Knowing what the first unit's hours were and what percentage learning curve performance has been realized, one can predict the expected unit hours (standard time plus realization factor).

$$RF_n = RF_1 \times \text{quantity (log LC/log 2)} \tag{3}$$

$$RF_{20} = 9.65 \times 20 \ (\log 0.831/\log 2)$$

$$= 9.65 \times 20 \ (-0.0804/0.3010)$$

$$= 0.95 \times 0.4493$$

$$= 4.33$$

See Fig. 8.13.

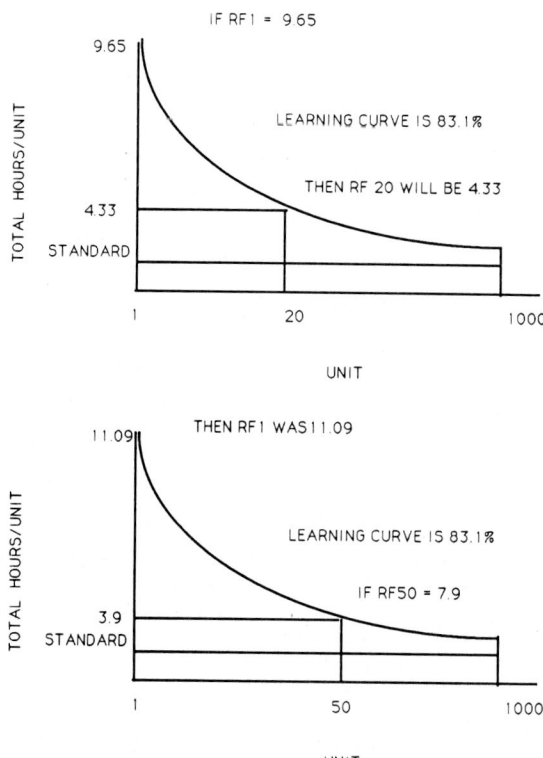

REALIZATION FACTOR FORECASTING

FIGURE 8.13 Realization factor forecasting.

Similarly, if the first unit realization factor is not known but historical costs indicate that product A is running a 3.9 realization factor at the fiftieth unit (assuming 0.831 learning curve), one can solve for what the realization factor at unit 1 was (11.09).

$$RF_1 = RF_n$$

$$n \ (\log LC/\log 2) = 3.9/50 \ (\log 0.831/\log 2)$$

$$= 3.9/50 - 0.804/0.3010$$

$$= 3.9/0.3518$$
$$= 11.09$$

See Fig. 8.13.

BIBLIOGRAPHY

Reference Books

"Manufacturing Time Forecasting Curves," chap. 16 in Frank W. Wilson, ed., *ASTME Manufacturing Planning and Estimating Handbook,* McGraw-Hill, New York, 1963.

Hartmeyer, Fred C., *Electronics Industry Cost Estimating Data,* The Ronald Press, New York, 1964.

"Current MTM Research-Learning Curves," chap. 20 in Delmar Karger and Franklin Bayha, eds., *Engineered Work Measurement,* 2d ed., Industrial Press, New York, 1966.

Jordan, Raymond, *How to Use the Learning Curve,* Cahners Publishing Company, Boston, Mass., 1973.

Hancock, Walton M., "The Learning Curve," in H. B. Maynard, ed., *Industrial Engineering Handbook,* 3d ed., chap. 5, McGraw-Hill, New York, 1971.

Cochran, E. B., *Planning Production Costs—Using the Improvement Curve,* Chandler Publishing Company, San Francisco, Calif., 1967.

Cochran, E. B., *The Learning Curve,* Industrial Education Institute, Boston, 1965.

Special Publications

Learning Curves—Theory and Application, American Institute of Industrial Engineers Publication No. 6 Monograph, Work Measurement and Methods Engineering Division, Comprehensive Bibliography of 6 Books, 98 Articles, and 55 Reports and Bulletins.

Learning Curve Program Using MTM, MTM Association for Standards and Research, 1985.

Learning Curve Research on Manual Operations: Phase II, Industrial Studies Report 113A; Short Cycle Operation Report 112, MTM Association for Standards and Research, 1969.

Smith, Jason, *Learning Curve for Cost Control,* Industrial Engineering and Management Press, 1989.

Hearings before the Subcommittee on Administration Practice and Procedure of the Committee on the Judiciary US Senate 99th Congress Examining of Agency Information Flow, July 23–24, 1985, Serial No. 5-99-42.

Reports and Bulletins

Guidelines for Application of Learning/Cost Improvement Curves, Dr. Leon M. Delionback Systems Analysis and Integration Laboratory, Marshall Space Flight Center, Alabama, October 1975, United States Department of Commerce NTIS Report N76-13970.

The Learning Curve: Historical Review and Comprehensive Survey, Louis E. Yelle, University of Lowell, *Decision Sciences,* vol. 10, no. 2, April 1979, pp. 302–328, Comprehensive Bibliography of 99 Articles, Reports, and Bulletins.

Articles

Malstrom, Eric M., and Richard L. Shell, "A Review of Product Improvement Curves," *Journal of Industrial Engineering,* May 1979.

Anderlohr, George, "Determining the Cost of Production Breaks," *Journal of Industrial Engineering,* September 1965.

Bump, Edwin A., "Effects of Learning on Cost Projections," *Management Accounting,* May 1974.

Abernathy, William J., and Kenneth Wayne, "Limits of the Learning Curve," *Harvard Business Review,* September/October 1974.

Ghenauwet, Pankaj, "Building Strategy on the Experience Curve," *Harvard Business Review,* March–April 1985.

INCENTIVE PROGRAMS

CHAPTER 1
MEASURED DAYWORK

Louis M. Kuh
Principal, Kuh & Associates
Management Consultant
Stamford, Connecticut

The term "daywork" implies nonincentive conditions, but "measured daywork" qualifies as an incentive program, although one that is not associated directly with pay rates. The term has been used in various contexts over the past sixty or so years, ranging from some form of pay rate adjustment to the simple provision of work measurement for other than labor control purposes. More recently, however, measured daywork refers to a basic system of labor control that plays a key role in the maintenance of a satisfactory level of productivity.

DEFINITION

Measured daywork is a form of managerial control involving (1) the development of time standards for the performance of work; (2) regular determination of worker and group performance, utilization, and productivity; (3) an hourly rate of pay that is not raised or lowered by the level of performance; (4) the mutual recognition that as a condition of employment the worker is responsible for maintenance of an acceptable level of performance, quality of production, attendance at work, and safety on the job; and (5) recognition that management has the responsibility to maintain working conditions, material supply, and machinery and equipment at levels that will permit the average worker to meet the established standards.

It is step 4, above, that provides the incentive application of a measured daywork program. Perhaps the only difference between measured daywork and other incentive programs is the actual determination of wage payments. Figure 1.1 illustrates the various characteristics of pay systems, and the applicability of each characteristic for unmeasured daywork, measured daywork, and other incentive systems of management control.

Time Standards. Proper labor time standards are an essential part of a measured daywork system. Depending on the type of work being done, any appropriate method of standards development may be used, such as stopwatch time study, group-timing technique, standard data, or engineered predetermined time data.

It must be realized that accurate labor standards are essential for labor control, as well

CHARACTERISTIC	UNMEASURED DAYWORK	MEASURED DAYWORK	INCENTIVE SYSTEMS
WAGE PAYMENT			
Predetermined	X	X	
Variable – Performance			X
WORKER PERFORMANCE			
Unmeasured	X		
Measured		X	X
SUPERVISORY CONTROL			
Subjective	X		
Objective		X	X
WORKER RESPONSIBILITY			
Undefined	X		
Defined		X	X
MANAGEMENT RESPONSIBILITY			
Undefined			
Defined	X	X	X
BASIS FOR MANAGEMENT CONTROL –			
OPERATING & FINANCIAL			
Estimates/History	X		
Work Measurement		X	X

FIGURE 1.1 Characteristics of daywork, measured daywork, and incentive systems.

as for work planning and control, cost determination, and any other effective management control. The fact that the time standard does not directly determine the workers' pay rate in no way permits the use or application of an inaccurate standard.

The correct labor standard includes all features such as the 100 percent of "fair day's work" leveling of the raw time, and proper application of PR&D (personal, rest, and unavoidable delay) factors.

Finally, standards must be maintained as rigidly as for any other type of incentive plan or any other effective application of work measurement.

Performance, Utilization, and Productivity.   The first-line supervisor, and all levels of management, require reports that keep them informed about worker performance, utilization, and productivity. That information is the basis for management action designed to maintain an acceptable level of product output.

All too often, performance is the only measurement made. Performance alone is not only deceptive, it may be manipulated to provide results favorable to the worker. By adding the measurement of utilization, management not only can detect attempts to manipulate performance data but will have a measure of those activities that reduce the amount of time actually spent on productive work. Productivity is the measure of overall effectiveness as determined by the product of performance and utilization.

Performance is determined by dividing the standard time required to do a task by the actual time used. The actual time used does not include any time spent on an activity not included in the time standard. It is essential that downtime be identified and coded for application within the timekeeping system in use. Typical downtime activities include machine breakdown, lack of material, reset machine for quality specification, lack of material handling equipment, or any other item that effectively stops production.

Utilization is a measure of the time spent working on productive tasks as a percent of the total time available for work. The total time available is the actual work time, plus downtime, plus time spent on assigned unmeasured work.

Productivity is measured by calculating earned time (from the time standard) as a percent of the total time at work. As an example, assume that a worker produced 650 items that earned 6.86 standard hours, in 7.0 actual work hours. Performance is $(6.86 \times 100)/$

7.00, or 98 percent. Assuming an 8-hour day, utilization is $(7 \times 100)/8.00$, or 87.5 percent. Productivity is $(6.86 \times 100)/8.00$, or 85.75 percent. The same value can be calculated by multiplying performance (as a decimal) by the percent utilization, or 0.98×87.5 percent, equals 85.75 percent.

Wage Payment Structure.   Any normal nonincentive wage structure is usually acceptable. However, it is desirable that the wage should be based on a wage scale derived from good job descriptions and a sound job evaluation plan, and one that is competitive in the area.

It is highly desirable that each wage level should be composed of at least three steps, with significant differentials between steps. The lowest step is used for entry to the wage level, whether by hire or by internal promotion. That step would be in effect for any designated ''trial'' period of employment.

The intermediate step is a learning step, to be paid when a worker has successfully completed the trial period but has not yet learned the full requirements of the position.

The final step is the full job rate for all jobs at that wage level. All those workers who have qualified for the job as described in the job description receive the job rate.

Conditions of Employment—Performance, Quality, Safety, and Attendance.   In order to maintain an acceptable level of production, the implementation of a measured daywork system and the labor standards must be accompanied by a set of guidelines for satisfactory individual performance. Obviously, one cannot require incentive level (over 100 percent) performance for a nonincentive wage. When the measured daywork program is replacing an incentive pay program, it may be expected that the worker can and should provide a fair day's work—100 percent performance—in return for the established wage. However, when the measured daywork program is the first introduction of time standards, a target performance of 85 or 90 percent is a more realistic expectation. In either situation, the following concept is applicable, although the acceptable performance levels will be different.

Although standards and a labor reporting system are designed to identify a worker's performance when work is available, it would be unrealistic to expect a constant 100 percent performance level. But it is reasonable to expect a 98 percent performance level based on a 2-week moving average. In fact, a range from 95 to 100 percent could be expected to occur.

Any 2-week period where the moving average drops below 95 percent, however, will require corrective action. Such action should be designed to improve operator performance. In addition to the normal procedure of finding out what the problem is, actions could include progressive penalties prior to definitive action. Such steps could include:

1. Time standard and method review, correction and worker retraining, if required, with results expected in 2 weeks.
2. Written warning by the foreperson, with results expected in 2 weeks.
3. A reduction to the learning level rate, or the entry rate if the individual is already at the learning rate, for a maximum period of 30 days. Rate restoration will occur when a 2-week moving average performance reflects a satisfactory performance.
4. A reduction to a lower-grade (and therefore simpler) job, or a layoff if no opening is available. This step is just short of dismissal, and may not be practical.
5. The third repetition of poor performance in the same job over a 6-month period, or continued inability to demonstrate satisfactory performance should result in dismissal.

A simpler alternative is for steps 2 and 3 to provide an opportunity for the worker to demonstrate ability on some other task. A fourth failure would be grounds for dismissal. The primary consideration is that each employee should be given the opportunity to show that there is a job which he or she can learn, and can perform satisfactorily.

Total satisfactory job performance will be determined by the foreperson in accord with the performance levels achieved, the workers' demonstrated ability to meet the essential requirements of attendance, quality, and safety, and any additional requirements identified in the appropriate job description.

Management Responsibilities.   Just as the worker has basic responsibilities for performance, attendance, quality, and safety, management has a basic responsibility to ensure that the worker has the opportunity to perform. Machinery and equipment must be in good operating condition, and not subject to frequent breakdown. Material of the proper quality and quantity must be available as required. Supervisory techniques must not be adversarial but should always be supportive. Manufacturing or assembly services should be available as required. In other words, all those facets of operations that are not the worker's responsibility must be provided and sustained by management.

MANAGEMENT REPORTS

Probably the most important part of a measured daywork program is the information provided to the operating managers. Timely information about worker performance and utilization is critical for the first-line supervisor, who must make key decisions about counseling the workers, possible disciplinary action, as well as requirements for management action.

Daily, weekly, and monthly reports may be issued to various levels of management. All reports are based on raw and/or corrected daily time reporting by individual workers and/or their forepersons when the worker is not present. The following paragraphs identify typical report levels and frequency, the report content, and some indication of the use to which the reports are put.

Level One—Daily.   The first-line supervisor should receive a basic report each day, issued as soon as possible after the day's work has been completed. The report provides detailed information as reported by each employee, and as calculated from the reported data. The typical report includes:

1. Worker clock number and name
2. Part number produced
3. Operation number with short work description
4. Time card clock hours
5. Quantity produced
6. Time standard
7. Earned hours and worker performance
8. Downtime description and time card clock hours
9. Indirect tasks and time card clock hours

A concurrent error report is also issued. That report identifies part numbers and operation numbers reported, but not on record.

The foreperson's review of the report should include a review of quantities produced as well as the performance level achieved by the operator. A check of quantities may reveal data input errors that would not be picked up by the computer. Review and correction of all items on the error report is essential.

It is very important that the foreperson observe the amount of downtime reported each day. An indication that a large amount of downtime is being reported while performance approaches maximum levels may be a sign that workers are fudging their reports to demonstrate high performance.

Level One—Weekly.   The weekly report is a compilation of the daily reports, *as corrected.* The report is used by the foreperson as a detail reference for problems revealed in the level two weekly report.

Level Two—Weekly.   Level two reports are issued for the use of the superintendent and the foreperson on a weekly basis. The reports consist of a weekly summary of individual operator and department activity daily and for the week. Daily operator performance is given as well as the 2-week moving average of performance. Total department performance is also provided daily, for the week, and a 2-week moving average.

A summary of downtime and/or indirect hours used by each worker is given, including vacation time, sick time, or other absentee data.

Finally, department or group utilization and productivity are reported. It is desirable to report a dollar index of progress, but an index that fluctuates only with operating indicators, excluding overtime pay and any pay deviation from the job rate. The "index dollar per earned hour" is discussed in more detail in the following section of this chapter.

A typical level two report is shown in Fig. 1.2a and b. It is the basis for supervisory

DEPT D PERFORMANCE SUMMARY   Week Ending: 2-13

NAME / CLOCK #	MONTH: FEBRUARY DAY:	8 DAY	9 DAY	10 DAY	11 DAY	12 DAY	13 DAY		TWO	
WOODS 2835	1 EARNED HRS	7.0	9.3	9.7	7.4	3.8	2.8		40.1	70.6
	CLOCK HRS	10.0	10.0	10.0	11.0	10.0	5.0		56.0	110.0
	PERF - %	70.0	93.1	97.3	67.2	38.4	56.4		71.6	64.2 *
SMITH 2841	1 EARNED HRS	7.0	7.4		7.1	4.7			26.1	61.7
	CLOCK HRS	9.0	8.5		8.5	7.9			33.9	80.7
	PERF - %	78.1	86.8		82.6	59.1			77.0	76.4 *
CONKLIN 2978	1 EARNED HRS	9.5	9.1	10.6	9.6	7.9			46.7	98.0
	CLOCK HRS	8.6	8.3	10.0	9.0	7.5			43.4	95.2
	PERF - %	110.0	110.0	105.8	106.7	105.1			107.5	102.9
JAMES 3737	2 EARNED HRS	9.1	11.7	12.8	8.7	8.0			50.3	94.7
	CLOCK HRS	9.0	10.0	10.0	10.5	9.0			48.5	99.7
	PERF - %	101.0	116.7	127.6	83.0	89.3			103.7	95.0
STAMAN 4012	1 EARNED HRS	6.6	7.0	10.8	6.6	6.7	6.1		43.8	86.1
	CLOCK HRS	10.0	11.0	14.7	10.5	10.0	5.0		61.2	115.2
	PERF - %	65.8	63.5	73.3	63.1	67.7	121.0		71.6	74.8 *
SOMERS 4528	1 EARNED HRS	13.8	10.4	10.9	8.9	9.4			53.4	105.9
	CLOCK HRS	15.0	11.0	9.5	8.9	10.0			54.3	103.6
	PERF - %	91.9	94.7	115.3	100.2	93.8			98.3	102.3
HODGES 6409	1 EARNED HRS	8.7	5.7	9.6	10.8				34.8	102.2
	CLOCK HRS	10.6	6.9	8.4	10.2				35.9	85.8
	PERF - %	82.7	83.1	114.6	105.8				96.7	119.1
HANCOCK 7106	1 EARNED HRS	11.2	10.8	7.0	9.1	10.0			48.1	109.4
	CLOCK HRS	7.8	7.5	8.5	8.1	8.0			39.8	85.3
	PERF - %	144.5	143.2	82.6	111.9	125.7			120.7	128.4
DEPARTMENT DIRECT - TOTAL	EARNED HRS	72.9	71.3	71.4	68.1	50.6	8.9		343.2	728.6
	CLOCK HRS	79.9	73.2	71.0	76.7	62.4	10.0		373.1	775.4
	PERF - %	91.2	97.5	100.5	88.8	81.1	88.7		92.0	94.0

* Counseling/Disciplinary Action Required

FIGURE 1.2a Level two—weekly report—performance.

DEPT D PERFORMANCE SUMMARY   Week Ending: 2-13

DOWNTIME

NAME	2/8	2/9	2/10	2/11	2/12	2/13	WEEK TOTAL
WOODS							0.00
SMITH		0.51 BM	SICK	0.46 BM	1.10 BM		2.07
CONKLIN	1.36 WM	1.72 NC			1.98 NS		5.06
JAMES							0.00
STAMAN							0.00
SOMERS			1.54 WJ	1.14 WF			2.68
HODGES	6.45 WM	0.90 RM	3.64 RM	2.84 RM	11.00 RM	5.00 RM	29.83
HANCOCK	1.23 BM	1.48 BT	0.40 WF	0.90 IC			4.01
TOTAL HRS DOWNTIME	9.04	4.61	5.58	5.34	14.08	5.00	43.65

DOWNTIME DISTRIBUTION

	2/8	2/9	2/10	2/11	2/12	2/13	WEEK TOTAL
BT-BROKEN TOOL		1.48					1.48
NM-NO MATERIAL							.00
BM-BAD MATERIAL	1.23	.51		.46	1.10		3.30
NC-NO CARRIER		1.72					1.72
NS-NO MAT'L SERVICE					1.98		1.98
RM-RESET MACHINE		.90	3.64	2.84	11.00	5.00	23.38
WM-WAIT - MAINT	7.81						7.81
WF-WAIT - FOREPRSN			.40	1.14			1.54
WJ-WAIT - JOB			1.54				1.54
IC-INDIRECT - CLEAN				.90			.90
IO-INDIRECT - OTHER							.00
SICK			8.00				8.00
VACA - VACATION							.00
UA - UNREPORTED ABSENT							.00
TOTAL DISTRIBUTION							43.65

	2/8	2/9	2/10	2/11	2/12	2/13	WEEK TOTAL
CLOCK HRS - PROD	79.9	73.2	71.0	76.7	62.4	10.0	373.1
CLOCK HRS - TOT	89.0	77.8	76.6	82.0	76.4	15.0	416.8
UTILIZATION - %	89.8	94.1	92.7	93.5	81.6	66.7	89.5
EARNED HOURS	72.9	71.3	71.4	68.1	50.6	8.9	343.2
CLOCK HOURS	89.0	77.8	76.6	82.0	76.4	15.0	416.8
PRODUCTIVITY - %	82.0	91.7	93.2	83.0	66.2	59.1	82.3
INDEX DOLLARS	972	1015	1018	979	925	534	5443
INDEX $/EARNED HR	13.32	14.23	14.26	14.39	18.28	60.15	15.86

FIGURE 1.2b Level two—weekly report—utilization and productivity.

review of department progress, and for review with the foreperson to determine what has caused problems and what corrective action has been taken or should be taken.

A set of curves may also be provided. The curves display the primary indicators of performance, utilization, productivity, and a cost index. The curves should be based on either weekly data or a specific period moving average and are usually effective when posted in the area of the group or department depicted.

Figure 1.3 is a typical set of curves, including targets.

Level Three—Weekly.   Level three reports are issued weekly to the factory manager and the superintendents. The reports summarize the key indicators for each department as reported weekly for a calendar quarter. Target values for each indicator may also be given.

A curve may be generated to show overall plant performance. Together, the report and the curve provide the factory manager with a quick evaluation of progress and a basis for regular review with the superintendents. Samples are shown in Figs. 1.4 and 1.5.

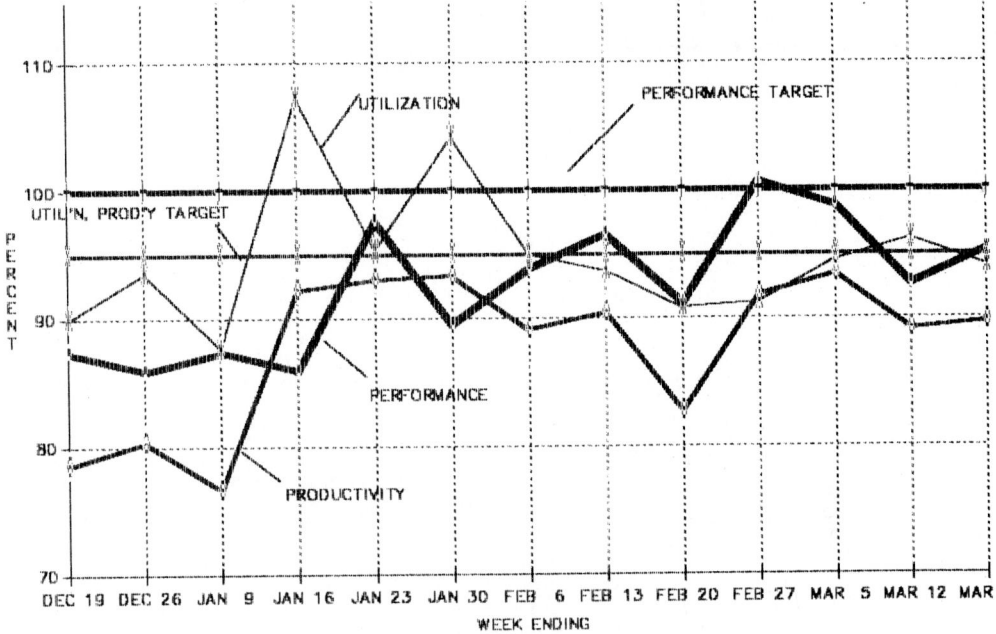

FIGURE 1.3   Department B—performance, utilization, and productivity.

Level Four—Weekly or Monthly.   Level four reports are issued weekly (or monthly if preferred) to senior management and operating management from the superintendent up. The report provides a measure of the benefits achieved weekly and on a cumulative basis in terms of benefits based on the index dollars per earned hour calculation.

The report covers a 13-week period (equivalent to a calendar quarter) on a continuing basis. The cumulative total benefit for the operating unit is given. The data may also be shown by a curve. Examples are shown in Figs. 1.6 and 1.7.

Each report level is designed to provide an appropriate level of information to specific managers, as well as providing each successive level with the type of information that may be required to respond to inquiries from above.

INDEX DOLLARS

Index dollars per earned hour (index $/EH) is similar to the conventional dollars per standard hour, but it differs in two very important ways. The index $/EH covers all clock hours charged to a department but does *not* add premium pay for overtime hours or for shift work. Overtime is a variable time, as is shift assignment. Incorporation of premium pay would incorporate a variable that is not a measure of either department performance or effectiveness.

Index dollars are calculated by multiplying the total clock hours charged to the department by the individual job rates of the workers, not the actual rate paid to the worker. The job rate is used because it avoids fluctuations from the entry rate to the learning rate to the job rate that are not related to improvements in productivity.

Figure 1.8 shows an actual comparison between the index dollar per earned hour and the normal accounting department's dollar per standard hour. The bulk of the gap between

MEASURED DAYWORK PROGRAM – OPERATING RESULTS: THIRTEEN WEEKS ENDING 3/19

DEPT	INDICATOR	UNIT	TARGET	DECEMBER		JANUARY				FEB				MARCH		
				19	26	9	16	23	30	6	13	20	27	5	12	19
A	PERFORMANCE – WEEK	%	100.0	95.0	93.8	90.9	91.2	91.3	91.4	91.5	91.9	94.4	96.6	95.6	91.0	91.2
	PERF – 2 WK MOVING AVG	%	100.0	93.2	94.6	92.6	92.1	91.9	91.8	91.4	91.7	93.1	95.6	96.1	93.4	91.4
	UTILIZATION	%	95.0	93.7	93.2	87.5	91.2	93.4	95.7	95.2	94.5	93.6	93.4	92.7	91.8	90.5
	PRODUCTIVITY	%	95.0	89.0	87.4	79.5	83.2	85.3	87.5	87.1	86.8	88.4	90.2	88.6	83.5	82.5
	INDEX $/EARNED HOUR	$	12.80	14.99	15.07	17.96	15.97	15.90	14.79	15.08	15.82	15.54	14.72	15.27	16.06	15.68
B	PERFORMANCE – WEEK	%	100.0	87.3	85.9	87.4	85.9	97.6	89.5	93.8	96.5	91.1	100.7	98.8	92.6	95.3
	PERF – 2 WK MOVING AVG	%	100.0	88.5	86.8	86.8	86.6	92.1	93.3	91.6	95.1	93.9	96.4	99.7	95.7	94.1
	UTILIZATION	%	95.0	89.9	93.5	87.6	107.3	95.3	104.2	95.0	93.7	90.8	91.2	94.5	96.2	94.1
	PRODUCTIVITY	%	95.0	78.5	80.3	76.6	92.2	93.0	93.3	89.1	90.4	82.7	91.8	93.4	89.1	89.7
	INDEX $/EARNED HOUR	$	12.15	15.70	13.60	15.53	16.46	14.79	15.87	14.74	14.42	15.33	13.48	12.64	12.88	12.58
C	PERFORMANCE – WEEK	%	100.0			96.0	93.2	93.8	97.0	93.5	91.3	91.3	92.4	89.5	91.2	91.3
	PERF – 2 WK MOVING AVG	%	100.0			96.0	94.4	93.5	95.3	95.1	92.4	91.3	91.9	90.8	90.3	91.2
	UTILIZATION	%	95.0			97.2	99.5	98.6	100.0	98.3	84.1	90.5	97.6	98.0	98.7	97.6
	PRODUCTIVITY	%	95.0			93.3	92.7	92.5	97.0	91.9	76.8	82.6	90.2	88.7	90.0	89.1
	INDEX $/EARNED HOUR	$	12.57			12.37	12.15	11.75	11.81	13.20	13.85	12.67	12.28	12.18	12.52	12.93
D	PERFORMANCE – WEEK	%	100.0							95.8	92.0	102.7	82.9	92.3	86.7	89.1
	PERF – 2 WK MOVING AVG	%	100.0							95.8	94.0	96.7	91.9	87.7	89.9	88.4
	UTILIZATION	%	95.0							93.0	89.5	80.0	87.5	84.9	80.2	90.5
	PRODUCTIVITY	%	95.0							89.1	82.3	82.2	72.5	78.4	69.5	80.6
	INDEX $/EARNED HOUR	$	12.68							14.12	15.54	15.79	16.13	14.69	17.62	14.59
E	PERFORMANCE – WEEK	%	100.0								62.3	91.0	87.8	85.4	101.3	79.9
	PERF – 2 WK MOVING AVG	%	100.0								62.3	77.5	89.4	86.7	92.0	90.6
	UTILIZATION	%	95.0								75.4	89.1	90.0	84.5	53.4	66.5
	PRODUCTIVITY	%	95.0								47.0	81.1	79.0	72.2	54.1	53.1
	INDEX $/EARNED HOUR	$	10.35								25.5	14.79	14.89	15.36	18.15	19.87
F	PERFORMANCE – WEEK	%	100.0										108.5	103.2	99.5	107.5
	PERF – 2 WK MOVING AVG	%	100.0										108.5	106.4	101.3	102.4
	UTILIZATION	%	95.0										94.7	90.9	78.9	79.3
	PRODUCTIVITY	%	70.0										102.7	93.8	78.5	85.3
	INDEX $/EARNED HOUR	$	13.65										13.42	16.28	18.28	15.77
PLANT PERFORMANCE FOR WEEK				90.2	90.7	91.1	90.2	90.9	92.0	93.3	91.1	94.0	96.4	94.9	92.3	93.9

FIGURE 1.4 Level three—weekly—operating results.

the two curves is overtime premium pay. That is reflected by the narrowing of the gap in January, when a major effort to reduce overtime was initiated.

It is interesting to note that the actual dollars per standard hour curve shows a more emphatic reduction over the 13 weeks than does the index dollar per earned hour curve. The index dollar curve is a direct reflection of movements in the department's productivity.

Another application of the index dollar per earned hour is the determination of the total benefits derived from the application of measured daywork. When relatively accurate production data (quantity, part number, operation number, and worker) are available from a period prior to installation of measured daywork, the new time standards can be applied to establish a base from which any improvement can be measured. That was the basis for the savings shown in Figs. 1.6 and 1.7.

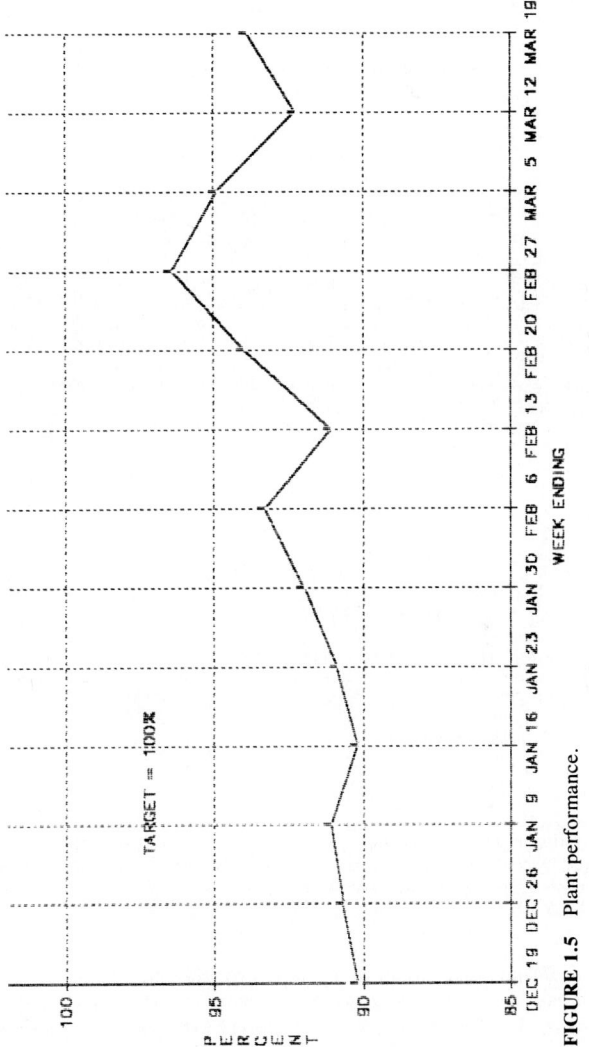

FIGURE 1.5 Plant performance.

MEASURED DAYWORK PROGRAM - BENEFITS REALIZED: THIRTEEN WEEKS ENDING 3/19
BASIS: INDEX DOLLARS PER EARNED HOUR

DEPT.	ITEM	VALUE $	DECEMBER 19	26	JANUARY 9	16	23	30	FEB 6	13	20	27	MARCH 5	12	19
A	19__ BASIS	16.96													
	TARGET	12.80													
	CURRENT INDEX $		14.99	15.07	17.96	15.97	15.90	14.79	15.08	15.82	15.54	14.72	15.27	16.06	15.68
	CURRENT EARNED HOURS		533	287	429	633	562	646	690	674	622	734	704	634	555
	REALIZED SAVE - WEEK	$	1050	542	-429	627	596	1403	1298	769	883	1644	1189	571	710
	CUM REALIZED SAVE	$	1186	1728	1298	1925	2521	3924	5221	5990	6873	8517	9706	10277	10988
B	19__ BASIS	18.24													
	TARGET	12.15													
	CURRENT INDEX $		15.70	13.60	15.53	16.46	14.79	15.87	14.74	14.42	15.33	13.48	12.64	12.88	12.58
	CURRENT EARNED HOURS		410	222	382	462	569	573	565	627	486	688	745	707	624
	REALIZED SAVE - WEEK	$	1040	1029	1036	822	1963	1358	1979	2395	1415	3276	4174	3788	3532
	CUM REALIZED SAVE	$	4528	5557	6592	7414	9377	10735	12714	15109	16523	19799	23973	27761	31293
C	19__ BASIS	16.25													
	TARGET	12.57													
	CURRENT INDEX $				12.37	12.15	11.75	11.81	13.20	13.85	12.67	12.28	12.18	12.52	12.93
	CURRENT EARNED HOURS				283	355	373	333	358	311	357	412	510	420	405
	REALIZED SAVE - WEEK	$			1099	1456	1678	1478	1092	746	1279	1634	2077	1565	1345
	CUM REALIZED SAVE	$			1099	2555	4233	5711	6803	7549	8828	10462	12538	14103	15448
D	19__ BASIS	14.47													
	TARGET	12.68													
	CURRENT INDEX $								14.12	15.54	15.79	16.13	14.69	17.62	14.59
	CURRENT EARNED HOURS								386	343	302	294	350	254	358
	REALIZED SAVE - WEEK	$							135	-367	-399	-487	-77	-801	-43
	CUM REALIZED SAVE	$							135	-232	-631	-1118	-1196	-1997	-2040
E	19__ BASIS	13.10													
	TARGET	10.35													
	CURRENT INDEX $									25.54	14.79	14.89	15.36	18.15	19.87
	CURRENT EARNED HOURS									81.4	133.6	128	113	96	75
	REALIZED SAVE - WEEK	$								-1013	-226	-230	-256	-483	-508
	CUM REALIZED SAVE	$								-1013	-1238	-1468	-1724	-2207	-2714
F	19__ BASIS	15.26													
	TARGET	13.65													
	CURRENT INDEX $											13.42	16.28	18.28	15.77
	CURRENT EARNED HOURS											450	295	315	342
	REALIZED SAVE - WEEK	$										827	-301	-951	-174
	CUM REALIZED SAVE	$										827	527	-424	-599
	CUMULATIVE REALIZED SAVINGS		5714	7284	8990	11895	16131	20369	24873	27403	30355	37019	43826	47514	52376
	CUM RLZD SVGS + FRINGE @ 59.8%		9131	11640	14366	19008	25778	32550	39747	43790	48507	59156	70033	75927	83696

FIGURE 1.6 Level four—weekly or monthly—benefits realized.

When data are not available from a previous period, the initial results at the time of measured daywork installation can form the base, and improvements or savings can be measured from that base. Obviously there must be some improvement, or there would not be a financial reason for installing a measured daywork program.

MANAGEMENT RESPONSIBILITIES

The definition of measured daywork identified a number of responsibilities that management must fulfill if the worker is to be able to achieve the targeted performance and productivity. Most of those responsibilities will be fulfilled when the various operating and

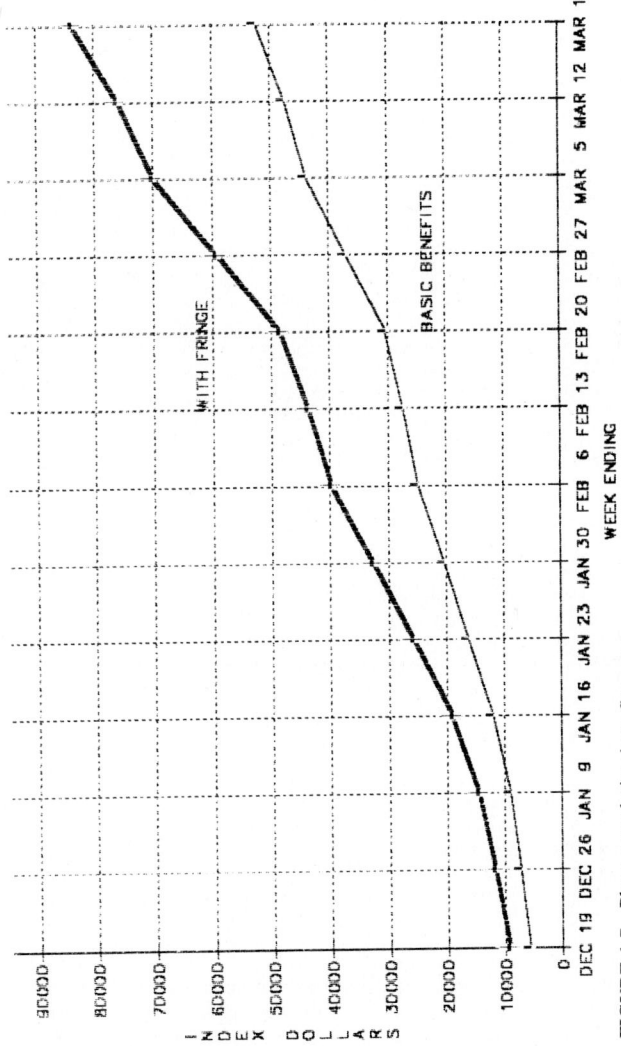

FIGURE 1.7 Plant cumulative benefits.

6.13

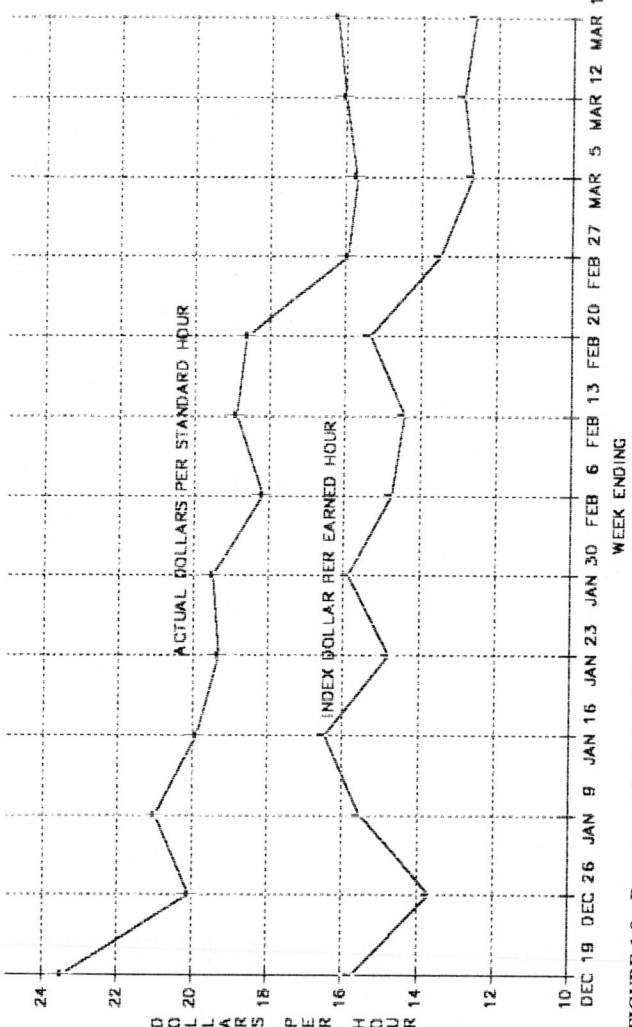

FIGURE 1.8 Department B—index dollar per earned hour comparison.

staff departments carry out their mandates. It is important to specifically review the foreperson's role.

The Foreperson's Role.   The foreperson has a vital role in assuring that a measured daywork program is effective. In fact, the foreperson's work should be the same no matter what incentive or nonincentive system is in use. Some 40 years ago, it was recognized that the foreperson was required to do more than any individual could be expected to do effectively. It was recognized that the primary task was working with and leading subordinate workers to achieve the output goals that had been established.

To provide the time required for the basic task, more and more staff services were provided for direct operations. The latest move has been toward reducing staff services and placing a greater task load on the foreperson. The introduction of material requirements planning (MRP) systems has encouraged nonoperating management to insist on the elimination of production scheduling and control departments. As a result, in many organization the foreperson has become a planner and an expediter. The amount of time available for reviewing worker performance and work methods and then working with the worker to make improvements has been greatly reduced. The amount of time spent in the foreperson's department has been greatly reduced, and in many cases there has been a growth of official or unofficial group leaders. In most cases, that task is not a desirable function, particularly when the group leader usually has little knowledge of or interest in company management objectives.

Each foreperson must be in position to review his or her level one reports each day and to be able to respond to indicated problems that interfere with improved productivity. The foreperson should be able to balance reported production time and reported downtime for accuracy and for correct indication of actual or potential problems. There should be time to review daily productivity results with each worker and to provide the proper and effective support required.

SUMMARY

Measured daywork is an incentive-type system of management control that differs from most incentive systems in that the incentive is not direct adjustment of pay but an agreement between management and the worker to maintain a given level of work volume and quality while maintaining a satisfactory attendance and working safely, as a basic condition of employment. Management is required to maintain working conditions that will accommodate the productivity and quality goals.

Management is provided the benefits of time standards as a basic management tool and is given the opportunity of establishing truly effective working relationships with the workers.

CHAPTER 2
CONVENTIONAL WAGE INCENTIVES

Albert F. Celley, Ph.D.
A. F. Celley & Associates
Toledo, Ohio

This chapter describes, discusses, and analyzes conventional financial wage incentive systems. Its purpose is to provide management, industrial engineers, and union leaders with an insight into how wage incentives motivate workers, how they are designed, and what principles and practices are required to assure equitable results to both the worker and the company. Such readers will then be better prepared to understand and cooperate with the design, the implementation, and the maintenance of wage incentives so that the incentive system will truly increase productivity for the company and afford increased earnings for the increased effort provided by the employees.

A well-founded wage incentive system leads to greater productivity, decreased unit costs, and the ability to become more competitive in a world market, and it returns a financial reward to the worker through higher pay and increased profits to the company.

GENERAL HISTORY OF WAGE INCENTIVE PLANS

The basic purpose of wage incentives throughout industrial history was intended to reward the employee with additional compensation above the base rate for exerting above normal skill and effort in the production of goods and services.

Incentives of one type or another have been applied since one individual began to labor for another. The motivation for an employee to perform at an above normal level of productivity consists of and includes many contributing factors, some psychological and some monetary. Successful wage incentive plans have always included elements of both factors.

Halsey Plan. Wage incentive plans in what is considered modern application began before the turn of the twentieth century in the era of Frederick W. Taylor and "scientific management." An early application was the Halsey plan, where emphasis was usually placed upon dividing the gains resulting from increased productivity between the employer and the employee. This early plan was used when standard times were estimated rather than established by formal work measurement techniques. Such estimated standards usually were more liberal than work measurement based standards, and thus made the potential for increased wage payment more acceptable to the worker as well as permitting a gain to the employer.

Bedaux and Haynes Plans. Two similar wage incentive plans that were widely applied during the period up to World War II were the Bedaux and Haynes plans. The Bedaux was described as a "point" plan and the Haynes as a "manit" plan. Under the typical application of the Bedaux plan the worker was paid 75 percent of the hours earned, while the remaining 25 percent of the earned hours went into a pool from which indirect workers and supervisors were also paid an incentive. The selling point of this incentive plan was that direct and indirect workers as well as first-line supervision must cooperate and work together to attain the extra compensation.

The Haynes manit plan was generally applied to production areas where operations, methods, and conditions were somewhat unstandardized. The participation in incentive earned was typically split as follows: 50 percent to the workers, 10 percent to supervision, and 40 percent to be retained by the company.

Gantt Plan. A well-known plan applied previous to World War II was the Gantt plan. While many plans such as the Bedaux and Haynes plans began to pay bonus at 100 percent performance in small regular increments as productivity increased over 100 percent, the Gantt plan paid 15 percent in a lump sum immediately upon attaining 100 percent performance. Thereafter a small marginal increase in incentive earnings was paid as productivity increased. This 15 percent wage incentive paid upon attaining 100 percent performance provided a strong motivational potential for the employee to reach the 100 percent goal. To be effective to all parties, this type of bonus payment plan required that excellent methods engineering and work measurement techniques be applied to establish and maintain equitable standards of measurement.

Needless to say there were an infinite number of variations of the described incentive plans. Their details of design and application can readily be found in many texts devoted to this subject. Also the First, Second, and Third Editions of Maynard's *Industrial Engineering Handbook* (McGraw-Hill, New York) are recommended as a concise source of information describing the major types of wage incentive plans and their variations.

Piecework. A popular wage incentive application that has existed from the first days of industrial-type production has been the "piecework" system. This system is most feasibly applied where production can be measured in discrete units of material, parts, subassemblies, or final product. Piecework earnings are determined by paying a fixed monetary price per piece produced by the operator. The larger the volume of units produced, the greater the earnings. The system is easy to understand and for that reason readily acceptable to the employees. In addition, because the piecework system deals directly in monetary terms, it is easy for all concerned to directly equate the value of the volume produced to wage earnings. The piecework system remains popular in many environments today as it has for many, many years.

The dollar value of the piece produced can be derived from estimates, from historical rates of productivity, from negotiation with the employees or union, or on a work measurement basis. Where a substantial volume of different piece rates exist, it can become arduous to change the dollar value of each piece rate when the base hourly rate of pay is changed for any reason. All too often the value of the piece rate is negotiated rather than derived by professional methods analysis and work measurement. The piece rate then is subject to political manipulation by the worker, the union, or the company. Therefore, the accurate labor cost per unit of task produced can differ substantially over time from the true and equitable labor cost of the work performed.

One-for-One Plan. A very popular wage incentive system found in many areas is defined as the "one-for-one" plan. This type of incentive system has in many areas replaced the true piecework plan because incentive standards are expressed typically in terms such as "standard hours per unit produced." The standard hours per unit are nominally determined by professional methods analysis and work measurement techniques. The incentive basis inherent in this plan pays a 1.0 percent incentive over the guaranteed base wage at 100 percent performance level for each 1.0 percent increase in performance. For example, at the point where standard hours earned (number of units produced times the standard hours per unit) divided by the actual hours worked equals 100.0 percent, no incentive is earned. When the performance level rises to 101.0 percent, 1.0 percent incentive is earned, and so forth.

Emerson Plan.   In a historical perspective the last major type of incentive system, the Emerson plan, paid an incentive that commenced at less than 100 percent performance. For example, when an employee attained a performance level of 81 percent, based on standard hours divided by actual hours, an incentive of 1.0 percent would be paid. When the employee reached 100 percent performance, 20 percent (or some other incentive level) bonus or incentive would be paid. Each 1.0 percent performance increase paid an additional incentive of 1.0 percent. Many variations existed in the application of this type of plan. Its advantages were that an employee began to earn a bonus at less than 100 percent performance, and as the performance level increased, the unit cost of production steadily decreased for the company.

Typical wage incentive payment plans based upon professional work measurement and methods analysis are not limited to the amount the worker may earn as long as working conditions remain unchanged.

Elimination of Incentive Plans.   In the past 20 years many firms have dropped or bought out of their incentive plans because the poor management of such incentive plans allowed them to deteriorate to the point where many systems became significant cost penalties, as well as led to poor employee and union relations. Some of the major reasons for the deterioration of incentive plans result from a poor understanding of the economics underlying successful incentive systems; the desire to avoid the constant difficult decisions incentive administration gives rise to; the increase in mechanization and automation of many former manual tasks; and the desire by industrial engineers and industrial engineering schools to place a low priority on classical work measurement techniques and incentive plan applications.

PREVALENCE OF WAGE INCENTIVE PLANS

After roughly a century of the continued growth and widespread application of incentive plans, Norma W. Carlson, writing in the *Monthly Labor Review,* May 1982, states that incentive wage plans continue to decrease in popular application in 37 major industries in the United States when comparing their incidence for the 1973–1980 period with that for 1961–1968. However, alternative behaviorally focused methods of motivating workers are attracting increased attention from both labor and management. Emphasis on machine-paced operations appears to be a major reason for the decreased incidence of financial wage incentive pay plans. Thus time-based pay rates have replaced many former incentive applications established on quantity-based measures. Financial incentives still tend to be concentrated in a restricted area of industries whereby individual employees have the potential to exert substantial influence on productivity. Incentive plans which establish a close link between output and earnings are designed to fill two roles: both to motivate employee productivity and also to provide a basis for wage compensation. Time-based payment plans compensate the employee on a fixed time period rate and rely on supervisory effort and skills or machine-paced processes to maintain a predetermined level of productivity. Highly automated industries typically rule out financial incentive wage systems because the operating crews have a minimum or no control over the pace of production and the quantity of output. The steel industry is the exception in the United States.

Table 2.1 illustrates the change in percentage of employees compensated on time-rated versus incentive pay plans in 37 industries for the two periods under comparison. These

TABLE 2.1 Percentage and Related Workers Covered by Time-Rated Wage Payment Plans in Selected Manufacturing Industries, 1961–1968 and 1973–1980

Industry	Percentage of time-rated workers 1961–68	1973–80	Percentage point change
Glass containers	62	88	+26
Meatpacking	66	90	+24
Other pressed or blown glass	64	79	+15
Cellulosic fibers	84	98	+14
Candy and other confectionery products	75	89	+14
Brick and structural clay tile	68	80	+12
Women's hosiery	25	37	+12
Corrugated and solid fiber boxes	64	75	+11
Ceramic wall and floor tile	58	68	+10
Hosiery, except women's	30	39	+9
Miscellaneous plastics products	87	95	+8
Wood household furniture, except upholstered	80	88	+8
Pulp, paper, and paperboard mills	90	97	+7
Gray iron except pipe and fittings	75	82	+7
Leather tanning and finishing	48	55	+7
Industrial chemicals	95	99	+4
Prepared meat products	92	96	+4
Fabricated structural metal	92	96	+4
Clay refractories	76	80	+4
Steel foundries	75	79	+4
Motor vehicle parts	69	73	+4
Paints and varnishes	98	100	+2
Gray iron pipe and fittings	75	77	+2
Wool yarn and broadwoven fabric	73	75	+2
Cotton and manmade fiber textiles	68	70	+2
Textile dyeing and finishing	89	90	+1
Cigarettes	100	100	0
Flour and other grain mill products	99	99	0
Petroleum refining	99	99	0
Motor vehicles	98	98	0
Nonferrous foundries	82	82	0
Men's and boys' suits and coats	25	25	0
Noncellulosic fibers	99	98	−1
Clay sewer pipe	77	76	−1
Men's and boy's shirts	23	22	−1
Leather footwear	30	27	−3
Basic iron and steel	33	20	−13

Source: Monthly Labor Review, Department of Labor, Washington, D.C., June 1983.

data reflect the selection between time rates and incentive pay that depends on technological and economic environments, managerial preferences, union philosophy, and traditions.

Nonfinancial Incentives. Carlson states that other approaches to employee motivation have resulted from management's concern over rising inefficiencies and decreased productivity. Questions are being raised as to the efficacy of financial wage incentive compensation as a motivator of increased productivity. Therefore, attention is being directed to alternate means of improving productivity. Some of these alternatives are directed toward better interpersonal relations, improved job interest, greater work satisfaction, reduced absenteeism, reduced turnover, reduced waste, and a strong focus on a high quality of goods and services. All these factors when appropriately implemented and administered result in increased productivity. This approach to improved productivity has been described generally as *quality of worklife*. Some are still experimental, some controversial, and on some the jury is still out. Many quality of worklife motivational techniques and practices combine the art of industrial and organizational psychology, such as the employee's need for personal fulfillment, recognition, and involvement in decisions along with management on issues that directly affect performance of the task. Worker participation in decision making has increased since the early 1970s even though the concept was practiced much earlier in a few organizations. This nonfinancial incentive approach to increased productivity has been often presented as labor-management committees and as quality circles, a practice borrowed from Japanese manufacturing. It appears that the adoption of financial incentives as the sole source of increased productivity will be much less in demand in the future, and nonfinancial motivators as just discussed will continue to become an ever greater source of management interest and practice.

PSYCHOLOGISTS' VIEWS OF WAGE INCENTIVE PLANS

There is a divergence of opinion among psychologists as to the contribution of wage incentives as a source of motivation for employees to increase their productivity. Mitchell Fein, writing about wage incentive plans in Maynard's *Industrial Engineering Handbook* (third edition, 1973), excellently characterizes the views of psychologists on this subject. Psychologists generally agree that, as expressed by Gellerman: "There is no doubt that money has an important effect on the thinking and behavior of production workers; but this effect is neither as simple nor as strong as management has often assumed."

Some psychologists believe that incentives will motivate people. Leavitt says: "Money incentives have come to occupy a central place because money is a common means of satisfying all sorts of diverse needs in our society and because money may be handled and measured. Money is 'real'; it is communicable. Many other means to need satisfaction are abstract and ephemeral. Moreover, money incentives fit with out culture's conception of what work means...." Vroom supports this approach: "Much of the evidence that is available concerning the effect of wages on performance is consistent with the assumption that people strive to maximize the amount of their wages."

Some psychologists see incentive pay as more than just money; it carries recognition with it which is a source of satisfaction to the employee. A high productivity level raises earnings and the standing of the employee on the recognition scale. Where incentives do not operate freely but are controlled by the workers, this obviously does not occur.

Whyte takes a more restrained attitude toward incentives: "Systems of financial incentive in industry yield a net gain in productivity, but most incentive plans will fail to release more than a small portion of the energy and intelligence workers have to give to their jobs." There are others who hold that money is not as strong a motivator as is commonly believed; that workers have nonfinancial needs which, if satisfied, will also motivate them to increase their productivity.

Numerous studies of the relation between job satisfaction and performance indicate that a high level of satisfaction does not necessarily motivate workers to expend increased energies directed toward improved performance. Simply put, happy workers may just continue to be happy workers, not necessarily high producers. Job satisfaction will improve the climate for increased productivity; it will reduce turnover and secure other desirable attitudes. There is no assurance, however, that job satisfaction without motivation will improve productivity.

Reporting on the role of financial compensation in industrial motivation, Opsahl and Dunnette, in a broad study of current research, state that: "Strangely, in spite of the large amounts of money spent and the obvious relevance of behavioral theory for industrial compensation practices, there is probably less solid research in this area than in any other field related to worker performance. We know amazingly little about how money either interacts with other factors or how it acts individually to affect job behavior. Although the relevant literature is voluminous, much more has been written about the subject than is actually known. Speculation, accompanied by compensation fads and fashions, abounds; research studies designed to answer fundamental questions about the role of money in human motivation are all too rare.''

A number of psychologists have reported on plants with wage incentives. Gellerman states that ''their firsthand accounts of what actually happens when incentives are introduced form the basis for the most comprehensive analysis yet of how the dollar affects the worker attitudes.'' This aspect of the behavioral scientists' findings is often missed and the writings are used to judge incentive plans, whereas the writings are not critiques of incentive plans but only report workers' reactions to incentives. These must be viewed for what they are: case histories reporting primarily the opinions of workers who obviously reacted to their environment. In many of the reported cases, the incentives appeared to be deficient. Inferences cannot be drawn in such plants of how workers react to incentives in general, but only to deficient incentives. The writings of the psychologists do not seem to take this into account, perhaps because they could not accurately identify deteriorating incentives.

To those not expert in the operation of incentive plans, the anecdotes and case histories of the incentive games played between workers and management reported by the behavioral scientists make interesting reading. Much of the behavioral literature depicts management as almost continually in trouble with incentives. To someone not experienced in incentives, reading this literature is comparable with a layman's reading medical case histories of sick people. After a while, it appears as though the world were filled only with sick people. Although a few successful plans are reported, many of the cases describe problem-ridden incentives. The behavioral writings on incentives can prove misleading where these are drawn from limited case histories.

Although some psychologists suggest that increased pay will not necessarily motivate workers because there are some needs which money cannot satisfy, it is generally believed by managers that money is the prime reason that people work and that people in general endeavor to maximize their earnings.

Fein concludes his discussion by stating that managers looking for clear recommendations from the psychologists will be disappointed; support can be found for and against incentives in their writings. But there is much to be gleaned from the behaviorists' findings by managers who understand the temper of their plants and who put together the right combinations of people relationships and motivation.

UNION VIEWS ON WAGE INCENTIVE PLANS

History reveals that the official view of many in the leadership of the AFL/CIO and also in several other large international industrial unions is opposition to wage incentive plans. This attitude arose from the misuse of incentive plans by the unilateral tightening of standards by management when no significant change could be demonstrated in methods, materials, equipment, tools, or other relevant working conditions. The tightening of these standards resulted in a cut in worker wages. Often the standards were not an accurate reflection of the defined task because the standards were not established on the basis of professional industrial engineering techniques. The result was that employees found ways to substantially increase their wages without a requisite increase in effort and skill. To counteract this condition, management would arbitrarily rein in the high earnings by tightening the standards. Naturally, this practice became a significant target in the union management battleground.

It is important to note, however, that the practices found in individual plants regarding the union's views of wage incentive plans vary widely. In essence it has been the author's experience that ''the local union is pretty much what local management makes it.'' This is true even when the local union is affiliated with an aggressive international union. It should not be overlooked that the local union may have a significant measure of autonomy to make decisions based on its own operating environment and strength of leadership.

The view of unions as expressed by practice can be observed from the most recent study conducted by the Bureau of Labor Standards regarding the prevalence of wage incentives in industry as summarized in Table 2.1. Incentive coverage varies markedly by industry. Some industries are traditionally oriented toward wage incentive plans, with such plans applied as the major source of productivity in largely manual operations. However, more and more highly mechanized and process industries do not utilize wage incentives because the pacing of the product output is designed into the plant process operations. In similar manner the low coverage in auto and appliance assembly results from the designed productivity of the line with its established crew sizes. On the other hand, auto parts suppliers apply wage incentives to achieve a high cost-effective rate of productivity.

The steel industry, although a basic process industry, utilizes incentive standards in large part to achieve a high output at a competitive quality and low cost per ton produced because of traditional practice for over a century. Normally the United Steel Workers union does not oppose incentives or the increased coverage of workers in their modern steel plants.

In general the local union and often its international does not actively oppose wage incentives that afford the employees an opportunity to increase their compensation as long as the employees' interests can be equitably safeguarded. Where good industrial relations exist between management and the union, wage incentive plans, their continuance and new implementation, can be established. However, labor standards and wage incentive practices and policies should never become a bargainable issue. Management sets the standards and designs and alters the incentive plan as necessary. Standards and incentive problems can be a basis for grievances and be adjudicated through the grievance machinery and possibly by arbitration; but the quantitative values of standards should always be determined by professional industrial engineering and work measurement analysis and never by negotiation.

In final analysis, whether unionized or not, the success of an incentive plan lies completely with the attitude of the employees who are impacted by it. Where such employees perceive that an incentive plan is equitably administered and their best interests are enhanced, they will support it. If the opposite attitude becomes tantamount, then they will not support it and the plan will fail to achieve the productivity objectives of management. Also see chap. 4 in this section for further information.

INCENTIVES PART OF TOTAL COMPENSATION

Incentive earnings generally are a part of total earnings. Incentives provide extra pay for extra performance in addition to the regular guaranteed base wage rate for a defined job. Because incentive earnings are a part of total earnings, it is essential to examine the various components of wages.

A minimum wage rate is an ultimate basic wage payment below which no hourly wage rate should fall. Over this minimum, it is generally agreed that a systematic relationship should exist among the values of jobs performed and their basic rates of pay. So that this relationship may be logical and equitable, an objective procedure for comparing the tasks that are included in the job is required. This procedure is called job evaluation. Job evaluation is an orderly and logical method for measuring and determining the relative worth of jobs. See Sec. 8, Chap. 4 for further details of job evaluation.

After job evaluation is performed on all jobs, job classifications are set up. Each job is grouped with other jobs of similar value, and a base pay rate is established for each group. Normally several pay ranges exist within each group so that employees can be rewarded with higher earnings based on merit or length of service. As a rule, merit rating is applied to guide the granting of wage-rate increases. This procedure when professionally applied should minimize bias and errors in judgment. All the foregoing should correlate to the requirements for the job and the payment of wages for satisfactory performance. Incentive earnings are generally paid in terms of the base wage rate for increased output.

The Base Wage Rate as a Stimulus of Productivity. The method of compensation is a very basic factor which causes people to be productive. The base hourly rate (that rate which is normally guaranteed) should be derived from a professionally established job evaluation system. The base rate itself does not primarily correlate to productivity, but it does represent the security of a constant rate of earnings to the worker. This income or measure of job security is, for the majority of employees, a powerful motivator to maintain a reasonable level of production. For those who desire a higher level of income, the knowledge that base rates increase with jobs of a higher evaluation score may be available to those workers who can meet the job requirements. This factor can act as a stimulus to higher than average production in the present job. However, the average rate of productivity where only a base hourly rate exists, with no output goals established by methods analysis and work measurement, generally will be quite low even with effective supervision. This is because neither the employee nor the supervision has a defined quantity of production as a goal. This type of wage payment system in and of itself does not provide a continuing motivation to increase output.

THE CASE FOR WAGE INCENTIVES

There is no question about the benefits to be attained from a soundly designed and administered wage incentive plan. Yet there are innumerable managements and labor leaders who oppose the application of wage incentives. There are several reasons for this opposition. In the early days of wage incentive applications unscrupulous managements abused the system by cutting wage rates when workers earned excessive wages because management had not performed a good job of methods engineering, work measurement, or incentive plan design. In latter years managements have bought out of many incentive plans because they felt that worker performance could be paced by mechanization of the process and often, more the truth, they did not want to continually go to battle with workers and their bargaining units over deteriorated wage incentive systems. Advocates of measured daywork plans often overlook the fact that, for measured daywork plans to operate effectively over a longer period of time, such plans must be administered as carefully as good wage incentive plans. If a comparison is made between a well-administered measured daywork plan and a well-administered incentive plan, the advantages of the wage incentive plan generally can be summarized as greater productivity, lower unit costs, and increased employee earnings.

CHARACTERISTICS OF SOUND WAGE INCENTIVE PLANS

The typical characteristics of a sound wage incentive plan are as follows:

1. There must be a direct relationship between something of value which is measured (frequently output) and the performance in terms of the standard measure.
2. It should be simple enough for employees to understand readily and to compute their own incentive pay with some training.
3. The standards upon which the wage incentive plan is based should be accurately established by thorough industrial engineering analysis.
4. The plan should provide for the changing of production standards whenever changes in methods, materials, equipment, or other controlling conditions are made in the operations represented by the standards.

5. Standards should be guaranteed unless changes occur which clearly alter the work measured.

6. To be effective, the plan should be sufficiently motivating to convince workers that they are being adequately repaid for turning out quality products in an increasing quantity.

7. The plan should be unrestricted as to the amount of earnings. There should be no fixed maximum amount of earnings. However, no wage incentive plan can continue to be effective very long if high earnings can be made without a high input of effort.

8. Under ordinary circumstances, management should guarantee that the employees' basic rates of pay which existed prior to the plan will be maintained as the minimum rates of pay under the plan.

9. In fairness to owners and to consumers, the plan should usually result in a reduction of the unit labor cost of manufacture.

10. The plan should be so established that it may be related readily to other management controls, such as quality control, production control, or cost and budgetary controls.

11. In general, the plan will be most effective when applied to individuals or small groups rather than to large groups.

12. The plan should have the continued attention of those directly responsible for its operation and the continued attention and support of top executives.

13. The plan should be equitable to employees, managers, and owners in its design and in its administration.

14. Definite written instructions covering policy application procedures, and method of performing the operation should be provided so that the plan may be carefully maintained.

15. Management and employees or their representatives should be in agreement as to the adoption or modification of the incentive plan.

Simple but Sound Plans.   One of the characteristics given for a sound wage incentive plan is that it should be simple enough for employees to understand and compute their own incentive pay, if so desired. Nevertheless, it is vital to put emphasis upon the phrase "sound incentive plan" and not sacrifice soundness to achieve simplicity. Inherently, the relationship between a measurable evidence of superior output and the input of labor effort is usually more complex for indirect labor than for direct labor. Therefore, desirable as it is to keep an incentive plan simple and direct, it is a fact that complex relationships sometimes cannot be measured in a simple manner which will avoid gross inaccuracies. Every effort must be made to find the simplest possible means, consistent with requisite accuracy, of measuring work and establishing incentive plan procedures.

Nature of Desirable Standards.   The chief requisite of a sound wage incentive plan is the establishment of standards which truly and continuously measure effort. If standards are set "too tight" or "too loose," difficulties arise. Standards which seem to be "too tight" are likely to result in discouraged efforts and decisions to hold back productivity until adjustments are made. Standards which appear to be on the "tight" side at least have the advantage of raising employee complaints where they can be investigated and either explained or rectified.

Many standards are, through carelessness or poor judgment, set "too loose." At other times, standards which were originally representative of a fair day's work deteriorate because of changes in methods, equipment, or material but are not changed to fit the new conditions.

In these cases, where for one reason or another standards become "too loose," and workers apply themselves and earn much more than was originally anticipated by either labor or management, jealousies among workers are likely to occur which tend toward an easing of effort upon the part of the worker. Thus, often limits are set by union edict, by groups of workers, and often by apprehensive forepersons or other supervisors; so that production which might have been attained with proper standards is never reached. Under these circumstances, management often never learns how much product could be turned out and what its true competitive unit cost might have been.

Incentive Unit of Measurement.   In taking steps to measure any job for purposes of incentives, the basic question is, "What is the unit of measurement?" Put another way, "What are we trying to accomplish with the job, and therefore what kind of yardstick is being set up to do the measuring?" It is important to keep in mind that, no matter what is selected as a basis for measurement, industrial engineers are concerned usually with a measurement of efficiency or effectiveness which relates the input of laborpower to the output of a product or service. This rule holds even when the minimum laborpower, one worker only, is clearly needed at an official work location to perform the work. In such cases, it is necessary to specify that the standards and the incentive apply only if one worker and no more work against the incentives.

Looking beyond time measures of how many pieces per hour or how many minutes are required to produce a unit of product, let us consider other measures used for incentives.

When materials are costly and the most effective utilization of material is important, a basis for incentives may be found in relationships between raw material and resulting finished products with proper consideration for and control of laborpower. The means of expressing this may sometimes be found in the amount of scrap and spoiled material.

In processing industries, many high costs are inherent in the processes. Labor cost, as such, may be relatively low, but labor and supervision also can keep costs low. Therefore, in a processing plant, the measure of the performance of the relatively few workers is to be found first of all in the material; that is, for every ton of raw material taken in how much finished product will be produced. The process may utilize electricity, steam, or heat. It is entirely possible to secure a measure in terms of these which tells whether a good job or a poor job is being performed.

Mechanical-equipment utilization and the measurement of utilization as a basis for incentives follow the same lines. Sometimes the effective utilization of mechanical equipment is of considerably greater importance to the successful operation of a business than the utilization of laborpower. Importantly, both may be brought into and made a part of the incentive measures.

There are many opportunities for greater utilization of incentives in the process industries. In many cases, measurements in such industries should be divided into various phases including labor, yield, power saving, material savings, and closer quality adherence. Each should be handled separately with proper attention given to the control of labor.

If savings can be made in various ways other than on labor, albeit labor contributes to the savings, it is clear that labor and labor organizations will be much more receptive than if the plan looks as though it will cut down the number of workers. Thus, while it is still desirable to control labor in relation to the general level of work volume, this can frequently be done in process-plant application merely by holding the specified labor force at its earlier level.

Incentive Period.   In practice, common periods of time used for calculating incentive earnings are a day, a week, 2 weeks, 4 weeks, or a month. Often the actual time for accomplishing a specific task or job order is used, so many hours, days, or weeks, and so forth.

Just as an incentive usually has a stronger appeal or greater motivation if applied to an individual rather than to a group of people, so an incentive is likely to have a stronger appeal if applied to a short period of time rather than to a longer period of time.

While this may be true, there is the weakness that, if employees get off to a bad start when a single day or a short task is a period, they may have a strong inclination not to try to recover lost time but to take it easy and accept guaranteed daywork pay for that day. Also, in anticipation of having an occasional bad day, employees have been known to hold back units of work (banking) to be turned in on a bad day, thereby building up a record of standard hours not actually earned on that day. But this distorts records and makes it difficult to control production in meeting shipping dates or building inventories.

It is to the advantage of the company and of the employees also, because their well-being depends upon the successful operation of the company, to maintain a steady, high level of output. Many believe that this steady, high level of output can be maintained most successfully if the incentive period is at least a week in length. There is less tendency for production to be up and down. Employees know that their incentive earnings depend on the level of production over this entire period of time, which is long enough to give them a chance to make up for unusual delays and take advantage of "breaks" to offset them.

MEASURED DAYWORK AS AN INCENTIVE

Measured daywork is examined in detail in Sec. 6, Chap. 1; therefore, only a brief outline of this subject will be covered here. Because of the cost and attention required to maintain a pure wage incentive system, many firms have made the decision to adopt measured daywork as a basis for control of their production operations. A measured daywork system provides a degree of motivation, but the motivation is not as effective as that a pure wage incentive system provides. Measured daywork programs allow management to establish production goals by which employees can be evaluated as to their effectiveness; but excellent supervision of all production factors (labor, material, waste, and the like) is essential to really motivate the workers to attain established output goals.

The measured daywork goals should be determined from standards established by good methods analysis and work measurement. Compared with wage incentive plans employees are normally paid a higher base rate to compensate for the lack of an incentive opportunity. The make-out level of production frequently is established below that of an incentive system but higher than the expected level of production of nonmeasured work. When properly administered, a measured daywork system can provide an incentive to greater productivity through the higher hourly pay rate. Data reporting the level of worker performance usually go to first-line forepersons and supervisors as well as to an employee.

The motivation for increased output through the application of a measured daywork system must arise from management in the form of closer control and improved planning, rather than from the incentive self-stimulus normally provided by a pure wage incentive plan. Successful application depends largely on effective supervision.

ECONOMIC COMPARISON OF TYPICAL MODERN INCENTIVE PLANS

Emerson 20 Percent Plan.   The following example, a typical modified Emerson-type plan, is based upon commencing the opportunity for incentive earnings at a level below 100 percent performance. The value of this lower than 100 percent opportunity for incentive earnings is that it motivates the employee to increase productivity earlier on the effort scale and continue such increased effort above the 100 percent performance make-out point. Taking these givens, a typical application would consist of:

1. Standard hours divided by actual hours equals percent performance.
2. Percent performance at 100 percent pays a base incentive or bonus of 20.0 percent.
3. Commencing at 76 percent performance, 0.8 percent bonus is paid for each increase in 1.0 percent performance.
4. The basic wage rate will be $10 per hour.
5. The incentive rate will be 0.200 standard hour per unit produced.
6. Incentive earnings will be calculated by multiplying the percent bonus earned by the actual hours worked.

TABLE 2.2 Example Modified Emerson Plan at 20.0 Percent Bonus Payout

Number units produced	Standard hours earned	Actual hours worked	Percent performance	Percent incentive	Incentive hours earned	Total hours earned	Total dollars earned	Unit labor cost, $
(1)	(2)	(3)	(4)	(5)	(6)	(7)	(8)	(9)
45	9.00	8.00	112.5	30.0	2.40	10.40	104.00	2.311
44	8.80	8.00	110.0	28.0	2.24	10.24	102.40	2.327
43	8.60	8.00	107.5	26.0	2.08	10.08	100.80	2.344
42	8.40	8.00	105.0	24.0	1.92	9.92	99.20	2.362
41	8.20	8.00	102.5	22.0	1.76	9.76	97.60	2.380
Base 40	8.00	8.00	100.0	20.0	1.60	9.60	96.00	2.400
39	7.80	8.00	97.5	18.0	1.44	9.44	94.40	2.421
38	7.60	8.00	95.0	16.0	1.28	9.28	92.80	2.442
37	7.40	8.00	92.5	14.0	1.12	9.12	91.20	2.465
36	7.20	8.00	90.0	12.0	0.96	8.96	89.60	2.489
35	7.00	8.00	87.5	10.0	0.80	8.80	88.00	2.514
34	6.80	8.00	85.0	08.0	0.64	8.64	86.40	2.541
33	6.60	8.00	82.5	06.0	0.48	8.48	84.80	2.570
32	6.40	8.00	80.0	04.0	0.32	8.32	83.20	2.600
31	6.20	8.00	77.5	02.0	0.16	8.16	81.60	2.632
30	6.00	8.00	75.0	00.0	0.00	8.00	80.00	2.667

Column		Column	
1. Production report.		6. Col. 3 times Col. 5	
2. Col. 1 times 0.200 standard hour per unit.		7. Col. 3 plus Col. 6.	
3. Payroll hours reported.		8. Col. 7 times $10.00 per hour.	
4. Col. 2 divided by Col. 3.		9. Col. 8 divided by Col. 1.	
5. Col. 4.			

From Table 2.2 it should be obvious that as the performance increases from 75 to 112.5 percent, the employee's hourly wage rate increases from $10 to $13 while the labor cost of production decreases from $2.667 to $2.311 per unit produced.

If the incentive base was reduced so that a 10.0 percent bonus was paid at 100 percent performance, the wage payment and unit cost would be as shown in Table 2.3.

After the industrial engineer presents the economics of such plans to management, it becomes incumbent upon management to decide the value of the trade-offs comparing the motivational pull of the incentive earnings potential and the unit labor cost to be achieved.

TABLE 2.3   Example Modified Emerson Plan at 10.0 Percent Bonus Payout

Number units produced	Percent incentive	Incentive hours earned	Total hours earned	Total dollars earned	Unit labor cost, $
(1)	(2)	(3)	(4)	(5)	(6)
45	15.0	1.20	9.20	92.00	2.044
44	14.0	1.12	9.12	91.20	2.073
43	13.0	1.04	9.04	90.40	2.102
42	12.0	0.96	8.96	89.60	2.133
41	11.0	0.88	8.88	88.80	2.166
Base 40	10.0	0.80	8.80	88.00	2.200
39	09.0	0.72	8.72	87.20	2.236
38	08.0	0.64	8.64	86.40	2.274
37	07.0	0.56	8.56	85.60	2.316
36	06.0	0.48	8.48	84.80	2.356
35	05.0	0.40	8.40	84.00	2.400
34	04.0	0.32	8.32	83.20	2.447
33	03.0	0.24	8.24	82.40	2.497
32	02.0	0.16	8.16	81.60	2.550
31	01.0	0.08	8.08	80.80	2.606
30	00.0	0.00	8.00	80.00	2.667

Standard hours, actual hours, and percent performance calculated in same manner as shown in Table 2.2.

One-for-One Plan.   The following incentive plan, the one-for-one plan, illustrates the incentive earnings potential and its effect on unit costs based on the same base wage rate as illustrated above, but paying no incentive until 100 percent performance is attained.

Examining Table 2.4, it should be noted that as the performance increases from 75 to 112.5 percent, the employee's hourly wage rate increases from $8.00 to $11.25 while the unit labor cost decreases from $2.667 to $2.00 at 100 percent efficiency and remains at $2.00 as the employee's performance increases above 100 percent. By comparing these two examples of major wage incentive application, we gain insight into the economic power of a wage incentive plan, assuming such a plan is capably administered and maintained.

The first example by the nature of its design is anticipated to increase the motivation to produce by beginning to pay bonus early on the performance scale with a constantly decreasing unit labor cost across the full range of the performance increase. The one-for-one plan does not begin to pay an incentive until a 100 percent performance level is attained, then pays only a minimum incentive increase beyond 100 percent performance while the unit labor cost remains constant. However, as the performance level increases up to 100 percent there is a substantial decrease in unit labor cost.

TABLE 2.4 Example One-for-One Plan Payout

Number units produced	Standard hours earned	Actual hours worked	Percent performance	Incentive hours earned	Total hours earned	Total dollars earned	Unit labor cost, $
(1)	(2)	(3)	(4)	(5)	(6)	(7)	(8)
45	9.00	8.00	112.5	1.00	9.00	90.00	2.000
44	8.80	8.00	110.0	0.80	8.80	88.00	2.000
43	8.60	8.00	107.5	0.60	8.60	86.00	2.000
42	8.40	8.00	105.0	0.40	8.40	84.00	2.000
41	8.20	8.00	102.5	0.20	8.20	82.00	2.000
Base 40	8.00	8.00	100.0	0.00	8.00	80.00	2.000
39	7.80	8.00	97.5	0.00	8.00	80.00	2.051
38	7.60	8.00	95.0	0.00	8.00	80.00	2.105
37	7.40	8.00	92.5	0.00	8.00	80.00	2.162
36	7.20	8.00	90.0	0.00	8.00	80.00	2.222
35	7.00	8.00	87.5	0.00	8.00	80.00	2.286
34	6.80	8.00	85.0	0.00	8.00	80.00	2.353
33	6.60	8.00	82.5	0.00	8.00	80.00	2.424
32	6.40	8.00	80.0	0.00	8.00	80.00	2.500
31	6.20	8.00	77.5	0.00	8.00	80.00	2.581
30	6.00	8.00	75.0	0.00	8.00	80.00	2.667

Column	
1. Production report.	5. Col. 2 minus Col. 3.
2. Col. 1 times 0.200 standard hour per unit.	6. Col. 2.
3. Job or payroll hours reported.	7. Col. 6 times $10.00 per hour.
4. Col. 2 divided by Col. 3.	8. Col. 7 divided by Col. 1.

Of course, there are many variations in the application of the Emerson wage incentive plan which will affect the economic payoff of each plan. This type of plan, in the author's experience, has a strong motivational pull to attain early and continuing higher rates of productivity throughout the total practical range of the worker's skill and effort input. The one-for-one plan in comparison, while simple in its design and application, has a much lower incentive motivation because no incentive earnings are available until 100 percent performance is attained. Clearly each firm considering the implementation of a wage incentive plan must carefully and professionally calculate the relative trade-offs between the motivational power and the economic factors of each plan.

ADMINISTRATION OF INCENTIVES

Emphasis must be placed on the need for proper administration of an incentive plan if it is to continue to function satisfactorily. Companies whose incentive plans are well maintained report high productivity, low costs, good morale, high quality, and a good competitive position.

Managements who have had improper incentive administration are quite concerned about rising costs, unreasonable union attitudes, grievances, arbitrations, "pegged" production, and business lost to competitors. Companies arrived at this state gradually by improperly or inadequately administering their incentive plans. This is a difficult and costly process to correct. It can be done, it takes time, but once the plan is reestablished on a sound basis, extreme care must be taken to see that incentive administration is sound in the future.

Management must adhere to several basic rules if the incentive plan is to be an asset to the company and its employees.

Rule 1. Clearly Define Incentive Policies and Commit Them to Writing. With all the varied forces at work in a company, it is essential to have an agreed-upon, recorded understanding of wage incentive policies. Otherwise, individuals such as the plant manager, whose prime responsibility is production, or the personnel manager, whose main goal is harmony, or the union leader, who wants greater pay for the people, will make some little change that sooner or later can have a mushrooming effect. One authority stated, "Most plans go haywire because, somewhere along the line, management replaces equity with charity." If policies are correct, clear, and in writing, management is much less likely to allow this to happen, and mistakes can be avoided in the future.

Rule 2. Provide Adequate, Trained, Competent Personnel to Administer the Plan. This applies to the industrial engineers who have the technical task of establishing or correcting standards or investigating potential trouble spots.

This rule also applies to the line organization which must make the plan work. The supervisor, for example, should cooperate with the industrial engineer and the worker to assure that equitable standards are developed. It is then up to the supervisor to see that workers follow the right method and that their material is as specified. If supervisors, because of lack of training, have no knowledge of the significance of their actions in giving undeserved extra allowances or in forcing loose standards, they may open the door to a deteriorated incentive situation.

Rule 3. Establish Accurate Standards Based on Good Methods. The standards, of course, are the heart of the incentive plan. If they are loose, the operator has no trouble in meeting them. The operator then either restricts production to keep earnings down or earns more than fellow operators for the same applied effort. If the standards are tight, there is insufficient incentive opportunity for the operators, and they will be inclined not to try to meet or beat the standards.

It is essential to develop the best possible method before standards are established. This will result in a sounder, more consistent earnings situation, and will avoid many future difficulties.

Rule 4. Record the Method and Adjust Standards When Methods Change. Failure to follow this rule is responsible for the deterioration of incentive plans more than any other single factor. Ordinarily, changes in method are more evolutionary than revolutionary. If a change is revolutionary, chances are good that the operation will be restudied and a new standard established. Most changes, however, are small, of the ''creeping'' variety. Any one of the small changes may have little effect on the plan. Cumulatively, however, small changes lead to inconsistencies and ''runaway'' earnings that will eventually destroy the incentive structure. Thus, it is essential that management have a record of the method on which the original standard was based and adjust the standard as the method is changed. This will require an auditing or reporting procedure that points out changes in methods and the need for standards adjustment. It ties in with rule 2 in that sufficient trained personnel must be available to perform this vital function.

Rule 5. Cover the Work with Standards as Completely as Possible. Complete incentive coverage of a department or a plant will avoid two major problems. First, where incentive and daywork are mixed, it is often difficult to get an accurate check of the time spent on each category. Therefore, time juggling can occur which will weaken the incentive plan materially and raise costs.

Second, when workers become accustomed to incentive earnings, they resist working at the regular guaranteed base rate. This leads to a strong demand for the payment of additional money, such as average hourly earnings, for time spent on jobs not covered by standards. The inevitable result of paying incentive earnings for a nonincentive performance is to weaken the incentive plan.

Rule 6. Train the Incentive Workers. The operators cannot meet the standard if they do not know the method. Their natural inclination is to bring pressure on management to loosen the standard. If they succeed, and they often do, the standard is relaxed. Sooner or later the operator will begin to find the better method and the standard will become loose.

This situation need never exist if proper attention is paid to training. The operators must be trained not only in proper methods but in the fundamental philosophies of incentive operations. In an area as sensitive as wage payment, it is important that the operators know the plan thoroughly as it affects them, and understand the reasons for the plan.

Rule 7. Provide the Incentive Workers with Sufficient Work and the Right Tools, Materials, and Instructions. One of the most discouraging things that can happen to an incentive worker is to work at an incentive pace, beat the standard, and then run out of work. People, in general, will work harder if they can see work piled up ahead of them. This is particularly true of incentive workers.

Every effort must be made to help the incentive worker eliminate delays. If the operators have to wait for a job, or wait to have materials corrected or tools changed, they will soon lose the desire to try to earn a bonus. When the conditions are correct, however, they will apply themselves diligently and produce at an excellent level of productivity.

GROUP INCENTIVE PLANS

The group wage incentive system has been developed to cover groups where it is difficult to measure the productivity of individual workers and their individual contribution to the total output. Natural groups that may fall into an incentive unit are exemplified by indirect labor crews such as forklift operators, material handlers, machine crews, progressive assembly line crews, total departments, and the like. In the group system each person shares in the earnings of the group according to the individual person's work time in the group. Normally the best results are attained where the group size is limited to some natural sphere of work interest. Cooperation, fellowship, joint interest, and supervisory effort and control are generally more easily achieved in smaller groups. As the size of the group increases beyond some nominal point, individuals tend to lose an awareness of their contribution, and the motivation to increase productivity decreases.

The advantages of the group system include:

1. Individual operators cooperate better.
2. Less supervision may be required.
3. New employees can be trained by experienced employees.
4. Nonproductive labor is reduced.
5. More accurate total costs prevail.
6. Product costing is simplified.
7. Timekeeping is simplified.
8. Employees work more conscientiously.
9. Earnings are more equitably shared.
10. Wages do not fluctuate greatly.
11. Operators are assigned work for which they are best suited.
12. The working environment is made more pleasant.

The major disadvantages of the group plan are:

1. It may be difficult to account for incomplete jobs.
2. No time allowances are provided for individual jobs.
3. No focus generally is made on individual performance.
4. It may be difficult to find a supervisor or natural leader to motivate the group.

The calculation of incentive earnings can parallel the individual incentive procedure calculations such as earnings based on total piece rates in proportion to each worker's time charged as part of the crew, or total standard hours divided by total actual hours to calculate a group percent efficiency.

All the same strong administrative factors required in the application of individual incentive plans are also required in the administration of group plans. It is not unusual to find earnings limits placed upon group plans because of the difficulty of assessing the true contribution of each member to the total task.

INCENTIVES FOR INDIRECT LABOR

Whenever possible, indirect labor should be measured and paid a bonus under the same type of wage incentive plan used for direct labor. Unfortunately, while possible, this is sometimes quite difficult.

To make incentives available to indirect employees, some firms have agreed to pay indirect labor the same bonus or some fixed percentage of the bonus paid to direct labor. This involves fallacious reasoning. Merely to pay all or a portion of the indirect workers a bonus equivalent to the bonus earned by a group of direct workers without any measure of the production of the indirect workers is not only an uncontrolled increase in pay but tends to destroy the motivation of the indirect workers' incentive plan.

In the absence of the possibility of a direct measure of shop indirect labor, it is sometimes possible to establish by analysis a definite number of hours of indirect labor required to service direct labor which is earning a definite number of standard hours. A measure may be established which is a standard series of ratios of shop indirect-labor hours to cost center, department, or plant standard hours at any level of capacity. This is a form of variable budget. Improvement in relation to the measures may result, according to a predetermined schedule in the payment of bonus to shop indirect labor.

A bonus payment to indirect workers for this increased productivity may, of course, be made in accordance with a wage incentive plan which is best suited to the conditions under control of indirect labor. Ordinarily, in such an approximate plan, where no studies have been made on an individual basis to determine the necessary work involved, it is generally agreed to be advisable and appropriate to pay as bonus a percentage of hourly rate not greater than one-half of the percentage increase in productivity.

It should be emphasized that, although expediency may demand such a solution as described, the soundest and fairest practice calls for a detailed study of individual indirect jobs with remuneration a function of the effort expended.

Therefore, such an expediency should be carefully planned so that it may be temporary and readily convertible to a properly engineered plan when time and conditions permit.

For methods of measuring indirect labor accurately, see Sec. 5, Chap. 5.

Maintenance Incentive Plans.   Many firms have measured maintenance and repair work so that such measurement can be used as the basis of an indirect type of incentive plan. All the maintenance crafts including carpenters, millwrights, bricklayers, welders, pointers, pipefitters, electricians, machinery and equipment repairers, and similar trades, as well as building maintenance operations, have been measured and standard times developed that can be applied across a wide range of industries and organizations.

In some circumstances standard times have been determined for repetitive tasks through the application of traditional work measurement techniques. In many other circumstances, the slotting system, whereby maintenance jobs are compared with typical jobs within a given time slot, is used as the basis of determining efficiencies of work performance.

Larger firms have taken the lead in such maintenance applications because it pays, as a rule, to measure diversified maintenance operations and set up standard data. The volume of maintenance work in a larger company lends itself more readily to work measurement than the limited volume of a smaller firm because the potential savings of having maintenance unit costs controlled are greater if a practical wage incentive plan can be derived from professional maintenance work measurement. Preventive maintenance is especially amenable to work measurement because such work normally consists of repetitive tasks.

The application of incentive systems should not be overlooked by management as a strong motivational force to improve the productivity of maintenance employees in the process and highly mechanized industries that require large maintenance forces.

Incentives for Office Work.   Methods of measuring this type of work are discussed in Sec. 5, Chap. 7. Standards and incentives have been applied to such work as typing, duplicating, filing, billing, timekeeping, payroll work, bookkeeping, and addressograph operation, as well as many other office operations. These incentive applications have been made in terms of standard times established from standard data. Also, it is possible to relate the volume of work in a small office to information which may be obtained readily, such as number of orders and number of invoices. For a successful application, it is necessary, as has been brought out previously, to control the input of laborhours in relation to the output of effective office work performed.

In all indirect labor and office application of incentives, it is natural that large companies lead the way. Small companies can learn much from the success of larger companies in applying incentives to nondirect labor operations. Often the standard data developed in the larger companies may be applied with modifications in smaller companies. This is particularly true of data for similar office operations. Emphasis must be placed in all cases on the word "modifications" because in general, standard data developed in one company should not be applied in another company until the data have been thoroughly checked as to suitability and any necessary modifications made.

SUPERVISORY INCENTIVES

Forepersons and other supervisors play a very significant role in the operation of any firm. If desirable bases for the measurement of their performance can be determined, an incentive system may be applied which can do much to motivate successful operation of the firm. If the firm has profit sharing or other overall incentive plans, forepersons may participate in these. Nevertheless, it is well to consider how measures may be set up which are designed to provide strong incentives for forepersons that are coupled to productivity, quality, waste, and the like.

Bonuses have been paid to supervisors in some companies in accordance with the performance of direct shopworkers. Although it is true that better than standard output by shopworkers can be attained because of good management by a foreperson, this is only one of a number of important bases for the incentive measurement of forepersons. A weakness in this procedure which should always receive careful consideration is that a foreperson may be stimulated to be neglectful or even to overlook loosened shop incentive measures. This is an example of the hazards that exist in supervisory incentives if such plans are not carefully designed, audited, and maintained.

General Characteristics of Supervisory Incentive Plans.   Supervisory incentives are usually set up to provide a financial bonus to supervisors for superior performance in controlling costs, productivity, quality, waste, turnover, and other factors for which the supervisor has primary responsibility. Such measures should be based upon factors over which the supervisor has control. Such plans should be kept as simple as possible. The duties of a foreperson are more complex than those of a worker in the shop; therefore, an incentive plan for forepersons can be expected to be correspondingly more complex and generally based upon multiple criteria.

It is important that an overall plant factor be included in each supervisory incentive plan. The best performance of supervisors is to be found when they cooperate and team up for the general good of the company. A supervisory incentive plan should never provide such a strong incentive for the operation of a single department that the proper operating relationships to other departments and to the company as a whole are neglected. The overall factor in each supervisory incentive plan should be such that it tends to assure that the supervisor keep in mind relationships with the rest of the plant.

Considering the positions in the line organization which are above a foreperson, it should be evident that a rearrangement of measures which have been established for forepersons, together with measures for overall plant performance, can be established for such positions as general forepersons, superintendents, and works managers so that persons in these positions may also participate in a supervisory incentive plan.

How far should supervisory incentive plans be extended above the level of forepersons as well as to specialized staff positions? This is subject to careful study. In general, this rule may be applied. No incentive plan should be considered unless the measures established are truly related to those things which management considers important that a supervisor do well. If measures cannot be found which relate to important aspects of a supervisor's work, supervisory incentive plans may not be useful.

Inherent Differences between Worker and Supervisory Incentives.   The wage incentive plans for production workers which have been described are based largely upon the rate at which work is produced. Such incentive plans are intended to stimulate employees to maintain a steady high level of output. Stimulus relates to greater expenditure of physical energy, within limits not detrimental to the health of workers, for the purpose of securing additional units of production per worker-hour worked.

Keyperson or supervisory incentives, on the other hand, relate much less to physical effort. Instead, planning and control of operations are emphasized. To a considerable extent this requires mental effort and cooperation with other supervisors. The supervisor must use thought to plan the work and to control the accomplishment of the work in terms of such plans. Also supervisors, whether direct-line or staff, should contribute ideas to bring about greater effectiveness of company operations. It is recognized that new ideas have some favorable influence on the level of achievement in terms of supervisory incentives. It is recognized also that supervisors, as a part of management, are expected to contribute ideas which will help to make a company more successful even though such ideas do not relate in a direct way to supervisory incentive performance.

ELEMENTS TO BE INCLUDED IN COMPREHENSIVE INCENTIVES

If incentives for both workers and supervisors tend only partially to stimulate, what elements are lacking to provide more comprehensive incentives? Certainly emphasis should be placed upon enlisting the constructive thinking of all company employees. Ideas and improved morale are the goals rather than a high level of physical effort alone or a limited level of mental effort. It is evident that a plan which satisfies these criteria will act to draw upon the vast mental reservoir of all employees. It will be supplemental to, rather than a substitute for, conventional incentive plans which have been discussed in the foregoing portions of this chapter. For indirect employees who may not be covered by existing incentive plans, it provides a means whereby sweepers, material handlers, maintenance workers, office employees, and others whose work is frequently difficult to measure in a direct way may also receive incentive payments for their contributions to the common good of the company. Thus a means is provided of bringing forth the thinking and ideas of all employees so that a company may be strengthened both by using new ideas and by giving to all employees a greater sense of participation.

Emphasis is given to the principle that valuable ideas may flow from many sources and that gains resulting from ideas should be shared. In this are involved all the elements of harmonious cooperation of employers and employees for their common good. The purpose is to minimize any limiting by employees of the most effective conversion of raw materials into finished products using all the combined abilities of employees and the facilities of a plant; and at the same time to eliminate any ceiling on the amount an employee may be allowed to earn as a result of individual and collective efforts. In the application of such a plan, it is severely limiting to believe that within management and its engineering department lies the only wellspring of ideas. It rejects any thought that the interests of employees and employers are antagonistic. Rather, it provides a positive cooperative way to share in progress.

COMPREHENSIVE INCENTIVE PLANS

An example of a comprehensive incentive plan for supervisors at the first line of management could include:

Incentive basis	Percent emphasis	Percent bonus (20% base)
Department efficiency	50.0	10.0
Quality	25.0	5.0
Waste	25.0	5.0
Total	100.0	20.0

Department efficiency could be calculated by dividing total standard hours earned by the employees supervised by a foreperson by the actual hours charged against the operations. Quality could be based upon a scale developed from percent rejects, rejects in following departments, or customer returns or reworks on a plantwide basis. Waste would be based upon a scale derived from the difference between good material leaving the department and material accepted into the department. A plant superintendent's incentive plan might be designed to reflect a stronger emphasis placed upon cost, quality, and delivery performance. For example:

Incentive basis	Percent emphasis	Percent bonus (20% base)
Budgeted cost	60.0	12.0
Quality	20.0	4.0
Percent on-time delivery	20.0	4.0
Total	100.0	20.0

It is not uncommon to find that incentive earnings for supervisors, similar to those for indirect labor, are limited to say 110 percent performance or 28.0 percent as in the example above. Industrial engineers take strongly opposed positions on the subject of limiting incentive earnings because such practice may limit performance. In the case of supervisors and indirect labor such limitations tend to be justified by management because they may be rewarded with excessive incentive earnings when such high earnings occur from factors beyond their control or contribution. Similar reasoning applies to indirect labor incentive earnings plans.

The opportunity for designing incentive emphasis is unlimited and should always emphasize the most critical elements to be attained over which the various levels of supervision have control. The incentive basis and the percent emphasis should be altered as the needs of the firm change so that the major strategic goals are fulfilled.

It is especially significant to reiterate that an incentive plan, whether for shop employees or the supervisory force, is not a substitute for good management. All too often forepersons take the approach that an incentive system will make their supervisory responsibilities less arduous. However, the wage incentive system will continue to be effective only as supervisory performance continues to be excellent and the nuts and bolts of the standards and incentive programs are excellently maintained. If anything, the application of a successful wage incentive program requires more intensive supervision, not less. It has its basis in a full recognition of the distinctive value of each individual to the success of the company and to the full recognition of the latent abilities of every employee no matter what their position in the company may be. It affords both an opportunity for and a demonstration of the concept that each individual in a company, given an opportunity, can develop into a stronger, more effective person.

INSTALLATION OF WAGE INCENTIVE PLANS

Successful installation of any new or revised program can best be summed up by the following five major points.

1. Tell them in advance.
2. Tell them the truth.
3. Tell them why.
4. Tell them the bad with the good.
5. Encourage them to speak up and question.

Installation. Installing a wage incentive program requires the same communication process described in Sec. 4, Chap. 1 of this handbook, which will be reoutlined here. It is important to remember that the communication task should be focused not only on the employees who will be affected by and work under the incentive plan, but also toward all levels of management from the very top down through the forepersons and all other first-line levels of supervision. Successful incentive plans based upon professional work measurement techniques of whatever type are best suited to the conditions surrounding the work to be performed under the incentive program. A good work measurement program should consist primarily of these functions: proper layout of the department and workplace, methods analysis and improvement, proper employee selection and training, ergonomic consideration, correct equipment and tools, effective communication throughout the standards setting process, and the professional application of whatever time measuring techniques are to be used to establish the standard time allowed to perform each defined task.

Implementation. Successful implementation and continuing validity of an incentive system commence early in the development of the system. To merely complete all the technical components of a work measurement and incentive program, then announce to all individuals in the organization by letter, a bulletin board notice, and a brief meeting or so just prior to the installation of such a program will not easily or readily accomplish good and lasting implementation. As with all human endeavors resistance to change comes to the fore here especially as wage incentive plans are a part of total wages paid, often a significant part.

Top management understanding and continuing support is essential from the very first consideration of any program. Top management exhibits its support by effectively communicating its support down through all other levels of management in the organization. It is as important for the industrial engineers to project the advantages and economics of an incentive plan to the total hierarchy of management as it is to sell the workers, who will be directly impacted.

Excellent communications must be reemphasized. Communication throughout the organization is the foundation of successful implementation. Good human relations result from good communications. Successful implementation is the result of good selling from the top manager down through the foreperson and work measurement analyst. Well-practiced persuasive skills are at a premium in the implementation process. All such individuals can benefit by instruction, training, and practice in this human art. Successful selling can be practiced and achieved by any individual, even those who do not consider themselves to be salespeople in the usual meaning of the term. No apology need be made for the term "selling" when applied in a professional area. It is a fact of management life that many worthwhile and expensive projects have failed because they were not properly and persuasively sold to all the affected members of the organization.

Before a credible incentive program can be established and implemented, the following activities and/or tasks generally must take place, or at least be considered to establish the proper environment for successful implementation.

1. The organization must have defined and communicated the long-range and short-range purpose for establishing and implementing a wage incentive program.

2. All members of the organization must understand this purpose and the long- and short-range goals.

3. All members must be afforded the opportunity to contribute to the establishment of such a program both at the beginning and during the process of implementation.

4. Continuous and appropriate communication and persuasion should be practiced to avoid, or at best to minimize, resistance to the wage incentive program.

5. Both the dollar and human cost should be recognized and kept in mind as the program develops and is implemented.

6. All the various industrial engineering and work measurement analysis and incentive design should be determined and performed by trained and experienced analysts.

7. Continuous follow-up communications, meetings, and explanations as appropriate should be maintained and scheduled at all times. If the organization is represented in whole or in part by a union, it is important to maintain good communication with the local union officers and their international representatives. The key to avoiding or minimizing resistance to change is enhanced by avoiding surprises, avoiding misunderstandings about the purpose and the techniques, and being sensitive at all times to good human relations.

8. The contribution of ideas, techniques, and better approaches should be welcomed and publicly acknowledged at all times.

9. Appropriate training of all individuals when required supports acceptance of change and new programs at all levels of the organization.

10. Any new or revised wage incentive program can best demonstrate its validity by selecting an area where it can be easily and successfully installed early in the total implementation program.

11. It is extremely important to document all policies, purpose, techniques, analysis, standards, proper reporting, and results for permanent record. One only has to have taken part in grievance, arbitration, or contract negotiations to realize how important documentation of all factors applying to a wage incentive plan becomes.

12. It is likewise important to build into any wage incentive program a reporting system—manual or electronic—that is easily understood at all levels in the organization. The results of incentive earnings should be made publicly available to the employees as well as to their supervisors.

13. An acceptable reporting system requires accurate reporting of units produced, who produced them, the labor standard applied, standard hours developed, actual hours charged, percent efficiency, and incentive earned in hours and dollars. What is actually contained in an incentive reporting system of course will depend on the type of system used as well as the needs of the organization. The variations can be numerous, but the significant issue is that adequate records be published and preserved to facilitate comparisons over subsequent time periods and various jobs run. It is also important that where a reporting and control system is to be computerized the computer program be designed to fit the program, not the reverse.

14. The reporting process should be designed to facilitate a proper audit trail. For example, total units reported as produced by single or multiple operators and/or equipment should be compared with total units scheduled, warehoused, or shipped with losses reconciled to such a total.

15. Trial runs should be scheduled along with the training activities to serve as a learning process for all concerned as well as to debug the program.

16. Never, I repeat, never discontinue using an old or present system before all the bugs have been eliminated or the necessary changes made in the system being implemented. Always run a dual or parallel system until the new wage incentive program is functioning satisfactorily. Such a debugging process is in itself a major training process at all levels in the organization.

Selling an Incentive Program.   The major steps in selling an incentive program include the following:

1. Avoid a threatening approach. Always emphasize the positive aspects of the program, but do not hide any negatives.

2. Be aware of and understand each individual's and each group's (management, union, worker, and the like) vested interest in the success or failure of a wage incentive pro-

gram. Insight into "What's in it for me?" is a basic step in selling any product, program, or idea.

3. The most significant single aid to the selling function is the ability to communicate ideas clearly, concisely, and logically. Of course, this is much more easily stated than accomplished; therefore, careful preparation and practice are essential.

4. Many types of media are available—letters, posted notices, handouts, house organs, and the like. However, in general, the one-on-one personal presentation has the greatest impact when practical. If a one-on-one is not feasible, small groups are next best. This approach is especially effective where the human "resistance-to-change" factor is expected to be encountered. The one-on-one approach allows the salesperson to garner insight into the recipient and tailor an approach to fit each individual's circumstance. Another multiple approach is to present the program by taking advantage of all benefits to be gained by the delivery of the message with a combination of the various media available. A one-on-one or a presentation to a small group helps avoid the problem of having dissident subgroups obstruct a presentation made in a large group—especially where the exhibition of union or dissident politics may be a factor within the group. Selling is essentially an implementive process and can in no way be underestimated when introducing and installing new programs.

5. Resistance to change because of fear of how a wage incentive program will impact on available work, crew sizes, workload, long-term employment and wages is a key factor that must be overcome as much as possible before the incentive program can commence. If a reduction in the work force is expected, the employees must be told up front.

Again, it cannot be emphasized enough that the implementation of a wage incentive system is the most critical part of the program because such a program can severely impact productivity, wages, and human relations.

WHAT TO AVOID IN WAGE INCENTIVES

The following pitfalls should be avoided in establishing and operating wage incentive plans:

1. Failure fully to inform employees and their bargaining agents regarding plans and proposed procedures in establishing incentives.

2. Failure to have supervision play a major role in the setting up of incentive plans.

3. Failure to recognize the high level of competency required to establish and maintain wage incentives properly.

4. Failure to take into account, analyze, and establish standards for materials and spoilage whereby there is a clear understanding that incentive payment applies only to the production of acceptable work.

5. Failure to maintain measured standards and wage incentives properly and continuously once they have been established. The operation of an incentive plan over a period of time is fully as difficult—probably more difficult—than its design and installation.

The practice of setting a temporary standard in new plants or on new operations should be kept to a minimum. In any event it should be clear to all that the standards are temporary for a defined period of time.

DETERIORATED WAGE INCENTIVE PLANS

Of the few major problems in modern management that can cause constant disruption in good human relations, and require a disproportionate share of effort and dollars, the deteriorated wage incentive plan probably ranks right up near the top. This problem is treated in detail in Sec. 6, Chap. 4 of this handbook, so only a brief outline of the subject will be included here.

Deteriorated wage incentive plans are a detriment in any organization. Management is constantly belabored because productivity and profits are inadequate. Competition is difficult because unit costs are high and often no accurate basis exists for establishing competitive labor rates and direct labor and product costs, and because of problems of scheduling production, calculating budgets, comparing performance levels, and evaluating alternate methods and equipment.

Employees become unhappy when their jobs are lost or plants must be closed because of a decreased ability to compete. In addition, wage inequities create serious and continuing human relations problems among management, labor, and their unions.

Action to rehabilitate a deteriorated plan is as follows if management makes the decision to continue with wage incentives. First some of the causes of a deteriorated plan include:

1. It appears to be too difficult to fight the union and the workers over the changes required to maintain the plan adequately.

2. Management does not fully understand the nature of an incentive system and the role it plays as a source of human motivation and potential for increasing productivity.

3. Management is unaware of the symptoms of a deteriorated plan.

4. The timing is not right to make a change.

Once the basic management practices that have led to the deteriorated plan are identified and accepted by management, a habilitation schedule should follow these steps:

1. Identify the cause (not to be confused with the symptoms): improper design of the plan, improper implementation, improper administration, or improper maintenance of standards.

2. To obtain and quantify the true causes, a series of specially designed analyses should be professionally conducted on a work sampling basis to assess the true percent performance and utilization of the employees under the methods being practiced.

3. Making a decision to rehabilitate an incentive plan is one action, but convincing the employees and the union to accept such a wholesale emotionally tinted revision, and then implementing it, is a management challenge of enormous proportions. However, the payoff in improved productivity, profitability, and competitive pricing can be as rewarding as was the original installation.

4. The revision must be wisely negotiated and bargained for after establishing revised representative labor standards based upon proper methods and training. The loss of workers' wages must be bought out on some basis equitable to both management and the workers, making sure that the employees are apprised of the potential negative economics that could result in the plant's being moved or shut down if the revised wage incentive plan is not accepted by them.

5. The implementation plan will follow the same general schedule as outlined in the implementation of a new plan described earlier in this chapter.

The most important aspect in rehabilitating a deteriorated wage incentive plan is, as it is with all other major decisions, the dedication of management at all levels in supporting the program to a successful conclusion, and assuring a continuing high level of incentive plan administration and maintenance. See Chap. 4 in this section for further details.

SUMMARY

It is important to emphasize that wage incentive plans are imperfect tools and require constant administration and maintenance. Wage incentive systems have a century of history—

both good and at times not so good. Good incentive experience requires the highest type of technical and human management; but the payoff where largely manual operations exist can be significant and return the cost of development, administration, and maintenance many times over through increased productivity.

There has been a decrease in the application of wage incentive plans over the last score of years because of the mechanization of processes, the attempt to avoid the cost of administering such plans, the human strife that occurs between management, the workers, and the union, and the attempt to experiment with behavioral factors as motivators to maintain or increase productivity.

Wage incentive systems should still be considered as significant motivators of increased productivity. When successfully applied, they will serve to exhibit the best in good management.

BIBLIOGRAPHY

Aft, Lawrence, S., *Productivity Measurement and Improvement,* Reston, Reston, Va., 1983.

Barnes, Ralph M., *Motion and Time Study,* 7th ed., Wiley, New York, 1980.

Carlson, Norma W., "Time Rates Tighten Their Grip on Manufacturing Industries," *Monthly Labor Review,* Bureau of Labor Statistics, Washington, D.C., May 1982.

Ferrell, Michael D., "Making Wage Incentive Plans Work," *Industrial Engineering,* November 1982.

CHAPTER 3
GAIN SHARING

Roger M. Weiss
Managing Partner
H. B. Maynard and Company, Inc.
Pittsburgh, Pennsylvania

Gain sharing is generally regarded as a concept that permits sharing with all employees a portion of the savings which result directly from their collective efforts to reduce costs, improve quality, and increase productivity. Gain sharing has also been referred to as a business proposition between the company and its employees. In return for a share of the success, more open communications, and an opportunity to participate in improving overall company performance, the company expects to receive higher productivity, lower costs, improved competitiveness in the marketplace, and the creation of an operating environment where teamwork and mutual trust are key factors in the success of the company. The employees' part of this business proposition is a commitment to the success of the company, with their continuing efforts to put their knowledge of problems and solutions related to their job activities to use in order to improve overall company performance. In return, the employees expect to receive monetary rewards, greater security, improved job satisfaction, and increased participation in those matters which most closely affect their jobs. This business proposition is clearly a win-win situation.

Gain sharing plans are generally characterized by the following:

- They are based on improvements in productivity.
- They are administered as group plans.
- Usually all personnel are at one particular site, and all but a few executives participate in the plan.
- Every plan has one or more base measures which represent the threshold for bonus payment.
- Plans require employee involvement to generate benefits in the form of bonuses which are shared between the employees and the company.

HISTORY AND BACKGROUND

History. The first formalized gain sharing plan was probably the one devised by Joe Scanlon around 1935. Because of a successful turnaround and saving of union jobs at the Empire Steel and Tinplate Company, where Joe Scanlon was the president of the local union, the United Steelworkers appointed him director of a newly established production engineering department. Scanlon's mission was to utilize the same employee involvement process that he had pioneered at Empire to work with other companies on the verge of collapse to save as many union jobs as possible. Unlike many of the companies with whom he had worked, the Adamson Company was a successful company desiring to increase its productivity in the face of World War II wage controls. Scanlon worked with the president of the company to add a critical element to his previous success with employee involvement. That element was a bonus system that would encourage the workers' assistance in finding ways to improve operations. The Scanlon model has been further developed into a comprehensive process which includes involvement and reward, and has subsequently been followed by a number of companies to this day.

In 1948, a similar sharing model was developed by Alan Rucker based on his extensive research to find a stable financial measure. The so-called value added measure became very popular because of its stability within industry sectors, its ease of understanding, and its ease of application requiring only a few years of historic data to develop a usable measure—in contrast to the Scanlon measure, which required extensive history and tailoring to establish a reliable base measure.

In 1974, Mitchell Fein advocated the use of sharing measure based on hours rather than dollars. Fein reasoned that a formula based on hours produced would be more understandable and acceptable to employees than measures derived from financial data. His system, called Improshare,[1] is used by approximately 200 companies in the United States.

In addition to these three best-known approaches to gain sharing, there are countless plans of custom design developed by all types of enterprises. Profit sharing is a form of sharing but is normally not considered a gain sharing program; it is covered in more detail in Sec. 4, Chap. 6.

Background.   In the fall of 1986, the American Productivity and Quality Center, along with the American Compensation Association, sponsored a nationwide survey entitled "People, Performance and Pay."[2] The survey examined the use of nontraditional reward systems and, in so doing, provided a basis for understanding the growth of gain sharing and its use as a reward system today. Approximately 1500 companies responded to the survey. They were divided equally between producers of goods and producers of services. At that time, 20 percent of the goods producers were using some type of gain sharing, but only 8 percent of the service companies had reported gain sharing plans. Those numbers are very likely proportionately higher at this writing because the use of gain sharing has been growing steadily and more rapidly since the time of the survey.

In 1986, 19 percent of the companies who had gain sharing plans had implemented them 6 to 10 years earlier. Seventy-three percent of the companies had implemented their plans within the last 5 years. When queried by the survey as to the reasons for using gain sharing, 92 percent of the companies indicated that productivity improvement was a major factor. Equally important for 60 to 70 percent of the companies were quality, employee relations, and labor costs.

TYPES OF PLANS

This section discusses the characteristics of four different types of gain sharing plans. This discussion is meant to provide the reader with an overview. More detailed information on the theory, characteristics, and design of each of these plans is available from several sources.[3-5] Generic, rather than specific, names have been given to the plans based on their characteristics in order to emphasize the fact that extensive variations are possible in all cases, and therefore they can logically be grouped under a generic name rather than the specific plan recommended by its original author.

Total Employment Cost Plan.   The total employment cost approach to gain sharing establishes the historic ratio of total payroll dollars of the participants to sales value of production. In this *single-ratio* application, monthly sales value of production is multiplied by the base percentage to determine allowed payroll costs. These costs are compared with actual payroll costs, and the difference becomes the gain sharing bonus. Seventy-five percent of this bonus pool is allocated to the employees, with the remaining 25 percent going to the company. It is quite common, as with most financially based plans, to place one-fourth to one-third of the employee share in an escrow account to offset those months when the bonus is negative (actual employment costs exceed the base ratio times the sales value of production). In addition to the single-ratio approach, there are several variations of the formula. The most common is the *split ratio,* which is utilized when there are major differences in the base ratio between product lines. In this case, a base ratio is calculated for each product line and the sales value of production for each line and its accompanying total payroll costs are calculated separately for each product. The results are then summed into one bonus pool which is divided as before. Other common applications of the total cost approach include the *multicost ratio,* where the base percentage is the ratio of total employment costs plus other costs that may be included to the sales value of production. A fourth variation is the *allowed labor approach,* which calculates the bonus pool by subtracting actual labor from a total allowed labor figure, which is usually a function of the standard cost system. Multicost ratios or value added approaches are normally used when single ratio or split ratio are found to be not consistent enough historically to be credible. A sample monthly bonus calculation is shown in Fig. 3.1.

```
                            TOTAL COST

                         MONTHLY CALCULATION

    1.   SALES                                          $2,000,000

    2.   LESS RETURNS, ALLOWANCES, DISCOUNTS            (    50,000)

    3.   NET SALES                                       1,950,000

    4.   CHANGE IN INVENTORY                            (   300,000)

    5.   VALUE OF PRODUCTION                             1,650,000

    6.   ALLOWED PAYROLL COSTS
              BASE RATIO X VALUE OF PRODUCTION             577,500
                   0.35 X 1,650,000

    7.   LESS ACTUAL EMPLOYEE COSTS                     (   530,000)

    8.   BONUS POOL                                        47,500

    9.   LESS COMPANY SHARE 25%                         (    11,875)

   10.   EMPLOYEES' SHARE                                  35,625

   11.   LESS RESERVE FOR DEFICIT MONTHS               (    10,688)
              30% X 35,625

   12.   BONUS DISTRIBUTION                               24,937

   13.   PARTICIPATING PAYROLL                           380,000

   14.   BONUS PERCENTAGE = 24,937/380,000                 6.6%
```

FIGURE 3.1   A sample monthly bonus calculation for a total employment cost plan.

The total cost calculation, though important, is really a secondary consideration in the original Scanlon philosophy. The key to the success of the total cost plan or any gain sharing plan is employee involvement. Scanlon stressed this above all other things.

Value Added Plan.   Like the total cost plan, the formula for the value added plan also uses a ratio; however, it is based on total payroll dollars divided by production value. Production value is sales value produced less outside purchases for things such as materials, supplies, and energy. The value added base measure is derived historically and has been found to be very consistent in most manufacturing environments. Rather then an arbitrary percentage split between employee and company, the employee share is measured by their historic share of the total production value. A typical calculation is shown in Fig. 3.2. The illustrations which follow provide a clearer explanation of how bonuses are earned and shared between the employees and the company utilizing the value added method.

VALUE ADDED

MONTHLY CALCULATION

1.	VALUE OF PRODUCTION		
	(Sales + Various Adjustments)		$1,500,000
2.	LESS OUTSIDE PURCHASES		
	MATERIAL AND SUPPLIES	$760,000	
	OTHER OUTSIDE PURCHASES AND		
	NON-LABOR COSTS	350,000	(1,100,000)
3.	VALUE ADDED		400,000
4.	ALLOWED EMPLOYEE COSTS		
	BASE RATIO X VALUE ADDED		172,000
	0.43 X 400,000		
5.	LESS ACTUAL EMPLOYEE COSTS		(152,000)
6.	EMPLOYEES' SHARE		20,000
7.	LESS RESERVE FOR DEFICIT MONTHS		
	33% X 20,000		(6,600)
8.	BONUS DISTRIBUTION		13,400
9.	PARTICIPATING PAYROLL		90,000
10.	BONUS PERCENTAGE = (13,400/90,000)		14.9%

FIGURE 3.2 A sample monthly bonus calculation for a value added plan.

Essential to any financially based sharing plan is the development of a historical measure which can be utilized to establish the base from which bonuses are calculated. In the case of value added, a history of the sales value of output is used. This sales value of output, illustrated by the circle, represents the sales value of what was produced in a given month, less discounts and returns and adjusted for inventory change (Fig. 3.3).

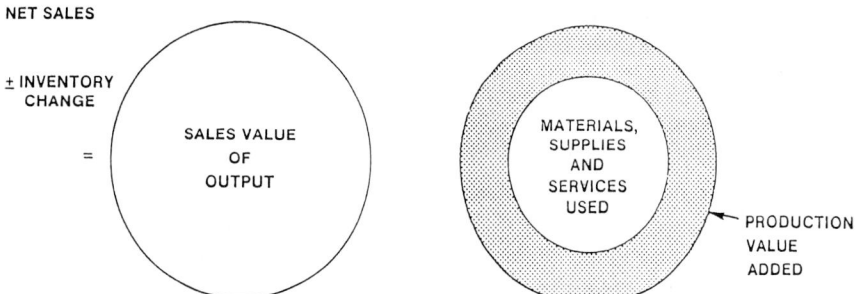

FIGURE 3.3 The circle represents the sales value of what was produced in a given month, less discounts and returns and adjusted for inventory change.

FIGURE 3.4 The hole in the doughnut represents outside purchases subtracted from sales value produced. The remaining doughnut is called production value, or value added.

The next step in the calculation is to subtract all those things that were purchased and used in producing that sales value. These outside purchases of materials, supplies, and services are subtracted from sales value produced as shown by the hole in the doughnut. The remaining doughnut is called production value, or valued added (Fig. 3.4).

The historical data are utilized to determine on average what percent the pay and benefits of all the employees participating in the plan represent of total value added. The shaded section of the doughnut represents that relationship (Fig. 3.5).

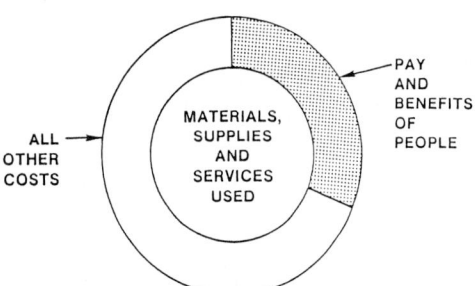

FIGURE 3.5 The shaded area of the doughnut represents the pay and benefits of employees participating in the plan.

The value added method creates gains to be shared by the company and the employees three ways. Typically, the total gain realized at the end of each month is a combination of all three of these improvements. The first is illustrated by looking at what happens if the amount of sales value produced through more shipments and better quality is increased. The dotted line represents this increase, and the shaded area represents the employee share of this increase, all other things being equal (Fig. 3.6).

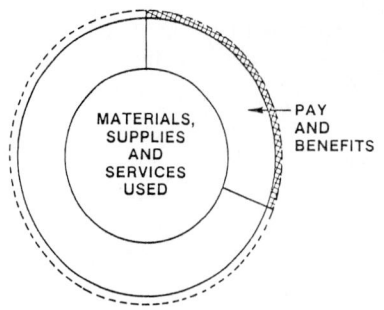

FIGURE 3.6 Example of three ways the value added method creates gains to be shared by the company and employees.

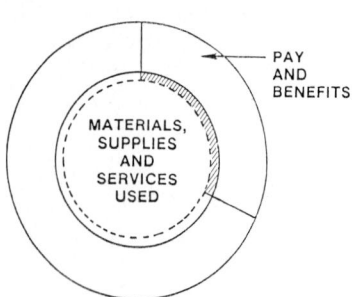

FIGURE 3.7 Example of gain created by reducing the material and supply costs through scrap reduction and better planning and purchasing.

A second way to create gain would be to reduce the material and supply costs through scrap reduction and better planning and purchasing. The dotted line inside the doughnut illustrates this reduction. The shaded area illustrates the portion of the gain that would go to the employees (Fig. 3.7).

Finally, if the same production value is turned out with reduced labor costs through better scheduling, methods improvement, labor productivity improvement, and so forth, the entire savings show up as extra bonus for the employees. This is illustrated by the shaded segment in the pay and benefits area (Fig. 3.8). Since the costs associated with producing sales value are dynamic, the interaction of all three of these situations ultimately produces gains when the allowed employee costs exceed the actual employee costs for the accounting period.

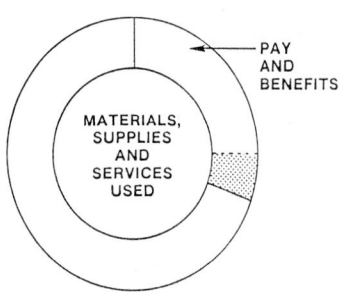

FIGURE 3.8 Shaded area illustrates extra bonus to the employees when same production value is turned out with reduced labor costs through better scheduling, methods improvement, and so on.

Typical of all gain sharing plans, a weekly, or in the case of financially oriented plans, monthly accounting must be made to the participants. An employee example of a monthly value added statement is shown in Fig. 3.9. Also typical of the financially based plans, the payoff percentage is the ratio of the bonus pool to the total gross compensation paid to the participants for the period. The balancing account, or reserve account mentioned previously, is credited with the monthly contribution so that the participants know at any time the amount of money in this account. This money will be distributed at the end of the plan year in the same way as the monthly bonuses have been distributed. If the amount of money in the account is negative, no bonus distribution will be made and the account will start at zero at the beginning of the next plan year.

In order to better understand the calculation of the company's share under the value added system, a management example of the monthly value added statement for the same month is shown in Fig. 3.10. The economic productivity standard represents the dollars of value added per dollar of payroll. The company share this particular month is the total gain, $27,289, minus the employee share of $9300, or $17,989, a very high percentage of which will contribute directly to margin. The consistency of the value added base measure

-Plant Employees-

1. What we sold and added to inventory	$582,473
2. Less Materials and Supplies we used up	261,784
3. ACTUAL PRODUCTION VALUE WE CREATED	$320,689
4. Your share @ 34.08%	$109,300
5. Less Regular Payroll and Benefits	100,000
6. Gross Bonus	$ 9,300
7. Less "Cushion" set-aside of 1/3	3,100
8. Cash Payoff now	$ 6,200

9. Total of four pay checks to all people at work or

eligible to work on last working day of the month

was $76,518.

10. "Payoff" Per Cent = $\dfrac{\$\,6{,}200}{\$\,76{,}518}$ = 8.10%

Each eligible employee receives 8.10% of the total of his four regular pay checks, including shift premium and overtime, for this month.

Balancing Account

Previous Balance	$ 18,245
Addition (Subtraction) this month	3,100
New Balance	$ 21,345

The Account is redetermined at the end of each fiscal year and the balance distributed to employees.

FIGURE 3.9   An employee example of a monthly value added statement.

over the years recommends its careful consideration where operating history indicates its consistency in past years.

Earned Hour Plan.   The concept of paying incentive on earned hours has been the basis of many conventional incentive programs. The difference in the gain sharing approach is that the basis for payment begins with the historical actual production hours required per unit of product produced, this in vivid contrast to what industrial engineers know as engineered standards. Like the financial measures, the historical hours contain all of the past inefficiencies and reflect average performance to date. Engineered standards, of course, reflect an overall performance of 100 percent, or the theoretical fair day's work level. The concept of sharing improvement, hence the name Improshare, is predicated on starting at

-Plant Employees-

1. Wages for the Month:
   Sum of weekly pay checks and odd days $80,000
   Accrued "Fringes"   <u>20,000</u>

   TOTAL REGULAR PAYROLL $100,000

2. Economic Productivity Standard   x $ 2.934

   - - - - - - - - - - - - - - - -

3. STANDARD PRODUCTION VALUE $293,400

4. ACTUAL PRODUCTION VALUE   <u>320,689</u>

5. GAIN (Loss) $ <u>27,289</u>

6. Employees Share of Gain @ 34.08% $ 9,300

7. Less Reserve set-aside of 1/3   <u>3,100</u>

8. Immediate Cash Distribution $ 6,200

9. "Payoff" Per Cent = $\dfrac{\$\,6{,}200}{\$\,76{,}518}$ = 8.10%

   - - - - - - - - - - - - - - - -

Each eligible employee receives 8.10% of the total of his four regular pay checks, including shift premium and overtime, for this month.

_____

Reserve Status

Previous Balance $18,245
Addition (Subtraction) this month   <u>3,100</u>
New Balance $21,345

- - - - - - - - - - - - - - - -

The Reserve is redetermined at the end of each fiscal year and the balance distributed to employees.

FIGURE 3.10   A management example of a monthly value added statement.

the present level of productivity and sharing on a 50/50 basis improvement over that level. Mitchell Fein, in his development of the details of this plan, has been very precise and comprehensive. The reader should review the referenced materials to gain a thorough understanding of the plan provisions and functions. For the purpose of this discussion, the plan calculations may be summarized as follows.

Employee hour standards are calculated for each product or major product line as the ratio of total production hours divided by units produced. The total standard value hours, for the purposes of calculating the base, are equal to the appropriate employee hour standards times units produced during the base period. A base productivity factor is then calculated by dividing the total employee hours for all employees to be included in the plan by the total standard value hours. The example in Fig. 3.11 illustrates this. A typical bonus calculation is shown as Fig. 3.12.

As mentioned previously, there are many conditions and stipulations for the development and administration of the plan too numerous to describe in this chapter. A major

PRODUCTION EMPLOYEES: 20 (A-8, B-12)
NON-PRODUCTION EMPLOYEES: 10
NUMBER OF HOURS WORKED BY EACH EMPLOYEE: 40
UNITS PRODUCED: A-80, B-60

EMPLOYEE-HOUR STANDARDS (EHS)
PRODUCT A = 8 X 40/80 = 4 HOURS
PRODUCT B = 12 X 40/60 = 8 HOURS

TOTAL STANDARD VALUE HOURS (TSVH)
PRODUCT A = 4 HOURS X 80 UNITS = 320
PRODUCT B = 8 HOURS X 60 UNITS = 480
 TOTAL TSVH   800

BPF = (20 + 10) 40/800 = 1.5

FIGURE 3.11 This shows the derivation of a base productivity factor (BPF).

UNITS PRODUCED:
 PRODUCT A = 100 UNITS
 PRODUCT B = 50 UNITS
 BPF = 1.5
 TOTAL EMPLOYEE HOURS = 950

BONUS:
 PRODUCT A = 4 HOURS X 100 UNITS X 1.5 =600
 PRODUCT B = 8 HOURS X 50 UNITS X 1.5   =600
  EARNED HOURS   1200
  ACTUAL HOURS   -950
  GAINED HOURS   250
  EMPLOYEE BONUS HOURS
  (0.5 X GAINED HOURS)   125
 BONUS = 125/950   13.2%

FIGURE 3.12 This shows a typical bonus calculation.

feature of this plan that exists in no others is the buy-back stipulation when the earned bonus share for the employees exceeds 30 percent. Above this level, hours may be banked and carried forward, or by mutual agreement, the standard may be adjusted back to the 30 percent level. The major difference between the earned hour type of plan and the other two financially derived bonus plans is the lack of a specific tie between overall company performance and the bonus pool when the earned hour concept is used. The earned hour approach more closely typifies conventional incentive programs in that it ignores those business factors that can create real earning performance problems that can only be worsened by the payment of bonuses based on measures not tied to overall performance.

Cost Factor Plan.   Cost factor plans have been devised where linking the plan performance to the marketplace was thought to be either inequitable or impossible, and the need to control specific cost areas was an overwhelming consideration. There are many varieties of cost factor plans. Basically they have the following characteristics.

A series of cost-related factors are developed and evaluated for their incremental value per each percent of improvement. Then on a weekly, or more commonly a monthly basis, the change in these selected cost indexes times the bonus factor per incremental change equals the bonus pool contribution. The sum of all the factor contributions is equal to the total bonus pool. The example calculation in Fig. 3.13 demonstrates this.

This example illustrates three different types of cost factors. The first four factors have sufficient available measures to establish a base from which to measure improvement. Unlike the plans previously discussed, this case may or may not be historically achievable. In fact, in most cases, it is an established goal. Also for these first four factors, an economic value for each percent improvement can be determined using accounting data. In many plans, an attempt is made to equalize these incremental dollars so that percentage improvement in all factors is worth the same. Our example more accurately defines the importance of each factor by showing the differential in incremental improvement value. The second type of cost factor has to do with the employee involvement program and suggestions submitted and implemented by the employees. Here the value for each submitted or implemented suggestion is an arbitrary value established by the company.

Finally, the so-called theme factor is something that can be added or subtracted at the will of management. The base is rather arbitrary as is the value for percentage improvement. In order to maintain some semblance of equity, the theme factor will not detract from the bonus pool should it be negative. Once the bonus pool is established, the calculation and distribution of bonus is usually similar to the other financial plans; that is, the company share, usually 50 percent, is subtracted, then the employee share is reduced by

	PERCENT CHANGE	$ EACH PERCENT	BONUS $
FACTORY PRODUCTIVITY			
BASE 65% ACT. 67%	+ 3%	2,100	6,300
QUALITY COST			
BASE 2% ACT. 1.8%	+ 10%	230	2,300
ON-TIME DELIVERY			
BASE 85% ACT. 88%	+ 3.5%	2,000	7,000
INVENTORY TURNS			
BASE 2.5% ACT. 2.55%	+ 2%	4,100	8,200
SUGGESTIONS			
SUBMITTED 20		100	2,000
IMPLEMENTED 5		300	1,500
24 HOUR SERVICE			
COMPLIANCE (THEME)			
BASE 90% ACT. 95%	+ 5.5%	1,000	5,500
		TOTAL BONUS	32,800

FIGURE 3.13   This shows an example of a cost factor plan calculation.

a reserve amount, and then the available bonus is paid as a percentage of the participating payroll. This reserve is used in the same manner as in other financial plans.

Several important differences characterize the cost factor plan. In addition to having the same drawback as the earned hour plan because it is not tied to overall company performance, the incremental value of the cost factor may vary with volume changes and will need to be periodically changed to reflect the impact of inflation or other cost changes. Also, any time a base measure does not have any historical significance, it is always subject to question and its acceptance by the participants is more difficult. Finally, and perhaps most importantly, because the cost factors focus only on specific areas, the overall improvement that can be realized is limited. Over time these measures will tend to max out, thus stifling future continuous improvement motivation.

Profit Sharing.   Profit sharing is not normally categorized as a gain sharing plan, though it has been in use at least as long. In the survey previously mentioned, 32 percent of the participants indicated they had some form of profit sharing. The survey participants also unanimously agreed that profit sharing plans have little or no motivational effect on daily operations. This experience is attributed to the fact that, generally speaking, employees feel profit is beyond their individual control and contribution. Bonus payment often lags performance by as much as a year or as little as 3 months. The level for sharing has been arbitrarily set by executive management, with no employee participation. These factors all taken together add up to lack of involvement and therefore motivation.

GUIDELINES FOR DEVELOPMENT

Management Commitment and Support.   To be successful, gain sharing requires both involvement and reward. Employee involvement requires management commitment and support at the highest level. Characteristic of successful programs has been a high level of senior management commitment to involvement, with the adoption of this commitment at all levels of the organization. This commitment can only be won through broad opportunities for participation, coupled with education. Commitment must be demonstrated by daily support manifested in personal involvement in the process of developing, implementing, and ultimately managing the gain sharing process.

A properly designed plan, correctly implemented, has the potential for substantial rewards for both the employees and the company. Conversely, an ill-conceived plan can be costly, not only in a financial sense but also in the context of damaging employee management relations to the extent that participation in other programs essential to improving productivity and quality will be difficult, if not impossible. Bearing in mind the importance of this development process, the following steps are offered as guidelines to this critical process.

Preliminary Plan Design.   The type of plan, and ultimately the plan measure, is the most critical step in the development process. In addition to the four well-known plans that were described previously, hundreds of customized plans are designed to fit specific company situations. Because of the complexity of this process and the potential for far-reaching benefits or problems, this task is best carried out by selected members of executive management working closely with an experienced professional. This combination usually provides the fastest determination of practical alternatives and allows management to determine the type of plan that will be acceptable.

Plan Type.   There are distinct advantages to the use of a financially oriented or economic measure, as opposed to a physical measure (such as hours, tons, or pieces). The economic measure can be related directly to company performance and to the monthly financial data. Economic measures can be tied to acceptable levels of profitability and provide dependable control over critical cost areas as they relate to the bonus. Physical measures, on the other hand, operate quite independently of company performance, and may be set at a level of productivity that is initially well below that which should be expected at current operating costs. The use of a physical measure may be considered where an economic measure is not feasible. Determining the type of plan provides specific direction as to what the plan measure will be. This determination can usually be made in a few weeks, depending on the skill of the preliminary design group.

Operating Procedures.   Once the plan measure has been determined, essential operating procedures should also be identified. These procedures are less critical than the selection of the plan type but are nevertheless important to accomplish some preliminary simulation. Identification of those employees who will be participating in the plan is essential. Determination of monthly fluctuations relative to seasonal swings and how they will be handled is also an important component of the initial operating procedure. In addition to these, the following should be determined:

How often will the bonus be paid?

How will the bonus be calculated?

What provisions will be made for accruals?

What percent of the bonus will be set aside for the balancing account?

What level of plant performance will be the threshold for bonus payment?

What rules of eligibility will apply?

What provisions for ratification and change will be necessary?

Simulation.   In order to assure that the plan will benefit both the company and the employees, a simulation of monthly variations through the base period should be made. This will enable the design team to predict what percentage of the bonus should be set aside in the balancing account, and will help assure that the year-end results, once resolved, will result in one additional payment. The importance of this step should not be underestimated, since negative year-end results are an indicator of poor plan performance, as well as a negative incentive for future plan initiatives.

Business plan pro formas can also be used to develop simulations for future years in order to anticipate the impact of major events affecting company performance, such as price increases, wage increases, market share changes, volume changes and new products. Simulation is also used to evaluate the extent to which additional improvements may be required in order to achieve an acceptable level of performance to implement bonus payment.

Implementation Requirements. During the course of the preliminary plan design step, it is possible that a number of issues or problems that would tend to impede the successful implementation of a plan may be identified. Typically, these include lack of efficient access to necessary data; an overall plant performance level that is well below that which should be expected for the existing levels of compensation; problems with other management systems such as planning, scheduling, and cost control; and finally, potential problems with management and employee attitudes relative to working cooperatively to achieve mutual benefits. Once identified, specific steps to correct these problems must be part of the conditions for implementing the proposed plan.

Management Approval. It is absolutely essential that the results of the preliminary design study are totally acceptable to all levels of management with approval authority. Along with this acceptance, necessary resources must be provided to correct any problems that might stand in the way of implementation and development of a comprehensive plan to move forward with the subsequent steps of the process on a timetable that is realistic and has the highest priority of company management.

Gain Sharing Plan Development Organization. The shared responsibility of developing and implementing the gain sharing plan from this point on is absolutely essential to its success. A steering committee of no more than eight people should be organized so that the interests of management and the participating employees are appropriately served. This committee should then be responsible for determining the objectives of the plan, based on the preliminary design study, and selecting the plan coordinator(s) who will ultimately chair the design and later the gain sharing administration committee. A ratio of one full-time plan coordinator for each 500 employees is a good rule of thumb.

The size of organization is normally not a detriment to gain sharing; however, the larger the organization, the more important an adequate communication structure will become. Gain sharing plans have been successfully implemented in plants employing from 50 to 15,000 people. In order to institute an appropriate communications network, the organization should be divided into groups of 20 employees, with each of those groups having one communicator or representative. These communicators are selected by a peer nomination process and later, through the same process, will select the members of the design team. Ideally the design team should not exceed 15 people.

Peer selection process is simply one of nominating three people from an individual group in order of their preference, that is, giving three votes to the first choice, two to the second, and one to the third. This pool of communicators then provides the resource for personnel to participate in the initial development and implementation process, and later in the process of administration.

The plan design team is selected by the communicators and therefore becomes representative of the participating employee group. In addition to the elected members, the controller, or equivalent position for the organization, becomes a design team member because of the need for access to essential data in order to administer the plan. The plan coordinator(s) will chair this committee.

The design committee is responsible for finalizing the design of the plan, developing the details of the employee involvement program, and gaining acceptance of the plan of involvement and reward from all employees.

Development and Implementation of an Employee Involvement Program. Since employee involvement is key to the success of the gain sharing plan, a suitable program should be organized and implemented under the direction of the design task force as soon as possible. The goals of the program should be clearly identified, along with short- and long-term objectives. Program organization will center around the small group communicators and first-line supervision. Appropriate procedures for idea development, submission, and acceptance should be developed. A computerized tracking system for all ideas should also be an integral part of the program design. Special training of the members of the design task force and the communicators or representatives must be done in order to familiarize them with the type of feedback charts and displays needed to capture the imagination of the plan participants in terms of tracking improvement.

The key to the involvement program itself is to establish an efficient procedure for processing individual and team developed ideas to their completion in the shortest period of time. Employee communicators or representatives will work with first-line supervision and the program coordinator(s) to establish specific procedures which will allow continuous development and implementation of these ideas. An example of a typical weekly report tracking completion of ideas is shown in Fig. 3.14.

The organization of the employee involvement program is critical to its success. A typical organization diagram is shown in Fig. 3.15.

Idea generation may be the product of a series of employee training programs, including basic work flow and methods improvement, facilities improvement, JIT, total quality management, and other training in techniques designed to improve overall quality and productivity. A formalized structure and tracking system is key to the success of the employee involvement program. Though the program may be implemented prior to the introduction of the gain sharing reward process, opportunities for rewards under the gain sharing system should be provided within 3 months of the involvement program introduction.

Development and introduction of the employee involvement program as a first step for the design task force provides an opportunity for achieving quick results through teamwork, and will set the stage for the more difficult task of finalizing the gain sharing plan design. In addition, the involvement program raises the sensitivity of all the employees as to what can be accomplished, and begins the improvement process prior to the beginning of plan accounting. In many cases, this may be necessary to reach an acceptable level of productivity prior to introducing bonus payment.

Final Plan Design.   Finalization of the plan design begins where a preliminary design process left off. The type of plan has been selected, a preliminary measure may have been developed, and preliminary determination of plan participants has been completed. All of the work done in the preliminary stage is reviewed with the design task force, and details regarding both the plan measure, participation, policies, and procedures are finalized or developed for the first time. Properly done, the preliminary design step does not usurp any of the prerogatives of the design task force, but rather provides a sound basis from which to complete the development process. To do otherwise could result in a plan design that would ultimately be unsatisfactory to management and take excessive time to develop.

Finalization of plan design must be done on a consensus basis with everyone on the task force in full agreement with the conclusions. As the design process progresses, frequent meetings between the design task force members and the communicator representatives for the employee participant groups are essential in order to communicate progress on the design and to receive input from the employees in response to what is being planned.

The final step of the plan design is to document the policies and procedures, and the calculation of the base measure to support future administration of the plan. In addition, an employee booklet describing the plan in sufficient detail and providing answers to most commonly asked questions should be developed for use in the employee orientation sessions which will precede the implementation of the program.

Presentation of the Plan to the Participants.   The plan brochure developed in the last step of the design process is utilized as the focus of small group meetings of all employee participants. It is essential to the success of the plan that there be a high level of understanding of how the plan functions and, more importantly, what actions will create a gain situation.

G A I N S H A R I N G

NUMBER	ORIGINATOR	GROUP	DESCRIPTION	INVESTIGATION			SUPERVISOR OR MANAGER	DECISION	ADOPTION		
				WHO	TARGET DATE	ACTUAL DATE			WHO	TARGET DATE	ACTUAL DATE
155	S. HUGHES 01/12/90	8	When receiving raw steel make sure both ends are clearly color marked.	E. SCRUGGS	01/15/90	01/15/90	E. SCRUGGS	CAN DO	A. PARKER	01/15/90	01/15/90
157	G. PARKER 01/12/90	4	Need to re-draw and recalculate wgts on snow plow sketches	B. SNADER	01/15/90	01/15/90	B. SNADER	CAN DO	B. SNADER	01/15/90	01/15/90
164	J. MUSSER 01/26/90	6	Install on-off switch on back west door light	B. SMITH	02/06/90	01/31/90	B. SMITH	CAN DO	D. ROWLES	02/02/90	02/01/90
166	R. MC CARTHY 01/26/90	1	Need another L shaped angle hook for unloading trucks at cupline	J. PRUITT	01/22/90	01/22/90	J. PRUITT	CAN DO	D. ROWLES	02/01/90	01/31/90
174	S. WIGGINS 01/26/90	2	Check on buying grease by the bulk instead of tubes	D. SPURLOCK	02/05/90	01/31/90	R. ROBINSON	CANNOT DO		01/31/90	01/31/90
178	D. DAUGHERA 01/26/90	6	Make a new stop bar w/ set screws in front on #211 punch	B. SMITH	02/05/90	01/31/90	B. SMITH	CAN DO	D. ROWLES	02/05/90	02/05/90
183	M. BLOOMFIELD 01/26/90	3	Each shift should set up for following shift on punches, burners, etc	D. GERHART	01/22/90	01/22/90	R. JONES	CANNOT DO		01/22/90	01/22/90

FIGURE 3.14 Example of a typical weekly report tracking completion of ideas.

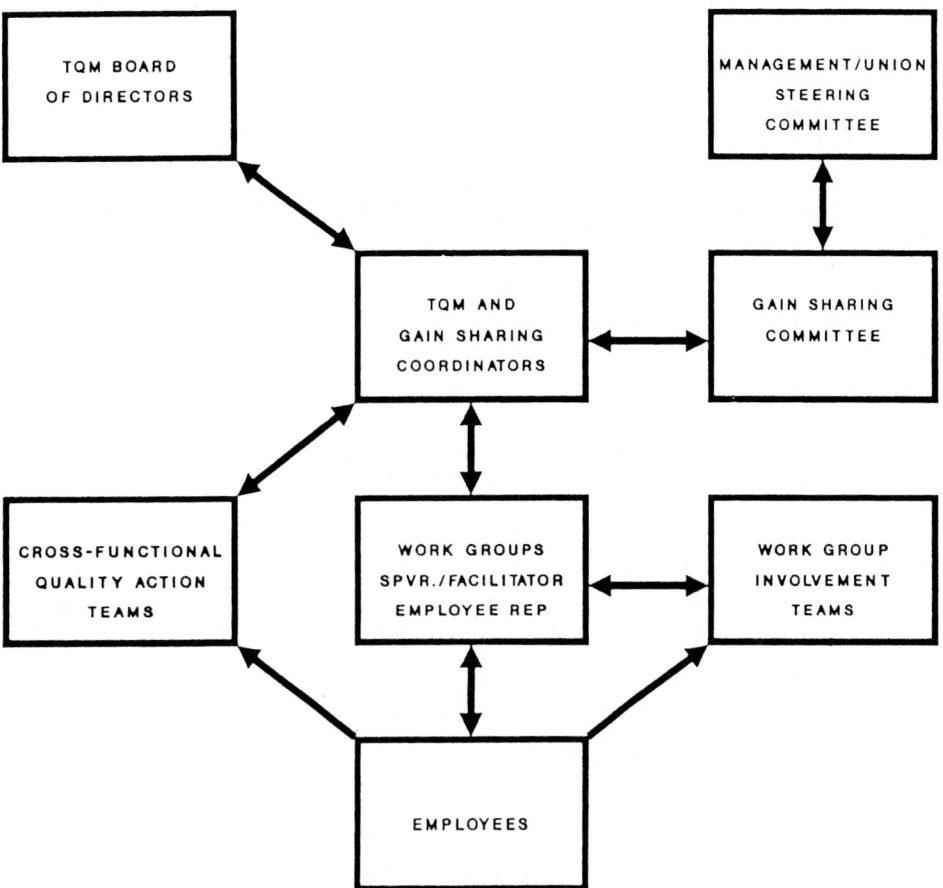

FIGURE 3.15   Example of a typical organization diagram.

Each functional group within the organization has a contribution to make. Even though these contributions are not necessarily equal, they are all important. Without total commitment of each function, the plan will not be successful. As a result of the employee orientation, a measure of the employee's desire to participate should be taken. In certain situations, an actual vote of acceptance may be appropriate. Unless a high percentage of employees indicate their desire to participate in the plan, implementation may be premature.

Gain Sharing Plan Implementation.   Timing is critical to the appropriate implementation of a gain sharing program. The beginning of accounting for bonus purposes requires that the employee involvement organization and program be in place and working, that necessary support systems for reporting essential data for the calculation of the bonus are functioning properly, that employee education has been completed, and that a high percentage of the employees have indicated a desire to participate. In addition to these prerequisites, the date for implementation should be predicated on the fact that an acceptable level of company performance has been reached as defined by the base measure, and that the near term (1-year) outlook for business improvement is positive.

In the event that negative situations relative to the business environment exist or are predicted to exist within the short term, consideration should be given to postponing implementation of plan accounting until such time that appropriate steps can be taken to assure bonuses in the early months. Once the plan has established its credibility through the payment of bonus checks and the increasing trend of positive performance, negative trends, occasioned by the marketplace or other unforeseen problems, can be taken in stride. Employees will focus more clearly on what governs the earnings of bonuses when there are bonuses and when monthly checks are distributed. Until that focus is established, acceptance of the plan in the early months is in peril.

The educational process that has occurred through the plan design process will develop an understanding for the need for proper timing, and therefore the ultimate decision as to whether to proceed with plan implementation should be made by the design task force. Once the plan has been implemented, the design task force will evolve into a gain sharing administration group and will be responsible for reporting monthly performance, as well as making recommendations for improving results. It is therefore incumbent on this group that they be responsible for the ultimate decision of implementation since they will be responsible for explaining the results in future months.

Reporting Results.   Analyzing and reporting results on a monthly basis is critical to continuous improvement. The first step in the process typically involves the company controller and the gain sharing coordinator. Together they review the results and prepare an explanation for what has occurred over the past month. These results are then explained to the executive management group and subsequently presented to the gain sharing committee. The gain sharing committee members are then charged with transmitting this explanation to the other communicators from the groups they represent so they in turn can hold short meetings with small groups of employees to explain the results to them. A one-page explanation, in addition to a one-page bonus calculation, should be posted to assure appropriate transmittal of information.

In addition to reporting plan and reward results, an analysis of the effectiveness of the employee involvement program over that month is also an integral part of this reporting process. Key indicators of the plan are:

Quantity, which is the number of suggestions submitted per 100 employees. This measure has been found to be directly related to the size of the bonus.

Quality, which is the percentage of submitted ideas that have results in positive change.

Speed, which is the number of working days that have transpired between the submittal of an idea and its appropriate disposition.

Participation, which is the percentage of employees who have submitted one or more ideas within the last 6 months.

As mentioned previously, ideas come from individuals, work group teams, and cross-functional teams, and are the key to creating bonus dollars. Typically, the gain sharing base measure is predicated on history. Unless history can be improved upon through innovative ideas effectively implemented, there can be no gain. In addition, innovative strategies for management to counter market trends and other influences that go beyond individual employee control are key to maintaining the viability and effectiveness of a successful plan and its comparable impact on continuing improvement in company performance.

CONCLUSION

This chapter has presented an overview of the gain sharing concept, its application, and guidelines for development. Though apparently simple in theory and application, successful development and implementation of gain sharing plans are historically a product of the combined experience of professionals in the field, and capable management and employee teams. Available expertise should be seriously considered in the first steps of the development process in order to assure a sound foundation for the plan and avoid the multiple pitfalls that can defeat the effort. Any enterprise wishing to explore the application of gain sharing should be able to find the resources to be successful. Because of the potentially far-reaching impact of this process, inexperienced personnel should be discouraged from experimentation in this critical area.

REFERENCES

1. Fein, Mitchell, "Improshare®: An Alternative to Traditional Managing," American Institute of Industrial Engineers, Norcross, Ga., 1981.
2. Odell, Carla S., "People, Performance and Pay," The American Productivity and Quality Center, Houston, Tex., 1987.
3. Doyle, R. J., "Gainsharing and Productivity," AMACOM, New York, 1983.
4. Freund, W. C., "People and Productivity," Dow Jones—Irwin, Epstein E. 1984.
5. General Accounting Office, "Productivity Sharing Programs: Can They Contribute to Productivity Improvement?" U.S. GAO, Gaithersburg, Md., AFMD 82-11, Mar. 3, 1981.

CHAPTER 4
AUDITING AND RESTRUCTURING INCENTIVE PLANS

Michael D. Ferrell
Vice President Process Improvement
IBP
Dakota City, Nebraska

George Matkov, Jr.
Attorney at Law
Matkov, Salzman, Madoff and Gunn
Chicago, Illinois

Wage incentive programs have played a vital role in improving productivity in a variety of applications. Unfortunately, in all too many cases, these incentive systems have deteriorated from lack of updating, erosion of basic principles, and lack of appropriate audit and top management review. The result is scores of deteriorated wage incentive plans which no longer provide appropriate motivation for their participants and result in high costs, falling levels of productivity, and lost jobs.

TRACING THE CAUSES

Companies that are in trouble with incentive plans can usually look to one or a combination of causes:

1. Improper plan design
2. Improper installation
3. Poor administration
4. Inadequate maintenance

If the plan was poorly designed, the original standards may have appeared accurate because they produced anticipated earnings. In reality, the workers may have "pegged" their earnings by setting output ceilings, a form of "price fixing" at the employee level.

Another design flaw is the absence of effective job evaluation and classification. With an incentive system, it is necessary to develop a consistent relationship between job requirements and dollars paid per hour. Without this feature, there is no way of ranking jobs reasonably to differentiate properly between skilled and unskilled workers.

In addition, without job evaluation, it is difficult to compare a company's wage scale with prevailing competitive wages in the area. A competitive wage should represent 100 percent work performance, that is, a fair day's work. Incentive or premium payments should then be earned and paid only for performance above this 100 percent level.

It is an important feature of plan design that the incentive earning potential be in a proper relationship with nonincentive earnings potential. Experience has shown that plan designs offering incentive earnings of less than 15 percent of total earnings usually do not motivate workers to incentive performance levels. Conversely, plan designs targeting incentive earnings above 30 percent of total earnings create tremendous pressures for abuse. These abuses take the form of hourly earnings guarantees and misstatement of time and production reporting. High incentive earnings opportunities also create inequities between incentive and nonincentive workers. Incentive opportunities in the 25 percent range can be defended on the basis of additional work effort requirements with their resultant additional output and lower overall unit product cost. Incentive opportunities above the 25 percent level cannot be logically defended on the basis of pay for performance requirements. This is because the *average* experienced and qualified worker cannot achieve performances against properly engineered standards much above the 125 percent performance level over the course of a production shift.

Performance requirements should also be equal between jobs. Proper work measurement techniques should be used to engineer the operation times according to the job method. There should be a direct relationship between what people are doing in terms of work requirements and what they are being paid. Plans that are not soundly based create inequities because they do not pay employees according to their achievements.

Proper plan design also requires that incentive earnings be paid only *for net, good output* within the control of the worker. Paying for total output can provide a motivation to produce scrap or rework, while maximizing incentive earnings. This situation obviously defeats a primary objective of incentives: lower unit product costs. A company must maintain control over scrap and rework in order not to pay more than once for its output.

An improperly installed system usually results from a desire by management to see an immediate payoff. The proper study of methods is often neglected in this process. Frequently, insufficient attention is also paid to training the workers to meet the new standards. When they cannot rapidly attain the standards, the response is often a management directive to loosen the standards.

This action results in inconsistencies which create pressures to loosen all standards to the level of the loosest (bootstrapping). As the installation progresses, methods and standards are not kept in a proper relationship. The result is improper documentation of the methods upon which the standards are based. Proper auditing to check for method changes, so that standards can be adjusted accordingly, is often overlooked.

This condition leads to unrealistic earnings in relation to base rates. In some companies, the average earnings by labor grade is more than 200 percent of the incentive base rate.

PROPER ADMINISTRATION

To administer a wage incentive plan properly, all levels of management and supervision must understand policies, procedures, methods improvements, work measurement, and other corporate guidelines. This is acquired knowledge that should be developed by adequate training on an organized basis, with continuing training provided to guard against slippage. Poor administrative practices can creep into any area of wage plan administration. A few examples will provide an idea of the type of abuse that can occur.

One very important administrative ingredient is accurate time and production reporting for both productive and nonproductive activities. To properly pay incentive wages, a company must know precisely how much production occurred and in what period of time this output was achieved. Otherwise, control is lost.

It is very tempting for supervision to authorize average earnings, daywork, delays, additional operations, or additional work. Whoever authorizes deviations should be responsible for the result of such authorization and not in a conflict-of-interest position.

Direct incentives should not be viewed as a right but as an opportunity that management will try to maximize in terms of the percent of time the incentive work force will have incentive *opportunity*. This requires that only required build schedules be produced and that production is sequenced properly. Occasionally, adherence to this principle will require work force balancing to work schedule requirements as well as removing incentive opportunity from incentive workers when build schedules have been achieved.

If job assignments are given out indiscriminately without regard to how work should flow through the plant, and if additional work is released to afford incentive opportunity, the incentive plan and work flow will also be hurt as the work-in-progress inventory begins to grow.

Finally, an incidence of grievances either high or low does not indicate the health of an incentive plan. Few grievances can be a sign that there is poor communication, a loose system, low utilization, or other inequities that are undefined. Conversely, a high incidence of grievances can be an indicator of tight incentive plan maintenance.

MAINTENANCE

No incentive plan should be installed and then forgotten. Management must not turn it over to the clerks to administer. An incentive program touches too many sensitive spots in the corporate body. It functions in a fluid environment as jobs and methods change and people move in and out of the organization. Consequently, it is vitally important that the maintenance of the plan be entrusted to adequate numbers of people who thoroughly understand the concepts and technical basis for it.

CONSTANT ATTENTION NEEDED

Management should devote constant attention to every detail affecting its wage incentive plan if it wants to enjoy lasting success. Thoughtful, planned action on a continuing basis consistent with the principles and fundamentals of sound incentives is desirable if an incentive wage program is to operate properly.

The same philosophy applies to an incentive plan that has gone bad. It is not enough to recognize the preceding symptoms. Management must act positively, consistently, and with a total commitment to convert the deteriorated productivity, pay, and cost situation. Dealing with the symptoms of the problem (that is, making some piecemeal changes) will not cure the illness.

Once incentive systems have deteriorated, they normally cannot be easily fixed. At some point in the deterioration process, management responsible for the company's operating health must make a decision. Since the cure is time-consuming and possibly painful, will it be worth the effort? It will require a longer-term program, carefully planned and carried out, to effect the cure.

There is an approach that will direct management to the right choice regarding incentive correction. It involves detailed analysis of the situation that includes three distinct steps:

- Fact finding
- Determining improvement potential
- Developing a plan for realizing that potential

Auditing the Problem. When dealing with a deteriorated incentive situation, it is extremely important that evaluation and actions be based upon factual information. Typically, significant misinformation is prevalent in a deteriorated incentive environment. This is due primarily to the exposure particular individuals have had to the incentive system over time and the misinformation generated by the system. For example, company-generated pay reports such as W-2 earnings statements are usually accurate indicators of what employees actually earned under the incentive system. However, performance, or productivity data generated from the plan, generally does not give an accurate reflection of

TABLE 4.1 Paid Incentive Earnings versus Incentive Employee Productivity

PRODUCTIVE DEPARTMENT	PRESENT INDICES -- INCENTIVE GROUP ONLY				PAID INCENTIVE EARNINGS % E	PAID EARNINGS AS A % OF PROD. (E ÷ D) 100
DEPARTMENT	UTILIZ. % A	PERF. % B	METHODS % C	PROD. D = A x B x C		
102	73	96	95	67	159	237
103	89	94	93	78	163	209
104	78	97	93	70	155	221
105	79	99	93	73	155	212
106	82	102	97	81	158	195
107	73	99	96	69	152	220
108						
109	64	99	93	59	150	254
110	76	92	93	65	153	235
ALL DEPTS.	**78**	**98**	**95**	**73**	**157**	**215**

true efficiency or productivity achieved. Under a properly designed and administered incentive plan, pay performance and productivity should be equal. However, as shown in Table 4.1, it is common to find pay performance of 150 to 200 percent for true levels of productivity below 100 percent.

The primary focus of this auditing activity should be centered in the following areas:

1. Employee earnings under the existing plan
2. True productivity levels achieved under the existing plan

Wage Payment. The most critical issue with the workers is individual earnings. Data are needed in order to make an objective evaluation in this area, and to assess the human relations impact of any proposed changes. Such data can be provided by the development of a scatter diagram as depicted in Fig. 4.1. This diagram shows the average hourly straight time earnings by individual by labor grade. These data point out not only what earnings expectation levels people have become accustomed to but also the earnings inequities within a labor grade, between labor grades, and between the incentive and nonincentive workers. In most cases of deteriorated incentive systems, the individual's earnings are not a primary function of his or her labor grade but are directly related to how lucrative his or her standards and administration situations have become.

It is common to find relatively unskilled incentive workers earning at levels equal to or above more highly skilled nonincentive workers. In many of these cases, the highly compensated incentive workers are achieving these payment levels at less than incentive levels of output. Such inequities traditionally have been dealt with using piecemeal approaches such as adding money to the base rates of the nonincentive workers and withholding money from the incentive base or calculation rate and paying such monies [cost of living adjustments (COLA) and wage increases] as an hourly adder. Over an extended period of time, this approach can significantly reduce the incentive motivation or pull of the plan. Some plans have approximately only 20 percent of the total hourly rate subject to incentive earnings. When this type of situation occurs, the incentive worker achieves a 1/5 of 1 percent increase in pay for every 1 percent increase in output. Under such conditions, the motivation is generally for looser rates rather than output increases.

The important feature of these data, however, is that they point out what you can expect in terms of reaction from the work force to proposed changes in the system. The data also graphically depict the situations that need to be addressed in order to arrive at equity

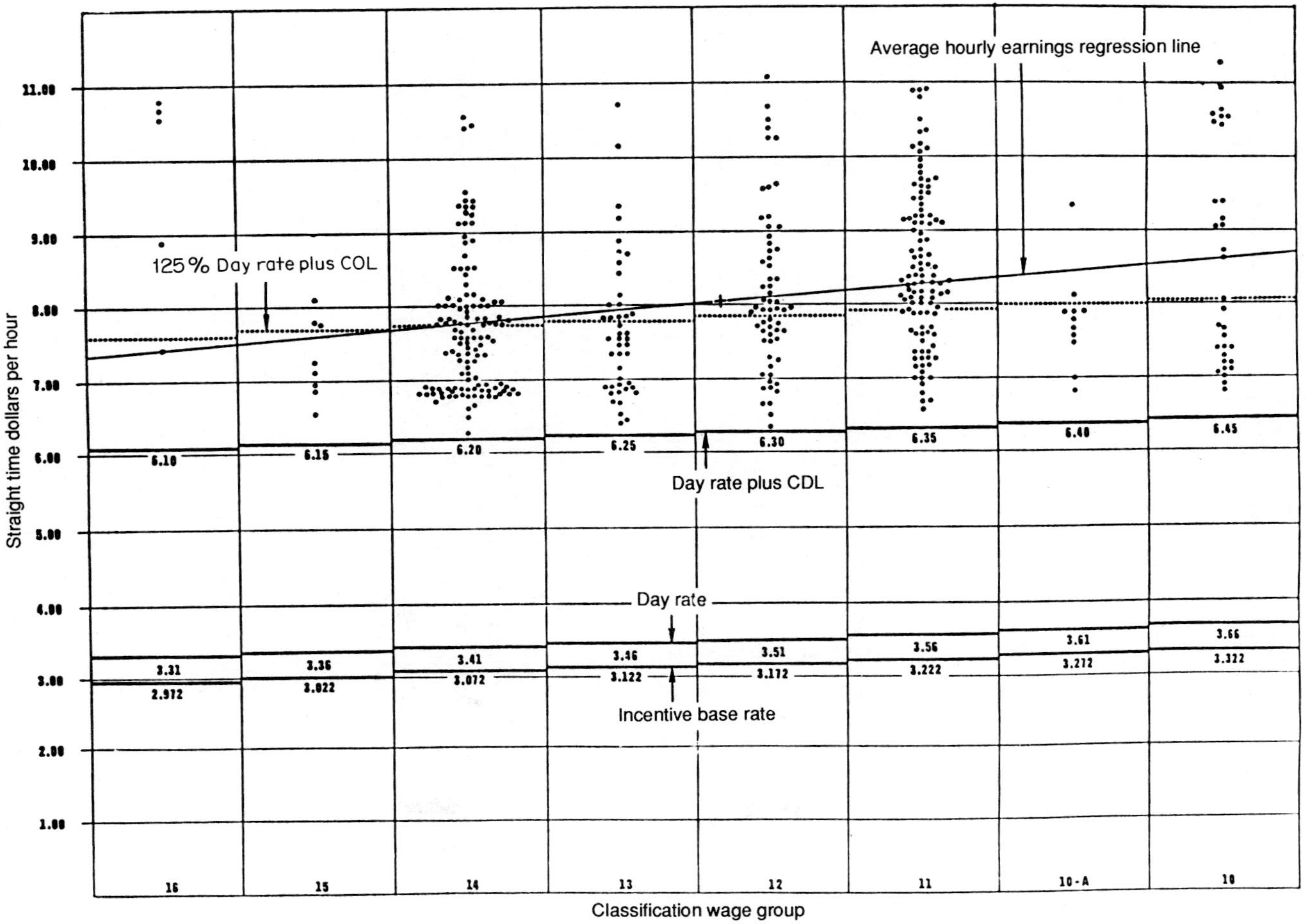

FIGURE 4.1 Scatter diagram of average hourly earnings by wage group.

in wage payment for performance achieved. These data can also point out the feasibility of utilizing various correction approaches. In the situation depicted in Fig. 4.1, it is likely that some form of earnings protection will be needed because of the very high earnings experienced by some individuals. Requiring these workers to take pay cuts will eliminate their support for the proposed plan. Additionally, if wage levels are high under the present plan, in most situations it is not practical to introduce gain sharing for the existing incentive workers. To do so usually will result in output reduction from the high performers. Also, to the extent that present wages may be noncompetitive, the company in a competitive environment cannot afford in the long run to share wages to higher levels—even for some reduction in unit product labor costs. There may be possibilities, however, to overlay gain sharing or other group motivational systems on top of a corrected direct incentive.

Generally, the earnings data, when coupled with the productivity data, will not economically support removing incentives entirely from an operation. In order to gain acceptance to remove incentives, companies are faced with paying somewhere near existing wage levels with a real danger that output levels will fall below existing levels. A more economically viable approach is usually to restructure the incentive plan to pay approximately existing wages for full measure of output and to tie wage payment directly into this realization. To motivate the nonincentive workers, other motivational plans can be incorporated to obtain the productivity gains possible from motivating the entire work force.

Determining Improvement Potential.   The wage payment data provide the information on what is being paid. These data need to be compared with what levels of true productivity are being realized for this wage payment level. To obtain the facts about a particular situation, a series of specially designed productivity studies should be conducted. One is a rating of work performance levels (the pace or effort level exerted by the worker when working productively). This is usually done on a sampling basis and covers both manual methods and process time. Table 4.2 shows the results of such an evaluation.

Another sampling study should be performed on worker utilization (the amount of time the worker spends in productive activities such as cycle time, setup, changeover, and the like). Table 4.3 shows the results of a work sampling study embodying over 17,000 observations.

TABLE 4.2   Productive Departments Observed Performance Levels (Manual Portion of Work)

DEPARTMENT		NUMBER OF OBSERV.	OBSERVED PERFORMANCE (%)		
NO.	NAME		LOWEST	HIGHEST	AVERAGE
	Incentive				
103		28	80	110	94
104		41	80	105	92
105		35	85	115	97
106		32	80	115	98
107		56	80	125	102
108		25	85	125	102
109		22	85	115	100
110		18	85	115	98
111		21	80	110	92
	Total Incentive	**278**	**80**	**125**	**97**
	Non-Incentive				
103A		20	80	100	91
105		9	85	95	90
107		12	80	100	93
108		16	75	100	88
109		9	85	105	94
110		28	75	105	88
111		40	75	100	89
	Total Non-Incentive	**134**	**75**	**105**	**90**
	PLANT PRODUCTIVE DEPARTMENTS	**412**	**75**	**125**	**95**

TABLE 4.3 Work Sampling Results

Number of observations and percent of total (17,603 total observations)

Dept.	Working	Idle	Waiting						Clean up	Early pass	Absent					Un-known	Other	Total
			Down-time	Tools	Crane	Mat'l.	Heat	Pull car			Union	Medical	Per-sonal	Get	Payroll			
101	1156	296	45	33	18	27	9	148	27	322		1	47	2	2	7	7	2,147
	53.9	13.8	2.1	1.5	0.8	1.3	0.4	6.9	1.3	15.0		—	2.2	0.1	0.1	0.3	0.3	
102	1571	304	20	7	4		9	353	26	339		3	5		1	12	12	2,666
	58.9	11.4	0.8	0.3	0.2		0.3	13.2	1.0	12.7		0.1	0.2		—	0.5	0.5	
103	1715	407	41	34	60	31	19		1	288	2	4	78	12		12	13	2,717
	63.1	15.0	1.5	1.3	2.2	1.1	0.7		—	10.6	0.1	0.1	2.9	0.4		0.4	0.5	
104	1114	127	11		14	7	108		23	325		2	31	1	1		10	1,774
	62.9	7.2	0.6		0.8	0.4	6.1		1.3	18.3		0.1	1.7	—	—		0.6	
105	1224	526	81	26		59	7	142	17	450		1				37	6	2,577
	47.5	20.4	3.1	1.0		2.3	0.3	5.5	0.7	17.5		—				1.4	0.2	
106	2550	568	46	1	3	9		330	31	554		3	13	1		30	14	4,153
	61.4	13.7	1.1	—	0.1	0.2		7.9	0.7	13.3		0.1	0.3	—		0.7	0.3	
107	998	186	5						4	353		22	1					1,569
	63.6	11.9	0.3						0.3	22.5		1.4	—					
Totals	10,328	2414	249	101	99	133	152	973	129	2631	2	36	175	16	5	98	62	17,603
Obs. %	58.7	13.7	1.4	0.6	0.6	0.8	0.9	5.5	0.7	14.9	—	0.2	1.0	0.1	—	0.6	0.4	
Maximum possible errors	0.7	0.5	0.2	0.1	0.1	0.1	0.1	0.3	0.1	0.5		0.1	0.1	0.1		0.1	0.1	

Finally, there is the consideration of methods. For example, what are process time practices with regard to internal wait time and multiple machine assignments? How effective are the workplace layouts and measurement procedures?

Productivity Evaluation.   The productivity evaluation evolves into a formula that can be expressed as

$$\text{Performance} \times \text{methods} \times \text{utilization} = \text{current productivity level}$$

Productivity, as we have defined it, is a combination of three indexes:

Utilization: Amount of work time spent in productive activity, as determined by work sampling

Performance: Skill and effort applied while productively working, as determined by performance rating during the work sampling

Method level: The current methods used, compared with accepted industrial practice, as determined by observation of the work area

The productivity results that can be achieved under various types of plans can then be compared with the existing situation. For example, if performance is 100 percent, methods level is 100 percent, and utilization is 60 percent, the effectiveness rating would be 60 percent for the existing level of productivity.

Under a sound standard hour, one-for-one incentive plan, normal expectation is for performances to average around 118 percent. The ideal numbers would be 125 percent for performance, 100 percent for methods, and 95 to 100 percent for utilization for a combined rating of 118 and 125 percent.

The present productivity level can then be compared with what can be reasonably expected under a revised system. An out-of-control incentive situation could have 100 percent performance, 60 percent utilization, and 80 percent methods for a combined rating of 48 percent. The differential, by area or department, between the current situation and the improved situation is where the savings will be realized. Table 4.4 presents the results of such an analysis.

When this differential is converted into dollars in terms of the number of hours in each of the studied areas, the picture can become even worse. For the low rating previously discussed, the company may be paying a substantial incentive bonus—sometimes in excess of 50 percent of total wages (Table 4.1). This kind of analysis provides the economic justification for determining whether incentive restructuring is worth the risks and the effort.

PRODUCTIVITY IMPROVEMENT PLAN

It is important to be aware of the economic trade-offs involved in switching from deteriorated incentives. The only real alternative to incentives is buying out of incentives, which means paying current rates and usually suffering a significant loss in productivity or output. That is the main reason for compiling the study data. The data indicate the present level of productivity, what incentives can do, and what would happen without incentives. These data can also be used as an educational aid in the negotiating process. Figures 4.2 and 4.3 give a graphic illustration of what can be accomplished with restructured incentives.

Wage Structure.   The plan should examine the present wage structure, determine requirements to put it on an equitable basis for the future, and indicate how this can be made acceptable to the work force. Workers find it extremely difficult to accept a significant loss in their standard of living. Therefore, management should provide earnings protection for

TABLE 4.4   Direct Labor Productivity Summary, Present versus Potential

DEPARTMENT	NO. EMPL.	PRESENT				EXPECTED				POTENTIAL PROD. GAIN %
		UTIL.	PERF.	METH.*	PROD.	UTIL.	PERF.	METH.	PROD.	
	A	B	C	D	E = BCD	F	G	H	I = FGH	L = I - E/J
100	22	82	100	92	82	87	120	100	104	
101	21	89	103	95	92	93	120	100	112	
102	7	96	101	100	97	96	104	100	100	
103	6	85	101	92	86	90	120	100	108	
104	7	84	98	100	82	90	116	100	104	
105	17	88	103	95	91	92	120	100	110	
106	9	87	104	95	90	90	120	100	108	
107	12	64	103	95	66	80	111	100	89	
108	5	82	106	95	87	90	120	100	108	
109	14	83	100	95	83	93	120	100	112	
110	17	79	100	95	79	93	120	100	112	
111	9	78	100	95	78	93	120	100	112	
TOTAL DIRECT INCENTIVE	**146**	**83**	**102**	**100**	**84**	**91**	**118**	**100**	**107**	**22**
112	17	71	93	100	66	85	93	100	79	
113	23	83	91	95	76	90	120	100	108	
114	27	84	92	95	77	90	120	100	108	
115	5	77	94	95	73	90	120	100	108	
116	2	72	101	100	73	90	100	100	90	
TOTAL DIRECT NON-INCENTIVE	**74**	**80**	**92**	**100**	**74**	**89**	**113**	**100**	**101**	**27**
TOTAL DIRECT	**220**	**82**	**98**	**100**	**80**	**90**	**117**	**100**	**105**	**24**

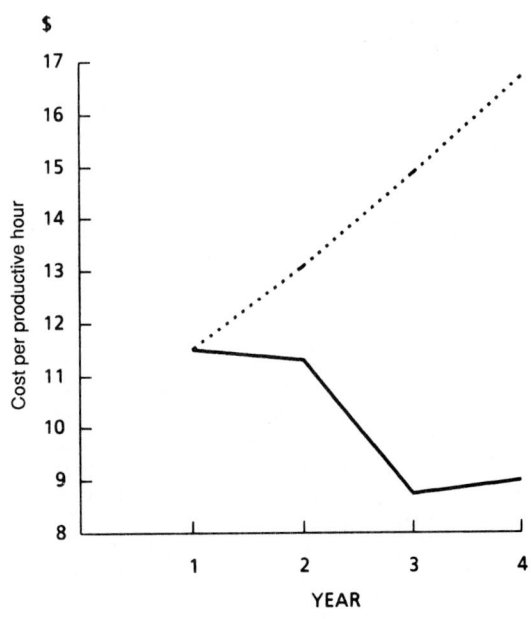

——— With restructured incentives

······ Without restructured incentives

FIGURE 4.2   Total cost per productive hour (150 workers).

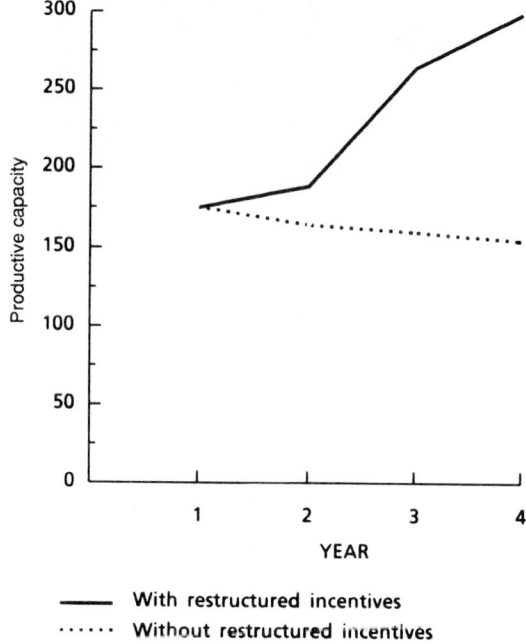

FIGURE 4.3   Productive capacity (150 workers).

them. If earnings protection is properly constructed, it does not cost the company additional wage payment dollars because it is money that is currently being paid out, and perhaps is even contractually required. The guidelines for earnings protection are that the money be paid to incumbents only, that the protection be either bought out or gradually phased out, and that the payments be directly tied to increases in productivity. Earnings protection provisions will be discussed in more detail under "Negotiating the Change."

PLAN STABILITY

After the data are assembled and the improvement potential is calculated, a stable plan for accomplishing the improvement goals should be developed. If improved management control is the objective, a wage incentive plan, a rated measured daywork plan, or a straight measured daywork system will provide detailed information on staffing requirements, utilization, performance, and job costing. These can be structured in individual or group plans and can cover both manual and clerical employees.

If one of the primary objectives is an increase in productivity and maximization of employee earnings opportunities, a variety of incentive bonus plan features as shown in Fig. 4.4 can be used to tailor a plan to fit specific situations. The important considerations, however, are to ensure economic feasibility of the approach and to ensure that human relation problems of acceptance have been recognized and dealt with properly. The plan should deal with eradicating the causes of the existing problems and not just its symptoms. It must deal with long-term issues and be as fair and equitable to all parties as possible under the circumstances. The plan must be believed in by management and worker groups alike and must be documented by negotiated contractual language that protects all parties.

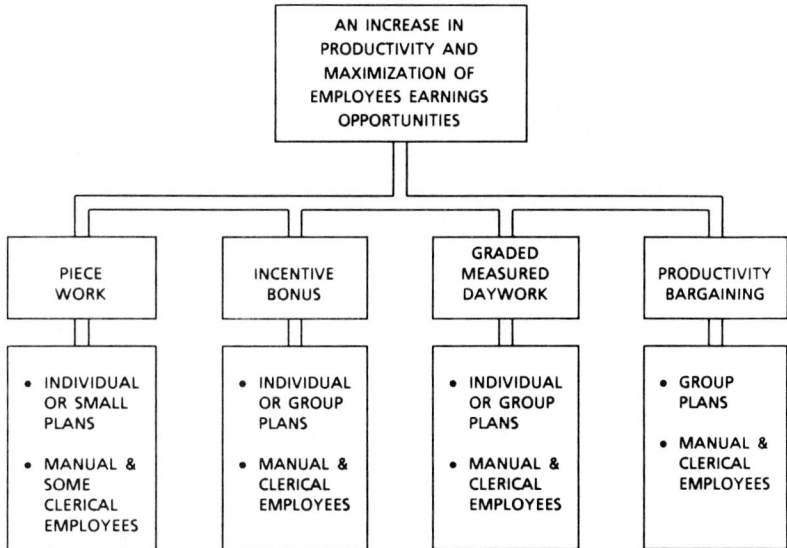

FIGURE 4.4 An increase in productivity and maximization of employees' earnings opportunities.

NEGOTIATING THE CHANGE

The entire area of negotiating the restructuring of wage incentive systems is very tough. It is probably the most emotional area of collective bargaining. It hits established and negotiated practices; it impinges the pocketbook nerve, and it affects the amount of work the employees have to do each day to earn their wages. Unions are very sensitive about all these issues. Therefore, any kind of negotiation involving incentive restructuring will require a very thorough job of preparation.

Union Attitudes.   One of the first things to do before deciding to negotiate a wage incentive system restructuring is to analyze what the attitude of the union is going to be. Unions do have different attitudes about wage incentive systems. For example, generally speaking, the United Steel Workers (USWA) are more receptive to wage incentive systems than the United Auto Workers. Therefore, the USWA has traditionally been more receptive to talking about a restructuring of a wage incentive system.

Other union attitudes will vary. In other unions the international officials may not have a fixed attitude about wage incentive systems, but the local officers may have gone to a training school and may have some very strong attitudes toward wage incentives and restructuring or against doing it in a particular way.

Timing of the Negotiation.   Another key consideration is the best timing. There are two philosophies. One is to wait until contract termination time to broach the subject with the union. The other is to approach it midcontract. At termination time, there is the advantage of having a contract expiration data, which can be the target for bringing the negotiations to a close. Incentive correction can be tied in with the wage and fringe benefit package and made a condition of agreement in order to get the wage improvements. The biggest disadvantage to this approach is that it takes a lot of time to explain and to familiarize a bargaining committee with the provisions and restructuring of a wage incentive system. If other important items are coming up in negotiations, many days may be required just on the wage incentive system issues. Other areas could get less attention than necessary, making the union unhappy that it can't get adequate discussion of its desired improvements in the contract. This could lead to a strike threat posture just because of lack of time to complete everything. Another possible disadvantage of bringing it up for the first time at contract termination is the emotional factors previously mentioned. The union might be impatient and unwilling to listen.

If it is approached at midcontract, the union knows they don't have to accept it if they are not convinced it is a good and/or necessary thing for them. Even if no agreement is reached, at least the big problems of the time factor and the emotions have been taken out of the contract expiration negotiations. The union may understand what it is that the company is trying to do. Generally taking up this problem is handled best substantially before a contract expires in order to get everything exposed at a time when people are willing to understand the issues. Perhaps this should be done only 90 to 120 days before the contract expires, but it at least should take place at a time when there is no pressure from the ordinary problems of negotiating a new labor contract.

Bargaining Leverage.   To obtain a new incentive wage plan, there has to be some kind of leverage. There must be something that will induce the union to be willing to agree on the restructuring. Sometimes this can be something fairly easy to provide. While it may not always be possible to find these windfalls, other aids or conditions can serve as leverage. One general rule for good leverage is an overwhelming desire of the company to get a restructured system. The company must realize how much their runaway program is costing them and the problems it is creating for them. If it comes to the critical part of a negotiation to revise a wage incentive system and the management negotiating team doesn't know how many hundred thousand dollars or how many million dollars this means to the company, they are likely to back away.

Top management personnel in the company must understand exactly what is at stake when negotiations begin. This determination may have to extend to the point of a willingness to take a strike in order to get the correction program. The leverage doesn't always have to involve a willingness to take a strike. Some other things can be brought to bear in a situation, which also involves decision making at the highest level of a company. This can be done on the basis of a fear of loss of jobs and even loss of a plant. Many times these issues can be far more effective even than willingness to take a strike.

If it is feasible to confront the union with an important decision in this area, it can be far less expensive and far more important than getting a good agreement for the restructuring of a wage incentive system. For a company involved in heavy fabrication, foundry-type operations, or a forge-type operation which is not readily moved, alternatives on continuing to operate or relying upon other facilities in the company can save a large amount of money in negotiations. Another factor which can be helpful is to point out that there has been a real loss of incentive opportunity in the existing program. Many companies have built up side pay, either through the cost-of-living provision or through annual negotiations that dictate the general increase is to go on as a side payment or as an additive. If a large buildup of side payments has watered down the incentive and if the employees begin to feel that the percentage of pay available as incentive earnings isn't worth anything to them, this may bring about a desire on their part to help get the system restructured.

There can be some additional arguments which the union can use when they go back to their membership for ratification. By agreeing to this wage structuring, they are going to help eliminate the inequities for the skilled employees and the other dayworkers where out-of-control incentives have submerged the earnings of those people by comparison with incentive paid employees in the plant. Also, the temporary transfers, which may be a current problem because of the big variations in earnings, won't be a concern under the restructured wage incentive system. All these things can help in getting ratification, but they aren't the real leverage that obtains the agreement.

Negotiation Ideas.   One thing that a union is going to be very concerned about in negotiating any restructuring is the possible loss of earnings that each individual or each classification of employees may suffer if they give up their existing program and go to a new one. In order to get around this problem, many companies have offered some kind of earnings protection or earnings indemnity arrangement. This can help decrease resistance to the program and can be extremely important in obtaining acceptance of the agreement.

A company's management should be willing to explain how standards are going to be set under the new program. A proposal to go to a new program without further definition will create problems in the workers' minds as to whether this will be done on a decent and sound basis. Being able to show some sample standards to the employees under the new program so they can actually see that it is fair helps in the acceptance process.

Once the general approach you are going to take in alleviating these fears has been decided upon, you must decide exactly how the process will be initiated. You certainly should consider whether to sit down with representatives of the national union in advance and tell them what you have in mind. Typically that's a good thing to do. However, it depends on the orientation of the international representative or the regional director. Also, you may want to consider meeting separately with the top officials of the local union to explain to them or maybe meet with them in combination with the international before meeting with the entire bargaining committee. Frequently these representatives appreciate being briefed in advance. Sometimes they provide immediate reactions which can be used to help shape the approach that should be taken in the negotiation.

Negotiation Documents.   What should you have in the way of documents in the bargaining? There are three parts or three documents which are used.

Right to Convert.   The first document is the agreement which gives the company the right to convert the old system to the new system. Many items must be included in this document. The most important item is called a zipper clause. This clause wipes out all past practices and side agreements concerning the old incentive plan. These past practices and agreements must be eliminated to properly correct the existing problems. Many companies have had not only loose or ineffective standards but other problems caused by supplementary agreement or established practices. One example of these bad practices is permitting forepersons to approve downtime without really knowing what it is, or authorizing a foreperson to approve downtime at the end of the day. In effect, this permits employees to figure their time cards backward in order to come out with the prescribed earnings which they normally enjoyed. Much of this type of practice will frequently not be included in your labor contract, but if you don't wipe it out with a zipper clause, you will end up fighting over whether or not you did negotiate to get rid of all those deleterious practices in your contract.

Earnings Protection.   The second document is an earnings protection clause. There are two major types of differentials. The first is called an "incentive differential"—or for short, an ID. The ID is computed as follows: First, a base period (such as the last 3 months preceding agreement on the new program) is selected and the employee's average earnings while on incentive is computed for this base period. You then inflate the incentive base rate for the new plan by the incentive expectancy of the new plan. If the figure is less than the base period average, the difference is the ID.

EXAMPLE:

Base period average incentive earnings	$ 6.50
Base rate ($5.00) times 120% incentive expectancy	6.00
ID =	$ 0.50

This ID should not be looked upon by the company as a "cost." All it does is to assure employees that they'll be able to earn what they had been earning before the restructuring. It's not a "cost" because it's already being paid. It is really a postponement of some of the savings that management intends to achieve out of the correction program.

The second type of differential was pioneered in the basic steel industry. When basic steel revised their incentive programs many years ago, they came up with what they call an "out-of-line differential," or an OLD. That's a well-known and well-accepted term in the area of wage incentive systems. What they did was to say that when a particular area of this plant is converted to the new wage incentive system, and it's done area by area, the old system survives until such time as the new method is installed.

Prior to the installation of a particular department or area on the new system, the new standards were developed and a comparison of payroll calculation of the new plan against the old standards was run in terms of old base rates, old guarantees, and the like, for a defined period. If the old standards yielded an average of $8.00 an hour and the new standards, new base rates, and new guarantees produced $7.60 an hour, the OLD would be $0.40 per hour. Now, that OLD preserves in it all the looseness of the old system. In other words, if the employees were producing under those old standards but not really at incentive pace, that looseness will still get preserved in the differential. As such, this is certainly not the most desirable way to proceed.

Another disadvantage of the OLD is that all standards have to be in existence in order to measure this differential at the time of conversion in a particular department. And there may be changes of equipment, and the like that are made about the time of the conversion. That creates significant problems. Therefore, the OLD is not recommended, but if it can't be negotiated any other way, it is a possibility.

Of the two earnings protection arrangements the ID is preferable compared with the OLD. The ID does not preserve the looseness of the old system, because it is based upon what the employees were actually earning during the base period. If they weren't performing right up to incentive potential, they don't get the advantage of that potential in the calculated differential. And the figure that is subtracted from the base period average does not assume that work is being done at an incentive pace in order to achieve that 25 percent earnings target. This results in a differential which when paid to employees for each hour they work on incentive will preserve their former earnings potential.

Under the ID calculation for equal effort workers are not going to get equal pay. This is because if they weren't working at incentive pace before the new system, they're going to be working at incentive pace afterward. However, by working at the required pace under the new program, they will be able to achieve their former earning potential. Expressing this differential as a percentage generally works out best. However, percentage application may extend the time period for reducing IDs to zero. As wage increases are added to base rates, a percentage ID will increase the differential for the future (in cents per hour), if it is preserved. However, a percentage ID is a way to keep an ID from subverting the real incentive under the program for employees who have very large IDs.

In the ordinary situation where it's just bargaining power against bargaining power, it will probably be necessary to work out some type of earnings protection differential. If so, guidelines should be spelled out for how and when this differential is paid. This payment should go to incumbents only; it shouldn't go to everybody coming into this classification in the future. Otherwise it will be difficult to get rid of it. It should be an additive to incentive earnings only. If it is applied to both daywork and incentive hours, it can have a tendency to dampen incentive effort on the part of the employees who may be receiving very large differentials. Therefore, it is best to calculate and pay it only on incentive time. There also should be provisions for what happens when someone who is entitled to an ID gets transferred to some other classification either temporarily or permanently.

Buyouts.   Differentials should not last forever and can be terminated with a buyout. A buyout typically is a certain number of hours, 2000 or 3000 or 4000 hours, times the amount of that differential. Buyouts can be compulsory. If so, the differential doesn't provide a strong argument at the bargaining table that it will protect earnings, but the buyout does get rid of the differential. It may be enough of a bargaining sop to gain union acceptance of the program. Many people like to get that large sum of money at the time of conversion.

It's far easier, however, to negotiate an optional buyout. It is very important to give the option for only a limited period of time, typically the first 2 weeks after the department is put on the new incentive program. If the employees have the option forever, obviously they're going to wait until it's time to move from the classification or retire or be laid off to exercise their options.

How many people elect a buyout when it's optional? In one situation with a 3000-member plant, approximately 75 to 80 percent of the people bought out, even though it was optional. It will depend somewhat on the attitudes of the employees. It's very tempting for them to select the buyout. The money can buy a new TV set or it can serve as a down payment for the next car they want to buy, and so on. People become opportunistic about this, and even though it doesn't necessarily make long-term economic sense, workers will buy out far more frequently than you might expect.

There are other ways to extinguish this differential. There can be a time limit; that is, it will only be paid for a certain number of years. Several companies have been successful with this approach. The differential will be paid for only 3 years or 2 years or some stated period. Any such limitation makes the attractiveness of the differential far less for the union than a continuing differential or one that lapses when the employee bids into a new classification.

Another way to eliminate differentials is to subtract a certain amount of money each time a general wage increase is granted. The steel industry, for example, has job evaluation. They agree to an across-the-board increase and then they agree to additions to the increments between labor grades. They subtract the amount of that increment from the existing differential for each labor grade. It's a very slow way to back those differentials down, but at least gradually it does erase them. If the differential agreement is in cents per hour and if inflation continues to increase, the differential gradually becomes a smaller and smaller percentage of the employee's final pay.

The New Incentive Plan.   The third document is the new plan, and it has several parts. The following points are directed toward replacing the deteriorated incentive plan with a new one-for-one standard hour plan as this option, although the most complicated, is usually the most economically advantageous.

There should be a clause that gives management the right to use any sound industrial engineering technique. It can be much more elaborate than that, but essentially this clause guarantees the right to the use of predetermined times.

The agreement should contain the earnings opportunity target, whatever it may be, such as 20 or 25 percent. This language should be couched in such a manner that not every standard has to meet the target.

There should be the right to revise standards for method changes. A percentage limitation such as 3 or 5 percent is not a problem. Elemental restrictions, that is, a clause saying only those elements of a standard which are affected by the method changes can be revised, creates a big problem. Such language should be resisted. If elemental limitations are agreed to, following are several conditions that should be stipulated to dampen the negative effect of this provision.

Exceptions.   First, there should be an exception for "revamped method jobs." These are jobs where very large changes are made. This can be defined in terms of change in time, that is, by a certain percentage of total time or changing a certain percentage of the elements in a job. For example, by exempting from elemental restrictions those changes exceeding stated limits, such as 20 or 30 percent, we avoid the problem of having to recapture tiny elements here and there where the job is completely changed. At the same time the union gains the protection which they want against the company's seizing upon minor method changes to justify large revisions in the standard.

There should also be an allowance for picking up accumulation of changes. Also, management must reserve the right to change standards for any method change. There should not be exceptions to the right to revise a standard for operator initiated changes. Nothing but problems comes out of that. If some substantial period of time passes between a change and the time the industrial engineering department discovers it, there is frequently a tremendous argument as to who initiated that change. Usually the union is going to win that argument, because they can parade operators who have been in this classification for years who will swear that "so and so did this" or "it came about this way." By this time the foreperson is either retired or has gone someplace else, and it's very difficult to rebut the union evidence. As such, any method change should be the basis for revising the standards.

It is strongly recommended, from the standpoint of administration, that earnings be calculated on either a daily or weekly basis and that the base rate should not be guaranteed by the job (job payoff). It is impractical for a company to administer a contract properly under which a base rate or day rate is guaranteed by the job.

Delays.   Another provision that should go into the standard hour plan, which again relates to administration, is called "the 6-minute exclusion rule" on delays. This language says that the only unavoidable delays which are paid are those delays which exceed 6 minutes (a tenth of an hour) and then only for that portion which is in excess of 6 minutes. In other words, the first 6 minutes of each delay is taken care of by the unavoidable delay allowance and is not again paid for as downtime.

An incentive system is not just a system to encourage improved productivity; it's a program which may also encourage employees to cheat. If delay reporting and payment is not defined, cheating can occur and the incentive plan will be circumvented.

Other Provisions.   Another area definitely to be covered is that only base rate is paid for nonincentive assignments. There may be some minor exceptions such as an assignment to make samples, which justify a step rate (a rate higher than base rate but less than incentive pay levels). But if the deviations are broadened to include temporary transfers for the convenience of the company, no rate situations, and the like, it is going to cause serious trouble. These exceptions should be as few as possible and *do not* pay average incentive earnings.

Provide for payment only for pieces produced to the quality standards. Monitor those pieces and have sufficient inspection so that the employee really has to correct his or her own errors. It shouldn't go to some repair center where somebody repairs it and the repair employee is paid for it after already having paid another worker who should have done it correctly the first time.

Finally, do not put the administrative manual into the agreement. That administrative manual for wage incentive systems does not belong in the collective bargaining agreement. There are many good instructions to industrial engineers that are terrible as contract language. It causes all the grievances concerning standards to concentrate on whether the industrial engineer who established that standard did it according to the manual. The real question is whether the standard is correct according to the test earnings opportunity stated in the contract.

For example, the administrative manual may say that in timing a job, the rater should not rate below 80 percent. This is generally a good industrial engineering instruction, but if that clause is in the contract, it creates a situation in which the employee who is being time studied deliberately slows down below 80 percent. Nobody will work above 80 percent, and the result is a performance rating of 80 percent even though the actual pace was 60 or 70 percent. This is done by the time study engineer in order to get a standard on the job.

The preceding documents cover the main points. A complete labor agreement will include a number of provisions in addition to these three basic documents. It is important that a clear understanding exists between the company and the union and all new and changed provisions. These provisions, plus diligent controls and dedication, should give you the basis for maintaining a sound system for many years.

IMPLEMENTING THE NEW PLAN

Properly implementing the new plan is a critical step. There are two basic objectives. First, projected benefits for the workers and the company must be realized within the established time and cost. Second, a positive management-employee relations attitude must be maintained throughout the installation.

The following 10 steps will help ensure effective installation of a new wage incentive plan:

1. Identify negotiated rights, policies, and restrictions that will be involved with the technical development.
2. Train company and union representatives in the techniques to be used.
3. Provide orientation training to management and union officials in the basic features of the program and foster an appreciation of the techniques to be used.
4. Prepare a master schedule of the overall program, identifying projects and milestones to be met.
5. Analyze methods and develop standard data and measurement approaches to be applied to the operations.
6. Develop time reporting and management control systems, procedures, and reports.
7. Apply standards to incentive operations.
8. Calculate incentive earnings and distribute to employees.
9. Provide follow-up maintenance procedures to keep the wage incentive plan up to date. Check administration to see that correct methods are followed, that counts have been reported correctly, and that standards have been properly applied. Check abnormally high or low performances.
10. Provide information by all effective means on how the system works and how everyone can benefit from it. Communicate, communicate, communicate. This step is conducted concurrently with the nine previous steps.

AN ATTAINABLE GOAL

Deteriorated wage incentive plans are not good for any business. Management suffers because productivity and profits are inadequate. Employees are hurt when jobs are lost or plants must close because unfair incentives drain the ability of companies to complete in the marketplace. Inequities in earnings opportunities create enormous industrial relations problems for management and employees alike.

Many companies have taken action to correct their wage incentive plans with tangible benefits to everyone. A great deal of attention to detail is required to correct a wage incentive program, but compared with the extraordinary results that can be obtained, the effort is very worthwhile.

The most important ingredient in correcting a deteriorated wage incentive plan, or any other major business problem, is management's dedication to seeing the program through to a successful conclusion and to ensure that the principles and fundamentals of sound incentives are adhered to on an ongoing basis.

BIBLIOGRAPHY

Bowey, Angela, *Payment Systems and Productivity*, New York: St. Martin's Press, 1986.

Goldman, Alvin L., *Settling for More: Mastering Negotiating Strategies and Techniques*, Washington, D.C.: Bureau of National Affairs, 1991.

Institute of Industrial Engineers, *Gainsharing: A Collection of Papers*, Atlanta/Norcross: Industrial Engineering and Management Press, 1983.

Loughran, Charles S., *Negotiating a Labor Contract: A Management Handbook*, Washington, D.C.: Bureau of National Affairs, 1984.

Matejka, Ken, *Why This Horse Won't Drink: How to Win—and Keep—Employee Commitment*, New York, N.Y.: AMACOM, 1991.

Weiss, Andrew, *Efficiency Wages: Models of Unemployment, Layoffs, and Wage Dispersion*, Princeton, N.J.: Princeton University Press, 1990.

MANUFACTURING ENGINEERING

CHAPTER 1
DEFINITION AND SCOPE OF MANUFACTURING ENGINEERING

Frank J. Riley
Senior Vice President
The Bodine Corporation
Bridgeport, Connecticut

Historically, The Society of Manufacturing Engineers (SME) has defined manufacturing engineering as:

> that specialty of professional engineering which requires such education and experience as is necessary to understand, apply and control engineering procedures in manufacturing processes and methods of production of industrial commodities and products and requires the ability to plan the practices of manufacturing, to research and develop the tools, processes, machines and equipment, and to integrate the facilities and systems for producing quality products with optimal expenditures.

It has also stated:

> Engineering involves a high degree of creative activity in the identification of theories and their development into practical applications. It is at this level that concepts are formulated and ideas are transformed into design and product engineering activities. The manufacturing process begins with the engineered product and ends with the final operation on the production line, in between product engineering and distribution lies the industrial function called manufacturing engineering.

The manufacturing engineer performs many tasks in a wide range of industries. An encompassing definition, if necessary, must be very general. A definition supplied by the U.S. Department of Labor defines a manufacturing engineer as one who:

> directs and coordinates manufacturing processes in industrial plants; determines space requirements for various functions, and plans or improves production methods including payout, production flow, tooling and production equipment, material fabrication, assembly methods and manpower requirements; communicates with planning and design staffs concerning product design and tooling to assure efficient production methods; estimates production times and determines optimum staffing for production schedules; applies statistical methods to estimate future manufacturing requirements and potential; approves or arranges approval for expenditures;

and reports to management on manufacturing capacities, production schedules and problems to facilitate decision making.

These historical definitions, which were accurate for many years, are now subject to dramatic change. Manufacturing today must operate in a world of global competitiveness necessitating greater flexibility of operation, more rapid communication, and redefined organizational structure.

The scope or breadth of knowledge and the depth of knowledge required are growing so rapidly that a single definition of manufacturing as a career profession must be seriously challenged. Manufacturing engineering will increasingly become a professional assignment for people with a variety of professional education and experiences. To fully comprehend emerging trends in manufacturing engineering, it is necessary to look at the historical development of manufacturing engineering.

HISTORICAL DEVELOPMENT OF MANUFACTURING ENGINEERING

North America

The present engineering specialty known as manufacturing engineering has its roots in two areas.

One area involved the plant layout activities normally associated with industrial engineers working in manufacturing enterprises. These plant layouts were usually of departments focusing on specific manufacturing processes and means of transportation of parts from one department to another until all necessary operations were completed, and lines for the components to be assembled in a finished product.

The second activity was the various forms of tool design done to produce locating jigs and fixtures, determine tool geometry, determine optimal speed and feed rates for metal cutting operations, and design punches and dies and molds for primary forming operations. While plant layout people usually had a college level of education, often in the specialty of industrial engineering, tool designers came both from academic sources including vocational schools, technical institutes, and correspondence schools or from a tool and die maker background leading to work in tool design departments. These activities were usually conducted under line management of industrial companies involved in discrete parts manufacturing.

During the entrepreneurial phase of most of America's major manufacturing corporations, senior management often had an engineering background, and often was personally involved in the product design and process development of the products they manufactured. Most of the entrepreneurial leadership had intense hands-on experience in the manufacturing arena. Mechanical engineering schools included substantial theoretical and practical education in the basic manufacturing processes as an integral part of the requisite courses for graduation with a mechanical engineering degree.

Professional Societies. During the Great Depression of the 1930s, a profession society for tool engineers was created in Detroit and spread rapidly throughout the industrial centers of the United States, primarily in the northeastern quadrant of the country. These tool engineers performed production miracles at the start of World War II in converting factories from civilian production to military production using tools and machines fundamentally unchanged from those used in World War I. Tool geometry was developed empirically and a great reliance put on formal apprenticeship programs to ensure knowledge of essential mathematical skills and basic understanding of metallurgical processes.

Following a severe postwar recession, industry began to regroup to meet large consumer demands for manufactured products. Organizational skills developed during World War II to produce new types of weapons systems, mathematical techniques such as operations research, and training developed by the military to develop effective skills rapidly led to a major restructuring of the organizational or administrative layers of large corporations. Increasingly, nontechnical managers rose to the control of companies to a degree unknown in any other industrial country. Top management was more concerned with market development, capital expansion, and addressing economic and legal issues, placing lesser emphasis on manufacturing skills in the selection of top management. As part of the organizational restructuring, industrial engineering groups were often broken into plant maintenance and services handled by industrial engineers, and plant layout planning and capital equipment appropriation preparation handled by manufacturing or production engineering departments, who also supervised the tool engineering or tool design activities within the plant.

The professional society, recognizing this changing scope of activity, changed its name to the American Society of Tool and Manufacturing Engineers, but its membership basically remained people involved in fabricating processes in discrete parts industries. The educational level of those people serving in tool or manufacturing engineering roles for the most part did not include graduate engineers. A rush of new capabilities in materials, machine design, and machine control demanded higher levels of education, but much of this was achieved through continuing education activities of professional societies, or by vendor sales activities rather than through formal training at the university or collegiate level.

Studying the Role.   To tailor its continuing education activity and the increasing scope of manufacturing engineering, the society changed its name to The Society of Manufacturing Engineers (SME) but the general membership remained people involved in the traditional tool design functions of companies or were employees of those companies supplying machinery and tools to tool designers. In the late 1960s SME commissioned Arthur D. Little to do a study entitled "The Manufacturing Engineer Today and Tomorrow." This study was designed to define the manufacturing engineer's role in a changing social and technological environment. When the report was published in 1969, the average engineer had yet to encounter such technological advances as direct numerical control, computer-aided design, and artificial intelligence, nor had they become familiar with microprocessors or even many computers. The manufacturing engineering environment had not had to react to social legislation involving affirmative action, job enrichment, product liability, or occupational safety and health consideration. The 1969 report, a good historical document of the *existing* role of manufacturing engineers, could not foresee the rapid technological changes which occurred, and hence was not completely useful in preparing SME's activities in the field of continuing education.

At the same time, universities preparing engineers, under enormous pressure to produce more rounded citizens, were forced to include in their core curriculum courses in the humanities, literature, and history. Under these pressures, and those from accreditation groups, courses in the theory and practice of fundamental processes were eliminated. In the late 1970s the SME board, noting the rapid use of computers and electronic controllers and the increasing demand for baccalaureate levels of education, engaged the Battelle/Columbus Laboratories to prepare a second report redefining the manufacturing engineer in the context of a rapidly changing social and industrial environment. This report was intended to guide SME in its continuing education activities during the 1980s and to equip it to effectively influence the curricula at engineering schools primarily through participation in accreditation bodies. The manufacturing engineer in the study was defined as a person who plans, develops, and optimizes the processes of production but continued to limit the scope of manufacturing engineering to the activities between the completion of product design and the beginning of the production phase. At that time, the majority of manufacturing engineers still did not have a baccalaureate level of education. It was determined, however, that the roles of planning occupied more manufacturing engineers than did the tool design function. Almost three quarters of those interviewed in the preparation of the study were involved in some form of continuing education specifically related to their manufacturing engineering assignment. The vast majority of those interviewed still regarded practical experience as the primary source of technical training.

Expanding Education.   Manufacturing engineers then began to feel the impact of social pressures on the work environment, and the impact of foreign competition became a significant consideration in management decisions, particularly in areas of labor-intensive operations. The need for computer competence had by then caused a further revision of mechanical engineering curricula to include courses in developing computer literacy. This totally eliminated any time for manufacturing process courses within the traditional mechanical engineering degree requirements. The Society of Manufacturing Engineers, by a variety of activities, including the Manufacturing Education Foundation, began to ask universities to create departments of manufacturing engineering as one of the engineering disciplines offered. States, recognizing the increasing education needs for technicians, generally developed a large number of associate degree level technology schools. In the very late 1970s and early 1980s, the issue of global competitiveness became a national concern while at the same time several potential technological tools such as machine vision and robots, industrial lasers, and others offered promise through major capital investment to regain America's competitive productivity edge.

Role of Other Engineering Disciplines.   Large expenditures were made in the manufacturing engineering area, totally changing the nature and scope of manufacturing engineering activities. Much of the promise of these new technological tools, and continued advancement of the more established manufacturing machinery and tooling never came to full fruition because of a shortage of trained manufacturing engineers to apply and implement these new opportunities properly. The increasing use of computer controls and computer interfacing meant that many electrical engineers were performing manufacturing engineering functions without fundamental knowledge of basic manufacturing processes. The level of education required to comprehend and implement these new technologies could not often be obtained in the more conventional forms of continuing education. Universities, particularly in the graduate levels of education, put major emphasis on manufacturing research.

At the end of the 1980s the Society of Manufacturing Engineers commissioned the A. T. Kearney consultancy group and the University of Michigan to do an in-depth study of the present role of engineers and their perceived role in the next century. This chapter deals with the present functional roles of manufacturing engineers and the perceived changes in these roles in the foreseeable future.

International

In a world of global competitiveness, nations without significant petroleum reserves recognize that success in manufacturing is the primary source for growth in national wealth and raising or maintaining standards of living. In almost every country other than the United State, active government programs encourage and support the growth of the manufacturing sector. Responsibility for manufacturing policies is often assigned to the highest level of government. Emerging economic powers generally have governmental agencies supportive of manufacturing education and manufacturing engineering research. In these countries, the manufacturing engineering has a professional and social standing unknown in North America.

In Germany, IPA is an organizational structure for professional manufacturing engineers as a constituent part of the overall engineering umbrella association. German universities having production engineering departments usually have Franhofer institutes dedicated to specific areas of manufacturing technology. These institutes allow universities to collaborate interactively with industrialists on specific projects involving production or manufacturing engineering content.

In other countries, manufacturing engineers are often referred to as production engineers, plasticity engineers, or precision engineers. The term production engineer, however, is falling out of favor as reflecting too narrow a scope of activity, since manufacturing engineers find their functional roles expanding into product design and product planning and into various levels of postdelivery support for products. The English Institute of Production Engineers has recently been granted permission by the Privy Council to change its name to reflect the word manufacturing rather than production engineering. In most countries, manufacturing engineers have a higher level of overall educational attainment than that commonly found in North America. As the depth and breadth of knowledge requirements increase, graduate degrees among manufacturing engineers become the norm rather than the exception. Emerging countries tend to look toward manufacturing engineering developments in Europe and in Japan rather than in the United States since these countries are more actively involved in advanced manufacturing engineering technology. The United States still holds a quantitative edge in manufacturing engineering systems activities but its leadership is under continuous attack.

Relationships with Other Engineering Disciplines.   The present manufacturing engineering profession emerged in the last 60 years in an environment in which industrial organizations and academic institutes compartmentalized various types of engineering disciplines into separate departments. As the functional scope and emerging roles for manufacturing engineers constantly grow, the span of manufacturing engineering practitioners reaches out into disciplines in a way not perceived by tool engineers 40 or 50 years ago. Some of these are worth examination.

Industrial Engineering.   Industrial engineering has long been deeply involved in many aspects of industrial or manufacturing activity. Industrial engineers have been involved in production monitoring and improvement through such activities as time and motion study. They have been deeply involved in material handling aspects, parts storage, and containerization. Industrial engineers have the responsibility to address power needs and environmental considerations critical to successful manufacturing, and are often responsible for implementation of industrial safety devices and conformance with governmental occupational hazard regulations. In the past, a significant portion of industrial engineers went into manufacturing concerns. Today, industrial engineers are finding employment opportunities in commercial buildings, hospitals, data processing centers, and other non-manufacturing-related functions. At the same time, many of the traditional roles of industrial engineers within industry are being absorbed by system planners or incorporated into overall system design. Since more and more industrial manufacturing equipment is highly computer integrated, the industrial engineer is sometimes unprepared to participate in a multidisciplinary team involved in the design or implementation of such systems.

Mechanical Engineering.   Mechanical engineers traditionally had courses in manufacturing processes as part of the core curriculum during their education. Increasingly, the demand for well-rounded citizens has placed a new emphasis on arts and communication skills, and the demand for computer literacy has placed new burdens on basic required curricula for accreditation of the institute and graduation of the student. In order to accommodate courses involving civics, sociology, literature, history, communication skills, and computer literacy, schools have had to abandon most of their process-oriented courses at the undergraduate level. While mechanical engineers are still qualified to do basic research programs, they are ill equipped for the most part to address process decisions. This has led to foreign criticism of American engineers based on the perception that Americans invent products but do not know how to make them reliably or cost-effectively.

Although there is a growing recognition of the need to reintroduce process engineering courses, few know how to fit this in with the ever-bulging requirements for graduation. Some schools are pursuing a 5-year program for a baccalaureate level of graduation. Other schools have pursued, with varying degrees of success, industrial internships and cooperative education programs. Most of these programs have met with great industry acceptance of their graduating students. Schools, however, find it hard to find committed industrial partners for internships or cooperative study in which corporations pledge commitments to the students for the entire period of their college education.

Electrical Engineering.   Electrical engineering and computer sciences seem to be those fields of engineering which attract the brightest and most capable students. The knowledge requirements for understanding the fundamentals underlying electrical engineering leave little room for manufacturing engineering courses. The electronics industry, in general, has avoided many of the manufacturing problems found in the discrete parts mechanical industries because of an ability to highly standardize component configurations and component packaging. Where the components do not adapt to standard packaging techniques, there remains a high level of human involvement in the manufacturing process, and often such manufacturing is transferred to geographical areas where plentiful and inexpensive labor is available. As the role of integrated circuits and microchips spreads to almost every type of product, manufacturing engineers are increasingly becoming involved in electronics manufacturing and assembly.

Materials Sciences. In general, materials engineers are well qualified for the production of raw manufacturing materials or continuous-process industries since their education concerning the chemical content or atomic structure of the materials is usually intertwined with a knowledge base of the manufacturing processes and environmental concerns. For the most part, manufacturing engineers have little involvement with most materials science engineers.

Quality Control. The competitive need for product quality relied in the past on identification of defective components and assemblies after manufacture through an inspection or quality control department. As the true costs of quality are increasingly recognized by manufacturing management, there is a growing understanding that almost all quality problems can be related to improper process selection, capital equipment incapable of manufacturing to necessary product specifications, or product design inappropriate for available state-of-the-art fabrication or assembly processes. As management begins to recognize the relationship of product-process relationships and process reliability, the responsibility for product quality is being switched from defect detection and product quality verification to process control. Manufacturing engineers are increasingly integrating quality control and quality verification into each incremental step of the manufacturing process, ensuring that no added value is contributed to a product or component already defective. Manufacturing engineers are becoming involved in activities formerly restricted to quality control professionals.

Production Control. The tendency in recent years to understand the cost of in-process and finished parts inventory has led to major physical and organizational restructuring of manufacturing companies. Most in-process inventory and finished goods inventory control relates to tighter integration of the manufacturing processes, and faster knowledge of production levels at each incremental stage of the production process. Since in-process and finished goods storage is so often related to manufacturing processes, conveyors, and buffer storage, manufacturing engineers are often involved in redesigning overall manufacturing systems to achieve production control goals.

This continuous outreach of manufacturing engineering into areas traditionally held by other engineering disciplines, within both academic institutions and industrial organizations, often leads to friction and a lack of collaboration so necessary in multidisciplinary teamwork. This topic is addressed below under Emerging Roles for Manufacturing Engineers.

Certification and Registration. Professional registration has long been critical in engineering disciplines in which engineering efforts have enormous social implications. Structural engineers, architectural engineers, industrial engineers, and others often make decisions exposing manufacturers to broad liability, both civil and criminal, and government censure. Societies have recognized the need for uniformly established credentials for engineers above and beyond the diplomas for undergraduate or graduate levels of education. Professional registration usually involves a determination of application skills and hands-on experience combined with a necessary base of professional knowledge. Most engineering disciplines have had comprehensive professional registration programs and carefully monitored examinations. Industrial practitioners, however, in many states are not treated as individual professionals but as members of manufacturing organizations, and many states provide industrial exemption to any need for professional registration for those engineers involved in manufacturing activities. Many states have not created a professional registration examination or division for professional registration as a professional engineer in the field of manufacturing engineering. Many engineers who have secured professional registration find that professional registration is not significant in advancement, promotion, or compensation in an industrial environment.

Certification is another means of recognition of achievement and application skill in specific areas of engineering discipline. The Society of Manufacturing Engineering created a certification program for manufacturing engineers both as generalists and in specific technologies. Certification is available in coating and finishing, robotics, machine vision, and other specialized disciplines within the generic field of manufacturing engineering. Certification seems to be a growing peer review form of recognition of application skills and seems to have a higher level of acceptance for promotion and compensation within industrial organizations.

MANUFACTURING ENGINEERING EDUCATION

Professional education for manufacturing engineers has faced two severe challenges. The first is the relative age of manufacturing when compared with more established engineering disciplines. The second is that of addressing the ever-changing educational needs of a profession at a time when more stable engineering professions find a declining interest in established educational institutions, which involves issues of tenure and budget restraint.

Engineering is defined by the Accreditation Board of Engineering and Technology (ABET) as:

> The profession in which a knowledge of the mathematical and natural sciences gained by study, experience, and practice is applied with judgment to develop ways to utilize, economically, the materials and forces of nature for the benefit of mankind.

The definition of manufacturing engineering adopted by the board of SME is given earlier in this chapter.

Manufacturing engineering technology is a part of this definition. Technology programs normally do not require as many math and science courses as engineering programs, even though they do involve application of both. More emphasis goes into laboratory work, skill development, and applied engineering. A modest but rapidly growing number of schools have been accredited in manufacturing engineering degree programs, on both the baccalaureate and the graduate level.

Engineering technology is defined by the Accreditation Board for Engineering and Technology (ABET) as:

> That part of the technological field which requires the application of scientific and engineering knowledge and methods combined with technical skills in support of engineering activities; it lies in the occupational spectrum between the craftsman and the engineer at the end of the spectrum closest to the engineer.

The critical phrase in this definition is "in support of engineering activities." This assumes that basic physical skills have been developed which will allow active involvement in the production process, and at the same time, basic theory and knowledge have been acquired which allows engineering types of decisions to be made.

Industrial technology, on the other hand, may be characterized by involvement in the area between engineering and management. The National Association of Industrial Technology (NAIT) defines industrial technology as:

> consisting of degree programs of study designed to prepare management-oriented technical professionals in the economic-enterprise system. Industrial Technology degree programs and professionals in Industrial Technology careers typically will be involved in: 1) the application

of significant knowledge of theories, concepts and principles found in the humanities and the social and behavioral sciences, including a thorough grounding in communications skills; 2) the understanding and ability to apply principles and concepts of mathematical and physical sciences; 3) the application of concepts derived from, and current skills developed in, a variety of technical disciplines including—but not limited to—materials and production processes, industrial management and human relations, marketing, communications, electronics and graphics, and may include; 4) a field of specialization, for example: electronic data processing, computer integrated design and manufacturing, construction, energy, polymers, printing, safety and transportation.

It should be noted that both manufacturing engineering technology and industrial technology have a common element of applied skills development. The primary difference, as identified in the two definitions, is in the distinction between ''support of engineering'' and ''management-oriented.'' Therein lies the reason for differences in program requirements between the two. Industrial technology programs are usually characterized by fewer requirements in math and science than engineering programs and more requirements in the area of management and business administration.

Engineering Technologist. The engineering technologist must be applications-oriented, building upon a background of applied mathematics through differential and integral calculus. Based upon applied science and technology, the technologist must be able to produce practical, workable results quickly; install and operate technical systems; devise hardware from proved concepts; develop and produce products; service machines and systems; manage construction and production processes; and provide sales support for technical products and systems.

Engineering Technician. With a minimum of 2 years of postsecondary education, ideally in engineering technology, with emphasis in technical skills, the engineering technician must be a doer, a builder of components, a sampler and collector of data. The technician must be able to utilize proved techniques and methods with a minimum of direction from an engineer or an engineering technologist. He or she should not be expected to make judgments which deviate significantly from proved procedures.

The technician should expect to conduct routine tests, present data in a reasonable format, and be able to carry out operational tasks following well-defined procedures, methods, and standards.

Technician training, of skills and knowledge in all fields of engineering, relies on familiarity with tools, equipment, and apparatus. It often provides specialized education and training that emphasize manipulative skills. Apprenticeship programs once were a common preparation, and although apprenticeships are still important in some specializations, the associate of applied science degrees seem to have become the more popular route. Students gain both skill and fundamental knowledge of math and science and in-depth knowledge in a technical specialty. By a combination of job experience and further education, technicians often assume increasing responsibility such as in supervision or become more adept in their technical specialty.

Undergraduate Levels. Manufacturing engineering courses at the baccalaureate levels are available in many schools. In some instances, degrees are granted specifically in manufacturing engineering, and in other cases as options in mechanical or industrial engineering. Active participation in developing ABET accreditation has led to a steady growth of schools offering baccalaureate level education in manufacturing engineering. Such courses may be traditional engineering courses while others offer technology-oriented courses as described above. Severe budget constraints and changing enrollment levels in specific engineering disciplines make it very difficult to develop new departments specifically for manufacturing engineering. Unfortunately, participation in education at the undergraduate level is not as professionally rewarding for the professors involved when compared with professional opportunities in graduate teaching.

Graduate School.   Relatively few schools offer graduate degrees in manufacturing engineering. Several major conferences have been held on the configuration of graduate levels of education in manufacturing engineering. Although accreditation criteria are available, it has proved to be extremely difficult to obtain the necessary grants or funding for high levels of activity for graduate degrees in manufacturing engineering without specialization.

Manufacturing Research.   Significant funding has been done in manufacturing-oriented research by the National Science Foundation and Department of Defense. Such research is often done in areas of specific interest of the professor with little active direction from industrial partners. Recent international conferences and bilateral agreements have emphasized the high level of applications orientation of manufacturing research in Japan and Europe when contrasted to the theoretical or generic research more commonly found in American universities. The proposed Japanese initiative IMS has put a heavy emphasis on industrial partnerships, and directed research to meet specific industrial needs. Reduction in government funding has led to more effective relationships between American industry and academia in directed multidisciplinary research.

Manufacturing System Engineering.   Many schools maintain departments or centers for graduate-level activity in the area of manufacturing system engineering. This pertains to the computer integration of the design in manufacturing processes. American schools have a clear leadership position in this area of manufacturing-related education and are working diligently to improve the process content of manufacturing systems engineering schools.

Vocational Schools and Technical Colleges.   Economic conditions and social expectations in the United States have worked to eliminate most of the traditional apprenticeship programs still prevalent in Europe for the development of engineering technologists. At the same time, the level of education necessary for today's sophisticated manufacturing equipment has produced strong public support for associate degree level vocational schools and technical colleges. Vast funds have been expended to provide technicians with sufficient education to understand, maintain, and operate today's sophisticated industrial machinery. Accreditation is available for these schools in specific areas of manufacturing engineering. Since these schools tend to be state funded, tuition is inexpensive and many students find that 2 years spent in an associate level technical college is an inexpensive alternative to the first two years in a more formal baccalaureate level engineering school. Many students of technical colleges therefore continue for higher levels of education and do not enter industry on the technologist's level. The opportunities for engineering technologists remain extremely high, and the shortage of personnel at this educational level has become extremely critical for many manufacturers.

FUNCTIONAL SCOPE OF MANUFACTURING ENGINEERING

Manufacturing encompasses the total spectrum of conversion of raw materials into products of higher social or economic benefit. As such, it includes all continuous process industries such as refineries, discrete parts manufacturing, and activities similar to those found in automotive production. It includes the preparation of foodstuffs and lumber products. It encompasses products manufactured for both civilian consumption and military use. This conversion of raw materials through manufacturing processes accounts for approximately two-thirds of all the increase in the wealth of the United States. Although the concept of manufacturing spans such corporations as oil refineries, steel mills, fish canneries, plywood plants, and soap and detergent manufacturers, it is more commonly thought of as the process of discrete parts manufacturing and assembly. The selection of equipment and plant design for continuous process industries has often been done by architectural engineering companies, metallurgists, chemical engineers, and petroleum engineers. With the exception of mechanical and electrical engineering, engineers trained for continuous process industries usually receive training not only in the product but in the processes of manufacturing or conversion of that product as part of their baccalaureate level of education.

Traditionally, manufacturing engineering is more commonly thought of as relating to discrete parts industries. Conversion from this limited approach will be long and difficult since discrete part industries usually make a structural or organizational distinction between product design and process planning. The functional roles of manufacturing engineering are therefore responsive to organizational structure, particularly that which evolved since World War II.

Present Functional Roles of Manufacturing Engineers in Industry

The areas described below reflect historical roles of manufacturing engineering in the period of World War II until the mid 1980s.

Tool Design. The primary role of many manufacturing engineers is that of tool design. Tool design encompasses the design of material cutting or forming tools, the design of jigs and fixtures used to locate parts properly for primary or secondary machining operations, the design of locating jigs for mass production of interchangeable parts, the design of molds for both metal and plastic material, and the design of transport containers used during the manufacturing process.

Tool design was, for many years, the primary occupation of manufacturing engineers since it involved the selection of processes and tools used in the conversion of raw materials to finished components. The focus was not entirely on tools and processes since tool design engineers often requested modification to product designs to provide locating surfaces or provide for internal rigidity to take the necessary torques and forces generated during machining or forming operations. These requests for product design changes were reactive and did not involve the concept known today as simultaneous engineering.

Tool design remains an essential and vital component of the manufacturing engineering function, but the nature of this activity has been radically changed by the advent of numerically controlled machines. Modern machine tools with numerical controlled capability and improved spindle and bearing design are capable of manufacturing many parts with a minimum of jigging and fixturing. Such machines require certified cutting tools which can be preset off the machine and machine parts to specific tolerances or dimensions. The role of tool design therefore has shifted into the area of machine control programming to achieve parts specifications. Modular jigs and fixtures designed specifically for numerically controlled machines have reduced the workload in tool design departments. Increased capability of advanced computer-aided design systems can be programmed to ensure the proposed component designs are capable of withstanding machining or forming pressures. Despite the increasing capabilities of computer-aided design tools, human review and input are often necessary in areas such as injection mold design, because of the variety of factors that may enter into material flow and proper heat transfer.

Systems Planning. From the earliest days of the industrial revolution, it has been traditional to structure manufacturing facilities by the manufacturing function performed. Companies traditionally had casting departments, turning departments, milling departments, welding departments, and the like. In more recent years, modern plants recognizing the cost of large in-process inventory have tended to rearrange plants into entire manufacturing systems. Such systems, or in smaller terms work cells, have the ability to perform all of the manufacturing operations necessary to produce a component in a single department. Manufacturing engineers today are deeply involved in system planning. Work cell configuration and production system planning have three goals. The first is to reduce in-process inventory. A second goal is a corollary of the first, namely, the reduction in floor space derived from in-process inventory reduction. The third, also related to the first, is the improvement in product quality achieved by a reduction in material handling and a reduction in the length of time between the primary, secondary, and tertiary operations. System engineering today involves not only the machine controls but input and output relating to quality measurement programs, production controls systems, and management information systems.

There is often confusion in nomenclature between systems planning and manufacturing systems engineering. System planning involves the overall layout of the manufacturing facility which may require computer simulation of work flow and probable production levels. Manufacturing systems engineering pertains in most cases to the computer data flow between machine tools and management information systems. System planning includes a determination of appropriate processes, whereas most manufacturing system engineering functions pertain to the control or measurement of these processes.

Capital Procurement.   Another major role of the manufacturing engineer in most industrial organizations is a preparation of capital procurement budgets for the necessary equipment to manufacture, assemble, and package industrial products. Such budgetary work involves expense accounting. In most cases, machine tool justification, or the decision to purchase capital investment, is based upon corporate targets for return on capital investment. The hurdle rate for such capital investment is generally higher than for any other form of expenditure in industrial corporations. In recent years, most corporations expect returns on investment in the range of 35 to 40 percent. Companies producing products with long product life may be willing to accept a lesser amount, while other companies with rapidly changing markets may look for significantly higher rates of return. The work of the manufacturing engineer is often inhibited by accounting systems that place a heavy premium on direct labor content in manufacturing and do not have quantitative measurements for many of the factors that should be of concern in major capital investment. Since manufacturing engineers rarely are able to carry their presentations to top management, it is imperative that their capital appropriations submissions be as accurate as possible, be self-explanatory, and anticipate questions from other parts of the management team. The preparation of capital appropriation requests is perhaps one of the major occupations of manufacturing engineers employed by large corporations.

Production Control.   Traditionally, manufacturing engineers, as part of the capital procurement process, have been expected to estimate accurately the production capabilities of capital equipment and specialized tooling. Since the net production, when compared with gross cyclic capability, varies widely from one type of production equipment to the other, vast experience in line operations performance is extremely critical to anticipation of probable overall production systems. Under the old style of organizational structure with large departments devoted to specific technologies and substantial in-process inventory, production control was often a clerical function with problems addressed by departmental supervisors.

In a world increasingly using integrated work cells, the buffer storage capacity of in-process inventory has lessened to the point where production control often means interactive communication between various machine tools in an integrated manufacturing system. Manufacturing engineers must often plan using the tools of computer simulation to determine the minimum buffer storage to ensure total production system throughput. Such simulation must provide not only for throughput under normal operating conditions but also for off-loading or on-loading at various points in the production line to ensure continued production in the event that one element of the work cell is down for significant repairs or alteration. Often this production control evaluation must be done on a team basis, but it has become an increasing responsibility for manufacturing engineers in industry.

Advanced Manufacturing Planning.   In the era between 1950 and 1980, many companies addressed the problem of lack of interaction between product designers and process planners by placing within product design departments advance manufacturing planning specialists. These specialists, as part of the product design team, were intended to examine proposed initial concepts in the light of the most efficient manufacturing processes. Many leading industrial firms had advanced manufacturing planning specialists in product design departments. The results of these advanced manufacturing engineering groups usually were beneficial, but sometimes the results were controversial. If one accepts the fact that there are optimal product designs for given processes, the decision by an advanced manufacturing engineer working in the product design department often had strong influence on product design without that advance manufacturing planning engineer being responsible for capital appropriation requests, machine productivity, machine performance, or profitability. Often the advance groups used emerging technology in which durability and production rates were not proved. There often was severe animosity between the manufacturing engineering groups in line operations and the advance manufacturing engineering groups who reported to corporate staff. The problems of advanced manufacturing planning have for the most part been overcome through the concept of simultaneous or concurrent engineering where that is effectively practiced. Only a limited number of engineers were employed in advanced manufacturing planning, and this type of interaction is no longer a significant area for manufacturing engineers.

Assembly Mechanization.   Until the late 1950s mechanized assembly of products was limited to those types of components that could be fed from unoriented bulk condition by existing feeder mechanisms. The development of the vibratory feeder in the late 1950s opened up an entire spectrum of possibilities for assembly mechanization at the same time that mature industrial countries were facing labor shortages or increased labor costs. Assembly of high-volume products was often transferred to third world countries where production costs were low but quality problems and transportation costs were enormous. The advent of the vibratory feeder led to increasing recognition of assembly as being a form of manufacturing susceptible to high levels of engineered mechanization.

In the late 1970s renewed interest in industrial robots and the potential of machine vision seemed to indicate an era where assembly mechanization could be applied to flexible production requirements and lower volumes of production. Many engineers were involved in the development aspects of robotics and machine vision in hopes of developing new capabilities in assembly. While these two techniques have found broad application in electronic circuit board manufacturing and in certain types of manual joining operations such as welding, these advanced technologies have not found broad-scale economically justifiable application in discrete parts manufacturing.

A common trend of major corporations to purchase low value added components or subassemblies from outside partners has placed a new emphasis on the importance of assembly engineering within major corporations. Assembly mechanization and assembly system planning are major functional roles for manufacturing engineers in many corporations. The new emphasis on product quality and reliability using test equipment integrated into the very manufacturing processes means that assembly mechanization involves much more than the actual placing of parts into specific spatial relationships. Assembly mechanization engineering is a growing and critical field in the present functional roles of manufacturing.

Total Quality Management.   Manufacturing engineers are becoming directly involved in total quality management. This concept, widely practiced in defense industries, is being applied in consumer products manufacturing to meet customer expectations of quality and reliability. Total quality management is directly concerned with manufacturing process reliability. This is an extension beyond quality verification commonly found in modern assembly and fabricating equipment. Manufacturing engineers are often leaders in total quality management teams because of the heavy interaction between appropriate process identification and process limitations in the light of product quality expectations.

It will be seen that the above major functional roles of manufacturing engineers in industry are major expansions of manufacturing engineering into areas formerly the domain of electrical and mechanical engineers, chemical engineers, production controllers, quality control specialists, and product designers. These extensions often mean that single individuals are limited in their ability to perform these functional roles which often must be handled by teams rather than individuals. All of these major functional roles will probably remain for the foreseeable future, but their weight and relative importance will change relative to new and emerging roles for manufacturing engineers.

EMERGING ROLES FOR MANUFACTURING ENGINEERS

Amid the dizzying array of sophisticated new technology, an increasingly competitive global marketplace, and widespread moves toward greater decentralization, the role of the manufacturing engineer is expected to change rapidly over the next few years. Against this backdrop in 1988 the Society of Manufacturing Engineers (SME) commissioned a study designed to determine what the manufacturing engineer's job description would look like by the year 2000, and to ascertain what additional steps, if any, ought to be set in motion to prepare him or her for the profound changes that are anticipated (see Fig. 1.1). This study was titled "Countdown to the Future: The Manufacturing Engineer in the 21st Century." More commonly referred to as "Profile 21," it identified several worldwide trends.

Today's manufacturing engineers are increasingly involved in a widening variety of tasks and activities inside and outside an enterprise. With the exploding environmental scope, no one individual will be able to be "all things to all people." Rather, manufacturing engineers of the year 2000 will play one of three alternative roles categorized as follows:

1. Technical specialist
2. Operations integrator
3. Manufacturing strategist

Each of these roles requires a different balance of depth and breadth skills. (See Fig. 1.2.) The depth skills include the traditional technical engineering disciplines. The breadth skills include nontechnical capabilities such as effective communications, foreign language fluency, teamwork, and the ability to deal with broader business issues.

Technical Specialist.   The first category of manufacturing engineers, technical specialists, will need focused, in-depth skills. These technologists will focus on a specific aspect of manufacturing in great detail. The traditional and new emerging technical disciplines will be their forte. Most manufacturing engineers fall into this category today. While concentrating on technical skills, specialists cannot afford to neglect nontechnical breadth skills.

Technical specialists will not be effective in isolation. They will be the "detail men" of the integrator and the strategist. While they will play a vital role and put out day-to-day fires, they will be only a "piece of the puzzle," one element in a broader equation. Many believe strongly that the importance of the technical specialist will decline as expert systems become more common. "Technical experts temporarily serve as resources, but once their knowledge has been depleted, and converted into systematic rules for decision-making, their usefulness is attenuated," states Shoshana Zuboff, in his book, *In the Age of the Smart Machine.*

Operations Integrator.   Global competition has focused corporations on the need for more functional integration. One study participant stated, "Integration is the key to future success. Integration of people, equipment, and information, all tied into one smooth operation." An integrated manufacturing system is one element of the broader aspect of functional integration. Today 27 percent of respondents to the study said they are involved in integrated manufacturing systems, and 57 percent thought that they will be involved in integrated systems by the year 2000.

Manufacturing engineers acting as operations integrators will possess a relatively equal

Technologies	Absolute Percent Increase	Percent Currently Required to Use	Percent Required to Use in 2000	Growth Multiple
▪ Expert systems, artificial intelligence and networking	36%	11%	47%	4.3
▪ Automated material handling	35	23	58	2.5
▪ Sensor technology, such as machine vision, adaptive control, and voice recognition	35	16	51	3.2
▪ Laser applications, including welding/soldering, heat treating and inspection	34	17	51	3.0
▪ Integrated manufacturing systems	30	27	57	2.1
▪ Advanced inspection technologies, including on-machine inspection and clean room technology	25	32	57	1.8
▪ Flexible manufacturing systems	24	32	56	1.8
▪ Simulation	23	17	40	2.4
▪ Composite materials	20	16	36	2.3
▪ CAD, CAE, CAPP, or CAM	13	56	69	1.2
▪ Manufacturing in space	11	2	13	6.5
▪ Bio-technology	7	1	8	8.0

FIGURE 1.1   Forecast of technologies used in manufacturing.

balance of breadth and depth skills. They will act as quarterbacks and call the signals for the entire manufacturing team. Those who function in this role are uniquely qualified because they develop and coordinate the entire manufacturing process from product design through after-sales service—they are the new manufacturing engineers in the role of operations integrators. In short, operations integrators will play a central role and interact with almost all functions within a corporation. No other discipline shares this multifaceted responsibility. Figure 1.3 illustrates the overlapping relationship with others.

Many characterizations were suggested to depict this new manufacturing engineering role. They ranged from Superman to decathlete to "the person who makes it holistic." One round-table participant summed it up best: " A Renaissance man, one acting as a technical, strategic, computer proficient, people-oriented, hands-on integrator."

No function will stand alone in the organization of twenty-first century companies. The need for integration across all business functions is driving out isolated, unilateral approaches. Interdependence is key and bound to grow. The manufacturing engineering function is at the center of much of this interdependence.

Of particular importance to manufacturing engineering is its relationship to product de-

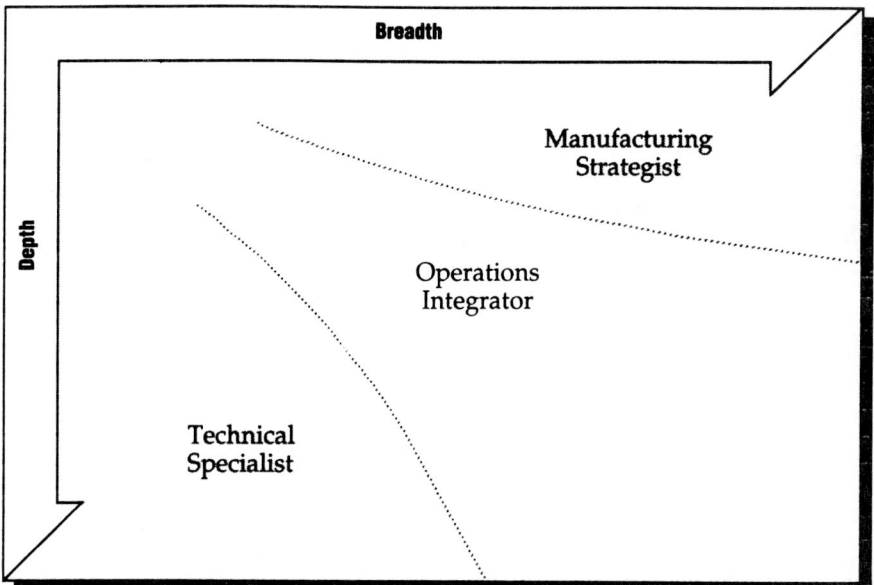

FIGURE 1.2   Multiple roles of the manufacturing engineer.

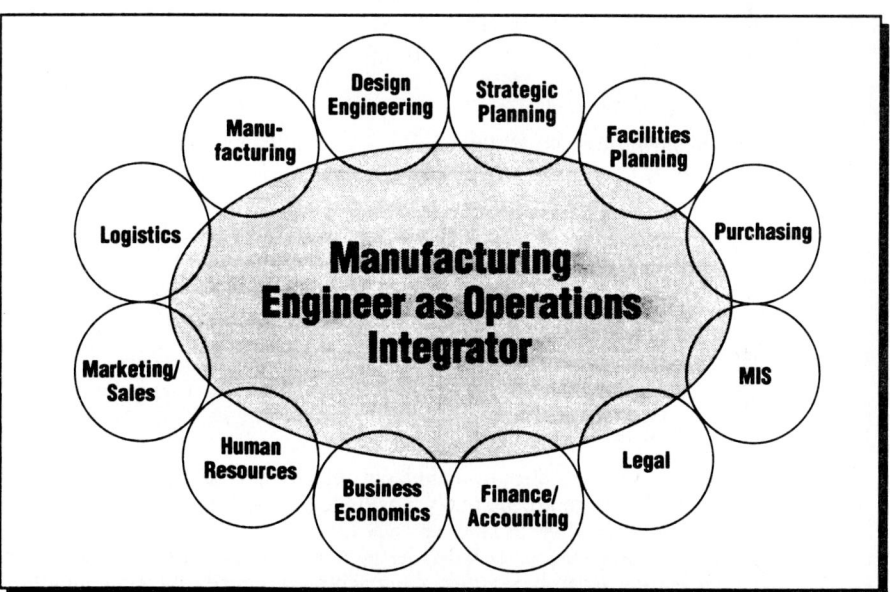

FIGURE 1.3   Role of the operations integrator.

sign. No longer can these be viewed as two separate activities divided by invisible walls. Many feel that the creation of the manufacturing engineering discipline has only served to build a wall between engineering disciplines, that is, manufacturing engineering and product engineering. All sources of information for the SME study—CEOs, round table and Delphi participants, and manufacturing engineers—stressed that a company's future success will depend ultimately on how well the various functional organizations work together. No longer will manufacturing engineers see a new product design and ask, "Can we make it?" "Today they are separated. Separated by buildings, separated by management, separated in time. To integrate design and manufacturing, it will be necessary to combine the product engineer and the manufacturing engineer into the same organization."

Manufacturing Strategist.   The third major category of manufacturing engineers, manufacturing strategists, will need breadth skills rather than significant depth skills. They will develop strategic manufacturing plans based on company business objectives. They will function as advisers to management and challenge the organization to motivate people, innovate, automate, or emigrate in search of the most cost-effective manufacturing strategy.

These strategists, in contrast to operations integrators, will be more general and will concentrate on the "what?" rather than the "how?" While lacking the technical capability of both the integrator and the technical specialist, they will play a valuable role to raise and address business issues directly impacting a company's ability to serve its customers through better manufacturing.

The role of the manufacturing strategist will be to influence the entire organization to implement worldwide strategies to stay in business well into the twenty-first century.

Lastly, the study identified key trends currently evolving in major industrial corporations.

Manufacturing engineers will function increasingly as operations integrators.

They will serve more as integration engineers, coordinating people, information, and technology within an organization.

Thus a significant number of manufacturing engineers will need to emphasize management and business skills over scientific and mathematical skills in their ongoing professional development.

They will have to understand new technologies and processes and be able to assess these from a broad managerial perspective in terms of the impact on their business.

Given the growing overall importance of manufacturing engineers' increasingly complex tools to everyday planning, logistics, and work flow, they must assume a greater role as strategists.

This is important from two perspectives:

1. Top management must recognize and facilitate the important advisory role of the manufacturing engineer—a role which may prove crucial to a company's competitive position. It would appear that the importance of this changing role is not yet widely recognized at this level.

2. A significant number of manufacturing engineers must be ready to grow into this role for which they are uniquely positioned, conceptualizing problems and solutions not only in technological and logistical terms but also in terms of the end user in a shifting global marketplace. Despite the apparent urgency of preparing for this broader role, there appears to be no formal curriculum in place to accommodate this need.

The strategic role of the manufacturing engineer will be important for another reason: Because of the increased costs of developing new technologies, there will be greater emphasis on collaborative and sharing agreement (limited partnerships, joint ventures, and the like) to share R&D resources for technological innovation.

This shift toward collaboration will radically reduce the level of vertical integration for many companies and will lead to a far different view of competitors. Teamwork and people skills will play an important role in the work of manufacturing engineers. As a result of sophisticated and complex technology, and the need for access to much more information about every phase of manufacturing, by the year 2000, manufacturing engineers will work much more in teams than individually.

So-called people skills will become of greater importance to the success of this profession. While technology has often been cited as the key to improving productivity, quality, and innovation, Profile 21 results indicate that effectively dealing with people and recognizing individual skills and points of view of employees at all levels will be critical to remaining competitive in the manufacturing environment of the future.

Some have interpreted extracts of Profile 21 as indicating a reduction in the roles of technical specialists. This may be due in part to the broad management perception in the late 1970s and in much of the 1980s that technology alone could solve many manufacturing problems and produce a new competitiveness for the organization. Much research effort and capital investment was made in new or emerging technologies without matching skills in the implementation of these new technologies on the factory floor and integration of these new technologies with gigantic existing capital investment.

The functional scope of manufacturing engineering described earlier in this chapter will not disappear. Instead it will no longer be limited to the small period of time between completed product design and shipment of the product. The emerging roles for manufacturing engineers will give them a greater responsibility and greater accountability for all activities within the manufacturing enterprise. The technical specialist will require increasing capability, and will require masters or doctoral levels of education.

The operations integrator, by nature, will be involved in capital equipment projects. Traditionally, established manufacturing enterprises have maintained a large in-house capability to handle these programs. With the ever-increasing sophistication of manufacturing technology, no individual can have a truly effective knowledge of all manufacturing technology options. Operations integrators will require teams of people whose collective knowledge will span the spectrum of manufacturing processes. It is difficult and expensive for corporations to maintain such integration teams on the corporate payroll unless there is a continuing series of major capital equipment programs. The role of the operations integrator and manufacturing engineer will be done by consultancy groups such as architectural engineering firms or major production equipment builders. These operation integrators will be expected to operate with a high level of professional ethics since they will have access to much confidential information concerning the strategic planning of major corporations. They will have to be able to interact with corporate manufacturing strategists and be able to draw on a large pool of technical specialists both within and without the corporation.

Most large corporations in the discrete part industry are grappling with the terms of engagement of such operations integrators. Continuous process industries are more acclimated to such outside engagement in the building of large continuous process plants. As noted above, today's engineering schools, particularly those on the graduate level, are capable of producing superbly trained technical specialists. Training people for the role of operations integrators or manufacturing strategists in a formal education mode is proving more difficult. Manufacturing strategists will probably require more business school training than technological training, and it is expected that many schools will soon offer degrees on the masters level in technical administration. Preparing future engineers for the role of operations integrator is being aggressively attacked in many schools and is putting great stresses on the traditional departmental structure of engineering and business schools. Many schools already require that masters theses be developed by interdisciplinary teams, and many major research grant giving agencies are insisting on a multidisciplinary approach for future research programs.

Figure 1.4 is an attempt to illustrate the trends of change from the established historical roles for manufacturing engineers to the emerging roles for them within the manufacturing enterprise. It is also clear that most individuals working in the manufacturing engineering function will have baccalaureate-level degrees in specific disciplines other than manufacturing engineering.

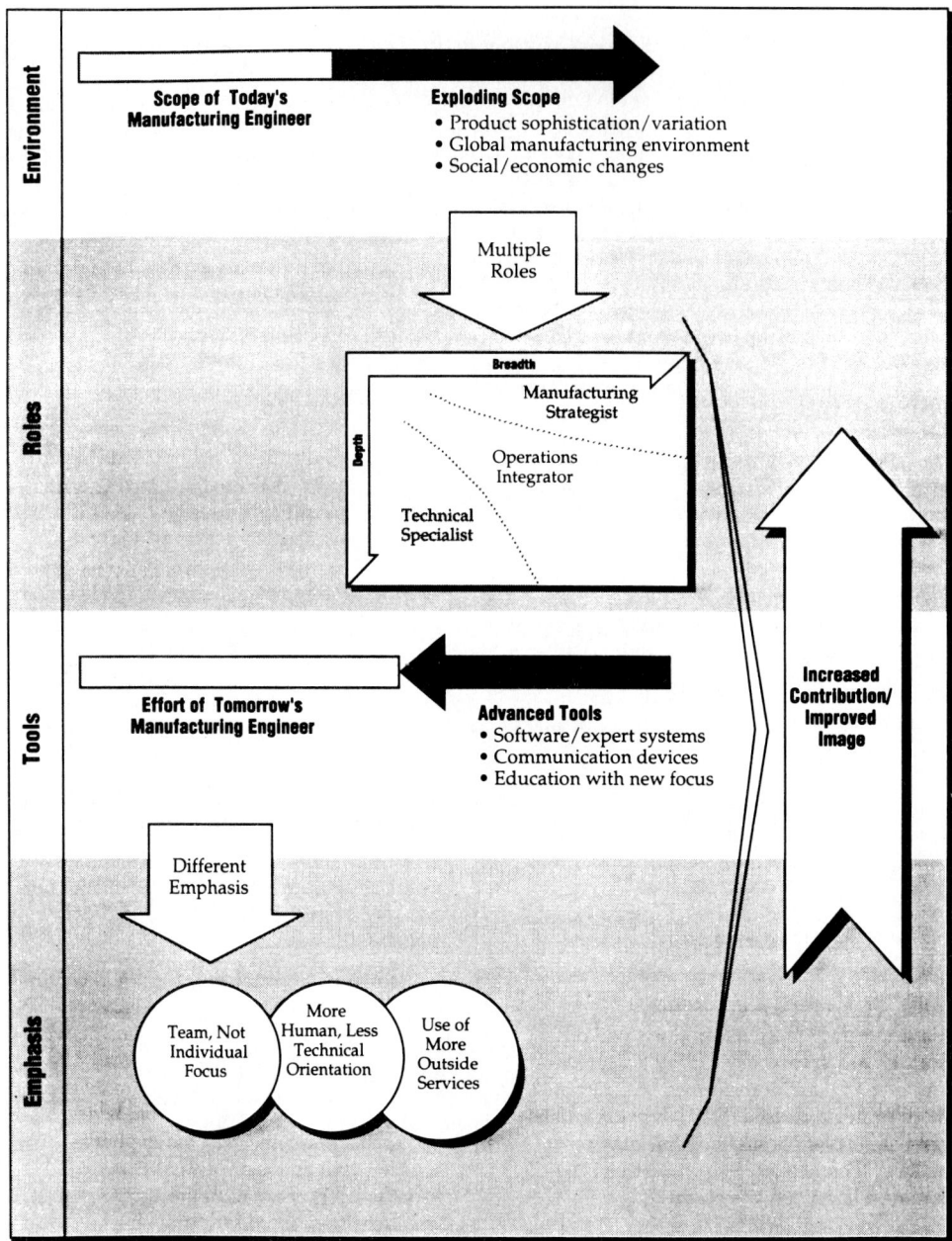

FIGURE 1.4   Manufacturing engineer in the year 2000—issues and implications.

THE EVOLVING ENVIRONMENT FOR MANUFACTURING ENGINEERING

Simultaneous Engineering.   In order to meet the ever-shortening demands for new or improved products, industrial organizations are increasingly attempting to use the concept of simultaneous or concurrent engineering. This concept involves interaction between product designers and manufacturing engineering groups to ensure an appropriate match between product design and process application, while at the same time eliminating the redundancy that occurred when manufacturing engineering was a function performed after the product design was complete. In Japan, almost all process engineering and, in fact, production equipment are built in-house, and manufacturing engineers, machine builders, and product designers report to the same management, thus ensuring a tight and effective cooperation among the three groups. What is achieved is interaction among these groups. In Germany and in other European countries, bridges have long been established between academic production engineering groups and industrial companies to ensure the most advanced manufacturing processes are planned with the development of new products. The strong and effective partnerships that have developed in Europe involve long-term relationships where price considerations are secondary to the development of an effective match between product design and process planning.

In the United States, to date, simultaneous engineering has not reached its full potential for several significant reasons. Perhaps the first and foremost is that there is not a significant pool of qualified manufacturing engineers capable of operations integration. Schools have produced superb capability in systems engineering, but there is a severe shortage of qualified operations integrators capable of matching information needs with process development. The second significant factor is that capital equipment procurement methodology has not been modified in the last 40 years to permit the freedom, flexibility, and timeliness of capital equipment purchasing decisions necessary in order to implement simultaneous engineering. The third factor that clouds North American use of simultaneous engineering concepts is the tendency of product design to report to marketing management, while manufacturing engineering usually reports to line or operating management. Manufacturing engineers will be forced to find more effective ways of implementing the simultaneous or concurrent engineering concept. It is perhaps unfortunate that the words simultaneous or concurrent are not sufficient. What is really intended in "interactive" engineering from the initial days of product design.

Multidisciplinary Teamwork.   Global competitiveness is causing many corporations to examine the hierarchical structure that has been a traditional pattern for industrial organizations in North America. Many industrial corporations have adopted organizational structures that reflect vertical reporting paths and horizontally structured collaboration lines. This leads to a very complex and difficult organizational structure but reflects the increasing recognition of the necessity for multidisciplinary teamwork in emerging manufacturing organizations. The enormous growth of manufacturing technology and computer capability in the 1970s and 1980s, as well as developments in robotics, machine vision, lasers, and advanced fabricating processes, have greatly diminished the effectiveness of vertically structured organization with each department maintaining skills and rights to specific technologies. Multidisciplinary engineering teams are increasingly indicated as the most effective means of meeting global competition. Such teamwork is highly dependent upon communication skills and interpersonal relationships, which are often ignored in formal engineering training programs. Many organizations are finding that the ability to develop interpersonal skills is as critical, if not more critical, than knowledge of technological processes.

Project Employment.   The constant acquisition and merger of large industrial corporations is part of an overall effort to reduce corporate overhead. In the past, manufacturing engineers were usually employed on a lifetime basis colored only by recessions and depressions. Corporations are now reluctant to hire people for a modest number of major capital equipment programs, and then maintain these highly skilled personnel on a lifetime employment basis, particularly when the fringe costs of such employment have become increasingly significant.

More and more industrial corporations are hiring engineers on a contract basis for specific projects. While these engagements are often done on an individual basis, many professional engineers find that various forms of social security, particularly in health and life insurance, as well as retirement funding are put at major risk. Professional manufacturing engineers are thus finding employment with architectural engineering and consulting engineering companies or with production equipment builders. This provides some form of economic security while allowing them to participate in the project employment mode that is increasingly being used by large corporations for professional engineering employees.

Collaborative Research.   Traditionally, industrial corporations have expended large funds on internal research and development activities to develop new products and new processes to maintain their competitive position. Collaborative research has been foreign to both industrial organizations and academic organizations for a variety of reasons. Not only are there competitive advantages to close control of research developments but for individuals, patents achieved through research can provide enormous economic rewards. Today, the cost of development of new products and processes often surpasses the financial capabilities of the largest organizations. There has been in the United States, also, a concern about violation of antitrust regulations concerning collaboration or collusion between industrial organizations to secure market advantages.

In the future, more and more companies will participate in collaborative, precompetitive research. This research may be done with organizations, academic institutions, or governmental bodies sharing different cultural attitudes, different ethical positions, and severe language and communication problems. Manufacturing engineers in the years to come must be equipped to handle and overcome difficulties in effective collaboration, while at the same time protecting the rights and future competitive position of their employers. Control of intellectual property will be an essential element of a manufacturing engineer's professional activities.

Engineering Education.   All professional practitioners feel the need for continuing education. Many professional bodies mandate continuing education to maintain licenses or certification. In no field, however, is the need for continuing education more important than in manufacturing engineering. While most professional educations have certain basic fundamentals that remain unchanged, the very basis of manufacturing engineering decisions changes radically and in increasingly shorter time frames. Often new manufacturing technologies do not supplement old but instead totally eradicate the function and economic viability of older technologies. The rapid change in cost-effective manufacturing technologies means that people who are negligent in maintaining their education will not slowly become less effective but can often find that their functional role in life is completely displaced by a new technology. In the past, the cost of education has often been borne by the employer of professional engineers. With project employment becoming a norm for many manufacturing engineers, the cost of continuing education will be borne by the individual and will take a substantial portion of available work time.

Prospective manufacturing engineers will have to be extremely careful about the curricula offered by the university-level institutes they attend and the relevancy of their curricula to a knowledge base that can be modified and that can produce skills in communications and problem solving which are increasingly seen as essential to personal professional development. Students must be careful about enrolling in educational institutes mired in strongly tenured vertically compartmented departments of engineering education. Instead, they must seek out institutions with strong capability in developing multidisciplinary teamwork and collaborative research. Manufacturing engineering is a growing professional function. It will offer challenging and rewarding education opportunity to those who have professional degrees in manufacturing engineering and also turn to graduates with degrees in other specific engineering disciplines. The evolving environment will probably, for the foreseeable future, have more professionals from fields outside manufacturing engineering than those with specific degrees in manufacturing engineering.

BIBLIOGRAPHY

This chapter quotes extensively from the following material developed under commission of the Society of Manufacturing Engineers.

"A Report to the American Society of Tool and Manufacturing Engineers Prepared by Arthur D. Little, Inc.," *The Manufacturing Engineer—Today and Tomorrow*, Dearborn, Mich., 1968.

Koska, Detlef K., and Joseph D. Romano, *Countdown to the Future: The Manufacturing Engineer in the 21st Century*, "A Research Study Conducted by A. T. Kearney, Inc., commissioned by the Society of Manufacturing Engineers," Dearborn Mich., 1988. Executive Summary; Issues and Implications; Questionnaire results.

Heytler, Peter G., *Countdown to the Future: The Manufacturing Engineer in the 21st Century*, Delphi Forecast, "The Delphi Forecast...Conducted by the University of Michigan, Industrial Development Division, as Part of the Profile 21 Research Study commissioned by the Society of Manufacturing Engineers," Dearborn, Mich., 1988.

CHAPTER 2
VALUE MANAGEMENT

R. Terry Hays
Vice President
Value Analysis, Incorporated
Newport Beach, California

Value management is a function-based analytical methodology used to improve businesses by identifying opportunities to remove unnecessary costs from their products and services while assuring that quality, reliability, product performance, and other critical factors meet customer expectations. The value management methodology can be applied to products, manufacturing processes, administrative procedures, organizational studies, and construction projects. Studies can be performed on existing projects, projects in process of being developed, or projects as they are being conceptualized. While many of the techniques used in value management can be applied by individuals, the most significant results are obtained through the application of value management by multidisciplined teams.

There are two essential aspects to the application of value management: (1) the unique method of function analysis must be applied, and (2) the organized job plan steps must be taken in proper sequence. It is these factors that differentiate value management from other analytical or problem-solving methodologies.

The terms value analysis (VA), value engineering (VE), and value control (VC), while having specific definitions, refer to the same basic methodology, and through general usage have become synonymous. Today, value management (VM) has become generally accepted as a term encompassing all the others.

Studies performed on existing projects are termed value analysis. In VA studies much is known about the current situation, and the VA team is taking a "second look" at the project to improve the situation. Often some of the cost savings identified through VA studies are used to offset the cost of change.

Studies performed on projects during the development phase are termed value engineering. In VE studies some critical information is known about the project and the VE team is analyzing the project before significant cost has been expended to set up the manufacturing for the project, thus significantly reducing the cost of change. Often people think only new designs use VE; however, new manufacturing processes, procedures, and organizational studies can and should be subject to VE studies prior to implementation.

Value control studies are performed at the very onset of a project to help set cost and quality targets and to assure the project has long-term viability with the customer. Value control studies are often followed by VE studies as the design progresses to further optimize the design and proposed manufacturing process.

Regardless of the type of study, the VA techniques are basically the same. The major difference between the various types of studies is how you prepare for the study.

Because of significant cost savings that generally occur as a result of value management studies, VM is often associated with cost reduction. However, cost reduction and value management are distinctly different. Cost reduction activities are part-oriented. This usually means altering manufacturing methods, relaxing tolerances, thinning or changing material, and so on. Normally, this will produce savings without an alteration of the design concept. Value management is function-oriented and generally leads to new or refined concepts that perform needed functions more simply with higher quality and more economical manufacturing processes or construction techniques.

THE ORIGIN OF VALUE MANAGEMENT

The process of VM is not new. It was developed as a result of the effect of World War II when shortages of strategic materials forced the adoption of alternative materials and methods. After the war, when industry adjusted back to peacetime standards, there was a tendency to return to the original concepts. However, at General Electric (GE), it was discovered that in a number of cases the new methods resulted in less expensive and better performing products than the original designs.

The concept of an approach to improved cost effectiveness and quality was the brain child of Harry Erlicher, GE's vice-president of purchasing and traffic. He assigned Larry Miles to develop a methodology that would *cause* these changes to occur. Between 1947 and 1952, Larry Miles developed the basic techniques of what he originally called value analysis. Larry realized that analyzing *functions,* not parts or materials, was the key to improving product value and managing costs.

This was a completely new concept. Previously when a company wished to improve profits or increase sales, they tried, often successfully, to reduce the cost of an object. Too often this resulted in reducing quality, performance, or customer satisfaction at the same time. This is not meant to be a criticism of traditional cost reduction methods, but we must realize that there is a point where cost reduction ends and waste begins.

Today, value management is one of the most versatile and effective business improvement tools available to management. Value management techniques are in use around the world, not only in heavily industrialized countries like Japan, France, and Germany but also in countries such as Brazil, India, and Hungary, where it is helping the national economies to grow.

CONCEPTS OF VALUE

The objective of any VM study is to improve the value of whatever is being studied. Unfortunately, people have differing opinions as to what affects the value of a product. It is also important to avoid confusing cost with value. Added material, labor, or overhead increases cost, but not necessarily value. If added cost does not improve the ability to perform the necessary functions, value is lessened. Three basic elements provide a measure of value to the user. They are performance, cost, and delivery.

Performance.   Appropriate performance requires that the product or service have a predetermined level of quality, reliability, interchangeability, maintainability, producibility, and marketability.

Delivery.   The customer requires acceptable delivery, usually at a specific place within a given time.

Cost.   In most cases, the customer is looking for cost that is competitive with similar products or services provided by other suppliers.

The relationship of these three elements, as shown below, is important to satisfy the customer and maximize value. From this relationship it is easy to see that value can be improved by either improving performance or delivery or reducing cost. While most VA studies have specific objectives such as quality improvement, cost reduction, or improved manufacturing throughput, making decisions based on the value relationships assures that one element is not improved at the expense of another element. Most VA studies result in the improvement of several elements of value simultaneously.

$$\text{Value} = \frac{\text{performance and delivery}}{\text{cost}}$$

From this value relationship the following definition for value has evolved:

VALUE: The most cost-effective way to reliably accomplish a function which will meet quality expectations of the customer.

JOB PLAN

In value work, there is an organized approach which must be followed if significant results are to be obtained. This system is called the "job plan." The job plan is comprised of seven major phases. Key to the success of the value management approach is following these steps in sequence and avoiding the temptation to try to reach a conclusion before the project has been thoroughly understood and analyzed. Typically, the information phase is performed prior to a value management workshop, and the reporting and implementing phases after the workshop. Listed below are the steps of the job plan.

Information.   Preparation is the key to any study. Gathering and reviewing the appropriate information prior to starting a project provides a team with a basic level of understanding. Depending upon the type of study, the information required will vary slightly. Each type of study requires that customer needs and requirements be understood, specific projects and goals be defined, and current costs be gathered and organized. Also, the team members should agree upon and specify the scope of the study so they can properly focus their efforts.

Function Analysis.   Function analysis is the heart of the value management methodology. The ultimate objective of function analysis is to identify the functions that are not providing good value and need to be improved. There are three steps in function analysis. The first is simply defining the functions occurring in the project and categorizing the type of function. Next, the functions are organized into a FAST diagram (function analysis system technique). The FAST diagram shows how the functions relate to each other and provides the team with a visual image of these function relationships. The final step of function analysis is developing cost-function relationships. This is accomplished by assigning part or process costs to the various functions. Through this process, the team can analyze which functions have excessive cost and do not provide good value to the customer. Typically, in a project three or four functions stand out as requiring improvement, either from the cost-function relationships that were developed or through the identification of a function(s) as a root cause of a quality, reliability, efficiency, or other problem.

Speculation.   In the speculation phase of the job plan, each function that was identified as needing improvement during the function analysis phase has its own creative session. During these creative sessions, any idea that can be associated with that function is recorded so that it may be evaluated at a later time. Typically, this is done utilizing brainstorming techniques and identifying numerous ideas on each function selected. Key to a successful creative session is to avoid the evaluation of the ideas during the creative process.

Evaluation. The evaluation phase consists of four distinct steps that help you to (1) eliminate some of the nonsense ideas that occurred during the brainstorming sessions, (2) organize the ideas into logical groupings so that as you evaluate the ideas you are focused in one area at a time, (3) evaluate the ideas, and (4) select the best ideas. The first two steps allow the detailed evaluation to occur more effectively. Step 3 is where most of the team's time will be spent in discussing, conceptualizing, and evaluating the ideas. The fourth step is an optional technique that is used when several competing ideas emerge from the evaluation process and a decision needs to be made as to which of these new concepts would provide the best value solution. At the end of the evaluation phase, the team has concepts that need to be further developed.

Development. Depending upon the type of study being conducted, the development phase may be completed as part of the workshop or as follow-up to the workshop. The objective of this phase is to develop the concepts identified during the evaluation phase into specific proposals. Before the team starts to develop the proposals, it is often necessary to establish an action plan to organize the team's efforts, as some of the work will be performed by individual team members and some by the team as a whole. With a good action plan in hand, the team can then develop and finalize each proposal.

Report. A value management study is concluded by a final report to the management group regarding the team's recommendations on the project studied. The objective of the management presentation is to get a decision on the team's proposals, so that any funding or action required to implement those proposals can be integrated into the company's plans. The team is not expected to implement the change; rather, the change is implemented as any other change would be implemented in the organization. Owing to the nature of their responsibilities, some team members may be part of the implementation effort.

Implement and Audit. While the implementation and auditing portions are not part of a value management study, they are crucial to total value management. As the objective of the study was to develop proposals that would improve upon the cost effectiveness, quality, and efficiency of the organization, nothing has been accomplished until the change has been implemented and the benefits received. Having a process to track the implementation status of the projects and audit the results improves the effectiveness of a company's value management effort, as it assures the implementation is completed in a timely fashion and verifies the final benefits.

INFORMATION PHASE TECHNIQUES

Value management projects are selected by the company's management staff. The rationale for selecting each project may vary slightly, but with each project management is looking to the VA team to provide specific recommendations that can improve the product. A clear understanding of what the project is and what situations management hopes to see improved should come first.

Gather Background Information. In order for the teams to thoroughly understand the project and be prepared to conduct the study, some pertinent data need to be gathered. Listed below is some of the key information that needs to be gathered and reviewed prior to any study.

1. A sample of the items to be studied.
2. The bill of material listing all the parts included in the project.
3. The assembly and part drawings.

4. The manufacturing routing for the items under study. These data should include detailed time and costs.

5. Marketing requirements, features, and annual or contract quantities.

6. Project specifications (material-manufacturing-purchasing-engineering).

7. Quality information (reject rate, scrap and rework costs, field problems, customer complaints, and the like).

Additional information beneficial for process and procedural studies is:

8. Any company policies and written procedures which define the present process or procedure.

9. Storage requirements for both raw material and finished goods inventory.

10. A layout showing the present flow of material and equipment locations of areas under study.

11. A list of the capital equipment currently used in the process.

Understand Customer Needs and Requirements. A key part of the project background information is customer information. Understanding the project from the customer's point of view is important. If the product has been on the market for several years, it may be necessary to update the customer information. Are the customer's wants and needs the same as when the product was originally introduced? Have the customer's requirements changed? Are there new features the customer would like or must have? Are there any product concerns or problems? How does the competition differ from your product? These are all questions that should be asked prior to your value study. Many times the perceived wants and needs of your customer vary considerably from their actual requirements.

Establish Objectives and Goals. The basic objectives and goals of the team are usually provided to the team by the management group or project initiator. After review of the project background, and based on the individual team members' knowledge of the project, objectives may be clarified and/or added to by the team. While many value studies traditionally focus on cost improvement, other objectives regarding quality, performance, and delivery are often important considerations. It is important that the team members understand not only what they are studying, but why, if they are to make recommendations that can best improve their product.

Determine Scope. In order to solve any problem, the parameters of the study must be clearly defined. While a large amount of documentation is gathered for a study, it is important to agree upon what is included in the study and where the study stops, the interface points of the study, so the team can properly focus its efforts.

The scope of design projects is specified by listing the major components or subassemblies included in the study. Manufacturing process studies are specified by first identifying the endpoints of the study and the major activities between those endpoints. To further clarify the scope, the team often specifies what is not included in the study. Examples of how projects have been defined are shown in Figs. 2.1 and 2.2.

Understand Current Costs. Cost is one of the most misunderstood items in business today. The "cost" of the product under study may vary greatly depending upon who you ask and what level of cost they are accustomed to using. Is the cost fully burdened? Does it include profit? Is it just material cost? Does the cost information reflect estimates, or are they based on quotes or other actual information?

Cost visibility techniques are used to organize the cost information, provide the team with a better understanding of the current cost situation, and identify which elements are driving the cost. First, the appropriate level of cost for your project is determined. Most organizations today use material, labor, and a variable portion of overhead for design stud-

---

PROJECT: Steering Gear - Hydraulic Assist

OBJECTIVES: Reduce Cost and Maximize Commonality with other Rack & Pinion Gears

PROJECT TYPE: **Design,**
(Identify one)   **Manufacturing Process,**
    Administrative Procedure

SCOPE: Steering Gear Housing, Rack, Hydraulic Valve Assembly, Tie Rods, Bellows,
    Housing Cover, Guides and Bushings.

SCOPE DOES NOT INCLUDE: Steering Column Coupling or Attachments to the Vehicle

ANNUAL COST OF PROJECT: $ 29,490,750

Figure 2.1   Example of project definition for a design study.

PROJECT: Compressor Wheel - LS 1200 Turbine

OBJECTIVES: Reduce Cost and Improve Quality

PROJECT TYPE: **Design,**
(Identify one)   **Manufacturing Process,**
    Administrative Procedure

SCOPE: From the receipt of the Casting thru Machining and Final Inspection, packed and
    ready to ship to the Customer

SCOPE DOES NOT INCLUDE: Casting the part or Assembly into the Turbine

ANNUAL COST OF PROJECT: $ 395,000

Figure 2.2   Example of project definition for a manufacturing study.

ies. In manufacturing and administrative studies, the cost includes only labor and variable overhead. Material cost is considered a function of the design and is not included unless it relates to such things as scrap, excess required for manufacturing, and material consumed in testing. In administrative studies, the cost of forms or other consumable items may also be considered.

By developing the cost in this fashion the team can focus on the product and process cost that they can impact in the study. The format for organizing the cost information varies slightly between design and process and procedural studies.

Product Design Studies.   These studies use a cost visibility worksheet to organize the cost data. Figure 2.3 shows a typical product cost breakdown.

Manufacturing Processes and Administrative Procedure Studies.   These studies use a sequence flowchart to organize their cost information and to graphically detail the process or procedure as it currently exists. To develop the chart, list the operations down the left side of the form and identify those performing the tasks across the top. Indicate the flow of operations by placing circles in the appropriate locations; then insert the time per event, cost per event, elapsed time, and miscellaneous costs in the appropriate columns. This document provides the team with easy-to-understand, current information regarding the study project. An example of a sequence flowchart is shown in Fig. 2.4. Listed below are several important items to consider as the cost data are analyzed.

DETERMINE TOTAL COST: Based on the costing ground rules, determine the total cost for your product. A product that "sells" to your customer for $18.29 may have a

PROJECT : STEERING GEAR - HYDRAULIC ASSIST   TEAM NO. : _____11_____   DATE: __NOV 3__

MATERIAL COST: $ __60.09__   LABOR COST: $ __34.44__   VARIABLE OVERHEAD: $ __31.54__

UNIT COST: $ __131.07__   ANNUAL VOLUME: __225,000__   ANNUAL COST: $ __29,490,750__

IDENTIFICATION NUMBER	DESCRIPTION	QTY.	PURCH PARTS $	RAW MATERIAL $	DIRECT LABOR $	VARIABLE O/H $	TOTAL $
1732458	VALVE ASSEMBLY	1	3.95	5.22	7.98	6.38	23.53
2257483	HOUSING ASSEMBLY	1	14.66		11.78	11.78	35.86
7745356	TIE ROD ASSEMBLY	2	15.22	6.01	3.48	2.78	27.49
1733339	TUBE ASSEMBLY	1		2.26	2.78	2.22	7.26
9547345	RACK	1		2.59	6.77	5.42	14.78
1233445	BELLOWS	1	1.90				1.90
1436459	ROD BUSHING	4	3.21				3.21
1000576	RACK GUIDE	1	3.59				3.59
1743855	HOUSING COVER	1		.29	.50	.40	1.19
	FINAL ASSEMBLY				4.90	3.92	8.82
	PACKAGING		.35		.27	.22	.84
	SCRAP		.89	.30	1.25	1.00	3.44
	TOTAL		43.42	16.67	39.44	31.54	131.07

Figure 2.3   Example of design project cost worksheet.

PROJECT: __COMPRESSOR WHEEL__   DATE: __OCT 20__   SHEET _1_ OF _1_

ANNUAL VOLUME: __1250__ $ __316.00__ PER _1_   ANNUAL COST: __$395,000__

WHAT? / WHO?	REC DEPT	TURN DEPT	GRIND DEPT	INSP DEPT				EVENT TIME	COST	TOTAL TIME	MISC
RECEIVE & INSP RAW MAT'L	O							.72	36.00	6.0	
MACHINE HUB		O						.63	32.50	1.7	
GRIND HUB			O					.56	28.00	1.9	
GRIND BLADES			O					.45	22.50	2.4	
REMOVE BURRS & WASH			O					.24	12.00	1.3	
INSPECT				O				.76	38.00	1.2	
BALANCE			O					.82	41.00	1.7	
SPIN TEST			O					.22	11.00	2.1	
INSPECT				O				.45	17.75	.9	
SHOT BLAST SURF & INSP			O					.22	11.00	3.1	
HAND BLEND PROFILE			O					.13	6.25	.6	
X-RAY (3 VIEWS)				O				.07	8.25	.3	45.00
IDENTIFY WHEEL & PACK			O					.06	6.75	3.0	3.75
							TOTAL	5.33	267.50	26.2	48.75

Figure 2.4   Example of manufacturing project sequence flowchart.

total *design* cost (material, labor, and some portion of overhead) of $12.79 or a total *manufacturing* cost (labor, some portion of overhead, and miscellaneous costs) of $7.79, for the purpose of the value management study.

DETERMINE ANNUALIZED COST: In order to establish a base for determining cost improvements, calculate the annualized cost of the study item by multiplying the unit cost by the number of pieces produced per year. The annualized cost will be used to determine annualized savings in the following fashion:

Annual project cost (current) minus annual project cost (proposed)

= gross annual savings less cost of change = net first year savings

In most studies, projects that show a payback in less than 1 year are easily accepted by management for implementation.

FUNCTION ANALYSIS TECHNIQUES

Function analysis techniques are used in defining, analyzing, and understanding the functions of a project, how the functions relate to one another, and which functions require attention if the value of a project is to be improved.

Definition of Functions.   All functions are defined by two words—a verb and a noun. To state what something does in two words is sometimes difficult, but it helps to simplify terminology, improve communications, and create better understanding. When choosing the words that define function, make them as broad and generic as possible. Avoid words that predetermine the way the function should be performed.

For a design study, electric motors produce torque, light bulbs generate light, a steering gear transmits force, and calrods produce heat. In manufacturing process studies, a machining or casting process is designed to shape material, a material handling procedure is designed to deliver material, and an inspection process is to verify product. In administrative procedure studies, a payroll system is designed to distribute payroll, an inspection report procedure is designed to identify status, and a project approval procedure is designed to authorize change. Simple statements such as these ensure clarity of thought and enhance project understanding.

The selection of the noun is also important. Try to select a noun that can be expressed as a measurable parameter. For instance, a shaft transmits torque, not turns a pulley. We can quantify torque and answer the question, "How much torque?"

There are two major types of functions within the scope of a study item—basic and secondary. The *basic function* is the specific work that a product, process, or procedure is designed to accomplish. *Secondary functions* are the other functions that the device performs and are subordinate to the basic function. They support the basic function and allow the product, process, or procedure to work and sell. Secondary functions may be categorized as either required, aesthetic, or unwanted. Required secondary functions are necessary to allow the basic function to happen or happen better. Aesthetic secondary functions improve the appearance of the product and make it more desirable to the customer. Unwanted secondary functions are generally undesirable by-products of either the basic or other secondary functions and often require additional cost to minimize their impact.

As examples of the various categories of functions consider:

- *An Overhead Projector:* Its basic function is to project images. In addition, the overhead projector has many required secondary functions, such as convert energy, generate light, focus image, enlarge image, receive current, transmit current, and support weight.

Unwanted functions such as generate heat and generate noise and the aesthetic function of enhance decor also exist.

- *A Manufacturing Process:* Its basic function is to produce product. In addition, the manufacturing process has many secondary functions, such as generate shape, move material, attach components, inspect product, store material, protect product, set up equipment, and smooth surface. Generate scrap is an unwanted function that plagues most manufacturing processes. Aesthetic functions are generally not found in manufacturing processes.
- *A Hiring Procedure:* Its basic function is to fill vacancy. In addition, the hiring procedure has many secondary functions, such as create announcement, interview candidates, prepare requisition, conduct orientation, evaluate application, and select candidate. While administrative procedures may have unwanted functions, aesthetic functions are rare.

FAST Diagram.   FAST is an acronym for function analysis system technique. The FAST diagram is a powerful value management technique which (1) shows the specific relationships of all functions with respect to each other, (2) tests the validity of the functions under study, (3) helps identify missing functions, and (4) broadens the knowledge of all team members with respect to the project. At first glance, FAST appears to be similar to a PERT chart or a flowchart. However, the basic difference between FAST diagraming and these other techniques is that FAST is function-oriented and not time-oriented. Figure 2.5 shows the basic ground rules for developing a FAST diagram.

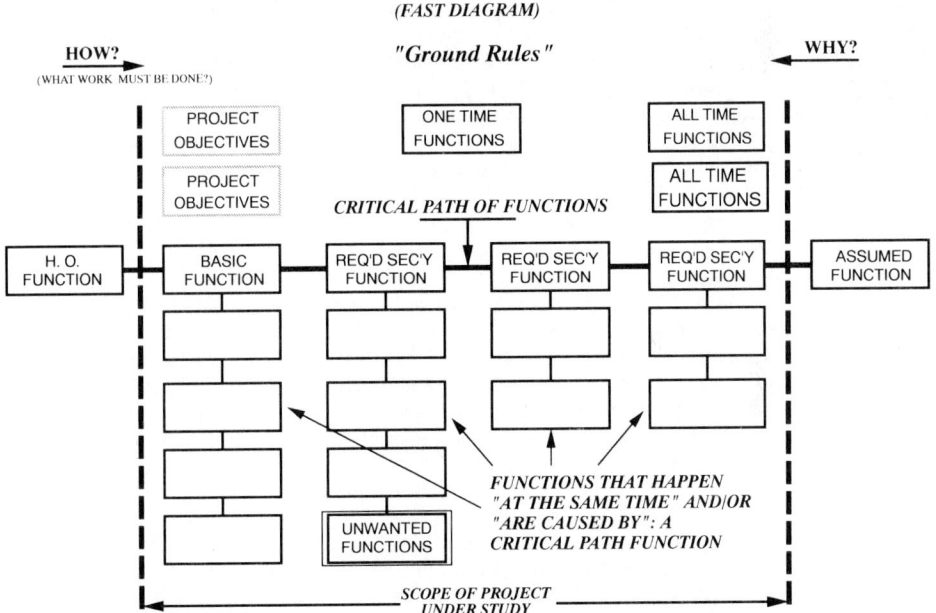

Figure 2.5   FAST diagram ground rules.

The most important aspect of the FAST diagram is constructing the critical path. This is accomplished by first placing the basic function just to the right of the left-hand scope line on the critical path. Then, ask the question, "Why is the basic function performed?" The function that answers this question is the higher-order function. This function is placed to the immediate left of the left scope line. In order to check the validity of the selection of the higher-order function, ask the question, "How is the higher-order function accomplished?" The answer should be the basic function.

All other functions on the critical path will lie to the right of the basic function. To determine the proper arrangement and relationships of the functions on the critical path, continue to ask the two basic logic test questions: How? Why? Using the "how" test, key on the basic function and ask the question, "How is the basic function accomplished?" The function answer should be placed to the immediate right of the basic function. The "how" question can also be asked, "What work must be done?" The second test of "Why" works in the opposite direction. Ask the question, "Why do I (verb) (noun)?" The answer should be the function to the immediate left on the critical path, in this case the basic function.

Continue building the critical path by asking "how-why" questions until you reach the right-hand scope line. The function which lies on the critical path and to the right of the right-hand scope line is called the "assumed function." The assumed function is outside of scope. The team must assume that this function is occurring properly and focus its attention on the functions that it can affect. Typically, the critical path between the scope lines includes no more than five functions. If there are more than five functions on the critical path, the team may have laid out a time sequence of functions and should reevaluate their "how-why" logic.

Once the critical path has been developed, the other secondary functions the product performs need to be distributed on the FAST diagram. If the unassigned functions "happen at the same time" and/or "are caused by" functions on the critical path, place these functions below the appropriate critical path function. If some of these functions are "unwanted secondary functions," they are highlighted by placing a double-lined box around that function.

When a function is directly related to more than one critical path function or happens "all the time," such as an aesthetic function, it is placed above the critical path functions in the upper right corner of the diagram. Functions not directly related to critical path functions that occur "one time," such as setup, assembly, and packaging functions, are placed in the middle of the diagram above the critical path. If there are specific "design objectives" the team wishes to keep in mind as the FAST diagram is constructed, place them above the basic function and place them in dotted boxes.

Cost-Function Relationships.   Cost-function relationships provide direction for the team related to areas of greatest opportunity. In other words, the team is able to identify which functions are not providing good value.

A cost-function worksheet is used to help develop this technique. The functions within scope are listed across the top of the form. Then the major cost groups (from the cost visibility worksheet or sequence flowchart) are listed down the left-hand side of the form with the associated incremental costs in the total cost column.

Next the function(s) impacted by each part or subassembly is identified. Once this is done, the team must estimate how much of the cost of each part or subassembly belongs to each function. This need not be a precise estimate.

Finally, all columns are added vertically to determine how much cost is allocated to each function. Typically, three or four functions will be responsible for 60 to 80 percent of the total cost. Figure 2.6 shows an example of a cost-function worksheet. The percentage of the total cost that each function represents is then transferred to the FAST diagram for further analysis. Examples of this are shown in Figs. 2.7 and 2.8.

Identify Functions Requiring Improvement.   Having completed this technique, the team is ready to key on specific functions and develop additional alternatives. Typically, less than 20 percent of the total cost is related to the basic function. However, this is what the customer wants to buy! The areas of potential cost improvement are generally obvious by simple review of the costed FAST diagram. It is equally obvious that several functions do not require any attention, as their cost contribution is relatively small.

Cost is not the only consideration used to determine functional areas needing improvement. Quality, reliability, customer satisfaction, and productivity are also critical criteria.

DIVIDE COSTS INTO FUNCTIONAL AREAS   **STEERING GEAR**   SHEET 1 OF 1

OPERATIONS, ASSEMBLIES, OR PARTS	TOTAL COST	FUNCTIONS (VERB-NOUN)											
		TRANS FORCE	CONV ENERGY	TRANS TORQUE	DIRECT FLUID	REC TORQUE	CONT PRESS	RESTR TRAVEL	DAMP SHOCK	ADJUST LENGTH	ALLOW ATTACH	SHIELD CONT	REC FLUID
VALVE ASSEMBLY	23.53		5.25	3.75	4.88	1.65	4.50				3.50		
HOUSING ASM.	35.86	20.45			11.85			1.80			0.80	0.10	0.86
TIE ROD ASSEMBLY	27.49	20.39						0.60		3.00	3.50		
TUBE ASSEMBLY	7.26	2.90	2.56								0.60	1.20	
RACK	14.78	5.33	9.45										
BELLOWS	1.90											1.90	
ROD BUSHING	3.21		2.90									0.31	
RACK GUIDE	3.59	2.75						0.70	0.14				
HOUSING COVER	1.19											1.19	
FINAL ASSEMBLY	8.82	1.75	2.20		1.37	1.80	0.20	0.20		0.25		0.15	0.90
SCRAP	3.44	0.15	1.93	0.75	0.05		0.18			0.30			0.08
TOTAL	131.07	53.72	24.29	4.50	18.15	3.45	4.88	3.30	0.14	3.55	8.40	4.85	1.84
%		41.0	18.5	3.4	13.8	2.6	3.7	2.5	0.1	2.7	6.4	3.7	1.4

Figure 2.6 Example of cost-function worksheet.

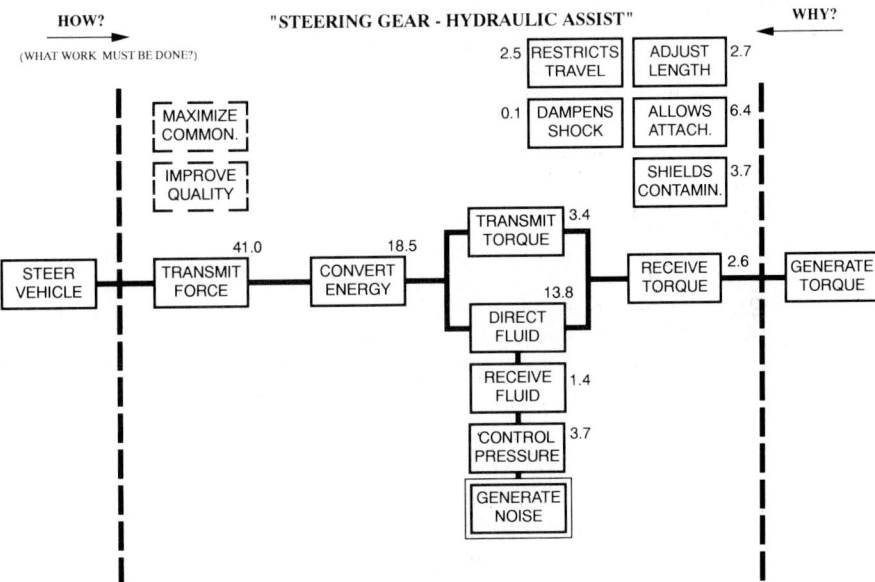

Figure 2.7 Example of design project costed FAST diagram.

Problems in any of these areas need to be related to the functions. The final decision regarding which functions the team should focus its attention on is based on a combination of the cost-function analysis, the relative importance of the functions as shown on the FAST diagram, functions that dominate in quality, reliability, or customer satisfaction concerns, and functions that are non-value-adding or unnecessary.

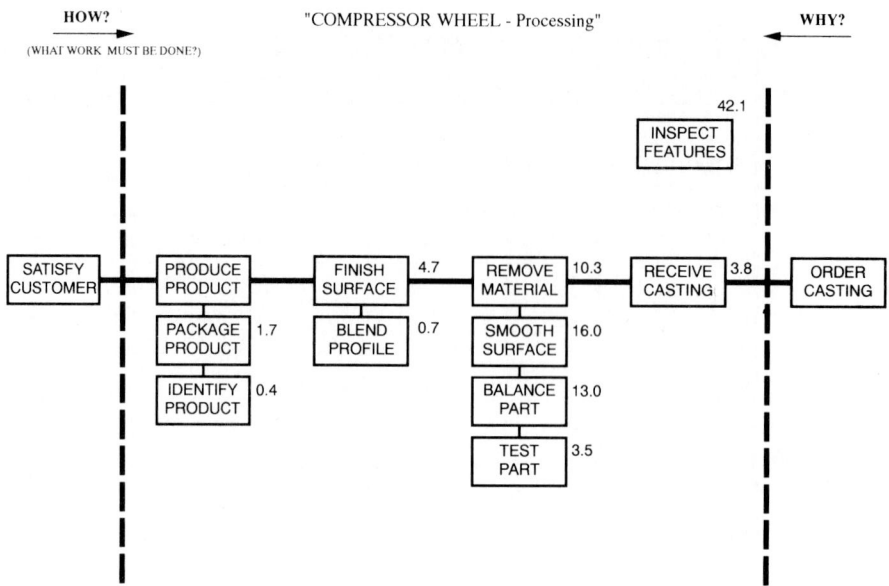

Figure 2.8 Example of manufacturing project costed FAST diagram.

SPECULATION

Several factors stimulate the need for creative processes. Every design is a compromise achieving only the best possible solution within a given time. New effects, new materials, new manufacturing processes, and new data communication technology are continuously being discovered and developed. Customer needs and desires are constantly changing. Thus creative opportunity exists and always will exist in every phase of industrial activity. The function analysis phase has provided the team with direction as to where new ideas need to be developed to improve the product or process. Now the team members must apply their creative ability.

An often misunderstood aspect of human behavior involves a person's ability to be creative. Many people believe creative ability is something you either do or do not possess. This is a misconception. The fact is that every person has the ability to be creative, but creativeness does not just happen. It must be given the opportunity to develop and grow. This will occur if problem understanding and hard work accompany a strong desire to succeed.

CREATIVE IDEAS ARE THE RESULT OF DILIGENT EFFORT: Creative people have often been portrayed receiving their ideas through some mysterious illumination or inspiration, sometimes referred to as a "flash of genius," when it is really the climax of exhaustive research. Thomas Edison is quoted as saying, "Genius is 99 percent perspiration and one percent inspiration."

Creative Thinking Defined.   Although the dictionary defines "create" as "causing to be or to come into existence," interpretations vary. For our purposes, we define *creative thinking* as:

A NEW COMBINATION OF THOUGHTS AND OF THINGS BASED UPON PREVIOUSLY ACQUIRED KNOWLEDGE, EXPOSURE, AND EXPERIENCES

Today there are tremendous opportunities for creative expression. Each new idea becomes an additional building block to be combined with any of the already existing blocks. Thus creativity is a regenerative process, with each new contribution permitting a whole series of additional combinations or creations.

The Creative Session.   The objective of any creative session is to generate a large quantity of ideas. By developing many ideas, you have the opportunity to select the idea(s) that best meet the criteria of your project.

During a creative session, the atmosphere must be open, positive, and receptive to the ideas being generated. Every idea needs to be verbalized and recorded. This may trigger another thought or idea in a teammate's mind and assures that no ideas will be lost during the idea-generation process. The fact that the idea may not be initially considered as a "solution" to the problem should not inhibit its inclusion on your list. The objective is to develop a long list of *ideas, not answers.*

To assure that you have not suppressed any ideas, defer judgment and evaluation. Not only must you refrain from judging ideas that are presented by others, but you must also refrain from judging your own ideas. Each idea serves one of two purposes: it is either *a potential solution* or *a stimulus for other ideas.*

Too often people will only suggest ideas that they consider as possible solutions. These "ideas" are generally not much more than the traditional answers to the problem. To reach beyond this myopic tendency, it is important that a minimum of 100 ideas be generated for each function selected.

Producing the quantity of ideas that is necessary requires the probing of the subconscious mind. The subconscious is stimulated by the association of ideas, which is the psychological basis of creative thinking. The basic principles of association are:

*SIMILARITY—or a like idea.*By this we mean that a bolt used to fasten parts might remind you of a screw, a rivet, or a spot weld; or a wire used to transmit current might cause you to think of a fiber or hairlike strand of material.

*CONTIGUITY—or adjoining idea.*By this we mean that a feather might cause you to think of a bird, a chair of a table, or an electrical wire of a connector, plug, or switch.

*CONTRASTING—or opposite idea.*Here, light might cause you to think of dark, hot of cold, short of tall, and high voltage of low voltage.

Creative Session Ground Rules.   Generating 100 ideas on any function or activity is made easier if you follow these four basic ground rules:

- Express the problem free from all specifications.
- Assume that every idea will work.
- Search for ideas with a competitive spirit.
- Encourage other team members to contribute their ideas.

In addition to these basic ground rules, your creative session will be even more productive if you remember to keep the creative session moving quickly. It should not take any more than 20 to 30 minutes to generate 100 ideas on any topic. If you exceed this time, any of the following situations could be occurring.

Judgment Is Occurring as the Ideas Are Being Presented.   Not only will this slow the process, but it will severely restrict others from contributing because of fear that their ideas will be instantly judged. This also impedes the spontaneity required during creative sessions.

Discussion of the Ideas as They Are Being Presented.   Too often team members try to explain their ideas rather than just state them. When discussion occurs during the creative session, evaluation of the ideas soon follows. The recorder should write down just enough of the idea so that the originator can recall the thought at a later time. The discussion of ideas should occur during the evaluation phase, not the speculative phase.

A Lack of "Off-the-Wall" Ideas. The creative process should be fun. Some of the seemingly "far-out" ideas are not only valuable as stimuli for additional ideas, but when combined with some more traditional ideas may lead the team to a new, truly innovative solution. If "off-the-wall" ideas are not being generated, the teams may need to relax and focus their thinking on ideas rather than answers.

Quality ideas will occur as a by-product of the quantity of ideas generated. By following the four ground rules for creative sessions, there should be no problem developing the quantity of ideas necessary to achieve the quality solutions ultimately desired.

EVALUATION TECHNIQUES

The purpose of the evaluation phase is to systematically reduce the large number of ideas generated during the speculation phase to a number of concepts that appear promising in meeting the project's objectives. During the evaluation phase, the obvious nonsense ideas that were developed during brainstorming sessions will be eliminated, the ideas will be organized into logical groupings and then analyzed with respect to project criteria, and the best combination of ideas will be identified.

Eliminate Nonsense. The first step in the process of transforming ideas to answers is to sort out and eliminate the "nonsense" and/or strictly "thought-provoking" ideas. These are the ideas that were generated during the speculation phase owing to the association of ideas, and while obviously nonapplicable, they may have contributed greatly to the creative process by stimulating thoughts that could lead to other, more meaningful ideas. However, caution must be used as the team starts eliminating the nonsense ideas, as it is too easy to remove ideas that may initially sound impractical but really could be developed into beneficial answers.

The team should slowly read each idea on the list. If any team member thinks the idea should be left on the list, it remains. Typically, from a list of 100 ideas, 40 to 60 will be left after this process.

Organize Ideas. The second step in the evaluation phase is to group the ideas surviving from the first step into similar categories. Ideally, having a number of small groupings of ideas helps to optimize the evaluation process and allows the team to focus on one specific area at a time.

Once the groupings are established, the team needs to determine which of the ideas within those groupings are independent ideas, and which of those are competitive. Independent ideas are those that can be implemented in conjunction with other ideas in that grouping. Frequently, several ideas from any grouping will ultimately result in recommendations. The remaining ideas in that grouping are competitive. Of the competitive ideas, only one or the other could be implemented; both ideas could not be implemented concurrently.

The competitive ideas in a group are typically evaluated first, to determine which of those ideas is best. Then the independent ideas are evaluated, one at a time, to determine which of those also will be beneficial. If there are more than five or six competitive ideas, it is often helpful to rank them based on which idea, if it would work, would be the most cost-effective. This does not include cost of change, only recurring manufacturing costs. This is often difficult, as the information at this point is very preliminary. However, the team is not trying to do a cost estimate, but merely evaluate cost on a relative basis to determine which ideas would be the lower-cost and which ideas would be the higher-cost and how they would rank in between.

Having these priorities set enables the team to start with the potentially lowest-cost idea, evaluate it, and if indeed it appears that it would meet all the criteria and work, the team members must ask if they have found "the lowest-cost alternative to reliably accomplish the function." If this is the case, the team needs to question whether they need to analyze the other alternatives. This prioritizing can help reduce the number of ideas evaluated, because if an option is obviously more expensive and a lower-cost alternative will work and meet the customer's requirements, it is probably not necessary to spend time evaluating the higher-cost alternatives. Of course, if the first or second ideas are found unacceptable, the team continues to evaluate the ideas until it either finds one that is acceptable or eliminates all the ideas in this category.

Once all the ideas have been organized into logical groupings and ranked where necessary, the team is ready to start the evaluation of the ideas. The category or group of ideas the team should start the evaluation with may be obvious. Often the result of analyzing one particular grouping will have significant impact on the other groupings.

Evaluate Ideas. This step in the evaluation process consumes the most time and energy. With the ideas organized, the team can focus on one group at a time to find which ideas can best improve the current situation. The ideas in these groupings are only a few words. In order to evaluate the ideas, the team must first conceptualize what the idea is and how it could affect their project. Often sketches are done, prints are reviewed, or the idea is just discussed by the team to allow all members to thoroughly understand the concept.

In some cases, after the initial discussion it becomes obvious that the idea is worth pursuing and further discussion or evaluation at this time would not be beneficial. The team identifies the idea as a "keeper," continues to the next idea, and repeats the process. If the idea is not an obvious "keeper," the team uses the knowledge gained through their discussion and develops a T-chart to aid in the evaluation of the concept (see Fig. 2.9).

+	-	What can be done?
Eliminate Chips, tabs and screws	Increased tooling lead time	Running change, with vendors, tolerable
Controls gap	Adds dimensional tolerance on pad	Vendor control, mold tools, OK
Matches vertical locations	Added gauge requirement	Rework gauge, OK for next model
Improves durability	Modifies nozzle tooling	Release by Eng., OK for next model
Reduces assembly time		
Improves service-ability		
Improves alignment		
Reduces squeaks and rattles		
Eliminate nozzle screws		
Snap-on nozzles		
Highest quality		

Figure 2.9 T-chart example.

The vertical line of the T separates the positive attributes of the concept from the apparent negative features. First, list all the positive features of the concept in the left column of the T-chart. The first advantage should be obvious, such as low cost, high quality, or whatever was the basis of the ranking. The list may include not only ''technical'' benefits but such considerations as acceptability by the customer, similarity to known or current processes or designs in use, degree of risk, or ease of implementation.

Once the team has listed all the positive ideas they can think of relating to the concept, they must challenge the concept, identify the potential problems or concerns with the concept, and list any negative aspects on the right side of the T-chart. Again, caution must be used. Just because you list some negative or potential problem features, do not take the attitude that you will automatically eliminate the idea for consideration. Rather, analyze the idea to see if the negative features can be overcome or at least reduced so they can be made tolerable. Typically, the negative features will be of two types—people or technical. By looking at them closely, you may be able to improve the concepts. Many teams like to add a third column to the T-chart describing the solutions to correct the negative to positive, eliminate the negative, or at least reduce them until they are tolerable.

For one or two categories of ideas, T-charting presents a relatively straightforward approach to selecting the most appropriate ideas to develop into final recommendations. However, when there are several categories of ideas, the process becomes more complicated. The final ''best'' answer(s) may involve combining ideas from several categories. Assuming the process of T-charting selected the best idea(s) within each category, theoretically combining these choices should represent the best total answer. However, there may be a conflict between the categories. As the team reviews all the ideas that can potentially work from the various categories, it often becomes apparent that all the ideas do not combine into one package of recommendations. When this occurs, the team needs to combine the related ideas into one concept and evaluate these competing concepts using the evaluation matrix as described below.

The majority of value management studies do not require evaluation beyond the T-charting process. The T-charting process organizes the evaluation data and allows the team to decide which idea or group of ideas should be carried through to the development phase.

Select Best Alternatives.   Traditionally, choices between potential alternatives are selected on personal subjective mental weighting of the criteria and risk. When this happens, one particular criterion often overpowers the rest and the ultimate decision is often not the best decision for the company or customer. When deciding whether or not to implement change, management is faced with substantial investments of money and time. Thus management needs to evaluate the alternative with respect to all the criteria in order to make the proper decision.

In order to respond to these needs, a matrix analysis based on utility theory for selecting the best value alternatives was developed. It is a mathematics-oriented process which reduces much of the subjectivity of comparative analysis. Further, it is capable of combining benefits, evaluating resource mixes, and comparing potential alternatives for implementation. This matrix has been refined to incorporate the definition of value representing the ratio of performance, delivery, and cost.

The matrix is a very effective tool in evaluating alternatives, since diverse criteria such as reliability, customer needs, quality, and implementation requirements can all be incorporated into the analysis and recommendations. This approach can provide management much better information to aid in their final decision.

Determine Performance Criteria.   The initial step is to recognize and define needs. Too often, product and service decisions are based on what an organization can best design and produce rather than what the customer needs. That situation causes a value mismatch and will result in loss of business. Determining the performance criteria will require gathering information from throughout the organization and, most importantly, from the customer.

Determine Relative Importance of Each Criterion.   Two steps are important to the determination of the final outcome—the selection of criteria and the relative weighting (importance) of these criteria. Of these two, selection of criteria presents fewer problems. Various management sources can decide which criteria should be used for internal reasons, and market research can determine which factors and features the customer wants. The basic question is "How important is each criterion in relationship to the other?" For instance, engineering might state that reliability, weight, and performance are essential; manufacturing may want the decision based on producibility and quality; marketing may favor appearance, delivery, and maintainability, and so on. The essential task is to determine the relative importance of the factors. Is reliability twice as important as weight? Is delivery half as important as quality? Is producibility equal to appearance?

While many different approaches can be used to determine the weighting factor, the "paired comparison method" is one of the most effective. This method is based on the assumption that the simplest and least emotional decision considers only two criteria at a time and determines which is more important. In essence, it only requires an answer to "is criterion A more important than criterion B?" rather than a judgmental "how much more important is criterion A than B?" choice. By comparing criteria against each other in this fashion, the relative importance of each criterion is easily established. The paired comparison matrix (Fig. 2.10) is an effective way to record and tally the decisions.

		DECISION					TOTAL	%
RELIABILITY	A	A	A	D	A	A	4	27
SERVICABILITY	B	C	D	B	F		1	7
PRODUCABILITY	C	D	C	F			2	13
QUALITY	D	D	D				5	33
WEIGHT	E	F					0	0
EASE OF USE	F						3	20

Figure 2.10 Example of a paired comparison matrix.

In this example the team identified six criteria as important to the final decision. The letter in the box relates to the letter code of the criteria that the team considered the most important in the decision between the related pair of criteria. The number of times each criterion was selected is totaled by counting the number of times A, B, and C, and so on occurred, and entering that quantity in the TOTAL column. Note that each letter can occur in only one row and one column. Once the totals for each criterion have been determined, percentages can be calculated and the contribution that each criterion has to the total decision can easily be determined. One criterion will always be zero. This does not mean that the criterion is not important, rather that it is just the least important, and all alternatives must meet the minimum acceptable level for this criterion.

When using this technique, it is essential that the customer's opinion of criteria and relative importance be considered in the process. This will help the team avoid a value mismatch between the producer and the customer.

Rate Alternatives.   Once the relative importance of the criteria has been determined, each alternative is rated for each criterion. The most effective way to rate the alternatives is to rate each alternative for one criterion before proceeding to the next criterion. To accomplish this, a consistent scale of measure should be used. In practice, any scale will work as long as the same scale is used for all the criteria. An easy scale to use is 1 to 10, with 10 being excellent and 1 being poor.

Compute the Total Score for Each Alternative.   The total score for each alternative is computed by multiplying the weighting established times the rating by criterion for each

EVALUATIVE CRITERIA	WEIGHT OF IMPORTANCE	BEST ALTERNATIVES FROM 'T' CHART WORKSHEETS RATE FROM 10 (EXCELLENT) TO 1 (POOR)			
		IDEA 'A'	IDEA 'B'	IDEA 'C'	CURRENT METHOD
RELIABILITY (MTBF)	27	4 / 108	7 / 189	7 / 189	5 / 135
SERVICABILITY (MTTR)	7	5 / 35	7 / 49	9 / 63	8 / 56
PRODUCABILITY (TTP)	13	9 / 117	7 / 91	8 / 104	6 / 78
QUALITY (Rej Rate)	33	4 / 132	8 / 264	9 / 297	7 / 231
EASE OF USE	20	5 / 100	8 / 160	6 / 120	7 / 140
TOTAL WEIGHTED CRITERIA	100	492	753	773	640
COST - ESTIMATED		13.50	15.00	20.00	22.00
VALUE RATIO = $\dfrac{\text{CRITERIA}}{\text{COST}}$		36.44	50.20	38.65	29.09

Figure 2.11 Example of evaluation matrix.

alternative and summing the totals. Figure 2.11 illustrates the process using three alternatives and the current method.

Estimate the Cost of Each Alternative. Based on the organization's normal estimating practices, the predicted cost of each alternative is determined. Cost estimates for the alternatives are often developed from historical data or variations to the current design. Implementation cost is not included other than how it might affect the recurring manufacturing costs.

Compute the Value Ratio. Value can be determined by the relationship:

$$\text{Value} \approx \frac{\text{performance and delivery}}{\text{cost}}$$

Thus dividing the total weighted factors for each alternative by their estimated cost will provide a ratio indicating the best total value alternative.

Select the Alternative That Offers the Best Total Value. Although alternative A has the lowest cost and alternative C has the highest criteria total, alternative B offers the best total value based on the ratio of performance delivery factors to cost.

Use of the evaluation matrix procedure provides an easily understood and administered process for orderly decision making. This technique avoids the potential of one criterion dominating the others. Allowing cost to dominate the decision might result in alternative A's being selected, or if quality dominated the decision alternative C might be selected.

Neither of these decisions represents the best overall decision for the company or the customer. This method not only reveals strengths and weaknesses of each alternative but also shows where improvements in individual factors could significantly impact the final decision.

DEVELOPMENT

The core of the value management study includes the function analysis, speculation, and evaluation phases of the job plan. At the completion of the evaluation phase, the team has concepts that need to be developed into proposals. The development phase consists of two steps: planning and execution.

Planning.   At this phase of the value management process, a number of promising concepts have been identified. It is now the responsibility of the team to identify the steps necessary to validate their concepts. The acceptance and implementation of the ultimate proposals will not occur without establishing a solid plan. When developing an action plan, several major tasks must be considered.

What Needs to Be Done?   Identify the tasks necessary to develop the concepts into solid proposals. This might include:

- Obtaining piece cost estimates
- Obtaining tooling quotes
- Developing implementation costs
- Reviewing your ideas with other affected groups
- Developing a preliminary layout or drawing
- Fabricating samples
- Testing samples, processes, or procedures
- Verifying equipment availability
- Verifying concepts and their benefits

Each task should be identified and listed separately so that specific assignments can be made.

Who Should Be Assigned the Task?   Identify the person on the team that could best obtain the required information and assign the tasks accordingly. Try to avoid overloading any one member of your team. Remember, this is a team project.

When Should the Task Be Completed?   Establish deadlines for verifying your concepts within 4 to 6 weeks after completing the evaluation phase. Assign a due date to each task.

Plan Future Team Meetings.   Plan meetings on a regular basis to review the project status, update your action plan, and make new assignments. Establish a regular day, date, and time for these meetings.

Are There Any Possible Roadblocks?   Some of the potential changes may encounter obstructions as the ideas are being sold or discussed. By anticipating potential "roadblocks," we can deal with them effectively.

Execution.   During the execution of the action plan, the team members individually work on their assigned tasks. This work is typically integrated into the everyday activities of the team members.

The team should target to have the development phase completed in 4 to 6 weeks. To achieve this, the team members must maintain open communications with each other. This is accomplished through a series of weekly team meetings, which average 1 to 2 hours each. During these meetings, the team reviews project progress, updates action plans, pools information, documents and finalizes the proposals, develops an implementation plan, and prepares the management report.

Once the items on the action plan have been completed in the development phase, the team needs to develop activity and timing required for implementation. This information will further aid management in their final decision.

REPORTING

In the reporting phase, the team presents specific recommendations for change to their management group and requests that action be taken on these proposals. The management group hearing the recommendations should have the authority to make the necessary decisions. This is not just an informational report; it should initiate action. To accomplish this objective, the presentation must be carefully planned and prepared.

Report Planning.   When preparing your report, key on the specific objectives of your presentation: selling the ideas and obtaining management approval for them. To attain the acceptance for the recommendations, the team must first examine why the ideas are worth implementing. Emphasize the performance and delivery benefits first. Most of these points are already highlighted on the T-charts and idea evaluation matrix from the study and can easily be summarized. It is also important to address any concerns that may exist regarding the proposals and identify what can be done to either minimize or overcome these problems.

Once the improvements have been detailed, the effect on cost should be documented. The cost change should also be presented on an annualized basis. The cost to implement can then be subtracted to determine the net first year impact of the change.

Presentation Format.   When developing the presentation, it is important to make sure that you are providing a clear and concise picture of your proposals, why they should be implemented, who should be involved in the implementation, and the timing required. All this information should be conveyed in 10 to 12 minutes. This can be accomplished with a well-organized and well-delivered presentation. The presentation should be structured into three sections: the introduction, body, and conclusion.

Introduction.   In the introduction phase of your report the following questions should be answered: "What is the topic and scope of the project?," "Who was involved?," and "Why was this topic studied?" Prepare your audience for the proposals that they will be hearing by identifying the number of proposals and whether they are long-term or short-term. This information serves as a transition into the body of your presentation.

Body.   The body of your report contains specific recommendations your team wants management to act upon. Each proposal presented to management should include a discussion of WHAT, WHY, WHEN, and WHO. These four elements of a proposal are detailed below.

What is the proposal? Explain in sufficient detail so your plan can be clearly understood, but avoid getting too detailed, as this will tend to make the proposal confusing. You can always answer questions if additional information is needed.

Why should the proposal be accepted? Describe the improvements over the current method and state the effects such as quality, ease of assembly, reliability, throughput, safety, and communications. Once these benefits have clearly been established, the effect on cost should be presented. Here the annual cost change (change in cost per unit times the annual volume) should be presented. From the annual savings the cost of change is subtracted so that a first year effect of the proposal can be determined. It is also important that concerns surrounding the proposal be addressed at this point.

When can these value improvements be realized? Before management approves any proposal, they will want to know how long it will take for the improvements to be realized. Will it be 6 months or 1½ years? An implementation schedule for the proposal showing the steps necessary to make the change should be prepared.

Who should implement the proposal? In most cases, with the approval of the proposals the involvement of the value study team ends and the implementation is assigned to the line organization. The departments involved should be reviewed by management to evaluate laborpower requirements and set priorities.

In the body of the presentation you are selling the proposals. Identify the important features of the proposals. Do not dwell on the present situation. Discuss other products or services that may also be impacted beneficially by the team's proposals but were out of the scope of your study.

Conclusion. After all the proposals have been presented, the objective is to get management to initiate action. In the conclusion of the presentation summarize the impact of all proposals, then ask the management group for their decision.

Presenting the Recommendations. The selling of your proposals is often dependent on not only the facts of your study but how the facts are presented. A clear and concise report is much more effective than a long, wordy, detailed one. Details can be provided in a written report that supplements the presentation. The written report should be provided to management one week prior to the presentations to allow management to familiarize themselves with the study and be in a better position to make a decision.

IMPLEMENTATION

The objective of any value management study is to recommend change that will improve the project that was studied, but the objective of value management is to implement change and facilitate the continuous improvement necessary to assure the long-term success of any organization. To assure that study results are being implemented and maximized and that value management is being effectively applied, an implementation plan customized to the management style of the organization is necessary. The implementation of value management focuses on: (1) the recommendation of potential cost and quality improvements through the application of the function-oriented methodology of VA and VE; (2) the establishment of a procedure to realize the timely implementation of these improvements; and (3) the development of a capability to reapply the value methodology on a continuous basis. In other words, "to manage change and manage costs on purpose."

While each industry is different, there are several keys to success that an organization should consider when implementing value management:

1. The understanding, support, and involvement of all functional organizations within the company. This must be properly organized and structured to blend smoothly with the operating philosophy of management.

2. Selecting someone knowledgeable and experienced to guide the development of this critical function.

3. Educating a significant percentage of key personnel from all line and staff functions within the organization in the techniques of VM. This is the base of trained people around which VM is structured and developed.

Education. To compete successfully in the changing business environment, management must allow their people to work smarter. The education of key people in the discipline of VM is necessary if the techniques are to be successfully applied. The education is best provided on real projects, where multiple-disciplined teams study such areas as product designs, manufacturing processes, and administrative procedures. These training studies should not be conducted on "canned" projects, rather on carefully selected projects from within the organization. This provides not only a learning experience for the participants, but real improvements from the projects that were studied. The workshop is conducted under the guidance of qualified VM professionals. To be truly effective the workshop must deal with two major aspects: methodology and people.

Once people have learned the value management techniques, further studies can be conducted. These studies are generally guided by a value manager. The combination of employees educated in value management, a value manager, and a management team committed to VM decision making is recognized by many in VM and is the model from which they have patterned their value activities. Figure 2.12 shows how the VM process functions.

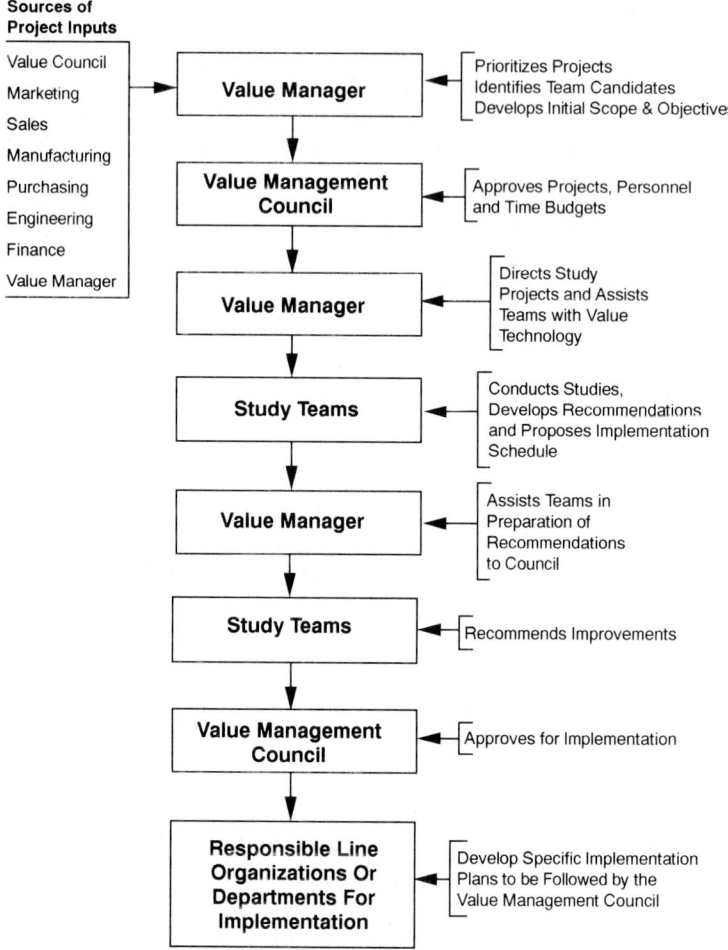

Figure 2.12 Value management project flowchart.

The Value Management Council.   The primary responsibility of this group is to provide the necessary guidance and direction to achieve the overall objectives of VM. The authority for decisions in planning, approving, and implementing projects rests with the value management council. Typically, this is the president or general manager and staff.

The Value Manager.   The value manager typically reports to one of the members of the value management council administratively but also reports monthly to the total value management council on matters relating to VM. This person is responsible for developing and maintaining VM in the company. In addition, the value manager keeps the VM council informed of all aspects of the VM progress.

Value Analysis–Value Engineering Study Teams.   Members of a typical value management team may include personnel from engineering, purchasing, reliability, production engineering, manufacturing, or finance. Team members are selected to best suit the individual projects and to make the best possible use of available personnel. Normally, a minimum of half of the members of the study team will have had formal VM training. The team is responsible to apply the VM techniques and recommend changes that will improve the project and the company.

Continuous Improvement.   Using the approach outlined above for implementing and applying value management in an organization, management can focus their people on the areas necessary to continually improve their business operations. Typically, companies actively using value management receive approximately $10 of cost improvement for every $1 invested in VM. Some companies have maintained ROI in excess of 30:1 for their VM efforts. While cost is not the only benefit received, it is the easiest to measure. In recent times, with the increased emphasis on quality, many companies have used the VM techniques as a major tool for accomplishing their quality objectives.

BIBLIOGRAPHY

Miles, Lawrence D., *Techniques of Value Analysis and Engineering,* 2d ed., McGraw-Hill, New York, 1961.

Osborn, Alex F., *Applied Imagination,* Scribner's, New York, 1963.

Society of American Value Engineers, *Adding Value with Value Management,* Society of American Value Engineers, Northbrook, Ill., 1990.

Society of American Value Engineers, *Proceedings of the International Conference,* Society of American Value Engineers, Northbrook, Ill., 1970–1990.

Value Analysis, Incorporated, *Value Analysis/Value Engineering Workshop Seminar Workbook,* Value Analysis, Incorporated, Newport Beach, Calif. 1990.

CHAPTER 3
PROCESS ANALYSIS AND OPERATION PLANNING

Joseph Metz
Vice President
Booz-Allen & Hamilton, Inc.
Beloit, Wisconsin

Joseph Peake
Senior Consultant
Booz-Allen & Hamilton, Inc.
Beloit, Wisconsin

A product is depicted by a drawing or package of drawings, pertinent specifications outlining the materials required, product configuration, and functional capabilities. The manufacturing activities necessary to accomplish the production of an end product must be processed or arranged in an orderly, workable sequence. This analysis and planning is the bridge between design engineering and product manufacturing. It encompasses every phase of industrial and manufacturing engineering by establishing a manufacturing plan which is economical and supplies a quality product.

Process analysis and operation planning are required for all manufactured products regardless of their size, material makeup, or type of construction. A fast-food hamburger, a computer chip, a bottle of perfume, and an airplane all follow process plans during their manufacturing process. An adequately defined operation plan can improve any task which adds value to a product with materials, laborpower, or equipment. The detail and complexity of the process plan will vary greatly from simple handwritten notes to detailed CAD/CAM-created prints and computer-generated procedures specifying parts, tools and equipment, gauges and their readings, and even which hand is employed for each movement.

This chapter presents the basic concepts of process analysis and operation planning for industrial applications. The main objective is to present techniques and a guide to generate effectively and efficiently the best operation plan for a product's current manufacturing life cycle.

FUNCTIONAL ORGANIZATION

Personnel and Duties. Personnel responsible for performing process analysis and operation planning will vary from plant to plant as will their duties. This is normally dependent upon the size and type of industrial activity.

Small Job Shops.   In small job shops, such as machine shops or tool and die businesses, quite often the shop foreperson is responsible for routing the parts to appropriate machines, utilizing the production workers' skill and a close relationship with them to eliminate detailed analysis and planning. Other than a job packet specifying quantities and containing drawings or sketches, no other data are furnished to the worker. Workers utilize tools, equipment, and methods they deem suitable to produce an acceptable part in a reasonable time frame. This works well with skilled production workers but is becoming risky with the current decline of the skilled work force.

Small- to Medium-sized Production Plants.   Small- and medium-sized plants, with product lines that have a relatively stable design and are manufactured repetitively, normally utilize a separate industrial or manufacturing engineering group to perform process analysis and operation planning.

Often the personnel will be former production workers who have a close knowledge of both the product and the current manufacturing equipment and practices. A drawback to this staffing practice is that the personnel may not be willing to learn about new techniques and therefore will not be receptive to incorporating changes in tooling, equipment, or methods. Process planning assignments vary from plant to plant. Normally, they include creating a detailed operation plan and applying work standards to the operations, which allows semiskilled workers to produce a quality part at a reasonable cost. Additional duties may include development of NC (numerical control) programs, tooling selection and design, and detailed work measurements, such as time studies or other standards development.

Large Production Plants.   Large plants with one or more repetitively manufactured products normally utilize specialists, each performing only a segment of the process analysis and operation planning. Personnel are generally a mix of former production workers and degreed industrial and manufacturing engineers. In some plants, an individual may be assigned to perform operation planning for only a specific process, such as primary machining or sheet metal fabrication. Additional associates will then create any remaining operation plans, apply standards, specify or design tooling, and create necessary NC programs, each according to their assigned specialty. The benefits of having specialists may be offset by the loss of continuity of process flow and the need for closer supervision. The more individual specialist groups utilized, the greater the problems of liaison with all other manufacturing support functions.

Organizational Level.   The process analysis and operation planning function is important because of its primary position in a company's ability to operate at a profit while producing a quality product. Personnel making the decisions of manufacturability, operation definition, work standards application, equipment allocation, and tooling design must be at a sufficient manufacturing hierarchy level to be able to enforce adherence to the defined plans. A plan which cannot be enforced or maintained leads to false manufacturing costs and incorrect allocations of resources. It is advantageous to have the supervisor reporting as part of the top tier of the plant or company organization. This ensures that top management has an awareness of the manufacturing effort and recognizes the importance of adhering to defined processes.

Organization Relationships.   The personnel performing process analysis and operation planning are in a pivotal position between product engineering and manufacturing operations. They receive product information from design engineering, interpret the data into terms of process requirements, and then create operation plans with work standards and tooling requirements for manufacturing to use. It is necessary for them to interrelate with a broad array of support personnel, including quality control, production control, plant engineering and maintenance, material handling, receiving and shipping, and industrial relations. A close association will be required between process engineering, tool design, and NC programming for efficient and effective methods and tooling to be developed. Therefore, the process engineer must be conversant with a wide variety of technical terms to adequately interpret and distribute information necessary to assist each support group in performing its functions. In actuality, the manufacturing or industrial engineer performing the analysis and planning serves as a coordinator among technical functions and between technical and administrative functions.

ANALYSIS AND PLANNING DATA REQUIRED

For anyone to perform a process analysis and create an operation plan, a certain amount of information must be available. Successful process engineers must be able to research existing and new techniques rapidly as the needs arise. They should constantly look for new approaches specifically applicable to their type of manufacturing. A wealth of reference material is available in the form of textbooks, technical papers, technical data supplied by the machine tool industry, and technical journals.

In-Plant Data.   The process engineer must have pertinent data about the plant and product to be able to specify operations appropriate for the available manufacturing equipment and production needs. This information should be provided and maintained on the user's computer system, which can be done quite economically. The amount of detail desired in the process analysis and operation plan will determine the volume and scope of data necessary. The following are examples of the type and detail of data required for effective operations plans.

Part Drawings and Bills of Material.   Detailed drawings of each component, subassembly, or assembly along with any corresponding bills of material and other specifications available are required for each analysis and plan to be completed. The drawings or specifications should list all dimensions, tolerances, surface finishes, and other general information about the part. Bills of material must specify all component parts and the required quantities that are used to build the assembly or subassembly. Based on known manufacturing parameters, changes may be suggested to allow manufacturing to make a more cost-effective and quality product.

Expected Product Life and Manufactured Quantities.   Both the anticipated life of a product and its expected magnitude of production have an effect on operations planning. A product with a short life span will probably be manufactured by a different method than a product with a long anticipated life span. Similarly, a part that is expected to be made only in minimal quantities will normally be produced on different equipment than those parts needed in large volumes. This is especially true if no similar parts are currently being produced to allow utilization of the same equipment.

Equipment List.   A current listing of production equipment with their associated workstation identifiers is necessary. This allows each operation task to be assigned with a specific work center for production loading and data collection. Equipment details such as size and tolerance capabilities, feed and speed ranges, and tooling available should also be shown.

Work Center Load Forecasts.   The sales forecast, exploded by workstation time required, can be used to guide process engineers to equipment with available open time and away from overloaded equipment. It can also be utilized to highlight areas where additional equipment is desirable before bottlenecks occur.

Material Specifications.   A number of material specifications are necessary to generate an operation plan. Material sizes, availability, and its normal stocked condition, such as bar stock, forging, or casting, are essential to ascertain preparatory or initial process operations. The "as stocked" hardness data determine if the operation plan has to allow for thermal treatments, make stock allowances, and perform special process operations. The specifications must also list qualified heat treatments and the resultant material conditions achieved. Tooling and tool grade selection depend on machinability ratings and material conditions, such as hardness and surface condition.

Speed and Feed Data.   Each company needs to develop speed and feed tables for use by personnel developing work standards or creating NC programs. These tables will be based upon proved values which produce quality parts at optimum production levels. Values from textbooks and vendor charts are normally rough averages which may not be what your plant can achieve in production. All types of processes and tooling utilized should be included in the data bank for best results.

Tooling and Gauging Standards and Inventory.   To prevent a proliferation of special nonstandard tools and gauges, it is necessary to standardize with functional but universal equipment. This does not prohibit special tools and gauges when they are essential for a manufacturing process. The list of standard equipment needs to specify application guidelines, with gauge information providing frequency of gauging recommendations. In addition to quantities, inventory data should include average life and frequency of use information. For gauges, certification records are also included. This information may not be used extensively in the preparation of the operation plans but will be required to maintain adequate tool and gauge supplies.

Work Measurement and Standard Data.   Personnel involved with applying time standards to operation plans must have standard data which are both current and of sufficient detail to allow development of accurate work standards. These work standards are normally used to determine product costs and to gauge worker efficiency. During the process development, these data can be utilized to verify that the most economical processes are being selected for equipment availability, production run quantities, and required deliveries.

Abbreviation Listing.   A manual listing acceptable abbreviations is indispensable in describing repetitive process terms in the shortest space. Without the manual, workers tend to create their own abbreviations, which may or may not be clear to everyone using the operation plan.

Scrap and Rework History.   This information, combined with statistical process control (SPC), can be used to determine if marginal processes are being utilized. If a specific process operation continually produces scrap or rework, a detailed analysis should be performed to determine the causes. During process operation development, this specific process may be avoided if alternatives are available. See Sec. 11, Chap. 3 for further information.

Cutting Fluid Applications.   Knowledge of cutting fluids and their application can be used to help produce optimum production rates. Improper specification may cause excessive tool failure rates, resulting in lost production time. Some product applications demand that certain chemicals not be used or applied during that part's manufacture.

ANALYSIS AND PLANNING STEPS AND CONSIDERATIONS

The sequence of events necessary to establish an effective manufacturing process varies from company to company and depends upon plant size and the product manufactured. Companies with large and specialized staffs normally require more steps and operate with greater detail. The following examples contain the most normally used sequences.

Preproduction Drawings.   Design drafters may not produce an initial drawing which is ready for production. They may have a limited manufacturing knowledge because their primary goal is to represent the functional requirements of a product by a series of detailed drawings of each component, subassembly, and assembly. The initial material selected may consider only structural needs, with no thought given to machinability. The net result is that the first release of drawings must be considered as preliminary or preproduction.

Manufacturing Feasibility Review.   Many organizations use a review board to assess manufacturing requirements and suggest design improvements. Regardless of whether a review board or a process engineer is utilized, all preproduction drawings should be analyzed for the following:

1. Are dimensioning and datum surfaces compatible with accepted manufacturing practices?
2. Are bills of material and casting or forging drawings available?
3. Are sufficient stock allowances provided on castings, forgings, and stampings to allow for anticipated mismatch or distortion in heat treatment?
4. Are sufficient clearances and access allowed for proper assembly of all components?
5. Are maximum allowable tolerances applied to nonfunctional characteristics?
6. Are tolerances and surface finishes on functional characteristics realistic, and is statistical tolerancing used where possible?
7. Are adequate clamping and locating surfaces provided for manufacturing?
8. Are value analysis suggestions for lower cost applicable? See Sec. 7, Chap. 2 for details.

The information on the above items is collected in the form of comments and is reviewed with the responsible design engineer. Acceptable suggestions or trade-offs are agreed upon and incorporated into each applicable drawing. these revised drawings become the production drawings.

Make or Buy Decisions.   The next step is to decide whether to make or buy. Information from design engineering, purchasing, production control, and processing is needed to formulate a qualified decision. Normally a team consisting of responsible individuals from each of these groups will prepare an analysis for deciding whether to produce an item in-plant, to purchase it complete, or to purchase it semifinished and complete it in the plant. Proper decisions are based upon true cost comparisons, in-plant workloads, lead time comparisons, and in-plant versus vendor capability. New equipment and processes being considered or purchased weigh heavily on the decision. A factual estimate of all parameters of producing each item in-plant must be presented. See Sec. 9, Chap. 6 for more information.

In-plant Production Considerations.   The two points to consider when preparing to establish process routings are the anticipated product life cycle and the required production quantities. If a product is to be made only for a short time or in limited quantities, it is more economical to use standard equipment, tooling, and methods wherever possible. However, if a long product life cycle or a high production volume is planned, then the time spent in selecting equipment, designing tooling, and choosing process methods is justifiable. A factor that can be used to offset low requirements and short life cycles is the ongoing production of one or more similar parts. Often the new product is a slight variation of one in current production. Equipment, tooling, and methods may well be interchangeable with only minor variations.

Production worker knowledge is another point to consider when attempting to develop operation plans. If the work force is stable and experienced, manual equipment can be utilized more effectively than if the work force is new and inexperienced. High production and specialized equipment can normally utilize an inexperienced work force more easily than manufacturing cells and conventional equipment. However, the use of NC equipment and knowledgeable programmers using qualified procedures can be used to improve the effectiveness of an inexperienced worker on more conventional machines.

Developing the Process.   The method of developing a process remains the same regardless of the variation in product nature, level of production, lead time allowed, and the like. The formal steps in this procedure are:

1. Create a general statement of the manufacturing operations to be performed.
2. Establish a provisional process.

3. Develop alternative processes.

4. Select a production process.

5. Communicate the selected process to other affected activities.

6. Perform detailed processing.

Each of these steps is not necessarily recorded on paper or entered into a computer. Many times they exist only in the mind of the personnel creating the operations plan. These steps or portions of them may also be worked on by more than one individual.

General Statement.   This step simply identifies the generic type of equipment needed to provide the design features of the product. Do the quantities dictate special high-production equipment or slower, but standard machines? Here experience, product knowledge, and existing similar process routings help in making the decision.

Provisional Process.   The in-plant data are used now to establish a provisional process consisting of operation sequences that will generate each design feature. Production quantities, required due dates, raw materials, equipment and tooling costs, and work standards are all considered in creating this process.

Alternative Processes.   Analysis of the provisional process may indicate a high manufacturing cost, excessive tooling cost, or areas where the confidence level of achieving the desired requirements is marginal. In such cases, it is advisable to develop alternative process methods for those operations presenting the problems.

Production Process.   By comparing the alternative processes with the provisional process, a compromise will emerge that will optimize the elements of cost, quality, flexibility, and risk. This will become the production process.

Communication of the Process.   With the production process selected, it now becomes necessary to communicate its requirements to other support areas, such as work standards, tool design, quality control, and NC programming. Hopefully, these groups have been involved in any critical decision making during the selection of the production process and are simply receiving a working copy.

Detailed Processing.   This step covers items such as process prints, tooling lists with recommended speeds and feeds, and other operation-specific information. The cost and effort required to complete all these elements prohibit their use in many plants, although segments may be done to assist production.

It must be remembered that the process is not complete until the specific tooling, machines, and time standards are defined.

Pilot Production Run.   The first production run of a part can be considered a pilot run, as any new, untried process may require some modifications. Ideally the person who created the operation plan will be responsible for ensuring the defined process is followed and making any necessary adjustments. It is imperative that the process be followed to the letter. If any approved deviations are not fully documented, it will be difficult to determine if they were caused by design or process problems. Adjustments should be performed only after a full analysis of the problem encountered has been concluded. The need to meet production schedules can cause additional labor to be added, which could have been avoided by proper analysis of the problem. For this reason, it is desirable to have some slack time allowed in any schedule which uses new and untried processes.

Process Review and Updates.   Once a production process has been proved, it might not be reviewed for a considerable period of time. New equipment, revised methods, and design engineering changes are generally the reasons operation plans need to be modified. Any process change after the first production run must consider the following:

1. Parts or assemblies in process and material on hand

2. Cost of the change, including effect on tooling, material, and delivery schedule

3. Anticipated savings or added cost

Although cost reduction is a constant philosophy of manufacturing, changes should be made with caution. Minor savings have been known to create problems and resultant losses. Too often in eagerness to show an improvement, consideration is not given to the cost of retraining, potential scrap and rework, and rebalancing machine loads and assembly lines. A pilot run may also have to be made to prove validity of the revised process. There is much to be said for maintaining stability in manufacturing.

PROCESS FORMAT AND CONTENT

Process format can vary greatly but tends to follow similar patterns in like industries. The document received by the production operator will depend on methods of duplication, distribution, and presentation.

Types of Forms.   Drawings with handwritten notes serve as a format in some job shops. Still pictures of elaborate operation setups or part assemblies are often utilized either to aid workers or even to be their only process document. At the other end of the spectrum are part files kept at each machine consisting of detailed process prints showing specific surfaces, dimensions, finishes, and tolerances to be machined for that defined operation step accompanied by computerized procedures detailing every movement made. All tooling and gauging will be designated with instructions for their usage. Audiovisual equipment such as video cameras and recorders are used to preclude the need for written documentation, particularly in assembly operations. Both still pictures and films can be easily and quickly changed by taking pictures or filming the new operation sequence.

Having details of each process printed, typed, or sketched onto master reproducible forms is still used by many companies. Figures 3.1 and 3.2 are examples of this method for assembly and fabrication processes.

This technique of developing a format and having preprinted master reproducible forms is giving way to the computer age. Currently the approach is to use a computer to store and reproduce the required process documentation. In this manner, the process engineer has remote access to all stored data. A format is developed and stored as a master, allowing the engineer to scan, change, add, or delete data by entering requests through a terminal and keyboard. The results may be viewed on the terminal screen or by requesting a printout from a high-speed printer. Figure 3.3 is an example of a computer process format. To generate a production order for the shop, a production control person simply has to enter an order number and the required quantities, then request documentation printout through the computer terminal

Process Content.   Every company has its own theories of how much or how little should be included in an operation plan or process document. For all but very small plants, the following items are normally mandatory for inclusion.

Part Identification.   A part number along with its description is necessary on all pages of a process document to identify the item as it proceeds through its production sequence. Normally the drawing number is used, as this aids in reducing the proliferation of numbers associated with a part and its production.

Process Revision Information.   Process routings like engineering drawings need to contain revision levels and a record of what has been changed. This information is necessary to maintain a running history of how parts have been made in the past. The finished product may require traceability not only to drawing revision level but also to the routing revision level.

Drawing Identification.   Where the part number is different from the drawing number or there are multiple drawings, it becomes necessary to include all drawing numbers on the first sheet of the process.

Bill of Material Information.   Bills of material are required for all subassembly and assembly processes. The required quantities and correlation of part identities between the

PART NUMBER [X]	OPER. NO. 40	PART NAME	PUMP	ASSEMBLY [X]	PAGE NO. 11	MODEL NO. 34F [X] 66-3W

SEQ. NO.	B.M. QTY NO.	PART NO.	PART DESCRIPTION	DESCRIPTION OF OPERATION
		647224	GUIDE ASM.	PRECLEAN & ULTRASONIC CLEAN IN FREON DEGREASER & PLACE IN WORK AREA
		472425	SIDE PLATE	
		648547	ROLLER BRG	
		648548	PIN	
1		647224	GUIDE ASM.	PICK UP GUIDE ASM & PLACE IN WORK AREA
4		872425	SIDE PLATE	PICK UP SIDE PLATES & ROLLER BRG & POSITION IN GUIDE ASM. & ASSEMBLE PIN
2		648547	ROLLER BRG	
1		648548	PIN	
4		872425	SIDE PLATE	PICK UP SIDE PLATES & ROLLER BRG & POSITION IN GUIDE ASM. & ASSEMBLE PIN
2		648547	ROLLER BRG	
1		648548	PIN	
2		890460-3	NUT	PICK UP NUTS & SCREWS & ASSEMBLE IN GUIDE ASM.
2		AN510C10-11	SCREW	
				PLACE GUIDE ASM. ON HOLDING FIXTURE
				PICK UP TORQUE WRENCH & TORQUE SCREWS TO 30-40 IN LBS-LAY ASIDE TORQUE WRENCH
				REMOVE GUIDE ASM. FROM FIXTURE & LAY ASIDE

FIGURE 3.1 Typical assembly operation detail sheet.

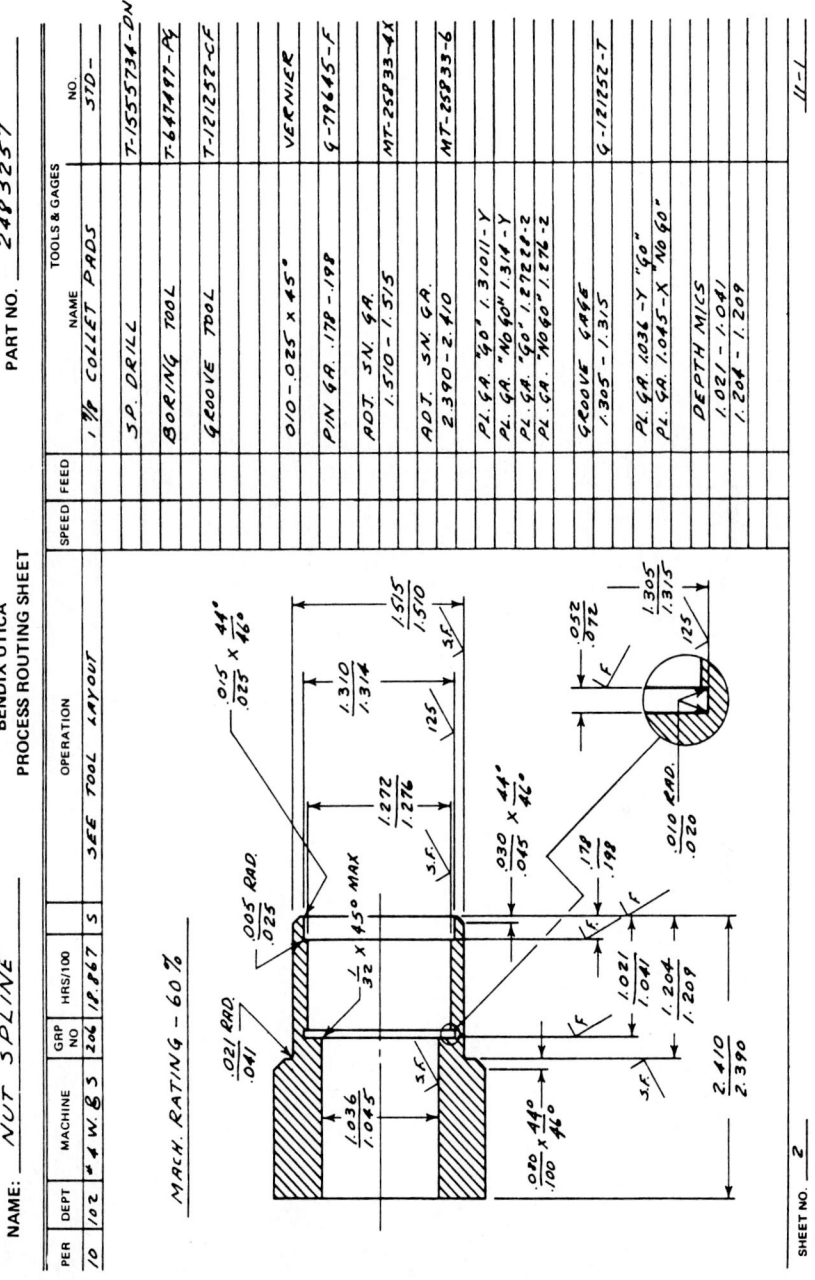

FIGURE 3.2 Typical fabrication process with pictorial view of operation.

SD1020 REQUESTOR : TJSILVA H A R D   R O U T I N G   R E P O R T PRIMARY 11.11.22 7/16/90 PAGE 1

PRODUCT	PRIMARY DESCRIPTION	SECONDARY DESCRIPTION	FLANNER	DAT/OBS	PUMP TYPE	LOT SIZE	FAMILY
9990304	SHAFT,MASTER ROUTE #4 (STEPPED	& TAPERED SHAFTS,HARD CHROME)		N N		1	

DRAWINGS REV/DESCRIPTION

MCN: NONE

E29'S:

COMPONENT ITEM

• •

WC#		SEQ #	OP.CODE	DESCRIPTION/COMMENT/TEXT	SETUP	RUN TIME
200	WAREHOUSE	10	099	- CUT TO LENGTH	.10	.10
035	E9	20	060	FACE TO LENGTH AND CENTER	.30	.42
021	E40 *NC*	30	099	- 1.TURN FOR GRIND, ALLOWING .C10 - .012" FOR FINISH 2.ROUGH TURN TO PRE-PLATING DIAMETER PER DRAWING. 3.FINISH THREAD AND/OR GROOVE, IF REQUIRED	.20	1.27
080	G1	40	099	- ROUGH GRIND TO PRE-PLATING DIAMETER PER DRAWING.	.20	3.00
057	M1	50	099	- MILL KEYWAY(S)	.20	.97
710	DIML INSP	60	070	INSPECT-TAG FOR CHROME PLATE	.00	.20
220	SUB-CONT	70	114	SUB-CONTRACT HARD CHROME PLATING	.00	.00
710	DIML INSP	80	073	INSPECT PLATING	.00	.20
080	G1	90	099	- FINISH GRIND OD AND DEBURR	.20	6.00
710	DIML INSP	100	053	INSPECT - MARK ITEM# AND CODE#	.00	.30
200	WAREHOUSE	110	059	STORE & RECORD LOCATION	.00	.00
					TOTAL SETUP 1.20	TOTAL LABOR 12.46

*** END OF REPORT ***

FIGURE 3.3 Typical computerized machining master route sheet.

drawings and routings is supplied with a bill of material. This may be a separate document, as is the drawing. If its identification number differs from the part number, it must also be specified on the process.

Material Identification.   The first sheet of a process should identify the material specification, configuration, and size. If a casting or forging, then their part numbers are required along with any previous heat treatment information. Where multiple parts are made from a certain size of raw material such as a bar, it may be necessary to specify the quantity of parts that will be produced.

Operation Number.   Each operation should be identified with a separate number. Consideration should be made to allow room in the numbering system for added operations.

Workstation Identity.   Each workstation requires its own individual identifying number. A workstation may be a machine, assembly station, or other work area. This number will be used for machine loading and data collection to calculate utilization. It is useful to have each workstation number linked to an operation-type identifier to associate all like workstations for machine loading. This extra number need not appear on the process document, although it can be helpful.

Operation Description.   For all operations, at least a basic description of the elements such as turn, face, drill, or mill is required along with the affected dimension. Other specific information such as tolerances, sequences, gauging procedures, and frequency may be included. Detail may extend to even specifying motion patterns as in the assembly process (see Fig. 3.1).

Production Time Standard.   Each operation should have a standard in allowed hours or minutes per piece or in hours per 100 pieces. It is also a good practice to show the standard and to include a standard for setup when applicable.

Tooling and Gauges Required.   While it is mandatory to identify only fixtures, special tools, and gauges, it is good practice to list all tools and gauges required for each operation (see Fig. 3.2). All operations routed to NC machines must have separate tool lists. This allows tooling personnel or the operator to obtain and set essential tooling and gauging in advance of actual needs and ensures that proper tooling will be used.

In larger plants or where conditions dictate, additional information may be needed for the operation plan. Even though these data are useful, the cost of documenting it may be prohibitive under many circumstances.

Speeds and Feeds to Use.   The speed and feed for each machining element may be predetermined and included on the process sheet. This is especially helpful when the work force is new and unskilled. It may even be advantageous to predict the number of cuts before a tool is to be changed because of wear. Operations routed to NC machines do not need to have speeds or feeds specified on the routing, as the NC program will contain appropriate values.

Tool Layouts.   A separate form may be developed for each machine using multiple tools, to define how the machine is to be set up. Machine tool builders have provided such forms for their machines in the past. Currently many plants with NC machines utilize tool matrices, whereby normally used standard tooling is resident in assigned machine pocket locations. Special or nonstandard tools are assigned to open pocket locations by the programmer during generation of the NC program. This information should be accompanied by drawings showing tool dimensions. These data are kept in catalogs with copies available in tool design, NC programming, and the tool presetting area. Normally each NC machine will have a catalog containing only the tool data specific to it.

Workplace Layout.   The optimum workplace layout can be shown. Normally reserved for assembly operations, it may be desirable for machining or other operations.

Process Operation Drawing.   A drawing of the part showing what surfaces are to be machined is helpful. The drawing can identify how to locate and grip the part as well as accurately pinpoint surfaces to be deburred. This separate document may be a fixture drawing showing the part and machine table relative to the fixture. Often NC programs are accompanied by drawings showing tool path movements in relation to the part configuration.

Handling Equipment.   Specialized lifting or handling equipment, their locating points, and instructions for their use may need to be defined for safety or quality reasons. Serious accidents have occurred by inappropriate use of standard equipment and failure to follow specified procedures.

Special Protection and Conditions.   The process may define proper part storage, oxidation protection, and cleaning requirements. Special storage containers can be necessary for a number of different reasons. Parts may require a plastic netting or oil-type coverage to prevent surface finish damage. The routing can also include cautions to the operator regarding potential dangers of chemicals when improperly used.

PROCESS CONTROLS

Well-managed manufacturing organizations establish strict process and manufacturing control procedures covering changes and allowable deviations. It is essential that these procedures be understood and simple enough that people will comply. To maintain control, adherence must be enforced. Consideration should be given to the following controls affecting operation planning and processing.

Request for Changes.   Processes should be considered essential blueprints for manufacturing. Only after requested changes have been properly reviewed for validity and approved by responsible supervision are any modifications made. Process change requests need to originate with the use of a form provided to responsible support personnel and supervision. The requester records the desired changes and the reasons for them. Process engineering evaluates the proposal, looking at its effects on quality, lead time, parts currently in production, equipment, tooling, and the like. Costs to initiate the change are weighed against the potential benefits and savings. Personally discussing the request with the originator is helpful in maintaining a close working relationship. Regardless of the decisions, the originator should receive documentation explaining the reasons for or against approval. If revisions are authorized, a sequence of steps and timing to institute the change must be established.

Alternate Methods.   The use of alternate processes or operations is normally considered poor practice. The tendency for them to become the rule rather than the exception increases production costs. Occasionally there is a justifiable reason to deviate from the standard process such as:

1. A part is obtained or manufactured with unacceptable deviations. To meet deliveries or because of cost, the part must be completed by the use of additional operations or alternate manufacturing methods.

2. A specified machine is not available or has an overload of work, and to meet schedules, other machines have to be substituted.

3. The absence of a skilled operator may require that multiple, less-skilled workers share in the workstation tasks to prevent bottlenecks.

4. Owing to limited lead times, requested tooling or equipment is not currently available and makeshift tooling or additional operations are necessary.

A form documenting the deviation and the recommended corrective action or procedure is prepared by the shop supervisor or support personnel. Process engineering verifies that the suggestions will correct the problem. Costs are calculated by appropriate departments, and depending upon the amount of cost, various levels of management need to approve alternate methods as a safeguard against their proliferation.

Disposition of Equipment and Tooling.   A form to request disposition of equipment or tooling is mandatory. It allows a thorough review to be made determining what process or product, if any, might be affected. Many decisions are made without consulting the process engineering section, resulting in disposal of tools or equipment essential for the manufacturing of current production parts. The originator of the request should be responsible for obtaining the various departmental approvals prior to disposition of the equipment.

AIDS AND SPECIAL TECHNIQUES

Group Technology (GT).   Consistency in processing can be significantly aided by utilizing group technology. The ability of obtain similar parts for use as reference will decrease processing time immensely by allowing the processor to make minor changes rather than developing a process from scratch. Master routings may be developed for part families to provide uniform methods by which parts can be manufactured. Comparable parts allow other functions such as design engineering, tool design, and NC programming to also perform their jobs in less time. Where a product has many variations, GT may even prevent duplicate designs. Using group technology, tool design can create tooling that will accommodate a family of parts, reducing the tooling cost per part. With the aid of GT, NC programming will be able to produce master programs capable of quickly generating similar tool paths for part families.

The classification system must be easy to understand and use, containing standard industry terms. The codes used should be based on visible attributes such as raw material, geometric shape, and manufacturing features. There are two types of code structure, hierarchical and chained.

The hierarchical code, also known as a monocode, is constructed so that each code digit is dependent on the preceding digit like branches of a tree. This type of code can be used for 90 percent of data retrieval needs in most organizations. It is preferred because it normally requires only four to eight digits to classify each part.

The second coding system, known as chained or polycode, is constructed such that each code digit signifies a set meaning. This requires that a position be reserved for each represented attribute, resulting in codes that may be 30 to 40 characters long. Although it provides easy search capability for a particular feature, the length of the code restricts its use.

Computer-Aided Process Planning (CAPP).   Most plants design parts utilizing CAD (computer-aided design), and many of them employ CAM (computer-aided manufacturing) to control their machine tools; however, a weak link has emerged between the two areas. This hurdle is the lack of computer-aided process planning. Process engineers receive CAD drawings containing all necessary manufacturing information from design engineering. They develop a routing using their experience and similar or master routings, employing a computer only for data retrieval and storage. Tool design creates needed tool drawings on the CAD system, utilizing the data of the CAD geometric part model. The routing is passed on to the NC programmer, who uses the computer system, the hard copy process, and the CAD part and tool drawings to create and load a program onto the CAM system for the machine tool. The part is now machined under computer control. It can be seen that the only person not fully utilizing the computer system is the process originator. The bridge over this hurdle is computer-aided process planning.

Computer-aided process planning is an expert system that captures the knowledge of individuals experienced in the manufacturing resources and practices of the company along with generic manufacturing engineering principles. Manual systems are dependent on people who will eventually retire and leave the organization, taking all their knowledge with them. CAPP will draw on the geometric model generated on the CAD system, match the characteristics and components of the part to the capabilities of equipment on the factory floor, and produce the process routing. There are two basic types of CAPP systems, variant and generative.

Variant Systems.   The variant approach is the most commonly used procedure today. This method establishes a set of standard process plans for all the part families that have been identified by group technology techniques. Each standard plan consists of the sequential set of manufacturing operations to complete a part, including work element descriptions, machine numbers, tooling, and work standards. These plans are stored on a computer system and are retrieved when appropriate for new or revised parts. To retrieve a plan, a part-family search is initiated using the applicable classification code. When a match is achieved, the standard plan is retrieved and the new part identities are entered. Additions, deletions, and changes may be made to any process field to obtain the specific operation requirements of the drawing. The process is electronically released for production once all modifications have been completed.

Generative Systems.   The generative approach, using the part's geometric model data and based on appropriate algorithms, makes the various technological decisions to generate a process plan without reference to previously created routings. Input from the process engineer is limited to monitoring and solving arbitrary decision conflicts. The computer database must contain the manufacturing logic, capacities of existing equipment, standards creation data, and the like. Because the decision logic may vary from industry to industry, the process parameters need to be modified for each company's manufacturing environment. Detailed hierarchical tree logic coding needs to be attached to every part's geometric model for effective process generation. Systems have been developed that allow design engineering to create a part's CAD model by defining surfaces according to the manufacturing method used to produce it. This places the responsibility of the manufacturing operation plan into the design engineering function. The process engineer and the design engineer need to work closely or become one person to effectively process the part as it is being designed. Few design engineers have the manufacturing knowledge and few process engineers have the design knowledge to perform this function efficiently.

With refinements in both coding technology and systems logic design, it appears the generative method may have the greatest growth and savings potential. Unfortunately it is not readily available for general industry usage at present.

Outside Processing Assistance.   It is common practice in industry to purchase outside tool design service. A well-trained tool designer has little difficulty performing to acceptable standards for a number of different companies. This is feasible because tool design practices are reasonably standard throughout industry. This is not the case, however, in processing and operation planning. The firm selling outside processing service must first learn the total capability of the plant to be serviced, including active machine inventory, standard tool inventory, conditions of machines and tools, material handling equipment, and the like. All in-plant accepted practices and drawing interpretations, as well as the skill level of the production operators, must be fully understood. When and if a need arises to consider outside processing assistance, careful, thorough selection and indoctrination of the outside firm will be required. It must also be realized that the outside firm may not be available to troubleshoot or make changes at later dates. Many companies cross-train process engineers and tool designers. When a peak load occurs, the tool designers are shifted to processing. Tool design rather than processing is subcontracted.

Tolerance Charts.   The process engineer must be able to establish in-process control to ensure that the end product will in fact meet the drawing dimensions within the tolerances specified. Simple parts with a limited number of machining operations present no serious problem. Complex parts that require a number of machining cuts involving rough, semifinish, and finish machining; heat treatment; or plating present a more difficult problem. Industrial scrap barrels have been filled over the years because of inadequate consideration of stock allowances for each operation required to produce a finished part. Haphazard allowances for distortion, shrinkage, or growth during heat-treating, and guessing at the effect of plating buildup have cost industry millions of dollars in delays, scrap, and rework.

One answer to this problem is the tolerance chart technique of manufacturing control.

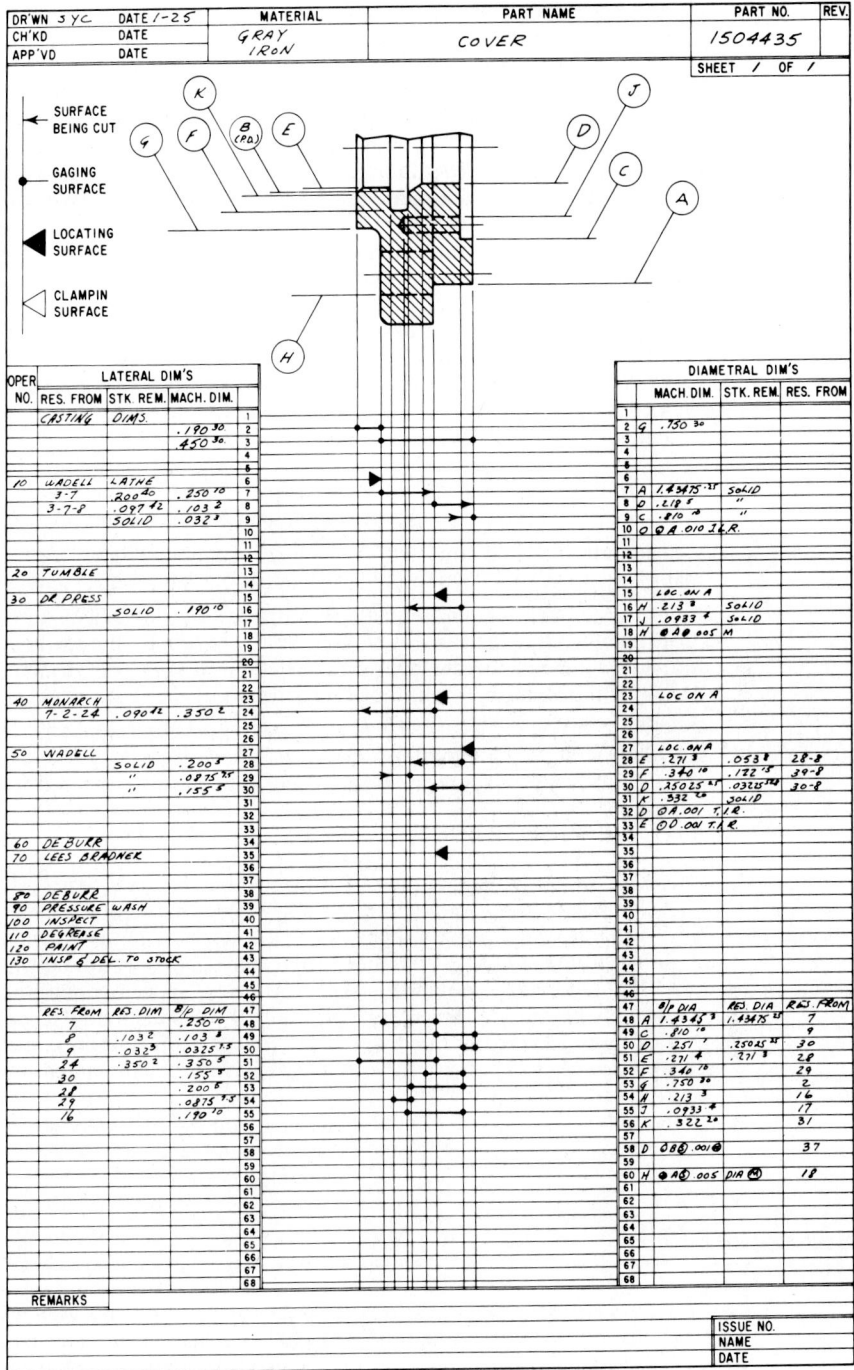

FIGURE 3.4 Typical tolerance chart for a machined part.

This technique involves the construction of a graphic breakdown of each plane involved, showing every dimension machined and the locating surface from which it is machined. All tolerances are shown as equal bilateral tolerances. This allows for adding or subtracting dimensions as required, while the mean tolerances are always added. A properly constructed tolerance chart will trace the stock removal of every operation, define the stock allowed for each subsequent operation, and accumulate the tolerances of every dimension machined (see Fig. 3.4). The tolerance chart is a tool which can be used by value engineering to effect sizable cost reductions. It will allow fast determination of where process dimensions are held tighter than is necessary. It will allow trade-off of tolerances so that the least expensive machining operations are applied. It can be used to establish statistical tolerancing and statistical process control (SPC), allowing tolerance relief with proper distribution control.

Line Balancing.   Any situation where a total manufacturing task is broken down into a series of operations which, by their nature, cannot be performed independently on a lot or batch basis requires a line balance analysis. The process engineer must arrive at the optimum approach, considering:

1. What portion of the assembly can be handled as subassemblies on an off-line basis?
2. What portion of the assembly must be handled on a continuous-line basis?
3. Considering the total time involved on the continuous-line portion of the task, how many operations are required to meet the scheduled requirements?
4. Can the total task be split equally into operations with reasonable break points? The problem of loading the early operations lightly to ensure that the units do not bottleneck in the beginning of the line must be considered. It is not unreasonable to have a differential of 10 percent between the allowed times of the first operation and the last operation.
5. At what point in the assembly sequence is it necessary to process inspection operations? This is largely dependent upon accessibility and expected frequency of rejects.

See Sec. 13, Chap. 5 for detailed information on line balancing.

Simulation.   An effective method of performing line balancing is to use computer simulation. A number of software packages are available that allow you to perform "what-if" scenarios to determine the best plans for your facility. Originally available only in a hard copy format from mainframe computers, personal computers are now capable of providing graphic simulation. By defining the part variables, such as quantities and cycle times, along with the planned systems, you can watch the products flow through the plant. Problem areas are highlighted or indicated as they occur. Adjustments can continue to be made via the computer until either all the problems have been eliminated or they have been reduced to acceptable levels. The ability to condense time frames through the use of the computer allows the testing of conditions that would take weeks to occur in real-life situations. See Sec. 12, Chap. 4 for more information on simulation.

BIBLIOGRAPHY

Group Technology at Work, ed., Society of Manufacturing Engineers, Dearborn, Mich., 1984.

Salvendy, Gaviel, *Handbook of Industrial Engineering,* Wiley, New York, 1982.

Tool and Manufacturing Engineers Handbook, desk edition, Society of Manufacturing Engineers, Dearborn, Mich., 1989.

CHAPTER 4
PROCESS CONTROL

Gregory L. Tonkay
Assistant Professor
Department of Industrial Engineering
Lehigh University
Bethlehem, Pennsylvania

Process control encompasses the principles and equipment used to control machines and processes in manufacturing environments. This chapter is divided into three sections. The first two sections discuss process control principles for two different types of manufacturing systems: (1) control of continuous processes and (2) control of discrete processes. The third section, programmable logic controllers (PLCs), discusses a physical implementation of the concepts presented in the first two sections.

CONTINUOUS PROCESS CONTROL

Fundamentals of Control Theory. Modern control theory deals with the principles and procedures used to control continuous processes. In this discussion, the equipment or process to be controlled will be referred to as the plant. In designing a control system, different and often conflicting goals must be considered. For example, sometimes the goal is to speed up the response of the plant so that the output of the system can respond more quickly and effectively to changes in the input. At other times, the goal is to maintain the output of the plant at a desired level while disturbances act to change the output.[1]

Open-Loop Control. Open-loop control is the simplest and most inexpensive method of control. A block diagram of an open-loop control system is shown in Fig. 4.1. In open-loop control the controller receives no feedback about the operation of the plant. Instead, a model is developed in the controller to predict the output of the plant under different conditions. One of the major difficulties with an open-loop control system is trying to design the controller to compensate for disturbances to the system. The controller must be able to accurately predict the effect of disturbances on the plant and then determine the desired signals to send to the plant to overcome the effects of the disturbances.

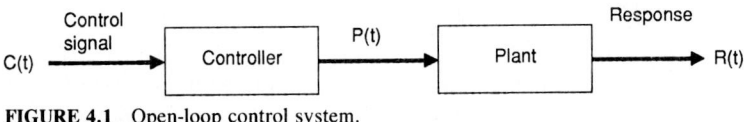

FIGURE 4.1 Open-loop control system.

An example of an open-loop control system is an educational robot that uses step motors to drive the joints. A step motor, also known as a stepper or stepping motor, is an electric motor that rotates a given angle—say 1.8 degrees (0.031415 radian)—whenever it receives a pulse from the controller. The resolution, or number of steps per revolution, varies with the manufacturer and model. The normal resolution is from less than 200 to over 50,000 steps per revolution, with the most common value being 200.[2] Hence the most common step angle is 1.8 degrees (0.031415 radian).

In order to move to a target position, the controller must first determine the angle of rotation required for the motor to reach the position and then calculate the corresponding number of pulses required to move the angle. In order to move at a constant velocity, the controller must send a series of pulses at a constant frequency to the motor. The acceleration and deceleration of the motor can be controlled by sending pulse trains that increase or decrease in frequency.

Disturbances to a step motor system might include a large countertorque applied to the motor by an overweight payload or collision with an obstacle. In an overweight payload situation, when a pulse is sent to the motor, the motor "slips" because it cannot overcome the inertia of the heavy payload. The controller assumes the motor has moved when in actuality it is in the same position. The controller cannot determine that an error has occurred, and all subsequent calculations will be in error by the number of slips that have occurred. If the controller had some feedback that told it when the motor slipped, it could try to correct the situation by sending extra pulses or alerting the operator that an error had occurred. This feedback to the controller would close the control loop and create a closed-loop system.

Closed-Loop Control.   By taking an open-loop control system and providing feedback to the controller about the response of the plant, a closed-loop system is created. A block diagram of a closed-loop control system is shown in Fig. 4.2. Typically, a closed-loop control system is more expensive because of the extra sensors required to provide the feedback. However, it is easier to design a closed-loop control system to compensate for disturbances because the controller can measure the response of the plant and modify the strategy if it is not achieving the desired results.

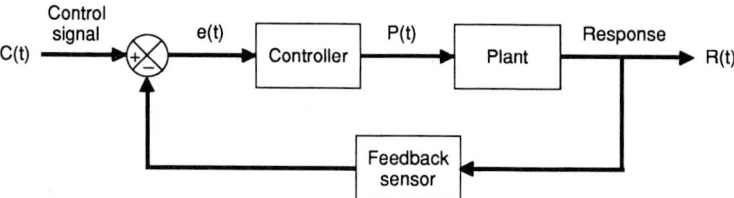

FIGURE 4.2   Closed-loop control system.

Types of Control Action.   Several distinct types of control action can be implemented to better control a process. Depending upon the characteristics of the process being controlled, the magnitude, or *gain,* of each of these actions can be independently set. The location of the controller can be seen in Fig. 4.2. The feedback signal taken from the output of the process is subtracted from the control signal or set point to form an error signal. This error signal is fed into the controllers discussed below. The output of the controller directly drives the plant or process.

Proportional Control.   With proportional control, the error signal is multiplied by a proportional gain constant. The gain could be greater or less than 1, depending on the magnitude of the error signal $e(t)$ compared with the plant input $P(t)$. The higher the gain, the faster the response of the plant to changing input conditions. However, too much gain can cause the system to become unstable. The equation for a proportional controller is

$$P(t) = K_p e(t)$$

where $P(t)$ = output of the controller
K_p = proportional gain constant
$e(t)$ = error signal input to the controller
t = time

With proportional control and direct feedback of the response, the controller will attempt to drive the plant to a state where system response $R(t)$ equals the control signal $C(t)$. When $R(t)$ exactly equals $C(t)$, the error will be zero and the controller will not attempt to drive the plant. Another characteristic of a proportional controller is that when the error signal $e(t)$ is large, the controller will attempt to drive the plant faster by sending a larger signal.

Integral Control.   With integral control, the error signal is integrated or summed over time to achieve an integrated error term. This error term is multiplied by an integral gain constant and fed to the plant as the control signal. The form of an integral controller is

$$P(t) = K_i \int_0^t e(t) \, dt$$

where K_i = integral gain constant

The primary purpose of integral control is to overcome the effects of sustained load disturbances such as friction in a mechanical system. Assume that a target tracking system is attempting to lock on a target that is at 90 degrees to a reference point. If the tracking system started at 0 degrees and only proportional control was available, $P(t)$ would be large as the system started to move. As the tracking system approached 90 degrees, the error signal would be reduced along with $P(t)$. However, because of friction, wind, or other forces acting on the tracking system, $P(t)$ might be too small to move the tracking system, thus causing it to stop moving and point at 88 degrees. A proportional controller would continue to send a control signal proportional to the 2-degree error but too small to overcome the disturbances, no matter how long the controller waited.

On the other hand, an integral controller would accumulate the error over time. As time went on, the integrated error term would increase, along with the control signal sent to the plant. Eventually, the integrated error term would be large enough to overcome the disturbances and move the tracking system to 90 degrees.

The disadvantage of integral control is that it tends to cause overshoot. If the integral error builds up too quickly or the gain constant is too large, it might cause the tracking system to swing beyond 90 degrees. This would cause a negative error that would eventually reduce the integral term to 0 and possibly even force the integral term to become negative. As the gain was increased, the magnitude of the oscillations would increase and the system would eventually become unstable.

Derivative Control.   With derivative control, the derivative or change in error signal is multiplied by a derivative gain constant and fed to the plant as the control signal. The form of the derivative controller is

$$P(t) = K_d \frac{de}{dt}$$

where K_d = derivative gain constant

Derivative control can provide the capability to respond more accurately to quickly changing signals. When a change is sensed in the error term, the output of the controller $P(t)$ is directly proportional to the amount of change. If the disturbances changed quickly, derivative control would quickly compensate, unlike integral control, which would take some time for the error to accumulate enough to move the plant. The disadvantage of derivative control is that if the error term is constant, no controller action is taken to change the output of the plant, even if the output is far from the requested input. For example, in the target tracking problem if the target remained at a constant bearing of 90 degrees and the disturbances remained constant, the error would not change and the controller would not attempt to move the tracking system from the zero position. For this reason, derivative control is rarely used by itself. Instead, it is combined with proportional or proportional and integral terms. Thus the proportional, integral, and derivative terms can be combined to form a PID controller with the form

$$P(t) = K_p e(t) + K_i \int_0^t e(t) \, dt + K_d \frac{de}{dt}$$

In physical implementations, sometimes the form is slightly changed so that the proportional gain is multiplied by all three terms

$$P(t) = K_p \left[e(t) + K_i \int_0^t e(t) \, dt + K_d \frac{de}{dt} \right]$$

In summary, in a PID controller the proportional term acts to force the error to 0, the integral term acts to overcome long-term disturbances that the proportional controller cannot overcome, and the derivative term acts to speed up the response in situations where the error term varies quickly. Further information on PID controllers can be found in several sources.[1,3,4]

Analog Control Loops.   Analog control loops are set up such that all signals in the control loop are analog. This implies that a digital device such as a computer cannot be part of the loop. However, computers are often used in the control loops to provide additional flexibility and capability. In these situations, analog-to-digital converters (ADCs) and digital-to-analog converters (DACs) must be used to interface the computer to the control system. Many sensors provide analog outputs and are easily integrated with analog control systems. Examples include resolvers, thermocouples, pH meters, tachometers, potentiometers, and strain gauges.

Digital Control Systems.   Digital control systems operate by passing all data in digital form. This eliminates the need for DACs and ADCs in the control system. However, it requires digital sensors or sensors that can convert their analog output to digital output. Examples of digital sensors include switches, encoders, and sensors that activate at predetermined levels of temperature or pressure without providing a measure of the variable, for instance, the oil pressure sensor which is connected to the dashboard warning light of a car. Any of the analog sensors previously discussed can be connected to ADCs to provide digital signals.

The PID algorithms that are implemented in a digital computer are similar to those presented previously except that time is changed from a continuous to a discrete variable. For digital systems, the sampling rate, or discrete time interval between samples, is an important factor. The sampling rate must be fast enough to ensure the control system does not miss changes in system variables. A discussion of digital PID control principles is found in Dorf and Bollinger.[3,5]

Combined Analog and Digital Control Systems.   Control systems can combine analog and digital elements so that digital signals are present in part of the loop and analog signals are present in the rest of the loop. Many of the control systems for robots and NC systems use this hybrid approach.

DISCRETE PROCESS CONTROL

Discrete process control deals with systems that process discrete information and parts. Generally, the information is binary; however, this is not a requirement. An example of discrete process control is a conveyor and packaging system for crayons. The machine dispenses the crayons, one of each color at the same time, into a box. Sensors detect if the crayons were damaged or the box was not filled. A conveyor carries the good boxes to a bundling unit and rejects the bad boxes for inspection and repackaging.

Sensors.   The sensors required for discrete process control are generally binary. Mechanical, optical, or proximity switches can detect the presence or absence of parts. Pushbutton switches can be used for operator input. Limit switches can detect position of equipment. Bar code readers can detect product ID codes or model numbers. Sensors are added to the process control design wherever errors are likely to occur or information is required for decision making.

Binary Logic.   Since most of the sensors used in discrete process control are binary, binary logic is a primary tool used by the system designer. Binary logic consists of the functions AND, OR, NOT, NAND, NOR, and XOR (exclusive OR). The truth table for these functions is shown in Table 4.1. Boolean algebra defines a list of identities, theorems, and proofs to manipulate logical expressions. A list of the basic identities is shown in Table 4.2. Further material on boolean algebra can be found in Booth and Greenfield.[6,7]

TABLE 4.1   Truth Table for Binary Logic Functions

		AND	OR	NOT	NAND	NOR	XOR
A	B	$A \cdot B$	$A + B$	$\overline{A}$	$\overline{AB}$	$\overline{A + B}$	$A \oplus B$
0	0	0	0	1	1	1	0
0	1	0	1	1	1	0	1
1	0	0	1	0	1	0	1
1	1	1	1	0	0	0	0

TABLE 4.2   Identities of Boolean Algebra

$$\overline{\overline{A}} = A$$
$$A \cdot \emptyset = \emptyset \cdot A = \emptyset \qquad 1 + A = A + 1 = 1$$
$$A \cdot 1 = 1 \cdot A = A \qquad 0 + A = A + 0 = A$$
$$A \cdot A = A \qquad A + A = A$$
$$A \cdot \overline{A} = 0 \qquad A + \overline{A} = 1$$
$$AB = BA \qquad A + B = B + A$$
$$A(A + B) = A \qquad A + AB = A$$
$$A(\overline{A} + B) = AB \qquad A + \overline{A}B = A + B$$
$$AB + A\overline{B} = A \qquad (A + B)(A + \overline{B}) = A$$
$$\overline{ABC} = \overline{A} + \overline{B} + \overline{C} \qquad A + B + C = \overline{\overline{A}\,\overline{B}\,\overline{C}}$$
$$AC + A'BC = AC + BC \qquad (A + C)(A' + B + C) = (A + C)(B + C)$$

Ladder Logic Diagrams.   Ladder logic diagrams provide a method to represent discrete process control logic. They were first developed for use with relay logic control circuits. The shell of a ladder logic diagram is shown in Fig. 4.3. The vertical lines on the left and right are the power lines to the system. Each horizontal line, or rung, is a separate circuit. Any rung that provides at least one complete path for current to flow will energize the output devices in that rung. By convention, the input devices are shown on the left and the output devices are shown on the right. Figure 4.4 shows the symbols of the most common elements found on ladder logic diagrams. Of these, the switch and the relay are two of the most important. Switches can be normally open, normally closed, or toggle switches that have no normal position.

The relay, often called a control relay, is an electromechanical device. An electrical schematic is shown in Fig. 4.5. The relay contains a coil which is connected in the ladder

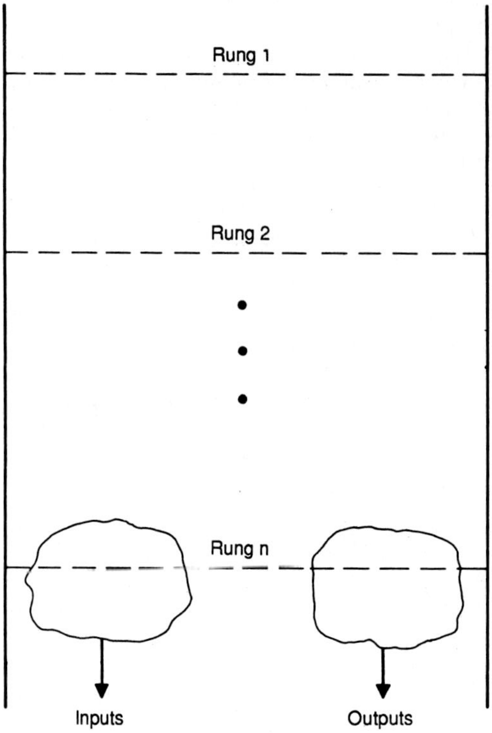

Inputs

——o⟋o—— Normally open switch

——o o—— Normally closed switch

——| |—— Normally open relay contacts

——⫫—— Normally closed relay contacts

FIGURE 4.3 Ladder logic diagram shell.

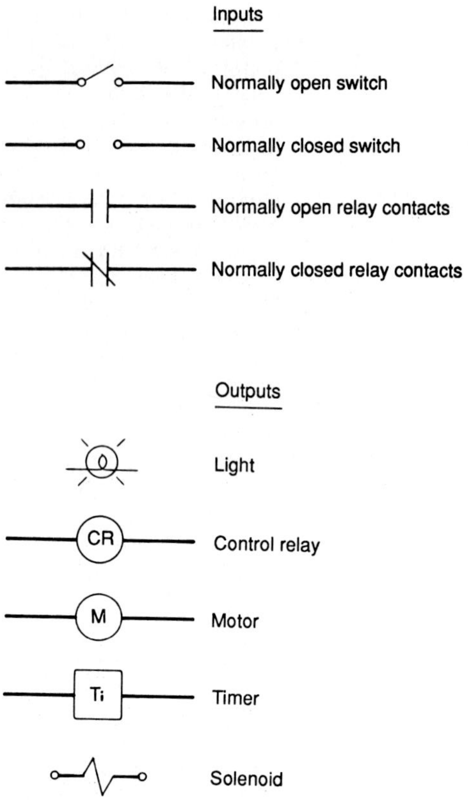

Outputs

Light

CR   Control relay

M   Motor

Tᵢ   Timer

Solenoid

FIGURE 4.4 Common symbols in ladder logic diagram.

Rung 1

Rung 2

Rung n

Inputs

Outputs

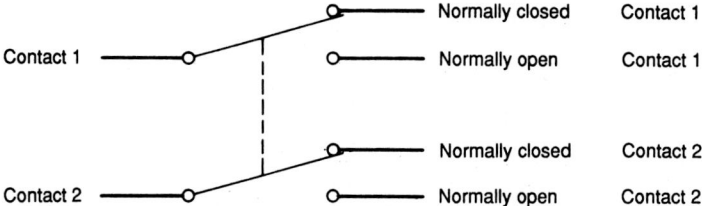

Contact 1 ——o   o—— Normally closed   Contact 1

                   o—— Normally open   Contact 1

Contact 2 ——o   o—— Normally closed   Contact 2

                   o—— Normally open   Contact 2

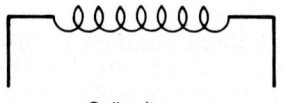

Coil voltage

FIGURE 4.5 Electrical schematic of a control relay.

as an output. When the coil is energized, the magnetic field it creates attracts a spring-loaded set of contacts. Depending on the relay, the contacts may close when the coil is activated or they may open. Generally, there are several sets of contacts that open or close at the same time. These sets of contacts are electrically isolated so that they can be used in different rungs of the ladder.

Generally, there is one output on a rung; however, there could be several. The inputs on a rung are grouped together to form logical expressions. Figure 4.6 shows the ladder logic representation for AND, OR, NOT, XOR, NAND, and NOR. Two contacts in series form a logical AND while two contacts in parallel form a logical OR. A normally closed contact represents a NOT function. To form a NAND or NOR requires two rungs of logic or simplification using boolean algebra theorems. The first rung performs the AND or OR function and controls a relay. The second rung inverts the output of the relay by using a normally closed set of contacts from the relay. Interconnections of these simple logic units can form rungs which represent complex logic.

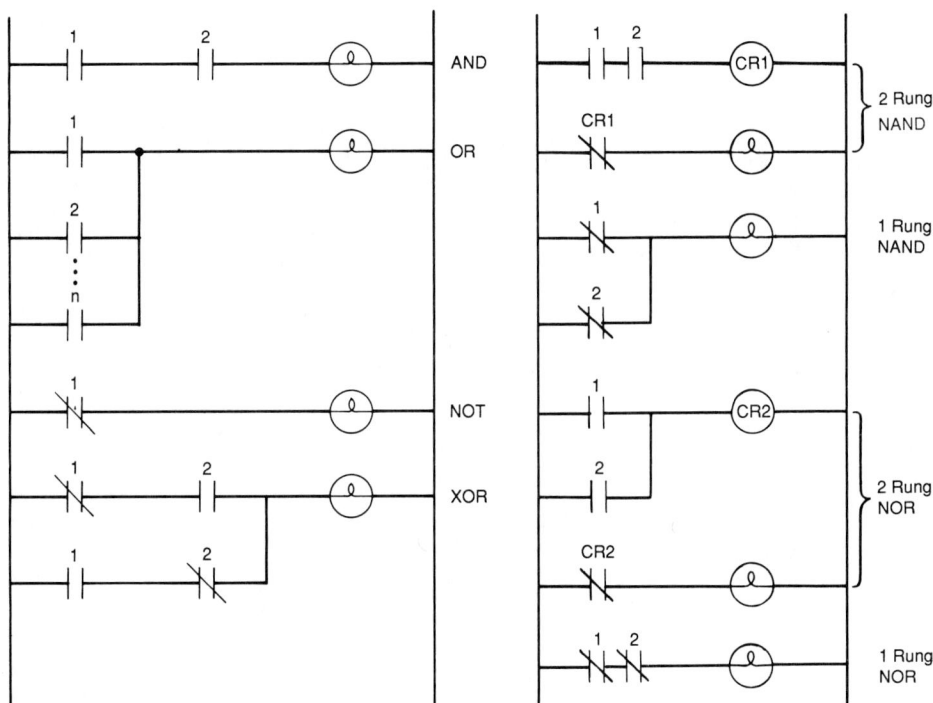

FIGURE 4.6 Ladder logic representation of logic functions.

Combinational Logic.   Combinational logic is a term given to the class of discrete control problems where the state of any output can be determined by combining the present state of other inputs and outputs in the system. In combinational logic, what has happened in the past does not matter; only what is presently happening matters. An example of a simple combinational logic problem is the control of a warning buzzer in a car. If the keys are left in the ignition and the driver opens the door, the buzzer will sound. The buzzer will also sound if the headlights are left on and the driver opens the door. The ladder logic diagram to represent this problem is shown in Fig. 4.7.

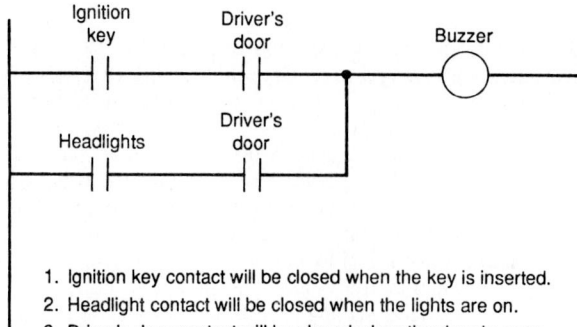

1. Ignition key contact will be closed when the key is inserted.
2. Headlight contact will be closed when the lights are on.
3. Driver's door contact will be closed when the door is open.

FIGURE 4.7 Ladder logic representation of car buzzer.

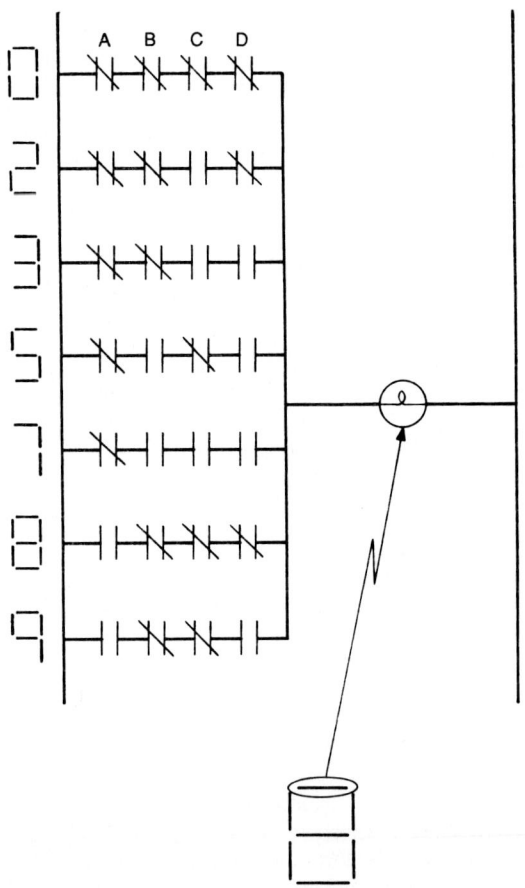

FIGURE 4.8 Ladder logic representation of top bar of seven-segment LED.

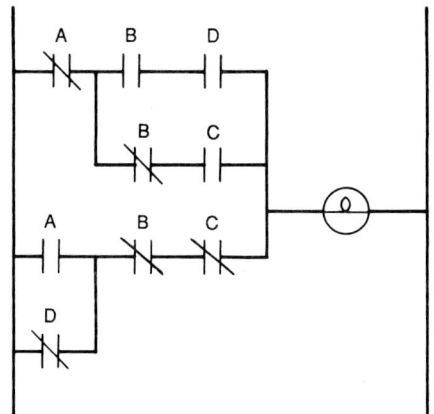

FIGURE 4.9 Simplified ladder logic representation of top bar of seven-segment LED.

A more complex combinational logic circuit is one which takes a four-bit binary number as four inputs (one input for each bit with "a" as the most significant bit and "d" the least significant bit) and attempts to light the appropriate bars on a seven-segment LED array similar to one found on most calculators. For simplicity, only the logic to control the top bar of the seven-segment LED array will be considered. Furthermore, only the binary combinations which represent the digits 0 to 9 are considered. The assumption will be made that the combinations which represent the hexadecimal numbers A to F will never occur. In this problem, the top bar should light for the numbers 0, 2, 3, 5, 7, 8, and 9. One straightforward method to solve this problem would be to provide a separate path to the output for each of the binary combinations which should light the bar. This solution is shown in Fig. 4.8. This circuit could be simplified using one of several techniques such as boolean algebra, karnaugh mapping, or the method of prime implicants.[6,8,9] Figure 4.9 shows an example of the circuit after it has been simplified using karnaugh mapping.

Sequential Logic. While some control problems fall into the category of combinational logic, many more fall into the category of sequential logic. Often the control system must know something about what has happened in the past in order to make correct decisions. Sequential logic circuits provide this capability. They use the present state of inputs and outputs in the system plus the states of memory devices that provide information about what has occurred in the past. If designers include memory devices in the control system, they must also provide the logic required to turn the memories on and off. This makes sequential logic circuits larger and more difficult to design than combinational logic circuits. In control circuits that use chip-level gate logic, flip-flops are usually used as memory devices. In relay control circuits, relays can be wired to act as memory devices.

Flip-Flops. There are many different types of flip-flops. However, they all have inputs to SET or turn the memory on, RESET to turn the memory off, and outputs to indicate th present state of the memory. Additionally, flip-flops are available with inputs to preset or clear the memory and synchronize the changing of the memory with other flip-flops and devices in the control system. A block diagram of the simplest type of flip-flop, the S-R (set-reset) flip-flop, is shown in Fig. 4.10. This flip-flop has two inputs: SET and RESET and two outputs: the state of the memory and the inverted state of the memory. The truth table for the S-R flip-flop is shown in Table 4.3. Notice that the output state is undefined when both the SET and RESET inputs are activated. Hence the control system logic should never allow both inputs to be activated at the same time.

Relays. Several circuits can be implemented to use a relay as a memory device. Two circuits are shown in Fig. 4.11.

In both circuits one set of relay contacts is connected as an input to keep the relay activated once it is SET. For the sake of this

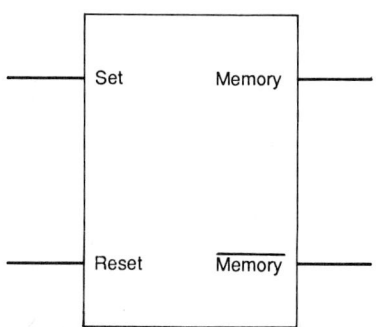

FIGURE 4.10 Block diagram of set-reset flip-flop.

TABLE 4.3   Truth Table for Set-Reset Flip-Flop

Set	Reset	M^{V+1}	$\overline{M}^{V+1}$
0	0	M^V	$\overline{M}^V$
0	1	0	1
1	0	1	0
1	1	Undefined	

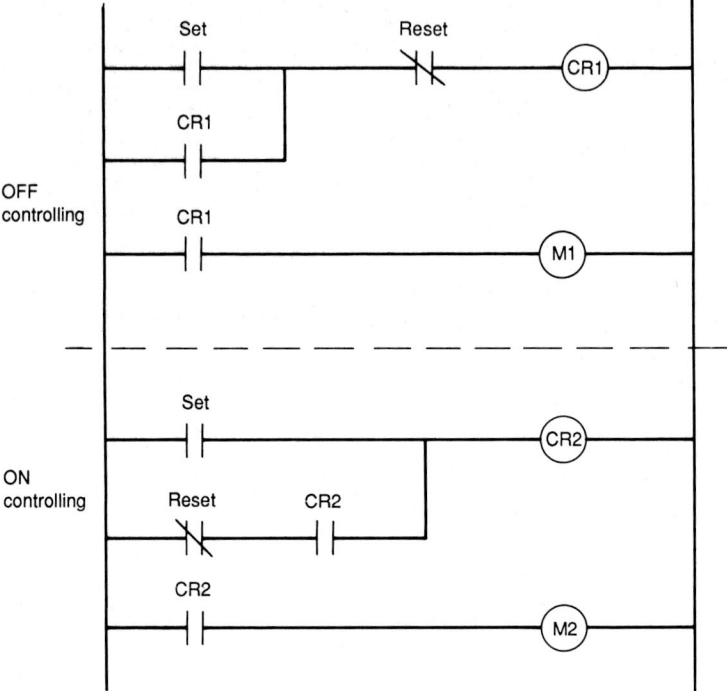

FIGURE 4.11   Relay memory circuits.

explanation, the relay is assumed to be RESET or off at the start. To turn the memory on, the SET input is activated, which completes a path through the rung and activates the relay. As soon as the relay is activated the relay contacts close and a second, parallel path is provided to keep the relay activated. At this time, if the SET input is removed, the second-ary path still exists through the contacts of the relay. Thus the relay remains energized and remembers that the SET input was previously activated. The relay will remain activated until the RESET input is activated. When this occurs, the path through the rung is broken and the relay is deactivated.

The difference in the two circuits shown in Fig. 4.11 lies in what happens when both the SET and RESET inputs are activated simultaneously. In the on-controlling circuit the relay will be activated. In the off-controlling circuit the relay will not be activated. Off-controlling memories are often used in motor control circuits utilizing separate push-button switches for the start and stop inputs. For safety reasons, if the operator hits both the start and stop buttons simultaneously, the relay turns off and stops the motor.

Sequential logic circuits are more difficult to design than combinational logic. If a pre-determined sequence occurs each cycle, the problem is somewhat simplified. For example, consider the punch-press operation (shown in Fig. 4.12) in which the inputs are defined as follows:

S   = 1 if two palm buttons are depressed, 0 otherwise
C   = 1 if clamp is extended (clamped), 0 otherwise
PT   = 1 if punch is at the top, 0 otherwise
PB   = 1 if punch is at the bottom, 0 otherwise

The outputs are defined as follows:

X   = 1 will extend the clamp, 0 will retract the clamp
Y   = 1 will extend the punch, 0 will retract the punch

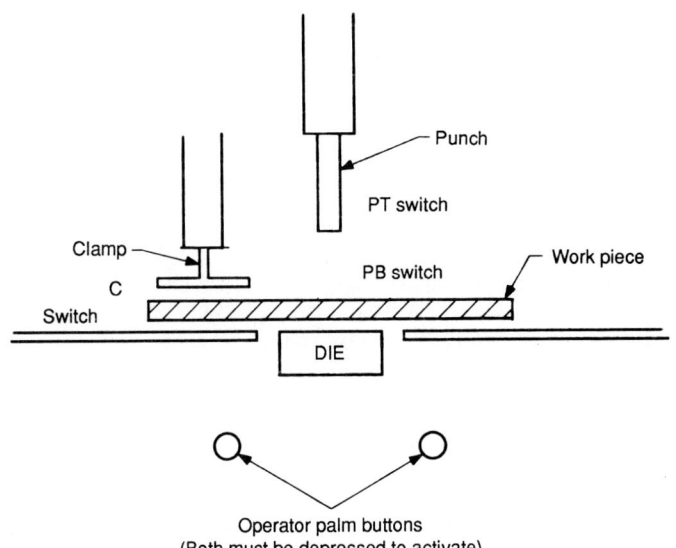

FIGURE 4.12   Punch-press example problem.

The predetermined normal sequence of events is shown in Table 4.4. The need for sequential logic arises because steps 4 and 6 require different outputs for the same inputs. The

TABLE 4.4   Normal Sequence of Events for Punch-Press Problem

	Inputs			Outputs		
Step	S	C	PT	PB	X	Y	Comments
1	0	0	1	0	0	0	Unclamped, punch retracted
2	1	0	1	0	1	0	Palm buttons pressed so clamp
3	1	1	1	0	1	1	Clamp extended so punch
4	1	1	0	0	1	1	Punch is extending
5	1	1	0	1	1	0	Punch extended so retract
6	1	1	0	0	1	0	Punch is retracting
7	1	1	1	0	0	0	Punch is retracted so unclamp
8	1	0	1	0	0	0	Done, wait for removal of hands
9	0	0	1	0	0	0	Same as step 1

same inputs in the first half of the cycle extend the punch, while in the second half they retract the punch. To remedy this problem, a memory device can be added to the system, the output of which can act as an extra input to eliminate the conflict between steps 4 and 6. However, additional logic is required to determine when to set and reset the memory device. Assuming the memory device starts the cycle in the reset state, the best place to set the memory device is between the steps where the conflict occurs (4 and 6) or at step 5. The modified sequence of events is shown in Table 4.5.

TABLE 4.5 Normal Sequence of Events Modified with a Memory Device to Eliminate Conflicts

	Inputs					Outputs			
Step	S	C	PT	PB	Mem	X	Y	Set	Reset
1	0	0	1	0	0	0	0	0	0
2	1	0	1	0	0	1	0	0	0
3	1	1	1	0	0	1	1	0	0
4	1	1	0	0	0	1	1	0	0
5	1	1	0	1	0	1	0	1	0
6	1	1	0	0	1	1	0	0	0
7	1	1	1	0	1	0	0	0	0
8	1	0	1	0	1	0	0	0	0
9	0	0	1	0	1	0	0	0	1

Using Table 4.5, an unsimplified ladder logic diagram can be easily constructed. A separate rung is required for each output, a total of 4 in this problem. The clamp, output X, should be activated during steps 2, 3, 4, 5, and 6. By ORing the input conditions in these steps, the rung is constructed as shown in Fig. 4.13. The same process can be applied to the other three outputs to form the complete ladder logic diagram shown in Fig. 4.14. Simplification techniques could be applied to this problem to find a better solution. In an actual control system, the designer would have to consider safety issues such as the occurrence of events not in the normal sequence. For example, if the punch was extending and the operator released the palm buttons, the punch should immediately retract. However, the control system should not instruct the clamp to retract before the punch is fully re-

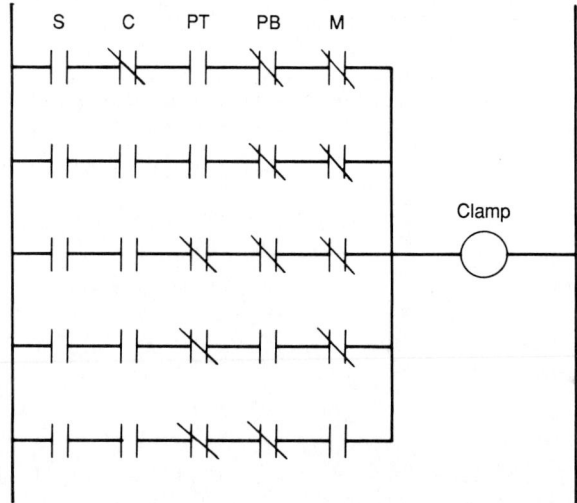

FIGURE 4.13 Rung that controls clamp in punch-press operation.

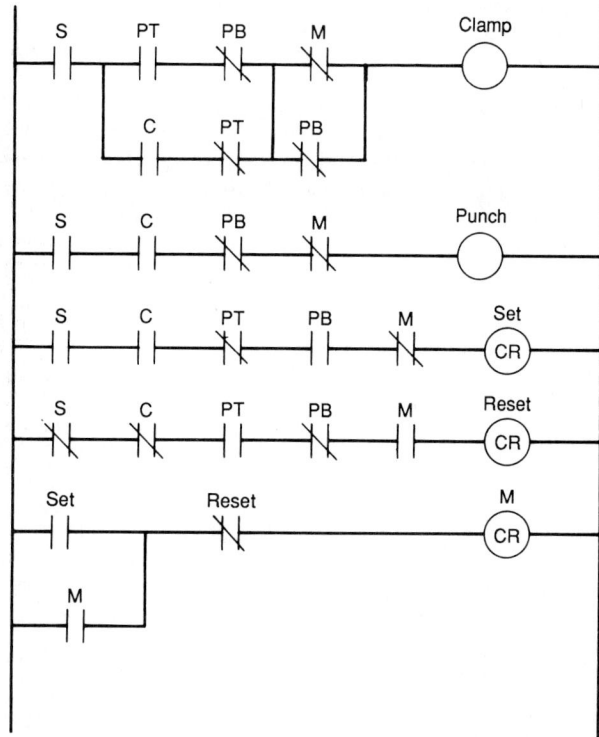

FIGURE 4.14 Ladder logic diagram to control punch-press operation.

tracted. Otherwise, the punch could stick in the part and foul the machine. Furthermore, operators cannot be allowed to remove their hands from the palm buttons in the middle of the cycle and then replace their hands to continue the cycle or tie one of the palm buttons down so that only one hand is required to operate the press.

A more difficult class of sequential problems deals with random changes in inputs instead of the predetermined sequence as discussed previously. These problems are not as simple to solve and require other techniques such as the Hoffman method presented by Pessen.[9] The Hoffman method uses flow diagrams, also known as state diagrams, to map all the possible inputs that can occur.

PROGRAMMABLE LOGIC CONTROLLERS

Programmable logic controllers (PLCs) are in widespread use for both continuous and discrete process control. Recent developments have provided the PLC with increased speed, programming, and communications capabilities. This section describes the development of the PLC, the components of the PLC, the operation of the PLC, and common methods of programming the PLC. A more complete discussion of PLCs can be found in Johnson and Petruzella.[10,11]

PLC Development. The design criteria for the first programmable logic controller were developed by the Hydramatic division of the General Motors Corporation in 1968. They had large numbers of discrete parts control systems based on relay control. Designing, fabricating, and debugging the control systems were expensive and changes usually occurred each model year. They believed that advances in solid-state electronics could be applied to solve this problem. Thus the first generation of PLCs was designed solely to replace relays.

With the advancement of microelectronics and microprocessors, the PLC achieved greater capabilities while becoming smaller in size. Math functions, communication capabilities, and high-speed counters became available. Specialized modules were developed to sense values of continuous variables rather than just a binary ON/OFF condition. This development opened the door to many continuous process control applications. Software enhancements provided the capabilities to use subroutines and interfaces to BASIC, C, and other high-level programming languages. Networking of PLCs by means of high-speed coax and fiber-optic cables provided alternative methods to control large or remote systems. Modules are now available which provide connection to other factory networks using MAP (manufacturing automation protocol).[12]

PLC Components.   Most PLCs are designed in a modular fashion. The advantage of modular construction is twofold. First, the unit can expand and change to meet the needs of the company. New modules can be added if more or different process variables must be controlled or monitored. Second, maintenance and repair is simplified. Modules can be maintained in inventory and swapped with a faulty unit in a matter of minutes, an important consideration when controlling multi-million-dollar production lines. The modules in a PLC can be divided into the following categories: power supply, processor/memory/logic, inputs and outputs (I/O), communication, and special-purpose modules. The relationship of these modules is shown in Fig. 4.15.

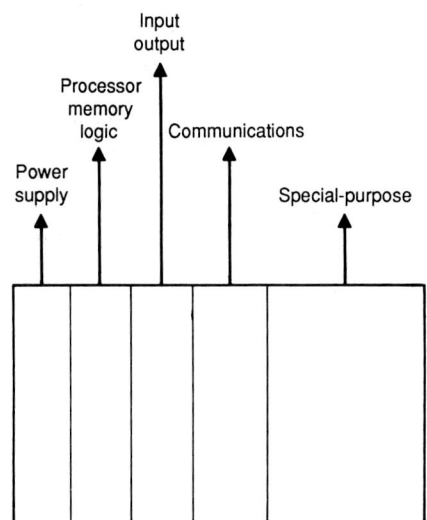

FIGURE 4.15   Modules in a programmable logic controller.

Power Supply.   The power supply provides the other modules in the unit with filtered and regulated power. Often, the power supply is contained in the base unit which houses the backplane, a series of connectors which the other modules plug into. If the system contains remote units, a power supply is usually located in each remote unit.

Processor/Memory/Logic Modules.   The processor is the heart of the PLC. It executes the program which causes the PLC to scan the inputs and update the outputs. Often, processor modules which execute at faster speeds are available at an additional cost for large or fast applications. This allows the user to upgrade capabilities with minimal hardware and programming changes. Since larger programs need more memory, additional memory can usually be added to the processor module or to a separate memory module. In larger and more specialized PLCs, logic and math modules are available to further decrease scan time in applications that require time-intensive numerical calculations or special logic functions. By off-loading these calculations from the processor, substantially higher overall speeds can be realized.

Input and Output Modules.   Input and output (I/O) modules are available to sense and control virtually any signal. Since the processor is a digital computer, the function of the input modules is to convert external signals into digital values that the processor can manipulate. Similarly, the function of the output modules is to convert digital data into control signals. The general structure of an input module is shown in Fig. 4.16. The input signal is filtered and debounced if it is binary. The signal is also isolated from the processor circuits so that incorrect input voltages or currents will not harm the processor or other I/O modules. Generally, optical isolation is used because it is less expensive. Input modules are available to read binary signals such as TTL, 5 VDC, 24 VDC, and 115/230 VAC; analog signals such as 0 to 10 VDC, -10 to $+10$ VDC, and 4- to 20-mA current loop; and thermocouples.

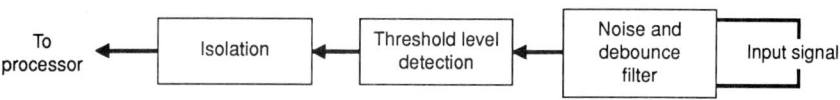

FIGURE 4.16   Structure of an input module.

The output modules are also isolated from the processor. The general structure of an output module is shown in Fig. 4.17. The output conversion device could be mechanical such as a relay or solid-state such as a triac or DAC (digital-to-analog converter). Output modules can be used to drive binary signals of various voltages such as TTL, 5 VDC, 24 VDC, and 115/230 VAC. They can also drive analog signals such as 0 to 10 VDC, -10 to $+10$ VDC, and 4- to 20-mA current loop.

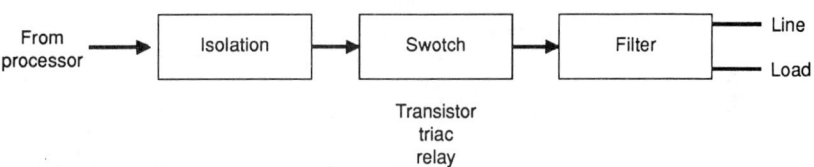

FIGURE 4.17   Structure of an output module.

Communication Modules.   Communications modules allow the processor to transmit and receive data and commands from other PLCs or computers. Modules are available for many different communications standards such as RS-232, RS-422, and MAP as well as proprietary links such as General Electric GEnet;[13] Allen Bradley Data Highway and Data Highway 2;[14] Texas Instruments TIWAY I and II; and Gould-Modicon Modbus and Modway.[15] Some PLCs can control remote units in distant locations. The remote units look identical to a PLC except there is no processor. Instead, a receiving module connects to the cable from the controlling PLC and passes the appropriate data to and from the I/O modules.

Special-Purpose Modules.   Special-purpose modules are those that do not fall into one of the previous categories. Examples include axis-positioning modules, PID loops, and high-speed counters.

PLC Operation.   Since the PLC was originally designed to replace relay logic, an attempt was made to make the control systems interchangeable. However, in relay logic each rung is processed continuously and independently of every other rung. Therefore, a parallel processing computer system would be required to identically replace a relay control system. To date no implementations using parallel processors have been reported. Instead, sequential processors are implemented in the pseudo-parallel method explained in the next section. The rungs are processed quickly in sequential order.

PLC Scan.   A scan consists of one check through all the rungs plus some input and output and possibly some communication. A diagram of what happens during a typical scan cycle is shown in Fig. 4.18. First, the processor reads all of the inputs that are attached to the PLC and stores them in a data table. Next, the processor checks each rung to see if a path is available to energize the outputs. During this check, when the program

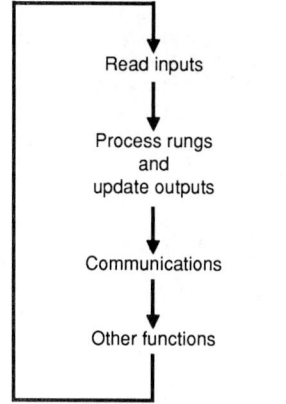

FIGURE 4.18 Typical scan cycle of a PLC.

references an input device, the processor looks for the value in the data table instead of checking the input module.

If an input changes after the processor has read the value at the beginning of a scan, it is ignored until the beginning of the next scan. Since all the rungs use the inputs that were read at the beginning of the scan, each rung is treated as if it is seeing the data at the same time as any other rung. This process makes it seem as if the rungs are scanned in parallel. However, a problem occurs when outputs are updated. Many PLC manufacturers update an output as soon as a rung is scanned and found to provide a complete path for current to flow. Herein lies the difference between a PLC and a relay control circuit. If the first and last rung each depend on the same input conditions, the first will be activated almost a full scan before the last. In most applications this makes little difference. However, in certain critical applications this idiosyncrasy must be addressed. If converting from a relay control system to a PLC control system, changes may be required in the ladder logic to accommodate the sequential scanning of rungs.

Many PLC manufacturers make a provision for critical inputs that must be checked when the rung is analyzed rather than at the beginning of the scan. These inputs are called fast I/O. However, they do increase the time required to scan the rung and are normally reserved for specialized applications.

PLC Modes of Operation. The scenario described above occurs when the PLC is in the RUN mode. This is the mode when the program is scanning normally and the process is being controlled. In this mode, the operator can monitor inputs, outputs, timers, counters, and data values.

Usually, there are several other operating modes. In the PROGRAMMING mode, operators can edit the program. They can add, delete, or change elements and rungs. On some PLCs, they can print out the program. In the SINGLE SCAN mode, operators can cause single scans to occur. This is a tool to allow debugging of programs. Often they can force inputs or outputs to be on or off. In this way, they can test the logic of the program without actually reading the signals in from the process. Other operating modes allow the operator to read and write programs from or to an external device.

PLC Programming. There are many different ways that PLCs can be programmed. Factors affecting programming methods include the manufacturer and the size of the PLC. The most common methods of programming include ladder logic, boolean, functional block, and high-level languages.

PLC programs offer some important advantages over relay circuits. The most important is that relays and other output devices can have an unlimited number of contacts. In contrast, most control relays are limited to a few sets of contacts and circuits must be designed with this limitation in mind. Generally, this results in few rungs with more complex logic. PLCs, on the other hand, use an electronic bit to represent each input and output. This single bit can be referenced as many times as required in the program without adversely affecting the scans of other rungs because there is no electrical interaction between the elements represented in the rungs.

Ladder Logic Programming. Ladder logic programming is one of the most popular methods of programming. With this method, the programmer enters the program using a graphical display on a computer or hand-held programmer. Editing keys provide the capability to add elements, delete elements, connect elements or parts of rungs, and change elements. The programmer composes a ladder logic diagram that looks the same as one drawn on paper. Each element is identified as input or output. For elements like timers or counters, additional data must be entered such as preset values, or conditions under which to reset.

Boolean Programming.   Boolean programming is very similar to ladder logic programming except it was developed before the advent of friendly, graphic interfaces. With this method, the programmer enters a sequence of mnemonics to describe what elements are adjacent to other elements and how they are connected together. An example of a boolean program is shown in Fig. 4.19. In this example, STR can be thought of as a left parenthesis used to start a grouping of elements. OR STR and AND STR can be thought of as a right parenthesis used to end a group. Complex logic can be represented. However, for most people it is easier to understand a graphic ladder logic interface than boolean mnemonics.

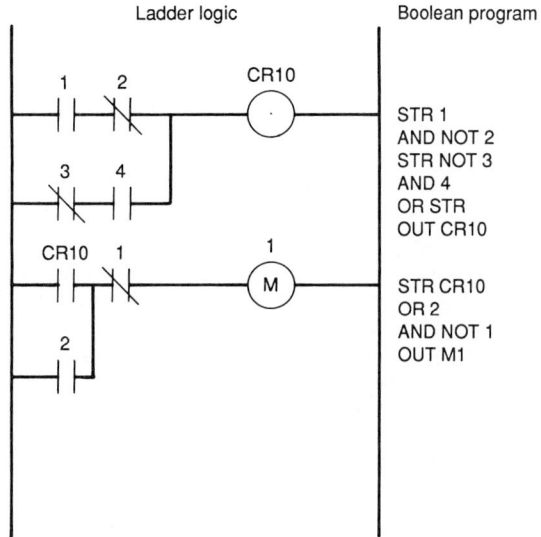

FIGURE 4.19   Boolean program for a PLC.

Functional Block Programming.   Functional blocks are high-level instructions that are incorporated into ladder logic diagrams. Each block is an instruction that is similar to a subroutine. The block can require several parameters, similar to a subroutine requiring arguments. During a scan, if the rung is energized the block is executed. Some of the functions that can be implemented using blocks are timers, counters, arithmetic operations, data transfers, logical shifts, and sequencers.[15] Note that many PLCs also implement these functions using methods other than functional blocks.

High-Level Programming.   High-level programming languages such as C, FORTRAN, and BASIC can be used to program some PLCs. It is convenient to implement some subroutines that require specialized functions or calculations in these languages. Specialized modules or programming are normally required to interface with a high-level language.

Computers versus PLCs.   Computers can also be used to control discrete processes. In fact, PLCs are nothing more than specialized computers that are hardened to operate under severe environmental conditions. Specialized circuits can be designed or purchased to interface computers to industrial processes. In the future the differences between PLCs and computers will be reduced. PLC capabilities will shift toward higher-level programming languages, graphical output, user-friendly input, and general data-processing capabilities. This will enable PLCs to interact better with automation systems in the factory.

REFERENCES

1. Seborg, Dale E., Thomas F. Edgar, and Duncan A. Mellichamp, *Process Dynamics and Control*, Wiley, New York, 1989.

2. *Programmable Motion Control*, Parker Hannifin, Compumotor Division, Rohnert Park, Calif., 1990.

3. Dorf, Richard C., *Modern Control Systems*, 5th ed., Addison-Wesley, Reading, Mass., 1989.

4. Koren, Yoram, *Computer Control of Manufacturing Systems*, McGraw-Hill, New York, 1983.

5. Bollinger, John G., and Neil A. Duffie, *Computer Control of Machines and Processes*, Addison-Wesley, Reading, Mass., 1989.

6. Booth, Taylor L., *Introduction to Computer Engineering: Hardware and Software Design*, 3d ed., Wiley, New York, 1984.

7. Greenfield, Joseph D., *Practical Digital Design Using IC's*, Wiley, New York, 1977.

8. Friedman, Sander B., *Logical Design of Automation Systems*, Prentice-Hall, Englewood Cliffs, N.J., 1990.

9. Pessen, David W., *Industrial Automation: Circuit Design and Components*, Wiley, New York, 1989.

10. Johnson, David G., *Programmable Controllers for Factory Automation*, Dekker, New York, 1987.

11. Petruzella, Frank D., *Programmable Logic Controllers*, McGraw-Hill, New York, 1989.

12. Stacy, Alan H., *The MAP Book: An Introduction*, Macmillan, New York, 1985.

13. *GEnet Factory LAN Series Six Network Interface*, GE Fanuc Automation, Charlottesville, Va., 1987.

14. *Allen-Bradley 1785-LT3, 1785-LT, 1785-LT2 PLC-5 Family Programmable Controllers*, Allen-Bradley, Milwaukee, Wis., 1987.

15. Jones, C. T., and L. A. Bryan, *Programmable Controllers: Concepts and Applications*, IPC/ASTEC, Atlanta, Ga., 1983.

CHAPTER 5
COMPUTER-CONTROLLED PROCESSES AND TOOLING

Mikell P. Groover
Professor of Industrial Engineering
Lehigh University
Bethlehem, Pennsylvania

The digital computer is widely used in industry today, not only to perform the business functions of a firm but also to control its manufacturing operations. The way in which control is accomplished varies with differences in production process and size of operation. Computer control is a key component in computer-integrated manufacturing (CIM), automation, and robotics—other subjects covered in this handbook. This chapter presents computer-controlled processes and tooling, including the hardware and software used in a computer control system, and how the technology has evolved over the years and been influenced by advances in computer technology. The coverage begins with two early control schemes that have contributed significantly to the development and importance of modern computer process control.

DIRECT DIGITAL CONTROL

Control of production operations by digital computers can be traced to the oil refinery and chemical processing industries in the late 1950s and early 1960s. These types of complex processes are characterized by large numbers of variables and associated control loops for each variable. They were traditionally controlled by means of analog devices, typically one or more dedicated to each feedback loop. The various loops had their own set-point values and in most instances operated independently of all other loops. Any coordination of the system was accomplished in a central control room, where workers made adjustments in the individual settings to attempt to achieve a semblance of stability and economy in the process. The hardware cost for all the control loops was substantial, and the human means of coordination was far from optimal.

With the commercial development of the digital computer in the 1950s, it seemed natural to replace some of the analog elements with a computer. Not all the components in each loop could be replaced, but some of them could: the analog controller which performed the conventional proportional-integral-derivative (PID) control functions, the recording and display units, and the comparator which related the measured output to the set point. If these devices could be replaced for all the control loops in the system, the cost of the digital computer might be justified. Operating on a time-shared, sampled-data basis so that each loop could be serviced, the computer would receive the measured values of the process variables, make the calculations associated with the controller functions, and send back commands to the various actuators and process interface devices. It would also maintain records of process performance.

This form of control was called *direct digital control* (DDC, for short). The central computer was "directly" linked to the production process, hence the name for this control type. Because of the high cost of mainframe computers of those early days, direct digital control could be cost justified only for very large systems—ones with numerous control loops so that the savings in analog hardware components would exceed the price of the mainframe.

DDC was originally conceived as a more efficient way to carry out the same types of control actions as the analog elements it replaced. However, the digital computer is capable of doing much more than simply imitate the operation of a group of analog controllers. Three enhancements of this early form of computer control have since been developed:

1. Wider variety of control algorithms. Analog controllers are somewhat limited in terms of the scope of the computations they can perform. Generally, their operations are restricted to some combination of proportional-integral-derivative (PID) control. The digital computer is considerably more versatile in the control calculations it can be programmed to execute. For example, optimal control algorithms can be accomplished on the control loops for which they are appropriate.

2. Changes in the control algorithm. It is relatively easy to change the control algorithm if that becomes necessary, simply by reprogramming the computer (perhaps not as easy to do as it is to say). Altering the analog control loop is likely to require hardware changes that are more costly and less convenient.

3. More than loop control. As long as the computer is connected to every loop, why not program it to integrate the information from all of the signals and use some optimizing algorithm to manage the entire process? The computer would thereby make the required adjustments in the set points for the individual feedback loops, rather than the crew of workers in the central control room. (This third enhancement will be examined in more detail in the later discussion of "supervisory control.")

These enhancements of direct digital control have rendered the original DDC concept obsolete. Today, computers perform all the functions described above and more, and computer technology itself has progressed dramatically so that much smaller and less expensive yet more powerful computers are available for process control than the large units available in the early 1960s.

NUMERICAL CONTROL

In the late 1940s and early 1950s, before digital computers were available, it was recognized in the manufacturing industries that a better method for controlling machine tools was needed. At that time metal-cutting machine tools were controlled exclusively by human operators, sometimes with the aid of templates, or patterns, that could be used to guide the path of the cutting tool. However, part shapes were becoming more difficult to cut, largely as a result of mathematically defined curved geometries of components used in newly developed high-speed aircraft. The concept of numerical control, originated by John Parsons and Frank Stulen, who did contract machining for the U.S. Air Force, was to use coordinate position data contained on punched cards to define the surface contours of airfoil shapes. A prototype three-axis milling machine was developed in one of the laboratories at the Massachusetts Institute of Technology under Air Force contract, and the machine was first demonstrated to the public in 1952. This prototype and subsequent machines that were developed and sold by machine tool companies during the 1950s and 1960s were not computer-controlled. Instead, they utilized the hardwired electronic controls that represented the technologies of those years.

Numerical control (NC) involves the control of a machine by means of numbers and other symbols, coded in an appropriate format that can be interpreted by the machine's control unit. It is a form of programmable automation because the numbers and symbols can be changed to alter the sequence of instructions that regulate the operation of the machine. When a new production job is released to the machine, the program is rewritten specifically for the new job. NC had its origins in the machine tool industry, and modern versions of NC are widely used today. In addition, many other applications have been developed for this type of positioning control. Table 5.1 presents a list of the important applications of numerical control.

TABLE 5.1 Applications of Numerical Control and Similar Positioning Systems

Machine tool applications:	
NC milling machines	Milling and related operations
NC drill presses	Hole drilling and related operations
NC turning machines	Turning and related operations
NC machining centers	Milling, drilling, automatic tool changing, and other functions
NC pressworking	Sheet-metal punching, some bending
NC flame cutting	Torch cutting of flat plates and sheets
Other applications:	
Electric wire wrap	Back-pin wiring of electrical boards
Coordinate measuring machine (CMM)	Measuring dimensions of mechanical parts in inspection
x-y plotter	x-y control of pen for plotting on paper
Drafting machine	Similar to x-y plotter, for drafting
Component insertion	Position and insert parts on flat plane, such as printed circuit board
Industrial robot	Positioning of mechanical manipulator

The common operating feature of all these applications, machine tool and other, is positional control of a processing element, such as a cutting tool, relative to an object being processed, such as the work part being machined. A requirement of a numerical control system, therefore, is that it must have a coordinate system by which the relative position of the "tool" can be defined with respect to the machine table on which the "work" is clamped. The coordinate system commonly used in most NC applications is shown in Fig. 5.1. It is based on the cartesian coordinate system (x, y, and z axes), with rotation (a, b, and c axes) defined about these linear axes. Not all six axes are used in every system. Some processes require control of only two axes, for example, an x-y plotter. Others need control of five or six axes, an example being an industrial robot whose coordinate system may be different from the cartesian system, depending on the anatomy of the robot. (It might be argued that an industrial robot goes beyond our definition of numerical control because its controls enable it to do more than merely position its manipulator. However, in terms of its position control capabilities, it operates as a numerical control system. Other NC systems are also capable of more than just position control, for example, the tool-changing capability of a machining center.)

It should be mentioned that in the positioning of the tool relative to the work, it is not only the tool that moves in absolute space. To achieve relative positioning, either the tool or the machine table can be moved, depending on the mechanical design of the system. In

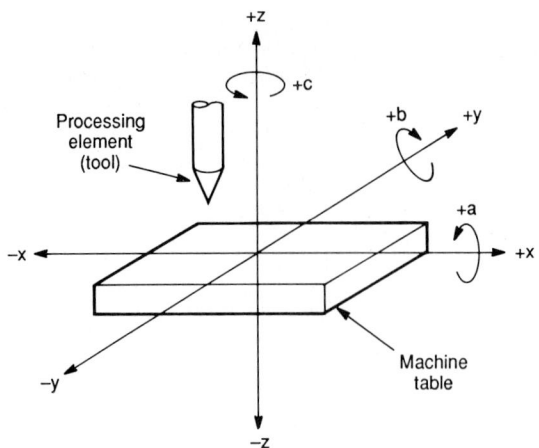

FIGURE 5.1 Coordinate system used for most numerical control applications.

many NC systems, it is the table that is moved in relation to a fixed tool, in order to accomplish positioning.

There are differences in the level of control exerted over the relative position of the tool in an NC system. Two categories can be defined: (1) point-to-point, and (2) continuous path. In *point-to-point* or *position* control, the objective of the control system is to move the tool to a position defined in the coordinate system, without regard to the path taken to achieve the position. Once the tool reaches the desired position, some operation is performed at that location; for example, a hole is drilled or a spot weld is made. In *continuous path* systems, called *contouring* systems in machining terminology, the path taken by the tool is controlled in real time. That is, the relative position of the tool is continuously controlled to achieve a desired motion trajectory. To accomplish this level of control, each axis must be controlled in terms of both position and velocity, and its movement must be coordinated with the movements of all other axes in the system. Continuous path NC systems permit the tool to follow mathematically defined paths, such as straight lines, circles, ellipses, and other complex curved shapes like those found in the design of airfoils, aerospace components, and automobile car bodies.

COMPUTER-ASSISTED PART PROGRAMMING

The digital computer was applied in numerical control applications almost from the beginning of the technology, but not as the controller unit in the NC system. The application was for *NC part programming,* which is concerned with the planning and coding of the sequence of programming steps to be performed on the numerical control machine tool. In the late 1950s and early 1960s, the *APT* part programming language was developed (APT stands for automatically programmed tooling). This language, still used today, permits a person to define the geometry of the work part and then specify the tool path required to machine the finished part, using simple English-like statements. The "part programmer" then enters the program into the computer, which performs the necessary calculations to generate a tool path code that can be interpreted by the machine tool controller. The medium used to input the program into the controller was punched tape—1-inch-wide paper tape with punched holes coded to represent the tool path commands. The controller therefore required a punched-tape reader in order to read the contents of the tape.

APT has been largely replaced in many installations by enhanced versions of APT—languages based either directly on APT or on APT concepts. In addition, CAD/CAM (computer-aided design and computer-aided manufacturing) began its rapid commercial development in the 1970s—also based on APT concepts. Modern CAD/CAM systems, equipped with advanced software, are capable of performing many of the NC part programming functions automatically, thus providing valuable savings in part programmer time.

DIRECT NUMERICAL CONTROL

As computer technology developed, the idea of establishing a communication link between a central computer and the individual NC machine tools in the factory became popular. Introduced in the late 1960s, *direct numerical control* (DNC) involves the use of a large computer to control the operations of a number of separate NC machines, as pictured in Fig. 5.2. (Note the use of the word "direct" in the term, perhaps borrowed from direct digital control.) In DNC, there is a direct connection between the computer and the machine control unit (MCU), and the control is accomplished in real time. The tape reader is omitted from each MCU in a DNC system, and the programs are communicated from the central computer to each MCU. The specified part program is called from mass storage (magnetic tape), and the required commands are sent to the individual MCUs as they are needed. The transmission of instructions is accomplished on demand, almost instantaneously, so that no machine tool has to wait for the computer. In principle, it should be possible for one computer to service more than a hundred machine tools operating in this mode.

In addition to transmitting programs to the NC machine tools, there was another objective in DNC: to receive data on operating performance back from the individual machines. The performance data included measures such as production rates, piece counts, tool changes, uptime and downtime statistics, and utilization. Thus direct numerical control involved a two-way communication between the computer and the shop floor. It controlled the machines in the factory through the downloading of part programs, and it collected performance data from the machines which helped managers control the overall operations of the plant. Advantages of DNC included: (1) elimination of tape and tape reader (which had proved to be one of the least reliable components of a conventional NC system); (2) improved part program storage; (3) greater computational capability in the central computer than in the individual MCU (for certain functions such as circular inter-

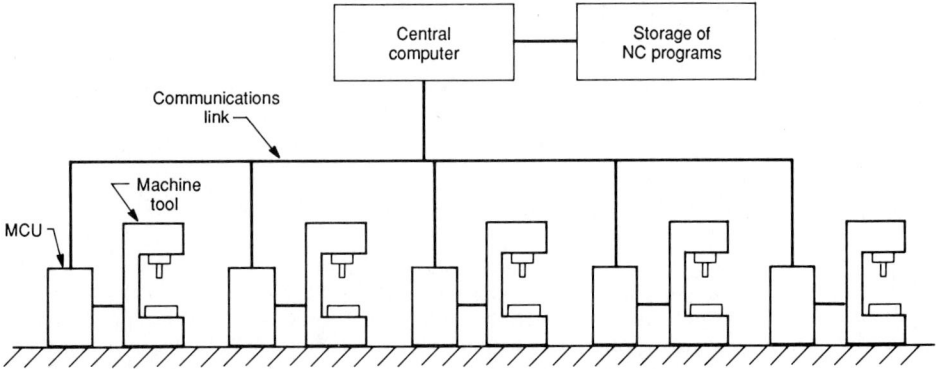

FIGURE 5.2 Configuration of a direct numerical control system.

polation during contouring); and (4) better management of the factory through current information about the operations.

Direct numerical control is still in use, but no new systems are being installed according to the original design. Since the time of the first DNC systems, computer technology has developed rapidly. The physical size and price of a computer have dropped dramatically at the same time that its speed and performance have increased significantly. Minicomputers and microcomputers have opened up new applications and opportunities, including the feasibility of using one computer for each machine tool.

COMPUTER NUMERICAL CONTROL

Computer numerical control (CNC) is an NC system in which a dedicated small computer serves as the machine control unit. The configuration of a CNC system is illustrated in Fig. 5.3. For a stand-alone CNC system, a tape reader is used for initial input of the part program. Once the program is entered into internal storage, the machine tool is driven from the contents of its computer memory. This differs from the operation of a conventional NC system, in which the NC tape must be cycled through the tape reader for every part that is machined. Some of the other features typically included in a computer numerical control system are:

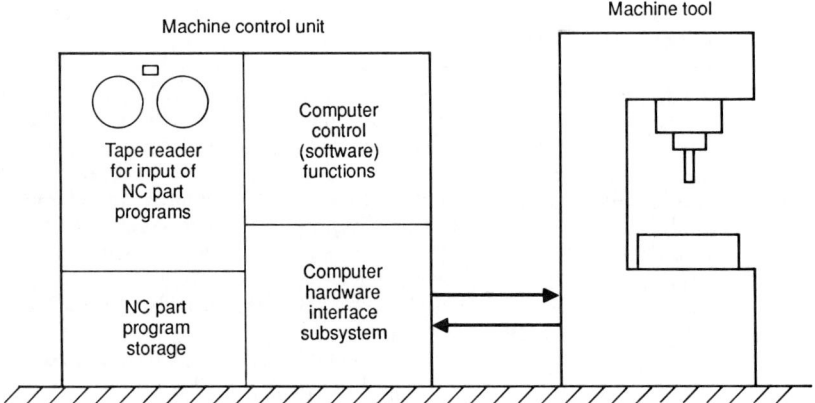

FIGURE 5.3   Key components of computer numerical control (CNC).

- *Program editing at the machine tool site.*Part programming mistakes are common, so that three or more versions of the part program must be tried out to correct programming errors before an acceptable tape is produced. CNC makes it possible to correct these errors at the machine site, without taking the tape back to the programming room.

- *Storage of more than one part program.*The newer CNC controllers possess sufficient capacity to store more than one part program, thus allowing commonly used programs to be retained at the machine site.

- *Fixed cycles and programming subroutines.*Frequently used machining routines, such as pocketing (a milling operation to create a shallow cavity in the surface of a part) and bolt hole circles (a series of holes equally spaced around the perimeter of a circle), can be made to reside in memory and be called as a subroutine from the part program by a MACRO-type statement.

- *Diagnostics.*CNC machines often include diagnostics capabilities, designed to monitor the operation of the system continuously and check for malfunctions or signs of impending breakdowns.
- *Communications interface.*Most CNC controllers are equipped with a standard communications interface to link the machine tool to other computer systems in the factory.

Modern NC machine tools and other NC-type devices (that is, systems that rely on mechanical positioning of a "tool" relative to a "work") are all computer-driven. The old-style hardwired electronics controllers are rarely used.

Many CNC machines are installed as stand-alone units, that is, with no interconnections to other computer systems in the plant. However, it is of interest to note that a growing number of CNC machines are connected to central computers in the factory very much like a direct numerical control hierarchy. However, whereas the original DNC systems consisted of machine tools with adapted NC controllers, the newer versions of this hierarchy use CNC machines. That is, the central computer is linked to other computers (the CNC units) rather than hardwired MCUs. This configuration, in effect, fits the definition of a distributed computer system. Accordingly, the new version of DNC is called *distributed numerical control*. It possesses all the advantages of the previous direct numerical control, plus the added benefit of being able to link computers to other computers.

TYPES OF COMPUTER PROCESS CONTROL

The preceding discussion indicates the variety of ways in which computers are used for controlling and observing the manufacturing process. We can identify three basic categories: (1) process monitoring, (2) preplanned control, and (3) supervisory control. The three categories will be described as pure control types, although various combinations of the three types represent the usual industrial practice.

Process Monitoring.   Computer process monitoring involves the use of a digital computer to collect and record data on the process and associated equipment. The data collected by the computer can generally be classified as follows:

- *Process data.*These are measured values of input and output variables that indicate process performance.
- *Equipment data.*This refers to measured or calculated values that indicate machine status, such as machine utilization, tool change statistics, and machine malfunction diagnostics.
- *Product data.*Measured or counted data relating to product quality and production rates.

This category is not really a type of computer process control, because the computer does not directly control the process. Control of the process remains the responsibility of human operators, who manage it using the collected data and computer-prepared reports for guidance.

Preplanned Control.   In this category, the computer is used for directing the process or equipment to perform a predetermined sequence of operating steps. The control sequence is defined by means of a program which must be prepared in advance. The program specifies the order in which the steps are to be carried out and includes IF-THEN statements to cover the various processing conditions that might be encountered. Preplanned control often uses feedback control loops to verify that each step in the operating sequence has been properly executed before proceeding to the next step. However, in some cases, feedback is not required, and such systems execute the processing sequence in an "open-loop" fashion. Preplanned control goes by different names in industry, and some examples will help to illustrate this control mode.

NC-Type Systems. NC, discussed previously, controls a sequence of movements of a tool relative to an object being processed. In our present context, this control program is executed as a sequence of preplanned movements that are accompanied by value-added work on the object.

Sequence Control. This refers to a form of preplanned control that usually includes a combination of sequencing and logic control. A *sequencing system* operates by means of internal timers to determine when changes in process parameters will occur. The changes are typically ON/OFF functions. *Logic control* refers to a control mode in which logical decisions and actions about the process are taken in response to events that have occurred in the production system. The output signals of the controller are determined by the signals indicating process status. Sequence control is widely used for industrial process control and is typically accomplished by *programmable logic controllers,* a special form of digital computer designed for convenient interfacing with industrial processing equipment.

Program Control. This control type, most commonly used in the processing industries, refers to applications in which the computer is used to start up or shut down complex industrial processes, or to change over the process from one product grade to another, or to control the process through a sequence of processing steps in chemical batch processing. In program control, the objective is to guide the process from one operating condition to the next by making appropriate changes in the process variables.

Supervisory Control. Supervisory control denotes a control system in which the computer is programmed to optimize some overall objective function for the process. The objective function, or index of performance, as it is sometimes called, is usually based on one or more economic criteria such as maximum production rate or minimum cost or optimum quality. Supervisory control is often superimposed on other control modes discussed. The other controls are connected directly to the process, while supervisory control directs the operations of these process-level control systems, as suggested by the diagram in Fig. 5.4.

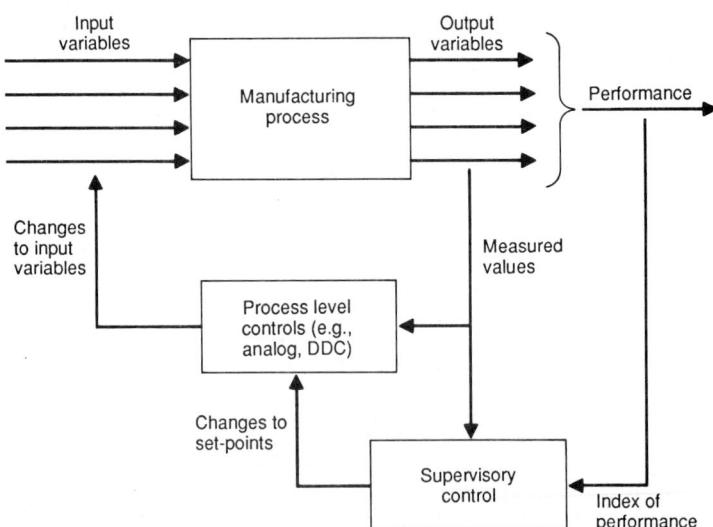

FIGURE 5.4 Supervisory control superimposed on other process-level control systems.

Several control strategies are used in supervisory control to optimize the process. They can be classified as follows:

Regulatory Control.   The objective here is to maintain process performance at a certain level or within a given tolerance band of that level. This is appropriate, for example, when some measure of product quality is the performance attribute, and it is important to keep the quality at the specified level or within a specified tolerance band. As indicated in Fig. 5.5, regulatory control is to the overall process what feedback control is to an individual control loop in the process.

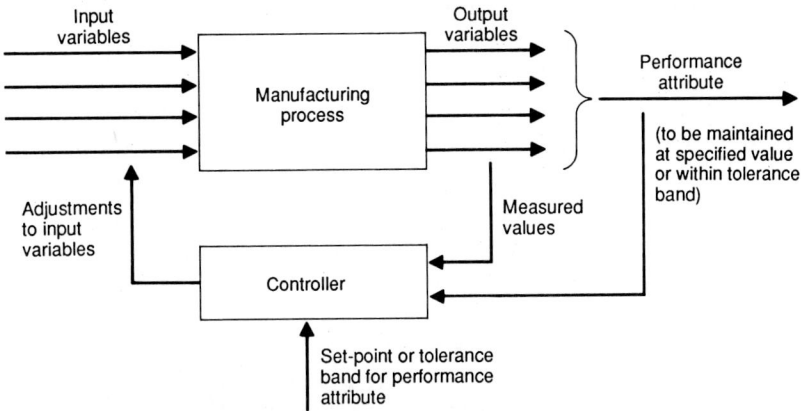

FIGURE 5.5   Regulatory control.

Feedforward Control.   Feedforward control attempts to anticipate the effect of disturbances that will upset the process by sensing them and compensating for them in advance of the upset. As shown in Fig. 5.6, the feedforward control elements sense the presence of a disturbance and introduce a compensating parameter that cancels any effect the disturbance will have on the process. In the ideal case, full compensation results. However,

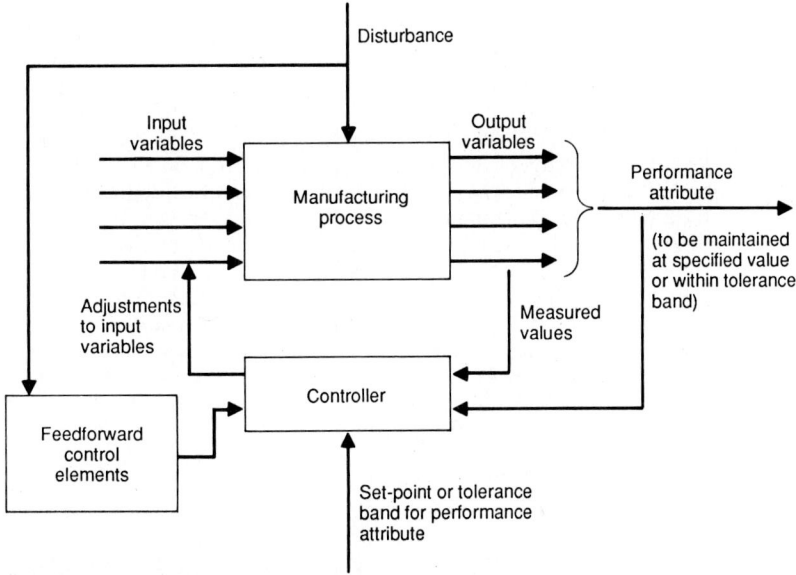

FIGURE 5.6   Feedforward control, combined with feedback control.

since this is unlikely, feedforward control is usually combined with feedback control, as shown in the block diagram.

Steady-State Optimization.   This term refers to a class of optimization problem in which it is assumed that there is perfect knowledge of the process operation in the form of a mathematical model; that there is a measure of system performance, called an index of performance, that can be applied to the model; and finally, that values of the system parameters which optimize performance can be mathematically determined. The control system operates in an open-loop manner, as indicated in Fig. 5.7. A variety of mathematical techniques exist for solving steady-state optimal control problems, including differential calculus, calculus of variations, linear programming, dynamic programming, and others.

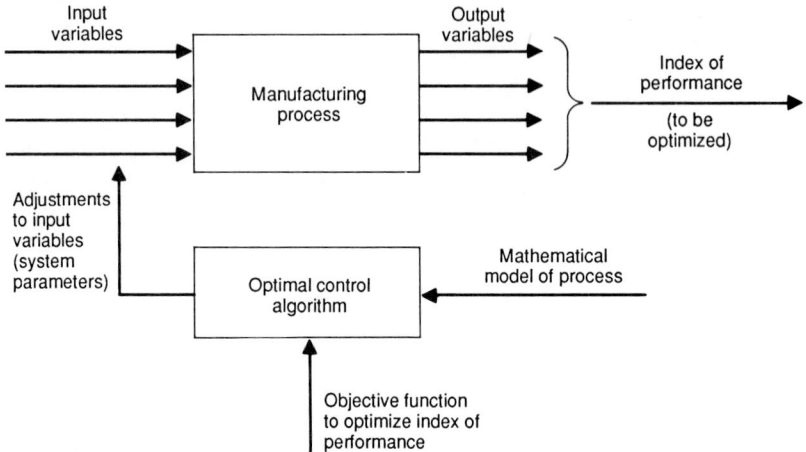

FIGURE 5.7   Steady-state (open-loop) optimal control.

Adaptive Control.   Steady-state optimal control operates as an open-loop system. Adaptive control can be considered as a combination of feedback control and optimal control. As in feedback control, the relevant process variables are measured; and as in optimal control, an index of performance is used. What distinguishes adaptive control from the other two types is its special capability to cope with a time-varying environment. In both feedback control and steady-state optimal control, the environment is assumed to remain constant. Otherwise, system performance is degraded. An adaptive control system is designed to compensate for its changing environment by monitoring its own performance and altering some aspect of its control mechanism to achieve optimal or near-optimal performance. The environment in a manufacturing process refers to the day-to-day variations in raw materials, tooling, atmospheric conditions, and the like. As depicted in Fig. 5.8, three functions characterize the operation of an adaptive control system: (1) *identification*—the current performance of the system is measured or computed based on data collected from the process; (2) *decision*—the control system figures out (computes by some preprogrammed logic) what changes in its input variables or internal parameters should be made to improve performance; and (3) *modification*—the decision is implemented by making the adjustments indicated. These adjustments are typically made by altering the internal control parameters of the process-level controller, as illustrated in the block diagram, or by changing the values of the inputs to the process.

Search Techniques.   These are used for a special class of adaptive control problem in which the decision function cannot be sufficiently defined. Specifically, the influence of the input variables on performance is not known, or not known well enough to utilize adaptive control as described above. Accordingly, it is not possible to determine what changes to make in the input variables or internal parameters of the system in order to effect the desired improvement in performance. Thus a search strategy is employed in

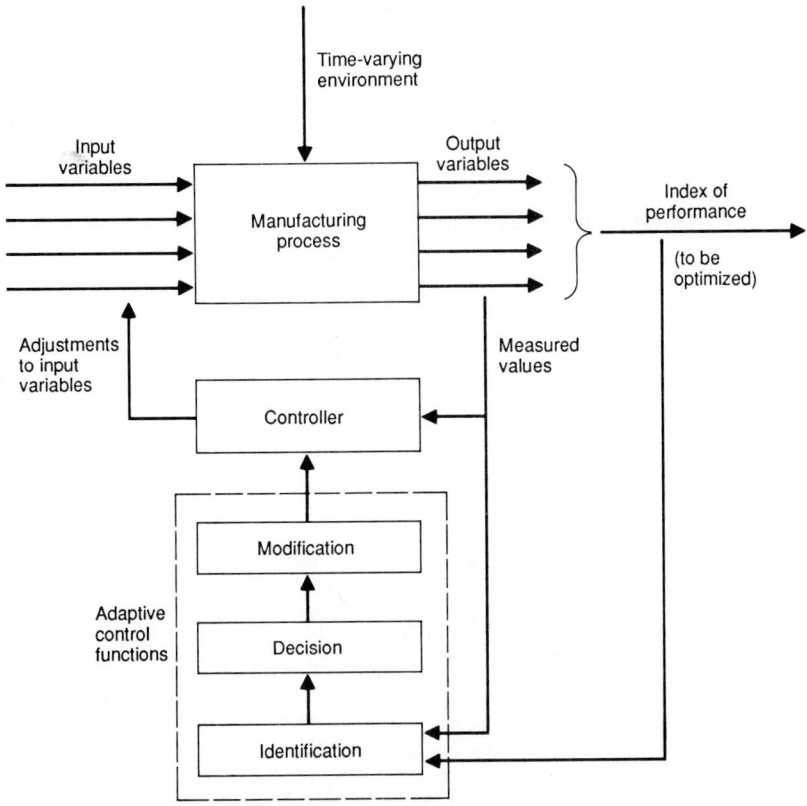

FIGURE 5.8 Configuration of an adaptive control system.

which systematic changes are made in the input variables to observe what effect these changes have on the output variables. Based on the results, the appropriate changes are made in the inputs so that system performance will be improved. The most familiar search strategies are those based on the use of gradients, which indicate which input variables will provide the maximum improvement in the index of performance for the process.

Other Specialized Techniques.   These include learning systems, expert systems, and other artificial intelligence approaches used in process control.

REQUIREMENTS OF PROCESS CONTROL SOFTWARE

Programming for computer process control is distinguished from data processing or engineering and scientific applications by several requirements, all concerned with the need to communicate with the process on a real-time basis. The software requirements can be stated concisely as the following:

- *Timer-initiated events.*The computer must be able to manage events that occur in clock time, some at regular intervals (such as sampled-data values of process variables) and others at distinct points in time (such as lowering the furnace temperature at the end of a heat-treat cycle).
- *Process-initiated interrupts.*The computer must be programmed to respond to incoming signals from the process, so that it interrupts its regular program execution to service the process.

- *Computer commands to process.* In addition to incoming signals from the process, the computer must be capable of transmitting control signals to the process, to actuate hardware devices or to adjust a set point in a control loop.

- *System- and program-initiated events.* The process control system rarely stands alone. It is becoming increasingly common for computer systems to be interconnected in networks, to transmit data and instructions between the system components; for example, data on machine utilization or instructions to begin a new production order.

- *Operator-initiated events.* Finally, the computer must be able to accept input from the (human) operator, for program editing, startup instructions, or emergency shutdown.

Most of the above requirements can be satisfied by designing the computer system and associated software so that its current operations (whatever software it is currently executing) can be interrupted by events which have higher priority. This kind of system is called an interrupt system. In the following paragraphs, this and other features of control software will be discussed.

Priority Interrupt System.   All computer systems have interrupt capability (if nothing else, shut off the power). But a more sophisticated interrupt system is required in process control applications. An *interrupt logic system* is a computer control feature which permits execution of the current program to be suspended so that another program or subroutine can be executed in response to an incoming signal indicating a higher-priority event. When the interrupt signal is received, control is transferred to a predetermined subroutine that is designed to deal with the specific interrupt. The location and status of the current program are remembered so that its execution can be resumed after the interrupt has been serviced. Interrupt conditions can be classified as internal or external. Internal interrupts are generated by the computer system itself, and include timer-initiated data recording and commands to the process. External interrupts are events that are external to the computer system, such as an out-of-tolerance process variable or an operator input. The reason why an interrupt system is required in process control is so that more important programs will be executed before less important ones—so that higher-priority functions can interrupt functions with lower priorities. Accordingly, the possible functions performed by the control computer must be prioritized. A typical ranking of functions might be as shown in Table 5.2.

TABLE 5.2 Possible Priority Levels in an Interrupt System

Priority level*	Computer function
1	Operator inputs
2	System interrupts
3	Timer interrupts
4	Commands to process
5	Process interrupts

*1 = lowest priority; 5 = highest priority.

The actual number of priority levels, and the relative importance of the functions, must be designed for the individual process control situation. For example, emergency shutdown of a process for safety reasons would be an operator input; yet it would occupy a very high priority level. Most operator inputs would have low priorities.

Diagnostics.   Another desirable feature of the computer system is the capability to identify the reason behind any failures or malfunctions that may beset the process. These occurrences can result in costly downtime and loss of production. The process must often be stopped and repaired by humans. The typical diagnostics system has three modes of operation: (1) status monitoring, in which the system observes and records the important process and equipment variables; (2) failure diagnostics, during which a malfunction or failure has occurred, and the sensed variables must be interpreted and the problem diagnosed; and (3) recommended repairs, in which the system provides a suggested procedure to be used by the repair crew. Recommendations are often based on the collective knowledge of experts on the particular equipment or process. In some applications, diagnostics systems are designed to anticipate the occurrence of failures by observing subtle changes and trends in the operating characteristics of the process and predicting possible malfunctions that may result from these changes.

Error Detection and Recovery.   The use of a diagnostics system usually presumes that human operators will help in diagnosing the malfunction, make the necessary repairs, and restart the operation. With the increasing use of computers for process control, the computer is being called on not only to detect the problem but also to perform the necessary steps to repair and restore the system—automatically, without human involvement. Error detection and recovery, as this capability is called, consists of two phases: error detection and error recovery. Error detection involves the use of sensors interfaced to the process and the proper interpretation of the sensor signals by the computer system to identify when an error has occurred and the nature of the error. Typical errors in process control include equipment and tooling failures, significant variations in raw materials, and human mistakes (for example, entering an incorrect identification code for the product). When an error in the process is detected, this interrupts the normal process control functions, so that the error recovery procedures can be executed. In error recovery, the required corrective actions are taken for the particular error so that the normal operating mode can be resumed. Some of the possible recovery procedures are automatic tool change when a tool breaks or wears out, abandon further processing of the work part if it has been damaged during the operation, and compensation for variables that have suddenly changed. If the same recovery procedure must be invoked on several successive cycles, it usually means that a systematic error is occurring, and a human operator must be called. The problems in designing an error detection and recovery system are (1) anticipating the variety of possible things that could go wrong in the process, (2) developing the appropriate sensor package to reveal these errors when they occur, and (3) devising appropriate procedures that will either correct or compensate for the errors. These problems are usually very specific to the individual process.

INTERFACE HARDWARE

For process control, the computer must collect data from and transmit command signals to the manufacturing operation. The types of data and signals communicated between the computer and the process vary, depending on the sensors and actuating devices used. Three types of data, illustrated in Fig. 5.9, can be distinguished: (1) continuous analog signals, (2) discrete binary data, and (3) discrete data not restricted to binary values. Continuous analog signals are variables which can assume any of a continuum of values (within some practical range) and which vary over time. Examples in process control applications include force, flow rate, temperature, pressure, and velocity. Discrete binary variables

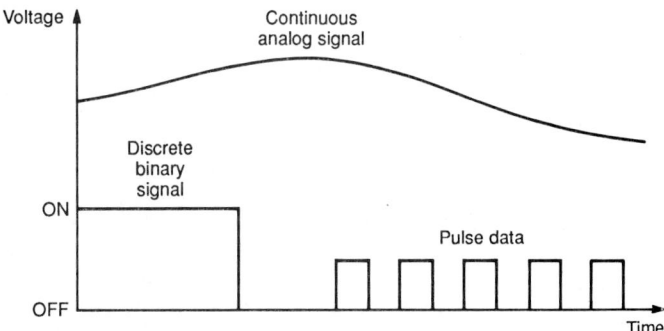

FIGURE 5.9   Three types of manufacturing process data.

can assume either of two possible values, usually on or off, opened or closed, etc. Examples include voltage to an electric motor on or off, a valve open or closed, and contact or no contact of a limit switch. Discrete data other than binary have multiple values, the values being limited to certain discrete, often integer, levels. These data sometimes take the form of an electrical pulse train, a series of pulses which can be counted to obtain the data value. Digital transducers (sensors) operate in this way.

The following paragraphs discuss many of the interface hardware devices used to communicate these data between the computer and the process it controls.

Sensors and Transducers.   These devices are used for collecting data from the manufacturing process. A *transducer* is a device that converts one type of physical quantity to another, such as, temperature to electrical voltage. The reason for making the conversion is to evaluate the signal more conveniently. Transducers are often referred to as *sensors* if they are used to measure a physical variable. Some of the desirable features of a sensor for process control include accuracy and precision, speed of response, reliability, minimum drift, and ease of calibration. Few measuring devices possess all these features, and the process control designer must usually select among a variety of available sensors for the given application. A partial list of common sensors and measuring devices is presented in Table 5.3.

TABLE 5.3   Common Sensors Used in Process Control

Sensor	Process variable measured
Accelerometer	Vibration
Ammeter	Electric current
Bourdon tube	Pressure, vacuum
Infrared sensor	Temperature
LVDT*	Angular or linear displacement
Limit switch	Contact or no contact
Manometer	Pressure
Ohmmeter	Electrical resistance
Optical endcoders	Rotational speed and displacement
Pyrometer	Temperature
Photometer	Illumination, light intensity
Pitot tube	Flow rate
Potentiometer	Voltage
Strain gauge	Force, pressure, torque
Thermistor	Temperature
Thermocouple	Temperature
Venturi tube	Flow rate

*Linear-variable-differential transformer.

Analog-to-Digital Converters.   As its name indicates, an analog-to-digital converter (ADC) is a device that converts a continuous analog signal into approximately equivalent digital data. Most process signals are continuous and analog in nature. In order for the computer to interpret the signals, they must be converted into digital form. Figure 5.10 illustrates the ADC process, showing how the continuous analog signal is converted into a series of individual data values capable of being interpreted by the computer. Operating features that distinguish ADCs include sampling rate (the rate at which the continuous signal is sampled), conversion time (the time it takes to convert the signal into digital form), and resolution (the precision with which the analog signal is evaluated).

Digital-to-Analog Converters.   A digital-to-analog converter (DAC) performs the opposite function of an ADC. It is utilized to send control signals to actuators that are driven by

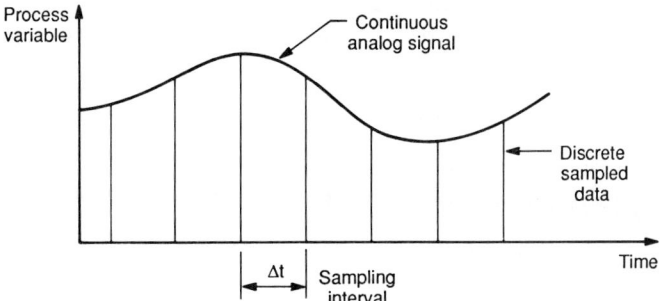

FIGURE 5.10 Analog signal converted into series of discrete sampled data by analog-to-digital converter.

continuous signals. The output of the computer control algorithms, in digital form, must be converted back to continuous (analog) form, and that is the function of the DAC. Digital-to-analog conversion consists of two steps: (1) decoding, in which the digital data output of the computer is converted into analog value, and (2) data holding, in which the analog value is made into a continuous signal (usually electrical voltage) that can be used by the analog actuator at the process. This two-step process is repeated at periodic intervals to obtain a stepwise analog signal as shown in Fig. 5.11.

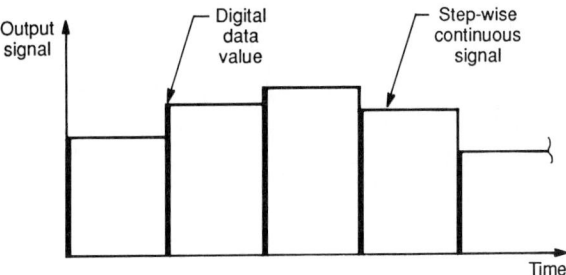

FIGURE 5.11 Digital output converted into stepwise analog signal.

Contact Interfaces.   Contact interfaces are of two types: input and output. A contact input interface is a device by which binary data are read into the computer. In consists of a simple contact device that can be either open or closed (on or off) to indicate the status of a limit switch, valve, motor, or other similar device. Scanning of these contacts by the computer is one of the main functions of a programmable logic controller.

With a contact output interface, on/off signals are communicated from the computer to the process. Based on the control algorithms programmed into the computer, the contact positions are set in either of two positions: on or off. These positions are maintained until changed by subsequent events in the process. Hardware controlled by the contact output interface includes alarms, indicator lights (on control panels), solenoids, constant-speed motors, and the like.

Pulse Counters and Generators.   Some of the data flowing from the process to the computer is in the form of pulse trains, most commonly electrical pulses, for example, from digital transducers. To illustrate a typical application, suppose the rotational speed of a motor is to be measured. The motor shaft is connected to an optical encoder, which generates a certain number of electrical pulses for each rotation. To measure rotational velocity, the pulse counter is used. It actually measures the time (time can be measured with great precision) during which a specified number of pulses are received from the encoder. To determine velocity, the number of pulses is divided by the measured time and by the number of pulses per revolution of the optical encoder.

Pulse generators are used by the control computer to produce either (1) a specified number of pulses or (2) a certain rate of frequency of pulses. In case 1, the specified number of pulses might be used to control the x axis or y axis of a positioning table. In case 2, the pulse rate (that is, the frequency of the pulse train) could be used to control the rotational speed of a stepping motor. A pulse generator operates by repeatedly closing and opening an electrical contact, to produce either a specified pulse count or a specified pulse rate.

SUMMARY

This chapter has provided an introductory treatment of the subject of computer process control. The historical evolution of this technology and some of the important applications have been reviewed. These applications are pervasive in high, medium, and low production and throughout the wide spectrum of modern industrial processes. The various hardware and software components described in this chapter, as well as the control strategies that have been discussed, have made computer process control an important tool in the automation of manufacturing operations.

BIBLIOGRAPHY

Bollinger, J. G., and N. A. Duffie, *Computer Control of Machines and Processes,* Addison-Wesley, Reading, Mass., 1988.

Groover, M. P., *Automation, Production Systems, and Computer Integrated Manufacturing,* Prentice-Hall, Englewood Cliffs, N.J., 1987.

CHAPTER 6
FLEXIBLE MANUFACTURING SYSTEMS: DESIGN AND OPERATING PROBLEMS AND SOLUTIONS

Kathryn E. Stecke
Graduate School of Business Administration
The University of Michigan
Ann Arbor, Michigan

In the metal-cutting industry, a flexible manufacturing system (FMS) consists of several computer numerically controlled (CNC) machine tools, integrated via an automated material handling system, all under the control of one or more computers. An FMS is capable of the concurrent manufacture of diverse part types in unit batch sizes. Both a good design and the efficient operation of FMSs involve some intricate operations research problems.

First, examples of existing FMSs are described. Various generic problems are defined. Then mathematical models that can or have been used in the design, planning, scheduling, and control of these systems are discussed. These models have been used in the *design* of an FMS to determine, for example, the appropriate number of machine tools of each type, the amount of flexibility desired, the capacity of the material handling system, and the size of buffers. They have also been used in the *planning,* or system setup, of an FMS to determine which part types should be selected for simultaneous machining over some upcoming time period, the optimal partition of machine tools into groups, allocations of pallets and fixtures to part types, and the assignment of operations to machine tools and hence the associated cutting tools of each operation among the limited-capacity tool magazines of the machine tools. Mathematical models have been used in the *scheduling* of an FMS to determine the optimal input sequence of parts and an optimal sequence at each machine tool given the current part mix. *FMS control* issues involve the real-time monitoring, to be sure that the system is performing as one thinks it is and that the production expected has been achieved.

This chapter also overviews many of the models that are useful in solving such problems. The references provide additional information on much of the research that has been done to address these problems.

INTRODUCTION TO FMSs

A flexible manufacturing system (FMS) is an integrated system of machine tools linked by automated material handling. Because of the versatility of the machine tools and the quick (seconds) cutting tool interchange capability, these systems are quite flexible with respect to the number of part types that can be produced simultaneously and in low (sometimes unit) batch sizes. These systems can be almost as flexible as, and more complex than, a job shop, while having the ability to attain the efficiency of a well-balanced assembly line.

An FMS consists of several computer numerically controlled (CNC) machine tools, each capable of performing many operations. Each machine tool has a limited-capacity tool magazine that holds all the cutting tools required to perform each operation. Once the appropriate tools have been loaded in the tool magazines, the machines are under computer control. During system operation, the automatic tool interchange capability of each machine allows no idle setup time in between consecutive operations or between the use of consecutive tools. When a new tool is required, the tool magazine rotates into position, and the changer automatically interchanges the new tool with the one that is in the spindle in seconds.

Each part type that is machined is defined by several operations. Each operation requires several cutting tools (say, about 5 to 20). All tools for each operation need to occupy slots in one or more machine tool's tool magazine.

Each cutting tool takes 1, 3, or occasionally 5 slots in a machine's magazine. Magazines can have, say, 40 to 160 slots. Sixty is typical. Tools wear and break, so a computer needs to track the lives of all tools. A tool that breaks during the cut can severely damage the part and sometimes the machine or spindle.

Tools can be delivered to the FMS either manually or automatically, for example, via automated guided vehicles (AGVs). The delivered tools are loaded into the magazines either manually or automatically. Examples will be given shortly.

FMSs have an automated materials handling system that transports parts from machine to machine and into and out of the system. These may consist of wire-guided AGVs, a conveyor system, or tow-line carts, with usually a pallet interchange with the machines.

The interfaces between the materials handling system and the part are usually pallets and fixtures. Pallets "sit" on the cart and fixtures hold and clamp the parts onto the pallets. Pallets are usually identical. Fixtures are usually of different types. Fixtures are different in order to be able to hold securely different types of parts. Fixtures also hold identical parts in different orientations throughout their processing. The number of pallets in the FMS defines the amount of work-in-process inventory in the system.

After some machining, parts are often checked at the machine by interchanging automatically a probe into the spindle. The probe does some at-the-machine inspection of the cuts that were made. After several operations, a cart may bring the part to a washing station, to remove the chips before either further machining, refixturing, or inspection.

After several operations, often some parts are brought to an inspection station for more precise and more complete checking of the cuts to see if all tolerances were met.

Some existing systems will now be described. Figure 6.1 is a schematic of an FMS (see Stecke and Solberg[53]) that was built by White Sundstrand machine tool company for Caterpillar Tractor Company in Peoria, Ill. It consists of four five-axis machine tools, Omnimills, three four-axis machine tools, Omnidrills, two vertical turret lathes, an inspection station, and a washing station.

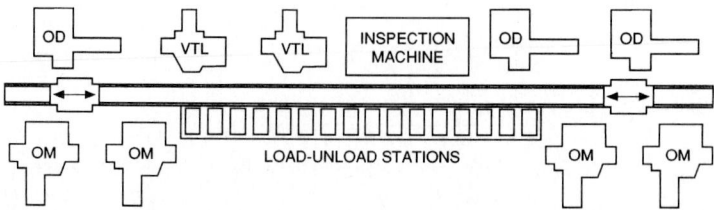

FIGURE 6.1 Sundstrand/Caterpillar flexible manufacturing system (FMS).

There's a 16-station load/unload (L/UL) area, where parts are manually palletized and fixtured for input into the system, where finished parts leave the system, and where refixturing takes place. Some of the stations are dedicated to the input of a particular part type, some dedicated to the output of a particular part type, and some dedicated to refixturing. The remainder provide a central buffer area for in-process inventory. There is no local storage at the machines. If a part's "next machine" is busy, that part is sent to an undedicated L/UL station to wait. After certain machining operations are performed on a part, a cart brings it to a particular L/UL station. The part is manually refixtured, that is, taken off its current pallet and fixture combination, rotated, and clamped on a fixture of a different type of another pallet on an adjacent dedicated L/UL station.

Two tow-line carts automatically transport parts from machine to machine to L/UL to washing to inspection. Once a part has been manually set up, the pallet-fixture-part combination is transported automatically and visits machines as a unit. All scheduling and control are under computer control unless manually overridden at a local station.

Two sizes of housings for automatic transmissions are machined. Each housing consists of a case and a cover. After machining operations are performed on each, a cover is manually put on a case at a particular L/UL station and the assembled part is sent out for additional machining operations.

There are two sizes of housings, so six different part types are machined. Each part (type) requires one or two manual refixturings before it is finished. Once a part is manually palletized and fixtured, the part and its processing are under computer control. The computer controls the machining operations (speeds and feeds of the CNCs) and the movement of parts (scheduling the operations), monitors tool wear, calculates remaining expected tool life of all of the many cutters that are used, ...

Each part consists of several operations per fixturing. An operation is a machine visit and is defined by its processing time at a machine tool and by the cutting tools required. For an operation to be performed at a particular machine tool, all of the cutting tools have to have been previously loaded into the tool magazine. Tool loading is manual here. Machines are idle when the tools are changed.

In 1984, the Sundstrand/Caterpillar FMS was replaced with a new Sundstrand system to machine (initially) the same six part numbers. The two Bullard turret lathes from the old line were used in the new system. The four Omnimills and three Omnidrills were replaced with updated Series 80 Sundstrand Omnimills.

Two other FMSs located in Torino, Italy, and built by Comau are described in Stecke.[45,46] The Comau equipment differs from the Sundstrand example in that there are local buffers at each machine tool. In particular, each machine tool has both an input and an output buffer.

In the latter FMS, several times during their processing, parts are batched for automatic transport out of the FMS for some manual processing and then brought back to the FMS for additional machining.

A schematic for the Comau-Torino FMS described in Stecke[45] is given in Fig. 6.2. It consists of six identical five-axis Comau CNCs, with three each on both sides of a washing and inspection unit and refixturing spot. Each machine tool has two primary tool magazines of 60 slots each. Primary means that the machine has direct access to both magazines.

Components for six different aluminum gearboxes are initially machined on the FMS. The requirements of each gearbox can vary by 5 to 10 percent daily. The FMS requirements for these six gearboxes consist of 2, 2, 2, 2, 3, and 4 components, respectively, to result in 15 part types to be manufactured on the FMS. To hold these 15 part types, four fixture types are required. Other components machined in a job shop as well as bought components will also feed the gearbox assembly line. Two carts transport parts.

Some other existing systems are described in Barash,[6] Cavaillé et al.,[12] and Ranky.[31] Most FMSs are devoted to the fabrication of metal parts, although there are other types, for example, flexible flow systems (FFSs) for fabricating circuit boards. See Ahmadi et al.[3]

Models are useful to analyze and sometimes solve many of the problems associated

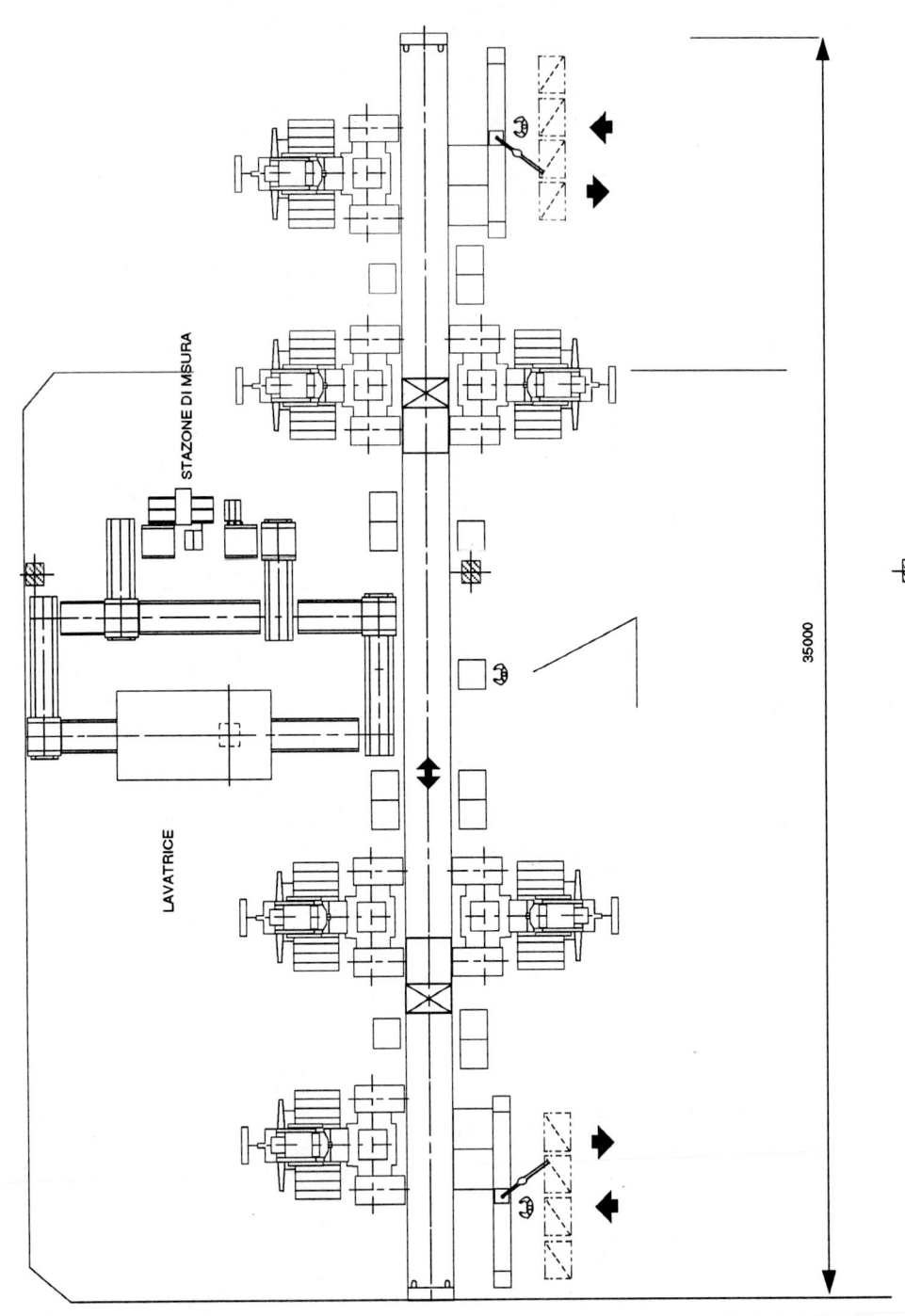

STAZIONE DI MSURA

LAVATRICE

35000

FIGURE 6.2 Layout of a Comau Torino flexible manufacturing system (FMS).

with the design, planning, and operation of these highly automated systems. This chapter next overviews these FMS problems that have to be addressed under the next heading "Overview of FMS Design, Planning, and Operating Problems." Various models that can and have been used to investigate aspects of these problems are presented later under the heading "Mathematical Models Useful to Analyze FMS Problems." Much current research addressing these is described. Finally, the scope of application of each model is discussed under "Model Applicability."

OVERVIEW OF FMS DESIGN, PLANNING, AND OPERATING PROBLEMS

In this section, we briefly list the many decisions that have to be made during the design, implementation, and subsequent operation of an FMS. Additional details and descriptions of these problems can be found in Stecke.[43]

A decision to automate should be based on both economic comparisons and strategic considerations. Assuming that management has decided that flexible manufacturing is appropriate for a particular application, perhaps to increase capacity in a certain department producing changing products or for new families of part types, the following design issues have to be addressed.

FMS Design Problems.   An initial consideration is to determine the families of part types that will be manufactured and/or assembled on the FMS or FFS. Then candidate combinations of machine tools and cutting tools can be generated that can take care of all of the operations that have to be performed as well as meet expected production requirements.

Group-technology-like concepts and methods can help with these decisions. One difference with traditional group technology applications is that in an FMS, it is desirable to have several identical machines that can perform the same operations (for alternative routing possibilities, for example). Another difference is the higher diversity of parts that the FMS will produce. This diversity in turn requires additional flexibility of certain types from the system.

The amount of flexibility that is needed or desired has to be decided, and this helps to determine the degree of automation and the type of FMS that is designed. Impacting the latter decision is the type of automated material handling system (MHS) that will move the parts from machine to machine. See Buzacott,[9] Browne et al.,[8] Stecke and Browne,[47] and Sethi and Sethi[34] for information on a spectrum of flexibility options. The required capacity of the MHS also has to be determined, in the case of tow-line carts or wire-guided vehicles. The type and size of buffers (to hold in-process inventory) has to be determined, for example, a central or local buffer area. The control structure hierarchy among the computers controlling machines, the MHS, and each other has to be designed and built. The FMS layout has to be decided. The numbers and designs of both pallets and fixtures of different fixture types have to be specified. Planning, scheduling, and control strategies to operate the system have to be created and developed. These will be discussed shortly.

There is a lot of iteration among all these *FMS design problems* as candidate solutions are suggested and then often determined to be infeasible when other factors are considered. Design specifications and needs change. These mandate that initial FMS designs vary widely. Efficient and accurate mathematical and other models are required to help narrow in on the appropriate FMS design. Following the development and subsequent implementation of the FMS design, models are also useful to help set up and schedule production through the system.

FMS Planning Problems.   Because of the quick automated cutting tool capability, negligible setup time is associated with a machine tool between consecutive operations as long as all the cutting tools required for that next operation have previously been loaded into the machine tool's limited-capacity tool magazine. However, determining which cutting tools should be placed in which tool magazine and then loading the tools into the magazine requires some "planning and system setup" time. Setup decisions that have to be made and implemented before the system can begin to manufacture parts are called *FMS planning problems*. When the system has been setup and can begin production, the remaining problems are those of *FMS scheduling*.

The first *FMS planning problem* is to decide which of the part types that have production requirements (either forecasted demand or customer orders) should be those next manufactured during the same time over the immediate time period.

This information can be used to help determine the amount of pooling among the identical machine tools that can occur. Pooling, or identically tooling all machines that are in the same machine group, has many system benefits. For example, alternative routes for parts are automatically allowed and also machine breakdowns may not cause production to stop. This is because all machine tools in a group, being tooled identically, are able to perform the same operations.

Another *FMS planning problem* is to determine the relative ratios at which the selected part types should be on the system, to attain a good utilization, say. The limited numbers of pallets and fixtures of each fixture type impact these production ratios. Also, determining the minimum number of pallets and fixtures that are needed to maintain these production ratios is required.

Finally, each operation and its associated cutting tools of the selected set of part types has to be assigned to one or more of the machine tools in an intelligent manner. Different loading objectives that can be followed are applicable in different situations.

When all these decisions have been made and the cutting tools loaded into the selected tool magazines, production can begin. Then the following *FMS scheduling problems* have to be addressed.

FMS Scheduling Problems.   Subsequent problems are concerned with the operation of the system after it has been set up during the planning stage. One problem is to determine an appropriate policy to input the parts of the selected part types into the FMS, or efficient means to determine which parts to input next. In some situations, a periodic input may be sufficient. (See Agnetis et al.[2]) In others, determining the next input to keep producing according to the calculated production ratios is appropriate. Parts have to be input so as to not exceed capacity.

Then, applicable algorithms to schedule the operations of all parts through the system have to be determined. Real-time scheduling is usually more appropriate for these automated systems, as opposed to a fixed schedule. Tool breakage, down machine tools, and the like would totally disrupt a fixed schedule. However, a fixed schedule is useful as an initial guideline to follow.

Potential scheduling methods range from simple dispatching rules to sophisticated algorithms having look-ahead capabilities. Scheduling might be helped by determining priorities among all part types for those parts waiting for a particular machine tool and perhaps also priorities among the machine tools. Machine breakdowns and the many other system disturbances should be considered when developing scheduling and control procedures. If the system is "set up" during the planning phase, with sufficient care and flexibility, the scheduling function will be much easier.

FMS Control Problems.   By *FMS control,* we mean the continuous monitoring of the system to be sure that it's doing what was planned for it to do and is meeting the expectations set up for it.

For example, during the *FMS design* phase, policies have to be determined to handle breakdown situations of many types. However, all breakdown types cannot be anticipated. If a machine tool goes down, what should be done with the present schedule? Is the breakdown expected to be long or short? Should the planned schedule be revised? Should a new schedule be developed? It might be appropriate to determine how to return to the original schedule as soon as possible. In any case, it is desirable to reallocate operations and reload the cutting tools (if they have to be) so that the tool-changing time is minimized.

System reliability is an issue. Maintenance policies have to be determined. Preventive maintenance should be performed, which could perhaps be scheduled on a regular basis. The number of repairpersons to have per shift has to be determined as well as priorities among their many maintenance tasks.

Inspection points (checks for part quality) and the frequency of inspection of in-process and finished parts must be determined. The appropriate inspection equipment has to be selected, interfaced with the existing machine tools, and implemented.

Monitoring procedures for both the processes and cutting tool lives has to be specified as well as methods to collect data of various types (monitoring and breakdown). Tool life estimates should be reviewed and updated. Reasons for process errors have to be found (that is, machine or pallet misalignment, cutting tool wear and detection, swarf problems, and so on) and the problems corrected.

Many of these procedures can and should be devised during the design phase. New control, monitoring, and breakdown problems that surface during implementation and operation require immediate attention. Some of these problems are new and system-specific and so may not be anticipated before their actual occurrence.

Hierarchical Approaches.   Because the planning and scheduling problems are complex and require a lot of data consideration, many of these problems have been framed and subdivided within a hierarchy. The solution of each subproblem provides constraints on problems lower in the hierarchy. The partition of FMS problems into *planning* (before time zero) and *scheduling* (after production begins) is one example of a hierarchy. The *FMS planning problems* are another hierarchical decomposition of a system setup problem. Stecke[42,44] suggests hierarchical and iterative approaches to several of these problems.

FMS scheduling problems have also been addressed hierarchically. Both the uncertainty in demand as well as machine breakdowns are accounted for. For example, Akella et al.[4] account for demand during any particular failure state at the higher level of a hierarchy of problems. A lower-level control determines which part should be input next and when, while also considering breakdowns.

MATHEMATICAL MODELS USEFUL TO ANALYZE FMS PROBLEMS

Models are useful to identify key factors that will affect system performance and to provide insight into how a system behaves and how the system components interact. Models should be applied to help determine the appropriate procedures to design and set up a system or strategies to help run a system efficiently.

In this section, we overview many of the models that can be and/or have been used to analyze the problems mentioned in the "Overview of FMS Design, Planning, and Operating Problems." In the next section, "Model Applicability," we indicate how these models can be and have been used and which problems each model might be applicable to. Relevant research on various approaches to date will be noted. A whole range of different models are required in order to address the hierarchy of various problems to be solved.

Some of the models that we now describe are simulation, queuing networks, perturbation analysis, mathematical programming, Petri nets, and artificial intelligence. We note that Buzacott and Yao[10] review the early FMS research that used analytic models, mostly queuing network models. Suri[58] provides a brief review of some models that can be used to evaluate suggested or candidate solutions to the problems described in "Overview of FMS Design, Planning, and Operating Problems."

Simulation.   Depending on the amount of information that is built into a particular model, simulation has the potential to be the most detailed and flexible model, allowing as much detail as desired or necessary to mimic reality. Simulation can also potentially be the most expensive and time-consuming to develop, debug, and run. Many computer runs may be required to investigate the possibilities before a decision is made. Of course, this need not be. With the aid of simulation languages and/or sufficient modeling capability, a person can efficiently and quickly capture the necessary detail in a particular FMS model. Usually, a simulation is tailor-made for the particular application at hand.

FMSs have been modeled using some of the existing general-purpose languages, such as SLAM II, GASP IV, GPSS/H, SIMAN, or SIMSCRIPT. Some special-purpose (but still general) *manufacturing system* simulation languages have been developed. Most of the following examples of these languages can be applied on a microcomputer. Many have graphics capabilities of various types.

MicroNET is an interactive network simulation language developed by Pritsker & Associates. MAP/1 is another product offered by Pritsker & Associates. SEE-WHY is one developed by British Leyland Systems, Ltd. SIMAN is another network language developed at Pennsylvania State University. CAPS/ECSL (computer-aided programming for simulation/extended control simulation language) had in part been developed in the U.K. SPEED was developed by Horizon Software, Inc. MAST was developed by CMS Research and is an outgrowth of GCMS, developed at Purdue University. GFMS of Draper Laboratories is another outgrowth of GCMS. These are not evaluated here. The main purpose is to acknowledge their (and others) existence and usefulness in modeling an FMS.

Queuing Networks. Both open and closed queuing networks (OQNs and CQNs) have been used to model an FMS at an aggregate level of detail. These models can take into account the interactions and congestion of parts competing for the same machines and the uncertainty and dynamics of an FMS. Most simple queuing networks require, as input, certain average values, such as the average processing time of an operation at a particular machine tool and the average frequency of visits to a machine. The outputs that are obtained and useful for evaluating the performance of a suggested system configuration are also average values and include the steady-state expected production rate, mean queue lengths, and machine utilizations. These models are very efficient, running in seconds of computer time. They also give adequate estimates of steady-state performance measures.

Solberg[40] was the first to suggest the use of a simple, single-class, multiserver, closed queuing network to model an FMS. His computer program, called CAN-Q, uses Buzen's efficient algorithm to analyze product form queuing networks. Dubois[18] describes a variation of this model to mimic the availability or lack of various amounts of in-process inventory.

In an OQN, parts arrive externally according to a Poisson process and leave when completed. OQNs have been used for FMS performance evaluation insights.

CQNs contain a fixed number of parts with no external arrivals or departures. The congestion due to parts competing for the same limited resources (machines) is captured. The required normalizing constant provides most of the relevant measures that can evaluate the performance of a particular configuration.

Mean value analysis (MVA) is an alternative means to Buzen's algorithm of providing steady-state mean performance measures for product form networks. MVA provides an easier means to analyze multiclass queuing networks that can model individual part types. Although they are more time-consuming to analyze, they also allow more modeling detail. However, these models are less flexible in that they cannot easily handle multiple servers. Its main extension to CAN-Q is the ability to explicitly model each part type, allocating a fixed number of pallets to each type. The production rates of each part are then provided. However, production ratios of each part type can't be set in advance as input.

Cavaillé and Dubois[11] apply MVA to systems that contain no machine groups. Suri and Hildebrant[59] allow the pooling of machines into machine groups in a package called MVAQ. Suri[57] demonstrated the robustness of these CQN models to provide reliable results even when the assumptions are quite different from reality.

Perturbation Analysis. Perturbation analysis is a technique developed at Harvard University. The analysis provides much more additional information to that normally provided by a simulation's output. (See Ho and Cao[24] and Suri,[58] for example.)

The analysis proceeds as follows. A discrete-event dynamic system of interest (such as an FMS) is simulated and a sample path of a trajectory is observed. The relevant data are collected. One event is perturbed. Perhaps another customer enters the system or the mean service time at a machine is changed. The perturbation is propagated throughout the system over time. The consequences of that perturbation and others caused by it are tracked over time along the initial sample path. Perturbations are added using some superposition rules to determine the net effect on a chosen performance measure. The net result is to obtain additional results from only one simulation run, but as if n simulations had been performed. From the observations of one sample path, a gradient vector of output is estimated. Sensitivity analyses can be performed on a number of parameters.

To apply this technique, much time is spent in analyzing the system of interest and developing system-specific perturbation equations to propagate the perturbations. The analysis is tailored to the particular application, just as a simulation would be.

Mathematical Programming.   Some of the problems that were mentioned in the "Overview of FMS Design, Planning, and Operating Problems" have been formulated mathematically, some as nonlinear integer programs (Stecke[42]), others using linear or integer programs (Rajagopalan[29]).

Depending on the problems formulated, some formulations are detailed and tractable (and hence immediately useful). Other formulations are detailed and untractable. (However, heuristic or other algorithms can or have been developed from the exact formulations to solve the problems. For example, see Wittrock.[63]) Some formulations are aggregate and do not contain all details of the problem being modeled.

Timed Petri Nets.   Petri nets are useful to model systems whose behavior can be described as interferences between asynchronous and concurrent processes (that is, FMSs). They can be useful for analyzing transient, steady-state, and real-time control issues.

For certain subclasses of timed Petri nets, in particular, those that are decision-free, very efficient algebraic techniques have been developed to analyze the performance of a Petri net. A decision-free, timed Petri net description is equivalent to linear state equations in a {max, +}-based algebra. Machine utilizations, information on the bottleneck machine, cycle time and hence production rate, and length of the transient period can all be obtained via efficient algorithms, based on graph-theoretic concepts for the analysis of cyclic event graphs. See Cohen et al.[13] and Dubois and Stecke.[19,20]

Artificial Intelligence.   Expert systems are being developed and applied to such diverse areas as medical diagnostics, oil or mineral exploration, computer system configuration, and production control. The usual method of encoding knowledge for expert systems is in the form of rules. Production rules are typically of the form: *If* (a series of conditions are satisfied), *then* (a set of consequences can be produced). The implementation process of an expert system consists of various ways of scanning lists of the *if* parts of rules in order to match them with the *then* parts. Expert systems can be developed to address some of the *FMS scheduling* and *control* problems described in the "Overview of FMS Design, Planning, and Operating Problems." See Shaw and Whinston,[38] Ranky,[30] Shaw,[37] Stecke et al.,[50,51] Villa and Rossetto,[61] and Devedžić.[16]

MODEL APPLICABILITY

In this section, we indicate the scope of applicability of each of the models described in "Mathematical Models Useful to Analyze FMS Problems." The usefulness of each model is discussed as well as which problems each model can address. References are provided.

Simulation.   Simulation can be and has been used for all problem types. It is the most widely used modeling tool. For example, Renault Machines Outiles used a simulation model to help design their FMS in Boutheon, France, and will use the simulation in future similar design problems. VUOSO has developed a simulation model to help solve some of their tool management problems (planning problems) at their FMSs at Celakovice and Olomouc, Czechoslovakia. The Vought LTV FMS in Dallas uses simulation to help schedule and control production.

Some existing flexible manufacturing installations have been simulated using the simulation languages mentioned under "Simulation." For example, Stecke and Solberg[53] investigated alternative loading and scheduling strategies (using GASP IV) for the system built by White Sundstrand for Caterpillar Tractor Company in Peoria, Ill. (see Fig. 6.1.) Cavaillé, Forestier, and Bel[12] simulated a proposed design for the Renault FMS in Boutheon, France. Schriber and Stecke[33] and Stecke and Kim[50] describe how GPSS/H can be used to model some of the unique aspects of an FMS, such as AGV movement and limited buffers. Graver and McGinnis[22] use simulation to solve a tooling design problem of determining how many tools of each tool type need to be commissioned for use in an FMS.

Because of machine breakdown and other random events, the nature of an FMS is stochastic, despite largely deterministic processing times. Expertise in simulation output analysis, in the design of experiments, and in determining appropriate confidence limits is needed.

At an advanced stage of the *FMS design process,* simulation is useful to get a view of the behavior of the system as a function of candidate scheduling and operating policies. Detailed questions can be analyzed and system parameters determined. However, simulation may be too detailed or expensive to use during initial design consideration. Initial designs can vary widely.

Queuing Networks.   Queuing networks are useful in providing quantitative answers to some *FMS design problems,* such as determining the necessary capacity (perhaps in terms of the number of carts) of the material handling system (see Schriber and Stecke,[32] determining the number of machine tools that are required of each type (see Lee and Johnson[25]), and locating bottleneck processes. Queuing networks are useful for narrowing in on appropriate "ballpark," preliminary designs, to suggest several possible configurations. Then models that allow more detail can be used to fine-tune actual design parameters.

Queuing networks have been used to determine the necessary number of pallets and fixtures (Solot[41]). They've been used to determine the allocation of buffer storages in an FMS (see Shanthikumar and Yao[36] and So[39]).

Closed queuing network models have proved useful in providing insight into how system components interact as parts compete for the same limited resources. They provide qualitative information about some of the *FMS planning problems,* in particular about machine grouping (see Stecke and Solberg[54]) and cutting tool loading. Most of the loading results pertain to workload balancing when there is no grouping (see Shanthikumar and Stecke[35]). For results on unbalancing machine workloads for unbalanced grouping, see Stecke and Solberg[54] and Dallery and Stecke.[14]

Since the most relevant queue discipline for an FMS (which retains product form solutions) is FCFS, which requires exponential servers, queuing networks are *not* very useful in studying breakdown situations or in scheduling and control problems. However, dynamic load control policies for FMSs have been constructed by Tenenbaum and Seidmann.[60]

Perturbation Analysis.   Perturbation analysis is a tool to help pull information from the data of one simulation run of a discrete event system, such as an FMS. Gradient vectors can be determined and sensitivity analyses performed on various parameters.

The perturbed event cannot be too large. For example, the addition of a new machine tool cannot be analyzed using PA techniques. The technique has been applied to queuing networks for which product form is not preserved, for example, to study finite buffer situations to help determine the correct size of the buffers.

Mathematical Programming.   Graves and Redfield,[23] Afentakis,[1] and Lee and Johnson[25] have used mathematical programming to address various design problems, such as equipment selection (Graves and Redfield) and machine layout (Afentakis and Lee and Johnson).

Mathematical programming has been used to both formulate, and sometimes solve, several of the *FMS planning* and *scheduling problems*. For example, the FMS machine grouping and tool loading problems have been formulated in all detail as nonlinear integer problems. (See Stecke.[42,44]) Branch and bound (Berrada and Stecke[7]) and implicit enumeration techniques (Mazzola et al.[27]) are useful for solving several of these problems. Arbib et al.[5] optimally balance machine workloads. Stecke and Talbot[55] have suggested some heuristics for the *FMS loading problem* and Wittrock[63] and Das and Khumawala,[15] for a scheduling problem.

Linear and integer programming has been used by Rajagopalan,[29] Stecke and Kim,[48–50] Agnetis et al.,[2] and Stecke and Toczylowski[56] to address the FMS part type selection problem. Rajagopalan[29] develops heuristics to address the problem.

Akella et al.[4] formulated the higher-level scheduling problem as a continuous-time dynamic program, considering both the rate of demand and the state of the system with respect to machine failures. Linear programming is used at a middle level to determine a suboptimal, but appropriate, production rate control. Rabinowitz et al.[28] schedules the activities of robots in assembly cells.

Escudero[21] and Stecke and Kim[48] have used mathematical programming to address input sequencing problems. Agnetis et al.[2] develop an optimal approach for input sequencing into a two-machine FFS that generalizes Johnson's algorithm.

Many of the other models described here are used to evaluate the candidate solutions that are generated by some of these mathematical programming techniques.

Timed Petri Nets. Petri nets, in conjunction with certain modeling conventions, appears to be a general modeling tool (see Wadhwa and Browne[62]). In particular, activities requiring many resources (such as a part requiring a machine tool, cart, robot, cutting tools, and so on) can be modeled. Many activities having durations (such as processing times, transportation times, and setup times) can also be modeled. See Dubois and Stecke[20] for a description and some examples of FMS applications of these modeling capabilities.

The most general means to *analyze* a general Petri net description is simulation. In general situations, Petri nets can also be useful in designing a well-defined simulation.

On the other hand, for certain subclasses of timed Petri nets, called decision-free, efficient algebraic techniques have been developed to analyze and evaluate their performance. See Cohen et al.[13] and Dubois and Stecke.[19]

These subclasses of timed Petri nets can be useful to investigate real-time, steady-state, and transient scheduling and control problems, such as the determination of relevant control rules to order operations waiting for each machine tool or the determination of the minimum number of pallets required to maintain the required production ratios. Information that is readily attainable from the analysis of a decision-free Petri net includes the cycle time and production rate, the period and the order of the period, the bottleneck machine, the utilizations of the bottleneck machine and all other machines, the critical resources, and the duration of the transient period that precedes steady state. This information can be used to address many of the *FMS design* and *scheduling* problems that were described earlier.

SUMMARY

A brief overview of FMSs and the various mathematical models that can be and have been used to examine and solve many of the design and operation problems associated with flexible manufacturing has been provided. Each model is useful under different circumstances and for different types of problems. For some problems, it is useful to use a hierarchy of models to solve them.

ACKNOWLEDGMENTS

The author's work was supported in part by a summer research grant from the School of Business Administration of The University of Michigan.

REFERENCES

1. Afentakis, Panos, "A Loop Layout Design Problem for Flexible Manufacturing Systems," *International Journal of Flexible Manufacturing Systems,* vol. 1, no. 2, pp. 175–196, April 1989.

2. Agnetis, Alessandro, Claudio Arbib, and Kathryn E. Stecke, "Optimal Scheduling in a Flexible Flow System," *Proceedings of the Second International Conference on Computer Integrated Manufacturing,* R.P.I., Troy, N.Y., May 21–23, 1990.

3. Ahmadi, Javad, Stephen Grotzinger, and Dennis Johnson, "Emulating Concurrency in a Circuit Card Assembly System," *International Journal of Flexible Manufacturing Systems,* vol. 3, no. 1, pp. 45–70, March 1991.

4. Akella, Ramakrishna, Yong Choong, and Stanley B. Gershwin, "Performance of Hierarchical Production Scheduling Policy," *Annals of Operations Research,* vol. 3, pp. 403–425, 1985.

5. Arbib, Claudio, Mario Lucertini, and Fernando Nicolò, "Workload Balance and Part-Transfer Minimization in Flexible Manufacturing Systems," *International Journal of Flexible Manufacturing Systems,* vol. 3, no. 1, pp. 5–26, March 1991.

6. Barash, Moshe M., "Computerized Manufacturing Systems for Discrete Products," chap. VII-9 in Gavriel Salvendy, ed., *The Handbook of Industrial Engineering,* Wiley, New York, 1982.

7. Berrada, Mohammad, and Kathryn E. Stecke, "A Branch and Bound Approach for Machine Load Balancing in Flexible Manufacturing Systems," *Management Science,* vol. 32, no. 10, pp. 1316–1335, October 1986.

8. Browne, Jim, Didier Dubois, Keith Rathmill, Suresh P. Sethi, and Kathryn E. Stecke, "Classification of Flexible Manufacturing Systems," *FMS Magazine,* vol. 2, no. 2, pp. 114–117, April 1984.

9. Buzacott, John A., "The Fundamental Principles of Flexibility in Manufacturing Systems," *Proceedings of the 1st International Conference on Flexible Manufacturing Systems,* Brighton, England, pp. 13–22, October 1982.

10. Buzacott, John A., and David D. W. Yao, "Flexible Manufacturing Systems: A Review of Analytical Models," *Management Science,* vol. 32, no. 7, pp. 890–905, July 1986.

11. Cavaillé, Jean-Bernard, and Didier Dubois, "Heuristic Methods Based on Mean-Value Analysis for Flexible Manufacturing Systems Performance Evaluation," *Proceedings of the 21st IEEE Conference on Decision and Control,* Orlando, Fla., pp. 1061–1065, December 1982.

12. Cavaillé, Jean-Bernard, J. P. Forestier, and Gerard Bel, "A Simulation Program for Analysis and Design of a Flexible Manufacturing System," *Proceedings of the IEEE Conference on Cybernetics and Society,* Atlanta, Ga., pp. 257–259, October 1981.

13. Cohen, Guy, Didier Dubois, Jean P. Quadrat, and M. Viot, "A Linear-System-Theoretic View of Discrete-Event Systems," *Proceedings of the 22nd IEEE Conference on Decision and Control,* San Antonio, Tex., Dec. 14–16, 1983.

14. Dallery, Yves, and Kathryn E. Stecke, "On the Optimal Allocation of Servers and Workloads in Closed Queuing Networks," *Operations Research,* vol. 38, no. 4, July–August 1990.

15. Das, Sidhartha R., and Basheer M. Khumawala, "An Efficient Heuristic for Scheduling Batches of Parts in a Flexible Flow System," *International Journal of Flexible Manufacturing Systems,* vol. 3, no. 2, pp. 121–148, March 1991.

16. Devedžić, Vladan, "A Knowledge-Based System for the Strategic Control Level of Robots in Flexible Manufacturing Cells," *International Journal of Flexible Manufacturing Systems,* vol. 2, no. 4, pp. 263–287, July 1990.

17. de Werra, D., and M. Widmer, "Loading Problems with Tool Management in Flexible Manufac-

turing Systems: A Few Integer Programming Models," *International Journal of Flexible Manufacturing Systems,* vol. 3, no. 1, pp. 71–82, March 1991.

18. Dubois, Didier, "A Mathematical Model of a Flexible Manufacturing System with Limited In-Process Inventory," *European Journal of Operational Research,* vol. 14, no. 1, pp. 66–78, January 1983.

19. Dubois, Didier, and Kathryn E. Stecke, "Dynamic Analysis of Repetitive Decision-Free Discrete Event Processes: Algorithmic Issues and the Algebra of Timed Marked Graphs," *Annals of Operations Research,* vol. 26, "Production Planning and Scheduling" (Maurice Queyranne, ed.), pp. 151–193, 1990.

20. Dubois, Didier, and Kathryn E. Stecke, "Dynamic Analysis of Repetitive Decision-Free Discrete Event Processes: Applications to Production Systems," *Annals of Operations Research,* vol. 26, "Automated Manufacturing Systems" (Joseph B. Mazzola, ed.), pp. 323–397, 1990.

21. Escudero, L. F., "An Inexact Algorithm for Part Input Sequencing with Side Constraints in FMS," *International Journal of Flexible Manufacturing Systems,* vol. 1, no. 2, pp. 143–173, April 1989.

22. Graver, Thomas W., and Leon F. McGinnis, "A Tool Provisioning Problem in an FMS," *International Journal of Flexible Manufacturing Systems,* vol. 1, no. 3, pp. 239–254, June 1989.

23. Graves, Stephen C., and Carol Holmes Redfield, "Equipment Selection and Task Assignment for Multiproduct Assembly System Design," *International Journal of Flexible Manufacturing Systems,* vol. 1, no. 1, pp. 31–51, September 1988.

24. Ho, Yu Chi, and Xiren Cao, "Perturbation Analysis and Optimization of Queuing Networks," *Journal of Optimization Theory and Applications,* 1983.

25. Lee, Heungsoon Felix, and Roger Vivian Johnson, "A Line Balancing Strategy for Designing Flexible Assembly Systems," *International Journal of Flexible Manufacturing Systems,* vol. 3, no. 2, pp. 91–120, March 1991.

26. Lin, Yuh-Jiun, and James J. Solberg, "Effectiveness of Flexible Routing Control," *International Journal of Flexible Manufacturing Systems,* vol. 3, nos. 3, 4, pp. 189–211, June 1991.

27. Mazzola, Joseph B., Alan W. Neebe, and Christopher V. R. Dunn, "Production Planning of a Flexible Manufacturing System in a Material Requirements Planning Environment," *International Journal of Flexible Manufacturing Systems,* vol. 1, no. 2, pp. 115–142, April 1989.

28. Rabinowitz, Gad, Abraham Mehrez, and Subhashish Samaddar, "A Scheduling Model for Multirobot Assembly Cells," *International Journal of Flexible Manufacturing Systems,* vol. 3, no. 2, pp. 149–180, March 1991.

29. Rajagopalan, S., "Formulation and Heuristic Solutions for Parts Grouping and Tool Loading in Flexible Manufacturing Systems," *Proceedings of the Second ORSA/TIMS Conference on Flexible Manufacturing Systems: Operations Research Models and Applications,* Ann Arbor, Mich., Elsevier Science Publishers B.V., Amsterdam, pp. 311–320, Aug. 12–15, 1986.

30. Ranky, Paul G., "A Real-Time, Rule-Based FMS Operation Control Strategy in CIM Environment," Part 1, *International Journal of CIM,* vol. 1, no. 1, pp. 55–72, 1988.

31. Ranky, Paul G., *Flexible Manufacturing Cells and Systems in CIM,* CIMware Ltd., Guildford, Surrey, England, 1990.

32. Schriber, Thomas J., and Kathryn E. Stecke, "Machine Utilizations Achieved Using Balanced FMS Production Ratios," *Annals of Operations Research,* vol. 15, pp. 229–267, 1988.

33. Schriber, Thomas J., and Kathryn E. Stecke, "Machine Utilizations and Production Rates Achieved by Using Balanced Aggregate FMS Production Ratios in a Simulated Setting," *Proceedings of the Second ORSA/TIMS Conference on Flexible Manufacturing Systems: Operations Research Models and Applications,* Ann Arbor, Mich., Elsevier Science Publishers B.V., Amsterdam, pp. 405–416, Aug. 12–15, 1986.

34. Sethi, Andrea Krasa, and Suresh Pal Sethi, "Flexibility in Manufacturing: A Survey," *International Journal of Flexible Manufacturing Systems,* vol. 2, no. 4, pp. 289–328, July 1990.

35. Shanthikumar, J. George, and Kathryn E. Stecke, "Reducing Work-in-Process Inventory in Certain Classes of Flexible Manufacturing Systems," *European Journal of Operational Research,* vol. 26, no. 2, pp. 266–271, August 1986.

36. Shanthikumar, J. George, and David D. Yao, "Optimal Buffer Allocation in a Multicell System," *International Journal of Flexible Manufacturing Systems,* vol. 1, no. 4, pp. 347–356, September 1989.

37. Shaw, Michael J., "A Pattern-Directed Approach to Flexible Manufacturing: A Framework for Intelligent Scheduling, Learning, and Control," *International Journal of Flexible Manufacturing Systems,* vol. 2, no. 2, pp. 121–144, December 1989.

38. Shaw, Michael J., and Andrew B. Whinston, "A Distributed Knowledge-Based Approach to Flexible Automation: The Contract Net Framework," *International Journal of Flexible Manufacturing Systems,* vol. 1, no. 1, pp. 85–104, September 1988.

39. So, Kut C., "Allocating Buffer Storages in a Flexible Manufacturing System," *International Journal of Flexible Manufacturing Systems,* vol. 1, no. 3, pp. 223–237, June 1989.

40. Solberg, James J., "A Mathematical Model of Computerized Manufacturing Systems," *Proceedings of the 4th International Conference on Production Research,* Tokyo, Japan, August 1977.

41. Solot, Phillipe, "A Heuristic Method to Determine the Number of Pallets in a Flexible Manufacturing System with Several Pallet Types," *International Journal of Flexible Manufacturing Systems,* vol. 2, no. 3, pp. 191–216, May 1990.

42. Stecke, Kathryn E., "Formulation and Solution of Nonlinear Integer Production Planning Problems for Flexible Manufacturing Systems," *Management Science,* vol. 29, no. 3, pp. 273–288, March 1983.

43. Stecke, Kathryn E., "Design, Planning, Scheduling, and Control Problems of Flexible Manufacturing Systems," *Annals of Operations Research,* vol. 3, pp. 3–12, 1985.

44. Stecke, Kathryn E., "A Hierarchical Approach to Solving Machine Grouping and Loading Problems of Flexible Manufacturing Systems," *European Journal of Operational Research,* vol. 24, no. 3, pp. 369–378, March 1985.

45. Stecke, Kathryn E., "Algorithms for Efficient Planning and Operation of a Particular FMS," *International Journal of Flexible Manufacturing Systems,* vol. 1, no. 4, pp. 287–324, September 1989.

46. Stecke, Kathryn E., "Planning and Scheduling Approaches to Operate a Particular FMS," *European Journal of Operational Research,* 1991.

47. Stecke, Kathryn E., and Jim Browne, "Variations in Flexible Manufacturing Systems According to the Relevant Types of Automated Material Handling," *Material Flow,* vol. 2, no. 2, pp. 179–185, July 1985.

48. Stecke, Kathryn E., and Ilyong Kim, "A Study of FMS Part Type Selection Approaches for Short-Term Production Planning," *International Journal of Flexible Manufacturing Systems,* vol. 1, no. 1, pp. 7–29, September 1988.

49. Stecke, Kathryn E., and Ilyong Kim, "Performance Evaluation for Systems of Pooled Machines of Unequal Sizes: Unbalancing Versus Balancing," *European Journal of Operational Research,* vol. 42, pp. 22–38, 1989.

50. Stecke, Kathryn E., and Ilyong Kim, "A Flexible Approach to Part Type Selection Using Part Mix Ratios in Flexible Flow Systems," *International Journal of Production Research,* vol. 29, no. 1, pp. 53–75, January–February 1991.

51. Stecke, Kathryn E., Ilyong Kim, and Moonkee Min, "A Knowledge-Based Approach to Part Type Selection Considering Due Dates in Flexible Manufacturing Systems," *Proceedings of the 12th IMACS World Congress on Scientific Computation,* Paris, France, July 17–22, 1988.

52. Stecke, Kathryn E., Moonkee Min, and Ilyong Kim, "A Hybrid Model-Based Approach for the Production Planning of FMSs," *Proceedings of the Third International Conference on Expert Systems and the Leading Edge in Production and Operations Management,* Hilton Head Island, S.C., May 20–24, 1989.

53. Stecke, Kathryn E., and James J. Solberg, "Loading and Control Policies for a Flexible Manufacturing System," *International Journal of Production Research,* vol. 19, no. 5, pp. 481–490, September–October 1981.

54. Stecke, Kathryn E., and James J. Solberg, "The Optimality of Unbalancing Both Workloads and Machine Group Sizes in Closed Queuing Networks of Multiserver Queues," *Operations Research,* vol. 33, no. 4, pp. 882–910, July–August 1985.

55. Stecke, Kathryn E., and F. Brian Talbot, "Heuristic Loading Algorithms for Flexible Manufacturing Systems," *Proceedings of the Seventh International Conference on Production Research,* Windsor, Ontario, Canada, Aug. 22–24, 1983.

56. Stecke, Kathryn E., and Eugeniusz Toczylowski, "Profit-Based FMS Dynamic Part Type Selec-

tion over Time for Mid-term Production Planning," *European Journal of Operational Research,* 1991.

57. Suri, Rajan, "Robustness of Queuing Network Formulae," *Journal of the Association for Computing Machinery,* vol. 30, no. 3, pp. 564–594, July 1983.

58. Suri, Rajan, "An Overview of Evaluative Models for Flexible Manufacturing Systems," *Proceedings of the First ORSA/TIMS Special Interest Conference on Flexible Manufacturing Systems: Operations Research Models and Applications, Ann Arbor, Mich.,* Aug. 15–17, 1984.

59. Suri, Rajan, and Richard R. Hildebrant, "Modelling Flexible Manufacturing Systems Using Mean Value Analysis," *Journal of Manufacturing Systems,* vol. 3, no. 1, pp. 27–38, January 1984.

60. Tenenbaum, Abraham, and Abraham Seidmann, "Dynamic Load Control Policies for a Flexible Manufacturing System with Stochastic Processing Rates," *International Journal of Flexible Manufacturing Systems,* vol. 2, no. 2, pp. 93–120, December 1989.

61. Villa, Agostino, and Sergio Rossetto, "Rule-Based Production Planning in Flexible Manufacturing Systems," *International Journal of Flexible Manufacturing Systems,* vol. 2, no. 1, pp. 5–24, October 1989.

62. Wadhwa, S., and Jim Browne, "Modeling FMS with Decision Petri Nets," *International Journal of Flexible Manufacturing Systems,* vol. 1, no. 3, pp. 255–280, June 1989.

63. Wittrock, Robert J., "Scheduling Parallel Machines with Major and Minor Setup Times," *International Journal of Flexible Manufacturing Systems,* vol. 2, no. 4, pp. 329–341, July 1990.

CHAPTER 7

CONCURRENT ENGINEERING: INTEGRATED PRODUCT AND PROCESS DESIGN

James L. Nevins
Arthur D. Little, Inc.
The Charles Stork Draper Laboratory Inc.
Cambridge, Massachusetts

To be competitive in today's environment, management must review all their basic strategies of operation, use of facilities, and of course how they invest their capital. In this chapter we look at new strategies to support operations, particularly those dealing with the design, manufacture, and fielding of products.

STRATEGIES

The goals generally are to reduce the cycle time from product concept to market by 50 percent, reducing costs by similar amounts while increasing quality—but particularly to give early visibility to many downstream options at the product concept level, thus reducing risk. The reason for doing this is usually extreme competitive pressure. To accomplish these goals, a team is assembled and applied on the next new or redesigned product that the company is planning.

The advantage of this method is that it can be implemented instantly with your present people without the necessity of investing in a single piece of new technology (computers, robots, FMSs, and the like). The investment is in people, that is, changing their awareness by training. Teams are then formed with designers, manufacturers, suppliers, and users, basically refocusing the entire company to meet the perceived competitive threat. Contrary to some claims, the issue is not simply communication between diverse groups or individuals. It is consideration of the many downstream options and supporting decision by analysis at the concept design point rather than serially.

One problem with this technique is that it presents many institutional-social problems to management: how individual team members resolved their career issues, what happens to them when they leave the team and return to their old organizations, the use of outside "facilitators," what types of training are needed for staff, for management, and so on.

This larger strategy attempts to apply all the typical downstream activities at the product concept level. It is called concurrent engineering (CE) or concurrent design (CD)[1] or integrated product and process design or simultaneous engineering or integrated system management. On a higher level, which seeks to integrate all the activities of the company, it is called EIF (enterprise integrated framework).[2] Much has been written about these techniques in the media, professional publications as well as government studies.[3-6]

Levels of Application.   What is it, why is it important, and how is it applied!

First Level.   There are three principal levels of application. The first level, as already described, involves a management decision to integrate all the normal serial operations into an integrated team. The approach is very effective, as indicated in the various references cited earlier. The down side to the first type of implementation is that eventually everyone will do it. *Therefore, there is no long-term competitive advantage.* Thus more advanced strategies—levels 2 and 3—are needed.

Second Level.   The second level of CD/CE implementation involves the use of new tools and methods (a number of them computer-based) to aid this process by giving quantitative visibility to options, trade-offs, and constraints due to economic-technological issues. It is generally the approach used by companies after they have tried a few projects using the first method and realized that new tools and methods may be helpful to alleviate some of the problems associated with the integrated team only concept.

For example, designers usually are looked on as the elite; they have the most computers or workstations, they have the analysis tools, and the like. The manufacturers or users have many years of experience, but not much in the way of quantitative results to buttress their viewpoint. Thus, when arguments take place with designers the manufacturing people find it hard to rebuff the computer-supported analysis results used by the designers to highlight their arguments.

Third Level.   The third level represents the integration of these various tools into a feature-based design system involving solid models, extensive databases with analysis tools for process planning, scheduling, economic modeling, and automatic system design.

At level 3, the free-standing software tools used in level 2 to move people away from subjective assessments are integrated into a system based on features. This system offers computer-aided design (CAD) solid models and extensive databases with analytical tools for process planning, scheduling, economic modeling, and automatic system design. The system is oriented toward engineering analysis, not geometry, which is the principal way designers interact with CAD systems whether they be two-dimensional or solid models. The system helps *define* the product as well as aid in its detailed design. Using the system, individual team members or the team as a whole can quickly explore innovations in product design or quality, technical options, or economic issues at the product concept level. We describe this in more detail later.

Importance of Strategies.   Figures 7.1, 7.2, and 7.3 indicate the effect early decisions have on the total life cycle cost of a product. Figure 7.1 relates how the life cycle cost of a product is determined during various phases of design. Figure 7.2 illustrates that the 5 percent point in cumulative cost for the product coincides with the point where 90 percent of the life cycle cost is committed. Figure 7.3 shows the cost impact of various design decisions. Both Figs. 7.2 and 7.3 emphasize the brief "window of opportunity" for leveraging decisions to control costs and reduce risks.

How Are Strategies Applied?   The primary focus in this chapter is complex electromechanical products. Our view is that in the world of integrated chips or PWA (printed wire assemblies) manufacture many tools exist and in a number of companies a high level of integration already exists. This is not true in the electromechanical world. The tool list is not complete and the integration of these tools is embryonic at best.

The discussion is illustrated by assembly, not because assembly is the most important function but because the problems of design, manufacturing process variations, as well as piece-part manufacturing tolerance, all come together on the assembly floor. To illustrate, piece-part design and manufacture is generally a unidirectional process; that is, ideal geometry is created in a CAD database which is then imposed on deformable materials with

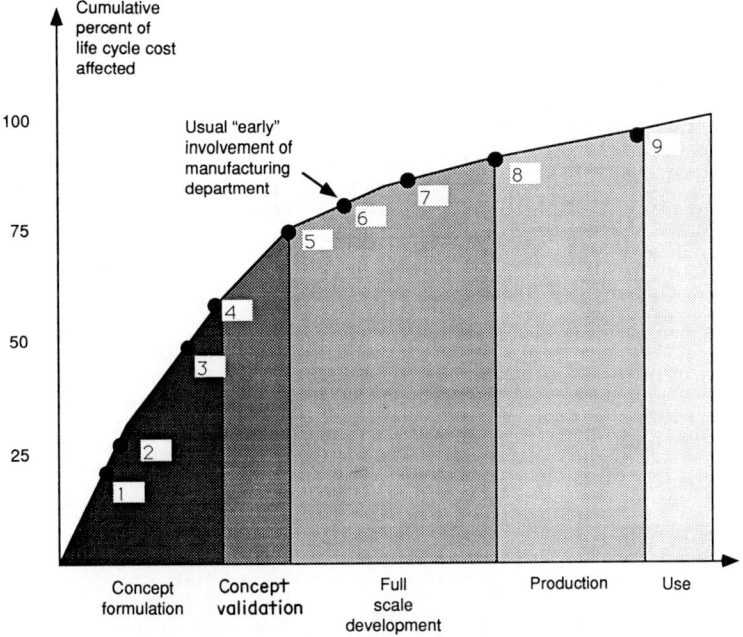

1. Define use patterns
2. Define alternatives
3. Develop alternatives
4. Freeze subsystems
5. Prove feasibility

6. Provide preliminary designs
7. Provide detail designs
8. Provide manufacturing plans
9. Product

FIGURE 7.1 Life cycle phases (Ref. 1).

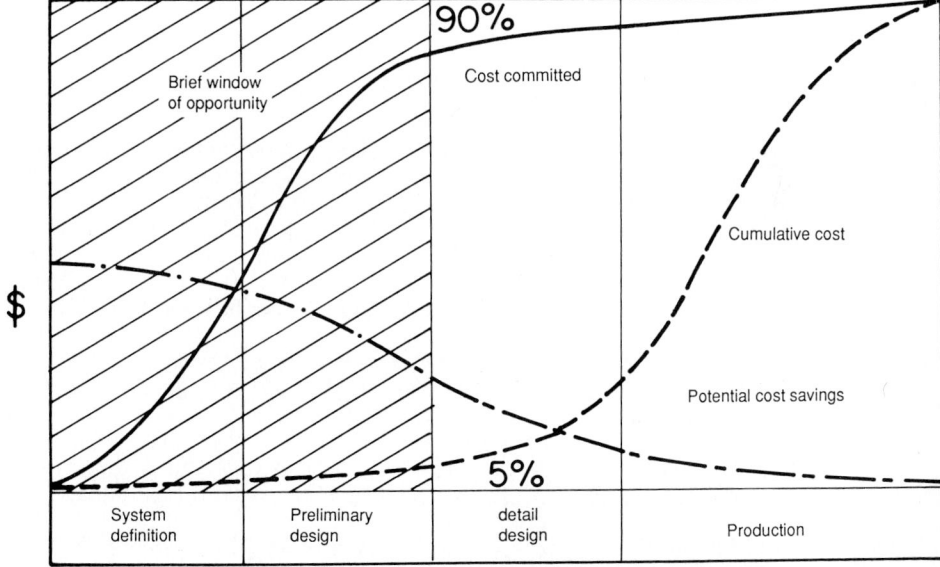

FIGURE 7.2 Decreasing leverage for cost savings (Manufacturing study, U.S. Air Force), integrated computer-aided manufacturing (ICAM study).

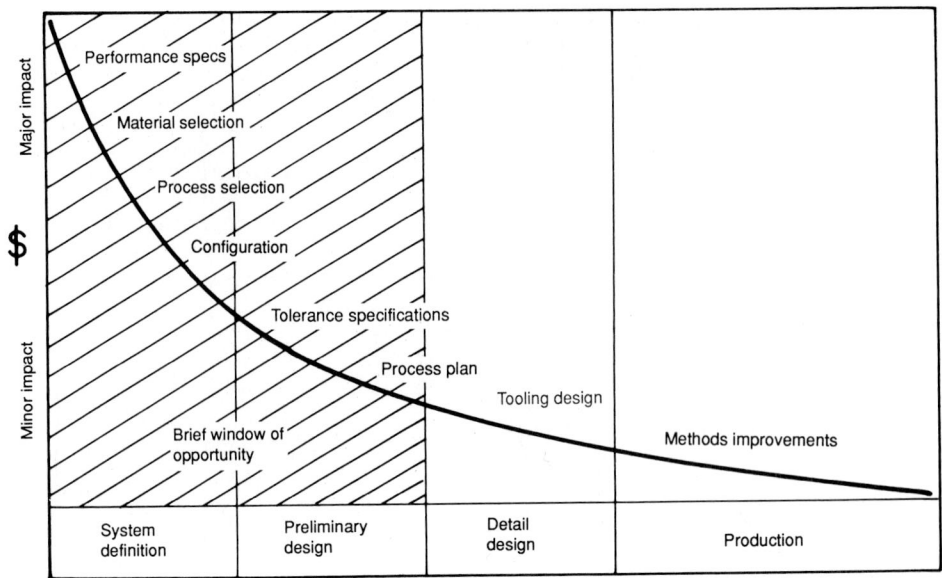

ICAM study

FIGURE 7.3   Cost impact of design decisions (U.S. Air Force ICAM study).

some tolerance. The process may be metal removal, or forming, or casting, or near-net shape operations like powder metallurgy. In any event the process tends to be one-directional with little feedback. A newer strategy using "engineered" materials is an integrated approach to piece-part manufacture from advanced materials. The functional activities are manipulation of the microstructures of materials, piece-part geometry, and performance requirements coupled to material issues and process modeling of machines and tooling.

TOOLS AND METHODS

Present Tools.   A number of tools, methods, and techniques have been developed to help assembly or manufacturing system designers. These tools range from DFA (design for assembly—Boothroyd and Dewhurst) to design axiomatics (Suh et al.) and the extensive design-aid handbooks created by the Fraunhoffer Institutes in Germany. Table 7.1 lists an initial set of references to this body of work. Since this work has been extensively published and is summarized in Ref. 1, we will not go into detail here. Instead we briefly describe some of the new tools that readers may not be familiar with.

New Tools—What Do They Offer?   Some of the new tools available are part-mating science, liaison sequence analysis, system synthesizing techniques, as well as programs capable of automatically generating process plans and the task-resource matrix needed to automate the generation of the system synthesizing technique. These tools were developed primarily to support assembly system design but their potential use is much broader, as will be illustrated. The following is a brief description of these tools. An example of how they are used will be given later.

Part-Mating Science.   This is an analysis technique for specifying the engineering requirements for physically mating piece parts, both rigid and compliant or springy parts. Figure 7.4 shows the basic geometry, insertion angles, and friction forces associated with the simple in-

TABLE 7.1   Present Analysis Tools Reference List

Salvendy, G., ed., "Value Engineering," chap. 7.3, *Handbook of Industrial Engineering*, Wiley, New York, 1982.
Taguchi, Shin, and Diane Byrne, "The Taguchi Approach to Parameter Design," *Proceedings 1986 ASQC Quality Congress*; also Taguchi, G., and Y. Wu, *Introduction to Off-line Quality Control*, American Supplier Institute, Romulus, Mich., 1986.
Boothroyd, G., and P. Dewhurst, *Design for Assembly Handbook*, Boothroyd and Dewhurst Associates, Kingston, R.I., 1985.
Mather, H., "Logistics in Manufacturing: A Way to Beat the Competition," *Assembly Automation*, vol. 7, no. 4, pp. 175–178, 1987.
Suh, N., et al., "On an Axiomatic Approach to Manufacturing Systems," *ASME Journal of Engineering for Industry*, vol. 100, p. 127, 1978.
Roth, K., "Design Models and Design Catalogs," *Proceedings 1987 International Conference on Engineering Design*, pp. 60–67, ASME, Boston, 1987.
Anonymous, "Systematic Approach to the Design of Technical Systems and Products," VDI Standard 2221, VDI-Verlag GmbH, Düsseldorf, FRG, 1986.

sertions of a peg into a hole for rigid part analysis. Figure 7.5 illustrates a class of compliant or springy parts and Fig. 7.6 shows that any desired insertion force or withdrawal force pattern can be created when these analysis techniques are applied to electrical connectors.[1]

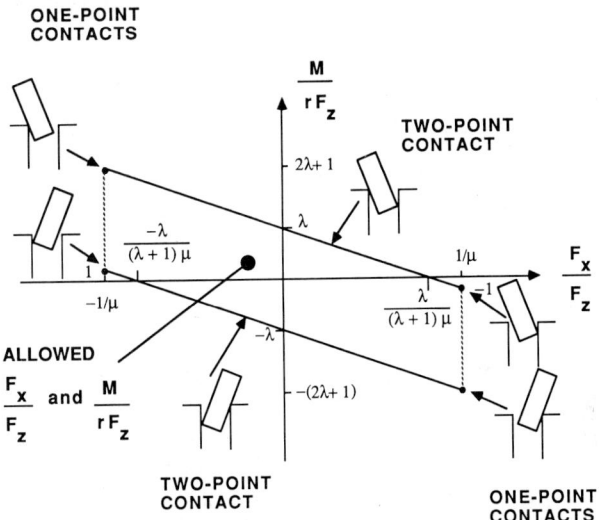

FIGURE 7.4 Single part mate. The theory of part mating is well established based on work at Draper over the last decade (Ref. 1). The criteria for single part mate success are captured in the diagram of this figure indicating how to avoid jamming. Thus any individual part mate can be evaluated and its difficulty reported to the designer.

ASDP (Assembly System Design Program). This is a computer program based on dynamic programming with heuristics capable of synthesizing manufacturing systems based on economic-technological issues created from a process plan, technical resources including people, and constraints (ROI, annual volume, cost-performance indexes, and so on). It does not simulate a system that you have given it—it creates completely new systems based on a detail description of the problem, resources, and constraints.

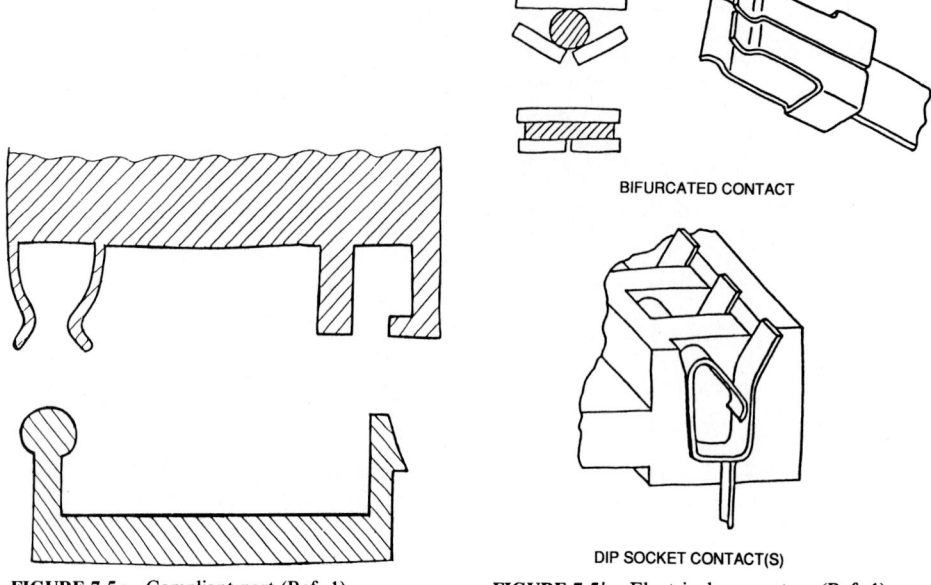

FIGURE 7.5a   Compliant part (Ref. 1).

BIFURCATED CONTACT

DIP SOCKET CONTACT(S)

FIGURE 7.5b   Electrical connectors (Ref. 1).

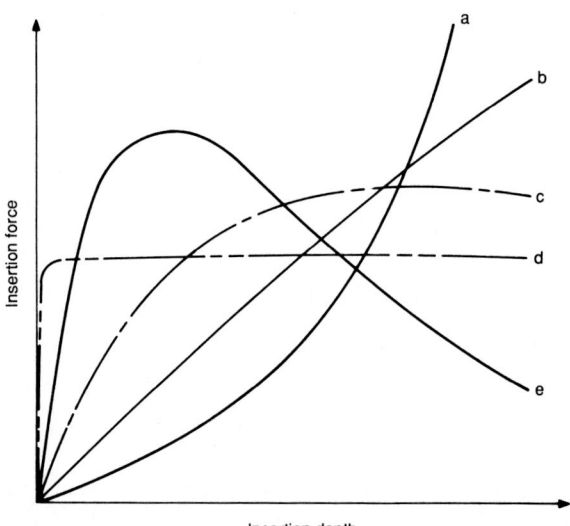

FIGURE 7.6   Insertion force versus insertion depth design space (Ref. 1).

   Figure 7.7 is a simplified block diagram of the system. Figure 7.8 shows how this type of information can be used to couple market volume issues to capital investment required to unit cost to assemble as a function of technology. Shown by the figure is a manufacturing system based on either completely manual operations or one based on what is called ''world class'' manufacturing. That is, someone in the world has used the particular technology in manufacturing, at about the same annual volume, and has received the desired

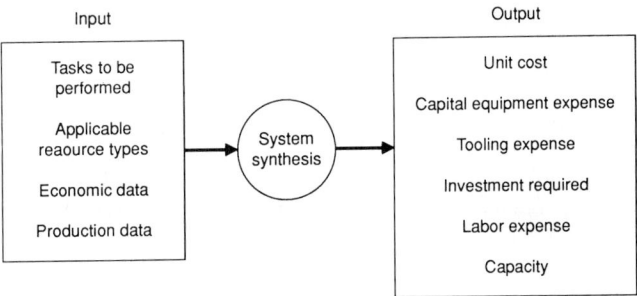

FIGURE 7.7 Simplified block diagram of an assembly system design program (ASDP)/(Ref. 1).

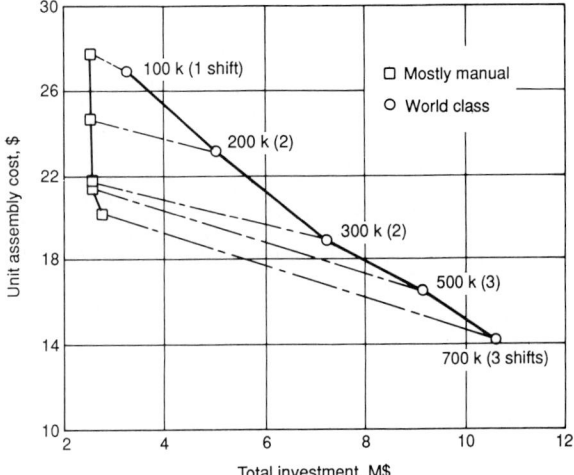

FIGURE 7.8 Final assembly options (Ref. 1).

performance, yield, and reliability at the specified cost. It is a method of identifying the amount of risk a corporation takes when it uses new technology. The manual system indicates that there is very little economy of scale with the manual system; that is, the unit cost reduces from $28 per unit at 100,000 units per year to $20 per unit at 700,000 per year. The world class system goes from $27 per unit to $14 per unit over the same volume range, but it takes almost five times the capital investment. If marketing can identify the range of possible market demand, management can then decide the best manufacturing capital investment strategy to suit the market uncertainties. All this information can be obtained at the concept design point. Other uses of the system will be shown later.

   SPM (Assembly Sequence, the Assembly Process Plan, and the Assembly Task-Resource Matrix).   This is a method for automatically generating the necessary process plan and task-resource matrix needed by ASDP starting from a simple bounding-box description of the individual piece part and data on the part mates. More detail on this system and how it is used can be found in Ref. 7.

   LSA (Liaison Sequence Analysis).   This is a method for identifying all possible ways to assemble a product starting from liaisons between parts, where a liaison may be a process, a test, or the physical mating of two parts. A typical industrial engineering study might identify one or at most two ways of assembling something; LSA will show all possible ways something can be assembled. The results of LSA can then be correlated with reliability modeling and economic-technological analysis to obtain the best product quality for a certain capital investment.[1]

An illustration of how LSA is applied is shown by Fig. 7.9 for the assembly of an automobile solid shaft rear axle. As mentioned earlier, a liaison may be a physical mating or joining of two piece parts, a process, or a test. For the rear axle shown, items K, L, and M, under parts, are tests. The liaisons as well as the derived precedence relationships are shown. Figure 7.10 is the "diamond" diagram which results. Also shown are three studies, one called the base-line system created by the client, one addressing the brake cable subassembly, and one addressing the option of automatic insertion of the C-washers that retain the axles inside the carrier.

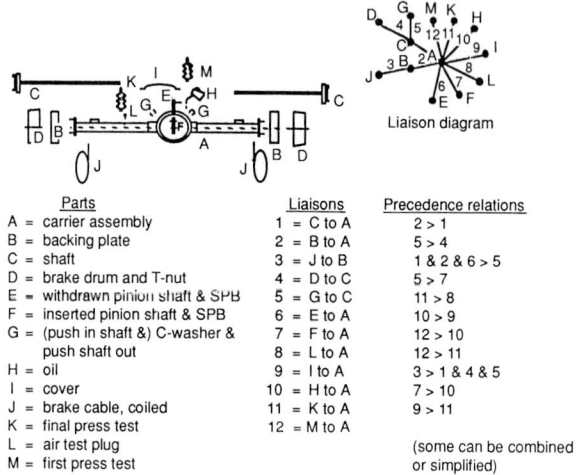

Liaison diagram

Parts	Liaisons	Precedence relations
A = carrier assembly	1 = C to A	2 > 1
B = backing plate	2 = B to A	5 > 4
C = shaft	3 = J to B	1 & 2 & 6 > 5
D = brake drum and T-nut	4 = D to C	5 > 7
E = withdrawn pinion shaft & SPB	5 = G to C	11 > 8
F = inserted pinion shaft & SPB	6 = E to A	10 > 9
G = (push in shaft &) C-washer &	7 = F to A	12 > 10
push shaft out	8 = L to A	12 > 11
H = oil	9 = I to A	3 > 1 & 4 & 5
I = cover	10 = H to A	7 > 10
J = brake cable, coiled	11 = K to A	9 > 11
K = final press test	12 = M to A	
L = air test plug		(some can be combined
M = first press test		or simplified)

FIGURE 7.9   Automobile Rear Axle Liaisons (Ref. 1).

Integration of Tools.   The tools listed can be used individually, as a group, or as an integrated set. They have been applied individually on studies ranging from precision mechanisms (piece-part clearances of the order of 1.5 microns or 60 millionths of an inch) to an automation strategy for shipyards. Basically the tools are generic and thus independent of the product domain.

Feature-Based Design (FBD) Systems.   At the time of this writing a system exists in the prototype stage that implements a form of feature-based design for assembly.[8] It does not automate the design process but instead is a decision and design aid for designers interested in integrated design. Feature-based design captures design intent (assembly topology, product function, manufacturing, or field use) while creating part and product geometry. Design for assembly, as used here, extends existing ideas about critiquing part shapes and part count to include assembly process planning, assembly sequence generation, assembly fixturing assessments, and assembly process costs. As used here, features are not restricted to the set required to accomplish the machining of an individual part; they may describe attributes to enable piece-part manufacture, or describe a process, or define some form of testing.

Figure 7.11 is a block diagram of the prototype assembly design system integrated with a manufacturing design system. Figure 7.12 defines the architecture of the prototype system. The system shown contains all the elements described earlier except for part-mating analysis. It contains a solid modeler as well as a common database for all the relevant analysis. The interface to the system is not the usual CAD designer interface, that is, pri-

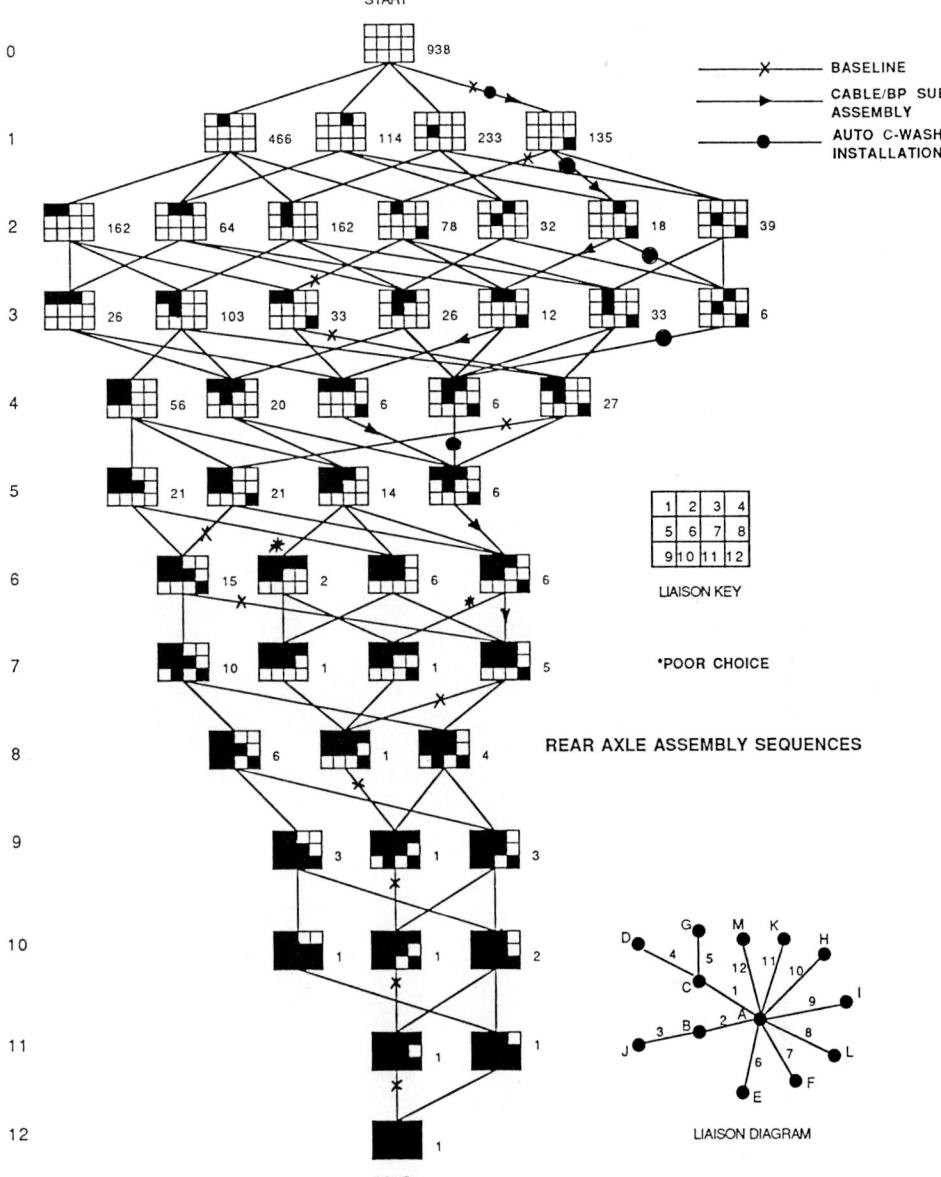

FIGURE 7.10 "Diamond" diagram of all possible ways of assembling axle represents all valid assembly sequences that result from the precedence relations. Each block represents a single unique state of assembly, with established liaisons between by marks in the (numbered) cells. Each line represents an assembly move, the establishment of a liaison. Full disassembly is represented by the top (empty) state, full assembly by the bottom (full) state (Ref. 1).

marily through geometry. Rather it is an engineering analysis interface. Thus the focus is not the solid model graphics but the engineering analysis information stored in the database.

This system is an example of the types of new CAD systems that CAD vendors will be offering to their customers over the next 3 to 5 years.

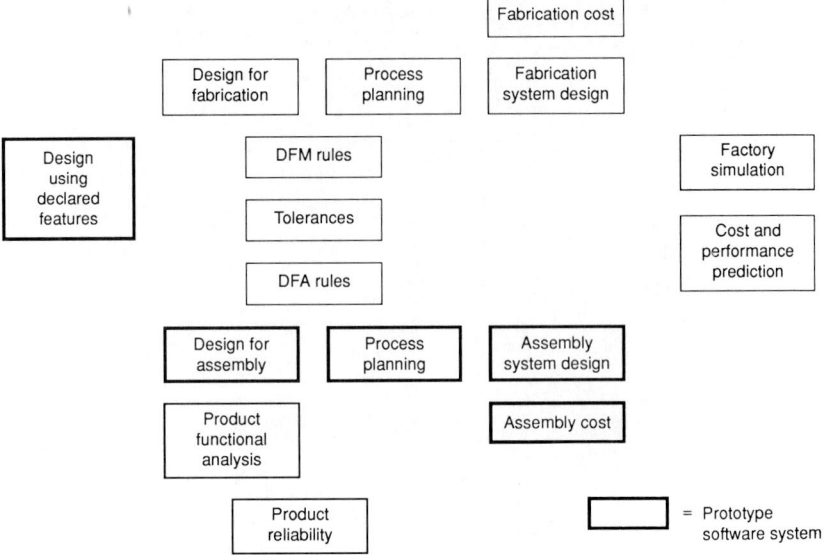

FIGURE 7.11   FBD block diagram (Ref. 8).

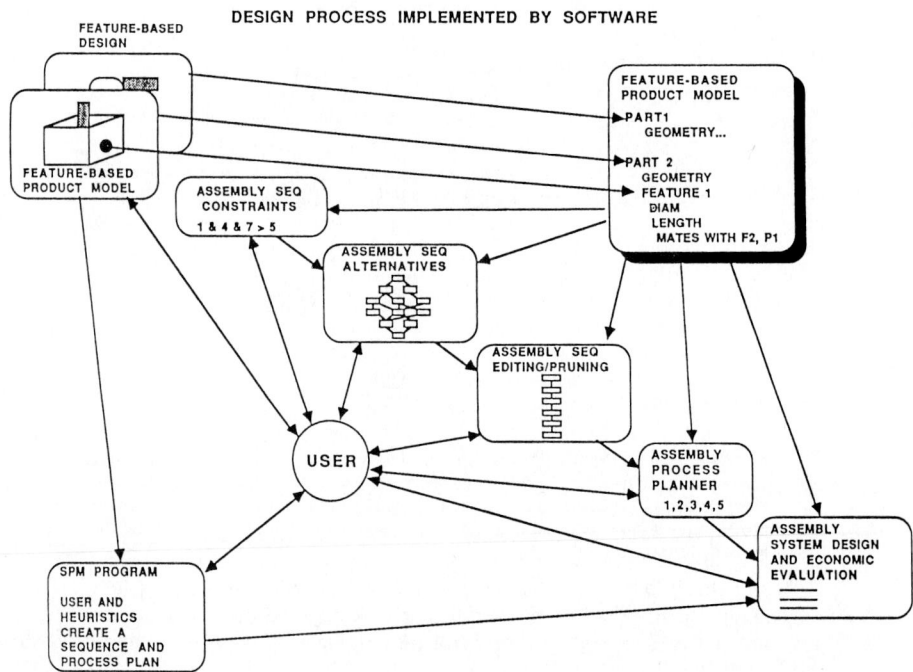

FIGURE 7.12   FBD software architecture (Ref. 8).

HOW DO YOU APPLY THE NEW TOOLS?

Introduction and General Approach.   This section summarizes a study of automobile rear-drive independent suspension rear axles made to determine the feasibility of automating all or part of their final assembly.* The axle is illustrated in Figs. 7.13*a* and *b*. The project involved all aspects of assembly, both technical and economic, including

- Assembly operations
- Assembly system design and floor layout
- Material flow timing and workstation timing
- Robots, tooling, and controls
- Parts handling and conveying
- Interface to management systems
- Investment cost estimates and financial justifications
- Computer simulations, transport studies, and throughput studies
- Suggestions for redesigns to improve automatibility

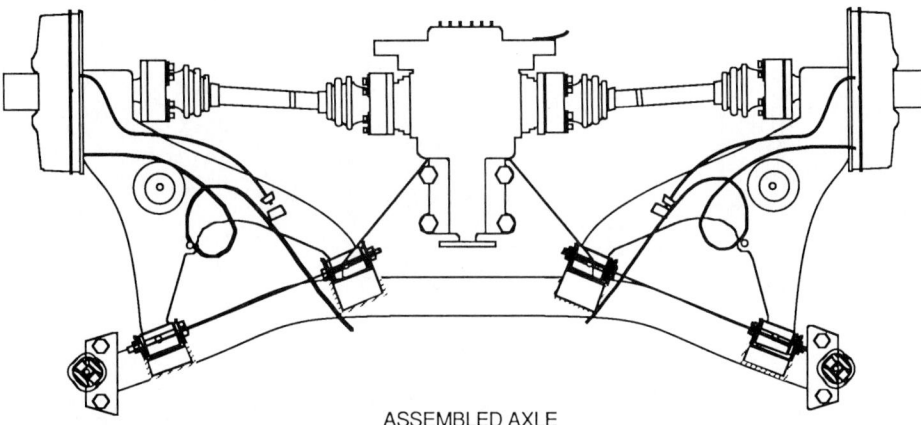

ASSEMBLED AXLE

FIGURE 7.13*a*   Independent rear suspension rear axle (Ref. 1).

The client's manufacturing engineers were heavily involved in the study.

The project began with a statement of the manufacturer's objectives for the product, dealing with model mix, production volume and cost projections, quality issues, and the desire to learn more about assembly automation. Engineering analyses based on part-mating theory[1] were used to predict difficult mates, which often arise from close clearances and tolerance stackups, including robot, jig, and gripper errors. Several alternate assembly sequences were considered, since these alternatives affect final system layout, cost, tolerances, jig and grip surfaces, deployment of people where necessary, and so on. Several redesigns and robot lab experiments were suggested to reduce assembly risk, obtain timing data, or make several operations feasible for automatic assembly.

Candidate assembly sequences were then paired with possible assembly technologies, including fixed automation, robots, and people. These were analyzed by computer simulation and economic justification programs to determine good candidate systems for presentation to the manufacturer. Vendors were consulted concerning feasibility and cost for var-

---

*This illustration is a composite of several actual studies.

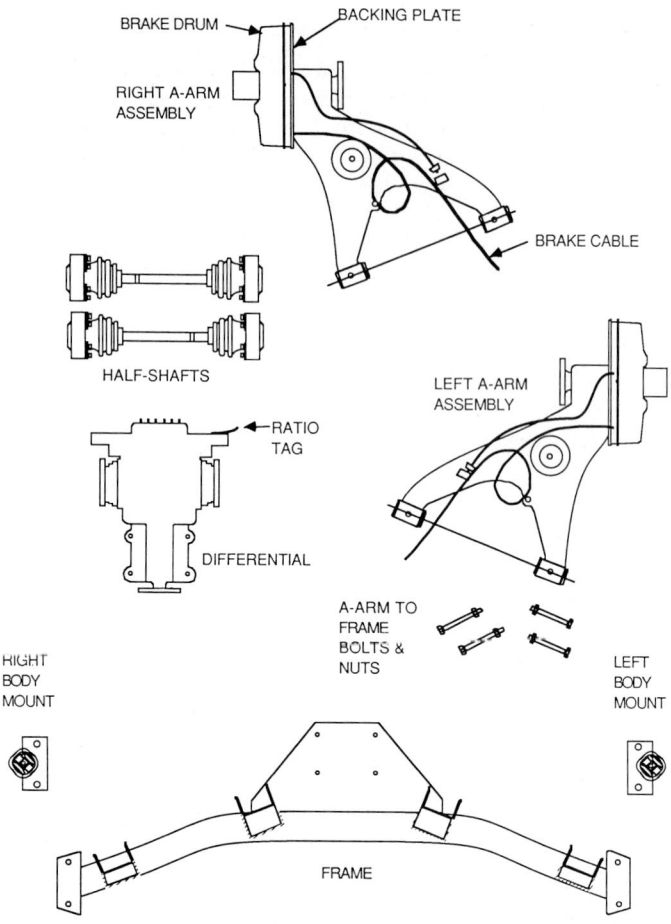

FIGURE 7.13*b*   Parts of the rear axle (Ref. 1).

ious robots, tools, and transport options. A final list of options was presented, along with recommendations.

Characteristics of the Rear Axle.   The axle is made in several models, but the differences are internal to the differential and the brake systems. These comprise different gear ratios and, possibly, disk rather than drum brakes. In the future the manufacturer may want the option to assemble other models on the proposed assembly system. These models would be geometrically similar and have the same part identity and count but would be larger by perhaps 5 to 10 percent. Production volumes are currently projected to be in the range of 250,000 to 300,000 per year, with future models adding about 20 percent.

This is a mature product, well tested in the field. Only minor redesigns were considered acceptable. These deal with part mates that are presently deemed difficult or unsuitable for robots or automatic machines: attachment of the "A-arms" to the frame and insertion and coiling of the emergency brake cables. Solutions require increased attention to tolerances on a welded assembly or cooperation with vendors.

Two other aspects of the product affect the feasibility of automation, namely, the accuracy of part fabrication and the method of presenting parts to the assembly system. In some instances, the probability of unsuccessful part mating is increased by parts that mate with tight tolerances and do not always have the generous chamfers shown on the drawings. As is typical in such situations, assembly workers have learned to overcome these problems, but automated systems are not yet that smart. The tooling and part mating recommendations sought to overcome these problems, but it is generally advisable to correct them where they occur.

Because redesign was not a major feature of this study, the emphasis fell on tooling and assembly system design. A major study point concerned material transport options and part provisioning. Provision of parts on kits was compared with more conventional feeding directly to the production line on conveyors and racks. Automated guided vehicles (AGVs) were compared with inverted monorails technically and economically, including counting how many vehicles were needed for carrying either a pallet fixture or a pallet plus parts kit(s). The interface between fabrication and assembly, which occurs at the kitting or conveyor loading area, was also examined carefully to determine feasible alternatives. These include structured dunnage ("dunnage" refers to either disposable or reused parts carriers, baskets, boxes, and so on), simple vision systems, and better parts handling in the fabrication areas.

Part-Mating Issues.   Every part mate involved in the axle final assembly was studied. Two involve risky mates, affect possible assembly sequences, or have other impacts on the final assembly system. These include attaching A-arms to the frame and mating half shafts to the A-arms and the differential. Each operation will be discussed in turn.

Major Parts.   The major parts are the *frame,* the *differential* (with its four mounting bolts, nuts and washers), two *body mounts* (with two bolts and nuts each), two *A-arms* (with wheels, brake cables, and brake drums already installed, plus their two hinge bolts and nuts), and two *half shafts* (with six bolts and three bridge washers at each end of each shaft). Each of these bolted joints is of critical safety importance, and the torque must be measured and certified to be within specifications.

Assembly of *A-arms* to the frame involves heavy parts. Two locations on the arm must pass between ends of brackets that are welded to the fráme and then must be positioned so that hinge bolts may be passed through and fastened with nuts. The lateral clearance available to fit the arm ends into the brackets ranges from 0.008 inch (0.2 mm) to 0.120 inch (3 mm). Bolt-to-hole clearances range from zero to 0.004 inch (0.1 mm). Assembly is aided by the fact that A-arm ends contain rubber grommets that mate to the frame brackets. However, the large chamfers at the open ends of the brackets that appear on the drawings are almost nonexistent on real parts. Widening these chamfers is essential to successful mechanical assembly. The alternative would be to greatly tighten the tolerances on the widths of all the mating parts as well as tolerances on location of the brackets. However, the brackets are welded to the frame, so their location is bound to be somewhat uncertain. Furthermore, tightening tolerances is expensive as well as unnecessary. Another recommendation was to provide tapered tips on the hinge bolts. These recommendations were accepted.

The differential is also heavy, but its mate to the frame is simple. Its mate to the half shafts is not simple, however, and must be considered when assembly of shafts to A-arms and of A-arms to frame are considered. One suggestion was further simplifying the fastening of the differential to the frame by making the nuts captive to the frame in some way. While not accepted immediately, this suggestion was later included in a major redesign of the frame in which the nuts were replaced by threaded holes in the differential itself.

Mating of half shafts to differential and A-arms involves several steps. People currently make a subassembly off-line consisting of a half shaft and an A-arm. The loose end of the shaft has the screws and bridge washers already installed. [The rubber boots around the constant-velocity (CV) joints keep the screws from falling out.] Because of the action of the CV joint, the length of a shaft can be changed over a range of 1.25 inches (3.175 cm) with considerable manual effort. The differential is added to the frame first. Then the shaft-arm subassembly is added. The assemblers compress the shaft to minimum length, support both the arm and the shaft in two hands, drop the arm ends into the frame, and push the shaft's cone end into the differential flange's cup. Attachment of the bolts involves rotating the shaft (or the brake drum) while pushing on the screws until the holes align and the screws drop in, then driving the screws. This is an involved and dextrous procedure which the analysts wished to avoid or simplify in mechanized assembly by changing the assembly sequence or method.

Brake Cables.   They present serious assembly difficulties that rippled through the entire study. They are awkward, and both ends must be dealt with at different times during assembly. The drum must be put on after one end of the cable is installed in the backing plate and before the other end is wrapped temporarily around the A-arm. The last operation seems unavoidably manual, forcing one of two alternatives: if brake drum installation were automated, a person would still be needed at the end of the line just to wrap cables around the A-arm; or brake drum installation could be manual and the person who installed them would also wrap the cables.

In this case, the manufacturer did not have sufficient space for an integrated final assembly system including assembly of the A-arms and chose to build A-arms manually as subassemblies. At final axle assembly, the brake cables must be unwrapped from the A-arms and threaded through holes in the frame. This awkward task will still be manual until extensive redesign is undertaken.

The *ratio tag* is an awkward part, hard to feed, easy to lose. A recommendation was that it be replaced with some other marking method, such as bar codes. The bar code could include model, date, and time information and could be used by the production control system to coordinate launching of kits or matching model-specific parts to each axle.

Other Parts and Tasks.   In general, other parts do not present great difficulties. Nuts can be fed down chutes to nut runners while they are in tool storage. Robotic pickup of kitted or conveyor-fed shafts, frames, and A-arms can be aided by attention to kit tolerances and gripper design. Simple vision systems above pallets or on grippers might be needed to take out gripping uncertainty, or grippers could be designed with wide enough throw and attention to closure motions so that parts are gripped correctly even if they are a bit off location in the kit. Such methods might eventually allow automatic kitting, depending on how much it costs to obtain good structured dunnage. As discussed below, some recommended assembly methods include fixtures with powered degrees of freedom of their own.

Part Fabrication.   It was noted above that some parts are not made as specified by the prints. This may not affect axle performance but could cause problems for assembly machine vendors. The analyst recommended that the parts be made to print or else a set of ''as built'' prints be prepared for issue to the vendors. This would help ensure that tools, grippers, and fixtures are designed to as accurate specifications as possible, reducing the risk of rework and delay in getting the assembly system operating at full capacity.

From a management point of view, one should strive for the level of quality that the product or the assembly system needs. The assembly system will ''inspect'' the parts and probably jam on ones that are too far out of tolerance. From an economic point of view, one must remember that the parts probably account for about 93 percent of an axle's in-plant cost, and final assembly accounts for only about 7 percent. Efforts to improve part quality could cost too much. Quality for quality's sake is justifiable only on the grounds that a spirit of quality will inevitably result in better axles, even if one cannot precisely trace specific warranty reductions or other benefits to specific investments. A spirit of quality could eliminate many poor practices that management could be unaware of without requiring that each be discovered and an individual cure prescribed.

Assembly Sequences.   The assembly of this product is fairly straightforward and there are not many assembly sequence options. An informal analysis quickly revealed the choices. One may build the axle ''right side up'' or ''up side down.'' If right side up is used, the A-arms and differential are on top of the frame while the body mounts are underneath it. These relationships are reversed if the axle is built upside down. In addition, there is one important subassembly option, that of adding half shafts to A-arms before assembling either to the frame.

Although the subassemblies remain an option, it was decided to pursue only right side up assembly. This orientation is better for ease of installing the A-arms to the frames.

Tooling, Jigs, Fixtures, and Workstations.   This section covers design of pallets, kits, and robot workstations. These designs required a decision as to how many axles to put on a pallet. Economic and system timing considerations (see below) showed that, depending on the speed of the pallet transport system, either one or four axles per pallet might be economically preferable.

Pallet design must conform to access needs of people, robots, and kits. For economy, one pallet design was used for all axle models. A suitable design for four axles per pallet places them on about 27-inch centers, leaving about 16 inches for a walk-in space to help a person attach the half shafts to the differential, if that were to remain a manual task. If longer pallets were used to create more space between axles, rotational-type robots would have difficulty reaching four axles. If space for people were a problem, the robot could be mounted on a linear axis alongside the pallet at an extra cost of $20,000 to $30,000 and about 3 seconds per move. Then a longer pallet could be used.

Robot access difficulties arise because of the unusable space around the center of the robot. This space has typically a 36 to 45 inch radius. The unusable space problem is not solved by mounting the robot upside down. It is eased, however, by using the powered linear axis parallel to the pallet.

An alternative to rotational robots would be gantries. Their reach is essentially unlimited. However, they are typically about twice as expensive as rotational robots. They were recommended only where their reach is essential.

The above studies used scale two-dimensional CAD based on manufacturers' drawings of reachable regions in their literature. Such regions may shrink when different wrist axis position combinations are used. About 25 reach analyses were done for this study. Figure 7.14 shows a typical reach analysis.

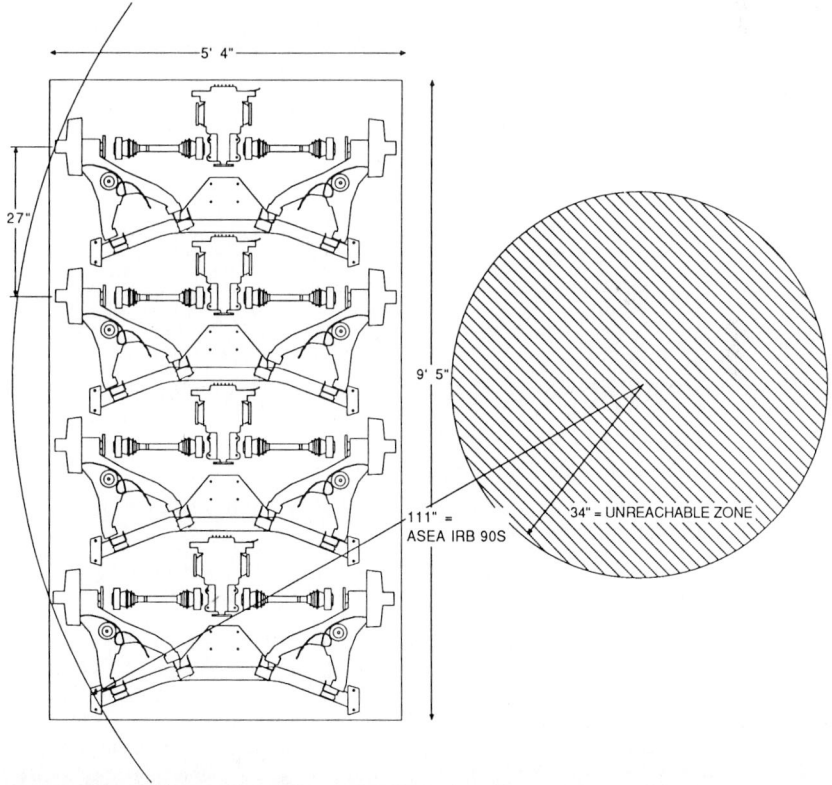

FIGURE 7.14 Reach analysis for an ASEA robot and pallet with four axles (Ref. 1).

Robots have about 0.02- to 0.05-inch repeatability. To avoid oscillations when carrying heavy tools and parts, they cannot be run at full speed. To compensate for repeatability and other errors (jigging, part fabrication, gripping), chamfers were recommended on grippers, grip surfaces of parts, and the entry faces of mating parts. The width of these chamfers must sum to dimensions larger than the largest contemplated errors. Reference 1 explains how these calculations are performed. To reduce contact forces during mating in the presence of error, it was recommended that tools contain engineered compliances such as remote center compliances.

If kitting were chosen, a parts kit would contain the half shafts, A-frames, and differential for one axle assembly. Kits would be approximately 5.5 feet long by 2 feet wide and would go on the pallet before the frame. Kits of the required size, however, would crowd the space needed for assembly unless either a separate kit pallet (and transport vehicle) were provided or as few as one or two axles rode a pallet. Extra vehicles would be costly and geometrically awkward. Kitting would thus rule out four axles per pallet.

Assembly System Design.   Assembly system design has six main aspects: capacity planning, equipment choice, assignment of tasks to equipment, transport choice, floor layout, and kitting. These interact heavily, and there are many alternatives. We identified several cases worth studying deeply. Computer synthesis, analysis, and simulation tools played a large role in the design. Table 7.2 lists the alternatives studied.

TABLE 7.2   Assembly System Options Studied

Production volume	No. of units per pallet	Unit-unit time, sec	Station-station time, sec	Tool change time, sec	Remarks
300,000 per year	1	3	5	5	Inverted monorail conveyor
	2	3	5	5	
	4	3	5	5	↓
	1	3	15	5	Automated guided vehicles
	2	3	15	5	↓
	4	3	15	5	↓

Source: Ref. 1.

Capacity Planning.   Capacity is influenced by several factors:

- The speed of individual workstations
- Time lost while robots change tools
- Time lost while people or robots shift attention from one axle on a pallet to the next (called unit-unit time)
- Time lost shifting from one pallet to another or waiting while transport takes a pallet away and brings another (called station-station time)

Robot tool changing is needed only if a robot is given more than one task and different tools are needed for some of them. Shifting from one axle to another arises if several axles are on a pallet versus only one. Finally, time is lost shifting pallets between stations due to their size; lost time can be reduced by double-buffering pallets at stations, although this requires purchase of additional pallets. Transport type also affects station-station time, with AGVs assumed to be slower than conveyors unless double buffering is used. (Double buffering involves providing two work locations for a robot; the robot works at one while transport brings a new pallet to the other. This technique works well for stations with one robot but is almost impossible to arrange at stations with one robot on each side of a work area or with fixed automation equipment. Two robots might be used for cooperative work to mate half shafts, the differential, and the frame, for example.)

In the present case, the economics favored a system with several fixed automation stations, severely limiting the opportunities for a parallel arrangement. The reasons for this conclusion are discussed in the next section.

A second choice related to the above discussion concerns the number of axles per pallet. Quantities from one to four were considered. Use of several axles per pallet has the beneficial effect of spreading over several axles the lost time due to tool changes and shifting pallets in and out of the station. More than four per pallet could cause robot reach problems.

On the other hand, as discussed in the next section, at the target production volume, manual stations in favored trial designs contained only one person. To install the body mounts on several axles on a pallet, for example, the person would have to walk around to the other side of the pallet to do half the task. This wastes a lot of time. At higher production volumes, several people would be needed to do the work and it would be natural to assign some of them to each side of the pallet.

Furthermore, ASDP analyses showed that a cost penalty was associated with fixed automation equipment that was designed to be large enough to deal with four axles. This cost was not offset by the timesaving associated with sharing station-station move time over four axles except in the case where AGVs were used (15 sec station-station time). This conclusion, too, emerged from the studies discussed in the next section.

Equipment Choice and Task Assignment. These two aspects of system design are discussed together because the ASDP algorithms do both at once in an optimal way that meets the required production volume while minimizing annualized cost per axle. The issue of serial versus parallel stations and number of axles per pallet can also be resolved economically at the same time. Number and type of transport vehicles can be decided as well if one iterates between ASDP analyses and scale floor layouts of the system.

Equipment choice required preparation of a database containing candidate "equipment" (people, robots, or fixed automation), giving purchase cost, operating cost, operating speed, need for human supervision, and other economic data. For each assembly step, candidates were listed, including estimated task time. The engineer's judgment is required when deciding which equipment types are suitable candidates for each task as well as when estimating task times. The relevant data for the axle appear in Table 7.3.

This table indicates that four assembly resource types were considered: manual ("MAN"), fixed automation ("FXD"), a small robot of the SCARA type ("RBS"), and a large jointed robot ("RBB"). All resource types were deemed adequate for all assembly tasks except that RBS was suitable only for tasks 2 and 3 due to its small size and load capacity, and task 7 was deemed manual. To account for the cost of monorails or AGVs, a transport resource called "TRN" was also provided.

The computer calculated the attainable production rate while allowing for necessary tool changes and dead time while pallets are shifted in and out. Each type of resource was assigned an uptime factor (ranging from 80 percent for people to 99 percent for transport) to allow for failures.

In this study, 5 sec was allowed for a tool change (10 sec for a large robot). It was assumed that an AGV would require 15 sec to remove a pallet while another AGV brought in a new one. An inverted monorail conveyor was assumed able to do this in 5 sec. Choosing four axles per pallet greatly reduced the effects of these time differences.

Table 7.4 gives the results of these analyses. This table shows that the best (that is, the lowest unit cost) combination is given by one unit per pallet and 5 sec station-station time, meaning use of monorail conveyors rather than AGVs. If AGVs are desired for other rea-

TABLE 7.3 Applicable Technology Chart for Rear Axle

TITLE _IRS Rear Axle_   DATE _6/4/88_

_235_ WORKING DAYS PER YEAR   _0.2395_ ANNUALIZED COST FACTOR

_2_ SHIFTS AVAILABLE   _$24_ AVERAGE LOADED LABOR RATE ($h)

_5 or 15_ sec STATION-TO-STATION MOVE TIME   PREPARED BY: _____

RESOURCE DATA SET NAME: _IRSDAT_   TASK DATA SET NAME: _IRSTSK_

	UNITS	DAYS
PRODUCTION BATCH DATA	200,000	_____
	250,000	_____
	300,000	_____
	600,000	

Task Number	MAN $C: 2000$ $\rho: 1.5$ $\varepsilon: 80$ $V: 4.00$ $t_c: 5$ $m_s: 0.83$		FXD $C: 0$ $\rho: 1.5$ $\varepsilon: 95$ $V: 6.00$ $t_c: 5$ $m_s: 4$		RBS $C: 40000$ $\rho: 2.5$ $\varepsilon: 90$ $V: 6.00$ $t_c: 5$ $m_s: 4$		RBB $C: 80000$ $\rho: 2.5$ $\varepsilon: 90$ $V: 6.00$ $t_c: 10$ $m_s: 4$		TRN $C:$ $\rho:$ $\varepsilon:$ $V:$ $t_c:$ $m_s:$	
1. Put frame on pallet	15 | 101	15000	10 | 201	75000			10 | 401	10000		
2. Attach body mounts to frame	25 | 102	5000	15 | 202	100000	15 | 302	20000	15 | 402	20000		
3. Subassemble shafts to A-arms	60 | 103	30000	15 | 203	300000	25 | 303	40000	25 | 403	50000		
4. Attach A-arms to frame	30 | 104	2000	15 | 204	300000			20 | 404	40000		
5. Place differential on frame	15 | 105	15000	10 | 205	150000			8 | 405	20000		
6. Mate differential shafts, and frame	75 | 106	15000	20 | 206	250000			35 | 406	50000		
7. Arrange brake cables, attach to frame	40 | 107	2000								
8. Transport: conveyor or AGV's									5 | 508	258000 or 963300

* For each resource:

C = hardware cost ($)
ρ = installed cost/hardware cost
ε = up-time expected (%)
V = operating/maintenance rate ($/h)
t_c = seconds in tool change time
m_s = maximum stations per worker

When a resource can be used on a task:

Operation time (sec)	Tool number
Hardware cost ($)	

Source: Ref. 1.0

sons, the recommendation is to use four axles per pallet. Single-shift operation results in the opportunity to choose a parallel station layout due to the large number of human resources needed. However, the unit cost penalty is evident.

As a result of these studies, a serial system layout consisting of three people, three fixed automation stations, and one each of the robot types was selected. It is shown schematically in Fig. 7.15a and in more detail in Fig. 7.15b. ASDP does not consider enough geometric information and constraints to permit a detailed station layout to be designed entirely by computer. Thus the engineers improved this design by

TABLE 7.4 Results of System Synthesis Studies (ASNP)

Each solution is based on providing 300,000 units per year. Some designs have reserve capacity above this level (Ref. 1).

No. of units per pallet	Unit-unit time, sec	Station-station time, sec	Tool change time, sec	No. of shifts	Unit cost, $	Solution* I J K M N
1	3	5	5	2	3.32	3 3 1 1 1
1	3	5	5	1	3.93	10 4 0 0 1
4	3	5	5	2	3.62	3 2 1 1 1
1	3	15	5	2	4.71	4 2 1 1 1
4	3	15	5	2	4.47	3 2 1 1 1

*Note: Solution type IJKMN means that the recommended system contains I manual workers, J fixed automation stations, K small robots, M large robots, and N transport resources. N is always 1.

```
IRS  REAR  AXLE-300K-CONV-1  AX/PALLET



         33.33 seconds   Usable Cycle Time
    90.0 %  Bottleneck up-time      108.00  Units/hr expected
         406080 units  Actual Capacity of this System

      1433000 ($)  Total Investment,  rho Factor =  1.58
         5.06 Workers at  24.00 $/hr required
      46.00 $/hr  System Operating/Maintenance Rate

      0.739 Year required for 2.0 Shift Operation
      235 Days required for 1.48 Shift Operation
```

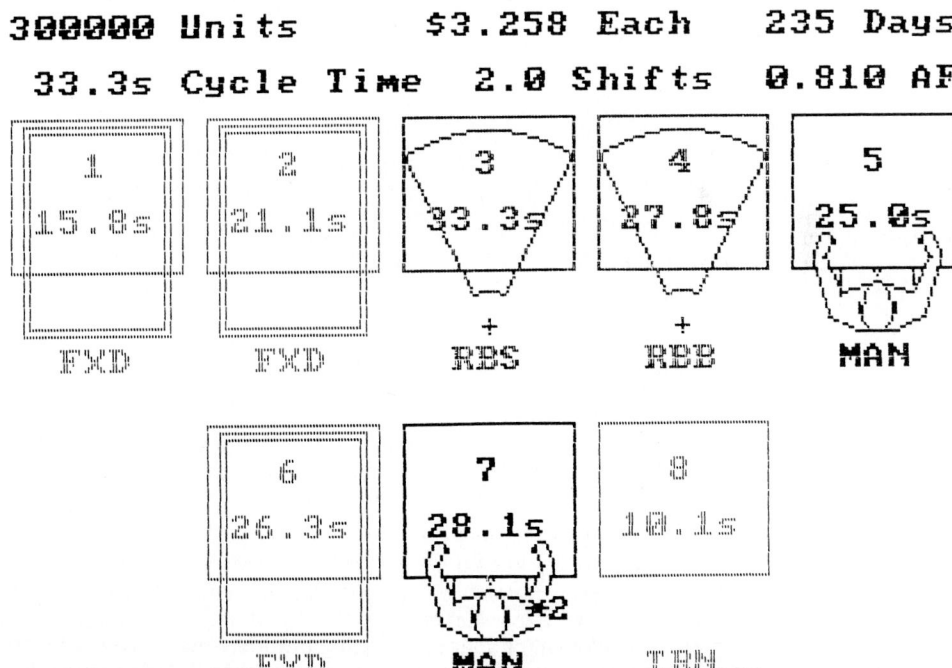

FIGURE 7.15a System synthesis output showing schematic design of assembly system (Ref. 1).

```
IRS  REAR  AXLE—300K—CONV—1  AX/PALLET
```

300000 units in Production Batch

Unused Resource Available Time Scale Factor   0.810

190.4 Days for 2.0 Shifts; 235 Days for 1.62 Shifts

Resources

*	MAN	FXD	RBS	RBB	TRN
Avltime *	24.2	29.7	27.9	27.9	31.2
Nsta *	7.0	4.0	2.0	5.0	1.0
Wrkr *	8.4	1.0	0.5	1.3	0.2
Vrate *	0.0091	0.0033	0.0033	0.0033	0.0024
Esttime *	41.43	18.33	22.50	23.00	5.00
Vrstlty *	1.0	1.0	1.0	1.0	1.0

Synthesized System

Task	Resource used	Resource Cost	Variable Cost	Operation Time	Tool Change	Tool Number	Station Cost
1	FXD- 1	0	10526	10.0	0.0	201	39982
2	FXD- 2	0	15789	15.0	0.0	1202	53309
3	RBS- 1	37069	27778	25.0	0.0	303	37069
4	RBB- 1	74138	22222	20.0	0.0	404	37069
5	MAN- 1	1066	51431	15.0	0.0	105	7996
6	FXD- 3	0	21053	20.0	0.0	2206	133272
7	MAN- 3	2132	137149	40.0	0.0	2107	2132 * 2
8	TRN- 1	0	3704	5.0	0.0	508	86758

Apparent System Cost = $866534

Resource	Total Cost	Number used	Time used	Unit Cost Fixed	Variable	Number of Tasks	Tools	Workers
MAN	287624	3	28.1	0.044	0.914	2	2	3.61
FXD	326562	3	26.3	0.755	0.333	3	3	0.75
RBS	107471	1	33.3	0.247	0.111	1	1	0.25
RBB	144540	1	27.8	0.371	0.111	1	1	0.25
TRN	111202	1	10.1	0.289	0.081	1	1	0.20

108.00 Units per Hour
33.3 seconds Cycle Time Expected
90.00 % Bottleneck Station Up-time Expected
406080 units Production Capacity of this System
46.00 $/hr System Operating/Maintenance Rate

977400 Cost ($) to produce 300000 units, with Unit Cost ($)   3.258
1433000 ($) Total Investment required
908000 ($) for required Hardware

FIGURE 7.15*b* System synthesis results (Ref. 1).

1. Reversing the sequence of tasks 1 and 2, putting the body mounts on the frames while the frames were in the feeder track before they are put on the pallet, thus employing the fastener-motivated subassembly discussed above

2. Using the large robot for part of its cycle to obtain and hold the swing arm so that the small robot can maneuver the axle shaft onto it and drive the screws

3. Using station 3 simply for manually placing the differential on the frame a few inches to the left of its final location, and aligning the half shafts to the differential laterally and rotationally in anticipation that final mating of differential, half shafts, and frame will occur at the next station

4. Providing that station with sliding action under the pallet to maneuver the differential and half shafts into their mated positions

This design is reasonable in that it gives fixed automation some straightforward assembly actions, reserving for a person or robot those tasks which require maneuvering in several degrees of freedom at once or possible search for correct mating and alignment. Figure 7.16 shows the major assembly operations arranged in four stations for the case of one axle per pallet.

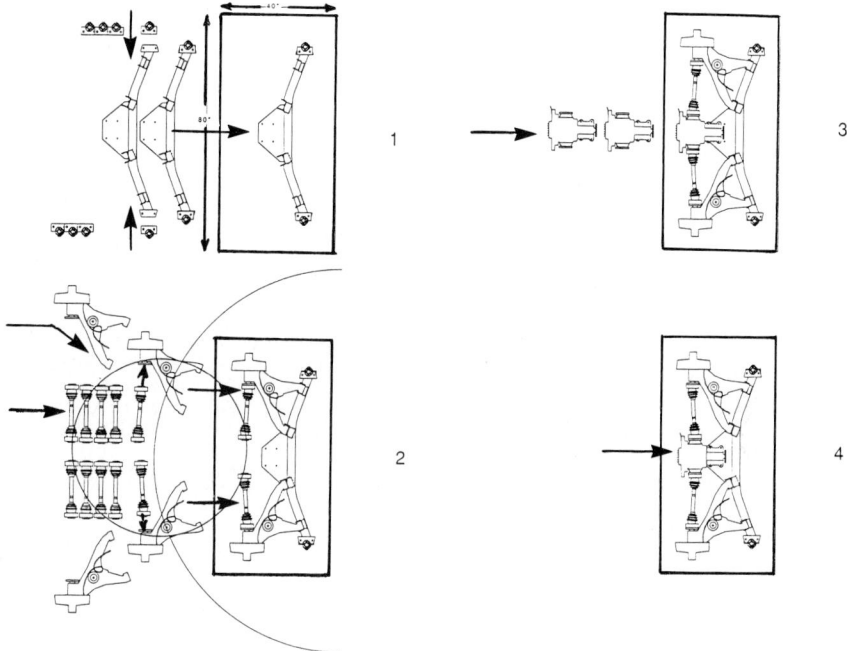

FIGURE 7.16   Arrangement of assembly operations into four stations (Ref. 1).

Choice of Transport Method.   Two types of transport were investigated, AGVs and inverted monorails. The cost of these was included in the ASDP studies. An iterative approach was required because the cost of transport method depends on floor layout and number of stations to be served, which depends on equipment selection and task assignment, which depends on cost of transport, among other factors. The analysis method consisted of using ASDP to obtain the required number of stations, making a rough floor layout, determining the length of transport track needed and approximate number of vehicles, then determining a cost for an inverted monorail system and an AGV system, and finally rerunning ASDP with TRN resources representing these costs.

The cost structures of these types of transport are quite different. The cost of inverted monorails for axle assembly was estimated to be 90 to 95 percent in the conveyor rails themselves, based on a cost figure by an experienced vendor ($600 to $800 per running foot). Vehicles are so inexpensive ($2000 for power-free and $5000 for self-powered) that doubling their number to support buffering would not seriously affect the system's final cost. Tracks appear to be modular and can be reused as long as they are not placed below the main floor level. Reuse would be especially easy if self-powered vehicles were used since there would be no drive chain to resize and restring.

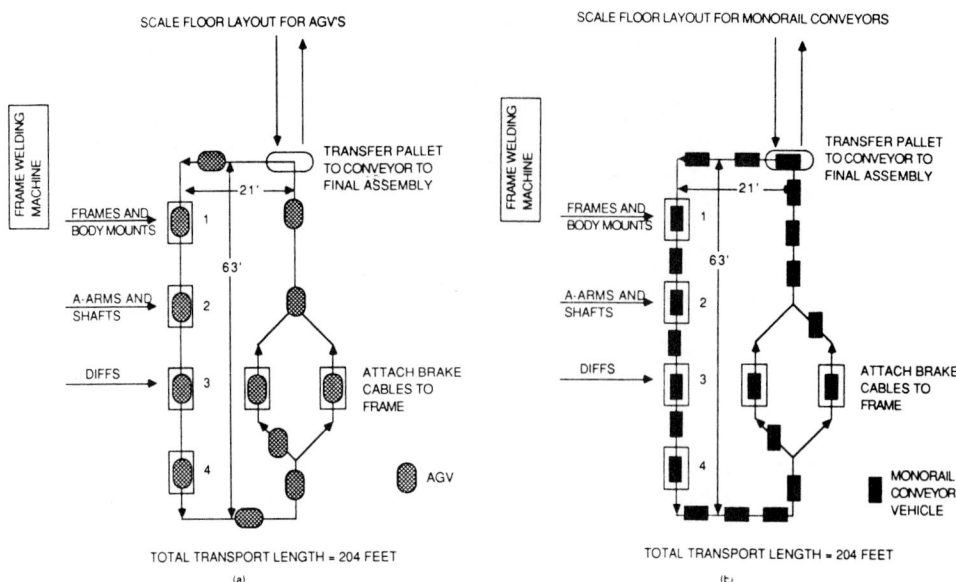

FIGURE 7.17   Scale drawing of assembly system for transport analysis purposes. (*a*) Arrangement for 12 AGVs based on placing one AGV at each station and approximately equally spacing others along the rest of the track. (*b*) Arrangement for 21 monorail conveyor vehicles based on placing one at each station, one between each assembly station, and equally spacing the rest. Total cost for AGVs = $79,000 × 12 + $75 × 204 = $963,300. Total cost for monorail conveyors = $5000 ×21 + $750 × 204 = $258,000. These costs were used for resource TRN in separate ASDP analysis.

On the other hand, the cost of an AGV system for axles would be dominated by the vehicles, at 70 percent, versus route-related costs of software, system integration, and guideway at about 25 percent (based on an estimate provided by another vendor, using the same floor plan that the conveyor vendor used). The vehicles could be reused if the route were changed, but the route costs would have to be paid again. A cost estimate of $79,000 per vehicle and $75 per foot of track was used. Figure 7.17 shows the scale floor layout used to estimate number of vehicles and track length, based on the four-station scheme evolved using ASDP. In the final analysis, because of the simple nature of this system and the economy of the serial arrangement, it was decided to use inverted monorails.

Floor Layout.   Floor layout is affected by the assembly sequence, serial-parallel structure of the system, size of pallets and equipment, safety zones, and the need to recirculate pallets, transport vehicles, and kit dunnage. If taxi mode is used for AGVs (AGV leaves pallet, goes elsewhere, and picks up another one), then the efficiency of dispatching and the speed with which a vehicle can respond are increased if extra pathways are provided.

All the floor layouts were laid out in scale via CAD. Robot reach areas were taken from manufacturers' literature. Each layout avoided material transport pathways that cross. The scale models were used to determine travel times for AGVs and conveyors for use in the ASDP analyses and dynamic simulations. The layout did not include kitting, which was finally ruled out by the awkward shapes of the parts and the increased pallet size needed to carry kits and provide assembly space.

Economic Analysis.   Economic analyses were carried out on several of the candidate assembly system designs to determine their rates of return. The current manual system was analyzed to obtain a reference variable assembly cost of $4.47 per axle. ASDP was used for this analysis, using resources MAN and TRN-conveyor and the same data as in Table 7.3.

Method of Analysis.   The economic analysis recognized three cost elements: labor, assembly system equipment, and transport system equipment. Each kind of equipment has allocated to it a purchase cost and an operating rate, and within the operating rate are some labor costs for supervision and maintenance. The cost of pallets is also included in equipment cost estimates. The rate of return is calculated by calculating a stream of savings in comparison with the current manual system and using those savings to pay back the equipment investment.

Labor cost is assumed to be $24 per hour, including fringe benefits. This is typical of the U.S. auto industry. Current Internal Revenue Service rules concerning tax rates and depreciation schedules are used (34 percent tax rate and 7-year MACRS depreciation schedule).

Results.   Table 7.5 gives the results of this analysis. The total investment of $1,433,000 is taken from Fig. 7.15*b*. The savings stream is obtained as the difference between the manual baseline system's variable unit cost of $4.478 and the proposed automated system's variable unit cost of $1.55 multiplied by the annual volume of 300,000 for a first-year savings of $878,400. The analysis shows a rapid payback slightly under 2 years with an internal rate of return of nearly 43 percent. There is an adequate cushion in these results for errors in estimating costs and times. In addition, the system as designed has a two-shift production capacity of over 400,000, allowing some excess for market growth.

TABLE 7.5   Result of Cash Flow Analysis to Determine Rate of Return on Proposed Automated Assembly System. (Ref. 1)

Year	Ratio	Tax rate	Depreciable	Savings	Depreciation	Tax rate
0	100%	34%	60.00%			
1				$878,400	14.29%	34%
2				$922,320	24.49%	34%
3				$968,436	17.49%	34%
4				$1,016,858	12.49%	34%
				Salvage value	31.24%	34%
		Total investment		$1,433,000		
		Depreciable investment		$859,800		
		Internal rate of return		42.93%		
		RHO		1.67		

Pro-forma cash flow

Year	Income	Depreciation	Taxes	Net income	Credits	Disc net income
0	($1,433,000)	$0	($194,888)	($1,238,112)	$0	($1,238,112)
1	$878,400	$122,865	$256,882	$621,518	$0	$434,830
2	$922,320	$210,565	$241,997	$680,323	$0	$333,002
3	$968,436	$150,379	$278,139	$690,297	$0	$236,392
4	$1,016,858	$107,389	$309,219	$707,639	$0	$169,541
4*	$268,602			$268,602	$0	$64,354
Gross totals	$4,054,616	$591,198	$1,086,237	$2,968,378	$0	$1,238,120
Net totals	$2,621,616	$591,198	$891,349	$1,730,266	$0	$8

Nominal capital recovery per year			$772,749	
Payback in about		1.85 years		

CONCLUSIONS AND RECOMMENDATIONS

The study concluded that automation of axle assembly appeared feasible. Without re-design, some operations would remain manual, and for some operations involving brake cables it is possible that no redesign would ever make them automatable. The recommended systems were mixes of people, fixed automation, and tool-changing ro-bots.

Automation was also determined to be economically attractive, exceeding the minimum required 21 percent internal rate of return, and using cost estimates for robots, tools, fixed automation, transport systems, and engineering. The original system assumed AGVs. On a cost basis, inverted monorail conveyors were clearly less expensive, even after including a complete ripout and rerouting (assuming AGVs, monorail vehicles, and most of the monorail track could be salvaged, and assuming that the monorails were placed on rather than below the floor).

SUMMARY

The new tools described earlier coupled with the illustration of how to use these new tools should allow the reader an outline of how to address similar problems. They, and the oth-ers mentioned, are the way products and manufacturing systems will be designed in the future.

REFERENCES

1. Nevins, J. L., and D. E. Whitney, et al., *Concurrent Design of Products and Processes; a Strategy for the Next Generation of Manufacturing,* McGraw-Hill, New York, 1989.
2. ESPRIT Research Reports, Project 688, AMICE, vol. 1, *Open System Architecture for CIM,* Springer-Verlag, New York, 1989.
3. "The Role of Concurrent Engineering in Weapons System Acquisition," Institute for Defense Analysis (IDA) Report R-338, December 1988. NTIS/DTEC Access no. AD-A203615.
4. "Findings of the U.S. DoD Technology Assessment Team (TAT) on Japanese Manufacturing Technology," CSDL Report R-2161, June 1987.
5. "GE Keeps Those Ideas Coming," *Fortune Magazine,* Aug. 12, 1991, pp. 41–49.
6. "The Best-Engineered Part Is No Part at All," *Business Week,* May 8, 1989, p. 150.
7. "SPM (Assembly Sequence, Assembly Process Plan, Assembly System Task/Resource Matrix), Version 1.1," 1990, available from CSDL Library, 555 Technology Square, Cambridge, Mass.
8. De Fazio, T. L., A. C. Edsall, R. E. Gustavson, J. A. Hernandez, P. M. Hutchins, H.-W. Leung, S. C. Luby, R. W. Metzinger, J. L. Nevins, K. K. Tung, and T. E. Whitney, "A Prototype for Feature-Based Design for Assembly, Rev. 1," July 1990.

CHAPTER 8
COMPUTER-INTEGRATED MANUFACTURING

Joel N. Orr
Chairman
Orr Associates, Inc.
Virginia Beach, Virginia

Computer-integrated manufacturing is the automation of the entire manufacturing process with computers. The coinage is generally attributed to the late Dr. Joseph Harrington, whose 1973 book of that name describes it as a "puzzle" whose pieces include computer-aided design, computer-aided manufacturing engineering, computer-aided materials management, numerically controlled machine tools, computer scheduling and production control, source data collection, and all the standard accounting functions. He noted that the new control and communication structure defined thereby was greater than the sum of its parts, and hence worthy of naming—although he explicitly eschewed creating a new acronym.

The acronym, despite Dr. Harrington's objections, has taken hold; "CIM" is "computer-integrated manufacturing." However, computer-integrated manufacturing is *not* the mere computerization of factory operations; it is something wholly new that takes advantage of the power of the computer to perform the functions of the factory more effectively than was ever possible before.

Designing a CIM system means applying systems thinking to the manufacturing enterprise; in brief, it means viewing the organization as a unit with certain inputs and certain desired outputs, and designing systems that are both computer-based and people-integrated to cause the inputs to be transformed into the outputs.

But the transformation of a manufacturing company from partly automated to computer-integrated is a complex and difficult task. Technological challenges are the least of the problems; they can generally be resolved by competent professionals with suitable budgets. Organizational, methodological, and philosophical obstacles to integration present themselves throughout the factory. It is easy to design a fully automated process, but difficult to design a series of steps leading from the present manual systems and islands of automation to CIM in an economical and humane sequence. For this reason, CIM cannot be sold as a packaged product or service; each factory requires entirely different considerations.

The benefits of CIM are many. The most obvious one is economic: a well-designed CIM-based factory can have a break-even point of about 30 percent of its operating capacity, while a conventional plant at best breaks even at 50 to 55 percent of its operating capacity. CIM-based plants can thus be run profitably at much lower loading factors than conventional plants. Consequently, in difficult economic times, CIM-based plants will continue to operate after conventional plants have been shut down.

The goal of "lot size one" as a profitable minimum can be realized with CIM, but not without it. Total automation opens vistas of flexibility that cannot be achieved in conventional manufacturing; thus CIM-based factories can also respond more quickly than conventional ones to the changing needs of the market.

Likewise, CIM's drawbacks are largely evident: What to do with superannuated workers, how to retrain factory personnel, how to restructure to exploit the benefits of CIM—and how to deal with additional technological complexity. The lack of explosive growth in CIM is due not to a poor benefit-cost ratio but to management's inability to visualize itself in the automated factory.

Finally, CIM is not a name for the total vision of the "factory of the future"; it is one leg of the tripod called by consultant Thomas Gunn, "world-class manufacturing," the other two being total quality control (TQC) and "just in time" (JIT). The benefits promised by CIM require the other two "legs" of the tripod for their realization.

OVERVIEW

A CIM facility, in concept, ought to be monolithic—or at least appear so. But owing to the historical development of the factory, seamless CIM is almost impossible. Moreover, the conversion of an existing factory to an automated one is a process that must perforce proceed on a piecemeal basis. It is thus reasonable to consider the components of CIM individually, while knowing that their full effectiveness is revealed only in the aggregate.

Management Issues. Management expert Peter Drucker said in a *Harvard Business Review* article: "Full realization of the systems concept of manufacturing is years away. It may not require a new Henry Ford. But it will certainly require very different management and very different managers. Every manager in tomorrow's manufacturing business will have to know and understand the manufacturing system. We might well adopt the Japanese custom of starting all new manufacturing people in the plant and in manufacturing jobs for the first few years of their careers."

For CIM to succeed, management must view the factory as a system, and apply systems thinking to its design, development, and maintenance. Recent history indicates that this process must also be informed by quality considerations, and structured within the context of the organization's suppliers and customers.

Technology Issues. One of the complications of CIM is that it is not a single technology that can be applied homogeneously to everything that occurs in a factory. It is, rather, a concept or principle—the use of computers to effect and monitor what goes on within the factory. But the automation of product design consists of different activities from the automation of product manufacturing. The technologies that support team design efforts, for example, are nothing like those involved in the hierarchical control of machining centers and automatic guided vehicles.

What these diverse activities have in common is that they employ computers and networking. Moreover, they can be arranged into "chains" of activities, in which the output of one is the input to the next. The computer and the network supply a context in which this can take place with a minimum of human intervention.

Some of the technologies of CIM include:

Computer-aided design

Numerical control

Robotics

Expert systems

Automatic guided vehicles

Computer-aided testing

Automatic assembly

Communication networks

Remote sensing

Digital instrumentation

Conceptual Issues. Successful CIM requires a shift in the understanding of "what does a factory make?" For CIM to work, planners must recognize that factories manufacture and manipulate *data,* and the actual products are simply one manifestation of the data. To correct a flaw in the product of an automated factory, it is far more effective to modify the data that control its manufacture than to change each product as it comes out of production.

Concomitantly, the automated factory is a collection of data-processing processes, some of which manifest as manufacturing operations. So the success of CIM relies on the success of the implementation of the underlying data-processing technologies. Most CIM systems have as an underlying assumption that the geometry of the product is the significant invariant data produced by the factory, on which all other data depend. Finishes, for example, are alphanumeric characters associated with particular surfaces; materials are similarly associated with solid parts.

COMPONENTS OF CIM

CADD. Geometric data originate in computer-aided design and drafting (CADD). Computer-aided design and drafting has evolved from drafting automation. The overt process of producing engineering drawings was not clearly separable from the underlying process of design. Initially, the automation of drafting—drawing production—was identified as "computer-aided design." Today, systems that are designed to automate the drawing production process differ in many subtle ways from those that are meant to empower designers in the process of ideation, sketching, and analysis.

The engineering drawing is a highly stylized product. Automated drafting systems simplify many aspects of its creation. For example, dimensioning can be done almost entirely automatically, with the drafter needing only to specify the type and location of the dimension text. Witness lines and other features are automatically generated by most systems.

However, engineering drawings, being only two-dimensional, are not sufficient to serve as the geometric fundament of CIM. For CIM requires models that are three-dimensional, in order to fully and unambiguously define products. So the engineering portion of a CIM organization must use computer-aided *design,* not computer-aided *drafting,* and must produce three-dimensional geometric data.

Three-dimensional geometric data can take three common forms: wireframe, surfaces, and solids, in order of progressively diminishing abstraction. Wireframe models model only the edges of objects, and lack "containment" information—that is, it is impossible in a wireframe to know whether a point is "inside" or "outside" an object. Surface models contain wireframe information to which have been added descriptions of bounding surfaces. While such models can be used to determine containment, it is possible in a pure surfaces system to generate models that contain geometric inconsistencies, such that they could not be made in reality. Only solid models, which require more computing power than the more abstract wireframe and surfaces, have sufficient information to make them accurate analogs of real objects.

In solid modeling systems, interferences among parts can be automatically detected. Moreover, sectioning and other "natural" operations can be performed on the models, and the results are realistic.

CAE. For CIM to be efficient, the model created in the computer-aided design process should be able to be digitally tested, through the design verification process that generally follows initial design. In a nonautomated environment, a model shop usually builds a physical prototype, and it undergoes tests, the results of which are used to redesign the product. Then the model and drawings are given to manufacturing engineers for design of the process by which it is to be manufactured.

CAE, or computer-aided engineering, is the term used to describe computer-based engineering analysis of digital models. The most popular form of CAE for mechanical engineers is finite-element analysis. In electronics, there are various forms of circuit analysis that are applied in this manner.

CAM. Computer-aided manufacturing, or CAM, usually refers to numerically controlled machine tools, the means of controlling them, and the associated computer-based systems within the factory. This category also includes other physical devices, such as flexible machining systems (FMS), robots, and automatic guided vehicles (AGVs), and data control systems, such as manufacturing requirements planning software (MRP).

In the CIM environment, CAM systems are able to receive data in digital form from the design phase. Digital descriptions of tool paths for numerically controlled machine tools, for instance, can be transmitted directly from the computer network to the machine tool controllers. (This arrangement is termed *direct numerical control—DNC*.)

DBMS (Database Management Systems). CIM is, in a sense, *databased manufacturing*. The databases for its support are therefore crucial elements. A robust CIM database must be equally efficient with graphical and nongraphical data.

Recent developments in database management technology have made the implementation of CIM easier than before. Relational database technology has expanded to include graphical data types. Moreover, there are now relational database management systems that can span nonhomogeneous computer systems over networks—a development that was necessary for the evolution of existing manufacturing facilities, which have many different computer systems in place, into computer-integrated facilities.

Networks. The most important technology for CIM is *networking*. The digital connectivity that allows the data to flow from design through production to field support and back is the key to the "integrated" aspect of CIM.

One of the most important computing trends of the 1990s is the move toward distributed databases. In the past, all databases were centralized. At first—in the 1960s—there *was* usually only one computer in a company. CADD, CAM, MRP, and other application terminals were connected to it, but all the data resided on the central system. Computer science yielded *timesharing* operating systems, which made it possible for many users to share a single computer while giving each the impression that they had a computer to themselves.

Much sophisticated programming was required to enable users to share a database without interfering with one another. "Record locking"—a way to signal all other users that a particular record in a file is in the process of being modified and cannot be accessed until that process is complete—became an important mechanism in database management systems.

In the large commercial timesharing systems of the 1960s, the first important stage of expansion was *hierarchical*. The big central computer system acquired smaller systems which handled part of the computational burden, under the control of the main system. Users experienced the system as a single computer. But the databases were, by and large, still centralized. See Fig. 8.1.

As graphics became more and more important to engineering computing, it became apparent that the timesharing approach had a serious limitation: communication rates. While a screenful of text contains about 2000 characters or bytes, a screenful of engineering graphics often takes 100,000 bytes or more. It was obvious that even with relatively high-

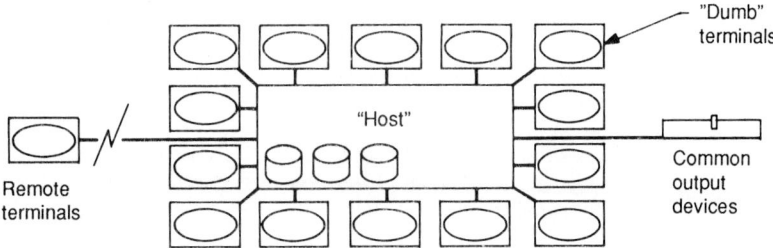

FIGURE 8.1   Centralized computing. (*Source: Joel N. Orr.*)

bandwidth communication lines—say, 9600 bits per second (bps)—a 100,000-byte picture would take almost 2 minutes to transmit. Such communication rates are about 200 times too slow for interactive work.

Developments like fiber optics, which allow much greater bandwidths, have developed more slowly than expected. The physical challenges of splicing fibers without major signal attenuation were difficult to overcome.

Nicholas Negroponte, founder of MIT's famous Media Lab, pointed out that the most inexpensive component of computer systems would ultimately be the computer itself. The Media Lab had a system in which each user had a computer, and all the computers shared an expensive array of disk drives.

At the Xerox Palo Alto Research Center, distributed computing took another step forward in the 1970s with the development of the independent workstation and the *client-server* model of systems. "Servers" are network-accessible data repositories; "clients" are workstations that access them.

With the advent of distributed computing systems, consisting of clients and servers, data can reside in any storage facility. The challenge of distributed databases is this: How to allow *anyone* on the network to update any part of the database, while keeping everyone informed as to the status of the files—*without* maintaining multiple copies of the database. For it is certain that multiple copies will wind up "out of synch" with one another. Today's answer to this quandary is to keep files on servers in networks, so that they are logically centralized, though they may be physically distributed.

A full discussion of networks is beyond the scope of this chapter. Salient characteristics of networks include topology, physical media, protocols, bandwidth, and cost. Popular factory networking schemes include *token ring* and *collision sensing* protocols, ring and star topologies, coax and twisted-pair physical media. See Fig. 8.2.

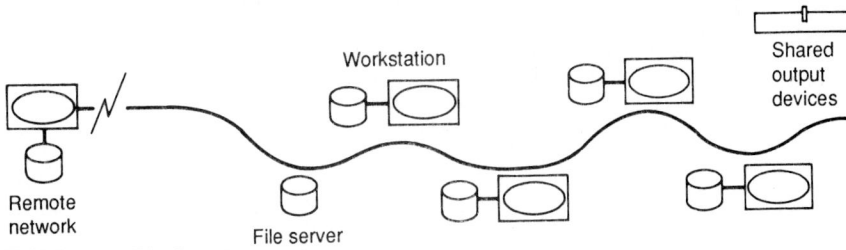

FIGURE 8.2   Distributed computing. (*Source: Joel N. Orr.*)

Networks are not generally homogeneous. What works well for general factory communications may prove too slow for "real-time" applications such as adaptive control of machine tools or AGVs, while the faster systems seldom have the error detection and correction capabilities of systems designed for handling large complex messages.

Computational Hierarchies. In addition to the engineering workstations and servers, the engineering data management network, and the accounting computers, many other computers serve the CIM organization. Most of them are *programmable logic controllers,* or *PLCs,* simple computers designed for integration into industrial systems and therefore offering a particularly wide range of input-output options, as well as a variety of robust (''hardened'') installation possibilities.

PLCs work with machine controllers and are generally tied into cell or area computers, which in turn report ''up the line'' to departmental systems. The complexity of systems with even a small number of cells and areas is great, and adds to the challenges of CIM implementation. Maintaining network integrity and system operations in the face of local malfunctions, with a minimum of human intervention, is a task worthy of the leading computer scientists. However, few are drawn to this field, and development is therefore perhaps slower than it ought to be.

CIM IN PROCESS ENGINEERING

In one sense, CIM has been more widely applied to process than to product engineering. Inserting sensors into flow lines, the output of which is used to regulate the flows, while other sensors monitor temperature, pressure, and other factors to control a process automatically—such ideas are familiar to the process engineer. It is in the surrounding aspects of CIM—arrangements with suppliers, the monitoring of quality—that product engineering has moved ahead more quickly.

While the typical product engineer is concerned with objects that must be produced in quantity, the process engineer's focus is usually on a single ''product'' at a time. For the product engineer, the consequences of the quality of the design are amplified by the number of units produced; for the process engineer, amplification occurs in effects on the project time line.

The actual concerns of the process engineer differ from those that occupy the thoughts of the product engineer. Pipes, pressure vessels, special symbolic drawings (P&IDs—process and instrumentation diagrams; PFDs—process flow diagrams; and electrical, hydraulic, and pneumatic schematics)—these are the stock in trade of the process engineering professional. Projects are generally large and long in duration. Environmental factors such as temperature, pressure, and corrosion figure significantly in designs; aesthetics, on the other hand, do not. But more than any other factor, it is the *size* of the typical process engineering project that differentiates its designers from those of consumer products or cars or machinery. That size dictates that most projects be designed by a well-coordinated team.

Since the objects that must be designed in this environment are large and unique to each project, mistakes can be very costly. Delays in construction can cost hundreds of thousands of dollars per week, or even more. So pilot projects, to test key concepts, realistic simulation, and project coordination are important.

WHAT THE COMPUTER OFFERS

Modeling a plant in the computer instead of in plastic has much to commend it. The process can be faster, cheaper, and more flexible, and can provide far more information for the subsequent refinement and construction processes. Graphics need not be copied from the model; they are automatically generated from it.

By housing all the data in a network server, the organization retains control of the project at all times. Data are always accessible to all who are working on the project. The model is always up to date. Developing model geometry in the computer makes it directly available to analysis programs, and the results of the analyses can be immediately used to modify the model. Concepts can thus be refined quickly.

Administrative and accounting data can be derived directly from the geometry of the model. Material take-offs, for instance, can be performed almost automatically. Shop and construction drawings that are guaranteed to be consistent with the design can be generated quickly and easily.

Finally, the networked workstation serves as a locus for all project communications. It makes it possible for groups working in different locations to be addressing the same database, without loss of synchronization. Even projects that are geographically dispersed can be centrally managed.

WHAT THE COMPUTER REQUIRES

Getting all the benefits from such a system necessitates major changes to the working processes of most companies. Working processes must be completely rethought in the light of the capabilities of computers. Not doing so leaves an organization with meager productivity increases, even if many processes are automated. The chain of data generation and control is only as efficient as its least efficient link.

CIM AND CONCURRENT ENGINEERING

"Concurrent" engineering is generally considered to be the same as "simultaneous" engineering—a manufacturing strategy in which design of a product occurs at the same time as the design of the process by which the product is to be produced. But another meaning of "concurrent" makes it the preferred term: "Being in or characterized by complete agreement." Agreement between the design and the manufacturing engineers is a commonly held objective, but it is unlikely to come to pass without some radical changes. For further information see Sec. 7, Chap. 7.

Concurrent engineering goes along with successful CIM. The automated communication environment within the CIM factory facilitates contact between manufacturing and design engineers, and the systems point of view demands that production factors be taken into account in the design process.

While no designer plans to design a product that cannot be easily manufactured, accounts of product designs that had to be completely redone in order to be manufactured effectively and economically abound in the business literature. Why? Designers know that the source of the information they need to avoid such straits is the counsel of manufacturing professionals. But cultural and organizational factors make it difficult for them to seek such help.

To attain "agreeable engineering," engineering education, manufacturing management, and engineering management must be addressed. Educators must research the power and implications of concurrence and develop useful methodologies for migrating quickly to such an environment. Manufacturing and engineering management must negotiate a quick end to the adversarial relationship that still pervades most factories, when it comes to the design-through-manufacturing process.

The Japanese have demonstrated the benefits of "continuous improvement": When operations are begun in a new factory, and a bottleneck is uncovered, workers gather around and applaud the discovery. A committee is then formed to deal with the bottleneck, and everyone returns to standard operating mode, in which the slogan is, "find the next bottleneck." Since all human creations are by nature imperfect, is is clear that there will be a "next bottleneck."

By contrast, the standard American approach is to consider the new factory perfect. When the first bottleneck turns up, as it inevitably will, the "guilty" are punished, the bottleneck is embarrassedly dealt with, and perfection is once again assumed. The engineering process is considered "perfect" in most American manufacturing organizations today; to get to "agreeable engineering," the concepts of "continuous improvement" must be applied.

An attitude that is a survival trait under some circumstances can be a challenge to survival under others. What prevails in most organizations is, "If it ain't broke, don't fix it." But since the performance measures in most organizations are not properly tied to the real world, internal issues are judged by criteria that may be irrelevant to reality. In particular, neither design nor manufacturing engineers are usually appraised on their ability to cooperate or on overall manufactured product results. But they must be, in order for concurrent engineering to be implemented.

Many technologies and adjuncts for the support of concurrent engineering have emerged. "Technologies" include special CADD packages and engineering database management systems created with concurrent engineering in mind. "Adjuncts" include networking and data translation capabilities in general.

More than any other development, the network has created an environment in which design and manufacturing engineers can get together on projects with minimal disruption of their existing work routines. Because of effective networking, concurrent engineering projects can be well executed even when the participating design and manufacturing groups reside in different time zones.

Some network-ready CADD programs allow a user in one location to "mark up" a design on the screen in such a way that others can easily distinguish the graphical "commentary" from the original document. Coordination can thus take place across great distances.

There is much more to concurrent engineering than technology. If "agreeable engineering" is to take hold, organizations must change, and engineering schools must prepare engineers who will help implement the new methods.

CIM AND QUALITY

Much attention is being paid to proactive, preemptive quality assurance today, under the name "total quality control" (TQC). Quality is enjoying considerable popularity as a measure of success for manufacturing firms.

Quality is defined, ultimately, by the customer. It is widely recognized today as an important competitive factor. CIM provides a manufacturer with a framework in which to measure quality and use the measurements to prevent future failure. Lot traceability, statistical techniques, and other methods that are gaining popularity today are much easier to implement in a CIM-based plant than in a conventional one.

It is not simply that computers provide a convenient mechanism for the implementation of quality tools, although that is certainly true. It is that the tight coupling and deliberate communications among the components of the CIM-based plant resonate with the fundamental notions underlying TQC.

OBSTACLES TO CIM

The problem in factory automation is *people,* not technology. Envisioning technological integration in the factory has been deceptively easy; achieving it has not. Babel is an inescapable aspect of our information processing heritage. It is proving difficult to attain to the uniformity of "language" demanded by the vision.

The difficulties of interfacing can be overcome by executive order, by simply sticking with a single vendor; this circumvents much of the babel problem for the computers, albeit at the price of using only what the chosen vendor has to offer. Technology for doing most of what the vision promises can be bought ''off the shelf'' today. But *people* are holding up progress, because they don't see themselves in the beautiful vision of CIM. So they resist change, leading the company in directions in which it didn't expect to go. To combat this, management must spread the message of CIM and its value, *translated into personal terms for each worker.*

In sum, thinking about CIM provides us with an excellent opportunity for examining, and changing, the way the people parts of organizations work. The solutions to the problems of the technological obstacles to CIM are in sight, but they will never be implemented unless we change the organizational framework to truly accommodate people, as individuals. The organismic nature of organization must emerge. CIM involves a great many functions within an organization. The complexity of this relationship is illustrated by Fig. 8.3.

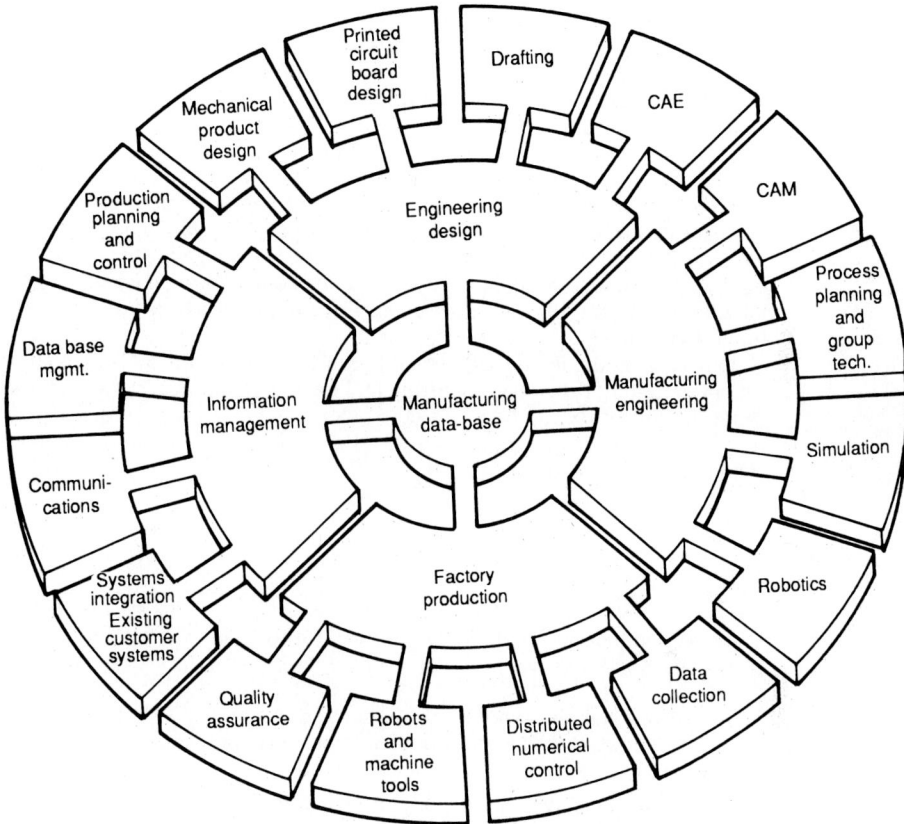

FIGURE 8.3 Generic CIM architecture. (*After E. Teicholz and J. N. Orr,* Computer-Integrated Manufacturing Handbook, *McGraw-Hill, New York, 1987.*)

SUMMARY: FROM CADD/CAM TO CIM

Fully automated factories are unlikely to become epidemic in the next 10 years. The social and organizational obstacles to such a radical change are proving more formidable than anyone expected. Technology is infiltrating engineering and manufacturing; bridges among the islands of automation now seem to be an inevitable intermediate step on the road to total integration.

This is a time of transition, and the successful products will be those that ease the transition. Automated drafting systems are a case in point: They do not seem to be a step toward CIM, but they provide measurable productivity improvement in the automation of a manual process. (Of course, by familiarizing users with computers, they do facilitate further automation. But that is only a side effect, not a measurable benefit.)

The greatest opportunities in CIM lie in the further automation of the design process. For this to occur, a new and broadly applicable design methodology must be developed. Until then, technology can only be applied to fragments of the process.

The challenges of engineering drawing, as implicit in centuries-old engineering practices, have been met by existing CADD packages; it is now possible to draw lines, circles, and splines, place and edit text and dimensions, and otherwise make and modify engineering drawings three to ten times as fast as before.

The next step will expand the metaphor of the engineering drawing far beyond its original usage to a new concept of design. This means design that creates a digital model that can be used by the rest of the manufacturing operation, design that incorporates manufacturing considerations, design that sheds the limitations of the engineering drawing.

It is yet to be determined what shape the next metamorphosis of the engineering drawing will take. The apparently obvious development of the three-dimensional solid model has been in existence for some time, in growing sophistication, but has yet to take hold. Is engineering tradition simply slow to change, or is the new object—the three-dimensional solid model—just not that useful? Most observers believe that engineers who were not trained to design in three dimensions have difficulty learning how to do so on a computer, but that engineers who are now using three-dimensional design tools in school will find them "natural."

Once the model is in digital form, it can be used to control the manufacturing process. The integration of the manufacturing and engineering processes with the material control, ordering, inventory, and accounting systems has been technologically feasible for some time. The widespread application of CIM awaits, more than anything else, a change in attitude on the part of management.

IMPLEMENTING CIM

Management decision is the only prerequisite to CIM. The prescription for implementing CIM is simple:

1. Use this handbook, and other publications and educational forms, to gain a thorough understanding of CIM and its implications at large. Equip your staff and your management with a similar education.

2. Thoroughly analyze existing operations. Flowchart the explicit, business-oriented flow of information and materials. Describe in detailed narrative form the implicit interpersonal forces that are at work.

3. Design the fully automated operation, with both detailed flowcharts and extensive narratives. Involve workers at all levels in the design process.

4. Consider the description of existing operations as the initial state of a dynamic system, and the design of the fully automated operation as the final state. Design a series of intermediate steps that will bring the organization from the initial state to the final state with minimum trauma to the organization and its people.

5. Set up an organizationwide implementation committee, and empower it to act.

This is not a simple process. It is difficult and time-consuming. But there is no other.

BIBLIOGRAPHY

Bray, O. H., *Computer Integrated Manufacturing: The Data Management Strategy,* Digital Press, Maynard, Mass., 1988.

Gerelle, E. G. R., *Integrated Manufacturing Strategy, Planning, and Implementation,* McGraw-Hill, New York, 1988.

Gunn, T. G., *Manufacturing for Competitive Advantage: Becoming a World-Class Manufacturer,* Ballinger Publishing, Cambridge, Mass., 1987.

Harrington, J., *Computer Integrated Manufacturing,* Academic Press, New York, 1973.

Nevins, J. L., D. E. Whitney, and T. L. De Fazio, *Concurrent Design of Products and Processes: A Strategy for the Next Generation in Manufacturing,* McGraw-Hill, New York, 1989.

Suzaki, K., *The New Manufacturing Challenge,* The Free Press, New York, 1987.

Teicholz, E., and J. N. Orr, *Computer-Integrated Manufacturing Handbook,* McGraw-Hill, New York, 1987.

CHAPTER 9
INDUSTRIAL CONTROLS AND SENSORS

Cecil Smith, Jr.
President
Cecil Smith Incorporated
Baton Rouge, Louisiana

Although digital computers were used in industrial control systems prior to 1960, the high cost of computers prevented their widespread use in industrial control systems until the advent of the microprocessor. Although introduced only in the early 1970s, by 1980 it was clear that microprocessor-based control products would eventually displace all pneumatic and electronic analog controls in industrial production facilities.

The development of industrial controls followed two routes, with surprisingly little interaction between the two. One route was via the manufacturing industries, such as automotive, aerospace, and the like. Another route was via the process industries, such as petroleum refining, foods, and so on. Very few industrial applications involved a significant overlap between these two technologies. One example of an application with significant overlap is injection molding equipment used to manufacture products such as plastic containers. The controls for the extruder (or extruders in a multilayer plastic product) are clearly process controls. However, the controls for the molding assembly are more typical of those required in the manufacturing industries.

As digital technology began to be used for automation, it was recognized that the various automation functions needed to be integrated, both between themselves and with other functions within the production facility. This led to a concept referred to as either hierarchical control or computer-integrated manufacturing. Although the manufacturing industries and the process industries proceeded along different routes, the final structure is surprisingly similar, with the differences being more in terminology and less in function.

COMPUTER-INTEGRATED MANUFACTURING

The best context in which to understand industrial controls and sensors is that of computer-integrated manufacturing. In some quarters, this concept was formerly called hierarchical control. This concept is applicable in both the manufacturing and the process industries. As illustrated in Fig. 9.1, the structure is very similar, although different terms are used at certain levels.

Manufacturing Industries	Process Industries
Corporate	Corporate
Plant	Plant
Cell	Supervisory
Machine	Regulatory
Sensors and actuators	Sensors and actuators

FIGURE 9.1 Levels of control.

Sensors and Actuators.   These devices provide the interface between the controller and the machine or process being controlled. Sensors translate some variable on the machine or process to a signal (usually electrical) that can be subsequently processed by the control equipment to obtain the current value of the variable of interest. Actuators accept a signal from the control system and in some fashion impose this signal on the machine or process being controlled.

Sensors can be either continuous or discrete. Examples of continuous sensors are measurement devices for temperatures, pressures, flows, etc. Such sensors translate the value of the variable being measured into a signal that can assume any value between two extremes, generally referred to as the lower range and the upper range. Examples of discrete sensors are proximity switches, limit switches, on/off level switches, and the like. Such sensors produce an output at one of the two possible voltage levels, such as 0 or 5 volts dc, 0 or 24 volts dc, 0 or 110 volts ac, and so on.

Actuators can also be either continuous or discrete. Examples of continuous actuators are variable-speed drives, pneumatic positioning valves, and the like. Examples of discrete actuators are on/off motors, solenoid valves, and so on.

Machine or Regulatory Control.   Functions at this level provide the basic regulation of the machine or process being controlled. For a machine, regulatory control is provided by controllers such as numerical controllers or programmable logic controllers. For processes, regulatory control is provided by single-loop or multiloop microprocessor-based controllers. In years past, machines and processes could usually be operated manually, the incentive to install automatic control equipment being higher throughput, better quality, and the like. But today, no machine or process can be operated without regulatory control equipment of some type.

Cell Control or Supervisory Control.   In manufacturing facilities, several machines are often grouped to form a cell. In process plants, several unit operations are interconnected to form the process. In order to realize the maximum performance from a cell with several machines or a process consisting of several unit operations, either or both of the following functions are required:

1. Coordinate the various machines or unit operations. Parts or materials usually flow from one machine or unit operation to another, which produces the need to coordinate the various components of the production facility.

2. Optimize conditions within the machines or unit operations. Regulatory controls are normally expected to do exactly as told. For example, a regulatory temperature controller does whatever is required to maintain the temperature at the specified target, even though this target might not be optimal. Supervisory control is responsible for determining the optimal value for the target supplied to the regulatory controller.

Plant Control.   At this level reside the various functions required for efficient operation of the entire plant. Functions at this level include keeping track of the raw materials and products, routing production activities throughout the plant, monitoring equipment utilization, and so on. The computers at this level communicate with the cell or supervisory computers via a plant local area network (LAN).

Corporate.   Normally implemented as part of the management information system (MIS) capabilities, functions at this level are directed to the long-term or strategic issues. An essential part of this function is the collection, analysis, and presentation of information to the corporate-level managerial personnel.

MEASUREMENT DEVICES

A sensor is an element that generates a signal (usually a low-level electrical signal of some form) that is related to the current value of a process variable of interest. This signal may assume either of two forms:

1. *Voltage (usually a low-level dc voltage).* Such signals are often referred to as analog inputs and can be digitized using A/D (analog-to-digital) converters. For most industrial measurements, a resolution of 12 bits or 1 part in 4000 is adequate, as most industrial measurement devices have an accuracy on the order of 0.1 percent (or 1 part in 1000). Where higher resolution is required, it is possible to obtain A/D converters with a resolution of 15 bits or 1 part in 32,000. Where an even higher resolution is required, some approach other than analog signals must be used.
2. *Pulse.*These are normally used on rotating elements and may assume a simple form such as an optical detector aimed to the spot where a bright band passes in front of the optical detector once for each rotation of the element. The output of the optical detector is a pulse, which may be converted to rotational speed via either of the following:
   a. Count the number of pulses received over a fixed time interval. For the above optical detector example, one pulse is received on each rotation, so a count of the number of pulses per minute is the rotation speed in revolutions per minute.
   b. Determine the time between pulses, or actually, the time from the leading edge of one pulse to the leading edge of the next pulse. For the above optical detector, the reciprocal of the time (in minutes) between pulses is the rotational speed in revolutions per minute.

Of these two options, counting the number of pulses received over a specified interval of time is more commonly provided by industrial controllers. Although the pulses could be counted by software routines, use of a hardware pulse counter input module reduces the load on the processor, which can become significant at high pulse rates (over 1000 pulses per second).

When the input signal is a voltage, it is most convenient if the electrical signal is directly proportional to the value of the process variable of interest. If so, such sensors are said to be linear. However, if the input signal is nonlinear, today's electronics technology permits linearizing circuits to be incorporated into the measurement device (with the swing to digital, the nonlinear signal is often digitized and the linearization accomplished via digital computations).

Where the control system is in close proximity to the measurement device, the signal from the sensor may be connected directly to the control system. However, when the control system is several hundred feet from the sensor, this approach is risky. For example, a thermocouple converts a temperature difference to a dc voltage of only a few millivolts (rarely does this voltage exceed 30 millivolts). Although the thermocouple wire could be several hundred feet long, twisted shielded pairs are required to avoid induced voltages from the electromagnetic fields invariably present in manufacturing facilities.

This risk can be avoided by installing a transmitter at or near the location of the sensor. In addition to linearizing the output of the sensor, the transmitter also converts the sensor output to a current signal over a range of 4 to 20 milliamperes. Adjustments on the transmitter specify the engineering span. The lower range of the engineering span is the value of the process variable corresponding to 4 milliamperes. The high range of the engineering span is the value of the process variable corresponding to 20 milliamperes. A milliampere (current) signal is used because such signals are the least influenced by the electromagnetic fields associated with electric motors and similar equipment within the plant.

With the continuing swing to digital technologies at all levels within industrial control systems, some transmitters are now capable of generating digital outputs. Instead of connecting such transmitters to control systems via 4- to 20-milliampere signals, the transmitters are connected via digital communications systems, with several transmitters connected to each communications link. Such transmitters are often referred to as "smart transmitters" because they include functions such as the following:

1. Checks on the internal electronics, such as verifying that the voltage levels of internal power supplies are within specifications.

2. Checks on environmental conditions within the instruments, such as verifying that the case temperature is within specifications.

3. Compensation of the measured value for conditions within the instrument, such as compensating the output of a pressure transmitter for the temperature within the transmitter.

4. Compensation of the measured value for other process conditions, such as compensating the output of a capacitance level transmitter for variations in process temperature.

5. Linearizing the output of the transmitter.

6. Configuring the transmitter from a remote location, such as changing the engineering span of the transmitter output.

7. Automatic recalibration of the transmitter. For example, an ultrasonic-level transmitter can recalibrate by directing its beam to a reflector at a known position, thus compensating for the effect of ambient temperature on sonic velocity.

Over the years, a wide variety of principles have been employed to measure the values of the variables of interest. In order to be usable within an industrial control system, the measurement device must output a signal acceptable to the control system. For example, the height of water flowing over a weir has been used for years to measure the flow rate of water in an open channel. This approach is very compatible with a human operator, since a scale can be provided at the weir to permit the height to be read visually. Today's industrial controllers cannot perform such visual readings. In order to provide a suitable signal to an industrial controller, an ultrasonic device (similar to sonar on ships) must be installed to provide an electronic signal indicating the height of the water over the weir.

In the above illustration of a weir, a measurement device most suitable for use by humans can sometimes be adapted for use by an industrial controller. Conversely, some measurement devices provide a signal directly acceptable to an industrial controller but do not inherently provide for a visual readout to a human operator. For example, the millivolt signal from a thermocouple can be accepted directly by a large number of industrial control products, but conversion equipment is required to provide a visual readout.

For some of the common measurement requirements, the following summarizes the measurement devices available for providing an input signal to an industrial controller:

Temperature. Today, the thermocouple is the most commonly used sensor for measuring temperature. The output of the thermocouple is a low-level dc voltage that is related to the temperature difference between the hot junction (at the temperature of interest) and the cold or reference junction (located in the control system or readout device). The voltage must be compensated for variations in the temperature at the cold junction (called cold-junction compensation) and then linearized (the relationship between voltage and temperature is moderately nonlinear). The thermocouple is the most economical measurement technology for temperature, especially when several thermocouples can be multiplexed to a single readout.

Another approach to measuring temperature is the resistance temperature detector (RTD), which is based on the principle that the resistance of a metal (usually platinum) is a known function of temperature. Basically, the resistance of a platinum resistance element is measured and then linearized to obtain the temperature. Although somewhat more expensive than thermocouples, RTDs usually retain their accuracy better over time than thermocouples.

The thermistor is a semiconductor element whose resistance varies with temperature. These elements are very sensitive and can be used to measure temperatures very accurately over a narrow span.

For high temperatures or where physical contact is not permitted (such as measuring the temperature of a plastic film from an extruder), the pyrometer is a noncontact temperature-measuring device based on the Planck radiation formula. Several variations of pyrometers are available, some designed for high-temperature applications (such as furnaces) and others for low-temperature applications (such as plastic films).

Pressure.   Meters are available to measure either static pressure or differential pressure. Within the meter, the sensitive element actually measures the difference in pressure across the element. When used to measure static pressure, one side of the element is exposed to atmospheric pressure.

The older versions of pressure meters used a bourdon tube or other mechanical mechanism to sense the pressure. But today these approaches have been almost totally replaced by instruments using strain gauge technologies, where an electrical property (usually capacitance) of the sensitive element varies with the differential pressure across the element.

Meters are available for measuring both very low pressures (a few inches of water) and very high pressures (thousands of psi). Furthermore, pressure measurement is also the basis for measuring other variables, notably flow and level.

Flow.   Measuring the flow of a liquid or gas is required in a large number of applications. The orifice meter is commonly used. An orifice is inserted into the flow line, and the pressure drop across the orifice is measured. The pressure drop is proportional to the square of the flow, so the measured differential pressure must be linearized by taking the square root. Although the orifice meter is very popular, it cannot be applied in situations where no pressure drop is available (gravity flow), where two-phase flow is present (such as slurries of solids in liquid), where the liquid is near its boiling point (the pressure drop causes flashing), and so on.

There are several alternatives to the orifice meter. The turbine meter can be used to measure the flow of clean fluids very accurately. The flowing fluid causes a rotor within the turbine meter to spin, and the speed of rotation is sensed using a magnetic pickup coil. Unfortunately, the bearings required for mounting the rotor have proved to be a common source of problems, sometimes even in clean fluids.

For applications such as slurries in water, the magnetic flowmeter is a possible solution. Electromagnets are used to create a magnetic field within the flowmeter. When the fluid flows through the magnetic field, an electrical potential is generated that is proportional to the velocity of the fluid flowing through the meter. This electrical potential is sensed via electrodes within the meter. The flow rate of the fluid can be determined from the electrical potential.

Other popular flowmeters include the vortex shedding meter, various versions of ultrasonic flowmeters, and the Coriolis flowmeter. The Coriolis is somewhat unique in that it is one of the few true mass flowmeters. All other meters are volumetric meters, and when used as mass flowmeters must be compensated for anything that affects the fluid density (temperature, composition, and if gas, pressure).

Level.   When the top of the vessel is open to the atmosphere, the pressure at a fixed point below the liquid surface is directly proportional to the level in the vessel. Thus a pressure meter located at this point can be calibrated in terms of liquid level. When the vessel is not open to the atmosphere, a differential pressure meter must be used.

Another approach to measuring level is the displacer. If the displacer is held at a fixed position, the upward force on the displacer increases as the level in the tank increases and decreases as the level decreases. The upward force is directly proportional to the volume of liquid displaced by the displacer.

Floats can also be used to measure level. However, the mechanical nature of these level measurement devices makes them relatively unattractive for industrial level measurement.

Weight.   Another approach to determining how much material is in a vessel is to place the entire vessel on load cells to determine the total weight. In effect, the empty weight of the vessel is subtracted from the total weight to determine the weight of the contents of the vessel. When the contents of the vessel is a solid, weight measurement is about the only reliable approach to determining how much material is in the vessel.

When load cells are used, considerable attention must be directed to the design and installation of the process vessel. The weight of the vessel must be uniformly distributed on the vessel supports, and extreme care must be used to avoid any horizontal stresses from the piping. Flexible connectors are preferred, but these in turn are objectionable when the material within the pipe is at high pressure, is highly corrosive, is highly toxic, and the like.

Today's load cells can give resolutions approaching 1 part in 100,000. That is, a vessel weighing up to 10,000 pounds can be weighed to a resolution of 0.1 pound. Unfortunately, today's A/D (analog-to-digital converters) are limited to a resolution of about 1 part in 32,000. Therefore, analog inputs cannot be used for interfacing the high-resolution load cells. The more common approach is to use a serial communications interface.

Physical Properties.   Depending on the application, various physical properties might be of interest. Common examples of physical property measurements include density and viscosity. Density can be measured with a variety of techniques, but no single approach has yet proved to be widely applicable. For liquids, most principles (such as differential pressure) used as the basis for measuring level can be adapted to measure density. However, where very high resolutions are required, instruments based on various types of U-tube arrangements are normally used.

Several instruments are also available for measuring viscosity. Industrial rotational viscometers similar to the laboratory rotational viscometers can be acquired. Industrial capillary tube viscometers are also popular, especially when capillary tube viscometers are used in the laboratory. However, several vibration or ultrasonic viscometers are available for industrial use, although no counterpart is available for use in the laboratory.

Composition Measurements.   Where the composition of materials must be determined, industrial versions of analytical instruments can be installed. Chromatography is a common approach for gases and liquids. Various versions of infrared and ultraviolet instruments are available. For solids, analyzers such as x-ray fluorescence are available.

Such measurement technologies are certainly expensive. However, the major problem with using such devices for on-line measurements is the sample system. Sample probes are available for withdrawing gases and liquids from process streams. Care must be taken to ensure that the probe withdraws a representative sample, and then care must be taken to ensure that the sample is not subsequently altered in some way (such as condensation of moisture). For solids, it is extremely difficult to ensure that a representative sample is acquired.

ACTUATORS

The actuator is the device that imposes the wishes of the control system on the machine or process being controlled. The actual outputs of digital control systems are electrical signals in some form. However, these signals are of relatively low power. The actual motive force for the actuator may assume either of the following forms:

1. *Electrical.* This may be in the form of constant-speed ac or dc motors, variable-speed ac or dc motors, solenoids such as those utilized on some two-position valves, and so on.

2. *Pneumatic.* Air has been an extremely popular source of power for valve actuators. Leaks do not contaminate the area in the same manner as leaks from hydraulic systems. When it may be possible for explosive vapors to be present in the vicinity of the actuator, pneumatic actuators have a clear advantage over electrical actuators with regard to safety. Another advantage is that on loss of power pneumatic actuators can be set to fail open or fail closed, which is difficult to achieve using motor-driven actuators. Finally, pneumatic actuators have proved both inexpensive and highly reliable.

3. *Hydraulic.* Where large forces are required, hydraulic actuators are often the only viable alternative.

Although some programmable logic controllers provide output modules sufficient for driving small- to medium-size solenoid valves, a transducer of some type must usually be inserted between the low-power output of the control system and the final actuator.

Although invariably electrical, the nature of the output from the control system may assume one of the following forms:

1. *Position signal.* When the actuator can position its output to any value between two extremes (often simply designated as 0 and 100 percent), the control system will calculate a position value, which can be considered to be a fraction between 0.0 and 1.0. This value is then converted to a voltage output, the usual signal range being 0 to 5 volts dc or 0 to 10 volts dc. In plants where the actuator is located some distance from the control system, the value may be converted to a current output over the customary 4- to 20-mA range. Note that upon loss of power at the control system, the position output to the actuator is 0 percent.

2. *Velocity signal.* Instead of accepting a position signal (for example, position to 74.2 percent), some actuators accept an incremental signal (for example, increase by 2.1 percent from current position). The output signal is a pulse signal of one of the following two forms:

   a. An on-off pulse of fixed duration that causes a stepper motor in the actuator to increment (or decrement) by one step (usually about 0.1 percent of full travel). The amount that the actuator is to be incremented is converted to a pulse count. For example, if each pulse increments the actuator position by 0.1 percent, then 21 pulses would be required to increment the actuator position by 2.1 percent.

   b. An on-off pulse of variable duration that is output to an integrating amplifier. The amount that the actuator is to be incremented is converted to a pulse duration time.

   Perhaps the primary advantage of the incremental actuator is that the final actuator retains its last position on loss of power. However, if such actuators must be positioned to a specific value, a position feedback signal from the actuator is required.

3. *Discrete (on/off).* These outputs drive actuators that can assume only one of two possible states. Examples are ac motors, solenoid valves, and electrical switches. In some cases, a positive feedback for the state of the actuator is required. For example, a valve can be outfitted with two limit switches, one that is closed when the valve is fully closed and another that is closed when the valve is fully open.

It is also possible to effectively convert a discrete actuator into a proportioning actuator via a technique called time proportioning control. With this technique, the final actuator is energized for a specified fraction of each consecutive period of time, called the cycle time. This approach is commonly used in electrical heater elements such as those used on plastic extruders. For example, if the cycle time is 10 seconds and the control system output is 74.2 percent, the heater is on for 7.42 seconds of each 10-second period of time. Note that the switch for the heater is subjected to one on-off cycle each 10 seconds, which translates to 6 per minute, or 360 per hour, or over 3 million per year. The desirability of solid-state switches such as triacs over mechanical switches should be obvious.

DISCRETE LOGIC

Discrete logic is required in virtually all types of industrial control systems. The controller for most manufacturing machines requires considerable discrete logic. For applications such as an automated warehouse, substantial discrete logic is required for controlling the conveyors. In process control applications, discrete logic is required for implementing the safety logic for protecting both personnel and equipment.

Traditionally, this discrete logic was implemented in a hardwired fashion, typically resulting in a cabinet with hundreds of relays coupled with coils, counters, timers, and the like. The complexity was normally measured by the number of relays required to implement the logic. The logic was represented by relay ladder diagrams. It was the responsibility of the electricians to troubleshoot and repair any problems with the discrete logic.

In the early 1970s, a digital alternative to the hardwired relay cabinets was introduced in the form of the programmable logic controller (customarily shortened to PLC and sometimes PC for "programmable controller," although the latter leads to confusion with "personal computer"). The early programmable logic controllers were limited to those functions required to implement discrete logic, namely, boolean logic, timers, counters, and the like. The logic for the programmable logic controllers was developed using the same relay ladder diagrams as those used for the conventional hardwired relay cabinets.

The programmable logic controller was the first digital technology to compete successfully on a cost basis with conventional approaches. In the early days, the number of relays that had to be replaced in order to justify a programmable logic controller started at around 500. However, this number quickly dropped. Today, very small programmable logic controllers (sometimes called micro-PLCs) with only 16 or so I/O points are on the market.

Programmers for the programmable logic controllers are capable of presenting the relay ladder diagrams directly on video displays and accepting modifications via the video display. Thus the logic can be developed in the same manner as for the relay cabinets. Furthermore, these same displays are available for troubleshooting problems. The "current flow" is displayed in high intensity on the video display, which conveys the same information that the electrician could determine with conventional test meters but in a much more vivid manner.

For the initial PLCs, the I/O was limited to discrete inputs and discrete outputs. In addition to interfacing to the plant equipment, the I/O also supported the panel housing the lights, switches, digital displays, thumbwheel switches, and other operator interface equipment. As time passed, analog I/O was added to the PLC, and CRT technology began to replace the panel-mounted equipment for the operator interface.

When PLCs initially appeared, they were used to perform essentially the same functions as the relay cabinets they were replacing. However, it was practical to implement far more complex logic in the PLC than in the relay cabinets. Three advantages of the PLC made this possible:

1. The software implementation of the logic allowed more complex logic than was practical to implement in the form of hardwired relays.
2. The modifications that inevitably arise during commissioning of the control system could be implemented much more quickly.
3. The CRT displays of the relay ladder logic make it possible for electricians to troubleshoot the more complex logic.

Without such advances, it would simply not be practical to implement the complex logic required for applications such as automated warehouses.

Given the popularity of PLCs, the manufacturers were anxious to add other capabilities in order to make their product suitable for a variety of additional applications. One of the functions quickly added was arithmetic capabilities. By using binary-coded-decimal interfaces to load cells, the PLC could now control the weighing systems found in food processing facilities such as flour blenders. The entire list of ingredients could be loaded into the PLC, which could then weigh and blend the necessary materials with little or no operator intervention (except possibly to weigh any materials required in very small amounts).

With arithmetic capabilities on board, the next logical step was to add analog I/O (voltage inputs and outputs) and continuous control functions such as the proportional-integral-derivative (PID) algorithm. While the PLC was easily capable of performing such functions, acceptance by the user community was less than astounding.

The addition of the above functions required that extensions be improvised for the relay ladder diagrams. Arithmetic computations and PID loops were never incorporated into the traditional relay cabinets. Continuous control configurations were represented in loop diagrams, not relay ladder diagrams. Furthermore, PID loops were normally serviced by instrument technicians, not electricians. It was not difficult to devise extensions to the relay ladder diagrams to accommodate the new functions, but their acceptance by the user community was another issue. On the other hand, the migration to a new methodology for programming PLCs is hampered by the fact that many labor contracts require that the logic be displayed to the electricians in the form of relay ladder diagrams.

Programming PLCs using a methodology called Grafcet was developed in Europe. This is a graphical programming approach with much in common with the structured programming concepts developed within the computer science community. Sometimes referred to as sequential function charts, Grafcet is much more convenient than relay ladder diagrams for implementing sequence logic (for example, add material A, add material B, stir for 5 minutes, and so on). The Grafcet approach has now been introduced into the U.S. market, although most initial implementations translate Grafcet into relay ladder diagrams, the usual explanation being that existing labor contracts can still be honored.

NUMERICAL CONTROLS

In order to produce a part with a milling machine, the proper sequence of machine positions, spindle rotations (speed and direction), tool positions, and the like, must be followed. At one time, human operators were responsible for making the machine follow the proper sequence. But as more complex parts had to be machined to ever tighter tolerances, the need for automatic sequencing arose.

The basis for a numerical control system is a numerical representation for each of the possible operations that the machine is capable of performing. For example, rotate spindle might be code 23, which would be followed by the desired rotational speed. All other operations that the machine could perform would be assigned a unique numerical code.

For any part to be produced, it is necessary to prepare a program in the form of a numerical representation for each operation in the sequence required to produce the part on the machine. When numerical control first appeared in the early 1970s, this sequence of operations was loaded into the machine's numerical controller from a punched paper tape. This program could be debugged by single-stepping through the sequence. Where necessary, the sequence could be modified until the part was produced correctly. The final program could then be punched onto a paper tape and saved.

Numerical control offered several major advantages. Production rates as number of parts produced per hour were far higher. The uniformity from part to part was much better. It became practical to produce more complex parts without incurring unacceptable reject rates.

With the early numerical controlled machines, parts still had to be produced in lots or batches. The machine would be outfitted with the proper tools to make the part, the numerical control tape for the part would be read into the numerical controller, and the machine would proceed to produce the required number of parts. The tools would then be changed, another numerical control tape loaded, and another part would be produced.

The availability of a numerical controller for a machine led to more capabilities being added to the machine. Specifically, much additional flexibility could be imparted to the machine by providing for the automatic changing of certain tools. This reduced the changeover time from part to part, thus making it possible to schedule production in smaller lots for each part.

Punched paper tape was never an especially convenient medium. With the ability to change tools more quickly, the time required to read the tape containing the numerical control program became significant. The paper tape could be eliminated and the time to read the program greatly reduced by storing the program as a file on a disk. The numerical controller essentially became a full-fledged computer, the result being known as computer numerical control (CNC).

If carried to the limit, the "just-in-time" philosophy of production means that a part would be made just before it is required in the next step of the production process. Keeping track of what parts are needed when is the responsibility of the cell controller. When the cell controller concluded that it is time for a computer numerical controlled machine to make a specific part, the program for making that part could be loaded via a local area network (LAN) and the part then made. The result is the ultimate in a flexible manufacturing facility, where any part can be made whenever needed. See Chap. 6 in this section for further information on flexible manufacturing systems.

LOOP CONTROLS

The basic objective of a control loop is to maintain a process variable such as a pressure, flow, level, and the like, at or near a target or set point. Figure 9.2 illustrates the schematic for a pressure control loop. The information flow in simple loops such as the one in Fig. 9.2 involves four major constituents:

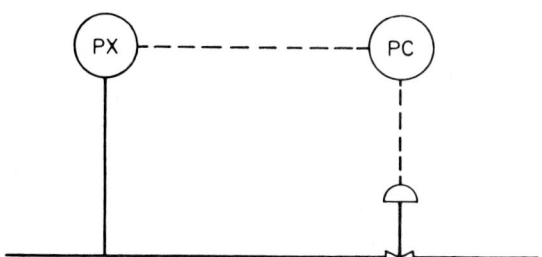

FIGURE 9.2   P and I diagram of a pressure control loop.

1. *Measurement device.* For the pressure loop in Fig. 9.2, this element generates a signal that indicates the current value of the pressure. The output of the measurement device is designated the measured variable, process variable, or *controlled variable*.
2. *Controller.* The difference between the set point and the controlled variable is the *error signal*. Industrial controllers first determine the error signal, and then apply logic to the error signal to generate the output to the process.
3. *Actuator.* The controller output is normally referred to as the *manipulated variable*. The actuator imposes the manipulated variable on the process. Positioning valves and variable-speed drives constitute the major category of actuators.
4. *Process.* The nature of the process determines how changes in the manipulated variable lead to changes in the controlled variable. Furthermore, it should be noted that the controlled variable may be influenced by the following:
   a. Manipulated variable. In order for the loop to have any hope of functioning, the manipulated variable must have a significant influence on the controlled variable.

b. Disturbances. This class of variables includes all variables other than the manipulated variables that influence the controlled variable.

Of all these constituents, the process is the key. Ultimately, the characteristics of the measurement device, the actuator, and the controller must reflect the characteristics of the process.

An industrial control application may require from as few as two or three to as many as several hundred control loops as illustrated in Fig. 9.2. A set of plant drawings called the process and instrumentation (P and I) drawings contain the schematic representation of the required control loops. Especially where hundreds of loops are involved, a convention for the symbology used to draw the loops is essential. The pressure control loop in Fig. 9.2 is based on the symbology from the Instrument Society of America (ISA). An alternative is the symbology from the Scientific Apparatus Manufacturers Association (SAMA). Industries such as petroleum refining, chemicals, and foods almost exclusively use the ISA symbology. Similarly, the fossil fuel electric utility industry uses the SAMA symbology almost exclusively.

In control system analysis, the block diagram is the preferred representation for the loop. Figure 9.3 presents the block diagram for the pressure control loop. Block diagrams are composed from only two basic elements:

1. *Summer or comparator.* The output of this element is the inputs added or subtracted according to their respective signs.
2. *Transfer function.* This "block" expresses the relationship between the input to the block and the output from the block.

Some physical components of the control loop require more than one element in the block diagram. For example, in Fig. 9.3 the controller is represented by a summer plus another block.

Similarly, the process is represented by two blocks and one summer. One block contains the transfer function $G_M(s)$, and represents the influence of the manipulated variable on the controlled variable. The other block contains the transfer function $G_D(s)$, and represents the influence of disturbances such as downstream pressure on the controlled variable. Although other avenues are available, the traditional use of transfer functions in the Laplace domain continues to be the preferred approach to representing the dynamic characteristics of the process being controlled.

The block diagram in Fig. 9.3 indicates that there exist two inputs to the control loop from the outside world. This fact leads to two distinct classes of control problems:

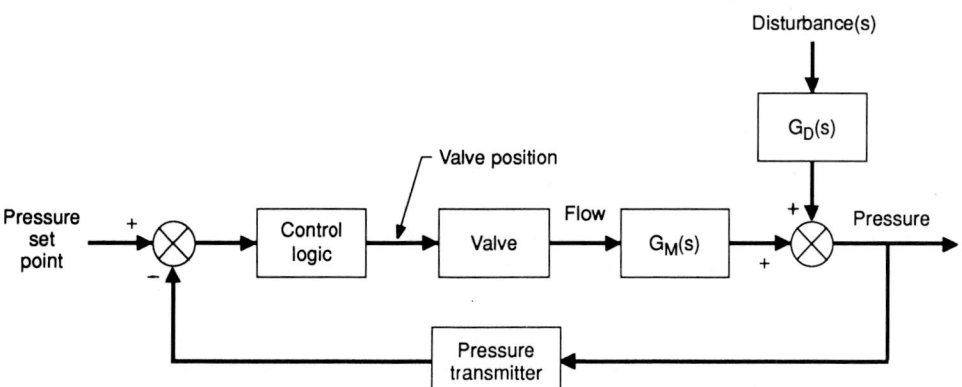

FIGURE 9.3 Block diagram of a pressure control loop.

1. *Servo.*This problem addresses how effectively the control loop follows or tracks changes in the set point or target. In applications such as motion control, this aspect of the control problem is of paramount importance. Most of the initial control theory was directed to this problem, and the term "servomechanisms" was often used synonymously with control theory.

2. *Regulator.*This problem addresses how effectively the control loop responds to changes in the disturbances. A good example of such a problem is the hot water heater in one's home. The set point for hot water temperature is rarely changed; instead, the control system must respond to whatever changes arise in hot water demand (that is, different quantities of hot water, but all at the same temperature).

In practice, most loops are exposed to both changes in the set point and changes in one or more disturbances. Fortunately, the algorithms normally used for regulatory control will respond to both.

The equation corresponding to the comparator within the control loop in Fig. 9.3 is:

$$E(t) = SP(t) - PV(t)$$

where $E(t)$ = error signal, as percent of span
$SP(t)$ = set point, as percent of span
$PV(t)$ = process variable, as percent of span

In most industrial controllers, the input to the control equation is the error signal $E(t)$, and the output is the manipulated variable $M(t)$. Although many algorithms are possible, the two most commonly used in industrial control systems are on/off and proportional-integral-derivative (PID). The following subsections consider both of theses.

On/Off Control.   For applications where precise control is not required, the simplicity of an on/off controller is appealing. The output of this control equation is to a discrete actuator, such as a solenoid valve or an electrical switch. Thus the final actuator is also simpler.

A simple example of on/off control is the temperature control logic within the thermostat (a combined measurement device and controller) in a residential heating system. The logic is simply as follows:

1. When the temperature is below the target or set point [$E(t) < 0$], turn the heater on.

2. When the temperature is above the target or set point [$E(t) > 0$], turn the heater off.

An algorithmic expression of this logic is as follows:

$$M(t) = 100\% \qquad \text{if } E(t) < 0$$

$$M(t) = 0\% \qquad \text{if } E(t) \geq 0$$

Actually, it is irrelevant which case carries the equality.

The response of an on/off controller is illustrated in Fig. 9.4. The controller variable does not line out at the target but instead assumes a never-ending cycle about the target. Such cycles are called limit cycles and are characterized by a period (time for one cycle) and amplitude (distance from peak to valley of the cycle).

While the on/off algorithm is quite simple, there are operational problems with the simple form stated above. The cases are as follows:

1. The time interval between switches is very short, leading to what is sometimes referred to as "chatter." This can be reduced by using the switching curve illustrated in Fig. 9.5. The error deadband E_{db} is an adjustable parameter. However, this is achieved at the expense of a longer period and a larger amplitude in the resulting limit cycle.

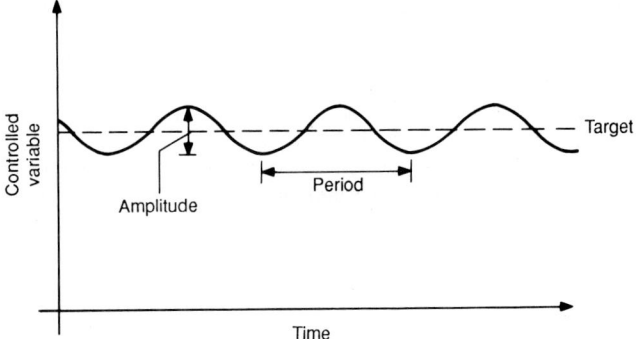

FIGURE 9.4 Response of an on/off controller.

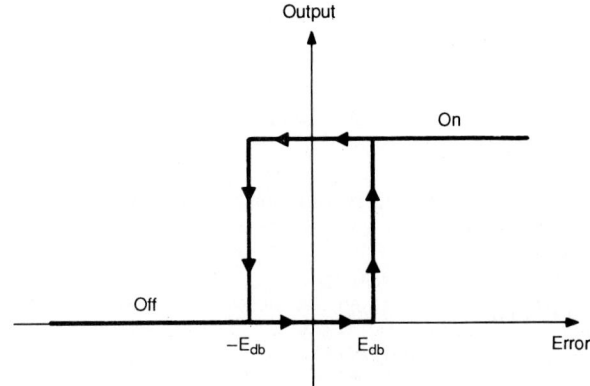

FIGURE 9.5 Switching curve for an on/off controller.

2. The time interval between switches is very long, resulting in a limit cycle with an unacceptable large amplitude. This can sometimes (but not always) be corrected by incorporating a term involving the rate of change of the process variable. That is, the algorithm becomes

$$M(t) = 100\% \qquad \text{if } E(t) - K_D \, dPV(t)/dt < 0$$

$$M(t) = 0\% \qquad \text{if } E(t) - K_D \, dPV(t)/dt \geq 0$$

The coefficient K_D must be adjusted to give the desired performance. If adjusted properly, the result is a limit cycle with a lower amplitude and a shorter period (which also means more actuator movements per day).

While the above approaches are available to address the problems, they result in a more complex controller with a parameter that must be adjusted (usually by trial and error) at the time the control loop is commissioned. Unfortunately, this at least partially offsets the on/off controller's advantage of simplicity.

PID Control. The proportional-integral-derivative (PID) control equation has now been used for over a half century. The early advocates for computer control systems promised to develop a control equation far superior to the PID equation. However, the equation implemented in today's microprocessor-based control systems is little different from the equation used in pneumatic controllers in the 1940s.

The classical form of the PID control equation is expressed by the following differential equation:

$$M(t) = K_c\left[E(t) + \frac{1}{T_i}\int E(t)dt + T_d\frac{dE(t)}{dt}\right] + M_R$$

where $M(t)$ = controller output, as percent of output span
K_c = proportional gain, %/%
T_i = reset time, minutes
T_d = derivative time, minutes
M_R = bias or manual reset, as percent of output span

This control equation involves the following three terms or modes:

1. *Proportional.* This component of the output is proportional to the current value of the error.
2. *Integral or reset.* This component of the output is proportional to the integral of the error.
3. *Derivative or rate.* This component of the output is proportional to the rate of change of the error.

The complete equation is sometimes referred to as the three-mode control equation. However, the following subsets are possible:

1. *Proportional only (P).* Responds quickly to changes in error but does not line out at set point.
2. *Integral only (I).* Responds very slowly, with main use in loops with a large noise component on the measured variable.
3. *Proportional plus integral (PI).* Most commonly used control equation.
4. *Proportional plus derivative (PD).* Rarely used.

The coefficients K_c (the proportional gain), T_i (the reset time), and T_d (the derivative time) are called tuning parameters. Their values must be adjusted so that the characteristics of the controller are "in tune" with the characteristics of the process. This procedure is called tuning. For the most part, tuning is done in a trial-and-error manner. Step-by-step tuning techniques are available, and some manufacturers now offer automated tuning facilities within their control products. However, for the most part, the traditional trial-and-error methods continue to be used.

When the controller is configured, one of the configuration options is direct or reverse action. A direct-acting controller will increase its output when the value of the measured variable is increasing. When the control equation is written in the form as above, this will occur when the numerical value of the proportional gain K_c is negative. A reverse-acting controller will increase its output when the value of the measured variable is decreasing. When the control equation is written in the form as above, this will occur when the numerical value of the proportional gain K_c is positive. In most industrial controllers, the tuning parameter K_c is always specified as a positive value, and the sign is specified separately in a configuration parameter called the action.

Whereas the on/off controller results in a limit cycle in the controlled variable, the PID controller will seek an actuator position that results in the controlled variable lining out at the target. Figure 9.6 illustrates three different responses to a change in the set point of the controller:

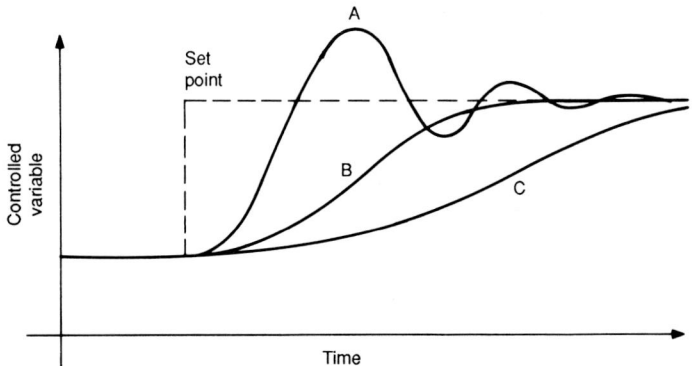

FIGURE 9.6 PID responses to a change in set point.

1. Response *A* gives an overshoot and oscillation with a decay ratio of ¼, the decay ratio being the amplitude of the second peak divided by the amplitude of the first peak.
2. Response *B* gives a minimal overshoot, with the response arriving at the set point as quickly as possible but without any significant excursion above the set point.
3. Response *C* is a very sluggish response that very slowly attains the set point.

For a given loop, the adjustments of the tuning parameters K_c T_i, and T_d determine which of these responses is achieved. Most control texts recommend response *A* as being a reasonable compromise between a rapid initial response and a short line-out. However, operators of industrial facilities are usually more comfortable with response *B*.

Although the classical form of the PID equation bases all modes on the error signal, most implementations depart from this approach with regard to the derivative mode. Instead of being based on the error signal, the derivative mode is based on the measured variable, as per the following equation:

$$M(t) = K_c \left[E(t) + \frac{1}{T_i} \int E(t)dt - T_d \frac{dPV(t)}{dt} \right] + M_R$$

In addition, some filtering or smoothing is usually incorporated into the derivative mode in order to reduce the impact of noise in the measured variable on the performance of the loop.

The term "mode" is actually used in two contexts:

1. Modes of the control equation. This refers to the proportional, integral, and derivative terms.
2. Modes of operation, which are manual and automatic.

In the manual mode of operation, the control calculations are not performed. Instead, the operator can directly specify the output of the controller. That is, in manual mode, the operator could set the output of the controller at 65 percent. In manual, the value of the set point is immaterial, since the control calculations are not being performed.

In the automatic mode of operation, the control calculations are being performed. As the result of these calculations is the output of the controller, the operator cannot directly specify a value for the controller output. Instead, the operator specifies the value for the target or set point, from which the error is calculated for use in the various terms of the control equation.

Another consideration in the application of the PID control equation is bumpless transfer. While in manual, suppose the operator set the output of the controller to 65 percent. Further suppose the operator now wishes to switch to automatic. This transfer will be bumpless provided the value for the controller output just after the switch to automatic is still 65 percent. That is, bumpless transfer means that there will be no change in the controller output over a manual-to-automatic transfer.

Basically, bumpless transfer is achieved by properly setting the value of the bias or manual reset M_R. The proportional mode equation is written as follows:

$$M = K_c E + M_R$$

Bumpless transfer is achieved by solving this equation for M_R:

$$M_R = M_o - K_c E$$

where M_o is the controller output during manual mode of operation. The bumpless transfer calculations must be either performed throughout the time the controller is in manual or performed as part of the manual-to-automatic mode transfer. In the PI or PID control equation, the bias M_R is normally implemented as the initial condition on the integrator in the integral or reset mode.

Whenever the control equation contains the integral term, logic must be incorporated to prevent the integral term from becoming very large, either positively or negatively. In essence, when the manipulated variable attains the limit of either 0 or 100 percent, the output in the controller is said to be "saturated" and further increases in the value of the integral term have no effect on the control actuation. The logic required to prevent the integral term from becoming unreasonably large either positively or negatively is referred to as *reset windup protection* and is incorporated into the commercially available industrial controllers.

The above forms of the control equations have been expressed in differential equation form. Technically, these forms are suitable only for implementation in analog electronic or pneumatic controllers. For digital systems, the corresponding discrete equation is used. When the derivative mode is based on the process variable, the difference equation is as follows:

$$M_n = K_c \left[E_n + \frac{T}{T_i} \sum_{i=1}^{n} E_i - \frac{T_d}{T} (PV_n - PV_{n-1}) \right] + M_R$$

where M_n = controller output at sampling instant n, in percent of span
E_i = error signal at sampling instant i, in percent of span
PV_n = process variable at sampling instant n, in percent of span
T = time interval between control calculations, minutes

In most control applications, the sampling time T is very short, and the performance of the discrete algorithm is equivalent to that of the continuous algorithm.

The above form of the discrete control equation is referred to as the position form, as the result M_n is the desired position of the final actuator. When the output is via an analog output module, this form is usually most appropriate. However, an alternate form of the control equation is as follows (obtained by differencing each term in the position algorithm):

$$\Delta M_n = K_c \left[(E_n - E_{n-1}) + \frac{T}{T_i} E_n - \frac{T_d}{T} (PV_n - 2PV_{n-1} + PV_{n-2}) \right]$$

This form of the control equation is called the velocity algorithm, as the result ΔM_n is the change in actuation position. This form of the control equation is more compatible with incremental actuators driven by pulse outputs. See Chap. 4 in this section for additional information on process control.

SUMMARY

As in most other areas, the impact of digital technology continues to be a dominant factor in industrial controls. At the start of the 1970s, digital technology was used very sparingly in industrial control systems. During the 1970s, industrial control equipment based on digital technology became the solution of choice for almost all requirements. However, this equipment was used largely for the same functions as would have been done with traditional control equipment. During the 1980s, the power of the microprocessor based control equipment began to be used to perform functions beyond what could reasonably be undertaken with traditional equipment. During the 1990s, the control functions will be integrated with the other information processing functions within industrial organizations to create a computer-integrated manufacturing environment.

CHAPTER 10
ROBOTICS AND AUTOMATION

Nicholas G. Odrey
Professor of Industrial Engineering
Lehigh University
Bethlehem, Pennsylvania

This chapter considers, in general, the field of automation and, in particular, the evolution of the field from simpler mechanized devices to the flexible automated systems that currently exist. Coupled with the concept of flexibility has been the advent of the robot over the past three decades. Robotics has emerged from its initial applications on the factory shop floor to sophisticated autonomous devices capable of a variety of tasks. This chapter considers the evolution of automation, robots within automated systems, and applications of robotics primarily within traditional manufacturing and assembly operations. Programming aspects pertaining to robotics are also reviewed. Implementation strategies for robotic work cells are discussed, as are the economic justification and quantitative and qualitative performance measures to consider prior to implementation.

Issues on systems integration are of prime concern to industrial engineers. Such issues contain the structure for modeling flexible manufacturing and flexible assembly systems and include the hierarchical reference models used to implement specific architectures for control of such systems. Hierarchical control development has led to the development of concepts pertinent to intelligent machines or robots.

Mechanization and Automation.   The precursor to automation is mechanization, which can be defined in its simplest sense as the transfer of skills and manual activities to machine operations. The primary difference between mechanization and automation is that automation includes feedback for controlling an automated system. We distinguish here microautomation from macroautomation in that microautomation pertains to the low-level control system commonly envisioned for industrial automation. As such, microautomation is primarily concerned with logical control focused on individual machines and the logical linkage between machines and devices. In particular, microautomation pertains to the servomechanisms, hydraulic and pneumatic devices, and associated low-level software used for physical movement of parts through a production system which typically results in "islands" of automation.

Macroautomation, as the term indicates, has a larger scope and deals with the coordination and supervision of the various smaller scoped islands of automation. This has led to various models and paradigms for communication and control of large-scale manufacturing (automated) systems. Pertinent issues include reference models and corresponding architectures for implementation of large-scale systems.

Automation and Production. Production volume and product variety are considerations to take into account when classifying automated production systems. As noted by Groover,[6] such systems can be classified into three basic types:

1. Fixed automation
2. Programmable automation
3. Flexible automation

The relationships of product variety and product volume to the type of automated system are indicated in Fig. 10.1. Fixed automation is characterized by having the sequence of operations necessary to manufacture or assemble a product fixed by the equipment configuration. As such, there is typically equipment which is inflexible to product changes. The advantage is high production rates. Conversely, programmable automation equipment is highly reprogrammable to accommodate high product variety but has low production rates relative to fixed automation. Parts are typically loaded into programmable automated production systems in batches. Each batch consists of a different part, and the machines comprising the system to manufacture the part are reprogrammed for each batch. Changeover from one batch to the next also requires a change in the physical setup of the machine tools, that is, their fixturing and tooling. Such changeover results in a loss of production time. Examples of programmable automation include numerically controlled (NC) machine tools and industrial robots.

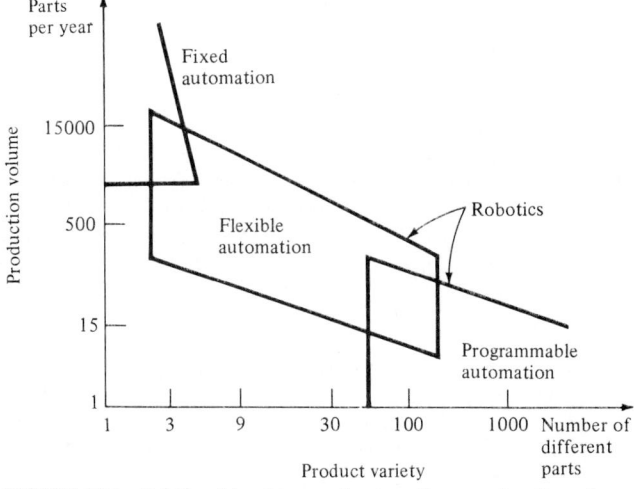

FIGURE 10.1 Relationship of types of automation as a function of production volume and product variety. (*Groover et al., Fig. 1-1, p. 5.*)

Flexible automation has evolved over the past two decades from its roots in NC machine tools. Conceptually, a flexible automated system has the capability of producing a variety of parts with minimal changeover time from one part to the next. The ability to change part programs and to change the physical setup of the production system with little or no loss of production time is the primary difference between flexible automation and programmable automation. To accomplish this, one strategy has been to formulate flexible manufacturing cells based on group technology principles. The overall objective is to increase productivity by properly designing a flexible system that offers advantages similar to fixed automation schemes of mass production but has the capability for handling part variety.

Mechatronics and Concurrent Engineering.   Closely related to the issues of robotics and flexible automation is the concept of "mechatronics," which was originated by the Japanese in the early 1970s. Mechatronics[11] has been described as "the union of mechanical and electronic engineering needed in producing the next generation of machines, robots, and smart mechanisms for applications such as manufacturing, large-scale construction, and work in hazardous environments." Even more so, it is a multidisciplinary approach using integrated teams of designers, manufacturing engineers, purchasing, marketing, and sales personnel to design both the product and the system to manufacture it. Complementary to the term mechatronics would be concurrent (simultaneous) engineering, which is a term more predominant in U.S. industries. The issues now are toward integration as opposed to the specialization and separation which were predominant in the study of work systems during the first half of the twentieth century.

Various terms have been used to describe the integrations of product design and process design such as "design for manufacturability," and "design for automation." The motivation for these efforts has been the drive to reduce market lead time, increase both the quality and reliability of products, and recognize reduced product life cycles. A consumer-oriented society has arisen in which homogenization of products has changed to multiple products with various options as evidenced, for example, by automobile manufacturers.

As noted by Kuo and Hsu,[8] the Japanese are focusing on four key areas in regard to concurrent engineering. These areas are, in many ways, fundamental to the existing discipline of industrial engineering and its future development and include:

- Intelligent manufacturing systems
- Intelligent computer-aided design
- Human-machine interface
- Maintenance

Particularly important to industrial engineers would be their contribution to the development of intelligent manufacturing systems. Issues in plant layout to achieve a systematic methodology for production configuration and implementation, the associated economic evaluation of different configuration alternatives, the algorithms and expert system techniques for production control, the effect of ergonomics and human factors in production systems, and the overall informational system infrastructure are fundamental to intelligent manufacturing systems. In addition, the issues of designing systems for maintenance management and long-term quality and reliability fall within the domain of industrial engineering.

The function of mechatronics, or in more general terms, concurrent engineering, is to provide a rational framework for the integration and implementation of the various subsystems which comprise the disciplines and organizational functions of an engineering and manufacturing facility. The primary intent is thus to transcend the boundaries, usually educationally based and artificially induced, which typically exist between design and manufacturing.

FUNDAMENTALS OF ROBOTS

The following definition of an industrial robot is given by the Robotics Industries Association (RIA):

An industrial robot is a reprogrammable, multifunctional manipulator designed to move materials, parts tools, or special devices through variable programmed motions for the performance of a variety of tasks.

This definition is compatible with the classification of programmable automation, but robots are also used in flexible automation and fixed automation systems. For example, an automation line using many robots to perform spot welds in which robot programs are downloaded from a computer to a specific robot controller could be considered a high-production flexible automation system.

Robot Classification.   To classify robots, one has to consider the various types. Two standard classifications suggested by Engelberger[2] are:

1. Mechanical configuration based
2. Control method based

The classification based on mechanical representation considers the various joints and links comprising the physical structure of the robot and their relationship to each other. Classification by control pertains to the type of technique implemented to control the robot. This classification considers the following subclasses: Non-servo-controlled and servo-controlled. Each of these classification techniques is considered in subsequent sections.

   Mechanical Configuration.   The majority of commercially available industrial robots can be grouped into four basic configurations:

1. Polar configuration
2. Cylindrical configuration
3. Cartesian coordinate configuration
4. Jointed-arm configuration

These configurations are depicted schematically in Fig. 10.2.

   The polar configuration (*a*) has a telescoping arm which pivots about a horizontal axis and also rotates about a vertical axis. The work volume, a term referring to the space within which the robot can manipulate the wrist end, thus defines a spherical volume in which the robot can perform its task. Cylindrical configured robots (*b*) use a vertical column with the robot arm attached to a side which can move up and down the column. Simultaneously, the arm can move radially with respect to the column. The cartesian or rectilinear robot (*c*), also termed a gantry robot, has three mutually perpendicular axis which define a rectangular work volume. The jointed-arm robot (*d*) most resembles a human arm and consists of a series of links connected by rotary joints which when referenced from the base are referred to as the shoulder, arm, and wrist joints. Obviously, different applications might dictate the most appropriate configuration. Regardless of configuration, the function of the robot arm configuration is to position a wrist assembly which orients an end effector to the proper position and orientation (jointly referred to as the POSE) dictated by the task at hand.

   Orientation of the robot wrist provides proper orientation of an attached end effector to the task to be performed. Various representational schemes can be used to describe orientation. One common method of specifying orientation in practice is by the wrist roll, pitch, and yaw angular deviations. This motion is illustrated in Fig. 10.3. Wrist roll, also referred to as wrist swivel, involves rotation about the arm axis. Wrist pitch or bend is the up and down rotation of the wrist, whereas yaw refers to the right or left rotation of the wrist. These motions typify the three degrees of freedom (DOF) associated with orientation. Individual joint motions are referred to as degrees of freedom, and a typical industrial robot has four to six degrees of freedom which provide the POSE for a specific task. Three joints provide positioning of the arm or body of the robot and two to three joints actuate the wrist motion.

   Control Classification.   Various techniques have been developed to control the various axes of a robot simultaneously. The simplest type of control is the non-servo-controlled or limited-sequence robot. For more complex control and greater flexibility, current industrial practice employs servo-controlled robots. Non-servo-controlled robots

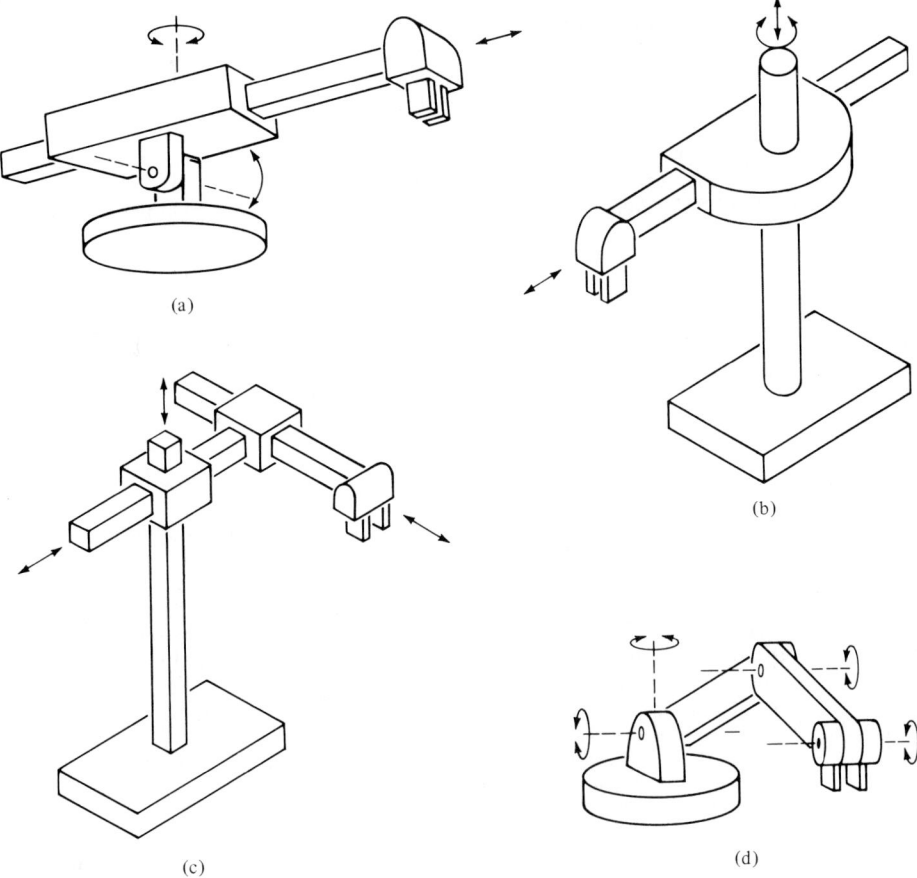

FIGURE 10.2 Four basic robot configurations: (*a*) polar, (*b*) cylindrical, (*c*) cartesian, and (*d*) jointed-arm. (*Groover, et al., Fig. 2-1, p. 22.*)

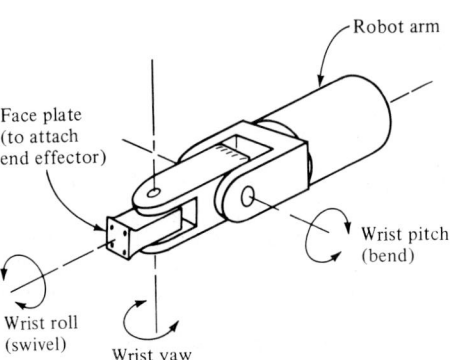

FIGURE 10.3 Degrees of freedom associated with a robot wrist. (*Groover, et al., Fig. 2-11, p. 20.*)

are also called limited-sequence robots, end-point robots, pick-and-place robots, or bang-bang robots. Such a robot is controlled by setting mechanical stops or limit switches to establish end points of travel of each joint. The mechanical setup to give the proper position and sequence of stops serves as a rudimentary programming approach rather than a computer-intensive robot programming language.

The servo control of an industrial robot is accomplished by comparing feedback information to the command input such that a desired trajectory will be followed, that is, a closed-looped system. Feedback information on position, velocity, or other physical variables is provided by continuously monitoring the variables of interest. Several sophisticated robot arm control techniques and algorithms have been developed for robot controllers. The point of view taken here is to consider how commands are input to the robot controller such that a robot follows a desired path. From this viewpoint, servo-controlled robots can be classified as playback robots with point-to-point (PTP) control or playback robots with continuous-path (CP) control.[7]

The term playback refers to the teaching of positions or motions with subsequent recording of these "taught" variables into the robot memory such that they may be played back during operation of the robot. Point-to-point (PTP) control of a robot consists of "teaching" a series of desired point locations which are recorded in memory. The robot moves in proper sequence from one point to another during operation, but the path to get from one point to another is not controlled by the programmer but rather is calculated by the computer using an interpolation scheme. Depending on the degree of sophistication of the controller it is possible to specify straight-line interpolation or circular interpolation in addition to the joint interpolation schemes commonly available in the majority of PTP robots. In joint interpolation, the controller determines for each joint how far each joint must move from the first defined point to the next. Based on a move time determined by the joint that requires the longest time, the motions of all joints are coordinated such that the start and stop times of all joints are the same.

PTP robots have many applications such as loading and unloading of machines and spot welding in automotive assembly lines and have proved quite adequate in industrial settings, particularly if high loads and long reaches are required. For complex paths and high end-of-arm speeds where loads and arm reach are not predominant, a continuous-path (CP) robot may prove more advantageous. Such applications include arc welding, spray painting, and grinding of complex surfaces. CP robots can perform motion cycles in which the trajectory followed is controlled. CP robots differ from PTP robots in that points are not taught manually in a CP robot. Instead, programming is accomplished by an automatic sampling routine which records points as an operator physically moves the arm along a complex path. The sampling rate is high (on the order of 60 to 80 recorded points per second) such that the motion cycle is divided into thousands of individual closely spaced points along the trajectory and the points are recorded into the controller memory. Playback of these recorded points results in an extremely smooth motion.

Response time is a function of the speed of motion of a particular robot and is fundamental in determining the work cycle time. To minimize production cycle time, one must consider the effect of various factors when setting robot speed. These include:

1. End effector positioning accuracy
2. The weight of the object
3. The total distance moved

Inertial effects for heavier objects imply that slower speeds should be set to move such objects. In addition, it should be noted that the size, construction and configuration, and type of drive system determine the load-carrying capacity of a robot. Rated weight-carrying capacities are specified without the end effector attached and can range from less than a pound to very heavy loads. For example, the Prab Versatran Model FC has a capacity of 2000 pounds.

The issue of accuracy is contained within the precision of movement of the robot arm. Precision can be defined as a function of (1) spatial resolution, (2) accuracy, and (3) repeatability. Spatial resolution (SR) is the smallest increment of movement within a robot work volume. Accuracy refers to the capability to achieve a desired target point whereas repeatability pertains to the capability to return to a previously programmed point. It should be noted that accuracy refers to the capacity to be programmed to achieve a given target point but the actual programmed point might differ from the target point. These differences are due to limitations of the control resolution (CR), which is a function of the bit storage capacity designed into the controller memory.

Mechanical inaccuracies result from the various errors that can occur in a robot's construction or operation. These include gear backlash, stretching of pulley cords, leakage of fluids, or material and structural imperfections. Such errors degrade the overall accuracy of the robot and are magnified when the arm is fully extended. Mechanical inaccuracies are principally responsible for repeatability errors. A robot manufacturer typically specifies repeatability as the radius of an idealized sphere and expresses the specification as a plus or minus value. The majority or robots have extremely high repeatability (can be ± 0.001 inch or less) but many of these same robots can have comparatively greater inaccuracy values.

Control Systems and Components. This section gives an overview of the control structure, actuation devices, grippers, and types of sensors that may be used with an industrial robot. The review is not all-inclusive but is intended to give some of the basic concepts relevant for robot and automation implementation. For details, the reader is referred to the various texts in the field.

Controller. A robot controller has the function to control the manipulator as programmed by the user to perform the prescribed task. A robot controller not only can be used to control the robot itself but through interfacing with various other equipment can function as a work-cell controller. Figure 10.4 is a schematic for a general robot controller. Typically, such controllers are hierarchically arranged with a microcomputer at the top level which serves as a supervisory computer. At the lowest level, there are microprocessors to serve as controllers for each degree of freedom of the robot, and this level also contains a power amplifier, a digital-to-analog (DAC) converter, and a joint encoder for feedback. Motion commands are executed by the controller from program memory or operator input via a teach pendant or computer terminal. Each motion command provides the information for the mathematical processor to perform the calculations necessary for joint motion. Once the coordinate transformation calculations are completed, the executive processor downloads the results to the joint controllers as individual position commands. Each joint controller then provides the signal to an amplifier to drive the actuator. For example, the Unimation PUMA 560 series robot arm has a DEC LSI-11/02 as the main

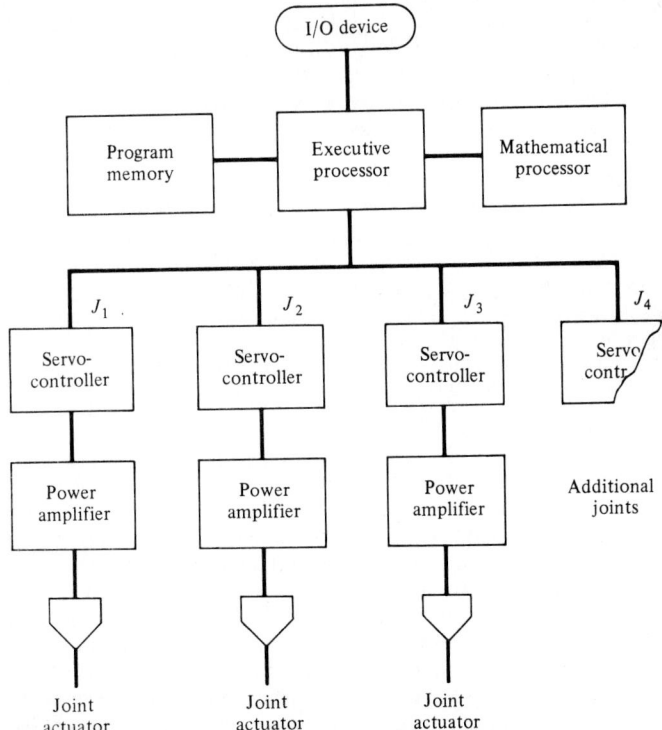

FIGURE 10.4 Elements of a general robot controller. (*Groover, et al., Fig. 14-13, p. 110.*)

supervisory computer and six Rockwell 6503 microprocessors arranged in a hierarchical structure. Each microprocessor communicates with the LSI-11/02 through an interface board. Such an architecture provides several advantages including expansion of the system to more joints and information flow between joint control elements.

Actuators. Actuators which provide the power to move the joints of a robot are typically pneumatic, hydraulic, or electric devices. Pneumatically activated devices consist of both linear cylinders and rotary actuators to provide motion. Pneumatic actuators are as a class less expensive and more reliable than other systems, but it is difficult to control speed or position due to the compressibility of the air which is used as the fluid medium. Accuracy can be enhanced by using mechanical stops, and pneumatically actuated robots are useful for light load applications involving pick-and-place operations.

For applications requiring heavy payloads (typically greater than 10 pounds and as high as 2000 pounds), the hydraulic drive is the system of choice. High power-to-weight ratios, greater accuracy, greater frequency response with smoother performance at low speeds, and a wide speed range are some of the advantages of hydraulic drive over pneumatic systems. Disadvantages include large floor space requirements and potential hazardous conditions from high-pressure fluid leaks.

Electric drive systems are becoming more predominant in industrial robot environments. Although they do not provide the speed or power of hydraulic systems, electric drives have better accuracy and repeatability, require less floor space, and consequently are very suitable for precise work such as assembly. Electric drive robots are typically actuated by dc stepping motors or dc servomotors. Currently, stepping motors are predominant on small "teaching" robots used in educational institutions or in laboratory automation environments. A stepper motor output consists of discrete angular motion increments initiated by a series of discrete electrical pulses. Stepper-motor-driven robots are used for light-duty applications since a heavy load may cause a loss of steps and subsequent inaccuracy. In addition, stepping motors are limited in resolution and have a tendency to be noisy, although it must be noted that progress is being made to overcome these inherent disadvantages. Still, the ability to have high torque at small angular velocities, to operate in an open-loop manner, and to be directly compatible with digital control techniques does make stepping motor robots advantageous in certain applications.

DC servomotors provide excellent controllability with minimum maintenance requirements. Torque control is possible by controlling the voltage or current, respectively, applied to the motor. DC servomotors using permanent magnets to generate the magnetic field are a common method of actuating a robot joint. Advantages offered by such motors include high stall torque, a small frame size and light weight, and a linear-speed curve which reduces computational effort. An excellent discussion on motor selection in the design of a robotic joint is provided in the text by Klafter et al.[7]

End Effectors. The end effector attaches to the end of the robot wrist and enables the robot to perform the tasks required. The end effector is considered as special-purpose tooling and is typically custom engineered. A wide variety of end effectors exist to perform different work functions. The various types can be classified[5] into two major groups:

1. Grippers

2. Tools

Grippers are used to grasp and hold objects whereas tools can be used to perform work on a part rather than merely grasp it. Examples of tool-type end effectors are spot welding tools, arc welding torch, water jet cutting tools, and rotating spindles for drilling, routing, and grinding.

Grippers consists of mechanical devices, magnets, suction devices, adhesives, or other devices to grasp and hold objects. Mechanical grippers can be classified as single or multiple grippers depending on the number of grasping members attached. Cases where more than two grippers (fingers) are used are rare because of cost and reliability issues associated with increasing the number of gripper devices. Examples of mechanical grippers are illustrated in Fig. 10.5.

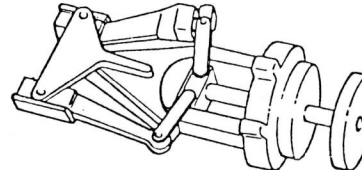

Standard hand
This is an inexpensive and all-purpose hand that will accept a virtually infinite variety of custom fingers. Fingers are tailored to the parts to be manipulated or moved. The parts should be of moderate weight. Simple linkages provide both the finger action and the force multiplication needed to grip the object sufficiently tightly. At the completion of finger closure, the fingers exert their maximum clamping force on the part.

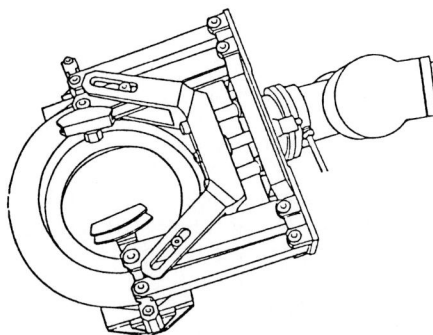

Cam-operated hand with inside and outside jaws
Assume that a part is re-oriented between the time when the part is placed in a machine and when it is removed. This special hand is one of those which will deal with this problem. When the part is oriented as shown, the hand can grasp it on the OD by employing the outer self-aligning pads. If the part is turned over, the inner pads will grasp the ID.

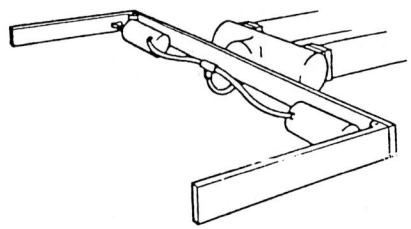

Special hand for cartons
The dual-jaw hand will open wide to grasp inexactly located objects of light weight. Lifting and placement of cardboard cartons is an application. Actuators and jaws can be re-mounted in any of several positions on the fixed back plate, making it practical for the same dual-jaw hand to move large cartons on one day and smaller cartons the next.

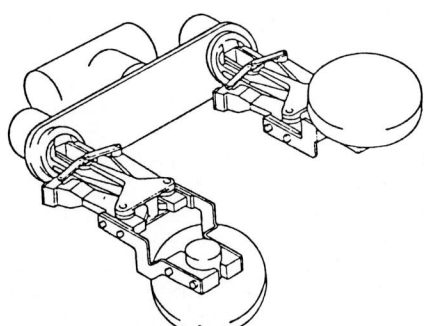

Double hand
Does a robot application call for the hand to remove a finished part from a machine and replace it with an unfinished part? A double hand with double actuators is a possible choice. It will pick a part out of the chuck of a machine, swivel, and place a new part back in the chuck, for instance. Thus, an industrial robot with this hand does not need to expend time to put one object down before it manipulates another: the hand seldom makes a trip while empty. Parts should not be of more than moderate weight when the double hand is used.

FIGURE 10.5 Examples of mechanical grippers for use on industrial robots. (*Engleberger, Fig. 3.2, pp. 45 to 48.*)

Various factors need to be considered in assessing gripper requirements. Consideration has to be given not only to the part surface reachability when attempting to grasp the object but also to size variations resulting from processing and any potential quality issues such as scratching and destroying the part during gripping. In addition to determine the grasping force, factors considered must be the part weight and shape, the robot arm speed and acceleration, the orientation of the part within the gripper relative to the motion of the gripper, and physical considerations such as the coefficient of friction, fabrication materials, temperature protection, and mounting connections. These factors contribute to sound gripper design.

Sensors.   Sensors are fundamental in the use of robots and other automated systems. (See Chap. 9 of this section for further information.) The primary use of sensors can be categorized as follows:[5]

1. Safety monitoring
2. Work-cell control interlocks
3. Part inspection for quality control
4. Position and related information on parts in a work cell

Safety or hazard monitoring is important to ensure the protection of workers in the vicinity of a robot or other equipment. The National Institute of Standards and Technology (NIST) defines three levels of safety sensor systems in robotics:

Level 1: Perimeter penetration detection
Level 2: Intruder detection inside a work cell
Level 3: Intruder detection in the immediate vicinity of the robot arm

Level 1 protection is typically accommodated by a fence around the work cell with simple switches to monitor any gates. Level 2 protection could involve pressure-sensitive mats within the work cell or light curtains. One implementation of level 3 detection has been the mounting of ultrasonic sensors on the wrist of a robot to detect any obstacle (including human presence) in the path of the robot arm.

Work-cell control interlocks pertain to signals sent to the controller with subsequent signals sent from the controller to the equipment within the work cell. The function of work-cell control interlocks is to coordinate the sequence of activities for different equipment in the work cell. Various types of sensors are used to send signals to the controller and are termed input interlocks. Output interlocks are the command signals sent from the controller. Interlocks are often implemented by means of simple limit switches or more advanced sensors.

Sensors are beneficial for determining part characteristics essential for quality control. The use of sensors enables 100 percent inspection of parts for specific characteristics as opposed to traditional manual techniques based on statistical sampling. Depending on the type of sensors used, there are limitations as to what characteristics or defects are being inspected. For example, a probe to measure part diameter would not detect flaws in the surface. An issue is the determination of the proper sensor strategy to employ for the task at hand. Sensor strategies, their placement, and their integration are important not only for quality control but also for overall work-cell control and safety.

Positional data and related information such as orientation, color, and size are important for part identification, determining any random position and orientation of parts in a work cell, and providing feedback information for improved control. All such information is typically processed to be either the robot controller or a separate work-cell control in real time for execution of the programmed work cycle. As with the other categories of sensor applications, the sensor systems chosen constitute a part of an overall control system whose function is to provide timely, accurate execution of the tasks to be accomplished.

Sensor use in robotics and automated systems includes a wide range of devices. These devices can be divided into the following categories:[5]

1. Tactile sensors
2. Proximity and range sensors
3. Miscellaneous sensors and sensor-based systems
4. Machine vision

Tactile sensors consist of either touch or force sensors to indicate contact with an object or contact plus the magnitude and direction of the contrasting force. Tactile sensing can be accomplished in various ways. For example, an IBM gripper with strain gauges to detect positive and negative forces in three mutually orthogonal directions is indicated in Fig. 10.6. The figure also indicates the use of an optical sensor to indicate whether the gripper has enclosed the object of interest. In addition, much use is made of force-sensing wrists which are mounted between the end of arm and the attached end effector. Such wrists provide three-component force information and three-component moment information at the end of arm.

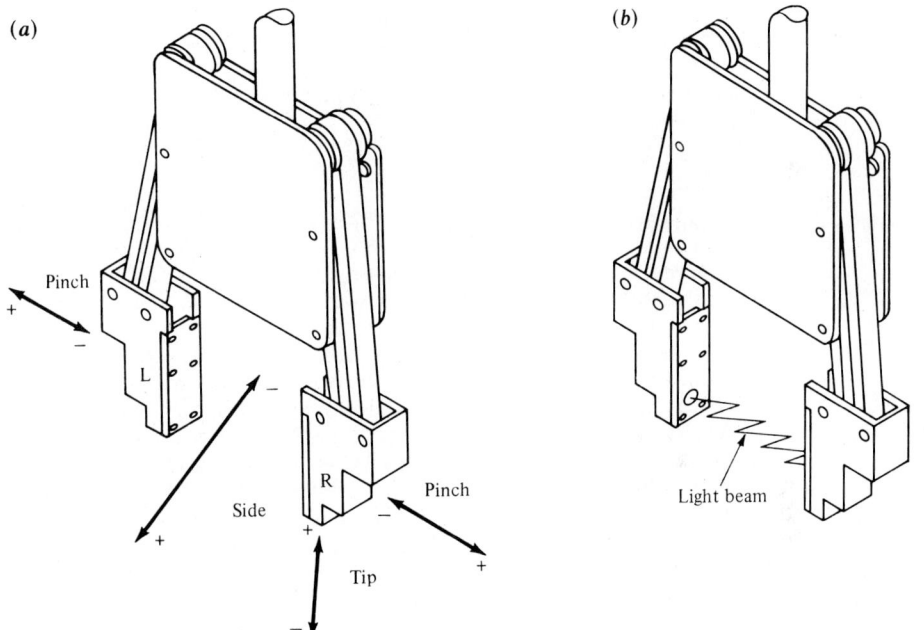

FIGURE 10.6 Sensored gripper used on IBM 7565 robot: (*a*) tactile sensing features, (*b*) optical sensing with light-emitting diode. (*Groover, et al., Fig. 90-P-2, p. 287.*)

Proximity and range sensors have a wide range of applications where it is necessary to determine closeness or distance from an object. Technologies used for designing such sensors include optical devices, acoustics, electric field techniques, and others. Miscellaneous sensors include those that might be used for interlocks and other purposes.

Machine vision has proved to be an important sensor technology and has found a wide variety of applications. Robotic applications of machine vision include inspection, identification, and more recently, visual serving and navigation. In addition, there are various special-purpose applications such as range finding via triangulation techniques, avoidance of obstacles, and automatic robot path planning. Machine vision, when coupled with force and torque sensors, provides a high degree of flexibility in automated manufacturing.

It should be noted that regardless of the choice of sensor there are certain desirable characteristics that all sensors should possess. These include high accuracy, high precision, a wide operating range, and a rapid response speed. In addition, a sensor should be easy to calibrate and have high reliability, and the cost to purchase, install, and operate it should be as low as possible.

Robot Programming. Programming language development for robots has occurred on an ad hoc basis so that the majority of robot manufacturers have their own robot programming language and operating system. This disadvantage is being addressed by various standards, groups, and research organizations. To date, robot programming may be classified into four categories on the basis of the level of user integration:[5]

1. Joint-control languages

2. Primitive motion languages

3. Structured programming languages

4. Task-oriented languages

First-Generation Languages. First-generation languages constitute the first two levels. Joint-control languages require that the user program in joint space and are found most frequently on small educational robots. Examples include ARMBASIC and RASP for use on Microbot or Rhino robots, respectively. Primitive motion languages have been common to industrial robots the last two decades and are divided into two basic methods:

1. Powered lead-through

2. Manual lead-through

Powered lead-through programming methods make use of a teach pendant to control the various robot joints. As previously noted, this point-to-point servo-control method entails powering the robot arm and wrist through a series of points in space where each "taught" point in space is recorded in memory for subsequent playback during the work cycle. For more complex curvilinear paths in space the manual lead-through (also called the "walk-through") method is used and employs continuous-path (CP) control.

Teach pendant control for commercially available robots has been developed to be used in conjunction with programming languages. These languages use instructions which can be input into a robot controller and include branching capability, wait and delay commands, and other signal controls which enable coordination with other devices in a work cell. The function of the teach pendant is to ensure that a point specified within a programming language is physically realizable in position and orientation and the taught point is repeatable once the program is run. Even with such extended capabilities, primitive motion language programming has certain limitations. These limitations include the difficulty in programming associated with increased program complexity, the issues pertaining to integration of robots to other computer-based systems, and the fact that the robot cannot be used while it is being programmed. The latter issue implies that the batch size must be large enough to justify the programming costs. To overcome these limitations, robotic programming has evolved into second-generation languages.

Second-Generation Languages. Second-generation robotic languages, also referred to as structured programming languages and evolving into task-oriented languages, possess structured control constructs as used in computer programming languages. Programming is thus very similar to regular computer programming. A teach pendant is still used to ensure proper robot path control at specified points. The features and capabilities of second-generation languages when compared with first-generation languages include:

1. Motion control for more complex geometry
2. Advanced sensory interactive capability
3. Increased communication and data-processing capability
4. Improved adaptive control capability

Commercially available second-generation languages include AML (a manufacturing language) by IBM, VAL II by Unimation, and MCL (machine control language) by McDonnell-Douglas.

Second-generation languages, in general, also have provisions for interfacing to other computers and provide enhanced control of activities within a work cell in addition to keeping records and generating reports. Access to the databases of CAD/CAM systems gives the geometry and physical properties of the parts to be manipulated, the data pertaining to the machine tool fixtures, and other devices with which the robot interacts. Such access also provides the information to enable off-line programming and simulation of a robot in a work cell.

Robot simulation utilizing the graphics capabilities of computer-aided design (CAD) systems is motivated by the need for off-line programming and debugging capability. Graphic simulation also benefits robot cell layout, training, and various implementation, optimization, and collision problems that might arise. Several available commercial packages include products such as PLACE (McDonnell Douglas Manufacturing Industry Systems Company) and Robot-SIM (General Electric's Calma Co.).

The evolution of structured programming techniques has led to the development of off-line programming and task-oriented languages. Task-oriented languages enable a user to work in the domain of the problem without a need to know the detailed mathematical computations necessary to accomplish a task or even to have a detailed knowledge of computer programming. A user need only specify a task in a natural language such as INSERT PIN IN HOLE. Task-oriented languages are still being researched and developed. One notable accomplishment is the Autopass language developed by IBM.

ROBOT APPLICATIONS IN MANUFACTURING

Robots have proved to be beneficial in many industrial and nonindustrial environments. This section focuses on the applications of robots within a manufacturing setting. Currently, such applications fall within three broad categories:

1. Material handling and machine loading and unloading
2. Processing
3. Assembly and inspection

Material Handling and Machine Loading and Unloading.   Applications in this category pertain to the grasping and movement of a work part from one location to another. Examples include load and unload operations for metal cutting operations, die casting, plastic modeling, and forging operations.

Positioning and orientation (POSE) information is important for proper grasping and subsequent motion, particularly if there are not sensors that provide such information prior to pickup. In many instances, specialized grippers are designed for grasping and holding. Obviously, weight-carrying capacity of the robot must not be exceeded, and the robot chosen should be of the correct configuration and associated work volume so as to allow for maneuverability of the gripper and be capable of reaching all points within the work cell. Certain applications might require a high accuracy and precision whereas others may not. It should be noted that such higher requirements result in more sophisticated drive mechanisms and controllers with associated increased costs.

Robot control requirements are typically unsophisticated for most material-handling operations. Two to four degrees of freedom might suffice in many tasks. More demanding operations such as pelletizing may require up to six degrees of freedom with stricter control requirements and more programming features. Material-handling applications also need to consider total distances moved. Minimization of such distances by proper cell design reduces overall cycle time and can increase production rates. It is also important to utilize all machines in the cell effectively including the robot to decrease and balance idle times that occur within the work cycle.

Processing Applications.   Processing applications in which the robot actually performs work on a part require that the robot's end effector is a tool. The specific processing application determines the type of tool. Examples include spot welding electrodes, arc welding, and spray painting nozzles. Typical processing operations currently making common use of robotics are given in Table 10.1. Many more processing operations have potential for increased robotic usage.

TABLE 10.1   Most Common Robotic
Applications in Manufacturing Processes

Spot welding	Grinding
Continuous arc welding	Deburring
Spray coating	Polishing
Drilling	Wire brushing
Routing	Riveting
Waterjet cutting	Laser machining

Welding.   Spot welding and arc welding represent two major applications, with spot welding probably the single most common application for industrial robots owing to their wide use in automobile assembly lines. Robots used in spot welding must meet the following requirements in order to perform the task:

1. Sufficient payload capacity to manipulate the welding electrodes.

2. A work volume of sufficient size for the product

3. Sufficient degrees of freedom for access to work areas

4. Memory capacity in the controller for the many positioning steps to accommodate all spot welds in a cycle

5. Programming flexibility to allow for different model changes or different product lines

Robots in spot welding improve weld quality and provide more consistent welds and better repeatability of weld locations. Safety is improved by removing the worker from potential electric shocks or burns.

Continuous arc welding, commonly used in fabrication industries, is a more difficult process for application of a robot than spot welding. A wide variety of sensors for robotic arc welding are commercially available and are designed to track the welding seam and provide feedback information to the controller for the purpose of guiding the welding path. Two basic categories of such sensors exist: noncontact and contact sensors. Contact sensors use an oscillating probe positioned ahead of and connected to the welding torch to provide position data as feedback for path corrections.

Noncontact arc-welding sensors consist of two basic approaches in today's commercial systems: arc-sensing systems and machine vision systems. Arc-sensing systems, also referred to as "through-the-arc" systems, use as feedback measurements taken on the arc itself. These measurements may be the current (constant-voltage welding) or the voltage (constant-current welding), which is taken by having the welding torch perform a weave pattern along the joint seam as programmed into the robot. The weave motion causes measurement differences which are interpreted as vertical and cross-seam position of the welding torch. Adaptive positioning may be accomplished by regulating arc length (constant-current systems) as irregular edges or gap variations are encountered during forward travel. Vision systems use a camera mounted on the robot to track the weld seam ahead of the torch. Deviations from the programmed weld seam path are detected and corrections are fed back to the controller to track the seam automatically.

A robotic arc-welding cell provides several advantages over manual welding operations. These advantages include higher productivity as measured by "arc-on" time, or the actual time that welding is occurring during a production shift. Manual arc-on time ranges from 10 to 30 percent, the higher figure reflecting batch production. Comparatively, arc-on time in batch production for robotic welding ranges from 50 to 70 percent. Factors contributing to this increase include elimination of worker fatigue, increased on-time with no need for rest breaks, and the use of parts positioners. A robot can perform operations at one position while a worker can load and unload at another position. In addition, safety and quality of work life are improved by removing a manual laborer from potentially hazardous and dangerous situations. Improved product quality results from the increased accuracy and repeatability provided by a robot.

Spray Coating.   Another major application of robots in manufacturing is spray coating in which paint or other coatings are applied to an object. Robots have proved suitable in such applications because of the many health hazards to human operators. These hazards include fumes and mist, nozzle noise which can result in hearing impairments, fire, and possible carcinogenic ingredients. Advantages include lower energy consumption, improved consistency of finish, and reduced amounts of paint used.

In general, the requirements for robots used in spray-coating applications can be summarized as follows:

1. *Continuous-path control:* The intent is to emulate the motion of a human operator.

2. *Hydraulic drive:* Provide a smooth motion without the potential danger of electric sparks that may occur with electric drive systems.

3. *Manual lead-through programming:* This method of teaching the robot duplicates the painting motion more closely.

4. *Multiple program storage:* Provides capability to access different programs quickly for any part changes.

Other Processing Applications.   Other processing operations using robots involve various machining or cutting operations. Machining operations employ end effectors which are powered spindles attached to the robot wrist. A tool such as a wire brush or grinding wheel is attached to the spindle, and the robot positions the rotating tool against a stationary workpiece and proceeds with the processing operation. It should be noted that such applications of robots have an inherent flexibility, and there is a disadvantage in that some operations lack accuracy as compared with a regular machine tool. In general, small robots tend to be more accurate, but large robots typically have the greater strength and rigidity to withstand the stresses resulting from the tool against the work part.

Assembly.   Automated assembly has become a major application of robots, particularly small high-speed robots. They have proved to be particularly suitable for microelectronics assembly. In the application of robots to assembly, two basic areas are considered, namely, parts presentation and assembly tasks.

Part presentation refers to the position and orientation of the part as it is presented to the robot for future assembly tasks. Various methods such as bowl feeders, magazine feeders, trays, and pallets are used to present the part properly. Of these devices, the most common is a vibratory bowl feeder for small parts.

Assembly operations consider two basic categories: parts mating and parts joining. Parts mating includes assembly situations such as peg-in-hole or hole-on-peg as might occur when a gear is placed on a shaft or multiple peg-in-hole as occurs when a semiconductor chip module is inserted into a circuit card. Parts joining considers not only the mating but also some type of fastening procedure to hold the parts together. Possible joining operations include fastening screws, press or snap fits, welding and related joining methods, adhesives, and crimping.

Applications of robots in parts joining typically utilize specialized tools attached to the robot wrist. For example, an attached powered screwdriver with self-tapping screws is commonly used. Glues and similar adhesives are applied via a dispenser attached to the robot wrist, and the robot can follow a complex path as in arc welding.

For parts mating such as peg-in-hole, remote center compliance (RCC) devices have proved to be an excellent solution. In peg-in-hole insertion, two possible positioning errors can occur: lateral position error and angular error. An RCC device, mounted between the robot wrist and its gripper, compensates for such errors by compensating for the contact forces and moments generated through the "give" or compliance achieved by having the RCC constructed with elastomer springs. Figure 10.7 indicates the reaction of this device for the two types of errors.

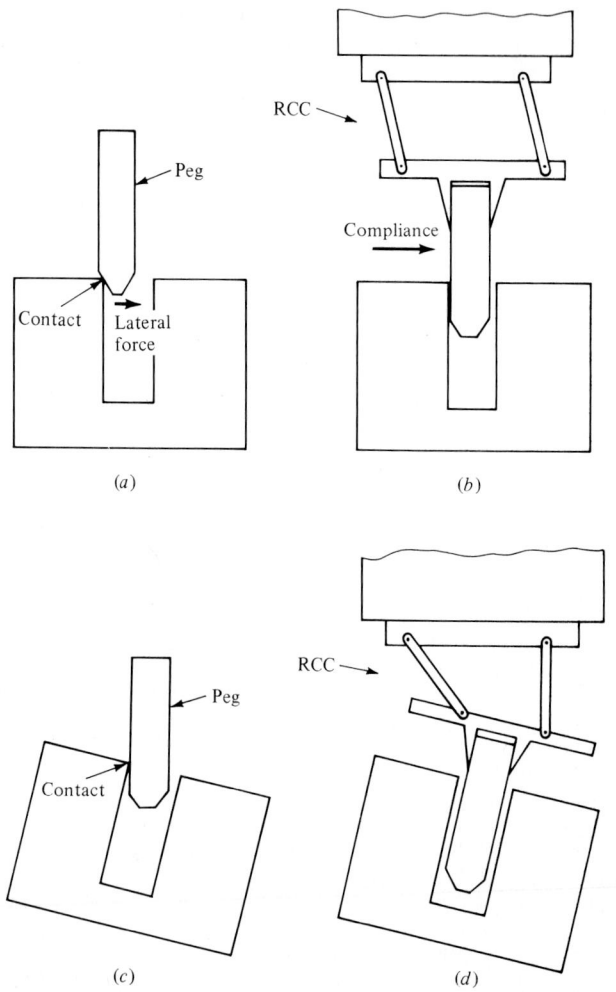

FIGURE 10.7   Action of remote center compliance device for peg-in-hole task: (*a*) and (*b*) action for lateral displacement, (*c*) and (*d*) action for angular error. (*Groover, et al., Fig. 15-17, p. 428.*)

Other approaches include instrumenting the RCC or providing the needed compliance in the robot itself. The latter approach was the impetus for the SCARA class of robots, where SCARA is an acronym for selective compliance arm for robotic assembly. This type of robot is stiff vertically but is relatively compliant laterally. An instrumented RCC, termed an IRCC, provides an indication of forces and moments acting at the wrist and allows high-speed part insertion due to the inherent compliance while allowing monitoring and data collection of the forces during operation.

Inspection.   With the growing emphasis on product quality, there is significant interest in automating inspection on a 100 percent basis. Machine vision systems, robot-manipulated active sensing for inspection, and robot loading and unloading operations with automatic test equipment are being integrated into total inspection systems. Inspection involves the checking of parts, assemblies, or products to verify whether they conform to the specifications of the engineering design.

Robotic application of vision systems includes part location, part identification, and bin picking. In inspection, machine vision systems can carry out tasks including dimensional accuracy checks, surface vision, flaw detection, and completeness and correctness of an assembled product. Current vision systems are predominantly two-dimensional systems capable of extracting feature information, analyzing these features, and comparing them with known objects previously trained into the memory of the system. Typical features and measurements for object identification include area, perimeter length, diameter, center of gravity, eccentricity, number of holes in the object, and first- and higher-order moments of the image of the object.

ROBOT CELL DESIGN

It is essential to integrate the robot properly into a production or assembly process. Integration refers not only to the design of the control system and communication devices which coordinate the activities among the various components in a cell but also to the physical layout of the work cell. In conjunction with the proper layout, the analyst must ensure that the different pieces of equipment do not interfere with each other by timing the equipment cycles properly. The amount of time required for the work cycle is an important overall consideration in the planning of the work cell, since the cycle time determines the production rate for the cell. Subsequent sections discuss the machine interference problem and robot cycle time analysis.

Machine Interference.   For robot cell layouts, one must be concerned about potential interference problems. Interference may be defined for two different situations: (1) when work volumes of robots overlap one another or (2) where multiple machines serviced by one robot have individual machines with too much idle time while the robot is servicing one machine. In the case of overlapping volumes, the robots can be physically separated, or if the tasks require both robots, they can be programmed to be coordinated such that there is no chance of collision.

The second case is termed machine interference. Machine interference is expressed as the percent of total idle time of all machines in the cell to the robot cycle time. This is identical to the definition for a human worker in place of a robot, but the machine interference expected for a robot cell would be lower. This is due to greater worker cycle times and lower effort level. In addition, human cycle time variation would be greater because of a fatigue factor.

Example.   A robot-centered cell has three machines serviced by one robot where the function of the robot is to load and unload the machines. Each machine's processing time and robot service time can be summarized as follows:

	Process time, sec.	Service time, sec.
Machine 1	30	20
Machine 2	15	10
Machine 3	20	10

A machine cycle time is the sum of the process time and service time. The cycle time of the robot is the total service time summed over all machines (travel time between machines is neglected in this example).

A robot and machine process chart is helpful to determine the best sequence of robot servicing times for the machines. For the above example, such a chart is depicted in Fig. 10.8. In general, if the robot cycle time is greater than the machine cycle time, interference will result. The converse indicates that the robot will be idle for part of the cycle.

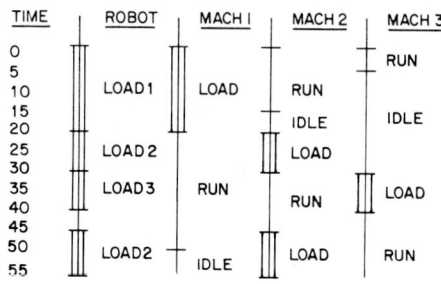

Machine interference = 35/55 = .6363 = 63.63 %
Robot idle time = 5/55 = 9.091 %

FIGURE 10.8 Robot and machine process chart for example.

Robot Cycle Time Analysis. Cycle time is basic to determining the production rate. An approach developed by Lechtman and Nof[9] for analyzing the cycle times of robots, called robot time and motion, or RTM, is similar to the methods time measurement (MTM) technique common to classical industrial engineering. Ten general categories of robot work cycle elements are categorized by four major groups:

1. *Motion elements:* These pertain to manipulator motion with or without a load.

2. *Sensing elements:* Sensor-related activities such as vision, force, and position sensing.

3. *End effector elements:* Relate to time taken by action of attached gripper or tool.

4. *Delay elements:* Based on any necessary need to account for potential waiting and processing conditions.

The reader is referred to the research of Nof and Lechtman[12] for complete details.

JUSTIFICATION AND IMPLEMENTATION

Criteria for Justification. Robotics and their implementation into systems provide the flexible automation capability within the framework of modern production systems. To evaluate such production systems, both technical and economic issues must be addressed. Typical technical issues apply to the reduction of throughput times with corresponding increases in production rates and the reduction of in-process inventories. The issues of economics which are addressed in the following section are from the traditional point of view, but it is instructive to note that there are other criteria which also should be evaluated before a final decision is made. As noted by Proth and Hillion,[16] these criteria are both quantitative and qualitative.

Quantitative criteria include not only the reduced throughput time and in-process inventory but also increased productivity of resources coupled with fewer resources. In addition, another measurable criterion is the reduction in management staff and monitoring staff. This is due not only to the smaller quantities handled as compared with classical mass production systems but also to the automatic monitoring by sensors. Quality improvement can be measured both quantitatively and qualitatively. Qualitative benefits from quality improvement result from improved customer satisfaction, increased competitiveness, simplified production management, and other factors. The potential benefits and cost reductions that may be achieved by installing an automated system are difficult to evaluate, and one should be aware that evaluation reflects on the long-term commitment of the company and impacts the various functions of the corporation as well as the environment. Table 10.2 summarizes some of the potential sources of cost and savings that are difficult to evaluate. In conjunction with these criteria, one then needs to consider the traditional economic analysis used to justify any proposed project.

Methods of Economic Analysis.   Three methods are commonly used in industry to analyze investments and comparing alternative projects. These methods are:

1. Payback period method
2. Equivalent uniform annual costs (EUAC) method
3. Return on investment (ROI) method

The payback method considers the initial investment cost (IC) balanced against the net annual cash flow (NACF) during the life of the project. The net annual cash flow is obtained for each year by obtaining the difference between the revenues and the operating cost. To determine a payback period of n periods, one can use the expression

$$0 = -(\text{IC}) + \sum_{i-1}^{n} (\text{NACF})$$

where costs are treated as negative values and revenues, savings, or profits are treated as

TABLE 10.2   Indirect Cost and Savings in a Robot Application Project

1. *In-process inventories:* The savings in in-process inventory result from a reduced manufacturing lead time with a robot installation. Shorter operation cycle times, use of the second and third shifts, and the possibility for combining separate operations into one robot cell are reasons why the manufacturing lead time is reduced.
2. *Finished inventories:* The technical feasibility of using robots in flexible, adaptable manufacturing cells and assembly systems provides the opportunity for reducing the production batch size. Smaller lot sizes translate into lower final inventories.
3. *Materials savings:* In some applications, robots use the raw materials more efficiently in the production process. This leads to a lower usage rate of these materials. Robotic spray painting operations are an example of these savings; the consistency with which the paint is applied by a robot allows a reduction in the total amount of paint consumed as compared with a manual spray paint operation.
4. *Less scrap and rework:* The avoidance of human error in the operation, the consistency of the robot cycle (in terms of both timing and positional repeatability) are some of the factors that contribute to a more uniform product and a reduction in the scrap and rework rates when robots are used.
5. *Equipment utilization:* When robots are used to automate an operation, the utilization of the existing equipment generally increases. The reasons for the increase include the opportunity to convert to multishift operation of the equipment when robots are integrated into the operation, fewer breaks in the shift as compared with the requirements in a manual operation.
6. *Material handling:* When several operations are combined into a single robot cell, the amount of material handling in the plant is reduced.
7. *Floor space:* A well-designed robot cell typically reduces the amount of floor space required for the operation. This is especially true when several operations, previously accomplished at separate workstations, are combined into a single robot work cell.

Source: Groover, Table 12-1, p. 359.

positive values. It is assumed that all cash flows occur at the beginning or the end of a year. Net cash flows are assumed to be end-of-year transactions whereas investments are considered to occur at the beginning of the year.

Many corporations today require relatively short (on the order of 1- to 3-year) payback periods to justify investment in a project. It is not unusual in the shortened product life cycle environment of the current market to see payback period requirements of no greater than 1 year. The difficulty with this technique is that it does not consider the time value of money and should be considered as a first pass-attempt at justification.

The equivalent uniform annual cost (EUAC) method does consider the time value of money and converts all investments, cash flows, salvage values, and any other revenues and costs into their equivalent uniform annual cash flow over the anticipated life of the project. It makes use of various interest factors associated with engineering economy calculations. These interest factors are tabulated for various interest rates. It is assumed the reader is familiar with engineering economy through various texts in the field such as the one by White, Agee, and Case[17] listed in the references. Also see Sec. 9, Chap. 1. The following example serves to illustrate the basic EUAC method.

Example.   A company uses a minimum attractive rate of return (MARR) of 30 percent as a criterion to decide whether to invest in a robot project. The project is expected to have a 5-year service life and annual operating costs and annual revenues of $20,000 and $65,000, respectively. The initial investment cost is $100,000.

The EUAC flow can then be determined from the expression

$$\text{EUAC} = -100{,}000 \, (A/P, \, 30\%, \, 5) + 65{,}000 - 20{,}000 = +\$3942$$

where $(A/P, \, 30\%, \, 5)$ is the discrete compounding interest factor symbol for determining an equivalent uniform series of cash flows A which would result from a present investment P at an interest rate $i = 30$ percent over a period n of 5 years. This factor is determined by

$$(A/P,i,n) = \frac{i(1 + i)^n}{(1 + i)^n - 1}$$

The positive value obtained (EUAC $= +\$3942$) indicates that the robotic project would be a sound investment.

In many situations, a company is interested in determining the rate of return on the investment based on the estimated costs and revenues and then comparing this rate with the minimal accepted rate for the company. This return on investment (ROI) method is similar in structure to the EUAC method, but now we determine the interest i in the above equations. For the example given, this would be equivalent to setting the EUAC equation to zero and determining the interest i. For this example $(A/P, \, i, \, 5) = 0.45$ and the corresponding return on investment is found by interpolation to be 34.94 percent, which justifies the investment since it is above the MARR of 30 percent.

One important factor is that a robot represents programmable automation such that the robot is reusable from one project to the next. To recognize the opportunity for subsequent use of a robot associated with shorter product life cycles, a salvage value is assigned to the robot at the end of a project. A robot may be reusable for three or four more projects over its expected life. Standard approaches include straight-line or accelerated depreciation techniques to determine the salvage value. If we consider in the previous example that the robot has a service life of 8 years for the project life of 5 years as stated, then the annual depreciation (straight-line) would be $12,500 and would result in a salvage value of $100,000 − 5($12,500) = $37,500. This salvage value would be treated as a positive cash flow for the project, and its amortized value would increase the ROI to 39.20 percent.

Another factor to be considered in automated production is the difference in production rates over a manual process that automation brings. An automated method (including a robotic implementation) would be expected to outproduce a manual method, and an analysis should reflect this. A robotic changeover from a manual method would thus have the potential to affect revenues for any project. The two alternatives (manual versus automated) can be analyzed if the value added by the operation is known per product. The difficulty is that in practice it may prove quite frustrating to obtain a firm value added for a given operation because the operation is one of a sequence of processing steps to make the product. Given that the value added can be determined, the additional annual revenues can be determined for each alternative, and a decision can be made on the basis of the economic analysis. Another potentially more feasible approach is to determine the EUAC for each alternative and, based on the production rate for each method, determine the unit cost for a product for each method. The alternative having the lower unit cost would then be chosen.

Strategies for Robotic Implementation.   A logical approach should be taken to implement robotics within a manufacturing firm. The approach presented here follows that of Groover et al.,[5] and it should be recognized that a particular company may have nuances that would give modifications to the approach presented. The steps for implementing robotics can be listed as follows:

1. Initial familiarization with the technology
2. Plant survey to identify potential applications
3. Selection of an application(s)
4. Selection of a robot for the application(s)
5. Detailed economic analysis and capital authorization
6. Planning and engineering the installation
7. Installation

It should be remembered that the underlying issue is one of systems integration and any potential robotic application should consider the impact on the total system and include the equipment, controllers, and other necessary hardware to have a fully functional and integrated system.

Technical Familiarization.   The first step is to ensure that the personnel have sufficient expertise to identify potential applications. Such information can be obtained from books, technical magazines and journals, robot manufacturing companies, consulting firms, and various conferences and trade shows.

A critical factor for the introduction of robotics technology within a corporation is management support. This support is necessary and crucial for success of a robotic or flexible automation project. The planning and installation of a flexible manufacturing project can take several years for completion, and it is essential that management provide continuing support, encouragement, and commitment over the total project length. Concomitant with management support is the participation of the production personnel who must ultimately accept the project. Without worker acceptance, it would become very difficult to install and operate a robot cell within a plant. To assess the potential for successfully implementing robotics, many companies have developed a work force acceptance checklist. One such checklist developed by the General Electric Company is given in Fig. 10.9.

Plant Survey.   To conduct a plant survey, two general categories of robot applications must be distinguished: (1) a project that involves the design of a new plant or (2) placing a robot project in an existing facility. In the first case, there is a higher degree of flexibility in not only incorporating robotics and automation technology but also in achieving work force acceptance. The second case is more common and usually involves replacing a human operator(s) by a robot(s) in a production environment. The focus here is on this second category.

The objective of a plant survey is to determine if existing operations are suitable for automation or robotic implementation. The survey for a robotic installation has the following general considerations:

1. Hazardous or uncomfortable working conditions
2. Repetitive operations

Checklist:

Item	Points to be Distributed	Driving	Restraining
1. Can workers be openly assured of job retention?	20		
2. Can workers displaced, but retained, be placed in equally rated jobs?	15		
3. Will the installations benefit the workers in terms of: a. Health? b. Safety? c. Relief from dehumanizing jobs? d. Relief from dirty, overly hot, back-breaking, onerous tasks?	15		
4. Is the present union–management climate favorable to open exchange? Disclosure of economic conditions? Labor unrest and frequent grievances? Usually distrustful? (If no union, assign points on similar issues for management–work force relations.)	15		
5. Is the present economic condition of the organization sufficiently healthy to guarantee that promises are kept?	5		
6. Have manufacturing engineering and other management units shown the ability to establish rapport with workers or does inordinate "social distance" exist?	5		
7. Is there management recognition and concern for the dehumanizing aspects of jobs to be performed by a robot? Or is the concern solely economic?	5		
8. Is there a plan to select and upgrade workers who will supervise or perform setups for the robot?	5		
9. Will workers on incentive rates be penalized by new rates or robot downtime that is not attributable to operator?	5		
10. Has management in the past demonstrated respect and regard for the talents, skills, and intelligence of the workers?	3		
11. Is the organization willing to share the results of this checklist with the work force and/or union?	3		
12. Will robot training be on organization time? Is there willingness to send the workers (if required) to the robot vendor's training school?	2		
13. Can workers express their concerns, apprehensions, and fears, without riducule?	2		

Scoring:

1. Total Driving Points = _____

2. Total Restraining Points = _____

   Net Score (1–2) = _____

Interpretation:

Range of Net Score	Probability of Acceptance
80–100	High. Implementation may proceed, assuming management acceptance conditions are equally high rated.
60–80	Proceed with Caution. After examination of the feasibility of changing strength of existing forces.
40–60	Insufficient. Reexamination of forces and management action required to increase probability.
Below 40	Failure More Than Likely. A score in this range indicates a poor probability of even modifying the forces.

FIGURE 10.9 General Electric work force acceptance checklist to assess robot implementation.

3. Difficult handling jobs

4. Multishift operations

Various hazards can occur in a workplace. These include physical or health hazards resulting from toxic chemicals, carcinogenic material, heat, flames, or other potentially dangerous workplace conditions. A meeting with personnel concerned with OSHA standards within the plant is beneficial to identify potential locations of such hazardous environments. Even if a job situation is not actually hazardous, it might be considered by the workers to be very uncomfortable or undesirable to work in. Current examples of such situations include spot welding and spray painting.

High- and midvolume production typically has many examples of repetitive operations. This involves a sequence of work elements that are performed repetitively. Robots are suitable for such applications because of their capability to be programmed for motion pattern from one cycle to the next with high repeatability. The basic requirement is to have the proper end effector and sufficient work volume to accomplish the task. It could prove helpful to investigate injury (particularly muscular injury) reports with medical personnel and ergonomics experts to aid in identifying potential job situations which could be alleviated with the aid of robots.

Many handling jobs can be identified as difficult. This could include difficult-to-hold parts or tools due to weight, shape, or temperature. Awkward shapes, such as plate glass, are amenable to some form of automated assistance to hold and manipulate the part.

Multishift operations associated with high demand of a product are likely candidates for robot applications. As compared with human workers, who typically have a high variable (labor) cost, an industrial robot substitution would have a high fixed cost but a relatively low variable cost. An advantage is that the high fixed cost of the robot can be distributed over the number of shifts. This would result in reducing the total operating cost for the robotic application.

In addition to attempting to identify the operations that possess the characteristics addressed above, the survey may also include looking for operations which require protective or safety clothing, ventilation systems or other special equipment to protect the workers, and special material handling or power-assisted devices (for example, hoists or cranes) which are currently being used. Plant supervision can provide information on operations requiring more than one shift.

Selection of an Application. After identification of potential robot applications, the problem is to determine which application is the best to pursue. In addition to the obvious economic criteria, there are also technical criteria to consider. Usually, a simple application which does not require a high degree of sophistication for a robot, its controller, and the applications end effector design, and which in addition is easy to integrate into the overall system is a good initial choice for a company. A fundamental rule would be to implement any straightforward applications to minimize the risk of failure. The General Electric Company has been successful in choosing robot applications by considering the technical criteria given in Table 10.3.

TABLE 10.3 Technical Criteria for Determining a Robot Application

- Operation is simple and repetitive
- Cycle time for the operation is greater than 5 seconds
- Parts can be delivered to the operation in proper location and with proper orientation
- Part weight is suitable (typical upper weight limit is 1100 pounds)
- No inspection is required for the operation
- One or two workers can be replaced in a 24-hour period
- Setups and changeovers are infrequent

Source: General Electric Co.

Robot Selection.   After selection of the application, a robot must be chosen from the many commercial models available. Various sources such as the *Robotics Product Database*[3] can aid in the selection. Selection needs to consider the appropriate combination of technical features suitable for the application. These features include the number of degrees of freedom, the type of drive and control systems, the programming capability, accuracy and precision, load capacity, and others such as interfacing capability.

Economic Analysis and Capital Authorization.   Management authorization of the project usually requires a detailed economic and technical analysis. The economic analysis estimates the financial benefits to the company as determined by such measures as the ROI and payback period previously discussed. Technical justification details the engineering and technical feasibility of the project. Application features, any required changes to existing equipment, acquisition of new equipment, and fixturing and tooling should be described in detail. Additional analysis on effects on labor, potential problem areas, anticipated production rates, and other similar characteristics should also be provided. Based on this analysis, a capital authorization decision is made as to whether or not to provide funds for the project.

Planning and Engineering the Installation.   Many of the analysis and design features discussed in this chapter are integral to the planning and engineering of the installation. Issues addressed include the operational methods, the work cell design, the control of the work cell, end effector and other fixturing and tooling designs, and safety considerations for the work cell. In effect the issues pertain to the systems integration topics within a planned work cell.

The study at this stage considers the basic purpose and function of the planned work cell. It should be noted that differences exist in the way a human performs a task and the way a robot is programmed to perform the same task. Sensing issues are critical to maximize performance and minimize potential collision damage. The issue of tooling, part, and fixturing needs to be thoroughly investigated. Fixturing and tooling considerations may be different in a robot application compared with a human operation. The RTM method may be useful in assessing cycle time for the operation. In addition, a work cell layout for a robot may be different than for a corresponding layout with human operators. Robots can be mounted overhead on gantry, directly to machine tools, or in other unusual locations. Safety should also be designed into the work cell, and maintenance aspects should also be considered.

Installation.   Installation involves the implementation of all the prepared detailed plans. Table 10.4 provides a checklist of the various activities included in the installation phase. Other aspects of installation include safety, training, maintenance, quality control of the product, and the normal start-up problems, debugging, and trial production runs before the work cell is fully operational. The time required to complete the installation is

TABLE 10.4   Check List of Activities Included in the Installation Phase

Purchase of the robot(s) and other equipment and supplies needed to install the work cell
Preparation of the physical site in the plant where the robot cell is to be located. This might include altering the foundation to support heavy machine tools in the cell and to fix the relative positions of the robot and other equipment. Also included would be any provisions for protection of the robot from its environment (for example, high temperatures, dangerous fumes or mist in the atmosphere, electrical noise, fire hazards)
Provision of electrical, pneumatic, and other utilities for the cell
Adaptation of standard pieces of equipment for use in the cell
Placement of robot and other equipment: installation of conveyors and other materials-handling systems for delivery of parts into and out of the cell
Installation, checkout, and programming of the work cell controller
Installation of interlocks and sensors, and integration with the work cell controller
Installation of safety systems
Fabrication of end effectors and other tooling

Source: Groover et al., Table 16-3, p. 467.

very difficult to estimate, and actual installation time may range anywhere from 3 months to 1 year or more.

SYSTEMS INTEGRATION

Systems integration, in a general sense, is a technological area which goes beyond robotics and shop-floor automation and their systems. Many technologies besides robotics are included, and it is the function of the industrial engineer to make them compatible with an overall factory planning and control system. These technologies include computer-aided design and manufacturing, computer-aided process planning, manufacturing information systems, flexible manufacturing systems, and other technical and management systems pertaining to factory production and automation. Reference models to elucidate the structure for large-scale manufacturing systems have been developed, and various such models exist. In general, a reference model serves as an abstract model of a structure of cooperating decision-making units and provides a reference for common terminology concepts and a framework for identifying generic manufacturing tasks. One example of a hierarchical reference model is that of the National Institute of Standards and Technology (NIST) which is being implemented within the Automated Manufacturing Research Facility at NIST. Such hierarchical models have served to initiate intelligent robot control and have been extended to overall factory control.

Intelligent Robots.   There is a continuing effort to develop control algorithms for reference model hierarchies. Even more intensive efforts have been applied to the development of intelligent controls at the lower levels of the hierarchy, particularly the equipment level. This has led to the concept of intelligent machines or robots. As with factory models, the pursuit of developing intelligent robots is a multidisciplinary area. Attainment of intelligent robot or machine control means that any goal must be reached autonomously without any intervention by a human operator. Fundamental issues that need to be addressed include not only the control structure but also sensor strategies, multisensor integration or fusion, learning and decision making, programming, mobility and navigation, trajectory planning, and overall systems integration.[14]

Intelligent robots, as with intelligent factory automation, will provide an abundance of problems and opportunities for industrial engineers. It is anticipated that there will be heightened activity in both theoretical and practical developments in the issues pertaining to the technology and the management of such technologies within automated systems, and such activity will continue well into the twenty-first century.

REFERENCES

1. Coiffet, P., and M. Chirouze, *An Introduction to Robot Technology,* McGraw-Hill, New York, 1982.
2. Engelberger, J. F., "Robotics in Practice," AMA COM: A Division of American Management Associations, 1980.
3. Flora, P. C., ed., *Robotics Product Database,* 6th ed., TecSpec, Orlando, Fla., 1989.
4. Friedman, S. B., *Logical Design of Automation Systems,* Prentice-Hall, Englewood Cliffs, N.J., 1990.
5. Groover, M. P., M. Weiss, R. N. Nagel, and N. G. Odrey, *Industrial Robotics: Technology, Programming, and Applications,* McGraw-Hill, New York, 1986.
6. Groover, M. P., *Automation, Production Systems, and Computer-Integrated Manufacturing,* Prentice-Hall, Englewood Cliffs, N.J., 1987.
7. Klafter, R. D., T. A. Chmielewski, and M. Negin, *Robotic Engineering: An Integrated Approach,* Prentice-Hall, Englewood Cliffs, N.J., 1989.

8. Kuo, W., and J. P. Hsu, "Update: Simultaneous Engineering Design in Japan," *IE Magazine,* October 1990, pp. 23–26.

9. Lechtman, H., and S. Y. Nof, "A User's Guide to the RTM Analyzer," *Technical Report,* School of Industrial Engineering, Purdue University, 1981.

10. Michel, M., "Justification Models for Flexible Manufacturing," *Robots' 10 Conference Proceedings,* SME publication, 1986.

11. Nevine, J., et al., "JTECH Panel Report on Mechatronics in Japan," *SAID Technical Report,* March 1985.

12. Nof, S. Y., and H. Lechtman, "The RTM Method of Analyzing Robot Work," *Industrial Engineering,* April 1982, pp. 38–48.

13. Odrey, N. G., "Integrating the Automated Factory," *Data Processing Management,* vol. 13, no. 5, 17 pp., October 1985.

14. Odrey, N. G., "Control Systems," *1992 Yearbook of Science and Technology,* McGraw-Hill, New York.

15. Pessen, D. W., *Industrial Automation: Circuit Design and Components,* Wiley, New York, 1989.

16. Proth, J. M., and H. P. Hillion, *Mathematical Tools in Production Management,* Plenum Press, New York, 1990.

17. White, J. A., M. H. Agee, and K. E. Case, *Principles of Engineering Economic Analysis,* Wiley, New York, 1977.

CHAPTER 11
TEAM MANAGEMENT CONCEPT

Yash P. Gupta
Frazier Family Professor
Department of Management
School of Business
University of Louisville
Louisville, Kentucky

In the past few years we have witnessed a fast erosion of U.S. competitiveness. In 1989, for example, Japan outspent the United States in modernizing and expanding its industries in spite of the fact that the U.S. economy is twice the size of the Japanese economy.[13] Japan is winning the economic competition by producing autos, TV sets, calculators, and cameras that are better, less expensive, and preferred by American buyers.[1] The impact of manufacturing can be felt in all service support industries as well as the U.S. economy as a whole, with a great many service jobs directly linked to their manufacturing counterparts. In fact, it is likely that up to seven service jobs will be lost for every manufacturing job lost.[9]

The U.S. organizations are under increasing pressure to respond to shortening product life cycles, increasing customer expectations on reliable delivery dates, and increasing sophistication of world class manufacturing plants. In response to these factors, a growing number of organizations have undertaken a major restructuring of production by introducing advanced manufacturing technologies. The flexibility offered by these technologies allows organizations to make design changes easily, retool quickly from one product to another, and make a variety of parts with the same equipment. Thus it enables organizations to lower the volume break-even point and adjust rapidly to changing markets. But the solution to fading competitive ability, sluggish productivity growth, and poor quality cannot be found exclusively in the mythical black box of a miraculous technology.[14] To realize the full potential of technology, American managers need to commit themselves to tapping the immense reservoir of competitive strength represented by the skills, insight, and energy of their organization's work force; that is, it is essential to focus attention on the human element in achieving an industrial renaissance.[18] One must view human resource as human beings rather than things because people must be considered as a resource and an opportunity rather than as a problem, cost, or threat.[11]

Since old-style confrontational approaches to managing the work force no longer work, such ways of managing need to give way to a greater emphasis on leadership, adaptability to change, and responsibility.[11,27] A variety of efforts which involve the work force more directly and more fully in the planning as well as the execution of production activities need to be adopted. Today's workers, in their search for a job, are choosing an organization whose total package matches their needs rather than opting for the highest salary as in the past. That shopping list includes more responsibility and commitment, the opportunity to do interesting and meaningful problem solving, and for self-actualization.[26] Workers are better educated, less accepting of traditional authority, and more willing to be involved in decisions that affect their day-to-day work lives.[20] In essence, an organization's attempts to increase worker involvement can take a variety of forms, including such practices as job enrichment, participative management, quality of work life programs, organizational development, organizational democracy, and employee problem-solving teams like quality circles, task forces, corrective action teams, and self-directed teams.

To begin with, teams encourage interaction, promote positive outcomes such as increased motivation, stimulation of ideas and relevant knowledge, more creative individual problem solving, and even a new framework to view organizational problems. The fundamental assumption supporting the use of teams is that a team's output will be superior to that of its average and, in some cases, to that of its best member.[17] In fact, a team approach has been used with increasing frequency to select R&D proposals for funding, plan innovative projects,[10] enhance the quality of organizational products or services,[22] and achieve self-management in innovative settings.[28]

The success of the team approach has been high for some organizations while others have floundered.[19] Honeywell's commercial flight division, for instance, owes its big edge in the flight-navigational market (80 percent market share) to teamwork. Virtually all plant functions are performed by teams, which account, in part, for 1988 profits coming in at 200 percent over projections. In addition, Ford uses this approach to cut development time of new models by one-third,[29] while AT&T also successfully employed this approach in developing its new 4200 cordless phone,[12] as did NCR with its recently introduced 2760 electronic cash register.[23] The Lockheed Corporation's Aeronautical System Group has even managed to reduce the time for designing and manufacturing sheet-metal parts from 52 to 2 days.[24]

The trend toward using teams and empowering employees to manage themselves has significant implications for an organization. For example, what factors contribute to the success of teams? What is the role of supervisors in this arrangement? How will their roles change? and What new skills will be needed by the employees to participate in teams? The purpose of this chapter, then, is to discuss these questions.

TEAMS

Teams and the potential they hold for producing significant results are being hailed as the productivity breakthrough of the 1990s. Organizations that have committed to the team approach have often labeled it as the unveiling of a "new paradigm" because this approach represents a fundamental shift away from scientific management's segmentation of work into dull, repetitive jobs. According to Reich,[25] in this paradigm, "entrepreneurship isn't the sole province of the company's founder or its top managers. Rather, it is a capability and attitude diffused throughout the company. Experimentation and development go on all the time as the company searches for new ways to capture and build on the knowledge already accumulated by its workers."

Traditional workplace design, based on scientific principle, entailed reduced discretion at each step of a process so that the jobs encompassing those steps could be learned fast, carried out accurately, and yield more. The employee's knowledge was limited to discrete, narrowly defined tasks which had very limited, if any, knowledge of the whole task. Under a team structure, on the other hand, team members are expected to focus their energies on the total business or project as if the business were their own and to assume full responsibility and accountability for business results.

The teams, then, create a cycle of positive dynamics in which each member plays an important role and has a positive impact on other members of the team. This environment of mutual reinforcement enables individual team members to reach higher levels of performance than if they worked individually. Consequently, teams stimulate member participation and involvement, enhance performance, increase job satisfaction, and lower absentee and turnover rates.[8] Teams also encourage team members to develop their sense of identity and loyalty to their organizations since under this arrangement their impact is more direct, noticeable, and appreciated (see Table 11.1). In a team setting, the team has a clearly defined mission to service the needs of equally well defined customer(s). According to Boyett and Conn,[7] in organizations where a team approach will be implemented:

> Teams, to a much greater extent than traditional organizational units, would have a clearly defined customer or a group of customers, either external or internal to the organization. Within teams, each member of the team would be expected to be flexible. Over time, each member of the team would be expected to learn, when the occasion demanded, to perform all of the jobs within the team and, most likely, would be paid extra for doing so. The need to learn a variety of jobs would mean that team members would be continuously learning new skills. In respect to day-to-day operations, teams would be largely autonomous. Team members would be provided with extensive information on team operations and would use that information to make most day-to-day operating decisions and to provide themselves with feedback on their performance (pp. 236–237).

TABLE 11.1   Benefits of Work Teams

- More sharing and integration of individual skills and resources
- Untapping and use of unknown member resources
- More stimulation, energy, and endurance by members working jointly than is usual when individuals work alone
- More emotional support among team members
- Better performance, in terms of quality and quantity, and more information
- More ideas for use in problem solving
- More commitment and ownership by members around the team goals
- More sustained effort directed at team goals
- More team member satisfaction, higher motivation, and more fun
- The sense of being a winner, greater confidence, and the ability to achieve more

Source: Bassin.[5]

Types of Teams.   Teams are known by several names, including self-directed teams, self-regulated work teams, self-managing teams, high-performance teams, and sociotechnical design teams. To some organizations self-managing is a misnomer for their teams, since these organizations do retain a management structure. On their journey from traditional to self-management, teams go through several predictable stages, requiring increasing levels of employee involvement. It is a journey that can take anywhere from 2 to 10 years to complete. In order to understand this concept, it helps to visualize these stages as stops along a continuum. Bassin,[5] for example, categorized the development of business teams into four stages: informal teams, communication teams, formalized business brand teams, and development and venture teams. At the same time, Boyett and Conn[7] argued that generally U.S. organizations initially adopt work-unit teams that gradually evolve to self-managed teams as the final stage of their evolution (see Table 11.2 for the differences between the two ends of the continuum).

Work-unit teams and self-managed teams have several dynamics in common. They both have shifted the responsibility of controlling performance and solving problems, traditionally a domain of supervisors and managers, to employees. In both cases the employees are expected to attend team meetings and participate in team activities. Under work-unit teams, managers and supervisors continue to perform traditional functions while maintaining old organizational structures, including functions like planning, budgeting, hiring, firing, and disciplining. Managers and supervisors may, however, seek greater input

TABLE 11.2   Differences Between Work-Unit Teams and Self-Managed Teams

Work-Unit Teams	Self-Managed Teams
Traditional organizational structure is retained with fewer levels of hierarchy.	Traditional organizational structures with divisions and departments along functional lines give way to teams of employees. They operate almost as small, independent business units.
Team develops key performance measures and establishes corresponding goals in collaboration with the management and supervisors.	Teams have "mission" to serve a customer(s)—internal or external.
Team meets with managers and supervisors to review performance and identify other areas needing improvement.	Team members assume the responsibility of managers and supervisors. Team elects a leader who facilitates team meetings.
Managers and supervisors may be reduced in number, but the positions remain intact. Managers and supervisors provide information and resources, facilitate team meetings, and coach employee problem-solving efforts.	Managers coordinate activities of various teams, ensure that they have the resources, advise teams on technical issues, and help resolve disputes.
Managers continue to perform traditional roles such as planning, budgeting, hiring, firing, and disciplining; however, with greater employee input into these decisions.	Planning, budgeting, hiring, firing, and discipline of team members is performed by the team itself.
	Teams have the responsibility, equipment, and other resources necessary to produce the entire product or service.

from employees in such decision making. In work-unit teams the members develop key performance measures for the team and establish goals to measure the performance of the team in association with managers and supervisors. The teams, however, are responsible for their own performance.

Under self-managed teams the traditional organizational structures drawn along functional lines may not exist. In this model, jobs are broadly defined, task responsibilities are rotated among its members, and the team runs as a business performing the various functions of the business including selection, promotion, and training of employees. Under this system the employees are remunerated, in the main, according to the number of tasks they have mastered, which, in turn, provides another incentive for an employee to continually learn new skills. Employees are supervised and evaluated by the team, thus eliminating the need for managers and supervisors (Fig. 11.1). The underlying assumption here is that peer pressure is a more powerful tool than the orders of a supervisor.[6] Supervisors, in some cases, are given significantly more responsibilities such as coach, trainer, and facilitator. Egalitarianism is promoted by an all-salaried work force and by eliminating distinctions between employee groups in dress codes, office space, parking, and eating facilities.

Characteristics of an Effective Team.   The implementation of a team approach is usually driven by a clear and pressing business reason such as the need to improve quality, productivity, and customer satisfaction. For effective team building, management must be convinced that the people closest to the work should be responsible for doing the job and deciding how it should be done. There are several characteristics of an effective team which top management must foster.[15]

Goal.   Each team must have a clear mission, goal, and purpose with well-ordered objectives. Thus team members must know what is expected from them and how their contribution is going to impact the organization as a whole. Tables 11.3 and 11.4 provide ex-

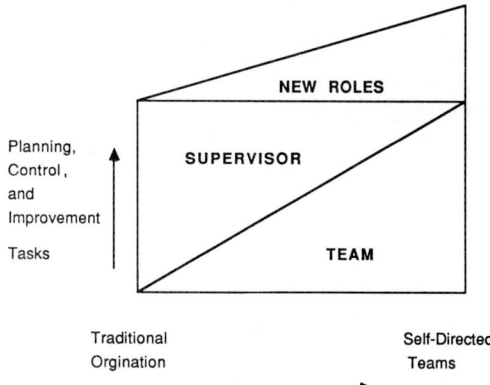

FIGURE 11.1 Changing role of supervisors and team members.

TABLE 11.3 An Example of a Set of Expectations from Members of a Team

- Work for consensus on decisions, objectives, and plans, and then to support the team's decision.
- Openly communicate with each other.
- Trust, respect, support, and have genuine concern for other team members.
- Listen and try to understand other team members' point of view and related background.
- Encourage and assist other team members to improve themselves and learn from them whenever possible.
- Strive for self-improvement but not at the expense of another team member.
- Provide the team with your honest opinion, not just what the team wants to hear.

TABLE 11.4 Team Responsibilities

- Meeting established production and customer requirements
- Maintaining safety standards
- Checking and maintaining established quality and environmental standards
- Maintaining and repairing their equipment
- Ensuring cleanliness, orderliness, and appearance of workplace and equipment
- Coordinating with other teams
- Providing feedback and exchanging information
- Educating and developing team members
- Keeping time for hours worked and balancing working hours
- Maintaining team effectiveness
- Selecting team members (replacement hiring)
- Self-scheduling the team
- Containing cost and controlling overtime
- Sharing responsibilities
- Optimizing the process
- Evaluating the team and team members

Source: A manufacturing organization.

amples of expectations and team responsibilities, respectively, drawn from the employee handbook of a manufacturing organization. At the same time, team members must be able to clearly understand priorities which are developed via discussion and negotiations.

Talent.   The team should have the talent necessary to fulfill its objectives. A team approach generally requires multidisciplinary skills and eagerness on the part of members to deliver that talent. The team members should be committed, then, to continuous personal improvement and professional development.

Role.   A team member's role must be carefully defined and understood by both the individual and the remainder of the team. Each member should recognize and appreciate how an individual's unique role contributes to the overall effectiveness of the team.

Procedures.   The team members should develop effective and efficient operating procedures which they must use to accomplish their respective responsibilities. The procedures must include issues related to planning, decision making, running meetings, and communicating effectively.

Interpersonal Relations.   The rules and norms of behavior should be openly discussed and agreed upon, establishing a climate of trust, mutual respect, excellence, and innovation. The members should be able to resolve conflicts as they arise in a constructive fashion and be able to challenge each other in a way not seen as personal confrontation or attack. Thus team members must have flexibility, objectivity, and humor.

Reinforcement.   Team members should express appreciation to each other for their contribution to the team and be able to reinforce each other's efforts for the team.

External Relations.   The team should be able to establish a constructive relationship with its broader environment such as customers or other managers who provide feedback, resources, or other assistance to the team. Members should be able to scan the environment to recognize threats and opportunities so that they can capitalize on opportunities and minimize the threats and concomitantly build their credibility.

Leadership.   The role of team leader is extremely important for a team to be effective. A team leader should ensure that the team goals are set and provide guidance for meeting these goals, help the team establish priority areas, develop plans and obtain needed resources to meet those priorities, and provide communication and coordination of efforts within the team and among other teams. At the same time, the leadership approach should be flexible to allow for changing needs of the team.

Reward and Recognition.   The team should obtain recognition and the appropriate authority in order to be able to complete their mission. An individual's contribution to the team should be respected, valued, and always rewarded.

IMPLEMENTATION OF TEAM CONCEPT

Traditional organizational structures do not lend themselves to work teams. Embedding teams in these hierarchical and political environments results in worker frustration and rejection of the concepts. To ease the implementation of new organizational structures, detailed planning and preparation must be undertaken well in advance. A prerequisite for an organization planning to implement a team approach includes the development of a strategic vision which must be crafted and articulated with clarity, continuity, and consistency; a clarity of expression that makes company objectives understandable and meaningful; a continuity of purpose that underscores their enduring importance; and consistency of application across business areas that ensures uniformity throughout the organization.[4]

It is also important that management pay attention to the choice of issues that must be considered by the teams. Local tasks-related issues where concrete agendas and clear technologies can be applied tend to work best. According to Kanter,[16] it is more difficult for employees to deal with distant, broad, and open-ended issues.

Planning for a team concept involves examining the purpose of this team effort and the conditions necessary for its success and studying the prevailing environmental factors.

The goals for a team approach are diverse and vary from one organization to another. One cannot, however, minimize the need to address this diversity through a commitment to the dual outcomes of productivity and human needs.[30]

A team concept can be successful in many different types of environments. It is important, though, to recognize the difference between implementing the team concept in existing facilities as opposed to new facilities. Walton[30] suggested that startups are more likely to have workers who lack technical and human skills as well as problem-solving capabilities. In such situations, it may be appropriate for supervisors to provide direct supervision rather than just facilitating. On the other hand, the implementation of this concept in "greenfield" plants is not hindered by existing cultures which may need to be revised. Similarly, it is essential that the presence or absence of unions must be considered in planning for implementation of the team concept since communication and close coordination with union officials are critical for any effort to change work systems.

The implementation of a team approach is a serious long-term undertaking for any organization. As a result, the team approach requires a fundamental change in management's thinking and philosophy, requiring significant adjustments for managers within the organization. Such a radical metamorphosis demands that obstacles or barriers to its success be overcome early.

Obstacles to the Team Concept.   Anthony[3] provided a framework in which these diverse barriers are categorized into four different types: organizational barriers, managerial barriers, subordinate barriers, and situational barriers. Organizational barriers include tradition, philosophy and values, quality of policy, procedures and personnel, structure, and lack of reward system. Managerial barriers have to do with habits, failure to understand team management, lack of security, and fear. The barriers for subordinates encompass lack of competency and desire to practice participative management, lack of content, knowledge, or expertise, fear, and lack of awareness of being expected to participate. Time, task, and environmental influences constitute situational barriers.

One of the prime responsibilities of a team is to evaluate the performance of their peers. This includes continuous monitoring, documenting the performance of the individual team members, and terminating services of an individual in the event of poor performance or lack of discipline. Generally, team members do not like to participate in this unpleasant task and often wish to revert to traditional management approaches for this particular issue and thus lose the essence of the team approach.

Since the team approach is so radically different from the traditional norms which the majority of American workers have experienced, these workers often have the tendency to resist accepting the new responsibilities entailed in this concept.

Most individual team members think of themselves as functional experts and identify their sphere of potential contributions within that area. In the team approach, however, members are not only expected to contribute in their functional areas of knowledge, but they are also required to think in a holistic fashion and to consider the total business. This thinking requires a significant change in the expectations regarding contributions and input on the part of the team members.

Most business organizations reward and recognize performance on an individual basis and have few, if any, mechanisms to do so on a group basis. According to Bassin,[5] considering an individual as a unit for performance evaluations could be an obstacle in obtaining the peak performance of teams:

> Business team peak performance is a group phenomenon; however, most business organizations are structured around individual performance and competition. Reward systems, promotions and a sense of identity are individually geared and generated. The individual and his or her contribution is the fundamental unit in work cultures, not the group.
> This is a different basic assumption than what is required for a peak performing team. In such a team, the group vision must be stronger than individual agendas. The opportunities for group members to help one another must be unencumbered by competition among them.
> If personal competition interferes with group cooperation, the possibilities for peak perfor-

mance diminish. As in a sports team, if one member values his or her personal performance (above the team's), team cohesion is lessened and team performance suffers.

Establishing a team culture which includes building a sense of trust, defining roles and procedures, and gaining a degree of confidence in the abilities of a team requires sustained effort and attention and considerable time with a continuity of participation from team members. If the continuity is broken, it causes a severe disruption, since it breaks the interdependence among members of the team.

In several organizations, perhaps the biggest obstacle to a successful team approach is the resistance from managers and supervisors. Managers and supervisors often perceive that teams erode their power, influence, and ultimately their importance. In fact, they recognize that their repertoire of management skills, often developed after years of struggle and experience, may become, at least in part, obsolete (see the next section for further discussion on this issue).

In traditional organizations career movements are determined by the performance of an individual in his or her functional area; in the team approach the focus is more on an individual's business acumen and his or her functional identity becomes secondary. The logical movement of those team members is to become team leaders and progress to general management. This arrangement may pose a threat to those employees more comfortable with the functionally dominant notion of career movement.

Team membership cannot be equated with equalization of power or elimination of hierarchy. In some cases, the external status of an individual team member may influence the action of others in the team discussions and, thus, the feeling of fear and retribution may maintain and sometimes entrench a hierarchy rather than eliminate it.

Individual members of the team bring different resources and personalities to a team; similarly they come with different needs and interests. The combination of these, in some cases, propels these individuals to attempt to exert more influence and power over team affairs than others, a factor that can be dysfunctional since the team may become an arena for individuals to flex their power muscles.[16]

There is a risk that a team may become so self-engrossed that it loses sight of the fact that it is only one entity of a large organization. Thus their goals and activities may not be in consonance with those of the organization as a whole. Similarly, they may spend their time and efforts in duplicating solutions which have already been arrived at previously by other teams.

In some cases management builds up unrealistic expectations from the team concept. Irrespective of how well this concept works, it will not eradicate all of an organization's malaise. Thus the team approach is more a way of broadening the skill base of employees and using their talents to solve problems and invent needed programs. As a result, management must manage expectations by setting realistic and attainable goals, providing opportunities for people to learn by redesigning their jobs, creating a learning environment, and providing appropriate learning programs rather than forcing people to participate against their will.

As a team matures and achieves success, members desire more responsibility and are often capable of assuming the role of the next layer of management, which could lead to confusion and frustration for both team members and management.

Role of Supervisors.   In several organizations, before implementing a team concept, management expected resistance from employees and unions. Experience has shown, however, that the real resistance to the team approach emanated from supervisors.

Supervisors play an important role in the implementation of team management. In the event that supervisors view this new design of organization as detrimental to their career, they may withhold their support and consequently may contribute to its failure. In a study, Klein[18] reported that over two-thirds of the supervisors had fears about their jobs and confusion over their role. Supervisors generally do not show open resistance to the change; the resistance does, however, manifest itself in several forms. For example, without the knowledge or influence over what has transpired at team meetings, supervisors perceive the team to be a threat, which can lead to innuendoes about the team and its competence in making decisions which, in turn, discourages employees from participating in the team. Another way in which supervisors show their resistance to the team is by withholding their assistance, that is, stressing that it is not their job, but the team's. However, when a problem reaches the point where the team cannot handle it, supervisors step in and resume their traditional position of authority.

There are three fears which supervisors experience. First, they feel a threat to their job security, that is, they fear that they will become redundant. Second, they are concerned about job definition, and third, they worry about the additional work generated by teams.

Supervisors tend to feel more unsure and are less supportive of teams when they are not fully aware of their role, what expectations are required of them, and how their performance will be measured. A lack of a well-defined set of responsibilities may add fuel to their fears caused by the prospect of losing their job.

With the implementation of team management, additional work such as coordination and follow-up for the program is generated. This extra work invariably falls on the shoulders of supervisors, who are, in most cases, not compensated for the additional work.

Management must examine the concerns of supervisors and develop a coordinated strategy which should make supervisors an integral part of the change process. According to Klein,[18] this strategy should include a wide variety of elements.

Training is an extremely important first step in mitigating the resistance of supervisors to the team management. In addition to the classroom and seminar-based training, training should include ongoing consultations with feedback and coaching from managers. The managers may become role models to modify supervisors' behavior and will also reinforce their commitment to the new way. Thus supervisors must be involved in the design and implementation of the team concept; this role must be explicitly designed. Such a plan may include training, team building, coaching, information sharing, joint goal setting and problem solving, linking teams to resources, customers, and suppliers, and conflict resolution.

Since day-to-day decision making is likely to become the domain of teams, managers should delegate increased responsibility and appropriate authority to make the supervisors' jobs more meaningful, allowing them to gain respect and prestige in the eyes of the employees. Another avenue to reduce the resistance involves encouraging supervisors to establish networks with other supervisors so that they can exchange their views and seek peer assistance in dealing with their fears.

Training Requirements. The implementation of team management requires that the organization provide training for its employees and managers. Anderson and Anderson[2] suggested that it is only when the employees accept the validity of the team concept that they will be willing to make the necessary changes, something which can be accomplished only through substantial training and development efforts.

The introduction of training programs should be gradual and should be completely explained to all employees. Training in a team management environment would be substantially different from that in traditional organizations. For example, it would not be acceptable simply to provide employees with skills which limit them to one specialized job. Similarly, it would not be appropriate for organizations to determine training needs unilaterally and prescribe training methods which have been done in a partnership between trainee and trainer. Thus training should be comprehensive in scope, systematic in delivery, job-related, fair in its evaluation, individualized, and classroom- and performance-based.

Numerous areas of skills and concomitant training needs important for the success of team management have been discussed in the literature.[2,3,21] Based on the several descriptions of skills needed for team management, it appears that the categories of communication, interpersonal relationships, managerial skills, team skills, and leadership and motivation skills may be used to define the broad areas of training that need to be addressed in team management organizations.

Communication skills include the ability to express ideas and convey information to others quickly and effectively. In team management, since much of the information required to perform the job, solve problems, and work effectively with others will come from verbal interaction, it is important that the employees develop extremely good verbal and listening skills. Similarly, interpersonal relationships skills include giving and receiving feedback, appropriate assertiveness, and conflict resolution skills.

Managerial skills usually encompass training in decision making and problem solving, performance evaluation, and career development. Problem solving requires knowledge of how to identify and select problems, develop an action plan, collect data and organize relevant data, analyze data, and analyze decisions. Since much of this problem-solving activity occurs in teams, employees need experience in group processes and consensual decision making. Managerial skills also involve an understanding of how individual behavior affects others, dealing with undesirable behavior in others and handling stress.

At the same time, leadership and motivation skills are extremely important in team management organizations. The training skills must deal with issues such as what organizations are, how they work, why they exist, what makes them effective or ineffective, how leaders influence members of teams, and the importance of ethical standards.

In addition to these skills employees must be trained in technical skills and cross-functional areas that would allow for job rotation within teams.

Organizational Structure.   In addition to the planning process discussed earlier, formal structures within the organization facilitate the implementation and operation of the team concept, including the organizational chart, operational procedures, and the framework for a team function.

An organizational structure should be relatively flat in order to enhance communications between various levels of hierarchy. Figure 11.2 shows the organizational structure of a large manufacturing organization with only four levels, that is, team members, team leaders, production manager, and plant manager. The figure shows that each level has a broad span of control and few intermediary positions between production and management. The organizational chart, then, reflects a formal structure designed to facilitate its team management approach.

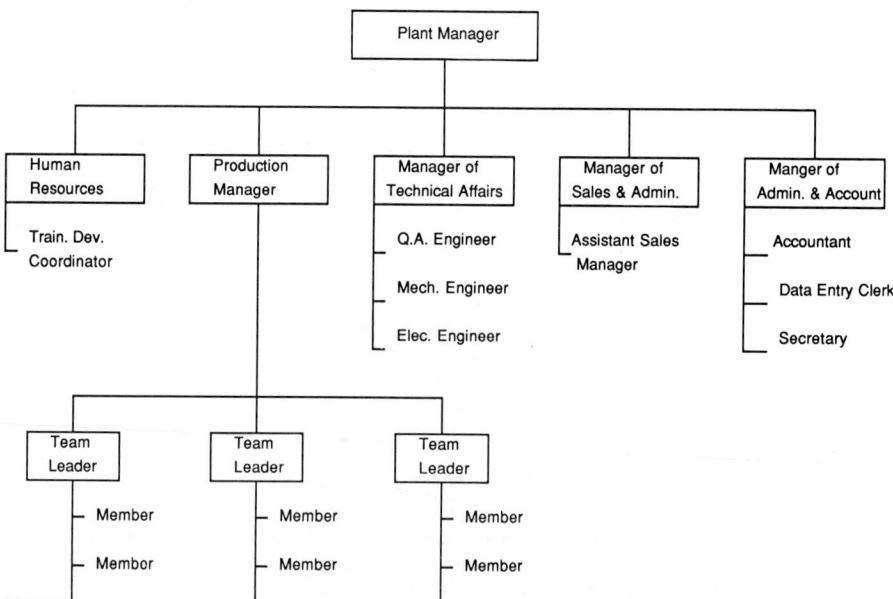

FIGURE 11.2   Organizational structure of a large manufacturing plant.

Operational Procedures.   Operational procedures may include a broad spectrum of activities that relate to the day-to-day work of an organization, such as job design, skill level of employees, renumeration plans, and accountability systems. Job design should revolve around the team as a basic unit for performing work. The team should complete a whole task within a functional area and the team members should develop the skills necessary to perform all jobs within their team's area of responsibility. This, in turn, should lead to an increase in the flexibility of an organization.

This skill level of employees has several implications for training and hiring policies. Hiring policy should be based on the premise that if an organization hires better, more committed people, the company would have a competitive advantage. The employee selection process therefore should be a rigorous but fair and objective procedure which identifies the individual capabilities and organizational needs, establishes high expectation levels, and creates high self-esteem for those selected. In one aluminum rolling mill, for example, the primary consideration in hiring was to determine those applicants who were best suited to work in a team management environment. For the applicant, the employment process involved eight steps, one of which was an interview by team members, who then participated in the final hiring decision.

Ideally, employee skill levels should be closely linked to how people are paid; that is, people should be paid according to their acquired knowledge and skills rather than just their performance. As team members master additional skills, they should be awarded additional compensation. Thus management should develop a comprehensive pay and progression plan which should specify the requisite skills and how they are to be demonstrated and measured.

Accountability systems, then, are an important element in team management which are sometimes viewed as having fewer controls or less accountability. Accountability systems should focus on the team as the basic unit not only in meeting production goals but also in such areas as a peer review system for performance appraisal and statistical process control system for monitoring product quality.

Team Structure.   Although an organization's operational procedures provide a framework for team management, a need exists to establish a structure which defines the team boundaries and outlines the responsibilities each team has for its own operations. For example, a large aluminum rolling mill uses the star concept in which each point of the star represents an area of responsibility within a team (Fig. 11.3). In this concept, all team members are assigned a point on the star. The star representatives are responsible for the task and duties on that point (for example, looking after planning, and operating systems) for a fixed period of time. At the end of that period the team members are rotated to a new position on the star.

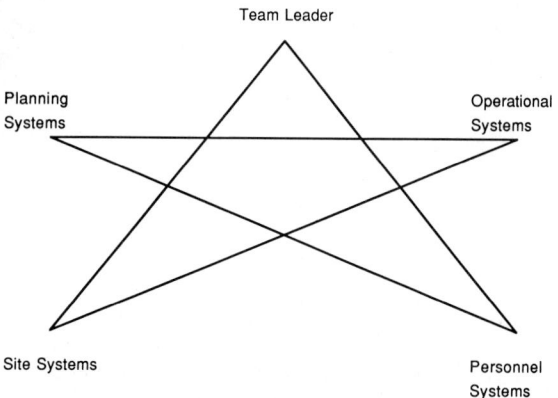

FIGURE 11.3   Team structure, a star concept.

The team leader runs the meetings, coordinates with other teams, completes performance appraisals, and resolves conflicts. The role of team members may change at various times to include responsibilities for performing needed tasks in the areas of operational systems, personnel systems, site systems, or planning systems. Team members themselves decide where employees will work, provided they have received training in the skill area to which they are assigned.

Operational systems include tasks like operating the equipment, quality control, and routine preventive maintenance. Personnel systems might include selecting and training new team members, developing and conducting peer appraisals of team members, and liaison with a companywide committee concerned with the review of personnel policies and procedures. Site system involves all aspects related to the work site, especially housekeeping and safety. Similarly, the planning system is concerned with keeping accurate records, scheduling and obtaining necessary supplies, and project personnel needs in order to plan effectively to meet production goals.

ARE TEAMS FOR EVERY ORGANIZATION?

Work teams may not be the ideal way to organize work for every organization or employee, or for every situation, but they offer a certain amount of promise as a basic pattern of approach in most organizations. In an organization where the team approach may be suitable, a related, and equally important issue regarding team formation concerns the degree of formalization. How formal and structured, or how autonomous are the teams to be? Teams can be molded along a continuum, ranging from work-unit teams to self-managed teams. Although the unique power and energy of teams increase as they move toward the self-managed model, this is clearly not appropriate for all settings. Thus matching the need and setting with the appropriate model is critical for the team's success.

An organization whose product or service requires a significant labor component is likely to benefit most from the team approach. Management needs to consider whether an increase in human efficiency will generate an increase in productivity. In a continuous-process chemical plant, for example, where two workers with very specialized skills watch over four million dollars worth of equipment, improving the productivity of those two workers may have little overall impact. Rather, the factor to consider is the content of individual jobs. If production jobs require only a single rudimentary skill, work teams may not provide any productivity gains. The same idea applies at the other end of the skill spectrum, since the team approach could be counterproductive when jobs are highly specialized. For example, a clothing fasteners producer found that given their technology, the most they wanted was better communication with workers. The equipment for making buttons, rivets, and clasps had not changed for generations, and each unique machine required an artisan to pass on his or her skill to an apprentice. Subtle skills like these requiring extended observation and practice may not lend themselves to the cross-training so essential to the team approach.

In the next section, a case drawn from Rohm and Haas Kentucky Inc. will be discussed. This case illustrates several issues discussed in the chapter so far.

ROHM AND HAAS KENTUCKY INC.—A CASE

The Rohm and Haas Kentucky Incorporated plant (RHK) employs 800 people to keep four production areas running and has an annual payroll of $31 million. The majority of these people are members of two unions, Oil, Chemical and Atomic Workers International Union (OCAW) and International Brotherhood of Firemen and Oilers. With an investment in the plant totaling $250 million, RHK is largely involved in changing a chemical called methyl methacrylate, along with other materials, into four major substances—Plexiglas molding powder, Paraloid additives, Rhoplex acrylic emulsions, and Acryloid coating resins.

In 1980, the employees represented by the OCAW went out on strike when the parties were unable to reach an agreement on a first contract. The strike lasted 10 weeks, and during this period the plant was run by management, plant forepersons, salaried employees, and the mechanics who were called upon to perform all the jobs required to keep production moving, with titles and classifications becoming largely irrelevant. People talked to each other about problems in order to find solutions as informal teamwork became the order of the day. With a high level of motivation among those who worked, demonstrating that people could adapt to change, everybody was interested in doing a good job. Except for a few delays during the first week of the strike, RHK never missed a shipment during this period. In fact, in spite of having fewer people they set production records. The employees' remarkable performance, played out under the restrictive conditions of the strike, led management to ponder how such motivation could be fostered in "normal times."

The 1980 strike and the lessons learned from it were, according to management and workers, precipitating factors in developing what was later called "a new way to work." The second major catalyst in bringing about change at Louisville was the downsizing taking place in some of the older Rohm and Haas plants, forcing all the plants to increase their efforts to improve productivity and quality to ensure long-term survival.

Immediately after the strike was over, management decided that the time had come to open the lines of communication and to push decision making down to the lowest appropriate level throughout the organization. New efforts to improve communication began with meetings between the plant manager and all employees. "Management by Walking Around" was resurrected and reinstalled as the apparent way of operating. A weekly plant newsletter and a series of communications meetings with supervisors were launched, early efforts which sowed the seeds for a quiet industrial revolution in the work and the work life of the RHK plant.

As a follow-up to initial efforts directed toward improving communication, the plant began to involve operators in revising both its utilities training program and some of its production training manuals. Management then established a plantwide steering committee to coordinate operator training.

In 1982, corporate and plant management provided assurances to all workers on questions of job security, training, local autonomy, and corporate support. The plant management stressed that management did not have a fully developed work plan to give to employees for implementation but that employees would be instrumental in creating a new work plan for themselves.

By 1983, Rohm and Haas Tennessee had established the use of self-regulating work teams. Based on their experience, RHK undertook a bold, full-scale experiment—the first design of a new production unit as a self-regulating work team. The unit, a Paraloid impact modifier plant called KV3, was still in the planning stages, so it afforded a rich opportunity to involve workers in the design of both the facility and the work that would be performed in it.

What Makes Teams Work at Rohm and Haas?   Members of a Rohm and Haas work team learn and rotate through all the jobs involved in the team's particular production responsibility. All members take turns at being a team leader so that a rotating leadership role replaces the traditional foreperson, or first-line supervisor position.

A team becomes self-regulating by preparing members to have a thorough working knowledge of everything needed to carry out the team's business, including regulating itself by reaching consensus on how the work can be performed most efficiently and by making many of its own decisions. The team regulates the quality of its production by having all its members well versed in statistical process control. Consequently, a work team understands what it is producing, how it will be used, and what standards the product has to meet to satisfy customers. (The customer may be another unit in the plant or another manufacturer which uses Rohm and Haas chemicals.)

Consequently, a Rohm and Haas work team is semiautonomous because it has nearly everything it needs to carry out its operation. First, it has a personnel mix tailored for results. Equipment in need of repair has the immediate attention of team mechanics. Previously, production units had to call upon a centralized maintenance department for all mechanical needs. A unit's request would go on a list with a host of others, its priority to be established by someone far removed from production. Now, a team's own production and quality goals determine its priorities, and its members' ready expertise enables it to work accordingly. Second, a team has the authority to make decisions and to take appropriate action. If SPC/quality control charting shows variations in a product or if safety questions arise, the team may choose to shut down the unit for adjustments. If a team needs supplies, permits, or work orders, it has the authority to obtain them.

Work Teams in KV3 Unit.   RHK's KV3 unit, the plant's first effort in work team organization, manufactures polyvinylchloride modifiers, or plastics additives, which go into shrink wrap, plastic food containers, molded bottles, and blister packages. The modifier plant was still on the drawing board when management decided to consider new ways to organize the work performed in it. A team of 10 hourly and salaried employees was formed to research various approaches and to make site visits to plants with innovative work methods. The research team then developed a proposal for a work team for the new plant, and interested workers volunteered to become members by bidding for the jobs.

The original training was intensive and extensive. Conducted on the job for 4 months, the training focused on sociotechnical systems (STS) to instruct employees on how to do the work and how to work together. In this process, workers learned how to reach consensus and make decisions as a group. In turn, the technical part of the training emphasized quality, consisting of learning new skills required to run the automated unit.

Since KV3 is self-supporting, the teams are responsible for the entire production process, from taking orders to handling customer complaints. Between these starting and finishing points, the team orders raw materials; schedules and conducts quality control; makes, coagulates, and dries its emulsion; packages the product; and ships it. A team member who today is involved in production may be performing completely different duties in the next job rotation. The KV3 work team is composed of a number of employees who work as four groups with the groups rotating through two 12-hour shifts to run the unit around the clock, 7 days a week. As each shift begins, team members in the group meet briefly to determine collectively how to accomplish the work at hand.

Once a month, the individual groups stay in the plant for an extra hour after their respective shifts to meet for more detailed discussion of the unit's operation. Every 5 weeks, the KV3 unit shuts down entirely for 3 hours so that all four groups can meet together. Communication among team members is also facilitated by a team steering committee made up of representatives from the four groups, which meets every few weeks to tackle various problems.

The operators in KV3 agree that the team approach is "a lot more work, more headaches and more responsibility." But, they quickly add, "95 percent of the unit's team members would want to keep it this way." "The company wanted the decision making process moved down, so everyone on the team has a part in it. We're making more decisions and more types of decisions (regarding) engineering, safety, and schedules," noted a team member. "It may take us longer to make a decision this way, but we may make a better decision by going through it." Conceding that there are "still a lot of decisions for management," he said that workers now see the management role as one of facilitator for the production crew. A worker acknowledged another positive aspect of shared decision making this way: "The team doesn't always come up with the right decision. But because everyone shares in it, they don't stand around pointing fingers. They move on. It eases up the burden on management, too."

While KV3 served as a greenfield plant for Rohm and Haas Kentucky, it was original in every way; other units have had to undergo redesign to adapt their ongoing businesses to new work methods.

Major Problems Encountered in Implementation.   Many of the serious problems encountered in implementing a team concept at Rohm and Haas Kentucky can be attributed to the broad uncertainty among management and workers alike about the results and ramifications of change. Compounding the overall problem of dealing with the unknown were difficulties with clear articulation and communication of the company's objectives.

The timing of team management created additional hardships. More time devoted to planning the effort might have lessened the articulation problems and produced a more lucid vision of outcomes for a worried work force. The timing of the introduction of team concept also proved troublesome as workers and managers tried to grapple with an escalating pace of change. An additional problem associated with the fast-paced change was the accelerating need for training and retraining to equip employees for participation in decision making and handling new responsibilities.

The role of the unions in the plant proved to be another significant problem in facilitating change. The workers' organizations were deliberately bypassed by management, which favored an approach involving employees, not their representatives. The unions, for their part, did not have a response to the new ways of working until long after the changes were well on their way to becoming institutionalized. Differences of opinion regarding the team concept among local members stood in the way of consensus and a unified response. Lack of direction from the international unions abetted the local problem.

Uncertainty about the effects of workplace changes was inevitable given the experimental and ambiguous nature of the changes. Early announcements of teams' management were exploratory rather than definitive since management was probing the work force for possible approaches to improved productivity and quality. Consequently, there was no strategic plan to guide the effort and no one could accurately predict the results.

The predominant concern of workers was job security. In spite of corporate pledges that no one would be laid off as a result of productivity improvements achieved through employee participation, hourly employees questioned the impact of increased productivity on plant laborpower. Furthermore, employees had reasons to fear the inevitability of fundamental changes in the way work was to be organized and in their job descriptions.

While many operators and mechanics welcomed the promised opportunity to have input into the plant's work, lower-level supervisors felt extremely threatened by the changes. Of all the employees at RHK, these supervisors harbored the most uncertainty about the effect the innovations would have on their work and their livelihood. They were told their jobs would change drastically, but no one seemed to articulate adequately how—since no one knew the anticipated role in the new work structure. Exacerbating the problem was the fact that the forepersons were not involved in the process of realigning their responsibilities early in the transformation. Instead of reaping the buy-in of the supervisors, which might have occurred had they been included, plant management gained their opposition and hostility. As implementation proceeded, these workers were the most profoundly touched. The supervisors lost their power over people, as surely as they lost shift differential pay, in the changing plant environment. Management, in retrospect, realized that the forepersons problem posed a major obstacle in bringing about change. But the forepersons, who were all reassigned to new jobs, were not the only personnel to have experienced loss of authority in the transition. Because of the importance of shared decision making, which had been driven down to the lowest possible levels, plant management also had to learn to work in a new way by forfeiting power, while retaining ultimate responsibility.

Better articulation and a corresponding reduction of uncertainty at all levels might have occurred if corporate and plant officials had devoted more time to planning the various efforts before implementing them. However, the company's decision to involve employees in the planning process made it necessary for employees to be brought in at an early stage when goals were vague and the steps required to meet those goals were largely unknown. Employees, many of whom had never been asked to participate in job-related decision making, were presented with questions, not answers. Consequently, doubt and confusion, along with resistance, were predictable outcomes.

Demanding training and retraining requirements created still more major problems for

Rohm and Haas Kentucky. At every level, employees had to be schooled in sociotechnical systems in order to function productively in the new environment. The company's quality drive necessitated broad training of workers in SPC. In discussing the problems associated with the training effort, employees continually cited the need for more learning opportunities. Several workers expressed the opinion that the forepersons should have been the first employees to be trained in the changing workplace as preparation for the leadership roles they would assume. Others complained that their training came too late in the process to adequately prepare them for their transformed jobs.

While no one in the company would quote the cost of the plant's commitment to training and retraining, company management agreed that the effort has been very expensive. The price tag would certainly be a major consideration for other plants which sought to replicate RHK's comprehensive training program. However, minimal investment in training can ultimately push a company out of business. In the next section, a discussion on the benefits accrued from the implementation of the team management approach is provided.

Results Obtained from the Implementation of Team Management. If a "cultural change" is what RHK attempted to bring about, there are some tangible signs that it has succeeded.

- This is a plant without time clocks—the clocks were banished in 1985 and replaced with time cards which first-line supervisors filled out for workers. Now, in most instances, workers are fully responsible for documenting their own time.

- Supervisors are rapidly being retrained in their new role as resource people, training coordinators, and facilitators, or referred to informally by some workers as coaches.

- Production operators sit in quiet carrels, writing and editing work manuals which are used to run everything from forklifts to chemical reactors.

- A look at the plant's parking lot reveals a striking symbol of egalitarianism with no reserved spaces. Early arrivals—hourly and salaried—get the shortest walk between their vehicles and their workstations.

The results of the implementation of team management at RHK can be tied directly to the major initiatives: the new emphasis on productivity and quality, the work structures and training designed to enhance productivity and quality, the unprecedented level of employee involvement, and improved labor-management relations. Results highlighted in this case study, therefore, concern changes in the plant's products and in its work force since the introduction of the team management.

Records of customers complaints bear out the positive impact of the quality emphasis. Since 1984, the first year for which complete data were recorded, the number of customer complaints has decreased by almost half from 2.5 percent of orders shipped to 1.3 percent in 1990. Plant records also provide evidence of increasing productivity at RHK. "Growth indexes," which reflect growth according to each year's absolute per pound numbers, indexed to 1982, reveal expansion in production and shipping volume, and in the size of the plant work force.

Figure 11.4 depicts a rapid growth in production from 1982 to 1983, followed by steady increases, which by 1986 had doubled the baseline year. With additional gains from 1986 to 1990, the growth index for production reflects 169 percent growth since 1982.

Similarly, the plant's growth index for shipments (Fig. 11.5) shows steady increases in pounds shipped from the facility in the first 3 years from 1982. From 1986 on, growth in pounds shipped rose sharply, reaching a new high of 1.74 in 1990, as compared with 1.0 in 1982.

A complement growth (Fig. 11.6) depicts expansion of the work force and includes all regular full-time Rohm and Haas Kentucky employees. The consistent growth represents the addition of some 300 hourly and salaried employees to the plant since 1982.

Further evidence of productivity growth is seen in comparisons of pounds shipped per employee over the 1982–1990 period. Again using an index keyed to 1982, Fig. 11.7 illustrates a continuous increase in pounds produced per employee through 1987. The dip in

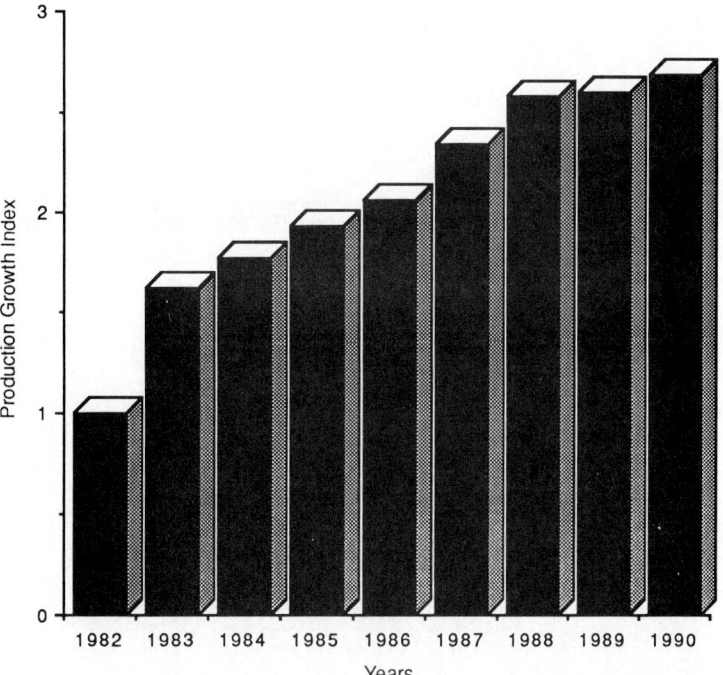

FIGURE 11.4   Rohm and Haas Kentucky production growth index (1982 = 1).

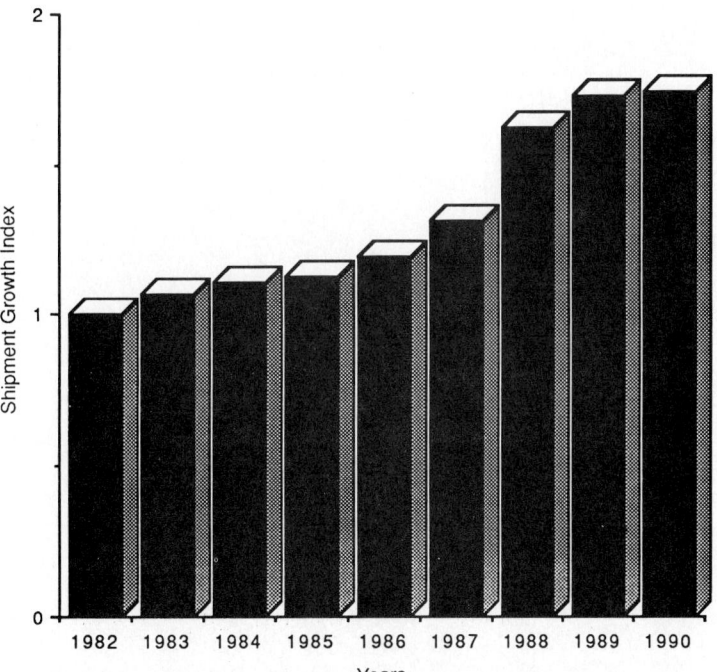

FIGURE 11.5   Rohm and Haas Kentucky shipment growth index (1982 = 1).

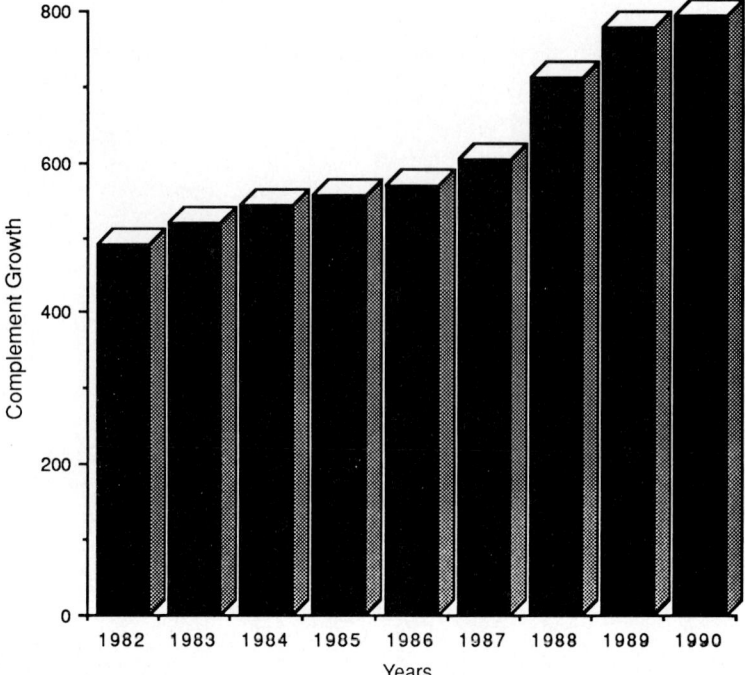

FIGURE 11.6   Rohm and Haas Kentucky complement growth.

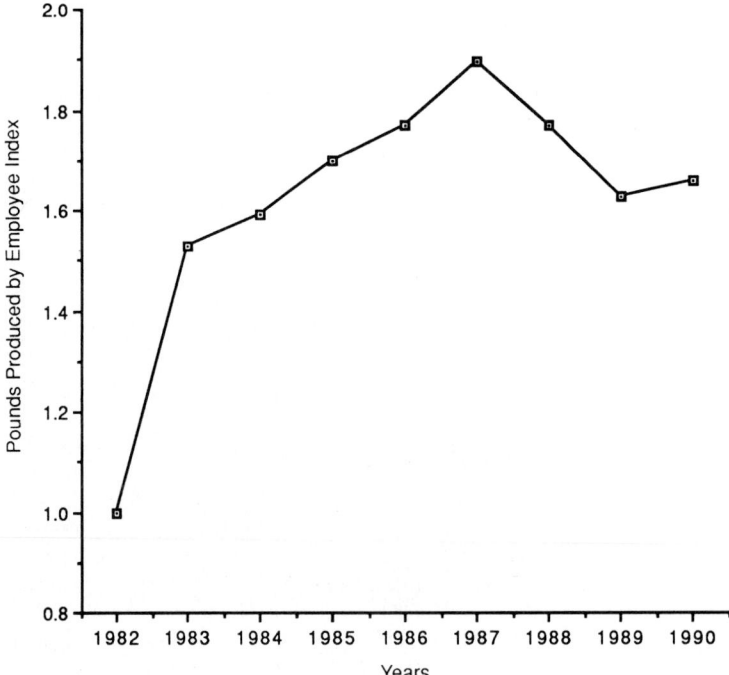

FIGURE 11.7   Rohm and Haas Kentucky pounds produced per employee index (1982 = 1).

productivity from 1987 to 1988 reflects the addition of employees for expansions which did not produce a product until 1989.

The cost of production per pound index has declined from 1.0 in 1982 to 0.939 in 1990. If one considers inflationary increases over this period, the cost of production per pound reflects a substantial decline.

Examination of data related to employee relations also reveals significant changes since the 1980 strike and the advent of workplace innovations in the plant. Grievances filed by members of the two unions at Rohm and Haas Kentucky dropped from a total of 121 in 1981 to 38 in 1990, a 70 percent reduction during the period.

Plant absenteeism records, shown as a percentage of scheduled hours, indicate reductions in all three employee classifications—hourly; salaried employees who are not exempt from overtime compensation regulations, such as clerical staff; and salaried employees who are exempt from overtime regulations, such as technical personnel. Between 1982 and 1986, hourly workers' average absenteeism declined from 3.25 to 2.76 percent absence from scheduled hours. Absentee records of nonexempt personnel were the most improved over the period, declining from an average of 4.75 percent of scheduled work hours in 1982 to 2.67 percent in 1986. Exempt staff experienced a reduction in absenteeism from an average of 2.84 percent of scheduled hours in 1982 to 2.32 percent in 1986. The plant discontinued absentee recordkeeping after 1986 because management did not think absenteeism was a significant problem.

Turnover, many plant employees acknowledged, has never been a problem at Rohm and Haas, a facility characterized by many as a benevolent, lifelong employer. But, even with its record of employee longevity, the plant has improved in this regard since 1980. Further evidence of the satisfaction of Rohm and Haas Kentucky employees surely lies in the fact that wages have continued to improve throughout the plant's transformation. A comparison of the growth index of the weighted average hourly rate in the plant with the consumer price index indicates that the hourly rate has increased consistently since 1980, the year to which growth is indexed.

In 1979–1980 Rohm and Haas Inc. had concerns about any further investment in RHK. During the 1986–1990 period, however, the parent company invested about $200 million in the plant, reflecting the healthy confidence the parent company has in RHK.

That RHK has seen remarkable improvements in quality, productivity, and employee relations since its introduction of the team concept seems more than coincidental. No data exist to measure and support a direct cause-effect analysis to the improvements. It is evident, however, that the combination of workplace changes, geared to creating value for the customer, has had a major, if unquantifiable, impact. In the end, the results of the efforts are due to pervasive improvements in everyone's performance.

REFERENCES

1. Alden, V., "The Trade Deficit: Stop Looking for Scapegoats," *Business Week,* Issue 2889, Apr. 8, 1985, p. 20.

2. Anderson, R., and K. Anderson, "HRD in Z Type Companies," *Training and Development Journal,* vol. 36, no. 3, pp. 14–23, 1982.

3. Anthony, W. P., *Participative Management,* Addison-Wesley, Reading, Mass., 1978.

4. Bartlett, C. A., and S. Ghoshal, "Matrix Management: Not a Structure, a Frame of Mind," *Harvard Business Review,* vol. 64, no. 4, pp. 138–145, 1990.

5. Bassin, M., "Teamwork at General Foods," *Personnel Journal,* vol. 67, no. 5, pp. 62–70, 1988.

6. Beer, M., B. Spector, P. R. Lawrence, D. Q. Mills, and R. E. Walton, *Human Resource Development: A General Manager's Perspective,* The Free Press, New York, 1985.

7. Boyett, J. H., and H. P. Conn, *Workplace 2000: The Revolution Reshaping American Business,* Dutton Books (Penguin Group), New York, 1991.

8. Bryan, L. A., Jr., "The Japanese and American First-Line Supervisors," *Training and Development Journal,* vol. 36, no. 1, pp. 62–68, 1982.

9. Cohn, S., and J. Zysman, "Why Manufacturing Matters: The Myth of the Post-Industrial Economy," *California Management Review,* vol. 29, no. 3, pp. 9–26, 1987.

10. Davies, G. B., and A. W. Pearson, "The Application of Some Group Problem-Solving Approaches to Project Selection in Research and Development," *IEEE Transactions on Engineering Management,* vol. 27, no. 3, pp. 66–73, 1980.

11. Drucker, P. F., *People and Performance: The Best of Peter Drucker on Management,* Harper, New York, 1977.

12. Dumaine, B., "How Managers Can Succeed through Speed," *Fortune,* vol. 18, Feb. 13, 1989, pp. 54–59.

13. Giffi, C., A. V. Roth, and G. M. Seal, *Competing in World-Class Manufacturing: America's 21st Century Challenge,* Business One Irwin, Homewood, Ill., 1990.

14. Gupta, Y., "Human Aspects of Flexible Manufacturing Systems," *Production and Inventory Management,* vol. 30, no. 2, pp. 30–37, 1989.

15. Huszczo, G. E., "Training for Team Building" *Training and Development Journal,* vol. 44, no. 2, pp. 37–43, 1990.

16. Kanter, R. M., Dilemma of Managing Participation, in M. Beer and B. Spector, eds., *Readings in Human Resource Management,* The Free Press, New York, 1985.

17. Kernaghan, J., and R. Cooke, "The Contribution of the Group Process to Successful Project Planning in R&D Settings," *IEEE Transactions on Engineering Management,* vol. 33, no. 3, pp. 134–140, 1986.

18. Klein, J. A., "Why Supervisors Resist Employee Involvement," *Harvard Business Review,* vol. 62, no. 5, pp. 87–95, 1984.

19. Kumar, S., and Y. P. Gupta, "Cross-Functional Teams Improve Manufacturing at Motorola's Austin Plant," *Industrial Engineering,* vol. 23, no. 5, pp. 32–36, 1991.

20. Lawler, E. E., *High-Involvement Management: Participative Strategies for Improving Organizational Performance,* Jossey-Bass, San Francisco, Calif., 1986.

21. Lee, C., "Beyond Teamwork," *Training,* vol. 27, no. 6, pp. 25–32, 1990.

22. Munchus, G., "Employer-Employee Based Quality Circles in Japan: Human Resource Policy Implications for American Firms," *Academy of Management Review,* vol. 8, no. 2, pp. 255–261, 1983.

23. Port, O., "The Best Engineered Part Is No Part at All," *Business Week,* Issue 3104, May 8, 1989, p. 150.

24. Port, O., "Smart Factories: America's Turn?" *Business Week,* Issue 3104, May 8, 1989, pp. 142–148.

25. Reich, R. B., "Entrepreneurship Reconsidered: The Team as Hero," *Harvard Business Review,* vol. 65, no. 3, pp. 77–83, 1987.

26. Rendall, E., "Quality Circles: A 'Third Wave' Intervention," *Training and Development Journal,* vol. 35, no. 3, pp. 29–31, 1981.

27. Schneider, W. E., "The Paradigm Shift in Human Resources," *Personnel Journal,* vol. 64, no. 11, pp. 14–18, 1985.

28. Sims, H. P., and J. W. Dean, "Beyond Quality-Circles: Self Managing Teams," *Personnel,* vol. 62, no. 1, pp. 25–32, 1985.

29. Taylor, A., "Why Fords Sell Like Big Macs," *Fortune,* vol. 7, Nov. 21, 1988, pp. 122–128.

30. Walton, R. E., Work Innovations in the United States, in M. Beer and B. Spector, eds., *Readings in Human Resource Management,* The Free Press, New York, 1985.

HUMAN FACTORS, ERGONOMICS, AND HUMAN RELATIONS

CHAPTER 1
MANUFACTURING ERGONOMICS

C. M. Gross
Biomechanics Corporation of America, Melville, New York

J. C. Banaag
Biomechanics Corporation of America, Melville, New York

R. S. Goonetilleke
Biomechanics Corporation of America, Melville, New York

K. K. Menon
Biomechanics Corporation of America, Melville, New York

The engineering of world class manufacturing facilities requires an unprecedented level of attention to the human factor. As product differentiation becomes more difficult to achieve solely on technological grounds, facility and process design must now be fused into the product development cycle. This is known as building quality into the process. The common denominator in this equation is the ubiquitous customer, with the concept of customer extending from the purchaser of the product to the individuals building the products. This sensitive, pragmatic, and somewhat egalitarian view toward the utilization of labor has developed into a survival tactic for sophisticated manufacturing organizations. Successfully addressing the human-equipment interface requires a broad and strong ergonomics knowledge base.

Ergonomics is a young science which grew out of the need to better accommodate military personnel during World War II. It is ironic that what was developed as a tool to make fighting more efficient is now the preferred technique for preventing musculoskeletal injuries in the workplace. As an interdisciplinary science, ergonomics draws its knowledge from several main tributaries (Fig. 1.1).

The practice of ergonomics began with the collection and use of anthropometric data. These data combined with observations were used to estimate the "goodness-of-fit" between equipment and personnel. Early ergonomists were concerned with accommodating variously sized individuals. As performance requirements became more critical, size considerations were expanded to encompass strength, reach, vision, cardiovascular capabilities, cognition, mission survivability, and most recently cumulative musculoskeletal injury. The last of these concerns addresses the question "Why ergonomics?" In traditional manufacturing environments, safety and health concerns are separated from manufacturing and design concerns, with one group being in personnel and the other in production or

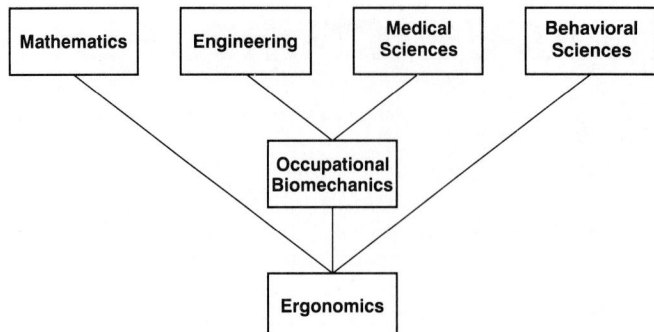

FIGURE 1.1   The science of ergonomics. (*Imai, 1986.*)

research and development. This suggests a fundamental problem indigenous to interdisciplinary sciences: addressing heretofore separate problems in a parallel manner. In a sense, biomechanically based ergonomics has taught us of the connectivity between product manufacturing and health management. Bridging the personnel gap between pure safety concerns and pure engineering concerns remains a major challenge in the practice of manufacturing ergonomics. The solution is to learn more. Both safety experts and industrial engineers need to become familiar with occupational biomechanics, the discipline which allows us to quantify forces on and within the human body at work. It is this measurement-based science which is largely responsible for the advances made in musculoskeletal injury prevention during the last decade. More to the point, biomechanics is a major tool for optimizing manufacturing facilities, addressing both economic and safety concerns.

Using biomechanics, it is possible to measure low-level, cumulative force exposures to the body. Using these measurement tools combined with necessary exposure limits it is possible to prevent musculoskeletal injuries from occurring. This measurement base also provides the quantitative basis for better utilization of human capabilities, or improved "fit."

The large current focus on quality improvement as a competitive advantage is a logical position once it's realized that well-distributed technology and capital require a new front on which to wage and acquire a strategic advantage. Though simplistic, process-driven advantages have been the mainstay of successful corporations in the 1980s. In the 1990s and beyond, process improvements will dominate competitive agendas. Ergonomics is a process for refining manufacturing systems and products through the study of how they interact with their users. Biomechanics is the decision tool. As such, ergonomics is a high-technology improvement process. These are the two basic differences between ergonomics and other process improvement strategies, that is, quality circles, Kaizen,* and the like.

As programs for continuous improvement (Kaizen in Japanese) have demonstrated both quality and cost advantages in the manufacturing and design process, they deserve careful study. The focus on the person-process, rather than the results is the main point of Kaizen. Kaizen is an extension of the Taoist philosophy applied to pragmatic business improvement. The results have been nothing short of astonishing.

According to Imai (1986), there are two main approaches to industrial progress, the relentless gradual effort and the great leap forward. As manufacturing becomes more sophisticated, differences in approach as contrasted with differences in technology may translate into significant differences in market penetration, quality, and customer satisfaction. In a sense the antonym for Kaizen is innovation. In the eastern countries (that is, Japan) the philosophy drives the process, whereas in the west the results feature the philosophy. Herein lies the difference: Kaizen, although originally an American idea (Deming et al.,

---

*Kaizen is a trademark of the Kaizen Institute, Ltd.

TABLE 1.1 Kaizen versus Innovation

	KAIZEN	INNOVATION
Effect	Long-term and long lasting but undramatic	Short-term but dramatic
Pace	Small steps	Big steps
Timeframe	Continuous and incremental	Intermittent and non-incremental
Change	Gradual and constant	Abrupt and volitile
Involvement	Everybody	Select few champions
Approach	Collectivism, group efforts, systems approach	Rugged individualism in ideas & efforts
Mode	Maintenance and improvement	Scrap and rebuild
Spark	Conventional know-how and state-of-the-art	Technological breakthroughs, new theories
Practical requirements	Requires little investment but great effort to maintain it	Requires large investment, but little effort to maintain it
Effort orientation	People	Technology
Evaluation criteria	Process and efforts for better results	Results for profit
Advantage	Works well in slow growth economy	Better suited to fast growth economy

Source: Imai, 1986.

1950s), is culturally better suited to an eastern approach to manufacturing as compared with innovation-driven industry. Imai's list comparing Kaizen qualities with those of innovation clarifies the distinction (Table 1.1).

From a competitive, industrial engineering perspective, how is the manufacturing advantage to be realized? The concept is to use ergonomics as a high-technology, Kaizen tool. Ergonomics is a process tool which naturally incorporates innovation. Rather than being individual-oriented continuous improvement (Kaizen), ergonomics is interface-oriented continuous improvement. The technology is a necessary component to this approach, as interface assessments require primarily quantitative rather than intuitive tools because of their complexity. Biomechanics is the tool.

In our experience, this approach to continuous improvement provides greater power and agility in the addressing of protean manufacturing problems when compared with Kaizen. Of course, ergonomics may be considered a Kaizen "type" tool. A systematic approach to bringing ergonomics into the manufacturing process is now termed total ergonomic quality (TEQ). This is a complete corporatewide program for ensuring ergonomic quality throughout design, production, and marketing (Fig. 1.2).

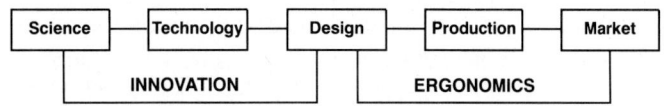

FIGURE 1.2 Innovation versus ergonomics. (*Imai, 1986.*)

TEQ starts with the training of management, engineers, and line employees in ergonomic techniques and responsibilities. TEQ ends with training and sharing the results of implementing ergonomics with the same individuals. Between these ends are process improvements utilizing ergonomic-engineering techniques. These techniques utilize biomechanics and statistical analysis tools (Table 1.2).

TABLE 1.2 Basic Statistical Analysis and Biomechanical Tools Utilized within a Manufacturing Ergonomics Program

STATISTICAL TOOLS	BIOMECHANCIAL TOOLS
Pareto diagrams	Dynamic Biomechanical Models
Cause-effect diagrams	Quantitative Muscle Measurement (EMG)
Histograms	Flexible Force Measurement Systems (FFM)
Control charts	Human Computer Aided Design
Scatter diagrams	Ergonomic Data Bases
Check Sheets	Cardiovascular measurements of fatigue
Relations diagram	Human motion tracking (video, sonar, infrared)
Affinity diagram	Musculoskeletal questionniares
Tree diagram	
Matrix diagram	
Matrix data-analysis diagram	
Process decision program chart	
Arrow diagram	

"What is ergonomics?" and "How is it used?" are the first topics this chapter addresses.

ERGONOMICS FOR PREVENTING WORKPLACE INJURIES

Back Injuries.   Apart from headaches, low back pain is the largest cause of pain in the United States (Khalil, 1984) and the major reason for physician contact (Mandell et al., 1989). Despite improved medical care, automation, and preemployment examinations, there has been little decline in back injuries (Ayoub and Mital, 1989).

There is little doubt that back injury is mainly caused by the nature of the individuals' work. Snook (1978) showed a high correlation between the specific act at work and the compensable back injury. Svensson and Anderson (1983) found heavy work to be strongly correlated to low pack pain. In this study, the heaviest tasks had the highest incidence of pain.

Back pain follows a highly remissive and recurring pattern. Attacks of back pain are self-limiting; regardless of the treatment used, most people feel better in 4 weeks and 90 percent spontaneously heal in 3 months. About 70 to 80 percent of low back pain attacks recur.

Nurses have among the highest incidence of occupational back injury (Jensen, 1985, 1986; Klein, 1984; Harber et al., 1985). The repeated lifting and moving of patients is suggested as the cause. A review of back injuries in 26 states (U.S. Department of Labor, 1982) showed that they occurred under the following conditions:

- The majority of movements at the time of injury were bending and twisting.
- The average duration an object was held at the time of injury was less than 1 minute.
- Of those who report a back injury, 30 percent lift over 100 times a day.
- The average weight of the object lifted at the time of injury was 40 to 100 lb (18.1 to 45.3 kg) in 70 percent of the cases and over 100 lb (45.3 kg) in 30 percent.
- The distance the load was carried in 80 percent of the back injury cases was less than 5 ft (1.52 m).
- The position of the load at the time of injury was on the floor in 50 percent of the cases.
- 35 percent of the workers felt that the loads they lifted were too heavy.
- About 50 percent of the workers who reported injuries had prior back injury.
- The position of the back at the time of the injury was fully or partially flexed in 83 percent of the injuries.

TABLE 1.3   Medical, Social, Psychological, and Other Risk Factors for Low
Back Pain

Constitutional:	Age, weight, height, back muscle strength, fitness, back mobility, genetic factors.
Medical:	Severe scoliosis, difference in length of legs, multi-level degeneration, disc resorption, disc herniation, severe arthrosis facets, spondylarthropathies, spondylolysis, spondylolisthesis, sacralisation/transitional vertebra, skeletal defects, fractures, neoplasmata, severe kyphosis, lumbar kyphosis, gravities.
Psychosocial:	Depression, anxiety, family problems, hypochondriasis, somatization, dissatisfaction with work and a high degree of responsibility.
Demographic:	Socio-economic and educational level, location of home.
Other:	Sports and level of activity, gardening, smoking, alcohol consumption, coughing.

Source: Hildebrandt, 1987.

Other risk factors include body constitution, postural problems, skeletal and medical problems, and psychosocial and demographic factors as shown in Table 1.3.

Hand, Arm, and Wrist Injury.   Injuries in the region between the hand and shoulder often belong to the category known as repetitive strain injuries (RSI). They are also referred to as cumulative trauma disorders or repetition injuries (Fraser, 1989; Sheng and Gross, 1988). RSIs are due to inflammation, pain, swelling or tenderness in tendons, nerves, synovial sheaths, synovial membranes, the lateral and medical epicondyles, and bursae. The most common RSIs are tendonitis, synovitis, tenosynovitis, epicondilytis, bursitis, and carpal tunnel syndrome.

Tenosynovitis affects the tendons and sheaths at the wrist; epicondilytis affects the muscles that originate at the epicondyles of the elbow; and carpal tunnel syndrome afflicts the median nerve as it passes through the carpal tunnel. It has been recognized that repetitiveness alone is not a good characterization of the condition; in many cases it may be caused by static forces. Though we use the term repetitive strain injury because of its familiarity, occupational overuse injury better explains the condition without necessarily linking it to repetitiveness.

More than one repetitive strain injury may occur simultaneously. The pattern of occurrence noted among a sample of women (Brown and Dwyer, 1984) is a useful pointer for the need for early intervention when symptoms first appear. Women with one RSI condition continued working and developed other conditions; the presence of one condition in a limb led to others in the same limb; resting one arm led to the appearance of problems in the other arm; inadequate rest following trauma or surgery led to the development of a new injury.

High-incidence patterns of RSI have been reported among certain types of work. A sample of these is shown in Table 1.4. Recently in the United States, the meat-packing industry has been focused on by the Occupational Safety and Health Administration (OSHA) for a "national special emphasis program" because of the high incidence of repetitive stress injuries (Bureau of National Affairs, 1990).

The risk factors for repetitive stress injuries have not been conclusively established, particularly because RSIs cover a wide range of symptoms. Some high-risk situations that have been suggested include tools and workplace equipment that cause extreme deviation of the hand and wrist, work surfaces that are too high or too low, tools or processes which require high hand forces, and the use of vibrating tools. Further, leisure activities which

TABLE 1.4   Work Types with High Incidences of RSI

Work Type	Specifics	Reference
Electrical and electronic	Winding wires and using small hand tools	Fraser, 1989
Poultry and meat fleshing boning and packaging	Use of knives and small tools and the lifting of carcasses.	Armstrong, 1982; Kivi, 1984
Manual sewing	Especially leather, canvas and heavy material.	Fraser, 1989
Word and data processing		Pinkham, 1988

require actions similar to those that caused the problem at work may contribute to the development of RSI in nonwork environments.

The Importance of Matching Workplaces to Human Capabilities.   If stressors in the workplace are reduced and operator comfort enhanced, then it is very likely that workplace injuries will be reduced. Physical and cognitive work involves the interaction between humans and machines. The elements of a human-machine system are:

1. Human
2. Interface
3. Machine
4. Environment

Whenever these four elements are in harmony, the injury potential is minimized, if not eliminated. In a more explicit sense and in a manufacturing context this implies that the human capabilities should match the task at hand. Any mismatches or poor fits are potential contributors to error and injury.

The Role of Ergonomics in Workstation Layout and Injury Prevention.   As defined before, ergonomics is the science dealing with work laws for humans. The key principles underlying this study encompass the man-machine system such that the human capabilities match the requirements of the job components, interface-machine-environment.

In the area of manufacturing, an inappropriate workstation layout may lead to a job stressor when its use exceeds the capabilities of the human operator. For example, when the reach and bending requirements or the frequency of each component task increases, back and hand injuries potential increases. The extent of the mismatch is best evaluated through a task analysis, which will be discussed later in this chapter. The accommodation of all individuals is achieved through adjustability. Ergonomics, when applied to layout and design, will help prevent injuries, improve performance, and improve operator well-being.

Workplace and Tool Considerations for Minimizing Wrist Injuries.   The reduction of stress to the wrist may be achieved by making sure that the following principles are adhered to:

Keep the Wrist Straight.   By maintaining a neutral wrist position during repetitive workplace activities, the amount of flexion, extension, and ulnar and radial deviation of the wrist is, by definition, minimized. As repetitive and forceful flexion or extension movements of the wrist are thought to gradually damage the median nerve, a wrist maintained in a nearly straight or neutral position will be less likely to sustain injury during repetitive motion activities.

Methods for ensuring that the wrist is maintained in as neutral a position as possible for a specific workplace activity include:

1. Biofeedback training

2. Task-specific tool selection

3. Task-specific workstation layout

Reduce the Transmission of Vibration to the Hand.   Low-frequency vibration (10 to 60 Hz) has been found to contribute to cumulative trauma disorders such as Raynaud's disease and vibration white finger. Thus, by reducing the transmission of vibration to the hand during the use of vibratory hand tools, it may be possible to reduce these vibration-related disorders. Methods for reducing vibration transmission include:

1. Vibration damping of the tool or workstation

2. Personal protective equipment use (warm, dry gloves with or without viscoelastic inserts)

3. Limiting daily exposure time

Keep Forward Reaches Short.   Acceptable limits of reaching should consider the reach frequency, body size of the worker population, and the distance and force required during each reach. Workstations and work methods should be designed so that workers are not required to perform extensive reaches on a repetitive basis.

Select Appropriate Hand Tools.   Ensure that the fit between the user and the tool is good and that the tool characteristics fit the required use. Factors which should be considered in handtool selection include weight, size, shape, ease of control, and surface texture. Tool characteristics should match the task requirements; for example, discrimination should be made between power and precision operations.

Evaluate the Mechanical Advantage of Tools.   Whenever possible, let the tool do the work instead of the user. This can be accomplished through semiautomation or by choosing a tool which provides adequate leverage to minimize force required during task performance.

Evaluate the Gripping Surfaces to Ensure High Coefficients of Friction and Smooth Edges.   The potential for tool slippage may be minimized by selecting a tool which provides a gripping surface and shape which matches the geometry of the hand posture required. Slip-resistant surfaces can be provided through granulated surfaces, rubber coatings, and rounded serrations.

Bend the Tool, Not the Wrist.   Tasks which require *frequent and similar* wrist deviations are best accommodated with tools bent in the direction of hand or wrist deviations. Modified hand tools are highly task-specific. A tool alone cannot be ergonomic. To have an ergonomic situation, you must carefully define the circumstance of use and the working population. The label "ergonomic" describes the interaction between the tool, the user, and the work process. The amount of bend in a handle is highly task-specific. As an example, a recent study by Konz (1986) indicates that for simple nail driving, hammers with a 10° bend are preferred over straight and highly bent models.

Stay Close to the Body's Center of Mass.   A well-designed hand tool should be usable as close to the center of mass as possible. This reduces bending and twisting movements as well as the magnitude of static muscle contractions (Fig. 1.3). Thus holding tools closer to the body improves control, comfort, and endurance. For many tools, shorter reaches also increase the amount of torque that can be applied.

Avoid Stress Concentrations.   The holding surface of the tool should be large enough and contoured in such a manner that it will distribute contact forces over the largest practical area of the hand (Fig. 1.4).

Maintain Sharpness of Cutting Tools at All Times.   Inadequately sharpened blades may significantly increase the amount of wrist force necessary to perform cutting operations. Blades which are easily changed or sharpened at the workstation should be utilized for repetitive cutting operations.

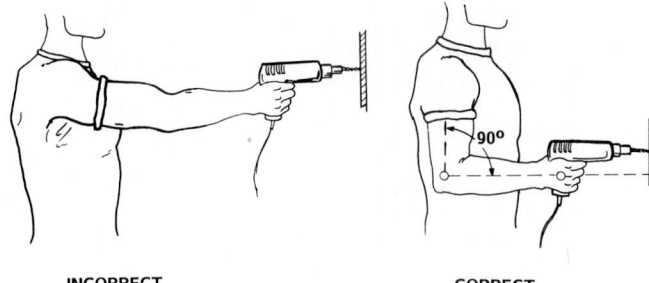

INCORRECT   CORRECT

FIGURE 1.3   Illustration of tool held at center of mass.

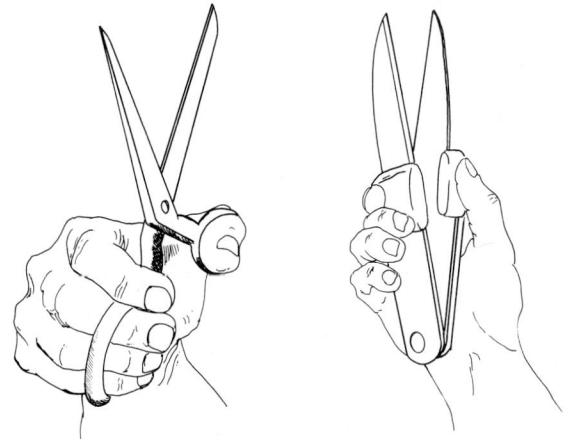

Traditional Scissors   **Modified Scissors with Spring Loaded Mechanism**

FIGURE 1.4   Distribution of contact forces through tool design.

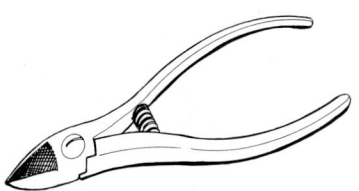

Traditional Wire Cutters with Spring

FIGURE 1.5   Spring-loaded mechanism in tool design.

Closing Tools Should Have Spring-Loaded Mechanisms.   This provides better tool control and requires less muscle force to operate (Fig. 1.5).

Tool Sizing.   To minimize flexor muscle activity and wrist force, a hand tool should fit the hand of the user (Fig. 1.6). The ability to grip a tool depends on hand size, surface contour, surface material, and handle width (separation).

Hand Tool Design.   Hand tools extend capability of the hand. The following principles will assist you in the selection of tools.

Design Tools to Be Used by Either Hand.   Without tools, the preferred hand is 5 to 15 percent faster and stronger than the nonpreferred hand. With tools, the advantage in performance of the preferred hand depends on the type of tool. With simple tools, such as an electric drill, the speed advantage is about 10 percent. The speed advantage for tools requiring some movement (hand saws, hammers) is 40 percent; for those requiring manipulation (scissors) it is 50 percent.

**Power drill with grip handle that
fits the shape of the hand**

FIGURE 1.6 Proper handtool sizing.

There are two main reasons why tools should be designed to be used by either hand. First, the preferred hand is not always the right hand. For about 10 percent of the population it is the left hand. Second, if a tool can be used by either hand it can be used alternatively—first with the preferred hand and then with the nonpreferred hand. This not only improves productivity since the hand can rest while the person continues working but also reduces repetitions on each wrist—very important for preventing carpal tunnel and tenosynovitis problems.

Power with Motors More than Muscles. As pointed out above, electrical or compressed air power is inexpensive. Power tools permit speeds, forces, and capabilities far in excess of that furnished by human muscles. Battery-operated power tools permit more safety (no danger from electrical shorts) as well as improved convenience.

Use a Power Grip for Power; Use a Precision Grip for Precision. Although there are a number of grips, the most important are the power grip and the precision grip. Force can be (1) parallel to the forearm (suitcase handle, electrical drill, pistol), (2) at an angle (hammer, pizza cutter, reverse-grip pliers), or (3) torque about the forearm (corkscrew with a T handle).

Precision grips have a pinch grip by the thumb versus the first finger (or first and second finger). Precise control is needed for the tool shaft. When the tool shaft is under the thumb, the grip is called internal. When the tool shaft is above the thumb, the grip is called external.

Make the Grip the Proper Thickness, Shape, and Length. Every tool has two ends—one working on the material, the other on the hand. Female hands tend to be smaller than male hands; thus hand tools designed for men tend to be too large for women.

For a power grip, the best diameter is 1 to 2 inches; the most common mistake is a too small diameter (say 0.5 inch) handle which cuts into the hand. To prevent rotation of the tool in the hand, a bearing surface helps. A spherical handle is another alternative; although there are no bearing surfaces, the large contact area permits lower pressure. If the tool is to rotate in the hand, the area within the pinch grip should be circular to permit simple finger movements. A change in the cross section along the tool grip axis (1) reduces movement of the tool in the hand, (2) permits more force to be exerted, and (3) may act as a shield. A pommel (a shield at the rear of the handle) permits greater force when the tool is pulled toward the body. Finger grooves are not recommended since they do not fit all size hands. A wedge shape (gradual change in diameter) is better. The length for a power grip should be at least 4 inches (5 may be better) so all four fingers can grip.

For an external precision grip, the length should be at least 4 inches so the tool can be supported by the base of the thumb and first finger. For an internal precision grip, a short shaft tends to dig into the palm, so the shaft should be long enough to extend beyond the palm.

Design the Grip Surface to Be Compressible, Nonconductive, and Smooth. A compressed grip is easier on the hand, giving a better coefficient of friction and reducing vibration. Use rubber or wood and avoid hard plastic and metal. Grips should not conduct heat or electricity. Grips which are not conductive gain heat more slowly and as a result do not get as hot. The grip should release slowly to the hand so that the user can let go before being burned. A knife works by exerting a force over a very small area. Sharp edges on the tool handle act as knives to the hand. Grind away sharp edges, cover them with tape, or dip the handle into plastic. A sign of a poor grip is a mark left on the hand.

Consider the Angles of the Forearm, Grip, and Tool.   The best position for the wrist is the "handshake" position. A contorted wrist leads to problems. Following the ergonomic philosophy that equipment should adjust to people rather than people adjusting to equipment: If a bend is needed, it should be in the tool, not the wrist.

Sufficient clearance also is needed to prevent burns and pinch points. It is desirable to have a spring keeping the tool normally open. Without the spring, there is the temptation to insert a finger between the handles to open them; the finger usually gets pinched. Tool handles, especially locking handles or toggle clamps, should have at least a 1-inch opening in their fully closed position.

Use the Appropriate Muscle Group.   The fingers should be used for manipulative work; let the strong muscles of the forearm do the power work. The force of the tool is furnished by the forearm muscles rather than finger muscles. Another common problem is "trigger finger," due to repetitive use of the first finger or thumb. A trigger strip actuated by two to four fingers is the preferred alternative.

Workstation Adjustment Checklist

In addition to tool considerations, the workstation must be evaluated. For seated work, review the workstation checklist in Table 1.5. The key principle is adjustability. Operators must be familiar with all adjustment features and principles of proper workstation and work method fit. Adequate adjustable support for the back, arms, and legs is necessary to reduce cumulative static muscle forces.

Table 1.6 presents an overview of ergonomic intervention strategies for managing cumulative trauma disorders.

HUMAN ANTHROPOMETRY

Overview

Anthropometry is a principal branch of anthropology which deals with the measurement of the human body. Anthropometric variables refer to the characteristics of the body that can be standardized. The science of human measurement drew increased interest with the development of ergonomics, primarily when designing jobs and machines to "fit" the human. The goal of ergonomics is to make workplaces, equipment, and products fit the potential human users' capacities for reach, grasp, and clearance.

A feeling of "ownership" is present whenever a good or product fits a person. In a man-machine environment, such ownership can be brought about by designing machines and equipment to accommodate the body measurements and capabilities of the potential user population. Within reasonably homogeneous populations, it is empirically true that anthropometric measures conform to a bell-shaped curve or a gaussian distribution (Fig. 1.7).

A common concept in the area of anthropometry is *percentile*. The xth percentile is defined as the point where x percent of the population has the same value or less than that value for a given measurement. For example, if stature is the variable of interest, a human with a 95th percentile height would be taller than 95 percent of a given population. Alter-

TABLE 1.5 Workstation Checklist

	YES	NO	N/A
1. Is the height of the worksurface adjustable to accommodate different operators?			
2. Is there sufficient space for temporary storage at the workplace?			
3. Are there inefficient work motions because of workspace layout?			
4. Is there sufficient clearance for handling or maintenance tasks?			
5. Is the worktable adequate for the smallest and largest worker for reach and clearance?			
6. Is there sufficient lighting at the workplace?			
7. Is there too much glare at the workplace?			
8. Is there too much noise at the workplace?			
9. Are there gaseous fumes in the area?			
10. Is the workplace temperature between 68° and 78°F.			
11. Is the air circulation good?			
12. Is the seat used suitable for the task?			
13. Is the chair height adjustable and easy to adjust?			
14. Is the chair backrest easily adjustable?			
15. Is a footrest provided?			
16. Is the footrest large enough to support both feet and allow a change of position?			

natively, one may say that only 5 percent of the people are taller than the 95th percentile individual.

The two parameters that describe the gaussian distribution are the mean (m) and the standard deviation (s). The pth percentile (Table 1.7) of a gaussian variable (X_p) can then be found by using $X_p = m + sz$, where z is the standardized normal deviation found in any statistics text. A few commonly used values in the design of man-machine systems are shown in Table 1.7.

Measurement Instrumentation. The standard anthropometric measurement instrument is the anthropometer. This is a rigid rod 2 m long, with two counterreading scales. Commonly, the rod is split into three or four sections which fit into one another. Elaborate anthropometers are fitted with mechanical or electronic reading devices. A stadiometer is a fixed anthropometer, primarily used for the measurements of stature. For transverse diameters, calipers may be used. A pelvimeter measures up to approximately 600 mm and the cephalometer up to approximately 300 mm. Skinfold thickness is measured with a constant-pressure skinfold caliper with a pressure of 10 g per mm^2. For arcs and girths, a flexible steel tape with a flat section may be used. It is undesirable to use self-straightening steel tapes. Goniometers are used to measure angles. Electrogoniometers (or elgon) are electrical potentiometers which can be calibrated to read angles. When a constant voltage is applied to the elgon, the resistance is a function of the angle.

TABLE 1.6   Ergonomic Intervention Strategies for Wrist Injury Prevention

PROBLEM:	POSSIBLE RECOMMENDATION:
1. REPETITIVENESS	1. Use mechanical aids 2. Enlarge work content 3. Rotate workers 4. Spread work uniformly across workshift 5. Automation
2. LARGE JOINT FORCES	1. Increase mechanical advantage of tools 2. Decrease the weight of tools, containers and parts 3. Increase the friction between handles and the hand 4. Optimize size and shape of handles 5. Select gloves to minimize effects on performance 6. Balance hand-held tools and containers 7. Use torque control devices 8. Fine tune work/rest ratios 9. Mechanical assists 10. Automation
3. LARGE HAND/ARM CONTACT FORCES	1. Enlarge corners and edges 2. Pads and cushions 3. Tool size selection
4. POOR POSTURE	1. Locate work properly 2. Orient work properly 3. Select tool dewsign for workstation 4. Change work method 5. Postural biofeedback training
5. VIBRATION	1. Select tools with minimum vibration 2. Select process to minimize surface and edge finishing 3. Mechanical assists 4. Use isolation for tools that operate above resonance point 5. Provide damping for tools that operate at resonance point 6. Vibration damping gloves 7. Adjust tool speed to avoid resonance

Source: Armstrong, T. J., and Lifshitz, Y., 1987.

TABLE 1.7   Values of Standard Normal Deviates

Percentile	z
2.5th	-1.96
5th	-1.64
50th	0.00
95th	+1.64
97.5th	+1.96

Adjustments to Measurement.   Most measurements reported are for humans who are minimally clothed. When these data are to be used for design purposes, such measurements should be adjusted to account for clothing and posture. Clothing adjustments as estimated by Eastman Kodak (1983) are as follows:

+ 2.5 cm for standing height

+ 0.5 cm for sitting heights

+ 0.8 cm for breadths

+ 3.0 cm for foot length

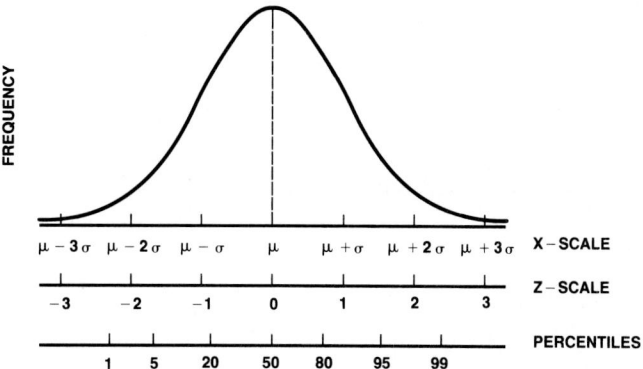

FIGURE 1.7   Percentiles and the z-statistic in a gaussian distribution.

Since measurements are made with the subject in an erect position, adjustments ought to be made also for postural slump. Estimated adjustments (Eastman Kodak, 1983) are

- 2.0 cm for standing height
- 4.5 cm for sitting height

Estimation.   The designer is frequently required to estimate the distribution of some measures which may not be reported directly for a particular population. Under such circumstances they need to be estimated. A few useful techniques are:

Correlation and Regression.   In the correlation and regression method, the assumption is that the correlation between two variables is the same for two populations (population 1 and population 2). The regression equation for population 1 is then used to estimate the unknown variables in population 2.

Sum and Difference Dimensions.   In the sum and difference method specialized equations are used. When an unknown dimension is equivalent to the sum of two known dimensions a and b, the

$$m_{(a+b)} = m_a + m_b$$

$$s^2_{(a+b)} = s^2_a + s^2_b + 2rs_a s_b$$

If the unknown dimension is the difference of two known dimensions, then

$$m_{(a+b)} = m_a - m_b$$

$$s^2_{(a-b)} = s^2_a + s^2_b - 2rs_a s_b$$

where m_a and m_b = the two means of the known variables a and b and s_a and s_b = the sample standard deviations of the same two variables
r = the sample correlation coefficient

Ratio scaling is commonly used (Barkla, 1961; Pheasant, 1986) for two populations (reference and target) similar in age, sex, and ethnicity. For two variables a and b

$$m_a/m_b \text{ (in population 1)} \sim m_a/m_b \text{ (in population 2)}$$

$$s_a/s_b \text{ (in population 1)} \sim s_a/s_b \text{ (in population 2)}$$

Anthropometric Data

Anthropometric data have been classified into two general classes: static and functional (dynamic) dimensions. Static data are obtained with the human in a standardized, static position, while functional (or dynamic) dimensions are obtained with the body being involved in some physical movement. Anthropometric literature of different countries has been published by many researchers. Two sources that list a few different countries are Garret (1971) and Pheasant (1986). A few others are presented in Table 1.8.

TABLE 1.8   Sources of Anthropometric Data

Population	Source
Egypt	Moustafa (1987)
Germany	Jurgens (1973), Kroemer (1964), Wagner (1988)
United Kingdom	Andrew and Manoy (1972), Davies et al. (1980), Haslegrave (1979, 1980, 1986), Montegriffo (1968), Murrell (1965), Pheasant (1982), Redgrove (1979)
France	Bouisset (1967), Bouisset and Monod (1961), Rebiffe et al. (1983), Wisner & Rebiffe (1963)
Hong Kong	Courtney (1984)
India	Eveleth & Tanner (1976), Goswami et al. (1987)
Japan	Yanagisawa (1974)
Korea	Fernandez (1989)
Netherlands	Molenbroek (1987)
Poland	Batogowska and Slowikowski (1974)
Sweden	Inglemark and Lewin (1968), Lewin (1969), Lindgren (1976), Thiberg (1965-1970)
Switzerland	Grandjean & Burandt (1962), Grandjean (1973)
United States	Armed Forces Personnel Clauser et al., (1972), Dempster (1955), Garret (1970, 1971), Hertzberg et al., (1954), Gordon et al (1989), NASA (1978), Van Cott and Kinkade (1972). Robinette and Fowler (1988) Civilian Personnel Diffrient et al. (1981), Kroemer et al., (1986), McFarland & Stoudt (1960), NASA (1978), Snyder et al. (1977), Van Cott and Kinkade (1972),
U. S. S. R.	Ermakova et al. (1985), U. S. Army Labcom (1986)

As may be evident, most of the data published are restricted to a few countries, making it very inconvenient for designers to quickly access information for global product design owing to the number of sources required. The anthropometric database ErgoBase™ (BCA, 1989) provides an alternative in the form of a computerized database compiling anthropometric data from many countries and special populations.

Uses of Anthropometric Data.   The main purpose of measurement and estimation of anthropometric data is to design "machines" which enhance the fit with the intended users. By enhancing, it is meant that the design improves operator comfort and reduces the strain experienced by the body while working in the environment. People vary in size and strength. Hence no one design will be optimal for all users. In the application of anthropometric data to workplaces, products, and tools, the following methods may be employed.

Design for the Average.   The design dimensions are based on the 50th percentile of the user population. Because of the variability, it is very desirable to design workplaces and equipment that are adjustable. Although adjustable features may be provided, they may not be utilized. Using the adjustability depends on the speed, training, and effort required and the degree of perceived benefit to the operator. The adjustments may not be required. However, the presence of adjustability will allow a larger population to work comfortably. The following two design methods are based on this concept.

Design for the Extreme Individual.   The designs are based on the extreme size of an anthropometric variable(s) that the potential current user population possess. With a new design, this would indirectly translate to 100 percent of the population. In some cases, such a design would not be financially viable. For this reason, the general tendency is to design for a range of individuals.

Design for a Range.   The range *normally* chosen in ergonomic practice is from the 5th percentile woman to the 95th percentile man. However, depending on the component or item being designed, the ergonomist may decide to use another suitable range. When designing for a range, the percentage of the population for which the item was designed is known as the *accommodated percentage* (McCormick and Sanders, 1982). Designing for a range is the most sensible approach for accommodating employees and customers.

Work Space.   An element which goes hand in hand with anthropometry is work envelopes. The *functional arm reach envelope* is a work-space envelope in three-dimensional space that represents the limits of convenient arm reach. Such an envelope may be traced through the use of a reach anthropometer (Roth, Ayoub, and Halcomb, 1977). The reach areas for a seated 5th percentile female and the standing reach areas of one arm and both arms are shown in Figs. 1.8 and 1.9.

Computer-Aided Man-Machine System Design.   Several computerized packages to evaluate human-machine fits have emerged over the past. The older models are BOEMAN (Ryan, 1971) and the crewstation geometry evaluator (CGE) (Katz, 1972). Both these models were not interactive and were batch-oriented, although the BOEMAN model did have the ability of producing a three-dimensional human on a screen.

A 2-D graphic aid was developed by the Australian Department of Defense (Hendy et al., 1984) for use with the RAAF Airtrainer CT-4A cockpit. The development of this program was based on the concept of fitting a given percentage of individuals for which adequate anthropometric data existed, rather than using the conventional approach based on pooled anthropometric data. For this reason, the simple 2-D model should not be used to define dynamic reach envelopes or volumetric relationships.

The SAMMIE (system for aiding man-machine interaction and evaluation) model (Bonney, 1979; Kingsley et al., 1981) includes the anthropometric tables necessary to generate a three-dimensional image of a user. This system has been used as a design and evaluation tool. Simpler and less sophisticated models are the computerized accommodated percentage evaluation (CAPE) model (Bittner, 1975) and the crewstation assessment of reach (CAR) (Edwards et al., 1976).

The COMBIMAN (Kroemer, 1972; McDaniel, 1976; Korna and McDaniel, 1985) system was developed to assist in the design and analysis of crew stations. This model

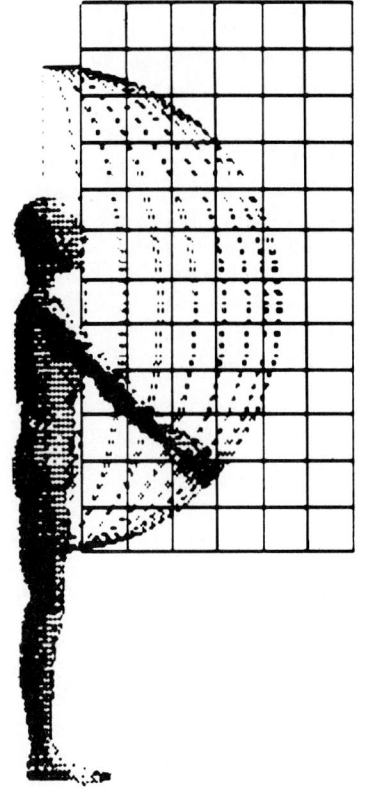

FIGURE 1.8 Standing reach envelope (one arm) for 5th percentile male. (*Courtesy of HumanCad Inc., Melville, New York.*)

is based on a 35-link skeletal system. The CREW CHIEF (Korna et al., 1988) simulation package is another program developed to analyze the interactions of the man-model's physical characteristics and capabilities with the workstation configuration. CREW CHIEF operates interactively with the CADAM* (computer aided design and manufacturing) package. This package was primarily designed to analyze the maintainability of designs, analyze the interaction of a maintenance technician with a system, and enable the user to evaluate limitations and capabilities in the areas of physical accessibility, strength, and visibility.

The main problem with the above packages has been the task specificity, specialized computer hardware requirements, and the level of user friendliness. Mannequin™ is a simpler yet powerful system ideal for man-machine systems design. The package is designed to run on the most popular type of computers, personal computers. The 3-D capabilities of Mannequin can be utilized to produce working drawings of workplaces. The anthropometric databases built into the package allow the user to evaluate and design "machines" with great ease. An example of a "machine" design in the presence of a human operator is shown in Fig. 1.10.

TASK ANALYSIS

Overview.   The most important methodology in ergonomic assessment is task analysis. A task may be described as one which comprises a set of human actions ultimately resulting in the output of the system. The term task analysis has had many meanings in this century. Task analysis in an ergonomic context refers to the process of identifying and describing subtasks and analyzing these for the successful performance of the job. The smallest subtask to be used in an ergonomic analysis is one which has no fixed boundaries but one which the ergonomists determine based on the job at hand. The elements of a task analysis are

1. Task description
2. Requirement specification
3. Analysis
4. Evaluation

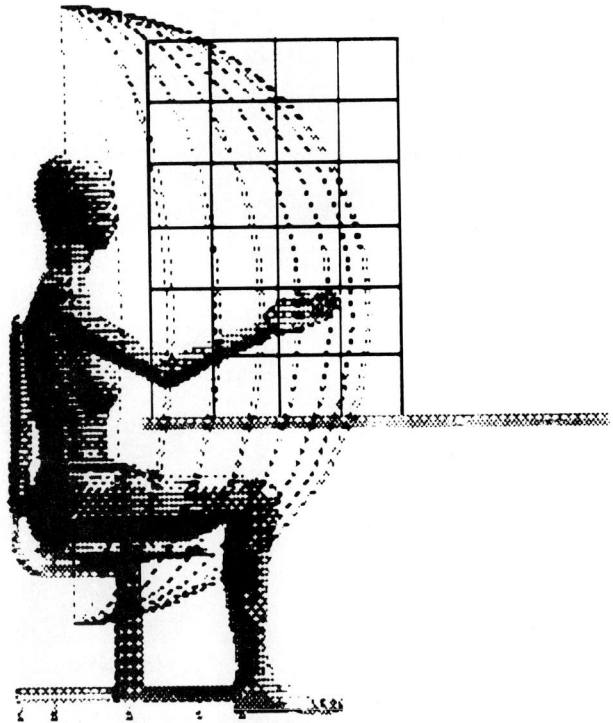

FIGURE 1.9   Forward reach envelope of seated 5th percentile female.
(*Courtesy of HumanCad Inc., Melville, New York.*)

The requirement specification of the job entails measuring the four primary components of each subtask, namely,

1. Force
2. Frequency
3. Posture
4. Environment

The following sections describe the different techniques used in current ergonomic practice.

Force Measurement.   In the study of biomechanics, force plays an important role. Force measurement is done through the use of force transducers. The simplest force transducers are mechanical, and these include spring dynamometers and cable tensiometers. Isokinetic dynamometers are another kind used for obtaining back strength. Electronic transducers, on the other hand, are used when accuracy is important (Fig. 1.11).

The most common types of electronic transducers are the strain gauge, piezoelectric transducers, and linear variable differential transformer (Eastman Kodak Company, 1983). For electronic transducers the strain induced as the result of the force (stress) varies the electrical characteristics and hence the voltage. This voltage can then be calibrated to indicate the force. When multiple force transducers are used, such transducers are referred to as a force plate or force platform (Morgan and Bhattacharya, 1984). Force plates are used to study ground reaction forces that resist the feet. Another application is for the measurement pressure between buttocks and seat surfaces when seated (*Autoweek*, 1990).

FIGURE 1.10   Human-machine design with HumanCad. (*Courtesy of HumanCad Inc., Melville, New York.*)

Ergonomic Seat Pressure Assessment of Automotive Seats

FIGURE 1.11   Flexible circuit force measurement system.

Muscle Strength.   Mechanical force transducers are used to measure isometric strength of human muscles. However, the most commonly used technique for measuring muscle load is electromyography (EMG). EMG is based on the electromechanical coupling of muscles and provides a means of quantifying the muscular strain. Rapidly fluctuating differences of potential within the tissues and across the skin are created when muscles are activated by signals from the nerves. A record of the potential changes (ranging from a few microvolts to a few millivolts) between two electrodes placed on the skin surface or embedded in the tissues is called an electromyogram. The electrodes used for EMG are of three types: surface electrodes, needle electrodes, and fine-wire electrodes. The surface electrode is the most popular type. They consist of a disk which is placed over the muscle being studied. The needle electrode is inserted into the muscle of interest, and even though this results in an invasive procedure the experimenter has better control in isolating the muscle. The fine-wire electrode methodology is similar to that of needle electrodes. The electrode here comprises two fine wires which are inserted into the muscle studied with the use of a needle. The needle is removed, leaving only the fine wires inside the muscle.

In ergonomic assessments, surface electrodes are common because of their ease of use and attachment. With surface electromyography different electrode arrangements are possible. Some are the unipolar, bipolar, tripolar, and concentric annular. The tripolar arrangement, for example, is comprised of three evenly spaced electrodes in a linear array. The potentials of the two outer electrodes are averaged, and the difference to the central electrode potential is recorded.

There has been disagreement in the exact relationship between the electrical signal and the tension generated on the muscle. Bigland and Lippold (1954) reported a linear relationship while Bouisset (1973) reported a nonlinear relationship under different experimental conditions. However, there is general agreement that a linear relationship exists in static conditions while a strong correlation is present with dynamic conditions. An example of an electromyogram is shown in Fig. 1.12. The standards for reporting EMG data are given by Winter et al. (1979).

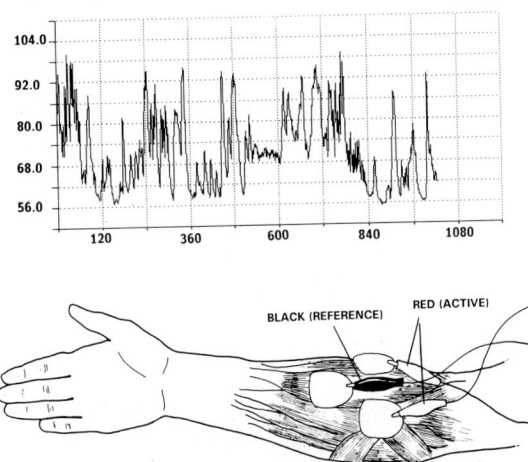

**Electrode Placement for Flexor
and Extensor Muscles**

FIGURE 1.12   Electrode placement and EMG of flexor and extensor muscles during load carrying.

EMG signals can be processed in many different ways. If overall muscle force is of interest, the signal must be integrated. Such integration is possible through the use of the peak value, the rectified average value, the root-mean-square (rms) value, the number of level crossings (for example, zero crossings) per unit time, the number of peaks per unit time, and so on. Through the use of spectral analysis techniques, center or median frequencies may be obtained. Chaffin (1973), Petrofsky (1979), and others have reported how EMG spectral density can be used to determine muscle fatigue. When muscular fatigue is present, the amplitude of the higher-frequency components decreases while the amplitude of the low-frequency components increases. The mean power frequency is a useful measure for detecting this change.

Prior to the processing of the signals it is customary to normalize the signals. One such normalization is done with respect to each subject's maximum EMG activity. The maximum amount of force that a muscle can exert is known as the maximum voluntary contraction (MVC). When the EMG activity is expressed as a percentage of the maximum

EMG activity, the measure is referred to as the percent maximum voluntary contraction (% MVC). Muscular strength is normally expressed in these units. As the % MVC approaches 100, the muscle is working near its maximum strength and can hold such force for only a few seconds. The muscular force and the maximum endurance time (MET) relationship was first developed by Rohmert (Fig. 1.13).

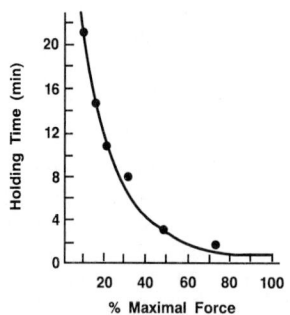

FIGURE 1.13   Force-time relationship for pull condition (Rohmert's 1968 curve). (*J. R. Wilson and N. C. Corlett, 1990.*)

Another normalization procedure is with respect to the resting EMG activity. An alternative measure is the relative muscle activity defined as follows:

$$\text{Relative activity} = \frac{\text{recorded EMG} - \text{resting EMG}}{\text{maximum EMG} - \text{resting EMG}}$$

In the normalization it is important that the correct posture be used. If the EMG signals of the wrist are being evaluated, the grip type and posture are both important. To take into account extreme variabilities in a work cycle Silverstein et al. (1986) have proposed the "adjusted force" measure defined as

$$\text{Adjusted force} = (\text{variance/mean force}) + \text{mean force}$$

Based on this measurement they concluded that the combination of high force and high repetitiveness (frequency) increases the risk of cumulative trauma diseases as opposed to any other combination of force and frequency.

Posture Measurement.   Researchers have used many methods for posture measurement. These can be categorized as

1. Direct measurement techniques
2. Indirect measurement techniques

Posture measuring instruments may be mechanical, eletromechanical, or optical. Electrogoniometers are frequently used to quantify joint motion. The criteria to bear in mind with the use of any angle measurement technique are

1. Accuracy
2. Compactness
3. Ease of use
4. Unconstrained movement of joint

The electrogoniometer is attached to the limbs of the body with the axis of rotation of the goniometer coinciding with the axis of rotation of the limb. When a constant voltage source is applied between the ends of the goniometer, the resistance of the potentiometer is a function of the joint angle. The electrogoniometer can thus be calibrated to measure the angle directly.

Researchers have developed many methods to measure the motion of the trunk. O'Brien and Paradise (1976) used strain gauges on the lower back to measure trunk flexion. Nordin et al. (1986) used a flexion analyzer for trunk movements in the sagittal plane. This instrument consisted of a pendular potentiometer, a five-level analog-to-digital converter, nine digital registers, and the appropriate control circuits. One of the drawbacks of this device is that the unit needs to be worn on the back of the subject studied.

The flexible curve method (Burton, 1986) is an inexpensive way to perform postural measurement. The disadvantage of this method, however, is that it is cumbersome and complex for many limb joints.

Optical instruments are based on imaging techniques such as cinematography, video recorders, multiple-exposure photographic techniques, or optoelectronic techniques. In cinematography, Cine or motion picture cameras are used. The most common and least expensive method that exists today is video recording. Another simple and economical approach is the multiple-exposure technique. Reflective markers placed on the body are tracked with a still camera with each illumination made using a strobe light. The resulting photograph shows a stick figure which is used for analysis. The disadvantage of this method is that it needs to be done in a dark room and where a strobe light will have no effect on the work performed.

One optoelectronic technique available today is the Selspot system (Selcom Inc., Sweden). This system uses infrared light sources, light-emitting diodes, or laser light sources. The light sources are attached to the body and are then sensed by the Selspot cameras. The camera output can be computer processed for precise position information. The capability of obtaining speed, acceleration, and rotation of limbs makes this a powerful method for motion analysis.

The Vicon system (Oxford Medical Systems, U.K.) is another system used for postural measurement. In the Vicon system, infrared retroreflective markers are detected by video cameras. From this point detection, 3-D coordinate files for body marker positions are produced.

Another system, the Spectron miniature inclinometer, uses single-axis electrolytic potentiometers. The output voltage of the unit is proportional to the angle from the vertical. Biplane inclinometers have been used by many researchers for head and spinal postures (Nordin, 1982; Weber et al., 1986). These instruments use miniature load cells mounted on two tumbling weights. The resultant force is then proportional to the angle of the tumbling weight with respect to the vertical.

The goal of these measurement systems is to quantify the position of the body in space. The analysis which follows this quantification is either observational, that is, intuitive, or biomechanical. There is a trade-off between one activity to accurately specify position and the time and encumbrance necessary to achieve this. As a result, most workplace posture measurement systems have given way to the simplicity and lower cost of video recording.

Environment.   The elements of concern in an ergonomic analysis related to environment are:

1. Noise
2. Illumination
3. Motion and vibration
4. Climate

Noise is measured in decibels. In the decibel scale, sound pressure (p) is expressed as a fraction of a reference pressure (p_{ref}) value. Sound pressure level (SPL) in decibels = 20 log (p/p_{ref}). In the measurement of sound the two important parameters are the sound pressure level and the frequency. The sound pressure level is usually measured with the use of a sound-level meter. Sound-level meters built to American National Standards Institute (ANSI) standards possess different weighting scales for the various frequencies. These scales are designated as A, B, or C. The D scale is used primarily as a measure of aircraft noise and is not of importance here. The C scale weights all frequencies equally. The Occupational Safety and Health Administration (OSHA) standards are based primarily on the A scale. The sound levels and their relation to exposure duration for human beings are shown in Fig. 1.14.

A more detailed description of noise measurement may be found in Jones and Chapman (1984). When the sound pressure levels cannot be reduced below "safe" levels it is necessary that the operators in such environments use ear protection. The reusable types of

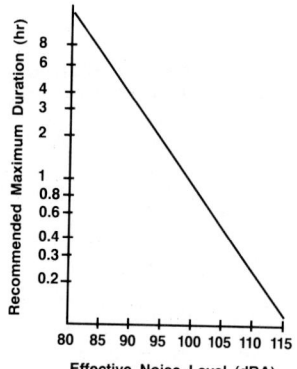

FIGURE 1.14 Noise level and maximum recommended exposure duration. (*Eastman Kodak Company, 1983.*)

devices are earplugs and earmuffs. Disposable forms such as cotton are used sometimes. Harris (1979) provides other means of reducing the environmental noise encountered in industry.

Illumination is the measure of light falling on a surface. In the SI system it is measured in lux. Illumination is also measured in footcandles (fc). Luminance is the amount of light emitted in a given direction by a luminous source or by an illuminated surface. In the SI system luminance is measured in candelas per square meter (cd/m^2). It is also expressed in footlamberts (ftL). A photometer is used to measure the amount of luminance. Many authors have reported studies where strong correlations exist between the level of lighting and operator behavior. The reader is referred to the IES Lighting Handbook (1982, 1983) and Kaufman and Haynes (1981a, 1981b) for a more comprehensive analysis.

Vibration on the human system can be whole-body or segmental (for example, hand-arm). ISO (1974, 1978, 1985) gives the effects of vibration duration on human comfort. The ISO (1979, 1984, 1988) provides guidelines and principles to measure the vibration transmitted to the hand.

The modern measurement of vibration involves the conversion of mechanical energy to electrical energy. The most common types of transducer used for vibration measurement are piezoresistive accelerometers (for whole-body vibration) and piezoelectric accelerometers (for hand-transmitted vibration). Accelerometers produce electrical signals proportional to the acceleration present.

The electrical signals from the transducer may be instantaneously processed or else stored on a tape recorder or computer for later analysis. The two common types of analysis are amplitude composition and frequency composition. Many measures have been used to quantify the amount of vibration, for example, mean, standard deviation, root-mean-square, skewness, kurtosis, root-mean-quad, vibration dose value, estimated vibration dose value, peak, crest factor, and so on. For vibrations which change with time the rms value may not be an appropriate measure. Under such circumstances the vibration dose value will be a more meaningful measure.

For whole-body vibration this measure will give a better indication of the total severity of the vibration magnitude. The vibration dose value (VDV) has been defined by

$$\frac{[T_s \Sigma x^4(i)]^{0.25}}{N}$$

where T_s = vibration duration

N = total number of samples with frequency f_s (samples per second)

$x(i)$ = sample data values where i = 1 to N

The estimated vibration dose value (*eVDV*) is defined by

$$[(1.4R)^4 \times T_s]^{0.25}$$

where

$$R = \text{root-mean-square value} = \frac{[\Sigma x^2(i)]^{0.5}}{N}$$

The simple measure when the peak value and the rms value are not appropriate is the crest factor, which is defined as

$$\text{Crest factor} = \text{peak value/rms value}$$

The crest factor provides an indication of the shape of the wave studied. For example, in sinusoidal vibration the crest factor is $(2)^{0.5}$, or 1.414.

It is important that vibration measurements be done at the human-machine interface. Measurements carried out at any other point will not represent a true picture of the value, since damping occurs with almost any surface. In an automobile seat, for example, if vibration is important, the measurement ought to be made between buttock and seat surface and not at the floor of the vehicle or at the cardiac region of the human. In every case the two important elements are mounting the transducer at the correct location and whether the characteristics of the transducer are influenced by the vibration being measured.

When measurements have to be made at such interfaces, it is necessary that the transducers be placed at the interface itself. However, with all such measurements it must be ensured that the transducer does not affect the human body or the "machine" surface characteristics. SAE (1974) has defined a device suitable for many such measurements. However, a few limitations such as the inability to mount existing accelerometer designs exist. The International Organization for Standardization (ISO) has provided many important characteristics and measurement procedures for vibration (1979, 1984, 1988). Whitham and Griffin (1977) provide a sufficient size platform for mounting using the "seat interface for transducers indicating acceleration received (SIT-BAR). This device has a contoured lower surface and a flat upper surface such that the transducer will compress similar to the human buttocks. Lawther and Griffin (1980) have described ways in which acceleration may be estimated using six translational accelerometers.

Climate.   The factors in the area of climate affecting human performance are temperature, air velocity, and humidity. The reader is referred to the numerous publications of the American Society of Heating, Refrigeration and Air Conditioning Engineers (ASHRAE) for more elaborate descriptions in this area (for example, ASHRAE, 1981).

Analysis and Evaluation.   Methodology in the area of analysis and evaluation is very limited. The effects of force and frequency on hand disorders were first quantified by Silverstein et al. in 1986. The authors concluded that a high odds ratio of cumulative trauma disorders were present with high force (in this case an adjusted hand force greater than 6 kg was considered high) and high repetition. A job with a cycle time of less than 30 seconds or a job where 50 percent of the time the operator performed the same fundamental cycle was categorized as high repetition. Drury (1987) used the measure *daily damaging wrist motions* to quantify wrist damage potential in industrial tasks. This measure was defined as the total frequency of the damaging wrist motions over one day. The damaging wrist motions were ones which exceeded 10 percent of the range of joint motion. Drury identified four zones for movement. Zone 0 was categorized as one with no exposure (up to 10 percent of range of motion), zone 1 as low exposure (10 to 25 percent of range), zone 2 as moderate exposure (25 to 50 percent of range), and zone 3 as severe exposure (more than 50 percent of range).

Both the above studies have primarily concentrated on two variables, either force and frequency or posture and frequency. However, in manual work force, frequency and posture are important. In redesign of workplaces it is necessary to eliminate all hazardous tasks. However, because of cost and technology constraints it may not be physically feasible to redesign all components of a job. In such cases a quantitative measure of value would be the force-frequency-posture index (FFPI) defined for each subtask, i (SFFPI$_i$) as

$$\text{SFFPI}_i = (\text{force} \times \text{frequency} \times \text{zone number})$$

and the FFPI index for job (JFFPI) as

$$\text{JFFPI} = \text{summation over all subtasks (SFFPI}_i)$$

The FFPI values and the biomechanical analysis can then be used to compare the job requirement specification with the operator capabilities. Redesigns of workplaces are based on this comparison such that the job demands are less than or equal to the operator capabilities.

BIOMECHANICS

Definition.   Biomechanics is an interdisciplinary field that fuses physics and engineering and medicine to analyze the forces and moments acting upon the body joints during the performance of daily activities, whether at rest or while in motion. The scientific basis for biomechanics is drawn from the disciplines of statics, dynamics, engineering anthropometry, kinesiology, and bioinstrumentation. Biomechanics is the key tool for quantitative ergonomic assessments.

Biomechanics deals with the areas of biostatics (the study of the forces acting on bodies at rest), biodynamics (the fundamental laws describing the kinematics and kinetics of body movement), and occupational biomechanics (the application of biomechanical principles toward improving working conditions). This section deals primarily with the concepts of occupational biomechanics and its role in the broader field of ergonomics. In essence, the human body is a kinematic linkage system. When a load is applied on the hands, reactive forces and moments are produced throughout the body joints in the linkage system.

Statics and Dynamics.   Mechanics is an applied science which aims to explain and predict physical phenomena and lay the foundations for most engineering applications. The basic concepts used are those of space, time, mass, and force. The concepts of space and time are used to define an event, specifying the location where and the instance when the event occurred. The concept of mass is used to characterize and compare bodies on the basis of certain fundamental mechanical experiments. The concept of force represents the action of one body on another. Statics deals with bodies at rest while dynamics deals with bodies in motion.

The forces acting on bodies may be separated into two categories: (1) external forces and (2) internal forces. External forces represent the action of other bodies on the body under consideration, either causing it to move or assuring that it remains at rest. Internal forces are those which hold the body together. Forces are vector quantities characterized by magnitude, direction, line of action, and point of application. The method of analysis is usually to draw a system of forces acting upon a body. This is commonly referred to as the free-body diagram.

The state of equilibrium of a body is determined from this free-body diagram. This means that the additive effect of all external forces acting on a body and the combined moment effects of the forces should be equal to zero.

For static conditions, the reactive force produced about a certain body joint to prevent motion can be calculated through the additive effects of the external load applied and the forces due to the weight of appropriate body segments. The reactive moment produced on the body joint can be determined in a similar manner, this time considering not only the forces but also the perpendicular distances of these forces (lever arm) from the body joint. Figure 1.15 shows a simplified example of the reactive force and moment produced about the elbow joint as a load is applied on the hands.

For dynamic conditions, inertial forces, in addition to gravity, act on a body segment as it is pivoted about a joint center. These inertial forces are the combined effects of the tangential force (tangent to the arc of motion) at the body segment's center of mass and the centrifugal force (perpendicular to the arc of motion). Aside from the mass of the body segment, the distance from the joint center to the body segment center of mass, the instantaneous angular velocity of the body segment and the instantaneous angular acceleration of the body segment need to be defined to calculate the tangential and the centrifugal forces. Please refer to Fig. 1.16 for an example of how these forces are calculated.

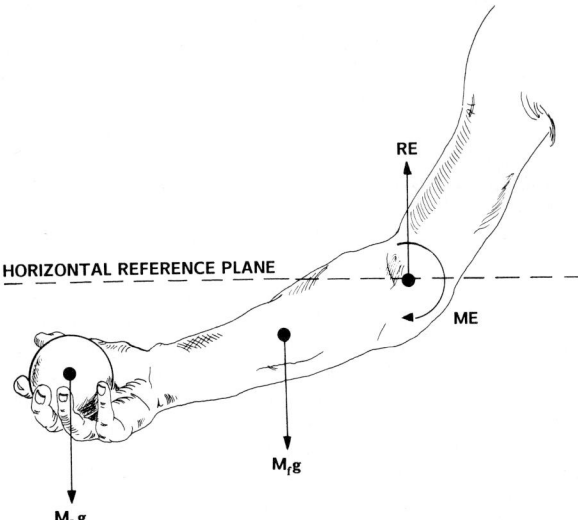

FIGURE 1.15   Reactive force and moment at the elbow joint.

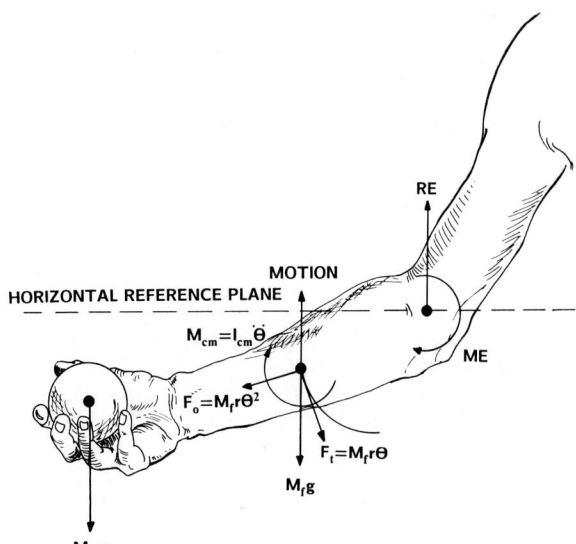

FIGURE 1.16   Calculation of tangential and centrifugal forces.

The NIOSH Work Practice Guide to Manual Lifting.   The National Institute for Occupational Safety and Health publication entitled *Work Practices Guide to Manual Lifting* (NIOSH, 1981) offers specific guidelines for evaluating human lifting capabilities intended for general industry use. The guidelines were meant to be applicable only to smooth, two-handed, symmetric lifting. It also calls for unrestricted posture, good handles on a moderately wide object, good environmental conditions, and no other significant manual materials handling tasks being performed during the course of the working day. The guide defines two limits: the action limit (AL) and the maximum permissible limit (MPL). The range between these limits is defined to consider the large variability in the general U.S. working population.

The action limit (AL) has been established as the maximum weight that produces 3400 N (approximately 750 lb) of compressive force on the L5/S1 disk, requires a metabolic load of 3.5 kcal per minute for a healthy young female, and can be sustained by over 99 percent of the males and 75 percent of the females. Epidemiological data have also indicated an increased risk of injury on lifting jobs exceeding the AL.

The maximum permissible limit (MPL) has been defined as three times the value of the action limit. Lifting at this limit produces 6400 N (approximately 1410 lb) of compressive force on the L5/S1 disk, requires a metabolic load of 5.0 kcal per minute for a healthy young male, and can be performed by only 25 percent of the males and less than 1 percent of the females. Epidemiological data also indicated that injury and severity rates are significantly higher for most workers who perform lifting jobs above the MPL.

Three ranges of stresses caused by lifting jobs have been defined using the AL and MPL. Lifting below the AL is presumed to create nominal stress and is of little risk to most people. Lifting above the MPL is considered unacceptable, and engineering controls (such as the use of hoists and similar materials handling mechanisms) should be implemented to eliminate the stresses produced by the lifting conditions. Lifting between the AL and MPL is also considered unacceptable without the implementation of engineering and/or administrative controls. Administrative controls include proper employee selection (functional capacity testing), employee training programs (proper work habits and overall fitness training), and job redesign.

The equations to determine AL and MPL (both in N) are as follows:

$$\text{AL} = 392 \times (15/H) \times [1 - (0.004 \times |V - 75|)] \times [0.7 + (7.5/D)] \times [1 - (F/F_{max})]$$

$$\text{MPL} = 3 \times \text{AL}$$

where H = the horizontal distance between the load center of mass at the origin of the lift and the midpoint between the ankles. Values of H range between 15 and 80 cm (6 to 31.5 in).

V = the vertical distance from the surface where the object to be lifted is placed to the load center of mass at the origin of lift. Values of V range from 0 to 175 cm (0 to 70 in).

D = the vertical distance of travel of the load. Values of D range from 25 to $(200 - V)$ cm $[10$ to $(80 - V)$ in]. If the vertical distance of travel is less than 25 cm, then 25 cm is used as the minimum value.

F = the average frequency of lifting. Values for F range from a minimum of 0.2 lift per minute (one lift every 5 minutes) to a maximum value F_{max} defined by the duration of the lifting task (1 hour or 8 hours). The values for F_{max} are as follows:

$$F_{max} = 12 \text{ lifts per minute if } V < 75 \text{ cm and task duration} = 8 \text{ hours}$$

$$F_{max} = 18 \text{ lifts per minute if } V > 75 \text{ cm and task duration} = 1 \text{ hour}$$

$$F_{max} = 15 \text{ lifts per minute if } V > 75 \text{ cm and task duration} = 8 \text{ hours}$$

$$\text{or if } V < 75 \text{ cm and task duration} = 1 \text{ hour}$$

It should be noted that as of this writing, the NIOSH guidelines are under review. The effects of twisting while lifting (asymmetric lifting) are currently being considered.

Computation of Dynamic Forces and Torques on the Human Body.   The NIOSH Guide applies only for calculating the spinal compressive force during symmetrical lifting in the sagittal plane. To describe and evaluate manual work, it is often necessary not only to analyze the reactive forces on the various body joints but also to compare load moments at these joints with population capabilities. In general, the forces acting on the hands are viewed as vector quantities acting separately on each hand. Biomechanical models then predict the moment load on each body joint.

Since the human body may be considered as a multilinkage system, a sequential set of analysis is performed to calculate the reactive forces and moments about a body joint. Most whole body biomechanical models divide the human body into seven links: hand and forearm, upper arm, shoulder to L5/S1 disk, shoulder to hip, L5/S1 disk to hip, thigh, and lower leg. Usually the calculations begin with the forearm-hand link and continue until reaching the stationary link (in most lifting tasks, the analysis ends with the foot). Hence the reactive force and moment about the elbow joint are added to the other forces and torques acting on the shoulder joint, and the reactive force and moment about the shoulder are added to the other forces and torques acting on the L5/S1 disk, and so on.

Recently, a two-dimensional dynamic biomechanical model (BackSoft®) to analyze the stresses imposed on the body during workplace tasks has been developed. Centers of gravity are calculated for the hand and forearm link, the upper arm, the torso, the head, the pelvis, the thighs, and the lower legs. Link lengths and center of gravity weights and locations are calculated as a percent of the height and weight of the person being evaluated. The compressive force produced on the L5/S1 disk is compared with the value of 3400 N [which is the compressive force produced on the L5/S1 disk when lifting at the action limit (AL) level] to determine whether administrative or engineering controls are required. The predicted strength of each joint is calculated as a function of the enclosed angle of the joint to determine the percent of the population capable of performing the task being analyzed.

Because of the complications of the calculations, certain assumptions were made in the calculation of body motion and torques for the BackSoft® program. These include

1. Constant angular velocity for each limb
2. Fixed center of gravity for each limb
3. Limb rotation about a joint in the third dimension is neglected
4. All forces and torques act at the joint center of rotation
5. Forces are applied by both hands
6. Both feet are flat on the ground in a similar position

The sign convention is: positive X axis to the right, positive Y axis upward, and positive moment counterclockwise. Consider a person lifting a load from the floor to a shelf. The free-body diagram of the linkage system is presented in Fig. 1.17. For purposes of simplicity, only the equations for calculating the reactive forces and moments about the elbow

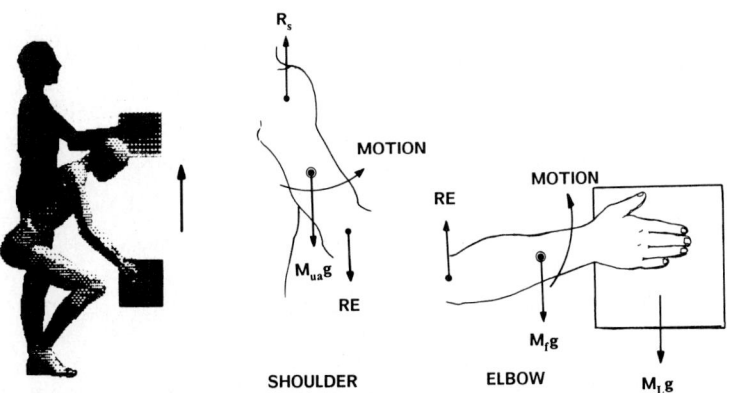

FIGURE 1.17   Free-body diagram of the linkage system.

and shoulder joints will be shown. The instantaneous reactive force on the elbow joint is computed by adding the component effects of the load and the gravity effects of the segment weights at this joint. The moment about the elbow joint is computed by adding the segment rotational moment of inertia, the linear acceleration effects, and the static effects due to gravity. The reactive forces and moment about the elbow are then calculated as follows:

$$F_{Ex} = 0$$

$$m_L a_{Lx} + m_f a_{fx} - R_{Ex} = 0$$

$$R_{Ex} = m_L a_{Lx} + m_f a_{fx}$$

$$F_{Ey} = 0$$

$$m_L g + m_L a_{Ly} + m_f a_{fy} - R_{Ey} = 0$$

$$R_{Ey} = m_L g + m_L a_{Ly} + m_f a_{fy}$$

$$M_E = 0$$

$$M_E - D_{EL} m_L g (\cos \beta_E) - D_{EL} m_L a_{Ly} (\cos \beta_E) - D_{EL} m_L a_{Lx} (\sin \beta_E)$$
$$- D_{Ef} m_f a_{fy} (\cos \beta_E) - D_{Ef} m_f a_{fx} (\sin \beta_E) - I_f \ddot{a}_f = 0$$

$$M_E = D_{EL} m_L g (\cos \beta_E) + D_{EL} m_L a_{Ly} (\cos \beta_E) + D_{EL} m_L a_{Lx} (\sin \beta_E)$$
$$+ D_{Ef} m_f a_{fy} (\cos \beta_E) + D_{Ef} m_f a_{fx} (\sin \beta_E) + I_f \ddot{a}_f$$

where R_{Ex} = force at the elbow joint in the x direction
R_{Ey} = reactive force at the elbow joint in the y direction
m_L = mass of load
g = acceleration due to gravity
a_{Lx} = instantaneous acceleration of the load in the x direction
a_{Ly} = instantaneous acceleration of the load in the y direction
m_f = mass of the forearm
a_{fx} = instantaneous acceleration of the forearm in the x direction at the center of mass
a_{fy} = instantaneous acceleration of the forearm in the y direction at the center of mass
M_E = reactive moment about the elbow joint
D_{EL} = distance from elbow joint to center of mass of the load
D_{Ef} = distance from elbow joint to center of mass of the forearm
β_E = forearm angle from the horizontal axis
I_f = moment of inertia of the forearm about the axis through the center of mass
$\ddot{a}_f$ = angular acceleration of the forearm about the elbow

The instantaneous reactive forces and moments on the shoulder joint are computed byadding the reactive forces and moment about the elbow joint and the gravity and moment of inertia effects of the upper arm at this joint. The reactive forces and moment are then calculated as follows:

$$F_{Sx} = 0$$

$$m_{ua} g + m_{ua} a_{uax} + R_{Ex} - R_{Sx} = 0$$

$$R_{Sx} = m_{ua} g + m_{ua} a_{uax} + R_{Ex}$$

$$F_{Sy} = 0$$

$$m_{ua} g + m_{ua} a_{uay} + R_{Ey} - R_{Sy} = 0$$

$$R_{Sy} = m_{ua}g + m_{ua}a_{uay} + R_{Ey}$$

$$M_S = 0$$

$$M_S - M_E - I_{ua}\ddot{a}_{ua} - D_{Sua}m_{ua}a_{uay}(\cos \beta_S) - D_{Sua}m_{ua}a_{uax}(\sin \beta_S) = 0$$

$$M_S = M_E + I_{ua}\ddot{a}_{ua} + D_{Sua}m_{ua}a_{uay}(\cos \beta_S) + D_{Sua}m_{ua}a_{uax}(\sin \beta_S)$$

where R_{Sx} = reactive force at the shoulder joint in the x direction
$\quad\;\; R_{Sy}$ = reactive force at the shoulder joint in the y direction
$\quad\;\; R_{Ex}$ = reactive force at the elbow joint in the x direction
$\quad\;\; R_{Ey}$ = reactive force at the elbow joint in the y direction
$\quad\;\; m_{ua}$ = mass of the upper arm
$\quad\;\;\;\; g$ = acceleration due to gravity
$\quad\; a_{uax}$ = instantaneous acceleration of the upper arm in the x direction at the center of mass
$\quad\; a_{uay}$ = instantaneous acceleration of the upper arm in the y direction at the center of mass
$\quad\;\; M_S$ = reactive moment about the shoulder joint
$\quad\;\; M_E$ = reactive moment about the elbow joint
$\quad\;\; I_{ua}$ = moment of inertia of the upper arm about the axis through the center of mass
$\quad\;\; \ddot{a}_{ua}$ = angular acceleration of the upper arm about the shoulder
$\quad D_{Sua}$ = distance from shoulder joint to center of mass of the upper arm
$\quad\;\; \beta_S$ = upper arm angle from the horizontal axis

The reactive forces and moments about the other body joints (for example, L5/S1 disk, hips, knees) are calculated in a similar manner. These are computed by adding the effects of the body segment mass and accelerations to the corresponding reactive forces and moments about the preceding body joint.

Reducing Musculoskeletal Stress through Workplace Redesign.   When redesigning working conditions, several factors need to be considered. These include the following:

1. Amount of force applied to manipulate the loads being handled
2. Size, shape, stability, and center of mass of the load
3. Design of handles, if any
4. Workplace layout
5. Cycle times and durations of the tasks
6. Environmental conditions
7. Personal protective devices and other work habits or methods

One of the primary factors included in the calculation of the NIOSH recommended action and maximum permissible limits is the horizontal distance between the center of mass of the load and person performing the manual materials handling task. In general, a smaller horizontal distance results in lower reactive forces and moments, particularly about the L5/S1 disk. Hence the amount of bending down or forward leaning should be minimized as much as possible. Also, any obstructions between the worker and the load should be eliminated. A common way of minimizing this horizontal distance is through the use of roller conveyors, gravity-fed slides, titled stock bins, and similar devices which bring the loads to the person performing the task instead of the person's having to reach for the load.

Another factor considered in the calculation of the NIOSH action and maximum permissible limits is the vertical location of the center of mass of the load with respect to the floor surface. Research studies have indicated that the lifting limits are 30 percent lower for lifting below the knuckle height than for lifting at knuckle height. Recently, the use of spring-loaded pallets or vertically adjusting support tables has been implemented to allow the person to maintain a more erect posture regardless of how many more loads are left on the pallet or table.

WORK METABOLISM

Principles of Energy Conversion in the Body. The *energy expenditure* of the human body can be determined from measurements on the respiratory system. The cells of the body get most of their energy from reactions that require oxygen, and finally eliminate carbon dioxide. Some basic definitions about the respiratory process follow (Astrand and Rodahl, 1986; Kroemer, Kroemer, and Kroemer-Elbert, 1986).

The *respiratory system* is made up of all the structures that move air in and out of the lungs. Oxygen in the air is taken into the bloodstream and carbon dioxide and water are removed from the body. The volume of air depends on the inspiratory and expiratory muscle activity. When the air from the lungs is forcibly expelled as much as possible, the lungs still contain a residual volume of approximately 1000 mL. A forced maximum inspiration of air adds a volume of about 5000 mL known as the vital capacity, a measure of the lung's fitness. During normal, quiet breathing a tidal volume of approximately 500 mL is moved in and out. The pulmonary ventilation is described by the minute volume and is calculated by multiplying the tidal volume by the frequency of breathing.

The *oxygen intake* V_{0_2} is the volume of oxygen absorbed per minute from the inspired air. At rest the value is about 0.2 liter/min. The heart rate is the number of times the heart's ventricles contract per minute. The resting rate is about 60 to 70 beats per minute. The pulse rate is an estimate of the heart rate.

The *metabolic processes* in the body release the energy contained in the nutrients that are taken in. The output is in the form of energy to maintain the body, as work and as waste.

The *basal metabolism* is the minimum amount of energy required for the body to function. It is measured under controlled conditions and is approximately equal to 4.9 kJ/min for a 70-kg person. The resting metabolism is measured at the beginning of the day with the subject resting. The resting metabolism is 10 to 15 percent higher than the basal metabolism. It is different from the working metabolism, which is the increase in the metabolic level between the resting level and the working level.

Metabolic Activity During the Work Cycle. Figure 1.18 shows how the oxygen uptake increases very little at the start of work. After this, the oxygen intake rises rapidly and then reaches a steady state at the level required by the body to perform the work. Thus at the start of the work the oxygen demand is more than the amount available, and an "oxygen debt" is incurred. The extent of the deficit depends on the intensity of the work. At the cessation of work, the oxygen demand drops rapidly and then tapers off gradually. The "oxygen debt" is repaid at this stage, with about double the amount as the deficit.

The reason for the slow increase in the oxygen demand at the start of work is that it takes time to increase the flow of oxygen-rich blood to the body. The metabolic processes at this stage are anaerobic, and lactic acid, a waste product, is produced; the oxygen present in the muscles is insufficient for the work level. When the work is over a prolonged period the oxygen supply matches the cardiac requirements and a steady state that can be maintained is reached. The workload at this stage should not exceed 50 percent of the maximal oxygen uptake.

When the work level requires more than half the maximal oxygen uptake, anaerobic processes are needed again. These result in lactic acid accumulation, and fatigue. To restore a balance between oxygen supply and demand, rest pauses can be inserted in the work period. Figure 1.19 shows the effect of the rest periods on the metabolic recovery.

Measurement of Energy Expenditure
Energy Expenditure Tables. Table 1.9 shows the energy expenditures for different

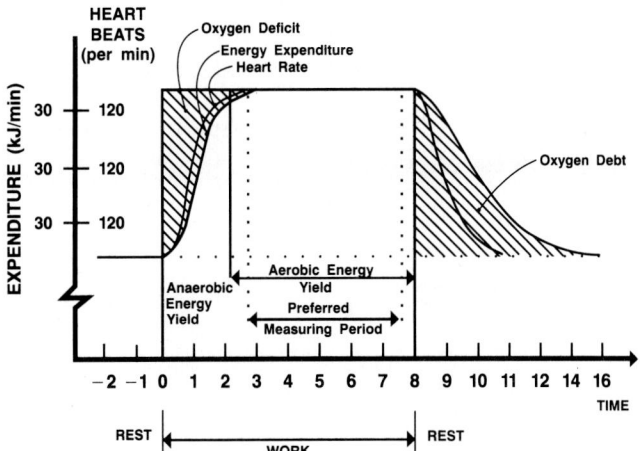

FIGURE 1.18 Energy expenditure during the work cycle. (*Kroemer et al., 1986.*)

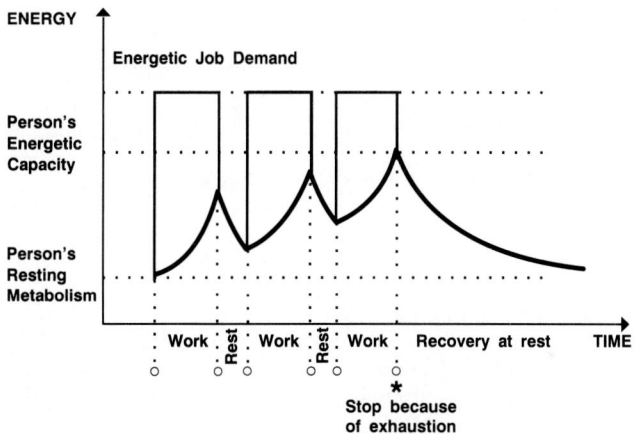

FIGURE 1.19 Effect of rest periods on metabolic recovery. (*Kroemer et al., 1986.*)

grades of work, and Table 1.10 for some selected activities. The energy determined by this method is an approximation, since the actual energy depends on the exact task, the fitness and skill of the individual, and the pace of work. The values from the table should be applied only to the time when the specific task is being performed.

Oxygen Consumption Method. The oxygen intake during work can be measured and used to estimate the energy expenditure. This method of estimation should be used when the levels of effort vary over the duration of the task. Oxygen intake is determined by measuring the volume of expired air and computing its oxygen content. The volume of expired air is multiplied by the difference in oxygen content between the expired air and the inspired air. The oxygen consumption in liters per minute is multiplied by 4.9 to give the energy in kilocalories. Except when the level of work effort is high, there is a linear relationship between oxygen consumption and energy expenditure.

TABLE 1.9   Classification of Work According to Energy Expenditure and Heart Rate

	Range for Class of Activity	
Class of Activity	Total Energy kcal/min	Heart Rate beats/min
Resting, sitting	1.5	60-70
Very light work	1.6-2.5	65-75
Light work	2.5-5.0	75-100
Moderate work	5.0-7.5	100-125
Heavy work	7.5-10.0	125-150
Very heavy work	0.0-12.5	150-180
Unduly heavy work	over 12.5	180+

Heart Rate Method.   The heart rate of an individual has a linear relationship to the oxygen intake (Kroemer et al., 1986). The heart rate can be used to estimate the workload if the relationship for an individual is known and if the general work conditions are similar. Figure 1.20 (Astrand and Rodahl, 1986) illustrates the relationship between heart rate and oxygen uptake. The measured maximal oxygen uptake is used to construct another parallel scale with the load expressed as a percentage of maximal aerobic power. The weighted mean of the heart rate is used to assess the approximate average oxygen uptake.

Analytical Methods.   Garg (1976) proposed a set of regression equations to predict energy expenditure. The equations give the energy expenditure for lifting in three different postures—standing, stooping, and squatting, and for carrying loads. A job is first divided into its elemental subtasks and their individual energy requirements computed. The total energy expenditure for the job is determined by summing up the energy expenditure for all the subtasks.

Asfour's (1980) method also uses regression to develop a set of equations that predict the energy for lifting and lowering tasks. The equations apply to the sagittal plane and to twisting movements. Intaranont's (1983) regression model predicts the threshold for anaerobic work and capacity for lifting weights.

Psychophysical Methods.   Psychophysical methods of measuring energy expenditure depend on one's perception of task effort. These are based on models of the relationship between a physical stimulus and the sensation of the stimulus. They were originally formulated by Weber (1834) and Fechner (1860). Weber's law states that the smallest change, or the just noticeable difference, that can be detected in a stimulus depends on the absolute magnitude of the stimulus. Fechner (1860) proposed a logarithmic relationship between the perceived magnitude of a stimulus and its absolute value.

Borg (1982) extended these laws to develop a scale of ratings of perceived exertion (RPE). The scale is shown in Table 1.11. The subject's perception of the intensity of the

TABLE 1.10 Average Energy Cost in kcal/min for Selected Activities

Body Position and Activity	Total Energy Cost (Kcal/min*) Typical	Range
1. Heavy activity at fast to maximum pace.		10.0-20.0
2. Jogging, level, 4.5 mph	7.5	
3. Lifting, 44 lb., 10 cycles per minute:		
-floor to waist	8.2	
-floor to shoulder	10.8	
4. Reclining at rest	1.3	
5. Running, level, 7.5 mph	12.7	
6. Shovelling, 18 lb load 1 yd with 1 yd lift, 10 times per min.	8.0	
7. Sitting at ease:		
-light hand work (writing,typing)	1.7	1.6-1.8
-moderate hand and and arm work (drafting, light drill press, light assembly, tailoring)		
-light arm and leg work (driving car on open road, machine sewing)	2.8	2.5-3.2
-heavy hand and arm work (nailing, shaping stones, filing)	3.5	3.0-4.0
-moderate arm and leg work (local driving of truck or bus)	3.6	3.0-4.0
8. Standing at ease:		
-moderate trunk and arm work (nailing, filing, ironing)	3.7	3.0-4.0
-heavy arm and trunk work (hand sewing, chiselling)	6.0	4.0-8.0
9. Walking, casual (foreman, lecturing)	3.0	2.5-3.5
-moderate arm work (stock room, sweeping)	4.5	4.0-5.0
-carrying heavy loads or with heavy arm movements (carrying suitcases, scything, hand lawn mowing)	7.0	6.0-8.0
-transferring 35 lb sheets 2 yd at trunk level, 3 times per min	3.7	
-pushing wheelbarrow, level, with 220 lb load	5.5	5.0-6.0
-level: 2 mph	3.2	
3 mph	4.0	
4 mph	5.9	
-up: 5° grade at 3 mph	8.5	
mailman climbing stairs	12.0	
-down: 5° grade at 3 mph	3.4	

*Multiply by 4 to obtain values in Btu/min.
Source: AIHA Technical Committee on Ergonomics, 1971.

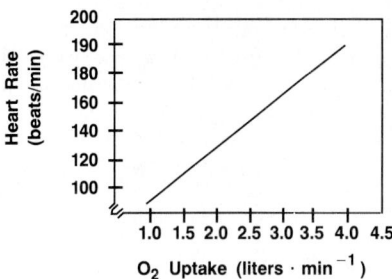

FIGURE 1.20 Relationship between heart rate and oxygen uptake. (*Astrand and Rodahl, 1986.*)

TABLE 1.11 The Borg RPE Scale

6	No exertion at all
7	Extremely light
8	
9	Very light
10	
11	Light
12	
13	Somewhat hard
14	
15	Hard (heavy)
16	
17	Very Hard
18	
19	Extremely hard
20	Maximal exertion

Source: Borg, G. A. V., 1962.

work, in RPE, is obtained using the verbal descriptions on the scale. The heart rate is 10 times the RPE.

Computation of Rest Pauses. Muller (1953) recommended rest pauses for carrying loads, carriage pulling, climbing stairs, walking, and bicycling. The net energy expenditure rate that could be performed over a day was taken as 4 kcal/min. The net energy expenditure was linearly related to the percentage of recovery time to working time. Thus for a task of duration 10 min and requiring 8 kcal/min, the rest time was shown to be 10 min.

Spitzer (1952) determined the rest pause as

$$R = \frac{M - 1}{4} \times 100$$

where R = resting time as a percentage of working time
M = net energy expenditure
= total energy cost − resting energy cost

Murrell (1964) determined the rest pause as

$$R = \frac{T \times (M - 4)}{M - 1.5}$$

where R = resting time, min
T = total working time, min
M = net energy cost, kcal/min

Mital and Shell (1985) developed a comprehensive energy model to determine rest allowances. The model is based on the worker's physical condition (age, weight, hours of sleep, fitness), the shift duration, the number of tasks performed during the shift, and the metabolic energy requirements for each task, the worker's aerobic capacity, and the energy requirement when not working. A program in BASIC to perform the computations is listed in Ayoub and Mital (1989).

WORKPLACE REDESIGN

Case Study 1: Wrist Intensive Assembly Line Operation
Problem. Over a 3-year period, reported incidences of carpal tunnel syndrome had been continually increasing. An analysis of injury data indicated that employees performing three repetitive assembly operations seemed to be particularly susceptible to these disorders and to other strains, sprains, inflammations, and muscle injuries. After consultation with the company's ergonomics task force, it was determined that a quantitative ergonomics assessment would be performed to identify potential causes of musculoskeletal injury and to develop recommendations for improving employee comfort, safety, and productivity during task performance.

Quantitative Assessment Methods.   A total of 27 employees (12 female, 15 male) at the plant were assessed on-site as they performed their normal job activities. Several data types were acquired, including anthropometric characteristics of the subject population, physical requirements for task performance, job characteristics (including the number and nature of tasks performed, rotation, schedules, and shifts), work method characteristics, workstation characteristics (including dimensions and constraining factors), tool characteristics (including weight, handle size, handle shape and configuration, and center of mass), throughput and production quality requirements and environmental factors. An assessment of employee ratings of task difficulty using the Borg scale was utilized as a comparison with the quantitative findings.

Results.   Owing to the nature of the tasks, the data analysis focused on the stressors place upon the shoulders, arms, and wrist during task performance. Computer analysis and digitization techniques were used to quantify the working postures, forces exertions, reach requirements, heart rate, and repetitive motions during task performance. These analyses provided an understanding of the percentage of time during which the employees maintained deviated or flexed wrist postures and to what degree. The type of grasp generally employed while gripping the tools was also determined. Figure 1.21 shows an example of the results of the postural analysis for one of the tasks evaluated.

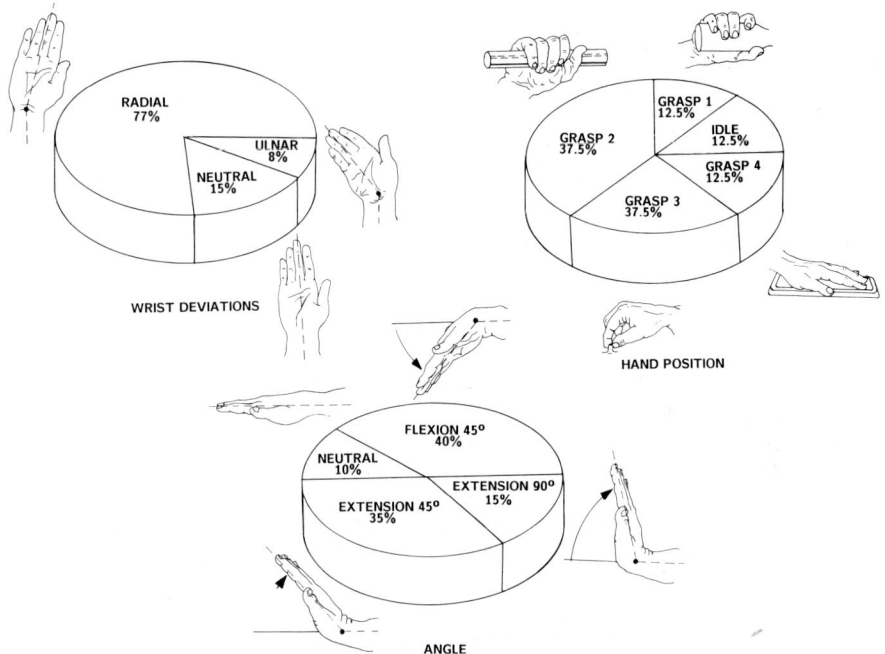

Analysis of Wrist and Hand Position

FIGURE 1.21   Postural analysis results.

An estimation of wrist injury likelihood was made by comparing the force and repetition characteristics of each task with the odds ratio suggested by Armstrong, Fine, and Silverstein (1985). According to this method, tasks which require high repetition and high force have a "high and significant" contribution to wrist injury likelihood. Tasks which require low repetition or low force have a "low and significant" contribution to wrist injury. The potential for wrist injury is generally considered "negligible" if both the force and repetition are low.

In this case, the analysis showed that the comfort and safety of the employees could be improved by modifying the design of one hand tool to improve the postural requirements to perform the task. The before and after postures required for task performance are shown in Fig. 1.22.

**Posture Before
Tool Redesign**

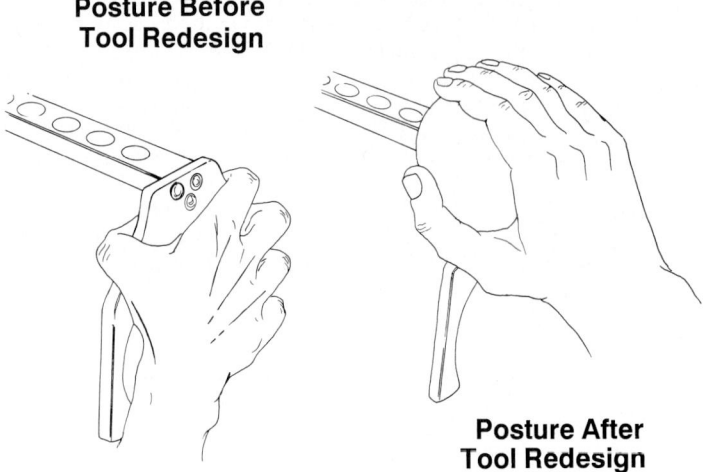

**Posture After
Tool Redesign**

FIGURE 1.22   Postures before and after ergonomic design solution.

The procedure which was followed during the implementation of this engineering design recommendation is shown in Table 1.12. Employees who utilize the new tool receive instruction in its proper use so as to minimize musculoskeletal stress. Additional recommendations included the implementation of a job rotation schedule such that approximately four different tasks are performed by each employee during a given shift.

TABLE 1.12   Ergonomic Solution Implementation Procedure

- Development of prototype modified tool.

- Testing of prototype on small number of experience operators.

- Follow-up evaluation of employee subjective rating of comfort after using the prototype for one month.

- Refine and approve new design based on user feedback.

- Manufacture modified tool in necessary quantities.

Case Study 2: Materials Handling Operation
Problem.   Increasing incidences of low back discomfort, pain, and injury in employees prompted an ergonomic assessment of the shipping dock loading job. The magnitude of the problem was further corroborated by the results of the musculoskeletal questionnaire completed during the assessment, which indicated that 80 percent of the shipping dock loaders had experienced low back discomfort during the preceding 12 months and 31 percent had been prevented from doing normal work by this discomfort.

An additional factor which had to be considered in the solution developed for reducing employee musculoskeletal stress and injury was that independent truckers were commonly used in this operation. This meant that the solution had to be mobile and located at the shipping dock as opposed to on each individual truck.

Quantitative Assessment Methods.   A total of 10 male shipping dock loaders were assessed on-site as they performed their normal task of loading packages from a pallet to a trailer truck. The data acquired were similar to those described in Case 1. The most informative data for a materials handling task such as this are the videotape recordings which provide information on the angles of each body joint during task performance, the physiological data which are used to classify the difficulty of the work, and information on the physical characteristics of the workstation and the materials being transferred.

Results.   The data analysis in this study focused on determining the musculoskeletal joint forces and torques on the elbow, shoulder, hip, knee, and L5/S1 spinal disk. Required strengths for the performance of the materials handling tasks were then compared with anthropometric data to determine the percentage of the normal population who would have the necessary strength to perform the tasks. The results indicated that the limiting factor for this operation was shoulder strength, with the required amount being possessed by only 38.1 percent of the U.S. male population. The spinal compressive forces upon the L5/S1 disk during this task were also found to exceed the current acceptable limit as specified by the NIOSH action limit. A summary of the data is given in Table 1.13.

TABLE 1.13   Sample Task Analysis Results

Task: Shipping Dock Loader
Working Heart Rate
Mean: 97.6 bpm S.D.: 6.6 Range: 91–105 bpm Classification: Heavy Work
Borg Rating of Perceived Difficulty
Mean: 13.8 S.D.: 1.2 Range: 12–15 Classification: Somewhat difficult to difficult
Maximum Compressive Force on L5/S1 disc = 1100 lb. (5057 N)[*] [*]exceeds NIOSH Action Limit (AL) of 740 lb. (3400 N)
% of Population with Necessary Elbow Strength to Perform Task = 38%

To reduce the musculoskeletal stress resulting from manual trailer loading into independent trucks, a mobile conveyor and package placement system was recommended. This system utilizes an overhead vacuum lifting mechanism to load the required packages from the pallet to the mobile conveyor system, which is then driven into the trailer for unloading. In addition to the mechanization of this previously manual task, training in proper lifting techniques and the use of materials handling equipment was also recommended for picking of special orders from the storage rack area.

The Cost Benefits of an Ergonomics Program

The benefits of a sound biomechanically based ergonomics program are myriad. Productivity is increased, product quality is improved, musculoskeletal injuries are removed, and costs are reduced. All these results are laudable. But for the sector of the business community that is bottom-line-oriented, a question always arises as to the special manner in which costs are reduced. The cost benefits of an ergonomics program are achieved by savings in both direct and indirect costs relating to employee injuries.

The direct costs include:

- Wages paid to injured employees
- Medical expenses for treatment of injured employees
- Rehabilitation costs
- Workmen's compensation and disability insurance costs
- Lost time

The indirect costs include:

- Reductions in productivity
- Cost of replacement
- Retraining costs
- Litigation arising from employee injuries
- Supervisory costs in assisting, investigating, and reporting injuries
- Administrative costs in processing injury claims
- In-plant medical and rehabilitation costs
- Wages paid to other workers during time their work was interrupted due to injury of their coworker
- Cost of spare capacity for emergencies
- Employee turnover
- Employee absenteeism
- Reduction in employee morale

The indirect costs of injury are difficult to measure in dollars. A rule of thumb often used is that indirect costs are approximately four to eight times direct costs. Therefore, if direct costs are $1 million, the indirect costs would be a minimum of $4 million and the total cost would be $5 million.

The cost benefits of a sound, biomechanically based ergonomics program are not theoretical. They are actual if the program effort and quality are sustained. A major U.S. chemical company recently implemented such a program to detect, evaluate, and correct ergonomic problems. The focus of this program was on back injuries. The preliminary results were dramatic. The company's total recordable injuries related to on-the-job back problems were reduced by more than 50 percent. The estimated cost savings for the first year exceeded $10 million. Now this is a major cost benefit.

The cost benefits that can be achieved as the result of a sound, biomechanically based ergonomics program are substantial. The cost benefit achievable in a single year can pay for the cost of the program many times over, and the cost savings may be like an annuity, paying off in cost savings for many years to come.

Acknowledgment   The authors thank A. Fein for his contribution to the cost-benefit section.

REFERENCES

AIHA Technical Committee on Ergonomics, "Ergonomics Guide to Assessment of Metabolic and Cardiac Costs of Physical Work," *AIHA Journal,* vol. 32, no. 8, pp. 560–564, 1971.

Armstrong, T. J., J. A. Foulke, B. S. Joseph, and S. A. Goldstein, "Investigation of Cumulative Trauma Disorders in a Poultry Processing Plant," *American Industrial Hygiene Association Journal,* vol. 43, pp. 103–115, 1982.

Armstrong, T. J., L. J. Fine, and B. A. Silverstein, *Occupational Risk Factors: Cumulative Trauma Disorders of the Hand and Wrist,* Final Report on Contract #200-82-2507, 1985.

Armstrong, T. J., and Y. Lifshitz, "Ergonomic Intervention Strategies for Wrist Injury Prevention," in American Conference of Governmental Industrial Hygienists, eds., *Ergonomic Interventions to Prevent Musculoskeletal Injuries in Industry,* MI: Lewis Publishers, pp. 73–75, 1987.

Asfour, S. S., Energy Cost Prediction Models for Manual Lifting and Lowering Tasks, Doctoral dissertation, Texas Tech University, 1980.

ASHRAE, Thermal Environmental Conditions for Human Occupancy (ASHRAE Standard 55-1981), American Society of Heating, Refrigeration and Air Conditioning Engineers, Atlanta, Ga., 1981. The standard for indoor thermal comfort.

Astrand, P. O., and K. Rodahl, *Textbook of Work Physiology,* McGraw-Hill, New York, 1986.

Autoweek, Should Your Car Seat Fit as Well as a Fine Suit? Mar. 12, 1990.

Ayoub, M. M., and A. Mital, *Manual Material Handling,* Taylor & Francis, London, 1989.

"Back Injuries During Lifting," Department of Labor, *Bulletin* 2144, 1982.

Barkla, D, "The Estimation of Body Measurements of the British Population in Relation to Seat Design," *Ergonomics,* vol. 21, pp. 123–132, 1961.

Bigland, B., and O. C. J. Lippold, "The Relation between Force, Velocity and Integrated Electrical Activity in Human Muscles," *Journal of Physiology,* vol. 123, pp. 214–224, 1954.

Borg, G. A. V., *Physical Performance and Perceived Exertion,* Gleerups, Lund, 1962.

Borg, G. A. V., "Physiological Bases of Perceived Exertion," *Medicine and Science in Sports and Exercise,* vol. 14, no. 5, pp. 377–381, 1982.

Bouisset, S., "EMG and Muscle Force in Normal Motor Activities," in J. E. Desmedt, ed., *New Developments in Electromyography and Clinical Neurophysiology,* vol. 1, Karger, Basel, pp. 547–583, 1973.

Brown, M. C., and J. M. Dwyer, "Repetition Strain Injury," *Proceedings of the Conference on Repetition,* New South Wales, Australia, 1984.

Bureau of National Affairs, "Job Safety: New Approaches and Heightened Expectations as 1990s Begin," *Special Report,* DER no. 12, 1990.

Burton, A. K., "Measurement of Regional Lumbar Sagittal Mobility and Posture by Means of a Flexible Curve," in N. Corlett, J. Wilson, and I. Manenica, eds., *The Ergonomics of Working Postures,* Taylor & Francis, London, 1986.

Chad, E. Y., "Experimental Methods for Biomechanical Measurements of Joint Kinematics, *CRC Handbook of Engineering in Medicine and Biology,* 385–411.

Chaffin, D., "Localized Muscle Fatigue—Definition and Measurement," *Journal of Occupational Medicine,* vol. 15.4, pp. 346–354, 1973.

Drury, C. G., "A Biomechanical Evaluation of the Repetitive Motion Injury Potential for Industrial Jobs," *Seminars in Occupational Medicine,* vol. 2, no. 1, March 1987.

Eastman Kodak Company, Human Factors Section, *Ergonomic Design for People at Work,* vol. 1, Van Nostrand Reinhold, New York, 1983.

Fechner, G. T., *Sachen der Psychophysik,* Breitkopf and Hertel, Leipzig, 1860.

Fraser, T. M., *The Worker at Work,* Taylor & Francis, London, 1989.

Garg, A., A Metabolic Rate Prediction Model for Manual Material Handling Jobs, Doctoral Dissertation, University of Michigan, 1976.

Garret, J. W., "The Adult Human Hand: Some Anthropometric and Biomechanical Considerations," *Human Factors,* vol. 13, pp. 117–131, 1971.

Gilad, I., D. B. Chaffin, M. Redfern, and S. N. Byun, "System Comparison of Two Advanced Methods for Measuring Angular Displacement of Torso," in F. Aghazadeh, ed., *Trends in Ergonomics/Human Factors* V, Elsevier, New York, pp. 847–856, 1988.

Harber, P., E. Billet, M. Gutowski, K. SooHoo, M. Lew, and A. Roman, "Occupational Lowback Pain in Hospital Nurses," *Journal of Occupational Medicine,* vol. 27, pp. 518–524, 1985.

Harris, C. M., *Handbook of Noise Control,* McGraw-Hill, New York, 1979.

Hildebrandt, V., "A Review of Epidemiological Research on Risk Factors of Low Back Pain," in P. Buckle, ed., *Musculoskeletal Disorders at Work,* Taylor & Francis, London, pp. 9–16, 1987.

Hult, L., "Cervical, Dorsal and Lumbar Spinal Syndromes," *Acta Orthopedical Scandinavical,* suppl. 17.

IES (Illuminating Engineering Society of America) Industrial Lighting Committee, Proposed American National Standard Practice for Industrial Lighting, *Lighting Design and Application,* vol 13, no. 7, pp. 29–68, 1983.

IES (Illuminating Engineering Society of America) Industrial Lighting Committee, Proposed American National Standard Practice for Office Lighting, *Lighting Design and Application,* vol. 12, no. 4, pp. 27–60, 1982.

Imai, M, *Kaizen, the Key to Japan's Competitive Success,* Random House, New York, 1986.

Intaranont, K., A Study of Anaerobic Threshold for Lifting Tasks, Doctoral Dissertation, Texas Tech University, 1983.

ISO, Guide for the Evaluation of Human Exposure to Whole-Body Vibration, ISO 2631 (E), International Organization for Standardization, Geneva, 1974.

ISO, Guide for the Evaluation of Human Exposure to Whole-Body Mechanical Vibration and Shock, ISO 2631, International Organization for Standardization, Geneva, 1978.

ISO, Principles for the Measurement and the Evaluation of Human Exposure to Vibration Transmitted to the Hand, DIS 5349.1, International Organization for Standardization, Geneva, 1979.

ISO, Guidelines for the Measurement and the Assessment of Human Exposure to Hand-Transmitted Vibration, DIS 5349.2, International Organization for Standardization, Geneva, 1984.

ISO, Evaluation of Human Exposure to Whole-Body Vibration, Part 1: General Requirements, ISO 2631/1-1985, International Organization for Standardization, Geneva, 1985.

ISO, Hand-Held Power Tools—Measurement of Vibrations at the Handle, Part 1: General, ISO 8662/1, International Organization for Standardization, Geneva, 1988.

Jensen, R. C., "Events That Trigger Disabling Back Pains among Nurses," *Proceedings of the 29th Annual Meeting of the Human Factors Society,* pp. 799–801, 1985.

Jensen, R. C., "Work Related Back Injuries among Nursing Personnel in New York," *Proceedings of the 29th Annual Meeting of the Human Factors Society,* pp. 244–248, 1986.

Johnston, R. C., and G. L. Smidt, "Measurement of Hip-Joint Motion during Walking," *The Journal of Bone and Joint Surgery,* vol. 51-A, no. 6, pp. 1083–1094, 1969.

Jones, D. M., and A. J. Chapman, *Noise and Society,* Wiley, Great Britain, 1984.

Kaufman, J. E., and H. Haynes, *IES Lighting Handbook,* reference volume, Illuminating Engineering Society of America, New York, 1981a.

Kaufman, J. E., and H. Haynes, *IES Lighting Handbook,* applications volume, Illuminating Engineering Society of America, New York, 1981b.

Kettelkamp, D. B., R. J. Johnson, G. L. Smidt, E. Y. S. Chao, and M. Walker, "An Electrogoniometer Study of Knee Motion in Normal Gait," *The Journal of Bone and Joint Surgery,* vol. 52-A, no. 4, pp. 775–790, 1970.

Khalil, T. M., S. S. Asfour, and E. A. Moty, "Case Studies in Low Back Pain," *Proceedings of the 28th Annual Meeting of the Human Factors Society,* pp. 465–470, 1984.

Kivi, P., "Rheumatic Disorders of the Upper Limbs Associated with Repetitive Occupational Tasks in Finland in 1975–1979," *Scandinavian Journal of Work, Environment, and Health,* 5, Supplement 3, 1984.

Klein, B. P., M. A. Roger, R. C. Jensen, and L. M. Sanderson, "Assessment of Workers' Compensation Claims for Back Sprain/Strains," *Journal of Occupational Medicine,* vol. 27, pp. 443–448, 1984.

Kroemer, K. H. E., H. J. Kroemer, and H. E. Kroemer-Elbert, *Engineering Physiology: Physiologic Bases of Human Factors Ergonomics,* vol. 4 in G. Salvendy, ed., *Advances in Human Factors/Ergonomics,* Elsevier, Amsterdam, 1986.

Lawther, A., and M. J. Griffin, "Measurement of Ship Motion," in D. J. Oborne and J. A. Lewis, eds., *Human Factors in Transport Research,* vol. 2, *User Factors: Comfort, The Environment and Behavior,* Academic Press, London, pp. 131–139, 1980.

Mandell, P., M. H. Lipton, J. Bernstein, G. J. Kucera, and J. A. Kampner, *Low Back Pain,* Slack International, New Jersey, 1989.

Mital, A., and R. L. Shell, "A Comprehensive Metabolic Energy Model for Determining Rest Allowance for Physical Tasks," *Journal of Methods Time Measurement,* vol. 9, pp. 2–8, 1985.

Morgan, R. P., and A. Bhattacharya, "Force Platform Technique for the Quantification of Postural Change Induced by Alcohol Indigestion," in A. Mital, ed., *Trends in Ergonomics/Human Factors* I, Elsevier, New York, 1984.

Muller, E. A., "Physiological Basis of Rest Pauses in Heavy Work," *Quarterly Journal of Experimental Physiology,* vol. 38, pp. 25–215, 1953.

NIOSH, *A Work Practices Guide for Manual Lifting,* Technical Report No. 81-122, U.S. Department of Health and Human Services, National Institute for Occupational Safety and Health, Cincinnati, Ohio, 1981.

NIOSH, Criteria for Recommended Standard–Occupational Exposure to Noise, U.S. Department of Health, Education and Welfare, National Institute for Occupational Safety and Health, Cincinnati, 1972.

Nordin, M., G. Hultman, R. Philipsson, A. Ortelius, and G. B. J. Andersson, "Dynamic Measurements of Trunk Movements during Work Tasks," in N. Corlett, J. Wilson, and I. Manenica, eds., *The Ergonomics of Working Postures,* Taylor & Francis, London, 1986.

Nordin, M., Methods for Studying Workload, with Special Reference to the Lumbar Spine, thesis, Goteborg, 1982.

O'Brien, C., and M. G. A. Paradise, "The Development of a Portable Non-invasive System for Analyzing Human Movement," *Proceedings of the Sixth Congress of the International Ergonomics Association,* Maryland, pp 390–392, July 1976.

Petrofsky, J. S., "Computer Analysis of the Surface EMG during Isometric Exercise," *Computers in Biology and Medicine,* vol. 45, pp. 83–95, 1979.

Pheasant, S., *Bodyspace: Anthropometry, Ergonomics, and Design,* Taylor & Francis, London, 1986.

Pinkham, J., "CTS Impacts Thousands and Costs are Skyrocketing," *Occupational Health and Safety,* August, 1988.

Rohmert, W., *Die Beziehung zwischen Kraft und Ausdauer bei Statischer Muskelarbeit,* Schrifteneihe Arbeitsmedizin Sozialmedizin Arbeitshygiene, Band 22, p. 118. A. W. Gentner Verlag, Stuttgart, 1968.

Schoenmarklin, R. W., and W. S. Marras, "Validation of a Hand/Wrist Electromechanical Goniometer," *Proceedings of the Human Factors Society 33rd Annual Meeting,* pp. 718–722, 1989.

Sheng, W., and C. M. Gross, Carpal Tunnel Syndrome: A Review of the Literature, 1988.

Silverstein, B., L. Fine, T. Armstrong, B. Joseph, B. Buchholz, and M. Robertson, "Cumulative Trauma Disorders of the Hand and Wrist in Industry," in N. Corlett, J. Wilson, and I. Manenica, eds., *The Ergonomics of Working Postures,* Taylor & Francis, London, 1986.

Snook, S. H., "The Design of Manual Handling Tasks," *Ergonomics,* vol. 21, pp. 963–985, 1978.

Society of Automotive Engineers, "Measurement of Whole Body Vibration of the Seated Operator of Agricultural Equipment (the Society of Automotive Engineers recommended practice)," *Society of Automotive Engineers SAE J1013 Handbook,* Part II, pp. 1404–1407, Society of Automotive Engineers, Detroit, Mich., 1974.

Spitzer, H., *Physiologische Grundlagen für den Erholungszuschlag bei Schwerarbeit,* REFA-Nachrichten, Darmstadt, 1952.

Svensson, H. O., and G. B. H. Andersson, "Low-Back Pain in 40–47 Year Old Men: Work History and Work Environment Factors, *Spine,* vol. 3, pp. 272–276, 1983.

Weber, J., A. van der Star, and C. J. Snijders, "Development and Evaluation of a New Instrument for the Measurement of Work Postures; in Particular the Inclination of the Head and the Spinal Column," in N. Corlett, J. Wilson, and I. Manenica, eds., *The Ergonomics of Working Postures,* Taylor & Francis, London, 1986.

Weber, E. H., De pulse, resorptione, auditu et tacticu, Kochler, Leipzig, 1834.

Whitham, E. M., and M. J. Griffin, "Measuring Vibration on Soft Seats," Society of Automotive Engineering Paper no. SAE 770253, International Automotive Engineering Congress and Exposition, Detroit, 1977.

Wilson, J. R., and N. C. Corlett, eds., *Evaluation of Human Work, a Practical Ergonomics Methodology,* Taylor & Francis, London, 1990.

Winter, D. A., G. Rau, and R. Kadefors, "Units, Terms and Standards in the Reporting of Electromyographical Research," in 4th Congress of the International Society of Electromyographical Kinesiology, Boston, 1979.

CHAPTER 2
A BIOMECHANICAL APPROACH TO ERGONOMIC PRODUCT DESIGN

C. M. Gross
Chief Executive Officer
Biomechanics Corporation of America
Melville, New York

Ergonomics is the science which combines medical and engineering measurements to solve problems at the interface of people and machines. In product design, ergonomics has come to mean harmony—the harmonic interaction, compatibility, or "fit" between people and products. Harmonious accommodation and fit are integral to product quality, safety, and ease of use. A good product or tool extends or enhances the capabilities of the human body—for example, a surgeon's scalpel, an artist's pen; ergonomic soundness maximizes the comfort, safety, and efficiency of its use.

An ergonomically sound product or tool provides the user with improved functionality and perceived value, resulting in higher levels of customer satisfaction and product demand, ultimately providing the producers a competitive market advantage. Thus ergonomic product design is becoming a significant economic issue in the sense that better-quality products can be instrumental in increasing a company's market share and, when designed efficiently, can ultimately impact a country's global market share.

There is more to designing ergonomic products than meets the eye. Scientific measurement of ergonomic fit is as important as the aesthetic and marketing considerations. Ergonomics concerns itself with creating a "transparent interface" between user and product, whereby the physical responses of the person to using the product can be measured and predicted. Creating the transparent interface is fundamental to the product development cycle and precedes the final industrial design of the product. Today there is technology for measuring and testing ergonomic fit that is more cost-effective than multiple prototypes and focus groups.

While the science of ergonomics uses both qualitative and quantitative research, this chapter focuses on biomechanical testing and modeling techniques which are used to quantitatively measure ergonomic fit, and presents one case study highlighting the application of some of these techniques. The way in which these techniques fit into the product development cycle is illustrated in Table 2.1.

TABLE 2.1 Ergonomic Product Development

1. Ergonomic design specification and development of prototypes	Define preliminary ergonomic guidelines to optimize product design for the intended uses and anticipated user population(s).
	Develop recommendations for embodying the guidelines in a cost-effective, practical product design; provide alternatives.
	Model on CAD system; perform human computer-aided design.
	Develop prototypes.
2. Establishment and testing of experimental design	Identify typical use scenarios for product class.
	Set up experimental station to simulate most common working postures during product use.
	Define hypothesis to be tested (characteristics of product, user and use situation likely to contribute to user perception of comfort, and actual physical and physiologic requirements of product use).
	Develop specific study protocol and data collection criteria.
	Identify required subject characteristics.
3. Biomechanical assessment of product use	Perform controlled subject trials of product.
	Acquire a variety of subjective and objective data.
	Reduce and analyze data to develop a clear understanding of the effect of various tool characteristics on user perception of comfort and physical and physiologic requirements during use.
	Document study methodologies and findings.
	Make recommendations for product improvement based on refined ergonomics guidelines.
	Repeat step 1.

BIOMECHANICAL ASSESSMENT

Biomechanics is the combination of medicine and engineering for studying the forces on the human body. Biomechanical tools are used to determine the physical response of the individual to using the product. The quantitative toolbox consists of an array of biomechanical and physiological monitoring techniques.

The major functional categories for the biomechanical assessment of products are

1. Quantifying motion
2. Measuring body force
3. Quantifying body stress
4. Measuring accommodation and fit
5. Quantifying fatigue
6. Evaluating comfort

For each of these measurement categories, specific tools have been developed (Table 2.2). A better understanding of the tools will help elucidate the ergonomic process in product design.

Quantifying Human Motion
Digital Video. The quantification of human motion is usually performed using digital video. This is a technique whereby the spatial location of joint centers may be extracted from multiple video images to facilitate the measurement of limb motion.

TABLE 2.2 Ergonomic Toolbox

Assessment category	Assessment tool
1. Motion	Digital video
	3D sonar
	3D infrared imagery
	Electrogoniometers
2. Force	Load cells
	Force-sensitive resistors
	Triaxial accelerometers
3. Stress	2D and 3D dynamic and static biomechanical models
	Digital electromyography
4. Accommodation and fit	Global ergonomic database (for example, Ergobase™)
	Human, computer-aided design (for example, Mannequin™)
5. Fatigue	Oxygen consumption meter
	Heart rate telemetry
	Ratings of perceived exertion (RPE)
	Metabolic assessment programs
6. Comfort	Questionnaire assessment of product characteristics
	Ranking questionnaires
	Statistical analysis relating comfort to product and user characteristics

3D Sonar and Infrared Imagery.   Two newer techniques allow for the rapid measurement of three-dimensional body trajectories under a wide variety of product-use scenarios. Three-dimensional infrared tracking is especially good for quantifying rapid motions, for example, analyzing the motion of a golf swing to improve the design of a club. Three-dimensional sonar digitization offers an advantage in measuring product-person interaction, by allowing the ergonomist to digitize the object's geometry and by combining this digitized object with the trajectory of body joints in three dimensions. This is done in real time in a CAD environment. The selection of motion measurement technique requires care, as the time and expense required to make these measurements are proportional to the detail and accuracy required of the data.

Electrogoniometers.   An electrogoniometer may be used to measure body joint motion. When placed over the axis of rotation, this type of electropotentiometer uses a charging voltage to continually measure joint angle changes. The signal produced may be charted or entered into a computer for processing and recording motion during a particular activity. While relatively inexpensive and easy to use, electrogoniometry has limitations in that each joint must be measured separately and the instrumentation may interfere with the performance of normal job tasks.

Measuring Body Force

Traditional Load Cells.   Following the measurement of human motion, it is necessary to quantify external forces acting on the body. These forces may arise from the weight of objects, by the use of a handle, the vibration of objects which is transmitted to the hand, the hand forces generated during the activation of tools and products, or by skin contact with supporting surfaces such as the floor, a chair, and the like.

Traditional load cells measure forces applied to inanimate objects. When you apply this technique to measure the loads on flesh, it deforms because of its viscoelastic nature. Therefore, traditional force measurement techniques offer little assistance in the quantification of forces acting on the human body.

Force-Sensitive Resistors.   To measure the force that a product places on the body, it is necessary to have a flexible transducer—a transducer which deforms upon contact with skin and yet accurately measures the applied forces (Fig. 2.1). These force-sensing resistors mounted in flexible mylar packages allow for the transparent measurement of forces to the hand and other areas of the body.

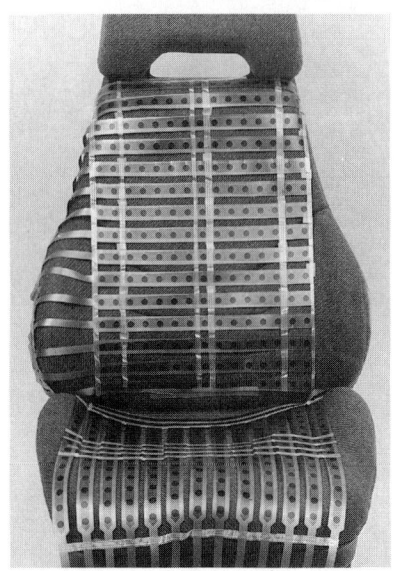

FIGURE 2.1   Force-sensitive resistors.

Triaxial Accelerometers.   In the development and assessment of certain products, particularly hand tools, it is necessary to quantify the levels of vibration transmitted to the hand during product use. Vibration is a vector quantity possessing both magnitude and direction. Obtaining comparative measures requires that the measuring transducers be placed at fixed locations. The International Standards Organization (ISO) specifies the hand-arm coordinate system used which is defined at the third metacarpal of the hand. Since it is not always possible to mount a transducer directly at this point on the hand, the transducer is often mounted on the tool handle, at the point where the hand grips the tool handle. During the collection of vibration data it is important to monitor the signal for overloading or poor response using an oscilloscope. The vibration signals can be processed using a dynamic signal analyzer in order to perform band analysis upon each product and/or different use scenario being assessed. By plotting the results of the analysis, comparisons may be made with ISO and ANSI standards to determine if the vibration level produced is acceptable.

Quantifying Body Stress

Biomechanical Models.   Assessing human stress helps to determine the likelihood of discomfort or injury which will be borne by the user. Currently there are two major techniques for evaluation of user stress. The first utilizes dynamic and static biomechanical models. These models are essentially predictors which allow the ergonomist to evaluate forces on joints and the magnitude of muscle tension developed. When these measurements are compared with cutoff scores for safe or efficient body force levels, an estimation of body stress can be achieved. For example, during the loading of paper into a high-volume copy machine, the user is required to pull out a tray at the base of the machine and insert a package of several thousand sheets of paper. To evaluate the force on the person's spine and the likelihood of discomfort or injury, a biomechanical model of the spine can be utilized to estimate task force on the L5/S1 disk and other joints (Fig. 2.2). Using these models, we can "see" the stress on the spine, much as utilization of a noise monitor allows us to "see" the potential for damage to the tympanic membrane and the resulting hearing loss. So the biomechanical model becomes a predictive tool for the ergonomist.

The National Institute of Safety and Health (NIOSH) in its "Work Practices Guide for Materials Handling" recommended the establishment of cutoff limits for spinal force. This was the first time a national research group put forward the concept of controlling back injuries by controlling force exposure to the spine. Prior to this, most effort to limit back injuries was focused on limiting the loads listed as opposed to forces applied to the spine. NIOSH recommended the establishment of a safe spinal load factor called the action limit. The action limit was recommended to be 3400 N, or 760 pounds. (Refer to Fig. 2.2 for a comparison between the forces acting on the spine during a heavy lift and this government-regulated limit of 3400 N.) Biomechanical models can be constructed for any

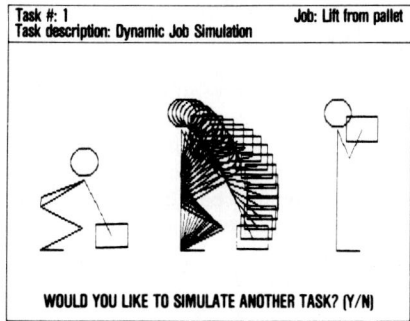
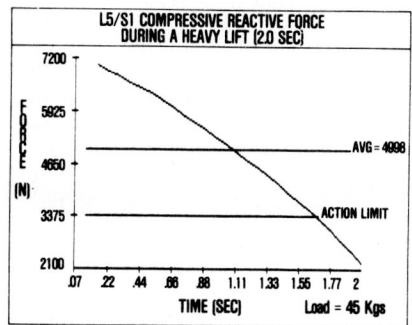

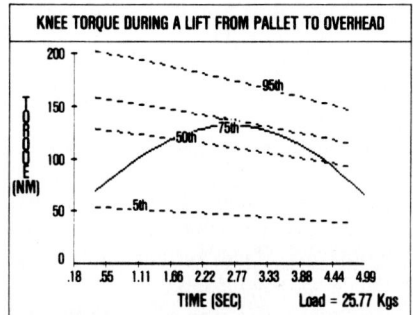

FIGURE 2.2 Output of biomechanical model.

joint of the body and serve as an experimental tool for first-level prediction of discomfort or injury in the performance of a task or utilization of a product.

Electromyography.   Another technique for measuring body stress, which is particularly useful for analyzing the forces exerted by various muscle groups (that is, wrist, shoulder, elbow), is the use of electromyography. Using electromyographic signals, which are produced as a result of muscle contraction, it is possible to record, in real time, an analog to joint force. As muscle tension increases, so does its electrical output in a proportional manner. Electromyography is a valuable tool that displays, during any activity being performed, which muscles are operating and to what extent they are operating under stress. It enables the ergonomist to understand the sequence of muscle use and to evaluate excessive muscle use required to operate particular tools or products.

Many products are developed which leverage the strength-to-weight ratio and the flexibility of the hand. To measure the stress on the wrist while operating a product with the hands, electrodes are placed over the forearm flexor and extensor muscles. The signals which are recorded by these electrodes may be amplified and calibrated to an independent measure of force or recorded as a percentage of maximal effort, with the voluntary maximum exertion being required of the user at the beginning of the test. In the case of excessive force requirements, it is fairly well understood that 6 kg of hand force, or greater than 20 percent of maximum voluntary contraction force (MVC), produces muscle fatigue and discomfort in the wrist and is to be avoided. In the less extreme cases, subtle electromyographical differences may be utilized to indicate the benefit of the product's mechanical interface to the user.

Measuring Accommodation and Fit.   The evaluation of accommodation and fit is of strategic importance in assuring customer comfort, performance, and product loyalty. Fit is the business of comparisons, and "global fit" requires the ability to assess the size, strength, and capabilities of potential customers throughout the product development process. It is impractical to perform fit assessment today without taking advantage of ergonomic databases and computer-aided design systems.

Global Ergonomic Databases.   Relevant information regarding human dimensions and capabilities may be accessed by ergonomists and designers in a variety of mediums. Traditional sources include textbooks compiling published civilian and military anthropometric measures. Currently, it is possible to access ergonomic data from large internationally available on-line databases, nationally based databases, general ergonomics databases, and databases on selected areas of ergonomics (anthropometry, biomechanics, workplace design).

Human Computer-Aided Design.   Computer-aided design systems have greatly automated and enhanced the efficiency of new product development programs. These systems have been very helpful in conceptualizing and rendering the product designs and, in some instances, in assessing manufacturability. Where they have been lacking is in the incorporation of human anatomical and anthropometric data with which the transparent interface between user and product can be made visible and the interactions can be measured.

The premise for human computer-aided design (HumanCAD™) is that a product can be designed in a shorter period of time and to better fit the consumer if articulated human bodies combined with ergonomic information are available in the computer-aided design environment. This enables questions such as the following to be answered: "Does the product fit the user?" "Does the user have adequate reach and strength to operate the product?" "Is the product within the user's visual field?" "Can users close their hands around the product?" These are questions which should be answered prior to prototype fabrication to reduce the product development cycle and make the design process itself more efficient. For example, to evaluate the fit of a person in a microscope workstation configuration, using HumanCAD™ it is possible to perform replacements of specific population percentiles (both male and female) in a simulated working environment (Fig. 2.3). Now that human modeling software is available for personal computers, these capabilities are accessible to any product designer and, over time, should contribute to the reduction of product development cycles and some of the costs inherent in them. The ability to present the human computer-aided designs as a solid model or rendered image also increases the designer's ability to conceptualize (and present) the implications of a product or workstation on the end user.

Quantifying Fatigue.   Assessment of fatigue are helpful in the evaluation of long-term customer satisfaction. Fatigue while using a product or tool leads to errors and issues of safety and comfort.

Oxygen Consumption Meter.   The amount of effort expended by a worker can be determined by measuring the amount of oxygen extracted by the lungs from the inhaled air and the amount of carbon dioxide exhaled. The greater the work capacity of a person, the larger the amount of oxygen that the person can extract from the inhaled air.

The maximal effort, or work capacity of a person, is expressed in maximal aerobic power. This is determined by the direct measurement of an individual's maximal oxygen uptake during exercise. The physical workload of a job may be assessed either by measurement of the oxygen uptake during the actual work operation or by indirect estimation of the oxygen uptake on the basis of the cardiac pulse rate recorded during the performance of the work. The validity of using oxygen uptake as a basis for determining energy expenditure has been established. This method is used to determine the energy cost of a wide variety of work activities.

Heart Rate Telemetry.   Cardiovascular fatigue (as opposed to mental fatigue) is most readily measured through the evaluation of heart rate. There is generally a linear relationship between oxygen uptake and heart rate. The use of recorded heart rate in the field, compared with the heart rate at known workloads on a bicycle ergometer, may be used as a basis for the estimation of the workload when the same large muscle groups are used.

Telemetric heart rate devices are easy to wear and do not disrupt, are unobtrusive for

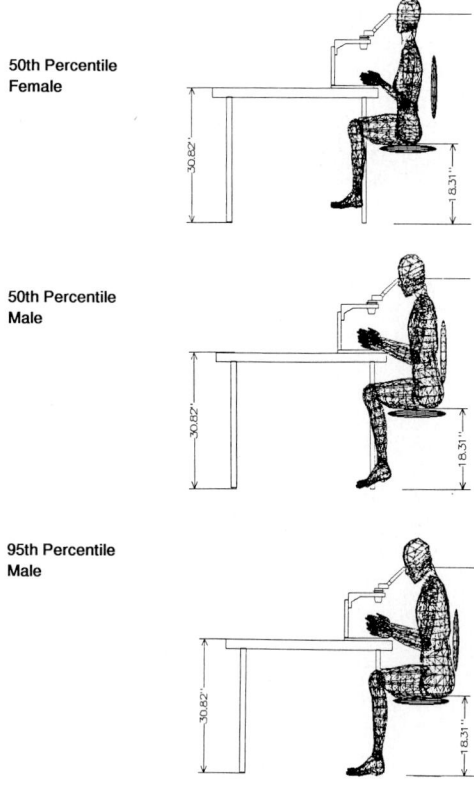

50th Percentile
Female

50th Percentile
Male

95th Percentile
Male

Fit of three U.S. population percentiles
in a microscope workstation

FIGURE 2.3   Human computer-aided design.

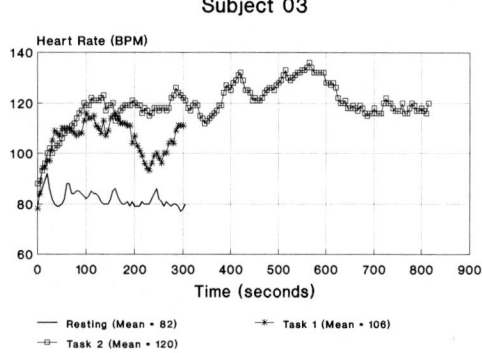

Subject 03

FIGURE 2.4   Output of heart rate assessment.

the user, and render time-coded heart rate data during product use (Fig. 2.4). Cardiovascular fatigue is an acceptable indicator of stress even for low levels of heart rate elevation. It can be measured by constructing a product-specific fatigue stress test. The goal of this test is to amplify the cardiovascular requirements for using a given product over a short period of time.

An example concerning the evaluation of several different automobile entry planes demonstrates this concept. A major car manufacturer wanted a system of evaluating the stress on the driver during ingress and egress of the vehicle. One of the variables found to be useful was the change in the heart rate above the resting level as a function of the difficulty of entering and leaving a specific automobile. As automobile entry or exit typically requires less than 3 seconds, no significant differences in heart rate activity were found during entry or exit under normal conditions. The test was modified to require each test subject to exit and enter the vehicle five times in a row at their normal pace. During this entire period of time, heart rates were recording using a telemetry system. It was found that the repetition in cycles made it possible to distinguish the stress contribution for the different entry planes for the vehicles in question.

Rating of Perceived Exertion.   Questionnaires may be used to evaluate perceived exertion during a particular task or product use trial and to classify the task in terms of effort required. A common approach to estimate this is to use the 15-point rating scale shown in Table 2.3. This scale, as developed by Borg (1962), is based on the range of heart rate levels during dynamic work. It was designed for obtaining subjective ratings of whole body effort. This scale was modified in 1980, as shown in Table 2.4. The modified scale is used to evaluate ratings of effort level for large muscle group activities. The ratings on this modified scale are multiplied by 10 to provide an estimate of the percent of maximum voluntary contraction (% MVC) that a given muscle effort requires. Subjective data determined from the large muscle group scale help the analyst identify on which body part fatigue may be likely to occur first.

TABLE 2.3 Rating of Perceived Task Difficulty (Whole Body)

6	
7	Very, very easy
8	
9	Very easy
10	
11	Fairly easy
12	
13	Somewhat difficult
14	
15	Difficult
16	
17	Very difficult
18	
19	Very, very difficult
20	

Source: Borg (1962).

Metabolic Assessment Programs. Metabolic energy expenditure rate models are one way of calculating the required rest for a given amount of task performance. Muller (1953) developed a rest-pause schedule based on the assumption that the net amount of energy expenditure which can be performed safely per day is 4 kcal/min and that rest pauses should be scheduled based on the percent of recovery time. Other models such as Spitzer (1952) and Murrell (1964) suggest that resting time be recommended as a percentage of total working time.

Evaluating Comfort

Questionnaire Assessment of Product Characteristics. During a biomechanical product assessment, it is useful to have the subjects complete a comfort questionnaire addressing various design features of the product relative to comfort during use. The questions should be very product-specific. The questions should refer to forces required during specific actions, control, comfort of various body parts involved in product use, product feedback, comfort of product dimensions (height, length, width), contours, overall product performance, and ease of use.

Ranking Questionnaire. When evaluating several products within the same class, it is also useful to have subjects rank the products in terms of several broadly defined characteristics (that is, fit, ease of use, overall performance). This information can then be correlated with other biomechanical measures to obtain insights into the product characteristics contributing to user satisfaction.

Relating Comfort to Specific Product Characteristics. Comfort assessments are performed by statistically analyzing the responses to the above questionnaires with respect to

TABLE 2.4 Rating of Perceived Task Difficulty
Scale (Large Muscle)

0	Nothing at all
0.5	Very, very weak (just noticeable)
1	Very weak
2	Weak (light)
3	Moderate
4	Somewhat strong
5	Strong (heavy)
6	
7	Very strong
8	
9	
10	Very, very strong (almost maximum)
	Maximal

Source: G. Borg, "A Category Scale with Ratio
Properties for Intermodel and Interindividual Compari-
sons," paper presented at the International Congress of
Psychology, Leibig, Germany, 1980.

specific subject and product characteristics. For example, how does the hand anthropometry of a subject affect the subject's rating of the hand comfort of the product being tested, or how does the product shape affect perceived ease of use.

As a more specific example, in the quantification of human comfort while sitting, it was necessary to first establish an analog for comfort based on physical measurement. Pressure distribution on the body was selected. A wide variety of seats were then instrumented using a tiny flexible transducer to measure how loads were distributed for different percentile men and women. After 1300 seat-subject evaluations, a pattern began to emerge as to the relationship between perceived comfort as measured on a Likert questionnaire scale and load distribution measured with the flexible circuits. From these two data sets, a multiple regression equation was established which enables the prediction of perceived comfort from the objective physical measure of load distribution. Figures 2.5*a* and *b* and 2.6*a* and *b* illustrate comfortable and uncomfortable seat pan pressure determined by this model.

A necessary procedure in the ergonomic design of a product is to establish the quantification basis for evaluating the product-customer interfacing characteristics. In the previous example, our goal was to design an ergonomically sound seat, but before we could achieve this goal, we needed to develop a measurement criterion for seat comfort.

SEAT PAN PRESSURE : VERY COMFORTABLE

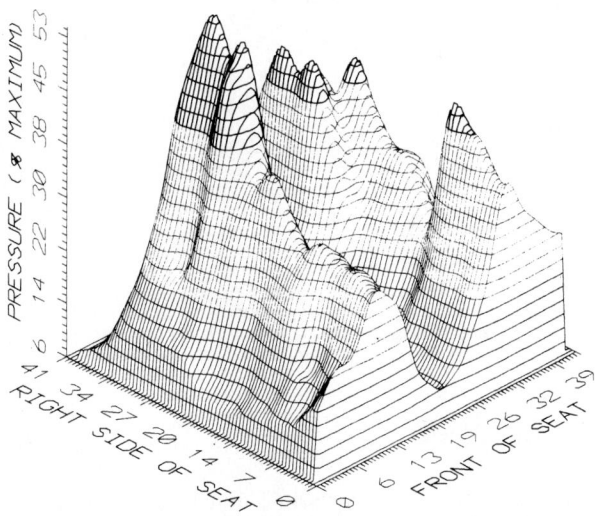

FIGURE 2.5*a* Relationship between pressure and comfort (contour
map): Seat rated "very comfortable."

SEAT PAN PRESSURE : VERY UNCOMFORTABLE

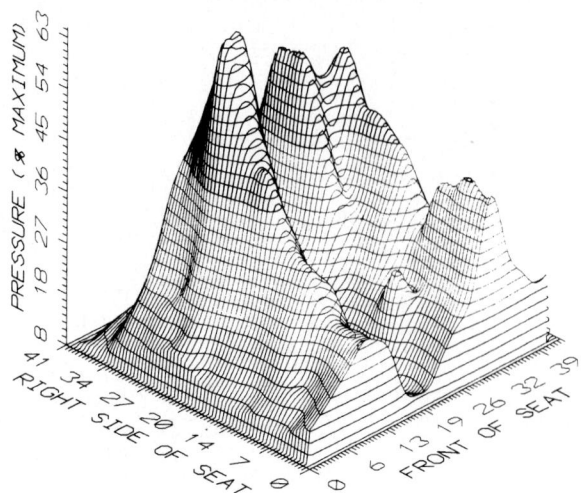

FIGURE 2.5*b* Relationship between pressure and comfort (contour map): Seat rated "very uncomfortable."

SEAT PAN PRESSURE : VERY COMFORTABLE

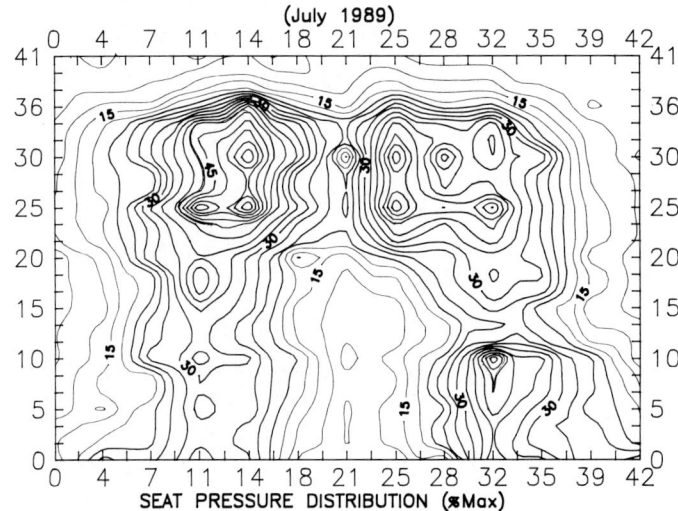

FIGURE 2.6*a* Relationship between pressure and comfort (distribution): Seat rated "very comfortable."

BEST-IN-CLASS ASSESSMENTS

Ergonomic assessments of products can also be used as a strategic device to probe for product weaknesses and identify specific areas for product improvement. The purpose of fusing ergonomics to product development is twofold: first, to have an ergonomically sound product, second, to assure that the product is the best it can be, in a specific class.

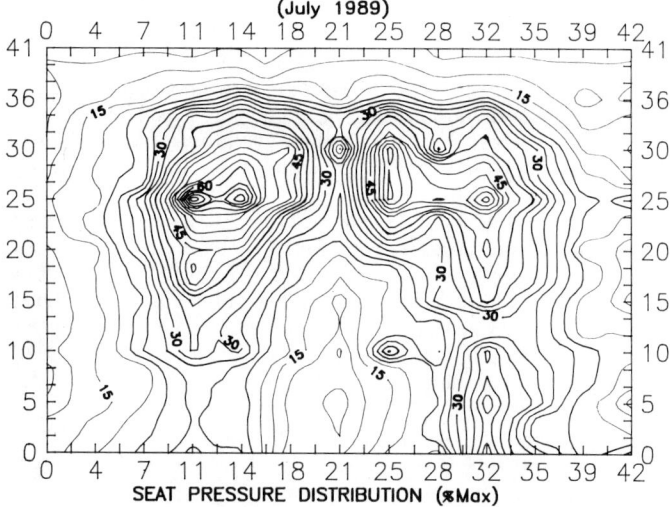

FIGURE 2.6*b*   Relationship between pressure and comfort (distribution): Seat rated "very uncomfortable."

The difference between very good and superior products is usually not accounted for by the technology content. Rather, the way the product is designed, the way the features are integrated, and the diligence with which user interface issues are addressed can make the biggest difference.

Best-in-class assessments generally include the preparation of an internal "consumer report" based on the results of the ergonomic assessments of the product and its competitors. The steps to best-in-class assessments are

1. Identify key products to be included in the best-in-class assessment.
2. Study product use under actual customer scenarios.
3. Develop a quantitative measurement approach for evaluating specific product features.
4. Conduct product analysis.
5. Use the quantitative measurement scale to develop an overall numerical rating of product performance.
6. Identify those features necessary to position the product best-in-class, or to improve the ergonomic content of the product being analyzed.

Using the aforementioned measurement tools, it is possible (under an aggressive program) to complete a best-in-class assessment in 1 to 3 months for many mass market products. The following case study is an example of a successful best-in-class assessment for the purpose of rating pneumatic drills. Following this rating, a new ergonomically sound drill line was designed and manufactured which incorporated ergonomically sound features.

In summary, if ergonomics is properly applied to the product design and development process, the result will be a product which is simple to operate, comfortable to use, and appropriate for the user and the conditions of use. Working together, designers and ergonomists equipped with biomechanics technology can develop high-quality products that will increase consumer acceptance and market share.

ERGONOMIC PRODUCT REDESIGN CASE STUDY: PNEUMATIC DRILL

Introduction. This case study is based on an assessment performed for the power tool division of Cooper Industries (CPT). This study was selected as an example of the application of biomechanical measurement technologies to product design. It also well demonstrates that ergonomics is a continuous process which may be used to enhance the value of existing products by making subtle, carefully defined refinements based on a thorough understanding of user capabilities and requirements.

The primary objectives of the project were (1) to identify specific product characteristics which contribute to user comfort and ease of use, and (2) to utilize these data to ergonomically redesign the product to maximize user comfort and ease of use and to decrease the musculoskeletal stress experienced by product users.

A best-in-class assessment of six competitive pneumatic drills was conducted, using biomechanical analysis. All the drills evaluated fall into the same tool class, that is, primarily utilized in an occupational setting to accomplish a variety of possible job requirements, and considered "substitutes" in a marketing sense. A discussion of each of the project phases is provided following the model set out in Table 2.1. As CPT had already produced a preliminary prototype drill, this project began at the second phase of the product development cycle.

Establishment and Testing of Experimental Design

Purpose. To define the data collection methods and test them to ensure the required data for developing ergonomic design, recommendations were collected in an accurate and consistent manner.

Method. Through discussions with CPT's product engineering and design staff, combined with a review and postural digitization of videotapes of drill use in a manufacturing setting, five postures were identified as typical use scenarios for the product class. A specialized workpiece was constructed to facilitate the collection of data in these varying postures during the user trials. Sample data collection sessions were then conducted to ensure that the setup optimized the data collection goals. Once the data collection method had been established, the experimental protocol was developed. This included the nature and size of the subject population, what data elements were to be collected, the sequence and randomization of user trials, and intended analysis techniques to be applied to the data acquired.

Six competitive tools were provided by Cooper Power Tools and 20 individuals with prior experience in the use of pneumatic and power tools were utilized in the biomechanical assessment as test subjects. The subjects were recruited through a classified advertisement in a regional paper and compensated $50 per 5-hour testing session.

Results. Figures 2.7 and 2.8 illustrate the postures chosen for analysis during the subject trials and the way the workpiece was configured to ensure consistent data acquisition. Table 2.5 details the data collection techniques applied during user trials for the six analysis categories defined in the protocol.

Biomechanical Assessment of Drill Use

Purpose. The study was conducted to analyze experienced drill users during the performance of product trials and develop a database, and apply statistical techniques to understand the relationship between product characteristics, subject characteristics, physical use requirements, and subjective product evaluations.

Method. Following the study protocol, data collection from the 20 subjects was conducted over a continuous 4-week period. Data reduction and analysis of the collected data began simultaneously to shorten the total length of the study.

Results. Table 2.6 summarizes the assessment results.

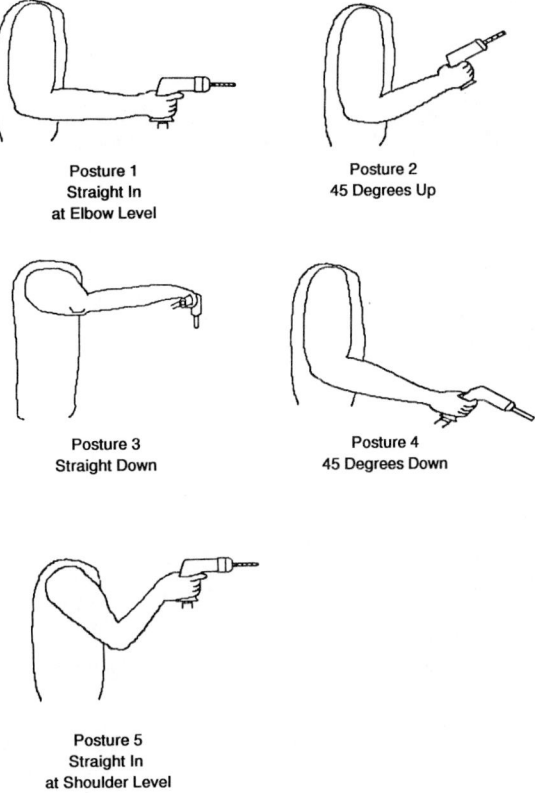

FIGURE 2.7   Sketch of five postures tested.

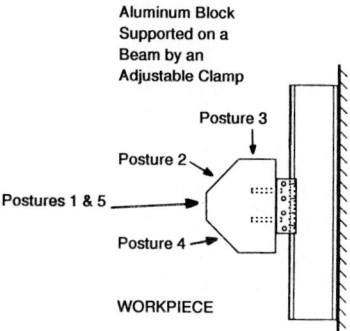

FIGURE 2.8   Setup for workpiece.

Ergonomic Design Specification and Prototype Development

Purpose.   To apply relationships identified during the data analysis to optimize drill characteristics which contribute to user comfort and preference and minimize undesirable tool characteristics.

Method.   From the above results, the following conclusions regarding the ergonomic design of drill handles were drawn:

1. Tool handle size is a distinct feature which greatly affects the comfort associated with tool use.

2. None of the tools tested accommodated all subjects comfortably. This may be attributed to the variation of hand anthropometry among the subject population. This variation is reasonably expected to be similar among normal tool users in an occupational setting.

3. Tool weight, effort required to grip handle, and the shape of the grip contours are factors which affect user comfort. If a tool does not fit comfortably at the onset of use, it is not likely to become more comfortable with extended, continuous use.

TABLE 2.5 Data-Collection Techniques

Tool characteristics: Tool weight, handle size, handle shape and configuration, trigger shape and configuration, trigger force, center of mass

Subject characteristics: Age, height, weight, dominant hand, anthropometric dimensions of the hand and other body segments gathered through questionnaires and standard measurement techniques

Postural analysis: Side view (sagittal plane) and top view (transverse plane) during drill use in each of five postures recorded on videotape

Hand force assessment: Forearm flexor and extensor EMG levels associated with tool grasping and trigger activation collected using digital electromyography

Vibration analysis: Hz level of vibration (frequency and amplitude) transmitted from the tool handle into the hand during tool use measured using low-mass accelerometers mounted on the drill handle

Subjective ratings of comfort: Ratings of specific drill characteristics, preference rankings, and free form comments elicited through written questionnaires

---

TABLE 2.6 Biomechanical Assessment Results

Subject characteristics:

1. 17 male/3 female
2. Average age 33.1 years (range 20–63 years)
3. Average height 69.3 in (range 62–77 in)
4. Average weight 170.3 lb (range 110–215)
5. Average anthropometric dimensions similar to 50th percentile U.S. population

Postural analysis:

1. Wrist extension during tool use ranged from 143 to 156° (180° is neutral)
2. Wrist ulnar deviations during tool use ranged from 146 to 172° (180° is neutral)
3. Included elbow angles during tool use ranged from 62 to 111°
4. Working posture was found to produce significant differences in elbow angle but not wrist extension and ulnar deviation

Hand force assessment:

1. No significant differences between the six tools for any of the EMG measures
2. Significant differences between postures for extensor muscle EMG as a percent of maximum grip exertion and mean force values
3. Significant differences between postures for flexor muscle EMG as a percent of maximum grip exertion and mean force values

Vibration analysis:

1. Significant difference between postures with posture 5 (drilling at shoulder level) producing the highest vibration values and posture 3 (drilling straight down) the lowest
2. Significant difference between tools
3. While all tools fell within acceptable levels of vibration according to the ANSI standard, the ISO standard was frequently exceeded during all of the tool trials

Subjective ratings of comfort:

1. One tool consistently had the best ratings for most of the features evaluated
2. This tool was also rated as the preferred drill based on ranking scale administered
3. The tool ranked least for overall comfort was also ranked the least preferred drill based on ranking scale administered
4. Highly significant differences were found between the six drills tested, particularly for attributes related to tool weight, effort required to grip, shape of grip contours, size of grip handle, and overall comfort rating
5. A strong correlation was found between perceived comfort and the inner dimension of the handle base to the top of the barrel on the hand
6. Perceived tool-to-hand fit was strongly related to the force required to depress the trigger and to the length of the user's hand. Lower trigger forces were preferred

---

Results.   The above conclusions, combined with generally accepted ergonomics guidelines for tool design, led to the definition of the following ergonomic features as a basis for proper pneumatic drill design.

Anthropometric Considerations of Handle Size

Recommendations.   Produce different tool handle sizes to accommodate the 50th percentile female, 50th percentile male, and 95th percentile male populations. Dimensions for each of the tool handle sizes were specified on the basis of published U.S. anthropometric data. User selection of the most suitable tool was to be accomplished through a simple wall chart (Fig. 2.9). To avoid the expense of producing completely separate tools to provide the different handle sizes, design a standard tool housing to accommodate interchangeable, varied sized handles.

Optimal Grip Biomechanics

Recommendations.   Subtle contours in the tool handle, particularly around the base of the thumb, will help distribute the grip forces to as large a pressure-bearing area on the fingers and palms as possible. This will lead to improved tool performance and a reduction in excessive hand pressure from inadequately distributed mechanical loads on the handle. The subtle curvatures running along the length of the handle from the base to the top will also allow users with varied anthropometry to feel that they have a good grip around the tool and prevent the tool from sliding down off the hand.

Additional recommendations regarding handle shape included the use of an oblong shape for the cross section of the handle. This shape corresponds better to the shape the palm makes as it wraps around objects than the traditional elliptical or cylindrical shape.

Low-Force Trigger Actuation

Recommendations.   Two new design concepts were proposed to achieve the goal of reducing trigger force during use. The first was the conventional index finger trigger with modifications, and the second conceptualized a thumb-activated trigger design.

1. *Index finger trigger design:*
   a. Recessed trigger to allow index finger to align with the other fingers on the grip.
   b. Flange between the trigger and lower part of the handle to prevent the drill from sliding down during trigger release and to prevent pinching by the trigger button.
   c. "Flattened out" button shape for the trigger to allow for quick activation of the drill regardless of the location of the force applied to the trigger.
   d. Better contour around the thumb area to provide a firmer grip around the drill surface.
2. *Thumb-activated trigger design:*
   a. Trigger button on side of handle, activated by thumb, will allow four fingers to maintain handle grip during activation.
   b. "Flattened out" button shape for the trigger to allow for quick activation of the drill regardless of the location of the force applied to the trigger.
   c. Would require modification of trigger mechanism to allow the necessary air flow into and out of the drill.
   d. Would require different tools for right-handed and left-handed users.

Minimal Vibration Transfer

Recommendation.   Place a vibration-damping material around the air flow cylinder prior to securing the handle to the tool housing. This will isolate the handle from the high-frequency vibrations (over 500 Hz). Other general recommendations included the minimization of tool weight to the extent feasible, and the consideration of thermal handle characteristics to minimize handle temperature drop as much as possible.

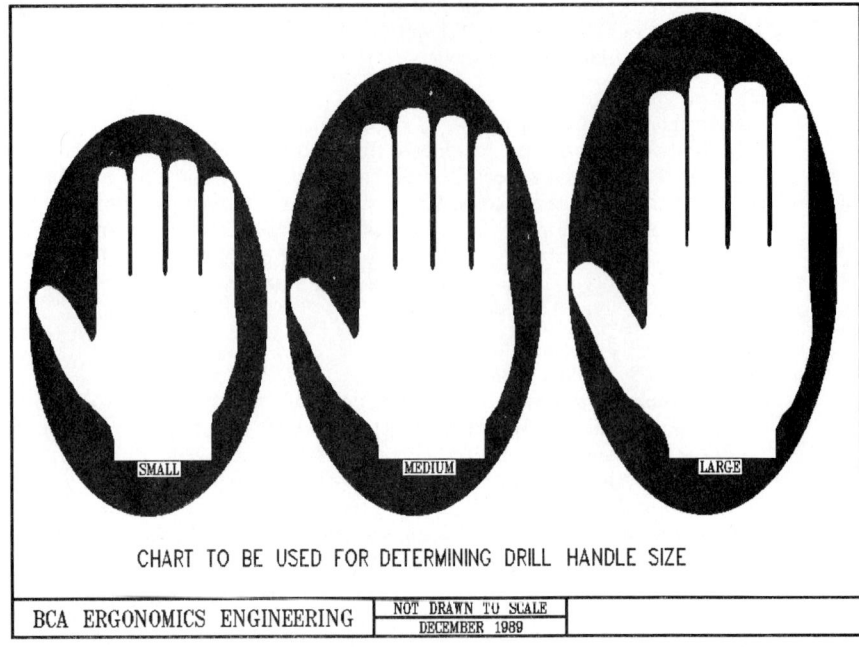

FIGURE 2.9   Wall chart for selecting proper drill size.

Study Outcome.   Conceptual prototypes for each of these two handle designs were fabricated by BCA and provided to Cooper Industries for evaluation. The outcome of the study was the production of a new ergonomic drill line which embodies the recessed index finger trigger design and the subtly contoured handle (Fig. 2.10). The tool is produced in two sizes (small and large). Feedback from corporate users of the new tool indicates that the

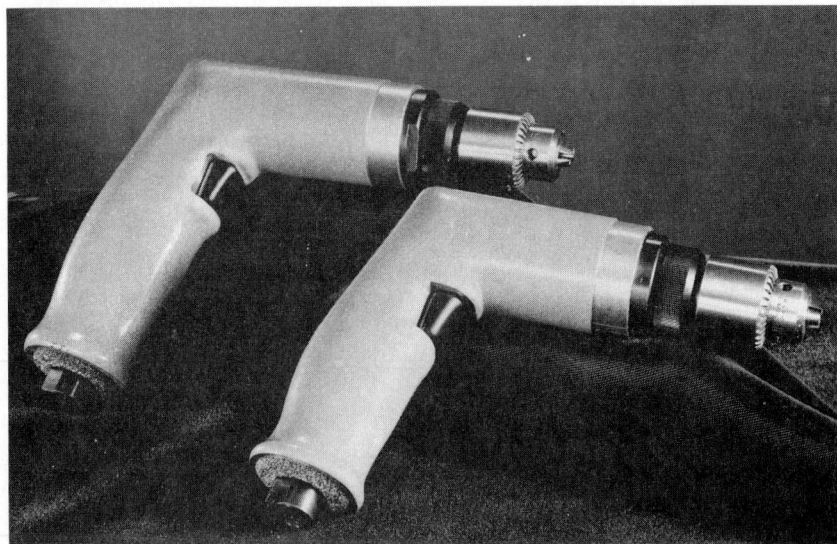

FIGURE 2.10   Ergonomically featured drill line.

new design has been well received. Systematic follow-up assessments of the tool users are being conducted to measure the efficiency of these ergonomic featured pneumatic drills over time.

Acknowledgment.   We thank Cooper Industries for their participation in this project and for permission to publish this case study.

REFERENCES

Ayoub, M. M, and A. Mital, *Manual Material Handling,* Taylor & Francis, London, 1989.

Biomechanics Corporation of America, *Principles of Ergonomics,* prepared for United States Department of Labor, OSHA, 1986.

Borg, G. A. V., *Physical Performance and Perceived Exertion,* Gleerups, Lund, 1962.

Gross, C. M., "Ergonomic Workplace Assessments Are the First Step in Injury Treatment," *Occupational Health & Safety,* May 1989.

Gross, C. M., "Ergonomic Aspects of Computer Terminal Operations," in *Success Factors for Implementing Change: A Manufacturing Viewpoint,* Society of Manufacturing Engineers, 1989.

Gross, C. M., and A. Fuchs, "Ergonomics Program Reduce Costly Workplace Injuries," *Occupational Health & Safety,* January 1990.

Eastman Kodak Company, Human Factors Section, *Ergonomic Design for People at Work,* vol. 2, Van Nostrand Reinhold, New York, 1983.

ISO, *Principles for the Measurement and the Evaluation of Human Exposure to Vibration Transmitted to the Hand,* DIS 5349.1, International Organization for Standardization, Geneva, 1979.

Megaw, E. D., "The Future Role of Ergonomics Databases," *Ergonomics,* vol. 33, no. 4, pp. 469–476, 1990.

Mital A., and R. L. Shell, "A Comprehensive Metabolic Energy Model for Determining Rest Allowance for Physical Tasks," *Journal of Methods Time Measurement,* vol. 9, pp. 2–8, 1985.

Muller, E. A., "Physiological Basis of Rest Pauses in Heavy Work," *Quarterly Journal of Experimental Physiology,* vol. 38, pp. 25–215, 1953.

Silverstein, B., L. Fine, T. Armstrong, B. Joseph, B. Buchholz, and M. Robertson, "Cumulative Trauma Disorders of the Hand and Wrist in Industry," in N. Corlett, J. Wilson, and I. Manenica, eds., *The Ergonomics of Working Postures,* Taylor & Francis, London, 1986.

Spitzer, H., *Physiologische Grundlagen für den Erholungszuschlag bei Schwerarbeit,* REFA-Nachrichten, Darmstadt, 1952.

Wilson, J. R., and N. C. Corlett, eds., *Evaluation of Human Work, a Practical Ergonomics Methodology,* Taylor & Francis, London, 1990.

CHAPTER 3
SELECTING, DEVELOPING, AND EVALUATING PERSONNEL

Robert Fitzpatrick
Consulting Psychologist
AMANSCO, Incorporated
Pittsburgh, Pennsylvania

This chapter deals with three major areas of human resources management or personnel administration. First is the evaluation of individual job performance, usually called performance appraisal. The second area is personnel selection, including (1) inferring the characteristics of people who are likely to succeed at a given job in an organization, (2) recruiting candidates, (3) obtaining job-relevant information about each candidate, and (4) applying appropriate decision rules to choose those who will be offered the job. Third is training and development of employees, to help them maintain old skills and master new ones.

In enterprises of all kinds, the foundation of organization has traditionally been the job. Each employee is assigned specific tasks and responsibilities. A job description defines the job requirements and describes the work environment.

In principle, management designs and defines the job for the employee. In practice, however, an explicit effort to design a job is seldom undertaken. Much of industrial engineering can be viewed as job design, but most jobs do not have the benefit of industrial engineering effort. It is usually the case that only the job incumbent (and perhaps the immediate supervisor) knows just what happens day to day on the job. Hence it is necessary in most cases to make a special effort to determine the precise nature of the job. Such an effort, or job analysis, may take any of a number of forms. Some of these are described in Sec. 3. Other sources can be recommended.[1] Some form of job analysis underlies all the methods and concepts described in the remainder of this chapter.

INDIVIDUAL JOB PERFORMANCE APPRAISAL

The evaluation of individual job performance, or performance appraisal, is normally carried out by each employee's supervisor. Occasionally, appraisals are done by technical specialists, by panels of supervisors and specialists, or even by the employees themselves.

---

[1]S. Gael, ed., *The Job Analysis Handbook for Business, Industry and Government*, Wiley, New York, 1988.

However, none of these variations has been adopted widely or supported by solid evidence.

It is possible and might often be desirable to evaluate the performance of task groups or teams rather than of individuals. Such group evaluations are seldom made, partly because organizational reward systems are focused on the individual.

Purposes of Performance Appraisal. The two major purposes of individual job performance appraisal are:

1. To provide information to management for use as a basis for decisions concerning pay increases, promotions, demotions, assignment to training courses, and other human resources management actions.
2. To help the individual and the supervisor plan the employee's further development and career progress.

For the first purpose, it is necessary to derive from the performance appraisal a single overall indication to designate the employee's performance. The more favorable this overall indication, the more likely is a desired management action such as a pay increase (and the larger the increase). For the second purpose, however, it is of little or no use to have an overall indication of performance merit. What is needed is a series of indicators showing what parts of the employee's performance have been relatively good and what parts have been relatively deficient. Then it is possible to plan to exploit the good performance and to remedy or ameliorate the deficient aspects.

Supervisors as Appraisers. It is generally assumed that the supervisor has ample opportunity to observe each employee's performance or the major outcomes of the employee's efforts. This assumption is not necessarily correct. Supervisors are often busy at tasks which conflict with the task of observing subordinates; the employees may do some of their work in locations different from that of the supervisor; many supervisors are responsible for more subordinates than they can keep systematic track of; and there are other circumstances which limit supervisors' ability to observe employee performance and products.

It is also assumed in most organizations that supervisors will be able to remember and properly weigh their observations of each employee when it comes time to fill out the appraisal form. But human memory is subject to distortion and loss. It helps if supervisors keep notes of their observation, but supervisors seldom do so consistently. In some cases, information may be recorded automatically or by a third party, as in the case of salespeople whose sales records are enshrined in a cash register computer tape or in an order book.

Usually, the supervisor is told to complete a written appraisal form periodically (typically, every 6 or 12 months). Such a form normally has spaces for descriptions or ratings of several aspects of the employee's performance. The completed appraisal form, in most cases, is submitted for final approval to the next higher level of management. Upon approval, it is made part of the employee's permanent file and is consulted when personnel decisions are made. It is normally expected that the supervisor will show the appraisal form to the employee and use it as a basis for discussion of the employee's past and future job performance. Supervisors do not always do this, since it takes time and may be an unpleasant experience for both participants.

Types of Performance Measures. A bewildering variety of performance appraisal methods and measures have been proposed and used. For convenience, measures are here discussed under four headings: production measures, behavioral performance records, rating scales, and other measures. In many cases, more than one type of measure is used.

Production Measures. A production measure is a measure of work outcome. Either quantity of satisfactory production, quality of production, or both quality and quantity of production may be measured. It is often necessary to make adjustments in a production measure to take account of factors which might make the measure unfair to some employees. For example, one worker might be assigned a more difficult task than another, might have a less reliable machine at which to work, or might work in more adverse environmental conditions. In any case, a production measure alone is seldom sufficient, since the way in which a task is done is typically as important as the production outcome of the task. For example, a worker may achieve high quantity and quality of production, but at the expense of a high level of scrap and a low level of cooperation among the group with which the worker is associated. If such factors as scrap and cooperation are important to the overall organizational objectives, a production measure alone is insufficient to characterize the worker's job performance.

Behavioral Performance Records.   A behavioral performance record is a record of job-relevant activities or behaviors of the employee. Rather than focusing on the products of the employee, such a measure is concerned with the process by which the employee achieves results. In some cases, the record is a simple narrative by the supervisor. However, since different supervisors write highly different narratives (and some will not write any narrative at all), it is desirable to provide some rules and procedures for making the record.

The critical incident technique[2] has been used with some success for this purpose. A *critical incident* is a report of an occurrence judged by a competent observer to be an example of especially effective or especially ineffective behavior. Thus the supervisor records only clearly important events and ignores trivial, routine, or ambiguous performance. Supervisors usually find that a record of incidents serves as a good basis for discussion with the employee and for planning ways to develop or improve performance. However, it is difficult to derive a meaningful overall score from a record of incidents, and most users of this type of system have resorted to some kind of supervisory rating to provide the overall score.

Rating Scales.   A rating scale presents two or more categories from which the rater chooses one as being most descriptive of the ratee or of some aspect of the ratee's performance. Generally, the description is evaluative: one of the categories represents the highest degree of employee merit, another the second highest, and so on.

A single general rating of overall job performance or merit is known as a *global* rating. When aspects or components of performance are rated, they are usually called *dimensions*.

Ranking may be defined as a type of rating in which each employee in a group is evaluated in relation to all the other employees in the group. Ranking is acceptable when a small number of employees is to be rated and when it is satisfactory to assume that differences between adjacently ranked employees are constant.

The number of rating categories can in principle range from two to infinity. In practice, the range is generally from four to ten. People are able to make appropriate distinctions among about seven categories and not many more. To proliferate categories far beyond seven is to pretend to a fineness of discrimination which is not achievable. On the other hand, to have only two or three categories is to fail to distinguish as well as can readily be done in most situations.

A binary scale may be useful in certain situations. Such a scale normally takes the form of a *checklist,* which consists of a series of statements or dimensions, each of which is to be designated by the rater as descriptive or not descriptive of the ratee's performance.

It is desirable to label at least some of the rating categories so that a common meaning for each category can be communicated. Such a label is called an *anchor.* Anchors may be numbers, adjectives, phrases, or sentences. A *behavioral anchor* is one which refers directly to behavior or performance of the employee. This type of anchor is preferable to others in most cases. Figure 3.1 illustrates some common types of rating scales.

Several performance appraisal rating systems have been developed through the use of critical incidents. The best known of these have been referred to as *behavioral expec-*

---

[2]J. C. Flanagan, "The Critical Incident Technique," *Psychological Bulletin,* vol. 51, pp. 327–358, 1954.

```
           0 __ 1 __ 2 __ 3 __ 4 __ 5 __ 6 __
```

```
      Unsatisfactory __ Poor __ Fair __ Good __ Excellent __
```

```
   Fails to meet  Barely meets  Meets standards  Clearly exceeds
     standards       standards      adequately        standards

        ☐            ☐             ☐               ☐
```

```
        Below Average       Average      Above Average
             ◯               ◯              ◯
```

```
   BOTTOM 10% __ NEXT 20% __ MIDDLE 30% __ NEXT 25% __ TOP 15% __
```

FIGURE 3.1   Examples of rating scales (illustrative of the range of possibilities; not necessarily indicative of best practice).

tation scales[3] or *behaviorally anchored rating scales*. An example of this type of scale is shown in Fig. 3.2. A variation is called a *behavioral observation scale* (see Fig. 3.3).

Other Measures.   A number of other ways of measuring employee job performance are applicable in certain cases, usually as only parts of a comprehensive appraisal system. Many organizations keep records of the training courses each employee has attended. Employees may be evaluated partly on the basis of the numbers and levels of courses taken. Evaluation may also consider the quality of employees' training performance, as reflected in such indexes as course grade, examination scores, and length of time needed to complete the course.

Records of absence and tardiness are maintained for some classes of employees. The use of this information in employee evaluation may have some unfortunate side effects, since it implies that the organization wants only the employee's presence and not the employee's best effort. Problems also arise in distinguishing between justified absences, due to illness or some other condition beyond the employee's control, and those absences which the employee could avoid.

Information received about an employee from sources other than the immediate supervisor may in some circumstances be used for evaluation. Commendations or complaints received from customers or suppliers are examples.

Records of work-related accidents, normally kept for safety and legal reasons, may also be used for individual evaluation. Of course, it is appropriate to use an accident for evaluation only when it may reasonably be attributed in substantial part to poor performance of the employee.

Some organizations have found it useful to ask the employee's peers or subordinates to participate in evaluation. Care must be taken in such a program to avoid invidious comparisons and possible erosion of cooperative activity.

---

[3]P. C. Smith and L. M. Kendall, "Retranslation of Expectations: An Approach to the Construction of Unambiguous Anchors for Rating Scales," *Journal of Applied Psychology*, vol. 47, pp. 149–155, 1963.

Responding to Customer Dissatisfaction

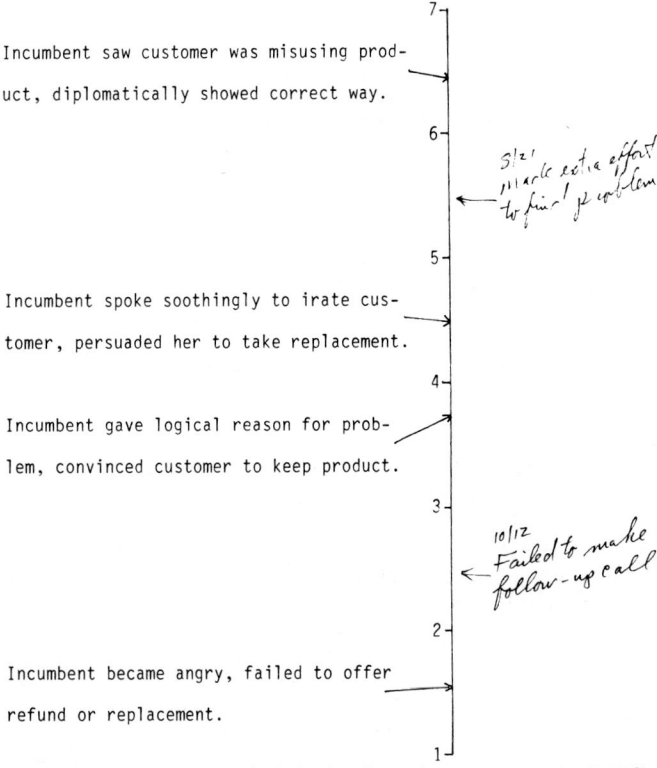

FIGURE 3.2   An example of a behaviorally anchored rating scale (BARS). The scale values of the illustrative incidents on the left would have been established in a preliminary study. In this version of BARS, the supervisor is to make notes of incidents as they occur, and to use those notes as the basis for an overall rating at the end of the appraisal period.

Maintaining Safe and Orderly Workplace

	Almost Never	1 2 3 4 5	Almost Always
1. Cleans up spills immediately and thoroughly.		_ _ _ _ _	
2. Puts scrap into designated container.		_ _ _ _ _	
3. Sweeps workplace area twice or more times daily.		_ _ _ _ _	
4. Wears eye protection when required.		_ _ _ _ _	

FIGURE 3.3   An example of a behavioral observation scale (BOS).

Desirable Characteristics of a Performance Appraisal System.   Several qualities should characterize a good set of performance appraisal procedures.[4] These qualities should be reviewed in developing a new system or evaluating an old one.

Relevance.   The performance appraisal system, as a whole and in each of its parts, should be relevant for the organization and for the job of the employee. For example, if the job does not call for the employee to generate novel solutions to problems, it would be irrelevant and harmful to include a rating of ingenuity as part of the performance appraisal.

Comprehensiveness.   Comprehensiveness refers to the degree to which the performance appraisal covers all dimensions of the job and all aspects of the work situation. The more comprehensive the system, the better. In many cases, comprehensiveness can best be achieved by using several types of appraisal measures: for example, a production measure, critical incidents, some rating scales, absence and tardiness records, and the like. An alternative approach is to combine several approaches in a unified system.

Discriminating Power.   A good performance appraisal system should distinguish among employees. If every employee receives the same appraisal, no information has been generated by the system. People's job performances and products do differ, and the appraisal should reflect the differences.

It is also desirable that the system should distinguish among rating dimensions for each employee. Raters sometimes commit the *halo error,* which consists of rating a given employee the same on all or most of the dimensions rated. Rater training can be helpful in reducing halo. However, halo is not the most important error and there is evidence to indicate that training aimed at reducing halo can actually produce less accurate ratings. Also, it is not clear that halo is in all cases an error, since dimensions may be positively intercorrelated; that is, for example, the employee who is high in productivity may also tend to be high in attendance.

Reliability.   Reliability here means the degree to which the scores resulting from appraisal are free from error. It is usually estimated by the product moment correlation coefficient between score distributions resulting from (1) application of the appraisal procedure on two occasions (test-retest reliability), or (2) application of two versions of the procedure on the same occasion (equivalent forms reliability). Perfect reliability is of course not possible; but the higher the reliability, the better.

Acceptability.   The supervisor, the employee, and higher management should all accept the appraisal system as a reasonable and fair way of measuring job performance merit. Otherwise, the process will be resisted and eventually eroded. One way to promote acceptability is to arrange for the participation of employees at all levels in the development and implementation of the appraisal system (see Sec. 8, Chap. 5). Acceptability tends also to be high when supervisors maintain frequent communication with subordinates about their job performance, so that the periodic written report is a review of matters already discussed rather than a revelation of stored-up complaints.

Fairness.   Each employee should have the same opportunity as any other employee to receive a favorable appraisal score when it is merited. If supervisors tend to denigrate the performance of some classes of employees without regard to their actual performance, the system will be dysfunctional and possibly illegal.

Usefulness for Development.   The performance appraisal system should encourage transmission of performance information to the employee. It should assist the employee and the supervisor in planning the employee's further improvement and career progression.

SELECTION

Worker Specification.   A worker specification (sometimes called a job specification) lists and describes the characteristics of a person who would be expected to succeed on the

---

[4]H. J. Bernardin, "A Performance Appraisal System," chap. 10 in R. A. Berk, ed., *Performance Assessment Methods and Applications,* Johns Hopkins University Press, Baltimore, 1986, pp. 277–304.

job. It should be based on the job analysis, along with general knowledge about people's characteristics in relation to their job success. Areas to be covered in a specification include knowledge and skill, cognitive abilities, health and fitness, psychomotor abilities, and temperament. When there are legal considerations, such as a requirement for a license or a minimum age, these should be included also.

Recruiting.   Recruiting is the process of seeking potential applicants for a job and inducing them to apply. Often, the applicants will seek the employer, who can then be the passive recipient of applications. However, it is generally effective and sometimes necessary for an employer to take active steps to recruit suitable candidates. The choice and implementation of recruiting procedures can have an important impact on an organization's overall success.[5]

In practice, recruiting includes the initial screening of applicants, since it is wasteful to continue dealing with an applicant who is unqualified for the job or who is highly unlikely to be selected.

Internal versus External Search.   Filling the job with people already employed by the organization is likely to help create or maintain a favorable organizational climate. Internal recruits have the additional advantage of familiarity with organizational policies, rules, and customs. On the other hand, an infusion of new styles and points of view from external recruits may be invigorating to an organization. It is often desirable to consider both internal and external sources of personnel.

Breadth of Search.   Relatively unskilled positions may usually be filled from the immediate neighborhood of the workplace. For other jobs, it is necessary to search on a national or even international scale. The cost obviously increases with the breadth. The recruiter must weigh costs versus benefits of extending the search.

Type of Appeal.   The recruiter may use a variety of arguments or information to attract applicants. It is important to determine first what positive features the job can be convincingly said to have. Then, the appeals should emphasize those features to which good applicants are likely to respond. Among the features and appeals are salary, potential for career progress and personal development, the reputation of the employing organization, location of the workplace (climate, convenience to transportation, and so on), and workplace comfort and safety. The recruiter should take care not to oversell the job; on the contrary, there is evidence to show that a realistic preview of job conditions and demands is often effective in attracting a preponderance of applicants who, if placed, will stay with the job.

Recruiting Media and Sources.   Recruiting is essentially a process of communication. The recruiter may use any communication medium and may consult a variety of sources of information. Common recruiting efforts include

1. Advertisements (newspapers, magazines, radio, television, billboards)
2. Telephone, letter, or in-person inquiries of people such as college professors likely to know of people who might be suitable applicants
3. Scheduled visits to schools, technical conventions, trade shows, and other institutions which can serve as intermediaries in bringing recruiter and potential applicant together
4. Retention of an employment agency or executive search firm
5. Inquiries to professional or union organizations which maintain lists of members or other people qualified in certain vocations
6. Requests to current employees to refer people who might be suitable applicants
7. Maintenance of recruiting offices or representatives in places where potential applicants are likely to be found

---

[5]S. L. Rynes, and A. E. Barber, "Applicant Attraction Strategies: An Organizational Perspective," *Academy of Management Review,* vol. 15, pp. 286–310, 1990.

8. Direct mail solicitations to groups of people, such as members of professional organizations, a substantial proportion of whom are qualified for the job

Affirmative Action.   Many organizations believe that they can benefit from having a work force which is broadly representative of the society or societies within which the organization operates. Minorities, women, older workers, handicapped people, and others may not be proportionally represented as employees, or may be employed disproportionately in low-paying and low-potential jobs.

To attain and retain a balanced work force, organizations must make special efforts to attract underrepresented groups to apply for appropriate jobs. They should target their recruiting efforts more strongly toward the groups who are underrepresented. For example, in recruiting for a job in which Hispanic people have not been employed, it may be advisable to place a large advertisement about the job in a Spanish-language newspaper and only a small one in an English-language newspaper. Thus non-Hispanic applicants are not excluded but Hispanics are especially encouraged to apply.

Various governmental jurisdictions have adopted policies to encourage affirmative action. The U.S. federal government, as a consequence of a presidential executive order, requires most of its contractors to develop and pursue affirmative action programs. Some have viewed affirmative action as a "numbers game" and have recommended hiring or promoting unqualified people to satisfy the governmental requirements. This is an expensive and ineffective policy, both for the organization and for the employees involved.

The targeted recruiting approach to affirmative action can confidently be recommended in the case of relatively low level jobs. However, it may require some modification for skilled and managerial jobs from which women, blacks, and other minorities have in the past been excluded. Since people in these groups have not had opportunities to accumulate the training and experience needed to function in high-level jobs, it may be appropriate to specify a relatively low level of entry skill for the job and to institute training programs which will help less experienced employees to gain skills as quickly as possible.

Selection Pocedures.   A *selection procedure* is any item or aggregation of information which influences a selection decision. A *selection decision* is one which involves choosing some people and (implicitly or explicitly) rejecting others to be offered a job, a salary increase, or a career advantage. Often, a selection decision is influenced in unknown, or at least unacknowledged, ways. In such a case, the selection procedure is informal and unofficial; the organization does not know how selection is done. Organizations should aim to define all selection procedures clearly and describe them publicly.

A variety of measures may be used as selection procedures. The major types are described in the following paragraphs.

Self-Reported Biographical Information.   Information about the background and experiences of the applicants is obtained, usually from the applicants themselves. Not all applicant information is completely accurate, so it may be desirable to check on claims of education and experience by contacting schools and former employers; however, this kind of checking is expensive.

Almost all organizations require applicants to fill out an application blank listing educational background and work experience. Other information may also be requested; it should be either job-related or needed for identification and communication (home address and telephone). It is possible to assign numerical values to the application blank responses and to derive an overall score intended to represent the relative value of the individual's education and experience as preparation for the job to be filled.

A biographical information blank (BIB) consists of a series of questions or items about the applicant's life history, attitudes, and preferences, usually in multiple-choice form. Each item is scored according to the expected job performance of applicants giving each possible answer.

Biographical information has been found to be an effective predictor of job success in a number of applications. However, careful and usually expensive development of the means of collecting information and the scoring system is necessary, and there remain questions of fairness to disadvantaged groups in the society.

Third-Party Information.   The employing organization may, and usually should, seek information about applicants from people other than the applicants themselves. Previous employers and schools may be asked to confirm and evaluate the experience or education of the applicant. Unfortunately, in the current climate, many organizations are reluctant to communicate more than dates of employment or enrollment. In the case of an internal applicant, records of performance appraisals may supplant or supplement supervisors' reports. If the previous job has different requirements from those of the job to be filled, it is necessary to evaluate the relevance of job performance information.

It is common to request or accept letters of recommendation and other communications from people who know the applicant. There is little evidence to support the usefulness of this type of selection procedure.

A *background investigation* involves interviewing neighbors and social acquaintances as well as former teachers and supervisors. It may also include a check of credit and criminal history. This technique has rarely been evaluated, but the available evidence suggests a moderate degree of effectiveness, especially when the background investigation is used in selecting people for jobs in which honesty and integrity are critical.[6]

Tests.   In a test, the applicant is presented with a standard set of cues or stimuli; the applicant's responses are noted and a score is derived. Normally, those applicants with relatively high scores are judged to be relatively suitable for the job. Tests, of course, should be carefully chosen or developed to be job relevant and appropriate for the candidates.

Most tests are designed to measure ability, temperament, or interests. Tests of ability are also called "maximum performance" tests; certain responses are clearly "correct," and the applicant aims to make as many correct responses as possible. Temperament tests are usually aimed at estimating "typical performance," that is, habits and preferences. The most common type of temperament measure is called an inventory or questionnaire rather than a test; it presents descriptive phrases and asks the examinee to indicate how well each phrase describes his or her own typical behavior or belief. Temperament inventories are scored partly on the basis of rational analysis or theory, and partly on the basis of correlational studies to determine (1) which responses tend to form clusters and (2) which clusters are associated with identifiable behavior patterns. Interest tests are similar to temperament inventories; generally, they are used for vocational counseling and career planning rather than selection. Examples of common types of test items are shown in Fig. 3.4. Normally, an ability test would contain at least 20 items of each major type to be measured, while a temperament or interest inventory would have at least 100.

The test cues are typically presented by verbal or numerical symbols (written or oral), and responses are required in similar form. Alternatively, in what are called *performance tests*, cues or responses may be more realistic. The cues may be presented by film, still pictures, or live actors. The responses in a performance test consist of actions which the examinee judges to be appropriate responses to the cues. Performance tests tend to be relatively expensive to develop and administer, and relatively difficult to administer and score consistently. Only in a few cases has the realism of a performance test been shown to compensate for its high expense and low consistency.

When the applicants for a job have training or previous experience relevant to the job, a test of job knowledge is appropriate. A good deal of evidence indicates that job knowledge is an essential element of job success, and that tests of job knowledge can be excellent selection procedures.[7]

Tests of health and fitness constitute a special class of tests. Medical examinations, in

---

[6]M. A. McDaniel, "Biographical Constructs for Predicting Employee Suitability," *Journal of Applied Psychology*, vol. 74, pp. 964–970, 1989.
[7]J. E. Hunter and H. R. Hirsh, "Applications of Meta-Analysis," chap. 10, in C. L. Cooper and Ivan T. Robertson, eds., *International Review of Industrial and Organizational Psychology 1987*, Wiley, Chichester, England, 1987, pp. 321–35.

(a) <u>Domicile</u> means about the same as (1) suit (2) home (3) vase

(4) will.

(b) Complete the series: 1 3 7 15   (1) 23 (2) 26 (3) 31 (4) 49.

(c) In the following alphabetical listing, <u>Schmidt</u> would come

immediately after (1) Scanlon (2) Schein (3) Schoonover

(4) Smidt.

(d) Would you prefer to take a guided tour of (1) a museum?

(2) a factory?

(e) When you approach a group of people, do you usually start a

conversation right away? (1) Yes (2) No.

FIGURE 3.4 Examples of test items. (*a*) A verbal knowledge item. (*b*) A quantitative reasoning item. (*c*) A clerical filing skill item. (*d*) An interest item. (*e*) A temperament or personality self-descriptive item.

particular, have been used extensively as selection procedures. Surprisingly, the scanty available evidence suggests that medical exams are not especially effective in personnel selection.[8]

Literally thousands of tests are available from commercial publishers and consultants. They cover a wide gamut of abilities and temperament factors. An organization may also develop its own tests. It is of course necessary to beware of a wide variety of pitfalls which may entrap the unwary test purchaser or developer. However, with appropriate technical advice, it should be possible to obtain tests for just about any selection situation. Listings and critiques of tests are available from several sources.[9]

The Interview.   An almost universal selection procedure is the interview. It seems natural and obvious to most people that useful information will result from an interview. Unfortunately, in many and perhaps even in most cases, the interview contributes little to the effectiveness of selection. Nevertheless, it can be useful under certain conditions.

Most interviews are left entirely to the discretion of the interviewer. The value of the interview can normally be increased by limiting the interviewer's options. Specific actions which are likely to improve the interview as a selection procedure are:

1. Standardize the interview. That is, specify what the interviewer is to ask and what judgments are to be made, and discourage idiosyncratic procedures by the interviewer.

2. Train the interviewers. Help each interviewer learn the procedures and practices which are most effective.

3. Emphasize the job. The interview should be aimed at finding out how well the applicant would perform on the specific job.

4. Make a record of the interview. Require the interviewer to take notes and use them to justify the judgments made about the applicant.

5. Require multiple judgments. The interviewer should not make a single overall judgment as to the suitability of the applicant but should provide judgments about each major factor which the interview is designed to explore.

---

[8]M. A. Rothstein, *Medical Screening of Workers,* The Bureau of National Affairs, Washington, D.C., 1984.
[9]Especially the Buros Institute of Mental Measurements at the University of Nebraska—Lincoln, and the Test Corporation of America, Kansas City, Mo.

Interviews which are relatively high in the above characteristics are called *structured*. Examples of systematically structured interview formats are given by Janz[10] and Latham.[11] The structured interview is a much more disciplined affair than the usual interview; in many respects, it may be viewed as an orally administered test of job knowledge or intelligence.

The applicant interview typically has purposes other then selection. The interviewer usually gives the applicant information about the job and the organization, and "sells" the organization to the applicant. Purposes such as these are legitimate reasons for an interview and may be important enough to influence the organization to retain the interview, even if it is not cost-effective as a selection procedure.

Decision Strategies in Selection.   For each selection procedure, a score should be generated to characterize each applicant. Then a decision strategy must be generated to determine how the scores will be used to decide which applicant or applicants should be offered the job. The major types of strategies include subjective decisions, multiple cutoff, weighted sum, and mixed strategies.

Subjective Decisions.   In typical personnel practice, selection decisions are made informally and unsystematically. A decision maker or committee reviews the information available on each candidate and simply makes a judgment as to which should be offered the job. This method has little to recommend it in most instances.

Multiple Cutoff.   In this method, a cutoff, or minimum acceptable score, is set for each selection procedure. To be selected, an applicant must meet all the cutoffs (or, when there are many selection procedures, all but one or two). This is a suitable method when there are many applicants and more than a few openings. It can be especially attractive when one or more of the selection procedures is relatively expensive. If the less expensive procedures can be administered first, some candidates will be eliminated without need for the expensive procedures. The organization will thus save money, but at the expense of reduced selection accuracy.

A serious problem with this method is the choice of cutoff points. There is no completely satisfactory method of making this type of choice. Methods developed in educational measurement are probably the best.[12]

Weighted Sum.   A composite score is derived for each applicant:

$$C = \text{sum } W_i X_i \qquad \text{from 1 to } n \tag{1}$$

where there are n selection procedures, X_i is the score on the ith selection procedure, and W_i is the associated weight. Weights may be set on the basis of subjective judgment but are more likely to function well if they are set on the basis of a previous study which allows the calculation of regression coefficients. In principle, the use of regression coefficients as weights produces the most accurate selection possible with a linear model. A linear model is appropriate for almost all selection situations.

Mixed Strategy.   It is common to use a mixture of the multiple cutoff and the weighted sum strategies. For example, it is frequently effective to use a four-stage strategy: (1) an initial rough screening based on résumés, application forms, and perhaps brief interviews; (2) paper-and-pencil tests for those who pass the screening; (3) interviews and performance tests for a group of those ranking highest on the tests; and (4) starting at the top of the ranking list of those remaining, a medical examination for each applicant in turn until a sufficient number of applicants has passed the exam and accepted a job offer.

---

[10]T. Janz, "The Patterned Behavior Description Interview: The Best Prophet of the Future Is the Past," chap. 11 in R. W. Eder and G. R. Ferris, eds., *The Employment Interview: Theory, Research, and Practice,* Sage Publications, Newbury Park, Calif., 1989, pp. 158–168.

[11]G. P. Latham, "The Reliability, Validity, and Practicality of the Situational Interview," chap. 12 in Eder and Ferris, op. cit., pp. 169–182.

[12]S. A. Livingston and M. J. Zieky, *Passing Scores: A Manual for Setting Standards of Performance on Educational and Occupational Tests,* Educational Testing Service, 1982.

Evaluating Selection Procedures. The key concept in evaluating selection procedures is that of *validity*. in the context of personnel selection, validity means the degree to which selection decisions are correct. Validity is a characteristic not of the selection procedure or procedures but rather of the decision process. Suppose that by some magical process we could determine how well each candidate would perform the job if given the opportunity; a perfectly valid selection procedure would predict the job performance of each candidate perfectly. Of course, it is not possible to know how the unsuccessful candidates would have performed on the job. All that can be done is to gather evidence bearing on the likelihood that the selection procedures will produce good decisions. Three major types of evidence bearing on validity are recognized: criterion-related, content-related, and construct-related.

Criterion-Related Validity Evidence. Evidence of criterion-related validity is obtained by studying, for a group of employees, the relationship between selection procedure scores and criterion scores representing job or training success. The usual index of the relationship is the product moment correlation coefficient, $r = C_{xy}/S_x S_y$, where C_{xy} is the covariance of the two sets of scores and S_x and S_y are the respective standard deviations.

It is conventional to attribute validity to a selection procedure if the correlation derived from a given study is high enough, given the number of cases in the study sample, to be significantly different from zero at the 0.05 level of confidence. However, this convention is a dubious one, since the coefficient is unstable unless the study sample is quite large and since other validity evidence may be available or obtainable. It is better practice to consider the results of a criterion-related validity study to be only one piece of evidence on which to base an overall judgment of validity.

The criterion measure should be selected or developed to represent job satisfactoriness. In most cases, a global measure of job performance is best. It is not necessary for the criterion to be highly reliable, as long as it is possible to obtain an accurate estimate of its reliability. The estimate of validity is reduced by criterion unreliability, but the coefficient may be adjusted to take this into account. For discussions of this and other technical problems in studies of criterion-related validity, see the volume on test validity edited by Wainer and Braun.[13]

Content-Related Validity Evidence. This type of evidence bears on the extent to which the content of the selection procedure is similar to the content of an important aspect of the job. By *content* is meant either knowledge or skill necessary for performing well on the selection procedure and on the job. Thus, to gather this type of evidence about a test of knowledge of arrest procedures as a selection procedure for police sergeants, one would analyze the test and the job to determine if the knowledge needed to get a high score on the test is also needed to perform the job well. For a typing test as a selection procedure for secretaries, one would perform a similar analysis in relation to the skills involved.

There is no accepted index of content validity. Rather, a judgment is made as to the strength of the content-related validity evidence. This type of evidence should not serve as the sole basis for evaluation of a selection procedure. It must be supplemented by other types of validity evidence, or by evidence showing how well the selection procedure functions in practice, or preferably by both.

Construct-Related Validity Evidence. This type of evidence consists of data or argument tending to show the degree to which (1) the selection procedure is an adequate measure of a defined psychological construct or idea, and (2) the construct is an important element in performing the job well. The construct would normally characterize people's habitual behavior; that is, it would be a *trait*. The evidence for construct-related validity tends to be complex and extensive. A series of studies, rather than one-shot research, is normally necessary.

Reliability. Reliability is defined as described above for performance appraisal measures. It may also be estimated in the same ways, and by some procedures which consti-

---

[13]H. Wainer and H. I. Braun, eds., *Test Validity,* Erlbaum, Hillsdale, N.J., 1988.

tute a special case of equivalent forms reliability. It is especially important to determine reliability for a selection procedure when much of the validity evidence is of the content-related type.

Other Indicators of Functioning.   Some other matters to be checked:

1. For a test or similar procedure, difficulty should correspond at least roughly to the standards which are applied on the job.
2. Procedures for collecting information should be clearly described in a set of instructions.
3. Scoring instructions should be clear and should produce consistent scoring.
4. Time limits for tests and similar procedures should be appropriate and adhered to strictly.
5. Test security should be rigidly maintained when it is needed (that is, when previous knowledge of test content would provide an applicant an unfair advantage).

Validity of a Composite.   The validity of multiple selection procedures has been studied only in the context of criterion-related validity, using a weighted sum strategy. Consider the equation connecting the selection procedure scores to criterion scores:

$$C = \text{sum } \hat{W}_i X_i + e \tag{2}$$

where e is the error term and the other terms are as given in Eq. (1). When e is minimized in the least squares sense, Eq. (2) is the regression equation, in which the weights produce the best linear composite. The validity coefficient is then the multiple correlation, that is, the correlation of the best linear composite with the criterion.

Three factors determine the weight to be accorded each selection procedure in the regression equation. The first factor, of course, is the criterion-related validity of the selection procedure; the higher that validity, the higher the weight. The second factor is the variability of the selection procedure; weights are inversely related to variability. Third is the pattern of intercorrelations of the selection procedures with each other; the lower the intercorrelations for a given selection procedure, the more weight is given to that procedure.

Regression weights tend to be unstable statistically, unless the study sample is very large. Hence the use of exact weights is normally unnecessary and multiple validity coefficients should be considered only as rough estimates.

Validity Generalization.   It has often been asserted that the validity of a selection procedure or composite is limited to the specific situation in which it has been studied. Recent meta-analytic studies, however, have shown that generalization across organizations and situations may normally be justified, if the job and the environment in the two organizations or the two situations are reasonably similar. Hence, before adopting or trying a selection procedure for a given purpose, one should survey validity studies of that selection procedure for similar jobs in other organizations. Confirmatory studies in the using organization are of course desirable, but they are not always necessary or definitive.

Utility.   The final determination of the usefulness of a selection procedure or program may, for many organizations, rest on the question of economic utility, or cost-benefit analysis. A general function expressing utility in selection is

$$U = VPS - C \tag{3}$$

where U = dollar gain per selectee per year
   V = criterion-related validity
   P = productivity variation, or the standard deviation of the criterion, expressed in dollars per year
   S = selectivity, or the average selection procedure score or composite score of those who are placed on the job, divided by the standard deviation of scores for all applicants
   C = cost per selectee per year

This is a stripped-down version of the formula. More complex statements have been proposed to take account of the time value of money, tax considerations, and other factors.

Fair Employment Considerations. Ethical employers take pains to assure that all applicants are given fair and impartial consideration. The ethical motive is reinforced in the United States by various laws, especially Title VII of the Civil Rights Act of 1964, under which it is illegal to discriminate in a personnel decision on the basis of race, color, religion, sex, or national origin. A distinction is made between *disparate treatment* and *disparate impact.* Disparate treatment occurs when the employer intends to discriminate and proceeds to do so. On the other hand, disparate impact may come about without an explicit intention on the part of the employer. It involves the use of a "facially neutral" selection procedure which, however, results in discrimination. Such discrimination is also illegal, *unless* the employer can show that its selection practices serve a business purpose; a showing of validity would normally be relevant.

Various legal complexities characterize the issue of employment discrimination, so that it is advisable to seek legal advice on such matters.

TRAINING AND DEVELOPMENT

It is traditional to distinguish between training and development, presumably on the basis that development is a more ethereal and idiosyncratic activity than training. But both words refer to the process of purposefully changing behavior, and there is little to be gained from positing two such processes.

People change whether or not they want to change. Unplanned change, however, is unlikely to constitute progress toward the goals of the individual or of the organization. It is important for organizations, in concert with their employees, to organize for planned change in the skills, knowledge, work habits, social interactions, and attitudes of employees. (Let us refer to all these as changes in *skills*.)

Psychological Ideas in Behavior Change. Psychological theory suggests a number of ideas for changing behavior. Some of the most immediately useful ideas are summarized in the following paragraphs.

Reinforcement. Given stimulus S, a person makes response R. If R is followed by a state or condition which changes the probability that the person will respond R to stimulus S next time, that state or condition is said to be a reinforcer. In less technical terms, a reinforcer is a pleasant outcome which causes the person to tend to repeat the response or an unpleasant outcome which causes the person to tend to avoid making the response in the future.

Logically, there are four possibilities after stimulus S:

1. The person responds R; the outcome is pleasant, so R becomes more probable on the next presentation of S.
2. The person responds not-R (any response other than R); the outcome is pleasant, so not-R becomes more probable next time.
3. The person responds R; the outcome is unpleasant, so R becomes less probable.
4. The person responds not-R; the outcome is unpleasant, so not-R becomes less probable and hence R may be more probable.

All these possibilities have been shown to be effective under some circumstances. However, the first one—positive reinforcement—is considered to be the most broadly useful.

Repetition of a stimulus-response sequence with no consequent reinforcement leads to extinction, or disappearance, of the response. However, it is not necessary to reinforce a desired response on each appearance. On the contrary, actions which are initially reinforced sporadically tend to be more long-lasting and resistant to extinction than those initially reinforced on each occasion.

For reinforcement to be effective in changing the behavior of an employee, it is critical that the reinforcer be clearly contingent on the desired actions of the employee. The employee must be convinced that the reinforcement will be delivered if, and only if, he or she responds as specified. This is a point often overlooked. Organizations typically make little effort to emphasize, for example, the relationship (if there is one) between the amount of a paycheck and the productivity of the worker during the corresponding pay period.

Many applications of reinforcement theory aim merely to encourage or discourage actions which are already in the employee's repertory. However, it is also possible to change behavior through reinforcement, using the idea of *shaping*. In shaping, a reinforcement is delivered when some approximation of the desired action is performed. Early in the process, the approximation may be a very poor one; as shaping proceeds, the reinforcement is applied only as the approximation becomes a little closer, until the action is as desired. For example, to teach a scowling salesperson to smile at customers, start by rewarding a facial expression which is less unpleasant than usual, then reward neutral expressions, and gradually proceed to reward more and more pleasant facial expressions.

Modeling.   Shaping is often a laborious process and is normally unnecessary with adult learners. We learn many of our skills by observing skilled performers and copying their actions. If a model of performance is to be the basis for learning, it is critical that the model be a good one. Often, a new worker is told to learn from watching experienced employees, chooses a less than perfect model, and learns some poor work habits as a result. Management should take pains to see that employees learn only from others who do the job well. Many modern training programs use films or videotapes to model correct behaviors, thus assuring that the model is correct and consistent.

Active Practice.   Learning is facilitated when the learner is active rather than passive. If the skill to be learned involves physical movement, training should be designed so that the learner is actually performing the motions or manipulations as soon as and as often as is feasible. If the skill involves knowledge or reasoning, the learner should spend as much of the training time as possible in producing the knowledge or reasoning. For example, instead of reading and rereading material in a textbook, the learner should put the book aside as soon as possible and try to recite or rehearse the essential content. Of course, the book must be consulted to check accuracy and fill gaps.

Goal Setting.   A powerful tool for improving performance or knowledge is goal setting. It is usually not especially effective merely to wish to be or to do better. However, when people set goals properly, they get results. Properly set goals are (1) specific and measurable ("I will be tardy no more than two times next month," rather than, "I will try not to be tardy so much."), (2) tied down by a time schedule when appropriate, (3) difficult enough to be challenging but easy enough to be feasible of achievement, and (4) accepted by the learner. The last point is often overlooked. Some managers and trainers persist in setting goals for learners, without regard for what the learners may think. If the goal is to be effective, the learner must accept it and make a commitment to work toward it. It follows that the learner must either participate in setting the goal or trust the trainer to set the goal fairly and reasonably.

Knowledge of Performance.   Practice does not necessarily make perfect. The learner may be practicing incorrect or inefficient skills, and hence learning the wrong things. The learner should have access to timely and accurate information about his or her performance. Sometimes, the learner can observe the performance or its outcome, and judge how good it is. In other cases, the learner needs the help of a device or another person to provide information about the performance.

Knowledge of performance, or *feedback,* is not always good. Often, it takes the form of criticism or is interpreted by the learner as criticism. People tend to become defensive when criticized, and to reject the information. Trainers and supervisors should (1) provide positive as well as negative feedback, (2) build a climate of trust before attempting extensive criticism, and (3) emphasize how the learner can do better rather than how he or she has done wrong.

Transfer. Transfer is the degree to which the learner will apply the newly learned skill on the job. Training and development programs are often carried out in classrooms or other contexts remote from the job. When the employee returns to the job, the new knowledge or skill may not be used or may not fit the job situation. To avoid such an outcome, the training and development planner should (1) make sure that the employee *intends* to use the new skill on the job; (2) design the training to be relevant to the job (for example, by simulating the job environment in the classroom, or by asking learners at frequent intervals how they might apply newly learned skills on the job); (3) teach principles which can be applied broadly to job situations, rather than teaching specific procedures which may not apply in some job situations; and (4) ensure that the learner's supervisor and co-workers are expecting the new behavior and that they will support and encourage it on the job.

Distribution of Practice. Given that x hours are to be devoted to a learning task, how many sessions should be scheduled and how long should each session be? There is extensive evidence indicating that, to maximize learning and remembering, there should be a relatively large number of relatively short sessions. Of course, some minimum session time may often be dictated by the need for setting up equipment, assembling a group of learners, and so on. Within reasonable limits set by practical considerations, however, it is best to minimize session length.

Sessions need not be of uniform length. If it is necessary for logistical or other reasons to schedule one or more long initial sessions, it would be desirable to have a number of short review sessions in succeeding weeks or months. For example, it is often the case that an employee leaves the normal workplace to attend a training course, which may range in length from an hour to a year. Whatever the length of the course, the employee and the supervisor would be well advised to find subsequent opportunities to review any of the course content which the employee does not get regular opportunity to practice. Review and rehearsal, in other words, are important parts of the training.

Training Environments. Training may take place in a variety of environments or contexts.

Classroom training. Most formal training is carried out in a setting similar to that of the school classroom. Such a setting allows easy control and organization of groups of learners, and many training methods and media are easily accommodated. The major disadvantage of the classroom is that it is remote from the job, not so much in distance as in overall atmosphere and context. It may be difficult to promote transfer in the classroom.

Simulated Job Environment Training. To encourage transfer, the training facility may be designed to simulate the job situation, or some part of it. The computer operator trainee, for example, may learn at a console similar to that used on the job. The trainee ideally practices job-relevant responses to job-relevant inputs.

On-the-Job Training. For maximum transfer, it may seem that training should be delivered on the job site. Sometimes this reasoning is justified. In most cases, however, on-the-job training is inefficient and ineffective. Instruction is typically informal and unplanned. The beginning trainee is likely to hinder productive activities. The supervisor and one or more experienced workers must take time away from regular duties to instruct and monitor the learner.

Even after initial learning of a job takes place, on-the-job training continues. It is generally assumed that employees learn as they earn, and that the more experience an employee gets, the more job-relevant knowledge and skill that employee accrues. No doubt, learning takes place; but it may not always be the kind which management would prefer. In many cases, what the experienced employee learns is how to make excuses, how to seem busy without actually doing much, or how many absences one can get away with.

Attitudinal Training.   Some organizations have determined that they will benefit if their employees display certain attitudes and styles of interaction with other people. Typically, it is desired that employees be more open, frank, and sharing with others. A number of training programs have been developed with the intention of modifying attitudes.

Most such programs are of the sort called T-groups, encounter groups, or sensitivity training. A group of people meets over a period of several hours or days, with the guidance of a facilitator. They carry out exercises aimed at clarifying their attitudes and feelings, and they discuss their feelings about themselves and each other, not necessarily in the context of work relationships. They are encouraged to develop an open and trusting style.

Many, though probably not all, such programs have an impact on the attitudes and behavior of participants. However, there are at least two problems with most sensitivity programs. First, since feelings are expressed openly, the emotional atmosphere in a group may become highly charged. Some people may find the tension so threatening that they withdraw physically or emotionally. The facilitator must be competent and alert to identify impending problems, and should be prepared to obtain professional counseling for those who may need it. The second problem is that transfer from a sensitivity group to the job is likely to be small. Attempts to be open and emotionally sharing on the job are not always welcomed by those who have not participated in the training. When this is the case, the new style of interacting is likely to be abandoned.

Training Needs.   Many training programs are designed on the basis of someone's intuitive notion of what is needed. Intuition is sometimes right, but the organizational batting average will be improved if formal and systematic procedures are used.

To identify training needs, it is logically necessary first to identify the set of knowledge, skills, and attitudes required by the job, second to inventory the set of knowledge, skills, and attitudes possessed by the job incumbents, and third to compare the two sets. The training needs are the difference; that is, the job requirements which the incumbents do not meet.

The job requirements are established in the job analysis and job description. However, many job descriptions do not deal with knowledge, skill, and attitude requirements. Hence it may be necessary to extend and expand the job description to cover these matters and to give consideration to the ways in which experts do the job.[14] Also, it is important to update job descriptions at intervals to take account of new methods, new equipment, and new policies.

Training Objectives.   Given a statement of training needs, the training developer can then formulate objectives for instruction. In most situations, this is a straightforward process. However, it may be necessary to specify two types of objectives: those which characterize the skill needed on the job, and those which are necessary for the development of the skill. Objectives of the second sort are called *enabling* objectives. For example, if the skill to be learned is estimating distance from a map, it may for some learners be necessary first to teach the ideas of ratio and proportionality.

There are some drawbacks to precisely specified objectives. Advanced and superior learners will often feel unduly constrained by them and may suffer a loss of motivation. Some subject matters are sufficiently complex that clear statements of learning objectives are difficult or impossible to formulate. One should be alert to the possibility that such a subject matter is too vague to be taught. However, it seems clear that some complex sets of ideas or other skills are simply resistant to easy formulation but can nevertheless be studied and learned with profit.

Once the objectives have been formulated, the training content can readily be specified. It consists of material which deals with the objectives and with nothing else. Some objectives are so broad that they cannot be covered completely. In such a case, it is necessary

---

[14]M. T. H. Chi, R. Glaser, and M. J. Farr, *The Nature of Expertise,* Erlbaum, Hillsdale, N.J., 1989.

to sample from the objective and to provide guidelines for generalizing to those points which are not explicitly covered.

Training Methods and Media. A training method is the general strategy used in training or development. A training medium is the physical means by which the method is implemented. For most purposes, it is not necessary to distinguish strictly between the two ideas.

Lecture and Discussion. In this method, a trainer presents material orally, answers trainees' questions, asks questions, and facilitates discussion. This is a very flexible method, often criticized but seldom shown inferior to other methods. A lecture may readily be updated or otherwise changed to fit changing circumstances. The success of the method is highly dependent on the skill and motivation of the trainer.

Film and Videotape. A film can present a wide variety of visual and auditory information. It can use animation, live performers, or a combination of the two. It is especially useful for modeling an activity in a standard form. The initial cost of making a film is high, but the cost per showing can be small over many showings. It is hard to keep a film up to date.

Programmed Instruction. In this method, a small amount of instruction is presented, a response by the learner is required, and immediate feedback is given. This sequence is repeated until all the material has been presented and mastered. Normally, programmed instruction is self-paced, allowing quick learners to go fast and slow learners to take the time they need. Programmed instruction may be used with a computer to allow branching from the mainstream of instruction, primarily to help learners who have misunderstandings or who do not understand a point completely.

Case Discussion. Under the direction of the trainer, a group discusses a case problem. The case problem is designed so that the trainees, if they are to respond to it effectively, should make use of the skills which the program is designed to teach. If the case is realistic and job-relevant, transfer is facilitated. However, the method encourages verbal gymnastics, often at the expense of job-related skill development.

Role Play. The learner is asked to play the role of a job incumbent meeting a situation which calls for some of the skills to be learned. Learners take turns playing the roles of others in the situation. This method is useful when the skills to be learned require communication and dealing with other people.

Simulation and Games. Some aspects of the job are simulated, along with a situation requiring the use of the skills to be learned. Learners are to act as if they were on the job. There may be competition between teams of learners, or the learner may simply practice the skills involved. This method allows for a high level of complexity in the application of skills, but it is typically difficult to control the situation and to evaluate the learner's performance.

Measuring Trainee Progress. An important but often ignored element of training design is that of measuring how well trainees are learning or have learned the intended skills. The measure may be a test of the usual classroom type using pencil and paper. An alternative is to simulate a job situation which would require application of one or more relevant skills, have each trainee play the role of a job incumbent dealing with the situation, and evaluate the effectiveness of the trainee's actions in relation to the training objectives. Ultimately, it is desirable to observe the employee's behavior back on the job, identify instances when the skill is appropriate, and record the degree to which the employee applies the skill effectively.

Evaluating Training Programs. Careful planning by experienced people can go a long way toward producing an effective training program. Unfortunately, however, there is no way to be sure that such a program will work. Hence it is essential to evaluate the effectiveness of each program. It is useful to categorize evaluation efforts according to the type of outcome considered, as in the following paragraphs.

Program Effectuation. A common problem with training programs is that they are simply not carried out as designed. A number of cases have been documented in which a training program was carefully planned, supervisors and trainers informed, and program materials made available, but the program was not implemented or was carried out in a distorted form. To check how well the program has been effectuated, the evaluator should seek information such as (1) number and job titles of people who have served as trainers and trainees, (2) distribution of training materials, such as textbooks or tools, (3) number and length of meetings or training sessions, (4) attendance at each meeting or training session, (5) objectives covered at each session, (6) adequacy of facilities used, (7) availability and use of teaching equipment, such as film projectors, and (8) number of trainees whose progress was evaluated, and the completeness of the evaluations. All this information is to be compared with that which was planned and to be interpreted in the light of the training needs and related organizational functioning.

Acceptance. If a training program is to be effective, it is generally necessary that the trainees approve of its plan and implementation, or at least find them acceptable. It is generally worthwhile to ask trainees how interesting and useful they found the program to be. In addition, trainees may be able to criticize and evaluate the program with some degree of objectivity and knowledgeability. Normally, trainee evaluative information is collected by means of a questionnaire or interview.

Change in Trainee Behavior. Since the training program is designed to change trainee behavior, it makes sense to check whether in fact such change has taken place as a result of the program. The measuring instrument would normally be the same as that used to evaluate individual trainee progress. Since change is to be measured, the evaluation test should be administered both at the beginning and at the end of the program. Alternate forms of the test should be used when the trainees' memory of specific test content might influence their responses to the postprogram test.

Change in Job Performance. The payoff for any training program is improved job performance. Change of this sort is typically difficult to observe and quantify. Some ways of doing this are:

1. Interview the supervisors of a sample of participants after an appropriate time. Ask them to report specific occasions when the participant has performed better in areas relevant to the program than would have been expected before he or she was trained.

2. Ask supervisors or participants to keep a log or diary of events indicative of progress or lack of progress in areas relevant to the training program.

3. Use interviews or questionnaires to third parties, such as customers, to estimate the extent to which new skills are being applied on the job.

4. Collect records of production before and after the training program; compare pre- and postprogram data in areas relevant to the training.

5. Collect pre- and postprogram performance appraisals for a sample of participants. Check differences relevant to the program.

Data collected by any of these methods are likely to be somewhat unreliable. In some cases, the data will be unacceptably poor. However, just as a strong signal overcomes noise, a strong program impact will show up in spite of unreliability of data. If the impact of the training program is not strong enough to be detected, it is possible that it is not strong enough to justify the cost of the program.

Utility of Training Programs. Utility may be estimated as

$$U = EP - C$$

where U = utility in dollars per trainee per year
E = effect size
P = productivity variation as in Eq. (3)
C = cost of the program per trainee per year

Effect size is the difference in job performance between the average training program graduate and the average comparable employee who has not yet participated in the program, in standard deviation units. It should normally be estimated through one or more studies of job performance change.

BIBLIOGRAPHY

Berk, R. A., ed., *Performance Assessment: Methods and Applications,* The Johns Hopkins University Press, Baltimore, 1986.

Eder, R. W., and G. R. Ferris, eds., *The Employment Interview: Theory, Research, and Practice,* Sage Publications, Newbury Park, Calif., 1989.

Gagné, R. M., ed., *Instructional Technology: Foundations,* Erlbaum, Hillsdale, N.J., 1987.

Goldstein, I. L., and Associates, *Training and Development in Organizations,* Jossey-Bass, San Francisco, 1989.

Muchinsky, P., *Psychology Applied to Work: An Introduction to Industrial and Organizational Psychology,* 3d ed., Brooks/Cole, Pacific Grove, Calif., 1990.

Smith, M., and I. Robertson, eds., *Advances in Selection and Assessment,* Wiley, New York, 1989.

CHAPTER 4
JOB EVALUATION[1]

George J. Matkov, Jr.,
Attorney at Law
Matkov, Salzman, Madoff & Gunn
Chicago, Illinois

Debra I. Danner,
Attorney at Law
Matkov, Salzman, Madoff & Gunn
Chicago, Illinois

Ideally, employees should be paid according to the nature of the job they perform and its value in relation to other jobs in the organization for which they work. The external competitiveness of a company's wage structure is also an important factor in its pay policies, since it must continue to attract and retain qualified applicants in a competitive employment market.

Establishing job values and internal pay differentials is one of the most important challenges an organization must face in developing and administering its pay plan. The larger the organization, the more important consistency and fairness are to the rules of job structuring and wage allocation among the many departments. The more people performing the same or similar tasks, the greater the demand for "equal pay for equal work." The more difficult, strenuous, or tedious the work tasks or conditions, the more employees will expect to be rewarded for their efforts.

Employee dissatisfaction with their employer's wage practices will have a definite impact on their performance and productivity, as well as on their relationship with company management and coworkers. Some may leave. Others will attempt to organize a union to protect their interests, file grievances with the existing union, or seek recourse through the panoply of administrative agencies and judicial bodies. Potential applicants and recruits may forgo the company's employment offer altogether in favor of better-paying, more attractive positions.

Responding to such concerns with random, across-the-board pay increases or ad hoc individual wage adjustments will inevitably serve only to aggravate the problem and perpetuate the perceived (or actual) inequities in the company's pay structure. More importantly, it will also undermine the credibility of the company's overall wage and human resource policies themselves. Instead, companies must address such problems head-on by periodically reviewing their job structure and wage system as a whole, and revising it when necessary. It is equally important that a common understanding be fostered among both employees and management as to how the entire pay structure operates, so that people's perceptions can run closer to reality.

---
[1] Grateful acknowledgment is due to Marty Nadis, CCP, a senior compensation analyst at The Management Association of Illinois, and Dennis Sulik, supervisor of Industrial Engineering at Sta-Rite Industries, Inc., in Delavan, Wisconsin, for their kind assistance in reviewing this chapter.

Job evaluation is the tool by which this overall review and adjustment can be accomplished. Its primary goal is to assist in establishing pay structures that are equitable and defensible—that is, those which ensure equal pay for equal work and reward employees for working in jobs which are of more value to the company and/or require greater effort, skill, or hardship.

JOB EVALUATION

Briefly stated, job evaluation is a formal process by which an organization determines the relative value of the various jobs within it in order to develop and maintain a sound basis for wage administration. For most organizations, job evaluation is no longer the novelty it was during the 1940s when many of the basic systems were developed. Instead, it has become a well-respected and widely used management tool in companies of all sizes around the world.

Job evaluation begins with an analysis and description of each job as it currently exists in the company. The contents of the job are then measured using a set of categories or criteria, in order to determine its relative value. Once the evaluation is completed, the jobs are ranked according to their scores, and then grouped into a hierarchy of grades which form the basis for the establishment (or revision) of the wage structure. For that point, the organization can set minimum and maximum pay rates for each grade based upon a variety of internal and external factors. The resulting pay system must then be checked for accuracy and consistency and then implemented and maintained with a fair and even hand.

There are three basic principles of job evaluation to keep in mind: (1) the job evaluation process establishes the *relative* rather than the absolute value of jobs; (2) it is the *job*, rather than the person performing the job, which is being evaluated; and (3) the results of a job evaluation are *just one factor* in the determination of the appropriate wage rate for a particular job or class of jobs.

THE MANAGEMENT DECISION

A job evaluation program is a major undertaking that requires a great deal of planning, time, and effort from start to finish on the part of everyone involved. The first major decision that must be made, therefore, is whether a job study should be undertaken at all. This decision is important not only because of the expense involved, but because the process and its results will have a substantial impact upon the personnel decisions and operations of the company for a long time to come. For this reason, the decision to embark on a job study must be made at the top level of an organization, and must receive the full support of its chief executives as well as the on-site management team. In order to make this commitment clear to all employees, the announcement of the study should also come from the company's top management rather than, for example, the personnel or industrial engineering department.

A job evaluation program is usually considered when one or more of the following circumstances exist:[2]

• Generalized internal dissatisfaction and frequent disputes have arisen concerning the wage structure, including claims that similar work does not result in equal pay, that equal pay is given for dissimilar work, or that differences in pay are not related to the work actually performed. (See discussion below.)

---

[2]Adapted from International Labour Organisation, *Job Evaluation*, Geneva, 1986, pp. 4–5.

- There is concern that certain groups of workers (such as women) are being underpaid as the result of unlawful discrimination.

- New equipment or new methods of work have been introduced which change the content of many jobs, resulting in the need for establishing a new basis for remunerating fairly those workers who have been affected.

- Organizational changes (such as a business consolidation) have been made which necessitate a revision of the pay scheme in order to harmonize diverse wage rates into a single, coherent structure.

- Changes in the nature of the company's work, machines, production methods, or products have left an unmanageable number of job descriptions to administer.

- The business plan of the organization calls for changes in resource allocation and labor cost management.

- New pay delivery vehicles (such as pay-for-performance, gain sharing, or other performance incentives) are being introduced which require that a sound base pay structure be in place.

- A high employee turnover rate and growing number of unsuccessful recruitment attempts indicate that better-paying employment opportunities are being offered elsewhere in the industry or area.

Job evaluation is more than simply a palliative for rectifying certain wage problems, however. The in-depth analysis that is required for both individual jobs and the organization's entire structure forms a solid basis for modern human resources management and policy making. In fact, job evaluation is most useful when it is part of a package of measures aimed at resolving a wider set of management issues.

EQUAL PAY FOR EQUAL WORK

One of the most important factors impacting the growth of job evaluation in the last 30 years has been the proliferation of federal and state legislation that mandates that employers give employees equal pay for equal work.

Discrimination in the form of unequal pay based on sex is made unlawful by the federal Equal Pay Act[3] and by Title VII of the Civil Rights Act of 1964,[4] as well as by Executive Order 11246 (covering government contractors) and state civil rights and equal pay acts. "Equal work" is defined under the federal Equal Pay Act as work that is substantially equal, but not necessarily identical.[5] The purpose of the federal Act has been described as follows:[6]

The Act was intended as a broad charter of women's rights in the economic field. It sought to overcome the age-old belief in women's inferiority and to eliminate the depressing efforts on living standards of reduced wages for female workers and the economic and social consequences that flow from it.

Specifically, the federal Equal Pay Act provides:

No employer having employees subject to any provisions of this section shall discriminate...between employees on the basis of sex by paying wages to employees...at a rate less than the rate at which he pays wages to employees of the opposite sex...for equal work

---

[3]29 U.S.C. § 206(d) (1963).
[4]42 U.S.C. § 2000e et seq. (1964).
[5]29 CFR § 1620.13(a).
[6]Schultz v. Wheaton Glass Co., 421 F.2d 259 (3d Cir. 1970), cert. denied 398 U.S. 905 (1970).

on jobs the performance of which requires equal skill, effort, and responsibility, and which are performed under similar working conditions, except where such payment is made pursuant to (i) a seniority system; (ii) a merit system; (iii) a system which measures earnings by quantity or quality of production; or (iv) a differential based on any other factor other than sex....

The act prohibits, for example, classifying jobs as "light" or "heavy" and then placing all or nearly all females in the "light" jobs at a lower rate of pay, when the females perform substantially the same work as males working in the "heavy" jobs such as maintaining two separate categories for "janitors" (males) and "cleaners" (females).

While the equality or inequality of jobs will generally be judged by the actual job requirements and performance, rather than job classifications or titles,[7] having a bona fide job evaluation system in place can go a long way toward defending the legitimacy of an employer's pay structure. Note that the Equal Pay Act specifically states that job comparisons under it will be guided by four factors: skill, effort, responsibility, and working conditions. Not surprisingly, these factors have been the basis for most job evaluation systems since their inception.

The potential for bias in the job evaluation process itself, however, cannot be ignored. While a job study may be more objective than ad hoc personal judgments made by company management, the process is still subjective in many ways. From the selection of the members of the job evaluation committee and the job analysts themselves, to the preparation of the job descriptions, the selection and weighing of the factors by which the value of each job will be determined, and the selection of benchmark jobs, an employer must make every effort to maintain the integrity of the evaluation process and ensure that whatever inequalities may have existed are not perpetuated in the job evaluation plan. If, for example, the job evaluation committee consists only of white males who have experience with a very narrow range of jobs in the plant, or greater weight is uniformly given to factors which are less prevalent in jobs usually occupied by a particular employee group (such as weight-lifting requirements rather than education), the process will likely be tainted and less defensible in charges of discrimination.

WHO SHOULD DO THE JOB STUDY?

Once the decision has been made to conduct a job study, it is necessary to determine who will do it. The answer will depend upon the type and scope of the project to be done and the resources available. The company may hire an outside consulting firm to do the entire job, or it may choose to do the job itself internally, utilizing qualified company employees. It may also elect to utilize a combination of these two approaches, using an outside firm on a consulting basis to assist company employees in performing the study.

Contracting with an outside firm may be the quickest way to accomplish the study because the analysts are already trained and experienced professionals whom the company will pay to do the job—and only that job—efficiently and correctly. Their work is more likely to be fair and impartial, since they presumably have no ax to grind or favors to be done within the company. Their impartiality and expertise also become important if the results of the study are challenged in a grievance or legal proceeding. Hiring an outside firm may also be desirable when a company does not have, or cannot commit, the internal resources necessary to get the job done. Many companies are reluctant to take employees away from their regular assignments in order to be trained and then participate in such a project.

The argument for doing the study internally is that company employees will take ownership of the project more readily and come to identify themselves more closely with it,

---

[7]29 CFR § 1620.13(e).

since they will be responsible for both its progress and its results. In addition, company employees are more familiar with the traditions of their workplace and may be better able to adapt the evaluation program in light of the job content, wages, and lines of authority that already exist. (This "familiarity" with the inner workings of the company does, of course, have its downside, since current employees may be less able to step away from their own subjective perceptions and experiences and view the company's overall job structure in a more objective light.) Using company employees to do the study also provides training for those whose job it will be to acquaint the rest of the work force with the study results, and who will be expected to police and maintain the wage structure after it has been established. Too often when an outside firm completes the study, management accepts the results as a finished product which should not (or cannot) be modified with the passage of time. However, as will be emphasized throughout this chapter, job evaluation is part of an *ongoing process* of keeping the company's job and wage structure balanced and up to date.

Using an outside organization of experts on a consulting basis combines the advantages and minimizes the disadvantages of using either internal or outside resources exclusively. Experts can be brought in to train and assist company management in setting up and then implementing a sound job evaluation system, and can be retained to make (or assist the company in making) periodic checks on the system with which they will already be familiar. While the up-front costs of this approach may seem high, they should be weighed against the costs of training and monopolizing the company's internal resources during the course of the job study, as well as the cost of rehabilitating and validating a system gone awry.

Job Evaluation Committee.   Regardless of which approach is used, the job evaluation project should be guided by a central committee composed primarily of company personnel. The membership of this committee is generally left to management discretion. Most committees have a permanent chairperson and permanent members, as well as temporary members such as supervisors, forepersons, industrial engineers, and quality control specialists, who may be brought in when matters concerning their particular area or expertise are being considered.

While management may choose to assume the bulk of responsibility for the project and conduct the job evaluations themselves, more accurate data and greater acceptance of pay decisions will result if both direct supervisors and employees are involved in the process. Some companies set up *joint job evaluation subcommittees* specifically for this purpose. Joint subcommittee members can participate in the study itself, or, perhaps more importantly, serve as a reviewing body for the project's results before they are finalized. They should also play a critical role in disseminating information about the final wage program and its underlying rationale when it is introduced to their fellow employees.

The selection of employee members of the joint committee is critical to its success. Like those from the management team, they should represent a wide variety of departments, have a broad range of work experience, and be well respected by their peers.[8] Employees with "hidden agendas" or particular grudges can subvert or, at the very least, delay the entire process.

The Role of the Industrial Engineer.   The role of a company's industrial engineering staff, be it one person or an entire department, in a job evaluation program will depend upon the approach the company decides to take. Industrial engineers may find themselves directing the entire job study from start to finish, or they may serve as analysts or consultants during the evaluation process itself. Regardless of their precise role, however, industrial engineers will provide valuable information and lend expertise to the job evaluation process. Moreover, once the study is completed, industrial engineers become an integral part of the implementation and administration of the company's pay system.

---

[8]In a union setting the employer may be obligated to include union representatives on the committee.

It is therefore critical that industrial engineers, like company management, become familiar with the various job evaluation systems and their methods.

SELECTION OF JOBS TO BE INCLUDED IN JOB STUDY

While it may be possible to include all the jobs in an organization in a job evaluation program, it is seldom desirable to group them all together into one study. Instead, jobs should be separated into large natural divisions or major groupings, which may be commensurate with existing or planned separate pay plans. For example, in a manufacturing company it is best to evaluate factory jobs separately from the office jobs (clerical, supervisory, and administrative). Jobs of high-level managers, officers, and executives are also commonly segregated for evaluation purposes, and are often left out of job evaluation programs entirely.

Regardless of which groups are selected for a particular evaluation project, the overall end result must be equity among *all* jobs in the company.

EMPLOYEE COMMUNICATIONS

One of the most important, though historically most neglected, aspects of a job evaluation program is the need to foster good employee communications. In order to achieve maximum employee acceptance and cooperation, company management must let the work force in on what is going on and why. Moreover, one of the goals of job evaluation is to promote understanding and agreement among all members of the work force as to what is expected and what is important. A good job evaluation program will lend a common language to promote this dialogue.

As mentioned above, the decision to embark upon a job evaluation program should be communicated from the company's top level of management. This will not only add legitimacy to the study itself, but will also serve to affirm the company's commitment to responding to employee concerns and facing the challenges of a changing marketplace. The company's communications efforts cannot stop there, however. As part of the initial planning process, a systematic program of employee communications should be devised to keep employees apprised of the progress of the study through announcements, meetings, and handouts. For example, group informational meetings should be held before the job analysis phase of the program begins to acquaint employees with the job evaluation process, and to enhance employee cooperation and acceptance of the work being done around them—and of the analysts themselves. Updates should then be given periodically during the course of the study, indicating its progress and the expected completion date. A company that neglects these important human relations aspects of a job study program may find itself making important wage decisions based on inaccurate and/or incomplete job data.

At the conclusion of the job study, employees should be given the opportunity to review their job descriptions and rankings before the monetary values are attached, and to discuss them with their supervisors and/or the job analysts who worked on the project. The evaluation committee should make every effort to provide justification for the decisions which have been made during the job evaluation process, and to take the time necessary to listen to employee concerns. As mentioned above, employee members of the job evaluation subcommittee can play an important role in disseminating the information and increasing employee acceptance and understanding of the program. An impartial appeals process or tribunal may be set up in order to resolve employee complaints before the final wage structure is put into place.

When the entire project is completed and the company's overall wage plan is set, employees should be presented with the results in a clear and understandable manner, and again be invited to ask questions of their supervisors or job evaluation committee members. A written guide to the company's pay system and job evaluation program should be prepared for future employee and management reference and use.

DEFINITIONS

Before beginning a more detailed discussion of the various job evaluation systems and their methods, it is important to define several of the basic terms that are commonly used in the evaluation process.[9]

Job.   According to the American Compensation Association, a job is "the total collection of tasks, duties and responsibilities assigned to one or more individuals whose work has the same nature and level."[10] While a job may, in fact, be a composite of the work performed by more than one individual, the analyst should always treat a job as being performed by a single worker in order to discount individual abilities and performance.

Task.   A task is defined as one or more elements that constitute a distinct activity that is a logical and necessary step in the performance of work by an employee. A *task element* is "the smallest step into which it is practical to subdivide any work activity without analyzing separate motions, movement, or mental processes." A *duty* is "a group of tasks that constitute one of the distinct and major activities involved in the work performed." A *responsibility* is "a duty or group of duties which describe the major purpose or reasons for the existence of a job."[11]

Position.   A position is "the total work assignment of an individual employee, comprised of a specific set of duties and responsibilities. The total number of positions in an organization equals the number of employees plus vacancies."[12] Analysis on the basis of positions, however, is undesirable because two or more positions might have exactly the same or very similar descriptions.

Skill.   A skill is that which (with whatever it is composed of) must already be possessed by the worker performing the job.

JOB ANALYSIS

Job analysis is the foundation for the entire job evaluation process. It is the process of systematically obtaining, categorizing, and documenting all pertinent information relating to the nature of a specific job, including such things as the tasks which comprise the job, and the skills, knowledge, abilities, and responsibilities required for successful job performance.[13] The object of a job analysis is to develop an accurate and concise job description which can then be used to evaluate the contents and "value" of that job. A thorough job analysis will also serve to highlight the relationships and hierarchies currently existing within the organization.

---

[9]Although these terms and definitions are widely accepted, particular job evaluation systems may have their own distinct definitions of these and other key terms.

[10]American Compensation Association, "Job Analysis, Job Documentation and Job Evaluation" (Certification Course 2), 1990.

[11]Ibid.

[12]Ibid.

[13]War Manpower Commission, Division of Occupational Analysis, "Training & Reference Manual for Job Analysis," p. 7, Government Printing Office, Washington, D.C., 1944.

Job analysis is a labor-intensive activity, demanding time and attention from jobholders, supervisors, managers, and human resources professionals alike. It takes time to determine what information is to be collected, and then more time to gather, process, analyze, and document the data. Moreover, because it is an intrusive process that inquires into things such as what people do in their jobs, how their jobs interact with other jobs, and the nature of their supervision, it can provoke anxiety among the work force if they are uncertain about management's intentions. Therefore, time must also be taken to establish a basis of trust and understanding among the management, the analysts, and the employees in order to ensure that the information that is gathered is accurate and complete.[14]

In addition, as Lange observed, because it is focused on an intangible thing called a "job," which exists in a dynamic, constantly changing environment called "the workplace," the information collected during a job analysis often has a limited useful life. Thus, even if the job is considered stable and unchanging, it is usually necessary to verify the information gathered when questions arise a year or more after the study has been completed.[15]

The fruits of a thorough job analysis are many. Aside from their use as a base measure during a job evaluation program, accurate job descriptions also serve to clarify the perceptions of both the company and the jobholder as to what is expected of him or her while working in that job. These articulated expectations form the cornerstone of an organization's performance appraisal program, and become the basis for both reward and discipline. Carefully written job descriptions are also critical to other personnel decisions, such as employee selection, training, promotion, and transfers.

In addition, the results of an accurate job analysis may be used by employers to validate their personnel decisions in the event such actions are challenged as illegal or discriminatory. For instance, this information will prove invaluable when issues and controversies arise over such things as exempt/nonexempt classifications, equal pay, comparable worth, employment discrimination, and the validity of preemployment tests. Under the Americans with Disabilities Act of 1990 (the "ADA"), for example, the agencies and courts will look to an employer's previously maintained job descriptions as evidence of the "essential functions" of a particular job from which a disabled worker was excluded in order to determine whether or not he or she could have performed those functions, with or without reasonable accommodation. It is therefore essential that these job descriptions be not only accurate, but *gender-neutral* and *nondiscriminatory* on their face.

Information Gathering.   The information for a job analysis is gathered from a variety of sources, including existing job descriptions, questionnaires, surveys, interviews, and direct observation. The more data the analyst is able to obtain in this information-gathering process, the more accurate his or her job assessment will be.

The first step in the job analysis process is to determine the specific type of information that is needed and the methods which will be used to collect it. The data collected should clarify the nature of work being performed (including principal tasks, duties, and responsibilities), as well as the level of work being performed. It should also include the extent and types of knowledge, skill, mental and physical effort, and responsibility required for the work. Details regarding work environment, hazards, and general physical conditions which affect the work may also be relevant.[16]

The extent of information to be included in each completed job description will depend in large part upon the type of job evaluation system selected, and the factors upon which the jobs will be evaluated during that process. For example, the point factor system developed by the National Metal Trades Association (NMTA) analyzes each job using 11

[14]Norman R. Lange, "Job Analysis and Documentation," in Rock and Berger, eds., *The Compensation Handbook* (3d ed.), McGraw-Hill, New York, 1991, p. 51.

[15]Ibid.

[16]American Compensation Association and American Society for Personnel Administration, *Elements of Sound Base Pay Administration* (2d ed.), 1988, p. 4.

TOOL AND DIE MAKER

Perform duties to repair and modify a variety of tools, dies, molds, gauges and fixtures and construct prototype parts and models.

Work from prints, sketches, drawings and instructions:

1. Plan and build close tolerance piercing, blanking and forming dies working from sketches and piece parts.
2. Tryout, check and debut new tools to ensure required past specifications are maintained.
3. Modify existing tools to meet new part manufacturing needs.
4. Construct welding, machine shop and assembly tools, fixtures and gauges.
5. Plan and build machine repair parts working from sample parts and sketches.
6. Work with engineering staff to construct prototype parts and machines, assemble units and recommend product design changes for ease of manufacturing.
7. Plan operation sequences and perform a variety of difficult machining setups.
8. Troubleshoot tool, die and mold production problems, analyze and diagnose nature of problem and take corrective action.
9. Keep supervisor informed of unusual equipment or tool problems.

Use a variety of machine tools, height gauge, surface plate, optical comparator, micrometers, veneer calipers, precision blocks and hand tools.

Follow safety rules and keep work area in a clean and orderly condition. Perform other related duties as assigned.

FIGURE 4.1   Job description for a tool and die maker. (*From National Metal Trades Association.*)

factors. If a job description does not already contain information regarding these factors, it must be revised and expanded. Figure 4.1 shows a job description for a tool and die maker done under a point factor system. A whole-job evaluation system may, on the other hand, require less detailed information.

Data Sources.   Regardless of which evaluation system is selected, it will generally be necessary to gather both qualitative and quantitative information about the jobs included in the study from a variety of primary and secondary data sources.

Primary data sources are the employees who actually are performing or supervising the work. Commonly used techniques for collecting job information from primary sources include:

- Questionnaires
- Interviews
- Logs or diaries
- Direct observation
- Work plans or manuals
- Process, setup, and method sheets

Because each of these techniques has its own advantages and disadvantages, they are best used in combination wherever possible.

All other sources of information about jobs are referred to as *secondary* data sources. Frequently used secondary sources of information about jobs include the U.S. Department of Labor's Dictionary of Occupational Titles and the Occupation Outlook Handbook, pay surveys, existing job descriptions, and information on similar jobs at other companies.

Most compensation plans use more than one method of acquiring data in an effort to achieve comprehensiveness, accuracy, flexibility, administrative efficiency, and uniformity. Each organization must weigh the cost, complexity, and availability of its resources to determine which approach will achieve the best results.[17]

TYPES OF JOB EVALUATION SYSTEMS

Once the data have been collected and organized through the job analysis process, and job descriptions have been developed, the "value" of each job must be determined using a particular method of job evaluation.

Job evaluation systems are generally divided into two categories: qualitative and quantitative.[18] While all job evaluation systems are to some extent 'quantitative" because they place jobs in hierarchical classes, *qualitative* systems generally compare jobs on a whole-job basis, whereas *quantitative* systems generally analyze their component parts. Moreover, in quantitative systems, the degree to which a job possesses a specific evaluative factor (such as level of responsibility) is measured using a finer unit of measurement, usually a numeric value. Qualitative systems, on the other hand, rely more upon rough estimations and gross measurement devices to achieve a job hierarchy. While qualitative systems are still in use, quantitative systems and their hybrids are by far the most popular in today's business world.

Traditionally, there have been four basic job evaluation systems: ranking, grade description, point factor, and factor comparison. The first two systems are classified as qualitative, the latter two as quantitative. With the rapid growth and pervasive use of job evaluation since the post–World War II era, however, have come new systems and hybrids which are almost as numerous as the consultants who work in this field. Figure 4.2 shows the relative popularity of various job evaluation systems in use today.

Therefore, choosing a job evaluation system is the first important step for an organization planning a job study. Each system has its advantages and disadvantages in terms of complexity, time, cost, and ease of comprehension and administration. Each company must choose the system that is most compatible with its needs and resources, both now and in the foreseeable future. Keep in mind that, while an inflexible or limited system may solve a short-term problem, it may not be flexible enough in the long run to accommodate the natural changes and fluctuations in organization. Companies should consider adopting a system which will grow with them and lend itself to constant updating and revision. If a company has used a particular system in the past, but its results have fallen into a state of disrepair, it should weigh the need to be consistent by using the same system with the value of implementing a method that is more flexible and easily administered, particularly if the company's needs have changed since the former plan was implemented.

The following section serves as an introduction to the features of the four basic job evaluation systems and their hybrids, including a brief discussion of the advantages and disadvantages of each.

JOB RANKING SYSTEM

Job ranking, a nonquantitative system, is considered the simplest form of job evaluation. It uses simple essay descriptions of jobs to determine the overall worth of a job by measuring it against all other jobs in the organization.

---

[17]Ibid.

[18]These two categories are also sometimes referred to as "analytical" and "nonanalytical," "qualitative" and "nonqualitative," or "quantitative" and "nonquantitative."

	Formal Job Evaluation Systems					
	Percent of companies					
	All companies	By industry			By size	
		Mfg.	Nonmfg.	Nonbus.	Large	Small
Number of companies	(197)	(65)	(84)	(48)	(87)	(110)
Organization has formal job evaluation system(s)	75%	77%	70%	81%	78%	72%
	(148)	(50)	(59)	(39)	(68)	(80)
Job evaluation methods:*						
Point-factor	57	66	49	56	60	54
Job classification	84	26	44	28	26	40
Ranking	25	30	25	18	24	26
Factor comparison	18	16	15	26	19	18
Other	10	14	10	5	10	10
Sources of job evaluation systems:*						
System created by own staff	47	50	51	38	49	46
Standard commercial system	26	34	17	31	25	28
System created by consultant	24	18	27	26	21	26
Customized commercial system	16	14	17	15	19	13
Other	10	14	10	5	10	10

*Percentages are based on the number of organizations with one or more formal job evaluation systems, as shown by the preceding row of numbers in parentheses.

FIGURE 4.2 Relative popularity of various job evaluation systems in use today. (*From* Compensation (BNA Policy and Practices Series), *Bureau of National Affairs, Inc., Washington, D.C., 1990, p.315:215*)

Job Analysis. Because the process of job ranking involves a whole-job, job-to-job comparison, rather than the breaking down of each job into its component parts, the degree of job analysis required before the evaluation process begins is minimal. At minimum, each job description should contain certain identifying information, a summary statement of the job, a listing of the tasks comprising the job, and the worker specifications.

Since the ranking will be done by overall job "worth," the meaning of this term should be discussed at the outset. Worth may be defined by criteria such as department size, job responsibility and/or complexity, necessary qualifications or skills, and importance to the company.

Simple Ranking. The initial ranking of jobs is often done at a departmental level by those well acquainted with the work in that department. Each evaluator or rater is given a job description for every job in that department and is asked to rank the jobs in a hierarchy from highest worth to lowest worth.

The task of ranking can be accomplished informally by giving raters a separate index card for each job title and asking them to place the cards in order to reflect ranking decisions. Ranking can also be done numerically: the most difficult job is given the rank of 1, the next most difficult the rank of 2, and so on until all the jobs have been ranked.[19] Evaluators are told to avoid giving the same rank to more than one job since this defeats the purpose of job evaluation—to differentiate between jobs.

[19]Under this system, if two jobs are given the same rank, each job is given the average of that rank and the following rank. For example, if two jobs are ranked 11, they both receive the rank of 11.5. The job ranking immediately below these two tied jobs is given the rank of 13. The same principle is followed if three or more jobs tie.

Benchmark Jobs.[20]   In order to facilitate the job ranking process, certain benchmark jobs are often selected in each department—or among the larger group of jobs included in the study—in order to provide a meaningful framework against which the remainder will be judged. The selection of these benchmark jobs—usually a number roughly equal to 15 to 20 percent of the jobs to be studied—should be done by the consensus of the evaluation committee as a whole as the first important step in the evaluation process. The selected jobs should be representative of the variety of the tasks and responsibilities found in the plant, so that the rest of the jobs in the study can be ranked by comparison with them.

Paired Comparison.   Another method used to facilitate job evaluation is the process of paired comparison. Using this method, evaluators compare each job separately with each of the other jobs in the study, so that only two jobs are being compared at a time. A sample worksheet for paired comparison is shown in Fig. 4.3. The pairs should be randomized so that there is no systematic presentation of one job in the list. That is, each job should be placed in the second position as often as in the first and so on. The results of each rater's comparisons may be summarized as shown in Fig. 4.4.

```
Instructions: In each of the pairs of jobs listed
below, underline the job which you believe is more
difficult and should receive the higher wage. Please
be sure to make a choice for each pair even though it
may be hard to distinguish between the difficulty
level of the two jobs.

Assembler                        Milling Machine Operator
Milling Machine Operator         Tool and Die Maker
Machinist, General               Shaper Operator, Spindle
Welder, Gas                      Assembler
Machinist, General               Milling Machine Operator
Tool and Die Maker               Machinist, General
Assembler                        Machinist, General
Shaper Operator, Spindle         Tool and Die Maker
Assembler                        Shaper Operator, Spindle
Welder, Gas                      Shaper Operator, Spindle
Milling Machine Operator         Welder, Gas
Welder, Gas                      Machinist, General
Shaper Operator, Spindle         Milling Machine Operator
Tool and Die Maker               Welder, Gas
Tool and Die Maker               Assembler
```

FIGURE 4.3   Paired comparison worksheet.

The paired comparison method has the advantage of simplifying the decisions that have to be made because the evaluator is comparing only two jobs at a time. However, while this is practical when the number of jobs involved is small, the number of comparisons which must be made increases very rapidly as the number of jobs increases. For example, for 10 jobs the number of comparisons is 45. For 40 jobs it is 780.

The formula for determining the number of comparisons is $[n(n-1)/2]$, where $n =$ number of jobs.

Integrating Department Rankings.   Once all the jobs have been ranked separately by department, the next step is to obtain a single set of rankings for all the jobs included in the

---

[20]The use of benchmark jobs is common among other job evaluation systems as well, as will be seen below.

JOB	NUMBER OF TIMES JUDGED MORE DIFFICULT	RANK
Assembler	0	6
Welder, Gas	3	3
Milling Machine Operator	2	4
Shaper Operator, Spindle	1	5
Machinist, General	4	2
Tool and Die Maker	5	1

FIGURE 4.4   Summary of paired comparison judgments.

job study. This step is usually accomplished by a committee which includes people whose knowledge of the jobs cuts across the various departments.

In this step, the jobs in one department are compared with the jobs in another. Then the jobs in a third department are compared with those in the first two departments and so on, until all the jobs have been considered. Worksheets similar to the one shown in Fig. 4.5 can be used by raters to accomplish this task.

Job Grades.   Once the ranking results have been assembled for the entire set of jobs which were studied, the jobs should be listed from highest to lowest worth. This list (or job hierarchy) must then be divided into job grades. Because this system of job evaluation is not done by any precise form of measurement, the divisions made at this point are somewhat arbitrary and guided largely by convenience and individual preference. For example, jobs which are roughly equal in responsibility and pay may be grouped together.

Advantages and Disadvantages.   The advantages of the job ranking system are that it usually takes less time than the other methods, requires less paperwork, and is easy for evaluators to understand (although the subjective nature of the evaluative decisions makes them difficult to explain). For this reason, it may be a good first step for an organization that has never undertaken a formal job evaluation project. It could also be an effective approach in smaller organizations and self-contained divisions or departments where jobs are already clearly defined and understood. Some critics also commend the subjectivity of the ranking approach for avoiding the hypocrisy of seeming to be scientific. In any event, it is clearly superior to the random, personalized rate-setting approach used by some organizations, which causes the problems enumerated earlier in this chapter.

On the other hand, job ranking is generally less accurate than other job evaluation methods, since it merely places jobs in a hierarchical order from high to low, without regard to degree and without reference to a standard unit of measure. The subjectivity of this approach also makes the exchange of information and opinion between rankers, as well as the demarcation of labor grades, more difficult. Raters may also be unduly biased by the current pay or status of the job incumbent. Another disadvantage is that the rankers should ideally be thoroughly familiar with all the jobs being evaluated. In a large organization, this is difficult to achieve.

Finally, since the ranking process is generally informal, intuitive, and undocumented, it does not provide managers or employees with any concrete justification of the results. This presents serious drawbacks in explaining and understanding the system, as well as in defending it against claims of bias.

Total rank	Department				
	Accounting	Clerical	Engineering	Administrative	Sales
1				President (1)	
2				Vice president (2)	
3					
4				General manager (3)	
5					
6	Controller (1)				
7					
8				Assistant general manager (4)	
9					
10	General auditor (2)		Senior mechanical engineer (1)		
11					
12			Mechanical engineer (2)		Sales manager (1)
13					
14	Auditor (3)			Executive assistant (5)	
15					
16	Senior accountant (4)		Chief designer (3)		Jobbing representative (2)
17					
18	Accountant (5)		Senior designer (4)		
19		Chief clerk A (1)			
20					
21	Junior accountant (6)	Chief clerk B (2)	Designer (5)		Senior sales representative (3)
22		Senior clerk A (3)			
23	Bookkeeper (7)	Senior clerk B (4)	Assistant designer (6)		Sales representative (4)
24					
25	Assistant bookkeeper (8)	Clerk A (5)	Junior designer (7)		Junior sales representative (5)
26			Senior drafter (8)		
27		Clerk B (6)	Drafter (9)		
28			Junior drafter (10)		
29		Assistant clerk (7)	Tracer (11)		
30					
31		Junior clerk (8)	Junior tracer (12)		

FIGURE 4.5 Integrated rankings of jobs in different departments. (Numbers in parentheses indicate departmental rankings.)

GRADE DESCRIPTION OR JOB CLASSIFICATION SYSTEM

Like the ranking method, the grade description (or job classification) system is another qualitative approach to job evaluation. It compares jobs on a whole-job basis, rather than on their component parts, and then places them in one of a predefined set of grade (or class) descriptions. The classification system is therefore the exact opposite of the ranking system: the latter starts by ordering jobs and then separates them into grades; the former begins with grades or classes and then slots jobs into them based on their characteristics.

The job classification method has been most widely used for salaried jobs, and is the system utilized by the federal government of the United States.[21]

Job Analysis.   The process of evaluating jobs using the grade description system begins with a selection and analysis of jobs to be included within the job study. Jobs are generally separated into "job families" first. A job family is generally defined as one with an identifiable continuum of jobs in a career progression—such as sales, engineering, accounting, and clerical. Next, the factors which will be considered in the classification process are selected, so that the job descriptions prepared during the job analysis step will be useful in slotting jobs into the defined grades.

As an example, the federal government of the United States uses the following eight factors in its job classification system: (1) difficulty and variety of work; (2) degree of supervision received or exercised by the jobholder; (3) judgment; (4) originality; (5) type and purpose of official contacts; (6) responsibility; (7) experience; (8) knowledge. Other organizations, of course, may use different factors in their classification systems.

Writing the Grade Level Description.   The crux of the grade description method obviously lies in the writing of the descriptions of the various grade levels into which jobs will be classified. Each of these descriptions must reflect some degree of the factors selected above.

For example, the first grade of the federal General Schedule (GS) classification system, GS-1, represents the lowest job classification and includes those jobs:[22]

> the duties of which are to perform, under immediate supervision, with little or no latitude for the exercise of independent judgment—(1) the simplest routine work in office, business, or fiscal operations; or (2) elementary work of a subordinate technical character in a professional, scientific, or technical field.

At the highest levels the definitions are more complex. For example, the description of grade GS-15 specifies:[23]

> Grade GS-15 includes those classes of positions the duties of which are—(1) to perform, under general administrative direction, with very wide latitude for the exercise of independent judgment, work of outstanding difficulty and responsibility along special technical, supervisory, or administrative lines which has demonstrated leadership and exceptional attainments; (2) to serve as head of a major organization within a bureau involving work of comparable level; (3) to plan and direct or to plan and execute specialized programs of marked difficulty, responsibility, and national significance, along professional, scientific, technical, administrative, fiscal, or other lines, requiring extended training and experience which has demonstrated leadership and unusual attainments in professional, scientific, or technical research, practice, or administration, or in administrative, fiscal, or other specialized activities; or (4) to perform consulting or other professional, scientific, technical, administrative, fiscal, or other specialized work of equal importance, difficulty, and responsibility, and requiring comparable qualifications.

---

[21]The Classification Act of 1949, Title VI, established the General Schedule (GS) for most white-collar positions in the federal government. The U.S. Office of Personnel Management (OPM) has recently undertaken a major revamping of its approach to publishing classification standards, experimenting with more simplified, "generic" standards.

[22]5 U.S.C. § 5104(1).

[23]5 U.S.C. § 5104(15).

The number of grade levels can be predetermined or can be selected after the level description process is completed. The federal GS system has historically contained 18 levels. Descriptions for the lowest and highest levels are generally written first, in order to define the limits of the grade description scale. These descriptions can then be tested by selecting some of the most simple and most difficult jobs to see if they are satisfactorily classified by the two extreme definitions. Descriptions for the second highest grade and the second lowest grade are then prepared and tested using jobs which would usually be classified at those levels. This process is continued until there are sufficient grade levels to encompass all the jobs in the study.

Grade level descriptions are not easy to write. Evaluators must select and include information regarding key job factors—such as the level of responsibility, skill, experience, and knowledge that are required—to guide them in their decision-making process. Each description must be general while at the same time specific enough to cover a variety of tasks and duties. This is often difficult to achieve because, for example, many jobs have some tasks which are at a low level of difficulty, and other tasks at a high level. In addition, each grade level must be distinct from the grade levels adjacent to it and at the same time must represent a logical step in a continuum without creating discernible gaps.

Moreover, it must be possible to assign each job under consideration to one of the grade levels without difficulty. If such difficulty is experienced with too many jobs, the definitions of the grade levels must be revised or new levels must be added. Grade levels may also be eliminated during this process if only a few jobs were assigned to a particular job level, and if such jobs approximated very closely the worth of jobs in adjacent levels.

Figure 4.6 illustrates a grade description plan for factory jobs.

Classifying Jobs.   Once grade descriptions are drafted, the evaluators begin the task of slotting jobs into those preset categories. This step should theoretically progress smoothly and rather quickly, since the majority of work has already been done. As with the ranking system, benchmark jobs may be selected for each grade level for ease of comparison and slotting. A *slotting matrix* is sometimes used to facilitate comparisons across organizational lines with grades on the vertical axis and departments or job families on the horizontal axis. See Fig. 4.7.

Advantages and Disadvantages.   Like the ranking system, the job classification system is relatively simple and inexpensive to implement. This is particularly true when employers take advantage of the many books and prepackaged classification descriptions which are commercially available. Moreover, the system is considered more objective and exacting than the job ranking system because of the precision required in the grade description process. New or restructured jobs can therefore be classified more readily. It is also viewed as more flexible than its analytical counterparts, particularly in the context of collective bargaining.

On the other hand, the nonanalytical nature of the job classification system carries with it many of the same drawbacks as the ranking system—namely, its subjectivity. Devising the grade descriptions can be an enormously difficult and contentious process, particularly in larger organizations where a broad range of jobs is to be included in the study. While narrow classifications prove unwieldy, expansive grade descriptions yield little differentiation between the various levels. Moreover, the placement of jobs into the predetermined classifications can be problematic where individual jobs contain characteristics of several different grade levels, or match most but not all of the minimum characteristics of a particular grade level.

In addition, appraising each job as a whole does not, by definition, allow for any detailed analysis of its component parts and cannot be expected to yield accurate, or even quantifiable, results. The assumed differential between adjacent grade levels is also deceptive, as there is no measure of the degree to which one classification is greater or lesser than the other. Finally, companies with less distinct job families—such as those where career ladders cut across job groups or are less susceptible to precise definition—may have

GRADE 1

Jobs included in Grade 1 are very simple, None requires over 1 month of experience and most can be learned satisfactorily in 1 week. Light laboring jobs such as janitor or sweeper and other light unskilled jobs are in this grade. Many jobs having to do with packing the product come in this category. Often the most difficult part of the job is that workers must be on their feet nearly all the time.

GRADE 2

This grade includes more jobs and more employees than any other. Most of these are concentrated on the numerous semiautomatic machine-feeding, sorting and inspecting jobs. Most of the laboring jobs in the plant also fall within this bracket. Helpers and service people as well as learners on machine jobs are for the most part included here, too. So also are operators of relatively simple equipment. As a rule the experience requirements for jobs in Grade 2 run between 1 week and 3 months. Responsibilities ; on these jobs are usually very small although they often rate high on effort.

GRADE 3

Almost as many jobs are included in Grade 3 as in Grade 2. Operators of machines of medium difficulty are included here. Inspection jobs involving responsibility and discretion are in this grade. Most of the jobs involving learning or helping to set up and operate complex machines are Grade 3 jobs, as are some maintenance jobs of semiskilled variety. This grade is definitely one covering semiskilled jobs.

GRADE 4

The setting up and operating of most of the plant machines are included in this grade. Grade 4 also covers many maintenance jobs and a variety of individual jobs involving considerable skill. Few of the9e jobs can be learned in less than 1 year and most require from 1 to 3 years of experience. Responsibilities on these jobs are usually substantial .

GRADE 5

Jobs in this grade are all of a high degree of skill and as a rule take up to 5 years to learn. Most of them also involve substantial responsibilities for products and materials and frequently considerable responsibility for the work of others. Skilled maintenance jobs, setting up complex machines, and some floorperson's jobs are included. These latter jobs include certain minor supervisory activities.

GRADE 6

Only jobs requiring a high degree of skill are in this grade. Most of the small number of jobs are floorperson's. On all these, both the experience and responsibility demands of the jobs are high. The most highly skilled maintenance department jobs are also in Grade 6, as are several machinist, toolmaker, and diemaker jobs. As a general rule, from 5 to 8 years of experience is required on these jobs. An ability to work independently with only a small amount of supervision is characteristic of most of these jobs.

GRADE 7

This grade covers jobs similar to those in Grade 6 except that these are a little more exacting. There are only three jobs in this grade. These are the pattern-maker, toolmaker, Grade B, and the most difficult floorperson's job in the plant. Eight to ten years of experience is required.

GRADE 8

Jobs in this grade are the most difficult and require the most skill of any jobs in the plant. The employees on these jobs are expected to be able to plan and carry out their work with little supervision. Only the top toolmaker and machinist jobs merit inclusion in this bracket, and these rank here largely because the employees on this work must almost be machine designers to carry out their work. Much of the equipment in the plant is specially built, or is very old, and parts must be designed and made by these employees. These jobs, calling for from 8 to 10 years of experience, are the top jobs in the plant.

FIGURE 4.6 Grade description plan for factory jobs. (*From J. L. Otis and Richard H. Leukart,* Job Evaluation, *2d ed., Prentice-Hall, Inc., Englewood Cliffs, N.J., 1954, pp. 98–99. Reprinted by permission.*) [*N.B.: Modified to be gender-neutral.*]

Benchmark Job Slotting Matrix				
Grade	Accounting	Purchasing	Personnel	Marketing
I	Accounting clerk I	Purchase clerk		
II	Accounting clerk II			Clerk typist
III	Payroll clerk	Senior purchasing clerk	Personnel assistant	
IV	Senior payroll clerk			Marketing analyst

FIGURE 4.7   Slotting matrix for clerical jobs. [*From American Compensation Association,* "Job Analysis, Job Documentation and Job Evaluation" (*Certification Course 2), 1990.*]

difficulty devising grade descriptions and conforming their job structure to a classification system.

POINT FACTOR OR POINT RATING SYSTEM

The point factor (or point rating) system—and its many variations—is the most commonly used job evaluation system in industry today. It is considered a quantitative method because it utilizes points and rating scales to measure specific job characteristics or factors which are common to many jobs—for example, skill, effort, responsibility, and working conditions.

The point factor system is a relatively sophisticated and multistep process which first entails selecting factors, dividing those factors into degrees, and assigning point values to those degrees. Each job is then evaluated and assigned a point value for each factor. For example, a particular job may receive 42 points for education, 16 points for physical effort, 60 points for experience, and so on. The points obtained by each job on the various factors are then combined, and a job hierarchy is devised based on those totals.

Selecting the Factors.   As its name implies, the selection of the factors to be used in the job study is the cornerstone of this method of job evaluation. While many factors can be utilized, certain factors appear again and again in point factor plans, although sometimes under different names. Figure 4.8 provides a sample list of those factors most commonly used in factory settings.

As a practical matter, the selection of factors is usually predetermined by the particular version of the factor system which is chosen. Two of the most popular and well-established systems in this country are those developed in the 1930s by the National Electrical Manufacturers' Association (NEMA) and the National Metal Trades Association (NMTA).[24] These point factor systems have since been adapted and revised many times over by these and other consulting firms around the world. Figure 4.9, for example, shows a comparison between the factors used in the NEMA and MNTA plans and an example of those used by H. B. Maynard & Co., a prominent consulting firm, for manual operations. Indeed, much of the appeal of the point factor approach to job evaluation lies in its adaptability, since the preselected factors and their weightings can be changed to suit the needs of a particular organization. An organization can even "start from scratch" and devise its own set of factors and point values.

---

[24]In 1963, the name of the latter organization was changed to the "American Association of Industrial Management" (AAIM). It is currently known as the Management Associations of America (MAA).

SKILL		
Education	Job knowledge	Resourcefulness
Education or mental development	Knowledge of machinery and dexterity with tools	Versatility
Trade knowledge	Knowledge of materials and processes	Job skill
Schooling		Manual dexterity
Experience	Mentality	Manual accuracy and quickness
Previous experience	Mental capability	Dexterity
Experience and training	Accuracy	Degree of skill and accuracy
Training time	Ingenuity	Physical skill
Training required	Initiative and ingenuity	Ability to do detailed work
Time required to learn trade	Judgment and initiative	Social skill
Time required to adapt skill	Intelligence	

EFFORT		
Mental effort	Fatigue due to eye strain	Muscular or nerve strain
Mental application	Physical effort	Fatigue
Mental or visual demand	Physical application	Monotony of work
Concentration	Physical demand	Monotony and comfort
Visual application	Physical or mental fatigue	

RESPONSIBILITY (FOR)		
Safety of others	Supervision of others	Protection of materials
Material or product	Supervision exercised	Physical property
Material and equipment	Cost of errors	Plant and services
Equipment or process	Necessary accuracy in checking, counting, and weighing	Cooperation and personality
Equipment		Coordination
Product	Effect on other operations	Details to master
Machinery and equipment		Quality
Work of others	Spoilage of materials	

WORKING CONDITIONS		
Unavoidable hazards	Occupational hazard disease	Surroundings
Hazards involved		Dirtiness of working conditions
Exposure to health hazard	Danger—accident from machinery or equipment	Environment
Exposure to accident hazard	Danger—from lifting	Job conditions
		Disagreeableness

FIGURE 4.8   Factors selected for factory point rating systems. (*From Jay L. Otis and Richard H. Leukart,* Job Evaluation, *2d ed., Prentice-Hall, Inc., Englewood Cliffs, N.J., 1954, pp. 118–119. Reprinted by permission.*)

When customizing a point factor plan for its own specific needs, an organization should keep certain basic principles in mind:

1. The factors should be *pertinent* to the type of positions included in the job study. If no supervisory positions are included, there is no point in having the factor "supervision of others." However, "supervision received" might be very appropriate.

2. Only *important* factors should be selected. Otherwise the final list will be unmanageable. It is suggested that 10 to 12 factors be used and that 15 be the maximum number. It should be possible to cover the more important factors and some of the less important job characteristics within these limits.

3. The factors selected *should not overlap in meaning*. For example, "physical skill" and

NEMA/ NMTA		H. B. MAYNARD	
General factors	Subfactors	General factors	Subfactors
Skill	Education* Experience Initiative and ingenuity	Skill	Education and intelligence Learning time Coordination
Effort	Physical demand Mental and/or visual demand	Effort	Physical effort Nervous effort Posture Eye strain
Responsibility	For equipment or process For materials or product For safety of others For work of others	Responsibility	Damage to materials or products Damage to equipment or tools Safety to others
Job Conditions	Working conditions Hazards	Working conditions	Accident hazard to self Water exposure location Temperature exposure Noise exposure Oil or dirt exposure Suspended dust exposure Fume exposure
		Investment	Tool expense Excessive clothing expense

*NMTA lists this factor as "knowledge."

FIGURE 4.9 Comparison of NEMA and NMTA factors with an example of H. B. Maynard and Co., Inc. factors for manual operations. (*From National Electrical Manufacturers' Association, "Guide for Use of NEMA Job-Rating Manual," New York, 1946; National Metal Trades Associates, "The National Position Evaluation Plan Manual," 1980; H.B. Maynard and Company, Inc., "Factor Evaluation for Manual Operations," 1980.*)

"dexterity" have practically the same meaning. An effort should therefore be made to select factors which are unique with respect to each other.

4. The factors chosen must be ones which lend themselves to *differentiation* in terms of "amount" of the job characteristics that they represent. This means that they must be *ratable* and that they can be described in terms of varying degrees. They must also be *quantifiable,* at least in terms of brief verbal descriptions which show a clear hierarchy from high to low. To take an easy example, "education" can be described in precise terms, such as "eight years of school," "high school graduate," or "Ph.D. in economics." On the other hand, a factor such as "truthfulness" would be much more difficult to quantify on a varying scale (although fortunately such a factor would not be included in a job study plan).

5. Similarly, factors should not be included on which all or most of the jobs would be given the same rating. This is equivalent to adding a *constant* to the value of every job and does not contribute to differentiating among the jobs. For example, if all the jobs are performed under the same physical conditions, a factor for this job characteristic is not required.

6. The factors should represent those for which the employer is willing to pay—that is, *compensable* factors.

7. The factors selected must be *acceptable* to both employees and management. If a sufficient number of jobs are extremely repetitive, omission of a "monotony" factor would lead to dissatisfaction, even though it could be demonstrated statistically that

this factor contributes little to the final point ratings of the job, or that it could easily be covered under another factor.

8. The factors should be *understandable* to both employees and administrators, both at the time the evaluation is done and in the future when the job evaluation and pay plan must be maintained.

9. The factors must be selected with *legal considerations* in mind. Race, gender, marital status, age, disability, and religion, among other things, may not be taken into consideration.[25]

Selecting factors to be included in the plan is not a simple matter. The guidelines enumerated above show some conflict among themselves, and careful judgment and common sense will obviously have to be exercised in making the final selection. Moreover, as a rule of thumb, a company should try to use as few factors as possible, while still ensuring that the range of jobs being evaluated will be adequately measured. Most factor plans include between 3 and 7 factors, though some companies use up to 15. The time required to evaluate the jobs and complete the job evaluation process will directly reflect to the number of factors selected.

As with other systems, *benchmark jobs* can be utilized in the factor selection process. They can be used to test the applicability of a preliminary list of factors and to suggest those which should be included and excluded in the final evaluative scheme. They also provide guidance to evaluators when rating the remainder of the jobs in the study group.

Defining the Factors.   The job factors which are chosen should be clearly defined and should mean the same thing to all persons involved in the job study project. Varying interpretations of ideas, concepts, words, and phrases should be anticipated as much as possible, and appropriate revisions made to the stated definitions. For example, if "training" is a factor, it should be carefully stated whether this includes formal schooling only or, in addition, a period of apprenticeship. Some factors which are very specific, such as "responsibility for funds," will be relatively easy to define. Other, more general factors, such as "complexity of duties," will be more difficult. An example of a definition for a specific factor is the following:[26]

> *Experience.* This factor measures the minimum length of time usually or typically required to attain quality and quantity performance standards under normal supervision. Do not include any knowledge considerations which have been evaluated under Knowledge [another factor] or any additional time after competency is reached.
>
> Experience is of two kinds: (1) previous qualifying experience on related work or lesser positions either within the organization or outside; and (2) the "breaking-in time" or period of adjustment or adaptation on the specific position itself. Both periods must be added together to properly reflect the evaluation.

An example of a definition for a more general factor is the following:[27]

> *Complexity of Duties.* Use this factor to appraise the job's requirements for independent action, exercise of judgment, and creative effort in devising new methods or new products. Rate a job high in this factor if it requires a great deal of judgment, and ability to resolve complex data or problems into units that can be evaluated and compared. Rate the job low in this factor if it is circumscribed by standard practice.

Defining Degrees for Each Factor.   Once the factors have been selected and defined, they must be divided into degrees. This step consists essentially of constructing a rating scale for each factor. Each level or category of the scale is a degree. For example, the factor

---

[25]For example, discrimination in employment on the basis of race, age, national origin, gender, marital status, and religion are among those prohibited by Title VII of the Civil Rights Act of 1964. The Americans with Disabilities Act of 1990 prohibits employment discrimination against persons with disabilities.

[26]NMTA, *The National Position Evaluation Plan Manual* (1980).

[27]Johnson, Boise, and Pratt, *Job Evaluation,* John Wiley & Sons, Inc., New York, 1946, p. 76.

"experience" cited above may contain five degrees, as shown in Fig. 4.10. The degree definitions for the general factor "complexity of duties" are illustrated in Fig. 4.11.

Experience

1st Degree

Up to and including 3 months

2nd Degree

Over 3 months up to and including 12 months

3rd Degree

Over 1 year up to and including 3 years

4th Degree

Over 3 years up to and including 5 years

5th Degree

Over 5 years

FIGURE 4.10   Degrees of experience. (*From NMTA*, "The National Position Evaluation Plan Manual," *1980*.)

The number of degrees within each factor should not be greater than are reasonably required to differentiate the jobs. There may be as few as two or three, if all the jobs can be grouped into that many distinct categories with respect to the factor concerned. All the factors included need not contain the same number of degrees.

Degree definitions should be stated in terms that are understandable to both analysts and employees. For example, terms common to the trade or industry may be more readily accepted than words which are not customarily used. Whenever possible, objective terms should be used rather than subjective terms: for instance, the phrase "machining operations involving tolerances below 0.001 inch" is more useful than "machining operations involving precise accuracy." Examples should be used to illustrate elements within the degree definitions and the degree itself.

Determining Relative Values of Job Factors (Weighting).   The next step in the point system of job evaluation is one of the most crucial: assigning relative values to the job factors selected. When totaled, these relative values will equal 100 percent.

Relative values of factors should be assigned based upon a consensus among those involved in the study as to how much each factor contributes to the total value of the jobs being studied. For example, while the level of skill may be more important to engineering positions, the level of responsibility may be primary in defining the value of managerial jobs. Evaluators must discipline themselves to think of the entire universe of jobs and not of one particular job or group of jobs when ranking and weighting factors. Important questions may include: "Should jobs be paid more for experience than for hazards?" and "Should education receive more consideration in pricing jobs than monotony?"

Figure 4.12 shows a comparison of sample factor weights for three employee groups. In most job evaluation plans, the major factor, "skill," receives the highest weight. The differences among the various plans, however, serve to emphasize the fact that, while other plans may be helpful as reference points, a point system can be tailor-made for one's own organization.

Assigning Point Values to Degrees.   After the relative weights of each factor have been determined, points must be assigned to the degrees in each factor. The first step is to decide upon the theoretical range of total points that could be obtained on all the factors. That decision must be reached upon the lowest score that any job could possibly achieve and the highest.

Degree	Definition	Typical jobs
1	Simple routine duties, requiring the use of only a few definite procedures and little individual judgment, the work either being performed under immediate supervision or involving little choice as to methods of performance.	Keypunch operator, messenger
2	Duties are clearly prescribed by standard practice but require the use of several procedures and the making of minor decisions requiring some judgment.	Clerk C, stenographer A, typist A.
3	Duties involve an intensive knowledge of a restricted field and require the use of a wide range of procedures and the analysis of facts to determine what action, within the limits of standard practice, should be taken.	Detail engineer, aircraft design; junior engineer, aircraft design; methods engineer B; major estimator A; major estimator B; executive secretary; placement assistant, personnel; librarian A.
4	Duties involve general knowledge of company policies and procedures and their application to cases not previously covered. Duties require working independently toward general results, devising new methods, and modifying or adapting standard procedures to meet new conditions. Decisions, however, are based on precedent and company policy.	Major engineer, layout, aircraft design; engineer, layout, aircraft design; lead technician A, motion picture laboratory; senior auditor A, internal auditing.
5	Difficult work on highly technical or involved projects, presenting new or constantly changing problems. Duties require outstanding ability to deal with complex factors not easily evaluated, or the making of decisions based on conclusions for which there is little precedent.	Assistant group engineer, aircraft design; lead engineer, aircraft design.

FIGURE 4.11 Degrees of complexity of duties. (*Modified from Forrest H. Johnson, Robert W. Boise, Jr., and Dudley Pratt,* Job Evaluation, *John Wiley & Sons, Inc., New York, 1946, pp. 77–78.*)

Because the theoretical range is an arbitrary matter, any two numbers sufficiently far apart to permit differentiation among jobs can be selected. A simple and convenient method is to allow the percentage weights selected for the factors also to represent the point values for the lowest degree in each factor. Figure 4.13 is an illustration of a completed chart with points and degrees assigned. Theoretically, a job could receive a minimum of 100 points under this plan. The values for the intermediate degrees in Fig. 4.13 have been assigned on the basis of an arithmetic progression. Some plans utilize a geometric or a variable progression to obtain the intermediate degree values.

Rating the Jobs. The next step involves rating the jobs based upon their job descriptions and assigning final point values to them.

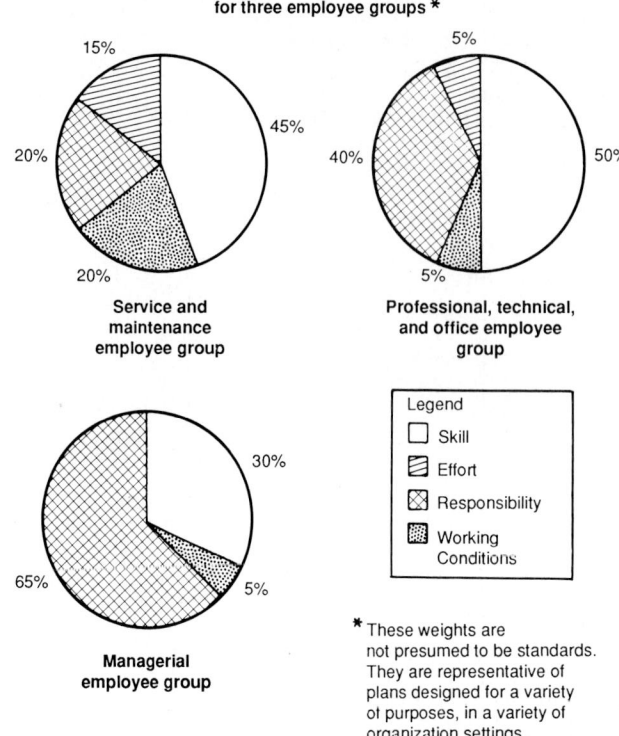

**Sample factor weights
for three employee groups ***

Service and
maintenance
employee group

Professional, technical,
and office employee
group

Managerial
employee group

Legend

☐ Skill

▨ Effort

▨ Responsibility

▨ Working
Conditions

* These weights are
not presumed to be standards.
They are representative of
plans designed for a variety
ot purposes, in a variety of
organization settings.

FIGURE 4.12   Comparison of sample factor weights. [*From American Compensation Association,* "Job Analysis, Job Documentation and Job Evaluation " (*Certification Course 2), 1990.*]

Factor	1st Degree	2nd Degree	3rd Degree	4th Degree	5th Degree
Skill					
1. Knowledge	14	28	42	56	70
2. Experience	22	44	66	88	110
3. Initiative and ingenuity	14	28	42	56	70
Effort					
4. Physical demand	10	20	30	40	50
5. Mental or visual demand	5	10	15	20	25
Responsibility					
6. Equipment or process	5	10	15	20	25
7. Material or product	5	10	15	20	25
8. Safety of others	5	10	15	20	25
9. Work of others	5	10	15	20	25
Job conditions					
10. Working conditions	10	20	30	40	50
11. Hazards	5	10	15	20	25

FIGURE 4.13   Points assigned to factor degrees; point ranges for labor grades. (*From NMTA*, "The National Position Evaluation Plan Manual," *1980.*)

Before starting this stage of the process, however, a set of guidelines, descriptions, and tables—preferably in handbook form—should be developed as a reference tool to promote consistency among evaluators. This *job evaluation manual* should then be tested by the evaluators on a handful of jobs to verify that the plan will yield consistent and satisfactory results.

After the plan has been verified, the evaluation process may begin in earnest. Each job description must be reviewed and evaluated on each factor separately and assigned a point of value for that category. The number of points assigned for each factor are then totaled to obtain the total point value for the job. Figure 4.14 shows the rating of the tool and die maker job described in Fig. 4.1.

It is sometimes useful to group similar jobs for analysis at the same time—for example, single and multiple drill press operators. In some cases, the job analyst may need to obtain additional information relating to the job description or to some particular factor in the specification in order to evaluate the job accurately.

Obtain Final Point Values for all Jobs.   The results of each evaluator's assessments are then compiled by the job evaluation committee in order to reach the final point value to be assigned. Jobs are then listed in order of their point values (the job hierarchy) to complete the picture. Even at this point, anomalies may appear, and points may need to be adjusted accordingly, based on the "big picture."

Job Grades.   Once the job hierarchy is established, the list must be divided by job grades based on the total scores. While the line of demarcation between some adjacent grades may be easy to determine—where there is a natural gap in the point totals—others will be set arbitrarily based on preference and convenience, among other reasons. Both the NEMA and NMTA systems include recommendations regarding the number of grades and the point ranges for each. The NMTA point ranges are illustrated in Fig. 4.15.

Advantages and Disadvantages.   The point system is more complicated than either the ranking or grade description system. It is more difficult to understand, requires more time to construct, and involves considerably more paperwork.

However, the fact that it is and has been the most popular job evaluation method would seem to indicate that it has advantages which outweigh the disadvantages. First and foremost, a point factor system minimizes the degree of subjectivity found in nonquantitative systems. It allows job comparisons to be made across job family or functional lines and can therefore include a much wider range of jobs under its umbrella. This makes it more useful for medium- to large-size organizations with a large number and/or broad range of jobs within them. While it may be more cumbersome for smaller companies, it is also a dynamic system which can grow (or contract) with the size of the organization over time. Once in use, both management and employees can understand the point factor system with ease and can usually reach similar point values for a job independently. Moreover, the fact that agreement among job analysts participating in the study tends to be high also adds credibility to, and promotes confidence in, the job rating results.

THE FACTOR COMPARISON SYSTEM

The factor comparison system, another quantitative method of job evaluation, was originally developed by E. J. Benge in the late 1920s in an effort to improve upon the point factor system.[28] It incorporates some of the elements of the point factor method and emulates parts of the ranking system as well. As a practical matter, however, its complexity has made its application and use limited.

---

[28]Eugene J. Benge objected to the point factor system on the grounds that: (1) it assumes all jobs are composed of the factors that are selected; (2) point values are assigned to degree levels of the factors in an arbitrary manner, especially the upper limits of the factors; (3) the unit of measurement, namely, the "point," is undefined; (4) factors are frequently not defined; and (5) the final value of a job is based on a job analysis rather than a comparison of the job with other jobs.

Tool and Die Maker		
FACTORS	DEGREE	POINTS
Knowledge:		
Use advances trade mathematics together with visualizing and the use of complicated prints, and use all varieties of precision measuring instruments such as micrometers, height gauge, precision blocks, and optical comparator. Equivalent to a complete apprenticeship.	4	56
Experience:		
Over 4 and up to 5 years	4	88
Initiative and ingenuity:		
Highly diversified work. Repair and modify a variety of tools, dies, molds, gauges, and fixtures and construct prototype parts and models. Exercise considerable judgment to modify existing tools to meet new manufacturing needs and troubleshoot tool, die, and mold production problems, where only general methods are available in making a broad decision requiring considerable initiative and ingenuity.	4	56
Physical demand:		
Light physical effort required to handle parts, material and tools, which is equivalent to frequently lifting or moving lightweight material or, occasionally, average weights.	2	20
Mental and/or visual demand:		
Concentrated mental and visual attention required to layout and machine close tolerance tools and die sections, to very close limits of high accuracy and quality.	4	20
Responsibility for equipment or process:		
Improper set up and operation of machine tools would result in damage to tool holding devices, requiring repair or replacement. Probable damage over $25 but seldom over $50.*	3	15
Responsibility for material or product:		
Improper construction of dies and molds would result in rework or scrapping of dies and molds. Probable loss over $100 but seldom over $200*	3	15
Responsibility for safety of others:		
Normal care required in performance of duties to prevent injury to others. An accident involving machine tools, should it occur, would probably result in lost time, loss of finger, or eye injuries	3	15
Responsibility for work of others:		
Responsible for own work.	1	5
Working conditions:		
Good working conditions with exposure to noise, dust, dirt, and oil, but with none present to the extent of being disagreeable.	2	20
Hazards:		
Injuries, should they occur, would result in lost time due to loss of finger or eye injuries while operating machine tools.	3	15
Total points:		325
Grade:		3

*1935–39 average price figures.

FIGURE 4.14 Rating of tool and die maker job. (*From National Metal Trades Association.*)

Score range	Grades	Score range	Grades
139 and under	12	250–271	6
140–161	11	272–293	5
162–183	10	294–315	4
184–205	9	316–337	3
206–227	8	338–359	2
228–249	7	360–381	1

FIGURE 4.15 Point ranges for labor grades. (*From NMTA*, "The National Position Evaluation Plan Manual," *1980*)

Under this system, jobs are ranked according to whether they contain more of a specific compensable factor than others.[29] Factor rankings for each job are then assigned numerical values and combined to form a total job score. This score is then translated directly into a dollar figure, fixing the wage rate of the job immediately.

Selecting the Factors.   Five basic factors are generally used in a job comparison scale: mental requirements, skill requirements, physical requirements, responsibilities, and working conditions. The precise definitions of these factors are set forth in Benge's manual.[30]

Once the jobs to be included in the study are selected, job descriptions must be written for them that take into account the various factors selected for the comparison.

Ranking Key Jobs.   The second step in the factor comparison system is the selection of 15 to 20 key (or *benchmark*) jobs among the larger group of jobs to be studied. The proper selection of these jobs is the most critical step of this job evaluation method, since the remainder of the process depends entirely upon it. The duties and current pay rates of these jobs must be undisputed among the evaluators participating in the study and must reflect an adequate sample of the range of the pay scale.

Once the key jobs have been selected, evaluators independently rank the chosen jobs separately for each factor using job descriptions. The key job possessing the lowest amount of one of the factors is assigned the rank of 1. Each key job is therefore ranked five times—once for each factor.

The results of the ranking are then compiled and presented to the entire group of evaluators for decisions upon a final ranking for each key job. Where unanimity or a clear majority cannot be achieved, the particular job involved is deleted from the list of key jobs.

Apportioning Pay Rates among the Factors.   The next step is to apportion the current wage or salary of each key job among the five factors. A convenient form to use for this step is illustrated in Fig. 4.16. The columns headed "Rank" show the final rankings assigned to the key jobs for each of the factors. The columns headed "Points" show the proportion of the current hourly rate that has been assigned to the five factors. For each job, the sum of the points must equal the current pay rate.

Apportioning the pay rates among the factors may be difficult. In actual practice, one job may not have a large enough current pay rate to permit the rank desired in one or more factors. Another may have too large a pay rate to apportion among the factors without changing the final rankings. Evaluators are asked to apportion wage rates several different times over the course of several weeks. The results of their efforts are then compiled and presented to the group for decisions upon a final ranking for each key job. Figure 4.17 illustrates a sample allocation of money values for five factory jobs. Persistent anomalies are noted and are generally deleted from the key job list.

---

[29]For this reason, the approach is sometimes referred to as the "weighted-in-money" or "direct-to-money" method.

[30]Eugene J. Benge, *Job Evaluation and Merit Rating*, National Foremen's Institute, Inc., New York, 1943.

Rater_____   Date_____

Key job	Current pay rate /	Mental		Skill		Physical		Responsibility		Working conditions	
		Rank	Pts	Rank	Pts	Rank	Pts	Rank	Pts	Rank	Pts
Assembler.........	2.43	4	51	4	45	3	78	3.5	42	3	27
Automatic screw- machine operator.	4.59	13	102	14	120	4.5	84	13	78	15	75
Brake operator.....	3.21	7.5	57	9	72	6	87	5.5	48	13	57
Carpenter.........	3.96	10	78	12	96	10	108	11	63	10.5	51
Expediter.........	3.54	15	117	6	57	1.5	72	14	84	1.5	24
Janitor...........	2.40	2	36	2	30	11	⑨⓪	2	24	5	36
Machinist.........	4.20	11	87	13	105	13	123	12	66	6.5	39
Material mover.....	2.49	1	27	1	24	15	150	1	18	4	30
Millwright.........	3.75	5.5	54	10	75	14	141	8	54	10.5	51
Painter...........	3.09	5.5	54	7	⑦⑧	4.5	84	3.5	42	14	69
Pipe fitter.........	3.30	7.5	57	8	69	7.5	99	7	51	12	54
Timekeeper........	2.85	12	96	3	33	1.5	⑤④	9.5	60	1.5	24
Tool and die maker.	4.98	14	111	15	135	12	120	15	93	6.5	39
Truck driver.......	2.94	3	42	5	51	9	105	5.6	48	8.5	48
Turret lathe operator.........	3.72	9	72	11	93	7.5	99	9.5	60	8.5	48

FIGURE 4.16   Key job data sheet for factor comparison system. (*Adapted from R.C. Smyth and M.J. Murphy,* Job Evaluation and Employee Rating. *McGraw-Hill Book Company, New York, 1946, p. 21.*)

Job	Wage rate in money units	Skill		Mental requirements		Physical requirements		Responsibility	
		Money value attributed	Ranking of job	Money value attributed	Ranking of job	Money value attributed	Ranking of job	Money value attributed	Ranking of job
Toolmaker	20	9.0	1	5.0	1	2.0	3	3.0	1
Machinist (grade 1)	18	8.0	2	4.0	2	1.0	5	2.0	2
Electrician (grade 1)	16	6.0	3	3.0	3	3.0	2	1.5	3
Assembler (grade 1)	14	4.0	4	2.0	4	1.5	4	1.0	4
Janitor	12	2.0	5	1.0	5	4.0	1	0.5	5

FIGURE 4.17   Allocation of money values to different factors; ranking of jobs under factor comparison method. (*From International Labour Organization,* Job Evaluation, *Geneva, 1986, p.55.*)

It is sometimes necessary to add additional key jobs to the job comparison scale for three reasons: (1) to fill the wide spread that may exist on any of the measuring sticks between two adjacent jobs; (2) to make up for the loss of original key jobs; and (3) to validate the selection of the remaining key jobs. Additional jobs are chosen as before, and all six steps described above are repeated. From 30 to 50 supplementary key jobs should be sufficient to construct the final form of the job comparison scale. Evaluation of the supplementary key jobs should be accompanied by reexamination of the original key jobs.

Setting Up the Job Comparison Scale. The remaining key jobs are set up as a job comparison scale, as shown in Fig. 4.18. This consists of five scales or "measuring sticks," one for each factor. Note that the jobs of painter, janitor, and timekeeper do not appear because they were eliminated as key jobs. Each measuring stick has a number of points from low to high. Various points along each scale are defined in terms of one of the key jobs.

Cents	Mental requirements	Skill requirements	Physical requirements	Responsibility	Working conditions
150			Material mover		
147					
144					
141			Millwright		
138					
135		Tool and die maker			
132					
129					
126					
123			Machinist		
120		Automatic screw-machine operator	Tool and die maker		
117	Expediter				
114					
111	Tool and die maker				
108			Carpenter		
105		Machinist	Truck driver		
102	Automatic screw-machine operator				
99			Turret lathe operator		
96		Carpenter	Pipe fitter		
93		Turret lathe operator		Tool and die maker	
90					
87	Machinist		Brake operator		
84			Automatic screw-machine operator	Expediter	
81					
78	Carpenter		Assembler	Automatic screw-machine operator	
75		Millwright			Automatic screw-machine operator
72	Turret lathe operator	Brake operator	Expediter		
69		Pipe fitter			
66				Machinist	
63				Carpenter	
60				Turret lathe operator	
57	Pipe fitter / Brake operator	Expediter			Brake operator
54	Millwright			Millwright	Pipe fitter / Millwright
51	Assembler	Truck driver		Pipe fitter	Carpenter
48				Truck driver / Brake operator	Turret lathe operator
45		Assembler		Assembler	
42	Truck driver				
39					Tool and die maker
36					Machinist
33					
30					Material mover
27	Material mover				Assembler
24		Material mover			Expediter
21					
18				Material mover	
15					
12					
9					
6					
3					

FIGURE 4.18 Job comparison scale for factor comparison plan. (*Adapted from R.C. Smythe and M.J. Murphy,* Job Evaluation and Employee Rating, *McGraw-Hill Book Company, New York, 1946, pp. 24–25.*)

Evaluating the Remaining Jobs.   Using the five job scales containing the key jobs and sup-plementary key jobs as guide points, the committee decides where on each scale the remaining jobs in the study should be placed. This determines the total wage rate for each job and completes the process.

Advantages and Disadvantages.   The construction of a factor comparison plan, although simple and logical in design, is a time-consuming and sometimes difficult process. Individ-ual job analysts or committee members are required to make many important decisions: selecting key jobs, ranking key jobs, rating key jobs, and making ratings consistent with rankings, integrating their judgments with those of their fellow committee members and resolving differences of opinion, and selecting and evaluating additional key jobs. Proba-bly the strongest criticism of the factor comparison method is that it establishes scales on the basis of monetary units. It is felt by many that job analysts will be unduly influenced by the current wages being paid for the jobs, and this in turn may prolong, rather than remedy, existing inequities.

However, the fact that the method has been used in industry indicates that its useful-ness may sometimes compensate for the difficulties involved in constructing the plan. One definite advantage of the factor comparison method is that it is tailor-made for each com-pany, since the primary basis for evaluating jobs is to compare them with selected "key jobs" in that particular company. On the other hand, key jobs, like any job, are subject to change, and as soon as this happens, the scale of measurement also changes.

Proponents of the factor comparison system also note that it obviates the need for the next step of most job evaluation systems—that of translating the rankings into money. Moreover, when the relative worth of the various occupational groups remains constant, a single multiplier can be used to keep the whole wage structure up to date. Finally, because it compares characteristics, it avoids what some consider the drudgery of using predeter-mined degree definitions and their narrow parameters.

HYBRID SYSTEMS

Given the limitations of each of the four traditional job evaluation methods, it is not sur-prising that various systems have been developed since the 1940s that attempt to combine the best features of some or all of the old systems while remedying their flaws. Others have adopted a new approach to job evaluation altogether.

In a broader sense, of course, every company creates its own hybrid system when it modifies an existing job evaluation plan to meet its own needs.

Regardless of the approach, these new systems tend to have two things in common: a search for greater accuracy and an effort to attain greater acceptability. The former is of-ten achieved through increased use of universal, nonoverlapping factors, benchmark jobs, multiple verification and consensus steps, and mathematical formulas designed to diminish the role subjectivity plays in the evaluation process. The latter generally involves in-creased employee participation and involvement. Among those hybrids which have expe-rienced some widespread use is the combination point factor system.

Combination Point Factor Comparison System.   This method combines features of the factor comparison system with the more flexible point factor system. In this hybrid, the same preliminary steps are followed as in the factor comparison system through the step of setting up the job comparison scale. One essential difference, however, is that the apportioned rates for each key job need not equal the current average pay rate for that job. That is, if the entire evaluation committee is in agreement, the current rate may be disregarded. In addition, the rates are doubled or tripled and then rounded off to multiples of 5 or 10. The purpose of this is to disguise the monetary aspect of the measuring sticks and thereby reduce the biasing effect of the job analyst's knowledge of current wage rate.

Once the job comparison scale is set up, the factors are weighted primarily on the basis of the range of points utilized for each. The remaining steps follow the point factor system: definitions are written for the factors; degree levels for each factor are defined and assigned point values; and a job evaluation manual is constructed and used for the remaining jobs and for new jobs in the same manner as the point system.

JOB COMBINATIONS

After the job analysis, rating, and initial job hierarchy are completed, the evaluation committee may identify several existing jobs which should be combined into a single job for administrative or other reasons.[31] The smaller the number of positions, the greater the company's flexibility in work assignments, and the simpler the wage system is to administer. Job combination may also provide employees with the opportunity for job enrichment by broadening their responsibilities and experience and thus making them more promotable. Companies can often reduce the number of job descriptions in their workplace by as much as 50 percent during a job evaluation study simply by eliminating job titles which are unpopulated or drastically underpopulated, and those which have become obsolete as the result of technological change.

Once the decision is made to combine jobs, each newly consolidated job must be analyzed and evaluated using the same job evaluation method applied to the rest. The positions must then be given their new place in the job hierarchy before the final rankings are issued.

While most employees whose jobs are combined will already have the necessary skill and ability to do the work required by their new job description, others will require job skill training before their proficiency will reach expected levels. The company should make clear its commitment to providing this training and should allow for it in setting standards and wage rates, as the learning process may take months or, in some high-skill areas, even years to complete.

CLASSIFICATION OF EMPLOYEES

After all the jobs have been described, identified, evaluated, and placed on a hierarchy, each employee must be matched with the title of the job which he or she performs. If the job descriptions are accurate, the task of classifying employees should be relatively easy, since job analysts have observed the work they perform and obtained information by interviewing workers and supervisors.

Problems may arise, however, when an employee's job title or job description has been substantially revised (or even eliminated) during the job study process. In these cases, the displaced employee will have to be slotted into the new job and may need to be trained for the new position prior to the implementation of a new pay system.

REVIEW OF JOB EVALUATION RATINGS

Before finalizing employee classifications, the company is well advised to allow at least its first-line supervisors to review the job study findings once again. To further promote acceptance of the new structure, employees should also be given the opportunity to see their job descriptions and rankings and to discuss any questions or points of disagreement with

---

[31]In some cases, an employer may have determined ahead of time that certain jobs should be combined and will treat the combined jobs as one job in the earlier stages of the job study, rather than after the preliminary hierarchy has been devised.

their supervisor or members of the job evaluation committee prior to implementation. In a union shop this privilege may also have to be extended to union officers and shop stewards, depending upon the terms of the collective bargaining agreement. In any case, one or more of the job analysts should be available to both supervisors and employees to answer questions and explain the principles underlying the job descriptions and the classifications. At this stage, however, no monetary values should be attached to the various jobs, grade levels, or classifications (except when a direct-to-money method of job evaluation is used).

While this review of job evaluation ratings may be time-consuming, it comes at a critical point in the process. It represents the last opportunity to change the point values or rankings of any jobs before these values are used as a criterion for evaluating the fairness and competitiveness of current wages and implementing the company's revised pay plan.

Misclassification of employees will inevitably foster employee morale problems and poor acceptance of the new job and wage program. Misclassified individuals will feel disgruntled if they are underpaid, and perhaps uneasy if they are overpaid. Their coworkers will attribute the existing injustice to the company's overall human resource policy or the job evaluation system itself, rather than to an error in the manner in which the system was applied.

Misclassification may also cause the wage curve to be inaccurate. If the number of misclassifications is large enough, the slope of the curve may be affected, as will the final classification structure which is superimposed on the curve.

THE WAGE CURVE

The next major step in achieving the company's final wage structure is setting the company's wage curve. This process may involve several steps, and draws on information derived from the job evaluation project, as well as from outside market surveys.

The *wage curve* is a graphic representation of the relationship between job values and wages paid for those jobs. When this relationship is a straight line, it is generally referred to as a *wage line*.

Diagnostic Wage Curve.   Before deciding on how (or whether) the company will change its existing pay plan as a result of the job evaluation project, it is necessary to get a clear picture of the company's current wage structure.

The company's present wage curve is obtained by setting up a scattergram in which the abscissa represents point values or labor grades and the ordinate represents current wages. Each employee's job is plotted on the chart, and a line of best fit is either drawn freehand through the plottings or, preferably, computed by statistical methods. While the usual, or at least preliminary, procedure is to determine the straight-line relationship, if the plottings appear to follow a curve, it is advisable to compute a second-degree curve for the data. The computational procedures for obtaining the straight-line trend and the second-degree curve may be found in most advanced statistics texts. Prepackaged computer software programs are also available to calculate and derive these curves quickly and accurately.

Figure 4.19 shows both the straight-line and the second-degree curves computed from the same wage data. It is readily apparent that a wage structure based on the second-degree curve pays higher pay rates for the jobs in the very high and very low labor grades, and pays lower pay rates for the intermediate labor grades, as compared with a wage structure based on the straight line.

When the job of each employee is plotted on the scattergram, as indicated above, the number of plottings is equal to the number of employees. As an alternative, wage curves are often derived in which each job is plotted only once for all the employees on that job. In this case, the median wage paid these employees is the ordinate value for the plotting, and the number of plottings is equal to the number of jobs. The two methods should yield practically the same curve.

Plotting the wage of each employee has the advantage of showing to what extent individual employees deviate markedly from the general trend for the entire organization. This

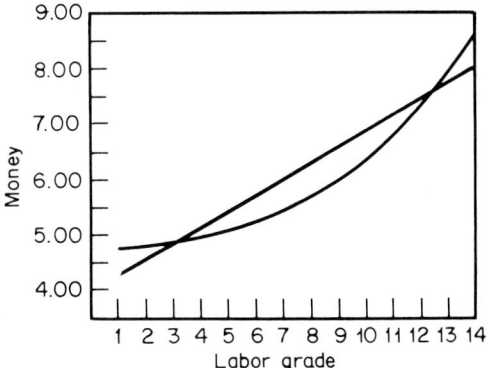

FIGURE 4.19 Straight line and secondary wage curves computed from the same data.

information will be needed to resolve individual pay inequities. Using median values, on the other hand, saves time in deriving the curve. This becomes particularly important when several wage curves have to be computed, as in a community wage survey (see below). The wage curve based on plottings of individuals will, however, be more accurate, if only because it is based on a greater amount of data and because the "pulling" effect of each employee on the wage curve is taken into account.

Regardless of the statistical method used, plotting the wage curve will bring the anomalies in the company's pay structure into sharp relief. For example, a point plotted for a job in a lower labor grade may appear significantly higher than those charted in a higher labor grade. This is a sure sign that something is amiss.

ESTABLISHING PAY RATES FOR JOB GRADES

The next step in the process is establishing the minimum and maximum pay rates for each job grade. For ease of administration, it is customary to establish a pay rate for each grade or class, rather than each individual job.

Basic Principles.   There are many ways of determining pay rates for job classes. Each method has its advantages and disadvantages. The method selected depends to a large extent upon the desired slope of the company's wage curve—that is, on the desired pay differential between low and high job grades. If the company believes that jobs in its higher grades should be paying a substantially higher wage than those in the lower grades—because they require highly skilled employees, for example, and/or those in short supply—the wage line will be steeply sloped. On the other hand, if there is not much differentiation between the jobs in the lower grades and the jobs in the higher grades—for example, in a plant where the difference in skill levels required for jobs in the higher and lower grades is not significant—the wage line will be relatively flat. These two types of wage lines are illustrated in Fig. 4.20a.

Other variables come into play as well, such as the size (or point range) of each labor grade, the breadth of the wage range within each job grade (that is, the difference between the minimum and maximum rates), and the degree of overlap between adjacent wage ranges. If, for example, it is felt that the point unit represents the same amount of job difficulty along the entire range of jobs, equal job classes are justified. However, many consultants feel that a point unit at the upper end of the scale represents more job difficulty than one at the lower end and therefore advocate job classes that are successively greater in size.

Similarly, wide wage ranges make it possible to grant more within-grade wage increases

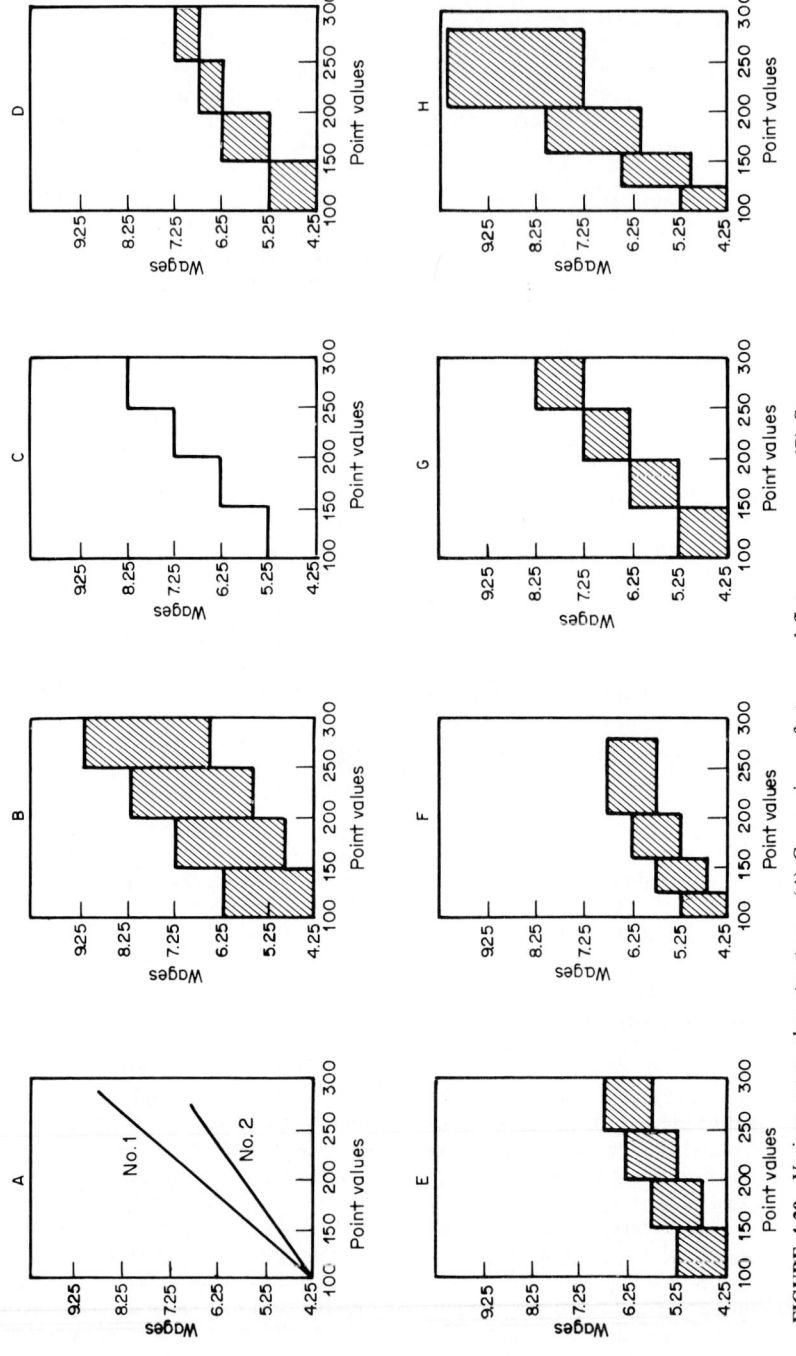

FIGURE 4.20 Various wage plan structures. (*A*) Comparison of steep and flat wage curves. (*B*) Steep curve, multiple grade overlaps. (*C*) Steep wage line, single rates. (*D*) Flat curve, no overlap. (*E*) Flat curve, overlap. (*F*) Flat curve, size gradually larger, overlap. (*G*) Steep curve, no overlap. (*H*) Steep curve, overlap, larger grades.

based on merit or tenure, which may increase employee morale. On the other hand, because wide wage ranges generally overlap, employees in one job class often receive lower wages than some employees in the lower job classes. Depending on range sizes, this overlap may extend over several grades. This is illustrated in Fig. 4.20b, where there is overlap over three grades. The trend in wage administration is to reduce overlap as much as possible.

For example, Fig. 4.20c illustrates a wage system with a steep wage line, and a single rate for each job grade (that is, no minimum or maximum rates and no overlap).

Figure 4.20d shows a relatively flat wage line where there is no overlap between wage ranges and therefore very narrow ranges within the grades. In contrast, Fig. 4.20e shows a flat wage line where adjacent wage ranges are permitted to overlap and can therefore be wider. Figure 4.20f shows a flat wage line where the size of the labor grades is progressively larger and wage ranges overlap. The philosophy behind this type of job classification is that, because it is easier to differentiate between job classes in the lower point values than in the higher point values, a wider range of points should be used as the jobs become more difficult.

Figure 4.20g shows a steep wage line where no overlap is permitted, but where wage ranges within labor grades can still be relatively wide because of the slope of the line. Figure 4.20h shows a steep wage line with overlapping wage ranges, each of which is equal to the preceding wage range plus 40 percent. The use of a percentage money spread makes it possible to have a wider wage range for the more difficult positions.

The decisions regarding each of these wage rate variables are a matter of company policy. There is no scientific solution to this problem because no matter how objective a job evaluation system attempts to be, the unit of job difficulty is abstract and merely a convenient way of quantifying a human judgment.

Number of Job Grades.   The number of job grades or classes in a company's job structure will depend in large part on the method of job evaluation used. Most systems recommend the use of a specific number, based upon the types of jobs included in the job study and the range of job difficulty values. For example, a hierarchy involving clerical jobs alone can be separated into 6 to 10 classes. If supervisory and professional jobs are included in the hierarchy, anywhere from 3 to 8 more classes will be necessary. For factory jobs the number of grades generally varies between 8 and 12, with an average of 10. In general, the tendency is toward *fewer* rather than many job classes.

The number of job grades is also subject to some manipulation based on the company's desired wage curve and financial goals. If, for example, there is a large employee population in a higher labor grade, the company may wish to adjust the point value cutoff for that grade so that the employees are spread over two grades, rather than one. The job descriptions, evaluations, and point values themselves, however, should never be changed.

RESOLVING WAGE INEQUITIES

After the job grades and wage ranges have been established, it will invariably be found that the existing wages of some employees fall above the wage maximum or below the wage minimum set for their job classes. The best way to show this is to plot each employee on the chart showing the job classes. Each job class, with its established wage minimum and wage maximum, will "box in" most of the workers, leaving a small number above and below each job class. Those above the maximum wage for each job class could be considered overpaid, and those below the minimum underpaid.

Those employees falling *below* the wage line should be brought up to at least the minimum rate either immediately or through a short, step-in progression program. (The difference between their current rate and the newly set minimum is sometimes referred to as their *green circle* rate.) While it is possible that the resulting increase in the payroll will be offset by a reduction in the wages of workers who are above the maximum, in practice this may be very difficult to do without jeopardizing the success of the job study.

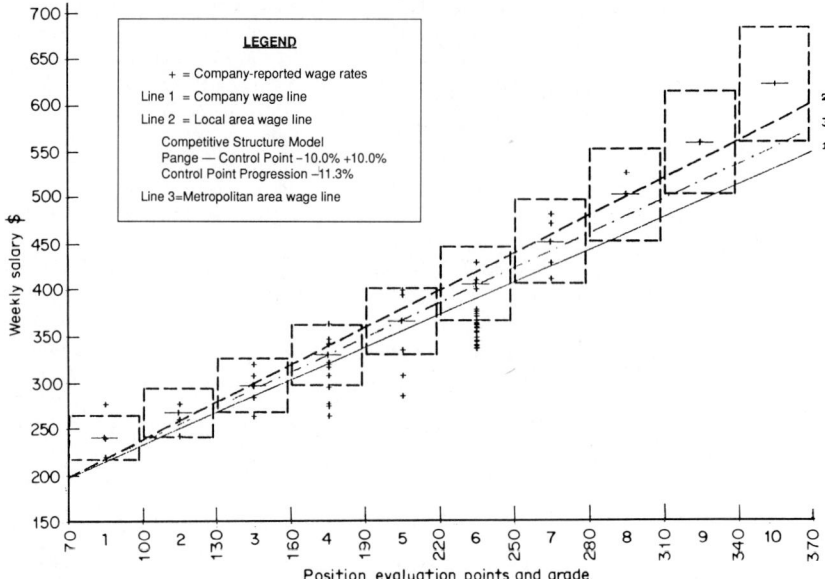

FIGURE 4.21   Comparison of company's existing wage rates with community trends. (*From Management Association of Illinois, "Sample Nonexempt Salary Survey Comparison," 1988.*)

This serves to illustrate a cardinal point about job studies. A company must be ready to sustain increased payroll costs in order to remedy the inequities within its wage structure and/or bring it within a wage range that is competitive in the industry. Among those benefits it can expect to see in return are lower rates of employee turnover (which means less time and money spent in recruiting and training), improved employee morale (which usually translates into higher productivity and greater cooperation), and a uniform wage system (which can significantly reduce administrative costs).

"Red Circle" Rates.   Workers whose existing wage is *above* the new maximum rate for their class should, whenever possible, be transferred to more difficult jobs in higher job classes with pay commensurate to their current earnings. If this cannot be done, the company may choose to carry them at their old pay rates indefinitely (so long as they perform their work efficiently) or, more commonly, for a fixed period of time. The amount these employees are paid above the newly set wage is often referred to as their "red circle" rate. For example, if the new set maximum rate for the job class is $8.75 per hour, but the incumbent employee was previously making $9.25 per hour, the red circle is $0.50 per hour.

As part of its red circle earnings protection program, a company may elect to offer these "overpaid" employees an opportunity to "buy out" their red circles. Under such a program, eligible employees are given a one-time-only opportunity to take (as a lump sum) their red circle amount multiplied by some preestablished number—such as, for example, the number of work hours in the succeeding 1 or 2 years. If they take this buy-out option, these employees would then work at the newly established rate for that job immediately. This type of program has the obvious advantage of cleaning up the company's payroll, since up to 75 to 80 percent of red circle employees will choose the buy-out option. It also allows the company to pay out in present, rather than future, dollars (cash flow permitting). Thus, the red circle buy-out option can actually save the company money in the long run.

Employees who elect to retain their red circle rate will lose their red circle at the expiration of the period fixed under the program (such as 2 years). Most programs also provide that the employees lose their red circles when they transfer to a different job classification. Their replacement is then automatically paid at the rate consistent with the established wage structure. The red circle rate can also be subtracted from any subsequent wage increase so that the employee's red circle amount steadily decreases over time.

The goal of any red circle program should be to eliminate pay anomalies from the company's wage plan as quickly as possible—and, in the meantime, to treat those aberrations uniformly.

COMMUNITY AND INDUSTRY WAGE SURVEYS

While an organization may derive its own wage curve as soon as it has compiled the wage data and job difficulty data required for the computation, the most important question concerning the current wage structure has not yet been answered. In its simplest form, this question is, "Is the general wage structure of the company good or bad?" The answer may be derived through the use of a community or industry wage survey which will show where the company's wages are in relation to those against which it is competing in the labor market.

Because wage surveys involve a large number of variables, a great deal of planning and effort is required to ensure that the information obtained will be accurate and useful.

Selecting Participants.   The *companies* to be included in the survey must be chosen carefully. Ideally, they should have each of the jobs which the sponsoring organization has selected for the study, and be competing in the same labor market for persons to fill these jobs geographically. Participating organizations will therefore generally be members of the same industry, unless the job study is based on jobs common to several industries—for example, clerical and office jobs—in which case the market can be defined geographically. While most surveys include up to 25 to 30 carefully selected companies, it is more important to obtain a good sampling of companies than to strive for a large number. Many excellent surveys have included as few as 5 to 10 companies.

Selecting Jobs.   Careful attention must be given to the *jobs* that will be included in the survey. Smyth and Murphy give five criteria for selecting jobs:[32]

1. The jobs should be distributed over the whole range of evaluated jobs.
2. The jobs should have remained relatively stable in recent years.
3. The jobs should exist in nearby or competing companies.
4. The jobs should be filled by as large a number of workers as possible.
5. There should not be an unusual shortage or surplus of workers qualified to fill the jobs.

The number of jobs surveyed in each participating company need not be large. Twenty-five to forty carefully selected jobs will yield practically the same wage curve as 300 to 400 jobs.

Getting Information.   Deriving the necessary *information* from each company also requires careful planning. The objective is to obtain sufficient information about the same job in each of the companies so that a fair comparison may be made of the wages paid for that job. The wages paid for similar jobs in two different companies may be equal, but further analysis may reveal that the worker in one company is actually "richer" because of a less demanding work schedule or the availability of items often described as supplemental in-

---

[32]R. C. Smyth and M. J. Murphy, *Job Evaluation and Employee Rating*, McGraw-Hill, New York, 1946, pp. 97–98.

come—for example, paid holidays, insurance benefits, profit-sharing pay-outs, and bonuses. For this reason, it is essential to differentiate among the base pay, earned pay, supplemental income, and hours worked of each employee. The usual practice is to focus on the base hourly rate or base weekly salary of each employee on the job at each participating company and then consolidate the results for all the companies into a single report.

The method by which a wage survey is conducted must also be determined. Wage surveys are generally made on the basis of job title, job description, or job evaluation.

Wage Survey by Job Title.   Using this method, an employer sends to other companies in the community or industry a list of job titles such as "receptionist," "machinist," or "janitor." The cooperating companies then enter their pay rates after each of the job titles and return the completed list. The wage information for each job title is then consolidated and returned to the cooperating organizations so that each can compare its pay rates with those of the others.

While this survey method may be the easiest to complete, it is the most inaccurate. Although a job title may be common to many companies, the content of the jobs it represents may vary considerably. For example, a tool and die maker may work under much poorer conditions, be required to have a higher skill level, or receive more supervision in one company than in another. The duties and responsibilities of foreperson in one company may also be quite different from those in another. For reasons such as these, the results of wage surveys conducted on the basis of job title may be very misleading.

Wage Survey by Job Description.   Wage surveys by job description represent a decided improvement over the job title method and are the most frequently used method of data collection. Using this method, the originating organization furnishes brief job descriptions which can be matched with the content of jobs in the cooperating organization, rather than just the job titles.

However, a wage survey by job description still leaves much to be desired. The job descriptions are usually brief "thumbnail sketches," so one can never be sure that the selected jobs have really been properly matched contentwise in each of the participating organizations.

Wage Survey by Job Evaluation.   The answer to the vagueness of either of these first two approaches is to extend the methods of job evaluation into the wage survey itself. The job evaluation method of wage surveying is therefore touted to be the most accurate and defensible, though it is probably the most difficult to conduct.

The essential difference between the job description and job evaluation methods of conducting a wage survey is that in the latter, both wage data and job evaluation data are obtained for the jobs surveyed in each participating company. The analyst completing the wage survey conducts his or her investigation armed not only with questionnaires and forms for recording wages and number of employees on each key job but also with a job evaluation manual so that point values for the jobs can be determined. This means the sponsoring company must obtain the specification for each job in the survey in terms of the factors used in the job evaluation manual. Although this seems like additional work, it may actually save time in the long run because it is more important for the job analyst to obtain a representative sampling of key jobs in each company than to attempt to match particular jobs which may, in fact, be apples and oranges. That is, there is no need to worry about whether the job of "receptionist" in one company is less or more difficult than in another company, since such differences will be reflected in the point values that are obtained.

Other Services.   Regardless of the survey method used, a company would be wise to consult both primary and secondary sources for information regarding wages in their target survey group. Secondary sources include the U.S. Department of Labor's Bureau of Labor Statistics, state departments of labor or industrial relations, and industry or local business associations that compile this type of information themselves and provide it to their members at no or low cost. Companies who use these sources, however, should be wary of the survey technique employed and its limitations. Moreover, regardless of the technique used, it is always recommended that an organization conduct its own follow-up interviews with cooperating companies whenever possible to verify the accuracy and recentness of the results, as well as to ask for additional clarification and comment.

Finally, it is also important that the organization which intends to use survey data—as well as the survey takers themselves—screens the information for anomalies which may be the result of unlawful discriminatory practices in the marketplace. The fact that *other* employers in the community or industry are unlawfully discriminating against women or other protected groups in their pay policies is no excuse for a company's perpetuating the problem in its own wage structure.

THE FINAL WAGE STRUCTURE

Once the results of the wage survey have been tallied, the company must analyze where its wages fall in relation to those of other organizations in the industry or community. To facilitate this comparison, wage curves should be derived from the survey information and compared—preferably in a single chart—with the company's own existing wage data. Figure 4.21 is an example of this type of chart, where local and metropolitan wage lines are plotted on the same graph as the company's existing wage data. The wage *ranges* of the local area wage line are also superimposed on this chart for additional comparison and references.

The next important question a company must ask of itself, of course, is "Where *should* our wages fall in the community?" The answer must be derivative of the company's fundamental wage policy. In essence, each company can choose one of three positions.[33]

1. The company can choose to be a leader among its labor market competitors and deliberately maintain a wage scale that averages 5 to 10 percent (or occasionally even more) *above the average for the industry*. Since this policy obviously increases the attractiveness of the com-pany to potential applicants, it is often used when a company's needs appear to outstrip the resources available in the labor pool. Other employers choose to remain in a leadership role because it is part of their business tradition.

   The long-term risks associated with taking this particular position are several. First, there is no guarantee that a higher than average wage scale will attract more productive employees. A solid applicant screening and performance evaluation system must be in place to ensure that the company sees a return on its investment. Second, a traditional wage leader sometimes finds itself in the unenviable position of having to drop back from the pole position because of its own financial situation or a change in the company's leadership and wage philosophy. This change can affect the morale of its current work force as well as its image and future recruiting posture.

2. In the alternative, an organization may decide (perhaps by default) to maintain a wage curve that is appreciably *lower than the industry average*. This position may be taken for purely financial reasons. The company may be in its formative stages, struggling for its existence on small capital. Or the company may be part of an industry that is forced to operate on extremely low profit margins because of the nature of its product, its gradual displacement by technological change, or other competitive problems peculiar to that industry. The organization may also be the victim of poor management, which has resulted in the dissipation of profits and depletion of capital. Or the company's management may (mistakenly) believe that a minimal wage scale will increase an already acceptable margin of profit.

---

[33]Adapted from Smyth and Murphy, op. cit., pp. 131–133.

Maintaining a low wage scale has a number of risks. It will usually make it difficult both to attract and to keep qualified employment candidates, as employees will be more apt to go elsewhere the moment a better-paying opportunity arises. Inability to secure an adequate labor supply and high employee turnover translate to increased operating expense and overhead costs, since time and money are spent in unsuccessful recruitment attempts and in continually training employees who then leave before their production levels compensate for their training costs. Lost productivity resulting from an untrained work force and low employee morale can also be by-products of a low-paying wage scale.

3. Most companies adopt the third choice—the policy of maintaining their wage scale *at or near the industry or area average*. An average wage structure is characterized by:[34]
   a. A base rate structure equal to the average of the industry or area as to both the absolute amounts of money paid and the relation of those amounts to one another (that is, the slope of the base rate curve).
   b. Average earned rates that are comparable with the averages of the industry and area as to both the absolute amounts of money paid and the relation of those amounts to one another (slope or curve).
   c. A minimum hiring rate that is equal to the average of the industry or area.
   d. An aggregation of miscellaneous additions to income or industrial relations policies (such as paid vacations, subsidized insurance coverage, holidays, or night-shift premium) that are reasonably comparable with the practices of a majority of the concerns in the industry or area.

Regardless of the approach taken, however, companies must continually monitor the relationship of their wages to those in the community or industry. When those wage rates fluctuate, companies must decide whether to maintain their same relative position or to alter it in light of their own business forecast.

Employee Communications. Once the desired modifications have been made to the company's current wage structure based on the job evaluation program and information gleaned from wage surveys, top-level management must approve the finished pay system and give the go-ahead for its implementation.[35] After that approval is secured, the new pay system must be presented to all company employees in a form that will be easily understood by everyone. Group meetings, handouts, and handbooks are all useful informational tools. Figure 4.22 is an example from a wage system presentation for production jobs

Labor grade	Wage range	Point value range
1	$10.88–13.60	320 and above
2	$10.44–13.05	289–319
3	$9.08–11.35	258–288
4	$8.44–10.55	227–257
5	$7.64–9.55	196–226
6	$6.68–8.35	165–195
7	$5.80–7.25	Below 165

FIGURE 4.22 Completed wage structure—production jobs.

based on the point method. This is a critical step in the employee communications program which, if done well, will pave the way for a smooth implementation of the new system and increased employee cooperation and acceptance of the company's overall pay practices.

---

[34]Smyth and Murphy, op. cit., pp. 131–133.
[35]In a unionized setting, of course, the company must negotiate with union representatives over employee wages before implementation.

KEEPING THE WAGE STRUCTURE UP TO DATE

It is not uncommon for an excellent wage structure to be set up and then become unsatisfactory within a relatively short period of time because it has been inadequately policed and maintained. For example, changes may be made to the content of jobs, which render job descriptions and specifications obsolete. New equipment, processes, or specialties may be introduced to the workplace, which require that new job descriptions be developed and placed on the job hierarchy. The level of wages in the community or industry or the tenor of the labor market may also fluctuate, causing the company's entire wage curve to become misaligned. A well-maintained job evaluation system will be periodically revised to reflect these changes (both internal and external), so that their cumulative effect does not subvert the entire system.

Because the administration of a company's wage structure and policy is such a major responsibility, it is often delegated to a committee composed of a chief analyst, the company's comptroller, and department heads, who are assisted by one or more full-time wage and job analysts. Members of the committee that conducted the job evaluation project should also be included to maximize the transfer of knowledge as well as achieve consistency in the system's application.

WAGE ADJUSTMENTS

There are many ways in which an employee's pay will change. Increases may result from a promotion or transfer to a higher-paying position, or merit or tenure bonuses. Conversely, pay may be cut as a result of a demotion, transfer, break in service, or poor performance. Wages may also be increased or decreased because of general wage adjustments.

General Wage Adjustments.   General wage increases usually take two forms: lump-sum increases or straight, across-the-board increases. Lump-sum increases or bonuses are generally one-time payments made to employees in lieu of (or supplemental to) a general wage increase. This type of increase has become more popular, particularly in concession bargaining, because employee base rates remain the same, increases are not compounded, and the benefits which are tied to base wages (for example, holiday and vacation pay, shift differentials, insurance and pension benefits) are not affected.[36] Lump-sum payments are also effective cost-controlling measures for employers since a 3-year contract, which pays three equal lump-sum bonuses, provides the equivalent of a wage increase in the first year, followed by wage freezes in the second and third years. Lump-sum increases are attractive to employees because they provide a significant sum of money for immediate consumption.

Across-the-board increases, on the other hand, will increase employees' base pay and corresponding benefits. This type of increase can be expressed in terms of dollars and cents or as a percentage. A portion of the increase may be recovered from a cost-of-living increase from the prior year or treated as an advance of future cost-of-living (COLA) increases. Some employers have adopted a practice of granting deferred increases, where the scheduled bonus payment is doled out in increments paid at various intervals.

Across-the-board pay cuts are generally meted out in a similar manner, using percentages or progressive step-downs or give-backs. General wage freezes can affect scheduled general increases, individual merit or tenure increases, or both.

Individual Merit or Tenure Increases.   The administration of individual increases based on merit or tenure is an important part of a company's wage program. In some organizations

---

[36]Also, because the wage base does not increase, FICA payments related to the wage base do not increase. (Note, however, that FICA must be paid on the lump-sum payment.)

department heads initiate recommendations for merit wage increases based on employee performance. In other organizations the amount and timing of increases may be dictated largely by union contract. In still others wage increases are automatically given at specified intervals, provided the employee's work record has been satisfactory. Whatever the method, it should be the responsibility of the wage committee to see that individual increases are granted when they are due and deserved.

With respect to wage administration, the following issues with regard to individual increases should be kept in mind:

1. Wages must be kept within established ranges and should not go above the maximum. This may be a very difficult principle to follow, especially when the labor market is tight. Making one exception inevitably leads to pressure to make similar exceptions for other employees. Avoid the temptation to pay an exceptional employee at a rate above the set maximum by encouraging the employee to take on more responsibility and post into a job in a higher grade (with higher pay).

2. When a merit increase is granted to one employee, the records of other, similarly situated employees should be reviewed to determine if they too are deserving of an increase. Accurate and fair methods of measuring job performance are essential in this area.

3. The wage range of a job class should be utilized fully and in an intelligent manner. Workers whose performance is barely satisfactory should be paid the minimum wage of the job class. Workers with exceptional performance records should receive the maximum. Workers whose performance is somewhere in between should receive an appropriate sum between the two extremes.

4. Increases granted solely on the basis of tenure on a job should be clearly identified as such and be handled separately from merit increases.

5. Because department heads will not always treat their subordinates alike in recommending merit increases, the committee should periodically review merit ratings and requests for wage increases of all employees.

6. Policies with respect to merit increases should be enforced in a consistent manner. If the maximum rate for one job class is raised, the maximums of other job classes should also be adjusted. If some workers with a certain amount of tenure are rewarded, other workers with the same amount of tenure should also be rewarded.

Two-Tier Wage Structure. Companies facing difficult financial times or a glutted labor market may be tempted to institute a tiered wage program for new hires. In a tiered system, newly hired and/or rehired employees are offered a training wage that is below the regular minimum rate established for certain job classes. In some tiered wage programs the lower-tier employees remain on a separate wage track indefinitely. Other tiered programs ensure that the wages of second-tier employees are gradually stepped up to the regular rates.

Not surprisingly, tiered wage systems are difficult to administer over the long haul, as employees are transferred, promoted, or laid off, and the labor market or economy continues to shift. Tiered wage systems have also been known to create a significant amount of tension in the workplace, since employees who are working side by side doing the exact same work are paid at sometimes radically different rates. In the airline industry, for example, "A-scale" pilots can make as much as $160,000 per year, while entry-level "B-scale" pilots may earn as little as $21,000 per year. It is not surprising, then, that the elimination of this two-tiered structure has become a primary sticking point in many contract negotiations in the airline industry.

Conversely, when the job market is tight, a company may have difficulty hiring employees at its established minimums if other companies begin to hire at accelerated rates. Under these circumstances, a company may consider instituting a new hiring wage for the lowest or the two lowest job classes that is substantially higher than its regular rate. This creates the same type of friction between the newly hired employees and the incumbents whose wages fall below these special rates. For this reason, a company electing to step up its hiring rates should plan to integrate the employees into the regular pay structure as soon as possible or make adjustments in the pay of other employees according to a fixed schedule.

Transfers.   When employees are transferred from one job to another within a job class, the company may continue their former rate of pay, increase it immediately, or increase it by a step progression over time. A problem may arise when transferred employees who were already receiving the maximum of their job class are kept at their former rate of pay but are then unable to perform as efficiently as other workers in the new position. In order to stem the resentment of lesser-paid incumbents toward the new arrivals, an employer can publicize and consistently enforce a policy that persons who are transferred will not be reduced in pay, so that all employees concerned will know what to expect when such a transfer occurs.

Promotions.   When an employee is promoted to a higher job class, he or she will generally receive the minimum rate of the new class. There is always the possibility, though, that the employee may not perform as well on the new job as on the old. Demoting the employee back to his or her old job is always embarrassing, especially if a wage increase was given with the promotion. One solution to this problem is to have a certain fixed "try-out" period, during which both the employee and his or her supervisor can assess the employee's ability to perform in the new position. If, at the end of this period, the employee is unable or unwilling to perform up to standard, he or she is returned to his or her former position and former pay rate. It is also possible to leave employees at their former rate of pay during this tryout period and increase them only after their performance has proved satisfactory. The length of the try-out period will vary with the type of work performed and can be specified in the job description.

Another solution is to have overlapping wage ranges and to promote only those individuals who have attained a wage in their job class which is at least at the minimum of the next job class. An employee who performs well during the try-out period may then be granted a pay raise within the new job class. If not, he or she can be placed back on the former job without a decrease in pay. This approach does, however, limit a company's ability to promote based on merit, since less senior employees may not have reached the top of a pay grade before they are promotable.

Demotions.   Demotions, which are often made necessary because of inefficiency, reductions in force, and slack business periods, raise the question of whether individuals whose rate is above the maximum of the class to which they are demoted will continue to receive their old wage or be paid some lesser amount. It is generally agreed that the new salary should be within the established wage range of the new position. Some companies have adopted the policy of setting the employee's new wage at the same point in the wage range of the new job as it was in the old. For example, an employee who was at the midpoint of the range when demoted would be placed at the midpoint of the lower wage range of the new job. Other companies allow employees to retain their former wage rate for a certain period of time before their pay is reduced—for example, for 90 days. This approach allows the company more flexibility for returning the employee to a former job in the event of an immediate business turnaround. Still other organizations provide for gradual step-down periods, in order to decrease the immediate impact of the pay cut on the employee's financial situation.

CONCLUSION

Job evaluation is not an exact science. It depends upon the subjective judgment of those who perform the analysis, and the cooperation and understanding of those whose jobs are being studied. Job evaluation is an ongoing process, rather than a static product, and requires a long-term commitment to implement and maintain its results. If done well, however, job evaluation can form a sound basis for a pay system which is logical, fair, and administrable, as well as cost-effective.

FURTHER READING

American Compensation Association, "A Bibliography of Compensation Planning and Administration Publications, 1975–1985" (3d ed.), Phoenix, 1987.

American Compensation Association, "Job Analysis, Job Documentation and Job Evaluation" (Certification Course 2), 1990.

American Compensation Association and American Society for Personnel Administration, *Elements of Sound Pay Administration* (2d ed.), 1988.

Bureau of National Affairs, *Job Evaluation Policies & Procedures* (PPF Survey No. 113), Washington, D.C., 1976.

International Labour Organisation, *Job Evaluation*, Geneva, 1986.

Johnson, Boise, and Pratt, *Job Evaluation,* John Wiley & Sons, Inc., New York, 1946.

Lange, Norman R., "Job Analysis and Documentation," in Rock and Berger, eds., *The Compensation Handbook* (3d ed.), McGraw-Hill, New York, 1991.

National Metal Trades Association, *The National Position Evaluation Plan Manual,* 1980.

Otis, Jay L., and Richard H. Leukart, *Job Evaluation* (2d ed.), Prentice-Hall, Inc., Englewood Cliffs, N.J., 1954.

Rock, Milton R., and Lance A. Berger, eds., *The Compensation Handbook* (3d ed.), McGraw-Hill, New York, 1991.

Smyth, R. C., and M. J. Murphy, *Job Evaluation and Employee Rating,* McGraw-Hill, New York, 1946.

U.S. Merit Systems Protection Board, "OPM's Classification and Qualification Systems: A Renewed Emphasis, A Changing Perspective," Washington, D.C., 1989.

War Manpower Commission, Division of Occupational Analysis, "Training and Reference Manual for Job Analysis," Government Printing Office, Washington, D.C., 1944.

CHAPTER 5
GRIEVANCES, ARBITRATION, AND COLLECTIVE BARGAINING

George J. Matkov, Jr., Attorney at Law
Matkov, Salzman, Madoff & Gunn
Chicago, Illinois

Patrick W. Kocian, Attorney at Law
Matkov, Salzman, Madoff & Gunn
Chicago, Illinois

Industrial engineers often affect the lives and livelihoods of many people. Their decisions can have a significant impact upon the wages, working conditions, and demands upon workers on the shop floor, particularly in companies using measured daywork standards or incentive compensation systems. Consequently, their actions are likely to be, at least indirectly, at the center of many employee complaints; indeed, industrial engineers themselves may even be seen by other employees as the source of lower wages, stricter standards, higher expectation, and other unwelcome changes.

It is inevitable that any industrial engineer who works on production processing methods and standards will become involved in an employee dispute. When this occurs in a unionized company, it is essential for the industrial engineer to be familiar beforehand with the process of resolving employee complaints under the unique procedures applicable there.

EMPLOYEE COMPLAINTS AND EMPLOYEE RIGHTS

Employers generally want to hear about employee complaints for two commonsense reasons: employee morale affects employee performance; and employees may, from their perspective, identify production problems and potential improvements that might be unseen by management.

But while an employer may choose to listen to employee complaints or to consider an employee's explanation for unsatisfactory performance, an employee's right to demand such a hearing is by no means self-evident. In particular, employee rights differ dramatically between nonunion and unionized workplaces.

Generally, in a nonunion company, employees are presumed to be employed "at will" unless they are hired under an individual employment contract. This means that the company can change their wages, benefits, or terms of employment whenever it wants; such employees can be demoted, fired, or disciplined for virtually any nondiscriminatory reason or no reason at all. Of course, employers may have statutory obligations to their employees, such as the obligation to pay minimum wages and overtime, to pay employees whatever wages are due on time, to observe safety regulations and procedures, and to not discriminate on the basis of race, sex, religion, age, national origin, disability, and occasionally other characteristics (such as, for example, employee sexual orientation or workers compensation history). Beyond these and other basic statutory restrictions on the employment relationship, however, nonunion employees who are not covered by an individual contract have no rights in the workplace except the right to quit.

In practice, of course, things are often different. Many nonunion employers recognize some procedure by which employees may make job-related complaints. Some of these procedures are formal, with different levels of "appeal"; more often than not, complaints are aired by way of an informal "open door" process under which employees are permitted to speak with any level of management at any time. Because courts in many states have begun to find—based on language in employee handbooks or other documents, or on some notion of the public interest—that even at-will employees may have implied rights to certain conditions of employment, or even to continue employment itself, absent good "cause" for the change or the discharge, many nonunion employers have chosen to investigate employee complaints and listen to employee explanations for misconduct and poor performance, if only to reinforce their subsequent actions if they are later challenged in court.

However, even where nonunion employers choose to listen to employee complaints, the procedure they adopt is almost always less formal than that observed by unionized companies—probably because less is at risk. Whatever the complaint, the employer in a nonunion setting ultimately can defend its actions to an at-will employee as long as it shows any nonarbitrary reason—such as efficiency, economy, or convenience—because there is no underlying contractual obligation to maintain any particular terms and conditions of employment or to treat the employee consistently or equivalently with others (assuming no illegal discrimination). In other words, assuming no express or implied contract is involved, the employee complaint procedure in a nonunion company is ultimately illusory as a dispute-resolution mechanism, because it is incapable of involuntarily binding the employer to any result.

The union setting is very different. Federal law guarantees the right of employees to engage in "concerted activities...for mutual aid and protection,"[1] and this includes the right to make complaints in a reasonable fashion.[2] More importantly, an employer is obligated to negotiate a collective bargaining agreement to govern the terms and conditions of the employees represented by a union. As a result, the unionized employer, unlike the nonunion company, cannot change employment terms or discharge an employee, merely "at will." Instead, it has to observe the terms of the labor agreement. The agreement can be enforced by a lawsuit in federal court,[3] and in addition, the union can threaten to strike to persuade employer compliance. Consequently, employee complaints under a collective bargaining agreement have to be taken seriously.

But lawsuits are costly, and strikes even more so. For this reason most labor contracts contain a mandatory, formal dispute-resolution procedure concluding with neutral arbitration; the company agrees to accept arbitration decisions as binding, in exchange for the union's agreement not to strike over an alleged breach of contract. In the lexicon of labor relations, complaints under this procedure are called "grievances," and it is settled federal policy to encourage the resolution of labor contract disputes through the grievance and arbitration process, rather than by resort to work stoppages or lawsuits.[4] In fact, courts

---

[1]National Labor Relations Act, § 7, 29 U.S.C. § 157 (1991).

[2]*N.L.R.B. v. City Disposal Systems, Inc.*, 465 U.S. 822, 104 S. Ct. 1505 (1984).

[3]Labor Management Relations Act, § 301, 29 U.S.C. § 185 (1991).

[4]*United Steelworkers of America v. Warrior & Gulf Navigation Co.*, 363 U.S. 574, 580–581, 80 S. Ct. 1347, 1352 (1960); see also, e.g., *AT&T Technologies, Inc. v. Communications Workers of America,* 475 U.S. 643, 648, 106 S. Ct. 1415, 1418 (1986).

will refuse to even hear a claim for the breach of a labor contract until the parties first complete the negotiated grievance process.[5]

GRIEVANCES IN A UNION SETTING

Types of Grievance.   A grievance generally must claim a violation of some section of the collective bargaining agreement. However, some labor contracts allow grievances over alleged departures from other terms of employment, such as nonbargained employer policies, shop rules and practices, or even alleged violations of public laws, such as those prohibiting race discrimination. Even when the scope of the grievance procedure is narrowly limited to enforcing only the terms contained within the contract itself, the contract often contains a general nondiscrimination, "fair treatment" clause which may suffice as a basis for grieving employer conduct which is not directly governed by the contract. In short, it is prudent to assume that almost any company policy or action which directly affects terms and conditions of employment, whether or not covered by the collective bargaining agreement, can be subject to a grievance.

While the scope of the grievance procedure may be broad, the range of grievance issues generally faced by industrial engineers is more narrow. The industrial engineer is involved in many activities which, although they ultimately affect the terms and conditions of employees, do so in an indirect way. For example, employee grievances are unlikely to involve such industrial engineering activities as plant location surveys, feasibility studies, cost estimations and controls, production planning. Industrial engineers are much more likely to be called to deal with grievances involving the design of, and payment for, particular jobs. One arbitrator has classified industrial engineering grievances within two basic categories: "job structuring" and "work performance or measurement" grievances.[6]

"Job structuring" grievances generally involve the categorization and "pricing" of the functions performed in a particular job classification. This involves not only a review of the formal job description but also an evaluation of the comparative skill and effort involved in each of the various tasks actually performed by occupants of the job classification, and the relative significance of each such task in the composite of duties that are regularly required of employees in that position. The industrial engineer is called upon to justify the relative wage rate determined for various jobs based upon their task components, as well as the division of labor between jobs. Job structuring grievances are often related to changes arising from the introduction of new technology or production methods, and the resulting elimination of some jobs and the realignment of tasks associated with others.

"Work performance or measurement" grievances, on the other hand, focus on the actual and projected performance of various job tasks. There are two basic types of performance or measurement grievances: those which challenge the production standard as an economic item, that is, the impact of the standard on incentive earnings; and those which challenge the standard as applied to a skill determination in a job qualification dispute or a discipline, demotion, or discharge action proposed for a substandard employee. Some of these grievances challenge the legitimacy of production standards and incentive rate structures, on grounds that they are inaccurate or were determined by a process which is improper under the contract, is inherently biased, or is based on inaccurate premises. Other grievances may accept the existing standards but contend that interim events or aberrations justify a revision to or deviation from them.

The Grievance Process.   The contractual grievance procedure usually consists of several levels, or "steps," of appeal. Grievances ordinarily must be presented first to the lowest

[5]*United Paperworkers Int'l Union, AFL-CIO v. Misco, Inc.,* 484 U.S. 29, 37–38, 108 S. Ct. 364, 370 (1987).
[6]Wiggins, Ronald L., *The Arbitration of Industrial Engineering Disputes,* Bureau of National Affairs, Washington, D.C., 1970.

level of management involved in the dispute—with respect to employee grievances, this is usually the line supervisor or foreperson. If unresolved, the grievance moves up the management chain of command at levels specified by the contract; for example, a typical second-step grievance is presented to a department head, a third-step grievance to the division or human resources director, and the final step to the company president or designee. At each step of the process, the company representative issues a written response, and the grievant must present any appeal within a contractually mandated time limit—otherwise the grievance is waived. In the absence of settlement at the final grievance level, either the union or the employer (but not the employee grievant) may decide to refer the case to binding arbitration, again within a limited time period and with proper notice to the other.

Grievance meetings tend to be fairly informal. They take place in offices or conference rooms, with the grievant and the union representative on one side (often literally one side of a table) and the deciding management official on the other, sometimes accompanied by a labor relations representative and/or a manager from the relevant company operation (such as the industrial engineer). The grievant usually has completed a formal grievance form and is expected to state the case (directly or through a union representative) to the employer representative. Ideally the grievant or the union will identify the contract provision alleged to be breached, the grievant's interpretation of it, and its proposed application to the facts of the grievance. However, there is no particular order of discussion following the grievant's initial statement. Generally a colloquy between the parties continues until each party is satisfied that they understand their respective positions and the relevant facts. Witnesses can be called if the contract or past practice of the parties permits, but there is no formal examination of evidence and no judicial protocol. Likewise, there is usually no recorded transcript, although both sides should take notes. In most cases, the company representative should withhold any response to the grievance until a written decision can be issued after the meeting.

Responding to a Grievance. The industrial engineer's involvement in the prearbitration stage of the grievance process may take several forms. First, the engineer may be consulted by the supervisor in preparing the first-step answer or by other company officials preparing statements of the company position at later steps. Some supervisors might be inclined to summarily deny a grievance in so many words, but that really defeats the dispute-resolution purpose of the grievance procedure. When invited to draft a grievance answer, the industrial engineer should prepare a concise but responsive explanation of the company's conduct. This should include an explanation of the pertinent engineering activities involved in the challenged action, and identify the contract provisions which permit them. The grievance response should be a genuine attempt to convince the grievant that the company acted correctly, without malice and in accordance with valid and reasonable procedures.

Second, the industrial engineer may be involved in the company's internal investigation of the grievance. The company can and should grant grievances which state a meritorious complaint; however, it generally will not fall to the industrial engineer to make that call, and other issues—such as bargaining leverage, pending arbitrations or other litigation, and union politics—may enter the equation. The role of the industrial engineer as a grievance investigator is to supply the company with *all* the relevant facts, and not just those which support the company's position. Whatever its response to a grievance, the company is much better off when it knows of facts which undermine its case early on, rather than being surprised at a later arbitration hearing.

Finally, the industrial engineer may be involved, at the request of the company personnel department, in personally participating in the grievance meeting. When asked to join the grievance meeting, the industrial engineer should insist on a thorough briefing on the grievance beforehand, an opportunity to investigate the matter, and a clear outline of what the engineer will be expected, and is authorized, to discuss. The industrial engineer must take care, of course, to leave the conduct of the meeting to the management representative and should avoid rendering any opinions on the case—unless these have first been discussed with the company.

The role of the industrial engineer in grievance processing can be very important, even essential. In most cases, however, it will not be the industrial engineer who is authorized to represent the company and bind it to a particular grievance resolution. The industrial engineer should never attempt to discourage an employee from filing a grievance—which would itself violate federal labor law—and must resist the temptation to deal directly with the grievant to settle the case after a grievance has actually been filed, unless asked to do so by the company and then only with the consent of the union. The industrial engineer can be a great asset to the company in addressing employee grievances, but only as an adviser and not the decision maker.

GRIEVANCE ARBITRATION[7]

While grievance processing is fairly inexpensive and routine, arbitration can be costly and is a more serious matter for both parties. It is costly because it requires more preparation and obligates the parties to pay for the time of the arbitrator, court reporter, and sometimes an outside advocate, that is, an attorney; it is more serious because arbitration decisions are binding on the parties in the matter at hand, and potentially binding in other disputes in the future. Consequently, most grievances are settled without going to arbitration, either by a mutually satisfactory result or by one party's conclusion that an unsatisfactory grievance decision is not worth the effort, cost, and risk of arbitration. Often the union will recognize that a grievance is meritless but will pursue it through the multistep grievance process for institutional or political reasons; but once grievance appeals are exhausted, these reasons must weight especially heavily before the union will go forward to arbitration on a case it knows it is likely to lose.

Arbitration of one kind or another is now accepted as a means of settling some or all contract grievances in almost all labor agreements. The historic purpose of arbitration was to fill the gaps in the contract unavoidably left by negotiators who could not foresee all possible disputes which could arise in the employment relationship. As such, arbitration was seen as a continuation of the collective bargaining process, and many arbitrators understood their mandate to be a matter of perceiving the deal which the parties *would have* made had the disputed issue been identified at the bargaining table. However, as the labor-management relationship between a company and a union matures over time and more and more factual scenarios are encompassed within a settled or imposed interpretation of the contract, the frequency of arbitration tends to decline, and over the years arbitrations have come to focus more on a legalistic contention over contract terms than on an equitable contract settlement. A corollary of this, however, is that the number of arbitrations will increase and, to a lesser extent, the authority presumed by an arbitrator may expand, when new conditions, new contract language, or new procedures—such as new job classifications or a new incentive compensation system—are implemented.

The foremost feature of grievance arbitration is the agreement to be bound by the decision of an impartial decision maker who is mutually selected by the parties. Selection of the arbitrator is thus a critical task, and there is no given method by which this must be done—the parties have to devise a procedure. The labor agreement is the touchstone of the process, and several methods are in common use. One of the most common is an agreement to use an arbitrator assignment service, such as the American Arbitration Association, a private concern, or the Federal Mediation and Conciliation Service, an inde-

---

[7]In addition to arbitration of employee grievances, in some cases unions and employers will agree to "interest arbitration" to settle contract terms which were not resolved in negotiations. This type of arbitration is really a feature of collective bargaining and not a method of resolving employee contract complaints.

pendent agency of the U.S. government. The service provides the parties with a list of arbitrators based on some limited qualifications (usually geographic), and each party in turn eliminates from the list that arbitrator it least prefers, until one remains. Other contracts provide for the parties themselves to suggest arbitrators until agreement is reached. Another method is to establish in advance a permanent panel of agreed-upon arbitrators who are known to the parties and eventually become familiar with their particular business and contractual relationship; these are then assigned to cases in some prearranged sequential order.

Selection of the arbitrator is especially important in cases involving industrial engineering issues. In these cases, general principles of labor contract interpretation, with which all arbitrators become familiar, often are secondary to very technical questions of process measurement. Some arbitrators are particularly experienced in time study, wage incentive, and other engineering issues—some, albeit too few, are themselves industrial engineers—and ideally these arbitrators should be sought for cases involving such issues. In companies where incentive compensation systems are widely used, and thus industrial engineering–type grievances can be expected to occur with some frequency, the parties may formally agree to seek arbitrators with special qualifications (either by their own selection or by instruction to the arbitrator selection service) or even to establish a separate permanent panel of specialized arbitrators for such cases. However chosen, the industrial engineer can assist the company in evaluating the qualifications of potential arbitrators in incentive or other engineering-intensive cases.

Management Rights.   Most arbitrations involving the industrial engineer are matters of contract interpretation and require the union, or the grieving employee, to prove that the company has violated the contract. Arbitrations involving the discharge or demotion of a poorly performing employee are an exception, insofar as the burden of proof in such cases is generally understood to lie with the employer to prove "just cause" for the adverse action. However, even in employee discipline cases, unless the accuracy of an individual performance measurement itself is in dispute, industrial engineering issues come into play principally by the union's challenge to the methods used by the company to establish performance standards. This ultimately turns on a definition of management rights.

In the past, labor arbitrators confronted the issue of management rights from the perspective of the employment-at-will doctrine which prevailed until recently; in other words, arbitrators viewed labor agreements as express exceptions to the legal presumption that a company could do pretty well what it wanted in business with little regard to the impact of its actions on employees. In that view, although employers might have a legal obligation to bargain with a union, once satisfied, any matters not directly encompassed within the terms of the resulting labor agreement were understood to be "reserved" to management's discretion. If the contract didn't prohibit it, the company's action was within its reserved management rights.

Some arbitrators still adhere to this "reserved rights" doctrine, but many more are inclined to broadly construe labor contract language to encompass unspecified management conduct or to impose an implied management obligation in areas where no express obligation is stated. In light of this, most collective bargaining agreements now contain an express enumeration of rights which are reserved exclusively for management unless expressly limited elsewhere in the contract. A good so-called *management rights* clause might read:

Except to the extent expressly abridged by specific provisions of this agreement or applicable laws, the company reserves and retains all of the inherent rights and authority to manage the business, provided that these rights will not be used for the purpose of infringing upon any rights expressly granted by this agreement. These include, but are not limited to, the following rights: to establish or continue policies, practices, and procedures not contrary to the terms of this agreement; to determine the number, location, and types of its plant or plants and its operations and the methods, processes, and materials to be employed; to discontinue processes or operations in whole or in part, or to discontinue their performance by employees of the company, and to contract out any or all such operations; to determine the times during which operations shall be continued; to select and determine the number and types of employees required to operate any equipment or machinery; to determine the amount of overtime to be worked; to assign work to such employees in accordance with the requirements determined by the company; to establish and change schedules and assignments; to transfer, promote, or demote employees, or to lay off, terminate, or otherwise relieve employees from duty for lack of work or other reasons according to the terms of this contract; to establish standards of work performance; to make, enforce, or change reasonable rules for the maintenance of discipline and protection of life and property; to suspend, discharge, or otherwise discipline employees for just cause; and otherwise to take such measures as management may determine to be necessary for the orderly and economical operation of the company's business.

Unfortunately, few agreements include such a comprehensive reservation of management rights; most state a much more abbreviated version. Moreover, after the initial collective bargaining agreement between the company and the union is negotiated, it is almost impossible to obtain union agreement to expand a deficient management rights clause. Consequently, most arbitration decisions involving industrial engineering issues and a company's right to make industrial engineering–based decisions must rely on a conjunction of an ambiguous management rights clause, a residual notion of reserved management rights, and an interpretation of express contract language.

A general principle of contract law is that the more specific contract language prevails over the more general. Thus any arbitration in which the company's defense is based upon management rights (rights either presumptively reserved or expressly withheld by contract language) must consider whether the dispute is addressed under more specific language in another section of the agreement. In making that determination, arbitrators are called upon to interpret the contract.

Arbitral Principles of Contract Interpretation.   In principle, arbitrators are not supposed to interpret a provision of a collective bargaining agreement unless it is ambiguous. If the words of a provision are plain and clear, the arbitrator should apply their clear meaning. Indeed, the contract itself often denies an arbitrator the power "to add to, subtract from, disregard, alter, or modify any of the terms of this agreement." In practice, however, those grievances which reach arbitration cannot be resolved by a simple reading of the contract. Occasionally, the meaning of a disputed contract provision can be clarified by reference to other contract language. More often than not, however, when management rights are involved, the arbitrator will have to determine—and perhaps settle for the future—the true meaning of certain contract terms by reference to evidence outside of, or "extrinsic" to, the agreement.

Contract interpretation, in essence, requires the arbitrator to divine the intent of the parties at the time of bargaining. Testimony at the arbitration hearing as to what one party "must have" surely intended when agreeing to certain contract provisions years earlier is obviously suspect as a rationalization after the fact, and of course is easily contradicted by the opposing party. To supplement such testimony and resolve conflicting recollections of the parties, the arbitrator will usually resort to three extrinsic sources to reveal intent: past practice, bargaining history, and prior arbitration awards or grievance settlements construing the same contract language or addressing a similar issue.

Past Practice.   Arbitrators will review the past conduct of the company and the union to determine contractual intent with the presumption that the parties themselves may have settled upon an unwritten interpretation of the contract language in question. Because the ultimate objective is to determine what the parties in fact have *agreed* to, rather than to impose an interpretation which is merely reasonable to an objective observer, past practice will not be held to bind the parties to future conduct unless it demonstrates a deliberate and mutual decision to proceed in a particular way in all like circumstances in the future. Consequently, arbitrators look for evidence of consistency in a past practice over a long period of time, before concluding that the practice demonstrates a contractual meeting of the minds; in other words, an ad hoc arrangement in one instance will not be held to commit either party to something more than a resolution of the matter then at hand. In addition to duration and consistency, an arbitrator may also look for evidence of mutuality, a quid pro quo, before concluding that a past practice of the parties suffices to constitute an unwritten agreement.

Bargaining History.   There are two types of bargaining history evidence, one of which is more useful to an arbitrator than the other. The first type attempts to show the parties' intentions and strategy during bargaining, from such evidence as daily bargaining notes (kept by each party) and testimony by the relevant negotiators. Ultimately, such evidence is too self-serving to be given much credence.

More telling, to the arbitrator, is evidence of actual bargaining proposals made, modified, and withdrawn by the parties at previous bargaining sessions. Arbitrators are loath— or should be, in any event—to award a party through arbitration what it failed to achieve through negotiation. Consequently, proof that the contract language now in dispute survived intact from a previous attempt to negotiate a change is particularly probative. An abandoned proposal by the union to modify the current contract language to obtain the very objective which would result from the interpretation it now asserts in arbitration indicates that the union believed at the time that the language did not truly support that interpretation. Absent another explanation, if the proposal was withdrawn in the course of bargaining, the arbitrator will conclude that the parties revealed their mutual, albeit reluctant, agreement about what the contract means. The same conclusion, to a lesser extent of certainty, can be drawn from bargaining proposals exchanged between the same parties for similar contracts governing different employees, as in a multiplant company, where the bargaining representatives for both sides were the same.

Grievance Settlements and Arbitration Awards.   The most compelling method to illuminate an ambiguous contract provision is to introduce evidence of previous grievance settlements or arbitration awards, construing the same language for the parties under similar circumstances. Although sometimes referred to as grievance or arbitral "precedent," this is a weighted misnomer, usually invoked by the party proffering the evidence. In fact, arbitrators will chafe at the suggestion they are absolutely bound by prior grievance settlements, or even arbitration awards, in future contract arbitrations. As a practical matter, however, arbitrators are often conclusively persuaded by evidence of like grievance or arbitration decisions. Because they usually recognize the need for stability and predictability in labor-management relations, arbitrators are inclined to sustain the interpretation of a prior arbitral award involving the same parties and contractual provisions, and to invite the parties to remedy any unsatisfactory results at the bargaining table instead. However, this is not an absolute rule, and arbitrators are free to disagree with another arbitrator's prior contract interpretation which they find unreasonable or unfounded, or which they may choose to distinguish on some basis from the circumstances involved in the case at hand.

Also relevant, but less persuasive, are arbitration decisions involving different parties contending the same issues and construing similar contract language. Because many labor contract clauses have become almost standardized in the industry, it can be plausibly argued that the parties must have shared the prevailing interpretation of such clauses. Published arbitration awards thus become part of a "common law" of labor relations, from which commonly used contractual clauses derive a settled, universal meaning that can be arguably presumed and imposed upon any agreement.

Arbitral Principles in Industrial Engineering Cases.   A summary of arbitral authority regarding industrial engineering issues would be undone by two problems. First, the resolution of any individual grievance is, to a very large extent, dependent on the language of the contract, which may differ greatly from language in other contracts. Second, the scope of potential arbitral issues in this area is quite broad and resists generalization.

Nevertheless, a discussion of arbitral approaches to the most commonly arbitrated industrial engineering issues will illuminate for the industrial engineer, as a witness or an assistant to the company's arbitration advocate, the nature of the arbitrator's inquiry.

Job Restructuring and Evaluation Issues. An employer may create new jobs or re-structure old ones as technology or production needs demand, and contractual job and wage provisions generally do not operate to restrict this right. However, whenever a job is created or changed, a question arises as to what the occupants of the job will be paid. This is often a function of the placement of the job on a contractual wage scheme, based upon an evaluation of the functions and requirements of the job. In other words, in most cases job evaluation questions are directly related to a determination of the appropriate hourly wage.

Collective bargaining agreements typically contain a provision that the company reserves the discretion to evaluate a new job which it has established, subject to the union's grievance. The arbitrator may make two inquiries in a case challenging a job evaluation: whether the employer used the proper method for evaluating the new job, and whether it applied that method correctly.

When the parties have expressly agreed to use a particular standard job evaluation method, there is no question about the propriety of the employer's right to use that method or the company's obligation not to deviate from it. Where no express agreement exists, the parties may dispute whether there is an implied agreement as to the proper job evaluation method, essentially a question of past practice. In either event, if there is an agreed-upon method or job evaluation manual, the arbitrator will look only to see if the company properly weighed the factors identified therein, according to its own terms. The company has no obligation to revise a settled but anachronistic evaluation system to reflect current conditions. For example, where an agreed job evaluation procedure includes an equation of jobs which are comparable in various evaluated job factors, the company will be obligated only to make the comparisons required by the evaluation manual, even if certain jobs are clearly no longer comparable because of intervening changes in job content.

In reviewing the industrial engineer's evaluation where there is no contractual or other guidance, arbitrators have considerably more discretion, although they are ultimately limited to determining whether the company's job evaluation is unreasonable, arbitrary, or capricious. In these cases, most arbitrators have adopted the approach used by most standard evaluation systems of comparing similar, already-evaluated jobs with the job whose evaluation is being arbitrated. In considering whether the new job, as evaluated, conforms to similar existing classifications, arbitrators look to the union to present evidence of higher-rated jobs containing similar characteristics.

A related job evaluation grievance arises when an employee contends that modifications to job duties warrant a job reevaluation and, derivatively, a wage increase. The company has no obligation to revise and upgrade a job evaluation unless the collective bargaining agreement requires it to do so, or unless it has revised the evaluations of other positions in response to similar changes in job duties. Often the contract will require that a new job evaluation will be undertaken by the company when the duties of the job are changed "significantly," "materially," or otherwise substantially. In such cases the union has the burden of proving that the job duties have changed fundamentally with respect to factors normally examined by the company in evaluating jobs.

Incentive Pay Issues. Grievances challenging wage incentive plans are among the most complex faced by arbitrators, and for that reason, it is preferable to select for these cases an arbitrator who is trained or experienced in incentive matters. Incentive challenges tend to be very specific to the contract and the operation involved, but several arbitral principles and practices have emerged.

When faced with a challenge to the validity of a production standard against which an incentive worker performs, arbitrators seek guidance in the terms of contractual or stipulated system for setting such standards. Where the contract thoroughly addresses the process, the scope of the arbitrator's inquiry is largely limited to a review of the company's compliance with those procedures in setting the standard. For example, such cases may dwell on the question of whether the company followed contractual or stipulated guidelines in time study measurements, permitted proper union observation of the study, or provided the union access to data from which the standard was derived. The arbitrator articulates the methods and principles provided in the contract or any incorporated system, and applies them to the facts presented at the hearing.

A more likely case is that the contract uses undefined and nonoperational words such as "reasonable," "normal," or "equitable." In those cases, an arbitrator must rely on testimony by the parties to understand the methods actually used to determine standards. The arbitrator will require substantial testimony about the measured operation, the nature of the production process, the obstacles and interruptions commonly experienced on the job, and the allowances for such nonproductive time in the standard. In addition the arbitrator will need to know the settled conditions under which a deviation from the standard is permitted, or when employees are presumed to be "off standard." So informed, the arbitrator will then determine if the company complied with generally accepted industrial engineering procedures to determine the standard, if the company's measurements under those procedures were accurate, and if the standard recognizes operational realities revealed in the testimony.

In an incentive standard arbitration, the parties can expect the arbitrator to spend a great deal of time observing and studying the operation. Some arbitrators will make time studies of their own—either themselves or by outside experts—and in some instances the parties may individually or mutually request an independent study. More often, however, arbitrators will make rough estimates of time requirements for use as a guideline and will observe other operations for comparative purposes. If a new time study is made, the parties may stipulate to an acceptable margin of deviation between the grieved standard and the new measurement, such as plus or minus 5 or 10 percent.

For the most part, arbitrators determine only whether the standard is adequate and will change it only if there are patent procedural irregularities or if the standard is unachievable by almost all employees. Arbitrators generally agree, however, that a mere reduction in incentive earnings for all or some employees following the otherwise proper implementation of a new standard will not, by itself, demonstrate that the standard is inappropriate. On the other hand, if the standard is valid, the company is generally free to increase line speed to effectively compel employees to achieve maximum incentive earnings.

Skill and Ability Determinations. A collective bargaining agreement typically contains a provision regulating the company's procedure for selecting employees to fill new or vacated bargaining unit positions; most contracts also describe procedures for allocating limited jobs in the event of a layoff. Often these mechanisms are heavily weighted toward seniority, but almost all contain some recognition of an employee's ability to perform the duties of the relevant job, particularly in the context of a competitive application for promotion. In some instances this involves an evaluation of the employee's performance after a period of time in the new position, a sort of job-specific probationary period. In other cases, particularly with respect to skilled trades jobs, the contract may permit the company to require applicants seeking the upgraded job to undertake a performance test.

Even where not specifically negotiated by the parties, arbitrators generally recognize the right of the company to unilaterally design and use tests to determine job qualifications. Such tests must be fair and must bear some rational relationship to the job or a demonstrated value in predicting success at the job. For example, a qualification test based on the performance of only one relatively small, discrete function of the job might not be fair, if at the same time all but a handful of the employees currently in the position do not perform that function well but perform other functions of the job sufficiently well to be successful in the position overall. In that respect, statistics showing job performance by other applicants, both successful and unsuccessful, will be persuasive to the arbitrator.

Poor Performance Discharge. Ability evaluations are particularly pertinent in arbitrations in which employees contest their dismissal or demotion from a job based on inadequate performance. Absent unusual language in the collective bargaining agreement, arbitrators almost universally recognize an employer's right to remove employees who are incapable of meeting reasonable performance standards, even where the deficiency results not from slothfulness but from the basic inability of a motivated, earnest employee to meet the standard. In these cases, the arbitrator is likely to uphold the demotion or discharge if

1. The performance standard is reasonable, that is, most employees can meet it.
2. The poorly performing employee had notice of the standard.
3. The employee's performance was measured over a representative period of time, usually a minimum of 2 weeks.
4. The employee was afforded some advance warning that his or her performance was deficient and could lead to discharge.

The standard for minimally acceptable performance does not necessarily bear any relationship to the base wage incentive standard. The latter is designed wholly to regulate compensation, not performance.

Because arbitrators recognize that the employer's right to expect a "fair day's work for a fair day's pay" is fundamental to the employment relationship and the company's ability to manage its business, the arbitral standard of deference to employer dismissals of poorly performing employees is higher than in cases involving discipline for employee misconduct. Although some arbitrators cite with approval an employer's attempt to offer training or other assistance to bring the substandard employee up to an acceptable level, that is not usually required. Similarly, employee intent or motive, often an issue in a conduct-based employee discharge, is irrelevant to a performance case, where poor performance is often a matter not of willful loafing but simply of inherent ability. Consequently, the arbitral penchant for progressive discipline—which looks for an escalating series of discipline as a way to correct employee misbehavior before discharge—is generally inapplicable to performance cases, where employee volition is not in issue. Provided most employees are capable of meeting the standard, and the poorly performing employee was notified of the standard and given a fair opportunity to achieve it, arbitrators will uphold performance-based discharges of even long-term, well-meaning employees with otherwise good employment records.

The Arbitration Hearing.   Arbitration is a more structured and formal process than the grievance procedure, largely because the arbitrator is neutral and must be presumed to have no knowledge of the underlying grievance or the relevant contract provisions involved. Because both parties deserve an equal opportunity to persuade the arbitrator of their case, formal procedures are observed in order to guarantee fairness and an opportunity for each side to introduce evidence under uniform and unbiased procedures.

Arbitrations follow a hearing format, requiring an ordered presentation of evidence. The party with the burden of proof—the union or employee in most cases, the company in discipline or discharge cases—proceeds first, usually with an opening statement of its case. Unlike the grievance meeting, where company and union positions are merely stated and explained, nothing may be presented to the arbitrator except by authenticated evidence of fact, either documents or testimony; unsupported conclusions generally are not welcome. In other words, an arbitration case must be made by introducing evidence from someone who is shown, or is stipulated by the parties, to have personal knowledge of relevant facts or documents. Only the evidence contained in the arbitration record, that is, the testimony given and documents submitted at the hearing, along with posthearing briefs if submitted by the parties, may be considered by the arbitrator in reaching a decision.

Unlike a court of law, the arbitration proceeding comes to the disputants on their own ground; that is, the arbitrator reports to a hearing site designated, and paid for, by the parties. This is often, but not always, a conference room on the company premises or an off-premise site, such as a hotel, which is near the company facility (thereby allowing witnesses to be called from work only when needed). Each side is represented by an advocate, often but not necessarily an attorney, who is sometimes supported by a technical adviser. In many cases the parties hire a court reporter to transcribe the arbitration hearing, especially when posthearing briefs will be filed; otherwise the arbitrator will merely take notes or record the proceedings for personal use.

Arbitration witnesses are usually sworn to tell the truth, and all evidence, documentary or testimonial, is subject to cross-examination by the opposing party. Often witnesses are "sequestered," or excluded from the hearing room when not testifying, in order to prevent their testimony from being influenced by evidence presented before them. Either party can object to any evidence offered by a witness, including testimony, on normal evidentiary grounds. However, the standard judicial rules of evidence are somewhat relaxed in arbitration, and most evidence ultimately will be received by the arbitrator to be "given the weight it deserves."

Preparation for the Hearing.   The sooner the industrial engineer becomes involved in the company's preparation for arbitration of a case involving industrial engineering issues, the better. Even experienced arbitration advocates may not be familiar with the technical aspects of a case involving industrial engineering processes, so the industrial engineer should be invited as early as possible to meet with the company advocate in order to shape the case strategy. Ideally the labor relations representative will notify the industrial engineer as soon as a grievance involving industrial engineering issues is designated for arbitration; however, industrial engineers should not hesitate to contact the labor relations director to inquire about the status of a case they know may proceed to arbitration, lest their necessary involvement in defending the company is overlooked until the last minute.

The first step in the industrial engineer's preparation for arbitration should be to dissect the elements of the disputed company decision which raises industrial engineering aspects. For example, in a work performance case, the industrial engineer should outline each step involved in the company's determination and implementation of the implicated performance standards. For each step that went into the process, the industrial engineer should describe (1) the relevant data (machine specifications, for example), (2) the elements of and basis for the decision formula, (3) the measurements taken, and (4) any documents, including personal notes, generated or used in the process. This outline will greatly assist the engineer in marshaling a narrative justification for the company's actions and will probably be of substantial help to the company advocate in organizing the presentation of the company's case.

It usually takes several months after the final step in the grievance process before an arbitration hearing can be scheduled. In the meantime, the industrial engineer should review all the documents identified in the case outline, recheck relevant measurements, and develop confidence in every aspect of the procedure involved in the engineering issue in the case. It is a good idea to check any machines which might be involved in the dispute, to determine whether their operations have changed since the particular standard was determined or measurement made.

Although industrial engineers are not usually asked to defend the company's interpretation of the relevant provisions of the collective bargaining agreement—they presumably act only at the direction of others—they should nonetheless review these provisions and other published policies (for example, a job evaluation guidebook) to confirm in their own minds the propriety of the company's procedures. Indeed, this review serves more than to reinforce the industrial engineer's confidence in the case—it also provides an important "reality check" of the company's position in the case. It cannot be presumed that the labor relations representatives, who may have little operations experience or technical aptitude, fully understand the pertinent industrial engineering procedures involved in the case. Their conviction in the merits of the company's position should not discourage the industrial engineer's own review of the contractual basis for the company's action. If there are any doubts that the actual procedures involved in the disputed action conform to the company's obligations under the contract or another published policy, the engineer should bring this to the attention of the company's director of labor relations or arbitration advocate without delay. It is better to learn this sooner rather than later, while settlement is still possible.

The industrial engineer should meet with the company advocate at least twice, the fist time early in the prehearing period, then again shortly before the hearing. At the first meeting the industrial engineer should alert the advocate to all relevant facts and documents, whether favorable or not, and make sure the advocate fully understands the industrial engineering procedures involved. If the case involves a piece of equipment or an operating procedure at the plant, the industrial engineer should also arrange to demonstrate and explain its operation to the advocate. Once again, the sooner this is accomplished the easier it will be for the advocate to formulate the presentation of the case and the defense of the company.

Shortly before the hearing, the industrial engineer should meet with the advocate again, this time to assist the advocate in preparation of exhibits, and if designated as a witness, to prepare his or her testimony and practice responses to sample cross-examination. By then the company will have identified and assembled all the documents it may use as arbitration exhibits, and the industrial engineer can practice explaining them. Even when someone is thoroughly familiar with potential exhibits, it can be surprisingly difficult, at first try, to succinctly summarize their meaning for the uninitiated. This prehearing review offers the industrial engineer an opportunity to formulate a clear explanation of the potential exhibits, so as to describe them better as a witness at the arbitration hearing.

Conduct at the Hearing.   The industrial engineer is likely to have one or both of two possible roles at the hearing: witness for the company, and adviser to the advocate.

Guidelines for the Adviser.   An arbitration hearing presents the arbitrator with two versions of the facts and the contract. Each side tries to describe its version, through witnesses and exhibits, in a manner that can be easily apprehended, is compelling in its logic, and if possible, is inviting in its simplicity. In short, each side sets out to tell its story of the case. As an adviser to the company's advocate at the arbitration hearing, the industrial engineer should be fully acquainted with the company's theory of the case and its view of the facts beyond the industrial engineering aspect; in other words, to be an effective hearing adviser—especially if serving as the only management representative assisting the advocate—the industrial engineer should understand the advocate's plan for telling the company's entire story.

The nature of the adviser's role will be defined in advance by the advocate, but the first responsibility almost certainly is to pay close attention to the testimony of the union witnesses and to alert the advocate when the testimony is false, misleading, or incomplete. An adviser should take notes of the testimony and pay close attention to the witnesses, in order to assist the advocate to keep track of potential cross-examination questions. The adviser can also help keep exhibits and other documents in order and readily available when the advocate needs them.

While advisers should not presume to represent the company's case themselves, most advocates will welcome their input of information bearing on the case as it develops, including proposed questions for cross-examination. This can be done by unobtrusively handing the advocate brief notes during the hearing, if the information is immediately relevant, or by bringing matters to the advocate's attention at one of the many breaks that usually occur in the course of the hearing. Unless asked beforehand, however, the adviser should avoid interrupting the advocate with whispered verbal comments unless urgently necessary.

Guidelines for the Witness.   There are some primary rules to guide any witness at an arbitration hearing or, for that matter, at any other hearing:

- Look the questioner in the eye.
- Never answer a question you do not understand.
- Be forthright and tell the truth.
- Talk clearly and slowly.
- Be positive, except when you are not.
- Do not guess unless asked to do so.

An arbitration witness will face questions from advocates for both parties: ''direct'' examination from the party who calls the witness and ''cross-examination'' from the opposing party, which is limited to the subject matter of the testimony given on direct examination. Generally the industrial engineer will be called first during the company's case and thus will face initial, direct examination from the company's advocate.

The difficulty inherent in testifying on direct examination is that the advocate generally is prohibited from asking ''leading'' questions; that is, the advocate cannot ask a substantive question by providing the answer and asking for assent to it. For example, instead of asking whether ''the performance standard against which the grievant was measured accounted for interruptions in material flow caused by new preproduction quality inspections,'' the advocate must solicit an explanation through open-ended questions such as, ''what interruptions to material flow are anticipated by the performance standard?'' That is why prehearing preparation is so important: the industrial engineer will know in advance which information the advocate is soliciting and which information should not be volunteered.[8]

On the other hand, a witness should be careful not to be too quick in answers given on direct examination. While prehearing preparation is expected—and can be admitted if inquired about by the arbitrator or the union advocate—testimony is supposed to be based upon the witness' own knowledge, not the repetition of facts or opinions provided by the company. If the testimony appears too rehearsed, the witness can lose credibility before the arbitrator. It is more believable for witnesses to concede that they cannot recall less important facts associated with a remote event—what time of day a certain bargaining session ended, for example—than to reconstruct the entire day with implausible exactitude.

Cross-examination, on the other hand, has its own perils for the witness. On direct examination, the witness is generally invited to give a full answer; on cross-examination, however, the union advocate will try to limit the company witness to specific agreement or disagreement with carefully constructed and weighted questions. It is a good idea for the industrial engineer to practice responding to cross-examination with the company advocate before the hearing.

In general, an industrial engineer should observe the following guidelines on cross-examination:

- *Don't argue and don't show anger or hostility.* Arbitrators are not impressed by impassioned testimony. If the union advocate is being unfair or rude, the arbitrator will notice or the company advocate, not you, will bring it to his or her attention.

- *Be patient.* The company will have its chance during additional, or ''redirect,'' examination to clear up any confusion resulting from ingenuous or crafty cross-examination questions.

- *Answer only the question asked, not the question that should have been asked.* Avoid volunteering information beyond that specifically sought by the question, unless necessary to explain an unfavorable ''yes'' or ''no'' answer. (The union advocate may successfully object to any unsolicited elaboration, but don't worry: the company advocate can give you the opportunity to explain any answer on redirect examination). When you're finished, sit still and wait for the next question.

- *Pause before answering any question.* Wait to see if the company advocate objects to a question before answering it. Although the arbitrator may claim to disregard an unfa-

---

[8]It is possible, but not common, for the union to call the industrial engineer first, as an ''adverse'' witness, in which event the union advocate will ask questions on direct examination, and the company's advocate will be limited to cross-examination about the issues raised on direct. When that happens, the rules with respect to leading questions are reversed: the union advocate can ask leading questions of an adverse witness on direct examination, while the company's advocate on cross-examination may not. In other words, the rule on leading questions recognizes and follows the true loyalties of the witness, not the order in which the witness is called. However, the union's use of the industrial engineer as an adverse witness does not prevent the company from recalling the engineer again during its own case, to ask questions about matters not raised during the union's direct examination.

vorable answer if elicited by an improper question, for all practical purposes, once the answer is out, the objection is lost.

- *Watch out for complex questions.* Ask the union advocate to restate a question which is confusing, ambiguous, or awkwardly stated. Don't try to answer a multiple question; make the advocate break the question into its component parts.
- *Don't speculate.* If you do not personally know or remember something, say so without apology and leave it at that.

In some cases the arbitrator may also ask questions of the witness. Answer them respectfully, but keep an eye on the company advocate as you proceed. This will allow the advocate an opportunity to object if the answer begins, unavoidably, to stray into dangerous territory.

Industrial engineers are called as witnesses in arbitration hearings to talk about matters which other company managers and labor relations representatives are not trained or experienced enough to talk about themselves. They bring a special, technical knowledge to the proceeding, and for that reason are treated, by the arbitrator at least, with some deference and respect. But to be an effective witness, the industrial engineer must successfully explain a technical issue to a nontechnical audience. It would be too difficult for the industrial engineer, and might be perceived as condescending by an arbitrator, to abandon technical jargon altogether. However, the industrial engineer must be prepared to take the time to explain the technical questions and facts in issue—the jargon—in lay terms. In this respect, the industrial engineer, as a witness, must take a cue from the arbitrator, who may or may not be experienced in the industrial engineering issues involved in the case.

Postarbitration Considerations. Following the hearing and the optional submission of posthearing briefs, the arbitrator will issue a written decision. The quality of arbitration decisions varies widely. Ideally, an arbitrator's decision should describe the grievance, set forth the relevant provisions of the labor contract and the positions of the company and the union, explain the reasoning leading to an award either granting or denying the grievance, and if the grievance is granted, set forth a rational remedy for the grievant.

Most labor contracts provide that arbitration awards are "final and binding" upon the parties. Consequently, courts are extremely reluctant to set aside arbitration awards, even poorly reasoned ones, except when they are so patently irrational as to have no discernible basis in the collective bargaining agreement.[9] Absent the rare decision which may be subject to judicial correction, the company and the union are stuck with the award, along with its persuasive, if not precedential, interpretation of the contract—at least until the parties negotiate new contract language to correct that interpretation.

The industrial engineer's involvement in the arbitration process does not end with the award, at least not with an unfavorable one. Industrial engineers should carefully review arbitration decisions and offer their own appraisal of the effect of such decisions on the company's operations and productivity objectives, both within the unit which was the subject of the grievance and elsewhere in the production facility. They should also suggest methods to temper the adverse impact of a bad arbitration award and propose procedures to correct the award which can be incorporated into proposals in later negotiations for a new labor agreement.

COLLECTIVE BARGAINING

Arbitration ultimately is an extension of the collective bargaining process. Not only is it itself a creature of the contract, it is a means, in the periods between full-scale negotia-

---

[9] *United Steelworkers of America v. Enterprise Wheel and Car Corp.,* 363 U.S. 593, 596–598, 80 S. Ct. 1358, 1360–1361 (1960); *see also United Paperworkers Int'l Union v. Misco, Inc., supra* note 5.

tions, to maintain the vitality and relevance of the contract in the face of constantly changing circumstances. If the industrial engineer has a role in arbitration, then for the same reasons the industrial engineer must be involved in the process of establishing the contractual terms which are interpreted in arbitration.

The Process of Collective Bargaining.   A company and a union can and do negotiate agreements at any time. However, their biggest contract, often called the "master" collective bargaining agreement, generally runs for a given term, commonly 3 years.

The term of the contract necessarily defines the cycle of the bargaining process. Early in the process—soon after the previous contract is signed and ratified by the union—a well-organized labor relations management will begin to define the company's goals for the next bargaining cycle. This should be an inclusive process: managers at all levels should be consulted for their identification of obstacles in the current agreement and suggestions for improvement. Soon into this assessment, the company should organize a broad-based negotiations committee to distill the best ideas from management input, and from that basis to formulate contract proposals.

Ideally the first draft of the company's "wish list" of contract revisions should be prepared a full year before bargaining begins. A winnowing-out process follows which eliminates or refines proposals based upon a more realistic appraisal of the union's anticipated response to them, and an evaluation of each proposal's importance to the company—high, medium, or low priority—in comparison with the trade-offs necessary to obtain union agreement. Finally an initial composite set of company proposals is exchanged for the union's proposals when negotiations begin at some agreed time before the expiration of the contract.

Negotiation meetings take place between bargaining "teams" of the company and the union. The composition of these teams will vary in each case, depending on the subject matter of negotiations and institutional political considerations—especially for the union but also, to a lesser extent, for the company as well. In many cases subcommittees of the respective bargaining teams will be formed to discuss esoteric or technical issues—for example, revised job classifications. In many other instances, the teams negotiate in only a pro forma sense, while true negotiations take place in so-called side bar sessions, which are private meetings between only one or two team members from each side.

Negotiations typically proceed slowly at first, with little substantive agreement until contract expiration draws near. An imminent contract expiration gives the parties an incentive to deal with the hard issues. When the contract expires, the union is free of its no-strike obligation and can stop work at any moment; at the same time, the company can replace strikers or lock out the workers and close the plant. Neither side wants to face this showdown, so the greatest part of bargaining occurs in the days before the deadline. Often the company and the union will agree to a brief extension of the contract expiration deadline if there is a good prospect of concluding negotiations soon; but the extension provides only a limited reprieve, because the union cannot afford to diminish its bargaining power by freely forestalling its threat of a strike.

The Role of the Industrial Engineer.   Too often, industrial engineers are not invited to take a significant part in the collective bargaining process. That is a shortsighted mistake. The industrial engineer can contribute tremendously to the formulation and presentation of the company's proposals at each step of the process.

In the development of the company's proposals, industrial engineers, because of their daily familiarity with current production practices and problems, should be able to identify provisions of the existing contract which obstruct efficiency and productivity. They can invoke concrete examples of problems arising in the past under the current contract language, and calculate their economic impact. In so doing, industrial engineers can clarify for the company the relative priority of recommended contract changes.

Those same qualities enable industrial engineers to articulate and justify company proposals to union representatives at the bargaining table. Their firsthand knowledge of work processes gives them, in the eyes of the union bargaining team, much more credibility than the labor relations representatives or other management officials in explaining the company's position on issues involving job structures and performance expectations. This is particularly true in the negotiation of new wage incentive systems, in which production standards and job classifications are revised, or in proposals for productivity-based gainsharing programs, which must establish production and quality benchmarks. By virtue of both their training and experience, industrial engineers are capable of comparing any existing productivity and incentive programs with proposed new systems, by calling upon their own ready knowledge of detailed shop-floor examples. This makes them well suited to represent management on bargaining subcommittees assigned to resolve such issues.

Finally, industrial engineers can provide an essential check against the company's concession to apparently innocuous proposals which have hidden adverse consequences. From their unique perspective, industrial engineers may be able to perceive a potential for mischief in contract language which might be overlooked by other members of the bargaining team.

There are times, of course, when contract ambiguity is deliberately sought by the parties precisely for its imprecision, allowing both to later argue their own interpretation of the contract—in other words, essentially deferring the question to another time and another forum, such as arbitration. There are other times when bargaining realities compel the company to accept loose contract terms. However, the company's entry into an ambiguous contract should be with a knowledge of the risks involved. The industrial engineer can identify some of those risks and can counsel against the use of subjective and vague contract language when the potential of an adverse arbitral interpretation is too great.

CONCLUSION

Industrial engineers have long played an important, but sometimes unacknowledged, role in the labor relations process. They continually guide the company to a more efficient use of resources but in doing so often change the nature of work for other employees. To the extent their analysis and recommendations form the empirical basis for a company's actions, industrial engineers are indispensable parties to the process of understanding and resolving the disputes which result from those actions.

Handling employee grievances and arbitrations is a natural part of an industrial engineer's job—as can and should be the industrial engineer's involvement in the process of shaping and negotiating the collective bargaining agreement. Both functions can be accomplished with the same professional skill which industrial engineers bring to the technical aspects of their work and can likewise be a source of as much professional satisfaction.

FURTHER READING

Elkouri, Frank, and Edna Asper Elkouri, *How Arbitration Works,* 4th ed., Bureau of National Affairs, Washington, D.C., 1985, and supplements.

Labor Law Course, 26th ed., Commerce Clearing House, Chicago, Ill., 1987.

Rubin, Milton, "The Arbitration of Wage Incentives," *Arbitration of Subcontracting and Wage Disputes, Proceedings of the Thirty-Second Annual Meeting of the National Academy of Arbitrators,* Bureau of National Affairs, Washington, D.C., 1979.

Wiggins, Ronald L., *The Arbitration of Industrial Engineering Disputes,* Bureau of National Affairs, Washington, D.C., 1970.

CHAPTER 6
OCCUPATIONAL SAFETY

Donald S. Bloswick
Assistant Professor, Department of Mechanical Engineering
Director, Ergonomics and Safety, Rocky Mountain Center for
Occupational and Environmental Health
University of Utah
Salt Lake City, Utah

During the last half of the nineteenth century, the industrial revolution made a profound impact on The United States. American production methods changed from craft shops to mechanized factories, first in New England in the textile industry but soon throughout the entire country. The National Safety Council (NSC, 1988) notes that these changes in production methods can be summarized as:

1. The substitution of steam and other inanimate power sources for animal sources
2. The substitution of machine power for human power and skills
3. The development of new processes for making iron, steel, industrial chemicals, and other materials
4. The reorganization of work from craft shops to large factories or mills with an efficient division of labor

These production methods greatly expanded the quantity and types of products available to the average American. They also expanded the magnitude and types of hazards present in the industrial workplace and resulted in the awareness of the need for industrial safety programs.

The National Safety Council (NSC, 1988) estimates that the work-related death rate has decreased from approximately 15 per 100,000 people in the U.S. population in 1930 to less than 5 per 100,000 in 1985. Some of this decrease must be attributed to the recognition of the importance of industrial safety and the implementation or enhancement of industrial safety programs. A disconcerting fact, however, is that the disabling injury frequency *rate* (cases per million person-hours worked) has increased from an average of 6.26 for the 5 years 1961–1965 to 10.68 for the 5 years 1971–1975 (DeReamer, 1980). (The severity rates were comparable.) While some of the increase in the disabling injury frequency rate may be due to the impact of the OSHAct, and resulting increased record keeping requirements, one still must be concerned with this trend. One must also be concerned when reviewing the absolute human and financial cost of work-related injuries, illnesses, and deaths in

1989. The Bureau of National Affairs (BNA, 1990) notes the following highlights of a Bureau of Labor Statistics survey of 1989 data:

1. There were 6.6 million work-related injuries and illnesses in 1989, or 8.6 per 100 full-time workers. This is nearly the same as 1988 but an increase from the 7.9 per 100 workers in 1986 and 8.3 in 1987.
2. Nearly half of these injuries resulted in lost work time or restricted work activity.
3. There were an estimated 3600 work-related deaths in establishments of 11 or more employees.

The National Safety Council (1989) notes that for 1988:

1. 10,600 workers were killed on the job.
2. Deaths and injuries in the workplace cost an estimated $47.1 billion.
3. Each death cost approximately $550,000 and each disabling injury $16,800. Each worker must produce goods or services in the amount of $410 just to offset these accident costs.

In summary, while progress has been made in reducing the human and dollar cost of occupational accidents, continued emphasis is necessary to protect the life and health of workers while accomplishing the organization's total performance objectives.

Definitions

Accident. An accident is an unexpected event which interrupts the work process and carries the potential for injury or damage. Accidents may or may not result in fatality, injury, or property damage but have the potential to do so (Firenze, 1988). An accident may be attributed to a human factor, a situational factor (operations, tools, equipment, and/or materials), or an environmental factor.

The following definitions, adapted from Hammer (1989), illustrate additional safety-related concepts:

Hazard. A hazard is a condition which has the potential to cause injury, damage to equipment or facilities, loss of material or property, or a decrease in the capability to perform a prescribed function.

Danger. The danger inherent in a situation is dependent on the relative exposure to a hazard. For example, a high-voltage transformer is a significant hazard but may present little danger if locked in an underground vault.

Damage. Damage is the severity of injury or magnitude of loss which results from an uncontrolled hazard. A worker on an unguarded beam 10 feet above the ground is exposed to a similar hazard (potential for fall injury) and is in the same danger (exposure to fall) as a worker on an unguarded beam 300 feet above the ground. The possibility of damage, however, is much greater in the latter case.

Risk. Risk is a function of the probability of loss (danger) and the magnitude of potential loss (damage). It can be thought of as (probability of loss) × (magnitude of potential loss).

Safety. Safety is the absence of hazards or minimization of exposure to hazards. Firenze (1991) also notes that safety is the control of hazards to an acceptable level.

OCCUPATIONAL SAFETY STANDARDS

History of Safety Standards. In about 1750 B.C., Hammurabi's code presented probably the first written safety guidelines (including a penalty clause). It notes that "If a builder constructs a house for a man and does not make it firm and the house collapses and causes the death of its owner, the builder shall be put to death" (Hammer, 1989). In the United

States, interest in safety was a logical accompaniment to the industrial revolution, and in 1867, Massachusetts became the first state to use factory inspectors. Ten years later it passed a state law requiring safeguarding of dangerous machinery.

The steel industry pioneered the safety movement in the United States. In 1892, the Joliet Works of the Illinois Steel Company established a safety department. In 1906, the United States Steel Corporation created a committee for inspection and accident prevention, and its president, Judge Elbert Gary, issued what might well be the first company safety policy:

> The United States Steel Corporation expects its subsidiary companies to make every effort practicable to prevent injury to its employees. Expenditures necessary for such purposes will be authorized. Nothing which will add to the protection of the workmen should be neglected (NSC, 1988).

The first national safety standards were attempts to deal with boiler explosions. In 1915, specifications established by the American Society of Mechanical Engineers (ASME) were adopted nationwide as a voluntary code for the development and maintenance of boilers.

In 1912, the first general industry safety conference was held in Milwaukee. The following year the National Council for Industrial Safety was organized in New York. Shortly thereafter, the organization was enlarged to include other types of safety and the name was changed to the National Safety Council. The American National Standards Institute (ANSI) evolved from the National Safety Council. The function of ANSI is to determine when a national consensus relative to voluntary standards has been reached, rather than to generate these standards. Consensus is reached by coordinating the development of standards by the national groups and organizations concerned. These consensus standards have often been adopted by states and federal agencies as the basis for government regulations.

Workers' Compensation.   Prior to the enactment of workers' compensation legislation it was necessary for the worker to sue the employer in a court of law and prove that the accident or injury was due solely to the employer's negligence. As a defense, the employer could plead that (1) the injured employee had contributed through his or her own negligence, (2) the injury was due to the negligence of a fellow employee, or (3) the employee had knowingly assumed the risk inherent in the job. These defenses were often successful. Even when the employers' negligence was clear, the immediate costs of the trial and medical bills worked against the employee. The primary purpose of workers' compensation legislation was to provide prompt and reasonable income and medical benefits to victims of industrial accidents (or their dependents) regardless of fault. It assured the workers of payment, and the workers gave up the right, except in rare situations, to sue the employer for work-related injuries.

In 1911, Wisconsin and New Jersey passed the first workers' compensation laws in the United States. By 1915, 30 states had passed some type of workers' compensation legislation. Initially, workers' compensation laws were declared invalid, a violation of the Fourteenth Amendment. Requiring an employer to pay damages without regard to fault was considered taking property "without due process of law." In 1917, the U.S. Supreme Court, ruling in *White v. the New York Central Railroad,* declared that such taking of property, because of the extreme degree of public interest involved, was within the states' police powers. This decision resulted in the remaining states quickly passing their own workers' compensation laws. At present there are 50 state workers' compensation laws, 3 within the U.S. federal government, and 4 more for the District of Columbia, Guam, Puerto Rico, and the Virgin Islands.

Government Involvement—OSHA.   During the first half of the twentieth century the federal government's involvement in safety legislation was largely limited to setting safety and health standards for its contractors. The Walsh-Healy Public Contract Act of 1936 provided that contracts in excess of $10,000 entered into by an agency of the Unites States prohibit the use of materials "manufactured in working conditions which are unsanitary or dangerous to the health and safety of the employees."

Federal legislation during the 1960s was aimed primarily at specific industries. The Construction Safety Act of 1969 required that all federal or federally financed or assisted projects in excess of $2000 comply with established safety and health standards enforced by the U.S. Secretary of Labor. The Federal Metal and Nonmetallic Mine Safety Act of 1966 and the Federal Coal Mine Health and Safety Act of 1969 also directed attention to occupational safety and health. The Federal Mine Safety and Health Act, promulgated in 1977, established a single mine safety and health law for all mining operations enforced by the Mine Safety and Health Administration (MSHA) within the Department of Labor.

Occupational Safety and Health Act. On October 29, 1970, Richard Nixon signed Public Law 91-596. This law, the Williams-Steiger Occupational Safety and Health Act of 1970 (OSHAct, 1970), became effective on April 28, 1971. Its stated purpose is:

> To assure safe and healthful working conditions for working men and women; by authorizing enforcement of the standards developed under the Act; by assisting and encouraging the States in their efforts to assure safe and healthful working conditions; by providing for research, information, education, and training in the field of occupational safety and health; and for other purposes.

The Occupational Safety and Health Administration, or "OSHA," was established to enforce the OSHAct. OSHA is located within the Department of Labor. Under the provisions of the OSHAct the National Institute for Occupational Safety and Health (NIOSH) was established within the Department of Health, Education and Welfare (currently the Department of Health and Human Services). While OSHA is primarily an enforcement agency, the primary functions of NIOSH are to perform safety and health research, develop and establish recommended standards, and facilitate the education of personnel qualified to implement the provisions of the OSHAct.

The regulations relating to the OSHAct are included in Parts 1900 to 1999 of Title 29 of the Code of Federal Regulations (CFR) and the operation of the Occupational Safety and Health Review Commission (OSHRC) are included in Parts 2200 to 2499. Unless specifically excluded, the OSHAct applies to every employer who has one or more employees and who is engaged in a business affecting interstate commerce. Major examples of specific exclusions of the OSHAct are state and local government employees, self-employed persons, farms at which only immediate members of the farm employer's family are employed, and workplaces already protected by other federal agencies under other federal statutes (operators and miners covered by the Federal Mine Safety Act of 1977, for example). Employers with fewer than 11 full- or part-time employees are excluded from the record-keeping requirements of the OSHAct.

OSHA Standards. There are three separate sets of standards: General Industry (29 CFR 1910), Construction (29 CFR 1926), and Maritime Employment (29 CFR 1915–1919). Summaries of major portions of the OSHA standards have been prepared by OSHA and are available in digest form. OSHA Publication 2201 is a summary of the OSHA General Industry Standards (OSHA, 1988b) and OSHA Publication 2202 is a summary of the OSHA Construction Standards (OSHA, 1989a).

Application of the General Duty Clause. In order to consider hazards not specifically included in OSHA standards, OSHA has turned to the provisions of Section 5(a) of the OSHAct, or the "General Duty Clause," which states that

> Each employer- (1) shall furnish to each of his employees employment and a place of employment which are free from recognized hazards that are causing or are likely to cause death or serious physical harm to his employees; (2) shall comply with occupational safety and health standards promulgated under this Act.

In summary, four governmental units have the primary responsibility to carry out the act:

1. *The Occupational Safety and Health Administration (OSHA)* is concerned with national, regional, and administrative programs for developing, and ensuring compliance with, safety and health standards. It also trains OSHA personnel. U.S. Department of Labor, Department of Labor Building, 200 Constitution Avenue, NW, Washington, D.C. 20210; (202) 523-8148.

2. *The Occupational Safety and Health Review Commission (OSHRC)* reviews citations and proposed penalties in enforcement actions contested by employers or employees. 1825 K Street, NW, Washington, D.C. 20006; (202) 634-7970.

3. *The National Institute for Occupational Safety and Health (NIOSH)* is a research, training, and education agency. U.S. Department of Health and Human Services, 1600 Clifton Road NE, Atlanta, Ga. 30333; (404) 329-3061.

4. *The Bureau of Labor Statistics (BLS)* conducts statistical surveys and establishes methods for acquiring injury and illness data. U.S. Department of Labor, 200 Constitution Avenue, NW, Washington, D.C. 20210; (202) 523-1092.

SAFETY MANAGEMENT

Accident Statistics and Record Keeping.   Accident statistics and record keeping are useful in evaluating the safety level of a facility or industry, to determine where to allocate safety resources, and to determine the effectiveness of control methodologies.

Accident Statistics.   Traditional accident statistics involve the calculation of frequency and severity rates. Accident frequency is the number of incidents which occur for a specific number of hours worked. The American National Standards Institute has traditionally used a base of 1 million person-hours worked. OSHA has established 100 person-years or 200,000 person-hours as the base for accident statistics (OSHA, 1989b). (An OSHA recordable incident is one which involves lost time, restricted work, or medical care other than minor first aid). For example, if there were 3 recordable accidents in a year when there were 400,000 hours worked, the OSHA accident frequency rate for the above situation would be

$$\frac{(\text{Number of accidents}) \times (200,000)}{\text{Hours of employee exposure}} \quad \text{or} \quad \frac{(3) \times (200,000)}{400,000}$$
$$= 1.5 \text{ accidents per } 200,000 \text{ hours worked}$$

The same procedure may be used to determine the number of a particular type of accident. For example, in the above example if two or three accidents resulted in lost workdays, the lost workday injury (or LWDI) rate would be

$$\frac{(2) \times (200,000)}{400,000} = 1 \text{ LWD injury per } 200,000 \text{ hours worked}$$

It is also possible to determine an injury *severity* rate. This is done by using a measure which involves days lost from the job or on restricted work activity. If in the above example the two lost workday accidents resulted in a total of 20 days away from work or on restricted work activity, one measure of severity would be

$$\frac{(20) \times (200,000)}{400,000} = 10 \text{ lost (or restricted) workdays per } 200,000 \text{ hours worked}$$

ANSI calculations of severity rates assign a fixed number of days to fatalities or permanent partial disabilities.

Record Keeping.   While record-keeping mechanisms may be established for any number of reasons, the record-keeping requirements of the OSHAct establish the minimum requirements for most employers. *All* employees subject to the OSHAct must keep records except:

1. Employers with a total of 10 or fewer full-time or part-time employees during the year at all the employer's work sites
2. Employers who conduct primary business in one of the standard industrial classifications specifically exempted by OSHA

Specific state or local requirements may require even those employers noted above to maintain similar safety records. Form OSHA No. 200 is the basic log and summary of occupational injuries and illnesses. This form includes information relating to the employer, employee name and work location, type of injury or illness, and the extent and outcome of the injury or illness. Essentially all work-related injuries and illnesses are required to be recorded on the OSHA No. 200 except those requiring only minor first aid (minor scratches, cuts, burns, splinters, and the like). Work-related injuries and illnesses which require any days of restricted work activity must be noted. Supplemental data relating to the injury or illness must be recorded on form OSHA No. 101. There are several publications which provide detailed information about record-keeping requirements including retention posting and so on (BLS, 1986).

Accident Causation Models.   Prior to the development of formalized approaches to accident control, accidents were viewed as chance occurrences or "acts of God." A variation on this theme is that accidents are inherent consequences of production. These attitudes or approaches yield no information about causation, and they limit solutions to those that mitigate the adverse consequences of the event. Perhaps the greatest handicap to the evolution of a systematic approach to safety is the willingness of people, even those with analytical training, to accept accidents as acts of God and to believe that "...the causal sequences involved were those of chance or luck that were incapable of any form of examination beyond mere tabulation" (Wigglesworth, 1972).

Accident Proneness.   During the 1920–1940 era the theory of accident proneness became popular. It was based on studies which indicated that certain individuals had a disproportionate number of accidents. Accident proneness assumes a relatively permanent, personal idiosyncrasy which predisposes the individual to have accidents. The assumed permanence of the idiosyncrasy is an important issue. The tendency to have an accident is actually a function of situational and personal factors including age, health, vision, fatigue, stress, mental state, and the like. The "proneness" to have an accident is really a function of many situational factors which increase the possibility of an accident during a particular period of time.

Studies purporting to identify accident-prone individuals have often employed incorrect statistical techniques which fail to compare distributions of accidents which happen to "accident-prone" people with distributions of accidents which would occur entirely by chance.

Biorhythm.   One behavioral theory which became popular in the 1970s was that individuals would be more likely to have accidents during critical days of their 23-day physical cycle, 28-day emotional cycle, or 33-day intellectual cycle. As with the concept of accident proneness, no hard scientific data have proved the theory's validity.

Circadian Cycle.   Experiments with the 24-hour biological cycle called the circadian rhythm show more promise than do biorhythms. Body functions which vary in an approximate 24-hour pattern include body temperature, blood pressure, respiration, blood sugar, hemoglobin levels, urine volume, coordination, sensory discrimination, and rate of cell division. When this rhythm is disturbed, fatigue and disorientation often result. Shift changes which disturb this rhythm may cause accidents. A partial solution may be employee screening so that alternative or rotating shifts are assigned to workers who are able to adjust quickly and safely.

Heinrich's Domino Theory.   In 1931, Heinrich (Heinrich, 1980), after reviewing a large number of insurance claims, noted that

1. Industrial injuries result only from accidents.
2. Accidents are caused directly only by the unsafe acts of persons or exposure to unsafe mechanical conditions.
3. Unsafe actions and conditions are caused only by the acts of persons.
4. Faults of persons are created by environment or acquired through heredity.

According to Heinrich, five factors in the chronological accident sequence occur in a fixed, logical order. These are (1) social environment and ancestry, (2) fault of person, (3) unsafe act, mechanical or physical hazard, (4) accident, (5) injury. Just as the fall of the first domino in a row causes the fall of the entire row, an injury is caused by the action of the preceding factors. The injury is inevitable unless the series is interrupted by the removal of a factor. If, for example, the unsafe act or mechanical hazard is removed, the accident and the injury will not occur. Heinrich's focus on unsafe acts and conditions affected the direction of industrial safety. Through the exploration of unsafe acts, safety professionals delved into psychology, medicine, biology, sociology, and communication skills. Through the exploration of unsafe conditions, safety professionals performed research in the areas of engineering, physics, and chemistry.

Critics have faulted Heinrich's lack of recognition of multiple causation. For example, a worker's fall from a defective ladder may be attributed to an unsafe act (climbing a defective ladder) or unsafe condition (defective ladder). These causes may result in disciplinary action against the worker for the unsafe act or getting rid of the unsafe ladder. Further investigation, however, might result in multiple solutions: an improved inspection procedure, improved training, better assignment of responsibilities, or prejob planning by supervisors. These root causes often relate to the management system and affect not only the accident under investigation but other operational problems which might cause accidents in the future.

It is important to determine not only the existence of an unsafe act or condition and how it can be corrected, but why it was permitted, and whether supervision and management have the knowledge and resources to prevent it.

Systems Engineering Approach.   System safety has been defined as "the total set of men, equipment, and procedures specifically designed to be imposed on an industrial system for the purpose of increasing safety" (Brown, 1973). These techniques can be applied to three areas:

1. The design of the end product for safety.
2. The design of the manufacturing process for safety.
3. The design of the safety system. This is the overall perspective, emphasizing the entire management system.

Where traditional industrial safety focuses on the operational phase, the systems approach also includes the conceptual, design, and disposal phases. This approach is based on the assumption that the system consists of an interacting set of discrete elements (human, mechanical, situational, environmental) and that controls can be developed so that the system can perform its function safely. Specific systems safety analysis techniques will be discussed later in this chapter.

Cooperation between Management and Workers.   To be effective, safety programs must involve both management and workers. A quote from the OSHA Safety and Health Guide for the Meatpacking Industry (OSHA, 1988c) illustrates this point:

An employer's commitment to a safe and healthful environment is essential in the reduction of a workplace injury and illness. This commitment can be demonstrated through personal concern for employee safety and health, by the priority placed on safety and health issues, and by setting good examples for workplace safety and health. Employers should also take any necessary corrective action after an inspection or accident. They should assure that appropriate channels of communication exist between workers and supervisors to allow information and feedback on safety and health concerns and performance. In addition, regular self inspections of the workplace will further help prevent hazards by assuring that established safe work practices are being followed and that unsafe conditions or procedures are identified and corrected properly. These inspections are in addition to the everyday safety and health checks that are the routine duties of supervisors.

Since workers are also accountable for their safety and health, it is extremely important that they too have a strong commitment to workplace safety and health. Workers should immediately inform their supervisor or their employer of any hazards that exist in the workplace and of the conditions, equipment and procedures that would be potentially hazardous. Workers should also understand what the safety and health program is all about, why it is important to them, and how it affects their work.

Accident Investigation. [This section is adapted from OSHA Publication 2288, *Investigating Accidents in the Workplace* (OSHA, 1977)]. Accident investigation involves the investigation of every factor relating to an accident in order to determine the events leading up to, and the cause(s) of, the accident. There are two primary goals of accident investigation: (1) determine the causes(s) of the accident, and (2) prevent the accident (or similar accidents) from happening again. The basic principles of accident investigation are noted below.

1. The investigator should have a basic familiarity with the equipment, operation, or processes involved and an understanding of the conditions or circumstances which might be associated with the type of accident which is under investigation.

2. Every effort should be made to get to the accident scene promptly. As time passes, it becomes increasingly difficult to gather the facts associated with the accident. (Prompt investigation also decreases the likelihood that the same procedures or conditions will cause additional accidents and injuries.)

3. Two important attributes are comprehensiveness and creativity. It is critical that one perform a thorough job of gathering facts through photography, interviews, accident reconstruction, and the like. The analysis of the facts to determine the accident cause(s) often requires creativity.

4. Recognize that accidents do not always have a single cause but are often caused by a combination of personal, environmental, procedural, physical, or other factors.

5. The primary purpose of accident investigation is to improve safety and health conditions in the workplace. It is important to determine if a violation of applicable safety standards was a factor in the accident, or, if not, if a standard (or interpretation of the standard) should be revised to cover a hazardous condition which contributed to the accident.

A list of the basic equipment required for accident investigations, based on that recommended for OSHA compliance officers, is included as Fig. 6.1.

Liberty Mutual Insurance Company (no date) suggests that every accident should be investigated for a cause and a result and notes that the result might be death, injury, damage, interruption of planned activities, or a combination of these. They also note that the investigation of an accident which results in minor injury or damage might provide as many constructive conclusions as the investigation of a more serious accident. The form used to record data during an accident investigation may vary from company to company, but Liberty Mutual suggests that, for accidents involving injuries, the general contents should include

1. All available information about the injured person

2. What the injured person was supposed to be doing at the time of the accident and what was actually being done

SAFETY EQUIPMENT	☐ ☐ ☐ ☐ ☐	Hard Hat Safety Glasses or Goggles Earplugs Safety Shoes/Rubber Boots Protective Clothing
PHOTOGRAPHY	☐ ☐ ☐ ☐ ☐ ☐ ☐	Camera Film—Black and White Color Video Camera, Tape, Extra Batteries Flash and Other Attachments Photography Log Sealing Devices Folding Rule (in fractions of feet)
SKETCHING	☐ ☐ ☐ ☐	Graph Paper Ruler, Protractor, Compass Pencils Carbon Paper
INTERVIEWING	☐ ☐	Interview Comment Sheet Handheld Tape Recorder
SAMPLING	☐ ☐ ☐ ☐ ☐	Sample Log Chemical/Reflective Detectors (Gases) Containers and Caps (Liquids) Plastic Bags and Envelopes (Solids) Tags/Labels
WRECKAGE EXAMINATION	☐	Parts Retrieval Log
OTHER	☐ ☐ ☐ ☐ ☐	Stop Watch Magnifying Glass Sticks of Chalk Compass 50-Foot Cloth Tape

FIGURE 6.1 Recommended basic equipment for accident investigation. (*Reprinted from* Investigating Accidents in the Workplace, *OSHA Publication 2288, 1977.*)

3. How the employee was performing the operation

4. What training the employee had received

5. The past accident record of the employee and work area

6. An examination or inspection of the tools, machinery, personal protective equipment, and other physical conditions existing at the time and place of the accident

Training.   Personnel injured at work often lack the information, knowledge, and skills required to protect themselves. OSHA (1987b) notes that various surveys by the Bureau of Labor Statistics have found that:

1. Of 724 workers injured while working with scaffolds, 27 percent indicated that they had received no information on the use of the scaffold they were using.

2. Of 868 workers suffering head injuries, 71 percent said that they had received no training regarding the use of hard hats.

3. Of 554 workers hurt while maintaining equipment, 61 percent indicated that they had received no training on lockout procedures.

OSHA suggests that employers be sure that safety training be made an essential part of new employee orientation and plant routine. OSHA standards require specific training in many types of hazardous work. An effective, comprehensive training program will result in increased efficiency and reduced absenteeism and workers' compensation costs.

Safety training material is available from a wide variety of commercial sources. Information on safety and health training from government sources is available as noted below:

1. A catalog of "OSHA Publications and Audiovisual Materials" is available from the OSHA Publications Office, Room S-4203, Washington, D.C. 20210; (202) 523-9655.

2. A six-part self-study program "Principles and Practices of Occupational Safety and Health" for first-line supervisors is available from the Superintendent of Documents, Government Printing Office, Washington, D.C. 20402; (202) 783-3238. Total program cost is $29.95, but parts must be purchased separately.

3. A book *A Resource Guide to Worker Education Materials in Occupational Safety and Health* (228 pages) lists training materials and publications on safety and health which are available from some public and private organizations. It may be purchased from the Superintendent of Documents, Government Printing Office, Washington, D.C. 20402; (202) 783-3238. $7.00.

4. The OSHA Training Institute provides training for industrial personnel in the areas of general industry and construction safety. Tuition is modest, but participants must register in advance. Information is available from the OSHA Training Institute, 1555 Times Drive, Des Plaines, Ill. 60616; (312) 297-4810.

SAFETY ENGINEERING

Safety engineering may be thought of as the application of engineering and management principles to systems consisting of workers, equipment, materials, and processes within a defined environment with the goal of reducing the probability and severity (risk) of injuries and property damage. In this section the basic principles relating to safety engineering are presented.

The Safety Professional.   Whether called a safety engineer, safety director, loss control manager, or by some other title, the safety professional normally functions as a specialist at the management level. The safety program should enjoy the same position or status as other established activities of the organization, such as sales, production, engineering, or research.

The safety program involves occupational health, product safety, machine design, plant layout, security, and fire prevention. The position of safety professional combines engineering, management, preventive medicine, industrial hygiene, and organizational psychology. It also requires a knowledge of systems safety and ergonomics. The safety professional must have a thorough knowledge of the organization's equipment, facilities, and manufacturing process and must be able to communicate effectively and work with all types of people.

The emphasis in this area is reflected in the growing membership of the American Society of Safety Engineers (ASSE). This organization now has nearly 25,000 members and added 4050 new members during the 12-month period of July 1989 to June 1990 (Broderick, 1990). In 1968, the ASSE was instrumental in forming the Board of Certified Safety Professionals (BCSP). The purpose of the BCSP is to provide professional status by certification, to qualified safety professionals who meet strict educational and experience requirements and who pass a series of rigorous examinations.

Construction Safety.   Workers in the construction safety industry must be protected from all normal industrial safety hazards and additional hazards more common to construction sites, such as open excavations, falling from elevations, falling objects, temporary wiring, excessive dust and noise, and heavy construction machinery. OSHA standards relating to construction are contained in the Code of Federal Regulations (29 CFR 1926). OSHA Publication 2202 is a summary of the OSHA construction standards (OSHA, 1989a).

Construction safety programs may be thought of as providing for worker safety during the processes of transportation, excavation, fabrication, erection, and demolition.

Transportation.   Care must be taken to prevent trucks and other transportation from colliding with or contacting power lines, other vehicles, or other facilities. Traffic patterns in a construction site are often unclear and may vary from day to day. Efforts should be made to establish clear traffic flow patterns and communicate this to all affected personnel. Vehicles, particularly those with obstructed views to the rear, should be equipped with a reverse signal alarm.

Excavation.   [This material is adapted from OSHA Publication 2226, *Excavating and Trenching Operations* (OSHA, 1982). The reader is referred to this document or Code of Federal Regulations, 29 CFR 1926, Subpart P, for further information.] Excavations and trenching cave-ins are estimated to cause approximately one hundred fatalities each year in the United States, and for each fatality there are estimated to be fifty serious injuries. The costs and safeguards for excavation projects are dependent on the traffic, proximity of structures, type of soil, surface and groundwater, water table, underground utilities, and weather.

OSHA requires that all trenches over 5 feet deep, except those in solid rock, be sloped, shored, sheeted, braced, or otherwise supported. Trenches less than 5 feet deep must be protected if hazardous ground movement is expected. Factors to be taken into account when determining the design of a support system are soil structure, depth of cut, water content of soil, weather and climate, superimposed loads, vibrations, and other operations in the vicinity. The approximate angle of repose for sloping on the sides of excavations ranges from vertical (90 degrees from horizontal) for solid rock, shale, or cemented sand and gravel to a 2:1 slope (26 degrees from vertical) for well-rounded loose sand.

Fabrication.   Fabrication processes at construction sites generally involve the same basic operations as in general industry. Electrical hazards, explosions, fires, hand and power tools, and the like are discussed later in this chapter.

Erection.   [This material is adapted from OSHA Publication 2202, *Construction Industry—Occupational Safety and Health Standards Digest* (OSHA, 1989a). The reader is referred to this document or Code of Federal Regulations, 29 CFR 1926, Subparts E, L, M, and R for further information.] The following items, often important issues during the erection of structures at construction sites, are briefly discussed: scaffolds, guardrails and toeboards, ladders, safety nets, steel erection.

Scaffolds should be able to accept at least 4 times the maximum intended load and should be erected on a sound footing able to accept the maximum intended load without settling.

Guardrails and toeboards should be installed on all open ends and sides of platforms located 10 feet or higher above the ground and on scaffolds 4 to 10 feet above the ground which have a minimum dimension of less than 45 inches in either direction. A standard railing consists of a top rail approximately 42 inches above the floor able to withstand a lateral load of at least 200 pounds, an intermediate rail approximately halfway between the floor and the top rail, and a toeboard. A standard toeboard should be of substantial construction, at least 4 inches high with openings not to exceed 1 inch.

Ladders should have the rungs uniformly spaced at 12 inches and extend at least 36 inches above the landing. Portable ladders should be erected so that the ratio of the working length (base to point of contact at top) to base is 4:1. Portable metal ladders should not be used for electrical work.

Safety nets should be provided when workplaces are higher than 25 feet above the floor if the use of scaffolds, temporary floors, or other safer procedure is not practical.

During *steel erection* a temporary floor should be maintained within two stories or 30 feet (whichever is less) directly below where work is being performed. A 1/2-inch wire rope or equivalent should be installed at a height of 42 inches around the perimeter of temporary floors. Except where structural integrity is maintained by the design of the building, a permanent floor should be installed so that there are no more than eight stories between the erection floor and the highest permanent floor.

Demolition.   Demolition should be performed by specialists who are familiar with relevant regulations and the procedures required to protect themselves, other workers, and the general public.

Electrical Hazards.   [This material is adapted from OSHA Publication 2236, *Controlling Electrical Hazards* (OSHA, 1985). The reader is referred to this document or Code of Federal Regulations, 29 CFR 1910, Subpart S for further information.]

Electricity is analogous to water flowing through a hose. The power-generating station is the "pump," the current (amperes) is the volume of "water" flowing, and the voltage (volts) is the "pressure." The resistance to the flow of electricity is measured in ohms and is a function of the type, cross-sectional area, and temperature of the material subject to the current flow.

Electrical Shocks.   Electricity must travel in a closed circuit through a material called a conductor. When the body is a part of this circuit, the electric current passes through the body from one point to another and a shock is the result.

The severity of the shock received by a person is a general function of the amount of current flowing through the body, the path of the current between the points of contact of the body with the circuit, and the duration of the contact. Other factors which may affect the severity are the frequency (Hz) of the current, the phase of the heartbeat, and the general health of the individual. There are no absolute levels of current which cause the same sensation in all individuals. Figure 6.2 indicates the general effect of a 60-cycle current of 1 second duration passing from the hand to the foot (a common route). Note that current above the 5- to 30-milliampere range may cause the loss of muscle control and may prevent the individual from voluntarily releasing the energized contact. This may cause a longer duration of exposure, resulting in severe injury or even death.

Injuries from Electrical Hazards.   The most common injury from electrical shock is burns. These may be electrical burns resulting from the electric current passing through the body tissue, arc or flash burns resulting from the high temperatures produced by an electrical arc or explosion, or thermal contact burns when the skin comes in contact with the hot surfaces of overheated electrical conductors or energized equipment or when clothing is ignited. Electrical shock may also cause secondary injuries (sometimes called body reaction injuries) due to involuntary muscle reaction and falls. Injuries and property damage may also result from fires caused by electrical arcing or explosions.

Correcting Electrical Hazards.   Electrical accidents are generally caused by unsafe equipment, unsafe environmental conditions, or unsafe work practices. Electrical hazards can be minimized through the use of insulation, guarding, grounding, mechanical safeguards, and safe employee work practices.

Insulation involves the covering of electrical conductors (or potential conductors) with a material which has a very high resistance to electric current flow. Some good insulators are glass, mica, rubber, and plastic. OSHA's general requirements are that circuit conductors be insulated with a material suitable for the voltage and existing conditions (temperature, moisture, contaminants, and the like) to prevent accidental contact.

Indoor electrical installations over 600 volts which are accessible to unqualified persons must be guarded by enclosing them in a lock-controlled area or in a metal case. *Guarding* of live parts of 50 volts or more may be done by

1. Location in a room or similar enclosure which is accessible only to qualified personnel

EFFECTS OF ELECTRICAL CURRENT IN THE HUMAN BODY

Current	Reaction
1 Milliampere	Perception level. Just a faint tingle.
5 Milliamperes	Slight shock felt; not painful but disturbing. Average individual can let go. However, strong involuntary reactions to shocks in this range *can* lead to injuries.
6-25 Milliamperes (women) 9-30 Milliamperes (men)	Painful shock, muscular control is lost. This is called the freezing current or "let-go"* range.
50-150 Milliamperes	Extreme pain, respiratory arrest, severe muscular contractions.* Individual cannot let go. Death is possible.
1-4.3 Amperes**	Ventricular fibrillation. (The rhythmic pumping action of the heart ceases.) Muscular contraction and nerve damage occur. Death is most likely.
10+ Amperes	Cardiac arrest, severe burns and probable death.

---

*If the extensor muscles are excited by the shock, the person may be thrown away from the circuit.

**Where shock durations involve longer exposure times (5 seconds or greater) and where only minimum threshold fibrillation currents are considered, theoretical values are often calculated to be as little as 1/10 the fibrillation values shown.

FIGURE 6.2   The effects of electric current on the human body. (*Reprinted from* Controlling Electrical Hazards, *OSHA Publication 3075, 1986.*)

2. Installation of permanent, substantial screens or other partitions to exclude unqualified personnel
3. Location of the parts on a balcony, gallery, or platform elevated and configured to exclude unqualified personnel
4. Elevation of at least 8 feet above the floor

Grounding is normally a secondary measure which provides a low-resistance path to the earth or ground so that any excessive voltages will use this path and not the body as the route to complete the circuit. This reduces the possibility that an individual will be shocked through contact with improperly energized parts such as the casing of an electrical hand tool. The "service ground" or "system ground" consists of one wire which is grounded at the transformer and at the service entrance of the building and is intended to prevent damage to machines, tools, and insulation. The "equipment ground" provides a path to ground from the specific tool or machine and protects the worker.

Mechanical safeguards automatically terminate or limit the electric current when a ground fault, overload, or short circuit occurs. Fuses and circuit breakers monitor the amount of current in a circuit and open the circuit when the current flow is excessive. They serve primarily to prevent or reduce direct damage to conductors and equipment. They do little, however, to protect operators from direct shock hazards. Ground-fault circuit interrupters (GFCI) are designed to terminate electrical power when there is a current loss (due to a short, for example) in the circuit which may be hazardous to operators. The GFCI senses when the current loss is as small as 0.005 ampere and terminates electrical power within as little as 0.025 second (OSHA, 1987a). GFCI are often used in high-hazard areas such as construction sites.

Employee safe work practices are required to minimize electrical hazards. They include

1. Deenergize electrical equipment before performing maintenance operations.
2. Use only electrical tools which are safe and properly maintained.
3. Use good judgment and follow applicable safety guidelines when working near energized lines.
4. Use adequate, properly maintained personal protective equipment.
5. Avoid (by 10 feet at least) overhead power lines with ladders, cranes, or other equipment.

An excellent resource is *An Illustrated Guide for Electrical Safety* (OSHA, 1983a), which is also available through the American Society of Safety Engineers (ASSE, 1983).

Fires, Explosions, and Pressure. [This material is adapted from *Accident Prevention Manual for Industrial Operations* (National Safety Council, 1988) and *Occupational Safety Management and Engineering* (Hammer, 1989). The reader is referred to these documents or Code of Federal Regulations, 29 CFR 1910, Subpart L for further information.]

Fires. For a fire to start there must be fuel, an oxidizer, and an ignition source. The fuel and the oxidizer must be in proper proportions, and an uninhibited combustion chain reaction must occur.

Generally fires pass through four stages. The *incipient* stage generates no visible smoke, flame, or significant heat but considerable combustion particles. Ionization fire detectors can be used to detect fires in this stage. As the amount of combustion particles increases and smoke is visible, the *smouldering* stage begins. Photoelectric fire detectors respond to this smoke. When the point of ignition is reached, the *flame* stage begins. The resulting infrared energy may be detected by infrared fire detectors. The flame stage usually becomes the *heat* stage very quickly. The resulting heat energy may be detected by thermal fire detectors.

Fires are generally divided into one of the following four classes according to the fuel involved:

Class A: Solids such as coal, paper, and wood which produce char or glowing embers

Class B: Gases and liquids which require vaporization for combustion

Class C: Class A or B fires which involve electrical equipment

Class D: Fires involving magnesium, aluminum, titanium, zirconium, or other easily oxidized metals

The *flash point* of a liquid is that temperature at which it will give off sufficient vapor to momentarily ignite and burn. The burning stops as soon as the vapors are consumed. If the temperature rises above the *fire point,* the burning will continue after ignition. Liquids are often classified as flammable if their flash point is below a certain temperature or combustible if their flash point is above that temperature. The ratings for flammable and combustible are different for different organizations.

Fires can be extinguished by (1) removal or isolation of the fuel from the oxidizer (usually air), (2) increasing the volume of inert gas in the oxidizer, (3) quenching or cooling the heat of combustibles, or (4) inhibition of the combustion chain reaction.

Explosions.   An explosion is a sudden, violent release or expansion of a large amount of gas and can be caused by sudden release of compressed gas as well as a chemical reaction. Explosions may cause damage and injuries through the resulting shock waves, material fragments, or body movement resulting from the shock wave. Explosion damage may be prevented by (1) minimizing the use and storage requirements of explosive or pressurized materials, (2) isolation of materials and processes which might explode from people or valuable equipment, (3) use of pressure release devices, valves, or "blowout" panels, (4) use of suppressants which inhibit the chain reaction involved in chemical explosions, and (5) control and elimination of explosive dust concentrations.

Hand and Power Tools.   [This material is adapted from OSHA Publication 3080, *Hand and Power Tools* (OSHA, 1986b). The reader is referred to this document or Code of Federal Regulations, 29 CFR 1910, Subpart P for further information.]

Hand Tools.   Hand tools are nonpowered and include anything from hammers to screwdrivers. Hazards from hand tools often result from misuse and improper maintenance. Saw blades, knives, and other tools must be directed away from areas where other employees are working. Knives, scissors, and other cutting tools must be kept sharp since dull tools are frequently more hazardous than sharp tools. Personal protective equipment such as mesh gloves, hand and arm guards, and protective aprons should be used when workers are using knives and other cutting tools. Spark-resistant tools should be used where sparks produced by iron or steel hand tools are a dangerous ignition source.

Power Tools.   Power tools may use electric, pneumatic, liquid fuel, hydraulic, and powder activation. General power tool precautions include

1. Never carry a tool by the cord or hose.
2. Never yank a cord or hose from the receptacle.
3. Keep cords and hoses away from heat, oil, and sharp edges.
4. Disconnect tools when servicing and changing accessories.
5. Keep bystanders at a safe distance.
6. Secure workpiece so two hands can use the tool.
7. Avoid accidental starting.
8. Maintain tools properly.
9. Maintain good footing and balance.
10. Do not wear loose clothing.
11. Properly tag damaged tools.

Hazardous Material.   [This material is adapted from OSHA Publication 3084, *Chemical Hazard Communication* (OSHA, 1988a). The reader is referred to this document or Code of Federal Regulations, 29 CFR 1910, Subpart H for further information.]

Approximately 32 million workers are potentially exposed to chemical hazards which may result in disorders such as heart ailments, kidney and lung damage, sterility, cancer, burns, and rashes as well as cause fires or explosions.

Each container in the workplace must be tagged, labeled, or marked with the identity of hazardous material it contains. It must include prominently displayed warnings in written, picture, or symbol forms that convey the hazards of the chemical.

Chemical manufacturers and importers must develop material safety data sheets (MSDS) which include the name of the hazardous chemical, its specific chemical identity, physical and chemical characteristics, known acute and chronic health effects, exposure limits, if the chemical is considered to be a carcinogen, precautionary measures, emergency and first-aid procedures, and the organization that prepared the MSDS. The MSDS for chemicals in a particular work area must be readily accessible to employees in that area. In addition employers must establish a training and information program for all employees exposed to hazardous chemicals.

Machine Guarding.   [This material is adapted from OSHA Publication 3067, *Concepts and Techniques of Machine Safeguarding* (OSHA, 1983b). The reader is referred to this document or Code of Federal Regulations, 29 CFR 1910, Subpart O for further information.]

Guarding is required at the point of operation, around power transmission apparatus, and around other moving parts. Hazardous motions include rotating, reciprocating, and transverse, and hazardous actions including cutting, punching, shearing, and bending.

Guarding mechanisms must, as a minimum, (1) prevent contact between the worker and dangerous moving parts, (2) be firmly attached to the machine and discourage tampering, (3) protect from inadvertent insertion or dropping of foreign objects, (4) create no new hazards, (5) create minimum interference with job performance, and (6) allow safe maintenance.

Machine guarding may be grouped into five general classifications: (1) guards, (2) devices, (3) location and distance, (4) feeding and ejection mechanisms, and (5) miscellaneous aids.

Guards.   Guards may be fixed, interlocked, adjustable, or self-adjusting. A *fixed* guard is a permanent part of the machine and generally encloses the entire point of operation. A fixed guard is often preferred because of its simplicity. Care must be taken, however, to allow safe access for inspection and maintenance. When an *interlocked* guard is opened or removed, the machine automatically shuts off or cannot cycle until the guard is replaced. *Adjustable* guards accommodate parts of different sizes or shapes. *Self-adjusting* guards adjust automatically to the movement of the stock or part which is being inserted.

Devices.   Devices may be presence-sensing, pullback, restraint, safety controls, or gates. *Presence-sensing devices* detect the presence of a foreign object (hand, for example) in the operating area and interrupt the operating cycle. Presence-sensing devices are generally photoelectric, radio-frequency, or electromechanical. *Pullback* devices include attachments to the operator's hands, wrists, or arms which withdraw the body member from the point of operation when the machine cycles. *Restraint devices* also include attachments, usually to the wrist, which keep the operator's hands away from the point of operation. *Safety controls* may be of several types. Safety tip controls (bars, tripwires, triprods) provide a quick means to stop the machine in an emergency situation. Two-hand controls require the concurrent use of two hands to cycle the machine, and a two-hand trip requires the concurrent use of two hands to start the machine cycle. A *gate* is a movable barrier which must be in place at the point of operation before the machine cycle can start.

Location and Distance.   The dangerous parts of the machine must be located so that they are not accessible or do not present a hazard to the operator during normal operation. Workers may be protected by a wall, or dangerous parts may be located high enough to be out of any possible reach. Operator controls may be located at a safe distance from the machine if the operator is not required to tend the process.

Feeding and Ejection Mechanisms.   Feeding and ejection mechanisms may protect the operator by eliminating the need for the operator to place his or her hands in the point of operation. Guards and devices may still be required to protect the operator.

Miscellaneous Aids.   These include awareness barriers which remind workers of dangers or dangerous areas, protective shields, and tools which may be used to insert and remove stock from the point of operation.

A list to remind the reader of important machine guarding issues is shown in Fig. 6.3.

Material Handling and Storage.   [This material is adapted from OSHA Publication 2236, *Materials Handling and Storage* (OSHA, 1986a). The reader is referred to this document or Code of Federal Regulations, 29 CFR 1910, Subpart N for further information.] Injuries may result when material is moved and stored by hand or when moved and stored with the assistance of machines.

Requirements for All Safeguards

	Yes	No
1. Do the safeguards provided meet the minimum OSHA requirements?	___	___
2. Do the safeguards prevent workers' hands, arms, and other body parts from making contact with dangerous moving parts?	___	___
3. Are the safeguards firmly secured and not easily removable?	___	___
4. Do the safeguards ensure that no objects will fall into the moving parts?	___	___
5. Do the safeguards permit safe, comfortable, and relatively easy operation of the machine?	___	___
6. Can the machine be oiled without removing the safeguard?	___	___
7. Is there a system for shutting down the machinery before safeguards are removed?	___	___
8. Can the existing safeguards be improved?	___	___

Mechanical Hazards

The point of operation:

	Yes	No
1. Is there a point-of-operation safeguard provided for the machine?	___	___
2. Does it keep the operator's hands, fingers, body out of the danger area?	___	___
3. Is there evidence that the safeguards have been tampered with or removed?	___	___
4. Could you suggest a more practical, effective safeguard?	___	___
5. Could changes be made on the machine to eliminate the point-of-operation hazard entirely?	___	___

Power transmission apparatus:

	Yes	No
1. Are there any unguarded gears, sprockets, pulleys, or flywheels on the apparatus?	___	___
2. Are there any exposed belts or chain drives?	___	___
3. Are there any exposed set screws, key ways, collars, etc.?	___	___
4. Are starting and stopping controls within easy reach of the operator?	___	___
5. If there is more than one operator, are separate controls provided?	___	___

Other moving parts:

	Yes	No
1. Are safeguards provided for all hazardous moving parts of the machine, including auxiliary parts?	___	___

Nonmechanical Hazards

	Yes	No
1. Have appropriate measures been taken to safeguard workers against noise hazards?	___	___
2. Have special guards, enclosures, or personal protective equipment been provided, where necessary, to protect workers from exposure to harmful substances used in machine operation?	___	___

Electrical Hazards

	Yes	No
1. Is the machine installed in accordance with National Fire Protection Association and National Electrical Code requirements?	___	___
2. Are there loose conduit fittings?	___	___
3. Is the machine properly grounded?	___	___
4. Is the power supply correctly fused and protected?	___	___
5. Do workers occasionally receive minor shocks while operating any of the machines?	___	___

Training

	Yes	No
1. Do operators and maintenance workers have the necessary training in how to use the safeguards and why?	___	___
2. Have operators and maintenance workers been trained in where the safeguards are located, how they provide protection, and what hazards they protect against?	___	___
3. Have operators and maintenance workers been trained in how and under what circumstances guards can be removed?	___	___
4. Have workers been trained in the procedures to follow if they notice guards that are damaged, missing, or inadequate?	___	___

Protective Equipment and Proper Clothing

	Yes	No
1. Is protective equipment required?	___	___
2. If protective equipment is required, is it appropriate for the job, in good condition, kept clean and sanitary, and stored carefully when not in use?	___	___
3. Is the operator dressed safely for the job (i.e., no loose-fitting clothing or jewelry)?	___	___

Machinery Maintenance and Repair

	Yes	No
1. Have maintenance workers received up-to-date instruction on the machines they service?	___	___
2. Do maintenance workers lock out the machine from its power sources before beginning repairs?	___	___
3. Where several maintenance persons work on the same machine, are multiple lockout devices used?	___	___
4. Do maintenance persons use appropriate and safe equipment in their repair work?	___	___
5. Is the maintenance equipment itself properly guarded?	___	___

FIGURE 6.3 Important machine guarding issues. (*Reprinted from* Concepts and Techniques of Machine Safeguarding, *OSHA Publication 3067, 1983b.*)

Moving by Hand. The weight of the load and bending, twisting, and turning of the body are often associated with injury during manual material handling. Injuries include (1) musculoskeletal strains and sprains from lifting or moving loads which are too heavy or too large, (2) fractures and bruises caused by dropped or moving material, or getting caught in pinch points, and (3) cuts and bruises caused by dislodgment of improperly stored material or incorrect cutting of ties or other securing devices.

General guidelines to minimize the musculoskeletal hazards associated with manual material handling include

1. Keep the load close to the body.
2. Use the most comfortable posture.
3. Do not twist while lifting or lowering the load.
4. Lift slowly and evenly (don't jerk the load).
5. Securely grip the load.
6. Use a lifting aid or get help.

Refer to the NIOSH *Work Practices Guide for Manual Lifting* (AIHA, 1983) for additional information on the analysis of manual material handling tasks.

Moving by Machine. When a powered industrial forklift is used to move material, the load must be squarely centered on the forks as close to the mast as possible. The lift truck must never be overloaded. Stacked loads must be correctly piled and cross-tiered whenever possible.

Conveyers must be equipped with emergency stop devices. Nip points and other hazards must be guarded and guards must be provided where a conveyer passes over work areas or aisles.

When stacking material, height and weight limitations, accessibility, and material stability must be considered. All materials stored in tiers should be placed on racks, interlocked, or secured in some way to prevent falling or collapse. Flammable or combustible liquid, liquefied petroleum gas, explosives and blasting agents, and other hazardous materials must be stored in accordance with applicable safety and health regulations.

Sufficient clearance must be provided at aisles, loading docks, doorways, areas where turns are required, or other locations where the movement of material by machines might present a hazard. Permanent aisles and storage locations should be marked.

Noise. Noise may be defined as unwanted sound. Excessive noise may result in (1) decreased hearing sensitivity, (2) immediate physical damage, (3) interference or masking of particular sounds, (4) annoyance, (5) distraction, and (6) contribution to other types of disorders (Hammer, 1989).

The reduction of the adverse effects of noise in the workplace may be accomplished through (1) early planning, (2) reduction of the noise at the source, (3) insulation against reflected noise, and (4) use of personal protective equipment (PPE). The implementation of early planning to reduce the potential for noise exposure is the preferred option. Reduction of noise at the source and insulation can be very costly, and the use of PPE may be difficult and cause worker discomfort.

OSHA requires that employers monitor noise exposure levels to identify employees who are exposed to noise at or above 85 dB(A) averaged over an 8-hour day (time weighted average, or TWA). Hearing protection must be available to employees who are exposed to an 8-hour TWA of 85 dB(A) or above, and hearing protection must be worn by employees who are exposed to an 8-hour TWA of 90 dB(A) or above. The hearing protection must attenuate employee exposure to an 8-hour TWA of 90 dB(A).

Personal Protective Equipment. [This material is adapted from OSHA Fact Sheet 86-08, *Protect Yourself with Personal Protective Equipment* (OSHA, 1986c). The reader is referred to this document or Code of Federal Regulations, 29 CFR 1910, Subpart I for further information.]

OSHA standards require employers to furnish personal protective equipment (or PPE) and require employees to use this equipment when there is a "reasonable probability" that the use of such equipment will prevent injury. Data from the Bureau of Labor Statistics indicate that 60 percent of workers with eye injuries were not wearing eye protection, 99 percent of workers suffering face injuries were not wearing face protection, 84 percent of workers sustaining head injuries were not wearing hard hats, and 77 percent of workers incurring foot injuries were not wearing safety shoes.

The type of eye and face protection should be based on the type of hazard present and the degree of exposure. Selection criteria include comfort, snugness of fit, durability, and maintainability. Head protection must be able to resist penetration and absorb the shock associated with a blow to the head. Some situations also call for protection against electric shock. Foot and leg protection is required to protect against falling or rolling material, sharp objects, molten metal, and hot, wet, and slippery surfaces. Safety shoes must be sturdy and have an impact-resistant toe.

Hearing-conservation programs require the use of ear protectors in some cases. Ear protectors may be preformed or molded, which are fitted to the individual, or waxed cotton, foam, or fiberglass, which are self-forming. Disposable earplugs should be worn once and discarded, and nondisposable earplugs should be properly maintained. A variety of PPE is available to protect the arms, hands, and torso from cuts, heat, splashes, impact, acids, and radiation. This equipment must be selected to fit the particular task. Respiratory protection is required when there is exposure to air contaminated with hazardous dusts, fogs, fumes, mists, gases, smokes, sprays, and vapors.

Employees must be trained in the proper use and maintenance of PPE. They must also be aware that the use of the PPE does not eliminate the hazard. If the PPE fails, harmful exposure may result.

Radiation. Radiation is energy transmitted through space. Ionizing radiation (x-rays, gamma rays, cosmic rays, for example) changes atoms into ions through the addition or removal of electrons. A radioactive material is generally considered to be a substance that emits ionizing radiation. Adverse effects of exposure to excessive radiation include cancer, birth defects in future children of exposed parents, and cataracts. There is no conclusive evidence of a cause-effect relationship between adverse health effects and current levels of occupational radiation exposure. It is advisable, however, to assume that some health effects may occur at some occupational exposure levels. The hazard associated with exposure to ionizing radiation can be reduced through the reduction of exposure time, increase in distance between the worker and the radiation source, and appropriate shielding between the worker and the radiation source.

Nonionizing radiation such as ultraviolet, visible light, infrared, microwave, and lasers may also be hazardous. The risks from these sources can be reduced through the use of appropriate glasses, skin creams, clothing, gloves, and face masks as well as the time, distance, and shielding measures noted above.

Robot Safety. [This material is adapted from *NIOSH Alert—Request for Assistance in Preventing the Injury of Workers by Robots* (NIOSH, 1984). The reader is referred to this document or *Safe Maintenance Guide for Robotic Workstations* (NIOSH, 1988) for further information.]

Robotic workstations significantly extend the capabilities of the worker. While workers may recognize the hazards associated with the working end of the robot arm, they may not recognize the dangers associated with robot maintenance or the movement of other parts of the robot. The safety of robotic systems must consider the actual design of the robot, worker training, and worker supervision.

Robot Design. Robot design should include

1. Physical barriers with interlocked gates
2. Motion sensors, light curtains, or floor sensors that stop the robot when a worker crosses the barrier

3. Barriers between robotic equipment and free-standing objects to eliminate "pinch points"
4. Adequate clearance around all moving components
5. Remote diagnostic instrumentation to facilitate troubleshooting away from the moving robot
6. Adequate illumination of control and operational areas
7. Marks on the floor and working surfaces to indicate the movement area of the robot

Worker Training. Training of operators and maintenance personnel should include the following items:

1. Familiarity with all working aspects of the robot including range of motion, known hazards, programming, emergency stop methods, and safety barriers
2. Importance of staying out of the reach of the robot during operation
3. Necessity of operating at reduced speed and awareness of all pinch points during programming

Worker Supervision. Supervisors of operators and maintenance personnel should

1. Assure that no one is allowed within the operational area of the robot while first shutting down or reducing the speed of the robot
2. Recognize that, over time, workers may become complacent or inattentive to the hazards inherent in robotic equipment

Slips and Falls. Slips and falls may be from an elevation or on the same level. Slips and falls from elevations were mentioned in the earlier discussion of construction safety. Falls on the same level tend to result from either a stumble, which is the contact of a foot or leg with an unexpected obstruction, or an actual slip between the shoe and walking surface.

Stumbles can best be prevented by good housekeeping, proper illumination and marking of walkways, and load-carrying techniques which do not overload or affect the visibility of workers. Slips can be prevented by proper maintenance of the work surface and the selection of footwear which optimizes the slip resistance between the footwear and the walking surface. While some standards suggest that adequate slip resistance in the workplace is defined by a coefficient of friction of 0.5, there is some lack of agreement as to how this slip resistance is to be measured. In some cases, such as walking on ramps or pushing and pulling heavy loads, a higher slip resistance may be required. Emphasis should be placed on the maximization of slip resistance under all expected operational conditions through the selection of optimum shoe and floor surface materials.

SYSTEMS SAFETY ANALYSIS

The general area of system safety analysis may be defined as a directed or systematic process for the acquisition, review, and analysis of specific information relevant to a particular system. This process is methodical, careful, and purposeful. The purpose is to provide information for informed management decisions. System safety analysis techniques can be categorized as either inductive or deductive (Firenze, 1988).

Inductive Methods of Systems Safety Analysis. Inductive methods use observable data to predict what *can* happen. These techniques consider systems from the standpoint of the component parts and determine how a particular mode of failure of component parts will affect the performance of the system. Major inductive methods are preliminary hazard analysis (PHA), job safety analysis (JSA), failure modes and effects analysis (FMEA), and systems hazards analysis (SHA).

Preliminary Hazard Analysis (PHA).   Preliminary hazard analysis is a qualitative study conducted during the conceptual or early developmental phases of a system's life. Its objectives are to

1. Identify known hazardous conditions and potential failures
2. Determine the cause(s) of these conditions and potential failures
3. Determine the potential effect of these conditions and potential failures on personnel, equipment, facilities, and operations
4. Establish initial design and procedural requirements to eliminate or control these hazardous conditions and potential failures

In some cases an additional step, the estimation of the probability of an accident due to the hazard, is performed between steps 3 and 4 above.

Figure 6.4 presents an amusing but illustrative PHA of the feathers, flax, and beeswax wings used by Daedalus and Icarus in Greek mythology (Hammer, 1989).

The PHA is often based on a limited number of hazards which are determined as soon as initial facts about the system are known. These are the basic hazards which must be dealt with even though there are many different circumstances which might lead to them. The design process may be monitored to determine if these hazards have been reduced or eliminated and, if not, if the effects can be controlled.

Job Safety Analysis (JSA).   Job safety analysis is a written procedure designed to review job methods, uncover hazards, and recommend safe job procedures. Smith (1980) notes the following four basic steps in making a JSA:

1. Select the job, usually basing selection on potential hazards or high incidence rates.
2. Break the job down into a sequence of steps. Job steps are recorded in their normal order of occurrence. Steps are described in terms of *what* is done ("lift," "attach," "remove"), not *how* it is done.
3. Identify the potential hazards. To determine what accidents can happen one should (*a*) observe the job, (*b*) discuss the job with the operator, and (*c*) check accident records.
4. Recommend safe job procedures to avoid the potential accident.

A basic JSA form should include the job steps, the hazards associated with these steps, and recommend safe procedures, but the form may be altered to meet specific organizational needs by including such information as the name of the person performing the analysis, the name of the operator and supervisor, or the name of reviewers or approvers of the analysis.

Failure Modes and Effects Analysis (FMEA).   Failure modes and effects analysis is both a system safety and reliability analysis used to identify the critical failure modes that seriously affect the safe and successful life of the system, and an analysis of failure modes that could prevent a system from accomplishing its intended mission. This technique permits system change in order to reduce the severity of failure effects. FMEA is organized around the basic question "What If?" The areas which are covered and the questions which are asked move logically from cause to effect.

1. *Component:* What individual components make up the system?
2. *Failure mode:* What could go wrong with each component in the system?
3. *System causes:* What would be the cause of the component failure or malfunction?
4. *System effects:* What would be the effect of such a failure on the system, and how would this failure affect other components in the system?
5. *Severity index:* Consequences are often placed into one of four reliability categories (Firenze, 1991):
   a. Catastrophic: May cause multiple injuries, fatalities, or loss of a facility
   b. Critical: May cause severe injury, severe occupational illness, or major property damage

PRELIMINARY HAZARD ANALYSIS

IDENTIFICATION _____ Mark I Flight System _____

SUBSYSTEM _____ Wings _____   DESIGNER _____ Daedalus _____

HAZARD	CAUSE	EFFECT	PROBABILITY OF ACCIDENT DUE TO HAZARD	CORRECTIVE OR PREVENTIVE MEASURE
Thermal radiation from sun	Flying too high in presence of strong solar radiation	Heat may melt beeswax holding feathers together. Separation and loss of feathers will cause loss of aerodynamic lift. Aeronaut may then plunge to his death in the sea.	Reasonably probable	Make flight at night or at time of day when sun is not very high and hot. Provide warning against flying too high and too close to sun. Maintain close supervision over aeronauts. Use buddy system. Provide leash of flax between the two aeronauts to prevent young, impetuous one from flying too high. Restrict area of aerodynamic surface to prevent flying too high.
Moisture	Flying too close to water surface or from rain	Feathers may absorb moisture, causing them to increase in weight and to flag. Limited propulsive power may not be adequate to compensate for increased weight so that the aeronaut will gradually sink into the sea. Result: loss of function and flight system. Possible drowning of aeronaut if survival gear is not provided.	Reasonably probable	Caution aeronaut to fly through middle air where sun will keep wings dry or where accumulation rate of moisture is acceptable for time of mission.
Inflight encounter	a. Collision with bird	Injury to aeronaut	Remote probability	a. Select flight time when bird activity is low. Give birds right-of-way.
	b. Attack by vicious bird	Injury to aeronaut	Remote probability	b. Avoid areas inhabited by vicious birds. Carry weapon for defense.
Hit by lightning bolt	Bolt thrown by Zeus angered by hubris displayed by aeronaut who can fly.	Death of aeronaut	Happens occasionally	Aeronaut should not show excessive pride in being able to perform godlike activity (keep a low profile).

FIGURE 6.4   Preliminary hazard analysis of flight of Daedalus and Icarus. (*Reprinted with permission from W. Hammer*, Occupational Safety Management and Engineering, *4th ed., Prentice-Hall, Englewood Cliffs, N.J., 1989, p. 553.*)

      c. Marginal: May cause minor injury or minor occupational illness resulting in lost workday(s) or minor property damage

      d. Negligible: Probably would not affect the safety or health of personnel but is still in violation of a safety or health standard

    6. *Probability index:* How likely is the event to occur under the circumstances described and given the required precursor events? These probabilities are based on such factors as accident experience, test results from component manufacturers, comparison with similar equipment, or engineering data. Probability categories may be developed by individual companies or analysts but are sometimes classified as (Firenze, 1991)

      a. Probable (likely to occur immediately or within a short period of time)

   b. Reasonably probable (probably will occur in time)
   c. Remote (possible to occur in time)
   d. Extremely remote (unlikely to occur)
7. *Action or modification:* After the failure modes, causes and effects, severity, and probability have been established, it is necessary to modify the system to prevent or control the failure.

    Firenze (1991) notes that the severity index, probability index, and a third index relating to personnel exposure may be used to determine the overall risk. A review of the above steps makes the objectives of FMEA clear. FMEA is intended to rank failures by risk (severity and probability) so that potentially serious hazards can be corrected.
    When the analysis includes the severity, probability, and criticality indexes, it is sometimes called a failure mode, effect and criticality analysis (FMECA).
    Systems Hazards Analysis (SHA).   Systems hazards analysis includes the human component, a strength of job safety analysis, and the hardware component, a strength of failure mode and effects analysis (Firenze, 1978).
    SHA concentrates on the worker-machine interface. What process is being performed on what equipment? What major operations are required to complete the process? What tasks or activities are required to complete an operation? The thesis of SHA is that failures (undesired events) may be eliminated by systematically tracking through the system, looking for hazards that may result in a failure situation.
    In the language of SHA the terms process, operation, and task have specific meanings. Process means the combination of operations and tasks that unite physical effort and physical and human resources to accomplish a specific purpose. An operation is a major step in the overall process (for example, drilling and countersinking stock on a drill press). A task is a particular action required to complete the operation (for example, placing a cutting tool in a holder prior to sharpening the tool on the grinder).
    Once the process to be analyzed has been identified, it is broken down into its operations and tasks. To do this the analyst must be familiar with the tasks involved in the operation and the interactions between and within the system being analyzed and associated systems and subsystems. Often a flow diagram is constructed to record what is taking place throughout the flow of operations and tasks that fulfill process demands. This enables the analyst to see the pertinent subsystems, methods, transfer operations, inspection techniques, and man-machine operations.
    One type of hazard analysis process (adapted from Firenze, 1991) includes the identification and recording of information relating to

1. *Process* (turning between centers on a machine lathe)
2. *Major unit operations* required to complete the process (rough-turning steel stock)
3. *Tasks* required to complete an operation (select cutting tool and place in holder)
4. *Variance* from safe practices with the potential to cause hazards (incorrect cutting tool used)
5. *Hazard* which has the potential to cause an accident (worker in close proximity to lathe when incorrect cutting tool used)
6. *Triggering event* causing hazards to result in accidents, brought about by human error, situational, or environmental factors (starting the lathe)
7. *Accident* resulting from effect of triggering event on hazard (stock comes off centers when lathe is running)
8. *Effect* indicating type of injury or damage resulting from the accident (eye injury)
9. *Hazard consequence classification* (called the severity index in FMEA)
10. *Hazard probability* (same as in FMEA)
11. *Procedural requirements* to eliminate or reduce hazards in the workplace

12. *Safety and personal protective equipment* requirements to reduce the possibility of injuries and illnesses while performing operations and tasks

13. *Instructions and recommendations* to ensure safety and health in the workplace

As was noted in the discussion of FMEA, Firenze (1991) notes that the severity index, probability index, and a third index relating to exposure may be used to determine the overall risk.

Deductive Methods of Systems Safety Analysis.   Inductive methods of analysis analyze the components of the system and posit the effects of their failure on total system performance. Deductive methods of analysis move from the end event to try to determine the possible causes. They determine how a given end event *could* happen (Firenze, 1978). One widespread application of deductive systems safety analysis is fault tree analysis.

Fault Tree Analysis (FTA).   Fault tree analysis postulates the possible failure of a system and then identifies component states that contribute to the failure. It reasons backward from the undesired event to identify all the ways in which such an event could occur and, in doing so, identifies the contributory causes. The lowest levels of a fault tree involve individual components or processes and their failure modes. This level of the analysis generally corresponds to the starting point in FMEA.

FTA uses boolean logic and algebra to represent and quantify the interactions between events. The primary boolean operators are AND and OR gates. With an AND gate, the output of the gate, the event that is at the top of the symbol, occurs only if *all* the conditions below the gate, and feeding into the gate, coexist. With the OR gate, the output event occurs if *any one* of the input events occur. Figure 6.5 (Hammer, 1989) illustrates the basic gates used and a simple FTA.

When the possibilities of initial events or conditions are known, it is possible to determine the probabilities of succeeding events through the application of boolean algebra. For an AND gate the probability of the output event is the intersection of the boolean probabilities, or the product of the probabilities of the input events, or

$$\text{Prob (output)} = (\text{Pr input 1}) \times (\text{Pr input 2}) \times (\text{Pr input 3})$$

For an OR gate the probability of the output event is the sum of the "union" of the boolean probabilities, or the sum of the probabilities of the input events minus all of the products. Where the probabilities of the input events are small (less than 0.1, for example), the probability of the output event for an OR gate can be estimated by the sum of the probabilities of the input events, or

$$\text{Prob (output)} = (\text{Pr input 1}) + (\text{Pr input 2}) + (\text{Pr input 3})$$

Management Oversight and Risk Tree Analysis (MORT).   Management oversight and risk tree analysis is defined as

A formalized, disciplined logic or decision tree to relate and integrate a wide variety of safety concepts systematically. As an accident analysis technique, it focuses on three main concerns: specific oversights and omissions, assumed risks, and general management system weaknesses (EG&G, 1984).

It is essentially a series of fault trees with three basic subsets or branches:

1. A branch that deals with specific oversights and omissions at the work site.

2. A branch that deals with the management system that establishes policies and makes them work.

3. An assumed risk branch that acknowledges that no activity is completely free of risk and that risk management functions must exist in any well-managed organization.

FAULT TREE SYMBOLS

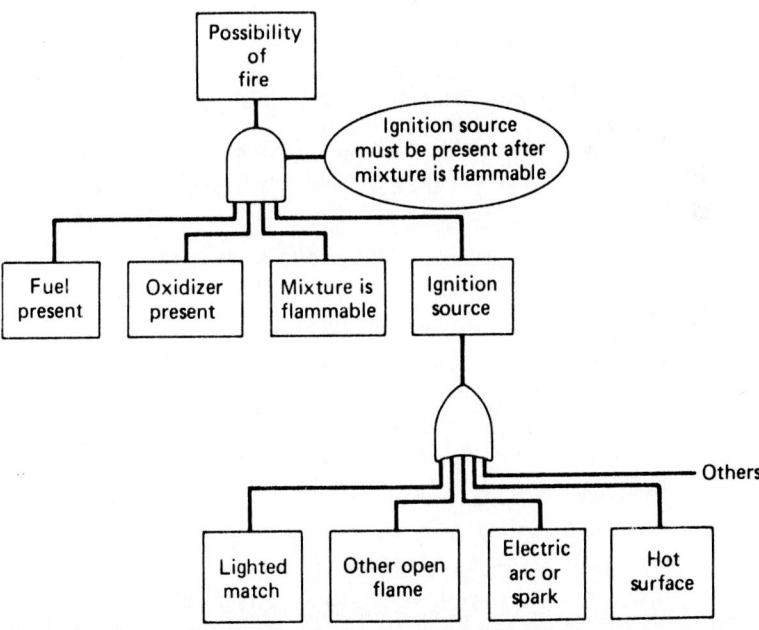

AND condition (or gate): *All* events leading into it from underneath must occur before the event leading out of it at the top can occur. In a Fault Tree Boolean equation, the AND condition is generally expressed by AB, A·B, or (A)(B) = C

OR condition (or gate): *Any* event leading into it from underneath will cause the event leading out of it at the top to occur. In a Fault Tree Boolean equation, the OR condition is generally expressed by +, such as A + B = C

A condition which must be satisfied before the event above the AND or OR can occur

Any event shown in such a "house" will always be present when the events in this tree occur

A transfer symbol: the events which will occur where this is shown will be the same as where the inverted triangle with the same identification is shown

FIGURE 6.5   Fault tree analysis symbols. Fault tree analysis of fire. (*Reprinted with permission from W. Hammer,* Occupational Safety Management and Engineering, *4th ed., Prentice-Hall, Englewood Cliffs, N.J., 1989, p. 558.*)

These assumed risks are those undesirable consequences that have been quantitatively analyzed and formally accepted by appropriate management levels within the organization.

MORT includes about 100 generic causes and thousands of criteria. The MORT diagram terminates in some 1500 basic safety program elements that are required for a successfully functioning safety program. These are elements that prevent the undesirable consequences indicated at the top of the tree. MORT has three primary goals:

1. To reduce safety-related oversights, errors, and omissions
2. To allow risk quantification and the referral of residual risk to proper organizational management levels for appropriate action
3. To optimize the allocation of resources to the safety program and to organizational hazard control efforts

MORT programs and their associated training courses place emphasis on constructing trees for individual program needs and on a set of readymade MORT trees which can be used for program design, program evaluation, or accident investigation.

EG&G Idaho, Inc., is one of the primary users and developers of MORT and offers training courses in MORT application.

SUMMARY

The importance of safety in the specification, design, and installation of production equipment, facilities, and systems cannot be overemphasized. Engineers must include safety considerations early in the design process to minimize the losses resulting from accidents and injuries in the workplace.

Some basic principles of occupational safety have been presented in this chapter. For further information relating to OSHA regulations the reader is directed to the OSHA General Industry Standards (29 CFR 1910) and Construction Standards (29 CFR 1926). The reader is also directed to the numerous standards and recommended practices developed by other organizations and referenced by or included by reference in the OSHA standards. These organizations include:

Agriculture Ammonia Institute—Rubber Manufacturer's Association

American Conference of Government Industrial Hygienists

American National Standards Institute

American Petroleum Institute

American Society of Agricultural Engineers

American Society of Mechanical Engineers

American Society for Testing and Materials

American Welding Society

Commerce, Department of

Compressed Gas Association

Crane Manufacturer's Association of America, Inc.

The Fertilizer Institute

General Services Administration

Health and Human Services, Department of

Institute of Makers of Explosives

National Electrical Manufacturer's Association

National Fire Protection Association
National Food Plant Institute
National Institute for Occupational Safety and Health
Public Health Service
Society of Automotive Engineers
Underwriters' Laboratories

REFERENCES

American Industrial Hygiene Association, *Work Practices Guide for Manual Lifting*, AIHA, Cleveland, Ohio, 1983.

American Society of Safety Engineers, *An Illustrated Guide to Electrical Safety*, ASSE, Des Plaines, Ill., 1983.

Broderick, J., "Maximizing Our Impact," *American Society of Safety Engineers Professional Safety*, October 1990.

Brown, D. B., "Systems Engineering in the Design of a Safety System," *American Society of Safety Engineers Professional Safety*, February 1973.

Bureau of Labor Statistics, *Recordkeeping Guidelines for Occupational Injuries and Illnesses*, 1986.

Bureau of National Affairs, *Occupational Safety and Health Reporter*, Nov. 21, 1990, p. 1036.

DeReamer, R., *Modern Safety and Health Technology*, Wiley, New York, 1980.

EG&G, *Glossary of Terms and Acronyms*, Publication 76-45/28, Idaho Falls, Idaho, 1984.

Firenze, R. J., *The Process of Hazard Control*, Kendall Hunt, Dubuque, Iowa, 1978.

Firenze, R. J., *Evaluation and Control of the Occupational Environment*, National Institute for Occupational Safety and Health, 1988.

Firenze, R. J., *The Impact of Safety on High-Performance/High-Involvement Production Systems*, Creative Work Designs, Inc., 1991.

Hammer, W., *Occupational Safety Management and Engineering*, 4th ed., Prentice-Hall, Englewood Cliffs, N.J., 1989.

Heinrich, H. W., *Industrial Accident Prevention*, McGraw-Hill, New York, 1980.

Liberty Mutual Insurance Company, *Investigation Is Important*, No date.

National Institute for Occupational Safety and Health, *NIOSH Alert—Request for Assistance in Preventing the Injury of Workers by Robots*, 1984.

National Institute for Occupational Safety and Health, *Safe Maintenance Guide for Robotic Workstations*, 1988.

National Safety Council, *Accident Prevention Manual for Industrial Operations*, 9th ed., National Safety Council, Chicago, Ill., 1988.

National Safety Council, *Accident Facts*, 1989.

Occupational Safety and Health Act, Public Law 91-596, Dec. 29, 1970.

Occupational Safety and Health Administration, *Investigating Accidents in the Workplace*, Publication 2288, 1977.

Occupational Safety and Health Administration, *Excavating and Trenching Operations*, Publication 2226, 1982.

Occupational Safety and Health Administration, *An Illustrated Guide to Electrical Safety*, Publication 3073, 1983a.

Occupational Safety and Health Administration, *Concepts and Techniques of Machine Safeguarding*, Publication 3067, 1983b.

Occupational Safety and Health Administration, *Controlling Electrical Hazards*, Publication 3075, 1986a.

Occupational Safety and Health Administration, *Materials Handling and Storage*, Publication 2236, 1985.

Occupational Safety and Health Administration, *Hand and Power Tools*, Publication 3080, 1986b.

Occupational Safety and Health Administration, *Protect Yourself with Personal Protective Equipment*, Fact Sheet 86-08, 1986c.

Occupational Safety and Health Administration, *Ground-Fault Protection on Construction Sites*, Publication 3007, 1987a.

Occupational Safety and Health Administration, *Improving Workplace Protection for New Workers*, Fact Sheet 87-07, 1987b.

Occupational Safety and Health Administration, *Chemical Hazard Communication*, Publication 3084, 1988a.

Occupational Safety and Health Administration, *General Industry—Occupational Safety and Health Standards Digest*, Publication 2201, 1988b.

Occupational Safety and Health Administration, *Safety and Health Guide for the Meatpacking Industry*, Publication 3108, 1988c.

Occupational Safety and Health Administration, *Construction Industry—Occupational Safety and Health Standards Digest*, Publication 2202, 1989a.

Occupational Safety and Health Administration, *Field Operations Manual*, 3d ed., Government Industries, Inc., Rockville, Md., 1989b.

Smith, L. C., "The J Programs," *National Safety News*, September 1980.

Wigglesworth, E. C., "A Teaching Model of Causal Mechanisms and a Derived Theory of Countermeasure Selection," *American Society of Safety Engineers Professional Safety*, August 1972.

ECONOMICS AND CONTROLS

CHAPTER 1
ENGINEERING ECONOMY*

G. A. Fleischer
Professor, Industrial & Systems Engineering
University of Southern California
Los Angeles, California

Engineering economy rests on the proposition that refusing to expend scarce resources is rarely, if ever, the wisest course of action. Rather, the problem is choosing from a variety of investment alternatives in order to best satisfy decision makers' immediate and longer-term objectives. The operative word is *economy,* and the essential ingredient in economy is *selection.* This chapter is dedicated to the principles and procedures of the selection process, especially when the economic characteristics of alternatives are of principal or significant concern.

FUNDAMENTAL PRINCIPLES

Before developing the mathematical models appropriate to evaluating capital proposals, it will be useful to identify the fundamental principles that give rise to the rationale of capital allocation. Moreover, some of these principles lead directly to the quantitative techniques developed subsequently.

1. *Only feasible alternatives should be considered.* The capital budgeting analysis begins with determination of all feasible alternatives, since courses of action that are not feasible, because of certain contractual or technological considerations, are properly excluded.

2. *Using a common unit of measurement (a common denominator) makes consequences commensurable.* All decisions are made in a single dimension, and money units—dollars, francs, pesos, yen, and so forth—seem to be most generally suitable. Of course, not all consequences may be evaluated in money terms. (See Principle 9 below.)

3. *Only differences are relevant.* The prospective consequences that are common to all contending alternatives need not be considered in an analysis, because including them affects all alternatives equally.

4. *All sunk costs are irrelevant to an economic choice.* A *sunk cost* is an expense or a revenue that has occurred before the decision. All events that take place before a de-

---

*Portions of the material included in this chapter have been adapted from G. A. Fleischer, *Engineering Economy,* PWS-Kent Publishing Company, 1984. It is reproduced here by permission of the publisher.

cision are common to all the alternatives, so sunk costs are not differences among alternatives.

5. *All alternatives must be examined over a common planning horizon.* The *planning horizon* is the period of time over which the prospective consequences of various alternatives are assessed. (The planning horizon is often referred to as the *study period* or *period of analysis.*)

6. *Criteria for investment decisions should include the time value of money and related problems of capital rationing.*

7. *Separable decisions should be made separately.* This principle requires the careful evaluation of all capital-allocation problems to determine the number and type of decisions to be made. For example, an analyst may be called on to make recommendations concerning a technological alternative (the equipment to be acquired, for example) as well as a financial alternative (the source of funds to finance the acquisition). If separable decisions are not treated separately, optimal solutions may be obscured in the analysis.

8. *The relative degrees of uncertainty associated with various forecasts should be considered.* Because estimates are only predictions of future events, it is probable that the actual outcomes will differ to a greater or lesser degree from the original estimates. Formal consideration of the type and degree of uncertainty ensures that the quality of the solution is evident to those responsible for capital-allocation decisions.

9. *Decisions should give weight to consequences that are not reducible to monetary units.* The irreducible as well as monetary consequences of proposed alternatives should be clearly specified in order to give managers of capital all reasonable data on which to base their decisions.

10. *The efficacy of capital-budgeting procedures is a function of their implementation at various levels within the organization.* Capital-allocation procedures must be clearly described and understood at all levels within the organization having responsibility, in whole or in part, for these decisions.

11. *Postdecision audits improve the quality of decisions.* The quality of decisions is directly affected by the analyst's ability to forecast the future with reasonable precision. The only way to judge predictive ability is to audit the results of the decision at a later date. In this way it is possible to determine the extent of an analyst's predictive bias.

EQUIVALENCE AND THE MATHEMATICS OF COMPOUND INTEREST

A central notion in engineering economy is that cash flows (that is, the receipt or payment of an amount of money) that differ in magnitude but that occur at different points in time may be *equivalent*. This equivalence is a function of the appropriate interest rate per unit time and the relevant time interval. Mathematical relationships describing the equivalence property under a variety of conditions are described in the remainder of this section.

Useful Conventions.   The following conventions will be used in this chapter.
   Cash Flow Diagrams.   In the literature of engineering economy, cash flow diagrams are frequently used to illustrate the amount and timing of cash flows. Generally, a horizontal bar or line is used to represent time, and vertical vectors (arrows) are used to represent positive or negative cash flows at the appropriate points in time. These cash flow diagrams are illustrated later in Fig. 1.1. The "shaded" arrows in the right-hand portion of the figure represent cash flowing continuously and uniformly throughout the indicated period(s).

Functional Notation.   As the algebraic form of the various equivalence factors can be complex, it is useful to adopt a standardized format which is easily learned and has a mnemonic connotation. The format which is in general use* is of the form

$$(X/Y,i,N)$$

which is read as "to find the equivalent amount X given amount Y, the interest rate i and the number of compounding or discounting periods N."

Discrete Cash Flows—End-of-Period Compounding.   In this section we assume that a cash flow A_j occurs at end of period j. Interest is compounded or discounted at the end of each period at rate i per period. The interest rate i is constant over $j = 1,2,\ldots,N$. The periods are of equal duration.

Single Cash Flows.   Consider a single cash flow P to be invested at the beginning of a time series of exactly N periods. Let F represent the *equivalent future value* of P as measured at the end of N periods hence, assuming that interest is compounded at the end of each and every period at interest rate i. Then,

$$F = P(1 + i)^N = P(F/P,i,N) \tag{1}$$

It follows immediately that, given a future amount F flowing at the end of N periods hence, the *equivalent present value* P is given by

$$P = F(1 + i)^{-N} = F(P/F,i,N) \tag{2}$$

The growth multiplier as shown in Eq. (1), $(1 + i)^N$, is known in the literature of engineering economy as the (*single payment*) compound amount factor. The discounting multiplier shown in Eq. (2) is known as the (*single payment*) present worth factor.

The cash flow diagrams, algebraic forms, and functional forms for these two factors are shown later in Fig. 1.1. Tabulated values for $i = 10$ percent and for various values of N are given in Table 1.7.

EXAMPLES.   A sum of $1000 is invested in a fund which earns interest at the rate of 1 percent per month, compounded monthly. To determine the value of the fund after 24 months, using Eq. (1):

$$F = \$1000(1.01)^{24} = \$1269.73$$

A certain investment is expected to yield a return of $100,000 exactly 8 years in the future. Assuming a discount rate of 10 percent per year, what is the equivalent present value? Using Eq. (2):

$$P = \$100,000(1.10)^{-8} = \$46,651$$

How long will it take a sum of money to double if interest is earned at the rate of 8 percent per period? From Eq. (1):

$$\$2 = \$1(1.08)^N$$

$$N = \ln 2/\ln 1.08 \approx 9 \text{ periods}$$

An investment of $10,000 yields a return of $20,000 five years later. What (annual) rate of return was earned? From Eq. (1):

$$\$20,000 = \$10,000(1 + i)^5$$

$$i = (\$20,000/\$10,000)^{1/5} - 1 = 14.87 \text{ percent}$$

*This is the functional notation recommended in *Industrial Engineering Terminology*, revised edition, Industrial Engineering and Management Press, Industrial Engineering Institute, Norcross, GA, 1991.

Uniform Series (Annuity).   Consider a uniform series of cash flows A occurring at the end of each of N consecutive periods. That is, $A_j = A$ for $j = 1,2,\ldots,N$. The equivalent future value F at the end of N periods is given by

$$F = A\left[\frac{(1 + i)^N - 1}{i}\right] = A(F/A,i,N) \qquad (3)$$

The factor in brackets is known as the (*uniform series*) compound amount factor. To find A given F:

$$A = F\left[\frac{i}{(1 + i)^N - 1}\right] = F(A/F,i,N) \qquad (4)$$

The factor in brackets is known as the *sinking fund factor*.

The equivalent present value of this uniform series is given by

$$P = A\left[\frac{(1 + i)^N - 1}{i(1 + i)^N}\right] = A(P/A,i,N) \qquad (5)$$

The factor in brackets is known as the (*uniform series*) *present worth factor*. To find A given P:

$$A = P\left[\frac{i(1 + i)^N}{(1 + i)^N - 1}\right] = P(A/P,i,N) \qquad (6)$$

The factor in brackets is known as the *capital recovery factor*.

As before, the appropriate cash flow diagrams, algebraic forms, and functional forms are shown in Fig. 1.1. Tabulated values for $i = 10$ percent are given in Table 1.7.

EXAMPLES.   (A 10 percent interest rate is assumed for all the following examples.) A sum of $1000 is invested at the end of each period for 15 periods. What is the amount in the fund after the 15th payment has been made? From Eq. (3):

$$F = \$10,000(F/A, 10\%, 15)$$

$$= \$10,000(31.772) = \$317,720$$

(Note that the value for the compound amount factor has been taken from Table 1.7.) How much must be invested at the end of each year for 15 years in order to have $20,000 in the fund after the 15th payment? From Eq. (4):

$$A = \$20,000(A/F, 10\%, 15)$$

$$= \$20,000(0.0315) = \$630$$

How much must be invested today in order to yield returns of $2500 at the end of each and every year for 8 years? From Eq. (5):

$$P = \$2500(P/A, 10\%, 8)$$

$$= \$2500(5.335) = \$13,337$$

Certain equipment costs $50,000, will be used for 5 years, and will have no value at the end of 5 years. What is the equivalent annual (end of year) cost? From Eq. (6):

$$A = \$50,000(A/P, 10\%, 5)$$

$$= \$50,000(0.2638) = \$13,190$$

Arithmetic Gradient Series.   Let $A_j = (j - 1)G$ for $j = 1,2,\ldots,N$, where G represents the amount of increase or decrease in cash flow from one period to the next. This results in an arithmetic series of the cash flows of the form $0,G,2G,\ldots,(N - 1)G$ for periods $1,2,\ldots,N$, respectively. Given the gradient G the equivalent present value is given by

$$P = G\left[\frac{(1 + i)^N - iN - 1}{i^2(1 + i)^N}\right] = G(P/G,i,N) \tag{7}$$

and the equivalent uniform series is given by

$$A = G\left[\frac{(1 + i)^N - iN - 1}{i(1 + i)^N - i}\right] = G(A/G,i,N) \tag{8}$$

Again, the appropriate cash flow diagrams, algebraic forms, and functional forms are shown later in Fig. 1.1. Representative tabulated values are given in Table 1.7 for $(P/G,10\%,N)$ and $(A/G,10\%,N)$.

With appropriate algebraic manipulation, it may be shown that

$$(P/G,i,N) = G(1/i)[(P/A,i,N) - N(P/F,i,N)]$$

and

$$(A/G,i,N) = G(1/i)[1 - N(A/F,i,N)]$$

It also may be shown that

$$(P/G,i,N) \rightarrow 1/i^2 \text{ as } N \rightarrow \infty$$

and

$$(A/G,i,N) \rightarrow 1/i \text{ as } N \rightarrow \infty$$

EXAMPLE.   Costs of manufacturing are assumed to be $100,000 the first year and to increase by $10,000 in each of the years 2 through 7. If interest is at 10 percent per year, determine the equivalent present value of these costs. Using Eqs. (5) and (7) and taking the appropriate factor values from Table 1.7:

$$P = \$100,000(P/A, 10\%, 7) + \$10,000(P/G, 10\%, 7)$$

$$= \$100,000(4.868) + \$10,000(12.763) = \$614,430$$

Note: This analysis assumes that all cash flows occur at end of year.

Geometric Gradient Series.   Consider a series of cash flows $A_1, A_2,\ldots,A_N$ where the A_j's are related as follows:

$$A_j = A_{j - 1}(1 + g) = A_1(1 + g)^{j-1} \tag{9}$$

where g represents the rate of increase or decrease in cash flows from one period to the next. With cash flows discounted at rate i per period, the equivalent present value of the geometric series is given by

$$P = A_1\left[(1 + i)^{-1}\sum_{j=1}^{N}\left(\frac{1 + g}{1 + i}\right)^{j-1}\right]$$

$$= A_1\left[(1 + g)^{-1}\sum_{j=1}^{N}\left(\frac{1 + g}{1 + i}\right)^{j}\right] \tag{10}$$

$$= A_1\left[\frac{1 - (1 + g)^N (1 + i)^{-N}}{i - g}\right]$$

As N approaches infinity, this series is convergent if $g < i$. Otherwise ($g \geq i$) the series is divergent.

Tables for the geometric series present worth factor $(P/A_1, i, g, N)$ may be generated using Eq. (10). However, the factor may also be computed by using the more readily available compound amount (F/P) and present worth factors (P/F) by noting that

$$P = A_1 \left[\frac{1 - (F/P, g, N)(P/F, i, N)}{i - g} \right] \tag{11}$$

EXAMPLE. Manufacturing costs are expected to be \$100,000 in the first year, increasing by 5 percent each year over a 7-year period. Find the equivalent present value of these cash flows assuming a 10 percent discount rate (per year) and end-of-year cash flows. Using either Eq. (10) or Eq. (11):

$$P = \$100,000(P/A_1, 10\%, 5\%, 7)$$

$$= \$100,000(5.5587) = \$555,870$$

Effective and Nominal Interest Rates.   An interest rate is meaningful only if it is related to a particular period of time. Nevertheless, the "time tag" is frequently omitted in speech because it is usually understood in context. If someone reports earnings of 6 percent on investments, for example, it is implied that the rate of return is *per year*. However, in many cases the interest-rate period is a week, a month, or some other interval of time, rather than the more usual year (per annum). At this point it would be useful to examine the process whereby interest rates and their respective "time tags" are made commensurate.

As before, let i represent the *effective* interest rate per period. Let the period be divided into M subperiods of equal length. If interest is compounded at the end of each subperiod at rate i_s per subperiod, then the relationship between the effective interest rates per period and per subperiod is given by

$$i = (1 + i_s)^M - 1 \tag{12}$$

The *nominal* interest rate per period r is simply the effective interest rate per subperiod times the number of subperiods, or

$$r = M i_s \tag{13}$$

Period and Subperiods: An Example.   It is often necessary to compare interest rates over a common time interval. Consider, for example, the case of consumer credit, say, a major oil company or bank "charge card" for which interest is compounded monthly at rate 1.5 percent of the unpaid balance. Here $i_s = 0.015$ and $M = 12$. The *nominal* rate per annum, by Eq. (13) is $12 \times 0.015 = 0.18$. The *effective* rate per annum, by Eq. (12) is

$$i = (1.015)^{12} - 1 = 0.1956$$

The nominal rate is only an approximation of the effective rate. Defining the *absolute error* as $(i - r)$, the *percent error* is $1 - (r/i)$.

$$\text{Percent error} = 1 - \left[\frac{M i_s}{(1 + i_s)^M - 1} \right] \tag{14}$$

For a given value of i_s, the percent error increases as M increases. And, for a given value of M, the percent error increases as i_s increases. The error can be considerable. For example, when $i_s = 0.02$ and $M = 12$, the percent error is approximately 10.5 percent.

Periods and Superperiods.   Consider a uniform series of cash flows A occurring at regular intervals. Specifically, the cash flows occur every M periods, with the first cash flow occurring at the end of period m and the last cash flow occurring at the end of period n, where $1 \leq m \leq n \leq N$. There are exactly $[(n - m)/M] + 1$ cash flows in the resulting uniform series. The time between cash flows, the *superperiod*, consists of M periods, where M is integer-valued, with the start of the first superperiod at the end of period $m - M$. The equivalent present value of this uniform series of cash flows is given by

$$P = A \left\{ \frac{(1 + i)^{n-m+M} - 1}{(1 + i)^n [(1 + i)^M - 1]} \right\} \tag{15}$$

For example, consider major overhaul expenses of $20,000 each occurring at the end of year 5 and continuing, every 2 years, up to and including year 13. ($A_j = -\$20,000$ for $j = 5,7,9,11,13$.) Assuming a 10 percent discount rate:

$$P = \$20,000 \left\{ \frac{(1.10)^{13-5+2} - 1}{(1.10)^{13} [(1.10)^2 - 1]} \right\}$$

$$= \$20,000(2.19834) = \$43,967$$

Continuous Cash Flows—Continuous Compounding.   Assuming that the number of subperiods at the end of which interest is compounded or discounted becomes infinitely large, the effective interest rate per period is

$$i = \lim_{M \to \infty} \left\{ \left[\left(1 + \frac{1}{M/r} \right)^{M/r} \right]^r - 1 \right\} \tag{16}$$

$$= e^r - 1$$

where e is the base of the natural (napierian) logarithm system and is approximately equal to 2.71828.

Assume that a total of $\overline{A}$ dollars flows over one interest period, with $\overline{A}/M$ flowing at the end of each and every one of the M subperiods within the period. As before, the *effective interest rate* is i per period and the nominal rate is r per period. Interest is compounded at effective rate $i_s = r/M$ per subperiod. Let A represent the equivalent value at the end of the period:

$$A = (\overline{A}/M)(F/A, i_s, M)$$

As the number of subperiods (M) becomes infinitely large, it may be shown that

$$A = \overline{A} \left[\frac{e^r - 1}{r} \right] \tag{17}$$

The value in brackets is known as the *funds flow conversion factor* because it has the effect of converting a continuous cash flow (during the period) to a discrete cash flow (at the end of the period). Using the relation $i = e^r - 1$, Eq. (17) can be restated in terms of the effective periodic interest rate:

$$A = \overline{A} \left[\frac{i}{\ln (1 + i)} \right] \tag{18}$$

The funds flow conversion factor is useful in modifying the end-of-period factors, previously discussed, to accommodate the "continuous" assumptions. To illustrate, consider the factor for determining the equivalent present value of a cash flow $(\overline{F})$ flowing continuously and uniformly during the Nth period hence. Combining Eqs. (2) and (18):

$$P = \overline{F} \left[\frac{i}{\ln (1 + i)} \right] (1 + i)^{-N}$$

$$= \overline{F} \, (P/\overline{F}, \, i, \, N) \tag{19}$$

Similarly, consider the factor for determining the equivalent present value of a uniform series of continuous cash flows $(\overline{A})$ flowing during each and every period through N periods. Combining Eqs. (5) and (18):

$$P = \overline{A} \left[\frac{i}{\ln (1 + i)} \right] \left[\frac{(1 + i)^{N} - 1}{i(1 + i)^{N}} \right]$$

$$= \overline{A} \, (P/\overline{A}, \, i, \, N) \tag{20}$$

The factors given by Eqs. (19) and (20) are included in Table 1.7, in the columns headed $P/\overline{F}$ and $P/\overline{A}$. (The factor $F/\overline{A}$ has been incorporated into the table.)

For ease of reference, all of the equivalence models described above are summarized in Fig. 1.1. Models for *discrete cash flows* are shown in Fig. 1.1(*a*) under the two compounding conventions: (1) *end-of-period* compounding at effective interest rate i and (2) *continuous* compounding at nominal interest rate r. Models for *continuous cash flows* are shown in Fig. 1.1(*b*) under the assumption of continuous compounding at (1) effective interest rate i and (2) nominal interest rate r.

METHODS FOR SELECTING AMONG ALTERNATIVES

A variety of methods are used to evaluate alternative investments. Associated with each is a *statistic*, that is, a "figure of merit," and a *decision rule* which is used to select from among alternatives on the basis of the statistics. These are presented briefly here for a set of evaluation methods which are most commonly used in engineering economy.

Present Worth (Net Present Value).   "Present worth" (PW) and "net present value" (NPV) are equivalent terms. The former is widely used in the literature of engineering economy; the latter is common to the literature of finance and accounting.

   Present Worth of a Proposed Investment.   The *present worth* is the equivalent present value of the cash flows generated by the proposed investment over a specified time interval (planning horizon N) with discounting at a specified interest rate i. One of several algebraic expressions for present worth (PW), assuming end-of-period cash flows A_j and end-of-period discounting at rate i, is

$$PW = \sum_{j=0}^{N} A_j (1 + i)^{-j} \tag{21}$$

The *planning horizon* represents that period of time over which the proposed project is to be evaluated. It should be of sufficient duration to reflect all significant differences between the project and alternative investments. The discount rate i is the *minimum attrac-*

END-OF-PERIOD COMPOUNDING EFFECTIVE INTEREST RATE (i)		CONTINUOUS COMPOUNDING NOMINAL INTEREST RATE (r)

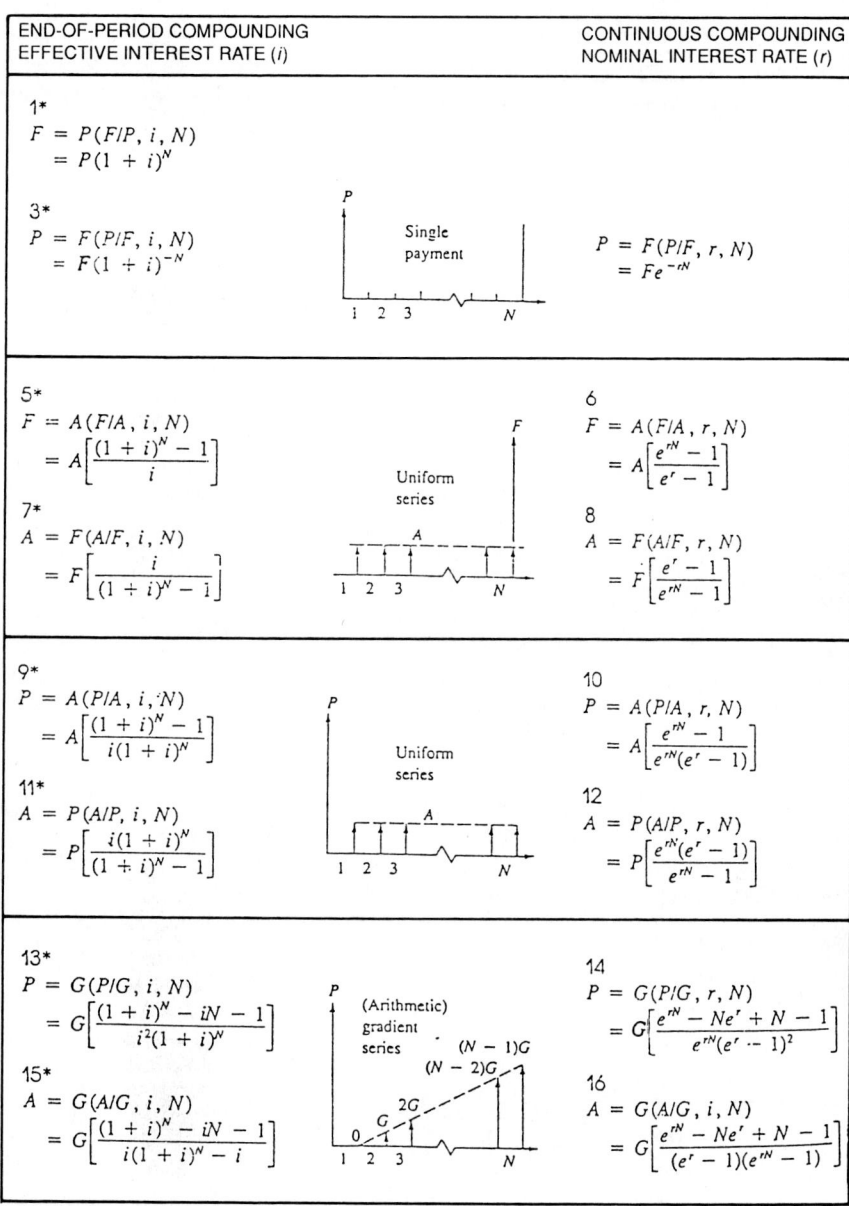

1*
$$F = P(F/P, i, N)$$
$$= P(1 + i)^N$$

3*
$$P = F(P/F, i, N)$$
$$= F(1 + i)^{-N}$$

Single payment

$$P = F(P/F, r, N)$$
$$= Fe^{-rN}$$

5*
$$F = A(F/A, i, N)$$
$$= A\left[\frac{(1 + i)^N - 1}{i}\right]$$

7*
$$A = F(A/F, i, N)$$
$$= F\left[\frac{i}{(1 + i)^N - 1}\right]$$

Uniform series

6
$$F = A(F/A, r, N)$$
$$= A\left[\frac{e^{rN} - 1}{e^r - 1}\right]$$

8
$$A = F(A/F, r, N)$$
$$= F\left[\frac{e^r - 1}{e^{rN} - 1}\right]$$

9*
$$P = A(P/A, i, N)$$
$$= A\left[\frac{(1 + i)^N - 1}{i(1 + i)^N}\right]$$

11*
$$A = P(A/P, i, N)$$
$$= P\left[\frac{i(1 + i)^N}{(1 + i)^N - 1}\right]$$

Uniform series

10
$$P = A(P/A, r, N)$$
$$= A\left[\frac{e^{rN} - 1}{e^{rN}(e^r - 1)}\right]$$

12
$$A = P(A/P, r, N)$$
$$= P\left[\frac{e^{rN}(e^r - 1)}{e^{rN} - 1}\right]$$

13*
$$P = G(P/G, i, N)$$
$$= G\left[\frac{(1 + i)^N - iN - 1}{i^2(1 + i)^N}\right]$$

15*
$$A = G(A/G, i, N)$$
$$= G\left[\frac{(1 + i)^N - iN - 1}{i(1 + i)^N - i}\right]$$

(Arithmetic) gradient series

14
$$P = G(P/G, r, N)$$
$$= G\left[\frac{e^{rN} - Ne^r + N - 1}{e^{rN}(e^r - 1)^2}\right]$$

16
$$A = G(A/G, i, N)$$
$$= G\left[\frac{e^{rN} - Ne^r + N - 1}{(e^r - 1)(e^{rN} - 1)}\right]$$

FIGURE 1.1   Cash flow models and mathematical models for selected compound interest factors: (*a*) Discrete cash flows.

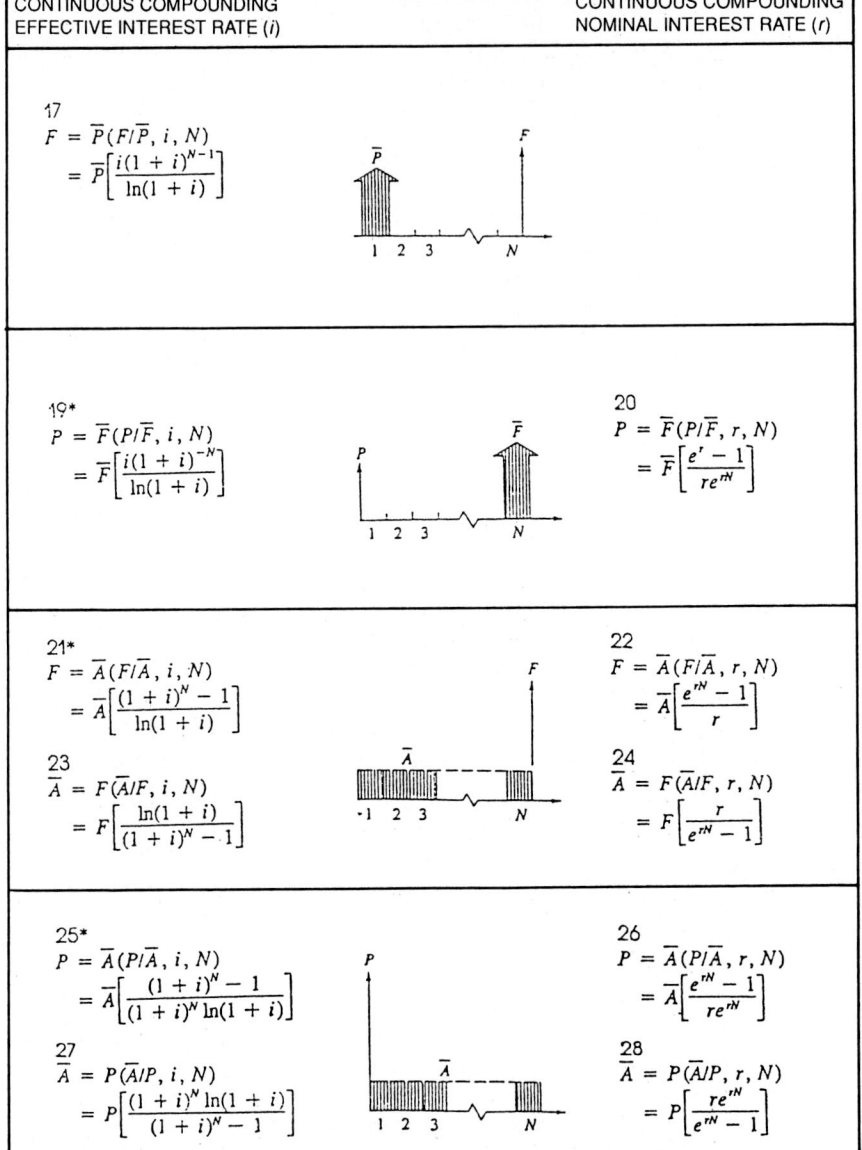

CONTINUOUS COMPOUNDING EFFECTIVE INTEREST RATE (i)	CONTINUOUS COMPOUNDING NOMINAL INTEREST RATE (r)

17
$$F = \overline{P}(F/\overline{P}, i, N)$$
$$= \overline{P}\left[\frac{i(1 + i)^{N-1}}{\ln(1 + i)}\right]$$

19*
$$P = \overline{F}(P/\overline{F}, i, N)$$
$$= \overline{F}\left[\frac{i(1 + i)^{-N}}{\ln(1 + i)}\right]$$

20
$$P = \overline{F}(P/\overline{F}, r, N)$$
$$= \overline{F}\left[\frac{e^r - 1}{re^{rN}}\right]$$

21*
$$F = \overline{A}(F/\overline{A}, i, N)$$
$$= \overline{A}\left[\frac{(1 + i)^N - 1}{\ln(1 + i)}\right]$$

22
$$F = \overline{A}(F/\overline{A}, r, N)$$
$$= \overline{A}\left[\frac{e^{rN} - 1}{r}\right]$$

23
$$\overline{A} = F(\overline{A}/F, i, N)$$
$$= F\left[\frac{\ln(1 + i)}{(1 + i)^N - 1}\right]$$

24
$$\overline{A} = F(\overline{A}/F, r, N)$$
$$= F\left[\frac{r}{e^{rN} - 1}\right]$$

25*
$$P = \overline{A}(P/\overline{A}, i, N)$$
$$= \overline{A}\left[\frac{(1 + i)^N - 1}{(1 + i)^N \ln(1 + i)}\right]$$

26
$$P = \overline{A}(P/\overline{A}, r, N)$$
$$= \overline{A}\left[\frac{e^{rN} - 1}{re^{rN}}\right]$$

27
$$\overline{A} = P(\overline{A}/P, i, N)$$
$$= P\left[\frac{(1 + i)^N \ln(1 + i)}{(1 + i)^N - 1}\right]$$

28
$$\overline{A} = P(\overline{A}/P, r, N)$$
$$= P\left[\frac{re^{rN}}{e^{rN} - 1}\right]$$

FIGURE 1.1   (*Continued*) Cash flow models and mathematical models for selected compound interest factors: (*b*) Continuous cash flows.

tive rate of return, that is, the rate of return which could be expected if the funds to be invested in the proposed project were to be invested elsewhere.

Present Worth of the "Do Nothing" Alternative. Let P represent the initial investment in a proposed project. If P were to be invested elsewhere, rather than in the proposed project, then this "do nothing" alternative would yield $P(1 + i)^N$, assuming compounding at rate i for N periods. The present worth of this course of action is zero, as can be seen by

$$\text{PW} = A_0 + A_N (1 + i)^{-N}$$

$$= -P + P(1 + i)^N (1 + i)^{-N} = 0$$

Comparing the proposed investment with the "do nothing" alternative, it follows that the investment is economically attractive (preferred to doing nothing) if its PW > 0.

The do nothing alternative is sometimes known as *alternative zero.* Here, PW($\emptyset$) = 0.

Multiple (More Than Two) Alternatives. We have seen that given two alternatives, (1) the proposed project and (2) do nothing, the "invest" decision is indicated if PW > 0. But suppose that there are more than two alternatives under consideration. In this case, the PWs of each of the alternatives are rank-ordered, and *that alternative yielding the maximum PW is to be preferred* (in an economic sense only, of course). To illustrate, consider the four mutually exclusive alternatives summarized in Table 1.1. Present worths have been determined using Eq. (21) and assuming $i = 10$ percent. As noted in the table, the correct rank ordering of the set of alternatives is IV $>$ II $>$ III $> \emptyset >$ I.

TABLE 1.1 Cash Flows for Four Mutually Exclusive Alternatives

End of period	Alternative I	Alternative II	Alternative III	Alternative IV
		Assume $i = 20\%$		
0	−$1000	−$1000	−$1100	−$2000
1–10	0	300	320	520
10	4000	0	0	0
Net cash flow	$3000	$2000	$2100	$3500
PW	−$ 354	$ 258	$ 242	$ 306
AW	−$ 85	$ 62	$ 58	$ 73
FW	−$2192	$1596	$1496	$1894

It is *not* necessary to adjust the PW statistic for differences in initial cost. This is so because any funds invested "elsewhere" yield a PW of zero. In our example, consider alternatives II and III. Initial costs are $1000 and $1100, respectively. Alternative II may be viewed as requiring $1000 in the project (yielding PW of $258) and $100 elsewhere (yielding PW of $0). The total PW(II) = $258. This may now be compared directly with alternative III: PW(III) = $242. Each alternative accounts for a total investment of $1100.

Annual Worth (Equivalent Uniform Annual Cost). The *annual worth* (AW) is the uniform series over N periods equivalent to the present worth at interest rate i. It is a weighted average periodic worth, "weighted" by the interest rate. Mathematically,

$$\text{AW} = (\text{PW})(A/P,i,N) \qquad (22)$$

If $i = 0$ percent, then AW is simply the average cash flow per period, that is,

$$\text{AW} = (1/N) \sum_{j=0}^{N} A_j$$

By convention, this is known as the *"annual..."* method, although the period may be week, month, or the like. This method is most often used with respect to costs, and in such cases it is known as the equivalent uniform annual cost (EUAC) method.

The decision rule applicable to PW is also applicable for AW (and EUAC). That is, a proposal is preferred to the do nothing alternative if AW > 0, and multiple alternatives may be rank-ordered on the basis of declining AW (or increasing EUAC). Given any pair of alternatives, say, X and Y, if PW(X) > PW(Y), then AW(X) > AW(Y). This is so because $(A/P,i,N)$ is a constant for all alternatives as long as i and N remain constant.

The annual worth method is illustrated in Table 1.1. Note that the ranking of alternatives is consistent with that of the PW method: IV > II > III > $\emptyset$ > I.

Future Worth.   In the *future worth* (FW) method, all cash flows are converted to a single equivalent value at the end of the planning horizon, period N. Mathematically:

$$FW = (PW)(F/P,i,N)$$

The decision rule applicable to PW is also applicable to FW. A set of mutually exclusive investment opportunities may be rank-ordered by using either PW, AW, or FW. The results will be consistent. The future worth method is illustrated in Table 1.1.

Rate of Return.   There are three ways of calculating rate of return.

Internal Rate of Return.   The internal rate of return (IRR), often known simply as the rate of return (RoR), is that interest rate i^* for which the net present value of all project cash flows is zero. When all cash flows are discounted at rate i^*, the equivalent present value of all project benefits exactly equals the equivalent present value of all project costs. One mathematical definition of the IRR is that rate i^* that satisfies the equation

$$\sum_{j=0}^{N} A_j (1 + i^*)^{-j} \equiv 0 \qquad (23)$$

This formula assumes discrete cash flows A_j and end-of-period discounting in periods $j = 1,2,\ldots,N$.

The discount rate used in present worth calculations is the opportunity cost—a measure of the return that could be earned on capital if it were invested elsewhere. Thus a given proposed project should be economically attractive if and only if its internal rate of return (IRR) exceeds the cost of opportunities forgone as measured by the firm's minimum attractive rate of return (MARR). That is, an increment of investment is justified if, for that proposal, IRR > MARR.

Multiple Alternatives.   Unlike the PW/AW/FW methods, mutually exclusive projects may *not* be rank-ordered on the basis of their respective IRRs. Rather, an incremental procedure must be implemented, as illustrated in Table 1.2. Alternatives must be considered pairwise, with decisions made about the attractiveness of each increment of investment. As shown in Table 1.2, we conclude that IV > II > III > $\emptyset$ > I. These results are consistent with those found by the PW/AW/FW methods.

TABLE 1.2   (Internal) Rate of Return Analysis of Alternatives from Table 1.1

Step	Comparison of alternatives	Cash flows (A_j) A_0	A_1–A_{10}	A_{10}	Incremental rate of return, %	Conclusion (MARR = 20%)
1	$\emptyset \to$ I	−$1000	0	4000	14.9	I < $\emptyset$
2	$\emptyset \to$ II	−1000	300	0	27.3	II > $\emptyset$
3	$\emptyset \to$ III	−1100	320	0	26.3	III > $\emptyset$
4	$\emptyset \to$ IV	−2000	550	0	24.4	IV > $\emptyset$
5	II $\to$ III	−100	20	0	15.1	III < ii
6	II $\to$ IV	−1000	250	0	21.4	IV > II

External Rate of Return.   When properly applied, the internal rate of return method yields the same solutions as those obtained from the annual worth and present worth methods, yet the algebraic structure is such that analysts may easily be led to incorrect solutions.

Consider the end-of-period model described by Eq. (23):

$$\sum_{j=0}^{N} A_j (1 + i^*)^{-j} \equiv 0$$

This expression may also be written as

$$A_0 + A_1 x + A_2 x^2 + \cdots + A_N x^N = 0 \tag{24}$$

where $x = (1 + i^*)^{-1}$. Solving for x leads to i^*, so we want to find the roots x of this Nth-order polynomial expression. Only the real, positive roots are of interest, of course, because any meaningful values of i^* must be real and positive. There are many possible solutions for x and, by extension, for the rate of return i^*.

The solution procedure requires the use of an auxiliary interest rate, the rate of return available from reinvested capital, to ensure that all positive cash flows occur at the end of the last period N. The revised problem has only one variation in sign, and thus there is a single *external rate of return (ERR)*. It is "external" in the sense that it is the result of the amounts and timing of cash flows of the original investment as well as of the influence of the auxiliary interest rate.

As before, let A_j represent the cash flow at the end of period j, for $j = 0,1,2,\ldots,N$. Further, let

$$A_j = \begin{cases} R_j & \text{if } A_j > 0 \\ -C_j & \text{if } A_j < 0 \end{cases}$$

The R_j's are the positive cash flows and the C_j's are the negative cash flows. Let k represent the auxiliary interest, assuming that k is also the MARR. The equivalent value of all positive cash flows at the end of period N is given by

$$\sum_{j=0}^{N} R_j (1 + k)^{N-j}$$

and the external rate of return i_e^* is the solution to the equation

$$\sum_{j=0}^{N} C_j (1 + i_e^*)^{N-j} = \sum_{j=0}^{N} R_j (1 + k)^{N-j} \tag{25}$$

In words, the external rate of return is that interest for which the equivalent value of all *negative* cash flows at the end of period N is exactly equal to the equivalent value of all *positive* cash flows at the end of period N, where the equivalent value of the latter is determined by the auxiliary interest rate. The investment is justified if $i_e^* > k$.

The external rate of return method is rarely used. It is of interest only in those cases where (1) the cash flows are such that multiple internal rates of return are obtained and (2) the decision maker desires a single meaningful rate of return. If there is only one change in direction of sign of the cash flows, at most one IRR will result. Even in those instances where multiple IRRs are obtained, it is recommended that the PW method, rather than the rate of return method, be used because of the possible misunderstanding of the significance of the ERR statistic.

Benefit-Cost Ratio.   The benefit–cost ratio method is widely used in the public sector.

Benefit-Cost Ratio and Acceptance Criterion. The essential element of the benefit-cost ratio method is almost trivial, but it can be misleading in its simplicity. An investment is justified only if the incremental benefits *B* resulting from it exceed the resulting incremental costs *C*. Of course, all benefits and costs must be stated in equivalent terms, that is, with measurement at the same point(s) in time. Normally, both benefits and costs are stated as "present value" or are "annualized" by using compound interest factors as appropriate. Thus,

$$B\!:\!C = \frac{\text{PW (or AW) of all "benefits"}}{\text{PW (or AW) of all "costs"}} \tag{26}$$

Clearly, if benefits must exceed costs, the ratio of benefits to costs must exceed unity. That is, if $B > C$, then $B\!:\!C > 1.0$. This statement of the *acceptance criterion* is true only if the incremental costs *C* are positive. It is possible, when evaluating certain alternatives, for the incremental costs to be negative, that is, for the proposed project to result in a reduction of costs. Negative benefits arise when the incremental effect is a reduction in benefits. In summary,

For $C > 0$, if $B\!:\!C > 1.0$, accept; otherwise reject.

For $C < 0$, if $B\!:\!C > 1.0$, reject; otherwise accept.

Multiple Alternatives. Like the rate of return method, the proper use of the benefit-cost ratio method requires incremental analysis. Mutually exclusive alternatives *should not be rank-ordered on the basis of benefit-cost ratios.* Pairwise comparisons are necessary to test whether increments of costs are justified by increments of benefits.

To illustrate, consider two alternative projects *U* and *T*.

| | Present Worths | | | |
Comparison	Benefits, $	Costs, $	B:C	Conclusion
$\emptyset \rightarrow T$	700,000	200,000	3.50	$T > \emptyset$
$\emptyset \rightarrow U$	1,200,000	600,000	2.00	$U > \emptyset$
$T \rightarrow U$	500,000	400,000	1.25	$U > T$

On the basis of benefit-cost ratios, it is clear that both *T* and *U* are preferred to the do nothing alternative ($\emptyset$). Moreover, the *incremental* analysis indicates that *U* is preferred to *T* since the incremental $B\!:\!C$ (= 1.25) exceeds unity.

It will be noted here that PW analysis would yield the same result: $\text{PW}(T) = \$500,000$ and $\text{PW}(U) = \$600,000$. It may be shown that this result obtains in general. That is, for any number of mutually exclusive alternatives, ranking based on *proper* use of the benefit-cost ratio method using *incremental* analysis will always yield the same rank order resulting from proper use of the present worth method.

The Net Benefit-Cost Ratio. There are several variations of the benefit-cost ratio, all of which, *properly used,* will lead to consistent, correct results. One such variation is the *net benefit-cost ratio,* which is of the form

$$\text{NB}\!:\!C = \frac{B - C}{C} \tag{27}$$

Since $\text{NB}\!:\!C = (B\!:\!C) - 1$, the acceptance criterion is $\text{NB}\!:\!C > 0$ when $C > 0$. All other observations concerning the benefit-cost ratio method apply here as well.

Payback. The payback method is widely used in American industry to determine the relative attractiveness of investment proposals. The essence of this technique is the determination of the *number of periods required to recover an initial investment*. Once this has been done for all alternatives under consideration, a comparison is made on the basis of respective payback periods.

Payback, or *payout,* as it is sometimes known, is the number of periods required for cumulative benefits to equal cumulative costs. Costs and benefits are usually expressed as cash flows, although discounted present values of cash flows may be used. In either case, the payback method is based on the assumption that the relative merit of a proposed investment is measured by this statistic. The smaller the payback (period), the better the proposal.

(Undiscounted) payback is that value of N^* such that

$$P = \sum_{j=1}^{N^*} A_j \tag{28a}$$

where P is the initial investment and A_j is the cash flow in period j. *Discounted payback,* used much less frequently, is that value of N^* such that

$$P = \sum_{j=1}^{N^*} A_j (1 + i)^{-j} \tag{28b}$$

The principal objection to the use of payback as a primary figure of merit is that all consequences beyond the end of the payback periods are ignored. This may be illustrated by a simple example. Consider two alternatives V and W. The discount rate is 10 percent and the planning horizon is 5 years. Cash flows and the relevant results are as follows:

End of year	Alternative V	Alternative W
0 (initial cost)	−$8000	−$9000
1–5 (net revenues)	4000	3000
5 (salvage value)	0	8000
Undiscounted payback	2 years	3 years
PW at 10%	$7164	$7337

Payback is a useful measure to the extend that it provides some indication of how long it might take before the initial investment is recovered. It is a helpful *supplementary* measure of the attractiveness of an investment, but it should never be used as the sole measure of quality.

Return on Investment (ROI).   There are a number of approaches, widely used in industry, that use accounting data (income and expenses) rather than cash flows to determine "rate of return," where income and expense are reflected in the firm's accounting statements. Although there is no universally accepted terminology, this accounting-based approach is generally known as *return on investment (ROI),* whereas the cash flow approach results in internal rate of return (IRR or RoR).

One variation of this approach is to define ROI as the ratio of the average annual accounting profit to the *original book value* of the asset. Another variation is the ratio of the average annual accounting profit to the *average book value* of the asset over its service life. In any event, such computations are based on depreciation expense, an accounting item which is not a cash flow and which is affected by relevant tax regulations. (See "Depreciation" below.) Therefore, the use of ROI is not recommended as an appropriate figure of merit.

Unequal Service Lives.   One of the fundamental principles, noted earlier, is that alternative investment proposals must be evaluated over a common planning horizon. Unequal service lives among competing feasible alternatives complicate this analysis. For example, consider two alternatives: one has life of N_1, the other has life of N_2, and $N_1 < N_2$.

Repeatability (Identical Replication) Assumption. One approach, widely used in engineering economy textbooks, is to assume that (1) each alternative will be replaced at the end of its service life by an identical replacement, that is, the amounts and timing of all cash flows in the first and all succeeding replacements will be identical to the initial alternative; and (2) the planning horizon is at least as long as the common multiple of the lives of the alternatives.

Under these assumptions, the planning horizon is the least common multiple of N_1 and N_2. The *annual worth* method may be used directly since the AW for alternative 1 over N_1 periods is the same as the AW for alternative 1 over the planning horizon.

Specified Planning Horizon. Although commonly used in the literature of engineering economy, the repeatability assumption is rarely appropriate in real-world applications. In such cases, it is generally more reasonable to define the planning horizon N on some basis other than the service lives of the competing alternatives. Equipment under consideration may be related to a certain product, for example, which will be manufactured over a specified time period.

If the planning horizon is *longer* than the service lives of one or more of the alternatives, it will be necessary to estimate the cash flow consequences, if any, during the interval(s) between the service life (or lives) and the end of the planning horizon. If the planning horizon is *shorter* than the service lives of one or more of the alternatives, all cash flows beyond the end of the planning horizon are irrelevant. In the latter case it will be necessary to estimate the salvage value of the "truncated" proposal at the end of the planning horizon.

AFTER-TAX ECONOMY STUDIES

Most individuals and business firms are directly influenced by taxation. Cash flows resulting from taxes paid (or avoided) must be included in evaluation models, along with cash flows from investment, maintenance, operations, and so on. Thus decision makers have a clear interest in cash flows for taxes and related topics.

Depreciation. There is a good deal of misunderstanding about the precise meaning of *depreciation*. In economic analysis, depreciation is *not* a measure of the loss in market value or equipment, land, buildings, and the like. It is *not* a measure of reduced serviceability. Depreciation is strictly an *accounting* concept. Perhaps the best definition is provided by the Committee on Terminology of the American Institute of Certified Public Accountants:

> Depreciation accounting is a system of accounting which aims to distribute the cost or other basic value of tangible capital assets, less salvage (if any), over the estimated life of the unit (which may be a group of assets) in a systematic and rational manner. It is a process of allocation, not of valuation. Depreciation for the year is the portion of the total charge under such a system that is allocated to the year.*

Depreciable property may be tangible or intangible. *Tangible* property is any property that can be seen or touched. *Intangible* property is any other property, for example, a copyright or franchise.

Depreciable property may be real or personal. *Real* property is land and generally anything erected on, growing on, or attached to the land. *Personal* property is any other prop-

---

*American Institute of Certified Public Accountants, *Accounting Research Bulletin no. 22* (American Institute of Certified Public Accountants, New York, 1944) and American Institute of Certified Public Accountants, *Accounting Terminology Bulletin no. 1* (American Institute of Certified Public Accountants, New York, 1953).

erty, for example, machinery or equipment. (*Note:*   Land is never depreciable as it has no determinable life.)

To be depreciable, property must meet three requirements: (1) it must be used in business or held for the production of income; (2) it must have a determinable life longer than 1 year; and (3) it must be something that wears out, decays, gets used up, becomes obsolete, or loses value from natural causes.

Depreciation begins when the property is placed in service; it ends when the property is removed from service.

For the purpose of computing taxable income on income tax returns, the rules for computing allowable depreciation are governed by the relevant taxing authority. An excellent reference for the U.S. government is *Depreciation, Publication 534,* published by the Internal Revenue Service (IRS), U.S. Department of the Treasury. Publication 534 is updated annually.

Currently three principal systems are used for the computation of allowable depreciation for federal income tax purposes in the United States. The *first* of these is a set of permissible methods—principally, straight-line declining balance and sum of the year's digits methods—for depreciable property placed in service *before 1981*. (These methods are still widely used by a number of taxing authorities other than the U.S. government.) *Second,* the accelerated cost recovery system (ACRS), introduced by the Economic Recovery Tax Act of 1981, is applicable to depreciable property placed in service *after 1980 but before 1987*. *Third,* the Tax Reform Act of 1986 introduced a number of significant changes, including a new modified accelerated cost recovery system (MACRS) for depreciable property placed in service *after 1986*. Each of these methods is described briefly below.*

Allowable Methods Prior to 1981.   Prior to 1981 the following four methods were in use.

1. *Straight-line method.* In general, the allowable depreciation in tax year *j*, D_j, is given by

$$D_j = \frac{B - S}{N} \qquad \text{for } j = 1,\ldots,N \tag{29}$$

where B is the adjusted cost basis, S is the estimated salvage value, and N is the depreciable life.

Allowable depreciation must be prorated on the basis of the period of service for the tax year in which the property is placed in service and the year in which it is removed from service. For example, suppose that $B = \$90,000$, $N = 6$ years, $S = \$18,000$ after 6 years, and the property is to be placed in service at midyear. In this case,

$$D_j = \frac{\$90,000 - \$18,000}{6} = \$12,000 \qquad \text{for } j = 2,\ldots,6$$

$$D_1 = D_7 = (6/12)(\$12,000) = \$6000$$

The *book value* of the property at any point in the time is the initial cost less the accumulated depreciation. In the numerical example above, the book value at the start of the third tax year would be $\$90,000 - \$6000 - \$12,000 = \$72,000$.

2. *Declining balance method.* The amount of depreciation taken each year is subtracted from the book value before the following year's depreciation is computed. A constant depreciation *rate a* applies to a smaller, or declining, balance each year. In general,

$$D_j = \begin{cases} \pi_1 aB & \text{for } j = 1 \\ aB_j & \text{for } j = 2,3,\ldots,N+1 \end{cases} \tag{30}$$

---

*The discussion of depreciation accounting is necessarily abbreviated in this handbook. The reader is encouraged to consult competent tax professionals and/or relevant publications of the Internal Revenue Service for more thorough treatment of this complex topic.

where π_1 = portion of the first tax year in which the property is placed in service $(0 < \pi_1 \le 1)$

B_j = book value in year j prior to determining the allowable depreciation

Assuming that the property is placed in service at the start of the tax year $(\pi_1 = 1.00)$, it may be shown that

$$D_j = Ba(1 - a)^{j-1} \tag{31}$$

When $a = 2/N$, the depreciation scheme is known as the *double declining balance method*, or simply DDB.

To illustrate using the previous example, suppose that we have DDB with $a = 2/6 = 0.333$. Since $\pi_1 = 6/12 = 0.5$,

$$D_1 = 0.5(0.333)(\$90,000) = \$15,000$$

$$D_2 = 0.333(\$90,000 - \$15,000) = \$25,000$$

Salvage value is not deducted from the cost or other basis in determining the annual depreciation allowance, but *the asset cannot be depreciated below the expected salvage value*. In other words, once book value equals salvage value, no further depreciation may be claimed.

3. *Sum of the years digits method.* The annual depreciation allowance is a declining fraction of the cost or other basis of each single-asset account reduced by the estimated salvage value. In particular, the depreciation in year j is given by

$$D_j = \left(\frac{N - j + 1}{\text{SYD}}\right)(P - S) \tag{32}$$

where SYD is the sum of the years digits, that is,

$$\text{SYD} = 1 + 2 + \cdots + N = N(N + 1)/2$$

This method results in a declining book value quite similar to that found by using the declining balance method, except that here we are assured that exactly $B - S$ is depreciated over N years.

4. *Other allowable methods.* The straight-line, declining balance, and sum of the years digits methods are specifically mentioned in Section 167 of the Internal Revenue Code of 1954. In general, the law also permitted "any other consistent methods" for federal income tax purposes if the total depreciation deductions taken during the first two-thirds of the asset's useful life are not more than the total allowable under the declining balance method. One cannot depreciate below salvage value, and "negative depreciation" is not allowed. A variety of methods have been developed—sinking fund method, units of production method, and so on. Please see any of the references at the end of this chapter for additional discussion.

Accelerated Cost Recovery System (ACRS). The Economic Recovery Tax Act of 1981 required that, with certain exceptions, property placed in service beginning in 1981 must be depreciated by a new method for the purpose of determining federal income taxes, the *accelerated cost recovery system (ACRS)*.

For the purpose of applying ACRS, all recovery property is divided into seven *classes* based on recovery period. For each class, statutory *depreciation percentages* are specified for each year during the recovery period. See Table 1.3 for applicable percentages for cer-

TABLE 1.3 Applicable Percentages for Computing Annual Depreciation under ACRS

Recovery year	Type of property		
	3-year	5-year	10-year
1	25	15	8
2	38	22	14
3	37	21	12
4		21	10
5		21	10
6			10
7			9
8			9
9			9
10			9

tain of the recovery classes (not included are percentages for low income housing and real property).

The annual depreciation expense in a given year under ACRS is the product of the unadjusted basis and the applicable percentage for that year.

$$D_j = B[p_j(k)] \qquad (33)$$

where D_j = depreciation expense in year j, B = unadjusted basis, and $p_j(k)$ = applicable percentage in year j for property class k. The *unadjusted basis* is generally the initial cost of the property; it is unadjusted in the sense that there is no reduction for prior depreciation.

The Economic Recovery Tax Act of 1981 permits taxpayers to elect an alternative recovery percentage determined in a manner similar to that of the straight-line method. Using the *alternate ACRS method,* for each property class (k), the applicable percentage for each year $p(k)$ is

$$p(k) = 1/n(k) \qquad (34)$$

where $n(k)$ is the recovery period selected by the taxpayer for that property class. (Note that *recovery period* under ACRS is comparable with *usable life* under the pre-1981 methods.) The three options available to taxpayers for each property class are

Recovery property (k)	Optional recovery period $(n(k))$
3-year	3, 5, or 12 years
5-year	5, 12, or 25 years
10-year	10, 25, or 35 years
15-year real 15-year public utility	15, 35, or 45 years

The alternate ACRS method, like the ACRS method, ignores salvage value when computing annual depreciation expenses.

MACRS and AMACRS. In this section, we refer to the modifications in the accelerated depreciation method available under the 1986 Tax Reform Act as MACRS (modified accelerated cost recovery system—which can be pronounced "makers") and the alternate straight-line method as AMACRS (alternate modified accelerated cost recovery system—pronounced "a-makers"). These methods apply to most depreciable property placed in service after Dec. 31, 1986.

1. *Property Classes.* The 1986 Tax Reform Act describes eight property classes, each with a defined class life. The classes of interest here are *3-, 5-, 7-, 10-, 15-, and 20-year properties.* (Two additional classes, *nonresidential real* property and *residential rental* property, will not be discussed further.)

2. *Depreciation Methods.* Under *MACRS,* the 200 percent declining balance (DB) method is used for the early years of depreciation for the 3-, 5-, 7-, and 10-year classes; the 150 percent declining balance method is used for the 15- and 20-year classes. Switch to the straight-line (SL) method occurs in the first tax year for which the SL method, when applied to the adjusted basis at the beginning of the year, will yield a larger deduction than had the DB method been continued. Zero salvage value is assumed.

AMACRS, the alternate to MACRS, uses the straight-line method with zero salvage value assumed. With certain exceptions, the recovery period is equal to the class life of the property.

3. *Conventions.* With certain exceptions, both MACRS and AMACRS assume that all property placed in service (or disposed of) during a tax year is placed in service (or disposed of) at the midpoint of that year. This is the *half-year* convention.

There is an exception to the above rule under MACRS. If more than 40 percent of the cost basis of all depreciable property (other than residential rental property or nonresidential real property) placed in service during the tax year is placed in service during the *last quarter* of the year, a midquarter convention must be used. That is, properties placed in service during any quarter are assumed to have been placed in service at the midpoint of that quarter. This "40 percent rule" does not apply to AMACRS; the half-year convention is used in any event.

4. *Depreciation Percentages.* The annual depreciation percentages under *MACRS*, assuming the half-year convention, are summarized in Table 1.4.

TABLE 1.4 Annual Depreciation Percentages under MACRS (Half-Year Convention)

Recovery year	Recovery Period (k)					
	3-year	5-year	7-year	10-year	15-year	20-year
1	33.33	20.00	14.29	10.00	5.00	3.750
2	44.45	32.00	24.49	18.00	9.50	7.219
3	14.81	19.20	17.49	14.40	8.55	6.677
4	7.41	11.52	12.49	11.52	7.70	6.177
5		11.52	8.93	9.22	6.93	5.713
6		5.76	8.92	7.37	6.23	5.285
7			8.93	6.55	5.90	4.888
8			4.46	6.55	5.90	4.522
9				6.56	5.91	4.462
10				6.55	5.90	4.461
11				3.28	5.91	4.462
12					5.90	4.461
13					5.91	4.462
14					5.90	4.461
15					5.91	4.462
16					2.95	4.461
17						4.462
18						4.461
19						4.462
20						4.461
21						2.231

For 3-, 5-, 7-, 10-, 15-, and 20-year properties, the depreciation percentage in year j for property class k under *AMACRS* is given by

$$p_j = \begin{cases} 0.5/k, j = 1 \\ 1.0/k, j = 2, 3, \ldots, k \\ 0.5/k, j = k + 1 \end{cases} \tag{35}$$

Other Deductions from Taxable Income. In addition to depreciation, there are several other ways in which the cost of certain assets may be recovered over time.

Amortization.   Amortization permits the taxpayer to recover certain capital expenditures in a way that is like straight-line depreciation. Qualifying expenditures include certain costs incurred in setting up a business (for example, survey of potential markets, analysis of available facilities), the cost of a certified pollution control facility, bond premiums, and the costs of trademarks and trade names. Expenditures are amortized on a straight-line basis over a 60-month period or more.

Depletion.   Depletion is similar to depreciation and amortization. It is a deduction from taxable income applicable to a mineral property, an oil, gas, or geothermal well, or standing timber. There are two ways to figure depletion, cost depletion and percentage depletion. With certain restrictions, the taxpayer may choose either method.

Cost depletion is determined by dividing the adjusted basis of the mineral property by the total number of recoverable units in the deposit and multiplying the resulting rate per unit by the number of units sold during the tax year. (*Timber* depletion may be determined only by the cost method. The cost does not include any part of the cost of the land.)

Percentage depletion is a certain percentage, specified for each mineral, of the taxpayer's gross income from the property during the tax year. The deduction for depletion under this method cannot be more than 50 percent of the taxable income from the property, figured without the deduction for the depletion. Even if the taxpayer has already recovered the full cost or other basis of the property, the deduction for percentage depletion may be taken.

Section 179 Expense.   The taxpayer may elect to treat the cost of certain qualifying property as an expense rather than as a capital expenditure in the year the property is placed in service. *Qualifying property* is "Section 38 property"—generally, property used in the trade or business with a useful life of 3 years or more for which depreciation or amortization is allowable, with certain limitations—and that is purchased for use in the active conduct of the trade or business.

The total cost that may be deducted for a tax year may not exceed $10,000. The expense deduction is further limited by the taxpayer's total investment during the year in Section 179 property: the $10,000 maximum is reduced by $1 for each dollar of cost in excess of $200,000. That is, no Section 179 expense deduction may be used if total investment in Section 179 property during the tax year exceeds $210,000. Moreover, the total cost that may be deducted is also limited to the taxable income which is from the active conduct of any trade or business of the taxpayer during the tax year. See IRS Publication 534 for more information.

The cost basis of the property must be reduced by the amount of the Section 179 expense deduction, if any.

Gains and Losses on Disposal of Depreciable Assets.   The value of an asset on disposal is rarely equal to its book value at the time of sale or other disposition. When this inequality occurs, a gain or loss on disposal is established and the transaction has certain tax consequences.

In general, the *gain* on disposition of depreciable property is the net salvage value minus the adjusted basis of the property (its book value) at the time of disposal. The *adjusted basis* is the original cost basis less any accumulated depreciation, amortization, Section 179 expense deduction, and, where appropriate, any basis adjustments due to investment credit claimed on the property. A negative gain is considered a *loss* on disposal.

All gains and losses on disposal are treated as *ordinary* gains or losses, *capital* gains or losses, or some combination of the two. The rules for determining these amounts are too complex to be discussed adequately here; interested readers should therefore consult a competent expert and/or read the appropriate sections in *Tax Guide for Small Business* (IRS Publication 334) or a similar reference. Currently (1992), the tax rate for capital gains is the same as that for ordinary income.

Federal Income Tax Rates for Corporations.   Income tax rates for corporations are adjusted from time to time, largely in order to affect the level of economic activity. Currently the *marginal* federal income tax rate for corporations is 34 percent for all taxable income greater than $75,000. However, for taxable income between $100,000 and $335,000, there is a 5 percent additional tax so as to bring the *average* tax rate up to 34 percent for taxable income in excess of $335,000.

When income is taxed by more than one jurisdiction, the appropriate tax rate for economy studies is a combination of the rates imposed by the jurisdictions. If these rates are independent, they may simply be added. But the combinatorial rule is not quite so simple when there is interdependence. Income taxes paid to local and state governments, for example, are deductible from taxable income on federal income tax returns, but the reverse is not true; federal income taxes are not deductible from local returns. Thus, considering only state (t_s) and federal (t_f) income tax rates, the *combined incremental tax rate* (t) for economy studies is given by

$$t = t_s + t_f(1 - t_s) \qquad (36)$$

Timing of Cash Flows for Income Taxes.   The equivalent present value of tax consequences requires estimates of the *timing* of cash flows for taxes. A variety of operating conditions affect the timing of income tax payments. It is neither feasible nor desirable to catalog all such conditions here. In most cases, however, the following assumptions will serve as a reasonable approximation.

1. Income taxes are paid quarterly at the end of each quarter of the tax year.
2. Ninety percent of the firm's income tax liability is paid in the tax year in which the expense occurs; the remaining 10 percent is paid in the first quarter of the following tax year.
3. The four quarterly tax payments during the tax year are uniform.

The timing of these cash flows can be approximated by a weighted average of quarter-ending dates.

$$0.225(1/4 + 2/4 + 3/4 + 4/4) + 0.1(5/4) = 0.6875$$

That is, the cash flow for income taxes in a given tax year can be assumed to be concentrated at a point 0.6875 into the tax year. (An alternative approach is to assume that cash flows for income taxes occur at the end of the tax year.)

After-Tax Analysis.   The following procedures are followed to prepare an after tax analysis.

1. Specify the assumptions and principal parameter values, including:
   - Tax rates (federal and other taxing jurisdictions, as appropriate).
   - Relevant methods related to depreciation, amortization, depletion, investment tax credit, and Section 179 expense deduction.
   - Length of planning horizon.
   - Minimum attractive rate of return—the interest rate to be used for discounting cash flows. *Caution:* This rate should represent the *after-tax* opportunity cost to the taxpayer; it will almost always be lower than the pretax MARR. The same discounting rate should *not* be used for both before-tax and after-tax analyses.

2. Estimate the amounts and timing of cash flows other than income taxes. It will be useful to separate these cash flows into three categories:
   - Cash flows that have a *direct* effect on taxable income, as either "income" or "expense." Examples: sales receipts, direct labor savings, material costs, property taxes, interest payments, state and local income taxes (on federal returns).

- Cash flows that have an *indirect* effect on taxable income through depreciation, amortization, depletion, Section 179 expense deduction, and gain or loss on disposal. Examples: initial cost of depreciable property, salvage value.
- Cash flows that do not affect taxable income. Examples: working capital and that portion of loan repayments that represents payment of principal.

3. Determine the amounts and timing of cash flows for income taxes.
4. Find the equivalent present value of cash flows for income taxes at the *beginning of the first tax year*. To that end, let P_j denote the equivalent value of the cash flow for taxes in year j, as measured at the start of tax year j.

$$P_j \approx T_j \, (1 + i)^{-0.6875} \qquad j = 1,2,\ldots,N + 1 \qquad (37)$$

where i is the effective annual discount rate and N is the number of years in the planning horizon.

The equivalent present value of all the cash flows for taxes, as measured at the start of the first tax year, is given by

$$P(T) = \sum_{j=1}^{N+1} P_j \, (1 + i)^{-j+1} \qquad (38)$$

5. Find the equivalent present value of the cash flows for taxes, where "present" is defined as the start of the planning horizon. For example, if the property is placed in service at the end of the third month of the tax year, the present value adjustment is $P(T) \times (1 + i)^{3/12}$.
6. Find the equivalent present value of all other cash flows estimated in step 2 above. Use the after-tax MARR. Here the "present" is defined as the start of the planning horizon.
7. Combine (5) and (6) to yield the total net present value (NPV), or present worth (PW).

Note: If it is desired to determine the *after-tax rate of return* rather than PW (or FW, EUAC, and so on), steps 4 to 7 above must be modified. With the appropriate present worth equation for all cash flows, set the equation equal to zero and find the value of the interest rate i^* such that PW = 0. This is the after-tax IRR for the proposed investment.

Numerical Example. Consider the possible acquisition of certain manufacturing equipment with *initial cost* of $400,000. The equipment is expected to be kept in service for 6 years and then sold for an estimated $40,000 *salvage value*. *Working capital* of $50,000 will be required at the start of the 6-year period; the working capital will be recovered intact at the end of 6 years. If acquired, this equipment is expected to result in *savings* of $100,000 each year. The timing of these savings is such that the continuous cash flow assumption will be adopted throughout each year. The firm's after-tax MARR is 10 percent per year. The *present worth* of these cash flows, *other than income taxes,* is

$$\text{PW} = -\$400,\!000 + \$40,\!000(P/F, 10\%, 6)$$

$$-\$50,\!000 + \$50,\!000(P/F, 10\%, 6)$$

$$+\$100,\!000(P/\overline{A}, 10\%, 6)$$

$$= \$57,\!800$$

Assume that there is no Section 179 expense deduction. The equipment will be placed in service at the middle of the tax year and depreciated under MACRS as a 5-year recovery property using the half-year convention. The incremental federal income tax rate is 0.34; there are no other relevant income taxes affected by this proposed investment. The

TABLE 1.5   Cash Flows for Income Taxes—Numerical Example

Tax year j	Depreciation rate $p_j(5)$	Depreciation D_j	Gain G_N	Other revenue R_j	Taxable income $R_j-D_j+G_N$	Income taxes T_j	PW factor $(1.10)^{0.3125-j}$	PW @ 10% P_j
1	0.2000	$80,000		$40,000	−$40,000	−$13,600	0.93657	$12,737
2	0.3200	128,000		80,000	−48,000	−16,320	0.85143	13,895
3	0.1920	76,800		80,000	3,200	1,088	0.77403	−842
4	0.1152	46,080		80,000	33,920	11,533	0.70366	−8,115
5	0.1152	46,080		80,000	33,920	11,533	0.63969	−7,378
6	0.0576	23,040		80,000	56,960	19,366	0.58154	−11,262
7	0	0	$40,000	40,000	80,000	27,200	0.52867	−14,380

PW measured at start of first tax year	$15,345
Adjustment factor	$\times(1.10)^{0.5}$
PW measured at start of planning horizon (rounded to closest $100)	−$16,100

PW of the effects of cash flows due to income taxes is summarized in Table 1.5. The total PW for this proposed project is as follows:

Cash flows other than income taxes	$57,800
Effect on cash flows due to income taxes	−16,100
Net present worth	$41,700

Spreadsheet Analyses.   A wide variety of computer programs are available for before-tax and/or after-tax analyses of investment programs. (Relevant computer software is discussed from time to time in the journal, *The Engineering Economist.*) In addition, any of several spreadsheet programs currently available may be readily adapted to economic analyses, usually with very little additional programming. For example, Lotus 1-2-3, Version 2.0 and above, include financial functions to find the present and future values of a single payment and a uniform series (annuity), as well as to find the IRR of a series of cash flows.

Table 1.6 is an illustration of one possible application of a computer-generated spreadsheet. Here Lotus 1-2-3 is used to evaluate the numerical example presented in the previous section. Columns (A) through (D) are self-explanatory. Further:

TABLE 1.6   Spreadsheet Analysis—Numerical Example

MARR = 0.1

Project year j A	Investment and salvage value B	Working capital C	Savings during year j D	PW of discrete cash flows E	PW of continuous cash flows F	Total present value G
0	($400,000)	($50,000)		($450,000)		($450,000)
1			$100,000		$95,382	$95,382
2			$100,000		$86,711	$86,711
3			$100,000		$78,828	$78,828
4			$100,000		$71,662	$71,662
5			$100,000		$65,147	$65,147
6	$40,000	$50,000	$100,000	$50,803	$59,225	$110,028
Total	($360,000)	$0	$600,000	($399,197)	$456,957	$57,759
	Present worth (NPV) of cash flows for taxes					($16,094)
	Net present worth					$41,665

$$\text{Col. (E)} = [\text{Col. (B)} + \text{Col. (C)}] \times (1.10)^{-j}$$

$$\text{Col. (F)} = \text{Col. (D)} \times (1.10)^{-j} \times (0.10/\ln 1.10)$$

$$\text{Col. (G)} = \text{Col. (E)} + \text{Col. (F)}$$

The net present worth is completed in Table 1.6 by subtracting the PW of cash flows for taxes from the total PW of the pretax cash flows in column (G).

INCORPORATING PRICE LEVEL CHANGES INTO THE ANALYSIS

When allocating limited capital resources among competing investment alternatives, the effects of price level changes can be significant to the analysis. Cash flows, proxy measures of goods, and services received and expended are affected by both the *quantities* of goods and services as well as their *prices*. Thus, to the extent that changes in price levels affect cash flows, these changes must be incorporated into the analysis.

The *Consumer Price Index* (CPI) is but one of a large number of indexes that are regularly used to monitor and report relative price changes. The CPI is by far the best known, yet it is not particularly useful for specific economic analyses. Analysts should be interested in *relative price changes of goods and service that are germane to the particular investment alternatives under consideration*. The appropriate indexes are those that are related, say, to construction materials, costs of certain labor skills, energy, and other cost and revenue factors.

General Concepts and Notation.   Let p_1 and p_2 represent the prices of a certain good or service at two points in time t_1 and t_2, and let $n = t_2 - t_1$. The relative rate of price change between t_1 and t_2, average per period, is given by

$$g = \sqrt[n]{p_2/p_1} - 1 \tag{39}$$

We have *inflation* when $g > 0$ and *disinflation* when $g < 0$.

Let A_j = cash flow resulting from the exchange of certain goods or services, at end of period j, stated in terms of *constant* dollars. (Analogous terms are *now* or *real* dollars.) Let A_j^* = cash flows for those same goods or services in *actual* dollars. (Analogous terms are *then* or *current* dollars.) Then

$$A_j^* = A_j(1 + g)^j \tag{40}$$

where g is the periodic rate of increase or decrease in relative price (the *inflation rate*).

As before, let i = the MARR in the absence of inflation, that is, the real MARR. Let i^* = the MARR required taking into consideration inflation, that is, the *nominal* MARR. The periodic rate of increase or decrease in the MARR due to inflation f is given by

$$f = \left(\frac{1 + i^*}{1 + i}\right) - 1 = \frac{i^* - i}{1 + i} \tag{41}$$

Other relationships of interest are

$$i^* = (1 + i)(1 + f) - 1 = i + f + if \tag{42}$$

and

$$i = \left(\frac{1 + i^*}{1 + f}\right) - 1 = \frac{i^* - f}{1 + f} \tag{43}$$

Models for Analysis.   It may be shown that the *future worth* of a series of cash flows $A_j{}^*$ $(j = 1,2,\ldots,N)$ is given by

$$FW = (1 + i^*)^N \sum_{j=0}^{N} A_j (1 + d)^{-j} \qquad (44)$$

where

$$d = \frac{(1 + i)(1 + f)}{1 + g} - 1 \qquad (45)$$

and i, f, and g are as defined previously. From Eq. (44) it follows that the *present worth* is given by

$$PW = \sum_{j=0}^{N} A_j (1 + d)^{-j} \qquad (46)$$

Note: In these models it is assumed that both the cash flows and the MARR are affected by inflation, the former by g and the latter by f, and $f = g$. If it is assumed that both i and A_j's are affected by the same rate, that is, $f = g$, then Eq. (46) reduces to

$$PW = \sum_{j=0}^{N} A_j (1 + i)^{-j} \qquad (47)$$

which is the same as the PW model ignoring inflation.

To illustrate, consider cash flows in constant dollars (A_j) of $80,000 at the end of each year for 8 years. The inflation rate for the cash flows (g) is 6 percent per year, the nominal MARR (i^*) is 9 percent per year, and the inflationary effect on the MARR (f) is 4.6 percent per year. Then

$$d = \frac{1 + i^*}{1 + g} - 1 = \frac{1.09}{1.06} - 1 = 0.0283$$

and

$$PW = \sum_{j=1}^{8} A_j (1 + d)^{-j} = \$80,000(P/A, 2.83\%, 8) = \$565,000$$

Multiple Factors Affected Differently by Inflation.   In the preceding section it is assumed that the project consists of a single price component affected by rate g per period. But most investments consist of a variety of components, among which rates of price changes may be expected to differ significantly. For example, the price of the *labor* component may be expected to increase at the rate of 7 percent per year, and the price of the *materials* component is expected to decrease at the rate of 5 percent per year. The appropriate analysis in such cases is an extension of Eqs. (44) through (46).

Consider a project consisting of two factors, and let A_{j1} and a_{j2} represent the cash flows associated with each of these factors. Let g_1 and g_2 represent the relevant inflation rates, so that

$$A_j^* = A_{j2} (1 + g_1)^j + A_{j2} (1 + g_2)^j \qquad (48)$$

It follows that

$$FW = (1 + i^*)^N \left\{ \left[\sum_{j=1}^{N} A_{j1}(1 + d_1)^{-j} \right] + \left[\sum_{j=1}^{N} A_{j2}(1 + d_2)^{-j} \right] \right\} \qquad (49)$$

and

$$PW = \left\{ \left[\sum_{j=1}^{N} A_{j1}(1 + d_1)^{-j} \right] + \left[\sum_{j=1}^{N} A_{j2}(1 + d_2)^{-j} \right] \right\} \qquad (50)$$

where

$$d_1 = (1 + i^*)/(1 + g_1) \qquad d_2 = (1 + i^*)/(1 + g_2) \qquad (51)$$

Interpretation of IRR under Inflation.   If *constant* dollars (A_j) are used to determine the internal rate of return, then the *inflation-free IRR* is that value of ρ such that

$$\sum_{j=0}^{N} A_j (1 + \rho)^{-j} = 0 \qquad (52)$$

The project is acceptable if ρ > i, where i is the inflation-free MARR as in the preceding section.

If *actual* dollars (A_j^*) are used to determine the internal rate of return, then the *inflation-adjusted IRR* is that value of ρ* such that

$$\sum_{j=0}^{N} A_j^* (1 + \rho^*) = 0 \qquad (53)$$

To illustrate, consider a project which requires an initial investment of \$100,000 and for which a salvage value of \$20,000 is expected after 5 years. If acquired, this project will result in annual savings of \$30,000 at the end of each year over the 5-year period. All cash flow estimates are based on constant dollars. It may be shown that, based on these assumptions, $\rho \approx 19$ percent.

It is assumed that the cash flows for this proposal will be affected by an inflation rate (g) of 10 percent per year. Thus $A_j^* = A_j (1.10)^j$, and, from Eq. (52), $\rho^* \approx 31$ percent.

The investor's inflation-free MARR (i) is assumed to be 25 percent. If it is assumed that the MARR is affected by an inflation rate (g) of 10 percent per year, then $i^* = 1.10(1.25) - 1 = 0.375$.

Each of the two comparisons indicates that the proposed project is not acceptable: $\rho(19\%) < i\ (25\%)$ and $\rho^*(31\%) < i^*\ (37.5\%)$.

TREATING RISK AND UNCERTAINTY IN THE ANALYSIS

It is imperative that the analyst recognize the uncertainty inherent in *all* economy studies. The past is irrelevant, except when it helps predict the future. Only the future is relevant, and *the future is inherently uncertain.*

At this point it will be useful to distinguish between *risk* and *uncertainty,* two terms widely used when dealing with the noncertain future. *Risk* refers to situations in which a probability distribution underlies future events and the characteristics of this distribution are known or can be estimated. Decisions involving *uncertainty* occur when nothing is known or can be assumed about the relative likelihood, or probability, of future events. Uncertainty situations may arise when the relative attractiveness of various alternatives is a function of the outcome of pending labor negotiations or local elections or when permit applications are being considered by a government planning commission.

A wide spectrum of analytical procedures is available for the formal consideration of risk and uncertainty in analyses. Space does not permit a comprehensive review of all these procedures. The reader is referred to any of the general references included in suggestions for further reading for discussion of one or more of the following:

- Sensitivity analysis
- Risk analysis
- Decision theory applications
- Digital computer (Monte Carlo) simulation
- Decision trees

Some of these procedures can be found elsewhere in this handbook. Other procedures widely used in industry include:

- *Increasing the Minimum Attractive Rate of Return.*Some analysts advocate adjusting the minimum attractive rate of return to compensate for risky investments, suggesting that, since some investments will not turn out as well as expected, they will be compensated for by an incremental "safety margin" Δi. This approach, however, fails to come to grips with the risk or uncertainty associated with estimates for specific alternatives, and thus an element Δi in the minimum attractive rate of return penalizes all alternatives equally.

- *Differentiating Rates of Return by Risk Class.*Rather than building a "safety margin" into a single minimum attractive rate of return, some firms establish several risk classes with separate standards for each class. For example, a firm may require low-risk investments to yield at least 15 percent and medium-risk investments to yield at least 20 percent, and it may define a minimum attractive rate of return of 25 percent for high-risk proposals. The analyst then judges which class a specific proposal belongs in, and the relevant minimum attractive rate of return is used in the analysis. Although this approach is a step away from treating all alternatives equally, it is less than satisfactory in that it fails to focus attention on the uncertainty associated with the individual proposals. No two proposals have precisely the same degree of risk, and grouping alternatives by class obscures this point. Moreover, the attention of the decision maker should be directed to the causes of uncertainty, that is, to the individual estimates.

- *Decreasing the Expected Project Life.*Still another procedure frequently employed to compensate for uncertainty is to decrease the expected project life. It is argued that estimates become less and less reliable as they occur further and further into the future; thus shortening project life is equivalent to ignoring those distant, unreliable estimates. Furthermore, distant consequences are more likely to be favorable than unfavorable; that is, distant estimated cash flows are generally positive (resulting from net revenues) and estimated cash flows near date zero are more likely to be negative (resulting from startup costs). Reducing expected project life, however, has the effect of penalizing the proposal by precluding possible future benefits, thereby allowing for risk in much the same way that increasing the minimum attractive rate of return penalizes marginally attractive proposals. Again, this procedure is to be criticized on the basis that it obscures uncertain estimates.

COMPOUND INTEREST TABLES (10 PERCENT)

Table 1.7 presents compound interest tables for the single payment, the uniform series, and the gradient series.

TABLE 1.7 Compound Interest Tables (10 Percent)

N	Single payment			Uniform series				Uniform series		Gradient Series		N
	Compound amount	Present worth		Compound amount		Present worth		Sinking fund	Capital recovery	Uniform series	Present worth	
	F/P	P/F	$P/\bar{F}$	F/A	$F/\bar{A}$	P/A	$P/\bar{A}$	A/F	A/P	A/G	P/G	
1	1.100	0.9091	0.9538	1.000	1.049	0.909	0.954	1.0000	1.1000	0.000	0.000	1
2	1.210	0.8264	0.8671	2.100	2.203	1.736	1.821	0.4762	0.5762	0.476	0.826	2
3	1.331	0.7513	0.7883	3.310	3.473	2.487	2.609	0.3021	0.4021	0.937	2.329	3
4	1.464	0.6830	0.7166	4.641	4.869	3.170	3.326	0.2155	0.3155	1.381	4.378	4
5	1.611	0.6209	0.6515	6.105	6.406	3.791	3.977	0.1638	0.2638	1.810	6.862	5
6	1.772	0.5645	0.5922	7.716	8.095	4.355	4.570	0.1296	0.2296	2.224	9.684	6
7	1.949	0.5132	0.5384	9.487	9.954	4.868	5.108	0.1054	0.2054	2.622	12.763	7
8	2.144	0.4665	0.4895	11.436	11.999	5.335	5.597	0.0874	0.1874	3.004	16.029	8
9	2.358	0.4241	0.4450	13.579	14.248	5.759	6.042	0.0736	0.1736	3.372	19.421	9
10	2.594	0.3855	0.4045	15.937	16.722	6.145	6.447	0.0627	0.1627	3.725	22.891	10
11	2.853	0.3505	0.3677	18.531	19.443	6.495	6.815	0.0540	0.1540	4.064	26.396	11
12	3.138	0.3186	0.3343	21.384	22.437	6.814	7.149	0.0468	0.1468	4.388	29.901	12
13	3.452	0.2897	0.3039	24.523	25.729	7.103	7.453	0.0408	0.1408	4.699	33.377	13
14	3.797	0.2633	0.2763	27.975	29.352	7.367	7.729	0.0357	0.1357	4.996	36.801	14
15	4.177	0.2394	0.2512	31.772	33.336	7.606	7.980	0.0315	0.1315	5.279	40.152	15
16	4.595	0.2176	0.2283	35.950	37.719	7.824	8.209	0.0278	0.1278	5.549	43.416	16
17	5.054	0.1978	0.2076	40.545	42.540	8.022	8.416	0.0247	0.1247	5.807	46.582	17
18	5.560	0.1799	0.1887	45.599	47.843	8.201	8.605	0.0219	0.1219	6.053	49.640	18
19	6.116	0.1635	0.1716	51.159	53.676	8.365	8.777	0.0195	0.1195	6.286	52.583	19
20	6.728	0.1486	0.1560	57.275	60.093	8.514	8.932	0.0175	0.1175	6.508	55.407	20

TABLE 1.7 Compound Interest Tables (10 Percent) (*Continued*)

N	Single payment Compound amount F/P	Single payment Present worth P/F	Single payment Present worth P/F̄	Uniform series Compound amount F/A	Uniform series Compound amount F/Ā	Uniform series Present worth P/A	Uniform series Present worth P/Ā	Uniform series Sinking fund A/F	Uniform series Capital recovery A/P	Gradient Series Uniform series A/G	Gradient Series Present worth P/G	N
21	7.400	0.1351	0.1418	64.003	67.152	8.649	9.074	0.0156	0.1156	6.719	58.110	21
22	8.140	0.1228	0.1289	71.403	74.916	8.772	9.203	0.0140	0.1140	6.919	60.689	22
23	8.954	0.1117	0.1172	79.543	83.457	8.883	9.320	0.0126	0.1126	7.108	63.146	23
24	9.850	0.1015	0.1065	88.497	92.852	8.985	9.427	0.0113	0.1113	7.288	65.481	24
25	10.835	0.0923	0.0968	98.347	103.186	9.077	9.524	0.0102	0.1102	7.458	67.696	25
26	11.918	0.0839	0.0880	109.182	114.554	9.161	9.612	0.0092	0.1092	7.619	69.794	26
27	13.110	0.0763	0.0800	121.100	127.059	9.237	9.692	0.0083	0.1083	7.770	71.777	27
28	14.421	0.0693	0.0728	134.210	140.814	9.307	9.765	0.0075	0.1075	7.914	73.650	28
29	15.863	0.0630	0.0661	148.631	155.945	9.370	9.831	0.0067	0.1067	8.049	75.415	29
30	17.449	0.0573	0.0601	164.494	172.588	9.427	9.891	0.0061	0.1061	8.176	77.077	30
31	19.194	0.0521	0.0547	181.944	190.896	9.479	9.945	0.0055	0.1055	8.296	78.640	31
32	21.114	0.0474	0.0497	201.138	211.035	9.526	9.995	0.0050	0.1050	8.409	80.108	32
33	23.225	0.0431	0.0452	222.252	233.188	9.569	10.040	0.0045	0.1045	8.515	81.486	33
34	25.548	0.0391	0.0411	245.477	257.556	9.609	10.081	0.0041	0.1041	8.615	82.777	34
35	28.102	0.0356	0.0373	271.025	284.361	9.644	10.119	0.0037	0.1037	8.709	83.987	35
40	45.259	0.0221	0.0232	442.593	464.371	9.779	10.260	0.0023	0.1023	9.096	88.953	40
45	72.891	0.0137	0.0144	718.906	754.280	9.863	10.348	0.0014	0.1014	9.374	92.454	45
50	117.391	0.0085	0.0089	1,163.910	1,221.181	9.915	10.403	0.0009	0.1009	9.570	94.889	50
55	189.059	0.0053	0.0055	1,880.594	1,973.130	9.947	10.437	0.0005	0.1005	9.708	96.562	55
60	304.482	0.0033	0.0034	3,034.821	3,184.151	9.967	10.458	0.0003	0.1003	9.802	97.701	60
65	490.372	0.0020	0.0021	4,893.715	5,134.514	9.980	10.471	0.0002	0.1002	9.867	98.471	65
70	789.748	0.0013	0.0013	7,887.483	8,275.592	9.987	10.479	0.0001	0.1001	9.911	98.987	70
80	2,048.405	0.0005	0.0005	20,474.045	21,481.484	9.995	10.487	0.0000	0.1000	9.961	99.561	80
90	5,313.035	0.0002	0.0002	53,120.348	55,734.168	9.998	10.490	0.0000	0.1000	9.983	99.812	90

SUGGESTIONS FOR FURTHER READINGS

Books

American Telephone and Telegraph, Engineering Department, *Engineering Economy*, 3d ed., McGraw-Hill, New York, 1980.

Au, Tung, and Thomas P. Au, *Engineering Economics for Capital Investment Analysis*, Allyn and Bacon, Boston, Mass., 1983.

Bierman, Harold, Jr., and Seymour Smidt, *The Capital Budgeting Decision*, 6th ed., Macmillan, New York, 1984.

Blank, Leland T., and Anthony J. Tarquin, *Engineering Economy*, 3d ed., McGraw-Hill, New York, 1989.

Canada, John R., and William G. Sullivan, *Economic and Multiattribute Analysis of Advanced Manufacturing Systems*, Prentice-Hall, Englewood Cliffs, N.J., 1989.

Clark, F. D., and A. B. Lorenzoni, *Applied Cost Engineering*, 2d ed., Dekker, New York, 1985.

Collier, C. A., and W. B. Ledbetter, *Engineering Cost Analysis*, 2d ed., Harper & Row, New York, 1987.

DeGarmo, E. P., W. G. Sullivan, and James A. Bontadelli, *Engineering Economy*, 8th ed., Macmillan, New York, 1989.

Fabrycky, Wolter, and Gerald J. Thuesen, *Economic Decision Analysis*, 3d ed., Prentice-Hall, Englewood Cliffs, N.J., 1984.

Fleischer, Gerald A., *Engineering Economy: Capital Allocation Theory*, Wadsworth, Boston, 1984.

Grant, Eugene L., W. Grant Ireson, and Richard S. Leavenworth, *Principles of Engineering Economy*, 8th ed., Wiley, New York, 1990.

Kleinfeld, Ira H., *Engineering and Managerial Economics*, Holt, Rinehart and Winston, New York, 1986.

Newnan, Donald G., *Engineering Economic Analysis*, 3d ed., Engineering Press, San Jose, Calif., 1988.

Sepulveda, J. A., W. E. Souder, and B. S. Gottfried, *Theory and Problems of Engineering Economics*, Schaum's Outline Series, Wiley, New York, 1984.

Smith, G. W., *Engineering Economy: Analysis of Capital Expenditures*, 4th ed., Iowa State University Press, Ames, Iowa, 1987.

Sprague, J. C., and J. D. Whittaker, *Economic Analysis for Engineers and Managers*, Prentice-Hall, Englewood Cliffs, N.J., 1986.

Swalm, R. O., and J. L. Lopez-Leautaud, *Engineering Economic Analysis: A Future Wealth Approach*, Wiley, New York, 1984.

Thuesen, H. G., and W. J. Fabrycky, *Engineering Economy*, 7th ed., Prentice-Hall, Englewood Cliffs, N.J., 1989.

Wellington, Arthur M., *The Economic Theory of Railway Location*, 2d ed., Wiley, New York, 1887. This book is of historical importance; it was the first to address the issue of economic evaluation of capital investments due to engineering design decisions. Wellington is widely considered to be the "father of engineering economy."

White, J. A., M. H. Agee, and K. E. Case. *Principles of Engineering Economic Analysis*, 3d ed., Wiley, New York, 1989.

Journals

Decision Science
The Engineering Economist
Financial Management
Harvard Business Review
IIE Transactions
Industrial Engineering

Journal of Business
Journal of Finance
Journal of Finance & Quantitative Analysis
Management Science

CHAPTER 2
PLANNING FOR PROFITS AND BUDGETARY CONTROL

Edmund J. McCormick, Jr.
Chairman
McCormick & Company
Summit, New Jersey

Robert L. Staehle
Chairman Emeritus
McCormick & Company
Summit, New Jersey

Profits don't just happen. They must be *planned*. The development of realistic plans for the company and for each of its major divisions and product lines is an essential function of management. The continuing comparison of actual operating results with the plan is one of the most important means of control available to management.

The *object* of profit planning is to make the most effective use of resources and thereby obtain the highest level of sustained profits. In almost every major industry there are companies with relatively modest resources that are making profits equal to or greater than those of competitors with far more capital. Nine times out of ten, the reason is that the more successful company is doing a better job of planning. It is using a carefully coordinated system to chart the course it means to follow.

The examples used in this chapter are concerned with manufacturing companies. However, the same basic principles apply to service and financial companies such as communications, data processing, utilities, transportation, insurance, banking, and similar organizations that are primarily delivering intangibles to end users.

Strategic planning is concerned with long-term goals and the changes in organization, investment, product mix, and financial resources required to attain these objectives.

Tactical planning is concerned with the period immediately ahead, where the problem is to make the most of resources available. Both types of planning are important, but they call for different approaches and require different kinds of decisions.

Strategic planning usually places its emphasis on the direction of movement rather than the amounts that will be involved. Its object is to highlight the general direction that the company will take over the long term. The time span involved will vary with the characteristics of the enterprise. For example, a forest products company may plan 75 years out in terms of raw materials; a high-fashion clothing enterprise will probably limit its strategic planning to 3 years at best. Detailed planning on a long-term basis is usually not realistic, but such plans require translation to hard numbers. This should be a relatively simple procedure concentrated on total sales, marginal income, period costs, and profit figures. Marginal income is the overall indicator that should be closely tied with long-term return on investment or capacity.

Tactical planning is usually a year out or, in a few cases, a quarter. It must deal with current realities, and as a result, the planning should be more detailed, particularly in terms of variable costs. However, the same basic measures used in strategic planning can be used as focal points for the tactical plan.

Assumptions are required in both cases. The future of any company can be significantly changed by such broad influences as inflation, material costs, shifts in supply and demand, or swings in the national economy. The impact of such changes will vary with the particular characteristics of the industry. Planners should have a clear idea of what will happen to key variables—costs, volume, marginal income, and most important of all, profits—if the assumptions do not pan out. One of the uses of the model of company operations that planners build is to analyze the effect that a change in assumptions would have on the profit plan. Using the model for this sort of sensitivity analysis is one of the major steps toward arriving at the final goals of the profit plan.

The planner needs a good cost control system that will provide the basis for an accurate analysis of profitability, product by product and business unit by business unit. It is impossible to make projections that mean anything if costs are swinging erratically up and down or if costs attributable to each profit or unit have never been determined.

The profit plan itself is a form of control in the sense that it sets goals (what should happen) and matches them against actual results. However, this is on a longer-term basis.

To back up the profit plan, the company needs a control-performance measurement system that can track and measure short-term performance. Controls should be developed that are simple, timely, and motivational in nature. The system should be carefully tailored to reflect the operating conditions as well as its management style.

Planning, however, should not be confused with a control system based upon standard costs. This mistake obscures many of the greatest values that a good planning system offers. Planning is concerned with costs, of course. But its real focus is on profits. To analyze the profitability of units or products, the planner must use a variety of tools besides costs.

To construct a *preliminary plan,* the planner's first estimate of profitability in the period for which the plan applies, the planner uses two sets of projections, one related to the profit and loss statement, the other to the balance sheet.

Projections keyed to the company's profit and loss statement include:

- Sales for each division, product, or product line.
- Variable manufacturing costs attributable to each.
- Variable costs of marketing and distribution attributable to each.
- The difference between the selling price and the total variable costs of manufacturing and marketing. This is the marginal income attributable to each unit or product.
- Period costs (common and distributed).

Projections keyed to the balance sheet include:

- Accounts receivable.
- Property, plant, and equipment.
- Inventories of raw materials and of finished goods.
- Investment in research and development.
- Investment in patents and other proprietary items.

Using these basic tools, the planner drafts a preliminary plan showing the established performances of each unit or product and the total profits this will generate for the company. The planner then constructs a model that compares the preliminary plan with targets for the same business unit or product line estimated on the basis of rate of return that management seeks to get on resources earmarked for that unit or product line. Such a comparison may reveal an opportunity to utilize ideal capacity, change the product mix, or reduce investment in some lines. It may also show the need to revise the targets to make them conform to realistic forecasts.

The model can be used to answer "what if" questions, showing what would happen to sales, costs, and profits under various assumptions about markets, prices, investment, and product mix. In many ways it is the most powerful tool available to top management for charting the future course of the company. At the same time, it is invaluable at lower levels of management where it can be used to improve profitability or to analyze the effects of a change in design.

The industrial engineer should be a key person throughout the planning process. Industrial engineers' understanding of the manufacturing process, the raw material requirements, and the plant and equipment needs in each product line is invaluable in determining realistic targets. Their continuing role in product evaluation puts them constantly in touch with all areas and divisions of the organization.

Unfortunately, many corporations have downgraded the role of the industrial engineer in recent years. They have created corporate planning offices with vaguely defined responsibilities. As a result, the industrial engineer has often been shoved out of the executive offices and consigned to life in the shop. This is a serious mistake. Effective planning must involve the entire organization. While the financial aspects of planning will be a primary responsibility of the treasurer, financial vice president, or controller, operational planning and substantial parts of organization planning call for exactly the kind of insight that the industrial engineer can contribute.

The ideal planning team consists of representatives of corporate planning, accounting, and industrial engineering. The corporate planner will bring an understanding of long-term goals and policy; an accountant will supply the data on past performance, broken down by divisions and products; the industrial engineer will contribute knowledge of what each division of the company does and how it works. The starting point for planning, of course, must be the sales forecast, broken down by divisions and products. The next item should be for the industrial engineer to convert the sales forecasts into manufacturing targets. This will reveal at once whether or not the forecasts are feasible, given existing capacity and manufacturing methods. More than anyone else in the company, the industrial engineer can bring realism to the planning function. Shut the industrial engineer out of the planning process, and the plans that result are likely to be pipe dreams rather than workable guides to the future.

CONCEPTS THE PLANNER USES

The end result of planning should be to maximize the ratio of profits to the equity capital employed, that is, the rate of return on the assets that the owners of the business have committed to it. One of the common mistakes that companies make is to concentrate on profit as a percent of sales rather than keeping a steady watch on what the owners are making on their investment.

Table 2.1 shows key items from the balance sheets and income statements of two manufacturing companies. Their profits to sales ratios compare as shown in Table 2.2.

On the face of it, Company B seems to be doing substantially better than its competitor. But take a look at what happens when the analyst refused to stop with the percent of profit and compares the two companies on the basis of the rates of return they earned on the resources used in their businesses (Table 2.3).

It is apparent that Company A, with a somewhat smaller investment, is making a sig-

TABLE 2.1 Company Comparison: Company A to B ($000)

	Plan	Revised	Difference	
BALANCE SHEET				
Assets	$110,000	$190,000	$80,000	Greater
Liabilities	45,000	90,000	45,000	Greater
Equity Capital Employed	65,000	100,000	35,000	Greater
INCOME STATEMENT				
Net Plan Sales	$225,000	$270,000	$45,000	Greater
Variable Mfg. Cost of Sales	135,000	135,000	0	No Diff.
Total Marginal Income	90,000	135,000	45,000	Greater
Period Mfg.	20,000	40,000	20,000	Greater
Cost of Sales	155,000	175,000	20,000	Greater
Period Selling & G&A	40,000	55,000	15,000	Greater
Operating Profit before Tax	30,000	40,000	10,000	Greater
Profit After Tax	15,000	20,000	5,000	Greater
RATIOS				
Marginal Income	40.00%	50.00%	10.00%	Greater
ROA Before Tax	27.00%	21.00%	−6.00%	Smaller
Profit to Equity Capital Employed After Tax	23.00%	20.00%	−3.00%	Smaller
Profit to Sales	13.30%	14.80%	1.50%	Greater
Margin of Safety	33.00%	29.60%	−3.40%	Smaller
Breakeven	$150,000	$190,000	$40,000	Greater

TABLE 2.2 Company Comparison of Profit to Sales (P/S) Ratios: Company A to Company B ($000)

	Sales	Profits	Profits before taxes as a percent of sales
Company A	$225,000,000	$30,000,000	13.33%
Company B	270,000,000	40,000,000	14.81%

TABLE 2.3 Comparison of Profit to Equity Capital Employed (ECE): Company A to B ($000)

	Equity capital employed	Profits	Profits before taxes as a percent of equity capital employed
Company A	$65,000,000	$30,000,000	46.15%
Company B	100,000,000	40,000,000	40.00%

nificantly better return on its assets than Company B. Obviously, Company A's profit planning is making the most out of what it has. Further evidence of this is the pretax return on total assets—27 percent for Company A, which compares with 21 percent for Company B.

The succeeding sections of this chapter outline the steps that Company B can take to upgrade its performance and improve its returns.

The profits to equity capital employed ratio (P/ECE) is a product of two other ratios: profit to sales (P/S) and sales to equity capital employed (S/ECE). The profit to sales ratio measures the number of cents that the company can keep out of each sales dollar. The sales to equity capital employed ratio measures the number of times equity capital employed turned over in terms of sales dollars. The two ratios multiplied together give the profit on equity capital employed ratio:

$$P/S \times S/ECE = P/ECE$$

Control of the P/S and S/ECE ratios—and through them the P/ECE ratio—is achieved by comparison of actual and target marginal income, the amount of the sales dollar that is left after costs generated by the product process. To estimate marginal income, the planner must forecast sales and identify the two major categories of cost—those that vary with the rates of production and those that are fixed for the period ahead regardless of output levels.

The planning process begins with the preparation of the marketing plan. An effective marketing plan will answer such questions as:

- What products will sell and in what volumes?
- At what prices?
- At what promotional costs?
- When and by what selling method?
- What product mix does this plan require?

But the answers to these questions will be useful only if their impact on profits can be determined. To make such a determination, the planner must turn to marginal income accounting. Because the unit profit contribution for a particular product stays constant in the marginal income approach, it is easy to pinpoint the effect of volume swings. Simple multiplication of the units to be sold by the constant rate of contribution will give the answer in terms of profits.

Marginal income estimates provide the benchmarks for making pricing decisions. Knowing how much each product contributes to total marginal income, sales executives can develop a price structure that maximizes the sales of the products that contribute most to company profits. Similarly, when open plant capacity exists, any price obtained above the standard direct cost of an item will generate marginal income to cover period costs and contribute toward profit.

As executives concerned with pricing know all too well, the marketplace is always shifting. Pricing policies must be reevaluated continually in response to shifts in competition, demand, and supply. The key to successful pricing is rapid and knowledgeable response to these hectic conditions in the actual marketplace. The marginal income approach provides the timely and reliable guidance the marketing executive needs to make a successful response designed to achieve the profit targets. With a rapid feedback of information on market conditions and direct cost variances, the marketing executive can readily detect departures from the planned targets and change the pricing and selling effort accordingly.

Marginal income is the most useful of all concepts available to the planner. It is a key that opens the way to reliable, scientific analysis of profit opportunities in spite of the uncertainties and breathtaking changes of the modern business world. Marginal income costing is particularly well suited to computerization because it is specifically designed to deal with variances, real and projected, in volume, prices, various kinds of costs, and capacity utilization. It provides investors with a clear picture of the results of operations.

The vital role of marginal income costing makes precise definitions essential.

Variable Costs.   Costs that go up or down in step with production of the product or performance of the service involved are variable costs. They are the specific costs of making or delivering a product or service.

Period Costs.   Costs that vary only gradually over time periods so long as operations remain within normal capacity levels are period costs. They are considered the costs of being in business and are not susceptible to control at production level. In most cases, variable costs are controlled at the line or production level, while period costs are controlled at the management level. In many cases they are determined by management policy, which can be evaluated on the basis of its economic impact on the company.

Marginal Income.   This is sales minus variable costs. It can be measured at two levels—manufacturing marginal income, which is the amount left out of the sales dollar after direct costs of production have been subtracted, and marketing marginal income, which is what remains after the direct sales and distribution costs have been paid.

In service and financial types of organization, variable costs can be identified in several zones or layers of activity. A bank or insurance company, for example, may have branch office, regional, and division variable costs. A marginal income figure can then be identified for each one depending on the routing of the service or product.

IDENTIFYING VARIABLE COSTS

In distinguishing between variable and period costs, the planner should not underestimate the power of the company to control its costs. If an expense cannot be clearly identified as period, it should be classified as variable. The decision should be made on the basis of what could be done, not on the defeatist assumption that nothing will be done.

Most costs won't present serious problems of classification, but a planner who goes through the process of analyzing costs should be on the watch for chances to lower costs without lowering output or quality.

Labor.   Assuming no major change in productivity, production line labor costs will vary directly with the rate of output. But all payroll costs are by no means variable. Some employees are, in effect, period cost workers, while the majority of line workers are variable cost workers. A maintenance crew, for example, will have little to do in slack times, but it will continue to draw wages. By contrast, production line workers can usually be furloughed when output is reduced, although the union contract may restrict management's choices in such cases. The important thing for the planner is to compare work actually done with the standards for such work, leaving the problem of costs incurred regardless of production levels to the company policymakers. Management might, for instance, put maintenance work out on contract with a service organization instead of keeping it in house. The need for such a decision might be revealed by the planning process, but until the change is actually made, the maintenance crew's wages will be a period cost. This can also be true of contract maintenance if specific tasks are ordered regardless of rates of operation. Generally, forcing a cost to a variable basis will better control profits.

Raw Materials.   So long as the specifications for the product remain the same, the cost of raw materials can be expected to vary with the rate of production. However, there may be opportunities here to change the specifications or find a new source of supply, thereby reducing the cost at each level of output. Here, again, the planning process highlights opportunities to increase profitability.

Distribution Costs. Though some marketing costs will be keyed to sales and production, others may depend on management decisions. This is especially true in marketing and distribution where different means of reaching the final consumer are likely to involve strikingly different costs.

Figure 2.1 shows four different channels of distribution available to a soft drink company. Although the product can remain unchanged as it passes through these channels, the profit content can change significantly. The various channels involve an almost infinite number of combinations of price, cost of distribution, and investment. Large deliveries to jobbers and supermarkets not only involve the lowest cost but require the lowest investment. Consequently, prices can be lower. Special events may involve high labor costs and demand special equipment, but they can build customer goodwill. In setting up targets for sales and profits, management will have to take account of all these considerations.

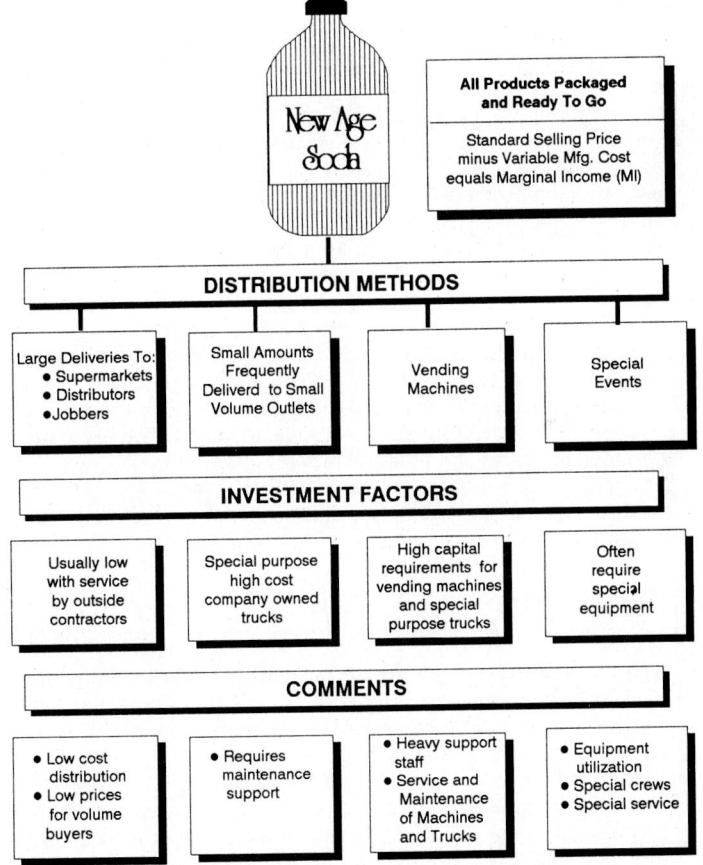

FIGURE 2.1 Typical distribution pattern for a soft drink company.

In many cases, it will be useful to analyze distribution and sales costs on the basis of sales regions. The costs of each region can be separated into variable (including commissions, commercial travel, automobile, and freight) and period (clerical and supervisory office) costs. A careful study of the marginal income generated by each distribution channel may suggest that the geographic areas covered should be reduced to generate a higher return by more intensive selling.

Some businesses have a choice in how a service may be rendered. A transportation company can effect a delivery by company vehicle, outside contractors, air, rail, and water carriers. Communications can be written, verbal, electronic, or any number of combinations.

Variable cost planning is a precision tool. It is far and away the best method available not only for profit planning but also for the entire cost system of the company. By associating only variable costs with a product, management eliminates the information fog caused by volume variance. Most of the budgeting and standard costs systems of the past were inflexible and became useless if output missed the target rate the plan set for it. This happened because in trying to assign whole costs to each product, companies used a device called under- or overabsorbed burden, and this was estimated on the basis of an assumed "normal" rate of operation. But as any plant manager knows, total costs per unit will come out to a predetermined figure only if production hits the assumed level on the nose. If it runs higher, per unit costs will be less than forecast. If it runs lower, they will be greater.

The variable costs system works on the sound assumption that period costs should not be assigned to the product until after marginal income has been estimated. Since variable costs are a function of output rates, cost per unit remains constant as long as output does not approach the limits of capacity and create inefficiencies that raise costs.

The marginal income approach also gives management a realistic basis for evaluating new products. It shows precisely what a proposed new product will add to costs, because it focuses on direct out-of-pocket costs incurred as a result of producing the new product. Unless it will entail the addition of salaried staff or an increase in investment, the marginal income of a new product can fairly be regarded as an addition to profit.

The variable costs estimates that emerge from the planning process should become the basis of an effective system of controls designed to make the goals of the plan a reality. The targets set by the plan should not be regarded as idle dreams. They should be treated as firm commitments by each division of the company or each product manager, and they should apply to each profit the company makes.

This means that the profit planning has to be coordinated with day-by-day supervision of operations. It will involve people as much as it involves production schedules. Regular variance meetings and monthly reports of variable costs on a product-by-product basis are essential parts of effective cost control. This is an area where the industrial engineer's understanding and experience can make a vital contribution.

MANAGING PERIOD COSTS

Accountants used to use the term "fixed" costs as though no change was possible. The modern term, period costs, reflects increasing recognition of the fact that these costs are fixed for a specific time frame only. They may be out of management's reach for months, but most of them can be managed over time.

It is true that some costs, such as insurance and depreciation on existing plant, are fixed for the foreseeable future and little can be done to change them. But others, such as heating, air conditioning, and snow plowing are seasonal. Still others, including a number of major staff costs, are determined by previous management decisions. For example, the number of lawyers, secretaries, and other support staff on the payroll at any time is fixed only in the sense that management has authorized them. A decision to use outside counsel or to set up a stenographic pool might cut these costs significantly.

Predetermined costs such as advertising programs and contributions are also the result of previous management decisions. They could be changed abruptly, as any advertising manager will testify, but if they were well conceived and designed to bring public support and interest, management should think twice before whacking at them. It is important to identify and classify each of these costs in terms of how it originated and how much change could be made if management were willing to review previous decisions.

This is a job for the planners, including the industrial engineer. It must be done by carefully weighing the advantages and disadvantages. It should never be done by issuing a blanket order to cut everything in a particular accounting category by a fixed percentage. As good accountants are quick to say, generally accepted accounting principles followed blindly will lead to total confusion and controversy.

Cross-Checking with Charts.   Once period costs have been determined, it is possible to cross-check the estimate by constructing the familiar break-even chart, which plots sales on the horizontal axis and profit or loss on the vertical. A trend line fitted to recent sales or loss figures will cross the dividing line between profit and loss at the break-even point, and it will intersect the vertical axis at the loss figure equivalent to total period costs. Figure 2.2 is the break-even chart for Company B. It plots the ratio of profits to sales. Sales are on the horizontal axis and profit or loss is on the vertical. With fixed costs of $95 million and sales projected at $270 million, Company B has a margin of safety (percent of sales over the break-even point) equal to 29.6 percent of sales. At the $270 million sales level, the chart shows that Company B will have a $40 million profit. Marginal income (MI) is $135 million ($95 million of fixed cost plus $40 million of profit). The chart demonstrates the effect that a change in cost or sales will have on profits.

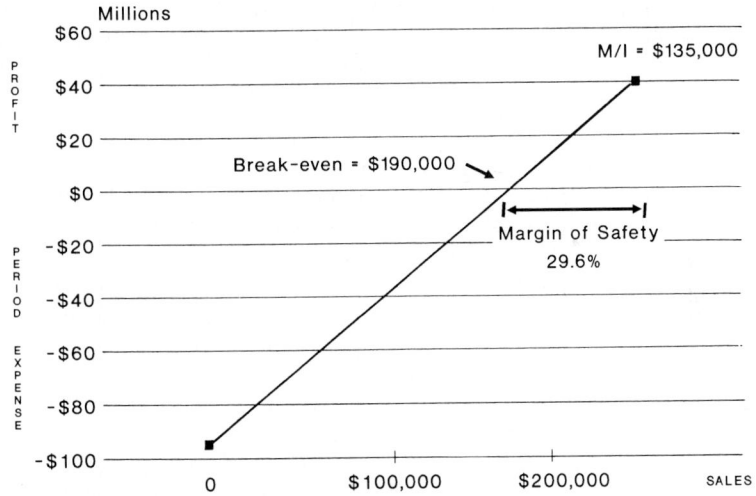

FIGURE 2.2   Break-even chart, Company B.

However, as in the comparison of Company A and Company B (Table 2.1), the ratio of profit to sales can conceal important differences between companies or products. If the charts are being used to estimate the effects of changes in volume or price, management should make a break-even chart that plots marginal income rather than profit or loss associated with each level of sales.

A sophisticated version of the break-even chart is the product analysis chart (Fig. 2.3), which plots the behavior of several marginal incomes generated by different products or divisions of the company rather than profit and loss.

This eliminates a number of distortions found when lumping diverse products into a sales total. Using the data from Table 2.4, four of Company B's product lines are plotted

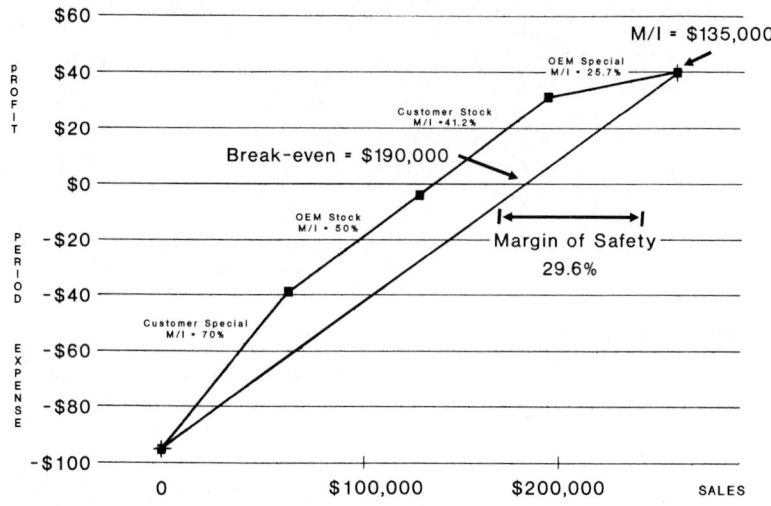

FIGURE 2.3 Product analysis chart, Company B.

in order of descending magnitude. The slope of each line represents the MI income ratio for each product. The product analysis chart can show management which products can bring the greatest profit per dollar of volume. Company B is engaged in four major activities. It is original equipment manufacturer (OEM) of devices held in stock until sold to other manufacturers who build them into finished products. It is OEM of specialty items made to order. It makes stock items sold directly to final customers. And it makes customer specialty products sold direct to final customers.

Table 2.4 is the table of values used in plotting the product analysis chart in Fig. 2.3 and the resulting MI ratios. Management would want to concentrate on those items which produce the greatest marginal income. However, they should be cautious not to make changes using marginal income alone. Instead, management's goal, and in fact duty to shareholders, is to seek maximum return on assets.

TABLE 2.4 Cumulative Product Marginal Income Analysis: Company B ($000)

Product	Sales	MI %	Cummulative MI
Customer special	$80,000,000	70.00%	$56,000,000
OEM stock	70,000,000	50.00%	91,000,000
Customer stock	85,000,000	41.20%	126,000,000
OEM special	35,000,000	25.70%	135,000,000

The product analysis chart can reveal some interesting opportunities to change the product mix and increase total profit. Both manufacturing and sales may be involved in the solution that finally emerges.

Manufacturing. By plotting each product on the product analysis chart, management can see the items bringing the greatest profit per dollar of volume. Armed with this knowledge, management can then consider changing the product mix to maximize profit. Other things being equal, it will want to concentrate on the items that generate the greatest marginal income even if this means keeping production of the remaining items at a minimum. In Company B's case, this strategy would mean diverting capacity from the OEM special order to OEM stock. Obviously, this would yield a larger dollar profit on the same sales volume.

At this point, however, a caution is necessary. Decisions should not be made on the basis of marginal income alone. As later discussion will demonstrate, management must always seek to maximize return on assets. If OEM special orders require comparatively little resources while OEM stock work calls for heavy use of plant and equipment, the higher profit associated with increased emphasis on OEM stock may actually represent a lower return on equity employed. The product analysis chart should suggest to Company B's planners that adjusting the resources required is worth looking into, but this should not be treated as a final answer.

Sales.   Knowing the marginal income that each product contributes, the sales manager can see where to concentrate his or her selling efforts and where an adjustment in prices will yield greater profit. The sales manager's skill in anticipating the way the markets will respond to price changes will then enable him or her to set up the pricing structure that yields the highest profits to sales ratio. The accounting department can help the sales manager by supplying information on the past relation of prices and volume for each product. But as any good executive knows, past experience can be misleading if it is blindly projected into the future. History is useful in decision making, but it is no substitute for accurate forecasting and a sound understanding of the company's markets.

SETTING THE CRUCIAL RATIOS

With the figures on costs, sales, and marginal incomes in hand, the planner can begin setting up profit targets. There are two yardsticks that can be used to do this. One is what has been earned in the past. The other is what could be earned with the most effective possible use of the resources available to management. These two figures establish a minimum and a maximum—a range of profit within which the company's projected performance should fall.

Perfection is rare, and a realistic final target is likely to fall short of the maximum. But it should be high enough to keep the entire company on its toes, leaving no room for complacency and self-satisfaction.

Realistic targets cannot be pulled out of a hat. They have to be established by careful analysis of projections for key items in the profit and loss statement and the balance sheet. These projections can be divided into any logical organization categories, such as:

- Business units
- Major divisions of the company
- Product groups
- Individual products or product lines

Whether a particular product should be singled out or lumped in with a group will depend on whether or not it makes a significant contribution to profit. It is a waste of time to go into lengthy analysis of operations that have no real impact on costs or profits. Too much detail clutters the final plan and impairs its ability to focus clearly on profit opportunities.

To develop a realistic final profit target, a planning team must go through a five-step process:

1. Develop preliminary projections of sales, variable costs, and marginal income for each planning unit; then estimate how much total profit will remain after period costs have been deducted.

2. Develop projections for the balance sheet items attributable to each unit.

3. Calculate the rate of return that management expects to make on each asset shown by the balance sheet.

4. Develop alternate profit targets for each unit designed to yield the required rate of return and cover direct costs in order to provide greater profitability for the company in total.

5. Compare the marginal income targets of the preliminary plan with marginal income driven by the rate of return method.

A careful study of differences between the two tentative targets for each unit will show where a change of prices, a shift in product mix, a reduction of costs, or change in product-related assets could increase the profits to equity capital employed ratio (P/ECE) and maximize total profitability of the company. The planners can use the model they have constructed to ask a variety of "what if" questions, assessing the impact of possible changes. In the end they can set up a final group of targets that will come close to the top of the profit range.

The Sales Forecast. Construction of the model begins with the preliminary plan, and construction of the preliminary plan begins with sales forecasts. The sales manager, by training, experience, and organizational responsibility is best qualified to prepare a sales forecast. The sales manager will, of course, be guided to a large extent by the expectations of the salespeople who work under him or her.

Most salespeople are enthusiastic by nature. This is one of the qualities that makes them valuable, but it means that usually their original estimates need a certain amount of deflation.

Take the case of the sales manager of a new and fast-growing company in the rubber products business, who submitted a volume budget for the coming year. Projected sales for one product line were at $15 million. Because the manager had only five salespeople, this would have meant an average of $3 million per salesperson. A careful inquiry revealed that except for the company's star performers, an excellent sales performance would be half that. Obviously, the sales manager's projection was unrealistic, but it suggested that the potential market was there. Instead of ditching the target entirely, the company moved it 1 year ahead and decided to hire five additional experienced salespeople. Then it prepared an interim plan on a more modest basis for the year just ahead.

There are several steps, involving very little labor, that will help make a sales target realistic. One is to take account of the growth trend of the company and its industry. Every industry has a rate of growth over a period of years. This growth curve can be found and projected. In many respects it is the same type of curve as the learning curves the industrial engineer constructs for hand labor.

Many industry curves fit the growth of population—for example, the confectionery industry. Others, such as electronics and various forms of packaging, have far outstripped the rate of population growth. Advancing technology, new applications, and widening acceptance of old applications give them added momentum.

Once a satisfactory growth curve for an industry has been established, it is easy to keep it up to date. But it is important to use good judgment and the best possible statistics in setting up the curve. More than one company has missed out on a potential expansion of its markets because it was working with statistics that lumped its products in with slow-growth items that pulled down the totals.

Usually, the trade association for an industry will be able to provide figures showing volume by years. Actual units are best, but dollar figures, corrected for price changes, will do if units are not available.

Another device for improving the accuracy of sales forecasting is to break total sales down into its components. This breakdown may be in terms of:

- Totals by salesperson
- Totals by product or product group

The industrial engineer can check the practicality of sales forecasts by plotting the projected sales against the available capacity. Table 2.5 shows the results for Company B. This analysis not only will show whether or not projected sales can be met with existing capacity; it will also give management a picture of what capacity it is using and how much will be idle.

TABLE 2.5   Capacity Analysis: Company B ($000)

Product	Planned Sales	Added sales at 100% capacity	MI %
Customer special	$80,000,000	$10,000,000	70.00%
OEM stock	70,000,000	4,500,000	50.00%
Customer stock	85,000,000	20,000,000	41.20%
OEM special	35,000,000	40,000,000	25.70%

Management, of course, still has the option of putting more effort into its sales program. If the profits yielded by the final sales volume target are inadequate, it may be possible to adjust the targets to achieve the profit goal. These adjustments may call for expansion of the company itself or changes in its basic structure as, for example:

- Additional capacity in certain departments
- Larger sales forces
- Alterations or additions to sales territories
- New products

A sales manager with a $2 million promotion budget can use Table 2.5 to work out a strategy. Heavy promotion of a customer special will be tempting because of the 70 percent marginal income, but the wide spread between planned MI and the ROA target MI developed later suggested that a price cut would be smarter. OEM stock with a 50 percent MI has only $4.5 million in unused capacity. The sales manager will scarcely get the promotion costs back before production hits the limit ($4.5 million × 50 percent = $2.25 million). This would give a gain of only $2,250,000 for all the sales manager's efforts. It will make better sense to concentrate on customer stock where there is $20 million potential capacity at 41.2 percent MI, provided there is room for market expansion.

When the sales forecast has at last been keyed into the profit target, it becomes the sales manager's target for the year. To allow for seasonal variations, it is usually best to set the target in terms of quarters. The sales manager will accept a quarterly breakdown; ordinarily he or she will balk at a monthly quota on the grounds that it is impractical. So long as the combined sales for the quarter hit the target, it does not really matter if individual months show deviations.

Manufacturing Costs.   Once the sales target is set, the planners can establish a target for manufacturing costs. Like the sales forecasts, the manufacturing cost targets should be stated briefly and simply. They should never be designed for use in the detailed variance analysis that effective cost control requires. The planning forecasts should stick to the most significant targets and the broad measures of performance. The purpose of planning is to give the company a basic outline of what it must do to maximize profitability. That purpose should never be buried in an avalanche of details.

In several major categories of cost, however, the planners will have to ask questions that the cost control system should answer. These include:

- The capacity utilization of equipment centers
- The price to be paid for production materials and supplies
- The target allowance for rework and scrap
- The rates to be paid for labor
- The allowance for maintenance

The standard cost system should pick up most of these items on a "should be" or target basis. The industrial engineer is best suited to address this array of attainable targets or standards.

The final manufacturing targets should be expressed in terms of volume and in terms of units, using the same categories as the sales forecasts. In addition, supplementary targets should be developed for projected material costs. These will be used by the purchasing department in preparing the schedule of materials procurement. Scrap and rework allowances should be approached with the idea of bettering the performance of the past. At this stage of the planning process, the variable costs of manufacturing will be summed up based on Fig. 2.4.

Cost	Expressed in	Source
Direct labor	Per unit of product; total by quarter, by year	Cost system standards
Material	Per unit of product; total by quarter, by year	Cost system standards for usage Purchasing forecast for cost
Indirect labor	Variable portion per machine or direct labor hour or other production measures	Graphic projection of standard crews at various volumes separating cost into fixed and variable portions
Expense	Same as indirect labor	Same as indirect labor

FIGURE 2.4 Typical variable cost elements.

A supplementary schedule can also be prepared to show the projected usage of each machine center, based on forecast volume. The back of this sheet can be used to present the sales department with a graphic demonstration of what utilizing the unused capacity could do for sales.

Before clearing the manufacturing cost forecasts, it is wise to make a simple check on the way the company's overall performance stacks up. This can be done by taking some unit of measure—pairs of shoes, yards of cloth, barrels of oil, tons of folding cartons—and dividing it into the total number of salaried and hourly workers involved in producing.

The resulting figure is a broad measure of productivity and a good common denominator for comparing a company's performance with that of competitors. This simple check may flag the planner that something is out of line in the operation and lead the planner to set up higher goals for the company.

The same principles that were used to separate variable costs from period costs for manufacturing can be applied to distribution costs. It is important for the planner to do a thorough job in this area. In some cases, distribution costs are more than the cost of manufacturing.

One of the major tasks of the planner is to determine the most effective distribution channels to use in delivering the company's products to the final consumers. As the soft drink company (Fig. 2.1) discovered from its analysis, different methods involve substantially different costs. As a result, the same product can have a wide range of marginal involves and returns on investment, depending on the distribution system used to move the merchandise to the consumers.

Service companies also have alternative methods of delivering their product. A communications company can serve its customers by written document, voice, telephone, radio, telex, fax, electronic mail, or in person. Each method has its own individual impact on revenue, costs, and marginal income.

With the profit targets and the variable costs clearly identified and the sales forecasts reviewed, the planners can then proceed to calculate the other important projections keyed to the profit and loss (operating) statement. These include:

- Preliminary marginal income targets derived by subtracting variable costs from sales projections for each business unit or product

- Estimated period costs that can be distributed to business units or product lines on the grounds that if production were discontinued entirely, there would be no such costs

- Estimated period costs that cannot be so distributed

- A profit target for each planning unit or product derived by subtracting attributed period costs from marginal income, and a profit target for the company as a whole derived by totaling these targets and subtracting undistributed period costs

Table 2.6 illustrates the brief P&L format needed to display these proposed targets in

TABLE 2.6   Preliminary Plan and P&L: Company B ($000)

	Income and Expense Category	OEM Stock	OEM Special	Customer Stock	Customer Special	Total Company
1	Net Planned Sales	$70,000	$35,000	$85,000	$80,000	$270,000
2	Less: Variable Labor & Expense	15,000	6,000	25,000	20,000	66,000
3	Variable Material	20,000	20,000	25,000	4,000	69,000
4	= Total Variable Cost	$35,000	$26,000	$50,000	$24,000	$135,000
5	= Mfg. Marginal Income	$35,000	$9,000	$35,000	$56,000	$135,000
6	Marginal Income (MI) Percent	50.00%	25.71%	41.18%	70.00%	50.00%
7	Less: Mfg. Period Costs – Distributed	6,000	2,000	13,000	9,000	30,000
8	Selling G&A – Distributed	9,000	3,000	12,000	11,000	35,000
9	Total Distributed Period Cost	$15,000	$5,000	$25,000	$20,000	$65,000
10	Net Margin before Common Costs	$20,000	$4,000	$10,000	$36,000	$70,000
11	Mfg. Period Costs – Common					10,000
12	Cost of Sales					$175,000
13	Common G&A					20,000
14	Total Common Period Cost					$30,000
15	= Operating Profit before Tax					$40,000
16	Operating Profit after Tax					$20,000
17	Average No. of Units	100,000	50,000	500,000	400,000	
18	Average Sales Price/Unit	$700.00	$700.00	$170.00	$200.00	

the most effective form for decision making and communication. Horizontally, this presentation can be developed for the major categories of the business, such as industrial, consumer, and other categories. Alternatively, the P&L can be set up in terms of the corporate divisions, reflecting the specific profit center organization of the company. This approach will be particularly useful to individual managers. When necessary, the diffuse P&L estimates can be further broken down by product groups or product lines.

Balance Sheet Projections.   In addition to the P&L targets, the planner must determine a group of targets based on the balance sheet. Table 2.7 shows the assets that Company B would require to meet the profit and loss statement targets of Table 2.6. To make such a forecast, the planners must work closely with the company treasurer and/or controller, who is primarily responsible for procuring the funds for the necessary investment.

TABLE 2.7   Preliminary Plan Balance Sheet by Business Segment: Company B ($000)

Asset	OEM Stock	OEM Special	Customer Stock	Customer Special	Replacement Value	Book Value
Accounts Receivable & Cash	$14,000	$5,000	$16,000	$15,000	$50,000	$50,000
Inventories						
Raw	2,500	1,000	4,500	2,000	10,000	
Work in Process (WIP)	12,000	3,000	7,000	3,000	25,000	
Finished Goods	7,000		8,000		15,000	
Total Inventories	$21,500	$4,000	$19,500	$5,000	$50,000	$50,000
Equipment & Buildings	45,000	4,000	51,000	60,000	160,000	90,000
Total Assests @ Replacement Value	$80,500	$13,000	$86,500	$80,000	$260,000	$190,000

The treasurer and controller can make an enormous contribution to efficient operation by arranging for the fastest turnover of assets. And the first requirement of effective planning is to recognize that it is not the ratio of profit to sales alone that counts; it is the percent of profit in combination with the turnover that makes a good profit performance. Although the company treasurer or controller performs many other services, watching inventories and receivables is one of the most important. Behind companies rising with spectacular growth curves is often an able treasurer and/or controller who has learned the value of placing inventory at the right point to service customers efficiency, who makes effective use of bank loans, and who keeps receivables at a low level.

Two of the balance sheet items that the planner must forecast are directly related to sales: accounts receivable and inventories. The average investment, or balance, in accounts receivable will also reflect such things as customer payment practice and customary practice in the particular line of business. Like the P&L items, accounts receivable can be classified by product line, product group, division, or business unit.

Inventories are usually divided into three natural categories: raw materials, work-in-process, and finished goods. Each will have a target return consistent with the level of investment risk. Company policy, lead times, and turnovers should be taken into consideration in determining what balance in each category will be consistent with projected sales.

The remaining balance sheet items are fixed assets—and for some companies, research and development, and proprietary investment—which are not directly related to sales. In forecasting for these items, the planner will have to begin by deciding on a method of valuation.

As Table 2.7 illustrates, valuation of the fixed assets—largely property, plant, and equipment—can be approached from two different directions, each of which is potentially useful. First, all fixed assets could be valued at replacement cost. In most cases this will be preferable because it leads to pricing decisions that reflect major economic changes. The established company will price its products at the same level as a new entrant seeking an adequate return on the higher investment it was forced to make.

The second approach is to use depreciated value of the original cost of the assets. The trouble with this is that it will not provide sufficient return for equipment replacement, modernization, or growth. In some cases, however, it may be more realistic than attempting to work with replacement costs. This may be particularly true of industries where technology has changed radically and new entrants use equipment so different from that of the older producers that costs cannot be compared. In this case, some companies use the replacement cost necessary to equal the best technology available to produce today's product. This is sometimes referred to as equivalent replacement cost. New production rates and labor crewing must also be reevaluated. This is a sophisticated but valuable approach in a high-technology industry.

Many companies do not carry a line for research and development on their balance sheets. Where the information is available, however, the planner can use it to recognize the short- and long-term R&D expenses applicable to each business unit. The return expected will then depend on the anticipated success ratio of the R&D projects. In most cases, the R&D expenses used in projecting costs for the preliminary plan will be entirely different from the investment figure produced by the balance sheet projection.

Proprietary investment presents problems of valuation similar to those encountered in R&D. The object of making this projection is to explore the possibilities of setting a target return on investment in proprietary products, which may or may not be patentable.

THE MODEL

Both parts of the *preliminary plan,* the R&L items and the balance sheet projections, are now complete, and the planner has a separate set of tentative sales and marginal income targets for each division of the company and for each product group that it sells. The next step is to put together a model, which will enable the planners to analyze the forecasts of the preliminary plan in terms of rates of return on assets as well as in terms of their contribution to total profits. To do this, the model compares the targets of the preliminary plan with targets developed by an alternate plan designed to yield a selected rate of return on each class of assets. The answer section of the model will be the final guide for construction of a revised plan aimed at maximizing the return of assets.

A computerized model allows instantaneous consideration of alternate price, volume, and capacity utilization structures to maximize the unit's profitability. The object of the exercise is to prepare a revised plan with the most profitable utilization of the company's capacity. Because profit center managers are involved in the final decisions, the revised plan will take into consideration actual market conditions, including price, demand, elasticity, and market share. At the same time, the model provides the profit center manager with extensive opportunities to see what might be achieved by altering various elements under his or her control. Such "what if" questions will lead to a realistic final decision that should put performance close to the top of the profit range.

To compute the alternate ROA-based targets, the planners first determine the rates of return they expect on each of the balance sheet items of the preliminary plan:

- Accounts receivable
- Inventories of raw materials
- Work-in-process
- Inventories of finished goods

- Property, plant, and equipment
- Research and development
- Proprietary investments

The expected rate of return shown in Table 2.8 will usually be different for each of these categories and could often be significantly different for two companies in the same line of business. For example, the expected rate on accounts receivable could vary from money market rates, to the prime lending rate of the banks, and on up to something much higher, depending on the credit rating of the customers involved. Target returns on property, plant, and equipment will be substantially higher because of the heavy long-term, low liquidity of the investment. Target returns on inventories will vary with the risk factors, such as shelf life and returns. The expected return on R&D and proprietary investments will depend on estimates of risk and assumptions about the useful life of the investment. Planners can often check their specified rates of return against the statistics, ratios, and return rates published by outside sources, such as trade associations, Robert Morris Associates, and Dun & Bradstreet.

TABLE 2.8   Alternate Plan by Business Segment: Company B ($000)

	Asset	OEM Stock	OEM Special	Customer Stock	Customer Special	Total Company	Selected ROA
1	Accounts Receivable & Cash	$1,680	$600	$1,920	$1,800	$6,000	12.00%
	Inventories						
2	Raw	300	120	540	240	1,200	12.00%
3	Work in Process (WIP)	1,920	480	1,120	480	4,000	16.00%
4	Finished Goods	1,820	0	2,080	0	3,900	26.00%
5	Total Inventories	$4,040	$600	$3,740	$720	$9,100	
6	Equipment & Buildings	13,500	1,200	15,300	18,000	48,000	30.00%
7	Total ROA Dollars	$19,220	$2,400	$20,960	$20,520	63,100	
	Calculate Additional ROA Required						Line Source
8	Total Distributed Period Cost	15,000	5,000	25,000	20,000	65,000	9 (Tbl. 2.6)
9	Add: Total ROA Dollars	19,220	2,400	20,960	20,520	63,100	7
10	Total ROA plus Period Costs	34,220	7,400	45,960	40,520	128,100	Total
11	Less: Marginal Income (MI)	35,000	9,000	35,000	56,000	135,000	5 (Tbl. 2.6)
12	Additional ROA Required	$780	$1,600	($10,960)	$15,480	$6,900	11 − 10
	Calculate Additional ROA Allocation						
13	Variable Labor & Expense	15,000	6,000	25,000	20,000	66,000	2 (Tbl. 2.6)
14	Variable Labor & Expense Percentage	22.73%	9.09%	37.88%	30.30%	100.00%	13/Total 13
15	Additional Return Required	1,568	627	2,614	2,091	6,900	14 x Total 12
16	Total ROA Required	$20,788	$3,027	$23,574	$22,611	$70,000	7 + 15

The next section of the model is a calculation designed to convert the rate of return targets into sales and profit targets comparable with those of the preliminary plan. The first step is to add the return on assets (ROA) assigned to each business unit or product to the distributed period costs allocated to it. The results are then added to give a total for the company. This total of ROA and distributed period costs is then compared with the marginal income that would be generated by the preliminary plan. The difference between the two represents the amount by which the targets based on ROA alone could fall short (as in the example, Table 2.10) of covering undistributed period costs and yielding the same profit as the preliminary plan.

The next step is to distribute this difference to the business units or products so that the final marginal income and profits of both plans will be the same. This distribution is necessary because marginal income, at the corporate level by definition, is the total of period costs plus profit. The purpose of the calculation, therefore, is to determine targets for each business unit's marginal income that will:

• Cover distributed period costs attributed to the unit.
• Fulfill the return on assets targets for each.
• Cover all common period costs.
• Yield the same profit objectives as the preliminary plan.

The distribution of the difference should be based on the criteria that best represent the value added to the product by each unit. One method is to distribute the difference in proportion to the variable labor costs and expenses. When the model is used at the levels of the corporate hierarchy which are not authorized to make capital investment, this method is probably the most appropriate. An alternative is to base the distribution, or part of it, on research and development or proprietary investment.

Once the distribution is completed, a final calculation gives the planners the answer section of the model. This shows the marginal income, sales, and variable costs for each unit as determined by the preliminary plan and the ROA targets. The total marginal income, sales, and profits of the company will be the same in both sets of figures, but the performance forecast for the individual business units or products will be significantly different.

Table 2.9 shows a simple version of the model built for Company B. The first section

TABLE 2.9 Target versus Actual Marginal Income Analysis by Product Group: Company B ($000)

		OEM Stock	OEM Special	Customer Stock	Customer Special	Total Company
P	Sales (Plan)	$70,000	$35,000	$85,000	$80,000	$270,000
&	Less: Variable Mfg. Cost of Sal	35,000	26,000	50,000	24,000	135,000
L	Distributed Period Cost	$15,000	$5,000	$25,000	$20,000	$65,000
	Common Period Cost					30,000
	Operating Profit/Tax					40,000
R	Accounts Receivable	$1,680	$600	$1,920	$1,800	$6,000
O	Inventories	4,040	600	3,740	720	9,100
A	Equipment & Buildings	13,500	1,200	15,300	18,000	48,000
	Selected Return	1,568	627	2,614	2,091	6,900
	Total ROA	$20,788	$3,027	$23,574	$22,611	$70,000
A	Distributed Period Cost	$15,000	$5,000	$25,000	$20,000	$65,000
N	Add: ROA	20,788	3,027	23,574	22,611	70,000
S	Total Dist. Period Cost and RO	$35,788	$8,027	$48,574	$42,611	$135,000
W	Less: Plan Marginal Income	35,000	9,000	35,000	56,000	135,000
E	Target vs Plan Difference	$788	($973)	$13,574	($13,389)	$0
R	Add: Plan Sales	70,000	35,000	85,000	80,000	270,000
	ROA Target Sales	$70,788	$34,027	$98,574	$66,611	$270,000
	ROA Target Marginal Income	50.56%	23.59%	49.28%	63.97%	50.00%
	Plan Marginal Income %	50.00%	25.71%	41.18%	70.00%	50.00%

consists of the forecasts of sales, variable costs, and marginal income for each business unit. The next section shows the target returns for each category of asset projected by the preliminary plan. This is followed by a calculation of the additional income required to bring the income estimated by the ROA method up to the levels of the preliminary plan. These additions to the ROA targets are distributed on the basis of value added by each planning unit. The answer section at the bottom of the page shows the two different approaches that would generate the same total profit for the company in distinctly different ways.

The model does not offer management an either/or choice between one approach and the other. It offers an opportunity to use comparative data for each business unit, product group, or product line to make choices that will provide the maximum possible benefit. For example, take the targets for two products shown by the model in Table 2.10.

TABLE 2.10 Target versus Actual Marginal Income Ratios by Product Group: Company B

	OEM Stock	OEM Special	Customer Special	Customer Stock
ROA Aprroach MI	50.56%	23.59%	49.28%	63.97%
Preliminary Plan MI	50.00%	25.71%	41.18%	70.00%
Marginal Income Differences	0.56%	-2.12%	8.10%	-6.03%

It is obvious that profits will be increased if management holds "customer special" at preliminary plan levels and attempts to increase the marginal income of "customer stock." But additional possibilities should be explored. Though "customer special" is a low market share operation, this may be because its products are overpriced. The figures show that there is room for a modest price cut, and if demand is price-elastic, this might stimulate a rise in volume that would increase total marginal income with a decrease in the marginal income ratio.

Similarly, with "customer stock," management should consider a simple increase in prices to raise marginal income per unit. However, if the market response will be to reduce unit volume so much that total sales, measured in dollars, are reduced, a price increase will not be the best answer. In any case, the manager of the business unit should explore such possibilities as changing costs, altering product specifications, or changing the assets devoted to each product line. The model will show all the components of costs and investments attributable to the manager's unit. In order to increase the return, the manager should ask whether any of the following could be reduced:

- Investment in the product line
  - Receivables
  - Inventory
  - Equipment
- Variable costs in the product line
  - Materials
  - Labor
  - Expense
  - Marketing or sales expense
- Distributional costs

The profit center manager can also see how product relates to other products or segments of the company. This may reveal opportunities to increase profits by altering the product mix. For example, where the same equipment is used in manufacturing two products, changing the priorities in equipment utilization may facilitate output of the product with the higher return.

Other measures can be added to the model to give the profit center manager a better understanding of the choices that are open. For example, when a company is operating at capacity, it is more realistic to think in terms of profits per hour than profits as a percent of sales. The necessary figures can be shown on the model as:

- ROA targets marginal income per hour
- Preliminary plan marginal income per hour

It can also be useful to compare the two approaches with profit planning in terms of marginal income per unit. This calculation requires measuring the performance of the various profit centers on the basis of a common unit or equivalent unit of output. In using all these measures, the important thing for the profit center manager to remember is that if the margin income set by the ROA method can be equaled or exceeded, total company profits and return on investment will be increased.

The model, however, should be used for realistic planning, not for dreaming. All the figures in it and all the conclusions drawn from it should be consistent with the company's resources and capabilities. The check of sales forecasts against capacity (Table 2.3), therefore, should now be expanded into a careful analysis of the load each target will put on facilities. Table 2.11 shows what such a comparison would reveal about the practicality of a proposed 20 percent expansion in the sales of OEM stock items.

TABLE 2.11   OEM Stock—Departmental Capacity Analysis: Company B ($000)

Production Activities	Total Hours Capacity	Planned Hours	Excess or (Deficit)
Casting	10,000	6,000	4,000
Machining	20,000	18,000	2,000
Assembly	12,000	13,000	(1,000)
Testing	9,000	8,000	1,000
Engineering	3,000	4,500	(1,500)
Total Hours	54,000	49,500	4,500

While total capacity utilization under this plan would be 91.7 percent (49,500 planned hours divided by 54,000 total hours), there obviously is not enough engineering time or assembly hours to support the proposed sales level.

Every organization has capacity limitations. Few businesses are an "on/off" operation. There are usually one or two elements in the manufacturing process that act as constraints, just as the engineering assembly operations did in Company B. In a process industry, the limit might come from a step in the process flow or possibly a material.

Intangible capacity considerations, such as the number of people available and their capabilities, frequently dictate the limits. How much the company can do will depend to some extent at least on what sort of staff it has for R&D, field service, and engineering. In some organizations, the limits are set by financial factors. For a bank, deposits and capital determine capacity. In other cases, cash flow sets the ceiling.

A comparison of all the model's projections with the capacity limits not only will avoid nasty surprises in the future; more important, it will show management where new investment or a change in the product mix could open up opportunities that were never before identified.

THE POWER OF THE MODEL

The model can be used at all marketing levels of the organization (corporate, business unit, product line). Goal setting can be done from the top down or from the bottom up—from corporate to product or from product to corporate. The president may look at a total business unit and ask, "How can this unit be made more profitable?" Or the product manager may ask, "What can be done to make this product more profitable?" In either case, the model is designed to deal with the chain of more specific questions that each of the broader questions generate. The answers, showing the effect on division profit, can then be transmitted to different levels of the organizational hierarchy.

Other typical questions that the model will answer include:

• How can I use my capacity more profitably?
• What effect will my mix of sales (by either division, plant, product line, or product) have on my profits?

The model is designed to quantify the considerations involved in this kind of evaluation. It gives the various profit centers a common set of measures for communication.

The first consideration in addressing such questions is price and market share. If product sales managers can tell whether a price increase or decrease will be accompanied by an increase, decrease, or no change in volume, the model can evaluate various strategies in terms of marginal income, capacity utilization, and the final effect on profits.

The "What If" Matrix.   In addition to listing all "what if" questions, the profit planners may construct a matrix with three major constraints and six major questions. Figure 2.5 shows a simple matrix. In a more complicated version with three constraints and six questions, there will be boxes on the grid where significant answers appear to proposed actions. (In three boxes one of the constraints rules out the action that one of the questions proposes, and no answer will be possible.) Of course, the model can also assimilate all combinations of "what if" questions under each constraint, and it can work on the assumption that profit can vary without constraint.

The "what if" matrix is one of the analytical tools that the model makes available. Its purpose is to take the blind gambles out of planning and base each decision on scientific analysis and full information about the company's operations. It can and should be used to develop a final plan for the period ahead that will bring company profits closer to their potential maximum than either the preliminary plan or the ROA approach alone.

PROFIT PLANNING IN ACTION

On the basis of the model it has constructed (Table 2.8) and various "what if" studies, Company B decides to make some important changes in the targets projected by the preliminary plan (Fig. 2.6).

• In customer stock items, the management decides to increase material specifications, even though this means a 10 percent increase in cost. This will lift the variable material costs from $25 million to $27.5 million. In addition, the company plans to add value to the product by tightening up on inspection and improving the quality of workmanship. This will raise labor cost 10 percent, from $25 million to $27.5 million. The company forecasts that improved quality will justify an increase of 8 percent in selling price. As a result, planned sales go up from $85 million to $91.8 million (8 percent increase).
• In customer special items Company B determines that it will cut unit sales prices by 15 percent. As the model shows, this will bring marginal income more in line with the mar-

PRELIMINARY PLAN	"WHAT IF" CABABILITY	REVISED PLAN
P & L	1. What if variable manufacturing costs increase or decrease?	P & L
	2. What if variable marketing costs increase or decrease?	
	3. What if distributed period costs increase or decrease?	
	4. What if common period costs increase or decrease	
Balance Sheet	5. What if asset balances are increased or decreased?	Balance Sheet
	6. What if the target return on assets employed are increased or decreased?	
	SELECTING THREE DIFFERENT CONSTRAINTS	
ROA Selection	1. Holding unit sales price and profit constant – but, unit volume varies.	Revised
MI% _____	2. Holding unit volume and profit constant – but, unit sales price varies.	MI% _____
ROA _____	3. Holding unit sales price and unit volume constant – but, profit varies.	ROA _____
P/ECE _____		P/ECE _____
P/S _____	RESULTING IN _____	P/S _____

FIGURE 2.5   Model plan elements—''what if'' analysis.

ginal income ratio derived by the ROA approach. Unit sales price will come down from $200 to $170 per unit, and as a result, Company B foresees an increase of 30 percent in volume. This will bring sales volume to $88.4 million. Labor and expense costs will go up from $4 million to $5.2 million, giving a total variable cost of $31.2 million.

The net result of these changes is to drop the marginal income ratio on customer stock from 41.3 to 40.1 percent and to lower the marginal income ratio for customer special from 70 to 64.7 percent. However, total marginal income of the company has increased from $135 million to $138 million, and profit before taxes is now $43 million instead of $40 million. In both cases, the marginal income ratio is lower, but volume has increased and assets are being utilized more efficiently. As a result, profit is higher. Table 2.12 shows how these moves have changed the targets and predicted profits.

The next step is to look at the balance sheet to see what changes can be made there and what changes the new sales forecasts will involve. Table 2.12 shows the results of this analysis. Management has decided to reduce the accounts receivable for customer stock items from $16 million to zero. In effect, it is making customer stock items a cash and carry business. At the same time, Company B proposes to increase the inventory turnover on customer stock items by 50 percent. This could be expected to reduce raw material inventories by $2.25 million, work-in-process by $3.5 million, and finished goods by $4 million. However, the planned upgrading of raw materials (resulting in a 10 percent cost

Product	Revised Plan Action	P&L Result	Balance Sheet Result
Customer Stock	Change product specifications	Increase variable cost 10% (From $25 to $27.5 million)	Increase inventory value 10% (From $19.5 to $21.45 million)
	Improve quality	Increase variable labor cost 10% (From $25 to $27.5 million)	Increase selling price by 10% (From $85 to $91.8 million)
	Increase inventory turnover	No change	Decrease inventory value 50% (From $21.45 to 10.725 million)
	Change price structure	Increase sales 8% (From $85 to $91.8 million)	Increase accounts receivable 8% (From $15 to $16.2 million)
	Change to cash and carry	No change	Eliminate accounts receivable (From $16.2 to $0 million)
Customer Special	Lower ROA to control competition	Decrease sales price by 15% (From $200 to $170 per unit)	Decrease accounts receivable value 15% (From $15 to $12.75 million)
	Increase market share	Increase sales 30% (From $80 to $88.4 million)	Increase accounts receivable 30% (From $12.75 to $16.575 million)
		Variable Labor and Material increase by 30% (From $24 to 31.2 million)	Increase inventory by 30% (From $5 to $6.5 million)

FIGURE 2.6   Revised plan—Company B—actions and assumptions by product.

increase) and workmanship in this area partially offset the reduction in quantities, and so the final inventory item on the balance sheet for customer stock will be $10.725 million.

In customer specialty items, Company B expects to increase accounts receivable by the same percentage as the expected increase in sales. The increase of $1.575 million brings total accounts receivable to $16.575 million. At the same time, it is necessary to increase total inventory value to reflect the higher total variable costs associated with larger volume. Applying the variable cost increase of 30 percent to raw material inventories of $3 million and work-in-process of $2 million brings total inventories for customer specialty business to $6.5 million. Table 2.13 shows the balance sheet entries associated with the revised plan. The changes have resulted in a decline in the replacement value of assets from $260 million to $238.3 million. If management chooses to use book value to estimate its investment, the decline will be the same dollar amount, bringing the total down from $190 million to $168.3 million. Either way, the amount of capital on which the company must earn a return is reduced by $21.7 million.

Putting it all together, Company B arrives at its revised plan. Table 2.14 shows how the new figures compare with the targets set by ROA analysis. No significant change has been

TABLE 2.12 Revised Plan and P&L Based on Assumptions ("What If Capability"): Company B ($000)

Income and Expense Category	OEM Stock	OEM Special	Customer Stock	Customer Special	Total Company
Net Planned Sales	$70,000	$35,000	$91,800	$88,400	$285,200
Less: Variable Labor & Expense	15,000	6,000	27,500	26,000	74,500
Variable Material	20,000	20,000	27,500	5,200	72,700
= Total Variable Cost	$35,000	$26,000	$55,000	$31,200	$147,200
= Mfg. Marginal Income	$35,000	$9,000	$36,800	$57,200	$138,000
Marginal Income (MI) Percent	50.00%	25.71%	40.09%	64.71%	48.39%
Less: Mfg. Period Costs – Distribut	6,000	2,000	13,000	9,000	30,000
Selling G&A – Distributed	9,000	3,000	12,000	11,000	35,000
Total Distributed Period Cost	$15,000	$5,000	$25,000	$20,000	$65,000
Net Margin before Common Costs	$20,000	$4,000	$11,800	$37,200	$73,000

Mfg. Period Costs – Common	10,000
Cost of Sales	$187,200
Common G&A	20,000
Total Common Period Cost	$30,000
= Operating Profit before Tax	$43,000
Operating Profit after Tax	$21,500

	OEM Stock	OEM Special	Customer Stock	Customer Special
Average No. of Units	100,000	50,000	500,000	520,000
Average Sales Price/Unit	$700.00	$700.00	$183.60	$170.00

TABLE 2.13 Revised Plan Balance Sheet by Business Segment: Company B ($000)

Asset	OEM Stock	OEM Special	Customer Stock	Customer Special	Replacement Value	Book Value
Accounts Receivable & Cash	$14,000	$5,000	$0	$16,575	$35,575	$35,575
Inventories						
Raw	2,500	1,000	2,475	2,600	8,575	
Work in Process (WIP)	12,000	3,000	3,850	3,900	22,750	
Finished Goods	7,000		4,400		11,400	
Total Inventories	$21,500	$4,000	$10,725	$6,500	$42,725	$42,725
Equipment & Buildings	45,000	4,000	51,000	60,000	160,000	90,000
Total Assests @ Replacement Value	$80,500	$13,000	$61,725	$83,075	$238,300	$168,300

made in OEM stock items or OEM specialty items because the original projections of marginal income were close enough to the ROA targets to indicate that the proposed prices would provide satisfactory yields.

The ROA target return for customer stock items has to be changed because of the decision to reduce the assets required for this part of the business. The target marginal income therefore drops from 49.2 percent of sales to 46.2 percent. At the same time, marginal income as projected by the revised plan drops from 41.2 to 40.1 percent. The gap between the ROA target and the planned performance remains, but it has narrowed by one quarter—from 8 to 6 percent.

In customer specialty items, the revised plan provides for a price cut that will lower marginal income from 70 to 64.7 percent. This is 4.4 percentage points ahead of the revised ROA target, but again the gap has narrowed. It was a full 6 points when the preliminary plan was compared with the first ROA target.

For the company as a whole, the revised plan drops marginal income from 50 to 48.4

TABLE 2.14 Revised Plan Move—Target Selling Prices and Target Marginal Income: Company B ($000)

		OEM Stock	OEM Special	Customer Stock	Customer Special	Total Company
P	Sales (Plan)	$70,000	$35,000	$91,800	$88,400	$285,200
	Less: Variable Mfg. Cost of Sales	35,000	26,000	55,000	31,200	147,200
&	Distributed Period Cost	$15,000	$5,000	$25,000	$20,000	$65,000
L	Common Period Cost					30,000
	Operating Profit/Tax					43,000
	Accounts Receivable	$1,680	$600	$0	$1,989	$4,269
R	Inventories	4,040	600	2,057	936	7,633
O	Equipment & Buildings	13,500	1,200	15,300	18,000	48,000
A	Selected Return	2,637	1,055	4,835	4,571	13,098
	Total ROA	$21,857	$3,455	$22,192	$25,496	$73,000
	Distributed Period Cost	$15,000	$5,000	$25,000	$20,000	$65,000
	Add: ROA	21,857	3,455	22,192	25,496	73,000
A	Total Dist. Period Cost and ROA	$36,857	$8,455	$47,192	$45,496	$138,000
N	Less: Plan Marginal Income	35,000	9,000	36,800	57,200	138,000
S	Target vs Plan Difference	$1,857	($545)	$10,392	($11,704)	$0
W	Add: Plan Sales	70,000	35,000	91,800	88,400	285,200
E	ROA Target Sales	$71,857	$34,455	$102,192	$76,696	$285,200
R						
	ROA Target Marginal Income %	51.29%	24.54%	46.18%	59.32%	48.39%
	Plan Marginal Income %	50.00%	25.71%	40.09%	64.71%	48.39%

percent. But since sales increase by $15 million, profit before taxes rises from $40 million to $43 million.

Table 2.15 shows how the revised plan compares with the preliminary plan. The return on assets increases from 21 to 25.5 percent, which brings it very close to the 27 percent that Company A reported in Table 2.1. The after-tax profit to equity capital employed ratio rises from 20 to 27.5 percent, which puts it ahead of the 23 percent ratio of Company A. This reflects the fact that Company B has maintained the same level of borrowing in relation to a lower equity capital employed value and has increased income at the same time. The margin of safety for Company B—the percentage by which operations will exceed the break-even point—rises from 29.6 to 31.8 percent. The break-even point itself is up from sales of $190 million to $196.28 million. Under the revised plan, Company B is leaner, more efficient, and more profitable than it would have been if planning had stopped with the preliminary plan.

Use of the model as a planning tool applies at all levels of a company. As Table 2.16 shows, the targets can be set not only by business segments or divisions but also by product groups and then by each individual product. Information flows up and down the corporate hierarchy, with everyone looking at the same kind of model and with each level making the decisions that are within the scope of its authority.

The flow of formatted information throughout the company is a vital part of planning. The decisions that make a final revised plan out of the projections of the preliminary plan cannot be pulled out of the air. They can be made only after people at all levels of the company have worked with the model, tried out a great variety of scenarios, and check the validity of all assumptions against what is known about the market and the cost structures of various products.

TABLE 2.15 Comparison Preliminary and Revised Plan: Company B ($000)

	Plan	Revised	Change	

BALANCE SHEET

	Plan	Revised	Change	
Assets	$190,000	$168,300	($21,700)	Decreased
Less: Liabilities	90,000	90,000	0	No Change
Equity Capital Employed	100,000	78,300	(21,700)	Decreased

INCOME STATEMENT

	Plan	Revised	Change	
Net Plan Sales	$270,000	$285,200	$15,200	Increased
Less: Variable Mfg. Cost of Sales	135,000	147,200	12,200	Increased
Total Marginal Income	135,000	138,000	3,000	Increased
Add: Period Mfg. & Variable Mfg.	40,000	40,000	0	No Change
= Cost of Sales	175,000	187,200	12,200	Increased
Add: Period Selling & G&A	55,000	55,000	0	No Change
= Net Operating Cost	230,000	242,200	12,200	Increased
Operating Profit before Tax	40,000	43,000	3,000	Increased
Profit After Tax	20,000	21,500	1,500	Increased

TABLE 2.16 Hierarchy of Model Applications

Level 1

CORPORATE BY BUSINESS SEGMENT OR DIVISION					
	Segment	Segment	Segment	Segment	Total
Marginal Income Distributed Cost P & L R O A					

Overall Planning and Control

Level 2

BUSINESS SEGMENT OR DIVISION BY PRODUCT GROUP		
	Product Group	Total
Marginal Income Distributed Cost P & L R O A		

P & L Responsibility
Marketing Strategy
Comparison to
  Industry Ratios
Capital Investment
Planning

Level 3

PRODUCT GROUP BY PRODUCT		
	Product Group	Total
Marginal Income Distributed Cost P & L R O A		

Capacity Defined
Product Pricing Target

The power of the model does not lie in comparison of projected sales and marginal incomes with targets that will yield a satisfactory ROA. Anyone can say, "I propose to sell my products at a price that will give me an equal return on each of the investments I made and yield a satisfactory ratio of profit to equity capital employed." It takes able people to examine the figures and determine what products are out of line and what can be done about it. The model is simply a device for bringing the experience, knowledge, and intelligence of people at every level into focus.

Selling the Idea of Planning. The industrial engineer or other expert who wants management to take this systematic approach to provide planning will often encounter these objections:

- It takes too much executive time.
- Our business is different; we cannot forecast sales.
- Our company is too small to afford this sort of planning.
- We are making good profits now, and we don't need to plan changes.

These statements may be true, but they tell more about weaknesses in the company's way of doing business than about its need for planning. Top executives and supervisors should be asked to make only key decisions. Detailed schedules prepared by the engineers and accountants should give them all the information they need. A properly designed and administered planning system will make far better use of executive time than a catch-as-catch-can system of decision making.

The company that cannot accurately forecast sales—a manufacturer of high-style products, for instance—can at least have a forecast as good as the toughest competitor's. And the executive needs, just as much as in any other company, to make the best possible use of resources.

Nor is planning a tool for the medium- and large-sized firm alone. Small companies need good planning and associated cost accounting as much as big companies. In fact, knowing their cost structure may make the difference between life and death. And with adequate cost identification, planning will not add materially to expenses.

Finally, the fact that a company is doing well does not mean that it could not be doing better or that it will always do well without any changes or review of its performance. This argument is likely to be heard from companies in a new, fast-growing industry where there is a temporary shortage of capacity. But yesterday's new industries are likely to be the scene of today's bloodletting. The electronics business, for instance, has been immensely profitable for some companies and fatal to others. The booming period is likely to be followed by hard times when strict control of costs and a clear view of the "what if" possibilities will be the key to survival.

Building the model and making the decisions it is designed to assist will call for hard thinking and a thorough study of a company's operations. But there is no company—large or small, prosperous or hard-pressed—that cannot benefit by the information and understanding that the model gives its management.

MAKING THE PROFIT PLAN WORK

Once a plan has been established, two groups of people must accept the responsibility for making it work. The administrator's group includes accounting, industrial engineering, and various planning groups. The operations' group includes marketing, distribution, manufacturing, and staff support.

The administrators should make sure that there will be:

- An early warning system
- Regular review of the data with operations people as the information becomes available
- A follow-up system to ensure corrective action when performance does not meet the targets of the plan

Reporting should be organized so that there is a flow of information to the top with diminishing detail at each level and clear emphasis on significant variances from the plan.

The plan will work only if people are willing to act on the information they receive. The signals must be clear, understandable, and timely. Follow-up should be automatic to ensure that remedial action is taken. More than anything else, however, the effectiveness of the profit plan will depend on the importance that the chief executive officer attaches to it.

Many techniques can be used to muster support throughout the organization:

- Publicizing the plan
- Rewarding outstanding performance
- Incorporating incentive systems when they are appropriate
- Generating participation by using peer pressure
- Using graphic aids and posters

In the end, the profit plan will be a success only if it becomes second nature to the entire organization and is continually highlighted by top management.

The Importance of Controls.   The best early warning system is a control structure that provides targets—keyed to the profit plan—for every performance level from the factory floor to the top executive office. Control systems should be developed for use and action. Everyone in an organization does not need complete information about all its activities. Each needs the information that will enable them to manage their own activities in a timely, efficient way.

A good control system will prescribe different approaches for the various components of the profit and loss statement. The control of production labor costs, for instance, may require daily or even shift control reports. Period expenses need not be reported more than monthly.

Sales controls will differ, depending on the product or services sold. Long lead time products, such as heavy machinery, do not require selling and market information as swiftly as high-fashion products, where daily reports are often the rule.

A well-designed control system should also cover distribution, inventory, and customer service. These areas, often neglected or completely forgotten, can spoil the final results of heads-up discipline in manufacturing and selling.

Assigning Responsibility.   The best way—in fact, the only way—to ensure continuity of action in making a plan work is to vest in a single individual the full responsibility for meeting a standard and correcting any deviations from it. Every level of supervision should be charged with maintaining the standards that apply to its operations. This policy will hold good for both line and functional organizations. The maintenance of planned standards should be delegated just as duties are delegated. And just as the authority to carry out duties and meet responsibilities is delegated, authority to enforce the standards should also be delegated. In this manner, deviations will be discovered and dealt with by the people closest to the problem and by those with a direct interest in keeping a spotless record.

Some companies use incentives to keep their supervisory personnel interested in meeting or exceeding standards. Badly designed incentives can be counterproductive; any company considering them should make sure that the following conditions are met:

1. Standards should be set so that they hold supervisors to the performance expected of them. The incentives should be used to reward supervisors who exceed the standards. They should not be used to pay for adequate but undistinguished performance.

2. The incentives should apply only to standards that are under the supervisor's full control.

3. The incentives and the conditions under which they are awarded should be carefully tailored to fit the conditions involved.

4. All the factors of supervisor controls should be given equal weight in determining the incentives.

Probably the surest method of keeping the supervisors interested in meeting standards is to encourage them to participate in setting the standards in the first place. When this method is used, it leaves little room for excuses or rationalization to explain failures. Most people in a supervisory position take pride in achievement. If they think the standards are fair, they will have a powerful psychological incentive to meet or exceed them. This applies not only to product but to sales as well.

While a keen feeling of personal involvement is essential to effective planning and control, deviations can be best analyzed and resolved when they are divorced from personalities. Supervisors must be trained to look at deviations as objectively as possible so that they can discover the real reason why a failure has occurred.

For example, a deviation might have occurred because a machine broke down. The supervisor could blame this on mechanical failure, which might be absolutely true. But if the supervisor stepped back and looked at the situation objectively, he or she might consider the fact that one duty was to check the machine periodically to make sure that a proper preventive maintenance program was being carried out. Identifying a failure to check the cause of breakdown will not recapture lost production time, but it may well prevent similar trouble in the future.

Highlighting the Variances.   Successful control depends on the comparison of actual results with planned results. The reporting of the results should be kept as simple and direct as possible. Actual and standard figures with the variance between them are the points to emphasize. Additional data may be interesting, but it is the variances that demand prompt action.

Table 2.17 shows a reporting format that matches the plan with the results of a control system utilizing standard costs. Vertically, it shows a profit and loss statement. Horizontally, the first column shows the P&L headings. The next column shows the initials of the person responsible for the account. The third shows the actual results—standard costs plus variances for variable costs and actual amounts for period expenses. The next matches the plan against the actual results, and the next after that shows the variance for each P&L heading. In essence, the variances measure the extent to which actual performance departed from the plan during the period covered by the reporting.

Analyzing the variance shown by this format is a major function of the industrial engineer. As a practical matter, the industrial engineer should concentrate on those that involve large dollar amounts or large percentage departures from target. It is important to begin with those that most affect the profit and be careful not to dilute the effort by attempting to analyze each and every variance.

Variance Administration.   A properly designed variance administration program is vital to making the profit plan work. It must be followed up by the variance meeting—the periodic meeting of the people charged with the immediate responsibility for meeting the standards that the profit plan sets. Variance meetings on the operational level, properly summarized and reported to the executives at higher levels of the organization, will go a long way toward avoiding surprises at the monthly reviews of the profit plan. Variance meetings should be short and to the point. They should concentrate on significant variances and not waste time nit-picking. But they should seek the answers to three basic questions:

- What caused the variance in this period?
- What are we going to do about it?
- What happened to the variances incurred in the last period?

All variances that show in the report should be studied. But not all variances reflect conditions that the company can do something about. The first thing the variance meeting must do is determine whether the cause of a deviation lies within the control of the business or outside it.

Among the variances that can be controlled from within, those that have the greatest

TABLE 2.17   Operating Statement ($000)

								Period _____	
DESCRIPTION	Respon-sibility	ACTUAL		PLAN		VARIANCE	PRIOR YEAR		
		Amount	% Sales	Amount	% Sales		Amount	% Sales	
<u>Sales</u>									
Discounts									
NET SALES									
<u>Standard Direct Cost of Sales</u>									
Raw Materials									
Labor & Expense									
Freight Applied									
TOTAL DIRECT COST									
STANDARD MFG. MI									
<u>Variable Mktg. Expense</u>									
Freight Out									
Other Variable Marketing Exp.									
Total Variable Marketing Exp.									
TOTAL VARIABLE COST									
MARKETING MI									
<u>Variances</u>									
Manufacturing Labor & Expense									
Materials									
VARIABLE CONTRIBUTION									
<u>Period Expenses</u>									
Manufacturing									
Materials Management									
Research									
Marketing									
Corporate Administration									
TOTAL PERIOD EXPENSES									
OPERATING INCOME (LOSS)									
<u>Other</u>									
Interest									
Billings to (From CIC)									
Misc. Income (Deductions)									
Pretax Profit (loss) Before									
Loss from Disc. Operations									
Loss from Disc. Operations									
Pretax Profit (Loss)									
Retained Earnings (Deficit)									
– Begining of Period									
– End of Period									

McCormick & Company * Summit * NJ

effect on profit should be considered first. An item such as material might deviate from the standard by a comparatively small percentage amount, but in relation to the profit objective, the effect might be very large. Conversely, the deviation from the standard on a maintenance item might be very large in relation to that same standard but very small in its impact on profits.

This rule of first things first means that variations in direct labor should almost always receive immediate attention. Some programs call for checking actual labor cost versus standard labor cost a number of times within the regular reporting period, so that deviations in this important item can be detected before they become too great.

The Glass House Approach.   An effective mechanism for the control of labor cost is found in the "glass house" approach. This approach recognizes that labor supply and labor demand within any given department or plant will not always match ideally. Because it is impractical to lay off people in many areas and on a daily or an hourly basis, a medium is provided to permit a supervisor to reduce the actual payroll cost by transfer of personnel. A control system cannot perform its functions if explanations of variances include statements such as "I could not meet my standard because I had to hold onto my people."

An additional line on the labor control report entitled "Management Policy Account" is the place for labor costs to be transferred whenever the supervisors cannot find productive work for their people in their own or any other department. The management policy account is the labor pool and should be the responsibility of the plant manager, who accepts the supervisor's transfers into the account through a written or verbal contract.

In this manner, the supervisors identify all excess labor and have no excuse for not attaining their individual efficiency standards. The supervisors also voluntarily identify for the plant manager the total excess people that are present in any given period of time within the plant. The plant manager now has several choices. One is to assign these people to varied "make work" projects such as painting, cleaning, or building maintenance, but still charged to the manager as a cost. Another is to make a permanent reduction in the payroll when sufficient management policy variance experience has been accumulated.

This variance is often called the "glass house," because it throws the spotlight on unutilized labor. There is a large plant in England where the glass house physically exists. Supervisors actually send unneeded people to a completely glass-enclosed room, centrally located in the plant. While there, the employee is permitted to rest while awaiting further assignment. It is the plant manager's responsibility to keep the glass house empty.

Sales Variance.   The first things first approach also means that prompt action should be taken on any variation from the most important standard of all: the sales quota for the period. If sales fail to reach standard, the rest of the standards may be achieved but the profit target will be missed. The existence of a sales volume variance of any size should be a matter of concern for all top management but especially the sales executive. Often, extra effort on the part of the sales force will stave off a threatened slump in income and profits.

A carton plant once came up with a serious sales variance. Under the former procedure, it would have been chalked off as a bad month. Under a well-constructed control system, however, the variance and its effect on overall objectives were immediately apparent. Top management applied heavy pressure on the sales division to increase sales for the next period so that quarterly standards could be met. This extra prod caused sales to contact a customer they had never been able to sell. Through the extra effort, they were able to land a substantial order, although ordinarily they would not even have called on this customer.

Beating Costs into Line.   If sales are meeting their quotas, the problem of achieving the profit targets resolves itself into a problem of keeping costs in line. Every company has certain costs, the control of which is the key to profits. These should be singled out early in the program and close watch should be kept on them.

It is in this area of cost control that the industrial engineer can make one of his or her greatest contributions. The industrial engineer is in touch with almost all phases of plant operations. An overall viewpoint, training, and experience help the industrial engineer determine the true reasons for cost variances and their consequences.

Often a particular variance will have several causes, and often the same variance will have multiple effects. Such cases require careful investigations and interpretation by someone whose knowledge is not confined to a single department. The industrial engineer is particularly well qualified to handle these cases.

In the course of a variance meeting, the industrial engineer who knows the company may find that he or she can make a wide variety of contributions:

- Interpretations of variances and the weight that each carried
- Suggestions to supervisors for remedying variances
- Analysis of major cost items and their components
- Suggestions for revisions of standards

In addition, the industrial engineer will probably conduct studies and investigations on particular problems—ranging all the way from personnel practices to plant utilities. For example, here are some of the situations that often need investigation to keep costs in line with the targets:

1. Deviations are often attributable to the human element. Investigations and studies of employee attitude, morale, and labor relations in general can be highly revealing. Absenteeism, lateness, grievances, and other personnel problems can influence labor standards greatly.

2. Standards, incentives, methods, and training procedures need periodic rechecking. The influence of time and new personnel can introduce subtle changes that go undetected but still influence established standards to a significant extent.

3. Methods employed in planning, scheduling, and controlling production need frequent restudy because they determine the utilization of people, materials, machines, plan, and time. A look at inventory and material controls is also important. It may reveal malfunctions affecting several different phases of operation. This same procedure should be applied to concerns that provide intangible products or services. Many times the basic operation involves processing information (or data) rather than material. Although the terms may change, the same examination is required.

4. Maintenance variations are becoming increasingly important as plants become more mechanized and automatically controlled. Negative and positive deviations from the standard have to be watched with equal alertness. In the long run, undermaintenance of facilities can be disastrous and overmaintenance very expensive. Deviations and maintenance requirements should be investigated by the industrial engineer.

5. In the average plant, a high percentage of direct costs is incurred by material handling operations. Handling does not enhance the value of an item. Deviations arising in this category can be curbed by studies of the layout, routing, and handling methods employed.

6. The industrial engineer can help the supervisor by ensuring that cost controls are being maintained and are doing the job they were set up to do.

7. Waste, scrap, and rework are key items. When they are traced to their sources and the reasons for their occurrence are determined, many other variations may be completely or partially explained. Industrial engineering will cooperate closely with quality control on such cases.

8. The industrial engineer can perform a great service to the supervisors and the plan as a whole by installing or maintaining a well-thought-out suggestion system. No one can see all the ways a job can be done. The person on the job should be encouraged to think about improvements and submit ideas for consideration. A well-run suggestion system can be a continuing source of new ideas.

Deviations Requiring New Equipment.   By making a complete survey of the physical and economic aspects of the problem, the industrial engineer can be especially helpful in the handling of deviations requiring new equipment. Improved methods through the installation of new tools and fixtures, improved equipment, and changes in material handling systems are all part of industrial engineering activities.

Take the case of a printing plant that had a large material handling variance. Investigation by the industrial engineer revealed that the addition of a forklift truck to the warehouse facilities would eliminate this variance. However, the truck would cost $20,000, and to justify such a capital expenditure, an increase of approximately $25,000 in gross profits would have to be achieved. But industrial engineering probed a little deeper into the problem and decided to recommend the rental of a forklift truck. The rental for each period was greater than the cost would have been if the company had purchased the truck, but the necessity of a large capital expenditure was avoided and immediate savings over the additional costs were realized.

One thing to remember in this connection is that equipment manufacturers are a valuable source of advice on the reduction of costs. Their representatives are usually available without cost and have a large fund of knowledge about similar situations. The advertisements and articles in pertinent trade publications are another source of real help.

Deviations in Materials and Supplies Cost. Frequently, in appraising the suitability of profit planning to its operations, management says, "The prices of our purchases jump around so much that any forecast would quickly lose its value." This objection can be met by providing for adjustment in the active planning period—from 1 year to 6 months and even down to 3 months if necessary. This will almost always do the trick. Few commodities have varied as much as wool, cocoa beans, and waste paper. Yet the erratic behavior of all three has been harnessed to the principles of planning.

One of the most effective steps in controlling purchase price variances is to free the buyer to buy. Surveys of purchasing practices reveal that few purchasing agents spend much time developing new and competitive sources of materials and supplies. Extending sources of supply and bringing into the spotlight items formerly given routine purchasing treatment pay handsome dividends.

In any case, the purchasing procedure used to purchase each material should be studied with these questions in mind:

1. Can costs be reduced by buying larger amounts?
2. Is it practical to lower quality?
3. Will long-term contracts be advantageous?
4. Can larger inventories be used to take advantage of off-season buying?
5. Are substitutes or other sources available?

Variations in the cost of materials from the planning period estimates may result from the way the materials are used rather than from changes in prices. For example, the real cause of a variance may be one of the following:

1. Excessive scrap and waste during processing
2. Introduction of new processes
3. Specification changes in products
4. Changes in materials used
5. Excessive quality specifications above needs
6. Damages during handling and storage

Deviations of this sort require investigation. Each factor may involve secondary ramifications that range all the way from a breakdown in communications to poor supervision. Reasons must be established, recorded, and corrected. Here again, the industrial engineer is qualified to handle the analysis of the problem.

COURSE CORRECTIONS

Plans are made in anticipation of future conditions, and when conditions change, it may be necessary to change the plans. Any good planning procedure provides for periodic revisions and updating. It is important, however, to distinguish between genuine revision and unnecessary changes designed to whitewash a failure in performance.

There will always be pressure for change that is rooted in emotion rather than rationality. The wide variety of "reasons" and "logic" that people can conjure up to support a change in the plan to justify their performance record is almost unbelievable. This can be the most difficult of the people-oriented problems the planner encounters, but it is not an acceptable reason for altering a plan that was soundly designed in the first place.

A typical company may encounter four circumstances that call for a review of the targets it has established:

- People performance
- Physical changes
- External influence
- Financial considerations

People performance can be a continuing problem, but plans and standards that are not met for this reason should not be changed except at profit planning time. Deviations from the plan during the period to which it applies should be highlighted to identify people performance, good or bad. Checking out the variances will also show how well the plan was conceived in the first place.

Physical revisions are likely to be part of a constant effort to improve operations efficiency. It can be argued that revised methods and procedures, new equipment, changes in layout, or alternation of processes call for an immediate change in planned targets. This assumes that the expected efficiencies will be realized on schedule and product costs can be lowered accordingly. A more sensible arrangement is to leave the plans and standards alone and measure the actual efficiencies as they are reported. This approach ensures that reliable information will be available before the company revises product cost standards and makes the accompanying changes in marketing programs and strategies.

Changes beyond the control of the organization should be handled judiciously. Raw material prices, in particular, should be constantly monitored so that significant savings in prices are reflected in adjustments of product or service costs. Violent changes in the economy, such as abrupt market breaks, natural disasters, or war, obviously warrant plan changes as soon as the planners can assess the consequences.

A major change in subcontractors' prices could force significant operating and marketing changes. Most companies, however, have learned the danger of making long-term commitments to suppliers without attaching stipulations about prices.

Periods of inflation and rising interest rates can be especially damaging if plans are not constantly reviewed to determine the effect on the company. Changes in interest rates, especially, will affect the company's ability to extend credit, maintain inventories, or make expenditures that consume working capital. Changes in cash flow can cause the plan and the original profit target to be outdated.

In general, however, a company will do better to stick with a well-designed, carefully constructed plan than to make wild guesses about the impact of changing circumstances. At a minimum, a good plan can provide a base line for measuring the changes caused by unexpected events. At best, the company's inherent strength and flexibility will enable it to achieve the targets it set in spite of difficulties it did not foresee.

SUMMARY

The purpose of profit planning is to maximize profits—not just dollar volume of profits but also the ratio of profits to equity capital employed. That is to say, profit panning is designed to ensure the largest possible return on investment to the owners of the business.

Effective profit planning is a comparatively simple and straightforward set of forecasts. The systems and procedures of planning should be designed to simplify and clarify the process of setting goals for the organization. A mass of uncoordinated forms and statistical reports is not planning; it is pointless busy-work.

This chapter has described the construction of a highly effective planning model and a variance control system to back it up. Beginning with a set of preliminary estimates of sales, costs, and marginal income, the model compares the results of this plan with goals developed by an alternate using a return-on-assets approach. Management can then use the model to explore the possibilities of changing the product mix, altering the price structure, reducing or increasing investment, or reducing costs by redesign of the product. The final plan that it adopts will offer the realistic prospect of yielding total profits for the organization close to the top of the range of possibilities.

Plans cannot be cast in bronze. Changing circumstances may cause the forecasts on which the plan was based to become out of date. Good planning procedure calls for periodic review and revision to take account of such developments.

It is a mistake, however, to change a plan simply because performance is not measuring up to expectations. The control system should identify every significant variance from planned targets and initiate immediate action to eliminate the causes.

Planning and ensuring that performance comes up to targets is management's first and most important function. But planning need not be a burden upon top management. The system can be set up so that the chief officers of the company make only the broad, fundamental decisions. Profit planning enables the top people to concentrate on the vital questions of how the company can make the best possible use of the resources at its command and how it can keep expanding and strengthening its position in the future.

Each separate planning system should be carefully tailored to the needs and structure of the company that uses it, and so each will be different in detail. The company that wants to install an effective profit planning system should turn the job over to an expert. It cannot expect to take a ready-made plan off the shelf and find that it fits.

All good planning systems, however, will have two things in common: They will set up realistic targets, and they will provide the machinery for making actual performance conform to these targets. Planning is not just a hopeful forecast of what the future will bring. It is a method of setting goals to maximize profits and guiding the operations of the company firmly toward these goals.

BIBLIOGRAPHY

McCormick & Company

McCormick, Edmund J., "Budgetary Control," chap. 8, in H. B. Maynard, ed., *Industrial Engineering Handbook, 3d ed.,* McGraw-Hill, New York, 1971.

McCormick, Edmund J., "Direct Costing," sec. 10, chap. 10, in H. B. Maynard, ed., *Handbook of Business Administration,* McGraw-Hill, New York, 1967.

McCormick, Edmund J., "Sharpening the Competitive Edge for Profits," *Financial Executive,* April 1975, pp. 22–27.

McCormick, Edmund J., Jr., "The Extraordinary Benefits of Direct Costing for Marketing Strategy," *Business Strategies,* February 1990, pp. 1–4.

McCormick, Edmund J., Jr., "Piercing the Volume Veil," *Business Strategies,* March 1990, pp. 1–4.

McCormick, Edmund J., Jr., "Super Charging Your P&L," *Journal of Bank Costing and Cost Management,* Summer 1990, pp. 1–8.

Other

Adelman, Richard L., "The Marginal Contribution Breakeven Point," *CPA Journal,* October 1983, p. 87.

Ames, B. Charles, and James D. Hlavecek, "Vital Truths about Managing Your Costs," *Harvard Business Review,* January–February 1990, pp. 140–147.

Arnstein, William E., and Frank Gilabett, *Direct Costing,* ANACOM, New York, 1980.

Christie, John, "Direct Costing—A System for Planning and Control," *Accountancy,* May 1979, pp. 83–84.

Cooper, Robin, and Robert S. Kaplan, "How Cost Accounting Systematically Distorts Product Costs," chap. 8, in William J. Burns and Robert S. Kaplan, eds., *Accounting and Management: Field Study Perspectives,* Harvard Business School Press, Boston, 1987.

Dudick, Thomas S., and Robert V. Gorski, eds., *Handbook of Business Planning and Budgeting for Executives with Profit Responsibility,* Van Nostrand Reinhold, New York, 1983.

Grinell, D. Jacque, "Product Mix Decisions: Direct Costing vs. Absorption Costing," *Management Accounting,* August 1976, pp. 36–42.

Kollaritsch, Felix P., *Cost Systems for Planning, Decisions and Controls,* Grid, Ohio, 1979.

O'Guin, Michael, "Focus the Factory with Activity-Based Costing," *Management Accounting,* February 1990, pp. 36–41.

Ostrenga, Michael R., "Activities: The Focal Point of Total Cost Management," *Management Accounting,* February 1990, pp. 42–49.

Salvary, Stanley C. W., "Profitability Analysis in the Decision-Making Process," *Journal of Systems Management,* March 1981, pp. 6–8.

Sandretto, Michael, "What Kind of Cost System Do You Need," *Harvard Business Review,* January–February 1985, pp. 110–118.

Tucker, Spencer A., *Profit Planning Decisions with the Breakeven Systems,* Thomond Press, New York, 1980.

Williams, B. R., "Measuring Costs: Full Absorption Cost or Direct Cost?" *Management Accounting,* January 1976, pp. 23–24, 36.

Wright, Norman H., Jr., "Comparison of Absorption and Direct Costing Methods," *Management World,* August 1976, pp. 16–17.

CHAPTER 3
ACTIVITY-BASED COSTING AND CONTROLS

Edmund J. McCormick, Jr.
Chairman
McCormick & Company
Summit, New Jersey

A revolution is taking place in factories and service organizations throughout the country. Technology is rapidly replacing traditional manufacturing methods and service delivery systems. The result has been dramatic change in ratios between fixed and variable costs which continue to surge at alarming rates. As fixed costs climb, they are wreaking havoc on profit margins, making them extremely vulnerable to competitive forces.

To compete in today's dynamic and rapidly changing global marketplace, our domestic firms need new leadership to understand and control their overhead costs as in no other time in our history. No professional is better equipped to provide this direction than the industrial engineer. Thus the gauntlet has been thrown down to our industrial engineering cadre to meet this critical challenge of the 1990s.

During the past decade, the industrial engineer has made enormous contributions to productivity through the introduction and installation of production, quality improvement, and waste elimination programs, like just-in-time (JIT), material resource planning (MRP), computer-aided manufacturing (CAM), and computer-integrated manufacturing (CIM). These offshore developments were adopted in this country to enable domestic firms to compete with overseas producers, especially Japan, but also the emerging European Community, as well. Unfortunately, even after the implementation of these techniques, the regaining of competitiveness has still been elusive. The question is, "If so many domestic manufacturers and service organizations have adapted advanced manufacturing and service delivery methods, why are they still languishing behind in the field of world competition?"

The Industrial Engineer as a Leader. The answer lies in the integration of technology with modern information and control systems. In spite of advanced manufacturing technology, many companies have failed to make their financial, managerial accounting, and costing systems conform with the changes in their manufacturing and service environments. Maintaining the financial costing system has traditionally been the responsibility of the cost accountant. Unfortunately, most cost accountants have been isolated from the changes that have been occurring on the factory floor. Those few who have understood their impact have been reluctant to introduce the needed system changes.

The academic training of the cost accountant is the culprit. Accounting education has remained virtually unchanged for decades, notwithstanding the changes in information needs throughout the organization. Such education might have been satisfactory in the less mechanized world of the 1960s and 1970s. But, as advanced technology was introduced and labor was displaced with fixed cost machinery and electronics, the cost accountants were left without the tools to make the transition. They did not have a clear understanding of how technology was invalidating their firm's cost systems. Without such knowledge, a severe gap developed, leaving cost systems outmoded, outdated—and worse. Worse because in many instances the old systems provided management information entirely inadequate for decision making.

It is now up to the industrial engineer (IE) to lead the way until cost accountants can regroup. With the IE's education and understanding of the new technologies, the need for informational changes, and the understanding of technology's impact on product costs, they are best equipped to make management aware of the dire need to update the organization's cost information systems.

This rapid growth of technology has left a void which needs to be filled. The IE is the ideal candidate to assume this new leadership role to integrate automation and new service delivery systems with the organization's information system. The output of a modern cost and control system isn't just dollars and cents statistics as in the old days; it includes important decision-making information such as machine hours, pounds, standards, variances, and other measures which have always been the responsibility of the IE. At this time, only the IE has the needed tools to assist in the modernizing of current costing systems—systems which have been harming domestic industry's ability to compete adequately.

Our old cost and control systems were designed around high-volume, lower-quality products, and for improving quarterly earnings at the expense of long-term company benefits. Such systems stymie efforts to achieve continuous improvement. These old systems need replacement, and the prime candidate to lead this revolution is not the cost accountant but the IE.

The product of all cost and control systems is *change*. Who is better equipped to handle this mission than the IE trained in the techniques of systems measurement, analysis, and productivity improvement—and, most importantly, change? One of the significant emerging changes in cost system integration is activity-based management. This chapter will familiarize the IE with this latest approach to cost information systems and their application.

What Is ABC? Activity-based costing (ABC) relates resources to their consuming activities. Put another way, it is the concept of managing all of a company's resources and the activities which consume them. Conventional wisdom states that the production of a product or service produces costs. However, it is actually the *activity* involved in the production of a product or service which creates the cost

If we can agree that an activity produces cost, then it follows that the actual cost of a product or service should be the sum total of the costs of each activity required to produce it. By breaking product cost down into various activities or events, costs can be controlled by managing the activities and the events that cause the cost-consuming activity.

The Purpose of Activity-Based Costing. The purpose of ABC is to remove the severe distortions caused by traditional costing systems, such as absorption-based and direct costing. These traditional systems, once adequate, when direct labor costs were a large percentage of product cost, rarely find an operating environment today where they can provide meaningful product or service costs. Activity-based management and costing takes the best attributes of absorption-based and direct costing and applies all indirect costs to products and services through an analysis of the activity that actually produced the particular cost. This method treats all costs as if they were *variable*.

Definition.   Activity-based costing attributes variable, fixed, and overhead costs directly to each product or service using the *activities* required to produce the product or service as the means of allocation. In ABC the cost of the product and service equals the cost of raw materials, plus the sum of all the costs of every activity used to produce the product or service. In that way, ABC costing is different from traditional costing which accumulates the cost of raw materials and direct labor, then applied overhead using an arbitrary allocation formula based on volume of production rather than activity. ABC uses a very different cost attachment process as a result of this new understanding of how products and services consume activities and in turn, activities consume resources.

Activity-based management (ABM) is a system utilizing activity-based costing plus a number of control elements. These consist of process value analysis, activity-based process costing, activity-based product costing, performance measurement, and responsibility accounting.

New England China Company Examples.   A good way to understand the application of ABC and ABM is to work through an actual example to compare information outputs between cost systems. In our comparison we use The New England China Company, a fictitious manufacturing company. It produces a full line of china, including dishes. This manufacturer has a diverse product line with over 1000 patterns produced in annual volumes ranging as low as 500 units. Figure 3.1 lists three of these patterns.

Assumptions for the New England China Company P&L Statements

	Blue	Floral	Gold	Total
Selling Price	$1.75	$2.00	$2.20	
Unit Volume	120,000	60,000	10,000	190,000
Product/Volume %	63.16%	31.58%	5.26%	100.00%
Direct Labor/Unit	$0.40	$0.40	$0.40	
Material/Unit	$0.60	$0.60	$0.60	
Average Run Quantity	4,800	2,400	400	7,600
Total Number of Setups	25	25	25	75
Variable Overhead				$20,000
Fixed Overhead				$80,000

FIGURE 3.1   These assumptions are used to construct the P&L statements in Figs. 3.2 through 3.4. The New England China Company is a manufacturer of a full line of china products. In this example three dish styles are shown, blue, floral, and gold patterns.

Although production volumes vary, direct labor and materials per unit are the same for each pattern. Volumes range from 10,000 to 120,000; however, total setups equal 25 regardless of production volumes. Variable overhead is $20,000, and fixed overhead is $80,000.

Using the data in Fig. 3.1, a profit and loss (P&L) statement can be constructed to show the effect of absorption-based costing, direct costing, and activity-based costing on margins. To keep this example simple, selling and general and administrative (G&A) expenses have been eliminated from the P&L and indirect cost is just setup cost.

Figure 3.2 shows the *absorption-based* P&L for the New England China Company. Sales are determined by multiplying unit selling prices ($1.75, $2.00, $2.20) by unit volume (120,000, 60,000, 10,000). Direct labor and material costs are calculated by multiplying the respective unit costs ($0.40 and $0.60) by the unit volume. Factory overhead was calculated by multiplying production volumes (63.16 percent, 31.58 percent, 5.26 percent) times the total overhead (O/H) of $100,000 (variable O/H $20,000 + fixed O/H $80,000 = $100,000). The resulting total expense was subtracted from sales to provide the gross mar-

Absorption-Based P&L for the New England China Company

	Blue	Floral	Gold
Sales $	$210,000	$120,000	$22,000
Direct Labor	48,000	24,000	4,000
Direct Material	72,000	36,000	6,000
Factory Overhead	63,158	31,579	5,263
Total Expense	$183,158	$91,579	$15,263
Gross Margin	$26,842	$28,421	$6,737
Gross Margin %	12.78%	23.68%	30.62%

FIGURE 3.2   This is a simplified P&L statement showing the effect of absorption-based costing for a high-, medium-, and low-volume product. Notice how the major portion of the factory overhead is applied to the blue pattern dishes. That is because overhead is allocated based on a percentage of direct labor. Therefore, as units of production increase, so does the allocated overhead to that product.

gins, which provides a margin variance of 12.78 and 30.62 percent between blue and gold pattern dishes.

This margin information indicates that the gold pattern plates are almost two and one-half times more profitable than the blue pattern plates. Management may now conclude that production of the blue plates should be reduced and production of gold plates with the greater margin should be increased.

In an uncomplicated example like this one, it can be seen that although the numbers report one set of circumstances, common sense may indicate the opposite is true. If the blue pattern plates were reduced or eliminated, the cost would be shifted to the gold pattern product with disastrous results. In a real-world environment with more product lines and more complex variables, common sense may not be sufficient to slice through the information fog created by these misallocations. In that case, management may be caught making decisions based on incorrect or misleading information—perhaps eliminating a profitable product or product line altogether.

The New England China Company decided to revamp their full-absorption costing system by implementing a direct-costing system, known also as a marginal income (MI) system, so that distortions caused by the arbitrary allocation of overheads would be reduced. Figure 3.3 shows the effect of this change. The first three lines of the P&L statement are identical to the absorption-based P&L statement. Line 4, the application of overhead, is

Marginal Income Based P&L for the New England China Company

	Blue	Floral	Gold
Sales $	$210,000	$120,000	$22,000
Direct Labor	48,000	24,000	4,000
Direct Material	72,000	36,000	6,000
Variable Overhead	12,632	6,316	1,053
Total Expense	$132,632	$66,316	$11,053
Variable Margin	$77,368	$53,684	$10,947
Variable Margin %	36.84%	44.74%	49.76%

FIGURE 3.3   The use of marginal income costing allows the removal of fixed overhead cost from the overhead equation. This has the effect of equalizing margins so that only unit-based costs are considered. These are costs that directly relate to production. Variable overheads are still allocated on the basis of a unit of production (for example, labor hours), rather than on the activity that originates the cost. However, the marginal income method still does not eliminate cost distortions, which arise from the arbitrary allocation process.

the only difference. The variable overhead of $20,000 is applied to each product based on the proportionate amount of direct labor used for the production of each pattern. The resulting margins do not contain the large differences seen in Fig. 3.2. Instead, they appear more equal, even though the blue pattern plate still appears to show the lowest profit margin while the gold pattern appears to be the most profitable.

Yet management might have good reason to suspect these numbers, too. The initial reaction is that it doesn't appear to make economic sense that the lowest-volume product should have the highest margin, when we know that all direct costs are equal. Overhead, though much less of it, is still being arbitrarily allocated instead of applied to the product which actually caused the cost. The problem is the remaining indirect costs are unassigned so that only partial product costs are seen. If those remaining costs could be assigned directly to the product based on how the product actually consumed them, management would have the "truest" product cost possible.

Figure 3.4 demonstrates the implementation of such a solution—an activity-based costing system. Lines 1 through 4 are identical to Fig. 3.3. Line 5 has been added to show the effect of directly assigning indirect overheads to the product. In this case the indirect cost is the setup cost. Since each of the three products has the same setup cost, it is an easy matter to assign one-third of this cost to each. The result is that the gold pattern plates are being produced at a loss, whereas the blue pattern plates (the highest volume) have the greatest margin.

Activity-Based P&L for the New England China Company

	Blue	Floral	Gold
Sales $	$210,000	$120,000	$22,000
Direct Labor	48,000	24,000	4,000
Direct Material	72,000	36,000	6,000
Variable Overhead	12,632	6,316	1,053
Activity-based Overhead	26,667	26,667	26,667
Total Expense	$159,298	$92,982	$37,719
Variable Margin	$50,702	$27,018	$(15,719)
Variable Margin %	24.14%	22.51%	(71.45)%

FIGURE 3.4   Activity-based costing significantly changes the gross margin when compared with the P&L statement in Fig. 3.2. Costs are now directly applied to the dishes based on how they are actually incurred. This method, said to provide the "true" cost, applies overhead based on direct labor hours or some other unit of production. Costs which are not volume-related, such as inspection, setup, and purchasing, are attached to those units of production which contain the highest number of direct labor hours. This distorts costs by reporting high volume, large batch size, and/or high-complexity products or services as *low-margin* items. It has just the opposite effect on low volume, small batch size, and/or low unit complexity.

Though these are simplistic examples, consider the impact on a P&L when all overhead costs are correctly attributed to product or services. It is not unusual to find triple-digit gross margin differences between the current costing system and an ABC system.

EVOLUTION OF ACTIVITY-BASED COSTING

Absorption-Based Accounting.   The majority of domestic firms continue to use absorption-based accounting systems for product costing. A recent study completed by the University of Rhode Island reported that 62 percent of the firms surveyed did not differentiate between fixed (period) and variable costs (the direct costing method). In addition. 93.7 percent still applied overhead on the basis of direct labor (the absorption method).

Why such is the case in the face of such compelling evidence against the absorption-based costing method is not easy to ascertain. A reason often offered is that absorption-based accounting is the method required for external financial reporting needed by the IRS, SEC, stockholders, and the like. Having an absorption-based system already in place, management may be reluctant to make significant changes or to run two systems (an internal system and an external system) at the same time. Management may have felt that it was more cost-effective simply to modify the external system for internal reporting purposes, unaware of the potential for many information system reporting inaccuracies.

As was noted earlier, in the past, manufacturing processes were labor-intensive, overhead was a small percentage of total cost, and the range of products (product diversity) was more limited. Therefore, a decision to modify the traditional system for costing purposes did not cause severe harm. However, with the move toward automation where labor costs are often less than 10 percent of production and overhead reaches toward 50 percent of the total cost or higher, direct labor can no longer be used as the method to allocate value-added to a product or service.

In organizations with considerable product diversity, product and service costs may be severely distorted. A high-volume run of a product will appear to cost more than a low-volume run. In reality, this is not true, but the inherent system flaws compel the numbers to be reported that way. Simple mathematics will demonstrate that though the cost of setup, material handling, purchasing, and the like may be the same on two production runs, the unit cost for a higher-volume run compared with a lower-volume run will be greater. However, the traditional method takes all these indirect costs, totals them, and then allocates the cost using labor hours or some other volume-based unit. The higher-volume run has more labor hours, so it gets more of the cost.

Since these costs do not vary on a per unit basis, they cannot be accurately accounted for in any system which assigns cost using production volume as the basis. That is why management reports will show (incorrectly) low-volume products as more profitable than a high-volume product. Whole product lines have been discontinued because of this faulty information.

Recently, Alcoa was prepared to close an important West Coast division after receiving negative margin reports from an outmoded management information system. Most of the senior management were in agreement—close it down. However, the vice president of engineering had a number of suspicions concerning the numbers. He convinced management to reconsider while a team was sent to review cost allocations. After a basic activity-based analysis was completed, the division showed margins that were surprisingly acceptable. Needless to say, the division is still in operation today doing well for Alcoa's bottom line.

The traditional method of allocation assumes that only volume-related bases such as labor hours, machine hours, and material dollars are used to allocate overhead to products or services. Allocations based on units of production (direct labor, machine hours, material), falsely assume that the cost of production varies in direct proportion to the number of items produced. This assumption may be true for direct cost such as certain labor, material, and supplies; but the cost of inspection, setup, engineering, and purchasing, which are not volume-related, varies with the number of inspections, setups, engineering changes, and purchase orders. Allocating non-volume-related costs requires the use of a cost base that is non-volume-related. Figure 3.5 demonstrates how a traditional system reports product margins, giving high-volume products the appearance of lower margins and low-volume products the appearance of higher margins.

Direct Costing. Over the past several decades, direct costing, also known as *marginal costing,* was installed by a number of leading-edge companies in an attempt to overcome some of the weakness of the absorption-based system. This method of costing separates cost, by behaviors, into fixed and variable components. By subtracting variable cost from sales revenue, a number, referred to as *marginal income* (MI), is obtained. By using direct

Absorption-Based Overhead Allocation Method

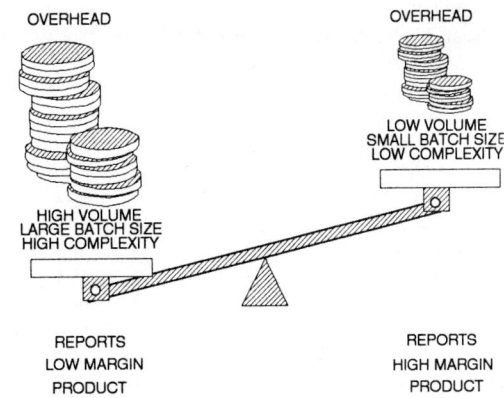

FIGURE 3.5   Absorption-based overhead allocation method.

costing for costing decisions and analysis, managers get a more realistic picture of relative profitability. Figure 3.6 shows a model of overhead allocation in a marginal income system. Overhead is treated by removing it from consideration from the variable cost of production. In this method overhead is said to be shown "below the line."

Direct Costing Overhead Allocation Method

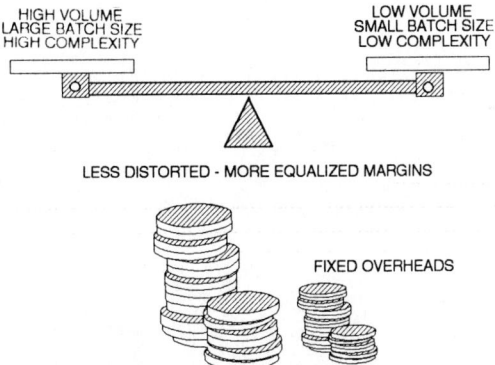

FIGURE 3.6   The direct costing or marginal income (MI) method removes all indirect costs from the product and services. By subtracting variable costs from sales revenues, the result will be MI. Relative profitability is made much clearer when the marginal income of each product or service is compared. Overhead is shown "below the line." MI is used to cover overhead. The remaining MI is profit. If there is not sufficient MI to cover the overhead, the remainder is negative, amounting to a loss.

Disenchantment with Marginal Costing.   However, many managers prefer to rely on full costing rather than direct costing. Some maintain that salespersons would be tempted to cut prices closer to margins if they knew their variable costs. The use of full costing actually eliminates this temptation by obfuscating true costs. Second, since fixed costs are not factored in, a number of managers feel that variable costs do not adequately reflect the demands that different products place on their fixed resources, such as plant and equipment. These managers are reluctant to drop products and services because of their concern with unabsorbed overhead, which would decrease total profits.

After all, product- and service-related decisions for introduction, pricing, and discontinuance should be long-term decisions, strategic in nature. Marginal income is based primarily on variable and incremental costs which by definition are short-term. Marginal costing made sense when variable costs such as labor, material, and certain overhead costs were a large proportion of the total manufacturing cost. We know that is not the case now.

Evolution of a ''New'' Cost System.   Although the latter part of the 1980s saw a surge of interest in activity-based costing, the principles have been available for 70 years. The concept for the ABC system was first advanced by C. Hamilton Church in the early 1920s. ABC never caught on, as it was difficult to implement owing to the large volume of transactions which needed to be recorded by hand. It was only after the widespread use of the personal computer and the popularization of spreadsheet programs that ABC could provide a cost-effective solution to the tediousness of manual recording.

Activity-based costing is certainly a novel way to look at costs, yet it has all the elements of an absorption-based system. The basic difference is that overhead is directly traced to the product or service instead of pooled and applied across product and services using an arbitrary formula (that is, percentages of labor hours, machine hours, and the like). Cost tracing first starts by identifying all the support activities required for production, then determining how the product actually consumes the various supporting activities. In that way, all overhead costs are attached directly to the product or service that consumes them. Figure 3.7 shows how overhead allocations are almost opposite in mag-

Acitivity-Based Overhead Allocation Method

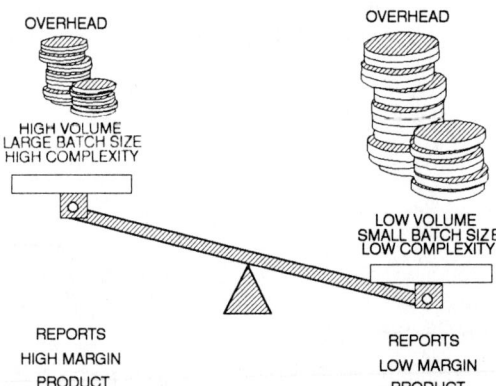

FIGURE 3.7   Activity-based costing significantly affects gross margins of the highest and lowest volumes, batch sizes, and complexity ranges of products and services. Those that fall within a midrange will also be affected, but not as radically. Comparing this with Fig. 3.5 demonstrates that ABC costing has the opposite effect in the assignment of overhead burdens. By correctly identifying which products receive these indirect costs, product and service costs taken on entirely different value compared with absorption-based and direct costing. These are the ''true'' product and service costs.

nitude from the allocations produced under the absorption-based system in Fig. 3.5. Figure 3.7 allocations are more accurately reflected, since they are based on how they are consumed by the product or service.

A demonstration of how fixed costs are assigned to a product is shown in the identification of *setup* and *engineering change* activities. Setup and engineering change costs depend on the quantity of each. Cost allocators are found by determining how the product or service is consuming setups and engineering changes. Therefore, assigning setup and engineering change costs to the product or service simply means determining the total setup and engineering change costs, then dividing each cost by the number of setups and engineering changes to determine the cost per unit. A product or service requiring two setups receives two units of setup costs, one requiring three setups would get three units, and so on. By identifying all activities and determining the relationship of how each activity is consumed in the production of the product and service, all indirect costs can be identified directly. Figures 3.8 and 3.9 show how engineering changes and setup cost might be assigned to two products, blue and gold pattern dishes. The method is compared with traditional unit costing to demonstrate the extreme cost distortions that full costing produces. The low-volume gold pattern dishes show that setup cost is almost 100 percent higher under the activity-based costing method, whereas the high-volume blue pattern plate is nearly 100 percent lower in cost.

ABC Systems Users.   Numerous ABC systems have been installed with highly successful results for the organizations that are using them. For example, Caterpillar, Inc., has had its system in operation for over 20 years. Hewlett-Packard first installed its ABC system as a pilot project in 1985. It was so effective that several years later they stopped tracking direct labor, which had been 2 percent of cost. Figure 3.10 lists a number of representative firms which have such systems in place.

Activity-Based vs Traditional Setup Cost

	GOLD PATTERN	BLUE PATTERN	TOTAL
Production Volume	5,000	1,000,000	
Cost Per Setup	$1,000	$1,000	
Number of Setups	2	200	
Batch Size	2,500	50,000	
Total Cost of Setups	$2,000	$200,000	$202,000
Direct Labor Hours Per Unit	2	2	
Total Direct Labor Hours	10,000	2,000,000	2,010,000
Setup Cost Per Direct Labor Hour			$0.10
($202,000/2,010,000)			
Activity-Based Costing Overhead	$0.40	$0.02	
(A=$1,000/2,5000: B=$1,000/50,000)			
Method. Convential Overhead	$0.20	$0.20	
($0.10 x 2 Direct Labor Hours)			
Traditional vs Activity-Based %	-99.01%	90.05%	

FIGURE 3.8 This example shows the setup cost for gold and blue pattern plates. Both setup costs are $1000. However, the blue pattern dish, with a total production of 1,000,000 units, has 200 setups, and gold, with only a 5000-dish production run, uses two setups. With two direct labor hours for each product, the total setup cost for each direct labor hour is $0.10. Using an activity-based approach, the unit cost of setup for the gold pattern is almost 100 percent greater than for the conventional method of allocation ($0.40 versus $0.20), whereas the activity-based unit setup cost for the blue pattern is almost 100 percent less ($0.02 versus $0.20).

Activity-Based vs Traditional Engineering Change Costs

	GOLD PATTERN	BLUE PATTERN	TOTAL
Production Volume	5,000	1,000,000	
Cost Per Engineering Change	$1,000	$1,000	
Number of Engineering Changes	1	2	
Total Cost of Engineering Hours	$1,000	$2,000	$3,000
Direct Labor Hours Per Unit	2	2	
Total Direct Labor Hours	10,000	2,000,000	2,010,000
Engineering Cost per Direct Labor Hour			$0.001
($3,000/2,010,000)			
Activity-Based Costing Overhead	$0.200	$0.002	
(A=$1,000/5,000: B=$2,000/1,000,000)			
Convential Overhead	$0.20	$0.20	
(A=$0.006 x 2 Direct Labor Hours: B=$0.006 x 2 Direct Labor Hours)			
Traditional vs Activity-Based %	-6,600.00%	33.00%	

FIGURE 3.9 In this example engineering changes are compared between gold and blue pattern dishes. Volumes and direct labor hours are identical to the example in Fig. 3.8. The differences relate to the number of engineering changes. In this example there is one change for gold and two for blue. The disparity of the cost distortion is even greater than in the previous example for setups, with a 6600 percent difference between activity-based costing and conventional costing for the gold pattern and a 33 percent difference for the blue pattern.

Leading Edge Organizations Using Activity-Based Costing

- Hewlett-Packard
- General Electric
- Caterpillar, Inc.
- Cal Electronic Circuit
- Siemen's Electric
- Tektronix

- John Deere
- AT&T
- Carolina Power & Light
- Schrader-Bellows
- Alexandria Hospital
- Union Pacific R.R.

- General Motors
- Harley-Davidson
- PPG Industries
- Warner Lambert
- Data Services
- Weyerhauser

FIGURE 3.10 Over 117 major organizations have adapted activity-based costing for internal cost management and control. The firms listed here are just a representative sample. They include manufacturing, transportation, financial, and service organizations.

These organizations installed the ABC system for a wide range of reasons. For instance, the Roseville Networks Division of Hewlett-Packard uses ABC to support the design for manufacturability of printed circuit boards. Northern Telecom employs ABC to eliminate "waste" and simplify manufacturing activities. Harley-Davidson uses ABC techniques in conjunction with its materials as needed program (MAN) to support a program of continuous improvement through the elimination of waste caused by non-value-added activities.

General Dynamics has an ABC system to identify cost at machine level to focus closely on the cost of individual parts and to record the cost benefits of smaller components in the manufacturing process. GM implemented an ABC system at an automotive stamping plant and uses it to better support engineering and management decisions. Cal Electronic Circuits, Inc., installed ABC, too, and uses it to identify process costs more clearly and the cost-volume-profit relationship for various products.

John Deere Component Works uses its ABC system to trace non-unit-volume activities to products to provide more accurate costing, which has resulted in an increase in bid awards. Siemen's Electric Motor Works and Tektronix Portable Instrument Division both installed a basic ABC system, which was deliberately kept simple so that managers could send a strong message using the new bases they selected.

ABC systems are a highly effective tool for departmental analysis, as well as for use in an entire facility. For example, the financial services department at Weyerhaeuser analyzed their activities by function (general accounting, invoicing, accounts payable and receivable, and database administration) to determine how these activities were utilized by the various divisions. The cost of this department was previously allocated by a percentage of a division's sales. Such an allocation had little relationship to the activities that caused administrative work within the financial department. A division with many customers and low sales volume versus a division with few customers and high sales volume would receive an incorrect allocation of how actual costs behaved. In fact, the cost reported would be just the opposite of the actual cost.

As of this writing, over 117 systems are installed and in use throughout the country. This number doesn't include hundreds of other organizations just beginning to install ABC systems ranging from regulated utility companies such as Carolina Power and Light to industry giants such as PPG Industries. During the last 24 months an extraordinary amount of interest in the subject has been demonstrated by the over 45 articles and 5 books published on the topic. It appears that ABC systems will be the costing and control system of choice in the next 5 years not only in high-tech firms but also in financial institutions, service, utility companies, and manufacturing companies.

BENEFITS TO THE ORGANIZATION

True Cost.   ABC systems benefit organizations in a number of ways. Such systems provide management with the true cost of strategic choices so that they do not have to rely on intuition. By segregating costs by activities, managers are able to approach cost reduction sensibly through cause and effect analysis. They are able to determine which are value- and non-value-added activities, perhaps for the first time. Thus, by focusing on the reduction on non-value-added activities, cost can be reduced without harming the long-term objectives of the firm.

Improves Decision Making.   Understanding the links (drivers) between resources and activities and, in turn, the activities producing the product or service will help managers make product decisions even without dollar figures. ABC focuses on product and process simplification to facilitate continued cost improvements and increased competitiveness. No longer will cost distortion lead to incorrect decisions on product additions or abandonment, as has been the case for many organizations.

Strategic Options.   Managers using the information provided by the ABC system can review a range of strategic options. They can now identify the truly unprofitable products and decide which steps should be taken. Is it in the best interest of the organization to abandon the product? Should prices be raised? Low-volume products and services tend to have much more price elasticity over high-volume items, which may allow pricing for a profit. By using ABC costing to shift indirect cost away from high-volume products or services, a manager will have the option to lower prices on products to increase market share. One of the most cost-effective ways to gain market share is by changing product mix. ABC develops the appropriate information to determine the best mix of products and/ or services.

Evaluates Technology.   ABC systems will provide a new tool to evaluate new process technologies. It provides the means to focus on the benefits of lowering material handling, improving quality to reduce inspections, reducing setups, improving process flow, and streamlining plant layout. Such costs can be readily identified on a product and product line basis, so adjustments in efficiencies can be made.

Encourages Product Redesign. Good managers are constantly encouraging their design engineers to modify products to use fewer components and to ease manufacturing cost. However, most organizations do not have a system in place that traces the benefits of doing so in concrete terms. The information supplied by an ABC system encourages this process by showing the cost and benefits of meeting this objective product by product. Every manager in the firm will understand, perhaps for the first time, the cost and benefit of designing for manufacturability.

Eliminates Traditional Standards. ABC systems do not rely on traditional fixed standards. Instead they use rolling standards, which continually compare prior periods with current periods. This promotes continued improvement toward perfection and eliminates the need to try to keep traditional standards current. The focus of such a system is not whether the standard or actual cost is right, but whether it is getting better. Thus the variances are used to improve the process—not to balance the ledger, as traditional systems do.

Highly Motivational. The use of ABC systems provides a highly efficient means to modify the entire organizational process. Just as important, it provides the method to judge how changes in the performance of activities affect overall cost. The ABC system not only provides a highly accurate method of costing but promotes activity efficiencies by exposing an activity that was once "buried" in an overhead pool. By separating the cost of these support activities, each department can directly trace the effect of their efficiencies on total product cost. Coupled with a responsibility accounting system, this information is highly motivational, since engineering can see the cost impact of their designs, purchasing will understand the impact of reducing or expanding vendors, and so on.

THE ABC SYSTEM MODEL

Similarities to Traditional Systems. The activity-based cost system has a number of similarities relating to traditional systems. It, too, is a two-stage allocation process. However, it is in the second stage of the allocation process that the two systems diverge. Figure 3.11 shows how a simple model ABC costing system might operate. As examples, two activity centers, customer service and purchasing, are illustrated for the New England China Company. Of course, in practice, an ABC costing system would contain many more activity centers.

The upper portion of the example shows the general ledger accounts (*resources*). At this early stage these accounts would be reviewed to determine which of these resources are consumed by which activities. In this example, the activities are customer service and purchasing. The next step is to determine the method of attributing these resources to the activities. Often called *first-stage cost drivers* in the literature, the term *resource drivers* is used here for clarity.

In the example, the resource, indirect labor, can be attributed to both activities based on the *resource driver*—labor hours. That is, the number of labor hours will determine the amount of indirect labor consumed by the activity of customer service and purchasing. This is the direct linear relationship we are looking for.

The allocation of utilities proves to be a bit more challenging. In the absence of metering, which of course is the best answer, a resource driver must be found which provides the linear relationship for the consumption of this activity. Head count or the number of labor hours could be used. However, they would not provide these two activities with the linear relationship that is needed and therefore would provide less than optimum results. Floor space is the best fit. Although it is not as precise as a meter, it is the most cost-effective resource driver for this purpose.

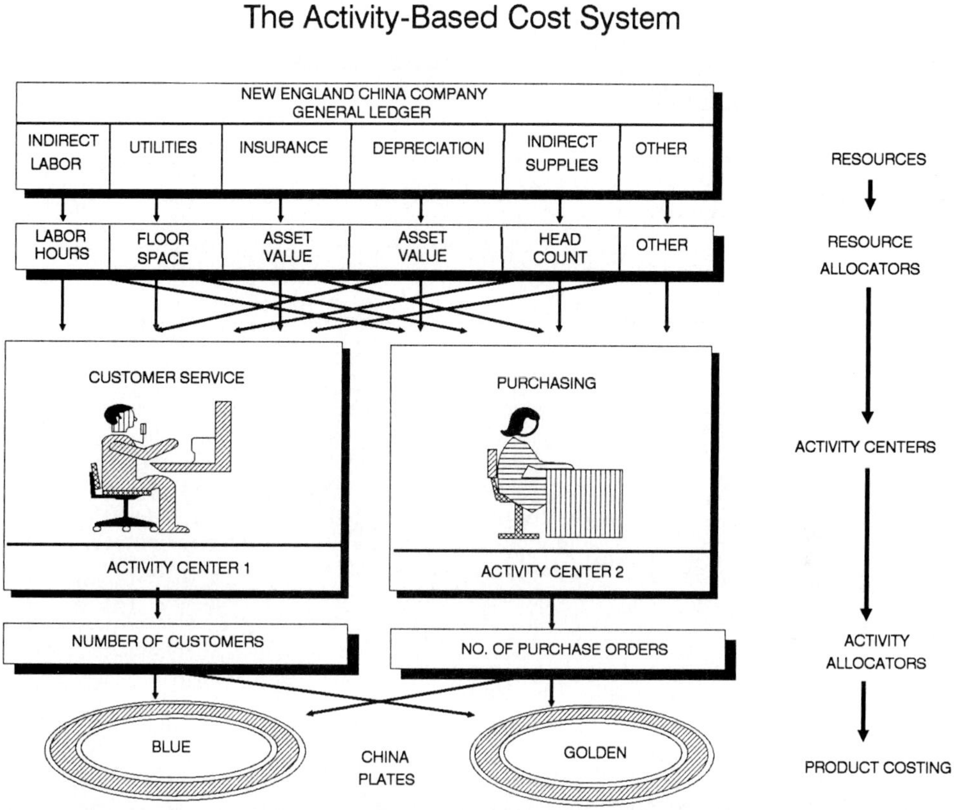

The Activity-Based Cost System

FIGURE 3.11   This is a schematic view of a partial ABC system and how it is developed. First, resources are identified from the general ledger. Then important activity centers such as purchasing and customer service are isolated. The method of allocating resources to the activity centers is determined through the selection of resource drivers. Activity drivers are then determined to accurately reflect the direct (linear) relationship of each activity center to each product or service.

The Second-Stage Difference.   Once these costs are assigned to the newly identified cost pool within the activity center, we are ready for the second stage. The second stage is the assignment of the amount of an activity consumed by the product or service. This process, often referred to as the *second-stage cost driver,* will be called *activity driver* here. In the case of the purchasing activity we have chosen an activity driver based on the number of purchase orders. That is, the amount of the purchasing activity a product, a china dish, consumes is directly related to the number of purchase orders generated to produce that pattern of china dishes. The customer service activity can be directly related in the same way by using the number of customers as the activity driver.

Unlike traditional costing, the second stage of the cost assignment process is not an arbitrary one. ABC does not allocate overheads based on one or two arbitrary methods, such as percentage of direct labor, material and/or machine hours, which have little or no relationship to how a product uses the overhead services. Instead, ABC systems identifies how these resources are consumed by each product or service and attaches according to this consumption pattern. There is very little indirect cost in an ABC costing system, since most costs can be "directly" attributed to the product or service.

FOCUS ON CONTROLS

Cost Control "After the Fact."   An ABC system has the desired benefit of being used as a highly effective control tool. As was stated earlier, traditional systems and thinking maintain that the production of a product or a service actually creates cost. That view examines cost after it has been incurred—it is easy to see that cost, once expended, cannot be controlled (modified). Traditional control systems rely on examining variances which are nothing more than a review of historical data. Such a system reveals only if a plan has been under- or overachieved. Control in these instances relies on making adjustments after the fact, in order to bring the potential for significant variances closer to plan at some time in the future.

A New Way to Control Cost.   ABC management (ABM) examines cost in a new and more "controllable" way. Earlier we learned that products and services consume activities and, in turn, activities consume resources. Therefore, the cost of all products and services is the total of all the activities consumed by the product or service.

In engineering terminology, control means to *regulate*. Typically, regulating is accomplished by a comparison of an actual or real-time statistic with a standard. The resulting variance between the actual and the standard determines the amount of adjustment required to bring a process back to the standard. Since cost is usually the prime target of control, the regulation of cost under traditional systems means that costs will exceed a standard at some point before the control system is able to spot it in order to take corrective action. By their very design traditional systems must encounter cost overages before any corrective action can actually be taken.

We have seen that the ABC costing system focuses on activities rather than costs. By organizing the work process into distinct activities a significant control advantage is gained. Controlling activity, rather than cost, is the objective of the ABM.

ABM systems control begins by separating activities into value-added and non-value-added categories. If a value-added activity is being consumed, costs may be increasing, but so is value. If a non-value-added activity is increasing, so are costs, with no added benefit. Therefore, an important aspect of the ABM system is to report both types of categories, so that managers can see how their outputs impact on the two basic activities. Control is almost automatic, as managers are provided with the opportunity to see activities in their areas of responsibility as value-added and non-value-added. They will place emphasis on the value-adding component.

DESIGNING THE ABC SYSTEM

Objectives.   ABC management information systems are typically more complex than traditional systems. Careful planning is needed to maximize the benefits. No two organizations will have the same information needs, because their company's cost drivers can be very diverse. However, every system has at least eight basic steps in common with each other which should be included in the design objectives. They are

- Determine design criteria
- Identify resource categories
- Identify activities
- Analyze and categorize activities
- Establish activity centers
- Determine cost pools within activity centers
- Determine resource drivers
- Determine activity drivers

Determining Design Criteria.   A number of important design choices should be made prior to attempting implementation of any ABC cost system. Four of the more important are included here. However, each organization will have different requirements, so use them as a guide. Careful consideration should be made to assure that the system will be designed to achieve both long- and short-term organization objectives. The four questions to be answered are:

- What are the strategic goals of the organization?
- How precise should the system be?
- Should the initial design be simple or complex?
- Should there be a pilot project first?

Strategic Issues.   A designer must never lose sight of the fact that all management information systems should be created to serve the long-term goals of the organization. Without knowing or even understanding what the strategic goals are, a system design could be fatally flawed. The designer must understand what information is required for each user in order to conduct the strategic mission. Having established this top layer of information needs, all subsequent information hierarchies will be much easier to identify.

Stand-Alone System.   The question of whether or not to integrate the ABC cost system into the organization's current accounting system or to install it as a stand-alone will depend on the objectives. There is no correct answer, since there are advantages to both. Having two systems tends to create controversy as to which system is correct. As the old saw goes, "A man with two watches will never know the correct time." More importantly, there is an extra cost attached to operating two databases instead of one. Information has to be rekeyed or downloaded into the new system, which could mean some data-collection delays or errors.

In spite of these disadvantages, many organizations have opted to run their ABC system on a stand-alone network. The reason offered most often is that it does not require the approval of the auditors; thus it can be up and running in a much shorter period of time. An integrated system requires a number of external and internal approvals, which could seriously delay or, worse, prevent the installation of the system altogether. Stand-alone systems are now operational in a wide variety of applications with excellent results in spite of the drawbacks of a dual system operation.

The Precision Question.   A number of issues must be decided regarding the precision provided by any cost system. With an ABC system, high precision is possible—but at a greater cost since there are many more variables to account for than in a traditional system. For that reason, precision may not be the primary objective. An ABC system does not have to be highly accurate to be highly effective. Since traditional systems have been *precisely* wrong for years, the ABC system that is *approximately* right will be a vast improvement.

To determine the amount of precision the organization can afford, a cost benefit analysis is conducted during the initial design phase. The 80/20 rule has application here. That is, 20 percent of an organization's products or services probably account for 80 percent of its cost. Once this relationship is established within the organization, selected costs and activities are chosen so that by very precisely controlling 20 percent (or whatever the ratio may be) of the activities, 80 percent of the costs are controlled. Of course, this will vary widely from organization to organization, but the concept remains valid.

As was seen in earlier examples, many traditional margins have been wrong by a factor of 100 percent or more. Choosing a less precise method of costing (interviews versus work measurement as an example) for establishing activity-based cost systems with margin variances of 10 to 20 percent may be perfectly acceptable—and more cost-effective.

The Complexity Factor.   The more activities that can be identified and the more cost drivers which can be related to these activities, the more precise the cost data will be. However, the system quickly becomes complicated, with a number of risks attendant to this design strategy. Cost is the most obvious. More importantly, users can be overwhelmed with excessive data, which would certainly discourage the use of the system over time. Ironically, the more data required for input, the more risk there is for error—precisely what system designers are seeking to avoid.

A compromise strategy may be to design a more complex system only in the early stages of the project. In doing so, the designers will be sure to discover all important activities and related allocators, which could be missed in a less comprehensive approach. Thus, by recognizing all the variables early in the design, a critical driver or activity has a much better chance of being caught prior to the implementation phase. Then, with the complex design completed, the designer has far greater options. The system can be installed as designed, installed in phases over time, or installed in a simplified format. It is far easier to pare down a design prior to installation than to regret having failed to include a crucial activity 6 months after installation.

THE PILOT PROJECT

Working Out the Design Flaws in Advance. In most situations the use of a pilot project prior to companywide installation is strongly suggested. A limited project scope will provide the design team a means to test, verify, and modify their work, as needed, without negatively impacting on the organization as a whole. It is here that major design flaws can be discovered and worked out. The importance of selecting the most representative area for a pilot project cannot be overemphasized.

Selecting the Pilot Project Area. Determining where to install the pilot project is fraught with risks. Giving careful consideration to the location of the installation within the organization, management support and the certainty of benefits will do much to improve the odds for a successful installation of the entire ABC system. One proven method of reducing these risks is selecting an area within the firm which is most representative of the organization. In that way the pilot project results can be transferred throughout the company without significant modification. However, the most important variable is that of a true management commitment within the area of the pilot.

Management Sponsorship. In considering in what area to conduct the pilot project, the greatest weight should be given to those managers who will provide the greatest support. The pilot project needs strong local management sponsorship to be successful, because those managers will later help sell the project to more tradition-oriented executives in the organization. Therefore, care should be taken so that every manager fully understands and supports the project.

The approach that produces the best results is an introductory management seminar where the benefits of the ABC system are explained. An open forum will quickly reveal who is the most enthusiastic manager. Have that manager champion the pilot. If a choice must be made between the best location and the area of strongest management support, choose the high level of support first. Enthusiastic management will make sure the pilot is a success, thus assuring the success of the roll-out throughout the entire organization.

Ascertaining Benefits. A third consideration in selecting the area for a pilot project is the speed and degree of success it will attain. It is not enough that the area under study be representative of the whole and that it has local management support; it must also be an area that can initially produce solid cost and benefit results. A project with high local management support may not succeed among more conservative organizational managers unless it shows more than modest cost benefits during this initial stage.

The choices made at this early stage will have the greatest impact on the success of the

ABC system than at any other milestone. It is imperative that system designers choose wisely at this critical junction.

IDENTIFICATION OF RESOURCE CATEGORIES

Resources.   The next logical step in the design process is the identification of the key resources. Resources include labor, material, utilities, and management. This step is not a difficult one because the resources we want are found in the general ledger accounts of the organization along with their dollar value.

A designer should not be inhibited by the current structure of the account categories. Such accounts can be combined as in the case of wages and benefits into one resource category, or additional accounts can be added to provide more detail showing more detailed categories of indirect labor. This decision is based on the needs of the organization.

ACTIVITIES

Definition.   Activities are a collection of actions made up of procedures and processes that cause work. Activities are incurred in direct response to the need to design, produce, market, and distribute products. Examples of activities include:

- Purchasing
- Receiving
- Disbursing
- Machine set up
- Machine operation
- Taking a customer order
- Redesigning a product
- Vendor negotiations

Identification of Activities.   To assist in determining the activities appropriate to a specific business, a model of the organization should be created. This model can be done by analyzing the product-line structure of the organization, then setting up the model using an organization chart format. The high-level value chain categories would include product introduction, operations, marketing, and postsales support. Within each of these functional categories, subcategories can be developed specific to the business. Perhaps the best way to determine subcategories is to develop an individual, specific bill of activities.

Bill of Activities.   A bill of activities is created to achieve the widest understanding of the relationship of activities to cost. Much like the bill of material found in manufacturing, this is a listing of the hierarchy which makes up each activity. As an example: designing a part is made up of a number of related activities and subactivities which might look like this:

1.0.0.0.0   Create part

  1.1.0.0.0   Generate product structure
    1.1.1.0.0   Design
    1.1.1.1.0   Prepare CAD drawing
    1.1.1.2.0   Assign part number

  1.2.0.0.0   Determine material sourcing

1.2.1.0.0   Cut requisition
1.2.2.0.0   Write purchasing specification
1.2.3.0.0   Qualify vendor

1.3.0.0.0   Procure material
1.3.1.0.0   Negotiate contract
1.3.2.0.0   Process order
1.3.3.0.0   Place order
1.3.4.0.0   Expedite

1.4.0.0.0   Internal processing
1.4.1.0.0   Receive order
1.4.2.0.0   Inspect
1.4.3.0.0   Add to inventory
    1.4.3.1.0   Set up bin location
    1.4.3.2.0   Move to stockroom
1.4.4.0.0   Price product
    1.4.4.1.0   Analyze and cost
    1.4.4.2.0   Research market
    1.4.4.3.0   Set price
1.4.5.0.0   Update catalog
    1.4.5.1.0   Design
    1.4.5.2.0   Write copy
    1.4.5.3.0   Photograph
    1.4.5.4.0   Typeset
    1.4.5.5.0   Paste-up

We can see that the generation of a new part triggers activities across the entire organization. Therefore, to cost this activity accurately, we need to develop the appropriate costs from engineering, purchasing, receiving, inventory control, accounting, marketing, and shipping.

Activity to Product Charting.   To fully understand the relationships between activities and costs, activities should be further separated between those which can be directly attributed to the product or service, and those that cannot. Figure 3.12 demonstrates how the New England China Company broke down its major activities into the appropriate subactivities. This activity and product matrix is very helpful to the ABC system designer, as such activity charting provides the needed understanding of the relationship of each cost with respect to the method of cost assignment to the product or service. At the New England China Company, four levels of cost attribution have been identified. The first, and preferred, is directly to the product. The next is through the product line itself. The third is through activity analysis. Finally, the company has elected to allocate administrative functions. This is the least desirable method; however, the designer has determined it to be the most cost-effective at this stage of the design.

TYPES OF ACTIVITIES

Another way of classifying activities besides the unit of work is to determine their functional levels within the organization. By developing a functional matrix, the system de-

Activity to Product Charting

Activities	Product Direct	Product Line	Activity	General
Engineering				
Product design	•	•		
Engineering change			•	
Product development				
Market research		•		
Material Acquistion				
Process purchase orders			•	
Qualify vendors			•	
Negotiate contracts			•	
Expedite			•	
Administration				•
Product Support				
Receive material			•	
Move material			•	
Store material			•	
Production planning	•			
Scheduling	•			
Setup	•			
Production				
Manufacture product	•		•	
Maintenence	•		•	
Marketing/Sales				
Advertising	•	•		
Promotion	•	•		
Selling	•	•		
Distribution	•	•		
Accounting				
Accounts payable			•	
Accounts receivable			•	
Payroll			•	
Credit/collections			•	
Human Resources				
Hiring	•	•	•	
Training	•	•	•	
Administration			•	
Corporate administration				•

FIGURE 3.12   The activities in the left-hand column are in high-level format. In practice subactivities which make up the high-level activity will be required in order to accurately make all cost attachments function in a linear manner. Each activity is categorized according to its ability to be directly attached to a product or the need for it to have to be attached indirectly through the use of product line, activity, or general allocation pools. A product pool summarizes all costs that can be directly related to the product. The product-line pool contains all costs relating to a particular product line. The activity pool contains support function costs such as engineering changes and setups. The general pool contains the costs that support the entire operation and which cannot be specifically related otherwise, such as the president, division management, and the like.

signer can readily see the impact of various activities at four key levels within the organization. Identification by unit level, batch level, product and service, and facility level of each of these activities will provide a clearer picture of how resources are being used.

Unit Level Activities. Some activities which are performed each time a unit is produced are

- Direct labor hours
- Machine hours
- Direct material hours
- Number of test hours

Batch Level Activities. Activities that are performed each time a lot or batch is produced are

- Setup hours
- Material handling
- Production order activity
- Issue purchase order for material

Product and Service Level Activities. These activities are performed, as needed, to support various product and service types, such as

- Product engineering change
- Parts administration
- Orders for supplies
- Sourcing material for a new product
- Cost of goods sold
- Establishing vendor relations

Facility Level Activities. These activities sustain the organization's general production process. A few of these activities could be directly related to the product and service levels; however, generally, they are not so clearly identified. These include

- Research and development
- Designing
- Processing customer orders
- Delivering product and services to customers

An example of how activities are identified at these various levels can be seen in Fig. 3.13. Here a detailed (low-level) list of activities demonstrates unit, batch, and product level activity relationships.

ACTIVITY CENTERS AND COST POOLS

Analyzing Activities. Since the ABC system is built around activities, it is important to understand clearly the major activities performed by the facility. Each overhead function must be identified to determine the activities that drive it. This determination is normally done by interviewing key managers and supervisors in the unit under analysis. In addition to determining activities performed, the interview will provide information on what drives those activities.

Functional Activity Matrix

Activity	Unit	Batch	Product
Setup hours		●	
Number of setups		●	
Avg. value of W.I.P.			●
Sales in $ and units	●		
Direct labor in a dept.	●		
Standard direct labor	●		
Cost of goods sold	●		
No. of:			
Shipments received			●
Shipments made			●
Parts purchased			●
Parts produced			●
Customer orders	●		
Purchase orders sent			●
Orders received			●

FIGURE 3.13 In addition to product charting, this component manufacturer categorized activities into three levels, unit, batch, and product. Thus a unit, batch, and product level pool was required.

Every salaried employee should be interviewed to determine the key responsibilities for each position and the activities performed. Separating employees into basic functional categories, such as product-related, equipment-related, and administration-related, facilitates the process.

Administrative employees cannot directly relate their activities to a product or service that caused their activity. But they can identify the individuals they supervise or support. Secretaries, administrators, and front office personnel will be in this category. The equipment-related employees will be able to identify equipment that causes their activities. Manufacturing engineering and maintenance supervision will be part of this category. The easiest group is the product and service-related individuals, who can directly trace their activities to the product and service itself, such as production and quality control.

Activity Centers.   Activity centers represent different stages of the production process. According to Robin Cooper, author of *Relevance Lost: The Rise and Fall of Management Accounting,* they can be machines, collections of machines, or entire departments. However, they are involved only indirectly in the actual assignment of costs to products or services. Once cost pools are established within each activity center, and resource and activity allocators are assigned, costs flow from the resource categories to each cost pool and then, in turn, to the products or services.

Cost Pools.   Cost pools are generated by splitting each resource category among activity centers. There is generally one cost pool for each resource category in each activity center. Defining cost pools for each organization ultimately affects the understanding of the way activities behave and the actual activity costs within each organization. Figure 3.12 shows the relationship of cost pools by product, product line, activity, and general allocation to a high-level list of activities.

Choosing Activity Centers.   Two considerations are to be taken into account when selecting activity centers. First, the designer must consider the effect of a separate activity center on product cost. The number of activity centers is not as crucial as how well the activity center reflects the important physical activities. The second consideration in selecting an activity center is the effect on the interpretation of activity cost. It must be easy for others to recognize the activity.

At this stage of development of the ABC system, there is no recommended optimal number of activity centers. However, it is suggested that in a process industry there should be a separate activity center for both plant and process level activities. The failure to distinguish between these two activities can lead to distortions and difficulty in interpreting data.

The type of activity centers will vary from company to company as well. The selection of activity centers depends upon the type of products produced, information needs, and company culture. For example, a chemical company may find it useful to have a "receiving raw materials" activity center. This center would be made up of a number of activities including inspection, material handling, and purchasing. Each cost pool within this activity center would be related to a product by a cost driver best suited for the purpose. In this instance it would be the number of raw materials used.

COST DRIVERS

Traditional Definition. A cost driver is an activity performed during the production or support of a product or service for which costs are associated. Using the term *cost driver* can create much confusion, as a number of definitions are currently in use for this term. For ABC costing purposes, it is used in this limited definition. However, a cost driver can be used in a much broader scale for operational purposes. In one definition cost drivers can be categorized into structural and executional drivers. A *structural* cost driver describes the economic makeup that drives the cost position for any given product and service group. These are strategic in nature and include scale, scope, experience, technology, and complexity. An organization with less experience will incur higher costs in comparison with a competitor of greater experience. Thus experience is said to drive cost.

Executional cost drivers determine an organization's cost position based on its ability to carry out an objective. These factors include the participation of the labor force, total quality management, capacity utilization, the efficiency of plant layout, production configuration, and the exploitation of the linkages with suppliers and customers through the organization's value chain.

The ABC Definition. For the purpose of this chapter, we confine a cost driver to the former definition, which simply means the *measure* of an activity. A cost driver implies which activities are taking place and how much each activity costs. It is therefore a method of *attributing* a cost to a product or service. With the correct cost driver, cost will increase in direct proportion to the level of activity, as measured by the activity driver. This is said to be *linear*.

Figure 3.14 illustrates how cost drivers bridge the gap between the first level of cost attribution—direct—and the last level and least preferred—allocation. The use of cost drivers provides the logical methodology to convert an indirect cost to a direct cost, thus providing a highly accurate and significantly more precise cost and control system.

Most academic accountants including Robert Kaplan, Robin Cooper, and Peter Turney use the term *cost driver* in describing the components of ABC. They go further and break down cost drivers into first- and second-stage components. However, many people find this terminology confusing because the term *cost drivers* does not provide sufficient hint of how they are used, without the added complication of trying to understand first- and second-stage drivers. Instead the terms *resource* and *activity drivers* will be substituted for the terms *first-* and *second-stage* cost drivers, respectively.

Figure 3.15 illustrates a number of cost drivers for overhead functions. In the example, activity drivers are related to various activity centers.

Cost Driver - The Missing Link

FIGURE 3.14 Traditionally, the choices for attaching costs to products and services were by either direct attribution, arbitrary allocation based on a non-volume-related formula, or both. The introduction of cost driver assigned attachments significantly improves the accuracy of all cost systems. It is the missing link in the chain of the cost attachment hierarchy bridging the gap between the most preferred and the least preferred method.

Determining Marginal Activity Cost. Since cost drivers measure the level of an activity, the ratio of the total cost of that activity to the level of that activity is the *marginal* cost of the activity.

$$\text{Marginal cost} = \frac{\text{total activity cost}}{\text{activity level}}$$

For example, if $1000 is the total cost for the production setup of floral pattern dishes (total activity cost) and 10,000 plates are produced (activity level), the setup cost (activity) of producing that lot of floral pattern dishes is $0.10 per dish produced (marginal activity cost).

Selecting the correct cost driver for each activity is not always easy. The correct cost driver will provide a cost behavior that increases in direct proportion to the level of activity, as measured by that cost driver. For example, if the activity being measured is setup, then the number of setups would be the appropriate cost driver assuming the cost of setups increases or decreases in direct proportion to the quantity of setups.

Using the Two-Stage Driver System. Just as traditional costing systems use two stages

Overhead Activity Drivers and Related Activities

Activity Center	Activity Drivers
Engineering	Engineering hours
Prototype/tooling	Piece-part volume
Administration	Number of workers
Human resources	Number of workers
Quality assurance	Number of setups
Waste treatment	Chemical dollar value
Purchasing	Number of purchase orders
Maintenance	Work hours

FIGURE 3.15 This table relates a number of activity drivers to their respective activities. As an example, engineering cost may be attached to a specific product using the number of engineering hours actually consumed by that product. This is a direct, or linear, relationship, as the cost of engineering varies directly with the number of engineering hours. Engineering hours will vary directly with the cost, whereas traditional costing systems lump engineering costs with a number of other support functions and allocate based on labor hours or some other arbitrary function having no linear relationship between engineering effort and a specific product.

of allocations, ABC utilizes a two-stage methodology, too. They are the resource driver stage and activity driver stage. Both serve as the method of unbundling costs. The benefit of this two-stage driver system is that we can understand which activity is taking place and how much each of those activities cost.

Resource Drivers.   This is the method of relating the cost of resources to be expended to the activity center that will consume the resource (first-stage cost driver). As an example, if we assume a manufacturing activity center to be a lathe, and the resource to be maintenance, we need to discover how maintenance cost relates to this activity center. One appropriate method of assigning the cost of this resource might be the number of maintenance hours. Thus maintenance hours becomes our *resource driver*. We can now show that there is a direct (linear) relationship between the activity center (lathe) and maintenance cost, because this cost will increase or decrease in direct proportion to the number of maintenance hours performed on this equipment. If head count is selected as the maintenance resource driver, such a relationship would be more difficult to establish because it is not a direct linear relationship. However, head count might be the appropriate resource driver to establish the relationship of administrative cost (resource) to customer service (activity).

Activity Drivers.   An activity driver provides the linear relationship (cause and effect) between activities within the activity center and the products and services which consume the activities. It traces cost from cost pools contained in each activity center directly to the product and service. An example of an activity driver might be "number of setups" and "number of inspections." It is at this level of the system that most of cost distortion is controlled.

Selecting the Best Activity Driver.   As with resource allocators, selection of activity drivers will be the ultimate determination of the accuracy of the ABC system. There are a number of choices in this regard. For example, Fig. 3.16 shows how we might categorize the activity, purchasing, into sublevel activities. We include negotiate contracts, qualify vendors, process purchase orders, expedite, and administrative support. Each activity has an appropriate activity driver which will directly relate to the variety (mix) of products and the product line. These include the number of suppliers, material classification, number of purchase orders, and head count.

Each one of the drivers can be used to provide a very accurate costing relationship. But, as was cautioned earlier, the complexity and subsequent cost of maintaining the information may outweigh the benefit. Therefore, the designer must determine which cost driver out of the group of drivers is the most representative of the cost of purchasing. In this example, the number of purchase orders provides the common link. Although the relationship is not entirely a linear one, because of activities such as administrative support and vendor qualification, it does reflect the predominant costs of processing and expediting.

Objective.   The basic objective in cost driver analysis is to identify the forces that influence cost behavior. There are three things to consider when identifying a driver: the cost of measurement, the degree of correlation, and its behavioral effect on the organization.

Selecting the appropriate cost driver for an activity is often not as simple as it would seem, though technically the methodology is straightforward. The sum of the unit allocators will equal the sum of the activity cost divided by the unit activity cost. If this relationship is valid, the activity and driver relationship has been isolated. However, often there is considerable ambiguity. When cost drivers do not capture this "cause and effect" relationship, confusion will result over the interpretation of a cost driver as a measure of an activity.

Cost of Measurement.   Will the benefit of determining the driver and reporting its effect be greater than the cost of adding it to the cost system? The law of diminishing returns is never more appropriate than when it is applied here. The more cost drivers, the more accurate the system will be—but the more costly it will be to operate. Therefore, consideration should be given to the marginal cost and benefit of including each additional driver in the system design.

Selecting the Best Common Activity Driver

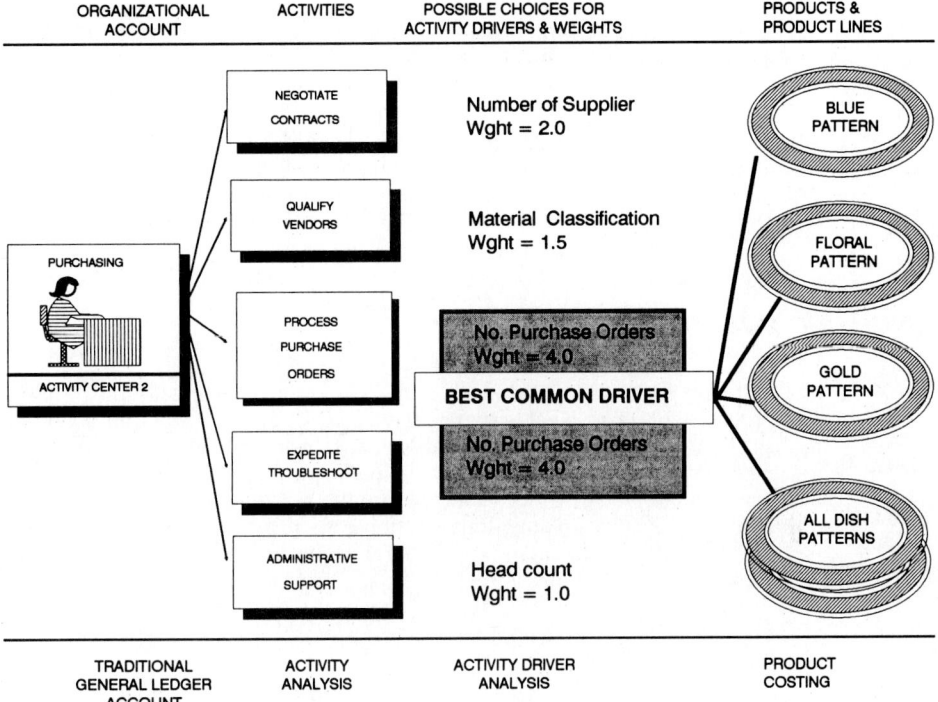

FIGURE 3.16 The system designer is faced with a wide variety of possible activity drivers. It is wise to keep an ABC system uncomplicated in the beginning. Therefore, the designer must look for the best common driver. In this example the number of purchase orders appears to be the best choice by using a spreadsheet and a system which weights drivers based on their impact on each activity. The outcome may not be an exact linear relationship with the specific activity of "negotiate contracts." However, for the group of activities, it has the highest weight. It is the best choice of a single driver for the majority of the purchasing activities.

Degree of Correlation. How each driver correlates to its respective activity is very important. The more linear the relationship the more accurate the assignment of cost. For example, if the actions of material handling and setup are combined to form one activity, then choosing a properly correlated cost driver becomes more complicated. Setup, by itself, has two drivers. One is time-based—number of hours. The other is unit-based—quantity of units. By adding in material handling cost, drivers relating to setups will have the effect of distorting the combination of material handling and setups. Using inspection hours (time-based) implies the longer the setup, the more costly material handling is—when material handling could be the same no matter how lengthy the setup time. Using number of setups (quantity-based) implies that the fewer the setups, the less material handling will cost. In the end, it still may be more cost-effective to combine both actions into one activity and use one driver that most closely approaches a linear relationship for both, even though some distortion will be built in.

Being practical, it may be more cost-effective to accept some distortion than to spend additional time and dollars to gather the needed data. Setting up the initial ABC system is fraught with a number of these potential inaccuracies. When the system is in place and underway, drivers can be more easily isolated as their effects on activities are fully understood. It is much more important to have the system framework in place than to be concerned with four-decimal accuracy.

Behavioral Effect. The third consideration in the selection of drivers is the behavioral effect they will have on the organization. A positive effect can be obtained by revealing the cost of activities in their direct relation to product costs, which were previously obscured or not available at all. As an example, engineering will now be able to understand how engineering changes impact product costs. There will be a direct measure of each change on a variety of products. By isolating purchasing and engineering activities for all to see, there will be a tendency to drive parts count down.

On the other hand, an ABC system will have a tendency to direct management toward high-volume, low-complexity production. Such an emphasis may technically maximize margins; however, low-volume specialized products may still be required to retain high-volume customers. A system designer must keep these behavioral tendencies in mind.

CHOOSING COST DRIVERS

Different cost drivers are required for different classes of activities. That is, unit, product, process, and plant level activities have distinct drivers. The unit level drivers might consist of a combination of direct labor hours, machine hours, or amounts of material. The process level consisting of depreciation, taxes, rent, process modification, and maintenance could include asset value, square footage, and maintenance hours, as an example. Plant level includes plant administration and the owning and maintaining of the plant. This highest level of resources would be assigned by head counts, maintenance hours, and square footage, to name a few.

Choosing cost drivers that have an obscure or misleading interpretation, such as using machine hours to assign building rent, increases the likelihood that activity costs will be interpreted incorrectly.

Criteria for Selecting Cost Drivers. There are a number of important criteria to consider when selecting cost drivers for inclusion in a system design. These include:

- Compatibility with operational measures used by manufacturing to monitor and control processes
- Meaningful parameters of operating controls
- Measurability in reasonable quantitative terms
- Ability to be used as the most convenient base for calculating indirect cost attribution to product and services

Other considerations which will aid in the selection of cost allocators include strategic initiatives, transaction volumes, administration and control, and complexity factors.

Strategic Initiatives. Part of the system design relies on the overall strategic plan of the organization. Using the strategic plan, the designer is able to focus on a number of critical activities, including marketing and sales expectations for the year. Training and development along with product, market research efforts, and special projects all impact on the type and amount of activity utilized in the ABC system.

Volume of Transactions. Volume is an important cost driver for a number of activities common to most businesses. These include customer service, purchasing, engineering, and service bureaus.

Administration and Control. Any center with the words "executive," "administration," or "control" in the titles falls into the administration and control driver category. These are typically head-count-driven.

Complexity Factors. Complexity tends to cause non-value-added costs to increase. Many overhead costs are driven by the complexity of production, not the volume of production. Therefore, nonvolume bases are required. A number of these costs are discretionary. Even though they vary with changes in the complexity of the production process, these changes are intermittent. Many overheads vary with transactions which are independent of the size of the order handled.

PULLING IT ALL TOGETHER

The implementation of an ABC system relies first on the identification of all costs which can be directly assigned to the product or service such as direct labor, material, and certain other costs. Second, costs which are not directly chargeable are isolated through a series of cost drivers. These costs are typically support costs. All remaining costs are then allocated through the traditional formulas. This procedure is used for both the first- and second-stage driver assignments, namely, resource and activity.

To minimize much of the confusion which this methodology creates, the use of a spreadsheet is very helpful. By listing all the potential drivers as in Fig. 3.17 and the key

Selected Manufacturing Processess and Cost Drivers

Key Processes

- Special Tooling
- Injection Molding
- Route
- First Firing
- Inspection
- Glaze
- Second Firing
- Inspection
- Place Decal
- Image Transfer
- Pin Stripe
- Clean
- Final Inspection
- Manufacturing Control

Key Cost Drivers

- Special Specifications
- Number of Pieces
- Size of Plates
- Batch Size
- Mix of Dish Sizes
- Number of Images/Dish
- Number of Setups
- Design Complexity
- Quality of Material
- Composition of Material
- Thickness of Dishes
- Fineness of Pin Stripe
- Customer Type
- Ambient Conditions

FIGURE 3.17 Fourteen basic manufacturing steps have been identified for the New England China Company. After identifying the key manufacturing processes, cost drivers need to be determined. The cost drivers listed here do not necessarily correspond to the adjoining process. These columns are random listings of processes and possible drivers. To correlate a driver or drivers to a process, a spreadsheet should be prepared using a system of weights to assign the appropriate drivers to the manufacturing processes.

manufacturing processes, the spreadsheet will facilitate the compilation of a table of weights for each driver under consideration. The designer may wish to make these determinations personally based on the results of the interviews. However, the most successful method is having each responsible manager provide an estimate of the relative importance of the driver for his or her area.

The designer can thus calculate the effect of each driver to the process or activity. If the cost effect is the strongest, that driver will be assigned the highest value; if it is the weakest it will have the lowest value. By going through all drivers in this manner, a total and average can be found for every driver. However, it may not be cost-effective to do every driver based on the considerations which were established earlier. An analysis similar to Fig. 3.16 should be made to consolidate drivers and determine the best common driver for a group of activities.

Reducing Drivers and Activities.   Prior to implementing the ABC system certain actions can be taken as a result of the activity and driver analysis. Cost drivers are determined by relating activities to events, circumstances, or conditions that create or "drive" the need for the activity and the resources consumed. These cost drivers should now be targeted for elimination or minimizing if they relate to non-value-added activities and optimized if they relate to value-added. Cost may be reduced in two ways: by the reduction of the activity itself or the reduction of the driver itself through process improvement.

It is important to understand that the reduction of drivers, which results in a reduced dependency on activities, does not lower costs until the excess resources are reduced or deployed into more profitable areas. The ABC system will provide all the relevant activities which make up the output of the organization. The system designer must then provide the reporting methodology so that management understands when resources are not being efficiently utilized.

At the conclusion of these efforts, the ABC system designer will have four categories of cost: direct, overhead driver costs, manufacturing or service process driver costs, and some minor left-over cost which must be arbitrarily allocated (Fig. 3.18). These costs can now be assigned to the organization's product and/or services with confidence that the new cost system will provide management with the most accurate costs possible.

The Final Results of ABC System Cost Assignments

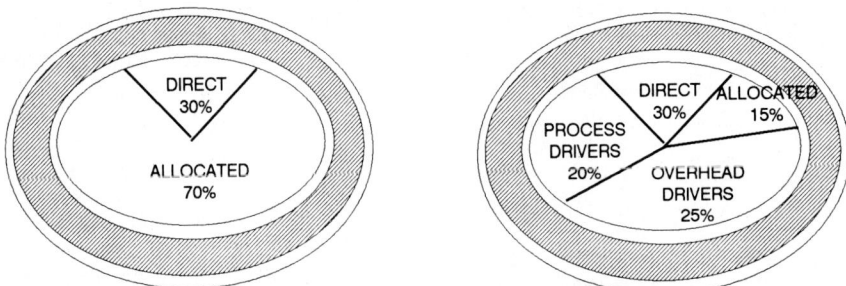

FIGURE 3.18   After all activities and the appropriate cost drivers are identified, costs can now be correctly assigned. In this example, the dish cost contained as much as 70 percent arbitrary cost assignment in the traditional cost system. Such arbitrary assignment meant that 70 percent of the costs did not vary directly with units of production. The ABC system leaves only 15 percent of costs independent of production volume. Later, designers may even wish to reduce this proportion if it is determined to be important and cost-effective.

FINAL IMPLEMENTATION

Once the pilot is operational, an in-depth analysis is conducted of the pro forma results of the new costs by a "top-level" comparison with existing cost system. Each cost pool is summarized listing costs from the most "direct to product" down. Costs pools which represent a summary of activities are explored to ascertain how well they represent the system design. Adjustments and changes are expected during this period.

Management Meetings.   Two important meetings should be held to inform management of the progress of the project. A results meeting should be held as soon as costs on the new system become available. This meeting should be limited to managers and engineers who are responsible for the product or service affected by the new data, because the data produced will be significantly different from the traditional company information system. Here the impact of the new product costs should be carefully explained. An objective of the meeting would be to reconcile product cost differences.

After the results, interpretive meetings should be held to instruct managers on how to interpret and use the new data. The objective of this meeting would be to explore what actions needed to be taken in light of the new data.

TRAINING AND DOCUMENTATION

Every stage of the ABC management project requires training to assure success. In the predesign stage, key managers must understand the benefits of an ABC project, because without their support, such an undertaking is not possible. Management must understand that designing a system that produces accurate product costs cannot guarantee the best system for controlling activities. Managers and users need to understand how to use and interpret the new information, which the ABC project will deliver—this means training. The implementation team must understand the theory and practice of ABC, so that they can design the appropriate system. Through the design and implementation processes there are a number of important milestones in which training plays a critical role in order to proceed to the next step.

The design plan should include at least the following:

- An ABC seminar for the plant management team
- A design seminar for the implementation team
- A progress meeting to be held at least biweekly to report findings
- An executive seminar for plant management to explain the final design
- A results meeting for select product and service line managers
- Interpretative meetings for industrial engineers and product line managers

THE BASIC REPORTING SYSTEM

An ABC reporting system does not need to be complicated to be effective. However, even the most basic system includes provisions for

- Cost control
- Product and service costing

Figure 3.19 provides a schematic model of an ABM reporting system. To accomplish its objectives the system needs to control and/or report on

- Activities
- Performance measures
- Responsibilities
- Use of resources

For a period to be useful, it is important for the user to be involved in the design process. This doesn't mean that an ABM system provides reports that only users want, but

A Complete ABM Reporting System

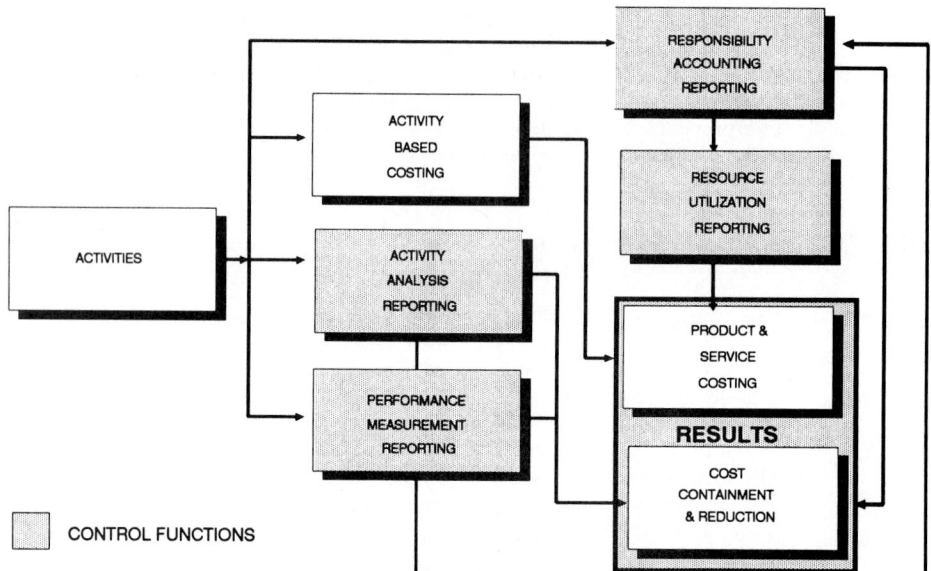

FIGURE 3.19 Putting together a complete cost management system utilizing ABC systems requires a means of performance measurement, a method of assigning responsibilities for results, and a method to focus on the efficient use of resources. This schematic model represents a comprehensive ABM system which incorporates all these functions.

the system will be ineffective if it is based only on what the designer feels is needed. There is a vast difference between a want and a need which the designer should understand.

Each report should contain only the relevant data for each manager. The simpler the presentation the more often the data will be used. For example, a report to purchasing should contain only information which relates to the person to whom it is submitted, and should show only information which is controllable by the person who is receiving it—in this case the director of purchasing. It is not necessary to report all data—only significant data.

The ABM system reports only by exception. Only items which are out of line with standards or budgets are reported—that is information. The report is useful only if it provides information. Information is that portion of data which informs the user. Providing just data for the sake of data wastes time and paper and is often confusing to the reader. Therefore, reporting by exception lets the reader know (provides information) that certain data are out of range relative to a standard. The user can then take corrective action.

ABM reports by level of responsibility in a pyramid design. First, there is a summary report which is then backed by a detail report. Operating management needs detail of individual transactions, middle management needs more of a summary, and upper management will receive trends, highlights, and exceptional items. Reports are provided more frequently at lower levels (that is, hourly to daily at the shop floor) and less frequently at senior levels (or weekly, monthly, quarterly). The ABM reporting structure parallels the organizational structure.

All variances which are unfavorable are shown in brackets. All positive variances are shown without brackets. This avoids confusion of terms.

The use of standards in the ABM system is very important. However, the standards used are historical rather than measured. This is because measured standards are difficult to keep current, if at all, whereas historical standards can be updated on each report based on a rolling average. In that way a comparison with perfection can be made as responsible managers work to reduce their rolling standards based on a set of management goals. As an example, the standards committee might set a group of standard targets each period which managers must accomplish in order to achieve their bonus.

Just as manufacturing techniques change through the manufacturing cycle, so will the need for information change. As significant changes take place, the ABM reporting system should be reviewed to determine what information should be added and what should be dropped. It is a good practice to review the entire reporting system regularly to keep current with new developments.

THE CRITICAL AUDIT

The costly effort of implementing an ABC system can be for naught unless a method is designed to monitor the effectiveness of the system. The moment any system is in place, external and internal forces are at work eroding its effectiveness. Such forces could include employee indifference, new manufacturing methods, employee turnover, management change, market conditions, and a host of other factors. The safest way to assure that the new ABC system is functioning as it should is to establish a regular audit procedure.

During the first year of operation there will be a need to make certain adjustments in order to assure that all users are satisfied. There should be extensive follow-up interviews at 3-, 6-, and 12-month intervals during the first year, which is the time when resistance is the highest. For the system to be successful, the design team must diffuse all objections by making fast corrections.

During the second year, an audit review at the 6- and 12-month periods should be sufficient. Thereafter, an annual systems' audit should be conducted to assure the system is modified to conform to operational changes during the year.

One international financial services firm installed systems three different times over 10 years because they failed to monitor their operations yearly. Had monitoring taken place, the time and expense of a completely new system could clearly have been avoided.

SUMMARY

The experiences of managers who have used ABC systems in a wide variety of manufacturing, administration, and service environments indicate that a properly designed ABC system provides a strategic and tactical advantage far superior to more traditional systems. ABC costing helps managers understand and eliminate complexity. It provides managers with "true" product costs and removes bad cost information from the management decision-making equation. Activity-based costing also helps managers understand the impact of sourcing decisions.

Activity-based costing can change the way managers determine the mix of their product line, price their products, and analyze the impact of new technology. The designer of an ABC system has the ability to choose cost drivers which can strongly influence behavior and are highly motivational. In addition, such a system will yield an inordinate amount of information which managers may choose to eliminate.

The challenge for system designers is to establish an ABC system that provides not only accurate product and service cost information but also information on activities that can be easily and correctly interpreted. ABM is a very useful control tool for any organization.

Through the use of rolling standards and the ability to control activities, ABM will be the management system of choice for those organizations wishing to compete in the global marketplace.

BIBLIOGRAPHY

Articles

Beaujon, George J., and Vinod R. Singhal, "Understanding the Activity Costs in an Activity-Based Cost System," *Journal of Cost Management,* spring 1990, pp. 51–72.

Borden, James P., "Review of Literature on Activity-Based Costing," *Journal of Cost Management,* spring 1990, pp. 5–11.

Cooper, Robin, and Robert S. Kaplan, "Measure Costs Right: Make the Right Decisions," *Harvard Business Review,* September–October 1988, pp. 96–103.

Cooper, Robin, "Implementing Activity-Based Costing at a Process Company," *Journal of Cost Management,* spring 1990, pp. 43–50.

Cooper, Robin, "The Rise of Activity-Based Costing—Part One: What Is an Activity-Based System?" *Journal of Cost Management,* summer 1988, pp. 45–54.

Cooper, Robin, "The Rise of Activity-Based Costing—Part Two: When Do I Need an Activity-Based System?" *Journal of Cost Management,* fall 1988, pp. 41–48.

Cooper, Robin, "The Rise of Activity-Based Costing—Part Three: How Many Cost Drivers Do I Need?" *Journal of Cost Management,* winter 1989, pp. 34–46.

Cooper, Robin, "The Rise of Activity-Based Costing—Part Four: What Do Activity-Based Cost Systems Look Like?" *Journal of Cost Management,* spring 1989, pp. 38–49.

Cooper, Robin, "You Need a New Cost System When...," *Harvard Business Review,* January–February 1989, pp. 77–82.

Frank, Gary B., Steven A. Fisher, and Allen R. Wilkie, "Linking Costs to Price and Profit," *Management Accounting,* June 1989, p. 22.

Goldhar, Joel D., and Mariann Jelinek, "Plan for Economies of Scope," *Harvard Business Review,* November–December 1983, pp. 141–148.

Harvey, Thomas W., "Cost Drivers: A Different Approach to Management Information in Banks," *The Journal of Bank Cost & Management Accounting,* spring 1990, pp. 5–28.

Johannson, Hank, "The Revolution in Cost Accounting," *P & IM Review and APICS News,* January 1985, pp. 42–46.

Johnson, H. Thomas, "Activity-Based Information: Blue Print for World-Class Management Accounting," *Management Accounting,* June 1988, pp. 23–30.

Jones, Lou, "Competitive Cost Analysis at Caterpillar," *Management Accounting,* October 1988, pp. 32–39.

Kaplan, Robert S., "One Cost System Isn't Enough," *Harvard Business Review,* January–February, pp. 61–66.

Kaplan, Robert S., "Yesterday's Accounting Undermines Production," *Harvard Business Review,* July–August 1984, pp. 95–101.

McCormick, Edmund J., Jr., "The Power P&L," *The Journal of Bank Cost & Management Accounting,* spring 1990, pp. 39–52.

McNair, C. J., "Interdependence and Control: Traditional vs. Activity-Based Responsibility Accounting," *Journal of Cost Management,* summer 1990, pp. 15–24.

Miller, J. G., and Vollman, "The Hidden Factory," *Harvard Business Review,* September–October 1985, pp. 142–150.

O'Guin, Michael, "Focus the Factory with Activity-Based Costing," *Management Accounting,* February 1990, pp. 36–41.

Ostrenga, Michael R., "Activities: The Focal Point of Total Cost Management," *Management Accounting,* February 1990, pp. 42–49.

Rotch, William, "Activity-Based Costing in Service Industries," *Journal of Cost Management,* summer 1990, pp. 4–14.

Roth, Harold, and A. Faye Borthick, "Getting Closer to Real Product Costs," *Management Accounting,* May 1989, pp. 28–33.

Sapp, Richard W., David M. Crawford, and Steven A. Rebischke, "Activity-Based Information for Financial Institutions," *The Journal of Bank Cost & Management Accounting,* spring 1990, pp. 53–62.

Shank, John K., and Vijay Govindarajan, "Strategic Cost Analysis: The Crown Cork & Seal Case," *Journal of Cost Management,* winter 1989, pp. 5–16.

Shank, John K., and Vijay Govindarajan, "Transaction-Based Costing for the Complex Product Line: A Field Study," *Journal of Cost Management,* summer 1988, pp. 31–38.

Troxel, Richard B., and Milan G. Weber, Jr., "The Evolution of Activity-Based Costing," *Journal of Cost Management,* spring 1990, pp. 14–22.

Turney, Peter B. B., "Ten Myths about Implementing an Activity-Based Cost System," *Journal of Cost Management,* spring 1990, pp. 24–32.

Turney, Peter B. B., "Using Activity-Based Costing to Achieve Manufacturing Excellence," *Journal of Cost Management,* summer 1989, pp. 23–31.

Books

Berlant, Debbie, Reese Browning, and George Foster, *Tomorrow's Accounting Today: An Activity Accounting System for PC Board Assembly,* CAM-I, Arlington, Tex., 1989.

Berliner, C., and James A. Brimson, *Cost Management for Today's Advanced Manufacturing: The CAM-I Conceptual Design,* Harvard Business School Press, Boston, Mass.

Bromwich, M., and A. Bhimani, *Management Accounting: Evolution Not Revolution,* CIMA, London, England, 1989.

Johnson, H. Thomas, and Robert S. Kaplan, *Relevance Lost: The Rise and Fall of Management Accounting,* Harvard Business School Press, Boston, Mass., 1987.

Lee, J. Y., *Managerial Accounting Changes for the 1990's,* McKay Business Systems, Artesia, 1987.

Porter, Michael E., *Competitive Advantage,* The Free Press, New York, 1985.

Staubus, George J., *Activity Costing and Input-Output Accounting,* Richard D. Irwin, Homewood, Ill., 1971.

Case Studies

Cooper, Robin, and Peter B. B. Turney, "Hewlett-Packard: Roseville Network Division," *Harvard Business School Case Series,* pp. 198–117.

Cooper, Robin, and Peter B. B. Turney, "Tektronix: Portable Instruments Division," *Harvard Business School Case Series,* pp. 188–142, 143, 144.

Cooper, Robin, and K. H. Wruck, "Siemens Electric Motor Works (A)," *Harvard Business School Case Series,* pp. 189–089.

Cooper, Robin, "Schrader Bellows," *Harvard Business School Case Series,* pp. 186–272.

Kaplan, Robert S., "Kanthal," *Harvard Business School Case Series,* pp. 190–002/003.

Kaplan, Robert S., "American Bank," *Harvard Business School Case Series,* pp. 187–194.

Kaplan, Robert S., "John Deere Component Works," *Harvard Business School Case Series,* pp. 187–107/108.

CHAPTER 4
COST ESTIMATING

J. E. Nicks

President
MiCAPP®, Inc.
Big Rapids, Michigan

Almost every manufacturing endeavor requires some form of estimating cost. Whether the activity is building a new skyscraper, designing an automobile, or developing a new bakery recipe, all require an estimate of cost. This chapter discusses every aspect of estimating cost and concludes with a listing of skills necessary for a cost estimator to possess.

Reasons Why Estimates Are Made. Cost estimates are developed for a variety of different reasons. The most important reasons are shown below.

Should the Product Be Produced? When a company designs a new product, a detailed estimate of cost is developed to assist management in making an intelligent decision about producing the product. This detailed estimate of cost includes an estimate of material cost, labor cost, purchased components, and assembly cost.

In addition to product cost, many other elements must be estimated. These include all tooling costs. A cost estimate must be developed for jigs, fixtures, tools, dies, and gauges. Also, the cost for any capital equipment must be entered into the estimate. These figures are usually supplied through quotations by vendors. An estimate of this nature will include a vast amount of detail, because if management approves the project, the estimate now becomes the budget.

Estimates as Temporary Work Standards. Many companies that produce product in high volume, such as automotive companies, will use estimates on the shop floor as temporary work standards. Temporary work standards are replaced with time studied work standards as rapidly as possible.

Cost Control. A job shop (contract shop) will use a cost estimate for cost control purposes because lot sizes are small and job shops seldom estimate work standards for what they produce. This use of an estimate for this purpose is different from temporary standards in that it uses the "meet or beat philosophy."

Make-or-Buy Decisions. When a company sets out to produce a new product, many components in the bill of materials are subject to a make-or-buy decision. A cost estimate is developed for comparison purposes. There are usually considerations aside from piece part cost. These may include tooling cost, vendor quality, and vendor delivery. See Sec. 9, Chap. 5 for further details.

Determine Selling Price. An estimate is used to determine selling price. The estimate is always a reflection of actual cost. In most organizations the marketing department has the responsibility of establishing a selling price, which can be substantially different from the cost estimate. There are many reasons for this. For example, a contract shop might be willing to sell the first order at something less than the estimate to develop a new customer.

Check Vendor Quotes (Purchase Analysis). An estimating function is often established for the sole purpose of checking vendor quotations on outsourced work. One automobile company has an entire department of cost estimators devoted to this task.

DIFFERENT TYPES OF ESTIMATES

Estimates can be developed in a variety of different ways depending upon the use of the estimates and the amount of detail provided to the estimator.

Importance of Understanding Estimating Methods. Every estimator should understand every estimating method and when to apply each, because no one estimating method will solve all estimating problems.

Guesstimates. Guesstimating is a slang term used to describe an estimate than lacks detail. This type of estimate relies on the estimator's experience and judgment. There are many reasons why some estimates are developed using this method. One example can be found in the tool and die industry. Usually, the tool and die estimator is estimating tooling cost without any tool or die drawings. The estimator typically works from a piece part drawing and must visualize what the tool or die looks like. Some estimators develop some level of detail in their estimate. Material cost, for example, is usually priced out in some detail, and this brings greater accuracy to the estimate by reducing error. If the material part of the estimate has an estimating error of plus or minus 5 percent and the remainder of the estimate has an estimating error of plus or minus 10 percent, the overall error is reduced.

Budgetary. The budgetary estimate can also be a guesstimate but is used for a different purpose. The budgetary estimate is used for planning the cost of a piece part, assembly, or project. This type of estimate is typically on the high side because the estimator understands that a low estimate could create real problems.

Using Past History. Using past history is a very popular way of developing estimates for new work. Some companies go to great lengths to ensure that estimates are developed in the same way actual cost is collected. This provides a way to use past history in developing new estimates. New advancements in group technology (explained later in this chapter) now provide a way for the microcomputer to assist in this effort.

Estimating in Some Detail. Some estimators vary the amount of detail in an estimate depending on the risk and dollar amount of the estimate. This is true in most contract shops. This level of detail might be at the operation level where operation 10 might be a turning operation and the estimator would estimate the setup time at 0.5 hour and the run time at 5.00 minutes. The material part of the estimate is usually calculated out in detail to reduce estimating error.

Estimating in Complete Detail. Where the risk of being wrong is high or the dollar amount of the estimate is high, the estimator will develop the estimate in as much detail as possible. Detailed estimates for machinery operations, for example, would include calculations for speeds, feeds, cutting times, load and unload times, and even machine manipulation factors. These time values are calculated as standard time and adjusted with an efficiency factor to predict actual performance.

Parametric Estimating. Parametric estimating is an estimating method developed and used by trade associations. New housing construction can be estimated on the basis of cost per square foot. There would be different figures for wood construction as compared with brick and for single-story construction as compared with multilevel construction.

Some heat-treating companies price work on a cost per pound basis and have different cost curves for different heat-treating methods.

Project Estimating.   Project estimating is by far the most complex of all estimating tasks. This is especially true if the project is a lengthy one. A good example of project estimating is the time and cost of developing a new missile. The project might take 5 years and cost millions of dollars. The actual manufacturing cost of the missile might be a fraction of the total cost. Major projects of this nature will have a PERT network to keep track of the many complexities of the project. A team of people with a project leader is usually required to develop a project estimate.

HOW ESTIMATES ARE DEVELOPED

Estimates are developed in a variety of ways depending on a number of different factors. These factors include the purpose of the estimate, how the company is organized, the amount of lead time to prepare the estimate, and the complexity of what is being estimated.

Estimating Accuracy and Consistency.   Another very important factor in how estimates are developed is the need to control accuracy and consistency of the estimate. One person making an estimate will be far more consistent than an entire department participating in the estimate's development.

Single Person.   In many companies, especially smaller ones, one person develops a cost estimate. This person usually has been in the business for several years and has had experience in both manufacturing whatever is being estimated as well as estimating experience. As pointed out above, a single-person estimate tends to be more consistent than a group estimate. If estimates are consistent, they can be made accurate by the application of math formulas such as learning curves, explained later in this chapter.

Committee Estimating.   Committee estimating is used especially where there is a lack of detail about the product being estimated. In developing a budgetary estimate for a helicopter transmission, for example, there are no detailed part drawings at this stage of the helicopter's development. Assembly sketches (loftings) are provided to illustrate the transmission size and weight. The collective judgment of the committee will provide a better estimate than the judgment of the individual. Also, parametric estimating is frequently used in this situation. The cost of spiral bevel gears can be estimated very accurately based on weight and number of teeth.

Department to Department.   Some companies develop estimates by moving the estimating paperwork through each department that can contribute to the estimate. Purchasing provides material cost, manufacturing engineering provides the process, industrial engineering provides the time values, and production control provides the machine loading. There are advantages and disadvantages to this procedure in developing an estimate. The chief advantage is that each person contributing information to the estimate is an expert in his or her field. The chief disadvantage is the amount of turnaround time to develop the estimate. Each person who makes a contribution to the estimate has other duties, and estimating new work usually is not a high priority.

Reporting Relationships.   Reporting relationships are very important, especially in manufacturing firms. The estimating function usually reports to the person in charge of manufacturing, typically the manufacturing manager. The theory behind this is that if people who report to the CEO of manufacturing contribute information to the estimate, they must live with it if the project is booked. Another reason for this reporting relationship is that marketing is usually given some authority over price but not over cost. Conventional organizational wisdom will rarely permit marketing to govern estimating.

COST DETAILS

In most companies, the elements of cost details making up the estimate are derived from the accounting department. Cost accounting is the function that collects actual cost, and there must be some correlation between how an estimate was developed and how actual cost is collected. The need for this correlation is based on improving estimating accuracy in the future.

Cost Details Descriptions.   Shown below is the standard industrial description of each major cost element.

Direct Labor.   Direct labor is that labor expended to add value to the product. This includes machine operators, assembly, and test people.

Indirect Labor.   Indirect labor is the labor required to support direct labor. While these people are required to support production, they add no value to the product being produced. They include crib attendants, material handlers, and shipping and receiving employees.

Direct Materials.   Direct materials are part of the product as it is shipped. These are both manufactured items and purchased finished components.

Indirect Materials.   Indirect materials are necessary to make and test the product but are not part of the product as it is shipped. One example of an indirect material is the fluid necessary to test a pressure vessel. The pressure vessel must be tested but is not shipped with the testing material. Indirect materials are part of product cost, but the material is used over and over again.

Overhead.   Overhead is an accounting term. This cost includes salary and management cost. In addition, overhead includes all other costs not listed above. Cost elements such as machinery cost, insurance cost, and office supplies are included. Overhead costs are usually expressed as a percent of direct labor cost for cost estimating purposes. See Sec. 9, Chap. 3 for a more accurate approach to allocating overhead cost.

General, Administrative, and Profit.   Some companies include general and administrative costs in overhead, while others add these in the estimate as an extra percent factor. Generally these costs include sales commissions, and administrative costs might include president and vice president wages. Overhead costs and general, administrative, and profit factors are determined by the accounting department, not the cost estimator.

THE SCREENING PROCESS IN A JOB SHOP

The screening process of what will and will not be quoted is a very important first activity in the estimating process.

The Sales Function.   In many job shops the sales force is paid on the basis of salary plus commission. This means the salesperson can make more money by "bringing in more paper" (items to quote). This creates a dilemma for the cost estimating function. The nature of this dilemma is explained below.

Quote Everything.   Some companies have a policy to quote everything that the shop has the capability of making. This can place a burden on the estimating function. As the volume of paperwork grows and turnaround times are held fast, less time is spent on any one individual estimate. This can lead to costly errors because of the lack of detail in the estimate and/or omissions in the estimates.

Committee Screening.   Other companies use a screening committee to be more selective of what is quoted. This committee is usually comprised of representatives from the following departments: marketing, estimating, manufacturing, accounting, and production control. Each member on the committee has a specific talent and can contribute a unique viewpoint. Whether a job is quoted or not is handled in two ways. The first is that all committee members have an equal vote. The second is that the representative from manufacturing has the final say.

The screening approach is very effective because the potentially unprofitable jobs are weeded out and the workload of the estimating function is more closely controlled.

Volume of Work versus Estimating Accuracy.   It is not necessarily true that the more time spent on an estimate the more accurate it becomes. Estimating accuracy depends on many factors. One of these is the volume of work expected from the estimating function. It is true that the less time spent on an estimate the more inaccurate the estimate will be. A balance must be found between the amount of time spent on an estimate versus the expected accuracy. In addition, it is important to remember that each estimator should spend about 30 percent of his or her time in reviewing past estimates and comparing these with actual performance. This review should include process lineups, time values, material costs, scrap rates, and outside service requirements.

STANDARD DATA

Standard data are defined as standard time values for all the manual work in an estimate. Standard data provide the opportunity for the estimator to be consistent in developing an estimate.

How Standard Data Are Developed.   Standard data are developed in a variety of ways depending on the industry that uses them. Experience shows that it is easier to develop standard data for a machining operation as compared with fabrication operations. This is because machining operations can be calculated by using speeds, feeds, and lengths of cut to determine time values. Most of the work content of a fabrication operation is manual effort rather than machine time, and for this reason reliable standard data for the fabrication industry are difficult to find. Listed below are the basic methods used to develop standard data.

Past History.   Many companies use past history or actual performance on jobs produced to develop standard data. Developing standard data this way rarely considers the best method of organizing work. This method is popular in smaller companies that do not have industrial engineers or time study engineers.

Time Study.   Larger, well-organized companies will develop standard data from stopwatch time studies. Time studies are used to establish rates of production. However, when time studies are also used to establish standard data, care must be taken in defining element content so work content can be isolated. Time study engineers must be taught how to establish the element content of their studies in a way that will permit the development of standard data.

Predetermined Time Standards.   Another approach in the development of standard data is to use one of the many predetermined time standards systems like MTM or MOST. (See Sec. 4, Chap. 5.) This method has its advantages and disadvantages. The chief advantage is consistency of the data, and the chief disadvantage is the amount of time necessary to develop the data. Some predetermined time standard systems are now computerized, which shortens the development time.

Standard Data Specific to a Shop and Lot Size.   It should be pointed out that all standard data are specific to a given shop and lot size. Standard data developed in a high-production shop under ideal methods are of little value to a job shop that runs lot sizes of 10 parts each. The reverse is also true. The use of efficiency factors or off standard factors can assist in using the same data for both conditions, but this is less than ideal. The reverse use of learning curves, that is, backing up the curve, is a better method of repricing work for small lot sizes. Using this method, the same standard data can be used for high and low production.

ESTIMATING MATERIAL COST

Many estimators will spend a great amount of time in developing the material content of an estimate as compared with the labor content. The theory behind this practice is that the material part of the estimate can be developed with a high degree of accuracy. The labor content in the estimate is subject to more error. When the material cost is added to the labor cost, it tends to make the entire estimate more accurate.

Wherever possible, if time permits, the estimator will get outside quotes for material cost. This is the best way to improve the overall accuracy of an estimate.

Bar Stock Estimating. Shown below is an example of a bar stock estimate for a part. The reader's attention is drawn to the amount of detail the estimator goes into to develop this part of the estimate.

Bar Stock Example. To develop an estimate for bar stock, we need to make some assumptions. The part length is 6 inches, the diameter is 1.875 inches, the width of the cutoff blade is 0.125 inch, the lot size is 200 parts, and the material is cold-rolled steel. The estimator will assume a 3 percent scrap factor; that is, enough material will be purchased to make 206 parts and the price per part will contain the scrap material cost. The calculations are shown below.

> 2-inch-diameter cold-rolled steel comes in 12-foot lengths and weighs 128.2 pounds per bar.
>
> Part length is 6 inches plus cutoff of 0.125 inch = 6.125 inches.
>
> 12-foot bar × 12 inches per foot = 144 inches long.
>
> 144 inches minus 2-inch chucking end = 142 inches.
>
> 142 divided by 6.125 = 23 parts per bar.
>
> 206 parts divided by 23 parts per bar = 8.8 bars or 9 full bars.
>
> 9 bars × 128.2 pounds per bar = 1153.8 pounds of material or 1154 pounds.
> $0.35 cost per pound = $403.90 total.
>
> 1154 × $403.90 divided by 200 parts = $2.02 per part.

This might seem like a great amount of detail to develop a material estimate, but it should have an accuracy of plus or minus 1 percent. Also, there are computer programs that will speed up the estimating process.

Other Considerations. Other considerations must be given to quantity extra cost, delivery, and packing extra cost as well. If the lot size is small, say 10 parts, the estimator must consider a sawing charge for part of a bar rather than purchasing one entire bar and charging the customer for a full bar. The reader will observe that there is not a great amount of skill in the development of this material but there is some detail.

Other Material. Other materials such as sheet stock estimates are handled in much the same way as bar stock. Typically, casting and forging estimates are developed by the supplier because the methods used to develop these costs are usually unique to the casting or forge shop.

ESTIMATING COMPONENTS OR ASSEMBLIES

The estimator must work through an array of details to develop an estimate. The amount of time spent and detail developed will usually be governed by the amount of dollar risk. If that risk is small, the estimator will rely on his or her judgment, but if the risk is large, the estimator will spend more time on the estimate. While reviewing the detail the estimator must work through, the reader should keep in mind the number of decisions the estimate must make. Each decision is critical to making a good estimate.

Bill of Material and Make or Buy.   If there is a bill of material, the first step is to review it and make decisions about what will be made and what will be purchased. Some of these decisions are not difficult. As an example, all fasteners might be purchased. The cost of each purchased component must be obtained from vendors, catalogs, or the estimator's judgment.

Labor Content.   For those items in the bill of material that are marked MAKE, the estimator must develop a process or operation lineup of how these components will be made. Again, the level of detail depends on the lot size and dollar risk.

The estimator may have access to past history on similar parts to include in the estimate. The estimator must estimate a setup and run time for each operation that each part must progress through.

Some operations, such as heat treat, are considered as overhead because there may not be a good way to collect actual cost per piece.

Work Center Rates.   The cost accounting department has the responsibility of developing work center rates to apply to the estimates. Most shops will have different work center rates for each different type of machine. This is necessary to preclude selling an hour of time on a drill press for the same amount as an hour of time on an NC machining center.

Cost versus Price.   All estimators understand that their job is to develop cost, not price. This cost is the estimator's best judgment of what the shop's actual performance will be. Most marketing departments are given some latitude to adjust cost into price if necessary. The shop's actual performance, however, is always compared with the estimate of cost, not price.

MATERIALS AVAILABLE TO DEVELOP AN ESTIMATE

Materials available for developing an estimate vary widely depending on what is being estimated. In most cases the quality of the estimate will depend on the amount of materials to make the estimate.

Estimating Materials.   Shown below is a listing of the materials available for making an estimate.

No Drawings.   In many cases there are no drawings of what is being estimated. One clear example of this is tool estimating. The estimator will develop an estimate for a progressive die, for example, by reviewing the piece part drawing. Some die estimators will develop a strip layout for the part and then estimate the die cost station by station.

Sketches.   Sometimes sketches of the part(s) represent the only information available. This is typically true for a budgetary estimate.

Line Drawings.   Loftings or line drawings are used for estimating in some industries. The pleasure boat industry represents an example. A full-scale lofting of a deck and hull is used to estimate both the material and labor for a new fiber glass boat.

Complete Drawings.   Complete drawings and specifications are available for estimating some work. The aircraft industry is one good example. Many times the estimator will spend more time reading the specifications than developing the estimate. This is necessary because the specifications will often determine the part process.

COMPUTER ESTIMATING

Computer estimating has become very popular in recent years primarily because of the advent of the microcomputer. Early efforts of computer estimating date back to the early 1970s but were cumbersome to use because they were on a mainframe and were card-driven. No less than 15 U.S. companies now offer estimating software for a microcomputer. Because the computer estimating industry is new, there are no real standards for estimating programs. Some programs are nothing more than a way to organize the calculations of an estimate, while others calculate all the details of the estimate.

Advantages and Disadvantages.   Shown below are some of the major advantages of computer cost estimating.

Accuracy versus Consistency.   Computer estimates are very consistent, provided they calculate the details of an estimate. Because these estimates are consistent, they can be made to be accurate. Through the use of consistent efficiency factors or learning curves, estimates can be adjusted up or down. This is one of the chief advantages of computer cost estimating.

Levels of Detail.   Some computer estimating systems provide different levels of estimating cost. The level of detail selected by the user depends on the dollar risk. Many estimators produce an estimate in more detail because the computer can calculate speeds and feeds, for example, much faster than an estimator can on a hand-held calculator.

Refinements.   Some computer estimating systems provide many refinements that would be impossible for the estimator to do in any timely manner. One example is to adjust speeds and feeds for material hardness. Typically, the harder the material the more slowly a part will be turned or bored. Another refinement is the ability to calculate a feed rate and adjust it based on the width of a form tool.

Source Code.   Some companies offer the source code uncompiled to their users. This is important because it affords the user the opportunity to customize the software. In addition, many companies have written their own software to do something that is not available on the market. If the source code is not compiled, the user can build upon a computer estimating system.

Disadvantages.   The chief disadvantage of computer estimating is that no one estimating system can suit everyone's needs. This is especially true if the source code is compiled and not customizable.

Another problem with computer estimating is that the estimator will, in all probability, have to change some estimating methods. Computer software for estimating cost is seldom written around one method of estimating.

GROUP TECHNOLOGY

Group technology is not new; it was invented by a Russian engineer over 30 years ago. Unfortunately, the subject is not taught in many of our colleges and universities, which partially accounts for the lack of use in American industry. Group technology (GT) is a coding system to describe something. Several proprietary systems are on the market. One such system, the MiCAPP system,™ uses four code lengths, a 10-, 15-, 20-, and 25-digit code. The code length selected is based on the complexity of the piece part or tool being described.

Uses for Group Technology.   Shown below are several uses for group technology along with several examples of use both internally and externally.

Cost Estimating.   GT can be used very effectively in estimating cost. Assume a company manufactures shaft-type parts. Also assume there is a computer database named SHAFT that contains a 10-digit code followed by a part number, that is, code—part number, code—part number, and so on. When an estimator must estimate the cost of a new shaft, the process starts by developing a code that describes the characteristics of the part. The first digit in the code might be assigned the part length, while the second digit is assigned the largest diameter, and so on. Next, the code is keyed in and the computer finds all the parts that meet the numeric descriptions and prints out the part numbers. The best fit is selected to be modified into a new part. All the details of each operation are retrieved. These include diameter, length of cut, number of surfaces, and the like. The estimator can alter these features and make the old part into a new one.

Actual Performance.   As the part is being produced, the estimated information is updated with actual performance and refiled. This gives the estimator the ability to improve estimating accuracy, because the next time the computer finds that part as one to be modified into a new one, the estimator is working with actual performance.

Other Uses for GT.   There are many other uses for group technology. One that is similar to estimating is variant process planning, in which a standard process plan is on file for each operation and can be modified into a new plan.

One carbide tool manufacturer produced a line of carbide drills and reamers and in their series 10 line had 758 different designs. After a matrix to describe these tools was developed, a code for each tool was developed and the database was established. The company conducted a redundancy search and found that 9 percent of the existing designs were either look-alikes or very similar. Now the company conducts a database search first when confronted with a new design.

PARAMETRIC ESTIMATING

Parametric estimating is the act of estimating cost or time by the application of mathematical formulas. These formulas can be as simple as multipliers or as complex as regression models. Parametric estimating, sometimes referred to as statistical modeling, was first documented by the Rand Corporation in the early 1950s in an attempt to predict military hardware cost.

Use of Parametric Estimating.   Many companies use some form of parametric estimating to develop sales forecasting. The four examples cited below will give the reader a good feel of how parametric estimating is used in a variety of different industries.

Construction Industry.   In developing a cost estimate for residential buildings, some cost estimators use a dollar value per square foot. The estimator constructs curves based on different construction such as wood or brick buildings and single- or multistory dwellings. These numbers can then be multiplied by the number of square feet in the building. Some construction companies have refined this process to provide additional detail. Carpeting, for example, could have a separate multiplier.

Heat Treating.   Most commercial heat-treating companies price their work based on a cost per pound and heat-treating method. Heat-treating costs are very difficult to define because many times more than one type of part is in the heat-treating furnace at the same time. It is difficult to think of a more effective way to estimate cost for this type of industry.

Tool and Die Industry.   As pointed out earlier, estimating cost for a progressive die can be very difficult because the estimator seldom has a die drawing to work from. Some tool and die shops have developed parametric estimating methods that take out some of the guesstimating. This method is known as the "unit value" method. Over a period of time, the estimator collects actual time values about dies being produced. Once the estimator is satisfied that the data are correct, they are averaged into usable hours. As an example, this might include 4 hours for every inch of forming or 3 hours for every hole under 2 inches in diameter. The unit value can stand for several meanings. For forming it is the number of inches being formed. For holes under 2 inches in diameter it represents the number of holes.

The estimator might establish a factor of 40 hours for a degree of difficulty. If the scrap cutter is "standard," the unit value is 1. If the scrap cutter is more difficult, the unit value might have a value of 1.5, where the hours allotted would be 60.

Helicopter Transmission.   A helicopter transmission is a large complicated assembly comprised of a planetary gear system, bevel gears, shafting, and housings. Budgetary estimates for a transmission are usually developed using a variety of parametric methods.

The housing costs are based on weight. The bevel gear cost is based on number of teeth, and the planetary gear cost is based on gear face width and number of teeth.

If methods like these were not employed, it would take several hundred man-hours to produce an estimate.

Collecting and Testing Data.   The single most important activity in parametric estimating is data collection and testing. Once the estimator develops the estimating method, enough sample data should be collected for a natural bell curve. Statistical testing of the curve is also very important. Once the parametric data are used for estimating, it is important to continually test them against actual performance and refine them as necessary.

LEARNING CURVES

Learning curves can be used for estimating cost of future production; this is well defined in Sec. 5, Chap. 8. Learning curves can also be used to adjust standard time for small lot sizes. If, for example, an estimator is estimating the cost of two lot sizes, a lot of 10 and a lot of 100 on a per unit basis, the estimator would not price a unit in the 10-lot size the same as a unit in the 100-lot size. This statement is made before any consideration for setup time. The estimator will usually make a mental judgment and add some percent or factor for each piece in the lot of 10.

Backing Up the Curve.   As pointed out earlier in this chapter, all standard data are set to some production level. That is, standard data developed in a high-production atmosphere must be adjusted for a shop producing lots of 10 units.

Estimates can be made by backing up the learning curve and adjusting standard time to better predict actual performance for small lot sizes.

Assumptions.   To adjust standard time by backing up the learning curve, two assumptions must be made. The first assumption is to determine at what quantity standard time can be reached. This can be determined mathematically by constructing a learning curve based on actual performance. There should be at least three or four points on the curve. The next assumption is to determine the learning rate for a given type of machine tool. This also can be calculated by testing standard time against actual. Typically, the more automatic a machine tool is the flatter the slope of the curve (learning rate).

Example.   In the example shown below, the time for the first piece is calculated by backing up the learning curve, and then this time value is adjusted for a lot size of 10 parts by repricing it forward on the curve. The assumptions are shown below.

T = standard time = 2.00 minutes.

10,000 = units produced to reach standard.

0.074 = exponent for a 95 percent curve.

T_1 = time for the first piece.

T_2 = average accumulated time each for a lot of 10.

$T_1 = (1/10{,}000)^{-0.074} \times 2.00 = 3.95$ minutes.

$T_2 = (T_1 \times 10)^{-0.074} = 3.33$ minutes.

Adjusting cost for small lot sizes using this method is far better than guessing because the math remains constant. Refinements in the two assumptions can be made until actual cost for small lot sizes can be calculated accurately. One company reported calculating actual performance within plus or minus 3 percent using this method. Caution must be exercised, however, when using this method of adjusting cost for manually operated machines because the learning rates are subject to fluctuation.

OTHER FACTORS THAT AFFECT COST ESTIMATING

There are other factors that affect the accuracy of a cost estimate. Several of these are cited below.

Project Estimating.   Inflation analysis and risk analysis come into play in project estimating. A multiyear estimate, such as many government contracts, is especially sensitive to both these factors.

Inflation.   When an estimate is being developed for future time periods, inflation rates are very important considerations. The three most popular measurements of inflation are the wholesale price index, the implicit price index, and the consumer price index, the last being the most quoted.

Because inflation rates are difficult to estimate accurately, most multiyear contracts have some provisions for reopeners to renegotiate. An after-tax evaluation of a multiyear project provides a more accurate assessment because it takes into consideration costs that are not sensitive to inflation. These costs might be loan repayment, leases, and depreciation costs.

Risk Analysis.   Risk analysis is a series of methods used to quantify uncertainty. Most of these methods are math models. Three broad classifications of risk associated with a project are cost, schedule, and performance. Some of the more popular methods of risk analysis are

1. Program evaluation and review technique (PERT).
2. Probabilistic analysis of network (PAN).
3. Risk information system and network evaluation technique (RISNET).

STATISTICAL ESTIMATING

The analysis of data through the use of statistical methods has been used for centuries. These data can be cost versus other information that leads to cost development. The practitioner must have a well-founded background in the use and application of statistical methods because an endless array of methods is available, several of which are described below.

Parametric Estimating.   Statistical estimating is another form of parametric estimating. The parametric methods discussed earlier in this chapter were industry-oriented whereas the methods discussed below are universal.

Regression Analysis.   The four most popular models of regression analysis are simple regression, multiple regression, log-linear regression, and curvilinear regression. Each math model is different and is designed for a specific use. Information can be regressed along a straight line or along a curve. Statistical estimating methods are very useful in parametric estimating. To use any of these methods also requires the user to have a sound knowledge of "goodness of data fit." Math models are available to determine how well data fit a straight line, curve, or log-linear relationship.

Computers.   Because of the complex nature of statistical estimating, the use of a computer is required. Fortunately, many good commercial programs, many of which are not expensive, are available on the market.

FEEDBACK SYSTEMS

A good feedback system to permit the estimator to review actual cost and compare it with the estimate is essential. The feedback system is the mechanism that permits the estimator to improve estimating accuracy. Shown below is one labor reporting system that permits actual performance to be reported in a way that permits this comparison.

Existing Systems. Most manufacturing organizations have some type of labor reporting system. A system can be as simple as filling in job cards or more automated, such as having terminals on the shop floor. The one described below is a simple one that permits data to be collected for cost accounting as well as cost estimating review.

Labor Reporting. Most good systems permit three types of direct labor entries to be made; direct labor setup time, direct labor run time, and direct labor nonstandard time. Direct labor nonstandard time is defined as time spent by direct labor people not working on standard elements. This is also sometimes referred to as nonproductive time. It may include performing machine maintenance, wait time, excessive tool trouble, and training.

There are major differences between nonstandard and off standard. Off standard is inefficiency associated in producing a part. The off standard costs should be included in the actual part cost. Nonstandard, or nonproductive, time reports are not associated with producing the part and should not be included in actual part cost. Most companies that use the nonstandard approach to developing part cost actually cross-charge direct labor time into an indirect labor account. The theory behind having a classification of time for nonstandard work is to separate out blocks of time over which the direct labor person has no control.

Another theory behind this system is to be able to compare true actual performance with the estimate. When the estimator estimates time, no regard is given to nonstandard elements such as machine downtime. An argument for keeping these times out of standard reporting can be made so that actual performance can be measured against the estimate. Most plants require the supervisor to approve a direct labor operator's request to go off standard and start measuring time for nonstandard. When a department starts reporting more than 5 percent of its time as nonstandard, the situation should be investigated.

Nonstandard and Nonproductive Time. Some systems make a distinction between nonstandard and nonproductive time. Nonstandard to an industrial engineer means an operation that has been added to the process, in addition to the standard operations, while nonproductive time is as defined above. If there is a substantial difference in the way a part is processed versus the estimate, the feedback system should include both nonstandard time for added or changed operations and the nonproductive time for wait time and the like.

It is not important for a plant to have a work standards system in place to develop and use a feedback system. If the facility does not have a work standards system, the estimate without off standard becomes the standard and all measurement for efficiency is made against the estimate. The feedback system is every bit as important as the estimator's estimating tools, and each estimator should spend at least 25 percent of the work time reviewing how close the estimate came to actual performance. The feedback system should be set up by operation for reporting purposes.

Management. When the estimating department encounters an abnormally high percentage of work that cannot be reconciled to the estimate because of processing differences, a red flag should go up to management. While some companies still use the meet or beat philosophy, this philosophy leads to redundant work and often makes comparing the original estimate with actual performance almost impossible because there is little resemblance between the process and the estimate. This situation is certainly counterproductive.

It is important for the estimator to be given feedback on other elements of the estimate, in addition to direct labor. These should include material cost, scrap and rework, and learning rates. While it is desirable to have this additional feedback on the same report as the direct labor feedback, it is not mandatory. If systems are already in place to produce a scrap report to explain why piece parts are scrapped, for example, it should not be necessary to be redundant for feedback purposes.

Computerized System. There are many computerized job costing systems on the market today, many of which are designed around a microcomputer. While it is not absolutely necessary for a small shop to have a computerized system of collecting and summarizing cost, it is highly desirable. Computerized systems have a wide range of capabilities and vary greatly in sophistication. Some systems are simply computer programs that collect actual cost as reported from the shop floor and produce a summary. Others permit the data to be entered via terminals from the shop floor as work is completed. Many of the latter systems do schedule updating and even reschedule workloads for work in queue at the next machine in the process. These systems, like MRP, can be costly to purchase and install, and usually require a computer that is larger than a microcomputer.

SKILLS NECESSARY TO BE A COST ESTIMATOR

The skills necessary to be a good cost estimator are wide and varied. Few companies train people to estimate cost; most people in the business gravitate into the position later in their career.

Skills and Intuition.   Estimating cost can be taught. As a matter of fact, it is taught in many colleges and universities, but the element of estimating cost that cannot be taught is intuition. A good cost estimator will have the ability to look at a completed estimate and make a judgment that the estimate is high, low, or just right. Intuition comes only with experience and the feedback referred to earlier.

Highlighted below are the important skills of a good estimator.

Knowledge of Estimating Methods.   A good cost estimator must have a good grasp of all the common estimating methods used in his or her field. These might include flash estimating, detail estimating, and parametric estimating. Having a knowledge of the common estimating methods places the estimator in a position of selecting the proper method for any situation.

Training and Education.   More and more colleges and universities are developing coursework in the discipline of estimating cost. These course offerings are usually part of an industrial engineering or manufacturing engineering curriculum.

In the past several years, the Society of Manufacturing Engineers has offered 3-day seminars on the subject of manufacturing cost estimating. These types of seminars are very useful for the person entering the field who does not have the opportunity to take course work at a college or university.

It is also very useful for a person in the field to have some basic education in cost accounting. While cost accounting will vary from company to company and industry to industry, the principles remain the same.

Experience.   Experience in the field of estimating cost can be achieved only by doing the work. Standard data forms and computer estimating assist in gaining estimating experience because both provide a regimented way of developing cost.

When people new to the discipline first start developing estimates, they are usually coached by a person who has been in the field for some years. This can be both good and bad. The "how to" part of estimating can be learned very well from an experienced estimator, but the young practitioner must be sensitive to the phenomenon of estimating bias. Every estimator has an estimating bias. This is the tendency to develop tight or loose estimates without understanding the bias. For example, if the shop has a high backlog of work, estimates may be on the loose side. If the shop needs work, the estimator's position might be to "sharpen the pencil."

Program Planning and Project Management.   It is very important for an estimator working on large projects to have some basic education in program planning and project management. Training in PERT and CPM is very useful. Indeed, most large government contracts require a PERT analysis to assist in assuring the program will remain on schedule. There are many good books on the subject and seminars offered to assist in the estimator's training.

BIBLIOGRAPHY

Nicks, J. E., *BASIC Programming Solutions for Manufacturing*, Society of Manufacturing Engineers, 1982.

Nicks, J. E., *Manufacturing Cost Estimating Software Manual*, Society of Manufacturing Engineers, 1986.

Ostwald, P., *Manufacturing Cost Estimating*, Society of Manufacturing Engineers, 1980.

Stewart, R., and R. Wyskida, *Cost Estimator's Reference Manual*, Wiley, New York, 1987.

Winchell, W., *Realistic Cost Estimating*, Society of Manufacturing Engineers, 1990.

CHAPTER 5
MAKE-OR-BUY DECISIONS

Robert I. Felch
Professor Emeritus
College of Business and Economics
Radford University
Radford, Virginia

An industrial firm has two sources for materials, components, and even its end products. It can produce them in-house from its own facilities, or it can obtain them by purchasing from outside suppliers. The decision process involved is known as "make or buy."

From the point of view of a purchasing manager, the process is one of selecting the right source. From a cost accounting point of view, every item acquired is considered as though it was purchased. From a philosophical point of view, the top management of a firm should consider its own production facilities as simply a source that must compete for the opportunity to be a supplier.

When a firm is producing an end product for the federal government, procurement jargon refers to the purchasing of supplies and components as subcontracting. But, in the private sector, the term subcontracting usually refers to the purchasing of items or services that normally are produced or provided in-house by a firm.

This chapter analyzes and discusses the make-or-buy decision process in the context of industrial firms in the private sector. However, since a firm in the private sector may be a supplier to a government agency, some special considerations are introduced for such circumstances.

To simplify the presentation, the chapter refers to the making or buying of materials and components. But services, and occasionally end item products, may become candidates for the make-or-buy decision process.

THE ORIGINS OF MAKE-OR-BUY DECISIONS

Past practices together with long-range plans create a kind of framework that appears to set a pattern for what will be made and what will be bought in a firm. But management is wise to be alert for fresh opportunities to benefit from what may seem to be nonroutine make-or-buy situations. These situations originate continually in any of a number of ways.

New Product. Ideally, in the development of a new product, all materials and every component are studied in advance of production. An analysis is then conducted to decide what the source will be for each of these. The logic behind this process dictates that, to be successful, a firm must include make-or-buy analysis as a regular function in product development. Criteria and the organization of these analyses are described later in this chapter.

Additionally, in an ongoing sense, make-or-buy decisions may originate in situations other than those of new product development as follows:

Value Analysis. A value analysis study may result in a changed component that deserves the same make-or-buy consideration as a new product component.

Unsatisfactory Supplier's Price or Performance. A supplier's behavior may trigger an investigation toward a different source that might be in-house production.

Supplier's Proposal. A supplier's proposal might introduce a new idea, improvements, or a price change that could call for reevaluation for the source for a component.

Labor. Changes in production schedules may result in either a shortage or an excess of labor resources in a firm. For example, a firm might revert from a pattern of buying an item to one of making it in-house in order to retain a skilled labor pool for a short run. However, consideration must then be given to the possibility of losing a valuable outside supplier that could be needed in the long run.

Organized labor may bring pressure to bear upon management to produce in-house what is planned for outside purchase. And of course, organized labor may jeopardize supply of an item by threat or introduction of a labor strike.

Product Revision. Like value analysis, product revision may introduce the need for a reassessment of make-or-buy relationships.

ILLUSTRATIONS

In studying the implications of, and processes for, make-or-buy decisions, it is helpful to have in mind some illustrations, particularly a few examples of extreme situations like the following. In past years, a number of firms like Ford Motor Company tended to integrate almost their total operations by acquiring, or investing in, all stages of manufacture even to the point of the extraction of raw materials. On the other extreme, many firms today operate essentially as assembly plants with little or no fabrication or finishing, acquiring practically all materials and components by purchase. The latter kind of operation underscores the trend toward more complex technology and specialization. Hence, with the passing of time, industrial firms on the whole purchase more and more.

In between the two extreme cases of making versus buying, one occasionally comes upon a startling example of what must have been a landmark make-or-buy decision. Take, for example, the Chrysler Corporation decision to incorporate Japanese-supplied engines in a number of their automobile models. There must have been a tangible impact on the organization, the facilities, and especially the labor pool at the home plant. Moreover, there was an intangible impact in the marketplace where the payoff for the decision depended in part upon buyers' perceptions of the value of a Japanese engine in what had been traditionally an American automobile.

CONSIDERATION OF THE OBJECTIVES AND CHARACTER OF THE FIRM

Contemplation of these simple examples of make-or-buy decisions leads directly to consideration of the underlying capabilities and objectives of a firm. Rationally, the owner(s) will evaluate the resources and capabilities of the firm and direct the management to achieve specific objectives. Ideally, the directive answers the question, "What kind of an enterprise are we?'' From the marketing standpoint, the firm may seek to acquire a complementary image for positive recognition by customers. Over time, the firm may develop a combination of qualities that distinguishes that firm and thus gives it character.

Make-or-buy decisions, as a matter of course, should conform with the objectives and character of the firm. But an overriding objective of almost every firm is viability, which is derived from ''the bottom line,'' return on investment (ROI). Hence, trade-offs will be made to adjust to changing economic conditions. At lower organizational levels in particular, decisions may assume some independence, as long as short-run benefits are seen to be accrued. With industries' evolution toward specialization, cost differentials between making or buying increase coincidentally. Therefore, make-or-buy decisions may not merely conform with the character of a firm, but they may serve to help shape that character over the long run.

THE FUNDAMENTAL FRAMEWORK FOR DECISION MAKING

Generally, each make-or-buy question falls into one of two categories, depending upon whether or not a substantial additional capital outlay is required for facilities, equipment, and tooling.[1] This is illustrated in Fig. 5.1.

For the category wherein production facilities are on hand, the decision process can be proceduralized more easily and may be helped by the use of computerized systems. More and better information is available than for the other category where a new start-up must be undertaken.

DECISION CRITERIA

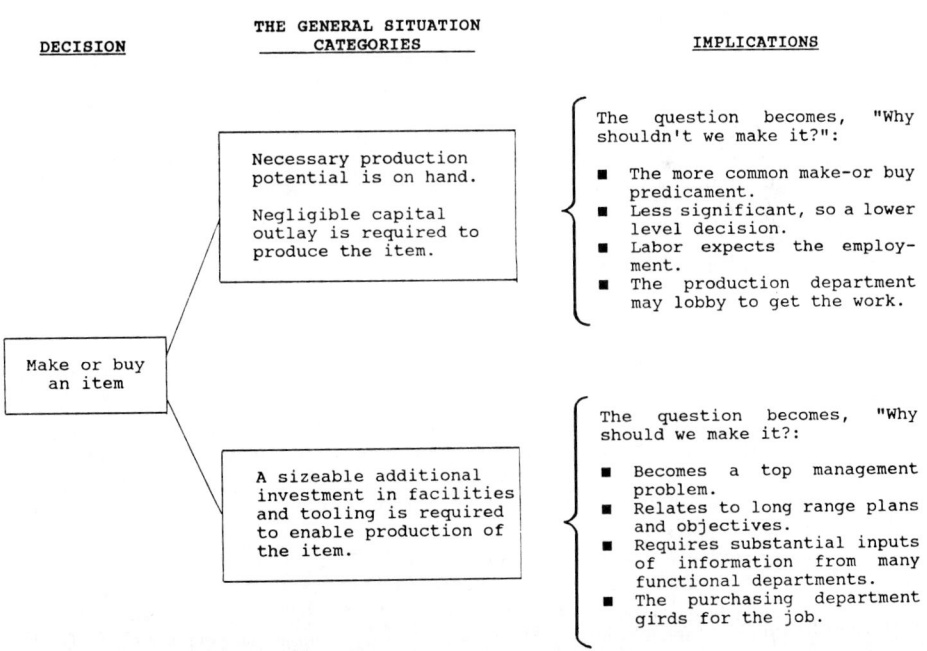

FIGURE 5.1 The fundamental framework for make-or-buy decision making (an oversimplified view).

The two criteria that stand out above all others are cost and availability of production facilities (the latter discussed above). These and other decision-making criteria that should be considered can be tabled in two groups to distinguish those that favor making versus those that favor buying. However, many of the criteria can fall into both groups, depending upon particular situations. For instance, it might cost either more or less to make a part than to buy it.

Important to the decision process is the need to consider *all* factors, because the omission of just one obscure but crucial consideration could upset an otherwise logical decision. For example, a seemingly good decision to make might have to be aborted because of an overlooked patent infringement.

A fairly comprehensive list of criteria follows together with brief comments. The more prominent criteria are listed first, but each decision case calls for its own arrangement of those criteria that are germane.

Cost.   A cost analysis calls for a determination of the cost of making an item which is then compared with the cost of buying that item. The trick is to assure that all the elements of each cost category are reconciled. For the make determination, direct costs of labor and materials can be reached straightforwardly. But many associated incremental costs may be subject to differing interpretations and misunderstandings or may even be overlooked. The cost to buy can be determined much more easily and usually consists of the purchase price, transportation and receiving, plus inspection. Of note is the fact that the cost to make is an estimate, whereas the cost to buy is known. Most firms prefer a known cost, even over a somewhat lower estimated cost, because of the validity of the cost structure.

Facilities.   The significance of already available facilities and capacity has already been emphasized. An additional investment in capital equipment and tooling for making in-house can result in a manufacturing momentum that is difficult to reverse. Labor can become entrenched, and there is a sunk cost for equipment and tooling.

Labor.   There usually is a tendency to stabilize the work force in order to realize efficiency in production. Make-or-buy decisions made with this objective may create purchasing problems when external suppliers are used as buffers. The purchasing manager should provide information for these kinds of make-or-buy decisions with regard to suppliers' abilities to absorb fluctuating order requirements.

When labor is organized, a union may attempt to influence managerial decisions by including clauses in labor contracts that would restrict a firm from buying items that could be made in-house.

Managerial and Specialized Abilities.   Beyond the consideration of operational labor is the need to consider the adequacy of managerial and specialized abilities for planning, controlling, and directing make work in-house.

Quality and Technical Expertise.   Unusual quality or technical requirements can influence a make-or-buy decision in either direction. Perhaps more commonly, a firm will elect to manufacture in-house an item that calls for unique or exacting specification. However, it is not unusual to find a situation in which a supplier is more capable in this regard. Occasionally, a firm will work closely with a particular supplier to overcome a special quality or technical requirement.

Quantity and Time.   Like quality, the quantity criterion can work in either direction. When only a small quantity of an item is required, the formidable job of tooling and setting up inevitably legislates a buy decision. On the other hand, when large and continuing requirements for an item are foreseen, an effort will usually be made to determine whether in-house manufacturing is feasible. But the criterion of quantity must be considered in relation to that of time. For instance, the need for a large quantity of a new item may legislate a buy for the simple reason that there is inadequate time for tooling up and preparing for in-house production. In this context of scheduling, time itself becomes the criterion for make-or-buy decisions.

The quantity and time criteria underscore the importance of reliable sales forecasting. Marketing management must be completely candid. When sales data are undependable, a contingent make-or-buy decision should be based upon conservative factors.

Assurance of Supply.   Predominant in the realm of services that a firm can realize from a supplier is the factor of reliable delivery. Usually, when reliable delivery of a purchased item cannot be assured, a firm will elect to make and thus achieve some control toward stabilizing the enterprise.

Make-and-Buy: Hedging and Managerial Control.   The notion of both making and buying has some fascinating features. For instance, a demanding top management might interject a buy decision for an item that is normally produced in-house just to check on the efficacy of its own production department. Or a firm that regularly produces an item could subcontract peak requirements. In emergency situations, especially for a critical item, a cooperative supplier may be agreeable to providing the hedge to assure adequate supply. In such circumstances, the firm's materials manager should be the leading individual to recommend an appropriate make-or-buy policy.

Design or Process Secrecy.   While this criterion may not be as commonly involved as others above, there may be justification for making (or less commonly buying) an item because of a secret or proprietary design or process.

Patent and Royalty Infringement.   The risk of exposure to lawsuit jeopardy in a case of patent or royalty infringement is clearly more prevalent when a new and innovative item is to be made in-house. To minimize the risk, a firm can (1) employ a patent attorney, or (2) make inquiry of the U.S. Patent Office, or perhaps most effectively, (3) apply for patent protection for the innovation. As mentioned earlier, adequate precautions must be taken, because the danger is obscure, and the risk is growing as high-tech industries multiply.

Return on Investment and Industry Expansion or Contraction.   When a firm's opportunity cost of capital lowers, and the industry involved is expanding, the risk of undertaking a make becomes more tempting. Such conditions serve to emphasize the importance of the recognition that underlying all make-or-buy decisions is the fundamental criterion of return on investment. Obviously, a company realizing a return on investment of 20 percent would have difficulty achieving a successful make of a new component in an industry where a 10 percent return prevails. Make-or-buy decisions must be tempered by good top-level financial and economic advice.

Other Criteria.   A number of other factors, depending upon less common circumstances, may have a decided impact on a make-or-buy decision: (1) More commonly with time, environmental and related political circumstances attend some production processes, especially in certain industries like chemicals; (2) even social pressures may interfere, as might be connected with a particularly distasteful kind of work which could be subcontracted without kickbacks; (3) changes in tax rates might upset a firm's cost-of-capital structure; (4) producers for government procurement operations perform in a complicated environment that receives some special considerations that are discussed in the following section.

CONSIDERATIONS FOR GOVERNMENT SUPPLIERS

A firm producing under a federal government contract operates in a highly regulated environment. The public's intrusion through politics and legislation into government procurement has resulted in a large unwieldy apparatus that reaches into nearly every aspect of supplier activity including make or buy. Even a subcontractor may find itself subject to governmental contract articles and provisions of a prime contract supplier.

The federal government's expressed policy is that it will not start or carry on any commercial activity to provide a service or a product for its own use if such a product or service can be procured from private enterprise through ordinary channels. Accordingly, the federal government is, by decree, very sensitive to make-or-buy situations.[3] This sensitivity is especially evidenced in cost-type contracts with suppliers in the private sector. In its zealous efforts to protect the public interest, the government controls a private contractor's make-or-buy process. The general nature of the government's procedure for this control is a requirement that each prospective supplier must provide a detailed plan of his or her make-or-buy program for any complex supply contract of large dollar value (currently the Federal Acquisition Regulations, referred to as the FAR, state "$5 million dollars or more"). Reasons must be given for categories of "must make" and "must buy." A supplier's "plan" may be incorporated in a contract, if a proposal is accepted. Cost-type contracts are regularly subjected to government audits. Good records must be kept.

It can be expected that the government's major consideration will be cost. Analyses of make-or-buy decisions may become very complex, with accounting technicalities and various costing issues.

The contingency or presence of a national emergency, such as war, may again bring about an incentive for suppliers to seek federal contracts. In the past, during emergencies, the federal government offered, in selected cases, provisions for accelerated depreciation of capitalized facilities needed to produce critical items. Under these provisions, selected suppliers would find it attractive to make, and both the public's and the suppliers' interests would be served.

The pervasive nature of government procurement, with its penchant for audits, coupled with public and political interest, require that a private firm that does business with the government must have its "house in order," so that its decision processes, including that of make or buy, will hold up under outside scrutiny.

VOLATILITY OF THE MAKE-OR-BUY SITUATION

Recognition of the composite effect of the criteria discussed above should lead to a conclusion that rigid formulas and rules of thumb are impractical for make-or-buy decisions. Each criterion is subject to change. Some, like changing materials costs, can turn a good decision into a bad one in a surprisingly short time. Larger firms may enjoy some cushioning benefits such as more cash, bigger labor pools, and more flexible facilities. Smaller firms must be especially prudent. But each make-or-buy decision is subject to the volatility of any one or more of the factors that are in the mix of the decision process. Still, the entrepreneur realizes that the price of profit is risk.

ORGANIZATION AND ADMINISTRATION OF THE DECISION PROCESS

It is commonly recognized that many, if not most, firms have no consciously expressed policy or systematic procedure for make-or-buy decisions.[1] In a young and up-and-coming firm, a production control clerk may decide alone. The purchasing manager may act only when a requisition is received from the production department.[1-3]

Experienced and progressive firms establish formalized organizational procedures for both make-or-buy decisions and their review. Because so many organizational elements are involved, a standing committee approach is popular. Coincidentally, an information system needs to be established to provide all relevant data to the decision-recommending body. Because of the breadth of the criteria and the implications of the decisions, each decision should have top management approval. Of course, the administration of the overall process can be systematized, and such techniques as management by exception may be implemented to facilitate the process. Finally, procedures should be introduced to assure the provision of adequate records, periodic reports to top management, and a plan for periodic reviews. In a firm that does business with the government, measures must be taken to assure that the procedures and records are responsive to the government's requirements.

CONCLUSIONS

Given a rational and cogent argument for systematic make-or-buy programs in manufacturing, it is puzzling to learn that many firms fail to have a recognizable policy and procedures for this decision process. In part, this may be because schools of business in higher education give little attention to the subject. Academic attention is prone to focus upon those managerial areas where quantitative methods can be applied, like queuing theory, linear programming, and networks. Make or buy thus far has eluded this kind of attention that is earmarked by deterministic and stochastic techniques. Any academic coverage is usually presented in a course for purchasing and materials management, and this kind of course is relatively rare in schools of business.

In conclusion, it seems that make or buy is a relatively underappreciated and unexplored managerial process. The industrial engineer is well suited for validating, introducing, or installing a make-or-buy program. The IE has the business sense, the technical skill, the aptitude needed for crossing organizational lines, the generalist's disposition, and the top-management point of view needed to cope with this kind of decision making.

REFERENCES

1. Donald W. Dobler, David N. Burt, and Lamar Lee, Jr., *Purchasing and Materials Management,* McGraw-Hill, New York, 1990.
2. Michiel R. Leenders, Harold E. Fearon, and Wilbur B. England, *Purchasing and Materials Management,* 9th ed., Irwin, Homewood, Ill., 1989.
3. Stanley N. Sherman, *Government Procurement Management,* Wordcrafters Publications, Gaithersburg, Md., 1981.

PLANNING AND CONTROL

CHAPTER 1
SYSTEMS ANALYSIS, DESIGN, AND OPERATING PROCEDURES

John R. Huffman, Ph.D., P.E.
President
John R. Huffman, P.E.
Los Angeles, California

SYSTEMS, SYSTEMS PROBLEMS, AND THEIR SOLUTIONS

Systems Concepts

A system is any set of interrelated and interdependent components which can be seen as satisfying a purpose, that is, performing a function, by processing input(s) into output(s) while attaining or attempting to attain one or more goals or objectives.[1]

The components of a system are

1. Its purpose
2. Its inputs, inputs in process, and outputs
3. The process performed
4. Any facilities and equipment required
5. Any people necessary to operate it
6. The information and instructions necessary for its operation
7. The environment within the system

These components are separated from the environment of the system by a boundary across which the inputs and outputs flow.

It is the relationships and dependencies between and among its components which enable a system to satisfy its purpose. When they are destroyed, the system cannot perform its function because it no longer exists. For example, if the operator, machine, and tool components of an operating workstation are removed from it and separated, each is still capable of performing its function; but no inputs can be processed because the relationships and dependencies between and among the components (and therefore the system) have been destroyed. If the relationships and dependencies are restored, the system is reconstituted; it will again process inputs.

From a broader viewpoint:[2]

1. Everything is a system.
2. Each system is composed of smaller systems.
3. Each system is paralleled by other systems.
4. Each system is part of at least one larger system.
5. Each system is part of at least one hierarchy of systems.

If a problem exists within system X, it is helpful to consider the next larger system in a hierarchy which includes system X to be its immediate environment and all larger systems in that hierarchy to be its general environment.[3]

The system boundary consists of all the interfaces with entities (in the immediate environment) which provide inputs to the system, receive outputs from it, or both. The general environment includes the factors which may influence system redesign in the long term. These can be classified as technology advances; cultural shifts; changes in the availability of labor, capital, and other resources; and changes in the political environment.*[3]

Systems Problems and Solutions

A systems problem arises when the performance of a system does not meet the needs, wants, or aspirations of one or more people or groups. Typically these "wants" (defined now to include needs and aspirations) are: (1) to restore the performance of a system to its previous level, (2) to improve the performance of a system, and (3) to create a system which does not now exist in any form. Performance is measured on scales as diverse as quantity or quality of output, profit, pollutants generated, and improvement in personnel attitude.

Every systems problem can be divided into two parts:

1. The physical, procedural, or other entity to be designed
2. The "wants" of *stakeholders,* that is, the individuals and groups affected by the changes incorporated into the solution system

The solution to a systems problem is a system design synthesized *and implemented* by a problem solver or problem-solving group.

SYSTEMS DESIGN (PROBLEM-SOLVING) METHODS

The procedures for designing a solution system constitute a problem-solving method. There is a variety of these ranging from the affective (or "gut feel") method and the copying of successful solutions, both used by managers, to rational procedures which include the research method, the engineering method derived from it, and systems engineering methods. All rational methods are iterative.

Rational Engineering Design Methods

Engineers have long used the engineering method, one version of which is shown in Fig. 1.1. The systems engineering method shown in the same figure, on the other hand, utilizes techniques developed since 1950. Those described in this chapter are modifications of the engineering method. They are based primarily on a 1981 modification first described in the text, *The Planning and Design Approach,*[4] and a subsequent restatement and expansion of

---

*Figure 1.3 shows the relationships among a system, its boundary, its inputs and outputs, and its immediate and general environment for the example problem.

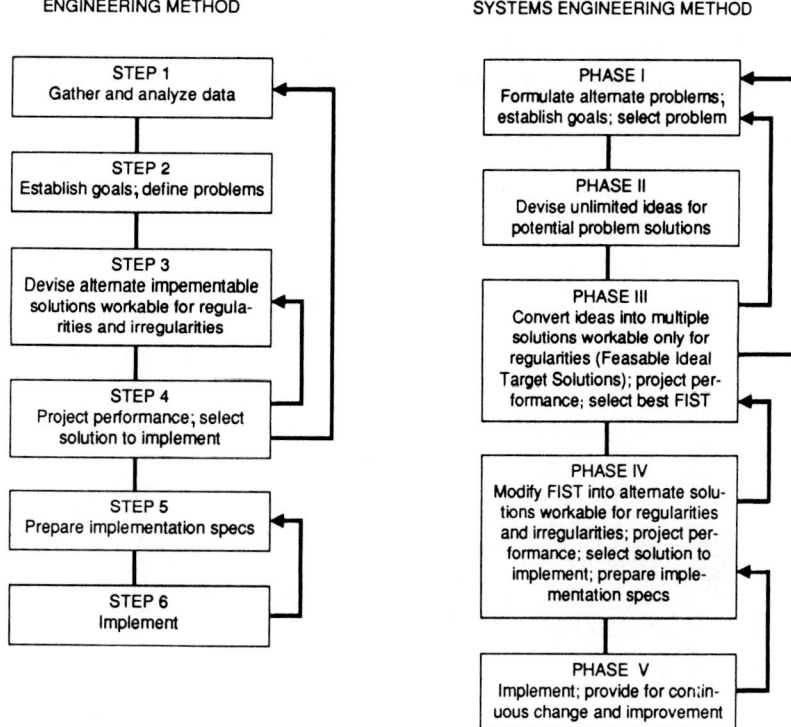

ENGINEERING METHOD   SYSTEMS ENGINEERING METHOD

FIGURE 1.1 The engineering and systems engineering methods.

its principles.[5] They include other changes, primarily amplifications, based on texts[6-8] published in 1989 and 1990, practice, and research.

The methods should be utilized with the understanding: (1) that they will be revised by future application and research and (2) that they should be adapted to the problem at hand. A suggested database (system matrix, Fig. 1.2) is provided to guide systematic recording of a description of the initial system, ideas, questions, setbacks, intermediate system descriptions, and specifications for the system to be installed.*

Comparison of Methods

The engineering and systems engineering methods employ different procedures and points of view: (1) to select or define the problem to be solved and (2) to devise possible solutions to recommend. They use essentially the same methods, but different philosophies, to select the solution to recommend, to implement it, and to monitor it thereafter.

Problem Definition.   The engineering method stresses: (1) gathering much information including what is wrong, (2) analyzing these data, and (3) establishing stakeholders' goals which become the bases for (4) selecting "the problem" to be solved.

The systems engineering method stresses:[9] (1) the creation of a hierarchy of possible purposes for the system to be redesigned, each purpose broader than its predecessor and corresponding to a larger system in which the problem lies, (2) the determination of goals,

---

*The system diagram of Fig. 1.3 depicts graphically the information that would be recorded in the components column of Fig. 1.2 and other information which would appear in the controls and interfaces columns.

Components \ Dimensions	Fundamental: basic or physical characteristics—what, how, where, who	Objectives: wants—motivations, global desires, ethical and moral precepts	Criteria, Measures of Effectiveness, and Goals; Design Specifications: how much, rates, when	Control: how to evaluate and modify an operating component or system	Interfaces: internal, with immediate environment	Future: projected or planned changes, research needs for all components
Purpose/Function						
Inputs: people, things, information to start process						
Outputs: desired (achieves ourpose) and undesired outcomes of process						
Process: unit operations, steps and flow to convert inputs into outputs						
Environment: physical and attitudinal—internal, immediate, general						
People: skills, responsiblities						
Facilities and equipment						
Information aids: manuals, instructions, rules, regulations						

FIGURE 1.2   System matrix. [*Adapted from Gerald Nadler,* The Planning and Design Approach, *Breakthrough Thinking Co., Los Angeles, 1981, Figure 1-5, p. 10 (copyright by the author).*]

and (3) the selection of the purpose to be satisfied, and therefore the system to be redesigned, on the basis of these goals. This approach increases the probability of selecting the "right problem"; it utilizes minimal data.

Solution Formulation and Selection.   From the start the engineering method seeks "the solution" which satisfies both the normal aspects ("regularities") and unusual aspects (exceptions or "irregularities") of the process and other components of the system. Practitioners employ a "doubting" attitude characterized by the statement, "You can't do that because..." and attempt to fit solutions into queuing, just-in-time, and other known solution formats. These approaches and the emphasis on defining "the problem" lead to innovative solutions only if the problem solvers are innovative.

On the other hand, the systems engineering method initially:[10] (1) seeks "ideal" solutions which satisfy only the regularities of system components and (2) employs a "believing" attitude characterized by the statement, "That solution could work now or will work when technology is available." It selects the best alternate ideal solution as a target and modifies it into other alternate solutions which accommodate system irregularities. It then selects the best of these modified ideal solutions for implementation. Both selection processes use known solution formats to test and evaluate solutions.

These steps inhibit the quick rejection of solutions excellent for regularities only; they stimulate conformists to devise innovative solutions.

Philosophy.   Practitioners of the engineering method traditionally have involved stakeholders only when necessary in the problem-solving process and anticipated that a solution is essentially "set" for a period of years. Systems engineers on the other hand: (1) involve stakeholders in every problem-solving step, including implementation, and (2) anticipate and provide for continuous change and improvement of the installed system.

Users of the engineering method often seek the "best" solution, which may be technically elegant but unusable;[11] to them, limitations on engineering time and cost are constraints. Systems engineering practitioners take a different point of view. They attempt to maximize: (1) the effectiveness of the recommended solution, (2) the likelihood of implementing it, and (3) the effectiveness of the engineering and other resources used to design it.[12] The systems engineering methods of this chapter increase the probability of approaching these goals while solving the "right" problem.

SYSTEMS ENGINEERING

The usual statement of a problem to be solved includes: (1) the wants of some or all stakeholders and (2) a description, usually partial, of the perceived system in which the problem exists. Sometimes it includes cost data. All are evident in this example problem assigned by the chief industrial engineer of the fictitious XYZ Co. To illustrate systems engineering methods it incorporates features from many real problems.

EXAMPLE: PROBLEM STATEMENT.   We receive covers for one of our products in cartons stacked in freight cars. Receiving personnel count the cartons and stack them on pallets; they prepare a pallet tag for each pallet and complete a receiving report for each shipment. Forklifts supply empty pallets and move the loaded pallets about 600 ft (183 m) to stores. Palletizers suffer frequent bruises and average a back strain every 3 months because the cartons are large and heavy.

The plant manager wants to reduce the $30,000 cost per year and the accidents inherent in these operations. Since the cover will be obsolete in 2 years, the return on any required investment must not only exceed our required minimum but also be the maximum possible over the entire 2 years.

Quick implementation is obviously essential. Don't take more than 90 days to reach that point or exceed the budget of $3000, which includes industrial engineering time.

Phase I, Selecting the Problem

The problem to be solved is often poorly defined when stated; many times it is selected incorrectly. Phase I addresses these possibilities by providing procedures which: (1) expose alternate, broader formulations of the problem stated and (2) make the problem selected from these alternates more likely to be the "right" problem to solve.

Establishment of the basis for evaluating alternate proposed solutions to this problem completes this phase.

Uncovering Broader Problem Formulations.   To expose broader versions of the stated problem, *with stakeholders:*

1. Observe the stated problem and review it briefly.

2. Create—with minimum information—a hierarchy of possible system purposes (functions), each larger than its predecessor, which define alternate problems and the corresponding larger systems in which the stated problem exists.

Observation and Review of the Problem. Use limited observation of the existing system and review of the problem with stakeholders to establish:

1. The unique aspects of the problem which differentiate it from similar problems
2. The data necessary: (*a*) to verify that the potential benefits justify the effort to solve the problem and (*b*) to plan the problem-solving effort and organization

The latter data should include a projection of solution life—the period which begins with installation of the solution to the current problem and ends when the next solution to the same problem is installed or the system is dismantled. Unintentionally, these data will also include: (1) possible problem solutions, (2) stated or unstated regularities and irregularities of the system, and (3) design specifications and constraints.

To promote creativity when devising problem-solving ideas and solutions, use the information obtained unintentionally only to the extent necessary to select the problem to be solved and to design the problem-solving effort.

Creating a Purpose (Function) Hierarchy.[13] A purpose hierarchy is a series of statements each of which defines a broader or larger function. The guidelines for creating a hierarchy are

1. Use the infinitive form of an action verb and an object to describe each function, as in "to make automobiles."
2. To minimize the influence of existing system components upon ideas for solutions, state each function as a "unit operation" by using a verb which describes the function in generic terms rather than operation by operation.
3. Be sure this unit operation adds value to the output of the system from the viewpoint of its user or consumer.
4. Make the first statement of a hierarchy the most specific one. Frame it to apply to the smallest system (usually the perceived existing one) which contains the problem.
5. Make each succeeding statement of the hierarchy larger (broader) than its immediate predecessor.
6. To ensure that the largest possible purpose is selected to define the problem to be solved, be sure the scope of the largest purpose far exceeds that of any one initially considered suitable for defining the problem.
7. To provide a maximum number of alternate potential problems to solve, and therefore systems to redesign, formulate successive purposes which define small steps between the smallest and largest purposes.
8. Do not reflect any "want" in a purpose as in, "to make automobiles at lower cost" because that focuses improvement efforts on stakeholders' wants instead of redesign of the system to meet them. If it is necessary to incorporate a want into a function statement, it may be stated as "to have lower cost automobiles"; such statements usually define the broadest purposes of a hierarchy.

To implement these guidelines start with the most specific purpose. Ask, "What is its purpose?" or "Why is it performed?" The answer to either question is the next larger purpose. Ask the same questions about it and each successively larger purpose to develop a complete hierarchy. This procedure will also arrange randomly suggested functions appropriately.

A hierarchy is valid if a review of it shows that the purpose of each function statement is the next larger function statement. Thus each purpose includes all smaller purposes, and solution of the problem in the system defined by one purpose solves the problem associated with each smaller function. When a problem is complex, its hierarchy may take the form of a tree with separate hierarchies (corresponding to the branches) combining to form a single hierarchy (corresponding to the trunk).

Selecting the Problem to Solve.   To select the problem more likely to be the "right" one:

1. Formulate the "wants" of stakeholders.
2. Select the purpose to be satisfied (the "focus" purpose), and therefore the problem to be solved and the system to be redesigned, by considering these wants. This problem will often be larger than the one posed originally.
3. Identify any essentially independent smaller system(s) included in the system corresponding to the focus purpose.

Establishing Stakeholders' "Wants."   Stakeholders' wants fall into two categories: (1) improvements of existing system performance and (2) design specifications and constraints (DS/C) for the solution system.

To establish wants:

1. Identify the stakeholders in solutions which might accomplish each purpose of the hierarchy.
2. Formulate each performance improvement desired as: (*a*) a broad objective, (*b*) a criterion which states specifically the performance measurement to be improved, or (*c*) a goal. These statements may or may not include a unit, called the measure of effectiveness (ME), which assesses the extent of the desired change.
3. Formulate any "go/no-go" design specifications and constraints (DS/C), including threshold (minimum or maximum) values for ME, applicable to the solution system. Questions to ask about threshold values and other areas to explore for DS/C may be suggested by these categories:

Source	Enforcement	Timing	Applicability	Type
Within system	Rigid	Now	All alternates	Technical
				Economic
Environment	May be relaxed	Later	Specific alternates	Psychological
		Always		Political

Source: Adapted from T. H. Athey, *Systematic Systems Approach*, Prentice-Hall, Englewood Cliffs, N.J., 1982 (Table 6.1, p. 90).

Selecting the Problem to Be Solved.   Select the function statement (or "focus" purpose) which defines the problem, and therefore the system which must be designed to solve it, by ascertaining for which purpose potential solutions:

1. Will satisfy the DS/C and threshold values of MEs.
2. Will satisfy to some extent the wants of stakeholders.
3. Will conform to applicable government, association, or other external controls and regulations.
4. Will meet any limitations imposed by management, the capabilities of those in the system or its immediate environment, and unusual implementation considerations.
5. May be possible, considering organizational politics and those individuals and groups in favor and opposed.

To ensure that the focus purpose (and the system associated with it) will be the largest possible, subject the broadest statement of the hierarchy to the questions first. Use such techniques as judgment, consensus, voting, the nominal group technique,[14] or a multiattribute evaluation model[15] to establish the focus purpose. If either of two functions could be the focus purpose, choose the larger one.

Identifying Independent Smaller Systems.   When the relationships between one or more smaller systems which are part of the problem system and the balance of the latter are weak or nonexistent, consider the smaller system(s) to be functionally independent of the problem system. Redesign each such system as a separate problem; integrate the solutions to each one with the solutions of the main problem in Phase IV.

Defining the Basis for Evaluating Problem Solutions.   To establish the basis for evaluating alternate potential solutions to the problem defined by the focus purpose, modify the wants used to select that purpose to include only those which are relevant to it. If a want is stated as a broad objective, derive the corresponding criteria, ME, and goal. If the want is stated as one or more of these four, derive the other(s).

ME and DS/C units may be quantitative (dollars per year, percent, and labor hours, for example) or qualitative (unacceptable, below average, average, above average, and very desirable, for example). Quantitative ME and DS/C are more desirable than qualitative ones because it is usually possible to measure the former routinely and at low cost while the latter must be evaluated by specific and costly efforts.

If performance of the existing system is not being measured in the units of the ME and DS/C chosen to evaluate solutions satisfying the focus purpose, measurements must be instituted promptly to provide a basis for assessing the change resulting from installation of the solution system.

Cautions.   Involve stakeholders in devising the purpose hierarchy, selecting the focus purpose, and devising the basis for evaluating solutions. Their purpose statements, some of which will usually include objectives or goals, should be accepted initially but revised by negotiation and persuasion to meet the guidelines. What stakeholders say, or don't say, as the focus purpose and evaluation basis are established will often reveal valuable ideas and hidden agendas.

The identification of smaller independent systems will require redesign of the problem-solving effort. It may require repetition of the prior steps of Phase I to define the purpose of each such system and the balance of the original problem. The time lost by iterating these steps may be regained subsequently by assigning additional personnel or specially qualified personnel to individual problems.

EXAMPLE: SELECTING THE PROBLEM TO SOLVE.   Observation of existing operations and discussions with possible stakeholders established that:

1. The unique aspects of the problem were: (*a*) the limited life of the solution, (*b*) the 2-year rate of return requirement, (*c*) the number of cars and cartons, each containing multiple covers, received per day, and (*d*) the size and weight of the carton which made it difficult to handle but resulted in pallet loads too high to handle two at a time.
2. The estimate of current costs was valid.

These observations and discussions also revealed that:

1. The "regularities" of the operation and any solution were no change in cover design or daily receipt volume.
2. The "irregularities" of existing operations were: (*a*) damage to cartons or covers in transit or during palletizing, (*b*) the creation of partial pallet loads, (*c*) the stoppage of palletizing pending the delivery of empty pallets or the removal of loaded ones, (*d*) incorrect, misapplied, or lost pallet tags and receipt summaries, and (*e*) space not available in stores.
3. Palletizers and forklift operators shared the plant manager's concern about accidents, wanted simpler procedures for handling receiving information, and thought palletized loads were an obvious solution to the problem. Management excluded any

solution requiring redesign of anything but cover packaging or revamping sales and marketing activities.

The projected life of any problem solution is 2 years.

The first column of Table 1.1 lists the purposes—stated as unit operations—comprising the hierarchy for the example problem; the second column shows the corresponding systems to be redesigned; and the last column lists solution ideas which will accomplish each purpose.

The first purpose in the hierarchy is the most specific one; it is the function of the corresponding (existing) system shown in the table. This system is defined by the system diagram* in Fig. 1.3, which shows the boundary and the immediate and general environments of the system in more detail than is usually necessary.

Function 2 of the hierarchy in Table 1.1 includes Function 1 and defines a larger physical system in which the problem also exists. Figure 1.3 would represent this larger system if: (1) the purpose were changed, (2) the transportation system (rail or truck) and the vendor's shipment loading area were transferred from the immediate environment into the system, and (3) other components of the system were augmented as necessary to reflect these changes. With additional, similar modifications Fig. 1.3 could represent each larger function of the hierarchy.

Table 1.2 summarizes the desired improvements of system performance and the DS/C of the stakeholders. The management restriction and rate of return DS/C will be enforced rigidly; the others may be waived to avoid eliminating a potential solution which apparently will perform well.

Purposes are listed in reverse order from largest to smallest to ensure evaluation of the largest purpose first. The estimate of the performance of the solution ideas for each purpose on each DS/C and objective listed across the top of the table is a consensus of stakeholders' judgments. They required limited additional information and assistance from staff specialists who also suggested more solution ideas.

After these evaluations were recorded in Table 1.2 as shown, stakeholders chose Purpose 3, "to move covers from vendor cover production to storage," as the focus purpose because ideas for solutions satisfying Purposes 5, 6, and 7 violated management restrictions** and ideas satisfying Purpose 4 would not satisfy three DS/C. While making this choice they—again with the help of staff specialists—concluded that: (1) the operations involving pallet tags and receipt summaries should be considered a functionally independent smaller system, (2) palletizing would be another smaller system once the physical characteristics of the unit load to be transported and stored had been determined, and (3) whether to adopt a just-in-time inventory policy would be another problem for a third, larger system.

Table 1.3 shows the stated and derived objectives, criteria, ME, and goals chosen to select the system (devised in Phase III) which best satisfies the focus purpose.

Phase II, Developing Ideas for Potential Solutions

The generation of *ideas,* or *concepts,* for problem solutions is the purpose of Phase II. Its goal is the largest possible number of such ideas—ideal or practical, innovative or conventional, major or minor in terms of required changes—which will satisfy *the focus or a larger purpose.*

Guidelines. To develop many solution ideas:

1. Stimulate the voicing of innovative ideas by asking questions such as: "How would I/we change this system if I/we could start from 'scratch'?" "What system design—innovative, ideal, or both—would accomplish the focus purpose?" "The next larger purpose?" "The next smaller purpose?"

2. Do not reject ideas that effect small improvements in the existing system, partial solutions, or details of solutions.

---

*This graphic description of a system and its immediate and general environment shows information that would be recorded in the components, controls, and interface columns of the system matrix.

**For this reason Table 1.1 does not include systems corresponding to these purposes.

TABLE 1.1 Purposes, Corresponding Systems, and Solution Ideas for the Example Problem

Purpose Hierarchy	Corresponding System	Solution Ideas
1. To move covers from transportation vehicle to storage	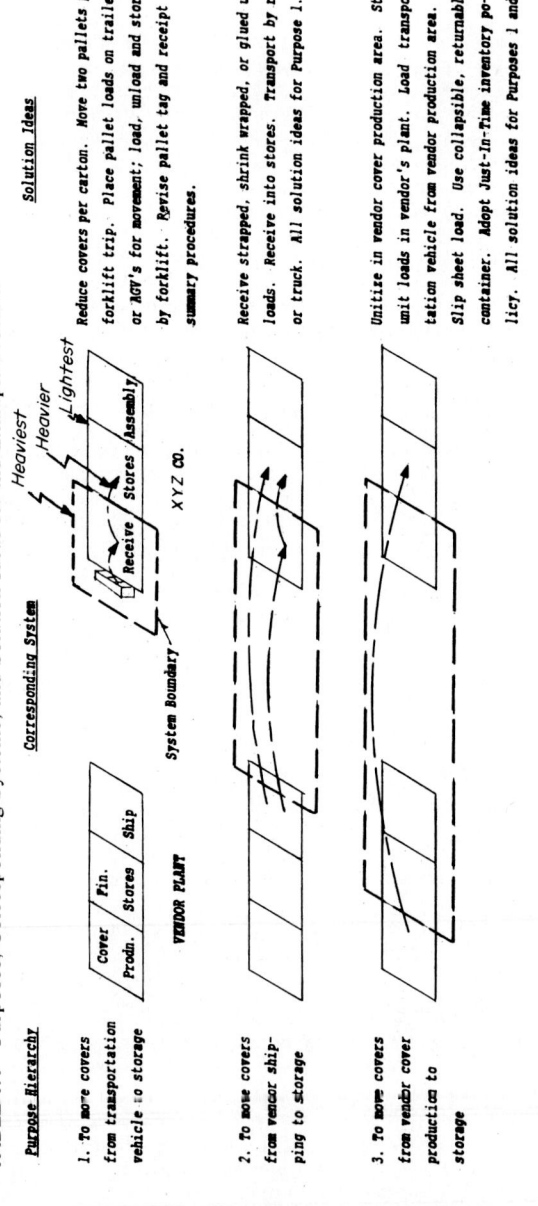	Reduce covers per carton. Move two pallets per forklift trip. Place pallet loads on trailers or AGV's for movement; load, unload and store by forklift. Revise pallet tag and receipt summary procedures.
2. To move covers from vendor shipping to storage		Receive strapped, shrink wrapped, or glued unit loads. Receive into stores. Transport by rail or truck. All solution ideas for Purpose 1.
3. To move covers from vendor cover production to storage		Unitize in vendor cover production area. Store unit loads in vendor's plant. Load transportation vehicle from vendor production area. Slip sheet load. Use collapsible, returnable container. Adopt Just-In-Time inventory policy. All solution ideas for Purposes 1 and 2.
4. To supply assembly line with covers		Move cover production to area adjacent to product assembly. Find closer vendor. Adopt Just-In-Time inventory policy.
5. To cover finished product		Redesign product to decrease cu. ft. per cover. Redesign product to eliminate cover.
6. To have product to ship		Revise plant layout to include cover production facilities. Buy finished product from vendor. All ideas for Purposes 4 and 5.
7. To have satisfied customers		Redesign product. Reduce price and/or time in-transit to customer.

(Corresponding System diagram labels:) Heaviest, Heavier, Lightest; Receive, Stores, Assembly; XYZ CO.; System Boundary; Cover Prodn., Fin. Stores, Ship; VENDOR PLANT

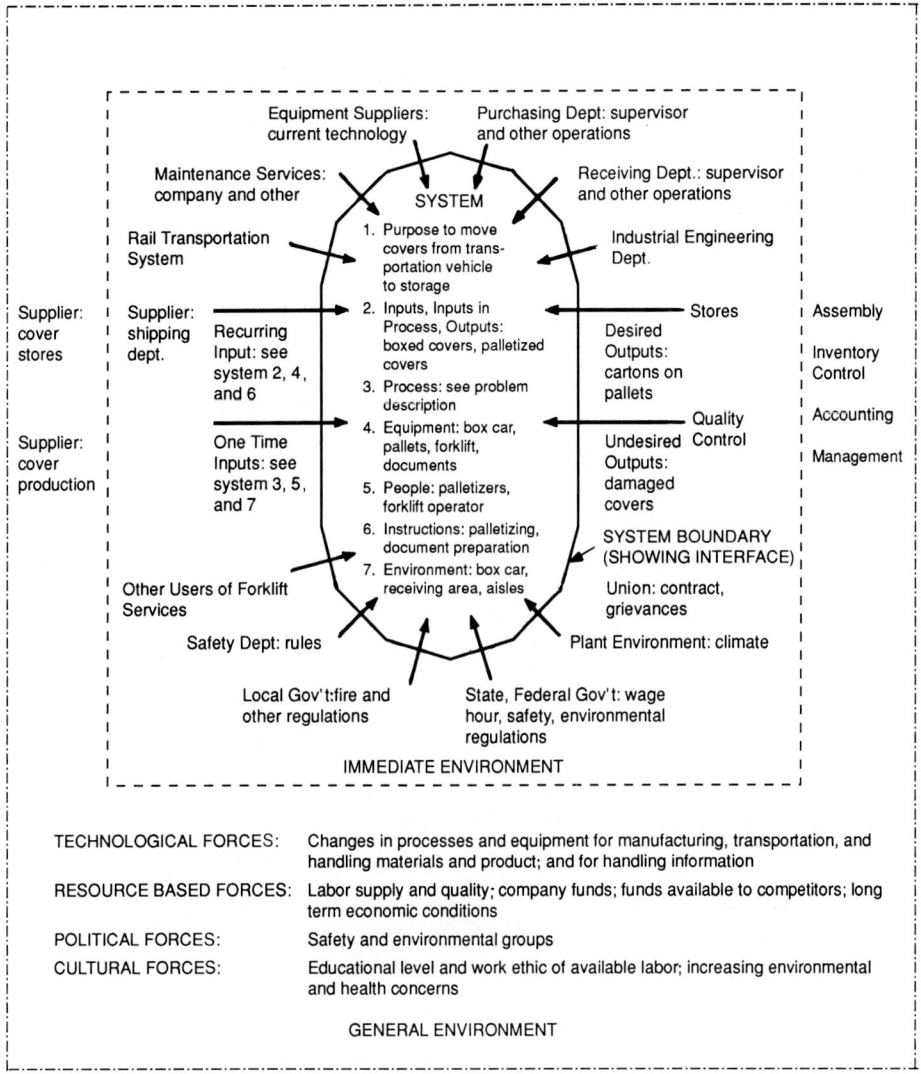

FIGURE 1.3 System diagram—Example problem existing system. (*Adapted from T. H. Athey*, System-atic Systems Approach, *Prentice-Hall, Englewood Cliffs, N.J., 1982, Figure 3.7, p. 35.*)

Adopt a "believing" attitude which suspends judgment of the feasibility of any concept. Never ridicule an idea or say, "That cannot be done because...". Encourage the statement of ideas in sufficient detail that their components are clear.

Techniques for Developing Solution Ideas. Methods of stimulating idea generation are based on: (1) the use of people as a source, (2) system components and the results of following systems engineering procedures, (3) reasoning processes, and (4) successful solutions to similar or analogous problems.

   Techniques Which Use People as a Source.   One of these techniques asks individuals and groups with stakes in the solution why each supports or opposes change; their responses will suggest solution ideas. Another solicits ideas from outsiders; their lack of

TABLE 1.2   Basis for Selecting the Focus Purpose for the Example Problem

Purpose	Possible solutions meet design specification and constraint					Possible solutions satisfy objective	
	Management restrictions	Implementation time ≤90 days	Total cost <$3M	Min % return exceeded		Improve worker safety	Reduce operating costs
7. To have satisfied customers	N						
6. To have product to ship	N	Reject					
5. To cover finished product	N						
4. To supply assembly line with covers	Y	N	N	N?		Y	Y
3. To move covers from vendor cover production to storage	Y	Y?	N?	Y		Y	Y
2. To move covers from vendor shipping to storage	Y	Y	Y?	Y		Y	Y
1. To move covers from transportation vehicle to storage	Y	Y	Y	?*		Y?	Y?*

Y = yes; N = no; ? = uncertainty, but belief.
*If the number of covers or cartons is reduced, worker safety will be improved but the cost savings may be negligible.

TABLE 1.3   Objectives, Criteria, Measures of Effectiveness, and Goals for the Example Problem

Stakeholder	Objective	Criteria	Measure of effectiveness	Goal
Plant manager, management	Reduce operating costs	Reduce labor costs	Wages + benefit costs per year	Max possible
	Quick implementation	Minimize time to implement	Days from assignment date until installed system operates as planned	Min ≤ 90 days
		or		
		Maximize savings	(90 − days above) × estimated saving per day	Maximum
	Return on investment	Maximize return over life of system	% return after taxes	Max ≥ min acceptable
Manager of industrial engineering	Implementation cost	Minimize implementation cost	$ expended	≤ $3000
Plant manager, workers	Improve worker safety	Minimize strains, other accidents	Manual handling accidents per year	0
		Minimize vehicle-related accidents	Vehicle-related accidents per year	0

knowledge about the problem ME and DS/C may lead to innovative suggestions. Questionnaires, brainstorming,[16] the nominal group method, and other techniques[17] may also be employed.

Techniques Based on System Components and the Use of Systems Engineering Procedures.   Solution concepts based on system components and the results of systems engineering procedures can be elicited by asking individuals and groups to:

1. State solution ideas that are suggested by words or phases of the function statement.

2. Indicate how projected system inputs, outputs, components, and environmental elements might be changed.

3. Devise solutions which satisfy one criterion.

4. Devise a solution that affects system performance negatively, then reverse the change.

5. Prepare a description of the ideal solution system.

Techniques Based on Reasoning Processes.   The application of engineering or other principles is a useful technique for devising solution ideas. For the industrial engineer, use of the methods improvement principles—eliminate, simplify, combine, or change sequence—and the associated operation chart and flow diagram is particularly effective.

Techniques Based on the Solutions of Similar or Analogous Problems.   Modifying the solution to a similar problem is a traditional engineering technique for generating solution ideas; it is extended here to include analogous problems. Since every problem is unique in some way, successful solutions to similar problems—particularly "off the shelf" solutions—should be examined carefully for applicability to the problem at hand.

Suggestion.   Further information about the idea-generating techniques described and others appears in Cross,[6] Hall,[7] Athey,[18] Sage,[19] and Nadler and Hibino.[20]

EXAMPLE: DEVISING ALTERNATE SOLUTION IDEAS.   Table 1.1 shows solution ideas which were devised for each purpose of the hierarchy to demonstrate how a purpose and its associated system trigger such ideas. Normally the ideas shown for Purposes 1 and 2 would be generated from consideration of the focus purpose and all ideas would be stated in more detail. In fact, the solution ideas for each purpose include all those for all smaller purposes.

The concept of handling cartons of covers in unit loads was suggested by the reduce accidents criterion, the reduce labor cost criterion, and the double reversal technique. The latter suggested degrading performance by handling covers individually instead of four per carton and then reversed that idea to suggest handling unitized loads of cartons.

The ideal solution idea, "move cover production to area adjacent to product assembly," satisfies the next larger (than focus) purpose "to supply assembly line with covers." As is often the case, it eliminates the operation which led to recognition of the problem.

Phase III, Devising Alternate Feasible Ideal Solution Targets (FISTs) and Selecting the Best FIST[21]

The Nature of a FIST.   The FIST is a *problem solution* which:

1. Performs the function of the focus (or a higher) purpose, is workable (feasible), meets all DS/C, and satisfies all threshold specifications for MEs—*but only for the regularities of the system.*

2. Provides the optimum balance—that is, performs best in the aggregate—on all measures of effectiveness.

A FIST which meets the first, or both, of these requirements is seldom an implementable solution because it is devised to accommodate only the relatively ideal conditions defined by the regularities of the problem system. The term FIST includes the word "ideal" for this reason.

A FIST which meets both requirements is a precursor of an implementable solution and a "target" solution because retention of the ideal aspects of this FIST is one objective when designing (in Phase IV) alternate implementable solutions based on it.

More than one FIST may be necessary to satisfy the regularities of a system.

Phase III Objectives.   The function and primary objectives of Phase III are

1. Conversion of the solution *ideas* generated during Phase II into the maximum possible number of alternate FIST candidates (potential FISTs), which: (*a*) satisfy only the performance requirements for the FIST and (*b*) are as mutually exclusive as possible
2. Selection from these candidates of the FIST which performs best in the aggregate

The secondary function of Phase III is the design of a "base system." It will be used when evaluating the implementable systems designed in Phase IV.

Devising FISTs.   These steps guide the design of FIST candidates:

1. Define the system "regularities" for which alternate potential FISTs will be devised.*
2. Classify each solution *idea* generated in Phase II as a complete FIST candidate, a component of a candidate, or a detail of a candidate.
3. Considering only regularities, create additions to the initial complete FIST candidates by using the components and details.

Defining System "Regularities."   Regularities are the characteristics of the components and environment of a system which are most common or most important and are not expected to change over the life of the solution system. Common characteristics fall into the 80 category of the "80/20 rule"; the emergencies for which airport fire fighting services are designed fall into the most important category.[22]

Devising potential FISTs to accommodate only the regularities of a problem enables a systems engineer to absorb the intricacies of a problem a few at a time and helps to avoid the confusion and diffusion of effort created by attempts to provide for irregularities in initial solution designs (as users of the engineering method do).

Classifying Solution Ideas as FIST Candidates, Components, or Details.   Because Phase II encourages the generation of as many solution *ideas* as possible, most ideas are components and details of potential FISTs; few are complete.

Categorize these ideas using these guidelines:

1. An alternate potential FIST is a solution idea which, if implemented, will perform the focus, or a higher, function for regularities even if its workability is credible but not certain.
2. A component is an idea which will satisfy this criterion only if it is combined with other components.
3. A detail modifies a component or a solution.

Devising Additional Alternate FIST Candidates.   The list of potential FISTs starts with any mutually exclusive candidates which fall in the first of these three categories. Modify these candidates and create additional ones by using the ideas classified as components and details.

The result of these efforts is a list of alternate potential FISTs, as mutually exclusive as possible, which—it is believed—satisfy the focus or a larger purpose for regularity conditions only.

Designing the "Base" System.   Phase IV evaluations which compare the performances of proposed problem solutions with that of the existing system may be biased toward the proposed solutions because it is often possible to improve existing system performance noticeably at a low cost. To eliminate this bias: (1) design alternate existing ("base") systems which improve performance but require minimum input of funds and other resources and (2) select the best performing alternate for inclusion with modified FISTs in the alternates from which the system to recommend is selected in Phase IV.

---

*Solution ideas may be devised during Phase II on the basis of these regularities if they do not hamper innovation.

The base system is not an ideal system.

Suggestions.   Retain the "believing" attitude during Phase III. Design as many FIST candidates as possible in each mutually exclusive category. Try to create systems which will provide a range of performance on each ME. Be alert for: (1) significant or sensitive system components or environmental factors and (2) any necessary revisions of the objectives, criteria, ME, and goals established during Phase I.

EXAMPLE: DEVISING ALTERNATE FISTS.   None of the solution ideas in Table 1.1 is a complete alternate FIST. This is apparent from Table 1.4, which classifies solution ideas for the focus purpose as components or details of potential FISTs on the basis of a process corresponding to that purpose.

TABLE 1.4   Examples of Components and Details of Solution Ideas for the Example Problem Focus Purpose

Process	Component	Detail
Unitize	People and equipment: manual palletizing, no equipment; no people, mechanical palletizer	Equipment: Conveyor to palletizer. Location: in cover production area, on shipping dock, elsewhere Output (unit load): Collapsible, returnable container; reusable pallet; disposable pallet; slip sheet. Strap, stretch wrap or glue load.
Transport to transportation vehicle	Process: transport to vehicle direct, via storage, or via storage and shipping. Equipment: fork lift; conveyor (direct only).	Equipment: forks, clamp or slip sheet attachment for forklift; type conveyor
Transport to Frammis Co. plant	Mode: rail, truck	Vehicle dimensions
Move covers from transportation vehicle to stores	Facilities: unloading location--receiving, stores. Equipment: forklift; forklift plus AGV; forklift plus trailer train	Equipment: forklift attachments; size trailers, trailers per train, tractor; AGV vehicle and guidance system specifications
Storage	Instructions: existing or just-in-time inventory policy	

The solution ideas for the focus and next large purpose shown in Table 1.1 can be combined into alternate potential FISTs which fall into two mutually exclusive categories:

1. Move cover production to an area adjacent to assembly.
2. Ship in unit loads and receive at the receiving dock or into stores.

Specific candidates in the first category would: (1) move individual covers from adjacent cover production to the assembly point by overhead conveyor or (2) unitize them for movement—from a less advantageously located cover production facility if necessary—and staging at the same assembly point by forklift or by company truck and forklift. Assuming the existing production facilities can be duplicated or an alternate process can be used (a significant factor), either is workable. The first of these two candidates eliminates all irregularities.

The feasibility of solutions in category 2 hinges on the willingness of the vendor to palletize cartons for shipment and (in some cases) to ship by truck (two additional significant factors). If that company will cooperate, almost any combination of the components and details in Table 1.4 will perform, for regularity conditions, the process shown in that table; each combination will be a workable solution now.

All workable FIST candidates in category 2 and the second potential FIST in category 1 will eliminate all existing irregularities except those related to pallet tags, receipt summaries, and a lack of space in the storeroom. They were relegated to functionally independent smaller systems.

Additional alternate FISTs providing a range of overall performances would be developed for each category.

The base system alternates would incorporate any workable solution ideas suggested by Purpose 1 in Table 1.1.

The steps of Phase III lead to the new set of objectives, criteria, ME, and goals shown in Table 1.5.

TABLE 1.5 Revised Objectives, Criteria, Measures of Effectiveness, and Goals for the Example Problem

Stakeholder	Objective	Criteria	Measure of effectiveness	Goal
Plant manager, management	Return on investment	Maximize return over system life	% return after taxes	Max ≥ min allowable
Manager of industrial engineering	Implementation cost	Minimize implementation cost	$ expended	≤ $25,000
Plant manager, workers	Worker safety	Minimize strains, other accidents	Manual handling accidents per year	0
		Minimize vehicle-related accidents	Vehicle-related accidents per year	0
Plant and quality managers, supervisors, and workers	Quality of covers manufactured	Equal or improve vendor quality	% rejects	≤ vendor %
Plant, production, and production control managers	Reliability of cover source	Minimize shutdowns due to late cover delivery	Shutdowns per year	≤ present 3 per year

The "return on investment" objective in it, for example, includes the cost and implementation objectives in Table 1.3. The plant manager, intrigued by moving cover production into the XYZ company plant, raised the maximum to the spending authorization limit. The quality and reliability objectives were needed because some alternates include a new source of covers—a production facility operated by the XYZ company.

Selecting the FIST with Optimum Performance. Before attempting to select the FIST which will perform best in the aggregate, eliminate any FIST candidate which will not satisfy the focus or next larger purpose. Then:

1. Provide the details necessary to assess the workability and performance of each candidate left; make these assessments.

2. Determine which of these FISTs performs best in the aggregate; it will be modified into possible implementable solutions during Phase IV.

Detailing Alternate Potential FISTs to Assess Workability and for Evaluation.   To determine the details necessary to assure the workability of each remaining candidate, visualize—in conjunction with stakeholders and experts—its operation *as it performs the system regularities*; consider expected, best, and worst case situations. If a potential FIST is not feasible, make it workable with minimal decrease of its ideal features. If it is not workable now but will be in the future, set it aside for inclusion in plans for continuous improvement of any installed solution. If it cannot be made workable now or later, set it aside. Then ascertain any "irregularities" each FIST candidate will accommodate without modification.

After establishing the potential FISTs that are workable, determine the data necessary: (1) to verify that each one, and each "base system" alternate, meets the DS/C's and threshold values for MEs and (2) to estimate the performance by each alternate of each ME. Plan collection of these data; they will be minimal because their specifications are known.

Selecting the Optimum Performance FIST.   Conventional engineering economy techniques are usually adequate to select the optimally performing FIST. When they require difficult judgments or when judgments are questioned, use more sophisticated techniques. Nadler[23] describes several of them—multiattribute decision theory, the decision tree, and a sequential ranking method—and discusses their defects. Hall,[24] Athey,[25] Cross,[26] and Huffman[27] provide examples illustrating different methods of applying multiattribute decision theory and alleviating its defects. Byrd and Moore[15] discuss this and other techniques.

If the number of FIST candidates is large, it may be useful to reduce the number by screening before employing any of the evaluation methods described. This technique employs estimates of workability and rough projections of alternate performance on DS/C and ME to remove from consideration FIST candidates that are not workable or cannot be made workable now or in the future and any feasible alternates that appear to perform poorly. Record the projections as in Table 1.6.

Screening for workability usually occurs as alternates are devised. It is not necessary to screen for overall performance if the number of FIST candidates is four or less.

TABLE 1.6   Screening Alternate Potential FISTs for the Example Problem

Alternate Potential FIST Description	Satisfies Purpose	Will Work	-----Alternate Potential FIST Satisfies DS/C or Is $\geq$ ME Threshold For------

			Can Implement Now	Max. Cost to Implement	Rate of Return	Worker Safety	Vehicle Accidents	Quality of Covers	Reliable Cover Source

Y = yes; N = no; ? = doubt but belief. Reject FIST candidate only for N.

-------Projected Performance (On a Scale of 0 to 10)-------

Cost to Implement	Rate of Return	Manual Accidents per Year	Vehicle Accidents per Year	Per cent Rejects	Shutdowns per Year

Cautions.   Involve stakeholders in the decision making required by Phase III. Hall[24] describes methods of group decision making and the problems associated with each one. Retain potential FISTs set aside or rejected during this phase for possible use if the problem solution (based on another FIST) recommended during Phase IV is rejected.

> EXAMPLE: SELECTING THE FIST.   After details were added, likely FISTs were identified by screening via Table 1.6. The FIST was selected by the multiattribute utility method; the relative weights it requires for criteria were determined by stakeholder consensus. They estimated the performance of each alternate on each ME with the aid of staff specialists who also calculated the aggregate performance of each alternate.
>
> The first two columns of Table 1.7 show the process and equipment and facilities components of the FIST selected.*

Phase IV, Recommended Solution Design, Approval, and Detailing

Converting the FIST selected during Phase III into an implementable solution is the function of Phase IV. Its objectives are: (1) the design of a solution (to recommend) which both accommodates irregularities with minimum loss of the ideal features of the FIST and increases its overall performance if possible, (2) approval of this solution, and (3) completion of the detailing necessary to plan its implementation.

Devising and Selecting the Solution to Recommend.   To create the solution to recommend:

1. Convert the FIST into multiple implementable solutions.
2. Project the performance of these solutions and select the solution to recommend.

Devising Alternate Solutions.   To design alternate implementable solutions which will accommodate all irregularities, modify the FIST by: (1) devising alternates for each component or new components which will handle the irregularities and (2) selecting from these alternates those which will improve system performance, if possible, and still maintain component relationships and dependencies.

Treat any remaining irregularities as regularities of functionally independent smaller systems; design systems which can be incorporated into the modified FIST to form an implementable solution. Then incorporate the designs of all such systems into each alternate. Throughout these efforts maintain the ideal aspect of the original FIST as much as possible.

Add, as additional alternatives to the implementable solutions devised, any "off the shelf" solutions or solutions that were successful elsewhere. This is a good point at which to consider potential difficulties in securing approval for any recommended system, acceptance by others, and possible implementation problems.

Projecting Performance and Selecting the Solution to Recommend.   Project performance and select the solution to be implemented by the methods employed (in Phase III) to select the FIST. The necessary performance projections will require more precision and less variability than those used at that time; prepare them.

Securing Approval.   When planning the steps, reports, and presentations to obtain approval of the recommendation, follow established practices but include not only the recommended solution but also others considered—including any "pet" ones—and the reason for their rejection. Consider not only those who must approve the recommendation but also such factors as: "Who can exert pressure for acceptance?" "Who will oppose

---

*This solution satisfies the focus purpose because vendor plant operations between the independent palletizing system and vehicle loading were considered to be another independent system. The physical solution system is therefore the one corresponding to Purpose 2 in Fig. 1.1.

TABLE 1.7   Example of the Design of an Alternate Implementable Solution for the Example Problem

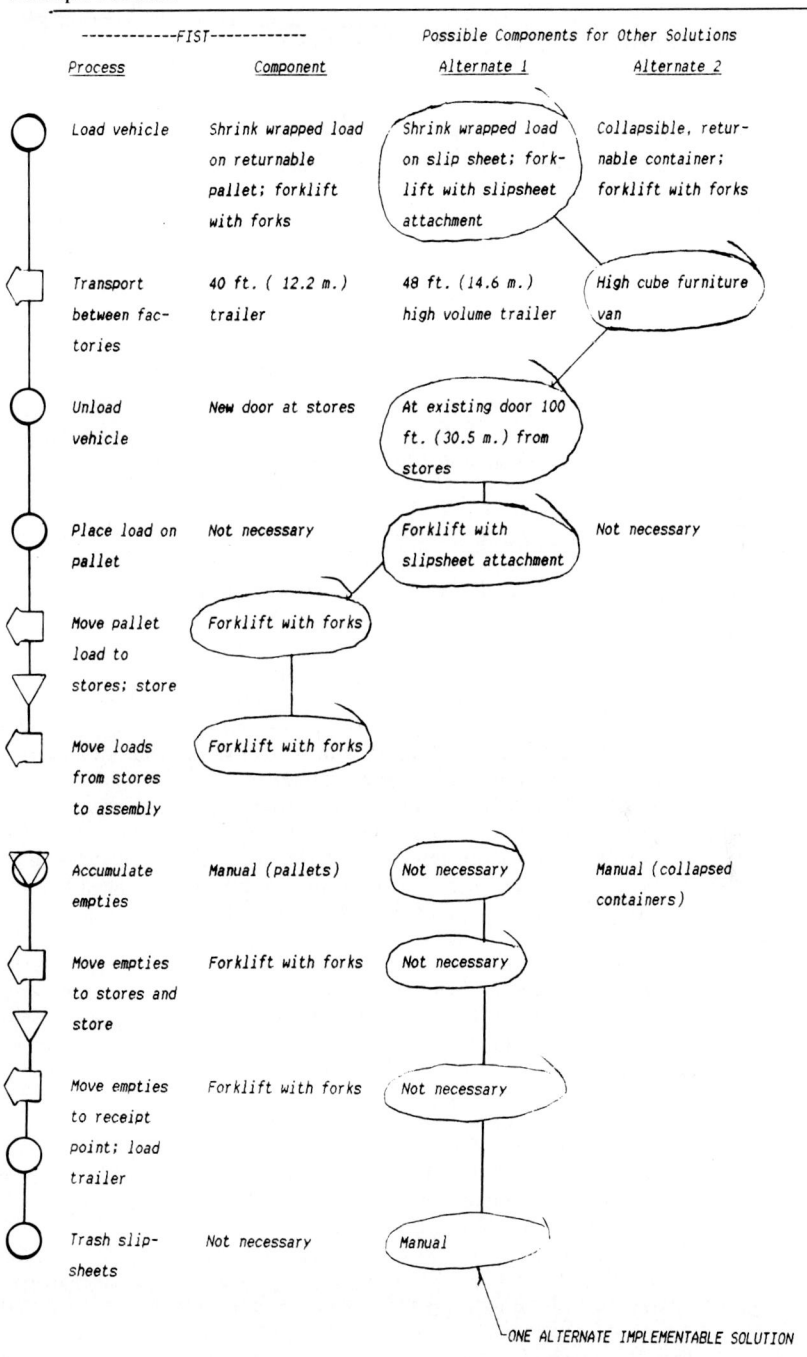

ONE ALTERNATE IMPLEMENTABLE SOLUTION

approval and how can their objections be overcome?" "Whose approval is not necessary but will be helpful?"

If the recommendation is accepted, detail it for implementation planning. If it is rejected, start over at the appropriate phase.

Detailing the Accepted Recommendation. Develop all the information necessary to plan implementation including such details as operations sequences, equipment specifications, detailed layouts, and predetermined time standards. The system matrix is very useful for establishing the data needed.

Suggestion. Review each step prior to and after presentation of the recommendation with stakeholders and others who can provide constructive suggestions.

EXAMPLE: DEVISING THE SOLUTION TO RECOMMEND. Columns 3 and 4 of Table 1.7 show possible alternates to the facilities and equipment component of the FIST of each step of the FIST process.

Alternate implementable solutions would be devised as shown by the example in the table. The solution to be recommended would be selected from these alternates, after detailing, by the evaluation method used in Phase III to select the FIST.

Phase V, Implementation Planning and Execution

Designing the Implementation Plan. Design a project management system to implement the approved recommendation. Provide not only for any necessary testing, installation, and start-up of the system but also for monitoring, modifying, and auditing it throughout its life. Focus plans to monitor and modify the system on achieving: (1) performance projections and (2) the acceptance of the system by stakeholders.

To foster acceptance, the implementation management system should: (1) permit people to adapt to new methods, equipment, and environments gradually, (2) allow them some leeway to adjust details of the recommended system, and (3) include steps to "sell" stakeholding groups represented but not involved in system design and to retain the support of others.

Incorporate continuous improvement into the plan by scheduling the implementation of systems rejected previously because they were considered feasible only in the future and by providing a mechanism for effecting other unexpected improvements that become apparent.

Use systems engineering techniques to design the implementation plan because it was not until the late 1980s that securing acceptance of the system was added to the traditional project management objectives of a system which meets performance specifications, on schedule, and at or below budget.

Implementation Follow-up

Monitoring and Modifying. Design the monitoring system as a feedback control system with two regularities: (1) detection and response to unsatisfactory performance or resistance to implementation or operation as planned and (2) detection and rapid response to signals indicating possible improvements.

Monitoring should compare actual against projected performance not only for each ME (in ME units) but also as aggregated for the solution overall because the latter was the basis for recommending the system. When overall performance is satisfactory, inadequate performance on one ME may not require quick correction; instead, the deviation may provide an opportunity to achieve higher than anticipated overall system performance.

Integrate any available suggestion system into the monitoring plan but provide a program of personal contact to detect improvements not suggested. Use other facets of monitoring discussed in project management texts.

Auditing Implementation and Operation.   Auditing includes a current evaluation of systems performance, but it stresses a review which will suggest how to improve execution of the systems engineering process. Questions such as: "Was the focus purpose correct?" "Were the objectives, criteria, MEs, and goals adequate to evaluate solution ideas, FIST candidates, and alternate implementable solutions?" "Were performance projections satisfactory?" "If not, why not?" provide the necessary information.

An audit at the time the system is turned over to line personnel for operation and another one 6 months or a year after that date are essential. Others at subsequent intervals of 1 year are desirable.

A BROADER VIEW OF SYSTEMS ENGINEERING

Applicability and Application

Systems engineering methods are applicable to any situation which requires restructuring (redesign) of an existing system or the structuring (design) of a currently nonexistent entity. The problem may involve a process or an inanimate object such as an ergonomically sound workstation. The methods may be applied to a system as a whole, to smaller systems which are a part of it, or to its components. They should be utilized to design all components of the system which solves a problem.

These procedures should not be expected to yield a break-through solution to every problem immediately. That solution can be the result of the continuous change and improvement provisions of every systems engineering problem solution.

The size and complexity of the problem determine the depth of the procedures used. Major ones justify use to the extent described in this chapter. For lesser problems abridged steps or thinking in terms of the methods and principles will suffice.

Limits to Application

Systems engineering procedures can be used to develop systems which: (1) present information in a manner that facilitates learning, (2) provide information for operating and supervising or evaluation, and (3) specify research procedures. They cannot be applied to the mental processes of these activities.

SYSTEM OPERATIONS

Make elementary systems engineering an integral part of operating practices by teaching its principles and its methods—except possibly implementation—to management, supervisors, and workers. Their applications of this knowledge will initiate continuous improvement without staff studies and provide a sense of accomplishment to individuals.

Conditions for Maximum Success

The success of systems engineering is fostered by a problem-solving environment in which:

1. Management and supervisors advocate systems engineering.
2. The actions of these groups support their words.

3. Communications both ways between the workers and management are open.

4. Personnel policies stress job security, competitive compensation, and recognition of achievement.

5. The data and resources necessary to evaluate employee suggested system improvements and to implement worthy ones are provided on a timely basis.

Supervisors and managers sometimes oppose the conditions because they see them as eroding traditional responsibility and authority. They are suspicious of systems engineering because solution systems are more likely than other methods to alter both. Systems engineering endeavors to overcome this resistance by involving those at all organizational levels in the systems engineering process.

THE INDUSTRIAL ENGINEER AND SYSTEMS ENGINEERING

Education at the undergraduate and graduate levels and experience uniquely qualify the industrial engineer for all aspects of systems engineering. Use of the procedures described in this chapter make it more likely that he or she will devise and implement innovative solutions to the "right" problems.

It is especially important that the industrial engineer participate actively in the implementation phase of systems engineering. This participation is a responsibility because no project is complete until successfully implemented; this participation is desirable because implementation provides valuable feedback called experience.

REFERENCES

1. Adapted from Gerald Nadler, *The Planning and Design Approach,* Breakthrough Thinking Co., Los Angeles, 1981, pp. 83–85 (copyright by the author).

2. Nadler, ibid., pp. 70–72.

3. Athey, Thomas H., *Systematic Systems Approach,* Prentice-Hall, Englewood Cliffs, N.J., 1982, pp. 32–35.

4. Nadler, Gerald, op. cit.

5. Nadler, Gerald, and Shozo Hibino, *Breakthrough Thinking: Why We Must Change the Way We Solve Problems, and the Seven Principles to Achieve This,* Prima, Rocklin, Calif., 1990 (copyright by the authors).

6. Cross, Nigel, *Engineering Design Methods,* Wiley, New York, 1989.

7. Hall, Arthur D. III, *Metasystems Methodology,* Pergamon, Elmsford, N.Y., 1989.

8. Dandy, G. C., and R. F. Warner, *Planning and Design of Engineering Systems,* Unwin Hyman, Boston, 1989.

9. Nadler, Gerald, and Shozo Hibino, op. cit., pp. 107–134.

10. Nadler, Gerald, op. cit., pp. 7–9.

11. "I Can't Work This Thing," *Business Week,* no. 3211, April 29, 1991.

12. Nadler, Gerald, op. cit., p. 1.

13. Adapted from Gerald Nadler, op. cit., pp. 135–142.

14. Delbecq, A. Van de Van, and D. Gustafson, *Group Techniques for Program Planning—A Guide to Nominal Group and Delphi,* Scott-Foresman, Chicago, 1975.

15. Byrd, Jack, and L. Ted Moore, *Decision Models for Management,* McGraw-Hill, New York, 1982 (discusses decision-making methods of all types).

16. Osborn, A. F., *Applied Imagination,* Scribners, New York, 1953.

17. For a more extensive list of these and other group processes which can be used to generate solution ideas, see Gerald Nadler and Shozo Hibino, op. cit., pp. 243–246.

18. Athey, Thomas H., op. cit., pp. 68–83.

19. Sage, Andrew P., *Methodology for Large Scale Systems,* McGraw-Hill, New York, 1977, pp. 170–176.

20. Nadler, Gerald, and Shozo Hibino, op. cit., pp. 148–152.

21. Adapted from Gerald Nadler, op. cit., pp. 154–161.

22. Nadler, Gerald, op. cit., pp. 155 and 156 provide specific suggestions for identifying regularities.

23. Nadler, Gerald, op. cit., pp. 159, 160.

24. Hall, Arthur D. III, op. cit., pp. 377–423.

25. Athey, op. cit., pp. 121–135, 206–214.

26. Cross, op. cit., pp. 101–124.

27. Huffman, John R., private publication, Los Angeles, Calif., 1988.

CHAPTER 2
PLANNING AND CONTROL OF MANUFACTURING SYSTEMS

Eric M. Malstrom
Professor and Head
Department of Industrial Engineering
University of Arkansas
Fayetteville, Arkansas

This chapter addresses both planning and control of manufacturing systems. It is important to note that entire books have been written about many of the topics covered in this chapter. The chapter therefore cannot provide complete "stand alone" coverage of many of the topics addressed. The approach used provides the reader an overview of a wide range of topics related to planning and control of manufacturing systems. Where more specific information is desired, the reader is referred to the references at the end of the chapter.

The chapter begins by overviewing time series forecasting methods. Forecasting is necessary to accurately anticipate future demand for products and goods. Master scheduling is discussed next. The master schedule reflects scheduled production for end items produced by those organizations that utilize explosion-based inventory systems.

A separate section is devoted to describing the steps in the purchasing process. Reorder point lot sizing methods are overviewed and compared with explosion-based inventory systems. The impact of just-in-time inventory philosophies is assessed on traditional explosion-based inventory procedures.

Process planning methods are next overviewed. Machine sequencing methodologies are summarized for single machines, job shops, and flexible manufacturing systems. Finally, movement and control of parts on the production floor is described by overviewing the procedures of both expediting and dispatching.

FORECASTING AND PROCUREMENT

Introduction. Before production may be either planned or scheduled, customer demand must be assessed over future time periods. This forecast usually is the basis for a scheduled plan of production regardless of whether reorder point, explosion-based, or just-in-time inventory systems are being used. The next step in the process is to develop a production plan. The nature of this plan depends on the type of inventory system that is being used. This section overviews forecasting methods, master scheduling concepts, purchasing procedures, lot sizing methods, and just-in-time inventory techniques.

Forecasting Methods.   In forecasting sales or demand levels, a variety of time series methods are commonly used. The term "time series" implies that the past history of demand levels will influence the future behavior of this demand over time.

   Ordinary Least Squares Regression.   Ordinary least squares (OLS) regression is used as a curve-fitting tool when a linear trend in data is present. Assume a scatterplot of historical data. A first-order equation of the form

$$Y_i = a + bX_i \qquad i = 1,2,\ldots,n \tag{1}$$

is fitted through the data. The intercept a and slope b of this equation are selected by the regression procedure to minimize the sum of the squares of the distances from each data point in the scatterplot to the straight line represented by the fitted equation. This minimization is illustrated in Fig. 2.1.

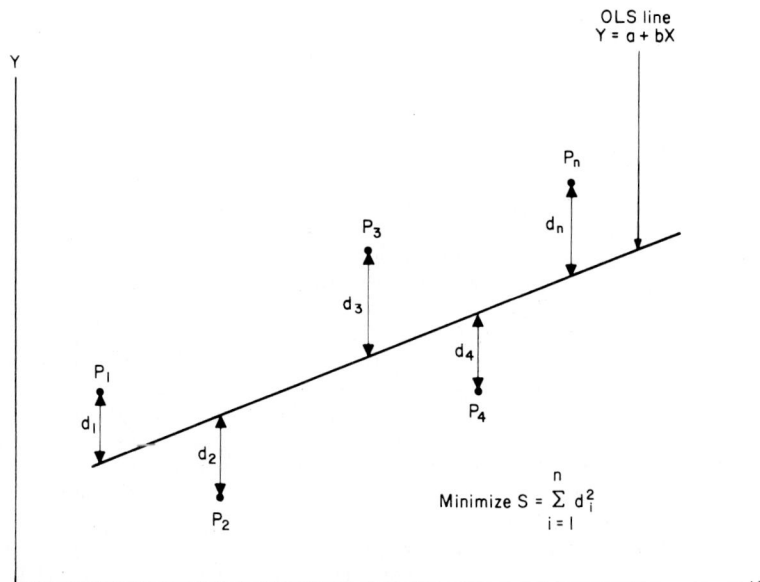

FIGURE 2.1   OLS minimization of the sum of squares.

   The quality of the "fit" of the straight line through the data scatterplot is generally measured by the coefficient of determination r^2. This parameter varies between 0 and 1, and represents the percentage of scatterplot data variation accounted for by the fitted equation. An r^2 value of 1.0 indicates that all data variation is accounted for by the fit. This is equivalent to saying that all of the data points in the scatterplot fall exactly on the fitted straight line. An r^2 value of 0 indicates that none of the data variation is accounted for by the first-order fit.

   In practice r^2 values in excess of 90 percent usually indicate a fit of good quality. Values of 50 percent or less suggest a poor fit and indicate that linearity was not present in the data scatterplot. Scatterplots indicative of both high and low r^2 values are depicted in Fig. 2.2.

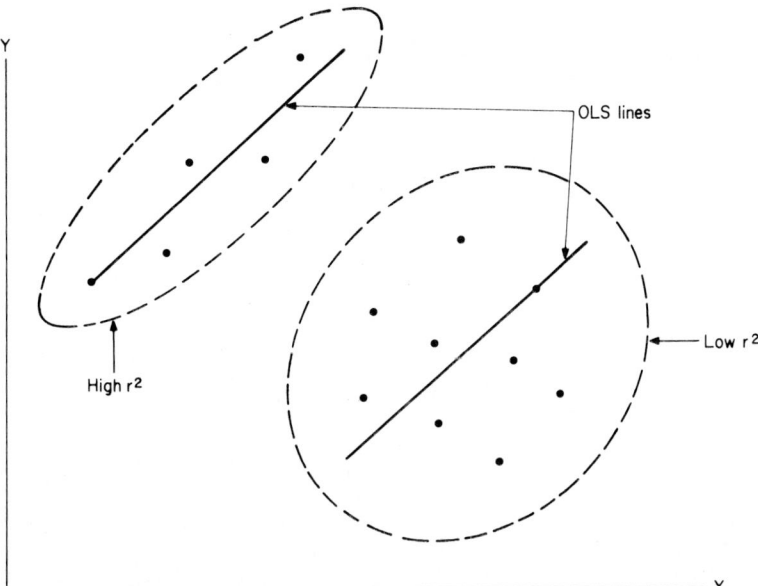

FIGURE 2.2   High and low r^2 values.

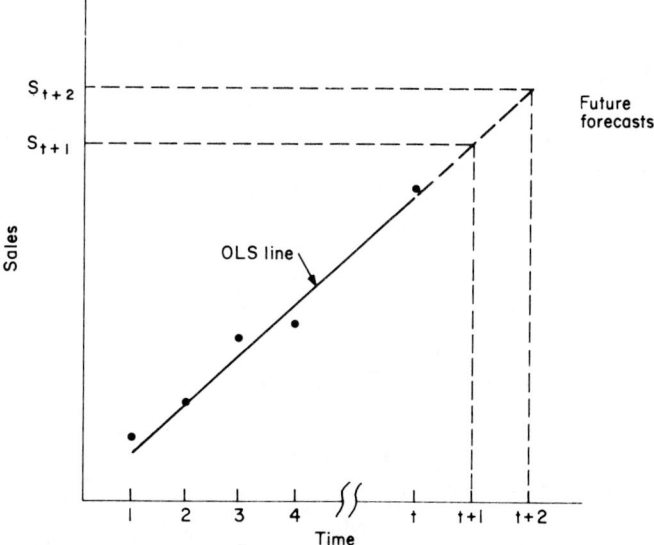

FIGURE 2.3   Extrapolation of OLS line.

Once a fitted line has been defined for the scatterplot, the forecasting procedure is straightforward. The fitted line is used to extrapolate demand values for future time periods. This procedure is illustrated in Fig. 2.3.

The OLS procedure is not confined to making linear or first-order fits. The method may be adapted to two types of exponential equations which are defined by the equations below.

$$Y_i = ab^X i \qquad i = 1,2,\dots,n \tag{2}$$

$$Y_i = aX_i^b \qquad i = 1,2,\dots,n \tag{3}$$

The OLS procedure is adapted by taking logarithms of Eqs. (2) and (3) to convert the equations to first-order. The OLS procedure is then modified by substituting logarithms of these values into the OLS normal equations. A good overview of OLS curve-fitting procedures has been presented by Draper and Smith.[17] Readers desiring more information on this subject are urged to consult this reference.

Other Curve-Fitting Methods. For scatterplots more irregular in form, other polynomial curve fits may be used. Orthogonal polynomials may be used to fit an nth-order polynomial through historical data. The equation is of the form

$$Y_i = a + bX_i + cX_i^2 + dX_i^3 + \cdots + mX_i^n \tag{4}$$

Caution should be used when using nth-order polynomials as a forecasting tool. It is possible to use this method to obtain a perfect curve fit if a polynomial order equal to the number of data points in the scatterplot is used. This perfect curve fit may lead to unrealistic expectations on the part of the user in terms of the accuracy of the forecasts that may be obtained by extrapolating future sales values from this curve.

Moving Averages. A variety of time series forecasting methods exist. These methods assume that the forecast for the next period is an average of previous historical sales values. With a simple moving average, the forecast for the next period is merely the average of the historical actual sales values for the n preceding periods. This is shown by Eq. (5).

$$F_{t+1} = A_t = (X_t + X_{t-1} + \cdots + X_{t-n+2} + X_{t-n+1})/n \tag{5}$$

where A_t = average over n periods through period t
F_{t+1} = forecast for period $t + 1$
X_t = sales or demand for period t

The term n reflects the number of periods in the moving average. For example, consider a four-period moving average. If it is now period 14, the forecast for period 15 would be the average of demands for periods 11, 12, 13, and 14.

A simple moving average assigns equal weights to all past data points comprising the forecast. With a weighted moving average, different weights are assigned to all past data as in Eq. (6).

$$F_{t+1} = A_t = w_t X_t + w_{t-1}X_{t-1} + \cdots + w_{t-n+1}X_{t-n+1} \tag{6}$$

where $i = \displaystyle\sum_{t-n+1}^{t} w_i = 1$
w_i = weighting factor corresponding to period i

For most applications, weights are generally selected to place more weight on recent data. As an example, suppose it is now period 22. A four-period weighted moving average might be written as

$$F_{23} = A_{22} = 0.4X_{22} + 0.3X_{21} + 0.2X_{20} + 0.1X_{19}$$

Exponential Smoothing. Exponential smoothing is a time series forecasting method that effectively "remembers" all past data. The technique is similar to the weighted moving average except that all previous demand values are contained in each forecast. A parameter called the smoothing constant may be adjusted to place more or less emphasis on more recent demand values.

Consider an actual demand $X_t = 120$. The corresponding forecast for period t, F_t, is known to be 100. Since the forecast is lower than the actual demand, it is apparent that the forecast for the next period, $t + 1$, should be higher. The question is, by how much? The exponential smoothing equation is defined as

$$\text{New forecast} = \text{old forecast} + \alpha \, (\text{present demand} - \text{old forecast}) \tag{7}$$

where α = the smoothing constant, $0 \le \alpha \le 1$

Substituting previously defined notation, we obtain

$$F_{t+1} = F_t + \alpha(X_t - F_t)$$

$$= F_t + \alpha X_t - \alpha F_t$$

$$= \alpha X_t + (1 - \alpha)X_t \tag{8}$$

where F_t = forecast for period t
$\quad F_{t + 1}$ = forecast for period $t = 1$
$\quad\quad X_t$ = actual demand during period t

Adjusting the smoothing constant allows the user to vary the extent to which the forecasting model is responsive to data fluctuations. The use of smoothing constants near 1.0 results in forecasts that place more weight on recent data. This causes the forecasts to be responsive to data fluctuations in recent historical demand values. The use of lower smoothing constants (generally less than 0.7) places more weight on older data. This tends to "damp out" the effects of irregularities in demand data.

Seasonality. A variety of methods exist to account for recurring irregularities in historical sales data. Trend components allow for automatic progressions in historical data. Cyclical components may be included in forecasting models to account for periodic recurring data fluctuations. Finally, seasonal components may be included to account for seasonal fluctuations in data.

A straightforward method for handling seasonal variation can be employed when historical sales data are plotted separately, by quarter. Consider the data points shown in Fig. 2.4. If the data from the first two quarters in each year are analyzed separately, it is apparent that sales have increased linearly over the past 3 years. This is not true for the data from the third and fourth quarters. Sales over the past 3 years for these quarters have been relatively constant.

Neither of these trends is apparent until the data are analyzed separately, by quarter. In this case, the best approach would be to use two different OLS fits, one for first and second quarter data and the other for third and fourth quarters.

Nothing requires that the forecasting methods for both data sets be identical. For example, if one of the two data sets was not linear, alternate curve fitting or time series techniques could be employed.

Model Validation. Which forecasting method to use is not always obvious. A good way to select the best method is to simulate how different techniques would have actually worked in practice when compared with one another. This may be accomplished in the following manner.

Model validation requires several periods of actual demand or sales data, usually over a period of 2 or more years. This historical database is then partitioned into two sections. The first section is used to develop the forecasting model. Models are developed on the first section of the database without the use or knowledge of any of the data values in the second section.

Forecasts are then made using the model. Forecast values are compared, period by period, with the actual demand values known to have occurred in the second section of the database. Deviations between forecast and actual values are tabulated for each forecasting method evaluated. The best method is the one that will yield the smallest sum of absolute deviations between forecast and actual values.

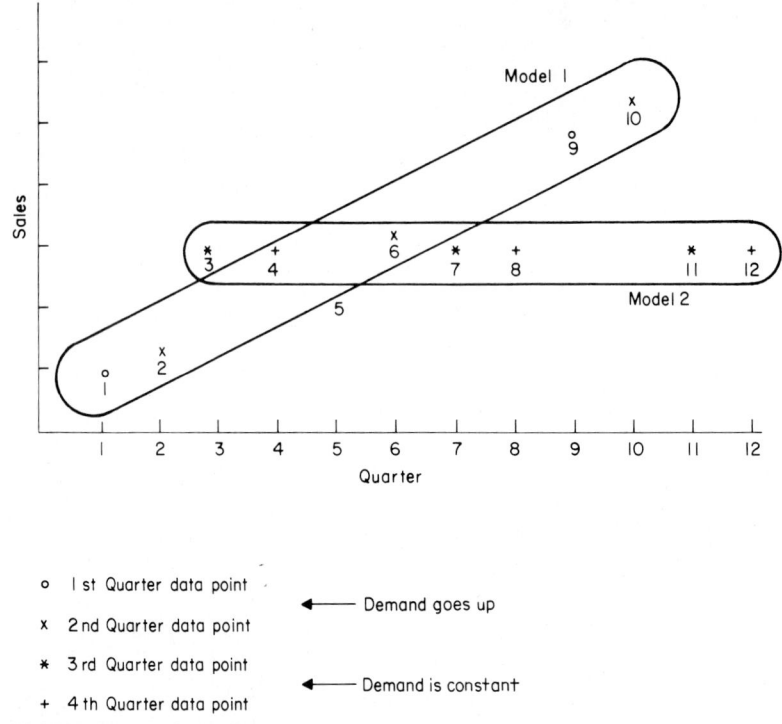

o	1st Quarter data point
x	2nd Quarter data point
*	3rd Quarter data point
+	4th Quarter data point

←———— Demand goes up

←———— Demand is constant

FIGURE 2.4 Analysis of demand data by quarter.

Summary. Curve fitting and time series forecasting methods both offer distinct advantages. Time series methods are usually easier to employ and are responsive to non-arithmetic changes in data periods. Time series methods are constrained in that the forecast is usually limited to one period into the future.

Curve fitting methods usually require more front-end work to set up. While not responsive to sudden changes in demand data, they offer the advantage of extrapolation, permitting forecasts to be made several periods into the future. The material in this section is intended to provide only an overview of forecasting techniques and methods. Readers desiring more detailed information on this subject are urged to consult Refs. 2, 3, 17, 20, 25, 41, and 47.

Master Scheduling. Vollman, Berry, and Whybark[48] define master scheduling as the anticipated build schedule for manufactured end products. The master schedule is not the result of a sales forecast. It is a statement of scheduled production as opposed to anticipated demand.

Sales forecasts may be regarded as critical inputs in determining master schedules. However, the master schedule also takes into account both limitations in factory capacity and the need to utilize such capacity as fully as possible.[48]

The master schedule indirectly determines the demands and related procurement schedules for all production components contained in the end items being produced. For example, a production schedule of 100 automobiles per month implies the need for prior procurement of 500 tires per month (four tires per car plus one spare). This procurement needs to be completed in advance of the car's being assembled so the tires are available to mount on the car at the time it is built. An explosion-based inventory system known as material requirements planning (MRP) facilitates the determination of procurement and production of production components. This system is the subject of a later section in this chapter. Also see Sec. 10, Chap. 4.

Vollman, Berry, and Whybark[48] define four master scheduling techniques. These methods are now separately described.

Cumulative Charting Model. The cumulative charting model is a straightforward graphical scheduling method. A scheduling time period is first selected. This is generally weeks or months. A scheduling period is generally selected. This is the interval of time over which the master schedule is to be constructed. This period is generally called the "planning horizon."

The scheduler graphs the cumulative units called for by the master schedule over the duration of the planning horizon. The cumulative sales demand over the planning horizon is next plotted on the same graph. The difference between cumulative production and cumulative demand is the inventory of end items expected to be in stock during any planning period.

Rolling through Time. Simple cumulative charting models fail to consider the difference between forecast and actual sales. As the schedule progresses, period by period, through the planning horizon, these two terms are likely to differ. The rolling through time method is a cumulative charting model that takes these differences into account.

Cumulative production and cumulative forecast sales are again plotted for the duration of the planning horizon. As each period is completed, actual sales are also plotted and compared with corresponding forecast values. When differences are noted, the master schedule is revised period by period. These revisions prevent shortages and excess inventory from occurring from period to period.

Order Promising. Order promising refines the rolling through time procedure. Again, cumulative production and the cumulative sales forecast are plotted over the planning horizon. Actual and forecast sales are compared with one another.

With this method, a backlog of orders must exist. In other words, the demand for end items of production must exceed the supply available in any period. The master schedule, forecast sales, and actual sales are compared with one another. This comparison permits end items to be committed for shipment to customers in future time periods in the planning horizon. It is this procedure that gives this technique its name.

The Time Phased Record. The time phased record is essentially a tabular presentation of the procedure described in the preceding section. Consider the table shown in Fig. 2.5. A 12-month planning horizon applies. It is now January 1. An on-hand balance in the amount of 40 units has been carried over from the preceding month.

	J	F	M	A	M	J	J	A	S	O	N	D
Forecast	10	10	10	10	10	20	20	20	20	20	20	20
Orders	5	5	5	5								
Available	25	10	55	40	30	10	50	30	10	50	30	10
Avail. to promise	10	0	50				50			50		
MPS			60				60			60		

On Hand 1/1 = 40

FIGURE 2.5   Time phased record.

The entries in the table are computed in the following manner. Orders in the amount of 5 units per month have been promised for each of the first 4 months. These orders must be satisfied in addition to the sales forecast. For January the total demand is the 10 units from the sales forecast plus the 5 units previously promised. This leaves $40 - 15 = 25$ available units at the end of January. Since 15 total units have been promised for February, March, and April, only $25 - 15 = 10$ units are available to promise at the end of January.

The quantity available at the end of any period can be determined from the relationship

$$A_i = A_{i-1} + MPS_i - F_i - O_{pi} \qquad (9)$$

where A_i = stock available at the end of period i
MPS_i = quantity scheduled for production in period i by the master schedule
F_i = forecast demand for period i
O_{pi} = order quantity of units previously promised for delivery during period i

The amount available to promise in any period is defined by Eq. (10).

$$ATP_i = A_i - \sum_{j=i+1}^{n} O_{pj} \qquad (10)$$

where ATP_i = amount available to promise in period i

Orders are generated on the master schedule in the following manner. A total of 10 units is carried forward from the end of February. The total demand for March is 15 units, 10 units from the sales forecast and 5 units that have been previously promised. The 10 units available is not sufficient to satisfy this demand. The master schedule therefore calls for an additional 60 end items to be available by the beginning of March. Applying Eq. (9) yields the amount available at the end of this month (55 units). Equation (10) yields a value of 50 units available to promise during March.

All other entries in the table are determined in this manner. The reader should note that the MPS quantity of 60 is arbitrary. Methods for determining actual lot sizes will be discussed in a later section. Readers desiring more detailed information on the subject of master scheduling should consult Refs. 25, 37, and 48.

The Procurement Process. One of the initial steps in the production process is that of procuring purchased parts. The "make or purchase" decision is not always straightforward. Malstrom[27,28] has indicated that purchased parts by definition are those that cannot be manufactured in-house. Occasions exist when parts that can be manufactured internally are also purchased from the outside. Under normal conditions, the least expensive alternative is selected.

Exceptions to the least cost procurement alternative do exist. It may be preferable to make parts in-house that can be procured at a lower cost from an outside vendor. In one situation "fill" workload is needed in the plant to keep from laying off employees who will be needed for other production tasks in the near future. In another situation the organization has a short lead time requirement and can produce the parts internally faster than they can be procured from an external vendor.

Several steps exist in the procurement process that make the purchase of component parts an expensive procedure. These steps are overviewed in the paragraphs that follow. Also see Sec. 9, Chap. 5 for more information on make-or-buy decisions.

Purchase Order Requisitions. The first step is the initiation of a purchase order requisition. This requisition is usually completed by the person who wishes the material to be procured. The purchaser selects the type of part or material, its specifications, and a desired vendor. The purchaser usually estimates the cost of the part(s) being ordered. Finally, an account number to which the cost of the order is to be charged is also supplied. Purchase order requisitions normally originate in the manufacturing departments of most production organizations.

Solicitation of Bids. The requisition is next forwarded to the purchasing department. If the cost of the order is high enough ($500 to $700 or more), the purchasing department will contact a variety of vendors to obtain bids on the order. Usually the lowest bid is selected. The part requested on the purchase order requisition is frequently replaced by a lower-cost "equivalent" part or material.

Purchase Orders. The purchasing department next initiates a purchase order (PO), which is forwarded to the vendor from whom the parts will be procured. The PO is a contractual document indicating the organization's intent to procure a specified quantity of a part or material at a preagreed unit cost. The intent to purchase is usually contingent upon the vendor's ability to deliver parts of an acceptable quality level by a specified date.

Inspection of Incoming Parts.   After the order is placed, follow-up procedures are initiated, particularly if the order is not delivered on time. Upon receipt, the parts are inspected to assure that they conform to defined quality levels. If the order does not conform to these levels, it is rejected and sent back to the vendor. A follow-up order for replacement parts is sometimes initiated. Alternately, the order may be canceled and a new order placed with a different vendor.

Issuing of Purchased Parts.   After the order has been successfully inspected, it is either placed into warehouse stock or issued directly to the production floor for insertion or use in a manufacturing operation. It is at this point that the account originally specified on the purchase order requisition is debited for the cost of the order.

Cost of the Procurement Process.   The cost of completing all of these steps is significant. In large organizations the cost can range from $60 to in excess of $300 per order. Many organizations that have adopted just-in-time (JIT) inventory procedures repetitively place successive orders for parts with the same set of preferred vendors. In such cases blanket purchase orders (BPOs) are generated. BPOs set up in advance agreements on cost, order quantity, quality levels, and delivery schedules. A significant "one-shot" cost is incurred in setting up a BPO with a preferred vendor. This "one-shot" cost is offset by significantly reduced costs of placing successive orders for the same part.

LOT SIZING METHODS

The concept of lot sizing addresses two questions with regard to parts that are either made or purchased. The first addresses when the order should be placed. The second addresses the determination of how many parts should be ordered.

Two families of lot sizing techniques exist. The first family is called reorder point lot sizing systems. Reorder point methods are used for parts whose demands are known to be independent of one another. Department stores, grocery stores, and stores selling replacement automobile parts are examples of organizations that would use reorder point inventory systems.

In a manufacturing environment, assembly relationships usually exist between a final assembly that is shipped to the customer and all of its component parts. If the demand for final assemblies is known, it defines corresponding demands for all component parts in the final assembly.

Reorder point inventory systems do not lend themselves to these types of production situations. Since the demands for end items and their components are functionally related, availability of component parts at the time they are needed in the manufacturing process cannot be ensured by reorder point systems. Explosion-based inventory systems are therefore used to address question of lot sizes.

Explosion-based inventory systems use lot sizing heuristics. Many of these heuristics are based on reorder point methods. The following subsections overview both reorder point and explosion-based lot sizing techniques. Just-in-time inventory systems are also overviewed. The impact of this philosophy on traditional lot sizing methods is examined.

Reorder Point Methods.   A variety of reorder point (ROP) lot sizing methods exist. Most of the mathematically straightforward models are based on a number of restrictive assumptions. Many of these assumptions are not true in practice. As these assumptions are relaxed, the computational complexity of the lot sizing models increases significantly. The assumptions are summarized below.

- Annual demand is constant and is known exactly.
- Orders are received instantly.

- Lead time is known and is constant.
- Order costs are known and are independent of order size.
- Purchase price is constant. Price may vary with the order size.
- Storage capacity is available to store up to one year's demand of an item.

Entire texts have been written on lot sizing models. It is therefore not feasible to cover all of them in detail in this section. The approach used will be to summarize popular methods in increasing order of mathematical complexity. The assumptions associated with each method will be summarized. Mathematical derivations of each approach will not be presented. However, lot sizing formulas will be included, where appropriate, to assist the casual reader in selecting the appropriate method.

Notation.   In describing lot sizing notation, it is necessary to address the concept of inventory cycles. This is best accomplished by reviewing the inventory stock level of a given part over time. This relationship is illustrated in Fig. 2.6. The first inventory cycle begins by assuming that an order in the amount of Q units has just been received. A constant demand is assumed so the stock level is depleted at a linear rate.

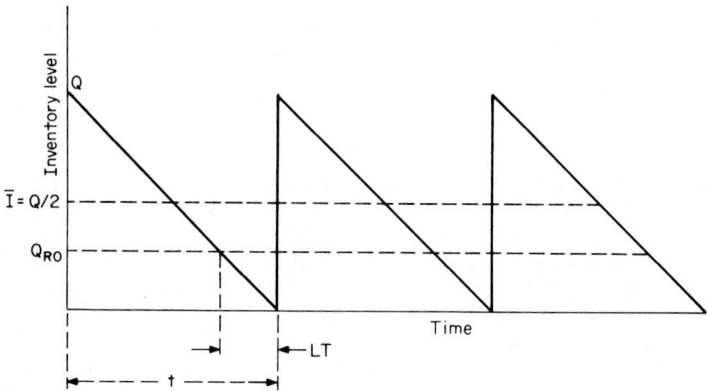

FIGURE 2.6 Inventory cycles and notation.

Initially, stock shortages are assumed not to occur. A second order is placed when the stock level reaches Q_{ro} units, the reorder point. This value defines the part's lead time LT since a new order must arrive exactly when the stock level for the part reaches zero. The maximum stock level is Q units; the minimum level is zero. It follows that the average stock level during the inventory cycle time t is $Q/2$ units.

A standardized set of notation for lot sizing has yet to be developed. Commonly used notation in many texts is similar to that presented below. This notation will be used in the subsections that follow:

TIC = total inventory cost
TIC_o = optimal or minimum TIC for a given lot size
Q = lot size or order quantity
Q_o = optimum lot size corresponding to TIC_o
R = annual demand in units per year
C_H = holding cost in dollars per unit-year
C_P = order cost in dollars per order
C_S = shortage cost in dollars per unit short-year
Q_{ro} = reorder point in units

LT = lead time
 B = buffer or safety stock level
 I = inventory level
 S = sales price in dollars per unit

Classical EOQ Model.   The classical economic order quantity (EOQ) model was first developed by Harris[23] in 1915. All of the restrictive assumptions listed at the beginning of this section apply for the EOQ model. In addition, part shortages are not allowed.

The model determines that order quantity which minimizes the sum of annual order costs and annual carrying costs for the part being ordered. The optimal order quantity is given by Eq. (11).

$$Q_o = (2RC_P/C_H)^{1/2} \tag{11}$$

The corresponding minimum total annual inventory cost is given by Eq. (12).

$$\text{TIC}_o = (2RC_PC_H)^{1/2} \tag{12}$$

EOQs with Shortages.   It is possible to adapt the previous model to allow it to address situations where stock shortages occur. Consider the inventory pattern shown in Fig. 2.7. In this illustration, the maximum inventory balance during any cycle is I_{max}. The period of positive inventory balance is t_1. During period t_2 a shortage in the amount of $Q - I_{max}$ units accrues. An order of size Q is needed to restore the inventory to its previous level of I_{max}. Q is the order size. Of this total, $Q - I_{max}$ units are effectively back-ordered. The optimal order size and corresponding minimum inventory cost are given by Eqs. (13) and (14).

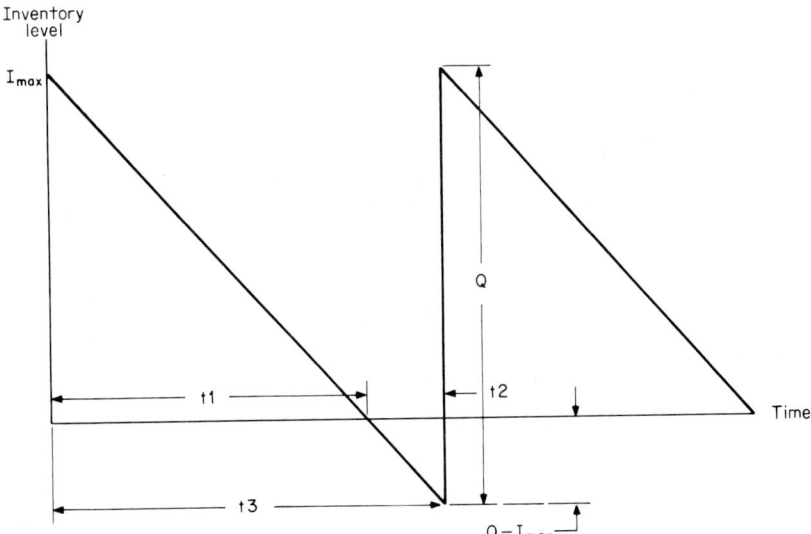

FIGURE 2.7   EOQ with shortages.

$$Q_o = (2RC_P/C_H)^{1/2} [(C_H + C_S)/C_S]^{1/2} \tag{13}$$

$$\text{TIC}_o = (2RC_PC_H)^{1/2} [C_S/(C_H + C_S)]^{1/2} \tag{14}$$

Readers should be advised that Eqs. (12) and (14) are valid only when $Q = Q_o$.

EOQs with Price Breaks. This model applies the EOQ methodology to situations where price breaks occur. Generally, vendors will offer products at discounted prices when larger orders are placed. For this model, it is necessary to define a new carrying cost parameter F_H. F_H defines holding costs as a fixed percentage of the annual inventory value of the part being stocked. The optimal and total annual inventory costs for this model are defined by Eqs. (15) and (16).

$$Q_o = (2RC_P/SF_H)^{1/2} \qquad (15)$$

$$\text{TIC} = C_P R/Q + SR + SF_H(Q/2) \qquad (16)$$

Equations (15) and (16) are applied in the following manner to solve for the optimal lot size. For an order situation with price breaks, each price must have a specific quantity interval. The quantity intervals may not overlap. The price per unit must decrease as the order quantity intervals increase in size.

Equation (15) is used to solve for Q for all values of S that apply for the quantity intervals in question. For each computation, the user must check to ensure that the value of Q obtained falls within the quantity interval for which the value of S used in the computation applies. Equation (16) is used to compute the total inventory cost associated with the quantity interval.

If Eq. (15) yields a value of Q lower that the lowest value of the quantity interval, the obtained value of Q is not used in the computation. Instead, the lowest value of Q in the quantity interval for which S applies is selected. This value is substituted in Eq. (16) to obtain the total inventory cost.

If the obtained value of Q is greater than the largest value in the quantity interval, the obtained value of Q from Eq. (15) is again not used. Instead, the largest value of Q in the quantity interval for which S applies is selected. This value is substituted in Eq. (16) to obtain the total inventory cost.

The preceding calculations are performed for all different values of S and their corresponding quantity intervals. An inventory cost associated with each value of S is determined. The optimal order policy is that quantity (and value of S) that has the smallest total inventory cost.

Economic Production Quantity (EPQ) Model. This model applies the EOQ logic to parts that are made, as opposed to purchased from an outside vendor. The production situation is depicted in Fig. 2.8. A part is produced internally at the rate of p units per day for a period of t_p days. If the daily demand for the part is r units per day, then the inventory balance increases by $(p - r)$ units for each day of production.

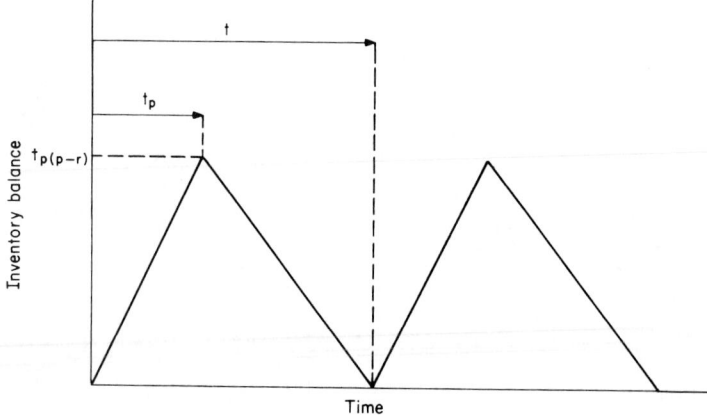

FIGURE 2.8 Economic production quantity stock levels.

At the end of the production period there exists an inventory balance of $t_p(p - r)$ units. This stock level is depleted at the rate of r units per day for the remainder of the inventory cycle. When the stock balance reaches zero, production of the part is again initiated, and the inventory cycle repeats. The optimal order size and corresponding inventory cost are given by Eqs. (17) and (18).

$$Q_o = \{2RC_P/[C_H(1 - r/p)]\}^{1/2} \tag{17}$$

$$\text{TIC}_o = [2RC_PC_H(1 - r/p)]^{1/2} \tag{18}$$

As before, Eq. (18) is valid only when $Q = Q_o$.

Variable Demand, Constant Lead Time Models. The preceding inventory models have all assumed that the demand for the product is constant during both the lead time and the total inventory cycle. This is rarely true in practice.

Consider the situation shown in Fig. 2.9. The demand from the beginning of each order cycle occurs at some average rate $\overline{D}$. While the stock is shown to be depleted at a constant rate during the inventory cycle, it will actually vary in accordance with some statistical distribution until the reorder point Q_{ro} is reached.

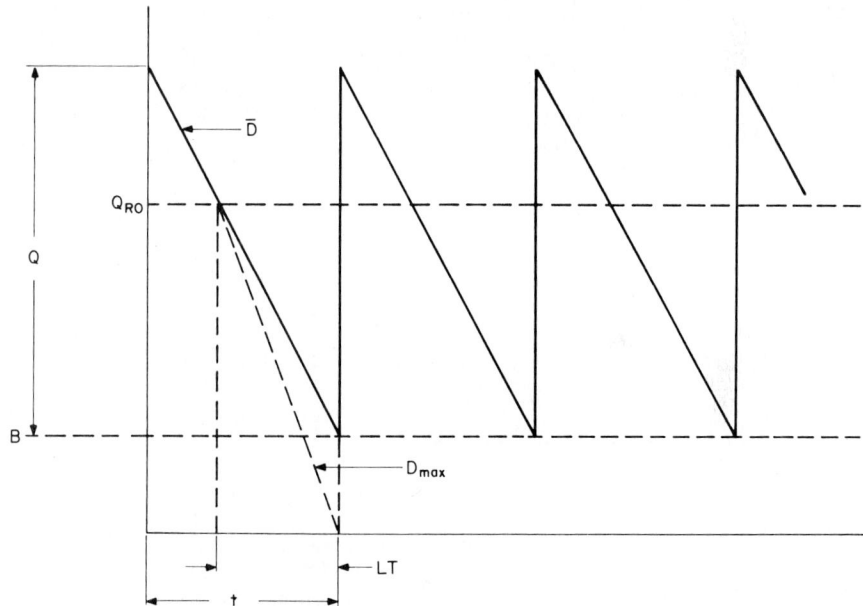

FIGURE 2.9 Variable demand constant lead time stock levels.

For analysis purposes, it is not necessary to know the demand variation prior to the time Q_{ro} is reached. The approach concentrates on determining demand variation during the lead time LT. The lead time is assumed to be constant.

The variance in the demand during the lead time is accounted for by carrying a buffer or safety stock. If the lead time demand continues at its average rate, the stock balance will be depleted exactly to zero by the time the next order arrives. The buffer stock is carried to satisfy lead time demand up to a rate of $D_{\max}$ units per day.

The level of $D_{\max}$ selected in determining the buffer stock level B determines the service level associated with the order policy. The service level is that percentage of the time during any order cycle that a stockout will not occur. The higher the service level, the higher the level of buffer stock.

Variable demand constant lead time models are of two types, back order and lost sales. Back-order models assume that when a shortage occurs, the product can be back-ordered, thus satisfying the demand at a later date. Lost sales models assume that when a shortage occurs, the demand for the units short is permanently lost.

The optimal lot size is again the one that minimizes total costs. In this case order costs and carrying costs again exist. However, there are now additional carrying costs associated with the buffer stock. The cost of back-ordering parts or lost sales also results when shortages occur during any inventory cycle.

Most models presented in the literature derive solutions corresponded to situations where the lead time demand is known to vary in accordance with normal or Poisson distributions. Product demand in practice rarely varies in accordance with these types of distributions. Discrete probability distributions are therefore recommended for such inventory situations. Readers desiring more information on this type of inventory model should consult the references at the end of this chapter on ROP models.

Constant Demand, Variable Lead Time Models. This model addresses the exact opposite of the situation described in the preceding section. Figure 2.9 again applies with the following changes. The demand is now constant at r units per day. The lead time varies in accordance with a known statistical distribution.

The demand that occurs during the lead time still must be determined. The concepts of buffer stock, back-order costs, and costs of lost sales still apply, as does the concept of service levels. Again, discrete probability distributions are recommended for use in describing lead time variation. A model accommodating discrete probability distributions has been presented by Riggs.[41] This model has been significantly refined by Lee, Malstrom, Vardeman, and Petersen[26] to address true average inventory levels when stockouts occur. Readers desiring more information on this type of model should consult these references.

Variable Demand, Variable Lead Time Models. This family of models imposes the fewest restrictive analysis assumptions but is also the most complicated set of inventory models. In this analysis situation, both the demand during the lead time and the lead time itself are allowed to vary. The concepts introduced in the preceding two sections again apply. The problem now becomes one of constructing a joint probability distribution in terms of both demand and lead time. This joint distribution will describe the lead time demand. Discrete probability distributions to describe both demand and lead time variation are again recommended.

Reorder Point Model References. Many of the models described in this section have been excerpted from Buffa and Miller.[6] However, a variety of newer texts also describe these models in greater detail. Interested readers desiring more information on this subject should consult Refs. 2, 3, 19, 20, 25, 40, 48, and 49.

Explosion-Based Inventory Models. Explosion-based inventory systems rely on requirements planning. Requirements planning may be defined as the management of raw materials, components, and subassemblies to ensure that these products are produced in sufficient quantity to satisfy the requirements for end items specified by the master schedule.

The most popular explosion-based inventory system is called material requirements planning (MRP). The most substantive treatment of MRP in early technical literature was by Orlicky.[37] MRP examines the assembly relationships between the component parts of an end item being produced. These relationships are used to generate both production and purchase schedules for "make" and purchased parts. These schedules ensure that sufficient components and subassemblies will be produced at the right time and in the right quantities to satisfy the forecast demand for end items.

Part Explosion Diagrams. A part explosion diagram is also called a bill of materials. The diagram indicates the "what goes into what" relationship of a manufactured end item. Each discrete part or subassembly is indicated by a separate node in the diagram. A sample node is illustrated in Fig. 2.10. As indicated in this illustration, the upper half of the

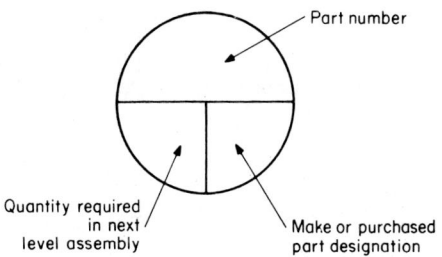

node contains the part number of the component or assembly. The lower left portion of the node indicates the number of the part or subassembly required at the next higher assembly level. Finally, the lower right portion of the node indicates whether the part is to be made or purchased.

A sample part explosion diagram is illustrated in Fig. 2.11. The diagram illustrates the component parts for an ax with a wooden handle. The reader should note that four distinct assembly levels exist, and that the nodes corresponding to each assembly level are arranged in columns on the diagram. Required make or purchase lead times are indicated to the right of each node, in most cases on the connecting assembly links of the diagram.

FIGURE 2.10 Notation for part explosion diagram. (*Adapted from Ref. 28.*)

Generation of MRP Tables.   The MRP logic is best illustrated with a detailed numerical example. All production schedules are a function of time. The time periods used in generating these schedules are called "time buckets." In practice, most commonly used time buckets are in either weeks or months.

Each production schedule for a component or assembly consists of a table with four sets of entries. These include the following:

> Gross requirements (GR): the amount of the parts or components required to satisfy the master schedule in any time bucket.

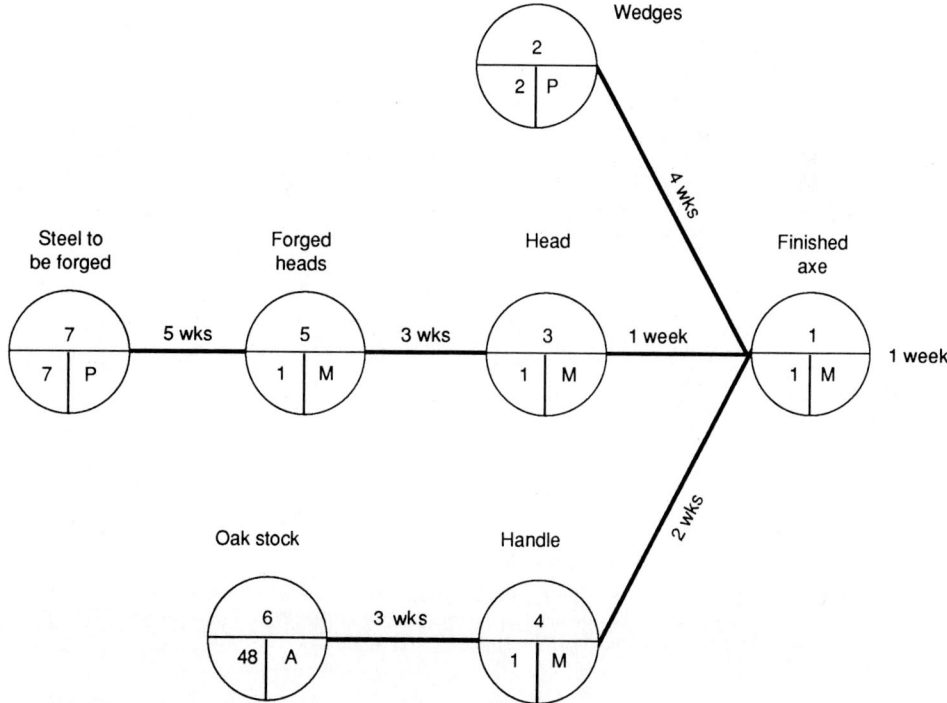

FIGURE 2.11 Sample part explosion diagram. (*Adapted from Ref. 28.*)

Scheduled receipts (SR): an order of quantity Q scheduled to arrive at the beginning of the time bucket. This order was placed LT time buckets ago, where LT is the lead time for the part or component.

On hand (OH): the on hand balance of the part or component that remains at the end of the time bucket. The on hand balance for any period t is given by Eq. (19).

$$OH_t = OH_{t-1} + SR_t - GR_t \tag{19}$$

where OH_t = parts on hand at the end of period t

SR_t = scheduled receipts that are to arrive at the beginning of period t

GR_t = gross requirements to be satisfied in period t

Planned orders (PO): an order of size Q initiated during period (time bucket) t. This order will arrive at the beginning of period $(t + LT)$ as a scheduled receipt.

Suppose the master schedule for the end item shown in Fig. 2.11 is as shown in Fig. 2.12. The MRP logic may be applied in the following manner. From Fig. 2.12, 25 end items must be available to ship at the beginning of period 21. From Fig. 2.11, it is apparent that the lead time for end items (part 1) is 1 week. This time allows for the final assembly of this part, which consists of part 2, part 3, and part 4.

Week	21	22	23	24	25	26	27	28	29	30
Demand	25	25	50	50	75	75	50	50	25	25

FIGURE 2.12   Example master schedule.

We begin by determining the gross requirements. Since a schedule for end items is desired, the gross requirements row of the table is merely the master schedule from Fig. 2.12. The result is shown in Fig. 2.13.

Week	20	21	22	23	24	25	26	27	28	29	30
Gross req.		25	25	50	50	75	75	50	50	25	25
Sch rec.		25	25	50	50	75	75	50	50	25	25
On hand		0	0	0	0	0	0	0	0	0	0
Pl. order	25	25	50	50	75	75	50	50	25	25	

FIGURE 2.13   MRP table for end items.

We assume that a separate order is placed for each period's demand. (This lot sizing decision is arbitrary, and specific MRP lot sizing methods will be presented in a later section.) The order schedule then becomes the planned order row of the table. Since the required lead time for the part is 1 week and a separate order is placed for each period, the planned order row is identical to the gross requirements row but starts 1 week earlier in time. (This example assumes a previous balance of zero end items from period 19.)

Refer again to Fig. 2.13. The scheduled receipts row depicts the arrival of each order 1 week after it is placed. Since the scheduled receipts exactly equal the gross requirements for each period, all end of period on hand balances are zero.

Consider now the generation of a procurement schedule for the wedges. These parts are driven into the end of the wooden ax handle to secure the ax head. From Fig. 2.11 it is apparent that the wedges (part 2) are purchased and require a lead time of 4 weeks.

We begin by determining the gross requirements for wedges. For any component part, the gross requirements may be determined by considering the relationship between parent and component nodes in the part explosion diagram. A "component node" is any node that goes into a node at the next higher assembly level in the part explosion diagram. The node that the component node goes into is called the "parent node."

Refer to Fig. 2.11. Part 1 is a parent node for the component nodes corresponding to parts 2, 3, and 4. Part 4 is a parent node for component node 6, and so on. The gross requirements for any component node may be defined as a function of the planned orders of the parent node in accordance with Eq. (20).

$$GR_c = PO_p \, (Q_g) \tag{20}$$

where GR_c = gross requirements of the component node
PO_p = planned orders of the parent node
Q_g = goes into quantity between the component node and the parent node

From Fig. 2.11 the goes into quantity between node 2 and node 1 is 2. This means that two wedges are required at the next higher assembly level. The gross requirements for wedges is then the planned orders for node 1 taken from Fig. 2.13 times 2. This is shown in Fig. 2.14.

Week	15 19	20	21	22	23	24	25	26	27	28	29	30
Gross req.		50	50	100	100	150	150	100	100	50	50	
Sch rec.						400				600		
On hand	400	350	300	200	100	350	200	100	0	550	500	
Pl. order		400				600						

FIGURE 2.14   MRP table for wedges.

Suppose a total of 400 wedges are in stock at the end of period 19. The on hand balances at the end of each period may be obtained by subtracting the gross requirements for each period from the initial stock balance. This is shown in Fig. 2.14. Completing these computations shows that a balance of 100 wedges is projected for the end of period 23. Further inspection shows that this balance is not sufficient to satisfy the scheduled demand for period 24, in this case, 150 units.

MRP does not allow part shortages to occur. It is therefore necessary to schedule an order of wedges to arrive at the beginning of period 24. Suppose this order size is 400. (Again, the selection of this lot size is arbitrary and is used for example purposes only.) The purchase lead time for wedges is 4 weeks. To arrive in period 24, the order would have to be placed in period 20. A planned order in this amount is therefore shown for this period.

Applying Eq. (19) yields the remaining on hand balances for periods 24 through 27. The on hand balance at the end of period 27 is zero. It is thus necessary to schedule another order to arrive at the beginning of period 28 to prevent a shortage from occurring. If this order size is arbitrarily selected to be 600 units, this quantity may be shown as a scheduled receipt in period 28. The corresponding planned order is shown in period 24.

This procedure may be used to generate production or procurement schedules for all of the remaining nodes in Fig. 2.11. Readers desiring a more detailed description of MRP should consult Refs. 2, 3, 19, 25, 37, 48, and 49). Also see Sec. 10, Chap. 4.

MRP Lot Sizing Heuristics.   Lot sizing in an MRP environment is equivalent to determining how many periods of gross requirements to combine into a planned order. When more than one level of the part explosion diagram is considered, it is usually not possible to prove the optimality of MRP lot sizing methods.

Instead, heuristics are used, many of which are based on the reorder point techniques that have been described previously. The effectiveness of these methods has been thoroughly investigated by simulating different types of rules with a variety of part explosion structures and demand patterns. The following subsections overview some typical MRP lot sizing methods. The comparative effectiveness of these rules in terms of total annual inventory cost is also described.

Lot for Lot Heuristic. The lot for lot (LFL) heuristic specifies that a separate order is placed for each period or time bucket. No periods of demand are combined. The order size is merely the gross requirement for the period in question.

The LFL method typically has high order costs, since separate orders are placed for each period with a nonzero demand. Carrying costs are minimized by this approach, since the stock is always used in the period in which it arrives. The LFL most closely represents the just-in-time (JIT) order philosophy which is described in a later section of this chapter.

Economic Order Quantity Heuristic. The economic order quantity (EOQ) heuristic applies the economic order quantity logic of reorder point inventory systems. This method assumes that demand from period to period is relatively constant. Each time it is necessary to place an order, Eq. (11) is used to determine the order size. If the order size happens to be less than the gross requirement for the period in question, the order size is increased to a level just large enough to prevent shortages from occurring. When an initial stock balance exists, the annual demand R in Eq. (11) is reduced by the amount of the initial stock balance. Carrying costs for the initial inventory are added to the total annual inventory cost.

Periodic Order Quantity Heuristic. The periodic order quantity (POQ) heuristic uses the EOQ logic to determine the optimal time interval between orders. An order is then initiated just large enough to cover the demand that is scheduled to occur over this time interval. Time periods in the interval are totaled in such a way that no order is scheduled for receipt during a period that has zero demand. This avoids incurring unnecessary carrying cost. This method responds well to demand patterns with wide fluctuations.

Least Unit Cost Algorithm. The least unit cost (LUC) algorithm computes for various order sizes the cost "per unit" chargeable to orders and setup and storage. That order size is selected that minimizes the total cost per unit.

Least Total Cost Algorithm. The least total cost (LTC) algorithm is also based on EOQ logic. It may be shown that the cost minimum corresponding to the optimal order size of Eq. (11) occurs at the point where the annual order costs and the annual carrying costs equal one another.

The LTC algorithm analyzes the gross requirements over a specified planning horizon. Various order quantities are evaluated. That order quantity is selected that makes the resulting order and carrying costs most closely equal one another.

Part-Period Balancing Algorithm. The part-period balancing (PPB) algorithm is very similar to the LTC approach to lot sizing. The primary difference between the two methods is an adjustment "look ahead/look back" routine. This feature prevents inventory intended to cover peak period demands from being carried in stock for long periods of time. It also helps orders from being keyed to periods with low requirements.

Silver-Meal Algorithm. The silver-meal (SM) algorithm is computationally more robust than the methods previously described. It is based on selecting the order quantity so as to minimize the cost per unit time over the time periods during which the order quantity lasts. This is a search on a time variable which is defined over the order quantity under the assumption that all inventory needed during a period must be available at the beginning of that period. This assumption of stock availability also holds for all of the previously described algorithms and heuristics.

Wagner-Whitin Algorithm. The Wagner-Whitin (WW) algorithm uses an optimizing procedure that is based on a dynamic programming model. It evaluates all possible combinations of orders to cover requirements in each period of the planning horizon. Its objective is to arrive at an optimal ordering strategy for the entire requirements schedule.

The algorithm does minimize the total cost of setup and carrying inventory, but only for the assembly level of the part being considered. It has the disadvantage of a high computational burden due to its mathematical complexity.

Comparative Performance of Heuristics and Algorithms. The comparative performance of lot sizing heuristics has been studied in detail over the last 15 years. Results of key studies have been documented by Choi, Malstrom, and Classen,[12,13] Choi, Malstrom, and Tsai,[9] and Taylor and Malstrom.[46]

Digital simulation has been used to simulate the heuristics under a variety of part explosion and demand conditions. More recent studies have evaluated larger part explosion product structures with increasing amounts of actual manufacturing data as inputs.

Of the heuristics previously described, the consistent best performer has been the periodic order quantity (POQ) rule. Other rules that have performed well include the least total cost (LTC) and least unit cost (LUC) rules. Marginal rules on the basis of cost have included the economic order quantity (EOQ), Wagner-Whitin (WW), and silver-meal (SM) methods. The lot for lot (LFL) rule has been consistently the worst-performing heuristic in all evaluations. Total annual inventory costs associated with this method are three to twenty times more expensive than the best-performing rules in the variety of simulation studies that have been conducted.[9,12,13,46]

The feature that has consistently distinguished good rules from the ones that are not cost-effective has been the order policy structured for end item products. Those rules that trigger frequent separate orders for end items incur annual setup or order costs that are extremely high. The increase in these costs is not completely offset by the corresponding lower carrying costs that are obtained.

This conclusion has some interesting implications for just-in-time (JIT) inventory systems. JIT order policies are most closely represented by the lot for lot (LFL) heuristic. This method has historically been the least effective of the rules evaluated. Malstrom[30] has determined that setup or order costs must be reduced to levels equal to 1/100 or less of the corresponding carrying cost for each node before the performance of the LFL heuristic begins to significantly improve relative to other lot sizing methods.

It is questionable whether this reduction in both setup and order costs is always attainable when JIT policies have been implemented. Burney[7] and Malstrom[29] have stated that such potential increases in order costs have the potential to negate many of the possible savings attainable with JIT policies. JIT inventory procedures are discussed in greater detail in the following section.

JUST-IN-TIME INVENTORY SYSTEMS

Just-in-time (JIT) inventory systems are known by a variety of names and terms. These include material as needed (MAN), minimum inventory production systems (MIPS), stockless production, continuous flow manufacturing, Kanban, and others. JIT has as its goal the elimination of waste. Waste is generally defined as anything other than the absolute minimum resources of material, machines, and labor required to add value to the product being produced.

JIT Benefits. In most cases, JIT results in significant reductions of all forms of inventory. Such forms include inventories of purchased parts, subassemblies, work-in-process (WIP), and finished goods. Such inventory reductions are accomplished through improved methods of not only purchasing but also scheduling of production.

JIT requires significant modifications to traditional methods by which parts are procured. Preferred suppliers are selected for each part to be procured. Special purchase arrangements are contractually structured to provide for small orders. These orders are delivered at exact times as required by the user's production schedule and in quantities small enough to be used in very small time periods.

Daily and weekly deliveries of purchased parts are not uncommon in JIT systems. Vendors contractually agree to deliver parts that conform to preagreed quality levels, thereby eliminating the need for the purchaser to inspect incoming parts. The arrival time of such deliveries is extremely important. If they arrive too early, the purchaser must carry additional inventory. If they arrive too late, parts shortages occur that can stop scheduled production.

Purchasers of such parts often pay increased unit costs to have parts delivered in this manner. While the "one-shot" costs of structuring the purchase agreement can be significant, the follow-on costs of procuring individual lots of parts each day or week can be reduced to near zero levels. Not having to inspect incoming parts can result in increased product quality and reduced inspection costs on the part of the purchaser.

Fabricated parts are scheduled for production so as to minimize work-in-process (WIP) inventory and stockpiles of finished goods. The JIT philosophy forces manufacturers to solve production bottlenecks and design problems that were previously "covered" by maintaining reserve inventory levels. Readers desiring a more detailed overview of JIT policies, procedures, and benefits should consult Refs. 7, 29, 43, 44, and 45. Also see Sec. 10, Chap. 4.

Cost Effectiveness of JIT Systems. The preceding benefits are realized after some significant investments of effort associated with JIT implementation. The cost of structuring blanket purchase arrangements with a variety of preferred suppliers can be significant. Large costs may also be associated with sophisticated tool design procedures to reduce setup costs for producing different products to near zero levels.

Such reductions are absolutely necessary if a lot for lot (LFL) order policy is to be applied as described in the preceding section on MRP. Setup costs must be reduced to at least 1/100 of the corresponding carrying costs for the part being produced.[30] If this reduction is not possible, the setup costs associated with an LFL like order policy may be significant enough to negate the benefits associated with JIT policies.[7,29]

Software to assess the cost effectiveness of JIT has been developed by Burney and Malstrom.[7] Written in the C Language, the software utilizes long sequences of pop-up screens. The screens contain sequences of tutorials that step the user through a detailed cost assessment procedure.

The procedure begins by helping the user to compile a detailed estimate of costs associated with the inventory system currently in use. JIT implementation costs are next estimated. Costs of blanket purchase agreements, setup reduction, personnel training, and lot sizing are all separately estimated. The user is also guided in estimating the costs of facilities modifications required by JIT.

JIT benefits are next assessed. The net change in inventory costs is estimated. The user is guided in ways to quantify the cost savings associated with better product quality and improved customer delivery. Readers desiring a more detailed description of the developed software should consult Ref. 7. For further information on inventory management and control procedures, see Sec. 10, Chap. 2.

PROCESSING

This section describes the fundamentals of the manufacturing processes associated with production. The areas of manufacturing engineering and related educational programs are overviewed, as is the need for integration of the design and manufacturing processes. Fundamentals of process selection are addressed as is the generation of process routings. These subjects are the topics of the subsections that follow.

Manufacturing Engineering. Manufacturing engineering is concerned with the design of manufacturing processes and process sequences to produce products of acceptable quality levels at minimum cost. Although a separately recognized engineering discipline, the development of separate 4-year programs in manufacturing engineering has long been ignored by most major colleges and universities. Only two programs existed nationwide in 1984 that culminated in an accredited bachelor's degree in manufacturing engineering.[31] There has been only a modest increase in the number of such programs since that time.

In light of the need in the United States to increase manufacturing competitiveness, this statistic may be somewhat surprising. The reasons for the lack of growth in 4-year manufacturing engineering programs have been examined by Malstrom and Kuper.[32] The authors state that much of the manufacturing emphasis in engineering curricula already exists in many programs of industrial and mechanical engineering.

The establishment of separate manufacturing engineering departments has historically been impeded by availability of qualified faculty, university reward systems for promotion and tenure, and the comparatively recent interest of the federal government in funding projects addressing applied manufacturing research.

To be effective teachers in this manufacturing engineering, it is desirable for incoming faculty to have significant amounts of prior industrial experience. University reward systems have historically failed to reward such experience in terms of starting salary, incoming rank, or credit toward either promotion or tenure.

Significant facilities and administrative costs are associated with the establishment of new academic departments and programs. Such costs are often offset by funded research opportunities. Manufacturing research has traditionally been done in-house in the manufacturing, as opposed to the academic community. Federal funded research opportunities for most universities have until recently been limited. At this writing, long overdue federal legislation is pending that will provide incentives to colleges and universities to undertake separate undergraduate curriculum development in this area.[15]

It should be noted that many programs in manufacturing technology exist at a variety of academic institutions. However, these programs focus primarily on metalworking principles and methods. The programs graduate students in many cases after the completion of only a 2-year curriculum. Students completing 4-year programs in manufacturing technology are not uncommon. Unfortunately, these individuals do not have professional parity with other graduate engineers and often do not rise to high enough professional positions to impact corporate decision making with regard to manufacturing procedures. Also see Sec. 7, Chap. 1.

Integration of Design and Manufacturing.   Product design and the procedures by which the products are manufactured have traditionally been performed separately by different groups of individuals. Design engineers are often now well versed in manufacturing procedures and processes. Individuals that design manufacturing processes have traditionally not been degreed engineers and are therefore not familiar with product design principles and procedures.

Concurrent engineering attempts to integrate the product design and its related manufacturing process design. For each product, teams are formed that consist of design, manufacturing, quality assurance, and purchasing personnel. The design of the product is performing concurrently with the selection of processes, equipment, and steps that will determine how it is to be built. Inputs by the quality assurance members of the team assure that the finished product can be manufactured to predetermined levels of quality. Inputs by the purchasing members of the team ensure that the product will be built with parts that are available and are economically procurable. See Sec. 7, Chap. 7 for further information.

Process Selection.   Developments in manufacturing automation over the past 30 years have increased the number of process alternatives available to manufacturing planners. The selection of the most cost-effective process is dependent upon part geometry, tolerance requirements, annual requirements, lot size, and other factors. This section overviews a number of automation alternatives available to process planners.

Numerical Control.   Numerical control (NC) has revolutionized machining process in recent decades. Numerically controlled machine tools have the ability to machine geometric contours not achievable with conventional equipment. There is the additional advantage of increased repeatability from part to part, without variations induced by a human operator.

Numerical control reduces tooling costs by eliminating the costs of sophisticated jigs and fixtures. Setup costs involved with the consecutive machining of different parts are usually small. If a general-purpose fixture can be used, the only cost of part changeover is the changing of the tape containing the part's program and the checking of the setup on the first part of the lot that is produced.

Numerical control is not a high-volume production process and may be considered a form of programmable automation. The purchase costs of NC machine tools are higher than their conventional, manual counterparts. Numerical control has been implemented to the extent that skilled operators capable of completing sophisticated, manual machining and gauging operations are becoming increasingly difficult to find.

Advantages of NC machining are partially offset by some overhead costs that are created by this form of automation. In addition to higher equipment costs, other costs are incurred for the programming, checkout, and storage of tapes containing the programs for parts to be manufactured. The topic of numerical control has been addressed in detail by DeGarmo et al.,[16] Neibel and Draper,[34] and Groover.[21] Interested readers are urged to consult these references. Also see Sec. 7, Chap. 5.

Flexible Manufacturing.   Flexible manufacturing involves the interconnection of groups of automated machine tools by an integrated and automated material handling system. Flexible manufacturing systems (FMSs) are used to complete sophisticated sequences of machining operations on families of manufactured parts. Part characteristics must be such that all machining operations can be completed on the set (or a subset) of the machine tools contained in the FMS.

Pallet fixtures are used to transport parts or groups of parts from one machine tool to another. Robots are often used to load and unload parts and/or pallet fixtures on and off the various machines. FMSs often contain provisions for automated part inspection. Hierarchical computer control is used to schedule and route parts through the system. Such control facilitates the tracking of machine tool usage and parts produced. This information is used to schedule periodic maintenance during which cutting tools and fluids may be changed.

Initial costs associated with FMS implementation are significant. System costs as high as $20 million are not uncommon. Product volume and projected labor savings must therefore be sufficiently high to justify these extremely high initial investment costs. Flexible manufacturing systems have been described by Groover.[21] This reference will provide additional information on this subject to interested readers. Also see Sec. 7, Chap. 6.

Robotics.   Industrial robots have been successfully utilized in a number of manufacturing applications in the past two decades. Typical applications include spray painting, the application of coatings, spot welding, arc welding, part transport during heat-treating operations, packing of finished parts, palletizing operations, and machine loading and unloading.

Robots eliminate repetitive, manual labor operations. They have successfully removed human workers from hazardous working environments and from physically stressful labor tasks. Like NC machine tools, robots have the requirement for the generation, checkout, and storage of various movement programs.

No standardization of programming languages for robots has yet occurred. It is not uncommon for different models of robots from the same manufacturer to be programmed in different languages. Languages used are often unique to a particular robot. This limitation has impeded the integration of robots into automatic manufacturing cells and systems.

Robots have achieved only limited use in complex assembly operations. Limitations in gripper design, tactile sensing, and vision have resulted in a need for complex fixturing for most assembly operations.[5] These requirements, together with the difficulty and cost associated with integrating vision systems into assembly operations, have kept robotic applications for complex assembly from becoming widespread.

A good overview of robotics has been provided by Engelberger[18] and Hall and Hall.[22] Readers desiring additional information on this subject are urged to consult these references. Also see Sec. 7, Chap. 10.

Fixed versus Programmable Automation.   Decisions to use fixed as opposed to programmable automation alternatives are usually based on the production quantity to be built. Fixed automation facilities are usually dedicated to the production of a single product, usually in extremely high volumes. Transfer lines used in the high-volume machining of engine blocks are an example of fixed automation.

Fixed automation is characterized by extremely high costs associated with equipment procurement. Any change in the product being manufactured results in a significant line changeover cost.

Examples of programmable automated equipment include both NC machine tools and industrial robots. These machines are capable of producing a variety of different parts, with comparatively modest setup and programming costs. Production output is much lower than dedicated, fixed automation equipment.

Cost relationships between conventional, programmable, and fixed automation equipment have been described by Groover[21] in terms of production quantity. An illustration adapted from this reference is presented in Fig. 2.15.

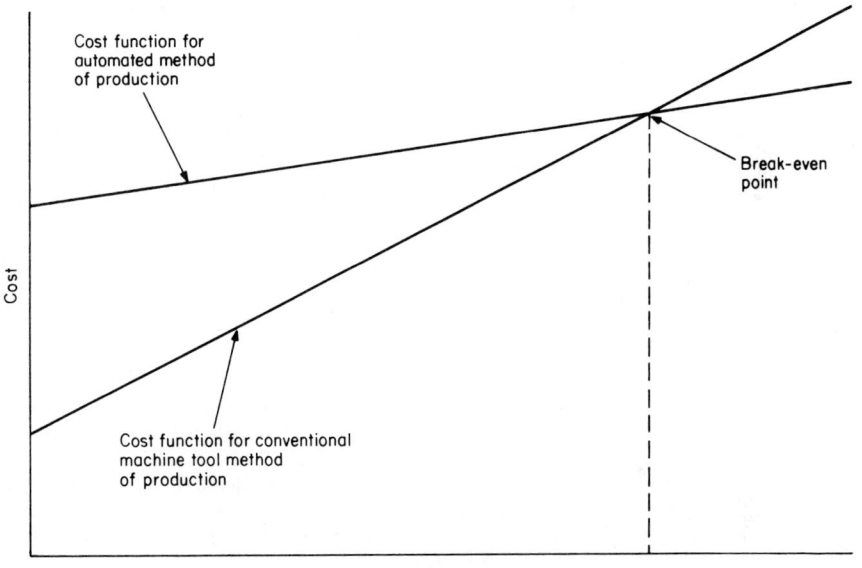

FIGURE 2.15   Cost trade-offs between conventional and automated production.

The Process Routing.   The process routing specifies that sequence of manufacturing operations that is required to produce a product or subassembly. Distinct process routings usually exist for each "make" part in the end item's bill of materials or part explosion diagram.

The process routing is divided into separate steps or operations. Each operation specifies the factory location or work center in which a set of manufacturing operations is to take place. The routing contains a description of these operations along with the set of machines or manufacturing facilities that is to be used.

If properly structured, the sequence of operations specified in the process routing will produce parts at minimum cost to prespecified tolerance and quality requirements. Performance standards are applied to estimate setup and run times associated with each routing operation. Inspection operations are also often defined as separate routing operations.

Where similarities exist with families of parts being manufactured, some progress has been made in using the computer to assist in process planning. Examples of computer-assisted process planning have been documented by Chang and Wysk[8] and McNeely and Malstrom.[33] Readers desiring additional information on this subject should consult these references. Also see Sec. 7, Chap. 3 and Sec. 12, Chap. 3.

SCHEDULING

This section addresses two different types of scheduling. At the macro level, master schedules (previously described) must consider both the capacity of the plant and its individual work cells. The master schedule must be continuously adjusted to match "per period" workloads with the capacities of machines, facilities, and available personnel. This goal is accomplished through the process of capacity planning.

At the micro level, the scheduling problem becomes one of determining a priority sequence for competing jobs or orders awaiting processing by a single machine or group of production facilities. The subsections that follow address these topics.

Capacity Planning.   Capacity planning is a method by which the master schedule is adjusted to balance the due dates of jobs or orders against the capacity of the plant and its individual work cells and facilities. This concept is perhaps best illustrated with the use of an example. Consider a hypothetical work center with one machine which is staffed by one worker. Let us suppose that a one-shift operation applies. Consider the 10-week production schedule shown in Fig. 2.16. For simplicity assume that each part requires 1 hour of processing time on the machine.

Week	20	21	22	23	24	25	26	27	28	29
Demand	25	25	50	50	75	75	50	50	25	25

FIGURE 2.16   Sample production schedule.

From Fig. 2.16 it is apparent that there is not enough work to fully occupy the machine and its worker during weeks 20, 21, 28, and 29. The demand in weeks 22, 23, 26, and 27 can be satisfied with the use of overtime. The demand in weeks 24 and 25 cannot be satisfied even if the worker completes six 12-hour days (a total of 72 hours).

The workload may be "smoothed" by adjusting the schedule as shown in Fig. 2.17. Suppose 25 units each from weeks 24 and 25 are moved to weeks 20 and 21. The result is shown in Fig. 2.17. The schedule for weeks 20 through 27 inclusive may now be satisfied with the use of 10 hours per week of overtime. While no full workload exists for weeks 28 and 29, it is likely that additional orders will arrive in the next several weeks to fully utilize the production facility during these time periods.

Week	20	21	22	23	24	25	26	27	28	29
Demand	(50)	(50)	50	50	(50)	(50)	50	50	25	25
					 25	25				

FIGURE 2.17   Schedule changes to smooth production.

In periods of work underload, the capacity planning procedure seeks to move orders back in time to match workload levels with existing capacities. In periods of work overload, orders are moved forward in time to reduce workload levels. When such schedule adjustments are not possible, it is necessary to hire additional people, add shifts, or lay off personnel. All three situations are undesirable because of the extra costs that are incurred.

Vollman, Berry, and Whybark[48] describe three separate types of capacity planning methods. The methods differ in the amount of production data used to afford increasing levels of detail in assessing workload levels. These methods are described in the following subsections.

Capacity Planning Using Overall Factors. Capacity planning using overall factors (CPOF) is a relatively simple approach that results in a "rough cut" capacity plan. The inputs come from the master schedule rather than from the MRP tables associated with individual parts in the bill of materials. Workload levels are derived from performance standards or historical data for end products only. Fabrication times for components included in the end item are embedded in these totals. These data are used to derive workload levels. The CPOF method does not consider the time shift associated with the lead times for all component parts in the end item.

Capacity Bills. This method provides a more direct linkage between different end products being produced and the respective capacities required by these different end items in various work centers. The method is responsive to changes in product mix of the end items produced.

Additional data are required to use this approach. Lot sizes for each end product and their respective components must be known. Setup and run times for each lot must be defined for each work center in which processing is required.

Resource Profiles. This approach further refines the capacity bills procedure. It considers the lead time requirements associated with each node in the parts explosion diagram. All data for the previous method are used, but are defined to occur in the specific period during which the work on a specific part or subassembly is scheduled to take place. This method is the most detailed (and time-consuming) of the three approaches that have been described.

More detailed information on each of these three capacity planning methods is presented in Ref. 48. The descriptions of each method are illustrated with detailed numerical examples.

Intra Work Cell Scheduling Methods. A variety of methods exist for the scheduling of jobs or orders within a given work cell. For most rules a notation of the form $n/m/C$ applies. In this notation the term n denotes the number of jobs or orders that are to be scheduled. The term m refers to the number of machines within the work cell. Finally, the term C refers to the objective or criterion addressed by the developed schedule.

Common scheduling objectives are to deliver or complete the orders by the due dates required by the customer. This is accomplished by minimizing the average or maximum lateness for a sequence of jobs or orders. Another common objective is to minimize the elapsed time that the order or job is "in process" within the work cell. This is equivalent to minimizing the average or maximum flow time for a sequence of jobs.

The mathematics involved in proving that job sequences derived from specific rules satisfy specific scheduling criteria are extremely complex. Most early work in analyzing scheduling methodologies therefore focused on work cells consisting of only one or two machines. A review of this early work is the topic of the subsections that follow.

Shortest Processing Time Rule. The shortest processing time (SPT) rule schedules jobs across a machine or set of production facilities in order of increasing processing times. For n jobs that are sequenced across a single machine, it may be proved that the SPT rule minimizes the mean flow time for all jobs. Flow time refers to the sum of the time the job spends in queue plus its processing time.

The primary disadvantage of the SPT rule is that jobs with long processing times usually are delayed in reaching the front of the queue. They are therefore often completed long after the due date required by the production schedule. This problem is addressed by using a truncated form of the SPT rule which forces jobs with long processing times to the front of the queue after they have awaited processing a specified length of time.

Due Date Rule. The due date (DDATE) rule sequences jobs across a machine or set of production facilities in ascending order of date by which the order or job is due to be completed. Those jobs with the earliest due dates are worked on first. For n jobs and one machine, it may be proved that the DDATE rule minimizes the maximum lateness for the sequence of jobs that are scheduled.

Slack Time Rule. The slack time rule (SLACK) sequences jobs across a machine or set of production facilities in order of increasing slack time. Slack time is the difference between a job's due date and its processing time. For any job i in a sequence of n jobs, the slack time is defined by Eq. (21).

$$t_i = d_i - p_i \tag{21}$$

where t_i = slack time for job i
d_i = due date for job i
p_i = processing time for job i

With the slack time rule, jobs with minimal slack have the greatest risk of being late. They therefore are placed first in the scheduling sequence. For n jobs and one machine, it may be proved that the slack time rule maximizes the minimum lateness for the sequence of jobs or orders that are scheduled.

Multiple Machine Rules.   Most scheduling applications involve the use of more than one machine. Conway, Maxwell, and Miller[14] overview two methods which address $n/2$ and $2/m$ scheduling problems. Johnson's algorithm is applicable for $n/2$ scheduling problems. Application of this procedure will yield a sequence that will minimize the maximum flow time of all n jobs across the two machines.

The authors also illustrate a graphical scheduling procedure for a $2/m$ scheduling problem. The goal again is to minimize the maximum flow time for the two jobs across the set of machines. Times at which both jobs will need a given machine are depicted on a two-dimensional graph as conflict areas. Scheduling paths are illustrated which pass around these regions and attempt to maximize the amount of time that both jobs receive simultaneous processing.

Readers desiring more information on either of these methods should consult either Conway, Maxwell, and Miller[14] or Baker.[1]

First Come, First Served and Random Scheduling.   These two methods are equivalent to doing no scheduling at all. In the first come, first served (FCFS) method, jobs are processed in the order in which they arrive at the machine or facility. With the random method, a completely arbitrary job sequence is randomly selected. The value of these methods comes from comparing them with other scheduling rules and heuristics. The FCFS and RANDOM rules serve as a comparison benchmark to show how much improvement can be obtained through use of other scheduling methodologies.

The RAND Simulations.   The RAND simulations were performed in the 1960s by the RAND Corporation. They have been described in detail by Conway, Maxwell, and Miller.[14] These studies are significant. They are one of the first large-scale digital simulation analyses that analyzed a variety of scheduling rules in a multiple-machine environment.

Much of the RAND evaluation focused on an $n/9$ scheduling environment. A variety of evaluation criteria were defined and included average number of jobs in queue, work hours remaining, work hours completed, average flow time, average job tardiness, and fraction of jobs tardy.

Job due dates were generated four different ways. These included a constant multiple of the job's processing time, a date proportional to the number of operations in the job, a constant due date for all jobs, and due dates that were randomly assigned.

A variety of scheduling rules were analyzed including SPT, DDATE, SLACK, RANDOM, and FCFS. Additional rules included those based on the amount of work in queue, the amount of work remaining, the number of job operations remaining, and those that prorated both due dates and slack time between a job's operations.

The SPT rule was consistently among the best performers for all evaluation criteria. SPT scheduled jobs were found to have the smallest average flow times. The SPT rule also performed best in terms of average tardiness and the number of jobs tardy.

The results of the RAND simulations have been confirmed in a number of subsequent simulation analyses. Because of the excessive lateness of SPT sequence jobs with large processing times, a truncated version of the SPT rule is generally recommended for use. Readers desiring additional information on the RAND simulations and scheduling rules in general should consult Refs. 1 and 14.

Scheduling in FMS Environments. Choi and Malstrom[10,11] constructed a working physical simulator to model an actual flexible manufacturing system. The system consisted of an automated storage and retrieval system (AS/RS), two NC lathes serviced by an industrial robot, and six identical multiaxis machining centers. Seven different families of parts were processed.

Two types of rules were studied. The first group was called part selection rules. These rules were used to determine the sequence of parts to be withdrawn from the AS/RS for processing. Seven part selection rules were analyzed including SPT, RANDOM, FCFS, DDATE, and SLACK. Two other rules were studied. The first prorated slack time over each job operation. The second was based on the part's dollar value.

The second group of rules was called machine selection rules. This rule set determined to which of the identical machining centers a part withdrawn from the AS/RS was routed. Four rules were studied. The first selected a machining center at random. The second minimized part travel time; in other words the closest machining center had the highest priority. The two remaining rules were based on work-in-process. The first routed the part to the machining center with the fewest number of parts in queue. The second routed the part to the machining center with the least amount of work hours in queue.

Evaluation criteria included total part traveling time, actual production output, total manufacturing throughput time, machine utilization, work-in-process inventory, and job lateness.

The best part selection rules included SLACK and SPT. The best machine selection rule was that based on routing the part to the machine with the least number of hours work in queue (WINQ). Combinations of these three rules were found by the authors to be superior to other rule combinations.[10]

DISPATCHING

Dispatching is a very important facet of the production control process. Dispatching involves the movement of parts, components, subassemblies, and end items so that they arrive at the appropriate work center exactly at the time they are needed in the production process. An alternate name for this procedure is shop floor control.

Three types of material are usually moved in the dispatching process. The first is the movement of a partially completed part or subassembly to the appropriate work center. The second is the movement of raw materials or components that are to be added at a particular process operation. The third is the movement of tooling, fixtures, gauges, and inspection equipment to the work center.

Not all parts "survive" all steps in the manufacturing process. For example, in metalworking, a part that has a dimension that is too large may often be reworked. In contrast, a part that has too much metal removed resulting in a small dimension is often scrapped.

A common procedure is for process planners to start greater quantities than are required at the beginning of the process sequence. Suppose historical scrap rates for a given set of processes are 10 percent. If 50 parts are required at the end of the sequence, a start quantity of 55 parts might be specified to begin the first process operation.

Those parts that can be reworked must be moved back through those work centers required to perform the necessary rework operations. Parts completing the rework process must be "rejoined" with the remaining parts on the order.

When the scrap rate for a set of processes is higher than expected, the start quantity may not be sufficient to satisfy the order requirements. Supplemental orders of sufficient quantity to account for excessive scrap rates may have to be placed. Moving these supplemental orders to and from the necessary work centers is also part of the dispatching process.

EXPEDITING

Greene[20] defines expediting as the "process of pushing shop orders that have fallen behind schedule." Expediting also encompasses the follow-up of orders for purchased parts that have not arrived on time.

The production scheduling process is dynamic. From data extracted from the initial master schedule, it is possible to establish required dates for both "make" and purchased parts through MRP and the product's explosion structure. Unfortunately, these dates are often subject to change. Capacity planning both compresses and stretches the master schedules for different end items. The result is a necessity to change previously established requirement dates for both purchased parts and fabricated subassemblies.

Other production difficulties result in the need for expediting. These include labor problems, equipment failures, and unanticipated rework or scrap. In JIT environments, "stretched" schedules may result in delayed requirement dates. Expediting in these cases may entail delaying the receipt dates for those parts and assemblies affected by the revised schedule.

SUMMARY

This chapter has sought to overview a number of principles and techniques of production planning and control. A wide variety of topics have been addressed. Several books have been written about each of the major topics that has been addressed.

The approach used has been to provide overviews of each subject. Where appropriate, numerical examples and mathematical notation have been used to illustrate concepts and procedures. None of the sections in this chapter are intended to provide "stand alone" coverage on any topic. References have been provided throughout the chapter that provide literature sources containing additional information on each major subject. Readers desiring additional information are urged to consult these references. Detailed bibliographic citations for all related literature follow.

REFERENCES

1. Baker, Kenneth R., *Introduction to Sequencing and Scheduling,* Wiley, New York, 1974.

2. Banks, Jerry, and W. J. Fabrycky, *Procurement and Inventory Systems Analysis,* Prentice-Hall, Englewood Cliffs, N.J., 1987.

3. Bedworth, David D., and James E. Bailey, *Integrated Production Control Systems: Management, Analysis and Design,* 2d ed., Wiley, New York, 1987.

4. Begeman, M. L., and B. H. Amstead, *Manufacturing Processes,* 6th ed., Wiley, New York, 1969.

5. Boothroyd, G., C. Poli, and L. E. Murch, *Automatic Assembly,* Dekker, New York, 1982.

6. Buffa, E. S., and Jeffrey G. Miller, *Production Inventory Control Systems,* Irwin, Homewood, Ill., 1979.

7. Burney, M. A., Eric M. Malstrom, and Sandra C. Parker, "A Computer Assisted Assessment of Just-in-Time Inventory Systems," *Proceedings of the Spring Annual Conference,* Institute of Industrial Engineers, San Francisco, May 1990.

8. Chang, T. C., and Richard A. Wysk, *An Introduction to Automated Process Planning Systems,* Prentice-Hall, Englewood Cliffs, N.J., 1985.

9. Choi, Richard H., Eric M. Malstrom, and R. D. Tsai, "An Extended Simulation of MRP Lot Sizing Alternatives in Multi Echelon Inventory Systems," *Production & Inventory Management,* vol. 29, no. 4, fourth quarter, 1988.

10. Choi, Richard H., and Eric M. Malstrom, "Evaluation of Work Scheduling Rules in a Flexible Manufacturing System Using a Physical Simulator," *Journal of Manufacturing Systems,* vol. 7, no. 1, 1988.

11. Choi, Richard H., and Eric M. Malstrom, "Physical Simulation of Work Scheduling Rules in a Flexible Manufacturing System," *Proceedings of the 8th Annual Conference on Computers and Industrial Engineering,* University of Central Florida, Orlando, March 1986.

12. Choi, Richard H., Eric M. Malstrom, and R. L. Classen, "Computer Simulation of Lot Sizing Alternatives in Three Stage Multi Echelon Inventory Systems," *Journal of Operations Management,* vol. 4, no. 3, May 1984.

13. Choi, Richard H., Eric M. Malstrom, and R. L. Classen, "Evaluation of Lot Sizing Alternatives in Multi Echelon Inventory Systems," *Proceedings of the Fall Systems Conference,* Institute of Industrial Engineers, Washington, D.C., December 1981.

14. Conway, Richard W., William L. Maxwell, and Louis W. Miller, *Theory of Scheduling,* Addison-Wesley, Reading, Mass., 1967.

15. Defense Manufacturing Engineering Education Act, S. 1331, in Authorization Bill S. 1066, United States Senate, Washington, D.C., 1991.

16. DeGarmo, E. Paul, J. T. Black, and R. A. Kohser, *Material and Processes in Manufacturing,* 6th ed., Macmillan, New York, 1984.

17. Draper, Norman, and Harry Smith, *Applied Regression Analysis,* Wiley, New York, 1966.

18. Engelberger, Joseph F., *Robotics in Practice: Management and Applications of Industrial Robots,* Avebury, London, 1980.

19. Evans, James R., D. R. Anderson, D. J. Sweeney, and T. A. Williams, *Applied Production and Operations Management,* 2d ed., West Publishing Company, St. Paul, Minn., 1987.

20. Greene, James H., *Operations Management: Productivity and Profit,* Reston, Reston, Va., 1984.

21. Groover, Mikell P., *Automation, Production Systems, and Computer Aided Manufacturing,* Prentice-Hall, Englewood Cliffs, N.J., 1980.

22. Hall, Ernest L., and B. C. Hall, *Robotics: A User Friendly Introduction,* Holt, Rinehart, and Winston, New York, 1985.

23. Harris, F., *Operations and Cost,* Factory Management Series, Shaw, Chicago, 1915.

24. Johnson, L. A., and D. C. Montgomery, *Operations Research in Production Planning, Scheduling and Control,* Wiley, New York, 1973.

25. Krajewski, Lee, and Larry P. Ritzman, *Operations Management: Strategy and Analysis,* 2d ed., Addison-Wesley, Reading, Mass., 1990.

26. Lee, Ted S., Eric M. Malstrom, S. B. Vardeman, and V. Petersen, "On the Refinement of the Constant Demand/Variable Lead Time Lot Sizing Model: The Effect of True Average Inventory Level on the Traditional Solution," *International Journal of Production Research,* vol. 27, no. 5, 1989.

27. Malstrom, Eric M., ed., *Manufacturing Cost Engineering Handbook,* Dekker, New York, 1984.

28. Malstrom, Eric M., *What Every Engineer Should Know about Manufacturing Cost Estimating,* Dekker, New York, 1981.

29. Malstrom, Eric M., "Assessing the True Cost Savings Associated with Just-in-Time Inventory Systems," *Proceedings of the Fall Annual Conference,* Institute of Industrial Engineers, St. Louis, November 1988.

30. Malstrom, Eric M., "Setup Cost Reduction Requirements for JIT Lot Sizing," Summary of Class Project Reports for IE 541, Advanced Production Control, Department of Industrial Engineering, Iowa State University, 1986.

31. Malstrom, Eric M., "Scarce Programs: Manufacturing," *IEEE Spectrum,* vol. 21, no. 11, November 1984.

32. Malstrom, Eric M., and George H. Kuper, "The Status and Needs of Manufacturing in American Engineering Programs," Unpublished Position Paper, Society of Manufacturing Engineers, Dearborn, Mich., 1987.

33. McNeely, R., and Eric M. Malstrom, "Generation of Process Routings: The Computer Takes Charge," *Industrial Engineering,* vol. 9, no. 7, July 1977.

34. Neibel, B. W., and A. B. Draper, *Product Design and Process Engineering,* McGraw-Hill, New York, 1974.

35. Nnaji, B. O., *Computer Aided Design, Selection, and Evaluation of Robots*, Elsevier, Amsterdam, 1986.

36. Oleston, Nils O., *Numerical Control*, Wiley, New York, 1970.

37. Orlicky, Joseph, *Material Requirements Planning*, McGraw-Hill, 1975.

38. Plossl, George W., *Manufacturing Control: The Last Frontier for Profits*, Reston, Reston, Va., 1973.

39. Pressman, Roger S., and John E. Williams, *Numerical Control and Computer Aided Manufacturing*, Wiley, New York, 1977.

40. Reinfeld, Nyles V., *Production and Inventory Control*, Reston, Reston, Va., 1982.

41. Riggs, James L., *Production Systems: Planning, Analysis, and Control*, 2d ed., Wiley, New York, 1976.

42. Roberts, A. D., and R. C. Prentice, *Programming for Numerical Control Machines*, McGraw-Hill, New York, 1978.

43. Schonberger, Richard J., *World Class Manufacturing*, The Free Press, Collier Macmillan, London, 1986.

44. Schonberger, Richard J., *Japanese Manufacturing Techniques: Nine Hidden Lessons in Simplicity*, The Free Press, Collier Macmillan, London, 1982.

45. Schonberger, Richard J., "Some Observations on the Advantages and Implementation Issues of JIT Production Systems," *Journal of Operations Management*, vol. 3, no. 1, November 1982.

46. Taylor, R. Bruce, and Eric M. Malstrom, "Simulation of MRP Lot Sizing Heuristics," Research Report, Department of Industrial Engineering, University of Arkansas, submitted to Northrop Aircraft Division, Los Angeles, Calif., March 1990.

47. Thomopoulos, Nick T., *Applied Forecasting Methods*, Prentice-Hall, Englewood Cliffs, N.J., 1980.

48. Vollman, Thomas E., William L. Berry, and D. Clay Whybark, *Manufacturing Planning and Control Systems*, 2d ed., Irwin, Homewood, Ill., 1988.

49. Wight, Oliver W., *Production Inventory Management in the Computer Age*, CBI Publishing, Boston, Mass., 1974.

CHAPTER 3
INVENTORY MANAGEMENT AND CONTROL

David W. Buker
Chairman
David W. Buker, Inc. and Associates
Antioch, Illinois

Improved inventory management and control is a key objective in every company's drive to control investment, improve cash flow, and increase profitability and return on investment. This chapter reviews the general principles of inventory management and discusses the planning, analysis, and control that are the foundation of a continuous improvement strategy for inventory management.

THE PURPOSE OF INVENTORY

Inventory is material or supplies that are held for future use or sales. Generally, it is finished goods waiting for a customer order. But it can also be goods or materials waiting for production or conversion into finished goods for the customer.

Not long ago, management thought inventory was a good thing; it was viewed as a valuable asset on the balance sheet. However, as business competition has intensified and costs have increased, inventory has come to be viewed somewhat differently.

Inventory has its costs of tied-up capital, storage space, handling, and obsolescence, all the costs of carrying inventory. There is a significant overhead cost or burden of carrying inventory or material, just as there is an overhead cost associated with labor costs. Inventory in the past has been carried as a cushion or safety stock to cover up for poor planning or poor performance, to protect against uncertainty in demand or variability in the supply process.

Companies can no longer afford the luxury of excessive inventory cushions or "safety stocks" if they are to be competitive in global markets. So while some inventory may be required in a company, managing and controlling it effectively has become a high priority. Inventory may be a necessary evil, but it carries a very high cost. And excessive inventory is cost added, a waste, a cover-up for poor planning. In fact, too much inventory may even be viewed as a liability.

Inventory is essentially a function of three things: the uncertainty of demand, the variability of the process, and the cycle time of the process.

Three types of variability or uncertainty may require inventory: (1) demand, (2) production, and (3) supply. These are important factors in the planning, control, and management of inventory.

Customer Demand. Depending on the industry and manufacturing environment, some inventory of finished goods is usually required to fill customer orders on a timely basis. The amount or type of inventory is dictated by the need to meet or beat the competitions' delivery lead time. Another factor that must be considered is the uncertainty of customer demand. So some cushion of finished goods may be planned to anticipate reasonable variations in customer demand. Remember, the better the planning and forecasting of demand for finished goods, the less inventory will be needed for uncertainty or variability of demand.

Production. The production process may have variability or uncertainty because of problems in quality, process reliability, tooling, and resource availability. An inventory of material in process provides a safety stock or cushion against uncertainties that can disrupt the production process. Proper work-in-process (WIP) inventory ensures the efficiency of a company's internal operations. Remember, the better the planning and scheduling and the shorter the cycle time, the less inventory will be needed for uncertainty or variability in production.

Supply Chain. Inventory is also required for smooth operation of the supply chain—vendor to manufacturer. Inventory in raw material may be required to protect against supply uncertainty or variability, such as vendor problems, transportation, and reliability of suppliers to allow smooth supply of raw materials and parts.

Remember, the better the supply relationships with the vendors, the less raw materials inventory will be needed for uncertainty or variability in the supply chain. See Fig. 3.1. If customer demand, production requirements, and supply chain requirements are known exactly, a company can plan requirements exactly for customer orders and will not require much additional inventory. Good inventory management means meeting customer demand with minimum inventory. Inventory investment is a function of (1) the accuracy of planning, scheduling, and execution; (2) the variability of demand, production, and supply; and (3) the cycle time of the process. Inventory investment can be used as a performance measurement tool to measure the quality of the planning and performance, the uncertainty or variability of the process, and the length of the cycle time. Remember, less is better.

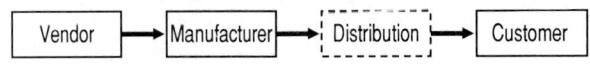

FIGURE 3.1 An illustration of the supply chain.

TYPES OF INVENTORY

Many types of inventory are found in the typical company, and they are classified and located according to their purpose or use. Three major categories apply to the inventory primarily related to a production process. See Fig. 3.2.

Raw Materials. Raw materials are acquired by the company in a raw form that needs further processing or conversion to make them part of an end product. Examples are basic

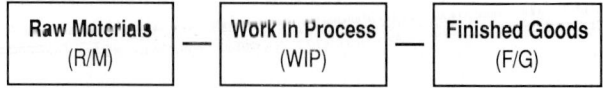

FIGURE 3.2 Inventory in the manufacturing production process.

raw materials to begin the production process, such as iron ore, crude oil, and lumber, or processed materials for general use—steel, wood, or chemicals, and the like. These are materials for primary operations, and this inventory is there to protect against variability in supply.

Work-in-Process.   This includes all production materials on which the company has performed some manufacturing, processing, or converting operations, but which are not yet in finished form. They are in the process of being worked on, and inventory protects against variability in the production process. Another category that can be considered part of this inventory is often referred to as finished parts, meaning that they are completed parts or components that are stored to be used in the final assembly of products or may also be sold as replacement parts.

Finished Goods.   This covers all completed products or finished goods produced and stored, awaiting sale or shipment to final customers. Finished goods inventory protects against variability of customer demand.

In addition to these three major classifications, there are additional classes of inventory, which can often reside at other locations.

Service Parts.   Parts commonly called service parts, spare parts, or spares are used to maintain the product or equipment the company sells or services. This inventory may be stored at the production location in finished parts or distributed and stocked with distributors, service locations, or locations closely involved in the repair or maintenance of the end product.

Distribution.   Finished goods as well as service parts are located, stored, or in transport in warehouses throughout the distribution network. These may include those owned by the company and located away from the central manufacturing plant in branch offices, company stores, and warehouses. They include goods shipped but not yet received or invoiced by distributors, retailers, or other customers and consignment stock, or goods belonging to the manufacturer but in possession of the prospective seller on consignment.

Supplies.   Items used to support or maintain operations either in the factory or in the office, but that do not become a part of the finished product, are classified by a variety of names, including general stores, maintenance repair, and operating (MRO) supplies. They include the nonproduction items regularly stocked by the company and either consumed in operations of the plant or office or needed to maintain its buildings or equipment. These are items for plant maintenance, machine repair, plant consumables, production consumables, office supplies, and so on. The items are usually expensed.

All of these inventories must be managed and controlled with the same disciplined objective to have the material available while minimizing the investment to achieve maximum efficiency in *all* areas of the business process.

What? How Much? When?   These three basic questions that drive inventory apply to all categories: raw materials, work-in-process, finished goods, and the like.

What to Order.   Forecasts of finished goods items determine replenishment orders for finished goods. The replenishment order determines what needs to be manufactured. Then this is broken down into what assemblies, subassemblies, components, and raw material are required to produce the product. These requirements are identified by a parts list, or bill of materials, that translates assembly requirements into the raw material requirements.

How Much to Order.   The objective in deciding how much to order is to focus on the material overhead cost—not merely the lowest purchasing cost, unit cost, or standard cost—to achieve the lowest total material cost. This requires establishing the most economical balance among the acquisition cost and the carrying cost.

Large order quantities enable orders to be placed infrequently and reduce acquisition and setup costs, but they increase the inventory carrying costs. Smaller quantities lower overhead and decrease the risk of obsolescence, but they require more frequent ordering and thus increase acquisition costs. For independent demand items having regular usage, the most economical balance can usually be obtained by calculating the economic order quantity (EOQ) for the item.

While there are a number of variations of the EOQ formula that apply to special situations, the simplest equation for determining the EOQ directly in pieces is

$$EOQ = \sqrt{\frac{2AS}{IC}}$$

where A = average annual usage, in pieces
S = setup and/or ordering costs
I = inventory carrying cost per year (as a decimal fraction)
C = unit cost of the item in dollars

This will give you the theoretical EOQ. The problem with this equation is that it assumes setup is fixed, which it is not. Setups can be worked on and reduced, and that will reduce the order quantity and the average inventory.

Thus, if we had a part with unit cost of $20 and annual usage of 3000 units, a setup cost of $50, and carrying costs of 0.5, the EOQ would be 173.

$$173 = \sqrt{\frac{2 \times 3000 \times \$50}{0.5 \times \$20}}$$

When to Order.   The question of when to order is: When is it needed? Forecasts of when a finished goods inventory item is needed can be used to calculate when assemblies, subassemblies, components, and raw materials are required. The bill of materials list and the lead time of each item can be used to determine when components, raw materials, or purchases are needed to meet the final production date. See Sec. 10, Chap. 2 for additional information.

COST OF INVENTORY

When considering the cost of inventory, one must look beyond the obvious—the purchased cost or standard cost of material. Inventory always carries an overhead cost, usually referred to as carrying costs; this overhead cost typically runs up to 50 percent of the purchase cost—and this total represents the total material cost (Fig. 3.3). In other words, items that cost $1, once in inventory, may really cost $1.50. Material overhead is an additional cost-added, and thus waste. Inventory management improvement should focus on reducing the total costs, which include these material overhead costs.

Eight major overhead costs are associated with inventory.

Acquisition Costs.   This administrative overhead includes the cost of requisitioning, sourcing, purchasing, shipping, receiving, and the like. Acquisition costs can add 5 percent cost to the value of the inventory per year.

Inspection.   This includes receiving inspection, in-process inspection, and finished goods inspection. Inspection costs can add another 5 percent cost to the value of the inventory per year.

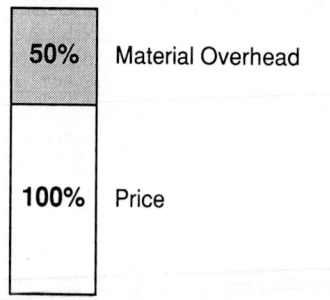

FIGURE 3.3   Chart representing the components of total material cost.

Storage. This is an obvious carrying cost, and it includes the cost of storage and warehouse space, security and related storage expenses, and taxes. Storage costs can vary widely, depending on the type and quantity of material and inventory stored and the kind of facility and space required. On the average, storage costs run at least another 5 percent cost to the value of the material stored per year.

Handling. All of the handling, moving, and transportation involved in controlling the inventory presents another obvious cost. It includes the wages and benefits of the personnel involved in these functions as well as all of the material handling systems and equipment that support their work. Handling customarily adds another 5 percent cost to the value of the inventory per year.

Interest. Inventory ties up one of a company's most versatile asset, cash. Since businesses have a limited amount of capital resources available to them from owners and creditors, capital invested in inventory carries a definite cost. It's the cost of capital. This cost is calculated as the cost of the money or the rate of return it could have earned were it invested in something else, such as government bonds or high-grade stocks. Interest costs, calculated on moderate estimates of what the capital could be expected to earn if wisely invested, add another 10 percent cost to the value of the inventory per year.

Obsolescence. Every business must face the grim fact of obsolescence to some degree. Parts in stock become obsolete because of a model change or a new product. This is particularly true in an engineered product, a high-tech product. Needs cannot be estimated with perfect accuracy, even with the most sophisticated computerized systems. Well-managed companies continuously work on surplus and obsolete inventory and dispose of it. A general rule is never to hold inventories for which there is no immediate need. Therefore, a part of the cost of inventory is an allowance to cover losses from obsolescence which may average up to 10 percent of the value of the inventory per year.

Depreciation. In accounting terms, depreciation is the reduction in value of a capital asset based on age or usage that often may or may not reflect any real loss of value. In the case of inventory, however, depreciation refers to damage and deterioration or loss due to storage, handling, weather, age, evaporation, or shrinkage. Depreciation varies with the type of inventory, but it normally represents about 5 percent cost of the value of the inventory per year.

Insurance. Insurance on inventory is a directly variable cost because it is normally paid at a rate directly proportional to inventory value. Another factor that affects insurance cost is the kind of facilities and security systems used for storing the inventory. The insurance costs average about 5 percent of the value of material stored per year.

Thus ordering, maintaining, and controlling inventory is expensive, as you can see. Adding the total carrying costs or material overhead costs:

Material Overhead Costs

	Percent per year
Acquisition	5
Inspection	5
Storage	5
Handling	5
Interest	10
Obsolescence	10
Depreciation	5
Insurance	5
Total overhead costs	50

These elements must be accurately calculated and analyzed to control the total material cost of all inventories.

GENERAL CONCEPTS OF INVENTORY

A number of general concepts of inventory need to be explored before we proceed.

Independent Demand.   Demand from the marketplace for end-product items, such as finished goods and service parts, is driven by factors which are independent of company decisions. This type of demand usually comes from relatively uniform customer orders, received continuously but also randomly throughout any time period. Forecasts of demand for these independent demand items are typically projections based on historical demand patterns that estimate the average usage rate, usage trends, and the pattern of demand variation. The demand generally draws down the inventory until the reorder point is reached, and then a replacement order is placed and received. See Fig. 3.4.

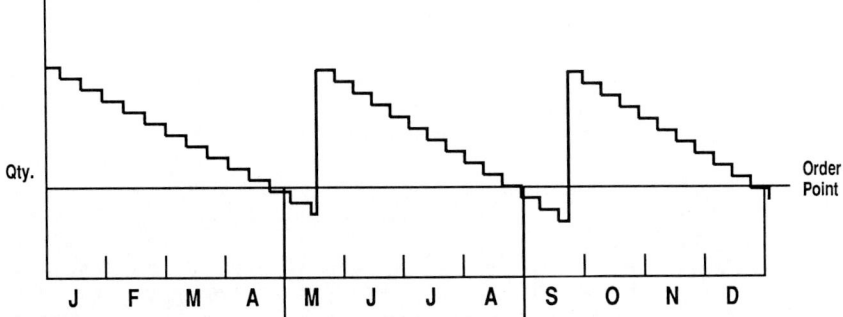

FIGURE 3.4   Graph showing independent-demand patterns over 1 year.

Dependent Demand.   Demand in manufacturing for materials needed to make finished goods is dependent on the demand of the end-product items. These dependent demand items are raw material, components, and lower-level subassemblies, and they are dependent on demand for the end-product item. This type of derived demand is usually intermittent, or dependent, since demand exists only at a time when the next higher level of assembly is being made. Requirements for the lower-level material requirements are dependent on the next higher level and can be calculated based on the assembly or production of the end product. Inventory is generally planned only to meet specific production requirements. See Fig. 3.5.

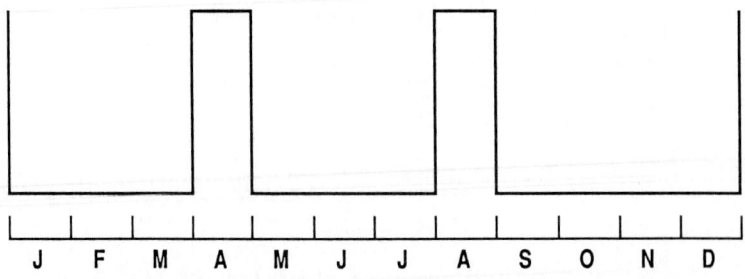

FIGURE 3.5   Graph showing dependent-demand patterns over 1 year.

Lead Time.   A concept essential to inventory planning is lead time, or the time that it takes to replenish an inventory item. It is how long it takes to purchase or manufacture the item.

Lead time is composed of many elements. Purchased lead time includes the time it takes to source, order, receive, and enter items into stock. Manufacturing lead time adds the elements of setup time, running time, queue time, and move or transit time. The lead time is important in knowing "when" to order, so that you have the time to order and get the item.

Lead time of the product can vary based on cycle time of the process and the inventory strategy of the company to deliver the product to the market. These inventory strategies are make to stock, assemble to order, make to order, or engineer to order. See Fig. 3.6.

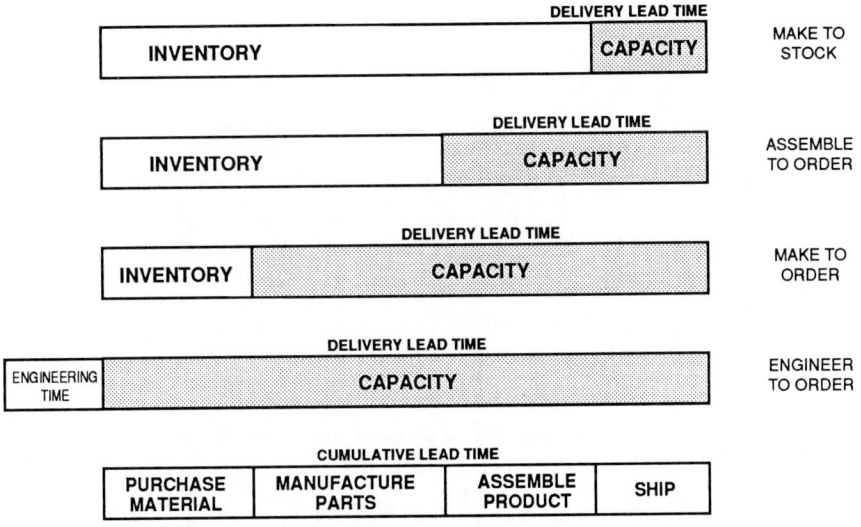

FIGURE 3.6   Bar chart illustrating the relationship of lead time to inventory strategy.

Make to Stock (MTS) Strategy.   The finished product is produced to stock and is in stock awaiting the customer order. This strategy produces a short customer lead time and a high level of customer service. But it requires better customer demand planning with a higher level of finished inventory investment.

Assemble to Order (ATO) Strategy.   The important parts or subassemblies are planned in inventory but not assembled. The customer delivery lead time and customer service are good. But it requires good customer demand planning and inventory investment in components for assembly.

Make to Order (MTO) Strategy.   The finished product is produced after the customer order is received. It allows for specific customer orders. This strategy requires a longer customer lead time but less investment in inventory.

Engineer to Order (ETO) Strategy.   The product is defined, designed, planned, and produced to meet the customer's specific requirements. This strategy requires a longer lead time but virtually no prior inventory investment.

Many companies pursue more than one of these strategies and sometimes all four at once, that is, different strategies for different product lines. Proper inventory planning strategy requires that the company has an understanding of the production lead time of each product and the competitive delivery lead time in the marketplace to implement the inventory planning strategy that will maximize customer satisfaction while minimizing inventory investment.

INVENTORY PLANNING

Inventory planning begins with projecting the company's needs or demands for inventory into the future. There are basically three future time frames that need to be considered: long-term, middle-term, and short-term. Each time frame is the primary focus of the three major planning levels in the company. See Fig. 3.7.

PLANNING LEVEL	PLANNING FUNCTION	LENGTH OF PLAN HORIZON
Top Management Planning	Finished Goods Demand Planning	Long Range 1 – 3 Years
Operations Management Planning	Product Mix Material & Capacity Availability	Medium Range 1 – 3 – 6 Months
Operations Management Execution	Detail Production & Delivery Schedules	Short Range Weeks Days

FIGURE 3.7 Chart relating planning levels and functions to time horizons.

Top Management Planning. Long-term planning deals primarily with forecasting the demand for finished goods or independent-demand inventory and the long-term production resources required to produce or supply the product to meet the company's market, product, and profit objectives. Top management planning should extend out for at least 1 year and be reviewed monthly to update the plan based on actual market and production performance.

Operations Management Planning. Medium-term planning translates top management's plans into the detail production resources or raw-materials inventory required to produce the required finished-goods inventory. Operations management planning is concerned primarily with material availability and capacity availability, and it should be reviewed weekly and updated based on actual performance.

Operations Management Execution. Short-term planning takes place at the front line in day-to-day execution of operations management's plan and the movement of raw materials, work-in-process, production supplies, and other dependent-demand inventories through the production process to produce or supply the finished product on a daily basis. Operations management execution typically implies daily or even more frequent targeting and review of performance.

The fundamental task in managing inventories is to translate the long-term into medium-term and short-term planning for finished-goods inventory, work-in-process, and raw-materials inventory to effectively meet customer demands with maximum operating efficiency and minimum inventory investment.

FINISHED-GOODS INVENTORY, INDEPENDENT-DEMAND PLANNING

Finished-goods inventories that are stocked to meet customer demand are usually found in finished-goods warehouses, distribution warehouses, stocking locations, or retail environments. These inventories characteristically include a large number of items. They are the end item stock-keeping units (SKUs) that are stocked for the customer, and their demand generally comes from many customers and is independent of other activities. Retailers can often count the number of SKUs in the hundreds and thousands when considering different variations such as sizes and colors.

Inventory management of these items relates directly to forecast demand and the level of customer service. Failing to order what is needed, in the quantity needed, at the time needed results in stockouts, poor customer service, and potential loss of sales. Conversely, ordering too much or ordering too soon will result in excessively high inventory and extra costs.

In many companies, independent-demand inventory control is heavily biased toward customer service at the cost of high inventory. Good customer service does not necessarily have to mean high inventory. Good customer service also depends on the accuracy of the demand plan (forecast) and the cycle time to replenish inventory. Another approach to excellent customer service without high inventory is to have accurate demand forecasts and short replenishment cycle times. Remember, what we want is the right amount of the right items, at the right time. Customer service is really a function of the accuracy of the forecast, the replenishment cycle time, and the inventory levels. The better the forecast, the shorter the cycle time, the better the service with lower inventories.

Customer service = forecast accuracy + cycle time + inventory levels

Order Point Systems.   For many years, people have relied upon historical data and past usage to develop demand patterns, set order points, and determine order quantities. These basic techniques involve setting reorder points for each inventory item (this reorder point depends on the replenishment cycle time), then when stock diminishes to the reorder point, placing an order for a specified order quantity.

Setting of the reorder point is influenced by four factors: the demand rate, the amount of uncertainty in the demand rate, the lead time required to obtain replenishment inventory, and management's policies regarding acceptable inventory shortages and customer service.

Reorder point = demand during lead time + safety stock

When there is no uncertainty in either demand rate or lead time for an item, determination of the reorder point is fairly straightforward. For example, if the demand rate for an item is exactly five units per day, and the replenishment lead time is exactly 1 day, the reorder point or trigger to launch the replenishment order is five units.

Yet constant demand rates and fixed replenishment lead times are rare in actual operations. Variations occur, not only from fluctuations in customer demand, but in replenishment lead times because of supply uncertainty. To provide protection when there is uncertainty in demand or replenishment time, the reorder point needs to be increased beyond the average demand during lead time to maintain some level of safety stock. But, remember, the objective is to have accurate forecasts with short lead times and reduced variability. This will result in lower inventories and higher-level customer service.

Order Rules.   Setting of inventory levels, order points, and order quantities is central to inventory planning and management for finished goods. These are stated in order rules. This process begins with forecasting average usage or demand. The order point is set at the demand during lead time. The safety stock is determined by the variability during lead time. The order quantity is determined by the economics of production or supply.

Once these order rules are determined, the inventory management process focuses on reviewing the available stock levels against the order point. Reviewing must be done frequently because of the constant rate of depletion of stock. Some items may be checked as frequently as every issue. This is particularly true with computer systems that make this possible. Real-time data can then be reviewed as frequently as appropriate to the business—a McDonald's store uses its point-of-sale (POS) inventory system to review inventory at least daily and usually every hour. Orders to replenish inventory items from central distribution or a nearby supplier can be launched daily or even several times a day in times of heightened demand or unanticipated traffic. However, the management goal is to forecast usage accurately so that the store can run on its regular shipment until the next scheduled shipment from the central warehouse or some other source of supply.

Order rules and order points should be reviewed and recalculated on a regular basis. Computers now make it easy to analyze rates of usage and recalculate order points and order quantities frequently to keep pace with changes in customer demand, replenishment cycle time, and other changes in the consumer environment.

There are several types of order rules and many variations of each, since any given situation or class of inventory may require different types of order rules involving differences in inventory levels and total inventory costs. The order rules most commonly applied in the distribution and retail environment are based on fixed order quantities and fixed order cycles or intervals.

Fixed Order Quantity. In some systems, the order point establishes when to order the inventory. When the stock on hand falls to the reorder point, the inventory planner places a reorder. How much to order is prescribed by a predetermined economic order quantity (EOQ). The EOQ is calculated as the quantity that will result in the lowest total costs of acquiring, producing, or carrying the item. Thus the order quantity is fixed and the time interval between orders may vary, depending on the rate of usage.

A good example of fixed order quantity inventory management can be seen in the maintenance of service parts at an auto dealer. When stock of a particular part, such as a gasket, reaches the reorder point for that item, it automatically orders an EOQ from the distribution center. The inventory control system may be as simple as a two-bin system. When one bin is depleted, the order is issued for replenishment inventory.

The advantage of the fixed order quantity for managing inventory is that it is flexible. Orders can be issued any time. When the order point is reached, an order is generated. Thus the order cycle may vary, but the order quantity does not. The nature of some businesses may also require that they develop separate order points and order quantities on items for different times of the year because of different (seasonal) usage rates.

Fixed Order Cycle. In some distribution systems, ordering takes place at a fixed interval or cycle. When this rule is used, items are ordered at regular fixed cycles such as every week. The specified interval indicates when to order. How much to order is determined by subtracting the stock on hand to determine an order amount to meet an established or target inventory level. In this case, the order quantity may vary, but the order review cycle is fixed.

A good example of fixed order cycle inventory management can be seen in the stocking of bread to retail outlets. Each retail site is visited by the bread distributor's representative on a predetermined schedule. The representative checks the loaves on the shelf, removes any outdated stock, and adds to what remains the quantity required to bring it to the desired inventory level. The replenishment quantity may vary from order to order (depending on the usage), but the interval at which the representative revisits, checks, and brings up the stock is fixed.

Fixed order cycle inventory management is most useful in situations where there are a large number of items that are ordered and delivered at one time, and there are no significant economies from ordering individual items in larger quantities.

Obviously, different inventory management policies apply to the different types of inventory found in different businesses. For example, running out of an expensive equipment service part is more likely to impose severe cost and operating penalties than running out of a general supply item. For that reason, the reorder signal will often be set at a relatively higher level on critical service parts.

Order point continues to be used in many companies for managing independent-demand finished-goods items, service parts, supplies, and stable usage items. While it may be executed manually, many companies now track finished-goods sales and inventories on computers, aided in the retail and warehouse environment by POS terminals, scanners, or bar-code readers for real-time data capture. The computers track on-hand inventory at all sales locations and provide automatic "reorder messages" to trigger replenishment orders and the proper distribution of inventory throughout the entire distribution system or supply chain.

DEPENDENT-DEMAND INVENTORY PLANNING

For many years, people relied upon the same order point method for managing both independent-demand finished goods and dependent-demand raw materials inventories. There were a number of problems, however, with using order point for these dependent-demand items. First, the approach is not oriented to future demand. It's based on historical usage. Second, and most importantly, it does not recognize the dependent-demand relationship. The lower-level components and raw materials are not needed in inventory until they are required for the next higher level. Their demand is dependent on the higher-level parent. With the advent of computers we are now able to plan more effectively and calculate dependent-demand inventories based on these dependent relationships.

The Closed-Loop Inventory Planning System.   First came computer-based material requirements planning (MRP) in the 1960s and 1970s, which shifted emphasis from order points and launching orders when an item appeared to be running out of scheduling and priority planning based on due dates for finished products and due dates for components and raw materials.

Next came manufacturing resource planning (MRP II) in the 1970s and 1980s, which expanded the degree of planning and control to encompass all functions of the company, including sales, manufacturing, engineering, purchasing, and production. This enabled management to integrate long-term, medium-term, and short-term planning into a total company closed-loop inventory planning system.

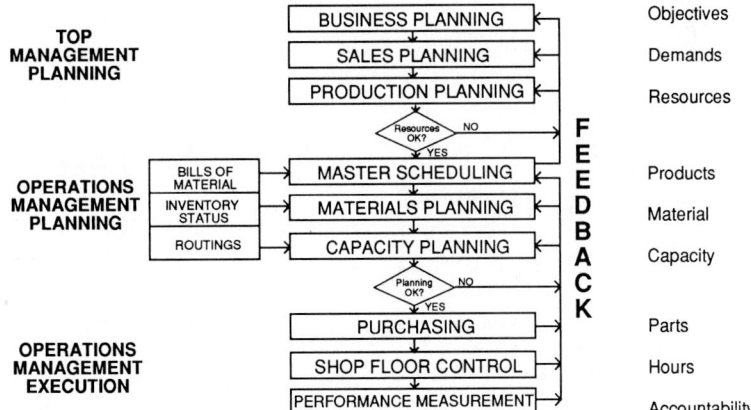

FIGURE 3.8   Chart presenting the closed-loop inventory planning system.

The closed-loop system translates top management's business plan, sales plan, and production plan for finished goods into rates of production that must be established and produced by operations management to meet customer demand. This is the "what, how much, and when" of the rates of finished product or finished goods by month that are needed for the company. See Fig. 3.8.

The next step is the heart of the computer-based inventory planning system. Operations management planning develops the master schedule of what, how much and when, which is the weekly detail statement of the mix of products to be produced, and then this schedule is exploded into the detail materials plan and capacity plan. Material planning is a time-phased priority planning system to schedule material to meet requirements. Capacity planning provides detail capacity requirements of labor and equipment to produce the product.

In the computer-based inventory planning system, these activities are supported by information in a database, which includes bills of material, inventory status, and routings. The bill of materials specifies the parts or materials needed to produce the final product. Routings specify the process or the operations in production. And the inventory status includes the on-hand quantity and location of the items in raw material, work-in-process, and finished goods that are available to produce the product.

Operations management execution is the final step which develops daily schedules of inventory and production for purchasing and production shop floor control. Purchasing then purchases the materials and parts required to support the inventory plan, and shop floor control moves the raw materials and subassemblies through the production process to meet daily schedules and produce the final product to meet the inventory plan. Performance measurement provides the monitoring device to review and communicate that all functions are performing to plans and to meet customer demand.

A computer-based closed-loop inventory planning system can provide much better planning and control of dependent-demand inventories. The computer takes over the multistep task of calculating requirements for all the parts through the bills of material, maintaining inventory record status, and projecting material and capacity requirements to produce the product.

ACCURATE INVENTORY RECORDS

Accurate inventory records are very important for a company's inventory management. They are important for many reasons:

1. They verify or attest to the physical inventory as an asset in determining the value of a company.
2. Customer orders for products can be accurately quoted and shipped from inventory.
3. Realistic production schedules can be developed and met because people can count on having the necessary parts and materials available in inventory when needed.
4. Production delays caused by unexpected shortages of critical materials can be eliminated and the need for costly, last-minute rush orders can be reduced.
5. Inventory levels can be reduced because "safety" stocks held to compensate for unexpected shortages or incorrect balance information are not needed.
6. Improved production efficiency, product quality, productivity, and customer service can result.

Effective inventory control depends upon accurate and timely inventory information. A key measurement of inventory control is inventory record accuracy.

Measuring inventory accuracy is a two-step process. First, the inventory items are physically counted. Next, the count is compared with the balances shown on the inventory records. When the counts match the balances shown on the inventory records exactly or within predetermined tolerance ranges, the inventory is accurate. Inventory accuracy of at least 95 percent is generally considered mandatory for effective inventory planning and control.

Inventory Transaction Processing System (TPS).   An inventory transaction processing system is required in a company to track the movement, location, quantity, and status of materials and parts as they physically move through the production process. The transaction processing system relies on people, processes, procedures, and computers to accurately account for the physical transfer of materials within the production process.

Inventory transaction systems should be simple and transparent and should reflect reality. A blueprint or layout of the facility is a handy starting point for identifying material flows and inventory control points. Inventory control points are anyplace where materials are transferred, such as the receiving dock, controlled stockroom areas, and the shipping dock. As material passes through an inventory control point, it should be documented and recorded by a transaction to maintain accurate inventory information.

Transaction information should include identification, or the part number of material being moved, quantity, its location, and the material's status. For example, when an item is received into the warehouse, a transaction should be recorded specifying the part number of the item and how many are being received into that stock location.

Controlled stockrooms can be helpful in improving inventory accuracy defined by physical barriers such as fences or by psychological barriers such as lines painted on the floor, signs on the walls, or other markings. Employees in each controlled stockroom are responsible for recording the appropriate transaction information any time materials move in or out of the area. The inventory accuracy of each controlled stockroom area should be measured, and performance results posted in each area for accountability.

Transaction recording can be facilitated through the use of bar coding, scanners, and optical character readers. This enhances both the accuracy and timeliness of the data capture. Paper forms may still be necessary under some circumstances, however. The following are guidelines for designing paper transaction forms:

1. Develop simple single-use forms. To minimize errors, a different form or paper color should be provided for each transaction type.

2. Make directions and field titles simple and easy to understand.

3. Minimize the amount of writing necessary to complete the forms. Preprint as much information as possible.

4. Organize the fields on the forms so that they can be completed in order as stockroom personnel receive or issue materials.

5. Make forms computer-scannable or clearly arrange the fields to match the computer system's input screens to facilitate accurate data entry.

To ensure that the inventory transaction system is up to date, transactions should be processed on a timely basis. Timely TPS is important for accurate inventory records and for audit reconciliation. Most inventory transaction systems generate a transaction audit trail for reconciling the inventory and identifying and correcting errors. The paper transaction documents completed by employees as materials move through the production process serve as a valuable, physical audit trail. These documents can be maintained in a file for reference whenever discrepancies are detected in the computerized records.

Transaction history reports can be generated detailing all transaction activity affecting on-hand balances. Reports by part number and location are also useful for identifying and correcting errors. Remember, the objective is to maintain accurate inventories, not to generate transactions and documents. Value added—not cost added. Less is more.

A number of methods can be used to verify the inventory. The two most frequently used procedures are the annual physical inventory and cycle counting.

Annual Physical Inventory (API).   Traditionally, manufacturers closed their plants for a number of days to physically count and verify their inventories. Discrepancies between the dollar value of materials counted on hand and of inventory on the books were reconciled. The books were adjusted, as necessary. The purpose of the annual inventory was to ensure that the company's books were accurate for accounting and tax purposes. The primary emphasis of this approach was on total dollar value of the inventory.

Today's competitive manufacturing environment requires a much different and higher degree of inventory accuracy. Managers require correct inventory information by item at all times to ensure the quality of their planning, scheduling, and control decisions. Inventory information that is reliable only once a year and only by total dollar value is of little use for modern production and distribution planning.

The accuracy of inventory balance information determined by an annual physical inventory is often questionable, even on the day that the inventory is completed. Because the annual inventory occurs only once a year, it is taken by numerous employees who are generally untrained in inventory taking. Counting and identification errors are likely to result, and the annual physical inventory has always been suspect. Despite the massive effort involved and loss of production time while the plant is shut down, the major problem with the API is that no ongoing problem-solving and accuracy improvements are likely to result. The API does not promote an ongoing process of continuous improvement and keeping the inventory accurate throughout the entire year.

Cycle Counting.   In place of the annual physical inventory, today's companies use a process called cycle counting, a proved method that helps maintain inventory accuracy over time. Cycle counting relies on continuous counts, or audits of the inventory, on a regular basis throughout the year. These counts are compared with the balances shown on the computerized inventory records. Any discrepancies are analyzed immediately to determine what caused the errors, and steps are taken to fix the error and prevent it from occurring again.

$$\frac{\text{Inventory}}{\text{record}} \atop \text{accuracy} = \frac{\text{number of records correct} \times 100}{\text{number of inventory items}}$$

Examples of accuracy tolerances for inventory items classified using the ABC method are shown in Fig. 3.9. Inventory records showing on-hand balances that match the physical count or fall within the tolerance range are considered accurate. Notice that 0 percent tolerance is allowed for high-dollar-value items (Class A). The wider tolerance range for class C items, 5 percent, reflects both their lower dollar value and the fact that weigh counting is routinely used instead of physical counting to determine both actual and transaction quantities for these items.

Inventory accuracy can also be measured by comparing the dollar value of inventory on hand, as shown by the inventory records, with the dollar value of the physical inventory. However, this measurement is not particularly useful for efficient manufacturing and improving inventory accuracy. As shown in the last column of Fig. 3.9, inventory accuracy measured in dollars produces a higher level of rate accuracy than count accuracy. For inventory control purposes, the count accuracy of on-hand balance by item by location is the important measurement, much more so than the dollar accuracy of the financial statements.

Remember, take care of the piece count accuracy, and the dollar accuracy will take care of itself.

Benefits of cycle counting in contrast to an annual physical inventory include

1. Efficient use of trained personnel
2. Regular error detection and correction
3. Minimal loss of production time
4. Improved inventory accuracy
5. Reduced inventory levels
6. Better productivity
7. Improved customer service

INVENTORY ACCURACY

Control		Count Analysis			Dollar Analysis		
Value Class	Tolerance	Parts Counted	Within # limit	Count Accuracy	On Hand (In Dollars)	Variance (In Dollars)	Accuracy (In Dollars)
A	0%	50	49	98%	7,500	75	99%
B	2%	75	72	96%	2,000	(100)	95%
C	5%	100	93	93%	500	(50)	90%
TOTAL		225	214	95%	10,000	(75)	99%

FIGURE 3.9   Examples of accuracy tolerances for inventory items classified by the ABC method.

Control Group Method.   Before implementing a full-scale cycle counting program for all inventory, it is a good idea to select a small sample control group of items for daily counting and reconciliation to prove the process and identify any problems. For example, a group of about 50 items, ranging in volume, price, and size from large to small, may be counted and reconciled. Differences between the counts and the balances shown on the computerized inventory records should be identified and corrected on a daily basis for this control group.

ABC Method.   This method maximizes dollar inventory accuracy while minimizing the effort and cost required for counting. Items are categorized as class A, B, or C, based on their dollar value. Class A items are counted most frequently, perhaps monthly, because they are usually about 10 percent of all inventory items that account for 60 to 75 percent of the total dollar value of inventory. Class B items are counted somewhat less often, perhaps quarterly. Class B items account for 20 percent of inventory items but comprise 20 percent of the inventory's total dollar value. Class C items are counted least often, perhaps only once or twice per year. Class C consists of the low-dollar-value items that make up the remaining 70 percent of the total inventory items.

Reorder Method.   The reorder selection method is designed to minimize the number of items that must be counted with each count. Using this technique, inventory items are counted whenever a reorder is issued, and the inventory is usually at the lowest level requiring the counting of the fewest number of items. Another advantage of this method is that when a quantity discrepancy is identified, there may still be time to prevent a stockout.

Free Count Method.   Using the free counts method, stockroom personnel count inventory items whenever they are servicing the inventory at a location, such as when a replenishment lot is received or when pulling the last item from a location—thus a "free" count.

Zone Count Method.   This is cycle counting by location or "zone." On a rotating basis, each zone's contents are counted. This method is used because zones keep the counting concentrated in one area. Also, inventory accuracy accountability is usually assigned by area. This is probably the best method.

Other Methods.   In addition to dollar value, some companies may use classification criteria such as how critical an item is to the finished product, length of procurement lead time, or amount of storage space required.

Items should also be counted whenever an error condition has been identified or a problem exists. For example, if the computerized inventory records show a negative balance on hand for a particular item or a quantity on hand without a valid stockroom location, the actual inventory status should be investigated and the records corrected.

Process of Continuous Improvement.   The objective is to fix the problems that are causing the discrepancies—which involves far more than just adjusting the numbers to bring them into balance. Once the problems are resolved, the records will begin to stay accurate on an ongoing basis.

Once inventory accuracy of at least 95 percent is consistently maintained for the small sample control group, then a full-scale cycle counting program can be launched that encompasses all the inventory. Various criteria can be used to select and schedule the inventory items that will be counted. These criteria include the ABC selection method, the reorder selection method, free counting, zone counting, and others.

Measuring Performance. Inventory accuracy is a measurement of performance indicating the accuracy of inventory balances on hand. Actual inventory should be compared with the balances shown on the inventory records at least once per year by physically counting the items. The measurement is expressed as the percentage of correct record balances. Correct on-hand balances are those that match, within preestablished tolerance ranges, the actual number of items on hand.

INVENTORY MANAGEMENT AND ANALYSIS

Inventory management and analysis are an important part of the management function. Timely inventory analysis enables managers to identify and control inventory investment problems. By monitoring and controlling inventory investment levels, turnover rates, lead times, and days of stock, many companies can significantly reduce their inventory investment and their total inventory investment costs.

Inventory Flow Model. Modeling has been used for analysis in many areas of management, and it is also an important tool for inventory management and analysis. Inventory models can be used to plan inventory levels and to highlight problem areas, such as inventory buildups or inventory imbalances The purpose of the inventory flow model is to model the present inventory against the inventory flow ratesto detect problems with the inventory. The information needed to construct the model is material, labor and overhead costs as a percentage of cost of sales, the annual volume, and the present inventory levels. This information is used to model each category of the present inventory—raw materials, work-in-process, and finished goods—and to establish inventory targets for each category and for the total inventory. See Fig. 3.10.

Materials Cost. As a percentage of cost of sales, figure generally 50 to 60 percent.

Labor Cost. As a percentage of cost of sales, figure generally 5 to 10 percent.

Overhead Cost. As a percentage of cost of sales, figure generally 30 to 45 percent.

Cost of Sales Dollar Volume. The annual total dollar volume in cost of sales (COS) divided by 12 gives the monthly total volume in COS rate, and that divided by 20 gives the daily COS rate.

Flow Percentage. The flow percentage for R/M is the percent that R/M is of the total cost of goods. In the example, 50 percent. The flow percent for finished goods is 100 percent. The flow percent for work-in-process is the R/M flow rate plus one-half the difference between the R/M rate and the F/G rate, or 50 percent plus 1/2 of 100 minus 50 percent = 75 percent.

Flow Rates per Day. Divide the flow percentage for each category into the total flow rate per day to determine the dollar flow rate per day for each category.

Inventory Dollars. The actual inventory dollars on hand by category of inventory and the inventory in total.

Raw Materials. The raw materials inventory divided by the raw materials flow rate per day equals the number of days of stock in raw materials, or $2500 divided by $125 per day equals 20 days of inventory.

Work-in-Process. The number of days of inventory are determined in the same way as raw materials. In this case $2500 divided by $187.50 equals 13 days.

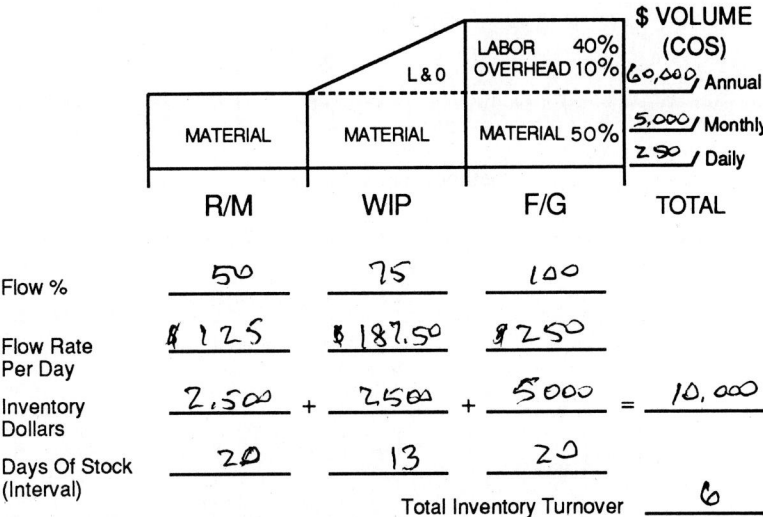

FIGURE 3.10   A basic inventory flow model.

Finished Goods.   The finished goods inventory is divided by the finished goods flow rate per day which is full cost (material + labor + overhead) to determine the number of days stock in finished goods, or $5000 ÷ $250 = 20 days.

Inventory Turnover.   Inventory turnover is the ratio of annualized cost of sales to inventory investment.

$$\text{Turnover} = \frac{\text{cost of sales}}{\text{total inventory}}$$

You can also calculate turnover of each category of inventory by dividing by the annualized cost of sales of the inventory for that category.

$$\text{Turnover} = \frac{\text{inventory cost of sales}}{\text{inventory category}}$$

This inventory flow model can be used to answer the following questions: What is the inventory by category and in total? What should the inventory be by category in total? What should the days of stock and cycle time be? What actions can be taken to reduce the cycle time, days of stock, and inventory for each category of inventory and the total inventory?

Inventory Model.   Inventory models can also be used by managers to construct inventory plans by category to monitor each category of inventory and the total inventory in the company. These inventory models make it easy to track planned to actual performance on a monthly basis to see if levels are being controlled and managed according to plan.

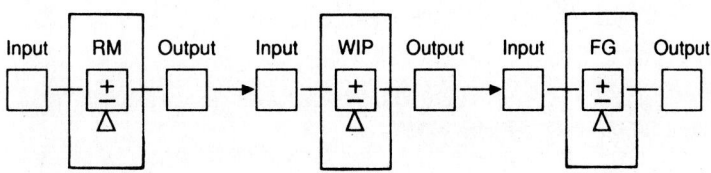

FIGURE 3.11   A flow model of the input-output process.

Input-Output Model. Input-output models are simple models for measuring the change in either input or output and the resulting change in that category being impacted. See Fig. 3.11.

Input-output models have many uses. This input-output concept can be used for your personal checkbook. If you spend more (output) than you put in (input), your bank balance will go down. It can be used for dieting. If you take in fewer calories (input) than you use up (output), you'll lose weight. If you don't, you won't.

In inventory, if we want a category of inventory to go down, we must put in (input) less than we take out (output). To reduce finished goods we must produce less than we sell. To reduce work-in-process we must plan to put in less material, labor, and overhead than we produce and ship out the door. To reduce raw materials we must plan to receive purchased material at a lower rate per day than we are using raw material in work-in-process, and so on. Input-output modeling can help to plan, control, and manage inventory flows. See Fig. 3.12.

	CATEGORY		JAN		FEB		MAR		APR		MAY		JUNE
SALES	Sales – $	P	6,000		8,000								
		A											
	COGS – $	P	4,000		5,000								
		A											
			$	TO	$	TO	$	TO	$	TO	$	TO	$
INVENTORY	Raw Materials	P	2,000	12	2,500	12							
		A											
	Work In Process	P	2,000	18	2,500	18							
		A											
	Finished Goods	P	4,000	12	5,000	12							
		A											
	Total Inventory	P	8,000	6	10,000	6							
		A											

FIGURE 3.12 Performance chart for tracking inventory flow.

By analyzing inventory models, managers can identify opportunities for cost savings, such as potential reductions in raw materials, work-in-process, and finished-goods inventories. Unfavorable trends, such as unplanned inventory buildups, can be detected, investigated, and resolved.

Opportunities for inventory reduction can be identified by developing understandable inventory models. These models plot actual inventory versus planned inventory and actual cycle times versus planned cycle time for each category of inventory. The models often reveal excess inventory that can be reduced or eliminated in order to reduce the total inventory.

For effective inventory management, performance objectives for inventory investment, turnover, cycle times, and days of stock should be established and monitored for each category of inventory. These measurements are a means for monitoring the effectiveness of inventory management and inventory improvement efforts.

Remember, inventory is a function of cycle time. The best way to reduce the inventory is to reduce the cycle time. Too much inventory is "cost added," not "value added." It is waste.

STRATIFICATION ANALYSIS—ABC

Vilfredo Pareto was an Italian economist who said that a small percentage of a group's items contribute the bulk of its costs, value, impact, and the like. This rule (Pareto's law) has many applications. For example, a small percentage of customers are often responsible for the largest percentage of sales volume. Similarly, a small percentage of out-of-stock items are generally the cause of the largest percentage of back orders.

In the typical manufacturing and distribution company, it is generally true that 20 percent of inventory items may account for 80 percent of inventory value while 80 percent of inventory items may account for only 20 percent of the value. Stratification analysis applies this rule in a time-honored tool of inventory management.

ABC Analysis.   Inventory items are categorized by their annual dollar volume. An ABC analysis lists inventory items in decreasing dollar-volume order and labels the high-dollar-volume items as "A," medium-dollar-volume items as "B," and low-dollar-volume items as "C." To generate an ABC analysis report:

1. Calculate annual dollar volume for each inventory item by multiplying the item's unit cost by its annual usage volume.

2. Generate a report in decreasing dollar-volume sequence showing item numbers, annual usage, unit costs, annual dollar volumes, and item counts.

3. Compute cumulative totals and percentages for item counts and annual dollar volumes on an item-by-item basis. The purpose of this step is to separate the three inventory classes. It is not necessary to perform these calculations for every item on the list—see step 4.

4. Delineate the A, B, and C categories based on cumulative item count and volume percentages. For example, a manufacturer may decide to place 10 percent of total inventory items in class A, 30 percent in class B, and 60 percent in class C. Because the items have already been sorted into decreasing dollar-volume order, the small percentage of items in class A will generally account for a large percentage of annual dollar volume, often as much as 80 percent or more.

Based on an ABC analysis of inventory, management can implement appropriate planning and control procedures for each class of inventory.

Class A items receive the most attention, since they account for the largest dollar volume yet are relatively few in number. By increasing control over class A items, fewer can be kept in stock. Managers can best leverage their time spent controlling inventory by concentrating on these high-dollar-volume items. In this way management can apply greater control over ordering and controlling costs to reduce the total material cost of the inventory.

Total Materials Cost and Turnover.   The information generated by an ABC analysis of inventory is a useful starting point for analyzing the effects on total material cost of different inventory investment and inventory turnover rates on different classes of inventory. Figure 3.13 shows a sample company's class A, B, C inventory classifications. It shows the turnover (TO) and order index using the normal approach and it shows the turnover and order index using the TMC-TO (Total Material Cost—Turnover) approach. The order index is determined by multiplying the percentage of total inventory items for a category by the turnover rate for the category. The order index is an indicator of the ordering and

% Dollars	% Item	Class	T.O.	Order Index	TMC-TO	Order Index
70%	5%	A	6	30	12	60
25%	25%	B	6	150	6	150
5%	70%	C	6	420	3	210
100%	100%		6	600	8.5	420

FIGURE 3.13   Chart showing total material cost as applied to inventory turnover by inventory class.

controlling costs for a category of products. The lower the index, the lower the ordering and controlling costs. Notice the contrasting effects on the order index. When the turnover for all three inventory items is six times per year, the order index is 600. Using the ABC concept of inventory control, the order index is reduced to 420, a reduction of 30 percent. The turnover of A items is increased to 12 times a year. This shows under the TMC-TO column. The overall TMC turnover is increased to 8.5 times per year, an increase of 40 percent.

Figure 3.14 shows how the savings in Fig. 3.13 were accomplished by introducing the dollar values for the three inventory categories. By focusing attention on the A items, it was possible to increase their turnover from 6 to 12 times per year. Because less attention was given to the C items, their turnover dropped to only 3 times per year. But since they constitute only a small dollar value, the total TMC-TO increased from an average of 6 to 8.5 times per year. The inventory levels are reduced by 30 percent.

% Dollars	Dollars	Class	T.O.	Inventory Levels	TMC-TO	Inventory Levels
70%	$10,500,000	A	6	$1,750,000	12	$875,000
25%	3,750,000	B	6	625,000	6	625,000
5%	750,000	C	6	125,000	3	250,000
100%	$15,000,000		6	$2,500,000	8.5	$1,750,000

FIGURE 3.14   Chart showing savings made in inventory levels by using the ABC categories of inventory controls.

This is balancing the inventory carrying costs with the ordering and acquisition costs by class of inventory to achieve lowest total material costs (TMC = acquisition costs + carrying costs).

The total result of applying the TMC-TO concept is to reduce the inventory and inventory carrying costs by 30 percent and also reduce the acquisition costs or ordering and controlling costs by 40 percent—working smarter, not harder.

SURPLUS AND OBSOLETE (CLASS D) INVENTORY

For inventory management and control purposes, there is another class of inventory. This is the surplus and obsolete inventory, sometimes called class D.

Surplus Inventory.   This is usable inventory, but extra or surplus stock that is on hand above the normal usage rate—usually for the next year. Because there is no forecast requirement for this inventory within the next year, keeping the surplus or slow moving items on hand merely increases total inventory carrying costs. Surplus items held for too long can eventually become obsolete.

Obsolete Inventory.   This consists of items that are no longer in demand. This may be because the items are perishable, have a shelf life, and have "expired" or because they are no longer stylish or have become technologically obsolete. They may no longer be used in products because of an engineering change.

Obsolescence is an acute problem for consumer products. By law, many products carry expiration dates. Such items include film, canned goods, drugs, and food and dairy products. Once these goods pass their expiration dates, they are unsalable, "obsolete," and must be written off and destroyed.

Stylish items and high-tech goods can become obsolete quickly because of changing styles and their short product life cycles. Examples of these items are clothing, cosmetics, electronics, and computers. Current sales of these types of products are not always accurate predictors of continued future demand.

Companies often write off as much as 10 percent or more of the value of their total inventories each year due to obsolescence. This comes right off the bottom-line profits. While some obsolescence may be unavoidable, many companies find they can reduce their write-offs to 5 percent or even less through better inventory control.

Many companies have accounting reserves for surplus and obsolete inventories in anticipation of writing off 25 percent per year for 4 years to provide for this total loss. The objective should be to reduce surplus and obsolete inventory by determining its causes and fixing the cause, whenever possible. Internal problems, such as inaccurate bills of material or engineering changes, and overproduction can be identified and corrected. Even unexpected changes in external demand should be identified and managed by prompt recognition and action. At a minimum, further acquisitions of materials for which there is no longer a forecast need should be stopped immediately.

Early warning systems that detect potential slow-moving and surplus items enable managers to take preventive action before items become surplus and obsolete. Management should carefully evaluate disposal alternatives for any items that do become surplus or obsolete to determine the most profitable recovery of the asset value.

Opportunity to Fix Causes.   An earnest effort should be made to identify obsolete and surplus inventory so that its causes can be analyzed and remedied. For independent-demand finished-goods items, differences between forecast and actual demand can cause surplus and obsolete inventory in finished goods. For dependent-demand items, differences between planned and actual usage can result in surplus or obsolescence in work in process or raw materials.

Analyzing significant variances of both types may reveal that some plans did not materialize, products changed, or the process was not managed properly, resulting in surpluses building up.

One of the major causes of surpluses and obsolescence is poor timing of engineering changes. Companies should time style, design, or package changes to make the best possible use of existing inventory. Whenever changes are planned, all of the affected functions should be consulted to determine optimal timing for the change. Change information and dates should be based on using up existing inventory and should be communicated accurately to all the relevant functions. Otherwise, surplus and obsolete inventory will result which will eventually have to be written off.

Another contributor to surplus and obsolete inventory is overbuying and overproduction. The benefits of placing or making large orders must be balanced against the risk of obsolescence. In this case, increasing order frequency and decreasing order quantity may lead to a decrease in total materials cost with less risk of surplus or obsolescence.

Solutions.   Unfortunately, Class D inventory is often identified after the fact. Surplus inventory becomes obsolete as expiration dates pass and demand drops. Companies can be in for an unpleasant surprise when they take an annual physical inventory and discover large quantities of unsalable and unsalvageable goods. The sizable write-offs that can result may be prevented by detecting surpluses earlier and timing changes to occur after existing inventories are depleted.

Once detected, surplus and obsolete items should be disposed of as profitably as possible. If an item can be used, as is, 90 percent or more of its former dollar value may be realized. For example, a part that is no longer needed because of an engineering change may still be salable to dealers as a pair or replacement part. If an item is unusable as is, reworking it for a different use or returning it to the vendor may net the company 75 percent of the item's former value. The next option is disposing of or selling the item at 25 to 50 percent of its former value. Offering a nearly obsolete item at a deep discount may make it attractive to customers or other manufacturers. The last resort is scrapping the

Surplus and Obsolete

Disposition	Value	
	Recovery	Loss
Use as is	100%	0
Rework Return to Vendor	75%	25%
Sell/Disposal	25-50%	50-75%
Scrap	10%	90%

FIGURE 3.15 Chart showing options and recovery values for disposing of surplus and obsolete inventory.

item. However, this is the least desirable alternative. Only 10 percent or less of the item's initial value may be recovered this way.

Early detection of surplus and obsolescence makes it possible to realize more of an item's original value upon disposal. If the problem is detected soon enough, the item can still be used, reworked, or returned. Timely action has been known to reduce write-offs from 10 down to 2 percent of an inventory's total value. See Fig. 3.15.

How can managers identify potential inventory problems while there is still time to react? One solution is to analyze the inventory against the projected or forecast usage to identify the items with no usage (potentially obsolete) and the items with inventory greater than 1 year's usage (potentially surplus). This analysis should be done at least once a year.

Remember, the objective is to eliminate or minimize the actions that cause surplus and obsolete inventory in the first place, and then dispose of the existing surplus and obsolete inventory for maximum value and minimum loss.

FUNCTIONS OF INVENTORY MANAGEMENT

The inventory management function can be in one of several places on a company's organization chart. Its exact location can depend on:

1. The size of the company
2. The financial condition of the company
3. The people involved

In a marketing-driven finished-goods-oriented environment, marketing may have the responsibility. In a financial-driven, cash-flow-poor environment, finance may have the responsibility. In a manufacturing environment, the inventory control function is generally placed under manufacturing management. It usually also interfaces with accounting since inventory represents a critical aspect of the company's assets and cash flow management.

Inventory management cuts across all areas of the company. With many different areas contributing to the management of inventory, inventory is really a total company responsibility.

The development of an organization plan begins with an understanding of the process to be performed. In the field of inventory control, the process consists of the following:

1. Determining *what* items are to be carried in inventory
2. Determining *how much* should be carried

3. Determining *when* to purchase or produce replenishments

4. Receiving, storing, and issuing inventory items as needed

5. Maintaining records of inventory quantities and values

6. Identifying and disposing of slow-moving, obsolete, or damaged inventories

7. Planning inventory investment levels

8. Organizing inventory planning functions

9. Controlling inventory flows and levels

10. Measuring inventory

The position responsible for performing each of these functions and the location of that position in the organization structure will vary from company to company. The basic responsibility is managing the inventory flows through the entire supply chain process. (*See* "Basic Questions" under "Summary" at the end of this chapter.)

IMPROVING INVENTORY MANAGEMENT

From an operating standpoint, one objective of inventory management is to provide the right materials, in the right quantities, at the right place, at the right time—just-in-time. A second objective is to minimize inventory investment while maintaining high levels of customer service. A third objective is to have sufficient inventory for efficient production operations. A fourth objective is to minimize losses resulting from inventory shrinkage, surplus, or obsolescence. A number of things can be done to improve the inventory management of a company.

Inventory Planning.   Effective inventory management relies on accurate inventory demand planning. The goal of inventory planning is to meet customer demands with maximum operating efficiency and minimum inventory investment.

The planning process begins at the top management level, with business, sales, and production plans for the next 12 months. Because these plans rely on long-term forecasts, they must be updated continuously (at least monthly) based on changes in customer demand and the competitive environment.

Top management's plans are translated into specific and detailed weekly operating plans by operations management. The weekly operating plans include the master production schedule, materials plan, and capacity plan. The master production schedule specifies the mix of what is to be produced. The materials and capacity plans detail the materials and resources required to support the production schedule and produce the product. The quality of the materials plan depends upon the accuracy of bills of material and inventory status information.

Execution of the production schedule is a daily, front-line management concern. The movement of raw materials, work-in-process, and production supplies through the production process must occur on an accurate, timely, and efficient basis in order to produce the product. The goal of improved inventory management can be achieved by measuring performance, identifying problem areas, and improving execution in the process on a daily basis. As the performance in the inventory planning process improves, the need for excessive inventory will be reduced. Remember, inventory is a cover-up for poor planning.

Performance Measurement.   Inventory management is a closed-loop process that consists of setting objectives and tolerance limits, developing action plans, allocating resources, assigning responsibilities, implementing plans, and finally, measuring performance in order to provide feedback for corrective action.

Today's computer-based inventory planning systems present companies with unprecedented capabilities for planning inventory and reporting conformance to plan on important

MANAGEMENT PROCESS

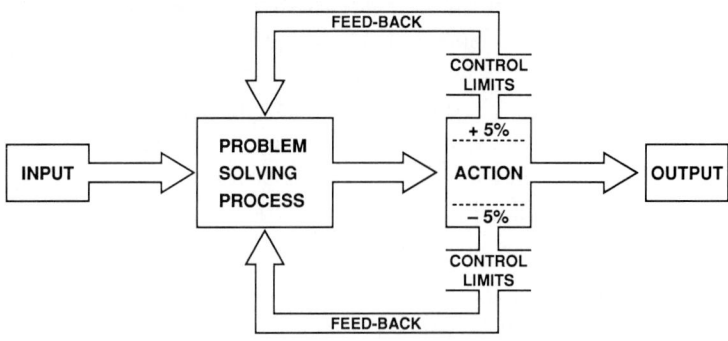

PROBLEM SOLVING PROCESS

1.) Define Problem	5.) Select Best Alternative
2.) Identify Facts	6.) Simulate or Pilot Alternative
3.) Determine Alternatives	7.) Implement Plan
4.) Evaluate Alternatives	8.) Feed-Back & Follow-Up

FIGURE 3.16

inventory performance measurements. See Fig. 3.16. Improving inventory management is a five-step process:

1. Establish measurable objectives.
2. Measure present performance.
3. Identify problem performance areas.
4. Develop an action plan with resources and responsibilities for solving the problem performance areas.
5. Measure performance on a regular basis and repeat steps 3 to 5.

Performance objectives should include a quantifiable performance target and the date when the target should be achieved. For example, a company may specify that it plans to increase inventory accuracy to 75 percent by June 1 and 95 percent by December 1 or reduce inventory levels by 25 percent by the end of the year.

There are a number of important performance measurement techniques for inventory. See Fig. 3.17.

Inventory Accuracy. This is a performance measurement indicating the accuracy of inventory balances on hand. The balances shown on the inventory records are compared with actual quantities on hand, determined at least once per year by physically counting the items. The measurement is expressed as the percentage of correct record balances. Correct on-hand balances are those that match, within preestablished tolerance ranges, the actual number of items on hand.

$$\text{Inventory accuracy} = \frac{\text{number of records correct} \times 100}{\text{number of inventory items counted}}$$

Most companies target for at least 95 percent inventory accuracy.

Inventory Investment. This compares actual inventory with planned inventory investment. The dollar value of each category of inventory and total inventory is tracked on a monthly basis, usually against a model, and a variance is calculated from the planned inventory investment.

$$\text{Performance to plan} = \frac{\text{actual inventory}}{\text{planned inventory}}$$

Inventory Turnover.   This is the ratio of annualized cost of goods sold to the cost of average inventory on hand for a given time period. Inventory turnover can also be calculated for each category of inventory or for the total inventory.

PERFORMANCE MEASUREMENT

FUNCTION	OBJECTIVE	MEASUREMENT
INVENTORY CONTROL	INVENTORY ACCURACY	PHYSICAL COUNT VS. INVENTORY RECORD
RESPONSIBILITY		
Materials	INVENTORY MANAGEMENT	INVENTORY INVESTMENT = INVENTORY PLAN

FIGURE 3.17   Chart showing functions and objectives for measuring performance of inventory management.

$$\text{Turnover} = \frac{\text{cost of goods sold}}{\text{average inventory on hand}}$$

Although the appropriate number of inventory turns varies by company and industry, a company's goal should be to continually increase the number of inventory turns while maintaining high-quality customer service. By increasing inventory turns, a company decreases its inventory investment and its total inventory investment costs. Inventory turnover can also be benchmarked against others in the industry to determine the performance in the industry.

Cycle Time.   This is another measure of the quality of inventory management. Cycle time can be measured in weeks or even days of stock. One way to measure cycle time is to simply measure the time a product is in the process. Another method of determining the cycle time is derived by dividing the number of days in the year by the number of inventory turns. For example, there are 360 days in a year so the company with 4 inventory turns has a cycle time of 90 days—3 months' worth of supply, or 90 days of stock. Cycle times for individual categories of inventory can also be calculated and reviewed.

Most companies strive to reduce cycle times, which is accomplished by improving the inventory flow and eliminating the wasted time in the process. Remember, inventory is a function of cycle time. If you want to reduce inventory investment, work on cycle times.

Inventory Modeling.   Inventory models are powerful tools for improving inventory management. Graphic representations of current inventory status are also useful for zeroing in on problem areas, such as raw-material, work-in-process, or finished-goods buildups, and identifying problems that cause unfavorable variances from performance objectives. Managers can use these models to determine and reevaluate optimal inventory levels and inventory policies to meet changing demands in the marketplace.

An inventory flow model shows the company's dollar investment in each type of inventory as a function of cycle time. The purpose of the flow model is to show the inventory investment by category of inventory both in total inventory dollars and in days of stock. The monthly inventory reports should show the actual inventory dollars by category of inventory and in total inventory against the plan. Based on this information, companies can make important inventory management decisions.

Another type of graphic inventory model is a cycle time bar chart showing the sequence and duration of the production process, materials movement, and queue time in the production of the product. This type of chart is useful for identifying steps that should be shortened or eliminated because they do not add value to the product. The goal of this process is to reduce the cycle time of the product and thus reduce the inventory.

A bar or line chart is another type of graphic inventory model which compares actual with planned performance. Measurements such as inventory investment, inventory turnover, and inventory accuracy can be charted and analyzed on a month-by-month basis to evaluate performance against plan and identify trends.

Performance measurements serve both diagnostic and motivational purposes. Once a company identifies areas of inventory that require improvement, the next step is to develop action plans for meeting performance improvement objectives. To provide necessary feedback to the employees and managers responsible for carrying out the plans, charts showing actual versus targeted performance should be posted on boards and in stockrooms. Managers should also receive detailed monthly reports and an analysis of problem areas for feedback to improve performance.

Improvements in inventory management can provide significant contributions to a company's bottom line. As long as adequate customer service levels are maintained, decreases in total inventory investment will immediately yield a higher return on a company's working capital investment.

Companies are increasingly recognizing the "opportunity costs" of carrying inventory. Dollars invested in inventory could provide a better return, in the long run, invested in new products or new processes to make the company more competitive. Managing inventories is an important part of managing a company. Remember, less is better. (*See* "Symptoms of Poor Inventory Management" under "Summary" at the end of this chapter.)

SUMMARY

Basic Questions.   Basic questions that need to be asked whenever dealing with inventory:

What should we stock?

How should we stock it?

What's our inventory strategy?

What are our inventory goals?

What is the total inventory?

What should it be?

What are the various categories of inventory?

What should they be?

What is the inventory level of each category of inventory?

What should it be?

What is our percentage of inventory accuracy?

How can we improve it?

What is the turnover of each category?

What should it be?

How many days of stock should we have on hand?

When should it be on hand?

Where are we overstocking?

Why? What can be done about it?

What can we do to improve our inventory performance?

Symptoms of Poor Inventory Management

1. Customer back orders
2. Missed ship dates
3. Continuously growing inventories, while order input remains constant or decreases
4. Lack of adequate storage space
5. Uneven production with frequent layoffs and rehirings
6. Frequent changes in production runs to meet changing sales requirements
7. Excessive machine downtime because of changeover and material shortages

8. Widely varying rates of inventory loss or turnover among branch warehouses and production facilities, or widely varying rates of turnover among major inventory items

9. Consistently large inventory write-downs because of distress sales, disposal of obsolete or slow-moving items, and so forth

10. Consistently large write-downs when physical inventories are taken

BIBLIOGRAPHY

APICS Dictionary, American Production and Inventory Control Society, Inc., Falls Church, Va., 1987.

Buker, David W., *Inventory Management,* Central Ohio APICS, 1976.

Buker, David W., and Thomas F. Ribar, "10 Steps to Inventory Record Accuracy," *P&IM Review with APICS News,* April 1989, p. 38.

Class A MRP II Performance Measurement, David W. Buker, Inc. and Associates, Antioch, Ill., 1988.

MRP II Newsletter, David W. Buker, Inc. and Associates, Antioch, Ill., 1981.

Martin, Andre J., *DRP Distribution Resource Planning,* Prentice-Hall, Englewood Cliffs, N.J., 1983.

Orlicky, Joseph, *Material Requirements Planning: The New Way of Life in Production and Inventory Management,* McGraw-Hill, New York, 1975.

Plossl, G. W., and O. W. Wright, *Production and Inventory Control,* Prentice-Hall, Englewood Cliffs, N.J., 1967.

Stickler, Michael J., "Database Accuracy: Getting It Right," *Systems/3X & AS World,* April 1988, p. 102.

Vollmann, Thomas E., William L. Berry, and D. Clay Whybark, *Manufacturing Planning and Control Systems,* Dow Jones-Irwin, Homewood, Ill., 1984.

Wright, O. W., *Production and Inventory Management in the Computer Age,* CBI Publishing Co., Inc., Boston, Mass., 1974.

CHAPTER 4
MATERIAL REQUIREMENTS PLANNING AND JUST-IN-TIME SYSTEMS

George Foo
Manufacturing and Engineering Director
American Telephone and Telegraph Co.
Little Rock, Arkansas

Larry Kinney
Operations Director
American Telephone and Telegraph Co.
North Andover, Massachusetts

Prior to the development of modern material management techniques, the most widely accepted technique for managing inventory used various reorder point models. These approaches were heavily influenced by techniques for calculation of economic order quantities. Traditional thinking taught that lower inventory investment could be obtained only at the expense of customer service levels.

Reorder point techniques required forecasts of demand over replenishment lead time. Replenishment lead time is the time it takes from recognition of need to receipt of material. Since both supply and demand have uncertainties, safety stocks were frequently used to buffer operations from risk. When it was projected that a component inventory would reach the safety stock level or reorder point at lead time, a replenishment order would be issued. This traditional approach suffered many frailties. Reorder point model forecasts generally assume uniform demand over time. This is a poor assumption, since very few manufacturers experience such smooth demand patterns. During this early period, with lead times generally averaging over 13 weeks and many extending to 20 weeks and more, each member of a supply chain used interval as an insulator from uncertainty. The net result was that backlogs of orders were gladly accepted since they provided a reassurance about short-term business volume. Furthermore, expediting was the rule rather than the exception.

Additionally, in an environment with many components and subassemblies, the demand for "number crunching" was astronomical. Without powerful computers, calculations were performed only at scheduled intervals. This, in turn, made the system unresponsive to the need for change. As with any system involving extensive manual calculation, errors were frequent.

As a result of errors, infrequent and inadequate planning, and long intervals, stockouts (shortages) were numerous and inventories were high. An "inventory management" cycle was all too common. Unacceptable stockout levels drove more inventory. Too much inventory then drove top management to edict reductions in stocks, resulting in more stockouts. Customer service deteriorated and overtime costs grew. Pressure to "eliminate" stockouts drove overordering behavior, with increases in inventory—and so forth.

Burgeoning inventories were only an irritant in the early sixties, since interest rates were running under 4.5 percent. They became a crisis as interest rates grew dramatically, increasing to double-digit percentages in the seventies. Also, inventory consumed cash that could otherwise have been used for purchases of capital equipment sorely needed to keep up with accelerating technology change. Better inventory management techniques became imperative. Material requirements planning (MRP) was the beginning of a new set of tools and techniques, facilitated by faster and less expensive computer resources.

MATERIAL REQUIREMENTS PLANNING (MRP)

MRP Systems. The fundamental difference between order point techniques and MRP is time phasing. In fact, an MRP system is a time-phased reorder point system for independent demands. Independent demand is unrelated to demand for other items. Customer orders and service part orders are examples of independent demands. Dependent demands are calculated by an MRP system as a result of its view of independent demands. Time phasing allows actual and forecast independent demands to drive demands for components at the right time, rather than using the continuous, smooth demand assumption used in order point models.

All MRP systems share a common objective,[1] which is to determine (gross and net) requirements, that is, discrete period demands for each item of inventory, so as to be able to generate information required for accurate inventory order action. This action pertains to procurement (purchase orders) and to production (shop orders). It is either new action or a revision of previous action. New action consists in the placing (release) of an order for a quantity of an item, due on some future date. The essential data elements accompanying this action are:

- Item identity (part number)
- Order quantity
- Date of order release
- Date of order completion

At a higher level, the MRP requires certain inputs to perform its calculations:

1. Master production schedule (MPS)
2. Bill of material (BOM)
3. Inventory status
4. Replenishment lead times
5. Manufacturing lead times

The last three inputs are frequently from a file called an item master file (IMF) that contains information for each part in the system. The MPS, BOM, and IMF are often considered part of an MRP system; however, in the strictest sense, they are simply inputs to the MRP process. For practical purposes, these functions as well as storeroom and procurement functions must be considered as integral subsystems for a working MRP system. Figure 4.1 is a schematic of a basic MRP system.

MRP examines time-phased demands from the master schedule and breaks these de-

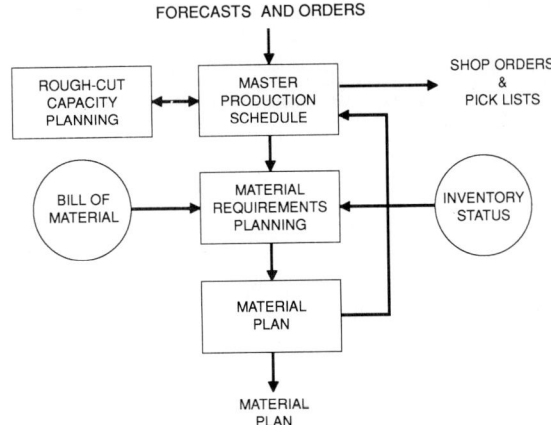

FIGURE 4.1 Basic MRP system.

mands into their component parts, using bills of material for this "explosion." These requirements are then "netted" by available inventories. The result is time-phased demands for components that must be satisfied by placing orders on vendors for the right quantities at the right time—this is the function called "material planning."

As a result of the MRP calculation, typical actions include

- Increase in order quantity
- Decrease in order quantity
- Order cancellation
- Advancement of order due date
- Deferment of order due date
- Order suspension
- Order placement

Major Modules of an MRP System.   The following sections will briefly describe some of the key modules of a modern MRP system. For further reading, there is a wealth of literature.[2-4]

Master Production Schedule (MPS).   The MPS is the driver of an MRP system. It receives all demands for product, and translates these demands into planned orders that are conveyed to the MRP logic section of the system. The format, features, and functions of the master production schedule vary widely among software systems—the MPS discussed below is an example.

The MPS is a time-phased schedule of end items and subassemblies. Each end item and subassembly in the master schedule has a unique part number and is called a planning unit. The demand for end items is a combination of independent demand orders (customer orders, distribution replenishment orders, and service part orders) and forecast. As demand orders are received and scheduled, they "consume" the forecast. Dependent demand is demand derived from higher-order demands (end items or higher-level subassemblies). Demand in the MPS for subassemblies is a combination of dependent demands calculated by MRP logic and independent demand (for example, service part orders).

Figure 4.2 shows a sample MPS grid that combines several important data elements on one "screen." As was pointed out earlier, there are as many variations of MPS as there are systems. Shown are 10 weekly "buckets"—current week 1 through week 10. The part shown is a child subassembly to a parent end item. It has independent demand from field technicians for repairs of the parent end item. This example is just a section of an MPS, which can have as many as 104 buckets which show the forecast over 2 years. Forecasting

PART NO: XYZ		MAJOR SUBASSEMBLY							
MLT: 2	CLT: 16		WIP: 403			BIN: 185			S/S: 25

	CURR	WK 2	WK 3	WK 4	WK 5	WK 6	WK 7	WK 8	WK 9	WK 10
IND FCT:	0	0	0	100	100	100	100	100	100	100
IND DMD:	89	103	61	65	22	3	0	0	0	0
DEP DMD:	43	269	304	280	300	300	300	300	300	300
PRJ AVL:	81	59	44	0	0	0	0	0	0	0
AVL PRM:	0	0	0	35	78	97	100	100	100	100
SCH REC:	53	350	350							
PLN ORD:	0	336	400	400	400	400	400	400	400	400

FIGURE 4.2   Sample master production schedule.

that far out is required to pass forecast volume changes to internal clients, such as engineering or finance, and to vendors.

Some terms need to be defined:

MRP Acronyms and Terminology

Acronym or abbreviation	Definition
MLT	Manufacturing lead time—the standard shop time for manufacture of the item
CLT	Cumulative lead time—MLT plus the lead time for procurement of the longest lead time component. Sometimes, planning and storeroom time is added. Given accurate lead time information, any increase to the MPS for this item inside this time frame will result in expediting, and a decrease will cause excess inventory or deexpediting
WIP	Work-in-process inventory—quantity of the item in process in the shop, regardless of the state of completion
BIN	Quantity of finished items in a storeroom
S/S	Safety stock—the quantity of the item required to buffer the plan from supply and demand uncertainty
IND FCT	Independent demand forecast
IND DEM	Actual independent demand orders
DEP DEM	Dependent demands calculated by the MRP system
PRJ AVL	Projected availability in the bin after supplies are received and demands are satisfied. Negatives indicate the plan is not doable. The master scheduler should consider replanning the material plan
AVL PRM	Available to promise—the quantity available to promise for independent demand orders each week. Whoever is responsible for assigning schedules to independent demand orders consistent with availability in the MPS must have this information to provide realistic schedules
SCH REC	Scheduled receipts. When a shop order is created for manufacture of the item, it has a schedule for delivery to bin after completion. This element shows the schedules for delivery from WIP to bin, based on the current open shop order schedules
PLN ORD	Planned shop orders. This shows the dates and quantities of the item, offset by MLT, required to fill demands not satisfied by BIN, WIP, and SCH REC. Some systems allow the planner to freeze (firm) planned orders, so that manual intervention is required to modify them. It is this line that is often referred to as the master schedule for the item, because these planned orders are furnished to the MRP logic section of the system for explosion into requirements for subassemblies and component parts

Using Fig. 4.2 as the basis, the calculation of the current week (CURR) planned order is the difference between supplies and demands through week 3 (current week plus 2 weeks manufacturing lead time), if demand exceeds supply.

$$\text{Supply} = \text{BIN (185)} + \text{SCH REC (753)} = 938$$

$$\text{Demand} = \text{S/S (25)} + \text{IND DMD (253)} + \text{DEP DMD (616)} = 894$$

Since there are 44 units available for WK 3, there are no planned orders for the current week.

Future week planned order calculation is illustrated for WK 4. Projected available supply going into week 4 is 44 units. In order to cover demands of 100 units IND FCT and 280 units DEP DMD, an additional supply of 336 units must be available WK 4. Therefore, the system plans to start (planned order) 336 units in WK 2.

Obviously, human analysis of the data for many active items over 2 years is quite a chore. Therefore, here, as with other areas of an MRP system, exception reporting is a virtual requirement. The system needs to identify problem areas (for example, negative PRJ AVL) to the planner for detailed analysis and corrective action.

Bill of Material.   The BOM (also known as the product definition master or product structure) defines the relationship between an end item and its component parts and subassemblies, or between a subassembly and its component parts and subassemblies. End items and subassemblies are called "parents," and the components and subassemblies used in their manufacture are called "children." A subassembly is a child to its parent end item or parent subassembly and a parent to its child subassemblies and components. Each hierarchical parent-child relationship is a "level" in the bill. Figure 4.3 is an example of a three-level bill of material.

Information contained in the BOM file must include the parent part number, child part numbers, the quantity of each part required to make the parent, and the date each child is

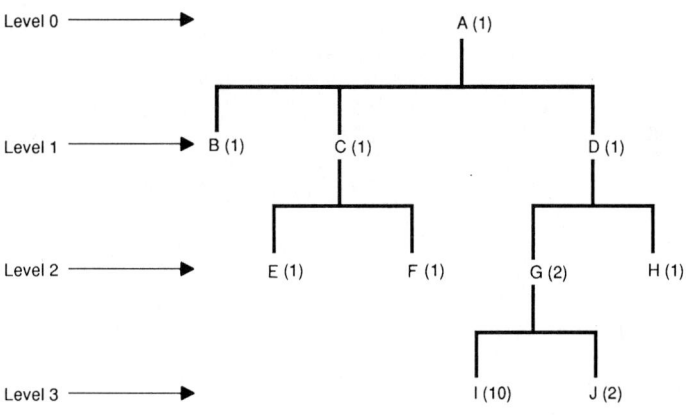

A = END ITEM

C, D, G = SUBASSEMBLIES

B, E, F, H, I, J = COMPONENT PARTS

EXAMPLE: AN ORDER FOR 10 PART A REQUIRES
10 SUBASSEMBLY D; 20 SUBASSEMBLY G;
AND 200 COMPONENTS I

( ) = QUANTITY

FIGURE 4.3 Sample three-level bill of material.

to become effective or to be removed from use in the bill (effectivity date control for scheduled engineering changes). Other elements may be present as well, such as dropout and yield percentages, shop floor delivery destination, and engineering revision level.

BOM Structuring for the MPS. The decision on what items should be included in the MPS is critical. According to Ref. 3 that decision should be based on the following criteria:

1. Have the minimum possible number of items in the MPS to improve management review and control.

2. Cover the maximum possible number of components in the MRP program driven by the MPS.

3. Generate the maximum possible amount of information on loads on manufacturing facilities.

Products generally fall into one of three categories:

1. *Make-to-Stock.* These are generally high-volume products with few, if any, configurations. Make-to-stock items are manufactured without end-customer orders and placed in storerooms until orders are received. Examples include telephone sets and personal computers. For these types of products, master scheduling at the end-item level is ideal. Forecasts are developed for the end items and, as orders are received, the forecasts are consumed by these orders. The concept of "consuming" a forecast means that the forecast, which is an estimate, is either partially, fully, or overrealized by actual orders.

2. *Assemble-to-Order.* These are products which have almost a limitless number of end-item configurations but are made up of a fixed set of components or subassemblies (for example, engine sizes). An example is an automobile.

Master scheduling at the end-item level, for assemble-to-order products, is unacceptable since the almost limitless number of end items are impossible to forecast accurately and master schedule with any hope of validity. Rather, the recommendation, for products of the complexity of an automobile or a mainframe computer, is to establish the MPS at the option or module level. An option or module is an intermediate product, subassembly, or group of parts which makes up an end item (for example, air-conditioning units for an automobile can be viewed as an option or module which is master production scheduled). Since the options or modules are simply manufacturable subassemblies, all that is required to support an MPS strategy for assemble-to-order products is to treat the subassemblies as end items. Thus the options become the highest level of the BOM for purposes of the MPS. This has the effect of designating them level 0 instead of level 1 in the bill of material structure.

For the assemble-to-order products, forecasts are developed for "typical" end items. Typicals are *planning* bills of material (PBOM) for an end item where component (in this case, buildable option) quantities are based on average usage over some time period.

3. *Engineer-to-Order.* These are products which are custom designed and manufactured per the customer's specifications. In general, engineer-to-order products are not forecast and are master scheduled at the end-item level for the purpose of material planning.

The differences between the master production scheduling strategy for a make-to-stock product and an assemble-to-order product are graphically illustrated in Fig. 4.4.

MRP Logic. Figure 4.5 illustrates the MRP logic process for the first two levels of the parent-child BOM stream A to D to G and H shown in Fig. 4.3. MRP logic uses the MPS as its time-phased source of demands—represented by planned orders in the MPS (end item A in Fig. 4.5). These demands are "exploded" through the BOM to generate gross requirements for each child of the parent (the gross requirements of subassembly D in Fig. 4.5 are an example of explosion of parent A into child D). Gross requirements are the time-phased quantities of children required to make the time-phased planned orders for a parent, without consideration for available inventories of the children.

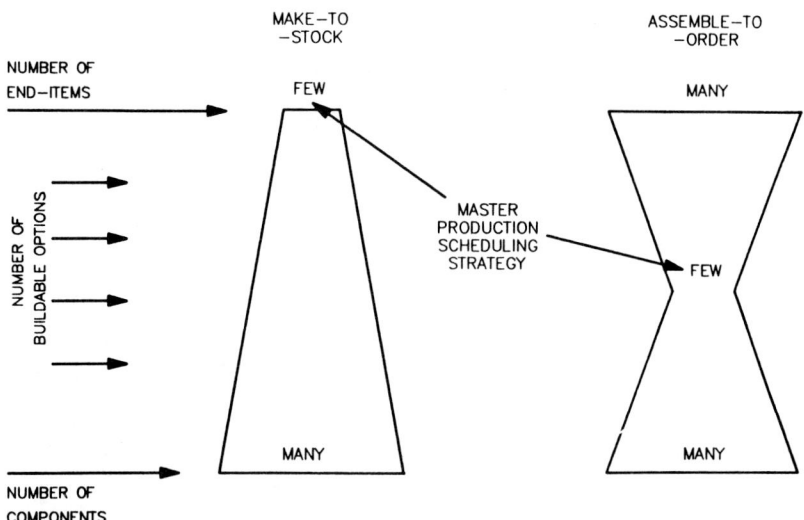

FIGURE 4.4   Master production scheduling strategy for various types of products.

Current on-hand inventory and scheduled receipts of the children are then subtracted from gross requirements to provide net requirements (in Fig. 4.5, there are no inventories of subassembly D on hand nor are there any expected scheduled receipts). At this point, if the children are component parts, they will receive no further attention from the MRP logic. However, children that are subassemblies (such as subassembly D) go through another iteration to generate gross requirements (subassembly G and component H in Figs. 4.3 and 4.5). This process continues until the bottom of each BOM is reached (for the BOM structure shown in Fig. 4.3, this process stops for component H but continues on for one additional iteration for subassembly G). Dependent demands for subassemblies are displayed on the MPS.

After the completion of the MRP explosion process, the component parts are aggregated from their various bill explosions, such that the material planner sees one stream of demands for each component. This process can be very time-consuming and demand substantial computer resources. Minimizing the number of levels in the BOM can save time and money. Furthermore, level proliferation has more serious consequences than just computation:

1. Each level requires a lead time to be assigned. Planned lead times are rarely accurate, and tend to be inflated. This is particularly true in systems that have weekly buckets—for example, if a subassembly has an actual manufacturing lead time of 2 days, you are forced to set the MLT to either zero or 1 week. In multilevel bills, inflated lead times are "stacked." As a result, the manufacturing lead time for the end item is dramatically overstated.

On the other hand, daily bucketed systems are not a panacea. Rather, they create the illusion of precision and cause unnecessary disruptions.[5]

2. Standard MRP logic requires shop orders (scheduled receipts from WIP) and storeroom receivals (to relieve open shop orders) for each level in the bill. The proliferation of shop orders adds overhead in their maintenance—they must be manually adjusted to reflect changes in demand. Storerooms are very expensive to operate—terminals, transactions, storekeepers, security, floor space, and storage facilities are required.

Material Planning.   Material planning is a term used to describe the activity of receiving gross and net requirement information for each component part from the MRP explo-

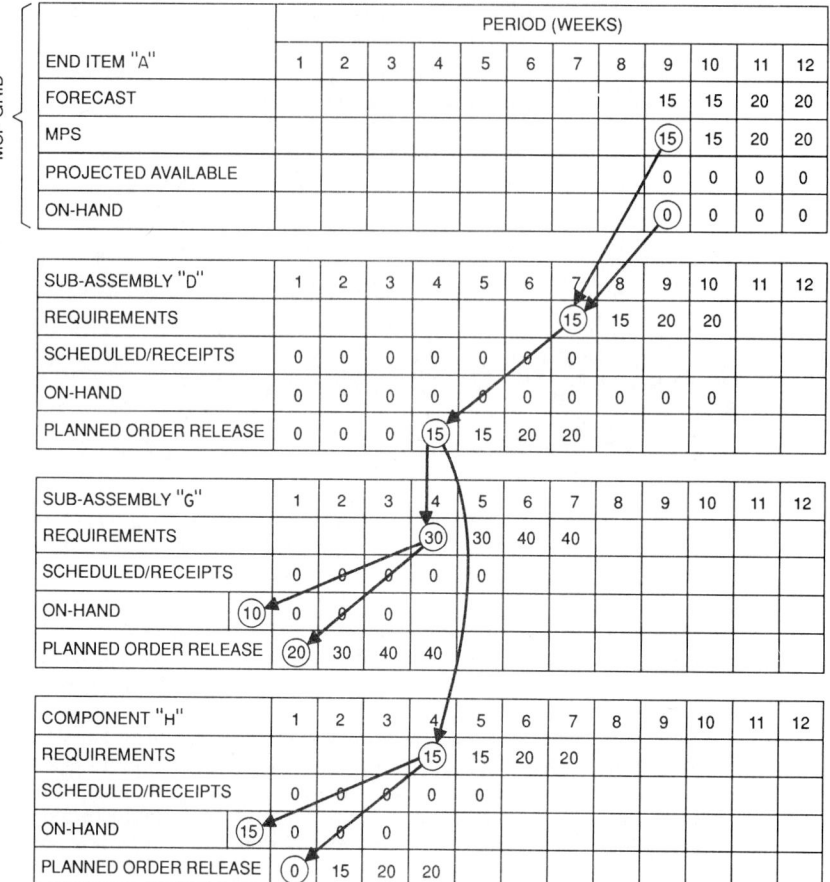

FIGURE 4.5   Materials requirements planning logic.

sion, comparing these time-phased dependent demand data with scheduled receipts from vendors, and deciding on what action to take, considering policies such as ABC classification.

ABC classification is the differentiation of components into three inventory classes, based on criteria such as unit cost, dollar volume, or lead time. A items receive more attention than B items. C items are sometimes on "automatic pilot," while the other classes receive much manual attention. A typical breakdown of items would have 20 percent in A, 40 percent in B, and the remainder in C. Other considerations that affect material planning include minimum and multiple order quantities, vendor reliability, and safety stock. The action will generally be to place orders (through purchasing, if a separate function), cancel orders, push orders, or pull orders. Placement of short interval orders and pulling orders up are known as expediting. Cancellation or pushing orders are deexpediting activities.

Material planning is similar to master scheduling in that it attempts to balance supplies and demands. Differences are that material planning generally deals with dependent demand, deals with far more items than the MPS, and interfaces with outside sources of supply.

Material planners must respond to changes in demand from the MPS in order to protect customer service (increased demands) and at the same time protect investment (decreased demands). They are constantly expediting or deexpediting as demands change. They are held responsible to resolve stockouts and to prevent them in the future. They are also held responsible for component overstocks.

Shop Order Release.   Planned orders in the master schedule represent future plans for release of orders to the shop as well as time-phased demands to be exploded by MRP to trigger component procurement. When the planned order is released by the master scheduler, it becomes a shop order (also called manufacturing order or work order), with a scheduled receipt date. In the MPS, it moves from the PLN ORD line to the SCH REC line in the above MPS grid example.

In addition, this release triggers generation of a "pick list" which contains a one-level explosion of the planning unit. The pick list contains information for the stock selectors in the storeroom such as part number, description, quantity, schedule for delivery to shop, delivery destination, and storeroom bin location.

Many manufacturers erroneously do not view their storerooms as work centers, even though all components (in a traditional MRP installation) are received into them and issued from them. The storeroom is a critical link between the material plan and its execution—the most advanced manufacturing process can be stalled by a storeroom that does not deliver the right parts to the right place at the right time.

Rough Cut Capacity Planning.   After a trial master schedule is created, a capacity check is made to detect possible bottleneck operations. The check requires a "load profile" for each end item in the MPS, which describes the resources required to make each item. The aggregate resources demanded by the MPS for each work center can be examined for smoothness and capacity bottlenecks.

Obviously, the accuracy of a capacity check depends on the accuracy of the information in the load profile for the MPS item—this information normally is developed from routing file data (discussed below with capacity requirements planning).

The master scheduler is expected to rearrange demand orders and/or forecasts to correct bottlenecks or to smooth work center loadings. Obtaining an optimal set of load profiles is an iterative process; sometimes sophisticated techniques are used to help.

Rescheduling demand orders in this process poses interesting problems if commitments have already been made to customers. If demand order scheduling is done on the basis of availability to promise, there should be no need to perform a capacity check, given that the production plan or forecast does not exceed capacity.

MRP II Systems.   As MRP logic replaced earlier techniques to manage manufacturing inventories, there was a recognition that MRP as described above was not enough. MRP assumed infinite capacity, both in the manufacturing facility and with vendor supply. In the real environment of finite capacities, additional tools were required to manage execution of the plan.

There was little or no provision for "closing the loops"—this is a term that means there is a process for feedback of "doability" of the plan provided by the higher-level functions. For example, if purchasing cannot provide a component requested by the MRP system, those data are provided to the master scheduler, who determines what end products are affected. As a result, customer orders, the production plan, and the business plan are changed to reflect realistic plans—this is called replanning.

In the early eighties, the term MRP II (manufacturing resource planning) was used to encompass MRP and additional techniques for correcting the shortcomings of MRP discussed in the previous section. Among these techniques are business and production planning, capacity requirements planning (CRP), and shop floor control (SFC). However, with the benefits of an MRP II system come the additional cost of establishing and maintaining the data required for CRP and SFC. Furthermore, CRP and SFC can drive excessive replanning, which is very disruptive to customers, shop operations, and vendors. Figure 4.6 illustrates the additional modules of a basic MRP II system when compared with a basic MRP system.

Business and Production Planning.   Every manufacturer must balance demand for its products against capacity according to its strategic objectives. Capacity consists of many factors, long-term to short-term. Long-term capacity may be new plants, plant expan-

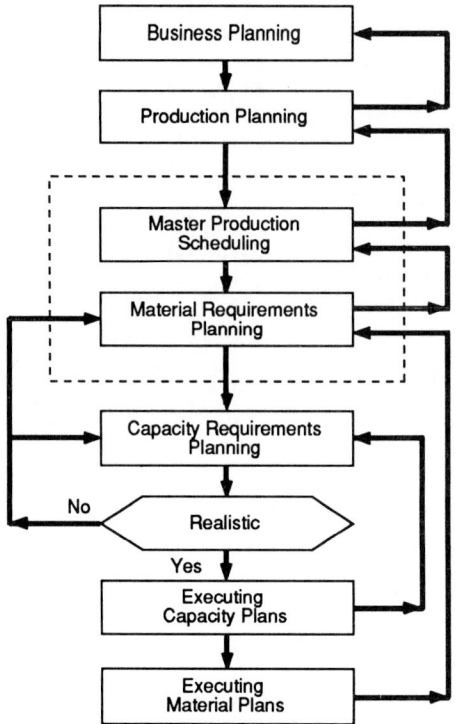

FIGURE 4.6   Basic MRP II system.

sions, or plant closings, which generally extend over 2 or more years. Medium-term capacity includes capital equipment acquisition over a 6-month to 2-year span. Finally, short-term capacity includes direct labor and material, for up to a 1-year period.

Business planning is responsible for communication of medium- to long-term capacity demands and should cover at least the next 5 years. Most business plans are updated annually, but significant changes may require interim updates for such things as technological breakthroughs, or disasters. The responsibility for preparation of the business plan varies, but marketing and top management are generally key players.

Aside from manufacturing operations, engineering, finance, human resources, and many other organizations must be aware of top management's future plans—the business plan must report these plans at product family levels, with at least quarterly granularity in the first 2 years. Vendors are critical partners in accomplishing capacity change, particularly expansion—MRP is the primary vehicle for communication of business volume plans to a manufacturer's vendor community.

Production planning provides more detail than business planning in terms of product groupings and time. This process should take place monthly (no less often than quarterly) to refresh the plan for the next 2 years, in monthly "buckets" the first year and quarterly the second year. The participants of production planning vary from enterprise to enterprise but frequently involve decision makers from manufacturing, marketing and sales, distribution, and finance.

The participants must be "decision makers" in that they must be empowered and capable of making tough choices—there is nothing less effective than a planning meeting that does not achieve a plan. A critical part of the process is understanding the reasons for variance of current actual demand from the previous forecast. This understanding, coupled with new information brought by the participants, will allow the best new production plan.

It is at this stage that some companies perform what is known as a rough-cut capacity check (also known as a resource requirements plan). The level of detail involved in this capacity check varies dramatically, depending on the relative magnitude of volume increase or mix change, the utilization levels of existing resources, and the intervals for capacity increases.

If it is determined that capacity should be checked, the chore is frequently handed to engineering for facility capacity, to operations for labor capacity, and to materials management for component capacity. Completion of the analysis may take several days or even weeks, in extreme cases. If there is insufficient capacity to support the proposed plan, it must be reworked in-line with capacity constraints.

It is important to differentiate the product plan described above and actual orders (for example, customer, service, distribution stocks) which are the "real" production plan within cumulative lead time. The production planning process recognizes "real" demand in recasting the plan, but the focus is on the forecast of future demands; therefore, the output of the production planning process is frequently called the "forecast."

Business planning and production planning may use some software tools such as statistical forecasting but are very judgmental in nature because of the myriad of factors influencing demand, including pending or new legislation, the state of the economy, new products, competition, sales force incentives, and pricing strategy.

Capacity Requirements Planning.   After the MRP process, each subassembly has its dependent demands refreshed, and the doability of the material plan, from a capacity standpoint, can be examined in detail. This is known as capacity requirements planning (CRP). The databases that CRP relies on are the routings file and the work center file—sometimes they are collectively referred to as routings.

The routings file includes the following data for each master scheduled item:

1. Manufacturing lead time based on a typical lot size
2. Operations to be performed
3. Planned move, queue, setup, and run times
4. The above information for alternate routings

The work center file includes the following data for each work center:

1. Number of shifts
2. Machine hours per shift
3. Labor hours per shift
4. Machine and labor efficiency
5. Machine and labor utilization
6. Effective daily capacity

The data elements above vary by company and system and are intended to illustrate the massive amount of data required to support CRP (as well as rough-cut capacity planning). This same set of data supports the MRP II execution system, shop floor control. Collection and maintenance of these data is a major undertaking. You will see later an alternative way of managing capacity that does not require this level of data.

When capacity bottlenecks are identified and cannot be overcome in time, the material plan must be replanned. This entails rescheduling either shop orders or independent demand orders or both. Once again, the impact on customer order schedules is of great concern, but the effects of confusion in the shop can be significant.

Shop Floor Control.   Shop floor control (SFC), also known as production activity control, is a system used by MRP II systems to orchestrate production activities to complete shop orders on time. Based on backscheduling techniques and lead time information, shop orders are released by the master production schedule, and represent quantities of subassemblies or end items, with specific schedules for completion. SFC communicates schedules and priorities to work centers via a dispatching process. Priorities are based on various algorithms. An example of such an algorithm is the *critical ratio:*

$$\text{Critical ratio} = (\text{LTC} - \text{TRDD})/\text{LTC} \qquad (1)$$

where LTC = lead time required to complete shop order
  TRDD = time remaining to complete shop order per the due date

The greater the critical ratio, the higher the priority of the shop order.

Dispatching requires the system to know what work centers operate on each order, what productive resources are consumed, and times involved with each operation. These data are contained in the routing file. Dispatching involves management of the queue of shop orders for each work center, using various techniques such as critical ratio and queue ratio. More details are discussed in Ref. 2. SFC, however, ignores two aspects of the real world of manufacturing:

1. The next operation often has limited capacity to store work in process.
2. The lead times or intervals used for backscheduling and establishing priorities are not static but are dynamic and depend on many factors, notably the loading on the shop. Actual timings such as move time, queue time, setup time, and run time in production almost always vary from those established in the routings database, regardless of attempts to make them accurate.

In addition, there are other factors which make the implementation of SFC systems cumbersome:

1. The cost of maintaining routing file data—or the alternative of inaccuracy
2. Lack of synchronization of customer order schedules and shop order schedules

As a result of the above limitations, SFC systems have seldom lived up to their promise. Too often judgmental manual intervention in scheduling shop orders is required (the position is known as "dispatcher"). Diligence must be observed when rescheduling shop orders, with respect to impact on customer order schedules.

Plossl (Ref. 3) has analyzed various SFC implementations and has defined the following prerequisites:

1. Simplify processes
2. Shorten intervals or lead times
3. Reduce setup times and lot sizes
4. Simplify bills of material
5. Level load the shop

These strategies are common with implementation of just-in-time (JIT) systems, which will be discussed later.

MRP and MRP II Performance Measurements. MRP II practitioners advocate performance measurements as a critical element in successful implementation. Metrics can serve as both goal setting and progress evaluation tools. One popular approach has been to establish classes of performance—A, B, C, and D—with class A as the pinnacle of success.

The following set of metrics, with a suggested class A level of performance, are characterized as the "basic pieces of data that must be accurate if the formal system is to work."[6]

1. Inventory (95 percent accuracy)
2. Bills of material (98 to 99 percent accuracy)
3. Routings (95 percent accuracy)
4. Master schedule (95 percent by item produced during the month)
5. Shipping dollars (100 percent shipped within the month)
6. Delivery performance (95 percent delivery against the original promise date each week)
7. Output by key work center (± 5 percent each month)
8. Shop delivery to schedule (95 percent each week)
9. Vendor delivery to schedule (95 percent each week)
10. Engineering delivery to schedule (95 percent each week)
11. Forecast accuracy (depends on product; 90 percent 60 days in advance is typical)

Few topics have as much controversy as performance measurements, which to adopt, their definition, how to measure, and what to do with results. One particularly dangerous use of measurements is comparison of one business with another. Many factors must be taken into account before a comparison has any meaning.

JUST-IN-TIME (JIT) SYSTEMS

Just-in-time (JIT) manufacturing is frequently described as a process for achieving continuous improvement through the systematic elimination of waste and variability. By definition, JIT programs encompass not only production control techniques, such as kanban systems (discussed later), but also total quality control programs, employee training and involvement initiatives, facility and line design, and the like. This section focuses on two aspects of JIT—production control and material provisioning.

JIT and Material Management.   There are many approaches for reducing waste, but JIT relies on physical material control to identify waste and eventually force the elimination of it. Figure 4.7 is an illustration of a JIT system applied to a production line consisting of several work cells and their respective output buffers. Within the JIT production line, not only are total inventory levels tightly controlled but so is the level of buffer inventories between the work cells. Production in and delivery of material to a work cell is triggered only when an inventory buffer is below a certain threshold as a result of being consumed by the succeeding operation. Furthermore, material cannot be delivered to a production line or a work cell unless an equal amount of material has left the line. This triggering signal for action can be an empty buffer, an empty container, a kanban card (more about this later), or other visible replenishment signals, all of which indicate that an item has been consumed and requires replenishment.

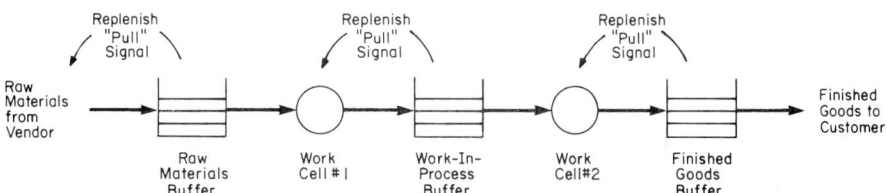

FIGURE 4.7   Illustration of a JIT/pull system.

JIT systems are reactive and, in contrast to MRP and MRP II systems, do no material planning. Rather, material is delivered to the line on a replenishment basis. Furthermore, JIT systems do not use computer-based shop floor control systems since work priority is determined by ascertaining what needs to be replenished based on what has been consumed.

Material Control to Identify Waste.   Some will argue that JIT systems are not too different from traditional reorder point replenishment systems. This is a gross simplification. Where JIT systems differ significantly from reorder point systems and why they are more effective is in their application to expose and to force the elimination of waste.

The basic premise of JIT systems is to continually lower inventories to surface problems which can then be solved. That is, in a JIT production environment, strict rules are imposed on the inventory levels and on the movement of material. By continuing to stress the manufacturing line, through the lowering of inventories, the production line becomes so lean that when a problem arises, the problem is visible since work has either slowed or has stopped. In a true JIT environment, that problem will trigger action, since work cannot resume, at the normal rates, until the problem has been adequately addressed. It is this exposure of waste and ultimate elimination that distinguishes JIT systems from traditional reorder point systems and, for that matter, MRP and MRP II systems which do not have these "built-in" behavioral incentives to improve performance.

A popular analogy for illustrating JIT philosophies is the "water-rocks" example shown in Fig. 4.8. According to the analogy, as the water (inventory) level is lowered, rocks (quality problems) are exposed. To be able to navigate freely, the rocks or quality problems have to be removed as they are exposed. Once those rocks are removed, the water level is lowered again until additional rocks are exposed. This process continues until all the rocks, or defects, are uncovered and removed.

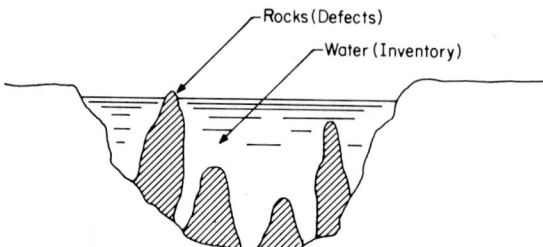

FIGURE 4.8 "Water/rocks" (inventory defects) analogy.

Reducing inventory levels not only lowers investment and makes quality problems more visible but also increases the velocity of the material through the production line, as predicted by Little's law:

$$I = WIP/D \qquad (2)$$

where I = manufacturing interval
D = total demand per unit of time
WIP = work-in-process

Well-managed JIT production lines thus have high material velocities, thereby earning the term "just-in-time." Further readings on JIT systems can be found elsewhere.[7-9]

Overall, JIT systems are simple and powerful techniques which when applied correctly will continuously improve quality and manufacturing performance. However, as with MRP and MRP II systems, JIT systems have their own limitations, and these will be discussed later.

Lot Size Reduction.   Another key parameter of JIT manufacturing is lot sizes. This parameter is especially important when a production line is required to make several items of which all are required to be assembled in order to deliver the finished product to a customer. For example, assume a finished product requires assembling widget A to widget B. Assume also that widget A and widget B are both made on the same production line. If the manufacturing lot size is 1000, that would imply that a finished product could not be assembled until 1000 widget A's and one widget B have been produced. However, if the lot size is 1, the finished product can be assembled and delivered after 1 widget A and 1 widget B have been produced. The inventory costs of the former scenario are considerably higher than the latter, thereby underlining the importance of small lot sizes.

From the perspective of minimizing inventory carrying costs, the ideal manufacturing lot size is one. However, "jumping" to a lot size of one without reducing setup times can be expensive from the standpoint of setup costs, production throughput, and asset utilization. Figure 4.9, based on the classical economic order quantity (EOQ) theory, is a simple illustration which shows the trade-offs between manufacturing setup costs and time and inventory carrying costs. According to Fig. 4.9, for a given level of production, increasing production lot sizes decreases setup costs but increases inventory carrying costs. From the perspective of total cost, the optimum lot size is where the setup costs curve

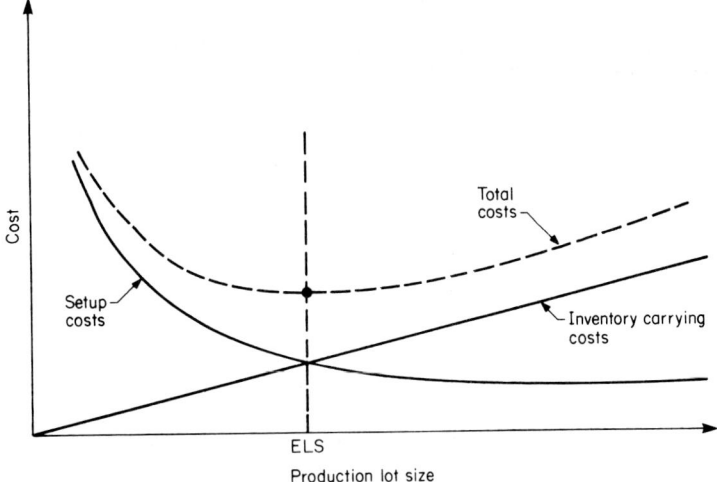

FIGURE 4.9   Impact of production lot size on production cost.

intersects the inventory carrying cost curve which is also represented by the minima of the total cost curve shown in Fig. 4.9. This point is known as the economical lot size (ELS). Reducing setup times will lower the setup cost curve and thus the economic lot size and the total cost, as shown in Fig. 4.10.

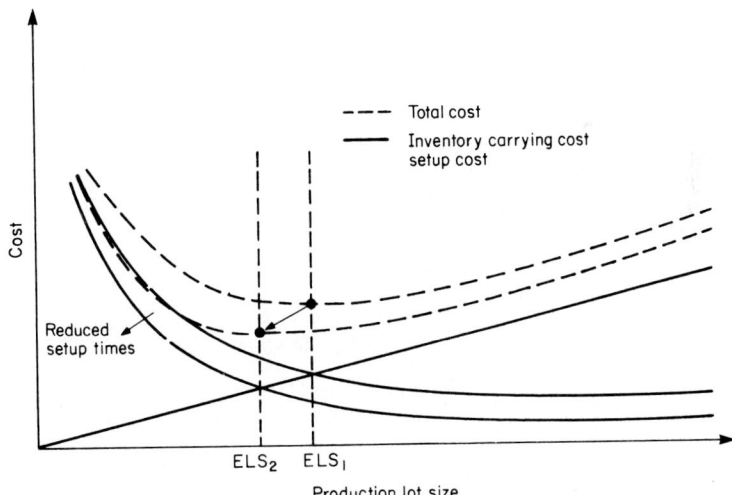

FIGURE 4.10   Impact of setup times on costs.

Besides the economic ramifications, smaller lot sizes enhance the JIT process. Smaller lot sizes result in lower overall inventories. Lower overall inventories cause greater exposure of problems when they arise and initiate actions which ultimately lead to improved manufacturing performance. Smaller lot sizes also have the benefit of lower scrap and lower rework costs since defects, if present, will be found more quickly and be smaller in total number. Additional information on reducing setup costs can be found in the literature.[10]

Kanban Systems.   A common JIT technique used to control the delivery and flow of material to and through a manufacturing line is the kanban, which means "visible signal." The kanban can be a card, a container, or an electronic signal. In a JIT manufacturing line, a kanban is associated with each manufacturing lot and ultimately the finished lot. When the material is consumed by either the succeeding operation or a customer sale, the kanban is dissociated from the consumed lot. This kanban then triggers action to replenish what has been consumed.

There are many variations of a kanban system, and each user is encouraged to design the system that matches his or her needs. One well-publicized variation is the dual kanban card system pioneered by the Toyota Motor Company.[11] The dual card kanban system consists of two types of kanban cards: conveyance and production kanban cards. As an example, consider the following process where there are kanbans between a producer's finished goods stock point and production line and kanbans between the producer and the raw materials vendor:

1. When the finished goods are consumed, a production kanban which was attached to the consumed lot is detached and delivered to the start of the production line, where the kanban triggers a production operation to replenish what has just been consumed.
2. The production, in turn, triggers the consumption of raw materials stock. The raw material lot has a conveyance kanban attached to it, and as the raw material is consumed, this kanban is detached and sent to the vendor, where the conveyance kanban initiates two simultaneous actions:
    a. The delivery to the producer of a fresh supply of raw material from the vendor's output stock with the producer's conveyance kanban attached to the delivered lot.
    b. The detachment of the vendor's production kanban card from the lot which has just been shipped. The detached production card then serves as the trigger that initiates action to replenish the vendor's stock that has been consumed.

Critical to the success of a JIT manufacturing line is maintaining control of the number of kanbans and their mix. Reducing the number of kanbans will reduce the level of inventory and increase production line speed as illustrated by Eq. (2). Caution, however, should be taken in reducing the kanbans to levels so low that the production line is "starved" for work and assets are not effectively utilized. A simple equation for estimating buffer or kanban inventories between two control points is as follows:

$$N = DI(1 + v)/C \tag{3}$$

where N = amount in kanban or buffer inventory
D = demand per unit of time
I = manufacturing interval
v = safety stock factor or service level variable
C = container capacity of kanban or buffer inventory

Note how kanban inventory levels are directly related to the manufacturing interval where the greater the interval, the greater the number of kanbans per inventory.

MRP/MRP II VERSUS JIT SYSTEMS

JIT is defined as a process which strives to eliminate waste which if implemented properly will result in "just-in-time" deliveries of material. An MRP system, on the other hand, is a computer-based material planning and production scheduling and control system which strives to deliver "just in time," and one result may be the elimination of excess material. An MRP system does not strive to identify and eliminate waste (other than perhaps material waste) and, by definition, is not a JIT process. Therefore, direct comparisons between MRP and JIT systems are valid only when the comparison is made between the material planning and production control aspects of MRP and the kanban system of a JIT process.

Push versus Pull Systems.   Material planning and production control of manufacturing operations can be classified as either "push" or "pull." In a "push" mode, an operation is delivering or building in anticipation of a future demand, and in a "pull" mode, the operation is building in reaction to a present demand, which can be an actual customer demand or a need to replenish stock. "Push" systems tend to plan, be proactive, and anticipative, and "pull" systems tend to expedite, be reactive, and be responsive. Kanban systems are representative of "pull" systems, and MRP systems and, in particular, MRP II systems are generally associated with "push" systems.

Limitations of Push Systems.   In a totally predictable environment, demand forecasts would always be realized; bills of material would be absolutely accurate; suppliers would ship on time and with total accuracy; nothing would be misplaced or miscounted; machines would never fail; all personnel would be present when expected; and all intervals in the process would be totally predictable. In such an environment, all manufacturing activities could be scheduled using the push/MRP system, every component would reach its destination just in time, and every manufactured item would be completed just in time.

However, no environment is totally predictable. To ensure product delivery in the face of the inevitable problems that arise throughout the chain from materials provisioning to product manufacturing, slack time is built into the schedule of an MRP system. Since slack is demanded by the system, there is little incentive to reduce it. Moreover, MRP and, in particular, MRP II systems suffer from incorrect lead times because they fail to recognize that lead times are not fixed but depend on congestion or load within an operation. Finally, acquiring and maintaining the computer-based MRP and MRP II systems can be expensive.

Limitations of Pull Systems.   When applied to expose quality problems, kanban systems become powerful tools in improving manufacturing performance. Kanban systems, in particular, are simple and cost-effective, do not require complex information systems or algorithms, and should be used where applicable. They are especially applicable to high-volume, constant-rate, repetitive flow manufacturing operations. However, kanban systems have some limitations. In particular, they tend to be inefficient and/or uneconomical for high-mix, high-demand volatility, dynamic environments which manufacture expensive items. This is because kanban systems require that all raw materials and buildable items within a production line as well as the finished goods storeroom be always in stock since delivery of the raw material and production of the buildable item are triggered only when the inventory (including work-in-process) is consumed and the stock falls below a certain threshold even when there is no actual customer demand. For products which have volatile demands and long production and delivery intervals, this can be expensive.

MRP/MRP II versus JIT Systems—When to Use.   General applicability of kanban systems and MRP systems is summarized in Fig. 4.11.[12] According to Fig. 4.11, for production processes which are dedicated to one or a few similar products and where the production is continuous and level and the production lead times are predictable and uniform, kanban is the system of choice for material provisioning, order release, and shop floor management. At the other extreme where the product is low-volume, complex engineered, or custom manufactured with no regularity in production patterns, MRP II for materials planning, order scheduling for order release, and shop floor control systems for shop management is the system of choice. However, most products and production systems fall between the two extremes, and generally the production control system of choice should be a hybrid that integrates the strengths of both approaches. An example of a hybrid system is one which uses MRP scheduling techniques for material planning but a JIT system for order release and for controlling the work flow on the manufacturing line.[13]

Ultimately, the choice between lead time based (look ahead and push) and kanban based production control systems depends on many factors,[14] including intervals and their variability, demand and its variability, forecast errors, and inventory requirements. For

	Material Planning	Shop Order Release	Shop Floor
Continuous Flow	Kanban	Rate Based	Kanban/Pull
Batch Repetitive Demand	Kanban or MRP	Pull or MRP	Pull
Batch Dynamic Demand	MRP	MRP	Pull or Order Scheduling
Custom Engineering	MRP	Order Scheduling	Operation Scheduling

(Left axis: Product Production Characteristics. Right axis: Lead-Time Variability — Low (top) ← → High (bottom))

FIGURE 4.11 Tailored production controls. (*Reprinted with permission from U.S. Karmarkar, Simon School, University of Rochester, Rochester, N.Y.*)

example, the inventory requirement under kanban-type policy is considerably larger than under the look ahead policy when the interval is relatively long but constant and the demands in the near future are known with near certainty but are quite variable. The difference in the two policies is even larger when the demands have high positive serial correlation. On the other hand, with small interval and/or small demand variability the difference is small and kanban is recommended because of its simplicity and low implementation cost. When the forecast error is large, both look ahead and kanban policies require large finished goods inventory and there is no appreciable difference in using one technique over the other. However, scheduling systems do not encourage continuous improvement and are usually costly to purchase and to implement. In many cases, commercial MRP and MRP II systems have many more features than usually required and can add considerable overhead to an operation. Further readings on choices and trade-offs in production control systems are available in the literature.[15–17]

SUMMARY

No single material planning or production control system is best for every production environment. MRP systems have their strengths in the planning of complex products while kanban systems are simple and best for low-option, repetitive manufacturing operations. Furthermore, kanban systems can be used as a technique to expose waste and help drive toward continuous improvement. More often than not, the best approach is a hybrid system which uses the strengths of both techniques. For example, the optimum approach could be using MRP for planning and JIT/kanban for execution.

In selecting a production control system, it is important to go through a series of steps before arriving at an efficient and cost-effective solution:

1. Analyze the products to be manufactured from the standpoint of demand and demand volatility, lead times (both production and procurement) and lead time variability, cost, and other production and product characteristics.

2. Based on the analysis, tailor a production control system to match the needs of the product.

3. Pilot or prototype the proposed system. Initiate corrections where appropriate.

4. Make a final selection and implement. The implementation should include not only the installation of the software (if required) and the population of the databases but more

importantly, education and training for the users. During the training, it is important to explain the need and rationale for the system, the basic concepts, and its usage.

5. Once implemented, an objective must be to continue to drive toward continuous improvement and higher levels of performance.

REFERENCES

1. Orlicky, J., *Material Requirements Planning,* McGraw-Hill, New York, 1975.

2. Fogarty, D. W., and T. R. Hoffmann, *Production and Inventory Management,* South-Western Publishing Co., Cincinnati, Ohio, 1983.

3. Plossl, G. W., *Production and Inventory Control: Principles and Techniques,* 2d ed., Prentice-Hall, Englewood Cliffs, N.J., 1985.

4. Orlicky, J., *Material Requirements Planning,* McGraw-Hill, New York, 1975.

5. Plossl, G., *Production and Inventory Control Applications,* George Plossl Educational Services, Inc., 1983, pp. 229–232.

6. Wight, O. W., *MRP II: Unlocking America's Productivity Potential,* 5th ed., Oliver Wight Limited Publications, Inc., Williston, Vt., 1983.

7. Schonberger, R. J., *Japanese Manufacturing Techniques—9 Hidden Lessons in Simplicity,* Free Press, New York, 1982.

8. Monden, Y., *Toyota Production System—Practical Approach to Production Management,* Industrial Engineering and Management Press, Institute of Industrial Engineers, 1983.

9. Hall, W. R., *Zero Inventories,* Dow Jones, Irwin, Ill., 1983.

10. Shingo, S., *Revolution in Set Up Time—Fundamental Approach to SMED System,* Productivity Inc., Stamford, Conn., 1984.

11. Shingo, S., *Study of Toyota Production System from Industrial Engineering Viewpoint,* Productivity, Inc., Cambridge, Mass., 1984.

12. Karmarkar, U. S., Simon School, University of Rochester.

13. Foo, G., and L. E. Kinney, "Integrated Pull Manufacturing—the Integration of MRP and Pull Systems," *Proceedings of the 1st International Conference on Systems Integration,* Apr. 23–26, 1990, IEEE Computer Society.

14. Krupka, D. C., and B. T. Doshi, "Integration of Planning and Execution Operations," *AT&T Technical Journal,* August–September 1990.

15. Karmarkar, U. S., Simon School, University of Rochester, "Getting Control of Just-In-Time," *Harvard Business Review,* September–October 1989, pp. 122–131.

16. Karmarkar, U. S., and I. M. Shivdasani, Simon School, University of Rochester, "Alternatives for Batch Manufacturing Control," *Quantitative Methods Working Paper Series,* QM 86-13, University of Rochester, June 1986.

17. Karmarkar, U. S., Simon School, University of Rochester, "Manufacturing Lead Times, Order Release and Capacity Loading," *Center for Manufacturing & Operations Management Working Paper Series,* CMOM 89-04, University of Rochester, 1989.

CHAPTER 5
ALTERNATIVE WORK SCHEDULES*

Richard E. Kopelman
Professor of Management
Baruch College
The City University of New York
New York, New York

Early in the nineteenth century, when it took days to travel from New York to Philadelphia by horse-drawn coach, President John Madison bemoaned the slowness of some information sources. More specifically, he lamented not hearing from his ambassador to Spain for more than 2 years. Reflective of his growing impatience, Madison declared that he was only willing to wait another 6 months; otherwise Madison would take action to find out what was going on.[1]

As we near the end of the twentieth century, products are regularly ordered and received cross-country in a day, and reliable information is available instantaneously from around the world. Yet, ironically, today people often feel there is too little time. Evidently, Toffler's "future shock"[2] has arrived.

Writing in 1925, Williams noted "the unity of life and labor—the complete impossibility of walling off the factory from the home... of dividing the hankerings of a man's working hours from those of his hours of leisure."[3] In the 1990s not only has the pace of life accelerated, the interrelatedness of work and life domains has increased, particularly in dual-career families. As noted in a feature article in *HRMagazine,* "Today's employees, forced to choose between work obligations and family responsibilities, often feel frazzled and guilty."[4] And when different shifts are worked, child care may entail a "baton pass" of children in a parking lot.[5]

Partly in response to these (and other) societal trends, work organizations have increasingly experimented with various alternative work schedules and employment arrangements, including (1) flexible work hours, (2) the compressed workweek, (3) permanent part-time employment, (4) peak-time employment, (5) job sharing, (6) leased employees, (7) telecommuting ("flexiplace"), (8) extended vacations, (9) sabbatical leaves, and (10) phased retirement.

This chapter focuses on two widely implemented alternative arrangements, flexible work hours (FWH) and the compressed workweek (CWW). After each intervention is defined, potential benefits and costs are identified, actual adoption rates are reviewed, and the effectiveness evidence is examined, with a particular emphasis on measures of productive efficiency. Factors influencing effectiveness (boundary conditions) are identified, and suggestions are offered with respect to possible implementation of both plans.

---

*This chapter updates and expands upon material in the author's book, *Managing Productivity in Organizations,* McGraw-Hill, New York, 1986.

FLEXIBLE WORK HOURS (FWH)

Although various specific plans exist, in broad terms FWH is a system of scheduling work that grants employees some choice as to when they work. Typically, there are hours when all full-time employees must work (the "core" hours) and hours over which employees can exercise some discretion (the flexible hours). For example, at Sandoz, Inc., the core hours are from 9:30 A.M. until 12:00 noon, and 2:00 to 4:00 P.M. The earliest allowed starting time is 7:30 A.M., and the latest stopping time is 6:00 P.M., yielding a 10½-hour "band width." Notwithstanding the existence of a 37½-hour workweek at Sandoz, employees are allowed to work up to five 8-hour days a week, banking up to 2½ hours per week. Thus, after three 40-hour weeks, the employee can take an extra day off with pay.[6] (Employees at Sandoz are not permitted to bank more than 2½ hours per week, as this would require paid overtime in accordance with the Fair Labor Standards Act; similarly, employees may bank only ½ hour daily, as the Walsh-Healey Act requires paid overtime after 8 hours.) Typically, companies with a 40-hour workweek permit employees to determine only when they will start (and stop) work each day.

There are eight key features in FWH plans that can vary: (1) the band width; (2) the core hours; (3) the flexible hours; (4) the length of the workweek; (5) whether banking is permitted—across days and/or weeks; (6) work schedule variability; (7) the method of monitoring time; and (8) the supervisor's role. The first five features have already been described. Work schedule variability pertains to whether employees are allowed to choose their work hours on a daily or less frequent basis (with weekly or monthly variability often requiring advance notice). Hours worked can be monitored in several ways—time clocks, sign-in sheets, honor systems—and the unit of measurement can range from ¼ to 1 hour. Supervisory roles can vary from granting the supervisor no control over individual schedules, to allowing an override in emergencies, to negotiating schedules in advance, to requiring prior supervisory approval of a work schedule.

In light of these eight key features of FWH plans, it is not surprising that more than 100 different variants have emerged in practice. The five most common plans, in order of increasing individual flexibility, are (1) staggered group hours (or "group flex," that is, the accounting department starts at 8:00, purchasing at 8:30); (2) staggered individual hours (or "flexitour"), whereby the individual chooses a daily schedule and must stay on that schedule for a period of time; (3) flexible hours (or "gliding hours"), with the individual setting his or her schedule daily, within certain parameters (such as core hours, number of days worked, maximum weekly hours); (4) flexible workdays (or "variable day" plans), which permit the individual to bank or owe days within the week; and (5) flexible workweeks (or "variable week" plans) that permit variations in hours across weeks or pay periods. Roughly a decade ago it was estimated that about 70 percent of extant plans consist of variants 2 and 3.[7] Perhaps the most liberal plan of all, "maxiflex," entails management merely defining the work to be accomplished; employees then choose their own work schedules.

Potential Advantages of FWH.   The literature on FWH has produced a prodigious list of potential advantages. For expository purposes, these benefits are classified below depending on whether the primary beneficiary is the employee or the organization. Listed first are the potential benefits to employees.

1. *Ameliorates interrole conflicts.* There are numerous examples of interrole conflict, such as employees who must care for children and/or elder parents, employees who want to work and study, and couples coping with marriage and dual careers. Indicative of the growing incidence of such concerns, the proportion of working mothers with preschool children (under age 6) was 12 percent in 1950, 32 percent in 1970, and 57 percent in 1990, and it should reach 65 percent by 1995. Even more dramatic is the rise in the proportion of working mothers with a child under one year of age: from 25 percent in 1978 to 50 percent

in 1990. The existence of FWH may permit a parent to see children off to school or be home when school lets out. While the growth of two-paycheck families has probably increased the incidence of interrole conflicts, such conflicts are particularly acute when both spouses are highly job involved, such as dual-career couples.

2. *Enhances feelings of autonomy.* Use of FWH gives many employees a measure of control over their jobs, and increased self-management; consequently, there may be heightened feelings of trustworthiness, autonomy, and general well-being. Indeed, implementation of FWH may "provide a new measure of self-respect for employees at every level."[8] Moreover, these concerns are especially salient to today's younger workers, many of whom seek some meaning and involvement at work.

3. *Reduces daily stress.* Clearly, there is less pressure on the individual if he or she does not have to arrive at work at a specific time. Concomitantly, the employee need not "build in" extra travel time to ensure against being late to work.

4. *Utilizes available work time.* With FWH the employee can start work upon arrival; it is not necessary to waste time waiting around for the "starting bell."

5. *Reduces travel time and travel costs.* By permitting travel during off-peak hours, potentially 1 hour of commuting time may be saved daily. If public transportation is used, though, there may be a cost savings coupled with a lengthier (off-peak) commute.

6. *Permits more daytime nonwork activities.* To the extent that individuals prefer to participate in daytime recreational activities, FWH may be advantageous.

It might be noted that some of the aforementioned advantages are unique to FWH, insofar as individuals have control over the hours they work. The advantages that follow might be seen as primarily beneficial to employing organizations.

7. *Increases work motivation and productivity.* To the extent that employees are able to work the hours they prefer, their motivation to perform should be heightened. Relatedly, it has been argued that FWH may match the employee's work schedule to his or her "bioclock." Hence the "late starter" need not sit around from 8 to 9 A.M. in a semisomnolent stupor.[9]

8. *Creates quiet periods.* By extending the hours of the workday, employees may experience relatively few interruptions during some time periods.

9. *Matches schedules to tasks.* By extending the hours of the workday, some difficult projects can be completed more efficiently. For instance, a task requiring several continuous hours need not be postponed if not started immediately. As one commentator put it, "the inherent logic of matching one's schedule to work-level variations...is hard to deny."[10]

10. *Improves customer service.* By extending the workday, the organization may be able to serve customers better (if adequate coverage is available).

11. *Virtually eliminates tardiness.* Provided that the employee arrives before the core period, FWH eliminates tardiness. As one employee noted, "It's hard to be late when you have a two-hour leeway."[11] Further, FWH obviates the unpleasant task of having to reprimand the late employee—an activity that often leads to anger, embarrassment, guilt, and fear—but not necessarily to improved punctuality.

12. *Reduces overtime costs.* Savings in overtime expenses are particularly likely to be realized if employees adjust their schedules to coincide with peak workloads, or if work can be assigned to multiple individuals. For instance, a last-minute work order can be processed by a clerk who ordinarily works until 6 P.M. This feature of FWH is generally seen as uncongenial by many union leaders.

13. *Facilitates employee retention and recruitment.* Because FWH is an attractive policy to many people—for some it is a new perquisite—the organization may be able to retain more (and better) employees and screen more applicants. A secretary and mother of two children commented, "I would not be able to keep this job if it were not for a flexible work schedule."[12] Similarly, Philip Kocher, an executive at Aetna, asserted that "instead of losing good people, we'd rather create work schedules so we can retain them."[13] Indeed, for some occupations, such as software engineers, flexible work hours are now viewed as a "nonnegotiable demand."[14]

14. *Reduces absenteeism, sick leave, and personal leave.* With fixed schedules, the employees who oversleeps or chooses to attend to personal business may decide to call in sick, and may even be paid in full. With FWH, employees need not have "imaginary illnesses" to obtain time off for personal reasons. As one plant manager put it: "With the advent of flexitime, our firm is no longer subsidizing dental appointments...[and the] extended lunch hour is no longer time lost by the company."[15] Instead, time not worked is made up by the employee.

15. *Increases employee skills.* FWH often leads to some cross-training of employees, so they can "cover" for each other.

16. *Improves external communications.* Some organizations, such as those on the west coast, may experience improved communication with the east coast, owing to an earlier starting time.

17. *More accurate performance appraisals.* Supervisors may be less concerned about punctuality, and possibly more attentive to more important aspects of job performance.

18. *Better resource utilization.* Scarce limited resources, such as data processing centers, may be utilized more fully with fewer bottlenecks and quicker turnaround times, owing to lengthier hours of operation.

As noted above, the potential advantages of FWH have been somewhat arbitrarily classified in terms of the principal recipient. Yet advantages to the individual generally also benefit the organization. For instance, reductions in interrole conflicts should enable employers to recruit better people, and employees to concentrate better at work. Indeed, human resource executives frequently cast the issue in terms of prudent organizational self-interest.

Potential Disadvantages of FWH. The literature of FWH has also yielded a lengthy list of possible disadvantages. These are reviewed first from the perspective of the individual employee.

1. *Employees responsible for lost time.* As noted above, the employee who is unavoidably (or avoidably) late for work still has to work the required number of hours. FWH may also result in the loss of absence privileges or short breaks. One union official noted that the worker who takes a couple of hours off to visit a doctor no longer is paid for that time. "With flexitime...you have to make up for the time you lost."[16]

2. *Reduced opportunity for overtime.* Because of the lengthier workday, and potentially greater management flexibility, overtime pay may be less available. The same union official stated: "There is no such thing as overtime with flexitime."[17]

3. *Potentially increased surveillance.* To keep track of actual hours worked, time clocks may be introduced (or reintroduced) or employees may be required to sign in and out. Supervisors may be overly conscientious in monitoring employee work hours, possibly causing tension and resentment.

4. *Unequal treatment.* There may also be resentment among those employees whose jobs exclude them from FWH. Internal service positions (telephone operators, security guards, and the like) will typically not be included in FWH plans since understaffing could be very detrimental. Those employees who formerly had exclusive claim to the prerogative of choosing their own hours may also feel a relative loss.

5. *Unwanted job enrichment.* Some employees may not want the added responsibility or variety that typically accompanies FWH.

A number of potential drawbacks have been noted from the organization's perspective. These are listed next.

6. *Inadequate staffing may occur.* The staff may be spread too thin, particularly during the early or late hours. One manager bemoaned the large number of employees who work early hours, leaving the organization understaffed for late afternoon work.[18]

7. *Increased overhead costs.* Given the longer hours of operation, there may be an increase in utility and internal service expenses. There are also likely to be extra outlays required for recording hours and keeping records.

8. *Problems in internal communication.* Internal communication and coordination may be more difficult, given that employees will not be available throughout the workday. Meetings and training sessions will be more difficult to schedule. One manager complained that the "new excuse around here is called *flex-out....* It used to be that if I couldn't find an inventory clerk, they would tell me the person was in the restroom. 'Flex out' is a better excuse. You can be gone away from your work station a lot longer."[19]

9. *Problems in external communication.* External communication and customer service may suffer, if clients or vendors are unable to contact appropriate personnel during noncore hours. As a consequence coworkers will have to take more messages, with the likely result being increased delays and errors.

10. *Increases difficulty of supervision.* There is wide agreement that FWH substantially complicates the supervisor's job. More careful planning will be needed to ensure adequate coverage, especially at times when some employees may choose to be absent, such as during peak load periods or on Friday afternoon. An additional burden results from having to plan and organize for workers who may arrive early (such as 7:00 A.M.). Consequently, many supervisors find that they have to work longer hours. One commented: "I know of no supervisor who takes advantage of the policy [FWH]. If we want to do an honest job of supervising our employees we have to work longer hours."[20] Moreover, supervisors may be forced to handle the abuses that occasionally occur, (such as employees who punch in or sign in for others, or simply do not put in a full day's work. Further, given the loss of authority and control (that is, to demand attendance at specific times), many supervisors will need to adopt a more participative approach to managing. Some will have difficulty making this adjustment, and may perform less effectively.

11. *Possible labor relations problems.* It is not unlikely that there will be strong union opposition to any changes in contractual agreements or business practices that affect overtime compensation or the conditions of work.

12. *Requires changes in personnel manual.* Not only must new policies be spelled out, but to the extent that jobs are altered, job descriptions and performance evaluation procedures may also need updating.

13. *Nearly irreversible decision.* Because of the popularity of FWH among employees, it may be a nearly irrevocable commitment—regardless of any effectiveness evidence. One manager compared FWH with stepping on a moving sidewalk, it "just keeps on going."[21]

Trends in the Adoption of FWH.   Beginning with the first known implementation of FWH in a West German aerospace company in 1967, FWH took hold rapidly in Europe. By 1974, there were some 3000 companies in West Germany with FWH; by 1977, there were 20,000. In the 1970s an estimated 25 percent of the Austrian, 30 percent of the French, and 40 percent of the Swiss work forces worked under FWH.

In the United States the adoption of FWH was energetic but less rapid than in Europe, for reasons discussed below. The first known adopting organization was Control Data in 1972. By 1974, roughly 4 percent of the U.S. work force had FWH; by 1978, the figure rose to 8 percent; by 1980 to 10 percent; and by 1985 to 12.3 percent.[22]

TABLE 5.1 Effects of Flexible Work Hours (FWH) on Productivity and Absenteeism

Study number and year		Setting and sample	Impact on productivity	Impact on absenteeism
1	1974	Smith Kline Corporation; 34 R&D employees	NR	15% (paid absences) −20% (single day absences)
2	1976	Berol Corporation; 7 locations; various occupations	No change	−50% −78% (personal leave days)
3	1976	State Street Bank; 4 departments; largely clerical employees	No change	No change
4	1977	Metropolitan Life Insurance Co.; 5 departments; 246 clerical employees	No "clear-cut impact"	No significant change in short-term absences
5	1978	Control Data Corporation, aerospace and microcircuit divisions; 386 managerial and nonmanagerial employees	"Slightly favorable trend"	NR
6	1979	U.S. Department of Labor, Office of Accounting	+3%	NR
7	1980	Mutual of New York; 22 clerical employees	+2.9%	−7.6%
8	1980	Insurance company claim processing department	+7.3%	−33.8%
9	1981	County welfare agency; 353 employees	+3.0%	−29.9% (paid absences) −143.9% (unpaid absences)
10	1981	Government agency; 64 clerical employees	−4.6%	NR
11	1981	U.S. Social Security Administration, Bureau of Data Processing; 350 employees	+12%	−3% (approx.)
12	1981	U.S. Department of the Interior, U.S. Geological Survey; 2230 employees	+10%	−7% (sick leave) −1% (annual leave)
13	1981	U.S. Office of Personnel Management, Bureau of Policy and Standards; 240 employees	Inconclusive results	No significant change

As would be expected, the adoption rate of FWH has varied depending upon sector, industry, and occupation. With regard to the most recent systematic sample (of 59,500 households in 1985) the following adoption rates were found: service sector, 14.5 percent; manufacturing, 9.8 percent; male employees, 13.2 percent; female employees, 11.1 percent. Adoption rates exceeded 15 percent in publishing, finance, insurance, real estate, and entertainment. Across occupations, adoption rates ranged from over 30 percent among math and computer specialists, 20 percent among salespeople, to only 4 percent among machine operators, assemblers, and inspectors. Although there has been steady growth in the adoption of FWH, the trend has largely reflected application among white-collar jobs in the service sector.

Effectiveness Evidence. Most of the early research on the effects of FWH was of poor quality, consisting largely of anecdotal and impressionistic accounts. Few studies obtained objective outcome data, such as physical productivity measures; the use of control (or comparison) groups was rare; and data were rarely collected for longer than a few months. Many of the early "studies" consisted of descriptive accounts that concluded, "We tried it and we liked it."[23] Indicative of the paucity of hard evaluative evidence, a survey of 78 organizations that adopted alternative work schedules reveals that in only 9 of the orga-

TABLE 5.1 Effects of Flexible Work Hours (FWH) on Productivity and Absenteeism (*Continued*)

Study number and year	Setting and sample	Impact on productivity	Impact on absenteeism
14 1981	Library of Congress; 150 employees	+10.5%	−43% (sick leave usage)
15 1981	Navy Finance Center; Cleveland, Ohio; 64 employees	NR	−50% (sick leave usage)
16 1981	Navy Finance Office; Long Beach, California; 55 employees	No change	No change
17 1981	U.S. Army Computer Systems Command; Fort Belvoir, Virginia; 82 employees	NR	No change (sick leave)
18 1981	U.S. Army Tank Automotive Command; Michigan; 400 employees	+2%	−29% (sick leave usage)
19 1981	U.S. Information Agency; 33 employees	+5%	No change (short-term absences) No change (sick leave usage)
20 1981	First National Bank of Boston; 125 employees	+10%	No change
21 1981	Occidental Life Insurance Co.; 700 employees	No change	−21% (full-day absences)
22 1981	Pitney Bowes, Inc.; 220 employees	No conclusive indication	−16%
23 1982	Large multinational corp.; approx. 250 secretaries, technicians, and drafters	NR	−71%
24 1984	Eastern plant; mfr. of business equipment; 455 operative workers and supervisors	+2½% (approx.)	−56%
25 1985	State government agency; 16 data entry operators; 57 programmers and analysts	0% 32% (input)	NR −20%
26 1986	Government agency; 75 accounting employees	3.2%	−18.9%
27 1987	New York State government, 245 employees in 2 agencies	NR	−16.5% (short-term sick leave) −10.5% (short-term annual leave)

Note: A complete list of references is available from the author on request.

nizations were objective effectiveness data obtained; and only 4 of the 78 organizations used this information to undertake a formal evaluation of the program.[24] It should be noted that although only a small *proportion* of extant studies has provided objective effectiveness evidence, the number of such reports is fairly sizable. Table 5.1 presents a survey of these reports.

Twenty-three studies have provided data concerning the impact of FWH on measured efficiency. In 13 cases the results were positive, the increases ranging from 2 to 32 percent; in 10 cases results were either inconclusive or negative. The median result was an increase of 2½ percent.*,†

---

*It might be noted that the study with the 32 percent increase actually did not measure output; rather the efficiency of computer programmers and analysts was estimated by the amount of central processing unit (CPU) time used (an input). The study is included here because a work sample found a high correlation between CPU time and lines of code produced—$r = 0.85$. The study reporting the next largest effect on productivity found an increase of 12 percent, however. In any event, exclusion of the questionable study would have had little overall impact, lowering the median result from 2½ to 2¼ percent.

Twenty-three studies have provided data concerning another objective criterion, absenteeism. Thirteen studies reported effects in terms of overall absenteeism; 11 reported results in terms of various component forms of absenteeism (for example, sick leave days used, short-term absences)—one study reported both overall and component absenteeism results. With regard to the more inclusive criterion, overall absenteeism, results ranged from no change to a reduction of 71 percent. The median effect was a decrease in absenteeism of 16 percent. Similarly, the median effects of FWH on component absenteeism indicators were reductions of 16.5 and 13.5 percent (across all indicators, and studies, respectively).‡

The impact of FWH on attitudes has generally been quite positive: typically 75 to 95 percent of respondents approve of FWH and express no interest in returning to fixed work hours. (In one study it was reported that all 125 employees were pleased with FWH, except one person who was described as "just an old grouch anyway."[27] Relatedly, about 85 percent of respondents have said that their job satisfaction (or morale) has improved as a result of FWH. Ronen, in a review of 27 studies, found that job attitudes improved in 26 cases.[28] However, findings pertinent to the effects of FWH on job satisfaction may be partly artifactual—they may reflect the desire to keep FWH. In one study where satisfaction was measured before and after FWH—but where *no reference was made to the possible impact of FWH*—there was no change in satisfaction. The researchers concluded that employees value FWH, not because it increases satisfaction per se "but because it makes working a bit easier. Thus, questionnaires used to gather employee attitudes about flexitime will yield favorable results if the link to flexitime is made obvious. When the link to flexitime is not salient, the traditional [positive] relationship may not occur."[29]

Boundary Conditions.   A number of factors have been identified that may affect the success of FWH.

Compatible Technology.   The task or technology should "fit" (be appropriate) for FWH to have positive effects. A suitable setting is one where individuals work independently on relatively isolated work modules (such as an R&D laboratory). FWH is also likely to be effective where longer hours of operation facilitate (1) the expanded use of a scarce resource (like a word processing center) or (2) improved external communication (possibly with customers and suppliers in other time zones). In contrast, unfavorable settings would seem to be those involving (1) sequential tasks (such as an assembly line with add-on or checking functions); (2) functionally narrow or fragmented work necessitating a high degree of internal communication, teamwork, and coordination; (3) multiple-shift, or 24-hour operations; (4) client-centered operations where relationships are critical; or (5) internal service departments that require continual readiness (such as reception, switchboard, and receiving). In view of these limiting conditions, it is not surprising that applications of FWH have been both more common and more effective in white-collar than in blue-collar jobs. When applied to both types of jobs in the same organization, reductions in overtime were 60 percent greater in one study and 100 percent greater in another.[30]

Organizational Size and Flexibility.   If it is necessary that a given work unit be fully staffed at all times, employees may need to be "borrowed" from other units. The cross-training of employees, therefore, may be a necessary boundary condition for the successful implementation of FWH. This likely will be of particular concern to small organiza-

---

†The exclusive reliance here on objectively measured efficiency reflects the considerable inaccuracy of perceptual estimates. For example, an early survey by Nollen[25] reported that the median impact of FWH on *subjectively perceived* productivity was 48 percent (p. 12); however, Nollen noted that observed increases in productivity ranged from 5 to 14 percent, and that such increases could be expected to occur one-third to one-half the time (p. 18)—yielding an average increase of slightly under 4 percent.

‡Although FWH has consistently yielded reductions in absenteeism, it might be noted that a nationwide survey of organizations found that use of FWH was associated with a slightly higher average absenteeism rate.[26] More specifically, the absenteeism rate among 207 organizations that had adopted FWH was 4.5 percent; in contrast, the rate among 780 nonadopters was lower, at 4.3 percent.

tions, where every member may play a vital role. Large organizational size, therefore, may provide a measure of flexibility, mitigating somewhat the need for cross-training.

Support of Involved Parties. It is important that union representatives and top management grant their imprimaturs. The commitment of first-level managers is probably of even greater importance, since they will face most of the difficulties in making FWH work. One supervisor, lacking enthusiasm for FWH, declared, "My employees can have flextime as long as they're here from 9 to 5."[31] Clearly, widespread cynicism will undermine successful implementation.

Employee Characteristics. Several individual difference variables may function as predisposing (rather than as necessary) boundary conditions; while they may not be crucial, they may improve the chances for success. FWH plans are more likely to be effective if employees value highly convenient work schedules or are

1. Young
2. Part of dual-career families (or single parents)
3. Receptive to increased responsibility and variety
4. Capable of working autonomously under limited supervision

To the extent that employees live in highly populated, traffic-congested areas, the benefits of FWH will be enhanced.

Supportive Organizational Climate. Evidence and logic indicate that FWH interventions are more likely to "work" where there is a climate with high trust; where delegation, freedom, and self-determination are encouraged; where experimentation is valued and failures are tolerated; and where there is a high degree of social cohesion (that is, people are willing to cooperate and "cover" for each other).

Clear and Appropriate Objectives. FWH should be introduced primarily to improve the quality of work life, and thereby to enhance the organization's ability to attract and retain good employees. If management expects to increase productivity, or decrease the number of paid hours not worked, disappointment will likely ensue, possible leading to discontinuance. Fortunately, evidence indicates that only about 15 percent of adopting organizations have implemented FWH with the expectation that productivity would improve.[32]

Guidelines for Implementing FWH. Organizations planning to implement FWH may want to consider a number of issues before adopting such a plan. Fixed work schedules have a long history of tradition and practice; hence there is likely to be some resistance to change. Certainly, the adoption of FWH is not a decision to be taken lightly, especially since the decision may be almost irrevocable—even if demonstrably unsuccessful from the organization's perspective. It is important, therefore, that several matters be carefully considered prior to implementation.

First, it should be determined who is responsible for the initial planning, policy creation, mechanics, and operational administration of the new plan. Initial planning should include the establishment of objectives and criteria for subsequent evaluation. Management should be clear as to *why* it is considering the adoption of FWH; some objectives are more realistic than others. A feasibility study should also be undertaken, possibly by an ad hoc committee, that examines (1) departmental (unit) functions and staffing requirements, (2) work flows plotted by time of day, (3) work units requiring unique work schedules (like cafeteria employees), (4) access to facilities and safety during extended hours, and (5) the degree of receptivity of union representatives (if pertinent). Related to the first four issues, consideration may be given to employee schedule preferences, possibly assessed via a questionnaire.

If FWH appears to be feasible, it is important that its implementation be planned carefully. As mentioned previously, eight design parameters need be decided upon in implementing FWH. It should be determined if, for example, the banking of hours will be permitted and if so whether across weeks as well as days. Meticulous attention must be paid to legal requirements and contractual obligations. The mechanics of tracking hours should also be considered, that is:

Will special monitors be needed?

Will employees be required to sign in and sign out?

If self-monitoring is employed, what provisions, if any, will be made to ensure accurate reporting?

Additionally, it is important to plan for an evaluation of the FWH plan. Effectiveness criteria should be identified in advance, and baseline measurements obtained (if possible). In brief, the introduction of FWH should not proceed impulsively. As one insurance company executive put it, "Just don't decide that next week in your company or government office you'll install some sort of flexible hours program."[33] The preliminary implementation steps undertaken by two companies that proceeded with considerable forethought and planning (Liberty Mutual and Control Data) have previously been described in published case studies.[34]

Several suggestions might be offered with regard to the initial phase of the implementation process. Implementation is likely to be successful to the extent that top management visibly and enthusiastically supports the change. One way to convey this support is by having high-level *line* executives conduct the orientation meetings. At such meetings the advantages and disadvantages of FWH should be explicitly identified. There should be a balanced and realistic presentation. Employees should be encouraged to express their concerns, a limited number of employees should attend each orientation meeting, and program parameters should be communicated clearly, repeatedly, and in writing. It is particularly important that orientation meetings be held with first-line supervisors, since they will handle most of the ensuing problems. Their commitment is crucial to the success of FWH. Management should also continue to consult with union representatives to ensure against misconceptions. The coordination of these efforts may require the appointment of a FWH project director, someone explicitly responsible for overseeing all aspects of implementation.

The intervention might be labeled an "experiment" or "pilot test" initially. By doing so, the costs associated with failure—and perhaps even the likelihood of failure—may be reduced. (One survey reported that 36 percent of such interventions were introduced on a trial basis.[35]) It may also be advantageous to conduct the initial test at a remote location, one with low visibility, or in a setting where employees are customarily given considerable autonomy.

Another way to reduce the risk of failure is to increase in stages the amount of flexibility granted to employees, for instance, gradually lengthening of the band width or adding a banking provision. It will be useful to hold meetings with first-line supervisors to discuss problems associated with the new role requirements and to provide feedback to supervisors so they can adjust their behavior accordingly. Finally, it is important to obtain accurate and extensive evaluation evidence from the trial run so that a sound basis exists for expanding or discontinuing the program.

THE COMPRESSED WORKWEEK (CWW)

The CWW entails scheduling the normal 36- to 40-hour workweek in fewer than 5 days, the most common form being the 4-day, 40-hour workweek (called the "four-forty"). There are, of course, numerous variations of the CWW. Two relatively prevalent variants are the 3-36 and, especially in recent years, the 9-80 (often called the "9/10 biweek").[36]

Potential Advantages of the CWW.   A number of potential benefits have been identified as possibly resulting from adoption of the CWW. For expository purposes, benefits are classified into two groups, depending upon whether the primary beneficiary is the employee or the organization. The former are listed first.

1. *Increased leisure time.*With the 3-36, 4-40, and 9-80 plans, the employee works 6 (or sometimes 7), 8, and 9 days every 2 weeks, respectively. Thus there is the opportunity to enjoy as many as 8 days off every 2 weeks instead of 4.

2. *Enriched jobs.*Job enrichment may occur if supervision is reduced (resulting in increased autonomy); this can happen if some supervisors do not work the longer hours. Enrichment may also result if 5-day coverage is retained while employees work on a compressed schedule; that is, having fewer employees on call may increase task variety and/or responsibility.

3. *Reduced commutation costs.*Not only will fewer days be worked, the longer workday may permit the use of public transportation during (less costly) off-peak hours. Traffic congestion may also be less troublesome.

4. *Increased opportunity for second job.*While this may be a benefit to some employees (those with a high need for earned income), it may be a drawback from the employer's perspective.
   It might be noted that the first and fourth potential advantages are relatively unique to the CWW (in comparison with FWH). The advantages that follow might be seen as primarily benefiting the organization.

5. *Improved employee recruitment.*Some organizations, particularly smaller ones, see the CWW as a valuable tool in attracting and retaining capable employees. The CWW may also serve to reduce the unattractiveness of evening and night shift schedules.

6. *Reduced absenteeism.*Because loss of a day's pay is more costly, absenteeism should be reduced. That employees have more personal time available should also contribute to lower absenteeism.

7. *Operating efficiencies.*The longer workday should contribute to lower setup and washup costs. Relatedly, plant and equipment may be used more efficiently if a second (that is, 3-day) mini shift is employed.

8. *Higher employee performance.*To the extent that better workers can be retained, that jobs are more motivating, that absenteeism is reduced, and that personal (nonwork) matters impinge less, performance quality and quantity should improve.

9. *Ease of scheduling.*Scheduling employees may be facilitated, particularly in settings with continuous process operations (like hospitals, petrochemical plants, and fire departments).

10. *Ease of implementation.*Given that employees work a fixed schedule, the CWW is a relatively simple intervention to implement.

Perhaps the most distinctive assessment of the advantages of the CWW was provided by Paul Samuelson and Riva Poor, who wrote:

> Progress comes from technical invention, and we shall be ever grateful to the discoverer of fire, the inventor of the electric dynamo, and the perfecter of hollandaise sauce. But there are also momentous *social inventions.* . . .
> The 4-day week is precisely such a social invention. Just as double entry bookkeeping may have done as much for modern life as the development of smelting, so will new ideas that enable mankind to find the good life in our present age of anxiety.[37]

Potential Disadvantages of the CWW.   Several potential drawbacks of the CWW have also been identified. These are reviewed, first from the perspective of the employee, then from the perspective of the employing organization.

1. *Heightened interrole conflict.*The extended workday may interfere with nonwork and family responsibilities. This is especially likely to be the case among single parents or dual-career couples.

2. *Lack of flexibility.*Employees will typically have to arrive at work on time, even if they attend a work-related meeting the night before.

3. *Heightened stress.*When the CWW is applied in organizations with 5-day (or longer) coverage, the staff may be thin at times. Understaffing, resulting from larger than anticipated absences—often on Mondays and Fridays—can be taxing.

4. *Fatigue.*Working a 10-hour day can be very tiring; when coupled with a 1- or 2-hour daily commute, the day can be exhausting.

     From the organization's perspective there are also several potential disadvantages. These are reviewed next.

5. *Problems in external communication.*Communications with vendors and customers may suffer if key employees are unavailable on regular workdays.

6. *Problems in internal communication.*Problems of coordination may arise, particularly if some parts of the organization operate 5 days a week.

7. *Performance problems.*Performance quality and quantity may decline if employees become fatigued toward the end of a 10- or 12-hour shift. One credit union found that the CWW was "zapping employees' energy levels and leaving member needs unserved in many instances."[38]

8. *Fewer hours worked.*There may be slippage in the actual hours worked due to fatigue and/or boredom.

9. *Problems in supervision.*Given the extended workday, supervision may be inadequate at times.

10. *Reduced flexibility.*It may be difficult to schedule overtime, even if required for urgent tasks. Along these lines, it may be too costly, or impossible, to set up a second (that is, 3-day) shift.

Trends in the Adoption of the CWW.   In the early 1970s the CWW had "a meteoric rise to fame and public attention."[39] In 1970 only about 40 companies had a CWW plan in use; by 1973 an estimated 3000 companies had implemented such a plan, and companies were converting to the CWW at a rate of about 150 per month. Evidently, 1973 marked the year of peak excitement about the CWW. By 1975 a *Wall Street Journal* article reported that "industry's love affair with the 'four-forty' had clearly cooled," and that many executives were not pleased with the results.[40] Indeed, a survey in 1975 reported that nearly one-third of the adopting organizations discontinued use of the CWW shortly thereafter.[41] Although the total number of organizations using the CWW continued to increase during the 1970s and 1980s, the rate of growth leveled off. Estimated participation rates for all U.S. employees in 1975, 1980, and 1989, respectively, were 2.2, 2.7, and 3.0 percent.

Effectiveness Evidence.   Most of the research on the CWW has been "highly impressionistic," to use the late William Glueck's term. His extensive review yielded 20 anecdotal and 10 "rigorous" accounts of the effects of the CWW.[42] However, even the "rigorous" studies are problematic because most relied on perceptions of productivity. Unfortunately, such perceptions are somewhat akin to contact lenses—in the eye of the beholder. That is, reports of perceived effects largely reflect attitudes toward the intervention (whether respondents want the program to continue) rather than actual changes. The present review examines only those studies which report objective evidence of effectiveness.

     Table 5.2 provides objective data concerning the effects of the CWW on productivity and absenteeism. Six studies reported evidence regarding productivity, with results ranging from a decrease of approximately 4.5 percent to an increase of 3.2 percent. The median

TABLE 5.2   Effects of the Compressed Workweek (CWW) on Productivity and Absenteeism

Study number and year	Setting and sample	Impact on productivity	Impact on absenteeism
1 1971	Small manufacturing company	+3.1%	+35% (hours lost) +4.9% (absence occurrences)
2 1973	Port Authority of N.Y. and N.J.; two departments	"No discernible trend"	NR
3 1973	Pharmaceutical company; 131 employees, various occupations	NR	−10%
4 1974	Industrial products company; 210 operative employees	NR	No change
5 1975	Accounting division of large multinational corporation; 474 clerical employees and supervisors	No significant change	NR
6 1975	Fabric mfg. company; 167 sewing employees	−1.1%	NR
7 1977	Industrial products company; 191 operative employees	NR	No change
8 1984	Midwestern plant; mfr. of business equipment; 559 operative workers and supervisors	−4½% (approx.)	−68%
9 1985	Large industrial corporation; 84 computer operations employees	NR	−33% (sick and personal leave hours)
10 1986	Government agency; 75 accounting employees	+3.2%	−18.9%

Note: A complete list of references is available from the author on request.

result was a change of 0.0 percent. Seven studies have provided absenteeism data: in four cases absenteeism declined, in two cases there was no change, and in one case absenteeism increased. Overall, the median effect was a decrease in absenteeism of 10 percent.

Surveys of employee attitudes have generally shown positive effects. Glueck[43] reported that job attitudes improved in 12 out of 18 studies; similarly Ronen and Primps reported that job satisfaction improved in 5 out of 9 (more rigorous) studies.[44] Yet even these results may be misleading because there is evidence that improvements in attitudes may not endure: "Initially the results are positive. Later, the results return to prior levels or decline relative to the 5-40 pattern. Perhaps there is a 'Hawthorne effect' in the early stages of the change."[45] Adoption of the CWW has frequently been followed by disenchantment and declining enthusiasm. Evidently, the diminished long-term effects reflect the cumulative impact of fatigue; indeed, for many companies the 4-day week has evolved into a "four-day headache."[46]

Boundary Conditions.   Several factors may affect the success of CWW interventions. The CWW is most likely to yield positive results under the conditions listed below.

Nontaxing Job Duties.   Because of the potential for fatigue, the CWW appears to be most effective on jobs that are neither physically nor mentally too demanding. Similarly, attitudes tend to be more positive among incumbents performing low-level jobs, those that offer little intrinsic satisfaction. Under these conditions the 4-day week may result in a perceived upgrading of status and responsibility; importantly, it may also allow employees to "get it over with" sooner.[47]

Minimal External Constraints.   The CWW is more likely to yield positive results in organizations—or organizational subunits—with minimal day-to-day contact with customers and/or suppliers. Such settings would include data processing, manufacturing, or accounting departments, possibly with a "skeleton" crew on hand for emergencies during the fifth day. Alternatively, the CWW is likely to be effective if the organization is able to operate with a "mini" (that is, 3-day) shift.

Adequate Organizational Slack. In small units, unanticipated absences and vacations can result in severe stress due to understaffing. This is especially a problem in settings where the organization operates 5 days a week but employees work on a compressed schedule. Large organizational (or subunit) size, therefore, may ameliorate this potential problem.

Employee Characteristics. Three individual difference variables have been found to reliably mediate the effects of the CWW. In general, young employees are more favorably disposed to the CWW, in part because of the long workday. Similarly, employees without family responsibilities tend to have more positive attitudes; single parents, though, often find the CWW to be burdensome. Employees with a desire for personal growth and achievement also tend to be more receptive to the CWW.

Enriched Jobs. The CWW tends to enhance satisfaction to the extent that it results in enriched jobs. If employees work four out of five workdays, it is likely that variety, autonomy, and authority will increase as people "cover" for each other and/or work with reduced or no supervision—if supervisors take days off. (In one study, however, only 14 percent of the supervisors took their regularly scheduled days off.[48])

Guidelines for Implementing a CWW. Organizations planning a CWW intervention should consider a number of issues prior to implementation. As noted in connection with FWH, it is important that management (1) anticipate and plan for some resistance to change; (2) assign responsibility for policy development, implementation, and evaluation; (3) conduct a feasibility study prior to implementation; (4) carefully review various design parameters and options (that is, 4-40 or 9-80); (5) demonstrate a high level of top management support; (6) elicit the opinions, and hopefully the support, of supervisors and key union representatives before implementation; (7) introduce the intervention as an "experiment" or "pilot test" so that it may be discontinued if unsuccessful; (8) conduct initial review meetings with key personnel to discuss problems; (9) provide supervisors with feedback early so that they can make any adjustments needed; and (10) collect extensive objective data on effectiveness.

Perhaps the biggest implementation issue concerns the choice of specific CWW plan features. In light of the problems found in the empirical literature with respect to the 4-40 plan—particularly the tendency for fatigue to set in—organizations might give extra consideration to the 9-80 plan. With this plan employees typically work eight 9-hour days and one 8-hour day every 2 weeks; about 30 percent of the time the extra day off can be taken on a Monday or a Friday. Clearly, this plan avoids the strain of 10-hour workdays while still offering the employee extra time off with pay.

CONCLUSION

Summarizing results, a comprehensive review of the evidence indicates that, on average, FWH plans yield an improvement in objectively measured productivity of 2½ percent, and a reduction in absenteeism of about 16 percent. FWH plans are generally quite well received, with approximately 80 percent of employees and 70 percent of supervisors *not* in favor of returning to fixed hours. Indicative of the widespread appeal, only 5 to 8 percent of FWH interventions have been subsequently terminated.[49]

The limited impact of FWH on organizational effectiveness is not surprising, given the relatively narrow nature of the intervention. Indeed, while it is logical to expect a (modest) relationship between work schedule and general job attitudes, there is little reason to expect a strong relationship between work schedule and job performance. As Dunham, Pierce, and Castañeda argued recently: "...why expect the work schedule to have a strong direct impact on worker effectiveness? Unless workers must be high performers in order to use a desired schedule, why would it motivate employee effectiveness?"[50] True, but as Nollen observed more than a decade ago, FWH may serve as the initial step in initiating a more comprehensive organizational development program.

A review of the empirical evidence concerning the CWW indicates that, on average, there is no effect in terms of productivity. CWW plans typically yield a modest reduction in absenteeism (a median decline of 10 percent), and roughly 70 percent of participants want the plan to continue. Attitudes toward the CWW have tended to sour over time, though, in part because of the cumulative effects of fatigue (and the nonobservance in practice of longer workdays). It is not surprising, therefore, that the discontinuance rate for the CWW has been roughly five times higher than for FWH. Moreover, in contrast to FWH, the CWW does not alter an employee's control over his or her work schedule. In place of increased discretion and autonomy, the worker is merely able to "escape" for longer contiguous periods.

All in all, future prospects seem far brighter for FWH than for CWW plans. An increasing segment of the work force is highly educated and desires "good" jobs: that is, jobs with more autonomy, discretion, and flexibility. FWH permits greater realization of these aims. Consider the comments of Rebecca Roloff, an executive at an insurance company:

> I love my job, but there are other things that are very important to me. To be as happy as possible in my work, I also need to be a good wife, a good mother, and a good friend. I think it is essential for companies to recognize, as mine already does, that people don't have two lives—one in the office and one at home. To maximize productivity, businesses must help their employees balance all their needs and responsibilities.[51]

While evidence indicates that the existence of FWH does not materially influence the amount of time working parents actually spend with children, it does permit working parents to have greater control in handling work and nonwork conflicts.[52] Furthermore, the effects of the FWH have been found to be greater among female in comparison with male employees.[53] This finding is salient, of course, given the large and growing number of working women. The typical CWW plan (the 4-40), however, often exacerbates the problem of work and nonwork conflicts. Because it is less severe, the 9-80 should witness increased popularity among CWW plans in future years.

All told, alternative work schedules will likely become increasingly prevalent, not so much because they improve individual job performance directly but because they enable organizations to compete more effectively for the most capable employees.

NOTES

1. Ulmer, Walter F., Jr., "Inside View," in the Center for Creative Leadership's newsletter, *Issues & Observations,* vol. 10, no. 4, p. 7, fall 1990.

2. Toffler, Alvin, *Future Shock,* Random House, New York, 1970.

3. Williams, Whiting, *Mainsprings of Men,* Scribners, New York, 1925, p. 3.

4. "Vying for Time," *HRMagazine,* vol. 35, no. 8, p. 37, August 1990.

5. Overman, Stephenie, "Plant Workers, Families, Find Not All Time Off Created Equal," *HRMagazine,* vol. 35, no. 8, p. 38, August 1990.

6. Ronen, Simcha, *Flexible Working Hours,* McGraw-Hill, New York, 1981, p. 181.

7. Nollen, Stanley D., "What Is Happening to Flexitime, Flexitour, Gliding Time, the Variable Day? And Permanent Part-Time Employment? And the Four-Day Week? *Across the Board,* vol. 17, no. 4, p. 9, April 1980.

8. Elbing, Alvar O., Herman Gadon, and John R. M. Gordon, "Flexible Working Hours: It's about Time," *Harvard Business Review,* vol. 52, no. 1, p. 155, January–February 1974.

9. Ronen, op. cit., pp. 58–59, 68.

10. Feuer, Dale, "In Favor of Flex-Time," *Training,* vol. 22, no. 7, p. 8, July 1985.

11. Silverstein, Pam, and Jozetta H. Srb, *Flexitime: Where, When and How?* Cornell University, Ithaca, N.Y., 1979, p. 13.

12. "Giving New Meaning to Clock-Watchers," *New York Times*, Oct. 22, 1985, p. A24.

13. Verespej, Michael A., "The New Workweek," *Industry Week*, Nov. 6, 1989, p. 12.

14. Easy Klein, "Beyond the Paycheck," *D&B Reports*, vol. 37, no. 2, p. 38, March–April 1989.

15. Curry, Talmar E., Jr., and Deane N. Haerer, "The Positive Impact of Flexitime of Employee Relations," *Personnel Administrator*, vol. 26, no. 2, pp. 63–64, February 1981.

16. Ronen, op. cit., p. 238.

17. Ibid.

18. DuBrin, Andrew J., *Contemporary Applied Management*, Business Publications, Inc., Plano, Tex., 1982, p. 199.

19. Ibid., p. 202.

20. Ibid., p. 202.

21. Elbing, Alvar O., et al., op. cit., p. 154.

22. It should be noted that 1980 estimates derived from the Bureau of the Census' Current Population Survey included data on self-employed individuals; in contrast the 1985 CPS data excluded self-employed individuals and hence are more pertinent. Thus the apparent leveling off in the incidence of FWH is partly a measurement artifact. Although two recent surveys of employers have reported higher adoption rates for FWH (in the range of 25 percent of all companies), it is important to note that the sample sizes are small and respondents not necessarily representative of all employers.

23. Newstrom, John W., and Jon L. Pierce, "Alternative Work Schedules: The State of the Art," *Personnel Administrator*, vol. 24, no. 10, p. 22, October 1979.

24. Burdetsky, Ben, and Marvin S. Katzman, "Evaluation of Alternative Work Pattern Applications," unpublished paper, George Washington University, 1981, pp. 1–11.

25. Nollen, Stanley D., "Does Flexitime Improve Productivity?" *Harvard Business Review*, vol. 57, no. 5, pp. 12, 18, September–October 1979.

26. Scott, Dow, and Steve Markham, "Absenteeism Control Methods: A Survey of Practices and Results," *Personnel Administrator*, vol. 27, no. 6, pp. 73–76, 81, June 1982.

27. Golembiewski, Robert T., Rick Hilles, and Munro S. Kagno, "A Longitudinal Study of Flexitime Effects: Some Consequences of an OD Structural Intervention," *Journal of Applied Behavioral Science*, vol. 10, p. 508, 1974.

28. Ronen, op. cit., pp. 137–139, 152, 192–193, 202–203.

29. Hicks, William D., and Richard J. Klimoski, "The Impact of Flexitime on Employee Attitudes," *Academy of Management Journal*, vol. 24, p. 340, 1981.

30. Ronen, op. cit., p. 73.

31. Hamilton, Patricia W., "Helping Out with the Kids," *D&B Reports*, vol. 37, no. 2, p. 19, March–April 1989.

32. Nollen, op. cit., p. 18.

33. Silverstein and Srb, op. cit., p. 23.

34. Gomez-Mejia, Luis R., Michael A. Hopp, and C. Richard Sommerstad, "Implementation and Evaluation of Flexible Work Hours: A Case Study," *Personnel Administrator*, vol. 23, no. 1, pp. 39–41, January 1978 (Control Data); Hermine Zagat Levine, "Alternative Work Schedules: Do They Meet Workforce Needs? Part 2," *Personnel*, vol. 64, no. 4, pp. 66–68, April 1987 (Liberty Mutual).

35. Levine, Hermine Zagat, "Alternative Work Schedules: Do They Meet Workforce Needs? Part 1," *Personnel*, vol. 64, no. 2, p. 60, February 1987.

36. Ahmadi, Mohammad, Farhad M. E. Raiszadeh, and William L. Wells, "Traditional vs Non-Traditional Work Schedules; A Case of Employee Preference," *Industrial Management*, vol. 28, no. 2, pp. 20–23, March–April 1986; Stu Newman, "Working Alternatives," *Supervision*, vol. 58, no. 7, pp. 11–13, July 1989.

37. Poor, Riva, ed., *4 Days, 40 Hours*, Bursk and Poor Publishing, Cambridge, Mass., 1970, p. 7. This quote appeared in the Foreword written by Paul A. Samuelson.

38. Donovan, Sharon, "New-Wave Workweeks," *Credit Union Management*, vol. 10, no. 6, p. 46, June 1987.

39. Asher, Jules, "Flexi-Time, Four-Day Week Thrive Despite Recession," *APA Monitor*, vol. 6, p. 5, April 1975.

40. Silverstein and Srb, op. cit., p. 10.

41. Calvasina, Eugene J., and W. Randy Boxx, "Efficiency of Workers on the Four Day Workweek," *Academy of Management Journal*, vol. 18, p. 605, 1975.

42. Glueck, William F., "Changing Hours of Work: A Review and Analysis of the Research, *Personnel Administrator*, vol. 24, no. 3, pp. 46–47, 66–67, March 1979.

43. Glueck, op. cit., p. 46.

44. Ronen, Simcha, and Sophia B. Primps, "The Compressed Work Week as Organizational Change: Behavioral and Attitudinal Outcomes," *Academy of Management Review*, vol. 6, pp. 66–67, 1981.

45. Glueck, op. cit., p. 46.

46. Bulkeley, William, "For Some Companies the Four-Day Week Is a Four-Day Headache," *Wall Street Journal*, Apr. 30, 1973, pp. 1, 36.

47. Latack, Janina C., and Lawrence W. Foster, "Implementation of Compressed Work Schedules: Participation and Job Redesign as Critical Factors for Employee Acceptance," *Personnel Psychology*, vol. 38, pp. 76–77. Interestingly, Latack and Foster found that although the CWW enriched jobs and made work more satisfying, the longer workday also made the desire for escape stronger.

48. Goodale, James G., and A. K. Aagaard, "Factors Relating to Varying Reactions to the 4-Day Workweek," *Journal of Applied Psychology*, vol. 60, p. 37, 1975.

49. Nollen, Stanley D., and Virginia H. Martin, *Alternative Work Schedules. Part 1: Flexitime*, Amacom, New York, 1978, p. 44; *National Productivity Review*, vol. 1, p. 244, spring 1982.

50. Dunham, Randall B., Jon L. Pierce, and Maria B. Castañeda, "Alternative Work Schedules: Two Field Quasi-Experiments," *Personnel Psychology*, vol. 40, pp. 237–238, 1987.

51. Davidson, Gail, "Changing Times: Flexible Work Options for Managers," *HBS Bulletin*, December 1990, p. 37. Reprinted with permission, © 1990 by the President and Fellows of Harvard College.

52. Presser, Harriet B., "Can We Make Time for Children? The Economy, Work Schedules, and Child Care," *Demography*, vol. 26, p. 534, 1989.

53. David A. Ralston and Michael F. Flanagan, "The Effect of Flextime on Absenteeism and Turnover for Male and Female Employees," *Journal of Vocational Behavior*, vol. 26, pp. 206–217, 1985.

CHAPTER 6
MAINTENANCE PLANNING AND SCHEDULING

Thomas A. Westerkamp
Contract Consultant
H. B. Maynard and Company, Inc.
Pittsburgh, Pennsylvania

This chapter is divided into five parts: organization, planning, scheduling, assigning work, and effective use of the computer in these maintenance functions. Measurement of maintenance activities is presented in Sec. 5, Chap. 6 of this handbook.

ORGANIZATION FOR MAINTENANCE PLANNING AND SCHEDULING

Organizing for maintenance activity is most successful when based on sound principles and when the organization form meets the tests of logic, communication, and balance.

Organization Principles.   The following are key principles to consider in organizing the maintenance function.

Customer-Service Relationship.   The natural justification for the existence of maintenance activity is that an enterprise requires facilities and equipment in good operating condition to carry out its responsibilities. These facilities must be maintained at a level that results in efficient and economical use for the purpose intended. The cost of construction and maintenance is part of the overall cost of operations, and funds for this activity are budgeted to the operating departments. In short, "the customer pays the bill." Operations is the chief customer or user of the construction and maintenance service. Maintenance provides the service required. This customer-service relationship is the foundation for assignment of authority and responsibility to the members of the organization team.

Optimum Crew Size Principle.   The optimum crew size is the smallest number of workers who can perform a job using a good representative method in a safe way.

Timeliness Principle.   Schedule control points must be at frequent enough intervals so that detection of problems occurs in time to bring about on-time completion of the job.

MAINTENANCE PLANNING AND SCHEDULING RESPONSIBILITY

KEY:
R = RESPONSIBLE   C = COORDINATES   A = ASSISTS

(NOTE: FOR A SPECIFIC WORK ORDER, ONLY ONE "R" APPLIES TO EACH ACTIVITY)

ACTIVITY DESCRIPTION	PRODUCTION REQUESTORS	ENG. & OTHER REQUESTORS	IND. ENG.	MRO STORES/ PURCH.	FOREMAN	GEN. FOREMAN	MAINT. SUPER.	MECH. ENG.	PLANNER COORD.	PLANNER	MAT'L. COORD.	MAINT. ENG.	TRADESMEN	CLERK TYPIST	MAINT. CLERK
1. DEFINE WORK REQUEST CONTENT	R	R							C	A	R				
2. INITIATE REPAIR WORK REQUEST	R	R													
3. INITIATE ENGINEERING WORK REQUEST		R													
4. APPROVE WORK REQUEST	R	R													
5. SET PRIORITY & COMPLETION DATE	R	R			A	C			A						
6. DELIVER WORK REQUEST TO MAINTENANCE	R	R													
7. FIELD CHECK. DETERMINE WORK CONTENT					A				C	R					A
8. DETERMINE MAT'L, TOOLS, EQUIPMENT & MANNING NEEDED		R			R	C			A						
9. FIND MAT'L, TOOLS & EQUIPMENT				A					C	R					
10. REQUISITION PURCHASED MATERIAL/TOOLS				A					C	R	A				
11. STAGE MAT'L, TOOLS, EQUIP. FOR DELIVERY				R					C	A	A				
12. DELIVER MAT'L., TOOLS, EQUIP. TO JOB SITE				R					C	A	A				
13. MAKE JOB SITE AVAILABLE	R	R													
14. PROVIDE PERMITS/ADMISSION TO WORK					R				C		R				
15. PREPARE WEEKLY WORK SCHEDULE	A	A			A	R			C	A			A		
16. PREPARE DAILY SCHEDULE										R			A		
17. APPROVE DAILY SCHEDULE															
18. ASSIGN TRADESMEN TO SCHEDULED JOBS					R								A	A	
19. LOAD JOB ASSIGNMENT BOARD					R					A					
20. VERIFY JOB STATUS, QUALITY & COMPLETENESS					R										
21. REPORT TIME, DELAYS, ADDITIONAL WORK DONE OR NEEDED - EACH JOB													R	A	A
22. RECORD EQUIPMENT I.D. & SPECIFICATIONS		A				C			A		R		A		
23. DESIGN PM TASKS & FREQUENCY												R			
24. APPLY PLANNED HRS. TO W.O.'S & ADJUST FOR CHANGES										R					
25. APPROVE TIME & DELAY REPORTING					R										
26. POST TIME & DELAYS TO WORK ORDERS DAILY									C					R	R
27. PREPARE WEEKLY CONTROL REPORTS									C					R	
28. ANALYZE WEEKLY CONTROL REPORTS			A		R	R	R	R							
29. ANALYZE EQUIPMENT HISTORY												R			
30. INITIATE PM CORRECTIVE ACTION					R	R	R	R				R			
31. AUDIT UMS DATA AND APPLICATION			R												
32. AUDIT SYSTEMS AND PROCEDURES			R												
33. SEQUENCE MULTI-CRAFT JOBS					R				C	A					A
34. SEQUENCE MULTI-SHIFT JOBS					R				C	A					A
35. DEVELOP OR REVISE UMS DATA			A						C	R		A			
36. APPROVE INTERNAL MAINTENANCE W.O.'S					R										

FIGURE 6.1 Activity responsibility chart.

Activity Responsibility.   Job control depends upon definite responsibility for each activity (Fig. 6.1) in the life of a work order. For example, the requester is responsible for delivering the work order to the maintenance department.

Organization Forms and Ratios.   There are two basic forms of maintenance organizations—central and area. In a central organization, all activity is controlled from a central maintenance shop. In an area organization, authority is delegated to various area shops, and maintenance crafts are assigned there. A central organization offers more control while an area organization offers better response due to shorter travel requirements. A third organization form, the area-central arrangement, is a combination of the first two forms. It is used in large, complex organizations and offers a compromise that achieves both high control and quick response. Trades used in an area every day are assigned there, while those not used often are dispatched from a central shop.

Organization Ratios.   Ordinarily, a good ratio of crafts to supervision is up to 15:1 while a good ratio of planners to crafts is 30:1. Therefore, a basic work unit is 1 planner, 2 supervisors, and 30 maintenance workers. General supervisors usually have four line supervisors reporting to them. One maintenance engineer and one material coordinator can support about 100 workers.

Effect of Other Factors on Ratios.   In establishing ratios, other factors such as scope of responsibility must be considered carefully. A supervisor with a crew of 15 who must spend half of the day obtaining materials is actually working with a craft ratio of 30:1 (15 at half time).

The skill of the maintenance supervision and work force and the nature of the work also affect ratios. A trained, experienced crew doing routine preventive maintenance assignments or working in a central machine shop may require only one supervisor for 20 to 30 workers provided that effective planning and material coordination functions exist.

PLANNING MAINTENANCE WORK

Communication is necessary in order to convert a request for service into a completed job. The first line of communication in all areas is the maintenance planner and the authorized requester. Others may be delegated to handle requests in certain well-defined situations. Nevertheless, the responsibility for maintaining good communications is theirs.

Using the Work Order to Communicate Maintenance Needs.   Since a large volume of these requests may occur in a relatively short time, they must be in writing. The standard form for this purpose is the work order, which transmits information between the requester and the planner, the planner and the foreperson, the planner and other crafts or shops, the foreperson and the craftsperson, and supervision and higher management. At each stage in the process the work order gains more complete information. As each job is completed, it is a constant, fresh source of feedback information that provides management with facts needed for corrective action. The objective of the work order is to promote optimum utilization of maintenance laborpower and to provide a written record of actual day-to-day work done.

The work order form shown in Fig. 6.2 is completed progressively by the requester, planner, worker, and foreperson. The requester fills in the work order information, records the request in the log, and forwards the work order to the planning office. The planner field checks the job, plans the work, and forwards the work order to the foreperson for assignment to the proper craft.

Work Orders for High-Frequency Tasks.   Many requests for maintenance service are completed in less than an hour. It is not unusual to find that 80 percent of the jobs account for as little as 20 percent of the workhours. The procedures that control this 20 percent of the workhours must be efficient and practical so that most of the foreperson and planner effort can be directed toward the 80 percent of the workhours. A separate work order for each of these short-range, repetitive jobs will soon have the planners and forepersons bogged down with unnecessary paperwork. On the other hand, any alternative procedure must retain for supervision a degree of control that promotes effective operations. Blanket work

WORK ORDER NO	DATE	TIME	REQUESTER	PHONE	DEPT/SO	EQUIPMENT NO	BUILDING	LOCATION

CLASS	PRIORITY	SHIFT	AVAILABLE	REQUIRED	PERMIT	APPROVED BY	RECEIVED	PLANNER

WORK NEEDED/PROBLEM

WORK PLAN
1.
2.
3.
4.

NO TRADE PLN HRS

MATERIAL DESCRIPTION

CAT NO UN QUANT

SPECIAL EQUIPMENT

OTHER WORK PERFORMED ☐   NEEDED ☐

	FOREMAN	DATE	TIME

workord 8-90

FIGURE 6.2 Sample work order form. (*Reprinted with permission of the publisher from Thomas A. Westerkamp, "Maintenance Manager's Standard Manual and Guide," Prentice-Hall, Englewood Cliffs, N.J., 1992.*)

orders with no specific, defined work content do not accomplish this. The alternative that best fits this situation is the use of weekly work orders (Fig. 6.3) issued to an individual worker and closed once a week. Two general groups of work can be profitably controlled with the weekly work orders: (1) specific routes traveled each day, and (2) grouped high-frequency assignments such as minor adjustments that occur frequently at random times.

All employees who are assigned checklists should also have other routine work orders assigned to them. They will need other work at times when no checklist work is required.

Checklist Work Order Benefits. Many benefits are gained by the maintenance worker, requester, and planner from use of the checklist work order. A checklist work order reduces the number of work orders, requires less writing for each job, reduces time waiting for assignments, reduces response time, and provides better communication. Preprinted planned hours are applied only once, and separate work order handling for each occurrence is not required.

Work Order Numbering. Volume of work orders processed in a year is very large even in fairly small plants. Therefore, success of the system will depend to a great extent on the careful control of the status of every work order from the time it is originated and received by the maintenance planner to the time it enters the closed order file (also see computer applications). A numerical control system using two different sets of numbers is used for this purpose. These sets of numbers are the requester's number and the work order number.

Numerical Control of In-Process Work Orders. Each of these numbers is listed in one or more registers along with the description of the work and information about the status of the job. Each register represents an information center for all work orders that have passed through that point of control. For easy retrieval of information, entries must be in numerical sequence. The actual sequence of requests from a single requester will be dif-

CHECKLIST NO. 1								
JOB	HRS. PER OCC.	FREQUENCY						TOT PLN HRS.
		M	T	W	TH	F	TOT OCC.	
ROLL AREA 475								
Adjust Turkey	0.7	I	II	I	III	₶	12	8.4
Adjust Side Plates	1.2							
Adjust Knife	0.7							
Adjust Hopper Door	0.6							
Adjust Brake	0.7							
Adjust Steam	0.8							
Adjust Flying Cutter	0.6							
Adjust Slitter Knife	0.7		I			I	2	1.4
Adjust Packing	0.3							
FINISHING AREA 476								
Adjust Trimmer	0.6							
Adjust Power Knives	0.9							
Adjust Saw	0.6							
Change Dowel Knife	1.1							
TOTAL PLANNED HRS.								9.8

FIGURE 6.3 Weekly checklist work order for specific daily assignments or grouped tasks. Hours per occurrence multiplied by number of tallies equals planned hours completed.

ferent from the sequence of receipt of work orders by the maintenance planning section from several different requesters.

Work Order Numbers for Emergency Work.   Since an emergency call is responded to in some cases before the paperwork is completed, the work order brought back to the shop by the mechanic may have no work order number on it. The work order is given to the planner who registers the job, assigns a number, and postapplies planned hours based on the work performed so that complete history records can be maintained.

Preventive-Predictive Maintenance.   As equipment becomes more sophisticated, the consequences of letting a breakdown happen become more severe. The foremost concern of management is the safety hazard presented when a machine breaks down while an operator is using it. In addition, the more automated the process, the more likely major machine damage will occur. Another serious effect is the disruption of the production or operations schedule, which can result in customer relations problems and even loss of orders.

Both preventive and predictive maintenance are advanced techniques that elevate maintenance above mere fire fighting and enhance management control.

Preventive Maintenance and Equipment History.   Implementing a preventive maintenance (PM) system is divided into five steps:

1. Establishing the PM system
2. Daily preventive inspection
3. Using the PM system
4. Equipment records
5. Revising the PM system

Definition of Preventive Maintenance.   Preventive maintenance is the systematic planning, scheduling, and completion on schedule of needed maintenance work that is designed to ensure highest availability of equipment and facilities, prolong the useful life of capital assets, and reduce costs. This work includes inspection, cleaning, lubrication, replacement, or repair and is scheduled annually to be done at regular, preplanned intervals.

Establishing the PM System.   To establish any preventive maintenance job, three factors must be determined. They are (1) work content, the description of the operations to be done, and the sequence of the operations; (2) frequency, the amount of calendar time or machine hours between successive repetitions of the job, and (3) schedule, the selected day(s) over a 12-month period when the job is to be done.

The preventive maintenance assignment form shown in Fig. 6.4 is used to describe the equipment or facility, the work to be performed, and material required. The amount of coverage included in one PM assignment varies widely. This will depend upon the nature of this work and the equipment involved. Inspection and lubrication jobs, for instance, are usually combinations of a large number of individual tasks performed at slightly different locations. Only a very short time is spent at one location (inspection or lubrication point). For this type of work, one PM assignment serves as a route sheet for a large number of short interval tasks.

Preventive maintenance assignments are established in the following manner:

1. Select a production area critical to the overall plant operation that is currently experiencing a high level of maintenance activity.
2. Starting with preventive inspection routes (see Fig. 6.4), define in detail the preventive maintenance required. Sources for this information are production and maintenance supervisors and engineering and vendor's operation and maintenance manuals.
3. Establish the frequency for repetition of the assignment.
4. Prepare the PM assignment.
5. Schedule the PM assignment annually.
6. After all the PM assignments for this area, machine, or department are identified and scheduled, go to the next area and repeat the five steps listed above until all areas of the plant are included in the PM system.

Applying Planned Hours.   Each PM assignment has a planned completion time assigned by a planner after the work description is completed. "Annual hours" are posted on the bottom of the form after calculating them as follows:

$$\text{Annual hours} = \text{planned hours} \times \text{frequency}$$

$$= 0.4 \text{ hour} \times 250 \text{ (daily)}$$

$$= 100 \text{ hours}$$

pni Written by	PREVENTIVE MAINTENANCE ASSIGNMENT	NO. PG. OF

1. EQUIPMENT NO. #1 and #2	2. EQUIPMENT NAME ALDRICH PUMPS			
3. DATE DUE Daily	4. LOCATION Extrusion Press – Bldg. 8472	5. SIZE 4		6. TYPE XCL
7. PLANNED HOURS 0.4	8. WORK TO BE PERFORMED (USE COLUMN HEADINGS ONLY FOR LUBE WORK. SEE BELOW)		9. FREQ. Daily	10. CRAFT Area Mechanic

11. JOB TYPE		R E F	LUBE LOCATION A.	FITTINGS NO. B	FITTINGS TYPE C	TYPE OF LUBE D	AMOUNT OF LUBE E	FREQ. (LUBE) ADD F	FREQ. (LUBE) REMOVE G
CLEAN	☐	1	Location			Condition			
REPLACE	☐	2	Oil Pressure for Pump Lube						
REPAIR	☐	3	System?			#1		#2	
CHECK	☐	4	a. At filter			35 psi		38	psi
LUBRICATE	☐	5	b. At outboard Bearing			42 psi		40	psi
12. MATERIAL		6	(Should be at least 30 psi)						
R E F	NO.	DESCRIPTION	7						
			8	Oil Level in Crankcase?			No Ok	✓	Ok
			9						
			10	Motor Pit Drain Clean?			✓ Ok	✓	Ok
			11						
			12						
			13						
			14						
			15						
			16						
			17						
			18	Is Water Flowing Thru Relief			✓ Ok		Yes	Ok
			19	Valve?						
			20	One Second						
			21	Delay for 3" By-Pass to Pump						
			22	After Being Energized?			✓ Normal		✓	Normal
			23						

REF	WORK REQUIRED
8	LOW – ADD OIL
18	SLIGHT LEAKAGE, REPAIR VALVE

A. Describe location of grease fitting or oil container.
B. Number of fittings at this lube location.
C. Grease or oil. Fitting or sup.
D. Lube code on product drum.

E. Number of shots or ounces.
F. & G. (W) Weekly; (2W) Every 2 weeks; (M) Monthly;
(3M) 3 Months; (6M) 6 Months; (A) Annual

Annual Hours	Date	Approved
100	5-20	P.M. Engineer

FIGURE 6.4 Daily preventive maintenance inspection.

Cost Approval for PM Assignments.   Each PM assignment schedule must be justified on the basis of improved quantity or quality of output, improved equipment availability, or improved asset life at optimum cost.

Predictive Maintenance.   In a predictive maintenance program, the engineer analyzes equipment condition while it is running and finds optimum repair intervals. This is always less costly and more reliable than the fixed frequency PM interval based on factors such as machine hours or calendar time.

Advantages of Predictive Maintenance.   Some key advantages of predictive maintenance are

- Warns of failure before it occurs
- Measures extent of substandard condition
- Identifies cause
- Portable or fixed
- Instant or continuous
- May not require shutdown
- Applies to existing equipment and acceptance tests for new equipment

Types of Instruments Used.   A wide range of diagnostic instruments can be applied to predictive maintenance. Some of these are listed in Fig. 6.5 showing the measurable characteristics used by engineers to identify maintenance problems and the detection instruments used. In all cases, these analytical techniques are extensions of human senses. They magnify normally undetectable physical characteristics so that impending quality problems or equipment failure, possibly leading to major machine damage or increased costs or downtime, can be avoided.

Comparison of Predictive Maintenance with Other Types of Maintenance.   The illustration in Fig. 6.6 shows the effect of predictive maintenance compared with emergency maintenance and preventive maintenance. Note that both downtime and cost are far lower using a predictive maintenance approach, in this case based on a differential pressure P limit to determine cleaning frequency in a heat exchanger. Clearly, emergency maintenance is the least effective and most costly approach.

Measurable characteristics	Type of instruments and tests
Light	Optical: microscopes, magnifiers, cystoscopes, cameras
Speed	Tachometers
Torque	Torque meter, prony brake
Tension and compression	Load cell, strain gauge, test specimen
Pressure	Pressure gauge, pressure recorder-controller
Volume	Flowmeters
Noise	Sound level meter
Vibration	Vibration amplitude and frequency analyzers
Dimension	Micrometers, electronic gauges, precision scale
Flaws	Ultrasonic, dye penetrant, magnetic particle, x-ray, radioactive isotopes
Leaks	Leak detectors, gas analyzers, soap test, freon tester
Electric current	Volt-ohm-ammeter, tong tester, continuity tester
Electrical	Megohmmeter
Corrosion insulation	In-line test specimen
Lubrication	Carbon residue, flash and fire, cloud and pour, spectrographic
Temperature	Thermometer, pyrometer, infrared ray
Liquid level	Pressure gauge, float gauge, liquid level meter

FIGURE 6.5   Measurable characteristics versus diagnostic instruments applied to predictive maintenance.

TYPE OF MAINTENANCE	REPAIR CYCLE	NUMBER OF REPAIRS	TIME TO REPAIR	ANNUAL REPAIR HOURS	ANNUAL REPAIR COST
EMERGENCY	AT FAILURE	6	16	96	$ 480,000
PREVENTIVE	4 WEEKS	13	2	26	130,000
PREDICTIVE	P = 50 psig	7	2	14	70,000

FIGURE 6.6 Comparison of cleaning heat exchanger tubes using emergency, preventive, and predictive maintenance shows that emergency maintenance is the most costly and results in longer downtime. The downtime is unscheduled.

Setting Up a Vibration Analysis Program.   Vibration of a running machine is the back-and-forth motion of its components from their position at rest. The total distance of movement is the peak-to-peak displacement, or amplitude of the vibration. The number of cycles of this displacement per unit of time is the frequency of the vibration. The speed of the machine movement changes depending on its position in the cycle. Since it reverses direction every cycle, the peak velocity occurs midway between reverses. A single vibration cycle is shown in Fig. 6.7.

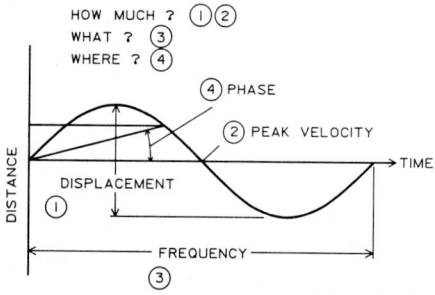

FIGURE 6.7 Characteristics of vibration. (*Courtesy of IRD Mechanalysis, Inc., Columbus, Ohio.*)

Each piece of moving or rotating equipment has a vibration tolerance limit. In a vibration analysis program, the vibration amplitude and frequency are measured periodically using either a fixed or portable vibration analyzer and recorded as in Fig. 6.8. When the preset tolerance limit is approached, the equipment can be scheduled for repair without a breakdown and without excessive repair cost due to shutting down more often than necessary. Five steps are required to set up a vibration analysis program:

1. Collect critical equipment identification data and prepare a history record card for each unit.

2. Determine the range of frequencies your equipment covers. Frequency is measured in cycles per minute (cpm). A range of 300 to 150,000 cpm covers most situations.

3. Select the vibration analyzer. This tool is a transducer pickup, connected to a meter that indicates amplitude of vibration in mils (0.001 inch) for a given frequency (cpm).

4. Set the tolerance limit for each piece of equipment established from a general machinery severity chart supplied by the analyzer vendor.

5. Train inspectors to take readings with the analyzer.

Using the System.   Initial readings and subsequent readings are taken at the same spots on the machine. Visible marks are made on each unit where horizontal, vertical, and axial measurements are to be taken to indicate condition. These measurements point to the exact part causing vibration close to the tolerance limit.

Common Vibration Causes.   Machine vibrations are caused by mechanical defects. Common causes and how they are identified are:

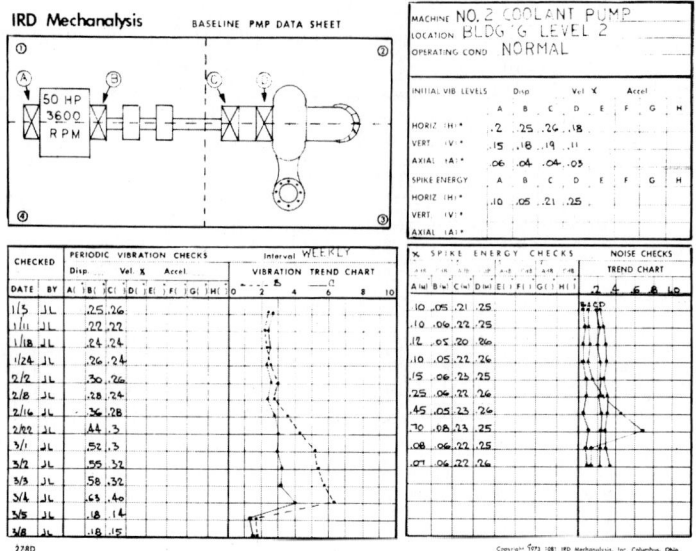

FIGURE 6.8   Vibration history. (*Courtesy of IRD Mechanalysis, Inc., Columbus, Ohio.*)

- Unbalance is the most common cause of vibration and is at the rotating speed of the machine.
- Misalignment, the second highest cause of vibration, is a high axial vibration at two times the rotating speed (rpm).
- Mechanical looseness of parts causes vibration at twice the rpm in the radial direction.
- Bent shaft vibration is similar to misalignment.
- Bad antifriction bearings will cause a vibration at very high frequency which is many times greater than the rotating speed.
- If the vibration cause is bad gears, the frequency will be the number of gear teeth on the bad gear times the rpm.
- Electrical trouble can cause electrical unbalance in rotors. The amplitude reading will instantly disappear when the power is shut off.
- If uneven or pulsating aerodynamic and hydraulic forces are present, vibration frequency will equal the number of blades times the rotation speed.

Other Advantages.   There are other advantages to vibration analysis. The number of spare parts can be greatly reduced if you can predict the need more accurately and in advance. The analyzer also lends itself well to witness tests prior to purchase of new equipment.

Equipment History Records.   The chief source of information for equipment history is the work order. In order to accumulate information about the repairs made on a single piece of equipment, the work performed is transferred from the work order to an equipment history record, or in simpler systems, the work orders are accumulated in a file by equipment or asset number. Also see discussion of computer use below for computerized work order and equipment record systems.

Development of the System.   To develop the record system, an equipment history record is prepared for each major unit. A property number tag is attached to the equipment and the same number is assigned to the record card for that unit. The nameplate information from the unit's nameplate is obtained for each piece of equipment. The accounting department notifies the maintenance department in writing each time a new asset is purchased, assigns a property number, and provides a property number tag. Complete nameplate information is entered on the heading of the record. This is the source for later identification of specifications when the supplier is contacted concerning the unit, when parts are requisitioned or quotations are requested, and when a replacement is made from stock spares.

Use of the Equipment Record System.   Periodically, a summary of specific equipment is prepared for analysis. This analysis includes the following:

1. Total occurrences of repair by type of repair or replacement
2. Total repair cost for the current period compared with past similar periods
3. Total downtime hours
4. Availability and condition of spares
5. Need for design improvements
6. Need for PM routines (total downtime hours or a history of repetitive repair are indicators)
7. Average time between repairs
8. Average time to repair

Using Equipment History for Statistical Analysis.   Accumulation of historical repair information is the means to valuable frequency-of-repair, repair-or-replace, and make-or-buy decisions.
    The most frequently used statistics are:

1. Time between failures (TBF)
2. Time to repair (TTR)

Under a program that is succeeding, TBF should be increasing and TTR should be decreasing. The charts in Figs. 6.9 and 6.10 illustrate the application of these data. The effect of design or method improvements should clearly show a reduction in time to repair, or an increase in time between repairs.

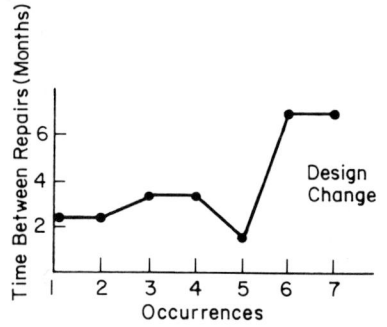
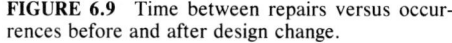
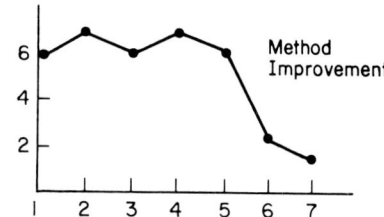

FIGURE 6.9   Time between repairs versus occurrences before and after design change.

FIGURE 6.10   Time to repair versus occurrences before and after repair method improvement.

SCHEDULING MAINTENANCE WORK

The objective of maintenance scheduling is to fulfill the most important responsibility of the foreperson—to schedule a full day's work for each worker the day before the work is assigned.

The Daily Maintenance Schedule. On the day before the work is to be done, the mainte-nance planner lists the work orders and planned hours for each on the daily schedule form (Fig. 6.11), and the maintenance foreperson forecasts the actual hours.

Role of the Backlog in Planning and Scheduling Work. Backlog is the approved mainte-nance repair and project workload. Backlog figures and trends are vital to determine how to distribute the work force by craft and by area among the various assignments. When the ready-to-work backlog is too low, the foreperson is forced to assign jobs which have not been planned, a handicap to improving productivity. When it is too high, jobs are inter-rupted frequently and too many are in progress for good control. A typical backlog range is 2 to 4 weeks' work measured as follows:

$$\text{Backlog weeks} = \frac{\text{total planned workhours}}{\text{number of workers} \times 40 \text{ hr/wk} \times \text{productivity}}$$

Example:

$$\text{Assume total planned workhours} = 400$$
$$\text{Number of workers} \qquad\qquad = \quad 7$$
$$\text{Productivity} \qquad\qquad\qquad = \quad 70\%$$

$$\text{Backlog weeks} = \frac{400}{7 \times 40 \times 0.70}$$

$$\text{Backlog weeks} = 2.0$$

To be comparable from week to week, backlog reporting must be consistent and com-plete. Two basic classes of work, unscheduled and scheduled, are included in computing the backlog as follows:

1. Unscheduled work—Types of work included in this class are (*a*) checklist hours—the planner obtains this number by totaling planned hours applied to checklist work orders each week; and (*b*) emergencies—here the planner estimates the hours on jobs which interrupt the foreperson's daily schedule.
2. Scheduled work—Includes the following types of work defined on work orders: (*a*) rou-tine repairs (not covered by checklist); (*b*) preventive maintenance scheduled jobs, such as inspection, cleaning, and adjustment; and (*c*) major projects, such as construc-tion, overhauls, and remodeling including engineering projects.

Backlog File. All pending work orders except those scheduled for today which should be in the job board and those which the planner is planning at the moment are kept in the backlog file. The sections of the backlog file are:

1. Emergency work orders should stay in this section only for the time required to notify the foreperson to assign the job to a worker or for postapplying planned hours to com-pleted emergencies.
2. To be planned.
3. Hold for material.
4. Hold for scheduling.
5. Next 30 days—The bulk of the work orders should be in this section.
6. Long-range—Projects not due to start for at least 30 days or jobs requiring outage of major equipment.

The planner who is successful at using the backlog file to control work orders will find many other tasks much easier, such as:

DAILY MAINTENANCE SCHEDULE

SUPERVISOR   SHIFT ☑1 □2 □3   AREA ELEC. SHOP   DATE 6-3-74

WORK ORDER NUMBER	JOB DESCRIPTION	J. SMITH	R. JONES	B. BROWN	J. DOE	HOURS PLAN	HOURS ACTUAL	COMPLETE (✓) (No)
1011		2.4				1.6		✓
1013		1.2				0.8		✓
1124		1.2				0.8		✓
1899	(Brief job description here)					0.8		✓
1079		1.8				1.2		✓
1067		0.9				0.6		✓
1103		1.5				1.0		✓
1105		1.1				0.7		✓
1085		2.2				1.5		✓
1955		3.0				2.0		NO
1956				8.1		10.8		✓
1020	REPLACE DIAPHRAGM, 6" STEAM REDUCING VALVE							

TOTAL HOURS   8.1   8.4   8.2   8.2   21.8

Note 1: Calculate forecast actual hours with nomograph as follows:
1. Current productivity = 66%
2. Job 1011 UMS Hours = 1.6
3. Forecast actual hours
```
   = 1.6
     .66
   = 2.4
   ====
```

Note 2: Double line indicates end of scheduled work. Emergencies are listed below.

Note 3: % Compliance = UMS Hours complete / UMS Hours planned
```
   = 19.8
     21.8
   = 91 %
   ====
```

FIGURE 6.11 Daily maintenance schedule.

10.135

- Daily schedule preparation
- Weekly backlog report preparation
- Answering requester inquiries
- Reporting on and following up on material
- On-time work order completion

All work orders are kept in the file and never stored, even temporarily, in any other location if the file is to be representative of the true workload. This includes interrupted work-in-progress. This work should be returned to the backlog file until it is ready to be completed.

Starting the Backlog File.   To set up the backlog file, five sources of information are used:

1. Operations and maintenance supervision
2. Planned projects
3. Facilities engineering
4. Preventive maintenance schedule and inspections
5. Vendor operation and maintenance manuals

Backlog Trend.   Since the workload is constantly changing, management must know whether the trend is increasing or decreasing. This trend is a deciding factor in determining the level and distribution of laborpower. It is displayed by the use of a backlog chart made up of information taken from the backlog file by type of work. The planner in each area reviews the backlog file weekly and totals the planned hours of work in the file plus estimated hours on unplanned jobs. These figures are posted to the backlog trend chart.

Maintenance Priorities.   A simple, effective maintenance priority system as shown in Fig. 6.12 is used to ensure that response to a request for maintenance service is made according to the urgency of the request. Maintenance priority policy is one of the most important considerations higher management faces in designing an effective maintenance program. If management decides that immediate response to all production-related repairs is essential, labor cost will be very high, as will unutilized time of workers between emergencies. If, on the other hand, a more reasonable expectation exists, say a response time of 1 or 2 weeks coupled with planned downtime for repairs, a much lower cost will occur but downtime will be lower. Compare the high- and low-priority systems below.

Safety Requirements.   Safety requirements are extremely important and should receive careful and continuing attention. These requirements will fall into all four of the priority categories shown in Fig. 6.12. Priorities should be established for safety work according to circumstances of each situation, and the work order should be plainly marked "SAFETY."

Preventive Maintenance Annual Schedule.   To use the PM system, maintenance supervisors are responsible for assigning the PM jobs to workers so that the jobs are completed on schedule. To accomplish this requires the combined efforts of the PM engineer, the maintenance supervisor, and the planner. The PM engineer and planner assist the supervisor by watching the annual schedule for occurrence of a PM requirement and preparing the work order for the work order backlog file (see Preventive Maintenance). The procedure for initiating a PM job is as follows:

1. A month before the scheduled date, the PM assignment numbers due are determined from the annual PM schedule calendar. A copy of each PM assignment that is due is prepared. For example, on April 1 the May assignments are prepared.

```
                  MAINTENANCE PRIORITY SYSTEM

The following four levels of priority are used to insure
that response to a request for maintenance service is made
according to the urgency of the request:

    Priority No. 1 - Emergency.  Includes emergency work
                     needed due to personnel safety, a
                     breakdown or possibility of major
                     machine damage.  Response must be
                     immediate.

    Priority No. 2 - Must be completed during this shift as
                     soon as a mechanic is available.

    Priority No. 3 - Service is needed within 24 hours.
                     Specify that the service is needed
                     before the end of a certain shift by
                     writing the number:

                         (1) for the day shift
                         (2) for the afternoon shift
                         (3) for the night shift

    Priority No. 4 - Scheduled work.  This work can wait
                     more than 24 hours.  It includes
                     routine repairs, P-M and project
                     installation and modification work.
                     The request must include a date when
                     completion is needed.
```

FIGURE 6.12 Maintenance priority system.

2. The PM clerk stamps the date due on each PM assignment. The stamped forms are forwarded to the maintenance planning supervisor for assignment to the planner.

3. The PM assignments are distributed and assigned by area or central shop supervisors. A supervisor is responsible for assigning PM work to the appropriate shift.

4. A central shop planner prepares a work order for each PM job.

PM Schedule Compliance Is Required for Success.   In the judgment of those who establish frequencies for the PM assignments, the schedule dates are essential for continued availability of equipment and facilities and for optimum life of capital assets. If the compliance is good and PM hours are increasing, breakdown hours will be reduced.

PM Compliance Trend Charts.   Two trend charts used to summarize current progress of the PM system, shown in Figs. 6.13 and 6.14, are total annual hours and percent of compliance. As annual hours of PM increase, percent compliance with the schedule should remain high, indicating good workload control.

ASSIGNING MAINTENANCE WORK

The maintenance supervisor's primary responsibility is to provide enough work to keep the crew busy all day and to direct the workers in the crew. The only one with the authority to assign work, the supervisor is responsible for authorizing completeness and quality of the work as well as the accuracy of work reporting.

Supervisor's Use of the Work Order for Assignment and Control.   Though making only brief entries on the work order, the supervisor's presence at the job site and close checking of

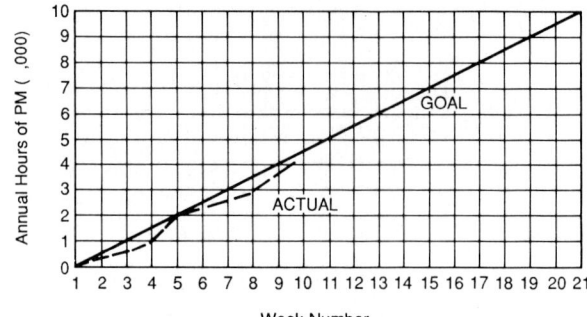

FIGURE 6.13   Trend of total annual PM hours scheduled.

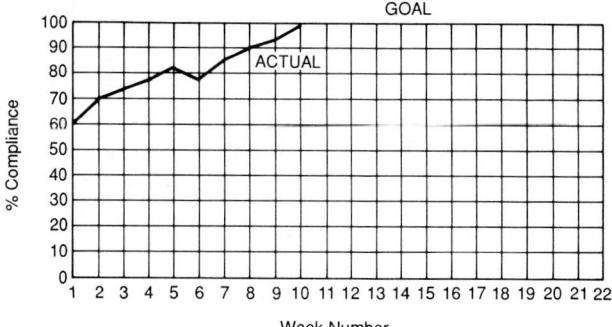

FIGURE 6.14   Trend of PM compliance with schedule.

the work in progress and completed work against the work assigned on the work order is absolutely essential to good control of maintenance repair and service work. Information provided by the supervisor includes date and shift work is completed, any additional follow-up work required, and the supervisor's signature approving completeness and quality of the work.

Guidelines for Using the Job Assignment Board.   The work order does not start action on a job until it is assigned to a maintenance worker. The foreperson uses the job assignment board in each shop to assign routine work at least one job ahead to each worker in the crew. Figure 6.15 shows a typical job assignment board. The foreperson puts the work orders into the appropriate slots in the "To Start" section arranged in sequence by priority, the highest-priority job in front and the lowest-priority in back. The worker takes them from the "To Start" slot in the same order as they were placed there by the foreperson, tears off a copy, places it in the "In Progress" slot in the job assignment board, and takes the copy to the job site. The worker then enters other work performed or needed, places both copies in either the "Interrupted" slot if the job is not finished or in the "Complete" slot if the job is finished. When both the "To Start" and "In Progress" slots are full, this indicates at a glance that each worker is now working on a job and has work planned ahead to start as soon as the in-progress job is finished. Completed jobs are picked up from the job assignment board during the day and at the end of the day by the foreperson.

Interrupted Jobs.   Work may be interrupted by reassignment to a higher-priority job, by nonavailability of materials or engineering information, or simply because the job was not completed by the end of the shift. As soon as the job is ready to reschedule, the same work order is returned by the foreperson to the "To Start" slot.

JOB ASSIGNMENTS				
Name	To Start	In Progress	Inter-rupted	Complete

Electricians 1st Shift

2nd Shift

3rd Shift

FIGURE 6.15 Job assignment board for work order assignment control.

Multicraft Jobs.   Maximum control is ordinarily gained by preparing a separate work order copy for each worker on the job for two reasons:

1. Visible evidence in the job board that the entire crew is fully scheduled. An empty "To Start" slot indicates that someone has run out of work.
2. Everyone knows where and with whom to work.

However, if this practice were always followed, the system would become very cumbersome. When assigning two or more workers to a crew of the same or different trades, paperwork is minimized by using one work order and an erasable plastic card for each additional worker showing:

1. Work order number
2. The name of the worker who has the work order
3. Any job instructions the foreperson wants to convey

In general, the number of multiworker work orders will be kept to a minimum and effectiveness will improve if the crew size principle is adhered to closely, that is: The optimum crew size is the minimum number capable of performing a job using a good, representative method in accordance with safe practices.

Maintenance Activity Time Reporting.   Time reporting is essential input to history records, backlog, and performance reports. Time reported is divided into two types—job time and nonjob time.

Job Time.   Job time includes planning and working time at the job site, job preparation and cleanup, travel time, personal and rest time, and minor unavoidable delays.

Nonjob Time.   Since all paid hours must be reported to maintain good labor control, it is necessary to provide a procedure to account for time spent on nonjob activities. Nonjob time includes both major delays not controllable by the worker and other nonjob activities. Typical nonproductive and delay categories are shown below.

Nonproductive Time Codes

- Vacation
- Holiday
- Union activity
- Funeral
- Jury duty
- Training
- Safety and security meetings

Delay Time

- Waiting for foreperson
- Waiting for other craft
- Waiting for operator
- Waiting for tools
- Waiting for equipment
- Waiting for work site
- Waiting for permits
- Waiting for materials
- Waiting for transportation

The Daily Time Distribution Card.   Daily reporting of actual hours and charges on the time distribution card by the workers and posting of time distribution cards to work orders provide the input data for the weekly maintenance control report. The relationship between a single work order and the time distribution card is shown in Fig. 6.16. Accumulating time and charges on a daily card is preferred to reporting this time on individual work orders. This daily accumulation enables the foreperson to account for all hours without the tedious paperwork required if work orders had to be sorted by employee to ensure that they totaled actual hours paid for.

DAILY TIME DISTRIBUTION		DATE 2-14	
NAME *Robert T. Smith*		CLOCK NO 1234	
TRADE CODE *MW*	DEPARTMENT *2*	SHIFT *1*	
WORK ORDER NUMBER OR DELAY CODE		HOURS CHARGED	
		REGULAR	OVERTIME
2801		2.5	
1124		4.2	
1020		0.8	
2		0.5	
TOTAL HOURS		8.0	

DELAY CODES

1 ASSIGNMENT	6 EQUIPMENT
2 PARTS	7 TRAINING
3 TOOLS	8 MEDICAL
4 PRODUCTION	9 MEETINGS
5 ASSISTANCE	10 UNION BUSINESS
	11 OTHER (EXPLAIN)

FOREMAN APPROVAL   *J. Foreman*   DATE   *2-14*

timecard 8-90

FIGURE 6.16 Daily time distribution card shows time charged to each job or delay.

Time Reporting for Emergency Service.   In order to control the maintenance workload, it is absolutely essential to have a written record of all work performed. An exception to this general rule is applied to extremely urgent requests such as immediate danger to personnel, a breakdown when critical equipment is involved and no alternate facilities are available, or major machine damage that will occur if repairs are not immediately made.

If a maintenance worker is already working in the area, authorized operations personnel advise the maintenance foreperson or planner by phone that the emergency exists and the worker's assistance is requested. A work order is provided after the fact for equipment records and performance control.

If no maintenance employees are working in the area, the authorized requester notifies the maintenance planner or foreperson that an emergency exists and the planner or foreperson gives the requester a work order number.

For further discussion of the use of job and time reporting information for management control reports and performance indexes, see Sec. 5, Chap. 6.

EFFECTIVE USE OF THE COMPUTER IN MANAGING MAINTENANCE PLANNING AND SCHEDULING

A growing list of maintenance management information systems are being automated through the use of data processing techniques. This trend highlights the information-intensive nature of the maintenance function. Earlier systems were batch mode—punch cards for work orders, time cards, inventories, and equipment histories updated in batches during off-hours, causing some delays. The display screen and keyboard have replaced the punched card, and the user interacts with the database storing, retrieving, and editing data instantly and getting on-screen or hard-copy verification and control reports in real time.

Wide Scope of Computer Uses in Maintenance. With the addition of programmable controllers, computers diagnose machine variations and automatically issue work orders and even adjust tolerances to bring them back into acceptable limits without human intervention. However, the labor-free maintenance department is even farther in the future than totally automated production. Meanwhile, a host of tedious information processing tasks are being automated, bringing us gradually closer to optimum automation and control. Figure 6.17 illustrates just a part of the scope of information handling activities the computer can productively perform.

Single-Station Systems. Small centralized maintenance organizations, where one or two planners plan the work for 50 or 60 tradespeople who work out of a central shop with central stores attached, can effectively use a single-station microcomputer system. Because of the large amount of data required in even the smallest plants, a typical single-station system has at least 640K (640,000 bytes or characters) of memory, a hard-disk drive with 20 MB (million bytes) of data storage, a monitor (CRT), 100-key keyboard, and an 80-character printer.

AUTOMAINT for Maintenance and Inventory Management. AUTOMAINT is an easy-to-learn system. At a major manufacturer, a technician was using AUTOMAINT and loading files within two days. To use AUTOMAINT, as with many user-friendly systems available, no prior computer experience is needed.

AUTOMAINT System Description. The AUTOMAINT system is usable on a wide variety of microcomputers, requiring only the MS DOS operating system. The preference of the majority of AUTOMAINT users is IBM/PC-XT (IBM is a trademark of International Business Machines, Inc.) or -AT or compatibles. Capabilities included are

- System and file password security
- Maintains large databases with fast retrieval of any work order
- Editing
- Data verification
- Report customization
- Menu and command driven to perform many information systems tasks quickly and automatically

AUTOMAINT Menus. The AUTOMAINT main menu is shown in Fig. 6.18. Creating work orders is accomplished with the function menu, which is selection (I) on the main menu.

PM Work Orders. By selecting (A) on the function menu, AUTOMAINT generates PM work orders. The system asks the week for which the orders are to be written, scans its file of PM schedules to find those that are to be completed this week, and then prints the orders and writes a record of the order in the computer backlog.

Activity or information	Input	Output
1. Audits	Specific information about a maintenance department in multiple-choice format	Present productivity Potential benefits Specific improvement areas Basis for future audit comparisons
2. Organization	Organization table and labor rates	Security system Craft assignments Labor cost Work schedules Vacation and overtime control
3. Training	All skills training programs	User practice
4. Stores, material, and tool control	Stores classifications, part numbering, location, cost, economic order quantities, withdrawal slip	Stores catalog, inventory, requisitions, purchase orders, vendor performance
5. Work order, planning, and scheduling	Work request, work priority	Work order Job plan and schedule Backlog status
6. Management controls	Time and work reporting, delays, budgets, standard time data	Performance coverage, delay cost, cost per planned hour Downtime Overtime Actual versus budget
7. Preventive and predictive maintenance and equipment history	Asset records, PM tasks and frequency, predictive inspection data	Material, labor cost by asset Mean time between failures Mean time to repair Highest repair cost items Highest downtime items PM routes Predictive inspection routes Repair priorities Repair work content
8. Work measurement, incentives	Work content, method, wage incentive pay formula	Standard time Performance Regular wages Bonus wages

FIGURE 6.17 Scope of handling computer activity or information.

Work Order Formats and Format Modifications.   AUTOMAINT files are structured to meet the widest possible variety of maintenance conditions. Most users require some modification to these standard formats in order to have the work order meet special needs or include a special code or appear more similar to existing work orders. For instance, a user may need very simple, one-line work orders, which are available in the short-form work order. Other users require a routine work order for multiple craft assignments. A work order can handle up to five crafts. A sample four-craft work order is shown in Fig. 6.19.

Work Order Support Files.   The PM system uses three files to prepare a work order:

1. Equipment information
2. Schedules
3. Tasks

All these files can be accessed through the menu. Equipment included in the PM program is entered into the equipment file. The scheduling file handles PM frequen-

```
┌─────────────────────────────────────────────────────┐
│                      MAIN MENU                        │
│                                                       │
│     Functions              Data Types                 │
│     ------------           ----------                 │
│                                                       │
│     CR   Create            EQ   Equipment             │
│     CH   Change            TA   Tasks                 │
│     DE   Delete            PL   Parts Lists           │
│     DI   Display           PI   Parts Inventory       │
│     PR   Print             EM   Employee              │
│     RE   Report            SC   Schedule              │
│                            NS   Non-Scheduled         │
│     PL   Plan              BA   Backlog               │
│     IS   Issue             OP   Open                  │
│     MO   Move              HI   History               │
│     CL   Close                                        │
│     UT   Utility                                      │
│                                                       │
└─────────────────────────────────────────────────────┘

       Function >              Data Type >
```

EX - Exit

FIGURE 6.18 AUTOMAINT main menu. (*Courtesy of H. B. Maynard and Company, Inc., Pittsburgh, Pennsylvania.*)

MULTI-CRAFT WORK ORDER

```
WORK ORDER #:          118      DATE INITIATED:      08/25/87
WORK ORDER TYPE:   NONSCHED     PRIORITY: 2      JOB# 22453M
DATE REQUESTED:    09/11/87     REQUESTOR:           WILSON

EQUIPMENT #:  AREA-328          DESCRIPTION: MAIN PACKAGING AREA
DEPT/AREA:                      FLOOR:  2
ROOM/SPACE:                     MANUFACTURER:

TASK DESCRIPTION:               REFURBISH MAIN ROOM
INSTRUCTIONS:      STRIP ROOM BARE, INCLUDING ELECTRICAL SERVICE
                   NEW ELEC DISTRIBUTION
                   NEW LIGHTING
                   NEW CEILING, PANELING, FLOOR
                   PAINT

PART NEEDED:


CRAFT1: MECH    CRAFT2:  ELEC    CRAFT3:   CARP    CRAFT4: PAIN
HOURS:    35             66                40              18

SPRV: WILSON    SCHED. START:  / /    ACT. COMPLETION  / /

COMMENTS:


MATERIAL $ BUDGET:              LABOR $ BUDGET:
MATERIAL $ ACTUAL:              LABOR $ ACTUAL:
```

FIGURE 6.19 Multicraft work order for four crafts. Additional instructions are given on supplemental work plans and drawings. (*Courtesy of H. B. Maynard and Company, Inc., Pittsburgh, Pennsylvania.*)

cies, which can be changed, for instance, from monthly to quarterly. The "Schedule Choice" number permits balancing the craft load from preventive maintenance work orders. Special schedules can also be created, for example, air conditioners in the fall and in the spring. PM tasks to be printed on the work order are written into the task file.

Work Order Backlog File.   When the user selects (B) on the open work order submenu, AUTOMAINT lists all incomplete work, grouped by craft. Other groupings are also possible.

Work Order Completion and History Analysis. AUTOMAINT permits closing out work orders, assigning labor and material costs, and moving work orders into history. Closing out work orders is accomplished via the AUTOMAINT menus. Work order history can be summarized by equipment and craft as in Fig. 6.20.

WORK ORDER HISTORY BY EQUIPMENT
EQUIPMENT #23

WO #	CRAFT	COMPL	SCHED	ACTUAL	PERF	TASK DESCRIPTION
66	ELEC	03/08/84	1.00	2.00	50.0	CONSOLE WON'T LIGHT
**	SUBTOTAL	**	1.00	2.00		
70	HVAC	04/06/84	1.00	2.50	40.0	STOP NOISE
71	HVAC	05/05/84	2.50	3.90	64.1	CONTROL MOTOR HOT
72	HVAC	05/19/84	4.00	4.00	88.8	REPAIR COOLER
115	HVAC	06/23/84	1.50	2.00	75.0	REPLACE BELT
**	SUBTOTAL	**	9.00	12.40		
***	TOTAL	***	10.00	14.40		

FIGURE 6.20 Work order history. (*Courtesy of H. B. Maynard and Company, Inc., Pittsburgh, Pennsylvania.*)

AUTOMAINT Material Files. AUTOMAINT also maintains parts files on equipment. The AUTOMAINT inventory record for individual parts is shown in Fig. 6.21. It indicates the balance, order point, order quantity, and minimum balance. It shows any parts allocated or on order and adjusts the on-order record when parts are received. Standard cost is included in the inventory record to permit inventory valuation. A purchase requisition for replenishing stock or procuring nonstock items is available.

PARTS INVENTORY

PART #: 129-984-01	DESCRIPTION:	VEE BELT SET 28 5/8 IN.	
MANUFACTURER: GATES	MANUFACTURER #: 129-984-01		
LOCATION: TOP RACK	COST: 8.00	UNIT: SET	CODE #1:
QUANTITY ON HAND	60	ALLOCATED	0
MINIMUM QUANTITY	20	REORDER POINT	40
ORDER QUANTITY	20	YTD USAGE	15
QTY ON ORDER 0	DATE ORDERED / /	PO #	
COMMENTS:			

FIGURE 6.21 Part record. (*Courtesy of H. B. Maynard and Company, Inc., Pittsburgh, Pennsylvania.*)

Time Reporting and Overtime Records. A custom feature of AUTOMAINT is the ability to record craft time and to produce up-to-the-minute reports on overtime and labor performance. A typical overtime report is shown in Fig. 6.22.

Multiple Station Systems. In larger facilities, or where several distant locations must be linked, multiple station systems are required because several planners, supervisors, material coordinators, warehouseworkers, and clerks are accessing a common information database. The concept of open system architecture is used. A variation of the AUTOMAINT system, a local area network (LAN) or wide area network (WAN) of CRT screens, keyboards, and printers connected to a common central processing unit (CPU) and disk drive(s) with a common information database is used in this application.

If two planners try to reserve the last 20-foot length of 1½ by 1½ by ¼ angle iron a few minutes apart, the first planner's order will be accepted and the second will see a stockout on

OVERTIME LISTING

NAME	CRAFT	SENIORITY	OVERTIME
WILSON, J.M.	ELEC	02/25/71	322.00
HERMANN, W.	ELEC	07/25/73	229.00
JONES, K.	ELEC	07/14/59	128.30
WOJNAR, V.	MECH	05/21/78	244.00
MONK, T.V.	PLUMB	11/05/49	623.80
BRUCE, W.K.	PLUMB	08/19/63	244.50
MILLER, H.H.	PLUMB	02/14/81	98.50
JENKINS, J.J.	ELEC	08/08/77	388.20
EDGAR, S. E.	ELEC	12/15/78	402.10
RISER, M.E.	ELEC	12/04/59	255.80
MICEY, P.B.	MECH	09/02/59	155.70
JILIAN, P.B.	ELEC	09/02/77	5.00
MOORE, W.W.	ELEC	10/11/88	366.00

FIGURE 6.22   Overtime listing. (*Courtesy of H. B. Maynard and Company, Inc., Pittsburgh, Pennsylvania.*)

the screen display. In a manual system, time delays may make it very difficult to keep up with minute-by-minute changes taking place. Even better, with good inventory control features, the computer can process purchasing documents before supplies get too low and emergency ordering is necessary, saving dollars not only on lower inventory levels but also on delivery, carrying costs, and maintenance craftsperson delays. For more about using the computer to produce management control reports and performance trend charts, refer to Sec. 5, Chap. 6.

CONCLUSION

This chapter has summarized the most modern and effective tools available for the maintenance manager's planning and scheduling toolbox. When a new tool is first used, it seems awkward and ineffective, it takes longer to use than familiar tools, and the user might tire of it. Persistence, however, will achieve results undreamed of. If some of the techniques described meet resistance, it is probably due more to this natural skepticism than to any flaw in the tool, because they have already stood the test of time in someone else's toolbox.

BIBLIOGRAPHY

Heintzelman, J. E., *The Complete Book of Maintenance Management,* Prentice-Hall, Englewood Cliffs, N.J., 1976.

Higgins, L. R., and L. C. Morrow, eds., *Maintenance Engineering Handbook,* 3d ed., McGraw-Hill, New York, 1977.

Westerkamp, Thomas A., *Maintenance Manager's Standard Manual and Guide,* Prentice-Hall, Englewood Cliffs, N.J., 1992.

Zandin, Kjell B., *MOST Work Measurement Systems,* 2d ed., Dekker, New York, 1990.

CHAPTER 7
NETWORK PLANNING AND CONTROL

Francis M. Webster, Jr.
Management and Computer Consultant
Cullowhee, North Carolina
Editor-in-Chief
Project Management Institute
Drexel Hill, Pennsylvania

Probably the two most significant developments in management in the last two decades have been in quality management and production scheduling. Quality management has resulted in major changes in operating methods, reallocation of responsibilities from management to operators, and substantial new training requirements. New approaches to production scheduling have moved through material requirements planning (MRP) to the current emphasis on just-in-time (JIT) concepts. These have required major changes in attitudes toward inventory, efficiency, utilization, and application of engineered time standards, and have created even more training requirements.

Although receiving less publicity, another "revolution" has been growing in importance: planning, scheduling, and control of project work. This new concept in management dates back to the mid-1950s and has been characterized by varying degrees of interest and understanding. Today, it is the basis for a major portion of economic activity: the management of change. It has developed from being a very specialized technique to a management philosophy supported by a range of techniques, concepts, and management theory. Indeed, project management has progressed to the point of being recognized as a distinct career and profession. Project managers are served by a professional society—the Project Management Institute (PMI)—which, among other things, has programs of education, certification of project management professionals, and accreditation of academic programs in project management.

Paralleling these developments in project management was the development of computer-based techniques for assembly- or progressive-line balancing. While the two techniques, project scheduling and assembly-line balancing, seem quite different, a more in-depth analysis of the tools and concepts reveals a striking similarity.

More recently, network planning has found its way into the executive suite as a convenient way in which to express corporate strategic plans, including alternative methods to achieve corporate objectives.

Preceding all these developments, dating back to 1936, is the line-of-balance technique. It was developed prior to the computer age and has apparently not yet been converted into an attractive computer-based technique.

These are the major areas of application of networks in planning and control in management. Others probably exist but have not been publicized because of their development for competitive advantage.

PROJECT MANAGEMENT CONCEPTS

The Concept of Projects.   Projects and project management may be one of the more misunderstood concepts in management. This is due in part to opinions of these new concepts developed in the early days and in part to the fact that projects have been a way of life throughout history.

The first developments in modern project management were sponsored by DuPont and the U.S. Navy, followed closely by the U.S. Department of Defense (DoD) and National Aeronautical and Space Administration (NASA). From these efforts, the impression was gained that the techniques were applicable in construction and maintenance (in the case of DuPont) and to very large scale defense and space systems development. This impression has been promulgated by authors of textbooks in production and operations management, management science, and operations research. With the advent of a variety of microcomputer-based project scheduling software and the availability and convenience of use of these packages, projects of all sizes have benefited from their application.

In the early years of modern project management, the cost of application was substantial for two reasons: user friendliness and training of users. The first systems were *not* user-friendly. Considerable computer expertise was required, and the methods for these techniques were inherently labor-intensive. Today, both of these deficiencies have been overcome. User friendliness has made it possible for novice users to use significant features of a package effectively with only a couple of hours of introduction, generally from the user's manual, help messages, or tutoring programs. Training in the use of project management techniques has become widespread, and many persons with project management responsibilities find it easy to plan their project in an on-line mode, with considerable benefit coming from a better understanding of the project and more accurate communication of their desired approach for the execution of the project.

Project efforts have been described in the earliest of recorded history. The change efforts of society are conducted by projects. Most managers and engineers have been involved in and many have managed projects. The degree and scope of responsibility has typically been rather limited, often with responsibility shifting from one individual to another as the project progresses through its various phases. Today, it is more common for an individual to be given responsibility for managing the project from its inception to its closure.

Thus many changes have taken place in the management of projects. While project management seemed to be almost entirely the application of network techniques, today the use of these techniques for planning, scheduling, and control can be considered 10 percent of project management, at the most. Behavioral considerations, contract management, risk management, and other concepts have been recognized to be of far greater significance. Perhaps the appropriate view is that the scheduling techniques have improved in both usability and usefulness to such an extent as to permit the project manager, and the project management team, to perform the planning and control functions with much less time and effort, thereby affording them the time to also perform the other functions far better.

Characteristics of Projects.   In its "Project Management Body of Knowledge"[26] PMI defines a *project* as any undertaking with a defined starting point and defined objectives by which completion is identified. This is indeed broad in scope, incorporating construction and defense and space systems but also including new product and/or process design and implementation and even cost reduction and methods improvement projects.

Projects are composed of activities, usually nonrepetitive, operating on an interrelated set of items which inherently have technologically determined relationships. One activity must be completed before another can begin. These relationships are described in the network diagram. This is further discussed below.

Projects involve multiple resources, both human and nonhuman, which require close coordination. Generally there is a variety of resources, each with its own unique technologies, skills, and traits. This leads to an inherent characteristic of projects—conflict. There is conflict *between* resources as to concepts, theory, techniques, and the like. There is conflict *for* resources as to quantity, timing, and specific assignments. And there are other sources of conflict in projects. Thus a project manager must be skilled in managing such conflict.

The "project" is not synonymous with the "product of the project." The word "project" is often used ambiguously, sometimes referring to the project and sometimes referring to the product of the project. This is not a trivial distinction, as both entities have characteristics unique to themselves. Some of the concepts apply to both. For example, the life cycle of a project includes the conceptual, development, implementation, and termination phases. The life cycle of a capital facilities product includes feasibility and acquisition (a project), operating it, major repairs or refurbishing (typically done as projects), and dismantling (often a project, if done at all). The project cost of creating the capital facilities product is generally a relatively small proportion of the life cycle cost of that product. See Fig. 7.1. Similarly, there is a life cycle of product development consisting of basic research, product research, design, and production. The first three are really projects. The objective of the production phase is typically to prolong its useful life, and this is often done through understanding the phases of the marketing product life cycle.

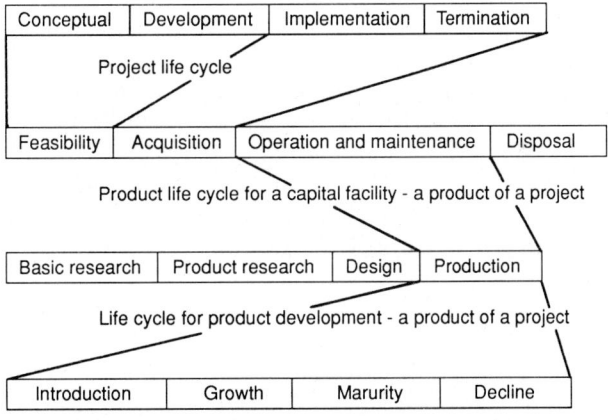

Marketing product life cycle for a mass produced product

FIGURE 7.1 Comparison of project and product life cycles. The lines are intended to relate the segments of one bar with related segments of another. That is, "Feasibility" on bar 2 is the "Conceptual" and "Development" phases of bar 1. "Acquisition" on bar 2 is the "Implementation" and "Termination" of bar 1. Similarly, on bar 3, "Production" is concurrent with "Operation" and "Maintenance" on bar 2 and also with the entire life cycle of the product from a marketing point of view (bar 4).

Managerial emphasis is on timely accomplishment of the project as compared with the managerial emphasis in other modes of work. Most projects require the investment of considerable sums of money prior to enjoyment of the benefits of the resulting product. Interest on these funds is a major reason for emphasis on time. Being first in the market often determines long-term market position, thus creating time pressure. Finally, a need exists for the resulting product of the project; otherwise the project would not have been authorized.

Thus time is of the essence. This time pressure, combined with coordination of multiple resources, explains why most "project management systems" have emphasized time management.

An Alternative Definition.   The above logic leads to an alternative definition.

> Project: A temporary process composed of a constantly changing collection of technologies and operations involving the close coordination of heterogeneous resources to produce one or a few units of a unique product or service.[30]

Given this, it is appropriate to examine the characteristics of the "process." Consider the following:

> The essential characteristic of the process by which a project is performed is the progressive elaboration of requirements or specifications.[30]

A project is initiated by a person (perhaps a member of an organization) recognizing a problem or opportunity about which some action is to be taken. That person, alone or in concert, develops an initial concept of the action to be taken in the form of a product, be it a product for sale, a new facility, or an advertising campaign. Much work needs to be accomplished to take this meager concept to the reality of the product. This work, though often not conceived as such, is accomplished by instituting a project.

The general concept is expanded into a more detailed statement of requirements. These are examined for feasibility...market, technical, legal, organizational, political, and the like, resulting in further refinement of the specifications. These are then the basis for general design, the products of which become the basis for detail design. The detail designs are followed by production designs, tooling, production instructions, and the like, each stage producing an elaboration on the specifications of the prior stage. Eventually, the product of the project takes shape, is tested, and is ready for operation. At this stage, give or take a few details, the project is completed.

Relation to Modern Quality Management.   This characterization of projects permits the adaptation of modern quality management concepts into the management of projects. Modern quality management begins with a new definition:

> Quality: Conformance to requirements and specifications.

The relevance of "conformance" can be made no more clear than as presented in the first nine pages of Chapter 4 in *Quality Is Free.*[6] To begin there must be a clear understanding of the customer's or client's requirements in a product. These then become the specifications for the next stage. This process continues throughout the project until the product of the project is completed.

Another important concept is "zero defects," or:

> Do the right thing right the first time.

Repeated experience has shown that more attention to doing jobs right the first time eliminates so much rework and scrap cost as to fully support the proposition that "quality is free." When these concepts are combined with another essential concept of modern quality management, a new perspective is available on the project as a process. This is:

The customer is the next person or operation in the process.

Thus, as the progressive elaboration of specifications proceeds, the "customer" is the next engineer, the tool builder, the ad layout person, and so on. If the product going to them has no defects, they can perform their tasks in the most efficient manner...and get it right the first time.

This concept of a project as a process is essential for the application of "process control," and more specifically, statistical process control, to the management of projects. One approach to this was presented in "Validating Technical Project Plans."[27] It involves procedures for reviewing the plans for a project in much the same manner as design or construction review teams analyze the design of a product for such things as structural integrity or constructability.

Another approach was outlined in "Reliability Maturity Index."[20] This concept identified events in a project network which marked the completion of activities which either measured or contributed to the reliability of the product of the project. Other concepts and techniques were identified in "Responsibility for Quality in a Project," including failure mode and effect analysis, fault tree analysis, and stratified random sampling.[18] Still others involve the development and application of methods and technologies which permit the operator on a project to perform inspections on the product immediately upon completion or even during the performance of an operation.

Project Management Techniques: A Taxonomy.   To establish the framework for discussion of the techniques available for managing projects, it is desirable to define a taxonomy of techniques (see Fig. 7.2). There are two classes: critical path techniques and other project techniques.

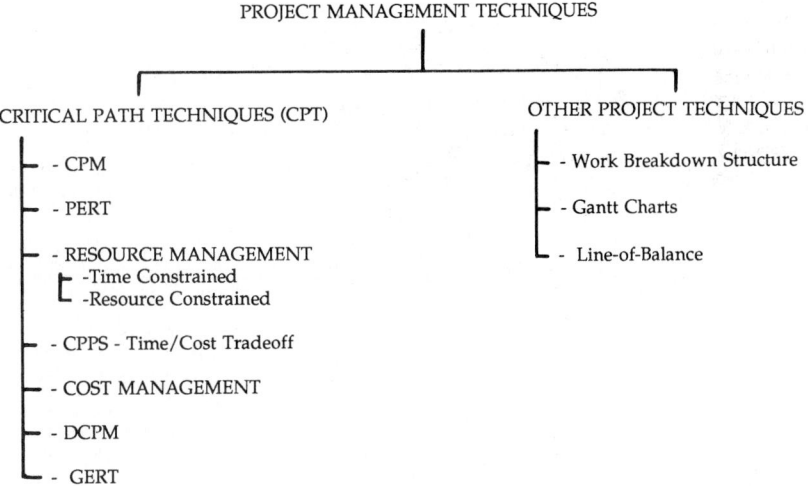

FIGURE 7.2   Taxonomy of project management techniques.

All of the critical path techniques have two things in common; they are network diagram–based and they involve the calculation of early and late times for the performance of each activity necessary for the performance of the project. These need to be understood before proceeding to the detail discussion of the techniques. Thus the next section presents some fundamentals of networks.

SOME FUNDAMENTALS OF NETWORKS

Introduction. A network diagram is a schematic display of the sequential and logical relationship of the activities which comprise a project. Generally, these technological relationships are very difficult to violate. For example, if getting oneself dressed is considered a project, it just does not make sense to put your shoes on before your socks. Whether to put on both socks and then both shoes or to complete the left foot before the right foot is generally a question of preference. In modern project management, a network diagram is used to portray these technological sequences. Figure 7.3 shows the use of networks to describe alternative ways of putting on socks (RSock and LSock) and shoes (RShoe and LShoe).

A.   RSock → LSock → RShoe → LShoe → Done

Serial network with preference for putting both socks on first and doing the right foot before the left foot.

B.   RSock → RShoe → LSock → LShoe → Done

Serial network with preference for putting sock and shoe on right foot first.

C.   RSock → RShoe ⟍
                            → Done
      LSock → LShoe ⟋

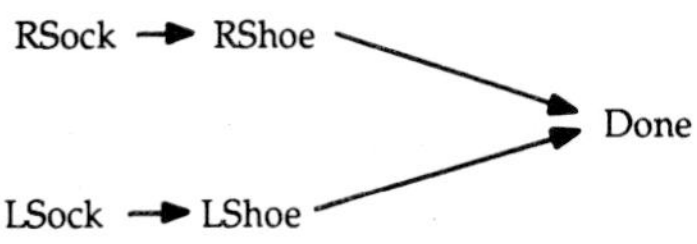

Parallel network with only technological relationships.

D.   RShoe → LShoe → LSock → LSock

A nonsensical network, in most instances.

FIGURE 7.3 Examples of network diagrams.

Example C, in Fig. 7.3, does not imply that both socks are put on simultaneously but, rather, it provides flexibility to determine the actual sequence based on other criteria. It is important for planners to focus on the technological relationships to prevent implicitly scheduling a project before really understanding the alternatives available. Example D in Fig. 7.3 would be nonsensical in most instances, for it implies not only putting the sock on over the shoe but also putting both socks on the left shoe.

Project Network Diagram Conventions. For several reasons a great deal of ambiguity has developed in the language of networking. PERT networks, CPM networks, and precedence diagrams are sometimes referred to as if they are unique and sometimes as if they are identical. It is helpful to develop a clear distinction which can be described by a three-dimensional matrix, with the dimensions being graphic convention, focus convention, and identification convention as shown in Fig. 7.4.

Graphic Convention. Project networks can be graphically portrayed in either "activity-on-the-node (AON)" or "activity-on-the-arrow (AOA)" notation. The examples in Fig. 7.3 are shown in AON notation. An example of AOA notation, based on C in Fig. 7.3, is

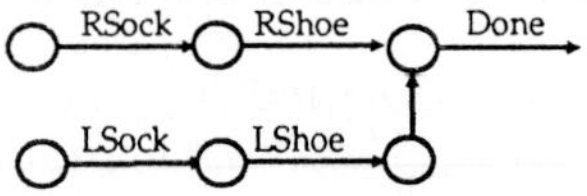

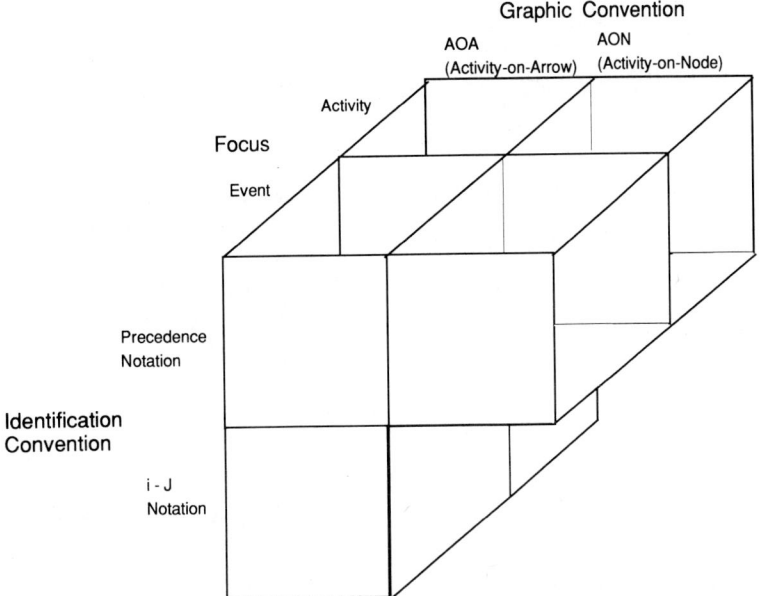

FIGURE 7.4   Networking conventions.

The original development of both PERT and CPM was based on AOA notation. At about the same time, John Fondahl, at Stanford University, was developing a comparable technique using AON.[13,14] Another development effort in Europe also used AON. In the United States, the impetus provided by the Department of Defense for use of PERT resulted in AOA's being the notation learned by most users. It was not until the PC versions of critical path techniques that AON started becoming really popular in the United States.

There are a number of reasons for preferring AON including ease of use, separating planning from scheduling, and ability to maintain the integrity of identification of activities as they are subdivided when adding more detail. Probably the greatest benefit to industrial engineers, however, is that it is exactly the same notation used in drawing process flowcharts, computer program flowcharts, and precedence (balloon) diagrams in assembly-line balancing. Thus AON requires less learning time and is less likely to result in errors from switching from one graphical mode to the other.

This does not mean that the AOA notation has no usefulness. After planning a project using AON, it is very useful to check the logic of the plan by plotting an AOA diagram on a time scale to see the temporal relationships between activities. Often it is possible to recognize that an activity is scheduled at what seems to be a totally illogical time. This can easily happen in a complex project because a precedence relation was inadvertently not specified or a relationship was specified which was really not necessary. Most CPT software packages today have the ability to plot the project in either AOA or AON modes.

Focus Convention.   PERT and CPM were developed for completely different purposes and therefore focused on the project in a different manner. PERT, as its name implies (program evaluation and review technique), was developed to assess the probability that a specific event could take place as scheduled. Thus the focus was on the "event." CPM was developed to improve the planning and control of the work on a project. Indeed, the original development resulted in the critical path planning and scheduling (CPPS) technique which focused on the time-cost trade-off inherent in scheduling the project in a shorter total duration. (This is discussed in more detail under "CPPS—Time-Cost Trade-off" below.)

Thus, from the beginning of these two techniques there was a difference in focus. Each has its place in modern project management. Top executives, in planning corporate strategy and approving projects, are primarily concerned with "When is it going to be completed?" Generally, they have too many other details on their minds to consider much more than the major milestones of the various projects ongoing in the organization. Indeed, there has been an axiom that no more than three milestones on a project should be reported on at a briefing of top executives lest they become confused from information overload. Recent interest in "prudency," especially in utility rate cases, has led top executives to take more interest in the major projects in their organization.

Identification Convention.   Both PERT and CPM, as originally developed, used the same method of identifying activities in a network, the *"i-j"* convention; that is, events were given identifiers, usually numbers, and an activity was denoted by the number of the predecessor event followed by the number of the successor event such as 110 - 130.

The systems developed using AON notation used what is commonly referred to as "precedence" notation. Early on, these systems permitted the use of alphanumeric codes to uniquely identify activities (instead of events). Thus, an activity can be given the code name "ABCDE" or "AESS." The relationship between two or more activities can be specified by simply listing the activity of concern followed by its predecessor(s) such as AESS - ABCDE, 34567, CM12P, which states in coded form that AESS cannot start until activities ABCDE, 34567, and CM12P are completed. An alternative also available permits listing the followers of an activity, and today some systems allow both follower and predecessor notation in the same project.

In the past, some knowledgeable persons argued for a nonmeaningful code, which led to the likes of "ABCDE." More recently, it has become accepted practice to use a meaningful code such as "AESS," which might stand for "erect structural steel in area A." Recently developed software provides several characters for identifying activities, making it quite easy to develop a standardized coding structure for activities commonly performed in an organization. This provides at least two benefits: ease of recalling the names of frequently used activities and the ability to compare like activities across projects to improve estimating and performance measurement.

Overlap.   In addition to the above conventions, a new logical relationship is now available. The original planning logic permitted only one basic logic between activities: the follower of an activity could not start until the preceding activity was finished. This proved quite frustrating in accurately portraying relationships where the follower could clearly start before the predecessor was finished. Thus four alternative logics were introduced and are popular today. They are:

Finish-to-start with overlap and delay—[B A(FS − 3 days)]—activity B can be started 3 days before activity A is finished or for [B A(FS + 3 days)] activity B cannot start until 3 days after A is finished. The former is useful to indicate that detail drawings can be started 3 days before the layout drawings are completed. The latter would be convenient for indicating that the forms cannot be removed from a concrete wall until 3 days after it was poured.

Finish-to-finish with overlap and delay—[B A(FF + 3)]—is convenient to indicate that it is all right to start activity B before A is finished; just do not expect to finish it until 3 days after B is finished.

Start-to-start with overlap and delay—[B A(SS + 3)]—is an alternative way to state that work can start on detail drawings 3 days after layout drawings have started.

Start-to-finish with overlap and delay—[B A(SF + 3)]—is perhaps somewhat less useful, but the logic is available in many systems today.

While these overlap capabilities are very useful in many situations, they can be overused. A network incorporating many of these relationships can be very confusing and thus diminish its ability to communicate. Manual calculations are much more involved, thus reducing the ability to analyze the network. Finally, it is easy to incorporate logic which has unintended consequences. Indeed, in the extreme, it has been shown that it is possible to construct a network which has a finish time which is before the start time. With these thoughts in mind, it should be easy to remember to use these relationships with discretion.

Example Project.   For purposes of explanation, a simple project is presented to illustrate each of the techniques. The explanation of each technique is presented in accompanying illustrations so the reader may skip it in reading the rest of the text. At each step only those concepts are introduced which are unique to that technique. The data are presented in AON, precedence, and work focus notation. There are many other sources which provide explanations of these and other variants. A description of the example project is provided in Example 7.1.

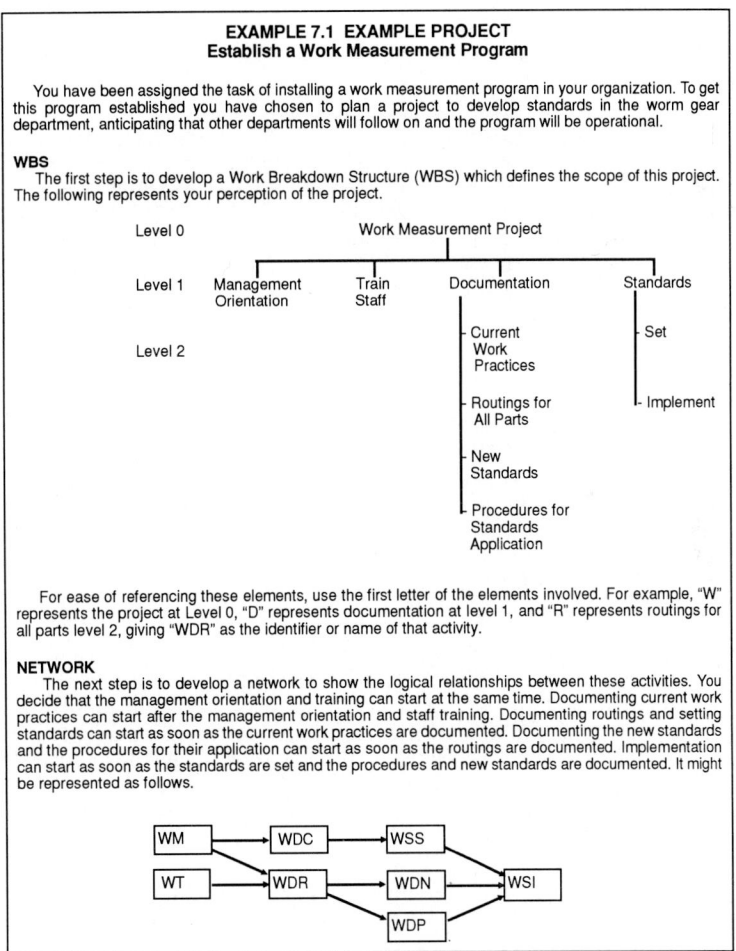

EXAMPLE 7.1 EXAMPLE PROJECT
Establish a Work Measurement Program

You have been assigned the task of installing a work measurement program in your organization. To get this program established you have chosen to plan a project to develop standards in the worm gear department, anticipating that other departments will follow on and the program will be operational.

WBS

The first step is to develop a Work Breakdown Structure (WBS) which defines the scope of this project. The following represents your perception of the project.

For ease of referencing these elements, use the first letter of the elements involved. For example, "W" represents the project at Level 0, "D" represents documentation at level 1, and "R" represents routings for all parts level 2, giving "WDR" as the identifier or name of that activity.

NETWORK

The next step is to develop a network to show the logical relationships between these activities. You decide that the management orientation and training can start at the same time. Documenting current work practices can start after the management orientation and staff training. Documenting routings and setting standards can start as soon as the current work practices are documented. Documenting the new standards and the procedures for their application can start as soon as the routings are documented. Implementation can start as soon as the standards are set and the procedures and new standards are documented. It might be represented as follows.

Estimating.   Estimating in projects is somewhat more involved and less precise than most estimating done by industrial engineers. For one thing, the "unique" nature of projects limits the usefulness of past data, makes time study of the job being done less applicable, and in general reduces the cost-benefit ratio of engineered time standards.

In addition, it is often necessary to differentiate between the work content of an activity and the actual duration which will be required. For one thing, it is generally not certain exactly who will perform the activity. One person may be very skilled at doing the activity but the person who ends up doing it may be much less skilled. Often, one person works on more than one activity at a time, and thus the work content must be divided by less than a full day to determine the actual duration. Much project work is performed with little supervision, resulting in varying degrees of efficiency on the same activity. Communications of exactly what is expected as the work product of the activity may be incomplete. Also, some activities push the state of the art of the technology and will therefore depend upon repeated trials, or even serendipity, to find the solution to a difficult problem.

The above tends to focus on estimating durations. Inherent in an estimate is an assumption of the methods and technologies that will be used. In addition, there are resource types to select, including not only human resources but the supplies and equipment required. Uncertainty is involved in all of these. From the above, cost estimates must be derived which could, if too high, result in the project's not being approved or, if too low, result in poor performance to budget, schedule, and/or technical requirements.

EXAMPLE 7.2 ESTIMATING

You estimate the activity durations to be as follows.

Activity	Days
WM	2
WT	5
WDC	3
WDR	2
WSS	4
WDN	4
WDP	5
WSI	2

Network Calculations. Network calculations are tedious but quite simple to master with a little practice. Any user of a computer program to perform these calculations is urged to invest the time and effort necessary to be comfortable with these manual calculations for at least three reasons. First, it is suggested that it is very risky to accept computer output as accurate without knowing the basis for the calculation of that output. Surprises can happen! Second, the computer output will be presented in tabular form and it is important that the user be able to read that output and interpret it properly. Third, many years of experience teaching the network approach to planning has indicated that most persons do not inherently think in terms of a complex network and require a learning experience like that involved in learning the calculation procedures well.

EXAMPLE 7.3 CALCULATIONS

It is necessary to add the predecessor information to the table to define the logic of the network. To calculate the feasible time frame for each of these activities, start with time zero for those activities having no predecessor. Time zero can be interpreted as the end of day zero. Add their durations to their start time to get the finish time. The finish time for WM can be interpreted as the end of day 2. WDR has two predecessors, WM and WT. WDR cannot start until both of the predecessors are completed, i.e., the end of the 5th day.

EXAMPLE 7.3 *(Continued)*

Proceed in a similar manner for the remaining activities. Note that WSI cannot start until after the end of the 12th day, after all three of its predecessors are completed. **The rule is: When calculating the early times, always choose the latest finish of the activity's predecessors.**

Act	Pred.	Days	ES	EF
WM	-	2	0	2
WT	-	5	0	5
WDC	WM	3	2	5
WDR	WM,WT	2	5	7
WSS	WDC	4	5	9
WDN	WDR	4	7	11
WDP	WDR	5	7	12
WSI	WSS,WDN,WDP	2	12	14

When calculating the late starts and finishes it is necessary to have a target completion date for the project. This target date could be arbitrarily chosen, perhaps by management, and could be either earlier or later than the earliest finish of the project, i.e., 14 days. On this project you have been given a target date by management of 13 days. Otherwise, ignore the early calculations while doing the late calculations.

From the network diagram it is clear that WSI is the ending activity. Give it the target finish time of 13. Subtract the duration to get the latest start time of 11. Post that 11 as the latest finish time of activities WSS, WDN, and WDP as they must be finished before WSI can start. Proceed up the list completing the calculation and posting of one activity at a time. After posting WDP's late start of 6 to WDR's late finish, WDN shows a late start time of 7. Since the 7 is later than the 6, ignore the 7 and proceed. **The rule is: When calculating the late times, choose the earliest start time of the activity's followers.**

Act	Pred.	Days	LS	LF
WM	-	2	2	4
WT	-	5	-1	4
WDC	WM	3	4	7
WDR	WM,WT	2	4	6
WSS	WDC	4	7	11
WDN	WDR	4	7	11
WDP	WDR	5	6	11
WSI	WSS,WDN,WDP	2	11	13

Completion of these calculations discloses a problem as activity WT's late start time is !1 which means it should have started yesterday if the target completion date is to be met. This calls for replanning.

EXAMPLE 7.4 REPLANNING

Reviewing the plan, we decide that the documentation of application procedures can be finished after the implementation has started so long as it is finished at the end of that activity, i.e., WDP and WSI can complete at the same time, the completion of the project. Combining the early and late calculations into one table gives the following. The only difference in these calculations is due to the removal of WDP as a predecessor to WSI.

EXAMPLE 7.4 *(Continued)*

Act	Pred.	Days	ES	EF	LS	LF	TOTAL SLACK	FREE SLACK
WM	-	2	0	2	2	4	2	0
WT	-	5	0	5	0	5	0	0
WDC	WM	3	2	5	4	7	2	0
WDR	WM,WT	2	5	7	5	7	0	0
WSS	WDC	4	5	9	7	11	2	2
WDN	WDR	4	7	11	7	11	0	0
WDP	WDR	5	7	12	8	13	1	1
WSI	WSS,WDN	2	11	13	11	13	0	0

Two columns have been added to this table to show total slack and free slack. Total slack is the difference between the time available and the time required to perform each activity. Using this definition, slack on activity WM is 2 (late finish of 4 minus the early start of 0) minus the duration of 2 for WM, i.e., 2. It can also be obtained by taking the difference LS – ES or LF – EF.

Free slack is the amount that an activity's completion can be delayed without delaying any of its immediate followers. For example, WSS with an EF of 9 is followed by WSI with an ES of 11. Therefore, WSS could be delayed up to 2 days with no impact on WSI. This information is useful when leveling resources or when asked to approve a slippage in an activity. If the slippage is less than the free slack, it will not delay any other activity.

CRITICAL PATH TECHNIQUES

CPM. For purposes of discussion, "CPM" will be used to refer to the simplest form of CPT involving the precedence relationships between activities, a single estimate of each activity's duration, and the calculation of early and late start and completion times and total slack. This distinction is important because of the ambiguity which exists in fact. If one refers to the typical textbook on management science and operations research, production and operations management, or industrial engineering, CPM will probably be found to refer to the time-cost trade-off algorithm. If one speaks with practitioners in the field, CPM will probably be used to refer to the simplest system as defined above. We prefer the vernacular of the practitioner. CPM is presented in the example above.

Program Evaluation and Review Technique (PERT). PERT was developed by a team representing the Naval Weapons Research Laboratory and Booz-Allen consultants to evaluate the status of the Polaris Missile Weapons System.[21] Admiral Rayborn credited PERT with taking a year out of that development project. It was originally developed to permit the expression of the uncertainty associated with moving from one event to another. Considerable mathematics was used in deriving the specific form of the distribution, the beta, and the practical simplification leading to three time estimates: a = optimistic or shortest reasonable time, m = most likely time, and b = pessimistic or longest reasonable time. For those interested, several alternative relationships have been developed based on arguments pertaining to the relative frequency of these times. The original PERT relationship assumed that a and b occurred about one time in a hundred each. The three time estimates were used to calculate a mean and standard deviation for each activity as follows:

$$t_e = (a + 4m + b)/6$$

where t_e = expected value of the duration of the activity as in expected value theory in probability and statistics

$$\sigma = (b - a)/6$$

where σ = standard deviation of the distribution of activity durations

The general logic of the calculations is identical to CPM except for the calculation of t_e for the duration and the calculation of the standard deviation of the ending event or activity. While the theory of PERT assumes a beta distribution, that is, one that may be skewed to either the right or the left, the central limit theorem nevertheless allows the addition of the activity distributions on the critical path with the resulting distribution assumed to approximate the normal curve of probability theory. Thus the mean of the ending activity or event is equal to the sum of the means of the activities on the longest path leading to that activity or event. Similarly, the variance of the ending activity or event is the sum of the variances (standard deviations squared) of the activities on that same path. The square root of the summed variances is the standard deviation of the completion time for that activity or event. Generally, this is done only for the latest activity or event in the project. Thus, substituting the three time estimates for each activity in the example network, the calculations are as shown below.

The resulting T_e of 13 is the expected completion time for the last activity or event on the critical path. There is a 50 percent chance it will take more time to complete the project but also a 50 percent chance that it will take less time.

EXAMPLE 7.5 PERT

To perform the PERT calculations it is necessary to have the three time estimates: a, m, and b. Had they been estimated originally they would have resulted in the durations used above except now we change the name of the durations to "t_e" to reflect the new meaning of the "expected value." The calculations of early and late times are the same as before.

Act.	Pred.	a m b	te	ES	EF	LS	LF	TOTAL SLACK	σ	σ^2
WM	-	1 2 3	2	0	2	2	4	2	2/6	
WT	-	3 5 7	5	0	5	0	5	0	4/6	
WDC	WM	1 3 5	3	2	5	4	7	2	4/6	
WDR	WM,WT	1 2 3	2	5	7	5	7	0	2/6	
WSS	WDC	1 3 11	4	5	9	7	11	2	10/6	
WDN	WDR	2 4 6	4	7	11	7	11	0	4/6	
WDP	WDR	3 4 11	5	7	12	8	13	1	8/6	
WSI	WSS,WDN	2 2 2	2	11	13	11	13	0	0/6	

In addition, the standard deviation (σ) and variance (σ^2) are shown. Since, in PERT, the variance of the end event, or activity, depends only on the activities on the critical path, sum the variance of the activities with zero total slack to get 36/36, or 1.0 days^2. Taking the square root of the variance, the standard deviation of the ending activity is 1.0 day. The probability of completing the project in 15 days, according to the PERT theory, is 97.5 percent.

It should be noted that the path WM–WDC–WSS–WSI has slack of 2 days. Yet it has a variance of 120/36 and therefore a standard deviation of 1.825 days. Thus, the probability of completing that path in 15 days is only a little better at 98.6 percent (Z = (15 – 11)/ 1.825 = 2.19). Thus, this path could be as, if not more, troublesome than the critical path.

The standard deviation for the critical path is a measure of the uncertainty associated with that activity or event. It permits the determination of the probability of completing the project on or before (or after) any specified time. For example, for the illustration network the standard deviation of the ending activity WSI is 1 day. Thus it can be said that the probability of completing the project in 12 days or less is 16 percent. Similarly, the probability of completing the project in 14 days or less is 84 percent. The probability of completing in 15 days or less is 97.5 percent. The probability is nearly 100 percent that the project will be done in 16 days or less.

Such information can be of considerable value. If, for example, a major meeting is dependent upon completing the project discussed in the illustrations, it would be desirable to schedule it no earlier than the 17th day. In more complex projects, it is not uncommon for several paths to converge just prior to a meeting or other high-cost high-visibility event. The more converging paths there are, the greater the probability that at least one of them will be late. Using some simple probability calculations, that probability can be calculated for alternative dates and the meeting scheduled accordingly.

Several papers have been written discrediting the uncertainty calculations of PERT. Some suggest that the optimistic and pessimistic times are not properly represented by 1 in 100 chances. Perhaps they ought to represent 1 in 20 chances, and so on. The user can select one of as many as four alternatives in this regard in some software. More serious is the error introduced by basing the entire calculation on only those activities which happen to be on the critical path in that particular calculation. As shown in the illustration, a less critical path can get into difficulty and take longer than the expected time for the critical path.

The best solution to this is to use one of the software packages which employ Monte Carlo simulation to more adequately develop the distribution of the ending activity or event by including the implications of all paths through the network. This capability has been commercially available for only a few years now, but its use is growing as risk analysis becomes of more pressing concern.

Resource Management.   Implicit in the above discussion has been the assumption that whatever resources are required in any given time period will be available. This is a tenuous assumption at best. Further, it is assumed that there is no limit on the number of resources which can be working on a given entity or in a given space at any one time. Such constraints can be incorporated in the network plan, but this is generally not a wise thing to do, as it may unnecessarily constrain other possible solutions.

Until resource implications are considered, the calculations discussed to this point merely provide time frames within which activities *may* be scheduled. There are two general classes of problems in this area, time constrained and resource constrained scheduling. See Examples 7.6 and 7.7.

If time is of the essence and whatever resources are required can be obtained, it is still desirable to bring the minimum necessary units of resource onto the project and avoid frequent hiring and firing. One experience by the author in using a program of this type resulted in a recommendation for the hiring of additional skilled tradespersons of, for example, 20 die setters and 10 electricians. At the same time, based on historical approaches to the same problem, the personnel department was attempting to fill requisitions for 10 die setters and 20 electricians. After examining our analysis, a quick revision was made to the hiring plans to match the project scheduling software's recommendation. Thus savings can be substantial both in not overhiring and in not hiring in the right pattern and incurring the resulting delays of the project.

Many software programs can handle this problem, albeit with varying degrees of sophistication. Research has been conducted on scheduling using job shop scheduling rules similar to the research familiar to most industrial engineers reported by Conway, Maxwell, and Miller.[5] In general, similar conclusions are reached. Nevertheless, most schedules derived in this manner can be improved on by experienced project managers.[8,9] Similarly, many project scheduling software programs can solve the resource constrained problem. There are two basic approaches: assigning resources to activities in accordance with one or more of the "job shop" scheduling rules and letting the finish time for the project go where it may or performing successive tries at the time constrained resource leveling procedure, lengthening the time available for the project at each try. The latter is demonstrated in Example 7.6.

EXAMPLE 7.6 TIME-CONSTRAINED RESOURCE LEVELING

Resource leveling is accomplished with the aid of a Gantt chart. Each activity is plotted with an "open bracket" marking the early start and a "close bracket" marking the late finish. The activity duration is plotted to show both the early (top line) and late (bottom line) alternatives. The resources required are shown on the left with the number in parentheses showing the resource days R-D.

It is useful to develop the early and late resource profiles by adding the resources required on any specific day. For example, for the early profile on day 1, WM and WT each require 1 unit for a total of 2 units of resource. On day 8, WSS requires 3 units, WDN requires 1 unit and WDP requires 3 units for a total of 7 units. Similarly, when determining the requirements for the late profile on day 8, WSS requires 3 units and WDN requires 1 unit for a total of 4.

	Resources Req'd		Day #	1	2	3	4	5	6	7	8	9	10	11	12	13	14	
	Units	R-D	Time	0	1	2	3	4	5	6	7	8	9	10	11	12	13	14
WM	1	(2)																
WT	1	(5)																
WDC	2	(6)																
WDR	4	(8)																
WSS	3	(12)																
WDN	1	(4)																
WDP	3	(15)																
WSI	1	(2)																
		(54)																
Early resource profile				2	2	3	3	3	7	7	7	7	4	4	4	1		54
Late resource profile				1	1	2	2	3	6	6	4	7	7	7	4	4		54
Average				4	4	4	4	4	4	4	4	4	4	4	4	4		52
Critical path				1	1	1	1	1	4	4	1	1	1	1	1	1		19
WM				1	1													2
WDP											3	3	3	3	3			15
WSS										3	3	3	3					12
WDC						2	2	2										6
Total				2	2	3	3	3	7	7	7	7	4	4	4	1		54

The average resource is determined by dividing the total resource-days required by the duration of the project, 13 days, to get 4 units per day. The critical path requirement is simply the resource requirements for those activities on the critical path. It makes sense to show those first because these activities cannot be moved without delaying the project completion. Below the critical path are shown those activities having slack. The sequence in which they are listed reflects the order in which the activities were considered for leveling. Activity WM was scheduled at its earliest time. Activity WDP was originally scheduled at its latest time . WSS and WDC were scheduled at their earliest times. When it became clear that 7 units were required on at least three days, WDP was moved forward one day to be in process on days 8 through 12, thus making the requirement for 7 units contiguous over 4 days. While there are other schedules which might be acceptable, this one has good characteristics in that it starts and ends with a small resource requirement and holds steady requirements of 3, 7 and 4 in between.

Two types of resources require different approaches to the problem. These can be characterized as homogeneous versus heterogeneous, common versus unique, or trade versus individual. The first of these, scheduling trades, assumes that all units of resource in a class have basically the same capabilities, that is, a carpenter is a carpenter is a carpenter. Most scheduling packages available today are of this type.

The other type of program, for scheduling individuals, assumes that each individual has unique capabilities. The author recognized this in attempts to design software for scheduling draftspersons and engineers in an engineering and drafting operation. One engineer could do bumpers very efficiently but was very slow on quarter panels (fenders). Another had comparable abilities of another mix. For lack of a better term, this was dubbed the "drafting room scheduling problem." It can be characterized as a three-dimensional assignment problem, where the resources and the activities form the usual dimensions but the problem is compounded by the seemingly random occurrence of the availability of the resources as well as the availability of the activities to which to assign the resources. It is unclear that this problem has been well solved to date in commercially available software. A similar problem occurs in many architect offices and software shops where each person is assigned several activities and works on more than one at a time. Thus the program should be able to monitor the hours allocated to each activity and the total commitment of the individual's time. Several packages have been developed to handle this problem; however, they tend to be tailored to the particular profession as opposed to being more general-purpose.

While analytical solutions have been proposed for these problems, they tend to be limited to small problems of little practical significance. It is likely that expert systems concepts and even artificial intelligence will provide a new class of solutions for these problems in the not too distant future.

EXAMPLE 7.7 RESOURCE-CONSTRAINED SCHEDULING

Resource constrained leveling proceeds in the same manner. In this case, the activities were scheduled in the sequence originally calculated, recognizing the limit of 4 units of resource per day. When that limit was reached, the next activity was delayed until resources were available. The project now requires 16 days.

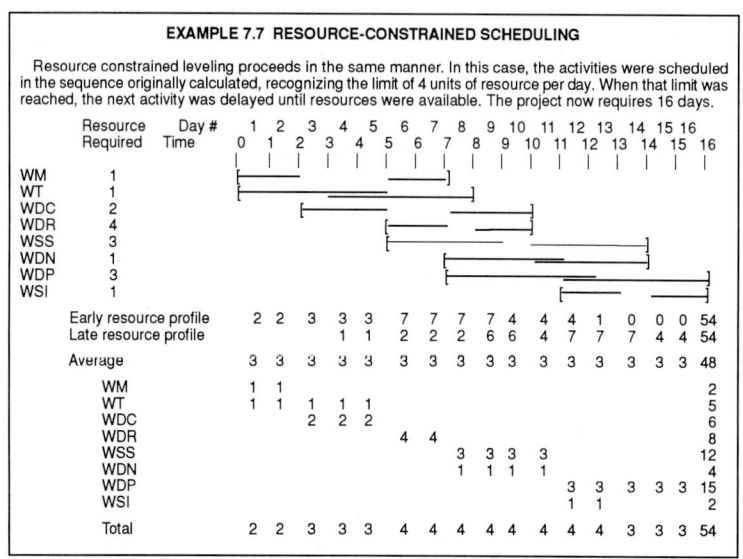

	Resource Required	Time
WM	1	
WT	1	
WDC	2	
WDR	4	
WSS	3	
WDN	1	
WDP	3	
WSI	1	

	1	2	3	4	5	6	7	8	9	10	11	12	13	14	15	16	Total
Early resource profile	2	2	3	3	3	7	7	7	7	4	4	4	1	0	0	0	54
Late resource profile			1	1		2	2	2	6	6	4	7	7	7	4	4	54
Average	3	3	3	3	3	3	3	3	3	3	3	3	3	3	3	3	48
WM	1	1															2
WT	1	1	1	1	1												5
WDC			2	2	2												6
WDR						4	4										8
WSS								3	3	3	3						12
WDN								1	1	1	1						4
WDP												3	3	3	3	3	15
WSI												1	1				2
Total	2	2	3	3	3	4	4	4	4	4	4	4	4	3	3	3	54

CPPS—Time-Cost Trade-off.   This was the technique that was developed as a result of the pioneering effort by Kelley (DuPont) and Walker (Sperry Univac) concurrent with the development of PERT.[16,17] The original solution technique was linear programming, and therefore the capacity was severely limited at the time. The problem, as originally characterized, was to determine the optimum project duration given that each activity in the network could be performed in a range of durations, one of which was least costly. This was desirable for, once a new process plant was authorized to be built or existing plant shut down for maintenance, every day the plant was not available, a loss was incurred due to the inability to sell the product which it might have produced were it on stream. The problem is akin to the classic inventory problem where one cost—carrying cost—decreases with smaller order quantities while another cost—ordering cost—increases. On the project, there may be opportunity costs—product which could have been sold if it had been available or interest costs, such as on the construction loan—which decrease with shorter project duration while other costs of performing the project increase with shorter project duration. Thus the objective is to determine the project duration for which the sum of these costs is the least as shown below.

To illustrate, consider only the activities on the critical path in our example problem. To shorten the time for the project, the duration of the critical path must be reduced. It is realistic in practice to analyze the critical activities because there will generally be only one or a few paths which require significantly more time than all the rest. These may consist of as few as 10 percent or less of the activities in the network. Further, some of these may not be amendable to time reduction while others may be very simple to reduce. For example, on a project to design a new engineering releasing system, a substantial reduction in negative slack was achieved when it was recognized that all interchange of documents was planned as activities handled through intracompany mail, each requiring 3 days. Those intracompany mail activities with negative slack were changed to hand-carry delivery, reducing the project duration by some 150 days.

EXAMPLE 7.8 TIME/COST TRADEOFF (CPPS)

To perform the crash calculations, two additional items of information are required for each activity, the crash cost and the number of days the activity can be crashed. The crash cost is the slope of the time-cost curve. Thus to reduce the project time in the least costly manner, it is necessary to reduce the durations of activities on the critical path and, later activities which were not originally on the critical path, crashing first those activities with the least slope or cost per day.

Act.	Pred.	Crash Cost ($/Day)	Crash Days	Normal Days	ES	EF	LS	TOTAL LF	SLACK
WM	-			2	0	2	2	4	2
WT	-	$100	2	5	0	5	0	5	0
WDC	WM	$ 75	1	3	2	5	4	7	2
WDR	WM,WT	$125	1	2	5	7	5	7	0
WSS	WDC	$ 50	1	4	5	9	7	11	2
WDN	WDR	$ 50	1	4	7	11	7	11	0
WDP	WDR	$ 80	2	5	7	12	8	13	1
WSI	WSS,WDN	$150	1	2	11	13	11	13	0

This is most easily accomplished by setting up the following table.

Paths			Crash Activity	Cost	Total Additional Cost
WT,WDR,WDN,WSI	WT,WDR,WDP	WM,WDC,WSS,WSI			
13 days	12 days	11days			
12 "	12 "	11 "	WDN	$ 50	$ 50
11 "	11 "	11 "	WT	$ 100	$ 150
10 "	10 "	10 "	WT & WSS	$ 150	$ 300
9 "	9 "	9 "	WDR & WDC	$ 200	$ 500

WDN is crashed first as it is the least costly activity on the critical path. This makes the first two paths equal in length at 12 days. WT is next as it is in each of these paths but is less costly than WDR. Now all paths are equal in length requiring that combinations of activities be crashed. WT and WSS combined have the lowest cost of $150. WDR and WDC are next at $200. The cost column gives the slope of the total cost curve for the project which can be weighed against the savings which might accrue for having completed the project earlier. For example, if the savings expected from implementing this project are $125 per day, there would be a net saving of $ 100 by performing the project in 11 days instead of 13. The important comparison , however, is the marginal cost. While there would be savings from performing the project in 10 days, it would only be $ 75. The marginal cost of the 10th day is $150 versus a savings of $125 resulting in a net marginal cost of $ 25.

This process can get very complex quickly but can be used on most projects to get considerable reduction of time with no more formal procedure than that shown above. For a more robust procedure, there are a few project management scheduling packages which incorporate this feature.

Cost Management.   Earned value analysis is the modern concept of what was originally known as PERT/COST, a system which was touted in the early 1950s but which was found to be too rigid in practice. Earned value is primarily a client-oriented cost control system in that it aids in ensuring that the client pays only for work that is done and materials actually delivered. It essentially recognizes costs at the time when the check is drawn to pay for the work or materials.

Earned value is very similar in concept to a standard cost approach for volume production operations, permitting analyses of cost differences due to schedule variance and spending variance. Because project work is essentially one-time in nature, the measurements must be based on estimated costs for each activity or work package.

Earned value analysis has three major cost measurements: budgeted cost of work scheduled (BCWS), budgeted cost of work performed (BCWP), and actual cost of work performed (ACWP). These are measured at the activity or work package level and summed progressively through the work breakdown structure to provide an overall measure for the project as a whole. Thus managerial analysis can begin by determining if there is a problem for the project and progress downward through the WBS to determine the contributing causes of a problem.

Cost management begins with a budget which, when approved, becomes the baseline for all further control. The first measure for an activity, group of activities, or the project as a whole is the budgeted cost of work scheduled (BCWS). It is the amount which should have been expended for work scheduled during the reporting period. It is accumulated with that for prior periods to obtain the amount which should have been expended to the reporting date. The cumulative cost curve to the end of the project is generally in the shape of an S and is called the S curve.

EXAMPLE 7.9 COST MANAGEMENT

For cost management purposes, consider the schedule developed in the Time Constrained Resource example as the official baseline schedule. Consider the cost of the resource to be $100 per day and the zeros will be dropped for convenience.

Day #		1	2	3	4	5	6	7	8	9	10	11	12	13	14
Time	0	1	2	3	4	5	6	7	8	9	10	11	12	13	14
WM		1	1												
WT		1	1	1	1	1									
WDC				⚡	2	2	2								
WDR							4	4							
WSS						3	3	3	3						
WDN								1	1	1	1				
WDP								3	3	3	3	3			
WSI												1	1		

Total Budgeted		2	2	3	3	3	7	7	7	7	4	4	4	1	54
Cumulative Cost (in $100)			2	4	7	10	13	20	27	34	41	45	49	53	54
Revised Budgeted		2	2	1	3	3	9	...							
Revised Cumulative Cost		2	2	4	5	8	11	20	...						
Actual Costs Incurred		2	3	4	4										
Cumulative Actual Cost		2	5	9	13										

If time now is at the end of day 4, the BCWS is $1,000. Assuming that activity WDC did not get started until day 4, the revised schedule shows it still taking 3 days. The revised cumulative cost now shows the BCWP to be $800, a schedule variance of $200 under spent. If the actual costs are as shown by period , the ACWP is $1300, a spending variance of $500 over spent. Thus, this project is, to date, in trouble on both schedule and and spending.

A forecast for the remaining activities would be necessary to project accurately into the final completion date and cost. WDC had slack of 1 day so its delay should not affect the completion date. Only detail knowledge of the causes of the cost overrun can aid in determining future costs. In the absence of such understanding, however, two assumptions are possible. First, it can be assumed that there will be no further overruns, in which case the EAC cost could be forecast as $5,900 ($5,400 + $500 variance to date). The second assumption might be that the remaining 8 days will incur cost overruns at the same rate, i.e., $125 per day ($500 / 4 days), in which case the EAC would be $6,900 ($5,400 + $500 + 8 x $125).

Not all work is performed to schedule, so the next measure is the budgeted cost of work performed (BCWP). This indicates the number of dollars which should have been expended during the reporting period, or cumulatively to the reporting date, in performing the work that was actually performed. The difference, BCWS – BCWP, is a measure of schedule variance, comparable with the volume variance in a standard cost system.

The work which is performed may not cost exactly what it was estimated to have cost. Thus the third measure is the actual cost of work performed (ACWP). The difference, BCWP – ACWP, is a measure of the spending variance, comparable with the rate variance in the standard cost system.

From these measures, then, the competent project manager can determine if actual expenditures are basically consistent with the actual progress on the project and determine what if anything can be done about it. Because of the pyramidal structure of the accounting system built into the project management software, a variance at the project level can be traced through the many levels of detail to the specific activity(ies) which is contributing to the problem. Thus it focuses the attention of the project manager on those areas most in need of diagnosis and direction.

Using various methods, based on different assumptions, a projection can be made of the estimate at completion (EAC) cost. The estimated completion date is more accurately projected based on the time and schedule calculations.

The earned value approach has been adopted by the Department of Defense (DoD), National Aeronautics and Space Administration (NASA), Department of Energy (DOE), and other federal agencies. The DoD version is called "Cost/Schedule Control Systems Criteria (C/SCSC)" and is well documented in Fleming.[11] Contractors performing work for the above agencies may be required to maintain cost and schedule records and controls and report in accordance with C/SCSC. It should be noted that certification of compliance with C/SCSC is based on the contractor's total management system, of which a part may be a standard project management schedule and cost system. While the vendor's package is not certified as such, it may be appropriate to determine if the package has been an integral part of a C/SCSC certified contractor's management system.

Project Strategic Analysis.   Two techniques developed in the early 1960s offer capabilities not found in the more familiar project management software packages. Because of the additional logic available in these, it is expected that they will become more popular as more deliberate strategic planning of projects develops and even more as project management concepts are used in corporate strategic planning.

Decision CPM (DCPM).   The techniques discussed above have some constraints which limit their flexibility in practical application. For example, there is no provision for identifying alternative methods of performing the same task or meeting the same requirement. The only way to accomplish this is to incorporate one alternative in plan A and the other in plan B. If the number of alternatives is appreciably greater than two *and* there is interaction between alternatives, the task of evaluating these can be formidable.

DCPM, a concept first published in 1967,[7] relaxes this constraint and allows alternative task sets to be defined, that is, mutually exclusive alternative methods for performing the task, one of which is ultimately to be selected. A task set is preceded by a decision node. The decision nodes are represented by a triangle while the tasks in a task set are portrayed by a circle, as are all ordinary tasks (or activities); that is, the activity-on-node graphic convention is used. The first task set might be identified as T_1 in the decision node (triangle) with the jobs belonging to the set being identified as $T_{1,1}$, $T_{1,2}$, and $T_{1,3}$. Based on some objective function, that is, minimum cost or minimum time, a decision can be made between the tasks in a task set. A sample problem is discussed in Wiest and Levy[32] in which there are two task sets composed of three alternatives each and one composed of two alternatives. It points out that there are 18 different ways of choosing from these and thus 18 different ways of performing the project. Clearly, these 18 alternatives could be portrayed as 18 different project networks.

The solution techniques include complete enumeration, a branch-and-bound algorithm, or a heuristic approach. While these techniques were documented in the early literature, the author is not aware of any currently commercially available software for solving this type of problem. It is discussed here because of the expectation that it will become available within the life span of this edition of this handbook. For more details on this technique see either Wiest and Levy[32] or Moder, Phillips, and Davis.[23]

Graphical Evaluation and Review Technique (GERT).   With the exception of PERT, the techniques discussed above are "deterministic" in nature; that is, they do not involve probabilistic elements. All of the above techniques, except DCPM, assume that all activities identified in the plan will be performed. None of the above techniques allow "loops" in the network, that is, an arrow which goes backward in the topology of the network, creating a path which eventually comes back to the same place in the network.

GERT portrays activities in the activity-on-arrow graphical notation, and nodes are either conventional deterministic events or decision nodes. Decision nodes have both an input side and an output side. The input side determines the conditions under which a job can be released, allowing following activities to proceed. For example, two or more alternatives may be planned to solve a particularly difficult problem. The first alternative finished might release subsequent activities and may, if properly coded, signal the completion of work on the preceding alternative activities. Four different logical alternatives are available on the input side including conditions relevant to the repeat of the node in the event of looping back. Output sides can be deterministic or probabilistic. If deterministic, the logic is the same as in the above techniques. If probabilistic, one of the emanating jobs is released depending on the probabilities assigned.

After being inculcated with "no loops" it may not be obvious why a loop would be useful in a network plan. Many times a part or assembly must go through a design-build-test sequence. If it fails on the test, it may have to repeat that same sequence. For example, automotive exterior lamp assemblies must pass stringent tests at the federal level and by some states in addition to tests in the company's own laboratories. Any of those tests may fail, resulting in a return to the design-build-test sequence.

GERT is also discussed in Wiest and Levy[32] and Moder, Phillips, and Davis.[23] For details of P-GERT, contact Pritsker & Associates, Inc., West Lafayette, Indiana.

OTHER PROJECT TECHNIQUES

Work Breakdown Structure.   One of the most fundamental tools in project management is the work breakdown structure (WBS). It is the tool by which a project, no matter how large, is subdivided into progressively smaller and smaller elements until the degree of detail is compatible with the needs of planning the project. A WBS is similar in appearance to an organization chart but it is quite different. The WBS identifies the work which must be performed in order to complete the project. The top level, level 0 (zero), is the total project. Level 1 typically identifies deliverables which, in total, represent all elements of the project. Further levels subdivide the work by a number of criteria such as components, responsibility or "trade," geographical location, and so on. Indeed, level 1 can be subdivided by any of these criteria unless specified by the client. For example, the U.S. Department of Defense requires level 1 to be by deliverables to ensure comparability of like elements across programs and vendors.

The WBS is an important tool in controlling the work content of the project. All work that is required by the scope statement in the contract for the project must be identified in the WBS. Conversely, no work should be included unless it is in the scope statement. This then becomes the baseline for controlling changes to the scope and negotiating for appropriate budget or charges. Not only is this important to ensure proper recognition of costs for the changes and appropriate charges to the client, it is vital in managing the project. *One of the anomalies of project management is that, while it is the management of change, it is changes to the project which are most likely to cause problems in the management of the project.*

The WBS is also important in clearly establishing the strategy of the project. The manner in which the project is subdivided determines what problems will have to be dealt with by the project manager as well as each lower-level member of the project team. Problems which are clearly within the responsibility assigned to the person responsible for a WBS element should be solved by that person. Problems which span two or more WBS elements require the involvement of the person who is responsible for all those elements. *One of the most important tasks of the project manager is managing the interfaces between WBS elements.*

Some basic principles must be followed in developing a WBS, including:

- *Uniqueness.* A unit of work should appear at one and only one place in the WBS.
- *Summative.* The work content of a WBS element is the sum of the WBS elements immediately below and related to that WBS element.
- *Unity of responsibility.* A WBS element is the responsibility of one and only one individual.
- *Consistency.* The WBS must be consistent with the way in which the work is actually going to be performed; that is, it should serve the project team first and other purposes only if practicable.
- *Motivation by involvement.* Project team members should be involved in developing the WBS (as well as the network plan), not only to ensure consistency but more to ensure "buy-in" to the project plan.
- *Documentation.* Each WBS element must be documented to ensure accurate understanding of the scope of work included, and not included, in that element.
- *Flexibility.* The WBS must be a flexible tool to accommodate the changes which are inevitable in a project while properly maintaining control of the work content in the project as per the scope statement in the contract.

Careful attention to developing a WBS according to these principles will reward the project manager and the organization with many benefits throughout the project.

Gantt Chart. The Gantt chart was developed by Henry Gantt about 1917 as a means for scheduling work in a job shop. It has become the most easily understood portrayal of a scheduled plan for a variety of types of work. While it was used extensively prior to computer methods for scheduling projects, it was often demeaned by the comment, "It conceals more than it reveals." Manual means for preparing Gantt charts were so time-consuming and expensive as to be close to "cast in bronze." As a result, the tendency was to delay redrawing them until the project was well off schedule.

Modern project management software programs prepare Gantt charts rapidly and inexpensively with a wide variety of graphical formats, as in Fig. 7.5. They are effective for communicating with clients, executives, middle managers, and line managers. Indeed, they are preferred by line managers, who generally want to know what they are supposed to do and when.

Gantt charts do have deficiencies. They typically do not show the relationship between

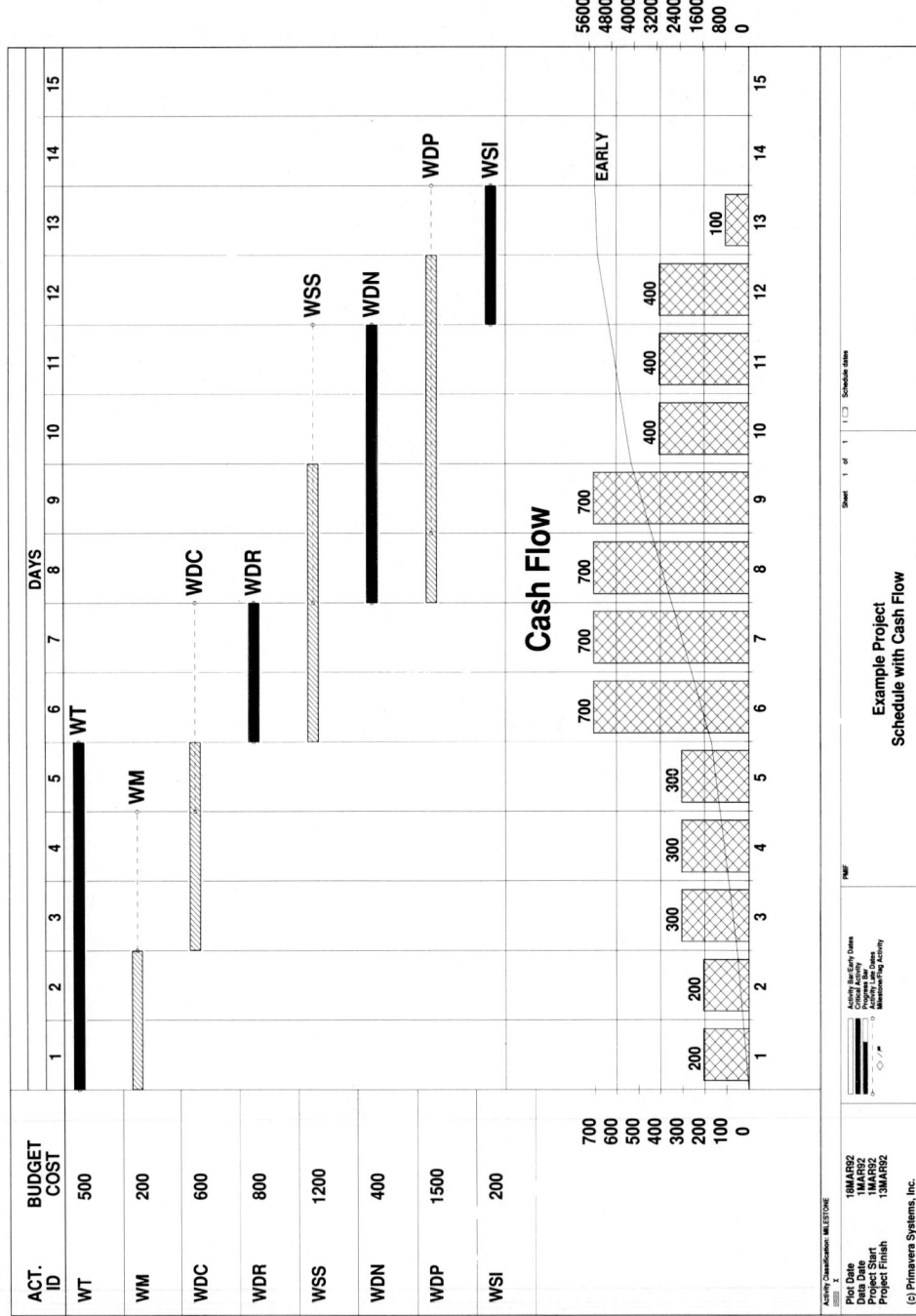

FIGURE 7.5 Example Gantt chart (based on example project). (*Courtesy of Primavera Systems, Inc.*)

activities in the network. A special version, the connected Gantt chart, overcomes this but in so doing becomes more complicated. In general, it is best to use the simple Gantt chart for most purposes, bringing out the connected version or even the network diagram for more complex problem solving.

OTHER NETWORK-BASED TECHNIQUES

Line-of-Balance.   Line-of-balance was developed in 1936 by the U.S. Navy and Goodyear as a technique for diagnosing the status of small-volume production efforts. It was used extensively by the armed forces procurement operations during World War II to diagnose contract status when deliveries fell behind schedule. It is used to integrate a production plan involving the production of a number of components resulting in assembly of multiple units of a final product. Several variations on the theme were developed, including the application to projects and new product start-up. There have been suggestions from time to time that line-of-balance should be computerized. To this author's knowledge, no such package has been developed which is commercially available. This is in spite of a conviction that there is a niche for a comprehensive and user-friendly package.

Line-of-balance (Fig. 7.6) consists of three graphic elements: a lead-time chart, an objective or schedule chart, and a progress chart. The lead-time chart resembles an AOA, event-oriented network, having multiple starts and a single finish. The objective chart shows the cumulative contract delivery schedule and the cumulative actual deliveries. The progress chart has a vertical bar for each event in the lead-time chart showing the actual number of units which have reached that point in the process. Combining information about the lead time for each event and the scheduled cumulative units which should have been delivered, a "line-of-balance" is derived which shows the number of units which should have reached each event at the time of the study.

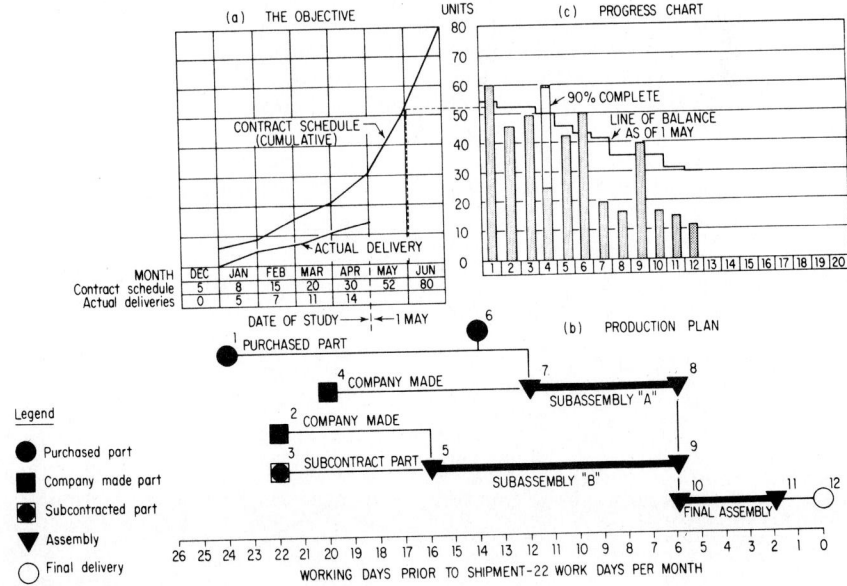

FIGURE 7.6 Line-of-balance technique diagrams.

Assembly-Line Balancing.   Maximum productivity on progressive lines (assembly and/or disassembly) can be achieved through proper assignment of work elements to work stations. Prior to, and even long after, the advent of computers this was a tedious task performed by industrial engineering staff. The earliest successful effort to use computers was by Tonge[28] in his heuristic system developed on white goods assembly lines at Westinghouse. Several formulations of the problem followed, many of which used analytical techniques, that is, techniques which guarantee an optimum solution in a finite number of steps. These analytical techniques, while theoretically useful, were limited in capacity and were therefore not of great practical use. Two development efforts were concurrently undertaken by General Motors and Chrysler in the 1960s, both of which were abandoned because new time-shared computer capabilities became available about the time the techniques were being completed. Private conversations between the author and his counterpart at GM confirmed that the approaches were almost identical.

The approach used at Chrysler was documented by Arcus as his Ph.D. dissertation and in subsequent articles on COMSOAL.[1] Arcus described the approach as it started at Chrysler in the early 1960s, relying on a Monte Carlo simulation approach. As the effort progressed, it moved more and more toward resembling what is now called "expert systems." By the mid-1960s, both efforts to approach assembly-line balancing were abandoned in favor of simpler approaches relying on analyst-computer interaction.

The analyst-computer interaction approach resulted in certain information about the work elements and their interrelationships, the line being assigned, and the production schedule being entered into the computer. The industrial engineer then arranged the work elements into workstations using all information available about alternative work methods, the work environment including location of critical facilities, and often the individual(s) who would be working at specific stations. The computer then performs all the arithmetic to allow the IE to try various alternatives in the process of assigning work elements, prevent the IE from doing something logically not permitted, prepare the finished data, and print out the completed assignments.

The essential graphic technique employed in assembly-line balancing is the precedence, or "balloon," chart. At first glance, it might be difficult to differentiate between a balloon chart and an AON project network diagram. The logic is similar in that a work element in a balloon chart cannot be assigned to a workstation until all its predecessors have been assigned to workstations. Figure 7.7 shows a portion of a typical balloon diagram.

Most discussions of assembly-line balancing assume as an objective function the minimization of idle time at each station for a given production volume. Based on this objective function, a set of heuristics (rules of thumb) are developed to use in selecting the work elements to go into a particular workstation. This may well produce an assignment which is suboptimal overall. One reason for the assignment being suboptimal is that it often leads to one or more stations at the end of the line having significant idle time. This leads to

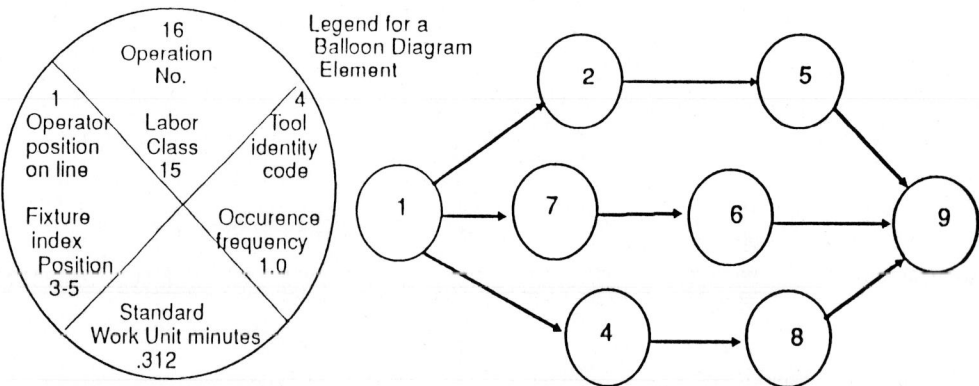

FIGURE 7.7   Example balloon diagram for assembly-line balancing.

complaints of inequity between operators. A more subtle problem is that assignments developed in this manner may have an appreciable amount of time spent in changing tools, changing position of the part or the operator, or other actions which do not change the nature of the product being worked upon. In the Chrysler effort, a more appropriate objective function was found to be to maximize the "do" work elements. "Do" work elements are those which are essential to change the nature of the product as it progresses from the initial workstation toward becoming a finished product. This substantially changed the logic which was incorporated into the computer program for assembly-line balancing.

While the above is presented in the context of a comprehensive computer program for assembly-line balancing, the logic is appropriate for either a manual approach or any degree of computerization of the process of developing assignments.

Corporate Strategic Planning.   The techniques of project planning and scheduling have not been widely used in the executive suite. Reports of interest in these capabilities for strategic planning are becoming more and more frequent. The most obvious application is to determine feasible target dates for implementing a strategic plan, often requiring the completion of several related projects. The next most obvious is to incorporate cost estimates to test the financial feasibility of the planned strategic move. Risks can be assessed on these two dimensions using the multiple time estimates to indicate the degree of uncertainty associated with individual activities and, with the proper logic, the impact on costs. At still another level, project networks might be used to test the feasibility of a strategic plan subject to limitations on the physical and human resources available. This might indicate the degree of need for additional resources.

These capabilities exist in many commercially available project management software packages today. The next capability that is likely to be introduced into the executive suite is the either/or logic of DCPM/GERTS. This will permit the definition of alternative plans to reach the same objective and more sophisticated analysis of the potential consequences and risks.

THE PRACTICE OF PROJECT MANAGEMENT

The Project Management Institute has published a "Project Management Body of Knowledge" (PMBOK) defining the essential terms and concepts which a person needs to understand to earn certification as a project management professional.[26] It is organized into eight basic areas: scope, quality, time, cost, risk, human resource, contract and procurement, and communications management. The following discusses the application of project management based on this framework.

Scope Management.   The first step in any project should be the identification and documentation of the objectives of the project, that is, the characteristics of the product of the project which is being contracted. A strategy is then chosen by which the project is to be undertaken, and it is expressed in a work breakdown structure (WBS). The WBS is an orderly, progressive detailing of the elements of the project resulting, at its lowest level, in identification of the activities which must be undertaken to achieve the project objectives. It is a more precise and disciplined way to identify the tasks in a project than the familiar "back of the envelope." There should be no attempt to portray the sequential relationships between the elements in a WBS. That is done later in the development of the network diagram. The WBS identifies the work content of the project, the work for which the contractor is to be paid. The approved scope statement, including the WBS, becomes the baseline against which technical and cost performance is measured.

It is important that the scope of a project be carefully defined for at least two reasons, omission and commission. Errors of omission result in the client's not being satisfied with the product of the project. Errors of commission, while perhaps leading to greater client satisfaction, probably result in work being performed for which proper compensation is not received.

Errors of omission may be recognized prior to completion of the project. If discovered early enough, their impact may be negligible. If not, they can result in significant tear-up and rework costs. They may result in extra costs, inadequate performance of the product of the project, and/or litigation to resolve disputes over the delays.

Errors of commission may enter the project in the scoping process by mistake or on purpose. In a large, complex project, miscommunication may lead to work being incorporated which is not really necessary or intended. Additional work may be incorporated in the project at lower levels in the organization to avoid risks. Even after the project scope is approved, additional work can creep into the project as a result of direct contact between the client and those responsible for performing specific activities. For example, in a software development effort, the client may persuade the programmer to introduce extra features in the software without proper recognition that this resulted in a change of scope. When this happens, the contractor is not likely to be compensated for the extra work. Change control procedures are a must to ensure that all changes to the project scope are documented, checked for adverse consequences, and approved, including payment if appropriate.

Quality Management.   Quality management has at least two major aspects: assurance and control.

Quality assurance in a project should include a thorough review of the network plan to ensure process capability. Just as designs for the product of the project are scrutinized by senior review, peer review, and structured walk-throughs, the project plan—the design for the project—should be subject to the same reviews. Activities should be identified for more in-depth review of the process by which those activities are to be performed. Such a formal review might have identified critical shortcomings in the project which resulted in the collapse of a building on opening day in a suburb of Vancouver, British Columbia.[31] At least it might have led to a recognition of those activities or contractors for which close quality control was essential.

Quality control in projects can be enhanced by the coordination of control efforts with the activities in the project. Having reviewed the project plans and identified appropriate activities, it is easy to follow progress on the project plan and ensure adequate steps are taken to measure and document critical variables and, even better, cause the work to be done right the first time.

Time Management.   Time management consists of four essential processes: planning, estimating, scheduling, and control.

Planning.   "Planning consists of identification of the intentions of the project management group with respect to the methods and procedures they intend to follow," that is, what they intend "to do, how it will be done, what will be used to do it."

Given the four steps in time management, it should be apparent that scheduling is a separate process. It may be less than apparent that planning and scheduling *should not* be done concurrently. This is a habit which has carried over from planning with a Gantt chart. That tool invited doing both, as well as estimating, concurrently. The risk of this approach is that, with no understanding of the interrelatedness of some activities, a schedule will be developed during planning which is inherently inefficient. It is better to wait until the planning and estimating are essentially completed, make the first computer schedule analysis run, and use the information which is developed to aid in proper scheduling. For example, until after the first schedule analysis, the planner does not know which path through the network is critical. On one system development project, the critical path was composed, in part, of some 50 activities which involved intracompany mail. Each activity was estimated at 3 days. It was an easy decision to have mail on such critical paths hand-carried. It would have been foolish to make that decision prior to knowing that this method, clearly the cheapest, was not appropriate for some activities. Generally, such revelations are more substantive in nature, but they often can be just as surprising to the project planner.

Estimating. Estimating, as it relates to the time management function, is the determination of the duration of an activity.[26] Inability to estimate accurately should never be an excuse for not using modern scheduling techniques for two reasons: First, the degree of accuracy is independent of the technique used; it is inherent in the projects we undertake. Second, the network approach to planning inevitably leads to a better understanding of the project, and therefore better estimating than comes with the use of any other planning and scheduling technique.

Owing to the unique nature of projects, the typical approaches to estimating used by industrial engineers in volume production situations are generally not applicable. For an organization which performs similar projects repeatedly, it may be useful to develop standard jobs in a manner similar to that used in setting standards for maintenance tasks. See Sec. 5, Chap. 6 for further details.

Inherent in the estimating process is the determination of the work content in resource hours, the units of resource which will be applied, the utilization and productivity of those resources, and in many instances, the skill level of the specific resource relative to the specific activity. Environmental circumstances affect these estimates, including the weather, hazards, and uncertainties associated with hidden elements such as underground conditions. The work climate of the organization can affect these estimates, so the estimator must be aware of factors contributing to organizational climate.

There are two theories as to who should do the estimating. One approach is for the person most directly responsible for performing the activity to make the estimates. This leads to greater psychological commitment to the estimates. A favorable work climate can result in consistently optimistic estimates which may have to be tempered by management. If the work climate is unfavorable, this will likely lead to "padded" estimates in order to avoid the unpleasant consequences of failing to complete the activity in the estimated time. The other approach is to have a professional estimator, who is more likely to provide an unbiased estimate. This is particularly appropriate when an incentive is associated with on-time completion of activities.

Scheduling. "A plan represents how one intends to execute a project without regard to when the project will be executed....the schedule applies the recognition of time and resource constraints."[26] The conflict between time and resource constraints is inherent and must be resolved. A project will not get done by management dictum if the time and/or resources are less than is required. The discussion of resource management techniques detailed the difference between time-constrained and resource-constrained scheduling. One will be controlling in any project and will determine the feasibility of completing the project.

Control. "Control contains, as its components: the measurement of what actually happened against what was expected to happen; what the results, or effect, will be; and if negative, the implementation of steps to prevent undesirable impacts and, if positive, the implementation of steps to ensure its continuation."[26]

The most popular method of measuring performance is by "percent complete." This is subject to an adage that "It takes 90 percent of the time available to complete 90 percent of the work and it takes another 90 percent of the time available to complete the last 10 percent of the work." Some practitioners have adopted a practice of placing a limit on the maximum duration of activities and then accepting reports of progress only in terms of completed activities. This becomes particularly poignant when payment for work accomplished is based on completed activities. Another approach is to have the responsible person simply answer "yes" or "no" to the question, "Are you on time?" It should be clear that the question applies to the activity's being started on time as well as to its being completed on time. Failure to start on time is the first indication that the activity will likely be late. Failure to complete an activity on time following a stream of "yes" answers can be very embarrassing. A "no" answer on intermediate progress reports should not be scorned, as it is an indicator that action may be required by management, an indicator which is received while there is still time for effective action to be taken. Precipitous reaction to unfavorable progress reports can have serious consequences, resulting in inaccurate reports of progress or excessive reliance on management to effect the solution.

Cost Management. Cost management is discussed under four headings: estimating, budgeting, control, and other applications.

Estimating. Cost estimating is the process of assembling and predicting costs of a project over its life cycle. The primary focus is on estimating the cost of the project itself. Techniques and practices used in justifying the product of the project such as payback and return on investment are also included.

Budgeting. This step involves relating the costs of the project to the activities in the work breakdown structure. From this is derived a cost plan, and when associated with the time schedule, a cash flow forecast can be obtained. Many project scheduling systems are flexible in recognizing the timing of cash flow. It can be assumed that the cash flows at the end of the activity or uniformly over the duration of the activity. Other assumptions are also possible in some systems.

When the schedule and costs are approved, the baseline plan is established against which future progress and expenditures are measured. One of the important uses of the information developed in this step is the measurement of expenditures over time. By accumulating the costs over all activities by time period, a histogram of expenditures is derived. If these costs are accumulated over all time periods, the result is an S curve of cash flow. If the project is performed to schedule, expenditures should flow along this curve.

Another use for the data developed in the budgeting process is cash management. This is particularly vivid on government projects where the funding is approved year by year, requiring speeding up the project one time and stretching it out the next. By changing the completion date of the project and scheduling the activities appropriately, the S curve will become steeper when accelerating the project and flatter when stretching the project out.

Control. "The process of cost control is the gathering, accumulating, analyzing, monitoring, reporting and managing the costs on an ongoing basis."[26] Cost control, or management, must be considered from four points of view.

1. The concerned party dimension
2. The time dimension of costs
3. Current cost control
4. Financial accounting systems compatibility

The Concerned Party Dimension. The major theory for cost management on projects is "earned value." Earned value as a concept can be characterized as the "standard cost" concept applied to projects. The implementation of earned value is well documented in publications by the Department of Defense and others who have officially adopted this theory. Such publications may be referred to as C-Spec, C/SCSC, or more completely, as cost and schedule control systems criteria.

C/SCSC presents the current requirements derived from what started as PERT/COST. For a variety of reasons, including its lack of flexibility, PERT/COST did not work in practice. C/SCSC was developed to provide more flexibility. Over time it has been adapted to ensure accurate and proper accounting for projects. This author's sardonic view of C/SCSC is that it is an attempt to close the loopholes by which contractors have, in the past, billed for costs which were not justified. This view is derived, in part, from the anecdotes recounted in a very good book on the subject by Fleming.[11]

C/SCSC is primarily a client's cost management system; that is, it provides assurances to the client that payment is not being made for work that is yet to be done. To the extent that project managers are clients of subcontractors, "earned value" is certainly a useful tool. On very large projects, it provides the project manager an overview of costs. No such system should be adequate for effective cost management to the exclusion of "managing by walking around."

The Time Dimension of Costs. If the project manager relies solely on the typical accounting system to manage the cost on the project, it will be too late and cost overruns will likely be the order of the day. The typical accounting system recognizes costs when they are actually incurred, that is, basically when the check is written. In non-project-oriented organizations this is often as much as 45 days after the performance of a given activity or the delivery of materials or equipment.

At minimum, the cost should be recognized when the work is performed or the delivery is received. Many project managers have resorted to collecting actual labor hours on a daily or weekly basis, extending them by some standard or average rates, and entering them in the project cost system immediately. This requires later reconciliation adjustments, but that is better than being in the dark for a considerable period of time. Similarly, costs on materials and equipment can be entered directly off the invoice rather than waiting for them to make the rounds through the accounting system.

The best management of project costs is achieved if they are recognized when they are committed. Many organizations manage costs based on a "committed cost control" concept. While specific implementations will vary, the basic concept is to recognize that cost is incurred at the time when the purchase order is released, a work order is authorized, or a similar event has occurred. The significance of this is easily seen when the lag between signing a purchase order and the writing of the check for pay for the purchased item is recognized. For major equipment items, this lag may be more than a year. If control is not exercised until the check is written, additional commitments can result in serious overruns.

In organizations which do not have a formal committed cost control system, managers often maintain supplementary records manually to ensure they do not overcommit authorized funds. The author is not aware of any project schedule and cost control systems commercially available as of this writing which provide true commitment cost control.

Current Cost Control.   The emphasis on earned value concepts could lead one to believe that such a system would be adequate to control costs on a project. A more realistic approach recognizes that costs are controlled by careful planning, communication, observation, and correction on a daily, if not an hourly, basis. Measurement of accomplishments each day and careful comparison with planned accomplishments provide information which can be used to make corrections before further efforts are expended. For example, in pouring a concrete retaining wall, the amount of concrete can be determined by measurement of the forms as well as recording the amounts received in each truckload as compared with the amount budgeted. Labor costs can be measured for the pour and compared with standards derived from similar work. Performance comparisons can thus be available before the next day begins. Most construction firms and other companies that normally work in the project mode have developed their own in-house systems for recording and measuring each day's production and having the results available the next day to permit corrective action as appropriate.

Effective cost management must be current. It must be done on a day-by-day basis just as it is on the shop floor of a manufacturing operation. This capability does not seem to exist in any commercially available software for project management.

Current cost control on projects can also be enhanced through the use of conventional industrial engineering techniques such as work sampling and methods studies. Work sampling can measure the frequency of delays of labor and equipment due to manageable causes such as lack of material or instructions, waiting on equipment, and breakdowns. Work methods can be applied at the planning stages of the work. For example, in constructing a wood frame house, there are many opportunities to introduce modern methods, jigs, and fixtures which can provide substantial reductions in work content to accomplish a specific task. Most skilled tradespersons lack training in the application of work methods analysis. Also see Sec. 4, Chap. 3 for further details.

Financial Accounting Systems Compatibility.   One of the most difficult problems involved in realizing adequate project cost control is achieving compatibility with the accounting system in the organization. Unless the organization operates primarily in the project mode, the accounting system will be oriented toward one of the other four modes of work efforts, that is, craft, job shop, progressive line, or continuous flow process. This will cause conflict. Many accounting systems are established primarily to accommodate the requirements of financial accounting. Many of the practices involved in financial accounting do not meet the needs of timely project cost control. Even in accounting systems with strong work order systems there is often a tendency to collect costs at a more summary level than is generally required for adequate project cost control.

If the above is true in your organization, special efforts may be necessary to achieve adequate control. For example, on some projects it has been found advantageous to make a copy of all labor distribution reports prior to submission to accounting. Labor hours are attributed to activities or work packages and priced at standard rates. Copies of all invoices are used to record payments to others. Other costs are dealt with in a similar manner. Monthly, a reconciliation is done and correcting entries are made to negate variances. It is extra work but sometimes necessary to achieve control on projects which are moving too fast to wait on the normal accounting data. Alternatively, control can be exercised over the resources used on the project and let the dollars fall where they may.

Other Applications.   There are other applications of cost control techniques, two of which are discussed below.

Life cycle costing has grown more important to ensure that decisions on the project are made with the proper perspective. Project costs are typically around 20 percent of the life cycle costs of the product of the project. Often a small marginal increase in project costs can actually reduce the total life cycle costs of the product. For example, a building's hallway floors might be covered with tile—perhaps a more expensive approach—in such a way as to allow a mechanical mopping machine to be used for cleaning the floor at a lower cost per year. Or landscaping can incorporate features which minimize maintenance.

Value engineering and analysis is a technique useful in identifying characteristics of the product leading to a lower life cycle cost. Also see Sec. 7, Chap. 2. Form and function are analyzed to ensure that all features incorporated in the project enhance the utility of the product. Many companies are implementing concurrent engineering and utilizing sociodiagram analysis to ensure that relevant functions and persons are involved in the design process. For example, companies have placed product engineering, purchasing, and manufacturing engineering in close proximity and assigned responsibilities to ensure consideration of production problems in the design of the product. Even maintenance is becoming involved as it is a significant part of life cycle costs. In concurrent engineering, efforts are made to involve the ultimate users in product design to ensure that the functions are performed in a "user-friendly" manner. Also see Sec. 7, Chap. 7.

Risk Management.   Risk analysis is becoming an integral part of project management. The most frequently used aspect of this seems to be measuring the time uncertainty in a manner similar to that developed in PERT. However, a more accurate approach is taken today based on Monte Carlo simulation. This approach considers the impacts of all paths through the network to the event in question. Thus reasonably valid statements can be made about the probability that a project or an intermediate milestone can be completed by a given target date. If the probability is too low, the target completion can be renegotiated to a later date or the plan revised to increase the probability to an acceptable level. Such considerations are of increasing relevance in multiproject programs where the results of the program cannot be realized unless all projects are completed per schedule.

In the PMI PMBOK, risk management is divided into two categories: identification and mitigation.

Identification.   Identification is subdivided into five categories. *Legal risks* involve licenses; patent rights; contractual failure; suits, both insider and outsider; and force majeure. *Technical risks* may involve state-of-the-art technology or new applications of proven technology or simply differences from expected quality, productivity, or reliability. *Internal nontechnical risks* can result from schedule delays, cost overruns, or cash flow problems. *External predictable risks* may involve economy and market forces, operational problems, or environmental or social impacts. *External unpredictable risks* include natural hazards; regulatory problems; and failure of other organizations, such as a governmental failure to complete necessary related facilities or services.

Mitigation.   Risk mitigation is subdivided into five categories. *Impact analysis* is used to assess the possible consequences of the various risks identified. Based on the impact analysis, *response planning* determines what action should be taken. One action is to spread the risk on *insurable* events including direct property damage such as auto collision, indirect consequential loss such as business interruption, legal liability such as property damage arising from another's negligence, and personnel including employee bodily injury. A *response system* should be put into place to ensure that appropriate action is taken should the risk be realized. Finally, *data applications* should be developed to establish both a historical and current project database to aid in assessing future risks.

Human Resource Management.   There are six subdivisions in human resource management, categorized under administration and behavioral.

Administration.   Some of the areas of general human resources administration with which a project manager should be familiar are *employee relations, compensation and evaluation,* and *government regulations and evaluation.* Included in these are recruiting, training, labor relations, performance evaluation, and the law pertaining to discrimination and equal opportunity. Most industrial engineers need this same knowledge, so it will not be discussed further here.

Behavioral.   Because projects deal with change, there is generally considerable interest by *individuals outside the project.* Ample evidence of this has appeared in the popular press relative to public concerns and resulting delays associated with nuclear facilities. Today the same concerns are being expressed with regard to waste disposal facilities and landfills.

Project management is probably more dependent upon knowledge of and skills at dealing with human behavior than any other mode of work effort. Because much of the work is done by professionals who tend to work independently, the role of supervisor is much less than it is in most volume production operations. Even skilled tradespeople work in an environment where supervision of methods and effort is often less strict than in most volume production operations. Thus efficient and effective utilization of resources is far more reliant on personal motivation of the individual operator. Knowledge and skill in dealing with individual *team members* and with the small and large groups who make up *the project team* are essential. Using power and influence with individuals helps project managers gain cooperation by using persuasion of others over which they have no direct authority. Team building and conflict management help the project manager create project team cohesiveness and maintain it in spite of the inevitable conflicts involved in creating change.

Contract and Procurement Management.   Certainly on large projects, most of the material and work efforts are obtained from other organizations. Thus contract and procurement management is a major element in the planning and execution of projects. Even small, in-house projects often involve many participants. Agreements must be obtained from each participant to ensure their commitment to the project. The concepts and processes involved are essentially the same as with external sources.

A clear understanding of the *objectives* of the project as manifest in the scope of work, the project strategy, the environment, and the risks is essential. Much of this can be reflected in the contract work breakdown structure which identifies relationships between WBS elements and their interfaces. There is no better documentation of this than in "Endicott Oil Field: First Offshore Arctic Oil Field Begins Production."[12] The construction of this offshore Arctic production and processing facility 20 miles east of Prudhoe Bay exemplifies the best use of contracting strategy to complete a $1.0 billion plus contract 9 months ahead of schedule and $600 million under budget. Oil production started only weeks after the arrival on a barge of a gas compression module (5160 tons, 96 feet wide, 195 feet long, and 105 feet high) from its erection site on the Mississippi River in Louisiana.

Information systems and methods of *procurement identification* are required to support the *acquisition process.* The latter includes knowledge of acquisition methods, source selection, contract types, procurement and tender documents, invitation to bid procedures, bid evaluations, and negotiations. *Contract administration* involves notice to proceed, performance control, financial control, change control, dispute resolution, and contract closeout. Finally, the cycle is complete only after *postcontract evaluations.*

Communications Management.   Communications management is gaining recognition as a necessity for smooth progress on projects. Networks are a major means of communicating. This should be uppermost in the minds of planners as they develop and document the network. If the planners are other than those who will execute the project, the plan must be communicated to the "doers" and their acceptance of the plan obtained. The plan must be communicated to upper management as an expression of the adequacy and completeness of the planning process. Conversely, it must be communicated in a manner which provides upper management with the information which is essential for them to know without overloading them with details relevant only at much lower levels in the project organization. Finally, the network must provide the lowest-level project participant the information essential for performing their individual activities in a timely and cost-effective manner. Thus, in many ways, developing a network is an art gained through experience as opposed to a science which can be simply taught.

Communicating schedules is essential if those responsible for performing are to start and complete their activities on time. Not only is it important to communicate the dates for start and completion, it is perhaps even more important that they be communicated in a manner which enhances acceptance of responsibility and commitment to the orderly progress of the project. This may be more of an art than developing the network plan, as it requires a careful assessment of the culture of the organization to determine what will be accepted in that organization at a particular time. Even in an organization which will accept the most rigorous discipline, *excessive pressures* to perform can lead to *excessive efforts* to perform and may reduce the acceptance of the system on later projects. On one project, the team was so motivated as to cause this author serious concerns about the physical and emotional well-being of team members due to excessive overtime and stress.

Gantt charts are still one of the preferred methods of communicating schedules. They help provide a perspective, showing project participants how their work relates to the overall project. The greatest objection to Gantt charts, that is, the cost of preparing them, has been substantially negated by modern project management scheduling and control software which generates Gantt charts automatically. Care must be used, however, to minimize changes to the schedule to avoid confusion as to the real schedule and eventually ignoring the schedule data altogether. One such approach to this is well documented in "Project Administration Methodology: Achieving Schedule Control on a Large Project—The Somers Project."[19] This project, involving consolidation of a number of regional operations within IBM, was characterized by volatility of the project plans and the absolute necessity that final moves and cutovers be flawlessly accomplished over weekends to ensure continued support of massive information systems.

A number of articles have appeared in the *PM NETwork* illustrating the importance of communicating with the various publics of a major project. One of these, "The Milwaukee Water Pollution Abatement Program: Its Stakeholder Management," shows how effective public relations solved many problems before they reached crisis proportions and gained public support for the project.[25] Another, "The Westlake Story: The Need for Coordination, Cooperation and Communication," describes how coordination and cooperation were achieved through effective communications.[22]

SOME OPPORTUNITIES AND CAVEATS

Opportunities.   Project management is the management of change. Change is increasingly the norm in business, industry, and government. The efficient management of change is essential to survival, and modern project management is the way to achieve that efficiency.

Many organizations that operate primarily in the project mode have adapted project management as a way of life. An excellent example of this is AT&T Business Communication Systems Division, which equips their project managers with laptop computers loaded with a standard set of software useful in managing projects.[24] Modern project management is used in the construction of the English Channel Tunnel and in developing new pharmaceuticals.[10] It is used in developing high-energy physics research capabilities at Sandia Laboratories[3] and on the "Super Conducting Super Collider."[2] It is used to get new products to the market more quickly and to manage the Voyager II trip to Neptune.[4] It is even used in running the Olympic Winter and Pan American Summer Games.[15]

It is becoming more popular for managers having responsibility for many small projects to simply keep track of their status and to understand the implications of both budget and resource availability. This area is one of the great opportunities for industrial engineers to contribute to the success of their organization. Modern project management can, and should, be applied at the plant level to manage all the non-volume-production work being performed in the plant. It should then be expanded, where appropriate, to include all such project activity in a multiplant organization. The author had such an application working long before either timesharing or microcomputer approaches to project management were available. Other than the slow turnaround it was well received. It included personnel studies as well as new product and facility preparations.

Caveats.   Care must be taken in applying modern project management in an organization not familiar with its application. It can drive a project team too hard and develop resistance. One plant manager favored allowing supervisors to spend all the money in work orders rather than risk having the unspent funds resulting from better project management returned to corporate. Care must be taken to ensure that benefits being enjoyed today are not lost with the introduction of new management tools.

Managers must understand that modern project management tools are more precise than previous tools. They cannot be used in the same manner as the tools of the past. *If they are misused, project team members will ensure that they are not accurate.* If used properly, they can support effective team building and lead to projects being completed in conformance with technical requirements, on time and, often, under budget.

One of the ways in which these tools can be misused is to put too much pressure on the team member to perform. People who are pushed to extreme will avoid using the systems if at all possible.

Finally, care must be exercised in using the tools to ensure the most appropriate means are used to improve performance. Following is a list, in a preferred sequence based on a general concept of the cost-benefit ratios, of ways to improve performance on projects.[29]

- *Managerial attention and involvement* to show interest and concern, but not to the point of interfering with those who have to accomplish the tasks.
- *Expediting* to ensure that obstacles to progress are removed and resources are available when required.
- *Improved methods* can reduce the work content, time required to complete a task, the resources used, and the resources wasted.
- *Reassigning resources* can move the most efficient resources to the activities which are most critical.
- *Reallocating resources* can move resources from noncritical activities, stretching their duration, to critical activities, shortening their durations.
- *Overlapping activities* can allow work to proceed more quickly but with some risk unless extra attention is given to adequacy of communications.
- *Defining activities in greater detail* can subdivide the work so more people can work on the project simultaneously.
- *Deleting certain activities* is often feasible after careful review with the client of what is really essential.
- *Changing the technology* applied is similar to methods improvement but may be more risky if it is a technology which is unfamiliar to the performing organization.

- *Subcontracting or buying* the goods or services can often result in economies as well as time savings, but with some loss in control and dependability.
- *Applying additional resources,* using the CPPS approach to time-cost trade-off, can reduce the total time required, but note that this is eleventh in this list.
- *Changing target dates* on the total project, still delivering essential parts of the project at dates acceptable to the client, can minimize perceived poor performance.
- *Changing the scope* of the project can permit timely delivery of essential elements and possibly avoid major failure on the project as originally conceived.
- *Changing the person responsible* often results in major delays while the new person becomes acquainted with the project *and* the team gets acquainted with the new person.
- *Changing the management information system* can lead to better performance but generally only after all participants have an opportunity to become familiar with it and it is adjusted to the peculiar needs of the organization.
- *Changing the organization structure* seems to be a quick fix, but often it takes considerable time for the project team members to begin to function as a team.

Modern project management is not a panacea, but used wisely it offers many opportunities for improving the performance of an organization's project work and, as a result, leads to better performance in other parts of the organization because projects are more likely to be completed on time and under budget.

SUMMARY

Project management has grown from an early conception that it was simply planning and scheduling, then to include cost management, and now to a comprehensive responsibility for seeing that all aspects of the project are carefully integrated and managed. One view of this progression is that the early focus on scheduling and cost management solved a major problem at the time—inadequate control of time and cost. The success in dealing with these two vital areas has substantially reduced the time required and improved the accuracy and utility of these functions, making more time available to manage the other elements of project success.

Project management experience has become recognized as one of the very important career steps on the ladder to general management. The project manager is the general manager of that project. The normal progression for successful project managers is on to larger, more challenging assignments.

The progressive industrial engineer can bring substantial benefits to an organization through project management. Purchase an inexpensive project management software package. Start with a small project in the department. Move to a larger project with another department which is receptive to change. By this time you should have a much better idea what is required in project management software for your organization. Then move up to a still larger project involving several departments. By this time, the experience gained and the successes recorded should build the acceptance to apply these concepts plantwide, if not corporatewide. Good luck!

REFERENCES

1. Arcus, A. L., "COMSOAL: A Computer Method of Sequencing Operations for Assembly Lines," *International Journal of Production Research,* vol. 4, no. 4, 1966.
2. Baggett, Neil, R. Aprile, T. Kozman, D. L. Pells, and E. J. Story, "The Super Conducting Supercollider," *PM NETwork,* vol. IV, no. 8, pp. 6–40, November 1990.

3. Barr, Gerald R., James P. Furaus, and Charles G. Shirley, "Particle Accelerator Research and Development at Sandia National Laboratories," *Project Management Journal,* vol. XIX, no. 1, pp. 29–47, February 1988.

4. Bartos, Ken, and William D. Brundage, "The Voyager 2—Neptune Encounter, A Management Challenge: Speeding toward Your Deadline at 42,000 Miles per Hour...Without Brakes," *PM NETwork,* vol. III, no. 4, pp. 7–23, May 1989.

5. Conway, Richard W., William L. Maxwell, and Louis W. Miller, *Theory of Scheduling,* Addison-Wesley, Reading, Mass., 1967.

6. Crosby, Philip B., *Quality Is Free,* McGraw-Hill, New York, 1979.

7. Crowston, W., and G. L. Thompson, "Decision CPM: A Method for Simultaneous Planning, Scheduling and Control of Projects," *Operations Research,* vol. 15, no. 3, pp. 407–426, May–June 1967.

8. Davis, E. W., and J. H. Patterson, "A Comparison of Heuristic and Optimum Solutions in Resource-Constrained Project Scheduling," *Management Science,* vol. 21, no. 8, pp. 944–955, April 1975.

9. Davis, E. W., and J. H. Patterson, "Resource-Based Project Scheduling: Which Rules Perform the Best?" *Project Management Quarterly,* vol. 6, no. 4, pp. 25–31, December 1975.

10. Engelhart, J., M. Malkin, and R. Rhodes, "From the Laboratory to the Pharmacy: Therapeutic Drug Development at Merck Sharp & Dohme Research Laboratories," *PM NETwork,* vol. III, no. 6, pp. 11–28, August 1989.

11. Fleming, Quentin W., *Cost/Schedule Control Systems Criteria: The Management Guide to C/SCSC,* Probus Publishing Co., Chicago, Ill., 1988.

12. Flones, Peter F., "Endicott Oil Field: First Offshore Arctic Oil Field Begins Production," *Project Management Journal,* vol. XVIII, no. 5, pp. 41–50, December 1987.

13. Fondahl, John W., *A Non-computer Approach to the Critical Path Method for the Construction Industry,* Department of Civil Engineering, Stanford University, Stanford, Calif., 1961.

14. Fondahl, John W., "Precedence Diagramming Methods: Origins and Early Development," *Project Management Journal,* vol. XVIII, no. 2, pp. 33–36, June 1987.

15. Holland, Robert G., Lee Peters, and Ray M. Shortridge, "The Story Behind the Games: The XV Olympic Winter Games and the X Pan American Games, *PM NETwork* vol. III, no. 8, pp. 5–25, November 1989.

16. Kelley, James E., Jr., and Morgan R. Walker, "Critical-Path Planning and Scheduling," *Proceedings of the Eastern Joint Computer Conference,* Boston, Dec. 1–3, 1959, pp. 160–173.

17. Kelley, James E., Jr., and Morgan R. Walker, "The Origins of CPM: A Personal History," *PM NETwork,* vol. III, no. 2, pp. 7–22, February 1989.

18. Kloppenborg, Tim, and Francis M. Webster, Jr., "Responsibility for Quality in a Project," *PM NETwork,* vol. IV, no. 2, pp. 25–28, February 1990.

19. Mahler, Ed, "Project Administration Methodology: Achieving Schedule Control on a Large Project—The Somers Project," *PM NETwork,* vol. V, no. 5, pp. 9–33, July 1991.

20. Malcolm, D. G., "Reliability Maturity Index (RMI)—An Extension of PERT into Reliability Management," *The Journal of Industrial Engineering,* January–February 1963, pp. 3–12.

21. Malcolm, D. G., J. H. Roseboom, C. E. Clark, and W. Fazar, "Applications of a Technique for R&D Program Evaluation (PERT)," *Operations Research,* vol. 7, no. 5, pp. 646–669, 1959.

22. Mask, Karen J., and Judith S. Kilgore, "The Westlake Story: The Need for Coordination, Cooperation and Communication," *PM NETwork* vol. IV, no. 5, pp. 13–18, July 1990.

23. Moder, Joseph J., Cecil R. Phillips, and Edward W. Davis, *Project Management with CPM, PERT and Precedence Diagramming,* 3d ed., Van Nostrand Reinhold, New York, 1983.

24. Ono, Dan, "Implementing Project Management in AT&T's Business Communications System," *PM NETwork,* vol. IV, no. 7, pp. 9–31, October 1990.

25. Padgham, Henry F., "The Milwaukee Water Pollution Abatement Program: Its Stakeholder Management," *PM NETwork,* vol. V, no. 3, pp. 6–18, April 1991.

26. Project Management Institute, "Project Management Body of Knowledge," *PM NETwork,* August 1987.

27. Thamhain, Hans J., "Validating Technical Project Plans," *Project Management Journal,* vol. XX, no. 4, pp. 43–50, December 1989.

28. Tonge, Fred M., *A Heuristic Program for Assembly Line Balancing,* Prentice-Hall, Englewood Cliffs, N.J., 1961.

29. Webster, Francis M., Jr., "Ways to Improve Performance on Projects," *Project Management Quarterly,* September 1981, pp. 21–26.

30. Webster, Francis M., Jr., "Integrating PM and QM," *PM NETwork,* vol. V, no. 3, pp. 24–32, April 1991.

31. Wideman, Max, et al., "An Analysis of a Failure: The Structural Collapse at the Station Square Development," *PM NETwork,* vol. IV, no. 2, pp. 5–30, February 1990.

32. Wiest, J. D., and F. K. Levy, *A Management Guide to PERT/CPM,* 2d ed., Prentice-Hall, Englewood Cliffs, N.J., 1977.

PROJECT MANAGEMENT SOFTWARE

The selection of project management software is not a simple task. While the basic features exist in many packages, some of the special features are important in determining the best for any given organization, project, and management objectives. The prospective user is advised to seek opinions from current users.

Allegro ($895 and up), The Allegro Group, Englewood, Colo. (303) 850-9854.

InstaPlan ($249–$549), Micro Planning International, Mill Valley, Calif. (415) 389-1420.

MacProject II ($495), Claris Corp., Santa Clara, Calif. (408) 987-7000.

On Target ($399), Symantec Corporation, Cupertino, Calif. (800) 441-7234.

Open Plan ($4200), Welcom Software Technology, Houston, Tex. (713) 558-0514.

Primavera ($4000), Primavera Systems, Inc., Bala Cynwyd, Pa. (800) 423-0245.

Project Director ($695), adRem Technologies Inc., Richmond Hill, Ontario (416) 886-7899.

Project Scheduler 5 ($685), Scitor Corp., Foster City, Calif. (415) 570-7700.

Sagacity ($1395), Erudite Corp., Burlingame, Calif. (415) 348-7714.

Schedule Publisher ($2000), Lucas Management Systems, Fairfax, Va. (703) 222-1111.

SuperProject 2.0 ($895), Computer Associates, San Jose, Calif. (800) 531-5236.

Texim Project ($1295), Texim Inc., St. Paul, Minn. (612) 290-9627.

Time$heet Professional ($200), Timeslips Corporation, Essex, Mass. (508) 768-6100.

QUALITY CONTROL

CHAPTER 1
Total Quality Management

Thomas P. Huizenga, Vice President
Juran Institute, Inc.
Wilton, Connecticut

Eric D. Dmytrow, Vice President
Juran Institute, Inc.
Wilton, Connecticut

I often hear this comment: "My company is in a different category of business, so it is difficult to engage in QC [Quality Control] or TQC [Total Quality Control]. We cannot do it." My answer has remained the same: "Instead of thinking about the reasons that you cannot do, why try to not discover what you can do?"

"TQC simply means that we do what we are supposed to do."[1]

What is TQM? The Japanese Connection. TQM, or total quality management, is a term often used today in many different circles. In industry, TQM is discussed frequently by those businesses which have been most severely impacted by quality of foreign manufactured goods, especially from Japan. In the U.S. federal government, TQM is a new buzzword being popularized primarily by the Department of Defense and defense contractors and is trickling over to other federal agencies. In the service sector, TQM techniques and methodologies traditionally used in manufacturing industries are being utilized by all types of service business such as banks, hospitals, and hotels.

The literature makes it clear as to the origins of TQM. In tracing these origins, we need to go back in time to the late 1940s and 1950s to postwar Japan. In order to survive as a nation, Japan had to build its ability to produce commercial goods which were salable in worldwide markets. "The initial step was taken in 1946 during the occupation (post–World War II) when W. G. Magil and H. M. Sarasohn of SCAP's Civilian Communications Sector undertook to instruct the Japanese telecommunications industry in quality control. Two years later, the Union of Japanese Scientists and Engineers (JUSE) set up a research committee of five members that was later to be known as the QC Research Group."[2] Most of the authorities on quality control in Japan are, or were, members of this group and include Professors Asaka, Ishikawa, Kogure, Mizuno, and Moriguchi. "The famed Deming came to Japan at the behest of SCAP in 1949 and again in 1950 as a consultant in statistical research."[3] In 1950, the Union of Japanese Scientists and Engineers invited Dr. W. E. Deming to lecture on statistical quality control for managers and engineers. Deming held a number of such lectures in the 1950s.

The use of statistical methods was found to be valuable for assigning causes of variation in manufacturing processes, clarifying the correlation between manufacturing conditions and product quality and reducing the work force needed for inspection.

According to Ishikawa, the early 1950s evolved as a period of overemphasis on statistical quality control. He cited the following lessons learned from that period.

1. It is true that statistical methods are effective, but we overemphasized their importance. As a result, people either feared or disliked quality control as something very difficult. We overeducated people by giving them sophisticated methods where, at that stage, simple methods would have sufficed.
2. Standardization progressed in the areas of product standards, raw material standards, technical standards, and work standards, but it remained pro forma. We created specifications and standards, but seldom made use of them. Many people felt that standardization meant using regulations to bind people.
3. Quality control remained a movement among engineers and workers in factories. Top and middle-level managers did not show much interest. A misconception was also that if a company started a quality control movement, it would cost money. In those days, we used to say, "Who is going to put a leash on the fat cat (top managers)?" Those of us who were members of the Quality Control Research Group tried to persuade top managers to join, but, perhaps because of our relative youth, our efforts were met with little visible success.[4]

As a result of these limitations, JUSE decided to invite Dr. J. M. Juran to Japan to conduct seminars for top- and middle-level managers to explain their role in QC activities. Dr. Juran's visit marked a transition in Japan's quality control activities from dealing primarily with technology based in factories to an overall concern for the entire management system. "There is a limit to statistical quality control which has engineers as its prime movers. The Juran visit created an atmosphere in which QC was to be regarded as a tool of management, thus creating an opening for the establishment of total quality control (TQC) as we know it today."[5] "Following the path shown in Juran's courses, JUSE created the QC middle management course in 1955 and the QC top management course in 1957."[6]

In the United States, the media have credited the Japanese success in quality almost totally to Deming. In contrast, the Japanese view Juran's teaching on quality as a tool of management to be the most significant factor in development of their version of TQM, which they refer to as total quality control (TQC).

Japanese TQM or TQC was developed by the Japanese quality experts named earlier and was developed from the teachings of Juran, Deming, and others. However, the real innovations of total quality management as a key part of the management system were developed by Japanese experts. The Deming prize today represents Japanese TQC as Japanese quality experts evolved it from its original beginning as an award for use of statistical methods. From the personal experiences of one of the authors in helping Florida Power and Light compete for and win the Deming Prize, it is noted that while many people believe this prize to be based on how well an organization implements Deming's 14 points, it is not. Again, it is based on Japanese TQC, which was innovated by Japanese quality experts.

Definition of TQM.   This brief review provides the background for which definitions of TQM can be formulated. Many abbreviations for total quality management exist in the literature and appear to be describing the same things. Since the history of modern TQM has its origin in Japan, we again need to refer to the development of quality there.

In Japan, quality activities are referred to as QC or quality control activities. "The traditional definitions of quality control include the following:"

Deming's Definition (1950).   "Statistical quality control is the application of statistical principles and techniques in all stages of production directed toward the most economic manufacture of a product that is maximally useful and has a market."[7]

Juran's Definition (1954).   "Quality control is the totality of all means by which we establish and achieve quality specifications, with statistical quality control that part of the means, for establishing and achieving quality specifications, which is based on the tools of statistical methods."[8]

In the third edition of his *Quality Control Handbook* in 1974, Juran revised this to: "Quality control is the regulatory process through which we measure actual quality performance, compare it with standards, and act on the difference."[9]

Japan Industrial Standards Definition (JISZ 8101). "A system of means whereby the qualities of products or services are produced economically to meet the requirements of the purchaser. Since modern quality control adopts statistical techniques, it is sometimes especially called *tokeiteki hinshitsu kanri* [statistical quality control (SQC)]."[10]

American National Standards Definition (ANSI ZI.7 1971). "Quality Control: The operational techniques and the activities which sustain a quality of product or service that will satisfy given needs; also the use of such techniques and activities."[11]

These particular definitions of quality control are ones which we in the Western world understand and use in our everyday language and beliefs. In Japan, the words in the definition of quality control are also closely aligned with statistical quality control as is demonstrated by the above Japan Industrial Standards definition. The meaning of *control* in Japanese quality control is considered as follows:

> To surely perform "plan-do-check-action," [PDCA] that is a series of procedure to achieve rationally and efficiently the aim of the enterprise, aims of divisions and sections. Since the action is taken for the "plan" and "do," the maintenance and improvement are promoted by repeating this procedure. To explain intelligibly, the term and the diagram called the circle of control or circle of PDCA are often used.[12] [See Fig. 1.1.]

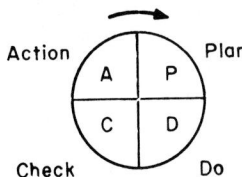

FIGURE 1.1 Circle of control (PDCA). (*Japanese Standards Association,* Guide to Quality Control and Company Standardization, *Tokyo, 1984, p. 22.*)

While the words in the Japanese definition of quality control appear to be close to the Western world's definitions, the meaning of quality control in Japan has broadened from the original use of statistical techniques in manufacturing to use of quality control throughout the company. In order to fully understand what the Japanese mean by quality control, more definitions are required. In JISZ 8101-1981, "Glossary of Terms Used in Quality Control," from the Japanese Standards Association, the term *quality control* is defined as follows:

> A system of means whereby the qualities of products or services are produced economically to meet the requirements of the purchaser.
> "Quality control" is sometimes called "QC" for short. In addition, since modern quality control adopts statistical techniques, it is sometimes specially called "statistical quality control," and "SQC" for short.
> In order to perform quality control effectively, throughout all phases of the enterprise activities such as market survey, research and development, planning of product, design, production readiness, procurement and subcontract, manufacture, inspection, sales and after sales servicing as well as finance, personnel affairs and indoctrination, whole personnel including from the executives down the managers, foremen and workers are required to participate and collaborate.
> The quality control activities conducted in such way is called "'company-wide' quality control" and "TQC" for short."[13]

Mizuno refers to QC in the context of Japanese-style TQC which means company-wide quality control that has been developed and implemented by Japanese companies. "Japanese-style TQC means integrated quality control in which the whole company—at every division at every level—is involved for the achievement of a common corporate goal, especially a product or policy goal. This is why Japanese-style TQC is better referred to as company-wide total quality control."[14] This is to be contrasted with quality control in the United States, where, still today, most companies and organizations believe quality to be the prime responsibility of the Quality Department, and quality control concentrates on product quality of manufactured products.

From Ishikawa, we also have a similar explanation of quality control. He believes that Japanese quality control is a thought revolution in management which represents a new way of thinking about management. His definition follows:

> To practice quality control is to develop, design, produce and service a quality product which is most economical, most useful, and always satisfactory to the consumer. To meet this goal, everyone in the company must participate in and promote quality control, including top executives, all divisions within the company, and all employees.[15]

Ishikawa provides insight into specific activities of Japanese TQC practice as it was evolving. Quality control activities in Japanese industries were expanded in the 1960s to include:

1. Establishing the top management policy on quality and the long-term QC plan of the entire company to realize the policy. The word *goal* is sometimes used as a substitute for the word *policy*.
2. Introducing the QC concept and techniques into new-product development.
3. Establishing the quality assurance system, which covers the whole company.
4. Conducting QC audits.
5. Expanding QC activities to include the sales and marketing activities of agents, trading firms, stores and shops, etc.[16]

Ishikawa stated

> ...that the following six characteristics were the ones which distinguished Japanese quality control from that of the West.
>
> 1. Company-wide quality control (CWQC); participation by all members of the organization in quality control
> 2. Education and training in quality control
> 3. QC circle activities
> 4. QC audits (Deming application Prize and presidential audit)
> 5. Utilization of statistical methods
> 6. Nationwide quality control promotion activities[17]

The actual meaning of TQM is the same as that which has evolved as Japanese TQC or CWQC. From the previously mentioned definitions of Mizuno and Ishikawa, we can define TQM as follows: TQM means the systematic approach which integrates quality-related activities throughout the entire organization, including market research, research and development, operational planning for production of goods or services, procurement, production, and service. Taken together, these activities produce products and services economically, meet the requirements of the customer, and are planned to achieve key organizational goals set by top management relative to quality.

The rest of the chapter will deal with implementing a TQM system within an organization. We will start with some definitions of how to think about quality.

HOW TO THINK ABOUT QUALITY

In order to understand TQM, it is useful to have some common viewpoints and understanding of key concepts in building a framework for a total quality management system. Discussion of a number of key concepts and definitions follows.

Quality Defined.   The word quality has multiple meanings. Two of those meanings are generally used and can be summarized as follows:

1. Quality consists of those product features which meet the needs of customers and provide product satisfaction.
2. Quality consists of freedom from deficiencies in products and processes.

To explain these it is first necessary to define some key words.[18]

- *Product* is the output of any process. Product consists mainly of goods and services.
- *Goods* are physical things: pencils, color television sets, office buildings.
- *Service* is work performed for someone else. Entire industries are established to provide services in such forms as central energy, transportation, communication, entertainment, and the like.
- *Product feature* is a property which is possessed by a product and which is intended to meet certain customers' needs.

Customers and Their Needs.   The following are definitions related to customers.

- *Customer.*A customer is anyone who is impacted by the product. Customers may be external or internal.
- *External Customers.*External customers are impacted by the product but are not members of the organization which produces the product. External customers include: clients who buy the product, government regulatory agencies, the public.
- *Internal Customers.*Internal customers are everyone within an organization who performs work in an organization. As work is completed and passed on, those receiving work from others are considered customers of those who supplied them with the work.
- *Customer Needs.*All customers have needs to be met, and the product features should be responsive to those needs. This applies to both external and internal customers.
- *Product Satisfaction.*Product features which respond to customer needs are said to provide *product satisfaction*. The degree of satisfaction can affect market share and product salability for suppliers of products.
- *Product Defects.*Defects take such forms as late deliveries, field failures of goods, errors in invoices, factory scrap or rework, and design changes. Each of such events is the result of some deficiency in a product or process. Each can cause the customer to become dissatisfied.
- *Product Dissatisfaction.*A consequence of product defects is that customers are dissatisfied. External customers express their dissatisfaction in such forms as complaints, returns, and claims. If the extent of dissatisfaction is too high, customers may stop buying the product, and lower sales, higher costs, and lower market share could result for the supplier.
- *Product satisfaction and product dissatisfaction are not always opposites.*While product satisfaction has its origin in product features, and product dissatisfaction has its origin in nonconformances, there are many products which give little or no dissatisfaction; the products do what the supplier said they would do. Yet the products are not salable because some competing product provides greater product satisfaction. The message is to make sure you keep in touch with your customers' ongoing needs through market research techniques.

Achieving Customer Satisfaction.   The *quality function* is the entire collection of activities through which we assure quality and achieve customer satisfaction no matter where these activities are performed. Juran's spiral of progress in quality depicts a typical progression of activities as carried out in industrial companies. Figure 1.2 is a depiction of this spiral. A company-wide quality function arises from the fact that product quality is the resultant

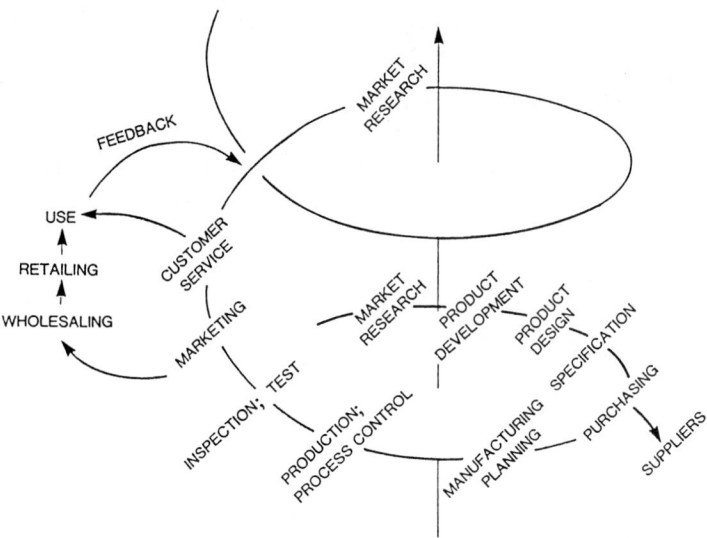

FIGURE 1.2 The spiral of progress in quality. *(Juran, J. M., and M. Gryna, Juran's Quality Control Handbook, 4th ed. McGraw-Hill, New York, p. 2.5, Fig. 2.1. Reproduced with permission of McGraw-Hill, Inc.)*

of the work of all departments around the spiral. Each of those departments has not only the responsibility to carry out its special function; it also has the responsibility to do its work correctly—to make its products fit for use. In this way, each department has a quality-oriented activity to carry out along with its main function. (These departmental quality-oriented activities are supplemented by other quality-oriented activities carried out by staff departments and by upper management.) Together they achieve customer satisfaction.

QUALITY MANAGEMENT

Managing for quality is done by use of the three managerial processes of planning, control, and improvement. Juran calls these the *quality trilogy*. The conceptual approaches are identical with those used to manage for finance. However, the procedural steps are special, and the tools used are also special. An overview of these three processes is as follows. Note that these three processes are fully expanded as shown in Fig. 1.4, and 1.7.

Quality Planning.   This is the activity of developing the products and processes required to meet customers' needs. It involves a series of universal steps after quality goals have been established:

1. Determine who are the customers
2. Determine the needs of the customers
3. Develop product features which respond to customers' needs
4. Develop processes which are able to produce those product features
5. Transfer the resulting plans to the operating forces

Quality Control.   This process is used by the operating forces as an aid to meeting the product and process goals. It is based on the feedback loop, and consists of the following steps:

1. Evaluate actual operating performance
2. Compare actual performance to goals
3. Act on the difference

Quality Improvement.   This third member of the quality trilogy aims to attain levels of performance which are unprecedented—levels which are significantly better than any past level. The methodology consists of a process—an unvarying series of steps:

1. Prove the need for improvement
2. Identify specific projects for improvement
3. Organize to guide the projects
4. Organize for diagnosis—for discovery of causes
5. Diagnose to find the causes
6. Provide remedies
7. Prove that the remedies are effective under operating conditions
8. Provide for control to hold the gains

The Juran Trilogy™.   The three processes of the quality trilogy are interrelated. Figure 1.3 shows this interrelationship. The Juran Trilogy is a graph with time on the horizontal axis and cost of poor quality (quality deficiencies) on the vertical axis. The initial activity is quality planning. The planners determine who are the customers and what are their needs. The planners then develop product and process designs which are able to respond to those needs. Finally, the planners turn the plans over to the operating forces.

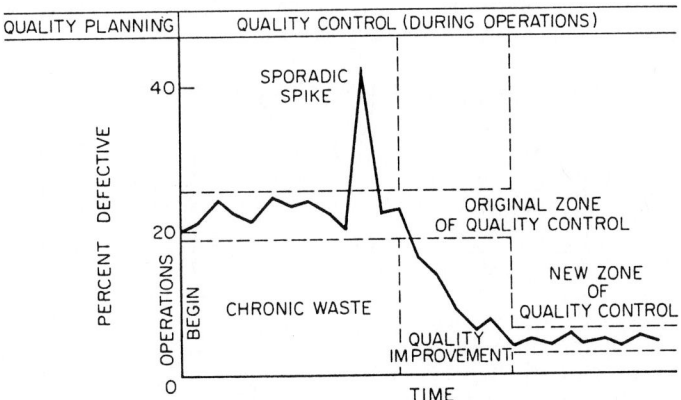

FIGURE 1.3 The Juran Trilogy *(Juran, J. M., and Gryna, F. M., Juran's Quality Control Handbook, 4th ed., McGraw-Hill, New York, 1988, p. 2.7, Fig. 2.2. Reproduced with permission of McGraw-Hill, Inc.)*

The job of the operating forces is to run the processes and produce the products. As operations proceed, it soon emerges that the process is unable to produce 100 percent good work. Figure 1.3 shows that 20 percent of the work must be redone because of quality deficiencies. This waste then becomes chronic because *the operating process was planned that way.*
Under conventional responsibility patterns, the operating forces are unable to get rid of that planned chronic waste. What they do instead is to carry out *quality control*—to prevent things from getting worse. Control includes putting out the fires, such as that sporadic spike in Fig. 1.3.
But the chart also shows that, in due course, the chronic waste was driven down to a level far below that planned originally. That gain was achieved by the third process in the

trilogy—*quality improvement*. In effect, it was realized that the chronic waste was also an opportunity for improvement, so steps were taken to seize that opportunity.

In most companies there exists a widespread situation in which:

1. Numerous operating processes are deficient. Each is an opportunity for quality improvement through the project by project approach.

2. The approach to quality planning is also deficient. This deficiency is the root cause of most of those deficient operating processes. That same deficient planning process continues to create new wasteful operating processes. The remedy is to improve the planning process.[19]

Quality Improvement

What is Improvement?   As used here, *improvement* means the organized creation of beneficial change; the attainment of unprecedented levels of performance. A synonym is *breakthrough*. The trilogy diagram (Fig. 1.3) shows graphically the nature of quality improvement and its relation to quality planning and quality control. Quality improvement is very different from "fire fighting." In the trilogy diagram, the removal of that sporadic spike is often called *fire fighting* (or *troubleshooting*). It restores performance to the prior chronic level, which was also the prior standard. In that same trilogy diagram, quality improvement takes the performance to unprecedented levels, such as improving a process yield from 60 to 95 percent. A complete picture of the quality improvement process appears in Fig. 1.4.

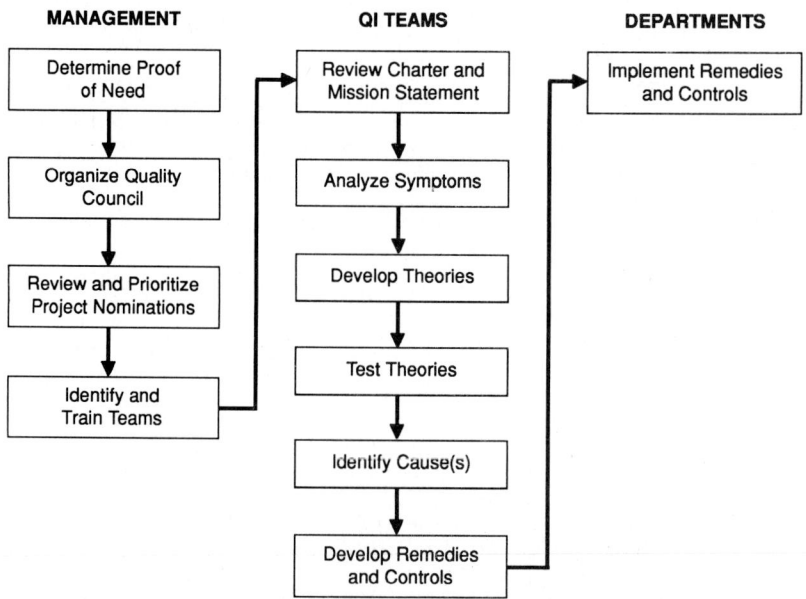

FIGURE 1.4   The Juran quality improvement process. *(Juran Institute.)*

Key Lessons Learned from Quality Leaders.   The following is a summary of key lessons learned relative to quality improvement from present-day quality leaders in Japan and the United States.

• The rate of quality improvement is vital in determining quality leadership. An organization with a lower rate of improvement can eventually be exceeded by a faster pace of improvement by its competitor. The United States and other Western nations have suffered significant losses in market share in many businesses because of the Japanese pace of improvement, allowing defect rates in parts per million compared with U.S. defect rates in parts per hundred.

- Improvement is made through completion of improvement projects. A project is defined as a problem scheduled for solution.
- Quality improvement applies universally to manufacturing and service organizations.
- The opportunity for improvement is big. The value of chronic waste (cost of poor quality) is on the average 20 percent of sales of an organization.
- Although quality improvement is not free, the return on investment for quality is among the highest available to managers and is not capital-intensive.
- To make quality gains at a substantial pace requires a new organizational infrastructure spearheaded by top management. Sometimes referred to as the *quality council* or *quality steering committee,* the quality council is made up of the top management team of the organization wishing to adopt TQM. The major common elements of responsibility usually include the following:

  Formulate the quality improvement policy, that is, priority of quality; need for annual quality improvement; mandatory participation.
  Estimate the major dimensions, that is, quality compared to competitors; cost of poor quality; length of the new products launch cycle.
  Establish the project selection process.
  Establish the team selection process.
  Provide resources: training; time for working on projects; diagnostic support; facilitator support.
  Assure that project solutions are implemented.
  Establish needed measures, that is, progress on quality improvement; performance versus competitors; managers' performance.
  Provide for progress review and coordination.
  Provide for recognition.
  Revise the reward system and business plans to include quality.

- Substantial training is required in order to teach teams proper problem-solving methodologies, as well as specific tools such as statistical tools to aid in problem solving. An example of a fundamental quality improvement methodology is Juran's famous breakthrough sequence which includes the following steps:

  Identification of a quality problem
  Understanding the problem symptoms
  Theorizing on potential causes
  Discovery of causes
  Developing and testing remedies to remove causes
  Establishing controls to hold the gains at the new level

Most organizations have some version of the above problem-solving process.

Cross-Functional Optimization.   In the traditional organization, the heads of the functions are given functional goals which on the diagram flow downward (Fig. 1.5) representing optimization within the function and the creation of functional boundaries. The diagram also shows that the major business processes (such as new product development) flow across the organization (cross-functional), with the end of the process flowing to the external customers. As the process flows internally, there are also internal customers within each function who are impacted by the process. The major impact of this type of functional organization for quality is the creation of many

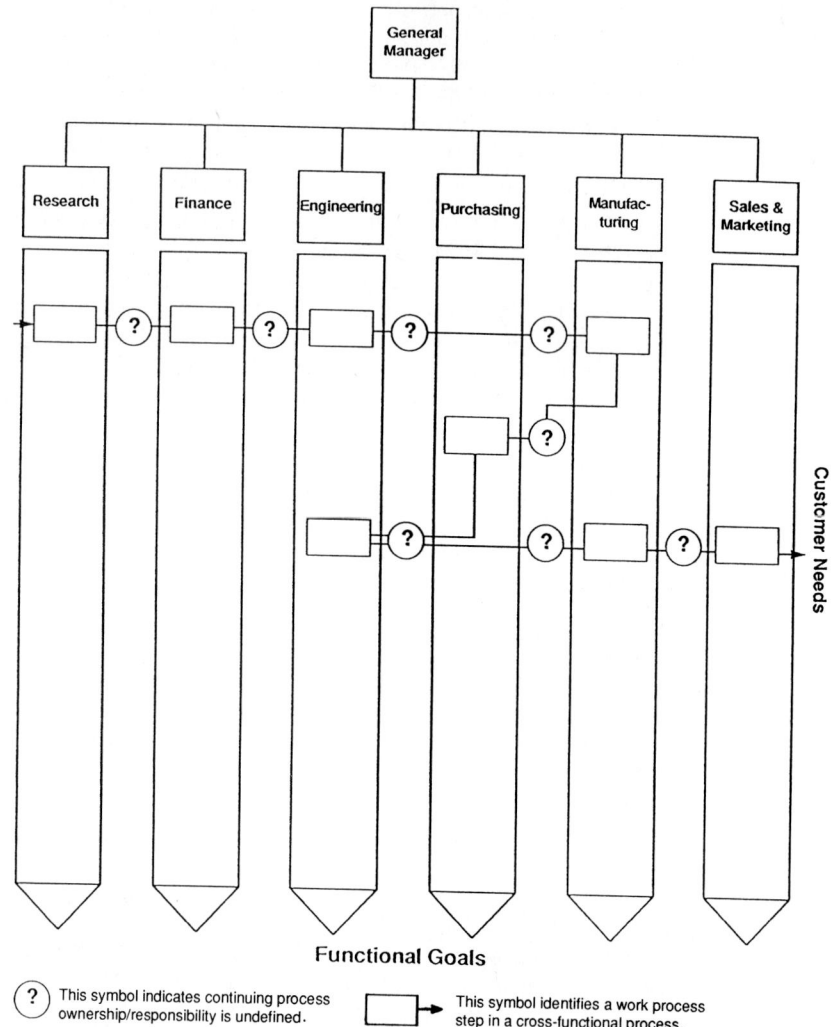

FIGURE 1.5   Traditional organization structure: functional versus cross-functional optimization. *(Thomas P. Huizenga.)*

chronic quality problems that go unresolved because of the lack of ownership of the problems that occur; no function owns the handoffs or "white spaces" between the functions as processes flow through the organization. These handoff points are denoted by the circles between the functions. The usual outcomes from such an organization design are:

- Functional optimization
- Cross-functional suboptimization
- Chronic quality problems and associated cost of poor quality
- Unplanned business processes
- Dissatisfied internal and external customers

The quality council must specifically address these issues in planning the TQM process. For further in-depth explanation of quality improvement, refer to Sec. 22 of *Juran's Quality Control Handbook,* 4th edition.

A summary of the key aspects for organizing the quality improvement process and infrastructure appears in Fig. 1.6.

MOBILIZING FOR QUALITY IMPROVEMENT

Establish quality council
Statement of responsibilities
Improvement policies, goals

Establish infrastructure	Projects collectively	Projects individually
Subcouncils	Strategic improvement goals	Nomination
Director of quality	Deployment	Screening
Quality improvement managers	Projects	Selection
Sponsors, champions	Resources	Mission statements
Facilitators	Progress review	Project teams
Structured improvement process	Recognition	Life cycle of a project: diagnosis, remedy, cloning
Training; methods; tools	Rewards	

FIGURE 1.6 Quality improvement interrelation of elements. *(Juran Institute, Inc., "Making Quality Happen.")*

Quality Planning

The Quality Planning Concept.   Quality planning is one of the three basic processes used in managing for quality. In its broadest sense, quality planning is the process of establishing quality objectives and developing the means (plans) for meeting those objectives.

The Quality Planning Road Map.   Within any organization, the quality plans are closely interrelated. At the top of the organization, the broad quality objectives require broad plans. Such broad plans then require establishment of subobjectives, each of which requires a subplan. In large organizations, this subdivision can involve many layers of objectives and plans.

Irrespective of level in the hierarchy, quality planning involves meeting the quality needs of customers. To meet these quality needs requires taking a universal series of actions. This series is shown graphically as the *quality planning road map* (Fig. 1.7), which consists of a structured, coherent sequence of steps. Each step is an activity whose ouput becomes an input to the next step in the sequence. The quality planning road map is a successful tool that helps new product development teams design and launch superior products and assists process improvement teams charged with planning or replanning processes. Examples of such planning projects include replanning of new product development and introduction processes or developing a successful new product.

The same infrastructure for improvement is also required for successful quality planning efforts. For further-in-depth explanation, refer to the section on quality planning in *Juran's Quality Control Handbook,* 4th edition.

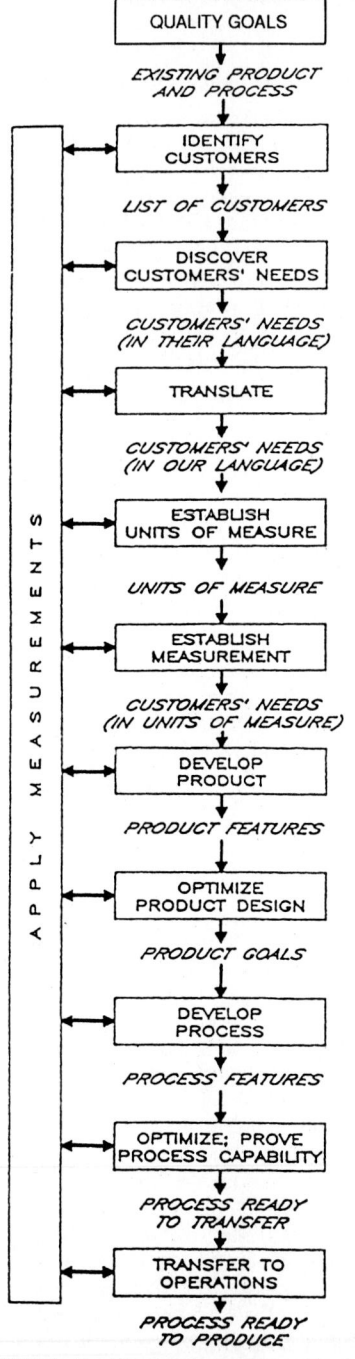

```
              ┌──────────────────────┐
              │    QUALITY GOALS     │
              └──────────────────────┘
                         ↓
               EXISTING PRODUCT
                 AND PROCESS
              ┌──────────────────────┐
              │      IDENTIFY        │
              │     CUSTOMERS        │
              └──────────────────────┘
                         ↓
               LIST OF CUSTOMERS
              ┌──────────────────────┐
              │      DISCOVER        │
              │  CUSTOMERS' NEEDS    │
              └──────────────────────┘
                         ↓
               CUSTOMERS' NEEDS
               (IN THEIR LANGUAGE)
              ┌──────────────────────┐
              │      TRANSLATE       │
              └──────────────────────┘
                         ↓
               CUSTOMERS' NEEDS
               (IN OUR LANGUAGE)
              ┌──────────────────────┐
              │     ESTABLISH        │
              │  UNITS OF MEASURE    │
              └──────────────────────┘
                         ↓
               UNITS OF MEASURE
              ┌──────────────────────┐
              │     ESTABLISH        │
              │   MEASUREMENT        │
              └──────────────────────┘
                         ↓
               CUSTOMERS' NEEDS
               (IN UNITS OF MEASURE)
              ┌──────────────────────┐
              │      DEVELOP         │
              │      PRODUCT         │
              └──────────────────────┘
                         ↓
               PRODUCT FEATURES
              ┌──────────────────────┐
              │      OPTIMIZE        │
              │   PRODUCT DESIGN     │
              └──────────────────────┘
                         ↓
               PRODUCT GOALS
              ┌──────────────────────┐
              │      DEVELOP         │
              │      PROCESS         │
              └──────────────────────┘
                         ↓
               PROCESS FEATURES
              ┌──────────────────────┐
              │  OPTIMIZE; PROVE     │
              │ PROCESS CAPABILITY   │
              └──────────────────────┘
                         ↓
               PROCESS READY
               TO TRANSFER
              ┌──────────────────────┐
              │    TRANSFER TO       │
              │    OPERATIONS        │
              └──────────────────────┘
                         ↓
               PROCESS READY
               TO PRODUCE
```

(vertical label at left: APPLY MEASUREMENTS)

FIGURE 1.7   Quality planning road map. *(Juran, J. M., and Frank M. Gryna,* Juran's Quality Control Handbook, *4th ed., "The Quality Planning Road Ma," McGraw-Hill, New York, 1988, p. 6.5. Reproduced with permission of McGraw-Hill, Inc.)*

Quality Control

Planning for Control.   Once planning is complete, the plans are put into operation. The responsibility of the operating personnel is mainly to meet the established goals. They do this through a planned system of quality *control.* Control is largely directed at meeting goals and preventing adverse change. This is in contrast to improvement, which is largely directed at creating beneficial change.

Much human energy is devoted to control. In organizations, this takes the form of meeting goals: delivery according to schedule, expenses according to budget, quality according to specification. This control *process* consists of a universal series of steps which, when applied to problems of quality, can be listed as follows:

1. Choosing the control subject, i.e., selecting what is to be regulated
2. Choosing a unit of measure
3. Establishing a goal for the control subject, i.e., specifying the target value for operating performance
4. Creating a sensor—a means of evaluating actual performance in terms of the unit of measure
5. Evaluating actual performance
6. Interpreting the difference between actual performance and goal
7. Taking action on the difference

The field of quality control is fairly well-established; traditional tools such as the Shewhart control chart were popularized throughout the 1980s. In Juran's trilogy model, note that quality control is the result of good practice in quality planning, with the establishment of capable processes as part of the planning effort. Quality control is also involved in the last step in the quality improvement sequence, which is to set up quality control to hold the gains.

The trilogy of quality processes is the foundation for TQM. An organization's success in TQM is directly related to the pace of quality improvement, the proper planning of products and replanning of processes to prevent chronic waste, the ability to establish capable processes, and the ability to control processes during operations.

For further elaboration on the subject of Quality Control, refer to *Juran's Quality Control Handbook,* **4th edition.**

MEASURING QUALITY

Quality measurement can be a motivator at all levels of the organization. Such measures should reflect organizational priorities as well as the customer's point of view. Another valuable aspect of quality measures is that they can communicate progress toward the achievement of a total quality management environment.

Quality measures should reflect the customer's point of view. Both internal and external customer satisfaction measures should be developed for desired quality.

Measurement should relate to the two major dimensions of quality:

- Product features
- Freedom from defects

Communication of quality-related features is best done through numbers. In order to say it in numbers, it is necessary to create a system of measurement consisting of:

- *Unit of measure.* A defined amount of some quality feature that allows for evaluation of that feature in numbers
- *A sensor.* A method or instrument which can carry out the evaluation and state the findings in numbers in terms of the unit of measure.

Freedom from Defects. Units of measure for product and process performance are usually expressed in technological terms; for example, fuel efficiency is measured in terms of distance traveled per volume of fuel; timeliness of service is expressed in minutes, hours, or days required to provide service.

Units of measure for product and process deficiency usually take the form of a fraction:

$$\% \text{ deficient} = \frac{\text{number of occurrences}}{\text{opportunity for occurrence}}$$

The numerator may be in such terms as number of defective units of product, number of field failures, or cost of warranty charges. The denominator may be in such terms as number of units produced, dollar volume of sales, number of units in service, or length of time in service.[20]

Product Features. While there is a universal formula for deficiencies, there is no such formula for measuring customer satisfaction. These measures must be derived out of the quality planning process. During quality planning, extensive use is made of the various tools of market research including customer surveys, focus groups, and the like.

CROSS-FUNCTIONAL QUALITY MANAGEMENT

This section deals with a vital part of TQM: the ability of the organization to plan major work processes to meet internal and external customer needs.

The Meaning of Quality Assurance. Mizuno believes that quality is not assured by inspection alone and that quality requires a rational design and correct implementation of quality control procedures and operations (preventive functions). "Quality assurance is investigating to make sure that quality inspection and quality control operations are being carried out correctly and checking design, manufacturing, and marketing divisions to see whether they are all working to maintain the targeted quality level. Quality assurance requires that the findings of these investigations be reported to management."[21]

Quality assurance requires that each function or division specify what it is doing in quality control and what it is expected to achieve. For example, Fig. 1.8 shows a partial listing of quality assurance activities for a sales function.

The kinds of items to be clarified by each function to develop quality assurance include:

1. Items requiring quality assurance
2. Work required to assure quality
3. People responsible for quality assurance
4. Data required for quality assurance control[22]

Some of the kinds of main functions and activities that should be documented include:

- New product development
- Personnel control
- Cost control
- Demand projection
- Product research
- Complaint processing
- Production planning[23]

Each of these main functions represents major business processes. Mizuno refers to quality assurance as the systems approach where an organization understands how the quality control activities and duties interrelate among the various divisions and functions. This is also referred to as *cross-functional management*. It is impossible to overemphasize the importance of taking a systematic approach for implementing quality control activities. "Quality assurance covers everything from product planning, maintenance and repair, and disposal. Quality assurance activities, therefore, must clearly define what is to be done at each stage to guarantee quality throughout the product's life cycle (in other words, define the quality functions)...."[24] These functions, of course, are identical to Juran's famous spiral of progress in quality; Mizuno's list of key functions and processes required to assure quality include:

1. Designing quality: deciding on the quality required for new products and new product types, and establishing, revising, and eliminating criteria
2. Purchasing and storing materials: materials control and inventory control
3. Standardization
4. Analyzing and controlling manufacturing processes
5. Inspection and processing of defective products
6. Complaint processing quality supervision and quality inspection
7. Equipment and installation management: constructing and installing equipment, preventive safety measures, measurement procedures
8. Personnel management: placement, education, and training
9. Management of outsourcing and subcontracting
10. Technology development: new product development, research management, and technology management
11. Diagnosis and supervision: auditing quality control activities and supervising quality control operations[25]

Company-Wide Total Quality Control

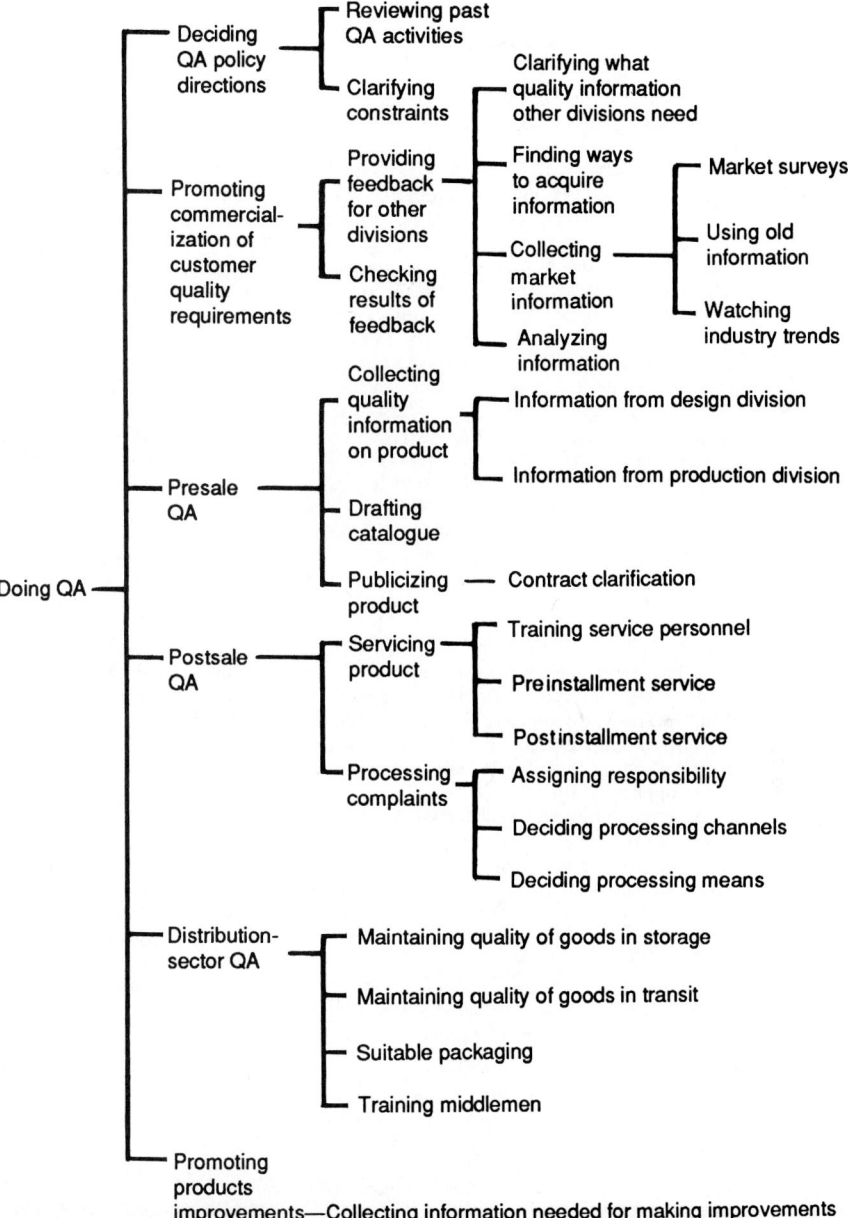

FIGURE 1.8 Partial listing of quality assurance activities for a sales function. *(Mizuno, S.,* Company-Wide Total Quality Control, *"Quality Assurance Duties of Sales Division," Asian Productivity Organization, Tokyo, 1988, p. 56. Distributed in the United States, Canada, and Western Europe by Quality Resources, White Plains, N.Y. Reprinted by permission of Quality Resources.)*

Cross-Functional Management. Ishikawa sheds some additional light on how to manage cross-functional activities.

> Japanese society is often described as a vertical society, and its industries share this structure. Industry has a strong top-to-bottom vertical bond while sectionalism hinders development of horizontal relations. For example, no matter how hard the division of quality assurance attempts to perform the function assigned to it, it cannot do so adequately within the existing organizational structure.
>
> Cross-function management which has cross-function committees for support can provide the woof to help the company run crosswise, making possible the responsible development of quality assurance.
>
> In textiles, the warp by itself remains a thread. Only when the woof is added, and when warp and woof are intertwined, will there be cloth. In a company, the analogy holds true. A vertical society resembling the warp is not an organization. It becomes a strong organization only when various functions such as quality assurance are intertwined with the warp.... Organizational management is possible only through the intertwining of the warp, which engages in management by divisions, and the woof, which engages in control by cross-function management.
>
> When we speak of cross-function management, many topics immediately come to mind, including quality assurance, quality control, cost (profit) control, new product development, control of subcontracting, sales control, and the list can go on and on. From the perspective of company goals, the main functions are the three functions of quality assurance, cost (profit) control, and quantity control. To these three may be added personnel control. All others are auxiliary functions defined by the steps to be taken or the means to be adopted.
>
> [**Teamwork and Ownership.**] In accordance with the functions to be managed, the company must establish cross-function committees. For example, a cross-function committee on quality assurance may be established. The chairman must be a senior managing director or a managing director who is in charge of that function. Committee members are selected from among those who hold the rank of director or above (if necessary, division heads may be included). The number should be around five. It is not desirable to select committee members only from among those who are directly connected with that specific function. Actually, it is better to have one to two persons from nonrelated divisions as committee members. Each cross-function committee must maintain a secretariat within the division that handles the function under consideration and appoint a secretary. The committee must be operated flexibly. When dealing with major functions, the committee must establish regularly scheduled monthly meetings which can engage in the audit of functions under study. The committee may also establish project teams under it.
>
> The committee then allocates responsibilities and authority for quality assurance to all affected divisions in concrete terms. It creates a viable system of quality assurance and establishes applicable rules.
>
> Every month the committee must study the conditions of quality assurance and determine if any claim has been registered against defective products. It must revise and redetermine the allocation of responsibilities periodically. At Toyota, the monthly meeting of the cross-function committee does all of these things efficiently. (Please keep in mind that the company has had about ten years of experience in cross-function committees before reaching this stage.) The committee's meetings are formal ones....
>
> The committee, however, does not implement quality assurance. Nor does it assume direct, day-to-day responsibility for quality assurance. That task is performed by each of the line divisions in this "vertical society." The responsibility of the committee is to let the woof be woven into the warp to strengthen the entire organization....[26]

In addition, Fig. 1.9 shows a conceptual diagram example of what Ishikawa means by cross-functional management.

QUALITY AND STRATEGY

Strategic Quality Planning Process
Describe the company's strategic quality planning process for short-term (1–2 years) and longer-term (3–5 years or more) quality leadership and customer satisfaction.

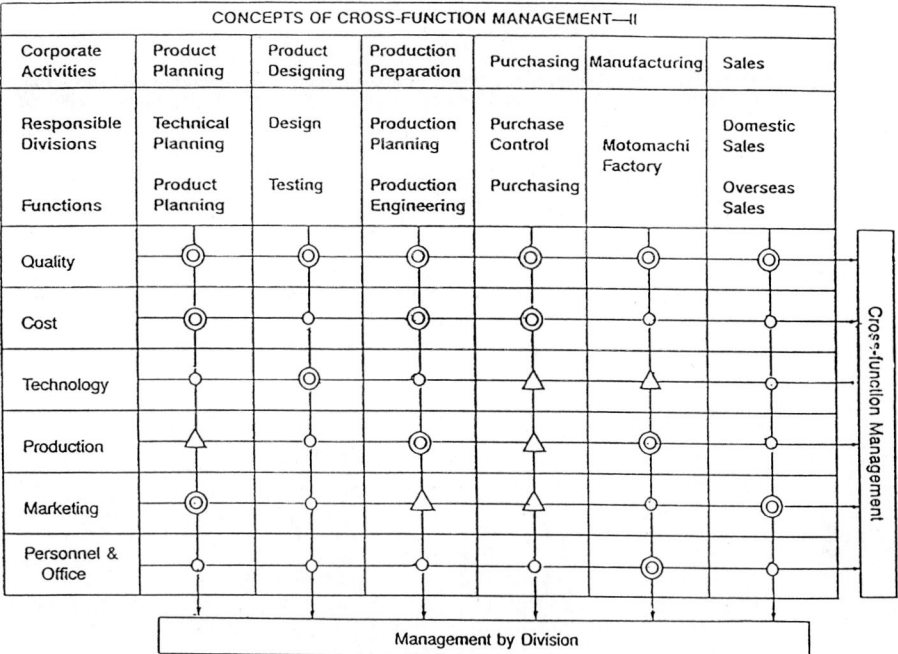

FIGURE 1.9 Conceptual diagram of cross-function management. *(Ishikawa, K.,* What is Total Quality Control, *Prentice-Hall, Englewood Cliffs, New Jersey, copyright © 1985, p. 115. Used by permission of the publisher, Prentice-Hall, a division of Simon & Schuster.)*

Some readers might recognize this as part of the introduction to Sec. 3.1 of the Malcolm Baldrige National Quality Award criteria. Management in many U.S. companies are sometimes puzzled that quality and business strategy are or could somehow be linked together. Others in these same organizations have been trying, perhaps for years, to get upper management's attention to focus on quality as a part of the business strategy.

The issue for some business leaders revolves around the question of whether there is a way to link quality as part of the business strategy, and if there is, what is involved in doing this. We will explore the vital steps to be taken to link quality and strategy, and set forth the means for doing this.

What is Strategic Planning?   Corporate planning and strategic planning models and tools have been used by business and organizations since the 1950s. Since we are trying to understand the link between quality and strategy, it is important to have some idea of what is meant by *strategic planning*. The following explanation of strategic planning will be used to guide the further discussion on this paper.

> *Strategic planning* derives from the military use of the word "strategy." It includes the definition of missions and objectives—how the company sees its purpose and where it wants to go—and determination of the best means to achieve those goals at a broad level. Strategic planning provides the basic direction and focus of the corporation, the so-called big picture. It is concerned with top management decisions about the company's businesses, asset acquisition/disposal, and product/markets. An assessment of the environment is needed, along with an appraisal of corporate strengths and weaknesses. Some of the company's basic strategic decisions might relate to questions such as: What business are we in? What business *should* we be in, now and in the future? What should be the geographic scope of operations? What are our research and development goals? How should products be sourced? What product or market is

most vulnerable to competition? What should the capital structure be? What is our distinctive competence? Where are we weakest? Should we diversify or prune our businesses?

Strategic planning is the top management decision process that focuses on the longer-range direction of the company and establishes the means by which that direction is reached.

Long-range planning gives the plan a time frame and can be used synonymously with strategic planning, which usually has a longer time horizon. A long-range plan should extend three to five years, as contrasted with a one- to two-year short-range time frame.[27]

Operational planning deals with the short term, the one- to two-year time frame. It should derive from strategic planning and is the next step in translating top management focus and strategy into a meaningful day-to-day guideline.[28]

It helps to see the process of strategic planning in a conceptual model. One such model is found in Fig. 1.10.

This model provides a way to understand key aspects of strategic planning as well as how quality planning fits with strategic planning. The key points include:

- *The plan to plan:* Deciding with top management how the system will work.
- *Planning inputs:* Outside interests, inside interests, databases, evaluations.
- *Formulating master strategies and developing program strategies for specific projects and resourcing those projects:* This planning represents long-range (5 + years) fundamental business planning for all aspects of the company.
- *Medium-range programs:* Specific functional plans are developed to show how strategies are to be carried out to achieve long-range objectives. The typical time frame is 5 years.
- *Short-range plans:* These plans generally represent the yearly operating plans that carry out the medium-range plans.
- *Implementation:* Carrying out the plans in actual operation.
- *Review:* Evaluation of results of implementation and execution of the plans.
- *Information flows:* Planning should include how information would flow throughout the planning process.
- *Decisions and evaluations:* Planning should also include any decision and evaluation rules, both qualitative and quantitative, throughout the entire planning process.

Can Quality be Part of Strategic Planning?   Referring to the strategic planning model in Fig. 1.10, we see that some of the issues involved in preparing the plan include the study of expectations of outside interests from such sources as customers, society, inside sources (all employees), performance factors, as well as evaluation of opportunities and threats and the analysis of an organization's ability to compete via its strengths and weaknesses.

Any organization undertaking such an exercise today, or as far back as the 1970s, would find it difficult not to have recognized quality as a competitive factor in some form, as having an impact on world markets and businesses within those markets. In fact, during the 1960s and 1970s, many U.S. companies lost their quality leadership to new, aggressive competition; the most obvious consequence was loss of market share. For example, here is a partial list of goods for which imports had gained a significant share of the North American market by 1980.

Stereo components	Athletic equipment
Medical equipment	Computer chips
Color television sets	Industrial robots
Hand tools	Electron microscopes
Radial tires	Machine tools
Electric motors	Optical equipment

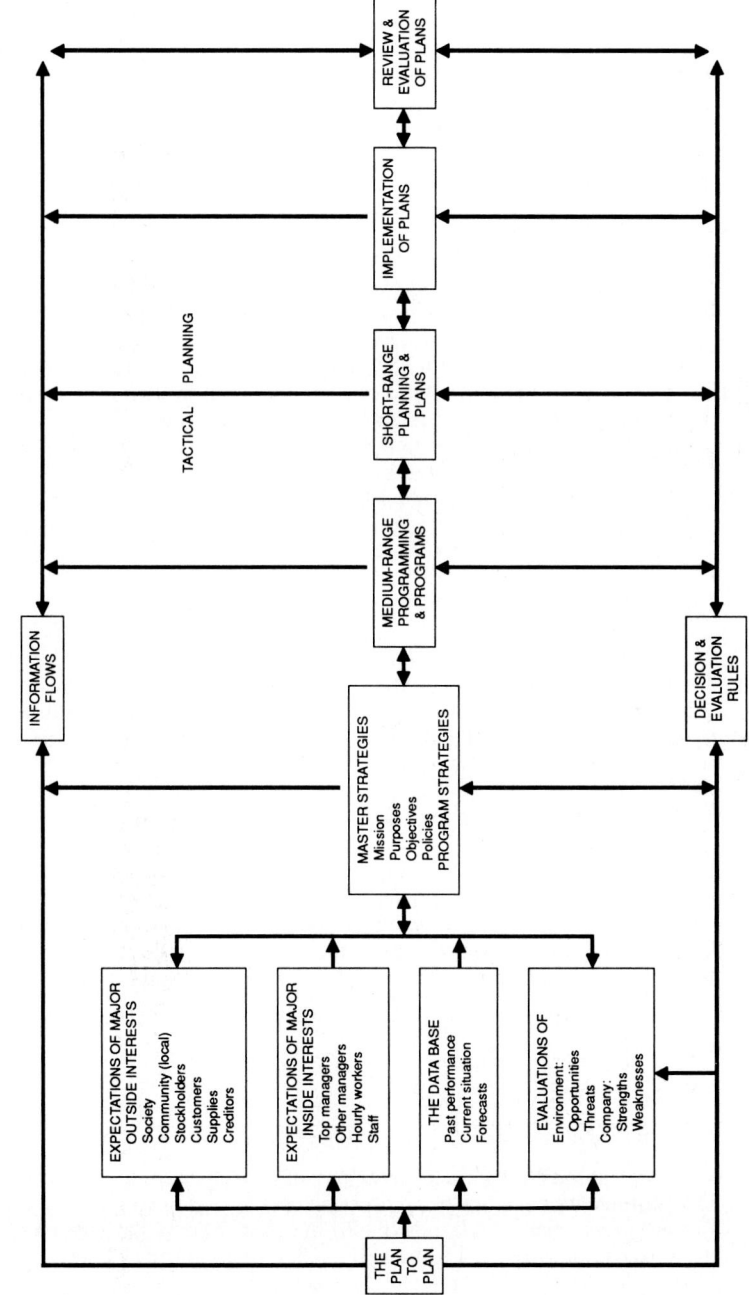

FIGURE 1.10 Structure and process of business company-wide planning. *(Steiner, G. A., Strategic Planning, The Free Press, a Division of Macmillan, Inc., New York, "Structure and Process of Business Companywide Planning," copyright © 1979, by The Free Press, p. 17)*

The reasons for the losses in market share were associated with quality in two respects:

1. The imports had quality features which were perceived as better meeting customer needs.
2. The imports did not fail in service as often as the domestic products.

Another major reason for recognizing quality in strategic planning is the growing awareness by companies and organizations that they have been living for many years with excessive costs because of chronic quality-related wastes. Juran states that about one-third of what we do consists of redoing work previously done. From the authors' collective experience of actual evaluations of the cost of poor quality in a variety of service and manufacturing organizations, the average cost of poor quality was found to be in the range of 20 to 40 percent of sales.

The specific link between strategic planning and quality planning is through the inputs for strategic planning. As we have seen earlier, planning for quality starts with the understanding of customer needs, of both internal and external customers; the understanding of the opportunity for quality improvement via the financial analysis of costs of poor quality; and through competitive benchmarking which evaluates the organization's strengths and weaknesses versus those of its competitors.

Now we have established a link between the strategic planning process and how quality can fit in. Strategic quality planning can be defined to be the structured process for establishing long-range quality goals and defining the means to reach those goals. This is to be accomplished at the highest level of the organization and should be included as part of the strategic planning process in an organization.

Of note here is an important part of business planning of any kind and certainly of quality planning. You do not plan in the abstract if your wish is to be successful in carrying out the plan. One can plan only after the goal has been established. Many companies have undertaken to implement both broad and narrow tactical quality strategies without the support of upper management. That is, the planning for getting better in quality via better yields, lower scrap, and rework was done in the abstract without being part of the business plan. Upper management and middle management typically pay little attention to any organizational activity not tied to the business strategy and goals. In addition, since the management reward structure (promotion and bonus) is usually tied to achievement of goals, any goal not tied to the business strategy is not perceived as legitimate and is therefore relegated to low or no priority.

The Origin of Strategic Quality Goals.   Strategic quality goals is a special category which has only recently entered the business plans of companies. Because the concept has a big impact on operations as well as on quality planning, we will examine the events which lead up to establishing strategic quality goals.

Some companies have adopted the word *vision* as an expression of what they would like to accomplish, or where they would like to be, sometime in the future. Statements of the vision take such forms as:

To be the low-cost producer

To be the market leader

To be the leader in innovation

To be the quality leader

These statements by themselves are not much more than a wish list. Publication of these statements does not tell the people in the organization what they should do that is different from what they have done in the past. The vision statement must be converted into a list of specific goals to be achieved, along with the plan to be followed in order to meet those goals. That conversion is accomplished by the planning process. Visions resemble wishes. They have little relation to reality until they are converted into specifics—into quantitative goals which are to be met within a specific time span.

When the Ford Motor Co. embarked on the Taurus model of automobile (in the early 1980s), one of the visions was to restore profitability. The quality-related goal became "best in class."

The new high priority assigned to quality has created a trend to expand strategic business planning to include strategic quality goals. Here are some actual examples of strategic quality goals which were established as part of companies' business plans:

Make the Taurus/Sable models at a level of quality which is best in class. (Ford Motor Co.)

Improve product and service quality 10 times by 1989. (Motorola's goal in January 1987.)

Reduce the cost of (poor) quality by 50 percent in 5 years. (3M Corp. in July 1982.)

Reduce billing errors by 90 percent. (Florida Power and Light Co.)

Note that all of the above goals relate to major processes, namely: new product launching, customer service, reduction in chronic waste, and billing.

Subject Matter of Strategic Quality Goals.   Despite the uniqueness of specific industries and companies, certain subjects for strategic quality goals are widely applicable:

*Product performance.*This goal relates to major performance features which determine response to customer needs: promptness of service, fuel consumption, mean time between failures, courtesy. These features directly influence product salability.

*Competitive performance.*This has always been a goal in market-based economies, but seldom a part of the business plan. The trend to make competitive quality performance a part of the business plan is new.

*Quality improvement.*This goal may be aimed at improving product salability and/or reducing the cost of poor quality. Either way, the end result after deployment is a formal list of quality improvement projects with associated assignment of responsibilities.

*Cost of poor quality.*The goal of quality improvement usually includes a goal of reducing the costs attributable to poor quality, such as a goal to reduce these costs by half over a 3- to 5-year period.

*Performance of major processes.*This goal has only recently entered the strategic business plan. The goal relates to the performance of major processes which are multifunctional in nature, that is, new product launching, billing, bidding for business, and purchasing.

Benefits of Strategic Quality Goals.   Establishment of strategic quality goals is a vital first step toward translating the vague vision into reality. Taking this first step also yields some major benefits relative to quality:

The process of selecting the goals helps to bring top management together to agree on key goals.

Since the goals must be approved at high levels, the upper managers must get involved personally.

Goals which are a part of the business plan are more likely to obtain the needed resources to meet the goals.

The reward system (for promotion, merit increases, or bonuses) associated with business plan goals makes it more likely that the goals will be met.

Deployment of Quality Goals.   A list of strategic quality goals is a wish list. To convert this list into realities requires getting into specifics: what actions need to be taken in order to meet the goals, who is to take those actions, and the like. The process for identifying these specifics is called *deployment of quality goals*.

As used here, *deployment* means subdividing the goals and allocating the subgoals to lower levels. Such deployment accomplishes some essential purposes:

- The subdivision continues until it identifies the specific actions to be taken.
- The allocation continues until it assigns clear responsibility for taking the actions.
- Those who are assigned to take the actions then respond by determining the resources needed and communicating this back to higher levels.

All of this is based on analysis of new and existing data. Through Pareto analysis, management identifies the key areas to focus resources at each level of the organization.

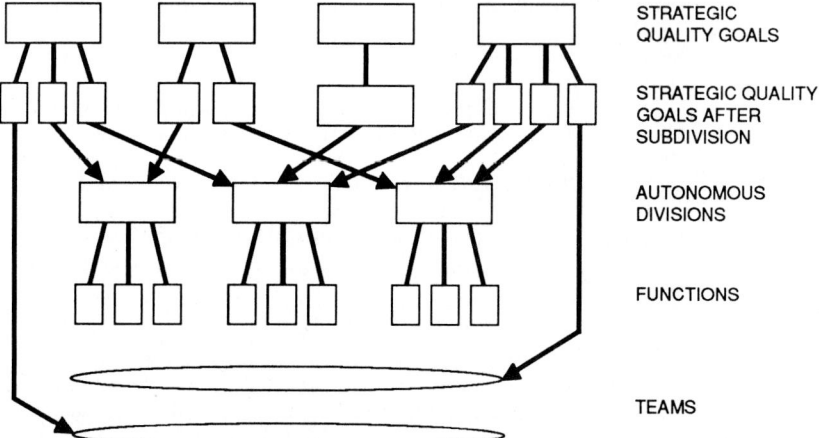

FIGURE 1.11   Goal deployment model. *(Juran Institute, Inc., "Planning for Quality," p. 9-9.)*

The strategic quality goals have already been established by the quality council. The need is to allocate these goals to lower levels. However, most strategic quality goals consist of broad-scope quality planning or improvement projects. The deployment process includes dividing up such broad goals into manageable pieces. For example, a quality goal to reduce the cost of poor quality by *Y* million dollars must first be subdivided and deployed to lower levels, and then broken down into manageable "bite-sized" projects that deal with specific problems such as scrap, rework, product returns, reliability, and the like, using the tools of the quality improvement process. See Fig. 1.11.

In some cases, the quality council has the information needed to do the subdividing. In other cases, the council may assign specific broad goals to lower organizational units, and leave it to them to do the subdividing (based on analysis of the data).

Identify the Needed Actions and Resources.   The allocations establish responsibility but do not make clear what are the actions to be taken. Only as these actions are identified does the wish list become real. What happens next is that each organization unit or project team examines its mission and comes up with conclusions such as:

According to the available data, the vital few error types (five of them) account for 80 percent of all errors. In our judgment, we should be able to bring the error rates of those five types down to levels which will result in an overall reduction of 50 percent. To do so we will need resources   as   follows....

At this stage, the deployment process has already accomplished a major purpose—to assign responsibility and to define the actions to be taken.

Provide Resources.   Now the direction of communication is reversed. The proposals from the action level flow upward to the quality council (top management), along with the requests for resources. The proposals are then talked out. There may be some negotiation and revision. What emerges is agreement on actions to be taken, resources to be provided, and results to be achieved.

The deployment process provides for participation by lower levels, as well as for communication both up and down. Strategic quality goals may be proposed at the top. The lower levels then identify the actions which, if taken, will collectively meet the goals. The lower levels also submit the plan for resources needed to achieve the goals. The subsequent negotiations between lower organizational units and top management then try to arrive at an optimum which balances the value of meeting the goals against the cost of doing so.

The two-way communication feature of the deployment process (a Japanese term is *catch ball*) has turned out to be an important aid to getting results. Feedbacks from companies using this process suggest that it is better than the process of one-way goal setting by upper managers.

The Price to Be Paid.   Quality is not free, and even though the investment in TQM activities has proven to be worthwhile, with good payback, there are many changes an organization must be willing to accept in order to gain benefits from a TQM system. The major changes are

- Establishment of broad quality goals as part of the company's business plan
- Adoption of cultural changes
- Rearrangement of business and organizational priorities
- Creation of a new infrastructure: quality council, quality teams, quality training
- Upper management participation and leadership in managing for quality
- Time to make the desired improvements
- Regular progress reviews by upper management

Strategy Issues.   One issue of adopting quality as a strategy is when to adopt it depending on the type of business you are in and the general life cycle state of the business. In a research report published by SRI International, four general stages of industry are considered for developing quality strategy: introduction, growth, maturity, and decline. Table 1.1 summarizes the role of quality and the considerations for developing quality strategy in each of these stages. The authors note that, ironically, a high-quality image is easiest to attain in the introduction and growth stages when product quality is less important to buyers; it is most difficult to attain in the maturity stage when product quality is of key importance to buyers. Therefore, companies in the introduction and growth stages should guard against assumptions that the current complacency of buyers will continue and should allow sufficient lead time in their quality strategies to ensure that they enter the maturity stage with a higher-than-average quality image.

As it turns out, it is well-known that it is also much cheaper to build quality into products and services and most expensive to try and fix broken things once the customer has experienced some dissatisfaction.

For all of those industries mentioned earlier affected by foreign competition, many found the price to be paid to get back to a competitive position was too high. Many businesses lost significant market share; some folded or were sold, and some decided to pay the price to remain competitive. Certainly the remaining auto companies are trying to make a comeback; Xerox and Motorola, both severely impacted by the Japanese, have also made impressive comebacks. So we know that it can be done in the United States. It seems that, again, the major issue facing us as a nation is the inability for us to remain

TABLE 1.1 Industry Evolution and Quality Strategy

Stage	Quality characterization	Considerations in developing quality strategy
Introduction	Overall quality image of the industry is not yet established. Quality is not a principal emphasis in buyer purchase decision.	Firms whose products have a higher than average quality reputation: • Attain profitability more rapidly • Tend to retain a larger market share during later growth stages • More easily meet threats of new entrants and rapid imitation of product designs by competitors
Growth	Technical and performance differentiations are key competitive areas.	In complex devices and systems, reliability becomes the key to widening the buyer group.
	Quality is improving but uneven: it is gaining importance in purchase decisions.	It is easier to establish or change quality image now than in the mature stage because processes are not yet fully standardized within the firm. The customer base is growing and the product is undergoing design changes.
Maturity	Quality tends to be highest and quality competition is substantial.	It is difficult to change quality image—buyers are experienced repeat customers. Market shares and prices are stabilizing.
	Service arrangements are an important element in quality competition.	Standardization of processes and techniques within firms is substantial. Costs are key to profitability. Quality improvement (fewer defects) can arrest loss of market share to competitors and contribute to cost reduction.
Decline	Quality is spotty. Price differentiation is declining.	High-quality innovative products can contribute to establishing a leadership position in a stable or slowly declining segment if they address the key attributes to the now sophisticated buyer and require substantial capital investment from competitors who would be reluctant to follow. Return on investment and product quality are most highly related in stagnant markets.

Source: Norman B. McEachron and Harold S. Javitz, *Managing Quality: A Strategic Perspective,* ©1981, SRI International, Research Report 658, p. 9, Table 3.

competitive in an industry after we have created it and dominated it in the early stages of development of the industry.

In a recent *Fortune* magazine article, one reason cited for companies choosing not to compete is that U.S. business spends too much money buying and selling operations and not enough in building them. Again, staying competitive requires a price to be paid in terms of investing in technology, human resources, and all the planning, time, and hard work needed to continually improve.

Another issue facing companies today is the type of strategic planning methodology employed by a company. Robert Hayes, a prominent business school professor from Harvard, has pointed out that the strategy of "portfolio management" has been employed widely by a generation of business school graduates. This methodology by its very nature focuses on "restructuring and rearranging the boxes" of the business portfolio, rather than getting inside the business and managing it better. For companies that would rather acquire and divest businesses than grow their own, "if it ain't broke, don't fix it" has a new corollary, "if it is broke—sell it."

As an alternative to existing strategies for managing businesses, time-based competitive advantage is becoming a new way to gain competitive advantage. A good explanation of using time to compete can be found in the book *Competing Against Time* by Stalk and Hout. The authors mention that the strategic implications of compressing time are significant. As time is compressed for things like manufacture and distribution of products, business process cycles, and developing and introducing new products, the following changes occur:

- Productivity increases
- Prices can be increased
- Risks are reduced
- Share is increased[29]

The authors note, though, that there is a price to be paid and that three tasks must be accomplished by the management of the company:

1. Make the value-delivery systems of the company 2 to 3 times more flexible and faster than the value-delivery systems of competitors
2. Determine how customers value variety and responsiveness, focus on those customers with the greatest sensitivity, and price accordingly
3. Have a strategy for surprising competitors with the company's time-based advantage[30]

As a summary, regarding the advantage of a time-based strategy, the authors conclude the following: Clearly, the time advantage is enabling time-based competitors to upset the traditional leaders of their industries and to claim the number one competitive and profitability positions. When a time-based competitor can open up a response advantage with turnaround times 3 to 4 times faster than its competitors, it will almost always grow 3 times faster than the average for the industry and will be twice as profitable as the average for all competitors. Moreover, these estimates are "floors." Many time-based competitors grow faster and earn even higher profits relative to their competitors.

Again, it is ironic to note that speed is, and always has been, an element of good quality planning and good quality in the form of process cycle time. The rub is that there is a price to pay in order to develop an organization that can compete against world-class competition, and that investment in resources and time is not one that many U.S. companies have made. When all the cards are on the table, adopting quality as a strategy is a business decision, since there is a significant price to pay.

Relationship to Japanese Policy Deployment.   Kondo provides insight into the Japanese approach to planning for and achieving quality goals company-wide. Kondo refers to this as *policy control,* or *hoshin kanri.*

> *Policy Control ("Hoshin Kanri") in Japanese Companies.* Quality control activity in Japanese industries is companywide, and top management personnel are in the position of leading and promoting the quality control activities of their companies. They are responsible for deciding top policy on quality of manufactured products and service and for establishing the long-term plan of companywide quality control in order to realize that policy. In addition, they evaluate whether the policy and the plan are being realized on schedule and whether any corrective actions need to be taken by top management. These activities are a form of companywide Deming's cycle and are called "policy control" ("hoshin kanri") in Japanese industries. Internal quality control audits by top management...are an effective way to evaluate the results as a basis for appropriate corrective actions. (This concept of policy control is followed in many but not all Japanese companies.)[31]

IMPLEMENTING TQM

Organizations wishing to implement TQM into the management system must develop a plan or a road map for doing so. TQM does not develop by itself. In addition, while TQM concepts are not difficult to grasp, the concepts do not bring themselves to life through human common sense or motivational strategies often expounded by management in the United States. What is required is a plan directed at achieving aggressive quality goals and a number of years of sustained effort to achieve the desired goals.

Figure 1.12 is an example of how an organization can drive toward a TQM environment. While the model does not demonstrate the complexity of attaining a TQM environment, it does explain essential elements of moving toward a total quality management system.

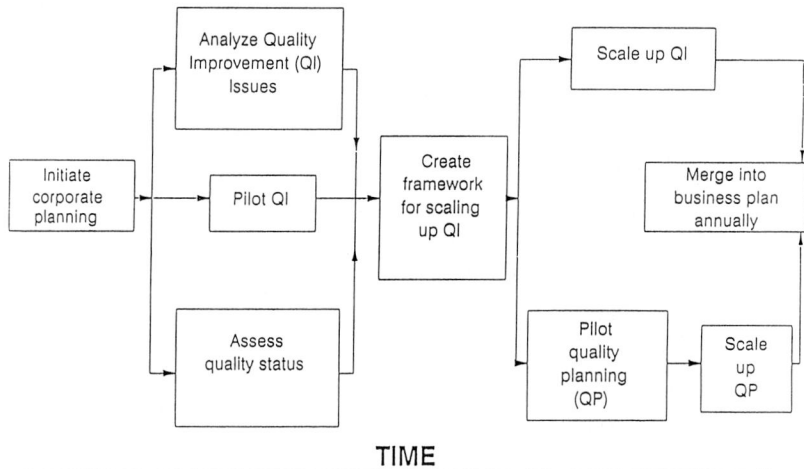

FIGURE 1.12   TQM planning model. (*Juran Institute, Inc., "Making Quality Happen—Upper Management's Role.℠"*)

Developing a Plan.   A brief description of the planning model items follows.

Initiate Corporate Planning.   In this step, an organization begins to evaluate quality as an element for future success. Initial discussions begin usually at high levels; studies are completed; visits to various companies are made; and reports of key findings are presented for management consideration and action.

Analyze Quality Improvement Issues.   At this step, evaluation of input from the initial planning efforts results in initial decisions by management to move forward with further quality activities. At this point, an organization makes a decision that quality is an important strategic element to pursue.

Pilot Quality Improvement.   Attempts are made by an organization to start a small-scale effort directed at quality improvement (QI). This effort starts to shape some fundamental elements of success for TQM, which includes organizing the infrastructure for TQM and carrying out a limited number of improvement projects directed at the elimination of chronic cross-functional problems. The success of these initial efforts is critical to the success of the TQM plan.

Assessing Quality Status.   Typical assessments undertaken include a study of cost of poor quality, an internal customer survey to determine employee perceptions about quality and company culture issues, market research to determine customer-perceived quality versus competitors and an internal assessment of quality systems and results within the organization. These assessments provide the organization with critical inputs in moving forward with TQM.

Framework for Scaling Up QI. Requirements for scaling up to achieve greater levels of improvements beyond the pilot phase include those key elements discussed in the section on quality improvement. Basically, this involves creating the organizational infrastructure and capability to achieve broad-scale improvement.

Scale-up QI. In this phase, an organization multiplies its improvement efforts with more teams working on more projects.

Pilot Quality Planning. Quality planning (QP) projects would be initiated to improve product development efforts and to replan or plan key organizational processes.

Scale-up QP. Just as with quality improvement, quality planning efforts would be expanded once success is proven by the initial projects.

Merge into Business Plan Annually. Quality improvement and planning activities should eventually become part of the business planning process mentioned earlier in the section "Quality and Strategy." Once business planning and quality planning are merged, the TQM environment is created.

Creating the Infrastructure. The primary infrastructure for TQM was explained in the sections "Quality Improvement" and "Quality Planning." As a brief recap, the essential items are the following:

- Top management quality council
- Subcouncils
- Cross-functional committees
- TQM staff
- Sponsors and champions
- Facilitators
- Structured processes for making improvements
- Training
- Teams
- Strategic quality plan

The most important aspect of creating this infrastructure is the direct participation and leadership of upper management.

Roles for Japanese managers in TQC are very similar to what has been advocated by Western TQM experts. Mizuno provides the following roles for top and middle management:

1. Establish and disseminate the corporation's quality policies
2. Identify priority quality problems and see that they are solved
3. Create an organizational plan for implementing quality control
4. Check and revise quality assurance activities as necessary[32]

To elaborate on the above, Mizuno offers the following detail:

1. Articulate the corporation's present circumstances and possible future problems, draw up business-year and long-term administrative plans to resolve these problems, and clarify these plans to everyone in the corporate structure.
2. Investigate the best ways to transmit the corporation's policy goals and set them in motion, and issue the necessary instructions for their implementation.

3. Keep track of the progress being made in implementing policies, diagnose the effectiveness of what is being done, and apply the lessons learned in setting new policies.

4. Check whether or not goals have been attained, and issue new instructions as necessary where problems remain.

5. Prepare and regularly update a policy manual that incorporates an in-depth analysis and evaluation of policy management to date, and appoint someone to head a policy management division handling the clerical functions required for top management policy audits (with the understanding that this TQC office or headquarters must work closely with the corporation's planning and supervisory divisions).[33]

An example of Japanese-style policy management in the United States was demonstrated by Florida Power and Light Co. (FPL). The following is a summary of policy deployment (policy management) at FPL.

Summary.   Policy deployment is the cornerstone of FPL's management system. It is a management process for achieving breakthroughs on major corporate problems and focuses on customer needs by deploying resources on a few high-priority issues. It is the primary means of achieving the corporate vision.

Prior to 1985, the company used management by objectives (MBO) as the principal method for achieving corporate objectives. MBO focused on the company's point of view without adequately considering the customer's point of view. Also, it did not provide a systematic process for ensuring that corporate objectives were met. Because of the difficulties that existed with the company's implementation of MBO, policy deployment was introduced in 1985.

In the introduction phase, policy deployment provided a management process for achieving corporate objectives. However, the company attacked too many problems and solved only a few. Also, the improvement objectives attempted were not focused on customer needs. Therefore, it was necessary for the company to take corrective action for the next cycle.

Corrective action was taken in each successive cycle. For example, in 1988, cross-functional committees were introduced to help the executives coordinate the major activities of the company from a corporate perspective.

The many improvements that were made clarified how corporate management activities were tied to an overall management system. This management system is based on a philosophy of continuous improvement of the quality of products and services that FPL provides its customers. It identifies both current and future customer needs.

Emphasis of Activities.   In developing FPL's policy deployment process, particular attention was given to the following to ensure the achievement of company objectives:

1. Establish policies focused on achieving customer satisfaction

2. Develop consistent policies and targets

3. Strengthen review of implementation activities

FPL Management System.   FPL's management system was enhanced with each policy deployment cycle. The following are four key characteristics of the FPL management system:

1. *Focus on customer needs.* Policies are identified and developed on the basis of quality elements, which are established through a combination of the customer's voice and the electric utility industry's obligation to serve.

2. *Management Reviews.* Improvement activities are reviewed by the president, responsible executives, and managers to check on the achievement of corporate policies, to take action when necessary, and to promote the quality improvement process (QIP). FPL has three levels of management reviews of improvement activities. Level 1 reviews, conducted by the president and executive vice presidents, are for vice presi-

dents who report to the chairman and the president. The business plan is reviewed with emphasis on corporate cross-functional issues and short-term plans. Level 2 reviews, conducted by appropriate vice presidents, are for district general managers, site vice presidents, plant managers, and department heads. The emphasis is on the business plan, with particular attention given to policy deployment activities and priority problems. Level 3 reviews, conducted by department heads, are for managers, superintendents, and supervisors. The major emphasis is on how well QIP has been implemented within their departments, on priority problems, and on the application of quality in daily work (QIDW).

3. *Cross-Functional Management.* Responsibility for improvement objectives is assigned to executives, although activities cross functional lines.

4. *Integration of Policy Development and Budget.* Through management consensus and catch ball, the resources are allocated to support annual improvement activities.

Since the time that policy deployment was introduced, several systems, such as the quality/delivery system and the cross-functional committees, have been added.[34]

Policy Deployment Flowchart.   Figure 1.13 outlines the three phases: establish policy, deploy policy, and implement policy.[35]

Training.   Organizations interested in establishing a TQM process will find that a substantial amount of training in the quality-related disciplines is required. Xerox reported that, during a 4-year period in the 1980s, it spent $125,000,000 on quality-related training. Motorola reported that it spends approximately 1 percent of sales on quality-related training. The reason for these significant levels of investment is that few educational institutions provide the type of quality-related training needed to implement TQM. It is therefore left to business and service organizations to fund.

Key aspects in planning the quality-training curriculum are to determine the following:

- Who is to be trained? Examples include top management, middle management, work force, and the like.
- What is the subject matter? This is determined by analysis of the types of problems and issues faced by an organization and the corresponding quality tools and techniques that deal with solving the problems.
- Will training be required on the job? If not, the training effort will contribute minimally toward achieving TQM goals.
- A training plan to meet organization needs, including costs, time, and schedule for carrying out the plan.

An example of a quality curriculum and training plan appears in Table 1.2, which describes FPL's quality-related training.

Resistance to Change.   For most organizations, a shift to TQM as a way of life represents a big change, since the focus of quality becomes company-wide. Key to overcoming cultural resistance will be aligning the organization to be receptive to change by fully communicating TQM goals; providing the organization with the time, training, and resources to meet goals; building a cross-functional cooperative environment with teams addressing chronic problems; aligning by management of quality plans and strategy and business plans and strategy so that quality is perceived as legitimate by the organization; and finally, changing the performance and reward systems to reflect achievement of quality goals.

THE FUTURE OF TQM

The U.S. National Quality Award.   From the experience of the authors, many forces are continuing to push quality as an ever-important part of global competition. The impact of poor quality has had a devastating effect on the United States in terms of market share

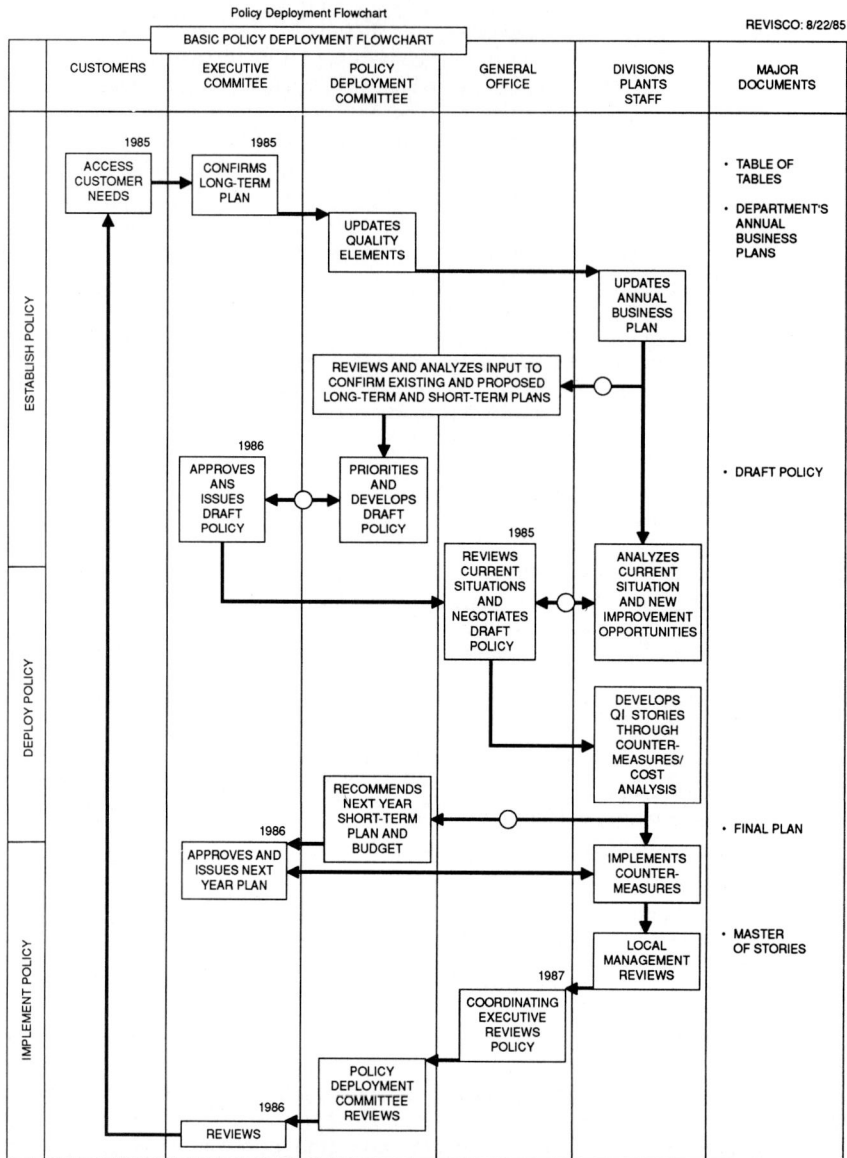

Policy Deployment Flowchart

BASIC POLICY DEPLOYMENT FLOWCHART

REVISCO: 8/22/85

FIGURE 1.13 Policy deployment flow chart. (*Dmytrow, E. D., "OIP—FPL's Continuous Improvement Process,"* 1989 ASQC Quality Congress Transactions, *Toronto, p. 354, Fig. 3.2. Reprinted with the permission of ASQC*)

losses for some key products and services. While the United States has a good level of quality in most industries and services, for the most part quality is still focused on product and operations. A total company-wide focus for quality (TQM) is emerging as a trend for the 1990s. A growing number of organizations are making progress toward TQM and world-class leadership. The Malcolm Baldrige National Quality Award (MBNQA) has been a big plus toward recognition of quality excellence for U.S. organizations. Addition-

TABLE 1.2   Implementation Status of QI Education and Training

Eligible population	Course title (year developed)	Course length, days	Content	Participants trained	
				December 1988 (est. actual)	% of target
Executives	Statistical Concepts for Executives (1988)*	2*	SQC and reliability tools	18	82
Managers and above	Orientation for managers (1983)	1	Introduction to QIP	168	31
	Leadership for Managers I (1984)	3	Managing QI teams	582	106
	Leadership for Managers II (1985)	3	Policy Deployment	605	110
	Leadership for Managers III (1986)	3	Quality in daily work	587	107
	Statistical concepts for Managers (1988)	5	SQC and reliability tools	280	51
Managers, Supervisors, Selected staff	Application expert (1987)	15	Statistical application	245	100
Supervisors	Supervisor-foreman awareness (1983)	1	Introduction to QIP	1021	55
	Supervising for Quality (1986)	5	Supervising teams, policy deployment, and QIDW	2468	133
	Supervising teams (1988)	2	Supervisory facilitation of teams	923	50
Facilitators, team leaders	Techniques (1985)	3	Selected SQC tools	224	NA
	Techniques (1988)	5	Scatter diagrams and control charts	713	51
Facilitators	Facilitator training (1983)	5	QIP administration and facilitating skills	976	100
Team leaders	Team leader training (1982)	5	QC tools, group dynamics	6122	100
Totals	13 courses developed by FPL	56	QIP	14932	NA

Source: Eric D. Dmytrow, "QIP—FPL's Continuous Improvement Process," *1989 ASQC Quality Congress Transactions,* Toronto, p. 364, Exhibit 6.1.
*Externally developed.

ally, the award criteria represents a framework for TQM. Figure 1.14 is a representation of the system with the titles of each of the main categories. This framework is providing U.S. organizations with knowledge of a system for TQM. The framework does not explain how to get to TQM, but can provide a means for assessment of an organization's quality system and can aid in development of improvement plans for any gaps discovered in the assessment.

Winners of this award are required to share information on how they achieved excellence in the award categories. This sharing has been of tremendous value to U.S. organizations, as well as to others, because it demonstrates that the United States can reach world-class quality levels in global markets. Japanese quality experts have also recognized winners as world class (authors' experience).

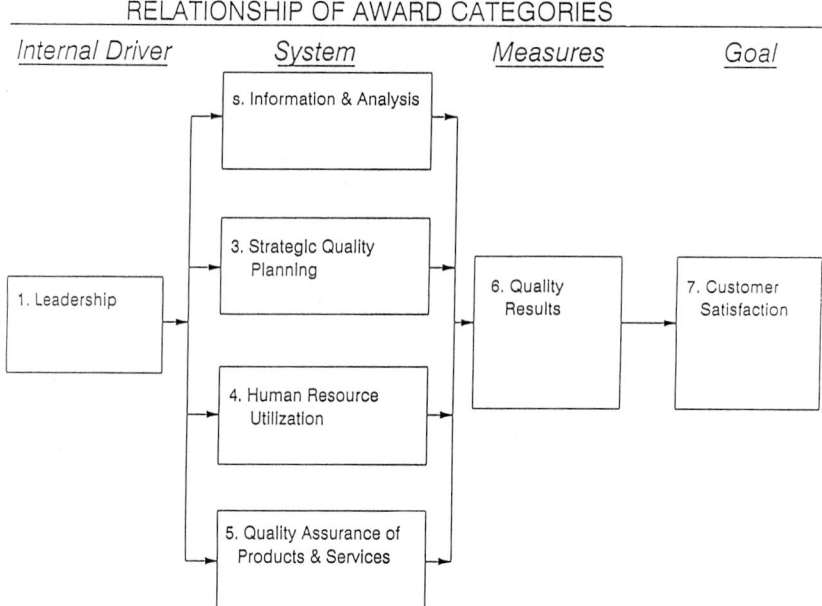

RELATIONSHIP OF AWARD CATEGORIES

| Internal Driver | System | Measures | Goal |

- s. Information & Analysis
- 3. Strategic Quality Planning
- 1. Leadership
- 4. Human Resource Utilization
- 6. Quality Results
- 7. Customer Satisfaction
- 5. Quality Assurance of Products & Services

FIGURE 1.14 Relationship of award categories. (*Juran Institute, Inc., "Planning to Win: Malcolm Baldrige National Quality Award."*)

Trends. The Baldrige award will have a continuing positive effect throughout the 1990s for promotion of TQM in the United States to help businesses compete in global markets with excellence in quality. Juran was so moved by the results achieved by the winners and the widespread interest among businesses in the United States that he changed his prediction of the United States catching up with Japan's pace of improvement sometime beyond the year 2000 to catching up in the 1990s. Juran's predictions on such matters have been remarkably accurate in the past, predicting major impact on U.S. businesses. We hope he is right about this new prediction.

NOTES

1. Ishikawa, K., *What is Total Quality Control? The Japanese Way,* translated by David J. Lu, Prentice-Hall, Englewood Cliffs, N.J., copyright © 1985, p. 7. Used by permission of the publisher, Prentice-Hall, a division of Simon & Schuster.

2. Mizuno, S., *Company-Wide Total Quality Control,* Asian Productivity Organization, Tokyo, 1988, p. 282. Distributed in the United States, Canada, and Western Europe by Quality Resources, White Plains, N.Y. Reprinted by permission of Quality Resources.

3. Mizuno, S., *Company-Wide Total Quality Control,* Asian Productivity Organization, Tokyo, 1988, p. 283. Distributed in the United States, Canada, and Western Europe by Quality Resources, White Plains, N.Y. Reprinted by permission of Quality Resources.

4. Ishikawa, K., *What is Total Quality Control? The Japanese Way,* translated by David J. Lu, Prentice-Hall, Englewood Cliffs, N.J., copyright © 1985, pp. 18–19. Used by permission of the publisher, Prentice-Hall, a division of Simon & Schuster.

5. Ishikawa, K., *What is Total Quality Control? The Japanese Way,* translated by David J. Lu, Prentice-Hall, Englewood Cliffs, N.J., copyright © 1985, p. 19. Used by permission of the publisher, Prentice-Hall, a division of Simon & Schuster.

6. Ishikawa, K., and K. Kondo, "Education and Training for Quality Control in Japanese Industry," *Quality,* no. 4, 1969, pp. 90–96. Reprinted with permission from QUALITY, a publication of Hitchcock Publishing, a Capital Cities/ABC, Inc., Company.

7. Deming, W., *Elementary Principles of the Statistical Control of Quality,* rev. 2d ed., JUSE, Tokyo, June 1952, p. 3.

8. Juran, J. M., *Planning and Practices in Quality Control—Lectures on Quality Control,* JUSE, Tokyo, July-August 1954, lecture 1, p. 2.

9. Juran, J. M., and F. M. Gryna, *Juran's Quality Control Handbook,* 3d ed., McGraw-Hill, New York, 1974, sec. 2, p. 11. Reproduced with permission of McGraw-Hill, Inc.

10. *Japanese Industrial Standard: Glossary of Terms Used in Quality Control* (JISZ 8101) 1981, prepared by Japanese Industrial Standards Committee, Tokyo, Japanese Standards Association, p. 2.

11. *American National Standard, Quality Systems Terminology,* approved January 18, 1979, revision of ASQC A3-1971 (ANSI ZI.7-1971), Milwaukee, Wis., p. 4. Reprinted with the permission of ASQC.

12. *Guide to Quality Control and Company Standardization,* Japanese Standards Association, Tokyo, 1984, p. 22.

13. *Guide to Quality Control and Company Standardization,* Japanese Standards Association, Tokyo, 1984, p. 21.

14. Mizuno, S., *Company-Wide Total Quality Control,* Asian Productivity Organization, Tokyo, 1988, p. 18. Distributed in the United States, Canada, and Western Europe by Quality Resources, White Plains, N.Y. Reprinted by permission of Quality Resources.

15. Ishikawa, K., *What is Total Quality Control? The Japanese Way,* translated by David J. Lu, Prentice-Hall, Englewood Cliffs, N.J., copyright © 1985, p. 44. Used by permission of the publisher, Prentice-Hall, a division of Simon & Schuster.

16. Ishikawa, K., "Recent Trends of Quality Control," *Reports of Statistical Applications and Research,* JUSE, vol. 12, no. 1, pp. 1–17.

17. Ishikawa, K., *What is Total Quality Control? The Japanese Way,* translated by David J. Lu, Prentice-Hall, Englewood Cliffs, N.J., copyright © 1985, p. 37. Used by permission of the publisher, Prentice-Hall, a division of Simon & Schuster.

18. DeFeo, J. A., et al., *Planning for Quality,* 2d ed., © Juran Institute, Inc., Wilton, Conn., 1990, G2–G15.

19. Juran, J. M., and F. M. Gryna, *Juran's Quality Control Handbook,* 4th ed., McGraw-Hill, New York, © 1988, sec. 2, p. 7. Reproduced with permission of McGraw-Hill, Inc.

20. Juran, J. M., and F. M. Gryna, *Juran's Quality Control Handbook,* 4th ed., McGraw-Hill, New York, © 1988, sec. 6, p. 35. Reproduced with permission of McGraw-Hill, Inc.

21. Mizuno, S., *Company-Wide Total Quality Control,* Asian Productivity Organization, Tokyo, 1988, p. 67. Distributed in the United States, Canada, and Western Europe by Quality Resources, White Plains, N.Y. Reprinted by permission of Quality Resources.

22. Mizuno, S., *Company-Wide Total Quality Control,* Asian Productivity Organization, Tokyo, 1988, p. 61. Distributed in the United States, Canada, and Western Europe by Quality Resources, White Plains, N.Y. Reprinted by permission of Quality Resources.

23. Mizuno, S., *Company-Wide Total Quality Control,* Asian Productivity Organization, Tokyo, 1988, p. 61. Distributed in the United States, Canada, and Western Europe by Quality Resources, White Plains, N.Y. Reprinted by permission of Quality Resources.

24. Mizuno, S., *Company-Wide Total Quality Control,* Asian Productivity Organization, Tokyo, 1988, p. 71. Distributed in the United States, Canada, and Western Europe by Quality Resources, White Plains, N.Y. Reprinted by permission of Quality Resources.

25. Mizuno, S., *Company-Wide Total Quality Control,* Asian Productivity Organization, Tokyo, 1988, pp. 71, 72. Distributed in the United States, Canada, and Western Europe by Quality Resources, White Plains, N.Y. Reprinted by permission of Quality Resources.

26. Ishikawa, K., *What is Total Quality Control? The Japanese Way,* translated by David J. Lu, Prentice-Hall, Englewood Cliffs, N.J., copyright © 1985, pp. 114–116. Used by permission of the publisher, Prentice-Hall, a division of Simon & Schuster.

27. Marrus, S. K., *Building the Strategic Plan,* John Wiley & Sons, New York, copyright © John Wiley & Sons, 1984, pp. 4, 5. Reprinted by permission of John Wiley & Sons, Inc.

28. Marrus, S. K., *Building the Strategic Plan,* John Wiley & Sons, New York, copyright © John Wiley & Sons, 1984, p. 5. Reprinted by permission of John Wiley & Sons, Inc.

29. Stalk, G., Jr., and T. M. Hout, *Competing Against Time: How Time Based Competition is Reshaping Global Markets,* The Free Press, New York, 1990, p. 31.

30. Stalk, G., Jr., and T. M. Hout, *Competing Against Time: How Time Based Competition is Reshaping Global Markets,* The Free Press, New York, 1990, p. 36.

31. Juran, J. M., and F. M. Gryna, *Juran's Quality Control Handbook,* 4th ed., McGraw-Hill, New York, 1988, p. 35F.12. Reproduced with permission of McGraw-Hill, Inc.

32. Mizuno, S., *Company-Wide Total Quality Control,* Asian Productivity Organization, Tokyo, 1988, p. 30. Distributed in the United States, Canada, and Western Europe by Quality Resources, White Plains, N.Y. Reprinted by permission of Quality Resources.

33. Mizuno, S., *Company-Wide Total Quality Control,* Asian Productivity Organization, Tokyo, 1988, pp. 37, 38. Distributed in the United States, Canada, and Western Europe by Quality Resources, White Plains, N.Y. Reprinted by permission of Quality Resources.

34. Dmytrow, E., et al., "QIP-FPL's Continuous Improvement Process," *1989 ASQC Quality Congress Transactions,* Toronto, pp. 352, 353.

35. Dmytrow, E., et al., "QIP-FPL's Continuous Improvement Process," *1989 ASQC Quality Congress Transactions,* Toronto, pp. 354.

CHAPTER 2
ON-LINE QUALITY CONTROL

Susan L. Albin
Department of Industrial Engineering
Rutgers, The State University of New Jersey
Piscataway, New Jersey

Elsayed A. Elsayed
Department of Industrial Engineering
Rutgers, The State University of New Jersey
Piscataway, New Jersey

Until recently the role of quality control activities has been limited to inspection and removal of defective units before shipment of the final product to the customer. In addition, the role of *statistical quality control* was limited to observing the variations in the quality characteristics of the product, and, as long as certain statistical tests were satisfied, the process was considered capable of producing "good quality" products.[9]

The increased competition among businesses and industries has altered the goal of quality control from ensuring the conformance of a product to specification to constantly improving the quality of the product (or service) during every phase of its life cycle. This has resulted in the implementation of a *total quality system* which encompasses two components: off-line and on-line quality control systems. The off-line system focuses on designing robust products and processes which are not sensitive to material and process parameter variations. The on-line system focuses on ensuring the quality of the product during actual production. The total quality control system has demonstrated its effectiveness in improving the quality of products and services while reducing quality costs.[10]

Definition of On-Line Quality Control. On-line quality control is defined as a set of activities that is applied during the actual production of a product to ensure that variations among products are reduced and that the desired quality characteristics are attained at minimum cost. These activities may include, but are not limited to, product inspection and sampling, process diagnosis, adjustment of the process parameters, monitoring of the process parameters, process stoppage and recovery, and feedback and feedforward process controls.

In discrete parts manufacturing, for example, an ideal on-line quality control system measures the dimensions of the part while the tool is cutting, and the measurement results at the point of cutting are fed back to the machine controls with automatic adjustment of tool tip position, taking into consideration the influence of temperature, clamping distortion, tool wear, and vibration.

Different manufacturing processes are influenced by different process parameters. In the machining example, the process parameters are the feed and speed of cutting, depth of cut, and the material being cut, whereas in the optical fiber drawing process, the parameters may include the speed of drawing, the temperature of molten material, and the material from which the fiber is drawn. Measuring and controlling the fiber coating concentricity are important quality characteristics which can be achieved by using an on-line quality control system.

Although the main objective of the on-line and off-line quality control systems is to reduce variation in products and processes, they do not, of course, eliminate it. This is due to different causes of variations which are difficult, if not impossible, to eliminate. The following section presents a brief discussion of causes of variation.

Causes of Variation. Factors that cause the functional characteristic of a product to deviate from its specified nominal (target) value are referred to as noise factors and are classified as follows:

1. External factors
   a. Variations in the operating environment from the design conditions such as temperature, humidity, vibrations, and voltage fluctuations.
   b. Human errors: improper use of the product because of human errors may cause the product characteristic to deviate from its target (nominal) value.
2. Internal factors
   a. Manufacturing imperfections: a major cause of manufacturing imperfection is the nonuniformity of product material. This factor has often led designers to introduce safety factors into their designs. Manufacturing imperfections may also be attributed to inadequate control of the manufacturing process.
   b. Product deterioration: products deteriorate (the quality characteristics deviate from the target values) with age and environment. Examples of product deterioration are increased resistance of a resistor, loss of spring resilience, change of the colors of painted surfaces, and change of the properties of plastic products with age. In general, the causes of product deterioration can be attributed to one or more deterioration-inducing effects, such as force, time, temperature, or reactive environment.

Some variation in products and processes must be anticipated. These variations can be reduced by using appropriate methods of off-line and on-line quality controls. In the following sections, methods are presented for evaluating and measuring product variation.

PROCESS CAPABILITY AND PRODUCT VARIATION

In a typical manufacturing environment, the capability of the process to produce products that conform to specified limits is determined in advance (during process design and baseline experimentation), and, when feasible, preliminary information is obtained regarding the degree of variation with respect to specified *nominals* (target values). Process capability studies provide a well-tried means of achieving these objectives. These studies are also used in monitoring processes for sudden drift or change.

Before describing the methods for evaluating process capability, we introduce the following definitions.

Process: A combination of machines, tools, methods, materials, and personnel employed to attain the desired quality and quantity of a product or service.

Process capability: An evaluation of the inherent precision and accuracy of the process; the quality-performance capability of the process under specified control conditions.

Process Capability and the Process Capability Index.   The process capability study is a comparison of actual process performance against engineering specifications for a part being produced or assembled. A measure of variation caused by manufacturing imperfections can be considered the capability of the process.

Conventionally, process capability is defined as 6 standard deviations that infer 99.73 percent of all the readings for a normal distribution to fall within the area bounded by 3 standard deviations from the mean. Thus, a process capability study mathematically constitutes a well-organized, carefully disciplined frequency-distribution analysis of the product's functional characteristic from its target. The formula for the process capability is given as follows:

$$\text{Process capability} = 6v = 6 \sqrt{\frac{\sum_{i=1}^{n}(x_i - m)^2}{n}} \tag{1}$$

where v = square root of the mean squared deviation (MSD) from target
$x_1, x_2, \ldots, x_n$ = individual measurements of the characteristic
m = target (nominal) value
n = number of individual measurements

The process capability index C_p is defined as

$$C_p = \frac{\text{tolerance}}{6v} \tag{2}$$

where tolerance is the difference between upper and lower specification limits of the characteristic being studied.

The C_p index, as defined above, is used as a quantitative measure of the variation of the process around the target (nominal or desired) value.

Example.   In a drilling operation, two holes must be drilled in sequence at two locations in the workpiece. The assembly operation of this workpiece with another part requires that the distance between the centers of the holes be 5.000 ± 0.001 in. (nominal distance is 5.000 in.). The distance between centers is measured for 10 successive pieces and the following observations are obtained:

4.998   4.990   5.001   5.000   4.997   5.001   5.004   4.995   5.000   5.000

What is the capability index of the drilling process?

$$v^2 = \frac{1}{10}[(4.998 - 5.000)^2 + (4.990 - 5.000)^2 + \cdots + (5.000 - 5.000)^2]$$

$$= 0.0000156$$

$$v = \sqrt{0.0000156} = 0.00395$$

$$\text{Process capability} = 6v = 0.0237$$

$$\text{Tolerance} = 5.001 - 4.999 = 0.002$$

From Eq. (2), the process capability index is calculated as

$$C_p = \frac{0.002}{0.0237} = 0.08439$$

The value of the C_p index is a measure of the process capability. The higher the value of C_p, the more capable the process of producing products with characteristics closer to their target values, and the proportion of the out-of-tolerance product becomes smaller. Obviously, a process with a low C_p, such as 0.08439 (as obtained from the above example) is a process that is not capable. As a result, the proportion of the out-of-specification product of such a process would be very high. In Japan, many manufacturers' minimum acceptable C_p is 1.33 (or 8/6), and some manufacturers are increasing the C_p value to 2.0 or higher.[11]

A process with sufficiently high C_p values effectively eliminates defectives and the need for inspection, consequently avoiding virtually all costs associated with inspection, scrap, and rework. For example, if the capability index of a process is equal to 2.0 (assuming that the observations follow normal distribution and the target is the mean value of the distribution) the defect incidence of this process is 0.000000002 percent or 2 parts per billion (ppb). This is obtained as follows: since $C_p = 2.0$, the probability that $X - m > 6v$ or $X - m < 6v$ represents the proportion of the production that falls outside the tolerance limits, that is, $P(X > m + 6v$ or $X < m - 6v)$, where X is a random variable, denotes the measurements. The proportion of the out-of-tolerance product above the upper limit is obtained by finding the value of Z, the standard normal random variable with mean equal to zero and standard deviation equal to one:

$$Z = \frac{X - m}{v} = 6$$

and $P(Z > 6v) = 0.000000001$. The proportion falling below the lower tolerance limit of $6v$ is also 0.000000001 (because of normal distribution symmetry). Therefore, the total proportion outside of the tolerance limits is $2 \times 0.000000001 = 0.000000002$ or 2 ppb.

The process capability index can be used to compare the capability of processes. In Fig. 2.1, process A has a C_p index less than 1.0 and the proportion of the product falling

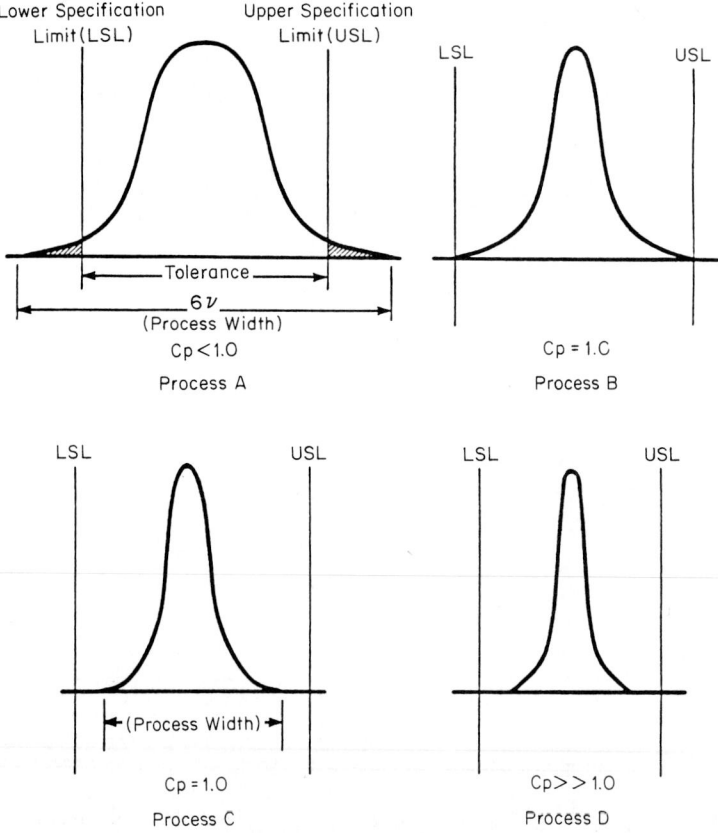

FIGURE 2.1 C_p index as a measure of process capability.

outside the tolerance limits is high. Process B, with a C_p index equal to 1.0, has a smaller proportion of defective product than process A. Although both processes C and D are capable, process D is a better process than process C because it produces the product with a much smaller variation from the target value. The C_p index can also be used to monitor the changes in the process capability from one period of time to another. It should be noted that a more accurate estimation of the C_p index is obtained when more measurements are taken. We recommend a minimum of 10 measurements.

For a process capability study to be meaningful, it should have measurements that are in a state of statistical control (discussed later in the process control section). Since the control charts are used, in addition to serving as a feedback mechanism, to determine whether a process is in a state of statistical control, they are essential prerequisites of a process capability study.

Upper and Lower Capability Index.   The upper and lower capability index C_{pk} is used when the specification limits are not symmetrical around the target. Consider the unilateral tolerance situation where only the upper specification limit is given. The C_p index for the upper specification limit is defined[16] as

$$\text{CPU} = \frac{\text{allowable upper spread}}{\text{actual upper spread}}$$

$$= \frac{\text{USL} - m}{3v}$$

where USL = upper specification limit
CPU = process capability index for unilateral upper specification limit
   m = target value of the process
   v = square root of the mean squared deviation of the process from target

Similarly, the process capability index for the lower specification limit is

$$\text{CPL} = \frac{m - \text{LSL}}{3v}$$

where LSL = the lower specification limit
CPL = process capability index for lower specification limit

Using the above equations, the CPU and CPL are related to C_p by

$$C_p = \frac{\text{CPU} + \text{CPL}}{2}$$

When the target value of the product characteristics is not the midpoint between upper and lower specification limits, the process capability index is measured by C_{pk} as follows:

$$C_{pk} = \min(\text{CPL}, \text{CPU})$$

where

$$\text{CPL} = \frac{m - \text{LSL}}{3v}$$

$$\text{CPU} = \frac{\text{USL} - m}{3v}$$

Example.   A cast-iron workpiece is to be face-milled on a numerical control machine using cemented carbide inserts. The specification of the principal dimension of the workpiece is set between 4.0000 + 0.0020 in. and 4.0000 − 0.0010 in. with a target value of 4.0000 in. The measurements of the principal dimension of 20 samples (sample size is 4) are given below. Calculate the C_{pk} index of the process.

Sample no.	Measurements of the principal dimension, in.			
1	3.9990	3.9980	4.0010	4.0015
2	4.0010	4.0015	4.0010	4.0010
3	3.9980	3.9990	3.9980	3.9990
4	4.0010	4.0015	4.0015	3.9990
5	3.9990	3.9980	3.9990	4.0010
6	3.9980	4.0015	4.0015	3.9980
7	4.0013	4.0012	3.9999	4.0000
8	3.9980	4.0016	3.9990	3.9989
9	3.9982	3.9989	4.0015	4.0011
10	4.0017	4.0016	4.0017	3.9981
11	3.9985	4.0010	4.0015	3.9988
12	3.9999	4.0020	4.0018	3.9980
13	4.0002	4.0013	4.0015	3.9990
14	3.9980	3.9989	3.9999	3.9987
15	3.9990	3.9980	4.0020	3.9981
16	3.9985	3.9980	3.9982	4.0018
17	4.0012	4.0019	4.0010	3.9999
18	4.0013	4.0005	4.0012	3.9980
19	3.9982	3.9991	3.9985	3.9996
20	4.0010	3.9990	4.0008	3.9992

Since the target value is not the midpoint between the specification limits of the workpiece, the capability index is then determined as follows. The C_p index for the upper specification limit is

$$\text{CPU} = \frac{\text{USL} - m}{3v}$$

$$= \frac{4.0020 - 4.0000}{3 \times 0.00133} = 0.5015$$

where v (mean squared deviation) is

$$v = \sqrt{\frac{1}{80}[(3.999 - 4.000)^2 + (3.998 - 4.000)^2 + \cdots + (3.9992 - 4.000)^2]}$$

$$= 0.00133$$

The C_p index for the lower specification limit is

$$\text{CPL} = \frac{4.0000 - 3.9990}{3 \times 0.00133} = 0.2506$$

and C_{pk} (the process capability index) is

$$C_{pk} = \min (0.2506, 0.5012)$$

$$= 0.2506$$

This implies that the current milling process is not capable of producing workpieces meeting the given specifications.

It should be noted that the C_p index does not take into account the centering of the process around the target value. It considers only the spread of the process. On the other hand, the C_{pk} index explicitly accounts for centering of the process by taking the distance from the nearest specification limit to the process center, then dividing by half the natural tolerance spread of the process (3σ). When the process is not in a state of statistical control, an estimate of the standard deviation is used and the measures corresponding to C_p and C_{pk} are replaced by the process performance indexes P_p and P_{pk} respectively.[7]

The process capability and the percent defective are statistical measures of the process and product quality. In many practical situations, it may be more appropriate to use economical measures instead. This is discussed under the loss function section.

STATISTICAL PROCESS CONTROL

The key instrument of statistical process control (SPC) is the control chart invented by Walter Shewhart in the 1920s. Its purpose is to alert operators and engineers of a change in a manufacturing process. At determined times, a sample of measurements is taken of the product or the manufacturing process and a sample statistic is computed from these measurements, an average or fraction defective, for example. The sample statistic is plotted on a graph where the y axis is the sample statistic and the x axis is some measure of time, hour of the day, or batch number, for example. If the sample statistic lies outside a band defined by upper and lower control limits, this signals a shift in the manufacturing process and the engineers and operators must identify and correct the problems.

Figure 2.2 shows a control chart constructed from data in Table 2.2, which is described later in this section. The x axis is batch number; the y axis is the sample fraction defective out of 150 items. The control limits are dotted lines. The centerline shows the overall average for this manufacturing process.

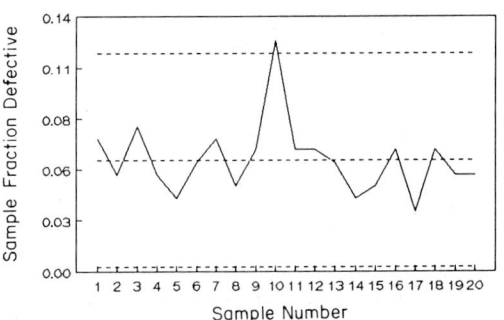

FIGURE 2.2 P chart for fraction defective.

Chance and Assignable Causes of Variation.   When the process is operating normally, the sample statistics vary due to chance causes although they fall within the control limits with high probability. Generally, many chance causes affect a process, each contributing a small amount to the outcome. When only chance causes are operating on a process, we say the process is in-control.

Manufacturing processes are also subject to assignable causes, the identifiable causes of change in a process. Examples of assignable causes are poor batches of raw materials, operator errors, and machines out of calibration. When assignable causes are operating on a process, we say the process is out-of-control.

Control charts serve as the alarm that sets off a feedback system. To close the feedback loop, there must be a procedure for identifying assignable causes consisting of technical instructions, such as a gauge specification, and organizational instructions, such as the position and name of the individual responsible under certain circumstances.

The control limits in a control chart are computed according to the variability of the process when it is in-control. They are not based on product specifications. Control limits can be moved closer together only when measurements show less variability resulting from changes in product design or process improvements.

Type I and II Errors.   There are two types of errors associated with SPC. The Type I error, or false alarm, occurs if a sample statistic falls outside the control limits when, in fact, no assignable cause is present. The false alarm results from sampling error; that is, the process is in-control but the sample selected is highly unlikely.

Process control charts are designed so that the chance of a false alarm is very small, on the order of 0.3 percent, to minimize unnecessary searches for assignable causes or shutdowns. The choice of control limits mainly determines the probability of false alarms.

The Type II error associated with SPC occurs if a sample statistic falls within the control limits when, in fact, the process is out of control. While the out-of-control condition remains undetected, some or all of the items produced may be of poor quality. Of course, there is a good chance the out-of-control condition will be detected at the next sampling point. This type of error is also a result of sampling error.

In most control charts the chance of the Type II error is typically much greater than the Type I error, on the order of 10 to 20 percent. To reduce the probability of the second type of error, increase the sample size used to compute each sample statistic. To minimize the consequences of this type of error, sample more frequently.

Selecting a Sample.   For SPC, samples should come from a homogeneous subpopulation; that is, the measurements come from products that have been subject to the same processing conditions, the same operators, the same machines, and the like. For example, if a product is assembled by two parallel lines it is better to have a separate control chart for each line than to average together measurements from the two lines on a common control chart.

To select the number and location of control charts in a manufacturing process, one must compromise between two goals. On one hand, it is best to sample from homogeneous subpopulations. On the other hand, it is unwieldy and expensive to make many measurements and manage many control charts. Often, there are constraints that help in the selection. For example, in many processes there are only a few points where product or process characteristics can be measured easily and quickly.

Interpreting Control Charts.   The sample data plotted on a control chart can give warning of a shift in the process by displaying a nonrandom pattern even when all sample points fall within the control limits. For example, a disproportionate number of points above the centerline may indicate the process mean is increasing. If every fifth point is high, there may be a systematic problem.

To detect nonrandom patterns, the runs test described in Ref. 8 is simple and useful. Count the number of points above the centerline and the number below. Let r be the smaller and s the larger. Now count N, the number of runs, that is, chains of points either above or below the centerline. For example, if all the points fall above the centerline, there is 1 run. If the first 10 points are above, then 2 below, then 5 above, there are 3 runs. Look up r and s in Table 2.1; find the critical number N^*. If $N \leq N^*$, there is evidence of nonrandom behavior. Apply the runs test to Fig. 2.2 to obtain $r = 9$, $s = 11$, $N = 12$, and $N^* = 6$ at the 0.05 level of significance. Since $N > N^*$, there is no evidence of nonrandom behavior. This test detects only large effects.

Control Charts for Attributes and Variables.   Control charts are classified as either *attribute* or *variable* charts. In attribute charts, including the P chart, the C chart, and the U chart, the data are integers, such as number of defectives or number of defects. A defective is an unacceptable item. In contrast, an item with one or several defects, such as a painted surface with a few small bubbles, may be perfectly acceptable. For variable charts, such as

TABLE 2.1 Testing Randomness—The Runs Test

							r								
s	6	7	8	9	10	11	12	13	14	15	16	17	18	19	20
6	3														
7	4	4													
8	4	4	5												
9	4	5	5	6											
10	5	5	6	6	6										
11	5	5	6	6	7	7									
12	5	6	6	7	7	8	8								
13	5	6	6	7	8	8	9	9							
14	5	6	7	7	8	8	9	9	10						
15	6	6	7	8	8	9	9	10	10	11					
16	6	6	7	8	8	9	10	10	11	11	11				
17	6	7	7	8	9	9	10	10	11	11	12	12			
18	6	7	8	8	9	10	10	11	11	12	12	13	13		
19	6	7	8	8	9	10	10	11	12	12	13	13	14	14	
20	6	7	8	9	9	10	11	11	12	12	13	13	14	14	15

s = cases on one side of average
r = cases on other side of average $\Big\}$ r always taken as the smaller number of cases; s the larger

Source: From Swed and Eisenhart, "Tables for Testing Randomness of Grouping in a Sequence of Alternatives," *Annals of Mathematical Statistics,* vol. 14, 1943, pp. 66–87, and Ref. 8, p. 970.

the $\bar{x}$ chart and the R chart, the data are continuous measurements of a dimension, for example.

When defectives or defects are counted for attribute charts, it is often possible to note the reason the item is defective or the type of defect that occurred. Then a Pareto chart can be constructed to identify the most important defect types, and a fishbone diagram can be used to trace the causes of the defects.[15]

P Charts for Fraction Defective.   P charts are attribute charts used when each item produced can be classified as defective or nondefective. P charts are applicable, for example, when controlling processes that produce glass containers, assemblies, and integrated circuits.

The statistic that is plotted on a P chart is the sample fraction defective. Count the number of defectives X in a sample of n items. The sample fraction defective is X/n.

To construct a P chart, assume the number of defectives in a sample of n has the binomial distribution. Assume the outcome of each item, defective or nondefective, is independent of all other items and, for each item, the probability it is defective is p.

To establish the centerline and control limits for a new control chart, obtain baseline data, design an initial control chart, and revise the chart iteratively until it is correct. The steps are given below:

STEP 1.   For the initial control chart, take m samples, each of size n. Denote the number of defectives in the m samples by $X_1, X_2, \ldots, X_m$.

STEP 2.   Compute an estimate $\bar{p}$ for the process fraction defective, based on all the baseline data, a total of mn items:

$$\bar{p} = \left(\sum_{i=1}^{n} X_j \right) \Big/ mn \qquad (3)$$

STEP 3.   Compute the centerline (CL), the upper control limit (UCL), and the lower control limit (LCL) of the P chart as follows:

$$CL = \bar{p}$$

$$UCL = \bar{p} + 3\sqrt{\frac{\bar{p}(1 - \bar{p})}{n}}$$ (4)

$$LCL = \bar{p} - 3\sqrt{\frac{\bar{p}(1 - \bar{p})}{n}}$$

where $\bar{p}$ is given in Eq. (3), n is the sample size, and 3 is the 99.87 percentile point on the standard normal distribution. If LCL is negative, set it equal to 0. So-called three-sigma control charts are most commonly used.[19]

STEP 4.   Construct the P chart and plot the m sample fraction defectives $X_1/n, X_2/n, \ldots,$ X_m/n. Investigate the points outside the control limits. If no assignable causes are found, use the present P chart and debug the process until it is in control. If assignable causes are found, revise the P chart.

STEP 5.   Revise the control chart by removing sample fraction defectives outside the control limits for which assignable causes have been identified. Go back to step 2 noting the sample size is n but the number of samples, m, is reduced.

Example.   A light bulb manufacturer wishes to construct a P chart for the fraction of defective bulbs. Twenty samples, each containing 150 bulbs, are tested. The data are shown in Table 2.2.

The total number of defectives in the 20 samples is 184. The initial estimates for the process fraction defective from Eq. (3) and the control limits from Eq. (4) are

$$\bar{p} = 184/(20)(150) = 0.0613 = CL$$

$$UCL = 0.0613 + 3\sqrt{\frac{(0.0613)(0.9387)}{150}} = 0.1201$$

$$LCL = 0.0613 - 3\sqrt{\frac{(0.0613)(0.9387)}{150}} = 0.0026$$

Figure 2.2 shows the P chart. Sample number 10 is outside the control limits. A review of that sample indicates a new operator was assigned to the machine for that shift. Remove sample number 10, which leaves $m = 19$, and revise the control chart, that is, CL = 165/(19)(150) = 0.0579, UCL = 0.1151, and LCL = 0.0007. In the revised control chart, no points fall outside the control limits. Put this P chart in place and monitor it.

TABLE 2.2   Data Set for Initial P Chart—Sample Size 150

Sample number	Number defective	Fraction defective	Sample number	Number defective	Fraction defective
1	11	0.073	11	10	0.067
2	8	0.053	12	10	0.067
3	12	0.080	13	9	0.060
4	8	0.053	14	6	0.040
5	6	0.040	15	7	0.047
6	9	0.060	16	10	0.067
7	11	0.073	17	5	0.033
8	7	0.047	18	10	0.067
9	10	0.067	19	8	0.053
10	19	0.127	20	8	0.053

Selecting the Sample Size for a P Chart.   Evaluate the choice of sample size n according to the two criteria given below and economic considerations. One criterion is to choose n such that UCL is positive; for three-sigma control charts,

$$n \geq 3^2(1 - \bar{p})/\bar{p} \tag{5}$$

Applying Eq. (5) to the example, find $n \geq 9(0.9421)/0.0579 \cong 147$.

Another criterion is to choose n such that there is a 50 percent chance of detecting a shift of magnitude Δ in the process fraction defective; for three-sigma control charts,

$$n \geq (3/\Delta)^2 \bar{p}(1 - \bar{p})$$

In the example, if $\Delta = 0.05$, then $n \geq 196$; if $\Delta = 0.02$, then $n \geq 1227$. The current sample size of 150 yields a control chart that is sensitive to large shifts in the process fraction defective.

Selecting the Number of Samples for the Initial P Chart.   To evaluate the choice of m, the number of samples used to construct the initial control chart, compute a 95 percent confidence interval for the estimate of the process fraction defective:

$$\bar{p} \pm 1.96 \sqrt{\frac{\bar{p}(1 - \bar{p})}{nm}}$$

where $\bar{p}$ is given in Eq. (3), 1.96 is the 97.5 percentile point of the standard normal distribution, and we assume $mn \geq 100$. In the example, the confidence interval is

$$0.0579 \pm 1.96 \sqrt{\frac{(0.0579)(0.9421)}{(19)(150)}}$$

If this interval is too wide, increase m.

C Charts and U Charts for Number of Defects.   C charts and U charts are attributes charts used when quality is measured by counting the number of defects on an item. Examples include the number of dust particles in a sample of a liquid chemical and the number of scratches on the painted finish of an item.

The sample statistic plotted on a C chart is the number of defects in the sample, D. The sample statistic plotted on a U chart is the sample number of defects per unit. To compute it, count the total number of defects D in the sample and divide by the number of units in the sample, n. Suppose there are 20 dust particles in a sample of 1.5 cm^3 of liquid chemical. On the C chart plot 20; on the U chart plot $20/1.5 = 13.33$ dust particles per cubic centimeter.

For the C chart and U chart, we assume the number of defects per unit is Poisson-distributed with mean u and the number of defects in a sample of n units is Poisson-distributed with mean $c = un$. Also, the number of defects in any sample is independent of the number of defects in other samples and the expected number of defects is equal for all samples of the same size. For a study of defects that are not Poisson-distributed see Ref. 11.

The iterative process for constructing and revising C charts and U charts is the same as for a P chart. Thus we give the formulas in the following sections without repeating the iterative procedure.

To construct a C chart, take m equal samples of material and denote the number of defects by $D_1, D_2, \ldots, D_m$. Compute an estimate $\bar{c}$ for the process mean number of defects per sample,

$$\bar{c} = \sum_{j=1}^{m} Dj/m \qquad (6)$$

The CL and three-sigma control limits UCL and LCL are

$$CL = \bar{c}$$

$$UCL = \bar{c} + 3\sqrt{\bar{c}}$$

$$LCL = \bar{c} - 3\sqrt{\bar{c}}$$

where $\bar{c}$ is given in Eq. (6) and 3 is the 99.87 percentile point on the standard normal distribution. Set LCL equal to 0 if it is negative.

To construct a U chart, take m samples of material, each consisting of n units and denote the number of defects by $D_1, D_2, \ldots, D_m$. Compute an estimate $\bar{u}$ for the process mean number of defects per unit,

$$\bar{u} = \sum_{j=1}^{m} D_j/mn \qquad (7)$$

The CL and three-sigma control limits UCL and LCL are

$$CL = \bar{u}$$

$$UCL = \bar{u} + 3\sqrt{\bar{u}/n}$$

$$LCL = \bar{u} - 3\sqrt{\bar{u}/n}$$

where $\bar{u}$ is given in Eq. (7) and 3 is the 99.87 percentile point on the standard normal distribution. Set LCL equal to 0 if it is negative.

To select the sample size for a C chart or U chart, we present two criteria. The first is to choose n such that UCL is positive. For three-sigma control charts, $n \geq 3^2/\bar{u}$. The second is to choose n such that there is a 50 percent chance of detecting a shift of magnitude Δ in the process mean number of defects per unit. For three-sigma control charts, let $n \geq (3/\Delta)^2 \bar{u}$.

$\bar{x}$ and R Charts for Variables. $\bar{x}$ Charts are used to control the mean dimension of items produced by a process, and R charts are used to control the standard deviation. (The R is for range.) These charts are used together.

We assume that all measurements made when the process is in control are independent and normally distributed with the same mean and standard deviation. Further, we assume the mean and the standard deviation are independent.

To use an $\bar{x}$ chart, measure a sample of n items at the required time. Plot the sample average on the $\bar{x}$ chart and the sample range on the R chart. If the measurements are $X_1, X_2, \ldots, X_n$, then, for sample j, the sample average is

$$\bar{X}_j = \sum_{i=1}^{n} X_i/n$$

and the sample range R_j is the difference between the largest and smallest of the n sample measurements. The iterative procedure for constructing these charts is the same as described earlier.

To construct $\bar{x}$ charts and R charts, one must obtain a table of factors for constructing variable control charts that is available in most quality control texts.[4,8,12,15,19] Take m samples each consisting of n measurements and denote the sample averages by $\bar{X}_1, \bar{X}_2, \ldots, \bar{X}_m$ and the sample ranges by $R_1, R_2, \ldots, R_m$. Compute an estimate $\bar{\bar{x}}$ for the process mean,

$$\bar{x} = \sum_{j=1}^{m} \bar{X}_j/m \tag{8}$$

The estimate for the process standard deviation σ is

$$\bar{\sigma} = \bar{R}/d_2 \tag{9}$$

where

$$\bar{R} = \sum_{j=1}^{m} R_j/m$$

and d_2 can be found in the table of factors for various sample sizes n between 3 and 25. [For control chart purposes, the above method of estimating the standard deviation is better than the usual estimate S in Eq. (16), since it is not necessary that the process mean remain constant.]

The CL and three-sigma control limits UCL and LCL for the $\bar{x}$ chart are

$$CL = \bar{x}$$

$$UCL = \bar{x} + 3\sqrt{\bar{\sigma}/n}$$

$$LCL = \bar{x} - 3\sqrt{\bar{\sigma}/n}$$

where $\bar{x}$ is given by Eq. (8) and $\bar{\sigma}$ is given by Eq. (9). For convenience, one can compute the control limits as follows:

$$UCL = \bar{x} + A_2\bar{R} \quad \text{and} \quad LCL = \bar{x} - A_2\bar{R}$$

where A_2 is in the table of factors, given n.

For the R chart, the centerline and control limits are

$$CL = \bar{R} \quad UCL = \bar{R}D_4 \quad LCL = \bar{R}D_3$$

where D_3 and D_4 are in the table of factors, given n.

Example. Dimension data has been collected for 10 samples, each consisting of 5 measurements. The sample averages are 11.3, 12.2, 11.9, 11.4, 12.0, 11.9, 11.5, 11.8, 11.9, 11.7. The sample ranges are 0.6, 1.5, 0.9, 0.7, 1.4, 0.4, 0.6, 0.4, 0.4, 0.6. The number of samples m is 10. The sample size n is 5. We have

$$\bar{x} = (11.3 + 12.2 + \cdots + 11.7)/10 = 11.76$$

$$\bar{R} = (0.6 + 1.5 + \cdots + 0.6)/10 = 0.75$$

The centerline and control limits for the $\bar{x}$ chart, with $A_2 = 0.577$ for $n = 5$ from the table of factors, are CL = 11.76, UCL = 11.76 + (0.577)(0.75) = 12.193 and LCL = 11.76 − (0.577)(0.75) = 11.327. For the R chart, the centerline and control limits, with $D_4 = 2.114$ and $D_3 = 0$, are CL = 0.75, UCL = (0.75)(2.114) = 1.585, and LCL = (0.75)(0) = 0. The first two sample averages are below and above the control limits, respectively. These points must be investigated and removed if necessary, and the control limits revised.

Assume, in this example, that the $\bar{x}$ chart with CL = 11.76, UCL = 12.193, and LCL = 11.327 is successfully in place. A customer requires that the dimension has a tolerance ±0.5. The customer reviews the control charts and sees that control limits and the sample points are all within 11.76 ± 0.4. The knowledgeable customer recognizes that only 84 percent of the parts would meet the customer's specification. Using elementary probability, compute the probability that a single part, with measurement denoted by the random variable X, falls within the specification, that is, the interval 11.26 to 12.26. We have

$$\Pr(11.26 < X < 12.26) = \Pr(-1.41 < Z < 1.41) = 0.84$$

where Z is the standard normal random variable with mean 0 and standard deviation 1.

One may also determine that the process is not adequate to meet the specification by computing the process capability index in Eq. (2). First estimate v with $\bar{\sigma}$ from Eq. (9), assuming the target mean equals the process mean; thus $\sigma = 0.75/2.326 = 0.32$ since $d_2 = 2.326$. Then compute $C_p = 1.0/(6)(0.32) = 0.52$.

A common error is to confuse the control limits and the specifications. Remember the points plotted on the control chart are averages of n measurements. Specifications refer to one part at a time.

Choosing k-Sigma Control Limits.   In the control charts described so far, we have used three-sigma control limits with a 0.3 percent chance of a false alarm. This is the most common choice in the United States. In certain applications where the consequence of a false alarm is not as great and the consequence of not detecting a shift are very great, it may be preferable to have a two-sigma limit, which corresponds to a false alarm chance of 5 percent, or a 2.5-sigma limit, which corresponds to a false alarm chance of 1.2 percent. Economic models in Ref. 19 describe equations to choose the optimal control limits but require detailed cost information.

In some applications two-sigma warning limits, in addition to three-sigma control limits, are added to the control chart. If two consecutive sample statistics fall between the two- and three-sigma limits, the process is considered to be out of control.

Detecting Small Shifts: The CUSUM Chart.   The Shewhart control charts are easy to understand and sensitive to large shifts in the process, say a shift of 2 standard deviations or more. However, they are not sensitive to small shifts in the process mean, and consequently such shifts may go undetected. The cumulative-sum (CUSUM) control chart overcomes this problem. On the other hand, CUSUM charts are less sensitive to large shifts and are harder to interpret.[19]

Problems in Control Charts.   If a control chart is not working, it is important to find out why. If a chart is constantly indicating false alarms, operators will soon ignore the control chart and the whole concept of process control loses credibility.

Problems are often caused by violations in the assumptions underlying the control charts. First, the sample statistics may be serially correlated, violating the assumption of independence stated for each of the control charts. Time-series models are sometimes used.[6,19] Second, the sample statistics may not have the assumed distribution. For work related to violations of normality assumptions, see Refs. 4 and 19.

LOT-BY-LOT SAMPLING

Lot-by-lot sampling is a decision-making tool to determine whether a lot or batch of material meets a standard. A sample of a specific size is randomly drawn from the lot and measured. The sample statistic computed from the measurements is compared to an acceptance number and the lot is accepted or rejected. Lot sampling is sometimes called *lot sentencing*.

Lot sampling can be used when a company receives material from an outside vendor or before a manufacturer releases material. It can also be used within a manufacturing process when one stage of production receives material from another stage or from work-in-process inventory.

Lot sampling should be kept to a minimum if SPC is implemented correctly and there is a program of constant improvement using off-line quality control. Compared to SPC and off-line quality control, lot sampling gives the smallest payoff in terms of improved product. Recently, many vendor certification programs require that the vendor provide evidence of good SPC practices in order to minimize the lot sampling by the consumer.

In lot sampling programs, vendors are informed of lots that are rejected. However, this information often cannot be directly used to correct manufacturing problems. It is unlikely that the reason for the problem can be identified after a considerable amount of time has elapsed and the product has been subjected to further handling and, possibly, further production steps.

One hundred percent inspection, an alternative to lot sampling, is necessary in some cases because of the critical nature of the product. Before implementing 100 percent inspection, one must be aware of the disadvantages: it is expensive; it encourages workers to depend on the inspection, rather than on their own work, for product quality; it extends the time that material spends on the factory floor; it increases work in process; and it adds another stage of handling. Because of inspection errors, there is no guarantee that the final result will be 100 percent nondefective. Machine vision and other computer-controlled devices for inspection may overcome some of the problems, as they become less expensive.

Producer's and Consumer's Risks.   There are two types of errors associated with lot sampling. The producer's risk α is the probability that a "good" lot is rejected. The consumer's risk β is the probability that a "bad" lot is accepted. Good lots come from a process where the fraction defective is equal to the acceptable quality level (AQL). The AQL is a satisfactory level for the purposes of lot sampling, and lots from such a process have a high probability of acceptance by the lot sampling plan, that is, in the range from 0.9 to 1.0. Bad lots have fraction defective equal to or lower than the lot tolerance percent defective (LTPD). In lots with this unacceptable level of quality, the probability of acceptance under a sampling plan is low, say less than 0.10. Other terms used for LTPD are *limiting quality level* (LQL) and *rejectable quality level* (RQL).

For a sampling plan to achieve the producer's and consumer's risks for which it was designed, samples must be randomly chosen from lots according to a carefully established procedure. The sampling can be done using random numbers.[12]

Knowledge of how the lot is formed can be helpful in selecting a sample. For example, if a lot consists of several identifiable production runs, it may be possible to stratify the sample, that is, choose randomly from each run.

Rectifying Inspection.   In rectifying inspection, a rejected lot is subjected to 100 percent inspection and the defective items are replaced by nondefective items. A measure of the performance of this type of sampling is average outgoing quality (AOQ), the average percent defective over all accepted lots including lots accepted after rectification. For example, suppose a process yields 100 percent defectives. The AOQ is 0 percent defective, since all lots will be rejected and then rectified. Given a sampling plan, AOQ can be computed for each process fraction defective. The maximum AOQ is called the average outgoing quality limit (AOQL).

Operating Characteristic Curves.   An operating characteristic (OC) curve is a graph in which the x axis is the process parameter and the y axis is the probability a lot is accepted for a particular sampling plan. The process parameter is usually a mean or a fraction defective. OC curves help evaluate the performance of sampling plans under different process means.

Figure 2.3 shows four OC curves corresponding to four sample plans with different sample sizes.[1] The probability of accepting a lot from a process with mean AQL is 0.95 for all the plans; the consumer's risks depend on the sample size.

Families of Lot Sampling Plans.   A family of sampling plans is designed with a particular objective. The families of attribute plans described here are (1) single and double sampling plans to ensure producer's and consumer's risks; (2) MIL-STD-105D; and (3) Dodge-Romig plans for rectifying inspection. We describe variable plans designed to ensure spe-

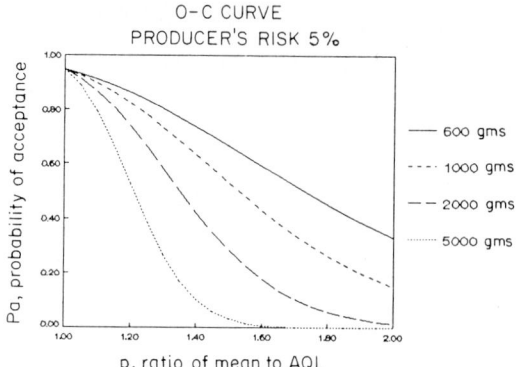

FIGURE 2.3   Operating characteristic curves for sampling plans with producer's risk of 5 percent. (*From S. L. Albin, "The Lognormal Distribution for Modeling Data When the Mean Is Near Zero,"* Journal of Quality Technology, *vol. 22, 1990, pp. 105–110.*)

cific producer's and consumer's risks and briefly mention MIL-STD-414. Finally we describe continuous sampling plans.

Single Sample Attribute Plans to Ensure Producer's and Consumer's Risks.   The objective of this family of sampling plans is to ensure that the producer's risk is α and the consumer's risk is β. If p is the process fraction defective, the requirements are

$$\text{Pr(accept lot}/p = \text{AQL)} = 1 - \alpha \tag{10}$$

$$\text{Pr(accept lot}/p = \text{LTPD)} = \beta \tag{11}$$

A lot is accepted if the number of defectives in a sample size n is less than or equal to c. We assume the number of defectives in a sample is binomially distributed and solve for n and c to satisfy Eqs. (10) and (11). (See Ref. 2 for a study in which the distribution is not binomial.) Since n and c are integers, it is often impossible to satisfy both requirements. Approximate solutions can be obtained conveniently from Table 2.3 or from a nomogram in Ref. 4.

Example.   The producer's risk is 0.05, the consumer's risk is 0.10, the AQL is 0.03, and the LTPD is 0.07. To find a single sampling plan, compute the ratio of the process fraction defective LTPD/AQL = 0.07/0.03 = 2.33. Check the last column of Table 2.3 to find the closest match, 2.31. From the first column, $c = 12$. The other columns give the product of n, the sample size, and p_f, the process fraction defective, such that the probability of acceptance is f. In column 3, the value of $np_{0.95}$ is 7.690. Since AQL = 0.03, $p_{0.95} = 0.03$ and $n = 257$. In column 6, the value of $np_{0.10}$ is 17.78. Since LTPD = 0.07, $p_{0.10} = 0.07$ and $n = 254$. The two solutions indicate there is no solution that satisfies both Eq. (10) and Eq. (11). One may choose according to which requirement is more important.

Double Sampling Plans for Attributes.   Double sampling plans reduce the number of items that must be inspected in the long run. Initially, a small sample is chosen, and, based on the results, the lot is accepted or rejected or an additional sample is taken to sentence the lot. Multiple and sequential sampling plans extend the idea of double sampling. These plans are each more difficult to administer than single sampling plans.

The double sampling plan is defined by four parameters, n_1, n_2, c_1, and c_2. The sample sizes for the first and second samples are n_1 and n_2. Accept the lot on the first sample if the number of defectives X_1 is less than or equal to c_1. Reject the lot on the first sample if X_1 is greater than c_2. If X_1 is greater than c_1 and less than or equal to c_2, take the second

TABLE 2.3 Values for Designing Single Sampling Plans for Attributes with α Approximately 0.05 and β Approximately 0.10

c	$p'n_{0.95}$	$p'n_{0.10}$	$p'n_{0.10}/p'n_{0.95}$ $= p'_{0.10}/p'_{0.95}$
0	0.051	2.30	45.01
1	0.355	3.89	10.96
2	0.818	5.32	6.50
3	1.366	6.68	4.89
4	1.970	7.99	4.06
5	2.613	9.28	3.55
6	3.285	10.53	3.21
7	3.981	11.77	2.96
8	4.695	12.99	2.77
9	5.425	14.21	2.62
10	6.169	15.41	2.50
11	6.924	16.60	2.40
12	7.690	17.78	2.31
13	8.464	18.96	2.24
14	9.246	20.13	2.18
15	10.04	21.29	2.12

Source: From F. E. Grubbs, "On Designing Single Sampling Inspection Plans," *Annals of Mathematical Statistics,* vol. 20, 1949, p. 256, and Ref. 8, p. 166.

sample. If the total number of defectives, $X_1 + X_2$, is less than or equal to c_2, then accept the lot. Otherwise reject.

One can compute the long-run average sample number (ASN) for a double sampling plan, given the process fraction defective. If the probability of making the decision to accept or reject on the first sample is P_1, then

$$\text{ASN} = n_1 + (n_1 + n_2)(1 - P_1) \tag{12}$$

Consider, again, the example described for the single sampling plan. To find a double sampling plan with $n_1 = n_2$, use Table 2.4. Other tables are available in Refs. 4 and 19. Compute $R = \text{LTPD/AQL} = 0.07/0.03 = 2.33$. Check the second column of the table to find the closest match, plan 15. Find $c_1 = 5$ and $c_2 = 13$ in the third and fourth columns. The fifth column gives the product of n_1 and p, the process fraction defective, such that the probability P of acceptance is 0.95. Thus $4.35 = n_1(0.03)$ or, equivalently $n_1 = n_2 = 145$. The seventh column gives the product of n_1 and p, the process fraction defective, such that the probability of acceptance is 0.10. Thus $10.08 = n_1(0.07)$ or, equivalently, $n_1 = n_2 = 144$.

Suppose, in the example, the process fraction defective is 0.028, a bit lower than the AQL, and $n_1 = 144$. Then P_1 is the probability there are less than or equal to 5 defectives or greater than 13 defectives out of 123 items in the first sample. From the binomial probability distribution, $P_1 = 0.786$. Then ASN equals 176 from Eq. (12). In contrast, the ASN for the single sampling plan was 254. For low and high values of the process fraction defective, the ASN is lower for double sampling plans. One may graph the ASN versus the process fraction defective.

Military Standard 105D for Attribute Sampling.   This system of sampling plans was developed during World War II, and the most recent revision, MIL-STD-105D, was adopted in 1971. The plans are known as ANSI/ASQC Z1.4 for civilian use and ISO/DIS-2859 in the international community. The full set of tables is available in Ref. 19.

MIL-STD-105D is based primarily on ensuring the producer's risk for a given AQL. The consumer's risk is protected largely through the choice of the inspection level. The

TABLE 2.4 Values for Designing Double Sampling Plans for Attributes with α Approximately 0.05 and β Approximately 0.10 and $n_1 = n_2$

| Plan number | R | Acceptance numbers | | Approximate values of p', n_1 for | | | Approximate ASN/n_1 for 0.95 point |
		c_1	c_2	$P = 0.95$	$P = 0.50$	$P = 0.10$	
(1)	(2)	(3)	(4)	(5)	(6)	(7)	(8)
1	11.90	0	1	0.21	1.00	2.50	1.170
2	7.54	1	2	0.52	1.82	3.92	1.081
3	6.79	0	2	0.43	1.42	2.96	1.340
4	5.39	1	3	0.76	2.11	4.11	1.169
5	4.65	2	4	1.16	2.90	5.39	1.105
6	4.25	1	4	1.04	2.50	4.42	1.274
7	3.88	2	5	1.43	3.20	5.55	1.170
8	3.63	3	6	1.87	3.98	6.78	1.117
9	3.38	2	6	1.72	3.56	5.82	1.248
10	3.21	3	7	2.15	4.27	6.91	1.173
11	3.09	4	8	2.62	5.02	8.10	1.124
12	2.85	4	9	2.90	5.33	8.26	1.167
13	2.60	5	11	3.68	6.40	9.56	1.166
14	2.44	5	12	4.00	6.73	9.77	1.215
15	2.32	5	13	4.35	7.06	10.08	1.271
16	2.22	5	14	4.70	7.52	10.45	1.331
17	2.12	5	16	5.39	8.40	11.41	1.452

Source: From Chemical Corps Engineering Agency, Manual No. 2, *Master Sampling Plans for Single, Dupli cate, Double and Multiple Sampling* (Army Chemical Center, Md., 1953), and Ref. 8, p. 189.

normal inspection level is II. More critical products are in level III and require samples twice as large as level II. Inexpensive products are in level I and require sample sizes one-half as large as level II. There are four additional levels for small sample sizes, S1 to S4, used for products where the testing is expensive or destructive and large sampling risks are allowable. The inspection level is a function of the product and the inspection test, not the producer's performance.

An important component of MIL-STD-105D is the switching rules to indicate whether a producer is subject to normal, reduced, or tightened sampling. According to the produc-er's recent performance, the level of sampling is changed according to specific rules. For example, if normal sampling is in place and two out of five lots have been rejected, then tightened sampling will be imposed. To return to normal sampling, five consecutive lots must be accepted.

MIL-STD-105D, in wide use, has been subject to some criticism.[8,19] One problem is that the producer's risk ranges from 1 to 9 percent for a given AQL, which can be expen-sive and unfair to a producer that meets the AQL. Also, there is some frequent, unjustified switching from normal to tightened to reduced sampling.

Dodge-Romig Plans for Attributes.   These plans are for applications where rectifying in-spection is in place. These plans are based primarily on ensuring the consumer's risk and the LTPD.

To use the plans, the process fraction defective and the lot size must be known. The plans minimize the expected number of items that must be inspected, given a requirement on either the AOQL or the LTPD. A full set of tables can be found in Ref. 19.

Variable Sampling Plans to Ensure the Producer's and Consumer's Risks.   The objective of these plans is to sentence lots where quality is defined as a specification on a measurement. An item that does not meet the specification is defective. The specification is either a lower specification limit, an upper specification limit, or both. The two requirements for the sampling plan are stated in the same terms as attribute sampling, as in Eqs. (10) and (11).

Variable sampling plans require smaller samples than attribute plans that achieve the same producer's and consumer's risks. The difference is especially great when the AQL is small. The disadvantages of variable sampling are the possible extra per-unit measurement cost and the need for a separate sampling plan for each dimension or feature of the product.

Variable sampling plans assume the measurements have a normal distribution. It is necessary to examine data to ensure that this assumption is reasonable. One method for checking the data is to plot the data on normal paper.[4]

The sample plan consists of the sample size n and the critical number k. A sample of n is measured and the sample average $\overline{X}$ is computed. Accept the lot if

$$\text{LSL only:} \qquad \frac{\overline{X} - \text{LSL}}{\sigma} \geq k \qquad\qquad (13)$$

$$\text{USL only:} \qquad \frac{\overline{X} - \text{USL}}{\sigma} \leq k \qquad\qquad (14)$$

$$\text{LSL and USL:} \qquad \frac{\overline{X} - \text{LSL}}{\sigma} \geq k \quad \text{and} \quad \frac{\overline{X} - \text{USL}}{\sigma} \leq k \qquad\qquad (15)$$

Given AQL, LTPD, α, β, and σ, the standard deviation of the normal distribution, the formulas for n and k are

$$n = \left[\frac{Z_\beta - Z_{1-\alpha}}{Z_{\text{AQL}} - Z_{\text{LTPD}}} \right]^2$$

$$K = \frac{k_1 + k_2}{2}$$

where

$$k_1 = \frac{Z_{1-\alpha}}{\sqrt{n}} + Z_{\text{AQL}} \qquad \text{and} \qquad k_2 = \frac{Z_\beta}{\sqrt{n}} + Z_{\text{LTPD}}$$

and Z_p is the point on the standard normal curve such that the area to the right is p.

If σ is unknown, compute the sample estimate S from the sample measurements X_1, $X_2, \ldots, X_n$:

$$S = \left[\frac{\sum_{i=1}^{n} (X_i - \overline{X}^2)}{n - 1} \right]^{1/2} \qquad\qquad (16)$$

Substitute S for σ in Eqs. (13), and (14), and (15) to obtain the acceptance rules. The formulas for k and n are

$$k = \frac{Z_\beta Z_{\text{AQL}} - Z_{1-\alpha} Z_{\text{LTPD}}}{Z_\beta - Z_{1-\alpha}}$$

$$n = \left[\frac{Z_\beta - Z_{1-\alpha}}{Z_{\text{AQL}} - Z_{\text{LTPD}}} \right]^2 \left[1 + \frac{k^2}{2} \right]$$

Note the sample size is larger by a factor $1 + k^2/2$ when σ is unknown.

Military Standard 414 for Variable Sampling.   This military standard, also adopted for commercial use as ANSI/ASQC Z1.9, is primarily based on protecting the producer's risk. It has similar features to MIL-STD-105D; tables are in Ref. 19.

Continuous Sampling for Attributes.   When the product is produced continuously, lot-by-lot sampling may be impractical. Therefore continuous sampling plans have been devised. These plans are useful for units produced on an assembly line, for example.

Dodge, in 1943, introduced the first continuous sampling plan, CSP-1, defined by i, the clearance number, and f, the fraction sampled. Initially, test all units until i nondefective units are found. Then test a fraction f of the units, selected randomly. Resume 100 percent testing if a defective is found. Tables in Ref. 19 give the CSP-1 plans and the associated AOQL.

An extension of CSP-1 is CSP-2, where 100 percent inspection is resumed when 2 defectives are found within a space of K sample units of each other. Usually $K = i$. Military standard, MIL-STD-1235B also gives continuous sampling plans.

THE LOSS FUNCTION

One of the common features among all statistical methods for evaluating process capability (that is, control charts and C_p index) is that these methods do not directly quantify the economic effects of process or produce deviation from the target values. In this section, the loss function method is presented. The loss function method is based on the concept that loss is incurred when a product's functional quality characteristic (denoted by y) deviates from its target (nominal) value (denoted by m), regardless of how small the deviation is. Figure 2.4 shows the relationship between quality loss and the amount of deviation from the target value. As shown in this figure, quality loss caused by deviation equals zero when $y = m$; the loss increases when the value of the functional characteristic moves in either an upward or downward direction from m. When the value of the functional characteristic exceeds either of the limits $m + \Delta$ and $m - \Delta$ (where 2Δ is the tolerance), the quality loss at these points is equal to the loss of the product's disposal or rework.

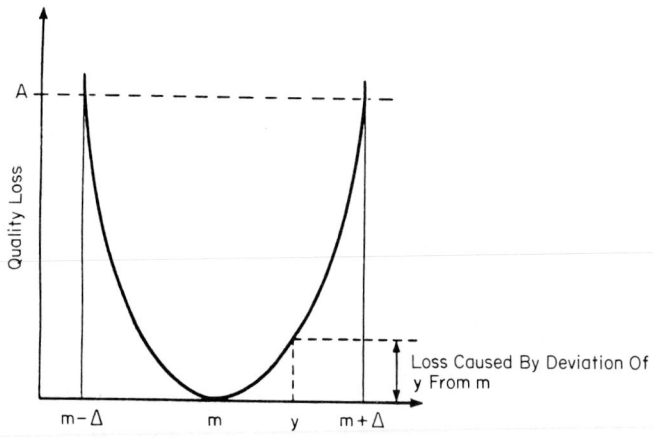

FIGURE 2.4   Relationship between quality loss and deviation from the target value m. (*Reprinted with permission from Ref. 22.*)

Derivation of the Loss Function.   Assume the loss of a defective part (because of discarding, repairing, or downgrading) to be A. Then denote the loss function by $L(y)$ and expand it, using Taylor's expansion. The following is obtained:

$$L(y) = L(m + y - m) \qquad (17)$$

or

$$L(y) = L(m) + \frac{L'(m)}{1!}(y - m) + \frac{L''(m)}{2!}(y - m)^2 + \cdots$$

Because $L(y) = 0$ when $y = m$ and the minimum value of the function is attained at this point (Fig. 2.4), its first derivative $L'(m) = 0$. The first two terms of Eq. (17) are equal to zero. If terms of $(y - m)$ with powers higher than 2 are neglected, Eq. (17) reduces to

$$L(y) = \frac{L''(m)}{2!}(y - m)^2 \qquad (18)$$

$$L(y) = k(y - m)^2$$

where k is a constant and its value is obtained as follows: When the deviation of a product's functional characteristic equals an amount Δ from the target value (m), the loss equals A. From Eq. (18), the following is obtained:

$$A = k\Delta^2 \qquad (19)$$

$$k = A/\Delta^2$$

As shown above, the loss function given by Eq. (18) is a quadratic relationship obtained by using Taylor's expansion. A different expression of the loss function can be used if such a function is available. The use of the loss function is now illustrated in estimating the loss caused by deviations from the target value.

Example.   Assume that the cost of repairing a failed product in the factory is 4 dollars. The specification for the target value is $m \pm 5$. Compare the losses caused by deviations from the target value in terms of two identical products, one is produced in factory A where deviations around the target follow the normal distribution, and the other is produced in factory B where the deviations follow a uniform distribution bounded by the lower and upper specification limits.

To calculate the losses caused by deviation from target, we need to determine the constant k in Eq. (18). Since $\Delta = 5$ and $A = 4$, from Eq. (19) we obtain

$$k = \frac{A}{\Delta^2} = \frac{4.0}{5^2} = 0.16 \qquad (20)$$

The expected loss L caused by deviation in the production of factory A is obtained by taking the expectation of $L(y)$ in Eq. (18). Thus,

$$L = kv^2$$

$$L_A = 0.16(10/6)^2 = 0.444 \text{ dollar}$$

The loss caused by deviation in the production of factory B is

$$L_B = 0.16\left(\frac{10}{\sqrt{12}}\right)^2 = 1.333 \text{ dollars}$$

where $v = 10/6$ when the product characteristic follows a normal distribution and $C_p = 1$, and $v = 10/12$ when the product characteristic follows a uniform distribution and $C_p = 1$.

A summary of losses caused by deviation for products produced in both factories is

TABLE 2.5   Quality Comparison between Two Manufacturers

Manufacturer	Average	Standard deviation	v^2	Expected loss L	Fraction Defective, %
Factory A	m	$\dfrac{10}{6}$	$\dfrac{100}{36}$	0.444	0.27*
Factory B	m	$\dfrac{10}{\sqrt{12}}$	$\dfrac{100}{12}$	1.333	0.00

*Obtained by using the standard normal distribution.

given in Table 2.5, which indicates that the losses caused by deviation in terms of products produced in factory B are approximately 3 times those of the same products produced in factory A, despite the fact that factory B has zero fraction defective.

Relationship between C_p and the Loss Function.   The loss function can be utilized in evaluating the effect of improving the process capability index on loss caused by deviation. For example, assume that factory A has improved the process capability index so that a new standard deviation of 10/8 is attained. What would be the losses caused by deviation from the target value?

With the use of Eq. (18) and $k = 0.16$ from Eq. (20), the following is obtained:

$$L = kv^2$$

$$L = 0.16\left(\frac{10}{8}\right)^2 = 0.250$$

The loss per unit of production would be decreased from 0.333 dollar (current process) to 0.250 dollar, resulting in 0.083 dollar savings per unit. This result can also be obtained by using the following relationship between the loss function and the process capability index:

$$L_1 = kv_1^2 \qquad \text{(loss of current process)} \tag{21}$$

$$L_2 = kv_2^2 \qquad \text{(loss of improved process)} \tag{22}$$

Divide Eq. (21) by Eq. (22) to obtain

$$\frac{L_1}{L_2} = \frac{v_1^2}{v_2^2} \tag{23}$$

But

$$C_{p_1} = \frac{\text{tolerance}}{6v_1} \tag{24}$$

and

$$C_{p_2} = \frac{\text{tolerance}}{6v_2} \tag{25}$$

By substituting Eqs. (24) and (25) into Eq. (23), obtain

$$\frac{L_1}{L_2} = \frac{C_{p1}^2}{C_{p2}^2} \tag{26}$$

which implies that the losses caused by deviation are reciprocally proportional to the squares of the C_p indices. It is sufficient to determine the loss and the C_p index at the beginning of the process. Then, the loss after the changes in the process capability can be determined at any production period thereafter by using the C_p index at the desired production period and by substituting it in Eq. (26).

Uses of the Loss Function.   The loss function can also be used to determine the optimal diagnosis period of the process, the optimal adjustment interval for the process parameters, and the acceptable deviation in the product's characteristic before a corrective action is taken. It can be used as a tool to determine the sensitivity of the product's characteristics to the parameters of the on-line quality control system.[12]

SENSORS AND SOFTWARE FOR ON-LINE QUALITY CONTROL

One of the most important steps in implementing on-line quality control systems is the ability to monitor and control the manufacturing process parameters during processing. Effective control requires accurate methods for measuring and evaluating the quality characteristics of the product under control. The information about the quality characteristics is fed back to the process control system (closed-loop systems are preferred) which adjusts the parameters of the process to ensure that the deviations of the product's quality characteristics from the target values are reduced to a minimum.

Sensors are the key element in providing the quality characteristic information to the process control system. Sensors are usually used in three general areas of monitoring processes. The first area is production monitoring, where sensors are utilized to determine the status of operations on the production floor, such as the total number of units produced, number of defective products, number of available machines, and so on. The second area of sensor application is machine and process monitoring, where information about the process parameters is obtained and used to make adjustments of the parameters. In the third area, sensors are used to monitor the environmental conditions of the process such as humidity, temperature, and ventilation.[2] On-line quality control requires the use of sensors in the three areas.

Choosing Sensors.   Performance requirements for sensors involve the need to detect the presence/absence, position, condition, or identity of an object. These broad categories may overlap with each other since a sensor may be applicable in multiple areas, depending on the manner in which it is employed.[2]

In general, sensors employed to evaluate an attribute quality characteristic, such as the presence or absence of a label from a can, a hole in a workpiece, or the color contrast on a product, are less complex than those sensors which provide measurements about the variable quality characteristic of a product. The output of the former sensors is usually a digital or on-off signal because there are only two conditions of interest in this situation. The latter sensors can be used to inspect a product or to measure the parameters associated with the production process. Examples of the product characteristics that might be measured include size, shape, chemical composition, mass, concentricity, flatness, surface finish, and color.[2]

It is important that the quality characteristics of the product and the parameters of the process are accurately measured. Therefore, in choosing a sensor, we recommend three steps. In the first step, the quality characteristic or process parameter must be classified as an attribute or a variable. Examples of the product characteristics that might be measured include size, shape, chemical composition, mass, concentricity, flatness, color, and surface finish. In the second step, appropriate characteristics of the sensors such as error, accuracy, sensitivity, reproducibility, and resolution need to be determined. We define these characteristics as follows:[7]

Error: The algebraic difference between the indicated value by the sensor and the true (or actual) value of measured variable. The smaller the error, the better the sensor.

Accuracy: The maximum overall error to be expected from a sensor. Accurate sensors have low overall error values.

Sensitivity: A measure of the change in output of a sensor (or instrument) for a change in input. Generally speaking, high sensitivity is desirable in a sensor because a large change in output for a small change in input implies that a measurement may be taken easily.

Reproducibility: The ability of the sensor or a device to repeatedly reproduce the same output value for a given input.

Resolution: The minimum measurable value of the input variable. This characteristic of the sensor can be changed only through the redesign of the sensor.

The third step to be considered in selecting a sensor is the data acquisition system to be used. The function of the data acquisition system is to provide the data (digital or analog) from sensors to the controller (usually a computer), which in turn transforms it to corresponding values of characteristics or process parameters. These values are then analyzed and the results are fed back (or fed forward) through the controller to the process where the parameters are adjusted when needed. Some sensors require a simple input/output interface with the controller while others require rather complex and elaborate interfaces. The acquisition system should also allow for multiple characteristics to be measured simultaneously.

Types of Sensors.   There is a wide range of sensors available for use in on-line quality control systems. They can be classified according to either the characteristic to be measured (force, dimension, viscosity, and so on), or the underlying principle of sensing (such as motion and force sensors, fluid sensors, moisture and humidity sensors, light and radioactivity sensors, temperature sensors, thickness sensors, proximity sensors, density and specific gravity sensors, and chemical sensors).[1,6,7,9,10]

The recent technological advances in computers, fiber optics, computer vision, and lasers have led to the development of highly accurate measuring devices capable of performing simultaneous measurements at high speeds. This has resulted in a wider use of such devices to perform on-line quality control. With the increase in automation and the demand for high-quality products, on-line quality control systems will be commonplace in production and manufacturing facilities.

Software.   The best source for current, commercially available software for all aspects of quality is the annual quality assurance/quality control software directory issue of *Quality Progress*. The March 1990 issue lists 170 suppliers and for each describes up to three software products. The products are for computers from PCs to mainframes. The prices range from under $100 to tens of thousands of dollars.

The categories of software listed in *Quality Progress* are calibration, capability studies, data acquisition, gauge repeatability and reproducibility, inspection, management, measurement, quality assurance for software development, quality costs, reliability, sampling, simulation, statistical methods, and "other."

Computer tools are helpful in achieving the goal of on-line process control. However, recording, manipulating, and storing data does not in itself improve quality. Managers, engineers, and operators must put their knowledge of their manufacturing systems to use. In statistical process control, for example, it is recommended that manual control charts be implemented before computerizing. Many Japanese firms use only manual SPC, believing that this enhances the feedback loop of control chart and system correction.

REFERENCES

1. Albin, S. L., "The Lognormal Distribution for Modeling Quality Data When the Mean is Near Zero," *Journal of Quality Technology,* vol. 22, 1990, p. 105.

2. Albin, S. L., and D. J. Friedman, "The Impact of Clustered Defect Distributions in IC Fabrication," *Management Science,* vol. 35, 1989, p. 1066.

3. Andreson, N. A., *Instrumentation for Process Measurement and Control,* Chilton Book, Radnor, Pennsylvania, 1980.

4. Banks, J., *Principles of Quality Control,* Wiley, New York, 1989.

5. Barkman, W. A., *In-Process Quality Control for Manufacturing,* Marcel Dekker, New York, 1989.

6. Box, G. E. P., and G. M. Jenkins, *Time Series Analysis, Forecasting and Control,* 2d ed., Holden-Day, San Francisco, 1976.

7. Case, K. E., and J. S. Bigelow, "Capability and Performance Indices: Proper Use in the Process Industries," *American Chemical Society Meeting,* Cincinnati, October 18–21, 1988.

8. Duncan, A. J., *Quality Control and Industrial Statistics,* 4th ed., Richard D. Irwin, Homewood, Ill., 1974.

9. Elmaghraby, S. E., "On Quality, Automation, Cultural Relativism, and Things like That," *The RJR Nabisco, Inc., Award, Distinguished Lecture Series,* North Carolina State University, Raleigh, N.C., 1987.

10. Feigenbaum, A. V., *Total Quality Control,* McGraw-Hill, New York, 1983.

11. Friedman, D. J., and S. L. Albin, "Clustered Defects in IC Fabrication: Impact on Process Control Charts," *IEEE Transactions on Semiconductor Manufacturing,* vol. 4, pp. 36–42, 1991.

12. Grant, E. L., and R. S. Leavenworth, *Statistical Quality Control,* 6th ed., McGraw-Hill, New York, 1988.

13. Herceg, E. E., *Handbook of Measurement and Control,* Schaevitz Engineering, Pennsauken, NJ, 1976.

14. Johnson, C. D., *Process Control Instrumentation Technology,* Wiley, New York, 1988.

15. Juran, J. M., and F. M. Gryna, Jr., *Quality Planning and Analysis,* 2d ed., McGraw-Hill, New York, 1980.

16. Kane, V. E., "Process Capability Indices," *Journal of Quality Technology,* vol. 18, 1986, p. 41.

17. Lenk, J. D., *Handbook of Microcomputer-Based Instrumentation and Controls,* Prentice-Hall, Englewood Cliffs, N.J., 1984.

18. Liptak, B. G., and K. Venczel (eds.), *Instrument Engineers' Handbook,* Chilton Book, Radnor, Pennsylvania, 1982.

19. Montgomery, D. C., *Introduction to Statistical Quality Control,* Wiley, New York, 1985.

20. Montgomery, D. C., "The Effect of Nonnormality on Variables Sampling Plans," *Naval Research Logistics Quarterly,* vol. 32, 1985, p. 27.

21. Sullivan, L. P., "Reducing Variability: A New Approach to Quality," *Quality Progress,* vol. 17, 1984, p. 15.

22. Taguchi, G., E. A. Elsayed, and T. C. Hsiang, *Quality Engineering in Production Systems,* McGraw-Hill, New York, 1989.

CHAPTER 3
OFF-LINE QUALITY CONTROL

W. J. Kolarik
Industrial Engineering Department
Texas Tech University
Lubbock, Texas

Off-line quality begins in the conceptual stage of product and process development. It is critical in the product and process design and development phases. Proactive or preventive off-line quality activities are absolutely essential; a heavy dependence on reactive quality activities is not competitive in today's marketplace.[1] Off-line quality includes the conceptualization, design, execution, and analysis of product and process experiments to assure that high performance (as perceived by the customer) is delivered to the customer at minimal cost. Off-line quality efforts also address product support in the field, to assure that all customers obtain a full measure of product performance. In addition, continuous product or service improvements result from diligent off-line quality efforts. This chapter stresses systematic questioning, experimentation, and thorough analyses to support continuous improvement action in products and services and the processes that create them.

SYSTEMATIC DESIGN AND DEVELOPMENT

There are clearly people as well as technical issues involved in off-line quality practices. People and technical issues come together through leadership and a sound application of available resources to identify and take advantage of critical quality improvement opportunities. Thorough and systematic strategies and thinking are critical in off-line quality efforts.

Systematic Thinking. A systems-level perspective is critical for studying quality-related issues. For example, the coordination between product concept, design, and manufacturing, relative to producing a high-quality product, requires a broad perspective that cuts across a number of departmental boundaries. Hierarchical models and analysis tools are useful in encouraging systematic and thorough thinking.

Quality Function Deployment. The quality function deployment (QFD) method of systematically moving from customer requirements to detailed product specifications[2] is useful in guiding both total quality management and off-line quality efforts.[3,4] Figure 3.1 illustrates the basic concept of QFD. It is important to focus on the customer's perspective and develop the product or service around the customer's wants and needs, moving one

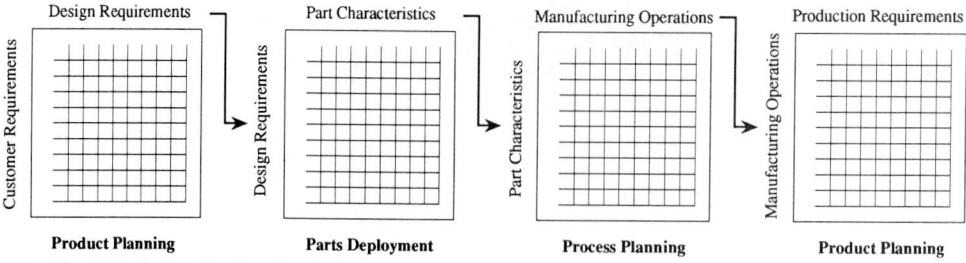

FIGURE 3.1   Quality function deployment concept.

step closer to product or service specifications in each step. For example, QFD seeks to systematically link the customer's requirements to systems-level design requirements, which are in turn linked to part characteristics, manufacturing operations, and finally production requirements. This linkage allows a better product (that customers want) to be developed faster and in a more cost-effective fashion, since product and production are considered simultaneously.

Goal Trees.   The goal tree is another hierarchical approach useful in off-line quality work. A goal tree approach is a positive (rather than a problem-driven negative) pragmatic method that can be used to provide a systems-level view and focus in off-line quality work. The goal tree is a logic-based success path methodology that has grown out of nuclear power plant research studies.[5]

A goal tree is simply a hierarchy of goals and functions, broken down (from the top down) to the technical means (success paths) available to accomplish the top-level goal. A success path refers to a bottom-up chain through the tree by which the goal can be accomplished. A simple goal tree structure is displayed in Fig. 3.2. The objective is to lay out a logical means (not necessarily unique) to accomplish the goal through accomplishing the subgoals, which are in turn accomplished through functional activities made up of success paths. Hence, the tree helps provide a systematic focus, and helps to avoid a trial-and-error approach.

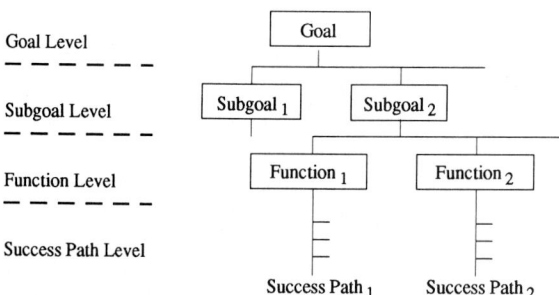

FIGURE 3.2   Goal tree concept.

Once a goal tree is developed, it can be pruned, or grown, to reflect changes over time. It is extremely helpful in assessing and communicating the "big picture." The tree structure helps to develop a feel for the issues involved in accomplishing the goal as well as the interactions between the issues involved.

Benchmarking.   A benchmark is defined as a reference point or a standard by which something can be measured or judged. Benchmarking was formally defined by D. T. Kearns, chief executive officer of Xerox:

> Benchmarking is the continuous process of measuring products, services, and practices against the toughest competitors or those companies recognized as industry leaders.

Benchmarking has evolved in recent years to represent a pursuit of the "best of the best" regarding both products and processes. Obviously, benchmarking, in a functional sense, is not a new concept. However, the formalization of benchmarking activities in a systematic and thorough fashion is relatively new.

The benchmarking strategy concentrates on both best practices (performance) and metrics (measurements). Successful benchmarking focuses on identifying and understanding best practices; metrics should be considered as important, but secondary. The best practices may be located somewhere in your own organization, in a competitor's organization, or in a functional area of a noncompetitor's organization. For example, if customer service through rapid order response is the focus, an organization in the fast food industry might be the best of the best and hence a good company to benchmark, even though your business happens to be sheet-metal fabrication. The point is that better practices may be in existence somewhere within your company, within your industry, or outside your industry. Camp[6] characterizes benchmarking as consisting of planning, analysis, integration, and action elements, all put together in a systematic fashion to improve both products and processes.

Robust Performance.   The term *robust performance* refers to a product's ability to deliver maximum performance (to the customer) over the entire range of anticipated applications, operating conditions, and environments. Off-line quality assurance is directed at providing robust performance through products and services that customers want, with a minimum amount of variation. Three general sources create this variation: (1) environmental, (2) operating, and (3) product. Environmental variables, including temperature, moisture, dust, usage patterns, and so forth create externally generated variation in performance. Internally generated variation results from deterioration or wear-out within a product. Unit-to-unit product variation creates yet another source of variation in product performance, when a group or population of items is considered. The point is that customers want a consistently high level of performance from each product unit they purchase, when operated under a variety of conditions.

Producing to Target.   One strategy that will help to assure consistent product and service performance is to assure the consistency of the product and service itself, through producing to targets, rather than within broad plus-or-minus specifications. A strategy of producing products to target requirements, rather than within broad plus-or-minus specifications, is clearly superior from both a technical perspective and a cost perspective.[7] Targets are generally of three types: lower is better, nominal is best, or larger is better. Lower is better would be applicable for production time, operating costs, warranty costs, defect counts and percentages, impurities, accident statistics, electrical noise, and so forth. Nominal is best refers to a specific point target such as a dimension, chemical content level, weight, and so forth. Larger is better is applicable to process yield, service life, time to failure, and so forth.

The point is to focus on the means to hit the target and reduce the dispersion about the target as long as such efforts are economically and technically feasible. A nominal case is shown in Fig. 3.3. Relying on inspection and assuming that all items between the specifications are equally effective not only increases production costs, it also decreases product performance. For example, tolerance stack-up in assemblies caused by encountering extremes in the plus-or-minus specification bands can degrade performance as well as create potential rework or hand-fitting and sorting problems, resulting in higher production costs. Inspection and rework as well as custom fitting are expensive alternatives to consistently producing to (or very near to) target.

Design Levels.   Genichi Taguchi,[8] a leading authority in the Japanese quality movement, delineates three levels at which quality is designed into products (these levels will also

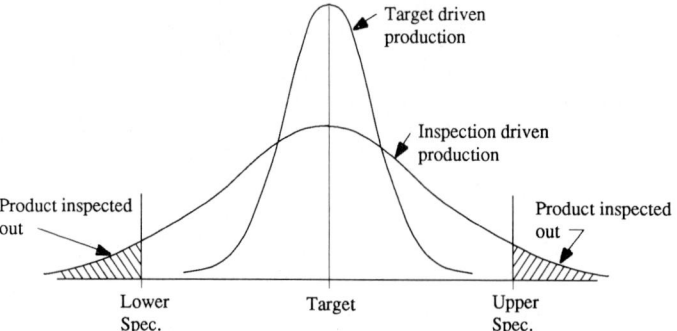

FIGURE 3.3   Target- versus inspection-driven production.

apply to services in general). The three levels are identified as (1) the system design level, (2) the parameter design level, and (3) the tolerance design level.

System Design Level.   System design applies to design at the functional level. At this level, relevant product or process technologies and approaches are identified. System level issues may include the total system view as the customer views and uses the product, or as the producer intends to build the product, or both. It may also reach within the product to include technical issues in subsystems, components, materials, interfaces (product, customer, service), production process technologies, assembly, maintainability, and so forth.

Parameter Design Level.   Parameter design is a secondary design level, within (or below) the system design level. Parameter design focuses on determining a "best level" of the design parameters identified and selected at the systems design level. The point is to meet the performance target (set at the systems level) with the least-expensive materials and processes and produce a robust product (insensitive to all three forms of variation).

Parameter design can be thought of as a process of optimizing the functional design with respect to both performance and cost. Taguchi's highly pragmatic approach to parameter design concentrates on designed experiments and specialized signal-to-noise-ratio measures. A sound and efficient experimental design program is mandatory to elicit an effective parameter design. A number of statistical tools are useful in parameter design analyses. Taguchi's signal-to-noise analysis (discussed later) is one of these tools. Classical ANOVA techniques, factorial arrangements of treatments, fractional factorials, regression, and response surface techniques are all potential aids in off-line quality work at the parameter design level. The point is to determine the best parameter levels (temperatures, pressures, materials, additives, purchased parts, and so forth) in the least amount of time and at the least expense, in order to gain performance and cost advantages in the marketplace. Hence, systematic experimentation, valid sampling techniques, and valid analyses (involving both location and dispersion measures) are required.

Tolerance Design Level.   Tolerance design is the final step in which the parameter tolerances are set. The parameter design step sets the best or optimal midvalues for the process parameters. The tolerance design step is a logical extension of parameter design to the point of a complete specification or requirement. In general, narrow tolerances should be given only to parameters where production variation will create critical performance problems. The point is to recognize that production costs escalate at nonlinear rates, as tolerances are decreased at linear rates. For example, a tight tolerance on a hole might call for a reaming operation in addition to a drilling operation, whereas the reaming operation may be eliminated if performance can be assured with a relaxed tolerance, or superior tooling, or different materials, or a different processing method. One should use target values and cost-performance calculations to guide tolerancing when possible.

Summary. The QFD, goal tree, and benchmarking methods are somewhat difficult to master and must be used with some discretion and patience. However, once developed, they can be reused, reviewed, and revised as conditions change. They also make excellent reference points, to serve as foundations for future continuous improvements efforts. The system-level thinking that they support and encourage also aids communications between functional areas and ultimately helps to create better products, from both a cost as well as a performance perspective. A product which exhibits robust performance under varied environmental and operational conditions is desirable. Unit-to-unit product variation must be minimized in order to assure customers consistently good performance. Using a three-level design strategy at the system, parameter, and tolerance levels helps to assure a systematic product design cycle which produces a product with robust performance characteristics at minimal cost.

EXPERIMENTATION, MODELING, AND DECISION

Customer, product, and process knowledge drive quality improvement. The primary source of knowledge is through empirical observations, either direct or indirect. Sometimes, knowledge can be acquired as the result of someone else's efforts through trade journals, technical papers, creative swiping,[9] and other means. At other times, one must put forth one's own efforts and resources to acquire knowledge through direct experimentation or as a by-product of ongoing operations.

Models. Once sufficient basic knowledge is available, one begins to formulate physical models to explain observed phenomena. These models may vary from simple rules of thumb to sophisticated statistical and stochastic models. The purpose is to leverage the knowledge available toward product and process improvement. Physical models may lead to the construction of physical simulation models which essentially represent a means of indirect experimentation. This experimentation may be useful in guiding parameter design and/or tolerance design efforts. However, physical confirmation experiments are typically run to confirm the simulation results (before major decisions or changes are made).

Most models and simulations represent physical phenomena in a highly abstract sense. Model validity is typically a matter of degree. For example, a good model should track the process it is modeling closely, but "perfect" agreement is typically beyond the capabilities of most models. Scaling and acceleration in physical experiments create additional challenges. Through modeling, one attempts to control the cost of knowledge and enhance the timeliness element required in effective decision making.

Critical Parameters. The experimentation necessary for effective off-line quality work involves three critical parameters: (1) physical parameters, (2) statistical parameters, and (3) economic parameters.

Physical Parameters. Physical parameters are critical parameters in a technical sense. They relate directly to product or process performance. Typically, one seeks to discover and/or calibrate a cause-effect relationship. Initially, a great deal of speculation may be involved in this discovery process. The cause-effect diagram is a useful tool to organize technical knowledge and experience. The point is to discover potential relationships involved, relative to the problem at hand (the effect), and then to obtain a feel for the relative influence each cause has on the listed effect. For example, the physical process by which tools wear and/or break can be conceptualized, organized, and displayed in a cause-effect diagram, as shown in Fig. 3.4. The resulting information is useful in developing action plans to ultimately improve tool life and product quality.

The incorporation of knowledge in an expert system is also a possibility. If a sufficient data or knowledge base exists, an expert system may be justified.[10] Commercial software

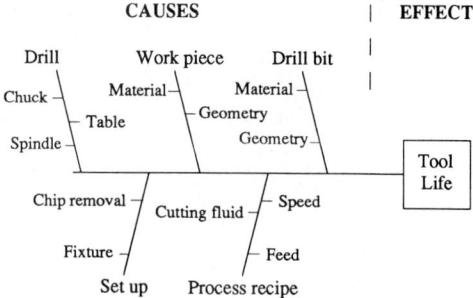

FIGURE 3.4 Identification of critical cause-effect relationships.

packages available for microcomputers and engineering workstations make this powerful tool feasible for off-line quality work.

Statistical Parameters. Most knowledge is gained from direct observation, through sampling studies. Very seldom does one have the opportunity to study a large number of product units or perform a complete census. In addition, most product units will perform somewhat differently when tested or used under the same conditions. Hence, statistics and statistical concepts are usually necessary in order to perform intelligent analyses and trigger informed decisions. The central issue in the Deming[11] and the Ishikawa[12] approach to quality is gathering sound statistics, properly interpreting the statistics, and then taking the proper management action regarding training or system modification.

Central tendency and dispersion are two fundamental statistical concepts which underlie most quality work. *Central tendency* is a location (to target) phenomenon; it refers to

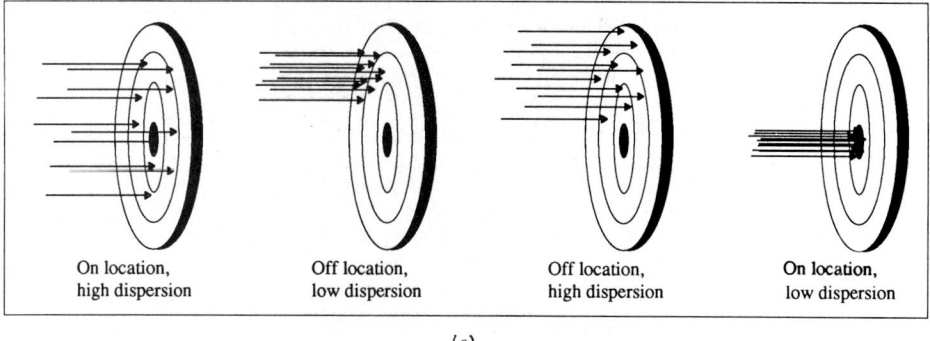

(a)

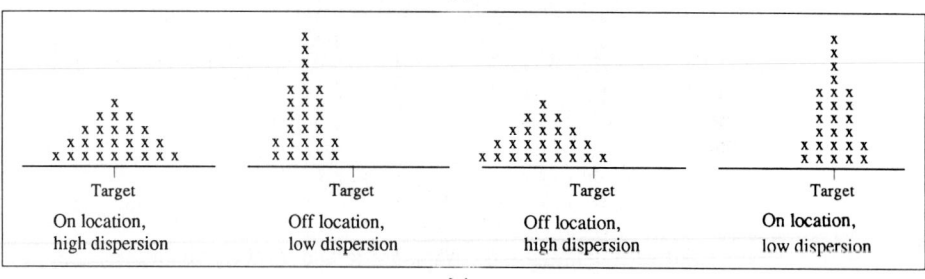

(b)

FIGURE 3.5 Location and dispersion illustration: (*a*) sporting targets (two-dimensional physical analogy); (*b*) check sheet targets (one-dimensional abstraction).

the center point of a population or set of data, relative to the zero point. *Dispersion* is a variation, spread, or scatter phenomenon; it is the tendency of a population or set of data to cluster about a center point. Both are critical. Figure 3.5 illustrates the concepts with respect to sporting targets as well as check sheets. Just like sporting targets, product and process targets must be hit consistently in order to gain competitive advantage. In other words, the most desirable outcome is one in which the target is hit in a repeatable fashion (on target with minimal dispersion).

Economic Parameters. The third critical parameter involves economics. Limited resources dictate that product and process efficiencies are an important determinant of product cost. However, field performance in terms of product life-cycle costs (including operation, maintenance, and failure and repair costs) must also be considered. Quality costs are typically classified as prevention, appraisal, and failure (internal and external) costs.[13] Crosby[14] takes an accounting view of quality costs and refers to the cost of nonconformance. Taguchi[8] chooses to view quality costs in terms of losses and quantify them through loss functions. A typical nominal, best loss function is shown in Fig. 3.6. Taguchi's concept of loss is termed a *loss to society*.

Designed Experiments. Experimentation is necessary to learn (or verify what we think we already know) about a product or process. Experimentation can be performed in a number of ways, ranging from trial and error "one knob at a time" approaches to carefully planned multifactor experiments. The former has given way to the last, because of its overall superiority in effectiveness (in isolating main effects and interaction effects and/or components of variation), efficiency (in smaller sample sizes), and timeliness (producing results faster).

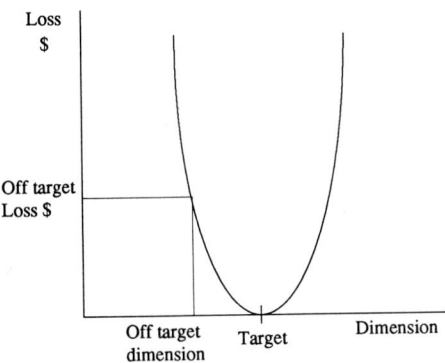

FIGURE 3.6 Loss function model (conceptual).

The classical body of knowledge known as design of experiments is extensive. Just like any speciality, experimental design has its own unique vocabulary, which must be understood to effectively utilize the body of knowledge. To introduce the concept of designed experiments, a number of concepts and terms must be defined and discussed. It is necessary to understand these concepts and terms in order to communicate with experts in the statistical field as well as to understand trade studies and reports based on experiments.

Objectives. A designed experiment is driven by experimental objectives and hypotheses. Objectives are structured in terms of the end results (knowledge) desired, expressed in the form of hypotheses. In order to clearly develop the objective and hypotheses, one must identify the experimental factors and responses.

Factors. Factors are independent variables (for example, surface speed, material). Factor levels are the physical levels of a factor (for example, 100 ft/min, 200 ft/min, steel, aluminum). Factors may be either quantitative (such as surface speed) or qualitative (such as materials). Sometimes, qualitative factors are termed *class* or *classification variables*. Typically, qualitative variables can take on only a few values (such as male, female) whereas quantitative variables can take on a great number of values (for example, any pressure from 0 to 400 lb/in^2).

Response. A response is the measured or observed *result* obtained at a given setting or *treatment* (combination of the factors). For example, thrust force in pounds, tool life in holes drilled, and so forth might be measured responses in a machinability experiment. Many times multiple responses are recorded in a single experiment. The results may be considered one response at a time (the usual case) or with respect to more than one response simultaneously (termed a multivariate analysis). Cause-effect diagrams and/or Pareto charts are very useful aids in helping one set objectives, identify factors and define meaningful factor levels. For example, tool life measured in linear distance drilled might be a meaningful response (the effect) while the factors of interest might be workpiece materials, speeds, feeds, and cutting fluids (the causes).

Hypotheses. Experimental objectives are typically addressed through stated hypotheses and hypothesis tests. The most common structure is to develop a null hypothesis H_0 (a hypothesis of no difference) and an alternative hypothesis H_1 (representing some sort of difference or deviation from the null hypothesis). The null hypothesis is typically assumed to be true until enough evidence is produced to reject it.

As an example of a hypothesis structure, one might set up the following set of hypotheses:

H_0: There is no difference in the average tool life (the response) with respect to the workpiece materials (factor levels) steel, copper, and aluminum.

H_1: There is a difference in the average tool life (the response) with respect to the workpiece materials (factor levels) steel, copper, and aluminum.

The interpretation (and the words chosen to express the interpretation) of experimental findings is critical. An interpretation resulting in the statement that "we will accept the null hypothesis" or "we will not reject the null hypothesis" is interpreted to mean "we do not see enough evidence to reject the null hypothesis," rather than "we believe the null hypothesis to be absolutely true." This point may seem rather trivial at first glance, but it is a critical point in the interpretation of results. The science of statistics and its usefulness in practice many times suffer from a lack of understanding and gross overinterpretation of results obtained from small samples.

Test of Hypotheses. A designed experiment consists of controlled physical conditions, with a *random* ordering of the treatments (factor level combinations). This control and randomization together constitute the *design*. The design results in a mathematical or statistical model. The model, along with a set of statistical assumptions, allows the hypotheses of interest to be assessed through formal statistical tests [usually t, F, or chi-squared (χ^2) tests]. For example, one can formally (statistically) test or assess the evidence provided by the response (tool life) in the form of sample means or averages pertaining to the treatments (steel, copper, or aluminum workpieces), relative to the dispersion present in the data. If the difference in the means is relatively large, compared to the dispersion associated with the means, evidence supports the rejection of the null hypothesis, and it is safe to conclude that all of the tool lives are not equal.

When dealing with small samples and drawing conclusions based on small samples, one must realize that statistical errors are possible. Table 3.1a summarizes the concept of statistical errors. The "absolute truth" is a "god's eye" view, which we mortals are not privileged to hold. Hence, we must draw conclusions relative to the sample indication, attempting to minimize errors by intelligent sampling, choosing sample sizes, and following careful experimental procedures. Walpole and Myers[15] provide a detailed, but practical, discussion of statistical hypotheses and inferences.

Table 3.1b, on the other hand, depicts the relationship between statistical significance and practical significance. Successes in off-line quality work come about through practical significance, relative to product and process improvements. But statistical significance provides support for drawing practical conclusions and making decisions.

Treatments. The terms *treatment* and *treatment combination* (of factors and factor levels) are used to describe what is done to the experimental material. The term *experimental unit* (eu) is used to refer to the largest collection of experimental material to which a single independent application of a treatment is made at random. An eu could be a piece of metal, a group of pieces, the volume of a chemical, an animal, a group or pen of animals, and so forth. An eu is defined in the context of an experiment, relative to the experimental objectives and resources available.

TABLE 3.1 Statistical and Practical Significance

(a) Statistical significance

		Absolute truth	
		H_0 true	H_0 false
Conclusion from sample	H_0 true	Correct conclusion	Type II error (β)
	H_0 false	Type I error (α)	Correct conclusion

(b) Practical significance

		Practical significance (physical dimension)	
		Significant difference	Insignificant difference
Statistical significance (statistical dimension)	Statistically significant difference (H_0 rejected)	Sufficient (and conclusive) physical results presented (decision support present)	Insufficient (but conclusive) physical results presented (decision support present)
	No statistically significant difference (H_0 not rejected)	Promising (but inconclusive) physical results presented (more evidence needed)	Nonpromising (and inconclusive) physical results presented (reevaluate experimental program)

Balance. Balance in an experiment implies the same number of observations for each treatment combination in the experiment. Balance is important in order to extract the same amount of information from each treatment combination.

Experimental Error. Experimental error is usually defined as the measure of variation which exists among observations (responses) on eu's treated alike. Hence, in order to obtain a measure of the experimental error directly, one must replicate (observe more than one response) for each treatment combination. A replication implies assigning multiple eu's to treatment combinations, treating the eu's independently, and observing the results. Multiple measurements on the same eu are considered a subsample, not a replication. Subsampling can be used to develop a measure of sampling error, which is the variation which exists within an eu along with instrumentation or measurement error.

P *Value.* A P value is a term used in computerized statistical analysis programs to represent an observed significance level (OSL). The *observed significance level* is defined as the smallest significance level at which one can reject the null hypothesis. In statistical terms, an OSL is the probability of observing a more extreme test statistic (t, F, χ^2, etc.) value given that the H_0 is true. Small P values support rejection of the H_0 in favor of the H_1. In practice, one rule of thumb of interpretation is to consider P values ≤ 0.01 as indicating very strong support for rejecting the H_0, $0.01 < P$ value ≤ 0.05 as indicating strong support for rejecting the H_0, $0.05 < P$ value ≤ 0.10 as indicating some support for rejecting the H_0, and P value > 0.10 as indicating little or no support for rejecting the H_0.

A statistical analysis and a resulting P value indicate the degree of support for rejecting the H_0, rather than an absolute yes or no answer. This degree of support concept results from the fact that small samples are not perfect indicators of population characteristics. Larger samples are better (yield more information) than smaller samples, but one must still avoid the absolute terms *always* and *never* and their mind-set when drawing conclusions from samples which are drawn from a population. The alternative is to examine every member of the population; only then can one be absolutely sure.

Main Effects and Interactions. Properly designed experiments are capable of measuring both main effects and interactions between factors. A main effect measures the failure of the factor levels of one factor (the factor in question) to respond the same (on the average) at all levels, averaged over all of the remaining factors. A two-factor interaction effect measures the failure of a factor to produce the same change in the average response (from level to level) at each level of another factor, all averaged over any remaining factors.

Graphical analyses can be very helpful to isolate and help explain main effects and interaction effects. Figure 3.7a depicts a drill diameter main effect which shows the average thrust force (averaged over all other factors) increasing as drill diameter increases. The curve would be flat (across all three diameter levels) as shown in Fig. 3.7a (upper right) if no diameter main effect were present. The points plotted are average responses, averaged over all factors (speed, feed, hardness) other than diameter. Figure 3.7b depicts a hardness × drill diameter interaction. In other words, the thrust response (averaged over speeds and feeds) is not constant from low hardness (level 1) to high hardness (level 2), across the three diameter levels. A similar plot with no interaction (parallel lines) is shown in Fig. 3.7b (upper right). In this depiction, one can observe a hardness effect (the hardness curves are separated) and a diameter effect; the average thrust increases as the diameter increases.

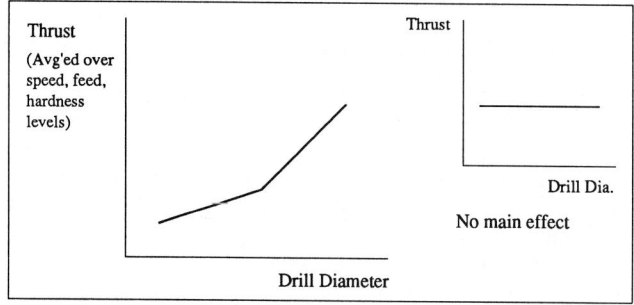

(a) Drill diameter main effect illustration.

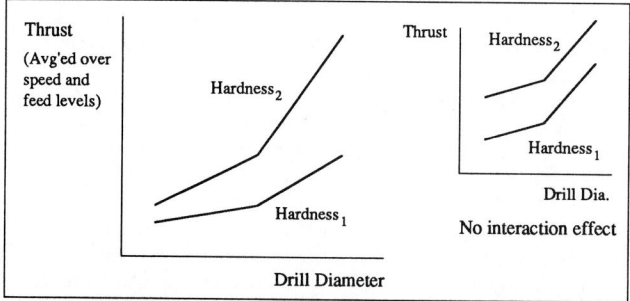

(b) Drill diameter and hardness interaction effect illustration.

FIGURE 3.7 Graphical interpretation of experimental data: (a) drill diameter main effect illustration; (b) drill diameter and hardness interaction effect illustration.

The critical rule of thumb is that, if significant interactions are present, then one cannot interpret each factor involved in the interaction independently. They must be interpreted simultaneously. For example, Fig. 3.7b shows that hardness effects should not be described independently of drill diameter effects. Rather, the response must be described at each hardness × drill diameter combination, taken as a pair. In other words, the thrust increases much faster at larger drill diameters than at smaller drill diameters when moving from hardness 1 to hardness 2. The knowledge of this dependence is expressed through the interaction effect. This same general rule of thumb pertains to interactions with more than two factors.

Records Analysis.   It is not always necessary to design and execute formal experiments in order to obtain product and process knowledge. Existing stores of data may be available from production, warranty, service, and even past experiments. It may be possible to use these data in lieu of performing an experiment.

Two basic conditions must be addressed before one should consider results from analyses of records data valid. First, the data should be relevant to the objective at hand. In this case, one proceeds to generate objectives and hypotheses as in a designed experiment. The second condition pertains to the integrity of the data. Ishikawa[12] suggests that one view any previously collected data set as suspect, if one cannot trace its source, conditions of collection, and so forth. His contention is that in many cases data are collected (and sometimes selectively screened) to show only what the collector desires to be shown. In other words, the data may produce a very biased picture of the product or process. Hence, decisions based on such data may lead to disaster. The safe procedure regarding records data is to doubt their integrity until established otherwise.

Summary.   Off-line quality work involves physical, statistical, and economic parameters. Each parameter plays a role in improving product and process performance and customer satisfaction. The use of designed experiments in off-line quality studies is critical in order to isolate the effects (main effects and interaction effects) of key variables, relative to the response of interest. The integrity of the data as well as the analysis method are important. Hypothesis testing and plots are helpful in assessing the extent of influence the variables have on the response. Finally, correct interpretation of the sample result is critical in order to support decisions.

EXPERIMENTAL METHODS AND MODELS

When product and process quality or performance are assessed, some performance parameters can be measured over a relatively short period of time: process yield, horsepower, package weight, purity, part dimensions, and so forth. Other measures require relatively long periods of time: product life, deterioration rates, time between failures, and so forth. In either case, sound experimental methods, analyses, and models are necessary to establish an unbiased picture of performance.

Classical Analysis of Variance.   Additive models and the classical analysis of variance concept are the foundation of experimental design and analysis. The ANOVA method is straightforward in principle. The elements involved include a structured randomization plan to average out possible inconsistencies and minimize biases (resulting from both controlled and uncontrolled variables) by providing independent observations of responses. The randomization plan includes a random allocation of the eu's to the treatment combinations. Analysis calculations can become complicated, but they are supported by a number of software packages.

Single-Factor Experiments.   In order to make effective use of the ANOVA method, a fundamental knowledge of the concept is necessary. In the simplest design, the completely randomized design (CRD), initially each eu has an equal chance of being assigned to any treatment. In this design, a treatment is defined as a level of the single factor. Each eu typically produces one response, unless a subsampling procedure is employed. Assuming a single response for each eu, one might obtain the data shown in Fig. 3.8. The data represent four treatments and three independent observations per treatment.

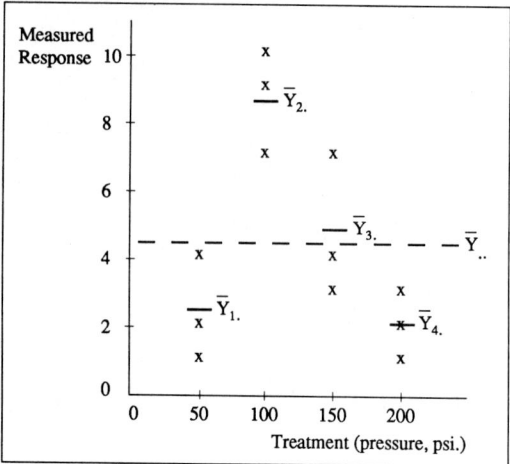

	Treatment			
	1	2	3	4
Obs.	(50 psi.)	(100 psi.)	(150 psi.)	(200 psi.)
1	4.0	9.0	7.0	3.0
2	1.0	7.0	4.0	2.0
3	2.0	10.0	3.0	1.0
$Y_{i.}$	7.00	26.00	14.00	6.00
$\overline{Y}_{i.}$	2.33	8.67	4.67	2.00

$$Y_{..} = 53.00 \qquad \overline{Y}_{..} = 4.42$$

(a) Experimental data.

(b) Scatter plot of experimental data.

FIGURE 3.8 Single-factor experiment: (a) experimental data; (b) scatter plot of experimental data.

Analysis. The analysis of the data (observations) primarily involves means and variances, developed according to the ANOVA technique. Each data point Y_{ij} is represented by a $\times$. The $\overline{Y}_{i.}$ symbol refers to the ith treatment average and the $\overline{Y}_{..}$ symbol refers to a grand average. The subscript dots indicate a summary over the subscript that it replaces; for example, $Y_{1.}$ is the total of all observations from the first treatment, $i = 1$.

Before this design is discussed in more detail, one should recognize that one typically desires to draw conclusions regarding both central tendency and dispersion. Central tendency is measured by the $\overline{Y}$'s in Fig. 3.8. Dispersion can be assessed by observing the scatter in the $\times$s within and between each treatment. This scatter is usually measured by computing sample variances.

The single-factor completely randomized design, or one-way ANOVA, can be represented in a model form as

$$Y_{ij} = \mu + \tau_i + \varepsilon_{ij} \tag{1}$$

Although abstract, model equations and their components are a critical part of the designed experiment and inferences that can be drawn. For example, the overall CRD mean μ is included as well as a treatment effect term τ_i and an experimental error term ε_{ij} in the CRD model. The error term serves as a residual term to represent natural variation between the eu's as well as any uncontrolled variables that may be affecting the responses Y_{ij}. An expansion, relative to the data, is shown in Eq. (2). Notice that the equality is preserved and the terms on the right-hand side correspond to the model terms in Eq. (1). The components of these terms can be identified in Fig. 3.8. The abstract model can be related to the empirical data, as shown in Eq. (3), by rearranging, squaring, and summing. Hence, the development of the term *sum of squares* is accomplished and the resulting partitioning concept is developed. This result is the basis for ANOVA techniques. The simplified ANOVA result is shown in Eq. (4).

$$Y_{ij} = \overline{Y}_{..} + (\overline{Y}_{i.} - \overline{Y}_{..}) + Y_{ij} - \overline{Y}_{i.}) \tag{2}$$

$$\sum_{i}^{a}\sum_{j}^{n}(Y_{ij} - \overline{Y}_{..})^2 = \sum_{i}^{a}\sum_{j}^{n}[(\overline{Y}_{i.} - \overline{Y}_{..}) + (Y_{ij} - \overline{Y}_{i.})]^2 \tag{3}$$

$$\sum_{i}^{a}\sum_{j}^{n}(Y_{ij} - \overline{Y}_{..})^2 = n\sum_{i}^{a}(\overline{Y}_{i.} - \overline{Y}_{..})^2 + \sum_{i}^{a}\sum_{j}^{n}(Y_{ij} - \overline{Y}_{i.})^2 \qquad (4)$$

The numerical results are typically displayed in an ANOVA table as shown in Fig. 3.9. This ANOVA table pertains to a balanced experiment with a treatment levels and n observations per treatment. The sum of squares shown are the partitioned terms of Eq. (4). The numerical results shown in Fig. 3.9 were obtained from the data shown in Fig. 3.8. Typically, computer software packages (or algebraically simplified forms for the sum of squares) are used in constructing ANOVA tables. Detailed calculations are described by Montgomery;[16] Hicks;[17] and Box, Hunter, and Hunter.[18]

Source	Degrees of Freedom	Sum of Squares	Mean Squares	F Statistic	P Value
Treatment	a-1	$n\sum_{i}^{a}(\overline{Y}_{i.} - \overline{Y}_{..})^2$	SS_{trt}/df_{trt}	MS_{trt}/MS_{err}	P_{trt}
(Between Trt's)	3	84.92	28.31	11.32	< 0.01
Experimental Error	a(n-1)	$\sum_{i}^{a}\sum_{j}^{n}(Y_{ij} - \overline{Y}_{i.})^2$	SS_{err}/df_{err}		
(Within Trt's)	8	20.00	2.50		
Total Corrected	an-1	$\sum_{i}^{a}\sum_{j}^{n}(Y_{ij} - \overline{Y}_{..})^2$			
	11	104.92			

FIGURE 3.9 One-way ANOVA.

Fixed Effects. Factors (and their resulting effects) involved in an experiment are classified as either fixed or random. A fixed effect is appropriate when one seeks to make an inference to a small set of hand-picked factor levels. For example, if process yield at four temperatures (100°F, 150°F, 200°F, and 250°F) is of interest, then one could use a four-level fixed effects factor (temperature) in the experiment. Statistical inferences, expressed through hypotheses, can be made only to the temperature levels selected. The hypotheses can be stated as:

H_0: All $\tau_i = 0$ or all μ_i are equal (all temperatures produce the same response level, on the average).

H_1: At least one $\tau_i \neq 0$ or at least one μ_i is not equal to the others (at least one temperature produces a different response level than the others).

An F statistic is appropriate for this test.

Support for the rejection of the H_0 implies that the average response is not the same at all treatment (temperature) levels. Data plots of response means are very good at helping to communicate fixed effects, or the lack of fixed effects, in the data. Even if the ANOVA detects no treatment effect, additional observations (data) or more highly controlled conditions in the physical experiment or better measurement equipment might eventually lead to the detection of a small effect. On the other hand, there may be no effect whatsoever (an unlikely case). This simple example points out the criticality of interpretation, wording, and communicating experimental findings. The interpretation is closely linked with the concepts of both statistical and practical significance described in Table 3.1.

The objectives in many off-line quality experiments focus on selecting a "best" method, additive, tool geometry, and so on. In fixed effects analyses, a number of methods exist to compare individual treatment means in pairs (two at a time) or in other combinations. Pairwise methods include the least significant difference, Duncan's multiple range, Tukey's honestly significant difference, and many others. Scheffé's method and single degree of freedom contrasts can be used to tailor many different types of treatment comparisons. References such as Walpole and Myers;[15] Montgomery;[16] Hicks;[17] Box, Hunter, and Hunter;[18] Moen, Nolan, and Provost;[19] and other basic statistics books describe these methods in detail.

Random Effects. The second basic statistical model, the random effects model, is appropriate if one seeks to estimate components of variation. Here, the situation involves a large number of possible factor levels. One then selects a of these levels at random. For example, one might want to estimate the treatment (temperature) variation involved in a production process. Hence, a temperature levels are selected at random (not hand-picked as in a fixed-effects case) from possible temperature levels occurring in the process. Then the experiment can be performed and an ANOVA table can be developed. Since the variance (assuming independence) of an observation, Y_{ij}, is

$$\text{Var}(Y_{ij}) = \sigma_\tau^2 + \sigma_\varepsilon^2 \tag{5}$$

where σ_τ^2 represents the treatment factor variation (temperature in this example) and σ_ε^2 represents a component of variation reflecting the natural variation in eu's and dispersion resulting from uncontrolled variables (other than temperature in this example). The σ_ε^2 term is estimated by the mean square error (MS_e). The mean square treatment (MS_{trt}) term estimates $n\sigma_\tau^2 + \sigma_\varepsilon^2$. The hypotheses tested, with the F test, in this case are

H_0: $\sigma_\tau^2 = 0$ (no temperature variation exists)

H_1: $\sigma_\tau^2 \neq 0$ (a temperature variation exists)

Hence, a significant result (rejection of the H_0) implies that we can safely conclude that $\sigma_\tau^2 > 0$ (that is, $\hat{\sigma}_\tau^2$, the estimated temperature variation, is large enough to detect in the presence of σ_ε^2 the random error or "noise" variation) and solve for an estimate of σ_τ^2 numerically. If one cannot reject the H_0, then the treatment variation is not detectable (in the presence of the error variation) and is assumed to be zero or nonexistent, according to the evidence at hand. More evidence (data) may result in a different conclusion. Typically, as the sample size increases, the ability to detect even a small treatment variation increases.

Blocking. Blocking on a single factor involves a restriction on randomization where the eu's are stratified by blocks and then randomized within each block, one block at a time. If the experiment is complete (each block contains each and every treatment one time), then the design is termed a randomized complete block design.

Blocking is a very powerful tool (when appropriate) which involves giving up a few degrees of freedom in the error term in exchange for an additional partition or block line in the ANOVA table. The physical requirements in blocking create some difficulties, since the eu's within each block should be more alike (with respect to the anticipated response) than eu's between different blocks. For example, in an experiment involving hand assembly methods, one might find it beneficial to block on operators (as individuals) in order to remove operator-to-operator variation from the experimental error mean square. This form of blocking could help isolate assembly method effects.

Single factor experiments blocked in two directions are termed *Latin square* designs. In a Latin square design, two restrictions on randomization are involved, typically called *row* and *column* effects. The Latin square restriction, that the number of treatments is equal to the number of columns (blocks in one direction) and also equal to the number of rows (blocks in the second direction), places rather harsh restrictions on the physical layout of the experiment. Experiments blocked in three directions are referred to as Graeco-Latin squares.

Multiple Factors. Single-factor experiments are very restrictive in the sense that conclusions must be drawn pertaining to one independent variable, with all others considered constant or controlled at a single level. In off-line quality there are typically many variables that serve to establish a process recipe or dictate product performance. Therefore, multiple factors are practical.

Multiple factors (two or more) can be designed into experiments in what is termed a factorial arrangement of treatments (FAT). The FAT concept requires that all possible treatment combinations (of factor levels) be developed. In a complete factorial, all treatment combinations are observed and their responses recorded and analyzed in an ANOVA table format.

In quality work, many times two or three levels of the selected factors are of interest. When two levels of each factor are studied, the arrangement is termed a 2^f, where f refers to the number of factors. When three levels are involved, the arrangement is termed a 3^f, and so on. One may also design FATs with different levels of different factors such as a four-factor FAT ($2 \times 3 \times 2 \times 4$) with 48 possible factor level combinations. The number of treatment combinations grows large very quickly. Hence, complete factorials with more than three factors and with three or more levels soon expand beyond the resources available to carry out the experiment. Fractional factorials, discussed later, are one answer to reducing the size of an experiment.

The typical form of a three-factor FAT model with fixed effects, in a completely randomized design, is

$$Y_{ijkl} = \mu + A_i + B_j + C_k + AB_{ij} + AC_{ik} + BC_{jk} + ABC_{ijk} + \varepsilon_{1(ijk)} \qquad (6)$$

where $i = 1, 2, \ldots, a$
$\quad j = 1, 2, \ldots, b$
$\quad k = 1, 2, \ldots, c$
$\quad l = 1, 2, \ldots, n$

Here, each Y_{ijkl} represents an observation. The A_i, B_j, and C_k terms are referred to as main effects. For instance, A_i measures the effect of the average response of the factor A levels to be the same, when averaged over all levels of factor B and factor C. The AB_{ij}, AC_{ik}, and BC_{jk} terms are referred to as *two-factor interactions* or *first-order interactions*. For example, AB_{ij} measures the failure of A (or B) to respond the same over all levels of B (or A), averaged over factor C. The ABC_{ijk} term represents the three-factor interaction or second-order interaction. This term measures the failure of the AB (or AC or BC) interaction to respond the same over all levels of C (or B or A). The $\varepsilon_{1(ijk)}$ represents an experimental error term, similar to the one discussed in the single-factor case. An ANOVA table for a three-factor FAT in a completely randomized design is shown in Fig. 3.10.

Care must be taken in interpretation when interaction is present. When significant interaction is not present, the factors can be interpreted independently of one another. But, when interaction is present, the factors involved in the interaction must be interpreted simultaneously. In other words, when significant interaction is present, one should describe the observed response relative to the levels of the other factors involved in the interaction. Plots are very helpful in multifactor interpretation. Figure 3.7 illustrates main effects and interaction plots.

If multiple observations are not made for each treatment combination, the error term cannot be estimated directly. Sometimes the higher-order interactions (three-factor and higher) are pooled and used to estimate the error terms. This use of the interactions to form an error term is discussed by Montgomery.[16] Sometimes this method is called *pooling up* (up the ANOVA table).

Random effects models can also be developed. They are useful to estimate components of variation. For example, if one was assessing product variation in a manufacturing process where machines, operators, and materials are involved, a 3^f random effects model could be used. It would be possible to determine the proportion of product variation created by each factor. Then, the most critical element could be addressed through a well-structured quality (variance reduction) project.

Source	Degrees of Freedom	Sum of Squares	Mean Squares	F Statistic	P Value
A	$a-1$	$\sum_{i} Y_{i...}^2 / bcn - CF$	SS_A/df_A	MS_A/MS_{err}	P_A
B	$b-1$	$\sum_{j} Y_{.j..}^2 / acn - CF$	SS_B/df_B	MS_B/MS_{err}	P_B
C	$c-1$	$\sum_{k} Y_{..k.}^2 / abn - CF$	SS_C/df_C	MS_C/MS_{err}	P_C
AB	$(a-1)(b-1)$	$\sum_{ij} Y_{ij..}^2 / cn - SS_A - SS_B - CF$	SS_{AB}/df_{AB}	MS_{AB}/MS_{err}	P_{AB}
AC	$(a-1)(c-1)$	$\sum_{ik} Y_{i.k.}^2 / bn - SS_A - SS_C - CF$	SS_{AC}/df_{AC}	MS_{AC}/MS_{err}	P_{AC}
BC	$(b-1)(c-1)$	$\sum_{jk} Y_{.jk.}^2 / an - SS_B - SS_C - CF$	SS_{BC}/df_{BC}	MS_{BC}/MS_{err}	P_{BC}
ABC	$(a-1)(b-1)(c-1)$	$\sum_{ijk} Y_{ijk.}^2 / n - SS_A - SS_B - SS_C$ $-SS_{AB} - SS_{AC} - SS_{BC} - CF$	SS_{ABC}/df_{ABC}	MS_{ABC}/MS_{err}	P_{ABC}
Exp. Error	$abc(n-1)$	$SS_{tot\,cor} - SS_A - SS_B - SS_C - SS_{AB}$ $-SS_{AC} - SS_{BC} - SS_{ABC} - CF$	SS_{err}/df_{err}		
Total Corrected	$abcn-1$	$\sum_{ijkl} Y_{ijkl}^2 - CF$			

where $CF = (\sum_{ijkl} Y_{ijkl})^2 / abcn$

FIGURE 3.10   Three-factor FAT ANOVA.

Confounding and Fractional Factorials.   A factorial arrangement of treatments is a very effective tool for experimental design in quality work, since there are typically a number of factors (variables) affecting even simple product or process performance. The amount of experimentation required in a full factorial (all possible treatment combinations) outstrips the available resources after about 4 or 5 factors at, say, 3 or 4 levels. For example, 4 factors at 3 levels require $3^4 = 81$ observations for each replication of the experiment. In the case of 5 factors and 4 levels, $4^5 = 1024$ observations would be required for 1 replication. The point is that for initial screening experiments, where one seeks to determine the critical variables affecting a given response (from a large group of variables), a full factorial experiment is not feasible. A fractional factorial experiment can be used to screen a number of factors and identify the most critical factors for further study through a more detailed follow-up experiment.

In order to reduce the amount of physical experimentation (increase the efficiency), but retain the information yield (sufficient effectiveness), factor confounding in blocks and fractional replications are useful for off-line quality studies. Confounding results in the loss of resolution or the ability to isolate and measure all effects (main effects and interactions), relative to the total resolution in a full factorial experiment. In practice, the measurement of some interaction effects may be of little, if any, importance. Therefore, one is usually willing to give up the ability to measure these effects in order to reduce the size of the experiment, without reducing the number of factors or factor levels. Typically, one attempts to confound an interaction effect that is not considered to be of critical importance (or highly significant) with another noncritical interaction effect. If two noncritical effects cannot be confounded together, a noncritical effect may be confounded with an effect of critical importance.

Resolution.   Fractional factorial experiments (FFEs) are classified by their resolution. Resolution refers to the "quality" of information that can be obtained from an experiment.[20] Resolution numbers are used to indicate the degree of resolution possible. Resolution numbers for a given FFE can be determined relative to the technical confounding structure.[16]

A full factorial of size $2^5 = 32$ treatment combinations will measure all 31 effects, from main effect A through the $ABCDE$ interaction. Figure 3.11 shows an example of 5 factors at 2 levels each in a machinability context. A full factorial consists of $2^5 = 32$ treatment combinations, represented by the squares in Fig. 3.11b and c, where the blacked-out squares represent a ½ and a ¼ FFE, respectively. In the FFEs, only a fraction of the treatment combinations are run. Hence, in the case of the ¼ FFE, 3 other ¼ FFEs could have been structured (other than the one shown). For example, the ¼ FFE for 5 factors at two levels would require 8 observations (represented by the black squares) rather than 32 observations.

Aliases. The confounding used in FFEs creates problems in interpretation. The columns shown in Fig. 3.11d represent the ANOVA rows and the aliases involved in the ¼ FFE illustrated. Aliases in an FFE are indistinguishable mathematically. For example, the A effect, the BE effect, the BCD effect, and the $ACDE$ effect sum of squares (in an ANOVA table) result in the same measure, since only a carefully selected fraction of the treatment combinations (represented by the black squares) was run. Therefore, in order to attribute the A row to the A effect, one must assume that the BE, BCD, and $ACDE$ effects are zero or nonsignificant. Otherwise, one is at a loss to interpret the results. The critical element in designing FFEs is to keep the resolution as high as possible by confounding higher-order interactions with each other or, if absolutely necessary, with main effects.

Interpretation. As long as orthogonality (balance) is preserved, the ANOVAs for FFEs are straightforward (similar to the full factorial ANOVAs). The problem of interpretation, however, may result in some confusion. The interpretation is straightforward if one assumes the aliases to be zero or negligible. In practice, physical insight is necessary to

Factor	Example	Level 1	Level 2
A	Speed	Slow (-)	Fast (+)
B	Feed	Low (-)	High (+)
C	Hardness	Soft (-)	Hard (+)
D	Cutting fluid	Present (-)	Proposed (+)
E	Geometry	Present (-)	Proposed (+)

(a) 2^5 machinability example.

(b) 1/2 FFE layout.

(c) 1/4 FFE layout.

ANOVA Aliases	ACDE BCD BE	BCDE ACD AE	ABCDE CD AB	ABCE ABD DE	ABDE ABC CE	BCE ADE BD	BDE ACE AD
Factors	A	B	E	C	D	AC	BC
Obs. 1	-	-	-	-	-		
2	-	-	-	+	+		
3	-	+	+	-	+		
4	-	+	+	+	-		
5	+	-	+	-	+		
6	+	-	+	+	-		
7	+	+	-	-	-		
8	+	+	-	+	+		

(d) 1/4 FFE treatment combinations, ANOVA rows and aliases.

FIGURE 3.11 Fractional factorial illustration (conceptual): (*a*) 2^5 machinability example; (*b*) ½ FFE layout; (*c*) ¼ FFE layout; (*d*) ¼ FFE treatment combinations, ANOVA rows and aliases.

justify ignoring the aliases, unless the resolution is extremely high (for example, V or better). In general, significant 3-factor interactions and higher are much more unusual than 2-factor interactions.

Resolution III designs can be challenging. For instance, the 2^5, ¼ FFE, of resolution III shown in Fig. 3.11 might show no significant factor A and factor B effects, but a highly significant factor E effect in the ANOVA. The result may be just as it looks, or, on the other hand (somewhat unlikely), the process may be showing a very high level of AB interaction (an alias of E) which might be important in a physical sense, and little if any factor E effect. Considering the number of aliases shown in this resolution III design, one could extend this hypothetical argument indefinitely. The point is that one's only protection in interpretation (outside of blind luck) is a sound grasp of the physical process and/or high resolution.

Additional Factor Levels.   The previous illustration represents a 2-level FFE. In some cases, 3- or possibly 4-level FFEs are useful, especially if one seeks to measure quadratic effects or zero in on "best" process settings or "best" product requirements. Additional levels (beyond 2) help to provide insight in explaining a response. FFEs of 3 or more levels are very useful in designing follow-up experiments, where a few driving variables have already been identified. Designing 3-level FFEs or higher is considerably more complicated than 2-level FFEs. In addition, it is possible to develop designs where some factors are set at 2 levels and others at 3 or 4 levels. Design guidance is available in a number of sources, including Refs. 16, 20, and 21.

Summary.   The use of designed experiments in off-line quality work provides a systematic structure for organizing and executing an experiment as well as an analysis method capable of measuring both main effects and interaction effects. Blocking and multiple-factor experiments are useful when extending an experiment beyond a single factor (variable). Fractional factorial designs are very useful in performing screening experiments where many factors (variables) are present and one desires to identify the most critical or driving variables. Typically, more detailed follow-up experiments are constructed after a screening experiment. The follow-up experiments usually contain only a few factors and are structured to provide high resolution. Experiments should always be designed in light of the available physical knowledge. Hypotheses and interpretations should be stated in terms of and related to the physical phenomena.

SPECIALIZED APPLICATIONS OF EXPERIMENTAL RESULTS

The ultimate objective of many off-line quality projects is to maximize a response such as process yield, minimize a response such as tool wear, or produce to a target value such as a purity level. Typically, the factors or variables influencing these responses will be of a quantitative nature. Hence, the recipe involves determining the "best" levels (from an infinite amount of possibilities) of the critical factors. Traditional experimental design approaches, as previously discussed, are certainly valid. However, specialized approaches have been devised that may enhance the effectiveness and efficiency as well as the timeliness involved in acquiring the product and process knowledge necessary to establish and maintain a competitive edge.

Regression and Response Surfaces.   Regression analysis is a general modeling approach where one seeks to develop a predictor $\hat{Y}$ in terms of independent variables X's. Typically, a least squares fitting approach is taken.[22] Data may be obtained through designed experiments or from historic databases. The usual model in multiple linear regression (MLR) is

$$Y = \beta_0 + \beta_1 X_1 + \beta_2 X_2 + \cdots + \beta_p X_p + \varepsilon \qquad (7)$$

where Y = response

X's = independent variables

β's = model coefficients

ε = error term, to compensate for the lack of a perfect fit

The error is said to contain dispersion resulting from natural variation as well as effects from variables or effects (possibly known or unknown) not included in the model.

For example, the pressure data shown in Fig. 3.8 has been modeled using the regression technique. The predictor developed is graphed in Fig. 3.12 and labeled *linear*. The predictor in equation is

$$\hat{Y} = 5.667 - 0.010(\text{pressure}) \qquad (8)$$

Hence, a predicted response can be calculated from Eq. (8) at a given pressure level.

Second- and higher-order models are useful in many quality studies. The MLR technique is useful in constructing quadratic models and higher-order polynomial models. In the example of Fig. 3.12, a quadratic model seems to be appropriate:

$$Y = \beta_0 + \beta_1 X_1 + \beta_2 X_1^2 + \varepsilon \qquad (9)$$

In this case, the fitted prediction equation is labeled *quadratic* in Fig. 3.12. The predictor is

$$\hat{Y} = -5.583 + 0.215(\text{pressure}) - 0.0009(\text{pressure}^2) \qquad (10)$$

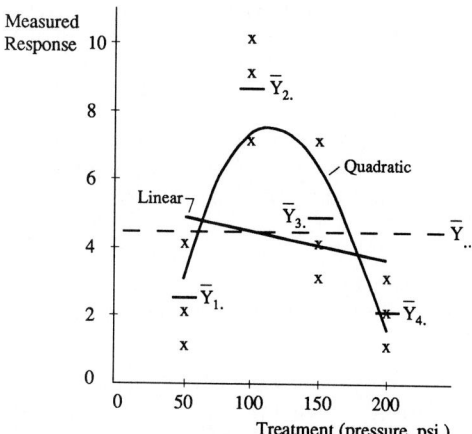

FIGURE 3.12 Linear and quadratic model illustration.

Response surface techniques involve a combination of experimentation and regression. Specialized forms of experimental designs are useful in response surface analyses. When a number of factors or independent variables, X's, are involved, classical experimental design becomes somewhat inefficient. Specialized designs, such as the central composite design, are used. The objective of such designs is to strategically select levels of the independent variables in patterns which produce more response data, Y's, in an area where the best response is expected. But, points are also placed around this selected area in a systematic pattern to broaden the scope of the experiment. Second-order or quadratic response surfaces are the most widely used. Quadratic models require at least three levels of each factor or independent variable. Response surface designs and analysis strategies are discussed in detail by Box and Draper[23] and Diamond.[20]

Regression and response surfaces offer powerful tools for the analysis and modeling of performance phenomena regarding quality measures and the variables which influence performance. However, care must be taken in the interpretation of the results. If the model does not adequately describe the data, or if one makes predictions beyond the observed data (used to construct the model), the predictions must be used with extreme caution.

Mixture Experiments.   A general assumption in most experimental designs is that the levels of one factor or variable are completely independent of the levels of all other factors or variables. For example, temperature levels are typically independent of cooking time. A number of quality-related experiments may not meet this criterion. If chemical composition is important, reducing or increasing the proportion of one ingredient will affect the proportion of another ingredient. For example, a two-solution mixture of 20 to 40 percent of chemical A and 80 to 60 percent of chemical B requires special treatment. Many alloy mixtures fall into the same category.

A number of specialized designs have been developed to address mixture-related studies. Typically, these designs are portrayed on variable scales from 0 to 100%; a triangular representation for three variables is common. Hence, a point somewhere on or within the figure represents a unique blend or mixture of the components. Diamond[20] and Cornell[24] provide discussions regarding useful mixture study strategies.

Summary.   Regression techniques and response surfaces are useful modeling tools. They are capable of relating a response of interest to a set of variables that can be controlled (or at least measured). In some applications, these models can be used to help improve, or optimize, a response of interest by studying the response (from the model) while the controllable variables are adjusted. The ultimate result sought is improved product or process performance. In other applications, regression and response surfaces can be useful for predicting performance levels which can be expected under a given set of input variables (product, process, or environment related). Mixture experiments call for specialized designs, because of the interrelationships between the mixture components, relative to the total makeup of the compound. If changing the level (concentration) of one factor automatically changes the level (concentration) of another factor, then the two factors are not independent and a mixture design is appropriate.

ROBUST PRODUCTS AND PROCESSES

Products and processes which produce performances that are robust (tolerant) to variation, resulting from the environment or application, will typically provide higher customer satisfaction than less robust products and processes. The degree to which products and processes are robust is determined by both design and production functions. Systematic and thorough methods are available to aid in both design and production related activities.

Taguchi Loss Functions and S/N Ratios.   The Taguchi loss functions and signal-to-noise (S/N) ratios[8,21,25,26] represent a pragmatic (and controversial with respect to theoretical statistics[27-29]) integration of classical statistics with the economic realities of manufacturing. These tools are useful for off-line applications, mainly at the parameter design state (described earlier). However, they can be extended into the tolerance design stage.

Loss Functions.   Three general cases of loss functions are applicable to quality work: (1) nominal is best, (2) lower (but greater than or equal to zero) is better, and (3) higher is better. The three loss functions are shown in Fig. 3.13. Taguchi describes loss as an economic loss to society by virtue of an off-target and/or highly variable production process. His loss functions are considered to be quadratic in nature. The loss functions are expressed mathematically as shown in Table 3.2, where k is a constant, y represents a part measurement, m represents a nominal (ideal target) value, and S^2 represents the variance

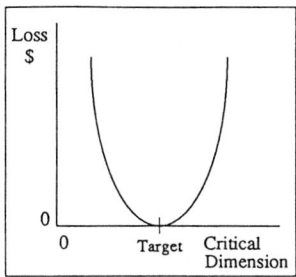 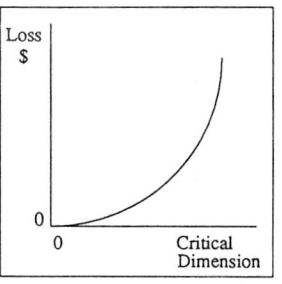 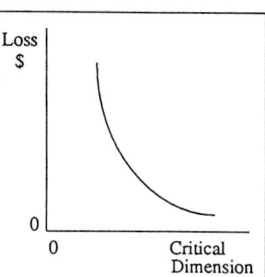

(a) Nominal is best.   (b) Lower is better.   (c) Higher is better.

FIGURE 3.13 Taguchi loss function (conceptual): (*a*) nominal is best; (*b*) lower is better; (*c*) higher is better.

TABLE 3.2 Taguchi Loss Functions

Characteristic	Individual loss/part	Average loss/part
Nominal is best	$k(y - m)^2$	$k[S^2 + (\bar{y} - m)^2]$
Lower is better	$k(y^2)$	$k[S^2 + \bar{y}^2]$
Higher is better	$k(1/y^2)$	$k(1/\bar{y}^2)[1 + (3S^2/\bar{y}^2)]$

about the average value of the y's, $\bar{y}$. Hence, the loss function structure includes both location and dispersion considerations, while the constant k calibrates the loss function to the application at hand.

A relatively high degree of variation, or noise, as measured by S^2, is generally an indication of a lack of off-line quality effectiveness. For example, poor tooling, fixtures, setups, operator training, maintenance, and so forth might lead to excessive product variation, thus adding to the loss. The point is to consider the loss both as a function of how close to target the average product measurement falls as well as how much variation exists from part to part, relative to the product mean.

Parameter Design Experiments. In general, parameter design is useful to improve quality without removing or controlling the cause of variation. In other words, parameter design aims to make the product or process robust to noise (or tolerant of variation) resulting from applications and environments as well as product nonuniformities. For example, one might seek to determine the least expensive materials, methods, and so forth, while maintaining adequate performance. In other cases, performance might be increased and costs lowered simultaneously when innovative thinking and technology are applied.

The *S/N* ratio method of parameter design proposed by Taguchi involves an FFE approach to experimental design (previously discussed) through orthogonal arrays and specialized response analyses, depending on whether nominal is best, larger is better, or smaller is better. The specialized responses represent special transformations of the observed data. Table 3.3 summarizes these fundamental transformations.

The usual procedure is to utilize orthogonal arrays for the inner design factors and the outer noise factors. These arrays are typically printed along with linear graphs and triangular tables,[8,21,25,30] which aid in designing the experiment. These experimental designs or layouts are also referred to as *matrix experiments*. The experimental design consists of assigning factors to array columns so as to maintain the desired degree of resolution (to isolate the desired effects) and minimize the number of trials. The linear graphs are useful in making column assignments and the triangular tables are useful in determining aliases (discussed earlier).

A brief example will serve to illustrate the *S/N* ratio procedure. The example is an extension of the previous fractional factorial machinability experiment, Fig. 3.11. The five

TABLE 3.3   Taguchi Signal-to-Noise (S/N) Ratios

Characteristic	S/N ratio	Comment
Nominal is best	$\eta = -10 \log V_e$	Measures only variation.
	$\eta = 10 \log \dfrac{V_m - V_e}{rV_e}$	Measures both variation and location.
Lower is better	$\eta = -10 \log \left[\dfrac{1}{r}\sum_{i=1}^{r} y_i^2 \right]$	The y_i refers to the ith observation from each design factor combination (treatment); $i = 1, 2, \ldots, r$. The r refers to the number of rows in the noise array.
Higher is better	$\eta = -10 \log \left[\dfrac{1}{r}\sum_{i=1}^{r} (1/y_i^2) \right]$	

$$V_e = \left\{ \sum_{i=1}^{r} y_i^2 - \left[\sum_{i=1}^{r} y_i \right]^2 / r \right\} / (r-1) \quad \text{and} \quad V_m = \sum_{i=1}^{r} y_i^2 / r = r(\bar{y}^2)$$

All S/N ratios are higher-is-better measures. S/N ratio measures are discussed in decibel (dB) units.

design factors A–E are assigned to L8 columns 1, 2, 4, 7, and 3, respectively, as shown in Fig. 3.14a and c. Columns 5 and 6 are not assigned main effects. Note that the L8 array in Fig. 3.14c and the 2^5 FFE developed earlier (Fig. 3.11d) are identical with respect to the factor effect rows and aliases (for example, the $+$ and $-$ assignments in the observations are identical). The alias structure is shown in Fig. 3.14c as well as Fig. 3.11d.

The outer array used in the example is an L4 with two noise factors at two levels each, Fig. 3.14b and d. Columns 1–3 represent Z_1, Z_2, and the $Z_1 \times Z_2$ interaction effects, respectively. Four experimental trials (set by the outer array rows) are required for each combination (the rows) of the interdesign array. Altogether, 32 observations (the y's in Fig. 3.14e) must be collected. Next, the S/N ratios (the η's in Fig. 3.14e) are computed for

Factor	Example	Level 1	Level 2
A	Speed	Slow (-)	Fast (+)
B	Feed	Low (-)	High (+)
C	Hardness	Soft (-)	Hard (+)
D	Cutting fluid	Present (-)	Proposed (+)
E	Geometry	Present (-)	Proposed (+)

(a) Design factors at 2 levels.

Factor	Example	Level 1	Level 2
Z_1	Work piece thickness	Thin (-)	Thick (+)
Z_2	Cutting fluid flow	Low (-)	High (+)

(b) Noise factors at 2 levels.

ANOVA Aliases	ACDE BCD BE	BCDE ACD AE	ABCDE CD AB	ABCE ABD DE	BCE ADE BD	BDE ACE AD	ABDE ABC CE
Factors	A	B	E	C	AC	BC	D
Columns	1	2	3	4	5	6	7
Obs. 1	-	-	-	-			-
2	-	-	-	+			+
3	-	+	+	-			+
4	-	+	+	+			-
5	+	-	+	-			+
6	+	-	+	+			-
7	+	+	-	-			-
8	+	+	-	+			+

(c) Inter L8 design array.

Factors	Z_1	Z_2	$Z_1 Z_2$
Columns	1	2	3
Obs. 1	-	-	
2	-	+	
3	+	-	
4	+	+	

(d) Outer L4 noise array.

Design Factor Trial	Observed Data Noise Factor Trial				Calculated S/N Ratios*
	1	2	3	4	
1	y	y	y	y	η
2	y	y	y	y	η
3	y	y	y	y	η
4	y	y	y	y	η
5	y	y	y	y	η
6	y	y	y	y	η
7	y	y	y	y	η
8	y	y	y	y	η

(e) Data and S/N ratio format.

* Regardless of nominal is best, lower is better or higher is better, the S/N ratio is judged on a higher is better criterion.

FIGURE 3.14   Taguchi signal-to-noise ratio experimental layout (conceptual): (a) design factors at 2 levels; (b) noise factors at two levels; (c) inter L8 design array; (d) outer L4 noise array; (e) data and S/N ratio format.

each row of four observations. The *S/N* ratio selected depends on the type of response desired (larger is better, smaller is better, nominal is best), as shown in Table 3.3.

Many possibilities exist in the analysis and interpretation of an experiment using both inner and outer arrays. Tentative conclusions can be drawn by direct comparisons of *S/N* ratios, followed by ANOVA techniques regarding the *S/N* ratios as responses. If replications of the inner array are made (with no outer array present), *S/N* ratios can also be computed and analyses may proceed as above. Use of the outer array forces variation from the outer array factors into the observations in a systematic fashion. If no outer array is used and replications are performed, variation will also be present, but not systematically forced into the experiment.

One may also disregard the *S/N* ratios and analyze the results in a classical fashion and still obtain the benefits of a systematic application of noise or environmental variables. Analyses may consist of direct response comparisons, plots of means, ANOVAs, and so forth. Hence, one should recognize the fundamental design as a means to systematically obtain data. Then, recognize the many forms of analyses available to help interpret the observations. The point is that the orthogonal arrays develop a systematic means for performing the physical experiment, supporting both effectiveness in gaining knowledge as well as efficiency in using company resources. However, one must typically give up the ability to measure all possible interaction effects.

Three of the four Taguchi *S/N* measures listed in Table 3.3 include both location and dispersion measures, relative to the treatment combinations observed. Classical fixed-effects ANOVAs typically assume that a homogeneity of variation exists over all treatment combinations. In other words, the ANOVA assumes that the variation within each treatment combination is equal. Thus, the classical ANOVA inferences are made relative to mean responses, with experimental error developed by pooling "within treatment" variances. The classical approach is very powerful in detecting differences in means. But that power is obtained by assuming equal variances across the treatment combinations. On the other hand, any method which can work effectively with both location and dispersion simultaneously is attractive for off-line quality work.

The validity of the Taguchi model, like all models (for example, ANOVA, regression, and so forth), requires a number of assumptions. As usual, one must be aware of the assumptions and ultimately assess the appropriateness of the model to the problem at hand. In most cases, it is advisable to obtain expert advice when developing an experimental program to assure that the proper tools are selected and tailored correctly, and the results are interpreted correctly.

True Zero Defects and Customer Satisfaction.   The typical consumer does not buy products or services in great quantities. Production statistics concerning the average number of defective items, defects per unit, and so on do not greatly interest the consumer. The performance of the one or two units the consumer buys determine product and service experience. For example, as a consumer, the performance you get out of the automobile you own influences your opinion of that automobile model much more than the performance of your neighbors' similar automobiles. In the short run, constant product and process improvement toward defect rates expressed in *x* defects per thousand, million, or billion are reasonable. However, long-run goals should seek "perfect" quality or "zero defects." Activities that support total customer satisfaction all the time must be aimed at understanding customer needs and expectations as well as technical issues. Otherwise, one might attain zero defects but produce a product that no one wants.

Off-line quality efforts must be aimed at total consumer satisfaction 100 percent of the time, not mere factory statistics. Shingo's[31] *poka-yoke* (mistake proofing) concept, beyond product and process statistics, makes a great deal of sense. The concept stresses the development of technical and organizational methods to aid in first preventing and second detecting mistakes in real time so that corrections can be made and product integrity maintained. Shingo places high priority on the development of innovative fabrication and assembly methods, fixtures, and mistake-sensing devices to help assure zero defects. For example, if an assembly is put together with four screws, arrange a device to always deliver four screws at a time. In this way, the assembler is unlikely to forget the fourth screw.

Summary.   A loss function concept focuses attention on the ability to produce to target, with respect to both location and dispersion measures. The Taguchi approach to parameter design utilizes inner and outer arrays and an S/N ratio transformation. The objective is to systematically introduce a realistic element of variation in an experiment and measure the robust characteristics of the response. The point is to design an experiment where the product (or process) will be subjected to noise or environmental variables at levels similar to those anticipated in field use by the customer (or within the manufacturing process). Hence, observations made under conditions likely to be encountered in actual practice, rather than highly "sterile" laboratory conditions, may lead to a more effective and efficient product and ultimately to a competitive advantage. The mistake-proofing concept and the robust design concept are highly compatible. The need to mistake-proof the entire system, from assessing customer needs, design, production, delivery, and finally to customer usage support, is clearly the direction that progressive companies are taking. This effort involves total quality management, on-line quality, off-line quality, reliability, and maintainability efforts all integrated within the production organization.

CONCLUSIONS

Customer satisfaction drives the pursuit of quality. Customers buy benefits and these benefits are discovered, developed, and delivered by serious competitors throughout the world. All serious competitors have access to the same pool of equipment, raw materials, and capital. The people resource is the key to a competitive edge. Through strategies, plans, and actions, people ultimately apply all resources, including human resources. Off-line quality assurance activities are in large measure responsible for discoveries and developments that lead to more effective and efficient products and services. But, effectiveness and efficiency are both relative measures and therefore are constantly changing. Hence, the timeliness element is essential. Table 3.4 summarizes

TABLE 3.4   Off-Line Quality Tool Summary

Check marks indicate relative usefulness

Product life phase	Quality function deployment	Goal tree	Classical designed experiments	Regression/ response surfaces	Loss functions	Taguchi S/N analysis
Customer needs	✓✓✓	✓✓			✓✓	
Research and development	✓✓	✓✓	✓✓✓	✓	✓	✓✓✓
Product design and development	✓✓✓	✓✓✓	✓✓✓	✓✓✓	✓✓✓	✓✓✓
Process design and development	✓✓✓	✓✓✓	✓✓✓	✓✓✓	✓✓✓	✓✓✓
Production	✓✓✓	✓✓✓	✓✓	✓✓	✓✓	✓✓
Delivery	✓✓	✓✓	✓✓	✓	✓	✓✓
Usage and support	✓✓✓	✓✓	✓✓✓	✓✓✓	✓✓	✓✓✓
Mainframe computer*			SAS, SPSSX, BMDP			
Microcomputer*	**	**	SAS, SPSS, SYSTAT, STATGRAPHICS, SOLO, MINITAB, JMP, STATVIEW		**	**

*Selected trade names of popular software.
**Emerging software.

the major off-line quality tools and current software support available. The number of check marks associated with each tool indicates its potential usefulness, relative to each product life phase.

Figure 3.15 depicts an off-line quality improvement cycle. The questioning element involves the formulation of conceptual hypotheses that have the potential to stretch or reach beyond the current state of knowledge. In quality work, the knowledge sought is primarily directed at cause-effect relationships, ultimately regarding customer satisfaction through products and services and the processes necessary to produce and deliver them. The experimentation element involves active exploration through efficient experimentation aimed at identifying and understanding the driving variables and variable interactions.

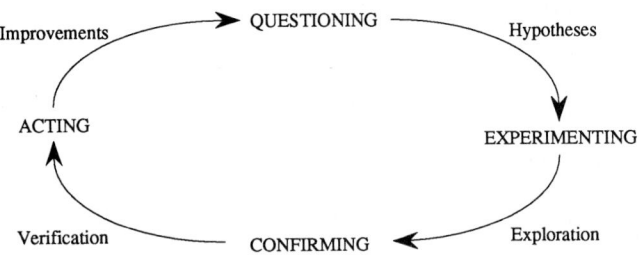

FIGURE 3.15 Off-line quality improvement cycle.

Experimentation, with the objective of finding the best materials or settings, typically involves measuring responses at fixed factor levels. It would be a rare coincidence if the best setting was one of the treatment combination levels selected. Hence, one attempts to get fairly close and systematically zero in on the best setting by developing additional experiments or using some type of response model.

Confirmation refers to the process of replicating results (usually tentative) from the exploratory work. The certification of exploratory findings is necessary to support major product and process decisions. The action element closes the loop. Adjustments are usually necessary to realize the benefits (improvements) of the knowledge gained from the other three elements in the off-line quality improvement cycle.

REFERENCES

1. Kolarik, W. J., and J. N. Pan, "Proactive Quality: Concept, Strategy, and Tools," *Proceedings—International Industrial Engineering Conference,* 1991, pp. 411–420.

2. Akao, Y., *Quality Function Deployment,* Productivity Press, Cambridge, Mass., 1990.

3. Ryan, N., *Taguchi Methods and QFD,* ASI Press, Dearborn, Mich., 1988.

4. Sullivan, L. P., "Quality Function Deployment," *Quality Progress,* June 1986, pp. 39–50.

5. Roush, M. L., et al., "Integrated Approach Methodology: A Handbook for Power Plant Assessment," Report SAND87-7138, Sandia National Laboratory, Albuquerque, N.M., October 1987.

6. Camp, R. C., *Benchmarking,* Quality Press, Milwaukee, 1989.

7. Sullivan, L. P., "Reducing Variation: A New Approach to Quality," *Quality Progress,* July 1984, pp. 15–21.

8. Taguchi, G., *Introduction to Quality Engineering: Designing Quality into Products and Processes,* Kraus International (Asian Productivity Organization), White Plains, N.Y., 1986.

9. Peters, T., *Thriving on Chaos,* Harper and Row, New York, 1987.

10. Mishkoff, H. C., D. Shafer, and D. W. Rasmus, *Understanding Artificial Intelligence,* 2d ed., Sams, Indianapolis, 1988.

11. Deming, W. E., *Out of the Crisis,* MIT Center for Advanced Engineering Study, Cambridge, Mass., 1986.

12. Ishikawa, K., *What is Total Quality Control? The Japanese Way,* Prentice Hall, Englewood Cliffs, N.J., 1985.

13. Feigenbaum, A. V., *Total Quality Control,* 3d ed., McGraw-Hill, New York, 1983.

14. Crosby, P. B., *Quality is Free,* McGraw-Hill, New York, 1979.

15. Walpole R. E., and R. H. Myers, *Probability and Statistics for Engineers and Scientists,* 4th ed., Macmillan, New York, 1989.

16. Montgomery, D. C., *Design and Analysis of Experiments,* 3rd ed., Wiley, New York, 1991.

17. Hicks, C. H., *Fundamental Concepts in the Design of Experiments,* 3d ed., Holt, Rinehart and Winston, New York, 1982.

18. Box, G. E. P., W. G. Hunter, and J. S. Hunter, *Statistics for Experimenters,* Wiley, New York, 1978.

19. Moen, R. D., T. W. Nolan, and L. P. Provost, *Improving Quality Through Planned Experimentation,* McGraw-Hill, New York, 1991.

20. Diamond, W., *Practical Experiment Designs,* Lifetime Learning Publications, Belmont, Calif., 1981.

21. Ross, P. J., *Taguchi Techniques for Quality Engineering,* McGraw-Hill, New York, 1988.

22. Draper, N. R., and H. Smith, *Applied Regression Analysis,* 2d ed., New York, Wiley, 1981.

23. Box, G. E. P., and N. H. Draper, *Empirical Model-Building and Response Surfaces,* Wiley, New York, 1987.

24. Cornell, J. A., *Experiments with Mixtures,* 2d ed., Wiley, New York, 1990.

25. Taguchi, G., *System of Experimental Design,* vols. I and II, Kraus International, White Plains, N.Y., 1987.

26. Kackar, R. N., "Off-line Quality Control, Parameter Design, and the Taguchi Method," *Journal of Quality Technology,* vol. 17, no. 4, October 1985, pp. 176–188.

27. Box, G. E. P., "Discussion, Off-line Quality Control, Parameter Design, and the Taguchi Method," *Journal of Quality Technology,* vol. 17, no. 4, October 1985, pp. 189–190.

28. Hunter, J. S., "Signal to Noise Ratio Debated," *Quality Progress,* May 1987, pp. 7–9.

29. Hendrix, C. D., "Signal-to-Noise Ratios: A Wasted Effort," *Quality Progress,* July 1991, pp. 75–76.

30. Phadke, M. S., *Quality Engineering Using Robust Design,* Prentice-Hall, Englewood Cliffs, N.J., 1989.

31. Shingo, S., *Zero Quality Control: Source Inspection and the Poke-Yoke System,* Productivity Press, Cambridge, Mass., 1986.

CHAPTER 4
RELIABILITY ENGINEERING

PART 1
RELIABILITY THEORY

S. M. Alexander
Professor
Department of Industrial Engineering
University of Louisville
Louisville, Kentucky

PART 2
RELIABILITY COMPUTATION AND EXPERT SYSTEMS

O. Geoffrey Okogbaa
Professor
Department of Industrial Engineering and Management Systems
University of South Florida
Tampa, Florida

PART 3
SOFTWARE RELIABILITY

Way Kuo
Professor and Chair
Department of Industrial and Manufacturing Systems Engineering
Iowa State University
Ames, Iowa

One of the implications of technological advances is the increased complexity of modern systems with competing design goals such as high performance, high reliability, low cost, and ease of maintainability.

This has created a need for detailed analyses of reliability, safety, and maintainability aspects of these systems at the design, operational, and disposal phases of the product or system life cycle. In addition, there is a growing interest in the development of standardized mathematical tools and procedures as well as computer-assisted design tools and software packages. Reliability analysis requires an understanding of the system and hence a correct description of the resulting topology, an understanding of the failure modes and the criticality of each mode, a capture of the characterization of the underlying process probability distributions, the use of the appropriate techniques and models to represent the system, and computation so as to extract estimates of the desired parameters. Furthermore, there is growing concern about the quality and reliability of software packages, since there is virtually no software package that can solve every reliability problem, and no industry standards with which to judge software.

Part 1 of the chapter deals with the issue of reliability theory, problem formulation, and parameter estimation and testing. Part 2 focuses on reliability computation with emphasis on complex systems (complex in terms of both topology and size). It also explores the use of expert systems as the technology of computing continues to advance. Part 3 looks at the issue of software reliability and its implication on the overall reliability and cost of systems.

PART 1
RELIABILITY THEORY

The commonly accepted definition of reliability is that "reliability is the probability that a device will satisfactorily perform its specified function for a specified period of time under a given set of operating conditions." The next section will include the development of a mathematical definition of reliability. In the final section, reliability prediction, design, and testing methods will be discussed. Examples will be provided to illustrate the methods discussed.

Mathematical Definition of Reliability.   As defined before, reliability is the probability that an item will not fail by a given time t, under a given set of operating conditions. The probability of failure by a given time t is referred to as the *unreliability* of the item. Mathematically, the unreliability is represented by

$$U(t) = \int_0^t u(t)\, dt \qquad t \geq 0 \tag{1}$$

where $U(t)$ = unreliability of product or system and $u(t)$ = probability density function of failure.

Therefore, the reliability R at time t of an item is

$$R(t) = 1 - U(t) \tag{2}$$

Hazard Rate.   The probability of failure of an item between time periods t_1 and t_2, given that the item has survived till period t_1, is given by

$$\frac{U(t_2) - U(t_1)}{R(t_1)} = \frac{R(t_1) - R(t_2)}{R(t_1)} \tag{3}$$

The failure rate is therefore

$$\frac{R(t_1) - R(t_2)}{R(t_1)(t_2 - t_1)} \tag{4}$$

The instantaneous failure rate is termed the hazard rate $z(t)$. It is obtained by taking the limit of the failure rate function as $h = (t_2 - t_1) \rightarrow 0$. This leads to the following:

$$Z(t) = \lim_{h \to 0} \left[\frac{R(t) - R(t + h)}{hR(t)} \right]$$

$$= -\frac{1}{R(t)} \left[\frac{dR(t)}{dt} \right] = \frac{u(t)}{R(t)} \tag{5}$$

which can also be written as

$$Z(t) = \frac{-d \ln R(t)}{dt} \tag{6}$$

From the above equation the product reliability can be defined in terms of the hazard rate as shown below:

$$R(t) = \exp \left[-\int_0^t z(t) \, dt \right] \tag{7}$$

Distributions for Failure, Reliability, and Hazard Rates.   The reliability, failure, and hazard distributions of a product can be estimated empirically from the following equations:

$$\hat{R}(t) = \frac{n(t)}{N_0} \tag{8}$$

$$\hat{u}(t) = \frac{1}{N_0(t_{i+1} - t_i)} \qquad t_i < t < t_{i+1} \tag{9}$$

$$\hat{z}(t) = \frac{1}{n(t_i)(t_{i+1} - t_i)} \qquad t_i < t < t_{i+1} \tag{10}$$

where N_0 identical items are put on test at time 0 and $n(t)$ survive at time t, and t_i represents the failure time of the ith item.

The distributions, most commonly used to represent unreliability, are the exponential and Weibull distributions. These and other distribution functions and the associated reliability and hazard functions are shown in Fig. 4.1a and b.

RELIABILITY PREDICTION

In the previous paragraphs we defined the functions related to the reliability of a part or component. Any product would be made up of a combination of components. The reliability of the product, or component system, can be estimated from the reliability of the individual parts. The components in the system could have a series, parallel, series and parallel, or other (complex) configuration.

System Reliability if the Components Have Constant Reliability
Series System.   In a series system, every component must function correctly for the system to function correctly. Schematically a series system can be represented as shown in Fig. 4.2. The reliability of the system, R_{sys}, in this configuration, if the component reliabilities are independent, is given by

$$R_{sys} = R_1 \cdot R_2 \cdot \cdot R_n \tag{11}$$

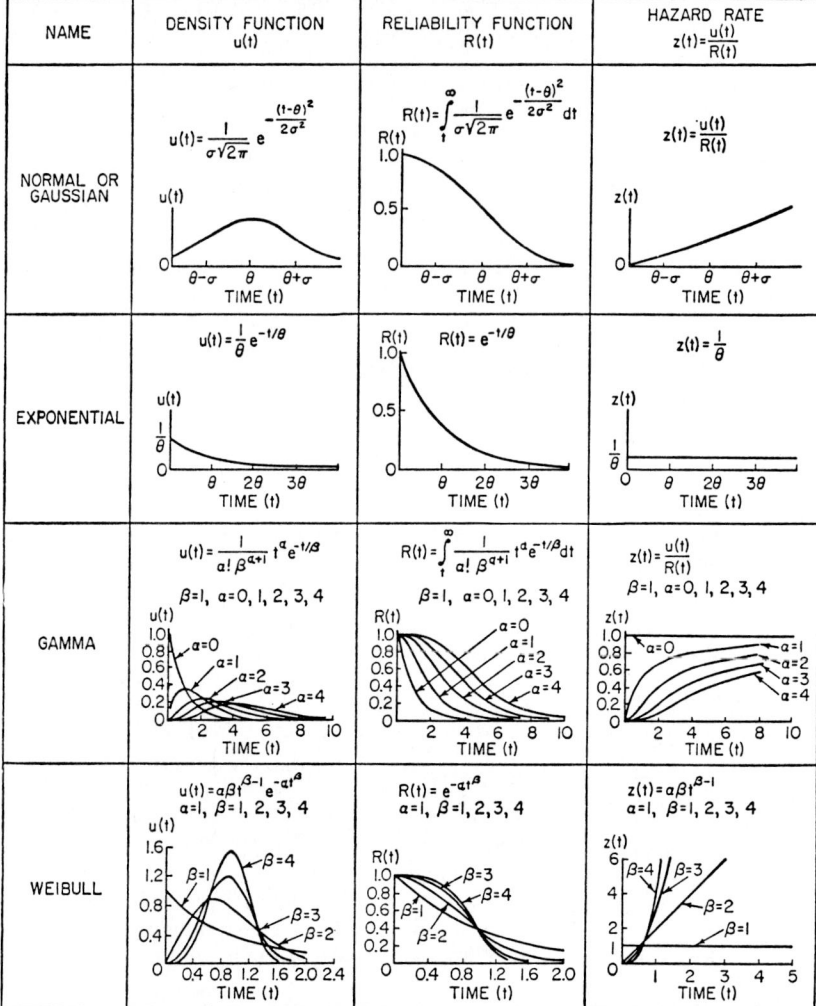

FIGURE 4.1 (*a*) Density and reliability functions and hazard rates of the normal, exponential, gamma, and Weibull distributions.

where $R_1, R_2, \cdots, R_n$ are the individual component reliabilities.

Parallel System.   A parallel system configuration is illustrated in Fig. 4.3. The reliability of the system with a parallel configuration is

$$R_{\text{sys}} = 1 - \prod_{i=1}^{n} (1 - R_i) \tag{12}$$

r-out-of-n system.   An *r*-out-of-*n* system functions correctly if *r* components out of the total number of *n* components function correctly. The system reliability for an *r*-out-of-*n* system, with identical components, is given by

$$R_{\text{sys}} = \sum_{k=r}^{n} \binom{n}{k} R^k (1 - R)^{n-k} \tag{13}$$

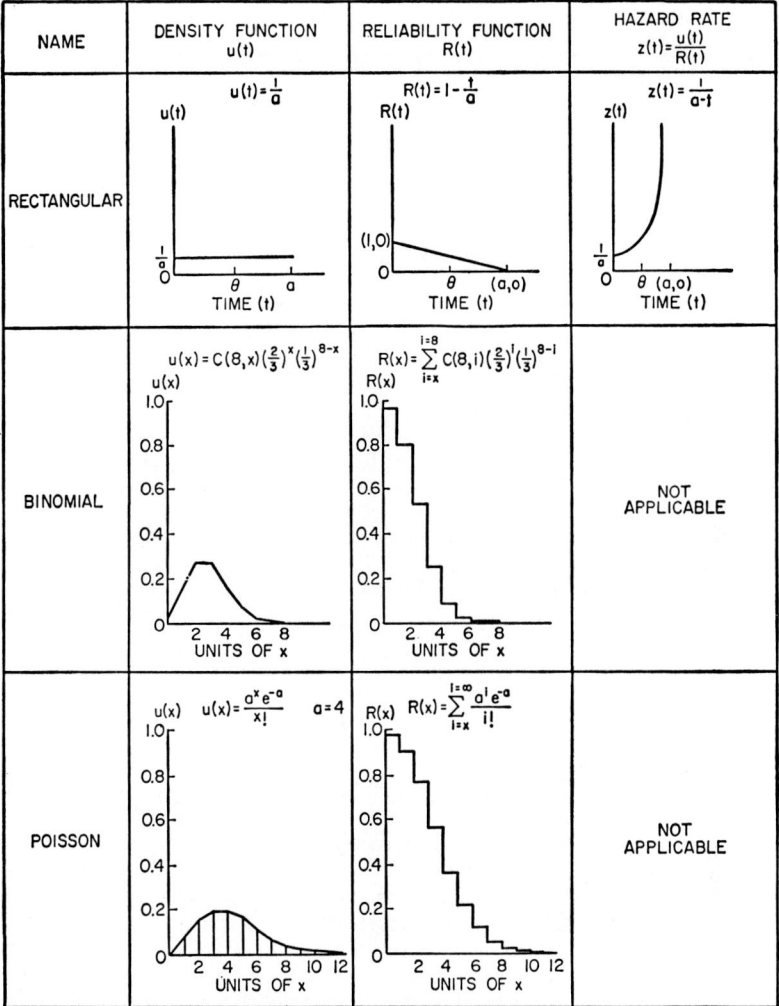

FIGURE 4.1 (*Continued (b)* Density and reliability functions of the rectangular, binomial, and Poisson distributions; hazard rate of the rectangular distribution.

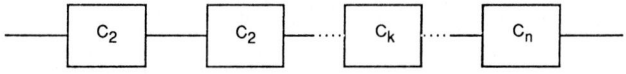

Non-Series-Parallel Systems. Some systems cannot be classified as either series or parallel systems. Such systems are defined as *complex systems*. An example of such a system is shown in Fig. 4.4.

The reliability of such a system can be obtained by enumeration, path tracing, or conditioning on a key element. All of these approaches look at all combinations of components that must function correctly for the system to function correctly. This approach to obtaining reliability is quite tedious and often intractable for practical problems, where

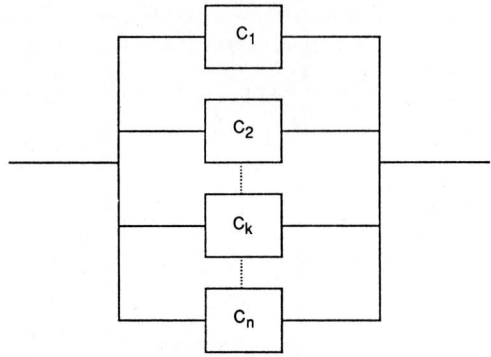

FIGURE 4.3   A parallel system.

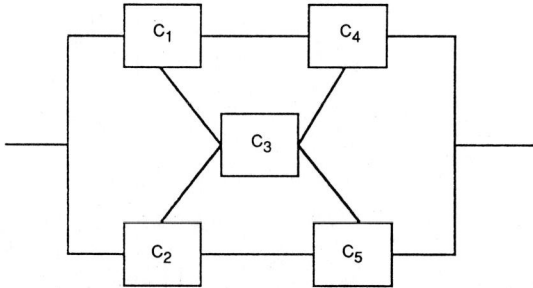

FIGURE 4.4   A complex system.

systems consist of a large number of components. Hence, in practice, approximation methods can be used. One such procedure is where one determines upper and lower bounds for the reliability of a complex system. A crude lower bound could be obtained by considering all the components to constitute a series configuration, and an upper bound could be obtained by considering a parallel configuration. Better bounds can be obtained by considering a series configuration of minimal cuts and a parallel configuration of minimal paths. A minimal cut is defined as a minimal set of components that must fail to cause system failure; a minimal path is a minimal set of components, which, if functioning, will ensure system operation. As an example, the complex network (Fig. 4.4) represented as minimal paths and minimal cuts is shown below:

Minimal paths	Minimal cuts
C1, C4	C1, C2
C2, C5	C4, C5
C1,C3,C5	C1,C3,C5
C2,C3,C4	C2,C3,C4

System Reliability if the Component Reliabilities are Time-Dependent.   If we consider components with constant hazard rates $z_i(t) = \lambda_i$, the reliability of a series system is

$$R_{sys}(t) = \prod_{i=1}^{n} R_i(t) = \prod_{i=1}^{n} e^{-\lambda_i t} \tag{14}$$

$$= e^{-(\lambda_1 + \lambda_2 + \cdots + \lambda_n)t}$$

The system reliability of a series system is like that of a single component whose hazard rate is the sum of all component hazard rates. The composite hazard rate of the series system is also constant.

For a pure parallel system, the system reliability is

$$R_{sys}(t) = 1 - \prod_{i=1}^{n} (1 - e^{-\lambda_i t}) \tag{15}$$

The hazard rate of a parallel system, composed of constant-hazard-rate components, is a monotonically increasing function of time. The mean time to failure (MTTF) of a parallel system of identical components is

$$\text{MTTF} = \frac{1}{\lambda}\left(1 + \frac{1}{2} + \frac{1}{3} + \cdots \frac{1}{n}\right) \tag{16}$$

where n is the number of components.

It is seen from the equation that increasing the number of components in a parallel system increases the mean time to failure at a declining rate. Hence, to increase the mean time to failure, redesign is a better option than redundancy.

RELIABILITY DESIGN

Reliability Allocation. In the preceding section we defined methods to predict the reliability of a system, given the configuration and the reliability values of the components. In the design process, however, we aim for a certain system reliability and we attempt to determine how to allocate the reliability specifications among the subsystems.

A simple approach would be to allocate the reliability to the subsystems according to the number of parts in the subsystems. For example, consider a subsystem consisting of three equipment items: a power supply, a receiver, and a transmitter. An MTBF of 200 hours is desired, which means a maximum allowable failure rate of 0.005 failures per hour.

A failure in any one of the three equipment items will cause a subsystem failure. The complexity of each equipment has been estimated as follows:

Power supply: 100 parts
Receiver: 255 parts
Transmitter: 560 parts

The weight assigned to each equipment is then

$$W_{(\text{power supply})} = \frac{100}{100 + 255 + 560} = 0.11$$

$$W_{(\text{receiver})} = \frac{255}{100 + 255 + 560} = 0.28$$

$$W_{(\text{transmitter})} = \frac{560}{100 + 255 + 560} = 0.61$$

The failure rates per equipment item can then be apportioned as follows:

Power supply: 0.11(0.005) = 0.00055
Receiver: 0.28(0.005) = 0.00140
Transmitter: 0.61(0.005) = 0.00305
System: = 0.00500

Dividing by the number of estimated parts per equipment will then provide the average part failure rate required. For example, for the power supply

$$\lambda_{ave} = \frac{0.00055}{100} = 5.5 \times 10^{-6} \text{ failures/h, or 5.5 failures/}10^6 \text{ h}$$

Another approach would be to use the formula based on the 1957 report of the Advisory Group on Reliability of Electronic Equipment (AGREE):

$$\theta_i = \frac{Nw_i t_i}{-n_i \ln R} \qquad \text{for } i = 1, 2, \ldots, k \tag{17}$$

where R = desired system reliability
  θ_i = required mean life of subsystem i; $\lambda_i = 1/\theta_i$
  t_i = time for which the ith subsystem is needed
  w_i = probability that the system will fail if the ith subsystem fails
  n_i = number of parts in subsystem i

$$N = \sum_{i=1}^{k} n_i = \text{total number of parts in subsystem}$$

Reliability Improvement.   If the designer is not able to meet the design specifications with the available components, then the following options remain:

1. Improve the reliability of the components.
2. Redesign, using fewer component parts in series.
3. Apply component derating techniques to reduce the failure rates.
4. Attempt to use redundancy if the other options are not feasible.

Selection of Components to Improve.   Consider a series system of five components with reliabilities $R_1 = 0.80$, $R_2 = 0.75$, $R_3 = 0.60$, $R_4 = 0.95$, $R_5 = 0.97$. Suppose the desired system reliability is 0.60.

$$R_{sys} = 0.80 \times 0.75 \times 0.60 \times 0.95 \times 0.97 = 0.33$$

is less than the desired system reliability. The question is to determine which components should be improved and by how much. The components are first ordered in increasing order of reliability, that is, $R_1 \le R_2 \le R_3 \le R_4 \le R_5$. Now each component is considered individually to determine whether it should be improved. Suppose each component were as good as the best; then the system reliability would be 0.85. Since this is better than 0.60, the reliability of C_5 is adequate. Similarly, we check the adequacy of the reliabilities of the other components. For example, $(0.95)^4 (0.97) = 0.79$, which is greater than 0.6. Therefore R_4 is adequate. But $(0.80)^3 (0.95) (0.97) = 0.47$; therefore R_3 is not sufficient.

The reliabilities R_1, R_2, and R_3 must be increased to $[0.60/(0.95)(0.97)]^{1/3} = 0.87$.

RELIABILITY TESTING

Life Testing
  Nonreplacement Test with Failure Times Not Recorded.   In order to obtain a point estimate of the reliability at time t^*, n items are put on test for the time t^*. At the end of this time period the number r that have failed is noted. The point estimate of the reliability is then

$$\hat{R}(t^*) = 1 - r/n \tag{18}$$

An interval estimate of reliability for the same time t^* is based on fact that the binomial parameter has confidence bounds based on the F distribution:

$$R_L(t^*) = \frac{1}{1 + \dfrac{r + 1}{n - r}F_2} \tag{19}$$

$$\hat{R}_u(t^*) = \frac{F_1}{F_1 + \dfrac{r}{n - r + 1}} \tag{20}$$

where $R_L(t^*)$, $R_u(t^*)$ = lower and upper $(1 - \alpha)$ confidence limits
$F_1 = F_{\alpha/2}(2n - 2r + 2, 2r)$
$F_2 = F_{\alpha/2}(2r + 2, 2n - 2r)$

If the products have a constant hazard rate, then the reliability estimates for times other than t^* can be obtained using the equation shown below:

$$\hat{R}(t) = \left(\frac{n - r}{n}\right)^{t/t^*} \tag{21}$$

Testing with Failure Times Noted.   When the failure times are noted, a point estimate of the mean time to failure is obtained from the equation

$$\theta = \sum_{i = 1}^{n} t_i/n \tag{22}$$

where θ = maximum likelihood estimate of the mean time to failure. The interval estimate is given by:

$$\theta_L, \theta_u = \left[\frac{2\sum_{i = 1}^{n} t_i}{X_{\alpha/2}^2(2n)}, \frac{2\sum_{i = 1}^{n} t_i}{x_{i - \alpha/2}^2(2n)}\right] \tag{23}$$

For example, suppose 10 items having constant hazard rates are tested till all fail. If their failure times are 75, 83, 97, 150, 200, 65, 98, 220, 175, and 167 h, then

$$\sum_{i = 1}^{n} t_i = 1330$$

and
$$\theta = 133.0$$

$$\hat{\theta}_L = \frac{2 \times 1330}{x_{.05}^2(20)} = \frac{2660}{31.41} = 84.69$$

where $\hat{\theta}_L$ is the 95 percent lower confidence limit of the mean time to failure.

Type II Censoring: Without Replacement.   When a number of components are tested, it is very possible that many would not fail for a very long time. In order to limit the amount of time for testing and obtaining an estimate of reliability, it is possible to limit the testing time to that when a predetermined number r of items fail. The rest of the items are censored, with the incomplete information being used to obtain a measure of reliability. The reliability is obtained by determining the maximum likelihood estimate of the parameter θ. This estimate is obtained from the likelihood function $L(\theta)$. The maximum likelihood estimator is

$$\hat{\theta} = \frac{t_1 + t_2 + \cdots + t_r + (n - r)t_r}{r} \tag{24}$$

where $t_1, t_2, \cdots, t_r$ are ordered values of failure times. It can be shown that this is also an unbiased estimate of θ.

Type II Censoring: With Replacement. Here failed items are immediately replaced with good items and the testing continues until a specified number r fail. The number of items available for testing must be at least $n + r - 1$ so that all n test stations are filled. Under this scenario,

$$\hat{\theta} = \frac{T}{r} \tag{25}$$

$$T = nt_r$$

Type I Censoring: With Replacement. Under this scheme the testing is truncated after a certain time interval t^*. The estimates of the parameter θ are

$$\hat{\theta} = \frac{nt^*}{r} \tag{26}$$

$$\hat{\theta}_L, \hat{\theta}_u = \left[\frac{2nt^*}{x_{\alpha/2}^2(2r + 2)}, \frac{2nt^*}{X_{1-\alpha/2}^2(2r)} \right] \tag{27}$$

Hypothesis Testing for CFR Components and Type II Censoring. Here we test components to determine whether they have an acceptable mean life. In the test procedure, a sample of n items are put on test. When r items fail, the testing is stopped, and the mean life of the sample is estimated. If the mean life is greater than a rejection constant C, the lot is accepted; otherwise it is rejected.

The rejection constant C is obtained from

$$C = \frac{\theta_a X_{1-\alpha}^2(2r)}{2r}$$

where θ_a = mean life considered to be acceptable.

Sequential Reliability Tests. The purpose of a reliability test is to establish in the shortest possible time, and at minimum cost, whether or not the reliability of a type of component or of a system is equal to or better than a specified minimum. The sequential probability ratio tests have been devised for this. This method enables us to make one of three decisions as each failure occurs: (1) accept, (2) reject, (3) continue testing. Essentially, what we do is establish two values of mean life θ_1 and θ_2. θ_1 is some minimum unacceptable value, and θ_2 is some chosen acceptable value. After r failures have occurred, we then compute the probability of r failures occurring for a mean life of θ_2. For example, in the case of the Poisson distribution, the probability of r failures in time t for an equipment whose times to failure are exponentially distributed is

$$P_0 = \frac{(t/\theta)^r e^{-t/\theta}}{r!} \tag{28}$$

where θ is the chosen mean life. Having computed this for θ_1 and θ_2, we take the ratio $P_{\theta 1}/P_{\theta 2}$ and compare this against two selected positive constants A and B which are based on previously agreed on risks—the consumer's risk β, or probability of accepting equipment with $\theta = \theta_1$, and the producer's risk α, or probability of rejecting equipment with $\theta = \theta_2$. These constants are given by

$$A = \frac{1 - \beta}{\alpha} \tag{29}$$

$$B = \frac{\beta}{1 - \alpha} \tag{30}$$

As each failure occurs, the ratio $P_{\theta1}/P_{\theta2}$ is computed, and the following decision rules are applied:
Accept if

$$\frac{P_{\theta1}}{P_{\theta2}} \leq B \tag{31}$$

Reject if

$$\frac{P_{\theta1}}{P_{\theta2}} \geq A \tag{32}$$

Continue testing if

$$B < \frac{P_{\theta1}}{P_{\theta2}} < A \tag{33}$$

For exponentially distributed times to failure,

$$p(r) = \frac{P_{\theta1}}{P_{\theta2}} = \left(\frac{\theta_2}{\theta_1}\right)^r \exp\left[-\left(\frac{1}{\theta_1} - \frac{1}{\theta_2}\right)t\right] \tag{34}$$

For example, let us assume that $\alpha = \beta = 0.10$ percent, so that $A = 9$ and $B = 0.111$, and that $\theta_1 = 100$ hours and we choose $\theta_2 = 200$ hours. Then

$$\frac{P_{\theta1}}{P_{\theta2}} = 2^r e^{-t/200}$$

If no failure occurs up to 200 hours, $P(r) = 2^0 e^{-1} = 0.368$. The value still lies between A and B; so no decision can be made. If no failure occurred to 440 hours, however, then $p(r) = 0.111 = B$, and we would accept the equipment. Remember that t is the sum of the operating times of all the equipment under test and r is the total number of failures. If five items of equipment were under test, and no failures occurred, the accept decision could be made in 88 hours of test time.

To ease the problem of computation, a graphical technique has been developed which enables one to determine instantaneously whether to accept, reject, or continue testing. It can be shown for the exponential case that Eq. (33) is of the form

$$a + bt < r < c + bt \tag{35}$$

where the left and right sides are equations of two parallel straight lines with equal slopes b. When these two lines are plotted on paper with t as the abscissa and r as the ordinate, the constants a and c are the intercepts of the lines with the ordinate. Numerically, a, b, and c are given by

$$a = \frac{\ln B}{\ln (\theta_2/\theta_1)} \tag{36}$$

$$b = \frac{(1/\theta_1) - (1/\theta_2)}{\ln(\theta_2/\theta_1)} \tag{37}$$

$$c = \frac{\ln A}{\ln(\theta_2/\theta_1)} \tag{38}$$

Figure 4.5 is a plot of the two parallel lines $a + bt$ and $c + bt$ in an r-t coordinate system; between these two lines, we plot the step function representing the cumulative number of failures versus the time t. This enables a rapid decision to be made without computation.

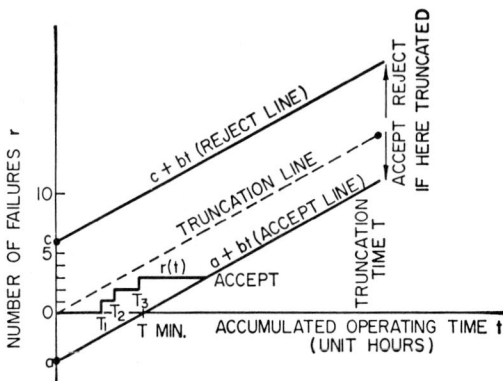

FIGURE 4.5   Sequential reliability test.

The parallel dashed line in the graph, which goes through the origin, represents the truncation line. If we must stop the test at some time T, and no prior accept or reject decision has been possible, the truncation rule says that an accept decision is made if at T the $r(T)$ value of the $r(T)$ step function is below the dashed line or touches it; a reject decision is made if at T the value of $r(T)$ is above the dashed line. If $T \gg T_{min}$, the effect of truncation on α and β will be small.

The graph for a sequential reliability test should always be prepared in advance. It requires a knowledge of α, β, θ_1, and θ_2, which are usually specified or agreed on. Its preparation is easy, and it simplifies the test procedure because no further calculations are necessary. The problem is reduced to computing the accumulated unit hours and entering in the graph the cumulative $r(t)$ line as the test proceeds.

The test described herein is an exponential test because the equations for the accept and reject lines are derived from the exponential function. It can be safely used for complex electronic equipment or for any complex device which has been debugged and in which no wear-out failures will occur during testing. Sequential probability ratio tests, based on Eqs. (31) and (32), also can be designed for distributions other than exponential. In those cases, the probability ratio $p(r) = P_{\theta 1}/P_{\theta 2}$ is no longer given by Eq. (34) and has to be calculated from the particular distribution involved. When this has been done, graphical procedures can be developed as in Eq. (35).

One of the most commonly used sequential test plans for equipment is that for which $\alpha = \beta = 0.10$ and $\theta_2/\theta_1 = 1.5$. The quantity θ_2/θ_1 is usually referred to as the discrimination ratio. Figure 4.6 portrays the graph and lists the accept-reject criteria based on the number of failures and the test time in multiples of θ_2. θ_n on the chart is the same as θ_2. MIL-STD-781A contains a number of sequential test plans for differing values of α, β, and discrimination ratio.

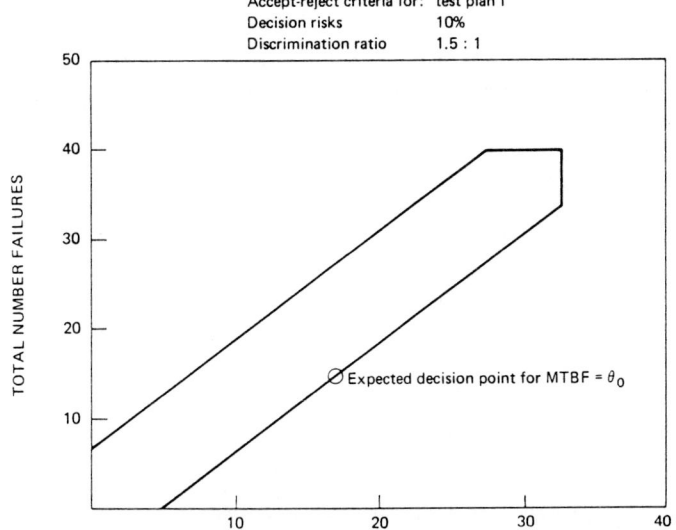

Accept-reject criteria for: test plan I
Decision risks   10%
Discrimination ratio   1.5 : 1

TOTAL NUMBER FAILURES

◯ Expected decision point for MTBF = θ_0

TOTAL TEST TIME (IN MULTIPLES OF SPECIFIED MTBF)

No. of failures	Total test time* Reject (equal or less)	Total test time* Accept (equal or more)	No. of failures	Total test time* Reject (equal or less)	Total test time* Accept (equal or more)
0	N/A	4.4	21	12.6	21.4
1	N/A	5.2	22	13.4	22.2
2	N/A	6.0	23	14.2	23.0
3	N/A	6.8	24	15.0	23.8
4	N/A	7.6	25	15.9	24.7
5	N/A	8.5	26	16.7	25.5
6	.5	9.3	27	17.5	26.3
7	1.3	10.1	28	18.3	27.1
8	2.1	10.9	29	19.1	27.9
9	2.9	11.7	30	19.9	28.7
10	3.7	12.5	31	20.7	29.5
11	4.5	13.3	32	21.5	30.3
12	5.3	14.1	33	22.3	31.1
13	6.1	14.9	34	23.1	31.9
14	6.9	15.7	35	24.0	32.8
15	7.8	16.6	36	24.8	33.0
16	8.6	17.4	37	25.6	33.0
17	9.4	18.2	38	26.4	33.0
18	10.2	19.0	39	27.2	33.0
19	11.0	19.8	40	28.0	33.0
20	11.8	20.6	41	33.0	N/A

*Total test time is total unit hours of "equipment on" time and is expressed in multiples of the specified MTBF

FIGURE 4.6   Sequential test plan. (*Source: MIL-STD-781A, "Reliability Tests, Exponential Distribution," U.S. Department of Defense, Washington, D.C., 1965.*)

Military Standards.   Many documents related to reliability are available from the U.S. Department of Defense (DOD). The documents cover diverse areas including sampling inspection, test procedures, and guidelines for the development of reliability growth programs. These publications can be obtained from:

Naval Publications and Forms Center
5801 Tabor Avenue
Philadelphia, PA 19120-5099

BIBLIOGRAPHY

Arinc Research Corp., *Reliability Engineering,* Prentice-Hall, Englewood Cliffs, N.J., 1964.

Barlow, R. E., and F. Proschan, *Mathematical Theory of Reliability,* Wiley, New York, 1965.

Bazovsky, I., *Reliability Theory and Practice,* Prentice-Hall, Englewood Cliffs, N.J., 1961.

Data Collection for Nonelectric Reliability Handbook, NEDCO I and NEDCO II, vols. I to III, AD-841-106, AD-841-107, and AD-841-108, June 1968.

Grosh, L. D., *A Primer of Reliability Theory,* Wiley, New York, 1989.

Hahn, G. J., and S. S. Shapiro, *Statistical Methods in Engineering,* Wiley, New York, 1967.

Ireson, W. G. (ed.), *Reliability Handbook,* McGraw-Hill, New York, 1966.

Lloyd, D. R., and M. Lipow, *Reliability: Management, Methods, and Mathematics,* Prentice-Hall, Englewood Cliffs, N.J., 1962.

MIL-STD-781A, *Reliability Tests, Exponential Distribution,* U.S. Department of Defense, Washington, D.C., 1965.

Naresky, J. J., "Reliability Engineering," *Maynard's Industrial Engineering Handbook,* 3d ed., McGraw-Hill, New York, 1971.

Pieruschka, E., *Principles of Reliability,* Prentice-Hall, Englewood Cliffs, N.J., 1963.

Rome Air Development Center, *Reliability Notebook,* RADC TR-67-108, vols. I and II, AD-845-304 and AD-821-640, November 1968.

PART 2

RELIABILITY COMPUTATION AND EXPERT SYSTEMS

During the past few decades, advances in technology and the increased complexity of systems have resulted in considerable demand for reliable, safe, and maintainable systems that can be produced at minimum cost. This need for product assurance has generated considerable discussions in the engineering and scientific communities regarding the role of concurrent engineering in all aspects of the design activity such as design for manufacturing (DM), design for assembly (DA), design for maintainability (DMT), and design for testability (DT). The increased awareness of the need for product safety and reliability clearly emphasizes the impact of reliability and design synchronization on product life-cycle consideration for enhanced product/system performance.

As the demand for systems with high levels of performance (less cost, high safety levels) increases, so does the need for standardized methods and techniques and associated computer-aided tools for determining the reliability of the resulting complex systems. Some of the reasons for the need include the following (Poucet, 1990):

1. The necessity to reduce effort expended by the user on some important but nonskilled activities such as drawing, data transfer, and the like.

2. The need to provide some degree of formalism to the types of analyses that are required for different problem situations.

3. The requirement for uniformity of modeling and solution approaches so as to enhance repeatability and completeness.

4. The paucity of the knowledge and science bases for modeling different classes and sizes of problems.

5. Cost reduction for modeling and analyses.

Thus the major objective of reliability design and analysis is to develop reliability criteria, based on specific functional requirements of the product/system, that are achievable with respect to other system parameters such as availability, profitability, safety, and so on (Mogos, 1989). The initial effort in this regard is the development of the functional relationships and/or interrelationships between the components and the parent system or product. In addition to the functional requirements consideration, there is the need to ensure that such reliability requirements are also compatible with the manufacturing processes. Based on this, possible system configurations in the form of reliability block diagrams are developed. These are then used to evaluate design configurations as well as the performance of the configurations with respect to the reliability goals, based on assessed component reliability values and/or failure rates. However, as the design progresses to higher levels of complexity and reality, more detailed analyses of the overall reliability is needed. If the product/system configuration or architecture and the resultant topology is such that the components can be decoupled easily into separate components without affecting the performance integrity of the system, then such decoupling could result in decomposition into independent subunits in which the components are either in parallel, series, or a configuration in which the system is made up of subsystems or components that have a basic topology that is both series and parallel. In this decomposed form, the reliability of the system is easily computed through analytical techniques by using well-known mathematical formulas (Sawyer & Foster, 1986; Mimouni, 1986; Kapur & Lamberson, 1977; Lewis, 1987). For the series configuration, the formula is

$$R_s = \prod_{i=1}^{n} R_i$$

where R_s = system reliability
$\quad\quad R_i$ = reliability of component i
$\quad\quad n$ = number of components in series

and for the parallel configuration, the formula is

$$R_s = 1 - \prod_{i=1}^{n} (1 - R_i)$$

where n = number of components or banks of components in parallel.

Several reliability evaluation techniques have been proposed. These include the binomial expansion approach (Dhillon & Singh, 1981), the network reduction approach (Kaufmann et al., 1977), the decomposition method (Kapur & Lamberson, 1977; Dhillon & Singh, 1981), the minimal-cut set and minimal-tie set method (Kaufmann et al., 1977; Dhillon, 1983; Shooman, 1968), the delta star method (Dhillon, 1983), the Markov process (state space) method (Dhillon, 1983), the path tracing method (Dhillon, 1983), and the event space method (Shooman, 1968; Sharma et al., 1982). All these methods are based on the assumption that the components are independent. Please consult the list of references for more detailed discussions about each method.

COMPUTATION APPROACHES FOR COMPLEX SYSTEMS

Because of the complexity of most practical systems and the nature of their failure modes, it is often not possible to decouple a system into independent subunits. This limits the ability to represent the system as strictly series, strictly parallel, or both. Thus, the computation of the reliability of such complex systems is often cumbersome and hence difficult to accomplish using analytical methods (Smith, 1976; Mimouni, 1986).

A few approaches have been proposed for addressing the computation problem for complex systems. The effectiveness of some of these methods has been greatly enhanced by the increased capabilities of different computing environments as well as new software and the associated platforms. These approaches can be grouped into two categories: the conventional approaches with varying levels of computer support, and artificial intelligence/expert system (AI/ES) approaches. A sample computation for a conventional approach is presented.

CONVENTIONAL COMPUTATION APPROACHES

One method of computing the reliability of complex systems that has been found to always be applicable is the method of *complete enumeration*. In this approach, all possible mutually exclusive ways in which the system could fail or survive are examined with a view toward characterizing the effect of each mode (failure or success) on the system operation (Kapur & Lamberson, 1977; Halpern, 1978). A major disadvantage of this method, however, is that the number of modes increases exponentially as the number of components in the system increases. Thus, for a system with n components, the number of mutually exclusive ways the system could operate could be as high as 2^n, with the actual number fixed by the system configuration.

Another method that is also used to analyze this kind of system is the *linked configuration* approach (Lewis, 1987). In this approach, one or more of the system components are assumed to provide the link to other components. The assumption of linkage, unless it is fairly obvious from the topology, can lead to different results and the possibility of nonunique solutions.

Decomposition/conditional probability (Shooman, 1968) is yet another approach that is used for studying this type of system. This method is based on the bayesian approach. In this method, the survival probability of the system is conditioned on the survival or failure of a component or group of components. First the system is decomposed into simpler subunits, and then the computations of the decomposed subsystems are performed separately and combined to obtain the reliability of the original system. As in the case of the linked configuration approach, the applicability of this method lies in the choice of the key component, that is, the component used to decompose the network. Once the component is chosen, the reliability computation becomes almost trivial. On the other hand, for a moderately sized complex system this method could be tedious since the choice of a key element may not be easily identified.

Monte Carlo simulation (Pritsker, 1986; Curry et al., 1989) is becoming the tool of choice by many researchers in the area of reliability analysis. Simulation is attractive to researchers because, instead of characterizing the system through analytical mathematical constructs, a descriptive, logical model of the system is developed and then used to capture the operational characteristics of the system. Estimates of the system parameters are then derived though statistical experimentation using the Monte Carlo approach of random number generation.

Sample Computation.   The following examples illustrate the computation of the reliability of a complex system using the methods described earlier. The system that is described by the reliability block diagram in Fig. 4.7 is considered complex because the system as shown cannot be reduced to a series or parallel configuration.

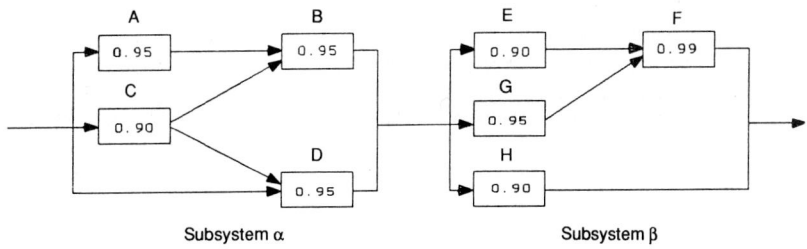

FIGURE 4.7 System configuration (Kapur & Lamberson, 1977).

Complete Enumeration Approach. An explicit enumeration of the mutually exclusive ways a system of n components would operate (survive) or fail is given by the binomial expansion of n, that is,

$$(p + q)^n = p^n + {}^nC_1 q^1 p^{n-1} + {}^nC_2 q^2 p^{n-2} + \cdots + q^n$$

where p_i = probability component i operative and q_i = complement of p_i.

While this expansion represents the mutually exclusive ways in which the system would survive or fail, the coefficient of each term of the expansion represents the maximum number of ways in which i components out of n would survive or fail. In other words, if i-out-of-n components are not operative, then the total number of ways this could happen is ${}^nC_i$. This number represents the upper bound of the number of possible combinations. However, the actual value is determined by the system architecture or configuration, but its value will be no more than ${}^nC_i$. For a given system, this number could be very large and is typically in the order of 2^n. Thus, rather than evaluate success—the number of ways a system would be operative given component failures—it is much less cumbersome to evaluate system failure.

Consider the system shown in Fig. 4.7 with subsystems α (with $n = 4$), and β (with $n = 4$). The following notation is used:

$$E_i = \text{event component } i \text{ is operative}$$

$$\overline{E}_i = \text{event component } i \text{ is not operative}$$

The probabilities associated with the failure of subsystem α because of the failure of the components are as shown on Table 4.1. From the table,

$$R(\overline{\alpha}) = P(\overline{\alpha}) = 0.0027375 \qquad R(\alpha) = 0.9976265$$

Note that in this subsystem, for $i = 1$, ${}^nC_1 = 4$. However, because of the system configuration, the number of possibilities is actually zero. Also for $i = 0$, the actual number is also 0.

TABLE 4.1 Probability of Failure for Subsystem α

i	${}^nC_i$	Actual failure number	Failure event	Probabilities
4	1	1	$\overline{D} \cap \overline{B} \cap A \cap \overline{C}$	$(0.05)(0.05)(0.05)(0.1) = 0.0000125$
3	4	3	$\overline{D} \cap \overline{B} \cap A \cap \overline{C}$	$(0.05)(0.05)(0.95)(0.1) = 0.0002375$
			$\overline{D} \cap \overline{B} \cap A \cap C$	$(0.05)(0.05)(0.05)(0.9) = 0.0001125$
			$\overline{D} \cap B \cap \overline{A} \cap \overline{C}$	$(0.05)(0.95)(0.05)(0.1) = 0.0002375$
2	1	1	$\overline{D} \cap \overline{B} \cap A \cap C$	$(0.05)(0.05)(0.95)(0.9) = 0.0021375$
				$P(\overline{\alpha}) = 0.0027375$

TABLE 4.2 Probability of Failure for Subsystem β

i	$^{n}C_i$	Actual failure number	Failure event	Probabilities
4	1	1	$\bar{H} \cap \bar{F} \cap \bar{E} \cap \bar{G}$	$(0.1)(0.01)(0.1)(0.05) = 0.000005$
3	4	3	$\bar{H} \cap \bar{F} \cap \bar{E} \cap \bar{G}$	$(0.1)(0.01)(0.1)(0.95) = 0.000095$
			$\bar{H} \cap \bar{F} \cap E \cap \bar{G}$	$(0.1)(0.01)(0.9)(0.05) = 0.000045$
			$\bar{H} \cap F \cap \bar{E} \cap \bar{G}$	$(0.1)(0.99)(0.1)(0.05) = 0.000495$
2	1	1	$\bar{H} \cap \bar{F} \cap E \cap G$	$(0.1)(0.01)(0.9)(0.05) = 0.000855$
				$P(\bar{\beta}) = 0.001495$

For subsystem β ($n = 4$, $i = 1, 2, 3, 4$), the probabilities of system failure given different combinations of component failures is as shown on Table 4.2. From the table,

$$R(\bar{\beta}) = P(\bar{\beta}) = 0.001495 \qquad R(\beta) = 0.998505$$

The resulting system configuration for the two subsystems is a series configuration whose reliability is easily computed as the product of the reliability of the two subsystems:

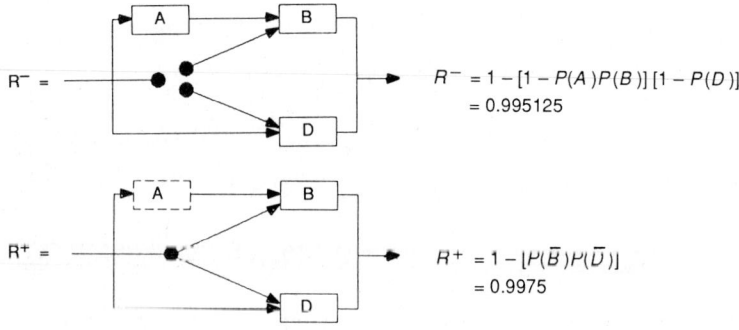

$$R(S) = P(S) = R(\alpha)R(\beta) = (0.9976265)(0.998505) = 0.99571159$$

Linked Configuration Approach

$$R = R^{-}(1 - R_L) + R^{+}R_L$$

where R_L = reliability of the link
R^{-} = reliability of system, assuming the link fails
R^{+} = reliability of system, assuming link works

For subsystem α, if the link is provided by component C, then the resulting configurations for both R^{+} and R^{-} are:

$$R^{-} = 1 - [1 - P(A)P(B)][1 - P(D)]$$
$$= 0.995125$$

$$R^{+} = 1 - [P(\bar{B})P(\bar{D})]$$
$$= 0.9975$$

For β, assuming the link is provided by component G:

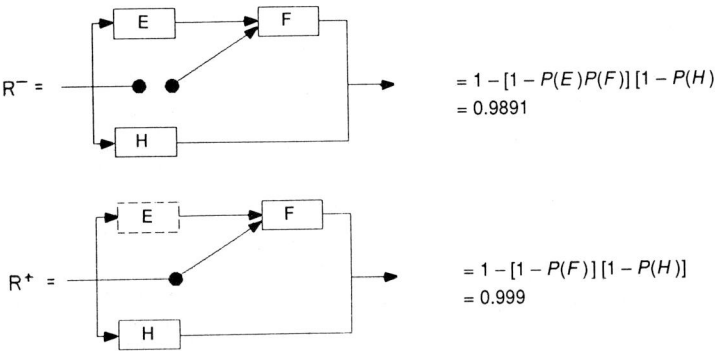

$R^- =$

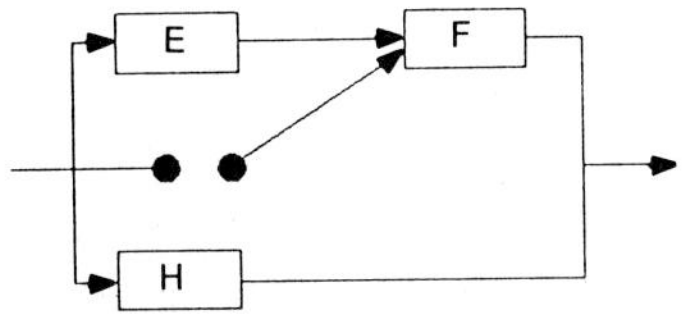

$= 1 - [1 - P(E)P(F)]\,[1 - P(H)]$
$= 0.9891$

$R^+ =$

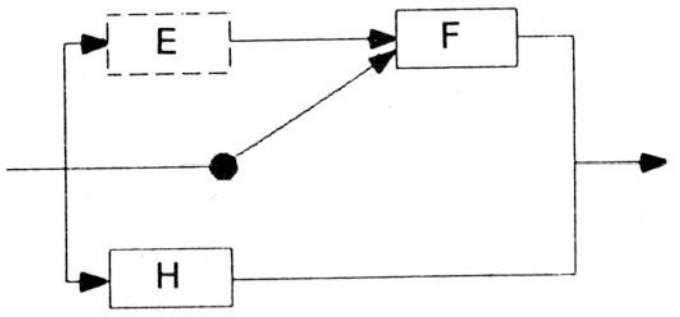

$= 1 - [1 - P(F)]\,[1 - P(H)]$
$= 0.999$

For the resulting configuration of the two subsystems, the system reliability is

$$R(S) = (R_\alpha)(R_\beta) = [(0.99125)(0.1) + (0.9975)(0.9)] + [(0.9891)(0.05) + (0.999)(0.95)]$$
$$= 0.99571159$$

Decomposition/Conditional Probability Approach. The conditional probability approach is based on the survival probability of the subsystem given the survival or failure of a component or group of components. From the concept of mutually exclusive events, the probabilities for both subsystems are

$$P(S_\alpha) = P(S_\alpha/C)P(C) + P(S_\alpha/\overline{C})P(\overline{C})$$
$$P(S_\beta) = P(S_\beta/G)P(G) + P(S_\beta/\overline{G})P(\overline{G})$$
$$P(S) = [P(S_\alpha)][P(S_\beta)]$$

where:

$P(S\alpha/C)$
$= 1 - [P(\overline{B})P(\overline{D})]$
$= 1 - (1 - 0.95)\,(1 - 0.95)$
$= 0.975$

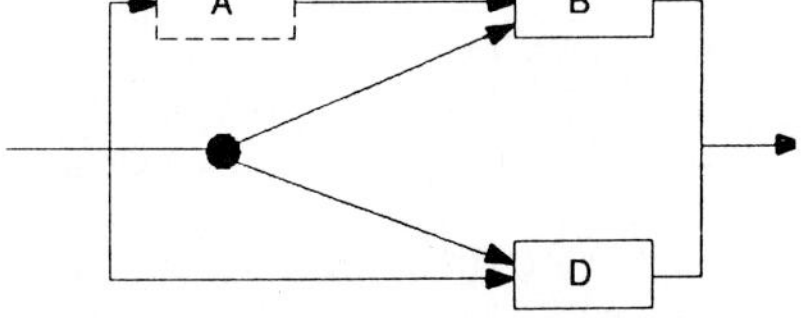

$P(S\alpha/\overline{C})$
$= 1 - [1 - P(A)P(B)]\,[1 - P(D)]$
$= 0.995125$

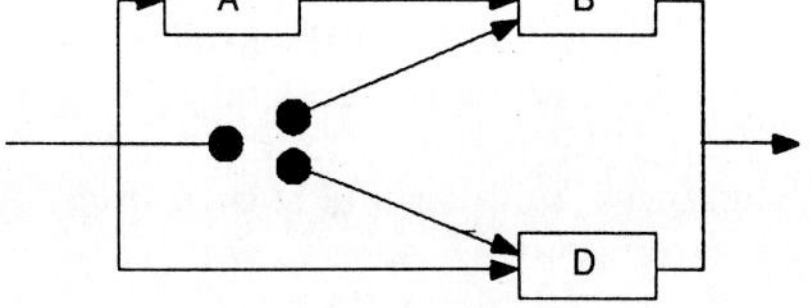

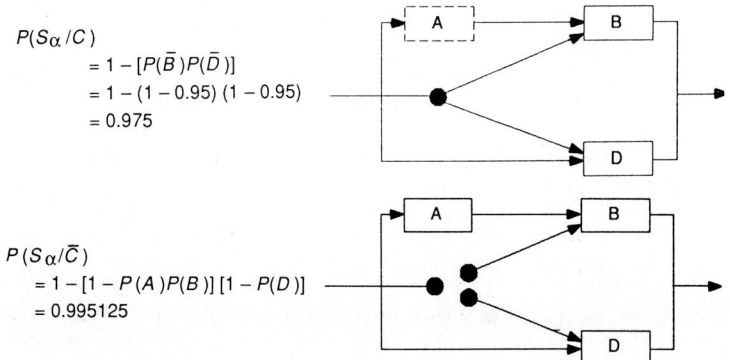

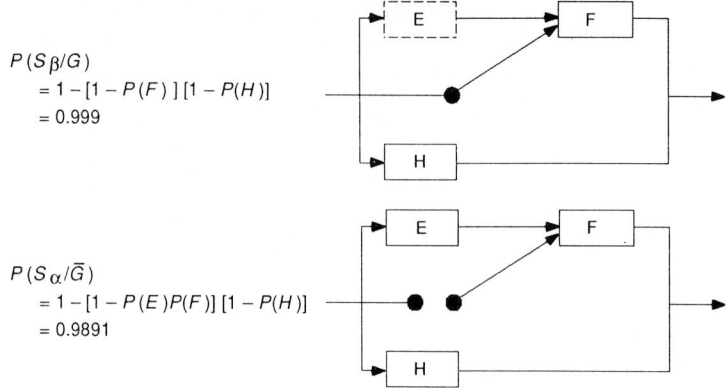

$$P(S_\beta/G)$$
$$= 1 - [1 - P(F)][1 - P(H)]$$
$$= 0.999$$

$$P(S_\alpha/\overline{G})$$
$$= 1 - [1 - P(E)P(F)][1 - P(H)]$$
$$= 0.9891$$

Substituting the appropriate values for the conditional probabilities and the component reliabilities gives the system survival probability value $P(S) = 0.99571159$, as before.

Computer Software Packages for Reliability Computation. Advances in computer technology have facilitated the development of software packages that incorporate some of the methods used to evaluate not only complex systems but the basic system configurations of strictly series and/or parallel. Table 4.3 is the result of a survey about some of the reliability software (including expert systems) currently in use.

In addition to that listed in the table, other software that has appeared in current literature includes:

CARDS. Computer-Aided Reliability Diagnostic System (CARDS) is an electronic design analysis expert system, developed by General Dynamics Space Systems Division, used primarily for the improvement of system reliability early in the design and development cycle. Developed on the Apollo DN330 platform, CARDS runs under the UNIX or AEGIS operating environment with windowing capability. CARDS is both rule-based and frame-based and uses LISP as the development language (Harbater et al., 1988).

HARP. Hybrid Automated Reliability Predictor (HARP) and its third-generation cousin CARE III (Computer-Aided Reliability Estimation) are reliability and availability predictor programs for large-scale, fault-tolerant, highly reliable digital-computer-based systems (Barvuso & Martensen, 1988). The model covers both transient and intermittent fault occurrences and offers the option of using a constant or nonconstant failure rate. It also has fault/error-handling parameters and uses the behavioral decomposition approach to handling the intractable explosion of the Markovian state space, a problem common to most stochastic modeling.

RAMCAD. Reliability and Maintainability in Computer-Aided Design (RAMCAD) is a U.S. Army initiative for an expert system for mechanical reliability prediction, software quality assurance, integrated design engineering analysis, and product assurance. RAMCAD/ULCE extends the RAMCAD concept to include a variety of relevant life-cycle costs and tradeoff design criteria (Cox, 1987; Naft & Petch, 1987).

MathCAD. These programs are mathematical "toolboxes" with symbolic manipulation capability to perform algebraic calculations. They can be used for step-by step sequential and logical reliability analysis. They are extremely useful in reliability analysis of simple and complex fault-tolerant systems. The program provides sensitivity and tradeoff analyses, with results shown in graphical format (Dowlatshahi, 1990).

RelDraw, MilStress, MainTain. RelDraw is a package that draws reliability block diagrams and calculates reliability. The probability of failure is calculated for all parts of the diagram and for the system as a whole by using failure rate data. The required hardware is an IBM PC (or compatible) with 640K random-access memory (RAM) and hard disk.

TABLE 4.3 Features of Packages for Maximum Likelihood Fitting

Date	Censor	Grafstat	Lindep	SAS 6.0	Star	Statpac	Survcalc	Survival	Survreg
Observed and right censored	Yes	Yes	Yes	Yes	Yes	Yes	Yes	Yes	Yes
Left censored	Yes	Yes	No	Yes	Yes	Yes	No	Yes	Yes
Interval	Yes	Yes	Yes	Yes	Yes	Yes	No	Yes	Yes
Transformations	Yes	Yes	Yes	Yes	Yes	Yes	Yes	Yes	Yes
Subset selection	Yes	Yes	Yes	Yes	Yes	Yes	Yes	Yes	Yes
Simulation	Yes	Yes	Yes	Yes	With effort	Yes	No	No	No
Distributions									
Exponential	No	Yes	Yes	Yes	Yes	Yes	Yes	Yes	Yes
Weibull	Yes	Yes	Yes	Yes	Yes	Yes	Yes	Yes	Yes
(Log)normal	Yes	Yes	Yes	Yes	Yes	Yes	Yes	Yes	Yes
(Log)logistic	Yes	Yes	Yes	Yes	Yes	With effort	Yes	Yes	Yes
Gamma	No	No	Yes	Yes	Yes	No	Yes	No	No
Extraneous value	No	Yes	No	No	Yes	Yes	No	Yes	Yes
General gamma	No	No	No	Yes	Yes	No	Yes	No	No
Other	No	No	Gompertz	No	Truncated	No	Gompertz	No	No
User-programmed distributions	No	No	Yes	Yes	No	Yes	Yes	Yes	Yes
Relationships									
Linear for location parameters	Yes	Yes	Yes	Yes	Yes	Yes	Yes	Yes	Yes
Linear without intercept	Yes	Yes	Yes	Yes	Yes	No	Yes	No	No
Log linear for scale parameters	No	No	No	No	Yes	No	No	No	No
Cox proportional hazards	No	Yes	Yes	Yes	Yes	No	Yes	Yes	Yes
User-programmed relations	No	No	Yes	No	No	Yes	Yes	Yes	Yes
Maximum likelihood fitting									
Stepwise	No	No	No	No	No	No	No	Yes	Yes
Hold coefficient/parameter Values fixed	No	No	Yes	Intercept And shape	Yes	No	Yes	Yes	Yes And constraint
Frequency count data (weights)	Yes	Yes	Yes	Yes	Yes	Yes	No	Yes	Yes
Fit output									
Estimates and asymptotes normal, confidence limits for parameter/coefficients	Yes	Yes	Yes	With effort	Yes	Yes	Yes	Yes	Yes
Percentiles	Yes	Yes	Yes	Yes	Yes	Yes	Yes	Yes	Yes
Fraction failing	Yes	Yes	Yes	Yes	Yes	Yes	No	Yes	Yes
User-programmed functions	No	No	Yes	No	No	Yes	No	Yes	No
Covariance matrix of estimates	Yes	Yes	Yes	Yes	Yes	Yes	Yes	Yes	Yes
Max. log likelihood	Yes	Yes	Yes	Yes	Yes	Yes	Yes	Yes	Yes
Likelihood ratio limits	No	Yes	No	No	No	No	No	No	No
Plot of fitted relations	No	No	No	Yes	Yes, percentiles	No	No	Yes	No
Probability plot of fitted cumulative density function	No	Yes	Yes	Yes	And confidence limits	No	Yes	Yes	Yes
Model evaluation									
Residuals	Yes	Yes	No	With effort	Yes	Yes	Yes	Yes	No
Probability plots (right censored)									
Exponential	No	Yes	No	With effort	Yes	Yes	Yes	Yes	Yes
Weibull	No	Yes	No	With effort	Yes	Yes	Yes	Yes	Yes
Extraneous value	Yes	Yes	No	With effort	Yes	Yes	Yes	Yes	Yes
Lognormal	Yes	Yes	No	With effort	Yes	Yes	Yes	Yes	Yes
Normal	Yes	Yes	No	With effort	Yes	Yes	Yes	Yes	Yes
Nonparametric	No	Yes	Yes	Yes	Yes	Yes	Yes	Yes	Yes

TABLE 4.3   Features of Packages for Maximum Likelihood Fitting (*Continued*)

Date	Censor	Grafstat	Lindep	SAS 6.0	Star	Statpac	Survcalc	Survival	Survreg
Model evaluation (*Continued*)									
Other	No	No	No	No	(Log)logistic	No	Gompertz	(Log)logistic	(Log)logistic
Peto/Turnbull cumulative density function estimate	No	No	No	No	Yes	No	No	Yes	Yes
Likelihood ratio tests	Yes	Yes	Yes	Yes	Yes	With effort	Yes	Yes	Yes
Cross-plots	Yes	Yes	Yes	Yes	With effort	Yes	No	Yes	No

Computer packages. Listed in the table are statistical packages with standard features suitable for fitting parametric accelerated testing models to data by maximum-likelihood methods. The table is partly based on the survey of statistical software for life data analysis by Wagner and Meeker (1985).

Censor is briefly described by Meeker and Duke (1981). Meeker and Duke (1982) provide the user manual. It runs on the IBM 370, VAX, and similar computers. Contact: Prof. William Meeker, Dept. of Statistics, Iowa State Univ., Ames, IA 50011, (515) 294-5336.

Grafstat features for fitting lifetime regression models to censored data are presented by Schatzoff and Lane (1986). It runs on IBM mainframe computers. Proprietary to IBM. Contact: Dr. Martin Schatzoff, IBM, 101 Main St., Cambridge, MA 02142, (617) 576-9235.

Lindep is described by Greene's (1986) 300-page manual. It runs on IBM PCs and compatibles. Contact: Dr. William Greene, 528 4th St., Brooklyn, NY 11215, (718) 499-8612.

SAS features for such fitting appear in the SAS (1985) user's guide under the Lifereg procedure and supplemental library procedures Coxreg and PHGLM. Version 6.0 now on PC's and UNIX workstations has all features in the table. Version 6.0 will soon be available on mainframes. Lefkowitz (1985) explains the basics of using SAS; this introduction is a model of simplicity and clarity for other manuals. SAS runs on most mainframe computers and PCs. Contact: SAS Inst. Inc., Box 8000, Cary, NC 27511, (919) 467-8000.

Star, developed by AT&T, has not been publicly released yet. Buswell, Meeker, Myers, and Gibson (1985) briefly describe it. Buswell, Meeker, and Myers (1984) provide the user's manual. It runs under UNIX. Contact: Dr. Jeff Hooper, AT&T Bell Labs, Crawford Corner Rd., Holmdel, NJ 07701, (201) 949-1996.

Statpac is briefly described by Strauss (1980). Nelson, Morgan, and Caporal (1983) provide a 35-page user manual. Nelson and Hendrickson (1972) provide the complete manual. It runs on the Honeywell 600/6000. Contact: Wayne Nelson, 739 Huntingdon Dr., Schenectady, NY 12309.

Survcalc is described in the user's manual by Lauchenbruch (1985). It runs on IBM PCs and compatibles. Contact: Wiley Professional Software, 605 Third Ave., New York, NY 10158, (212) 850-6788.

Survival is described in the 200-page manual by Steinberg and Colla (1988). It is a fully integrated module of Systat. It runs on the MacIntosh PC, VAX, NCR Tower, and Data General computers. Contact: Systat, Inc., 1800 Sherman Ave., Evanston, IL 60201, (312) 864-5670.

Survreg is described in the user guide by Preston and Clarkson (1980, 1983). It runs on IBM mainframes, PCs, and compatibles, Data General, Sun, and many others with Fortran 77 compilers. Contact: Dr. Douglas Clarkson, IMSL Inc., 2500 Park West Tower One, 2500 City West Blvd., Houston, TX 77042-3020, (713) 782-6060.

Source: Adapted from Nelson, W., *Accelerated Testing: Statistical Models, Test Plans and Data Analyses,* Wiley, New York, 1990, Table 1.1, p. 238. © Reprinted by permission of John Wiley & Sons Inc.

MilStress calculates failure rates for electronic components in accordance with the reliability prediction standard MIL-HDBK-217. Modules of up to 32,000 components can be used. The required hardware is an IBM (or compatible) with 640K RAM plus hard drive.

The MainTain package calculates mean time to repair (MTTR) at any level of an equipment or system. An IBM PC (or compatible) with 640K RAM and hard disk is required. The package is offered by ITEM Software Solutions, Item Software Ltd., 1990.

EXPERT SYSTEMS

Even with recent advances in computing technology, the design and implementation of reliability prediction programs for modeling complex systems (especially highly reliable fault-tolerant systems) is still a slow, cumbersome, and costly task and presents a major challenge for analysts and designers. Thus, there is the need for a computer-aided tool that would quickly and cost-effectively assist the design engineer and analyst in the use of domain-specific knowledge related to system structure, component failure rates, hazard functions, fault identification, and fault criticality in analyses of system reliability and availability, including capacity for fault-tolerance. Recent developments in artificial intelligence (AI), expert systems (ESs), and vastly enhanced (memory and speed) computing platforms hold great promise in the effort to determine the reliability of complex systems as well as that to achieve the overall goal of better quality assurance.

An *expert system* is computer software that has a wide base of knowledge in a restricted domain and uses complex inferential reasoning to perform tasks or reach conclusions (Giarratano & Riley, 1989; Nebendahl, 1988). ESs typically store and manipulate knowledge on a specific subject domain and solve problems by employing a large number of domain-specific facts and heuristics and by making logical deductions, sometimes with a high degree of skill, in complex and conflicting heuristic areas that would normally require the abilities of a human expert. They are also able to justify the inferences of experts or reproduce dialogues (as well as the final results) of experts in a specific problem area. Through AI reasoning and problem-solving techniques, ESs are able to draw conclusions that were not programmed into their knowledge base. This is in contrast to conventional programming techniques which require inputs that are certain, use primarily numeric computation in well-defined algorithms, and, when programmed correctly, produce correct results. Expert systems, on the other hand, use information that is not entirely complete, manipulate it by symbolic reasoning techniques without following a numeric model, and still product satisfactory answers.

The main features of expert systems that distinguish them from conventional computer programs include the ability to (Poucet, 1990; Mogos, 1989):

1. Manipulate and manage increasingly complex information processing needs
2. Handle and resolve conflicts for opposing reliability goals and criteria, besides arriving quickly at solutions or accommodating new knowledge and experience from a better understanding of system behavior and interactions

ESs are used whenever conventional data processing (DP) solutions are not possible or economical because of the complexity of the problem, the dynamics of the problem, or the combinatorial explosive nature of the problem, in which case knowledge-based solution methods are preferred (Nebendahl, 1988). Knowledge-based solution methods typically involve the use of rules or other structures that contain the knowledge and experience of experts, logical inferences, interpretation of ambiguous facts, and the manipulation of imprecise knowledge, that is, knowledge that is affected by certainty factors. ESs typically use only heuristics and are data-driven rather than procedure-driven.

The Structures of Expert Systems.   ES structures consist of several parts, namely the knowledge base, the inference engine, the working memory, and the user interface. Figure 4.8 is the

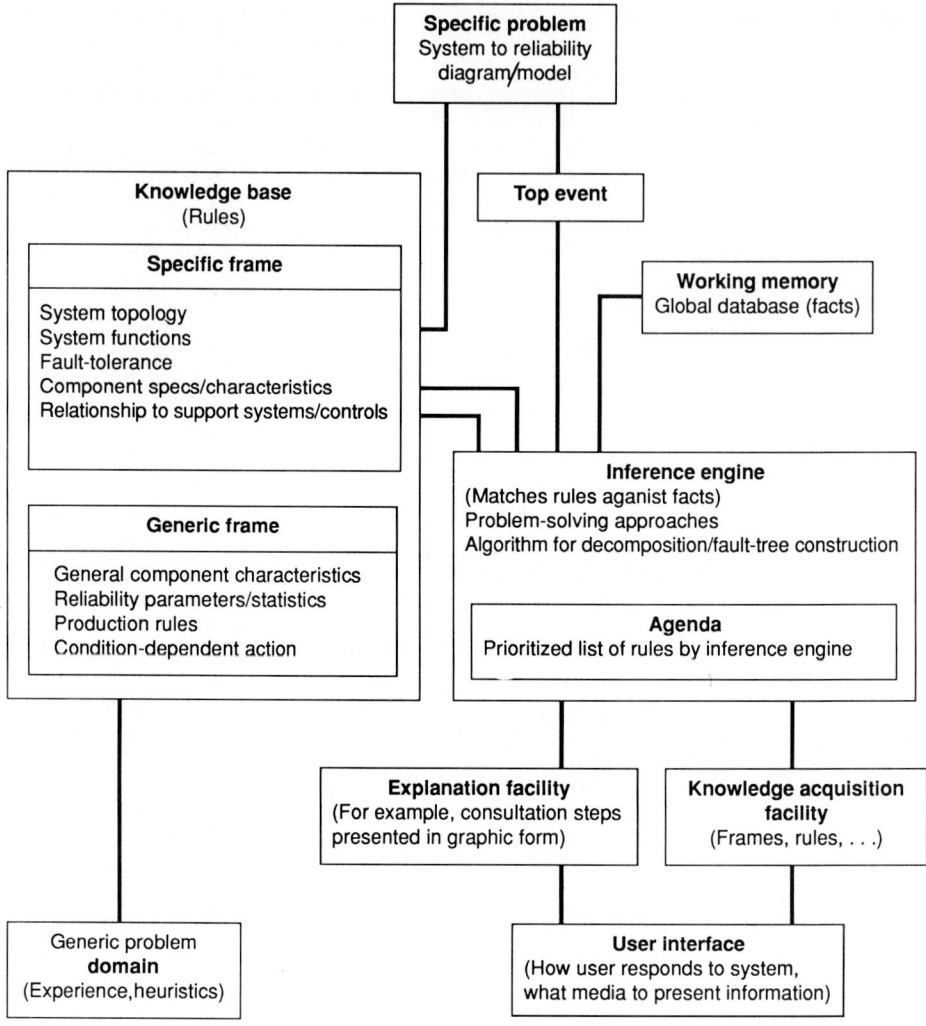

FIGURE 4.8 Structure of rule-based expert system.

architecture of the basic elements of an expert system for a reliability/fault-tree diagnostic system and their interactions (Giarratano & Riley, 1989; Poucet, 1990). Not shown is the database which, even though necessary, is not usually a standard feature of ESs. It is also possible for ESs to interact or interface with other software and equipment for control and monitoring purposes.

Knowledge Base.   The knowledge base (KB) contains a symbolic representation of the facts (data and their relationships), complex objects and their attributes, relationships between objects, and rules for processing knowledge and for deriving new knowledge from existing knowledge, that is, heuristics for a specific problem domain. The form in which the domain-specific knowledge is represented in the knowledge is used to characterize ESs. These are

- Semantic networks
- Frames
- Production rules (IF-THEN rules)
- Logic-based representation via predicate calculus

For reliability problems, the semantic is sometimes preferred to the forms since it is a collection of objects (nodes) connected by links (arcs). To represent component connectivity, two schemes for describing the graph structure or topology are widely used: the adjacency matrix (for the number of components) and the linked list (for connections that are non-zero) (Mogos, 1989).

Inference Engine.   Through the inference mechanism, the system knows which knowledge is required, when, and how. The inference mechanism is a complex computer program that handles the symbolic language processing. It controls the actions taken by the system and provides the problem-solving or the reasoning-strategy methods by which the rules, networks, or frames are processed. The most commonly used inference mechanisms are *forward chaining* and *backward chaining. Forward chaining* is a data-driven approach in which processing starts with the set of known facts, and then all the hypotheses in which these facts play a part are tested. *Backward chaining,* a goal-driven approach, starts with a set of goals, solutions, or hypotheses and then propagates to rules that support the goals or conclusions. An inference mechanism can also implement a combination of these methods (Mogos, 1989; Nebendahl, 1988).

Working Memory.   The working memory stores information about the current state of consultation or run of the expert system. It contains all the known facts, assertions, and conclusions reached by the system up to the present state. The contents of the working memory is used to justify system conclusions (Nebendahl, 1988; Giarratano & Riley, 1989).

User Interface.   The question and response dialogue between the system and the user during a consultation or a run of the ES is conducted via the user interface through a computer (personal computer, mainframe, minicomputer, or specialized AI machine). The user typically responds to questions or queries by using special symbols or natural language. The expert system also has an explanation component which provides considerable support to the user during a consultation (Nebendahl, 1988; Giarratano & Riley, 1989).

Knowledge Acquisition.   A knowledge engineer works with experts to structure and formalize their knowledge and make it available to the knowledge base. Empirical knowledge is the hardest to obtain because the experts themselves often do not recognize this as useful knowledge. The knowledge engineer has to develop a method for finding the quickest way to solve a problem and learn how to determine which rules to keep and which rules to disregard under certain circumstances. The *acquisition component* in an expert system is used by the knowledge engineer to implement the knowledge base. Expert systems are developed by an iterative method called *rapid prototyping.* This method includes steps such as concept development, knowledge acquisition and implementation, testing and verification, and analyses. Rapid prototyping also generates the possibility to create functional prototypes with relatively small development effort and then gradually perfect it.

Development Support and Programming Languages.   Development support is obtained through symbol-processing languages, such as LISP and PROLOG, program development environments, and shells. Symbol processing languages are used primarily for formalized representation of knowledge. They are dialogue-oriented and can easily be interpreted. Program development environments are an extension of programming languages providing complex functions and resources such as structure-oriented editors, compilers, and debuggers. Shells are tools which support the work of the knowledge engineer. Some shells are simple tools for structuring knowledge, that is, programs for representing knowledge relationships. Others provide one or more knowledge-representation mechanisms and, in some cases, an inference mechanism. Furthermore, some shells offer more extensive control options, or support the creation of the interface. Shells contain all the components of an expert system, except for the knowledge base. This enables the knowledge engineer to concentrate more on the major task, namely, the implementation of knowledge.

A Prototype Expert System—STARS.   STARS (Software Tools for the Analysis of Reliability and Safety) is a knowledge-based expert system implemented under the UNIX environment (Poucet, 1990). The knowledge representation is frame/rule-based, organized with a hierarchical structure. Its main use is for developing an integrated set of computer-aided reliability and analysis tools for the various tasks involved in system safety and reliability evaluation, including hazard identification, quantitative analysis [using HAZOP (Hazard and Operability Analysis) and failure mode and effect analysis (FMEA) tools], and logic model construction and evaluation (for example, sequence and system modeling, event-tree, fault-tree, and Markov process modeling tools). Construction of the fault tree is implemented via the frame/rule approach in which the deductive (goal-driven) reasoning and the heuristics are applied during the construction.

The tools (hazard identification and so on) obtain their knowledge through the implementation of four system KBs, the plant/unit KB, a substance KB, the matrix KB, and a component KB. The component KB contains characteristics for different types of components and knowledge about the way components behave under normal and nonnormal conditions. The inference engine uses production rules and backward chaining for control.

STARS runs on a UNIX workstation which makes it easy to develop an advanced user interface with windows and graphics capabilities. The implementation of the fault tree includes a CAD tool for describing system topology and for editing the fault tree.

CONCLUSION

In today's complex technological global environment, the need for component and system assurance has generated considerable interest about the design process, its parameters, and the overall impact of design concurrence or synchronization on product assurance. As systems become more complex, integrated, and cost-intensive, there is increased awareness of the implications of safety and availability on system performance and the need to develop adequate methodologies and procedures to aptly address these issues to achieve total system assurance. It is thus imperative to include reliability and safety parameters in all the phases of the product life cycle decisions. This means the synchronization of design decisions so that the impact of each element in the life cycle may be determined and given proper weight.

To do this will require a different way of thinking and evaluation of information needs, since there is much ambiguity at the initial phase of the design process. New and improved methods of reliability prediction will have to be developed to cope with the seemingly ambiguous initial phase of product and system development. Expert systems will play an important role in this regard. However, conventional computation approaches still have their role (at least for the foreseeable future) in helping define and determine reliability requirements. Furthermore, as we continue to depend very heavily on computer software, there is a strong need to establish guidelines and criteria with which to evaluate software, especially with respect to their reliability and availability.

REFERENCES

Bartee, T., *Data Communications, Networks, and Systems,* Sams, Inc., Indianapolis, 1985.

Barvuso, S., and A. Martensen, "A Fourth Generation Reliability Predictor (HARP, CARE III)," *Proceedings, Annual Reliability and Maintainability Symposium,* IEEE, Piscataway, N.J., 1988.

Berman, S. M., *The Elements Of Probability,* Addison Wesley, New York, 1969.

Brook, R. H. W., *Reliability Concepts in Engineering Manufacturing,* Wiley, New York, 1972.

Cox, J., "US Army Reliability and Maintainability in Computer Aided Design (RAMCAD) Initiatives," *Proceedings, Annual Reliability and Maintainability Symposium,* IEEE, Piscataway, N.J., 1987.

Curry, G., B. Deuermeyer, and R. Feldman, *Discrete Simulation: Fundamentals and Microcomputer Support,* Holden Day, Oakland, Calif., 1989.

Dhillon, B., *Systems Reliability, Maintainability and Management and Operation,* PBI, New York, 1983.

Dhillon, B., and C. Singh, *Engineering Reliability: New Techniques and Applications,* Wiley, New York, 1981.

Dowlatshahi, J. G., "Reliability Analysis Based on a Mathematical Tool," *Proceedings, Annual Reliability and Maintainability Symposium,* Piscataway, N.J., 1990.

Eckert, P. T., and P. E. Janusz, "Software Quality Assessment Measure," *1986 Proceedings Annual Reliability and Maintainability Symposium,* The Institute of Electrical and Electronics Engineers, Piscataway, N.J., 1986.

Giarratano, J., and G. Riley, *Expert Systems; Principles and Programming,* PWS-Kent Series in Computer Science, Boston, Mass., 1989.

Graham and Jones, *Expert Systems Knowledge, and Uncertainty and Decision,* Chapman and Hall, 1988.

Halpern, S., *The Assurance Sciences,* Prentice-Hall, Englewood Cliffs, N.J., 1978.

Harbater, S., W. Tonelli, and A. Belski, "Computer Aided Reliability System (CARD)," *Proceedings, Annual Reliability and Maintainability Symposium,* Piscataway, N.J., 1988.

Item Software Limited, RelDraw, MilStress, MainTain, Hampshire, England, 1990.

Kapur, K. P., and L. R. Lamberson, *Reliability in Engineering Design,* Wiley, New York, 1977.

Kaufmann, A., D. Grouchko, and R. Cruon, *Mathematical Models for the Study of the Reliability of Systems,* Academic Press, New York, 1977.

King, D., and P. Harmon, *Expert System: An Artificial Intelligence in Business,* Wiley, New York, 1985.

Lewis, E., *Introduction to Reliability Engineering,* Wiley, New York, 1987.

Mimouni, C., "Simulation and Antithetic Variates Techniques for Computing and Validating the Reliability of Complex System," unpublished master's thesis, University of Cincinnati, 1986.

Mogos, G., "The Determination of the Reliability of Complex Systems Using Expert Systems," unpublished master's thesis, University of Cincinnati, 1989.

Naft, J., and M. Petch, "A RAMCAD/ULCE Workstation Shell Structure," *Proceedings, Annual Reliability and Maintainability Symposium,* IEEE, Piscataway, N.J., 1987.

Nebendahl, D. (ed.), *Expert Systems: Introduction to the Technology and Applications,* Wiley, New York, 1988.

Nilson, J. N., *Problem-Solving Methods in Artificial Intelligence,* McGraw-Hill, New York, 1971.

Organick, E., A. Forsythe, and R. Plummer, *Programming Language Structures,* Academic Press, New York, 1978.

Parsaye, K., and M. Chignell, *Expert Systems for Experts,* Wiley, New York, 1988.

Poucet, A., "STARS: Knowledge Based Tools for Safety and Reliability Analysis," *Journal of Reliability Engineering and System Safety,* vol. 30, 1990.

Pritsker, A. A. B., *Introduction to Simulation and Slam II,* 3d ed., Halstead, New York, 1986.

"RelDraw, MilStress, MainTain," Item Software Limited, Hampshire, England, 1990.

Rubinstein, F. M., *Tools for Thinking and Problem Solving,* Prentice-Hall, Englewood Cliffs, N.J., 1986.

Savell, D. V., R. A. Perez, and S. W. Koh, "Scheduling Semiconductor Wafer Production: An Expert System Implementation," *IEEE Expert,* Fall 1989, pp. 9–15.

Sawyer, B., and D. Foster, *Programming Expert Systems in Pascal,* Wiley, New York, 1986.

Shooman, M. L., *Probabilistic Reliability: An Engineering Approach,* McGraw-Hill, New York, 1968.

Smith, O. C., *Introduction to Reliability in Design,* McGraw-Hill, New York, 1976.

Watanabe, M., et al., "CL: A Flexible and Efficient Tool for Constructing Knowledge-Based Expert Systems," *IEEE Expert,* Fall 1989, pp. 41–50.

Wolfgram, D. D., T. J. Dear, and C. S. Galbraith, *Expert Systems for the Technical Professional,* Wiley, New York, 1987, p. 5.

PART 3
SOFTWARE RELIABILITY

In the 1970s, the cost of software surpassed the cost of hardware as the major cost of a system.[1] In addition to the cost of developing software, the penalty costs of software failures are even more significant. The failure of software sometimes involves very high costs, human lives, and a social impact. Therefore, devising ways to measure and predict the reliability of software becomes an important issue.

In the past 30 years, more than 400 papers have been published in the areas of software reliability modeling, model validation, measurement, and practices. Since software engineering is an interdisciplinary subject, software reliability models are also developed from different perspectives of the software and different applications of the model. In order to pave the way for the future development and evaluation of highly reliable software and of systems involving software and hardware, a detailed taxonomy of the existing software reliability models and the assumptions behind those models is of value. This part presents a classification scheme for software reliability models. It is not our intention to review all published papers. Instead, we will highlight development of software reliability engineering in this part.

CHARACTERISTICS OF SOFTWARE RELIABILITY MODELS

In analyzing hardware reliability, reliability engineers are interested in the failure process because they emphasize the analysis of failure data and the design of the experiment. In software reliability, on the other hand, engineers are interested in the failure mechanism. Most software reliability models are analytical models derived from assumptions of how failures occur. The emphasis is on the model's assumptions and the interpretation of parameters.

In order to develop a useful software reliability model and to make sound judgments when using the models, engineers need an in-depth understanding of

- How software is produced
- How errors are introduced
- How software is tested

- How errors occur
- What types of errors occur
- What environmental factors can help us justify the reasonableness of the assumptions, the usefulness of the model, and the applicability of the model under a given user environment

General Description of Software and Software Reliability Software.   Software, often called a *computer program* or simply a *program,* is a collection of instructions or statements in computer language. Upon execution of a program, an input state is translated into an output state. Hence, a program can be regarded as a function mapping the input space to the output space, where the input space is the set of all input states and the output space is the set of all output states. An input state can be defined as a combination of input variables or a typical transaction to the program. The definition of software reliability is similar to that of hardware reliability; *time-domain software* reliability is defined as the probability of failure-free operation of software for a specified period of time under specified conditions.[1]

Computer programs are designed to perform some specified functions. When the actual output deviates from the expected output, a *failure* occurs. Since the definition of failure differs from one application to the next, it should be clearly delineated in the specifications. A *fault* is incorrect logic, incorrect instructions, or inadequate instructions that in their execution will cause a failure. In other words, faults are the sources of failures and failures are the realization of faults. Whenever a failure occurs, there must be a corresponding fault in the program, but the existence of faults may not cause the program to fail. Note that *error* and *bug* are loosely used by many authors and practitioners to represent fault and sometimes failure.

Bug-Counting Concept.   The bug-counting model assumes that conceptually there are a number of faults in the program. Given that faults can be counted as an integer number, bug-counting models estimate the number of initial faults at the beginning of the debugging phase and the number of remaining faults during or at the end of the debugging phase. Bug-counting models use per-fault failure rate as the basic unit of failure occurrence.

Depending on the type of models, we assume the failure rate of each fault to be either a constant, a function of debugging time, or a random variable from a distribution. Once the per-fault failure rate is determined, the program failure rate is computed by multiplying the number of faults remaining in the program by the failure rate of each fault.

During the debugging phase, the number of remaining faults changes. One way of modeling this failure process is to represent the number of remaining faults as a stochastic counting process. Similarly, the number of failures experienced can also be denoted as a stochastic counting process. If we assume perfect debugging (that is, a fault is removed with certainty whenever a failure occurs), the number of remaining faults is a nonincreasing function of debugging time. Under an imperfect debugging assumption, the number of remaining faults may increase or decrease. This bug-counting process can be represented by the binomial model, Poisson model, compound Poisson process, Markov process, or doubly stochastic process. Although the bug-counting concept is controversial, it is valuable in many applications.

User Environment.   Software reliability is subject to the user environment. Operational profile and system load are two environmental factors.

Operational profile is the distribution of input state execution. It is a discrete probability distribution function. Depending on the application, an input state could mean a typical transaction of daily operations, a partition of input space, or a combination of input variables. Since the relationship of input, fault, and failure is deterministic, how inputs are selected determines how failures occur. In other words, if the assumption is that all faults are equally likely to be detected, it implies that input states are selected randomly. During testing, the test cases should be generated randomly according to the operational profile, so that the testing strategy will conform with the assumptions of the model. In the operational phase, some input states are executed more frequently than the others and this occurrence must be considered when evaluating the reliability of the software.

The system load consideration is derived from the phenomenon that software is more likely to fail at peak hours than at the normal operational hours. In other words, the failure rate is not only a function of time (CPU time or operational time) but also a function of system load. This observation leads to a correction factor added to the software reliability model.[2]

Time Index.   Since materials for hardware deteriorate over time, calendar time is a widely accepted index for the reliability function. In software, however, failures will never happen if the program is not used. In the context of software reliability, time is more appropriately interpreted as the stress placed on or the amount of work performed by the software. The following *time units* have been suggested as indexes of the software reliability function.

- *Execution time:* CPU time
- *Operational time:* time the software is in use (usually estimated as 8 working hours per day)
- *Calendar time:* index used for software running 24 hours a day
- *Run:* a job submitted to the CPU
- *Instruction:* number of instructions executed
- *Path:* the execution sequence of an input

Models based on execution time, operational time, calendar time, and instructions executed belong to the time-domain model. Models based on run and path belong to the input domain model. Other time units such as development time, simulated execution time, and number of tests are also used as needed.

Although it seems that software reliability models do not have a unified index, the unification can be achieved through unit conversion. For example, Musa et al.[2] have proposed methods of converting their execution time model to the calendar time model. The input domain model can also be converted into a time-domain model through a factor of number of runs or paths executed per unit time.

Modeling Software Reliability.   Like hardware reliability modeling, software reliability modeling by a probability and stochastic approach has been successful. Although software and hardware reliability are quite different in several ways, as outlined in Table 4.4,[5] software reliability has been modeled by following the philosophy and pattern of hardware reliability. In fact, reliability of software is quite similar to that of hardware if the former is treated as if it were at the infant mortality stage of the latter.

The same terminology and definitions used with hardware reliability have been adopted for the prediction, estimation, and calculation of software reliability. In addition to convenience and tradition, the following considerations provide explanations for quantifying software reliability and modeling it just as we have modeled hardware reliability:

1. We need a unified performance measure so that two or more software systems can be compared to each other.

2. A quantitative measure, such as a probability measure, can help predict software survivability at different design, development, and maintenance phases. The probability measure used in hardware reliability has been well-known and widely adopted by engineers.

3. A system performance, which usually considers hardware, software, and human com-

TABLE 4.4 Hardware Reliability versus Software Reliability[5]

	Hardware reliability	Software reliability
Fundamental concept	Hardware fails because of physical effects	Software fails because of program error
Life-cycle causes:		
Design	Incorrect physical design	Incorrect program design
Development	Quality control problems	Incorrect program coding
Operation	Degradation and failure	Undetected program errors
Use effects	Hardware fails or wears out	Software does not fail or wear out
Function of design	Physics of failure	Programmer skill
Domains	Time	Time and data
Time relationship	Bathtub curve	Decreasing function
Mathematical models	Theory well-established	Theory established but not well-accepted
Time domain functions	$R = f(\lambda, t)$, exponential (constant λ), Weibull (increasing λ), normal (wearout)	$R = f(errors, t)$; no agreed-on time function models; various models proposed
Data domain	No meaning	Errors $= f$(data, tests)
Growth models	Several models exist	Several models exist
Metric	λ, MTBF, MTTF	Failure rate, number of errors detected or remaining
Growth application	Design, prediction, TAAF	Design, prediction, TAAF
Prediction techniques	Block diagrams, fault trees	Path analysis, complexity, simulation
Test and evaluation	Design and production acceptance	Design acceptance
Design	MIL-STD-781C (exponential); other methods (nonexponential)	Path testing, simulation, error seeding, bayesian
Operation	MIL-STD-781C	None
Use of redundancy:		
Parallel	Can improve reliability	Need to consider common cause
Standby	Automatic fault detection and switching	Automatic error detection and correction
Majority logic	m out of n	Impractical

MTBF = mean time between failures; MTTF = mean time to failure.

ponents, can be readily evaluated if performance of each of the individual components is modeled consistently.

4. Software performance varies as a function of time, which is also typical of the way hardware behaves. In addition, software maintenance, like hardware maintenance, requires debugging and correction staff.

5. Software quality and reliability are commonly perceived as an engineering subject; hence a quantitative measure is necessary. We would like to build reliability into the software design at an early stage.

System Tradeoffs. Simply achieving desired software reliability is not always possible. Hence, tradeoffs among functional capability, schedule, cost, and software reliability become very important. These tradeoffs involve further training of software engineers, marketing pressure, opportunity cost, and other economic considerations.

When evaluating software reliability for a specific system, software engineers often consider the degree of severity with which a fault may degrade the system. Examples include basic service interruption, basic service degradation, inconvenience with correction not deferrable, and minor tolerable effects with correction deferrable. When overall software reliability is considered, estimated failure intensity should be weighed against its degree of failure severity to get the overall failure intensity and failure rate.

CLASSIFICATION OF SOFTWARE RELIABILITY MODELS

Software reliability models can be classified into deterministic and probabilistic models. The deterministic model studies (1) the elements of a program by counting the number of operators, operands, and instructions; (2) the control flow of a program by counting the branches and tracing the execution paths; (3) the data flow of a program by studying the data sharing and data passing; and (4) other deterministic properties of a program.

Performance measures of the deterministic model are obtained by analyzing the program texture and do not involve any random events. The deterministic models include methods of software science, information content, software complexity, and software quality attributes. In general, these models empirically measure the qualitative attributes of software and are used either in the early phases of the software life cycle to predict the number of errors in a program or in the maintenance phase for assessing and controlling the quality of software.

The probabilistic model represents the failure occurrences and the fault removal as probabilistic events. It can be further divided into several models: error seeding, curve fitting, reliability growth, execution path, program structure, input domain, failure rate, nonhomogeneous Poisson processes, Markov processes, and bayesian inferences.

The error-seeding model estimates the number of errors in a program by using the capture-recapture sampling technique. Errors are divided into indigenous errors and introduced errors (seeded errors). The unknown number of indigenous errors is estimated from the number of introduced errors and the ratio of the two types of errors obtained from the debugging data.

The curve-fitting model uses regression analysis to study the relationship between software complexity and the number of errors in a program, the number of changes, the failure rate, or the time between failures.

The reliability growth model measures and predicts the improvement of reliability through the debugging process. A growth function is used to represent the progress. The independent variables of the growth function can be time, number of test cases, or testing stages, and the dependent variables can be reliability, failure rate, or cumulative number of errors detected.

The execution path model estimates software reliability on the basis of the probability of executing a logic path of the program and the probability of an incorrect path. This model is similar to the input domain model because each input state corresponds to an execution path.

The program structure model views the program as a reliability network. A node represents a module or a subroutine, and the directed arc represents the program execution sequence among modules. By estimating the reliability of each node, the reliability of transition between nodes, and the transition probability of the network and by assuming independence of failure at each node, the reliability of the program can be solved as a reliability network problem.

The input domain model uses *run* (the execution of an input state) as the index of reliability function, as opposed to *time*, for the time-domain model. The reliability of each run is defined as the number of successful runs over the total number of runs. Emphasis is placed on the probability distribution of the input state or the operational profile.

The failure rate model studies the functional forms of per-fault failure rate and the program failure rate at the failure intervals. Since mean time between failures is the reciprocal of failure rate, models based on time between failures also belong to this category.

The Markov model is a general way of representing the software failure process. The number of remaining faults is modeled as a stochastic counting process. When a continuous-time, discrete-state Markov chain is adapted, the state of the process is the number of remaining faults and time between failures is the sojourning time from one state to another.

When a nonstationary Markov model is considered, the model becomes very rich and unifies many of the proposed models. The nonhomogeneous Poisson process model is one extension. The nonstationary failure rate property can also simulate the assumption of nonidentical failure rates of each fault.

The bayesian model assumes a prior distribution of the failure rate. This model is used when the software reliability engineer has a good feeling about the failure process and the failure data are rare.

There are at least 20 specific models realized from the above general models. Each specific model is used for one type of application. In fact, none of them is definitely superior to others. For details of these models, see Refs. 2 and 4. The probabilistic model is the mainstream of the software reliability study because it can be integrated with the hardware reliability theory. As systems get more complex, the system functions will involve both hardware and software components. This common framework makes it possible to evaluate the reliability of a hardware-software system.

SOFTWARE LIFE-CYCLE COSTS

Studies of software cost have concentrated on development cost. However, life-cycle cost is the more appropriate one to be studied.

For hardware, the life-cycle cost is usually studied from a buyer's standpoint. Life-cycle costs can be divided into procurement, maintenance, and disposal costs. Since software development and maintenance are normally performed by the same organization, the software life-cycle cost is usually studied from the developer's point of view and is divided into costs of design and development and costs of operation and maintenance.

The software design and development process can be broken down into several phases:[1,3] requirement and specification, design, coding, and testing. Among these phases, testing (including unit tests, integration tests, and field tests) accounts for 40 percent or more of the development cost.

Operation and maintenance costs make up about 60 percent of software life-cycle costs. Included in maintenance costs are such activities as preventive maintenance, corrective maintenance, adaptive maintenance, enhancement, and growth. Since testing is part of each of these operation and maintenance activities, it becomes a major factor in the cost of maintenance as well as of design and development.

For common software projects, the reliability cost is mainly incurred by testing. To produce highly reliable software, we incur additional reliability cost at every phase of the software life cycle. Indeed, a large portion of the software life-cycle cost is devoted to achieving high reliability. Table 4.5 compares reliability costs incurred at each phase of the software life cycle for common and highly reliable software.

A number of time-domain software reliability models have been proposed. A categorical limitation of the existing models is that they treat all software failures equally, merely counting failures and not taking into account the widely differing severities of these failures. What is needed is a software reliability modeling technique that enhances the existing models to take into account the penalty cost associated with each failure.

TABLE 4.5   Reliability Cost and Software Life-cycle Phases[1]

Phase	Reliability cost of common software	Additional reliability cost for highly reliable software
Design and development		
Requirements and specifications	Basic requirement and specification walkthrough	Parallel development of requirements and specifications, and detailed validation
Design	Basic design walkthrough	Parallel design, fault-tolerant design, and detailed verification
Coding	Basic code walkthrough	Parallel coding of critical modules, fault-tolerance codes, and detailed code walkthrough
Testing	Basic testing	Extensive testing, stress test, and reliability assessment
Operation and maintenance		
Preventive maintenance	Totally devoted to reliability	Higher frequency of preventive maintenance
Corrective maintenance	Totally devoted to reliability	Immediate correction and extra testing
Adaptive maintenance	Testing	Extra testing
Enhancement	Equivalent to a development subcycle	
Growth	Equivalent to a development subcycle	

CONCLUSIONS

1. Embedding quality and reliability concepts into the software design process can decrease software life-cycle costs.
2. Considering software reliability models as design tools, one should be able to evaluate reliability-related software cost from a system viewpoint.

REFERENCES

1. Lin, H. H., and W. Kuo, "Reliability Related Software Life Cycle Cost Model," *Proceedings, 1987 Annual Reliability and Maintainability Symposium,* Piscataway, N.J., 1987, pp. 364–368.
2. Musa, J. D., A. Iannino, and K. Okumoto, *Software Reliability: Measurement, Prediction, Application,* McGraw-Hill, New York, 1987.
3. Chi, D., and W. Kuo, "Optimal Design for Software Reliability and Development Cost," *IEEE J. Selected Areas in Communications,* vol. 8, no. 2, February 1990, pp. 276–282.
4. Shooman, M. L., *Software Engineering: Design, Reliability, and Management,* McGraw-Hill, New York, 1983.
5. Kline, M. B., "Software and Hardware R & M: What are the Differences," *Proceedings, 1980 Annual Reliability and Maintainability Symposium,* Piscataway, N.J., 1980, pp. 179–185.

USE OF COMPUTERS

CHAPTER 1
COMPUTER FUNDAMENTALS

Mary T. McKinney, Director
Small Business Development Center
Duquesne University
Pittsburgh, Pennsylvania

The computer has been the core of the information age. Continued technological developments of the past two to three decades have propelled society into an era whereby the computer is an integral part of daily life. Once a machine relegated to special rooms and physically accessible only to data processing experts, the computer is now commonplace not only in factories and offices, but also in homes, automobiles, and in much of the equipment the average individual encounters daily.

This chapter is devoted to an overview of digital computer fundamentals. Its purpose is to serve as an introduction to the principles of the computer, which has become a major tool of the industrial engineering—and virtually every other—profession. The following chapters detail specific applications used by the modern manufacturing and industrial engineer which depend on the use of the computer.

Since the history and operation of the computer and its various components can fill volumes, it is necessary to narrow the scope of the subject. The pages that follow discuss coding and how data are represented to the computer. Subsequent comments discuss the basic hardware of components of the computer. The discussion then briefly reviews the technological developments of the past two decades and the emergent computer systems. Subsequent comments include a discussion of wide-area and local-area computer networking and centralized versus decentralized processing, followed by a discussion of systems and applications software. Finally developments expected in the future are reviewed.

Since fast-paced changes still characterize this dynamic field, each of the sections focuses on generic principles, in contrast to specific vendor products, and trends, rather than specific quantitative detail.

DATA REPRESENTATION

A basic concept in understanding how a digital computer functions is understanding how data are represented to the computer. Distinguishing between the code and the machine is not as difficult as it may appear at first. For example, the holes in a punched card or a punched paper tape are physical spots on a magnetic tape or disk, light beams in a fiber-optic cable, and electrical circuits on a silicon chip are representations of codes.

The one common characteristic of all the above examples is that they are based on a system which recognizes two possible, or binary, states. An electric current, a magnetic spot, or a light beam is either present or absent. The code itself can be represented on paper as a combination of 0's and 1's. Because 0 and 1 are used as digits in binary arithmetic (an arithmetic requiring only two symbols), they are usually referred to as *bits* (binary digits).

The translation of a decimal to a binary number is illustrated in Table 1.1. In any number system, the position of each digit in the number represents the base number raised to a power (exponent) of that position. In a decimal system (base 10), the first position going from right to left represents 10^0, which equals 1. Similarly, the second position represents 10^1, or 10, the third position represents 10^2, or 100, and the fourth position, 10^3, or 1000. Thus the number 1000 is equal to $(1 \times 10^3) + (0 \times 10^2) + (0 \times 10^1) + (0 \times 10^0)$. In a binary (base 2) system, there are only two digits, but the same logic is true. Hence, the decimal number 1000, when converted to base 2, is represented by 1111101000 (which is derived from $(1 \times 2^9) + (1 \times 2^8) + (1 \times 2^7) + (1 \times 2^6) + (1 \times 2^5) + (1 \times 2^4) + (1 \times 2^3) + (0 \times 2^2) + (0 \times 2^1) + (0 \times 2^0)$.

True binary representation is limited, however, by its cumbersomeness and inability to handle characters. These limitations are overcome by using standard-sized binary-based codes, grouped in fixed lengths. Typical lengths are 8 bits, called a *byte,* and a larger grouping of 16, 32, or 64 bits, called a *word.*

The two most popular coding systems for representing numbers and characters to the computer are ASCII (American Standard Code for Information Interchange) based on a 7-bit byte with a parity checking bit, and EBCDIC (Extended Binary-Coded Decimal Interchange Code) based on an 8-bit byte, shown in Table 1.2.. Each system represents the complete alphabet, decimal numbers, and punctuation marks, and utilizes a parity checking bit to lessen the chance of a transmission error. With the ASCII system, 2^7, or 128 symbols can be represented, and 2^8, or 256, with EBCDIC. An enhanced version of ASCII, ASCII-8, allows for 8 bits and 256 possible characters and numbers.

TABLE 1.1   Decimal-to-Binary Conversion

Decimal	Straight binary
Base 2 number derivation	Base 10 number derivation
0	0
1	1
2	10
3	11
4	100
5	101
6	110
7	111
8	1000
9	1001
10	1010
11	1011
...	...
100	1100100
...	...
1000	1111101000

TABLE 1.2   Examples of ASCII and EBCDIC Codes

Symbol or character	ASCII bit representation	EBCDIC bit representation
0	0110000	11110000
1	0110001	11110001
2	0110010	11110010
3	0110011	11110011
...	...	...
A	1000001	11000001
B	1000010	11000010
C	1000011	11000011
...	...	...

COMPUTER HARDWARE

Although the invention of the microprocessor in 1971 has greatly altered the architecture and construction of computer systems, the basic functional principles of the computer have remained the same. A computer is an electronic machine which, by means of stored instructions and information, performs rapid, often complex instructions, and compiles, correlates, and selects data. Applications are limited only by the creativity of the human beings who use computers and have been expanding exponentially over the past several years.

The Central Processing Unit (CPU).   This is the heart of the computer, interpreting and executing instructions and acting on data. The CPU also communicates with the input, output, and storage devices that are integral to a functioning computer system. Figure 1.1 illustrates the primary elements of a computer system: input device, CPU, secondary or auxiliary storage device, and output device. Technological developments continue to rapidly alter the components, capabilities, and performance of each segment of the computer and have enabled computers to network with each other far more readily than was possible less than a decade ago. Each of the elements is discussed below.

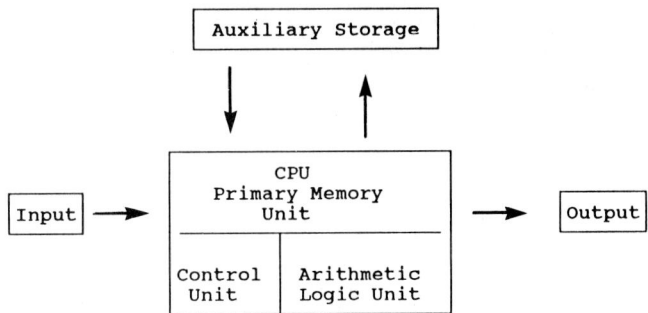

FIGURE 1.1   Basic components of a computer.

The CPU consists of an arithmetic/logic unit (ALU), primary or main memory, and a control unit. Electronic impulses from an input device are processed by the CPU under program control and transferred to the output device. Since memory is limited, for most applications the computer also accesses secondary storage. The CPU varies in size, depending on the computer system. A large mainframe CPU consists of hundreds of circuit boards, and on a basic microcomputer, the CPU is on a chip [sometimes called a microprocessor unit (MPU)]. The CPU contains various registers (devices that hold instructions and data), memory addresses, and an accumulator. The control unit directs the flow of program instructions and data between the various components. Since the contents of the registers can be acted on more quickly than the contents of main memory, the control unit loads program instructions and data to the registers from memory prior to processing by the ALU. The ALU performs both arithmetic and logical operations at great speeds and is kept continually supplied with instructions and data by the control unit.

An internal clock synchronizes the CPU components. The number of clock pulses per second, typically expressed in terms of megahertz (MHz, millions of cycles per second), play a major role in how fast an instruction is processed by the CPU. There are several other determining factors of computer speed, including word size (the number of bits processed simultaneously), memory capacity, and access time, as well as the presence or absence of special functions such as floating-point processing, pipelining, and memory cache. Figure 1.2 depicts the typical machine cycle for processing (fetching and executing) an instruction.

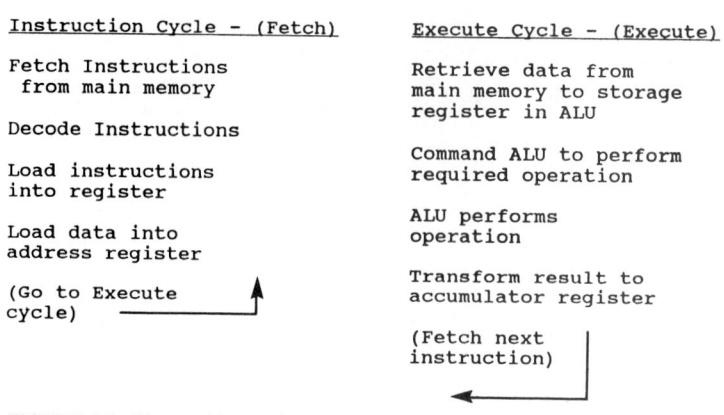

FIGURE 1.2 The machine cycle.

All machine cycle operations, except when the ALU actually performs the operation, are conducted by the control unit. A single program may involve millions of machine cycles in order to run. The time to process a single instruction, typically expressed in nanoseconds (ns, billionths of a second) or microseconds (μs, millionths of a second) is important when one considers machine performance. Current high-end machines perform in picoseconds (ps, trillionths of a second).

Memory. Data initially come into a computer through the input device and/or auxiliary storage and are placed into storage or memory until fetched by the control unit. Computers, depending on the model, have varying amounts of main memory, which is measured in terms of megabytes (MB, millions of bytes) or kilobytes (KB, thousands of bytes). Each bit of memory has a unique address which is accessed at rapid speeds by the control unit. The addresses are accessed directly, in contrast to a serial or sequential search, and, therefore, memory is commonly referred to as random-access memory (RAM). Bits are accessed in parallel from memory in word groupings. Therefore, the larger the word size, the faster data will be fetched from memory. Memory access is generally slower than processing speed and, therefore some models have a high-speed buffer, called *cache,* to hold data and instructions intermediately between storage and processing.

Many modern computers utilize a bus structure to transfer data between devices such as memory and I/O (input/output) devices and the processor. A bus is simply a group of circuits that transfers bits in parallel and, therefore, contains more circuits as word size increases. The size of the bus plays an important role in the data transfer speed. As data are processed, cache and RAM become overwritten, thereby making it necessary to have auxiliary storage.

Some applications which do not require data to be rewritten utilize hardware with read-only memory (ROM). Within ROM modules, the program instructions are contained in the hardware and supplied by the manufacturer or subsequently purchased and plugged into a board within the computer. Since ROM contains program instructions (typically referred to as software), it is sometimes known as *software-in-hardware,* or *firmware.*

Technological developments have increased the capacities and speed of memory. The most common memory used in modern computers is semiconductor memory. Composed of circuits on silicon chips, semiconductor memory, called MOS (metal-oxide-semiconductor), utilizes electrical currents and is dependent on a constant power source. If the power source fails and is not backed up with an emergency system, data are lost. Another technology is bubble memory, consisting of magnetized spots which retain magnetism indefinitely. Because of the complexity of material and cost, bubble memory has been limited to special applications requiring this technology.

Auxiliary Storage. Internal memory, no matter how large the capacity, is limited and is utilized for processing purposes. Internal memory usually does not contain sufficient space to run the operating system and applications programs, let alone the volume of data files that often must be accessed during processing. The role of auxiliary or mass storage is to provide large volumes of data that can be directly accessed by the control unit during processing and then stored again to be preserved for future use. Disks are the most readily used form of on-line auxiliary storage.

Retrievability of the information from the storage medium is very important. Access time is critical, yet must be balanced with efficiency in storage methods when dealing with large amounts of data. Random or direct access allows the control unit to directly address the data to locate and update, without reading all of the preceding records on file. This is in contrast to sequential access, normally associated with tape, wherein records must be read one after another until the data are located.

Storage Technology. Magnetic disks store data in the form of magnetic spots on a cylindrical surface, coated on both sides with magnetic material. Magnetic disks range in size from the floppy disks made popular with the introduction of microcomputers to disk packs containing several platters connected by a shaft. Data are accessed by a read/write head which glides so close to the surface as the disk spins that a human hair or smoke particles, if present on the surface, would create damage. Storage capacity ranges from 1 MB on floppy disks to several gigabytes (GB, billions of bytes) or even terabytes (TB, trillions of bytes) on the platter disk.

Optical disk technology, a development of the 1980s, uses a high-energy laser to burn in a permanent image and a low-energy laser to read the image. The technology is based on coding of laser pulses to correspond to the digital bit stream. Marks on the disk surface are approximately 1 μm in size ($\frac{1}{40}$ the thickness of a human hair), with track spacing at approximately 1.6-μm intervals. Initial applications were for read-only data.

Recent advances in erasable, or rewritable, optical disk technology holds much potential. The NEXT computer system, introduced in 1989, was the first computer system to incorporate a rewritable optical disk. Technological research and development during the 1990s will determine the future role of this technology.

CD-ROM (compact disk read-only memory) and WORM (write once, read many times) are optical disk systems for storing large automated files. Since the data in CD-ROM and WORM units are read only, applications are limited to situations where files are accessed, but not changed by the user. Examples are medical-record storage, service manuals, computer-aided design (CAD) graphics libraries, electronic publishing systems, and reference databases. Storage capacity is as high as several gigabytes for one CD-ROM and terabytes for a multiplattered WORM system (often referred to as a *juke box*).

Although access speed of optical disks is slower than magnetic media (an average speed of 80 ms as compared to a range of 17 to 50 ms), advantages are the greater amounts of data that can be stored and randomly accessed in compact space.

Although disk media are the most frequently used and convenient storage system for data, magnetic tape still plays a major role. Smaller systems utilize cassettes and tape cartridges (150 to 200 feet long), and larger systems utilize magnetic reels typically wound in lengths of 2400 feet in 10½-inch-diameter reels. Magnetic tape provides high-capacity storage at a relatively low cost.

Magnetic tapes are primarily used for backup data as well as for record storage. Both are applications which require relatively infrequent access by the user, and the economical, reliable magnetic tape medium provides a very suitable alternative to disk storage. Another advantage to having a tape storage and retrieval system available is that it provides redundancy in communications channels to the computer in case of disk failure.

COMPUTER INPUT/OUTPUT

Input and output devices make it possible for humans to communicate with the computer, and vice versa. Input equipment translates the normal media of human communication (written symbols and, to a degree, voice) into digitized symbols understandable to a computer. Output devices translate the digitized bit strings into symbols understandable to humans.

Devices such as video display terminals (VDTs), also known as cathode-ray tubes (CRTs), are used for both input and output. The user types information via a keyboard or manipulates the cursor via a keystroke, mouse, trackball, light pen, or touch of the screen, and the data are sent to the computer for processing. After processing, output can be returned to the screen for user viewing. CRTs are often characterized as "intelligent" or "dumb," depending on whether or not they have storage and sometimes processing power. In recent years, the field of ergonomics has greatly enhanced CRT design.

Often users prefer to have a hard copy of the output, and, therefore, printers are very important output devices. Output speed, desired format, and quality of print are three important considerations when choosing what type of printer to use. Printers are usually classified according to impact and nonimpact methods of affixing characters to the paper. An impact printer utilizes a mechanical printing element to physically strike the paper. Speeds vary depending on the resolution of the printer character.

Nonimpact printers which utilize laser heat or photographic technology have become very commonplace in recent years. A line-oriented laser printer can output approximately 20,000 lines per minute, providing characters of excellent quality. As printers become commonplace in offices, the readability of characters gains importance. Attention has turned to improving the quality of output, and the result has been the development of a wide variety of printers which produce "letter quality" output at relatively high speeds.

Several input devices utilize optical mark or character recognition. These techniques utilize reflected light to communicate the input signals to the computer. The familiar barcode analyzers, postal machines, and billing processors utilize this technology.

Other frequently utilized input devices include digitized tablets, voice-recognition devices, and robot sensing devices.

Both input and output may be routed from and to storage devices such as tape or disk, voice-controlled devices, and the control unit for a machine or a piece of equipment.

Graphics is a very important element of input and output which requires special mention. Graphics translates images into digital form and outputs them to a display screen and/or plotter. Graphics displays require a high-performance machine with significant memory in order to accommodate the bit map display, a copy of the image which refreshes the screen continually during the interactive session.

I/O devices are typically connected to the CPU via a bus through a front-end processor, a controller specifically designated for handling communications. Because I/O devices are slower than the processor itself, buffers are used to temporarily store data that is waiting to be printed and to assemble data for the processor.

TECHNOLOGICAL DEVELOPMENTS

In the 1950s and 1960s, mainframe computers, primarily manufactured by less than a dozen computer companies, were the mainstay. The early vacuum-tube-module-based mainframe processed about 10,000 operations per second. Both hardware and software technological developments, particularly the development of the transistor followed by the integrated circuit, improved performance steadily (processing increased to 1,000,000 instructions per second), but use was limited to relatively few applications when compared to what occurred after the invention of the microprocessor.

The first microprocessor, introduced in 1971, consisted of the control and ALU components of the CPU engraved on a silicon chip, the size of a thumbnail. This unit executed 60,000 operations in 1 second, contained 2300 MOS transistors, and literally paved the way to a computer revolution. Capacities and performance of chips developed rapidly throughout the 1970s and 1980s. By 1990, chips were available that incorporated 1 million transistors and contained 32 bit positions. Projections are that by the year 2000, a 64-bit chip containing up to 100,000,000 transistors will be available. This miniaturization of computer circuitry that has characterized the last quarter of the twentieth century has been referred to as very large scale integration (VLSI).

Prices have had an inverse relationship to performance, because of mass production techniques and surging market demand. Figure 1.3 illustrates the price-performance relationship for processing 1 million instructions per second (MIPS) during the rapid-development state of the microprocessor. In the 10-year period from 1980 to 1990, costs to process 1 MIPS decreased by a factor of 10 for the microcomputer and minicomputer and by a factor of 5 for the mainframe, with further reductions projected for the future. The differences between the personal computer, workstations, and microcomputer have rapidly become blurred. In terms of processing power, the powerful 32-bit chips of the 1990s approach mainframe capability.

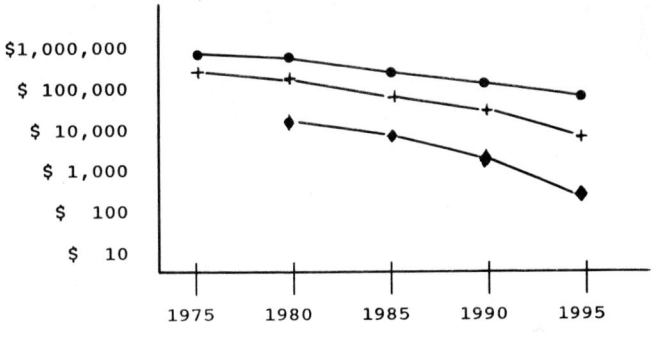

FIGURE 1.3   Price/performance trends to process 1 MIPS.

The technological advances created new applications and new markets for computers. By 1990, there were over 150 manufacturers of computers, in contrast to the 12 mainframe manufacturers of 1960. In 1970 there were fewer than 50,000 computers in use. Twenty years later, the figures are closer to 50,000 computers produced daily. Today's computer market is differentiated by size, capacity, speed, processing sophistication, data storage, and networking capabilities. Figure 1.4 demonstrates the trends that have occurred during the maturation of the computer industry.

VLSI technology, which decreased size and costs, greatly enhanced the computer's power, capacity, and reliability. All of these developments have caused sales to soar. The following describes the classifications of computer hardware that serve the user of the 1990s.

Mainframe.   Mainframe computers have been the dominant machine in the computer industry. The mainframe industry experienced double-digit growth until the latter half of the 1980s, when the rate of growth slowed to single digits, reflecting the emerging midsized and small machines. Although there are several models with varying capabilities, the typ-

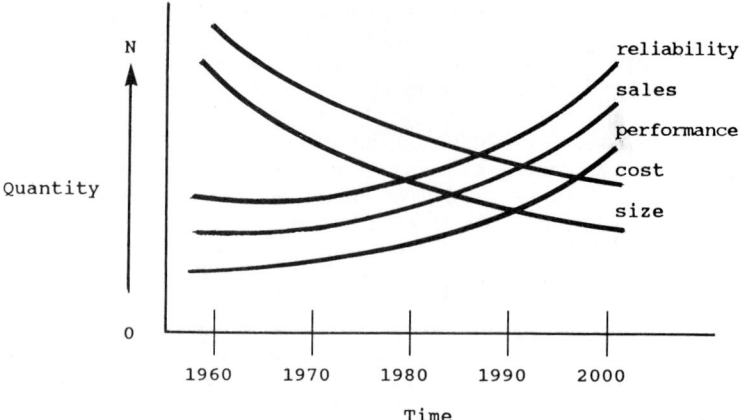

FIGURE 1.4   Technological trends.

ical mainframe system ranges from several megabytes to 2 gigabytes of memory; provides several mega- to gigabytes of storage; contains a 31-, 32-, 36-, or 64-bit address word; and costs in a range from $200,000 to $3 million. Many terminals, perhaps hundreds, are connected to the mainframe, many from outlying locations. The processor can handle hundreds of millions of floating-point instructions per second. Large mainframe systems usually employ multiple CPUs and link more than one machine together with special processors handling terminal and communications interfaces. Recent developments in disk technology have raised the storage limit to 16 TB. Mainframe computers are usually the workhorses of the large-scale organization, performing the accounting functions and other numerical calculations. The mainframe also contains the databases, processes orders, and handles the other functions that have been traditionally labeled data processing tasks. The mainframe requires a controlled environment, usually a room-sized clean facility with air-conditioning equipment.

Supercomputers.   The supercomputer was developed with the scientific community in mind. Scientific applications often require multidimensional calculations and computations of complex mathematical formulas that consume hours of computer time. Supercomputers introduced vector processing, which operates on an array of numbers simultaneously, in contrast to the normal mode of sequential processing of instructions. The CRAY-1 supercomputer, introduced in 1976, executed 32 operations simultaneously. Performance was rated at 150 million floating-point operations per second (MFLOPS).

The supercomputer utilizes reduced instruction set computer (RISC) technology, which eliminates hundreds of instructions representing a host of unnecessary contingencies, thereby increasing speed. Developments in the supercomputer field have increased performance to 16 billion floating-point operations per second (GFLOPS), with projections of up to 160 GFLOPS by the mid-1990s, and perhaps 1 trillion FLOPS (TFLOPS) by the year 2000. Each generation of supercomputer has improved performance by at least a factor of 10. Although early supercomputer technology depended on silicon, projections are that future computers will rely on gallium arsenide (GaAs) technology. The development of GaAs has been difficult because of the fragility of the material, but, once perfected, it will improve CPU performance dramatically because it is faster than silicon and does not require as much cooling. The cost of a supercomputer ranges from $8 to $20 million.

Minisupercomputers, also called *near-supercomputers,* which emerged in the latter part of the 1980s, can provide one-fourth to one-third of the capacity of a supercomputer at one-tenth the cost.

Minicomputers.   The distinction between mainframes and minicomputers has blurred over the past several years. The minicomputer, initially used by scientists, engineers, and others for special applications, introduced the concept of distributed data processing. Rather than being connected to a large installation, sharing the computer resources with the entire organization, the minicomputer with its simpler technology and less rigid environment, served just a few users simultaneously. Initially minicomputers were fairly small, 8- to 16-bit-word machines with limited capabilities. Today's minicomputer configuration is a processor with 16-bit to 32-bit addressing, several megabytes of memory, and one or more disk drives containing several megabytes and sometimes gigabytes of storage, with tape backup systems. Typically minicomputers use a network to communicate with other computers as part of a system. Minicomputers range in price from $15,000 to $250,000. As technology has developed, increased performance has allowed the introduction of faster and larger minicomputers, approaching the capacity of the mainframe. High-end microcomputers have been called superminicomputers and are typically characterized by faster processing, more memory, and faster disk-access times.

Microcomputers and Workstations.   A wide range of microcomputers, which are dependent on microprocessor technology, are available with 8-bit to 32-bit capacity, several megabytes of memory, and built-in hard disks with floppy disk and tape backup systems. The typical microcomputer usually consists of a single board (referred to as the motherboard), often with slots to allow for expansion.

Workstations are distinguished by their high-resolution graphics capability, greater performance, and use of the UNIX operating system (see below). The workstation's high rated performance is a result of its RISC architecture, delivering over 2 MFLOPS of performance. Microcomputers and workstations are typically single-user systems, often linked into a network. Prices and performance within this market vary dramatically, as has the multiplicity of users.

On the low end, a simple microcomputer system with peripherals capable of running basic user applications ranges in price from $500 to $10,000. Workstations range in price from $5000 to $300,000. Workstations are often used by engineers to design complex circuits and parts. They are single-user computers, with emphasis on high-resolution graphics and performance. The higher-end workstation approaches the performance of a minicomputer or a low-end mainframe, with an operating speed of at least 10 MIPS and 1 MFLOPS with expandable main memory.

Microprocessors, Robots, and Special-Purpose Machines.   Microprocessors are the basic controlling device of modern electronic and electromechanical equipment. A typical modern factory employs microprocessors in virtually every aspect, from production scheduling to manufacturing and quality control. When the microprocessor is embedded in a programmable machine that performs the repetitive, complex mechanical tasks in a humanlike way, the device is a robot. A robot is programmed to interpret data from sensors that recognize such patterns as movement and sound and, in turn, manipulate mechanical devices to perform actions in place of human beings. Standard as well as specially designed programming languages are utilized to control the robot's motors and activities.

Microprocessors have also been used in special-purpose equipment. When an application is specifically tailored, often a ROM program is developed that increases speed. The disadvantage of specially designed microprocessors is their inflexibility if change becomes necessary.

NETWORKING AND COMMUNICATIONS

As technological developments increased the use and capabilities of computers, the need likewise occurred for communications from peripherals to the computer and between computers themselves. Advances in the telecommunications industry have kept pace and interfaced with technology enhancements in the computer field, making the two industries very compatible. Early computer networking involved connecting terminals to nearby host computers. The deployment of the modem and other communications devices which enabled the conversion of digital-based computer codes to analog signals carried over telephone lines brought computer capabilities closer to the user. Networks linking computers and peripherals over great distances are referred to as *wide-area networks*.

The burgeoning of computer opportunities and the development of midsize and microcomputers created a demand for interfacing multiple computers within a short range. Local-area networking (LAN) provides short-range interconnectivity between the devices in the network, normally within one site. Data transmission on a local-area network is typically in digital form and does not require conversion.

Major companies with branches usually have the need for both wide-area and local networking. Networking has enabled companies to choose between centralized and decentralized computer systems. A decentralized system, known as a *distributed data processing* (DDP) system, utilizes several linked computers to facilitate data sharing. In a DDP system, multiple computers perform specialized tasks and communicate with other CPUs and/or devices as needed. In a centralized system, user terminals connect to the mainframe at headquarters.

Figures 1.5, 1.6, and 1.7 illustrate three alternatives for local-area networking. The star configuration (Fig. 1.5) consists of a hub computer that interfaces with other computers (nodes) which could in turn have a star network of interfaced computers. Communication between node 1 and node 2 occurs through the hub. The star network is the most widely used system. A disadvantage to the star system is that failure within the hub causes all system communication to fail. Typical applications of the star network are branch offices, or multiple departments, which have their own mini- and/or microcomputer installations and access the central computer database. The ring network depicted in Fig. 1.6 removes the potential failure barrier of a central hub. Each computer is connected to two others, and data are communicated throughout the ring. A special bit pattern signals which nodes are to receive the data. A variation of the ring system is to interconnect all computers in the loop. In a bus system, depicted in Fig. 1.7, all the nodes share a single cable. Messages are broadcast to each node at the same time, with each recognizing its own messages.

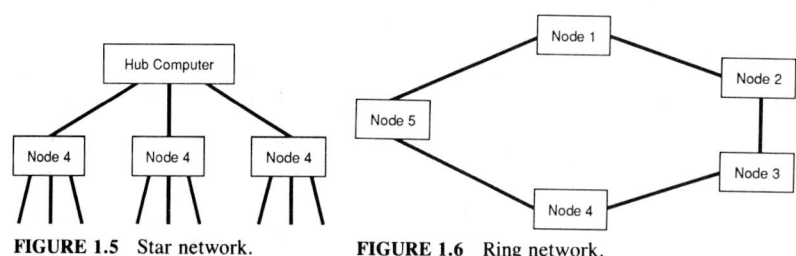

FIGURE 1.5   Star network.   **FIGURE 1.6**   Ring network.

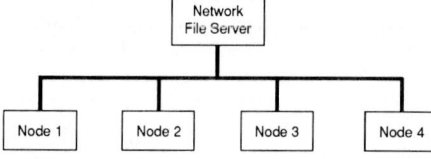

FIGURE 1.7   Bus network.

Although these network configurations are typically associated with local-area networks, some of the nodes may be linked by wide-area networks.

The major differences between local-area and wide-area networks are the communications medium and subsequent need to use or not use the signal conversion equipment. Coaxial cable is used in the vast majority of LANs. Twisted copper line, fiber-optic, microwave, and satellite connections are also used in wide-area networks.

Fiber optics, a thin, clear plastic cable free from electromagnetic interference, is one of the most promising technological developments, transmitting data at speeds up to 565 million bits per second (Mb/s).

The choice of connection usually depends on the speed required. Voice-grade telephone lines can handle speeds up to 9600 bits per second (b/s) and are less expensive than broadband channels which transmit large volumes of data at high speed. Coaxial cables, microwaves, and laser beams transmit data at rates up to 120,000 b/s.

Because computer terminals vary in the types of codes used (ASCII versus EBCDIC), as well as forms of transmission (parallel versus serial and asynchronous versus synchronous), the communications equipment has been adapted to recognize varying data transmission formats. Protocols are standards that have been developed and published by national and international groups to standardize transmission. Although there are still many models that cannot talk to each other, the communications industry has made significant strides in linking multiple-vendor terminals with CPUs as well as multiple-vendor CPUs with each other.

In addition to hardware and communications linkages, networking involves the use of sophisticated software. Software handles the monitoring functions of telecommunications, manages the flow of communications traffic, and interfaces with the operating system. Typical tasks performed by software are network performance monitoring of multivendor equipment, providing gateways to other host systems or networks, controlling network security, performing record and file locking, and providing diagnostics. Although communications network management and control systems (NMCS) suffer from the same vendor differentiation as the hardware and software industries, some standardization has begun to emerge to guide network system development.

COMPUTER SOFTWARE

A computer operates according to a sequence of instructions detailing specific actions it must perform that are outlined in a computer program. The collection of programs, together with the data used by the programs and documentation describing them, is known as *computer software*. Data on the computer are organized into files which consist of various records, typically organized by fields. Several systems of file organization and access methods have been identified. The direct-access approach mentioned earlier in the discussion of disks likewise has accommodated techniques for accessing data quickly. Programmers can write data into files and specifically create files for an application, or use a system database manager which coordinates system resources for the most efficient access method and compatibility with the particular installation. Software is typically categorized into two basic types: systems software and applications software.

Systems Software.   Systems software is the collection of programs that coordinate the various components of the computer system. The basic language of the computer is machine language, where the coding medium is binary digits. Since machine language is complex, intermediate assembler language and high-level programming languages such as COBOL, FORTRAN, Pascal, BASIC, C, LISP, and a host of others have been developed. Programs are characterized by a defined set of procedures which instruct the computer how to proceed.

Programming.   To illustrate some of the features of a typical computer language and how a program is written, a simple example to calculate the mean value of a number of observations follows. Assume the formula is

$$M = S/N$$

where M = mean
S = sum of values
N = number of observations

Assume also that there are nine observations whose values are 67.5, 82.0, 94.7, 86.2, 95.1, 35.0, 44.3, 72.0, and 66.7.

Figure 1.8 illustrates what the program looks like. For convenience, the lines are numbered in steps of 10, which allows for the insertion of additional instructions between any two statements if it is discovered later that something was overlooked in the original instructions given.

```
10    LET N = 1

20    READ X

30    LET S = S + X

40    LET M = S/N

50    LET N = N + 1

60    IF N < 10 THEN 20

70    DATA 67.5, 82.0, 94.7, 86.2, 95.1, 35.0, 44.3,
      72.0, 66.7

80    PRINT "THE MEAN IS"; M
```
FIGURE 1.8 Computer program for solving $M = S/N$.

Line 10 states LET N = 1. N will be used both as a counter to make sure all the values of the observations are added and in the formula $M = S/N$.

Then READ X is given on line 20. Note that on line 70 the data are listed. The first time the computer reaches this READ instruction, it will take the first value out of the list on line 70 and set it aside for whatever computation is to be performed. The numeric value of 67.5 is effectively removed from the list; the next time any READ statement calls for a value, the following number, 82.0, is selected. This process is repeated until the list is exhausted.

On line 30, the instruction LET S = S + X is given. On the first pass, this is equivalent to S = 0 + 67.5.

On line 40, the instruction LET M = S/N is given. This is equivalent to M = 67.5/1 on the first pass. Line 50 advances the counter N by 1, and prepares for a second pass through the loop. However, line 60 states IF N < 10 THEN 20; this means that if N is less than 10, go back to line 20 and repeat lines 20, 30, 40, and 50. On the ninth pass through this loop, M = S/N becomes M = 643.5/9 = 71.5. Line 50 advances N to 10 and line 60 says not to return to 20, but to continue to the next instruction, which is line 80. (If there had been 17 observations instead of 9, the number used for comparison would have been 17 + 1, or 18). Line 80 tells the computer to print out the statement THE MEAN IS 71.5.

In order for the computer to process a program written in a higher-level language, it must be converted to machine language. The BASIC program above is converted one line at a time by an interpreter. Compilers translate the entire program prior to running.

The programmer writes the program by using a set of utility programs and development tools consisting of editors, diagnostics, sorting and merging routines, and library maintenance routines. Also, automatic code generating programs have been developed and used to enhance programmer productivity and efficiency.

Operating System.   The operating system is the main program that allocates the system resources and manages and controls the computer's activities. The operating system schedules and performs input/output, allocates space and resources, provides monitoring and security functions, and governs the execution and operation for the various system programs and applications.

By necessity, the operating system is closely married to the hardware, although the UNIX operating system is constructed to make portability among various hardware systems a not-too-difficult task. Interface routines to the hardware are unique to each machine. For microcomputers, the clones of the IBM PC, which by definition use the same architecture, made possible the migration of the MS-DOS operating system to these machines.

Operating systems which handle several users simultaneously are more complicated than those which process jobs serially or serve only a single user. Time-sharing and multiple-programming-based systems differ in the way they allocate time to one of several programs, but both accomplish multiple user access by allocating short periods of uninterrupted time to each program. Real-time systems prioritize jobs, guaranteeing response time for the top-priority application.

The operating system also manages the movement of program segments to and from both primary and auxiliary storage, and provides interfaces to the input and output devices. The system's overall performance depends on the operating system's ability to manage these functions efficiently, as well as on the hardware's capabilities. Features such as cache memory and paging identify program portions most likely to be accessed and allow them to be processed faster by moving them to memory sooner than other portions, according to the operating system rules.

Parallel processing is a feature that will become more widely used in the future. Parallel, or multi-, processing segments programs to be run simultaneously on more than one processor. Great gains in system performance can obviously be accomplished as parallel processing becomes more developed.

Software Tools.   Another type of software is a group of programs referred to as software tools. Database managers, spreadsheet packages, graphics packages, and word-processing systems are among the most frequently utilized software tools. An argument can be made to classify these programs in either the systems or applications category. These programs are included with the systems software routines and certainly provide a major interface between a user application and the system. On the other hand, they are written in a high-level language, can be purchased as an application program, and often are the major application for the user environment. The focus of each of these programs is to create a prescribed format and set of rules to process very diverse data. The combined usage of these general-purpose systems in the modern office changed the nature of office and administrative work.

Application Programs.   Application programs are programs which are written in a programming language to solve a specific problem. Although hardware and systems software provide the major access point to problem solving by computer, neither would be of much use without applications programs. The term *user friendly* is used to connote that an application program's query and output features are user-oriented. The capabilities of modern computers have created a host of new applications for virtually every profession and industry. Typically used industrial engineering applications are: computer-aided design (CAD), computer-aided manufacturing (CAM), computer-aided process planning (CAPP), cost estimating, routing and scheduling, time standards, machine processes, materials resource planning (MRP), plant layout, and quality control. In addition, simulation and modeling routines enable engineers to interactively evaluate various alternatives. These programs work either individually or in combination with each other to aid the engineering profession to improve productivity at all phases of the manufacturing cycle—design, planning, management, production, and control.

Companies have reported productivity gains as high as an 80 percent reduction in design time when using CAD and a 50 percent reduction in setup and throughput times when using the various components of CAM. In addition, quality has improved, since progress can be monitored as production occurs rather than after it is finished. Inventory costs have been reduced substantially through better scheduling and monitoring provided by the computer.

Many companies utilize prepackaged software instead of writing customized programs to perform these functions. There are hundreds of manufacturing packages available to choose among, many providing interfaces for linkages to company databases.

Computer-integrated manufacturing (CIM) is the interfacing of all of the various manufacturing modules into a unified system, networked to the company's other major systems (finance, distribution, etc.). A fully operational CIM system is extremely complex because of its interfaces, and requires a very large investment. While complete CIM systems have been installed by relatively few corporations, CIM offers an excellent model or goal. The harmonizing of the various components affords cost savings in virtually every aspect of the production and delivery processes. The computer has also greatly assisted companies in the implementation of flexible manufacturing systems (FMS) and just-in-time (JIT) inventory systems. These combinations of computer and human resources will be a major factor in aiding companies to meet competitive challenges into the twenty-first century.

Artificial Intelligence.   Traditional computer programming employs an algorithmic approach which requires specific step-by-step input to solve predeterministic problems. A major limitation of programming has been its inability to solve problems that require humanistic heuristic reasoning. Artificial intelligence (AI) concentrates on emulating human reasoning in order to solve problems. AI algorithms analyze the methodology of how a human being solves a problem and translate the thought process to the computer. The computer then approaches the human reasoning process to solve the problem in contrast to executing an ordered set of instructions. Expert systems, neural networks, and fuzzy logic are alternative approaches within the artificial intelligence field.

The expert systems approach narrows the problem scope to a limited field and concentrates on defining the rules applying to this domain. Expert systems have been used as the basis for robot control, speech recognition, and manufacturing applications such as monitoring processes, diagnosis of equipment failure, and process planning. Neural networks broaden the scope of the problem through more direct simulation of human brain processes. This approach bypasses the need to assemble a basis of expert knowledge, because neural networks can learn directly by interfacing with the domain. The basis for a fuzzy logic approach is defining a means for the computer to analyze the imprecise information which human beings draw on for decision making. The computer then approximates the human logic for reaching a conclusion.

A more detailed discussion of artificial intelligence is presented in Sec. 12, Chap 5.

The Software Cycle.   The computer revolution has brought forth developments in both systems and applications software. Hardware technological developments require modifications in systems software so that applications can take advantage of the enhancements. Likewise, developments spawn new ideas for applications; this has kept the field dynamic over the past two decades. The typical software cycle consists of systems development, programming, testing, maintenance, documentation, and user training. Each of these stages requires a planned and systematic approach for a successful implementation.

The emergence of packages supplied by specialist vendors has eased the burden on programmers for in-house development of software. The vendor assumes responsibility for physically performing the various stages of the cycle, but the purchaser should ensure that the full cycle has been covered.

Linkages to system databases and network interfaces are also critical. With the proliferation of networking, security measures to protect the integrity of the data have become even more important.

CONCLUSION

One of the most interesting developments of the information age revolution is that the computer is no longer the sole domain of the computer expert. As more users gain familiarity with the knowledge and productivity gains made possible by using the computer as a tool, applications continue to burgeon. This democratization of computing has resulted in a growing trend among vendors toward creating systems with greater user compatibility, software portability, networking, data sharing, and user friendliness.

Although the everyday user can manage without knowing the details of system architecture, the field still has not evolved to the point where such basics as operating and networking system rules and major hardware capabilities are transparent. A knowledge of computer basics facilitates productivity for the computer user. For the computer systems purchaser, an understanding of computer fundamentals is essential to ensure the procurement of a unit containing appropriate hardware, software, and expansion capabilities for current and future use as well as interfaces for existing and potential data-sharing environments.

Although current technological developments in hardware, networking, and software have brought phenomenal gains in capabilities and performance in just one decade, the future promises to be as productive. Research and development to date in new materials, processes, artificial intelligence, and software algorithms indicates that there are still many advances forthcoming. As capabilities grow, new application opportunities will likewise emerge, which will make the computer field dynamic for many years to come. There is a central difference between the computer field and other industrial products whose growth has stabilized after reaching maturity. Rather than an end product in itself, the computer is also a tool which has spurred development of new products and applications. It has become the core of a new revolution of innovations, products, and applications which will be emerging for years to come.

This chapter has offered an overview of the computer's basic features. Additional details on specific areas of importance are presented in following chapters. The professional who wishes to keep abreast of the field is presented with many opportunities, including newspaper columns on computers, computer stores specializing in hardware and software, bookstores with sections devoted to computer topics, electronic bulletin boards, professional organizations, and university as well as professional development classes and seminars. The field is no longer reserved for specialists; it is now the era where knowledge of the computer is important for everyone.

GLOSSARY

Analog transmission   The transmission of continuous signals of data.

Artificial intelligence   The ability of a computer to perform humanlike intelligent actions, such as learning and reasoning.

ASCII   American Standard Code for Information Interchange, a 7-bit code used to represent data.

Assembly language   A low-level programming language that uses mnemonic codes instead of numeric instructions.

Asynchronous transmission   Transmitting data one character at a time over a line.

Batch processing   A technique by which transactions or other data are processed in groups.

Binary system   A system of numbers whose base is 2. The binary system uses only two symbols: 0 and 1.

Bit   A binary digit, 0 or 1. The smallest possible unit of data recognized by the computer.

Buffer   A high-speed, low-capacity temporary storage location used when data are to be transmitted.

Bus   Parallel conductors used to transmit data.

Byte   A sequence of binary digits that is acted on as a whole by the computer.

C   A programming language that combines high-level statements with low-level machine control to produce easy-to-use and highly efficient software.

COBOL   Common business-oriented language, a high-level program language developed for applications in business data processing.

Compiler   A computer program that translates a source program into machine language.

Controller A device that converts signals from the CPU to an input/output device.

CPU Central processing unit, the part of the computer system that interprets and executes instructions.

CRT Cathode-ray tube, an electron tube that displays information on a video screen.

Database An integrated collection of data that allows access to information of interest.

Data processing Computer manipulation of large quantities of data.

Digital transmission Transmitting data as discrete impulses.

Disk A flat, round storage medium on which data are organized in concentric tracks that may be accessed randomly.

EBCDIC Extended binary-coded-decimal interchange code, an 8-bit code used to represent characters and symbols.

Expert systems Computer programs designed to simulate an expert's decision-making process to come to a conclusion on a specific domain.

Field A group of related characters.

File A group of related records.

Floating point Number representation in which quantities are multiplied by a power of the number base.

Flowchart A diagram that uses symbols connected by lines to show the logic and sequence of an entire program and of each operation.

FORTRAN Formula translator, a high-level program language used to perform engineering, mathematical, and scientific applications.

Giga 10^9 (1,000,000,000, or 1 billion). Generally used in conjunction with storage capacity or processor speed.

Input Data entered into a computer.

Interpreter A computer program that translates a source program into machine language. The program executes the translation one statement at a time.

Kilo 10^3 (1000, or 1 thousand). Used in conjunction with storage capacity or processor speed.

Local-area network A system linking together several computers and peripheral devices that are in relatively close proximity to each other.

Machine language A programming language that requires no further interpretation by the computer.

Magnetic tape A reel or a cassette of tape whose plastic backing is covered by a thin film of magnetizable material used as a data storage medium.

Mainframe A very large computer that was originally manufactured in a modular fashion.

Mega 10^6 (1,000,000, or 1 million). Generally used in conjunction with storage or processor speed.

Memory The part of a computer where data are stored and implemented.

Micro One-millionth (1/1,000,000). Used in conjunction with time (seconds).

Microcomputer A small, self-contained computer whose CPU is contained on a single silicon chip.

Milli One-thousandth (1/1000). Used in conjunction with time (seconds).

Minicomputer A small, compact computer that is larger and has higher performance than a microcomputer and is smaller than a mainframe.

Modem Modulator/demodulator, a device that converts analog data into a digital signal that can be transmitted over a data phone line.

Multiplexer A device that transmits different sequences of signals down the same channel.

Multiprocessing The simultaneous execution of two or more program sequences by multiple computers that are under common control.

Multiprogramming The execution of two or more programs at the same time on the same computer.

Nano One-billionth (1/1,000,000,000). Used in conjunction with time (seconds).

Operating system   A collection of software that controls the operations of the computer and implements its functions.

Optical fiber   A fiber constructed of glass or clear plastic that is used to transmit coded light signals.

Output   Signals that are emitted from a computer system.

PASCAL   A high-level structured program language used for engineering, scientific, and business applications.

Peripheral   Any hardware device that can be attached to, and used with, a computer system.

Pico   One-trillionth (1/1,000,000,000,000). Used in conjunction with time (seconds).

Program   A set of coded instructions that cause a computer to perform particular actions.

Real-time processing   Processing data as soon as it is entered into the computer.

Record   A group of related fields.

Silicon chip   A microelectronic circuit built on a sliver of silicon.

Software   A general term referring to computer programs, procedures, and documentation.

Spreadsheet   A software package that allows numbers to be entered and calculated.

Supercomputer   An extremely powerful and very large computer. The fastest type of mainframe.

Synchronous transmission   Transmitting data one block at a time over a line.

Tera 10^{12} (1,000,000,000,000, or 1 trillion). Used in conjunction with storage capacity or processor speed.

Timesharing   A method of processing in which a CPU is shared by several users for different purposes at the same time.

Wide-area network   A system linking several computers and peripheral devices that are not in close proximity to each other.

Word processor   A computer system or software package that allows words and text to be input, edited, and output.

The preceding are examples of commonly used terminology in the computer field. Many computer textbooks contain a glossary of computer terms and definitions. Refer to textbooks and/or computer dictionaries for additional definitions.

BIBLIOGRAPHY

Aseltine, J. A., et al., *Introduction to Computer Systems: Analysis, Design, and Applications,* Wiley, New York, 1989.

Birnes, W. J. (ed.), *Microcomputer Applications Handbook,* McGraw-Hill, New York, 1990.

Bodner, M. S., *Micro to Mainframe Data Interchange,* TAB Professional and Reference Books, Blue Ridge Summit, Pa., 1987.

Data World Infodisk, Faulkner Technical Reports, Pennsauken, N.J., 1990.

McKenna, R., *Who's Afraid of Big Blue,* Addison-Wesley, Reading, Mass., 1989.

Parker, C. S., *Understanding Computers and Data Processing: Today and Tomorrow,* CBS College Publishing, New York, 1984.

Rabbat, G. (ed.), *Hardware and Software Concepts in VLSI,* Van Nostrand Reinhold, New York, 1983.

CHAPTER 2
DATABASE MANAGEMENT*

Charles W. McNichols, Dalton Professor of Business Administration
Radford University
Radford, Virginia

A database is a complete collection of data, or raw facts, which represents an organization's information resource. Like other resources, such as plant, equipment, and personnel, the information resource must be managed and maintained if it is to provide effective operational support. Since most medium- to large-scale organizations capture and store their data in computer-based systems, database management activities are usually associated with computer processing. However, effective database management requires the classical human manager's activities of planning, controlling, organizing, and communicating. A database administrator's position is often created to centralize these functions for the automated data resource.

THE INDUSTRIAL ENGINEER AND DATABASE MANAGEMENT

The elements of database management include recognition of data management problems; design activity leading to collection, storage, and retrieval of data in a manner which effectively supports the organization's needs; and use of computer software called a *database management system* (DBMS) to facilitate the automated support of data storage and access activity. In this chapter, we will examine organizational and management requirements for various levels of automation, the implications of these needs for data management requirements, and the specific capabilities and advantages of the DBMS approach to data processing. Before we address these elements in detail, we examine some of the specific areas in which industrial engineers (IEs) are likely to have a need for database management skills.

An industrial engineer's work activities often involve the gathering, interpretation, and communication of data describing the organization's operations. It is this generation or use of elements of the information resource that makes knowledge of database management principles important for an IE. Analysis of quality control data, project management activity, and functions related to methods time measurement (MTM) implementation provide three examples of IE tasks which require data management.

---

*Adapted from McNichols, C. W., and S. F. Rushinek, *Data Base Management: A Microcomputer Approach,* copyright 1988, pp. 87–122. Reprinted by permission of Prentice-Hall, Inc., Englewood Cliffs, N.J.

Effective data management can enhance professional productivity by decreasing the time required to produce analysis results and by increasing the confidence associated with such results. Many industrial engineers make extensive use of microcomputers to support the data-oriented part of their jobs, and microcomputer-level DBMSs may make this use more efficient.

With microcomputers, the historical position of the IE as a passive computer-system end user is often altered to make the individual the system designer and implementer. The motivation for an IE to know about database management thus has two facets. Technical knowledge of database techniques will make communication with data processing professionals more effective when they develop large-scale systems which support industrial engineering activities. And there is a growing need for end users, such as IEs, to be able to do their own database development and implementation work.

The advent of material requirements planning (MRP), manufacturing resource planning (MRP-II), and computer-integrated manufacturing, where the integration is provided by MRP or MRP-II, provides a strong incentive to establish an integrated database to support manufacturing activities. Some of these specific activities, which may involve industrial engineers whether an integrating MRP system exists or not, are listed below, along with a discussion of the nature of the activity that makes knowledge of database management important.

Inventory Control.   Relevant elements of the database would describe historical sales and order lead times and costs, as well as the current status of the inventory. The quality of these data will impact the effectiveness of models used to establish inventory policy. In an integrated MRP system, data in inventory files would have to be accessible to software supporting production planning activities and the order entry system.

Bill-of-Materials Processing.   The complexity of the data which must be stored in a bill-of-materials (BOM) system often provides the incentive for implementing such a system with database management software. The hierarchical structure of assemblies and subassemblies must be maintained in such a way that it can be rapidly accessed by many different applications programs which produce cost estimates and where-used lists as well as the exploded bill-of-materials output. It should also be possible to rapidly modify the structure of the BOM as changes in product design occur.

Supplier Evaluation.   To perform either routine monitoring or a special analysis of the impact of differences among vendors on the quality and timeliness of a manufactured product, data from a variety of functional sources may have to be brought together. Reference 1 presents an example showing how data would be drawn from the vendor history file (delivery time, back-order frequency), the inventory file (component description, cost, vendor identification), and the quality control file (scrap rates, failure and inspection data) to prepare such an analysis. The ability to tie data together from all of these files through the automated system may be critical to allowing such monitoring or special studies to be accomplished.

THE DATABASE AS A CORPORATE RESOURCE

Corporate databases represent resources which are expensive to obtain and maintain, yet are critical to the successful operation of every aspect of the business. In the manufacturing realm, likely to be of greatest concern to the IE, the database supports decisions requiring knowledge of the current status of orders, inventory, and work in progress. The relevant database maintenance activity includes correct and timely entry of data representing product design, receipts and issues from inventory, and status of products in the manufacturing cycle. Major difficulties in the successful operation of any level of

MRP or MRP-II type system can usually be traced to poor design of the database and/ or the software used to maintain it, or poor employee discipline in maintaining database accuracy.

Although conceptualized as a single physical activity, the automated part of an organizational database is unlikely to be stored in that fashion. Many years' business experience has demonstrated the impracticality of completely centralizing the computer database. Today's corporate database is likely to be a conglomeration of what Martin[2] refers to as *subject databases:* one for accounting, one for product management, one for manufacturing management, and so forth. The modern approach to database management emphasizes the use of common data management procedures and database software techniques to facilitate the maintenance and interchange of information within the organization.

BASIC TERMINOLOGY: THE DATA HIERARCHY

To introduce some of the terminology used to describe computerized databases, let us begin with a manual data processing example. Suppose, as industrial engineer for a small company, you have been tasked with establishing an inventory management system. To organize data describing the stock on hand, you might construct a manual record-keeping system organized around index cards and a card file. A logical approach would be to use an index card to represent a single line item that your company sells. With the index card as a departure point, the following elements of the data hierarchy can be defined:

Fields (Data Elements, Data Items).   A first step in setting up the system would be to define the specific pieces of data to be recorded about each stocked item. Some examples would include stock number, description, vendor name, number of units on hand, and the wholesale and retail prices. Each of these items is called a field (or, alternatively, a data item or data element). As suggested by Fig. 2.1, descriptive names would probably be assigned to each field, and preprinted cards including these field names might even be prepared to simplify the job of recording the data.

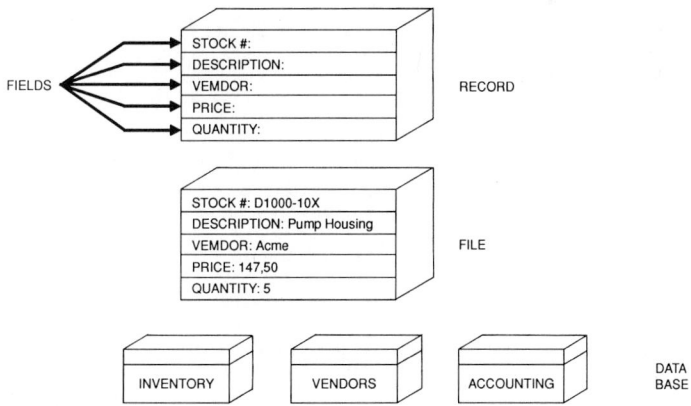

FIGURE 2.1   A hierarchy of data.

Records.   The complete collection of fields describing an entity defines a record. In our example, each index card will represent a record. The term *record* is used in two contexts in the example, and in the language of database management. On the one hand, our structure for the data as recorded on the index card defines the inventory record. In this sense, the record represents a design. When the cards are filled in for the actual inventory items, each card will be called a record. In this sense, the record represents the collection of physical items in the warehouse which have the same stock number.

Files. When a collection of our index cards has been used to record data about inventory items, a file has been created. A file is a collection of records of the same type: that is, records consisting of the same set of fields. It is important to distinguish between the names of fields (for example, *Stock Number* or *Price*) and the value the field may take on in a particular record (for example, F1000-26D or $295.37). The collection of values stored according to the record design is called an *instance* of the record.

Databases. When data that might be stored in individual, perhaps functionally oriented, files are treated as an information resource, the collection of files defines a database. Although an organizational information resource would technically include manual as well as automated records, for our discussion we will concentrate on a database defined as a collection of files which can be tied together for access purposes by computer software.

CONVERTING DATA INTO INFORMATION: BASIC DATA PROCESSING STEPS

While the terms *data* and *information* are often used interchangeably, there is a formal distinction between the two. Data are the raw materials of an information system: specific facts and figures like those represented by the values stored in the fields defined in our data hierarchy. Information is data that have been transformed to make them relevant to an end user, usually for decision-making purposes. This transformation process is performed by various combinations of the following data processing steps. These steps are often performed through the facilities of a database management system, and the efficiency with which specific DBMSs perform each activity represent important parameters for selecting DBMS software.

Storage. With the index card system we have been using as an example, the storage medium is the card itself, and the storage facility is a box in which the cards are kept. Computerized systems use magnetic tapes, disks, or some other electronic medium for storage. A DBMS can provide standardized mechanisms for entering data into these storage devices.

Retrieval. Data to answer managers' questions must be rapidly accessible if an automated system is to support decision making. The type of storage device used impacts the speed and flexibility with which data may be retrieved. Generally, the values stored in fields are used to define a retrieval request. For example, a quality manager might want to examine the inventory record for every item supplied by a specific company, if problems have been found with some of that company's components. Searching through every inventory record to make such a list could be expensive, even if the records are computerized. However, if the records are stored on a device permitting direct access (such as a magnetic disk), and if they are indexed by vendor name, the list could be rapidly prepared. Database management systems generally include capabilities for building indices and updating them as records are added or modified.

Selecting or Classifying. Selecting means extracting those records from a database which satisfy a set of conditions defined by values stored in the records—for example, all inventory records for items which have a wholesale price over $100.00. Classifying separates records into categories: inventory items might be classified by order lead-time intervals. Both selection and classification are important data processing steps in preparing printed reports.

Sorting. This is another data processing operation which is important in preparing easy-to-use printed reports. Sorting involves arranging records in ascending or descending order of the values in some field, which represents the *sort key*. For example, an inventory listing might be ordered by ascending stock number, and stock number would thus provide the sort key. An inventory listing printed without ordering the records would be almost useless, since a user would have to search through most of the report line by line to find a stock item of interest.

When many records contain the same value of the sort key, it is desirable to sort records on more than one field. To prepare a listing of vendors by state, the inventory file would be ordered by state. All of the records in the same state would be further sorted by vendor name. In this situation, the state is called the *major sort key,* and the vendor name is called the *minor sort key.*

Computing. Arithmetic calculations are often an integral part of the process of transforming raw data into information. In an inventory valuation report, the quantity on hand would be multiplied by the wholesale price in each record to obtain the value of that line item. If quantity on hand and price are stored in the file, it is unnecessary to store the product of the two, since it can be calculated when a report is prepared.

Displaying. None of the fundamental data processing activities are useful unless their results can be presented in an easily understood format. The printed report is the bread-and-butter product of computer systems at all levels, but there is growing interest in presenting numeric data in a graphical form.

DATA MANAGEMENT PROBLEMS: ENHANCING THE INFORMATION CONVERSION PROCESS

In theory, the collection and storage of data in the data hierarchy defined in the last section, combined with the basic data processing steps just listed, should provide access to an organization's automated information resource. In practice, a common lament in organizations with heavy investments in computer hardware and data collection is that data which have been collected are not accurate, and cannot be retrieved in a timely fashion or in a format which is really useful for dealing with nonroutine problems. Some of the data problems which lead to this situation, and which database management approaches hope to overcome, are the following:

Redundancy. Storage of the same data in multiple records or files represents redundancy, and leads to a host of data management problems. Suppose, for example, that a vendor's name and address are stored in each inventory record representing one of that vendor's products. The same data would have to be repetitively entered as each inventory record for a given vendor is constructed. Further, multiple storage of the same data wastes space in the files, and this costs money, both for the storage medium (tapes, disks, etc.) and for the extra computer processing time required to deal with these larger files. If a vendor moves, every record representing a product from that vendor will have to be updated. If any records are missed, or if some records are updated incorrectly, the quality of informational outputs from the inventory file suffers.

A better approach to file management is to store a vendor identification number in each inventory record and the vendor names and addresses in a separate vendor file, where each vendor would be represented only once. To prepare an inventory report which includes vendors' names and addresses, the computer software would use the vendor number to access the appropriate record in the vendor file as each inventory record is processed. If a vendor moves, updating a single record in the vendor file effectively corrects the address in all inventory records associated with that vendor.

Integrity.   Integrity refers to the reliability of data, and the redundancy problem just discussed is a major cause of data integrity problems. As an example, the redundant storage of vendor addresses in the inventory file leads to an integrity problem if the vendor moves and not all of the records are properly updated.

Data Access for Application Development.   Data needed to respond to a special information request may exist in a collection of files, but the expense and lead time involved in developing the computer programs to tie the data together may be prohibitive. Suppose the vendor reliability analysis used as an example earlier in this chapter is to be prepared, and data sources include a vendor file, inventory file, and quality control file. Preparation of an effective report will require a connection between the quality control file and inventory file on the basis of a stock number or combination of stock number and vendor number. A link will also be required between the inventory file and the vendor file, using some type of vendor identifier. If the database in which these files exist is maintained with a DBMS, the time and expense involved in preparing a specialized report will be minimized.

Even when an application does not require a complex integration of data from different sources, the programming time required to deal with data input and output can be substantial. When data are maintained with a database management system, standardized software is available to perform the data manipulation operations, thus speeding application development. The DBMS effectively isolates programmers from the details of data storage on physical devices, allowing them to access data needed for an application through the DBMS.

LOGICAL DATABASE STRUCTURES

The Logical versus the Physical View of Data.   To overcome the fundamental data problems just introduced, database management software will have to provide for the integration of data from different files. An example discussed earlier required linking the vendor number in an inventory file to the vendor's name and address information stored in a vendor file. There are a variety of ways to visualize this linking of separately stored data. These alternative models for relating data are called logical structures. While any of the models to be described could be implemented in computer programs developed from the ground up, a specific database management system generally deals with only one logical structure. There is no globally best model for dealing with data: each has its advantages and disadvantages. These generally are reflected in tradeoffs among flexibility, ease of use, and the speed with which data may be accessed and processed.

A common element of recent trends in database management software which permits for rapid application development, perhaps by nonprogrammers, is the high degree of separation of the user of data from the details of its storage. This is a concept called *data independence.* The idea is to let the user manipulate data from a logical viewpoint, concentrating on the logical relationships in the data. The database management software deals with the physical storage of data in records and files, and provides mechanisms for efficiently retrieving and organizing data. Thus, the application developer may not have to worry about the details of integrating fields which may be physically stored in different files, or even on different storage devices: tapes and disks, for example. A logical design for the overall data base, called a *schema,* and corresponding logical designs for subsets of the database of interest to specific users, called *subschemas,* provide enough details of database content and organization to let developers and end users retrieve needed information without knowledge of physical storage details.

Most database management software manipulates a database by using one of the following three data structures:

The Hierarchical or Tree Model.   Figure 2.2 provides an example of a hierarchical file

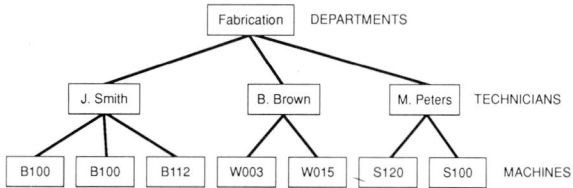

FIGURE 2.2   The hierarchical structure for a maintenance database.

structure which might be used in a maintenance record-keeping system. The records at each level (department, maintenance technician, machines) might be stored in separate files. Each record would contain a number of fields describing the entity at that level. For example, each department record might contain data about the department name, location, supervisor, and so forth. The technician record might include the maintenance worker's name, social security number, and skills. To prepare a report which includes data about the machines, their maintainers, and the department, data from all three files would be linked at the time the report is prepared.

In this example, maintenance technicians are assigned to specific departments, and each machine is assigned to a specific individual for maintenance purposes. Each record below the highest level belongs to a single record at the level immediately above it. These relationships are referred to as *parent-child* or *owner-member* relationships. A record can be both a parent and a child: maintenance technicians' records fit into this category. When a parent record has only one child record, the data relationship is said to be *one-to-one,* abbreviated 1:1. When more than one child or member record is associated with the parent, the relationship is *many-to-one,* abbreviated m:1. In the hierarchical model, a child or member record can be associated only with a single owner or parent record. Therefore, the hierarchical model would not be appropriate if a machine could be maintained by more than one individual.

The Network or Plex Model.   There are two kinds of logical complexity for which the hierarchical model is inadequate. In the simplest of these situations, a child record belongs to more than one parent. Figure 2.3 illustrates this situation with the maintenance database example modified to allow joint responsibility for technicians and their machines by two departments. A listing of machines by department would now show some of the same machines under two departmental headings.

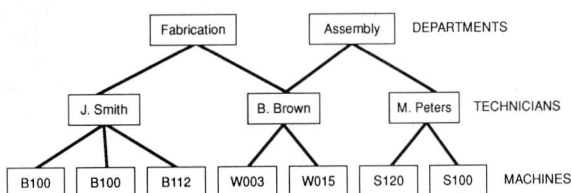

FIGURE 2.3   Network model—multiple parents for a record.

A more involved situation is illustrated in Fig. 2.4. In this extension of the maintenance database example, maintenance technicians may be assigned to multiple machines. There is also a need to view the maintenance personnel records as either parent or child records depending on the type of report desired. To prepare a report showing which machines each technician works on, the machine records would be the child records. To prepare a report showing the responsible maintenance technicians for each machine, the machine

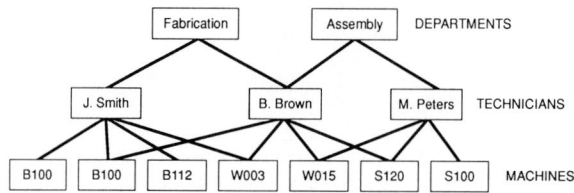

FIGURE 2.4   Network model—records viewed as either parent or child.

records would be the parents, and the personnel records, the children. This type of logical structure is called a *many-to-many* (or *m:n*) relationship, and is difficult to implement in a hierarchical database.

The Relational Model.   The two data models just discussed have as a common property the need to define relationships among records at the time the database is designed, and before records are actually entered. Thus, they are particularly appropriate for electronic data processing (EDP) and management information system (MIS) applications in which data retrieval requirements for reports or terminal displays can be predefined. Both of these models can be implemented in ways that provide very rapid access to the data, even though fields stored in diverse physical locations will have to be assembled into the logical records that the end user will see in a report.

For nonroutine data retrieval requirements, such as those that occur in decision support systems and, with increasing frequency, in management information systems, another logical structure is often applied. This is the relational data model. It is becoming the most popular of the three discussed in this chapter, particularly in microcomputer- and minicomputer-level DBMSs and in large DBMSs which are used to provide nonprogrammers with interactive data access. Although data retrieval is often slower in these systems than in those using hierarchical or network models, the relationships among records can be defined after data have been entered. The relational model is the most relevant for industrial engineers, because IEs are likely to make use of time-sharing or microcomputer-based interactive data retrieval software in support of their work activities. For this reason, we will examine the relational model in more detail than the hierarchical or network structures.

A relational database is viewed logically as a collection of two-dimensional tables (Fig. 2.5). Each of these tables corresponds roughly to our earlier description of a file, where each row in the table is one record, and each column represents one field. The association among fields forming one row is the relation that gives the relational model its name. In Fig. 2.5, the fields are all associated with a specific department in the department table and with a specific individual or machine in the tables representing those entities. The DEPT# field in the tech-

Department

DEPT#	LOC_CODE	DEPT_NAME
SHP01	BLDG20	SHIPPING
SHP01	BLDG15	SHIPPING
FAB20	BLDG15	FABRICATE
ASM01	BLDG20	ASSEMBLY

Technician

TECH#	NAME	SKILL_CODE	DEPT#
111221111	PETERS, MIKE	L010	FAB20
222334444	SMITH, MARY	L125	SHP01
123992000	BROWN, BOB	RP25	FAB20
888221111	ABLE, JOAN	L125	SHP01
777223222	SMITH, JOHN	L010	FAB20

Machine

MACHINE#	MAINT_CYCL	MAINT_HRS	MFG_MODEL	TECH#	DEPT#
B100	30	1.0	AMG1000	777223222	FAB20
B112	30	1.0	AMG1000	777223222	FAB20
W003	7	3.5	CIN123-2	123992000	FAB20
S120	14	2.0	ACM2411	111221111	FAB20

FIGURE 2.5   Relational tables representing the maintenance database.

nician table, and the TECH# and DEPT# fields in the machine table allow data from all three tables to be combined by the relational operations discussed later in this section.

When data are retrieved from a relational database structure, new tables might be temporarily constructed by combining elements of existing tables. An advantage of the relational model over the models discussed earlier is that the operations for combining tables can be defined mathematically, allowing a precise, unambiguous description of retrieval steps and results. For these operations to work properly, the following rules must be satisfied for data in each table:[3]

1. Each table may contain only one kind of record. In our example, the department, technician, and machine records would all be stored in separate tables.
2. Each row in each table has the same fields, and each field has a unique name.
3. Each field occurs only once in a record. There can be no repeating fields or groups. This means that we could not design a department record that included the social security numbers of five maintenance technicians assigned to the department. Following this rule avoids many problems in data updating and record access.
4. There is no predetermined sequence for the records in a table.
5. New tables can be produced by combining and/or extracting subsets of data from other tables.
6. Each row in a relational table is unique. That is, no two rows may contain exactly the same values in all matching columns. Each row in a relational table represents some entity: a department, technician, or machine in our maintenance application example. There would be no logical reason for including the same entity in the table more than once, and doing so would lead to confused and erroneous retrieval results.

One of the important features of the relational model is that it is possible to reproduce the hierarchical or network logical data relationships with this structure. Figure 2.6 shows how the hierarchical example of Fig. 2.2 could be converted to a sequence of two-way tables which satisfy the relational requirements. The actual tables would contain other fields describing departments, technicians, and machines, but the figure shows how multiple records would be created to represent the network linkages.

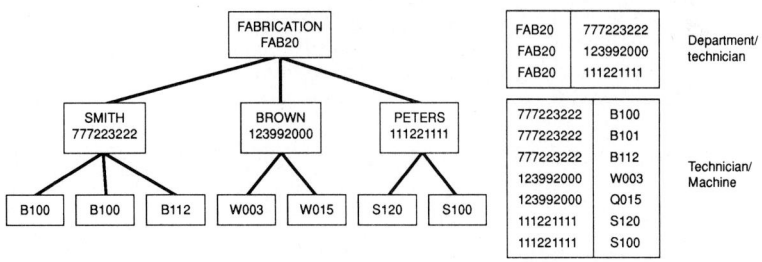

FIGURE 2.6   Conversion of a network logical view to a sequence of relational tables.

There are three fundamental operations performed on relational tables as part of a data retrieval process. Relational database management systems usually provide a mechanism for accomplishing these operations, although they are not always identified under their formal names in the data access strategies defined for a specific DBMS.

Join.   The relational join provides the logic for temporarily combining rows from different tables. Rows are combined by matching field values between the two tables which are joined. Figure 2.7 illustrates this operation using the maintenance database example. The objective is to prepare a table (perhaps to output as a printed report, or on a video terminal screen) which contains data from the maintenance technician records and the machine records.

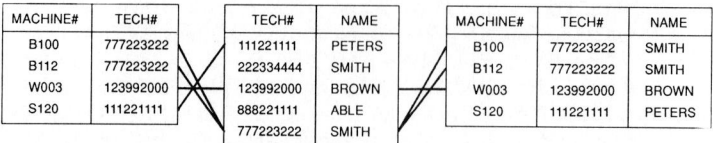

FIGURE 2.7   Joining relational tables.

In theory, the join operation is performed in the following manner. Beginning with the first row in the first table, all rows in the second table are examined to see if the logical condition defining the join is satisfied. In the example of Fig. 2.7, this would be a match between the social security number in the technician record and the machine record. When a match is found, a new record, consisting of fields from both the technician and the machine records, is output. When the end of the technician records is reached, the next row in the first (machine) table is compared with all rows in the second (technician) table. The process ends when every row in the first table has been examined.

For large tables, the join operation could be very time-consuming. If there are N records in the first table, and M records in the second, then $N \times M$ comparisons are performed. Many relational DBMSs provide faster techniques for performing joins when records are to be linked on the basis of fields which are indexed. In some cases, relationships among tables may be defined before data are entered, and the DBMS generates information that allows these joins to be rapidly performed. However, a true relational DBMS allows the join operation to be performed after data have been entered in tables, and on the basis of an arbitrary relationship between tables.

Selection.   The selection operation is relatively simple. It implies the construction of a new table by extracting a subset of the rows (records) from an existing table. If we wanted to prepare a maintenance report for only departments A and C, the first step in the retrieval process would be to prepare a new table consisting of only the A and C rows of the department table. Then, this table could be used as an input to a join to attach information from the technician and/or machine tables.

Projection.   Since not every field will be required in the new tables constructed from our relational database, the projection operation is needed to allow the extraction of a subset of the columns of a table, just as the selection operation allows extraction of a subset of the rows. There is an additional complicating factor with the projection operation. No two rows in a relational table can be identical, but extracting a subset of the columns in a table might lead to this condition, as illustrated in Fig. 2.8. Full implementation of the projection operation requires elimination of all but one of any redundant rows resulting from the elimination of columns in the new table.

MACHINE#	MAINT_HRS	MFG_MODEL	MAINT_HRS	MFG_MODEL	MAINT_HRS	MFG_MODEL
B100	1.0	AMG1000	1.0	AMG1000	1.0	AMG1000
B112	1.0	AMG1000	1.0	AMG1000	3.5	CIN123-2
W003	3.5	CIN123-2	3.5	CIN123-2	2.0	ACM2411
S120	2.0	ACM2411	2.0	ACM2411		

FIGURE 2.8   Projection with elimination of redundant rows.

DATA MODELING

The powerful data management techniques available in modern database management systems are of no value if the database is not properly designed. If the data (representing the raw material of an information system) are ill-defined, no amount of processing will be able to transform them into useful information. Even with microcomputer-level applications, some effort devoted to database design before data are collected and entered is important if the maximum benefit is to be obtained from the database.

In presenting the rules defining a relational table, the idea that each row in a table represents some real-world entity was introduced. In this sense, a database represents a model of objects in the real world. As with any modeling effort, a balance must be found between too much abstraction (which leads to inaccurate conclusions about the modeled process) and too much detail (which may be prohibitively expensive).

We will emphasize the relational data model in our modeling discussion, since an IE is most likely to be personally involved in data modeling for this data structure. However, the techniques described are generally applicable to the other data structures as well.

The approach to data modeling is impacted by the issue of whether the database is to support EDP, MIS, or decision support system (DSS) applications. Databases which support operating management are likely to be designed around the needs of a specific application, while those which support higher-level management requirements should model the entities independent of specific data processing applications.

EDP-level applications have a strong clerical orientation, and output products of these computer systems are well-defined. Thus, the content of the database can be determined by the output reports and displays which must be produced. Any field which is to appear in an output must either exist in the database or be capable of being calculated from fields in the database. If the computer system is designed solely for clerical support purposes, there is little justification for collecting and storing data beyond that needed for the standard reports. Thus, design effort can be concentrated on the format of the fields and efficient allocation of the data to specific files or tables.

When the design is for a management information system, the design process will still emphasize predetermined retrieval strategies and report formats. Since MIS products are used for analysis and decision-making purposes, there is a good chance that output products beyond those included in the original design will be requested. For this reason, extra effort should be expended on the design of a database for MIS purposes, to allow future access to its content. It is this evolutionary nature of MIS needs that has accelerated the interest in DBMS applications.

Design of the database is more difficult when it will become an integral part of a decision support system. Because the DSS will be applied to unstructured decision problems, database content and relationships cannot be determined solely on the basis of predefined output products. In this situation, likely to be a common one for industrial engineering applications of database management, the design should emphasize accurate representation of real-world relationships as well as retrieval flexibility. The relational model will generally provide the most effective logical view of data for DSS purposes.

KEYS

As part of the data modeling process, fields (or groups of fields) which will be used as keys should be identified. Two types of keys will be defined. *Primary keys,* which are unique identifiers of a record or field group, and *secondary keys,* which may occur in many field groups. A primary key provides a path to a group of fields in an unambiguous fashion, and would typically be used for updating purposes. In the maintenance example, a technician's social security number and a machine serial number represent primary keys. A *composite key,* consisting of the combination of department number and location code, would be an appropriate primary key for the group of departmental fields. An employee's name would not represent a good choice for a primary key because there are liable to be several employees with the same name, and alternative ways of representing the same name: Jack Brown and John Brown may or may not be the same individual, but social security number 123-44-5555 uniquely identifies a technician.

Secondary keys are established to provide rapid retrieval access to groups of fields for report preparation or terminal screen display. Skill code in the technician record might be used as a secondary key to permit rapid retrieval of information about all of the technicians with a particular code.

Next, we will want to formalize the grouping of fields, and we will refer to these groupings as records, again emphasizing the fact that the fields which define a record may or may not be physically stored together in the computerized representation of the database.

NORMALIZING DATA STRUCTURES

The objective for grouping fields into records is to provide efficient access to database contents, both from the standpoint of computer processing time and logical correctness of the information products produced. Record definition will have a major impact on minimizing the data redundancy and integrity problems introduced earlier. Unfortunately, it is possible to define records which work adequately for an initial application but lead to major problems when new applications are built or ad hoc retrievals are attempted. For this reason, a formal application-dependent approach should be followed for grouping fields.

The key to properly associating fields into records is a process called *normalization*. Records which have been designed according to this process are said to be in *normal form*. The process can be approached in steps, beginning with the first normal form and proceeding to the second and third normal forms. Additional levels of normalization may be defined, but, for most applications, records designed to meet the conditions of the third normal form will provide an effective database design.

Normalization is often discussed in conjunction with the relational database structure. It is particularly important to work with records in the third normal form with a relational DBMS to avoid erroneous information retrieval because of the retrieval flexibility inherent with this model. However, the process should be applied to record design with any of the data structures.

To illustrate the normalization process, we will employ the maintenance database example. Let us assume that a preliminary record design has been performed. We have picked a subset of the fields we would probably want in a real maintenance database to keep the example as uncomplicated as possible. The initial objective is to prepare a maintenance report sorted by department, technician number, and machine number which summarizes the routine maintenance requirements for the company's machines. As is often the case when records are designed for a specific application, all of the relevant fields have been grouped into a single record. Our starting point is a deliberately poor design, but we will see how the normalization process can convert the poor design into an effective one.

Figure 2.9 shows the initial record design as it might have existed as a paper form. Also presented is the bar format for a computer record. The bar is split across several lines because of its length.

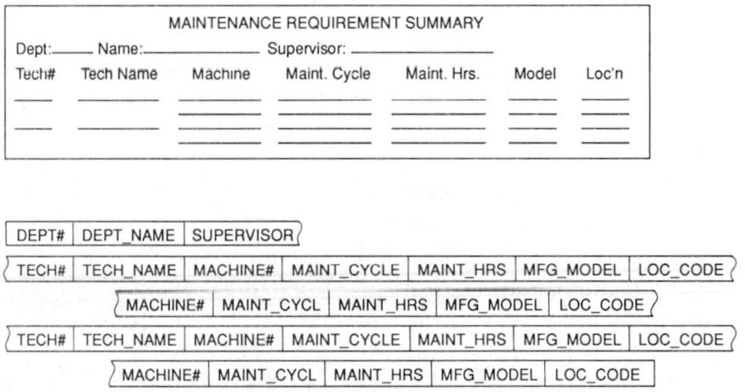

FIGURE 2.9 Unnormalized maintenance record.

Multiple associations which occur in the entities modeled by this record have led to the use of *repeating groups*. For example, space is provided in each record for data about two maintenance technicians, and the data about each maintenance technician can describe up to two machines. Elimination of these repeating groups represents the first step in the normalization process.

First Normal Form.   The record design presented in Fig. 2.9 really represents the hierarchical structure shown in Fig. 2.2. Conversion of this record design to first normal form is accomplished by replicating records so that a set of the records can be presented as a two-dimensional table, without repeating any fields (columns) of the table. Figure 2.10 shows this new record design, which is called a *flat file:* data represented as a two-dimensional table with no repeated fields or groups of fields.

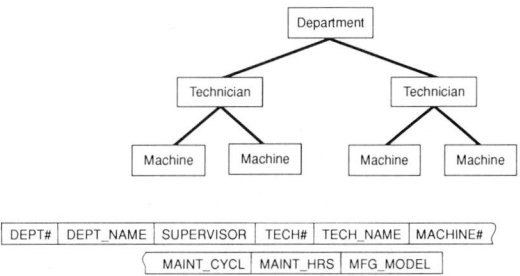

FIGURE 2.10   Implicit field hierarchy and first normal form representation of the maintenance record.

On first consideration, it may seem inefficient to convert one record into many to avoid repeated fields or groups. However, the presence of repeated data in a record creates several serious data management difficulties.

An obvious problem is that the record designer must make decisions about how many times each field or group of fields is to be repeated. In Fig. 2.9, the record design allows data to be recorded for two machines for each of two maintenance technicians. Suppose a department expands to the point where three technicians are needed. The record could be redesigned to accommodate a third repeated group, but the entire existing database would have to be modified to convert the records to the new format. In most cases, special computer programs would have to be written to accomplish this. Then, every application program which uses the file would have to be changed to work properly with the new record format. If the initial record design allows for several extra repeated groups to avoid the possibility of this later modification, the space for these data will be carried in all of the records in the file. This is expensive, both in terms of computer storage space (usually on magnetic disks) and processing time, since the time to process data in files is related to the number of characters of data stored in the file.

A more subtle problem with nonflat files is related to data retrieval. The computer program written to produce the maintenance report which provides the example application will be more complex to write with the nonflat file. To output the desired format, each record will have to be accessed, then the program will have to step through the repeating groups of fields describing technicians, and finally step through the repeating groups of fields describing machines. There is no simple way to output the records in technician number sequence unless the data have been entered in each record in this order, or a program is written to sort these fields within the records. The flat file presents a much simpler structure for report preparation.

For ad hoc data retrieval, the absence of repeating groups in our first normal form representation again offers many advantages. Suppose a manager wants to prepare a list of maintenance technicians who maintain a specific machine model number. With the flat file, the programming logic to accomplish this retrieval simply involves checking the MFG MODEL field, stored in the same place in every record, then listing the records which contain the desired value of the code. To accomplish this retrieval with the original form of the file, every repeating group would have to be checked, then data output from just the fields in the groups satisfying the model number condition. This is a more complex search situation, and one that few interactive data retrieval programs will support.

Second Normal Form. The next step in normalizing a record is only relevant when a record has a key which is defined by two or more fields in the record: a composite key. The appropriate primary key for the first normal form record is one which combines TECH# with MACHINE#. This key will be represented as follows: TECH# + MACHINE#, where the plus sign implies concantenation of the two fields. The primary key must provide unique identification of a record; no two records may have the same value of the primary key. Thus, to specify the primary key we must find a combination of fields that will have a unique value in each record. There will be multiple records with the same TECH#, since each technician maintains multiple machines. And there will be multiple first normal form records with the same MACHINE#, since several technicians can perform maintenance on the same machine. However, the combination of the two fields identifies a unique record.

No fields beyond those needed to uniquely identify the record should be included in the primary key. For example, it is not appropriate to add the DEPT# field to the primary key.

Second normal form is obtained when fields are allocated to records so that *every* field in a record is dependent on the whole key for that record. Each nonkey field in the record must be examined to see if it is associated with the entire key. If it is not, a new record should be created with a primary key which is a subset of the composite key. In our example, there are three possible dependencies. A nonkey field could be dependent on the entire key TECH# + MACHINE#, only the TECH#, or only the MACHINE#. Figure 2.11 illustrates the conversion of the first normal form records of Fig. 2.10 into the second normal form. Note that the fields which describe the technician are in a record which has TECH# as its primary key, the fields which describe the machine are in a second record which has MACHINE# as its primary key, and a third record linking technicians and machines has been created which has the composite primary key TECH# + MACHINE#. This is called an *intersection record,* and creating such a record is the proper way to capture the many-to-many relationship between technicians and machines. In our simplified example, there are no nonkey fields in the intersection record. However, additional fields specific to the combination of machine and technician would properly be stored here.

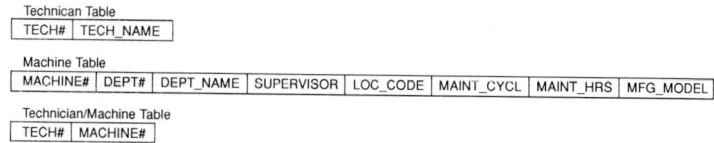

FIGURE 2.11 Conversion of records to second normal form.

The data management problems caused by records with composite keys which are in first but not second normal form are not likely to be as serious as those caused by nonflat files. However, data redundancy is a natural result of failure to implement the second normal form. In the first normal form maintenance example, data unique to the machine and data unique to the technician will be stored repeatedly in records describing the technician-machine combination. If a technician is replaced, records for every operating location will have to be updated with the new technician's identification. Missing a record, or erroneously updating one of the records, destroys the file's integrity. In a similar fashion, data about each machine will be stored redundantly if more than one technician maintains that location, again providing the potential for updating and retrieval problems.

Third Normal Form.   The final data problem we will examine results when the value stored in a field depends directly on the value stored in another field which is not a key. In the maintenance example, there is a unique DEPT_NAME and SUPERVISOR for each department, identified by the DEPT# field. In the current design, the DEPT_NAME and SUPERVISOR information will be stored redundantly, because there are multiple machines in the same department; therefore, there will be many records containing the same value in the DEPT# field. However, there is no second normal form violation, because the DEPT# field is not part of the primary key. The DEPT_NAME and SUPERVISOR fields are said to be *transitively dependent* on DEPT#. The heart of the problem created by this transitive dependence is again redundant data storage. A similar problem exists in the machine table. If the company owns several machines having the same MFG_MODEL, and the MAINT_CYCL and MAINT_HRS are determined by MFG_MODEL, these values are stored redundantly. The cure for these problems is to create additional department and maintenance tables, as shown in Fig. 2.12. The primary keys_ for these two tables are DEPT# and MFG_MODEL.

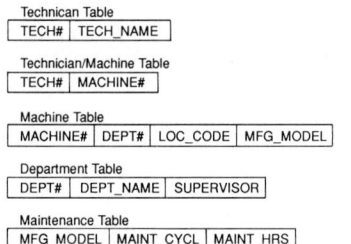

Technican Table
| TECH# | TECH_NAME |

Technician/Machine Table
| TECH# | MACHINE# |

Machine Table
| MACHINE# | DEPT# | LOC_CODE | MFG_MODEL |

Department Table
| DEPT# | DEPT_NAME | SUPERVISOR |

Maintenance Table
| MFG_MODEL | MAINT_CYCL | MAINT_HRS |

FIGURE 2.12   Converting the machine record to third normal form.

Data from several records could be required to prepare printed reports or produce screen displays. The logical associations necessary to reassemble records for output purposes are provided by *foreign keys:* fields in a record which match the primary keys in another record. The MFG_MODEL field in Fig. 2.12's machine table is a foreign key for the MFG_MODEL value, which is the primary key in the maintenance table. In the technician-machine table, the TECH# field is a foreign key for the same field in the technician table, and the MACHINE# field provides a similar association with the machine table. As these examples indicate, a foreign key can be either a nonkey field or part of a composite key.

If these rules are carefully applied, a database which avoids most logical data management problems will be obtained. The key to the normalization process is avoiding data redundancy. The benefit of a database in the third normal form is that minimizing redundancy enhances data integrity and increases the chances of obtaining logically sound information when data retrievals are performed.

SELECTING A DATABASE MANAGEMENT SYSTEM

Database management software is complex and expensive to develop. Because DBMSs deal with very general applications, few business organizations would consider developing such software in-house. Instead, a commercially available package will be obtained. The cost of a DBMS could range from a few hundred dollars for a microcomputer package to several hundred thousand dollars for a comprehensive mainframe implementation. In this section, the major parameters to consider in obtaining a commercial DBMS will be presented. These will not be of equal importance for all organizations, so it is important to begin the search for a DBMS with some consideration of the range and sophistication of the applications to which it will be applied. Because of the cost involved, an organization's data processing professionals will usually have a major input to the selection of a main-frame DBMS. The industrial engineer is more likely to control selection of micro-computer-level software, so the presentation of parameters will be slanted toward the small-system end of the spectrum.

Database Definition. The mechanism which provides the data description language and data dictionary function should be easy to use and capable of producing a printed representation of the database contents: field names, types, and sizes (measured in characters) at a minimum. Microcomputer DBMSs generally require that a database be defined by specifying the fields to be stored in each record. A question/answer dialogue might ask the user to input each field name, type (character or numerical data, for example) and length. Some DBMSs allow specification of field format, such as how many decimal places to carry in a numeric field. It may also be possible to input editing information such as a range of legal values for the field. This will prevent entry of out-of-range values when data are input.

It is important to consider the ease of modifying the database definition after data have been entered. As needs change, it may become necessary to add or delete fields, change field lengths, or change editing criteria. A well-designed DBMS will allow this to be done without requiring reentry of data already in the files.

Screen Display Design. Most data will be entered in a database through a cathode-ray-tube (CRT) terminal, and many data retrievals will be performed using this device. Most DBMSs provide a mechanism for interactive screen design, which allows a screen to be laid out on the CRT by using keyboard controls. Labeling information and locations for database fields to be presented will be identified. The screen design is then stored for later use in conjunction with a database application. Many of the software packages have a default CRT display mode for quick-and-dirty data entry. Knowing only the description of a record, the DBMS generates a screen display that lists all the field names and provides the operator the capability to key in the data. This capability is very attractive when end users want to implement their own databases.

Availability of an easy-to-use screen design facility will greatly reduce application development time and will encourage the provision of effective CRT displays for data entry and retrieval.

Report Design. Printed reports represent the major output from most database management systems. A report generation capability avoids the need to write programs in a language such as COBOL to retrieve data in a custom format. Report generators usually allow specification of fields to be included in the report, formatting information such as page and column headings, and sorting and totaling instructions. In many DBMSs, the report can be designed interactively, much the way a CRT screen is designed. As with the screen design capability, availability of a good report generator will make the data resource represented by the database more accessible to support the needs of analysts and managers.

Processing Design. There are two general approaches to data processing through a DBMS. Most large-scale systems use a host programming language. This means that data processing is accomplished with software written in a programming language such as CO-BOL that specifies a schema, then calls on the DBMS for data input and output. The advantage of the host language approach is that the full power and flexibility of the compiled language is available for software development.

Microcomputer DBMSs often provide an embedded, or self-contained, programming language. These languages are generally less comprehensive than the standard languages (such as COBOL), and result in slower processing of data. On the other hand, the embedded languages are generally easier to use for nonprogrammers, and may take special advantage of unique features of the specific DBMS for which they were designed.

The programming language, whether embedded or not, provides the capability to do arithmetic calculations with database contents. For industrial engineering applications, the computational limitations of an embedded language should be examined. For example, many of these embedded languages cannot compute square roots, powers, trigonometric functions, and so forth. Less expensive microcomputer DBMSs may contain no programming capability at all. Unless intended for very simple applications, this software is probably a poor investment for IE work.

Data Editing or Validation. Most DBMSs, even for microcomputers, will prevent the entry of nonnumeric data in numeric fields, and will limit the length of entered values to the size specified for a field at the time of its definition. More sophisticated software allows specification of editing parameters. For example, the edit check could require a positive number to be entered in a field representing hourly pay rate, and a legitimate date to be entered in a date field. It is almost always possible to implement this type of checking in systems with embedded or host programming language support.

Access to Records. Most of the data available to a DBMS will be stored on magnetic disk: floppy disks for inexpensive personal computers, and hard disks with capacities in the hundreds of millions of characters for bigger computers. To retrieve records and organize data for reporting purposes, sorting and indexing schemes are usually implemented. Sorting means rearranging all or a subset of the records in a file before output. Indexing is a means of providing a direct access path to records on disk, so that a record can be quickly found by specifying the value in one of its key fields.

In most cases, a DBMS with the ability to construct indexes is very desirable. This will allow a user to interrogate or update record contents from a CRT terminal or microcomputer keyboard without searching the entire file. The DBMS should have the ability to automatically update indexes when new records are added to a file, making the data immediately available for indexed retrieval and avoiding the processing time required to regenerate the index. And, it should be possible to maintain several different indexes for the same file. For example, we might want to index machine records in our maintenance database by both model and serial number.

Database Capacity. The maximum number of records which can be accommodated by mainframe DBMSs is usually very large, but the decreasing price of high-capacity storage devices for microcomputers makes the capacity issue an important one to examine when considering a microcomputer system. Three capacity measures are relevant: the maximum number of records which may be stored in a file, the maximum number of fields which may be stored in one record, and the maximum length of each type of field. The maximum length of a field represents the number of characters which may be stored in an alphanumeric field, and relates to the range of numeric values that can be stored in a numeric field.

Performance. Part of the price paid to obtain the advantages of database management through a DBMS is in the computer processing resources required to operate the system. For mainframe DBMSs, sophisticated performance evaluations are often performed by running benchmark tests with competing DBMS products or by using computer simulation techniques. Because these products are most often used in a multitasking environment, the mix of jobs typically run will impact performance. With microcomputer products, sorting speeds and indexed record access time are the most relevant performance parameters. Some software manufacturers publish test data for these processing features. If these data are not available, vendor demonstrations can be requested.

Security and Privacy. Access to records stored in a DBMS is usually restricted by implementing password protection. A potential user must enter a secret password to gain access to the software which displays data on a CRT or runs a report. This protection might be provided down to the field level, implying that different levels of users could have access to different subsets of the database contents. And separate access limits might be established for examining and changing data. For example, a payroll section manager might be allowed to examine and change an hourly pay rate, while a payroll clerk's password might allow display, but not modification, of the pay rate's value.

In some DBMSs, data are encrypted before they are stored on disk or tape. This prevents individuals with knowledge of the computer system from bypassing the passwords to directly examine or change records.

The relative importance of security features of a DBMS will depend on the nature of its applications. In general, mainframe DBMSs should have extensive data security and privacy features because they will be applied to a wide range of organizational applications, some of which are sure to need this protection. Databases implemented on a multiuser system of any size are likely to require security provisions because of the difficulty of controlling access to terminals. In many cases, microcomputer data security can be provided by using removable media such as floppy disks or cartridge hard disks for the database. The media can then be stored in a secure area when not in use.

Data Backup and Recovery. Hardware and operator failures are facts of life that must be planned for when a database management system is implemented. The disks or tapes on which data are stored may be destroyed through mechanical problems, software problems, or operator mistakes. The principal defense against loss of data is to provide a backup and recovery mechanism. At the simplest level, usually used with microcomputer DBMSs, this means routinely making copies of disks. If a master data disk is destroyed, the last backup can be used—with reentry of the most recent data to bring its contents up to date.

More sophisticated DBMSs provide a data logging and recovery capability. Backup copies of databases are still made at routine intervals, perhaps at the end of each work day. However, as data are entered which add to or modify the database, this information is recorded on a disk or tape separate from the primary database. If there is a system failure leading to loss of the primary database, the backup version of the database is brought into use, then updated with data from the data logging file. This avoids the need to manually reenter data, which could be a massive operation for a mainframe-level system.

Availability of Fourth- and Fifth-Generation Retrieval Languages. To provide the most comprehensive access to database content, some direct end-user retrieval capability is important. In many organizations, an information center provides terminals and software instructions to allow managers to perform their own data retrieval. The software which supports this activity is usually in the category of fourth- and fifth-generation retrieval languages. These are often optional features of a mainframe DBMS. They provide a more or less natural language user interface for data retrieval.

A great deal of research is currently under way in developing easier-to-use retrieval capabilities. The term *fifth generation* is often applied to the integration of artificial intelligence techniques with natural language. This leads to software capable of carrying on a dialogue with the user, allowing the retrieval language to interpret the intent of the user's question and convert it to a form amenable to database response.

Integration Capabilities. When a DBMS is used to support MIS and DSS activities, there will often be a need for more sophisticated processing of retrieved data than can be carried out within the scope of the DBMS itself. The ability to extract information from the database and input it directly to other software packages can extend the value of the data resource for decision-making purposes. Some integration samples, applicable to both the microcomputer and larger-scale computer environments, include:

Business Graphics. The message in numeric data can often be conveyed to management more effectively through pie, bar, or line charts than in a tabular form. A typical business graphics software package, integrated with a DBMS, would allow extraction and summarization of data from the DBMS, then presentation of the data in graphical form on a CRT, printer, blotter, or slide-producing device.

Spreadsheets. Electronic spreadsheets, available on both microcomputers and time-sharing systems, emulate a paper form with ruled rows and columns. Titles, numbers, and formulas may be entered in the spreadsheet, and powerful commands are available to operate on sets of values. Trigonometric functions and interest rate calculations are available to spreadsheet users. If data can be easily moved from a DBMS to the spreadsheet, processing limitations of the DBMS may be overcome. Most of the integrated spreadsheet packages have built-in business graphics capabilities.

Communications. Two forms of communications are worth considering. First, the ability to transmit data from a mainframe or minicomputer DBMS to a microcomputer DBMS or electronic spreadsheet allows local manipulation of the content of the corporate database without the cost of reentering data already stored on the larger computer. To provide this capability, the microcomputer might have special hardware installed which makes it appear to the mainframe to be a time-sharing terminal. In some cases, special software is available for both the mainframe and microcomputer to allow the small-computer user to request data from the mainframe through the keyboard.

Since mainframe database management systems are often operated in a distributed processing environment, where many computers are interconnected to share data and processing, the large-scale DBMS will have to provide support for data communications or integrate with other data communications software. A second form of DBMS communication support is needed when the database is physically distributed among computers, but is accessible as if it existed in a single location.

SUMMARY

Database management activities involve the design, maintenance, and use of collections of data describing an organization's activities. These comprehensive collections of data, called databases, are usually computerized. Software called a database management system is used to enhance the effectiveness of database management by standardizing data description and maintenance procedures, and by simplifying development of application software which uses the database.

The requirement for database management activities becomes greater as computer systems are asked to support the needs of analysts and decision makers at higher levels in a business organization. This is so because of the unstructured nature of many of the decisions made at this level, and because of the need to draw data together from different functional elements of the organization.

Successful database management begins with effective database design. This design activity represents an effort to model organizational entities through data describing their activities and relationships. Once fields have been defined, they must be grouped into logical records. A three-step record normalization process, which minimizes data redundancy and logical retrieval problems, is recommended as the approach to this grouping.

Database management software is available for all levels of computer sophistication, and is almost always purchased as a commercial product. Parameters for selecting an appropriate product must be evaluated in terms of the needs of the purchasing organization. These parameters describe methods for defining the database, its output products and processing, its ease of use, and the degree to which the software can be integrated with other software products.

REFERENCES

1. Senn, J. A., *Information Systems in Management,* Wadsworth, Belmont, Calif., 1982.
2. Martin, J., *Managing the Data-Base Environment,* Prentice-Hall, Englewood Cliffs, N.J., 1983.
3. Sandberg, G., "A Primer on Relational Data Base Concepts," *IBM Systems Journal,* vol. 20, no. 1, 1981, pp. 23–40.

CHAPTER 3
CAD/CAM AND COMPUTER GRAPHICS

Tien-Chien Chang, Associate Professor
School of Industrial Engineering
Purdue University
West Lafayette, Indiana

Sanjay Joshi, Assistant Professor
Department of Industrial and Management Systems Engineering
The Pennsylvania State University
University Park, Pennsylvania

Walter Hoberecht, Graduate Student
Department of Industrial and Management Systems Engineering
The Pennsylvania State University
University Park, Pennsylvania

This chapter reviews hardware and software technologies for CAD/CAM and computer graphics. Since computer graphics is the basis for CAD/CAM, it will be discussed first. The application of such new technologies in the field of industrial engineering and their implications will be emphasized.

INTRODUCTION

Computer-aided design (CAD) and computer-aided manufacturing (CAM) have become promising tools in improving product quality and increasing productivity in modern industry. CAD can be defined as any system that uses a computer to assist in the creation or modification of a design. CAM, as defined by CAM-I (Computer-Aided Manufacturing— International, a nonprofit organization based in Arlington, Tex.), is "the effective utilization of computer technology in the management, control, and operations of the manufacturing facility through either direct or indirect computer interface with physical and human resources of the company."

Design and manufacturing are two inseparable activities. Whatever is designed must be capable of being manufactured. Information generated in the design stage is used later in the manufacturing stage. The manufacturing operation also feeds information, such as producibility and cost, back to the design process. The design is then modified according to the feedback from manufacturing. The cyclic nature of the interdependence of these activities is shown in Fig. 3.1.

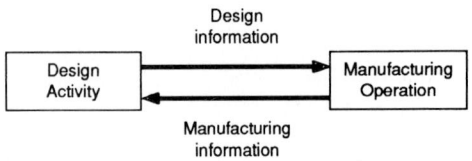

FIGURE 3.1   Interaction of design and manufacturing.

Applying computer technology to the link between the design system and the manufacturing system provides faster feedback and better communication. This helps lower costs and assists in providing better product quality. In addition, shorter product lead time and better documentation can be realized.

The integration of CAD and CAM is normally called CAD/CAM. Nowadays CAD/CAM is widely accepted as a CAD system having the capability to link with CAM functions. In this chapter our discussion is based on this narrower definition of CAD/CAM. Those who are interested in learning more about CAM and computer-integrated manufacturing (CIM) should see Sec. 7, Chap. 8.

In any design and manufacturing process, there are three major functions: synthesis, analysis, and presentation. In a traditional system, all three functions are carried out manually with the aid of drafting instruments, calculators, handbooks, and tables. A computer-aided system substitutes a computer for these traditional aids. In addition, a computer can assist in decision making, information storage, and automated information presentation. Since a picture is worth a thousand words, graphics is the best medium to transfer information. Computer graphics plays a major role in CAD/CAM. During the synthesis stage, interactive computer graphics assists humans in representing their ideas. The analysis of results can also best be presented pictorially by computer graphics. Therefore, computer graphics has become the most important part of CAD/CAM.

Although the major functions of CAD/CAM systems are design modeling, design analysis, and manufacturing control, there are applications in plant layout and business graphics too. Many industrial engineering jobs can be made easier with CAD/CAM systems. We will discuss both the technical aspects of CAD/CAM and some representative applications.

INTERACTIVE COMPUTER GRAPHICS

Interactive computer graphics may be defined as human interaction with a computer-controlled graphics input/output (I/O) device to create, manipulate, store, and retrieve pictorial data. It emphasizes the human-machine dialogue, which is conducted in real time. A typical interactive computer graphics system consists of a graphics display, a keyboard, and either a mouse, a light pen, or a graphics tablet. Pictorial data are input through either keyboard, mouse, light pen, or graphics tablet, and viewed on the graphics display.

The development of interactive computer graphics dates back to the early 1960s, when Ivan Sutherland developed the SKETCHPAD system at MIT.[12] Because of the cost of computer hardware, interactive computer graphics was not popular until the mid 1970s. With the drastic drop of computer hardware cost, and the development of user-friendly software, interactive graphics has gained wide acceptance. In the 1980s, the development of personal computers enabled nearly everyone to have access to some kind of interactive computer graphics application (computer games, business graphics, and so on).

Graphical I/O Devices.   Interactive graphics requires that the user be able to interact with the system. Some of the graphical I/O devices to input information to the system are:

Light Pen.   Light pens are photooptical devices capable of interacting directly with the image on a cathode-ray tube (CRT). A user can use a light pen to pick up a graphics element from the CRT by pointing the light pen at the image and clicking a button. The light pen can be used as a tracing or sketching instrument that permits direct drawing of lines and points on a CRT screen.

Tablet.   A data tablet is a graphical input device for inputting of point coordinates. It consists of a flat writing surface and a stylus which the user moves over the tablet. Most tablets in CAD systems are used to set up a menu of commands to make the user's task easier. Tablet menus typically include standard geometric construction and editing commands but may also include special commands created by the user for a particular application. The tablet is found by most users to be ergonomically superior to the light pen.

Mouse.   A mouse is used for applications that need to locate the relative position of a cursor on the screen. It has a small ball rolling on two orthogonal wheels attached to an encoder to convert it to coordinates on the screen. Push buttons may be mounted on the top of the mouse to enable a user to select commands. It can be rolled on any flat surface.

The mouse is cheap and simple. It has an advantage that the user need not pick it up in order to use it. The mouse has other unique properties. If the mouse is picked up and put down somewhere else, the cursor position will not change. Also the coordinates delivered by the mouse wrap around when overflow occurs. Its real disadvantages are that it cannot be used for tracing data from paper, since a small rotation or a slight loss of contact will cause cumulative error, and it is very difficult to hand-print characters for recognition by computer. For such applications, a tablet is essential.

Keyboard and Function Keys.   An alphanumeric keyboard is a standard device for input of commands and numbers used to communicate with the operating system and utilities. Keyboards may sometimes be extended by function keys. Some of the function keys may be programmable with respect to their effect, while some may have a fixed meaning. They can be used to select menu items and move a cursor.

CRT Display.   The cathode-ray tube is a device to display data and graphics. It is the standard CAD output device. The CRT uses electric fields to generate and deflect a finely focused beam of electrons to various parts of a phosphor-coated screen. The phosphor, when excited by the electron beam, glows to produce a visible spot or trace for a short moment.

A fundamental characteristic of the basic CRT is that the glow produced by the electron beam dies soon after the beam has moved away, depending on the persistence of the phosphor. This presents a problem for preserving an image on the screen for any duration, but it also provides the opportunity to generate fast-moving dynamic displays.

Various display techniques combining different deflection schemes and methods of preserving or rewriting the image have developed over the years. The three basic schemes are the direct-view storage tube (DVST), vector refresh, and raster scan.

Direct-View Storage Tube.   The DVST is an adaptation of the CRT with an extremely long-persistence screen. The image drawn by the electron beam stays on until it is erased by being reset. The DVST provides a flicker-free image, regardless of image size. Usually it can provide very high resolution. The main disadvantage is that it cannot erase parts of the picture selectively; it is therefore unsuitable for dynamic graphics.

Vector Refresh.   In vector-refresh displays, the image is vectorized and stored in display lists. Each element in the display list is a vector (pair of points). Any point drawn on the screen fades quickly. To overcome the fading of the phosphor, the picture is maintained on the screen by repeated regeneration, or *refreshing,* from the stored display list. The rate at which the image is refreshed must be faster than the flicker fusion rate of the eye (at least 30 times per second), for the image to appear steady. Complex images (in excess of 4000 vector inches) may require a refresh cycle longer than the phosphor can stay ignited, thus causing flicker. By changing the contents of the display list, an image can be partially altered. The advantage is a high degree of dynamics, high resolution, good image brightness, and high speed of interaction. A disadvantage is that the extra hardware and/or memory for the refresh cycle increase cost and system complexity.

Raster Scan. Raster scan displays differ in how the displayed data are represented. Raster scan eliminates the problems associated with DVST and vector-refresh methods. Raster scan implies that the image is scanned onto the screen in a raster sequence, that is, as a succession of equidistant scan lines. Scan lines are composed of *pixels* (smallest addressable dot on the display, a picture element). Pictures are drawn to the display from memory (frame buffer). Each pixel has a corresponding memory location in the frame buffer, specifying its on/off condition, color, and brightness. The beam scans across the lines and energizes pixels in accordance with the pattern stored in memory.

The picture is refreshed constantly at a fixed scanning rate and is not dependent on the complexity of the image. The resolution is determined by the number of pixels, and each pixel requires a separate memory element. A high-resolution (1000 × 1000 pixels) color display requires more than 1 megabyte (MB) of memory. Raster scan displays have good brightness, accuracy, and opportunity for unlimited color.

Non-CRT Displays. There are several types of non-CRT-based displays in computer terminals and workstations.

Plasma Panel Display. This display uses a panel with a raster of gas discharge channels, one for each representable point. The screen is flat, unlike the CRT. Its main disadvantage is its relatively poor resolution and its complex addressing and wiring requirements.

Laser Displays. Laser displays are one of the few high-resolution, large-screen display devices. A laser display is capable of displaying an image 3 feet by 4 feet and still has a relatively small spot size. It has been used in displaying maps, high-quality text, and elaborate circuit diagrams.

Other Non-CRT Displays. Some other display devices are liquid-crystal displays (LCDs) (including color LCDs), light-emitting diodes (LEDs), and electroluminescent-powder layers. These have not gained broad use in CAD installations.

Hard-Copy Devices. A CAD system terminal can only temporarily display an image. Hard-copy devices are used for:

- Creating check plots for off-line editing
- Making permanent copies of CRT displays
- Producing final drawings and documentation on paper, film, or microfilm

Two basic methods of plotting are vector plotting and dot matrix or raster plotting.

Vector Plotting. This is similar to displaying vectors on a vector-refresh CRT. Plotting can be done by using incremented or absolute coordinates. Two vector-plotting devices are drum plotters and flat-bed plotters.

The drum-type plotter uses special chart paper rolls and can produce continuous plots up to 120 ft in length. Several pens can be mounted in the carriage to obtain color and different line widths.

For flat-bed plotters, the paper is mounted on a flat bed and the pen movement is in both *x* and *y* directions. They generally provide higher resolution and a continuous display during plotting, as opposed to drum plotters.

Dot Matrix or Raster Plotting. This method of plotting is analogous to displaying information on a raster CRT. The presence or absence of dots is used to formulate the lines or other information on the document. Four types of raster plotters are electrostatic plotters, laser plotters, ink-jet plotters, and impact plotters/printers.

Electrostatic plotters are fast compared to line-drawing plotters. Time to output a picture is independent of the complexity of the picture. Color cannot be produced, but area shading is possible. This type of plotter requires special paper. Very high resolution (up to 400 dots per inch) can be obtained.

Laser plotters are similar to electrostatic plotters. They are very fast and best-suited for combining alphanumeric and graphic output. Both black-and-white and color plotters are available and have become very popular for graphics output.

Ink-jet plotters make use of electrically charged droplets of ink to produce the output. These can also produce color as well as black-and-white output.

Impact plotters/printers operate on a mechanical basis. Needles or flexible hammers are hit against ribbons of different colors while being moved along a line to be printed. They are very slow, and the resolution is low.

Other Plotters. Photoplotters operate like pen plotters except that a light source replaces the pen and film replaces the paper. Light beams with different apertures and speeds can be used to control line width and gray scales. Photoplotters are used mainly for integrated-circuit (IC) layout or for printed circuit (PC) board applications, such as preparing photomasks for PC board etching.

For microfilm plotters, the picture lines are either drawn directly on the film by electron or laser beam or projected on the film. Combined alphanumeric output is possible. Most of the computer microfilm-generating devices have the disadvantage of making positive microfilm, which is not as easily read as negative microfilm on screen. These plotters are particularly suited for space-saving archival storage.

GRAPHICS TECHNOLOGY

Some of the graphics technologies related to hardware have been discussed in the previous sections. Others which are related to software and/or general technology are discussed in this section.

Figure 3.2 depicts the procedure for plotting an object on a CRT terminal. Before one can see the picture on the screen, several steps have to be executed by the computer and the graphics terminal. First, the pictorial data for the object must be stored in an orderly manner. This is so that it can be easily manipulated and plotted. The data structure is built in such a way that it can accommodate all the graphics primitives (symbols) and/or text characters to be used. It also allows application programs to delete, add, and change the picture. A data structure manipulation language is needed to manipulate the data. Such a language (or set of subroutine calls) consists of commands such as add primitive after current primitive, delete current primitive, change parameters of current primitive, and update link list.

A picture is transformed before it is plotted. The operator may want to view the picture from a different angle and different distance. Transformation matrices for rotation, translation, scaling, and mirror imaging are used. Every endpoint of the picture has to be transformed to the new coordinate system. This can be done by applying the matrix to the data structure.

To display solid objects, the hidden parts have to be removed from the images of the solid objects. Hidden-line or hidden-surface algorithms are used to remove the hidden parts to create a more realistic image. These algorithms use some form of geometric sorting to distinguish visible parts of an object from those that are hidden. The algorithms work either in object space or image space.

A graphics application program gives the user the impression of looking through a window at a very large picture. Clipping is used to determine the visible and invisible portions of a picture element, and allows the invisible portion to be discarded.

Clipping and viewing transformations are used in conjunction to provide general viewing capabilities. The use of transformations gives us the ability to define pictures in coordinate systems of our choice, and then convert them to screen coordinates for display. The term *world coordinate system* is used to describe the space in which the picture is defined. The viewing transformation forms the bridge between the world coordinate system and screen coordinates. A window is specified in world coordinates surrounding the information to be displayed. In addition to the window, a view port can be defined on the screen where the contents of the window are displayed. Special effects such as panning and

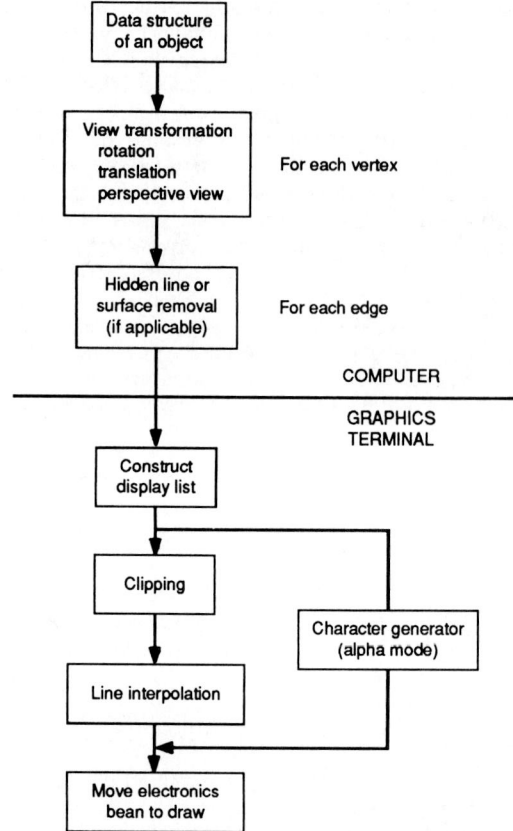

FIGURE 3.2   Procedure for displaying an object on a CRT.

zooming can be obtained by changing size and position of the window. Panning is the ability to scan a picture, and is obtained by linear translation of a window. Zooming is the ability to focus in on a small portion of the picture and produce an enlarged image. Zooming is obtained by changing the size of the window.

To display the picture on a raster scan device, the lines must be converted to a pixel representation. If the lines are inclined, then converting to pixel representation results in jagged lines. Interpolation algorithms convert a straight line into pixels for display.

CAD

The meaning of computer-aided design has changed several times in its relatively short history.[4] In the beginning CAD was synonymous with finite-element analysis. Later the emphasis in CAD changed to computer-aided drafting. Recently CAD is associated with the design modeling of three-dimensional objects.

CAD can be defined as a process which uses a computer to assist in the creation or modification of a design. It is a discipline which provides the methodology for specifying, designing, implementing, introducing, and using computer-based systems for design purposes. CAD concerns the utilization of computer systems for the design and communication of design information. The individual techniques used are not unique to CAD.

Design comprises synthesis, analysis, evaluation, and presentation of results aided by the intuitive creation of new information by the designer. These are the essential constituents of the iterative process leading to a final design. Synthesis of the design, in particular, requires some intelligence. Analysis is basically the repeated application of a set of rules. Evaluation deals with extracting pertinent information from the analysis performed and requires intelligence. Finally, presentation of results deals with preparing final documents and drawings to assist the user in viewing the design. Humans are better-suited for the task of synthesis and evaluation, since these processes require intelligence. Artificial intelligence techniques are being developed and introduced into CAD to make the computer suitable for these tasks. Analysis and presentation are tasks well-suited for computers, since these can be cast into algorithmic form.

The basic configuration of a CAD system consists of:

- Central processing unit, which can be a microcomputer, engineering workstation, minicomputer, or a large mainframe computer
- Graphics processing unit (graphics board or a separate graphics processor) for handling all interactive graphics
- Main memory for CPU and graphics processing unit
- Hard-disk storage unit for mass storage of programs
- Magnetic tape unit for system backup
- I/O devices

The CAD workstation provides the interface between the user and system. A user can design a part, put together an assembly, and conduct an engineering analysis directly and interactively on the CAD system.

A turnkey CAD system (which is any computer configuration that includes all hardware and software necessary for the system to function and produce results for a given application) has the following basic components:

- System hardware (computing equipment and peripherals)
- Graphics hardware (CRT displays, digitizers, plotters, and other related devices)
- Operating system software to control computers and peripherals
- Graphics software to produce all graphics primitives and commands
- Applications software to use the graphics software for specialized functions

Characteristics of CAD Systems.   CAD systems can be characterized by some aspects referring to the hardware and software components as well as their application areas (Table 3.1). Figure 3.3 represents the characteristics of CAD systems.[7]

Advantages of CAD.   CAD systems provide users with several advantages, such as:

- Increased productivity of the design process
- Improved design quality and accuracy
- Reduced development and testing time
- Reliability of design data
- A common database for use by other systems [CAM, material requirements planning (MRP), and others]
- Better design documentation

CAD Operating System Software.   Any CAD system, whether it be a large central system or a small stand-alone system, needs an operating system (or a limited subset of the op-

TABLE 3.1 Classification of CAD

By application area

1. Mechanical engineering
   - Two-dimensional mechanical part drawing
   - Three-dimensional mechanical part drawing
   - Assembly drawing
   - Detailing
   - Simulation capabilities, FEM, solid modeling
   - Tooling design
   - NC tool path
   - Manufacturing control
   - Process control
   - Quality control
   - Robotic simulation
2. Electronics
   - IC and PC board layout
   - Very large scale integration (VLSI) design
   - Logic and circuit design
   - Design analysis
   - Tooling analysis
   - Wire harnessing layout
3. Architectural engineering
   - Layout and structural design
   - Structural analysis
   - Piping and duct layouts
   - Space planning
   - Site planning
   - Producing plans, elevations, sections, details
4. Cartographic applications
   - Digital terrain modeling
   - Map plotting and generation
5. Other applications
   - Business graphics
   - Photo composition
   - Image processing

By hardware

1. Microcomputer-based
2. Minicomputer-based
3. Mainframe-based

By software used

1. Drafting
2. Three-dimensional wireframe
3. Three-dimensional solids
4. Surface modeling
5. Analysis software

erating system) to control the communication between the operator and CAD system hardware and software.

CAD Applications Software. Applications software performs specific tasks for users rather than for the computer. Some of the applications software available in a CAD system are:

- Drafting
- Wireframe models

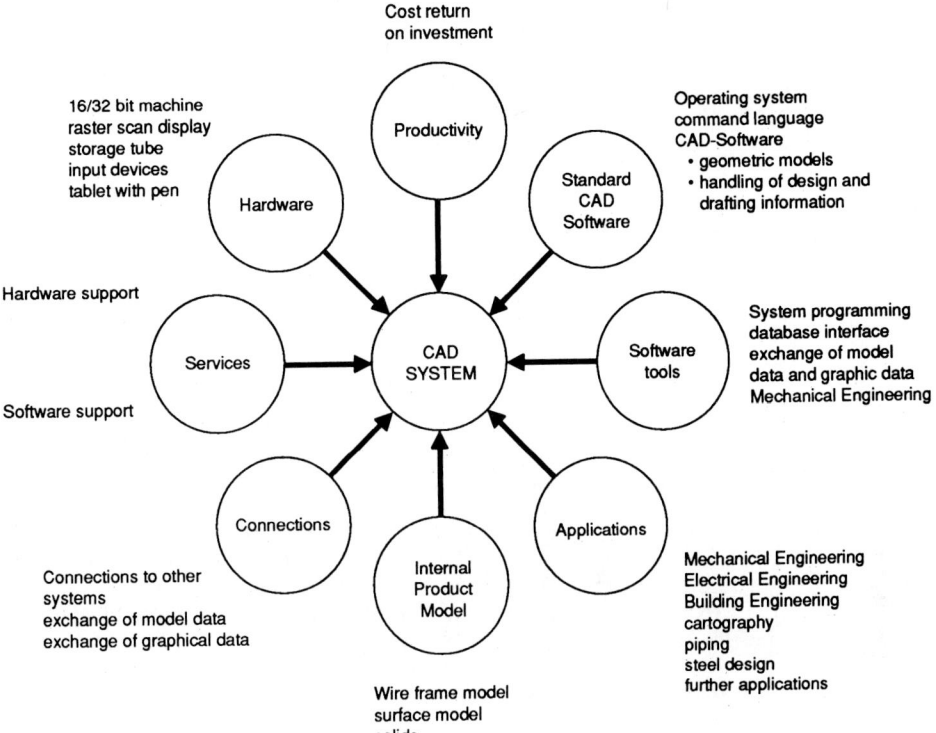

FIGURE 3.3 Characteristics of CAD systems. (*H. Brabowski and R. Anderl, "CAD-Systems and Their Interface with CAM," in* Lecture Notes of the Advanced Course on Computer Integrated Manufacturing (CIM '83), *U. Rembold (ed.), University of Karlsruhe, Federal Republic of Germany, 1983.*)

- Modeling of three-dimensional solids
- Sculptured surfaces
- Analysis software

Drafting. Two-dimensional drafting applications are a substitute for conventional drafting methods. The CAD system becomes a graphics editor in this case. Basic geometry such as points, lines, circles, arcs, cones, polygons, and spline curves can be defined interactively. Other drafting symbols such as tolerancing, cross-hatching, dimensioning lines, standard threads, and chamfers are also available to the operator.

A drafting package should also be capable of supporting overlay drafting and displaying a reasonable number of overlays simultaneously so that composite drawings can be made. The system should allow the user to retrace and restore steps taken in creation of a drawing, to aid correction. Some systems can support automatic or semiautomatic dimensioning. Zooming or panning functions are also available.

Using two-dimensional drafting systems has its limitations. A solid object cannot be represented as a solid, hence engineering analysis cannot be performed. The user cannot see the three-dimensional image on a display. Also there is difficulty in representing objects which are smoothly curved in two directions, for example, a sphere.

Wireframe Models. Wireframe models are a direct extension of the concept of edges used in two-dimensional drafting to three-dimensional space. A wireframe model describes only the edges and outlines of curves (see Fig. 3.4). Wireframe models are easy to generate and work with, simple to store and manipulate in computers, and useful as visual aids.

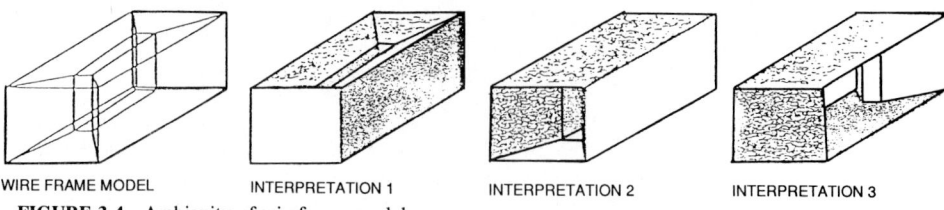

WIRE FRAME MODEL   INTERPRETATION 1   INTERPRETATION 2   INTERPRETATION 3

FIGURE 3.4   Ambiguity of wireframe model.

However, they do not capture enough of the shape properties of physical objects to convey the notion of solidity.

In a pure wireframe model there is no concept of surface. Surfaces are undefined entities whose intersections form edges which are explicitly represented as space curves. The shapes of surfaces whose boundary edges are defined by such curves are left to be determined by the user's imagination. Curved surfaces cannot be dealt with in a pure wireframe model. For example, the only "wires" that can be associated with a finite cylinder are two parallel circles, which is the same representation for two hemispheres. In the case of a sphere, there are no edges to represent it at all. To remedy this situation, wires are added where there are no real edges in the object. This is in order to approximate shapes of curved surfaces. These unreal or fake edges can be thought of as contours resulting from hypothetical planar sections of objects.

Another consequence of the absence of surfaces in wireframe models is that the notion of inside and outside is lost. Certain wireframe models are ambiguous; that is, the same representation corresponds to more than one solid (see Fig. 3.4). Absence of surfaces also makes it difficult to support a fully automatic elimination of hidden lines or computation of sectional views in a pure wireframe model. Hence, it is not sufficient for engineering applications requiring calculation of mass properties (volume, weight, center of gravity, moment of inertia, and so on).

In practical systems, wireframe models are often augmented by relatively simple surface definition facilities to provide a more complete geometric model (see Fig. 3.5). This augmentation may occur with or without extensive drafting functions extended to three dimensions. Extending two-dimensional drafting to three dimensions is done in many ways. One way is to work directly with three-dimensional coordinate values, instead of

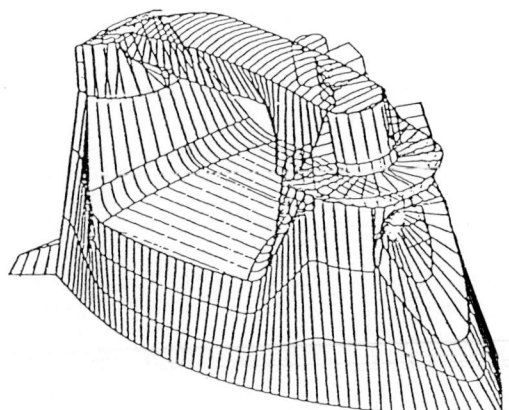

FIGURE 3.5   Wireframe model with ruled surfaces.

two-dimensional. However, this is not always convenient, since people are used to two-dimensional systems.

Another way is to provide a mode of operation where one coordinate can be factored out dynamically from the user-system communication. This is done by introducing the concept of a *current work plane* which permits users to carry on a two-dimensional dialogue with the system relative to a local coordinate system on the work plane. The local coordinates are converted to three dimensions automatically.

Wireframes are used frequently as a CAD tool to generate projection views of an object for drafting. There are also certain applications, such as semiautomatic generation of finite-element models, where wireframes provide adequate representations.

Modeling of Three-Dimensional Solids.   Several methods exist for constructing three-dimensional objects within a CAD modeling system.[9] The three-dimensional objects are used in CAD systems to build a model, modify a model, generate a projective display for line drawings, generate projective displays for hidden-line removal, evaluate collisions between separate solids, and compute geometrical and internal properties.

Representation schemes for three-dimensional objects should satisfy the following criteria: validity, completeness, uniqueness, conciseness, ease of creation and modification, and efficacy.

The commonly used methods in solid modeling systems are:

- Pure primitive instancing
- Spatial occupancy enumeration
- Cell decomposition
- Sweeping
- Constructive solid geometry (CSG)
- Boundary representation (BREP)
- Feature-based design

Pure primitive instancing is based on the notion of families of objects, each family being distinguished by a few parameters. Individual objects within a family are called *primitive instances,* and objects can be designed using a combination of instances. These schemes are unambiguous, easy to validate, concise, and easy to use, but limited to very small domains. The main problem is the difficulty of writing algorithms for computing properties of such objects.

Spatial occupancy enumeration is a representation using a sequenced list of spatial cells (*spatial arrays*) occupied by the solid. The cells are fixed-size and lie in a fixed grid. Spatial arrays are unambiguous, easy to validate, and unique, but require large storage. They are not very well-suited for modeling of the mechanical objects used in finite-element analysis.

Cell decompositions are general cases of spatial occupancy enumeration. The cells are tetrahedra which must be either disjoint or meet at a common face. Cell decompositions are unambiguous but nonunique, and validity is difficult to establish. They are used primarily in finite-element methods.

Sweeping is a technique for forming a solid object by moving a two-dimensional shape along a trajectory. Sweeping is of two types, translation and rotation (see Fig. 3.6). These schemes are unambiguous but not unique and are limited to objects which have rotational or translational symmetry. Sweeping is used as a means for entering object descriptions in geometric solid modelers.

Constructive solid geometry is a method of constructing objects from simple solids called *primitives.* The primitives may be blocks, cones, cylinders, spheres, or other geometric shapes. The technique used to perform the construction is the union, intersection, or difference of the primitives to form a solid. The construction procedure is represented internally as a tree, and is shown in Fig. 3.7. The CSG scheme is unambiguous but not unique. Validity of CSG trees can be easily ensured. It is concise and very easily creates a representation of unsculptured mechanical parts. However, CSG representation is not

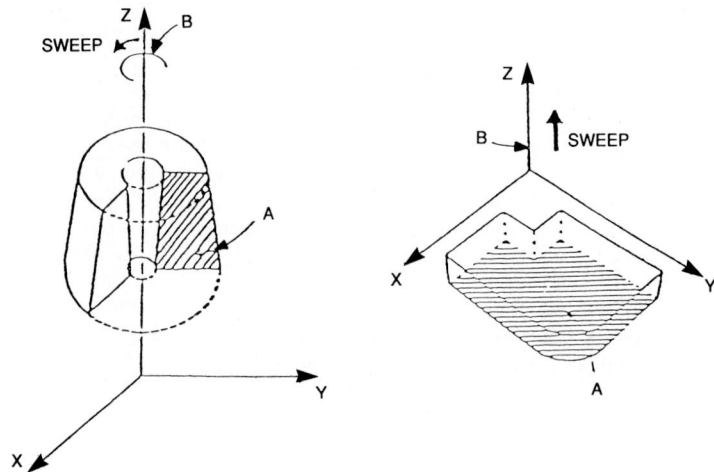

FIGURE 3.6 Translational and rotational sweep.

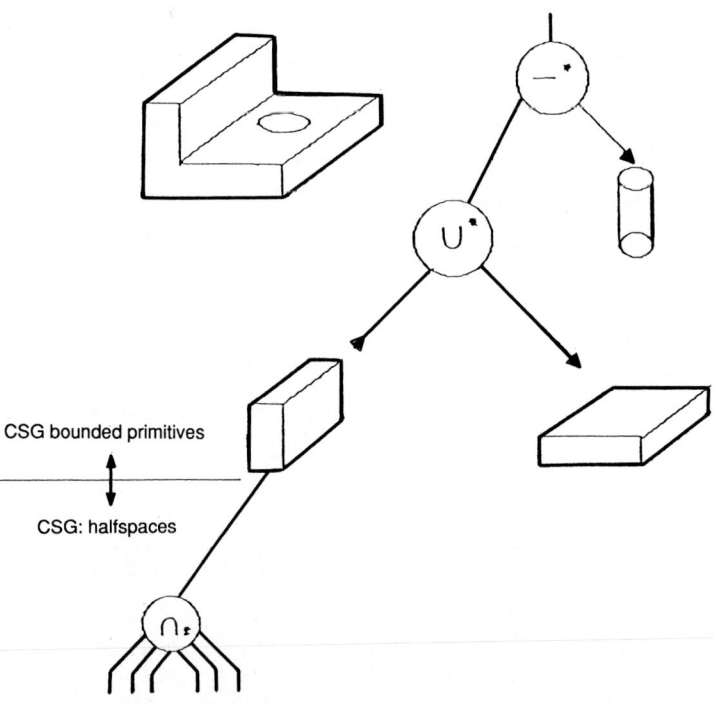

CSG bounded primitives

CSG: halfspaces

Six planar half spaces

FIGURE 3.7 Constructive solid geometry tree.

suitable for producing line drawings, and graphic interactions such as "pick an edge" are difficult to support directly. Current trends indicate CSG will be of considerable importance for manufacturing automation.

Boundary representation is a method of representing a solid object using its boundary surfaces (see Fig. 3.8). The faces of the solid are represented by its boundary, composed of edges and vertices. Boundary representation schemes are unambiguous if faces are rep-

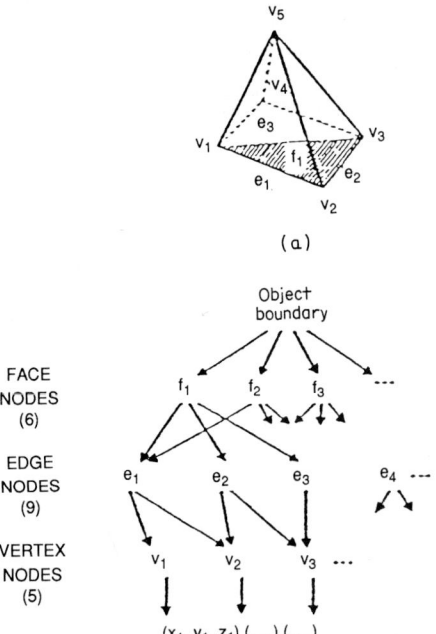

FIGURE 3.8   Boundary representation of a solid.

resented unambiguously, but generally not unique. Validity of BREP has to be established both by geometry and topology validity. BREP requires considerably more storage than CSG, and the volume of required data and the complexity of establishing validity make it difficult to construct without the help of computers. The main advantage of BREP is that the information between faces, edges, and relations between them is easily available and makes it suitable for generating line drawings, graphic displays, and graphic interaction.

Feature-based design is a method whereby a designer creates a part by selecting the features that the part should have. The concept of feature-based design[2] is based on providing the designer with a library of features along with a set of operators that can be used to create and modify designs. The library of features is similar to the primitives in a CSG system. The feature-based design representation can also be used to create a boundary representation for the part. The advantage of feature-based design is that it records as much information about a part as possible as early as possible. Proponents argue that when the designer starts the design process, most of the part features are already known. During the process of entering the design into a CAD system, this information is lost, only to be recreated later by feature recognizers. Additional benefits are also realized by feature-based design systems with regard to analysis, such as design for manufacturability, and functional information about the part (see Fig. 3.9).

Hybrid schemes are designed by combining the methods discussed above. The most common hybrid system is CSG/BREP, in which CSG is used as a basis for input and the object is stored in BREP, thus combining advantages of both.

Sculptured Surfaces.   In CAD, curves and surfaces principally serve the purpose of describing the shape of technical objects. CAD software for sculptured surfaces deals with the entire process of developing the description of such shapes with the aid of computers, from initial concept formalization to final determination. Many applications, such as vehicle body design and design of aircraft and ships, require modeling of surfaces.

In traditional design practice, the design of *free-form* curves (that is, the design of curves and surfaces by nonanalytical means, especially drafting methods) consists of two distinct phases: the *generation* phase, which results in preliminary shape description, and the *fairing* phase, which leads to final shape definition of appropriate quality. In the automotive industry, for example, a clay model of a car body shape is produced from a draft of the shape design, modified according to the stylist's instructions until the form can be digitized and faired to toolmaking tolerances.

The current trend in most industries is to use the computer to assist in the fairing phase as well as the generation phase. The two methods used for representing a surface in the computer are:

- Polygon mesh
- Parametric representation

A polygon mesh is a set of planar surfaces, each of which is bounded by a polygon and defined by a plane equation. The plane must be small enough to accurately represent the

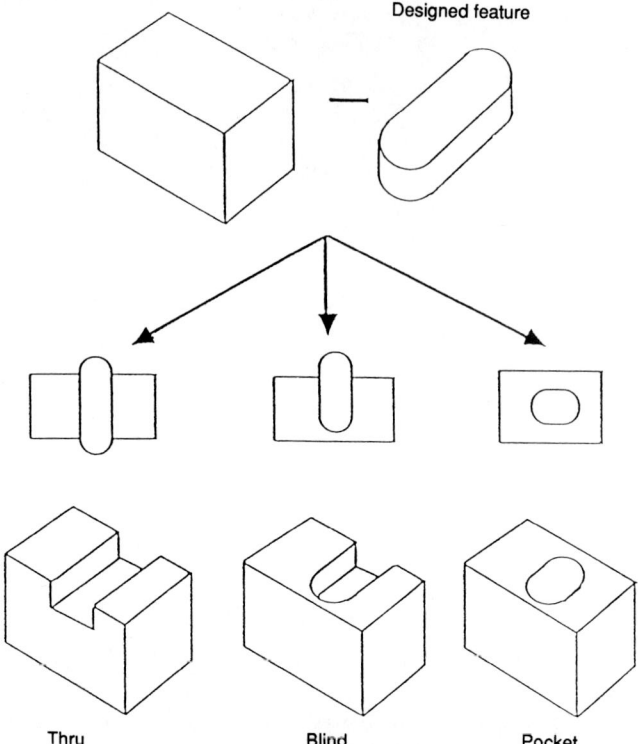

FIGURE 3.9   Design feature and manufacturing features.

actual curved surface. The amount of storage required is very large and the resolution is fixed, independent of the eventual display. Further, even with the most sophisticated continuous-shading models, polygonal techniques generally result in visually objectionable images. For some applications, polygonal representation may be quite sufficient and an exact surface description is not needed.

The parametric representation of sculptured surfaces was pioneered by S. Coons. P. Bezier introduced the use of nonlinear parametric equations for representing segments and patches to form surfaces. The points on the curved surface are defined by three equations, one each in x, y, and z, and each equation has two parameters u and v, up to the third order. The paths defined in this way can be made to satisfy the desired order of continuity.

When creating joints, the designer often wants to control the order of continuity at the joint. Zero-order continuity means that the two curves meet; first-order continuity requires the curves to be tangent at the point of intersection; second-order continuity requires that the curvatures be the same.

Four well-known methods for parametric representation of surfaces are:

• Coons surfaces
• Bezier surfaces
• B-spline surfaces
• Nonuniform rational B-spline surfaces (NURB)

The Bezier surface (Fig. 3.10) is defined as a *characteristic polyhedron,* which is specified in terms of the position vectors of its 16 vertices (called control points).

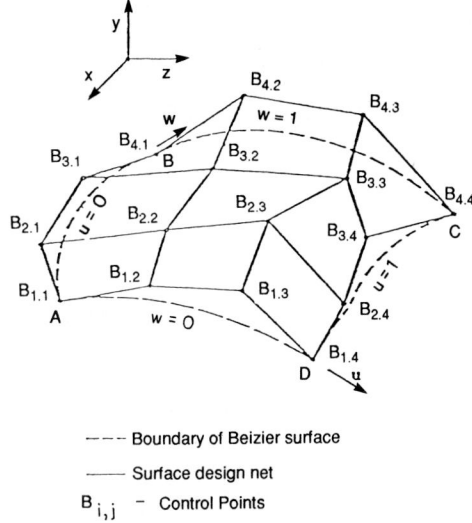

- - - - Boundary of Beizier surface

———— Surface design net

$B_{i,j}$ — Control Points

FIGURE 3.10   Bezier surface.

The patch is in a sense a smoothly blended approximation to the control polyhedron. Only the corner points lie on the curve. The configuration of the polyhedron gives the designer a good indication of the general shape of the corresponding patch and modification of the points alters the patch in a predictable way.

Bezier curves and surfaces exhibit global control, thus the movement of a single vertex affects the entire curve or surface. Pieces of separate Bezier's surfaces can be put together to form a composite surface. The method chosen depends on the degree of continuity desired (for details see Ref. 5).

Splines are named after the devices used by drafters and ship builders to draw curves. A physical spline is used like a French curve to fair a smooth curve between specified data points. A spline can be mathematically defined as a set of polynomials over a knot vector. B splines have the advantage of local control and have greater flexibility for controlling continuity. NURB is a more general form of B spline that can be used to describe nearly every kind of surface.

Analysis Software.   The most widely used group of methods for analysis is finite-element methods (FEM). Practical designs inevitably involve complicated geometric shapes and different materials. In order to quantitatively assess a design, it is necessary to decompose a complicated structure into basic elements, each being of geometric shape and made of a single material. Detailed analysis of each element will then be possible. By preserving the physical interrelationship of the elements, an approximate but fairly accurate evaluation of the structure can be obtained. It is widely used in:

- Static and dynamic analysis of complex structures such as airplanes, bridges, buildings, and cars
- Fluid flow, diffusion, and consolidation problems
- Lubrication problems
- Heat conduction and thermal stresses

Before a structure is analyzed by FEM, it is preprocessed to partition it into discrete elements. The common shape of a discrete element, called the *shape function,* is a triangle for a two-dimensional structure and a tetrahedron for a three-dimensional one; however, other shape functions can be used. CAD systems can be used to design the structure and help in automatic mesh generation (Fig. 3.11). The ability of CAD systems to assist in mesh refining and adding geometric details is best done in an interactive graphics environment. Presentation, interpretation, and evaluation of results is aided by computer graphics. Other analysis programs included in CAD systems are simulation of dynamic processes and kinematic analysis of mechanisms. Such capabilities provide the user with the ability to check interference of mating parts, collision between moving objects, and other dynamic processes. Analysis programs can also be used for automatic verification of numerical control (NC) tapes and to simulate the machining process.

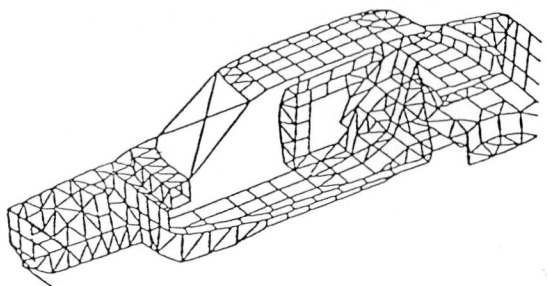

FIGURE 3.11   Idealization of a car body using finite elements.

Standardization of CAD-System Interfaces.   CAD systems have a number of hardware and software components which have to exchange data. The principal idea underlying the quest for standards was that the main body of the software should be device-independent. Another major concern which led to development of standards was the need to communicate with databases of drawing information in different CAD/CAM systems. Standards have been developed on all levels. In this chapter we discuss only the drawing information database level standards.

IGES.   The Initial Graphics Exchange Specification (IGES) is a part of American National Standards Institute (ANSI) Standard Y414.26M. It allows different CAD/CAM systems to exchange data. Since each CAD/CAM system has its own data format, it is not possible to transfer the design on one CAD/CAM system to another directly without translation. Under IGES, translations are performed by programs called pre- and postprocessors. Two-dimensional drafting and three-dimensional wireframe and surface models can be converted to IGES format. Currently IGES is available in all CAD systems.

PDES.   The Product Data Exchange Specification (PDES) was developed to overcome the weaknesses of IGES (file size, transfer time, sensitivity of geometric definitions). The PDES model is intended for complete graphical model representation. PDES is under continual development.

APPLICATIONS FOR INDUSTRIAL ENGINEERING

Several traditional industrial engineering functions can be assisted by CAD/CAM systems. Some of them are discussed in this section. However, the applications are not limited to the ones discussed.

Facility Layout.   There are several steps in the facility layout process. Flow data collection, overall layout planning, flow analysis, detailed layout, and layout installation are major steps. Collecting material flow information and preparing activity relationship charts are steps in the data collection stage. Overall layout planning and flow analysis consider only departmental layout. Computer programs such as CRAFT and CORELAP handle problems on this level.

Traditionally, detailed layout is done with the aid of two-dimensional or three-dimensional templates. Scaled-down templates representing machines, robots, carts, operators, and other shop entities are placed on a grid layout board. Besides aiding flow and other relation analyses, the visualization of the layout helps layout planners to understand the plan and hence produce a sound layout. This can help layout planners to determine conflicts and see the overall plan. After the layout is finished, a drawing is prepared from the template layout.

In a CAD/CAM system, templates can be modeled and stored in the system database. When needed, the graphics templates can be retrieved and placed anywhere on the CRT screen by using a menu-driven program. The CRT screen becomes an electronic layout board. Since CAD/CAM systems provide superior graphics creation and editing capability, the floor plan and device templates can be easily drawn and moved around. When the layout is finalized, a ready-to-use drawing can be easily plotted. The drawing can be used for the construction of the building or installing the layout.

Although any CAD/CAM system can be used for layout application, some systems come with a library of ready-to-use templates which make the task of the layout planner easier. Usually architectural CAD systems are better for layout application. Besides architectural symbols, such systems usually provide three-dimensional modeling capability. The layout can be viewed from different angles, and even light sources can be placed in the model to determine the effect on lighting conditions. Color is provided by all CAD/CAM systems and can be used to highlight different equipment and room partitions.[6]

Workstation Design.   Facility layout deals with the entire plant or shop. On the other hand, workstation design deals with humans or robots and the working environment surrounding them. It deals with detailed design of a working environment. In designing a workstation for human operators, human factors are always important. Can an operator reach all objects in the workstation with ease? Is the seating properly designed so that fatigue is minimized? These are some of the questions involved in workstation design. The industrial engineer usually follows the human body model to determine the height of chair and table and the ability to reach objects. Testing of the design may show that modification is necessary.

There are CAD systems which provide three-dimensional graphical human body models. The workstation designer can design the workstation around the model and simulate the human operator directly on the screen. One such model is called SAMMIE (System for Aiding Man-Machine Interaction Evaluation). SAMMIE is a 19-link computerized human model which is used for assessment of the operator's reach, vision, and fit for any given workstation.

Robotic Workstation Design.   Many CAD systems have mechanism animation and kinematic analysis modules. One of the areas where these tools are applied is in the modeling of robotic workstations. The workstation (parts, fixtures, and tools) and robots can be modeled and a three-dimensional simulation of the robot can be done to evaluate the design. With the advances in solid modeling, it has become easier to check for interference and collisions during motion of the robot links. Complex interiors and exteriors of the work cell can be viewed with color. Once the final decision is made, the drawing can be produced automatically.

Additionally, the model may be used as part of an automated path planning and collision avoidance system. Collisions can be detected as more than one object trying to occupy the same space.

Operation Sheet Preparation.   Operation sheets, sometimes called *process plans* or *work instructions,* contain the detailed instructions for producing a product. For manual operations, a detailed description of each operation and a sketch of the workpiece layout needs to be given. Such documents are traditionally prepared by industrial engineers. Preparation is a tedious job, especially when graphical instructions are involved. By using a CAD/CAM system, an operation sheet can be prepared and stored in the system. Workpiece and fixture models can be created once and used in different operation sheets for different workstations. Also see Sec. 7, Chap. 3.

Automated Process Planning.   Computer automated process planning (CAPP) is concerned with the preparation of a process plan for a part without human intervention, beyond the creation of the required databases. As such, CAPP fits properly in the broad definition given for CAM at the beginning of the chapter. Typically CAD/CAM provides the foundation on which a CAPP system is built. Also see Sec. 7, Chap. 8 for further information.

The fully automated generation of process plans from the CAD database is not a trivial task. This is because process planning requires a combination of mathematics, engineering, geometric modeling, computer science, artificial intelligence, and experience in manufacturing. Beyond the design of the part, information is required about the available machine tools, tooling, fixtures, and materials to prepare a viable process plan.

One stumbling block to process planning is interpreting those features of a design that require machining. There are two major classes of techniques for automatically converting a design from a CAD database into a process plan. The first attempts to recognize part features (hole, slot, and so on) from the geometric data that were saved when the part was designed. The second stores a representation of part features as the part is designed. There are several alternative techniques within each of these classes.

Group technology (GT) coding is just one example of how part features can be identified and interpreted automatically. It can further be used as a basis for structuring the CAD database. GT assists both the design engineer and the process planning system.

NC Programming and Tool Path Verification.   With the advent of numerically controlled machine tools came the development of programming languages such as Λ Programmed Tool (APT) and COMPACT II, suited to the description of machining motions. These languages offered programming productivity improvements over manually calculating the geometry of cutter motions. With ever more available computer power, graphics-assisted programming systems are now in widespread use. Where special-purpose languages provided a 10-to-1 improvement over manual programming, graphics-assisted techniques show a 20-to-1 improvement. Graphics systems also provide the user with immediate feedback about the motions that the machine tool will execute. This allows for faster program debug time, and improved communication between the shop and engineering.

There are two basic types of graphics-assisted tool path generation systems. The first is self-contained, providing graphics creation and editing capabilities within the package. The second type is designed to operate within an existing CAD package.

BREP systems that utilize sculptured surfaces or solid modeling are best-suited to NC tool path generation. In applications such as mold design, a model can be developed that assists the engineer during design and provides data for manufacturing. In addition to the data necessary to produce a part, the CAD system can also generate data to drive coordinate-measuring machines (CMMs). This assists in quality control and inspection activities.

Rapid Prototyping.   It is often desirable to have a physical rendering of a part to analyze or otherwise make use of. Frequently, parts will go through the design, prototype, and fabrication sequence several times. Decreasing the amount of time spent in the prototype generation phase helps to lower the cost of the design cycle and reduce the engineering lead time required for a new design. One technique that has successfully been used to this end is stereolithography. The stereolithography apparatus (SLA) uses a three-dimensional surfaced or solid model as its input data. SLAs produce parts through photopolymerization, transforming a liquid plastic monomer into a solid polymer by exposing it to ultraviolet light.

To actually produce prototypes, the control computer divides the CAD model of the part into several cross-sectional slices. By scanning the cross sections in sequence and lowering the elevator table prior to starting the next one, a solid rendering of the part is produced.

An alternative process similar to stereolithography is selective laser sintering, using lasers to heat a powder and cause it to fuse together. The CAD model is used for the generation of cross-sectional slices and for the control of the laser.

Prototypes that might take from several days to a few weeks to prepare with traditional techniques can be generated within a single day on systems such as these.[3,8,13]

Maintenance Plan. CAD/CAM systems can be used to maintain updates on the maintenance status of all equipment and facilities. Any change in the equipment and facility can be modified on the CAD/CAM database with ease. This is especially important in the chemical processing industry. When the facility is changed, it is difficult to update all the related hard-copy drawings. However, when a CAD/CAM system is used, a single database for all facilities is maintained. Every drawing is based on the master database. Maintenance plans become current and consistent.

CONCLUSION

In this chapter CAD/CAM and computer graphics are introduced. Some of their applications to industrial engineering are also briefly discussed. Although some specific applications are mentioned, the applications are not limited to the ones discussed. CAD/CAM is still an evolving technology, and only the initial development is visible. Far more advanced features will be available in the near future.

In the future CAD/CAM will be integrated into a total design and manufacturing environment. More user-friendly software, capable of natural language understanding, will be developed. Intelligent CAD systems will be able to design from the functional specifications provided by the humans. Lower costs of higher-resolution color dynamic workstations will allow wider use of CAD/CAM functions.

The developments in CAD/CAM will affect not only the designers or manufacturing engineers, but other users also. Many industrial engineers' analytical tools, such as optimization algorithms, will be interfaced with the graphics packages to produce easily understandable results. Some of these developments are being accomplished now and are in different stages of development. This is an exciting field for us to explore and its applications can be limited only by our imagination.

REFERENCES

1. Chang, T. C., and R. A. Wysk, *An Introduction to Automated Process Planning Systems,* Prentice-Hall, Englewood Cliffs, N.J., 1985.

2. Cunningham, J. J., and J. R. Dixon, "Designing with Features: The Origin of Features," *Proceedings of the ASME International Computers in Engineering Conference,* vol. 1, San Diego, 1988, pp. 237–243.

3. Deitz, D., "Stereolithography Automates Prototyping," *Mechanical Engineering,* vol. 112, no. 2, February 1990, pp. 34–39.

4. Encarnacao, J., and E. G. Schlechtendahl, *Computer Aided Design,* Springer-Verlag, Berlin, 1983.

5. Faux, I. D., and M. J. Pratt, *Computational Geometry for Design and Manufacture,* Ellis Horwood, Chichester, England, 1979.

6. Filley, R. D., "A Survey of Software for Facilities Planning and Design," *Industrial Engineering,* vol. 16, no. 5, May 1984, pp. 71–79.

7. Grabowski, H., and R. Anderl, "CAD-Systems and Their Interface with CAM," in *Lecture Notes of the Advanced Course on Computer Integrated Manufacturing (CIM '83),* U. Rembold (ed.), University of Karlsruhe, Federal Republic of Germany, 1983.

8. McKee, K., "Rapid Prototyping," *Manufacturing Review,* vol. 3, no. 2, June 1990, p. 131.

9. Mortenson, M. E., *Geometric Modeling,* Wiley, New York, 1985.

10. Newman, W. M., and R. F. Sproull, *Principles of Interactive Computer Graphics,* 2d ed., McGraw-Hill, New York, 1979.

11. Smith, B. M., "IGES: A Key to CAD/CAM Systems Integration," *IEEE Computer Graphics and Applications,* November 1983, pp. 78–83.

12. Sutherland, I. E., "SKETCHPAD," *Proceedings of AFIPS,* vol. 23, 1963.

13. Weiss, L. E., et al., "A Rapid Tool Manufacturing System Based on Stereolithography and Thermal Spraying," *Manufacturing Review,* vol. 3, no. 1, March 1990, pp. 40–48.

14. Wilson, P. R., "A Short History of CAD Data Transfer Standards," *IEEE Computer Graphics and Applications,* vol. 7, no. 6, June 1987, pp. 64–67.

CHAPTER 4
COMPUTER SIMULATION[*]

Mark O. Presnell
Engineer Consultant
Electronic Data Systems Corporation
Warren, Michigan

Simulation is one of the fastest-growing areas in the computer-based decision support marketplace. There have been three significant contributors to this explosive growth during the last decade. The first contributor was the addition of graphical output and animation. This development in technology has taken simulation from the back computer room into the boardroom. Instead of being presented as reams of computer output with some summary charts and graphs communicating the findings, the results are displayed by lively, animated media. A significant amount of information is displayed on the computer screen. The audience can absorb this information much more quickly because of the familiar, life-like representation of the problem under investigation.

The second significant contributor has been the increase in computer speed and drop in cost of hardware systems. Corporate mainframes are still used for running simulation models today, but only for the largest models. The growing trend is toward the personal computer and the engineering workstation as the hardware platforms of choice. The most transportable, powerful simulation tools have software offerings that span the hardware continuum from the personal computer to the mainframe. There are also many simulation tools developed for running only on the personal computer.

The third significant contributor is the development of easy-to-use simulation tools. Many simulation tools now have features such as custom modules to address material handling requirements, pull-down menus for improved user interface, and very complete standard output statistics.

As simulation continues to grow into the twenty-first century, the tools will be further enhanced. Computer-aided design (CAD) system interfaces, scheduling capabilities, process control logic development, computer links to maintenance systems, and other interface issues along a typical project life cycle will continue to be improved. This growth and improvement will be focused on shortening the total development time required to get products to market, lowering the amount of human effort required to do the analyses, and improving the quality and accuracy of the information provided to the decision-making process.

TRENDS IN THE INDUSTRY

Hardware. Much has been written about the trends in the hardware industry and it is not the author's intent to spend a great deal of time reviewing this subject. The simulation industry has been impacted by the changes in the hardware industry and it will continue to

[*]Special thanks go to Daxus Corporation, the author's former employer, for giving the author the time and support to write this chapter.

be influenced. Simulation tools will continue to take advantage of the increase in the ratio of computer speed to cost. In addition, changes in graphics capabilities will allow simulation software developers to enhance the graphics display features offered with their packages. More information will be dynamically displayed in real time with less impact on the overall speed and performance of the displayed graphics. As the cost continues to be driven down, more companies will be able to afford the required hardware for doing simulation modeling and analysis.

Software. During the last decade, software enhancements have focused on graphics improvement, ease of use, transportability across hardware platforms, and general flexibility and capability of the simulation tools. Moving into the twenty-first century, software enhancements will continue to be added to these areas. An area of great potential growth is the integration of simulation tools with the other computer-based tools required during the project life cycle. An example of this situation is the information provided by the facility and product design engineers in computer-aided design, manufacturing, and engineering (CAD/CAM/CAE) systems. These data are electronically pulled into the model because they provide a significant amount of information required for the simulation effort. Any additional information required for the simulation model is entered into the model by the modeler. After the simulation modeling and analysis is completed and an alternative has been selected, the detailed logic in the simulation model is electronically transferred to the process control development team and other computer system development teams. This ensures that the logic is captured and included in these systems in exactly the same manner as it was modeled, analyzed, and determined to be a working solution.

People. As with all technology, one of the most important pieces of an equation for success is people. The number of people who are knowledgeable and capable of providing simulation models and analyses will continue to grow. The simulation modeler will no longer be the computer programmer tucked away in a corner, perfectly happy to interact as little as possible with other people in the organization. The modeler will be a very visible, versatile person. Typical backgrounds include: operations research, industrial engineering, manufacturing engineering, computer network design, operations planning, and facility engineering. People with a good analytical problem-solving background, a thorough understanding of the area being modeled, a solid working knowledge of statistics, an understanding of design of experiments techniques, and the ability to communicate effectively with other members of the organization will be the norm. The importance of communication skills of the simulation modeler cannot be overstated because, as models become more comprehensive, the people providing information to the decision-making process will be going across many departments and areas in an organization. Part of the job requirement will be the facilitation of the entire experimental and analytical process to ensure accurate and unbiased results.

Uses. The use of simulation tools will continue to broaden as more people learn how simulation modeling can be effectively used. Manufacturing planning has been one of the largest application areas for simulation in both the discrete and process industries. Computer network analysis will grow in those areas where response time is particularly important to a computer application's success. The opportunities are endless. Simulation modeling tools and techniques are essential for evaluating complex environments which have nonpredictable, random events affecting the systems performance.

Integration with the System's Operation. Simulation tools will be used on an ongoing basis further down the project life-cycle curve. The simulation model of a manufacturing operation can be used as part of the continuous improvement process. This will allow the manufacturing personnel to evaluate alternatives before making changes on the shop floor. The true bottleneck of the system can be identified by using the simulation model. The plant management can then focus resources on the bottleneck area. Improvement in this area will typically have the largest direct impact on throughput. This longer-term use of simulation models will become more of a norm in the future. Companies will realize the benefit of keeping the models current and active in the decision-making process. Integration with other computer systems and sources of operating information will become more widespread. Material requirements planning (MRP-II) systems, finite scheduling systems, maintenance planning systems, worker planning systems, and inventory tracking systems will be integrated with simulation models as part of the overall corporate enterprise model.

INTEGRATION WITH OTHER TOOLS

Simulation will no longer be an island in the area of decision support technologies. Information from decision support tools will become integrated with information from many other sources. This integration of information will take place in a manner similar to the information planning and integration which is currently taking place for the entire enterprise model of many organizations. The integrated information will be from the ongoing operations and the advanced planning activities. The examples discussed below are not meant to be exhaustive, but only samples of what has been achieved in a few cases. The examples are not meant to describe today's industry norm.

Computerized Layout Planning Tools.   There are numerous facility layout planning tools (LPTs) on the market today. Many of these tools focus on analytical, relationship-diagramming techniques. An LPT allows the facility designer to quickly evaluate various layout options. The goal is to develop a facility layout that satisfies the constraints while keeping the non-value-added movement of material or product at a minimum. A substantial amount of information in this system is also used in simulation models. Examples of transferable information are distances, process routes for materials, size of containers being transported, method of transportation for the material, and flow rates between the different areas in the manufacturing operation.

Statistical Analysis and Statistical Process Control Tools.   Both input and output data can be analyzed with these tools before they are used in the simulation model. On the input side, data can be analyzed and fit to a distribution or combination of distributions. Particular data parameters include items such as downtime, variability in processing time, and reject rates. On the output side, the information collected during the execution of a simulation model can be analyzed. The analysis can include both the single simulation run and the multiple simulation runs. The analyses are being made to thoroughly understand the decision-making confidence level for the experimental results. As computers become more powerful, design of experiment tools and statistical analysis tools will be more directly linked. The experimental procedure will become more integrated and automated.

Computer-Aided Drawing (CAD) Packages.   The formal development of the simulation and CAD integration link will allow scale drawings being developed in the CAD package to supply the layout and detailed dimensional information to the simulation electronically. This information will be used in both the simulation and the animation of the system under investigation. The electronic link will yield a very accurate method for bringing these two technologies together. It will also decrease the development time for the simulation engineer during the animation development and material flow route development phases of modeling projects.

Material Requirements Planning Packages.   MRP-II packages contain a significant amount of information regarding the timing, quantity, and location for what will be produced in a manufacturing operation. This information is an important part of a manufacturing-related simulation model. It is the driver for the facility's intended manufacturing plan. Many MRP-II packages today have infinite capacity scheduling and planning capability. The output from these modules of the MRP-II package can be used by the simulation models. This production plan will still need to be modified by a finite scheduling module. In an ongoing operations environment, MRP-II data will need to be manipulated by some type of sequence-scheduling module or tool before being introduced into the simulation model. The planned sequence will become the simulation model driver, allowing an analytical review of the plan before it is released to the shop floor. The opportunity to test the schedules with the simulation model allows the scheduling rules to be improved.

Finite Scheduling Packages.   These packages are a rapidly developing set of tools that assume capacity is finite. Finite capacity is defined as a group of constraints that attempts to take into account the characteristics of the manufacturing operations being scheduled. The information contained in these systems can be used to drive the simulation model of the manufacturing facility. The simulation models can be used to evaluate the scheduling rules being implemented in the scheduling system. There are numerous tradeoffs to be made for any system. Simulation modeling provides the environment and opportunity to make quantifiable comparisons of the alternatives using the appropriate performance measures.

Maintenance Systems.   There are several areas where the integration of simulation models and preventive maintenance systems can be leveraged. The impact of different maintenance plans and strategies on production can be evaluated on a quantifiable basis. This will allow the identification of the best tactical plan for providing system maintenance. Issues to be evaluated can be the skill set, physical location, and number of maintenance people employed by the company. The simulation models can be used to look not only at strategies for the planned preventive maintenance downtime but also the unplanned random equipment failures that occur. Analyses can be performed on the overall system to determine the potential benefits of decreasing the amount of downtime on a particular company resource. If this piece of equipment is heavily utilized, it may in fact improve the facility's capabilities. If, however, the equipment is not heavily utilized, reducing the downtime on this particular resource may have no positive effect on the company's bottom line.

Downtime Databases.   One key piece of information going into simulation models is resource unavailability due to planned and unplanned downtime of the facility's resources. This is one area where simulation has been criticized in the past, namely for the lack of good historical data. As simulation models are used more frequently on an operational basis, the need and ability to use actual facility data increases. The opportunity exists for the facility's downtime databases to be accessed by the simulation models on a regular basis. This will ensure that the best available information is being used in the simulation model. This information will become more directly linked to the simulation model in the future.

Cost Management/Tracking Systems.   This area will become an increasingly important area for integration with simulation models. The simulation models will provide information in terms of throughput, time in system, inventory levels, and due-date performance. By accessing good cost tracking data, the models will be able to provide information regarding the cost of producing a given product or filling a particular customer's order. Much has been written about the inability of the cost-tracking systems to provide good information. The introduction of a properly implemented activity-based costing (ABC) program can provide the right information for use in the simulation models. See Sec. 9, Chap. 3 for more information.

Rapid Modeling Tools.   Rapid modeling tools (RMTs) are being developed to augment the tool kit of facility designers and planners. These tools are used upstream of simulation models to analyze a wide range of options. Many of these tools include queuing theory, optimization, linear programming, and spreadsheet-type capabilities. See Sec. 14 for more information. These tools require much of the same information eventually required by the simulation models. The integration of RMT tools with simulation tools will allow the information to be included in the simulation model in a faster, more accurate manner. There have been some attempts made at having the rapid modeling tools actually write the code for the simulation model through the use of a code generation module. The attempts have been limited in number and have been focused on a particular subset of problem types.

Time Standard and Methods Analysis Systems. The information contained in these systems provides some of the important requirements for a good simulation model. The information is used on both an upfront, planning basis and an ongoing, operational basis. The methods tools are used to look at ways to decrease the amount of effort required to pick up and assemble several parts, for example. The simulation model is used to look at the particular improvement activity from a total system perspective. It will assist in determining whether the proposed improvement will add anything to the company's bottom line or just contribute to idle time for the employee. The simulation model can help focus the detailed methods analyses on the areas that will have the most immediate impact. A significant amount of information contained in the time standard systems is needed in the simulation model. It is important to have both the time standard and the actual performance for a particular activity measured and collected over time, because not every employee can do a task in the same manner or length of time. This variability can be captured in the time standard system and then used by the simulation model on an ongoing basis.

Spreadsheet Tools. These powerful tools are highly utilized by facility planners and operations personnel. Two areas of integration with simulation models can be used. They will be discussed here. (There are undoubtedly other opportunities because of the seemingly limitless uses for spreadsheet tools.)

First, a significant amount of upfront facility planning is accomplished with the help of spreadsheets. The spreadsheets contain information about cycle time, aggregate efficiency, and product distribution. These data are used by the first, rough-cut, simulation models. The development of a common data interface between the spreadsheet and the simulation model will allow the planners to look at a larger number of alternatives posed by the marketing organization. The planners will be able to analyze the alternatives both from a deterministic point of view (spreadsheet) and probabilistic point of view (simulation) in a very rapid fashion.

The second opportunity is to use the spreadsheet tools as the data interface to the simulation model. Customized, user-friendly spreadsheet applications can be developed which allow operations engineering people to change parametric data in the spreadsheet when it requires updating. The engineer saves this information and imports it into the simulation model. The simulation model can then be executed while the data file created by the spreadsheet application is accessed. This gives the engineer the opportunity to use the simulation model application without having to know how to write a simulation model. The engineer will need to understand what the simulation output means, what is statistically significant, and what problems to look for when analyzing the simulation results, but will not be required to write simulation code.

USES OF SIMULATION TOOLS

As simulation tools become more popular and their benefits become more recognized, the uses and applications of these tools will continue to grow. A wide range of applications is found below. More detailed examples are found later in this chapter.

Material Handling Systems. Simulation has been used extensively on material handling systems. The nature of automatic storage and retrieval systems (AS/RS), automated guided vehicle systems (AGVS), power and free (P&F) conveyor systems, and electrified monorail systems (EMS) requires a substantial investment in computer control logic. The logic needs to be developed and proven before it is installed on the shop floor. Simulation models provide valuable insight into the number of pieces of equipment needed for the material handling system to meet the design objectives (P&F carries, AS/RS cranes, AGVS vehicles). In addition, the dispatching rules and control logic required by the material handling system are analyzed. This helps to ensure that the systems provide the appropriate level of support to the manufacturing or distribution system. It is advisable to do a sensitivity analysis around the variable number of pieces of equipment being considered for purchase. This helps ensure you are purchasing only what you need and not what will work under a given set of conditions.

Paper Flow in an Office Environment.   Paper flow in an office environment is treated much like product flow in a job shop or other manufacturing environment. Typical areas of analysis are

1. Number of people required
2. Number of steps in the overall process
3. Time required for the piece of paper to be in the system
4. Ability to process paper by the promised due date
5. Impact of serial versus parallel processing of the paper
6. Benefits and economies associated with individual versus batch processing of paper

It has been well-documented that a company's chances of success are enhanced by its ability to reduce lead time and meet due dates. The flow of paper adds to the company's total response time to customers. Through the use of simulation tools, companies benefit from determining ways to streamline their paper flow.

Computer Network Design.   Application of computer simulation provides benefits to the design of computer network systems. The greatest benefits are realized when response time is a critical measure of success. The ability to look at different alternatives for the distribution of software throughout the various levels of a typical seven-layer computer architecture, or across the nodes on the same level in the architecture, adds significant benefit to the design process. The architecture, like many problems, is first addressed using analytical, deterministic techniques to arrive at a rough solution. When a significant amount of random interactions are occurring, simulation provides a more complete set of information for making the necessary design tradeoff decisions. Several niche products addressing this problem are available in the market today. Special care should be taken to ensure that the assumptions used in these products are true for the particular application under investigation.

Flexible Manufacturing System (FMS) Design.   Many niche tools have been developed for the design of flexible manufacturing cells. These tools typically handle the global design of the systems quite well. But if the study being performed is detailed—that is, if it includes tool management in the various machines in the cell, manpower interfaces with the cell, tool wear tracking, and scheduling of product into the cell—it may be better to develop a simulation model using a more general-purpose simulation language. The general-purpose simulation language allows more detailed analyses. Simulation provides the experimental framework to evaluate:

1. Number of machines in the cell
2. Required capability of each machine in the cell
3. Schedule for introducing various products into the cell
4. Quantity of tools on each machine in the cell for a given product mix
5. Sequencing logic to be used inside the cell

Numerous opportunities exist for simulation to be used in understanding how FMS systems work and in ensuring the benefits are realized when an FMS system is implemented.

Material Handling with Overhead Cranes.   Frequently there are more overhead cranes in a facility than are actually needed. Excess crane capacity usually occurs because the existing facility is either overdesigned or the amount of work required by the cranes is reduced. Simulation tools are used when evaluating alternatives to handle more work with the existing cranes, to reduce the number of cranes being used, or to balance the work load across the cranes. The variability in times for when work arrives at the cranes, the potential interference and resultant blocked or idle time, and the response time of the cranes to manufacturing operations need to be considered. Simulation tools provide the ability to address the random events in these systems.

Retail and Consumer Outlet Design.   The classic area of analysis is the supermarket checkout line. Simulation is used to analyze the different alternatives for arranging and managing the checkout lines. The ability to get the customers through the checkout line and out the door is very important. Tradeoffs need to be made between installing or staffing additional checkout lines with their associated costs and losing sales when the customer decides to shop somewhere else because the line is too long.

Another area is drive-through windows at fast-food restaurants. On the basis of the physical arrangement and the amount of business, a manager needs to know when to take money from the customer and give food to the customer in return. The manager needs to know if it is better to do it at the same window or to accomplish these activities through separate windows.

Many examples can be given, but the issue really comes down to a tradeoff between the ability to satisfy the customer's requirements as well as or better than the competition does, in the most cost-effective, flexible way possible.

Warehouse and Distribution System Design.   Simulation is a valuable tool for understanding warehousing and distribution systems. A few potential areas of use for simulation modeling analysis are

1. Number and location of loading docks in a facility
2. Number of loading/unloading operators and strategy for deploying them at the docks
3. Number and assignments given to the material handling equipment operators throughout the operation

Many of the issues previously discussed for material handling systems hold true for warehousing and distribution systems. This is because of the significant amount of material handling equipment utilized in these operations.

Continuous Improvement Strategies.   Simulation models for a given facility allow the people working in the facility the opportunity to evaluate the impact of different improvement alternatives on the facility's performance. Simulation models focus the activity of improvement on the bottleneck or constraining operation. Focusing on the constraint provides the company with the largest return for its effort. Simulation also allows the total integrated view of the facility to be taken into account. This total system view permits the evaluation of a change on the entire, integrated system instead of only a specific area of the facility.

Operations Training.   The use of simulation tools for training grows as their models are no longer viewed as throwaway items. The models are maintained and enhanced for use during the operation of the facility. When new management of line supervision is brought into a facility, they are educated on what things work, what things do not work, and how the facility is currently being operated. These new people will invariably run into situations where they need to make operational decisions about running the facility. The operations people try to improve a problem area or correct a particular situation. Alternative decisions are tested on the computer to see if they yield the anticipated results. The testing is completed before a decision is made and tried out on the facility, customers, and employees.

WHY USE SIMULATION TOOLS?

Simulation packages are very powerful tools. They add significant insight and understanding about the planned or existing facilities under investigation. Simulation tools have a very solid place in the overall design and evaluation process. Simulation must, however, be used at the appropriate time during the project life cycle. Simulation does not replace the fundamental engineering that must take place long before simulation is used. A great deal has been said and written about why a project or person should employ simulation tools and techniques. Some of the primary reasons are described below.

Concept of Randomness.   Random events occur in nature, in manufacturing, in the service industry, and just about everywhere we look. If the world around us worked "on the average," simulation tools would not be necessary. The world could be analyzed with spreadsheets and calculators. In complex systems, where many sources of variability and randomness usually exist, simulation tools are valuable in analyzing the complex interactions of the numerous random variables. Simulation is used by the analyst to evaluate and determine what the expected outcome or results are most likely to be. This is a very important distinction requiring further exploration. Simulation does not predict what happens next in a system, but it provides some valuable insight about what will most likely happen and what the results will likely be.

Payback.   One concern that frequently arises during the process of trying to decide to use simulation tools on a project is the high cost of such an endeavor. Data collection and analysis, simulation software and hardware costs, and manpower requirements can add up to a large expense. Simulation is not cheap, but the benefits of use of these tools by a knowledgeable person frequently far outweigh the cost of the study. Industrial and manufacturing engineering trade magazines are flooded with success stories relating to the use of simulation tools. Frequently the payback is 10 times greater than the cost of the simulation study, with a breakeven point of less than one year. Savings are in the form of capital expenditure reduction or avoidance, operating cost reduction, reduction in required inventory levels, increase in throughput, and a decrease in the number of people required.

Insurance.   As previously discussed, the use of simulation tools probably will yield a very large payback if implemented properly. Even if a study yields savings approximately equal to its cost, it may be worthwhile because companies cannot afford risking lost production and delays when launching a new product or service. The use of simulation tools provides the comfort and security of knowing the facility design will provide the anticipated results. It is always better to eliminate guesswork beforehand. Waiting until start-up to discover problems can be disastrous. Many companies are betting their future on a major modernization program and cannot afford to make any mistakes. Simulation ensures that an effective solution is planned. Of course, a proper implementation of the plan is still required for success.

Communication Improvement.   One area directly related to the improvement in communication during the last decade is graphics. The saying "a picture is worth a thousand words" applies here. By using animated graphics depicting the key areas under investigation, all members have the opportunity to see and better understand what is contained in the simulation model and what the results mean. Animation is especially important when the results are not intuitively obvious. In addition to the animation, a thorough simulation modeling methodology ensures that information and assumptions are understood by all pertinent people and departments. This improved understanding greatly reduces the number of things "slipping through the cracks." A thorough methodology can also provide an environment for consensus building across the various groups or departments.

Goal Setting for Vendors Supplying Equipment.   Equipment vendors need to be evaluated. The evaluation should include not only what they say their product can deliver, but also what is required from their equipment if the overall system is to meet its objectives. The evaluation of the vendor's equipment and its interaction with other systems focuses on determining whether the facility meets it goals and objectives. The simulation model is used to evaluate the speed or throughput rate of the equipment, the setup time, the repair time, the frequency of unplanned downtime, and the scheduling rules for loading the equipment. In addition, a sensitivity analysis is done on various parameters. The analysis determines which parameters have an adverse impact on the system's performance. Parameters found to be sensitive are investigated further. The vendor-supplied information needs to be evaluated for accuracy and applicability. The sensitive requirements may even be written into a performance clause in the purchase order. It should be noted, however, that you may have a difficult time getting a vendor to agree to this type of term and condition because they are not in control of your facility. The vendors would be putting themselves in a very risky situation.

Continuous Improvement.   The continuous improvement philosophy and process is a rapidly growing strategy. Simulation is a very valuable tool for focusing a company's resources during this improvement process. Although the philosophy's goal is to empower the entire organization in this process, there is a need to prioritize and focus the effort in the areas yielding the largest gains. In the manufacturing arena, simulation assists in the evaluation of improvement strategies and provides information in terms of throughput, on-time delivery rate, inventory reduction, worker utilization, and lead-time requirements. This information is very useful in evaluating a proposed improvement alternative and trying to determine whether it will yield the desired benefits.

Evaluation of Operating Philosophies.   Simulation provides an experimental test environment for evaluating different operating philosophies. Before committing your organization to the current trend in the industry, use an appropriately developed simulation model of your facility to compare the new way of doing business against the current method and determine what the expected impact should be. The alternatives are quantified and compared to each other, and the probability of the best solution being selected and implemented is increased.

Test Environment.   The usefulness of simulation as a test environment cannot be overstated. If the area under evaluation does not exist, simulation provides a very valuable way to test the ideas and plans. This is accomplished before any capital investment is made. There is less risk in testing the planned operations on the computer. After the concrete is poured and the facilities are built, it becomes very expensive to make changes. These changes can be very costly to implement in terms of lost production, missed customer due dates, and decrease in worker morale, particularly in attitudes toward management.

Another issue is the ability to provide a cause-and-effect experimental environment. Many times in complex manufacturing operations, it becomes nearly impossible to determine the true impact of a change on the shop floor. There are too many events occurring simultaneously. Another occurrence may have actually caused the observed results, not the planned change. Any proposed change needs to withstand the test of time. Determining the change's impact on the system requires things to remain fixed for a period of time. Simulation models greatly compress the time window. In minutes, the simulation model provides feedback about what will most likely happen when the proposed change is made. It takes days and significantly more effort to acquire the same amount of information from a physical implementation on the shop floor.

GETTING STARTED

After a company or individual decides to proceed with using simulation modeling tools and techniques, it is time to determine what to do and where to start. The following section is a practical view of things to be considered during this process to ensure success. The following topics will be discussed:

1. Developing an understanding
2. Determining short- and long-term strategies
3. Make versus buy
4. Selecting simulation tools
5. Available simulation tools
6. Building internal capabilities

Developing an Understanding.   This step is very important to the overall success of the first simulation endeavor by an individual or company. The individual is encouraged to read as much as possible on the subject and talk to professionals in this area. You will probably find many consultants and vendors eager to assist you. Do not allow yourself to get caught up in the demonstrations, technologies, past project descriptions, or the capabilities of the organizations with whom you are talking to at this point. This evaluation will come later in the process. The emphasis is not on model building and analysis at the early stages but rather on the application of simulation to the area being analyzed. The individual is working toward the development of a clear set of goals and objectives for the first project as well as the longer-term simulation strategy.

Determine Short- and Long-Term Strategies.   A clear set of short- and long-term strategies needs to be developed. The focus depends on the size of the organization, the amount of simulation analysis needed, and the position or location where the champion resides in the organization. If there is not a significant, long-term need for simulation within a company, (enough to keep one or more people busy full-time), it probably makes more sense to purchase consulting services as required. When the need appears to be significant enough to keep one person busy full-time, long-term relationships with consultants or development of internal resources makes the most sense.

Make versus Buy.   This decision is always difficult. Its long-term impact to a company can be significant. The most important thing the company needs to do is be honest with itself. An accurate evaluation regarding the quantity of simulation modeling needs to be completed. The company must be committed to making a substantial investment if it decides to do the modeling internally. A company should decide on the development of long-term internal capability only if it is willing to make the appropriate commitment. The primary commitment is making simulation a full-time assignment. It is very difficult to do simulation modeling on a part-time basis. Using simulation tools, like using any sophisticated computer software language or package, falls into a ''use it or lose it'' situation.

This full-time focus is needed during the execution of a particular project for several reasons. First, it is very difficult for the simulation modeler to continually be putting down and picking up the simulation model while keeping track of the particular model's details. In addition, the faster a model is completed and results provided to the organization, the higher the level of acceptance and credibility the simulation study has. If simulation studies drag on too long, there is a tendency for assumptions and data to change in the model. This creates a general negative image of the simulation study because it is not utilizing current information and assumptions. If the anticipated volume of work is not significant enough to warrant development of internal capabilities, there are many qualified consultants available to address your simulation requirements. See Sec. 9, Chap. 5 for further details on make versus buy decisions.

Benefits and Costs.   There are relative benefits and costs associated with buying simulation consulting expertise and with developing your own capabilities. There are many significant factors to be considered (Fig. 4.1).

	INTERNAL STAFF	OUTSIDE CONSULTANT
BENEFITS	KNOWLEDGE OF COMPANY KNOWLEDGE OF SUBJECT BETTER UNDERSTANDING OF SYSTEM LOWER PURCHASING COSTS	HIRED ONLY WHEN REQUIRED MORE TOOLS AVAILABLE MULTIPLE RESOURCES AVOID PURCHASE OF SOFTWARE WIDER RANGE OF KNOWLEDGE
RISKS AND COSTS	UNTIMELY RESULTS INACCURATE RESULTS POOR TOOL SELECTION LOSS OF CAPABILITIES	MORE EXPENSIVE SERVICES AREN'T DELIVERED SPEND TIME EDUCATING CONSULTANT LESS UNDERSTANDING OF PROGRAM WRONG TOOL IS SELECTED

FIGURE 4.1   Make versus buy considerations.

The two primary ways to structure a purchase agreement with a simulation consultant are *time and material* and *fixed price*. If you are constrained by time and have little choice, a time-and-material contract can be arranged. This is best done when it includes development of the scope of work, the deliverables, and the project implementation plan. The consultant should then be able to provide a fixed-price bid for executing the plan. The other method is to fix-price-purchase the project from the beginning. You will need to develop a fairly detailed specification up front for this method to be successful. If the specification is not clear, you will receive a fixed-price contract that may be very specific in what it will deliver but may not be exactly what is needed. This can potentially lead to a significant amount of change orders. Two rules of thumb to follow when evaluating proposals are (1) if it sounds too good to be true, it probably is, and (2) don't make decisions on just price.

Defining Criteria for Decision Making.   The criteria for determining whether to develop or purchase simulation capabilities is not a black-and-white issue. In fact, the decision may be a hybrid decision. Your organization may want to start working with the appropriate consultant on the first project to "see how it is done by an expert." The longer-term plan has an organization doing additional models with internal capabilities. Management commitment, available resources, quantity of work, a simple problem for the first model, and reasonable timing requirements all need to be present when an internal strategy is pursued for the first time. If there is any doubt regarding any of these areas, it may be better to hire a consultant and expect significant involvement from your organization.

Evaluation of Options.   The evaluation process is subjective. There is no magic formula for the correct decision. The people providing input to the evaluation of the options need to be honest in the process. This helps ensure the decision that is eventually arrived at will lead to a successfully implemented solution. A well-documented evaluation methodology should be used to facilitate the decision-making process.

Selecting Simulation Tools.   If your organization has decided to do the simulation modeling work with in-house personnel, it becomes important to begin the selection process for a simulation tool or tools. This is not an easy task. The selection sets the future direction for your company. Your organization may want to go through this selection process even if the first models are going to be done by outside consultants. The consultant should supply the model in the simulation tool your organization has selected. The organization can then use and modify this model when ready to take on such an endeavor.

Defining Criteria for Decision Making.   The criteria used here are similar to those for any software evaluation task. There are the common issues to be addressed. In no specific order, Fig. 4.2 contains a partial list of criteria to be considered.

HARDWARE PLATFORM	GRAPHICS CAPABILITY
NUMBER OF LICENSES IN CIRCULATION	ERROR CHECKING CAPABILITIES
REGIONAL SALES/TECHNICAL SUPPORT	FLEXIBILITY
EASE OF USE	CUSTOM OUTPUT REPORTING
INITIAL AND ONGOING COSTS	CUSTOM INPUT DATA PROCEDURES
NO. OF RANDOM NUMBER STREAMS	OUTPUT ANALYSIS CAPABILITIES
ABILITY TO RUN IN BACKGROUND	SIZE OF ORGANIZATION
ABILITY TO RUN IN A BATCH MODE	QUALITY OF DOCUMENTATION
ABILITY TO RUN INTERACTIVELY	STANDARD FREQUENCY DISTRIBUTIONS
ABILITY TO CONTROL EXPERIMENTS	ABILITY TO COMPUTE STATISTICS
REQUIRED SUPPORTING SOFTWARE	CUSTOM FREQUENCY DISTRIBUTIONS
FREQUENCY OF TRAINING COURSES	PHYSICAL LOCATION OF COMPANY
NATIONAL USERS GROUP ESTABLISHED	PUBLISHED NEWS LETTER
FUTURE PRODUCT DIRECTIONS	QUALITY OF REFERENCE MANUALS
ABILITY TO OUTLINE AND COMBINE ALTERNATIVE SCENARIOS	OPERATING SYSTEMS USED BY THE SOFTWARE
ABILITY TO COMPUTE CONFIDENCE INTERVALS	

FIGURE 4.2   Criteria for simulation software selection. This figure can be made more exhaustive. Refer to the reading list at the end of the chapter for additional sources of criteria requirements.

Evaluation of Options.   The evaluation of simulation software tools takes a considerable amount of time and contact with each of the companies under evaluation. The first-pass evaluation focuses on developing a short list of simulation tools to investigate in greater detail. After a short list is made, the evaluation team visits the software supplier's facilities. The team needs to see first-hand the support organization and the training facilities and meet the people they would be working with. There are several scenarios for the selection and purchase of the simulation tool.

First, if the time, resources, and money are available, it is wise to attend a training course or a very detailed demonstration for each of the tools on the short list. This allows a look "under the hood." You need to determine whether the product is what the salesperson claims it is and how much effort is required to develop one of the supplier's demonstration models. Second, make arrangements with the software vendor to use the tool for 30 to 60 days after receiving the initial training. This allows for a thorough evaluation of the software and the organization standing behind the software. Frequently this can be arranged without charge, at the supplier's cost, or for a small monthly rental fee. The fee usually can be credited toward the purchase. Third, have the software supplier provide some initial consulting on a small project. The simulation modeler works very closely with the consultant. The modeler gains a good understanding about the use of the simulation tool and the methodology used in approaching the simulation project. Fourth, hire a nonaffiliated consultant to support the evaluation process. The consultant or consulting organization should have used many of the tools available on the market today. This knowledge is put to use in developing a good knowledgeable evaluation and selection of simulation software for your company.

Available Simulation Tools.   There are several good sources of information regarding available simulation languages. The list of simulation tools in Fig. 4.3 can be used as a starting point. The simulation tools are separated into three categories according to the type of problem and environment. (There is another category, which could be called "*specialty*

PHYSICAL	CONTINUOUS	DISCRETE
AutoBots	ASCL	AutoMod II
CimStation	C-Simscript	CADMotion
IGRIP	CSMP III	GENETIK
McAuto	CSSL III	GPSS/PC
	CSSL IV	GPSSH,PROOF
	DARE-P	MAST
	GASP IV	PCModel
	GENETIK	PROMOD
	SEEWHY	SEEWHY
	SIMPLE_1	SIMAN,CINEMA
	SLAMII	SIMFACTORY
		SIMSCRIPT II.5
		SLAM II, TESS
		SLAMSYSTEM
		WITNESS

FIGURE 4.3   Partial list of simulation tools.

simulators,'' but that is not addressed here; it would include tools such as NETWORK II.5 for the design of communications networks.)

It is important for those interested in simulation tools to do a thorough search to determine what is available in the market at the time. Many of the professional journals periodically publish reference and comparison articles regarding the different simulation tools on the market. The simulation tools and industry are in a high state of flux and will continue changing for some time. There is a continuous flow of new products coming into the marketplace. A literature search to identify current articles and information regarding simulation tools is advisable. The list of tools in Fig. 4.3 provides a starting point.

Building Internal Capabilities.   Internal capabilities are not built overnight. A sizable expenditure in terms of people, resources, and dollars needs to be made. This entry cost is an investment in your business. The capabilities are built while looking at both the short- and the long-term requirements.

Selecting People.   The selection of people from inside and outside the organization is a key requirement to developing successful simulation capabilities. People with the following backgrounds are typically successful in the simulation modeling environment: manufacturing engineers, operations research professionals, industrial engineers, facility planning engineers, mechanical engineers, chemical engineers, mathematicians, and other related technical professionals. The people need to have a familiarity with the use of a computer, statistical analysis techniques, and the business environment being modeled. Problem-solving skills, reasoning skills, people skills, and communication skills are also required for best results. For the typical simulation consulting project, it is frequently easier to teach people how to build simulation models and do the analysis than it is to teach them the business they are in and what is actually being modeled.

Training Requirements.   How to use the simulation tools is an obvious training requirement. This training is frequently offered by the software vendor. There are some very good third-party training classes available, and they should be considered. There are several other skills needed. If they are not present, it is necessary to provide training in those areas. People should be skilled in statistical analysis techniques, design-of-experiment methodologies, and the computer operating system and the line editor being used. Many of these skills are typically part of the people's backgrounds, described above. An assessment is required to determine whether formal training beyond the simulation-language training course is necessary.

Time to Become Proficient.   This should not be underestimated. On a relative scale, even ''easy to use'' simulation tools take some time to master. It may be easy to build a very simple model containing a minimal amount of detail. This is not the skill level required for solving complex problems. The general-purpose simulation languages are not as easy to use and require significantly more time to master. The benefits of using general-purpose tools are flexibility and expansibility. When an individual is getting started for the first time and is the only person in the organization with the experience in simulation, the learning and training period is conservatively 6 months for an easy-to-use package or simulator and 12 months for a general-purpose simulation language. These time estimates allow the individual to get experience in completing several small projects.

Start Small.   It is important to start with a project of reasonable size and difficulty. It takes time to become proficient in the entire spectrum of simulation modeling and analysis. A smaller problem allows the simulation modeler to become familiar with the simulation tool, with the implementation of a sound methodology, and with the other problems that invariably come up on the first project.

Simulation models, especially those with graphical animation, are attention-getters in an organization. This attention makes it particularly important for the first simulation project to be successful. The simulation modeler and the management involved in supporting simulation have their careers on the line. In general, the value and reputation of simulation is placed at risk inside the organization. Defining a problem that is difficult enough to warrant interest yet simple enough to ensure success is an art. Selecting the first project to be tackled deserves significant thought.

SIMULATION PROJECT METHODOLOGY

The importance of good, sound, reasonable methodology cannot be overstated. The success (or failure) of many projects has been attributed to discipline (or lack of it) in running the project. A good project planner has a clearly defined set of milestone deliverables, identifies who is responsible for the deliverables, and establishes the required dates for the deliverables. A good plan helps ensure the project is implemented in an organized manner. The overall success or failure of the project starts with a clear definition of the goals and objectives of the simulation study.

Benefits.   The benefits of using a structured project management methodology for simulation studies are basically the same as those of using a methodology for any other project. In addition to a schedule for the various milestones of the project, the most important benefit is the assignment of responsibility—particularly to the people who are required to contribute information and knowledge to the project but are not mainstream contributors. Supporting people generally have the tendency to receive other hot assignments; without a schedule, they may not find time to do a good job collecting the required information for the simulation project. Another benefit is status reports with which management can be provided an update on the project's progress. It is always better to start this discussion when things are going well and on schedule instead of waiting until the crisis mode of operation has been reached.

A Proven Methodology.   A proven methodology is described here. This methodology is not intended to be the "best" methodology but rather one that works. There are numerous variations in methodologies that can be applied to simulation projects. The decision to use a methodology is more important than the methodology chosen. It should be noted there is some overlap in the different phases. There is not always a distinct line between the phases because of the iterative nature of simulation studies.

The eight phases of the simulation project life cycle methodology are

1. Information gathering and process flow diagram development
2. Model development
3. Model verification and validation
4. Development of alternatives for analysis
5. Running experiments and analyzing results
6. Final report and presentation of results
7. User customization and user guide documentation
8. User training

Importance to Success.   The use of a proven methodology does not guarantee the success of a project, but not following any methodology at all is likely to damage the overall success of the project. Pulling the proper members of the company into the project is a great asset in determining whether the proper constraints and assumptions are included in the model. This makes the model more realistic and credible. The danger of bringing too many people into the project is loss of control. Frequently, people are not supplying the information in a timely fashion, but, more commonly, the project loses track of the original goals and objectives. Loss of the original intent and focus turns the project into one that never seems to be completed. There is a tendency for the participating members of the team to start coming up with ideas to test with the simulation model. At times, these ideas have little to do with the original scope of the project. Project milestones, defined deliverables, and a sense of timing and urgency keep things on the right track.

PITFALLS AND THINGS TO AVOID

The following issues are some of the more frequently encountered pitfalls. Whether the simulation modeling and analysis are done by in-house resources or by hiring a consultant, the issues identified here should be reviewed by the people responsible for the project.

Data Requirements.   One frequently raised criticism of simulation modeling is the lack of good data. The argument is that, without good data, the value of the simulation modeling effort is questionable. This idea has merit, and anyone involved in a simulation modeling project should not underestimate the time required for data collection. In addition to the physical collection of the data, time is required for evaluating the data and fitting them to a distribution.

It should be pointed out, however, that a great deal of benefit is realized even when some of the parametric data are not perfectly accurate and representative. A communication process, as previously described herein, occurs during a properly conducted simulation study and is very valuable. The process helps pull things together and identifies issues frequently forgotten or overlooked. Simulation requires a very detailed look at the information when the model is built. Another valuable result of simulation modeling activity with respect to data is focusing on the important areas. By doing some of the modeling with less than perfect data, the simulation modeler gains insight about what data are really important to the performance of the system. This allows the data collection effort to be focused on the critical areas and not on the areas having little impact on the system's performance.

Level of Detail.   A seasoned simulation modeler does not want to do any more coding than is absolutely necessary. The key issue is building the appropriate level of detail into the simulation model to meet the objectives of the study and to answer the questions required to support the decision-making process. Frequently, simulation modelers fall into the trap of building a very sophisticated model with more detail than required. When the level of detail increases in a model, many things occur. First, the time required to provide answers to the project team increases, and the answers may be too late. Second, the effort required to support the data collection effort becomes extremely expensive and time-consuming. Third, the computer time required to make one simulation run increases. This frequently leads to a longer response time for answering each question. Fourth, the cost to modify the model is greater than that of an appropriately detailed model. It is better to look at a large number of alternatives in less detail early in the planning process and narrow the field of options from there. The modeler may find the best alternative is not considered when looking at only a few alternatives in detail. This usually occurs because too much time is spent on the few detailed analysis alternatives and there is not time to look at the others.

Length of Project Milestones.   Another key issue is the size and amount of work associated with each task in the project plan. If the milestones are too long, there is a tendency for the simulation engineer to be working on the project with bad data or assumptions. Another result of long activities is the loss of enthusiasm and momentum within the organization. Short milestones and frequent contact with the engineers and managers in the organization enhance the level of interest in supplying data and other inputs. The engineers and managers are able to see some of the fruits of their labor in a timely fashion. Short milestones also make it easy to have an accurate picture of the project's status. The short milestones allow for the appropriate actions to be taken, keeping the project on track.

Communication.   Communication is an important part of practically any activity or project. Poor communication is a pitfall to be avoided in simulation modeling and analysis activities as well. The advent of graphics has greatly facilitated the communication process, but is has also created a new set of problems. As simulation grows and becomes more visible in an organization, the simulation modeler is required to work with a deeper vertical segment of a typical organization. In the past, it has been common for the simulation modeler to be explaining the results and findings to the project and process engineers in a typical manufacturing organization. Now, it is becoming more commonplace for the simulation modelers to be having discussions at the vice-president level in the manufacturing organization. As the discussions move up to that level in an organization, the manufacturing team needs to communicate with the product development team, the marketing team, and the materials management team. This upward movement of communication broadens the perspectives and issues requiring consideration by the simulation modeler. The simulation modeler must be effective in communicating with these people. Poor communication may lead to a significant amount of work and findings not being understood or not being used in the decision-making process.

Assumptions.   Simplifying and optimistic assumptions can have a negative impact on the quality of the information being provided to support the decision-making process. Assumptions need to be made and addressed at various times along the simulation project life cycle. Early in the process, when the possible options are numerous and the model is not detailed, more assumptions can be included in the analysis. As the model becomes more detailed and the information quality requirements for the decision-making process go up, the assumptions need to be removed by adding additional detail to the model. If the assumption cannot be eliminated, it needs to be tested. The modeler must determine the assumption's impact on the answers being provided to support the decision-making process. If answers and decisions are sensitive to the assumption, it needs to be eliminated or further substantiated. If this is not possible, the decision makers need to be aware of the potential solution's sensitivity to the assumption. Assumptions need to make good common sense at each phase of the project. The ability of the simulation modeler or the constraints of the software package should not influence the assumptions.

Tool Selection.   There are a large number of simulation modeling tools on the market today. The choice of what simulation tool to use can be a very difficult task. Both the short-term and long-term issues need to be considered before making a purchase. The primary driver is to pick the right tool for the job. Do not let the selected simulation tool drive the model building and analysis processes. Assumptions and modeling techniques should be used because they make sense for the analysis, not because of the limitations of the simulation software. The size of the problem, the type of data requirements to be tracked, the hardware platform, the people resource, the ease of use, and the ongoing vendor support all need to be considered. Some of the common problems encountered are

1. Model is too big to run on computer platform
2. Model execution is slow and timely results are not obtained
3. Data collection to support decision making cannot be accomplished
4. Number of attributes assigned to the entities or jobs is limiting
5. Allocated computer memory is not capable of handling the problem
6. Good, timely support from the software company is not provided
7. Debugging features are not adequate for model verification
8. Models are not transportable between different software versions
9. Software revisions are released with bugs in it by software vendor

Simulation software selection is a very complicated decision that needs to be made. Your organization may find it needs to develop capabilities in more than one simulation tool; however, your company may choose to not support simulation expertise in all languages. Instead, it may be better to establish relationships with consulting companies for providing expertise in the use of the required tools.

Poor Output Analysis. The output analysis is equally important to any other part of the project. The model needs to be run long enough and frequently enough to ensure that the observations are truly representative of what can be expected. In addition, the true cause-and-effect relationships need to be established. It is possible to review the data and find an appearance of a cause-and-effect relationship between one parameter and a performance measure. On further investigation, the modeler finds it is an entirely different parameter actually causing the change in results.

Looking at the data over time is another area to be considered. Do not fall into the trap of looking only at the average values for the simulation period. Inventory levels may be growing throughout the simulation model run, showing a certain average inventory level, when in fact the system under investigation is not in balance. Remember the world does not work on averages, and this is why simulation is important in the first place. Always include information regarding the ranges and confidence intervals. This helps reinforce the idea that the answers are not pure predictions of what will happen but rather some indication of what will probably happen. Refer to the list of additional reading material at the end of this chapter for more information on this subject.

No Clear Objective. Avoid doing simulation modeling efforts where there is not a clearly defined objective. These projects are almost never completed, rarely successful, and usually do not provide a good payback to the individuals and company involved in the project. Without clear objectives, the modeler has difficulty selecting a suitable simulation tool and determining the appropriate level of detail to be modeled. Examples of objectives are

1. Evaluate the proposed manufacturing layout to determine if it will meet the throughput objectives
2. Provide a validated, baseline model for testing future proposed changes to the existing operations
3. Determine the best way to design the drive-through window area of a fast-food restaurant and then use the model for training of operations personnel in managing the staffing of the drive-through windows

No Clear Questions to Be Answered. Clear questions are similar to the clear objectives issue raised above. They are very important to the success of projects. Without clear, specific questions to be addressed, it becomes unclear as to what data to collect, what experiments to run, and why the simulation modeling is being done in the first place. Examples of clear, specific questions are

1. What is the throughput of the manufacturing facility in jobs per hour?
2. What is the peak and average wait time required for people in line at the drive-through of a fast-food restaurant in minutes?
3. What is the utilization level of the eight injection molding machines?
4. How much time (minimum, average, maximum) does it take to process the customer order from the time when it is received until it is shipped to the customer?

Overuse of Simulation. Use the right tool for the right job at the right time. Too often, especially with its increased popularity, simulation is used when a simple spreadsheet and good engineering can solve the problem. The level of detail and sophistication required to support the decision-making process may not require simulation at that point in time. Remember, simulation should be used where there is a significant amount of random events occurring in a fairly complex system. If the system is simple and reasonably predictable, other analytical techniques such as spreadsheets, linear programming, optimization, and queuing analysis can be used.

Ease of Misuse. As simulation tools become easier to use, the opportunity for misuse grows as well. It is important for the simulation modeler to understand and correctly model the underlying logic and detail. The modeler needs to do more than build an animation or pretty picture of the system under investigation. Many of the "easy to use" products on the market today have given up something in order to be easy. Flexibility is frequently lost when making them easy to use. Simplifying assumptions are also made. It is critical for the simulation modeler to understand these assumptions and determine their impact on the situation under investigation. The assumptions can significantly alter the information being supplied to the decision-making process.

APPLICATION EXAMPLES

The examples that follow are included to show where simulation modeling and analysis can be applied. The opportunity to benefit from the use of simulation tools is seemingly endless. If the system is significantly complex and there are unpredictable, random events occurring, simulation provides important information to the analytical decision-making process. The information helps to determine how the system works, how it is planned to function, and how it can be improved.

Assembly System Utilizing AGVS.   Parallel workstation assembly systems appear to have many benefits over traditional, in-line assembly systems. These systems have been used in automotive assembly and major subsystem assembly operations. Like most alternatives, there are also a few disadvantages to be considered. The described example focuses on the use of simulation tools and methodologies in providing valuable information regarding the performance, strengths, and weaknesses of these systems.

   System Description.   Jobs are delivered to any one of the parallel lanes in the assembly area. The delivery system utilized in this example is an automatic guided vehicle system. Any other flexible, smart material-handling system can be substituted for AGVS. When the jobs reach the entry point (Fig. 4.4), they need to determine which lane to go into for processing. The system keeps track of past production as well as the system's current status. This information is very important for routing the next incoming job to the appropriate workstation. The job goes to the assigned lane and travels to the upstream stop point in the respective lane. If the primary workstation is free, the job travels directly to the primary workstation for processing. If the station is currently occupied, the job waits in the upstream stop point until the primary workstation is available. After the job has the appropriate amount of work completed, it goes to its available downstream stop point. The job is held at this point until the output rules are satisfied and the physical space is clear for the job to travel downstream.

   Technical Issues.   Technical issues are addressed by using simulation models of the existing or proposed parallel assembly system. The following partial list of technical issues for this system is where simulation can add value to the decision-making process.

1. What are the best input rules for jobs going into assembly area? (Fig. 4.5)
2. What output rules for jobs leaving the assembly area are the best? (Fig. 4.5)
3. What impact does the number of upstream stop points have on the assembly area's performance?
4. What impact does the number of downstream stop points have on the assembly area's performance?
5. What is the expected throughput of the system in terms of jobs per hour?
6. Is the work done by each team/lane balanced over time?
7. What is the impact of dedicating different lanes for specific types of jobs?
8. How many tools are required for each lane?
9. How much resequencing can be accomplished with the output rules while still meeting the throughput requirements?

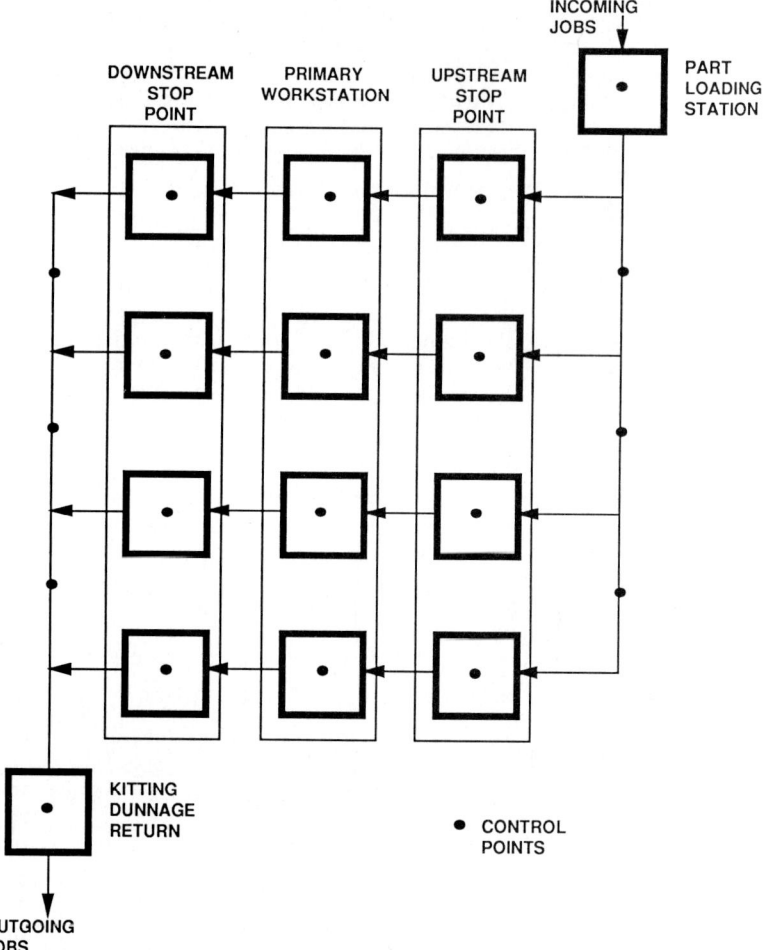

FIGURE 4.4   Parallel-workstation assembly system.

One significant benefit to using simulation is the ability to test the applicability, flexibility, and robustness of the proposed material handling and assembly system. Figure 4.5 contains examples of input and output rules for evaluation with the simulation model. This is not meant to be a complete set of rules but, rather, a good starting point.

Business Issues.   Business issues can also be addressed by using the simulation model of the assembly system. A partial list of business issues where simulation models provide information to support the decision-making process is:

1. What is the impact on system throughput when the jobs are held in the primary workstation until the workers have ensured that the jobs are 100% good quality?

2. How can the people in each lane be fairly rewarded according to the amount of work they do per shift?

3. How flexible is the assembly system in terms of jobs per hour and job content?

4. How much impact does the variable work rate from lane to lane and team to team have on the performance?

INPUT RULES	OUTPUT RULES
·ROUND ROBIN LANE SELECTION	·SAME SEQUENCE (FIFO)
·LEAST NUMBER OF JOBS TO DATE	·OLDEST JOB FIRST
·MOST NUMBER OF JOBS TO DATE	·RELEASE JOBS AS COMPLETED
·PRE-SELECTED LANE ASSIGNMENT	·HOT JOB FIRST
·SHORTEST EXPECTED FINISH TIME	·RELEASE TO SEQUENCING RULES FOR DOWNSTREAM OPERATIONS
·GREATEST AMOUNT OF SLACK TIME	
·FIRST AVAILABLE STATION FOUND	
·DEDICATED STATIONS ASSIGNED	

FIGURE 4.5 Input and output rules for a parallel-lane assembly system.

Other areas to be considered, but not specifically addressed with simulation, are the quality of work life, ergonomics, length of work cycle, ownership of work area, and flexibility in reacting to fluctuating sales volumes.

Machine Cell Design and Scheduling.   Much has been written about the use of group or cell manufacturing technologies, including their flexibility, ability to reduce inventory costs, and capability to support the just-in-time manufacturing strategy. While these benefits can be realized, a thorough understanding of what is being purchased and installed is necessary to ensure the desired level of success for the entire manufacturing operations, not just the cell being investigated. This example focuses on the use of simulation tools in understanding, planning, installing, and operating this type of manufacturing technology.

System Description.   Jobs or parts enter the cell or system at the load station and leave at the unload station (Fig. 4.6). This machining cell could be producing any number of items. Examples are gears, shafts, manifolds, valves, pump housings, and gear reducer housings. The jobs or parts enter the system where they are loaded on to a fixture either manually or automatically. This decision to load manually or automatically is not considered in this example. After the parts are loaded, the system's routing controller takes over. The fixture is routed to one of four machines to have the first set of operations performed. The conveyance method between the machines and the load and unload points is typically an AGVS or some type of mechanical shuttle system. The parts are routed to the appropriate machines in the desired sequence until all operations are successfully performed. The selection of this material-handling device depends on the vendor, other material handling systems in the plant, and many other factors. In this example, the material-handling device is a mechanical shuttle system. The parts are unloaded from the fixture and placed in a tray or container. The completed parts are then sent to their next location. This could be another in-process work area, a shipping area, or an in-process storage area.

Technical Issues.   Many technical system design and operational issues can be analyzed by using simulation models of current or planned systems. Simulation can be utilized to support and improve the decision-making process. Typical areas include:

1. What is the best set of input rules to be utilized in starting different job types into the system?

2. What strategy for tool allocation across four machines will yield the highest degree of flexibility and throughput capacity while addressing the issue of cost?

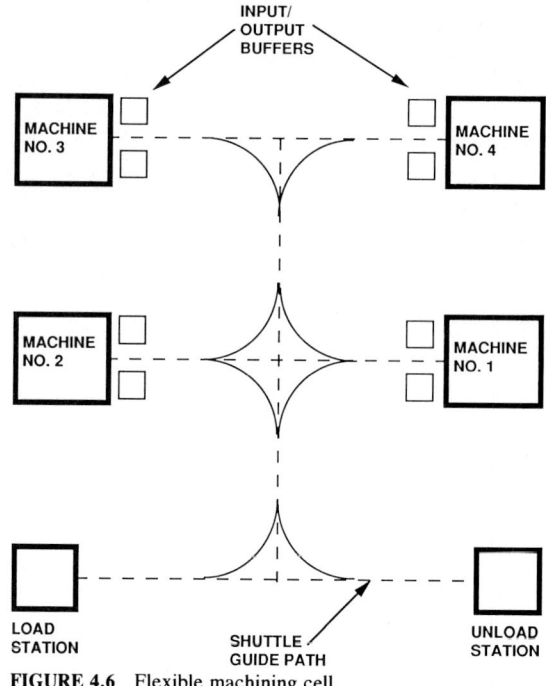

FIGURE 4.6   Flexible machining cell.

3. What is the expected throughput in terms of pieces per hour?

4. What is the best strategy to operate the cell when one or more of the machines is experiencing planned or unplanned downtime?

5. How are poor-quality raw materials and parts identified and removed from the system?

6. How many operations should be performed in each machine?

7. What is the impact of dedicating machines to producing certain parts while other machines are more flexible?

8. What tool change strategy should be implemented?

9. What should the size of the tool holder magazines be for each machine to support the system's throughput and flexibility requirements?

10. How many part fixtures should be in the system?

11. How much throughput can be realized by placing an input and output buffer next to each machine?

12. Should the system or cell be purged at the end of the shift or the end of a production lot before starting the next shift or batch of parts?

The true benefit of group or cell technology systems can be realized through an extensive information gathering and analysis activity. Simulation models can be very beneficial in providing good analytical information about this type of system. The information will support the decision-making process, whether the decision is to purchase the cell or to determine how best to operate the cell.

Business Issues.   Business issues relating to group or cell technology systems can be analyzed by using simulation tools and models of the system. A partial list of issues to be used as thought starters is

1. When using manual loading and unloading operations, how many operators are required to maintain throughput?

2. Can the load/unload operators be used for system maintenance and still maintain production?

3. Can the operators effectively load/unload more than one cell if the cells are located in reasonable proximity to each other?

4. How flexible is the cell in supporting the market-driven response time requirements, product volume demands, and the company's internal scheduling requirements?

5. How much is the throughput flexibility and capability worth in developing a business case to justify the program?

Other issues to be considered and evaluated are

1. The response time to the market for any particular purchase order of current product or for the introduction of new products

2. Employee ownership of product and work areas

3. Preventive and reactive maintenance strategies for the technology

4. Selection of manual versus automated loading and unloading of parts

5. Part inspection requirements to ensure that quality parts are being produced

6. Expansion plans for additional manufacturing cells or additional machines in a given manufacturing cell

7. Scheduling issues regarding multiple cells

This is not a complete list, but rather some things that should be considered. A thorough literature search will yield a large volume of information written on the subject and should definitely be pursued by those considering the use of manufacturing cell technologies.

Just-in-Time Manufacturing Systems. An industry buzz word in the manufacturing sector is *just in time*. Concepts include minimizing waste, reducing levels of inventory, balanced production rates and kanban cards. Proper use of simulation models allows a thorough evaluation of just-in-time manufacturing operations and the benefits and costs of implementing such a system in a given manufacturing environment.

System Description. This simple example will focus on a five-machine system shown in Fig. 4.7. These machines produce part A and part B. In addition, machine 5 produces part C, which is an assembled part made from part A and part B. Machines 2 and 3 are identical and can perform the required operations on parts A and B. Only one part type can be manufactured at a time on machines 2 and 3. A tool change is required when machine 2 or 3 changes the part type being produced. Machine 4 performs operations on the incoming parts, and, when the operations are complete, either part A or part B has been produced. Machine 4 can produce only one part type at a time. A machine setup is required each time machine 4 changes the part it is producing. All five machines are in close proximity to each other and parts are delivered to the machines by the machine operators using small dollies. Completed parts A and B leaving machine 4 and part C leaving machine 5 are sent to final packaging and shipping. The parts are then shipped to the distribution warehousing network. All machines are required to have periodic preventive maintenance operations performed. An addition, machines do require time for repair when an unplanned downtime occurs.

Technical Issues. The system-level technical issues need to be evaluated, quantified, and communicated to the operators and shop floor management. Just-in-time manufacturing is counterintuitive to many of the past practices followed in manufacturing companies. Through the use of simulation and animation tools, traditional, push-type production systems can be compared to a just-in-time, pull-type production system. Typical areas of evaluation include:

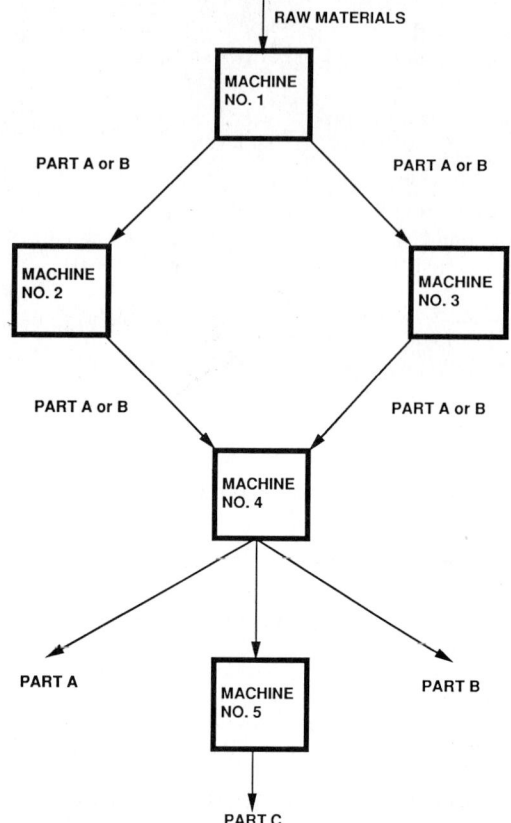

FIGURE 4.7   Just-in-time manufacturing system.

1. How much work-in-process inventory is required between machines?
2. How frequently should machine 4 be converted to produce part A and part B?
3. How long should machine 1 keep producing when either machine 2 or machine 3 unexpectedly stops producing?
4. What is the best transfer batch size between machines for each piece being produced?
5. Should machines 2 and 3 be dedicated to produce parts A and B, respectively?
6. Should machines 2 and 3 be operated differently when either of these machines is unavailable for manufacturing?
7. What is the expected amount of production throughput for this system?
8. How much time is required for a batch of parts to get through the system?
   Business Issues.   A partial list of business issues relating to a just-in-time manufacturing system is

1. How many operators are required to effectively run this system?
2. How predictable is the system in producing a customer's order on time and without delay?
3. What is the priority for building parts A, B, and C?
4. How are "hot" orders handled?

5. How many kanban cards should be in the system?

6. How much flexibility does the equipment and system have when trying to meet shifts in the market demand for the three part types?

Simulation can provide valuable insight about these systems. A good animation will allow new and existing manufacturing systems to be evaluated and then managed in a manner that focuses on minimizing waste.

CONCLUSION

The material covered in this chapter is provided to give the reader a better understanding and appreciation for simulation tools and their use from a practical standpoint. Substantial literature exists for further education in areas such as statistical analysis, data sampling and refinement, and experimental design. More research and knowledge is required beyond the information presented here and should not be overlooked before proceeding with the first simulation experience.

Tremendous improvements have been made to simulation tools, and this should be recognized. These advances, coupled with the increased capability of computer technology, have paved the way for a significant growth in the use of simulation tools. Properly executed simulation studies can provide meaningful information to support the decision-making process. The use of simulation models will continue to increase as the number of trained users grows. This increase is attributed to the number of academic courses available, the decrease in the price of the software, the increase in global competition, and the general increase in awareness and use of computer-based decision support tools. Like most engineering and decision support tools, they need to be applied to the right type of problem at the right time in the development cycle. Proper use of these sophisticated tools can provide lasting value and impact for an organization and should not be overlooked. Simulation tools will continue to become more of a mainstream part of the business environment as integration with other business tools and systems proceeds. In short, the recent past has been very bright for simulation and continues to look very promising in the foreseeable future.

FURTHER READING

Cox, W. A., "Eaton-Kenway Uses Simulation to Cut Costs and to Offer Solutions to Customer Needs," *Industrial Engineering*, February 1991.

Diaz, I., and S. Lezman, "Material Handling Simulation: Minimizing Bottlenecks and Improving Production Flow Using Lotus 1-2-3," *Industrial Engineering*, June 1988.

Johnson, M. E., "Selling Engineering Solutions Using Computer Animation," *Manufacturing Systems*, January 1990.

Law, A. M., and S. W. Haider, "Selecting Simulation Software for Manufacturing Applications: Practical Guidelines and Software Survey," *Industrial Engineering*, May 1989.

Law, A. M., and W. D. Kelton, *Simulation Modeling and Analysis*, McGraw-Hill, New York, 1982.

Pritsker, A. B., *Introduction to Simulation and SLAM II*, Systems Publishing, 1984.

Ramsey, M. L., and K. Tadibzadeh, "Push, Pull and Squeeze Shop Flow Control with Computer Simulation," *Industrial Engineering*, February 1990.

Schriber, T. J., *Simulation Using GPSS*, Wiley, New York, 1974.

Suri, R., "RMT Puts Manufacturing at the Helm," *Manufacturing Engineering*, February 1988.

Vasilash, G. S., "Simulation: Technology at the Edge," *Production*, December 1989.

CHAPTER 5
ARTIFICIAL INTELLIGENCE AND EXPERT SYSTEMS

Alan J. Rowe, Professor
Department of Management and Organization
University of Southern California, Los Angeles

Artificial intelligence and expert systems have been applied to many problems, including ones that have far-reaching implications for industrial engineers. In the past, computers have been used primarily for processing data, whereas artificial intelligence and expert systems are most appropriate where problems require access to a large knowledge base. Artificial intelligence increasingly incorporates model-based reasoning, theoretical knowledge, and simulation as the basis for finding solutions to complex problems.

Because computers can be made to behave in increasingly "intelligent" ways, and "expertise" can be incorporated into computer programs, more powerful solutions can be found than in current industrial engineering approaches. Artificial intelligence can couple the reasoning and judgment of the human mind with the power, speed, and memory of the computer to supply answers to problems that otherwise could not be solved efficiently.

AN OVERVIEW

A classification of artificial intelligence is shown in Fig. 5.1. Structured problems—those that are rule-based—are differentiated from unstructured problems that require reasoning and judgment. Artificial intelligence is shown on the horizontal axis. Expert systems are shown as an extension of computer systems that deal with complex problems.

Using this classification, we can represent alternative ways used to solve problems. The lower left-hand quadrant represents the more conventional applications of computers to process information. The upper left-hand quadrant covers expert/decision support systems which provide meaningful information and help focus on decision implications. Increasingly, decision support systems are incorporating elements of expert systems. The lower right-hand corner covers those systems that rely on well-defined data, knowledge, and inputs from experts. The final quadrant in the upper right deals with unstructured and complex problems. This area has the potential for extending the application of artificial intelligence to complex, interdependent, stochastic problems often confronting industrial engineers.

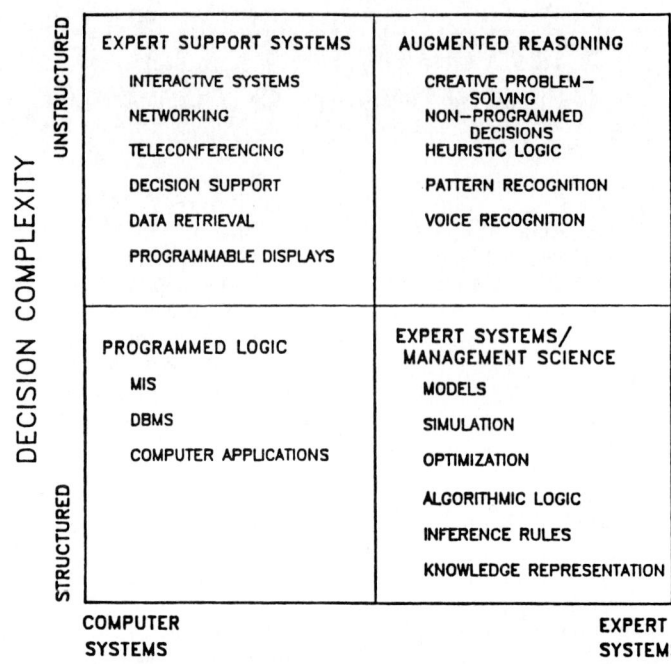

FIGURE 5.1   Artificial intelligence classification.

DEVELOPING EXPERT SYSTEMS

Knowledge engineering is the discipline used to capture the expertise of persons who have experience or know how to perform a complex, ill-defined task. This knowledge is typically represented in terms of rules. Unfortunately, human experts do not analyze how they think about problems they solve, nor are they aware of the rules that structure their knowledge. An important issue is who are the experts and how good are they in decision making? They have a tendency to simplify complex problems. Because of this, errors can be introduced such as:

1. Suboptimization
2. Wrong tradeoffs
3. Not considering time lags and variability
4. Not dealing with system interdependencies
5. Not being aware of the effects of transients
6. Introducing conflicting goals

This list is indicative of possible errors that experts incorporate in their solutions. Other problems are deficiencies in calibration (subjective-objective comparability), resolution (refinement), and coherence (internal consistency).

Heuristics and Expert Systems.   There are many definitions of heuristics, including rules of thumb, intuition, educated guesses, and application of experience or prior knowledge. For our purpose, heuristics are considered as the use of any cue, rule, or technique that reduces the need for extensive search for the solution to a problem. Heuristics can also be considered as intelligent search strategies for problem solving based on simple search criteria to discriminate among choices.

Heuristic reasoning utilizes intuition, judgment, and the ability to evaluate and integrate diverse kinds of information. Although solutions only approximate the optimum solution, heuristics can be described as the intelligent search of a complex situation based on "intuitive logic" that uses the logic of the left brain and the intuition of the right brain.

Cognition and Intelligence.   Fundamental to the field of artificial intelligence is *real intelligence*. Alan Turing proposed the definition "imitation of human thinking" as the basis for artificial intelligence. However, only a limited portion of human intellectual activity can be reproduced by computers. Although artificial intelligence will enable computers to replace some of the tasks humans perform, it is a long way from substituting entirely for human decision making. Computers do not have intuition, imagination, learning, emotion, analogical inference, common sense, or judgment. Without the aid of human input, computers are not yet capable of fully understanding language or pattern recognition.

Extremely complex problems involve the acquisition and exchange of knowledge and are bound up with pragmatic issues and semantics. These require insight, imagination, and a deep understanding of both computers and the application.

Artificial intelligence utilizes special-purpose computers and supercomputers to work on problems that previously could not be solved. While the speed of early computers was measured in thousands of floating-point operations per second (FLOPS), in which the decimal point is moved in very large and small numbers, today's is measured in gigaFLOPS, or billions of operations a second. Tomorrow's will be measured in teraFLOPS, trillions of operations a second. A single supercomputer going at teraFLOPS speed will have the power of 10 million personal computers.

APPLICATIONS OF EXPERT SYSTEMS

One of the most valuable assets of a company is its expertise. When Campbell Soup Co. knew that the expert in repairing cookers for sterilizing cans was about to retire, managers decided to build an expert system to translate the expert's knowledge into computer rules. Using interviews and analyzing the expert's troubleshooting heuristics, they developed a computerized adviser called Cooker. This expert system now solves almost all of the company's cooker malfunctions.

Another example of an expert system is XCON (Expert Configurer) that was applied by Digital Equipment Corp. during the early 1970s. Faced with the prospect of having to spend millions of dollars for new facilities, Digital developed XCON, an expert system to check sales orders and designs for each computer ordered. The system allowed Digital to ship components directly to the customer for final assembly and avoid building new assembly plants.

Schlumberger's geological analysis expert system enhances professional decision making. Schlumberger also uses this expert system to determine the probable value of oil wells from measurements taken during drilling. Petroleum geologists normally perform these analyses, and large sums of money are at risk.

Another example of a successful expert system application was developed by American Express to help its credit authorization staff, using 13 computerized databases, to determine the credit level for each customer. When a customer makes a large purchase that is outside the normal buying pattern, the Authorizer's Assistant expert system makes a recommendation regarding the authorization decision in a matter of seconds.

Westinghouse Electric Corp. had a problem with turbine failure and developed an expert system to monitor turbines. Because turbine failure can cause thousands of dollars of damage and loss of the use of the turbine, the system developed by Westinghouse continuously diagnoses data from monitors on the turbine and makes recommendations for maintenance.

Coopers and Lybrand uses the expert system Expertax to help accountants offer better tax advice. It provides experience to junior accountants for performing more accurate tax planning.

The Underwriting/Lending Advisor-System does risk assessment for bank loan officers and insurance underwriters by providing the following:

1. Qualitative and quantitative reasoning to analyze a customer's application to avoid incomplete and inaccurate information

2. Multiattribute assessments to compute interdependent factors such as financial terms and previous financial performance

3. Reference information such as ratings for different occupations and financial reports

The Underwriting/Lending Advisor is used to assist the underwriting of commercial insurance covering worker's compensation, inland marine, commercial autos, and computer facilities.

An expert system called PEOPL (Programmed Evaluation of Personnel) is used to evaluate personnel. This can be an expensive procedure and objective measures contribute only a small part to prediction of success. An expert system can improve personnel evaluation because supervisor ratings often are biased. PEOPL improves the selection and assignment of personnel and facilitates predictive and evaluative measures within selected subdomains. Many current measures contribute little to evaluation and are based on nonstandardized supervisory ratings that have considerable error.

Expert systems, in effect, are artificial experts, represented by a computer program. Relying on facts about a given problem and heuristic rules that experts use to solve such problems, an expert system appears to be able to reason.

THE LOGIC OF REASONING

Reasoning can be considered analytic (deductive) or heuristic (inductive). Model-based reasoning has focused principally on deductive logic, while human-based reasoning focuses on inductive or heuristic logic. Barr and Feigenbaum use five categories to describe reasoning:

1. *Formal reasoning* that deals with syntactic manipulation of data structures based on following specified rules in order to ''deduce'' new rules of inference

2. *Procedural reasoning* that examines heuristic approaches to answering questions and problem solving.

3. *Analogical reasoning* that compares similar objects or behaviors in order to predict behavior of an unknown phenomenon.

4. *Generalization and abstract reasoning* that deals with the ability to recognize cues to draw conclusions from these cues or inferences based on an individual's cognitive complexity.

5. *Metalevel reasoning* that provides inferences about outcomes based on knowledge about what one knows.

These five categories provide a framework for knowledge representation for many complex decisions.

A model representing the manner in which cognitive complexity determines how humans reason, make judgments, and understand information, and their preference for problem solving, is shown in Fig. 5.2. The figure points out that experts differ in their approaches to problem solving, decision making, reasoning, and judgment. With this

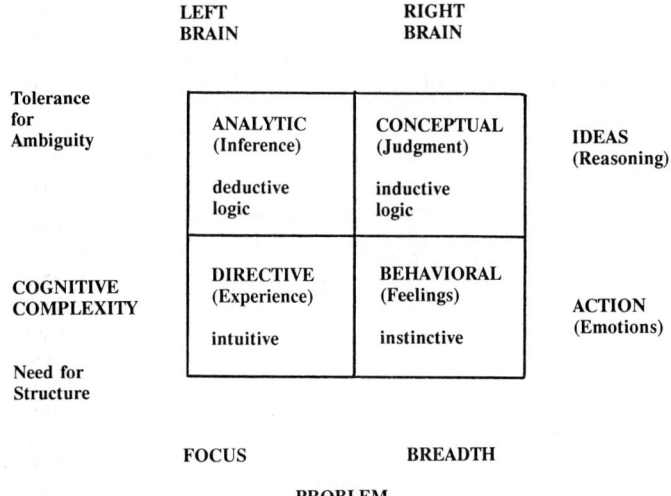

FIGURE 5.2   Cognitive model of decision making.

understanding, one can evaluate how well an expert is able to use heuristic rules for the solution of problems.

Many experts operate in the action domain and therefore "satisfice" their solution to problems. Simon suggested that this approach to using limited search for solutions was based on humans' having "bounded rationality." This approach is comparable to the action orientation of the cognitive complexity model. Analytic individuals do not necessarily follow limited search; rather, they tend to do extensive search, often involving extremely large amounts of data, and continue to do so over long periods of time. The conceptual individual is adept at concept formation and explores many options, considers constraints, and utilizes search as a stimulus for creative problem solving.

Model-Based Reasoning.   Model-based reasoning has a number of advantages over rule-based approaches. In general, rule-based or frame-based reasoning involves enumeration of all possible relevant conditions under consideration. Model-based reasoning deals at a more aggregate level because events or conditions can be represented by a single equation that is a close approximation of actual conditions. Building an expert system is also facilitated where model-based reasoning is utilized. It requires less time, fewer constraints or limits, and assures consistency from one application to the next. Interpretation is often easier because of knowledge about the models used. Heuristic rule development is also facilitated when applying appropriate models, for example, use of the Pareto law to categorize the relative importance of items or use of queuing theory to partition items in a queue into different priority classes without changing the expected delay.

Analogical Reasoning.   Analogical reasoning is considered a powerful approach used by experts to solve problems based on prior experience or relations to similar problems. It involves drawing on prior experiences, learning by example, learning by being told specifics, and empirically observing events. Encoding can describe an object and its properties, for example, temperature with a thermometer.

Representativeness is a heuristic that relies on intuitive judgment and affects analogical reasoning because:

1. Judgment is used to determine in which category an object belongs.

2. Judgment determines whether an observed event is significant or merely random.

3. Judgment is used to project general features from sample features.

4. Judgments are made about events that have not actually been observed, but on the basis of prior observed effects.

5. Predictions are made on the basis of observation of prior events.

The representativeness heuristic varies with each individual because of cognitive complexity and judgments made regarding perceived salient features. *Judgmental fixations* and *anchoring* describe how people make estimates given some initial value. Individuals tend to be remarkably resistant to new or further information even where the new data show that a prior conclusion was wrong.

JUDGMENT AND INFERENCE

The cognitive process involves understanding and evaluating information, cues, and constructs perceived. The judgmental aspect considers whether or not the information is similar to or resembles features that can be put into a particular category and an estimate of either consequences or causality in relationships. For example,

1. *Judgment* often depends more on preconceptions than relevant new information perceived.

2. *Intuitive judgments* based on normative decision models about subjective probabilities can be misleading.

3. *Representation* that attempts to classify concepts or objects can be illusive.

4. *Availability* describes the individual's intuitive judgment about the frequency of events and proportion of objects. These are subject to possible errors. Biased judgments result from those that are made without considering adequate data.

5. *Judgmental fixation* describes the anchoring of an individual's thinking regarding subjective probability and often is misleading.

Value judgments form a critical aspect of selection among alternatives, and preconception strongly influences intuitive judgments because:

1. People often search selectively for data that support their existing beliefs.

2. People give little weight to new evidence that opposes existing beliefs.

3. People have a primacy affect in which they stick to their first impression for drawing inferences that affect new information.

4. Beliefs persevere even though evidence may discredit strongly held values.

It is virtually impossible to separate an individual's values from judgments made, especially in complex, ill-structured decisions. Judgment thus combines both factual data and individual values. Furthermore, because many decision situations involve considerable uncertainty, value judgments still remain the principal means for evaluating, classifying, and selecting from among possible choices.

HOW EXPERTS USE HEURISTIC RULES IN INDUSTRIAL ENGINEERING PROBLEMS

Finding a workable solution can be based on a heuristic approach of separating problems into more important and less important parts. For example, a small percent of customers account for most sales. Likewise, a small percent of workers have the most accidents, and so on. If we formalize this intuitive logic, we have one of the building blocks of heuristic rule formulation.

Although there are a large number of possible alternative solutions for combinatorial problems, most of the solutions can be separated into categories of payoff. As shown in Fig. 5.3, the bounds range from equal payoff for any solution (trivial case) to the Pareto distribution of payoffs.

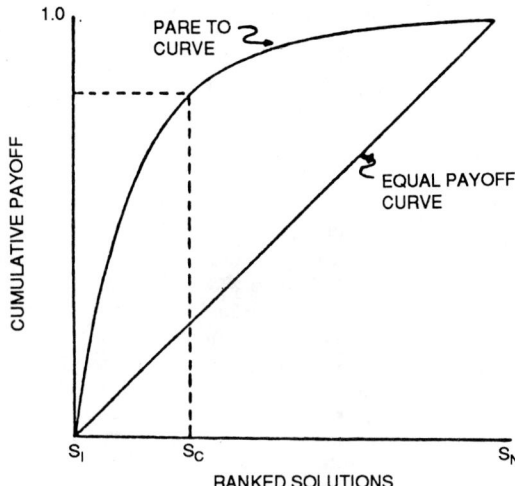

FIGURE 5.3   Comparison of possible solutions.

The Pareto curve shows that the cumulative payoff can be related to a small subset of the solution S_1 to S_c. The subset of solutions that falls within these limits is often what experts use as their heuristic approach.

In many cases, the payoff is obvious from an examination of the characteristics of the problem itself. For example, a typical problem that has a large number of possible solutions is the job assignment problem. Here one wants to find the assignment of jobs to machines that minimizes the total setup cost. The data in Table 5.1 show the setup costs for each possible assignment.

TABLE 5.1   Setup Costs in a Sample Job Assignment Problem

From machine	To machine				
	1	2	3	4	5
1-	-0-	9	6	3	2
2-	4	-0-	4	7	6
3-	5	9	-0-	3	10
4-	6	7	4	-0-	3
5-	8	4	7	2	-0-

The Machine Assignment Problem.   On the basis of the Pareto law, the lowest costs are circled, as shown in Table 5.2, to find the most efficient combinations. If the costs are ranked from lowest to highest, they would follow a Pareto curve. A heuristic rule for determining which are the lowest-cost assignments is to use a maximum cost difference of one dollar. The low-cost combinations, then, are 1-5, 1-4, 2-1, 2-3, 3-1, 4-5, 4-3, and 5-4, where the first number designates the row and the second, the column.

The machine assignment problem requires that all jobs in a factory be processed by available machines. This assumes that the sequence in which a job is performed on a machine affects the setup cost for tooling and setting of fixtures. Thus, in the example shown in Table 5.1, if a job is processed first on machine 1 and then on machine 2, the setup cost is $9, whereas, if job 1 is sent to machine 5 as the next stage, the setup cost is only $2. By examining the costs resulting from all

TABLE 5.2 Identification of the Minimum-Cost Solutions

From machine	To machine 1	2	3	4	5	Count
1-	-0-	9	6	③	②	2
2-	④	-0-	④	7	6	2
3-	⑤	9	-0-	8	10	1
4-	6	7	④	-0-	③	2
5-	8	4	7	②	-0-	1
Count	2	0	2	2	2	

possible sequences, it is seen that some are higher and some are lower. Using the Pareto law, we can see which pairs of sequences have the lowest-cost setup.

In order to find the low-cost pairs, a heuristic rule is used that assumes where there is only a difference of $1 between low-cost pairs, all jobs between would be included; for example, job 1 to 5 is $2 and job 1 to 4 is $3. Both pairs meet the heuristic cutoff. On the other hand, job 5 to 4 is $2 and 5 to 2 is $4, so the difference is $2 and does not meet the $1 criterion.

The next problem is to develop a heuristic rule for selecting which jobs to assign. This heuristic rule counts the number of low-cost alternatives in each row and column and then chooses the combination with the least number of low-cost alternatives (column 2 has zero low-cost assignments and rows 3 and 5 each have only one low-cost alternative). The next rule is to choose one of these assignments (choose 5-2 which has a zero count and is the lowest cost in column 2). The heuristic rule chooses either a low-cost or critical assignment. If this set of heuristic rules is followed, and 5-2 is chosen as the initial assignment, the combination shown in Table 5.3 turns out to be the optimum solution. If the sequence started with another alternative (1-5), the solution would not be optimal but would provide a very good approximation to the optimum.

TABLE 5.3 Optimum Solution in a Sample Job Assignment Problem

Assignment	Reason	Explanation
5-2	Critical	Because of low count in column 2
2-3	Low cost	Next assignment from machine 2
3-1	Critical	Because of low count in row 3
1-4	Low cost	Next assignment from machine 1
4-5	Remainder	Only assignment left

This same approach can be applied to the assignment problem of finding which jobs best match which machine, as shown in Table 5.4. The heuristic rule is the same as the one previously described. The lowest costs are circled for each machine. By observation, it can be seen that there was only one conflict for low-cost assignment. This is for machine 3. The obvious solution, thus, is to choose 1-10, 2-3, 3-8, 4-15, and 5-21, which turns out to be the optimum. This problem introduces an interesting variant on the application of

TABLE 5.4 Transformed Setup Costs in an Order-Matching Problem

Number of Machines	Number of orders 1	2	3	4	5	6	7	8	9	10	11	12	13	14	15	16	17	18	19	20	21
1	30	60	㉓	70	30	80	45	40	30	25	100	35	90	80	35	40	30	45	60	70	45
2	85	50	㊵	90	70	100	105	70	100	80	90	75	85	75	80	80	110	120	75	140	135
3	55	60	75	70	85	80	95	㉚	100	120	125	100	130	100	60	80	70	35	90	75	80
4	70	70	40	80	75	70	75	80	85	90	90	100	120	125	㉟	95	115	120	135	130	65
5	100	95	70	90	80	130	135	85	110	115	80	100	150	160	75	160	140	130	125	130	㊺

heuristic rules. Intuitively, the optimum was obvious without actually having to follow the rules once the low-cost alternatives were circled.

Ratio Delay Studies.   Another example is ratio delay studies used to determine the amount of time that a machine is idle. Samples are taken at random to determine whether the machine is running or is idle. This situation can be described statistically by a binomial distribution. Unfortunately, if the machine is idle only 1 percent of the time, and one wants a 95 percent confidence level, then a sample size of 156,000 is required. This is a totally unreasonable number of observations for most industrial situations. If one is willing to relax the requirements, and, for example, say that a 1 percent delay can be called either 0 or 2—that is, one is willing to tolerate a 100 percent error rather than a 5 percent error—the sample size reduces to approximately 1500, a significant reduction from the original sample size. On the average, however, the idle time still will be recorded as 1 percent even with the smaller sample size.

CLASSIFYING HEURISTICS

We can describe heuristics according to either the kind of problem being considered or the cognitive complexity of the decision maker. That is, we can state problems either as fitting generic categories or as fitting the number of variables that need to be considered. The generic categories can be described as follows:

1. *Deterministic:* These problems are assumed to be well-defined and not subject to variability.
2. *Stochastic:* These problems are subject to variability and can be considered either as risky or uncertain, depending on the information available.
3. *Combinatorial:* These are ones where either the sequence or the combination of factors determines the optimum solution.
4. *Organizational:* These cover the broad spectrum of problems that deal with people, politics, behavior, and so on.

The second classification scheme considers the number of variables in a problem:

1. *One-variable problem:* Because of its restrictiveness, this category almost always guarantees that the solution will be suboptimal.
2. *Two-variable problem:* This category covers those situations where it is possible to show a causal relationship between two variables or to compute tradeoff analyses such as cost versus benefit.
3. *Four-variable problem:* This class covers double balance (2×2) and represents equilibrium in a given situation. It also relates to the magic number 3 plus or minus 1.
4. *The N-variable problem:* This category covers all those problems that involve large numbers of variables that are highly interdependent and are not amenable to rigorous solution.

While any classification is arbitrary, the two described above have been found useful for application of heuristic methods.

A logical approach to applying heuristic rules incorporates:

1. A classification scheme which introduces structure into a problem
2. Analysis of the characteristics of the problem elements
3. Rules for selecting elements for each category to achieve efficient search strategies

4. Rules for successive selections, where required

5. An objective function that is used to test the adequacy of the solution at each stage of selection or search

A Complex Example Using Heuristic Methods. The branch-and-bound algorithm used to solve the "traveling salesman" problem can be described as follows:

1. Intelligently structure the search of the space of *all* feasible solutions.

2. The space of all feasible solutions is repeatedly partitioned into smaller and smaller subsets, and a lower bound is calculated for the cost of the solutions within each subset.

3. After each partitioning, those subsets with a bound that exceeds the cost of a known feasible solution are excluded from all further partitions.

4. Partitioning continues until a feasible solution is found such that its cost is no greater than the bound for each subset.

5. The branch-and-bound algorithm requires a heuristic rule that determines which of the currently active bounding problems is to be chosen for branching, as well as the method for deriving new bounding problems.

6. Branching strategies can be
   a. Branch from the lowest bound
   b. Branch from the newest active bound

In steps 3 and 4 above, it is possible, by using an efficient heuristic search algorithm, to avoid the iterative approach needed for many large combinatorial problems. What is proposed here is an approach based on a selection or search procedure that leads to a good, but not necessarily optimal, solution. The advantages of such a heuristic selection procedure are

1. Selection can often be done with a single iteration and is thus very efficient

2. Selection can be applied to large problems often without a computer

3. If the problem can be solved by using an expert system, it facilitates decision-maker interaction

4. Involvement by a decision maker in the solution leads to a greater likelihood of implementation

5. Good solutions generally can be used as initial solutions for more formal computational procedures

By using a heuristic approach, the traveling salesman problem can be solved with a single iteration. The result is always a good solution or the optimum solution. This eliminates the need to search *all* feasible solutions. This solution is shown in Fig. 5.4 as traveling the exterior path, with no back tracking and no crossovers.

This solution demonstrates how combining the properties of the left and right hemisphere helps to find a heuristic approach to the problem. The right hemisphere has the broad vision to look ahead and determine which direction to pursue. The left hemisphere ensures that the algorithm is followed, with no backtracking, no crossovers, and no skipping of cities that are to be visited.

The traveling salesman problem is difficult because of the large number of possible paths that can be followed in order to return to the city of departure. The number of possible paths $(n - 1)(n - 2)(n - m)$. Thus, for a 10-city problem, the number of paths is $9 \times 8 \times 7 \times 6 \times 5 \times 4 \times 3 \times 2 \times 1 = 362,880$. For larger problems, the number of paths increases rapidly; for example, for an 11-city problem there would be 3,628,800 paths.

As in machine assignment, a heuristic rule can be used to find either a very good solution or, in many cases, the optimum solution. The heuristic rule applied to this example is always travel an exterior path starting at any city, then move to the next adjacent city.

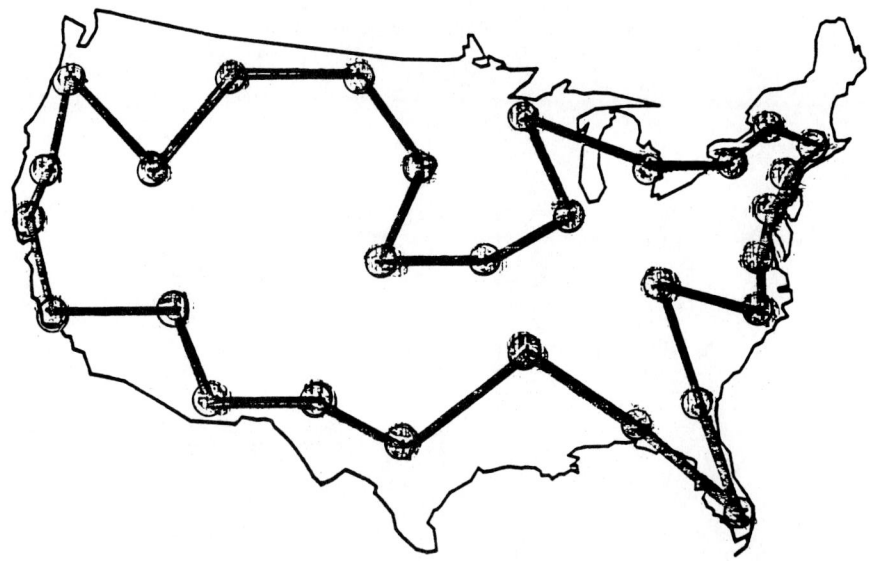

FIGURE 5.4   Traveling salesman solution.

Figure 5.4 shows all the cities that need to be visited and how, by applying this heuristic, one can find an excellent path.

AN EXAMPLE OF MODEL-BASED REASONING

Model-based reasoning combines heuristic rules and models, such as queuing theory, to facilitate developing diagnostics or decision aids. To illustrate this approach, a job shop scheduling problem will be described. A job shop is one where there is a fixed number of machines and an unknown number of jobs. Each new job that arrives in the shop requires differing amounts of work on some of the machines. The sequence of work to be done is also generally different for each job. With the problem stated in this way, the next step involves developing several heuristic rules:

1. Total expected time that a job would spend in the factory. This defines the promised schedule date for delivery.
2. A priority ranking rule that determines the sequence in which the jobs in a queue would be chosen.
3. The dollar value of the jobs and machine utilization also is taken into account.

To start, the time the job is in the factory is partitioned into three categories. First is the actual time needed for fabrication. Second is the time needed to move the job from one machine to another. Finally, there is the time spent waiting at each machine before the work can be done. Waiting time, or delay, is the one most difficult to estimate for the typical job shop. Thus, the heuristic rule that was developed directly considers the expected delay at each machine.

Starting with the average machine utilization, an estimate of the expected delay at each workstation, based on a queuing model, is determined. The dollar value of the jobs is then used as the basis for determining the overall schedule time. It is reasoned that assigning a higher priority to high-value jobs would reduce the average inventory in the plant. To accomplish this waiting time reduction, a flow allowance model is developed, as shown in

Fig. 5.5, that relates the time spent in the plant to the value of the job. An heuristic rule that assigns very low flow allowances thus could speed high-value jobs through the shop.

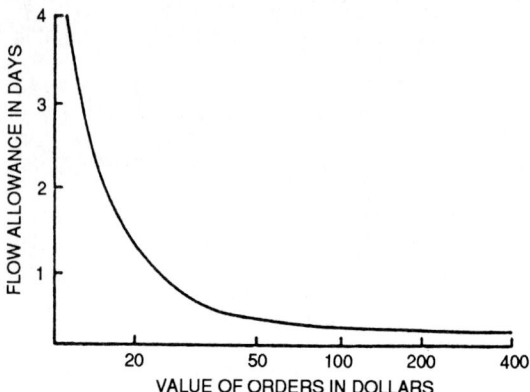

FIGURE 5.5   Flow allowance.

Next, the expected delay at each machine is used as the basis for assigning jobs in any given queue. This heuristic establishes a priority for each job and helps the factory manager determine which one to assign. The manager can override the rule if there is a special urgency such as shortage of material, missing tools, missing machine operators, or similar reasons for altering the priority sequence.

Finally, a heuristic rule for determining the date on which a job can be promised is needed. This rule considers the time for fabrication, the move time, the expected delay, the dollar value of the job, and any special urgency factors. The critical aspect of this rule is a model of queuing that allows jobs to have differing flow times without affecting the average delay. In addition, another heuristic rule is used to calculate the desired start date that would provide a specified assurance that at least 90 percent of the jobs would meet their promised delivery date.

An Heuristic Scheduling Decision Rule.   Allowances, when added to machining and transit time, establish the manufacturing cycle for each job. To determine the start date for each operation of a job, the following heuristic scheduling formula is used.

Value categories are established for each machine group such that the upper 10 percent of the jobs are considered high-value, the next 20 percent are considered medium-value, and the remaining 70 percent are considered low-value. With these percentages established, mean value for each category is determined by using the functional relation

$$F = kV^{-b}$$

as the basis for the flow allowance. The waiting time for each value category is then determined in the following manner (assume $b = 1$):

$$F = \frac{k}{V}$$

where k is an empirically determined constant and V represents the average value of jobs in each specified category.

The scheduling heuristic is next developed for the start date of each job:

$$S_i = D - (P_i + T_{i-1} + F_{ij}) + C$$

where S_i = planned start date for the ith operation
    D = promised delivery date for completion of the job

P_i = processing time for the ith operation

T_{i-1} = transit time from operation $i - 1$ to i

F_{ij} = flow allowance for operations $i = m$ to n, based on value and the delay for the ith operation at the jth machine group

C = cycle correction for critical jobs or special machine groups

Job Assignment Heuristic Decision Rules.   After jobs are scheduled and released to manufacturing, a heuristic decision rule is required which will enforce the planned cycle time. Dispatchers use this rule to choose which job to assign when a machine becomes available for work by using a priority index to compare the various jobs in a given queue awaiting assignment.

A heuristic rule that has been tested on many problems and has led to good solutions uses the planned flow allowance established during scheduling to establish a priority index:

$$\text{Priority}_i = \frac{D - S_i - F_{ij}}{N^{-b}}$$

where Priority$_i$ = priority index for a given job at the ith operation

D = fiscal day number when the priority among jobs is being compared

S_i = planned start date for the ith operation

F_{ij} = factor based on the flow allowances established during scheduling

N = number of operations remaining from i to m

b = empirically determined constant

The job with the largest algebraic priority index is assigned to a machine when there is more than one job competing. The deviation from start date, $D - S_i$, provides a measure of the lost time as a result of waiting for assignment or late delivery. When this deviation is compared with the flow allowance F_{ij}, the difference indicates residual time. Thus, when $D - S_i - F_{ij}$ is positive, it indicates that the job has exhausted its current flow time.

The residual flow time, $D - S_i - F_{ij}$, is divided by the number of remaining operations when the job is early. That is, the constant b is made positive. If the residual time is positive, the constant b is made negative, which results in multiplication by the number of operations. An alternative to dividing the residual flow time by the number of remaining operations is to divide by the expected remaining waiting time.

In studying the application of heuristic rules, Patterson compared them with the optimal solutions for resource-constrained project schedules. The problem involved scheduling activities in a project network in order to minimize project duration under conditions of multiple limited resources. The experiment consisted of comparing eight heuristic sequencing rules applied to 83 different problems. A bounded enumeration procedure was used to determine the optimum schedule duration. The results showed that the minimum slack heuristic scheduling rule, such as the one described above, produced the optimum in 24 out of the 83 problems and on the average was only 5.6 percent higher than the optimum solution.

By combining the heuristic rules described above with model-based reasoning, a factory manager at one of the General Electric factories was able to improve deliveries from 65 percent on time to 92 percent. Hughes Aircraft extended the use of this approach, and IBM applied it to scheduling its factories.

The job shop scheduling problem is a good illustration of the need for model-based reasoning as an adjunct to heuristic reasoning. Model-based reasoning adds a formal structuring to the analysis of problems.

CONCLUSION

The question sometimes asked is "Can expert systems be used alone without heuristic reasoning?" The answer is "Only for very well-structured problems with clearly defined objectives." In most complex problems, these are not sufficient by themselves to cover all aspects. While expert systems can be used for many problems, they still have limitations where there is lack of structure. Another consideration is that heuristic reasoning is based primarily on the intuition and logic used by experts. Here again, application of approaches such as the Pareto law as a guide to partitioning problems or the steepest ascent rule (highest priority) help formulate heuristic rules that go beyond IF-THEN statements. Thus, we can see that heuristics can be purely experiential, or they can be a combination of experience and formal heuristic approaches such as model-based reasoning. Where it is applicable, it is obvious that the last combination offers the most powerful constructs and greater likelihood of finding significantly improved solutions for industrial engineering problems.

FURTHER READING

Andriole, S. J., *Applications in Artificial Intelligence,* Petrocelle, Princeton, N.J., 1985.

Brule, J. F., *Artificial Intelligence, Theory, Logic and Application,* Tab, Blue Ridge Summit, Pa., 1986.

Barr, A., and E. A. Feigenbaum, *The Handbook of Artificial Intelligence,* William Kaufman, Inc., Los Altos, Calif., 1981.

Charniak, E., and D. McDermott, *Artificial Intelligence,* Addison-Wesley, Reading, Mass., 1985.

Chorafas, D., *Applying Expert Systems in Business,* McGraw-Hill, New York, 1987.

Elliot, L. B., "Analogical Problem-Solving and Expert Systems," *IEEE Expert,* vol. 1, no. 2, 1986.

Goldsmith, T. E., "Decision Making and Cognition: An Artificial Intelligence Perspective," *Proceedings Expert Systems in Production and Services,* Thomas Bermold, ed., Chicago, Sept. 13–15, 1988.

Green, E. P., and A. M. Krieger, "Models and Heuristics for Product Line Selection," *Marketing Science,* Vol:4, Iss:1, Winter 1985.

Harmon, P., et al., *Expert Systems,* Wiley, New York, 1988.

Hertz, D. B., *The Expert Executive,* Wiley, New York, 1988.

Hogarth, R., *Judgment and Choice,* Wiley, New York, 1980.

Holsapple, C. W., and A. B. Whinston, *Business Expert Executive,* Irwin, Homewood, Ill., 1987.

Koslov, A., "Rethinking Artificial Intelligence," *High Technology Business,* May 1988, pp. 18–25.

Michaelsen, R. H., et al., "The Technology of Expert Systems," *Byte,* April 1985, pp. 303–311.

Paterson, J. H., "A Comparison of Exact Approaches for Solving the Multiple Constrained Resource, Project Scheduling Problem," *Management Science,* Vol. 30, Iss. 7, 1984.

Pearl, J., *Heuristics,* Addison-Wesley, Reading, Mass., 1988.

Rowe, A. J., et al., "A Heuristic Approach to Complex Problem Solving," *Journal of Business and Economics,* Cal. State Univ., Los Angeles, 1977.

Rowe, A. J., "A Modeling Approach to Developing Heuristic Decision Rules for Use in Expert Systems," *Proceedings, First International Conference on Artificial Intelligence and Expert Systems,* AMK, Berlin, 1987.

Rowe, A. J., and F. R. Bahr, "A Heuristic Approach to Managerial Problem Solving," *Journal of Economics and Business,* vol. 25, no. 3, 1969, pp. 153–163.

Rowe, A. J., et al., "Management Use of Artificial Intelligence," *Applied Expert Systems—Trends and Issues,* Elsevier, New York, 1988.

Schank, R., et al., "The Quest to Understand Thinking," *Byte,* April 1985, pp. 143–145.

Shannon, R., "Expert Systems and Simulation," *Simulation,* vol. 44, Jan. 1985, pp. 275–284.

Simon, H., *The Science of the Artificial,* MIT Press, Cambridge, Mass., 1982.

Stewart, S. D., and G. Watson, "Applications of Artificial Intelligence," *Simulation,* June 1985.

FACILITIES AND MATERIAL FLOW

CHAPTER 1
FACILITIES PLANNING AND UTILIZATION

William E. Fillmore
Senior Consultant,
Richard Muther & Associates, Inc.
Kansas City, Missouri

Facilities are generally defined in the context of the fixed, or capitalized, assets of an organization. These include land, buildings, and equipment.

Planning is the act of establishing an intended method of accomplishing something. When applied to facilities, planning is used to define configuration and expected methods of operation for those facilities.

Utilization, as applied by industrial engineers, is the method by which something is turned to profitable use. Usually, this includes measuring effectiveness of such use.

Facilities often account for a large portion of the invested capital of a company. This investment is normally less liquid than other assets, such as inventory. Properly planned and effectively utilized facilities have a positive impact on operating costs and capacities. These facts taken together indicate that the effectiveness of planning and utilizing facilities can significantly affect return on assets. This chapter will focus on basic concepts underlying facilities planning, the process of planning facilities, and how measures of utilization can be used to enhance effectiveness of an implemented plan.

BASIC CONCEPTS

Facilities Components. Facilities have five physical components:

1. *Layout:* The physical arrangement of the facilities
2. *Materials handling:* The way that materials are moved within the facilities
3. *Communications:* Systems that transmit information to appropriate places in a timely manner
4. *Utilities:* Distribution of substances such as heat, light, power, and waste, as required
5. *Buildings:* Structures which house the facilities

The relative importance of each component varies from one facility to another. In most industrial applications, consideration of flow of materials tends to give major emphasis to layout and materials handing. In an office, communications and building components are dominant. A facility devoted to paint application would be more sensitive to utilities considerations. In planning facilities, all five components must be considered; one of them, however, is normally selected as the lead, or dominant component, on the basis of the nature of the facilities being planned.

Planning Fundamentals.   Each component involves three planning fundamentals, as shown:

Component	Fundamentals
Layout	Relationships
	Space
	Adjustment
Materials handling	Materials
	Moves
	Methods
Communications	Information
	Transmission
	Means
Utilities	Substances
	Distribution
	Conductors
Buildings	Form
	Materials
	Design

Layout involves definition of *relationships* between activity areas, such as buildings, departments, or workplaces; *space* required for each activity area, in amount, kind, and shape; and *adjustment* of these into an acceptable arrangement.

Materials handling involves *materials* being moved; *moves* between each origin and destination, along with conditions of the routes; and *methods* (route systems, equipment, and transport units) of moving the materials.

Communications involves *information,* such as facts, figures, ideas, instructions, and requests; *transmission* of information from one group or individual to others; and *means* (physical and procedural) of transmitting the information.

Utilities involves *substances,* such as power, air, heat, light, gas, sewage, and waste; *distribution,* accumulation, or dispersal of these substances; and *conductors* to be used for distributing the substances.

Buildings involves the *form* or shape required to accomplish the function; *materials* with which to build; and *design,* or resolving form and materials into a safe, economic, harmonious structure.

Integration of these five components and their respective fundamentals is basic to developing effective facilities plans.

Phases of Facilities Planning.   Facilities planning projects are organized into four planning phases, along with a preplanning and a postplanning phase:

Phase	Name	Activity
0	Preplanning	Prerequisite information gathered from business plans, strategies, and forecasts
I	Orientation	Definition of the project in terms of its scope, requirements, physical location, and external conditions
II	Overall plan	Solution in principle—block layout, overall handling and communication methods, primary utilities, and preliminary building plans
III	Detail plans	Solution in detail—detailed layouts for machinery and equipment, workplace-to-workplace handling, specific information on equipment and procedures, finite piping or ductwork, and detail construction drawings
IV	Implementation planning	Planning specific action steps for construction, rehabilitation, installation, and start-up
V	Execution	Actual physical activity required to make plans a reality

When arrayed against time, these phases come in sequence, and generally overlap. In very large projects, Phase II may be subdivided into overall plans within overall plans, or subphases. Conversely, in very small projects, Phases II and III may be combined.

There are several reasons for structuring projects into phases. Project organization is facilitated because different types of work are emphasized in each phase. Thus, focus and staffing can be logically adjusted as the project proceeds. The five physical components are integrated within each phase, so that overall plans are not mixed with detail plans. Phases provide logical management review points. And each phase provides specific results that can be distributed to and coordinated with people and functions concerned with the planning.

PROCEDURE FOR DEVELOPING FACILITIES PLANS

The planning phases, physical components, and planning fundamentals form the basis of a procedure sufficiently comprehensive to embody almost any project being undertaken. Steps less relevant to the project at hand may be covered by project assumptions, if properly documented, communicated, and approved.

Preplanning (Phase 0).   Outputs from preplanning are clear documentation of project parameters, identification of issues to be resolved, mission statements for the facilities being planned, identification of project organization, and approval for planning to begin.

Project Parameters.   A review of company business and strategic plans is undertaken to ensure that facilities plans developed match forecast requirements for the company. Of particular interest to facilities planners are current and planned products (or services) and markets; methods and technologies employed to produce products (or services); service activities required to support facilities; and time considerations, including operating times, seasonality, and rate of change of planning parameters. A summary document of these parameters, projected into the future, forms the basis for the planning work, and should be carefully reviewed and approved by management.

During the course of this effort, some issues may surface. For instance, do Engineering's staffing projections match Marketing's new product introduction schedule? Will the computer-aided design (CAD) system support new process technologies being planned? Will tightening environmental regulations force a change in the handling of scrap. These and other such issues must be clearly defined, resolved, and approved during preplanning.

Mission Statements.   Facilities being planned must support overall company plans. Accordingly, a mission statement should be developed which includes: existing conditions of facilities, suitability for intended uses, any longer-range plans committing them to a designated use, business units to be housed, products and operations to be involved, and posture on facilities investments and future expansion or growth. Typically, alternative scenarios will become apparent. During Phase 0, these are identified, documented, and presented, along with parameters and issues, for management resolution.

Organizing the Planning Project.   The magnitude and scope of the planning project will indicate project organization. A simple project involving minor rearrangement may be handled by an industrial engineer, in conjunction with area supervision and Maintenance. A major project involving extensive rearrangement, major expansion, or new construction might have a designated formal organization, complete with steering committee, full-time project manager, full-time project teams, and outside consultants, architects, and engineers. Most facilities-planning projects fall somewhere between these two extremes.

Organizational elements required for all projects include:

1. *Approval authority:* The specific individual (or group) with authority to approve plans and authorize necessary expenditures for implementation.

2. *Project manager:* An individual responsible to lead the project and take primary responsibility for budget and schedule.

3. *Project personnel:* People responsible for performing analyses and developing plans.

4. *Reviewers:* People who ensure compliance with policies and regulations, verify financial calculations and projections, and ensure compatibility with other organizational efforts, both current and planned.

These elements may include people both within and outside the organization. It is important to identify all people and their respective roles at the onset of the project, and to clearly document and communicate this information.

Phase 0 concludes with management approval of planning parameters, resolution of issues, facilities mission statement, and project organization.

Orientation (Phase I). Phase I outputs include: clarification as required of project scope and objectives, documentation of location and external conditions, review of existing facilities, determination of facilities requirements, and the master project schedule.

Location and External Conditions. Location includes identifying boundaries and characteristics of the facilities being planned. External conditions are things outside the area being planned which will affect the facilities planning and over which planners exert little or no control. These are defined in the context of the components of facilities attendant to the location.

External conditions related to *handling* are concerned with movements of material to and from the location. These include: relation to suppliers and customers (or distributors); access ways, such as rail, highways, waterways, pipelines, and air; transportation services and associated trip generation; and access for employee and visitor traffic, public transportation, fire protection equipment, contractors, and others.

External conditions related to *communications* include inbound and outbound transmissions of information. Examples of such conditions are: mail service; telephone lines; walkways for personnel and visitors; and broadcasting restrictions.

External conditions related to *utilities* address capacities, load variabilities, and locations of external sources of supply, along with disposal of effluent. Typically included are: water, electricity, gas, steam, and fuel oil; discharge services for effluent, along with required treatment and controls; easements; existing embedded lines; and tie-in points for services.

External conditions related to *buildings* are concerned with physical character of surrounding properties and applicable land-use restrictions. These may involve: ownership; topography, drainage, water table, and soil conditions; climate; prevailing wind patterns; zoning and building codes; existing buildings and structures; and area development plans. Once examined, external conditions are compared to facilities requirements, to identify any inherent discrepancies. Adjustments or acceptances based on this comparison should be clearly documented, and any resulting action approved by management.

Surveying Existing Facilities. A survey includes: description of each facility, its ownership, and location; space occupied and current use; age, physical condition, and construction limitations; suitability for current or other specific use; cost to relocate the facility or its current occupants; an inventory of machinery and equipment; capacity of production equipment and utility services; personnel assignments; and any other information deemed relevant to the project. Many of these data are available from company records. Other data may require special engineering studies or a review of legal documents. In all cases, a personal tour of the facilities is recommended for familiarity and verification.

Facilities Requirements. Requirements for capacity, staffing, support, and space are established for the project at hand and within the context of longer-range development. Decisions involving basic infrastructure, or flexibility and versatility in future facilities utilization, must consider the long term in order to be effective.

Capacity. Business plans are analyzed to establish desired output of products or services per period of time, as projected over the expected life of the plan. Next, planned operating methods are analyzed to determine capability of existing facilities. This analysis is done on an area-by-area basis, so that both overall capability and balance of capabilities are determined. It is essential to consider potential effects of changes in process technologies, management systems, and basic organizational structure. Any such changes can have a profound impact on operating methods and facilities. Comparing projected capability to desired output will establish increased capacity required.

Staffing. From the analysis of capacity, estimates are made of projected staffing magnitude and distribution. Consideration is given to such factors as degree of vertical integration, major or rapid product changes, and implementation of alternative technologies. These factors impact number of people required, associated job descriptions, and configuration, relative location, operating procedures, and utilities requirements of workstations. All such changes must be considered, along with consequences (training, learning curve, disruption, and so forth) of implementation, in determining requirements for facilities. Current and projected organization charts are used to document resultant staffing.

Support Systems. Facilities include utilities, transportation facilities, communications equipment, fire protection, security, personnel services, and maintenance. Support systems have physical and procedural aspects. Both are analyzed and documented, along with interaction among support systems. For example, a network installed for transmission of management information may also carry data about energy usage, security monitoring, equipment operating times, and maintenance requirements. A summary matrix of support systems versus time frame is used to document these analyses.

Space. Space has attributes of amount (area), kind (physical features required), and any mandatory shape or configuration required. There are several methods to determine space requirements for facilities.

1. *Calculation method:* Generally most accurate, it involves determining amount, kind, and shape for each space element or piece of equipment, then totaling these, along with any nonapportionable space, to get overall requirements. It includes operator work areas, maintenance areas, and material set-down areas.

2. *Conversion method:* This involves establishing space currently occupied; adjusting to actual current requirements; and converting to projected requirements for the proposed facilities, based on forecasts of operating parameters.

3. *Space standards:* These may be preestablished, or may be established as part of the project. Standards are typically used in office planning, where space assigned to an individual is determined by privacy requirements, equipment requirements, and in some cases by hierarchal position.

4. *Rough layout method:* This is often used on projects involving large equipment with fixed shapes. A layout which includes all required equipment, but is not necessarily optimal, is used to estimate space required for a given area.

5. *Ratio trend and projection method:* Particularly applicable to long-range planning, this method uses ratios of space to operating data (sales, employees, shipments, etc.) to forecast space requirements on the basis of projections of the operating data. Usually, several ratios are used and the results compared.

6. *Site saturation method:* In this method, which is frequently used in master site planning, existing space is apportioned into classifications, both underroof and yard. Such apportionment is used to guide division of total space available when the site is fully occupied.

Establishing facilities requirements, in practice, involves some variation or combination of the previously described methods for determining capacity, staffing, support, and space. The nature, size, and complexity of each project generally indicates which methods are most applicable.

Project Schedule.   Once appointed, members of the project organization develop a project schedule. All involved disciplines participate in the scheduling process, to ensure that all activities are included, and to give a sense of project ownership and responsibility. Project schedules vary. On a very small project, the schedule may be no more than a memo from the planner to the area supervisor, documenting purpose, scope, and timing of activities planned. Many projects are scheduled with a Gantt chart, which arrays tasks against a time scale. Progress on each task can be shown on the chart, allowing review and communication of overall project status. On a very large project, critical-path network analysis may be used. Output of the analysis is converted to project calendars or milestone charts or both, which typically are more understandable to less technically oriented people.

General Planning Model (Phases II and III).*   Facilities planning uses a general planning model for both overall (Phase II) and detailed (Phase III) planning. The model involves five major steps (see Fig. 1.1).

Step 1   Investigate inputs and influences, and clarify parameters.   Inputs include specific projections and probable forecasts of product and capacity requirements, operating and support technologies, and data relating to the external conditions and nonphysical influences previously documented. Resulting parameters are the project plan and assumptions, requirements and dominant considerations affecting the site, and definition of the activity areas involved. In addition, the lead (most important) component, usually layout, is selected.

Step 2   Interact major elements, and establish the conceptual or ideal plan for the lead component.   This interaction takes place four ways: (1) among all five of the physical planning components, (2) between the first two fundamentals of each of the other components, (3) between the three fundamentals of the lead component, and (4) with other entities associated with the physical and nonphysical influences involved in the prior (first) step.

To illustrate this second step, consider a facility with layout as its lead component. It is first recognized that there must be interaction with the other four components: handling, communications, utilities, and buildings. This interaction draws on analyses of the first two fundamentals of each of the other components to establish relationships and space, and for input to adjustment. For example, analyses of handling materials and moves will yield data for flow-based relationships, studying information and transmission for communications provides data for nonflow relationships, and all four other components impact adjustment. The resulting conceptual layouts are reviewed with others to ensure feasibility and capture benefits of alternative perspectives.

Step 3   Integrate the conceptual plan for the lead component into plans for each component, and develop these into preliminary facilities plans.   The planners share the conceptual plan with those who are planning the other four components and with other related parties. The conceptual plan becomes input to the third fundamental of other components, allowing those components' best plans to be developed. Then, these plans are sent back to the lead component for integration into preliminary facilities plans. This integrates handling methods, communications means, utilities conductors, and building design with the layout adjustment.

Step 4   Modify the preliminary facilities plans and refine them into specific alternative plans.   This step is primarily coordination and communication, drawing on review of preliminary plans for improvements. The number of variations is usually condensed into two to five distinctly different alternative plans. Impacts on operating and invest-

---

*Adapted from *Systematic Planning of Industrial Facilities,* R. Muther and L. Hales, Management and Industrial Research Publications, Kansas City, Mo., 1980.

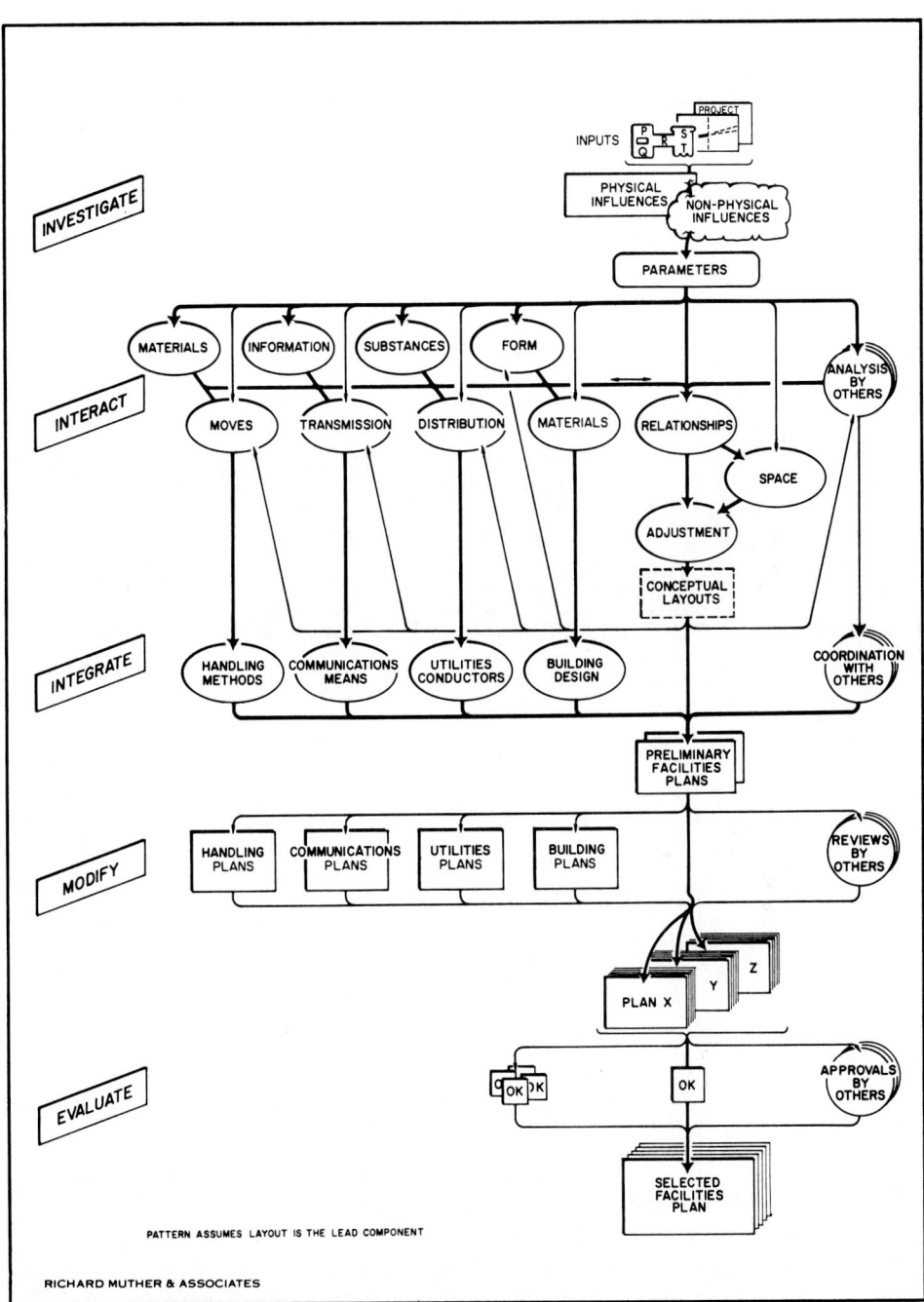

FIGURE 1.1 General planning model.

ment costs are estimated for each alternative. Plans are clearly visualized and described, in preparation for subsequent selection and approval. During this step, operating people should scrutinize the plans to understand potential impacts of each, because they will have to make the facilities work in the real world.

Step 5   Evaluate alternatives and approve a selected facilities plan. Evaluation generally involves three elements: cost comparisons and justifications, analysis of intangibles, and hidden factors. While planning itself may not involve requests for capital funds, approvers should have a clear understanding of financial implications of each alternative. Evaluation based on intangible factors draws from analyses of business plans and subsequent requirements developed in step 1 and input received during reviews. Hidden factors may involve confidential information not available to the planners. All three aspects are significantly aided by clear visualization and understandable presentation.

As stated previously, the planning model is used in both overall and detailed planning. In Phase III, the facilities of Phase II are divided into subfacilities, either by area or by component. Most commonly, the activity areas of Phase II each become subprojects to which the planning model is applied. Alternatively, divisions may be by component, if one component is heavily dominant at the detail level. In some other cases, a combination of division methods may be used.

Implementation Planning (Phase IV).   This involves planning the implementation organization, schedules, and budgets and determining what role, if any, outside contractors will play. Types of functions and often the makeup of the project team will change to reflect this different emphasis.

Organizing for Implementation.   The project manager establishes a clear definition of responsibility for each activity within the overall project, and clearly communicates this to individuals responsible, other members of the project group, and management. Such definition clarifies each person's role, averts later confusion, and helps to ensure that all items are covered. On large projects, subproject managers may be appointed.

Scheduling Implementation.   This is concerned with timing for physical work and related activities. For example, a project involving physical rearrangement tied to implementation of new assembly methods requires that operator training, materials procurement and delivery, tooling, and other requirements be synchronized with rearrangement activity. All projects should have a single master schedule, controlled by the project manager and distributed on a regular basis. Individual schedules must be coordinated with the master schedule. In addition, a plan should be developed for regular meetings to review schedule and budget status, including frequency, outline agenda, roster, chairperson, minutes distribution, and other points deemed pertinent.

Project Budgets.   A detailed cost estimate for each element of the total project, along with methods of tracking, reporting, and reviewing expenditures should be established. Typically, expenditures are classified by accounting professionals as capital, expense, or tooling, consistent with organizational accounting methods. Allocation of funds to project items should be consistent with subdivisions of the project defined by schedule and managerial assignments, giving each subproject manager the ability to coordinate budget with schedule. Budgets should be reviewed frequently, and current or projected variances resolved immediately, to prevent large unexpected overruns.

Outside Contractors.   On small projects, in-house maintenance people may perform all the necessary work, with limited review and follow-up by the planners involved. On major projects, outside contractors, in addition to many internal entities, may be involved. In this case, coordination and control become major efforts. Contracting for new construction requires that issues of project organization, delivery approach, contracting approach, and pricing approach be resolved. Planners must also consider the degree to which in-house personnel will be utilized on physical and managerial activities, and relationships of such personnel with contractors. Selection of appropriate contractors for the work is normally based on price, delivery, experience, and reputation.

Once contracting, schedules, budgets, responsibilities, and review meeting plans have been developed, they are reviewed and approved by the project team and management.

Executing the Plan (Phase V).   Phase V includes installing equipment, relocation, rearrangement, renovation, and/or new construction. The planner's role in this phase usually involves monitoring the progress of implementation, following up to ensure that items are implemented as planned, assisting resolution of problems or questions that arise, and making sure that the project is properly wrapped up.

Monitoring Progress.   During the development of implementation plans, budgets and schedules are established and approved. These are monitored to determine progress. Action is taken when deviations are encountered, either to adjust budget or schedule or to improve remaining tasks. Holding regular, often daily, progress meetings with all responsible parties is a practical method to identify deviations early on, before problems have grown to overwhelming magnitude. This also allows early identification of potential problems, so that prevention measures may be taken.

Following up on Implementation.   Executing the plan includes following up to see that the affected facilities are functioning as planned. Problems encountered typically arise from one of three sources: (1) operating people misunderstand the intended workings of the plan, (2) installation is not according to plan, or (3) the plan is inherently flawed from the start. In all cases, rapid corrective action should be taken. Such problems are avoided by involving operations people throughout the planning process, carefully managing installation, and using well-organized, comprehensive planning methods.

Wrap-up Activities.   Sometimes overlooked in the execution process are wrap-up activities. These include auditing costs and savings estimates, making as-built drawings, and organizing the project's file. Such activities are important not only to the project at hand, but to future projects as well.

FACILITIES UTILIZATION

Facilities contribute to organizational goals. Measuring and monitoring utilization can increase effectiveness of this contribution, through performance measurement, problem and opportunity identification, and trend analysis. In addition, facilities utilization data, properly organized, can serve as a timely and accurate database for planning.

Measuring and Reporting Utilization.   A general approach to utilization includes:

1. Establish objectives for facilities. These objectives should tie in to mission statements and planned operating parameters.
2. Identify information which is or could be available. The most common sources are company records maintained by Accounting, Facilities Engineering, or Industrial Engineering.
3. Select data which best describe performance against objectives. In this step, managers who will receive the information are an excellent resource to aid in selection.
4. Gather data and compare to objectives. It is best to combine current information with historical and projected trends for a broader perspective.
5. Summarize and report results, along with any specific recommendations for improvement. The fewest key informational points should be sent to top management, with some added detail to operating managers. Recognize that facilities utilization information is based on other data already received, and therefore reporting should be concise and brief.

Types of Data.   Specific data relevant to utilization are of three primary types:

1. *Output data:* Products, quantities produced, product mix, capacities, profits generated—frequently measured per unit of time.

2. *Operating data:* Financial information, such as costs, employment figures, and quality levels, and functional items such as utilities consumption and waste disposal.

3. *Asset data:* Inventory, equipment, and tooling, and amount and type of occupancy.

Essentially, output data indicate how well given facilities are performing with a given consumption of operating and asset items.

Ratios and Indexes.   Frequently, these data are used to derive ratios and indexes, which will be more statistically stable over time and allow simpler presentation to management. Ratios commonly used include:

1. Output versus operating item consumption, or how well the facilities are performing in the current environment

2. Output versus assets employed, or how well the facilities are achieving a return on capital

3. Operating data versus assets, or how well facilities support day-to-day operations

Like selection of data, determination of which ratios to monitor is dependent on the nature and characteristics of the organization involved. Historical and current information should be balanced with trend analysis and projections to give perspective.

Reporting Utilization.   Reports fall into three main categories:

1. *Status reports,* which keep management posted on current conditions and trends. These typically consist of a monthly fact sheet distributed with financial reports, along with an update of key indexes for top management.

2. *Problem reports,* which are issued as required to alert appropriate members of the organization to a facilities situation which will require action. Often, proposals will accompany such reports.

3. *Action reports,* which review the results of actions taken, frequently with recommendations on further activity. Action reports should always be issued as a follow-up to problem reports, but may also be used to communicate results of independent actions taken.

In all cases, it is essential to keep such reports succinct and readable.

Improving Utilization.   There are three primary approaches to improving facilities utilization:

1. *Increasing capacity without new construction:* Many utilization data suggest opportunities for improvement. Among approaches used are increasing working hours with added shifts or overtime; improving methods, processes, and equipment; redesigning products, often to simplify or standardize; adjusting inventory policy; subcontracting; and improving housekeeping.

2. *Rearranging for improved utilization:* Examining utilization trends may point out an activity or area whose performance indicates that change is necessary. This is not unusual—layouts which are static may not support dynamic change in the operating environment. Periodic rearrangement, often minor, can allow these areas to be more responsive to changing requirements.

3. *Effective facilities planning:* Building flexibility into each facilities plan can often defer the cost and disruption of rearrangement. Making certain that facilities are based on long-term as well as current requirements, carefully researching and clarifying parameters on which facilities plans are based, and using effective planning methods to ensure that plans developed support operational requirements all contribute to improved facilities utilization.

Conversely, a properly executed utilized program is invaluable for planning. Not only are the data used during planning, but often the awareness of facilities utilization helps management to understand, properly time, and support planning efforts.

SUMMARY

This chapter has addressed concepts and resultant methods for planning facilities; approaches to measuring utilization; and how effective planning and utilization complement each other. Both are part of an organization's total facilities management effort.

From a broader industrial engineering perspective, facilities planning and utilization activities provide added benefits. The structured nature of the planning process helps transform business and operational planning endeavors into tangible action strategies. An organized program for measuring and reporting utilization provides useful feedback to managers on the effectiveness of planning and on operational performance. Accordingly, these efforts make substantial contributions to overall organization management.

FURTHER READING

Hales, H. L., *Computerized Facilities Planning,* Industrial Engineering and Management, Norcross, Ga., 1985.

Lewis, B. T., *Management Handbook for Plant Engineers,* McGraw-Hill, New York, 1977.

Molnar, J., *Facilities Management Handbook,* Van Nostrand Reinhold, New York, 1983.

Muther, R., and L. Hales, *Systematic Planning of Industrial Facilities,* Management and Industrial Research, Kansas City, Mo., 1979 (vol. I) and 1980 (vol. II).

Tompkins, J. A., and J. A. White, *Facilities Planning,* Wiley, New York, 1984.

CHAPTER 2
SITE SELECTION

Hubert King Hardin
Manager, A. T. Kearney
Management Consultants
Chicago, Illinois

Expanding global competition and the need to be more responsive to customer requirements is giving the site selection process special emphasis in today's world market. The
important issues of earlier years—labor costs, logistics, and productivity—have to be considered in relation to quality of life and environmental concerns. This is not to say cost
issues and logistics are not primary selection factors, but the ability to assess location criteria becomes more difficult as quality of life and environmental needs attain higher levels
of importance.

This chapter's objective is to place in the hands of the industrial engineer an approach
for assessing both cost-related and quality-of-life issues as an aid in determining the most
suitable location for a new facility.

The site location team needs to understand there are no perfect communities; therefore,
selection of one versus another generally means comparing each against the other using
criteria developed at the project start. Once these criteria are established, the location
team will probably be able to find many locations that fit their profile. Differentiation of
the location criteria between *essential* and *desirable* often allows the search effort to be
appreciably reduced to the one or two best locations within specific geographic areas. A
winnowing process to reduce 48 contiguous states and their communities to a single recommendation can appear to be a monumental task, but undertaking it through an orderly,
proven approach will eliminate much of the trauma.

The methodology in this chapter outlines an approach for a new manufacturing facility.
Site searches for headquarters locations are significantly different. High-priority factors
identified for a manufacturing location, such as trainable and available labor, utilities,
transportation, and environmental regulations will be replaced by other considerations.
Important factors for a headquarters will be focused more on telecommunications, accessibility, livability, and the ability to recruit qualified personnel. While the major emphasis
for a manufacturing plant is generally cost-related, the major emphasis for a headquarters
location will be more concerned with quality-of-life issues. These qualitative factors become especially important when those conducting the evaluations and making the selection decisions are the ones being relocated to the new community.

Proper site location decisions are driven by many criteria, including strained capacities,
outmoded facilities, restrictive work rules, and shifts in customer markets. Some questionable reasons for considering relocation might include poor management practices, re-

duced productivity, and other correctable problems. Assuming that the reasons for deciding on a new location are proper, this chapter will provide you with a step-by-step approach allowing you to collect and analyze community data and make cogent recommendations to your management.

SIX STEPS TO A SUCCESSFUL SITE SELECTION

Kearney experience, gained from numerous site selection projects, has shown there are six proven steps which, if followed, allow intelligent location decisions to be made.

Step 1—Development of Site and Facility Location Criteria.   Location criteria are data developed to evaluate communities on a comparative basis. These criteria range from *essential* to *desirable*. Other information needed for this first step will be a summary of facility and operating requirements including facility area, work force projections by skill type, utility consumptions, and logistics needs.

Step 2—Selection and Evaluation of Benchmark Communities.   The objective of this step is to develop a listing of representative, or benchmark, communities in various states against which preliminary analyses are conducted to narrow the site search and winnow out undesirable locations.

Selection of these locations may be heavily influenced by management constraints, raw material sources, energy requirements, or customer locations. One way of selecting benchmark locations is to determine the sensitivity of transportation costs for alternative geographic locations. Evaluations can also be conducted by making macro assessments on a state-by-state comparative basis. Each state is evaluated, ranked, and compared against others included in the preliminary analysis. Data points for the evaluation might include only transportation costs or could include factors related to wage levels, union activity, manufacturing climate, and population characteristics. Communities identified in this process step are known as *benchmark locations* and are identified to facilitate the transportation analysis and provide points of reference for management.

Step 3—Selection and Evaluation of Candidate Communities.   Step 2 identified the states and/or communities most compatible for concentrating the search effort through the use of representative, or benchmark, locations. Actual identification of communities to be considered for relocation is done with the assistance of the various state industrial development agencies, which will have intimate knowledge of how well their communities match up with the location criteria developed in the first step. Each state representative is asked to provide the name of a single community which best fits the location criteria and facility requirements. These identified locations are known as *candidate* locations.

Having identified the candidate locations, the Site Location team can now begin to collect and analyze cost data for each. Data for this step may be obtained directly from the communities or from state industrial development agencies, site location specialists, utility companies, and others engaged in collating specific locational information. Analysis of qualitative data at this point in the process is generally less important than it will be during later visits to the finalist candidate locations.

Cost summaries are prepared for the candidate locations and a recommendation is made to conduct on-site visits to the finalist communities.

Step 4—Field Evaluations.   This step involves visits to each finalist location. During these visits, data are verified against those used for the evaluations in the previous step. In addition, potential sites are viewed, competitive-type industries are interviewed, community leadership is assessed, supports are identified, and quality of life is evaluated.

Quality-of-life factors are most often associated with livability and cultural amenities. These include historical attractions, festivals, museums, symphonies, theater groups, golf and tennis clubs, boating, fishing and other recreation areas, parks, libraries, churches, educational facilities, and overall attractiveness of the area.

Step 5—Analyze Comparative Data.   Following the community visits, each finalist location is summarized both from an operating cost perspective and qualitatively. This step can be detailed to whatever level is required by management to support the planned relocation.

Step 6—Recommendation.   The final step of the location process involves making a report to management of the recommended community. Following the recommendation, much work will still need to be accomplished prior to start-up of the new operation. Undoubtedly, management will want to visit the top recommended communities. Also, many other activities will need to be undertaken, such as negotiating for land acquisitions, obtaining permits, deciding environmental issues, reaching agreements reached with state and local governments, arranging financing details, and negotiating contracts.

The rest of this chapter will look in greater detail at these required steps for a relocation project to a new community.

CASE EXAMPLE—GOOD TOOL COMPANY

A company that we will call the Good Tool Company decided to relocate its manufacturing facilities to improve its competitive position. Having no previous experience with relocations, the company employed a consulting firm to assist it with this project.

The Good Tool Company formed an advisory steering committee made up of senior operations managers to monitor and direct the efforts of an appointed site location team. The site team consisted of the plant manager of the new facility and the industrial engineering manager.

Primary reasons for undertaking the site search included the noncompetitive position of Good Tool within its industry. Its major problems were outmoded facilities, a labor contract with restrictive work rules, and a shift in customer markets resulting in excessive distribution costs.

This example concerned a site location conducted in the United States. Non-U.S. site selection studies should consider the assistance of foreign nationals familiar with the countries where there is interest in relocating. Names of individuals involved in foreign industrial development can be obtained through their embassies. These specialists will be able to assist with local customs, regulations, logistical issues, trade restrictions, tariffs, permit requirements, labor skills, work force differences, and cautions related to foreign expansion.

Development of Site and Facility Location Criteria (Step 1).   Management inputs suggested the site location team should concentrate its effort in right-to-work states located in the center of the major customer markets. All functions except Sales, Marketing, and Engineering were to be considered for relocation and 10 salaried employees were to be transferred.

Facility Location Criteria.   With management's assistance, the location criteria in Table 2.1 were developed.

Site and Facility Data.   The site and facility data for the planned facility were also developed and are shown in Table 2.2.

Utility requirements for Good Tool were modest, which allowed the site location team to make the benchmark community selections relatively easily. However, an estimated one in seven manufacturers looking to relocate will be heavy users of water, sewage, electricity, and natural gas and other fuel resources. Some preliminary work to ascertain availability and reserves of these resources should be undertaken as an aid in putting together the benchmark community listing.

TABLE 2.1   Location Criteria

Factor	Essential	Desirable
Right-to-work state	X	
Industrial revenue bonds	X	
Vocational-technical school		X
Progressive community leadership	X	
Good minority relationships	X	
Out-of-doors/recreational amenities versus cultural/professional attractions	X	
Good labor climate with nonmilitant union history	X	
Favorable industrial climate (nondominance by major employer)	X	
High employee work ethics/productivity	X	
Accredited school system	X	
Favorable tax structure		X
Competitive wage rates/benefits	X	
Authority for interstate motor carriers	X	
Access to raw materials, supplies, services (assume available in metro area)		X
Rural versus urban site		X
Headquarters separate from manufacturing		X
1–1.5 hours from major metro area (50–75 miles)		X
Acreage 40–50, minimum 20	X	
Rail access to site		X
Access to interstate		X
All utilities to site	X	

TABLE 2.2   Site and Facility Data for the Planned Facility

Manufacturing area (masonry construction, 20-feet clear height in factory, light manufacturing, 40 × 40 foot bay size, 6-inch reinforced concrete floors, four truck docks, average density fire sprinkler system)	160,00 square feet
Office area, air-conditioned	10,000 square feet
Site size	25 acres
Workforce	25 salaried, 165 hourly
Electric power:	
Processing demand	2000 kW
Use/month	600,000 kWh
Electric air conditioning use/month (5 month/year usage)	250,000 kWh
Natural gas	
Heating use/year	36,000 MCF*
Processing	None
Parking spaces	125 vehicles

*MCF = million cubic feet.

Selection and Evaluation of Benchmark Communities (Step 2).   Because of the sensitivity of transportation costs to geographic location, the site selection team identified 15 states and representative communities for whom freight data were available and comparatively evaluated each, using a computerized data program to determine inbound and outbound freight costs. Locations were selected primarily on the basis of being within the major customer markets of Good Tool. Data needed for the analysis included raw material sources, customer locations, material commodity types, typical shipment sizes, annual tonnages, and freight rates. Inbound and outbound freight costs were calculated for each of the 15 benchmark locations. (An analysis at this level generally will provide adequate results by looking only at major vendor purchases and the largest customer shipments.)

It should be remembered that freight rates are dynamic and competitive. New rates generally can be negotiated to meet changed requirements brought about by new industry relocating to a community. One of Good Tool's objectives was to minimize freight advantages enjoyed by its competitors while at the same time reestablishing a competitive position in its cost of manufacturing.

The Good Tool Company received all of its raw and purchased materials and made all outbound shipments via commercial motor carriers. Many manufacturers compete today in a global market and will need to consider shipments by rail, water, and air. Each mode of required transportation will greatly influence the selection of benchmark communities. Table 2.3 summarizes the team's inbound and outbound freight analysis for the 15 benchmark locations.

TABLE 2.3 Inbound and Outbound Freight Analysis

Benchmark locations	Thousands of dollars		
	Inbound	Outbound	Total
Charleston, W. Va.	31	139	170
Saginaw, Mich.	29	157	186
Raleigh, N.C.	31	161	192
Louisville, Ky.	33	159	192
Nashville, Tenn.	33	162	195
Columbia, S.C.	32	166	198
Montgomery, Ala.	34	165	199
St. Louis, Mo.	47	153	200
Richmond, Va.	31	183	214
Kansas City, Kan.	47	184	231
Little Rock, Ark.	51	184	235
Dallas, Tex.	54	205	259
Salt Lake City, Utah	76	243	319
Atlanta, Ga.	35	164	199
Orlando, Fla.	51	200	251

After the transportation analysis, the site team dropped four states and added one for state-by-state evaluations of 14 factors covering wages, union activity, manufacturing climate, and population characteristics. These data are readily available from Department of Labor and Bureau of Census tables.

Examples of various available data are shown in Tables 2.4 to 2.7.

State surveys and reports are available through a number of organizations. One very good report is available through *Area Development Magazine,* which offers information on all the states, Puerto Rico, and the Canadian provinces. Their reports contain information on population, work force, unemployment, basic business taxes, and financial incentives.

The site location team, with inputs from the steering committee, assigned weighted values to each factor evaluated in Table 2.8:

Wage levels   40%

Union activity   35%

Manufacturing climate   15%

Population characteristics   10%

Table 2.8 depicts the factors and weighted values assigned for state-by-state evaluation.

The location team recommended that eight states be considered for in-depth evaluations. Table 2.9 lists the eight states best meeting the weighted-factor criteria agreed to by the steering and site teams.

TABLE 2.4   Union Membership

STATE	MANUFACTURING—UNION MEMBERSHIP (1,000)			PERCENT OF EMPLOYED [1]			STATE	MANUFACTURING—UNION MEMBERSHIP (1,000)			PERCENT OF EMPLOYED [1]		
	1984	1987	1988	1984	1987	1988		1984	1987	1988	1984	1987	1988
U.S.	5,285.9	4,883.6	4,771.1	27.3	25.8	24.9	MO	179.2	141.2	136.5	41.3	34.1	32.4
							MT	5.6	5.1	5.2	24.9	23.8	25.0
AL [2]	72.2	59.8	57.4	20.1	16.7	15.3	NE [2]	11.1	9.0	9.1	12.3	10.6	9.9
AK	4.3	4.0	3.7	38.1	29.6	25.5	NV [2]	1.6	1.6	1.5	7.6	6.8	6.2
AZ [2]	9.2	7.6	7.2	5.3	4.1	3.8	NH	9.9	8.6	8.2	8.0	7.2	6.7
AR [2]	28.6	28.4	27.3	13.4	13.0	12.0	NJ	184.9	168.1	165.6	25.4	24.7	24.8
CA	510.5	494.9	488.4	24.7	23.8	22.6	NM	5.6	4.2	4.2	15.3	10.8	10.4
CO	21.2	19.6	18.8	10.9	10.7	10.0	NY	668.2	612.8	588.0	50.4	50.3	48.2
CT	70.2	60.9	58.7	16.9	15.6	15.6							
							NC [2]	41.4	41.5	39.7	5.0	4.9	4.6
DE	13.8	14.3	14.1	19.6	20.7	20.5	ND [2]	2.1	1.8	1.7	13.6	11.5	10.4
FL [2]	43.3	47.6	48.2	8.6	9.1	8.9	OH	469.0	460.1	450.4	41.6	42.1	40.9
GA [2]	77.3	70.9	67.8	14.1	12.4	11.9	OK	30.6	27.2	26.9	17.5	17.1	17.1
HI	9.0	9.5	9.1	41.3	43.0	41.4	OR	49.9	46.3	45.2	24.8	23.0	21.4
ID [3]	7.3	5.0	4.9	13.3	9.3	8.6	PA	457.1	434.8	428.3	40.7	41.9	40.7
IL	407.7	317.8	314.8	40.9	34.2	33.3	RI	13.0	13.2	12.8	10.7	11.0	11.1
IN	248.7	250.9	237.8	40.1	41.3	37.6	SC [2]	14.6	13.2	11.8	3.9	3.6	3.1
IA [2]	65.3	48.9	46.8	30.8	23.1	20.8	SD [2]	1.1	.9	.8	3.8	3.1	2.7
KS [2]	22.6	21.1	20.6	12.8	11.8	11.4	TN [2]	78.4	69.6	67.2	15.8	14.0	13.5
KY	69.6	63.9	62.6	27.0	24.8	23.0	TX [2]	159.8	144.0	142.6	15.9	15.0	15.1
LA [2]	36.8	34.9	34.5	20.2	20.8	20.4	UT [2]	6.8	5.1	4.9	7.2	5.5	5.0
ME	23.8	20.4	20.1	21.5	19.8	18.7	VT	5.5	4.4	4.1	11.2	8.9	8.3
MD	71.2	62.9	62.0	32.5	30.3	30.0	VA [2]	48.5	53.1	52.1	11.5	12.5	12.2
MA	126.8	117.1	115.8	19.2	19.4	19.9	WA	99.7	94.1	93.6	34.6	30.0	28.2
MI	487.8	502.2	500.6	50.7	51.9	53.6	WV	34.9	29.1	25.7	38.1	33.6	29.8
MN	96.2	71.8	70.1	25.7	19.3	18.1	WI	161.4	138.9	133.3	31.1	26.7	24.3
MS [2]	19.0	19.8	19.0	8.7	8.8	8.1	WY [2]	1.6	1.5	1.3	20.0	18.5	15.9

[1] Employed in manufacturing.

[2] Right-to-work state.

[3] Right-to-work state beginning 1987.

Source: U.S. Department of Commerce, Bureau of the Census, "1990 Statistical Abstract of the United States," U.S. UNION MEMBERSHIP IN MANUFACTURING, BY STATE: 1984 TO 1988. (Data represent annual average, dues-paying full-time equivalent membership derived from financial records. Excludes unemployed members. In general, annual per capita revenues received by the parent organization were divided by the per capita rate to yield membership. For unions with multiple dues structures or other structures, other methods were used. See source for details. A right-to-work state has laws which prohibit collective bargaining contracts from including clauses requiring union membership as a condition of employment. Data for the District of Columbia not available.)

Selection and Evaluation of Candidate Communities (Step 3).   Once the states ranked one through eight were agreed to for possible relocation, the site team contacted representatives from the Departments of Commerce and Industry and others responsible for their state's efforts in attracting new industry. Each was asked to provide the name of the single community which best fit Good Tool's location criteria and facility requirements identified in Step 1. Table 2.10 is a list of the recommended candidate communities.

Candidate location selection having been completed, the site team next contacted community representatives identified by their state contacts. The team requested cost information which would allow it to make direct comparisons between the eight locations. This information was to be used to cost the facility requirements identified in Step 1. Included in their requests were the following:

• Average hourly labor rates, fringes

• Salaried labor, fringe costs

• Site costs

• Land development and construction costs

• Taxes (income, property, franchise, unemployment)

• Utility rates and costs

In instances where the community representatives did not have the available data, they acted as contacts and solicited the requested information through the appropriate utility

TABLE 2.5 Average Residential Prices for Utilities

Area, region and population size class	Average price per therm of utility (piped) gas		Range of therm consumption for Aug. 1990		Average price per KWH of electricity		Range of KWH consumption for Aug. 1990	
	July 1990	Aug. 1990	Low	High	July 1990	Aug. 1990	Low	High
U.S. city average	$0.581	$0.583	1	2,800	$0.087	$0.087	5	7,512
Region and area size [1]								
Northeast urban	.707	.706	1	2,800	.108	.108	26	7,246
Size A - More than 1,200,000	.721	.720	1	697	.114	.113	88	3,928
Size B - 500,000 to 1,200,000	.697	.697	1	252	.100	.103	26	7,246
Size C - 50,000 to 500,000	.636	.635	2	515	.101	.099	54	3,680
North Central urban	.484	.478	2	1,292	.088	.088	5	6,708
Size A - More than 1,200,000	.486	.477	10	1,292	.102	.101	5	6,708
Size B - 360,000 to 1,200,000	.520	.523	2	443	.070	.071	29	3,287
Size C - 50,000 to 360,000	.484	.481	38	676	.076	.076	39	3,626
Size D - Nonmetropolitan (less than 50,000)	.434	.433	16	429	.081	.082	21	3,749
South urban	.622	.622	2	1,026	.081	.081	30	7,512
Size A - More than 1,200,000	.623	.624	6	442	.084	.084	30	7,512
Size B - 450,000 to 1,200,000	.625	.626	2	1,026	.083	.084	112	5,260
Size C - 50,000 to 450,000	.627	.624	3	589	.081	.081	94	4,002
Size D - Nonmetropolitan (less than 50,000)	.604	.606	5	252	.074	.075	32	5,536
West urban	.604	.625	6	731	.083	.083	57	7,152
Size A - More than 1,250,000	.600	.629	6	731	.077	.077	75	7,152
Size C - 50,000 to 330,000	.523	.524	25	366	.065	.065	82	5,432
Size classes								
A	.590	.593	1	1,292	.095	.094	5	7,512
B	.614	.615	1	1,026	.084	.085	26	7,246
C	.558	.556	2	676	.080	.080	39	5,432
D	.501	.500	5	2,800	.080	.080	21	5,536
Selected local areas								
Chicago-Gary-Lake County, IL-IN-WI	.441	.435	23	635	.121	.118	169	6,708
Los Angeles-Anaheim-Riverside, CA	.719	.719	7	393	.108	.108	75	3,107
N.Y.-Northern N.J.-Long Island, NY-NJ-CT	.840	.841	1	697	.125	.123	88	2,929
Phil.-Wilmington-Trenton, PA-NJ-DE-MD	.775	.774	12	474	.127	.127	216	3,845
San Francisco-Oakland-San Jose, CA	.699	.699	6	191	.111	.111	108	2,457
Baltimore, MD	.642	.634	9	238	.087	.087	197	3,510
Boston-Lawrence-Salem, MA-NH	.698	.698	25	300	.095	.096	140	2,658
Cleveland-Akron-Lorain, OH	.478	.473	46	406	.127	.127	167	2,351
Miami-Fort Lauderdale, FL	.917	.920	8	92	.087	.087	304	2,961
St. Louis-East St. Louis, MO-IL	.469	.493	46	1,292	.098	.098	172	2,744
Washington, DC-MD-VA	.596	.602	6	442	.088	.089	30	3,763
Dallas-Fort Worth, TX	.727	.721	12	46	.071	.070	275	6,349
Detroit-Ann Arbor, MI	.549	.549	21	356	.098	.098	63	3,174
Houston-Galveston-Brazoria, TX	.494	.494	30	170	.085	.085	224	7,512
Pittsburgh-Beaver Valley, PA	.603	.602	31	426	.095	.095	240	3,257

[1]Regions are defined as the four Census regions.
Source: U.S. Department of Labor, Bureau of Labor Statistics, "CPI Detailed Report," August 1990: Table P2. Average residential unit prices and consumption ranges for utility (piped) gas and electricity for U.S. city average and selected areas.

company or other sources. (The use of a personal computer and spreadsheet software is an excellent way to sort and store relevant quantitative data collected in this step.)

After the significant data were collected, the Good Tool team developed a format for presenting a summary of the relative operating costs for each candidate location. All costs were annualized for clarity, and the information in Table 2.11 was prepared for the steering committee.

When the preliminary findings were reviewed with the steering committee, it concurred with the team's recommendation to perform in-depth field evaluations for the four communities having the lowest operating costs.

TABLE 2.6 Employment and Earnings

State and area	Average weekly hours			Average hourly earnings			Average weekly earnings		
	May 1989	Apr. 1990	May 1990°	May 1989	Apr. 1990	May 1990°	May 1989	Apr. 1990	May 1990°
Alabama	41.2	40.4	41.0	$9.08	$9.38	$9.38	$374.10	$378.95	$384.58
Birmingham	40.5	40.1	41.4	9.47	9.54	9.49	383.54	382.55	392.89
Mobile	41.6	41.1	42.2	10.77	11.28	11.04	448.03	463.61	465.89
Alaska	43.9	47.0	44.5	12.45	12.86	13.84	546.56	604.42	615.88
Arizona	40.8	40.3	41.0	9.91	10.09	10.00	404.33	406.63	410.00
Arkansas	40.8	39.5	40.4	8.20	8.49	8.53	334.56	335.36	344.61
Fayetteville-Springdale	42.3	38.9	41.5	7.26	7.62	7.63	307.10	296.42	316.65
Fort Smith	38.8	38.1	37.6	8.66	8.84	8.69	336.01	336.80	326.74
Little Rock-North Little Rock	40.8	40.3	41.2	8.80	9.03	9.10	359.04	363.91	374.92
Pine Bluff	43.2	42.3	44.9	10.33	10.98	10.92	446.26	464.45	490.31
California	40.5	39.6	40.5	11.13	11.35	11.43	450.77	449.46	462.92
Anaheim-Santa Ana	41.6	40.2	41.2	11.30	11.47	11.56	470.08	461.09	476.27
Bakersfield	39.9	39.9	39.7	12.12	12.21	12.27	483.59	487.18	487.12
Fresno	38.8	39.3	39.8	9.33	9.45	9.57	362.00	371.39	380.89
Los Angeles-Long Beach	40.9	39.4	40.6	10.55	10.77	10.83	431.50	424.34	439.70
Modesto	40.2	39.1	39.5	10.53	10.37	10.80	423.31	405.47	426.60
Oakland	41.0	40.2	40.7	13.39	13.44	13.54	548.99	540.29	551.08
Oxnard-Ventura	40.5	39.8	40.9	10.51	10.93	11.04	425.66	435.01	451.54
Riverside-San Bernardino	40.3	39.1	40.1	9.89	10.40	10.47	398.57	406.64	419.85
Sacramento	39.9	39.8	40.5	11.40	11.70	11.79	454.86	465.66	477.50
San Diego	41.3	39.7	40.5	11.18	11.44	11.52	461.73	454.17	466.56
San Francisco	37.9	40.3	40.1	13.09	12.91	13.00	496.11	520.27	521.30
San Jose	39.9	40.3	40.2	13.19	13.45	13.49	526.28	542.04	542.30
Santa Barbara-Santa Maria-Lompoc	39.6	40.3	40.4	11.34	11.81	11.89	449.06	475.94	480.36
Santa Rosa-Petaluma	37.1	38.1	37.8	10.60	11.30	11.27	393.26	430.53	426.01
Stockton	41.0	40.2	41.0	10.97	10.88	11.01	449.77	437.38	451.41
Vallejo-Fairfield-Napa	39.3	39.1	37.7	13.07	13.23	13.11	513.65	517.29	494.25
Colorado	39.8	41.1	41.5	10.51	10.73	10.68	418.30	441.00	443.22
Denver	40.3	40.6	41.3	11.82	12.21	12.19	476.35	495.73	503.45
Connecticut	42.0	41.3	42.6	11.16	11.37	11.45	468.72	469.58	487.77
Bridgeport-Milford	42.3	39.6	41.6	11.52	12.14	12.12	487.30	480.74	504.19
Hartford	42.5	41.4	41.3	11.73	12.34	12.24	498.53	510.88	505.51
New Britain	41.8	39.7	41.4	11.73	12.09	12.00	490.31	479.97	496.80
New Haven-Meriden	40.5	39.4	39.8	10.72	10.80	10.85	434.16	425.52	431.83
Stamford	43.2	40.5	40.9	11.91	11.81	12.19	514.51	478.31	498.57
Waterbury	42.3	40.2	41.7	9.94	10.23	10.36	420.46	411.25	432.01
Delaware	41.2	40.5	42.4	12.17	13.34	12.69	501.40	540.27	538.06
Wilmington	42.6	41.7	43.6	14.94	15.81	15.28	636.44	659.28	666.21
District of Columbia:									
Washington MSA	39.3	38.2	38.8	11.76	12.48	12.46	462.17	476.74	483.45
Florida	40.5	40.0	40.7	8.61	8.91	8.91	348.71	356.40	362.64
Fort Lauderdale-Hollywood-Pompano Beach	40.8	39.6	41.1	8.51	8.99	8.89	347.21	356.00	365.38
Jacksonville	40.1	39.0	39.8	9.13	9.50	9.46	366.11	370.50	376.51
Miami-Hialeah	39.4	39.1	40.3	7.19	7.35	7.52	283.29	287.39	303.06
Orlando	41.3	40.7	41.5	9.64	10.17	10.25	398.13	413.92	425.38
Pensacola	44.4	45.4	43.8	11.17	11.58	11.48	495.95	525.73	502.82
Tampa-St. Petersburg-Clearwater	40.7	39.6	40.9	8.67	9.35	9.37	352.87	370.26	383.23
West Palm Beach-Boca Raton-Delray Beach	40.4	39.2	39.6	9.36	9.51	9.57	378.14	372.79	378.97
Georgia	40.5	39.7	40.7	8.76	9.19	9.12	354.78	364.84	371.18
Atlanta	41.1	39.9	41.3	10.01	10.40	10.34	411.41	414.96	427.04
Savannah	43.3	47.4	47.3	11.55	11.24	11.12	500.12	532.78	525.98
Hawaii	39.3	39.0	39.8	10.38	10.88	10.86	407.93	424.32	432.23
Honolulu	39.3	39.1	40.9	10.44	11.13	11.02	410.29	435.18	450.72
Idaho	37.9	38.7	39.0	9.96	10.32	10.46	377.48	399.38	407.94

Source: U.S. Department of Labor, Bureau of Labor Statistics, "Employment and Earnings," July 1990. Table C-8. Average hours and earnings of production workers on manufacturing payrolls in states and selected areas.

Field Evaluations (Step 4). Arrangements were made and itineraries established for the Good Tool site selection team to visit each of the four finalist communities. The purpose of these visits was to verify and update previously collected data, assess the livability aspects of each, and determine community interest and support for new industry. Questionnaires were prepared in advance to allow collection of all community and state data necessary for the team to make a final recommendation. A survey questionnaire was forwarded to the community contacts prior to the team's scheduled visits. An explanation of the survey was included in an accompanying letter. (Inclusion of the preliminary facility requirements and a description of the project will aid those having to fill in the survey. One to two weeks should be allowed for the survey completion. The team should plan to com-

TABLE 2.7   State and Area Employment*

State and area	Manufacturing			Transportation and public utilities			Wholesale and retail trade		
	May 1989	Apr. 1990	May 1990p	May 1989	Apr. 1990	May 1990p	May 1989	Apr. 1990	May 1990p
Alabama	384.8	382.4	381.3	80.0	79.8	80.2	348.0	346.8	348.1
Birmingham	57.7	57.8	57.6	32.2	32.2	32.1	100.6	101.0	101.8
Huntsville	32.7	33.1	33.3	2.9	2.9	2.9	24.9	25.2	25.3
Mobile	25.9	26.9	27.3	9.9	9.7	9.7	45.9	46.7	46.8
Montgomery	18.9	17.9	18.1	5.3	5.1	5.2	30.1	30.4	30.3
Tuscaloosa	10.3	10.7	10.8	2.0	2.2	2.2	12.7	12.8	12.8
Alaska	16.4	15.4	16.7	22.2	18.5	19.5	44.5	43.9	45.6
Arizona	187.3	187.3	188.0	76.4	80.4	80.3	365.0	376.5	377.3
Phoenix	138.7	139.3	139.6	53.2	56.1	56.3	246.5	253.5	253.9
Tucson	28.1	26.7	27.0	9.6	9.5	9.6	58.1	57.6	59.9
Arkansas	230.3	229.6	230.4	54.0	54.9	55.0	204.1	209.3	211.4
Fayetteville-Springdale	13.2	13.7	13.6	5.1	4.9	4.9	12.1	12.3	12.5
Fort Smith	25.5	25.6	25.7	4.3	4.3	4.2	16.8	16.3	16.6
Little Rock-North Little Rock	33.3	33.1	33.3	16.5	16.3	16.2	59.6	60.8	61.1
Pine Bluff	6.6	6.3	6.3	2.7	2.6	2.6	7.4	7.4	7.5
California	2,158.9	2,132.1	2,130.8	600.8	604.1	608.2	2,942.9	2,993.8	3,009.7
Anaheim-Santa Ana	259.2	258.8	257.5	34.5	35.3	35.6	300.6	305.6	307.3
Bakersfield	10.6	10.5	10.5	8.0	8.3	8.4	39.6	39.6	39.8
Fresno	23.1	24.4	24.5	11.8	12.0	12.1	55.0	57.6	58.7
Los Angeles-Long Beach	895.5	883.9	882.9	212.5	215.4	216.3	959.6	971.7	975.1
Modesto	22.0	22.7	22.4	4.3	4.7	4.7	29.2	29.9	30.1
Oakland	113.4	115.6	116.0	58.2	60.7	60.9	213.6	219.7	220.9
Oxnard-Ventura	30.6	30.7	30.8	11.8	11.4	11.5	55.9	57.5	57.9
Riverside-San Bernardino	87.8	89.1	89.2	32.6	33.5	33.7	167.6	174.0	175.0
Sacramento	43.0	44.0	45.2	26.5	26.7	27.0	140.1	143.3	143.7
San Diego	134.6	135.3	134.9	35.6	36.0	36.3	232.7	240.0	242.2
San Francisco	81.2	81.2	80.9	75.9	78.6	79.3	216.1	215.9	216.5
San Jose	268.3	264.1	264.3	21.5	22.2	22.3	169.2	168.5	169.0
Santa Barbara-Santa Maria-Lompoc	22.4	22.8	22.7	5.3	5.4	5.4	34.5	34.3	34.6
Santa Rosa-Petaluma	20.5	21.2	21.3	6.2	6.6	6.6	33.7	36.4	36.5
Stockton	24.3	23.6	23.7	8.7	8.3	8.5	35.3	35.5	35.9
Vallejo-Fairfield-Napa	12.5	12.5	12.5	4.6	5.1	5.1	31.9	33.8	34.1
Colorado	192.3	194.3	194.5	92.3	92.8	93.1	361.5	359.6	362.1
Boulder-Longmont	30.5	30.9	30.6	2.4	2.4	2.4	23.9	23.7	23.8
Denver	95.5	95.9	96.4	66.8	67.9	67.8	200.4	200.0	201.1
Connecticut	361.4	352.9	352.1	73.2	71.6	72.0	379.5	376.8	379.1
Bridgeport-Milford	54.2	51.5	51.3	8.7	8.6	8.8	45.5	44.7	44.7
Hartford	85.9	86.3	86.4	18.5	19.0	19.2	104.4	102.7	102.2
New Britain	19.8	19.5	19.4	3.2	3.5	3.7	13.8	14.2	14.2
New Haven-Meriden	46.6	45.5	45.3	16.8	16.2	16.3	60.2	57.1	57.5
Stamford	21.1	19.6	19.4	5.7	5.9	5.9	29.6	29.6	29.8
Waterbury	23.3	22.3	22.6	3.5	3.4	3.3	17.9	17.3	17.0
Delaware	73.0	71.8	71.7	14.9	14.8	15.0	74.4	73.6	74.4
Wilmington	62.6	61.0	60.9	15.7	16.3	16.7	61.0	62.1	62.9
District of Columbia	15.8	15.9	16.1	24.9	24.9	25.0	63.2	64.0	63.7
Washington MSA	89.3	89.1	88.7	108.8	111.5	112.1	432.8	438.9	442.0
Florida	541.5	541.6	541.5	260.3	278.6	279.7	1,431.5	1,489.8	1,485.8
Daytona Beach	12.1	12.2	12.1	3.9	4.1	4.1	35.9	37.4	36.9
Fort Lauderdale-Hollywood-Pompano Beach	46.2	46.2	46.2	22.6	24.1	24.0	150.2	155.7	154.1
Fort Myers-Cape Coral	6.2	6.3	6.3	5.1	5.8	5.8	36.6	40.8	39.9
Gainesville	5.7	5.4	5.4	1.8	1.8	1.8	21.3	21.9	21.8
Jacksonville	38.0	38.1	38.3	28.6	29.0	29.2	111.2	113.1	112.9
Lakeland-Winter Haven	23.7	23.1	22.8	6.8	7.2	7.2	43.1	44.8	44.4
Melbourne-Titusville-Palm Bay	29.7	30.2	30.4	4.2	4.2	4.3	37.8	39.3	39.1
Miami-Hialeah	92.9	91.3	91.7	65.4	72.2	72.5	236.9	241.1	244.3
Orlando	55.5	54.4	53.9	27.4	28.5	28.7	142.3	146.4	147.8
Pensacola	11.3	11.5	11.3	6.4	6.4	6.4	32.8	33.1	33.4
Sarasota	8.5	8.7	8.6	3.5	3.6	3.5	33.9	35.1	34.5
Tallahassee	4.8	5.0	5.0	3.0	3.1	3.1	25.5	27.2	27.2
Tampa-St. Petersburg-Clearwater	95.7	95.7	95.8	39.7	42.3	42.4	238.8	247.9	246.6
West Palm Beach-Boca Raton-Delray Beach	34.1	34.4	34.2	13.5	14.2	14.3	95.0	102.1	100.1

*In thousands.
Source: U.S. Department of Labor, Bureau of Labor Statistics, "Employment and Earnings," July 1990. Table B-8. Employees on nonagricultural payrolls in states and selected areas by major industry.

plete any unanswered parts during its visits.) Survey questions and detailed data collected during each community visit included the following:

Labor.   Costs, availability, characteristics, and climate were assessed. Interviews were conducted with all major and potentially competing manufacturers within a 25-mile radius. During these interviews, usually held with the plant managers, data were requested concerning wage rates and fringes, work rules, labor turnover, labor productivity, employees, skills employed, training programs, and other inputs relevant to a manufacturing operation.

TABLE 2.8   Site Location Factors

Weight	Factor
	Wage levels (40%)
20	A. Average hourly earnings
20	B. Percent change in hourly earnings in last 5 years
	Union activity (35%)
3	C. Right-to-work law
14	D. Union membership percent
4	E. Percent change in union membership
4	F. Certification elections/100 establishments
8	G. Certification elections won
2	H. Work stoppages (to percentage of hours worked)
	Manufacturing climate (15%)
10	I. Percent change in manufacturing employment
5	J. Percent change in manufacturing establishments
	Population characteristics (10%)
3	K. Percent change in population
2	L. Persons in 18–44 age group as percent of total population
2	M. School population (5 to 17 years old) as percent of total population
3	N. Quit rate per 100 employees

TABLE 2.9   State Evaluations

State	Raw score	Ranking
South Carolina	77.6	1
Florida	74.6	2
North Carolina	73.8	3
Georgia	63.6	4
Virginia	57.3	5
Arkansas	51.1	6
Texas	49.0	7
Tennessee	48.0	8
Oklahoma	36.1	9
Kansas	31.3	10
Missouri	26.3	11
Alabama	25.1	12

TABLE 2.10   Candidate Communities

State	Baseline community	Metropolitan area
Arkansas	Pine Bluff	Little Rock
Florida	Panama City	Tallahassee
Georgia	Cordele	Columbus, Macon
North Carolina	Goldsboro	Raleigh
South Carolina	Florence	Columbia
Tennessee	Athens	Chattanooga, Knoxville
Texas	San Antonio	Dallas, Houston
Virginia	Blacksburg	Roanoke

TABLE 2.11   Candidate Location Financial Summary (Thousands of Dollars)

	Good Tool	A	B	C	D	E	F	G	H
						Location			
Labor costs	6,451	3,354	3,536	3,551	3,501	3,607	3,851	3,839	3,998
Transportation	186	198	235	199	214	251	192	259	195
Land/building	673	548	565	598	561	580	548	599	575
Taxes	331	68	73	145	164	269	160	238	433
Utilities	393	242	263	244	332	241	275	253	256
Total	8,034	4,410	4,672	4,737	4,772	4,948	5,026	5,188	5,456

Assessments were made in each community of the total number of people in the labor force and manufacturing, annual entrants into the labor force, unemployment rates, unionized employees, and labor disturbances during the past 5 years. Profiles were drawn for all major employers in the area by number of employees, products produced, unionization, and skills employed. Tables 2.12 and 2.13 show types of reports prepared during the community visits.

TABLE 2.12   Industrial Climate—Major Employers

Company	Union	Product	Employment
Kemp Furniture	None	Furniture	945
Barry	None	Slippers	575
A.P. Parts	None	Auto parts	400
Stevens	None	Textiles	376
Young-Squire	ACTWU	Garments	350
Americal Corp.	None	Garments	350
Borden Manufacturing	None	Textiles	336
Mt. Olive Pickle	None	Pickles	325
Burlington House	None	Draperies	320
Dewey Brothers	None	Foundry	300
Hevi-Duty Electric	CWA	Transformers	260
Celotex	P&PW	Roofing products	175
			4724
Total area manufacturing employment			8284

TABLE 2.13   Industrial Climate

- There are nine companies in the county which employ over 250 people. The largest employer is Wiscassett Mills Co., which employs 1800. The two largest nontextile manufacturers are Alcoa and Federal Pacific Electric Company, each employing 750.
- Unioned plants are Alcoa (Steelworkers), Federal Pacific (IBEW), and Flame Refractories (IBBM).
- Manufacturing employment is 11,098.
- Several firms with skills coinciding with Good Tool's needs include:
  1. Consolidated Fabricators (pressure vessels)—welders, fitters, machinists, form/bend
  2. Aeroquip Corp. (rigid hydraulic hose)—limited skills
  3. Metal forge (pressure vessels, tanks)—welders, fitters
  4. Federal Pacific (switchgear, electrical components)

Transportation.   Summaries were made for each location, denoting the number of motor carriers and terminals, availability of commercial air service and freight, interstate highways and interchanges, and rail service. Table 2.14 is an example of a transportation summary.

Training.   Communities were evaluated with respect to their available vocational training facilities and programs. Each state offered no-cost preemployment training programs tailored to meet Good Tool's needs.

Industrial Incentives.   Key incentives any site location team might expect could include preemployment training, extension of utilities to specific plant sites, and special financing assistance. Additional incentives are offered by states and the federal government, depending on the type of relocation program and skills to be employed.

Industrial Sites and Available Buildings.   Most important for the site team is to gain an understanding of what industrial sites and buildings are available, whether they are under the control of the local industrial authority, what improvements are planned, how good is the access, what restrictions exist, and other relevant information. Many communities offer "spec" buildings, which are partially finished buildings designed for quick occupancy. The team should also plan to meet with area contractors to gain further understandings of costs for various types of construction and to view first-hand examples of typical construction.

Operations Concerns.   Each community was analyzed with respect to fuels and utility availability, reserves, and costs. State and local tax rates and assessment practices were

TABLE 2.14   Transportation Analysis

	Motor carriers		
	Major interstate carriers	Average days to	
		Chicago	New York
Albemarle	10	3–5	3–5
Florence	27	2	2
Goldsboro	30	2–3	2
Orangeburg	38	2	5

	Airlines		
	Commercial	Freight	Nearest commercial service
Albemarle	No	No	Charlotte
Florence	Yes	Yes	Florence, Columbia
Goldsboro	No	No	Raleigh-Durham
Orangeburg	No	No	Columbia

	Highway system		
	Interstate	Distance	Conditions
Albemarle	I-85	35 miles	Fair
Florence	I-20, I-95	1 mile	Excellent
Goldsboro	I-95	20 miles	Good
Orangeburg	I-26	4 miles	Excellent

	Rail		
	Line	Average days to	
		Chicago	New York
Albemarle	N&W, Southern	6–8	5–7
Florence	Seaboard	6	4
Goldsboro	Southern, Seaboard	2–3	2–3
Orangeburg	Southern, Seaboard	3	4

reviewed. Meetings were held with local and state environmental agencies to determine permit requirements and to determine whether any harsh area regulations existed.

Quality of Life. Since quality of life means different things to each of us, the team made a comprehensive assessment of many factors, including area attractiveness and cleanliness; number and types of residential areas; school ratings; cultural and civic offerings; colleges and universities in the area; medical facilities; sporting, outdoor, and recreational pursuits; shopping centers; theaters; newcomer acceptance; and unique area attractions. Plant managers interviewed in each community were especially good contacts in that many communicated their feelings about how well they and their families had been received and how well their company fitted into the community.

Quality of life is often measured in part by using consumer price indexes. These indexes are published monthly by the U.S. Department of Labor. A portion of one of these tables is shown in Table 2.15.

Other. Various other data and information useful in comparing communities may include the type of local government, fire and police protection, fire rating for insurance purposes, number of hotel and meeting rooms, hospitals, and medical services.

Analyze Comparative Data (Step 5). The Good Tool team had reams of information in its files following its community visits. Each visit had required an average of 3 days as they met with local industrial development and Chamber of Commerce representatives, talked with the tax assessor, shared lunches with city and county government officials, talked with utility representatives, interviewed plant managers, and received inputs from educators and realtors. Other information was provided by various state agencies. On return to their office, previously developed operating costs were updated and summarized along with other information collected by the team. The updating included adjustments to the initial data and the application of credits for any expected productivity improvements. Also, start-up expenses were calculated to cover extraordinary charges expected during the construction period and for personnel relocations. Types of start-up expenses to consider might include:

- Personnel relocations
- Temporary living expenses
- Community visits
- Telephone and fax services
- Temporary inventory buildups
- Equipment disconnections, relocations, and reconnections
- Inventory and material transfer charges
- Duplicated operating expense during transition period
- Sale or disposition of old facilities
- Productivity losses, on-the-job training
- Recruiting and hiring expenses
- Severance payments

Using a 10-year economic life, discounted cash flow analyses were performed for the four finalist communities. An example of the cash flow summary is in Table 2.16.

Qualitative assessments were made for the communities and values assigned for each factor consistent with the location criteria developed during Step 1. An example of this assessment is shown in Table 2.17.

Recommendation (Step 6). For their final report presentation, the site team prepared an overview summary for each of the finalist locations and then ranked each of the charac-

TABLE 2.15 Consumer Price Index (CPI)*

	Monthly cities and pricing schedule 2 [1]											
	U.S. city average			Chicago-Gary-Lake County, IL-IN-WI			Dallas-Fort Worth, TX			Detroit-Ann Arbor, MI		
Group	Index	Percent change from—		Index	Percent change from—		Index	Percent change from—		Index	Percent change from—	
	Aug. 1990	Aug. 1989	June 1990	Aug. 1990	Aug. 1989	June 1990	Aug. 1990	Aug. 1989	June 1990	Aug. 1990	Aug. 1989	June 1990
Expenditure category												
All items	129.9	5.4	1.2	129.3	5.6	1.1	125.4	4.7	1.8	126.5	6.1	1.4
All items (1967 = 100)	386.9	–	–	379.7	–	–	386.6	–	–	372.6	–	–
Food and beverages	132.4	5.7	.7	131.2	7.1	.9	131.0	5.1	.9	127.2	5.3	.9
Food	132.7	5.7	.7	131.1	7.4	.9	131.3	5.2	.8	127.1	5.5	1.0
Food at home	132.4	6.3	.8	135.2	8.3	.8	128.4	5.4	1.1	127.7	7.2	1.3
Cereals and bakery products	141.3	5.4	.9	142.8	7.7	-1.4	134.5	.6	1.0	138.6	3.0	4.1
Meats, poultry, fish, and eggs	131.2	7.5	.9	137.8	11.7	.9	127.9	6.4	2.7	130.1	9.4	.7
Meats, poultry, and fish	131.9	7.6	.5	139.6	12.0	.4	128.8	6.4	2.2	130.9	9.8	.4
Dairy products	127.3	11.5	2.0	136.1	12.3	3.3	136.2	16.1	3.3	121.9	16.2	4.8
Fruits and vegetables	145.6	5.1	-.7	143.1	7.4	1.4	128.4	4.2	-1.4	133.1	6.0	-.5
Other food at home	124.2	3.8	.9	123.7	3.0	.5	121.5	1.9	-.6	118.3	3.9	.3
Food away from home	134.1	4.8	.7	123.6	5.6	1.2	137.0	4.9	.5	126.9	2.7	.2
Alcoholic beverages	129.8	4.7	.5	133.6	3.9	.4	129.3	3.4	.9	128.6	3.0	-.1
Housing	127.9	4.5	1.3	126.6	4.4	.9	116.1	2.6	1.9	122.1	5.0	1.0
Shelter	138.7	5.9	1.9	135.8	7.1	2.7	117.4	3.3	1.1	132.4	7.8	1.1
Renters' costs [2]	132.7	5.4	2.7	135.0	5.4	2.3	107.6	4.4	.4	131.5	3.8	1.2
Rent, residential	138.8	4.4	1.0	145.4	5.4	2.0	113.4	3.2	.1	137.3	4.6	.4
Other renters' costs	167.9	10.5	11.4	152.3	5.1	3.9	125.9	15.5	3.1	138.0	1.6	3.8
Homeowners' costs [2]	133.5	6.1	1.5	135.6	7.8	3.0	113.8	2.5	1.5	130.1	9.2	1.1
Owners' equivalent rent [2]	133.7	6.2	1.6	134.9	7.8	3.0	113.5	2.6	1.5	139.7	9.3	1.2
Fuel and other utilities	112.4	2.6	.4	116.7	.5	-3.0	115.7	2.9	4.6	114.7	-.1	1.2
Fuels	105.1	1.5	.1	107.3	-9.1	-8.8	116.2	-1.0	8.5	106.5	-2.8	.1
Fuel oil and other household fuel commodities	91.6	16.2	7.9	94.5	19.3	10.7	105.7	.2	.2	101.3	10.3	3.5
Fuel oil	91.4	18.7	10.4	98.8	22.7	14.6	NA	–	–	97.6	18.9	6.9
Other household fuel commodities [3]	115.3	11.9	3.1	109.7	12.4	3.0	110.7	.2	.2	123.7	7.7	2.3
Gas (piped) and electricity	111.3	.3	-.7	110.8	-9.3	-9.0	115.3	-1.0	8.7	109.1	-3.3	-.1
Electricity	121.8	1.2	-.4	136.0	-12.5	-14.0	112.6	-.8	14.7	131.8	4.0	-.1
Utility (piped) gas	95.0	-1.8	-1.5	87.8	-4.5	-.8	124.1	-1.4	-1.1	93.3	-9.9	.0
Household furnishings and operation	112.5	1.5	.2	110.2	-.9	-1.4	113.9	-.2	1.6	102.0	-.8	-.6
Apparel and upkeep	121.3	5.9	-.9	123.6	4.6	-.2	131.2	8.3	3.0	133.4	10.2	5.2
Apparel commodities	119.0	5.9	-1.2	124.5	4.6	-.2	126.2	8.0	3.0	133.6	10.4	5.4
Men's and boys' apparel	118.0	3.6	-.8	113.5	-1.2	-2.3	107.8	-4.8	6.2	122.9	5.9	1.1
Women's and girls' apparel	118.1	8.4	-1.4	122.0	5.9	3.6	137.3	13.5	.1	151.6	16.3	9.1
Footwear	116.8	3.3	-1.3	125.3	.4	-7.7	119.5	4.5	6.4	105.0	11.0	-3.2
Transportation	120.3	5.3	2.2	117.5	6.0	2.1	117.7	5.2	2.3	124.2	7.1	2.1
Private transportation	119.1	5.1	2.3	116.5	5.9	2.1	117.7	4.8	2.3	124.4	7.0	2.1
Motor fuel	103.4	13.6	9.2	110.6	17.5	8.1	99.6	10.3	4.2	104.5	16.2	8.5
Gasoline	103.3	13.3	9.0	110.1	16.5	7.0	99.6	10.2	4.1	104.3	15.8	8.2
Gasoline, leaded regular	112.3	18.1	12.0	130.7	23.0	9.6	NA	–	–	NA	–	–
Gasoline, unleaded regular	101.2	13.6	9.3	107.8	16.9	7.1	96.7	9.6	4.0	102.4	15.6	7.7
Gasoline, unleaded premium	105.3	11.9	7.7	111.9	15.6	7.0	102.5	12.0	4.4	111.8	16.5	11.8
Public transportation	140.0	8.4	.3	125.8	5.7	-.2	120.5	15.8	3.2	125.1	7.8	-.3

*1982–1984 = 100, unless otherwise noted.

Source: U.S. Department of Labor, Bureau of Labor Statistics, "CPI Detailed Report," August 1990. Table 24. Consumer Price Index for Urban Wage Earners and Clerical Workers: Selected areas, by expenditure category and commodity and service group.

teristics evaluated. This analysis was designed to provide a broader perspective for review by the steering committee. Excerpts from the summary are shown in Table 2.18.

On the basis of the quantitative results for the four finalist communities, the site location team determined that community A would have a present-worth advantage in excess of $13 million over a 10-year period. This finding, coupled with the fact that the quality-of-life assessment also favored location A, made the team feel most comfortable in recommending community A for its new manufacturing location. Selected members of the steering committee confirmed this recommendation through subsequent follow up visits to each of the finalist locations.

TABLE 2.16   Comparison of Alternatives—Cash Flow Summary (Thousands of Dollars)

Expense	A	B	C	D
Land and building	17,054	16,914	17,224	13,678
Machinery and equipment	38,341	38,341	38,341	38,341
Hourly labor	131,489	153,751	151,585	154,469
Salaried labor	44,120	44,129	44,120	46,440
Transportation	31,510	31,414	30,761	28,292
Productivity improvement	(39,444)	(46,128)	(45,473)	(23,048)
Property tax on building	1,705	896	1,950	780
Property tax on equipment	1,045	585	1,235	
Inventory tax	1,497			
Electric utility	3,899	4,078	3,668	4,256
Depreciation on equipment	28,048	28,048	28,048	28,048
Depreciation on building	5,426	5,376	5,476	4,344
Tax saved on depreciation of equipment	(11,219)	(11,219)	(11,219)	(11,219)
Tax saved on depreciation of building	(2,148)	(2,148)	(2,188)	(1,741)
Start-up	3,415	3,415	3,415	3,415
Total	$253,221	$268,940	$266,943	$286,055

TABLE 2.17   Qualitative Assessment of Communities

Criteria	Weight	A Rating	A Points	B Rating	B Points	C Rating	C Points	D Rating	D Points
Transportation	10								
Interstate access	7	4	28	5	35	2	14	4	28
Domestic airline service	2	3	6	4	8	1	2	1	2
International airline service	1	3	3	2	2	1	1	1	2
Labor Training	30								
Availability	8	5	40	5	40	4	32	3	24
Skill	7	5	35	4	28	3	21	3	21
Vocational-technical	15	5	75	5	75	3	45	3	45
Union	40								
Union climate	40	5	200	4	160	4	160	2	80
Livability	20								
Local college	3	3	9	3	9	1	3	1	3
Area attractiveness	6	4	24	3	18	3	18	3	18
Recreation	6	4	24	5	30	3	18	3	18
Housing	3	4	12	4	12	4	12	3	9
Newcomer acceptance	2	5	10	4	8	3	6	2	4
Total Points			466		425		332		254

Rating range: 1 to 5.

Next Steps.   Much work still awaited the Good Tool Company beyond that required to reach a consensus on a community to which they could relocate their manufacturing operations. Additional activities which were undertaken included:

- Negotiating the purchase of a plant site
- Development of equipment layouts
- Finalization of facility specifications
- Obtaining necessary local and state permits

TABLE 2.18 Community Overview

	A	B	C	D	Most attractive
Access					
Distance from	Charlotte	Columbia	Raleigh	Columbia	
Mileage	45	80	50	40	
Interstate	I-85/35 mi	I-95, I-20	I-95/20 mi	I-26	A/C
Civic leadership interest	Good	Average	Above average	Above average	D
New industry activity (countywide)	4 new industries in past 5 years	3 new industries since 1978	3 new industries in past 5 years	2 new industries in past 5 years	A/C
Housing, $50,000 to $100,000					
Number	80	100 +	134	30	
Cost per square foot	$35 with lot	$30–$35 with lot	$30 with lot	$30 with lot	A/B
Apartments/condos	Few available	Good supply	Good supply	Limited	
Use of job service office by area industry	Limited	Good	Limited	Good	B/D
Hotels					
Number	5	25	13	N.A.	
Rooms	142	2000	770	420	A/B
Schools (K–12)					
Tests in relation to national average	Above	Below	Above	Below	A/C
Percent going to higher education	55	67 (highest in state)	71	57	A/B
Schools—post high school	Stanly Tech., Pfeiffer College, Wingate College, Catawba College	Outstanding technical school, Coker College	Wayne Comm. College; graduate programs available at Seymour-Johnson Air Force Base; Duke; N.C. State	Outstanding technical school, S.C. State, Southern Methodist	A

13.31

Population:					A/B
City	38,600	32,000	16,500	17,000	
County	98,000	101,000	52,000	79,300	
Work force:					A (metalworking skills)
Total	39,380	46,900	22,400	34,370	
Manufacturing	8,284	12,900	11,098	13,900	
Metalworking	1,700	1,100	900	1,600	
Est. labor avail. within 25 miles	8,461	10,000	3,545	3,854	
Local:					
Unemployed (registrants)	2,475	2,716	1,166	3,854	
Mechanical trades	348	836	449	295	D
Rate	8.1%	7.7%	6.5%	9.9%	
Rate, lowest past 12 months	4.9%	5.8%	3.0%	N.A.	
Average hourly rate for general trade skills	$6.00	$6.75	$6.00	$6.00	D
Fringes	30.0%	30.0%	30.0%	25.0%	
Minority:					All have good relations
City	48.0%	36.0%	14.0%	47.0%	
County	33.0%	36.0%	11.0%	54.0%	
Unions	USWA (Alcoa), IBEW (FedPac), IBBM (Flame Refractory)	Textile Workers (Wellman Industries), Furniture Workers (La-Z-Boy), Paper Workers (So. Car Industries)	Paper Workers (Celotex), Comm. Workers (Heavy Duty), Textile (Young-Squire)	ACTWU (Ambler, Jeansville), Garment Workers (Orangeburg Garment), Teamsters (Ethyl), IWA (Champion), IAM (Utica Tool)	A/B
Percentage of manufacturing	9.6	10.0	13.5	19.0	All
Community encourages non-union environment					

TABLE 2.18 Community Overview (*Continued*)

	A	B	C	D	Most attractive
	Charlotte	Florence	Raleigh-Durham	Columbia	
Commercial air transportation					A/B
Livability	Outdoor activities prevail; two rivers, lake, national forest, state park, two country clubs, large municipal parks; YMCA; small shopping malls; many downtown stores empty; new tech. school campus; new 130-bed hospital	Ballet, concert, little theatre; YMCA; several large shopping malls, good downtown shopping; three country clubs; four hospitals with 604 beds; commercial air service; two rivers; reservoir; Darlington Track; Myrtle Beach 70 miles	Major shopping center with large malls; civic center; two country clubs; public 18-hole golf; Seymour Johnson SAC base; Morehead City/Wilmington-90 miles; state park; nine municipal parks; 340-bed hospital	Large city parks (Edisto Gardens); new 300-bed hospital; country club; public 18-hole golf; civic center; two area lakes; community theatre; 1 hour to ocean	A/B
Site size and cost per acre	57 ($3500); 65 ($3500)	Approx. 25 sites of 100 + acres; Choice @ $7500	122 ($7500); 122 ($7500); 200 ($4000, no water or sewer)	80 ($4500); 350 ($3500, no sewer)	D

- Development of architecture and engineering drawings
- Letting of construction contract
- Relocation of key personnel
- Hiring and training of new work force
- Equipment, material, and tooling relocations
- Disposition of old facilities

Today, the Good Tool Company is a respected member of the new community, a status attributable in large measure to the careful approach employed by the site location team.

BIBLIOGRAPHY

S. Freed, "Factories of the Future," *Site Selection Magazine,* February 1989.

S. Freed, "Locating Japanese Manufacturing Facilities in the United States," *Site Selection Magazine,* June 1990.

"U.S. Union Membership in Manufacturing by State: 1984–1988," 1990 Statistical Abstract of the United States, U.S. Department of Commerce, Bureau of the Census.

"Average Residential Unit Prices and Consumption Ranges for Utility (Piped) Gas and Electricity for U.S. City Average and Selected Areas," CPI Detailed Report, August 1990, U.S. Department of Labor, Bureau of Labor Statistics.

"Average Weekly Hours and Earnings," *Employment and Earnings,* July 1990, U.S. Department of Labor, Bureau of Labor Statistics.

"State and Area Employment," *Employment and Earnings,* July 1990, U.S. Department of Labor, Bureau of Labor Statistics.

"Consumer Price Indices," CPI Detailed Report, August 1990, U.S. Department of Labor, Bureau of Labor Statistics.

CHAPTER 3
PLANT LAYOUT

Richard Muther
President,
Richard Muther & Associates, Inc.
Kansas City, Missouri

Plant layout embraces the physical arrangement of industrial facilities. This arrangement, either installed or in plan, includes the spaces needed for material movement, storage, indirect laborers, and all other supporting activities or services, as well as for operating equipment and personnel.

The term *plant layout* sometimes means the existing arrangement, sometimes the proposed new layout plan, and often the area of study or the work of making a plant layout. Hence, plant layout may be an actual installation, a plan, or a job. The term also applies to the arrangement of office, laboratory, and service areas.

OBJECTIVES

Making a layout plan is not the end result, even of those responsible for its planning. Rather, improved operations, increased output, reduced costs, better service to customers, and convenience and satisfaction for company personnel are likely to be the chief objectives. It is important to target on these real aims; they are the only accomplishments required.

A planner should aim at certain general objectives in the layout. These include:

1. *Integration:* An integration of all pertinent factors affecting the layout
2. *Utilization:* An effective utilization of machinery, people, and plant space
3. *Expansion:* Easy to expand
4. *Flexibility:* Easy to rearrange
5. *Versatility:* Readily adaptable to changes in product, design, sales requirements, and process improvements
6. *Regularity:* A regular or straight division of areas especially when separated by building walls, floors, main aisles, and the like
7. *Closeness:* A practical minimum distance for moving materials, supporting services, and people

8. *Orderliness:* A sequence of logical work flow and clean work areas with suitable equipment for scrap, trash, and wastes

9. *Convenience:* For all employees, in both day-to-day and periodic operations

10. *Satisfaction and Safety:* For all employees

Basic requirements of any layout include the capability to produce the product called for in adequate quantity and at proper quality.

TYPES OF ARRANGEMENTS

The classic types of layout are three in number (Fig. 3.1).

Layout by fixed position, or by fixed material location, is the first. This is a layout where the material or major component remains in a fixed place. It does not move. All tools, machinery, workers, and other pieces of material are brought to it. The complete job is done or the product is made with the major component staying in the one location. Workers may or may not move from one assembly location to the others. Advantages are

1. Handling of major assembly unit is reduced (though increased parts handling to assembly point).
2. Highly skilled operators are allowed to complete their work at one point, and responsibility for quality is fixed on one person or assembly crew.
3. Frequent changes in products or product design and in sequence of operations are possible.
4. The arrangement is adapted to variety of product and intermittent demand.
5. It is more flexible in that it does not require highly organized or expensive layout engineering, production planning, or provisions against breaks in work continuity.

Layout by process, or layout by function, is the second. Here all operations of the same process or type of process are grouped together. All welding is in one area, all drilling in another, all stitching in the stitching room, and all painting in a paint shop. This layout has these advantages:

1. Better machine utilization allows lower machine investment.
2. It is adapted to a variety of products and to frequent changes in sequence of operations.
3. It is adapted to intermittent demand (varying production schedules).
4. The incentive for individual workers to raise the level of their performance is greater.
5. It is easier to maintain continuity of production in the event of:
    a. Machine or equipment breakdown
    b. Shortages of material
    c. Absent workers

Line production, or layout by product, is the third. Here one product or one type of product is produced in one area. But unlike layout by fixed position, the material moves. This layout places one operation immediately adjacent to the next. It means that equipment used to make the product, regardless of the process it performs, is arranged according to the sequence of operations. Advantages of this layout generally include:

1. Reduced handling of material
2. Reduced amounts of material in process, allowing reduced production time (time in process) and lower investment in materials
3. More effective use of labor:

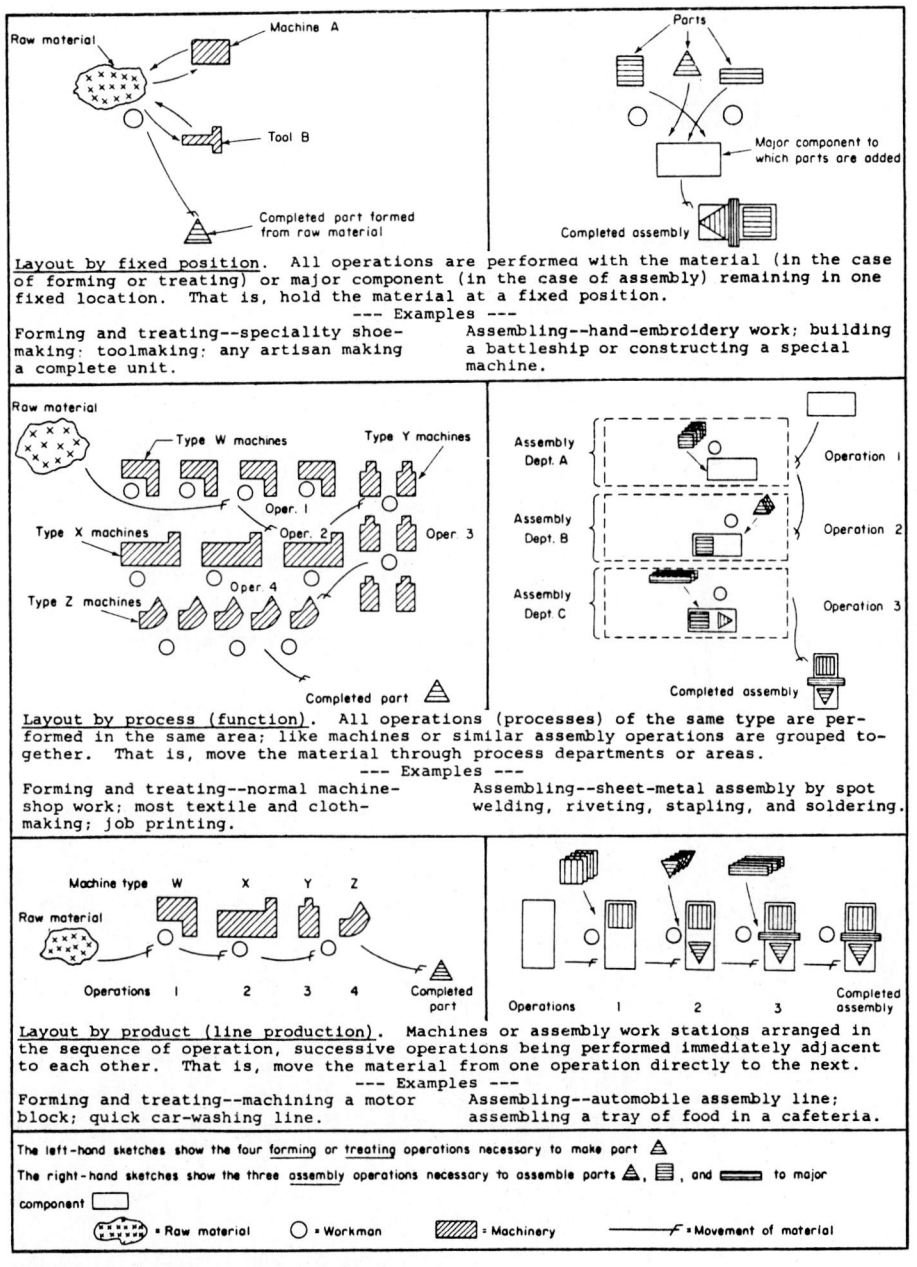

Layout by fixed position. All operations are performed with the material (in the case of forming or treating) or major component (in the case of assembly) remaining in one fixed location. That is, hold the material at a fixed position.

--- Examples ---

Forming and treating--speciality shoe-making; toolmaking; any artisan making a complete unit.	Assembling--hand-embroidery work; building a battleship or constructing a special machine.

Layout by process (function). All operations (processes) of the same type are per-formed in the same area; like machines or similar assembly operations are grouped to-gether. That is, move the material through process departments or areas.

--- Examples ---

Forming and treating--normal machine-shop work; most textile and cloth-making; job printing.	Assembling--sheet-metal assembly by spot welding, riveting, stapling, and soldering.

Layout by product (line production). Machines or assembly work stations arranged in the sequence of operation, successive operations being performed immediately adjacent to each other. That is, move the material from one operation directly to the next.

--- Examples ---

Forming and treating--machining a motor block; quick car-washing line.	Assembling--automobile assembly line; assembling a tray of food in a cafeteria.

The left-hand sketches show the four <u>forming</u> or <u>treating</u> operations necessary to make part △

The right-hand sketches show the three <u>assembly</u> operations necessary to assemble parts △, ▤, and ▭ to major component ▭

(⋯) = Raw material   ○ = Workman   ▨ = Machinery   ⌐ = Movement of material

FIGURE 3.1 The classical types of plant layout.

    a. Through greater job specialization
    b. Through ease of training
    c. Through wide labor supply (semi- and unskilled)
  4. Easier control:
    a. Of production; allows less paperwork

b. Over workers, with fewer interdepartmental problems; allows easier supervision

5. Reduced congestion and floor space otherwise allotted to aisles and storage

Which Type to Use. Use *layout by fixed position* or fixed material location when:

1. Material forming or treating operations require only hand tools or simple machines.
2. Only one or a few pieces of an item are made.
3. The cost of moving the major piece of material is high.
4. A high level of workmanship is needed, or it is desired to fix responsibility for product quality on one worker.

Use *layout by process* or function when:

1. Machinery is highly expensive and not easily moved.
2. A variety of products are made.
3. There are wide variations in times required for different operations.
4. There is a small or intermittent demand for the product.

Use *line production* or layout by product when:

1. There is a large quantity of pieces or products to make.
2. The design of the product is more or less standardized.
3. The demand for it is fairly steady.
4. Balanced operations and continuity of material flow can be maintained without difficulty.

In practice, most layouts are a combination of these classic types, or variations of them. Group technology layouts and flexible manufacturing systems are popular variations.

GUIDING CONCEPTS TO EFFECTIVE LAYOUT PLANNING

There are 10 concepts that guide effective planning of plant layouts (Fig. 3.2). The principles embodied in the concepts are stated below:

1. Every layout rests on relationships, space, and adjustment.
   - *Relationships:* The relative degree of closeness desired or required among things
   - *Space:* The amount, kind, and shape or configuration of the things being laid out
   - *Adjustment:* The arrangement of things into a realistic best fit

2. Basic input requirements for planning plant layouts are: product, quantity, routing, support, and time (P, Q, R, S, and T).
   - *P Product (or material or output):* What is to be made or produced
   - *Q Quantity (or volume):* How much of each item is to be made
   - *R Routing (or process):* How the product is to be made or the material converted
   - *S Support (or supporting services):* What backup will be used to convert the material into a product
   - *T Time (or timing):* When and for how long the product will be made

3. The closer in sequence one places the necessary operations, the less are the problems of material movement.

BASIC CONCEPTS OF LAYOUT PLANNING

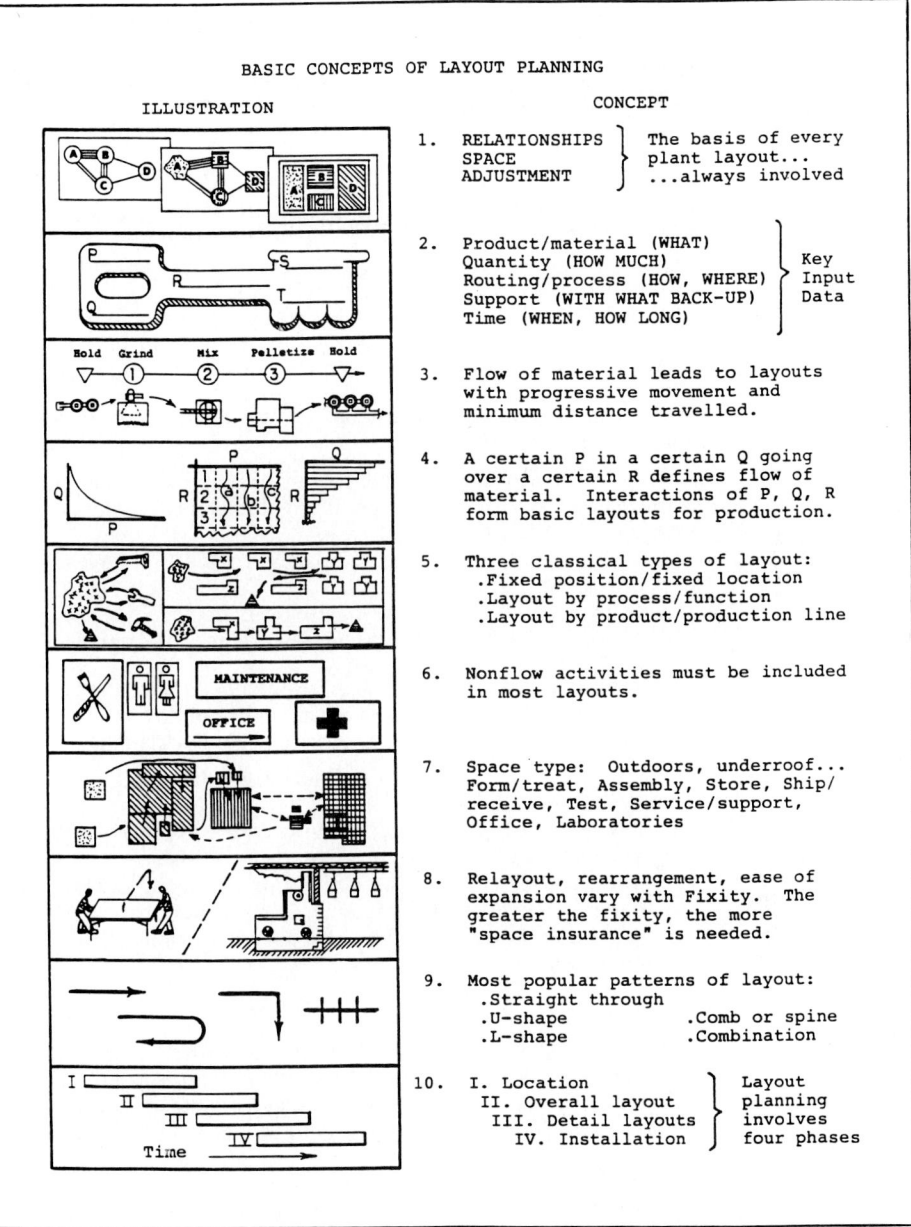

ILLUSTRATION   CONCEPT

1. RELATIONSHIPS ⎱ The basis of every
    SPACE   ⎰ plant layout...
    ADJUSTMENT   ...always involved

2. Product/material (WHAT)
    Quantity (HOW MUCH)   Key
    Routing/process (HOW, WHERE)   Input
    Support (WITH WHAT BACK-UP)   Data
    Time (WHEN, HOW LONG)

3. Flow of material leads to layouts
    with progressive movement and
    minimum distance travelled.

4. A certain P in a certain Q going
    over a certain R defines flow of
    material. Interactions of P, Q, R
    form basic layouts for production.

5. Three classical types of layout:
    .Fixed position/fixed location
    .Layout by process/function
    .Layout by product/production line

6. Nonflow activities must be included
    in most layouts.

7. Space type: Outdoors, underroof...
    Form/treat, Assembly, Store, Ship/
    receive, Test, Service/support,
    Office, Laboratories

8. Relayout, rearrangement, ease of
    expansion vary with Fixity. The
    greater the fixity, the more
    "space insurance" is needed.

9. Most popular patterns of layout:
    .Straight through
    .U-shape   .Comb or spine
    .L-shape   .Combination

10. I.   Location   ⎱ Layout
    II. Overall layout   ⎰ planning
    III. Detail layouts   ⎰ involves
    IV. Installation   ⎰ four phases

FIGURE 3.2 Ten basic concepts that underlie layout planning.

- Shortened distances
- Reduced incidence of pickup and set-down
- Lowered amount of work in process
- Quicker discovery of defects and oversights
- Diminished effort in material scheduling and control

4. Analyses of P, Q, and R lead to basic divisions and arrangement of industrial layouts where flow of materials is involved.

- *P-Q—product-quantity analysis:* Various products are plotted in descending order of their quantities. High-volume items will be mass-produced with layouts by product. Low quantities of large varieties call for a job-shop, custom-order, or made-to-order type of layout.
- *P-R—product-routing analysis:* Types of products (for example, wets and drys, large and small, ferrous and nonferrous) are arranged or cross-oriented against the sequence of operations required. Layouts based on this analysis get the benefits of togetherness by grouping both the similar types of operations and the similar types of products.
- *Q-R—quantity-routing analysis:* Quantity or intensity of material moved is plotted against each route (origin or pick-up to destination or set-down). High-quantity routes should be short and have more complex handling equipment.

5. The classical types of layout develop from the relative dominance of product, routing (process), and quantity.

- Layout by fixed position (where the major material remains fixed and the workers and working tools or machines are brought to and taken from it); employed when the product is physically dominant.
- Layout by process (where like equipment is together); typically used when the process or routing is dominant.
- Layout by product (where one operation is immediately adjacent to its preceding and succeeding operations); typically used when the quantity is dominant.

6. Relationships or closeness desired for reasons other than flow of material are basic to layout planning. Every layout must appraise the nonflow relationships and incorporate them into the layout plan. There are three general situations:

- Nonflow relationships between activity areas where no flow of material is involved, as in offices and laboratories.
- Nonflow relationships between activity areas supporting production, as with tool rooms, maintenance, and lunchrooms.
- Nonflow relationships between activity areas that also have flow of material between them, as with two production departments sharing the same equipment or same worker skills; or, negatively, keeping instrument calibration away from heavy presses.

7. Space may be classified according to its occupants.

- Different occupants need different types of space.
- Different types of space cost different amounts to provide, build, or lease.
- Different types of space cost more or less to maintain and to serve.
- Similar types of space have convertibility or ease-of-rearrangement benefits.

8. The greater the fixity of any operating or support equipment, the more space insurance it should have.

- The more costly to move equipment is likely to stay; the less-fixed equipment can be rearranged more easily.
- Space that will be highly fixed should be given special layout consideration. Do not let it block the way for growing operations.
- Provide extra or added space around fixed activity areas as insurance that it can stretch or grow.

9. Where products or materials are large or awkward and/or quantities are high, flow of material is important and the four dominant flow patterns are basic.

- *Straight-through:* In one end (side), out the other, with generally straight movement of materials
- *U-shaped flow:* Materials, fixtures, mobile handling equipment return to starting point, with in (receiving) and out (forwarding) on the same aisle or using the same dock doors
- *L-shaped flow:* In one side and out one end, or in one end and out the side, with accommodation to congestion or constraints in surrounding or external areas
- *Comb, spine, or dendridic flow:* Central concourse or back-to-back comb, with flexible, two-way flow assisting irregular or changing sequences of operations.

10. There are typically four phases of each layout-planning project.

- *Location:* Determine where the area to be laid out is to be located.
- *Overall layout:* Plan the overall or block layout (the gross or general arrangement), together with its major features.
- *Detail layouts:* Plan the detailed layouts of each department or subarea, including proposed placement of each specific piece of machinery or equipment.
- *Installation:* Plan the implementation of the layout and direct or coordinate the physical placing and hooking up of the machinery and equipment.

METHODS OF APPROACHING PLANT LAYOUT PROJECTS

Instinct and Intuition.   This is often fast, direct, and timesaving, but it is limited generally to simple or emergency situations or when the planner has deep experience and a record of sound decisions.

Find One Ready-Made.   Magazine articles, visits to other plants, discussions with planners from other companies, trade shows, and professional society meetings may lead to finding a layout—one that is spoken of enthusiastically and could be "just the thing." New ideas and methods are essential in this day of rapid change and certainly should be sought out, but what is good for someone else is not necessarily suitable for a different situation, and without at least some modifications, is not likely to be.

Team or Full Participation Approach.   This involves the democratic process. Get ideas from everyone, discuss them, and translate them into a visual presentation. Then call the group together for comment, make changes, and again solicit agreement of the group. This gives everyone involved a chance to participate, to have a bit of "ownership" and therefore to support the ultimate plan. But this approach may draw only on past experience, it can be time-consuming, and it may not take advantage of the analytical techniques so important to moving the company forward at the very time it has the opportunity to do something progressive and constructive. It can be highly effective when the team must stay and make the layout operate.

Flow of Materials.   Centuries ago, engineers discovered that moving material directly from each operation to the next afforded a logical sequence for control and reduced the cost of handling materials. By analyzing the sequence of necessary moves and arranging the layout accordingly, these benefits are gained. This is the approach most frequently taught. It is ideal for process-type industries such as oil refineries or flour mills. But this approach is limited to those situations where there are dominant patterns of material flow, for it does not fully recognize that relationships other than flow of materials may be equally or more important.

Organized Systematic Methodology.   Systematic layout planning (SLP) is a universally applicable approach. It incorporates the benefits of the other approaches and organizes the whole planning process into a rational system. It is generally recognized as the most realistically analytical of any approach yet developed. As a result, it develops plans more soundly and gets approvals faster.

Systematic layout planning is equally applicable to office, laboratory, service, warehouse, and manufacturing operations. It is also equally applicable to minor and major rearrangements, to existing or new buildings, and to new plant site planning.

SYSTEMATIC LAYOUT PLANNING*

SLP consists of a framework of phases, a pattern of planning procedures, and a set of conventions (Fig. 3.3).

The Four Phases of Layout Planning.   As each layout project runs its course from initially stated objective to its installed physical reality, it passes through the four phases of layout planning.

Phase I: Location. Here must be decided where the area to be laid out will be. Not necessarily a new-site problem, Phase I is more often one of determining whether the new layout or re-layout will be in the same place, in a present storage area which can be made free for the purpose, in a newly acquired building, or in a similar potentially available area.

Phase II: Planning the general overall layout. This establishes the gross or block arrangement and the basic flow patterns for the area. It indicates the size, relationship, and configuration of each major activity, department, and area.

Phase III: preparation of detailed layout plans. It includes planning where each piece of machinery and equipment will be placed.

Phase IV: Installation. This involves both planning the installation and physically placing and hooking up the equipment.

These phases come in sequence, and for best results, they should overlap each other. Phases I and IV are frequently not part of the layout planning engineer's specific project, even though the project must in every case pass through these first and last phases. Therefore, the layout planner concentrates attention on the strictly layout planning phases: II, general overall layout, and III, detailed layout planning.

Basic Input Data for Layout Planning.   Before looking at Phases II and III more closely, the basic input data or factors on which facts and information will be needed should be recognized. These are easy to remember when keyed to the alphabet of the facilities planner: PQRST. Practically every layout plan relies on these elements as a basis for its planning.

The Pattern of Procedures—Phase II, Overall Block Layout.   The analytical part of planning the overall layout begins with the study of the input data. Refer to the SLP capsule summary, Fig. 3.3, which shows the four phases, the three fundamentals, the expansion of them into both a short-form application of SLP (called *simplified systematic layout planning*) and a five-section planning pattern at the right. Also in the figure is a set of conventions, or sign language symbols, approved by the American Society of Mechanical Engineers (ASME) and International Materials Management Society (IMMS), and adopted as basic working tools of SLP.

---

*This portion of this chapter is condensed from current updatings of the SLP methodology. *Systematic Layout Planning* (see Bibliography) was initially published in book form in 1961, has been translated to nine foreign-language editions, and adopted worldwide as the most organized approach to layout planning.

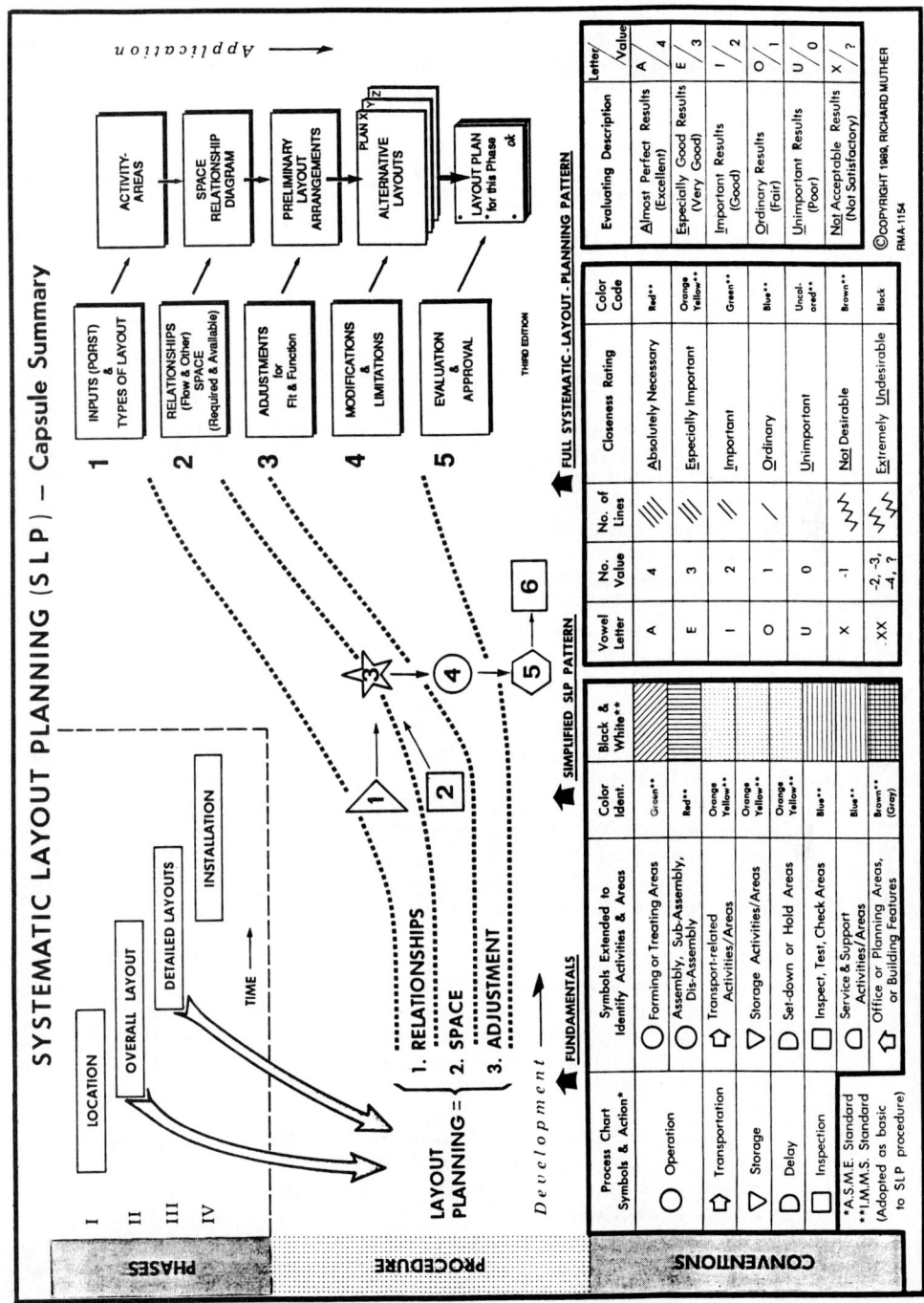

FIGURE 3.3 Capsule summary of systematic layout planning (SLP).

First comes analysis of the input data and the possible types of layout. From this, the division of the total space being laid out is clarified. The output of this section of the application is a list of activity areas (departments, work groups, product breakouts, and physical features such as shipping docks and main entrance).

The second section involves tying together the two primary fundamentals: relationships and space. It establishes: (1) the closeness, that is, desired relationships based on flow of material and nonflow considerations; (2) the space to plan for based on space needed and space available; and (3) a combination of the two, usually in a space-relationship diagram.

Determining relationships between activity areas must deal with both flow and nonflow factors. Heavy industry is based virtually entirely on flow; offices and laboratories almost entirely on non-flow-of-material considerations. The typical manufacturing layout involves both.

The flow and other-than-flow investigations are combined and visualized in an activity relationship diagram. In this process, the various activity areas or departments are geographically diagramed without regard to the actual floor space each requires.

To arrive at the space requirements, analysis must be made of process machinery and equipment necessary and of the service facilities involved. These area requirements must be balanced against the space available. Then the area allowed for each activity will be "hung on" the activity relationships diagram to form a space relationships diagram.

Relationships and space are essentially married at this point. The space relationship diagram almost becomes a conceptual layout. But it is not an effective layout until it is adjusted and manipulated to ensure good fit and function effectiveness. Practical shaping of the activity areas, deciding how to handle material, and adjusting to allow for main aisles and building features are part of this third section of the planning pattern. This leads to several preliminary layout arrangements.

In the fourth section, further modifications are considered for reasons of safety, control, convenience of operation, and so forth. As each potentially good idea is proposed, it must face the challenge of practicality. Modifying considerations and practical limitations are worked out as one idea after another is probed and examined. The ideas that have practical value are retained, and those that do not stand the test are discarded. Finally, two, three, four, or maybe five viable alternative layout proposals may remain. Each of these will work; each has value.

The problem, in the fifth section, lies in deciding which of these alternative layout plans should be selected. At this point, a cost justification should be made, together with an evaluation of intangible factors. As a result of this evaluation, a choice is made in favor of one alternative or the other, although in many cases the evaluation process itself suggests that a new, even better layout could be a combination of two or more of the alternative layouts.

The Pattern of Procedures—Phase III, Detailed Layouts.   The third phase, detailed layout, involves the spotting of each specific piece of machinery and equipment, each working aisle, and each storage rack for each of the activity areas or departments which have been blocked out in the previous overall plan. Phase III overlaps Phase II. This means that before the general overall layout is selected, certain details will have to be looked into. For example, the orientation of a major conveyer, a specific piece of equipment, may affect two activity areas and therefore the overall layout. This means that some adjustment may have to be made between departmental blocks as the detailed areas are being planned; that is, some readjustment of the general layout may be called for.

Note that a detailed layout plan must be made for each departmental activity area. In planning each detailed layout, the same pattern of procedures used in Phase II is repeated. However, the flow of materials now becomes the movement of materials within the department. The department relationships now become relationships among the equipment within the department. Similarly, the space requirements now become the space required for each specific piece of machinery and equipment and its immediate supporting area. And the space relationships diagram now becomes a rough arrangement of templates or other replicas of machinery and equipment, workers, and material set-down space. As in

Phase II, several alternative layouts may result. This leads to an evaluation to select the most satisfactory layout of machinery and equipment within the department.

Planning a manufacturing cell or a flexible manufacturing system is typically a detail-phase project. While layout is usually a dominant component, putting together an integrated cell is more than just layout planning. See Sec. 13, Chap. 5, "Production Line Techniques."

This SLP pattern of procedures provides a basic planning discipline and, at the same time, allows for logically different content of the PQRST input data. And just as the flow-of-materials analysis will become less important and the nonflow relationships more important in office or laboratory areas, so the entire pattern has the flexibility to be modified for the needs of any layout project. It becomes a matter of adjusting the importance of each step, or section, rather than changing the sequence or arrangement of them.

Set of Conventions.   A set of conventions is utilized to aid in planning, understanding, and communicating. The conventions are used throughout each step of the previously described pattern of procedures for diagraming, rating, visualizing, analyzing, and evaluating. The conventions are shown in the lower portion of Fig. 3.3. They consist of seven symbols, seven letters, seven line ratings, and five colors plus black and white. These are cross-integrated for multiple use in any application employing SLP.

Examples of SLP.   Figure 3.4 shows an example of the five steps (or sections) in the planning pattern for a company making plastic bags of various kinds. The planner followed these steps in developing an overall (block) layout. The planner then followed the same sequence—with different emphasis and different data, of course—to develop the layout for each departmental area.

Figure 3.5 shows a conceptual example of an SLP project. It shows, in simplified form, first the Phase I problem of location, then the Phase II overall layout followed by the Phase III detailed layouts of each department, and finally the Phase IV installation. Note the five sections of planning in Phase II, the repeating of the planning pattern for layouts within each activity area, and the use of shorthand conventions.

LAYOUT PLANNING TECHNIQUES

Determine the Flow.   The sequence of operations as the basis for the flow of materials is the heart of many layout plans. As a result, the process chart in its various forms is the most useful of all layout planning devices.

One-Product Analysis of Flow.   Figure 3.6 shows how the operation process chart practically leads to the layout plan. The horizontal material-feeding lines on the chart become delivery racks or conveyers in the layout, and the operations charted become subassembly benches or equipment in the layout. When a process involves forming or treating only, the operations and the information can be listed on a columnar worksheet without the necessity of chart symbols. As a general rule, however, start with an operation process chart when analyzing flow. Even if making a half-dozen different products, begin with charts for each one. A separate area and a separate layout may be needed for each product. Or a combined layout may be needed for all of them. This cannot be determined until the data are assembled in a form convenient for analysis.

Multiple-Product Analysis of Flow.   When there are a number of products or parts, use a combined, multiproduct, or gang process chart. The problem here is to combine products or classes or groups of products so that together they will give volume enough to justify an effective flow of material. Figure 3.7 shows an example of such a chart for the five different products of a factory producing small metal sign plates and shields. The first and last operations are shaded for quick spotting. Operations out of normal sequence are indicated by a slant line joining them with the preceding operation. Volume of business

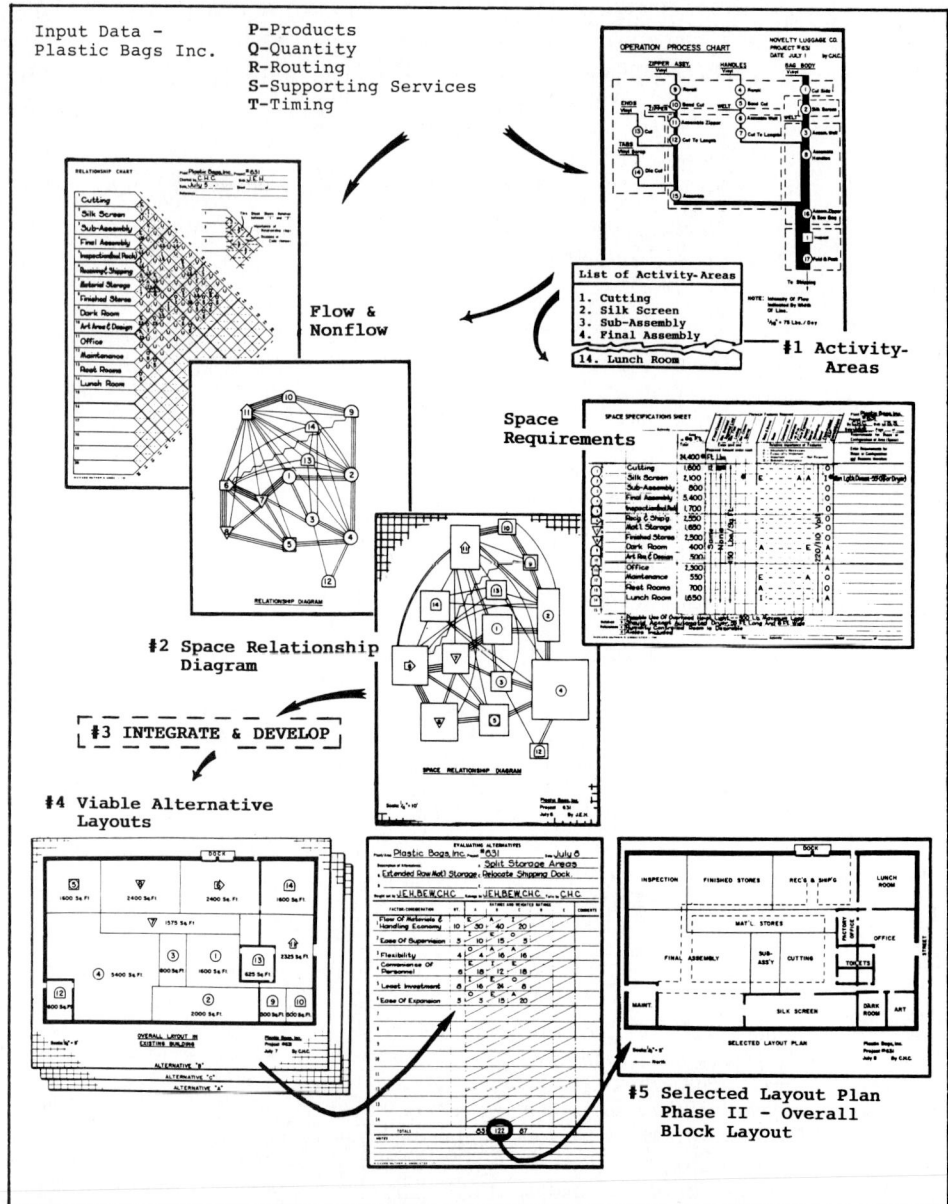

FIGURE 3.4 An example of planning an overall block layout for a plant making plastic airline bags and tennis racket covers and following SLP Phase II.

was determined from past records and sales forecasts, and is used to arrange the operations in the best flow of material.

If the products do not fall into natural groups, as in this example, juggle many different combinations to get a right grouping.

In classifying various products for flow possibilities, look for the following:

1. Products requiring similar machinery

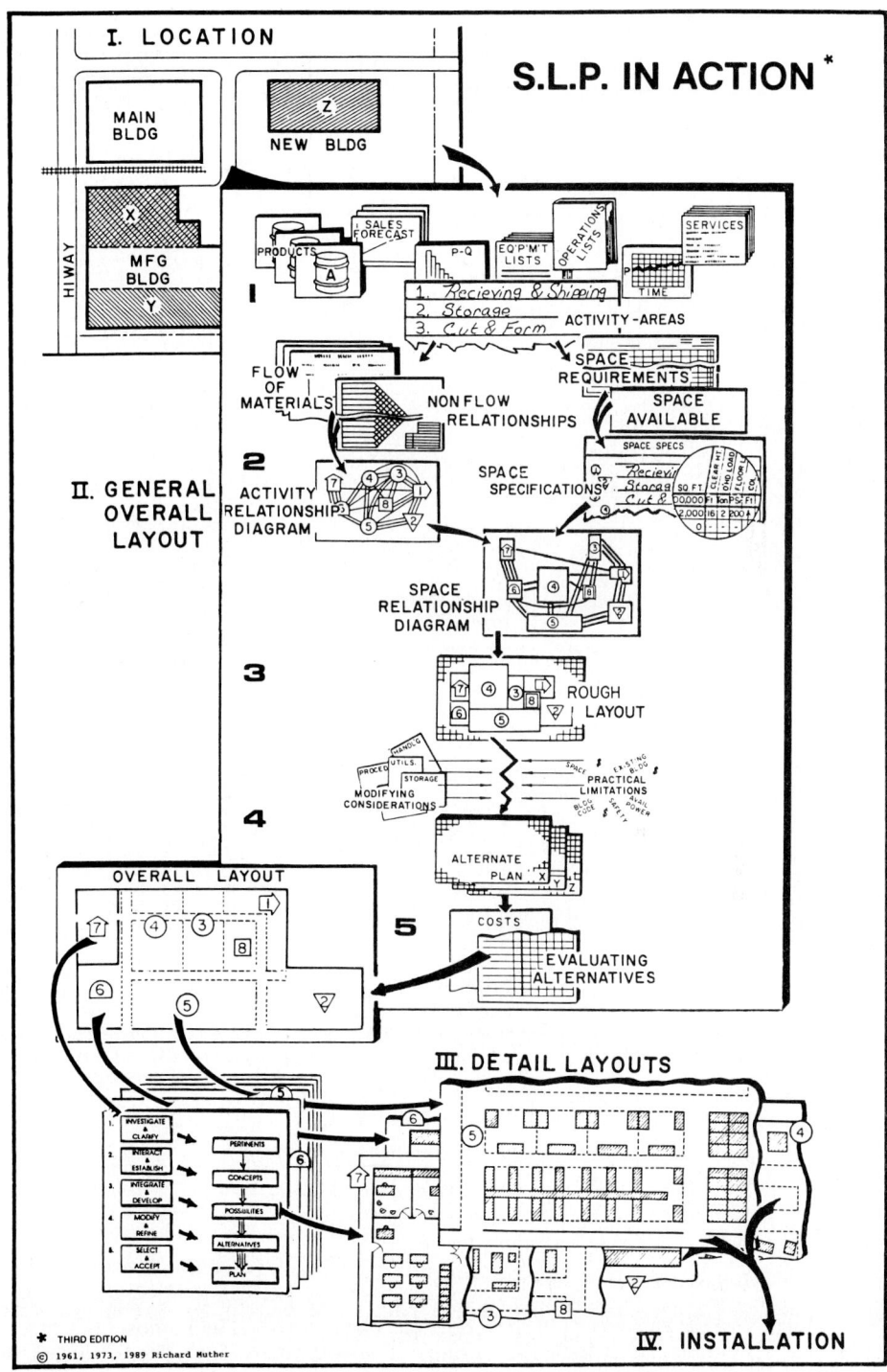

FIGURE 3.5   An example of a layout project applying SLP in its full four phases.

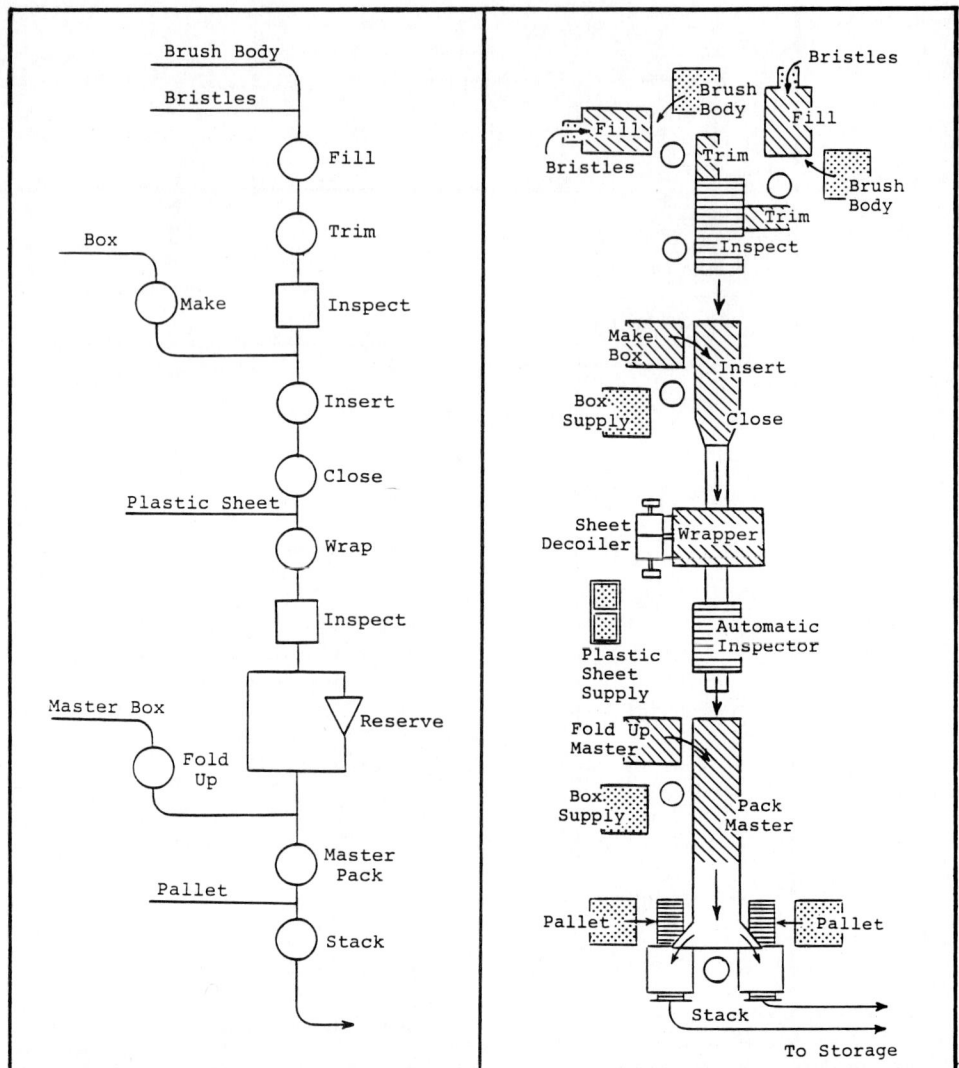

FIGURE 3.6 The process chart (left) as a direct basis for a layout.

2. Products requiring similar operations
3. Products requiring similar sequences of operations
4. Products requiring similar operation times
5. Products of similar shape, size, or purpose
6. Products requiring a similar degree of quality
7. Products of the same material

Figures 3.8 and 3.9 show how one plant combined into a flow pattern all its production of shafts. It is a classic example of group technology. The engineers on this project grouped certain products together to take advantage of flow and minimum distance. First, they gathered together operation sheets showing operations, sequence, machine type, and

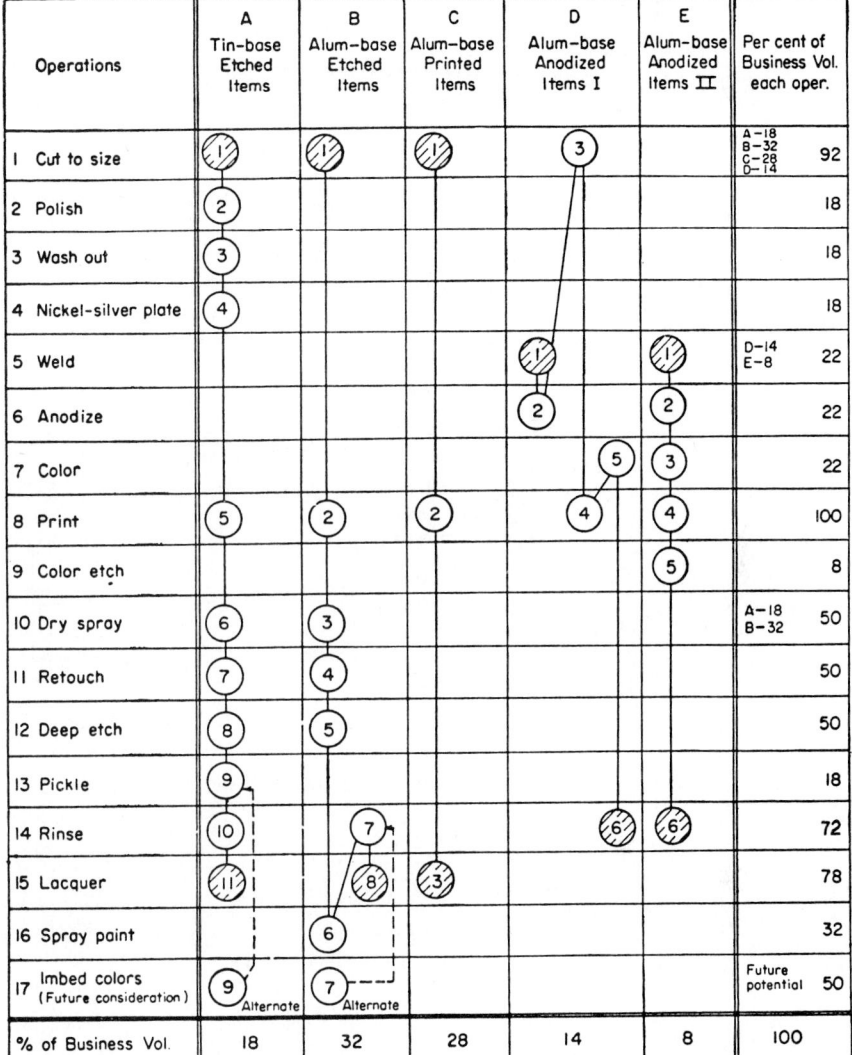

Operations	A Tin-base Etched Items	B Alum-base Etched Items	C Alum-base Printed Items	D Alum-base Anodized Items I	E Alum-base Anodized Items II	Per cent of Business Vol. each oper.
1 Cut to size	①	①	①	3		A-18 B-32 C-28 D-14 92
2 Polish	2					18
3 Wash out	3					18
4 Nickel-silver plate	4					18
5 Weld				①	①	D-14 E-8 22
6 Anodize				2	2	22
7 Color				5	3	22
8 Print	5	2	2	4	4	100
9 Color etch					5	8
10 Dry spray	6	3				A-18 B-32 50
11 Retouch	7	4				50
12 Deep etch	8	5				50
13 Pickle	9					18
14 Rinse	10	7		6	6	72
15 Lacquer	11	8	3			78
16 Spray paint		6				32
17 Imbed colors (Future consideration)	9 Alternate	7 Alternate				Future potential 50
% of Business Vol.	18	32	28	14	8	100

FIGURE 3.7 Multiproduct process chart showing combined operations for five different types of products.

machining time. Then they classified each one of the parts using the following considerations:

1. Parts completed in one machine, such as automatically turned parts
2. Parts almost completed in one machine but requiring simple additional machining, such as parts turned on automatic or turret lathes with subsequent simple drilling or milling
3. Parts of similar nature, especially main components, such as shafts, gears, flanges, and levers
4. Other parts requiring similar operation sequences, such as parts turned, milled, and drilled and parts milled and drilled

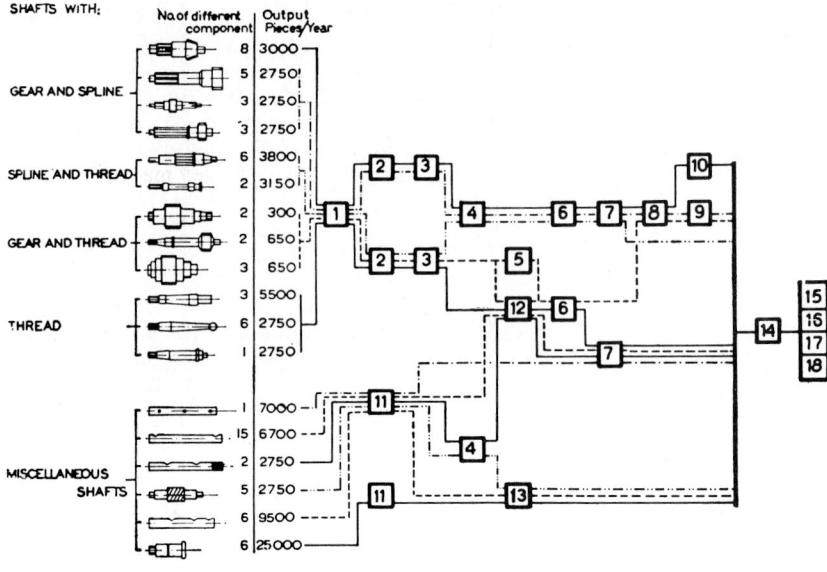

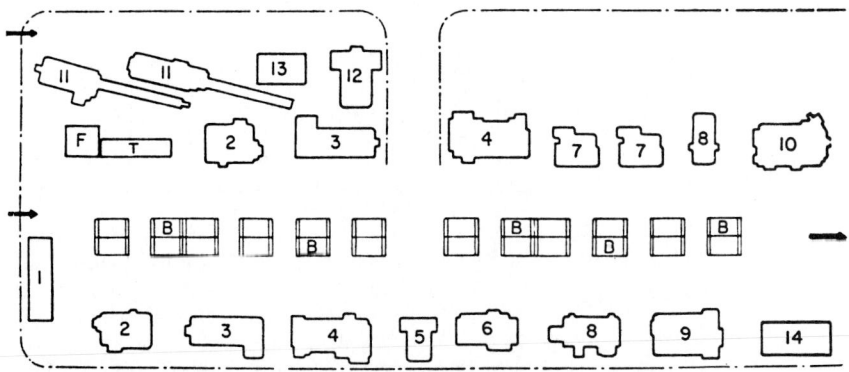

1 CENTERING LATHE.	7 3 sp. DRILLING PRESS.	13 HYDR. PLAIN MILLING MACHINE.
2 HYDR. CONTOUR LATHE.	8 CYL. GRINDER.	14 INSPECTION.
3 FINISH ENGINE LATHE.	9 GEAR MILLING MACHINE.	15 HARDENING DEPARTMENT.
4 SPLINE MILLING MACHINE.	10 GEAR CUTTING MACHINE.	16 GRINDING DEPARTMENT.
5 KEY MILLING MACHINE.	11 TURRET LATHE.	17 FINAL INSPECTION.
6 THREAD MILLING MACHINE.	12 UNIV. MILLING MACHINE.	18 ASSEMBLY STOCK.

FIGURE 3.8 Shaft production group flowchart. (*Courtesy of A. B. Scania-Vabis.*)

T – TOOLS AND FIXTURE • F - FOREMAN'S STAND • B - TRANSPORT BOXES

FIGURE 3.9 Shaft production group layout. (*Courtesy of A. B. Scania-Vabis.*)

In establishing subdivisions of these four classes, the engineers brought together parts which:

1. Required special-purpose machines for some of the operations
2. Were of similar size
3. Had to be machined to similar accuracy

Their chief aims were to obtain full utilization of the key machines in each group and to get a sequence of operations that was nearly the same for each part in a group. To do this, they compiled planning sheets showing the yearly production demands and the corresponding hours required on each type of machine. From this, they determined how many of each machine they should allocate to each group. Figure 3.8 shows the flow chart for the shaft production group. This group produces 80 different shafts, including fairly complicated ones with gears, splines, and threads. The key machines are the hydraulic contour lathes (2) and spline and gear milling machines (4 and 9). Figure 3.9 shows the layout of the shaft group.

A similar approach is to pick out representative parts and draw individual or multiproduct operation process charts for them. In doing this, select say, five of the most costly parts, five of the most fragile parts, five with the highest production requirements, and five with the greatest manufacturing difficulty or number of rejects. Compare the operation process charts of each by laying them parallel to one another, and from this, develop a pattern of flow.

Many-Parts or Many-Products Analysis of Flow.   When the products become too many to chart conveniently into some pattern of flow, the multiproduct process chart is inadequate. Because of the number of route paths involved, a cross-chart, or from-to chart, is better, especially in plants having a variety of nonstandardized products.

The from-to chart takes several forms and can be used in different ways. The main idea is to determine the amount of movement between each combination of two operations or areas. This is done by referring to operations lists or route sheets. The chart can be built for all parts that are involved or a representative selection of parts. Each move is recorded on the appropriate *from* line under its *to* column. The moves from each activity to each other activity are then tallied and totaled. Figure 3.10 shows a typical from-to chart in which unit loads per day is the charted value. Note that the total two-way flow includes moves both from-to and to-from, from example, 3 on line 4 and 13 on line 5.

Flow Alone Not Best Basis for Layout.   There are several reasons why the traditional flow of material—as determined predominantly by the sequence of operations—cannot be the sole basis for layout arrangements.

1. The supporting services must integrate with the flow in an organized way. The maintenance crib, the superintendent's office, the locker and rest rooms, and the transformer bank all have a relatively preferred closeness to each of the producing areas. They are all part of the layout; they must be planned into it, yet they are not part of the flow of materials.

2. Flow of materials is often relatively unimportant. In some electronic and jewelry plants, only a few pounds of material will be transported during an entire day. In other industries, materials are piped, or one skid load lasts a worker all week.

3. In completely service industries, office areas, or maintenance and repair shops, there is often no real or definitive flow of materials, even if paperwork, equipment, or people are regarded as the materials that flow.

4. Even in heavy-material-movement plants, where the influence of material flow will dominate the layout, there are many other reasons for certain operations to be close or kept apart. For example, the routing may call for this sequence: form, trim, treat, subassembly, assembly, and pack. For best flow of materials, treating should lie between trimming and subassembly. But treating is both a very dirty and dangerous operation. Therefore, it should be kept away from the delicate subassembly area and its high concentration of workers.

Some systematic way is needed of relating service activities to each other and of integrating supporting services with the flow of materials. The relationship chart is the best method of meeting this need.

FROM–TO CHART

Item(s) Charted: ALL ITEMS
Basis of Values: UNIT LOADS/DAY
Plant: ABC MECH. WORKS
By: RMH Date: JAN. 9
Project: NEW LAYOUT
With: DL
Page: 1 of 1

FROM \ TO	1 PRESS	2 WELD	3 MACHINE	4 ASSEMBLY	5 PAINT	6 STEEL STOR.	7 PARTS & SUPPLY	8 FINISHED STOR	9 DRIVEWAY	TOTALS
1 PRESS		15	12							27
2 WELD					11			2		13
3 MACHINE		3		15			2	12		32
4 ASSEMBLY					3			18		21
5 PAINT			7	13						20
6 STEEL STOR.	12	5	1							18
7 PARTS & SUPPLY		1	4	3	1					9
8 FINISHED STOR			2						20	22
9 DRIVEWAY						8	11			19
10										
11										
12										
13										
14										
15										
16										
17										
18										
19										
20										
TOTALS	12	24	26	31	15	8	13	32	20	181

NOTES:

RICHARD MUTHER & ASSOCIATES

FIGURE 3.10 From-to chart showing the flow of materials between each origin point and each destination.

The Relationship Chart.   The relationship chart is a semimatrix form where the relationship between each activity (or function or area or machine) and all other activities can be recorded. The basis of the form is shown in Fig. 3.11.

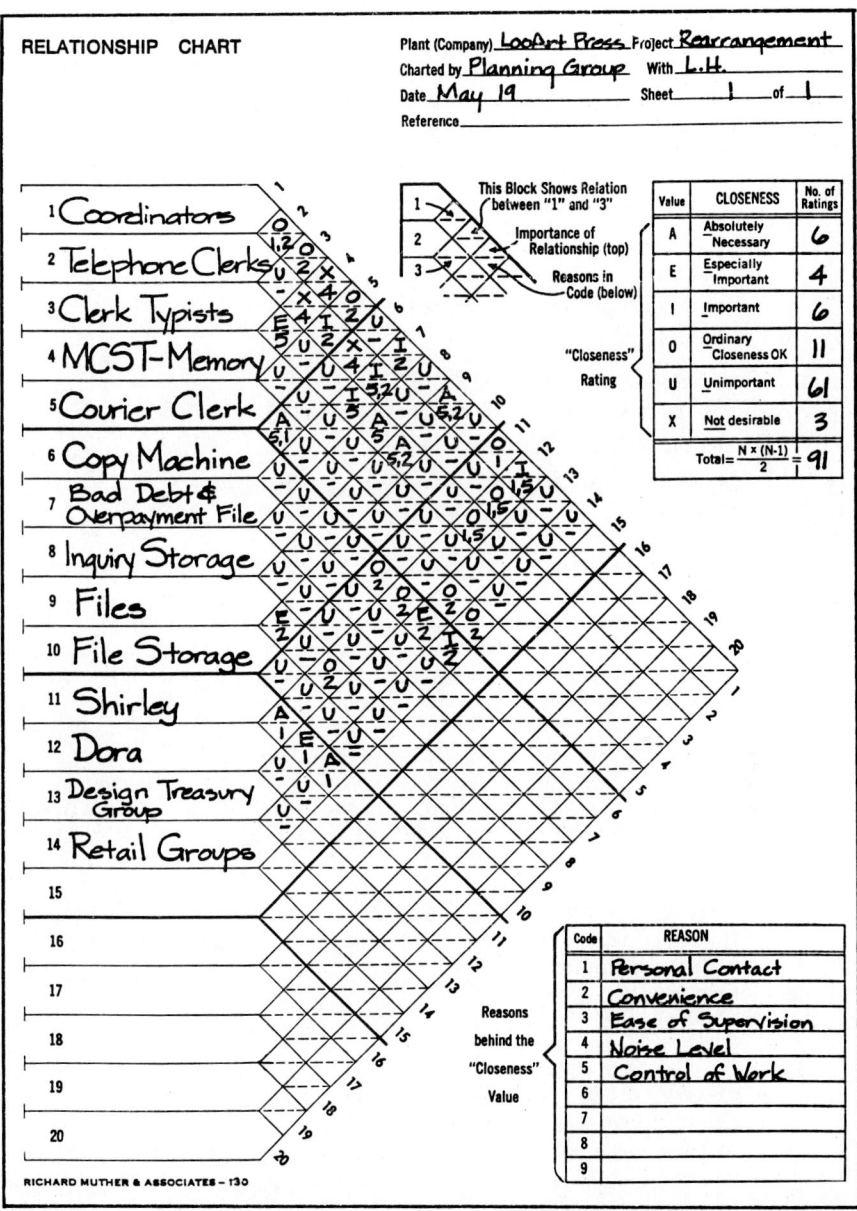

FIGURE 3.11   The relationship chart is extremely effective for planning all activities not tied together with a significant flow pattern. This chart was prepared for an office of clerical employees. It indicates that the coordinators must be near the files (A) and kept away from the MCST-Memory (X). Reasons are filled in and recorded by code number in lower half of the appropriate boxes.

The chart itself is almost self-explanatory. Where the activity on downward-sloping line 1 intersects the activity represented by upward-sloping line 3, the relationship between activity 1 and activity 3 is recorded. In this way, there is an intersecting box for each pair of activities involved. The basic idea is to show which activities should be located close together and which far apart, with all in-between relationships also rated and recorded.

Note that each box is divided horizontally. The upper part is for the closeness rating (A, E, I, O, U, or X). The lower half is for recording the reason causing the designated closeness value. This gives a rating and reason for each relationship.

Typical reasons supporting relationship ratings include the following, although many terms may be used and many other reasons are possible:

1. Flow of materials
2. Degree of personal contact
3. Degree of communicative or paperwork contact
4. Use of same equipment or facilities
5. Use of common records
6. Sharing of same personnel
7. Specific management desires or personal convenience
8. Supervision or control
9. Noise, dust, dirt, fumes, hazards
10. Distractions or interruptions

Combining Flow and Other-than-Flow Relationships.   One reason the relationship chart is so effective is that it records relationships based on flow of materials and other-than-flow reasons. Calibrating these two and getting a common denominator rating for the combination demands rational analysis. Figure 3.12 shows a work sheet example of how this can be kept track of during the sequence of calibrating flow to vowel letters, developing ratings for other-than-flow reasons, and combining the two into a net resultant relationship.

Diagraming the Relationships.   A diagram is a visual drawing of charted data. It can be made on a floor plan of an existing layout or on a blank sheet. In the former case, the flow will be tracked on a scale drawing of the area involved. In the latter, conceptual diagrams can be developed and analyzed. Figure 3.13 shows a flow diagram of the former type for a detailed layout. Figure 3.14 shows a quantified flow diagram for a metal furniture plant.

When flow and other-than-flow are combined in the same relationship data, the diagram is usually made first as a best-fit concept. This is usually done progressively by working with the high-rated relationships until all are included. Then the diagram is redrawn for final best fit.

The *activity* relationship diagram is drawn with symbols only, independent of space. After space requirements are established, they can be added and the diagram redrawn as a *space* relationship diagram. Figure 3.4 shows an activity relationship diagram and the subsequent space relationship diagram. Note that the relationships have been drawn in the number-of-lines convention, according to SLP procedure. Figure 3.15 shows a similar conversion from activity relationship diagram to space relationship diagram, but for a detailed layout situation. Here the space relationship diagram becomes a rough arrangement of templates.

Space Requirements.   There are at least five ways to establish space requirements.

1. *Calculation.*Determine the amount of space required for each piece of machinery or equipment, including areas for workers, maintenance service, material setdown, and

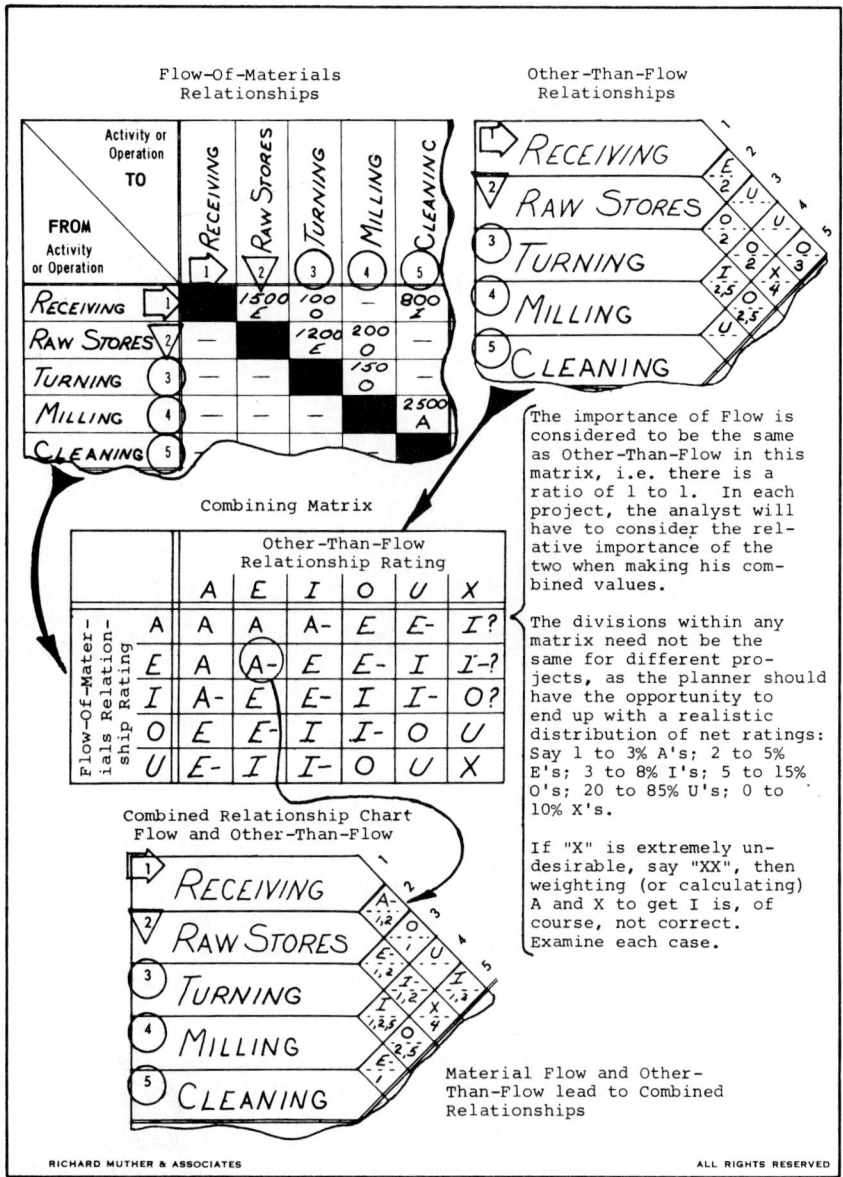

The importance of Flow is considered to be the same as Other-Than-Flow in this matrix, i.e. there is a ratio of 1 to 1. In each project, the analyst will have to consider the relative importance of the two when making his combined values.

The divisions within any matrix need not be the same for different projects, as the planner should have the opportunity to end up with a realistic distribution of net ratings: Say 1 to 3% A's; 2 to 5% E's; 3 to 8% I's; 5 to 15% O's; 20 to 85% U's; 0 to 10% X's.

If "X" is extremely undesirable, say "XX", then weighting (or calculating) A and X to get I is, of course, not correct. Examine each case.

Material Flow and Other-Than-Flow lead to Combined Relationships

FIGURE 3.12 Combining flow and other-than-flow relationships.

access to aisle; extend this by the number required for each piece of machinery, and add in space allowances for aisles and general or support areas.

2. *Conversion.*Determine the amount of space now used for each machine, machine group, or activity area; adjust this to what should be used to do the job efficiently at present; then convert this by some factor or multiplier to determine what will be needed for the new requirements.

3. *Rough layout.*Prepare a rough detailed layout plan to scale of a proposed or at least a

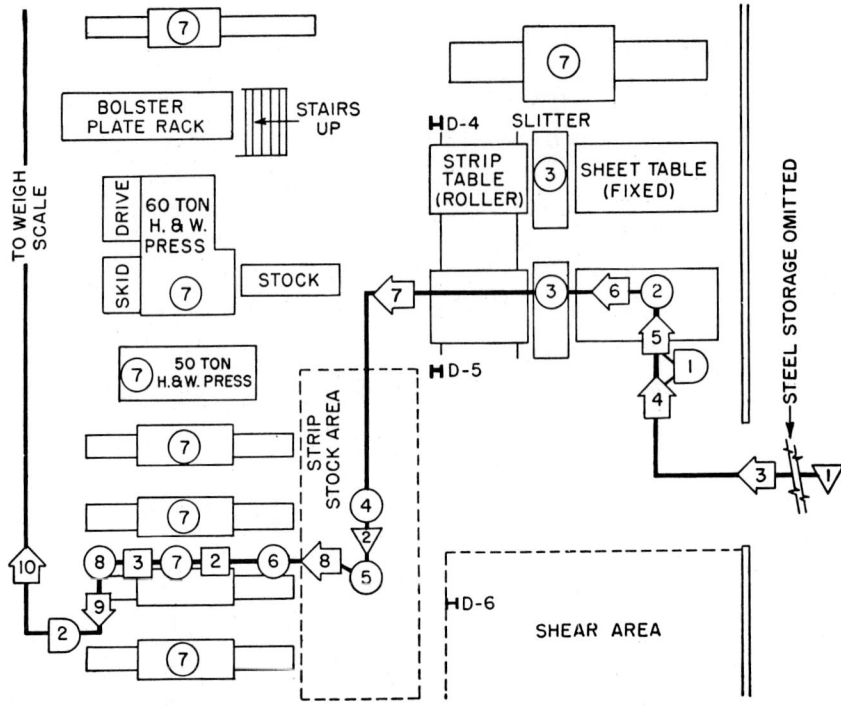

FIGURE 3.13 Portion of a detailed flow diagram showing what happens and where it happens to the sheet steel, strip stock, and laminations produced in a plant making small electrical transformers.

possible arrangement. It will in all likelihood not be the final, approved layout, but it will indicate approximate spacing between the equipment involved and will allow measuring the rough plan for total area requirements.

4. *Space standards.* In cases where certain types of areas are subject to repetitive layout planning, it is practical to develop standard amounts of space. This is particularly applicable for office areas or standard assembly bench layouts. But there is danger in using any standard if it is not understood.

5. *Ratio trend and projection.* There are a number of ratios that can be of value. From a plot of each ratio against time, a trend of that ratio is noted. This in turn can be projected into the future. Then, by knowing the projected ratio, the square feet required for any projected denominator can be calculated. For example, if 135 gross square feet per office employee is projected, 135 times a 5-year-plan figure of 100 office employees means that 13,500 square feet of office will be required to meet the five-year plan.

In practice, space requirements are not established quite this simply. In fact, several of the five methods may be employed on the same project. Moreover, space requirements must be balanced against space availability. Here, it may be most helpful to rate each of the activity areas on the relative importance of maintaining its space requirement. The same vowel-letter rating can be used effectively: A = absolutely necessary to honor the requirements, and so on. The areas rated O and U are squeezed the most when area requirements are reduced for a smaller available area. In industrial plants, these easily squeezed areas usually end up being storage, office, and flexible service areas, rather than production areas or fixed equipment areas.

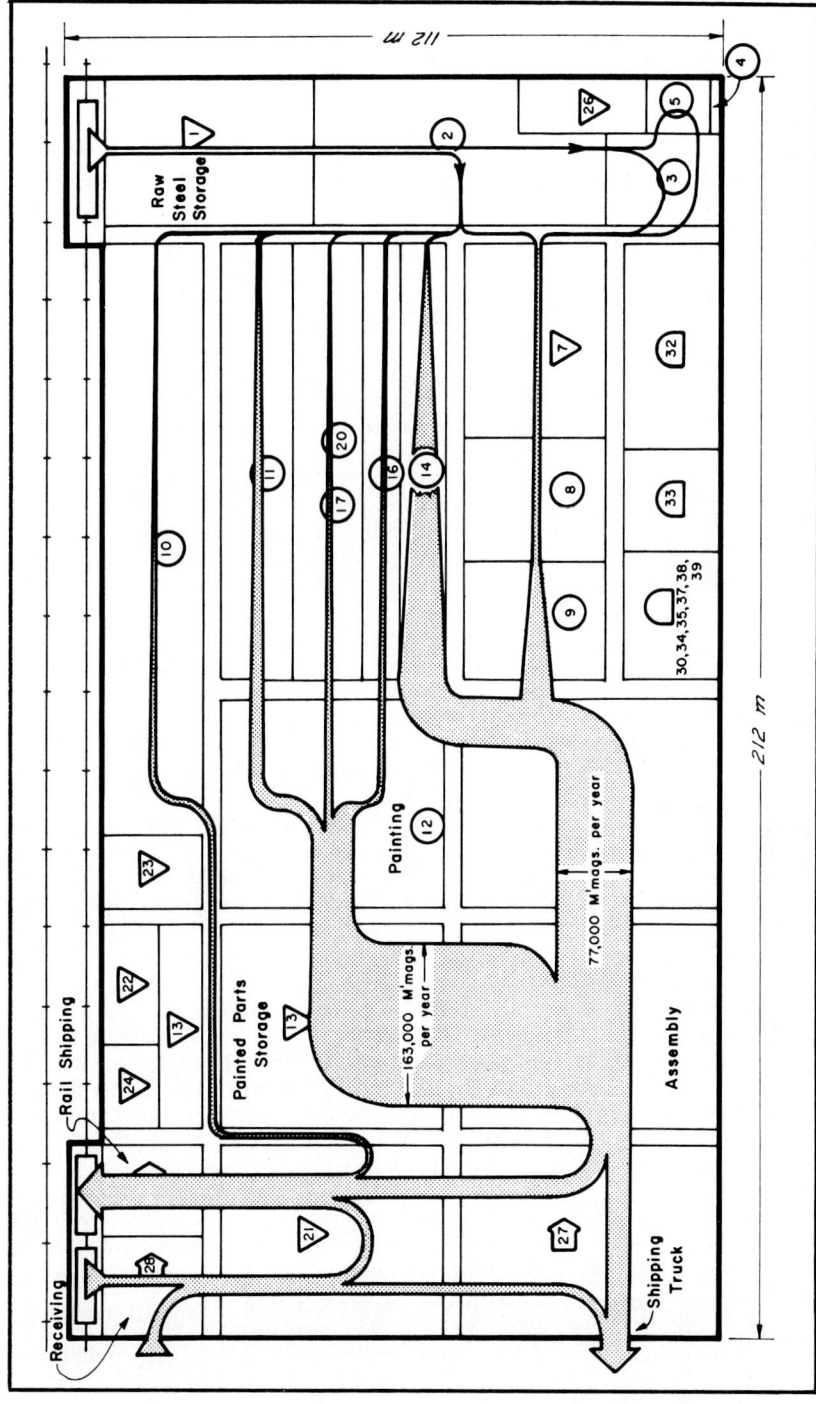

FIGURE 3.14 A flow diagram quantified by width of line indicating the intensity of material flow. The line widths increase as the product becomes more bulky and difficult to handle. This shows that it is most important, in order to reduce costs, to locate assembly close to painting and/or painted parts storage. This is a metal furniture factory.

13.57

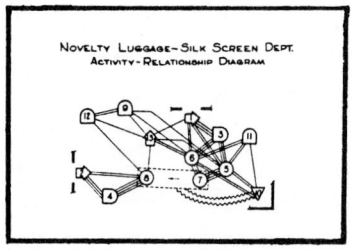
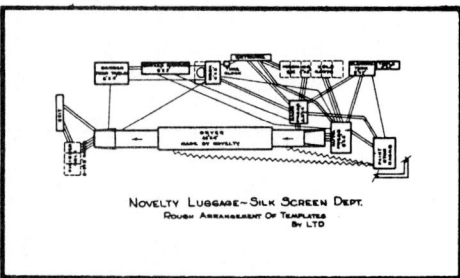

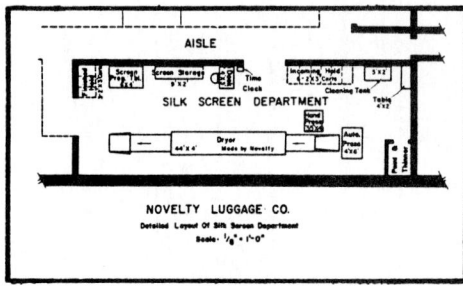

FIGURE 3.15 Conversion from activity relationship diagram to space relationship diagram for a detailed layout situation.

In any case, it is important to summarize the total planned space figures. Space comes in three basic forms: amount, kind or nature, and shape or configuration. Most experienced planners want to know, early in their projects, about all three of these aspects of the space with which they are working. As a result, the activities area and features sheet is recommended. Figure 3.16 shows how these data can be recorded.

Space Relationship Diagram. Working from the activity relationship diagram, each activity symbol is converted into its specific area. This may be done on cross-section paper at a convenient scale. Each activity will continue to be identified by symbol, number, and possibly name, but, in addition, it will be diagramed to scale and actual square feet will be shown. This way, the space is recorded both in actual numbers and in relative size.

Many refinements can be made in space relationship diagrams to show specific information pertinent to the project at hand—refinements involving existing buildings versus new construction, number of employees involved, need for space relief, or cost to relocate. These can all be coded into the diagram by use of colors, symbols, letters, and the like. In complex layout problems, it may be better to prepare the diagram on a reproducible print and, using several copies, show different significant information on each copy.

To get the full potential of planning the layout to the theoretically ideal conditions without the limitations of existing columns, walls, rail sidings, and the like, it is best to hang or superimpose the space as directly as possible right on the final redrawn activity relationship diagram. Later, when the space diagram is adjusted to the modifying considerations and their practical limitations, there will be ample opportunity to bring the theoretical diagram back into the constraints of existing buildings or other fixed features.

On the other hand, if it is known that certain fixed building features such as walls, columns, or floor loads definitely cannot be changed as part of the layout planning project, it may seem an unnecessary step to go all the way toward the fully ideal space diagram. So long as it is certain that there is no possibility of missing a real improvement, the fully ideal space diagram can be bypassed, but it should be remembered that many potential savings have been lost at this very shortcut.

Activity No.	Activity Name	Area in Sq. Ft. (Total 40,100, Aisle Incl.)	O'head Clearance Ft.	Max Overhead Supported Load	Max Floor Loading	Min Column Spacing	Water & Drains	Steam	Compressed Air	Foundations or Pits	Fire or Explosion Hazard	Special Ventilation	Special Electrification	Requirements for Shape or Configuration of Area (Space) and Reasons therefore
1	Receiving	450	15				-	-	I	-	-	-	-	Min. of 2 Dock Doors Width: 10 Ft. Min.
2	Material Std.	11,500	18				-	-	-	-	-	-	-	
3	Machining	3,000	15				-	-	E	-	-	-	-	
4	Wire Stringing	400	"				-	-	-	-	-	-	-	
5	Small Parts Subassembly	500	"				-	-	E	-	-	-	-	
6	Fluorescent Assembly	1,800	"				-	-	E	-	-	-	-	Long, Narrow Areas Suitable for Assembly Lines
7	Mercury Vapor Assembly	1,500	"				-	-	E	-	-	-	-	
8	Facade Light Assembly	2,100	"				-	-	E	-	-	-	-	Long, Narrow Areas Suitable for Assembly Lines
9	Finished Fixture Storage	3,500	18				-	-	-	-	-	-	-	
10	Pipe Receiving & Storage	2,800	"	3 Ton			-	-	-	-	-	-	-	Crane Way not Less than 30' Wide. 1 Truck Space-Underoot.
11	Pipe Bending	1,000	15				-	-	-	-	-	-	-	
12	Welding	5,500	"				-	-	I	-	E	E	A	
13	Paint	800	"				I	-	-	-	E	I	-	Must Accept Paint Dip Tank 30' Long.
14	Outside Pole Storage	-	-				-	-	-	-	-	-	-	
15	Shipping	450	15				-	-	-	-	-	-	-	Min. 2 Dock-High Doors- Must be conv. for loading poles to yd.
16	Tooling, Maint. & Test Lab.	700	12				O	-	A	-	-	-	O	
17	Employee Services	900	10				A	-	-	-	-	-	I	
18	General Offices	3,200	10				I	-	-	-	-	-	I	

RMA - 192-P COPYRIGHT 1963 - RICHARD MUTHER & ASSOC. K.C. MO.

FIGURE 3.16 Space specifications as recorded on an activities area and feature sheet.

Adjusting the Diagram. The space relationships diagram is adjusted and manipulated to create various possible arrangements. At this point, the operating and service managers should be brought back into the project, for there is now something for them to visualize. Furthermore, much of the adjustment must come as a result of the desires or practices of these people.

The space relationship diagram is almost a layout plan. It is not a very good one, in all likelihood, because the modifying considerations and the practical limitations—which hold within bounds the ideas for possible modifications—have not yet been incorporated into it.

There are many modifying considerations. Typical ones include:

1. Method of handling
2. Storage facilities
3. Personnel requirements
4. Location of main aisles
5. Utilities and auxiliaries
6. Procedures and controls

For each modifying consideration, a set of practical limitations must be weighed against it. This is a process of making many compromises. The objective is to get an arrangement of activities which will give the most practical overall combination of all considerations and limitations.

Typical limiting considerations include:

1. Project budget or costs
2. Floor loading and ceiling heights
3. Must-live-with existing features

4. Safety requirements

5. Personal preferences

By integrating the modifying considerations with the space relationship diagram and discarding all impractical ideas, the planner usually arrives at two to five alternative plans. Any of these can be made to work. The next problem is to decide which layout alternative to adopt.

Visualize the Layout.   Experienced layout planners know that a clear understanding of the plan they are making is the only way to get a sound layout. They have to visualize how the layout will look and how it is going to work. They also have to have some clear picture or reproduction of their layout to discuss with other people. They must have something others can see clearly.

The common methods of visualization are

1. Drawings and diagrams

2. Templates and layout boards

3. Three-dimensional models

4. Computer-aided design (CAD) screen and hard copy

Of these, drawings and diagrams are the most basic. They are readily made, easily altered, and inexpensive. In actually putting together a reproduction of a detailed layout, templates are the most valuable. There are many kinds, and they can be used in many ways. Essentially, they permit reproduction of as many different layout proposals as desired merely by rearranging the templates on the board. The same is true of CAD equipment, which goes much faster once the equipment templates are loaded.

To prove and check the layout plan or to help others visualize what is planned, the three-dimensional model comes into play. For these purposes, the model is supreme. But the great misunderstanding is that some people think models themselves are a substitute for layout planning; they are not. They glamorize the layout job; they help sell the layouts proposed; they develop interest; they help in training workers, supervisors, and staff personnel; they show suppliers of equipment what is wanted from them; they indicate clearness and prospective interferences in complex three-dimensional situations; and they act as a check on the planning of the layout planner. But, for their own visualization or planning, on most layout projects, layout engineers do not need them.

The most common pitfall into which the inexperienced layout planner falls is jumping into the use of these visualizations too early in the project, before the necessary facts to evaluate the various layout proposals have been analyzed.

The purpose of visualization is to assist in developing a sound layout. After getting information about the various input data involved, determining and diagraming the relationships, and conceiving the various ways to arrange these physical features, visualize them physically to see what they actually look like. Reproduce a likeness of each arrangement to see if it is as good as it first appeared. Adjust and change the arrangements; this is the time to move equipment without cost. These adjustments—still in the paper stage—lead to the arrangement that gives the best layout compromise. Planners use a physical means of visualizing each logical improvement to check their thinking; they do not start right in with sketching or templates and merely cut and try until, by trial and error, they come up with a layout that looks as if it will work.

The device of visualizing layouts should be readily changeable so it can record quickly every idea and proposed plan. In addition, the layout planner should make provision to keep track of ideas as they develop. Otherwise, changes may be made without capturing the suggestions. Some layout planners periodically supply each interested department head and management person with copies of their layout proposals as these are worked out. Copies do not cost much compared with the cooperation and valuable suggestions they stimulate. Caution: date every working copy.

Actually, once drawings, templates, hard copies, or models have been made up, they should be kept on hand if the plant anticipates any layout changes in the future. They should be properly identified and filed away, ready for use again. This is all part of good plant layout practice.

When planning is finished and the detailed layout is installed, be sure to bring the permanent layout record up to date. Keep the existing layout record untouched while planning a new layout. If no other record is available, changing the existing plan will destroy the record. This is why most experienced layout planners have a layout board, model, or hard copy that acts as the current record. They do all new layout planning on other copies which they can easily move about and change as necessary.

A scale model has to be kept horizontal. Some template layout sheets are also safer if kept flat. Most drawings or hard copies can be supported on the wall, which takes less floor space. But whether flat or on a wall, if the unit is large, it should be made in sections. Each section should be small enough to remove and work on conveniently and transport to the reproduction room or the president's office, if necessary.

Evaluate Alternative Layouts.   The best layout is always a compromise of the various factors, considerations, layout objectives, and types of layout. To select the best compromise, plan alternative layout proposals and eliminate those, or the portions of them, that compare unfavorably. Evaluating alternative plans should determine which proposal offers the best layout.

Various techniques of evaluation have been used successfully. Here are several:

1. Ranking based on selected considerations
2. Tally of gains and losses expected
3. Value rating of pros and cons
4. Rating of alternatives versus objectives
5. Rating alternatives versus total distance materials are moved
6. Rating alternatives for extent of honoring relationships
7. Audit of alternatives against established check questions

Perhaps the most frequently used evaluations, however, are the following:

1. *List the advantages and disadvantages.* This is the simplest way to evaluate alternatives: merely write down the advantages and disadvantages of each layout being evaluated. It is surprising how often such a listing quickly clarifies which alternative should be selected when the layout is not complex or costly.
2. *Factor analysis.* This method selects the factors or considerations on which the decision will be made. Each factor is given a weight value according to its importance (10, 9, 8, . . .). All the alternatives are then rated on one factor at a time. The rating (if the SLP vowel-letter ratings are used) is converted to a number and multiplied by the weight value. The weighted ratings are totaled for each alternative, and a numerical comparison is made. This increases the objectiveness of what can be a very subjective decision-making process. Moreover, it offers an excellent way of involving management in the selection and weighting of the factors, and the operating and support supervisors in rating the alternatives on each factor.
3. *Cost justification.* In important projects, costs will nearly always become a basis for selecting the best alternative. This means everything that goes into the cost of installation and operation. In establishing costs, the layout planner should consider the following list and charge against the installation every one that should be included. Moreover the planner should make a comparison or justification in accordance with methods

of cost analysis approved by the company's accounting or financial officials. Costs to be considered include:

a. Investment:
   (1) Initial cost of new facilities of all kinds: handling equipment, building changes, and so forth
   (2) Accessory equipment costs
   (3) Installation or occupancy costs
   (4) Depreciation, obsolescence, fees, permits, rentals, and so forth
b. Operating costs:
   (1) Material
   (2) Labor
   (3) General and burden

No matter how many layouts are investigated, none of them will have everything. There must be a tradeoff somewhere to get a practical solution. As a result, develop from the theoretical layout two or three practical solutions. Evaluate these, pick one that looks good, then develop its details. Otherwise, too much time may be spent debating which solution is best, and insufficient time will be left for developing the details.

Engineers are notorious for wanting to weight meticulously every scrap of fact or influencing detail. Be accurate and sound, but do not deliberate so long that the next phase of the layout is held up.

INSTALLING THE LAYOUT

Installing the layout is the fourth phase of layout-planning work. It follows location of area to be laid out, general overall layout, and detailed layout plan. The person planning the layout is sometimes responsible for seeing that the layout is properly installed. More often, that person is an advisor or coordinator, and the installation work rests with the plant engineer or maintenance department. As a very minimum, the layout planner is called on to supply the details of what the new installation should be like.

Information needed for layout installation usually includes:

A list of all new machinery and equipment to be installed or existing equipment to be moved or changed in location

A layout print, drawing, hard copy, or photograph explaining details of new locations

A schedule of moves

A specification sheet to show how each machine should be disconnected, moved, and hooked up

Figure 3.17 illustrates the practice used by one automotive supplier plant where almost every weekend is moving day. Because interference with production schedules cannot be tolerated, all moves must be carefully planned and scheduled. The procedure shown here makes sure that floor spaces are empty before trying to put machines into them.

1. The proposed new layout is shown in black on the diagram delivered to the plant engineer. The plant engineer marks existing locations (crosshatched) of machines to be relocated and shows route to follow and any intermediate locations (broken crosshatch).

2. A list of moves is issued by the plant superintendent. It tells what is to be done, but not how.

3. A moving order tells how and in what sequence moves shown on the proposed layout are to be made.

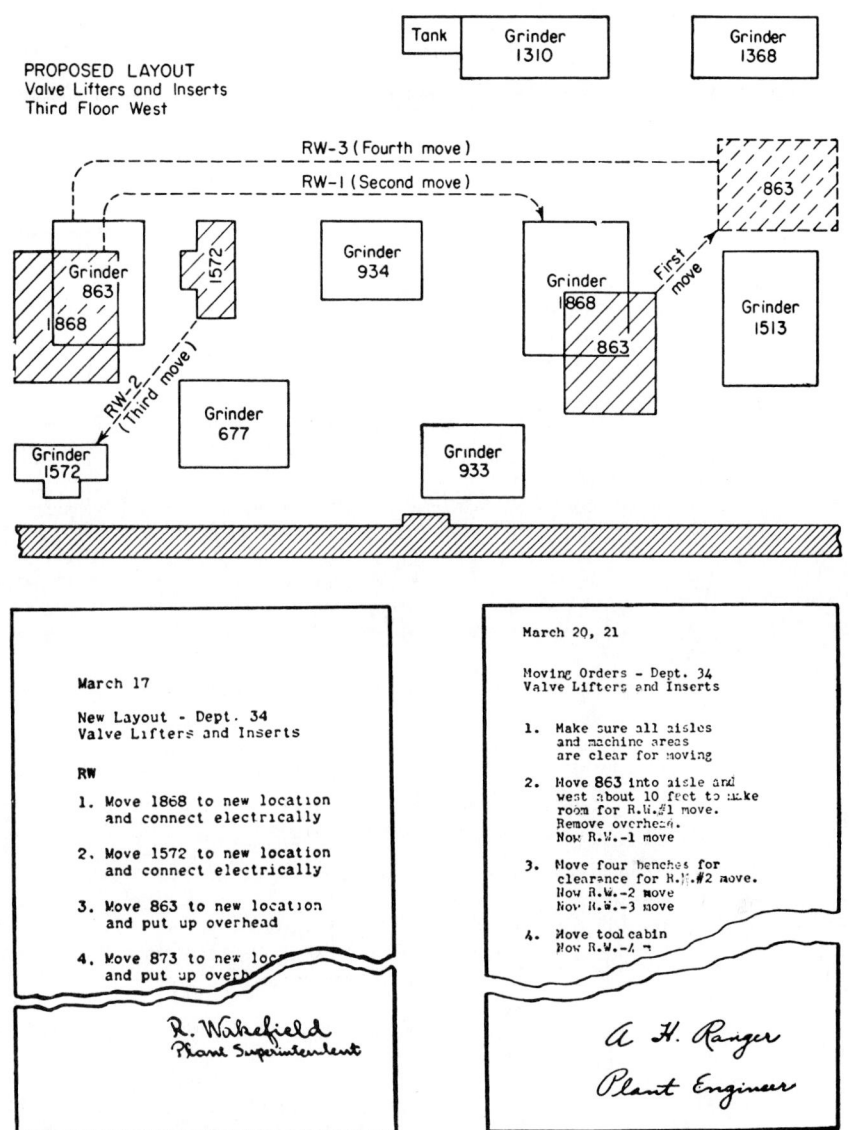

FIGURE 3.17 Method of planning small layout changes used by an automobile engine plant.

R. H. McCarthy of Western Electric Company recommended seven stages in layout installation: plan, provide, prepare, move, install, start up, and clean up. These seven might include the following advice:

1. Plan:
   a. Start planning early. Sound planning saves time during the fast-action stages.
   b. Determine the sequence of moves, weighing practical operating problems.
   c. Make up an inventory of everything to be relocated. Get the disposition of the balance.
   d. Schedule the moves in detail. Set up a timetable with specific dates and times.
   e. Assign a move number to each item, mark it on the inventory sheet, and check it against the machine tag.

2. Provide:
   a. Consider the use of outside moving and installation contractors and get bids from several.
   b. Have adequate help. Set up key persons in each department or area involved.
   c. Get ample moving equipment. Consider renting equipment to assist moving or to keep operations running during the installation.
   d. Provide good communications. Have telephones and lead persons at both ends of the move.
3. Prepare:
   a. Prepare new location. Foundations, partitions, cleaning, painting, and auxiliary service lines should be ready.
   b. Broadcast the plans. Let everyone know what is going on and take advantage of new ideas or suggestions.
   c. Tag every item to be moved. Use color and coding, and mark identification, move date, and destination.
   d. Notify employees what to do and when, where, and how to do it.
   e. Get equipment ready to move, and check it out before releasing it to movers.
4. Move:
   a. Keep the move on schedule. Post accomplishments each day.
   b. Move equipment intact. Try to keep equipment together to reduce reassembling time before it can operate.
   c. Move as close to the installation point as practical to reduce handling time by skilled installation crews.
   d. Keep movers coordinated by notification and frequent briefing.
5. Install:
   a. Expect to have last-minute changes. Do not get upset if the plan does not work perfectly, for it never will.
   b. Make use of temporary hookups, with permanent service connections to come later.
   c. Flag equipment ready for installation inspection, and have installation crews post accomplishments daily.
6. Start up:
   a. Check the installation. Be sure the placement and hookup are right.
   b. Release equipment for maintenance tryout and for supervisor's acceptance.
7. Clean up:
   a. Inspect installation, and note any loose ends.
   b. Set deadline for cleanup. Otherwise the installation remains temporary and production output suffers from the same attitude.

Who Does the Installation? Most firms do their own layout installation work. The following reasons indicate why:

- The cost is likely to be less when the company has its own maintenance crew.
- The maintenance personnel become familiar with the installation when it is put in and therefore find it easier to maintain and repair.
- There is less need for elaborate paperwork on contracts, prints, specifications, installation drawings, and the like.
- Where speed and time are important in hurried changeovers, it is frequently not practical to wait for an outside contractor.
- By having the company's own people do the installation work, the presence of maintenance personnel is subsequently assured in case of emergency.

There are, however, many advantages to employing outside contractors to make the installation:

- Contractors are often highly skilled and familiar with layout installation work and techniques; they will have the proper equipment available and can do a safe, efficient job.

- Frequently, the company does not have a large enough plant engineering staff to handle the infrequent task of layout installation.
- For new layouts, there may be no company staff at all.
- A company's own construction and maintenance crews will have a lot of other details to attend to during a re-layout, and they do not have time to handle the installation itself.

Where outside contractors are employed, it is practical to have one or more company people work closely with them. Be sure to specify for both groups what details of the installation each group will do. Thus, if there is an omission, one group will not assume that the other is handling it.

Other significant points in installation are

1. *Condition employees for the change.* The layout installation is a disrupting time for employees. Give employees the details about the new layout in writing and with diagrams. Talk enthusiastically and ask for comments. Brief them.
2. *Basic re-layout problem.* Making a re-layout is like a game of checkers—one move is made into the spot presently occupied by some other piece of equipment. Here the sequence of all moves must be planned so the movers do not try to move something into space already occupied. This problem, together with that of keeping up production during the change, is frequently a major obstacle limiting the design of the layout itself.
3. *When to install.* Although the time to install the layout is important, it is generally a matter of selecting the least inconvenient time rather than finding one that is fully acceptable to everyone. Frequently preferred times include:
   a. Times of annual changes in product design
   b. Time of plant shutdown for vacation
   c. Slack season
   d. Weekends or extended weekends
4. *Identify locating points.* Before beginning to move anything, experienced layout planners get their major aisles marked in. Otherwise, installation crews will set down equipment temporarily in the aisles, and by clogging them, make it necessary to move through areas where equipment is being installed. This leads to much shuffling. Also, columns should be identified before the installation if they are not already marked. Most equipment is going to be located from these aisle lines and columns. Whenever floor space is clear, the exact location of major equipment in the new layout should be marked on the floor.
5. *Coordinating the installation.* When a major move is involved and plans for its execution are completed, conduct a conference of the head of each function concerned. Following this meeting, advance written notification of the move schedule should be posted, again allowing sufficient time for possible conflicts to appear. When all are ready, the actual moves are usually initiated by a work order or an equipment move notice, together with an accompanying list of equipment.
   Perhaps the easiest way to schedule and control the installation of a new layout is the Gantt chart. This shows on the same sheet both the plan and the accomplishment to date (Fig. 3.18). For installations with a large number of tasks, those covering a very tight period of time, or those involving many suppliers or several contractors, a critical-path network diagram and schedule is more helpful. With either technique, the duration of each task should be estimated by those responsible for its completion. This promotes realistic estimates and a commitment to them.
6. *Layout planner should be on hand.* Layout planners should be ready to make any necessary changes during the installation. No matter how good the layout engineering has been, there will be adjustments at installation time. The planner should expect such problems and anticipate any necessary changes.
7. *The follow-up check.* There will always be bugs, regardless of how thorough the planning has been. The layout planner should check the actual layout—as installed and operating—against the approved plan. The planner should recognize any differences and

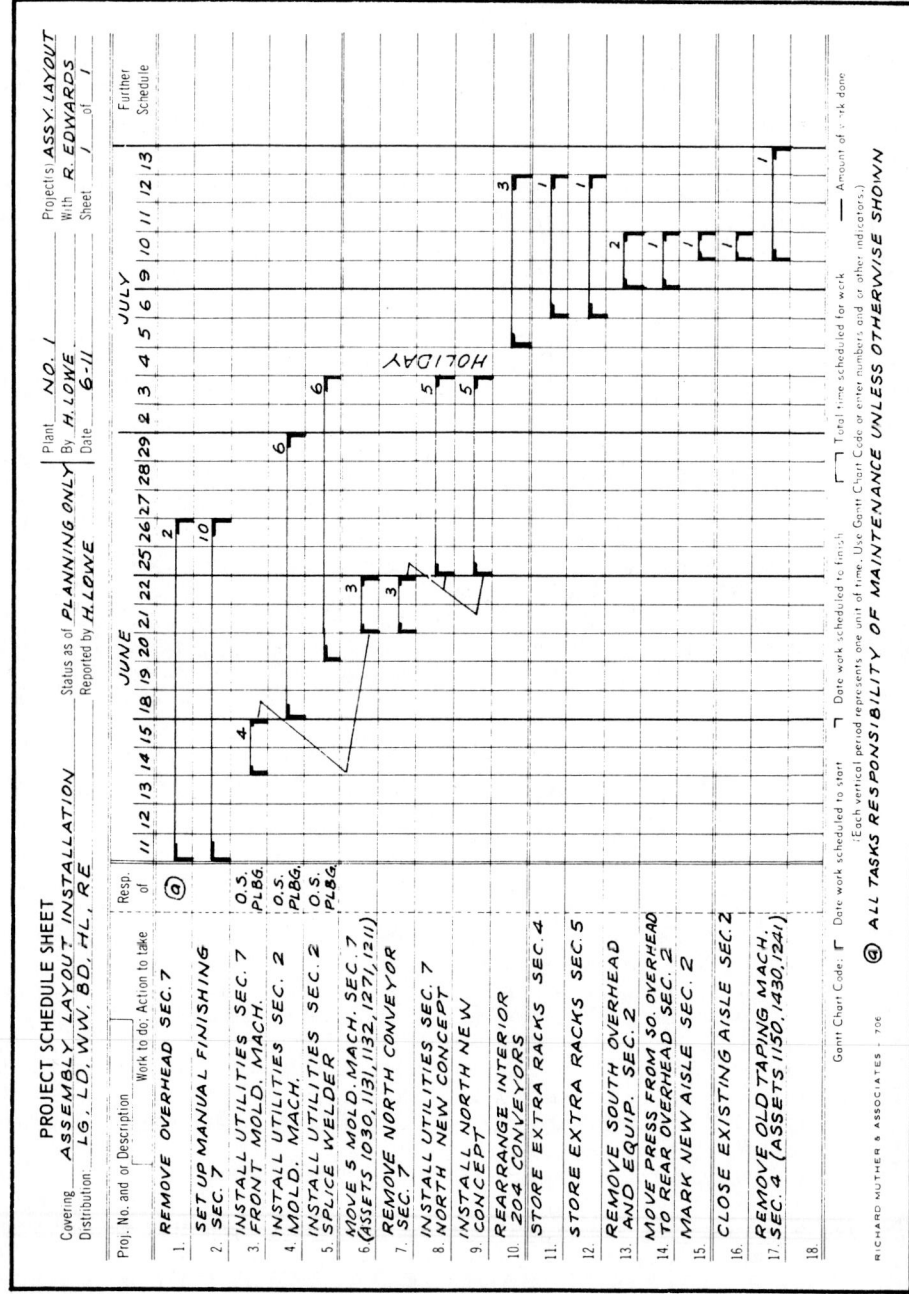

FIGURE 3.18 Gantt chart used to plan and schedule rearrangement of an assembly department. It shows what is to be done (down left side), who is responsible (in the next column), and the earliest date to start and latest date to complete (by brackets with connecting "duration" line). Note workdays for each task above completion-date bracket and zigzag lines indicating that precedence is required. These features can be added to the conventional Gantt chart to give more meaning to the schedule. As work progresses, the percent completion of each task's bracketed time is filled in, and a V is placed on the date scale to indicate the date of the latest status posting.

either accept them as satisfactory or make arrangements with the installation crews to reset equipment as called for.

8. *The layout record.* Even if for no other reason, a check of the installed layout is needed to bring the layout records up to date. Only in this way will a plan of the existing layout be ready and available for future reference.

OFFICE LAYOUTS

A simplified version of systematic layout planning is particularly adaptable to office and laboratory layouts. Figure 3.19 shows an example of simplified SLP applied to a small office layout. Although the simplified version is especially suited to small areas not having a dominant flow of materials, many analysts have found it valuable for large office layouts. They can "subcontract" smaller, less-complicated areas to the area supervisors by training them with the booklet, *Simplified Systematic Layout Planning.** The supervisors follow a simple, six-step procedure to lay out their own areas. Tremendous time savings can be achieved on involved projects for the less vital areas—and without default on the part of the total project leader.

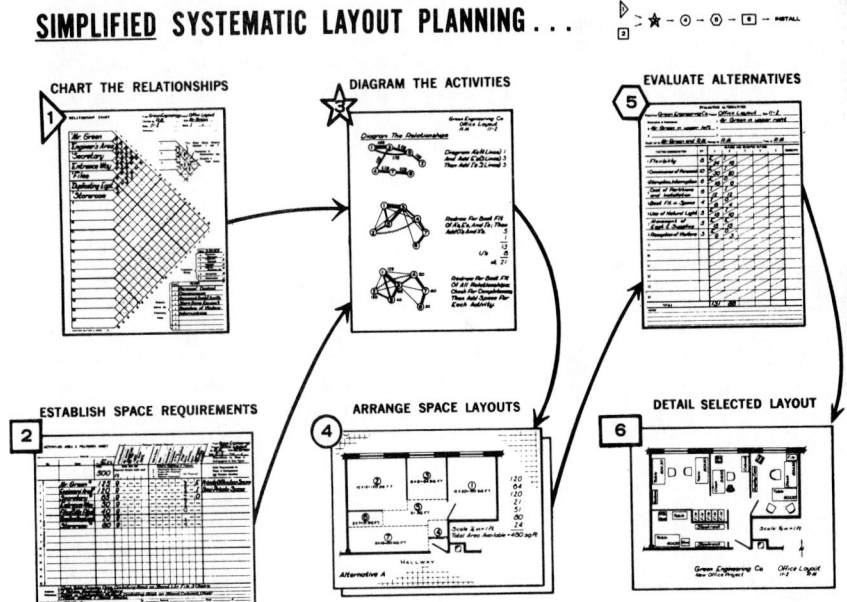

FIGURE 3.19   The six steps of simplified systematic layout planning. (*Source: R. Muther and J. D. Wheeler, "Simplified Systematic Layout Planning," Management and Industrial Research Publications, Kansas City, Mo.*)

The six-step pattern is shown in the center of the capsule summary in Fig. 3.2. The six steps are

1. Chart the relationships
2. Establish space requirements

---

*R. Muther and J. D. Wheeler, *Simplified Systematic Layout Planning,* Management and Industrial Research Publications, Kansas City, Mo.

3. Diagram activity relationships
4. Draw space relationship layouts
5. Evaluate alternative arrangements
6. Detail the selected layout plan

Although simplified SLP rests on the fundamentals of relationships, space, and adjustment, it is applicable to only a limited number of activities. As a result, it is used frequently in Phase III detailed layouts, or where the area involved is generally not more than 5000 to 8000 square feet.

COMPUTER-AIDED LAYOUT PLANNING

Computers in their various types and many models are very helpful in layout planning. They are most used in the following activities:

1. Processing input data and transferring data to graphic plots, graphs, diagrams, and tables
2. Calculating values of certain measurements for alternative proposed arrangements and scoring the relative worth of each
3. Visualizing quick alternative arrangements on display screens, by layout algorithms for block layouts or by interactive placing for both block layouts and individual equipment arrangements
4. Integrating various elements (utility distribution, building features, material flow intensities, telephone lines, and so forth) with the base plan layout by means of CAD overlays or layers
5. Preparing and changing layout drawings for review, approval, installation and as-installed records
6. Simulating alternative proposals and comparing them mathematically

Processing Support Data. Almost all layout projects of any size today employ some form of computer to organize, calculate, model, sort, rank and then print out, plot, or graph the information in more usable, readable, or interpretable form. Usually, this involves so-called small computers: personal and minicomputers. And there are a myriad of commercially available programs to support them. Often these are termed *decision support systems* or *computer-aided analysis systems*.

Typically, the following outputs of value to layout planners can be developed with these systems:

1. Sorting of product forecast requirements
2. Plotting of product-quantity curves (volume-variety analysis or product mix determinations)
3. Calculation of flow-of-material intensities by item from each origin to each destination, such as a printout of a from-to chart
4. Development of space requirements—historically, by activity area, class of space, floor load or ceiling height, current expansion need, or anticipated future growth
5. Ratio-trend plots with various calculated methods of projection, distance-intensity plots, communications matrixes, pairings or relationships by rating value, flow-of-materials arrays, department/machine adjacency tallies, and so forth
6. Sorting of questionnaire responses and grouping by machine loading

7. Space utilization reports, equipment affinity maps, and so forth

8. Alternative block layouts and their related space summaries

Layout Planning Algorithms.   The interest in mathematical algorithms for layout planning began in 1964 with introduction of the CRAFT routine. Since then, over 50 programs have been developed for block layouts. But their merits over sound SLP application have yet to be substantiated despite the seemingly convincing logic of each. Perhaps this is because, of necessity, many algorithms have limitations in their application which are not clearly pronounced beforehand for users, who then lose patience when they cannot make the programs work. Another fact often not appreciated is that algorithms do not produce an optimum or clearly best layout. And almost all require some graphical application to be done manually. Still, progressive planners should keep in touch with these developments, for sounder outputs are bound to come along.

Algorithms typically fall into two types:

1. Construction routines, which start with a relationship analysis and space determinations, and generate a theoretically best layout. CORELAP and ALDEP are the best known.

2. Improvement routines, which start with an existing or proposed layout and make changes of activity areas that reduce movement cost or improve net desired closeness of all relationships. Here the best known are perhaps CRAFT and COFAD.

CAD—Computer-Aided Design.   So-called CAD systems are really methods of, or aids for, visualizing layouts and their related elements and for manipulating templates or space replicas, quickly; reproducing drawings or changes in them more readily; and storing data, drawings, and documents. CAD is a big saver of time, particularly in Phase III, detail layout planning, and in preparing and maintaining installation documents and inventories or measurements related to the phase.

CAD provides the ability to create and manipulate drawings and diagrams on a screen, to interact personally with these graphics, and to make a hard copy almost immediately. And it can do so with equipment of reasonable cost and with readily available software.

The versatility of CAD equipment allows all manner of applications. CAD outputs can generally be classed as follows:

1. Working copies of the display screen image, scaled to the most appropriate working size

2. Finished diagrams or layout drawings, enlarged and reproduced at the appropriate size

3. Aids for presentation or installation, such as slides or transparencies; these are particularly helpful for three-dimensional images with color or shading identifications of various features

4. Documents or reports including charts, tables, and inventories of space, people, and classes of space, often with graphical records of various database information

As with any equipment, someone has to provide the dimensions and specifications and/or draw a master from which copies can be made. CAD is no exception here. But once such an inventory or menu is made, most CAD equipment can produce it almost immediately on the display screen; the user simply identifies what should be shown (usually by coded index) and where it is to be placed. Thus, one can very quickly call up the basic floor outline; lay onto it the major building features (such as columns, major openings, fixed walls, main aisles, and stairwells); and spot in the production and support equipment on the basis of previously CAD-produced process space flow, relationship diagrams, and equipment inventory. These are adjusted to best fit and may be checked for comparison as to prospective performance. Layers or overlaps of electrical distribution; underfloor sewer lines; heating, ventilating, and air conditioning; telephones; and material-handling equipment can all be tied to a base layout plan (Fig. 3.20).

Obviously the planner has to spend some time learning to become proficient with the equipment. Because of this, many planners find that simply overlaying onto a basic floor plan or building outline a must-be-located "glob" of space, to see what it would look like if placed there, is a practical first use of CAD. For companies having changing contracts for different quantities, just finding a place to do the work often becomes more important than planning it into a logical sequence of material flow. This is especially so if no long-term strategic plan has been established for the building or the site as a whole.

Simulation.   In planning, simulation is a technique that uses a computer to compare relative plans. The analyst proposes or speculates a change in conditions, converts it into a mathematical representation or replica, and computes against a base plan (existing or proposed) the relative benefits or losses of the change proposed. For example, one might test mathematically simulated alternative proposals for the arrangement of machinery and equipment or for a particular product mix.

Simulation modeling can be used in every phase of layout planning: to test candidate locations, to evaluate various layout concepts, to compare various options for layout of equipment and services, and to develop the most-preferred installation schedule. The most costly and complex the layout plans, the more value there is in reducing, with simulation in planning, the likelihood of risk in the installed operations.

WHO LAYS OUT THE PLANT?

To find out who does industry's layout planning and to identify the problems that plague planners, the American Material Handling Society (now MHMS) conducted a 3-year survey of American and European firms.

Perhaps the most notable fact uncovered is that plant layout engineering is indeed a staff function. More than half of the people who do this work are staff specialists.

Organizationally speaking, the layout function is evenly split between industrial and plant engineering. However, as might be expected, there are many other groups or individuals who have on occasion been given the responsibility for this function. And more recently, planning teams of regular employees are popular, especially for re-layout or expansion-and-rearrangement projects.

Managers were most critical of layout planners because:

1. They take too much time
2. They are not conscious enough of costs
3. They do not have the ability to do planning

Planners, on the other hand, were most critical of managers because:

1. They do not plan ahead
2. They do not share business plans with facilities planners
3. They do not give the layout planners enough time

MANAGEMENT INTEREST IN PLANT LAYOUT

Most managers realize that processes and methods are constantly changing and that product improvement is essential to a plant's existence. These conditions lead to continuous

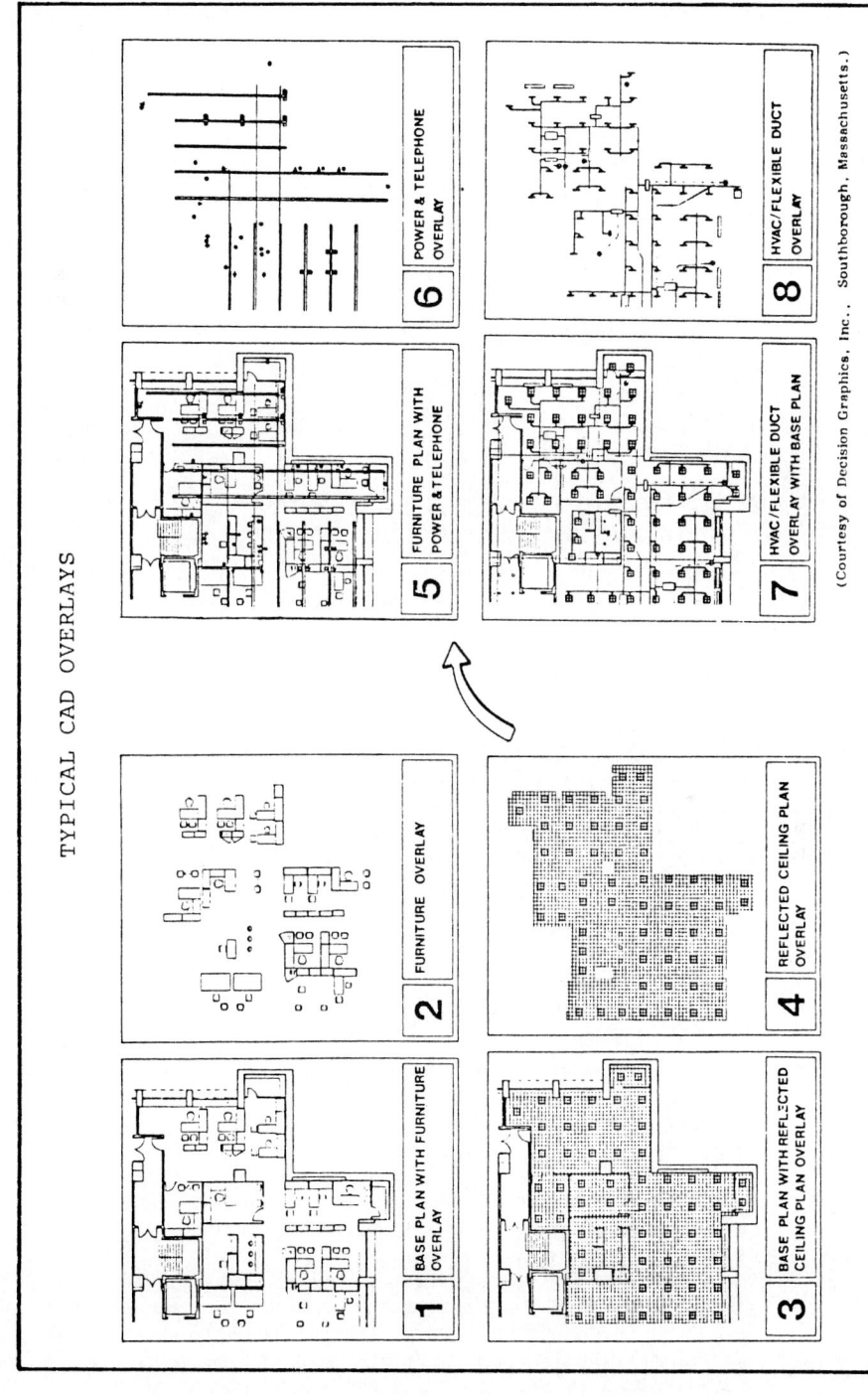

TYPICAL CAD OVERLAYS

(Courtesy of Decision Graphics, Inc., Southborough, Massachusetts.)

FIGURE 3.20 CAD equipment give the planner visual capability, especially with color on each overlay. Most CAD equipment pays off in ease of making changes or improvements and reproducing drawings.

projects in re-layout and in adjustments to existing layouts. Thus, a layout planned and installed is not necessarily completed and can be considered complete only when it is entirely replaced by a new layout.

When Is a New Layout Needed?   Because plant layout is a compromise of many factors, there will always be something about each layout that is imperfect. For this reason, one can criticize something in every layout. The most logical time for a layout change is when changing products, quantities, and process equipment. The more frequent or more extensive these changes, the more layout planning activity there will be.

Because changes in other things can be made at the time of a re-layout, management should note three further points:

1. A request for layout change may be made to permit a change to the just-in-time method of material control or for a less visible change in organization or production control.
2. A layout change gives management the opportunity to make other changes it may have been holding back.
3. A suggested change in layout may not always solve the real problem; a bad situation may be the result of several factors of which layout is but a part.

MANAGEMENT SUPERVISION OF LAYOUT PROJECTS

Management should conduct its assignment, supervision, and execution of the layout projects in ways that will tend to ensure success.

1. *Appoint a group or individual to head up the layout project.* Fix the responsibility in someone who knows layout problems and techniques. Other alternatives include:
   a. Assign the project to a team of various people closest to the areas involved
   b. Employ an outside specialist
   c. Develop layout engineers from within the firm
   d. Plan visits to other companies
2. *Supervise the layout work as necessary*
   a. Follow the guiding fundamentals of good layout
   b. Make a clear statement of the project
   c. Schedule the project
   d. Plan and check progress as each phase of the layout work nears completion
      Management should review progress in these four areas:

      Review Phase I (location): When the area to be laid out has been selected.

      Review Phase II (overall layout): When approval is requested for the general overall layout plan.

      Review Phase III (detailed layout plan): When approval is requested for the detailed layout plans.

      Review Phase IV (installation): When the layout is installed and ready to release to production personnel.

3. *Balance forcing accomplishment with getting good results.* Managers will lose out in the long run if they press their layout engineers for some visible plan too quickly. At the same time, endless refinement is not wanted. All too often, meticulous perfection of details is the chief cause of delay.
4. *Plan for the future.* Management must consider the long-range plans of the business and question whether the layout suggested will be as good at the halfway point in its life as it looks at the beginning. In developing layout plans, the wise manager will call for each layout to be integrated with a larger, longer-range plan for the facilities involved.

Indeed, no layout should be approved without consideration of its fit with the larger surrounding plan. Managers may be interested to know that one of the most difficult problems faced by a layout planner is to get management to look far enough ahead when planning facilities.

5. *Approve the layout.* Executives or department heads who are asked to approve layout plans normally consider three questions:

   a. What will be gained from this layout?

   b. What are the risks in this layout?

   c. How does this affect me personally and my group?

   The manager will certainly want to see analyses of the costs required and of the savings expected. The manager should watch for hidden costs in terms of increased maintenance, greater demands on service equipment, and the like. Four other factors should be checked: safety, convenience for personnel, conformance to laws and regulations, and changes in organization structure or procedures.

   There are two ways for management to be sure that all groups agree on the layout:

   Hold a meeting of all personnel involved to go over the layout plans together.

   Make sure all department heads, or those on a predetermined list, have initialed the layout plan.

Besides a good layout being more or less guaranteed by these methods, there is the important factor of group participation. In the eyes of department heads and supervisors, the layout engineer represents change. And with every change comes fear—fear that the new will mean less importance in the organization, more difficult operation of the department, greater dependence on others, or more pressure and rush service. Much of this can be avoided by building from the start the idea that the layout is a participative company project requiring their support and involvement.

6. *Build morale with a new layout.* By encouraging and developing cooperative activity in the layout project, a manager is building up the team. Be sure operating and service heads are included, and let employees in on the plans. Besides these internal activities to build employee enthusiasm, favorable publicity outside the plant can help swell employees' pride in "their" new layout.

BIBLIOGRAPHY

Hales, H. L., *Computerized Facilities Planning,* Industrial Engineering and Management, Norcross, Ga., 1985.

Holland, J. R., *Flexible Manufacturing Systems,* Society of Manufacturing Engineers, Dearborn, Mich., 1984.

Muther, R., *Systematic Layout Planning (SLP),* 2d ed., Management and Industrial Research, Kansas City, Mo., 1973.

Muther, R., and Hales, H. L., *Systematic Planning of Industrial Facilities—SPIF,* Management and Industrial Research, Kansas City, Mo., two volumes, 1979 and 1980.

Tompkins, J. A., and White, J. A., *Facilities Planning,* Wiley, New York, 1984.

CHAPTER 4
MATERIALS HANDLING

E. Ralph Sims, Jr., P.E.
Associate Professor
Industrial and Systems Engineering
Ohio University
Athens, Ohio
Chairman
The Sims Consulting Group, Inc.
Lancaster, Ohio

Materials-handling problems are business management problems. However, they normally require engineering solutions. The materials and goods which flow from the farm and mine to the factory, and on through the physical distribution system to the consumer, represent the flow of money through the enterprise and the economy. Every box, piece, cubic foot, or pound of product or material represents labor, material, and overhead dollars. *The movement of goods through the enterprise and the economy is the physical manifestation of cash flow.*

Likewise, every pallet, box, or pile of inventory, in storage or on hold, represents capital locked up in the business and unavailable for payroll, accounts payable, or investment. The economic impact of material flow varies with the *dollar density,* or value per unit, of the material. High-value merchandise has a greater financial impact on the enterprise, and "high-cube" goods have a greater impact on space and materials-handling equipment requirements.

Management of Materials Flow Is Management of the Business. Handling system design first requires a very complete analysis of the business enterprise and its management. Management's objectives, policies, and practices represent the operation parameters of the logistics system. The firm's product line and marketing pattern determine the material flow pattern. The product line's uniformity or diversity of handling characteristics, the transaction rate, the volume of movement, and the inventory profile at each stage of the operation define the physical parameters of the business and the materials-handling system. Materials-handling solutions require more than just equipment. They also require a logical and effective management system.

A Well-Engineered System Which Supports a Bad Management Situation Can Result in an Economic Disaster. If materials must be handled, the engineer must determine where, why, and how far they must be handled. The analyst must ask if the materials-handling work load and system characteristics can be altered by changing marketing policy, packaging design, quantity price breaks, unitization, or product design. One must also ask whether the process or layout can be changed and the moves simplified or eliminated. Can changes in materials management policy eliminate materials or reduce the volume of the movement? In short, since "the best materials handling is no handling," the *business* of materials flow must be dealt with before the *technology* or tooling of the materials flow is considered.

Materials Handling Is Time *and Time Is* Cost. The only "material" of production which earns money when it is idle is *money*. It earns interest when "stored" in a bank or other investment. All other materials in flow through the system, or in storage awaiting action and disposition, generate *cost*. And the generation of cost is directly related to *time*.

Materials in flow and in storage consume time, and at the very least, they generate cost because of the interest on the money they represent. In addition they require space, equipment, handling labor, insurance, and in many instances, incur damage and obsolescence losses. All of these expense factors are time-related. Therefore, since reduction of inventory and acceleration of movement will reduce the total amount of materials and goods residing in the system, these actions will reduce operating costs by:

- Increasing material turnover and return on inventory investment
- Reducing the dollar lockup in the system
- Reducing interest expense
- Increasing cash flow

This is the essence of the just-in-time concept of operations management.

Materials Handling Is an All-Cost Activity. It adds nothing to the usefulness, customer appeal, or salability of a product. It increases manufacturing and distribution cost without changing the product's function or appearance.

There are those who speak of time and place utility value as a reason for spending money on materials movement. However, this is a philosophy which is subject to argument. Time and place utility often reduces cost, but it is difficult to logically believe that it is adding any intrinsic value to a product. In addition, modern computer-based material control, real-time voice and data communications, rapid transportation, and just-in-time management tend to eliminate the need for preplacement of inventories to provide access convenience. In most cases, on-site prepositioned inventories only add cost to the system by increasing overhead and raising inventory levels. This cost can often be eliminated by centralized operations, rapid response, and smoothly controlled materials flow.

In manufacturing operations, the majority of materials-handling expense is involved in the movement and storage of work in process (WIP) between operations. In a modern just-in-time environment, or in a mechanized or automated operation, the materials-handling system is often an integral part of the manufacturing equipment design and the materials management system. Materials-handling cost can often be minimized by integrating materials-handling functions into the manufacturing machinery, building the handling operations into each manufacturing cell's design, and reducing WIP inventory to a bare minimum through implementation of a just-in-time management philosophy.

Materials-Handling Cost Reductions Are Pure Profit. Changes in materials-handling operations do not affect the design, function, or marketability of the product. These system changes do reduce the cost of the product's manufacture and distribution. If the market price is stable, the cost reduction increases the gross margin and is therefore all profit. These cost reductions can also permit price reductions for a competitive advantage.

THE MATERIALS-HANDLING EQUIPMENT MARKET

Equipment Manufacturers and Their Product Lines. The most important component of the materials-handling engineer's professional body of knowledge is a broad knowledge of the materials-handling equipment market. Because of its very fragmented development, the materials-handling equipment industry is technologically segmented and organizationally complex. Some vendors are beginning to offer system integration services, but most continue to sell only their own specialized lines of equipment.

The materials-handling engineer's primary tasks are to define the system's operating specifications, to design a well-integrated system, to select the required equipment, and to put the system into operation. The materials-handling engineer must be a business analyst, logistics system designer, applications and specification engineer, equipment selector, system integrator, and an implementation and installation coordinator.

The market is continuously expanding and changing, with new equipment and services appearing daily. The materials-handling engineer must learn about the vendors and their product lines and keep up with the state of the art. There are at least four distinct classes of materials-handling equipment vendors and there is a great deal of overlap in their channels of distribution. In an abbreviated form, the market structure is as follows:

System Integrators.   They design complete, and usually complex, systems involving conveyers, automatic storage and retrieval systems, in-process handling, robots, automatic guided vehicles and tow lines, and often the buildings to house the warehouse or handling system. These vendors often manufacture the primary components of the system, but they usually purchase the standard elements and subsidiary systems from other specialized vendors. They function as system designers, prime contractors, and purchasing agents for the ultimate owner of the system. This is known as the *turnkey* approach, in which the vendor takes responsibility for delivering an operating system to the user. System integrators also serve as a very important channel of distribution for the specialist manufacturers of many system components and subsystems. These system integrators often receive the original system concept design from the owner or the owner's consultant and then perform the detail design function. In other cases they perform the whole task on the basis of a user-defined requirement.

Equipment Manufacturers.   They usually produce a specific line of products such as industrial trucks and materials-handling vehicles, overhead handling systems, cranes and monorails, conveyers, pallet racks, storage and retrieval systems, automatic pallet loaders, or bulk-handling systems. They sell their products through multiple channels, directly to the user via the company sales and engineering staff, or through system integrators who include the products in their overall systems designs. They also market through consultants who seldom buy the equipment, but will design and specify systems using their equipment.

Wholesalers and Dealers.   They buy materials-handling equipment such as forklift trucks, standard conveyers, shelving, pallet racks, and hand trucks and resell to the industrial community. They often maintain showrooms, provide maintenance, and keep inventories of spare parts and frequently sold equipment. They also often maintain fleets of rental forklift trucks and other industrial vehicles and equipment. These dealers sometimes offer system design services and try to market through system consultants. Their main channel for sales is directly to the operating executives and plant engineers in the manufacturing and warehousing industries.

Mail Order.   Materials-handling equipment sales are becoming a more common channel for distribution to the small manufacturer and small-business owner. These companies sell manual handling equipment, storage furniture, conveyers, and various handling tools. They often maintain inventories and provide rapid delivery. They seldom offer design assistance or product service.

System Integrators and Consultants.   The owner or materials-handling engineer can also seek the advice and counsel of professional specialists. These sources include materials-handling engineering consultants and the engineering staffs of system integrators and materials-handling equipment manufacturers. In the case of the consulting engineers, the service is performed for a fee. In most cases, the ethical professional consultant will prepare an operating specification for issue to competing vendors. Most consultants will not buy for the client and have no fiduciary interest in the selection of the vendor. They fully represent the interests of the client in their effort to design an optimum system at the least possible cost. They will often assist in vendor selection and system implementation as the client's representative.

If a system integrator or manufacturer is used as the source of professional assistance, it is obvious that some bias will be involved. In most cases, adequate and competent professional support will be available. The manufacturer's professional staff usually can and will give system development assistance. However, the client will either pay a service fee or be expected to buy some or all of the equipment from the assisting firm. The cost of the vendor's engineering service will be included in the price of the equipment. A hazard in this case is the possibility of being sold the type of equipment that the manufacturer builds, even when a different system or type of equipment would better serve the application.

Materials-Handling Equipment Classification for Market Research.   The engineer must thoroughly survey the materials-handling equipment market. The purchasing department can help. The engineer must review and evaluate all available equipment and system options. In order to deal with the complexity of the market, it is desirable to classify materials-handling equipment and vendor literature in a research-oriented structure.

The most common classification pattern is by function and role. For example, all wheeled vehicles would be grouped together and then subdivided by function as forklift trucks, tractors, pallet transporters, automated guided vehicles, nonpowered vehicles, and so on. Forklift trucks might be further broken into cantilever, reach, straddle, turret, and walkee machines.

In the case of conveyers, the obvious first cut would be to separate package conveyers from bulk conveyers. In the case of package conveyers, the next breakdown would be by power and gravity, and then by roller, skate wheel, belt, and chute. In the live roller group, a further division might be according to type of drive, for example, belt, shaft, chain, and cable.

There are many ways to classify this complex galaxy of materials-handling equipment in order to expedite the market/vendor search and assist data retrieval. However, the individual engineer will probably develop a system to suit the project environment and the range of equipment under consideration.

Finding the Right Equipment.   This is the objective of the market search. The materials-handling engineer develops a system concept or a series of alternative system configurations and then uses standard market search procedures to find vendors. The tools used in this search usually include the *Thomas Register,* the annual trade journal directories published by *Materials Handling Engineering* and *Modern Materials Handling,* the Yellow Pages, and regional industrial directories. Employed materials-handling engineers usually enlist the assistance of their purchasing departments for the market/vendor search. Selection of the right equipment is a by-product of the feasibility study and the industrial engineering analysis of the various alternative system configurations under consideration.

THE MATERIALS-HANDLING SYSTEM DESIGN AND FEASIBILITY STUDY

Collecting the Required Data.   This is the first step in gaining a quantified knowledge of the project's business requirements. The required data include:

Product Data.   These include the item identity system, the physical shape and dimensions of the product, the individual and shipper package characteristics, product and package unit weight, pieces per package, pieces or packages per pallet and/or tote box, product and pallet stackability, product perishability and fragility, dollar density (dollars per cubic foot), control factors (that is, controlled substances and pilferables), environmental factors (such as refrigeration or cooling), and hazards.

Transaction Data.   These cover the number of units or the bulk quantity of each item in each move, order, or transaction; the number of moves for each item at each step of the flow system on a time basis (moves per hour, minute, day, week, or month, and so forth); the distance and the route traveled in each move; the method of movement in the present or proposed system; and the seasonal or periodic variations in the movement pattern or rate.

Inventory Data.   These include the number of units, bulk quantities, and/or unit loads of each item in the inventory at each stage or hold point in the system and in the system as a whole, on an instantaneous basis and maximum basis, for each control period (day, week, month, year) during a business cycle. The resulting information should be analyzed and modified to accommodate planned changes in inventory policies and marketing and manufacturing practices.

Following the development of one or several alternative system configurations, the engineer should be able to apply operations research and/or simulation techniques to test alternative system configurations. The alternative systems should be discussed with the operating executives who are to be the users of the system in order to identify operating problems which may be built into the system designs and surface objections and/or resistance to change in the user groups. The proposed designs can then be adjusted to accommodate these problems and defuse any operating objections. These actions may result in the elimination of some of the proposed alternatives because of practical operating problems.

Selecting Equipment Types and Identifying Sources.   The design engineer must survey the equipment market and select the materials-handling hardware which will support each of the proposed alternative system configurations. The engineer will then contact selected vendors to develop proposals and budget prices for use in estimating the capital cost and operating economics of each of the proposed alternative system proposals. Based on these budget estimates, the system designer will select the most economical alternatives from among the proposed system configurations and develop a preliminary system design. The design documentation should include preliminary layout and elevation drawings of the system, a flowchart of the proposed operation, a flowchart of the supporting information system, a staffing table, and an operating narrative describing all functions and operations in the system.

Use of simulation to test the system's function is a desirable next step. Each of the proposed alternative system configurations should be developed to the point where sufficient design and operating information is available to test the system's behavior with a standard simulation program. The simulation can be in the form of a mathematical model or a three-dimensional animated drawing on the computer screen. This procedure will focus the designer's attention on interferences and possible malfunctions before detail engineering is undertaken. This step will help to verify the validity of the system concept, refine and test its performance, and compare the functioning of alternative system configurations.

Preparation of Preliminary Operating Specifications.   This is the next step in the feasibility study process. The preliminary system design documents should include scale layouts and elevation drawings showing the building envelope; new and existing equipment locations; fire, utility, and security features; interfering "monuments" (immovable structures and operations) in any existing buildings; truck and rail docks; and site arrangement.

The preliminary equipment operating specifications should include a description of each process, handling, and storage operation and its performance requirements, a statement of the type of equipment needed for the operation, workstation layouts, and a statement of supporting utility requirements, materials-handling interfaces, throughput volumes, and transaction data.

The system narrative should describe the operation and be supported by process and materials movement flowcharts and an organization and staffing table.

A bid package should be developed which includes a cover letter or formal request for quotation (RFQ) which states the objectives of the project and the time, performance, and cost parameters. The bid package will include the drawings, operating specifications, operating narrative, and an RFQ for issue to vendors for the solicitation of firm quotes on equipment and/or turnkey bids. These documents become a part of the purchase contract between the vendor and the user. It is therefore essential to write the specifications as *operating specifications* and *not* design specifications. The vendor is obligated to provide a functioning system or equipment unit which will meet the operating requirements of the specification. The system designer must not limit the vendor's equipment or system design options in selecting or designing equipment which will comply with the operating requirements of the system. The user may require alternatives and optional features as a part of the operating specification. However, the mechanical and electrical design of the hardware and the basic configuration of any software should be the responsibility of the vendor, if the vendor is to be held responsible for fulfilling the specified performance requirements and complying with warranty agreements.

The vendor will submit preliminary, or budget, quotations for review with the user and for use as the basis for technical and contract (price) negotiations. These bids will usually include preliminary drawings of any vendor-suggested alternatives to the proposed design, product literature, and operating specifications. The price quotations and delivery terms will usually be in some detail, but also subject to negotiation.

These initial budget quotations will be used to rate and select the preferred vendor. The selected vendor and user will then negotiate a final design configuration, project schedule, and price.

The Economic Feasibility Issues.   The economic feasibility decision has several levels for evaluation. In some cases, the requirement for a materials-handling system is driven by process, safety, or facility configuration requirements and the feasibility decision is a technical one. In such cases, the most effective option is often chosen regardless of price in order to assure fail-safe operation. If technically comparable alternatives are available, the capital investment required is the usual basis for choosing one of them.

In many situations, the basic decision is whether to automate, mechanize, or do nothing and continue on a labor-intensive basis. Labor costs are becoming a shrinking component of operating expense and a well-managed manual or mechanized system seldom allows automation to be justified on the basis of labor savings alone. The capital-intensive nature of automation often precludes the commonly used 1- to 3-year return on investment. In some cases, further perceived savings can be shown in the safer handling and better control of material, and in the ability to implement just-in-time methods and reduce total inventory lockup. Capital-intensive systems must be justified on the basis of corporate performance and competitive advantage.

MATERIALS-HANDLING EQUIPMENT AND ITS APPLICATION

Bulk-Materials-Handling Equipment.   Bulk-materials-handling equipment breaks down into several basic classes, as shown in Fig. 4.1: conveyers, feeders, vehicles, and storage devices, with an infinite variety of designs and features. The following samples characterize this equipment.

Belt Conveyers.   These are usually of the trough type and in most cases are roller-supported. The rollers can be either rigid or a series of wheels on a flexible cable shaft. Slider belts in sheet-metal troughs and dust-proof enclosed belts are also available for handling bulk materials. Belt feeders are usually short conveyers with precisely variable speeds and stop/start controls which allow delivery of materials from hoppers or bins to processes in a controlled manner based on weight or flow rate. In the case of open belts, almost any material which is not too dusty can be conveyed at high speed and on slopes up to a few degrees below the angle of repose of the material. In the case of coal, ore, and other heavy bulk materials, the belts are often many miles long and deliver material at rates of hundreds, and even thousands, of tons per hour. In all bulk-handling systems, the moisture content, stickiness, and flow characteristics of the material dominate the flexibility of the equipment design.

Bucket Elevators and Bucket Conveyers.   These are used for vertical and steep-slope movement of bulk materials. There are several designs used in the construction of bucket

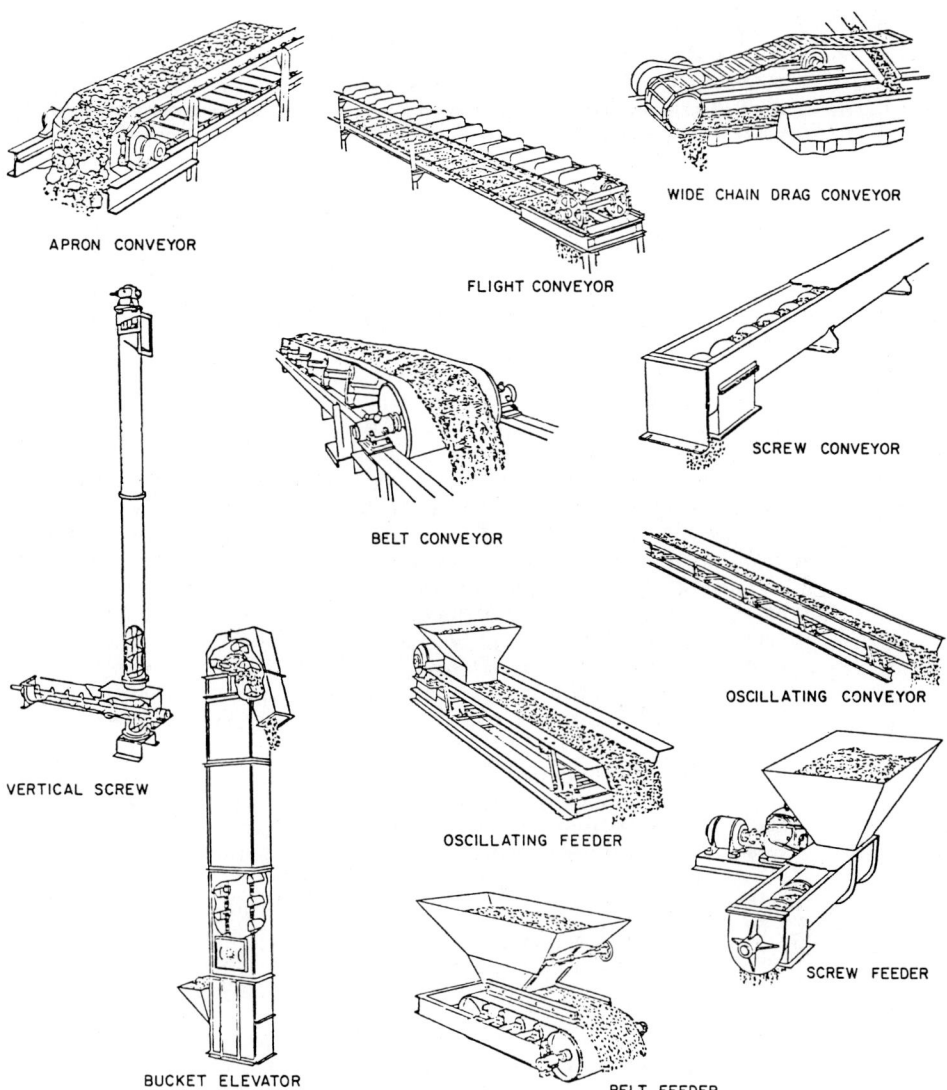

APRON CONVEYOR

FLIGHT CONVEYOR

WIDE CHAIN DRAG CONVEYOR

BELT CONVEYOR

SCREW CONVEYOR

VERTICAL SCREW

BUCKET ELEVATOR

OSCILLATING FEEDER

OSCILLATING CONVEYOR

SCREW FEEDER

BELT FEEDER

FIGURE 4.1 Typical bulk conveyer equipment.

elevators. Some have the buckets hung on chains, and others have them mounted on belts. In most cases, the buckets are loaded through a trough or chute at the bottom of the elevator casing and they dump their loads over the top of a head shaft pulley at the top of the system. They can go to great heights and can move very fast with great load-handling capacity. They are commonly used for delivery of bulk materials from below-ground rail and truck unloading conveyers to the top of bins and silos in the handling of grain, coal, glass sand, and other free-flowing bulk materials.

Bucket conveyers are usually flexible belts with pockets formed in them. These conveyers have the ability to move materials in horizontal, sloping, or vertical paths and around turns. They are used for ship unloading and in bulk systems where space and arrangement are a design problem.

Screw Conveyers.   These are usually quite short, and they are used for the movement of a variety of bulk materials, ranging from dry grain to wet clay in ceramic plants. Screw conveyers can have solid or segmented blades or flights and can serve in many roles, such as feeders, transporters, and mixers. The flights can have varying pitch to compact or spread the materials in transit. Screw conveyers with segmented blades are used in mixer applications. A screw conveyer's speed and volume can be precisely controlled. Screw conveyers make ideal feeders.

Vibratory Conveyers, Feeders, and Screens.   These can be either mechanically or electrically driven. They move the materials by making them hop along a pan-type surface. When used as a screen, the pan is made of either wire cloth or perforated sheet steel. In the feeder mode, precision of control is an important advantage of the vibratory machine. The frequency and amplitude of the motion is closely controllable, and this controls the speed of flow.

Drag Chains and Cableways.   These are used in heavy-movement installations to handle very rough materials or to move over long distances. Drag chains are used to haul logs in paper mills, to move castings and scrap iron in foundries, and to push ore and lump coal up steep slopes. Cableways are most often applied in the movement of coal or mineral ore over long distances in rough country. They can be single- or multiple-bucket and single- or multiple-cable. In many cases they travel for miles over hills, valleys, and rivers where a fixed belt installation would be too expensive. They are also used in the construction of dams and port facilities where crossing water or rough terrain is required.

Bulk Bins, Hoppers, Big Bags, and Unitized Bulk Operations.   These combine storage and handling within the bulk-materials-handling system. Almost all bulk systems require some form of surge capacity. Bulk bins, hoppers, silos, and other fixed holding devices provide this capacity. They are usually loaded by bucket elevators, screw conveyers, tipples, or trough belt conveyers. In most cases they are emptied by belt, vibratory, or screw feeders to deliver their contents to conveyers or containers which in turn feed processing operations. In the design of hoppers and other bulk containers, the most critical issues are the slope of the bottom and flowability and angle of repose of the material. The emptying of the hopper sometimes requires vibrators, thumpers, or air jets to loosen the material and make if flow from the hopper to the feeder or conveyer.

The big bag is a combination container for transport of bulk materials and portable hopper for storing and dispensing these materials.

Unitized bulk materials are in palletized bags or pallet boxes. In this kind of system, the bulk material behaves like packaged goods until it is emptied into the system by dumping the bags or using vacuum or gravity to remove it from the pallet box.

Vehicle Bulk-Handling Systems.   These utilize many common types of equipment. The most widely used device is the front-end loader which is usually a four-wheel machine and comes in very small to very large versions. Front-end loaders are used in such applications as feeding materials into processes, handling scrap and trash, loading and unloading vehicles, and feeding coal into power plants. These machines are sometimes mounted on tracks for rough-terrain operations and construction work. A companion type of bulk-handling machinery is the power shovel. These machines come in the form of backhoes, shovels, draglines, and clamshell cranes. They can be mounted on wheels or tracks and in some cases on railroad cars or barges. They can also be in fixed installations as at port-side bulk-loading facilities.

The most common type of bulk-handling vehicle is the dump truck. These come in many sizes and in both fixed-chassis and trailer form. They can be used for short hauls or over-the-road movements.

Marine Bulk-Materials Handling.   This is a very big segment of the flow of products in the world economy. Grain, coal, chemicals, ore, minerals, and some recyclable animal products are bulk-handled on and off ships and barges. In many cases the vessels are equipped with on-board loading and unloading gear. These include conveyer booms and traveling gantry cranes. In other cases the products are loaded from shore-based conveyers and unloaded by shore-based cranes. Pneumatic conveyer systems are often used to

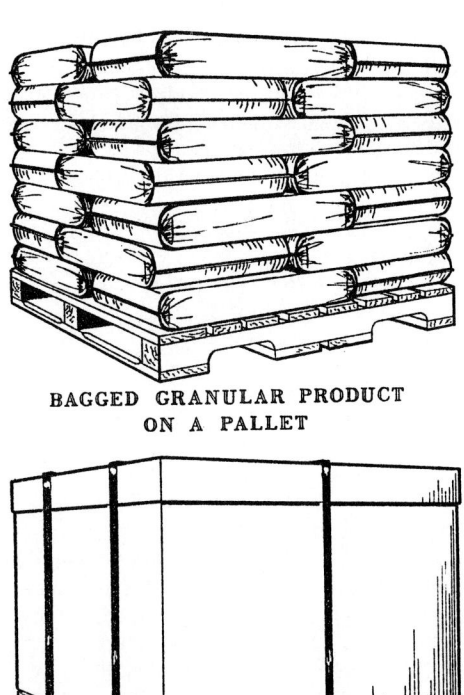

BAGGED GRANULAR PRODUCT
ON A PALLET

PALLET BOX
(GRANULAR PRODUCT)

FIGURE 4.2 Palletized bulk materials.

unload light materials such as grain and powdered minerals. Specially designed, high-volume systems are used to handle coal from conveyer-loaded barges into conveyer systems which feed power plants.

Unit- or Packaged-Materials-Handling Equipment.   The great majority of materials-handling operations in the modern economies of the world involve unit or discrete-item handling operations. In most cases the product being handled falls into the category of loose parts or products, cartons and packages, tote boxes, or pallets. In general, the first aim of the materials-handling system designer should be to homogenize the materials movement by applying the common denominator concept, which neutralizes the individuality of the product and simplifies the handling system. A pallet, carton, or tote box is a common denominator which hides the shape, mix, and quantity of the product from the prime mover or handling device, thus homogenizing the handling system and simplifying its design. This also increases the flexibility of the materials-handling system by minimizing the effect of product design changes or variations in batch quantities.

Pallets and Palletizing Operations.   These are the heart of unit- and package-handling operations. The pallet is the primary common denominator in commerce and industry. The design of the pallet pattern must create a stable and, if possible, stackable unit load. If interlocking is prevented by the shape of the package or part, auxiliary means such as strapping, gluing, stretch wrapping, shrink wrapping, or box palletizing must be applied to assure a stable unit load. Pallets are usually made of wood in a double-faced design with four-way access. The most common dimensions are 48 × 40 feet (1200 × 1000 millimeters) and 48 × 48 feet (1200 × 1200 millimeters). Other sizes are in use in private and military systems. Some pallets are made of plastic or fiberglass for use in food and chemical plants to permit washing and sterilization. Steel and aluminum pallets are also available.

FIGURE 4.3   A standard four-way, double-faced wood pallet.

Paper pallets and slip sheets are often used to eliminate the pallet return cost from the loading of palletized shipments. These one-way pallets are much less expensive than the least costly wood pallets. However, the use of slip sheets requires special equipment, and slip sheets are often used in conjunction with wood pallets to expedite internal operations while gaining the advantages of low-cost unit load shipping.

Palletizing operations can be manual or automated. Most automated pallet loaders receive the cartons or sacks via conveyer, arrange them in patterns by tiers, and stack them on the pallet to a predetermined load quantity. See Fig. 4.4. The pallet is then ejected onto a conveyer for transport or forklift pickup. More recently, robot palletizing machines pick up the carton or bag by vacuum, or with an end-effector clamp, and place the item on the pallet in a predetermined pattern. All of these machines can vary the pallet pattern for different items in random sequences based on computer control, often by reading the bar

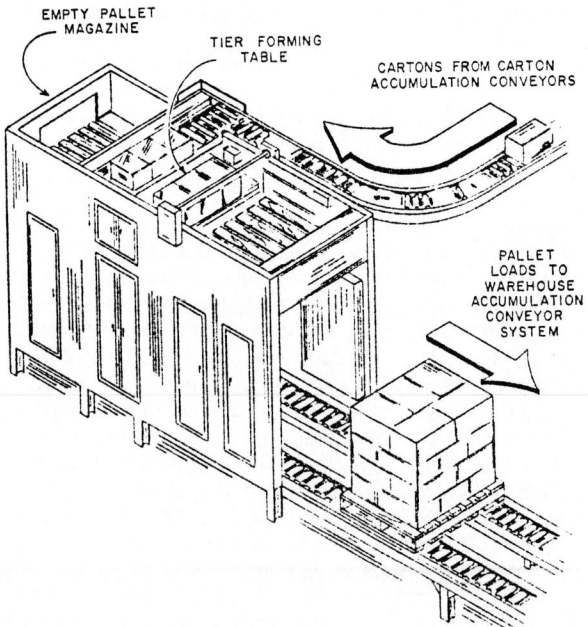

FIGURE 4.4   An automatic carton palletizer.

FIGURE 4.5 Good interlocking pallet patterns and stable unit loads.

code on the incoming package. Most have stacking speeds of 30 to 60 cartons or bags per minute. See Fig. 4.5.

Tote Boxes. These are used as the common denominator for internal movement of parts and materials in a factory and for movement of open-stock materials between warehouses and retail stores. The typical tote box system includes modular sets for modularizing the tote system with the overriding pallet system of the plant. Tote boxes are generally made of plastic or fiberglass and can be number-coded and bar-coded for tracking and identity. They also can be equipped with hinged or separable, lockable lids. Tote boxes used for kitting prepicked parts for assembly operations often double as part bins at the workstation. See Fig. 4.6.

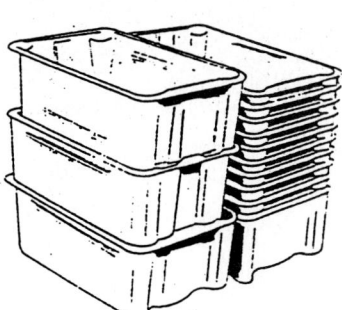

FIGURE 4.6 Typical tote boxes.

Package and Unit Conveyer Systems. These include a wide variety of both powered and nonpowered equipment designs and system configurations. Conveyer systems and their design requirements are too often treated as a hardware issue and dealt with in isolation from other elements of the materials handling system. Conveyers are the physical manifestation of material flow. They should be used when the flow is over a fixed route and stable in volume and material characteristics. The conveyer system should be used when the routing is stable and fairly continuous, and the vehicle should be introduced to accommodate variable routing and capacity requirements.

Gravity Conveyer. These include roller and skate-wheel types, ball tables, and chutes. See Fig. 4.7. Each type has both advantages and limitations in relationship to the types of cargo to be handled. Roller and wheel conveyers require a fairly hard, flat-bottomed package or product to avoid draping between the rollers or wheels. Roller conveyers can also handle items with multiple straight rails like a child's sled if the rails are parallel to the flow and lie across the axis of the rollers. For free and smooth flow, gravity conveyers are usually sloped at about 2 inches in every 10 feet. Inertia brakes or slide brakes are often used to slow acceleration, and booster belts are used to recover height in long runs. Ball tables are used in a level posture to allow easy manip-

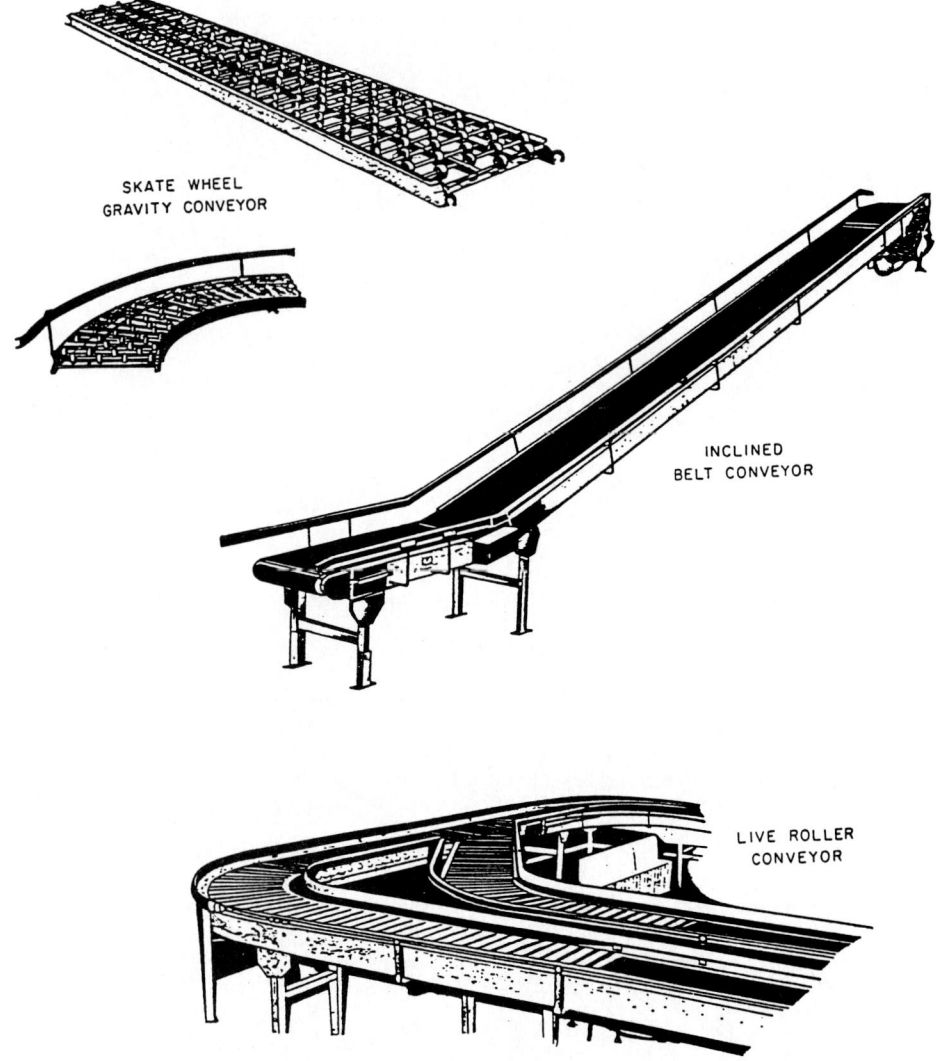

SKATE WHEEL
GRAVITY CONVEYOR

INCLINED
BELT CONVEYOR

LIVE ROLLER
CONVEYOR

FIGURE 4.7   Typical package conveyer equipment.

ulation of heavy boxes or other items in assembly and packing situations. When heavy
plates are handled in welding or shearing operations, the balls are often placed on the tops
of posts so the operator can move among them and maneuver the plates into the working
position.

Belt Package Conveyer.   These can handle most regularly shaped, hard, soft, and odd-
shaped parts, products, and packages. They can neutralize the shape of the cargo because
the belt functions as a table. Belt conveyers can usually carry loads up slopes in the 15 to
20° range. If rough belting is used, or the belts are slat-equipped, they can sometimes negotiate up to 30°. Belt spirals and turns are also available to make the design of these systems more flexible. One of the basic characteristics of belt conveyer systems is their ability to maintain product registration with the belt. This feature also prevents the

accumulation and grouping of batches. Accumulation operations are accomplished by inserting live roller or gravity conveyers into the belt conveyer system.

Belt conveyers are usually roller-supported and handle heavy loads at speeds of from a few inches per minute to hundreds of feet per minute. The slider belt variety is usually used for lighter loads where the belt friction on the slider bed is not a critical power factor. Slat-type belts are used for very heavy or difficult loads, such as handling hot castings in foundry operations. Small and narrow segmented plastic belts, which are really flat-topped chains, are used in food processing to handle bottles and cans through packing lines at very high speeds. These belts often have pockets in them to assure precise positioning or registration of the product with the belt for filling-machine timing.

A belt-tightening take-up mechanism is located in each belt conveyer unit. The return strand of the belt is usually supported by idler rollers under the conveyer frame, and, in some cases, is also used to carry cargo in the opposite direction.

Power-Roller Conveyer.   These are available in many design configurations. Figure 4.8 illustrates a loading system using both forklift trucks and extendable conveyer fed by conveyer accumulation lines. The most distinctive feature of a live roller is its ability to accumulate loads, and, in some designs, to vary the speeds for grouping and separating packages en route. In the most common type, the rollers are driven by friction from a belt pressed against the bottom of the load rollers by idler rollers. Another type is shaft-driven by grooves in the ends of each roller, with small belts connecting pulleys on the shaft to the rollers. Other variations are the cable-driven live roller, which uses a plastic-coated steel cable in place of the belt, and the chain-drive type, which has sprockets on each roller. The chain type is commonly used for handling very heavy loads. Power-roller conveyers are capable of the same high-speed movement as belt conveyers, but live roller systems usually cannot negotiate inclines of more than a few degrees of slope. This weakness can be reduced by using rubber-coated or rough-surface rollers with high-friction cargo. However, this design change will also reduce the live roller's capability for package accumulation.

FIGURE 4.8   A loading system using both forklift trucks and extendable conveyer fed by conveyer accumulation lines. (*Courtesy, The Sims Consulting Group, Inc.*)

Conveyer Turns and Switches.   These are an integral part of both power and gravity systems. See Fig. 4.9. In their general design, gravity and live roller turns are very similar. Both often use tapered rollers and guide rails for turns of various angles. Skate-wheel

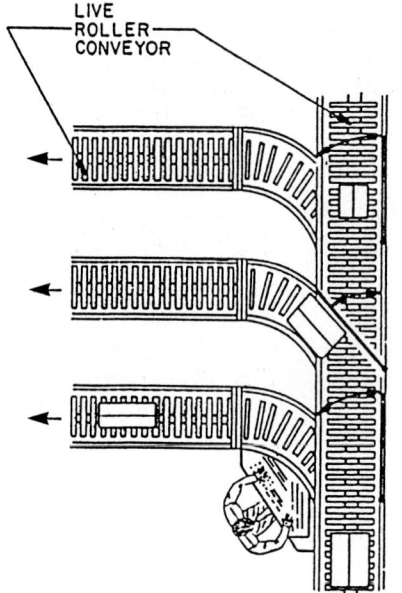

FIGURE 4.9   Typical conveyer switches.

turns usually depend on angling of the wheels and guide rails to route the packages around the bends. As stated above, belt conveyers use power belt turns which carry the product smoothly around bends. Although most turns are smooth and curved, some conveyer systems also use right-angle turns and switches to route the cargo. These vary from swing-arm sweeps and track shifters to pop-up right-angle transfers using lateral belts. In all cases, the key design issues are the timing of the switch to match the package flow and the avoidance of jamming. In most cases the packages must be separated by speed changes to produce gaps in the stream in order to safely switch a selected unit onto a branch line.

Conveyer Sortation and Accumulation Systems.
These are a natural outgrowth of switching capability and the accumulation ability of powered-roller and gravity conveyers. Sortation can also be accomplished on belt conveyer systems, with some technical limitations. Sortation systems are basically scheduled switching systems. The earlier methods used human operators to read labels and manually trigger downstream switches as the package passed them. Later versions used magnetic encoding of the belt and tipping tray systems, plunger-type pushers, or air blasts to eject the selected package from the stream into the accumulation line. These systems have been largely replaced by live roller systems using slide blocks to form smooth switch guides for packages identified by bar-code readers or preprogrammed distribution instructions. Most systems deliver the packages to live roller accumulation conveyers which feed workstations, pallet loaders, or loading and shipping operations. Large systems often have many accumulation lines leading to truck docks with extendable loading conveyers which project into the trailers. The extendable conveyers can move from dock to dock and can be used in conjunction with forklift trucks.

Pallet Conveyers.   These are heavy-duty variations of powered-roller or chain-type belt conveyers with multiple silent chain strands. Pallet conveyers are most often used to service automated storage and retrieval machines and pallet loaders. They are also used to deliver pallets of goods to high-volume order-picking stations and function as part of pallet racks in flow-through storage systems.

Monorail Conveyer Systems.   These are appropriate when it is desirable to utilize the overhead air rights in a plant to save floor space or provide grade separation for cross traffic in the handling system. They can also be used effectively when multilevel operations require direct materials movement between floors or in continuous in-line processes such as assembly, painting, plating, or washing. Monorail systems can also provide for in-process or in-transit storage between coordinated operations. See Fig. 4.10. Monorail systems can be built to provide a wide range of cargo capacities and offer both fixed-interval and power-and-free designs. Monorail systems can also be designed with independently self-propelled carriers which function like a railroad. Inverted monorail systems are also available, and they have many advantages over other types of load-supporting conveyer.

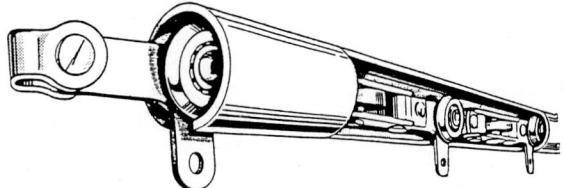

FIGURE 4.10 A typical tube-type light-duty monorail conveyer design. (*Courtesy Pacline Overhead Conveyer Corp.*)

Light-Duty Chain and Cable Systems. These are used in such light industries as needlecraft, toys, shoes, plastics, department stores, and electronics. In most cases the chain or cable has equally spaced support rollers and is housed in a square or round tube with a slot on the bottom or side to provide for the cargo carrier. The carrier is usually a universal or common-denominator ring or rod to which the specially designed product-carrying racks or trays can be attached. These light-duty systems are usually designed for loads of 100 pounds or less per carrier. These conveyers can make tight turns, go up or down steep inclines, and move at variable speeds. They can also be designed to automatically load and unload at workstations or interface with a belt or roller conveyer.

Heavy-Duty Chain-Type Monorail Systems. These perform the same type of functions as the light-duty versions. See Fig. 4.11. However, because of the need to design them for heavy loads, and often rough conditions, they are configured differently. Most heavy-duty systems are based on an I-beam-type structure. In most designs, the carrier is supported by a trolley which rolls along the bottom flange of the I beam and is connected to the other trolleys by a roller chain which provides the power and maintains the spacing of the trolleys. As in the case of the light-duty system, the trolley has a hole or rod which serves as a common denominator load rack or carrier connection. These systems often carry loads in the half-ton or 1-ton range and perform the same functions as the light-duty systems.

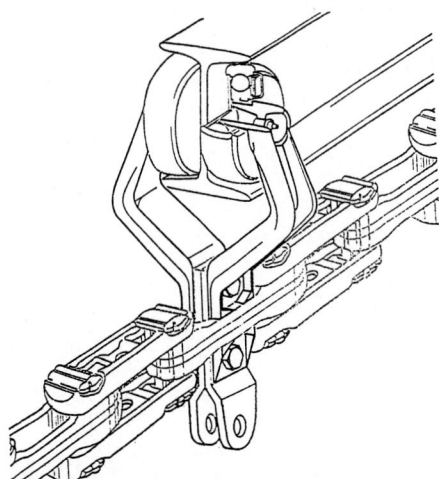

FIGURE 4.11 Typical construction of a heavy-duty monorail conveyer. (*Courtesy Cleveland Tramrail—Shepard Niles, Inc.*)

Power-and-Free Monorail Conveyer Systems.
These have similar characteristics to the chain or cable-driven monorails. However, in these systems the carrier can be separated from the driving chain or cable by an operator or by an automated control device. The power chain or cable continues to move, but the load carrier can start, stop, accumulate, or be switched to another track or sidetrack. This allows the load to be routed to a variety of destinations, accumulated on a storage track and later recalled, stopped to allow work to be performed on the cargo item, or accumulated to feed an operation or workstation. There are many designs for these systems, but most of them use a toggle or gripper mechanism to attach the load trolley to the power chain or power cable. Routing can be directed manually or by equipping the carrier with an on-board code device which

instructs the toggle. In some systems a bar-code card, a set of mechanical or magnetic buttons, or a specially designed lever informs the switch of the carrier's intention to turn off the main track. In computer-managed systems, the carrier's number or bar code is read to identify it and the computer instructs the switch. In accumulation systems, the holding track can be separately powered and controlled or gravity can be used to flow the disconnected carriers to the end of the holding track, where the actuator can attach them to the power chain or cable on command.

Inverted Monorail and Power-and-Free Systems.   These have many of the same design features as overhead systems, but they have some distinct operating advantages and are frequently used in assembly line operations. See Fig. 4.12. From a structural point of view, their main advantage is their floor-supported configuration and the elimination of extra roof-loading requirements. The floor installations also have the disadvantage of blocking cross traffic. However, inverted systems can also be installed in an overhead arrangement. Inverted systems can make sharp turns and with long loads mounted on dual carriers, this permits very tight turns for such large cargos as auto bodies in process. These systems can also go up slopes better than belt conveyers.

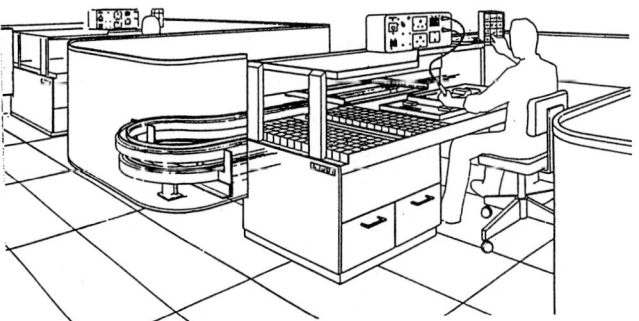

FIGURE 4.12   An inverted power-and-free system in an assembly operation. (*Courtesy Jervis B. Webb Co.*)

Inverted systems are most often applied in a power-and-free configuration in manufacturing operations. One of their main advantages is the ability of workers and machinery to access the cargo without interference from conveyer supports. In manufacturing operations, they often function in the power-and-free mode to deliver and pick up product at workstations.

Powered-Carrier Monorail Systems.   These are most often used in heavy-duty applications to move large and heavy cargo such as ladles of molten metal in a foundry, sand mixtures in a glass factory, or engines in a tractor plant. They are functionally similar to a power-and-free system in that each carrier can be independently routed to a destination and switched to other tracks and holding lines. They usually cannot negotiate steep slopes. However, instead of being attached to, and released from, a chain or cable, each powered carrier is equipped with a motor drive and gets its electricity from sliding contact with power rails on the track.

In some systems the carrier also receives instructions from track contractors. In other cases, the routing instructions are either coded into a controller or mechanical device on the carrier or into a computer which reads the bar code or other identity of the carrier and controls the switches along the route. In simpler systems, an operator follows the carrier and controls it with a pendant or radio controller. The powered carrier may be a single multiton lift unit on either a very short or an extensive track system. In some cases the system may consist of many carriers on a complex, computer-controlled railroad-type installation which routes cargo throughout an entire department or plant.

Vehicular Unit-Handling or Packaged-Materials-Handling Equipment.   This is the most commonly recognized area of materials-handling technology, and, at the same time, it is the most complex and labor-intensive in its application. Most modern vehicular handling machinery is battery-powered, although many forklift trucks are powered by internal combustion engines. See Fig. 4.13. Manually operated vehicles are also normally a part of the vehicular handling system. Most vehicular equipment requires an operator, although automated guided vehicle (AGV) technology is rapidly spreading throughout industry.

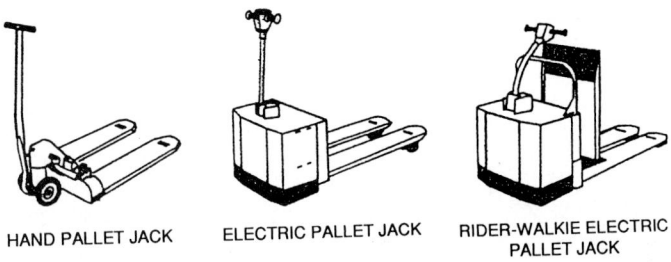

HAND PALLET JACK   ELECTRIC PALLET JACK   RIDER-WALKIE ELECTRIC
                                                                              PALLET JACK

FIGURE 4.13   Examples of manual and powered pallet transporters. (*Courtesy The Sims Consulting Group, Inc.*)

Materials-handling vehicles are used for package and unit handling, bulk materials handling, maintenance and construction, and both long- and short-distance movements. As a general rule of thumb, conveyers and other fixed materials-handling machinery should be applied when the route and volume of movement is predictable, stable, and fairly uniform in handling characteristics. By contrast, wheeled vehicles are most effective in variable-routing and variable-volume situations and usually for short-distance movements. Wheeled vehicles combine the flexibility and versatility of the human brain with the speed and muscle of the machinery. More than conveyers and other fixed materials-handling machinery, they are usually an extension and enhancement of the human worker. The automated guided vehicle has begun to capture the memory and decision-making capability of the human, and releases the operator from the system by introducing interacting or preprogrammed computer capability as a substitute for the human in the control loop.

Pallet Transporters and Materials-Handling Tools.   These are designed to enhance human capabilities in the local movement of materials. The hand pallet transporter, or *pallet jack,* is the most common tool and often a necessary one. Whenever pallets are in use, the materials are effectively immobilized unless a pallet transporter or forklift truck is available. The manual pallet transporter is a low-cost tool for short (under 100 feet) moves of pallets of up to 4000 pounds. This device can also be equipped with a fold-down frame for handling printer skids and other specialized unit loads. Pallet transporters are also commonly built with battery power and, in some configurations, accommodate a rider-operator. These machines travel at forklift speeds and can negotiate dock levelers to load trucks and railcars. The rider-type machines can also be built with long forks to carry two pallets. In this two-pallet configuration, the transporter is a more efficient transport device than a forklift truck, which has the primary function of stacking loads. Pallet transporters can also be equipped with special devices for handling slip sheets and unique cargoes.

Another handling tool is the two-wheel hand truck, which is very effective for moving loads of up to 500 pounds over short distances. These devices can be specially configured to handle barrels, drums, crates, furniture, refrigerators, and other major appliances. They are very useful for loading trucks which cannot carry the weight of a drive-on forklift truck. Other handling tools include such devices as roller peavey bars, roller pads, and air-cushion pads. These are generally found in the maintenance and millwright area or in the assembly of very heavy products such as machine tools and airplanes. Air-cushion devices are capable of supporting extremely heavy loads for maneuvering and short movements.

Counterbalanced Forklift Trucks.   These are the most common vehicular materials-handling machines. See Fig. 4.14. They are available in several basic configurations and with an almost infinite variety of design variations. The basic machines have two drive wheels in the front and two steering wheels in the rear. Other forklifts are designed with two front drive wheels and one center steering wheel in the rear. The three-wheel versions require less aisle to turn, but they are less stable in lifting operations. Forklift trucks come

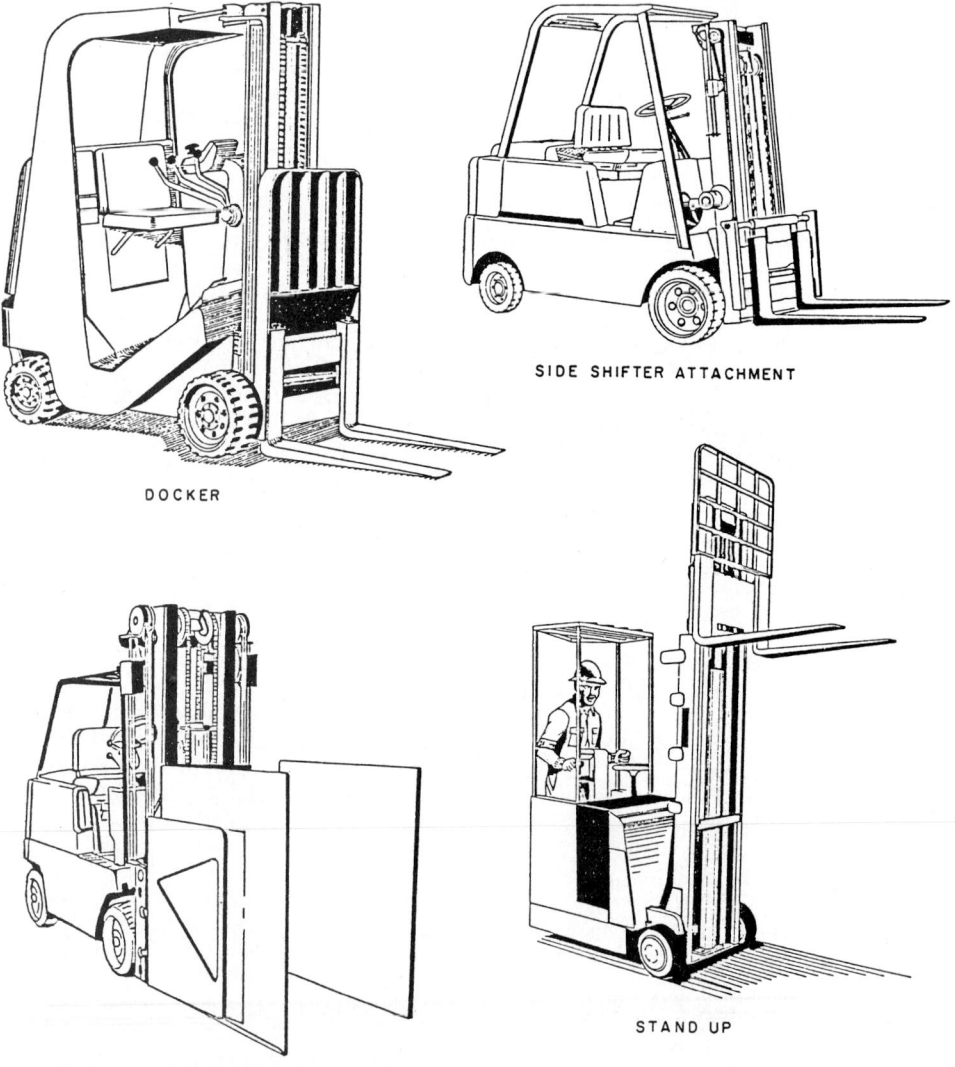

DOCKER

SIDE SHIFTER ATTACHMENT

CARTON CLAMP ATTACHMENT

STAND UP

FIGURE 4.14   Typical counterbalanced forklift trucks. (*Courtesy The Sims Consulting Group, Inc.*)

in standard capacities of 1000, 2000, 3000, 4000, 5000, 6000, and 10,000 pounds with 24-inch load centers. In most cases, the basic mast lift height is 240 inches. Many of the machines are equipped with a 54-inch-high free lift capability. Vehicles designated for truck-loading operations usually have a maximum lowered mast and overhead guard height of 83 inches. Some of the machines are designed specifically for truck-loading operations. These "dockers" are generally limited in their lift capability but have a very tight turning radius. The aisle-turning requirements vary from 10 feet to over 15 feet, depending on the machine's load capacity. Few forklift trucks are purchased in the basic design. The most common variations are in the mast height and lift height capability. The use of multiple masts can extend the machine's capability for extremely high lift heights at the cost of reduced load-lifting capacity. These machines can also be equipped with various attachments to rotate loads, clamp loads, side-shift loads, handle slip sheets, handle drums, and use center-hole rods for handling steel coils, carpets, or paper rolls. The attachment also reduces the lifting capacity of the machine.

Reach-Type Narrow-Aisle Forklift Trucks.   These are an aisle-space-saving variation of the basic forklift truck concept. See Fig. 4.15. They are normally battery-powered and are the descendent of the straddle fork truck. The operator stands on a rear platform which is equipped with a "dead-man" foot-pad switch. The straddle truck has outriggers to reduce footprint weight and aisle width. The outriggers on the straddle machine go around the pallet, and the load is carried within the wheelbase of the fork lift. This reduces the total weight of the machine and narrows the aisle required to turn into the load. However, it also forces wider gaps between pallets.

The reach-type machine is similar in basic design to the straddle-type except that it is equipped with a pantograph extension system which moves the forks and backrest out from the mast to a position where the heel of the forks is over the outrigger wheels. This effectively converts the straddle machine into a counterbalanced forklift configuration for stacking and picking up loads. In this mode, the reach truck behaves like a counterbalanced machine and it has the same range of attachment and height variations.

The main advantage of the reach forklift truck is its ability to retract the load to a point within its wheelbase and thereby reduce the turning-aisle requirements and the footprint floor loading. When equipped with large outrigger wheels, these machines are as versatile as the counterbalanced units. These units are seldom built with capacities in excess of 6000 pounds at 24 inches. They have become the standard forklift for conventional warehousing operations because of their space-saving feature and operating versatility.

The reach truck is also available with a double-extension pantograph to permit two-deep rack stacking of pallets.

Narrow-Aisle, Turret-Type Forklift Trucks.   These are capable of working in aisles which are only about 6 inches wider than the load or the machine. See Fig. 4.16. They have become quite common in high-rise pallet-rack-type warehousing operations and can operate in racks up to 40 feet in height. Most of these machines have rigid masts and are too large and cumbersome for general forklift applications. They are restricted to the storage phase of the materials-handling system. However, they can compete effectively with both counterbalanced and reach-type forklifts and with automated storage and retrieval systems in high-rise palletized warehouse applications.

The key feature of these machines is a transverse track on the mast which allows a fork turret to move laterally while a turret-mounted set of forks rotates through a 180° arc to reach either side of the aisle for placement and removal of pallets. This feature allows the machine to service both sides of the aisle in a high-rack pallet storage system without turning in the aisle. Because of the close fit, these machines are guided in the aisle by either a wire system or side rails. The minimum clearance between the racked pallets and the load or machine is 3 inches on each side. With a 48-inch-square pallet, a typical turret truck can operate tightly in a 60-inch aisle. However, in most installations it is more practical to use a 72- or 78-inch rack aisle to accommodate pallet and load overhand and permit rapid operations.

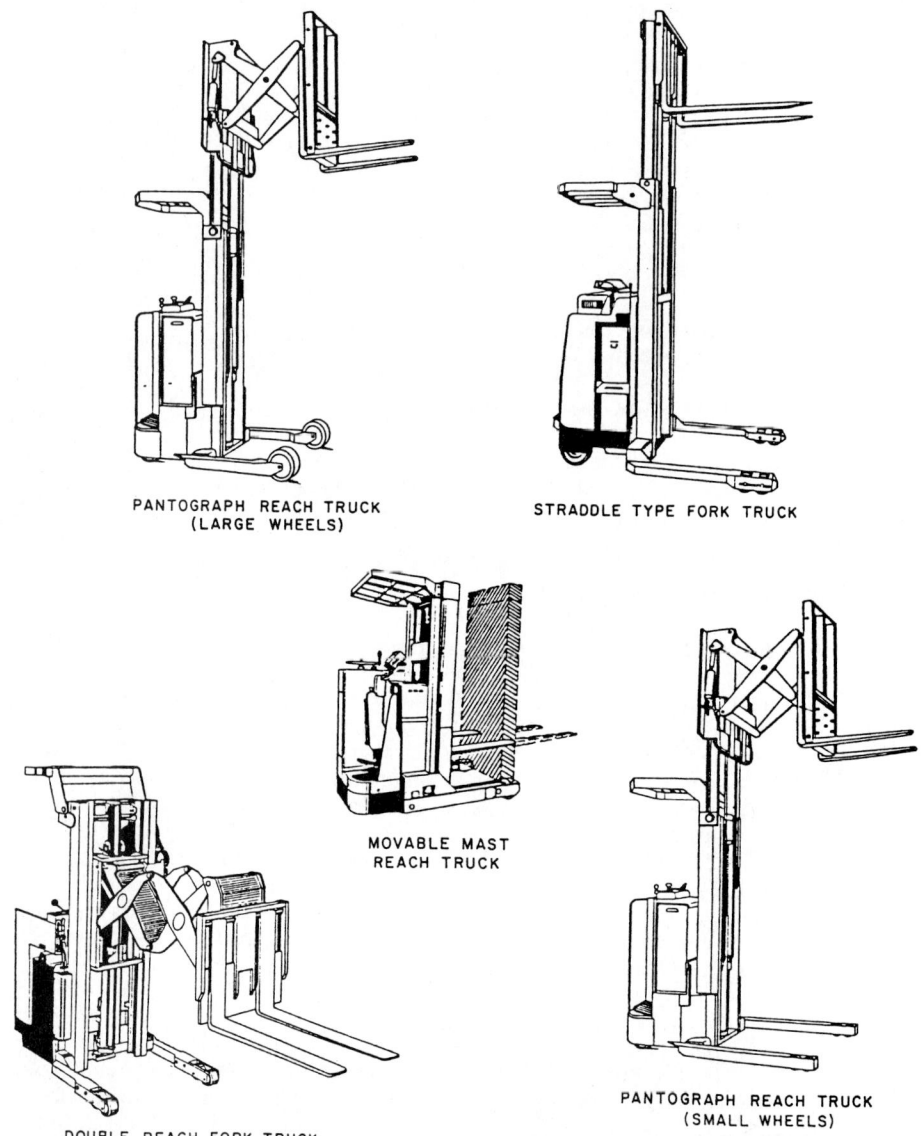

PANTOGRAPH REACH TRUCK
(LARGE WHEELS)

STRADDLE TYPE FORK TRUCK

MOVABLE MAST
REACH TRUCK

DOUBLE REACH FORK TRUCK

PANTOGRAPH REACH TRUCK
(SMALL WHEELS)

FIGURE 4.15   Examples of reach-type forklift trucks. (*Courtesy The Sims Consulting Group, Inc.*)

A variation on the basic turret truck is the hybrid vehicle. On these machines, the operator rides up and down in a mast-mounted cab, and there is a transverse track and a small auxiliary mast on the front of the cab. The fork turret is mounted on the auxiliary mast. This configuration permits the operator to more accurately manipulate the forks and to pick orders from the same vehicle used to place and remove pallets. In all other features, these machines perform like a basic turret truck.

Another variant on the basic turret truck is the swing-mast machine. This unit is based on a heavy-duty standard counterbalanced forklift truck chassis with the mast mounted on a transverse track carriage and hinged to swing 90° to the right. This configuration allows

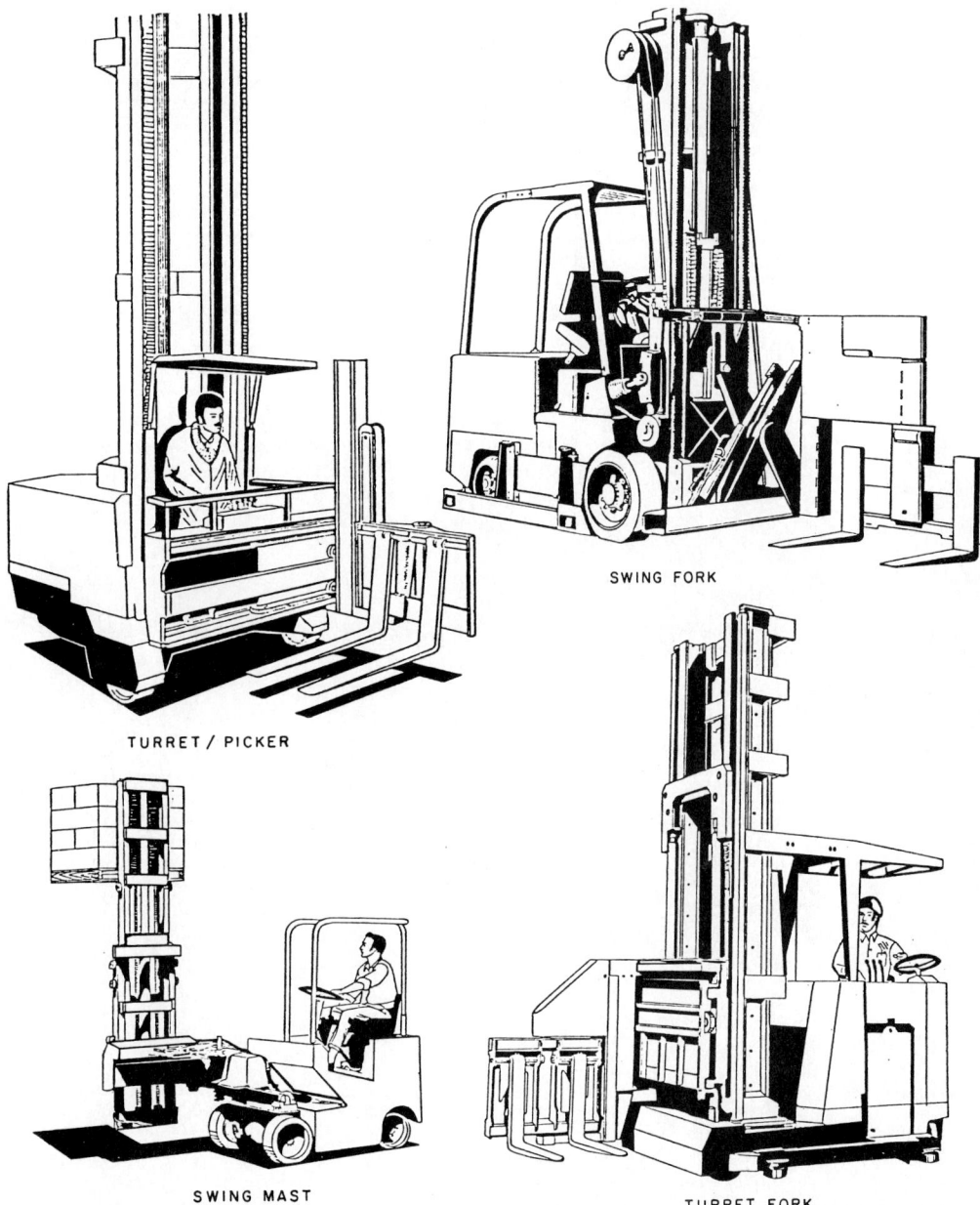

TURRET / PICKER

SWING FORK

SWING MAST

TURRET FORK

FIGURE 4.16 Typical narrow-aisle, high-lift turret-type forklift trucks. (*Courtesy The Sims Consulting Group, Inc.*)

the machine to pick up and drop pallets on its right side without turning in the aisle. To service the other side of the aisle, the machine must leave the aisle and reenter in reverse. These swing-mast machines are particularly useful in industrial plants where they can function as a standard counterbalanced forklift in block stacking and truck loading while also servicing extremely narrow aisles (58 inches with a 48- × 48-inch pallet) in rack storage operations. They can also haul long loads by leaving the forks in the turned position and carrying the pipe or other long items on the side of the vehicle. These machines can handle loads of up to 6000 pounds and lift pallets to 30-foot elevations.

Side-Loading Forklift Trucks.   These are being partially replaced by the turret-type machines. However, the large side loaders are still in use for handling pallets, large loads such as cargo containers, and long loads such as lumber and bundles of steel or pipe which are to be stored in cantilever racks. These machines usually have a track-mounted mast which moves across the chassis to pick up and place the loads. They can also be equipped with pantograph-mounted forks so that they can store two pallets deep in racks.

Automated Guided Vehicles.   AGVs are becoming more common in manufacturing and warehousing operations. See Fig. 4.17. The basic guidance capabilities are being greatly enhanced by new technology and different guidance methods. The functions are also being expanded to include forklift operations, tooling functions, robotics, and many different types of transport activities. The automated guided vehicle is no longer simply a robotic transport device. It has been integrated into automated manufacturing cells and assembly operations. It is used in offices, hospitals, factories, warehouses, and laboratories. It replaces manned vehicles and conveyers, and in many cases serves as an in-transit workbench for assembly and tool loading.

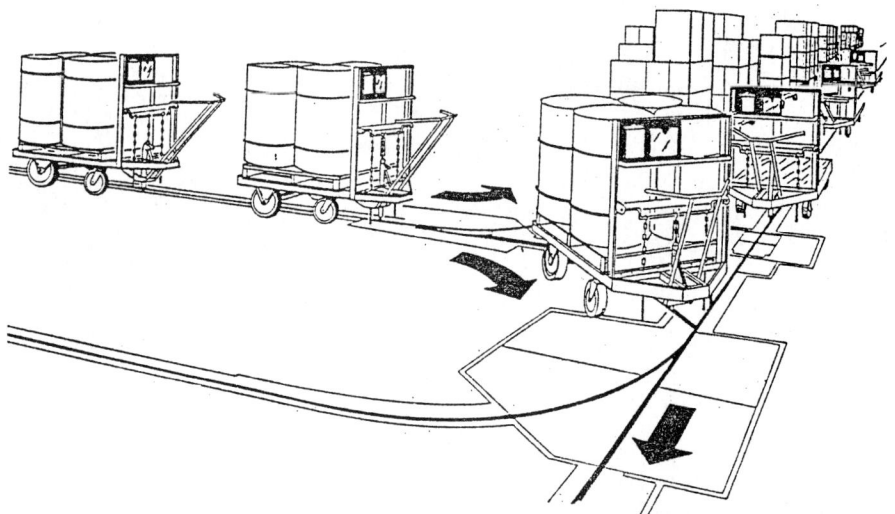

FIGURE 4.17   A typical towline system operation. (*Courtesy The Sims Consulting Group, Inc.*)

There are several methods for the guidance of AGVs, but three are the most common: wire guidance, optical guidance, and on-board computer control and navigation with en route fixes. The earliest and most common method uses wire guidance. In this case, a wire is embedded in a small slot in the floor, and the machine tracks the wire by sensing the alternating-current field around the wire. The wire can also be used as a medium for transmitting information and instructions to the AGV and thereby control its stops and route in a complex system. This same method is used to guide turret trucks and order-picking machines in pallet-rack aisles. In such cases, the wire can transmit instructions to the driver by presenting data on a digital or cathode-ray tube (CRT) readout on the machine.

Optical tracking of a stripe on the floor is another technique which has been used for a long time. In its basic form, optical tracking is less reliable than wire guidance because of the dirt and erasure factors. Recently, however, this method has been enhanced by the use of ultraviolet light and fluorescent paints which activate color-sensitive readers on the machine and transmit signal codes for guidance and instructions. When coupled with radio or infrared instructions and communications from a computer, this has proven to be a very effective system.

A variation on this system uses a steel ribbon located under a carpet or linoleum floor covering in an office or hospital. In this case the vehicle senses the metal and tracks the ribbon or electrical energy transmitted from the ribbon. The AGV receives instructions either from a preprogrammed on-board computer or by en route radio or infrared signals from the computer system or a stop station.

In all these systems, prepositioned floor markers can be used to instruct the AGV to stop, turn, load, unload, or return to base. These can be physically, optically, or electronically transmitted signals. Bar-code technology can also be used to control some systems.

In some very sophisticated systems, the AGV has an on-board computer and navigation system. In some cases, these systems are based on inertial guidance, as used in aircraft and missiles. These machines do not follow a wire or path. They are either preprogrammed or taught their work path and functions. The instructions are stored in the on-board computer and the machine starts and stops on either operator or computer command. As it progresses, the navigation system is checked by the machine's reading of a bar code fix on the wall of the building or a column with a laser scanner, or by receiving a signal via infrared or radio to correct its position information or change its programmed instructions. In some cases, these machines are used in hostile environments, clean rooms, or complete darkness.

The AGV can also serve as the platform, or "bench," to hold a tool pallet and fixture for loading parts into a manufacturing cell. It then travels to the cell and transfers the loaded fixture and tool pallet into the machine. In assembly applications, the AGV can carry components and parts to the workstation and, by using on-board lifts or robots, place the parts into the product being assembled.

Towline Systems and Tractor-Trailer Trains.   These have mass movement capability and a great deal of flexibility. The towline cart is drawn along an in-floor chain-powered track and can be attached at any point or diverted to a destination on instructions. In each case, the trailer or towline cart functions as a unit load or common denominator. It is capable of moving loads as an element or part of the system or, when off the system, as a manual cart and individual unit.

The trailer train simply tows a series of four-wheeled carts and functions like a highway truck. It picks up and delivers carts on a schedule or at random and is controlled by the tractor driver. In some cases, when repetitious paths are used, driverless tractors follow a wire like an AGV and perform in the same computer-controlled or preprogrammed manner. Both driverless and operator-driven trailer trains can vary their routings, cargo, and number of carts.

THE INTEGRATED MATERIALS-HANDLING SYSTEM

Automated Guided Vehicles and Their Application.   An integrated handling system requires an understanding of AGV abilities, limitations, and cost. The very sophisticated machines can do almost any task, at a price. See Fig. 4.18. The more standard AGVs can be an economical replacement for a manned vehicle by eliminating labor cost. This is especially true over long and repeated (or variable) travel paths or in situations where waiting time is a factor. When many and variable routings and delivery decisions are required, a manned vehicle with two-way radio communications and dispatching often has the advantage over the AGV.

Use of Robots in Materials-Handling Systems.   The use of robots is expanding. Robot-based automated pallet loaders are made by several manufacturers, and vehicle-mounted robots

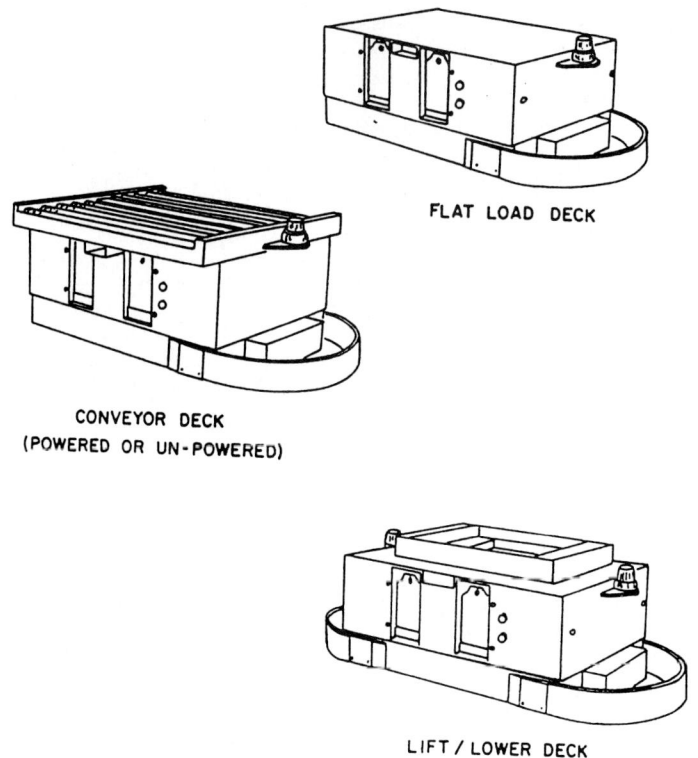

FLAT LOAD DECK

CONVEYOR DECK
(POWERED OR UN-POWERED)

LIFT / LOWER DECK

FIGURE 4.18   Typical automated guided vehicle configuration. (*Courtesy The Sims Consulting Group, Inc.*)

are being developed for order picking and clean-room manufacturing and assembly operations. Robots are almost commonplace in the loading and unloading of machine tools and in light assembly operations. Robots can replace workers in any fixed-location, limited-area handling situation. If mounted on an automated vehicle, a robot can provide mobile handling. In many cases, robots interface with conveyers, are mounted on tracks or AGVs, or are used to load and unload in plant vehicles, pallets, and tote boxes. The primary limitations to robot applications are the need to design sufficiently flexible mechanical movements and end effectors to deal with the variability of the tasks. In most systems, the integration is achieved by the use of automated guided vehicles, robots, and random-movement work pallet conveyers.

Coordination of Equipment and Vendors to Make a System.   This requires the materials-handling engineer to write a clear and complete operating specification and have a broad knowledge of both the technology of the equipment and the nature of the supplier market. The system designer must interpret the business and operating requirements into a system concept, prepare an overall operating concept, develop a preliminary system design, and screen the market for the best available suppliers to provide the equipment. After selecting the suppliers and jointly developing the detailed plans with them, the materials-handling engineer must mediate any technical or administrative conflicts or overlaps in the system and coordinate the designs to assure compliance with the operating specifications, smooth equipment interfaces, and proper installation scheduling. As a part of this task, the designer must be sure that the contracts and warranties are valid and nonconflicting. In most cases, either the owner or a prime contractor is assigned the task of managing the implementation and coordinating the vendors.

SUMMARY

Materials handling is the physical manifestation of the flow of money through an enterprise or an economy, and the control of materials flow is the control of the enterprise. It is therefore essential that we recognize this all-important and *all-cost* function as a primary target of industrial engineering attention. Every penny saved in materials handling is a penny earned in profit because materials handling does not change or improve the product or its value. To deal with materials handling, one must know the business issues, have the business facts, know the equipment market, and apply good industrial engineering and mechanical engineering principles. The computer has become an integral part of the materials-handling system, and, in many cases, the use of information techniques can reduce the need for and the cost of materials handling. In addition, the computer is essential to the management and control of the materials-handling system. Materials handling is an integral part of every operation in every enterprise. Materials handling is an all-cost activity which must be reduced, mechanized, or eliminated.

BIBLIOGRAPHY

Books

Ballou, R. H., *Business Logistics Management, Planning and Control,* 2d ed., Prentice-Hall, Englewood Cliffs, N.J., 1973.

Bowersox, D. J., E. W. Smykay, and B. J. LaLonde, *Physical Distribution Management—Logistics Problems of the Firm,* rev. ed., Macmillan, New York, 1968.

Francis, R. L., and J. A. White, *Facilities Layout and Location—An Analytical Approach,* Prentice-Hall, Englewood Cliffs, N.J., 1974.

Guelzo, C. M., *Introduction to Logistics Management,* Prentice-Hall, Englewood Cliffs, N.J., 1986.

Johnson, J. C., and D. F. Wood, *Contemporary Physical Distribution and Logistics,* 3d ed., Macmillan, New York, 1986.

Kulwiec, R. A., *Materials Handling Handbook,* Wiley, New York, 1985.

Muther, R., *Systematic Layout Planning,* Cahners, Boston, 1974.

Robeson, J. E., and R. G. House, *Distribution Handbook,* Free Press, New York, 1985.

Salvendy, G., *Handbook of Industrial Engineering,* Wiley, New York, 1982.

Sims, E. R., Jr., *Planning and Managing Materials Flow,* Industrial Education Institute, Boston, 1968.

Sims, E. R., Jr., *Planning and Managing Industrial Logistics Systems,* Elsevier, Amsterdam, 1991.

Sule, D. R., *Manufacturing Facilities Location, Planning, and Design,* PWS-Kent, Boston, 1988.

Tompkins, J. A., and J. D. Smith, *The Warehouse Management Handbook,* McGraw-Hill, New York, 1988.

Tompkins, J. A., and J. A. White, *Facilities Planning,* Wiley, New York, 1984.

Warehouse Modernization and Layout Planning Guide, NAVSUP Publication 529, United States Navy, Naval Supply Systems Command, Washington, D.C., 1985.

Journals and Periodicals

Distribution, Chilton, Radnor, Pa.

Journal of Business Logistics, Council of Logistics Management, CLM Publications, Oak Brook, Ill.

Material Handling Engineering, Penton, Cleveland.

Modern Materials Handling, Cahners, Newton, Mass.

Traffic Management, Cahners, Newton, Mass.

Transportation and Distribution, Penton, Cleveland.

CHAPTER 5
PRODUCTION LINE TECHNIQUES

Richard Muther
President
Richard Muther & Associates, Inc.
Kansas City, Missouri

A production line is the chief way to produce large quantities of standardized items at low cost. It is basically an arrangement of work areas. See Sec. 13, Chap. 3.

In its most refined state, line production is an arrangement of work areas where related operations are located immediately adjacent to each other, where the material moves continuously and at a uniform rate through a series of balanced operations which permit simultaneous performance throughout, and where the work moves toward completion along a reasonably direct path. These complete refinements, however, are not necessarily required.

Line production may be applied to operations other than production. Disassembly, inspection, repair and rework, overhaul, and salvage operations all may use lines. A cafeteria, physical examinations for soldiers, and even an after-party family dishwashing project may take advantage of the production line concept.

PREREQUISITES

Although the popular concept is a straight, conveyerized assembly line, these characteristics are not required. Prerequisites that must be present include:

Quantity. The quantity or volume of production must be sufficient to cover the cost of setting up the line. This depends, then, on the rate of production and the length of time the job will last. The quantity must be of a single standardized part or product or family of basically standardized products.

Balance. The required times for each operation on the line must be approximately the same. Times should be available for each operation and the equipment and personnel should be synchronized to a common balancing factor—usually expressed as time per workstation. This is typically termed: avoidance of bottlenecks.

Continuity. Once started, the line must continue to flow, for a stop at one point starves the rest of the operations. So precautions must be taken to assure a reliable supply of material, parts and subassemblies, and freedom from breakdown of equipment.

HOW TO PLAN A LINE

Further, no two companies operate the same. The nature of the product, requirements, and facilities vary with every line. Still, there are three basic determinants. These will be covered in the following discussion.

Product or Material.   Every plant layout—including those for production lines—begins with analysis of the product or material. The product itself should be checked to be sure it is designed for ease of production, not merely for functioning. Can the parts be made and assembled according to the drawings and specifications? Can they be changed so that they can be made and assembled more easily?

Quantity.   The quantity scheduled must be obtained and developed into a specific rate of production. From product and quantity, the engineer begins to plan the operations, their sequence, and the equipment required.

In dividing the product into components, generally try to keep as many operations as possible off the final assembly line. It is easier to do operations as part of the forming or subassembly operations, where the material is more accessible. The fewer the operations, the shorter is the line and the lower the investment in expensive assembly fixtures.

Process and Equipment.   Lining up the operations required, the capacity or number of pieces of each type of equipment selected, and the sequence of operations is an early step in planning a line. Right from the beginning, tie this back to the rate of production required:

Units per hour = (quantity per month + scrap)

÷ (number of working days per month

× number of hours per day the line will operate)

This is the rate which all workstations and equipment must meet if the line is to function as a unit.

Many times it is better to supply a line with equipment that is slower than the most efficient machine might be; there is little sense in planning overcapacity for one operation. Many plants slow down machines to the capacity of the line rather than have an operation that is so out of phase that it results in rehandling.

One major point: in tooling for line production, it is fundamental that the product design be fixed. To get tooled up and then have design changes come along is far more serious than in layout-by-process plants. With line production, just one change may upset an entire sequence of highly synchronized operations.

MOVEMENT OF MATERIALS

Movement of material is a fundamental part of planning the production line. It is the thing that ties all operations together. It is relied on to maintain the continuity of the line, and it must assure the delivery of parts and subassemblies as needed. The handling device must therefore be planned right into the line. Some lines, in fact, are built entirely around a basic handling device. Figure 5.1 illustrates the construction of a house on a conveyerized assembly line.

Conveyers form a useful part of many lines, but simple hand-passing, chutes, or portable wheeled fixtures are often more economical and usually more flexible.

Handling devices serve several purposes that should be recognized:

Transporting: Movement to, from, and along the line

Pacing: Maintaining a steady, uniform output

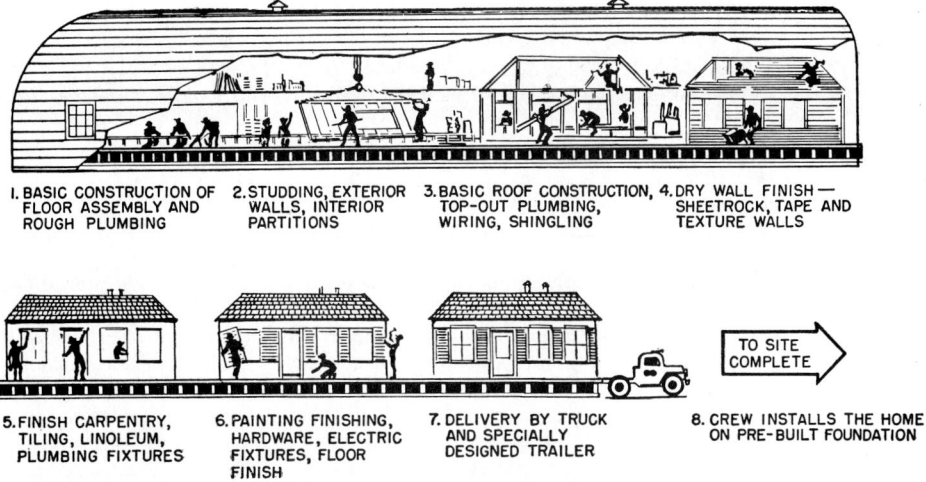

1. BASIC CONSTRUCTION OF FLOOR ASSEMBLY AND ROUGH PLUMBING
2. STUDDING, EXTERIOR WALLS, INTERIOR PARTITIONS
3. BASIC ROOF CONSTRUCTION, TOP-OUT PLUMBING, WIRING, SHINGLING
4. DRY WALL FINISH — SHEETROCK, TAPE AND TEXTURE WALLS
5. FINISH CARPENTRY, TILING, LINOLEUM, PLUMBING FIXTURES
6. PAINTING FINISHING, HARDWARE, ELECTRIC FIXTURES, FLOOR FINISH
7. DELIVERY BY TRUCK AND SPECIALLY DESIGNED TRAILER
8. CREW INSTALLS THE HOME ON PRE-BUILT FOUNDATION

TO SITE COMPLETE

FIGURE 5.1 House production.

Holding the work: Convenience and reduction of nonproductive manual handling

Storing: Especially for temporary reserves or cushions between operations

The selecting of the handling equipment will depend on the characteristics of the item and the moves to be made. Where the equipment is a portable rack or a holding fixture, do not overlook planning for its return to the head of the line.

Certain handling options are involved:

• Move a single piece or a lot (batch)

• Remove work or leave it on the conveyer

• Continuous or intermittent movement

• Recognition of work by color, bar code, position on conveyer, tags on product, automatic or timed diverters, electric eyes, readers, sensors, and so forth

In distributing the work along a line from one group of workers to another, automatic sweep-offs or diverters can be used to feed the parts to an accumulation area close to the point of use for the following operation. See Fig. 5.2.

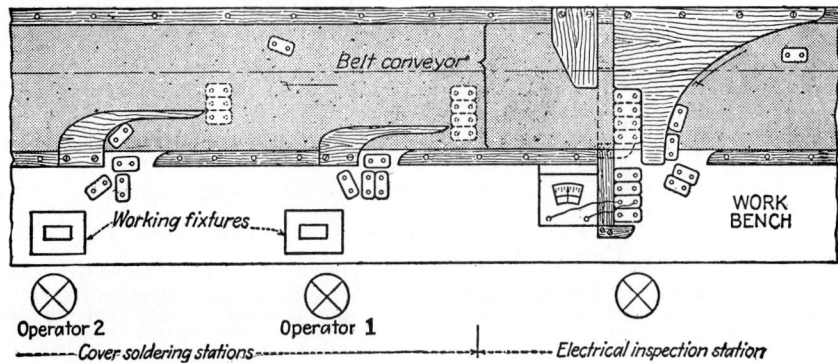

FIGURE 5.2 Simple distribution device.

Note that some conveyer speeds are so high that an operator must perform each cycle exactly on time or miss the operation. This leaves no allowance for bound-to-happen delays caused by hang-ups, dropped tools, or irregularity in parts, for example. More work can often be attained by slowing the conveyer and placing work on the conveyer closer together.

LINE BALANCE

Ensuring that all operations consume equal amounts of time and that that time is sufficient to meet the desired rate of production is the problem of line balancing. The desired rate of production is converted to a time per piece; this is called by different names: the *balancing factor, balancing time, cycle time,* or *station time.* This balancing factor is equal to the reciprocal of the rate of production; that is, it equals 1 divided by the rate of production.

Perfect balance is rarely achieved; there is always some extra time in at least one operation. However, an operator with idle time to balance may often be assigned additional work such as handling material to the line; extra inspections of work; applying lubricant, labels, or tags; and even more lengthy operations when the work is allowed to bank up at the workstation. Because of the difficulty of dividing machine operations, it is far harder to balance forming or fabrication lines than assembly lines, where the assembly time can be split at many places and the workers moved accordingly.

The term *bank* (or buffer) is commonly used for the accumulation of material waiting for an operation. Banks are common where parts have irregular amounts of work to be done on them. Banks are also used where parts are handled as a lot or batch on the line, where materials are delivered to or removed from the line at irregular intervals, or where a given machine or workstation is also used for some other part or product periodically.

Banks may be located directly in the line. Two examples are hoppers between punch presses, into which parts from the previous operation are elevated, and additional lengths of chute or roller conveyer, on which parts can accumulate but index themselves forward by sliding. Or the bank may be beside the line on a shunt, over-the-line shelf, or spare conveyer. These arrangements are far better than removing the bank to a separate storage point.

Where banks are held for protection rather than close balancing, they must be adequate to protect the operations subsequent to the point of interruption of flow. This means that the anticipated interruption or delay time divided by the station time or balancing time per piece will determine the number of pieces in the bank:

$$\text{Bank size (no. of pieces)} = \text{rate or production (pieces/h)} \times \text{time of interruption or delay (h)}$$

$$= \frac{\text{time of interruption or delay (h)}}{\text{balancing factor or station time (h/pieces)}}$$

LINE SPEED AND LENGTH

Speed of flow bears a direct relation to rate of production and space per workstation:

$$\text{Speed of line (ft/h)} = \text{rate of production (pieces/h)} \times \text{station length or space per piece (ft/piece)}$$

$$= \frac{\text{station length or space per piece (ft/piece)}}{\text{balancing factor or station time (h/piece)}}$$

The station length or space is dependent on the size of the part or unit, the room required by workers and equipment, and the amount of work to be performed there:

Length of line = station length or space $\times$ no. of stations
= station length or space (ft) $\times$ overall time for
completing piece or unit (total line h/piece)
$\div$ balancing factor or station time (h/piece)

METHODS OF GETTING BALANCE FOR FORMING OPERATIONS

These include:

- Improve the slow operations
- Change machine speeds
- Bank material and operate the slower machines overtime or on an extra shift
- Divert excess pieces to other machines not in the line
- Combine or group different items on combination lines

METHODS OF GETTING BALANCE FOR ASSEMBLY OPERATIONS

These include:

- Divide operations and apportion the elements
- Combine operations and balance groups
- Have operators move
- Improve operations
- Bank material and do slower operations on extra time
- Improve operator performance, particularly at the bottleneck operation

MECHANICS OF BALANCING

Balancing a line involves establishing:

1. The rate or production
2. The operations necessary and their required sequence considerations
3. The time necessary to perform each operation and preferably each element of it

These become prerequisites. For the line to function, it should be designed for a given rate of production; for it to operate as a unit, the operation times should be such as to let the material flow evenly. Time study, predetermined motion times such as those from methods-time measurement, standard time data, and machine capacity come into their own here. They are fundamental to establishing a balanced line, except in those cases where an overall time is estimated and the workers are left to balance out the operations themselves. Worker balancing can be done on simple, worker-paced operations when there is room for banking material between operations and when there is some form of group incentive. Although it works in many situations, it usually results in overly large protective banks and cannot be considered a precise method of balancing the line.

The steps in balancing are

1. Investigate the products, quantities, types of processes, related supporting services and time factors involved, and clarify the operations to be accomplished on the line.

2. Interact the work content (sum of the operations) with the sequence (precedence restrictions) of the operations and establish a quantified precedence diagram. In a small project, this may be as simple as a precedence diagram with times written next to the operation symbols. In larger, more complex projects, a computer algorithm may be employed. Such an algorithm is especially helpful when it is necessary to rebalance the line frequently.

3. Integrate the allocation of operations to workstations or groups and develop preliminary work assignments. Typically, several alternative assignments will be developed. During this integration process, it may be appropriate to adjust operations by combining or dividing operations or changing the work methods.

4. Modify and refine the preliminary assignments into alternative line-balance plans. It is important in this step to use whatever charts, diagrams, and other graphical tools are available to be sure the alternatives are clearly understood by the operating supervision.

5. Evaluate the alternative plans, on both tangible and intangible bases, to get a selected line-balance plan. Recognize that it is necessary that the plan be accepted—by both management and the operating people who will implement it.

This line-balance plan should be worked out before the line is installed. After installation, rebalancing or adjusting the balance is common, and the effectiveness of these procedures depends on how accurately the original balancing was done.

Figure 5.3 shows a line-balance work sheet. The product, quantity, and target cycle times are recorded at the top of the form. The operations are listed, together with the time required per unit for each operation. Sequence requirements are recorded for each operation.

From this, a precedence diagram is drawn (Fig. 5.4) and operation times are written next to each respective operation. The operations are grouped into four workstations of 8, 9, 10, and 10 minutes. In this alternative, both stations 3 and 4 are *pacing,* having the greatest amount of work time allocated to them. The amount of time lost to balance for each station is shown as a negative number. For this alternative, a total of 3 minutes per unit is lost to balance.

Alternative assignments of operations to stations are listed on the line-balance work sheet in the alternative assignment columns. The actual cycle time and time lost to balance is shown at lower left. Each alternative can then be compared to others to determine the preferred plan.

In balancing with a paced conveyer, care should be taken to arrange the line so an operator who occasionally cannot finish in time does not starve the following operator. To overcome this, some lines hold in each workstation a small bank of pieces that have been previously completed. When there is a delay, one of these pieces is placed on the conveyer. Where large assemblies are involved, utility workers may step in and help on operations that are temporarily taking more time than the station time allowed. Still other plants have warning signals, time-indicating lights, bells, clocks, markings on the conveyer, and the like to keep operators aware of the time they have left to complete their cycle.

Further points in balancing include:

1. Save *light* operations—those with lost or idle time to balance—for breaking-in or training stations for new operators.

2. Keep the first operation light to be sure the head of the line always feeds in the work on time.

3. Balance into the line the inspection, material handling, and other supporting or service operations so they are synchronized with the line.

LINE BALANCE WORK SHEET

Plant _Woodinville_   Project _New Assy. Line_
By _EFW_   With _MRN_
Target Cycle Time _11 Minutes_ /Unit   Date _8 May_   Sheet _1_ of _1_

Product _New Electronic Conversion Device_
Quantity _10,000 units/year_

No.	Description	Must Be Preceded By	Time / Unit	A	B	C	D	E	F
					STATION / GROUP WORK ASSIGNMENT ALTERNATIVE				
1	Get base & place in fixture	—	5	1	1	1	1		
2	Affix label / interlock to base	1	4	1	2	4	2		
3	Assemble bracket to base	1	3	2	1	1	1		
4	Assemble module #1 to bracket	3, 1	5	2	2	3	2		
5	Assemble insulator pad to bracket	3, 1	10	3	3	2	1		
6	Assemble module #2 to bracket	5, 3, 1	3	4	4	3	2		
7	Install connection buss to modules	6, 5, 3, 1, 4	3	4	4	4	2		
8	Assemble top cover to unit	7, 6, 5, 3, 1, 4, 2	4	4	4	4	2		
9								ⓐ	
20									

Alt.	Actual Cycle Time	Time Lost to Balance		Referenced Notations
A	10 minutes / unit	3 minutes / unit	a	Two lines with 2 stations each
B	10 minutes / unit	3 minutes / unit	b	
C	11 minutes / unit	7 minutes / unit	c	
D	9.5 minutes / unit	1 minute / unit	d	
E			e	
F			f	
			g	

WORK ALLOCATION SUMMARY

Richard Muther & Associates 560

FIGURE 5.3 Line balance work sheet.

4. A U-shaped line often allows operators to help each other and to combine the first and last operations for better balance.

5. Build in team motivation by shortening lines and having more of them, instead of using lengthy, less flexible lines.

PREPLANNED BALANCING

Preplanned sheets for determining the best line balance are helpful. For example, in a conveyerized filling-and-packing line, a sheet like the work/speed balance work sheet can be applied (Fig. 5.5). The operations and tasks are listed at left, the time data are recorded in the center, and alternative assignments of staffing are made at the right.

We assume a suitable methods study of the operations has been done and the task times have been determined from time study, predetermined times, standard data, and so forth. The standard minutes per unit are entered in the fifth column. Maximum speed per operator (in units per minute) is the standard minutes per unit divided into 1.

In the example shown, the first two and the last two tasks are combined into one per-

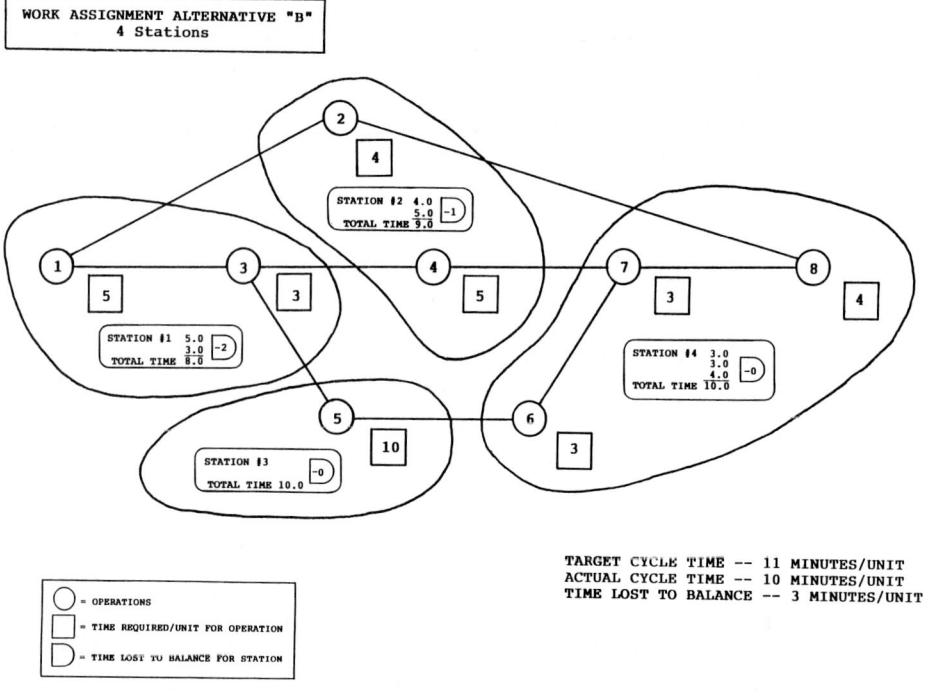

FIGURE 5.4 Work assignment alternative B.

son's assignment. The person with the lowest speed will limit the overall line output. That speed is indicated at the top of the column (55 units/minute in the first column). The limiting standard minutes per unit is recorded directly above (0.0182 minutes per unit).

At the bottom of the column, the total minutes per unit allowed for the work crew assigned (0.0727) is determined by dividing the total people assigned (4) by the speed (55). As each person is added to the work crew (usually at the previous limiting task and shown by a circle on the figure), the new limiting task is determined, and its time and speed are indicated at the top of the next column and the total minutes per unit allowed are indicated at the bottom. The labor utilization is calculated for each indicated speed and the "optimum" crew size is noted (in this case, 93 percent at 9 persons).

AUTOMATIC LINES AND COMPUTER SIMULATION

Characteristic of such industries as food, beverage, and toiletries are automatic production lines. Here unit packages (bottles, cans, jars, boxes) are cleaned, filled, capped, labeled, and packed for final shipment. The operations are performed with automatic equipment, typically running independently of each other, but linked in a production line by unit conveyers. Outputs of such lines are controlled by overall line speed, usually paced by the slowest machine in the line. Production workers assigned to the line are placed to monitor the continuous flow and to restore operating conditions after interruptions by package jams, machine malfunctions, missed cycles, and the like. The conveyer typically slides under the package when a machine goes down, allowing an accumulation bank right on the conveyer.

Computer simulation is a modern way of studying ways of improving such automatic

WORK/SPEED BALANCE						Plant _Rimville_ Project _9-2872_						
						By _JRL_ With _-_						
						Date _9-16_ Sheet _/_ of _/_						

Product No. _12-70_	Product Description _Hair Conditioner - sample_											
Operation _Fill, cap, label and pack_									Dept. _9_		Line/Cell No. _2_	

Line No.	Operation/Task Description	Control	Max Speed Per Oper.	Std. Min Per Unit	No. of People for indicated speed. Mins/Unit & Units per Minute							
					.0182 55	.0106 94	.0103 97	.0091 110	.0061 165	.0053 189	.0052 192	.0051 195
1	Supply carton & caps	M	200	.0050	⎱1	①	2	2	2	②	3	3
2	Feed bottles	M	180	.0056	⎰							
3	Fill & cap	A	195	.0051	/	/	/	/	/	/	/	①
4	Label	S	55	.0182	①	2	2	②	③	4	4	4
5	Pack (Bottles in carton)	M	120	.0083	⎱1	/	①	2	2	2	②	3
6	Palletize Cartons	M	500	.0020	⎰							
7												
18												

	.0442								
Total minutes/unit required →									
Total People →	4	5	6	7	8	9	10	11	
Total minutes/unit allowed →	.0727	.0532	.0619	.0636	.0485	.0476	.0621	.0564	
Labor utilization: Total minutes/unit required / Total minutes/unit allowed →	61%	83%	71%	68%	91%	93%	85%	78%	

Controls: Manual (M) Semi-automatic (S) Automatic (A)

Richard Muther & Associates 562

FIGURE 5.5 Work/speed balance work sheet.

production lines. A simulation model of the entire line's configuration, including the performance characteristics of each machine therein, can be developed to test improvement options via the computer without interruption of the physical line. That is, the model is used to evaluate potential changes, instead of tampering with the real line.

Figure 5.6 shows a block flow diagram of an automatic line for filling and packing jars of apple sauce. Empty jars are first rinsed with water, then filled with the sauce in a rotary multihead filler. The jars are then conveyed to a machine that seals a metal screw cap to each jar. They proceed to a roll-through labeler, where paper product or brand labels are affixed to them. The jar stream (flow) splits, after the labeling machine, via a conveyer diverter, into two case packers that place 12 jars into a partitioned corrugated shipper which becomes the finished-goods case after the flaps are sealed. The entire production line is paced by the filling machine that operates at a nominal speed of around 450 units/minute.

Performance of the filler is affected by one of these three conditions:

1. Downtime of the upstream (preceding) jar rinser causing the empty jar stream into the filler to be stopped
2. Downtime internal to the filler itself caused by such things as jams at the in-feed, irregular jars, and sauce flow interruptions

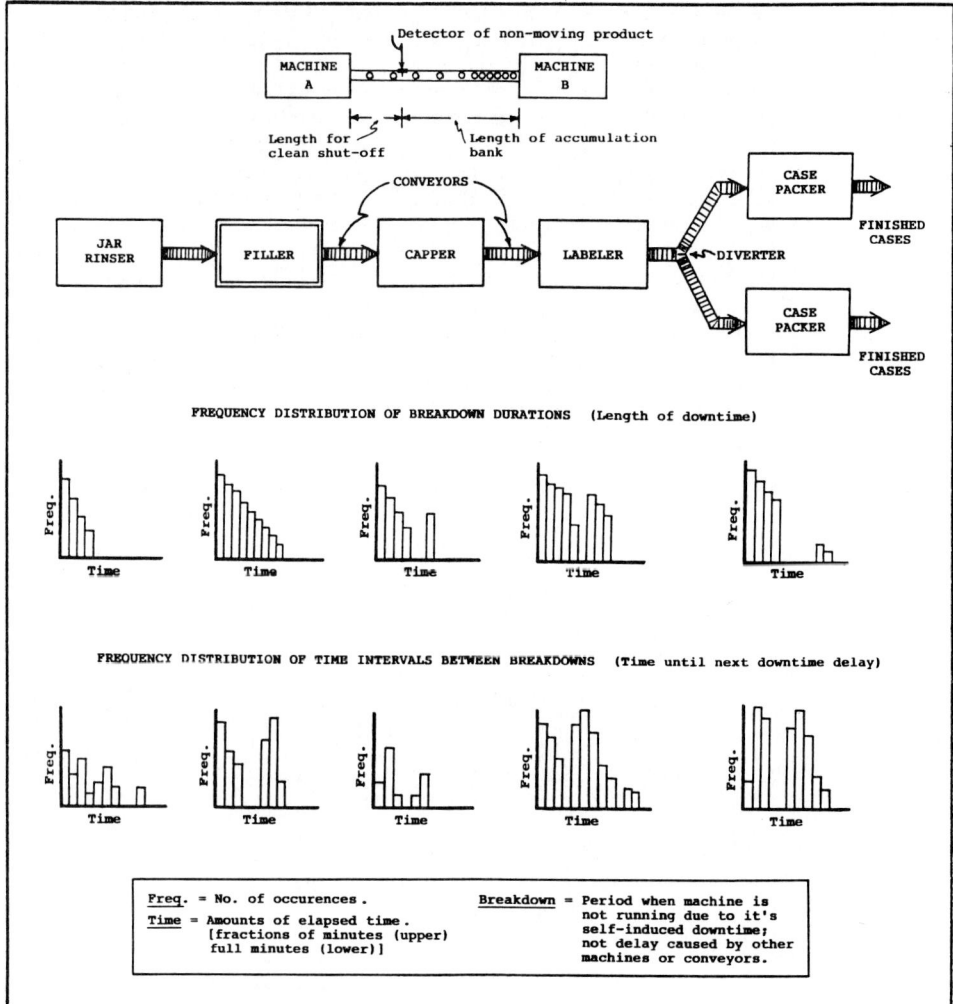

FIGURE 5.6 High-speed line computer simulation.

3. Downtimes at any of the downstream (following) machines that would fill up the conveyers and cause automatic shutoffs of the filler.

Stoppages of the filler are not instantaneous for conditions 1 and 3 because the between-machine conveyers have certain lengths on which product (jars) will accumulate, or bank. The accumulation capacity depends on the length of conveyer and the location of the detector or sensor triggering the machine shutoff along the conveyer.

The simulation model creates a representation of the line's operation, machine by machine. As a result it involves certain methodical measurements such as length of conveyer and placement of the sensor, as well as the operating characteristics of each machine (uptime, how long the machine is down, when the downtimes occur, and how long before it goes down again). Base data providing a true representation of these characteristics is collected, validated, and verified with actual line performance.

For the filling line described here, an engineering study team was assembled to record and compile line downtime statistics continuously for 2 operating days. The most important of these for model-building purposes were the individual machine downtime profiles. The lower portion of Fig. 5.6 diagrams the key elements of independent machine downtime, that is, frequency distributions of (1) breakdown duration times and (2) time intervals between breakdowns. The remaining line operating statistics collected during the study period were then used for model validation and verification.

The basic construct of the model—which recreates the functioning of the line machine by machine—is a stream of "transactions" (jars) entering and exiting blocks (machines) based on the internal (within machine) or external (in the conveyer) status of the blocks. (Product cannot enter a machine if it is full and cannot exit a machine to enter the accumulating conveyer if the conveyer bank is full.) The internal status of the blocks are predicated on random draws against the two types of frequency distributions. The computer model is run during a base-condition data-gathering study period, until steady-state conditions (full line with typical flows) are realized.

Then operating statistics are collected by the computer program until the run results match actual line outputs during the study period. The computer-generated statistics from the model are compared to the actual study-period performance results to see whether the model is a valid representation. Only then can hypothesis ("what-if") testing begin. These hypotheses generally take the form of changing machine speed, using different conveyer lengths (capacities), adding accumulation conveyers between machines, and altering speed balances between a machine and its upstream and downstream equipment.

For example:

1. What happens to line output if the conveyers between machines are modeled as infinite in length? How long (in terms of accumulating capacity) do the conveyers become when this is done?

2. What happens to overall performance if all equipment downstream from the filler is set at speeds 15 to 20 percent faster than they are now?

3. If, through improved maintenance, the mean downtime duration of the labeler could be reduced by 30 percent, what is the impact on total line output?

Using the computer simulation model along with what-if hypothesizing like that above can yield dramatic improvements with subsequent changes in equipment. Indeed, the data-collection phase preparatory to simulation modeling—by itself—often provides meaningful insights into line-operating problems that might not otherwise have been quantified.

FLOW PATTERNS

The most popular patterns of material flow are straight, U-shaped (or circular), L-shaped, and spine (or comb) (Fig. 5.7).

Straight lines are easiest to plan, establish, feed, and maintain. The U shape is actually the most popular. The spine accommodates operations that are not in a consistent process sequence.

PLANNING MANUFACTURING CELLS

Small production lines are often termed *manufacturing cells*. A cell, as in line production, is a series of operations that are physically located together and usually dedicated to working only on certain designated parts or products.

FLOW PATTERN	PLAN	BENEFITS
Straight-Through	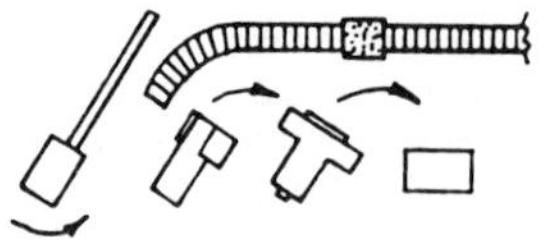	1. Easy to schedule, to follow, and to control. 2. Allows straight, inexpensive handling methods. 3. Easy access on two sides.
U-Shape or Circular	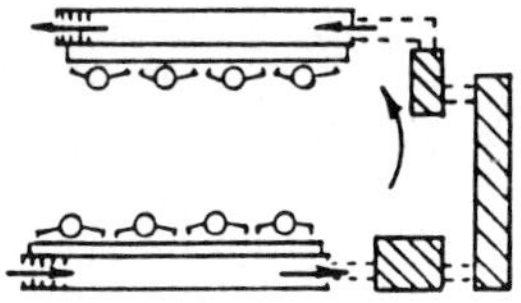	1. Fixtures and/or containers automatically returned to start, probably on same aisle. 2. Better workforce use, by balancing first and last operations and having load and off-load together. 3. Workers in center can help each other.
L-Shape	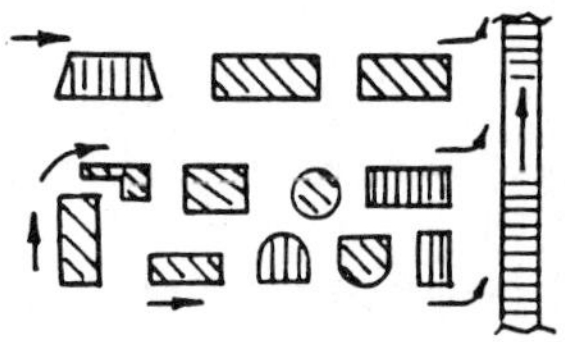	1. Allows fitting lengthy series of operations into limited space. 2. Lets feeding line start on aisle and end at point of use. 3. Easy to split in-flow and/or out-flow of physically different materials, products, supplies, special services.
 Comb or Spine	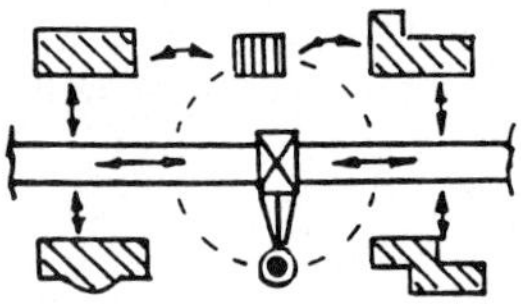	1. Well suited to sequences of operations that change or vary from job to job or part to part. 2. Permits multiple routings with automated integration of process, handling and controls. 3. Supports flexible manufacture with two-level conveyors, return-load dispatching, or robotic handling.

FIGURE 5.7 Basic flow patterns.

- *A product line cell* forms a progressive sequence of successive operations devoted to basically one item.
- *A group technology cell* forms a grouping of different operations dedicated to a group or family of parts or products.
- *A functional cell* forms a cluster or small functional department of similar operations set together to work as an integrated unit.

When the *quantity* (number of pieces per day) is high, and the product and process are conventional, a production line cell is preferred. When the *product* itself has special features or constitutes a group or family of items, the group technology cell is called for. When the *process* (operations and equipment) is dominant—large equipment, costly utilities, special building requirements—the functional cell is likely.

A number of cells working together in sequence form a product-layout department or macrocell, usually arranged as a modified production line.

Planning a manufacturing cell involves four phases (Fig. 5.8):

I. Orientation
II. Overall cell planning
III. Detail cell design
IV. Implementation

In Phases I and III, the planning pattern usually calls for the following five steps:

1. Clarify the parts, their classification, and their features.
2. Establish the process, its operations, sequence, equipment, and number of pieces of equipment required.

I. ORIENTATION

Location, external conditions
and planning the planning

The surrounding physical and
non-physical conditions/
considerations; plus the
objective, situation and
plan for planning the cell.

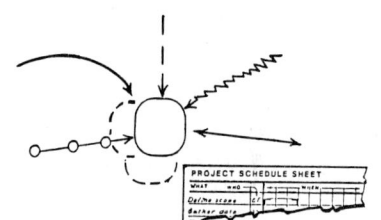

II. OVERALL CELL PLAN

The general plan for the
total cell

The overall (gross) arrangement
of machinery and equipment to
produce parts or assemblies in
an integrated sequence of
operations.

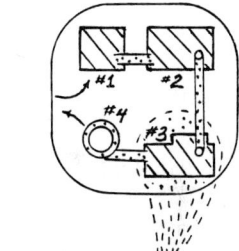

III. DETAIL CELL DESIGNS

Detail designs for each element

Detail (fine) plans or designs
(including documents or specifications) for each element
(machine, tool or control)
within the total cell.

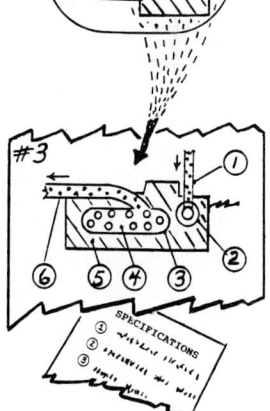

IV. IMPLEMENTATION

Do; take action on the plans

Schedule, provide, procure,
train, install, clean-up,
debug...

FIGURE 5.8 Planning manufacturing cells.

3. Integrate the handling methods, balance and utilization of workers or machines, communications and controls, and any adjustments in utilities or building features.

4. Modify and refine the rough or preliminary plans into viable alternatives.

5. Select and accept, or get acceptance of, the plan.

In the detail cell designs of Phase III, this sequence of steps is applied to each machine and its tooling, controls, and arrangements.

FLEXIBLE MANUFACTURING SYSTEMS

A flexible manufacturing system (FMS)—sometimes called a *versatile manufacturing layout*—involves an arrangement of operating and handling equipment that can readily accommodate changes in the sequence of operations. It provides materials-handling (internal transport) equipment that can move to any machine in any order programmed. Usually this involves preestablished tracking paths which can be wired in, or marked on, the floor so that transport vehicles can be directed individually to the next operation. Vehicles can be controlled by computer.

This automates the handling so that the sequence of material is a less dominant consideration. And it allows layouts by process (function) or by fixed position (fixed location) to be closely synchronized to production-line schedules. In a sense, an FMS lends itself to the classical types of layout much as group technology adapts families of parts and products to production-line types of layout.

Because the material handling is not tied to a fixed path, as in most production lines, an FMS provides a high degree of flexibility for changes in operation sequence, production quantities, variety mix, product size, lead times, machine utilization, machine availability because of breakdowns, and work in process.

STAFFING THE LINE

With line production, an operator generally stays at one workstation. Usually this work is on but one product. This means that (1) a less skilled operator can be used, (2) operators can be trained in a short time, (3) the labor market is widened, (4) production can be gotten under way quicker, and (5) waste motions can be reduced by keeping the work moving to the operator and taking it away.

The number of operators required for any line equals the total time (in standard hours allowed) required to perform all operations on one unit multiplied by the rate of production scheduled (in units per hour) divided by the length of the shift or period worked (in hours per worker). This assumes no idle time to balance. All calculations of this type, therefore, yield minimum or target values rather than actual values.

Some operators prefer not to work on a line. Their reasons include the following:

1. Their output is limited to the speed of the slowest operator on the line.

2. Their opportunity for variety of work and therefore for self-improvement is restricted.

3. They can easily be replaced, so they feel less secure.

4. There is pressure of compulsion, for they must keep pace with the speed of the line.

5. They have less personal freedom.

Other workers feel that line production offers certain advantages:

1. The work is easy to learn, and they can pay their way earlier or earn a bonus sooner.
2. The job becomes more or less automatic, rhythmic, and free from problems or mental concerns. The work is brought in and taken away with little planning, print reading, or judging of tolerances on the part of the worker.
3. Because less attention is required and operators are usually located close to each other, there is more sociability among them.
4. There is greater uniformity in the work, in the effort and skill required of the operator, and in the length of the working day for all workers on the same line.
5. There are fewer nonproductive delay and irritating interruptions.

To free workers from boredom or fatigue, fixed rest periods are generally recommended. Healthy competition and group self-direction gives workers greater interest. Rhythmic patterns of work and rotation or shifting of operators also can be used.

ASSURING CONTINUITY OF LABOR

- Avoiding or reducing absenteeism in the first place by a program of team motivation such as total quality management and/or self directed work groups.
- Transferring workers or borrowing personnel from other departments.
- Shifting workers into a new balance.
- Using alternative predetermined speeds and preplanned worker assignments.
- Using buffer banks between operations or an extra large float of in-process material.
- Grouping workers into teams and encouraging them to adjust the work and the balance and to suggest improvements.
- Scheduling rest periods that synchronize all personnel so the line can flow more uniformly.
- Employing relief operators who generally substitute for regular operators. They relieve each operator on the line at specified times for a specified period. This avoids congestion in rest rooms often caused by rest periods. Where processes must continue to run when once started, as in conveyerized paint booths, the use of relief operators is preferred.
- Calling on utility operators to fill in for absences, to pick up any mistakes or omissions, to train new workers, to check stock of material, to help out on a troublesome operation, and to assist the group leaders.
- Using other methods of overcoming the operator-away-from-the-line problem, such as offering attendance bonuses or imposing tardiness penalties, providing adequate training for extra operators, instituting a daily early check of worker attendance, encouraging social pressure from others in the pay group, and setting up duplicate lines.

Many plants feel that a wage incentive is of little value on production lines. The pace of a conveyer or the inherent pull of the line, they say, holds output at the normal amount. Besides, the increased effort or skill of one individual worker can seldom alter the output.

If an incentive is used, it should generally be a group incentive. In a group, workers have an interest in helping each other, in breaking in new workers, and keeping leaders, supervisors, or stockers notified of impending delays. Group incentives make timekeeping and administrative costs less than for individual incentives, and workers will exert pressure on slow operators.

ASSURING CONTINUITY OF MATERIAL

To keep a line flowing, the material must be right—in quantity, in quality, and in location. The functions of planning, purchasing, material control, tool maintenance, product engineering, and inspection assume great importance in line production plants.

Product Engineering must be sure it has supplied a design that can be readily produced and that will remain constant during the life of the run. Planning must synchronize and schedule the work so everything flows together as needed. Purchasing must ensure the reliability, quality, and dependability of suppliers. Material Control and Production Control must be sure the material is at the right place at the right time. Quality Assurance must guarantee material is current and will not cause delays. Tool Maintenance must ensure that the tools produce correctly.

All these functions must be planned with close ties to suppliers who understand the importance of continuity, and they must operate in a highly coordinated way to get full advantage of just-in-time feeding of parts or subassembled components. The can-making plant directly tied by conveyer to the cannery, with the latter actually scheduling the former, is an example.

Timing and continuity are more important than speed. This applies to both the preproduction planning to get the line into production and the planning and scheduling of the actual operations. And selecting suppliers with proven ability to supply the correct parts when needed is more important than price.

The important relationships in scheduling any material requirements are

$$\frac{\text{Monthly production}}{\text{working days per month}} = \text{daily rate of production}$$

$$\text{Usage per unit} \times \text{daily rate of production} = \text{daily requirements}$$

$$\text{Daily requirements} \times \text{days between deliveries} = \text{quantity per delivery}$$

Some techniques associated with efficient production lines include the following:

1. Store material at the point where it is used rather than in centralized storerooms.
2. Predetermine and identify all storage places along the line, the normal and minimum quantity of each item that should be held there, and the normal delivery quantity and frequency.
3. Use automatic controls or stock checkers to inventory the stock along the line and make sure that it is provided when the minimum bank is reached.
4. For small parts, replenish material along the line each night, being sure that enough is issued for a full day's run.
5. Manage your procurement so suppliers are synchronized closely with your line schedules.
6. Synchronize delivery conveyers to the speed of the final line, but for undependable operations hold a protective bank near the point of use.
7. Hold safety banks in the line or adjacent to it.
8. Have an adequate communication system available for the stock suppliers, with automatic signals for low stock and/or outages, or bar codes and readers to select the correct item.
9. Issue the stock to the line in sets, kits, or lists so that all parts for one unit or lot are supplied.
10. Make the material handlers lean over backward to take proper care of material; otherwise production operators are encouraged to be wasteful of material and careless with their work. Have material handlers report organizationlly to the planning or material control supervisor.

Where process departments feed parts to the line and the operations are faster than line speed, the technique of cycling is used. This procedure schedules the faster process department machines each to produce enough parts for, say, two weeks, but to run several parts, each on that two-week cycle. Quick tool changers, with tools prepositioned for ready use, frequently overcome the need for this type of cycling.

One thing is vital: the production planning and control function must have the authority to tell operating departments what, when, and how much to produce.

QUALITY

On an assembly line, there is usually little time for fitting and adjusting. The parts must be interchangeable.

With line production, there will be no centralized inspection point; inspectors must be spread throughout the plant. Centralized inspection increases handling, the very thing line production aims to reduce. But most important is to train operators to be responsible for quality as their number one job requirement.

A big benefit is that parts and subassemblies produced on a line tend to reveal errors or oversights more quickly.

MAINTENANCE

Proper maintenance of tools and equipment is essential to achieve continuity of operation. If the flow is interrupted, the line can no longer function as a unit. As a result, maintenance is more important, is more costly, is more decentralized, and is more dependent on preventive measures. Good line maintenance prevents breakdowns.

ENGINEERING CHANGES

Indiscriminate engineering changes can upset the whole line; a methods change at one point may throw the balance out of phase. Take extra care, therefore, to be sure that the design is right before tooling up. Then be sure that the tooling and methods are right before releasing the line to production.

Precautions and techniques to employ here include:

1. Thorough laboratory and field tests of the product before it is released for tooling.
2. Complete tool tryout and assembly of parts produced from these tools.
3. Pilot lot or trial run under the direction of industrial or manufacturing engineers, with adequate attention to debugging.
4. Station-by-station checkout with sign-off before release to production.
5. Freeze design for a full run, block, or period and then incorporate all engineering changes at one time.
6. Accept the fact that it is often more costly to make a change than to live with a design that is not perfect or a line that is not ideally balanced.
7. Maintain a small-order department where new products are tried out and leaders trained, the product later being moved to a regular production line.

LINE FLEXIBILITY

Besides machine utilization and labor balance, the chief limitation to the use of line production is variety of product. When there are many products, it is difficult to make them on the same line. This can be overcome in several ways:

1. Build a basically standardized product part of the way down the line. Then divide the line into spurs or distributaries to pick up the special features, or send the product to a small-order department or other modification point, or enclose the special attachments with the main unit to be assembled in the field.
2. Build a basic product but offer variety through selection by customers of standardized optional attachments, colors, trim, or other features.
3. Combine all similar parts into groups or families of parts, and create enough quantity to justify a line or manufacturing cell.
4. Make the line serve several products by changing over from one to the other but holding one design constant at any one period.
5. Freeze the basic design fundamentals, but periodically change the outward appearance.
6. Schedule special workers into the line to pick up special features only.
7. Equip each workstation or machine with quick tool changers or prepositioned jigs.
8. Block the line into a series of cells (or sections), and let each cell provide the variety of work called for by each specific product.
9. Establish a modified line and accommodate changes with programmed automatic guided vehicles.

Changes in volume or output may be taken care of by:

1. Altering the period of time the line operates by adding overtime or second shifts or reducing working hours.
2. Changing the number of workers in combination with changing the speed of the conveyer and spacing of the units on the line or altering the station assignment.
3. Leaving room for expansion.
4. Providing parallel lines that make different products but which can be converted to other products.

Major line changeovers usually come at a low point in the production season or time of month. Weekends are also used. Simple lines that are changed over frequently may be so handled at any time during the working day, the line operators themselves serving as stock handlers and tool adjusters during the change. When the change involves an entire plant and several days, the operators are usually given their vacations at that time and the changeover is made by maintenance and machine repair groups, often supplemented by outside contractors.

In restarting a line, the group leaders and utility operators are usually brought into the picture at the beginning. Aim for correct quality and a high rate of production right from the start, or there may be plaguing difficulties.

VARIATIONS AND MODIFICATIONS

There are several other types of line variations:

Progressive Groupings.   Here a few machines of each type are set up together and the work progresses from one group of machines to the next. Or clusters or cells of assembly operations are connected in progressive flow, cell to cell.

Broken or Extended Line.   This exists whenever a line is broken by movement of material to a storeroom or process-controlled area (like heat treating or plating). As long as the flow is broken and there must be a special handling operation, it makes little difference where the line begins again.

The Common Machine and Multiproduct Line.   Where there is an expensive machine that can serve two or more lines, the layout should feed the product to this operation so it can be a part of each line.

Group Production or Group Technology Cell.   When a grouping of machines common to several products is set up, the layout is termed group production (Fig. 5.9).

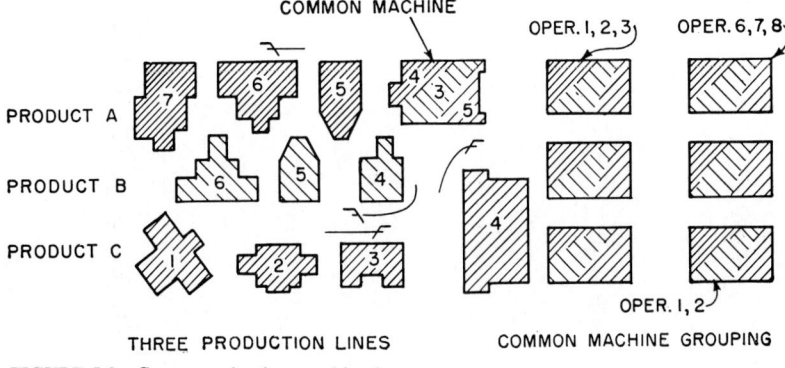

FIGURE 5.9   Group production combinations.

Flexible Manufacturing.   Programmed automatic guided vehicles route the work for special sequences of operations. Such vehicles can be used right in the line as the primary line conveyance.

Moving Workers through Fixed-Position Assembly.   Instead of the product being moved to the operators, the operators move progressively from one fixed-position assembly station to the next. Each performs the same assigned job on the several products. This saves the necessity of setting up a costly device to move the major component. Sometimes the workers move during the day and the product is moved up several stations at night.

BIBLIOGRAPHY

Gallagher, C. C., and W. A. Knight, *Group Technology,* Butterworth, London, 1973.

Holland, J. R., *Flexible Manufacturing Systems,* Society of Manufacturing Engineers, Dearborn, Mich., 1984.

Muther, R., *Production Line Technique,* McGraw-Hill, New York, 1944.

Tompkins, J. A., and J. A. White, *Facilities Planning,* Wiley, New York, 1984.

CHAPTER 6
WAREHOUSING AND DISTRIBUTION

Herbert W. Davis
Herbert W. Davis and Company
Englewood Cliffs, New Jersey

During the first half of the twentieth century, industrial engineering practice tended to concentrate on the manufacturing process—a process costing on average about one-half the selling price of the goods.

Since 1950, however, industrial engineering techniques have played an increasing role in the nonmanufacturing segment of this total cost framework. A major part of these activities involves the warehousing of product and its delivery to the manufacturer's customer—perhaps another company or plant, a wholesale distributor, a retailer, or a consumer. This chapter describes the warehousing and distribution function and the role of industrial engineering in its design, operation, and control.

ROLE OF THE DISTRIBUTION SYSTEM

Prior to World War II, most manufacturing was done at plants assigned either of the following roles:

1. *Geographic:* To serve a territory that might be the world, the United States, or a smaller geographic region. Location was a function of the economies of raw material availability, transport cost, and the market area served.
2. *Product:* To produce a specific product line that was the company's entire output or a portion of it.

In this environment, product could be shipped directly from the plant of manufacture to the customer. In large segments of business, however, plants were distant from customers, and reasonable, timely service required some intermediate storage. This storage system was the forerunner of the modern distribution center and network, with its computer control systems and sophisticated product flows.

Typically, a company with a high-volume, nationwide sales pattern might have had 100 or more warehouses, be supplied with product by one or more product-line-specialized facilities, and used water or rail transportation for the primary (plant-to-warehouse) leg. Local drayage to the customer was done by motor truck. Two things changed this pattern:

1. During World War II, modern concepts of logistics analysis, materials-handling systems and equipment, and mechanisms to time and control the total flow emerged.

2. The National Defense Highway System was authorized in the early 1950s, and it evolved by the 1960s into an extensive, easy-to-use national express highway network. This led to the growth and importance of major national highway motor carriers able to compete in price with the older water and rail systems, and to offer better, faster service.

These changes took place within the framework of developments in the sales and marketing function. Product diversity, reasonably prompt, complete delivery, and national pricing and promotional practices led to an explosion in the number and variety of products and styles offered to customers. Offering these more-sophisticated lines in an efficient manner led to the development of the modern distribution system based on decentralized, stand-alone regional warehouses tied together by an information and transportation system. Thus, by the mid-1960s, business had the need for sophisticated distribution systems and the conceptual physical designs, the materials-handling technology, and the transport system to make them work.

Today, that physical distribution system is the major link between the manufacturing plant, the customer, and the marketing/sales function. It is complicated, uses a very high level of information and materials-handling technology, and is a major operational area for industrial engineering.

DEFINITION OF PHYSICAL DISTRIBUTION

Physical distribution is the group of activities concerned with the control, movement, and storage of materials. These activities may take place within a single manufacturing facility or be played on a worldwide stage. The scope may include activities that occur prior to, during, or after the manufacturing process. In some companies, physical distribution includes purchasing of finished and raw materials, inbound transportation, plantwide storage and materials handling, shipping, and outbound transportation.

During the past decade, physical distribution has come to be considered primarily as the functions that occur after the manufacturing process, serving as the link between the manufacturing plant and the customer. Its assigned functions tend to be the physical, or product, aspects of marketing. Industrial engineering techniques have wide application in analyzing and improving all aspects of this physical process.

FUNCTIONS INCLUDED IN DISTRIBUTION

The physical distribution system used to control, move, and store products on the path from the manufacturing line to the customer is complex. A typical system has manufacturing plants, which may produce all or some of the product line, warehouses that are supplied products by the plants, and customers that are supplied by any of the plants or warehouses. Figure 6.1 illustrates such a system. Note that the physical facilities (plants and warehouses) are connected by transportation links. Orders for the plants and warehouses come into the system from the sales department through customer service and are directed to the plants or distribution centers for order fulfillment. Inventories are usually controlled by an administrative function that may be responsible also for the system design and control.

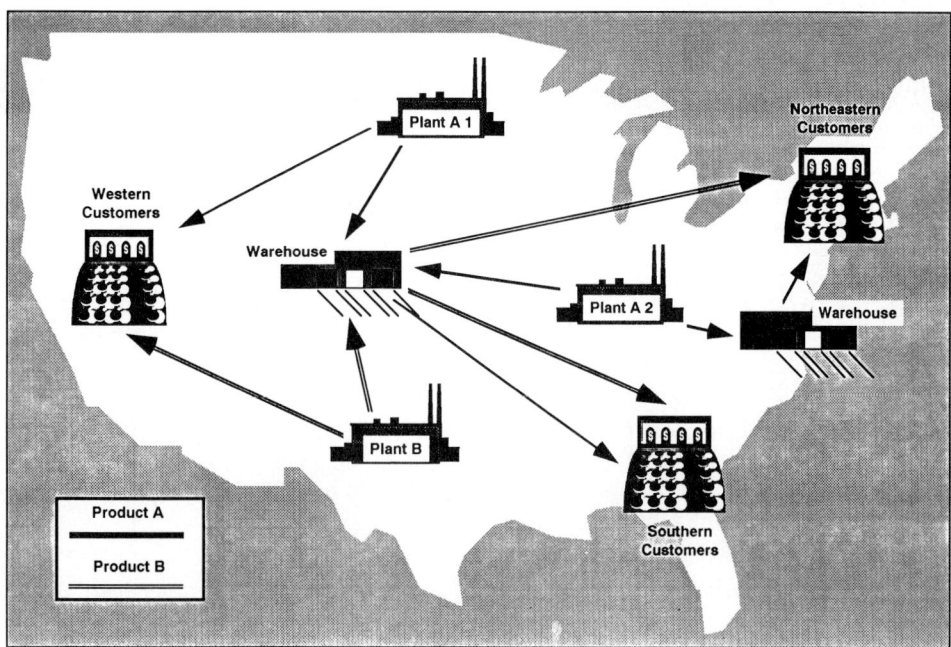

FIGURE 6.1   Two-product, three-plant, two-warehouse system.

Typically, there are five major functions assigned to physical distribution to manage:

Order Entry and Customer Service

1. Receive orders from customers and sales by telephone, fax, electronic data interchange, mail, or hand delivery.
2. Enter and edit the information, usually in a computer system.
3. Apply pricing.
4. Select shipping point and transfer information for picking, packing, and transport.
5. Track order and product status.
6. Report status to sales and customers.
7. Answer customer and sales inquiries on status.
8. Solve problems relating to these activities.

Warehousing

1. Receive materials from vendors, plants, and other facilities.
2. Verify material input and resolve discrepancies.
3. Place materials into storage, awaiting instructions.
4. Manage the physical quantities on hand.
5. Pick and pack materials for outgoing orders to customers or other warehouses.

Transportation

1. Route, rate, and control the use of freight carriers.

2. Transport goods from the plant or vendors to distribution centers and redistribute between multiple centers.

3. Transport goods from distribution centers and plants to customers. Many different modes may be used, including rail, motor truck, barge, ship, and aircraft. Shipment sizes may range from small parcels through containers or truckloads up to full bulk shiploads.

4. Receive, audit, and arrange payment for outside for-hire carriers.

5. Manage the company truck, rail, air, and water fleets.

Inventory Management

1. Determine how much material is needed to achieve inventory turnover, customer service, and cost objectives.

2. Order materials from vendors, plants, and warehouses.

3. Track materials flow and status.

4. Consider the cost to carry inventory, customer satisfaction, and warehouse and plant capacities in deciding when and how much material to order.

Distribution Administration

1. Determine and allocate funds and resources to the various distribution activities.

2. Design and manage the functional activities assigned to distribution.

3. Develop and manage the appropriate control systems.

WHAT DISTRIBUTION COSTS?

Physical distribution is one of the largest costs in the manufacture and sale of merchandise. Costs have tended to rise over the years when measured by product units (cases, pieces, or weight). Important factors in this cost increase have been:

1. A long-term decline in unit weight corresponding to the substitution of plastics and electronics for structural metals and mechanical controls, and the proliferation of protective and decorative packaging material.

2. An increase in the number of different items offered the customer, resulting in the distribution of fewer pieces per catalog number.

3. Refined inventory control and purchasing practices so that customers purchase fewer pieces spread over more-frequent ordering patterns.

When measured as a ratio to sales, however, distribution costs have been cyclical. Costs respond to a large number of external influences such as energy rates, service levels, interest rates, transportation costs and tariffs, competitive pricing, and company policies. Currently, distribution costs average about 7.2 percent of a manufacturer's sales revenue. The cost pattern from 1962 to 1990 is shown in Fig. 6.2. Figure 6.3 shows the decline since 1980 of the three largest cost elements: transportation, warehousing, and inventory. The three broad long-term trends are

1. *1962 to 1973:* In the United States, this was a period of steady growth marked by heavy price inflation from 1969 onward. It was a period of American economic dominance. U.S. companies' operations in Europe alone constituted the third largest world economy. In distribution, it was the period when companies adopted the distribution concept. Consolidated authority over the entire distribution budget resulted in the ability to

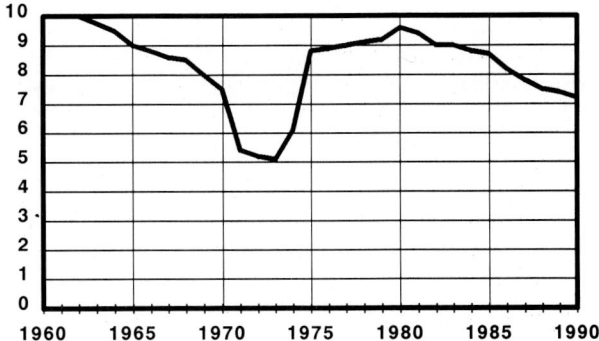

FIGURE 6.2 Distribution costs as percent of sales.

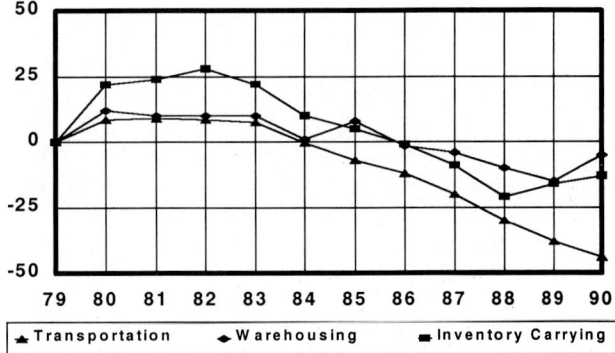

FIGURE 6.3 Annual percent cost change.

employ better-trained personnel supported by improved information systems. Distribution cost declined steadily from 10 percent in 1963 to a 1973 level of 5.5 percent of sales, a striking testimony to the power of the physical distribution concept.

2. *1973 to 1980:* The oil embargo ended the first distribution era. This second period was one characterized by the energy crunch, inflation, declining productivity, and a growing foreign presence in domestic U.S. markets. Preoccupation with cost containment pushed many companies into ignoring product quality and customer service issues. All of the gains of the 1963 to 1973 era were lost, and costs once again hit 10 percent of sales by the end of the 1970s.

3. *1980 to 1990:* The period started with transportation deregulation as an attempt to deal with costs through market forces rather than government regulation. The most important external factors in the period were the decisive changes wrought by the corporations in dealing with foreign competition, inflation, and productivity. This era was the time of corporate restructuring, offshore sourcing, manufacturing consolidation, capacity reductions, and centralization of major activities. It was a period of sustained, profitable growth in the domestic economy. In distribution, there was a new emphasis on productivity, a drive for customer service and quality excellence, and much better computer support. The result has been a significant and steady reduction in all of the major distribution costs. The current level of total cost is about 7.2 percent of sales and 38 cents per pound.

Distribution costs vary by product, product value, and company size:

1. *Product:* Distribution cost as a percent of sales has a strong, central tendency across a broad range of products (Fig. 6.4). The reasons for cost similarity are the relative importance of freight tariffs, interest on capital invested, wage rates, building rents, and energy. Companies tend to compare their product distribution costs to those of competitive companies and thus miss a major opportunity to learn from other industries.

Average Distribution Cost by Industry

	% of Sales	$/CWT
All companies	7.20%	$ 33.00
Industrial non-durable	6.63	7.26
Chemicals	6.77	6.86
Industrial durable	6.75	80.65
Industrial durable < $10/pound	8.01	25.67
Industrial durable > $10/pound	4.87	163.12
Consumer non-durable	6.76	30.78
Grocery	7.17	11.15
Food and beverages	7.04	6.21
Dry and packaged food	6.72	7.51
Canned and processed food	8.34	5.25
Temperature controlled food	6.00	6.38
Non-food in grocery channel	7.38	19.50
Consumer household products	8.15	12.95
Health and beauty aids	7.72	20.20
Pharmaceuticals	2.85	136.77
Consumer durable	6.54	31.88
Replacement parts	11.01	63.32
Retail companies	5.39	60.76

FIGURE 6.4 Average distribution cost by industry.

2. *Product value:* There is an important inverse relationship of distribution cost with product value per unit weight (Fig. 6.5). Small, lightweight products of high value such as jewelry, pharmaceuticals, and electronics tend to have low freight cost compared to bulky, heavy materials such as foods, machinery, and consumer appliances. This advantage is partly offset by large, more expensive inventories and by costly order-handling procedures associated with high-value products.

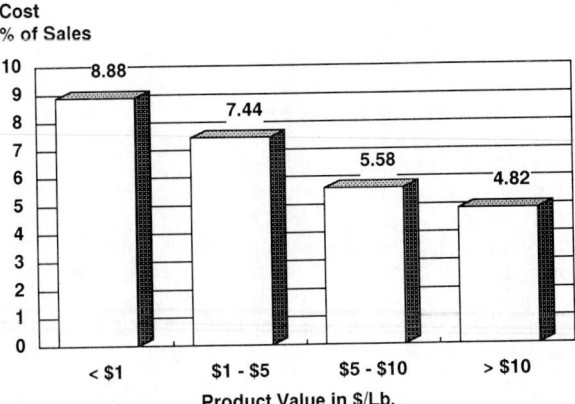

FIGURE 6.5 Product value.

3. *Company size:* Size is a complicated factor. Many large companies tend to have wage rates higher than small businesses. However, very large shippers tend to have strong negotiating leverage when dealing with carriers and other suppliers. This tends to reduce freight and material costs. Figure 6.6 shows that costs tend to be similar except for the two extremes of company size—very large and very small.

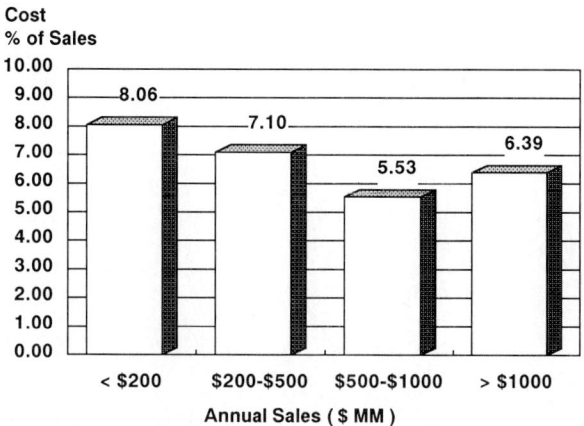

FIGURE 6.6   Company size.

Finally, a most interesting aspect of distribution cost is the similarity of total cost across products, continents, and company size. Probably this results from the almost universal application of common industrial engineering techniques.

DISTRIBUTION SYSTEM DESIGN

The most basic function of the physical distribution system is to efficiently and effectively move merchandise from the end of the production line to the consumer. This flow frequently involves a number of independent companies: the manufacturer, wholesale distributor, retailer, and the like. These channels of distribution are usually specified by the marketing organization, and their use is built into the basic organizational and material fabric of the company.

For the industrial engineer, therefore, the important segments of the distribution system are those under the direct control of the manufacturer. These channels are different for different products. There is a major difference between consumer and industrial products and durable versus nondurable goods.

Despite the complexity, however, there has developed a body of knowledge governing the design of an efficient product distribution system. Figure 6.1 shows a typical multiplant, multiproduct distribution system.

The mission of the system is to deliver product to the customer when the customer wants it, in the proper product mix, and at a reasonable cost. The design of a system, then, requires consideration of:

1. *Costs:* The costs usually consist of primary and secondary freight charges, warehousing expense, inventory carrying cost, and the expenses related to booking and processing the order.

2. *Customer service:* The key factors are prompt and complete fulfillment of the customer order.

The essence of the design problem involves meeting a prescribed set of customer service requirements, such as delivery of a complete order within a prescribed time frame, while simultaneously minimizing the distribution costs for freight, warehousing, and inventory. All of this needs to be done within the physical capacity constraints of the facilities involved.

This physical distribution system design is done most often using one or another of a competitive group of logistics models. It is seldom worthwhile to develop single-purpose models for a company. Such models are very costly or they oversimplify the problem to be addressed and do not use the best technology available.

Figure 6.7 describes the general steps in assembling and using a model to design the physical distribution system. The technique described is applicable to most commercially available models. The figure lists the types of data needed to support the model. The three basic types of data are:

1. *Sales model:* Defining volumes delivered by product group and by shipment size to each geographic market area. A universal geographic coding system is used, frequently the postal ZIP code.

```
             CHECKLIST FOR DISTRIBUTION SYSTEM DESIGN


    I    SALES MODEL
         A.  Obtain weight shipped by destination ZIP code
         B.  Identify major market centers for each product group
         C.  Obtain shipment size information
         D.  Characterize the market place by size and shipment size
         E.  Determine customer service requirements
         F.  Validate volumes to shipping point throughputs
         G.  Review with sales and marketing management


    II   SHIPPING POINT MODEL
         A.  Prepare list of all finished goods storage and shipping
             facilities
         B.  Conduct a cost analysis
             1.  Fixed/variable/other cost breakdown
             2.  Allocation of cost to product lines
             3.  Before tax/after tax issues
             4.  Customer and intra-company shipping
             5.  Building space, equipment, labor, inventory
         C.  Flowchart finished goods material flows
             1.  Nominal flow from plant/vendor through warehouse to
                 customer
             2.  Other flows such as returns and reworks
             3.  Chart costs, weight, and mode of transportation for each.
         D.  Summarize capacity limitations of each faciity
         E.  Validate data and volumes with accounting figures


    III  TRANSPORTATION MODEL
         A.  Determine modes to be studied.
         B.  Obtain base rates and discounts for common carrier
         C.  Obtain rates for other modes (for example, air)
         D.  Validate transportation cost modeled to accounting figures


    IV   RUN STRATEGY
         A.  Present system
         B.  Design year baseline
         C.  Baseline optimum
         D.  Study of best number and location of facilities
         E.  Service level study
         F.  Capacity study
         G.  Engineering economic study of results
```

FIGURE 6.7 Checklist for distribution system model.

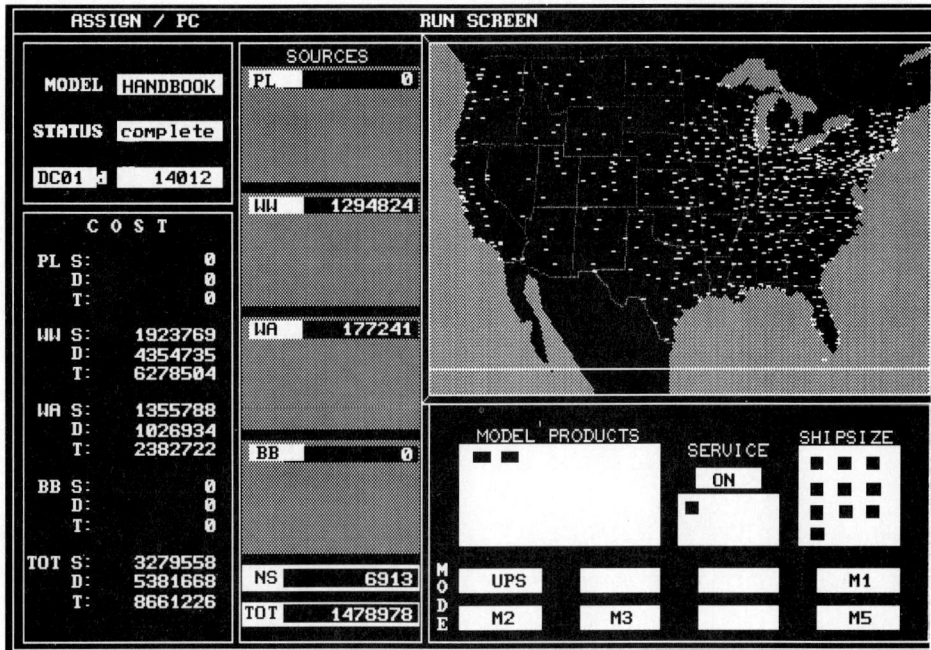

FIGURE 6.8   Model output graphics.

2. *Shipping point model:* Describing the materials flow and defining the costs for storage and handling at each facility considered. This is an important area for industrial engineering analysis, since the best results will be attained if the costs are carefully developed. Simple accounting-type allocations can lead to errors in the model output. It is important to consider a significant number of potential warehouse or plant locations, well beyond the current system configuration.

3. *Transportation model costs:* The means whereby the logistics model can cost out different ways of meeting the sales demand. This area requires considerable experience with the transportation system rate structure, the highway and rail networks, and the availability of transportation capacity at the shipping points tested.

Typical modeling practice is to run a series of tests of the data, to validate the model, and then to find the appropriate solution. This solution should be developed within the framework of economic justification, customer service requirements, and capacity constraints.

Commercially available models normally have an extensive set of graphic output to aid in analysis of the results and, if necessary, in specifying further runs. Figure 6.8 shows some typical output graphics.

Finally, pushpin-type graphic displays are available to help the analyst in all stages of the distribution system design. Such a display (Fig. 6.9) is useful in visualizing the largest and most important market areas—areas where warehouses might be indicated. Warehouses are frequently put in large market areas.

WAREHOUSING LEVELS

The storage and handling of materials is an important function in manufacturing and distribution. Storage levels normally used in the industrial process are

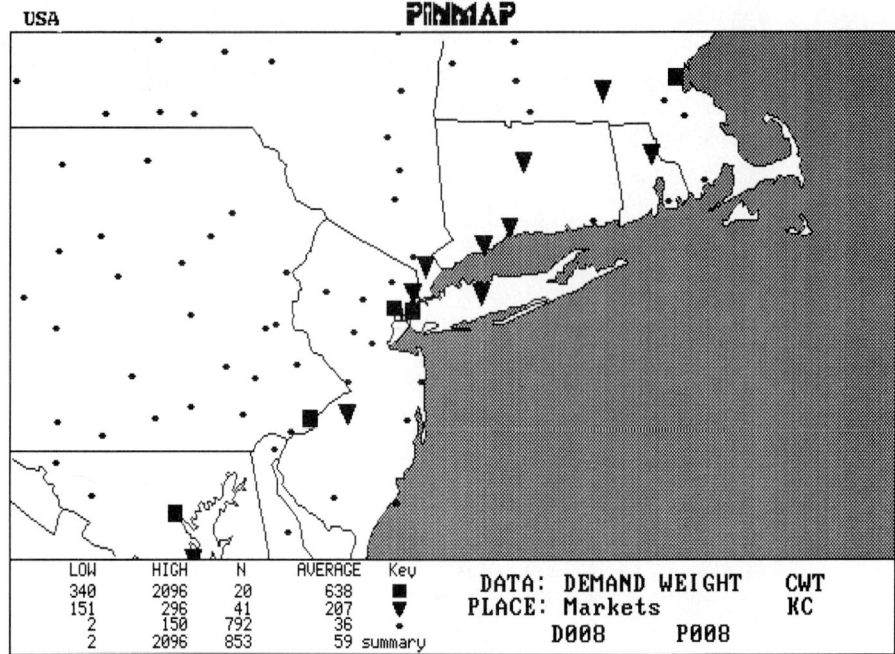

FIGURE 6.9 Market area graphics—pushpin-type display.

1. Raw material stores—chemicals, bar stock, component parts
2. Tool cribs—molds, dies, cutting tools
3. Maintenance supplies—paper, oils, electrical and plumbing repair parts
4. In-process materials—items stored between manufacturing operations
5. Plant finished goods warehouses
6. Public distribution centers
7. Private distribution centers
8. Bonded warehouses—usually for import goods held awaiting the payment of customs charges or for transfer to another country. May be for products on which local or federal taxes have not yet been paid.

In the general case, storage and warehousing occurs in or near either the plant or the market. Seldom are warehouses located between plants and markets.

Plant-located facilities either serve the plant operations (raw materials and tool cribs, for example) or are a major customer shipping point. The plant warehouse may also be the backup point to resupply a field distribution system.

Market-located facilities are positioned to supply customers with the company's products. These distribution centers may store the output from a number of plants. The customer can order products made by several plants and vendors, and receive a single shipment from the distribution center. Proper location planning can result in fast, complete delivery of a customer's order, tending to increase satisfaction and future volume.

Most warehouses are operated privately by companies for their own materials and products. There are many public warehousing companies, however, that offer space and labor on a for-hire basis. During the past three decades, the public warehouse industry has increased in size, complexity, and range of services offered. The warehouse, for example, might contract to do price ticketing, assembly and repacking, labeling, inbound material consolidation, outbound customer freight consolidation, and order receipt and entry. Public facilities with a tie-in to transportation carriers can also offer product tracking and status reporting. These services, added to an already high level of warehouse productivity, have resulted in a public warehousing growth rate higher than that of company-operated facilities.

WAREHOUSE DESIGN

The methods used to design the materials flow, handling, and storage activities and to control labor productivity in a modern distribution center are similar to industrial engineering practice in a manufacturing plant. There are a number of special conditions, however, in distribution facility design and operation:

Building Considerations.   Many warehousing facilities are located inside manufacturing plants. In such cases, it is common to find that the building is constructed to the plant specifications. Stacking heights and floor storage arrangements are set up to fit the manufacturing facility specifications. This practice results from the common use of space by both activities. Manufacturing frequently expands into the space occupied by warehousing.

In free-standing distribution centers, and on a few plant sites, the warehousing facility is designed to fit the unique characteristics of the distribution system. For example, modern stacking equipment can economically operate at heights of 40 to 85 feet or more. Some equipment can right-angle stack in a 5-foot-wide aisle. Other equipment may be secured to the building structure or the storage racks. Such dense storage patterns result in the design and construction of special-purpose buildings. In designing the modern distribution center, the industrial engineer must consider the following factors.

Material Flow.   The building can have a straight-through flow with receiving on one end and shipping on the other. Another popular approach is a U-shaped flow with common receiving and shipping areas. This method concentrates most of the building employees and activity for better control. Both methods are effective; the best can be determined on the basis of economic analysis and site configuration.

Levels.   Older facilities, and some very modern distribution centers, are frequently multilevel. Storage, however, is most efficient when concentrated on one floor level with a high stack height. Receiving, shipping, and packing operations, on the other hand, seldom require high ceilings. Normally, horizontal travel is less costly than vertical, leading to the current interest in single-level warehouses. The industrial engineer must reconcile these factors in preparing the design.

Bay Dimensions.   The storage pattern is a crucial factor in distribution center design. The buildup of storage spots and access aisles dictates the bay dimensions. Proper design can result in efficient or optimum bay dimensions. A bay is the floor area bounded by the building support columns. Thirty years ago, it was not uncommon to work with 3-foot-diameter concrete columns on 20-foot centers. In this situation, storage patterns were relatively inefficient. Current construction allows about 8 to 12 inches for steel columns, spaced 30 to 60 feet on centers. Figure 6.10 is an example of how pallets and pallet racks and the associated forklift access aisles are accumulated to determine bay dimensions. Note that the storage pattern is determined first. Then the column spacing is calculated to locate columns within the rack or storage structure. The final spacing may be any multiple which minimizes column space loss while providing a lower-cost steel frame roof structure. The final dimensions are decided by building cost calculations designed to balance the costs of lost space and extra-long steel members.

Ceiling Heights.   The distance between floor and lowest structural obstruction in a modern distribution center is determined by the storage stack height and the clearance needed for water dispersion from sprinkler heads. The storage area may contain storage racks on which palletloads of material are placed. There may be bulk stacks where

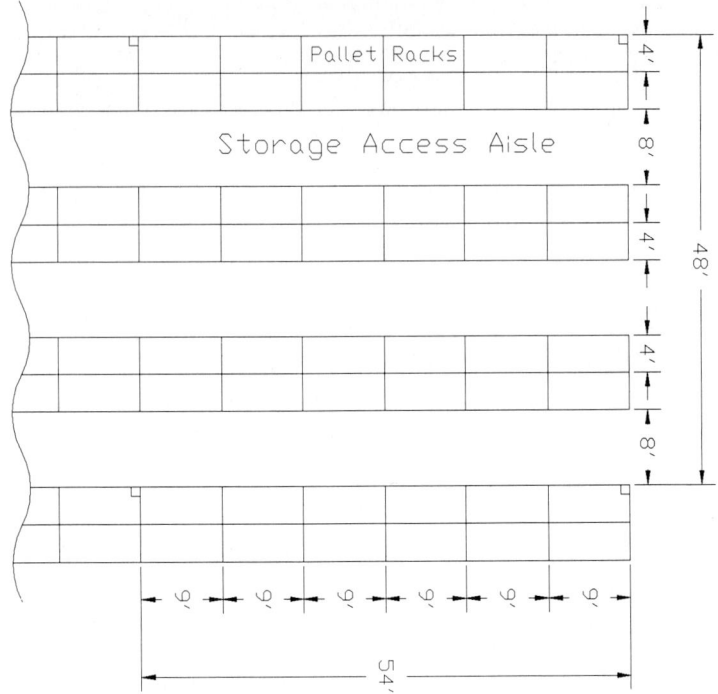

FIGURE 6.10 Typical bay dimensions.

palletloads are continuously stacked to the crushing limit. Pallet racks, however, normally are used in buildings with very high stack heights, because current lift equipment is capable of safely stacking much higher than crushing limits would permit. A typical ceiling height derivation is shown in Fig. 6.11.

Mezzanines. Because most modern distribution centers are constructed on a single level, the use of temporary and/or permanent mezzanines is an important building option. Mezzanines may be constructed with steel grating supported by storage racks, special columns, or building columns. They are used to more fully utilize the cubic space in a building. Typically, a warehouse may have storage covering 50 to 75 percent of the floor area. The other operations such as receiving, counting, marking, packing, and staging may total 50,000 square feet or more, but not effectively utilize the warehouse height of 30 or more feet. Thus, two or three overhead levels might be constructed to house these activities more efficiently.

Number of Truck Docks. Doors are expensive in terms of construction cost and energy loss. Determining the right number of truck and utility doors is complex, frequently requiring the use of simulation. Doors may be single-purpose (receiving, shipping, over-the-road trailer, and so on) or multipurpose, to fill all needs. Most warehouses are built with the floor 48 inches above grade and pavement. This provides for forklift access to typical highway trailers. Special-purpose docks for vans (24 inches) and ground level access for inside loading may be provided.

A method to accurately estimate the number of doors needed requires accumulating a record of truck arrivals (or unloading) and a separate record of outbound loads. The engineer needs to measure the average loading or unloading time for a sample time period. Given the average arrival and departure frequency and the average load/unload service

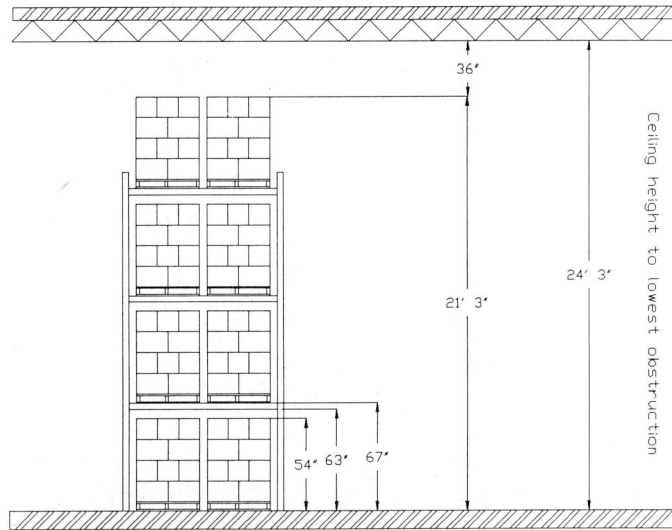

FIGURE 6.11   Ceiling height derivation.

time, queuing theory can be employed to determine the appropriate number of docks. Queuing tables are available to simplify calculation.

Length-to-Width Ratios.   In many cases the available land dictates the general configuration of the warehouse building. For unlimited sites, however, the ratio of building length to width is a useful design element. The selection depends on the desired materials flow and the handling and storage methods used:

1. *U-shaped flow:* The docks may be on one common wall to maximize control and cross-utilization of personnel. Buildings tend to be constructed square or to a 3:2 length-width ratio in these circumstances to minimize internal movement. Expansion is usually on the back wall opposite the door wall. This provides for low-cost additions, since the expansion need only provide lighting and minimal support services. Everything else is in the original section. It is also easy to expand on the other two walls if appropriate.

2. *Rectangular:* Straight-through materials flow buildings have docks at opposite ends with storage rack aisles parallel to the flow so that an item can move in a straight line from receipt to storage, picking and shipping. The building width is a function of the number of doors, which will be on about 12-foot centers. Thus, if 10 doors are needed for shipping, the building may be 120 to 150 feet wide. The long dimension is calculated to provide sufficient area for staging, storage, and operations. Typical ratios range from 1:2 to 1:5. Expansion of straight-through-flow buildings is on the long side to provide for additions to all the operations in a proportion roughly equivalent to the original space allocations. Straight-flow buildings have an inherent operating disadvantage: all material must traverse the entire long dimension.

3. *Hybrid:* Some warehouses have a large number of quite different activities dictated by product or corporate circumstances. Examples are cool and frozen-material storage, unit repacking or packaging, and hazardous materials handling. These special circumstances result in buildings that do not meet the general types described. A common hybrid today occurs when the building storage area is designed for very high storage. Stacker cranes can store products 85 or more feet in height and typically require aisles only 5 to 6 feet in width. In these cases, unique building specifications may be used to control access and environmental conditions in the storage module.

Warehouse Equipment.   Most warehouses use conventional equipment for the storage and movement activities. Some conventional items are

Pallet Racks.   These are used to store palletloads of product at multiple levels, making better use of floor space. Figure 6.11 shows a typical arrangement. Conceptually, racks are storage structures of formed steel with uprights fitted with movable bars set at appropriate heights to accommodate palletloads. Racks are usually strung in long lines with access aisles between them. A typical arrangement has modules consisting of a row of racks holding 4-foot-deep pallets, an 8- to 12-foot access aisle, and another row of racks. Other types of pallet racks are for double-deep storage or for drive-through to store pallets deeper. Finally, racks may be fitted with steel or plywood shelves to accommodate cases and small parts.

Storage Bins.   Usually of steel, bins are short sections of shelving designed to hold small lots of material. Many configurations are used, including drawers, slotted dividers, differing shelf heights, and reinforcing bars.

Flow Racks.   Picking individual items and small cases from bins or pallet racks may become laborious. For some high-volume operations, flow racks are used. A flow rack is usually a rack 8 to 10 feet wide and as deep, or deeper. Slide- or roller-equipped angle frames permit loading a case at the rear of the rack so that it will flow down the lane to the picking face. A few to a dozen cases may be contained in a flow lane. Each rack may be six or eight lanes wide and three to five high—a total capacity of perhaps 20 to 30 different items, each supported by a continuous feed of 10 cases or more. This gives a dense, usable storage pattern to support high-volume order-picking activities. In this arrangement, the picking face presents many more items to the picker per foot of access aisle, as compared to conventional bin or pallet rack storage.

Conventional Forklifts.   The oldest type of mobile pallet-moving equipment is the four-wheel industrial truck equipped with an elevating mast. Drivers may sit down, stand, or walk along, depending on the design. Power may be from battery, propane, or gasoline. Conventional forklifts are used in a wide array of missions, since they can travel great distances, carry loads up to several tons, maneuver in 12- to 15-foot aisles, and enter highway trailers safely. They are used for large bulk-storage areas where palletloads may be double- or triple-stacked, and rows of pallets may be 10 to 15 deep. Thus, a conventional forklift might service blocks of many hundreds of palletloads.

Narrow-Aisle Lift Trucks.   The typical narrow-aisle truck has two outriggers to straddle a pallet, providing a noncounterbalanced base on which to operate. The driver usually stands to operate the vehicle. Narrow-aisle vehicles are in wide use, right-angle stacking in 7- to 10-foot aisles, and stacking to heights of 30 feet or more. This gives dense storage patterns, usually based on concepts of random access to any pallet in the storage block. Narrow-aisle equipment usually cannot enter highway trailers, although some special designs with large front caster wheels are available.

Reach Trucks.   An important variation of the narrow-aisle truck is the use of special masts and forks that extend mechanically in the direction of travel. This allows the vehicle to stack materials closer together by eliminating the straddle outrigger. Other versions can reach out a full pallet depth to deposit loads in an inside rack. This increases storage density.

Very Narrow Aisle Trucks.   Special vehicles have been designed that can rotate their forks or forks and masts, called *swing-reach,* or *turret,* trucks. Because they do not have to turn to right-angle stack into a rack, they can operate in aisles only a little wider than the pallet. Aisles of 60 to 72 inches are common. Another characteristic is that the vehicles have to be very large and heavy to accommodate the complex mast equipment and provide a stable platform from which great pallet elevation heights can be achieved. These classes of equipment can store material safely at 40-foot elevations in aisles under 72 inches. The size and tight quarters usually require electronic or mechanical guidance to prevent contact and damage to the rack structure.

Stacker Cranes.   Stackers are manufactured in a wide range of configurations. Their basic purpose is to operate from the top of a storage stack on rails mounted to the building or rack structure. Heights are essentially limited only by economics, and stack heights of close to 100 feet are reasonably common. Stackers are usually operated by computers, fitting into highly mechanized or automated activities. In these cases, without operators, the building structures may have only minimal lighting and heating to preserve the product's life. Energy savings can be significant.

The facility designer should note that all narrow-aisle equipment, such as stackers and very narrow aisle swing-reach equipment, loses time when changing aisles. Appropriate facility layout, then, usually requires fairly long aisles with few occasions to turn into adjacent aisles. The typical facility is long and narrow—ratios like 5:1 or 10:1 for length and width.

Floor Tractors.   These units are used to pull trains of floor trailers over great distances in a warehouse. A frequently accepted rule is that elevating trucks should not travel more than 200 feet from their base. For greater distances, it is more efficient to load a pallet on a trailer and haul multiple loads to the destination. Floor tractors can pull trains of 10 to 12 trailers, each with two or more pallets aboard.

Automatic Guided Vehicles.   Essentially, the electric floor tractor can be equipped with computer and sensing devices to permit the vehicle to deliver and pick up goods throughout a warehouse. Installations of 50 to 100 automatic guided vehicles (AGVs) operating in multimillion-square-foot buildings are found today. The loading and unloading of the AGV is normally automated, and a master control computer directs the entire flow.

Conveyers.   Warehouse conveyers are used to move product within and between operations. The conveyers may be belt, roller, roller with over- and underbelt, skate-wheel-powered, or free. Typical applications are combined with flow racks (for picking operations or for long-distance movement of pallets or cases from storage), docks, and ancillary operations. Very complex conveyer systems, combined with scanners and reading devices, flow gates and computers, can result in extremely efficient, modern distribution centers.

Computer Control Systems.   An important productivity and control technique used in the modern distribution center can be characterized as a warehouse management system. Systems in use today are based on mini- or microcomputers. The computer receives customer orders, sequences them usually in batches or waves, and then controls all of the warehouse activities. The system may be equipped with lights and readout devices to indicate the item to be picked, quantity, and next operation, and to direct disposition. It usually controls the merging of the various components of an order, checks that all items are available, and prepares shipping manifests and carrier bills of lading. The system keeps track of orders and personnel and assures productive operation.

Work Standards, Incentives, and Cost Control.   Control of productivity in a warehouse presents different problems to the industrial engineer than those encountered in manufacturing activities. First, warehouse personnel are usually spread sparsely over hundreds of thousands of square feet of floor area. In manufacturing, there is normally a dense, concentrated population. Second, warehouse personnel are mobile—the essence of the operation is rapid physical movement in three dimensions. Finally, the work tends to be diverse and long-cycle, not paced by machinery.

Nevertheless, work standards have been applied in many distribution centers. Penetration is highest in warehouses closely allied with manufacturing facilities.

Standards.   Standards are set using the same techniques as in manufacturing:

1. Stopwatch studies of well-documented, short-cycle activities.
2. Elemental standard time data developed within specific industries and for the materials-handling function as a whole.
3. Higher-level standard data for long-cycle operations developed to aid in staffing decisions. These are widely used in industries such as grocery products, in associations such as public warehouse groups, and in government.

4. Ratio-delay-type studies to determine the total time spent in a warehouse divided among many functions. The studies are widely used as a starting point in determining which activities are large enough to warrant standards.

Incentives.   Monetary incentives may be attached to individual or group performance levels above standard output. Perhaps 30 to 50 percent of warehouses have some form of incentive compensation.

Cost Control.   Staffing requirements for warehouses frequently vary through the day, week, month, and season. Variable work loads are a vexing problem. Traditionally, most warehouses were staffed for a reasonably high level of activity—perhaps the 75th percentile. Overtime was used to reach the peaks, and layoffs, make-work, postponable work, and the like were used in low-volume periods. Recent expansion in the use of management control techniques and work standards have resulted in much better control of staff levels. Current methods use radio-frequency transmissions of work requirements, feedback loops, standards, and piece counts to control productivity. Part-time employees and interdepartmental transfers for temporary periods have facilitated productivity control.

PLANNING THE DISTRIBUTION CENTER

Given that a company either has or intends to set up a distribution center, the design project will require a high level of detailed information and data. The following outlines the design process:

Determine Functions to Be Included.   What functions will be contained in the warehouse? This can be a very long, complex list of activities. For example:

1. Receiving, counting, verifying, and accepting inbound materials and finished product
2. Transporting and storing the products in appropriate storage locations and equipment
3. Maintaining a control system to locate all materials and paperwork within the facility
4. Receiving and handling shipping orders
5. Picking, packing, and assembling outbound materials and marking them for accurate delivery
6. Routing outbound goods by carrier, calling the carrier, and staging and loading product onto the outbound vehicle
7. Checking outbound materials for accuracy and adjusting internal stock records.

Determine Initial Space Allocations.   Preliminary estimates are frequently made to determine the total space needed and to allocate it to the listed functions. This is called a *block layout*. At this stage, provision is made for utilities and support services, offices, staging areas, and so forth to estimate the building dimensions to a reasonable accuracy.

Develop Data on Volumes and Flows.   There are five basic types of data needed:

Inventory

1. Number of different items will be stored
2. Quantities expected for each item
3. Item dimensions and storage characteristics
4. Activity (picks and turnover), by item
5. Forecast of growth of the items or item groups and of new items expected
6. Nature of the items—fragile, hazardous, liquid, and so forth

7. Number of cases and pallets or other units to be stored
8. Normal ratios of items per case, cases per pallet, pallets per truck, weight per pallet

Receipts

1. Number per time period
2. Lot sizes
3. Need to segregate lots of an item
4. Seasonality

Shipping Orders

1. Number by time period
2. Seasonality
3. Types of orders
4. Characteristics—items per order, orders per shipment, and so forth

Order Analysis

1. Line items per order
2. Pieces
3. Cartons
4. Frequency distributions of pertinent data

Service Requirements

1. Timeliness in shipment
2. Accuracy requirements
3. Special markings
4. Promotional and regular materials

Observe Operations.   The engineer has to be knowledgeable about current warehouse methods in the existing facilities. Regular observation of each function performed, supplemented by flowcharts, current work standards, and lists of questionable practices, needs to be completed. The results of this work are normally discussed with operating managers to assure a full understanding of the current operation—its performance and requirements, special conditions, and problem areas that need to be addressed.

Establish Alternative Methods and Equipment.   In any warehousing function, there are a number of ways in which the work can be done. A new facility may have been accepted because more space is needed for expansion, or it may provide the room and the environment for major productivity or service improvements, provided that:

1. The job to be done has been described
2. The current problems and opportunities have been isolated
3. The current methods have been identified
4. Objectives for improvement have been established

The industrial engineer then has to describe a number of feasible alternative plans. The different plans usually involve an increasing level of mechanization or automation. Higher levels frequently have a high capital expense, but may have low operating labor cost. Higher stacking, for example, uses less floor space, but requires more expensive equipment.

The typical design study is done in two steps:

1. Individual operations are examined; for example, how high to stack. General answers are reached for each activity such as picking, order assembly, and storage.
2. Elemental results are aggregated into 5 to 10 feasible building layouts.

Evaluate the Alternative Designs.   These are evaluated for:

1. Feasibility and applicability to the facility mission
2. Operating cost
3. Investment requirement
4. Maintenance
5. Flexibility to suit changing needs in the future
6. Risk involved in achieving the desired results and savings
7. Implementation time

All this information is then evaluated by traditional engineering cost techniques, such as discounted cash flow and its variations. A decision can then be taken as to the best alternative for the circumstances evaluated.

Prepare Detail Designs.   Following acceptance of the basic facility conceptual design, a much more detailed plan needs to be prepared. This plan usually involves:

1. Contact with equipment vendors for additional ideas and constraints in the functional areas.
2. More detailed data in some areas to support elements of the design. For example, how many packing stations are needed? What conveyer speeds are most effective? How do you staff the lines for various load levels?
3. Simulation—modern computer simulation methods yield sound, operationally correct answers to many detail design questions. In particular, conveyer systems and staffing levels are sensitive to short-cycle volume and product-mix shifts. A simulation of the system in operation is a sound investment in achieving a problem-free facility start-up. The simulation can later be used for operator and supervisory training.

Prepare Written Recommendations.   At the conclusion of the design process, it is normal to prepare a complete written report on the project. The report may be needed to get internal or external financing. On another level, it should serve as an operating manual for the managers of the warehouse operation. The report typically includes:

1. *Equipment specifications.*Sketches, catalogs, prices, special requirements, numbers of units and operating speeds and conditions.
2. *Staffing.*The number of people needed at each function for varying volume levels should be specified. This can include job descriptions and reporting relationships.
3. *Operating narrative.*A written description of how the facility functions. The narrative starts at receiving, and traces the entire material flow, including storage and put-away, order picking, packing and assembly, and shipment loading.
4. *Facility layout.*The floor plan for fixed equipment showing all operating areas, staging, utilities, support functions and offices.
5. *Work standards.*Each repetitive job should have a standard that can be applied to measure and control productivity and to establish the building's staff requirement.
6. *Economic feasibility.*The initial budget level costs for construction, equipment, staffing, and implementation need to be refined. The final report should then present the economic and operational basis for approval of the warehouse investment.

BIBLIOGRAPHY

Bowersox, D. J., E. W. Smykay, and B. J. LaLonde, *Physical Distribution Management,* Macmillan, New York, 1971.

Jenkins, C. H., *Complete Guide to Modern Warehouse Management,* Prentice-Hall, Old Tappan, N.J., 1990.

McGee, J. F., W. C. Copacino, and D. B. Rosenfield, *Modern Logistics Management,* Wiley, New York, 1985.

Robeson, J. F., and R. G. House, *The Distribution Handbook,* The Free Press, New York, 1985.

Tompkins, J. A., and J. D. Smith, *The Warehouse Management Handbook,* Tompkins Associates, Raleigh, N.C., 1988.

CHAPTER 7
PRODUCT IDENTIFICATION AND TRACKING

Herbert W. Davis
Herbert W. Davis and Company,
Englewood Cliffs, New Jersey

In both manufacturing and distribution, the accurate, rapid identification of products and the use of this information in controlling the entire process have been key factors in productivity and service. Computers have had the ability to track products and control machining processes for many years. Recent advances in automated identification techniques have improved accuracy and timeliness. The result has been new levels of achievement in reducing manufacturing and distribution cycle times and more automated processes in these two related activities. The combination of product identification systems with computer technologies has been a key factor in the development of today's modern factories and distribution centers.

The purpose of this chapter is to describe the various technologies that are available to the industrial engineer and to illustrate them with some current applications. The field is developing very rapidly, so many of the detail methods will become obsolete, but the underlying concepts and technologies will be in place for many years.

TWO KEY ELEMENTS

A product identification and tracking system consists of two elements, or subsystems. First, it is necessary to have a technology that can identify the product or entity to be controlled. This technology is typically purchased, and there is a wide variety of equipment and methods available. Second, the identification information has to be sent through a communications system to a controller or computer that will interpret the information, update records, and trigger suitable actions—that is, the tracking system. It is very important to recognize that these two systems are quite separate. The industrial engineer can adopt any one of a myriad of identification technologies. This decision is almost wholly distinct from the decision on the computer processing system that will act on the acquired identification data. The processing system can be modified many times in the future, but it will be much harder to change the identification technique.

PRODUCT IDENTIFICATION TECHNIQUES

The simplest identification system, visual inspection by an operator, was the traditional input to factory systems until about 1960. It is still the most widely used product identification technique. It is, however, subject to considerable error, both in the identification process and in recording the information. The operator may not identify the item correctly and/or may make an error in writing or keying the proper nomenclature into the file. This inaccurate input then results in wasted effort and costly problems.

During the 1960s, a whole series of technologies were developed which have been refined into a fairly extensive list of optional approaches to automatically identify a part or to read information on a document. Each separate technology has some advantages that have led to its adoption in specific applications. Those technologies that are in general use today are described in the following paragraphs.

Bar Code Scanning. This is the method in widest use today. A bar code is a group of vertical solid lines that are printed together on a label. The width of the space between the lines can be varied to create a unique code; that is, the width of the spaces and their arrangement can be used to denote a letter, number, or symbol. Figure 7.1 shows a typical bar code.

00039391

FIGURE 7.1 Typical bar code.

The bar code is read by a scanner that moves a beam of intense light across the label. The light is reflected back by the spaces between the bars, interpreted by decoders into useful information, and transmitted to a computer or controller for receiving and action. Figure 7.2 illustrates the reading of the bar code label, decoding, and transmission to a process controller or computer for action.

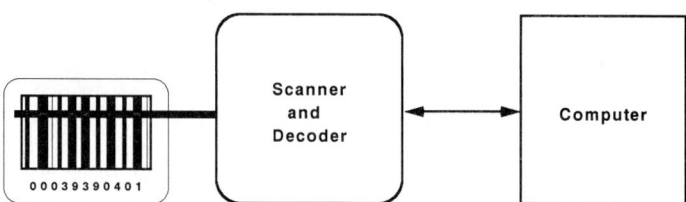

FIGURE 7.2 Identification—the key to the system.

There are a wide range of scanners available, from hand-held to fixed. The scanning technology also is extensive, with at least three different methods in use today:

1. *Helium-neon laser:* These have the longest scanning range and fast reading capability, making this method suitable for fixed stations.
2. *Laser diode:* Less power, high durability, and long life expectancy result in this technology being applied in hand-held, portable scanning installations.
3. *Infrared:* Low power usage, low cost, and small size are important factors. Infrared can read labels through grease, dirt, and opaque coverings, making the technique particularly useful on the shop floor.

Scanning today can be done at distances from as little as an inch to as much as 18 feet. The scanned information, in the form of digital signals, is transferred by wire or radio-frequency transmission to a decoder. The decoder senses the light intensity, differentiates between the spaces and bars, and assigns an alphanumeric character to the signal. The stream of signals is reduced and interpreted into a data set. This set can then be stored or transmitted, as required by the application.

Radio-Frequency Identification. This system is based on an identification tag attached to the item or lot. The tag responds to an inquiry from a reading device, transmitting on a specific radio frequency a stream of data which is picked up on an antenna wired to the control computer. The tag, an important system characteristic, can contain a fixed message (such as product code), or the computer can write new information on it.

The tag is the key element in radio-frequency identification. It is a fairly complex but usually inexpensive device. The tag contains a coil or plate antenna to receive and broadcast signals, a transceiver, control logic, and a nonvolatile memory. Some tags have an integral battery to serve as a power source; others draw their power from an induction coil activated by the scanner.

The other key unit is the scanner, or reader. This device may have an antenna, a transmitter-receiver, and an electronic processor to handle the data. The actual design varies with the requirements of the installation.

The tag can be preprogrammed at the manufacturing stage, or it can be reprogrammed in the field installation as a part of the application. A third option, programming during use, requires that the tag have read/write ability so that data can be added or deleted and status entered as the product moves through a process.

Scanning distances vary widely, from a few inches to 20 feet or more depending on the tag and scanner type and the application.

There are four levels of radio-frequency identification techniques in common use:

1. *Level I:* These are simple, one-bit tags used in retail surveillance systems to detect removal of an item from a retail store.

2. *Level II:* These tags store from 8 to 128 bits, are used in product identification systems, and are the entry to computer-stored information about the item. Thus, the process control computer may have information in its memory about the product, its specifications, and its last reported status. When product arrives at a workstation, the tag identifies the piece, and the process control computer can then supply the information to the machine to guide its actions.

3. *Level III:* This level contains up to 512 bits so that the product number, routing, and status information can be stored. Again, the information can be used as a key to the data stored in a controller.

4. *Level IV:* These are complex tags containing a large amount of information for use in a data-based application.

The higher-level tags can be used in automated manufacturing systems, notifying the machine controller of a product identification, which then triggers the machine setting data. Similar technology is used in directing product movement through a complex conveyer system at a distribution center.

Optical Character Recognition. The major difference between optical character recognition (OCR) and bar coding is that the OCR equipment can read a stylized character that is also readable by the human operator. The system requires a light source and a scanning device coupled to a converter that changes the scanned character into electronic impulses for transmission to the control computer.

Optical scanning has found wide application in postal facilities. Most pieces mailed have typed addresses readable by OCR equipment. The five- plus four-digit ZIP code on mail can be quickly read, and the electronic control system can then set sorting gates, diverters, and collators to complete the mail processing automatically. The result is extremely high-speed letter-sorting.

Magnetic Character Readers. In these systems, human-recognizable letters and numbers are imprinted on a document with magnetic ink. The system is best known for processing bank checks.

Voice Data Entry.   These systems are based on a technology that provides the ability to recognize voice patterns and decode them into a limited vocabulary. The operator speaks directly into a microphone, in a hands-free work environment. For example, the operator may, while unloading a truck, call out the case or product number, name, and other limited information. This voice-entered data is picked up by the system.

The system recognizes the words and numbers, converts these to electronic impulses, and transmits the information to a computer. Similar technology can be used to broadcast information back from the computer to the operators. Voice systems are also used in automated order entry, processing, and feedback applications. The technology is developing rapidly, and voice data entry is expected to have a large role in manufacturing and distribution operations.

Magnetic Stripe Reading.   This method is widely used for plastic credit cards and employee identification systems. Information is electronically recorded on a magnetic stripe which can be read by a decoder and transmitted or interpreted electronically. The technology is applied in manufacturing-floor data collection systems.

PRODUCT IDENTIFICATION APPLICATIONS

In its elementary form, product identification is used to identify and record an item, part, or material at a workstation. The identification technology may be any of the methods currently available to the industrial engineer, as described in the first part of this chapter. Rapid, accurate identification is a key element in the operation of modern retail, manufacturing, and distribution systems.

In retailing, the bar code system has been widely applied. Some large retailers and trade groups have standardized the format and nomenclature that they will use, resulting in a universally recognized product identification number. Machine-readable codes are imprinted by the manufacturer on a product or its package. This code is scanned at the retail store checkout counter, a customer charge is calculated, and the sales information is recorded. Typically, the cash register scanner transmits the information to a processor. Thus, information on sales by item can be aggregated to the store, division, and national level. The sales record can then be used by the retailer to manage inventory and cash. It can also be shared with the manufacturer as an aid in sales forecasting and inventory control. The data is used to calculate product movement, to generate market share information, and to trigger stock replenishment. This integrated information system is much faster and more accurate than the earlier manual systems. Further, information can be accumulated by item rather than by product group, and it is available rapidly at all management levels. The improved accuracy and detail of product movement data has resulted in significant reduction in inventory at all levels in the supply chain. The more efficient use of a retailer's inventory investment has been a major factor in better cost performance and service to the customer.

In manufacturing, various identification techniques are used to record the movement of an item or lot through the facility and to provide status, cost, control, and expediting information.

One of the key factors in manufacturing automation is the ability to identify a part or work order, because that provides access to any data or information stored in a computer or machine controller. Thus, knowing the item on which the machine will be working can lead directly to automation of machine settings and controls. This is well beyond the mere tracking of the status of an item or work order.

In the modern distribution center, a range of identification technologies are used to determine the items received from vendors; to maintain accurate stock location systems; to direct order picking, packing, and assembly; and to manifest, route, and control outbound orders. While bar coding is in the widest use, there are many examples of voice, radio-frequency, and other systems in use. A good example is shown in Fig. 7.3.

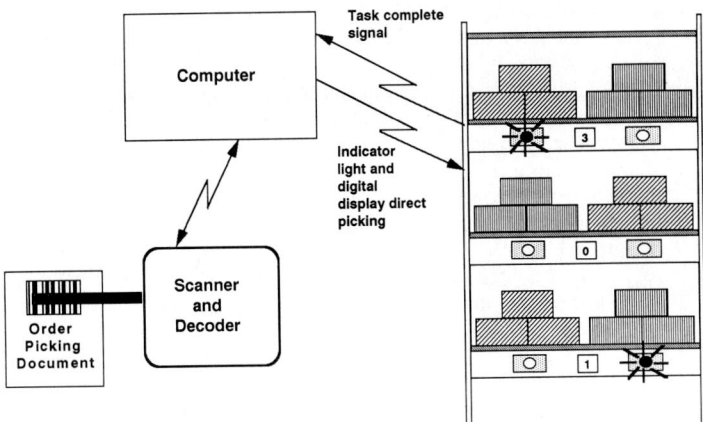

FIGURE 7.3   Computer-aided picking.

A modern automated order-picking system starts with a scanner to identify the customer order at the workstation. The scanned information signals a computer transmission to turn on lights to direct the order picker's attention to the correct item. A digital display notes the quantity and disposition of the pieces needed.

PRODUCT TRACKING

Product tracking is a logical development that stems from the combining of product identification technology with the extensive record keeping, analytical, and data processing capabilities of electronic computers. Basically, a product or a work order can be accurately identified when it arrives at a workstation. This information is then transferred automatically to a computer that records the arrival and adjusts related records to reflect the information. A product tracking software program then can process and utilize this information for a wide range of applications. Of primary interest are tracking systems in manufacturing, distribution, and in freight transportation.

Manufacturing.   There are three broad types of application of product identification and tracking on the shop floor. In the first, the company wants information on product status as material flows through the plant. Typically, the system is based on a shop floor data-collection system. These collectors accept scanned data announcing the arrival of a work batch at the station. Usually, additional information concerning labor time, operator number, machine time, quality level, and the like is entered at the same point through a keyboard, punched cards, tags, or some other automated data entry system.

The computer receives and processes the data and sends information to various departments and individuals to guide their actions. As the material moves from station to station, the tracking system can supply useful information to reduce the time cycle, avoid lag time in the flow, assure that labor and machine resources are available downstream, and provide status information for customer service and/or expediting.

The second major use of product identification and tracking in manufacturing is to keep records of material receipts, usage, and inventory. Normally, these reflect item, location, quantity on hand and on order, and so on. By maintaining these records and comparing them to requirements, the system can release and schedule work orders and provide information on bottleneck items, machines, and workstations. Inventory record keeping is an important element in the success of just-in-time assembly operations.

In the third major application, product identification techniques combined with stored data (in the computer or on a tag) can be used to set and control machining operations. This capability is a key factor in the rapidly developing automation of manufacturing.

Distribution.   Product identification and tracking have been widely applied in warehousing and distribution systems. Modern warehouses typically store thousands of different items and deal with hundreds or thousands of individual receipts and shipments in the course of a business day. Keeping track of orders, materials, and personnel in the modern distribution center is a complex activity. Bar coding is the most-used identification technique. The information scanned is transmitted to a tracking software program that can transmit control information and instructions back to the data terminal.

Figure 7.4 illustrates how product identification combined with a computer control system is used to control flow in a modern distribution center.

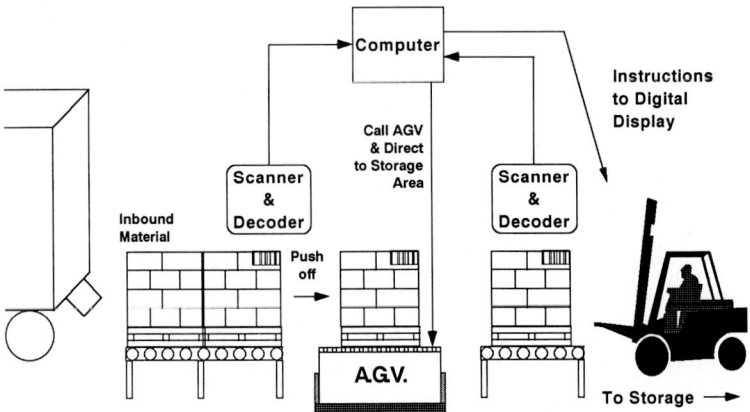

FIGURE 7.4   How product identification is used to control movement.

Typically, materials shipped to a facility are labeled by their manufacturer with bar-coded or other data. The data include company, purchase order number, product name and number, quantity, and the like. At the receiving dock, the label is read by a fixed or hand-held scanner.

The scanned data are verified by a blind count entered by the receiving operator. Both sets of data are used to access the computer records of purchase orders and related information. After verification, the computer directs the disposition of the materials received. Normally, this is done by automatic printing of an internal routing and identification tag or label which is put on the material. The printing is controlled by the tracking computer. Typically, the palletized load with its label is then picked up either automatically by a computer-guided vehicle or by a manually operated forklift truck. Again, the vehicle will have been scheduled or controlled by the tracking program.

The computer will select the storage or assembly line location to which the material is to be delivered and direct the vehicle and its movement. When product arrives at the designated location, the operator scans both the routing label and the identifying label at the destination storage location or workstation. This is verified by the computer, and the status information is adjusted in the computer file.

Following the completion of the manufacturing operation or the picking of materials for movement within the warehouse, the material, operator, and status are scanned and key-entered to continue the tracking process. Step by step, the computer can direct operations, select delivery locations, call and direct automatic and manually driven materials-handling equipment, and record status. The final operations typically involve loading of completed manufacturing and customer shipping orders onto transportation carriers. Figure 7.5 illustrates how a vertical scanner identifies an outbound order, combines this information with automated weight data, prints truck manifests, and sets conveyer gates to direct the order into the right truck.

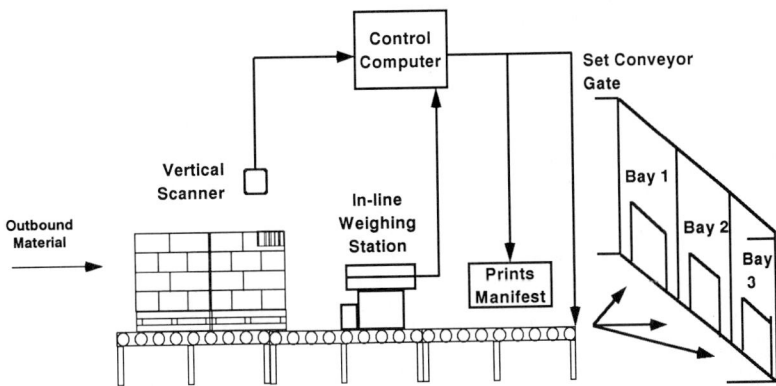

FIGURE 7.5 Elementary identification operations.

The materials are then handed off into the next tracking system. All of this depends on the existence of a product identification technology and a computer tracking and status software and hardware package. The assembly of these different technologies into a single coordinated flow and system are key elements in current automated manufacturing and distribution systems.

To illustrate: a very large central distribution center operated by a major U.S. manufacturer uses bar codes, scanners, and process control computers to manage the entire materials handling and product flow.

Receiving.   Materials are received in palletloads containing one or more items. Each pallet or case of an item has a manufacturing ticket identifying the number of cases of each item, quantity, date, and time. The pallet is removed from the delivery truck and deposited on an output conveyer after adjustment of quantity, load size, and the like to make sure it fits the physical system. The manufacturing ticket is wanded, variable data are entered, and a put-away ticket is automatically produced, showing the assigned location and quantity to be stored. The computer then calls an automatic guided vehicle (AGV) to pick up the palletload. It automatically delivers the pallet (either full or part) to the storing location receiving conveyer. The system next assigns a forklift truck to pull the pallet and deposit it in its designated location. The forklift operator wands the put-away ticket and a bar code label at the rack location. The computer receives and verifies the transaction and then updates inventory in the storage location.

Order Picking and Assembly.   The warehouse process control computer receives shipping orders from the company's mainframe computer. The processor then determines which items are needed from each storage zone in the warehouse. The local-zone forklift truck operators receive information by radio frequency, which is displayed as the next location and item to pick. The operator selects the correct number of cases, wands their bar code, and moves the product to an outbound conveyer. The process controller can verify the picked item identity and quantity, and signal necessary corrections. The controller then calls an AGV to pick up the pallet of material and move it to shipping.

Shipping.   On arrival at the shipping dock, the AGV deposits the pallet on a feed conveyer. Dock handlers scan the item and pallet, the computer signals the appropriate truckline, and the handler removes the pallet from the conveyer and drops it on the proper floor lane designated for the truckline. Priority "must ship" items are dropped close to the door. Multiple pallet orders are marshaled in the truckline drop spots, since part of an order can come from many locations in the distribution center. The shipping team leader calls in trailers and arranges for loading. The loader enters data into a computer at the dock face desk terminal, then wands each pallet as it is loaded into the truck. This relieves the dock area inventory in the warehouse computer.

Thus, the product is tracked at every movement stage through the facility. At any time, management personnel can inquire to determine the status of any item or order. Exactly the same system can be used in each stage of the manufacture, warehousing, and delivery of materials. All depends on the product identification technology.

Transportation.   During the 1980s, product identification and tracking techniques were extended beyond the building walls to include transportation systems. It is now feasible to transmit information from trucks via satellite and telephone communication systems linked to central control computers. The purpose is to supply the tracking computer with data about the receipt and loading of cargo onto the truck. Then, periodic transmissions can be sent both to direct the truck operator to the delivery point and to signal back significant information about current location, problems, progress, and costs. Thus, the shop floor tracking system can follow the manufactured product all the way to the customer destination.

A number of large motor carriers have designed their entire pickup, sorting, and delivery systems around the use of bar-code identification systems.

A typical over-the-road tracking system is illustrated in Fig. 7.6.

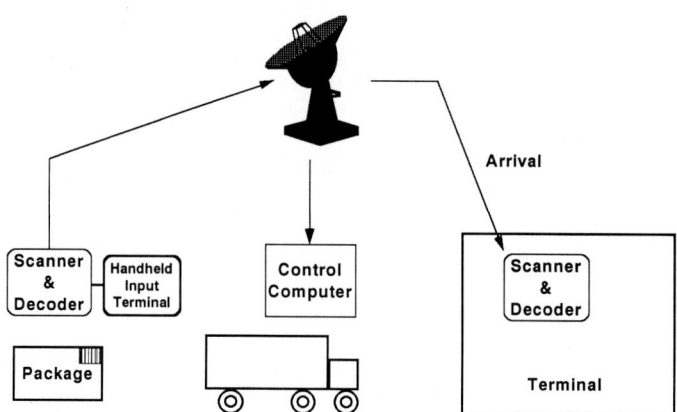

FIGURE 7.6   How product is tracked in an over-the-road carrier system.

In the example, a unique identification code number is assigned to each client tendering freight to the company. The client's shipping department receives bar-coded labels containing the client's unique number plus a sequential series of shipment numbers. The labels are supplied in rolls to facilitate application to the package or pallet.

Once the bar-code label is assigned to a specific shipper, all information pertaining to that particular shipper is input and stored in the carrier's main computer system. When the shipper sends a package, the warehouse applies a label. The driver will pick up the package and take it back to the local carrier terminal. Prior to shipping the package to the main sorting hub, the package is scanned by a laser gun. When it is scanned, the package is also weighed and the destination ZIP code is punched into a hand-held computer. This information is later used to generate the customer's shipping bill. The information is then transferred over telephone lines to the main sorting hub.

Thus, a file has been created that is unique to each shipment or package; as each package enters the hub, it is automatically scanned and identified so it can be sorted and moved to its next destination automatically. Each time it moves to a different location, it is scanned again, and its progress is tracked as it moves toward its final destination. Anyone within the carrier system can access the file to find out who sent the package, how much it weighed, when it was sent, where it was going, where it has been and when, where it is now, and when it is finally delivered.

The shipper's customer service department can access this information file, frequently from its own computer, using telephone lines. The system enables the shipper's customer, frequently a retail store operator, to monitor the flow of inbound goods. The higher level and accuracy of data permits all of the companies involved in the supply chain to operate more efficiently.

Increasingly, transportation companies are adopting this tracking technology to improve their service to shippers and, thus, to gain a competitive advantage. Typical arrangements include assignment of a unique shipper code to each company served, use of bar coding and hand-held scanners to read the shipper's label, and radio-frequency transmission to update carrier terminal records. The status of each shipment is tracked as the packages are picked up at the shipper's dock, transferred through the carrier's terminal, and delivered at the consignee's facility. Other pertinent status information such as dock delivery appointments, weather delays, en route progress reports, and schedule changes are reported and tracked. This same file can often be interrogated directly by the shipper to enhance customer communication.

Such systems are used today to coordinate outbound warehousing and delivery operations as well as the inbound flows of materials to assembly lines in just-in-time systems. The overall tracking network may involve computer coordination of the activities of dozens of companies. Their use aids accuracy and control to the overall system. This combination of product identification technology with product tracking and control software using radio-frequency transmission capabilities confers an obvious advantage to the user. Use of such linked intercompany systems, while now relatively rare, should become widespread quite rapidly.

BIBLIOGRAPHY

Bowersox, D. J., J. J. Closs, and O. K. Helferich, *Logistical Management: A Systems Integration of Physical Distribution, Manufacturing Support and Materials Procurement,* Macmillan, Riverside, N.J., 1986.

Burke, H. E., *Automating Management Information Systems: Principles of Bar Code Applications,* Van Nostrand Reinhold, Florence, Ky., 1990.

CHAPTER 8
ENERGY AND ENVIRONMENTAL CONSIDERATIONS

Nikhil Gandhi
Vice President, XENERGY Inc.
Burlington, Massachusetts

After the oil embargo in 1973, people all over the world watched with concern the dramatic increase in oil prices, the resulting disruption of economies in the Third World and the stagflation phenomenon seen in industrial nations. Growing awareness of the impact of energy costs brought about a fundamental change in the way energy consumption was viewed. Conservation efforts were launched, more efficient equipment was developed, and better housekeeping practices were implemented. In the United States, the amount of energy required per unit of gross national product declined by 28 percent.

Just as energy efficiency was being internalized, further price shocks were felt in 1979 that reinforced the wisdom of using energy efficiently. The oil cartel broke up in the 1980s, but different problems related to energy consumption emerged. Environmental considerations of energy production and consumption became important, as scientific evidence showed the disastrous impacts of excesses of using natural resources.

In the mid- to late '80s, the greenhouse effect, NO_x emissions, ozone depletion, and acid rain were the most frequently debated environmental issues. Carbon dioxide emissions from burning of fossil fuels is the primary cause of the greenhouse effect (warming of the earth's atmosphere). Refrigerants used in cooling cars and office buildings—popularly known as chlorofluorocarbons (CFCs)—depleted the ozone layer, allowing harmful radiation from the sun to reach the atmosphere. Similarly, NO_x emissions produced smog and burning of sulfur-rich fuel caused acid rain.

According to Jose Goldenberg, contributors to the greenhouse effect are energy production (57 percent), CFCs (17 percent), industry (3 percent), agriculture (14 percent), and deforestation and changes in the land-use pattern (9 percent). Thus, 77 percent of the greenhouse effect is from energy production and consumption.

As new research emerges and sensitivity to environmental considerations increases, the role of an industrial engineer will change from that of a manager of energy resources responsible to use them to increase profits to one who does the same in an environmentally responsible manner. The energy manager now has to integrate environmental issues in the analysis.

In addition to the environmental concerns, competitive pressures in the world marketplace began to dictate the need to be more energy-efficient. The energy intensity of the United States has declined from 25,000 British thermal units (Btu) per dollar of gross domestic product (GDP) in 1972 to 18,000 Btu per dollar of GDP in 1988. While this is a significant improvement, the energy intensity in the United States is nearly twice that of Japan and Germany. Despite the differences in the size of these countries, weather, demographics and other factors, it is sufficient to conclude that significant opportunities are available to improve the energy efficiency and reduce the energy intensity. Numerous studies have estimated the potential energy savings that are still feasible. These estimates include those prepared by the Department of Energy, Rocky Mountain Institute, American Council for an Energy-Efficient Economy, and many utilities and consulting firms and range from 15 to 50 percent of current energy consumption. The variations are due to the differences in assumptions, estimation methodology, energy costs, and regional characteristics. However, the fact is that significant improvement in energy efficiency is possible with existing technologies and more could be attained through technologies under development.

COMMERCIAL ENERGY USE

According to statistics compiled by the Energy Information Agency (EIA), commercial buildings in the United States consumed 5 quads of energy in 1986. Electricity accounted for 48 percent of energy consumption measured in Btu and natural gas accounted for 34 percent. The total expenditure on commercial energy consumption in 1986 was $60 billion, of which $47 billion was spent on electricity (78 percent) and only $8 billion (13 percent) on natural gas. Although electricity has a 48 percent market share in terms of Btu, higher electrical energy costs cause more dollars to be spent on electrical energy.

In the commercial sector, energy is used for a variety of end uses such as lighting, space heating, cooling, hot water, office and miscellaneous equipment, and refrigeration in grocery stores. The shares of the end uses vary depending on the type of building, region, and fuels used. For example, energy used for space cooling in hot-climate regions is higher than cooler regions. In office buildings, lighting, heating, ventilating, and air conditioning (HVAC), and office equipment are the major end uses, whereas, in restaurants and grocery stores, cooking and refrigeration are the predominant end uses.

Typically, 35 to 45 percent of electrical energy is used for the lighting end use in commercial buildings. Approximately 25 to 40 percent energy is used for heating, air conditioning, and ventilation; the remainder reflects other end uses. A commonly used method by which to compare energy use in buildings or building types is the energy use index (EUI), calculated per unit of floor area, for total energy used, or for each end use and fuel combination. The EUI for a facility is the total energy consumption of all fuels divided by the facility area in square feet. When the calculation is done for a specific fuel such as electricity, the result is the electric EUI. A more detailed calculation at the end-use level will yield the lighting EUI or HVAC EUI for a specific fuel.

INDUSTRIAL ENERGY USE

The industrial sector consumed 23.8 quads of energy in 1980, which declined to 21.1 quads in 1984. This represent roughly one-third of the total energy consumption in the United States. The energy intensity in the industrial sector has also declined by 1.5 to 2 percent per year from 1971 to 1986. The major end uses in the industrial sector are electric motors, thermal processes, welding, electrolytic separation, and lighting. Electric motors account for nearly 67 percent of industrial electrical consumption. Lighting at 12 percent is the next largest end use, followed by thermal processes, welding, and electrolytic separation; miscellaneous processes account for the remainder.

The comparison of energy usage between industrial plants or industry types can also be compared by calculating EUIs. However, the EUI calculated on a per square foot basis may not be appropriate for all industrial facilities. Some processes such as air separation and concrete preparation are energy-intensive but require less floor space. In such cases, EUIs calculated on a per employee or per unit output may be more appropriate measures of comparison.

PLANNING AND IMPLEMENTING AN ENERGY MANAGEMENT PROGRAM

To a large extent, the success of an energy management program depends on the commitment of top management. The organization must consider energy costs as an important component of the operating cost. The commitment to an energy management program can be expressed in many ways. First, a responsible person (energy manager) must be designated to monitor, control, and reduce energy costs. Second, energy-efficiency improvements or energy cost reduction projects must be evaluated, funded, and staffed as appropriate. Another way to ensure success of an energy management program, especially in industrial plants, is to involve manufacturing managers. Traditionally, the energy management function is assigned to an energy manager who reports to a facility manager or a building manager. A staff role of this nature without any involvement of production personnel restricts the implementation of energy efficiency measures to building operations. Opportunities to reduce lighting and HVAC energy costs exist in buildings, but much greater opportunities to reduce process energy costs may be overlooked without the involvement of production personnel. The energy manager often lacks the necessary backing to educate colleagues in manufacturing about the energy consumption aspects of process operations. As a result, high-efficiency motors may not be specified, or an adjustable speed drive may not be evaluated for a manufacturing application. By involving manufacturing personnel in the evaluation and decision-making process, all opportunities to reduce energy costs can be explored. An effective way to make manufacturing personnel sensitive to energy costs is to submeter plant, buildings, and even major equipment, and to apportion energy costs at this level.

Bill Analysis.   Once the energy management program is in place and top management is committed, the energy manager's job begins with understanding how much and what forms of energy are being used. Energy bills provide many clues to the consumption pattern in facilities. A minimum of the past 12 months' energy consumption should be plotted graphically to look for peculiarities. The plot may show high seasonal use from October to March, indicating space-heating usage. A similar hump in electrical energy use during the summer months indicates heavy air-conditioning usage. Extremely high demand during one month may be attributable to a malfunction or high production output, and a low demand may reflect a plant shutdown. A uniform billing demand indicates fairly consistent operation with little opportunity for load management, as opposed to a facility with high variations in monthly demand. Computing load factor by the formula

$$\text{Load factor} = \frac{\text{kWh used in a period}}{\text{peak demand} \times \text{hours in period}}$$

helps in understanding the variations in electricity usage. The higher the load factor, the more uniform the facility's usage.

The major components of an electric utility's rate structure are energy charge, demand charge, service charge, fuel adjustment, and taxes. The energy manager has more control over energy and demand costs, whereas other charges are not always controllable. The energy charge, measured in kilowatt-hours (kWh), is for the energy used during a specified period. If 1 kW power is used for 1 hour, then 1 kWh of energy is used. A utility may apply different charges for kilowatt-hours consumed during different time periods. For example, energy consumed during the day may cost more than the nighttime usage, if the utility has a time-of-use rate. Multiplying kilowatt-hour consumption by the cost per kilowatt-hour gives the electrical energy costs.

The demand charge depends on the rate of power used and is based on demand in kilowatts. Utilities charge for demand because they have to provide for the capacity that may be used. If a plant has two 20-kW heaters and only one of them is used for 10 hours per day, the utility will charge for 20-kW demand and 200 kWh per day for energy consumption. If, instead, the plant uses both heaters for 5 hours per day, the demand charge will be based on 40 kW demand and the energy charge will still be for 200 kWh. In this case, the utility must have sufficient reserves to meet 40 kW demand, and the cost of providing the reserve capacity is recovered by way of the demand charge, even though this capacity is used for only 5 hours, not 10 hours, per day. Demand charge is defined as dollars per kilowatt-month, and the customer pays for the highest demand registered in a given month. The demand, however, is measured as an average over a 15-minute or 30-minute interval. Unless the rate of use is sustained over that period, the customer will not pay higher demand charges. A popular misconception—that simultaneous starting of large motors creates a new demand peak—is not true because the motor starting is not sustained over 15 minutes.

The demand charge may vary by time of day or season. Utilities also differentiate between measured demand and billed demand. These differences arise because of poor power factor or a ratchet clause (explained later in this chapter).

Fuel oil is billed on a per-gallon basis. Natural gas is sold on a per-therm or million Btu (mmBtu) basis. Large customers transporting natural gas over a utility's pipelines pay only a service charge to the utility, whereas the natural gas cost is paid directly to the supplier.

A monitoring and tracking technique used frequently by energy managers is the comparison of current utility bills with the previous year's bills in the identical period. Such comparison shows trends in energy consumption and allows the energy manager to analyze changes. The comparison can be performed by the fuel unit and its cost. For example, fuel oil consumption can be compared in gallons as well as dollars. A common denominator for comparison, however, is the Btu of energy. Table 8.1 shows the conversion factors applicable to commonly used fuels.

TABLE 8.1   Conversion Factors

Electricity - 3,413 Btu/kWh
Fuel Oil No. 2 - 138,700 Btu/gallon
Fuel Oil No. 6 - 145,000 Btu/gallon
Natural Gas - 100,000 Btu/therm

Energy Audit.   The billing analysis identifies only potential problem areas. The alternatives available to increase the energy efficiency are identified only during the energy audit. The term *energy audit* has a negative connotation, but it is the most commonly used term in the industry. Other terms used to describe a comprehensive effort to understand how and where energy is used, identify energy cost reduction opportunities (ECROs), and evaluate ECROs are *plant energy study, energy survey, energy analysis,* and *energy utilization survey.*

The first step in performing an energy audit is to obtain a full inventory of energy-using equipment. Typically, a building sketch is prepared specifying the characteristics of walls, windows, and roof. In an industrial facility, a process flow diagram is also prepared. A plant tour is taken to survey the equipment: the motors, furnaces, boilers, lighting fixtures, HVAC equipment, and so on. A detailed inventory simplifies the evaluation of opportunities and the preparation of a maintenance schedule. The motor application, horsepower rating, motor type, and speed should be noted. The lighting fixtures should be inventoried by source type (fluorescent, incandescent, and so forth), lamp type, ballast, voltage, number of lamps per fixture, and fixture type, as applicable.

The second step in an energy audit is a survey of operating practices: How many hours of operation per day and when is equipment used? What controls are available? What are the temperature set points? Is the equipment fully loaded? Is the equipment idling? How well is the equipment maintained? This survey, particularly as related to operating hours of equipment, should occur both during regular hours and after business hours. A survey during nonworking hours shows lights and equipment inadvertently left on.

With full knowledge of equipment inventory and operating practices, an energy manager can estimate energy consumed for different end uses and identify opportunities for reducing energy costs. The evaluation can be performed manually or by computer programs. The building energy performance can be simulated using software programs such as DOE2, ASEAM, or the equivalent. The simulation software allows modeling of the building energy performance for different technology and control combinations. These programs do not perform technical and economic analyses of all measures. Computer programs are available that model the building's energy performance, perform technical and economic evaluations, and prepare a full professional report. One such software, XenCAP℠, marketed by XENERGY Inc., Burlington, Mass., performs all these tasks. XenCAP has been used to prepare energy audit reports for over 90,000 commercial and industrial facilities. Figure 8.1 shows the ECROs evaluated by the XenCAP software.

The depths of energy audits may vary substantially. One-day walk-throughs highlight ECROs for further study and are used as a screening tool. The plant energy study or feasibility study involves taking detailed measurements over several weeks. In this case, the analysis provides inputs for the design engineering phase of the project.

Economic Analysis.   The ECROs identified during the energy audit need to be evaluated for cost-effectiveness. The measures of cost-effectiveness are simple payback period, internal rate of return, net present value, and life-cycle benefit-cost ratio. Organizations use different methods to evaluate cost-effectiveness and have different acceptability criteria. Most prefer a simple payback period to evaluate investments because it is a simple and easy-to-use concept. A simple payback period of 2 to 3 years seems acceptable to most organizations, but a shorter or longer payback period may be used depending on the organization's policy. The concepts of engineering economic analysis are described in Sec. 9, Chap. 1. These concepts and methods are required to rank and compare ECROs with other investment opportunities.

All methods require estimated installed cost, annual operating costs, savings, and a discount rate, if time value of money is considered. The total cost has several components, that is, costs of a feasibility study, design engineering, construction supervision, equipment, and installation labor (which should include the cost of removal and disposal, if any, of existing equipment). Not all energy-using equipment requires special disposal, but certain items such as ballasts may require compliance with the state guidelines for hazardous-waste disposal.

Often ignored in the evaluation of ECROs is the annual maintenance cost or operation cost. For example, replacing a dc motor with an ac adjustable-speed drive will reduce the annual maintenance by eliminating carbon brushes, which are not used in an ac motor. The differential maintenance cost is often attributable to different equipment life. A 100-watt high-pressure sodium (HPS) lamp costs approximately $50 and is expected to last 24,000 hours, whereas a 500-watt incandescent lamp lasts 1000 hours and costs $6 to replace. If a 500-watt incandescent lamp is replaced with a 100-watt HPS lamp for an installation operating 3000 hours per year, then only $6.25 ($50 × 3000/24,000) would have to be spent replacing the HPS lamp. Compare that number with the $18 ($6 × 3000/1000) cost of changing the incandescent lamp three times in that period. The lower annual maintenance cost means higher annual savings.

The calculation of energy savings is fairly straightforward—the reduced energy consumption multiplied by the marginal cost of energy. The energy manager must capture all fuel savings and assign appropriate penalties for increased energy consumption, if applicable. For example, replacing lighting fixtures in an office building by more efficient equivalents means electrical energy savings for the facility. In addition, demand charges will be lower if the lighting fixtures were contributing to the facility's peak demand. Now, if these fixtures were located in conditioned spaces, the cooling requirement will be reduced because less heat is dissipated by more efficient fixtures. Similarly, the space heating requirement will increase because less heat is generated internally. Depending on the

XenCAP™
List of Measures

Lighting - Interior and Exterior

(a) Reduce Operating Hours by:
  • manual means
  • installation of switches
  • installation of time control devices
  • occupancy sensors
  • photocell control
(b) Reduce Light Levels by:
  • disconnecting fixtures
  • removing 2 lamps from multi-lamp fixtures
  • specular reflectors (all types)
  • installing no-light lamps
  • installing "Thriftmate" type lamps
  • current limiters
  • using lower wattage lamps
  • using ER-type lamps
  • replacing high output lamps
  • redesigning lighting system
(c) Refixturing with High Efficiency Lamps
  • incandescent to fluorescent
  • incandescent to HPS or metal halide
  • fluorescent or mercury vapor to HPS
  • installation of LPS
(d) Replacement of Standard Lamps with Low Energy Fluorescents (34 or 32 watts) or Incandescents
(e) More Efficient Lamp-Ballast Combination (core coil, electronic or hybrid)
(f) Upgrade Fluorescent Fixtures with 2nd generation lighting (T8-T10) technology

Temperature Control

(a) Reduce Winter Temperature/ Increase Summer Temperature by:
  • manual means
  • locking thermostats
  • thermostat control valves
  • dead-band thermostat
  • rezoning
(b) Modify Setback/Setforward Schedules by:
  • manual means
  • timeclocks
  • energy management systems
  • control modifications

Heating

(a) Service Heating System
(b) Reduce Burner Nozzle Size
(c) Install Flue Damper
(d) Install Turbulators
(e) Extinguish Pilot Lights in Summer
(f) Install Hot Water Reset System
(g) Replace Burner
(h)* Replace Boiler or Furnace or Heat Pump

Cooling

(a) Improve Maintenance of Equipment:
  • clean coils and filters
  • repair/replace belts
  • check refrigerant levels
(b) Install Economizer Cooling:
  • dry bulb
  • enthalpy

  • wet side (strainer cycle)
  • increase ventilation (low tech)
(c) Compressor Demand Controller
(d) Chiller Water/Condenser Water Reset
(e)* Replace Air Conditioner

Ventilation

(a) Reduce Outside Air
  • adjust or repair dampers
  • install new dampers (low leakage type)
  • system redesign
  • use energy management systems for duty cycling
(b) Reduce Fan Operating Time (Night Shutoff)
  • manual means
  • install time control
  • use energy management system
(c) VAV Conversion
(d) Exhaust Waste Heat Recovery

Distribution

(a) Add Pipe Insulation to:
  • domestic hot water pipes
  • heating water pipes
  • steam pipes
  • condensate pipes
(b) Add Tank Insulation Covers to:
  • domestic hot water tanks
  • condensate tanks
  • Process Tanks
(c) Add Duct Insulation
(d) Repair Steam, Hot Water or Duct Leak, Repair or Replace Faulty Steam Traps
(e) Install Destratification Fans

Hot Water

(a) Reduce Delivered Water Temperature
(b) Reduce Requirements with Low Flow Showerheads and Sink Aerators
(c) Shut Off or Cycle Circulator Pumps
(d)* Install New Stand-Alone Unit
(e) Install Vent Damper
(f) Install Swimming Pool Covers
(g) Install Heat Pump

Building Shell

(a) Modify Windows and Doors
  • add storm window/door
  • weatherstrip window/door
  • caulking
  • add custom glazing
  • reduce or modify window areas
  • add plastic storms
  • add night curtains
  • tinted film
(b) Replace Windows and Doors
(c) Add Wall Insulation:
  • add cavity wall insulation
  • install interior insulation
  • add exterior insulation
(d) Improve Ceiling Insulation
  • add insulation in available space
  • install interior insulation

  • add hung ceiling
  • add exterior insulation when reroofing
(e) Reduce Solar Gains
  • install reflective film on windows
  • install reflective roof materials
  • install window shading devices

Refrigeration

(a) Maintain/Repair Refrigerated Cases
(b) Defrost Control
(c) Add Case Covers or Doors
(d) Improve Compressor Efficiency
(e) Reclaim Compressor Waste Heat
(f) Limit Anti-Condensate Heater Operation

Cooking

(a) Improve Maintenance
(b) Improve Utilization
(c) Recover Heat from Exhaust
(d) Install Low Temperature Dishwasher

Transportation

(a) Add Radial Tires
(b) Install Fan Clutch
(c) Install Wind Deflectors
(d) Improve Maintenance
(e) Driver Efficiency

Process Energy

(a) Compressed Air System Maintenance/Modification
(b) Energy Efficient Motors
(c) Process Steam and Hot Water
(d) Process Tank Covers
(e) Variable Speed Drives

Load Management

(a) Energy Management Computer
(b) Time-of-Use/Signal Control Strategies
  • short-term lighting reduction
  • HVAC interrupts
  • process load interrupts
  • water heater shutoff

Alternate Energy Sources & Misc. Measures

(a) Solar Hot Water
(b) Swimming Pool Covers
(c) Solar Pool Heating
(d) Lower Pool Water Temperature
(e) Ceiling Fans
(f) Vestibules

Auditor Controlled Text

(a) Auditor Generated Text
(b) Customized Pre-formatted Text
(c) Full Free-Form Measures

* Fuel Switching Allowed

FIGURE 8.1   A list of energy cost reduction opportunities.

efficiency of the HVAC system and fuel costs, the reduced cooling load may not offset increased space heating equipment. Appropriate savings or penalties should be calculated to estimate net savings.

Another factor important in determining cost savings is the equipment life. Energy-efficient equipment may have a longer life than the standard equipment they replace. Electronic ballasts and high-efficiency motors last approximately 5 years longer than standard ballasts and motors. Economics based on life-cycle cost benefits is more attractive than a simple payback period comparison.

Implementation Plan.   After evaluation and screening of ECROs, the selected opportunities are to be implemented. The energy manager prepares a proposal for top management's review, and after approval is obtained, the design engineering and installation phase begins. The justification for energy projects is facilitated by innovative programs available from some utilities that view the improvement in energy efficiency of their customers as equivalent to meeting energy needs by building more power plants. These utilities may offer energy audits, cash rebates for installing energy-efficient equipment, and installation assistance, all of which reduce the total cost of installation and improve project economics.

Some states offer financial assistance, mostly reduced interest loans, to not-for-profit institutions. Energy Service Companies (ESCOs) offer installation services and also finance energy efficiency projects. The project cost is recovered according to a predetermined schedule as energy bills decrease. Funding and project implementation assistance reduce demand on organizational resources and improve the chances of obtaining top management's approval for projects.

UTILITY RATES

The most expensive energy used in a building or a plant is electrical energy. Electric utilities offer a variety of rate options to customers. Typically, large and small customers are offered different rates, differentiated by service class. Electricity rates for special applications may be available, for example, for electric space heating. Some utilities offer discounts for electric service purchased at a higher distribution voltage, in which case customers purchase and maintain their own transformers. The utility representative, frequently called the *account representative* or *customer service engineer,* is the best source of help in understanding and evaluating the most appropriate rate options for a specific situation.

Time-of-Use Rate.   Capacity-constrained electric utilities charge higher rates during the day, summer or winter, to discourage consumption or reflect true production cost. These are called on-peak or seasonal rates. For example, the on-peak period may be defined as 8 a.m. to 8 p.m. or 9 a.m. to 6 p.m. Electricity is much cheaper during the off-peak hours. The rate differential usually applies to both the energy and demand charge. Some utilities create more than two time periods and also offer different rates during the summer and winter. A facility manager should evaluate and restructure building or process operations to reduce energy consumption when the electricity is more expensive. Several strategies such as shifting the operation to off-peak hours, increasing production during the off-peak hours for use during on-peak hours, adding off-peak production capacity, and turning off unnecessary equipment would help in reducing the electric bill.

Curtailable and Interruptible Rate.   The curtailable rate motivates the customer to reduce peak demand when requested by the utility. This is a special rate offered by some electric utilities to customers contracting to shed load to a predetermined level. For example, a customer's contracted peak demand is 375 kW; on the day the utility calls requesting a reduction in peak demand, the registered demand is 425 kW; this customer, then, must shed 50 kW load to meet the contractual obligation. The utility pays a minimum bonus for the contract performance and possibly a higher incentive to exceed the minimum. This rate usually carries a penalty for nonperformance, but allows a sufficient notice period to comply. By activating emergency generators and shutting off noncritical equipment for a short time, customers can earn incentives for helping the utility meet its needs.

Interruptible rate service is a similar option. However, in this case, the utility controls specific equipment connected to a feeder controlled by the utility. When the utility needs additional capacity, it turns off the power supply to this equipment, instead of the customer shedding the load in curtailable rate service. The incentive is a lower year-round demand charge, and penalties are not applicable because the performance is assured. The number of interruptions per year is limited, which makes this rate attractive for the large customer that can allow noncritical equipment to be disconnected occasionally.

Real-Time Pricing.   The real-time pricing rate is still in an experimental stage, but is being considered more and more by electric utilities experiencing large weather-sensitive system peaks. As the name implies, electric rates are determined every hour on a real-time basis. On a hot summer day in the afternoon, the generation cost is maximum; therefore, the energy and demand charges are highest. At other times, these charges are much lower, providing an incentive to users to reduce electrical loads at peak hour.

Ratchet Clause.   Utilities often provide a disincentive or a penalty in the rate structure which may result in higher electric bills for some customers. One form of such a penalty is the ratchet clause. The ratchet charge links the demand charge a customer pays to the highest demand registered in the past 12 months. If a customer registers a very high demand in a given month and has lower demand in the following 11 months, the demand charges for these months will not correspond to the actual demand. Instead, the customer will be billed a minimum demand charge based on the higher monthly demand previously registered. Thus, using more equipment once for a short duration raises electric bills for the remainder of the year. The plant manager must strive to level loads and achieve a higher load factor in order to reduce electric bills.

Power-Factor Penalty.   Another penalty that a utility may impose for inefficient operation is the power-factor penalty. Power factor is defined as the ratio of actual power being used to the power apparently drawn from the line. When the power supplied by the utility equals that being used by electrical loads, the power factor is unity. In other words, the efficiency is maximum. This can happen only when the electrical circuit is resistive, consisting of loads such as resistance heating or incandescent light bulbs. The power factor for motors, transformers, and other inductive devices is always less than one. In such cases, the utility must supply more power to meet actual demand. Typically, the power factor penalty is levied when the power factor drops below 85 percent. The penalty may be shown separately in the electrical bill or may be charged directly for the kilovolt-amperes (kVA) or kilovolt-amperes reactive (kVAR) used. Power factor can be improved by installing correcting capacitors for the entire plant, or still better, for major inductive loads.

ENERGY-EFFICIENT TECHNOLOGIES

Technology developments over the past decade or so have been responsible for making innovative energy-efficient products available in the market place. For example, electronic ballasts, now available at an attractive price, have a savings potential of up to 40 percent of lighting energy and demand compared with the first generation of fluorescent lamps and ballasts. Technological innovations have succeeded in breaking the price barrier for technologies such as adjustable-speed drives which could save 30 to 45 percent of motor energy consumption, but until recently, were not used widely because of high capital cost. More efficient processes have been developed, and the use of alternative fuels has been explored for the benefit of energy consumers.

Lighting Technologies. Because lighting typically accounts for 30 to 50 percent of peak demand and consumption in the commercial sector, lighting retrofits have become the cornerstone of many utility- and state-sponsored energy conservation programs.

Since most lighting systems operate throughout the business day, reductions in lamp, ballast, and fixture wattage generally result in decreases in the peak demand of utilities, as well as substantial energy savings. Six important techniques will reduce lighting energy consumption:

- Energy-efficient lamp substitutions
- Energy-efficient lamp conversions
- Energy-efficient ballast replacements
- Automatic dimming devices
- Specular reflectors
- Occupancy sensors

Within each of these techniques is a host of individual technologies, some of which apply to the same lighting system and, hence, are mutually exclusive. For example, an inefficient incandescent lamp can be replaced with an energy-efficient incandescent *or* an even more efficient compact fluorescent lamp. The choice will be influenced by the decision maker's weighting of first costs, operating costs, color rendition, rated life, controllability, reliability, appearance, and other factors. A matrix of these factors is shown in Table 8.2.

TABLE 8.2 Applications Light Source Selector Guide

Characteristics of Sources	High Color Fidelity (Color Rendering)			Efficiency (Lumens/Watt)			Lumen Maintenance (Mean Lumens)			Rated Avg. Life (Hours)			Degree of Light Control			Input Power Required (for Equal Light)				System Operating Cost (For Equal Light)			Initial Equipment Cost (For Equal Light)			Total Owning & Operating Cost		
Relative Ratings of Sources	Very Important	Important	Unimportant	Highest (80 Up)	Medium (50-80)	Lowest (19-50)	Highest (85 Up)	Medium (75-85)	Fair (65-75)	Shortest (5000 or less)	Intermediate (5000-15000)	Longest (5000-25000)	Highest	Intermediate	Lowest	Highest	High	Intermediate	Lowest	Highest	Intermediate	Lowest	Highest	Intermediate	Lowest	Highest	Intermediate	Lowest
Incandescent	•					•	•			•			•			•				•							•	•
Tungsten Halogen	•				•	•	•			•			•			•				•							•	•
Fluorescent	•				•		•				•				•			•			•			•			•	
Clear Mercury					•	•	•				•	•		•				•			•			•			•	
Coated Mercury		•			•	•	•				•			•				•			•			•			•	
Clear Metal Arc	•				•				•		•		•										•		•		•	•
Coated Metal Arc	•				•				•		•			•									•		•			•
Clear Lumalux		•	•					•				•											•		•	•	•	•
Coated Lumalux		•	•					•				•											•		•	•	•	•
Clear Unalux		•	•					•			•	•											•		•	•		•
Coated Unalux		•	•					•				•											•		•	•		•

Note: Dot indicates that the light source exhibits the listed characteristics.
Metalarc = mercury vapor.
Lumalux = high-pressure sodium.
Unalux = low-pressure sodium.
Source: XENERGY training materials.

Energy-Efficient Lamp Substitutions. Energy-efficient (EE) fluorescent lamp substitutions are commercially available for virtually all standard fluorescent lamps. Most of the major lighting manufacturers offer such lamps and provide substitution guides. These lamps generally save 10 to 20 percent of the input wattage and generally result in a concomitant 5 to 15 percent decrease in light. Low-wattage lamps also have rated lifetimes of 12,000 to 20,000 hours, equal to their standard counterparts. The most common low-wattage lamps are rated at 25, 34, 60, and 95 watts, and replace 30-, 40-, 75-, and 110-watt standard lamps, respectively. Of these, 4-foot (40- or 34-watt) lamps are the most common type of fluorescent lamp in the commercial sector, especially in office buildings, where they may account for over 70 percent of all fluorescent lamps purchased.

Recently, even lower-wattage, 32-watt, 4-foot lamps have been introduced. These replacements increase lamp efficiency from 79 to 91 lumens/watt in comparison with standard lamps. An even lower-consumption 28-watt, 4-foot lamp (T8) is available for use with high-frequency electronic ballasts.

Because of its ubiquity, the 4-foot fluorescent lamp provides the best example of the energy-efficient lamp retrofit. A standard 40-watt F40T12/CW lamp sells for about $1.50 while its 32-watt replacement costs $3.00 and reduces the wattage by 20 percent, including ballast energy. If such a lamp operated for 3000 hours per year in a service territory where the average commercial price of electricity was $0.075/kWh, with a demand charge of $7.00 per kW-month, the simple payback from lighting energy savings alone would occur in less than 1 year.

The energy savings of a lamp-and-ballast combination differs somewhat from the rating of the lamps themselves. This interaction is described in the ballast section. The lower cooling requirement of reduced lighting energy consumption will improve the payback further.

Over the past few years, fluorescent lamps have been developed that are compatible with, or directly replace, lamps with Edison-type sockets. Improvements in these integral screw-in, modular compact and circline fluorescent lighting systems have opened the door to vast energy and power savings that occur when they replace the highly inefficient standard incandescent lamps.

The most common lamps, the PL-7 (7-watt), PL-9 (9-watt), and SL-18 (18-watt), replace standard 40-, 60-, and 75-watt incandescents, respectively. Besides having greater efficacy, these lamps last approximately 10 times as long as incandescents, thus saving on the costs of replacement lamps and labor over their rated life. The typical replacements shown above produce 10 to 20 percent lower light output, which may be unacceptable only in rare cases. However, the next higher compact fluorescent option will have higher light output as well as lower power requirements.

The potential savings from replacing incandescent lamps with fluorescents is, depending on the specific lamp conversion, in the range of 60 to 80 percent. For example, replacing a 75-watt incandescent lamp with an SL-18 (18-watt) fluorescent lamp would save 75 percent in fixture wattage and provide 570 kWh in total energy savings over the life of the new lamp. An SL-18 retails for about $15 and a standard 75-watt incandescent for about $1. The payback period will be less than 1 year.

High-Intensity Discharge Conversions. High-intensity discharge (HID) lamps include mercury vapor, metal halide, and high-pressure sodium lamps. Although not technically high-intensity discharge lamps, low-pressure sodium lamps are often discussed and associated with this family of lamps. Traditionally deployed for exterior lighting, HID lamps have recently been considered for more interior applications because of their high efficacy. These efficacies and other lamp characteristics are shown in Table 8.3.

Despite their energy savings potential, however, HID lamps have achieved only limited acceptance as a retrofit to existing interior lighting because of their poor color rendition. With their monochromatism, low-pressure sodium lamps make most colors appear gray or yellow. This severely limits their indoor applications, and hence, any contribution they might make to reducing summer peak loads.

High-pressure sodium (HPS) lamps provide color rendition that allows most colors to

TABLE 8.3   Comparison of Lamp Characteristics

Lamp Characteristics	Incandescent including Tungsten-Halogen	Fluorescent	Mercury Vapor	Metal Halide	High-Pressure Sodium	Low Pressure Sodium
Efficacy (Lumens per watt) incl. ballast	13-24	63-95	24-60	69-115	51-130	62-158
Lumen Maintenance	Fair to Excellent	Fair to Excellent	Very Good	Good	Excellent	Excellent
Color Rendition	Excellent	Good to Excellent	Poor to Good	Very Good	Fair	Poor
Light Direction Control	Very Good to Excellent	Fair	Very Good	Very Good	Excellent	Fair
Source Size	Compact	Extended	Compact	Compact	Compact	Extended
Relight Time	Immediate	Immediate	3-10 Minutes	10-20 Minutes	Less than 1 Minute	Immediate
Comparative Fixture Cost	Low-Simple Fixtures	Moderate	Higher than Incandescent & Fluorescent	Generally Higher than Mercury	High	High
Comparative Operating Cost	High-Short Life and Low Efficacy	Lower than Incandescent	Lower than Incandescent	Lower than Mercury	Lowest of HID Types	Low

Source: National Electrical Contractors Association.

remain generally recognizable. Nonetheless, the slight color shift that does occur may be unacceptable where exact color identification is needed. HPS lamps are suitable for many warehouse and industrial applications where color identification is not critical.

Metal halide lamps provide good color rendition and do so at relatively high efficacies. These lamps are good replacements for less efficient mercury vapor lamps. Metal halide lamps have more potential than HPS lamps for indoor applications in commercial and industrial buildings with high-bay fixtures such as warehouses and gymnasiums.

Mercury vapor lamps are the least efficient of the HID lamps and are generally the target of replacement by one of the lamp types discussed above.

Savings are generally large because HID lamps are 2 to 4 times more efficient than incandescents. When refixturing is not necessary, HID retrofits can pay back in 1 to 2 years. However, in many cases, refixturing is required for incandescent-to-HID conversions, and the resulting average paybacks are on the order of 4 to 6 years.

High-Efficiency Ballast Conversions.   Electromagnetic (core-coil) ballasts for fluorescent lamps provide four important functions for lighting system operation: they limit the current during lamp operation, improve the power factor, provide start-up voltage, and help suppress radio interference. Energy-efficient electromagnetic ballasts provide these same services while consuming significantly less energy than standard ballasts. The higher efficiency is obtained through the use of larger iron cores and substitution of copper for aluminum wirings. These ballasts are well proven and have recently become the federal standard.

Electronic ballasts perform the same service as conventional ballasts (starting and operating fluorescent lamps) but at much higher efficacies. Electronic ballasts have additional advantages as well. They convert the operating frequency of standard ballasts from 60 Hz to about 25 kHz. This higher frequency eliminates the flicker and hum associated with electromagnetic ballasts. There are generally two types of electronic ballasts: dimmable and nondimmable. Dimmable ballasts offer new opportunities for energy- and peak-saving lighting control strategies.

Because of the wide range of lighting system configurations that are possible with electronic ballasts, paybacks are highly site-specific. Although electronic ballasts consume about 12 watts less than standard ballasts, their savings are even greater than that because their higher frequency operation improves lamp efficacy as well. For example, a two-lamp, 4-foot fixture with an electronic ballast saves 23 watts in comparison with the same fixture powered by a standard ballast.

Automatic Dimming Equipment.   Automatic fluorescent lighting dimming is a relatively new technology, and equipment for its implementation is just now starting to be used commercially. Automatic dimming involves the sensing (with a photosensor) of the indoor light level to provide a feedback signal for a controller or dimming device. The power going to the fixture is then varied (reduced) to produce a preselected light level. Typically, the power level can be reduced from 100 percent to approximately 30 percent. This technology is used only on the perimeter of buildings to utilize the daylighting available during both cloudy and sunny days.

Specular Reflectors.   Recent progress in the development of specular reflectors offers significant improvement over traditional delamping. Specular reflectors increase luminaire efficiency by improving the reflectance of the luminaire's interior. Conventional luminaires have efficiencies of about 55 percent; that is, 55 percent of the light output of the lamps is reflected out of the luminaire. Luminaires retrofitted with specular reflectors have been shown to have efficiencies of 80 percent. There are two principal types of specular reflector retrofits: (1) silver-film-coated and (2) highly polished, anodized aluminum reflectors. The reflectivity of a specular reflector is very high compared with that of a fixture having a standard enamel-painted reflecting surface.

A recent Electric Power Research Institute (EPRI) report shows that, in laboratory tests, both types of specular reflectors increased luminaire efficiency by 43 percent. This improved efficiency allows up to two lamps to be removed from the luminaire. That is, if two lamps and one ballast are removed from a retrofitted four-lamp fixture, the demand is decreased by 53 percent but the reduction in illuminance is only 28 percent. Concerns about specular reflectors include light-level reduction and spotty light distribution, if fixtures are improperly designed. The simple payback generally occurs in about 2 to 3 years.

Occupancy Sensors.   Occupancy sensors sense the presence of people in the controlled area by infrared heat or ultrasonic sound sensors. These are excellent control devices to turn off lights automatically when no one is present. A time-delay setting delays the automatic switch-off until after a preset time. The lights are turned on instantaneously when occupancy is sensed. Most common applications are conference rooms, private offices, warehouses, stockrooms, and the like. Relatively inexpensive at $70 to $125 per unit, occupancy sensors pay back in under 2 years.

Although the paybacks of varying run times are generally 2.5 to 8 years, when considered in terms of burnout, the paybacks are often much shorter.

Energy-Efficient Motors.   Electricity use by motors accounts for a large percentage of commercial and industrial energy consumption and peak demand. Motors themselves are not an end use, but rather are the means by which other end-use services, such as cooling, ventilation, refrigeration, and industrial processes, are provided. Like other energy-consuming technologies, new motors have been designed that are more efficient than the units composing the existing motor population. High-efficiency motors obtain their greater performance by using thinner steel laminations in the stator and rotor core, minimizing the gap between the stator and rotor, and using more copper in the stator windings. These improvements lower operating temperatures (less waste heat generated) and, consequently, increase motor life.

High-efficiency motors are almost completely interchangeable with standard motors. They are available for use in virtually any application where standard motors are used, for example, for fans, pumps, compressors, and refrigeration, and in sizes from ½ to 300 horsepower. Some applications such as machine tools and hoists require high starting torque, special mounting, and/or frequent starts and stops and will not be suitable for high-efficiency motors.

TABLE 8.4 Comparison of Motor Efficiencies

Motor HP	Standard Efficiency (%)	High Efficiency (%)
1	70.5	87.0
2	75.2	87.3
3	77.4	89.5
5	81.7	89.5
7.5	85.0	91.7
10	85.7	91.0
15	86.7	92.4
20	87.8	93.0
25	88.3	93.6
30	88.8	94.1
40	89.8	94.5
50	90.4	94.5
60	90.7	95.4
75	91.2	95.4
100	92.0	95.4
125	92.1	95.4
150	92.7	95.8
200	93.0	96.2

As can be seen in Table 8.4, the improvement in efficiency is greatest in the lower horsepower motors; however, the larger motors generally offer greater demand and energy savings and have better paybacks. Typical efficiency gains are about 5 to 10 percent for smaller motors and 3 to 4 percent for larger units. The added cost premium of high-efficiency motors is about 15 to 30 percent over standard models.

Several variables combine to affect the payback of these units: hours of operation, electricity prices, motor size, and the point at which the standard motor is replaced. Motor replacement can occur during three points in a standard motor's service life: before rewinding, at rewinding, and at complete burnout. The cost of using a high-efficiency motor decreases from the full price of the new efficient motor to only the incremental difference between the new efficient motor and a new standard motor.

Adjustable-Speed Drives.   AC induction motors operate at a constant speed irrespective of the loading on the motor. The efficiency of an electric motor is highest at near full load, but drops off rapidly at loads less than 55 to 60 percent of the rated capacity. For applications such as fans, pumps, material-handling equipment, and air compressors, where the load varies substantially, energy can be saved by reducing the motor speed to match the load. The motor speed requirement varies directly with flow or volume, but the power requirement is proportional to the cube of the flow. Therefore, at 50 percent flow or load, the power requirement is only 13 percent of the full-load power.

AC adjustable-speed drives vary the frequency of power supplied to the motor to reduce speed to match the load requirement. The savings vary, for example, depending on the method of control, load factor, and hours of use. A recent study by the American Council for an Energy-Efficient Economy (ACEEE) assumed adjustable-speed drives would result in average savings of 22.5 percent of motor energy use in industrial applications.

Solid-State Devices.   Many industrial applications require variable-speed dc motors for process control. The use of dc motors for commercial buildings is limited mostly to elevators. The source of dc power for dc motors is most often a motor-generator (MG) set. Either the MG set is dedicated to a dc motor or the dc generator feeds into a dc bus and more than one dc motor then taps into the bus for dc power.

In a typical situation, the MG set consists of an ac motor driving a dc generator, which is a source of dc power. MG sets have been extremely reliable and rugged in construction, but the efficiency of conversion suffers because of the two rotating pieces of equipment involved. In most cases, it is feasible to replace an aging MG set with a solid-state variable-voltage rectifier. The rectifiers have an efficiency of approximately 96 percent or better, including the losses in the line transformer, which is considerably better than the 72 to 81 percent efficiency of the MG set.

The solid-state rectifiers can be used for replacing any MG set that drives a dc motor. Typical target installations are line shaft drives, helper drives, winders, supercalenders in paper mills, fans and pumps, feed conveyers for saw mills, winches, corrugated-board machinery, coaters, slitters, machine tools, and elevators.

Motor Controls.   Motors consume a large portion of the electrical energy used in industrial processes. In many situations, motors are left running even when not required by the process. Utilization of simple controls and instrumentation will provide a means to shut down idling motors. Typical equipment that can benefit from such controls includes conveyor systems, paint booth fans and pumps, and process exhaust fans. These frequently operate continuously, regardless of the intensity of use. In a conveyor system, for example, the motors run nonstop even though there may be periods during which no product is being transported. Shutting down the motors by automatic switching mechanisms can result in considerable cost savings.

Air Compressors.   The energy required by an air compressor depends on the state of the air (pressure, temperature) at both the intake and exhaust of the compressor. Typical compressed-air operations utilize air intakes located in or near the compressor, where the ambient temperature is typically 70 to 80°F. The efficiency of the compressed-air systems can be improved by utilizing intake air from the coolest possible locations. Cooler air is already more dense and requires less energy to bring it to the required pressure.

Air Leaks.   All compressed air systems, particularly those operating at high pressure, may have leaks which can waste as much as 20 percent of the compressed air produced. Although air leakage cannot be eliminated entirely, a careful survey of the system can reduce the waste to approximately 2 percent. While the air-using equipment runs only periodically, leaks discharge air continuously, causing the compressor to run longer. Leaks can best be detected during periods of low noise level (such as coffee breaks, lunch, and after normal working hours), when most equipment is off. After the leaks have been located, the defective seals, gaskets, or hoses should be repaired as part of normal operating procedure.

HVAC Equipment and Controls.   Equipment to provide space cooling is available today that is much more efficient than older cooling units. And even among the new technologies, there is an upper echelon of high energy efficiency ratio (EER) models that significantly outperform state and federal minimum standards. The EER is a measure of cooling efficiency representing the cooling effect in Btu divided by the energy consumption in watt-hours required to achieve that effect under designated operating conditions.

Among the ways to achieve high EER are the use of longer condenser and evaporator coils, more efficient motors for fans and compressors, and better low-temperature refrigerant line insulation than that used on standard models. Other techniques include reducing airflow-path rates in order to reduce fan energy consumption. Almost all manufacturers currently offer more efficient equivalents for all types of chillers, rooftop units, window units, and heat pumps.

Average EERs for room air conditioners (5000 to 20,000 Btu/h) are around 7.5, yet brands are available with EERs of 9 and, a few, above 10. New central air conditioners (20,000 to 40,000 Btu/h) average an EER of 8.8, but many manufacturers sell units with EERs above 10, with the best models reaching 12 to 15. In larger commercial and industrial applications, chillers are generally used to provide space cooling. New centrifugal chillers have EERs of 13.5 to 16. Reciprocating and screw chillers have slightly lower EERs of 11 to 13.

High-EER air-conditioning equipment provides significant energy and demand savings when compared with standard air-conditioning units. For example, a 12,000 Btu/h window unit with an EER of 9 will use 1000 kWh less energy over a 1500-hour cooling season than a unit with the same capacity and an EER of 6. Under full-load conditions, the high EER unit would also have a 0.68-kW lower demand than the less efficient model. Much larger savings occur when commercial and industrial chillers are replaced.

Absorption Chillers.   Absorption chillers use thermal energy for space cooling. The main difference between a mechanical cooling system and absorption cooling system is that the electrically driven refrigeration compressor is replaced by an absorber. The refrigerant vapor compressed by the compressor and released to the condenser is now absorbed by the absorbent solution. The absorbed refrigerant weakens the absorbent solution; therefore, to recover the refrigerant and increase the effectiveness of the absorbent solution, the weakened mix is pumped to a concentrator where heat is applied to recover the refrigerant. A stronger absorbent solution is then returned to the absorber. The heat can be applied by a direct-fired gas or oil system or by steam produced by a boiler. If the heat recovery is completed in one stage, the systems are called single-stage absorption chillers. If the heat is available at higher temperature, a second recovery stage is added, increasing the system efficiency. Such systems are called two-stage absorption chillers. The coefficient of performance (COP) of a single-stage absorption chiller is about 0.5, which is nearly half that of a two-stage system. By eliminating the refrigeration compressor, the electrical consumption in an absorption machine is limited to pumping energy. The economics is primarily driven by the difference between the cost of electricity and that of fossil fuels. In colder climates, absorption machines can be used for space heating as well, in which case the installation economics look much better. Currently, absorption chillers are more expensive than electric chillers, and paybacks are site-specific.

Variable-Air-Volume Systems.   A variable-air-volume (VAV) system delivers varying amounts of air as required by the conditioned spaces. The air volume is varied by controlling the fan speed or by discharge dampers. By providing the right amount of air, the VAV system eliminates the need for such wasteful systems as dual-duct heating and cooling systems, which maintain hot and cold air all the time and mix them as needed. The VAV system saves about 15 to 30 percent of HVAC energy and 25 to 35 percent of fan energy.

HVAC Controls.   A number of simple HVAC controls are available to save energy. The most commonly used are the dry-bulb economizer, enthalpy-controlled economizer, wet-side economizer, and chiller modulation controls. A dry-bulb economizer senses the outside temperature and allows more outdoor air for cooling when feasible. Colder outdoor air requires less energy and may even be used discretely for space conditioning. The limitation of dry-bulb control is that it senses only the sensible heat of outdoor air. The total heat in the air is also determined by the relative humidity. The enthalpy controller senses both dry-bulb temperature and humidity to determine how much outdoor air should be allowed. The use of *free cooling* is maximized by enthalpy controllers. A wet-side economizer performs a similar function for the chilled water system. When outside temperature is sufficiently low, the condenser water extracts heat from the chilled water, allowing the refrigeration compressor to be shut down. The chilled-water modulation controls raise the chilled-water temperature as the cooling load decreases, thus increasing the efficiency of the air-conditioning system.

Energy Management Systems.   An energy management system (EMS) is a computer-based system for automatically monitoring and controlling building lighting, HVAC, and other systems such as elevators, emergency generators, and water heaters. An EMS can be programmed for setforward (raise temperature during unoccupied hours in the summer) and setback (lower heating temperature during unoccupied hours in the winter), equipment shutoff, temperature settings, and lighting on/off controls from a single remote location. In addition, an EMS system can also perform load shedding and demand control functions. An EMS system is ideal for large facilities, especially when complex optimal start/stop controls, air and chilled water flows and sequencing are required. Simplified systems can be used to control medium-size facilities.

Waste Heat Recovery.   Outdoor air for ventilation imposes an appreciable load on the HVAC systems of most buildings. Because an equal amount of air has to be exhausted for the ventilation air supplied, engineers have sought to recover the energy in the exhaust air before it is rejected to the outdoors. This continues to be so, even though ventilation rates are being scaled downward as requirements in building codes are reevaluated, and as more is learned about how much ventilation air is necessary to maintain freshness in the indoor environment.

Some methods for reclaiming the energy in exhaust air, such as the runaround cycle and rotary heat exchangers, have been known and used for many years, but their economic viability is now much enhanced. Heat wheels have been developed for smaller applications; they have an edge on overall efficiency compared with the runaround cycle. Less costly, but also slightly less efficient, are static air-to-air exchangers. Engineering economics and maintenance factors need to be evaluated to determine the most cost-effective approach.

In industrial plants, significantly more opportunities exist to recover energy from waste heat sources. Energy costs can be reduced by recovering heat from flue gases, boiler blow-downs, condensate, air compressors, refrigeration compressors, and process exhaust. In order to save energy, there must be opportunities to use waste heat for space heating, hot water, or processes such as drying and curing. Care must be exercised to ensure that the temperature of the waste heat source is not lowered too much to cause condensation (flue gases) and that contaminants do not harm the quality of air (furnace or oven exhaust). In addition to the heat wheel, runaround system, and heat exchanger, industrial waste heat recovery equipment includes recuperators and heat pumps.

Industrial Processes.   The Electric Power Research Institute has identified seven factors affecting the choice of technology for industrial processes. These are product quantities and quality, plant siting, process design, equipment selection, sizing, operation, and fuel choice.

Deciding on the right factors is critical to the success of a business. In a competitive environment, the ability of an organization to respond to customer needs rapidly and cost-effectively determines profitability and survival. The technology selection decision must not lose sight of these fundamentals. Energy required by a specific technology or process is an important consideration in evaluating alternatives that meet business needs. In industries such as aluminum smelting and pulp and paper, energy costs as a percent of operating cost are very high, making the evaluation of energy efficiency an integral part of the technology selection process. In less energy-intensive operations, such evaluations are often considered unimportant, although they ought to be important. EPRI has identified technologies that are not necessarily the most energy-efficient, but flexibility, productivity, quality, and production output considerations may favor them overall. Table 8.5 summarizes these technologies and their major applications.

Boilers.   Boilers are used in commercial and industrial facilities to provide steam or hot water for space heating or process needs. The boiler efficiency can be improved in several ways to save money. Boilers are often found to use air in excess of that required for the most efficient combustion. The combustion process requires just the right amount of air. More air does not help the combustion process, but still needs to be heated, only to go out of the stack without any useful work. Excess air can be detected by measuring the oxygen level in the flue gases. A high oxygen level indicates excess air, a situation that can be corrected by adjusting the mix ratio of air and fuel. Monitoring kits and flue-gas analyzers are available to measure excess air so that adjustments can be made to achieve the highest efficiency.

Installing a flue gas economizer to recover heat from the flue gases to preheat the boiler feed water saves money, as does reducing the boiler blow-down and recovering heat from the blow-down. Similar savings result by returning condensate as makeup water. Finally, a boiler operating at a pressure higher than that required by the process wastes energy by heat loss from the boiler and distribution system and the higher stack temperature. These losses can be avoided by maintaining the appropriate boiler pressure.

Cogeneration.   Cogeneration—the simultaneous production of thermal and mechanical energy that can be used to generate electricity—is not a new technology. It has been used by the paper industry since the early 1900s. If a facility has sufficient year-round thermal load, cogeneration is an attractive way to meet the thermal load requirement while generating electricity. Most cogeneration installations are designed to produce electricity in this manner. If the electricity generated is insufficient to meet the electrical demand, the remaining requirement is supplied by the utility. In the very rare cases where more electric-

TABLE 8.5 Electrotechnology Applications

Technology	Definition	Application
Electrolytic separation and electrochemical synthesis.	Dissolution or synthesis of a chemical compound(s).	To produce chlorine, caustic soda, nylon.
Industrial Heat Pump	Absorb low grade heat and improve the heat quality through compression.	Evaporators
Microwave Heating	Heating by electromagnetic waves.	Food and chemicals industries requiring short processing time.
Direct-Arc Melting	An arc passing through the charge to melt it.	Scrap steel production.
Direct Resistance Melting	Resistance heating element transferring heat to the material.	Glass
Induction Melting	Eddy currents produced by induction process melt the charge.	Metal melting.
Electron Beam Heating	Heating by focusing electron beam under vacuum conditions.	Welding in automotive industry.
Flexible Manufacturing Systems	Assembling of one or more automated machine tools.	Flexible production of machined parts.
Induction Heating	Eddy currents produced by induction costs to heat the metal.	Ferrous and non-ferrous metal heating for heat treatment, forging, forming and paint drying.
Infrared Drying and Curing	Infrared radiation.	Textile, paper, paint and other coatings.
Laser Processing	Light amplification by stimulated emission radiation.	Cutting, drilling, melting surface treatment in metal fabrication.
Ultraviolet and Electron Beam Curing	Ultraviolet radiation.	Coating steel, aluminum or other applications where solid coating is needed.

ity is generated than is required, the producer may be able to sell excess power to the utility, subject to local rules and regulations.

Cogeneration systems are available from as low as 25 kW to hundreds of kilowatts capacity. Low-end systems target specific uses, mostly hot-water production. Higher-capacity systems are used to produce steam for process purposes. A topping-cycle cogeneration system produces electricity first, and the rejected heat goes to plant process operations. Bottoming cycle systems produce electricity by using rejected process heat. Both steam and gas turbine systems are available. Cogeneration may not be attractive for all because its economics depends on the difference in price between purchased electricity and cogeneration fuel, thermal load profile, maintenance cost, and projected fuel costs.

Cool Storage Systems.   In the last 4 years, cool storage systems have become popular for retrofits as well as for new construction. Cool storage systems produce chilled water at night during off-peak hours when electricity is cheaper. These systems may store the chilled water itself or convert it into ice, which requires less storage capacity. During the day, when cooling is required, the ice is melted to supply chilled water.

If the system is designed to meet the entire cooling demand of the building, it is known as a full storage system. More often, the cool storage system is used only to meet the peak cooling requirement, in which case the system is known as a *partial storage system*. The latter is more popular because the installation is less expensive, requires less space and saves demand charges by controlling the building's peak demand. Cool storage systems merely shift the production of chilled water to off-peak hours, resulting in lower energy costs. The payback period for cool storage systems depends on the demand charge differential and cooling requirement. Typical installations show a 5- to 10-year payback.

OPERATING AND MAINTENANCE PRACTICES

Approximately 5 to 10 percent of a facility's energy can be saved by observing good housekeeping practices and implementing a routine maintenance program. While very obvious, the golden rule of energy cost reduction—turn it off when not needed—is too often not observed. Energy is wasted in many ways:

- Leaving lights on after business hours or during business hours when no one is present
- Allowing equipment to idle while waiting for materials or products
- Keeping ventilation fans and air conditioners running longer than necessary
- Failing to set back or set forward temperatures during unoccupied hours

Most of these practices can be corrected by maintenance and operating personnel. To avoid human error, automatic controls can be installed to accomplish these tasks. For example, occupancy sensors automatically shut off lights when no one is present, paint booth exhaust fans shut off when the paint gun is not in use, an energy management system or a 7-day programmable thermostat monitors and changes the temperature settings as programmed, and automatic controls can turn off air conditioners.

Another excellent opportunity to capture energy savings arises when the existing equipment burns out or is at the end of its useful life. In many cases, replacing perfectly working equipment is shunned by energy managers even if financially attractive. The philosophy of "if it ain't broke, don't fix it" pervades, and concerns for possible disruptions when replacing working equipment override sound economic judgment. These concerns are no longer valid when the equipment is to be replaced anyway. Now, the difference between choosing the same equipment and its energy-efficient alternative is the incremental cost and incremental improvements. A standard electric motor can be replaced with a high-efficiency motor; a standard fluorescent lamp ballast can be replaced with an electronic ballast; a more efficient chiller can be installed or gas air-conditioning may be considered. An opportunity not seized at this time is probably forgone for the life of the replacement equipment, which in some cases is 15 to 20 years. An industrial engineer must evaluate more efficient alternatives in advance to help make better decisions when equipment needs replacement. A viable strategy is to specify the purchase of more efficient models on burnout of existing equipment.

Finally, a good maintenance program can save large amounts of money. Preventive maintenance not only eliminates premature breakdowns and resulting expensive repairs, but also maintains and improves the efficiency of equipment. Routine maintenance includes lubrication, checkups, cleaning, and even painting to improve the wall reflectance. Neglect, or the absence of a preventive maintenance program, means money lost through compressed-air leaks, leaking valves, failed steam traps, deteriorated insulation, loose-fitting doors, and scale deposits in boiler. Similarly, dirty walls and unwashed fixtures reduce light output; ballasts left in the circuitry after delamping consume energy; and replacing bulbs as they burn out instead of by group relamping every 2 to 3 years costs more in the long run. A good maintenance program has four steps:

- Taking an inventory of equipment and its condition
- Preparing routine maintenance tables that specify tasks to be performed and their frequency
- Defining a regular maintenance schedule
- Monitoring the program

A detailed maintenance schedule supplemented by careful measurements tracks the efficiency of equipment, and indicates when corrective action should be taken. See Sec. 10, Chap. 6 for more information on preventive maintenance programs.

BIBLIOGRAPHY

American Council for an Energy-Efficient Economy, *The Most Energy-Efficient Appliances*, Fall 1986.

Atwood, T., "Refrigerants of the Future: Facts and Facilities," *ASHRAE Journal*, February 1991.

California Energy Commission, *Review of Commercial Availability of Nongeneration Technologies*, Draft Report, 1985.

Clark, E. M., et al., "Retrofitting Existing Chillers with Alternative Refrigerants," *ASHRAE Journal*, April 1991.

"Commercial Buildings Consumption and Expenditures," EIA, DOE/EIA-0318, 1989.

Demand-Side Management, vol. 5, EPRI, EM-3597.

"Electrotechnology Reference Guide," EPRI, EPRI report EM-4527.

Environmental Impact Report, Non-Residential Building Energy Efficiency Standards, California Energy Commission, March 1991.

Epstein, G., and S. Manwell, "An Assessment of Environmental Tradeoffs Between CFC Use and Low Efficiency Cooling with Alternative Refrigerants," *Proceedings of Conference on Energy and Environment*, Alexandria, Virginia, April, 1991.

Fickett, A. et al., "Efficient Use of Electricity," *Scientific American*, September 1990.

Gandhi, N., "Computerized Energy Audit," IIE Fall Conference, Chicago, 1985.

Gandhi, N., "Energy Savings Potential of Selected Lighting Technologies," *Energy Management Division Newsletter*, IIE, 1987.

Gandhi, N., "Industrial Energy Management Success Using IE Techniques," *Proceedings of the Spring 1989 IIE Conference*, Toronto, May, 1989.

Gandhi, N., "Understanding and Estimating Industrial Energy Use for DSM Program Design," Fourth National Demand-Side Management Conference, EPRI, May 1989.

Geller, H. et al., "Acid Rain and Electricity Conservation," American Council for an Energy-Efficient Economy (ACEEE), June 1987.

Goldenberg, J., "Energy and Environmental Policies in Developing Countries," *Conference Proceedings, Energy and the Environment in the 21st Century*, vol. I, March 1990.

Grey P., J. W. Tester, and D. Wood, "Energy Technology: Problems and Solutions," *Conference Proceedings, Energy and the Environment in the 21st Century*, vol. I, March 1990.

GTE Sylvania, "Large Lamp Price Ordering Guide" and "Large Lamp Price Schedule," January and February 1986.

Lighting Technology, Inc., "Luminaire Retrofit Performance," prepared for the Electric Power Research Institute, EPRI EM-5094, Final Report, March 1987.

Lindsay, J., "Specular Reflectors for Fluorescent Troffers," *AEE Conference Proceedings*, Atlanta, Ga., October 1987.

Reddy, A. K., and J. Goldenberg, "Energy for the Developing World," *Scientific American*, September 1990.

Ross, M. H., and D. Steinmeyer, "Energy for Industry," *Scientific American*, September 1990.

Ross, M., "Emissions in U.S. Manufacturing," *Conference Proceedings, Energy and Environment in the 21st Century,* vol. I, March 1990.

Ross, M., "Modeling the Energy Intensity and Carbon Dioxide Emissions in U.S. Manufacturing," *Conference Proceedings, Energy and Environment in the 21st Century,* vol. I, March 1990.

"Superior Office Lighting—an Unusual Approach," *Electrical Construction and Maintenance Magazine,* November 1983.

Turner, W. C., *Energy Management Handbook,* Wiley-Interscience, New York, 1982.

Usibelli, A., et al., "Commercial-Sector Conservation Technologies," Lawrence Berkeley Laboratory, LBL-18543, February 1985.

XenCAP® RISE setup file, XENERGY, Burlington, Mass.

XENERGY internal reports, XENERGY, Burlington, Mass.

XENERGY Training Manual, XENERGY, Burlington, Mass.

XENERGY, *Service Life of Energy Conservation Measures,* prepared for the Bonneville Power Administration, Final Report, July 1987.

MATHEMATICS AND OPTIMIZATION TECHNIQUES

CHAPTER 1
MATHEMATICS FOR THE INDUSTRIAL ENGINEER*

Deborah Mitta
Assistant Professor
Department of Industrial Engineering
Texas A&M University
College Station, Texas

Many problems encountered by the industrial engineer are best solved by the application of mathematical, statistical, or programming procedures. There are a large number of such procedures available to the industrial engineer who has a sufficient grounding in mathematics and statistics to be able to apply them. This chapter and the other chapters of Sec. 14 assume that the industrial engineer has this grounding, for a handbook cannot serve as a text for teaching the fundamentals which are customarily taught in our educational institutions. Instead, the chapters of this section will describe, largely in mathematical terms, a number of the procedures which have been found useful by industrial engineers in solving the kinds of problems they may be expected to be confronted with in their never-ending search for improved ways of doing things. They will thus provide a useful source of information to the industrial engineer seeking the optimum procedure to apply to the specific problem at hand.

LOGIC

Proposition. A proposition is a sentence which we can assert is either true or false. In a proposition, a property of a certain object is mentioned.
   Example

1. The population of the United States is 200,000. This proposition is obviously false.
2. Neil Armstrong was the first man on the moon. As far as we know, this proposition is true.
   Notation. The usual notation for a proposition is a lowercase letter such as p or q.

---

*This chapter is a revision of the chapter by the same name in the third edition which was authored by Burton V. Dean and Maria Altschul.

Operations between Propositions.   There are a number of basic operations between propositions. By means of these and their combinations, we obtain complex propositional statements. These operations are the following:

1. And ($\wedge$): Given two propositions p and q, we define a new proposition $p \wedge q$ as the one which combines the statements given by the original propositions, and which will be true if and only if both are true. If at least one of them is false, then $p \wedge q$ is false. (See truth tables below.)
Example:
Let p be the proposition: Jones is a man.
Let q be the proposition: Jones is a college graduate.
Then $p \wedge q$ is the statement that Jones is a male college graduate.

2. Or (inclusive) ($\vee$): Given two propositions p and q, we define a new proposition p or q, which will be true if and only if at least one of the original propositions is true.
Example:
p: Jones went to a ball game.
q: Jones went to the movies.
$p \vee q$: Jones went to the movies or to a ball game.

3. Or (exclusive) (Δ): Given p and q, we define the new proposition $P\Delta q$ as a combination of p and q, which will be true if and only if exactly one of the original propositions is true.
Example:
p: Jones is in Africa.
q: Jones is in Asia.
$p\Delta q$: Jones is in Africa or in Asia.
Note: In spoken language, we make no distinction between the two different or's. The difference in logic is clear from the examples given above. When we say Jones went to the movies or Jones went to a ball game, we are not specifying time. Jones could have been at the movies and at the ball game, but not at both places at the same time. The proposition Jones went to the movies or to the ball game will then be true if Jones went to the ball game, or if Jones went to the movies, or if Jones did both activities. When we say Jones is in Africa, we mean now. Jones is in Asia also means now. It is then clear that Jones is in Africa or in Asia, and the composite proposition will be true if Jones is in Africa or if Jones is in Asia; but the third possibility which was present before, the one saying Jones is both in Africa and in Asia, is absurd in this case.

4. Negation of a proposition ($\sim$): Given a proposition p, the negation of p is a proposition $\sim p$, which affirms the contrary to what p affirms. It is true if and only if p is false.
Example:
p: It is raining.
$\sim p$: It is not raining.

5. Implication ($\Rightarrow$): Given two propositions p and q, we say that $p \Rightarrow q$ if p implies q. Intuitively, it means that the truth of p implies the truth of q. For a clearer idea, see the truth tables below.

Truth Tables.   Truth tables state the truth or falsehood of a proposition, given the truth or falsehood of its components. Furthermore, we can use these truth tables to define formally the operations listed above.

	p	q	$p \wedge q$	$p \vee q$	$p \Delta q$	$p \Rightarrow q$	$\sim p$	$\sim q$
1.	T	T	T	T	F	T	F	F
2.	T	F	F	T	T	F	F	T
3.	F	T	F	T	T	T	T	F
4.	F	F	F	F	F	T	T	T

T = true, F = false.

All these are intuitively clear, except maybe the column for $p \Rightarrow q$. Cases 1 and 4 are straightforward. Case 2 is easily understood because a true statement cannot imply a false one. Case 3 is not intuitively obvious and should be considered as a logical definition.

In the same way as we construct a truth table for two simple propositions p and q, we can construct one for the more complex propositions that result from applying the operations defined above.

Equivalence ($\Leftrightarrow$). We say that two complex propositions are equivalent if their truth tables are the same. For example, consider the propositions

$$(\sim p)\lor q \quad \text{and} \quad p \Rightarrow q$$

p	q	$\sim p$	$(\sim p)\lor q$
T	T	F	T
T	F	F	F
F	T	T	T
F	F	T	T

This is the truth table for the proposition $(\sim p)\lor q$. If we check on the column $p \Rightarrow q$ of the previous table, we see that it coincides with the column for $(\sim p)\lor q$ in this one. Then $[(\sim p)\lor q] \Leftrightarrow [p \Rightarrow q]$.

SET THEORY

Notion of a Set. A set is a collection, conglomerate, or group of objects. Traditionally, in the literature, capital letters $(A, B, X, \ldots)$ are used to represent sets, and the objects or elements which form these sets are represented by lowercase letters $(a, b, x, \ldots)$.

Notion of Belonging. Given a certain element or object x and a set A, there are two possibilities. Either the element forms part of the set or the element does not form part of the set. To represent this idea, we use the following notation:

$$x \in A \qquad \text{if } x \text{ belongs to } A$$
$$x \notin A \qquad \text{if } x \text{ does not belong to } A$$

A set which has no elements is said to be an empty set and is denoted by $\varnothing$.

Notation. A set is completely determined once we know all the elements that form it. Thus, we can characterize a set by enumerating or writing down all elements that determine it. That is, if X is formed by the elements x, y, and z, we write $X = \{x, y, z\}$.

Examples

1. Let A be the set of all digits; then

$$A = \{0, 1, 2, \ldots, 9\}$$

2. Let B be the set of all positive numbers divisible by 2; then

$$B = \{2, 4, 6, 8, \ldots\}$$

A more convenient way to characterize a set is by giving a property that the elements of the set, and only those elements, satisfy. That is, if A is the set of elements which satisfy property P, we write

$$A = \{x|x \text{ satisfies } P\}$$

Examples.   Using the sets stated above,

1. $A = \{x|x \text{ integer and } 0 \le x \le 9\}$
2. $B = \{x|x \text{ divisible by 2 and } x \text{ positive}\}$

Notion of Inclusion.   Given two sets X and Y, if all the elements of X are also elements of Y, we say that X is included (or contained) in Y. We also say that Y includes (or contains) X, or that X is a subset of Y. We write this as $X \subset Y$.
   Example.   Let

$$A = \{x|x \text{ integer}\}$$
$$B = \{x|x \text{ even}\}$$

then $B \subset A$.
   If both $X \subset Y$ and $Y \subset X$ occur, we say the sets are identical, and we write

$$X = Y$$

Operations between Sets

1. *Union:* Given X and Y, we define the union as

$$X \cup Y = \{x|x \in X \text{ or } x \in Y\}$$

To illustrate this and the following operations, we will make use of the Venn diagrams. Consider the set U of all possible objects (called universe or universal set). We represent it by a rectangle as shown below. The union of X and Y is given by the shaded area.

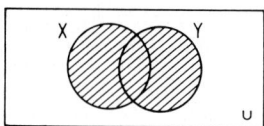

2. *Intersection:* The intersection of X and Y is defined as

$$X \cap Y = \{x|x \in X \text{ and } x \in Y\}$$

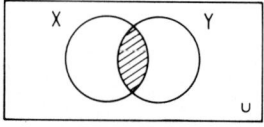

3. *Difference:* The difference between X and Y is defined as

$$X - Y = \{x|x \in X \text{ and } x \notin Y\}$$

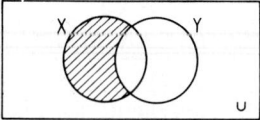

In particular, when X is the universal set, $\sim Y$ is denoted as $-Y$ and called the complement of Y; other common notations for the complement of Y are Y', Y^c, $\overline{Y}$.

Properties

1. $A \cap A = A \cup A = A$
2. $A \cap B = A$ if and only if $A \subset B$ if and only if $A \cup B = B$
3. $\left.\begin{array}{l} -(A \cup B) = (-A) \cap (-B) \\ -(A \cap B) = (-A) \cap (-B) \end{array}\right\}$ De Morgan's laws
4. $A \cap (B \cup C) = (A \cap B) \cup (A \cap C)$
   $A \cup (B \cap C) = (A \cup B) \cap (A \cup C)$
5. $A \cap (B \cap C) = (A \cap B) \cap C = A \cap B \cap C$
   $A \cup (B \cup C) = (A \cup B) \cup C = A \cup B \cup C$
6. $A \cap \emptyset = \emptyset$
   $A \cup \emptyset = A$
   $A \cap U = A$
   $A \cup U = U$
   $(-U) = \emptyset \qquad (-\emptyset) = U$

CALCULUS

Functions.   A function is a rule that assigns to each element of a set X an element of a set Y. X is called the domain; Y is called the range of the function.

Example.   Let $f(x)$ be the function that to each real number assigns its square. Then $f(x) = x^2$ and

$$X = \{\text{real numbers}\}$$

$$Y = \{x|x \text{ real and } x \geq 0\}$$

Limits of Variables.   Suppose we have a quantity x that varies, for instance, with time. If x approaches a constant value a in such a way that the difference $x - a$ gets eventually to be less than any preassigned number, we say that a is the limit of the variable x (in this case, as time approaches a certain value).

Note: If $x - a$ is a negative number, we consider $a - x$, which will then be positive. This is the definition of $|x - a|$ (absolute value or modulus of $x - a$):

$$|x - a| = \begin{cases} x - a & \text{if } x - a > 0 \\ a - x & \text{if } x - a < 0 \end{cases}$$

Properties

1. The limit of a finite sum of variables is the sum of their limits.
2. The limit of the product of a finite number of variables is the product of their limits.
3. The limit of the product of a constant times a variable is the constant times the limit of the variable.
4. The limit of a quotient of two variables is the quotient of the limits, if this quotient is defined, that is, if the limit of the denominator is different from zero.

Definition.   We say that a variable is "bounded from above" if there is a number b such that the variable is never greater than b.

The same definition holds for "bounded from below," replacing greater by smaller.

Property.   If a variable is bounded from above and never decreasing, it approaches a limit that is never greater than the bound.

The same holds true for a variable bounded from below and never increasing, replacing greater by smaller.

Limits of Functions.   We shall illustrate the idea of limits of functions by giving some examples.

Example 1.   Consider the function $y = x^2$ whose graph is shown below.

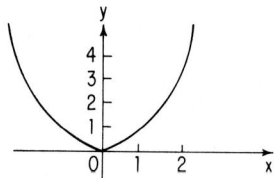

We can see that the lowest value that y takes is zero, when $x = 0$. As we move to the right, y increases indefinitely. We say that y tends to infinity as x tends to infinity, and we write

$$\lim_{x \to \infty} y = \infty$$

Example 2.   Consider the function $y = 1 + 1/x^2$.

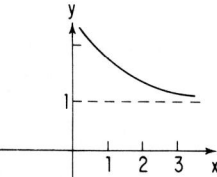

As x increases, y decreases and can be made arbitrarily close to 1. We say that y tends to 1 when x tends to infinity, and write

$$\lim_{x \to \infty} y = 1$$

When x decreases to zero, y increases indefinitely, and we write that as

$$\lim_{x \to 0} y = \infty$$

Consider now a general function $y = f(x)$, and a fixed value a of the variable. If the function $f(x)$ takes values very close to a value L when the values of x are close to a, then we say that $f(x)$ has limit L when x tends to a, and write it

$$\lim_{x \to a} f(x) = L$$

For the case of an infinite limit, if the function $f(x)$ takes values as large as we wish as the value of x gets closer to a, we say that $f(x)$ has limit infinity when x tends to a, and write

$$\lim_{x \to a} f(x) = \infty$$

Continuity.   The concepts of continuity and discontinuity of a function can be better understood by considering the following example.

Let $y = f(x)$ be the function whose graph is shown below.

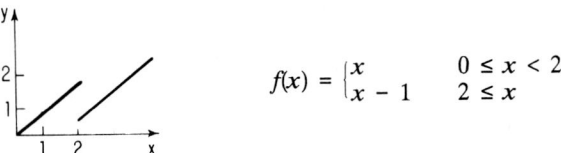

$$f(x) = \begin{cases} x & 0 \le x < 2 \\ x - 1 & 2 \le x \end{cases}$$

Intuitively, we see that at the point $x = 1$, the function is continuous, and at $x = 2$ it is discontinuous. That is, at $x = 1$ it continues increasing smoothly, whereas at $x = 2$ it decreases from the value 2 to the value 1. We are able to give the following definition.

Definition.   A function $f(x)$ is continuous at a point $x = a$ when

1. $f(x)$ is defined for $x = a$

2. $\lim_{x \to a} f(x) = f(a)$

If this is not the case, we say that the function is discontinuous at $x = a$.

In the previous example, $f(x)$ is continuous at $x = 1$, and discontinuous at $x = 2$. Note that although $f(2) = 1$, $\lim_{x \to 2} f(x) = 2$ when we approach $x = 2$ from the left.

There are four possible types of discontinuity. We shall illustrate them with examples.

Case 1.   In the case of the previous function, when

$$\lim_{x \to a^+} f(x) \ne \lim_{x \to a^-} f(x)$$

where $x \to a^+$ simply means that we approach x from the right and $x \to a^-$ that we approach it from the left.

Case 2.   Let

$$f(x) = \begin{cases} x & 0 \le x < 1 \\ -x + 2 & 1 < x \\ 2 & x = 1 \end{cases}$$

whose graph is shown below.

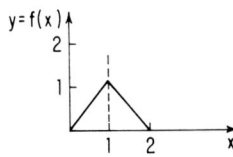

In this case, $\lim_{x \to 1^+} f(x) = \lim_{x \to 1^-} f(x) = 1$ but they are not equal to $f(1) = 2$. Formally, $\lim_{x \to a} f(x) = L$ exists, but $f(a) \ne L$. In this case, $f(a)$ is defined, but the same type of discontinuity exists when $\lim_{x \to a} f(x)$ exists but $f(a)$ is not defined.

Case 3.   Let $y = f(x) = 1 + 1/x^2$, whose graph was shown on p. 14.8, example 2. In this case, $\lim_{x \to 0} f(x) = \infty$, and the function is discontinuous at $x = 0$. Formally, $\lim_{x \to a}$.

Case 4

$$\lim_{x \to a} f(x) = -\infty$$

As an example of this case, take $f(x) = -1 - 1/x^2$. So far, we have considered continuity only at a given point. We now give the following definition.

   Definition.   If $f(x)$ is continuous at every x such that $a < x < b$ for some given a, b, we say that $f(x)$ is continuous in the open interval (a,b). If, furthermore, $\lim_{x \to a^+} f(x) = f(a)$ and $\lim_{x \to b^-} f(x) = f(b)$, then the function $f(x)$ is continuous in the closed interval $[a,b]$. If $f(x)$ is continuous at x for all values of x, we say that it is continuous.

Derivative of a Function.   When in industry we represent a relation between two variables in the form of a graph, we are often interested in the rate of change of the function represented, and especially in the range of values of the independent (or control) variable for which the rate of change may be positive or negative. The most important points are usually the extremes, where the rate of change is zero.

   To illustrate this, consider the curve indicating total cost shown below.

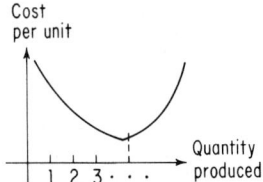

Production costs per unit decrease as we increase the production quantities, but stocking cost increases. The total unit costs are given in the figure. We are interested in that value of the production quantity that yields no change in the total costs. The value where the rate of change in costs with respect to production quantity is zero is the optimal amount to produce.

   We shall now introduce the concept of a difference operator applied to a function so that the concept of a rate of change can be developed. Then, by use of the theory of limits of functions, we shall define the derivatives of operations. As was seen in the example above, the concept of derivative is used in problems of production to find an optimal decision rule.

   Differences.   Let u be a function of n ($u = f(n)$). It is common to use the sequence notation when the independent variable is a positive integer number ($n = 0, 1, 2, \ldots$). Instead of writing $u = f(n)$, we write u_n when we want to indicate that u varies with n. The change in the function u when n increases in value of one is called "first difference," and is written as

$$\Delta u_n = u_{n+1} - u_n$$

and is still a function of n. Note that Δu_n is not a product, it is a single entity; the Δ in this case is not a number. It is called an *operator* because it stands for an operation or rule. Once we have the first difference Δu_n, we can compute the second difference.

$$\Delta^2(u_n) = \Delta(\Delta u_n)$$

In general, we can say

$$\Delta^k(u_n) = \Delta(\Delta^{k-1}u_n)$$

and the following properties hold:

1. $\Delta^r(\Delta^s u_n) = \Delta^r \Delta^s u_n = \Delta^{r+s} u_n$

2. $\Delta(u_n \pm v_n) = \Delta u_n \pm \Delta v_n$

3. $\Delta a u_n = a \Delta u_n$

4. $\Delta u_n v_n = u_{n+1} \Delta v_n + v_n \Delta u_n$

5. $\Delta \dfrac{u_n}{v_n} = \dfrac{v_n \Delta u_n - u_n \Delta v_n}{v_n v_{n+1}}$

Derivatives.   We have defined the difference by finding $u_{n+1} - u_n$. We have assumed that our scale was such that it was convenient for us to count unit by unit. This, however, can be extended to a more general concept of differences, and so we define

$$\Delta u_n = u_{n+h} - u_n$$

where h is any arbitrary interval, positive or negative.

Definition.   The rate of change of u_x over the interval x to $x + h$ (note that we consider x as the variable, instead of n, where x does not have to be integral) is defined as

$$\frac{\Delta u_x}{h} = \frac{u_{x+h} - u_x}{h}$$

This quotient gives us an approximation to how much our function is changing in a given interval x to $x + h$. Now we have to decide on the size we want our intervals to have. It seems only natural to think of intervals as small as we can possibly get them. For instance, when we talk about the speed of a vehicle, we are actually considering the change in distance divided by the change in time, and although we say the speed at a given instant, what we mean is the distance traveled by the vehicle in a very short interval of time, divided by the duration of the interval.

Even if we have limitations given by the measuring instruments as to how short an interval can be, theoretically we have already developed a weapon, the concept of a limit, which allows us to work with intervals as short as we want them. Consider the following example. Let

$$u_x = x^2$$

$$\Delta u_x = u_{x+h} - u_x = (x + h)^2 - x^2 = 2hx + h^2$$

The rate of change is

$$\frac{\Delta u_x}{h} = \frac{2hx + h^2}{h} = 2x + h$$

The smaller h is, the closer the rate of change gets to $2x$. We cannot say that for $h = 0$ the rate of change is $2x$, because for $h = 0$, $\Delta u_x = 0$, and the operation $0/0$ is not defined. But we can say that, as h tends to zero, the rate of change tends to $2x$. In symbols,

$$\lim_{h \to 0} \frac{\Delta x^2}{h} = \lim_{h \to 0} \frac{2xh + h^2}{h}$$

$$= \lim_{h \to 0} (2x + h) = 2x + \lim_{h \to 0} h$$

$$= 2x$$

Graphically, we can give an interpretation of the rate of change as shown. Let P be fixed. Let Q be a point that moves toward P; and $Q_1, Q_2, \ldots$ the positions Q takes.

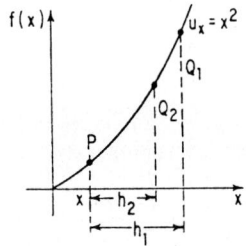

The slope of the chord PQ_1 is $[(x + h_1)^2 - x^2]/h_1$; for PQ_2 it is $[(x + h_2)^2 - x^2]/h_2$. As Q moves toward P, the interval h becomes smaller and smaller, and the chord PQ becomes the tangent line at P. We can now say that the rate of change at P, given by $\lim_{h \to 0} (\Delta x^2/h)$ is the slope of the tangent of the curve at P. At the minimum point of the curve, this slope is zero, and so is the rate of change. We give the following definition.

Definition.   If $f(x)$ is a function, we say that

$$f'(x) = \lim_{h \to 0} \frac{f(x + h) - f(x)}{h}$$

is the derivative of $f(x)$. The limiting process is called differentiation. $f'(x)$ is sometimes denoted as df/dx, $(d/dx)(f(x))$, $Df(x), \ldots$, where d/dx and D can be thought of as operations, as Δ was before. Just as with differences, we define the second derivative of a function as

$$f''(x) = \frac{d}{dx}(f'(x)) = \lim_{h \to 0} \frac{\Delta f'(x)}{h}$$

Operations.   Using the properties of the Δ operator given before and using limiting operations, we have the following:

1. $\dfrac{d}{dx}[u(x) \pm v(x)] = \dfrac{d}{dx}u(x) \pm \dfrac{d}{dx}v(x)$

2. $\dfrac{d}{dx}[au(x)] = a\dfrac{d}{dx}[u(x)]$

3. $\dfrac{d}{dx}[u(x) \cdot v(x)] = u(x) \cdot \left[\dfrac{d}{dx}v(x)\right] + \left[\dfrac{d}{dx}u(x)\right] \cdot v(x)$

4. $\dfrac{d}{dx}\left[\dfrac{u(x)}{v(x)}\right] = \dfrac{\left[\dfrac{d}{dx}u(x)\right] \cdot v(x) - u(x) \cdot \left[\dfrac{d}{dx}v(x)\right]}{[v(x)]^2}$

Optimization.   (Application of derivatives to find rules to minimize or maximize functions.) The problem of finding a maximum or a minimum of a function appears so often in industry that it is important to have a general procedure to find it. Note that finding a maximum of a function f is the same as finding a minimum for $-f$; so we need concern ourselves with solving only one of these problems. We shall consider only the minimizing problem. For example, if we are concerned with profits, our problem will be a maximization one; if we are concerned with costs, we try to solve a minimization problem.

Consider the case illustrated.

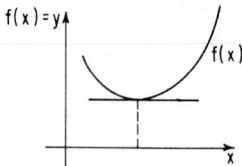

The function f starts decreasing, reaches a minimum, and then increases. Before the minimum is reached, $f'(x) = dy/dx < 0$; after it, $dy/dx > 0$. The conclusion then is that, for a minimum, it is necessary that $dy/dx = 0$.

Note: This, of course, holds if the derivative is a continuous function; otherwise, $f'(x)$ might not exist at the minimum.

Similarly, let us consider the necessary condition for determining the maximum of a function.

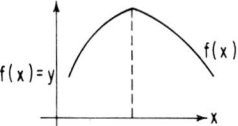

In the case illustrated, there is a maximum, and we have $dy/dx > 0$ before the maximum, and $dy/dx < 0$ after it. We conclude that here also it is necessary to have $dy/dx = 0$ at the maximum.

Now let us consider the following function:

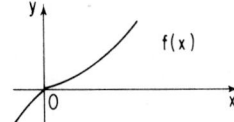

This function is $y = x^3$. If we compute dy/dx at the point $x = 0$, we find that

$$\left.\frac{dy}{dx}\right|_{x=0} = 3x^2\big|_{x=0} = 0$$

$\left.\dfrac{dy}{dx}\right|_{x=0}$ reads $\dfrac{dy}{dx}$ at $x = 0$.

So in this case, we have $dy/dx = 0$ at a point which is clearly not a minimizing value. This is called an inflection point.

Summarizing, we can state that a necessary condition for a minimum or a maximum of a function $f(x)$ at a point x_0 is that

$$\left.\frac{d(f(x))}{dx}\right|_{x=x_0} = 0$$

It is not, as was seen for $y = x^3$ above, sufficient to have $dy/dx = 0$ in order to have a maximum or minimum.

Conditions for Maximum and Minimum. Let $f(x)$ be a function and x_0 be a point at which $f'(x_0) = 0$. We wish to know if we are at an extreme point. If $f(x_0)$ is a minimum, then if h is small, we must have

$$f'(x_0 - h) < 0 \qquad \text{and} \qquad f'(x_0 + h) > 0$$

Then

$$\frac{f'(x_0 - h) - f'(x_0)}{-h} > 0$$

because $f'(x_0) = 0$, and

$$\frac{f'(x_0 + h) - f'(x_0)}{h} > 0$$

if we let $h \to 0$ in both inequalities, we get—provided it exists—$f''(x_0)$ as a second derivative, and analogous to the second difference Δ^2 defined before. So, if x_0 is a minimum, we conclude that $f''(x_0)$ has to be >0. The same reasoning shows us that for x_0 to be a maximum, $f''(x_0)$ has to be <0. If both $f'(x_0)$ and $f''(x_0)$ are zero, we cannot say anything. We might have a minimum as is the case with $f(x) = x^4$; we might have a maximum $f(x) = -x^4$; or we might have a shoulder or inflection point, as in $f(x) = x^3$. In these cases, then it is usually necessary to find higher order derivatives.

Before finishing our discussion about extreme points, we shall discuss functions with more than one extreme point, as the one whose graph is shown below.

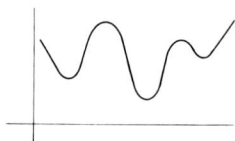

This function has three local minima and two local maxima. We usually want to find the global maximum or minimum. The way to do this is to examine the value of the function at each of these local extremes, and to select that one which gives us the maximum or minimum that is desired. However, in the discussion beginning on page 14.31, we see that if a function is convex (concave), it is guaranteed to have at most one minimum (maximum).

Integration. Suppose we have defined the operation sum ($+$) between two quantities a and b, and we get as a result

$$a + b = c$$

This operation of addition is clearly applied to two factors. Consider now the operation square root, applied to a quantity whose square root is defined. Assume we have

$$\sqrt[2]{a} = b$$

and suppose we want to determine the value of the original quantity a. What we do is apply to the result b the inverse operation to square root; that is, we raise it to the exponent 2. We have

$$a = (\sqrt[2]{a})^2 = b^2$$

In a very similar fashion, we can think of the operation differentiation and define its inverse, which we shall call *integration*.

For example, if we have a relation between profit P and advertising expenditure x given by $P = f(x)$, we showed before that, to get the optimal value of x, we set

$$\frac{dP}{dx} = f'(x) = 0$$

where dP/dx was the rate of change. Suppose now that by some experimental means we can determine the rate of change. The question is, can we use this rate of change to determine the original relationship? The answer depends on having an operation which acts as the inverse of differentiation. We could apply it to the rate of change and obtain as a result the original function. We define the following.

Definition. If $f(x)$ is a given function, an integral of $f(x)$ is a function $y = F(x)$ whose derivative, $F'(x)$, is equal to $f(x)$, and we write

$$y = F(x) = \int f(x)\, dx \qquad y \text{ is the integral of } f(x)$$

where

$$\frac{dy}{dx} = f(x) = F'(x)$$

The symbol $\int$ indicates the operation being performed. The symbol dx tells us to which variable it is applied where there is more than one variable involved. Note that in the definition, we said an integral of $f(x)$, and not the integral for $f(x)$. This can be made clear by noticing that if $F(x)$ is an integral of $f(x)$, then so is $F(x) + c$, where c is any constant. This is because

$$\frac{d(y + c)}{dx} = \frac{dy}{dx} = f(x)$$

Integration, however, is not to be considered only as the inverse of differentiation.

   Integration as the Limit of a Sequence of Sums.   In the previous discussion, we defined differences over an interval h as

$$\Delta f(x) = f(x + h) - f(x)$$

Then we said the rate of change was given by $\Delta f(x)/h$, and defined the derivative $f'(x)$ by letting h tend to zero:

$$f'(x) = \lim_{h \to 0} \frac{f(x + h) - f(x)}{h}$$

Now, suppose we want to find the results of the following sum:

$$S = \Delta f(x) + \Delta f(x + h) + \Delta f(x + 2h) + \cdots + \Delta f(x + nh)$$

Because

$$\Delta f(x) = f(x + h) - f(x)$$

$$\Delta f(x + h) = f(x + h + h) - f(x + h) = f(x + 2h) - f(x + h)$$

$$\cdots\cdots\cdots\cdots\cdots\cdots\cdots\cdots\cdots\cdots\cdots\cdots\cdots\cdots\cdots\cdots$$

$$\Delta f(x + nh) = f(x + (n + 1)h) - f(x + nh)$$

we get, when we cancel all the other terms,

$$S = f(x + (n + 1)h) - f(x)$$

That is, formally,

$$\sum_{r = 0}^{n} \Delta f(x + rh) = f(x + (n + 1)h) - f(x)$$

Suppose now that we wish to find the sum of a function $g(x)$ between the limits $x = a$ and $x = b$. Then

$$S = g(a) + g(a + h) + \cdots + g(b - h) + g(b)$$

In this case, $nh = b - a$. If we know a function $G(x)$ such that $\Delta G(x) = G(x + h) - G(x) = g(x)$, then we have

$$S = G(b + h) - G(a)$$

That is, to find the sum of the function $g(x)$ between a and b, find a function $G(x)$ whose difference is $g(x)$. Note that if we take limits, we get an interpretation very similar to the definition of integral stated above. The function $G(x)$ is called an indefinite sum of $g(x)$—indefinite because if $\Delta G(x) = g(x)$, then $\Delta[G(x) + k] = g(x)$ also. There is an infinite number of suitable functions $G(x)$.

Areas under Curves. Suppose we have the graph of the function $y = g(x)$ and we want

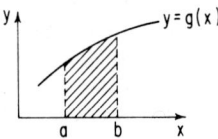

to find out what is the area under this curve between the values $x = a$ and $x = b$. To solve the problem graphically, we may divide the area in small rectangles, and add up the areas of the rectangles. If the rectangles are narrow enough, the shaded area which is ignored

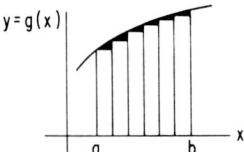

when adding would be small enough to give a negligible error. If the width of the rectangles is h, then the area is

$$S(b) = h[g(a) + g(a + h) + g(a + 2h) + \cdots + g(b - h)]$$
$$= h[G(b) - G(a)]$$

Now, we let h tend to zero, which will make the neglected areas in the above figure tend to zero. To do this, let us increase the right-hand boundary by h; the new sum is $S(b + h)$, and we get

1. $S(b + h) - S(b) = h[(G(b + h) - G(a)) - (G(b) - G(a))]$
$$= h[G(b + h) - G(b)] = hg(b)$$

Similarly, we get

2. $S(b - h) - S(b) = -hg(b - h)$

Dividing equations 1 and 2 by h and $-h$, respectively, we get

1′. $\dfrac{S(b + h) - S(b)}{h} = g(b)$

2′. $\dfrac{S(b - h) - S(b)}{-h} = g(b - h)$

When we let $h \to 0$, we get

$$S'(b) = g(b)$$

But, when $h \to 0$, S tends to the area under the curve $g(x)$, between a and b. We thus obtain the following results:

1. If $S(b)$ denotes the area under the curve $y = g(x)$ between $x = a$ and $x = b$, then $S'(b) = g(b)$.
2. To find the area, we must find a function $G(x)$ whose derivative is $g(x)$, and then compute $G(b) - G(a)$.

Such a function $G(x)$ is called an indefinite integral of $g(x)$. It is an indefinite integral, again, because if $G(x)$ is such a function, then $G(x) + k$ also is an indefinite integral of $g(x)$.

The process of finding $G(x)$ is called integration, and when we define the limits a and b, we have a definite integral, which we write

$$\int_a^b g(x)\, dx$$

Properties

1. $\displaystyle \int_a^b g(x)\, dx = -\int_b^a g(x)\, dx$

2. $\displaystyle \int_a^c g(x)\, dx + \int_c^b g(x)\, dx = \int_a^b g(x)\, dx$

LAPLACE TRANSFORM[1]

Let $f(t)$ be a given function which is defined for all positive values of t. We multiply $f(t)$ by e^{-st} and integrate with respect to t from zero to infinity. Then, if the resulting integral exists, it is a function of s, say $F(s)$:

$$F(s) = \int_0^\infty e^{-st} f(t)\, dt$$

The function $F(s)$ is called the Laplace transform of the original function $f(t)$, and will be denoted by $\mathscr{L}(f)$. Thus

$$F(s) = \mathscr{L}(f) = \int_0^\infty e^{-st} f(t)\, dt \qquad (1)$$

The described operation on $f(t)$ is called the Laplace transformation. Furthermore, the original function $f(t)$ in Eq. (1) is called the inverse transform or inverse of $F(s)$ and will be denoted by $\mathscr{L}^{-1}(F)$; that is, we shall write

$$f(t) = \mathscr{L}^{-1}(F)$$

Example. Let $f(t) = 1$ when $t > 0$. Then

$$\mathscr{L}(f) = \mathscr{L}(1) = \int_0^\infty e^{-st}\, dt = \left. -\frac{1}{s} e^{-st} \right|_0^\infty$$

Hence, when $s > 0$,

---

[1]Adapted partially from E. Kreyszig, *Advanced Engineering Mathematics*, Wiley, New York, 1967.

$$\mathcal{L}(1) = \frac{1}{s}$$

Properties of Laplace Transformation

Property 1.   The Laplace transformation is a linear operation; that is, for any functions $f(t)$ and $g(t)$ whose Laplace transforms exist and where a and b are constants, we have

$$\mathcal{L}[af(t) + bg(t)] = a\mathcal{L}(f) + b\mathcal{L}(g)$$

Property 2.   If $\mathcal{L}(f) = F(s)$ when $s > \alpha$, then

$$\mathcal{L}[e^{at}f(t)] = F(s - a) \qquad s > \alpha + a$$

That is, the substitution of $s - a$ for s in the transform corresponds to the multiplication of the original function by e^{at}.

We shall now see that differentiation and integration of $f(t)$ correspond to multiplication and division of the Laplace transform $F(s) = \mathcal{L}(f)$ by s. The significance of this property of the Laplace transformation is obvious, because in this way the operations of the calculus may be replaced by simple algebraic operations on the transforms.

Property 3.   Suppose that:

1. $f(t)$ is continuous for all $t \geq 0$.
2. $|f(t)| \leq Me^{at}$ for all $t \geq 0$ and for some constants α and M.
3. $f(t)$ has a derivative $f'(t)$ which is piecewise continuous on every finite interval in the range $t \geq 0$.

Then the Laplace transform of the derivative $f'(t)$ exists where $s > \alpha$, and

$$\mathcal{L}(f') = s\mathcal{L}(f) - f(0)$$

Property 4.   Let $f(t)$ and its derivatives—$f'(t), f''(t), \ldots, f^{(n-1)}(t)$—be continuous functions for all $t \geq 0$, satisfying line 2 of property 3 for some α and M, and let the derivative $f^{(n)}(t)$ be piecewise continuous on every finite interval in the range $t \geq 0$. Then the Laplace transform of $f^{(n)}(t)$ exists and is given by the formula

$$\mathcal{L}(f^{(n)}) = s^n\mathcal{L}(f) - s^{n-1}f(0) - s^{n-2}f'(0) - \cdots - f^{(n-1)}(0)$$

In this way, we obtain

$$\mathcal{L}(f'') = s^2\mathcal{L}(f) - sf(0) - f'(0) \tag{2}$$

Properties 3 and 4 may be used for determining transforms.

Example.   Let $f(t) = t^2/2$. Find $\mathcal{L}(f)$. We have $f(0) = 0$, $f'(0) = 0$, $f''(t) = 1$. Because $\mathcal{L}(1) = 1/s$, we obtain from Eq. (2)

$$\mathcal{L}(f'') = \mathcal{L}(1) = \frac{1}{s} = s^2\mathcal{L}(f)$$

or

$$\mathcal{L}\frac{t^2}{2} = \frac{1}{s^3}$$

Differentiation of $f(t)$ corresponds to multiplication of $\mathcal{L}(f)$ by s. Integration of $f(t)$ corresponds to division of $\mathcal{L}(f)$ by s.

Property 5.   If $f(t)$ is piecewise continuous and satisfies an inequality of the form 2 of property 3, then

$$\mathcal{L}\left[\int_0^t f(t)\, dt\right] = \frac{1}{s}\, \mathcal{L}[f(t)] \qquad s > 0, s > \alpha$$

Transformation of Ordinary Differential Equations. Ordinary linear differential equations with constant coefficients can be reduced to algebraic equations of the transform. For example, consider the equation

$$y''(t) + w^2 y(t) = r(t) \tag{3}$$

where $r(t)$ and w are given. Applying the Laplace transformation and using Eq. (2), we obtain

$$s^2 y(s) - sy(0) - y'(0) + w^2 Y(s) = R(s)$$

where $Y(s)$ is the Laplace transform of the (unknown) function of $y(t)$. This algebraic equation is called the subsidiary equation of the given differential equation. Its solution is clearly

$$Y(s) = \frac{sy(0) + y'(0)}{s^2 + w^2} + \frac{R(s)}{s^2 + w^2}$$

Note that the first term on the right is completely determined by means of given initial conditions, $y(0) = k,\ y'(0) = k^2$.

The last step of the procedure is to determine the inverse $\mathcal{L}^{-1}(Y) = y(t)$, which is then the desired solution of Eq. (3).

Example. Find the solution of the differential equation

$$y'' + 9y = 0$$

satisfying the initial conditions $y(0) = 0, y'(0) = 2$. The subsidiary equation is

$$s^2 Y(s) - 2 + 9y(s) = 0$$

Solving for $y(s)$, we obtain

$$Y(s) = \frac{2}{s^2 + 9}$$

From this and the table of some Laplace transforms (pp. 14.19–20), we find

$$y(t) = \mathcal{L}^{-1}(Y) = \tfrac{2}{3} \sin 3t$$

$f(t)$		$F(s) = \mathcal{L}[f(t)]$
1		$\dfrac{1}{s}$
t		$\dfrac{1}{s^2}$
t^n	$n = 1, 2, \ldots$	$\dfrac{n!}{s^{n+1}}$
t^a	a positive	$\dfrac{\Gamma(a + 1)}{s^{a+1}}$
$\dfrac{1}{\sqrt{\pi t}}$		$\dfrac{1}{\sqrt{s}}$

$f(t)$	$F(s) = \mathscr{L}[f(t)]$
$2\sqrt{\dfrac{t}{\pi}}$	$\dfrac{1}{s^{3/2}}$
$\dfrac{t^{a-1}}{\Gamma(a)}$ $a > 0$	$\dfrac{1}{s^a}$
e^{at}	$\dfrac{1}{s - a}$
te^{at}	$\dfrac{1}{(s - a)^2}$
$\dfrac{1}{(n - 1)!}t^{n-1}e^{at}$	$\dfrac{1}{(s - a)^n}$ $n = 1,2,\ldots$
$\dfrac{1}{a - b}(e^{at} - e^{bt})$	$\dfrac{1}{(s - a)(s - b)}$ $a \neq b$
$\dfrac{1}{a - b}(ae^{at} - be^{bt})$	$\dfrac{s}{(s - a)(s - b)}$ $a \neq b$
$\dfrac{1}{w}\sin wt$	$\dfrac{1}{s^2 + w^2}$
$\cos wt$	$\dfrac{a}{s^2 + w^2}$
$\dfrac{1}{a}\sin hat$	$\dfrac{1}{s^2 - a^2}$
$\cosh at$	$\dfrac{s}{s^2 - a^2}$
$\dfrac{1}{w}e^{at}\sin wt$	$\dfrac{1}{(s - a)^2 + w^2}$
$e^{at}\cos wt$	$\dfrac{s - a}{(s - a)^2 + w^2}$

DIFFERENCE EQUATIONS

The First Difference

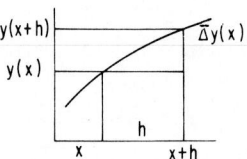

If a function $y(x)$ is given and h is a constant such that $x + h$ is in the domain of y, then Δy, the first difference of y, is a function whose value at x is given by

$$\Delta y(x) = y(x + h) - y(x)$$

Δ is called the difference operator and h the difference interval, which will be assumed to be constant (we are always using the same h) unless otherwise specified.

Example.   If $y(x) = x + 1$ and $h = 1$, then

$$\Delta y(1) = y(2) - y(1) = 3 - 2 = 1$$

$$\Delta y(2.5) = y(3.5) - y(2.5) = 4.5 - 3.5 = 1$$

In general, we can get the value of Δy for every value of x, using the formula

$$\Delta y(x) = y(x + 1) - y(x) = (x + 2) - (x + 1) = 1$$

Second and Higher Differences.   If a function y and its first difference are given, then the second difference of y, $\Delta^2 y$, is the difference of the first difference:

$$\Delta^2 y = \Delta(\Delta y)$$

or

$$\Delta^2 y(x) = \Delta y(x + h) - \Delta y(x)$$

Similarly, the third difference, $\Delta^3 y$, is the difference of the second difference:

$$\Delta^3 y = \Delta(\Delta^2 y) = \Delta(\Delta(\Delta y))$$

In general,

$$\Delta^n y = \Delta(\Delta^{n-1}(y))$$

The identity operator I is that operator which, when applied to any function y, produces a new function Iy identical with y. That is, for any x,

$$Iy(x) = y(x)$$

The symbol Δ^0 is defined as the identity operator, that is,

$$\Delta^0 y = Iy = y$$

By using this last identity, we can state now that

$$\Delta^n y = \Delta(\Delta^{n-1} y)$$

is valid for every $n \geq 1$. (Consider $\Delta^1 = \Delta$.)

Example.   Let $y(x) = x^2$.

$$\Delta y(x) = \Delta x^2 = (x + h)^2 - x^2$$

$$= x^2 + h^2 + 2xh - x^2$$

$$= h^2 + 2xh$$

Now

$$\Delta^2 y(x) = \Delta(\Delta y(x))$$

$$= \Delta(h^2 + 2xh)$$

$$= h^2 + 2(x + h)h - h^2 - 2xh$$

$$= 2h^2$$

and

$$\Delta^3 y(x) = \Delta(\Delta^2 y(x)) = 2h^2 - 2h^2 = 0$$

In general, if $y = x^2$, $\Delta^m y = 0$ for $m \geq 3$.

The Operator ε.   If y is a given function and x a constant, then we define $εy$ as the operator for which

$$εy(x) = y(x + h)$$

So we can now write $\Delta y(x) = εy(x) - y(x)$, and the following property is true:

$$ε^n y(x) = ε(ε^{n-1}y(x)) = y(x + nh)$$

where $ε^0 y(x) = y(x)$; $ε^1 = ε$.

Properties

1. $\Delta[cy(x)] = c\Delta y(x)$
2. $\Delta[y_1(x) + y_2(x)] = \Delta y_1(x) + \Delta y_2(x)$
3. If y is a polynomial of degree n, that is,

$$y(x) = a_0 + a_1 x + a_2 x^2 + \cdots + a_n x^n \qquad a_n \neq 0$$

then
$$\Delta^n y(x) = n!h^n a_n$$

and
$$\Delta^p y(x) = 0 \qquad \text{if } p > n$$

4. If u and v are two functions, then

$$\Delta[u(x) \cdot v(x)] = εu(x) \cdot \Delta v(x) + v(x) \cdot \Delta u(x)$$

Equivalence of Operations.   Two operations 0_1 and 0_2 are equivalent ($0_1 \equiv 0_2$) if, when applied to a function, the functions $0_1 y$ and $0_2 y$ are equal. We can say then that $\Delta \equiv ε - I$ because $\Delta y(x) = εy(x) - Iy(x)$. Because the last identity can be written as $εy(x) = \Delta y(x) + Iy(x)$, we can say that

$$ε \equiv \Delta + I$$

We are assuming then that we can manipulate operators as we do algebraic quantities, which is indeed true. With this new notation, we have

$$\Delta^0 \equiv I \qquad ε^0 \equiv I$$

$$\Delta^2 \equiv ε^2 - 2ε + I \qquad \Delta^3 \equiv ε^3 - 3ε^2 + 3ε - I$$

We define now $ε\Delta$ as the operator, so that

$$ε\Delta y(x) = ε[\Delta y(x)]$$

We can define $\Delta ε$ as $\Delta εy(x) = \Delta[εy(x)]$. In general, the order in which we take two operators to form this product is not irrelevant, but for the case of the operators $ε$ and Δ, it is; so we have

$$ε\Delta y(x) = \Delta εy(x) \qquad \text{or} \qquad ε\Delta \equiv \Delta ε$$

Another property of Δ and $ε$ is the following: If m and n are nonnegative numbers,

$$\Delta^m \Delta^n \equiv \Delta^n \Delta^m \equiv \Delta^{m+n}$$

$$ε^m ε^n \equiv ε^n ε^m \equiv ε^{m+n}$$

The Inverse Operator $(\Delta^{-1}$ and $ε^{-1})$ If Y is a function whose first difference is y, then Y is called an indefinite sum of y and denoted by Δ^{-1}. If

$$\Delta Y(x) = y(x)$$

then
$$\Delta^{-1} y(x) = Y(x)$$

Property.   If Y_1 and Y_2 are indefinite sums of y_1 and y_2, respectively, and c_1 and c_2 are arbitrary constants, then

$$\Delta^{-1}(c_1 y_1 + c_2 y_2) = c_1 \Delta^{-1} y_1 + c_2 \Delta^{-1} y_2$$

$$= c_1 Y_1 + c_2 Y_2$$

In the same way as we defined Δ^{-1}, we can define ε^{-1} as an operator such that if

$$\varepsilon Y(x) = y(x)$$

then
$$\varepsilon^{-1} y(x) = Y(x)$$

We see that
$$\varepsilon^{-1} y(x) = y(x - h)$$

We can also define the operators Δ^{-2}, Δ^{-3}, ... and ε^{-2}, ε^{-3}, ...

Difference Equations.   An equation relating the value of a function y and one or more differences Δy, $\Delta^2 y$, ... for each x value of some set of numbers S for which all these functions are defined is called a difference equation over the set S.

We are concerned, given a difference equation, with finding the function y for which the equation holds.

Example of a Difference Equation. $\Delta^2 y(x) + 2\Delta y(x) + y(x) = 0$, over the set of real numbers.

We are going to assume that our S set is either a finite or an infinite set of successive integers. If this is the case, instead of using $y(x)$ for the values of the function y over the set S, we can use y_k, where k is a subindex which indicates which value of x we are using. In any case, we must always specify the range of values of the x or of the index k.

Solutions of a Difference Equation.   Suppose we are given the following equation:

$$y_{k+1} - 2y_k = 0 \qquad k = 0, 1, 2,...$$

What is meant by the solution of the difference equation?

The equation is a relation between the values of y at the points k and $k + 1$. Is there a function y which makes this equation a true statement for every one of the k values over which the equation is defined?

The function $y_k = 2^k$, $k = 0, 1, 2,...$ is such a function. We can see that it satisfies the equation

$$y_{k+1} - 2y_k = 2^{k+1} - 2 \cdot 2^k = 0 \qquad k = 0, 1, 2,...$$

The function y_k defined before is then said to be a solution to the difference equation: $y_{k+1} - 2y_k = 0$. Note that it is a solution, but by no means the only one. As a matter of fact, all functions of the form $y_k = c(2^k)$, where c is any constant, are solutions to the equation. Any one of the solutions is called a particular solution, and the one containing the arbitrary constant c is called the general solution.

In general, a function y is a solution of a difference equation over a set S if the values of y reduce the difference equation to an identity over S.

Example.   The function given by

$$y_k = 1 - \frac{2}{k} \qquad k = 1, 2, 3,...$$

is a solution for the equation

$$(k + 1)y_{k+1} + ky_k = 2k - 3 \qquad k = 1, 2,...$$

To prove it, we substitute in the previous equation. We have

$$(k + 1)\left(1 - \frac{2}{k + 1}\right) + k\left(1 - \frac{2}{k}\right) = (k + 1) - 2 + k - 2$$

$$= k + 1 - 2 + k - 2 = 2k - 3$$

so the function y satisfies the equation.

Initial Conditions.   Given a difference equation with infinite solutions, we can determine which solution we want to choose by giving initial conditions that must be satisfied. To have a uniquely determined solution, we need as many initial conditions as different constants appear in the general solution.

  Example.   Consider the following solution:

$$(*) \qquad y_k = c(2^k) \qquad k = 0, 1, 2, \dots$$

which satisfies the equation

$$y_{k+1} - 2y_k = 0 \qquad k = 0, 1, 2, \dots$$

for every value of the constant c.

  If we want to find a solution which satisfies the initial condition $y_0 = 3$, we find which one of all the possible $y_k = c2^k$ will satisfy our extra equation.

  We see that $y_0 = c(2^0) = c$, replacing k by 0 in equation (*). So if we choose $c = 3$, we have

$$y_k = 3(2^k)$$

which satisfies both the equation $y_{k + 1} - 2y_k = 0$, $k = 0, 1, \dots$, and the initial condition $y_0 = 3$.

Some Examples of Problems That May Be Solved Using Differences

  Economic Dynamics.   One of the classical economic models gives national income Y as a function of consumption C and investment I. We have

$$(*) \qquad Y_t = C_t + I_t \qquad t = 0, 1, 2, \dots$$

We assume consumption to vary linearly with Y, that is,

$$(**) \qquad C_t = c + mY_t \qquad t = 0, 1, 2, \dots$$

We have that $c \geq 0$, $0 < m < 1$. We assume also that there exists a growth factor $r > 0$ such that

$$(***) \qquad \Delta Y_t = Y_{t+1} - Y_t = rI_t$$

Using equations (*), (**), and (***), we can state

$$Y_{t+1} - Y_t = rI_t = r(Y_t - C_t) = rY_t - r(c + mY_t)$$

or

$$Y_{t+1} = [1 + r(1 - m)]Y_t - rc \qquad t = 0, 1, 2, \dots$$

which is a first-order difference equation. The solution to this equation is

$$Y_t = [1 + r(1 - m)]^t\left(Y_0 - \frac{c}{1 - m}\right) + \frac{c}{1 - m} \qquad t = 0, 1, 2, \dots$$

where Y_0 is given as an initial condition.

Inventory.   Consumer goods are produced for sales and for maintaining inventory levels. Assume, for simplicity, that they are produced only for sales. Let u_t = number of units produced for sale in period t; let v_0 = net investment. The total income y_t produced in period t is equal to the total production of consumer goods plus net investment. Assume u_0 is given.

$$y_t = u_t + v_0 \qquad t = 0, 1, 2, \ldots$$

We assume that $u_t = \beta y_{t-1}$, $t = 1, 2, \ldots$, and β = marginal propensity to consumer. We have the following difference equation:

$$y_t = \beta y_{t-1} + v_0 \qquad t = 1, 2, \ldots$$

which is a first-order difference equation, whose solution is

$$y_t = \beta^t \left(y_0 - \frac{v_0}{1 - \beta} \right) + \frac{v_0}{1 - \beta} \qquad t = 1, 2, \ldots$$

where y_0 is given by the initial conditions.

LINEAR ALGEBRA AND CONVEXITY

Matrices.   A rectangular array of numbers is called a matrix. The notation is as follows. If

$$A = \|a_{ij}\| = \begin{bmatrix} a_{11} & \cdots & a_{1n} \\ \cdots & \cdots & \cdots \\ a_{m1} & \cdots & a_{mn} \end{bmatrix}$$

then A is an m by n matrix ($m \times n$). The number a_{ij} is called an element of the matrix.

Operations and Properties

1. Two matrices A and B are equal if all corresponding elements are equal; that is, if $A = \|a_{ij}\|$ and $B = \|b_{ij}\|$, then $A = B$ if $a_{ij} = b_{ij}$ for all i, j.
2. Given matrices A and B, we define

$$C = A + B$$

as the matrix whose elements c_{ij} are given by

$$c_{ij} = a_{ij} + b_{ij}$$

Thus, two matrices cannot be added unless they have the same number of rows and the same number of columns.

3. $A + B = B + A$
4. $A + (B + C) = (A + B) + C = A + B + C$
5. Given a matrix A and a real number λ, then

$$\lambda A = \|\lambda a_{ij}\|$$

is the product of A and the real number λ (λ is often called a scalar).

6. $\lambda A = A\lambda$
7. The product AB of matrices A and B is defined only if the number of columns of A is equal to the number of rows of B. In that case, we define $C = AB$ as the matrix whose elements c_{ij} are

$$c_{ij} = \sum_{k=1}^{n} a_{ik}b_{kj} \qquad i = 1,\dots,m \text{ and } j = 1,\dots,r$$

where A is $m \times n$ and B is $n \times r$. C will be an $m \times r$ matrix. *Note:* BA is not necessarily defined when AB is defined.

8. $(AB)C = A(BC) = ABC$
$A(B + C) = AB + AC$
Note: In general, matrix multiplication is not commutative even when both AB and BA are defined.

We shall now define some special matrices.
Identity Matrix. A square matrix is of order n, having ones along the diagonal running from upper left to lower right, and zeros elsewhere, that is,

$$I = \begin{bmatrix} 1 & 0 & 0 & \cdots & 0 \\ 0 & 1 & 0 & \cdots & 0 \\ 0 & 0 & 1 & \cdots & 0 \\ & & \cdots & & \\ 0 & 0 & 0 & \cdots & 1 \end{bmatrix}$$

If we write it in element notation, we have

$$I = \|\delta_{ij}\|$$

where

$$\delta_{ij} = \begin{cases} 1 & i = j \\ 0 & i \neq j \end{cases}$$

This symbol δ_{ij} is called the Kronecker delta.
Properties

1. $I^n = I$
2. If A is $m \times n$, then $I_m A = A I_n = A$.
3. $S = \|\lambda \delta_{ij}\| = \lambda I$ is a scalar matrix.
4. $D = \|\lambda_i \delta_{ij}\|$ is a diagonal matrix.

Null Matrix. A null matrix is a matrix whose elements are all zeros. It does not have to be a square matrix.

$$0 = \begin{bmatrix} 0 & \cdots & 0 \\ 0 & \cdots & 0 \\ & \cdots & \\ 0 & \cdots & 0 \end{bmatrix}$$

Properties. Whenever the operations are defined, we have

1. $A + 0 = A = 0 + A$
2. $A - A = 0$
3. $A0 = 0$
4. $0A = 0$
Note: The matrix equation $AB = 0$ does not imply that either A or B is equal to 0.

Example

$$\begin{bmatrix} 1 & 2 \\ 0 & 0 \end{bmatrix} \begin{bmatrix} -2 & 0 \\ 1 & 0 \end{bmatrix} = \begin{bmatrix} 0 & 0 \\ 0 & 0 \end{bmatrix}$$

Transpose Matrix.   The transpose of a matrix $A = \|a_{ij}\|$ is the matrix A', which has for columns the rows of A and for rows the columns of A. that is

$$A' = \|a_{ij}'\| \qquad \text{where } a_{ij}' = a_{ji}$$

Properties

1. $(A + B)' = A' + B'$
2. $(AB)' = B'A'$
3. $I' = I$
4. $(A')' = A$

Symmetric Matrix.   A symmetric matrix is a matrix A such that $A' = A$. A symmetric matrix must be square, and $a_{ij} = a_{ji}$ for all i and j.

Determinants and Inverse Matrix.   Given any square matrix A, we can find a number which is called its determinant $|A|$. To give a useful method for computing the determinant of a square $n \times n$ matrix, we shall first give a rule to find the determinant for a 2×2 matrix.
  Rule 1:

$$\text{If } A = \begin{bmatrix} a_{11} & a_{12} \\ a_{21} & a_{22} \end{bmatrix} \text{ then } |A| = a_{11}a_{22} - a_{21}a_{12}$$

We define now the cofactor A_{ij} of an element a_{ij} of the matrix A as the determinant of the matrix formed by the rows and columns of A, except for row i and column j, multiplied by $(-1)^{i+j}$. We have the following rule.
  Rule 2: To find the determinant of a square 3×3 matrix A,

$$A = \begin{bmatrix} a_{11} & a_{12} & a_{13} \\ a_{21} & a_{22} & a_{23} \\ a_{31} & a_{32} & a_{33} \end{bmatrix}$$

We compute the cofactors A_{ij} of the elements of any row (say i), multiply them by the corresponding a_{ij}, and add. For example, using row 1,

$$|A| = a_{11}A_{11} + a_{12}A_{12} + a_{13}A_{13}$$

$$= a_{11} \begin{vmatrix} a_{22} & a_{23} \\ a_{32} & a_{33} \end{vmatrix} (-1)^{1+1} + a_{12} \begin{vmatrix} a_{21} & a_{23} \\ a_{32} & a_{33} \end{vmatrix} (-1)^{1+2} + a_{13} \begin{vmatrix} a_{21} & a_{22} \\ a_{31} & a_{32} \end{vmatrix} (-1)^{1+3}$$

This is called an expansion in cofactors of row 1.
  Note: It is possible to expand in cofactors of a column instead of a row.
  Now we are able to give a rule to compute the determinant of any square matrix. Let A be an $n \times n$ matrix. Then, to find $|A|$, we expand in cofactors of any row, say row i:

$$|A| = a_{i1}A_{i1} + \cdots + a_{in}A_{in}$$

where $A_{i1}, \ldots, A_{in}$ are the cofactors of order $n - 1$, which we compute by the same method. Because we are reducing the size of the matrices at each step, we shall arrive finally to a number of 2×2 matrices, whose determinants are easy to compute using Rule 1.
  Definition.   The adjoint A^+ of matrix A is the transpose of the matrix obtained by replacing each element a_{ij} by its cofactor A_{ij}.
  Now we can define the inverse A^{-1} of a square matrix A as

$$A^{-1} = \frac{1}{|A|}A^{+}$$

A matrix A such that $|A| \neq 0$ is called nonsingular, and if $|A| = 0$, singular. Only nonsingular matrices have inverses, and every nonsingular matrix has an inverse.

Vectors.   We can define an n vector either as a row matrix with n elements (or column matrix) or as a point in the Euclidean n space. Just as (a_1, a_2, a_3) can be considered a point in a three-dimensional space $(a_1, a_2, \ldots, a_n)$ may be taken as a point in an n-dimensional space. We shall use row or column vectors according to notational convenience.

A vector $(a_1 a_n)$ is a row vector.

A vector

$$\begin{bmatrix} a_1 \\ \cdot \\ \cdot \\ \cdot \\ a_n \end{bmatrix}$$

is a column vector. We will use lowercase letters with a bar over them to indicate vectors.

Definition 1:   A unit vector $\bar{e}_i$ is a vector with 1 as the value of its ith component and all other components equal to 0.

Definition 2:   The null vector $\bar{0}$ is a vector whose components are zero.

Definition 3:   The identity vector has all its components equal to 1.

We say that $\bar{a} \geq \bar{b}$, where $\bar{a}$ and $\bar{b}$ both have n components, if $a_i \geq b_i$ for all i. In the same way, $\bar{a} \leq \bar{b}, \bar{a} < \bar{b}, \bar{a} > \bar{b}, \bar{a} = \bar{b}$ are true if $a_i \leq b_i, a_i < b_i, a_i > b_i$, or $a_i = b_i$, for all i, respectively.

Definition 4:   The scalar product of $\bar{a}$ and $\bar{b}$, both n component vectors, is $\Sigma_{i=1}^{n} a_i b_i$. $\bar{a}$ and $\bar{b}$ are both row vectors, the scalar product is denoted by $\bar{a}\bar{b}'$; if both are columns, by $\bar{a}'\bar{b}$; if $\bar{a}$ is row and $\bar{b}$ is column, by $\bar{a}\bar{b}$. Note that any vector a can be written in terms of the unit vectors, $\bar{e}_i$ as

$$\bar{a} = a_1\bar{e}_1 + \cdots + a_n\bar{e}_n$$

We shall think, then, of the unit vectors as the unit coordinates of our n-dimensional space. In this coordinate system, a_i is the ith coordinate of the point $\bar{a}$.

Definition 5: n-dimensional Euclidean space (E^n) is a collection of vectors (points) $\bar{a} = (a_1 \cdots a_n)$, for which addition and multiplication by a scalar are defined by the rules of matrix operations. Associated with any two elements of the space is a nonnegative number, called the distance between them, where, if

$$\bar{a} = (a_1, \cdots, a_n) \quad \text{and} \quad \bar{b} = (b_1, \cdots, b_n)$$

then

$$\text{Distance} = |\bar{a} - \bar{b}| = [(\bar{a} - \bar{b})'(\bar{a} - \bar{b})]^{1/2} = \left[\sum_{i=1}^{n} (a_i - b_i)^2 \right]^{1/2}$$

Linear Dependence.   *Definition* 1:   A vector $\bar{a}$ from E^n is a linear combination of vectors $\bar{a}_i, \ldots, \bar{a}_n$ from E^n if $\bar{a}$ can be written as

$$\bar{a} = \lambda_1\bar{a}_1 + \cdots + \lambda_n\bar{a}_n$$

for some set of scalars $\{\lambda_i\}$.

Definition 2:   A set of vectors $\bar{a}_1, \ldots, \bar{a}_m$ of E^n is linearly dependent if we can find scalars λ_i not all zero such that

$$\lambda_1 \bar{a}_1 + \cdots + \lambda_m \bar{a}_m = 0$$

If the only set of scalars for which this equality holds is the set $\lambda_1 = \lambda_2 = \cdots = \lambda_m = 0$, the vectors are linearly independent.

Basis. A set of vectors $\bar{a}_1, \ldots, \bar{a}_n$ is a basis for E^n if (1) all the vectors are linearly independent and (2) every vector in E^n can be written as a linear combination of $\bar{a}_1, \ldots, \bar{a}_n$.

Note: This representation of a vector in terms of the basis is unique.

Vector Spaces, Subspaces, Rank. *Definition* 1: A vector space is a collection of vectors which is closed under the operations of addition and multiplication by a scalar. It is denoted by V_n.

Example: E^n is a vector space.

Definition 2: A subspace S_n of V_n is defined as a subset of V_n which is itself a vector space.

Example: E^3, the three-dimensional Euclidean space, is a vector space. E^2, the plane, is a subset of E^3, and a vector space; thus it is a subspace of E^3.

Rank: Given an $m \times n$ matrix A, its n columns can be considered as vectors of E^m (each has m components). We define the rank of A, written $r(A)$, as the maximum number of linearly independent columns in A.

Properties

1. $r(AB) \leq \min [r(A), r(B)]$
2. The product of a matrix of rank k by a nonsingular matrix has rank k.

Simultaneous Linear Equations. Let a_{ij}, b_i be known constants, $i = 1, \ldots, m$ and $j = 1, \ldots, n$. Then, the following set of equations

$$a_{11}x_1 + \cdots + a_{1n}x_n = b_1$$
$$\cdots\cdots\cdots\cdots\cdots\cdots\cdots\cdots$$
$$a_{m1}x_1 + \cdots + a_{mn}x_n = b_m$$

is called a system of m simultaneous linear equations in n unknowns. We can write this system in matrix notation by letting

$$A = \|a_{ij}\| \qquad \bar{x} = (x_1, \ldots, x_n)' \qquad \bar{b} = (b_1, \ldots, b_m)'$$

The system now takes the form of

$$A\bar{x} = \bar{b}$$

In the particular case where $m = n$, if $r(A) = n$, so that $|A| \neq 0$, we obtain a unique solution $\bar{x} = \bar{A}^{-1}\bar{b}$.

METHODS FOR SOLVING A SYSTEM OF SIMULTANEOUS LINEAR EQUATIONS

Cramer's Rule. Let us have the following system of n simultaneous linear equations with n unknowns:

$$a_{11}x_1 + a_{12}x_2 + \cdots + a_{1n}x_n = b_1$$
$$\cdots\cdots\cdots\cdots\cdots\cdots\cdots\cdots\cdots\cdots$$
$$a_{n1}x_1 + a_{n2}x_2 + \cdots + a_{nn}x_n = b_n$$

To solve it, first compute $|A|$, the determinant of the coefficients a_{ij}. If $|A| \neq 0$, we proceed with the rule. If $|A| = 0$, there is either no solution or an infinite number of solutions.

Replace now the first column of A by the column of the b's, and compute the determinant of this new matrix A_1. Then x_1 is given by

$$x_1 = \frac{|A_1|}{|A|}$$

In general, to find x_i, replace the ith column of A by the column of the b's, and if we call A_i the matrix thus obtained, then x_i is given by

$$x_i = \frac{|A_i|}{|A|}$$

This method, although feasible for the cases $n = 2$ and $n = 3$, is not very efficient in finding numerical solutions for large values of n.

Gaussian Reduction Methods.   Given the system ($n \times n$):

$$a_{11}x_1 + \cdots + a_{1n}x_n = b_1$$

$$a_{n1}x_1 + \cdots + a_{nn}x_n = b_n$$

Assume $a_{11} \neq 0$. If $a_{11} = 0$, number the variables again, so that the one that takes the place of a_{11} is different from zero. This can always be done. Divide the first equation through by a_{11} and use this to eliminate x_1 in the other equations $2, \ldots, n$. We get

$$x_1 + a_{12}'x_2 + \cdots + a_{1n}'x_n = b_1'$$

$$a_{22}'x_2 + \cdots + a_{2n}'x_n = b_2'$$

$$\cdots \cdots \cdots \cdots \cdots \cdots$$

$$a_{n2}'x_2 + \cdots + a_{nn}'x_n = b_n'$$

where

$$a_{1j}' = \frac{a_{1j}}{a_{11}} \qquad\qquad j = 2, \ldots, n$$

$$a_{ij}' = a_{ij} - \frac{a_{1j}}{a_{11}}a_{i1} \qquad\qquad i = 2, \ldots, n \text{ and } j = 2, \ldots, n$$

$$b_1' = \frac{b_1}{a_{11}} \qquad b_i' = b_i - \frac{a_{i1}}{a_{11}}b_1 \qquad i = 2, \ldots, n$$

Now in the new system of equations, divide equation 2 by a_{22}' (if $a_{22}' = 0$, renumber equations and variables) and eliminate x_2 in equations 3 to n.

If $|A| \neq 0$, in a finite number of steps we arrive at a set of equations having the following form:

$$x_1 + h_{12}x_2 + \cdots + h_{1n}x_n = g_1$$

$$x_2 + \cdots + h_{2n}x_n = g_2$$

$$\cdots \cdots \cdots \cdots \cdots \cdots$$

$$x_n = g_n$$

We have from the nth equation

$$x_n = g_n$$

Substituting backward, we obtain the values of the other unknowns.

Basic Solutions.   We wish to study a set of m equations $Ax = b$ in n unknowns. If the rank of A is k, we can select k linearly independent columns from A, and assign arbitrary values to the $n - k$ variables not associated with these k columns. We give the following definition.

Definition.   If the rank of A is m (the number of equations in the system), if we select any $m \times m$ nonsingular matrix from A, and if all the $n - m$ variables not associated with this $m \times m$ matrix are set equal to zero, the solution to the resulting system of equations is called a basic solution and the variables associated with it are called basic variables.

Note: This concept of basic solution and variables is the basis for the development of the theory of the simplex method, applied for solving linear programming problems.

Example.   Let us have the following system:

$$x_1 + 2x_2 + x_3 = 4$$

$$2x_1 + x_2 + 5x_3 = 5$$

Setting $x_3 = 0$, we get $x_1 = 2$, $x_2 = 1$. We can set x_1 or x_2 equal to zero and obtain values for the other two. Either of the solutions obtained is a basic solution.

Convex Sets.   *Definition* 1:   In E^n, we define a line through the two points x_1 and x_2 to be the set of all points x such that the relation

$$x = \lambda x_1 + (1 - \lambda)x_2$$

is verified for some real number λ. The set $X = \{x | x = \lambda x_1 + (1 - \lambda)x_2, \lambda_{real}\}$ then represents a line through x_1 and x_2.

If we restrict λ to be between 0 and 1, we can state the following definition.

Definition 2:   In E^n, the line segment joining x_1 and x_2 is defined to be the set of points x such that, for some λ, $0 \leq \lambda \leq 1$:

$$x = \lambda x_1 + (1 - \lambda)x_2$$

Formally, if X is the line segment joining x_1 and x_2, X is of the form

$$X = \{x | x = \lambda x_1 + (1 - \lambda)x_2, 0 \leq \lambda \leq 1\}$$

Definition 3:   A set A is convex if, for any two points x_1 and x_2 belonging to A, the line segment joining these two points is contained in the set A. That is, if A is convex, given x_1 and x_2 in A, the point $x = \lambda x_1 + (1 - \lambda)x_2$, for $0 \leq \lambda \leq 1$, is also in A.

The expression $\lambda x_1 + (1 - \lambda)x_2$, $0 \leq \lambda \leq 1$ is referred to as a convex combination of the points x_1 and x_2.

Examples

1. A circle in E^2 is a convex set.
2. Regular polygons in E^2 are convex sets.
3. The set drawn in the following figure is not a convex set.

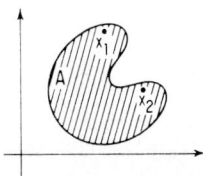

The line segment joining x_1 and x_2 is not included in A.

Definition 4:   A point x is an extreme point of a convex set if there do not exist any two points x_1, x_2, with $x_1 \neq x_2$ in the set, such that $x = \lambda x_1 + (1 - \lambda)x_2$ for some λ between 0 and 1, $\lambda \neq 0$ and $\lambda \neq 1$. Intuitively, an extreme point cannot be on a line segment joining two other points that belong to the set.

An extreme point is a boundary point of a convex set, but not all boundary points need be extremes. In example 1 above, all boundary points are extremes. In example 2, the vertices of the polygons are extremes, but any point on the boundary between two vertices is not an extreme point.

Properties

1. The intersection of two convex sets is convex.

2. Every vector space is convex. (Remember that a vector space is a set of all scalar combinations of vectors, and as such, it is convex.)

Definition 5:   The set of all convex combinations of a finite number of points is called the convex polyhedron spanned by these points.

Definition 6:   A cone C is a set of points with the following property: if x belongs to C, then μx belongs to C for all $\mu \geq 0$.

Definition 7:   A convex cone is a cone that is also a convex set.

Examples

1.

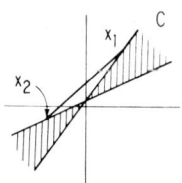

The shaded area is a cone, but is not a convex cone (we can see that the line segment between x_1 and x_2 is not in C).

2.

The shaded area is a cone, and is also a convex cone.

In the same way that the theory of basic solutions is the fundamental weapon used to develop the theory of the simplex method, the theory of convex sets, convex cones, and convex polyhedra is the basis for the geometrical interpretation of the same simplex methods.

It can be proved that there is a very well defined relation between systems of simultaneous linear equations (linear programming problems) and their basic solutions and convex polyhedral sets (the graphic representation of a system of simultaneous linear equations) and their extreme points.

The basic theory of the simplex algorithm consists in using the fact that all solutions to a system of simultaneous linear equations (linear program) are graphically the extreme points of a convex set that represents these equations, and thus only extreme points need be considered. But extreme points are the graphic equivalents of basic solutions, and then, of course, the problem of finding an optimal solution is simplified by the fact that only basic solutions need be considered as candidates.

Convex Functions.   *Definition* 1:   A function $f(x)$ is said to be convex over a convex set X in E^n if, for any two points x_1 and x_2 in X, and for all λ, $0 \leq \lambda \leq 1$:

$$f(\lambda x_1 + (1 - \lambda)x_2) \leq \lambda f(x_1) + (1 - \lambda)f(x_2)$$

Definition 2:   A function $f(x)$ is concave over a convex set X in E^n if, given any two points x_1 and x_2 in X, and λ, $0 \leq \lambda \leq 1$:

$$f(\lambda x_1 + (1 - \lambda)x_2) \geq \lambda f(x_1) + (1 - \lambda)f(x_2)$$

Examples

1.

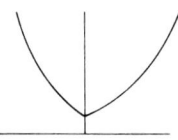

The function shown in the above figure is convex.

2.

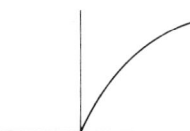

The function whose graph is shown above is concave.

Graphically, a function is convex if the segment joining any two points on the curve lies entirely on or above the curve; it is concave if that same segment lies entirely on or below the curve.

Properties

1. If $f(x)$ is convex then $-f(x)$ is concave.

2. A linear function is both convex and concave.

Definition 3: $f(x)$ is strictly convex over convex set X if, given any two points x_1 and x_2 in X, and λ, $0 < \lambda < 1$:

$$f(\lambda x_1 + (1 - \lambda)x_2) < \lambda f(x_1) + (1 - \lambda)f(x_2)$$

and is strictly concave if, with the same hypotheses,

$$f(\lambda x_1 + (1 - \lambda)x_2) > \lambda f(x_1) + (1 - \lambda)f(x_2)$$

Example

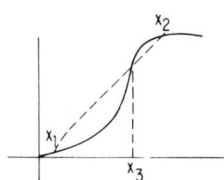

The function whose graph is shown in the above figure is neither convex nor concave. The segment joining x_1 and x_2 is above the curve to the left of x_3 and below the curve to the right of x_3.

This example shows that not all functions are convex or concave. In fact, most functions are neither convex nor concave. We ask, if the number of functions which have either of these properties is such a small number compared with the total number of functions we can find—both in theory and in real life—then why bother mentioning them? The answer to this question is given by the following.

Property.   Let $f(x)$ be a convex (concave) function over a closed set X in E^n. Then, any local minimum (maximum) of $f(x)$ in X is also the absolute or global minimum (maximum).

On page 14.12, we talked about extremes, and we gave a method for finding minima (or maxima) of functions in E^2. We also mentioned the fact that if a function has more than one minimum (maximum), the method for finding the optimum global extremes is to compute the value of the function at each of these local extremes and select the desired value.

It is clear that if we know a function is convex (or concave), we can apply the methods given to get a minimum (or maximum) and then be absolutely sure that this min (or max) is the optimal solution to our problem.

As an example of the application of this, we can say that most of the theory of stochastic inventory is based on convexity, or properties associated with convexity, of the cost function.

GRAPHS

To illustrate a relation between certain situations, we can use a geometric representation, called a graph.

A graph consists of points, called nodes or vertices, and line segments, called arcs or edges.

Example

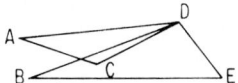

Many special graphs are used in graph theory:

1. A graph with no arcs is called a *null graph*.
2. A graph for which each node pair is connected is called a *complete graph*.
3. If we have a graph G, with some arcs joining some nodes, then G', the complement of G, consists of the same nodes as G and all the arcs joining them that did not appear in G. That is, G and G' together would form a complete graph.
4. Two graphs G_1 and G_2 are isomorphic if, whenever A_1 and B_1 are joined by an arc in G_1, there are corresponding nodes A_2, B_2 in G_2, which are also joined by an arc in G_2.

Planar Graphs.   Planar graphs are graphs that can be drawn in a plane in such a way that the edges have no intersections other than the vertices.

Example.   Maps of roads, if no bridges appear.

Number of Edges (or Arcs) in a Graph.   Where several arcs connect two vertices A and B, we say that the graph has multiple arcs. In general, instead of drawing all the arcs, we can draw a single one and assign a number of multiplicity to it to indicate how many times the arc should be repeated.

At every nonisolated node A in a graph, there will be some arcs having A as an end point. These arcs are incident to A. The number of such arcs is denoted by $\rho(A)$ and called the local degree at A.

Example

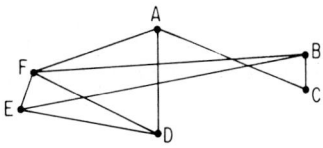

$$\rho(A) = \rho(B) = \rho(D) = \rho(E) = 3$$

$$\rho(F) = 4 \qquad \rho(C) = 2$$

The number of arcs in the graph is obtained as the sum of the number of arcs at each node, divided by 2 (because they are connected at both ends), that is, in this case,

$$\tfrac{1}{2}[\rho(A) + \rho(B) + \rho(C) + \rho(D) + \rho(E) + \rho(F)] = 9$$

In general, if a graph G has nodes $A_1,\ldots, A_n$ with local degrees $\rho(A_1),\ldots, \rho(A_n)$, then N = number of arcs in G is given by

$$N = \frac{1}{2}\left[\sum_{i=1}^{n}\rho(A_i)\right]$$

As a consequence, $\sum_{i=1}^{n}\rho(A_i)$ = even number.

Odd and Even Nodes.   The odd nodes A' are those for which $\rho(A')$ is an odd number. When $\rho(A'')$ is even, we say A'' is an even node.
  Example.   In the graph given above, $A,\ B,\ D,$ and E are odd; F and C are even.
  Property.   Any graph has an even number of odd vertices (nodes). It is easy to prove this property. Because the sum of all the local degrees is even, then the sum of all the odd local degrees has to be even also. Hence, there must be an even number of odd local degrees.
  A graph where all local degrees are the same (say r) is called regular of degree r. The number of its arcs is then $N = \tfrac{1}{2}\,nr$, where n is the number of nodes.

Connected Graphs.   When each node of a graph is connected to every other node by a sequence of arcs, the graph is connected.
  A circuit is a sequence of arcs such that no node is touched more than once, and the ending node is the same as the starting node.
  A tree is a connected graph with no circuits.
  A forest is a graph without circuits which has connected components.

Properties

1. A tree with n nodes has $n - 1$ arcs.

2. A forest with k components and n nodes has $n - k$ arcs.

  A graph G is bipartite if its node set N can be partitioned into two subsets N_1 and N_2 such that every arc in G connects a node in N_1 to a node in N_2.

Directed Graphs.   A graph G where a direction is indicated for every arc is called a directed graph.

CHAPTER 2
PROBABILITY AND STOCHASTIC PROCESSES

Ralph L. Disney
*Professor, Department of Industrial Engineering,
Texas A & M University, College Station, Texas*

The purpose of this chapter is to present some concepts in the areas of probability and random processes which are of practical value to the industrial engineer. Many of these uses appear (sometimes in disguised form) in other sections of this handbook. See, for example, Sec. 11, Chap. 4, and Sec. 14, Chap. 8. But the idea lurks behind some of the concepts in Secs. 4, 7, and 13 as well.

From the earliest studies of industrial processes, it has been apparent that random behavior is more often than not the milieu within which the industrial engineer must work. The point is most clearly seen when one views the almost immediate, favorable response of industrial engineering to the Shewhart studies in statistical quality control. Hardly an issue of the journals in fields of direct concern to the industrial engineer appear without several articles on topics such as statistical quality control, queuing applications, Markov chain applications, probabilistic inventory and material handling applications, and time measurement applications. It has become clear that probability theory and random process theory are part of the foundation of industrial engineering.

PROBABILITY THEORY

Probability theory is concerned with a concept called a *random experiment*. It is supposed that the outcome of the experiment is not known with certainty prior to performing the experiment.

The set of all possible experimental outcomes is called the *sample space S* of the experiment. Any one possible outcome is a member of the sample space and is called a *sample point s*.

If the number of sample points in S is countable (finite or countably infinite), the sample space is said to be countable or discrete. If the number of points in S is noncountable, S is said to be continuous.

For most engineering studies, any subset of S is an *event*. Hence, for our purposes, an event is a subset A of S. It is supposed that S is an event and the null set $\emptyset$ consisting of no sample points is an event, respectively called the *certain event* and the *impossible event*. The set of all events is called an *event space*.

Algebra of Events

1. If A_1 and A_2 are two events, then $A_1 \cup A_2$ is an event, and in general, finite unions, $A_1 \cup A_2 \cup A_3 \cup \cdots \cup A_n$ are events. It is supposed that all countable unions are events. Hence

$$\bigcup_{i=1}^{\infty} A_i$$

is an event. The event $A_1 \cup A_2$ is the set of points contained in A_1 or A_2 or both.

2. If A_1 and A_2 are events, then $A_1 \cap A_2$ is an event and

$$\bigcap_{i=1}^{\infty} A_i$$

is an event. The event $A_1 \cap A_2$ is the set of points contained in both A_1 and A_2. If $A_1 \cap A_2 = \varnothing$, A_1 and A_2 are said to be *mutually exclusive events*.

3. Because $A \subset S$, $\overline{A} = S - A$ is an event, called the *complementary event* of A. $\overline{A}$ is the set of sample points not in A.

Probabilities. To each event in the event space, one supposes that there can be attached a real number Pr[A] called the *probability* of the event A. These probabilities are determined by the *axioms of probability*:

1. $0 \le \Pr[A] \le 1$
2. $\Pr[S] = 1$
3. $\Pr[A \cup B] = \Pr[A] + \Pr[B]$ if $A \cap B = \varnothing$

Thus probabilities are nonnegative real numbers not greater than 1. The probability of the set S, the sample space, is 1. Hence the probabilities are normed. From axiom 3, one sees that the probability of A or B is the sum of the probabilities of the separate events A and B if A and B are mutually exclusive.

In general, one needs one more axiom which extends axiom 3:

4.
$$\Pr\left[\bigcup_{i=1}^{\infty} A_i \right] = \sum_{i=1}^{\infty} \Pr[A_i]; \; A_i \cap A_j = \varnothing \text{ for } i \ne j$$

Dependent and Independent Events. If $A_i \cap A_j \ne \varnothing$ for $i \ne j$, one is interested in $\Pr[A_i \cap A_j]$. This probability is called the *joint probability* of the events A_i and A_j. The events A_i and A_j are *independent events* if and only if

$$\Pr[A_i \cap A_j] = \Pr[A_i]\Pr[A_j] \tag{1}$$

If A_i and A_j are not independent, then

$$\Pr[A_i \cap A_j] = \Pr[A_i|A_j]\,\Pr[A_j] \tag{2}$$

where $\Pr[A_i|A_j]$ is the conditional probability of A_i, given A_j. If $A_iA_j \ne \varnothing$, then axiom 3 becomes

$$\Pr[A_i \cup A_j] = \Pr[A_i] + \Pr[A_j] - \Pr[A_iA_j]$$

Equation (1) generalizes immediately to many events. If

$$\Pr[A_1 \cap A_2 \cap \cdots \cap A_n] = \Pr[A_1]\,\Pr[A_2]\cdots\Pr[A_n] \tag{3}$$

then the events $A_1, A_2, \ldots, A_n$ are said to be *mutually independent*.

A collection of events may be independent in the sense of Eq. (1) but not in the sense of Eq. (3), as the following example illustrates. If the events are independent in the sense of Eq. (3), they are independent in the sense of Eq. (1).

Example 1.   Two coins are tossed, and one notes the occurrence of heads or tails on each toss. The sample space S for this simple experiment is

$$S = \{HH, HT, TH, TT\}$$

Let A_1 be the event "the first coin tossed is a head."
Let A_2 be the event "the second coin tossed is a head."
Let A_3 be the event "the coins match."
One sees that

$$A_1 = \{HH, HT\}$$

$$A_2 = \{HH, TH\}$$

The event

$$A_1 \cup A_2 = \{HH, HT, TH\}$$

The event

$$A_1 \cap A_2 = \{HH\}$$

Hence, A_1 and A_2 are not mutually exclusive events. Let $E_1 = \{HH\}$, $E_2 = \{HT\}$, $E_3 = \{TH\}$, $E_4 = \{TT\}$. All pairs of events E_i are mutually exclusive. For obvious reasons, let $\Pr[E_j] = \frac{1}{4}$ for $j = 1, 2, 3, 4$. Notice that

1. $0 \le \Pr[E_i] \le 1$
2. $\Pr[E_i \cup E_j] = \frac{1}{2}$ for every pair E_i and E_j
3. $S = E_1 \cup E_2 \cup E_3 \cup E_4$

hence

$$\Pr[S] = \Pr[E_1 \cup E_2 \cup E_3 \cup E_4]$$

and, because the events are mutually exclusive,

$$\Pr[S] = \sum_{i=1} \Pr[E_i] = 1$$

Hence, the assignment given for the $\Pr[E_j]$ satisfies the axioms of probabilities.

$$A_1 = E_1 \cup E_2$$

$$A_2 = E_1 \cup E_3$$

$$A_3 = E_1 \cup E_4$$

Also, because the E_j are mutually exclusive,

$$\Pr[A_1] = \Pr[E_1] + \Pr[E_2] = \frac{1}{2}$$

$$\Pr[A_2] = \Pr[E_1] + \Pr[E_3] = \frac{1}{2}$$

$$\Pr[A_3] = \Pr[E_1] + \Pr[E_4] = \frac{1}{2}$$

But

$$A_1 \cap A_2 = (E_1 \cup E_2) \cap (E_1 \cup E_3) = E_1$$

$$A_1 \cap A_3 = (E_1 \cup E_2) \cap (E_1 \cup E_4) = E_1$$

$$A_2 \cap A_3 = (E_1 \cup E_3) \cap (E_1 \cup E_4) = E_1$$

$$A_1 \cap A_2 \cap A_3 = (E_1 \cap E_2) \cup (E_1 \cap E_3) \cup (E_1 \cap E_4) = E_1$$

and

$$\Pr[A_1 \cap A_2] = \Pr[E_1] = \tfrac{1}{4} = \Pr[A_1] \Pr[A_2]$$

$$\Pr[A_1 \cap A_3] = \Pr[E_1] = \tfrac{1}{4} = \Pr[A_1] \Pr[A_3]$$

$$\Pr[A_2 \cap A_3] = \Pr[E_1] = \tfrac{1}{4} = \Pr[A_2] \Pr[A_3]$$

One sees immediately that A_1, A_2, and A_3 are independent when taken in any pairs. The events are pairwise independent. But

$$\Pr[A_1 \cap A_2 \cap A_3] = \Pr[E_1] = \tfrac{1}{4} \neq \Pr[A_1 \Pr[A_2] \Pr[A_3] = \tfrac{1}{8}$$

Hence, even though the events are independent in pairs, they are not independent when considered three at a time.

Bayes' Theorem.   Suppose one has events $A_1, A_2, \ldots, A_n$ that are mutually exclusive and

$$A_1 \cup A_2 \cup A_3 \cdots \cup A_n = S$$

(The events are "exhaustive.") Suppose one has another event, B. Bayes' theorem can then be stated as

$$\Pr[A_j|B] = \frac{\Pr[B|A_j] \Pr[A_j]}{\Sigma \Pr[B|A_j] \Pr[A_j]}$$

Example 2.   Two machine centers are manufacturing identical parts. Center 1 has been running about 5 percent defective parts, and center 2 has been running about 1 percent defective parts. Seventy-five percent of the total production has been coming from center 2. (Perhaps center 1 is an old machine used only for standby operation.) At final inspection, a part is found to be defective. What is the probability that it came from center 1?

Let B be the event "the part is defective" and let A_j be the event "the part came from center j for $j = 1, 2$." Then $\Pr[A_1|B]$ is the probability sought. Using Bayes' theorem, one finds

$$\Pr[A_1|B] = \frac{\Pr[B|A_1] \Pr[A_1]}{\Pr[B|A_1] \Pr[A_1] + \Pr[B|A_2] \Pr[A_2]}$$

$$= \frac{0.05 \times 0.25}{0.05 \times 0.25 + 0.01 \times 0.75}$$

$$= \frac{0.0125}{0.0125 + 0.0075}$$

$$= \frac{0.0125}{0.0200}$$

$$= 0.625$$

Random Variables.   In many applications, the outcomes of an experiment are nonnumeric. It is convenient to assign real numbers to sample points. One calls a function $X(s)$—a real valued function which maps the sample space into the real line—a random variable. Hence, a random variable is a function whose domain is the sample space and whose range is some subset of the real line.

Distribution Functions.   From the definition of a random variable, sample points are assigned real numbers. Collections of sample points are subsets of S and hence are events. Events have probabilities. Hence, one says that the random variable induces probabilities on the real line as follows. Let x be the set of points s such that $X(s) \leq x$ for a real number x. A is an event and has some probability assigned to it. We denote this probability by the following expression:

$$\Pr[A] = \Pr[X \leq x]$$

Notice that we have suppressed the variable s in this shorthand notation. $\Pr[X \leq x]$ is a function whose domain is the real line and whose range is the unit interval. We denote this function by $F(x)$ and call it the distribution function for the random variable X. Where no confusion is likely to arise, we use the abbreviated notation $F(x)$ for this function. For numerical work with these functions or various moments, it seems no longer to be necessary to carry tables of these. Most computers have available well-performing commercially available software. A perusal of issues of *Industrial Engineering* shows several such packages for microcomputers.

Some Properties of Distribution Functions

1. $F(x) = \Pr[X \leq x]$
2. $0 \leq F(x) \leq 1$
3. $F(-\infty) = 0$
4. $F(+\infty) = 1$
5. If $x_1 < x_3$ then $F(x_1) \leq F(x_2)$
6. $\lim\limits_{x \to a^+} F(x) = F(a)$

For many applied problems, $F(x)$ is either a differentiable function of x for almost all x or $F(x)$ is a step function with jumps at the integers. If $F(x)$ is differentiable, then let

$$dF(x) = f(x)\, dx$$

is called the probability density function of the continuous random variable X. Some useful probability density functions are shown in Table 2.1. In terms of probabilities, it is taken as

$$f(x)\, dx = \Pr[x < X \leq x + dx]$$

Note that $f(x)$ itself is not a probability. If $F(x)$ is a step function with jumps at the integers, then

$$F(x + 1) - F(x) = p_{x+1}$$

is called a *probability density function* or a *probability mass function* of the discrete random variable $\overline{X}$. In terms of probabilities

$$p_x = \Pr[X = x]$$

for x, an integer. Because both $f(x)$ and p_x are defined in terms of probabilities, one has

TABLE 2.1   A Short Table of Useful Probability Density Functions

Name	Density function	
Bernoulli	$p_j = \begin{cases} q & \text{if } j = 0 \\ p & \text{if } j = 1 \\ 0 & \text{otherwise} \end{cases}$ $p + q = 1$	
Binomial	$p_j = \begin{cases} \binom{n}{j} p^j q^{n-j} & \text{if } j = 0,1,2, \ldots, n \\ 0 & \text{otherwise} \end{cases}$ $p + q = 1$	
Geometric	$p_j = pq^j \qquad j = 1,2, \ldots$ $p + q = 1$	
Poisson	$p_j = \dfrac{\lambda^j e^{-\lambda}}{j!} \qquad j = 0,1,2, \ldots, \text{ and } \lambda > 0$	
Uniform	$p_j = \dfrac{1}{a + 1} \qquad j = 0,1,2, \ldots, a$	
Normal	$f(x) = \dfrac{1}{\sqrt{2\pi}\,\sigma} \exp\left[-\dfrac{(x - \mu)^2}{2\sigma^2} \right]$	$-\infty < x < \infty$
Gamma	$f(x) = \dfrac{a^n}{\Gamma(n)} x^{n-1} e^{-ax}$	$0 \le x < \infty$
Exponential	$f(x) = ae^{-ax}$	$0 \le x < \infty$
Beta	$f(x) = \dfrac{\Gamma(n + m)}{\Gamma(n)\Gamma(m)} x^{n-1}(1 - x)^{m-1}$	$0 \le x \le 1$
Weibull	$f(x) = \dfrac{b}{\theta - x_o} \left(\dfrac{x - x_o}{\theta - x_o}\right)^{b-1} \exp\left[-\dfrac{x - x_o}{\theta - x_o} \right]^b$	$x_o \le x < \infty$

1. $0 \le f(x)$

2. $\displaystyle\int_{-\infty}^{\infty} f(x)dx = 1$

and similarly for p_x with the integral replaced by a sum over all x. In addition, $p_x \le 1$.

Expectations.   If g is a function of a random variable X whose probability density function is $f(x)$ or p_x, then it can be shown that the expected value of the function is given by

$$E(g(X)) = \int_{-\infty}^{\infty} g(x)f(x)\, dx$$

if X is continuous, or

$$E(g(X)) = \sum_x g(x)p_x \qquad (4)$$

if X is discrete. A few properties of these expectations are given in Table 2.2.
   Table 2.3 gives the means and variances for some of the distributions given in Table 2.1.
   Two Other Useful Expectations

1. If X is a discrete random variable, one defines

$$E(z^X) = \sum_{i=0}^{\infty} z^i p^i$$

as the probability, generating function.

TABLE 2.2   Some Special Expectations*

$g(X)$	Name	Formula
X	Mean value or 1st moment about 0	$\mu = E[X] = \int x f(x)\,dx$
X^n	nth moment about 0	$\mu_n' = E[X^n] = \int x^n f(x)\,dx$
$[X - E(X)]^2$	Variance $V(X)$	$\mu_2 = E(X - E(X))^2$ $= \int (x - \mu)^2 f(x)\,dx$
$[X - E(X)]^n$	nth central moment	$\mu_n = E[(X - E(X))^n]$ $= \int (x - \mu)^n f(x)\,dx$
$aX + b$ a, b constants		$E(aX + b) = \int (ax + b) f(x)\,dy$ $= aE(X) + b$ $= b \quad$ if $a = 0$

*In this table, we have presented only the formulas for the case of a continuous random variable. The corresponding formulas for a discrete random variable are apparent from Eq. (4).

TABLE 2.3   Means and Variances for Some Distributions

Distribution	Mean	Variance
Bernoulli	p	pq
Binomial	np	npq
Geometric	q/p	q/p^2
Poisson	λ	λ
Uniform	$a/2$	$\dfrac{a(a + 2)}{12}$
Normal	μ	σ^2
Gamma	n/a	n/a^2
Exponential	$1/a$	$1/a^2$

2. A function used extensively in some work is the exponential form

$$E(e^{izX}) = \int e^{-izx} f(x)\,dx$$

if X is continuous, or

$$E(e^{izX}) = \sum_{j=0}^{\infty} e^{izj} p_j$$

if X is discrete. Throughout, $i = \sqrt{-1}$. To abbreviate notation, let us agree to put

$$\phi(z) = E(e^{izX})$$

in either case. This expectation is called the *characteristic function*. (See Table 2.4.) Some properties of $\phi(z)$ that are useful in probability are:

1. For any purely discrete or purely continuous random variable, $\phi(z)$ exists.

2. $\phi(0) = 1$

3. $\left. \dfrac{d^n \phi(z)}{i^n dz^n} \right|_{z=0} - E(X^n)$

4. If $\phi_X(z)$ and $\phi_Y(z)$ are the characteristic functions for independent random variables X and Y and if $Z = X + Y$, then

$$\phi_Z(z) = \phi_X(z)\phi_Y(z)$$

5. $f(x)$ or p_j can be retrieved from $\phi(z)$ using ordinary Fourier inversion methods. An important theorem regarding characteristic functions is the characterization theorem. If X and Y are two random variables with distribution functions $F(x)$ and $G(y)$, respectively, and if

$$\phi_X(z) = \phi_Y(z)$$

then $F = G$.

Chebyshev's Inequality.   Knowing only the first and second control moments, one can obtain some limits on the probabilities for a random variable. There are several types of theorems of this kind. We state only one here, called Chebyshev's theorem:

$$\Pr[|X - \mu| > t] \le \frac{\sigma^2}{t^2}$$

In many cases, Chebyshev's theorem is quite rough and forms a very loose bound on the probability. The theorem does lend some credence to the use of σ^2 as a measure of dispersion or spread of a density function, however.

TABLE 2.4   Probability Generating Functions or Characteristic Functions for Some Distributions*

Name	Probability generating function	Characteristic function
Bernoulli	$q + pz$	
Binomial	$(q + pz)^n$	
Geometric	$\dfrac{p}{1 - qz}$	
Poisson	$e^{-\lambda(1-z)}$	
Normal		$e^{-i\mu z + 1/2(\sigma_2)z^2}$
Gamma		$\left(\dfrac{a}{a - iz}\right)^n$
Exponential		$\dfrac{a}{a - iz}$

*See Table 2.1.

Multivariate Distributions.   If X and Y are two (or a countable number in general) random variables defined on the same sample space, then one is interested in the event

$$[X(s) \le x] \cap [Y(s) \le y]$$

This event will have a probability associated with it as

$$\Pr[(X(s) \le x) \cap (Y(s) \le y)]$$

As before, we define a real function F, called the joint probability distribution function, by

$$F(x,y) = \Pr[(X(s) \le x) \cap (Y(s) \le y)]$$

Then $F(x,y)$ determines the probability of the joint event $X(s) \leq x$ and $Y(s) \leq y$. If

$$F(x,y) = F_1(x)F_2(y)$$

where

$$F_1(x) = \Pr[X(s) \leq x]$$

$$F_2(y) = \Pr[Y(s) \leq y]$$

then the random variables are *independent random variables*. In general, if

$$F(x_1, x_2, \ldots, x_n) = F_1(x_1)F_2(x_2), \ldots, F_n(x_n)$$

the random variables $X_1, X_2, \ldots, X_n$ are independent random variables.
    If $F(x, y)$ is differentiable, then $f(x, y)$, where

$$\partial F(x, y) = f(x, y) \, \partial x \, \partial y$$

is the *joint density function* of the random variables. In the discrete case, one defines the joint density function

$$p(x,y) = \Pr[X = x; Y = y]$$

The marginal distribution function of X is defined by

$$F_X(x) = F(x, \infty)$$

where the subscript X indicates a distribution function for the random variable X. Similarly, for Y,

$$F_Y(y) = F(\infty, y)$$

The corresponding marginal density functions can be defined as differences of F_X or F_Y or by

$$F_X(x) = \sum_y p(x,y)$$

$$F_Y(y) = \sum_x p(x,y)$$

Replacing the sums by integrals with respect to the indicated variables would yield corresponding formulas for continuous $f(x,y)$. These definitions, of course, generalize to more than two random variables defined on the same sample space.
    If X and Y are not independent, one defines the conditional probability density function by

$$\frac{f(x,y) \, dx \, dy}{f_X(x) \, dy} = f_{Y|X}(y|x) \, dy$$

and similarly,

$$\frac{f(x,y) \, dx \, dy}{f_Y(y) \, dy} = f_{X|Y}(x|y) \, dx$$

The probability meaning of $f_{X|Y}(x|y)$ is that

$$f_{X|Y}(x|y) \, dx \, dy = \Pr(x \leq X \leq x + dx | Y = y)$$

Often the subscript on the conditional densities is dropped and one writes $f(x|y)$ or $f(y|x)$. The conditional density functions are density functions in that

$$0 \le f_{X|Y}(x|y) \, dx$$

$$\int f_{X|Y}(x|y) \, dx = 1$$

Similar definitions hold for $f_{X|Y}$ in the cases where X and Y are discrete random variables.

Moments of Multivariate Distributions.   Just as in the case of a single random variable, one can define the expectation of a function of the random variables X, Y, say $g(X,Y)$, as

$$E[g(X,Y)] = \int\int g(x,y) \, f(x,y) \, dx \, dy$$

Of particular importance in many statistical studies is the covariance of X, Y. The value

$$E[(X - E(X))(Y - E(Y))] = \int_y \int_x (x - E(X))(y - E(Y))f(x,y) \, dx \, dy$$

is called the *covariance of X and Y,* Cov $(\overline{X},Y)$. If X and Y are independent random variables, then the covariance of X and Y is 0. The converse is not true.

The value

$$\rho = \frac{\text{covariance of } X \text{ and } Y}{(\text{standard deviation of } X)(\text{standard deviation of } Y)}$$

is called the *correlation coefficient* and has the properties:

1. $-1 \le \rho \le +1$
2. If X and Y are independent, $\rho = 0$.
3. If $\rho = \pm 1$, then X and Y are two random variables related by $Y = aX + b$.

The Bivariate Normal Density Function.   Of considerable importance to the study of statistics is the bivariate normal density function.

The bivariate normal density is defined by

$$f(x,y) = (2\pi\sigma_z\sigma_y\sqrt{1 - \rho^2})^{-1}e^{-Q/2}$$

where

$$Q = \frac{1}{1 - \rho^2}\left[\left(\frac{x - \mu_x}{\sigma_x}\right)^2 - 2\rho\left(\frac{x - \mu_x}{\sigma_x}\right)\left(\frac{y - \mu_y}{\sigma_x}\right) + \left(\frac{y - \mu_y}{\sigma_x}\right)^2\right]$$

and

$$- \infty < x < \infty, - \infty < y < \infty$$

The means and variances are

$$E[X] = \mu_x \qquad V[X] = \sigma_x^2$$

$$E[Y] = \mu_y \qquad V[Y] = \sigma_y^2$$

The parameter ρ is the *correlation coefficient*. The probability density function is completely specified by μ_x, μ_y, σ_x, σ_y, ρ. This is one of the few cases where $\rho = 0$ implies that X and Y are independent.

The Hazard Function.   In many reliability studies, one is interested in the hazard function, defined as

$$\Pr[x < X \le x + dx | X > x]$$

The hazard function is given in general by

$$\frac{\Pr[x < X \le x + dx]}{\Pr[X > x]} = \frac{f(x)\, dx}{1 - F(x)}$$

For the Weibull density, for example, the hazard function is given by

$$\frac{b}{\theta - x_0}\left(\frac{x - x_0}{\theta - x_0}\right)^{b-1}$$

a polynomial in x. In the case $b = 1$, the random variable is exponentially distributed, and the hazard function is independent of x. This property is unique to the exponential distribution (for continuous random variables) and is called the *forgetfulness property* of the exponential.

Sums of Two Random Variables.   A common problem in probability is finding the distribution of a sum of independent random variables:

$$S_n = \sum_{j=1}^{n} X_j$$

One can show that

$$E(S_n) = \sum_{j=1}^{n} E(X_j)$$

"the expected value of a sum is the sum of the expected values." Furthermore, the variance of a sum—abbreviated Var(X)—is given by

$$\mathrm{Var}(S) = \mathrm{Var}(X) + \mathrm{Var}(Y)$$

if X and Y are independent random variables. If X and Y are not independent variables, one has the more general formula

$$\mathrm{Var}(S) = \mathrm{Var}(X) + \mathrm{Var}(Y) + 2\, \mathrm{Cov}\,(X,Y)$$

The density function of S can be computed (though not always in closed form) by using convolution arguments. For the case where $N = 2$ and X_j is discrete, one has

$$p_s = \sum_{j=0}^{s} p_j p_{s-j} \tag{5}$$

where

$$p_s = \Pr[S = s] \qquad s = 0, 1, 2, \ldots$$

$$p_j = \Pr[X_1 = j] \qquad j = 0, 1, 2, \ldots$$

$$P_k = \Pr[X_2 = k] \qquad k = 0, 1, 2, \ldots$$

Equation (5) is called the *convolution* of p_j and p_k. This can be generalized for any number of random variables in the sum S. If X_j is continuous,

$$f(s) = \int_0^s g(x_1)h(s - x_1) \, dx_1 \qquad 0 \le s < \infty \qquad (6)$$

where

$$f(s) \, ds = \Pr[s < S \le s + ds]$$

$$g(x_1) \, dx_1 = \Pr[x_1 < X_1 \le x_1 + dx_1] \qquad 0 \le x_1 < \infty$$

$$h(x_2) \, dx_2 = \Pr[x_2 < X_2 \le x_2 + dx_2] \qquad 0 \le x_2 < \infty$$

The characteristic function gives a useful way to determine the characteristic function of the sum of two independent random variables. For if

$$Z = X + Y$$

then

$$\phi_Z(z) = \phi_X(z)\phi_Y(z) \qquad (7)$$

This result generalizes to more than two independent random variables immediately.

Example 3. Let X_1 = the time to set up one of two machines and X_2 = the time to set up the other machine. Then $S = X_1 + X_2$ would be the time required for a work crew to set up the two machines. Suppose X_1 and X_2 are each exponentially distributed random variables with the same density functions

$$f_{X_1}(x) = f_{X_2}(x) = ae^{-ax} \qquad 0 \le x < \infty$$

Then, from Eq. (6),

$$f_S(s) = \int_0^s ae^{-ax}ae^{-a(s-x)}dx$$

$$= a^2 x e^{-ax} \qquad 0 \le x < \infty$$

From Table 2.1, it is apparent that S is a gamma distributed random variable with $n = 2$. Using Eq. (7) and Table 2.4, one has

$$\phi_{X_1}(z) = \frac{a}{a - iz} = \phi_{X_2}(z)$$

and

$$\phi_s(z) = \left(\frac{a}{a - iz}\right)^2$$

But if S is a gamma distributed random variable, one sees from Table 2.4 that

$$\phi_s(z) = \left(\frac{a}{a - iz}\right)^n$$

Hence, from Table 2.4, one concludes that S has gamma density function with $n = 2$.

A Central Limit Theorem. One of the remarkable properties of the normal density function is summarized in the central limit theorem, which can be stated as:

Let

$$X_1, X_2, \ldots, X_n$$

be a sequence of independent random variables, each with the same density function (not necessarily normal) with finite mean μ and finite variance σ^2. Further, let

$$S_n = X_1 + X_2 + \cdots + X_n$$

be the nth partial sum of the X's. Then

$$\lim_{n \to \infty} \Pr\left[a \le \frac{S_n - n\mu}{\sigma\sqrt{n}} \le b\right] = \int_a^b N(z; 0, 1)\, dz$$

where $N(z; 0, 1)$ is the normal density function given in Table 2.1.

In many applied statistics problems, one "takes a random sample of size n" (that is, one observes n independent random variables) from a "population" (that is, each random variable has the same distribution). It is reasonable to assume that the population has a finite mean and variance. One then determines

$$\frac{\sum_{i=1}^{n} X_i}{n}$$

called the *sample mean value*. From this, one wants to make statements about whether a process which produces the observations is "in control"; or one wants to estimate a reasonable range of values that could be expected to contain the process true mean μ; or one wants to test a hypothesis regarding the behavior of the means of two processes or one process under two conditions. In such cases, one can make rather precise comments about the probabilities associated with the mean values by using the central limit theorem. Notice that the central limit theorem allows one to make rather precise statements about the probabilities associated with the mean values knowing relatively little about the probabilities associated with the X's themselves. Hence, even though one might be able to say little about the probabilities of the X's, one can say a great deal about the probabilities of the mean of the X's by using the central limit theorem.

SPECIAL RANDOM PROCESSES

In many engineering studies, one is concerned with the time behavior of a process. Thus one is interested in $X(t)$ = the position of a particle at time t, or X_n = the time to produce a part in the nth day of production. In the study of random experiments, one is also interested in the time behavior of the process. Conceptually, one is concerned not with a single random variable X, but rather with an entire family of random variables, perhaps a sequence of random variables. Thus, one might define the sequence of daily demands for an item, $\{X_n : n = 1, 2, 3 \ldots\}$. Here n serves as a parameter which denotes which day we are concerned with, and X_n is the amount demanded on day n. The sequence $\{X_n\}$ is called a random sequence, or a discrete parameter (n) random process, or a stochastic process.

More generally one defines the family of random variables $\{X_t, \in T\}$ as a *stochastic process* or a random process. The set T, the set of all possible parameter values, is called the *parameter space*. Often, in application, the parameter set is taken to be a set of times. We shall be concerned only with the cases $T = \{0, 1, 2, \ldots\}$ or $T = (0, \infty)$. In the first case, $\{X_t\}$ is called a discrete parameter process, and in the second, it is called a continuous parameter process. Formally, $\{X_t\}$, or more properly $\{X_t(s)\}$, is a function X which maps points in the product space $S \times T$ into the real line, where S is the sample space for the experiment and T is the parameter space. One notes that if t is held fixed, then $X(s)$ is merely a random variable, as discussed under Probability Theory. On the other hand, if s is held fixed, $X_t(\cdot)$ is just a real valued function X whose domain is T and whose range is some subset of the real numbers.

The range of $\{X_t\}$ is called the *state space* of the random process. The state space may be continuous, as occurs in some reliability studies, or it may be discrete, for example, the integers. In the following discussion, we are concerned only with integer state random processes. If for some n (or t) $X_n = j$, $(X_t = j)$, we say that the "process is in state j at time n (or time t)."

Because $\{X_t\}$ is a collection of random variables, one must be able to determine the joint probability for all subcollections to make definite probability statements about the process. Hence, to completely define the probability structure of a random process, one must determine

$$\Pr[X_{t_1} = j_1; X_{t_2} = j_2; \ldots, X_{t_i} = j_i; \ldots; X_{t_n} = j_n] \tag{8}$$

for every value t_i, for all i, and all j_i in their respective spaces.

If the probability structure of the random process has the special property that

$$\Pr[X_1 = j_1, X_2 = j_2, X_3 = j_3, \cdots, X_n = j_n]$$
$$= \Pr[X_1 = j_1] \Pr[X_2 = j_2 | X_1 = j_1] \Pr[X_3 = j_3 | X_2 = j_2] \cdots \tag{9}$$
$$\Pr[X_n = j_n | X_{n-1} = j_{n-1}]$$

then one says the random process is a *discrete parameter Markov process*. A similar property must hold for a continuous parameter Markov process, that is, if for every t and n

$$\Pr[X(t_1) = j_1, X(t_2) = j_2, X(t_3) = j_3, \cdots, X(t_n) = j_n]$$
$$= \Pr[X(t_1) = j_1] \Pr[X(t_2) = j_2 | X(t_1) = j_1] \Pr[X(t_3) = j_3 | X(t_2) = j_2] \tag{10}$$
$$\cdots \Pr[X(t_n) = j_n | X(t_{n-1}) = j_{n-1}]$$

then $\{X(t)\}$ is a *continuous parameter Markov process*.

Discrete Parameter Markov Process.   In the structure (9), one defines $\Pr[X_1 = j_1] = p_{j_1}$, the initial state probability for state j. The vector $\bar{p} = (p_{j_1}, p_{j_2}, \ldots)$ is called the *vector of initial state probabilities*. The probability

$$\Pr[X_n = j | X_{n-1} = i] = p_{ij}(n - 1, n)$$

is called a *one-step transition probability* of the process.

In most applications, the conditional probabilities in Eq. (9) do not depend on the particular values of n. Hence,

$$p_{ij} = \Pr[X_m = j_m | X_{m-1} = j_{m-1}] = \Pr[X_n = j_n | X_{n-1} = j_{n-1}]$$

for every $m - 1, m, n - 1, n$. In this case, the process is said to have a *stationary transition mechanism*.

By Eq. (9), the entire set of joint probabilities for the process can be obtained from the initial state probabilities and the set of transition probabilities, in the stationary transition mechanism case. We will discuss only stationary transition mechanism chains in what follows. Hence, the entire probability structure of the process is determined in terms of a vector $\bar{p}$ of the initial state probabilities and a matrix $\bar{P}$ of the transition probabilities in this case. The vector $\bar{p}$ is a probability vector. That is,

$$0 \leq p_j \leq 1 \qquad \text{for every element } p_j$$

and
$$\sum_j p_j = 1$$

The matrix $\bar{P}$ is a stochastic matrix. That is,

$$0 \leq p_{ij} \leq 1 \qquad \text{for every element } p_{ij}$$

and
$$\sum_j p_{ij} = 1$$

that is, each row of $\bar{P}$ sums to 1.

n-*step Transition Probabilities and* nth-*step State Probabilities.*   The probability structure of a stationary transition mechanism, discrete parameter, Markov chain is completely determined by the vector of initial state probabilities and the matrix of one-step transition probabilities. One often needs to know, however,

$$p_{ij}^{(n)} = \Pr[X_{m+n} = j | X_m = i] \tag{11}$$

called the *n-step transition probabilities*. Because of stationarity of the transition probabilities, Eq. (11) is equivalent to

$$p_{ij}^{(n)} = \Pr[X_n = j | X_0 = i]$$

The *n*-step transition probabilities satisfy the *Chapman-Kolmogorov equations*

$$p_{ij}^{(n+m)} = \sum_k p_{ik}^{(n)} p_{kj}^{(m)} \tag{12}$$

In matrix form, Eq. (12) is given by

$$\overline{P}^{(n+m)} = \overline{P}^{(n)} \overline{P}^{(m)} \tag{13}$$

It follows that the matrix of *n*-step transition probabilities can be obtained from $\overline{P}$ as

$$\overline{P}^{(n)} = \overline{P}^n \tag{14}$$

That is, the matrix of *n*-step transition probabilities is simply the *n*th power of the matrix of one-step transition probabilities. $\overline{P}^{(n)}$ is a stochastic matrix.

The probability

$$p_j^{(n)} = \Pr[X_n = j] \tag{15}$$

is called the *nth-step state probability*. These probabilities are obtained from

$$p_j^{(n)} = \sum_i p_i p_{ij}^{(n)} \tag{16}$$

using Eq. (14) and defining $\overline{p}^{(n)}$ as the vector whose elements are those probabilities given by Eq. (15). The vector $\overline{p}^{(n)}$ whose elements are the $p_j^{(n)}$ is obtained from

$$\overline{p}^{(n)} = \overline{p} \overline{P}^n$$

Passage Probabilities.   One defines the probabilities

$$g_{ij}^{(n)} = \Pr[X_n = j, X_m \neq j \text{ for } m < n | X_0 = i] \tag{17}$$

and $$g_{ii}^{(n)} = \Pr[X_n = i, X_m \neq i \text{ for } m < n | X_0 = i] \tag{18}$$

as the *first passage probabilities* (17) and *first return probabilities* (18).

The $g_{ij}^{(n)}$ and $p_{ij}^{(n)}$ are connected by the relation

$$p_{ij}^{(n)} = g_{ij}^{(n)} + \sum_{m=1}^{n-1} p_{jj}^{(m)} g_{ij}^{(n-m)} \qquad n > 0 \tag{19}$$

The value

$$g_{jj} = \sum_{n=1}^{\infty} g_{jj}^{(n)} \tag{20}$$

is the *probability of ever returning to j.*

The value

$$\mu_j = \sum_{n=1}^{\infty} n g_{jj}^{(n)} \tag{21}$$

is called the *mean return time to state j*.

A Classification of State.   States of the chain can be classified in several ways.

1. If, for some $n > 0$, $p_{ij}^{(n)} > 0$, then state j is *reachable* from state i.
2. If state j is reachable from state i and state i is reachable from state j, then states i and j communicate.
3. If all states communicate, the chain is said to be an *irreducible chain*.
4. If, for some set of states C and for every $i \in C$ and $k \notin C$, $p_{ik}^{(n)} = 0$ for every $n > 0$, then C is said to be a *closed set of states*.
5. If C is a closed set containing one state i, then i is called an *absorbing state*.
6. If return to state i has probability 0 except perhaps at times n, $2n$, $3n$, $\dots$, then state i is called a *periodic state* with *period n*. Otherwise the state is *aperiodic*.
7. If $g_{jj} = 1$, then j is a *recurrent state*. Otherwise, j is a *transient state*.
8. If j is a recurrent state and $\mu_j < \infty$, then j is *positive recurrent*. Otherwise, j is a *null recurrent state*.
9. If state i is aperiodic, recurrent, and non-null, then it is called an *ergodic state*.

Three Important Theorems.   The following are three important theorems for classifying states of a chain.

1. In an irreducible Markov chain, all states are of the same type. If one state is recurrent, they are all recurrent. If one is transient, they are all transient. If one is null recurrent, they are all null recurrent. If one is periodic with period n, they are all periodic with period n.

2. In a Markov chain with a finite number of states, not all states can be transient and no states can be null recurrent.

3. In a finite, irreducible Markov chain, all states are positive recurrent.

Limit Theorems.   In a considerable number of engineering applications, one is concerned with the long-run behavior of a system. Quite often, systems are designed to optimize their behavior after such things as start-up effects have become negligible. In the study of Markov chains, the question of the limiting behavior of $p_{ij}^{(n)}$, or $p_j^{(n)}$, is of importance. We restrict attention here to aperiodic chains (every state is aperiodic).

A vector $\bar{x}$ is said to be a stationary probability vector if $\bar{x}$ is a probability vector satisfying

$$\bar{x} = \bar{x}\bar{P} \tag{22}$$

It can be shown that if the chain is irreducible and aperiodic, either of the two following conditions prevail:

1. All states are transient or null recurrent, in which case, $\lim_{n \to \infty} p_{ij}^{(n)} = 0$ for every pair i,j, and there is no stationary probability vector $\bar{x}$.

2. All states are positive recurrent and $\lim_{n \to \infty} p_{ij}^{(n)} = \pi_j$ exists. The π_j's are the unique stationary probability vector satisfying Eq. (22) above. In this case, $\pi_j = 1/\mu_j$. When π_j exists, one has $p_j^{(n)} \to \pi_j$ for $n \to \infty$.

Condition 2 above is often encountered in practice, and one is concerned with the elements π_j. These probabilities are called the *steady-state probabilities*.

Continuous-Parameter Markov Chains.   In the structure of Eq. (10) for a continuous-time Markov chain, one defines $\Pr[X_t = j] = p_j(t)$, the probability of being in state j at time t. In particular, $p_j(0)$ is the initial state probability for state j. The vector $\bar{p} = (p_0(0), p_1(0), \ldots)$ is called the *vector of initial state probabilities*.

The probability

$$\Pr[X(\tau_2) = j \mid X(\tau_1) = i] = p_{ij}(\tau_1, \tau_2)$$

is called the transition probability. In many applications, one finds that the transition probability depends only on the difference $\tau_2 - \tau_1 = t$. In this case, one uses the abbreviated notation $p_{ij}(t)$ for the transition probabilities, and the resulting process is said to be *time homogeneous* or to have a *stationary transition mechanism*.

By Eq. (10), the entire set of joint probabilities for the process can be obtained from the vector of state probabilities and the matrix whose elements are transition probabilities, in the time homogeneous case. We will discuss only this case in what follows.

Chapman-Kolmogorov Equations.   If one is concerned with an interval of length $t + s$, the transition probabilities $p_{ij}(t + s)$ are given by the Chapman-Kolmogorov equations.

$$p_{ij}(t + s) = \sum_k p_{ik}(t) p_{kj}(s) \tag{23}$$

The Instantaneous Transition Rates.   One defines the instantaneous transition rates

$$\lambda_{ij} = \lim_{s \to 0} \frac{p_{ij}(s) - p_{ij}(0)}{s} = \frac{dp_{ij}(0)}{dt} \tag{24}$$

$$\lambda_{ii} = \lim_{s \to 0} \frac{p_{ii}(s) - 1}{s} = \frac{dp_{ii}(0)}{dt}$$

which we assume exist and are finite.

The λ_{ij}.   The λ_{ij} are defined above. From their definition, it is clear that when they exist, the λ_{ij} are the derivatives of the $p_{ij}(t)$ functions evaluated at 0. There is another interpretation of this point that is extremely useful in building Markov models. From the definition of the λ_{ij}, one can argue that for some Δt near zero and sufficiently small,

$$p_{ij}(\Delta t) = \lambda_{ij} \, \Delta t + o(\Delta t)$$

That is, one can assume that, if the λ_{ij} exist in the sense of Eq. (24), then the transition probability function in the neighborhood of zero can be approximated by a straight line with slope λ_{ij}. The term $(o\Delta t)$ is the error of approximation and

$$\frac{o(\Delta t)}{\Delta t} \to 0 \qquad \text{for } \Delta t \to 0$$

It follows from the definition of λ_{ij} and λ_{ii} that

$$\lambda_{ij} \geq 0 \qquad \text{for } j \neq i$$

$$\lambda_{ii} \leq 0$$

For a *conservative Markov chain*,

$$\sum_j \lambda_{ij} = 0$$

and thus λ_{ii} can be obtained in any conservative process from

$$\lambda_{ii} = -\sum_{j \neq i} \lambda_{ij}$$

The Kolmogorov Differential Equation.   By using Eq. (23), it is shown that the transition probabilities satisfy the forward (25) and backward (26) Kolmogorov differential equations.

$$\frac{dp_{ij}(t)}{dt} = \sum_{k} p_{ik}(t)\lambda_{kj} \tag{25}$$

$$\frac{dp_{ij}(t)}{dt} = \sum_{k} \lambda_{ik} p_{kj}(t) \tag{26}$$

with initial conditions

$$p_{ij}(0) = \begin{cases} 1 & \text{if } i = j \\ 0 & \text{otherwise} \end{cases}$$

In matrix form, Eqs. (25) and (26) become

$$\frac{d\overline{P}(t)}{dt} = \overline{P}(t)\overline{\Lambda} \tag{27}$$

$$\frac{d\overline{P}(t)}{dt} = \overline{\Lambda}\overline{P}(t) \tag{28}$$

with initial conditions

$$\overline{P}(0) = \overline{I}$$

In this form, $\overline{I}$ is the identity matrix, $\overline{P}(t)$ is the matrix whose elements are the transition probabilities, $d\overline{P}(t)/dt$ is the matrix whose elements are the derivatives of those of $\overline{P}(t)$, and $\overline{\Lambda}$ is the matrix of the λ_{ij}.

It is clear from Eqs. (25) and (26) and the initial conditions that the transition probabilities are known in terms of the instantaneous transition rates. Hence, the entire set of probability needed to define $\{X_t\}$ is known when $\overline{p}(0)$ and $\overline{\Lambda}$ are known.

The State Equations.   In many cases, one is primarily concerned with the state probability $p_j(t)$. These state probabilities can be obtained from

$$p_j(t) = \sum_{i} p_i(0)p_{ij}(t)$$

If for some fixed i, $p_i(0) = 1$ and $p_k(0) = 0$, $k \neq i$, then $p_j(t) = p_{ij}(t)$, and the Kolmogorov Eqs. (25) and (26) can be used to determine the state probabilities merely by suppressing the subscript i to obtain, for example,

$$\frac{dp_j(t)}{dt} = \sum_{k} p_k(t)\lambda_{kj} \qquad j = 0, 1, 2, \ldots \tag{29}$$

This system of equations is often used in queuing theory to obtain the equation of state.

An Important Special Case—The Poisson Process.   In the special case

$$\lambda_{kj} = \lambda \qquad \text{if } j = k + 1$$

$$\lambda_{jj} = -\lambda$$

$$\lambda_{kj} = 0 \qquad \text{otherwise}$$

one has

$$\overline{\Lambda} = \begin{pmatrix} -\lambda & \lambda & 0 & 0 & \cdots \\ 0 & -\lambda & \lambda & 0 & \cdots \\ 0 & 0 & -\lambda & \lambda & \cdots \end{pmatrix}$$

The state Eqs. (29) yield for $p_0(0) = 1$, $p_k(0) = 0$, $k \neq 0$,

$$\frac{dp_0(t)}{dt} = -\lambda p_0(t)$$

$$\frac{dp_n(t)}{dt} = \lambda p_{n-1}(t) - \lambda p_n(t)$$

The solutions to the equations are

$$p_n(t) = \frac{(\lambda t)^n e^{-t}}{n!} \qquad n = 0, 1, 2, \ldots$$

and the Markov process is called a *Poisson process*.

Limiting Behavior. In an irreducible, time-homogeneous Markov chain with a continuous parameter, the limits

$$\lim_{t \to \infty} p_{ij}(t)$$

always exist. If in addition the chain is positive recurrent, then

$$\lim_{t \to \infty} p_{ij}(t) = \pi_j > 0 \qquad \Sigma \pi_j = 1 \tag{30}$$

That is, the limits exist; they are independent of the initial state and they form a probability density function.

The limiting probabilities π_j, when they exist, are the unique solution to the steady-state equation

$$0 = \overline{\Pi\Lambda} \tag{31}$$

or in component form,

$$0 = \sum_j \pi_j \lambda_{jk} \tag{32}$$

Equation (32) is a useful working tool. Under the stated conditions, these equations either have no solution that has the properties of a probability distribution or they have precisely one. The π_j are the *steady-state probabilities*, and it can be shown that

$$\lim_{t \to \infty} p_j(t) = \pi_j$$

is satisfied, as well as Eq. (30).

Birth-Death Processes. Of considerable interest in queuing theory applications is the special Markov process whose instantaneous transition rates are given by

$$\lambda_{i,i+1} = \lambda_i \qquad i = 0, 1, 2, \ldots$$

$$\lambda_{i,i-1} = \mu_i \qquad i = 1, 2, \ldots \tag{33}$$

$$\lambda_{ii} = -(\lambda_i + \mu_i) \qquad i = 1, 2, \ldots$$

$$\lambda_{ij} = 0 \qquad \text{otherwise}$$

A Markov process with this particular set of rates is called a birth-death process.
The state equations are given by

$$\frac{dp_0(t)}{dt} = -\lambda p_0(t) + \mu_1 p_1(t)$$

$$\frac{dt_j(t)}{dt} = \lambda_{j-1} p_{j-1}(t) -)\lambda_j + \mu_j)p_j(t) + \mu_j + p_{j+1}(t) \qquad j = 1,2,\dots$$

The *steady-state equations* are given by

$$-\lambda_0 \pi_0 + \mu_1 \pi_1 = 0$$

$$\lambda_{j-1}\pi_{j-1} - (\lambda_j + \mu_j)\lambda_j + \mu_{j+1}\pi_{j+1} = 0 \qquad (34)$$

The general solutions to the steady-state equations are given by

$$\pi_j = \frac{\lambda_0 \lambda_1 \cdots \lambda_{j-1}}{\mu_1 \mu_2 \cdots \mu_j}\pi_0 \qquad j = 1,2,\dots \qquad (35)$$

π_0 is determined from

$$\pi_0 = \left(1 + \frac{\lambda_0}{\mu_1} + \frac{\lambda_0\lambda_1}{\mu_1\mu_2} + \cdots + \frac{\lambda_0\lambda_1 \cdots \lambda_{j-1}}{\mu_1\mu_2 \cdots \mu_j} + \cdots\right)^{-1}$$

These steady-state solutions exist if

$$\left(1 + \frac{\lambda_0}{\mu_1} + \frac{\lambda_0\lambda_1}{\mu_1\mu_2} + \cdots\right) < \infty$$

(a stronger condition than is necessary but one that is useful in application).

Four Special Cases of Birth-Death Processes.   We note here four special cases of the birth-death process.

1. *The M/M/1 queue.*   In the *M/M/1* queue, one supposes that there is a "server" whose service times are exponentially distributed random variables with parameter μ and that successive service times are independent. "Arrivals" to the server form a Poisson process with parameter λ. Arrivals which occur when the server is busy form a single waiting line for their turn at service. It follows that the rate structure (33) is

$$\lambda_{i,i+1} = \lambda_i = \lambda \qquad \text{for } i = 0, 1, 2,\dots$$

$$\lambda_{i,i-1} = \mu_i = \mu \qquad \text{for } i = 1, 2,\dots$$

$$\lambda_{ii} = -(\lambda_i + \mu_i) = -(\lambda + \mu) \qquad \text{for } i = 1, 2,\dots$$

$$\lambda_{ij} = 0 \qquad \text{otherwise}$$

The state equations are given by

$$\frac{dp_0(t)}{dt} = -\lambda p_0(t) + \mu p_1(t)$$

$$\frac{dp_i(t)}{dt} = \lambda p_{i-1}(t) - (\lambda + \mu)p_i(t) + \mu p_{i+1}(t) \qquad i = 1, 2,\dots$$

The steady-state probabilities are given by

$$\pi_j = \rho^j(1 - \rho) \qquad j = 0, 1, 2,\dots$$

where $\rho = \lambda/\mu$ (the *traffic intensity*) upon using Eqs. (35) and (36). These steady-state solutions exist for $\rho < 1$.

   2. *The M/M/1 queue with finite capacity.*   If in case 1 above we retain the transition rate structure but require $\lambda_j = 0$ for $j \geq N$, and $\mu_j = 0$ for $j > N$, we are led for some $N > 0$ to the *M/M/1* queue with finite size N capacity. In this case, the steady-state probabilities exist for every ρ and are given by

$$\pi_j = \frac{\rho^j(1 - \rho)}{1 - \rho^{N+1}} \qquad j = 0, 1, 2, \ldots, N$$

$$\pi_j = 0 \qquad\qquad \text{otherwise}$$

   3. *The multichannel queue.* In the multichannel queue, one supposes that there are R servers sharing a common waiting line. Arrivals to the service system form a Poisson process with parameter λ. Each server performs service with service times exponentially distributed with parameter μ. (The servers are "identical.") It follows that

$$\lambda_{i,i+1} = \lambda \qquad \text{for } i = 0, 1, 2, \ldots$$

$$\lambda_{i,i-1} = i\mu \qquad \text{for } i = 1, 2, \ldots, R - 1$$

$$\lambda_{i,i-1} = R\mu \qquad \text{for } i = R, R + 1, \ldots$$

$$\lambda_{ij} = 0 \qquad \text{for } i \neq j \text{ and } i \neq i + 1, i - 1$$

The steady-state Eqs. (34) are

$$-\lambda\pi_0 + \mu\pi_1 = 0$$

$$-(\lambda + i\mu)\pi_i + \lambda\pi_{i-1} + (i + 1)\mu\pi_{i+1} = 0 \qquad \text{for } i < R$$

$$-(\lambda + R\mu)\pi_i + \lambda\pi_{i-1} + R\mu\pi_{i+1} = 0 \qquad \text{for } i \geq R$$

The steady-state probabilities are

$$\pi_i = \frac{\pi_0\rho^i}{i!} \qquad \text{for } i < R$$

$$\pi_i = \frac{\rho^i}{R!R^{i-R}} \pi_0 \qquad \text{for } i \geq R$$

and π_0 is obtained from the condition

$$\Sigma\pi_i = 1$$

The π_i exist if $\lambda < R\mu$.

   4. *The machine repair problem.* In a machine repair problem, we suppose that there are M identical machines, each subject to breakdown, that are repaired by one of R repair crews. Running times of each machine are random variables, exponentially distributed with parameter λ. The machines are identical. The repair structure is the same as in case 3. From this, it follows that if i machines are running.

$$\lambda_{i,i-1} = (M - i)\lambda \qquad i = 0, 1, 2, \ldots$$

$$\lambda_{i,i+1} = i\mu \qquad i = 1, 2, \ldots, R - 1$$

$$\lambda_{i,i+1} = R\mu \qquad i = R, R + 1, \ldots, M$$

$$\lambda_{ij} = 0 \qquad i \neq j, i - 1, i + 1$$

The steady-state equations are

$$-M\lambda\pi_0 + \mu\pi_1 = 0$$

$$(M - i + 1)\lambda\pi_{i-1} - [(M - i)\lambda + i\mu]\pi_i + (i + 1)\mu\pi_{i+1} = 0 \qquad i < R$$

$$(M - i + 1)\lambda\pi_{i-1} - [(M - i)\lambda + R\mu]\pi_i + R\mu\pi_{i+1} = 0 \qquad i \geq R$$

$$\lambda\pi_{M-1} + R\mu\pi_M = 0$$

Explicit formulas for π_i are complex but can be evaluated numerically. The π_i always exist in this case for finite M, R.

BIBLIOGRAPHY

Cinlar, E., *Introduction to Stochastic Processes,* Prentice-Hall, Englewood Cliffs, N.J., 1975.

Clarke, A. B., and R. L. Disney, *Probability and Random Processes: An Introduction with Applications* (2d ed.), Wiley, New York, 1985.

Feller, W., *Introduction to Probability Theory and Its Applications,* 3d ed., Wiley, New York, 1967.

Kemeny, J. G., and J. L. Snell, *Finite Markov Chains,* Van Nostrand, Princeton, N.J., 1959.

Morse, P., *Queues, Inventories and Maintenance,* Wiley, New York, 1958.

Parzen, E., *Modern Probability Theory and Its Applications,* Wiley, New York, 1960.

CHAPTER 3
PRACTICAL STATISTICS

John W. Adams

Associate Professor, Department of Industrial Engineering,
Lehigh University, Bethlehem, Pennsylvania

This chapter is a compendium of information on the subject of using statistical methods to solve practical problems. It is divided into sections with titles listed: (1) Random Variables and Probability Distributions, (2) Estimation of Parameters, (3) Tests of Hypotheses, (4) Analysis of Variance, and (5) Regression.

Each of the sections, after the first, is devoted to a single application. The first contains a discussion of the basic ideas and concepts of the theory of statistics. It is included for the benefit of users who are not familiar with the terminology and notations of applied statistics.

RANDOM VARIABLES AND PROBABILITY DISTRIBUTIONS

The purpose of this section is to explain and describe the use of two statistical ideas: *random variable* and *probability distribution*. These ideas are important to engineers because they provide a means for expressing engineering problems in statistical form, which can then be solved by using the methods of statistical analysis.

Engineering problems are solved by analyzing data, and the most effective methods of analysis are statistical because deterministic methods cannot cope with data that are contaminated with errors and uncertainty. Almost any piece of engineering data—whether in the form of an observation, a measurement, or an experimental outcome—has been, at least to some degree, affected by unidentified and unanalyzed causes. In such cases, repetition of the observation, the measurement or the experiment does not produce the same results over and over again. Instead the outcomes are scattered, haphazardly, without order or pattern, in ways that can only be described as random.

Statistical theory was invented to cope with randomness. It does so by defining a random variable to be a variable whose values, when observed repeatedly, are distributed over a spectrum of values according to a mathematically specified law of distribution. The value of a random variable cannot be computed prior to its observation, but its law of distribution can be used to compute the probability that the random variable will have a specified value when observed.

The simple expedient of regarding data as a random variable, obeying a law of distribution, has the effect of transforming an ordinary engineering problem into a mathematical problem. By means of this transformation, data acquire the mathematical properties of a random variable, and can, therefore, be manipulated and analyzed just like any other mathematical variable. Moreover, if the random variable is an accurate model of the process which generates the data, the analysis will produce useful results. The key to accuracy is the law of distribution. The model will be accurate if the law of distribution accurately describes the empirical distribution that an observer sees in the data.

Conventions and Definitions.   Some of the notation and terminology used in this chapter may not be familiar to all users. For the convenience of such users, a list of notational devices is given below followed by a brief explanation of each item in the list.

1. $n!$
2. $\binom{n}{j}$
3. $P(u < X < v)$, $P(X \leq x)$
4. $P(A|\text{condition})$
5. $\mathscr{E}(X)$ and $\text{Var}(X)$
6. Probability density function
7. Failure rate function

1.   The notation $n!$ Is read "n factorial," and it is shorthand for the product of the first n integers:

$$n! = 1 \cdot 2 \cdot 3 \cdots n, \, n = 1, 2, 3, \ldots$$

$$0! = 1$$

For example 4! is $1 \cdot 2 \cdot 3 \cdot 4 = 24$.

2.   The notation $\binom{n}{j}$ is read "n choose j," and it is shorthand for the number of different samples, each containing j objects, which can be selected from a set containing n distinct objects. Its value can be computed as follows:

$$\binom{n}{j} = \frac{n!}{j!(n-j)!} \qquad n = 1, 2, 3, \ldots; j = 0, 1, 2, 3, \ldots, n$$

3.   The notation $P(X \leq v)$ denotes the probability that the value of the random variable X, when observed, does not exceed v. If X is a continuous random variable,

$$P(X \leq v) = P(X < v)$$

and if X is a discrete random variable,

$$P(X \leq v) \geq P(X < v)$$

However, in both cases—discrete and continuous—it is true that

$$P(u \leq X \leq v) = P(X \leq v) - P(X < u)$$

4.   The notation, $P(A|\text{condition})$, denotes the probability of the event A, evaluated under the assumption that the condition is true. In this chapter conditions are always specifications of parameter values.

5.   The notations, $\mathscr{E}(X)$ and $\text{Var}(X)$, denote the expectation and the variance of the random variable X. For every random variable discussed in this chapter, there are simple formulas, which can be used to determine the values of $\mathscr{E}(X)$ and $\text{Var}(X)$. These formulas are included in the descriptions of the probability distribution functions.

6.   In this chapter, the notation $f(x)$ denotes a probability density function, which is any function possessing the following two properties:

- $f(x) \geq 0, \, -\infty < x < \infty$

- The region between the curve generated by plotting $f(x)$ as a function of x and the horizontal axis has area 1

Every continuous random variable has a probability density function, and if X is a continuous random variable, having the probability density function $f(x)$, then the value of $P(u < X < v)$ is the area of the region under the curve, generated by plotting $f(x)$ as a function of x, and above the interval (u, v), which is located on the x axis.

7. In this chapter, the notation $r(t)$ denotes a failure rate function, which is used in the study of the duration of life of a part or a component. The variable t denotes time, and the value of $r(t)$, at time t, is the instantaneous rate of failure of surviving components of age t. For additional details on the determination and use of failure rate functions see Ref. 1.

Types of Random Variables.   There are two distinct types of random variables:

- Discrete random variables
- Continuous random variables

The possible values of a discrete random variable form a discrete set—in many cases the possible values are integers. The possible values of a continuous random variable form a continuous set, usually in the form of an interval.

As a general rule, random variables representing data, which are in the form of counts, are discrete, and random variables representing measurements are continuous. The distinction between discrete and continuous is of more interest to mathematicians than to users. Nevertheless users will notice that the analysis of discrete random variables uses elementary mathematics, whereas the analysis of continuous random variables uses calculus.

The random variables, which are widely used in applications, are the four discrete random variables,

1. Binomial
2. Hypergeometric
3. Negative binomial
4. Poisson

and the four continuous random variables,

1. Normal
2. Exponential
3. Weibull
4. Uniform

In addition there are three important random variables whose probability distribution functions are derived from the normal distribution. These are

1. Student's t
2. Chi-squared
3. F

Each of the 11 random variables listed above has a probability distribution function, and each of these is described in the following sections.

Binomial Distribution Function.   Consider a manufacturing process which produces a stream of product, each unit of which is either defective or not defective when produced, and define the variables n, p, and X as follows:

- n is the number of units of product produced
- p is the probability that any given unit is defective
- X is the number of units which are defective

Then X is a discrete random variable, having the binomial distribution, and the probability that X has the value j can be computed as follows:

$$P(X = j) = \binom{n}{j} p^j (1 - p)^{n-j} \qquad j = 1, 2, 3, \ldots, n \tag{1}$$

and the probability that X does not exceed j can be computed as follows:

$$P(X \le j) = \sum_{k=0}^{j} \binom{n}{k} p^k (1 - p)^{n-k} \qquad j = 0, 1, 2, \ldots, n \tag{2}$$

Equation (1) is the binomial probability distribution function, and Eq. (2) is the cumulative binomial probability distribution function. The name *binomial* is used to emphasize the fact that the value of X is the result of n separate classifications, each one having only *two* possible outcomes—defective and not defective.

Parameters.   The binomial distribution has two parameters—n and p:

- n can be any positive integer,
- p can be any number in the interval (0, 1)

Evaluation of Binomial Probabilities.   In principle, Eq. (1) or (2) can be used to compute the value of any binomial probability, but hand calculations are impractical for values of n exceeding 10. Alternative methods of evaluation include tables, approximations—described in the discussion of the normal distribution—and computer programs.

The National Bureau of Standards[2] and the Harvard University Computation Laboratory[3] have both published extensive tables of the binomial distribution.

Table 3.1, which appears at the end of the chapter, is a tabulation of the cumulative binomial probability distribution function—$P(X \le j)$ for selected values of n, p, and j. It was computed by the author, using International Mathematical and Statistical Library (IMSL) routine BIND.[4]

Use of Table 3.1.   Values of n, p, and j appear in the margins of the table, and the corresponding entries are the probabilities. For example, the entry corresponding to the values $n = 15$, $p = .4$, $j = 5$ is .4032. Therefore,

$$P(X \le 5|n = 15, p = .4) = .4032$$

Also, if n is 15 and p is .4, as in the previous example,

$$P(5 \le X \le 6) = P(X \le 6) - P(X \le 4) = .3925$$

and

$$P(X = 7) = P(X \le 7) - P(X \le 6) = .1771$$

Table 3.1 contains no entries for values of p exceeding .5. To evaluate a binomial probability when p exceeds .5, use the following identity:

$$P(X \le j|n, p) \equiv 1 - P(X \le j'|n', p') \tag{3}$$

in which $n' = n$, $p' = 1 - p$, and $j' = n - j - 1$.

If n is 15, p is .6, and j is 5, it follows from Eq. (3) that

$$P(X \le 5|n = 15, p = .6) = 1 - P(X \le 9|n = 15, p = .4)$$

The probability on the right—value .4032—is in the table; therefore, the probability on the left, which is not in the table, has the value .5968.

Expectation and Variance.   A random variable X, having the binomial distribution with parameters n and p, has expectation and variance

$$\mathcal{E}(X) = np$$

$$\text{Var}\,(X) = np(1 - p)$$

The Hypergeometric Distribution.   Suppose that a sample, containing n items, is *randomly* selected from a lot containing N items of which k are defective. If X is the number of defective items found in the sample, when all of the items in the sample are tested, then X is a discrete random variable having the hypergeometric distribution, and the probability that X has the value j can be computed as follows:

$$P(X = j) = \frac{\binom{k}{j}\binom{N-k}{n-j}}{\binom{N}{n}}, \quad \max\,(0,\, n - N + k) \le j \le \min\,(n, k) \qquad (4)$$

Equation (4) is the hypergeometric probability distribution function.

Parameters.   The hypergeometric distribution has three parameters—N, k, and n:

- N can be any positive integer
- k can be any positive integer not exceeding N
- n can be any positive integer not exceeding N

Evaluation of Hypergeometric Probabilities.   In principle, Eq. (4) can be used to compute any hypergeometric probability. For example, if $N = 10$, $k = 3$, $n = 4$, and $j = 1$, Eq. (4) yields

$$P(X = 1) = \frac{\binom{3}{1}\binom{7}{3}}{\binom{10}{4}} = .5$$

However, hand calculations become tedious when N exceeds 20.

Extensive tables—over 700 pages—of the hypergeometric distribution were published by Lieberman and Owen.[5] Other methods for evaluating hypergeometric probabilities include binomial approximation.

Binomial Approximation.   When n is small relative to N—say n does not exceed .05N—the binomial distribution can be used to compute approximate values of hypergeometric probabilities. The approximation formula is

$$\frac{\binom{k}{j}\binom{N-k}{n-j}}{\binom{N}{n}} \approx \binom{n}{j}\left(\frac{k}{N}\right)^{j}\left(1 - \frac{k}{n}\right)^{n-j} \qquad (5)$$

Evaluation of the expression on the left of Eq. (5) would yield the exact value of a hypergeometric probability, and evaluation of the expression on the right will yield an approximation of the hypergeometric probability. However, the expression on the right is a binomial probability, which is relatively easy to evaluate.

The relation between the hypergeometric variables—N, k, n, and j—and the binomial variables—n', p', and j'—is $n' = n$, $p' = k/N$, $j' = j$. For example, suppose that $N = 100$,

$k = 25, n = 5, j = 3$. Then $n' = 5, p' = .25, j' = 3$, and the approximating binomial probability is the difference of the two binomial probabilities:

$$P(X \leq 3 | n = 5, p = .25) - P(X \leq 2 | n = 5, p = .25) = .0879$$

The exact probability, as calculated by use of Eq. (4) is

$$\frac{\binom{25}{3}\binom{75}{2}}{\binom{100}{5}} = .0848$$

The approximation, at least in this example, provides an answer which is good enough for practical purposes.

Expectation and Variance. A random variable X, having the hypergeometric distribution, with parameters N, k, and n, has expectation and variance

$$\mathcal{E}(X) = \frac{nk}{N}$$

$$\text{Var } (X) = \frac{nk(N - k)(N - n)}{N^2(N - 1)}$$

Negative Binomial Distribution. Consider a manufacturing process in which each unit of product has probability p of being acceptable when produced. Then suppose that the manufacturer has an order for r acceptable units and that production stops when the rth acceptable unit is produced. If X is the total number of units produced, then X is a discrete random variable having the negative binomial distribution, and the probability that X has the value j can be computed as follows:

$$P(X = j) = \binom{j-1}{r-1} p^r (1 - p)^{j-r} \qquad j = r, r + 1, \ldots \tag{6}$$

and the probability that the value of X does not exceed j can be computed as follows:

$$P(X \leq j) = \sum_{k=r}^{j} \binom{k-1}{r-1} p^r (1 - p)^{k-r} \qquad j = r, r + 1, \ldots \tag{7}$$

Equation (6) is the negative binomial probability distribution function, and Eq. (7) is the cumulative negative binomial distribution function.

Parameters. The negative binomial distribution has two parameters, r and p:

- r can be any positive integer
- p can be any number in the interval $(0, 1)$

Evaluation of Negative Binomial Probabilities. Equation (6) or (7) can be used to compute any negative binomial probability. For example, if $r = 5, p = .8$, and $j = 8$, Eq. (7) yields

$$P(X \leq 8) = \sum_{k=5}^{8} \binom{k-1}{4}(.8)^5(.2)^{k-5} = .9437$$

Tables for the negative binomial distribution are not needed because G. P. Patil[6] discovered a relation between the negative binomial and the binomial, which makes it possible to use the binomial probability formula to compute negative binomial probabilities. According to Patil,

$$\sum_{k=r}^{j} \binom{k-1}{r-1} p^r (1-p)^{k-r} \equiv 1 - \sum_{k=0}^{r-1} \binom{j}{k} p^k (1-p)^{j-k} \qquad (8)$$

To evaluate the sum on the left, which is a negative binomial probability, evaluate the sum on the right, which is a binomial probability. The negative binomial variables—r, p, and j—are related to the binomial variables—n', p', and j'—as follows: $n' = j$, $p' = p$, and $j' = r - 1$.

If r is 5, p is .8, and j is 8, as in the previous example, the binomial probability, corresponding to the case $n' = 8$, $p' = .8$, and $j' = 4$, is .0563, and when .0563 is subtracted from 1, as indicated in Eq. (8), the result is .9437. This is the same result obtained by use of Eq. (7).

Expectation and Variance.   A random variable X, having the negative binomial distribution, with parameters r and p, has expectation and variance

$$\mathcal{E}(X) = \frac{r}{p}$$

$$\text{Var}(X) = \frac{r(1-p)}{p^2}$$

The Poisson Distribution.   There are many manufacturing processes in which the units of product have flaws located within the product—along a length, over a surface, or throughout a volume. If in such a case it happens that the flaws have a constant average density—say λ flaws per unit length, or per unit surface area, or per unit volume—and if X is the number of flaws in a unit, then X is a discrete random variable, having the Poisson distribution, and the probability that X has the value j can be computed as follows:

$$P(X = j) = e^{-\lambda}\left(\frac{\lambda^j}{j!}\right) \qquad j = 0, 1, 2, \ldots \qquad (9)$$

and the probability that X has a value not exceeding j can be computed as follows:

$$P(X \le j) = \sum_{k=0}^{j} e^{-\lambda}\left(\frac{\lambda^k}{k!}\right) \qquad j = 0, 1, 2, \ldots \qquad (10)$$

Equation (9) is the Poisson probability distribution function, and Eq. (10) is the cumulative Poisson probability distribution function.

Parameters.   The Poisson distribution has one parameter, λ, which can be any positive number.

Evaluation of Poisson Probabilities.   In principle, Eq. (9) or (10) can be used to evaluate any Poisson probability. For example, suppose that the average number of flaws in a square yard of carpet is .015, and that X is the number of flaws found in a sample of 100 square yards. Then $\lambda = 1.5$, and the probability that X has the value 3 is

$$P(X = 3) = e^{-1.5}\left(\frac{1.5^3}{3!}\right) = .1255$$

and the probability that the value of X does not exceed 3 is

$$P(X \le 3) = e^{-1.5}\left(\frac{1.5^0}{0!} + \frac{1.5}{1!} + \frac{1.5^2}{2!} + \frac{1.5^3}{3!}\right) = .5578$$

Tables of the Poisson distribution are tabulations of the cumulative probability distribution—Eq. (10)—for selected values of λ. Unfortunately, the range of values of λ which occur in applications is very wide, and a generally useful table, therefore, has to be large. Extensive tables of the Poisson distribution were published by the General Electric Company.[7]

Expectation and Variance.   A random variable X, having the Poisson distribution, with parameter λ, has expectation and variance

$$\varepsilon(X) = \lambda$$

$$\text{Var }(X) = \lambda$$

The Normal Distribution.   Consider a manufacturing process in which the units of product each have a measurable property, which varies from unit to unit. If X is the value of a measurement on a unit of the product, then, prior to measurement, X is a continuous random variable, and a common assumption is that X has the normal distribution, whose probability density function is

$$f(x) = \frac{e^{-(1/2)[(x-\mu)/\sigma]^2}}{\sqrt{2\pi\sigma^2}} \qquad -\infty < x < \infty \tag{11}$$

Figure 3.1 is a drawing of the curve, generated by plotting $f(x)$, defined in Eq. (11), as a function of x. Examination of the drawing shows that the normal distribution is symmetric about its center, which is located at the point x equals μ. Not evident from examination of the drawing, but nevertheless true, is the fact that the region between the curve and the horizontal axis has area 1. The area of the shaded region in Fig. 3.1 is $P(u < X < v)$.

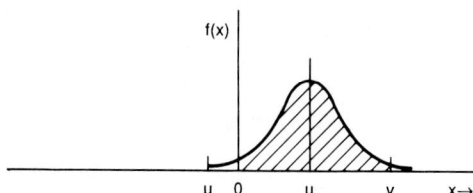

FIGURE 3.1   Normal density function.

Parameters.   The normal distribution has two parameters, μ and σ:

- μ can be any number, positive, negative, or zero
- σ can be any positive number

Evaluation of Normal Probabilities.   In the case of continuous random variables, probabilities are areas, and integral calculus provides general methods for evaluating the areas of regions bounded by analytic curves.

If X is a random variable, having the normal distribution, with parameters μ and σ, then, by definition,

$$P(X < v) = \frac{1}{\sqrt{2\pi\sigma^2}} \int_{-\infty}^{v} e^{(1/2)[(x-\mu)/\sigma]^2} dx \qquad -\infty < v < \infty \tag{12}$$

and, by means of the following transformation,

$$z = \left(\frac{v - \mu}{\sigma}\right)$$

Equation (12) can be reduced to the standard form

$$P(X < v) = \frac{1}{\sqrt{2\pi}} \int_{-\infty}^{z} e^{-u^2/2} \, du \qquad -\infty < z < \infty \tag{13}$$

The advantage of the standard form is that a probability, depending on the three variables μ, σ, and v, is expressed as a function of the single variable z.

In practice, the evaluation of normal probabilities is carried out by consulting tables whose entries are evaluations of the standard normal integral—the right-hand side of Eq. (13)—for selected values of z. Table 3.2, at the end of the chapter, was constructed in this way, using IMSL routine ANORDF[4] to carry out the numerical evaluations.

Structure of Table 3.2. Table 3.2 is based on the equation

$$\frac{1}{\sqrt{2\pi}} \int_{-\infty}^{z_\alpha} e^{-u^2/2} \, du = 1 - \alpha$$

in which α and z_α are variables; if the value of α is specified, the value of z_α is determined and vice versa.

Table 3.2 consists of pairs (α, z_α), with values of z_α in the margin of the table and the corresponding values of $1 - \alpha$ in the body of the table. For example, examination of Table 3.2 reveals that 1.96 is the margin entry corresponding to the table entry .9750. Therefore, the following is true:

$$\frac{1}{\sqrt{2\pi}} \int_{-\infty}^{1.96} e^{-u^2/2} \, du = .9750$$

and $\alpha = .0250$, $z_{.025} = 1.96$, and $1 - \alpha = .9750$.

Also useful is the identity

$$z_\alpha \equiv -z_{1-\alpha} \qquad 0 \le \alpha \le 1 \tag{14}$$

which is derivable from the fact that the standard normal distribution is symmetric about zero.

Use of Table 3.2. Suppose that X is a random variable, having the normal distribution, with parameters, μ and σ, and that, for specified values of u and v, it is required to evaluate the probability

$$P(u < X < v)$$

The first step is to express the probability in the form of a difference:

$$P(u < X < v) = P(X < v) - P(X < u) \tag{15}$$

The second step is to transform each of the probabilities on the right side of Eq. (15) into the standard form, thereby obtaining a value of z_α for each of the two probabilities as follows:

$$z_\alpha = \left(\frac{v - \mu}{\sigma}\right) \qquad \text{and} \qquad z_{\alpha'} = \left(\frac{u - \mu}{\sigma}\right)$$

Third, use Table 3.2, and, if necessary, Eq. (14), to find the values of α and α', which determine the value of the probability:

$$P(u < X < v) = \alpha' - \alpha$$

The value of α corresponding to a positive value of z_α can be found directly from Table 3.2, but if the value of z_α is negative, it is necessary to use Eq. (14). According to the identity, if z_α is negative, then $z_{1-\alpha}$ is positive, and the positive value of $z_{1-\alpha}$ can be used to find the value of α in Table 3.2.

For example, suppose $v = 9$, $u = 4$, $\mu = 6$, and $\sigma = 2$. Then

$$z_\alpha = \left(\frac{9 - 6}{2}\right) = 1.5 \quad \text{and} \quad z_{\alpha'} = \left(\frac{4 - 6}{2}\right) = -1$$

and, according to Eq. (14),

$$z_{1-\alpha'} = 1$$

The entry in Table 3.2 corresponding to the margin entry 1.5 is .9332, and, therefore, the value of α is .0668. The entry corresponding to the margin entry 1 is .8413, and, therefore, the value of α' is .8413. Hence,

$$P(4 < X < 9 | \mu = 6, \sigma = 2) = .8413 - .0668 = .7745$$

Expectation and Variance.   A random variable X, having the normal distribution, with parameters μ and σ, has expectation and variance

$$\mathscr{E}(X) = \mu$$

$$\text{Var }(X) = \sigma^2$$

Normal Approximation to Binomial Probabilities.   Suppose that X is a random variable having the binomial distribution with parameters n and p, and that Y is a random variable having the standard normal distribution with parameters $\mu = 0$ and $\sigma = 1$.

Values of binomial probabilities can always be obtained by application of Eq. (2) or by consulting tables such as Table 3.1. In some cases the normal distribution, Table 3.2, can be used to obtain accurate approximations of binomial probabilities. For an excellent discussion of the normal approximation to the binomial, including its limitations, see Ref. 8. According to Ref. 8,

$$\sum_{k=i}^{j} \binom{n}{k} p^k (1-p)^{n-k} \approx P\left(\frac{i - np - .5}{\sqrt{np(1-p)}} < Y < \frac{j - np + .5}{\sqrt{np(1-p)}}\right) \tag{16}$$

in which the expression on the left is the binomial probability to be approximated and the probability on the right is the normal approximation, which can be evaluated by using Table 3.2.

Accuracy depends in a complicated way on the values of n, p, i, and j. For a fixed value of n, accuracy is highest when $p = .5$ and deteriorates as p moves toward the tails—either 0 or 1. In general, accuracy increases as the value of n increases; if $p = .5$, acceptable results can be obtained, even when n is as small as 10, but if p is .01, n must be at least 1000 to get acceptable results. In all cases, accuracy is greatly diminished if $(i - np)$ and $(j - np)$ both have the same sign—either both negative or both positive.

Consider the following examples:

1. $p = .5$, $n = 10$, $i = 4$, $j = 6$
2. $p = .01$, $n = 500$, $i = 2$, $j = 6$
3. $p = .01$, $n = 1000$, $i = 7$, $j = 12$

The solutions are given below. The first number is the approximate value, as computed by the normal approximation, and the second is the true value as computed by the IMSL routine BINDF.[4]

1. .656, .657

2. .723, .692

3. .664, .654

The Exponential Distribution.   Suppose that X is the waiting time between successive breakdowns of a machine, and that the rate of breakdowns is constant, say λ breakdowns per unit of time. Then X is a continuous random variable, and if the breakdowns are random occurrences, having nothing to do with the aging of the machine, it is plausible to assume that X has the exponential distribution, whose probability density function is

$$f(x) = \lambda e^{-\lambda x} \qquad 0 \leq x < \infty \tag{17}$$

and whose cumulative probability distribution function is

$$P(X < x) = \int_0^x \lambda e^{-\lambda y}\, dy = 1 - e^{-\lambda x} \qquad 0 \leq x < \infty \tag{18}$$

Figure 3.2 is a drawing of the curve, generated by plotting $f(x)$, defined in Eq. (17), as a function of x. The region between the curve and the horizontal axis is 1, and this can be verified by substituting ∞ for x in Eq. (18). The area of the shaded region in Figure 3.2 is $P(u < X < v)$.

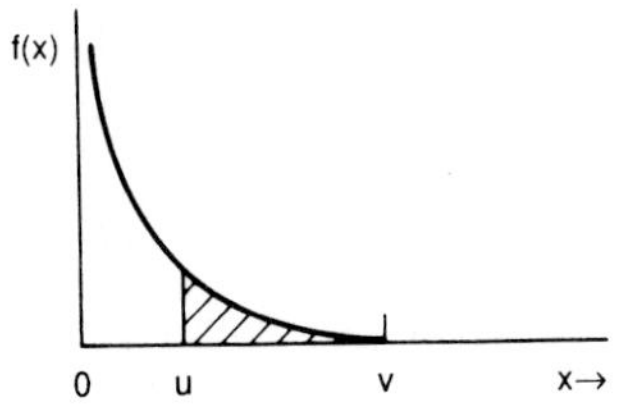

Parameters.   The exponential distribution has one parameter, λ, which can be any positive number.

Evaluation of Exponential Probabilities.   Equation (18) can be used to evaluate any exponential probability. There is no need for tables of the exponential distribution function, because, as indicated in Eq. (18), the exponential cumulative distribution function is an elementary function, which can be evaluated on a hand-held calculator.

FIGURE 3.2   Exponential density function.

Expectation and Variance.   A random variable X, having the exponential distribution, with parameter λ, has expectation and variance

$$\mathscr{E}(X)\frac{1}{\lambda}$$

$$\text{Var}\,(X) = \frac{1}{\lambda^2}$$

Use of the Exponential Distribution.   The exponential distribution is occasionally used in the analysis of failure data. For example, X could be the duration of life of some component, such as a valve or a motor, and it could be assumed that X has the exponential distribution. To judge the appropriateness of such an assumption, it is necessary to consider the failure rate function $r(t)$ of the component being studied. For information on failure rate functions see Ref. 1.

The value of $r(t)$, at time t, is the instantaneous rate of failure of surviving components at age t, and components which wear out must have failure rate functions which increase as the component ages. The failure rate function is determined by the probability distribution function which is being used as the model for the duration of life of the component. In the case of the exponential distribution,

$$r(t) = \lambda \qquad 0 < t < \infty$$

Since the failure rate function of the exponential distribution is constant over all time, it is inappropriate to use the exponential distribution in the study of components which wear out.

The Weibull Distribution. Suppose that X is the waiting time between successive breakdowns of a machine. Then X is a continuous random variable, and it is plausible to assume X has the Weibull distribution, whose probability density function is

$$f(x) = \lambda \alpha x^{\alpha-1} e^{-\lambda x^\alpha} \qquad 0 < x < \infty \qquad (19)$$

and whose cumulative probability distribution function is

$$P(X < x) = \int_0^x \lambda \alpha y^{\alpha-1} e^{-\lambda y^\alpha}\, dy = 1 - e^{-\lambda x^\alpha} \qquad (20)$$

Figure 3.3 is a drawing of the curve generated by plotting $f(x)$, defined in Eq. (19), as a function of x. The region between the curve and the horizontal axis is 1, and this can be verified by substituting ∞ for x in Eq. (20). The area of the shaded region in Fig. 3.3 is $P(u < X < v)$.

Parameters. The Weibull distribution has two parameters, λ and α:

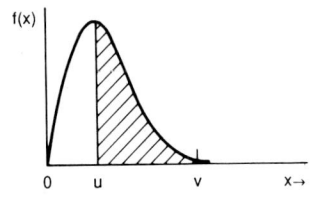

- λ can be any positive number
- α can be any positive number

FIGURE 3.3 Weibull density function.

Evaluation of Weibull Probabilities. Equation (20) can be used to evaluate any Weibull probability. There is no need for tables of the Weibull distribution, because, as indicated in Eq. (20), the cumulative Weibull probability distribution function is an elementary function, which can be evaluated on a hand-held calculator.

Expectation and Variance. A random variable X, having the Weibull distribution, with parameters λ and α, has expectation and variance

$$\mathscr{E}(X) = \left(\frac{1}{\lambda}\right)^{1/\alpha} \Gamma\left(1 + \frac{1}{\alpha}\right)$$

and

$$\operatorname{Var}(X) = \left(\frac{1}{\lambda}\right)^{2/\alpha} \left[\Gamma\left(1 + \frac{2}{\alpha}\right) - \Gamma^2\left(1 + \frac{1}{\alpha}\right)\right]$$

The symbol $\Gamma(x)$ denotes the gamma function. Beyer[9] has published a table which can be used to obtain values of $\Gamma(x)$ for selected values of x.

Use of the Weibull Distribution. The Weibull distribution is frequently used in the analysis of failure data. For example, X could be the duration of life of some component, such as a valve or a motor, and it could be assumed that X has the Weibull distribution. To judge the appropriateness of such an assumption, it is necessary to consider the failure rate function $r(t)$ of the component being studied. For information on failure rate functions see Ref. 1.

The value of $r(t)$, at time t, is the instantaneous rate of failure of surviving components at age t, and components which wear out must have failure rate functions which increase as the component ages. The failure rate function is determined by the probability distribution function which is being used as the model for the duration of life of the component. In the case of the Weibull distribution,

$$r(t) = \lambda \alpha(t)^{\alpha-1} \qquad 0 < t < \infty$$

The failure rate function of the Weibull distribution is an increasing function of t when α exceeds 1. Hence the Weibull distribution is a plausible choice of model for the study of components which wear out.

The Uniform Distribution.   Suppose that X is a random variable whose value is randomly selected from the specified interval (a, b). Then X is a continuous random variable, having the uniform distribution on the interval (a, b), and its probability density function is

$$f(x) = \frac{1}{(b - a)} \qquad a \leq x \leq b \tag{21}$$

$$= 0 \qquad \text{otherwise}$$

and its cumulative probability distribution function is

$$P(X < x) = \frac{x - a}{b - a} \qquad a \leq x \leq b \tag{22}$$

The most important use of the uniform distribution is in generating simulated values of random variables which have specified probability distribution functions.

Figure 3.4 is a drawing of the curve generated by plotting $f(x)$, defined in Eq. (21), as a function of x. The area of the region between the curve and the horizontal axis is 1, and this can be verified by substituting b for x in Eq. (22). The area of the shaded region in Fig. 3.4 is $P(u < X < v)$.

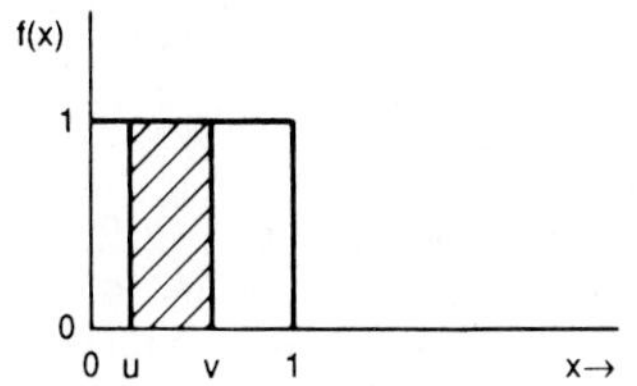

Parameters.   The uniform distribution function has two parameters, a and b:

- a can be any number not exceeding b
- B can be any number

FIGURE 3.4   Uniform density function.

Evaluation of Uniform Probabilities.   Equation (22) can be used to evaluate any uniform probability. Tables of the uniform distribution function are not needed because, as indicated in Eq. (22), the uniform cumulative distribution function is an elementary function, which can be evaluated on a hand-held calculator.

Expectation and Variance.   A random variable X, having the uniform distribution on the interval (a, b), has expectation and variance

$$\mathscr{E}(X) = \frac{a + b}{2}$$

$$\text{Var}(X) = \frac{(b - a)^2}{12}$$

Simulation of Continuous Random Variables.   The procedure for simulating continuous random variables is the following: if $P(X \leq x)$ is the cumulative distribution function of the random variable X, which is to be simulated, and if U is a random variable, having the uniform distribution on the interval $(0, 1)$, then the simulated value of X is the value of x which satisfies the equation

$$P(X \leq x) = U \tag{23}$$

Simulation is carried out by using a computer program, for example IMSL routine RNUN,[4] to generate simulated values of U, and then solving Eq. (23) for each simulated value of U.

For example, if X has the exponential distribution, with parameter λ, Eq. (23) becomes

$$1 - e^{-\lambda x} = U$$

and it is easy to verify that

$$x = -\frac{1}{\lambda} \ln (1 - U)$$

If λ is .01 and three simulated values of U are $U_1 = .67124$, $U_2 = .01842$, $U_3 = .29061$, the simulated values of X are $X_1 = 111$, $X_2 = 1.86$, $X_3 = 34.3$.

Equation (23) is difficult to solve when X has the normal distribution, and several alternate procedures have been proposed. However, a procedure discovered by Box[10] is easy to carry out. According to Box, simulated values of a random variable X, having the normal distribution, with parameters μ and σ, can be generated in pairs as follows:

$$X_1 = \mu + \sigma\sqrt{-2 \ln (U_1)} \sin (2\pi U_2)$$

and

$$X_2 = \mu + \sigma\sqrt{-2 \ln (U_1)} \cos (2\pi U_2)$$

If μ and σ have the values 10 and 2, and if two simulated values of U are $U_1 = .39865$, $U_2 = .84423$, then the two simulated values of X are $X_1 = 7.75$, $X_2 = 11.51$.

Simulation of Discrete Random Variables.   When a discrete random variable which has integer values is simulated the equivalent of Eq. (23) is the inequality

$$P(X \le j - 1) < U < P(X \le j) \tag{24}$$

The simulated value of X is the value of j which satisfies Eq. (24). For example, if X has the binomial distribution, with parameters n and p, $n = 2$, $p = .4$, then $P(X \le -1) = 0$, $P(X \le 0) = .36$, $P(X \le 1) = .84$, and $P(X \le 2) = 1$; and if three simulated values of U are $U_1 = .10734$, $U_2 = .90771$, and $U_3 = .55646$, then the three simulated values of X are $X_1 = 0$, $X_2 = 2$, and $X_3 = 1$.

The t Distribution.   The t distribution is derived from the normal distribution. Suppose that $X_1, X_2, \ldots, X_n$ are n independent random variables, each having the normal distribution, with parameters μ and σ. Suppose also that

$$\overline{X} = \frac{1}{n} \sum_{j=1}^{n} X_j \tag{25}$$

and that

$$S^2 = \frac{1}{n-1} \sum_{j=1}^{n} (X_j - \overline{X})^2 \tag{26}$$

Then the random variable

$$\frac{\sqrt{n}(\overline{X} - \mu)}{S} \tag{27}$$

has the t distribution, with parameter ν, and $\nu = (n - 1)$.

The probability density function of the t distribution is not reproduced here because it is of little interest to practitioners. However, Fig. 3.5 is a drawing of the curve generated by plotting the density function of the t distribution—say $f(x)$—as a function of x. Examination of the drawing shows that the t distribution is symmetric about its center, which is located at the point $x = 0$. Not evident from the drawing, but nevertheless true, is the fact that the region between the curve and the horizontal axis has area 1. The area of the shaded region in Fig. 3.5 is $P(u < t < v)$.

Parameters.   The t distribution has one parameter, $\nu = (n - 1)$, which may be any positive integer. Sometimes ν is called *degrees of freedom.*

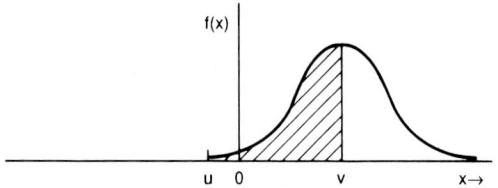

FIGURE 3.5 t distribution function.

Tables of the t *Distribution.*   Entries in a table of the t distribution are values of $t_{\alpha,\nu}$ which satisfy the equation

$$P(X < t_{\alpha,\nu}) = 1 - \alpha \tag{28}$$

in which X is a random variable, having the t distribution, with parameter ν, and α is a specified value in the interval (0, 1).

Table 3.3, at the end of the chapter, was computed by the author, using IMSL routine TIN.[4] Each entry is a solution of Eq. (28) for values of ν and α, specified in the margins of the table. For example, the entry corresponding to the case $\nu = 10$, $\alpha = .05$ is $t_{.05,10} = 1.182$.

There are no entries in the table for values of α greater than .1. For such cases, the required value can be found by application of the identity

$$P(X < t_{\alpha,\nu}) \equiv 1 - P(X < - t_{\alpha,\nu}) \tag{29}$$

in which X is a random variable, having the t distribution, with parameter ν.

For example, suppose that $\nu = 10$, $\alpha = .95$. Application of Eq. (29) yields

$$P(X < -1.182) = .05$$

and the required value of x is -1.182.

Use of the t *Distribution.*   Applications of the t distribution always occur in conjunction with the normal distribution. Assuming that $X_1, X_2, \ldots, X_n$ are independent random variables, each having the normal distribution, with parameters μ and σ, it follows from Eqs. (25), (26), and (27) that

$$P\left(-t_{\alpha/2,\nu} < \frac{\sqrt{n}(\bar{X} - \mu)}{S} < t_{\alpha/2,\nu}\right) = 1 - \alpha \tag{30}$$

in which the value of $t_{\alpha/2,\nu}$ is found in Table 3.3.

The probability statement in Eq. (30) can be used to construct a confidence interval for the parameter μ, or to test a hypothesis about the value of μ. Procedures for constructing confidence intervals and testing hypotheses are described later in the chapter.

The Chi-Squared Distribution.   The chi-squared distribution is derived from the normal distribution. Suppose that $X_1, X_2, \ldots, X_n$ are n independent random variables, each having the normal distribution, with parameters μ and σ. Suppose also that

$$\bar{X} = \frac{1}{n} \sum_{j=1}^{n} X_j \tag{31}$$

and that

$$S^2 = \frac{1}{n-1} \sum_{j=1}^{n} (X_j - \bar{X})^2 \tag{32}$$

Then, the random variable

$$\frac{(n-1)\,S^2}{\sigma^2} \tag{33}$$

has the chi-squared distribution, with parameter $v = (n-1)$.

The probability density function of the chi-squared distribution is not reproduced here because it is of little interest to practitioners. However, Fig. 3.6 is a drawing of the curve generated by plotting the density function of the chi-squared distribution, say $f(x)$, as a function of x. Not evident from the drawing, but nevertheless true, is the fact that the region between the curve and the horizontal axis has area 1. The area of the shaded region in Fig. 3.6 is $P(u < X < v)$.

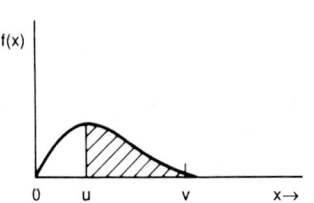

FIGURE 3.6 Chi-squared density function.

Parameters. The chi-squared distribution has one parameter, v, which can have any positive integer value. Sometimes v is called *degrees of freedom*.

Tables of the Chi-Squared Distribution. Entries in a table of the chi-squared distribution are values of $\chi^2_{\alpha,v}$ which satisfy the equation

$$P(X < \chi^2_{\alpha,v}) = 1 - \alpha \tag{34}$$

in which X is a random variable, having the chi squared distribution, with parameter v, and α is a specified value in the interval $(0, 1)$.

Table 3.4, at the end of the chapter, was computed by the author, using IMSL routine CHIIN.[4] Each entry is a solution of Eq. (34) for values of v and α, specified in the margins of the table. For example, the entries corresponding to the cases $v = 30$, $\alpha = .05$, $\alpha = .95$ are $\chi^2_{.05,30} = 43.773$, $\chi^2_{.95,30} = 18.493$.

There are no entries in the table for values of v exceeding 30. However, an approximation, discovered by Wilson and Hilferty,[11] provides adequate answers in most practical cases. Explanations and examples are given below.

Chi-Squared Approximations. Equation (34) has two unknowns, x and α. It is possible to specify the value of α and then to solve the equation for x, or to specify the value of x and then to solve the equation for α. The approximations of Wilson and Hilferty, mentioned above, provide approximate solutions to both problems.

If the value of α is specified, the formula given below can be used to find an approximation of the value of $\chi^2_{\alpha,v}$ which satisfies Eq. (34).

$$\chi^2_{\alpha,v} \approx v\left(z_\alpha\sqrt{\frac{2}{9v}} + 1 - \frac{2}{9v}\right)^3 \tag{35}$$

The value of z_α corresponding to the specified value of α, can be found in the margin of Table 3.2.

If the value of $\chi^2_{\alpha,v}$ is specified, an approximation of the value of α which satisfies Eq. (34) can be found by first using the formula given below to find an approximation of z_α, and then using the normal table—Table 3.2—to find the approximate value of α.

$$z_\alpha \approx \left(\left(\frac{x}{v}\right)^{1/3} - 1 + \frac{2}{9v}\right)\sqrt{\frac{9v}{2}} \tag{36}$$

The approximate value of α is the entry in Table 3.2, corresponding to the marginal value z_α, as obtained in the approximation.

The approximation can be used for values of v which exceed 30, and it steadily increases in accuracy as v increases. As the following examples show, it is quite accurate even when v is as small as 30.

Examples of the Use of the Approximations.   Two examples are presented. The first demonstrates the use of Eq. (35), and the second the use of Eq. (36).

To find the approximate values of $\chi^2_{.05,30}$, $\chi^2_{.95,30}$, first use Table 3.2 to determine the values of z_α, $z_{.05} = 1.645$, $z_{.95} = -1.645$, and then use Eq. (35) to determine the approximations

$$\chi^2_{.05,30} \approx 43.77 \qquad \chi^2_{.95,30} \approx 18.49$$

The approximations agree with the entries in Table 3.4, which shows that the approximation is good even when v is as small as 30.

Approximations of the values of α which satisfy $\chi^2_{\alpha,30} = 43.77$, $\chi^2_{\alpha,30} = 18.49$ can be found by use of Eq. (36). According to the formula, the two approximations of z_α are

$$z_\alpha \approx 1.645 \qquad z_\alpha \approx -1.645$$

and according to Table 3.2, the two approximated values of α are

$$\alpha \approx .05 \qquad \alpha \approx .95$$

Again the approximation agrees with Table 3.4 to a high degree of accuracy.

Uses of the Chi-Squared Distribution.   Applications of the chi-squared distribution frequently occur in conjunction with the normal distribution. Assuming that $X_1, X_2, \ldots, X_n$ are independent random variables, each having the normal distribution, with parameters μ and σ, it follows from Eq. (34) that

$$P\left(\chi^2_{1-\alpha/2,v} < \frac{(n-1)S^2}{\sigma^2} < \chi^2_{\alpha/2,v}\right) = 1 - \alpha \tag{37}$$

in which the values of $\chi^2_{1-\alpha/2,v}$ and $\chi^2_{\alpha/2,v}$ can be found in Table 3.4.

The probability statement in Eq. (37) can be used to construct a confidence interval for the parameter σ, or to test a hypothesis about the value of σ. Procedures for constructing confidence intervals and testing hypotheses are described later in the chapter.

The F Distribution.   The F distribution is derived from the normal distribution, under the following assumptions:

- $X_{11}, X_{12}, \ldots, X_{1n}$ are independent random variables, each having the normal distribution with parameters μ_1 and σ_1
- $X_{21}, X_{22}, \ldots, X_{2m}$ are independent random variables, each having the normal distribution with parameters μ_2 and σ_2

The F ratio is defined in terms of the following quantities:

$$\bar{X}_1 = \frac{1}{n} \sum_{j=1}^{n} X_{1j} \tag{38}$$

$$\bar{X}_2 = \frac{1}{m} \sum_{j=1}^{m} X_{2j} \tag{39}$$

$$S_1^2 = \left(\frac{1}{n-1}\right) \sum_{j=1}^{n} (X_{1j} - \bar{X}_1)^2 \tag{40}$$

$$S_2^2 = \left(\frac{1}{m-1}\right) \sum_{j=1}^{m} (X_{2j} - \bar{X}_2)^2 \qquad (41)$$

as follows: the ratio

$$\frac{S_1^2}{\sigma_1^2} \cdot \frac{\sigma_2^2}{S_2^2} \qquad (42)$$

has the F distribution, with parameters $v_1 = n - 1$ and $v_2 = m - 1$ and

$$\frac{S_2^2}{\sigma_2^2} \cdot \frac{\sigma_1^2}{S_1^2} \qquad (43)$$

has the F distribution, with parameters $v_1 = m - 1$ and $v_2 = n - 1$.

The probability density function of the F distribution is not reproduced here because it is of little interest to practitioners. However, Fig. 3.7, is a drawing of the curve generated by plotting the probability density function of the F distribution—say $f(x)$—as a function of x. Not evident from the drawing, but nevertheless true, is the fact that the region between the curve and the horizontal axis has area 1. The area of the shaded region in Fig. 3.7 is $P(u < X < v)$.

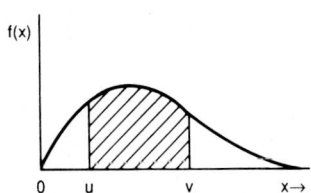

FIGURE 3.7 F-distribution function.

Parameters. The F distribution has two parameters, v_1 and v_2, which can have any positive integer values. Sometimes v_1 and v_2 are called *degrees of freedom.* In any case v_1 is associated with the numerator and v_2 with the denominator in the F ratio, as shown in Eqs. (42) and (43).

Tables of the F *Distribution.* Entries in a table of the F distribution are values of $F_{\alpha, v_1, v_2})$ which satisfy the equation

$$P(X < F_{\alpha, v_1, v_2}) = 1 - \alpha \qquad (44)$$

in which X is a random variable having the F distribution, with parameters v_1 and v_2, and α is a specified value in the interval $(0, 1)$.

Table 3.5, at the end of the chapter, was computed by the author, using IMSL routine FIN.[4] Each entry is a solution of Eq. (44) for values of v_1, v_2, and α, specified in the margins of the table. For example, the entry corresponding to $v_1 = 10$, $v_2 = 12$, $\alpha = .05$ is 2.75.

There are no entries in the table for values of α greater than .05. However, the identity given below can be used to find values of x satisfying Eq. (44) for large values of α—.99, .975, and .95. The identity is

$$P\left(\frac{S_1^2}{\sigma_1^2} \cdot \frac{\sigma_2^2}{S_2^2} < \frac{1}{x}\right) \equiv 1 - P\left(\frac{S_2^2}{\sigma_2^2} \cdot \frac{\sigma_1^2}{S_1^2} < x\right) \qquad (45)$$

As an example of the use of the identity, suppose $n = 11$, $m = 13$, $\alpha = .95$. First use Table 3.5 to find x such that the probability on the right-hand side of the identity is .95. In this step, the F distribution parameters are $v_1 = 12$, $v_2 = 10$, and, according to Table 3.5, the value of x is 2.91. The F probability on the left-hand side of the identity has the value .05 and its parameters are $v_1 = 10$, $v_2 = 12$. Hence,

$$P\left(\frac{S_1^2}{\sigma_1^2} \cdot \frac{\sigma_2^2}{S_2^2} < \frac{1}{2.91}\right) = .05$$

Uses of the F *Distribution.*   Applications of the F distribution always occur in conjunction with the normal distribution. Assuming that $X_{11}, X_{12}, \ldots, X_{1n}$ and $X_{21}, X_{22}, \ldots, X_{2m}$ are two independent sets of independent random variables, each having the normal distribution, with parameters μ_1 and σ_1, and μ_2 and σ_2, it follows from Eqs. (42) and (43) that

$$P\left(\frac{1}{F_{\alpha/2, m-1, n-1}} < \frac{S_1^2}{\sigma_1^2} \cdot \frac{\sigma_2^2}{S_2^2} < F_{\alpha/2, n-1, m-1}\right) = 1 - \alpha \tag{46}$$

in which the values of $F_{\alpha/2, n-1, m-1}$ and $F_{\alpha/2, m-1, n-1}$ can be found in Table 3.5.

The probability statement in Eq. (46) can be used to construct a confidence interval for the ratio σ_1/σ_2, or to test a hypothesis about the value of σ_1/σ_2. Procedures for constructing confidence intervals and testing hypotheses are described later in the chapter.

ESTIMATION OF PARAMETERS

One of the advantages of using probability distributions in the modeling of engineering process is that the distribution parameters serve as surrogates for important process parameters. In many practical cases, a complex process can be characterized in terms of one or two parameters, and in such cases the values of the parameters need to be estimated with as much accuracy as possible.

The estimate of the value of a parameter is always a function of observed data. The formulas used to compute estimates vary from case to case, but appropriate formulas are known for most practical cases. It is also possible to evaluate the accuracy of the estimate. The accuracy evaluation consists of an interval, which is likely to contain the unknown value of the parameter. Such intervals are sometimes called confidence intervals and sometimes interval estimates. Estimates consisting of a single number are called *point estimates.*

This section contains the formulas that are used to compute point and interval estimates for (1) the binomial parameter p and (2) the normal parameters μ and σ. It begins, however, with a precise definition of the interval estimate.

Point and Interval Estimates.   A point estimate of the value of a parameter is simply a number, determined by a specified formula. The usefulness of a point estimate is limited unless the user has some way of estimating its accuracy.

An *interval estimate* of a parameter is an interval whose end points are numbers, determined by specified formulas. The formulas are chosen so that the assertion—the true, but unknown, value of the parameter is included in the interval—has a specified probability of being true. Another name for an interval estimate is *confidence interval,* and the probability that the interval does contain the true value of the parameter is called the *confidence level* of the interval.

The usefulness of an interval estimate depends on its length—the shorter the better—but, in some practical cases, the problem of finding intervals of shortest possible length is not completely solved. Intervals determined by methods described in this chapter are not always the shortest possible, but they are close enough to be useful to practitioners.

Estimation of the Binomial Parameter **p.**   Let X be the number of successes (or failures) in n independent trials, each trial having probability p of being a success (or failure). The point estimator of the parameter p, say $\hat{p}$, is

$$\hat{p} = \frac{X}{n} \tag{47}$$

An interval estimate—confidence interval—of p can be determined either by use of the binomial distribution or by use of the normal approximation to the binomial distribution. Intervals determined by use of the binomial distribution have the advantage of accuracy and have the disadvantage of being hard to compute. On the other hand, the approximate intervals are inaccurate, especially when n is small, but easy to compute. Both methods are affected by the discreteness of the binomial distribution.

Effect of Discreteness.   There are two sources of discreteness in the binomial distribution; the sample size and the observed number of successes are both integers. The consequence of this discreteness is that it is impossible to find an interval, with confidence level exactly equal to a specified value, say $1 - \alpha$, for every possible value of α. For most values of α, the best that can be done is to construct an interval whose confidence level is not less than $1 - \alpha$.

Interval Estimate of p *Using the Binomial Distribution.*   Three pieces of information are needed to determine the interval estimate of p: the sample size, say n; the significance level of the estimate, say $1 - \alpha$; and the data, in the form of the observed number of successes (failures), say X.

In practice, the values of n and α are specified first, and the value of X is then observed. Since the observed value of X has only $n + 1$ possible values, it is possible to compute, for specified values of n and α, all $n + 1$ possible $(1 - \alpha)$-level interval estimates, prior to the observation of X. Then, when the observed value of X is known, the appropriate interval is selected from the precomputed list.

These $n + 1$ intervals are computed in a two-step procedure. First, for specified values of n and α, find the values of p_j which satisfy

$$\sum_{k=0}^{j} \binom{n}{k} p_j^k (1 - p_j)^{n-k} = \frac{\alpha}{2} \qquad j = 0, 1, 2, \ldots, n - 1 \qquad (48)$$

$$p_n = 1$$

In the second step, the $(1 - \alpha)$-level interval estimate for p is constructed, using the observed value of X, as follows: if the observed value of X is j, the $(1 - \alpha)$-level interval estimate of p is

$$(1 - p_{n-j}, p_j) \qquad j = 0, 1, 2, \ldots, n \qquad (50)$$

The shortcoming of this method is that Eqs. (48) are hard to solve. Computer solutions to this kind of problem are becoming available, and in the future such calculations will be routine, even on hand-held calculators.

Example of Interval Estimate of p *Using the Binomial Distribution.*   When n is 3 and α is .05, the values of p_j, which satisfy Eqs. (48) are $p_0 = .70760$, $p_1 = .90569$, $p_2 = .99160$, $p_3 = 1$, and the four possible .95-level interval estimates, constructed according to the rule in Eq. (50), are

- If X is 0, the .95-level interval estimate is $(0, .7076)$
- If X is 1, the .95-level interval estimate is $(.0084, .9057)$
- If X is 2, the .95-level interval estimate is $(.0943, .9916)$
- If X is 3, the .95-level interval estimate is $(.2924, 1)$

The observed value of X will be 0, 1, 2, or 3, and the .95-level interval estimate for p is selected from the above list. No matter which interval is chosen, the user can assert that the interval contains the true value of p, and the probability that the assertion is true is at least .95.

Approximate Interval Estimate of p.   Three pieces of information are needed to determine the approximate interval estimate of p: the sample size, say n; the significance level of the estimate, say $(1 - \alpha)$; and the data, in the form of the observed number of successes (failures), say X.

One of the standard textbook computations for the end points of an approximate $(1 - \alpha)$-level interval estimate of p is

$$\frac{2X + z^2_{\alpha/2} \pm z_{\alpha/2}\sqrt{4X\left(1 - \frac{X}{n}\right) + z^2_{\alpha/2}}}{2(n + z^2_{\alpha/2})} \tag{51}$$

in which $z_{\alpha/2}$ is the entry in the margin of Table 3.2, corresponding to the table entry $1 - \alpha/2$.

Example of an Approximate Interval Estimate of p. Suppose that $n = 30$, $\alpha = .05$, and that 12 successes have been observed in the 30 independent trials.

From Eq. (51), the approximate, .95-level interval estimate is (.246, .578), and the .95-level interval estimate—computed by the IMSL routine BINES[4]—is (.227, .594).

The IMSL interval is based on the binomial distribution, and a user can assert that it contains the true value of p, knowing that the probability that the assertion is true is at least .95. Since the approximate interval is shorter, a good guess is that the true level of confidence in the approximate interval is less than .95. In general it can be said that percent error in the approximation is smallest near the center of the binomial distribution ($p = .5$) and larger at the tails of the distribution (p near 0 or near 1). And, in all cases, the percent errors decrease as n increases.

Estimation of the Normal Parameters μ ***and*** σ. Suppose that $X_1, X_2, \cdots, X_n$ are independent random variables, each having the normal distribution, with parameters, μ and σ. The point estimator of μ is

$$\overline{X} = \frac{1}{n}\sum_{j=1}^{n} X_j \tag{52}$$

the point estimator of σ^2 is

$$S^2 = \frac{1}{n-1}\sum_{j=1}^{n}(X_j - \overline{X})^2 \tag{53}$$

and the point estimator of σ is S.

Interval estimates of μ depend on whether or not the value of σ is known. When the value of σ is known, the interval estimate of μ is based on the standard normal distribution, and, when σ is not known, the interval estimate is based on the t distribution. In all cases, the interval estimate of σ is based on the chi-squared distribution.

Hand evaluations of S^2, using Eq. (53), can be simplified by application of the following identity:

$$\sum_{j=1}^{n}(X_j - \overline{X})^2 \equiv \sum_{j=1}^{n}X_j^2 - \frac{1}{n}\left(\sum_{j=1}^{n}X_j\right)^2 \tag{54}$$

Either side of the identity can be used to compute the value of S^2, but using the right side is arithmetically simpler, because fewer subtractions are required. However, the advantage of simplicity is somewhat offset by the fact that the expression on the right, being a difference, is sensitive to rounding. A large error can result from what many users might regard as insignificant rounding. To avoid such errors, carry the maximum number of decimals through to the end of the calculation.

Interval Estimate of μ—σ *Known.* When the value of σ is known, the $(1 - \alpha)$-level interval estimate of μ is derived from the following probability statement:

$$P\left(-z_{\alpha/2} < \frac{\sqrt{n}(\overline{X} - \mu)}{\sigma} < z_{\alpha/2}\right) = 1 - \alpha \tag{55}$$

in which $z_{\alpha/2}$ is the entry in the margin of Table 3.2, corresponding to the table entry $1 - \alpha/2$.

The statement in Eq. (55) is true because the quantity between the inequality signs has the standard normal distribution. Moreover, it remains true for any legitimate rearrangement of the inequality, and it is possible, by means of such rearrangements, to isolate μ between the inequality signs, thereby obtaining the following $(1 - \alpha)$-level interval estimate of μ:

$$\left(\bar{X} - z_{\alpha/2} \frac{\sigma}{\sqrt{n}}, \bar{X} + z_{\alpha/2} \frac{\sigma}{\sqrt{n}} \right) \tag{56}$$

Examination of Eq. (56) reveals that the length of the confidence interval decreases as n, sample size, increases. This implies that the precision of estimation increases as sample size increases. Also the length of the interval increases as σ increases. This implies that if there is large variation in the data, large samples may be needed to obtain acceptable precision of estimation.

Example of an Interval Estimate of μ—σ Known. Suppose that σ is 1.5, and that a .95-level interval estimate of μ is to be computed, using $n = 15$, $\bar{X} = 20$. Since α is .05, the value of $z_{\alpha/2}$, found in Table 3.2, is $z_{.025} = 1.96$, and the .95-level interval estimate of μ is

$$\left[\frac{20}{15} - \frac{(1.96)(1.5)}{\sqrt{15}}, \frac{20}{15} + \frac{(1.96)(1.5)}{\sqrt{15}} \right] = (.574, 2.092)$$

Interval Estimate of μ—σ Unknown. When the value of σ is unknown, the $(1 - \alpha)$-level confidence interval for μ is derived from the following probability statement:

$$P\left(-t_{\alpha/2,v} < \frac{\sqrt{n}(\bar{X} - \mu)}{S} < t_{\alpha/2,v} \right) = 1 - \alpha \tag{57}$$

in which $t_{\alpha/2,v}$ is the entry in Table 3.3 corresponding to the column heading $\alpha/2$ and the row heading v.

The statement in Eq. (57) is true because the quantity between the inequality signs has the t distribution, with parameter $v = n - 1$. Moreover, it remains true for any legitimate rearrangement of the inequality, and it is possible, by means of such rearrangements, to isolate μ between the inequality signs, thereby obtaining the following $(1 - \alpha)$-level interval estimate of μ:

$$\left(\bar{X} - t_{\alpha/2,v} \frac{S}{\sqrt{n}}, \bar{X} + t_{\alpha/2,v} \frac{S}{\sqrt{n}} \right) \tag{58}$$

Since the t value in Eq. (58) decreases as n increases, it follows that the length of the confidence interval decreases as n, sample size, increases. This implies that the precision of estimation increases as sample size increases. Also the length of the interval increases as S increases. This implies that if there is large variation in the data, large samples may be needed to obtain acceptable precision of estimation.

Example of an Interval Estimate of μ—σ Unknown. Suppose that a .95-level interval estimate of μ is to be computed from these four observations: $X_1 = 8.2$, $X_2 = 8.3$, $X_3 = 8.4$, $X_4 = 8.2$.

First,

$$\bar{X} = \tfrac{1}{4}(8.2 + 8.3 + 8.4 + 8.2) = 8.275$$

and

$$S^2 = \tfrac{1}{3}[8.2^2 + 8.3^2 + 8.4^2 + 8.2^2 - \tfrac{1}{4}(33.1)^2] = .00917$$

Second, since v is 3 and α is .05, it follows from Table 3.3 that $t_{.025,3} = 3.182$, and the .95-level interval estimate of μ is

$$\left(\frac{33.1}{4} - 3.182\sqrt{\frac{.00917}{4}}, \frac{33.1}{4} + 3.182\sqrt{\frac{.00917}{4}} \right) = (8.12, 8.43)$$

The formula for evaluating S^2 is, as previously mentioned, sensitive to rounding. If, in the previous example, the number 33.1—the sum of the four observations—were rounded to 33.0, the result of the calculation would be 1.68 instead of .00917. This is a gross error, but it can be avoided by carrying a maximum number of decimals through to the end of the calculation.

Interval Estimate of σ. The $(1 - \alpha)$-level interval estimate of σ is derived from the following probability statement:

$$P\left(\chi^2_{1-\alpha/2,v} < \frac{(n-1)S^2}{\sigma^2} < \chi^2_{\alpha/2,v} \right) = 1 - \alpha \tag{59}$$

in which $\chi^2_{1-\alpha/2,v}$ and $\chi^2_{\alpha/2,v}$ are the entries in Table 3.4 corresponding to the row heading v and the column headings $1 - \alpha/2$ and $\alpha/2$.

The statement in Eq. (59) is true because the quantity between the inequality signs has the chi-squared distribution, with parameter $v = n - 1$. Moreover, it remains true for any legitimate rearrangement of the inequality, and it is possible, by means of such rearrangements, to isolate σ between the inequality signs, thereby obtaining the following $(1 - \alpha)$-level confidence interval:

$$\sqrt{\frac{(n-1)S^2}{\chi^2_{\alpha/2,v}}}, \sqrt{\frac{(n-1)S^2}{\chi^2_{1-\alpha/2,v}}} \tag{60}$$

The $(1 - \alpha)$-level interval estimate of σ^2 is obtained by squaring the end points of the interval estimate of σ in Eq. (60).

Since the χ^2 value in Eq. (60) increases as n increases it seems plausible, and is in fact true, that the length of the confidence interval decreases as sample size n increases. This implies that the precision of estimation increases as sample size increases. Also the length of the interval increases as S increases. This implies that if there is large variation in the data, large samples may be needed to obtain acceptable precision of estimation.

Example of an Interval Estimate of σ. Suppose that a .95-level interval estimate of σ is to be computed from the four observations $X_1 = 8.2$, $X_2 = 8.3$, $X_3 = 8.4$, $X_4 = 8.2$. First,

$$S^2 = \frac{1}{3}[8.2^2 + 8.3^2 + 8.4^2 + 8.2^2 - \frac{1}{4}(33.1)^2] = .00917$$

Second, since v is 3 and α is .05, it follows from Table 3.4 that $\chi^2_{.025,3} = 9.348$, $\chi^2_{.975,3} = .216$, and the .95-level interval estimate of σ is

$$\left[\sqrt{\frac{3(.00917)}{9.348}}, \sqrt{\frac{3(.00917)}{.216}} \right] = (.054, .357)$$

The .95-level interval estimate of σ^2 is obtained by squaring the end points of the interval for σ: (.003, .127).

Interval Estimate of the Ratio σ_1/σ_2 or σ_2/σ_1. In this case there are two normal populations, the first having the parameters μ_1 and σ_1, and the second having the parameters μ_2 and σ_2. The $(1 - \alpha)$-level interval estimates of the ratios σ_1/σ_2 and σ_2/σ_1 are both derived from the probability statement

$$P\left(\frac{1}{F_{\alpha/2,v_2,v_1}} < \frac{S_1^2}{\sigma_1^2} \cdot \frac{\sigma_2^2}{S_2^2} < F_{\alpha/2,v_1,v_2}\right) = 1 - \alpha \tag{61}$$

in which $v_1 = n - 1$, $v_2 = m - 1$, and $F_{\alpha/2,v_2,v_1}$ and $F_{\alpha/2,v_1,v_2}$ are entries in Table 3.5.

The statement in Eq. (61) is true because the quantity between the inequality signs has the F distribution, with parameters $v_1 = n - 1$ and $v_2 = n - 1$. Moreover, it remains true for any legitimate rearrangement of the inequality, and it is possible, by means of such rearrangements, to isolate either σ_1/σ_2 or σ_2/σ_1 between the inequality signs, thereby obtaining the following $(1 - \alpha)$-level interval estimate of σ_1/σ_2:

$$\left(\sqrt{\frac{1}{F_{\alpha/2,v_1,v_2}} \cdot \frac{S_1^2}{S_2^2}}, \sqrt{F_{\alpha/2,v_2,v_1} \cdot \frac{S_1^2}{S_2^2}}\right) \tag{62}$$

and the following $(1 - \alpha)$-level confidence interval for σ_2/σ_1:

$$\left(\sqrt{\frac{1}{F_{\alpha/2,v_2,v_1}} \cdot \frac{S_2^2}{S_1^2}}, \sqrt{F_{\alpha/2,v_1,v_2} \cdot \frac{S_2^2}{S_1^2}}\right) \tag{63}$$

Although it is not immediately obvious from examination of Eqs. (62) and (63), it is nevertheless true that increasing the values of n and m does shorten the length of the confidence interval. This implies that the precision of estimation increases as the sample sizes increase.

Example of Interval Estimates of the Ratios σ_1/σ_2 and σ_2/σ_1. Suppose that .95-level interval estimates of σ_1/σ_2 and σ_2/σ_1 are to be computed from the two sets of observations given below:

- $X_{1,1} = 8.2$, $X_{1,2} = 8.3$, $X_{1,3} = 8.4$, $X_{1,4} = 8.2$
- $X_{2,1} = 5.4$, $X_{2,2} = 5.9$, $X_{2,3} = 6.1$

Then

$$S_1^2 = \tfrac{1}{3}[8.2^2 + 8.3^2 + 8.4^2 + 8.2^2 - \tfrac{1}{4}(33.1)^2] = .00917$$

and

$$S_2^2 = \tfrac{1}{2}[5.4^2 + 5.9^2 + 6.1^2 - \tfrac{1}{3}(17.4)^2] = .13$$

Since v_1 is 3, v_2 is 2, and α is .05, it follows from Table 3.5 that $F_{.025,2,3} = 16.04$, $F_{.025,3,2} = 39.17$, that the .95-level confidence interval for σ_1/σ_2 is

$$\left[\sqrt{\left(\frac{1}{39.17}\right)\left(\frac{.00917}{.13}\right)}, \sqrt{(16.04)\left(\frac{.00917}{.13}\right)}\right] = (.042, 1.064)$$

and that the .95-level confidence interval for σ_2/σ_1 is

$$\left[\sqrt{\left(\frac{1}{16.04}\right)\left(\frac{.13}{.00917}\right)}, \sqrt{(39.17)\left(\frac{.13}{.00917}\right)}\right] = (.940, 23.6)$$

TESTS OF HYPOTHESES

A statistical hypothesis is a claim that the value of a parameter in a statistical model lies in a specified interval, or has a specified value. A test of a hypothesis is a formal procedure, often a simple formula, which always yields a conclusion: the claim is accepted or the claim is rejected. In all cases, the tests depend on data, often collected for the express purpose of testing the hypothesis.

Test procedures have been developed for many special cases. However, effective use of the procedures is not possible unless the user understands the concepts and is familiar with the terminology. Both concepts and terminology are explained and described in the following sections: (1) Specification of Hypotheses, (2) The Two Kinds of Error, (3) Tests about the Binomial Parameter p, (4) Tests about the Parameter μ—Normal Distribution, and (5) Tests about the Parameter σ—Normal distribution.

Specification of Hypotheses. The claim is called the *null hypothesis* and the converse of the claim is called the *alternate hypothesis*. The notations H_0 and H_a are used to denote the null and the alternate hypotheses respectively.

There are three standard forms of hypothesis tests. These are given below, where θ denotes a parameter of unknown value and θ_0 denotes the hypothesized value of θ:

$$H_0: \theta \leq \theta_0;\ H_a: \theta > \theta_0 \tag{64}$$

$$H_0: \theta = \theta_0;\ H_a: \theta \neq \theta_0 \tag{65}$$

$$H_0: \theta \geq \theta_0;\ H_a: \theta < \theta_0 \tag{66}$$

These hypotheses are called left-sided (64), two-sided (65), and right-sided (66). Sometimes the left-sided and right-sided tests are called *one-sided tests.*

In practical cases, tests of hypotheses are used to select courses of action. For example, suppose that the cost of installing a new process is justified only if the value of a certain parameter, say θ, does not exceed a specified value, say θ_0. In such a case, a left-sided test is used to make the decision.

The Two Kinds of Error. Statistical tests do not eliminate the possibility of false conclusions, and there are in fact two distinct ways in which a test of a hypothesis can lead to a false conclusion:

1. H_0 can be rejected when H_0 is true
2. H_0 can be accepted when H_0 is false

The error described in 1 is called an *error of the first kind,* and the error in 2 is called an *error of the second kind.*

Tests should be constructed to provide protection against both kinds of error. However, in most applications the consequences of error are not the same for both kinds of error, and the user may wish to take account of this fact in the construction of tests. Suppose, as in the example mentioned previously, it would be profitable to install a new process only if the value of a certain process parameter exceeds a specified value. In such a case, the two kinds of error would be as follows: the new process might be installed when the old is as good or better, or the old process might be retained when the new is better. If either kind of error occurs, a loss will be incurred, but it is unlikely that the losses will be exactly the same for both kinds of error.

Protection against Errors. It is always possible to construct tests which satisfy the condition

$$P(H_0 \text{ accepted}|\theta = \theta_0) = 1 - \alpha \tag{67}$$

and any test which satisfies Eq. (67) protects against an error of the first kind, because α can have any specified value.

A one-sided test protects against an error of the second kind if it satisfies the condition

$$P(H_0 \text{ accepted}|\theta = \theta_1) = \beta \tag{68}$$

in which the value of θ_1, specified by the user, is a value of θ for which H_0 is false. Similarly a two-sided test protects against an error of the second kind if it satisfies the two conditions

$$P(H_0 \text{ accepted}|\theta = \theta_1) = \beta_1 \tag{69}$$

$$P(H_0 \text{ accepted}|\theta = \theta_2) = \beta_2 \tag{70}$$

in which the values of θ_1 and θ_2, both specified by the user, are values of θ for which H_0 is false.

θ_1 and θ_2 are threshold variables whose values are specified by the user, depending on the application. For example, a user contemplating a left-sided test might decide that an error of the second kind has no significant consequences unless the true value of θ exceeds θ_0 by a threshold amount, say $\theta_1 - \theta_0$. After specifying the value of θ_1, such a user could construct a test satisfying Eqs. (69) and (70), and, because the values of α and β are both specifiable, the test would protect against both kinds of error.

Tests about the Binomial Parameter p. The tests described in this section are constructed by using the normal approximation to the binomial distribution. This is done for the sake of simplicity, but there is a risk that tests constructed by approximation do not satisfy all specifications exactly.

In general, the accuracy of the approximation increases as the sample size, say n, increases. However, for a fixed value of n, accuracy is greatest when p is .5 and least when p is near 0 or 1. A rule of thumb is that n should be at least 30 when p is .5, and at least 100 when p is as small as .01 or as large as .99.

One-Sided Tests on p. The one-sided hypotheses are

$$H_0: p \leq p_0; \ H_a: p > p_0 \tag{71}$$

$$H_0: p \geq p_0; \ H_a: p < p_0 \tag{72}$$

Data used in the testing of hypotheses are acquired by observation. In the case of binomial tests, there is an event whose probability of occurrence is p, and the user counts the number, say X, of occurrences of the event in a series of n independent trials. The number X is the data needed to carry out the test.

A left-sided test does not exist unless

$$p_0^n \leq \alpha \tag{73}$$

However, for all values of n, α, and p_0 which satisfy Eq. (73) the left-sided test is the following:

$$\text{If } X \leq \lfloor np_0 + z_\alpha \sqrt{np_0(1 - p_0)} \rfloor, \text{ accept } H_0; \text{ otherwise accept } H_a \tag{74}$$

The notation $\lfloor y \rfloor$ denotes the largest integer which does not exceed y, and the value of z_α can be found in Table 3.2. This test also satisfies Eq. (67), and, therefore, protects against errors of the first kind.

A right-sided test does not exist unless

$$(1 - p_0)^n \leq \alpha \tag{75}$$

However, for all values of n, α, and p_0 which satisfy Eq. (75), the right-sided test is

$$\text{If } X \geq \lceil np_0 - z_\alpha \sqrt{np_0(1 - p_0)} \rceil, \text{ accept } H_0; \text{ otherwise accept } H_a \tag{76}$$

The notation $\lceil y \rceil$ denotes the smallest integer which is not less than y, and the value of z_α can be found in Table 3.2. This test satisfies Eq. (67), and, therefore, protects against errors of the first kind.

Protection against Errors of the Second Kind.   The tests defined in Eq. (74) and (76) exist for all possible values of n. They protect against errors of the first kind no matter what the value of n, and they will protect against errors of the second kind if the value of n is sufficiently large. The equations given below can be used to determine a value of n large enough to protect against an error of the second kind.

$$\frac{\lfloor np_0 + z_\alpha\sqrt{np_0(1 - p_0)} \rfloor - np_1 + .5}{\sqrt{np_1(1 - p_1)}} = z_{1-\beta} \tag{77}$$

$$\frac{\lceil np_0 - z_\alpha\sqrt{np_0(1 - p_0)} \rceil - np_1 - .5}{\sqrt{np_1(1 - p_1)}} = z_\beta \tag{78}$$

If the test is left-sided, substitute the specified values of α, β, p_0, and p_1 in Eq. (77), and then solve for n. If the test is right-sided, substitute the specified values of α, β, p_0, and p_1 in Eq. (78), and then solve for n. If, in either case, it happens that the determined value of n is not an integer, round up to the nearest integer.

The value of z_β, given β, or the value of β, given z_β, can be found in Table 3.2, provided that the value of z_β exceeds zero. Negative values of z_β can be converted to positive values by use of the fact that $z_\beta = -z_{1-\beta}$.

Example of a One-Sided Test for p.   Suppose that a manufacturer claims that a unit of product, produced on a certain machine, has probability .01 of being defective. A test of the claim could be in the form of a left-sided test of hypothesis as follows:

$$H_0: p \leq .01; H_a: p > .01$$

Data for the test consist of the number, say X, of defective items found in a production run of n items, and, if the user has specified that $n = 200$ and $\alpha = .05$, then according to Eq. (74), the test is constructed as follows:

$$z_{.05} = 1.645$$

$$\lfloor 200(.01) + z_{.05}\sqrt{200(.010)(.99)} \rfloor = \lfloor 4.31 \rfloor = 4$$

If $X \leq 4$ accept H_0;otherwise accept H_a

This test is approximate in the sense that it was constructed by using the normal approximation to the binomial distribution. The true probability of an error of the first kind, computed using IMSL routine BINDF,[4] is .052, which slightly exceeds the specified value of .05.

Example Continued.   Suppose that the manufacturer specifies that p_1 is .02, and that β should be .05. In effect the manufacturer is demanding that the test be powerful enough to distinguish between two processes, one producing 1 percent defective and the other producing 2 percent defective. Intuition suggests that a sample of size 200, as used in the first part of the example, is not large enough to distinguish between values of p that are so close together. Intuition is verified by substituting the specified values of n, α, p_0, and p_1 in the left side of Eq. (77), and deducing that $z_{1-\beta} \approx .25$, $\beta \approx .60$. Protection against an error of the second kind is inadequate, because the probability of accepting H_0 when $p = .02$ is approximately .60.

To find a value of n sufficiently large to provide protection at a specified level, substitute the specified values of α, β, p_0, and p_1 in Eq. (77), thereby obtaining the equation

$$\frac{\lfloor n(.01) + 1.645\sqrt{n(.01)(.99)} \rfloor - n(.02) + .5}{\sqrt{n(.02)(.98)}} = -1.645$$

It is easy to verify that the value of n satisfying the equation is between 1465 and 1466. Therefore, a sample of size 1466 is needed to provide protection against an error of either kind at the specified level.

The revised test is constructed by using Eq. (74),

$$Z_{.05} = 1.645$$

$$\lfloor 1466(.01) + 1.645 \sqrt{1466(.01)(.99)} \rfloor = \lfloor 20.92 \rfloor = 20$$

If $X \leq 20$ accept H_0; otherwise accept H_a

The IMSL routine BINDEF can be used to determine that the true probability of an error of the first kind is .07 and that the true probability of an error of the second kind is .04. The true error of the first kind is slightly higher than specified, but the test would still be usable in a practical case.

Two-Sided Tests for p.   The two-sided hypothesis is

$$H_0: p = p_0; \; H_a: p \neq p_0 \tag{79}$$

Data used in the testing of the hypothesis are acquired by observation. In the case of binomial tests, there is an event whose probability of occurrence is p, and the user counts the number, say X, of occurrences of the event in a series of n independent trials. The number X is the data needed to carry out the test.

No meaningful two-sided test exists unless

$$p_0^n \leq \alpha \qquad \text{and} \qquad (1 - p_0)^n \leq \alpha \tag{80}$$

However, for all values of n, α, and p_0 which satisfy Eq. (80), the two-sided test is

$$\text{If} \; \left| \frac{(X - np_0)}{\sqrt{np_0(1 - p_0)}} \right| \leq z_{\alpha/2}, \text{ accept } H_0; \text{ otherwise accept } H_a \tag{81}$$

The inequality in Eq. (81) can also be written in the form

$$\lceil np_0 - z_{\alpha/2}\sqrt{np_0(1 - p_0)} \rceil \leq X \leq \lfloor np_0 + z_{\alpha/2}\sqrt{np_0(1 - p_0)} \rfloor \tag{82}$$

The notation $\lceil y \rceil$ denotes the smallest integer which is not less than y, the notation $\lfloor y \rfloor$ denotes the largest integer which does not exceed y, and the value of z_α can be found in Table 3.2. This test satisfies Eq. (67) and, therefore, protects against errors of the first kind.

Protection against Errors of the Second Kind.   The test defined in Eq. (82) exists for all possible values of n. It protects against errors of the first kind no matter what the value of n, and it will protect against errors of the second kind if the value of n is sufficiently large. The two pairs of equations given below can be used to determine a value of n large enough to protect against an error of the second kind at any specified level.

$$\frac{\lfloor np_0 + z_{\alpha/2}\sqrt{np_0(1 - p_0)} \rfloor - np_1 + .5}{\sqrt{np_1(1 - p_1)}} = z_{\eta - \beta_1} \tag{83}$$

$$\frac{\lceil np_0 - z_{\alpha/2}\sqrt{np_0(1 - p_0)} \rceil - np_1 - .5}{\sqrt{np_1(1 - p_1)}} = z_\eta \tag{84}$$

and

$$\frac{[np_0 + z_{\alpha/2}\sqrt{np_0(1 - p_0)}] - np_2 + .5}{\sqrt{np_2(1 - p_2)}} = z_\eta \tag{85}$$

$$\frac{[np_0 - z_{\alpha/2}\sqrt{np_0(1 - p_0)}] - np_2 - .5}{\sqrt{np_2(1 - p_2)}} = z_{\eta + \beta_2} \tag{86}$$

The value of n can be determined in two steps. First, substitute the specified values of α, β_1, p_0, and p_1 in Eqs. (83) and (84), and then solve for n and η. Second, substitute the specified values of α, β_2, p_0, and p_2 in Eqs. (85) and (86), and again solve for n and η. Each pair of equations is solved separately, and the solution values of n and η need not be the same for both pairs of equations. However, the value of n to be used in the test is the larger of the two solution values.

It is also helpful to remember the following:

$$\text{If } z_\eta \leq -3.5, \eta \approx 1; \text{ if } z_\eta \geq 3.5, \eta \approx 0$$

And in all cases, the value of z_β, given β, or the value of β, given z_β, can be found in Table 3.2, provided that the value of z_β exceeds zero. Negative values of z_β can be converted to positive values by use of the fact that $z_\beta = -z_{1-\beta}$.

Example of a Two-Sided Test for p.   Suppose that a manufacturer claims that a unit of product, produced on a certain machine, has probability p of being defective, and that the value of p is in the interval (.005, .015). A test of the claim could be in the form of a two-sided test of hypothesis as follows:

$$H_0: p = .01; H_a: p \neq .01$$

Data for the test consist of the number, say X, of defective items found in a production run of n items, and, if the user has specified that $n = 300$ and $\alpha = .05$, then according to Eq. (82), the test is constructed as follows:

$$z_{.025} = 1.96$$

$$[300(.01) - z_{.025}\sqrt{300(.010)(.99)}] = \lceil -.37 \rceil = 0$$

$$[300(.01) + z_{.025}\sqrt{300(.010)(.99)}] = \lfloor 6.37 \rfloor = 6$$

$$\text{If } 0 \leq X \leq 6 \text{ accept } H_0; \text{ otherwise accept } H_a$$

This test is approximate in the sense that it was constructed by using the normal approximation to the binomial distribution. The true probability of an error of the first kind, computed using the IMSL routine BINDF,[4] is .032.

Example Continued.   In this example, the manufacturer has specified that p_1 is .015 and p_2 is .005, and could specify that β_1 should be .02 and β_2 should be .03. Intuition suggests that a sample of 300, as specified in the first part of the example, is not large enough to distinguish between values of p so close together. Intuition is verified by first substituting the specified values of n, α, p_0, and p_1 in the left side of Eqs. (83) and (84) and deducing that $z_{\eta - \beta_1} \approx .95$, $\eta - \beta_1 \approx .17$; $z_\eta \approx -2.37$, $\eta \approx .99$, $\beta_1 \approx .72$, and second substituting the specified values of n, α, p_0, and p_2 in the left side of Eqs. (85) and (86) and deducing that $z_\eta \approx 4.09$, $\eta \approx 0$, $z_{\eta + \beta_2} \approx -1.64$, $\eta + \beta_2 \approx .95$, $\beta_2 \approx .95$. The large values of β_1 and β_2 show that distinguishing between values of p ranging from .005 to .015 is almost impossible with a sample size of only 300.

To find a value of n which is sufficiently large to provide the specified protection, substitute the specified values of α, β_1, p_0, and p_1 in Eqs. (83) and (84), and the specified values of α, β_2, p_0, and p_2 in Eqs. (85) and (86), thereby obtaining the four equations

$$\frac{\lfloor n(.01) + 1.96\sqrt{n(.01)(.99)}\rfloor - n(.015) + .5}{\sqrt{n(.015)(.985)}} = z_{\eta - .02} \tag{87}$$

$$\frac{\lceil n(.01) - 1.96\sqrt{n(.01)(.99)}\rceil - n(.015) - .5}{\sqrt{n(.015)(.985)}} = z_{\eta} \tag{88}$$

and

$$\frac{\lfloor n(.01) + 1.96\sqrt{n(.01)(.99)}\rfloor - n(.005) + .5}{\sqrt{n(.005)(.995)}} = z_{\eta} \tag{89}$$

$$\frac{\lceil n(.01) - 1.96\sqrt{n(.01)(.99)}\rceil - n(.015) - .5}{\sqrt{n(.015)(.985)}} = z_{\eta + .03} \tag{90}$$

The values of n can be determined by trial and error. If, in the first pair of equations, n is 7768, the value of η is immediately determined from Eq. (88): $z_{\eta} = -5.22$, implying that $\eta \approx 1$. Then Eq. (87) reduces to

$$-2.06 = z_{.98} = -z_{.02}$$

which, according to Table 3.2, is very nearly true. Similarly, if, in the second pair of equations, n is 4158, the value of η is immediately determined from Eq. (89): $z_{\eta} = 7.41$, implying that $\eta \approx 0$. Then Eq. (90) reduces to

$$1.92 = z_{.03}$$

which, according to Table 3.2, is very nearly true. Therefore, a sample of size 7768 is needed to provide protection against an error of either kind at the specified levels.

The revised test is constructed by using Eq. (82):

$$z_{.025} = 1.96$$

$$\lfloor 7768(.01) + 1.96\sqrt{7768(.01)(.99)}\rfloor = \lfloor 94.86\rfloor = 94$$

$$\lceil 7768(.01) - 1.96\sqrt{7768(.01)(.99)}\rceil = \lceil 60.49\rceil = 61$$

If $61 \le X \le 94$ accept H_0; otherwise accept H_a

The IMSL routine BINDEF can be used to determine that the true probability of an error of the first kind is .06, that the true value of β_1 is .02, and that the true probability of β_2 is .00.

Tests about the Parameter μ—Normal Distribution.   The normal distribution has two parameters, μ and σ, and tests about the value of μ depend on the value of σ. If the value of σ is known, the tests are based on the normal distribution, and if the value of σ is not known, the tests are based on the t distribution. The dependence on σ makes it difficult to provide protection against errors of the second kind when the value of σ is unknown.

In all cases there are three forms of test: left-sided, two-sided, and right-sided. The one-sided tests are described first.

One-Sided Tests of μ.   The one-sided hypotheses are

$$H_0: \mu \leq \mu_0; \; H_a: \mu > \mu_0 \tag{91}$$

$$H_0: \mu \geq \mu_0; \; H_a: \mu < \mu_0 \tag{92}$$

Data used in the tests consist of n observed values of a random variable, say X, having the normal distribution, with parameters μ and σ. Estimates of μ and σ^2, computed from the data, are

$$\bar{X} = \frac{1}{n} \sum_{j=1}^{n} X_j \tag{93}$$

$$S^2 = \frac{1}{n-1} \sum_{j=1}^{n} (X_j - \bar{X})^2 \tag{94}$$

If the value of σ is known, the one-sided tests—first left-sided, then right-sided—are

$$\text{If } \bar{X} \leq \mu_0 + z_\alpha \frac{\sigma}{\sqrt{n}}, \text{ accept } H_0; \text{ otherwise accept } H_a \tag{95}$$

$$\text{If } \bar{X} \geq \mu_0 - z_\alpha \frac{\sigma}{\sqrt{n}}, \text{ accept } H_0; \text{ otherwise accept } H_a \tag{96}$$

If the value of σ is not known, the one-sided tests—first left-sided, then right-sided—are

$$\text{If } \bar{X} \leq \mu_0 + t_{\alpha,\nu} \frac{S}{\sqrt{n}}, \text{ accept } H_0; \text{ otherwise accept } H_a \tag{97}$$

$$\text{If } \bar{X} \geq \mu_0 - t_{\alpha,\nu} \frac{S}{\sqrt{n}}, \text{ accept } H_0; \text{ otherwise accept } H_a \tag{98}$$

All four tests satisfy Eq. (67) and therefore protect against errors of the first kind. Values of z_α can be found in Table 3.2, and values of $t_{\alpha,\nu}$ (ν is $n-1$) can be found in Table 3.3.

Protection against Errors of the Second Kind—σ Known.   The tests defined in Eqs. (95) and (96) exist for all possible values of n. They protect against errors of the first kind no matter what the value of n, and they will protect against errors of the second kind if the value of n is sufficiently large. The equations given below can be used to determine a value of n large enough to protect against an error of the second kind at any specified level.

$$(\mu_0 - \mu_1) \frac{\sqrt{n}}{\sigma} + z_\alpha = z_{1-\beta} \tag{99}$$

$$(\mu_0 - \mu_1) \frac{\sqrt{n}}{\sigma} - z_\alpha = z_\beta \tag{100}$$

For a left-sided test, substitute the specified values of α, β, μ_0, and μ_1 in Eq. (99), and then solve for the value of n. For a right-sided test, substitute the specified values of α, β, μ_0, and μ_1 in Eq. (100), and then solve for the value of n. If, in either case, it happens that the solution value of n is not an integer, round it up to the nearest integer.

For all cases, the value of z_β, given β, or the value of β, given z_β, can be found in Table 3.2, provided that the value of z_β exceeds zero. Negative values of z_β can be converted to positive values by use of the fact that $z_\beta = -z_{1-\beta}$.

Protection against Errors of the Second Kind—σ Unknown.   The tests defined in Eqs. (97) and (98) exist for all possible values of n greater than 1, and they protect against errors of the first kind no matter what the value of n. However, when the value of σ is unknown, there is no way to provide protection, at a specified level, against an error of the second kind. This happens because the equations that determine the value of n depend on σ.

The upper limit of a confidence interval, based on an estimate of σ, provides a number which probably exceeds the true value of σ. This upper limit can be used in Eqs. (99) and (100) to determine a value of n large enough to provide protection, at or better than a specified level, provided the true value of σ does not exceed the upper limit. Of course if the true value of σ is less than the upper limit, the sample size, determined by using the upper limit, may be much larger than is necessary.

Example of a One-Sided Test on μ—σ Known. Suppose that a manufacturer claims that the average weight of a unit of product, produced on a certain machine, is 100 pounds, and that it is known that the value of σ is 1.5 pounds. A test of the claim could be in the form of a left-sided test of hypothesis:

$$H_0: \mu \leq 100; \quad H_a: \mu > 100$$

Data for the test consist of the average, say $\overline{X}$, of n observed weights, and, if the manufacturer has specified that n should be 25 and that α should be .05, then, according to Eq. (95), the test is constructed as follows:

$$z_{.05} = 1.645$$

$$\text{If } \overline{X} \leq 100 + \frac{1.5(1.645)}{\sqrt{25}}, \text{ accept } H_0; \text{ otherwise accept } H_a$$

Example Continued. Suppose that the manufacturer has specified that μ_1 should be 101 and that β should be .01. It may be that a sample of size 25 is not large enough to provide protection, at this level, against an error of the second kind. To find a value of n which is sufficiently large, substitute the specified values of α, β, μ_0, and μ_1 in Eq. (99), thereby obtaining the equation

$$(100 - 101) \frac{\sqrt{n}}{1.5} + z_{.05} = z_{.99}$$

However, $z_{.05} = 1.645$, $z_{.99} = -2.33$, and it is easy to verify that the value of n satisfying the equation is between 35 and 36. Therefore, a sample of size 36 is needed to provide protection against an error of either kind, at the specified levels.

Example of a One-Sided Test on μ—σ Unknown. Suppose that a manufacturer claims that the average weight of a unit of product, produced on a certain machine, is 100 pounds. A test of the claim could be in the form of a left-sided test of hypothesis:

$$H_0: \mu \leq 100; \quad H_a: \mu > 100$$

Data used in the test consist of the average, say $\overline{X}$, of n observed weights, and of the estimate of σ, say S, computed by using Eq. (94). If the manufacturer has specified that n should be 25 and that α should be .05, then, according to Eq. (98), the test is constructed as follows:

$$z_{.025} = 1.96$$

$$\text{If } \overline{X} < 100 - 1.96 \frac{S}{\sqrt{25}}, \text{ accept } H_0; \text{ otherwise accept } H_a$$

Two-Sided Test on μ. The two-sided hypothesis is the following:

$$H_0: \mu = \mu_0; \quad H_a \neq \mu_0 \tag{101}$$

Data for testing the hypothesis consist of n observed values of a random variable, say X.

having the normal distribution, with parameters μ and σ. Estimates of μ and σ^2, computed from the data, are

$$\overline{X} = \frac{1}{n} \sum_{j=1}^{n} X_j \tag{102}$$

$$S^2 = \frac{1}{n-1} \sum_{j=1}^{n} (X_j - \overline{X})^2 \tag{103}$$

If the value of σ is known, the two sided test is

$$\text{If } \mu_0 - z_{\alpha/2}\frac{\sigma}{\sqrt{n}} < \overline{X} < \mu_0 + z_{\alpha/2}\frac{\sigma}{\sqrt{n}}, \text{ accept } H_0; \text{ otherwise accept } H_a \tag{104}$$

If the value of σ is not known, the two-sided test is

$$\text{If } \mu_0 - t_{\alpha/2,v}\frac{S}{\sqrt{n}} < \overline{X} < \mu_0 + t_{\alpha/2,v}\frac{S}{\sqrt{n}}, \text{ accept } H_0; \text{ otherwise accept } H_a \tag{105}$$

Both of these tests satisfy Eq. (67), and therefore protect against errors of the first kind. Values of z_α can be found in Table 3.2, and values of $t_{\alpha,v}$ ($v = n - 1$) can be found in Table 3.3.

Protection against Errors of the Second Kind—σ Known.   The test defined in Eq. (104) exists for all possible values of n. It protects against errors of the first kind no matter what the value of n is, and it will protect against errors of the second kind if the value of n is sufficiently large. The equations given below can be used to determine a value of n large enough to protect against an error of the second kind at any specified level.

$$(\mu_0 - \mu_1)\frac{\sqrt{n}}{\sigma} + z_{\alpha/2} = z_{\eta-\beta_1} \tag{106}$$

$$(\mu_0 - \mu_1)\frac{\sqrt{n}}{\sigma} - z_{\alpha/2} = z_\eta \tag{107}$$

and

$$(\mu_0 - \mu_2)\frac{\sqrt{n}}{\sigma} + z_{\alpha/2} = z_\eta \tag{108}$$

$$(\mu_0 - \mu_2)\frac{\sqrt{n}}{\sigma} - z_{\alpha/2} = z_{\eta+\beta_2} \tag{109}$$

The value of n can be determined in two steps. First, substitute the specified values of α, β_1, μ_0, and μ_1 in Eqs. (106) and (107), and then solve for n and η. Second, substitute the specified values of α, β_2, μ_0, and μ_2 in Eqs. (108) and (109), and again solve for n and η. Each pair of equations is solved separately, and the solution values of n and η need not be the same for both pairs of equations. However, the value of n to be used in the test is the larger of the two solution values.

It is also helpful to remember the following:

$$\text{If } z_\eta \le -3.5, \eta \approx 1; \text{ if } z_\eta \ge 3.5, \eta \approx 0$$

And in all cases, the value of Z_β, given β, or the value of β, given z_β, can be found in Table 3.2, provided that the value of z_β exceeds zero. Negative values of z_β can be converted to positive values by use of the fact that $z_\beta = -z_{1-\beta}$.

Example of a Two-Sided Test on μ—σ Known.   Suppose that a manufacturer claims that the average diameter of a unit of product, produced on a certain machine, is 1 inch, and that it is known that the value of σ is .01 inch. A test of the claim could be in the form of a test of a two-sided test of hypothesis:

$$H_0: \mu = 1; \; H_a: \mu \neq 1$$

Data for the test consist of the average, say $\overline{X}$, of n observed diameters, and if the manufacturer has specified that n should be 50 and that α should be .05, then, according to Eq. (104), the test is constructed as follows:

$$z_{.025} = 1.96$$

$$\text{If } 1 - \frac{(.01)}{\sqrt{50}} 1.96 < \overline{X} < 1 + \frac{(.01)}{\sqrt{50}} 1.96, \text{ accept } H_0; \text{ otherwise accept } H_a$$

Example Continued.   Suppose that the manufacturer has specified that μ_1 is 1.003, μ_2 is .997, β_1 should be .02, and β_2 should be .03. Intuition suggests that a sample of 50, as specified in the first part of the example, is not large enough. This is verified by substituting the specified values of n, α, μ_0, and μ_1 in the left side of Eqs. (106) and (107), and deducing that $z_{\eta - \beta_1} \approx -.16$, $\eta - \beta_1 \approx .56$, $z_\eta \approx -4.08$, $\eta \approx 1$, $\beta_1 \approx .44$, and by substituting the specified values of n, α, μ_0, and μ_2 in the left side of Eqs. (108) and (109), and deducing that $z_\eta \approx 4.08$, $\eta \approx 0$; $z_{\eta + \beta_2} \approx .16$, $\eta + \beta_2 \approx .44$, $\beta_2 \approx .44$. The large values of β_1 and β_2 show that distinguishing between values of μ ranging from .997 to 1.003 is almost impossible with a sample size of only 50, when σ has the value .01.

To find a value of n which is sufficiently large to provide protection at specified levels, substitute the specified values of α, β_1, μ_0, and μ_1 in Eqs. (106) and (107), and the specified values of α, β_2, μ_0, and μ_2 in Eqs. (108) and (109), thereby obtaining the two pairs of equations

$$-.3\sqrt{n} + 1.96 = z_{\eta - .02} \tag{110}$$

$$-.3\sqrt{n} - 1.96 = z_\eta \tag{111}$$

and

$$.3\sqrt{n} + 1.96 = z_\eta \tag{112}$$

$$.3\sqrt{n} - 1.96 = z_{\eta + \beta_2} \tag{113}$$

The values of n can be determined by trial and error. If, in the first pair of equations, n is 180, the value of η is determined by Eq. (111): $z_\eta = -5.98$, implying that $\eta \approx 1$. Then Eq. (110) reduces to

$$-2.06 = z_{.98} = -z_{.02}$$

which, according to Table 3.2, is very nearly true. Similarly, if, in the second pair of equations, n is 164, the value of η is determined by Eq. (112): $z_\eta = 5.80$, implying that $\eta \approx 0$. Then Eq. (113) reduces to

$$1.88 = z_{.03}$$

which, according to Table 3.2, is very nearly true. Therefore, a sample of size 180 is needed to provide protection against an error of either kind, at the specified level of protection.

The revised test is constructed by using Eq. (104):

$$z_{.025} = 1.96$$

$$1 - 1.96 \frac{(.01)}{\sqrt{180}} = .9985$$

$$1 + 1.96 \frac{(.01)}{\sqrt{180}} = 1.0015$$

If $.9985 < X < 1.0015$, accept H_0; otherwise accept H_a

Example of a Two-Sided Test on μ—σ *Unknown.*   Suppose that a manufacturer claims that the average diameter of a unit of product, produced on a certain machine, is 1 inch. A test of the claim could be in the form of a two-sided test of hypothesis:

$$H_0: \mu = 1;\, H_\alpha: \mu \neq 1$$

Data used in the test consist of the average, say $\overline{X}$, of n observed diameters, and of the estimate of σ, say S, computed using Eq. (103). If the manufacturer has specified that n should be 25 and that α should be .05, then, according to Eq. (105), the test is constructed as follows:

$$t_{.025,25} = 2.06$$

If $1 - 2.06 \dfrac{S}{\sqrt{25}} < \overline{X} < 1 + 2.06 \dfrac{S}{\sqrt{25}}$ accept H_0; otherwise accept H_a

Tests about the Parameter σ—Normal Distribution.   Tests about σ are based on the chi-squared distribution. There are three forms of test—left-sided, two-sided, and right-sided. The one-sided tests are described first.

One-Sided Tests on σ.   The one-sided hypotheses are

$$H_0: \sigma \leq \sigma_0;\, H_a: \sigma > \sigma_0 \tag{114}$$

$$H_0: \sigma \geq \sigma_0;\, H_a: \sigma < \sigma_0 \tag{115}$$

Data for testing the hypotheses consist of n observed values of a random variable, say X, having the normal distribution, with parameters μ and σ. Estimates of μ and σ^2 computed from the data are

$$\overline{X} = \frac{1}{n} \sum_{j=1}^{n} X_j \tag{116}$$

$$S^2 = \frac{1}{n-1} \sum_{j=1}^{n} (X_j - \overline{X})^2 \tag{117}$$

The left-sided test is

If $S^2 < \chi^2_{\alpha,\, \nu} \dfrac{\sigma_0^2}{n-1}$, accept H_0; otherwise accept H_a \tag{118}

and the right-handed test is the following:

If $S^2 > \chi^2_{1-\alpha,\, \nu} \dfrac{\sigma_0^2}{n-1}$, accept H_0; otherwise accept H_a \tag{119}

Values of $\chi^2_{\alpha,\nu}$ and $\chi^2_{1-\alpha,\nu}$ ($\nu = n - 1$) can be found in Table 3.4, provided n does not exceed 30. However, for larger values of n, the chi-squared approximation, described in the section on the chi-squared distribution, can be used to approximate values of $\chi^2_{\alpha,\nu}$ and $\chi^2_{1-\alpha,\nu}$.

Protection against Errors of the Second Kind.   The tests defined in Eqs. (118) and (119) exist for all possible values of n. They protect against errors of the first kind no matter what the value of n, and they will protect against errors of the second kind if the value of n is sufficiently large. The equations given below can be used to determine a value of n large enough to protect against an error of the second kind.

$$B = \frac{\sigma_0^{2/3} z_\alpha - \sigma_1^{2/3} z_{1-\beta}}{\sigma_1^{2/3} - \sigma_0^{2/3}} \qquad (120)$$

$$B = \frac{\sigma_0^{2/3} z_{1-\alpha} - \sigma_1^{2/3} z_\beta}{\sigma_1^{2/3} - \sigma_0^{2/3}} \qquad (121)$$

$$n = 1 + \left[\frac{8}{9(-B + \sqrt{B^2 + 4})^2} \right] \qquad (122)$$

Equations (120) and (122) are used when the test is left-sided, and Eqs. (121) and (122) are used when the test is right-sided.

Example of a One-Sided Test on σ.   Suppose that a manufacturer claims that the value of σ in a certain process does not exceed 1.5. A test of the claim could be in the form of a test of a left-sided hypothesis:

$$H_0: \sigma \le \sigma_0; H_a: \sigma > \sigma_0$$

Data used for the test consist of the estimated value, say S, of σ, computed from n independent measurements by using Eq. (103). If the manufacturer has specified that n should be 30 and that α should be .05, then, according to Eq. (118), the test is constructed as follows:

$$\chi^2_{.05,29} = 42.557$$

If $S^2 < (42.557) \dfrac{1.5^2}{29}$, accept H_0; otherwise accept H_a

Example Continued.   Suppose that the manufacturer has specified that σ_1 should be 2.0 and that β should be .05. Intuition suggests that discrimination between σ values of 1.5 and 2.0 is not possible with a sample of 25. To find a sufficiently large value of n, use Eqs. (120) and (122) as follows:

$$z_{.05} = 1.645, z_{.95} = -1.645$$

$$B = \frac{(1.5)^{2/3}(1.645) + (2.0)^{2/3}(1.645)}{(2.0)^{2/3} - (1.5)^{2/3}} = 17.207$$

$$n = 1 + \left[\frac{8}{9(-17.207 + \sqrt{17.207^2 + 4})^2} \right] = 68$$

A sample of size 68 is sufficiently large to provide protection against both kinds of error at the specified level of protection.

Two-Sided Test on σ.   The two-sided hypothesis is

$$H_0: \sigma = \sigma_0; H_a: \sigma \ne \sigma_0 \qquad (123)$$

Data for testing the hypothesis consist of n observed values of a random variable, say X, having the normal distribution, with parameters μ and σ. Estimates of μ and σ^2, computed from the data, are

$$\bar{X} = \frac{1}{n} \sum_{j=1}^{n} X_j \tag{124}$$

$$S^2 = \frac{1}{n-1} \sum_{j=1}^{n} (X_j - \bar{X})^2 \tag{125}$$

The two-sided test is

$$\text{If } \chi^2_{1-\alpha/2,\nu} \frac{\sigma^2_0}{n-1} < S^2 < \chi^2_{\alpha/2,\nu} \frac{\sigma^2_0}{n-1}, \text{ accept } H_0; \text{ otherwise accept } H_a \tag{126}$$

Protection against an Error of the Second Kind. The test defined in Eq. (126) exists for all values of n exceeding 1. It protects against errors of the first kind, no matter what the value of n, and it will protect against errors of the second kind if n is sufficiently large. Unfortunately, however, the equations that determine the sample size are difficult to solve.

Example of a Two-Sided Test on σ. Suppose that a manufacturer claims that the value of σ, for a certain process, is in the range .8 to 1.2. A test of the claim could be in the form of a two-sided hypothesis:

$$H_0: \sigma = 1; H_a: \sigma \neq 1 \tag{127}$$

Data used for the test consist of the estimated value, say S, of σ, computed from n independent measurements, by using Eq. (117). If the manufacturer has specified that n should be 25 and that α should be .05, then, according to Eq. (126), the test is constructed as follows:

$$\chi^2_{.975,24} = 12.401, \chi^2_{.025,24} = 39.364$$

If $12.401 \left(\dfrac{1}{24}\right) < S^2 < 39.364 \left(\dfrac{1}{24}\right)$, accept H_0; otherwise accept H_a

ANALYSIS OF VARIANCE

In statistics, methods used to analyze data acquired in planned experiments come under the generic heading *analysis of variance*, which is often abbreviated ANOVA. This section begins with a general explanation of the terminology and concepts and continues with descriptions, including examples, of three widely encountered forms of ANOVA: (1) Terminology and Concepts, (2) One-Way ANOVA, (3) Two-Way ANOVA, and (4) Factorial Experiments—2^n.

Terminology and Concepts. A statistical experiment is a set of several different experiments, each carried out under controlled conditions. Controlling conditions means setting the experimental variables at specified levels and holding them constant during a single experiment, and then resetting them at different levels for the next experiment, and so forth.

The experimental variables are called *factors*, and there are two different kinds of factors, quantitative and qualitative. All factors are varied over a set of specified levels. The purpose of the set of experiments is to determine which factors influence the response.

Quantitative Factors. A factor is said to be quantitative if its value varies continuously. Variables like temperature and pressure are examples of quantitative factors. Of course the range of a variable—for example, temperature—can be restricted, but it is possible to do an experiment at any temperature within the allowable range.

Qualitative Factors. A factor is said to be qualitative if its values vary discretely. Variables such as operator, machine, or vendor are examples of qualitative factors. The levels of a qualitative factor are the names of the operators or the machines or the vendors used in the experiment.

One-Way ANOVA. Consider a production process in which a certain raw material, supplied by k different vendors, is used to produce a certain product. Suppose that the engineer responsible for the production of the product suspects that the raw material is affecting a measurable characteristic of the product. In such a case, the engineer could carry out an experiment, consisting of using material from each vendor to produce several units of product. The data would consist of a measurement on each unit of product, and the analysis would attempt to answer the question: does the average value of the product characteristic depend on the vendor that supplies the material?

The experiment described above is called a *one-way classification*, because vendor is the only factor. The data are divided into k groups, one for each vendor, and the data value Y_{ij} is the measurement on the jth unit, produced using material from the ith vendor. For the purposes of analysis, it is assumed that each measurement Y_{ij} is a random variable, having the normal distribution, with parameters $\varepsilon(Y_{ij}) = \alpha_i$; Var $(Y_{ij}) = \sigma^2$; $i = 1, 2, \ldots, k; j = 1, 2, \ldots, n_i$.

In this model α_i is a constant which may vary from vendor to vendor. However, the null hypothesis is that changing vendors does not affect the average response:

$$H_0: \alpha_1 = \alpha_2 = \cdots = \alpha_k; \, H_a: \text{at least two } \alpha\text{'s not equal}$$

Computations. It is convenient to use the following notation:

$$n = n_1 + n_2 + \cdots + n_k \tag{128}$$

$$Y_{i\cdot} = \sum_{j=1}^{n_i} Y_{ij} \quad i = 1, 2, \ldots, k \tag{129}$$

$$Y_{\cdot\cdot} = \sum_{i=1}^{k} \sum_{j=1}^{n_i} Y_{ij} \tag{130}$$

The formulas for computing the quantities used to test the null hypothesis are

$$SST = \sum_{i=1}^{k} \sum_{j=1}^{n_i} Y_{ij}^2 - \frac{Y_{\cdot\cdot}^2}{n} \tag{131}$$

$$SSC = \sum_{i=1}^{k} Y_{i\cdot}^2 - \frac{Y_{\cdot\cdot}^2}{n} \tag{132}$$

$$SSE = SST - SSC \tag{133}$$

The computations are summarized in the ANOVA table displayed in Table 3.6.

The null hypothesis—that changing vendors does not affect the product—is accepted if the F ratio does not exceed F_{α, v_1, v_2} when

- α is the probability, specified by user, of the first kind of error
- $v_1 = k - 1$, $v_2 = n - k$

TABLE 3.6   Analysis of Variance

Source of variation	Degrees of freedom	Sum of squares	Mean square	F ratio
Vendors	$k - 1$	SSC	$SSC/(k - 1)$	MSC/MSE
Error	$n - k$	SSE	$SSE/(n - k)$	
Total	$n - 1$	SST		

TABLE 3.7   Vendor Data

V_1	V_2	V_3	V_4
29	27	24	30
30	26	23	31
31	28	22	34
33	29	20	

Example.   Suppose that the material used to produce a product is supplied by four different vendors, and that units of product have been produced, using material from all of the vendors. Further suppose that all units have been weighed, and that it is suspected that the average weight of the product depends on the vendor that supplied the material. The column headings in Table 3.7 identify the vendors, and the entries in the columns are the weights. The computations are

$$n = 4 + 4 + 4 + 3 = 15$$

$$Y_1. = 123, \ Y_2. = 110, \ Y_3. = 89, \ Y_4. = 95$$

$$Y.. = 417$$

$$SST = 11,827 - \frac{417^2}{15} = 234.4$$

$$SSC = \frac{123^2}{4} + \frac{110^2}{4} + \frac{89^2}{4} + \frac{95^2}{3} - \frac{417^2}{15} = 203.233$$

$$SSE = 234.4 - 203.233 = 31.167$$

The computations are summarized in the ANOVA table in Table 3.8.

TABLE 3.8   Analysis of Variance

Source of variation	Degrees of freedom	Sum of squares	Mean square	F ratio
Vendors	3	203.23	67.74	23.94
Error	11	31.17	2.83	
Total	14	234.40		

In applications, the user specifies the value of α—here specified as .05—and the value of $F_{.05,3,11}$, found in Table 3.5, is $F_{.05,3,11} = 3.59$. The conclusion is determined by comparing the F ratio, computed in the ANOVA, and the F value from Table 3.5: $3.59 < 23.94$, and the vendor effect is significant.

Two-Way ANOVA.   Consider a bulk-annealing process in which time in oven and oven temperature are both controllable variables. Suppose that an engineer wishes to investigate the effect of the two variables on the hardness of the treated material. In such a case, the engineer could carry out an experiment, consisting of annealing material at different combinations of time and temperature. The data would consist of a hardness measurement on each unit of product, and the analysis would attempt to answer the questions: Does oven temperature affect hardness? Does time in oven affect hardness? Is the combined effect of the two factors the same as the sum of the individual effects?

The experiment described above is called a *two-way classification,* because there are two factors. The experimenter must specify the number of levels for each factor—say l levels for time and m levels for temperature. A complete replicate of the experiment consists of $m \cdot l$ experiments, and if the experiment is replicated r times, the total number of experiments is $r \cdot m \cdot l$. The notation Y_{ijk} is used to designate the kth hardness measurement, observed when time is at the ith level and temperature is at the jth level.

Two possible models are described here. In the first, the number of replicates is one, and in the second, the number of replicates exceeds one. The first model is simpler and requires fewer experiments, but it cannot detect the presence of interaction between the factors. The second model can detect the presence of interaction between the factors, if such interaction exists, but more experimentation is required.

First Model.   In the first model it is assumed that Y_{ij} has the normal distribution, with parameters

$$\mathscr{E}(Y_{ij}) = \alpha_i + \beta_j \qquad i = 1, 2, \ldots, l; j = 1, 2, \ldots, m \tag{134}$$

$$\mathrm{Var}\,(Y_{ij}) = \sigma^2 \tag{135}$$

The index k is omitted because it always has the value 1.

In this model, the α_i, $i = 1, 2,, l$, are a set of constants associated with the levels of the factor time, and the β_j, $j = 1, 2, \ldots, m$, are a set of constants associated with the levels of the factor temperature. If the α_i's all have the same value, it is evident that time has no effect on the expected response. Similarly, if the β_j's all have the same value, the factor temperature has no effect on the expected response. Two hypotheses can be tested:

$$H_0: \alpha_1 = \alpha_2 = \cdots = \alpha_l; H_a: \text{at least two } \alpha\text{'s not equal}$$

$$H_0: \beta_1 = \beta_2 = \cdots = \beta_m; H_a: \text{at least two } \beta\text{'s not equal}$$

Second Model.   In the second model it is assumed that Y_{ijk} has the normal distribution, with parameters

$$\mathscr{E}(Y_{ijk}) = \alpha_i + \beta_j + \lambda_{ij} \qquad i = 1, 2, \ldots, l; j = 1, 2, \ldots, m; k = 1, 2, \ldots, r \tag{136}$$

$$\mathrm{Var}\,(Y_{ijk}) = \sigma^2 \tag{137}$$

This model reduces to the first model if the equations in Eq. (136) have a solution when the λ_{ij} are all zero. Whether or not such a solution exists depends on the values of $\varepsilon(Y_{ijk})$, all of which are unknown. Nevertheless, if the entire experiment is replicated at least twice, it is possible to test the hypothesis that the reduced model fits the data. If the reduced model is rejected, the analysis has detected the presence of an interaction between the two factors.

As before, the α_i, $i = 1, 2, \ldots, l$, are a set of constants associated with the levels of the factor time, and the β_j, $j = 1, 2, \ldots, m$, are a set of constants associated with the levels of the factor temperature. If the α_i's all have the same value, it is evident that time has no effect on the expected response. Similarly, if the β_j's all have the same value, the factor temperature has no effect on the expected response. Three hypotheses can be tested:

- $H_0: \alpha_1 = \alpha_2 = \cdots = \alpha_l; H_a:$ at least two α's not equal
- $H_0: \beta_1 = \beta_2 = \cdots = \beta_m; H_a:$ at least two β's not equal
- $H_0:$ reduced model fits; $H_a:$ reduced model does not fit

Computations. It is convenient to use the following notation:

$$Y_{ij.} = \sum_{k=1}^{r} Y_{ijk} \qquad i = 1, 2, \ldots, l; j = 1, 2, \ldots, m \tag{138}$$

$$Y_{i..} = \sum_{k=1}^{r} \sum_{j=1}^{m} Y_{ijk} \qquad i = 1, 2, \ldots, l \tag{139}$$

$$Y_{.j.} = \sum_{k=1}^{r} \sum_{i=1}^{l} Y_{ijk} \qquad j = 1, 2, \ldots, m \tag{140}$$

$$Y_{..k} = \sum_{i=1}^{l} \sum_{j=1}^{m} Y_{ijk} \qquad k = 1, 2, \ldots, r \tag{141}$$

$$Y_{...} = \sum_{k=1}^{r} \sum_{i=1}^{l} \sum_{j=1}^{m} Y_{ijk} \tag{142}$$

The formulas for computing the quantities used to test the null hypothesis are

$$SST = \sum_{k=1}^{r} \sum_{i=1}^{l} \sum_{j=1}^{m} Y_{ijk}^2 - \frac{Y_{...}^2}{l \cdot m \cdot r} \tag{143}$$

$$SSR = \frac{1}{m \cdot r} \sum_{i=1}^{l} Y_{i..}^2 - \frac{Y_{...}^2}{l \cdot m \cdot r} \tag{144}$$

$$SSC = \frac{1}{l \cdot r} \sum_{j=1}^{m} Y_{.j.}^2 - \frac{Y_{...}^2}{l \cdot m \cdot r} \tag{145}$$

$$SSS = \frac{1}{r} \sum_{i=1}^{l} \sum_{j=1}^{m} Y_{ij.}^2 - \frac{Y_{...}^2}{l \cdot m \cdot r} \tag{146}$$

$$SSD = \frac{1}{l \cdot m} \sum_{k=1}^{r} Y_{..k}^2 - \frac{Y_{...}^2}{l \cdot m \cdot r} \tag{147}$$

$$SSI = SSS - SSR - SSC \tag{148}$$

$$SSE = SST - SSR - SSC - SSI - SSD \tag{149}$$

and the computations are summarized in the ANOVA table displayed in Table 3.9.

TABLE 3.9 Analysis of Variance

Source of variance	Degrees of freedom	Sum of squares	Mean square	F Ratio
Time	$l - 1$	SSR	MSR	MSC/MSE
Temperature	$m - 1$	SSC	MSC	MSC/MSE
Interaction	$(l - 1)(m - 1)$	SSI	MSI	MSI/MSE
Replications	$r - 1$	SSD	MSD	MSD/MSE
Error	$(ml - 1)(r - 1)$	SSE	MSE	
Total	$lmr - 1$	SST		

$$MSR = \frac{SSR}{(l-1)} \qquad MSC = \frac{SSC}{(m-1)} \qquad MSI = \frac{SSI}{(l-1)(m-1)}$$

$$MSD = \frac{SSD}{(r-1)} \qquad MSE = \frac{SSE}{(ml-1)(r-1)}$$

The null hypothesis that changing time in oven does not affect the hardness of the product is accepted if the F ratio does not exceed F_{α, ν_1, ν_2}, where α is the probability, specified by user, of the first kind of error and $\nu_1 = l - 1$, $\nu_2 = (ml - 1)(r - 1)$.

The null hypothesis that changing oven temperature does not affect the hardness of the product is accepted if the F ratio does not exceed F_{α, ν_1, ν_2}, where α is the probability, specified by user, of the first kind of error and $\nu_1 = m - 1$, $\nu_2 = (ml - 1)(r - 1)$.

The null hypothesis that there is no interaction between the factors is accepted if the F ratio does not exceed F_{α, ν_1, ν_2}, where α is the probability, specified by user, of the first kind of error and $\nu_1 = (l - 1)(m - 1)$, $\nu_2 = (ml - 1)(r - 1)$.

Example. In the experiment described above, the two factors are temperature and time, and the response is hardness. In this example, temperature has two levels—1000°F and 1100°F—and time has three levels—4 hours, 5 hours, and 6 hours. The data (made up for the purpose of illustration) are hypothetical measurements of hardness on specimens produced at the six possible combinations of the levels two factors. In addition each experiment was replicated three times. The simulated data are displayed in Table 3.10.

TABLE 3.10 Hardness Readings

	4 hours	5 hours	6 hours	
	54	41	37	
1000°F	55	43	38	
	53	42	36	
	162	126	111	399
	42	39	35	
1100°F	40	38	34	
	41	36	33	
	123	113	102	338
	285	239	213	737

The sums of squares are computed as follows:

$$SST = 54^2 + 55^2 + 53^2 + \cdots + 33^2 - \frac{737^2}{18} = 752.94$$

$$SSR = \frac{1}{9}(399^2 + 338^2) - \frac{737^2}{18} = 206.72$$

$$SSC = \frac{1}{6}(285^2 + 239^2 + 213^2) - \frac{737^2}{18} = 443.11$$

$$SSS = \frac{1}{3}(162^2 + 126^2 + 111^2 + 123^2 + 113^2 + 102^2) - \frac{737^2}{18} = 738.28$$

$$SSD = \frac{1}{6}(248^2 + 248^2 + 241^2) - \frac{737^2}{18} = 5.44$$

$$SSI = SSS - SSR - SSC = 88.44$$

$$SSE = SST - SSR - SSC - SSI - SSD = 9.22$$

The results of the computation are summarized in the ANOVA table in Table 3.11.

TABLE 3.11   Analysis of Variance

Source of variation	Degrees of freedom	Sum of squares	Mean square	F ratio
Temperature	1	206.72	206.72	224.69
Time	2	443.11	221.56	240.83
Interaction	2	88.44	44.22	48.07
Replications	2	5.44	2.72	2.96
Error	10	9.22	.92	
Total	17	752.94		

Degrees of freedom are computed as follows:

- Total: number of observations less 1, $18 - 1 = 17$
- Temperature: number of levels of temperature less 1, $2 - 1 = 1$
- Times: number of levels of time less 1, $3 - 1 = 2$
- Interaction: product of temperature degrees of freedom and time degrees of freedom, $1 \times 2 = 2$
- Replications: number of replications less 1, $3 - 1 = 2$
- Error: by difference, since the sum of all degrees of freedom equals total degrees of freedom, $17 - 2 - 1 - 2 - 2 = 10$

In applications the user specifies the value of α, specified as .05 in this example, and the values of $F_{.05,1,10}$ and $F_{.05,2,10}$, found in Table 3.5, are $F_{.05,1,10} = 4.96$, $F_{.05,2,10} = 4.10$.

Conclusions are determined by comparing the F ratios, computed in the ANOVA, and the F values from Table 3.5, as follows:

- $225 > 4.96$ and the temperature effect is significant
- $241 > 4.10$ and the time effect is significant
- $48 > 4.10$ and the interaction effect is significant
- $3 < 4.10$ and the replication effect is not significant

Factorial Experiments—2^n.   The notation 2^n denotes an experiment in which there are n factors, each at two levels. Hence, a complete replicate of the experiment consists of 2^n different experiments. If the experiments are replicated r times, the total number of experiments is $r \cdot 2^n$. A simple procedure for computing the ANOVA for any 2^n experiment, with r replications, was developed by Yates.[12]

Each of the 2^n different experiments is replicated r times, and, in the first step of the Yates method, the r results are summed, thereby forming 2^n numbers, each a sum of r numbers. Second, these 2^n numbers are arranged in a column of length 2^n in a specific order—the Yates order. The column of 2^n numbers is then processed as follows:

1. Starting at the top, divide the column into blocks, each block consisting of two consecutive numbers.
2. Form a new column whose top half contains the sums of the two numbers in the consecutive blocks in the same order as the blocks, and whose bottom half contains the

differences—obtained by subtracting the first number in the block from the second number in the block—in the same order as the blocks.

3. Repeat steps 1 and 2, using the previously formed column, until a total of n columns—not counting the original column of data that started the process—have been formed.

4. Modify the last column by squaring each number and then dividing by $r \cdot 2^n$.

The numbers in the last column are the sums of squares associated with the effects of the factors and all of the possible interactions of the factors. In addition, if there is replication—r greater than 1—it is possible to compute a sum of squares for error, as well as a sum of squares for differences in replications.

Yates Order in a 2^n Factorial Experiment. Table 3.12 contains a list of the eight experiments in the 2^3 factorial experiment. The list is in Yates order.

The symbols A, B, and C are used to denote the three factors, and the symbols + and − are used to denote the high and low levels of the factors.

Each row in the table corresponds to an experiment. For example, the first row corresponds to the experiment in which all three factors are set at the low level, as indicated by the three minus signs in the first row. This is the Yates order, and, prior to analysis, the experimental results must be arranged in the Yates order. This does not imply that the experiments must be performed in Yates order; in fact, the order of performance should be random.

TABLE 3.12 2^3 Factorial Experiment

A	B	C
−	−	−
+	−	−
−	+	−
+	+	−
−	−	+
+	−	+
−	+	+
+	+	+

The structure in Table 3.13 shows how to construct a table for the case of n factors. For every value of n, the signs in the first column alternate in a cycle of length one, and the signs in the jth column alternate in a cycle of length 2^{j-1}.

TABLE 3.13   Yates Analysis

Effect	A	B	C	Rep 1	Rep 2	Total	(1)	(2)	(3)	(4)
1	−	−	−	15	16	31	68	122	238	
A	+	−	−	19	18	37	54	116	70	306.25
B	−	+	−	6	7	13	68	34	−34	72.25
AB	+	+	−	20	21	41	48	36	46	132.25
C	−	−	+	15	16	31	6	−14	−6	2.25
AC	+	−	+	19	18	37	28	−20	2	.25
BC	−	+	+	4	5	9	6	22	−6	2.25
ABC	+	+	+	20	19	39	30	24	2	.25
				118	120	238				

Example—2^3 Factorial. The data in Table 3.13 are in Yates order, and all computations have been carried out by using the previously specified rules. The numbers used as data were made up for the purpose of illustration.

The values of SSD and SSE are computed as follows:

$$\text{SST} = 15^2 + 19^2 + \cdots + 5^2 + 19^2 - \frac{238^2}{16} = 519.75$$

$$\text{SSS} = \frac{1}{2}(31^2 + 37^2 + \cdots + 9^2 + 39^2) - \frac{238^2}{16} = 515.75$$

$$SSD = \frac{1}{8}(118^2 + 120^2) - \frac{238^2}{16} = .25$$

$$SSE = SST - SSS - SSD = 519.75 - 515.75 - .25 = 3.75$$

The results are summarized in the ANOVA table displayed in Table 3.14.

TABLE 3.14   Analysis of Variance

Source of variation	Degrees of freedom	Sum of squares	Mean square	F ratio
A	1	306.25	306.25	572.
B	1	72.25	72.25	135.
C	1	2.25	2.25	4.2
AB	1	132.25	132.25	247.
AC	1	.25	.25	.25
BC	1	2.25	2.25	4.2
ABC	1	.25	.25	.47
Reps	1	.25	.25	.47
Error	7	.54		
Total	15	519.75		

Degrees of freedom are computed as follows:

- Total: number of observations less 1, $16 - 1 = 15$
- Replications: number of replications less 1, $2 - 1 = 1$
- Main effects: number of levels of factor less 1, $2 - 1 = 1$
- Second-order interactions: $(2 - 1)(2 - 1) = 1$
- Third-order interactions: $(2 - 1)(- 1)(2 - 1) = 1$
- Error: by difference, since the sum of all degrees of freedom equals total degrees of freedom, $15 - 8 = 7$

In applications the user specifies the value of α, .05 in this example, and the value of $F_{.05,1,7}$, found in Table 3.5, is $F_{.05,1,7} = 5.59$. Conclusions are determined by comparing the F ratios, computed in the ANOVA, and the F value from Table 3.5, as follows:

- $572 > 5.59$ and the A effect is significant
- $135 > 5.59$ and the B effect is significant
- $247 > 5.59$ and the AB interaction effect is significant
- No other effects are significant

REGRESSION

Engineers know from experience that the values of measurements on characteristics of manufactured products are affected by the settings of process variables such as temperature and pressure. The regression model is a systematic way of exploiting this empirical fact.

The basic assumption in regression is that the expectation of an observable random variable, say Y, is a function of one or more controllable variables, say X_1, X_2, and so on. In applications, the user chooses the function to be used in the regression model from a list of simple functions. In this chapter the list consists of three choices:

1. Linear functions of one controlled variable
2. Linear functions of two controlled variables
3. Quadratic functions of one controlled variable

Each of these functions is determined by a small number of coefficients, and regression analysis provides methods for using data to estimate the values of the coefficients. Moreover, since the coefficients may be regarded as parameters of unknown value, it is possible to find interval estimates of the coefficients and to test hypotheses about their values.

The three cases are described and illustrated in the material which follows under the headings, "Linear Regression," "Multiple Regression," and "Quadratic Regression," respectively. Also included is an explanation of hypothesis testing in each case.

Linear Regression.   The linear regression model is

$$\mathcal{E}(Y|X) = \beta_0 + \beta_1 X, \ \text{Var}\ (Y) = \sigma^2 \tag{150}$$

and data used to estimate the values of β_0, β_1, and σ^2, have the form shown in Table 3.15.

TABLE 3.15   Linear Regression

i	X_i	Y_i
1	X_1	Y_1
2	X_2	Y_2
.	.	.
.	.	.
.	.	.
n	X_n	Y_n

The values of the Y_i in Table 3.15 are determined by observation, and the values of the X_i are controlled at levels specified by the user. Computations are simplified, if the model in Eq. (134) is replaced by the equivalent model

$$\mathcal{E}(Y|X) = (\beta_0 + \beta_1 \overline{X}) + \beta_1(X - \overline{X}), \ \text{Var}\ (Y) = \sigma^2 \tag{151}$$

in which $\overline{X}$ is the average of the n values of X_i in Table 3.15.

The estimators of $(\beta_0 + \beta_1 \overline{X})$ and β_1, say B_0 and B_1, are chosen to minimize the total sum of squares of deviations, say SSR,

$$\text{SSR} = \sum_{i=1}^{n} [Y_i - B_0 - B_1(X_i - \overline{X})]^2 \tag{152}$$

and the values of B_0 and B_1, which minimize SSR, must satisfy the following pair of linear equations,

$$\sum_{i=1}^{n} [Y_i - B_0 - B_1(X_i - \overline{X})] = 0 \tag{153}$$

$$\sum_{i=1}^{n} (Y_i - B_0 - B_1(X_i - \overline{X}))(X_i - \overline{X}) = 0 \tag{154}$$

Solutions to these equations exist, provided that there are at least two distinct values of X_i.

To compute the solutions, first carry out the preliminary computations shown

$$SXX = \sum_{i=1}^{n} (X_i - \bar{X})^2$$

$$SXY = \sum_{i=1}^{n} (X_i - \bar{X})Y_i$$

Then compute the values of the estimates, using

$$B_0 = \frac{\sum_{i=1}^{n} Y_i}{n} \qquad (155)$$

$$B_1 = \frac{SXY}{SXX} \qquad (156)$$

In summary,

- B_0 is an unbiased estimator of $(\beta_0 + \beta_1 \bar{X})$
- B_1 is an unbiased estimator of β_1
- $B_0 - B_1 \bar{X}$ is an unbiased estimator of β_0

Also, assuming that the Y_i are independent random variables, each having the normal distribution with parameters given in Eq. (151), it follows that B_0 and B_1 both have the normal distribution, with parameters as follows:

$$\mathscr{E}(B_0) = (\beta_0 + \beta_1 \bar{X}), \text{ Var } (B_0) = \frac{\sigma^2}{n} \qquad (157)$$

$$\mathscr{E}(B_1) = \beta_1, \text{ Var } (B_1) = \frac{\sigma^2}{SXX} \qquad (158)$$

$$\mathscr{E}(B_0 - B_1 \bar{X}) = \beta_0, \text{ Var } (B_0 - B_1 \bar{X}) = \left(\frac{1}{n} + \frac{\bar{X}^2}{SXX}\right)\sigma^2 \qquad (159)$$

Finally, an estimator of σ^2, say S^2, is

$$S^2 = \frac{1}{n-2}\left(\sum_{i=1}^{n} Y_i^2 - B_0 \sum_{i=1}^{n} Y_i - B_1 \cdot SXY\right) \qquad (160)$$

Hypothesis Testing.   It is possible to test hypotheses about the values of β_0 or β_1. For example, a user might wish to test the following hypothesis:

$$H_0: \beta_k \geq \beta_{k0}; H_a: \beta_k < \beta_{k0}, k = 0, 1$$

in which β_{k0} is the hypothesized value of β_k.

The test is based on the fact that the quantity

$$\frac{B_k - \beta_k}{\sqrt{\text{Var } (B_k)}} \qquad k = 0, 1 \qquad (161)$$

has the standard normal distribution, provided that the Y_i—values of observed response—have the normal distribution. Hence Eq. (96) can be used to derive the test when the value of σ^2 is known. If the value of σ^2 is not known, Var (B_k) is replaced with its estimated value in Eq. (161), and the modified quantity has the t distribution with parameter v, in this case, $v = n - 2$. Hence the modified version of Eq. (161) can be used to derive the test when the value of σ^2 is not known.

When the value of σ^2 is known, the test is

$$\text{If } B_k \geq \beta_{k0} - z_\alpha \frac{1}{\sqrt{\text{Var }(B_k)}}, \text{ accept } H_0; \text{ otherwise accept } H_a \qquad (162)$$

And, when the value of σ^2 is not known, convert Var (B_k) to an estimate, say $\hat{\text{V}}\text{ar }(B_k)$, by replacing the unknown value of σ^2 with its estimated value S^2. The test is then

$$\text{If } B_k \geq \beta_{k0} - t_{\alpha,v} \frac{1}{\sqrt{\hat{\text{V}}\text{ar }(B_k)}}, \text{ accept } H_0; \text{ otherwise accept } H_a \qquad (163)$$

In both cases the value of α, specified by the user, is the probability of an error of the first kind. The value of z_α is found in Table 3.2, and the value of $t_{\alpha,v}$ is found in Table 3.3. The value of v is the degrees of freedom—the number of observations less the number of location parameters estimated, which in this case is $n - 2$.

TABLE 3.16   Data for Linear Regression

i	X_i	Y_i
1	150°	40
2	150°	39
3	175°	43
4	200°	44
5	200°	46

The test shown above is right-sided, but left-sided tests and two-sided tests are easily constructed. To get a left-sided test change the symbol $\geq$ to $\leq$ and the sign $-$ to $+$ in Eqs. (162) and (163). To get a two-sided test, change α to $\alpha/2$ in Eqs. (162) and (163), and accept H_0 only if both one-sided hypotheses are accepted.

Example.   Suppose that the average weight of a unit of product is assumed to be a linear function of the temperature at which the product was produced, provided that the temperature is in the range 150 to 200°F. Suppose further that the data displayed in Table 3.16 are available for analysis. The computations are as follows:

$$\overline{X} = \tfrac{1}{5}(150 + 150 + 175 + 200 + 200) = 175$$

$$\text{SXX} = (-25)^2 + (-25)^2 + 0 + (25)^2 + (25)^2 = 2500$$

$$\sum_{i=1}^{5} Y_i = 40 + 39 + 43 + 44 + 46 = 212$$

$$\sum_{i=1}^{5} Y_i^2 = 40^2 + 39^2 + 43^2 + 44^2 + 46^2 = 9022$$

$$\text{SXY} = -(25)(40) - (25)(39) + (0)(43) + (25)(44) + (25)(46) = 275$$

$$B_0 = \frac{212}{5} = 42.4$$

$$B_1 = \frac{275}{2500} = .11$$

$$B_0 - B_1\overline{X} = 42.4 - (.11)(175) = 23.15$$

$$S^2 = \tfrac{1}{3}[9022 - (42.4)(212) - (.11)(275)] = .9833$$

A test of the hypothesis

$$H_0: \beta_1 \geq 0; \; H_a: \beta_1 < 0$$

is constructed, by using Eq. (162) when the value of σ^2 is known and Eq. (163) when σ^2 is not known. According to Eq. (158),

$$\text{Var } (B_1) = \frac{\sigma^2}{\text{SXX}} \qquad \hat{\text{V}}\text{ar } (B_1) = \frac{S^2}{\text{SXX}}$$

Hence, when the value of σ^2 is not known, the test is

$$\text{If } B_1 > -t_{\alpha,\nu} \sqrt{\frac{.9833}{2500}}, \text{ accept } H_0; \text{ otherwise accept } H_a$$

In this example n is 5, ν is $5 - 2 = 3$, and if α is specified to be .05, Table 3.3 is used to determine the value $t_{.05,3} = 2.353$. Hence,

$$\text{If } B_1 > -(2.353)(.0198) = -.05, \text{ accept } H_0; \text{ otherwise accept } H_a$$

Since B_1 is .11, H_0 is accepted.

Multiple Linear Regression—Two Independent Variables.   The multiple linear regression model, for two independent variables, is

$$\mathscr{E}(Y|X_1, X_2) = \beta_0 + \beta_1 X_1 + \beta_2 X_2 \qquad \text{Var } (Y) = \sigma^2 \qquad (164)$$

and data used to estimate the values of β_0, β_1, β_2, and σ^2 have the form shown in Table 3.17.

TABLE 3.17   Multiple Linear Regression

i	X_{1i}	X_{2i}	Y_i
1	X_{11}	X_{21}	Y_1
2	X_{12}	X_{22}	Y_2
.	.	.	.
.	.	.	.
.	.	.	.
n	X_{1n}	X_{2n}	Y_n

The values of the Y_i in Table 3.17 are determined by observation, and the values of the X_i are controlled at levels specified by the user. Computations are simplified, if the model in Eq. (164) is replaced by the equivalent model:

$$\mathscr{E}(Y|X_1, X_2) = \beta_0 + \beta_1\overline{X}_1 + \beta_2\overline{X}_2) + \beta_1(X_1 - \overline{X}_1) + \beta_2(X_2 - \overline{X}_2) \qquad (165)$$

in which $\overline{X}_1$ and $\overline{X}_2$ are the averages of the n values of X_{1i} and X_{2i}, and

$$\text{Var } (Y|X_1, X_2) = \sigma^2 \qquad (166)$$

The estimators of $(\beta_0 + \beta_1\overline{X}_1 + \beta_2\overline{X}_2)$, β_1, and β_2, say B_0, B_1, and B_2, are chosen to minimize the total sum of squares of deviations, say SSR,

$$\text{SSR} = \sum_{i=1}^{n} [Y_i - B_0 - B_1(X_{1i} - \overline{X}_1) + B_2(X_{2i} - \overline{X}_2)]^2 \qquad (167)$$

and the values of B_0, B_1, and B_2 must satisfy the following three linear equations:

$$\sum_{i=1}^{n} [Y_i - B_0 - B_1(X_{1i} - \overline{X}_1) + B_2(X_{2i} - \overline{X}_2)] = 0 \qquad (168)$$

$$\sum_{i=1}^{n} [Y_i - B_0 - B_1(X_{1i} - \overline{X}_1) + B_2(X_{2i} - \overline{X}_2)](X_{1i} - \overline{X}_1) = 0 \qquad (169)$$

$$\sum_{i=1}^{n} [Y_i - B_0 - B_1(X_{1i} - \overline{X}_1) + B_2(X_{2i} - \overline{X}_2)](X_{2i} - \overline{X}_1) = 0 \qquad (170)$$

Solutions to these equations exist, provided that there are at least two distinct values of X_{1i} and two distinct values of X_{2i}.

To compute the values of the estimates, first carry out the preliminary calculations:

$$\text{SX}_1\text{X}_1 = \sum_{i=1}^{n} (X_{1i} - \overline{X}_1)^2$$

$$\text{SX}_1\text{X}_2 = \sum_{i=1}^{n} (X_{1i} - \overline{X}_1)(X_{2i} - \overline{X}_2)$$

$$\text{SX}_2\text{X}_2 = \sum_{i=1}^{n} (X_{2i} - \overline{X}_2)^2$$

$$\text{SX}_1\text{Y} = \sum_{i=1}^{n} (X_{1i} - \overline{X}_1)Y_i$$

$$\text{SX}_2\text{Y} = \sum_{i=1}^{n} (X_{2i} - \overline{X}_2)Y_i$$

Then compute the values of the estimates, using

$$B_0 = \frac{\sum_{i=1}^{n} Y_i}{n} \qquad (171)$$

$$B_1 = \frac{\text{SX}_2\text{X}_2 \cdot \text{SX}_1\text{Y} - \text{SX}_1\text{X}_2 \cdot \text{SX}_2\text{Y}}{\text{SX}_1\text{X}_1 \cdot \text{SX}_2\text{X}_2 - \text{SX}_1\text{X}_2 \cdot \text{SX}_1\text{X}_2} \qquad (172)$$

$$B_2 = \frac{\text{SX}_1\text{X}_1 \cdot \text{SX}_2\text{Y} - \text{SX}_1\text{X}_2 \cdot \text{SX}_1\text{Y}}{\text{SX}_1\text{X}_1 \cdot \text{SX}_2\text{X}_2 - \text{SX}_1\text{X}_2 \cdot \text{SX}_1\text{X}_2} \qquad (173)$$

In summary,

- B_0 is an unbiased estimator of $(\beta_0 + \beta_1\overline{X}_1 + \beta_2\overline{X}_2)$
- B_1 is an unbiased estimator of β_1
- B_2 is an unbiased estimator of β_2

• $B_0 - B_1\overline{X}_1 - B_2\overline{X}_2$ is an unbiased estimator of β_0

Also, assuming that the Y_i are independent, each having the normal distribution with parameters given in Eq. (165), it follows that B_0, B_1, and B_2 all have the normal distribution, with parameters

$$\mathcal{E}(B_0) = (\beta_0 + \beta_1\overline{X}_1 + \beta_2\overline{X}_2),\ \text{Var}\ (B_0) = \frac{\sigma^2}{n} \tag{174}$$

$$\mathcal{E}(B_1) = \beta_1,\ \text{Var}\ (B_1) = \frac{SX_2X_2}{SX_1X_1 \cdot SX_2X_2 - SX_1X_2 \cdot SX_1X_2}\sigma^2 \tag{175}$$

$$\mathcal{E}(B_2) = \beta_2,\ \text{Var}\ (B_2) = \frac{SX_1X_1}{SX_1X_1 \cdot SX_2X_2 - SX_1X_2 \cdot SX_1X_2}\sigma^2 \tag{176}$$

$$\mathcal{E}(B_0 - B_1\overline{X}_1 - B_2\overline{X}_2) = \beta_0 \tag{177}$$

$$\text{Var}\ (B_0 - B_1\overline{X}_1 - B_2\overline{X}_2) \tag{178}$$

$$= \left(\frac{\overline{X}_1^2 \cdot SX_2X_2 - 2\overline{X}_1\overline{X}_2 \cdot SX_1X_2 + \overline{X}_2^2 \cdot SX_1X_1}{SX_1X_1 \cdot SX_2X_2 - SX_1X_2 \cdot SX_1X_2} + \frac{1}{n}\right)\sigma^2$$

Finally, an estimator of σ^2, say S^2, is

$$S^2 = \frac{1}{n-3}\left(\sum_{i=1}^{n}Y_i^2 - B_0\sum_{i=1}^{n}Y_i - B_1 \cdot SX_1Y - B_2 \cdot SX_2Y\right) \tag{179}$$

Hypothesis Testing.   It is possible to test hypotheses about the values of β_0, β_1, and β_2. For example, a user might wish to test the following hypothesis:

$$H_0: \beta_k \geq \beta_{k0};\ H_a: \beta_k < \beta_{k0},\ k = 0, 1, 2$$

in which β_{k0} is the hypothesized value of β_k.

The test is based on the fact that the quantities

$$\frac{B_k - \beta_k}{\sqrt{\text{Var}\ (B_k)}} \qquad k = 0, 1, 2 \tag{180}$$

each have the standard normal distribution, provided that the Y_i—values of the observed response—have the normal distribution. Hence Eq. (162) can be used to derive the test when the value of σ^2 is known. If the value of σ^2 is not known, Var (B_k) is replaced with its estimated value in Eq. (180), and the modified quantity has the t distribution with parameter ν. Hence the modified version of Eq. (180) can be used to derive the test when the value of σ^2 is not known.

When the value of σ^2 is known, the test is

$$\text{If } B_k \geq \beta_{k0} - z_\alpha\frac{1}{\sqrt{\text{Var}\ (B_k)}}, \text{ accept } H_0; \text{ otherwise accept } H_a \tag{181}$$

And, when the value of σ^2 is not known, convert Var (B_k) to an estimate, say $\hat{\text{Var}}\ (B_k)$, by replacing the unknown value of σ^2 with its estimated value S^2. The test is then

$$\text{If } B_k \geq \beta_{k0} - t_{\alpha,\nu}\frac{1}{\sqrt{\hat{\text{Var}}\ (B_k)}}, \text{ accept } H_0; \text{ otherwise accept } H_a \tag{182}$$

In both cases the value of α, specified by the user, is the probability of an error of the first kind. The value of z_α is found in Table 3.2, and the value of $t_{\alpha,\nu}$ is found in Table 3.3. The value of ν is the degrees of freedom—the number of observations less the number of location parameters estimated, which in this case is $n - 3$.

The test shown above is right-sided, but left-sided tests and two-sided tests are easily constructed. To get a left-sided test, change the symbol $\geq$ to $\leq$ and the sign $-$ to $+$ in Eqs. (181) and (182). To get a two-sided test, change α to $\alpha/2$ in Eqs. (181) and (182), and accept H_0 only if both one-sided hypotheses are accepted.

Example.   Suppose that the average weight of a unit of product is assumed to be a linear function of the temperature and pressure at which the product was produced, provided that the temperature is in the range 150 to 200°F and pressure is in the range 10 to 20 lb/in². Suppose further that the data displayed in Table 3.18 are available for analysis.

The computations are as follows:

TABLE 3.18   Data for Linear Multiple Regression

i	X_{1i}	X_{2i}	Y_i
1	150	10	152
2	175	10	174
3	200	10	199
4	150	20	143
5	175	20	165
6	200	20	191

$$\overline{X}_1 = \tfrac{1}{6}(150 + 175 + 200 + 150 + 175 + 200) = 175$$

$$\overline{X}_2 = \tfrac{1}{6}(10 + 10 + 10 + 20 + 20 + 20) = 15$$

$$SX_1X_1 = (-25)^2 + 0 + (25)^2 + (-25)^2 + 0 + (25)^2 = 2500$$

$$SX_1X_2 = (-25)(-5) + 0 + (25)(-5) + (-25)(5) + 0 + (25)(5) = 0$$

$$SX_2X_2 = (-5)^2 + (-5)^2 + (-5)^2 + (5)^2 + (5)^2 + (5)^2 = 150$$

$$\sum_{i=1}^{6} Y_i = 152 + 174 + 199 + 143 + 165 + 191 = 1024$$

$$\sum_{i=1}^{6} Y_i^2 = 152^2 + 174^2 + 199^2 + 143^2 + 165^2 + 191^2 = 177{,}136$$

$$SX_1Y = -25 \cdot 152 + 0 + 25 \cdot 199 - 25 \cdot 143 + 0 + 25 \cdot 191 = 2375$$

$$SX_2Y = -5 \cdot 152 - 5 \cdot 174 - 5 \cdot 199 + 5 \cdot 143 + 5 \cdot 165 + 5 \cdot 191 = -130$$

$$B_0 = \frac{1024}{6} = 170.67$$

$$B_1 = \frac{150 \cdot 2375}{2500 \cdot 150} = .95$$

$$B_2 = \frac{2500(-130)}{2500 \cdot 150} = -.867$$

$$B_0 - B_1\overline{X}_1 - B_2\overline{X}_2 = 170.67 - .95 \cdot 175 + .867 \cdot 15 = 17.43$$

$$S_2 = \tfrac{1}{3}[177{,}136 - (170.67)(1024) - (.95)(2375) - (.867)(130)] = .32$$

A test of the hypothesis

$$H_0: \beta_1 \geq 0; \quad H_a: \beta_1 < 0$$

is constructed by using Eq. (181) when the value of σ^2 is known and Eq. (182) when σ^2 is not known. According to Eq. (175),

$$\text{Var}(B_1) = \frac{SX_2X_2}{SX_1X_1 \cdot SX_2X_2 - SX_1X_2 \cdot SX_1X_2}\sigma^2$$

$$\hat{\text{Var}}(B_1) = \frac{SX_2X_2}{SX_1X_1 \cdot SX_2X_2 - SX_1X_2 \cdot SX_1X_2}S^2$$

Hence, when the value of σ^2 is not known, the test is

If $B_1 > -t_{\alpha,\nu}\sqrt{\left(\dfrac{150}{150 \cdot 2500}\right)}(.32)$, accept H_0; otherwise accept H_a

In this example n is 6, ν is $6 - 3 = 3$, and if α is specified to be .05, Table 3.3 is used to determine the value $t_{.05,3} = 2.353$. Hence,

If $B_1 > -(2.353)(.0113) = -.03$, accept H_0; otherwise accept H_a

Since B_1 is .95, H_0 is accepted.

Quadratic Regression.   The quadratic regression model is

$$\mathcal{E}(Y|X) = \beta_0 + \beta_1 X + \beta_2 X^2, \quad \text{Var}(Y) = \sigma^2 \tag{183}$$

and data used to estimate the values of β_0, β_1, β_2, and σ^2 have the form shown in Table 3.19.

TABLE 3.19   Data for Quadratic Regression

i	X_i	Y_i
1	X_1	Y_1
2	X_2	Y_2
.	.	.
.	.	.
.	.	.
n	X_n	Y_n

The values of Y_i are determined by observation, and the values of X_i are controlled at levels specified by the user. Computations are simplified if the model in Eq. (183) is replaced with the equivalent model

$$\mathcal{E}(Y|X) = (\beta_0 + \beta_1\overline{X} + \beta_2\overline{X^2}) + (\beta_1 + 2\beta_2\overline{X})(X - \overline{X}) + \beta_2(X-)^2 \tag{184}$$

in which $\overline{X}$ is the average of the n values of X_i in Table 3.19, and

$$\text{Var}(Y|X) = \sigma^2 \tag{185}$$

The estimators of $(\beta_0 + \beta_1\overline{X} + \beta_2\overline{X^2})$, $(\beta_1 + 2\beta_2\overline{X})$, and β_2, say B_0, B_1, and B_2, are chosen to minimize the total sum of squares of deviations, say SSR,

$$\text{SSR} = \sum_{i=1}^{n}[Y_i - B_0 - B_1(X_i - \overline{X}) + B_2(X_i - \overline{X})^2]^2 \tag{186}$$

and the values of B_0, B_1, and B_2 which minimize SSR must satisfy the following three linear equations:

$$\sum_{i=1}^{n}(Y_i - B_0 - B_1(X_i - \overline{X}) + B_2(X_i - \overline{X})^2) = 0 \tag{187}$$

$$\sum_{i=1}^{n}[Y_i - B_0 - B_1(X_i - \overline{X}) + B_2(X_i - \overline{X})^2](X_i - \overline{X}) = 0 \tag{188}$$

$$\sum_{i=1}^{n}[Y_i - B_0 - B_1(X_i - \overline{X}) + B_2(X_i - \overline{X})^2](X_i - \overline{X})^2 = 0 \tag{189}$$

Solutions to these equations exist, provided that there are at least two distinct values of X_i.

To compute the values of the estimates, first carry out the preliminary calculations

$$SXX = \sum_{i=1}^{n}(X_i - \overline{X})^2$$

$$SSXX = \sum_{i=1}^{n}(X_i - \overline{X})^4$$

$$SXY = \sum_{i=1}^{n}(X_i - \overline{X})Y_i$$

$$SSXY = \sum_{i=1}^{n}(X_i - \overline{X})^2 Y_i$$

Then compute the values of the estimates, using the formulas

$$B_0 = \frac{SSXX \cdot \sum_{i=1}^{n} Y_i - SXX \cdot SSXY}{n \cdot SSXX - SXX \cdot SXX} \tag{190}$$

$$B_1 = \frac{SXY}{SXX} \tag{191}$$

$$B_2 = \frac{n \cdot SSXY - SXX \sum_{i=1}^{n} Y_i}{n \cdot SSXX - SXX \cdot SXX} \tag{192}$$

In summary,

- B_0 is an unbiased estimator of $(\beta_0 + \beta_1\overline{X} + \beta_2\overline{X}^2)$
- B_1 is an unbiased estimator of $(\beta_1 + 2\beta_2\overline{X})$
- B_2 is an unbiased estimator of β_2
- $B_0 - B_1\overline{X} - B_2\overline{X}^2$ is an unbiased estimator of β_0
- $B_1 - 2B_2\overline{X}$ is an unbiased estimator of β_1

Also, assuming that the Y_i are independent, each having the normal distribution with parameters given in Eq. (184), it follows that B_0, B_1, and B_2 all have the normal distribution, with parameters

$$\mathcal{E}(B_0) = (\beta_0 + \beta_1\overline{X} + \beta_2\overline{X^2}), \ \text{Var}\ (B_0) = \frac{\text{SSXX}}{n \cdot \text{SSXX} - \text{SXX} \cdot \text{SXX}}\sigma^2 \tag{193}$$

$$\mathcal{E}(B_1) = \beta_1 + 2\beta_2\overline{X}, \ \text{Var}\ (B_1) = \frac{\sigma^2}{\text{SXX}} \tag{194}$$

$$\mathcal{E}(B_2) = \beta_2, \ \text{Var}\ (B_2) = \frac{n}{n \cdot \text{SSXX} - \text{SXX} \cdot \text{SXX}}\sigma^2 \tag{195}$$

$$\mathcal{E}(B_0 - B_1\overline{X} - B_2\overline{X^2}) = \beta_0 \tag{196}$$

$$\text{Var}\ (B_0 - B_1\overline{X} - B_2\overline{X^2}) = \left(\frac{\text{SSXX} + \text{SXX} + n\overline{X^4}}{n \cdot \text{SSXX} - \text{SXX} \cdot \text{SXX}} + \frac{\overline{X^2}}{\text{SXX}}\right)\sigma^2 \tag{197}$$

$$\mathcal{E}(B_1 - 2B_2\overline{X}) = \beta_2 \tag{198}$$

$$\text{Var}\ (B_1 - 2B_2\overline{X}) = \left(\frac{4n\overline{X^2}}{n \cdot \text{SSXX} - \text{SXX} \cdot \text{SXX}} + \frac{1}{\text{SXX}}\right)\sigma^2 \tag{199}$$

Finally, an estimator of σ^2, say S^2, is

$$S^2 = \frac{1}{n-3}\left(\sum_{i=1}^{n}Y_i^2 - B_0\sum_{i=1}^{n}Y_i - B_1 \cdot \text{SXY} - B_2 \cdot \text{SSXY}\right) \tag{200}$$

Hypothesis Testing. It is possible to test hypotheses about the values of β_0, β_1, and β_2. For example, a user might wish to test the following hypothesis:

$$H_0: \beta_k \ge \beta_{k0}; \ H_a: \beta_k < \beta_{k0}, \ k = 0, 1, 2$$

in which β_{k0} is the hypothesized value of β_k.

The test is based on the fact that the quantity

$$\frac{B_k - \beta_k}{\sqrt{\text{Var}\ (B_k)}} \qquad k = 0, 1, 2 \tag{201}$$

has the standard normal distribution, provided that the Y_i—the values of the observed response—have the normal distribution. Hence Eq. (162) can be used to derive the test when the value of σ^2 is known. If the value of σ^2 is not known, Var (B_k) is replaced with its estimated value in Eq. (201), and the modified quantity has the t distribution with parameter v. Hence the modified version of Eq. (201) can be used to derive the test when the value of σ^2 is not known.

When the value of σ^2 is known, the test is

$$\text{If } B_k \ge \beta_{k0} - z_\alpha\frac{1}{\sqrt{\text{Var}\ (B_k)}}, \text{ accept } H_0; \text{ otherwise accept } H_a \tag{202}$$

And, when the value of σ^2 is not known, convert Var (B_k) to an estimate, say $\hat{\text{Var}}\ (B_k)$, by replacing the unknown value of σ^2 with its estimated value S^2. The test is then

$$\text{If } B_k \ge \beta_{k0} - t_{\alpha,v}\frac{1}{\sqrt{\hat{\text{Var}}\ (B_k)}}, \text{ accept } H_0; \text{ otherwise accept } H_a \tag{203}$$

In both cases the value of α, specified by the user, is the probability of an error of the first kind. The value of z_α is found in Table 3.2, and the value of $t_{\alpha,v}$ is found in Table 3.3. The value of v is the degrees of freedom—the number of observations less the number of location parameters estimated, which in this case is $n - 3$.

The test shown above is right-sided, but left-sided tests and two-sided tests are easily constructed. To get a left-sided test, change the symbol $\geq$ to $\leq$ and the sign $-$ to $+$ in Eqs. (202) and (203). To get a two-sided test, change α to $\alpha/2$ in Eqs. (202) and (203), and accept H_0 only if both one-sided hypotheses are accepted.

Example. Suppose that the average weight of a unit of product is assumed to be a quadratic function of the temperature at which the product was produced, provided that the temperature is in the range 100 to 200°F. Suppose further that the data displayed in Table 3.20 are available for analysis.

TABLE 3.20 Data for Quadratic Regression

i	X_i	Y_i
1	100°	23
2	100°	24
3	150°	16
4	200°	27
5	200°	28

The computations are

$$\overline{X} = \tfrac{1}{5}(100 + 100 + 150 + 200 + 200) = 150$$

$$SXX = (-50)^2 + (-50)^2 + 0 + (50)^2 + (50)^2 = 10{,}000$$

$$SSXX = (-50)^4 + (-50)^4 + 0 + (50)^4 + (50)^4 = 25{,}000{,}000$$

$$\sum_{i=1}^{5} Y_i = 23 + 24 + 16 + 27 + 28 = 118$$

$$\sum_{i=1}^{5} Y_i^2 = 23^2 + 24^2 + 16^2 + 27^2 + 28^2 = 2874$$

$$SXY = -50{\cdot}23 - 50{\cdot}39 + 0{\cdot}16 + 50{\cdot}27 + 50{\cdot}28 = 400$$

$$SSXY = 2500{\cdot}23 + 2500{\cdot}24 + 0{\cdot}16 + 2500{\cdot}27 + 2500{\cdot}28 = 255{,}000$$

$$B_0 = \frac{25{,}000{,}000 \cdot 118 - 10{,}000 \cdot 255{,}000}{5 \cdot 25{,}000{,}000 - 10{,}000 \cdot 10{,}000} = 16$$

$$B_1 = \frac{400}{10{,}000} = .04$$

$$B_2 = \frac{5{\cdot}255{,}000 - 10{,}000{\cdot}118}{5{\cdot}25{,}000{,}000 - 10{,}000{\cdot}10{,}000} = .0038$$

$$B_0 - B_1\overline{X} = 42.4 - (.11)(175) = 23.15$$

$$S^2 = \tfrac{1}{3}[2874 - (16)(118) - (.04)(400) - (.0038)(255{,}000)] = 1$$

A test of the hypothesis

$$H_0: \beta_2 \geq 0; \quad H_a: \beta_2 < 0$$

is constructed by using Eq. (202) when the value of σ^2 is known and Eq. (203) when σ^2 is not known. According to Eq. (195),

$$\text{Var}\,(B_2) = \frac{n}{n \cdot SSXX - SXX \cdot SXX}\sigma^2$$

$$\hat{V}ar\,(B_2) = \frac{n}{n \cdot \text{SSXX} - \text{SXX} \cdot \text{SXX}}\,S^2$$

Hence, when the value of σ^2 is not known, the test is

If $B_2 \geq -t_{\alpha,\nu}\sqrt{\dfrac{5}{25{,}000{,}000}} \cdot 1$, accept H_0; otherwise accept H_a

In this example n is 5, ν is $5 - 3 = 2$, and, if α is specified to be .05, Table 3.3 is used to determine the value $t_{.05,2} = 2.920$. Hence,

If $B_2 > -(2.920)(.0004) = -.001$, accept H_0; otherwise accept H_a

Since B_1 is .04, H_0 is accepted.

REFERENCES

1. Barlow, R. E., and Proschan, F., *Statistical Theory of Reliability and Life Testing Probability Models,* Holt, Rinehart Winston, New York, 1975.

2. National Bureau of Standards, *Tables of the Binomial Probability Distribution,* Applied Mathematics, Series 6, U.S. Government Printing Office, Superintendent of Documents, Washington, D.C., 1950.

3. Harvard University Computation Laboratory, *Tables of the Cumulative Binomial Probability Distribution,* Harvard University Press, Cambridge, Mass., 1955.

4. IMSL, *Statistics/Library User's Manual,* Houston, Texas, 1987.

5. Lieberman, G. J., and D. B. Owen, *Tables of the Hypergeometric Probability Distribution,* Stanford University Press, Stanford, Calif., 1961.

6. Patil, G. P., "On the Evaluation of the Negative Binomial Distribution with Examples," *Technometrics,* vol. 2, 1960, pp. 501–505.

7. General Electric Company, Defense Systems Department, *Tables of the Individual and Cumulative Terms of The Poisson Distribution,* Van Nostrand, Princeton, N.J., 1962.

8. Feller, W., *An Introduction to Probability Theory and its Applications,* vol. 1, Wiley, New York, 1950.

9. Beyer, W. H., *CRC Standard Mathematical Tables,* 28th ed., CRC Press, Boca Raton, Fla., 1987.

10. Box, G. E. P., and M. P. Muller, "A Note on the Generation of Random Normal Deviates," *Annals of Mathematical Statistics,* vol. 29, no. 2, 1958, pp. 610–611.

11. Wilson, E. B., and M. M. Hilferty, "The Distribution of Chi-Square," *Proceedings of the National Academy of Sciences,* vol. 17, 1931, pp. 684–688.

12. Yates, F., "Complex Experiments," *Journal of the Royal Statistical Society,* vol. B2, 1935, pp. 181–223.

TABLE 3.1 The Binomial Distribution Function

$$F(n,j,p) = \sum_{i=0}^{j} \binom{n}{i} p^i (1-p)^{n-i}$$

See pp. **14.62–14.63** for instructions on the use of this table.

Entries in table are values of F(n, j, p).

n	j	p=.01	p=.05	p=.10	p=.15	p=.20	p=.25	p=.30	p=.35	p=.40	p=.45	p=.50
2	0	0.9801	0.9025	0.8100	0.7225	0.6400	0.5625	0.4900	0.4225	0.3600	0.3025	0.2500
	1	0.9999	0.9975	0.9900	0.9775	0.9600	0.9375	0.9100	0.8775	0.8400	0.7975	0.7500
3	0	0.9703	0.8574	0.7290	0.6141	0.5120	0.4219	0.3430	0.2746	0.2160	0.1664	0.1250
	1	0.9997	0.9928	0.9720	0.9393	0.8960	0.8438	0.7840	0.7183	0.6480	0.5748	0.5000
	2	1.0000	0.9999	0.9990	0.9966	0.9920	0.9844	0.9730	0.9571	0.9360	0.9089	0.8750
4	0	0.9606	0.8145	0.6561	0.5220	0.4096	0.3164	0.2401	0.1785	0.1296	0.0915	0.0625
	1	0.9994	0.9860	0.9477	0.8905	0.8192	0.7383	0.6517	0.5630	0.4752	0.3910	0.3125
	2	1.0000	0.9995	0.9963	0.9880	0.9728	0.9492	0.9163	0.8735	0.8208	0.7585	0.6875
	3	1.0000	1.0000	0.9999	0.9995	0.9984	0.9961	0.9919	0.9850	0.9744	0.9590	0.9375
5	0	0.9510	0.7738	0.5905	0.4437	0.3277	0.2373	0.1681	0.1160	0.0778	0.0503	0.0313
	1	0.9990	0.9774	0.9185	0.8352	0.7373	0.6328	0.5282	0.4284	0.3370	0.2562	0.1875
	2	1.0000	0.9988	0.9914	0.9734	0.9421	0.8965	0.8369	0.7648	0.6826	0.5931	0.5000
	3	1.0000	1.0000	0.9995	0.9978	0.9933	0.9844	0.9692	0.9460	0.9130	0.8688	0.8125
	4	1.0000	1.0000	1.0000	0.9999	0.9997	0.9990	0.9976	0.9947	0.9898	0.9815	0.9688
6	0	0.9415	0.7351	0.5314	0.3771	0.2621	0.1780	0.1176	0.0754	0.0467	0.0277	0.0156
	1	0.9985	0.9672	0.8857	0.7765	0.6554	0.5339	0.4202	0.3191	0.2333	0.1636	0.1094
	2	1.0000	0.9978	0.9842	0.9527	0.9011	0.8306	0.7443	0.6471	0.5443	0.4415	0.3438
	3	1.0000	0.9999	0.9987	0.9941	0.9830	0.9624	0.9295	0.8826	0.8208	0.7447	0.6563
	4	1.0000	1.0000	0.9999	0.9996	0.9984	0.9954	0.9891	0.9777	0.9590	0.9308	0.8906
	5	1.0000	1.0000	1.0000	1.0000	0.9999	0.9998	0.9993	0.9982	0.9959	0.9917	0.9844
7	0	0.9321	0.6983	0.4783	0.3206	0.2097	0.1335	0.0824	0.0490	0.0280	0.0152	0.0078
	1	0.9980	0.9556	0.8503	0.7166	0.5767	0.4449	0.3294	0.2338	0.1586	0.1024	0.0625
	2	1.0000	0.9962	0.9743	0.9262	0.8520	0.7564	0.6471	0.5323	0.4199	0.3164	0.2266

n	x											
	3	0.5000	0.6083	0.7102	0.8002	0.8740	0.9294	0.9667	0.9879	0.9973	0.9998	1.0000
	4	0.7734	0.8471	0.9037	0.9444	0.9712	0.9871	0.9953	0.9988	0.9998	1.0000	1.0000
	5	0.9375	0.9643	0.9812	0.9910	0.9962	0.9987	0.9996	0.9999	1.0000	1.0000	1.0000
	6	0.9922	0.9963	0.9984	0.9994	0.9998	0.9999	1.0000	1.0000	1.0000	1.0000	1.0000
8	0	0.0039	0.0084	0.0168	0.0319	0.0576	0.1001	0.1678	0.2725	0.4305	0.6634	0.9227
	1	0.0352	0.0632	0.1064	0.1691	0.2553	0.3671	0.5033	0.6572	0.8131	0.9428	0.9973
	2	0.1445	0.2201	0.3154	0.4278	0.5518	0.6785	0.7969	0.8948	0.9619	0.9942	0.9999
	3	0.3633	0.4770	0.5941	0.7064	0.8059	0.8862	0.9437	0.9786	0.9950	0.9996	1.0000
	4	0.6367	0.7396	0.8263	0.8939	0.9420	0.9727	0.9896	0.9971	0.9996	1.0000	1.0000
	5	0.8555	0.9115	0.9502	0.9747	0.9887	0.9958	0.9988	0.9998	1.0000	1.0000	1.0000
	6	0.9648	0.9819	0.9915	0.9964	0.9987	0.9996	0.9999	1.0000	1.0000	1.0000	1.0000
	7	0.9961	0.9983	0.9993	0.9998	0.9999	1.0000	1.0000	1.0000	1.0000	1.0000	1.0000
9	0	0.0020	0.0046	0.0101	0.0207	0.0404	0.0751	0.1342	0.2316	0.3874	0.6302	0.9135
	1	0.0195	0.0385	0.0705	0.1211	0.1960	0.3003	0.4362	0.5995	0.7748	0.9288	0.9966
	2	0.0898	0.1495	0.2318	0.3373	0.4628	0.6007	0.7382	0.8591	0.9470	0.9916	0.9999
	3	0.2539	0.3614	0.4826	0.6089	0.7297	0.8343	0.9144	0.9661	0.9917	0.9994	1.0000
	4	0.5000	0.6214	0.7334	0.8283	0.9012	0.9511	0.9804	0.9944	0.9991	1.0000	1.0000
	5	0.7461	0.8342	0.9006	0.9464	0.9747	0.9900	0.9969	0.9994	0.9999	1.0000	1.0000
	6	0.9102	0.9502	0.9750	0.9888	0.9957	0.9987	0.9997	1.0000	1.0000	1.0000	1.0000
	7	0.9805	0.9909	0.9962	0.9986	0.9996	0.9999	1.0000	1.0000	1.0000	1.0000	1.0000
	8	0.9980	0.9992	0.9997	0.9999	1.0000	1.0000	1.0000	1.0000	1.0000	1.0000	1.0000
10	0	0.0010	0.0025	0.0060	0.0135	0.0282	0.0563	0.1074	0.1969	0.3487	0.5987	0.9044
	1	0.0107	0.0233	0.0464	0.0860	0.1493	0.2440	0.3758	0.5443	0.7361	0.9139	0.9957
	2	0.0547	0.0996	0.1673	0.2616	0.3828	0.5256	0.6778	0.8202	0.9298	0.9885	0.9999
	3	0.1719	0.2660	0.3823	0.5138	0.6496	0.7759	0.8791	0.9500	0.9872	0.9990	1.0000
	4	0.3770	0.5044	0.6331	0.7515	0.8497	0.9219	0.9672	0.9901	0.9984	0.9999	1.0000
	5	0.6230	0.7384	0.8338	0.9051	0.9527	0.9803	0.9936	0.9986	0.9999	1.0000	1.0000
	6	0.8281	0.8980	0.9452	0.9740	0.9894	0.9965	0.9991	0.9999	1.0000	1.0000	1.0000
	7	0.9453	0.9726	0.9877	0.9952	0.9984	0.9996	0.9999	1.0000	1.0000	1.0000	1.0000
	8	0.9893	0.9955	0.9983	0.9995	0.9999	1.0000	1.0000	1.0000	1.0000	1.0000	1.0000
	9	0.9990	0.9997	0.9999	1.0000	1.0000	1.0000	1.0000	1.0000	1.0000	1.0000	1.0000
11	0	0.0005	0.0014	0.0036	0.0088	0.0198	0.0422	0.0859	0.1673	0.3138	0.5688	0.8953
	1	0.0059	0.0139	0.0302	0.0606	0.1130	0.1971	0.3221	0.4922	0.6974	0.8981	0.9948

n	k											
	2	0.0327	0.0652	0.1189	0.2001	0.3127	0.4552	0.6174	0.7788	0.9104	0.9848	0.9998
	3	0.1133	0.1911	0.2963	0.4256	0.5696	0.7133	0.8389	0.9306	0.9815	0.9984	1.0000
	4	0.2744	0.3971	0.5328	0.6683	0.7897	0.8854	0.9496	0.9841	0.9972	0.9999	1.0000
	5	0.5000	0.6331	0.7535	0.8513	0.9218	0.9657	0.9883	0.9973	0.9997	1.0000	1.0000
	6	0.7256	0.8262	0.9006	0.9499	0.9784	0.9924	0.9980	0.9997	1.0000	1.0000	1.0000
	7	0.8867	0.9390	0.9707	0.9878	0.9957	0.9988	0.9998	1.0000	1.0000	1.0000	1.0000
	8	0.9673	0.9852	0.9941	0.9980	0.9994	0.9999	1.0000	1.0000	1.0000	1.0000	1.0000
	9	0.9941	0.9978	0.9993	0.9998	1.0000	1.0000	1.0000	1.0000	1.0000	1.0000	1.0000
	10	0.9995	0.9998	1.0000	1.0000	1.0000	1.0000	1.0000	1.0000	1.0000	1.0000	1.0000
12	0	0.0002	0.0008	0.0022	0.0057	0.0138	0.0317	0.0687	0.1422	0.2824	0.5404	0.8864
	1	0.0032	0.0083	0.0196	0.0424	0.0850	0.1584	0.2749	0.4435	0.6590	0.8816	0.9938
	2	0.0193	0.0421	0.0834	0.1513	0.2528	0.3907	0.5583	0.7358	0.8891	0.9804	0.9998
	3	0.0730	0.1345	0.2253	0.3467	0.4925	0.6488	0.7946	0.9078	0.9744	0.9978	1.0000
	4	0.1938	0.3044	0.4382	0.5833	0.7237	0.8424	0.9274	0.9761	0.9957	0.9998	1.0000
	5	0.3872	0.5269	0.6652	0.7873	0.8822	0.9456	0.9806	0.9954	0.9995	1.0000	1.0000
	6	0.6128	0.7393	0.8418	0.9154	0.9614	0.9857	0.9961	0.9993	0.9999	1.0000	1.0000
	7	0.8062	0.8883	0.9427	0.9745	0.9905	0.9972	0.9994	0.9999	1.0000	1.0000	1.0000
	8	0.9270	0.9644	0.9847	0.9944	0.9983	0.9996	0.9999	1.0000	1.0000	1.0000	1.0000
	9	0.9807	0.9921	0.9972	0.9992	0.9998	1.0000	1.0000	1.0000	1.0000	1.0000	1.0000
	10	0.9968	0.9989	0.9997	0.9999	1.0000	1.0000	1.0000	1.0000	1.0000	1.0000	1.0000
	11	0.9998	0.9999	1.0000	1.0000	1.0000	1.0000	1.0000	1.0000	1.0000	1.0000	1.0000
13	0	0.0001	0.0004	0.0013	0.0037	0.0097	0.0238	0.0550	0.1209	0.2542	0.5133	0.8775
	1	0.0017	0.0049	0.0126	0.0296	0.0637	0.1267	0.2336	0.3983	0.6213	0.8646	0.9928
	2	0.0112	0.0269	0.0579	0.1132	0.2025	0.3326	0.5017	0.6920	0.8661	0.9755	0.9997
	3	0.0461	0.0929	0.1686	0.2783	0.4206	0.5843	0.7473	0.8820	0.9658	0.9969	1.0000
	4	0.1334	0.2279	0.3530	0.5005	0.6543	0.7940	0.9009	0.9658	0.9935	0.9997	1.0000
	5	0.2905	0.4268	0.5744	0.7159	0.8346	0.9198	0.9700	0.9925	0.9991	1.0000	1.0000
	6	0.5000	0.6437	0.7712	0.8705	0.9376	0.9757	0.9930	0.9987	0.9999	1.0000	1.0000
	7	0.7095	0.8212	0.9023	0.9538	0.9818	0.9944	0.9988	0.9998	1.0000	1.0000	1.0000
	8	0.8666	0.9302	0.9679	0.9874	0.9960	0.9990	0.9998	1.0000	1.0000	1.0000	1.0000
	9	0.9539	0.9797	0.9922	0.9975	0.9993	0.9999	1.0000	1.0000	1.0000	1.0000	1.0000
	10	0.9888	0.9959	0.9987	0.9997	0.9999	1.0000	1.0000	1.0000	1.0000	1.0000	1.0000
	11	0.9983	0.9995	0.9999	1.0000	1.0000	1.0000	1.0000	1.0000	1.0000	1.0000	1.0000

n	k											
	12	0.9999	1.0000	1.0000	1.0000	1.0000	1.0000	1.0000	1.0000	1.0000	1.0000	1.0000
14	0	0.0001	0.0002	0.0008	0.0024	0.0068	0.0178	0.0440	0.1028	0.2288	0.4877	0.8687
	1	0.0009	0.0029	0.0081	0.0205	0.0475	0.1010	0.1979	0.3567	0.5846	0.8470	0.9916
	2	0.0065	0.0170	0.0398	0.0839	0.1608	0.2811	0.4481	0.6479	0.8416	0.9699	0.9997
	3	0.0287	0.0632	0.1243	0.2205	0.3552	0.5213	0.6982	0.8535	0.9559	0.9958	1.0000
	4	0.0898	0.1672	0.2793	0.4227	0.5842	0.7415	0.8702	0.9533	0.9908	0.9996	1.0000
	5	0.2120	0.3373	0.4859	0.6405	0.7805	0.8883	0.9561	0.9885	0.9985	1.0000	1.0000
	6	0.3953	0.5461	0.6925	0.8164	0.9067	0.9617	0.9884	0.9978	0.9998	1.0000	1.0000
	7	0.6047	0.7414	0.8499	0.9247	0.9685	0.9897	0.9976	0.9997	1.0000	1.0000	1.0000
	8	0.7880	0.8811	0.9417	0.9757	0.9917	0.9978	0.9996	1.0000	1.0000	1.0000	1.0000
	9	0.9102	0.9574	0.9825	0.9940	0.9983	0.9997	1.0000	1.0000	1.0000	1.0000	1.0000
	10	0.9713	0.9886	0.9961	0.9989	0.9998	1.0000	1.0000	1.0000	1.0000	1.0000	1.0000
	11	0.9935	0.9978	0.9994	0.9999	1.0000	1.0000	1.0000	1.0000	1.0000	1.0000	1.0000
	12	0.9991	0.9997	0.9999	1.0000	1.0000	1.0000	1.0000	1.0000	1.0000	1.0000	1.0000
	13	0.9999	1.0000	1.0000	1.0000	1.0000	1.0000	1.0000	1.0000	1.0000	1.0000	1.0000
15	0	0.0000	0.0001	0.0005	0.0016	0.0047	0.0134	0.0352	0.0874	0.2059	0.4633	0.8601
	1	0.0005	0.0017	0.0052	0.0142	0.0353	0.0802	0.1671	0.3186	0.5490	0.8290	0.9904
	2	0.0037	0.0107	0.0271	0.0617	0.1268	0.2361	0.3980	0.6042	0.8159	0.9638	0.9996
	3	0.0176	0.0424	0.0905	0.1727	0.2969	0.4613	0.6482	0.8227	0.9444	0.9945	1.0000
	4	0.0592	0.1204	0.2173	0.3519	0.5155	0.6865	0.8358	0.9383	0.9873	0.9994	1.0000
	5	0.1509	0.2608	0.4032	0.5643	0.7216	0.8516	0.9389	0.9832	0.9978	0.9999	1.0000
	6	0.3036	0.4522	0.6098	0.7548	0.8689	0.9434	0.9819	0.9964	0.9997	1.0000	1.0000
	7	0.5000	0.6535	0.7869	0.8868	0.9500	0.9827	0.9958	0.9994	1.0000	1.0000	1.0000
	8	0.6964	0.8182	0.9050	0.9578	0.9848	0.9958	0.9992	0.9999	1.0000	1.0000	1.0000
	9	0.8491	0.9231	0.9662	0.9876	0.9963	0.9992	0.9999	1.0000	1.0000	1.0000	1.0000
	10	0.9408	0.9745	0.9907	0.9972	0.9993	0.9999	1.0000	1.0000	1.0000	1.0000	1.0000
	11	0.9824	0.9937	0.9981	0.9995	0.9999	1.0000	1.0000	1.0000	1.0000	1.0000	1.0000
	12	0.9963	0.9989	0.9997	0.9999	1.0000	1.0000	1.0000	1.0000	1.0000	1.0000	1.0000
	13	0.9995	0.9999	1.0000	1.0000	1.0000	1.0000	1.0000	1.0000	1.0000	1.0000	1.0000
	14	1.0000	1.0000	1.0000	1.0000	1.0000	1.0000	1.0000	1.0000	1.0000	1.0000	1.0000
16	0	0.0000	0.0001	0.0003	0.0010	0.0033	0.0100	0.0281	0.0743	0.1853	0.4401	0.8515
	1	0.0003	0.0010	0.0033	0.0098	0.0261	0.0635	0.1407	0.2839	0.5147	0.8108	0.9891
	2	0.0021	0.0066	0.0183	0.0451	0.0994	0.1971	0.3518	0.5614	0.7892	0.9571	0.9995

n	x											
	3	0.0106	0.0281	0.0651	0.1339	0.2459	0.4050	0.5981	0.7899	0.9316	0.9930	1.0000
	4	0.0384	0.0853	0.1666	0.2892	0.4499	0.6302	0.7982	0.9209	0.9830	0.9991	1.0000
	5	0.1051	0.1976	0.3288	0.4900	0.6598	0.8103	0.9183	0.9765	0.9967	0.9999	1.0000
	6	0.2272	0.3660	0.5272	0.6881	0.8247	0.9204	0.9733	0.9944	0.9995	1.0000	1.0000
	7	0.4018	0.5629	0.7161	0.8406	0.9256	0.9729	0.9930	0.9989	0.9999	1.0000	1.0000
	8	0.5982	0.7441	0.8577	0.9329	0.9743	0.9925	0.9985	0.9998	1.0000	1.0000	1.0000
	9	0.7728	0.8759	0.9417	0.9771	0.9929	0.9984	0.9998	1.0000	1.0000	1.0000	1.0000
	10	0.8949	0.9514	0.9809	0.9938	0.9984	0.9997	1.0000	1.0000	1.0000	1.0000	1.0000
	11	0.9616	0.9851	0.9951	0.9987	0.9997	1.0000	1.0000	1.0000	1.0000	1.0000	1.0000
	12	0.9894	0.9965	0.9991	0.9998	1.0000	1.0000	1.0000	1.0000	1.0000	1.0000	1.0000
	13	0.9979	0.9994	0.9999	1.0000	1.0000	1.0000	1.0000	1.0000	1.0000	1.0000	1.0000
	14	0.9997	0.9999	1.0000	1.0000	1.0000	1.0000	1.0000	1.0000	1.0000	1.0000	1.0000
	15	1.0000	1.0000	1.0000	1.0000	1.0000	1.0000	1.0000	1.0000	1.0000	1.0000	1.0000
17	0	0.0000	0.0000	0.0002	0.0007	0.0023	0.0075	0.0225	0.0631	0.1668	0.4181	0.8429
	1	0.0001	0.0006	0.0021	0.0057	0.0193	0.0501	0.1182	0.2525	0.4818	0.7922	0.9877
	2	0.0012	0.0041	0.0123	0.0327	0.0774	0.1637	0.3096	0.5198	0.7618	0.9497	0.9994
	3	0.0064	0.0184	0.0464	0.1028	0.2019	0.3530	0.5489	0.7556	0.9174	0.9912	1.0000
	4	0.0245	0.0596	0.1260	0.2348	0.3887	0.5739	0.7582	0.9013	0.9779	0.9988	1.0000
	5	0.0717	0.1471	0.2639	0.4197	0.5968	0.7653	0.8943	0.9681	0.9953	0.9999	1.0000
	6	0.1662	0.2902	0.4478	0.6188	0.7752	0.8929	0.9623	0.9917	0.9992	1.0000	1.0000
	7	0.3145	0.4743	0.6405	0.7872	0.8954	0.9598	0.9891	0.9983	0.9999	1.0000	1.0000
	8	0.5000	0.6626	0.8011	0.9006	0.9597	0.9876	0.9974	0.9997	1.0000	1.0000	1.0000
	9	0.6855	0.8166	0.9081	0.9617	0.9873	0.9969	0.9995	1.0000	1.0000	1.0000	1.0000
	10	0.8338	0.9174	0.9652	0.9880	0.9968	0.9994	0.9999	1.0000	1.0000	1.0000	1.0000
	11	0.9283	0.9699	0.9894	0.9970	0.9993	0.9999	1.0000	1.0000	1.0000	1.0000	1.0000
	12	0.9755	0.9914	0.9975	0.9994	0.9999	1.0000	1.0000	1.0000	1.0000	1.0000	1.0000
	13	0.9936	0.9981	0.9995	0.9999	1.0000	1.0000	1.0000	1.0000	1.0000	1.0000	1.0000
	14	0.9988	0.9997	0.9999	1.0000	1.0000	1.0000	1.0000	1.0000	1.0000	1.0000	1.0000
	15	0.9999	1.0000	1.0000	1.0000	1.0000	1.0000	1.0000	1.0000	1.0000	1.0000	1.0000
	16	1.0000	1.0000	1.0000	1.0000	1.0000	1.0000	1.0000	1.0000	1.0000	1.0000	1.0000
18	0	0.0000	0.0000	0.0001	0.0004	0.0016	0.0056	0.0180	0.0536	0.1501	0.3972	0.8345
	1	0.0001	0.0003	0.0013	0.0045	0.0142	0.0395	0.0991	0.2241	0.4503	0.7735	0.9862
	2	0.0007	0.0025	0.0082	0.0235	0.0600	0.1353	0.2713	0.4797	0.7338	0.9419	0.9993
	3	0.0038	0.0120	0.0328	0.0783	0.1646	0.3057	0.5010	0.7202	0.9018	0.9891	1.0000

n	x											
	4	0.0154	0.0411	0.0942	0.1886	0.3327	0.5187	0.7164	0.8794	0.9718	0.9985	1.0000
	5	0.0481	0.1077	0.2088	0.3550	0.5344	0.7175	0.8671	0.9581	0.9936	0.9998	1.0000
	6	0.1189	0.2258	0.3743	0.5491	0.7217	0.8610	0.9487	0.9882	0.9988	1.0000	1.0000
	7	0.2403	0.3915	0.5634	0.7283	0.8593	0.9431	0.9837	0.9973	0.9998	1.0000	1.0000
	8	0.4073	0.5778	0.7368	0.8609	0.9404	0.9807	0.9957	0.9995	1.0000	1.0000	1.0000
	9	0.5927	0.7473	0.8653	0.9403	0.9790	0.9946	0.9991	0.9999	1.0000	1.0000	1.0000
	10	0.7597	0.8720	0.9424	0.9788	0.9939	0.9988	0.9998	1.0000	1.0000	1.0000	1.0000
	11	0.8811	0.9463	0.9797	0.9938	0.9986	0.9998	1.0000	1.0000	1.0000	1.0000	1.0000
	12	0.9519	0.9817	0.9942	0.9986	0.9997	1.0000	1.0000	1.0000	1.0000	1.0000	1.0000
	13	0.9846	0.9951	0.9987	0.9997	1.0000	1.0000	1.0000	1.0000	1.0000	1.0000	1.0000
	14	0.9962	0.9990	0.9998	1.0000	1.0000	1.0000	1.0000	1.0000	1.0000	1.0000	1.0000
	15	0.9993	0.9999	1.0000	1.0000	1.0000	1.0000	1.0000	1.0000	1.0000	1.0000	1.0000
	16	0.9999	1.0000	1.0000	1.0000	1.0000	1.0000	1.0000	1.0000	1.0000	1.0000	1.0000
	17	1.0000	1.0000	1.0000	1.0000	1.0000	1.0000	1.0000	1.0000	1.0000	1.0000	1.0000
19	0	0.0000	0.0000	0.0001	0.0003	0.0011	0.0042	0.0144	0.0456	0.1351	0.3774	0.8262
	1	0.0000	0.0002	0.0008	0.0031	0.0104	0.0310	0.0829	0.1985	0.4203	0.7547	0.9847
	2	0.0004	0.0015	0.0055	0.0170	0.0462	0.1113	0.2369	0.4413	0.7054	0.9335	0.9991
	3	0.0022	0.0077	0.0230	0.0591	0.1332	0.2631	0.4551	0.6841	0.8850	0.9868	1.0000
	4	0.0096	0.0280	0.0696	0.1500	0.2822	0.4654	0.6733	0.8556	0.9648	0.9980	1.0000
	5	0.0318	0.0777	0.1629	0.2968	0.4739	0.6678	0.8369	0.9463	0.9914	0.9998	1.0000
	6	0.0835	0.1727	0.3081	0.4812	0.6655	0.8251	0.9324	0.9837	0.9983	1.0000	1.0000
	7	0.1796	0.3169	0.4878	0.6656	0.8180	0.9225	0.9767	0.9959	0.9997	1.0000	1.0000
	8	0.3238	0.4940	0.6675	0.8145	0.9161	0.9713	0.9933	0.9992	1.0000	1.0000	1.0000
	9	0.5000	0.6710	0.8139	0.9125	0.9674	0.9911	0.9984	0.9999	1.0000	1.0000	1.0000
	10	0.6762	0.8159	0.9115	0.9653	0.9895	0.9977	0.9997	1.0000	1.0000	1.0000	1.0000
	11	0.8204	0.9129	0.9648	0.9886	0.9972	0.9995	1.0000	1.0000	1.0000	1.0000	1.0000
	12	0.9165	0.9658	0.9884	0.9969	0.9994	0.9999	1.0000	1.0000	1.0000	1.0000	1.0000
	13	0.9682	0.9891	0.9969	0.9993	0.9999	1.0000	1.0000	1.0000	1.0000	1.0000	1.0000
	14	0.9904	0.9972	0.9994	0.9999	1.0000	1.0000	1.0000	1.0000	1.0000	1.0000	1.0000
	15	0.9978	0.9995	0.9999	1.0000	1.0000	1.0000	1.0000	1.0000	1.0000	1.0000	1.0000
	16	0.9996	0.9999	1.0000	1.0000	1.0000	1.0000	1.0000	1.0000	1.0000	1.0000	1.0000
	17	1.0000	1.0000	1.0000	1.0000	1.0000	1.0000	1.0000	1.0000	1.0000	1.0000	1.0000
20	0	0.0000	0.0000	0.0000	0.0002	0.0008	0.0032	0.0115	0.0388	0.1216	0.3585	0.8179
	1	0.0000	0.0001	0.0005	0.0021	0.0076	0.0243	0.0692	0.1756	0.3917	0.7358	0.9831

2	0.0002	0.0009	0.0036	0.0121	0.0355	0.0913	0.2061	0.4049	0.6769	0.9245	0.9990
3	0.0013	0.0049	0.0160	0.0444	0.1071	0.2252	0.4114	0.6477	0.8670	0.9841	1.0000
4	0.0059	0.0189	0.0510	0.1182	0.2375	0.4148	0.6296	0.8298	0.9568	0.9974	1.0000
5	0.0207	0.0553	0.1256	0.2454	0.4164	0.6172	0.8042	0.9327	0.9887	0.9997	1.0000
6	0.0577	0.1299	0.2500	0.4166	0.6080	0.7858	0.9133	0.9781	0.9976	1.0000	1.0000
7	0.1316	0.2520	0.4159	0.6010	0.7723	0.8982	0.9679	0.9941	0.9996	1.0000	1.0000
8	0.2517	0.4143	0.5956	0.7624	0.8867	0.9591	0.9900	0.9987	0.9999	1.0000	1.0000
9	0.4119	0.5914	0.7553	0.8782	0.9520	0.9861	0.9974	0.9998	1.0000	1.0000	1.0000
10	0.5881	0.7507	0.8725	0.9468	0.9829	0.9961	0.9994	1.0000	1.0000	1.0000	1.0000
11	0.7483	0.8692	0.9435	0.9804	0.9949	0.9991	0.9999	1.0000	1.0000	1.0000	1.0000
12	0.8684	0.9420	0.9790	0.9940	0.9987	0.9998	1.0000	1.0000	1.0000	1.0000	1.0000
13	0.9423	0.9786	0.9935	0.9985	0.9997	1.0000	1.0000	1.0000	1.0000	1.0000	1.0000
14	0.9793	0.9936	0.9984	0.9997	1.0000	1.0000	1.0000	1.0000	1.0000	1.0000	1.0000
15	0.9941	0.9985	0.9997	1.0000	1.0000	1.0000	1.0000	1.0000	1.0000	1.0000	1.0000
16	0.9987	0.9997	1.0000	1.0000	1.0000	1.0000	1.0000	1.0000	1.0000	1.0000	1.0000
17	0.9998	1.0000	1.0000	1.0000	1.0000	1.0000	1.0000	1.0000	1.0000	1.0000	1.0000
18	1.0000	1.0000	1.0000	1.0000	1.0000	1.0000	1.0000	1.0000	1.0000	1.0000	1.0000

TABLE 3.2 The Normal Probability Distribution Function

$$F(z) = \int_{-\infty}^{z} \frac{e^{-t^2/2}}{\sqrt{2\pi}}\, dt$$

See pp. **14.**67–**14.**68 *for instructions on the use of this table.*

z	.00	.01	.02	.03	.04	.05	.06	.07	.08	.09
0.0	.5000	.5040	.5080	.5120	.5160	.5199	.5239	.5279	.5319	.5359
0.1	.5398	.5438	.5478	.5517	.5557	.5596	.5636	.5675	.5714	.5753
0.2	.5793	.5832	.5871	.5910	.5948	.5987	.6026	.6064	.6103	.6141
0.3	.6179	.6217	.6255	.6293	.6331	.6368	.6406	.6443	.6480	.6517
0.4	.6554	.6591	.6628	.6664	.6700	.6736	.6772	.6808	.6844	.6879
0.5	.6915	.6950	.6985	.7019	.7054	.7088	.7123	.7157	.7190	.7224
0.6	.7257	.7291	.7324	.7357	.7389	.7422	.7454	.7486	.7517	.7549
0.7	.7580	.7611	.7642	.7673	.7704	.7734	.7764	.7794	.7823	.7852
0.8	.7881	.7910	.7939	.7967	.7995	.8023	.8051	.8078	.8106	.8133
0.9	.8159	.8186	.8212	.8238	.8264	.8289	.8315	.8340	.8365	.8389
1.0	.8413	.8438	.8461	.8485	.8508	.8531	.8554	.8577	.8599	.8621
1.1	.8643	.8665	.8686	.8708	.8729	.8749	.8770	.8790	.8810	.8830
1.2	.8849	.8869	.8888	.8907	.8925	.8944	.8962	.8980	.8997	.9015
1.3	.9032	.9049	.9066	.9082	.9099	.9115	.9131	.9147	.9162	.9177
1.4	.9192	.9207	.9222	.9236	.9251	.9265	.9279	.9292	.9306	.9319
1.5	.9332	.9345	.9357	.9370	.9382	.9394	.9406	.9418	.9429	.9441
1.6	.9452	.9463	.9474	.9484	.9495	.9505	.9515	.9525	.9535	.9545
1.7	.9554	.9564	.9573	.9582	.9591	.9599	.9608	.9616	.9625	.9633
1.8	.9641	.9649	.9656	.9664	.9671	.9678	.9686	.9693	.9699	.9706
1.9	.9713	.9719	.9726	.9732	.9738	.9744	.9750	.9756	.9761	.9767
2.0	.9772	.9778	.9783	.9788	.9793	.9798	.9803	.9808	.9812	.9817
2.1	.9821	.9826	.9830	.9834	.9838	.9842	.9846	.9850	.9854	.9857
2.2	.9861	.9864	.9868	.9871	.9875	.9878	.9881	.9884	.9887	.9890
2.3	.9893	.9896	.9898	.9901	.9904	.9906	.9909	.9911	.9913	.9916
2.4	.9918	.9920	.9922	.9925	.9927	.9929	.9931	.9932	.9934	.9936
2.5	.9938	.9940	.9941	.9943	.9945	.9946	.9948	.9949	.9951	.9952
2.6	.9953	.9955	.9956	.9957	.9959	.9960	.9961	.9962	.9963	.9964
2.7	.9965	.9966	.9967	.9968	.9969	.9970	.9971	.9972	.9973	.9974
2.8	.9974	.9975	.9976	.9977	.9977	.9978	.9979	.9979	.9980	.9981
2.9	.9981	.9982	.9982	.9983	.9984	.9984	.9985	.9985	.9986	.9986
3.0	.9987	.9987	.9987	.9988	.9988	.9989	.9989	.9989	.9990	.9990
3.1	.9990	.9991	.9991	.9991	.9992	.9992	.9992	.9992	.9993	.9993
3.2	.9993	.9993	.9994	.9994	.9994	.9994	.9994	.9995	.9995	.9995
3.3	.9995	.9995	.9995	.9996	.9996	.9996	.9996	.9996	.9996	.9997
3.4	.9997	.9997	.9997	.9997	.9997	.9997	.9997	.9997	.9997	.9998
3.5	.9998	.9998	.9998	.9998	.9998	.9998	.9998	.9998	.9998	.9998

TABLE 3.3   Student's t Distribution

See p. **14**.73 *for instructions on the use of this table.*

Entries in table are values of $t_{\alpha, v}$

v	$\alpha = .4$	$\alpha = .3$	$\alpha = .2$	$\alpha = .1$	$\alpha = .05$	$\alpha = .025$	$\alpha = .010$	$\alpha = .005$
1	0.325	0.727	1.376	3.078	6.314	12.706	31.821	63.657
2	0.289	0.617	1.061	1.886	2.920	4.303	6.965	9.925
3	0.277	0.584	0.978	1.638	2.353	3.182	4.541	5.841
4	0.271	0.569	0.941	1.533	2.132	2.776	3.747	4.604
5	0.267	0.559	0.920	1.476	2.015	2.571	3.365	4.032
6	0.265	0.553	0.906	1.440	1.943	2.447	3.143	3.707
7	0.263	0.549	0.896	1.415	1.895	2.365	2.998	3.499
8	0.262	0.546	0.889	1.397	1.860	2.306	2.896	3.355
9	0.261	0.543	0.883	1.383	1.833	2.262	2.821	3.250
10	0.260	0.542	0.879	1.372	1.812	2.228	2.764	3.169
11	0.260	0.540	0.876	1.363	1.796	2.201	2.718	3.106
12	0.259	0.539	0.873	1.356	1.782	2.179	2.681	3.055
13	0.259	0.538	0.870	1.350	1.771	2.160	2.650	3.012
14	0.258	0.537	0.868	1.345	1.761	2.145	2.624	2.977
15	0.258	0.536	0.866	1.341	1.753	2.131	2.602	2.947
16	0.258	0.535	0.865	1.337	1.746	2.120	2.583	2.921
17	0.257	0.534	0.863	1.333	1.740	2.110	2.567	2.898
18	0.257	0.534	0.862	1.330	1.734	2.101	2.552	2.878
19	0.257	0.533	0.861	1.328	1.729	2.093	2.539	2.861
20	0.257	0.533	0.860	1.325	1.725	2.086	2.528	2.845
21	0.257	0.532	0.859	1.323	1.721	2.080	2.518	2.831
22	0.256	0.532	0.858	1.321	1.717	2.074	2.508	2.819
23	0.256	0.532	0.858	1.319	1.714	2.069	2.500	2.807
24	0.256	0.531	0.857	1.318	1.711	2.064	2.492	2.797
25	0.256	0.531	0.856	1.316	1.708	2.060	2.485	2.787
26	0.256	0.531	0.856	1.315	1.706	2.056	2.479	2.779
27	0.256	0.531	0.855	1.314	1.703	2.052	2.473	2.771
28	0.256	0.530	0.855	1.313	1.701	2.048	2.467	2.763
29	0.256	0.530	0.854	1.311	1.699	2.045	2.462	2.756
30	0.256	0.530	0.854	1.310	1.697	2.042	2.457	2.750

TABLE 3.4 Chi-Squared Distribution Function

*See p. **14**.74 for instructions on the use of this table.*

Entries in table are values of $\chi^2_{\alpha,v}$.

v	$\alpha=.995$	$\alpha=.990$	$\alpha=.975$	$\alpha=.950$	$\alpha=.050$	$\alpha=.025$	$\alpha=.010$	$\alpha=.005$
1	0.000	0.000	0.001	0.004	3.841	5.024	6.635	7.880
2	0.010	0.020	0.051	0.103	5.991	7.378	9.210	10.597
3	0.072	0.115	0.216	0.352	7.815	9.348	11.345	12.838
4	0.207	0.297	0.484	0.711	9.488	11.143	13.276	14.861
5	0.412	0.554	0.831	1.145	11.071	12.833	15.086	16.749
6	0.676	0.872	1.237	1.635	12.592	14.449	16.812	18.547
7	0.989	1.239	1.690	2.167	14.067	16.013	18.475	20.277
8	1.344	1.646	2.180	2.733	15.507	17.535	20.090	21.955
9	1.735	2.088	2.700	3.325	16.919	19.023	21.666	23.589
10	2.156	2.558	3.247	3.940	18.307	20.483	23.209	25.188
11	2.603	3.053	3.816	4.575	19.675	21.920	24.725	26.757
12	3.074	3.570	4.404	5.226	21.026	23.337	26.217	28.299
13	3.565	4.107	5.009	5.892	22.362	24.736	27.688	29.820
14	4.075	4.660	5.629	6.571	23.685	26.119	29.141	31.320
15	4.601	5.229	6.262	7.261	24.996	27.488	30.578	32.801
16	5.142	5.812	6.908	7.962	26.296	28.845	32.000	34.267
17	5.697	6.408	7.564	8.672	27.587	30.191	33.409	35.719
18	6.265	7.015	8.231	9.390	28.869	31.526	34.805	37.157
19	6.844	7.633	8.906	10.117	30.144	32.852	36.191	38.582
20	7.434	8.260	9.591	10.851	31.410	34.170	37.566	39.997
21	8.033	8.897	10.283	11.591	32.671	35.479	38.932	41.401
22	8.643	9.542	10.982	12.338	33.925	36.781	40.289	42.796
23	9.260	10.196	11.689	13.090	35.173	38.076	41.638	44.182
24	9.886	10.856	12.401	13.848	36.415	39.364	42.980	45.558
25	10.519	11.524	13.120	14.611	37.653	40.647	44.314	46.928
26	11.160	12.198	13.844	15.379	38.885	41.923	45.642	48.290
27	11.807	12.878	14.573	16.151	40.113	43.195	46.963	49.644
28	12.461	13.565	15.308	16.928	41.337	44.461	48.279	50.994
29	13.121	14.256	16.047	17.708	42.557	45.722	49.588	52.336
30	13.787	14.953	16.791	18.493	43.773	46.979	50.892	53.672

TABLE 3.5a The F Distribution Function

See p. 14.76 for instructions on the use of this table.

Entries in table are values of $F_{.01, \nu_1, \nu_2}$

ν_2	$\nu_1=1$	$\nu_1=2$	$\nu_1=3$	$\nu_1=4$	$\nu_1=5$	$\nu_1=6$	$\nu_1=7$	$\nu_1=8$	$\nu_1=9$	$\nu_1=10$	$\nu_1=12$	$\nu_1=15$	$\nu_1=20$	$\nu_1=24$	$\nu_1=30$	$\nu_1=40$	$\nu_1=60$	$\nu_1=120$
1	4052	4999	5403	5625	5764	5859	5928	5981	6022	6056	6106	6157	6209	6235	6261	6287	6313	6339
2	98.50	99.00	99.17	99.25	99.30	99.33	99.36	99.37	99.39	99.40	99.42	99.43	99.45	99.46	99.47	99.47	99.48	99.49
3	34.12	30.82	29.46	28.71	28.24	27.91	27.67	27.50	27.34	27.22	27.03	26.85	26.67	26.60	26.50	26.41	26.32	26.22
4	21.20	18.00	16.69	15.98	15.52	15.21	14.98	14.80	14.66	14.55	14.37	14.19	14.02	13.94	13.84	13.75	13.65	13.56
5	16.26	13.27	12.06	11.39	10.97	10.67	10.46	10.29	10.16	10.05	9.89	9.72	9.55	9.46	9.38	9.30	9.20	9.11
6	13.75	10.92	9.78	9.15	8.75	8.47	8.26	8.10	7.98	7.87	7.72	7.56	7.40	7.31	7.23	7.15	7.06	6.97
7	12.25	9.55	8.45	7.85	7.46	7.19	6.99	6.84	6.72	6.62	6.47	6.31	6.16	6.07	5.99	5.91	5.82	5.74
8	11.26	8.65	7.59	7.01	6.63	6.37	6.18	6.03	5.91	5.81	5.67	5.52	5.36	5.28	5.20	5.12	5.03	4.95
9	10.56	8.02	6.99	6.42	6.06	5.80	5.61	5.47	5.35	5.26	5.11	4.96	4.81	4.73	4.65	4.57	4.48	4.40
10	10.04	7.56	6.55	5.99	5.64	5.39	5.20	5.06	4.94	4.85	4.71	4.56	4.41	4.33	4.25	4.17	4.08	4.00
11	9.65	7.21	6.22	5.67	5.32	5.07	4.89	4.74	4.63	4.54	4.40	4.25	4.10	4.02	3.94	3.86	3.78	3.69
12	9.33	6.93	5.95	5.41	5.06	4.82	4.64	4.50	4.39	4.30	4.16	4.01	3.86	3.78	3.70	3.62	3.54	3.45
13	9.07	6.70	5.74	5.21	4.86	4.62	4.44	4.30	4.19	4.10	3.96	3.82	3.66	3.59	3.51	3.43	3.34	3.25
14	8.86	6.51	5.56	5.04	4.69	4.46	4.28	4.14	4.03	3.94	3.80	3.66	3.51	3.43	3.35	3.27	3.18	3.09
15	8.68	6.36	5.42	4.89	4.56	4.32	4.14	4.00	3.89	3.80	3.67	3.52	3.37	3.29	3.21	3.13	3.05	2.96
16	8.53	6.23	5.29	4.77	4.44	4.20	4.03	3.89	3.78	3.69	3.55	3.41	3.26	3.18	3.10	3.02	2.93	2.84
17	8.40	6.11	5.18	4.67	4.34	4.10	3.93	3.79	3.68	3.59	3.46	3.31	3.16	3.08	3.00	2.92	2.83	2.75
18	8.29	6.01	5.09	4.58	4.25	4.01	3.84	3.71	3.60	3.51	3.37	3.23	3.08	3.00	2.92	2.84	2.75	2.66
19	8.18	5.93	5.01	4.50	4.17	3.94	3.77	3.63	3.52	3.43	3.30	3.15	3.00	2.92	2.84	2.76	2.67	2.58
20	8.10	5.85	4.94	4.43	4.10	3.87	3.70	3.56	3.46	3.37	3.23	3.09	2.94	2.86	2.78	2.69	2.61	2.52
21	8.02	5.78	4.87	4.37	4.04	3.81	3.64	3.51	3.40	3.31	3.17	3.03	2.88	2.80	2.72	2.64	2.55	2.46
22	7.95	5.72	4.82	4.31	3.99	3.76	3.59	3.45	3.35	3.26	3.12	2.98	2.83	2.75	2.67	2.58	2.50	2.40
23	7.88	5.66	4.76	4.26	3.94	3.71	3.54	3.41	3.30	3.21	3.07	2.93	2.78	2.70	2.62	2.54	2.45	2.35
24	7.82	5.61	4.72	4.22	3.90	3.67	3.50	3.36	3.26	3.17	3.03	2.89	2.74	2.66	2.58	2.49	2.40	2.31
25	7.77	5.57	4.68	4.18	3.85	3.63	3.46	3.32	3.22	3.13	2.99	2.85	2.70	2.62	2.54	2.45	2.36	2.27
30	7.56	5.39	4.51	4.02	3.70	3.47	3.30	3.17	3.07	2.98	2.84	2.70	2.55	2.47	2.39	2.30	2.21	2.11
40	7.31	5.18	4.31	3.83	3.51	3.29	3.12	2.99	2.89	2.80	2.66	2.52	2.37	2.29	2.20	2.11	2.02	1.92
60	7.08	4.98	4.13	3.65	3.34	3.12	2.95	2.82	2.72	2.63	2.50	2.35	2.20	2.12	2.03	1.94	1.84	1.73
120	6.85	4.79	3.95	3.48	3.17	2.96	2.79	2.66	2.56	2.47	2.34	2.19	2.03	1.95	1.86	1.76	1.66	1.53

TABLE 3.5b The F Distribution Function

Entries in table are values of $F_{.05, \nu_1, \nu_2}$

ν_2	$\nu_1=1$	$\nu_1=2$	$\nu_1=3$	$\nu_1=4$	$\nu_1=5$	$\nu_1=6$	$\nu_1=7$	$\nu_1=8$	$\nu_1=9$	$\nu_1=10$	$\nu_1=12$	$\nu_1=15$	$\nu_1=20$	$\nu_1=24$	$\nu_1=30$	$\nu_1=40$	$\nu_1=60$	$\nu_1=120$
1	161	200	216	225	230	234	237	239	241	242	244	246	248	249	250	251	252	253
2	18.51	19.00	19.16	19.25	19.30	19.33	19.35	19.37	19.38	19.40	19.41	19.43	19.45	19.45	19.46	19.47	19.48	19.49
3	10.13	9.55	9.28	9.12	9.01	8.94	8.89	8.85	8.81	8.79	8.74	8.70	8.66	8.64	8.62	8.59	8.57	8.55
4	7.71	6.94	6.59	6.39	6.26	6.16	6.09	6.04	6.00	5.97	5.91	5.86	5.80	5.77	5.74	5.72	5.69	5.66
5	6.61	5.79	5.41	5.19	5.05	4.95	4.88	4.82	4.77	4.73	4.68	4.62	4.56	4.53	4.50	4.46	4.43	4.40
6	5.99	5.14	4.76	4.53	4.39	4.28	4.21	4.15	4.10	4.06	4.00	3.94	3.87	3.84	3.81	3.77	3.74	3.70
7	5.59	4.74	4.35	4.12	3.97	3.87	3.79	3.73	3.68	3.64	3.57	3.51	3.44	3.41	3.38	3.34	3.31	3.27
8	5.32	4.46	4.07	3.84	3.69	3.58	3.50	3.44	3.39	3.35	3.28	3.22	3.15	3.12	3.08	3.04	3.00	2.97
9	5.12	4.26	3.86	3.63	3.48	3.37	3.29	3.23	3.18	3.14	3.07	3.01	2.94	2.90	2.86	2.83	2.79	2.75
10	4.96	4.10	3.71	3.48	3.33	3.22	3.14	3.07	3.02	2.98	2.91	2.85	2.77	2.74	2.70	2.66	2.62	2.58
11	4.84	3.98	3.59	3.36	3.20	3.09	3.01	2.95	2.90	2.85	2.79	2.72	2.65	2.61	2.57	2.53	2.49	2.45
12	4.75	3.89	3.49	3.26	3.11	3.00	2.91	2.85	2.80	2.75	2.69	2.62	2.54	2.51	2.47	2.43	2.38	2.34
13	4.67	3.81	3.41	3.18	3.03	2.92	2.83	2.77	2.71	2.67	2.60	2.53	2.46	2.42	2.38	2.34	2.30	2.25
14	4.60	3.74	3.34	3.11	2.96	2.85	2.76	2.70	2.65	2.60	2.53	2.46	2.39	2.35	2.31	2.27	2.22	2.18
15	4.54	3.68	3.29	3.06	2.90	2.79	2.71	2.64	2.59	2.54	2.48	2.40	2.33	2.29	2.25	2.20	2.16	2.11
16	4.49	3.63	3.24	3.01	2.85	2.74	2.66	2.59	2.54	2.49	2.42	2.35	2.28	2.24	2.19	2.15	2.11	2.06
17	4.45	3.59	3.20	2.96	2.81	2.70	2.61	2.55	2.49	2.45	2.38	2.31	2.23	2.19	2.15	2.10	2.06	2.01
18	4.41	3.55	3.16	2.93	2.77	2.66	2.58	2.51	2.46	2.41	2.34	2.27	2.19	2.15	2.11	2.06	2.02	1.97
19	4.38	3.52	3.13	2.90	2.74	2.63	2.54	2.48	2.42	2.38	2.31	2.23	2.16	2.11	2.07	2.03	1.98	1.93
20	4.35	3.49	3.10	2.87	2.71	2.60	2.51	2.45	2.39	2.35	2.28	2.20	2.12	2.08	2.04	1.99	1.95	1.90
21	4.32	3.47	3.07	2.84	2.68	2.57	2.49	2.42	2.37	2.32	2.25	2.18	2.10	2.05	2.01	1.96	1.92	1.87
22	4.30	3.44	3.05	2.82	2.66	2.55	2.46	2.40	2.34	2.30	2.23	2.15	2.07	2.03	1.98	1.94	1.89	1.84
23	4.28	3.42	3.03	2.80	2.64	2.53	2.44	2.37	2.32	2.27	2.20	2.13	2.05	2.00	1.96	1.91	1.86	1.81
24	4.26	3.40	3.01	2.78	2.62	2.51	2.42	2.36	2.30	2.25	2.18	2.11	2.03	1.98	1.94	1.89	1.84	1.79
25	4.24	3.39	2.99	2.76	2.60	2.49	2.40	2.34	2.28	2.24	2.16	2.09	2.01	1.96	1.92	1.87	1.82	1.77
30	4.17	3.32	2.92	2.69	2.53	2.42	2.33	2.27	2.21	2.16	2.09	2.01	1.93	1.89	1.84	1.79	1.74	1.68
40	4.08	3.23	2.84	2.61	2.45	2.34	2.25	2.18	2.12	2.08	2.00	1.92	1.84	1.79	1.74	1.69	1.64	1.58
60	4.00	3.15	2.76	2.53	2.37	2.25	2.17	2.10	2.04	1.99	1.92	1.84	1.75	1.70	1.65	1.59	1.53	1.47
120	3.92	3.07	2.68	2.45	2.29	2.18	2.09	2.02	1.96	1.91	1.83	1.75	1.66	1.61	1.55	1.50	1.43	1.35

CHAPTER 4
APPLICATIONS OF OPERATIONS RESEARCH

Michael Geurts
J. Darwin Gunnel Professor of Business Marketing
Marriott School of Management
Brigham Young University
Provo, Utah

Kenneth Lawrence
Department of Industrial Engineering and the Graduate School of
Mangement
Rutgers University
Piscataway, New Jersey

Heikki Rinne
Professor, Scaggs Institute
Marriott School of Management
Brigham Young University
Provo, Utah

J. Patrick Kelly
Kmart Professor of Retailing
Wayne State University
Detroit, Michigan

This chapter looks at some of the operations research methods used by decision makers in the following areas:

1. Resolving conflicting new product forecast parameters
2. Space allocation to departments in retail stores
3. Queuing theory and the number of sales stations
4. Operations research methods in price setting
5. Site selection models

CASE 1: RESOLVING CONFLICTING NEW PRODUCT FORECAST PARAMETERS

Forecasting sales for a durable product prior to the product's introduction is a necessary but difficult task. Most new product forecasting models require the estimation of parameters for use in the forecasting model. For a specific parameter, the forecaster is often faced with estimated parameter values that are not in agreement. A methodology of dealing with conflicting parameter values is applied to a diffusion forecasting model. Diffusion models are widely used to forecast new product sales for durable products.

When a new durable product is introduced into a market, the sales of the product will follow a pattern called the *life cycle of the product*. A prior knowledge of the shape of the life cycle would give managers very useful information to use in making manufacturing decisions. The area under the curve represents the total sales of the product, the peak height gives some indication of the peak manufacturing capacity required, and the width of the curve gives information about the useful life of the product.

The product life cycle concept can be an important concept in product planning since it shows a likely future performance of the product in terms of both sales and profits.

There are two problems in predicting the life cycle curve of a product prior to the existence of sales data. The first problem is the development of a model that allows the forecaster to generate an entire product life cycle using a limited number of parameters. With such a model, the problem of predicting the entire life cycle is reduced to that of predicting two or three meaningful parameters.

The second problem is the development of methods to estimate these key parameters prior to the existence of any sales data. The forecaster can use research reports to obtain estimates of the key parameters, or if research is not available, the forecaster can use an "expert" to estimate the value of the parameter.

However, there are many cases in which the research reports or the expert estimates of these parameters have conflicting values. The problems of conflicting research findings are quite common.

The Diffusion Model Derivation.   Assume that the new product is introduced at time $t = 0$ and let $S(t)$ be the number of cumulative sales of the product up to time t. The expected number of sales in an initial time period Δt would be $N \times P$ (buying in time Δt) where N is the total number of units purchased in all time periods and P (buying in time Δt) is the probability that an individual buys the product in a time interval Δt. But P (buying in time Δt) is itself the product of the probability of buying, given contact with the product in use and the probability of contact with the product in use during the time Δt; that is,

$$P(\text{buying in time } \Delta t) = P(\text{buying/contact}) \times P(\text{contact in } \Delta t) \tag{1}$$

The probability of buying, given contact, is the contagion parameter p, so that

$$S(\Delta t) = pNP(\text{contact in } \Delta t) \tag{2}$$

Now assume that the probability of contact is proportional to both the length of time interval Δt and the fraction of the population effectively using the product (that is, both carriers N_0 and previous purchasers), then

$$P(\text{contact in } \Delta t) = K \frac{N_0}{N + N_0} \Delta t \tag{3}$$

where k is a proportionality constant, N_0 is the number of carriers, or effective users, and N is the market size. Thus, Eqs. (2) and (3) yield at $t = 0$

$$S(\Delta t) = kp \frac{N_0}{N + N_0} N \Delta t \tag{4}$$

Once others among the susceptibles have made a purchase, they too become carriers capable of converting other susceptibles who have not purchased. Hence, at any future time t, when the cumulative purchasers is $S(t)$, the expected number buying in the next time period Δt, $S(t + \Delta t) - S(t)$, ought to be proportional to the number of susceptibles who have not yet purchased, $N - S(t)$, and the fraction of effective carriers that now equals $[N_0 + S(t)]/[N + N_0]$. So

$$S(t + \Delta t) - S(t) = kp \frac{N_0 + S(t)}{N + N_0} [N - S(t)] \Delta t \tag{5}$$

Dividing both sides of Eq. (5) by Δt and taking the limit at $\Delta t,\ T \to 0$ yields

$$\frac{dS(t)}{dt} = kp \frac{N_0 + S(t)}{N + N_0} [N - S(t)] \tag{6}$$

Equation (6) is then the basic differential equation that describes the relationship between the number of buyers up to time t and the basic parameters N, p, and N_0. Let $p^* = kp$, again a measure of contagion, be the basic rate parameter that is proportional to the probability of buying, given contact.

It should be noted that if

$$A = kp \frac{N_0}{N + N_0} \qquad \text{and} \qquad B = kp \frac{1}{N + N_0} \tag{6a}$$

and the ratio of A/B is equal to N_0, the result is a special case of the Bass model,[1] a widely known model in marketing and sales literature which has given rise to several other models. (In the Bass model, A is the coefficient of innovation and B is the coefficient of external influence.) Thus, the overall model would be given by

$$\frac{dS(t)}{dt} = A[N - C(t)] + BS(t)[N - S(t)] \tag{6b}$$

Returning to the original formulation, the solution to the differential equation in Eq. (6b) is well-known and has the form

$$S(t) \frac{N + N_0}{1 + (N/N_0)e^{-p^*t}} - N_0 \tag{7}$$

where $S(t)$ is an S-shaped curve. If the time t is in years, the sales in year t are simply given by $S(t) - S(t - 1)$. Given annual sales data, a forecaster can seek those parameters N, p^*, and N_0 that give the best agreement between the observed sales in year t and the predicted sales $S(t) - S(t - 1)$ in year t. Iterative methods for finding these "best" parameters are readily available and involve the use of nonlinear least squares techniques. Other authors have recommended using linear least squares techniques to fit the difference equation corresponding to Eq. (6b). Such a linearized approach violates the assumptions of the error structure in linear regression and will lead to poorer fits and highly biased estimates.

A forecaster faced with a new product introduction who wishes to use the model as a means of forecasting the timing of sales must estimate the three key parameters N, p^*, and N_0. Once these three values are estimated, the forecaster can plot product sales as a function of time for the entire life cycle. Generally the overall market size N is dependent on pricing, advertising, and distribution, as well as on inherent product features. For a given market strategy and a static marketplace without changes in competition, N has some "true" but unknown value. Direct market surveys will allow, at least conceptually, an estimate of N prior to product introduction. But what about the estimation of the remaining two parameters?

N_0, the "effective" number of prior users at time $t = 0$, created through promotion, is more difficult to interpret and estimate beforehand. A forecaster can, however, introduce a more easily interpreted substitute parameter by means of a mathematical manipulation. Given N and p^*, the value of N_0 is completely determined by the number of sales in the first year. Using Eq. (7) with $t = 1$ and solving for N_0 results in

$$N_0 = \frac{NS_1 e^{-p^*}}{N(1 - e^{-p^*}) - S_1} \tag{8}$$

where $S_1 = s(1)$, the sales in year 1. What remains is to estimate N. One way of estimating N is to use marketing research.[2] A second method would be to ask an expert what N would be. Finally, the forecaster could do both. Similar procedures could be used to develop an estimate of S_1. If a method exists for estimating p^*, the parameter N_0 can be obtained from Eq. (8). The basic forecasting model can then be viewed either as a function of N, p^*, and N_0 or as a function of N, p^*, and S_1.

Converging Conflicting Research Findings. The above section shows that the forecast of N_0 is based on two unknown parameters N and p^*. The value of both of these parameters can be estimated from marketing research. N (total market size) could be estimated as previously mentioned from a simple survey and/or from market demographic data, which is a common practice with new products. The parameter p, the proportion of total sales purchased during the time period Δt, can be estimated by using a number of diffusion forecasting models.

The problem is that the calculation of p^* from each model may yield different values. Often the values are very close, but in some cases the calculated p^*'s have taken on widely diverse values. The question is, "Which p^* should be used?"

Also, the estimates of N can take on various values for the new product.

The De Groot Model. De Groot[3] has developed a bayesian stochastic process in which subjective probability distributions are constructed to reflect the opinion a researcher has concerning research findings which conflict with the researcher's own findings.

The model may be viewed as a quantification of the delphi forecasting procedure in which probability distributions, rather than point estimates, are used to determine a final estimation, and a weighted linear combination process, rather than simple averaging, is used in combining end point estimates.

Initially, individuals read through the conflicting research results for the parameter estimation and then construct a probability distribution for various values of the unknown parameter. The individuals are then apprised of each of the other participants' subjective distributions and the additional information each of the other individuals possessed.

The revision continues until a steady state is arrived at or until $F^{(M+1)} \approx F(M)$.

The subjective distributions of the k individuals will converge if and only if a distribution F^* exists such that

$$\lim_{n \to \infty} F^{iM} = F^* \quad i = 1, \ldots, k \tag{9}$$

Geurts and Wheeler[4] have suggested an extension of De Groot's procedure to use when the researcher's distributions do not converge. This process is to elicit opinions from a panel of n outside experts. The n experts will be asked to make two sets of weights:

1. Weights indicating their relative confidence in each other
2. Weights indicating the relative confidence in the original reports

For example, consider the n experts' weightings of each other first. Let P_{ij} be the weight that individual i assigns to individual j's estimates. Assume that $P_{ij} \geq 0$ for every value of i and j, and that

$$\sum_{j=1}^{n} P_{ij} = 1 \tag{10}$$

for every value of i. Let P denote the $n \times n$ matrix consisting of the elements P_{ij}, where

$i = 1, \ldots, n$ and $j = 1, \ldots, n$. Then P is a stochastic matrix, since each element P_{ij} is nonnegative, and for each row the sum of the elements is 1. As De Groot showed, if there exists a positive integer r such that every element in at least one column of the matrix p^r is positive, than a consensus is reached. Another way of looking at the same condition is to consider the elements of P that are equal to zero. If it is possible through a series of row interchanges and corresponding column interchanges to write P as a matrix that can be partitioned into four submatrices, two of which are diagonal to one another and are null, then there will be no consensus.

If P is a stochastic matrix and $V = (v_1, \ldots, v_n)$, then v is said to be a *stationary probability vector* if $VP = V$ and the v_i are all nonnegative numbers whose sum is 1. De Groot showed that a common theorem of Markov chains leads to the conclusion that if P meets the criteria for convergence, then the vector V that is a stationary probability vector for P exists, is unique, and is the solution for consensus. P is calculated by solving the linear equations $VP = V$ together with the equation

$$\sum_{i=1}^{n} v_i = 1 \tag{11}$$

The result of this phase, then, is the consensus weighting of each of the n experts' opinions about each other.

Weighting the Conflicting Reports.   Suppose a high-technology company is introducing a new product that is an add-on extension of a product currently being sold. The company needs to forecast sales and uses the formula given as Eq. (8). The company needs to estimate p^* and N to make the forecast. The company uses a forecasting consultant who favors the calculation of p^* using one model. He feels p^* will be 0.46. Also he feels N will be 4000. The company forecaster has estimated N to be 7500; and using another procedure of estimating $P(t)$, she has estimated the value of p^* as 0.31. Suppose further that the two researchers are at odds with one another's research and construct a nonconvergence probability matrix of

$$P_1 = \begin{matrix} 1 & 0 \\ 0 & 1 \end{matrix} \tag{12}$$

Each feels there is no chance that the other researcher is right.

Since reaching convergence is impossible using the original researchers' estimates, two outside forecasting experts, A and B, are used to evaluate the p^* and N and each other's probability of being right.

In regard to A's and B's estimates of the reliability of each other's estimates, the following P matrix was constructed for the two variables.

$$P = \begin{matrix} 0.3 & 0.7 \\ 0.5 & 0.5 \end{matrix}$$

Individual A feels individual B is 0.7/0.3 or 2.3 times as likely to be right as himself. Individual B feels he is as likely to be right as individual A. When the previous discussion on convergence is applied, it is apparent that convergence is possible.

With the stochastic matrix P and the probability distributions F available, we now can establish the consensus distribution F^* for N and P^*. First, we must find the stationary probability vector V. Since P is an ergodic process, this is done by solving $VP = V$ together with the equation

$$\sum_{i=1}^{m} v_1 = 1$$

In our instance, $VP = V$ becomes

$$(v_1 v_2) \begin{matrix} 0.3 & 0.7 \\ 0.5 & 0.5 \end{matrix} = (v_1 v_2)$$

Carrying out the matrix multiplication gives:

$$(v_1 v_2) \begin{matrix} 0.3 & 0.7 \\ 0.5 & 0.5 \end{matrix} = (0.3v_1 + 0.5v_2)\,(0.7v_1 + 0.5v_2)$$

Therefore

$$v_1 = 0.3v_1 = 0.5v_2 \quad \text{and} \quad v_2 = 0.7v_1 + 0.5v_2$$

But

$$v_1 = 0.3v_1 + 0.5v_2 = 0.7v_1 - 0.5v_2 = 0$$

and

$$v_2 = 0.7v_1 + 0.5v_2 = 0.7v_1 - 0.5v_2 = 0$$

These two equations are identical. Thus, we need the additional equation, $v_1 + v_2 = 1$, to solve the problem. Solving this set of equations, we find

$$0.7v_1 - 0.5v_2 = 0 = 1.4v_1 - v_2 = 0$$

$$\underline{v_1 + v_2 = 1 = v_1 + v_2 = 1}$$

$$2.4v_1 \qquad = 1$$

$$v_1 = \frac{1}{24} \text{ or } \frac{5}{12}$$

$$v_2 = \frac{12}{24} \text{ or } \frac{7}{12}$$

Therefore

$$V = \frac{5}{12} \cdot \frac{7}{12}$$

The consensus distribution F^* is found by multiplying: $F^* = VF$. Thus,

$$F^* = \frac{5}{12} \cdot \frac{7}{12} \; \begin{matrix} F_1 \\ F_2 \end{matrix}$$

$$F^* = \frac{5}{12} F_1 + \frac{7}{12} F_2$$

To evaluate F^* at any point, one simply evaluates F_1 and F_2 at that point and then forms the weighted average of those two values. Figure 4.1 illustrates for each area of conflict the two experts' distributions and the consensus distribution.

What remains, then, is to construct the F vectors (the outside experts' probability distributions over the parameter p^* and N). The above cumulative distributions give maximum likelihood values of $N = 6230$ and $p^* = 0.37$. The parameter S, the number of adopters during the first time interval, has been estimated as 150 units. This estimate is based on the latest marketing research forecasts and has not resulted in any contradictory findings, as was the case with the other model parameters.

These values are then substituted into Eq. (8), and the following value for N_0, the number of carriers, is developed:

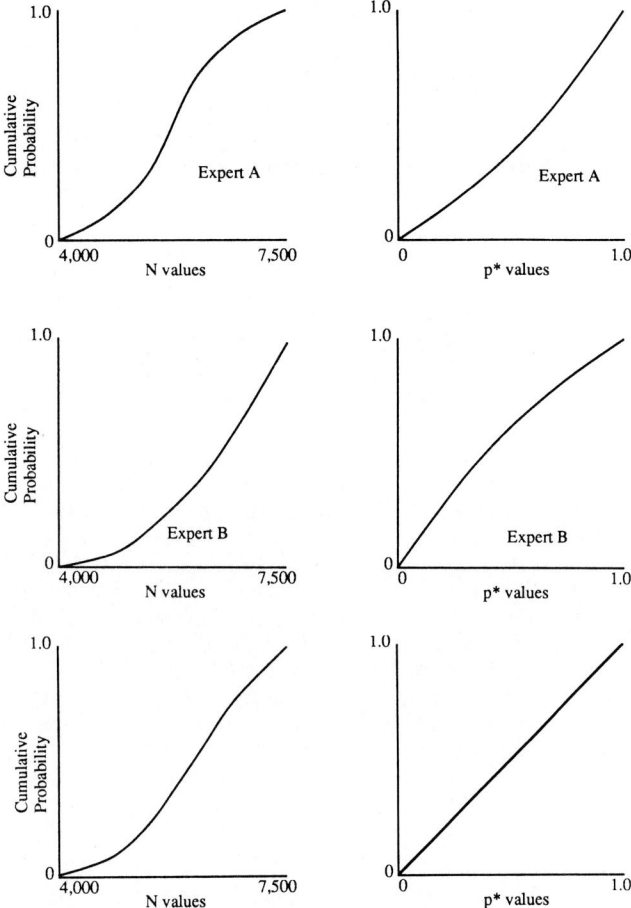

FIGURE 4.1 (*a*) Expert A's distributions for N and p^*. (*b*) Expert B's distributions of N and p^*. (*c*) Cumulative distributions.

$$N_0 = \frac{6230\ (150)e^{-0.37}}{6230(1 - e^{-0.37}) - 150}$$

$$= 334$$

Now using Eq. (7), a sales projection of the product can be made over the period 1980–1990. See Table 4.1.

CASE 2: SPACE ALLOCATION TO DEPARTMENTS IN A RETAIL STORE

A recent trend is the increased cost of building retail space. As retail space has become more expensive, retailers have looked at ways of managing the space better. The objective is to maximize profit by using existing space more efficiently. We propose a model for allocating space to departments in a department store. The proposed model uses managerial judgment, linear programming, and cross-correlation of sales between departments to specify the depart-

TABLE 4.1 Sales Projection for 1980 to 1990

	Year	$S(t)$, cumulative sales
1	1980	138
2	1981	324
3	1982	581
4	1983	904
5	1984	1311
6	1985	1811
7	1986	2292
8	1987	3049
9	1988	3416
10	1989	4131
11	1990	4794

ment location and the optional space to be allocated to the department. In addition, the model allows the departments to fluctuate in size monthly to take advantage of seasonal demand. The simulated effects of using this procedure predict significant increases in profits for the store. An in-depth treatment of this model is found in the *International Journal of Retailing,* vol. 2, no. 2, 1987.

The problem of increasing profitability by effectively using the gross selling space is a major goal of retail management. The problem has become increasingly important as the costs of building retail space have increased. Department store managers have three continuing decisions to make in managing the geography of departments in their stores: (1) What departments should the store have; that is, what mix of departments does the management want? (2) Where in the store should the departments be located? (3) How large should these departments be? The first decision gives rise to the second and third decisions. As a department store adds a new department, a decision must be made on where to locate the department and how large the department should be. Also, as a department store decides to eliminate departments, it has to decide if new departments should take up the same amount of space as the old department or if some of the freed space should go to existing departments. The model focuses on the second and third problems of where to locate departments and how much space should be allocated to these departments within a department store. Total store profits may be increased by reallocating space to departments in the store.

Prior research has been published on allocating shelf space for grocery or drugstore operations. However, past research does not address the issue of how space should be allocated to the departments in a department store setting; prior research ignores where departments should be located or how much space to allocate to departments. Our model starts to resolve the problem by specifically developing a space allocation model for departments in a department store.

Limits to Previous Research. The prior researchers who developed models for allocation of shelf space in grocery or drugstore operations solved the problem of how much space or how many facings should be allocated to a specific brand or a specific product class. In general, a limited space reserved for a department or a product class is allocated between products or facings to maximize or increase profits. This problem is not unlike the problem in a department store where the total department store space is allocated to different departments. However, in a department store setting, there are at least four additional considerations which have not been covered by the past space allocation research. First, past space allocation research does not focus on the decision of where to locate departments, products, or brands. As noted earlier, in a department store one must consider, in addition to allocating space, where departments should be located. Department stores frequently have different traffic patterns in different parts of the store and departments benefit differently from heavy traffic; impulse items benefit more from increased traffic than planned purchase items. Past space allocation research has given little attention to this location issue, probably because it is not as significant an issue for individual brands or products in a grocery store as it is for departments in a department store.

Second, department stores frequently have several floors with support walls and partitions to divide the store into selling areas. The boutique approach to merchandising within honeycomb walkways is a popular floor plan configuration. Therefore, the problem in a department store becomes how much space can be given to each department so that all fit into the physically constrained selling areas. This problem has not been addressed in past space allocation research.

A third problem not addressed in prior research is the need to use managerial insight to determine what departments should be next to each other to increase the aesthetics of the store, gain efficiency in shipping and stocking, and enhance management strategies. For aesthetic reasons, dresses probably should not be next to lawn mowers. Alternatively, a good strategy may be to place the women's dress department next to women's shoes or cosmetics in an effort to increase impulse purchases.

The art supply department and the school supply department are often located next to each other because (1) both departments may carry several of the same products, (2) sales help may have expertise in both areas and can be scheduled to work in either department, (3) there may be periods of the year in which the space allocated to the school supplies department should be expanded and that of the art supplies department reduced, and (4) there may be a cross-elasticity of demand between the two departments in that people who come for school supplies may be induced by the close proximity of art supplies to make purchases in the art supplies department. Thus, it becomes important in allocating floor space in department stores to look at managers' desires for department arrangements and alignment. This managerial input has not been included in prior research.

A fourth and final area of neglect in prior research on shelf space allocation is that it ignores the seasonality of sales of various products. It may be that the size of the gift department should be larger in December than in February. America's retail department store sales peak around consumption holidays such as Christmas and Mother's Day. The sales in most departments are very seasonal. Interestingly, some departments have contrasting seasonal sales patterns in contrast with other departments. The allocation of floor space is not a static problem. In determining space allocation needs, seasonal demand needs to be examined so that some departments expand during summer while others contract during summer.

What follows is a model to overcome these four limitations of past space allocation research. In addition to determining how much space to allocate to a department, this model attempts to locate departments on the basis of store traffic patterns; determines how to allocate space, given the physical building constraints; uses managerial insight to determine which departments should be next to each other; and allows seasonal fluctuation in space based on seasonal demand. The following section describes the model, after which we present an application of the model.

The Model.   There are three major procedures required in using the model to determine space allocation to departments.

Procedure 1.   The objective of the first procedure is to identify an approximate location in the store for each department. This assignment is based on traffic patterns within the store and managers' insight. Store managers feel that departments should be in certain geographic areas to increase both impulse buying and length of time in the store. This approximate assignment of a department to high-, medium-, or low-traffic areas depends on the degree of impulse purchasing in each department. Departments whose products have a high proportion of impulse buying are placed in high traffic areas, whereas departments whose products are shopping goods are placed in regions with low natural traffic flows. The specific steps for this first level of decisions are:

1. Divide the store into high-, medium-, and low-traffic areas.
2. Rank (or group) departments according to their potential for impulse buying.
3. Place departments in high-, medium-, and low-traffic regions on the basis of managers' judgments of impulse buying in each department. High-impulse departments should be assigned to high-traffic areas, while low-impulse departments are assigned to lower-traffic areas.

Procedure 2.   The objective of the second procedure is to allocate specific square footage for each department on a monthly basis. Given the approximate location of the department determined in the previous procedures, the specific space allocation for each department is determined by a linear programming routine which maximizes the total gross profit margin for the store on a monthly basis. The steps in determining square footage for each department are

1. Obtain *monthly* sales and profits for each department for the last 3 years on a monthly basis.

2. Get management estimates for the minimum and maximum square feet required for each department. These ranges should be chosen so that the minimum is the size that management feels is necessary for the department to exist as a department rather than a part of a department or selling area. The maximum constraint should reflect management's feel as to how large the department could be without detracting from the store image as a department store. For example, department store management may wish to limit the appliance or tool department in size to make sure the store image remains that of a department store rather than that of an appliance or tool outlet. On the other hand, it may be necessary to allocate some minimum square footage to the women's ready-to-wear department so that customers perceive it as a department rather than just "some clothes in the corner" of the store.

3. Run linear programs to maximize profits, given the above minimum and maximum constraints. This is done for each department on a monthly basis.

Procedure 3.   The objective of the final decision is to determine the specific location of each department. The departments are located next to each other according to two criteria: (1) the management criteria on which departments should be next to each other and (2) the cross-correlation of sales between departments. This allows the expansion and contraction of adjacent departments to be based on the variations of seasonal demand and profitability. The following steps determine the specific location of each department:

1. Calculate the correlations between the sales of all departments on a monthly basis.
2. Run a multiple dimensional scaling program using the correlations as similarity measures to determine departments that have the most similar and the most different yearly sales patterns.
3. Determine which complementary departments should be located next to each other on the basis of managerial considerations.
4. Determine the location and the amount of space allocated to each department on a monthly basis. The expansion and reduction of space is determined by the monthly linear programming solutions.

The Application.   The authors were able to obtain the support and cooperation of a large two-level retail store. The management of the store furnished the monthly sales and gross margin figures for the store's 10 major departments:

1. Women's Ready-to-Wear and Accessories
2. Books
3. Health and Beauty Aids, Candy and Foods
4. Records, Radio, and TV
5. Camera
6. Men's Clothing
7. Sports
8. Gifts and Small Appliances
9. School Supplies, Computers, and Stationery
10. Art Supplies

Management was anxious to increase impulse buying by placing frequently purchased shopping goods at the back of the store or in the basement level of the store so shoppers would spend more time in the store and also travel by what management considered impulse merchandise (novels and candy). Impulse items were to be placed near the checkout stand or at high-traffic points in the store. The first task was to identify the approximate location for each department. In doing this, the traffic and impulse-buying considerations were the most important factors. The departments were grouped into three groups based on the degree of impulse buying. On the basis of management judgment, the departments were ranked on the importance of impulse buying for each department. The rankings are

1. Health and Beauty Aids, Candy and Foods
2. Gifts and Small Appliances
3. Books
4. Women's Ready-to-Wear
5. Records, Radio, and TV
6. Sports
7. Men's Clothing
8. School Supplies, Computers, and Stationery
9. Camera
10. Art Supplies

Figure 4.2 shows the layout of the two floors, including the high-, medium-, and low-traffic areas within the store. On the basis of the above rankings and the traffic regions, the departments were grouped into the following three regions:

High-traffic region

1. Health and Beauty Aids, Candy and Food
2. Gifts and Small Appliances

Medium-traffic region

3. Books
4. Women's Ready-to-Wear
5. Radio, Records, and TV
6. Sports
7. Men's Clothing

Low-traffic region

8. School Supplies, Computers, and Stationery
9. Camera
10. Art Supplies

The second task was to allocate the available space to the departments on a monthly basis. This was based on the monthly profitability of each department. Linear programming models were developed to optimize the space allocation for the departments. The linear programming model is shown in Fig. 4.3. The monthly linear programming solutions are shown for all departments in Table 4.2, which also shows the difference in the actual gross margin and the gross margin predicted by the model.

As can be noted, the department space allocations have different patterns. For the most profitable departments (Health and Beauty Aids, Candy and Food), the maximum square

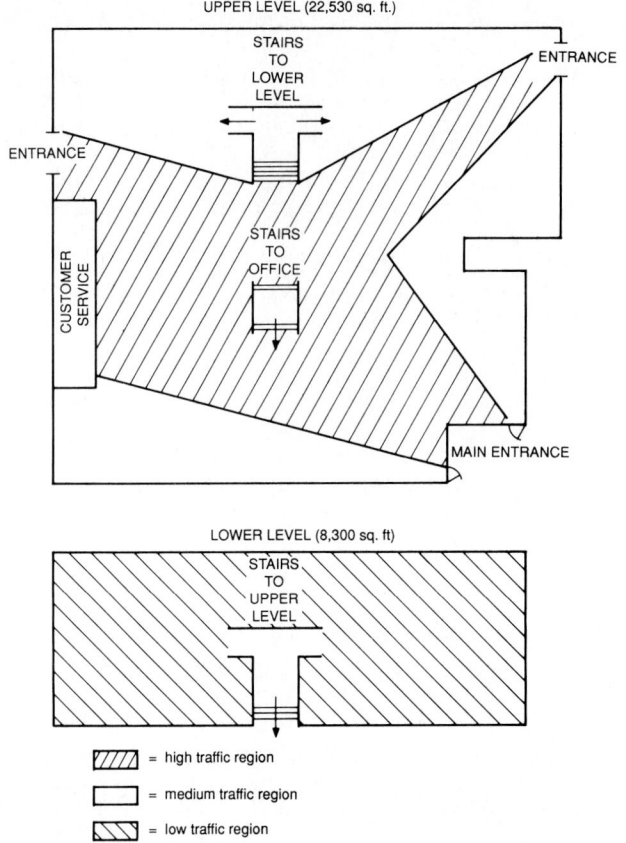

FIGURE 4.2   Traffic patterns and regions.

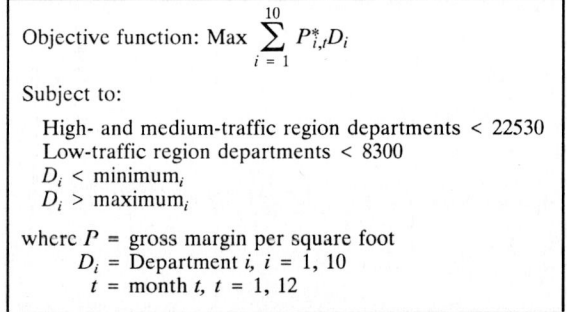

Objective function: Max $\displaystyle\sum_{i=1}^{10} P_{i,t}^* D_i$

Subject to:

  High- and medium-traffic region departments < 22530
  Low-traffic region departments < 8300
  D_i < minimum$_i$
  D_i > maximum$_i$

where P = gross margin per square foot
    D_i = Department i, i = 1, 10
    t = month t, t = 1, 12

FIGURE 4.3   Linear programming model to allocate floor space.

footage is allocated to each department throughout the year. For most departments the allocated space varies across the months between the maximum and the minimum allowed space. For a few less profitable departments, the allocated space is at minimum most of the months (Books, Men's Clothing, Women's Ready-to-Wear, and Art Supplies). Figure 4.4 compares the actual monthly profits and the predicted monthly profits based on the

TABLE 4.2 Linear Programming Solution for Floor Space Allocation, Square Feet for Departments by Month

					Department							
Month	Books	Women's Ready-to-Wear	Gifts and Small Appliances	Men's Clothing	School Supplies, Computers, and Stationery	Health and Beauty Aids, Candy and Food	Camera	Art Supplies	Radios, Records, and TV	Sports	Predicted gross margin	Actual gross margin
Nov	9000	2000	2400	2240	6472	2300	528	1300	1590	3000	378,930	349,303
Dec	9000	2000	2730	1500	6472	2300	528	1300	2000	3000	593,387	556,937
Jan	9000	2000	3730	1500	6472	2300	528	1300	2000	2000	301,235	258,582
Feb	9000	2000	4140	1500	6472	2300	528	1300	1590	2000	147,958	118,829
Mar	9000	2000	3730	1500	6100	2300	900	1300	2000	2000	251,094	225,901
Apr	9000	2000	2730	1500	6100	2300	900	1300	2000	3000	340,626	311,684
May	9000	2000	4140	1500	6100	2300	900	1300	1590	2000	176,568	159,891
Jun	9000	2000	3140	1500	6100	2300	900	1300	1590	3000	210,253	190,878
Jul	9000	2000	3140	1500	6472	2300	528	1300	1590	3000	223,725	199,324
Aug	9740	2000	2400	1500	6472	2300	528	1300	1590	3000	534,505	452,667
Sept	9000	2000	3140	1500	6472	2300	528	1300	1590	3000	532,531	445,346
Oct	9000	2000	3140	1500	6472	2300	528	1300	1590	3000	371,530	323,028
											4,062,342	3,592,370

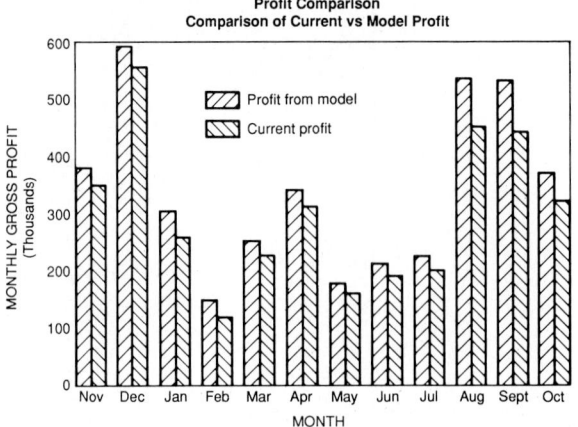

FIGURE 4.4   Profit comparison.

model space allocation scheme. In simulated results, the model predicts a 13 percent increase in gross margin for the year. Understandably, the actual results could be lower than these estimated results.

The last task was to determine the specific location and size for each department for each month. This decision was based on managerial judgment and the cross-correlation of sales. The correlation between the monthly sales for the departments in the high- and medium-traffic regions are shown in Table 4.3. These correlations were used as the similarity measures in multidimensional scaling. The two-dimensional map of these departments is shown in Fig. 4.5. (The departments in the low-traffic regions are located on the lower level and were not included in this analysis.)

The specific location of a department and the determination of which departments should be adjacent to each other were determined so that both departments with complementary products (high cross-elasticities) and departments with very different seasonal sales patterns were next to each other. The importance of having departments with com-

TABLE 4.3   Correlation Matrix of the Monthly Sales for Departments in the High- and Medium-Traffic Regions

	Books	Women's Ready-to-Wear	Gifts and Small Appliances	Health and Beauty Aids, Candy and Food	Radios, Records, and TV	Sports	Men's Clothing
Books	1.000						
Women's Ready-to-Wear	0.882	1.000					
Gifts and Small Appliances	0.023	−0.041	1.000				
Health and Beauty Aids, Candy and Food	−0.320	−0.274	0.706	1.000			
Radios, Records, and TV	0.425	0.551	0.474	0.393	1.000		
Sports	−0.134	−0.199	0.869	0.546	0.211	1.000	
Men's Clothing	−0.203	−0.193	0.896	0.705	0.575	0.762	1.000

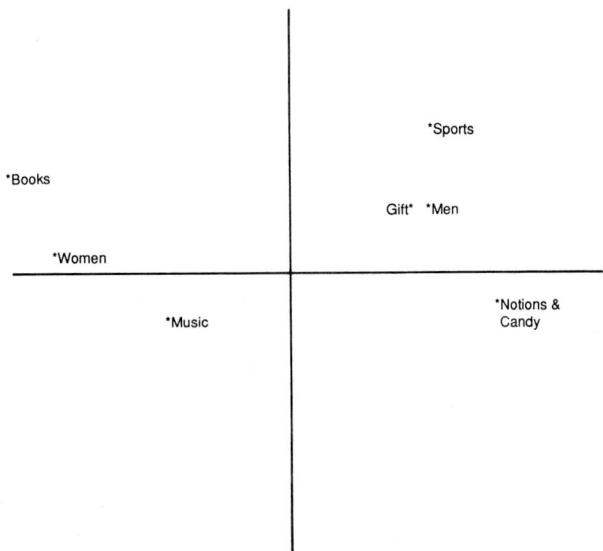

FIGURE 4.5 Multidimensional scale of departments.

plementary products next to each other is that it encourages impulse and complementary sales. On the other hand, placing departments with very different seasonal sales patterns next to each other allows the store to expand and contract the space allocated to each department monthly, as determined by the linear programming solutions.

The final location, the minimum and maximum square footage, and the direction of expansion and contraction of the departments are shown in Fig. 4.6.

Limitations of the Model.   A possible limitation of the model is the use of managerial judgment as a model input. If managers do not take the project seriously or are inept at estimating minimum or maximum department sizes, the profit can be less than optimal.

Using managerial judgment can be a mixed blessing. Managers can bring to the solution very valuable insight and wisdom. Also, involving managers can break down their resistance to the use of a model. Though managerial judgment may not always be accurate, it has been widely used in marketing models. For example, managerial inputs have been used in marketing mix decisions, new product introductions, retail management, and coupon promotion evaluation. The use of management judgment as input has been an accepted practice in modeling marketing decisions.

Also, the implementation of models with managerial input is likely to be smoother when management has had a significant impact on the decision process and model input.

The assumption of linearity is another possible limitation to this model. However, the maximum and minimum limits for each department were selected with this problem in mind. The ranges selected for each department were such that profit growth within these ranges was believed to be linear. Furthermore, the specific store in question had been renovated just a few years earlier, and some departments had been expanded and others reduced. The data for the different sizes of departments also supported the assumption that the profit-space relationships were near-linear within the specified ranges. The assumption of linearity in the model is not as serious a limitation as it may initially seem. Obviously, a researcher who felt that the relationships were not linear would use nonlinear programming instead of linear programming. Since the research indicated that the functional relationship between space allocation and profits was linear over the relevant range, linear rather than nonlinear programming was used.

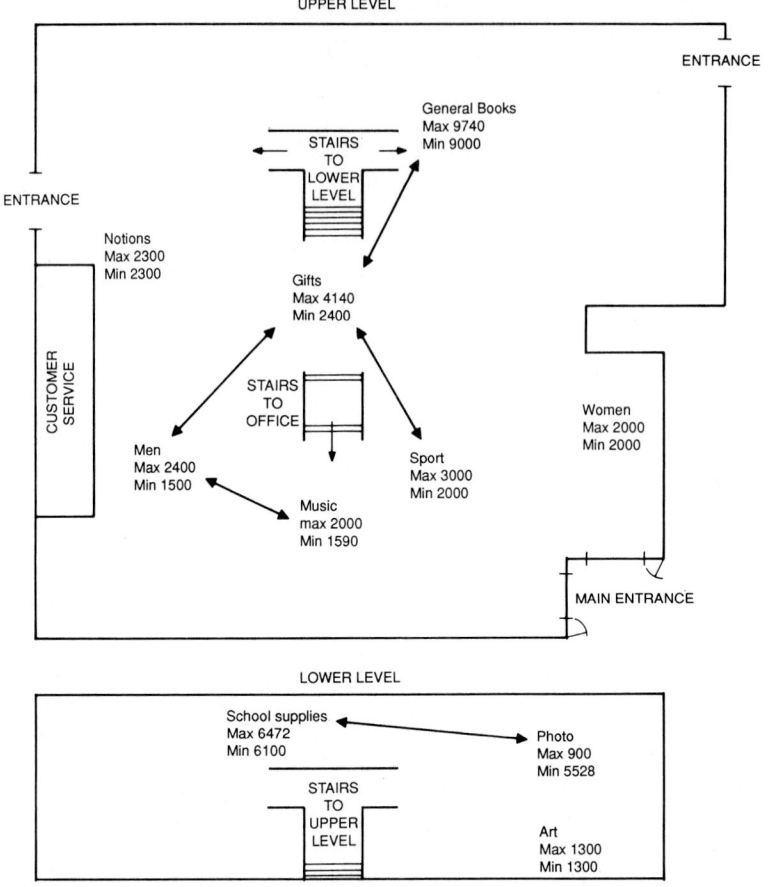

FIGURE 4.6 Final location for departments, with maximum and minimum square footage and patterns of expansion (arrows indicate the direction of expansion).

CASE 3: QUEUING THEORY AND THE NUMBER OF SALES STATIONS

It is very important in retail stores to provide the service level that keeps customers happy. In a retail store, customers are very service-conscious. Queuing theory is often used to determine how many checkout stands there should be to maintain a predetermined service level. The process is to identify the arrival rates at various times of the day for various days of the week, then to model service rates. The schedule for how many of the stations (checkout stands) to open at various times is determined by the model. In fast-food restaurants, the planning is done in 15-minute intervals.

CASE 4: OPERATIONS RESEARCH METHODS IN PRICE SETTING

Pricing is a very difficult managerial task for executives. Several models have been developed to set prices. One model used to set price in a bidding situation is the bidding bayesian model.

Bayesian Model.   Anderson and Sweeney[5] give an illustration of the bayesian model. Assume that the XYZ Company is bidding for exclusive manufacturing rights on a patent for widgets. There are three decision possibilities:

$$d_1 = \text{bid high}$$

$$d_2 = \text{bid at a medium price}$$

$$d_3 = \text{bid low}$$

The best decision will depend on the market's acceptance of the newly designed widget. For simplicity, suppose that the market acceptance will be in the form of one of only two possible states:

$$s_1 = \text{high market acceptance of widgets}$$

$$s_2 = \text{low market acceptance of widgets}$$

In accordance to the corresponding bid possibilities and states, the XYZ Company develops the following profit estimates:

	States of nature	
Alternatives	High acceptance (s_1)	Low acceptance (s_2)
Bid high, d_1	400	-50
Bid medium, d_2	300	40
Bid low, d_3	200	100

The company then develops probability estimates for the two possible states of nature (s_1 and s_2). Suppose that the company determines

$$P(s_1) = .4$$

$$P(s_2) = .6$$

The company is now in a position to determine the *expected monetary value* (EMV) for each decision alternative:

$$\text{EMV}(d_1) = 0.4(400) + 0.6(-50) = 130$$

$$\text{EMV}(d_2) = 0.4(300) + 0.6(40) = 144$$

$$\text{EMV}(d_3) = 0.4(200) + 0.6(100) = 140$$

Other things being equal, then, the middle bid (d_2) would be the best course of action.

Now, in order to make certain that the decision is a good one, the firm may want to do marketing research, which can then be used to update the original probability estimates of the two different states of nature. The following steps will help to decide how much the company should be willing to pay for the research.

Suppose that the marketing research will fall into one of two possible categories:

I_1 = favorable market report (those surveyed expressed much interest in widgets)

I_2 = unfavorable market report (those surveyed expressed little or no interest)

From here, the company will generally rely on historical data to determine the relationships between the two indicators and the two states of nature. Specifically, they must determine: given that a certain state of nature eventually turns out to be the actual, what is the probability that the marketing research would have predicted it as such? In other words, the following must be estimated:

$$P(I_1/s_1) \qquad P(I_1/s_2) \qquad P(I_2/s_1) \qquad P(I_2/s_2)$$

This is really a measure of the reliability of the marketing research.

Finally, the firm must determine the probabilities which correspond to the likely outcome of their marketing research $[P(I_1)$ and $P(I_2)]$, as well as the probabilities of the different states of nature, given the marketing research result:

$$P(s_1/I_1) \qquad P(s_2/I_1) \qquad P(s_1/I_2) \qquad P(s_2/I_2)$$

In determining the updated probabilities of the different states of nature, Bayes' theorem can be used:

$$P(s_j/I_k) = \frac{P(I_k/s_j)P(s_j)}{P(I_k)} \qquad (13)$$

These revised probabilities can be multiplied by the value of the state of nature to give revised expected monetary values. Finally, the company can estimate the cost of the research to determine if the research increased the EMV enough to justify the research being conducted.

One interesting expansion of the bayesian approach to pricing is to track competitors' bids on past projects and to incorporate that in the bidding process. Suppose, for example, that XYZ has one competitor, ABC Company. XYZ has tracked three states of nature for ABC. These are: (1) busy, (2) somewhat busy, and (3) desperate for work. When ABC is in state 1, it, on the average, bids 42 percent over XYZ's estimate of costs. In state 2, historically ABC has bid 24 percent over XYZ's estimated costs. Finally, when ABC is in state 3, it generally bids 7 percent over the costs that XYZ has estimated for the job. (*Note:* ABC does not know XYZ's estimated costs, but XYZ knows ABC's past bids and states of nature and so can develop the percent over estimated costs.)

We are now in a position to draw our decision tree, which is shown in Fig. 4.7. XYZ can then estimate ABC's state of nature and then incorporate this information into the bayesian model to determine its bid.

Other Pricing Procedures.   Price is a very easy variable to change. Thus, marketing managers often will make a best guess as to what price to charge within a range of acceptable prices. The range's upper bound is determined by what consumers view as the worth (value) of the product. The lower bound is total per-unit cost. Costs are determined from accounting records and worth is determined from marketing research. Morgenroth[6] has developed a technique for simulating a manager's price-setting behavior between these limits. The technique looks at evaluating the probabilities of different quantities sold at various prices within the range. Baumol[7] developed a model for pricing when only costs are known and a manager wants to change prices. His model is

$$\Delta PR = P_2 Q_2 - P_1 Q_1 - MC(Q_2 - Q_1)$$

where $P_1 Q_1$ = current revenue
$\qquad P_2 Q_2$ = new revenue
$\qquad Q_2 - Q_1$ = change in quantity sold
$\qquad MC$ = change in cost

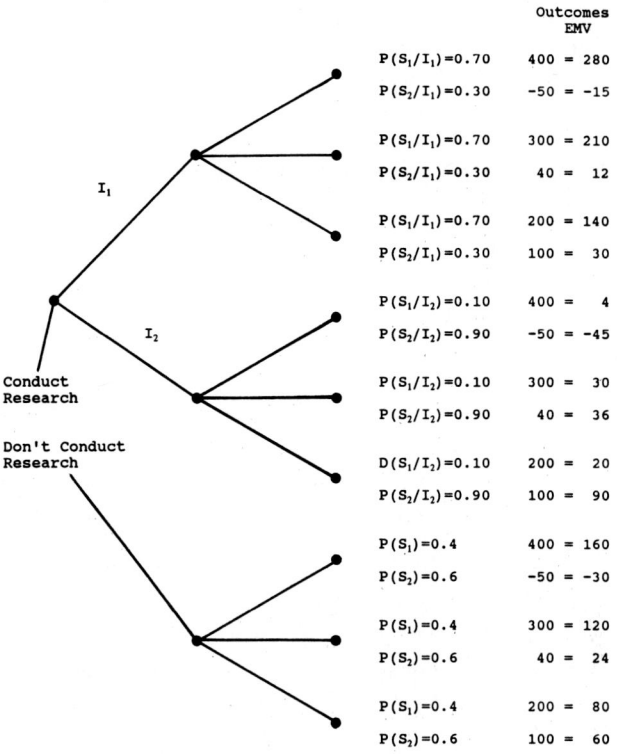

Outcomes
EMV

$P(S_1/I_1)=0.70$ 400 = 280
$P(S_2/I_1)=0.30$ −50 = −15

$P(S_1/I_1)=0.70$ 300 = 210
$P(S_2/I_1)=0.30$ 40 = 12

$P(S_1/I_1)=0.70$ 200 = 140
$P(S_2/I_1)=0.30$ 100 = 30

$P(S_1/I_2)=0.10$ 400 =   4
$P(S_2/I_2)=0.90$ −50 = −45

$P(S_1/I_2)=0.10$ 300 = 30
$P(S_2/I_2)=0.90$ 40 = 36

$D(S_1/I_2)=0.10$ 200 = 20
$P(S_2/I_2)=0.90$ 100 = 90

$P(S_1)=0.4$ 400 = 160
$P(S_2)=0.6$ −50 = −30

$P(S_1)=0.4$ 300 = 120
$P(S_2)=0.6$ 40 = 24

$P(S_1)=0.4$ 200 = 80
$P(S_2)=0.6$ 100 = 60

I_1

I_2

Conduct
Research

Don't Conduct
Research

FIGURE 4.7   Bidding decision tree.

A manager can estimate various profits or changes in profits by using the above model
and thus maximize profits. Basically, the model is a modification of the classical economic
theory of pricing, where MC = MR.

CASE 5: SITE SELECTION MODELS

Consistently, convenience has been found by marketing researchers to be a major factor in
selecting a product and where to buy that product. As a result, much effort has been devoted to using quantitative tools to select sites for retail establishments.

A very simple model is the *checklist method*. This requires the evaluation of the following by yes/no answers and tabulating the number of yeses.

1. Trading area potential
2. Accessibility
3. Growth potential
4. Business interception
5. Cumulative attraction potential
6. Compatibility
7. Competitive hazards
8. Site economics

Often this model is modified by weighting the variable and giving it a value of 1 to 10 rather than yes/no answers.

The *analog method* looks at people and sales potential as follows:

Zone radius, miles	Population in zone	Estimated per capita sales	Estimated weekly sales	Computed drawing power, %
0–0.25	4,700	2.00	9,400	28
0.25–0.5	12,900	0.76	9,804	29
0.5–0.75	23,000	0.22	5,060	15
0.75–1	36,300	0.12	4,356	13
Beyond			5,051	15
			33,671	100

This is done for each possible site, and the sales potentials of all sites are compared.

The *Reinitz model* looks at sales potential somewhat like the analog method. The procedure for a gas station would be:

1. Choose a local-area radius, usually 1 mile. Obtain car population and gasoline use.

2. Obtain a census of existing outlets, and rate them by a number of attributes; r_{ij} = rating of outlet i along attribute j.

3. Obtain (from consumers) importance weights of attributes: w_j = average importance weight of attribute j.

4. Now estimate local potential:

$$f_i = \frac{\Sigma_j w_j r_{ij}}{\Sigma_i \Sigma_j w_j r_{ij}} GL \tag{14}$$

where G = annual product consumption in area (gasoline); L = fraction of sales average customer buys locally; i = index covering all outlets in local area.

5. Now estimate transient sales potential. Determine how many potential transient customers there are and how many of these will stop. In the equations below, k = the number of ways to pass the site, L = road length (usually a few miles), R = set of road legs among transient trading routes, L_0 = average distance between refuelings, and q = average miles per gallon. Assuming that the amount of gas left in the tank is a random variable,

$$p \text{ (needing fuel)} = \frac{L}{L_0} \tag{15}$$

and the average quantity purchased is

$$Q = L_0 q \tag{16}$$

If traffic flow along route k is T_k cars per day, then the amount of gas bought per day is

$$G_k = T_k Q p \text{ (needing fuel)}$$

Along each route going by the site, this site would share the potential according to its attractiveness:

$$g_{ik} = \frac{\Sigma_j W_j r_{ij}}{\displaystyle\sum_{i \varepsilon R_k} \Sigma_j W_j r_{ij}} T_k L_0 q \qquad (17)$$

where g_{ik} = potential along route k and R_k = set of indexes of stations along route k. So the transient sales potential for the site is

$$g_i = \sum_k g_{ik} \qquad (18)$$

where g_i = total transient potential for site i and k = ranges over all transient routes that are associated with outlet i.

6. The potential for any site is then $f_i + g_i$ (local and transient).

A major problem in marketing is to determine the number of outlets that a firm should have. Empirical studies show a relationship between the number of outlets and market share. This is generally S-shaped.
Measure aggressiveness:

$$a = \frac{\text{number of recent outlets/total outlets}}{\text{number of recent industry outlets/total industry outlets}}$$

$$\text{Market share} = (a, s)$$

where a = aggressiveness and s = outlet share.
The challenge is to maximize the net present value (NPV) of outlets. Maximize:

$$z = \sum_{i=1}^{m} \sum_{t=1}^{y} \sum_{j=1}^{Xrt} V_{ijt} \qquad (19)$$

where m = markets, y = years, and X = outlets in market. Try different alternatives and combinations, and find the maximum. Basically, the procedure is to maximize the *incremental* NPV.
Build a table with markets and the incremental NPV of an added outlet in a given market:

Market	Outlets						
	1	2	3	4	5	6	7
1	5	4	5.6	3.2	4.3	2	3.4
2	3	2	6	7	3.2	4	5.4
3	5.5	3.2	6.5	3.4	5	6	2.2

In this example, NPV is maximized by adding 4 outlets to market 2. So those outlets are added, and the table is reconstructed, and the process is repeated.
The adding of new outlets in markets is subject to constraints such as site availability.

REFERENCES

1. Bass, F. M., "A New Product Growth Model for Consumer Durables," *Management Science,* January 1969, pp. 215–227.

2. Juster, F. T., "Consumer Buying Intentions and Purchase Probability: An Experiment in Survey Design," *Journal of the American Statistical Association,* vol. 61, September 1966, pp. 658–696.

3. De Groot, M. H., "Reaching a Consensus," *Journal of American Statistical Association,* vol. 69, March 1974, pp. 118–121.

4. Geurts, M., and G. Wheeler, "Converging Conflicting Research Findings: The Oregon Bottle Bill Case," *Journal of Marketing Research,* November 1980.

5. Anderson, D. R., and D. J. Sweeney, *An Introduction to Management Science,* West Publishing, St. Paul, 1982, pp. 414–434.

6. Morgenroth, W., "A Method for Understanding Price Determinants," *Journal of Marketing Research,* vol. 1, August 1964, pp. 17–26.

7. Baumol, W., *Economic Theory and Operations Analysis,* Prentice-Hall, Englewood Cliffs, N.J., 1961, pp. 396–397, 406–407.

CHAPTER 5
LINEAR PROGRAMMING

Hugh E. Warren
Professor
School of Business and Economics
California State University, Los Angeles, California

Linear programming deals with the efficient use of resources. Surveys have shown it is one of the most widely used mathematical techniques.[1] Early commercial applications were made in petroleum refining and in the manufacture of livestock feeds. The use of linear programming quickly spread to transportation of commodities, to scheduling of work crews and equipment, to location of facilities, and to coordination of large-scale manufacturing operations.

MODEL DESCRIPTION

There are similarities between linear programming and other quantitative models. Linear programming is a deterministic, optimizing model. This means the model assumes that its parameters are fixed numbers and that given input produces a unique answer that maximizes or minimizes a quantifiable objective. A common objective in maximizing problems is to increase some measure of the profit derived from the operation being modeled. Conversely, decreasing cost is a common objective in minimization problems.

The answer produced by an optimizing model consists of the optimal value of the objective and the values of so-called decision variables, which express how the optimal objective is achieved. For example, the objective might be to maximize the profit from the manufacture and sale of a variety of different products. The decision variables would include the quantity of each product to be manufactured. Linear programming has one more general feature. It is a constrained optimizing model; that is, it makes explicit allowance for adding constraints on the decision variables.

The distinguishing feature of the linear programming model is the requirement that the objective be a linear function of the decision variables and that the constraints be linear equations or linear inequalities. In algebraic notation the linear programming problem takes the form

Choose decision variables $x_1, x_2, \cdots, x_n$

to maximize or minimize $c_1 x_1 + c_2 x_2 + \cdots + c_n x_n$

subject to $a_{11}x_1 + a_{12}x_2 + \cdots + a_{1n}x_n \leq b_1$

$$\cdots$$

$$a_{m1}x_1 + a_{m2}x_2 + \cdots + a_{mn}x_n \leq b_m$$

There are two more technical requirements for a linear program. The first is that the decision variables be nonnegative, that is,

$$x_1 \geq 0, x_2 \geq 0, \ldots, x_n \geq 0$$

The second technical requirement states that, subject to nonnegativity, the decision variables can potentially take on any numeric value. This is called the *divisibility requirement* because it says in particular that fractions are acceptable answers.

In any mathematical model, assumptions and approximations are made in quantifying the objective, the decision variables, and the constraints. The question to be asked with respect to a potential application of linear programming is whether reasonable assumptions will satisfy the requirements for linearity, nonnegativity, and divisibility. An objective function which measures the temperature or yield of some physical process may well be a nonlinear function of the input variables. Nevertheless, linearity in the objective function is common when the objective is profit or cost and the decision variables measure units of output or input. The coefficients of the objective function are then the unit profits or unit costs. Usually the accounting systems that define costs and profits are themselves linear models.

Many functions that show nonlinear behavior at a large scale are approximately linear over restricted ranges. Consumption of resources as a function of output is typically nonlinear at the scale considered by economists. In the context of scheduling 1 week's production on a specific machine it might be very accurate to say that each of several products requires a fixed time per unit and that the constraint imposed by machine time is

$$a_1 x_1 + a_2 x_2 + \cdots + a_n x_n \leq b$$

where the a_i coefficients are the time required for one unit of each product and b is the total time available.

The nonnegativity requirement is rarely a matter of difficulty. Many measurements are by nature positive. The requirement of divisibility deserves more attention, since in practice many resources are bought or used in large standard batches and units of output may be individually significant. Dropping the divisibility requirement leads to the special area of integer programming, which is discussed in a separate section below.

The versatility of linear programming comes about because reasonable assumptions and approximations often lead to objective functions and constraints that are linear in the decision variables. The term *reasonable* is used to mean that the benefits from linear programming regularly exceed the costs of the analysis.

MAJOR APPLICATIONS

A few examples are briefly described here to indicate the scope of linear programming.

Product Mix. The objective is maximization of profit. The decision variables are the quantities of each product that should be produced in some coming period of time, such as the next day, week, or month. The objective function is commonly total contribution margin, which is defined as revenue minus the variable costs, that is, costs which vary in direct proportion to the quantity of each output. Final profit is contribution margin minus so-called fixed costs, which will not change over the period considered. The constraints include limitations on raw materials, labor, and plant capacity. Other constraints express required ratios between various products, such as between components and final assemblies. Still other constraints embody market conditions, for example, existing contracts or total demand.

Least-Cost Material Mix. A special application is finding low-cost diets, in which the decision variables are the quantities of various foods. The objective function is the total cost of the diet over a period of time. The basic constraints are nutritional requirements: so many calories, so much protein, so much of each vitamin and mineral. Additional constraints express factors such as palatability and variety. This problem has many refinements and extensions in practice. For example, a manufacturer of livestock feeds can combine the purchase of grains and other commodities over several periods of time for dozens of final feed products. The solution in this case consists of target purchase quantities by commodity and time period as well as recipes and a production schedule for the final products. More generally, material mix for any production process can be considered. An application to the use of wood products is given by Carino.[2]

The Transportation Problem.   The objective is to move some commodity, for example, wheat or crude oil, from various source locations to various destinations at least total cost. The primary constraints are the supplies available at each source and the demands at each destination. The decision variables define a shipping plan, that is, the quantities to be shipped from each source to each destination. Refinements and extensions of the basic problem are continually being introduced. For example, see Denardo.[3]

The Scheduling Problem.   The problem is to provide a service at stated levels with least cost or at levels to be determined with greatest profit. The objective might also be stated in terms of maximizing utilization of resources. The decision variables link equipment and personnel to tasks and periods of time. The variables are typically binary, that is, of value 1 if a particular assignment is made and of value 0 if the assignment is not made. Common constraints involve the number of successive time periods that can be worked and the locations or conditions associated with different tasks. Practical applications have been made in plant scheduling and the transportation industry. For recent examples see Frendewey,[4] Potts,[5] and Abara.[6]

Warehouse or Facilities Siting.   The objective is to minimize construction and operating costs by advantageous location of warehouses or other facilities. The decision variables specify locations, capacities, and ultimate operating levels. The constraints express available locations and the operating requirements. Links have been made between this general problem and the requirements of just-in-time manufacturing.[7]

The power of linear programming lies in its ability to simultaneously consider many possible courses of action in the context of many underlying constraints. In many applications the decision variables and constraints number in the hundreds. Moreover, some problems can contain aspects of all of the above examples. Advances in computer hardware and the refinement of the numerical methods for solving linear programs have increased the importance of the maintenance of databases for large problems and the presentation and interpretation of the results.

ELEMENTARY EXAMPLES

Although modern applications of linear programming tend to be very large, there are important concepts in the formulation and interpretation of linear programs that are best learned from simple examples.

Feed Example with One Constraint.   For a first brief example consider the production of a livestock feed that is formulated from corn and soybeans. Each kilogram of feed is required to provide 3600 calories of food energy. Current cost and calorie content per kilogram of each ingredient are

	Corn	Soybeans
Cost, cents	12	26
Calories	3600	4200

A formulation of this situation as a linear program is

Word description	Algebraic description
Decision variables:	
kg of corn per kg of feed	CORN
kg of soybeans per kg of feed	SOYB
Objective:	
Minimize cents per kg of feed	12 CORN + 26 SOYB
Constraint:	
Calories per kg of feed	3600 CORN + 4200 SOBY ≥ 3600

The minimal value of the objective function can be found by computing the cost per calorie from each ingredient:

	Corn	Soybeans
Cents per calorie	0.0033	0.0062

Since calories from corn are cheaper, and since corn has the minimal calorie content per kilogram, the optimal mix is CORN = 1, SOYB = 0, and the cost of a kilogram of feed is 12 cents. In one sense that is the answer to the problem.

A major value of linear programming is that it provides more information than the answer to the basic optimization question. Consider in this example what would happen if the factory ran out of corn and had to substitute soybeans to some extent. The cost of the feed would increase. The rate of cost increase would be of interest if, for instance, management were comparing the cost of stock outages to the cost of larger inventories. This rate change in the objective function is called a *reduced cost* and is one of the many extra numbers provided by linear programming. Reduced costs will be calculated in the examples given below.

Maximization problems can also occur where there is a single major bottleneck, possibly a material in short supply or limited time available on some machine. This means there is a single constraint in the problem formulation. Suppose profit is being maximized, and the variables represent quantities of various products produced. The problem is solved by computing the profit per unit of the constrained resource for each of the products. The product with greatest profit per unit of constrained resource is made exclusively, and none of the other products are made. If one unit of a suboptimal product were produced, say for an important customer, then the total profit would decline by the reduced cost for that product.

Feed Example with Two Constraints. Analysis by cost or profit per unit of constraint works in any situation having just one pressing constraint. Suppose, however, that a second nutritional requirement is added to the recipe for livestock feed, namely that a kilogram of feed must contain at least 150 grams of protein. Given that the contents per kilogram are

	Corn	Soybeans
Calories	3600	4200
Grams of protein	90	370

the problem is to find the recipe that minimizes the cost of the feed. A formulation of the linear program is as before, with the addition of one constraint:

Word description	Algebraic description
Constraint	
Protein, grams per kg of feed	90 CORN + 370 SOYB ≥ 150

The modified problem can be graphed by using cartesian coordinates. This is shown in Fig. 5.1, in which the horizontal axis measures kilograms of corn and the vertical axis,

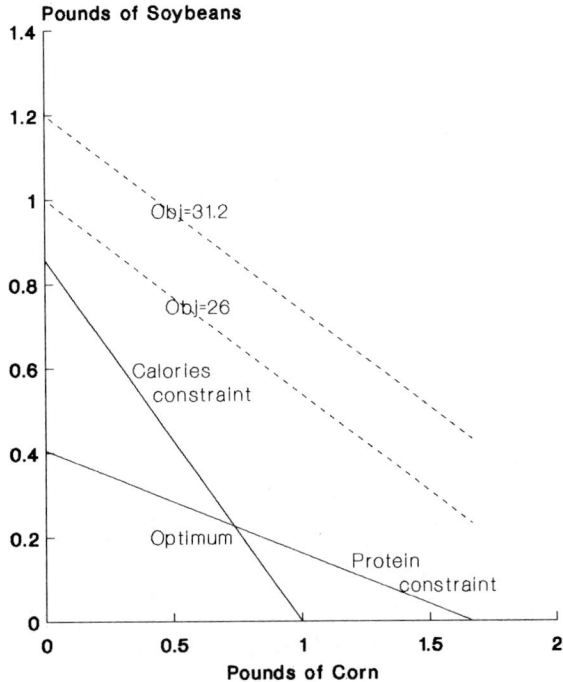

FIGURE 5.1 Graph for the feed example.

kilograms of soybeans. The solid slanting lines represent the lower limits of the calorie and protein constraints. The dashed lines represent lines of equal total cost. The least cost within the stated constraints occurs at

$$CORN = .736 \text{ kg and } SOYB = .226 \text{ kg}$$

and the total cost of the corn and soybeans is 14.7 cents. The method that obtains the solution will be described below. In practice, solutions are provided by well-tested computer programs. It is important even in simple problems to understand that the formulation must be an accurate picture of reality and that the various parameters in the formulation must often be gathered with some effort. In the feed problem, one might note that the corn and soybean recipe adds up to only 0.962 kg. Perhaps the formulation should have allowed for low-cost inert ingredients or included a constraint that the recipe total 1 kg.

Kiln Example.   The next example traces the development of a linear program from engineering and accounting data to specification of decision variables, an objective function, and constraints. It then proceeds to a solution and to interpretation of the solution. Standard terms will be defined in the course of the presentation.

Data for a linear program often come from several sources. Identifying, gathering, and organizing the relevant data are the first steps in solving the problem. In this example, a kiln operates around the clock in 6-hour cycles. A main ceramic housing and a ceramic insert are both fired in the kiln, but at different temperatures. Each main housing is sold with one insert. In the near term, there is high customer demand for the housing and for additional replacement inserts. Management wants to operate as profitably as possible.

A kiln cycle begins with loading the empty kiln, which can hold either 120 main ceramic housings or 240 ceramic inserts. The kiln is raised to the firing temperature for a specified period, then cooled and unloaded. Heat comes from natural gas. A cycle for main housings requires 150 therms of gas. A cycle for inserts requires 200 therms. Seasonal restrictions in the supply of gas limit the kiln to consumption of 10,000 therms in the next 2 weeks.

Firing the inserts produces a toxic by-product at the rate of 10 grams per insert. Disposal of the toxic by-product is currently limited to 30 kg per week. There is no such disposal problem related to the main housings, but a contract to provide 1200 housings must be filled from the next 2 weeks of production.

Revenue and cost information is given in the following table. Amounts are per unit.

Product	Main ceramic housing	Ceramic insert
Revenue	$25.52	$14.68
Incremental costs:		
Unfired unit	$11.680	$4.200
Gas	0.525	0.350
Toxic disposal	0	1.000
Packaging	1.745	0.860
Contribution margin	$11.57	$8.27

The incremental costs are those which will change in the near future as production volume changes. Incremental costs are also referred to as variable costs. Contribution margin is defined as revenue minus variable costs and is a form of profit suitable for short-term decision making. This example will specify the objective to be maximization of contribution margin from production in the next 2 weeks.

The choice of decision variables often involves the stage at which we measure activity, whether at input or output or some intermediate stage. Input and intermediate variables are convenient for personnel who carry out production. For example, the number of kiln cycles for main housings would be convenient for production scheduling. Different measures can be incorporated as separate decision variables so long as equations expressing the conversion factors are included as constraints. In this example the decision variables will be the number of output units of each product. A formulation of the kiln example appears in Fig. 5.2.

```
Word Description                              Algebraic Description

Decision variables:

 units of main ceramic housing
  produced and sold in next two weeks         HOU

 units of ceramic insert
  produced and sold in next two weeks                     INS


Objective:

 maximize contribution margin

  in next two weeks                           11.57 HOU + 8.27 INS


Constraints:

 kiln time, hours in 2 weeks                  .05 HOU + .025 INS ≤ 336

 gas supply, therms in 2 weeks                1.25 HOU + .833 INS ≤ 10000

 toxic disposal, in kg per week                         .005 INS ≤ 30

 contract for housings                        HOU                ≥ 1200

 at least one insert per housing              HOU -       INS ≤ 0
```

FIGURE 5.2 Formulation of the kiln example.

The algebraic description shows the linearity of the objective function and the constraints. A number of computations have been done to derive the coefficients from the original data. The coefficients of the objective function were previously obtained by subtracting variable costs from revenue. Note also, for example, that the kiln time coefficient of HOU is

$$0.05 \text{ hours/unit} = \frac{6 \text{ hours/cycle}}{120 \text{ units/cycle}}$$

and the gas supply coefficient of INS is

$$0.833 \text{ therms/unit} = \frac{200 \text{ therms/cycle}}{240 \text{ units/cycle}}$$

Deriving the coefficients of the objective functions and the constraints from more fundamental data is usually automated, and that derivation process is called *matrix generation.* The algebraic description is arranged in columns by variable so that the coefficients do indeed form a matrix.

The matrix of coefficients is the input to the linear programming solution algorithm. For the moment, the algorithm can be regarded as a black box, and the solution to the problem can be briefly stated as

    Earn $92,660 in contribution margin

    With the production of

Main ceramic housings	3720 units
Ceramic inserts	6000 units

All of the linear programming routines now in use actually provide a great deal of additional information about the nature of the constraints and objective function.

In this simple example, the constraints can be plotted by using algebraic geometry and coordinate axes corresponding to the two decision variables. In Fig. 5.3 note that each constraint defines a half-plane. Replacing the inequality sign in each constraint with an equality gives the equation for the straight line that bounds the half-plane. The coordinate axes themselves are considered boundaries for the so-called natural constraints HOU $\geq$ 0 and INS $\geq$ 0.

The intersection of all the half-planes is called the *feasible region,* because the coordinates of the points it contains correspond to production plans that fulfill all the stated constraints. The feasible point with the greatest value for the objective function must be one of the corners. This can be seen by observing that the objective function is constant along each of the dashed lines. In fact, plotting any line of the form

$$11.57 \text{ HOU} + 8.27 \text{ INS} = \text{constant}$$

will produce a straight line parallel to the two shown. Such lines of equal profit are called *isoprofit lines.* The maximum profit in the feasible region occurs at a corner because (1) the isoprofit lines are straight, (2) profit continually increases in one direction perpendicular to the isoprofit lines, and (3) the feasible region is a convex polygon. Note that a region is convex if, for each pair of points it contains, it also contains the straight line segment between those points.

The picture suggests a number of questions. For example, what is the role of the gas constraint? If this constraint were removed, the feasible region would remain the same. A constraint with this property is called *redundant.* If, however, the quantity of gas available were decreased from 10,000 therms, then the limiting line would move down and to the left, and eventually the gas constraint would affect the feasible region. For future planning it would be useful to know at just what lower gas supply this impact is first felt. This is an example of *right-hand side* analysis, so called because it investigates changes in the right-hand column of the algebraic description.

Another right-hand side question naturally arises from consideration of the toxic dis-

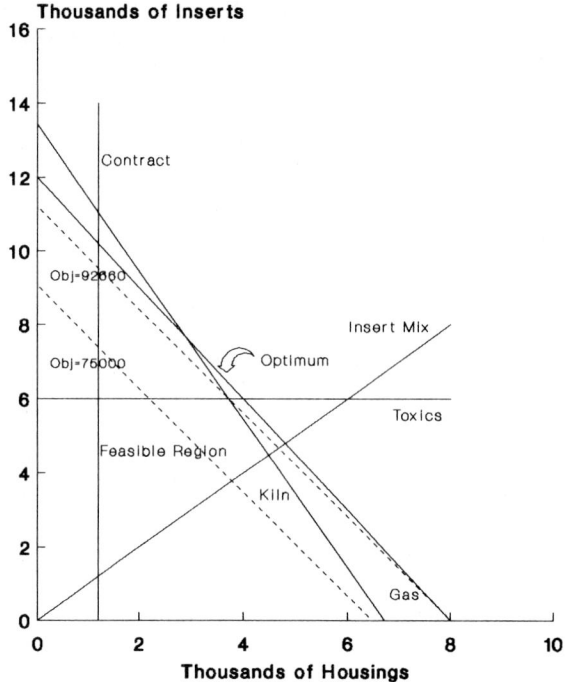

FIGURE 5.3 Graph for the kiln example.

posal constraint. If the limit of 30 kg per week could be increased, then the optimal profit would increase. What would increased disposal capacity be worth, and how much capacity would be useful? The diagram suggests the following qualitative scenario. Steadily raising the 30-kg limit will geometrically raise the horizontal limit line and increase profit at a steady rate. The constraints whose intersection defines the optimal values of the decision variables are called *active*. At 30 kg the toxic disposal and the kiln constraints are active and will continue to be active until the toxic disposal limit line reaches the point (2880, 7680), which is the intersection of the kiln and gas constraints. After that point the kiln constraint becomes inactive, the gas constraint becomes active and the profit function continues to increase, but at a slower rate.

Another set of questions revolves around the objective function. Suppose competition from a country with less demanding environmental laws forces a reduction in the selling price of inserts. Profit per unit would also decrease. Intuitively it would seem that profit could be increased with a mix of more main housings. Geometrically a decrease in the profit coefficient of INS causes the dashed isoprofit lines to rotate clockwise. The optimal profit will steadily decrease, but the optimal product mix will stay the same until the isoprofit lines become parallel to the limit line of the kiln constraint. Under that condition there are *alternative optimal points*; that is, any point on the boundary of the feasible region and on the kiln constraint will produce the optimal profit. A further reduction in the profitability of inserts will switch the optimal mix to the intersection of the kiln constraint and the need for one insert for each housing. In addition, the toxic disposal constraint becomes inactive.

These questions are answered quantitatively by what is called *sensitivity*, or *post-optimality*, analysis. Determining the limits where major changes occur is called *ranging*. The kiln example will be discussed further in the section on sensitivity analysis below.

The solution algorithms for linear programming by their nature produce answers for many of the sensitivity questions. The algorithms proceed in two parts: (1) addition of new variables to convert all the inequality constraints to linear equations and (2) repetitive solving for one group of variables in terms of others.

Quantitative answers to the sensitivity questions that have been raised require an algebraic approach. Just enough of the technique will be given here to aid in the interpretation of the output from the computer algorithms. The first step is to introduce variables which measure the distance to the limiting value of the inequality constraints. For less-than-or-equal-to constraints, these new variables are called *slack* variables. In the kiln example, the first three constraints become

Constraint	Equation with slack variable			
Kiln	.05 HOU + .025 INS + SKIL			= 336
Gas	1.25 HOU + .833 INS	+ SGAS		= 10000
Toxics	.005 INS		+ STOX	= 30

The constraint requirement for main housings is a greater-than-or-equal-to constraint. Subtracting a *surplus* variable expresses this type of constraint as an equation. In summary, the inequality constraints become linear equations by the addition of a slack variable or the subtraction of a surplus variable. The original variables are referred to as the *decision variables* or *structural* variables.

Constraint	Equation with surplus or slack variable			
Contract	HOU	− SCON		= 1200
Insert per housing	HOU	− INS	+ SMIX	= 0

The geometry of Fig. 5.3 shows that the optimal point is the intersection of the kiln and toxics limit lines. Algebraically this point is defined by

$$\text{SKIL} = 0 \quad \text{and} \quad \text{STOX} = 0$$

Substituting these zero values into the five equations that represent the constraints, one can solve for the remaining five variables. The solution values are

Main ceramic housings	HOU = 3720 units
Ceramic inserts	INS = 6000 units
Slack in gas constraint	SGAS = 350 therms
Surplus in contract	SCON = 2520 units
Slack in insert constraint	SMIX = 2280 units

In general, solving five equations in five unknowns would be onerous if done by hand. The job is easy in this example because each equation contains only a few of the variables. Systems of equations with this property are called *sparse*. The equations for real-world linear programs are almost always sparse, and the commercial computer algorithms take advantage of that fact.

The slack in the gas constraint answers the question about how much further the gas allotment could be curtailed before it affected the optimal production mix. The amount, 350 therms, is only 3.5 percent of the current 10,000-therm allotment. A small further curtailment or a small overrun in gas usage would reduce the possible profit.

Instead of setting SKIL and STOX explicitly equal to zero, it is possible to solve for the other variables in terms of SKIL and STOX algebraically. In particular,

$$\text{INS} = 6000 - 200 \text{ STOX}$$

and

$$HOU = 3720 - 20 \text{ SKIL} + 100 \text{ STOX}$$

Substituting these expressions into the objective function yields

$$Profit = 11.57 \text{ HOU} + 8.27 \text{ INS}$$

$$= 92{,}660.4 - 231.4 \text{ SKIL} - 497 \text{ STOX}$$

This last form of the objective function makes it clear that the maximum value occurs when there is no slack in the kiln and toxics constraints. The coefficients of SKIL and STOX also show the value of capacity in these constraints. An hour of kiln time is worth $231.40 and a kilogram per week of toxic disposal is worth $497. In each case the units of measurement are the same as for the right-hand side of the original constraint. The values are called the *dual prices* of the constraints. They are also called *shadow prices* or *marginal values* or *opportunity costs.*

In general, the active constraints have positive dual prices and the dual prices of the inactive constraints are zero. A so-called degenerate condition can occur when an active constraint shows a zero dual price. This would happen in the kiln example if the gas allotment were lowered by the critical 350 therms to 9650. Geometrically the limit lines for the kiln, gas, and toxics constraints would intersect in the single point (3720, 6000). In computer output, degeneracy is signaled by a constraint with zero slack or surplus and zero dual price.

The essence of degeneracy is that removing just one of the constraints that actively limit the objective function will not lead to any improvement. This is rare in small practical problems, but is more likely in very large problems with many variables and constraints.

SOLUTION METHODS

Simplex Method. The computer programs that solve the general linear programming problem use an algorithm called the *simplex method.* Many mathematical techniques foreshadowed the discovery of the simplex method by George Dantzig in the 1940s.[8] The method has been refined over the years by many workers. The refinements in the algorithm and advances in computer hardware have produced commercial programs that can rapidly solve sizable problems even on desktop computers.

The simplex algorithm begins by adding slack and surplus variables as in the kiln example above. It adds an additional variable, called an *artificial* variable, to each greater-than-or-equal-to constraint and each equality constraint. The artificial variables are added as a means to test whether there is any feasible region at all. Recall that a feasible point is one at which all the variables, including added variables, are nonnegative. If there were originally m constraints, and if variables are added so that the total number of variables is n, then n is made large enough so that $n - m$ variables can be set equal to zero, or other given bound, and the remaining m variables will be nonnegative.

To summarize, the simplex method deals with structural variables, which quantify the decisions to be made, and with *logical* variables—slack, surplus, and artificial—that serve the logic of the algorithm.

Solving for m variables in terms of the remaining $n - m$ is fundamental to the simplex method. The m that are solved for are called the *basic variables,* or simply *the basis.* The remaining variables are called *nonbasic.* Choosing a basis corresponds to choosing a point at the intersection of the limits of $n - m$ constraints. If the point is feasible, it is an extreme point of the feasible region. In essence, the simplex method finds an initial extreme point of the feasible region and then moves to an adjacent extreme point. Moving to an adjacent extreme point corresponds to exchanging a variable in the current basis for a current nonbasic variable. The nonbasic variable that enters the basis is chosen so as to rapidly increase the objective function in a maximization problem, or decrease it in a minimization problem. Different versions of the algorithm use varying criteria for choosing the new basis.

In the kiln example the simplex method would proceed as follows. Slack and surplus variables are added as before to express the constraints as equations. An artificial variable ART is added to the contract constraint. The number of constraints is $m = 5$. The total number of variables, structural and logical, is $n = 8$. The objective function and constraints are then expressed in following matrix form, which is called the *initial tableau*.

	HOU	INS	SKIL	SGAS	STOX	SCON	ART	SMIX	RHS
Objective	11.87	8.27	0	0	0	0	$-M$	0	
Kiln	1.25	0.833	1	0	0	0	0	0	336
Gas	0.05	0.025	0	1	0	0	0	0	10,000
Toxics	0	0.005	0	0	1	0	0	0	30
Contract	1	0	0	0	0	-1	1	0	1,200
Insert	1	-1	0	0	0	0	0	1	0

The column headings are the algebraic abbreviations of the variable names. The numbers below the variable names are the coefficients of the variables in the objective function and the constraints. There is an implied equals sign before the right-hand side (RHS) column, so that the initial tableau is equivalent to a matrix equation $Ac = b$. Variations of this format correspond to different conventions for writing equations.

The coefficients of the objective function appear immediately below the variable name. Initially, the slack and surplus variables have zero coefficients in the objective. The letter M stands for a very large number. If the problem is feasible, the artificial variable will have the value zero, canceling out the large coefficient.

The original kiln problem was two-dimensional. Constraints appeared as half planes with limiting lines, and the intersection of two lines defined a point. The artificial variable adds a third dimension perpendicular to the plane of the structural variables. In three dimensions, constraints are half spaces with limiting planes, and the intersection of three planes defines a point. In turn, each plane is defined by setting one of the variables equal to zero. The initial feasible point is at the intersection of the planes

$$HOU = 0 \quad \text{and} \quad INS = 0 \quad \text{and} \quad SCON = 0$$

Another way of saying this is that the initial basis consists of the slack variables and the artificial variable. In particular the value of the artificial variable is ART = 1200.

The feasible region in three dimensions can be visualized as an infinitely high prism whose base is the original two-dimensional feasible region. The sides of the prism are all vertical planes, except for the limiting plane of the contract constraint, which slants left to intersect the plane HOU = 0. The initial feasible point has coordinates HOU = 0, INS = 0, ART = 1200. In its first few iterations the simplex method will respond to the large negative coefficient $-M$ by removing ART from the basis. From then on each set of nonbasic variables will correspond to an extreme point of the original feasible region in the plane ART = 0.

In the general linear programming problem, the simplex method changes the set of basic variables to improve the value of the objective function computed by setting the nonbasic variables equal to zero, or other given bounds. Geometrically this corresponds to moving from one extreme point to an adjacent extreme point. An exception to this interpretation occurs at degenerate points that are the intersection of more than the minimum number of constraint limits. In such a case, the basis can change without a change in the extreme point.

The iterations of the simplex method stop when it is impossible to improve the objective function further. In maximization problems this occurs when all the coefficients of the variables in the reformulated objective are zero or negative, as in the kiln example. For minimization problems the condition for optimality is that the reformulated coefficients are zero or positive. If the optimality condition is reached while the basis still includes artificial variables, then the problem is infeasible.

Implementations of the simplex method include special techniques to speed computation and reduce the requirements for computer memory. Each step of the simplex method involves changing the tableau, which is the same as recalculating the matrix that represents the tableau. It turns out that the matrix can be partitioned so that only part of it needs to be stored in memory and recalculated. This modification is referred to as the *revised* simplex method.

The procedure described above for finding out if there is indeed a feasible region is called the *big-M* method. In practice, most routines use a different approach called the *two-stage* method. This method is based on the fact that the feasible region does not depend on any particular objective function. In the first stage, the objective function is set equal to the sum of the artificial variables and then minimized. If the minimal value is positive, then the problem is infeasible. If the minimal value is zero, then a feasible point has been found and the second stage can begin with this point and the original objective function.

Sparse Matrices.   A major group of refinements to the simplex method makes use of the fact that for most problems the matrix of the initial tableau is sparse, that is, it consists mostly of zero values. Hand computations applied to the kiln example show that major efficiencies can be gained from sparseness.

The reason that sparseness occurs is that, typically, individual constraints apply to only small subsets of the structural variables. Consider an extended feed-blending example in which corn and soybeans are used in four different products. Let C_i and S_i be the amount of corn and soybeans in 1 kg of the ith product. Let k_i and p_i be the respective requirements for calories and protein. The constraint matrix for this problem looks like

C_1	S_1	C_2	S_2	C_3	S_3	C_4	S_4	RHS
3600	4200	0	0	0	0	0	0	k_1
90	370	0	0	0	0	0	0	p_1
0	0	3600	4200	0	0	0	0	k_2
0	0	90	370	0	0	0	0	p_2
0	0	0	0	3600	4200	0	0	k_3
0	0	0	0	90	370	0	0	p_3
0	0	0	0	0	0	3600	4200	k_4
0	0	0	0	0	0	90	370	p_4

For most users of commercial linear programing packages, the significance of sparse matrix structure lies in reduced time for computer solution and reduced computer memory requirements.

Bounding Techniques.   Special techniques have been added to the simplex algorithm to capitalize on the form of certain constraints that are commonly seen in practical problems. One such technique incorporates bounds on individual variables. The simplex method automatically assumes each variable x satisfies $x \geq 0$. Bounds such as

$$x \geq L \quad \text{and} \quad x \leq U$$

could be incorporated as ordinary constraints. The simple form of such lower and upper bounds makes it more efficient to modify the simplex algorithm directly.

Another special form of constraint is called a generalized upper bound (GUB). In such a constraint, all the coefficients are either $+1$ or -1. A balance equation that says the volume of inputs to a process equals the volume of outputs will give this kind of constraint. The variables that occur in a generalized upper bound are called a *GUB set.* Any number of GUB sets can be handled simultaneously as long as the sets do not overlap. The part of a tableau that expresses the GUB constraints has a form similar to the following matrix:

$$\text{GUB 1} \quad 0 \quad 0 \quad \cdots \quad 1 \quad 1 \quad -1 \quad 0 \quad\quad 0 \quad 0 \quad 0 \quad \cdots \quad g_1$$

$$\text{GUB 2} \quad 0 \quad 0 \quad \cdots \quad 0 \quad 0 \quad\quad 0 \quad 1 \quad -1 \quad 0 \quad 0 \quad \cdots \quad g_2$$

$$\text{GUB 3} \quad 0 \quad 0 \quad \cdots \quad 0 \quad 0 \quad\quad 0 \quad 0 \quad\quad 0 \quad 1 \quad 1 \quad \cdots \quad g_3$$

Most commercial linear programming packages include both bounding for individual variables and generalized upper bounding. Solution times using these refinements are comparable to the times for problems with the number of constraints reduced by the number of single-variable and GUB constraints.

SENSITIVITY ANALYSIS

Marginal Costs and Ranging. The examples above have introduced the major areas of sensitivity analysis, which is also called *postoptimality analysis.* These areas are

Reduced costs, also called marginal costs or opportunity costs: These measure the cost of substituting a positive value for a structural variable which has optimal value zero.

Ranges for coefficients of the objective function: A range is given for each coefficient. As long as the coefficient stays in the range, the optimal values of the decision variables do not change, although the value of objective function usually will.

Dual prices, also called marginal values or opportunity costs: If the right-hand side of a constraint is changed, the optimal value of the objective function will change at the rate of the dual price.

Ranges for right-hand sides: The dual price is valid as long as the right-hand side of the corresponding constraint varies within the given range.

Formats for presenting sensitivity analysis vary widely. Reduced costs and dual prices are often presented in parallel with optimal values of the structural variables and slack or surplus values for the constraints. Sensitivity analysis for the kiln example might appear as in Fig. 5.4. Some of these numbers were mentioned earlier in a limited context. All of the numbers will be examined now in relation to each other.

The structural, or decision, variables in the kiln problem are the number of main housings and the number of inserts to produce during the coming 2-week period. The coefficients of the objective function are the per-unit profit figures for each product. As the per-unit profit figures change, the optimal product mix does not change continuously. Instead, the optimal mix is constant as each per-unit profit varies over a range. This is the range shown. When a coefficient reaches the end of its range, the old optimal mix and a new mix are both optimal. In the graph of the kiln problem, the end of the coefficient range corresponds to the isoprofit lines being parallel to a side of the feasible region. The new mix becomes the unique optimum as the coefficient moves beyond the end of the original range.

In the general linear programming problem, the optimal values of the decision variables do not change so long as a coefficient of the objective function is moved within its given range. This information has practical value when revenues and costs are known to fluctuate. Since coefficients for several variables may well change at once, the ranges should be taken as indicators of what will happen. Technically, the range applies when just the one corresponding coefficient varies. For example, if all revenues and costs were multiplied by a positive factor, the optimal decision variables would remain constant no matter how large or small the factor. If substantial changes are made to two or more coefficients, then the linear programming problem should be resolved.

The reduced costs in the kiln example are both zero, since both housings and inserts are

Results for the Kiln Example

Name: Kiln Example

Optimal Value of the Objective: 92,660.40 dollars

Decision Variables

Description	Value at Optimum	Reduced Cost	Measure Units
main ceramic housing	3,720	0.00	units per 2 weeks
ceramic insert	6,000	0.00	units per 2 weeks

Objective Coefficient Ranges in Which Above Values are Optimal

Decision Variable	Value at Optimum	Low End of Range	Current Coef.	High End of Range
main ceramic housing	3,720	0.00	11.57	16.54
ceramic insert	6,000	5.785	8.27	+infinity

Constraints

Description	Dual Price	Slack or Surplus	RHS Measure Units
kiln capacity	231.4	0	hours per 2 weeks
gas supply	0.0	350	therms per 2 weeks
toxic disposal	497.0	0	kg per week
contract for housings	0.0	2,520	housings per 2 wks
insert per housing	0.0	2,280	units per 2 weeks

Right Hand Side Ranges in Which These Dual Prices are Valid

Constraint	Dual Price	Low End of Range	Current RHS	High End of Range
kiln capacity	231.4	210	336	350
gas supply	0.0	9,650	10,000	+infinity
toxic disposal	497.0	22.4	30.0	38.4
contract for housings	0.0	-infinity	1,200	3,720
insert per housing	0.0	-2,280	0	+infinity

FIGURE 5.4   Results for the kiln example.

being produced. In large product-mix problems, some of the decision variables will commonly have optimal value zero, meaning that production of the corresponding items will reduce the profit as defined in the formulation of the linear program. Considerations not captured in the formulation may dictate production of some of these items. In general, the reduced cost measures the per-unit cost of introducing a variable that is shown with optimal value zero.

It can happen that a decision variable is shown with zero for both its optimal value and its reduced cost. Since a new positive variable can be introduced with no change in the objective function, there is a unique optimal value for the objective function but not a unique way to achieve it. This situation is also described by saying there are alternative optima. The practical advantage of this is that one can pursue secondary, perhaps nonquantifiable, objectives while still optimizing the objective of the linear program.

The available computer algorithms differ in the extent to which they analyze alternative optima. Some routines present only the first optimal solution found, and which is first may depend on the order in which variables and constraints occur in the database. Ideally, the output would explicitly state that alternative optima exist and also list all the optimal extreme points of the feasible region. In problems with many decision variables, such a list may be too long to be useful.

A situation for constraints, analogous to alternative optima, occurs when both a dual price and its matching slack or surplus are zero. This is called a *degeneracy*. The practical effect has to do with how changes in the right-hand sides of the constraints change the optimal values of the decision variables. When there is no degeneracy, either an increase or a decrease in the right-hand side of a constraint changes the dual price at the rate of the dual price. In case of degeneracy, a right-hand side change may have an effect on one direction but not the other.

When there are no degeneracies, the constraints with positive dual prices are the active constraints, that is, the ones which currently affect the optimal value. This is the case in the kiln problem. In that problem, the dual price of the constraint on kiln hours is $231.40 per hour available in the 2-week period. The effective range of this rate is 210 to 350 hours. The meaning is that, as kiln time increases or decreases, the optimal value of the objective function increases or decreases at the rate of $231.40 per kiln hour. One should note that the formulation of the kiln problem up to now does not capture the fact that the kiln is used only in 6-hour cycles and that the products are made only in full-kiln batches. Discrete jumps in the decision variables can be handled by the technique of integer programming.

The toxic disposal constraint can be used to illustrate the incremental nature of the dual price. Disposal of the toxic substance has a cost of $1 per insert. Each insert generates 10 grams of the substance, so the disposal cost per kilogram is $100. Capacity for disposing of one more kilogram per week would increase profit by the dual price, $497. The profit calculation has already accounted for the normal disposal cost of $100 per kilogram. Any premium of less than $497, in addition to the $100 base cost, could be paid, and overall profit would still increase. Careful attention to units of measurement is critical to the proper interpretation of dual prices in this example, and in general.

In nondegenerate cases, a zero dual price indicates an inactive constraint. In the kiln example, the constraints for natural gas, for the housing contract, and for making at least one insert per housing are all inactive. The range for the corresponding right-hand side shows how close the constraint is to being active. The original discussion of the kiln example pointed out that the natural gas constraint is not far from being active. The lower limit for the gas available is 9650 therms. Note that the difference between this lower bound and the current value of 10,000 is the same as the slack in the constraint, 350 therms. The upper bound of $+\infty$ means that additional gas is not currently of any value.

At the end of the effective range for a dual price, it is usually true that a new constraint becomes active. In the kiln example, when kiln hours decrease to 210, the optimum mix becomes 1200 housings and 6000 inserts, and at this point the contract for housings becomes an active constraint. When kiln hours increase above 350 it is the natural gas constraint that becomes active. Another way of saying that a constraint becomes active is to say that its slack or surplus variable leaves the basis; that is, the variable becomes zero.

Commercial computer routines often identify the newly active constraints under the heading LIMITING PROCESS. This heading can have different interpretations, depending on context. When used with an active constraint, a limiting process is one whose variable leaves the basis at the end of a range. For a variable corresponding to a constraint, that means the constraint becomes active. When the term *limiting process* is used with a currently inactive constraint, it refers to a variable that enters the basis. In the case of a slack or surplus variable entering the basis, it means that the corresponding constraint becomes inactive. The limiting processes for the constraints in the kiln example are given below.

Constraint	Limiting processes at ends of range	
	Low end	High end
Kiln capacity*	Contract	Gas supply
Gas supply	Kiln capacity	None
Toxic disposal*	Insert	Gas supply
Contract for housings	None	Toxic disposal
Insert per housing	None	Kiln capacity

*Active constraints.

The term *limiting process* is also used with respect to structural variables. In that case it refers to the variable that enters or leaves the basis at the ends of the range for the corresponding coefficient of the objective function.

The Dual Problem.   One often speaks of getting a new point of view on a problem, meaning that a new approach is taken even through the underlying problem remains the same. In linear programming, an analog to getting a new point of view is solving the so-called *dual problem*. Every linear programming formulation has a dual formulation. The original formulation is called the *primal* problem. If the primal is a maximization problem, then the dual is a minimization problem. The constraints of the primal correspond to variables of the dual, and the variables of the primal correspond to constraints of the dual.

The relation between the primal and the dual can be written succinctly in matrix notation. If the primal is

Minimize $\mathbf{c}^T\mathbf{x}$

with constraints $\mathbf{Ax} \geq \mathbf{b}$

then the dual is

Maximize $\mathbf{b}^T\mathbf{y}$

with constraints $\mathbf{A}^T\mathbf{b} \leq \mathbf{c}$

Here **c** is the vector of coefficients in the primal objective function. The vector **b** gives the bounds on the right-hand side of the primal constraints. The matrix **A** contains the coefficients of the variables in the constraints. The superscript indicates the transpose operation, which changes a column vector to a row vector and interchanges columns and rows in a matrix. Forming the dual of the dual gives back the primal problem. Figure 5.5 expands on the relation between primal and dual.

Primal	Dual
Structure	Structure
Maximization	Minimization
Minimization	Maximization
Variables	Constraints
Constraints	Variables
Objective coefficients	Limits in constraints
Limits in constraints	Objective coefficients
Optimal Values	Optimal Values
Objective value	Objective value
Variables	Dual prices
Dual prices	Variables
Reduced costs	Slacks or surpluses
Slacks or surpluses	Reduced costs

FIGURE 5.5   Relation between the primal and dual problems.

The mechanics of forming the dual can be illustrated with the livestock feed example discussed earlier. Formulations of the primal and the dual are compared in Fig. 5.6. Recall the meaning of the dual prices. Reducing the calorie constraint by 1 calorie per kilogram of feed will reduce the minimal cost of the feed by 0.0022 cents. Increasing the calorie requirement by 1 calorie increases the cost by the same amount. Likewise, increasing the protein requirement by 1 gram increases the cost of 1 kilogram of feed by 0.0453 cents.

Primal Formulation

Dual Formulation

Variables:

Variables (worth of one):

```
corn          CORN                     calorie       WCAL
soybeans                 SOYB          gram protein            WPRO
```

Objective:

Objective:

```
minimize    12 CORN +    26 SOYB      maximize   3600 WCAL + 150 WPRO
```

Constraints:

Constraints:

```
calories  3600 CORN + 4200 SOYB > 3600    corn     3600 WCAL +  90 WPRO < 12
protein     90 CORN +  .370 SOYB > 150    soybeans 4200 WCAL + 370 WPRO < 26
```

Optimal values

Optimal values

```
Objective:     14.72 cents per kg       Objective:    14.72 cents
```

Variables:

Variables:

```
CORN          .736 kilograms            WCAL         .0022 ¢ per calorie
SOYB          .226 kilograms            WPRO         .0453 ¢ per gram
```

Reduced Prices:

Reduced Prices:

```
calories    .0022 cents per calorie     corn       .736 ¢ per ¢ per kg
protein     .0453 cents per gram        soybeans   .226 ¢ per ¢ per kg
```

FIGURE 5.6   Primal and dual problems.

The dual problem can be described as maximizing the total worth of the calorie and protein mixture subject to the known worth of a kilogram of corn and a kilogram of soybeans. In light of the solution to the primal problem, the reduced prices of corn and soybeans in the dual can be quickly verified. For example, since the optimal primal mix calls for 0.736 kg of corn, a 1 cent per kilogram increase in the price of corn will increase the price of the feed by 0.736 cents per kilogram.

SPECIAL APPLICATIONS

The Transportation Problem.   The transportation problem involves moving a commodity from a number of sources, such as warehouses, grain elevators, or mines, to a number of destinations, such as stores, factories, or ports. The objective is to minimize the total transport cost. The formulation as a linear program is

Word description	Algebraic description
Decision variables:	
Cost per unit shipped from ith source to jth destination	x_{ij}
Objective:	
Minimize total transport cost	$\Sigma_{ij}\, c_{ij} x_{ij}$
Constraints:	
Supply at ith source	$\Sigma_j\, x_{ij} \leq s_i$
Demand at jth destination	$\Sigma_i\, x_{ij} \geq d_j$

The coefficients c_{ij} are the shipping costs per unit on each route. In general, the number of sources is not equal to the number of destinations. The solution is the minimal shipping cost and an optimal shipping plan given by the matrix of values x_{ij}.

The ideal circumstance in which total supply equals total demand is referred to as the *balanced problem.* If actual supply exceeds demand, a so-called dummy destination is created. Shipments to the dummy destination are interpreted as amounts left at the source. The cost of leaving material at each source could be set to zero. If certain sources are to be preferentially emptied, then an artificial cost per unit can be assigned to shipments from those sources to the dummy. The artificial cost can be large or small in comparison with the real shipping costs, depending on the urgency of emptying a source.

When total supply falls short of total demand, a dummy source is created. A shipment from the dummy means a shortage at the receiving destination. As with surpluses, shortages can be preferentially allocated by assigning nonzero costs to the dummy shipping routes.

For small problems the input data for the transportation problem can be represented in matrix form. For example, suppose wheat is to be shipped from three grain elevators to two deep-water ports. The right-hand column in the table below shows the supply at each elevator. The row at the bottom shows the demand at each port. Supply and demand are in tons. Shipping costs, in dollars per ton, are shown in the body of the matrix.

Matrix of Shipping Costs

Sources (elevators)	Destinations (ports)			Supply
	1	2	Dummy	
A	10	8	0	300
B	9	7	0	400
C	12	9	0	200
Demand	300	500	100	900 total

Shipping to the dummy destination represents leaving wheat at an elevator. No preference is expressed as to where the excess of 100 tons should be left.

The balanced transportation problem always has a feasible solution. Several heuristic methods are used to produce an initial feasible plan. The simplest is called the *northwest corner* method, in analogy to directions on a map, because it assigns amounts to be shipped starting in the upper left corner of the matrix. As much as possible is shipped from the first listed source to the first listed destination. If any supply remains at the first source, it is shipped to the second destination up to the amount of demand there, then to the third destination, and so on until all the supply at the first source is used up. Supply at the second source is then shipped to the destinations having unfilled demand, in the order they appear in the matrix. The balance of supply and demand assures that the shipments come out even by the time the procedure reaches the lower right, or southeast, corner. For the example, one has

Shipping Amounts by Northwest Corner Method

Sources (elevators)	Destination (port)			Supply
	1	2	Dummy	
A	300	0	0	300
B	0	400	0	400
C	0	100	100	200
Demand	300	500	100	900 total

Total shipping cost = 6700

In this case, the northwest corner shipping plan yields the least cost possible, although in general it does not. Note that the method completely ignores cost and depends only on the order in which sources and destinations are written.

Selecting the cheapest remaining route first is a fast method of producing a feasible shipping plan that takes cost into account, but it ignores possible tradeoffs between groups of routes. The common algorithm for the transportation method, called the *modified stepping stone* method, starts with any feasible shipping plan and evaluates all possible tradeoffs if the plan is varied slightly. A new shipping plan is developed using the best tradeoffs. That plan is evaluated in turn, and the method proceeds until no advantageous tradeoffs exist.

The transportation problem has many extensions to problems that do not involve physical shipment. As an example, consider a company that wishes to schedule production in one of its plants in light of the following projections for demand and capacity for its product.

Month	1	2	3	4	5	6
Demand, tons	40	45	50	65	70	75
Capacity, tons	50	40	30	60	90	90
$ cost per ton	100	120	130	100	80	80

The variations in capacity and cost per ton reflect an expansion and modernization program that will begin in the second month. During construction, production will be curtailed and operations will be less efficient. The product can be inventoried at a cost of $5 per ton per month. Customers will accept a delay in delivery of up to 1 month if given a discount of $8 per ton. Management wants to set a production schedule that will minimize costs over the 6-month period.

This production problem can be interpreted as a transportation problem in which the sources are producing months and the destinations are selling months. A formulation is

Effective Costs per Ton

Source (selling month)	Destination (selling month)						
	1	2	3	4	5	6	
1	100	105	110	115	120	125	50
2	128	120	125	130	135	140	40
3	M	138	130	135	140	145	30
4	M	M	108	100	105	110	60
5	M	M	M	88	80	85	90
6	M	M	M	M	88	80	90
Demand, tons	40	45	50	65	70	75	

M represents some very large cost per ton that effectively forbids back orders of more than 1 month.

As in the general linear programming problem, the input data for very large transportation problems commonly result in a sparse matrix, and printing the whole matrix becomes unwieldy. The same sensitivity analysis does apply. This means in particular that effective ranges can be given for the costs on each route, and reduced costs can be given for routes not included in the optimal shipping plan. In addition, dual prices measure the impact of changes in the supply and demand constraints.

The basic assumptions of general linear programming apply to the transportation problem. The divisibility assumption does take on a special character in the transportation context. If the supply and demand numbers have integer values, then the optimal decision variables, that is, the amounts shipped, automatically have integer values. The large quantities associated with shipping many commodities lessen the significance of integer amounts on the commodities themselves. The issue of integer values can be important in counting whole truckloads or shiploads.

The Assignment Problem.   An important special case of the transportation problem occurs when the demand and supply numbers are all equal to 1. This means that exactly one source is assigned to each destination. Although transportation may be part of what is going on, the one-to-one assignment is the essential feature. Examples are the assignment of machines to production work orders and the assignment of crews to various jobs. One commonly speaks of resources and jobs rather than of sources and destinations.

The data for the assignment problem are the costs associated with each pairing of resource and job. The common algorithm for solving the assignment problem is called the Hungarian-Flood method. The seemingly fanciful name reflects the names of the people who developed and refined the method. The output of the algorithm is an assignment matrix. For example,

	Assignment costs					Optimal assignment			
	Job					Job			
Machine	1	2	3	4	Machine	1	2	3	4
1	12	14	11	15	1	1	0	0	0
2	13	12	10	13	2	0	0	1	0
3	14	12	12	14	3	0	1	0	0
4	18	16	14	16	4	0	0	0	1

The total cost of the least-cost assignment is 50. The 1's and 0's in the assignment matrix are a consequence of the supply and demand numbers all being 1 in the transportation interpretation of the problem. A shipment of one unit from a machine to job means the machine is assigned to the job. Alternatively, the 1's and 0's can be thought of as binary switches: 1 means make the assignment, and 0 means do not.

Sensitivity analysis in the assignment problem can be done on the cost coefficients. Since the constraints all have right-hand sides fixed at 1, there is no practical interpretation for dual prices.

Integer Programming.   In many practical problems, the divisibility assumption is unrealistic. Integer programming was developed to deal with situations in which decision variables must be whole numbers. There are pure integer problems, in which all variables take integer values, and mixed integer problems, in which only some of the variables must be whole numbers. An important special case of an integer variable is one that represents a yes or no decision rather than a physical quantity. Such binary variables traditionally have the value 1 or 0.

To illustrate integer programming, consider a business that is planning to expand its facilities. The capital budget is $1,500,000 for new milling machines and building expansion. Two types of machines are being considered, the KJ42 and the NP18. Each KJ42 will generate $65,000 in annual contribution margin, a form of profit. The KJ42 has a cost of $200,000 and requires 1500 square feet of factory space. If model KJ42 is bought at all, at least two will be bought. Each NP18 will generate $78,000 in annual contribution margin, has a cost of $250,000, and requires 2000 square feet of factory space. At present there is 7000 square feet of space available for the machines. An additional 3500 square feet of factory space could be built at a cost of $420,000. The formulation in Fig. 5.7 presents this as an integer programming problem.

Note the roles of the binary variables. If $X = 0$, then the factory is not expanded. The whole capital budget is available for buying machines, but the floor space is limited to the existing amount. If $X = 1$, then the factory is expanded. The expansion takes up part of the capital budget, but floor space is available if the number of machines increases.

The M in the next-to-last constraint is any large positive number. The purpose of the constraint is to make sure that if $B = 0$, then KJ42 is not bought. The computer routines that solve the problem must have everything specified that is not explicitly done by de-

Word Description	Algebraic Description

Decision Variables:

  integer valued

    number KJ42 models purchased   K

    number NP18 models purchased   N

  binary valued, 1 = yes, 0 = no

    expand floor space   X

    buy any KJ42 models   B

Objective:

  maximize

    contribution margin
    in thousands of dollars   65 K +   78 N

Constraints:

  capital budget
  in thousands of dollars   200 K + 250 N + 420 X $\leq$ 1500

  floor space, in square feet 1500 K + 2000 N - 3500 X $\leq$ 7000

  B = 0 means no KJ42 models   K   - M B $\leq$ 0

  at least two KJ42 if any   K   - 2 B $\geq$ 0

FIGURE 5.7   Integer program for facility expansion.

fault. If $B \geq 0$, then the last constraint reduces to $K \geq 0$, which is no more than the constraint of nonnegativity, which *is* included by default in linear programming routines. If $B = 1$, then at least two of the KJ42 model must be bought.

The standard algorithm for integer programming uses the branch-and-bound technique. The procedure is to first solve what is called the *LP relaxation,* which is the integer problem without the restriction that the variables be integers. For a maximization problem the resulting optimal value in the LP relaxation is an upper bound for the integer optimal solution. The optimal decision variables in the LP relaxation are examined, and one with a noninteger value is chosen as the *branching* variable.

Suppose the branching variable is Y, its optimal value in the LP relaxation is y, and n is the greatest integer less than y. Two new relaxation problems are created; call them problems 2 and 3. Each is like the first LP relaxation, except with an additional constraint:

$$Y \leq n \text{ in problem 2} \quad \text{and} \quad Y \geq n + 1 \text{ in problem 3}$$

This divides the feasible region into two subregions. Each of the new problems is solved, and new branching variables are chosen. The dividing process continues, and additional constraints with *integer* bounds are added. It becomes more and more likely that a daughter problem will have optimal decision variables that are integers. Once that happens, the corresponding optimal value becomes a lower bound on the maximal objective value in the original integer problem. With the lower bound, other subproblems can be eliminated from the search. A number of different strategies for searching among the subproblems are discussed in Ref. 9.

Sensitivity analysis in integer programming problems requires caution. The essential difficulty is that integer variables are discontinuous variables, and small changes in parameters can produce big jumps. As computer time becomes even cheaper, the most cost-effective sensitivity analysis will often be to adjust the parameters of interest and run the program over again.

COMPUTER PACKAGES

The computer packages that handle linear programming problems usually have three modules. The first performs the functions of data management and provides input in standard form to the second module, which carries out the simplex algorithm and its refinements. The third part creates customized reports from the output of the second module. These modules are available from a variety of vendors, either as complete packages or as separate modules that interface with modules marketed by others.

Matrix Generators.   The first module is often referred to as the *matrix generator,* because one of its tasks is to create the matrix used in the simplex method. This module can include:

Routines to capture data. The source can be keyboard entry, engineering or accounting files on a mainframe computer, or spreadsheet files on small computers.

Database management. Practical use of linear programming demands efficient methods for adding, changing, sorting, and selecting data. Lists and tables are common data structures that must be maintained.

A computational language. The language describes how data elements are to be combined to provide the numbers required by the simplex algorithm. These computer languages are reminiscent of FORTRAN, but have syntax to operate with the data structures of linear programming.

Mathematical Programming Systems.   The second module contains the mathematical core of linear programming. Commercial names for this module often contain the initials MPS, standing for *mathematical programming system.* There is a standard format for the input to these routines. The basic features of this format are illustrated in Fig. 5.8. The numbers are from the earlier kiln example.

```
NAME            Kiln Example
ROWS
 N   PROFIT
 L   KILNHR
 L   NATGAS
 L   TOXICS
 L   INSMIX
COLUMNS
        HOU     PROFIT    11.57
        HOU     KILNHR    .05
        HOU     NATGAS    1.25
        HOU     INSMIX    1.0
        INS     PROFIT    8.27
        INS     KILNHR    .025
        INS     NATGAS    .833
        INS     TOXICS    .005
        INS     INSMIX    -1.0
RHS
        LIMITS  KILNHR    336.0
        LIMITS  NATGAS    10000.0
        LIMITS  TOXICS    30.0
BOUNDS
 LO  DEMAND     HOU       1200.0
ENDATA
```

FIGURE 5.8   MPS standard input.

The required sections of MPS input are titled NAME, ROWS, COLUMNS, and RHS. The line with ENDATA signals the end of the input data. The NAME section uses one line to assign a name to the problem. The ROWS section lists a name for the objective function and names for each constraint. The leading codes are N for no bound, L for a less than or equal bound, G for greater than or equal, and E for equal. The first N row is the objective function. The COLUMNS section pairs variables with the rows. The number to the right is the coefficient of the variable in the paired row. Unspecified coefficients are zero by default.

The RHS section gives the right-hand sides of the constraints. A right-hand side that is not specified, such as for the INSMIX constraint, is assumed to be zero. The name LIMITS is assigned by the user. Alternative sets of right-hand side bounds can be distinguished by these names. Such sets are used for sensitivity analysis.

The BOUNDS section is optional. It is used for upper or lower bounds on single variables. In this case it is used for the minimal demand created by the contract for main ceramic housings. The code LO stands for lower bounds. The code UP would be used for an upper bound. The name DEMAND is used in the same way as set names in the RHS section. A different set of single-variable bounds would have another name. Other options exist in the standard MPS format, but most users will take advantage of the options provided by matrix generators.

MPS output, that is, output directly from the various mathematical programming modules, is less uniform than MPS input. A typical format for elementary output is shown in Fig. 5.9. The heading lines identify the function being optimized and the set names of the right-hand side values and bounds that were used. The number of iterations refers to the steps in the simplex method and is a guide to the difficulty of the problem. The objective value is the optimum attained.

```
Problem Name: Kiln Example

Maximize
Objective: PROFIT
     RHS: LIMITS          at iteration   4
   Bounds: DEMAND       Objective Value = 92,660.40


COLUMNS Section
```

NO.	COLUMN	AT	ACTIVITY	LOWER BOUND	UPPER BOUND	REDUCED COST	OBJ.COEI
1	HOU	BS	3720.00	1200.00	+ INFINITY	.	11.57
2	INS	BS	6000.00	.	+ INFINITY	.	8.27

```
ROWS Section
```

NO.	ROW	TYPE	AT	ACTIVITY	SLACK ACTIV.	RHS	DUAL PRICE
1	PROFIT	N	BS	92660.40	-92660.40	.	.
2	KILNHR	L	UL	336.00	.	336.00	231.40
3	NATGAS	L	BS	9650.00	350.00	10000.00	.
4	TOXICS	L	UL	30.00	.	30.00	497.00
5	INSMIX	L	BS	-2280.00	2280.00	.	.

FIGURE 5.9 Typical MPS output.

The column labeled AT is sometimes given the more descriptive name STATUS. The symbol BS in the columns section means the variable is basic; that is, the variable is strictly between its upper and lower bounds. The upper and lower bounds are zero and plus infinity, unless overridden in the BOUNDS section of the input. Zero values, such as in the LOWER BOUND and REDUCED COST columns are often shown in MPS output as bare decimal points, as here, or as blank spaces.

Figure 5.10 shows excerpts from typical MPS output for sensitivity analysis. Two lines of output are used for each variable (COLUMN) and constraint (ROW). The pair of numbers shown for each variable in the column labeled OBJ.COEF/RANGE are the high and low values of the range corresponding to no change in the optimal decision variables. The pair of numbers in the column labeled OBJ AT OBJ/RANGE are the values of the objective function at the high and low. In the ROWS section the range for each right-hand side is shown by UPPER ACTIVITY and LOWER ACTIVITY, and the corresponding values of the objective function are shown under OBJ UPPER and OBJ LOWER.

Report Writers. Like matrix generators, the report writing modules usually contain what amounts to a programming language. The purpose of this language is to produce reports that will be readily understood by people in various parts of the organization. Four common audiences are marketing, production, finance, and central management. Each audience requires reports on different aspects of a problem and at different levels of detail.

```
COLUMNS Section
```

NUMBER	NAME TYPE	ACTIVITY STATUS	OBJ.COEF REDUCED COST	OBJ.COEF RANGE	OBJ AT OBJ RANGE
1	HOU	3720.00	11.57	16.4	110628.00
		SLACK	.	.	49620.00
2	INS	6000.00	8.27	+ INF	+ INF
		SLACK	.	5.785	43040.40

```
ROWS Section
```

NUMBER	NAME TYPE	ACTIVITY STATUS	RHS REDUCED COST	UPPER ACTIVITY LOWER ACTIVITY	OBJ UPPER OBJ LOWER
1	KILNHR	336.00	336.00	350.00	95900.00
	LE	BINDING	231.40	210.00	63504.00
2	NATGAS	9650.00	10000.00	+ INF	92660.40
	LE	SLACK		9650.00	92660.40
3	TOXICS	30.00	30.00	38.40	96835.20
	LE	BINDING	497.00	22.40	88883.20

FIGURE 5.10 Excerpts from MPS sensitivity analysis.

The features commonly included in customized reports are explanatory titles and captions, selection and summarization of data, and formatting into lists and tables. Most report writers could, for example, create Fig. 5.4 from the information in Figs. 5.9 and 5.10. Detailed sensitivity analysis such as limiting process information could also be added. The monograph by Palmer[10] includes examples of customized output from a report generator.

FURTHER READING

The brief list of books and articles in the references below is intended as an introduction to the literature of linear programming and to some of the principal journals. The four books cited range from the theoretical to the applied. The book by Schrijver[8] concentrates on the mathematical aspects of linear programming. It gives over 1000 references to the theoretical literature and contains many comments on the history of the underlying methods. The book by Gass[11] is an advanced textbook that gives a detailed exposition of the mathematics of linear programming and is notable for its organized bibliography. Over 700 books and articles published between 1939 and 1983 are classified by application and by industry.

The Murtagh book[12] contains both a concise treatment of solution methods and much practical advice. An extended example illustrates data organization and input to a commercial computer package. The monograph by Palmer et al.[10] describes a large-scale working system. It makes clear the need for careful data management.

Journal articles continually describe new techniques and applications of linear programming. Major outlets for such work have been *Decision Sciences, Interfaces, The Journal of the Operational Research Society, Management Science, The Naval Research Logistics Quarterly,* and *Operations Research.*

Computer packages for linear programming are available from many sources. The Engineering and Scientific section of the *Datapro Directory of Software*[13] lists vendors of packages for general use and for applications in particular industries. An increasing number of applications can be run on personal computers. These have been surveyed by Sharda.[14]

REFERENCES

1. Harpell, J. L., M. S. Lane, and A. H. Mansour, "Operations Research in Practice: A Longitudinal Study," *Interfaces,* vol. 19, no. 3, May–June 1989, pp. 65–74.

2. Carino, H. F., and C. H. LeNoire, Jr., "Optimizing Wood Procurement in Cabinet Manufacturing," *Interfaces,* vol. 18, no. 2, March–April, 1988, pp. 10–19.

3. Denardo, E. V., U. G. Rothblum, and A. J. Swersey, "A Transportation Problem in Which Costs Depend on the Order of Arrival," *Management Science,* vol. 34, no. 6, June 1988, pp. 774–783.

4. Frendewey, J. O., and R. T. Sumichrast, "Scheduling Parallel Processes with Setup Costs and Resource Limitations," *Decision Sciences,* vol. 19, no. 1, Winter 1988, pp. 138–146.

5. Potts, C. N., and L. N. van Wassenhove, "Algorithms for Scheduling a Single Machine to Minimize the Weighted Number of Late Jobs," *Management Science,* vol. 34, no. 7, July 1988, pp. 843–858.

6. Abara, J., "Applying Integer Linear Programming to the Fleet Assignment Problem," *Interfaces,* vol. 19, no. 4, July–August, 1989, pp. 20–28.

7. Das, C., and S. Heragu, "A Transportation Approach to Locating Plants in Relation to Potential Markets and Raw Material Sources," *Decision Sciences,* vol. 19, no. 4, Fall 1988, pp. 819–829.

8. Schrijver, A., *Theory of Linear and Integer Programming,* Wiley, New York, 1986.

9. Johnson, R. V., "Efficient Modular Implementation of Branch and Bound Algorithms," *Decision Sciences,* vol. 19, no. 1, Winter 1988, pp. 17–38.

10. Palmer, K. H., et al., *A Model-Management Framework for Mathematical Programming,* an Exxon Monograph, Wiley, New York, 1984.

11. Gass, S. I., *Linear Programming Methods and Applications,* 5th ed., McGraw-Hill, New York, 1985.

12. Murtagh, B. A., *Advanced Linear Programming,* McGraw-Hill, New York, 1981.

13. Datapro Research, *Datapro Directory of Software,* McGraw-Hill, Delran, N.J., 1990.

14. Sharda, R., "The State of the Art of Linear Programming on Personal Computers," *Interfaces,* vol. 18, no. 4, July–August 1988, pp. 49–58.

CHAPTER 6
NONLINEAR PROGRAMMING

Jayant Rajgopal
Department of Industrial Engineering
University of Pittsburgh
Pittsburgh, Pennsylvania

Nonlinear programming (NLP) is a subset of the more general area of mathematical programming, which has traditionally been part of the broad fields of operations research (OR) and management science (MS). However, in terms of its theory, methodology and applications, NLP overlaps into a number of other fields such as engineering, mathematics, and economics in addition to OR and MS.

The birth of mathematical programming is usually identified with the development of the simplex method in 1947 by George Dantzig. Although much of the mathematical theory of linear programming (LP) relates directly to the theory of linear inequalities and convex sets which have been formulated over the past century, it was the introduction of the simplex method and the advent of high-speed digital computers that stimulated interest in practical applications, as well as further research in mathematical programming.

The development of nonlinear programming began around the same time, with the Fritz John conditions in 1948, and the famous Kuhn-Tucker conditions.[1] Later Rockafeller, Wolfe, Cottle, and others developed the theory of NLP and extended the notions of duality. Commercial applications were begun in the 1950s by Charnes and Cooper among others; soon thereafter practical applications began to dominate. Today NLP is a well-developed area with a wealth of different methods and applications.

In the remainder of this chapter, nonlinear programs will first be defined and practical applications of NLP will be briefly discussed. Next some of the basic theoretical concepts will be outlined. Following this an overview of several algorithmic approaches to solving NLP problems will be provided. Finally, several special forms of nonlinear programs that arise in practice and for which special procedures have been developed will be presented.

DEFINITION AND EXAMPLES

In this section we define a nonlinear program, and then provide the reader with some idea of the types of problems to which NLP has been successfully applied, along with a couple of illustrative examples of practical nonlinear programs.

Nonlinear Programming Mode. We define a nonlinear program as follows:

Program 1

$$\text{Minimize } f(\mathbf{x})$$

$$\text{subject to} \quad g_j(\mathbf{x}) \leq 0 \quad j = 1, 2, 3, \ldots, m$$

$$h_j(\mathbf{x}) = 0 \quad j = 1, 2, \ldots, p$$

$$\mathbf{x} = (x_1, x_2, \ldots, x_n) \in \mathfrak{R}^n$$

where the functions g_j and h_j of the vector $\mathbf{x}$ are referred to as the *constraint functions* while the function f is called the *objective function*. Although there are a number of different types of nonlinear programming problems, they are all special cases of Program 1 above. The linear programming problem is a special case where the functions f, g_j, and h_j are all linear; such problems are relatively easy to analyze and solve via the simplex method. The general problem on the other hand presents a number of difficulties, as we shall soon see, and unfortunately there is no single procedure (like the simplex method for LP) that is capable of solving any general nonlinear programming problem.

Applications. Nonlinear programming has been applied widely to solve a number of different problems encompassing a wide variety of application areas such as engineering, mathematics, statistics, management science, economics, and computer science. Some specific examples include process equipment design, structural optimization, power transmission, capacity planning, inventory optimization, electromagnetic systems, wastewater treatment, tool engineering, mechanical systems design, reliability maximization, traffic analysis, nuclear systems, integral estimation, and petroleum refining.

This list is by no means comprehensive; for the interested reader, some books geared toward specific applications include those by Morris,[3] Levary,[4] Arora,[5] and Vanderplaats.[6]

We now demonstrate the use of nonlinear programming through two simple but illustrative examples. The objective is not to provide an in-depth analysis of a specific problem, rather it is to show how a typical NLP model would be developed.

Example 1: Consider the following problem adapted from Biegler et al.[7] A chemical process produces three products through the following steps:

STEP 1: Product is pumped from storage by a pump P_1 into a batch reactor R.

STEP 2: After the reaction, the output from the reactor is pumped by a second pump P_2 through a heat exchanger H, into a feed tank F.

STEP 3: From the feed tank the product is pumped by a third pump P_3 into a centrifuge C. Here the solids are separated and sent to a tray dryer D, from where the dried solids are sent to packaging.

Each product is produced in discrete batches, with a new batch being able to enter a specific stage in the process only after all of the previous batch has been completely cleared out. Thus there is a limiting cycle time associated with each product, as determined by the time required at the stage that takes the longest, and a batch of the product is produced at intervals of this cycle time.

The design problem is to find the appropriate batch sizes for each product, along with the capacities of the various stages, so that the total equipment cost (which is a function of capacities) is minimized. The design is constrained by the technological relationship between batch sizes and equipment capacities, and annual production requirements for the three products.

In order to formulate this as a nonlinear program, let us index the three products by $i = 1, 2, 3$. It should be noted that the reactor, the feed tank, and the tray dryer are batch stages, since an entire batch has to be collected in these before they start to operate. Their capacities are characterized by the volumes of product that they handle. On the other hand the three pumps, the heat exchanger, and the centrifuge are semicontinuous stages, since material flows through each of these in a stream. The capacities of these five stages are characterized by flow rates as opposed to volumes. Let us index the batch stages R, F, and D by $j = 1, 2, 3$ and the semicontinuous stages P_1, P_2, P_3, H, and C by $k = 1, 2, 3, 4, 5$. We define the decision variables as follows:

B_i = batch size of product i (lb), $i = 1, 2, 3$

V_j = volume of the batch stage j (ft^3), $j = 1, 2, 3$

R_k = capacity of semicontinuous stage k (ft^3/h), $k = 1, 2, 3, 4, 5$

t_{ik} = time required to process a batch of product i at semicontinuous stage k (h), $i = 1, 2, 3, k = 1, 2, 3, 4, 5$

T_i = limiting cycle time for product i (h)

The inputs into the NLP model will consist of the following parameter values:

S_{ij} = size factor for batch stage j (the volume of i processed at stage j for a unit mass of final product i)

D_{ik} = duty factor for semicontinuous stage k (the volume of i processed at stage k for a unit mass of final product i)

Q_i = quantity of product i to be produced per year

H = hours available for production per year

P_{ij} = fixed time associated with processing a batch of i at stage j

V_j^{max}, V_j^{min} = maximum and minimum capacities for batch stage j

R_k^{max}, R_k^{min} = maximum and minimum capacities for semicontinuous stage k

$C(j)$ = cost associated with batch stage j

$C(k)$ = cost associated with semicontinuous stage k

The cost functions are typically of the form $C(j) = a_j(V_j)\alpha_j$ and $C(k) = b_k, (V_k)\beta_k$, where a_j, b_k, α_j, and β_k are known values. The problem may then be formulated as follows:

$$\text{Minimize total cost} = \sum_{j=1}^{3} a_j(V_j)^{\alpha_j} + \sum_{k=1}^{5} b_k(V_k)^{\beta_k}$$

subject to:

1. Batch equipment capacity constraints

$$V_j \geq S_{ij}B_i \qquad i = 1, 2, 3; j = 1, 2, 3$$

2. Semicontinuous equipment constraints

$$R_k \geq (D_{ik}B_i)/t_{ij} \qquad i = 1, 2, 3; k = 1, 2, 3, 4, 5$$

3. Cycle time constraints

$$T_i \geq t_{ik} \qquad i = 1, 2, 3; k = 1, 2, 3, 4, 5$$

$$T_i \geq t_{ia} + P_{ij} + t_{ib} \qquad i = 1, 2, 3$$

where a represents the semicontinuous stages from which material flows into batch stage j and b represents the stage into which it flows from j.

4. Production time constraint

$$\sum_{i=1}^{3} (Q_iT_i)/B_i \leq H$$

5. Equipment capacity bounds

$$V_j^{\text{max}} \geq V_j \geq V_j^{\text{min}} \qquad j = 1, 2, 3$$

$$R_k^{\text{max}} \geq R_k \geq R_k^{\text{min}} \qquad k = 1, 2, 3, 4, 5$$

For the system defined here, the NLP is an optimization problem with 29 variables, and 9 constraints in (1), 15 in (2), 24 in (3), 1 in (4), and 18 in (5), for a total of 49 technological constraints, and 18 bounding constraints. This particular problem can be reformulated as a geometric program (a special type of NLP to be discussed later) and solved quite efficiently.

Example 2:   Our second example is from economic production lot sizing, an area that is very familiar to industrial engineers. This example is adapted from Smith.[8a]

Suppose we are producing N products at a facility and the following data are available:

D_i = annual demand for product i, $i = 1, 2, \ldots, N$

h_i = carrying cost for product i ($/unit/year)

r_i = replenishment cost for a production run for product i ($).

The problem is to determine an economic lot size (say Q_i) for each product so as to minimize the total annual cost C, which is given by $C = \Sigma_i\, h_i Q_i/2 + r_i D_i/Q_i$. The well-known economic lot sizing formula specifies that the optimal Q_i values that minimize C are given by $Q_i = (2r_i D_i/h_i)^{1/2}$.

Consider now the common situation where all products have to go through a bottleneck machine in the course of their manufacture. The values of Q_i calculated above could be infeasible if the total time available at the bottleneck machine is insufficient for all the setups for a production run that are scheduled at the machine. Suppose we also have the following data:

t_i = time required to set up the bottleneck machine for a production run for product i (h),

T = total time available at the bottleneck machine (h),

s_i = cost of one unit of time spent in setups for product i at the bottleneck machine ($/h),

f_i = cost of a replenishment for product i, exclusive of the cost of setup at the bottleneck machine ($).

Note that f_i could include the cost of initiating replenishment and setup costs at other machines before and after the bottleneck machine. Thus the actual replenishment cost r_i is the sum of f_i and the actual setup cost at the bottleneck machine.

The constrained optimization can now be formulated as the following nonlinear program:

Minimize total annual cost $C = \displaystyle\sum_{i=1}^{N} (h_i Q_i)/2 + (r_i D_i)/Q_i$

subject to $\displaystyle\sum_{i=1}^{N} (D_i t_i/Q_i) \leq T$

$$r_i = s_i t_i + f_i \qquad i = 1, 2, \ldots, N$$

In summary, it should be emphasized that these two simple examples are provided only to give the reader a flavor of what a typical NLP looks like, and to illustrate the variety of problems to which NLP is applicable. The interested reader is urged to refer to the books listed earlier in this section to get more details on different types of practical NLP problems.

BASIC THEORETICAL CONCEPTS

In this section we outline some important theoretical concepts that need to be understood in order to study NLP.

Convexity. The notions of convexity and concavity play a big role in nonlinear optimization. A set $\mathcal{S}$ is said to be convex if, and only if, for any two points x and y belonging to $\mathcal{S}$, and any λ between 0 and 1, the point $z = \lambda x + (1 - \lambda)y$ also belongs to $\mathcal{S}$. Figure 6.1 shows some examples of convex and nonconvex sets; a simple way of looking at it is that the line segment joining two points in a convex set lies entirely within the set.

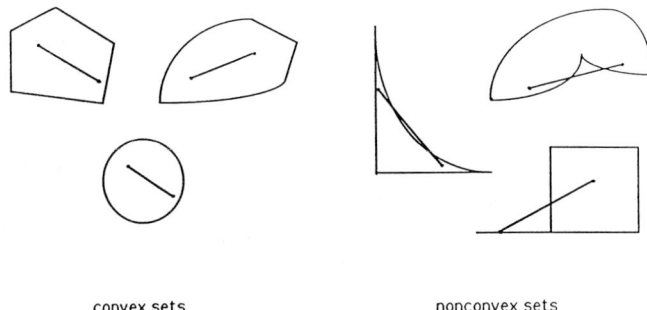

convex sets   nonconvex sets

FIGURE 6.1   Examples of convex and nonconvex sets.

A function $f(\mathbf{x})$ is said to be convex if, and only if, for all $\lambda \in (0, 1)$, $\lambda f(\mathbf{x}_1) + (1 - \lambda)f(\mathbf{x}_2) \geq f[\lambda \mathbf{x}_1 + (1 - \lambda)\mathbf{x}_2]$; for a function to be concave the direction of the inequality is reversed. A simple way of looking at it is that the chord joining two points on a convex (concave) function lies completely on or above (below) the function. Figure 6.2 shows examples of functions that are convex, concave, and neither convex nor concave. By definition, a linear function is both convex and concave.

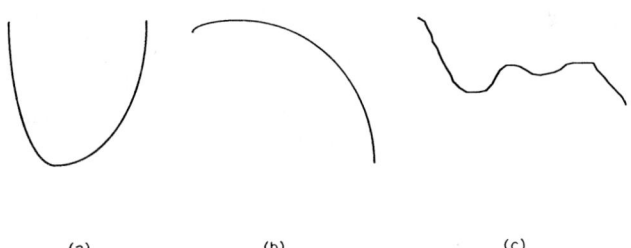

(a)   (b)   (c)

FIGURE 6.2   Examples of (*a*) convex, (*b*) concave, and (*c*) neither concave nor convex functions.

In general, it is relatively easy to minimize a convex function (or maximize a concave function) over a set of points that is convex; such a problem is referred to as a *convex programming problem.* Furthermore, a number of algorithms are guaranteed to work only under suitable assumptions of convexity and hence, it is useful to check these assumptions before attempting to use the algorithms; we shall briefly discuss procedures for doing so a little later. Finally, the conditions under which the existence of optima may be guaranteed are considerably simpler to check when certain convexity conditions are met. Thus convexity plays an important role in NLP.

The Difficulties of Nonlinear Programming. Perhaps the best way to understand the problems that arise from nonlinearity is to contrast NLP with linear programming. Most of the difficulties can be traced to the absence of one or more features found in LP.

Nonlinear Constraints. In LP a convex set is formed by the intersection of a finite number of half-spaces, with each half-space being determined by a linear constraint. Only a finite number of extreme points exist, and the optimum solution, when it exists, must be among these. However, with nonlinear constraints, it is possible for an infinite number of extreme points to exist, and convexity is no longer guaranteed for the feasible region.

Consider the feasible regions shown in Fig. 6.3; each one is determined by a single constraint function along with nonnegativity. Figure 6.3a has a linear constraint function while Fig. 6.3b and c have nonlinear constraint functions. Assume, for simplicity, that the objective in each case is linear.

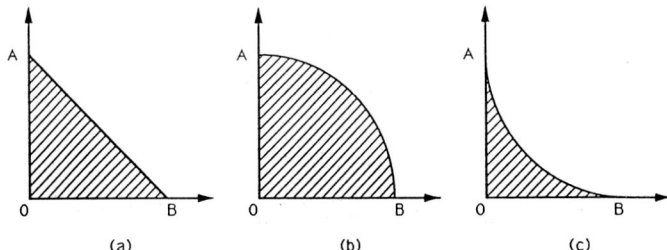

FIGURE 6.3   Feasible regions. (*a*) With linear constraint functions; (*b*) and (*c*) with nonlinear constraint functions.

In Fig. 6.3a we have a convex set with three extreme points 0, A, and B, one of which must be the optimal solution. In Fig. 6.3b, while the feasible region is convex, there are infinitely many extreme points; in fact any point on the boundary of the curve A-B is an extreme point. Thus there are an infinite number of candidate points for the optimum solution. This of course rules out an iterative scheme such as the simplex method. Finally in Fig. 6.3c we have a nonconvex region which, as we shall see, causes further problems.

Nonlinear Objective Function. Now consider a problem with a nonlinear objective; assume for simplicity that the constraints are all linear. The main problem here is that the optimal solution could occur at some point other than an extreme point.

Consider Fig. 6.4a and b, where the curves represent isoprofit contours, z_0 being the highest profit, and $z_0 > z_1 > z_2 > z_3$. However, in Fig. 6.4a, z_0 is located outside the feasible region, and the optimum solution for the constrained problem is at point D, which is

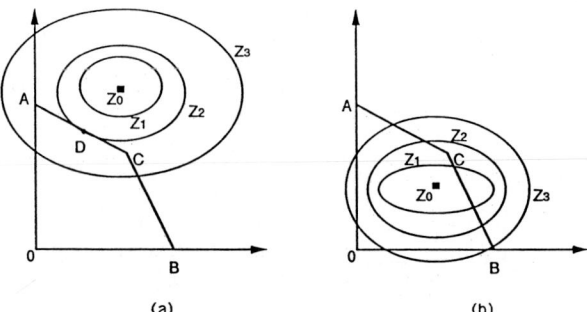

FIGURE 6.4   Isoprofit contours. (*a*) z_0 outside feasible region; (*b*) optimum solution at an interior point.

obviously not an extreme point. For the profit function of Fig. 6.4*b,* the optimum solution is located at an interior point. Once again, a straightforward simplex like approach is ruled out, although it should be pointed out that modifications of the simplex method have been made for such problems where the constraint set is a convex polyhedron.

Convexity and Concavity.   As mentioned earlier, it is relatively easy to maximize a concave function or minimize a convex function over a feasible region that is convex. With LP, these conditions are readily met, since the linear objective could be viewed as either convex (when minimizing) or concave (when maximizing). This is not always the case for the general NLP problem.

When the feasible region is nonconvex, a major problem is encountered in obtaining the global optimum as opposed to a local one. Figure 6.5*a* depicts a convex feasible set with a linear objective; if maximizing, the optimum is at *D.* This is a global optimum; regardless of the size of a move from *D,* the value of the objective can never be greater. Now contrast this with Fig. 6.5*b,* where the feasible set is nonconvex. At point *B* an optimum solution exists, since as we move toward *A* or *C* the value of the objective decreases; however, at the vertex *D* another optimum solution exists, and the value of the objective is greater than that at the previous optimum *B.* Since no other solution exists which yields a larger value for the objective function, the solution at *D* is called a *global* (absolute) optimum. The solution at *B* is called a *local* (relative) optimum.

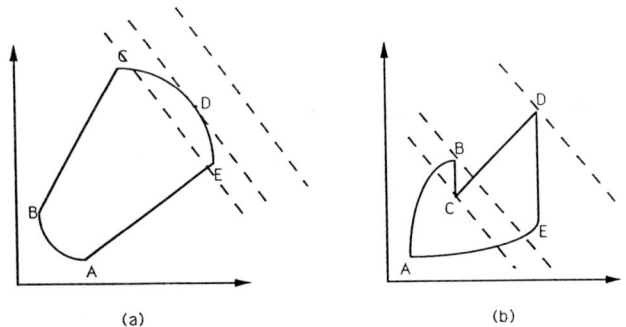

FIGURE 6.5   (*a*) Convex feasible set; (*b*) nonconvex feasible set.

Most algorithms for NLP problems which do not have a convex region give only local optima. However, by changing the starting point for the initial solution, we may be able to determine other local optima and then consider the local optimum solution that yields the best value for the objective function as the global solution. This may not necessarily be true, but in practice such solutions are frequently acceptable.

A similar problem is encountered if the nonlinear objective function does not meet certain criteria. Assuming a convex feasible set, a global solution is obtained for a minimizing (maximizing) problem if the objective function is convex (concave). Figure 6.6*a* shows a concave objective subject to a single linear constraint, and Fig. 6.6*b* shows a convex objective subject to a single linear constraint. The convex (concave) function exhibits increasing (diminishing) marginal returns.

Since a convex (concave) function is valley-shaped (hill-shaped), regardless of the starting point, all paths lead to the bottom (top). Thus, any computational scheme which obtains new solutions which have a lesser (greater) value for the objective than the previous one will eventually lead to the global solution, subject to the convex set of constraints. Obviously, no such statement can be made for the general case where the correct set of conditions does not exist.

In summary, all of the nice features that made LP so convenient to analyze are no longer guaranteed when we get into NLP; everything is much more fuzzy, and we must

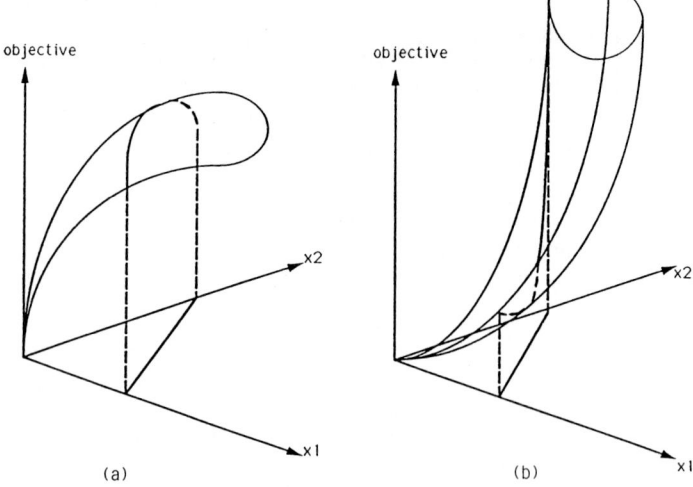

FIGURE 6.6 (*a*) Concave objective subject to a single linear constraint; (*b*) convex objective subject to a single linear constraint.

reconcile ourselves to the fact that it is impossible to develop a solution approach that is completely problem-independent.

Optimality Conditions. In this subsection we examine some of the basic theoretical concepts associated with nonlinear programming. Most of them have to do with being able to characterize an optimum solution in some fashion. In general, the term *optimum* used here refers only to a local (relative) maximum or minimum; in certain cases where special conditions are met, the local optimum will also be a global optimum. We will also, for the most part, restrict our discussion to minimization problems. This is because any problem where the objective is to maximize $f(\mathbf{x})$ is equivalent to the problem of minimizing $-f(\mathbf{x})$.

We now introduce some preliminary definitions and concepts. Assuming that the function $f(\mathbf{x})$ is continuous, real-valued and differentiable for $\mathbf{x} \in \mathcal{R}^n$, the *gradient vector* at the point $\mathbf{x}$ is defined as

$$\nabla f(\mathbf{x}) = \left[\frac{\partial f}{\partial x_1}(\mathbf{x}) \; \frac{\partial f}{\partial x_2}(\mathbf{x}) \cdots \frac{\partial f}{\partial x_n}(\mathbf{x}) \right]^T$$

If $f(\mathbf{x})$ is also twice continuously differentiable then we define its *Hessian* as the ($n \times n$) symmetric matrix $\mathbf{H}(\mathbf{x})$ which has the quantity $\partial^2 f(\mathbf{x})/\partial x_i \partial x_j$ as its entry in the ith row and the jth column.

The Hessian is an important matrix. Given an arbitrary $\mathbf{w} \in \mathcal{R}^n$, consider the quantity $\mathbf{w}^T \mathbf{H} \mathbf{w}$. If for all $\mathbf{w} \neq 0$, this quantity is nonnegative (nonpositive) the Hessian is said to be *positive (negative) semidefinite*. If the quantity is strictly positive (negative) for all nonzero $\mathbf{w}$, the Hessian is said to be *positive (negative) definite*. If the Hessian is not positive, negative definite, or semidefinite, it is said to be *indefinite*.

These are important concepts because a function is convex (concave) if, and only if, the Hessian matrix is positive (negative) semidefinite at all points. It is *strictly* convex (concave) if, and only if, the Hessian is positive (negative) definite. Thus the Hessian provides a ready way of checking if a function is convex or concave. The Hessian is also used to verify certain necessary and sufficient conditions; these are explained in the next paragraph.

The most common way of characterizing a solution is in terms of *necessary* conditions.

These are conditions that *must* be met by any optimum solution; however, if a point meets these conditions there is no guarantee that it is an optimum. In order to guarantee this, the point must also meet an additional set of conditions known as *sufficient* conditions. Given a candidate point, the sufficient conditions are in general more difficult to check than the necessary ones. Luckily, when a point satisfies the necessary conditions it *will* in many cases be an optimum solution. Most practical algorithms therefore look only for points that satisfy the necessary conditions.

We now examine optimality conditions for several specific types of NLP problems.

Unconstrained Optimization Problems. Let us first consider the problem of minimizing a function $f(\mathbf{x})$, where $\mathbf{x} \in \mathcal{R}^n$, that is not subject to any constraints.

A point $\mathbf{x}'$ is called a *stationary point* if $\nabla \mathbf{f}(\mathbf{x}') = 0$. For the single-variable function shown in Fig. 6.7, the points $A, B, C, D,$ and E are all stationary points. Such points are important because a necessary condition for a point $\mathbf{x}^*$ to be a local minimum is that it should be a stationary point. This is called the *first-order necessary condition.* Another necessary condition (called the *second-order necessary condition*) is that the Hessian at $\mathbf{x}^*$ should be positive semidefinite. A sufficient condition for $\mathbf{x}^*$ to be a minimum is that, in addition to the first-order necessary conditions, the Hessian at $\mathbf{x}^*$ should be positive definite; this is equivalent to saying that the function should be strictly convex in some neighborhood of $\mathbf{x}^*$. Thus the necessary conditions for a local minimum are also sufficient if the function f is convex; moreover every local minimum is also a global minimum.

To clarify these concepts, consider Fig. 6.7. Points B and E satisfy both the first- and second-order necessary conditions for a minimum, as well as the sufficient conditions for a minimum. Point C, on the other hand, satisfies the first- and second-order necessary conditions for a minimum but not the sufficient condition; this point is referred to as a point of inflection.

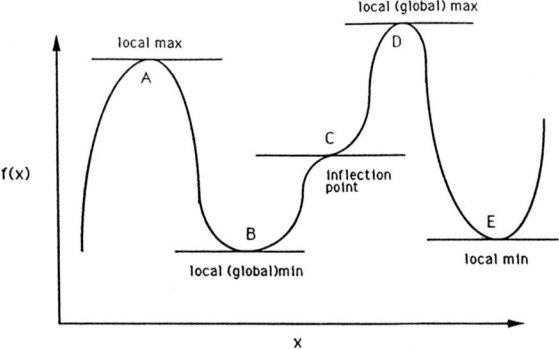

FIGURE 6.7 Clarification of concepts.

Equality-Constrained Problems. Now consider the case where the NLP has only equality constraints of the form $h_j(\mathbf{x}) = 0, j = 1, 2, \ldots, p$. To develop the necessary conditions for a minimum, we define the Lagrangian function

$$L(\mathbf{x}, \mu) = f(\mathbf{x}) + \sum_{j=1}^{p} \mu_j h_j(\mathbf{x})$$

Note that there is one μ_j for each constraint; collectively these are referred to as *Lagrange multipliers.* They are analogous to the simplex multipliers for LP and have a sound economic interpretation, in addition to being critical to the development of duality concepts in NLP. Duality concepts are beyond the scope of this chapter, and the interested reader is referred to one of the reference texts listed at the end of this chapter.

It may be shown that when certain regularity conditions (or *constraint qualifications*) are met, the equality-constrained minimization problem is equivalent to the unconstrained minimization of the Lagrangian function. Assuming that f and h_j are real-valued and continuously differentiable, the necessary conditions for optimality for the Lagrangian (from the last subsection) are that $\nabla \mathbf{L}(\mathbf{x}, \mu) = 0$. This is equivalent to saying that for a vector $\mathbf{x}^*$ to be a minimizer for the constrained problem, the necessary conditions are that there exist a vector $\mu^* \in \mathcal{R}^p$ such that†

1. $h_j(\mathbf{x}^*) = 0 \qquad\qquad j = 1, 2, \ldots, p$

2. $\dfrac{\partial f}{\partial x_i^*} + \displaystyle\sum_{j=1}^{p} \mu_j^* \dfrac{\partial h_j}{\partial x_i^*} = 0 \qquad i = 1, 2, \ldots, n$

The necessary conditions are sufficient if (1) and (2) above hold, and for every nonzero $\eta \in \mathcal{R}^n$ which satisfies $\eta^T \nabla h_j(\mathbf{x}^*) = 0$, for $j = 1, 2, \ldots, p$, it is also true that $\eta^T \nabla_x^2 L(\mathbf{x}^*, \mu^*) \eta > 0$. Moreover, the minimum in that case is also a strict minimum. Again, these sufficiency conditions are difficult to verify and in practice they are hardly, if ever, used. However for the special case where f is convex and each h_j is linear, the necessary conditions automatically imply the sufficient conditions, and the minimum is also a global minimum.

Problems with Both Equality and Inequality Constraints.   Finally, we look at the general NLP problem (Program 1) with both equality and inequality constraints and generalize the notion of the lagrangian to this case. The Lagrangian for the general problem is given by

$$L(\mathbf{x}, \lambda, \mu) = f(\mathbf{x}) + \sum_{j=1}^{m} \lambda_j g_j(\mathbf{x}) + \sum_{j=1}^{p} \mu_j h_j(\mathbf{x})$$

We make the usual assumptions that all functions are real-valued and continuously differentiable, and that the constraint qualification is met. Then the necessary conditions state that if the point $\mathbf{x}^*$ is a constrained minimum for $f(\mathbf{x})$, there must exist a vector $\lambda^* \in \mathcal{R}^m$ and a vector $\mu^* \in \mathcal{R}^p$ such that

1. $\nabla \mathbf{f}(\mathbf{x}^*) + \displaystyle\sum_{j=1}^{m} \lambda_j^* \nabla \mathbf{g}_j(\mathbf{x}^*) + \sum_{j=1}^{p} \mu_j^* \nabla \mathbf{h}_j(\mathbf{x}^*) = 0$

2. $\lambda_j^* g_j(\mathbf{x}^*) = 0 \qquad j = 1, 2, \ldots, m$

3. $\lambda_j^* \geq 0 \qquad\qquad j = 1, 2, \ldots, m$

4. The original constraints are satisfied at $\mathbf{x}^*$

Condition 1 states that $(\mathbf{x}^*, \lambda^*, \mu^*)$ is a stationary point for the Lagrangian function. Condition 2 (which is called *complementary slackness*) states that, at the optimum, if an inequality constraint is nonbinding, the corresponding multiplier is zero; conversely, if a multiplier is not zero, then the corresponding constraint must be binding. Condition 3 states that the Lagrange multipliers for the inequality constraints must be nonnegative. Collectively, these are referred to as the Karush-Kuhn-Tucker (K-K-T) necessary conditions for optimality.

Problems with Nonnegative Decision Variables.   Many real-world applications of nonlinear optimization (especially in engineering) call for nonnegative design variables. One could, of course, treat each restriction of the form $x_i \geq 0$ as an explicit constraint of the form $-x_i \leq 0$, define Lagrange multipliers for each of these, and then apply the K-K-T

---

†Constraint qualifications are primarily of theoretical interest, and have the purpose of verifying that problems are "well-behaved." Normally most practical problems will meet these conditions, and algorithms for solving NLP problems rarely, if ever, check to see whether constraint qualifications are met.

conditions of the previous subsection to check whether a point meets the necessary conditions for optimality. However, a more elegant approach is to use the following extension of the K-K-T conditions for this situation. The necessary conditions for Program 1, with the additional restriction that $\mathbf{x} \geq \mathbf{0}$, are

1.
$$\nabla \mathbf{f}(\mathbf{x}^*) + \sum_{j=1}^{m} \lambda_j^* \nabla \mathbf{g}_j(\mathbf{x}^*) + \sum_{j=1}^{p} \mu_j^* \nabla \mathbf{h}_j(\mathbf{x}^*) \geq \mathbf{0}$$

2.
$$\left[\nabla \mathbf{f}(\mathbf{x}^*) + \sum_{j=1}^{m} \lambda_j^* \nabla \mathbf{g}_j(\mathbf{x}^*) + \sum_{j=1}^{p} \mu_j^* \nabla \mathbf{h}_j(\mathbf{x}^*) \right]^T \mathbf{x}^* = 0$$

3. $\lambda_j^* g_j(\mathbf{x}^*) = 0 \qquad j = 1, 2, \dots, m$

4. $\lambda_j^* \geq 0 \qquad j = 1, 2, \dots, m$

5. The original constraints are satisfied at $\mathbf{x}^*$.

This concludes our brief discussion of the basic theoretical concepts behind nonlinear programming. It should be emphasized that by no means is this comprehensive; the interested reader is referred to one of the references for detailed discussions on the theory of NLP.

ALGORITHMS

The following two sections are devoted to algorithms for solving NLP problems. Literally hundreds of algorithms have been developed over the years for NLP; we will only provide an overview of the different classes of algorithms along with some representative examples.

It should also be emphasized that what we present here is only the essential logic of the algorithms; their actual implementation on a digital computer is a whole different issue, and the interested reader should refer to the original source of the algorithms for details and issues that relate to robustness, rate of convergence, and computational effort, among others. Robustness is the ability of the algorithm to reach an optimal solution from any starting point, and for all problems which it is designed to address. The rate of convergence relates to the speed with which we approach the optimum. Computational effort relates to storage requirements and CPU time used to compute all required quantities. An excellent general discussion of practicalities is contained in Gill, Murray, and Wright.[8b]

Algorithms for Unconstrained NLP

We begin with a section on algorithms for unconstrained optimization.

Line Search Procedures. The simplest algorithms are the ones used for finding the minimum of a univariate function $f(x)$, over some interval $[a_0, b_0]$, that is, to minimize $f(x)$, subject to $a_0 \leq x \leq b_0$. These procedures are commonly used within a number of iterative algorithms designed for general constrained NLP.

Line search procedures normally assume that the function to be minimized is unimodal, that is, that it has a single minimum value in the interval of interest. At the ith iteration the function is evaluated at discrete points within the current interval of uncertainty $[a_i, b_i]$, and a portion of the interval is then eliminated.

For instance, in Fig. 6.8a, the function is evaluated at a_i, b_i and three intermediate points p, q, and r inside the current interval of uncertainty. For the values shown, the new interval of uncertainty (a_{i+1}, b_{i+1}) would thus be (q, b_i). In Fig. 6.8b, we use only two intermediate points p and q, and the new interval would thus be (a_i, q). The procedure is halted when the interval of uncertainty is smaller than some prespecified tolerance.

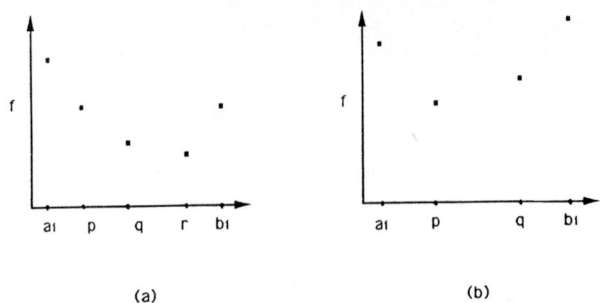

FIGURE 6.8   Line search procedures. (*a*) Function evaluated at three intermediate points; (*b*) function evaluated at two intermediate points.

The methods differ in how many intermediate points are chosen, and how these points are specified. Common algorithms include the three-point equi-interval search, the dichotomous search, the golden section search, the Fibonacci search, and polynomial interpolation. One of the better procedures is the golden section search which we describe here; the interested reader is referred to one of the references for details on the other procedures.

Golden Section Search.   This method uses two intermediate points. The location of these points is done in such a way that, at each iteration, one of the intermediate points from the previous iteration is used at the current iteration. Thus only one *new* point needs to be located at each iteration, and thus only one additional function evaluation is performed at each iteration.

To achieve this objective, suppose that the current interval is of length $b_i - a_i = l_i$. Let us locate p_i and q_i so that (1) $p_i = a_i + \alpha l_i$ and (2) $q_i = b_i - \alpha l_i$. We need to find the value of α.

At the next iteration, suppose $a_{i+1} = a_i$, $b_{i+1} = q_i$ as shown in Fig. 6.9a (the other possibility is that $a_{i+1} = p_i$, $b_{i+1} = b_i$). Then we want q_{i+1} to coincide with p_i as shown in Fig. 6.9b (so that we only need to locate p_{i+1} and evaluate the function at this point). Thus, when $a_{i+1} = a_i$ and $b_{i+1} = q_i$, we want $q_{i+1} = p_i$. Noting that $q_{i+1} = b_{i+1} - \alpha l_{i+1} = b_{i+1} - \alpha(b_{i+1} - a_{i+1})$, this implies that $p_i = q_i - \alpha(q_i - a_i)$. If we now substitute for p_i and q_i from (1) and (2) above, and simplify the expression, we obtain $\alpha^2 - 3\alpha + 1 = 0$. This quadratic equation yields $\alpha = 0.381$ (the other root is greater than 1 and hence may be ignored).

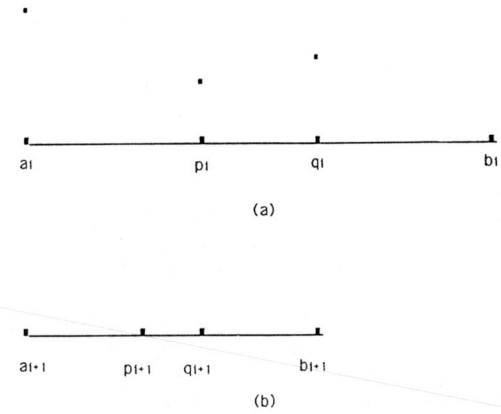

FIGURE 6.9   Golden section search.

It is easy to show that, if the second possibility occurs, then the value of α is identical. Thus at each iteration we locate a new point so that it is at a distance of 38.1 percent of the interval length from an end point, and this will ensure that at the next iteration only one new point will need to be located and the function will be evaluated at this one point only. The procedure continues till l_k is "sufficiently small" for some k.

Optimization of Multivariate Functions. The next class of algorithms consists of those that treat the problem of minimizing a function of several variables. This is also a very important class of algorithms, both in its own right, and because a number of methods for constrained problems first convert the constrained problem into an equivalent unconstrained one and then use an unconstrained optimization algorithm to solve the latter.

While there are a few procedures that search for the minimum without making use of derivative information, most of the better procedures are designed for differentiable functions. Recall that the necessary conditions require that the gradient vector vanish at the optimal point. Most algorithms in this class therefore try to search for a stationary point where the gradient vector is equal to the zero vector. A common criterion used to determine this is to check whether $\|\nabla f(x)\| < \varepsilon$, where ε is some small user specified tolerance. Other stopping criteria could also be used; refer to Gill, Murray, and Wright[8] for details on these.

Almost all procedures are descent algorithms which typically use the following logic at the ith iteration:

STEP 1: Is $\|\nabla f(x^i)\| < \varepsilon$? If yes, then accept x^i as the optimum solution. If not, go to Step 2.

STEP 2: Determine a search direction d^i such that the function f decreases continuously for at least some distance along d^i from x^i; that is, $f(x^i + \alpha d^i) < f(x^i)$ for all $\alpha \in (0, \delta)$, where $\delta > 0$.

STEP 3: Determine a step size α^* to be taken along the direction d^i [perhaps by minimizing $f(x^i + \alpha d^i)$ over all α in some interval $(0, \delta)$]. Let $x^{i+1} = x^i + \alpha^* d^i$, and return to Step 1.

The algorithms are called *descent* methods because they successively reduce the function value by moving along directions that will enable a reduction. They differ primarily in the specific procedures used to determine the search direction and the step size. We now turn to some common descent methods.

The Method of Steepest Descent. This procedure is based on the fact that the gradient vector points in the direction of maximum increase, so that the direction of steepest descent from x^i is $-\nabla f(x^i)$. Thus, at each iteration we move from x^i to some point $x^{i+1} = x^i - \alpha^* \nabla f(x^i)$, with α^* being determined by a line search.

While this method is guaranteed to converge, it is not the best procedure to use, since its convergence becomes very slow after the first few iterations. The successive directions generated are orthogonal to each other, and thus the method tends to zigzag around a "valley," or in the vicinity of the optimum. There are two major reasons for this. First, each search direction is a "cold start" which ignores all information from previous steps, and second, the procedure makes use only of the gradient vector and ignores curvature (second derivative information).

Algorithms Based on Conjugate Search Directions. Many of the better descent methods use successive search directions that are conjugate. Suppose we are given a square, symmetric matrix $\mathbf{Q}$. Two directions $\mathbf{v}^1$ and $\mathbf{v}^2$ are defined to be conjugate (or more precisely, Q-conjugate), if $(\mathbf{v}^1)^T \mathbf{Q} \mathbf{v}^2 = 0$. Geometrically, conjugate directions may be interpreted as follows: suppose we have a circle with two orthogonal radius vectors, and we consider the directions $\mathbf{d}^1$ and $\mathbf{d}^2$ as the directions along which these radius vectors point. Now, if the circle is stretched into an ellipse as shown in Fig. 6.10, the directions $\mathbf{d}^1$ and $\mathbf{d}^2$ will then get stretched into two conjugate directions $\mathbf{v}^1$ and $\mathbf{v}^2$.

The primary improvement that conjugate directions provide over the steepest descent

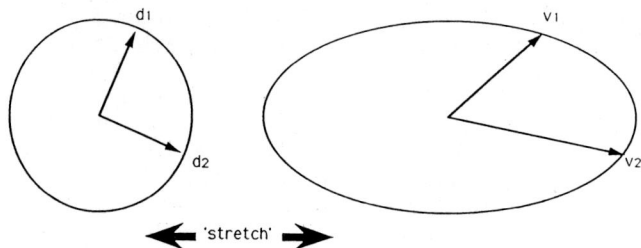

FIGURE 6.10   Circle stretched into ellipse.

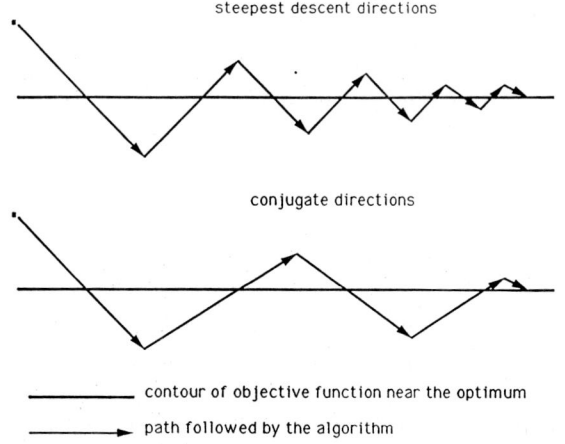

steepest descent directions

conjugate directions

—————— contour of objective function near the optimum

——————▶ path followed by the algorithm

FIGURE 6.11   Deflection of orthogonal directions.

procedure is that successive search directions are no longer orthogonal. Thus the effect of these procedures is to deflect the orthogonal directions so that the zigzagging effect is reduced. This is depicted in Fig. 6.11.

A second important feature of conjugate directions is that for a quadratic function of $\mathbf{x} \in \mathcal{R}^n$ with a positive definite Hessian, the minimum may be found from *any* starting point in at most n steps, by doing a series of successive minimizations along n conjugate directions. In fact, conjugate direction methods are designed for quadratic function forms where the Hessian does not change from one iteration to the next; however, in practice they tend to work quite well for general functions as well.

An example of a conjugate directions approach is the conjugate gradient method which follows the general scheme outlined for descent methods, and differs from the steepest descent method only in the procedure used to specify the search direction. It starts off with the steepest-descent direction at Iteration 1, and computes the search direction at Iteration k as $\mathbf{d}^k = -\nabla \mathbf{f}(\mathbf{x}^k) + \beta_k \mathbf{d}^{k-1}$, where $\beta_k = [||\nabla \mathbf{f}(\mathbf{x}^k)||/||\nabla \mathbf{f}(\mathbf{x}^{k-1})||]^2$. Thus the search direction at each iteration is the steepest-descent direction, plus some multiple of the previous search direction.

Quasi-Newton Methods.   A problem with the steepest-descent and the conjugate-gradient methods of the previous section is that in both cases we do not have any second-order derivative (curvature) information. If the latter were available, we could represent the surface of the objective more accurately, and thus obtain better search directions.

The classical Newton method does this by using the Taylor series expansion for the function f around a point $\mathbf{x}$, namely $f(\mathbf{x} + \delta\mathbf{x}) = f(\mathbf{x}) + \delta\mathbf{x}^T \nabla \mathbf{f}(\mathbf{x}) + (0.5)\delta\mathbf{x}^T \mathbf{H}(\mathbf{x})\delta\mathbf{x}$, where $\mathbf{H}(\mathbf{x})$ is the Hessian evaluated at $\mathbf{x}$, and $\delta\mathbf{x}$ is some perturbation vector. In order to minimize f as a function of $\delta\mathbf{x}$, we set $\partial f/\partial(\delta\mathbf{x}) = 0$, to obtain $\nabla \mathbf{f}(\mathbf{x}) + \mathbf{H} \, \delta\mathbf{x} = 0$, that is,

$\delta \mathbf{x} = - \mathbf{H}^{-1}(\mathbf{x}) \nabla \mathbf{f}(\mathbf{x})$. Thus Newton's method is equivalent to a descent method that chooses the search direction as $\mathbf{d} = - \mathbf{H}^{-1}(\mathbf{x})\nabla\mathbf{f}(\mathbf{x})$, and a step size of 1.0.

Although Newton's method uses second-order derivative information, it has several drawbacks. First, it is very sensitive to the starting point. Second, it requires that the function be twice differentiable. Third, it assumes that the Hessian is nonsingular at all points. Finally, it needs to actually compute second derivatives and then invert the Hessian, which often involves a lot of computational effort.

Quasi-Newton methods were developed in order to overcome some of these drawbacks. Rather than explicitly compute the Hessian and then invert it, these procedures try to build up an approximation to the inverse of the Hessian as the iterations proceed. These approximations are developed using only first-order derivatives.

While Quasi-Newton methods were also developed initially for quadratic functions with positive definite Hessians, they usually work very well for general problems as well. Essentially, they use the general formula $\mathbf{d}^k = - \mathbf{A}_k\nabla\mathbf{f}(\mathbf{x})$ to compute the search direction at the kth iteration, where the matrix $\mathbf{A}_k$ may be viewed as an approximation to $\mathbf{H}^{-1}$ (note that if $\mathbf{A}_k = \mathbf{I}_k$ this reduces to the steepest-descent procedure, while if $\mathbf{A}_k = [\mathbf{H}(\mathbf{x}^k)]^{-1}$ it reduces to Newton's method). Specific algorithms differ in the procedure used to update $\mathbf{A}_k$ and build up the approximation.

An example of a Quasi-Newton method is the Davidon-Fletcher-Powell algorithm.[9] It starts with some symmetric positive definite matrix $\mathbf{A}_0$ as an initial estimate of $\mathbf{H}^{-1}$ (in the simplest case, $\mathbf{A}_0 = \mathbf{I}$). At iteration k, it computes the search direction $\mathbf{d}^k = - \mathbf{A}_k\nabla\mathbf{f}(\mathbf{x}^k)$ and the step size α^* to minimize $f(\mathbf{x}^k + \alpha\mathbf{d}^k)$ in the usual way, so that the new design is $\mathbf{x}^{k+1} = \mathbf{x}^k + \alpha^*\mathbf{d}^k$. Suppose we now define $\mathbf{p}^k = \mathbf{x}^{k+1} - \mathbf{x}^k$ (the change in the design), and $\mathbf{q}^k = \nabla\mathbf{f}(\mathbf{x}^{k+1}) - \nabla\mathbf{f}(\mathbf{x}^k)$ (the change in the gradient vector). We compute the pair of correction matrices $\mathbf{B}_k = \mathbf{p}^k(\mathbf{p}^k)^T/(\mathbf{p}^k)^T\mathbf{q}^k$ and $\mathbf{C}_k = - \mathbf{A}_k\mathbf{q}^k(\mathbf{q}^k)^T\mathbf{A}_k/(\mathbf{q}^k)^T\mathbf{A}_k\mathbf{q}^k$. The matrix $\mathbf{A}_k$ is then updated for the next iteration as $\mathbf{A}_{k+1} = \mathbf{A}_k + \mathbf{B}_k + \mathbf{C}_k$.

While the updating scheme appears messy, it is actually quite elegant, and it may be shown that when applied to a positive definite quadratic form, the $\mathbf{A}_k$ converge to the Hessian of the quadratic form. Furthermore, $\mathbf{A}_k$ is positive definite for each k and the method will always converge to a local minimum.

Algorithms for Constrained NLP

We now turn our attention to algorithms that solve constrained minimization problems. We will look at general schemes here as opposed to algorithms developed for specific problem types. Constrained optimization algorithms can be broadly classified into two categories: *transformation methods* and *primal methods*. The former refers to algorithms that convert the constrained problem into an equivalent unconstrained one (or a sequence of unconstrained ones), and then apply one of the methods from the previous section to the unconstrained problem. Primal methods are algorithms that directly attack the constrained problem. We shall examine each class in order.

Sequential Unconstrained Minimization Techniques (SUMT). These are transformation algorithms that convert the constrained problem into a sequence of unconstrained ones. Essentially, they are one of two types: (1) exterior and (2) interior penalty function methods.

Assuming minimization, we define a penalty function as one that takes on a large positive value if a constraint is violated, and zero otherwise. Thus the penalty function penalizes the objective at points that are infeasible. As an example, for the constraints $g_j(\mathbf{x}) \leq 0, j = 1, 2, \ldots, m$, and $h_j(\mathbf{x}) = 0, j = 1, 2, \ldots, p$, we could define the penalty function

$$p(\mathbf{x}) = \sum_{j=1}^{p} \left| h_j(\mathbf{x}) \right|^2 + \sum_{j=1}^{m} \{ \text{Max} \left[0, g_j(\mathbf{x}) \right] \}^2$$

It is easy to see that $p(\mathbf{x}) = 0$ as long as all the constraints are satisfied; otherwise it takes on a positive value.

If we now define an *auxiliary* function $F(\mathbf{x}) = f(\mathbf{x}) + \mu p(\mathbf{x})$, where μ is some penalty parameter, we could do an unconstrained minimization of $F(\mathbf{x})$ as opposed to the constrained minimization of $f(\mathbf{x})$. The method will of course try to find a point that is in the feasible region [so that $p(\mathbf{x})$ is equal to zero and no penalty is incurred], while minimizing $f(\mathbf{x})$.

The solution to the unconstrained problem can be made arbitrarily close to that of the original problem by choosing μ sufficiently large. In practice, rather than solving a single problem with a large value for μ, we solve a series of unconstrained problems where the value of μ is successively increased, and the optimal point at any iteration becomes the starting point for the next one. This is because, if μ is chosen too large initially, too much emphasis is placed on feasibility and the step sizes become very small, causing the algorithm to stop at a suboptimal point.

The procedure may be generalized as follows: choose a starting point $\mathbf{x}^1$, a tolerance ε, an initial penalty parameter μ_1, and an *increase factor* $\beta > 1$. At iteration k, minimize $f(\mathbf{x}) + \mu_k p(\mathbf{x})$, using $\mathbf{x}^k$ as the starting point (usually), and let the optimal solution to this be $\mathbf{x}^{k+1}$. If $\mu_k p(\mathbf{x}^{k+1}) < \varepsilon$, we stop and accept $\mathbf{x}^{k+1}$ as the optimum solution, otherwise we let $\mu_{k+1} = \beta \mu_k$, and start iteration $k+1$.

The above procedure is referred to as an *exterior* penalty function method, since it begins with an infeasible point and generates a series of such points that converge to the optimum solution. Another SUMT algorithm is the *interior* penalty function method (or the *barrier* function method). These are methods that transform the original problem into a sequence of unconstrained ones, by using a barrier function that ensures that we never leave the feasible region. Thus, these methods generate a sequence of *feasible* points which converge to the optimal solution. A requirement here is that the interior of the feasible region be nonempty (thus equality constraints cannot be handled).

Ideally, we would like the barrier function to be zero in the interior and ∞ on the boundary; however this discontinuity causes problems. Therefore we choose a function that is continuous and nonnegative over the interior of $[\mathbf{x}|g_j(\mathbf{x}) \le 0)]$, and *approaches ∞ as the boundary is approached from the interior.* An example of a barrier function is $b(\mathbf{x}) = \sum_{j=1}^{m}[-1/g_j(\mathbf{x})]$. The auxiliary function to be minimized, $F(\mathbf{x})$, is now given by $F(\mathbf{x}) = f(\mathbf{x}) + (1/\mu)b(\mathbf{x})$. As with exterior penalty function methods, it may be shown that as μ approaches ∞, the minimum of $F(\mathbf{x})$ approaches the minimum of $f(\mathbf{x})$.

Barrier function methods work exactly like the interior penalty function method, except that we start with an initial $\mathbf{x}^1$ in the *interior* of the feasible region, and never leave the interior as we converge to the optimum.

Penalty function methods were among the earliest ones developed for constrained NLP, and many variations of these two schemes have been used. However, they all have several inherent problems, primarily the fact that they are ill-behaved near the boundary of the feasible region where the optimum usually lies. Furthermore, the choice of the increase factor β is also critical. Finally, the Hessian matrix (if used in the unconstrained minimization) often becomes ill-conditioned as the optimum is approached.

Another class of multiplier methods that alleviates some of the problems of SUMT, are the *augmented Lagrangian* (or *multiplier*) methods. In these methods, the penalty parameter μ does not have to go to infinity for the optimum to be approached, and thus the auxiliary functions tend to be somewhat better conditioned. Furthermore they also converge to a local minimum, and at rates that are faster than SUMT. The auxiliary function for these methods may be given by

$$F(\mathbf{x}) = f(\mathbf{x}) + \frac{1}{2}\left\{\sum_{j=1}^{m}\mu_j\left[\text{Max}(0, g_j(\mathbf{x}) + q_j)\right]^2 + \sum_{j=1}^{p}\mu_j'\left(h_j(\mathbf{x}) + q_j'\right)^2\right\}$$

where $q_j, \mu_j > 0$. In these methods, we start with some initial values for the parameters q_j, μ_j, q_j', and μ_j' associated with each constraint, and minimize $F(\mathbf{x})$. The parameters are then adjusted at each iteration, until the optimum is reached. A survey of these methods may be found in Bertsekas.[10]

Primal Methods.   This is the other broad class of methods for constrained optimization where the original problem is treated directly. Conceptually, the algorithms here are similar to unconstrained methods. The major difference is the fact that one has to keep the constraints in mind when deciding on both the search direction and the step size. Most algorithms try to linearize the objective and the active constraints at the current point, and then use these to find a search direction. The step size is chosen to reduce the value of a descent function which often includes penalty terms for violated constraints in addition to the objective function itself.

Perhaps the most straightforward procedure is the method of *feasible directions.* Suppose the feasible set is given by $\mathscr{X}$. A feasible direction at $\mathbf{d}$ is one for which there exists a $\delta > 0$ such that $\mathbf{x} + \alpha \mathbf{d}$ is in $\mathscr{X}$ for all α between 0 and δ. This merely says that we should be able to move along $\mathbf{d}$ from $\mathbf{x}$ for some finite distance and stay within the feasible region. In Fig. 6.12, $\mathbf{d}^1$ is a feasible direction at $\mathbf{x}$ while $\mathbf{d}^2$ is not. A feasible direction $\mathbf{d}$ is called an *improving* feasible direction if the objective decreases for some finite distance along $\mathbf{d}$. Under differentiability this implies that $\mathbf{d}^T \nabla f(\mathbf{x})$ should be negative.

A general strategy then is to start at a feasible point $\mathbf{x}^1$. At iteration k, we locate an improving feasible direction $\mathbf{d}^k$ from our current design $\mathbf{x}^k$, do an unconstrained minimization along $\mathbf{d}^k$ to obtain a step size α_k (perhaps by a line search), and move to $\mathbf{x}^{k+1} = \mathbf{x}^k + \alpha_k \mathbf{d}^k$. Methods differ in the exact procedure used to find an improving feasible direction at $\mathbf{x}^k$, and usually use only the constraints that are binding (or near-binding) at $\mathbf{x}^k$ when solving the direction-finding subproblem. The algorithm terminates when such a direction cannot be found.

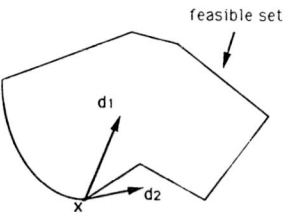

FIGURE 6.12   Feasible and nonfeasible directions.

Another well-known method that maintains feasibility at each iteration is the *gradient projection* method of Rosen.[11] Recall that the negative of the gradient vector at $\mathbf{x}$ points in the direction of steepest descent. However, with constrained problems, moving in this direction may destroy feasibility. Rosen's method uses a projection matrix $\mathbf{P}$ to deflect the steepest descent direction $-\nabla f(\mathbf{x})$ on to a hyperplane that is tangential to the binding constraints at $\mathbf{x}$, that is, $\mathbf{d} = -\mathbf{P} \nabla f(\mathbf{x})$. A step size is then computed along this direction. With linear constraints, one would simply move along the (linear) boundary, and thus feasibility is easily maintained. However, with nonlinear constraints, the new point (say $\mathbf{x}'$) will be infeasible, and therefore a series of correction steps will have to be made to return to the feasible region at the point $\mathbf{x}^{k+1}$. The procedure is depicted in Fig. 6.13.

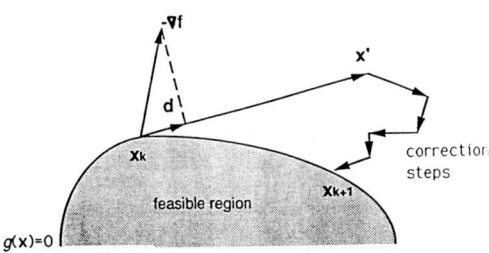

FIGURE 6.13   Correction step procedure.

This method has been applied successfully to many problems. However it is not without its drawbacks, the primary one being the fact that after the correction steps, $f(\mathbf{x}^{k+1})$ may not be less than $f(\mathbf{x}^k)$. In such cases one must reduce the step size, and execute the correction procedure from the new $\mathbf{x}'$ all over again. Thus, ensuring convergence is rather inconvenient.

Another successful algorithm is the *generalized reduced gradient* (GRG) method of Abadie and Carpenter.[12] It bears a nice analogy to the simplex method of LP, and is similar in spirit to the gradient projection method.

Suppose we have a nonlinear program with equality constraints only (perhaps by addition of suitable slacks). The direction-finding subproblem begins by partitioning the (feasible) vector of design variables $\mathbf{x}$ into an independent vector $\mathbf{x}_I$ and a dependent vector $\mathbf{x}_D$. Similarly the various gradient vectors at $\mathbf{x}$ are partitioned into $\nabla_I \mathbf{f}$ and $\nabla_D \mathbf{f}$, and $\nabla_I \mathbf{g}_j$ and $\nabla_D \mathbf{g}_j$. Suppose now that $[\delta_D \; \delta_I]^T$ is a vector of small changes in the current design. Then the corresponding changes in the objective (Δf) and constraint functions (Δg_j) are approximately $\delta_I^T \nabla_I \mathbf{f} + \delta_D^T \nabla_F \mathbf{f}$ and $\delta_I^T \nabla_I \mathbf{g}_j + \delta_D^T \nabla_F \mathbf{g}_j$ respectively.

In order for us to remain feasible after the changes, we must have $\Delta g_j = 0$, that is, $\mathbf{J}_D \delta_D + \mathbf{J}_I \delta_I = 0$, where $\mathbf{J}_D$ and $\mathbf{J}_I$ are matrices with columns containing the gradient vectors of variables in $\mathbf{x}_D$ and $\mathbf{x}_I$ respectively. Assuming that $\mathbf{J}_D$ is nonsingular, this implies that $\delta_D = - \mathbf{J}_D^{-1} \mathbf{J}_I \delta_I$. Thus, *given* a change δ_I in the independent variables, this is the corresponding change required in the dependent variables to maintain feasibility. Substituting this value for δ_D in the expression for Δf and simplifying, we have $\Delta f = \gamma^T \delta_I$ as the corresponding change in the objective, where the vector γ is called the *generalized reduced gradient* and is given by $\gamma = \nabla_I \mathbf{f} - [(\nabla_D \mathbf{f})^T \mathbf{J}_D^{-1} \mathbf{J}_I]$.

The search direction $\mathbf{d}$ is therefore chosen as the negative of the vector γ for the independent variables (in order to maximize the rate of decrease of f), and $\mathbf{J}_D^{-1} \mathbf{J}_I \gamma$ for the dependent variables. A line search is now conducted along this direction to determine the step size α. Typically, the procedure is to check whether the new point $\mathbf{x} + \alpha \mathbf{d}$ is feasible for the optimal step size. If not, the independent variables are fixed at their new values, and an iterative correction procedure is applied to the dependent variables to restore feasibility. If the new point is better than the previous one, we proceed to the next iteration; if not, the step size is reduced and the procedure repeated (like the gradient projection method).

GRG has been extensively tested on many applications and is, in general, considered one of the better NLP algorithms. Its main drawback is in the computational effort required during the line search step when the gradient vectors need to be reevaluated and the matrix $\mathbf{J}_D$ inverted for each iteration of the Newton-Raphson method. However, several clever schemes have been employed in practice to update $\mathbf{J}_D^{-1}$ without recomputing the gradient vectors.

Another class of algorithms consists of those that work by linearizing the constrained problem. Assuming that our current guess is $\mathbf{x}^k$, we may rewrite the problem of minimizing $f(\mathbf{x})$, subject to $g_j(\mathbf{x}) \le 0$ by using the Taylor series expansion around $\mathbf{x}^k$ for the functions involved. The problem then reduces to minimizing $f(\mathbf{x}^k) + (\mathbf{x} - \mathbf{x}^k)^T \nabla f(\mathbf{x}^k)$ subject to $g_j(\mathbf{x}^k) + (\mathbf{x} - \mathbf{x}^k)^T \nabla g_j(\mathbf{x}^k) \le 0$, for all j. If $(\mathbf{x} - \mathbf{x}^k)$ is denoted by $\mathbf{d}$, this is equivalent to minimizing $\mathbf{d}^T \nabla f(\mathbf{x}^k)$, subject to $\mathbf{d}^T \nabla g_j(\mathbf{x}^k) \le - g_j(\mathbf{x}^k)$, which may be easily seen to be a direction-finding subproblem that is a linear program in $\mathbf{d}$. Thus at each iteration of a *sequential linear programming* (SLP) algorithm we solve an LP to find the search direction. In the simplest case, $\mathbf{x}^{k+1}$ is chosen as $\mathbf{x}^k + \mathbf{d}^k$, while in others, some searching may be done along $\mathbf{d}$ for an optimal step size.

The convergence of SLP algorithms is usually very slow; worse yet, they often cycle between two points if the optimum solution is not at a corner point of the feasible region. A logical extension is to use second-order derivatives to incorporate curvature information and approximate the constraint and objective functions more accurately. In practice, it has been found that it is usually enough if only the objective is approximated by a quadratic function of the form $\mathbf{d}^T \nabla f(\mathbf{x}^k) + \frac{1}{2} \mathbf{d}^T \mathbf{N}_k \mathbf{d}$. The constraints remain the same, and at each iteration we solve a quadratic programming (QP) subproblem to find a search direction. A number of efficient algorithms are available for quadratic programming (which is discussed in the next section), and thus the QP subproblem is relatively easy to solve.

There have been numerous versions of this basic approach, with the difference being in the way the matrix $\mathbf{N}_k$ is specified at each iteration, the procedure for updating it, and in the specification of a descent function. The interested reader may refer to Pshenichnyj and

Danilin[13] for details. It may also be mentioned that the modern class of *sequential quadratic programming* (SQP) algorithms work on the same sort of logic.

This concludes our discussion of algorithms for constrained NLP. One of the points made in the literature (for example, Belegundu and Arora[14]) is that even some of the better algorithms described thus far often prove to be unusable for very large problems (such as structural optimization problems involving finite-element calculations) where function and gradient evaluations are computationally prohibitive and storage requirements are high. Parallel computing promises to be a way to address some of these issues. However, the final word is that what is good for one situation may not be good for another, and with NLP one has to often use a lot of intuition, as well as trial and error, to come up with the correct algorithmic approach to solve a problem.

SPECIAL FORMS OF NONLINEAR PROGRAMS

Unlike linear programming and the simplex method, one of the primary difficulties with nonlinear programming is that there is no unique methodology which is capable of solving any general NLP problem. Indeed, it is fair to say that there is no method that is even guaranteed to give *good* results with *all* NLP problems, and a general-purpose algorithm that works beautifully on one problem could fail miserably on another. Furthermore, we are talking only of local optima here; the author is unaware of the existence of any method that is guaranteed to find the global optimum of *any* general NLP problem.

This being the case, it is natural to try to classify nonlinear programs in some fashion and to then develop special algorithms for each class. The key word here is *structure,* and almost always, it is better to develop an algorithm that specifically exploits the structure of a problem, as opposed to using a general-purpose method. In this section, we therefore describe several special types of NLP problems for which more efficient solution procedures have been developed.

NLP with Linear Constraints.   In these problems, the constraint functions of the problem, namely $g_j(\mathbf{x})$ and $h_j(\mathbf{x})$, are all linear and of the form $\Sigma_{i=1}^{n} a_{ij}x_i$, so that the NLP may be represented as

$$\text{Minimize} \quad f(\mathbf{x})$$

$$\text{st} \quad \mathbf{A}_1\mathbf{x} \leq \mathbf{b}^1, \mathbf{A}_2\mathbf{x} = \mathbf{b}^2$$

This makes the feasible region a convex polyhedron (similar to LP) and considerably simplifies the problem, especially when all constraints are equalities. Several efficient algorithms have been developed for such problems, for example, Fletcher[15] and Gill and Murray.[16]

Quadratic Programming.   Quadratic programming (QP) constitutes a second class of nonlinear programs for which efficient algorithms have been developed. Several algorithms that solve general NLP problems proceed by solving QP subproblems at each iteration, and hence QP forms an important area for study. A quadratic program has constraints that are linear and an objective that is a quadratic function; in its general form it may be stated as follows:

Program 2

$$\text{Minimize} \quad f(\mathbf{x}) = \mathbf{c}^T\mathbf{x} + \tfrac{1}{2}\mathbf{x}^T\mathbf{Q}\mathbf{x}$$

$$\text{st} \quad \mathbf{A}\mathbf{x} \geq \mathbf{b}$$

where $\mathbf{A}$ is some constant matrix, $\mathbf{Q}$ is a constant square symmetric matrix, and $\mathbf{c}$ and $\mathbf{b}$ are constant vectors.

Obviously Program 2 is a special case of an NLP with linear constraints, and methods of the previous section could apply to these too; however, the special structure of the objective has been exploited (especially for the case where f is convex) to develop several algorithms exclusively for QP such as Fletcher[17] and Bunch and Kaufman.[18]

Convex Programming. An important class of NLP is convex programming. A convex program may be stated as follows:

Program 3

$$\text{Minimize} \quad f(\mathbf{x})$$

$$\text{subject to} \quad g_j(\mathbf{x}) \geq 0 \quad j = 1, 2, \ldots, m$$

$$h_j(\mathbf{x}) = 0 \quad j = 1, 2, \ldots, p$$

$$\mathbf{x} = (x_1, x_2, \ldots, x_n) \in \mathcal{R}^n$$

where f is convex, all g_j are concave, and all h_j are linear.

It is easily shown that the feasible region for the above program is a convex set, and thus we are minimizing a convex function over a convex set. Convex programming is one of the few categories of NLP for which global optima are available, and for which the K-K-T necessary conditions are also sufficient. Furthermore, if the Hessian at the optimum point is positive definite, then the optimum is also unique.

Obviously, the two types of NLP problems mentioned earlier in this section are also convex programs if the objective functions are convex. There are some algorithms, such as the cutting plane algorithm of Kelley,[19] that have been developed specifically for general convex programming problems. However, what is more important is that convex programs have several algorithmic implications (for instance, one does not have to guard against unboundedness or against the Hessian being indefinite) which result in the simplification and efficient use of general-purpose algorithms for NLP.

It should be emphasized here that for larger problems with reasonably complex objective and constraint functions, there is no straightforward, readily usable procedure for determining convexity. This being the case, convex programming techniques should be used only if it is *known beforehand* that the program is convex, (for example, a QP with a convex objective, or a posynomial geometric program).

Separable Programming. Separable programs are NLP problems where the objective and constraint functions are separable; a function $f(x_1, x_2, \ldots, x_n)$ is said to be separable if it can be written as $\Sigma_{i=1}^n f_i(x_i)$. Thus the separable programming problem is stated as

Program 4

$$\text{Minimize} \quad \sum_{i=1}^n f_i(x_i)$$

$$\text{st} \quad \sum_{i=1}^n g_{ij}(x_i) \leq 0 \quad \text{for } j = 1, 2, \ldots, m$$

The strategy for solving such problems is to replace each f_i and each g_{ij} by a piecewise linear approximation over some interval of interest (a, b) as shown in Fig. 6.14. This is done by choosing a set of k grid points $x_1, x_2, \ldots, x_k$, and approximating the value of the

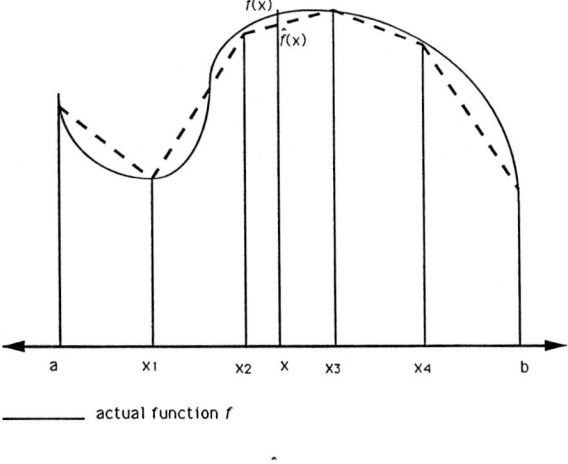

_____ actual function f

_ _ _ _ linear approximation $\hat{f}$

FIGURE 6.14   Strategy for solving NLP problems where objective and constraint functions are separable.

function [say $f(\mathbf{x})$] at a point x between two adjacent grid points x_m and x_{m+1} by the quantity $\hat{f}(\mathbf{x}) = \alpha f(x_m) + (1 - \alpha) f(x_{m+1})$, where $x = \alpha x_m + (1 - \alpha) x_{m+1}$ and $\alpha \in (0, 1)$.

By using this strategy, the original NLP can be converted to a linear program with the α (weights assigned to each grid point for each function) taking on the role of the decision variables. The simplex method (with the *restricted basis entry* rule) may then be applied to solve this LP; the reader is referred to Bazaraa and Shetty[20] for details of this procedure.

Geometric Programming   Perhaps the most elegant and interesting class of NLP problems is geometric programming (GP). GP has found wide applicability in a variety of areas, especially in engineering design. The derivation of GP is based on the classical arithmetic-geometric mean inequality. It was developed in the early sixties by Duffin, Peterson, and Zener, who also authored the first book on the subject.[21] Another book that is somewhat easier to read and has numerous practical examples, as well as extensions to GP is Beightler and Phillips.[22] Avriel later edited a volume that contains a collection of numerous important papers in the area.[23]

The objective and constraint functions in GP are *posynomials*. A posynomial is a generalized polynomial of the form $\Sigma_i c_i \Pi_{j=1}^{m} x_j^{a_{ij}}$ where the exponents a_{ij} may take on any sign, but the coefficients c_i are restricted to be strictly positive. The posynomial geometric programming problem may be stated as

Program 5A

Minimize $g_0(\mathbf{x})$

st $g_k(\mathbf{x}) \leq 1 \qquad k = 1, 2, \ldots, p$

             $x_j > 0 \qquad\quad j = 1, 2, \ldots, m$

where

$$g_k(x) = \sum_{i \in [k]} c_i \prod_{j=1}^{m} x_j^{a_{ij}} \qquad k = 0, 1, \ldots, p$$

The index set I numbers the n terms in the $(p + 1)$ posynomials, while the index subset $[k]$ numbers the terms in posynomial k; $I = 1, 2, \ldots, n$; $I = \cup_{k=0}^{p} f[k]$; and $[k] \cap [l] = \phi$ for $k \neq l$, and we assume that $c_i > 0$ for all i.

It is easy to show that this problem may be restated as an equivalent convex programming problem, and therefore if it possess an optimal solution, it must also be a global optimum. This is a very desirable characteristic, since this is one of the rare instances where a nonlinear program is guaranteed to yield a global minimum.

Program 5A is often called the *primal* GP problem. Closely associated with this primal is another problem called the GP *dual* problem. Just as in LP, the minimum value of the GP primal is exactly equal to the maximum value of the GP dual, and a well-defined relationship exists between the primal and dual variables at the optimum. This implies that either of these programs could be solved to obtain the optimum value and the optimum solution vector. The dual problem may be stated as

Program 5B

$$\text{Maximize} \quad V(\lambda, \delta) = \prod_{i \in I} (c_i/\delta_i)^{\delta_i} \prod_{k=0}^{p} \lambda_k^{\lambda_k}$$

$$\text{subject to} \quad \sum_{i \in [k]} \delta_i = \lambda_k \quad k = 0, 1, \ldots, p$$

$$\sum_{i \in I} \delta_i a_{ij} = 0 \quad j = 1, 2, \ldots, m$$

$$\lambda_0 = 1$$

$$\lambda_k \geq 0 \quad \delta_i \geq 0, \text{ for all } i \text{ and } k$$

The most striking feature of the dual program is that, unlike the primal, it is linearly constrained. Moreover, the objective function to be maximized is concave. Thus any of the specialized procedures developed for NLP problems with linear constraints could be applied to solve this problem; indeed the development of algorithms for the GP dual has been a rich area of research.

After the dual is solved the optimal values of the primal variables need to be recovered. The optimal primal and dual variables are related as follows:

$$\delta_i = \frac{\left\{ c_i \prod_{j=1}^{m} x_j^{a_{ij}} \right\}}{V(\lambda, \delta)} \quad \text{for } i \in [0]$$

$$\delta_i = \left\{ c_i \prod_{j=1}^{m} x_j^{a_{ij}} \right\} \lambda_k \quad \text{for } i \in [k], \text{ such that } \lambda_k \neq 0$$

On taking logarithms of both sides of the above equalities, we get a system of equations that are linear in $\log x_j$. Knowing the values of δ_i and λ_k, one may then solve this system for $\log x_j$ and thence for x_j. In the above system, it may be noted that an equation is always available for $i \in [0]$, but one is available for $i \in [k]$, where $k > 0$ *only if* $\lambda_k > 0$ at the optimum dual solution.

Solving the dual is not without its own difficulties. Noting that λ_k is a Lagrange multiplier for the kth primal constraint, if constraint k is slack at the optimum then $\lambda_k = 0$ and $\delta_i = 0$ for all $i \in [k]$. Thus quantities such as $(c_i/\delta_i)^{\delta_i}$ become undefined for such terms (even though, in the limit, they tend toward 1). In practical terms this means that algorithmic implementations on a computer will run into computational difficulties in the presence of slack constraints, unless some special provisions are made to handle such situations.

Indeed, all algorithms that directly approach the dual incorporate such procedures. A second (but somewhat infrequent) problem is that with a number of slack primal constraints, the rank of the log-linear system mentioned above may be insufficient to recover an optimal primal vector from the dual solution. One must then resort to the solution of so-called subsidiary problems; the reader is referred to Dembo[24] or Rajgopal and Bricker[25] for details.

These computational difficulties have led some researchers to conclude that solving the primal directly is, in general, a preferable approach. Several computational studies have been performed but it appears that there is no conclusive evidence that one approach is better than the other. However, the consensus seems to be that methods designed specifically for GP outperform general-purpose NLP algorithms.

While Program 5A was the original problem of Duffin et al.,[21] the assumption that all c_i have to be positive was later relaxed and the right-hand side in each constraint was also allowed to take on any sign. The resulting GP is referred to as *signomial* GP. Unfortunately, the signomial GP is not a convex program, and thus there is no way of guaranteeing global optima. Furthermore, many of the attractive structural properties of posynomial GP are also lost. All the same, attempts have been made to develop algorithms for this case; one of the better ones is the algorithm of Avriel, Dembo, and Passy.[26]

FURTHER READING

The references cited in this chapter constitute only a small number of relatively recent (for the most part) papers and books of importance. The interested reader is urged to use these for references to many classical papers in NLP. In addition, there is a wide range of excellent books in the general areas of optimization and nonlinear programming. A short list of highly recommended reading would include the books by Zangwill,[27] Avriel,[2] Bazaraa and Shetty,[20] Gill, Murray, and Wright,[8] Luenberger,[28] Evtushenko,[29] and Minoux.[30]

REFERENCES

1. Kuhn, H. W., and A. W. Tucker, "Nonlinear Programming," *Proceedings of 2d Berkeley Symposium on Mathematical Statistics and Probability,* J. Neyman (ed.), University of California Press, Berkeley, Calif., 1951.

2. Avriel, M., *Nonlinear Programming,* Prentice-Hall, Englewood Cliffs, N.J., 1976.

3. Morris, A. J., (ed.), *Foundations of Structural Optimization: A Unified Approach,* Wiley, New York, 1982.

4. Levary, R., (ed.), *Engineering Design, Better Results through Operations Research Methods,* North-Holland, New York, 1988.

5. Arora, J. S., *Introduction to Optimum Design,* McGraw-Hill, New York, 1989.

6. Vanderplaats, G. N., *Numerical Optimization Techniques for Engineering Design and Applications,* McGraw-Hill, New York, 1984.

7. Biegler, L. T., I. E. Grossman, and G. V. Reklaitis, "Application of Operations Research Techniques in Chemical Engineering," *Engineering Design, Better Results through Operations Research Methods,* R. Levary (ed.), North-Holland, New York, 1988.

8a. Smith, S. B., "Economic Lot Sizing with a Restriction on Setup Hours," *Production and Inventory Management,* vol. 11, 1970, pp. 82–88.

8b. Gill, P. E., W. Murray, and M. H. Wright, *Practical Optimization,* Academic Press, London, 1981.

9. Fletcher, R., and M. J. D. Powell, "A Rapidly Convergent Descent Method for Minimization," *The Computer Journal,* vol. 6, 1963, pp. 163–168.

10. Bertsekas, D. P., "Multiplier Methods: A Survey," *Automatica,* vol. 12, 1976, pp. 133–145.

11. Rosen, J. B., "The Gradient Projection Method for Nonlinear Programming, Part II: Nonlinear Constraints," *SIAM J. Applied Mathematics,* vol. 9, 1961, pp. 514–553.

12. Abadie, J., and J. Carpenter, "Generalizations of the Wolfe Reduced Gradient Method to the Case of Nonlinear Constraints," *Optimization,* R. Fletcher (ed.), Academic Press, New York, 1969.

13. Pshenichnyj, B. N., and Yu. M. Danilin, *Numerical Methods for Extremal Problems,* Nauka, Moscow, 1975.

14. Belegundu, A. D., and J. S. Arora, "A Study of Mathematical Programming Methods for Structural Optimization," *International J. Numerical Methods in Engineering,* vol. 2, 1985, pp. 1583–1624.

15. Fletcher, R., "An Algorithm for Solving Linearly Constrained Optimization Problems," *Mathematical Programming,* vol. 2, 1972, pp. 133–161.

16. Gill, P. E., and W. Murray, "Newton-Type Methods for Unconstrained and Linearly Constrained Optimization," *Mathematical Programming,* vol. 28, 1974, pp. 311–350.

17. Fletcher, R., "A General Quadratic Programming Algorithm," *J. Institute of Mathematics and Its Applications,* vol. 7, 1971, pp. 76–91.

18. Bunch, J. R., and L. C. Kaufman, "A Computational Method for the Indefinite Quadratic Programming Problem," *Linear Algebra and Its Applications,* vol. 34, 1980, pp. 341–370.

19. Kelley, J. E., "The Cutting Plane Method for Solving Convex Programs," *J. Society for Industrial and Applied Mathematics,* vol. 8, 1960, pp. 703–712.

20. Bazaraa, M. S., and C. M. Shetty, *Nonlinear Programming,* Wiley, New York, 1979.

21. Duffin, R. J., E. L. Peterson, and C. M. Zener, *Geometric Programming,* Wiley, New York, 1967.

22. Beightler, C. S., and D. T. Phillips, *Applied Geometric Programming,* Wiley, New York, 1976.

23. Avriel, M., (ed.), *Advances in Geometric Programming,* Plenum Press, New York, 1980, pp. 333–342.

24. Dembo, R. S., "Dual to Primal Conversion in Geometric Programming," *J. Optimization Theory and Applications,* vol. 26, October 1978.

25. Rajgopal, J., and D. L. Bricker, "On Subsidiary Problems in Geometric Programming," Technical Report 90-12, Dept. of Industrial Engineering, University of Pittsburgh, 1990.

26. Avriel, M., R. S. Dembo, and U. Passy, "Solution of Generalized Geometric Programs," *International J. Numerical Methods in Engineering,* vol. 9, 1975, pp. 149–169.

27. Zangwill, W. I., *Nonlinear Programming: A Unified Approach,* Prentice-Hall, Englewood Cliffs, N.J., 1969.

28. Luenberger, D. G., *Linear and Nonlinear Programming,* Addison Wesley, Reading, Mass., 1984.

29. Evtushenko, Y. G., *Numerical Optimization Techniques,* Optimization Software, New York, 1985.

30. Minoux, M., *Mathematical Programming,* Wiley, New York, 1986.

CHAPTER 7
MULTICRITERIA DECISION MAKING

Gary R. Reeves
Department of Management Science
University of South Carolina
Columbia, South Carolina

Kenneth Lawrence
Department of Industrial Engineering and the Graduate School of
Management
Rutgers University
Piscataway, New Jersey

Multicriteria decision making (MCDM) is concerned with the processes and techniques whereby multiple objectives are incorporated into the decision-making process. The field of MCDM can be subdivided into multiobjective decision making (MODM) and multi-attribute decision making (MADM). MODM is concerned primarily with deterministic problem situations involving a large (possibly infinite) number of feasible alternatives. Solution strategies involve the use of mathematical programming-based techniques. MADM is concerned with problems involving a relatively small number of alternatives often in an environment of uncertainty. Keeney and Raiffa[1] provide a thorough treatment of MADM. Surveys of MADM and MODM can be found in Hwang and Yoon[2] and Hwang and Masud,[3] respectively. The major focus of this chapter will be on MODM.

The field of MCDM has evolved rapidly over the last four decades. Early contributions to the field were made by Koopmans,[4] Kuhn and Tucker,[5] and others in the 1950s. The 1960s saw the introduction of goal programming by Charnes and Cooper.[6] The first specialized text on goal programming appeared in the early 1970s[7] and was followed by other texts and numerous application articles in the literature. The first international research conference devoted entirely to MCDM took place in 1972.[8] The proceedings of numerous subsequent conferences have appeared in the literature. Texts on MODM began to appear in the late 1970s, including those of Cohon,[9] Goicoechea, Hansen, and Duckstein,[10] Zeleny,[11] Chankong and Haimes,[12] Yu,[13] and Steuer.[14] Since that time there has been an explosive growth in the research literature on the theory and application of MCDM. Surveys and bibliographies of MCDM include Evans[15] and Stadler,[16] respectively.

BASIC CONCEPTS AND TERMINOLOGY

Multicriteria decision making involves multiple attributes, multiple objectives, or both. Decision alternatives have attributes. *Attributes* are the characteristics or qualities of the alternatives. Multiattribute decision making involves the selection of a "best" alternative from among a set of alternatives on the basis of their attributes. *Objectives* represent directions of improvement of attributes; *goals* establish desired target levels of attributes. While a characteristic of a decision alternative is an attribute, maximizing or minimizing that characteristic is an objective and achieving a certain target level for that characteristic is a goal. Multiobjective decision making is concerned with determining an alternative which optimizes or "best achieves" the objectives of the decision maker (DM). Multiple objectives are often conflicting and/or not measurable in the same or similar units (incommensurable). Thus, the concept of a "best" alternative is likely to be imprecise in a multiple-objective context, since there may be no feasible alternative which simultaneously optimizes all objectives.

A single-objective constrained optimization problem can be stated as:

$$\text{Max } [f(\mathbf{x}) = z] \tag{1}$$

$$\text{st} \quad \mathbf{x} \in S$$

where $\mathbf{x}$ is the vector of decision variables, $f(\mathbf{x})$ is the objective function and S is the feasible region. This framework can be extended to include multiple objectives as follows:

$$\text{Max } \{f_1(\mathbf{x}) = z_1\}$$

$$\text{max } \{f_2(\mathbf{x}) = z_2\}$$

$$.$$

$$. \tag{2}$$

$$.$$

$$\text{max } \{f_k(\mathbf{x}) = z_k\}$$

$$\text{st} \quad \mathbf{x} \in S$$

This problem is also referred to as a vector maximum problem (VMP). The decision variables can be continuous or discrete and the objective functions and constraints can be linear or nonlinear. If all the decision variables are continuous and all the objective functions and constraints are linear (the most important special case of MODM), the problem becomes a multiple objective linear programming (MOLP) problem:

$$\text{Max } \{\mathbf{c}^1\mathbf{x} = z_1\}$$

$$\text{max } \{\mathbf{c}^2\mathbf{x} = z_2\}$$

$$.$$

$$. \tag{3}$$

$$.$$

$$\text{max } \{\mathbf{c}^k\mathbf{x} = z_k\}$$

$$\text{st} \quad \mathbf{x} \in S$$

where $S = \{\mathbf{x} \in R^n | \mathbf{Ax} \leq \mathbf{b}, \mathbf{x} \geq 0, \mathbf{b} \in R^m\}$ denotes the feasible region in *decision space* and $Z = \{\mathbf{z} \in R^k | \mathbf{z} = \mathbf{Cx}, \mathbf{x} \in S\}$ denotes the feasible region in *criterion* or *objective space*.

An optimal solution to a MODM problem is one which optimizes all of the objective functions simultaneously. A point $\mathbf{x}^*$ is an *optimal* solution to the MOLP problem in Eq. (3) (usually referred to as an *ideal* solution in a multiple-objective context) if and only if $\mathbf{x}^* \in S$ and $\mathbf{Cx}^* \geq \mathbf{Cx}$ for all $\mathbf{x} \in S$. Since MODM problems often have conflicting objectives, an optimal solution may not exist (the ideal solution may be infeasible).

The concept of optimality in a single-objective context is replaced by the concepts of efficiency and nondominance in MODM. A point in decision space, $x \in S$, is *efficient*, if and only if there does not exist another $x \in S$ such that $Cx \geq Cr$ and $Cx \neq Cr$. Correspondingly, a point in objective space, $z \in Z$ is *nondominated*, if and only if there does not exist another $z \in Z$ such that $z \geq z$ and $z \neq z$. In words, a point is efficient or nondominated, if it is not possible to improve the value of any individual objective without worsening the value of at least one other objective. A complicating factor in MODM is that there can be many efficient (nondominated) solutions. A solution eventually selected by a DM is sometimes referred to as a *most preferred*, or *best-compromise*, solution.

Consider the following example:

$$\text{Max } z_1 = 4x_1 - x_2$$

$$\text{max } z_2 = -2x_2 + 3x_2$$

$$\text{st} \quad 2x_1 - x_2 \leq 8$$

$$x_1 + x_2 \leq 9 \tag{4}$$

$$x_1 \leq 5$$

$$x_2 \leq 6$$

$$x_1 \geq 0, x_2 \geq 0$$

The feasible region for this problem in decision space and objective space is given in Figs. 7.1 and 7.2, respectively. The extreme points of this problem are described numerically in Table 7.1. Since the objectives are conflicting, the problem does not have an optimal solution in the ordinary, single-objective sense. However, extreme points C, D, E, and F are efficient in decision space (nondominated in objective space). The boundary of the feasible region in objective space from extreme point C to extreme point F constitutes the *efficient frontier*. The set of points along the efficient frontier form the *efficient set*. For points in the efficient set, it is not possible to improve the value of either objective without worsening the value of the other objective. Extreme points A and B are nonefficient (dominated), as are all other feasible points not along the efficient frontier. Thus, although it makes sense to restrict the search for a most preferred solution to the efficient set, the set of efficient solutions will be infinite in most instances, and even the subset of efficient extreme point solutions can be quite large in practice.

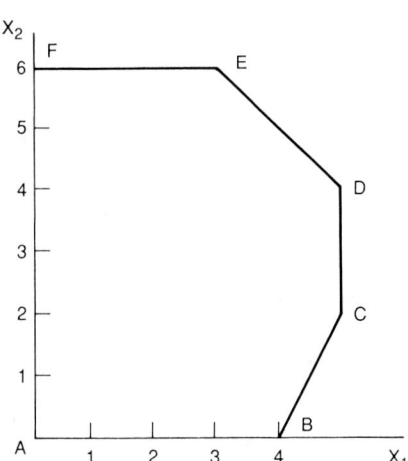

FIGURE 7.1 Example problem—decision space.

TECHNIQUES FOR MULTIPLE-OBJECTIVE DECISION MAKING

Techniques for multiobjective decision making can be classified according to when preference information is requested from the decision maker in order to help guide the overall solution process toward the identification of a DM's most preferred solution. Preference information can be requested from a DM by an analyst after, before, or during the solution process. These techniques will be referred to as posterior, prior, and progressive (or interactive) preference techniques, respectively.

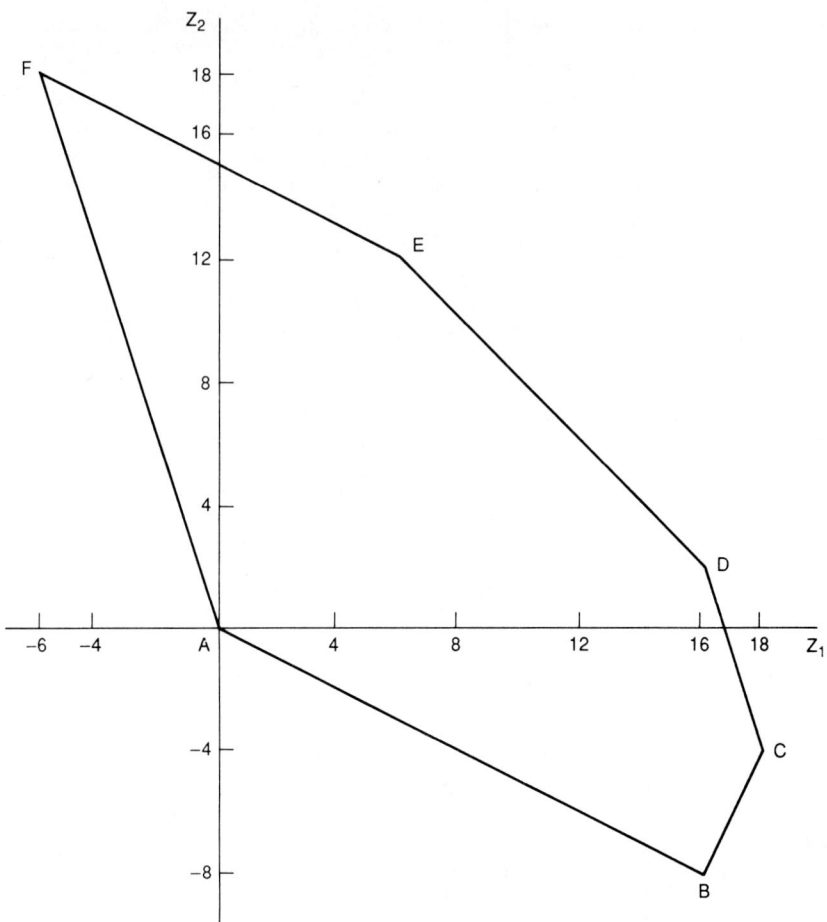

FIGURE 7.2   Example problem—objective space.

TABLE 7.1   Values of the Decision Variables and Objectives

Extreme point	Variables		Objectives	
	x_1	x_2	z_1	z_2
A	0	0	0	0
B	4	0	16	−8
C	5	2	18	−4
D	5	4	16	2
E	3	6	6	12
F	0	6	−6	18

Posterior Preference (Generating) Techniques. Posterior preference techniques attempt to generate an exact or an approximate representation of the set of efficient solutions to a MODM problem before requesting any preference information from a DM. These techniques have the advantage of minimizing required inputs from DMs prior to solution. However, the two major disadvantages of generating techniques are (1) they can require massive amounts of computational effort to generate potentially large numbers of efficient solutions during the solution process and (2) it can be difficult for DMs to process the resulting large numbers of efficient solutions down to a single preferred solution. Major categories of generating techniques include the weighting method, the constraint method and the multiobjective simplex method.

Weighting Method.   The weighting method is one of the oldest and conceptually simplest methods of MODM. It consists of forming a composite objective as a weighted linear combination of the original problem objectives, solving the resulting single objective optimization problem, and then varying the weights.

$$\text{Max } Z = w_1 z_1 + w_2 z_2 + \cdots + w_k z_k$$
$$\text{st} \quad \mathbf{x} \in S \tag{5}$$

An optimal solution to the weighted problem of Eq. (5) will be an efficient solution to the underlying multiobjective problem of Eq. (2), provided that all the weights are positive. The method follows directly from the Kuhn-Tucker conditions for an efficient solution.[5] Zadeh[17] was the first to recommend the use of weights to generate the efficient set. For MOLP problems involving only two objectives, Gass and Saaty[18] showed that all efficient extreme points can be generated by parametrically varying the weights.

For problems involving more than two objectives, however, the complete generation process becomes more complicated. For this reason the weighting method is most often used to generate an approximate, rather than an exact, representation of the efficient set. The major issue in implementing the weighting method is how to vary the weights. If too many weight combinations are chosen, computational effort increases and some efficient extreme points may be generated more than once. If too few weight combinations are chosen, some efficient extreme points may not be identified at all. In the example problem of Eq. (4), assuming $w_1 + w_2 = 1$, the weight ranges which would result in each efficient extreme point are given in Table. 7.2.

Constraint Method.   The constraint method treats all but one of the objectives of a MODM maximization problem as constraints subject to lower bounds and then optimizes the remaining objective over the revised feasible region,

TABLE 7.2 Weight Ranges for Efficient Extreme Points

Extreme point	Weight ranges	
	w_1	w_2
C	(.75, 1)	(0, .25)
D	(.5, .75)	(.25, .5)
E	(.33, .5)	(.5, .67)
F	(0, .33)	(.67, 1)

$$\text{Max } z_l$$
$$\text{st} \quad \mathbf{x} \in S \tag{6}$$
$$z_i \geq L_i \quad i = 1, 2, \ldots, l-1, l+1, \ldots, k$$

where the L_i are lower bounds on the achievement levels of the constrained objectives. By systematically varying the bounds and resolving Eq. (6), an approximate representation of the efficient set can be generated. Marglin[19] was one of the first to suggest this approach.

The constraint method, like the weighting method, is most practical as an approximate, rather than an exact, generating technique. An optimal solution to the constrained problem, Eq. (6), will be an efficient solution to the underlying multiobjective problem, Eq. (2), provided that the bounds on the constrained objectives are specified properly, although not necessarily an efficient extreme point. The major issues in implementing the constraint method are which objectives to constrain and how to vary their bounds. In the example problem Eq. (4), if the first objective is constrained and L_1 is varied in increments of 6 from −6 to 18, the resulting efficient solutions are given in Table 7.3. The first, third, and fifth solutions in Table 7.3 correspond to efficient extreme points F, E, and C, respectively, while the second and fourth solutions are nonextreme point (boundary) solutions.

Multiobjective Simplex Method..   The multiobjective simplex method is an extension of the simplex method for single-objective linear programming (LP) problems to handle multiple-objective linear programming problems of the form of Eq. (3). It deals directly with the individual problem objectives in a simplex-based environment without requiring the specification of any weights or bounds on the objectives. Several researchers have made important contributions to the theory of MOLP, including Philip,[20] Evans and Steuer,[21] and Zeleny.[22]

TABLE 7.3   Constraint Method Efficient Solutions

	Variables		Objectives	
Bound L_1	x_1	x_2	z_1	z_2
-6	0	6	-6	18
0	1.5	6	0	15
6	3	6	6	12
12	4.2	4.8	12	6
18	5	2	18	-4

An MOLP algorithm consists of three phases:

Phase I: Find an initial basic feasible extreme point.

Phase II: Find an initial efficient extreme point.

Phase III: Find all remaining efficient extreme points.

Phase I in MOLP is identical to Phase I in ordinary single-objective LP.

Phase II is based on the fact that, if an MOLP problem has an efficient solution, then there will be at least one extreme point which is efficient. Phase II consists of moving from an initial basic feasible extreme point to an initial efficient extreme point along the boundary of the feasible region through a series of simplex pivot operations. There are several methods for finding an initial efficient extreme point, including the weighting method mentioned previously and lexicographic or sequential optimization. In the weighting method a single weighted objective is formed as a linear combination of the original problem objectives and optimized over the feasible region. Lexicographic optimization considers objectives sequentially by optimizing the current objective and then constraining it to its optimal value before considering the next objective.

Phase III, which consists of finding all remaining efficient extreme points, is based on the fact that the set of all efficient extreme points is connected. Thus, given an initial efficient extreme point, it is possible to identify all remaining efficient extreme points through a series of simplex pivot operations. Additional subproblem tests beyond the basic simplex method are necessary at this stage to determine whether extreme points adjacent to the current efficient extreme point are themselves efficient. In addition to the previously mentioned MOLP references, Isermann,[23] Ecker and Kouada,[24] and Zionts and Wallenius[25] have proposed subproblem tests for efficiency.

Although the multiobjective simplex method can be used to generate all efficient extreme point solutions to a MOLP problem in theory, the increased computational, bookkeeping, and other overhead requirements of Phase III limit the size of problems for which an exact representation of the efficient set can be attained in practice. Steuer[26] has developed a computerized MOLP package called ADBASE. To lessen the computational requirements of Phase III, weight intervals can be specified for some or all objectives prior to solution.[27] This represents a compromise between a pure weighting approach which produces only a single efficient solution and a pure unweighted MOLP-based generating approach which produces all efficient extreme point solutions. In addition, ADBASE contains postsolution filtering routines to process large numbers of efficient solutions down into a more manageable number to lessen the information processing requirements for DMs.[28]

Prior Preference Techniques.   Prior preference techniques attempt to elicit DM preferences with regard to problem objectives prior to the solution of the underlying MODM problem and to incorporate this information into the solution process. DMs may be unwilling or unable to select a most preferred solution from among those provided by generating techniques which ignore DM preferences entirely. At the same time, DMs may not be able to specify complete and accurate preference information prior to problem solution. However, if DM preferences can be articulated with some degree of accuracy prior to solution, this information can be incorporated into the solution process to reduce the number of efficient solutions identified. This, in turn, should facilitate the selection of a final, most preferred solution. Major categories of prior preference techniques include utility function methods and goal programming.

Utility Function Methods.   Utility function methods seek to replace the MODM problem in Eq. (2) with

$$\text{Max } U(\mathbf{Z}) = U(\mathbf{z}_1, \mathbf{z}_2, \ldots, \mathbf{z}_k)$$ (7)

$$\text{st} \quad \mathbf{x} \in S$$

where $U(\mathbf{Z})$ is the DM's multiattribute utility function (MUF) defined over the problem objectives. The major advantage of utility function methods is that if the DM's utility function U can be assessed correctly, then solving Eq. (7) will result in the DM's most preferred solution. The major disadvantage of utility-based methods is, of course, that the correct assessment of the DM's utility function can be difficult, if not impossible. Some of the underlying assumptions of utility theory—that the DM is rational and consistent, for instance—can be of questionable validity in many practical situations. Correct utility function assessment also requires DMs to articulate these rational and consistent preference judgments prior to receiving any information on the problem solution space. Despite the questionable assumptions and practicality of utility function methods, there is a large body of literature on them, including books[1] and survey articles.[29]

The utility function $U(\mathbf{Z})$ can be of many forms. The most common form is the additive

$$U(\mathbf{Z}) = \sum_{i=1}^{k} U_i(\mathbf{z}_i)$$ (8)

followed by the multiplicative

$$U(\mathbf{Z}) = \prod_{i=1}^{k} U_i(\mathbf{z}_i)$$ (9)

The most important special case of the additively separable form Eq. (8) utilizes weights, $U_i(\mathbf{z}_i) = \mathbf{w}_i \mathbf{z}_i$, to indicate the importance of each objective, as in Eq. (5). For this special case, if the underlying MODM problem is linear, Eq. (3), then solving Eq. (7) reduces to solving a single-objective LP problem. If the multiplicative form Eq. (9) is used, the resulting optimization problem remains nonlinear.

Goal Programming.   Goal programming (GP) is clearly one of the most popular and widely utilized MODM techniques. Entire texts have been devoted to GP, including Lee,[17] Ignizio,[30] and Schniederjans.[31] Literally hundreds of application articles have appeared in the literature. Surveys of GP applications include Lin[32] and Zanakis and Gupta,[33] while Hannan[34] has surveyed advances in GP methodology.

An MODM problem in general, Eq. (2), or an MOLP problem in particular, Eq. (3), can be converted into a GP format as follows. The constraints in an MOLP context ($\mathbf{x} \in S$) are referred to as *system* or *hard* constraints in GP and remain unchanged, as do the decision variables. The MOLP objectives are converted into *goal* or *soft* constraints in GP through the addition of *deviational variables* and *goal target levels*

$$\mathbf{c}^i \mathbf{x} + d_i^- - d_i^+ = t_i \quad i = 1, \ldots, k$$ (10)

where t_i is the target level for the ith goal constraint. The deviational variables d_i^- and d_i^+ measure the underachievement and overachievement of the target, respectively. The negative deviational, or underachievement, variables d_i^- are analogous to slack variables in LP, while the positive deviational, or overachievement, variables d_i^+ are analogous to surplus variables. Contrary to LP, however, both deviational variables can appear in the same goal constraint, although at most one will be positive. Thus there are two types of constraints, system and goal, and two types of variables, decision and deviational, in a GP problem. To be feasible, a solution must satisfy the system constraints. In addition, a most preferred solution comes as close as possible to satisfying the goal constraints in terms of the overall GP objective function.

The objective in a GP problem is to minimize some function of the deviational variables, possibly prioritized and/or weighted. Thus a complete GP model can be formulated as

$$\text{Min } f(\mathbf{d}^-, \mathbf{d}^+)$$

$$\text{st} \quad \mathbf{c}^i\mathbf{x} + d_i^- - d_i^+ = t_i \quad i = 1,\ldots, k \tag{11}$$

$$\mathbf{x} \in S$$

$$\mathbf{d}^-, \mathbf{d}^+ \geq 0$$

With respect to individual goals, a DM can attempt to (1) come as close as possible to, (2) not underachieve, or (3) not overachieve a particular target level t_i by minimizing (1) $d_i^- + d_i^+$, (2) d_i^- or (3) d_i^+, respectively.

There are two primary forms of the GP objective in Eq. (11): the weighted sum and the prioritized. In the weighted-sum form, the DM seeks to minimize a single weighted sum of the deviational variables

$$\text{Min } \sum_{i=1}^{k} (w_i^- d_i^- + w_i^+ d_i^+) \tag{12}$$

resulting in a single-objective LP problem. The weighted-sum form assumes that all problem objectives are measurable in some common unit and that weight ratios represent marginal rates of substitution of units of one deviational variable for another (that a DM would be willing to worsen the achievement of one goal to improve the achievement of another).

The more popular of the two GP objective forms in practice has come to be the preemptive priority form

$$\text{Min } \sum_{j=1}^{r} \sum_{i=1}^{k} P_j(w_{ij}^- d_i^- + w_{ij}^+ d_i^+) \tag{13}$$

where the P_j's are the preemptive priority levels ($P_j >>> P_{j+1}$) and the w_{ij}'s are the weights assigned to the deviational variables of goal i at priority level j. This preemptive priority form assumes that some goals may be incommensurable and ranks the goals in order of decreasing importance. Note that the number of priority levels, r, is not necessarily the same as the number of objectives, k. Deviational variables from the same goal constraint can appear at different priority levels and deviational variables from different goal constraints can appear at the same priority level.

The solution process in preemptive GP operates by first achieving the highest-priority (most important) goal as completely as possible. Then it tries to achieve the remaining goals as completely as possible in order of importance with the added condition that it will not improve the achievement of any lower-priority goal if it would worsen the achievement of any higher-priority goal. Thus, the preemptive priority GP form assumes that a DM would not be willing to worsen the achievement of any higher priority (more important) goal by any amount, however, small, in order to improve the achievement of any lower-priority (less important) goal by any amount, however large.

To illustrate the preemptive GP solution process, consider the following example

$$\text{Min } P_1d_1^- + P_2d_2^+ + P_3d_3^- + P_4d_4^-$$

$$\text{st} \quad x_1 + x_2 \qquad\qquad\qquad \leq 12$$

$$x_1 + 3x_2 + d_1^- - d_1^+ \quad = 18 \tag{14}$$

$$x_2 + d_2^- - d_2^+ \qquad = 9$$

$$2x_1 + 2x_2 + d_3^- - d_3^+ = 30$$

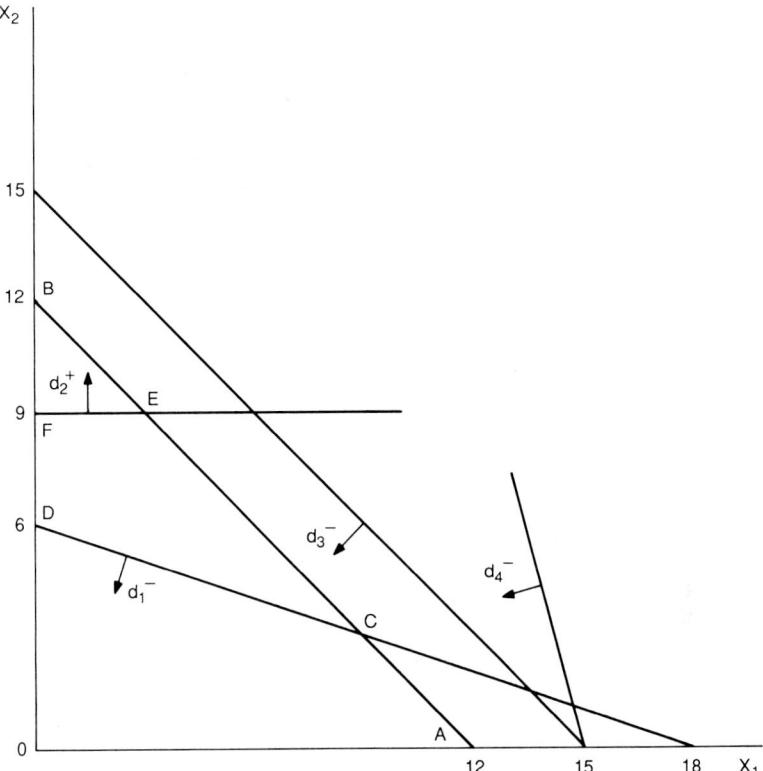

FIGURE 7.3 Goal programming example.

$$4x_1 + x_2 + d_4^- - d_4^+ = 60$$

$$\mathbf{x, d, d^+ \geq 0}$$

The solution space for this example is illustrated in Fig. 7.3. The set of points which satisfies the system constraints, including nonnegativity (the feasible region), is given by *0AB.* Considering the goals in order of stated importance, the points which satisfy the first priority objective lie on or above line segment *CD.* Thus, the feasible region is reduced to *BCD.* Similarly, the points which satisfy the P_2 objective lie on or below line segment *EF,* further reducing the feasible region to *CDFE.* Any point in this region meets both of the first two objectives completely $(d_1^- = d_2^+ = 0)$. With respect to P_3, it is not possible to achieve this objective completely without leaving *CDFE,* still it is desired to come as close as possible to achieving P_3 without worsening either P_1 or P_2. Since the third goal constraint happens to be parallel to the system constraint, this occurs everywhere along the line segment *CE* where $d_3 = 6$. Finally, it is not possible to achieve the P_4 objective completely either, but the point which comes closest is point *C* with $d_4^- = 21$. Thus, the best solution to this preemptive GP problem is $x_1 = 9, x_2 = 3, d_3^- = 6, d_4^- = 21$, and all other variables equal to zero.

This example problem illustrates the preemptive GP solution process graphically. It operates through a sequential reduction of the feasible region. As long as the best solution at the current priority level is not unique, then lower-ranking (less important) objectives can have an impact on the eventual solution. But as soon as the best solution becomes unique, any remaining objectives are, in effect, ignored.

Several options are available for the computerized solution of GP problems. As mentioned previously, a weighted-sum linear GP model can be solved as an ordinary single-objective LP problem. An LP solver can be customized to automate the preemptive GP solution process. Alternatively, the GP problem can be solved sequentially by solving a series of single-objective LP problems, one for each preemptive priority level, without necessitating any internal modification of the LP solver. After the LP problem is solved for a given priority level, the deviational variables at that level are constrained not to worsen from their current values, the objective function is replaced with that of the next priority level, and the process is repeated. Finally, the preemptive priority levels could be approximated by replacing them with numeric coefficients of different orders of magnitude and the resulting model solved as a single-objective weighted-sums LP problem.

As mentioned earlier in this section, GP is one of the most popular and widely utilized MODM techniques. The advantages of GP include the fact that it is conceptually simple and easy to understand, especially in its more popular preemptive form. This tends to make GP an appealing MODM technique for practitioners. The underlying behavioral philosophy of satisfying rather than optimizing which plays such an important role in GP is also appealing to many DMs. GP models are easily solvable with existing single-objective software. Many applications involving the use of GP have appeared in the literature.

GP has not been without its detractors, however, and some criticism of GP has appeared in the literature. Hannan,[35] Zeleny,[36] and others have addressed some of the controversy surrounding GP. Criticisms of GP include (1) the necessity of prior definition of goals, (2) the use of preemptive priorities, and (3) the possibility of dominated solutions. It may be difficult to select goals that are representative of a DM's true aspirations, as well as goal target levels, prior to model solution. There are theoretical difficulties in attempting to represent a DM's true but unknown underlying utility function using preemptive priorities. Finally, if goal target levels are set too loosely, the GP solution process may terminate with a dominated solution, leaving the DM unaware of the existence of improved solutions. Some of these difficulties can be alleviated, however, by converting GP from its strict prior preference form into an interactive solution process.

Progressive Preference (Interactive) Techniques.   Interactive techniques represent a compromise between the two extremes of the prior preference techniques and the generating techniques. Prior preference techniques assume that complete and accurate DM preference information is available prior to problem solution. Given this information, they return a single, most preferred solution. Generating techniques assume that no DM prior preference information is available and attempt to generate a complete or partial representation of the set of efficient solutions to a MODM problem. Obviously, there are difficulties associated with either of these two extreme approaches. It may be unreasonable to expect a DM to be able to supply complete and accurate preference information about a problem prior to solution. Similarly, the computational requirements of generating techniques and the information processing requirements on DMs after problem solution can be overwhelming. Interactive techniques represent a compromise between these two extreme approaches in that they attempt to involve the DM during, rather than exclusively either before or after, the MODM solution process.

Interactive procedures began to appear in the literature in the early 1970s. Since that time, a number of different procedures have been developed. Descriptions of many of the more well-known interactive procedures can be found in texts such as Refs. 2, 10, and 14. Aksoy[37] has published a recent bibliography on interactive MODM. Research on comparing interactive MODM procedures has begun to appear also (Wallenius,[38] Buchanan and Daellenbach,[39] Reeves and Gonzalez[40]).

Interactive MODM procedures share several characteristics. They explore the feasible region of a problem through a series of iterations, searching for an optimal or satisfactory, near-optimal solution. The procedures alternate between computational and decision-making phases. Intermediate solutions are generated, and DMs are asked to react to them in terms of tradeoffs, comparisons, rankings, and the like. New solutions are based on DM responses, and the process continues. As the DM learns more about the MODM problem and its solutions, DM feedback is used to guide the procedure toward improved solutions and eventual termination.

Given the current multiplicity of interactive MODM procedures, it is unlikely that any one approach will emerge as a clear favorite, because different procedures may be better suited for different types of DMs, for different types of decision situations, or even for different stages in the decision-making process. Reeves and Franz,[41] Steuer[14] (p. 362), and others have suggested important criteria or performance measures for evaluating and selecting an interactive MODM procedure, including:

1. Quality of solutions generated.

2. Simplicity/user friendliness. Ease of DM interaction with the procedure; amounts and types of inputs required from the DM; DM familiarity with the decision process.

3. Software and computational resource requirements.

4. Types of solutions generated. Efficient/nonefficient; extreme point/nonextreme point.

5. Convergence. Number of iterations necessary for the DM to identify a satisfactory solution.

6. Flexibility. Does the procedure allow the DM to change his or her mind, learn, or backtrack during the course of the solution process?

A description of several interactive MODM procedures follows.

STEM. STEM (Benayoun et al.[42]) is one of the earliest interactive MODM procedures. It consists of three phases: initialization, calculation, and decision:

1. *Initialization.* The ideal solutions are generated by optimizing each objective individually over the feasible region. Let z_i^* *denote the optimal value of the* ith objective and $\mathbf{z}^*$ the optimal criterion vector. A payoff table is constructed giving the range of each objective over the ideal set.

2. *Calculation.* At each iteration, minimize the maximum (weighted) deviation of the individual objective values from their ideal values by solving

$$\text{Min } \alpha$$

$$\text{st} \quad \alpha \geq \{z_i^* - z_i\}^* \mathbf{w}_i \; i = 1, \ldots, k \tag{15}$$

$$\mathbf{x} \in S$$

where S includes the original feasible region plus any constraints added at previous iteration, and the weights $\mathbf{w}_i$ give the relative importance of the deviations of the individual objective values from their ideal values. These weights are calculated as a function of the range of each objective value over the ideal set.

3. *Decision.* The compromise solution to Eq. (15) is presented to the DM. If all objective values are satisfactory, the process terminates. If not all objective values are satisfactory, then the values of some objectives which are satisfactory must be relaxed (allowed to worsen) in order to improve the unsatisfactory objective values. Constraints which allow this to happen are added and the calculation phase is repeated.

STEM is a semistructured approach for generating a sequence of improved solutions. Even if the DM's utility function were known, it would not be clear which objectives to relax and by how much. STEM requires only conventional single-objective software. It can identify nonextreme point solutions, but the possibility exists that compromise solutions could be nonefficient.

Compromise Programming—Displaced Ideal. Compromise programming (CP)[43] begins, as does STEM and other interactive methods, by determining the ideal solution (initialization phase). Assuming that the multiple objectives are in conflict and that, therefore, the ideal solution is infeasible, the calculation phase of CP assumes that the DM would still like to come as close as possible to the ideal solution. This criterion is used as a surrogate for the DM's true, but unknown, utility function. A subset of efficient solutions is determined whose elements are closest to the ideal solution according to various distance measures. Commonly used measures include, but are not limited to, minimizing (1) the sum of the deviations, (2) the sum of the squared deviations, and (3) the maximum deviation of the individual objective values from their corresponding ideal values.

In the decision phase, the compromise set is presented to the DM. If the set is small enough for the DM to pick a satisfactory solution, the process terminates. If not, the ideal is redefined, or displaced, over the compromise set, and the process is repeated. Compromise solutions are efficient, and the information processing requirements on the DM are modest.

Interactive Sequential Goal Programming.   The prior preference technique of goal programming in its pure form returns only a single solution to a MODM problem. In many instances, it would be helpful for DMs to be able to explore the impact of varying priority levels and weighting factors on problem solutions before having to select a single, final, most preferred solution. Since this is not a formal part of the basic, noninteractive GP methodology, however, it is often done on a random or ad hoc basis. The purpose of interactive sequential goal programming (ISGP)[44] is to allow for the progressive or interactive determination of DM preferences as alternative solutions are explored within a GP framework.

ISGP assumes that a DM can adjust desired goal levels iteratively as additional information is gained about the solution space of the problem in Eq. (2). Initialization consists of solving for the ideal solutions to bound each objective over the ideal set. On the basis of this information the DM establishes initial target levels for each goal.

In the calculation phase, a principal solution (PS) and a set of alternative solutions (AS) are generated. In the PS, the primary objective is to minimize the underachievement of the goal target levels, with a secondary objective of maximizing their overachievement to ensure that the resulting solution is nondominated. The AS satisfy the individual goals serially. A tradeoff table containing current solution information is formed.

In the decision phase, the DM is asked if any current solution is satisfactory. If so, the method is terminated. If not, the DM is asked to modify the goal target levels according to the information in the tradeoff table, and the method returns to the calculation phase.

ISGP is similar to the non-GP-based STEM and CP interactive MODM methods. It is more flexible than strict prior preference GP, and the possibility of dominated solutions is avoided. ISGP, along with STEM and CP, are less structured than some of the other interactive methods, but they are also simpler and more user-friendly.

Geoffrion-Dyer-Feinberg Method.   The Geoffrion-Dyer-Feinberg (GDF) method[45] is based on the Frank-Wolfe (FW) gradient-based method of nonlinear programming. In a single-objective context, and given an initial feasible solution, the FW method calculates the gradient of the objective function at the current solution as a locally optimal direction in which to search for improved objective values, and then determines how far to move in that direction before the objective function begins to worsen. The process is then repeated about the new candidate solution. Even though the DM's global utility function over multiple objectives is not known with certainty, the GDF method elicits local information about DM preferences to approximate local gradient directions for searching for improved solutions.

The GDF method does have a sound theoretical basis and should converge to a DM's most preferred solution, provided the direction and step size approximations are determined accurately enough based on DM feedback. However, this places a heavy, and some would argue an excessive, burden on the DM to supply accurate preference information at each iteration; moreover, a large number of iterations may be required for convergence, and intermediate solutions may not be efficient.

Zionts-Wallenius Method.   The Zionts-Wallenius (ZW) method[46,47] is a reduced-weighting space method. It begins by solving a weighted-sum problem [Eq. (5)] with arbitrary positive weights to find an initial efficient extreme point solution to the underlying

MODM problem. Those extreme points adjacent to the current solution, which is also efficient, are determined, and the resulting tradeoffs in objective function values are presented to the DM for consideration. For each such adjacent efficient extreme point, the DM determines whether the tradeoffs are desirable, undesirable, or neither. Depending on DM responses, future weights on each objective are constrained consistent with stated DM tradeoff preferences, a new set of weights is constructed, and the process is repeated. If weight constraints become inconsistent over time, the oldest such constraints are removed first. The process terminates when no new adjacent efficient extreme points are identified or when no adjacent efficient extreme point represents a desirable tradeoff for the DM.

The ZW method should converge to a DM's most preferred efficient extreme point solution, provided the DM is able to assess the tradeoffs between adjacent efficient extreme points accurately. However, some DM's may find the required tradeoff assessments difficult. The method tends to reduce the set of allowable objective function weights rapidly at early iterations. There is a built-in error-correcting capability in that the oldest objective function weight constraints are deleted first when inconsistencies occur. The method is more complex than previous methods, however.

Tchebycheff Method.   The Tchebycheff (or Chebyshev) approach[48] uses a weighted Tchebycheff (minmax) distance metric to sample the set of efficient solutions to a MODM problem in a progressively more concentrated way until a final solution is chosen. While other interactive methods also work with weighted combinations of original problem objectives. Tchebycheff weights are not intended to reflect the relative importance of problem objectives to the DM, but merely serve as a sampling vehicle. The procedure generates a fixed number of solutions per iteration for a fixed number of iterations.

Initialization consists of calculating the ideal criterion vector z^*. The calculation phase consists of generating a dispersed set of objective function weighting vectors and solving the corresponding weighted Tchebycheff problem[15] for each set of weights. The resulting solutions constitute a sample chosen from among the entire set of efficient solutions. In the decision phase, the DM is asked to select a single most preferred solution from among the current sample subset of efficient solutions. At each subsequent iteration, the weights are more concentrated and centered about those which resulted in the DM's most preferred solution at the previous iteration.

The Tchebycheff procedure is simple and requires a minimal amount of information from the DM at each iteration (choosing a single most preferred solution from a set of candidate solutions). Solutions generated are efficient and are not limited to efficient extreme points. The basic procedure does not afford the DM any flexibility for a change of mind during the solution process, however.

SIMOLP.   The simplified interactive multiple-objective linear programming (SIMOLP) procedure of Reeves and Franz[41] is also ideal-based. At each iteration, a single efficient extreme point solution is generated by solving a weighted-sum problem.[5] The weights are derived from the currently active solutions, but, as in Tchebycheff, they are merely a vehicle for sampling from the efficient set and are not designed to reflect DM preferences. Contrary to Tchebycheff, the DM is asked to select a least preferred solution from among the current set of active solutions (initially the ideal solutions plus the first weighted-sum solution). The least preferred solution is dropped, and the process continues.

SIMOLP is simple and user-friendly, requiring a minimal amount of information from the DM at each iteration (choosing a single least preferred solution from a set of candidate solutions). Whereas Tchebycheff operates by moving toward a most preferred solution, SIMOLP operates by moving away from a least preferred solution. Although SIMOLP generates only efficient extreme point solutions, it is more flexible than Tchebycheff in allowing the DM a change of mind and backtracking during the course of the solution process.

EXAMPLE

Consider the following multiple-objective blending problem from Hwang and Masud:[3]

$$\text{Min } z_1 = 0.225x_1 + 2.2x_2 + 0.8x_3 + 0.1x_4 + 0.05x_5 + 0.26x_6$$

$$\text{min } z_2 = 10x_1 + 20x_2 + 120x_3$$

$$\text{min } z_3 = 24x_1 + 27x_2 + 15x_4 + 1.1x_5 + 52x_6$$

$$\text{st} \quad 720x_1 + 107x_2 + 7080x_3 + 134x_5 + 1000x_6 \geq 5000$$

$$0.2x_1 + 10.1x_2 + 13.2x_3 + 0.75x_4 + 0.15x_5 + 1.2x_6 \geq 12.5$$

$$344x_1 + 460x_2 + 1040x_3 + 75x_4 + 17.4x_5 + 240x_6 \geq 2500$$

$$18x_1 + 151x_2 + 78x_3 + 2.5x_4 + 0.2x_5 + 4x_6 \geq 63$$

$$0 \leq x_1 \leq 6 \qquad 0 \leq x_2 \leq 1 \qquad 0 \leq x_3 \leq 0.25$$

$$0 \leq x_4 \leq 10 \qquad 0 \leq x_5 \leq 10 \qquad 0 \leq x_6 \leq 4$$

The decision variables represent the amounts of six basic food groups included in a diet: pints of milk, pounds of beef, dozens of eggs, ounces of bread, ounces of lettuce, and pints of juice, respectively. The constraints represent minimum daily requirements on vitamins (5000 units), minerals (12.5 milligrams), calories (2500) and protein (63 grams), respectively. There are also upper bounds on the amounts of each of the six basic food groups that can be included in the diet. The objectives are to minimize cost, cholesterol, and carbohydrates, respectively.

This problem is simple enough from an application standpoint to be easily understood, and yet sufficiently complex to possess a diverse and interesting set of potential solutions. Solution approaches to this problem will be illustrated using all three major categories of MODM techniques: posterior, prior, and progressive.

With respect to posterior preference or generating techniques, the problem is small enough that all efficient extreme point solutions can be generated with reasonable amounts of computational resources by using a multiobjective simplex code. There are a total of 20 efficient extreme point solutions to this problem. The corresponding objective (nondominated criterion vector) values for these solutions are displayed in Table 7.4. Costs range from \$2.26 to \$3.95, cholesterol from 17.9 to 100.4 units, and carbohydrates from 150 to 413.1 grams over the efficient set of solutions.

For larger, more complex problems, it may not be practical to attempt to generate the entire set of efficient extreme point solutions. Weight intervals could be specified for some or all objectives prior to solution to reduce the number of efficient extreme points generated.[27] For example, if the weight intervals $0.2 < w_1 < 0.7$, $0.3 < w_2 < 0.6$, and $0.1 < w_3 < 0.4$ were chosen for the cost, cholesterol, and carbohydrate objectives, respectively, where $\Sigma w_i = 1$, a reduced set of only six efficient extreme points would be generated. These six solutions correspond to nondominated criterion vectors 11, 14, 15, 17, 18, and 19 in Table 7.4. Other weight intervals would, of course, result in different subsets and numbers of efficient extreme point solutions.

The weighting and/or constraint methods could be applied to the nutrition prob-

TABLE 7.4 Nondominated Criterion Vectors for Example Problem

Extreme point	z_1 (cost)	z_2 (chol.)	z_3 (carb.)
1	2.26	67.8	281.7
2	2.30	71.5	264.7
3	2.36	72.9	253.5
4	2.40	70.3	253.3
5	2.50	79.0	234.0
6	2.52	70.0	245.0
7	2.53	24.7	404.5
8	2.54	23.8	409.5
9	2.56	23.0	413.1
10	2.56	64.2	256.1
11	2.78	28.8	356.9
12	2.80	55.5	263.2
13	2.82	23.0	389.0
14	2.93	39.9	305.2
15	2.94	17.9	412.0
16	3.19	100.4	152.2
17	3.31	94.4	151.8
18	3.55	63.9	207.4
19	3.88	74.1	168.3
20	3.95	96.7	150.0

lem as partial generating techniques. The solutions generated by the weighting method for various weight combinations are shown in Table 7.5. The first three entries, where the entire weight is given to each objective in turn, are the ideal solutions. Extreme point 1 minimizes cost at \$2.26; extreme point 15 minimizes cholesterol; and extreme point 20 minimizes carbohydrates. The next three entries weight each objective in turn twice as heavily as the others, while the final entry weights all three objectives equally. As the table illustrates, different weights can result in the same solution, while other solutions may be overlooked entirely. It can be difficult to choose representative weights, especially when the objectives are measured in different units or when the values of the objectives are of different magnitudes.

TABLE 7.5   Weighting Method for Example Problem

w_1	w_2	w_3	Ext. pt.	z_1	z_2	z_3
1.0	0.0	0.0	1	2.26	67.8	281.7
0.0	1.0	0.0	15	2.94	17.9	412.0
0.0	0.0	1.0	20	3.95	96.7	150.0
0.5	0.25	0.25	19	3.88	74.1	168.3
0.25	0.5	0.25	19	3.88	74.1	168.3
0.25	0.25	0.5	20	3.95	96.7	150.0
0.33	0.33	0.33	19	3.88	74.1	168.3

To illustrate the constraint method, the cholesterol and carbohydrate objectives could be assigned upper bounds and treated as constraints, leaving the cost objective to be minimized. The minimum cost solutions for several combinations of cholesterol and carbohydrate bounds are contained in Table 7.6. If the bounds are set too loose, as in the first entry in Table 7.6, the resulting solution may not be efficient. This solution is dominated by extreme point 1 in Table 7.4, which has the same cost but less cholesterol and carbohydrates. If the bounds are set too tight, as in entries 4, 6, and 7, there may be no feasible solution. The remaining entries in Table 7.6 are efficient, although not necessarily extreme-point, solutions to the original nutrition problem.

With respect to prior preference techniques, utility function methods and goal programming are similar to the posterior preference weighting and constraint methods, respectively, except that the weights and target levels are supplied by the DM prior to model solution. They are assumed to be an accurate reflection of DM preferences, and as a result, only a single, most preferred solution consistent with those prior preferences is generated.

In goal programming, the objectives would be converted into goal constraints through the addition of deviational variables and target levels established for each goal:

TABLE 7.6   Constraint Method for Example Problem

	Upper bound on		
	Chol., z_2	Carb., z_3	Min. cost, z_1
1	75	300	2.26
2	75	250	2.38
3	75	200	3.20
4	75	150	Infeasible
5	50	300	2.62
6	50	250	Infeasible
7	25	300	Infeasible

$$0.225x_1 + 2.2x_2 + 0.8x_3 + 0.1x_4 + 0.05x_5 + 0.26x_6 - d_1^+ \geq t_1$$

$$10x_1 + 20x_2 + 120x_3 - d_2^+ \geq t_2$$

$$24x_1 + 27x_2 + 15x_4 + 1.1x_5 + 52x_6 - d_3^+ \geq t_3$$

The goal-programming objective would be to minimize some function of the deviational variables (to minimize some function of the overachievement of the cost, cholesterol, and carbohydrate targets), possibly weighted and/or prioritized. Given such an objective reflecting DM preferences, goal programming would return a single solution. As with the posterior preference constraint method, if target levels are set too loose, the resulting solution might not be efficient. Once a target level is satisfied completely, for example, there is not necessarily any incentive to continue to move in a direction of improvement toward an efficient solution. Targets of $4.00, 110, and 415 for cost, cholesterol, and carbohydrates, respectively, are all achievable completely, but the resulting solution would be dominated by every efficient extreme point in Table 7.4. If, on the other hand, one or more targets are set too tight, the resulting solution will still be feasible, but some goals might not have any impact at all on the solution generated. In a preemptive model, once the solution at a given priority level is unique, remaining priorities have no impact on the solution process. For example, if the highest priority is to keep the cost of the diet at or below $2.25, which is not completely achievable, cholesterol or carbohydrate levels would have no impact on the solution generated.

There are a variety of progressive or interactive approaches for solving the example problem. Under the STEM method,[42] for example, the following constraints would be added:

$$\alpha \geq \{z_1 - 2.26\}^* w_1$$

$$\alpha \geq \{z_2 - 17.9\}^* w_2$$

$$\alpha \geq \{z_3 - 150.0\}^* w_3$$

where the numeric values are the ideal values for each objective from Table 7.4 and the w_i are weights representing the relative importance of the deviations of the individual objective values from their ideal values. For this example the initial weights would be given by

$$w_1 = 0.003 \qquad w_2 = 0.823 \qquad w_3 = 0.174$$

The objective in STEM is to minimize the maximum weighted deviation of the individual objective values from their ideal values. Minimizing α over the feasible region yields

$$z_1 = 3.07 \qquad z_2 = 45.44 \qquad z_3 = 282.62$$

If all objective values are satisfactory to the DM, the process would terminate. If not, one or more objectives must be allowed to worsen in order to be able to improve the values of one or more remaining objectives, and the process would continue.

EXTENSIONS

Most of the discussion in this chapter has focused on MOLP [Eq. (3)] as a subset of MODM [Eq. (2)]. This is because LP problems with either single or multiple objectives tend to be much easier to solve than their nonlinear or integer programming counterparts. MODM techniques can be and have been applied to problems involving discrete decision variables and nonlinear objective and/or constraint functions, although these generalizations do tend to create additional difficulties and challenges in the solution process. Another logical extension of the topics presented in this chapter is to consider the impact of multiple DMs on the MODM process. Other challenges include how to present the outputs of the MODM solution process for effective processing by DMs.

REFERENCES

1. Keeney, R. L., and H. Raiffa, *Decisions with Multiple Objectives*, Wiley, New York, 1976.
2. Hwang, C. L., and K. Yoon, *Multiple Attribute Decision Making*, Springer-Verlag, Berlin, 1981.

3. Hwang, C. L., and A. Masud, *Multiple Objective Decision Making,* Springer-Verlag, Berlin, 1979.

4. Koopmans, T. C., *Activity Analysis of Production and Allocation,* Wiley, New York, 1951.

5. Kuhn, H. W., and A. W. Tucker, "Nonlinear Programming," in J. Neyman (ed.), *Proceedings of the Second Berkeley Symposium on Mathematical Statistics and Probability,* University of California Press, Berkeley, 1951.

6. Charnes, A., and W. W. Cooper, *Management Models and Industrial Applications of Linear Programming,* vol. 1, Wiley, New York, 1961.

7. Lee, S. M., *Goal Programming for Decision Analysis,* Auerbach, Philadelphia, 1972.

8. Cochrane, J. L., and M. Zeleny (eds.), *Multiple Criteria Decision Making,* University of South Carolina Press, Columbia, 1973.

9. Cohon, J. L., *Multiobjective Programming and Planning,* Academic Press, New York, 1978.

10. Goicoechea, A., D. R. Hansen, and L. Duckstein, *Multiobjective Decision Analysis with Engineering and Business Applications,* Wiley, New York, 1982.

11. Zeleny, M., *Multiple Criteria Decision Making,* McGraw-Hill, New York, 1982.

12. Chankong, V., and Y. Y. Haimes, *Multiobjective Decision Making,* North-Holland, New York, 1983.

13. Yu, P. L., *Multiple Criteria Decision Making,* Plenum, New York, 1975.

14. Steuer, R. E., *Multiple Criteria Optimization,* Wiley, New York, 1986.

15. Evans, G. W., "An Overview of Techniques for Solving Multiobjective Mathematical Programs," *Management Science,* vol. 30, no. 11, 1984, pp. 1268–1282.

16. Stadler, W., "A Comprehensive Bibliography on MCDM," in M. Zeleny (ed.), *MCDM: Past Decade and Future Trends,* JAI Press, Greenwich, Conn., 1984.

17. Zadeh, L., "Optimality and Non-Scalar-Valued Performance Criteria," *IEEE Transactions on Automatic Control,* vol. 8, 1963, pp. 59–60.

18. Gass, S., and T. Saaty, "The Computational Algorithm for the Parametric Objective Function," *Naval Research Logistics Quarterly,* vol. 2, 1955.

19. Marglin, S. A., *Public Investment Criteria,* MIT Press, Cambridge, Mass., 1967.

20. Philip, J., "Algorithms for the Vector-Maximization Problem," *Mathematical Programming,* vol. 2, no. 2, 1972, pp. 207–229.

21. Evans, J. P., and R. E. Steuer, "A Revised Simplex Method for Multiple Objective Programs," *Mathematical Programming,* vol. 5, no. 1, 1973, pp. 54–72.

22. Zeleny, M. *Linear Multiobjective Programming,* Springer-Verlag, Berlin, 1974.

23. Isermann, H., "The Enumeration of the Set of All Efficient Solutions for a Linear Multiple Objective Program," *Operational Research Quarterly,* vol. 28, no. 3, 1977, pp. 711–725.

24. Ecker, J. G., and I. A. Kouada, "Finding All Efficient Extreme Points for Multiple Objective Linear Programs," *Mathematical Programming,* vol. 14, no. 2, 1978, pp. 249–261.

25. Zionts, S., and J. Wallenius, "Identifying Efficient Vectors: Some Theory and Computational Results," *Operations Research,* vol. 28, no. 3, 1980, pp. 785–793.

26. Steuer, R. E., "ADBASE Operating Manual," University of Georgia, Athens, 1989.

27. Steuer, R. E., "Multiple Objective Linear Programming with Interval Criterion Weights," *Management Science,* vol. 23, no. 3, 1976, pp. 305–316.

28. Steuer, R. E., and F. W. Harris, "Intra-Set Point Generation and Filtering in Decision and Criterion Space," *Computers and Operations Research,* vol. 7, no. 1, 1980, pp. 41–53.

29. Farquhar, P. H., "Utility Assessment Methods," *Management Science,* vol. 30, no. 11, 1984, pp. 1283–1300.

30. Ignizio, J. P., *Goal Programming and Extensions,* Heath, Lexington, Mass., 1976.

31. Schniederjans, M. J., *Linear Goal Programming,* Petrocelli, Princeton, N.J., 1984.

32. Lin, W. T., "A Survey of Goal Programming Applications," Omega, vol. 8, no. 1, 1980, pp. 115–122.

33. Zanakis, S. H., and S. K. Gupta, "A Categorized Bibliographic Survey of Goal Programming," Omega, vol. 13, no. 3, 1985, pp. 211–222.

34. Hannan, E. L., "Goal Programming: Methodological Advances in 1973–1982 and Prospects for the Future," in M. Zeleny (ed.), *MCDM: Past Decade and Future Trends*, JAI Press, Greenwich, Conn., 1984.

35. Hannan, E. L., "An Assessment of Some Criticisms of Goal Programming," *Computers and Operations Research*, vol. 12, no. 6, 1985, pp. 525–541.

36. Zeleny, M., "The Pros and Cons of Goal Programmming," *Computers and Operations Research*, vol. 8, no. 4, 1981, pp. 357–359.

37. Aksoy, Y., "Interactive Multiple Objective Decision Making: A Bibliography," *Management Research News*, vol. 13, no. 2, 1990, pp. 1–8.

38. Wallenius, J., "Comparative Evaluation of Some Interactive Approaches to Multicriterion Optimization," *Management Science*, vol. 21, no. 12, 1975, pp. 1387–1396.

39. Buchanan, J. T., and H. G. Daellenbach, "A Comparative Evaluation of Interactive Methods for Multiple Objective Decision Models," *European Journal of Operational Research*, vol. 29, 1987, pp. 353–359.

40. Reeves, G. R., and J. J. Gonzalez, "A Comparison of Two Interactive MCDM Procedures," *European Journal of Operational Research*, vol. 41, 1989, pp. 203–209.

41. Reeves, G. R., and L. S. Franz, "A Simplified Interactive Multiple Objective Linear Programming Procedure," *Computers and Operations Research*, vol. 12, no. 6, 1985, pp. 589–601.

42. Benayoun, R., et al., "Linear Programming with Multiple Objective Functions: Step Method (STEM)," *Mathematical Programming*, vol. 1, no. 3, 1971, pp. 366–375.

43. Zeleny, M., "A Concept of Compromise Solutions and the Method of the Displaced Ideal," *Computers and Operations Research*, vol. 1, no. 4, 1974, pp. 479–496.

44. Masud, A. S., and C. L. Hwang, "Interactive Sequential Goal Programming," *Journal of the Operational Research Society*, vol. 32, no. 5, 1981, pp. 391–400.

45. Geoffrion, A. M., J. S. Dyer, and A. Feinberg, "An Interactive Approach for Multicriterion Optimization with an Application to the Operation of an Academic Department," *Management Science*, vol. 19, no. 4, 1972, pp. 357–368.

46. Zionts, S., and J. Wallenius, "An Interactive Programming Method for Solving the Multiple Criteria Problem," *Management Science*, vol. 22, no. 6, 1976, pp. 652–663.

47. Zionts, S., and J. Wallenius, "An Interactive Multiple Objective Linear Programming Method for a Class of Underlying Nonlinear Utility Functions," *Management Science*, vol. 29, no. 5, 1983, pp. 519–529.

48. Steuer, R. E., and E. U. Choo, "An Interactive Weighted Tchebycheff Procedure for Multiple Objective Programming," *Mathematical Programming*, vol. 26, no. 1, 1983, pp. 326–344.

...nt

...ʒ Department

...necks, and various other congestion-related phenomena
...out virtually all types of production and service systems.
...at intersections, toll booths, bank teller stations, plant in-
...erable other settings in which randomly occurring demands
...vice mechanism. This chapter provides a framework and
...nd methods for analyzing queues.

...ework for general queuing systems including measures of
... Part 2 contains what might be considered as the elemen-
...Parts 3 and 4 follow with the more advanced models. The
...uing theory relies heavily on Markov chain models and
...utions. Part 5 provides some useful approximations for
...ult and not-so-fundamental conditions.

...R QUEUING MODELS

...ized over time t by an arrival process $\{A(t):t \geq 0\}$, a service
...of environmental conditions which includes the queuing dis-
...raints that might apply to the system components.
...nts of a queuing system are the customers, servers, and
...le in most applications these components are people, there
...e or more are machines.

...rson associated with a queuing system is generally trying to
...rs awaiting service, feeling that their time could be better

```
04-06-94  05:40p
Free:     684,032

<Current Directory>
<Parent Directory>
M595SYL.51
MB555SYL.51
MB595SYL.51
MBA595.
MBA596TQ.51
WP{WPC}.TRE
```

(handwritten: M580554.51)

```
Directory: B:\*.*

    .         Current   <Dir>   04-06-94 05:36p
    ..        Parent    <Dir>   04-06-94 05:24p
M595SYL  .51            2,230   04-06-94 05:33p
MB555SYL .51            8,827   04-06-94 05:31p
MB595SYL .51            5,194   04-06-94 02:10p
MBA595   .              7,345
MBA596TQ .51           17,906
WP{WPC}  .TRE              35
```

spent elsewhere, are normally discontent with prolonged waits. They therefore want to minimize waiting time and unless the service is pleasurable they want to minimize their total time in the system. Servers are bounded on the one side by boredom when task demands are low and on the other side by stress when they are high. Ideally, the server workload is coordinated to some suitable pace or rate. The service facility manager usually benefits from maximizing the amount and quality of service provided by the facility. When the queue length becomes large, the manager becomes concerned about adverse effects on quality, throughput, and customer satisfaction.

With these several points of view that exist in any queuing system, several measures are necessary in order to specify overall system effectiveness. The following are the common measures used to characterize queuing systems:

1. Number of customers in the system
2. Number of customers in the queue
3. Total waiting time in the system
4. Waiting time in the queue
5. Server idle time
6. Utilization of the system

In each case the performance measure will be some function of the arrival and service processes and the environmental conditions. Average values are typically used to represent the expectations of these measures. The most basic measure is the average number of customers in the system over time, to which we associate the system state. Transitions among these states, that is, the number of customers at various points in time, will change because of the randomness of the arrival and service processes and the environmental conditions. In general, both $A(t)$ and $B(t)$ are stochastic processes, and hence, for fixed t, each has an underlying probability distribution.

Classification of Queuing Systems. A standard classification and notation for queuing systems has been adopted originating from Kendall (1953) which specifies a system according to elements labeled a/b/c/d/ and e. Each of these elements is defined in Table 8.1. Elements a and b correspond to the arrival and service processes, respectively, c represents the number of parallel service channels, d the constraint on the number of customers permitted to enter the queue, and e is the queue discipline. For the common cases where d is infinite and e corresponds to the first-in–first-out (FIFO) order of service, they are typically suppressed in this notation.

PART 2: BASIC MARKOV QUEUING MODELS

Notation

c	= number of parallel servers
K	= system capacity limit
L	= long run average number in the system
L_q	= long run average number in the queue
λ	= mean arrival rate
λ_n	= mean arrival rate given n in the system
μ	= mean service rate
μ_n	= mean service rate given n in the system
$o(\Delta t)$	= order Δt, $\lim_{\Delta t \to 0} o(\Delta t)/\Delta t = 0$
p_n	= steady-state probability of n in the system

TABLE 8.1 Classification for Queuing Systems

Characteristic	Symbol	Definition
Interarrival time distribution (a)	M	Exponential
	D	Deterministic
	E_k	Erlang-k, $k = 1, 2, \ldots$
	H_k	Hyperexponential-k
	PH	Phase type
	GI	General independent
Service time distribution (b)	M	Exponential
	D	Deterministic
	E_k	Erlang-k
	H_k	Hyperexponential-k
	PH	Phase type
	G	General
Number of parallel servers (c)	$1, 2, \ldots, \infty$	
System capacity (d)	$1, 2, \ldots, \infty$	
Queue discipline (e)	FIFO	First-in–first-out
	LIFO	Last-in–last-out
	SIRO	Service in random order
	PR	Priority
	GD	General discipline

$p_n(t)$ = probability of n in the system at time t

W = long run average waiting time in the system

W_q = long run average waiting time in the queue

General Birth-Death Model. Many queuing situations can be modeled by a birth-death process which is a special Markov process for which transitions among states are restricted to nearest neighbors and hence jumps are not allowed. Associating arrivals as births and service completions as deaths and letting $X(t)$ represent the number of customers in the system at $t \geq 0$,

$$p_n(t) = P\{X(t) = n | X(0) = i\}$$

For convenience we suppress the first subscript i from $p_{in}(t)$. For an interval $(t, t + \Delta t]$ with $n \geq 0$, the state transitions can be described by way of the following postulates:

1. $P\{\text{arrival in } \Delta t | n \text{ in system}\} = \lambda_n \Delta t + o(\Delta t)$
2. $P\{\text{service completion in } \Delta t | n \geq 1 \text{ in system}\} = \mu_n \Delta + o(\Delta t)$
3. $P\{\text{more than 1 arrival or service in } \Delta t | n \text{ in system}\} = o(\Delta t)$
4. Arrivals and services are independent of each other

It follows from the Chapman-Kolmogorov equations that

$$p_n(t + \Delta t) = [1 - (\lambda_n + \mu_n)\Delta t + o(\Delta t)]p_n(t)$$
$$+ [\lambda_{n-1}\Delta t + o(\Delta t)]p_{n-1}(t)$$
$$+ [\mu_{n+1}\Delta t + o(\Delta t)]p_{n+1}(t) + o(\Delta t)$$

from which it follows that

$$p_n'(t) = \lim_{\Delta t \to 0} \frac{p_n(t + \Delta t) - p_n(t)}{\Delta t}$$

or

$$p'_n(t) = -(\lambda_n + \mu_n)p_n(t) + \lambda_{n-1}p_{n-1}(t) + \mu_{n+1}p_{n+1}(t) \tag{1}$$

with initial condition

$$p_n(0) = \begin{cases} 1 & n = i \\ 0 & n \neq i \end{cases}$$

Steady-State Balance Equations.   The differential equations of Eq. (1) can be solved in principle for various sets of values (λ_n, μ_n), $n = 0, 1, \ldots$. For most queuing systems these solutions are typically quite complicated. The steady-state solutions, however, provide results based on equilibrium or long-run average conditions, and these can generally be obtained without much difficulty.

For conditions such that steady-state probabilities exist; that is, as $t \to \infty$, $p'_n(t) \to 0$, and $p_n(t) \to p_n$, we obtain from Eq. (1) the balance equations:

$$0 = -(\lambda_n + \mu_n)p_n + \lambda_{n-1}p_{n-1} + \mu_{n+1}p_{n+1} \tag{2}$$

Solution Procedure.   Given a particular set of values (λ_n, μ_n), $n = 1, 2, \ldots$, we solve Eq. (2) to find the steady-state probabilities $(p_n : n = 1, 2, \ldots)$ from which we can compute the following steady-state performance measures:
Probability of idle system:

$$p_0 = 1 - \sum_{n=c}^{\infty} p_n \tag{3}$$

Average number in system:

$$L = \sum_{n=0}^{\infty} n p_n \tag{4}$$

Average number in queue:

$$L_q = \sum_{n=c}^{\infty} (n - c)p_n \tag{5}$$

The waiting times can be obtained simply by applying Little's formula, $L = \lambda W$. This gives

Average waiting time in system:

$$W = \frac{1}{\lambda} L \tag{6}$$

Average queuing time:

$$W_q = \frac{1}{\lambda} L_q \tag{7}$$

Little's formula applies to most systems provided the average arrival rate reflects the *effective arrival rate* that actually flows into service.

Single-Server Poisson Models

M/M/1—Basic Model.   Interarrival times are iid (that is, independent and identically distributed) exponential with mean $1/\lambda$, service times iid exponential with mean $1/\mu$, and a single server. The birth-death coefficients are

$$\lambda_n = \lambda \qquad n \geq 0$$

$$\mu_n = \mu \qquad n \geq 1$$

Substituting into the balance equations of Eq. (2) gives

$$0 = -\lambda p_0 + \mu p_1 \qquad n = 0 \tag{8}$$

$$0 = -(\lambda + \mu)p_n + \mu P_{n+1} + \lambda p_{n-1} \qquad n \geq 1$$

from which we obtain

$$p_1 = \frac{\lambda}{\mu} p_0$$

$$p_{n+1} = \frac{\lambda + \mu}{\mu} p_n - \frac{\lambda}{\mu} p_{n-1} \qquad n \geq 1$$

Solving recursively gives

$$p_1 = \left(\frac{\lambda}{\mu}\right) p_0$$

$$p_2 = \left(\frac{\lambda}{\mu}\right)^2 p_0$$

$$\cdots$$

$$p_j = \left(\frac{\lambda}{\mu}\right)^j p_0$$

Since the steady-state probabilities, p_j, $j = 1, 2,\ldots$, sum to 1, letting $\rho = \lambda/\mu$ gives

$$1 = p_0 \sum_{j=0}^{\infty} \rho^j = \frac{p_0}{1 - \rho} \qquad \rho < 1$$

since ρ^j is an infinite geometric series; thus $p_0 = 1 - \rho$ and the steady-state distribution is

$$p_n = (1 - \rho)\rho^n \qquad n = 0, 1, 2,\ldots \tag{9}$$

The expected number in the system under steady-state conditions is

$$L = \sum_{n=0}^{\infty} n(1 - \rho)\rho^n = (1 - \rho)\rho \sum_{n=1}^{\infty} n\rho^{n-1}$$

$$= (1 - \rho)\rho \frac{d}{d\rho}\left(\frac{\rho}{1 - \rho}\right)$$

or

$$L = \frac{\rho}{1 - \rho} = \frac{\lambda}{\mu - \lambda} \tag{10}$$

For the expected number in the queue,

$$L_q = \sum_{n=1}^{\infty} (n - 1)p_n = L - \sum_{n=1}^{\infty} p_n = L - \rho$$

or

$$L_q = \frac{\rho^2}{1 - \rho} = \frac{\lambda\rho}{\mu - \lambda} \tag{11}$$

Alternatively, we note that

$$L_q = L - E \text{ [number in service]} = L - \rho$$

From Little's formula, the expected waiting times are

$$W = \frac{1}{\lambda} L = \frac{1}{\mu - \lambda} \tag{12}$$

$$W_q = \frac{1}{\lambda} L_q = \frac{\lambda}{\mu^2 - \lambda\mu} \tag{13}$$

M/M/1/K—Finite Capacity. Interarrival times iid exponential with mean $1/\lambda$, service times iid exponential with mean $1/\mu$, a single server, a FIFO discipline, and no more than K customers are permitted in the system. The birth-death coefficients are

$$\lambda_n = \begin{cases} \lambda & n \leq K - 1 \\ 0 & n > K \end{cases}$$

$$\mu_n = \mu \qquad n \geq 1$$

Formulating the steady-state balance equations from Eq. (2) and solving recursively, we have

$$p_n = \rho^n p_0 \qquad n = 0, 1, \ldots, K \tag{14}$$

and to find p_0,

$$1 = \sum_0^K \rho^n p_0 = p_0 \left(\frac{1 - \rho^{K+1}}{1 - \rho} \right)$$

therefore

$$p_0 = \begin{cases} \dfrac{1 - \rho}{1 - \rho^{K+1}} & \rho \neq 1 \\ \dfrac{1}{K + 1} & \rho = 1 \end{cases}$$

Computing the other performance measures directly gives

$$L = p_0 \sum_{n=0}^K n\rho^n = \frac{\rho[1 - (K + 1)\rho^K + K\rho^{K+1}]}{(1 - \rho)(1 - \rho^{K+1})} \tag{15}$$

$$L_q = \sum_{n=1}^K (n - 1)p_n = L - \frac{\rho(1 - \rho^K)}{1 - \rho^{K+1}} \tag{16}$$

To apply Little's formula, we need the effective expected arrival rate

$$\lambda' = \lambda P(\text{system available}) + 0P(\text{system unavailable})$$

$$= \lambda(1 - p_K)$$

or

$$\lambda' = \frac{\lambda(1 - \rho^K)}{1 - \rho^{K+1}} \tag{17}$$

and

$$W = \frac{1}{\lambda'} L = \frac{\rho[1 - (K + 1)\rho^K + K\rho^{K+1}]}{\lambda(1 - \rho)(1 - \rho)^K} \tag{18}$$

$$W_q = \frac{1}{\lambda'} L_q = \frac{\rho^2[1 - \rho^{K+1} - (1 - \rho)(K + 1)\rho^{K-1}]}{\lambda(1 - \rho)(1 - \rho^K)} \tag{19}$$

Example.   The interarrival times for customers seeking service at a beauty salon are distributed exponentially with a mean of 20 minutes. Service is provided on a first-come–first-served basis by a beautician whose service time is exponentially distributed with a mean of 15 minutes. The shop has five chairs for waiting customers (not counting the customer being served) and it is assumed that potential customers enter only if a chair is available.

Here we have $K = 6$ with

$$\lambda = \left(\frac{1}{20}\right) 60 = 3 \text{ customers/hr}$$

and

$$\mu = \left(\frac{1}{15}\right) 60 = 4 \text{ customers/hr}$$

therefore, $\rho = \frac{3}{4}$.

So in Eq. (14), the steady-state distribution is

$$p_n = \frac{\rho^n(1 - \rho)}{1 - \rho^{K+1}} = \frac{(0.75)^n(1 - 0.75)}{1 - (0.75)^7}$$

$$= 0.289(0.75)^n$$

Applying Eqs. (15) and (16), we compute

$$L = \frac{0.75[1 - 7(0.75)^6 + 6(0.75)^7]}{0.25[1 - (0.75)^7]} = 1.92 \text{ customers}$$

and

$$L_q = 1.92 - \frac{0.75[1 - (0.75)^6]}{1 - (0.75)^7} = 1.21 \text{ customers}$$

The effective arrival rate, from Eq. (17), is

$$\lambda' = \frac{3[1 - (0.75)^6]}{1 - (0.75)^7} = 2.85 \text{ customers/h}$$

therefore, from Eqs. (18) and (19),

$$W = L/\lambda' = 1.92/2.85 = 0.67 \text{ h} = 40.4 \text{ min}$$

and

$$W_q = L_q/\lambda' = 1.21/2.85 = 0.42 \text{ h} = 25.5 \text{ min}$$

M/M/1—Finite Population. Interarrival times are iid exponential with mean $1/\lambda$, service times are iid exponential with mean $1/\mu$, a single server, and arrivals come from a finite source of size m. The birth-death coefficients are

$$\lambda_n = \begin{cases} (m - n)\lambda & 0 \le n \le m \\ 0 & n \ge m \end{cases}$$

$$\mu_n = \begin{cases} \mu & 1 \le n \le m \\ 0 & n > m \end{cases}$$

The resulting balance equations are

$$0 = -m\lambda p_0 + \mu p_1$$

$$0 = -[(m - n)\lambda + \mu]p_n + [m - n + 1]\lambda p_{n-1} + \mu p_{n+1} \qquad 1 < n < m - 1 \tag{20}$$

$$0 = -\lambda p_{m-1} + \mu p_m$$

Solving recursively, it follows that

$$p_n = \frac{m!}{(m - n)!} \rho^n p_0 \qquad n = 1, \ldots, m \tag{21}$$

where (to be a distribution, p_n, $n = 0, 1, \ldots$, must sum to 1)

$$p_0 = \frac{1}{\displaystyle\sum_{n=0}^{m} \left[\frac{m!}{(m - n)!} \left(\frac{\lambda}{\mu}\right)^n \right]}$$

To compute the performance measures it is convenient to first determine the steady-state expected queue length

$$L_q = \sum_{n=1}^{m} (n - 1)p_n = m - \frac{\lambda + \mu}{\mu}(1 - p_0) \tag{22}$$

and then compute

$$L = \sum_{n=0}^{m} np_n = L_q + (1 - p_0) \tag{23}$$

The effective arrival rate is

$$\lambda' = \sum_{n=0}^{m} (m - n)\lambda p_n = \lambda(m - L) \tag{24}$$

from which we can compute

$$W = \frac{1}{\lambda'} L \tag{25}$$

$$W_q = \frac{1}{\lambda'} L_q \qquad (26)$$

Multiple-Server Poisson Models

M/M/c *System.*   Interarrival times are iid exponential with mean $1/\lambda$, and c parallel servers, each with service time iid exponential with mean $1/\mu$, FIFO discipline, with birth-death coefficients:

$$\lambda_n = \lambda \qquad n \geq 0$$

$$\mu_n = \begin{cases} n\mu & 1 \leq n \leq c \\ c\mu & n \geq c \end{cases}$$

Substituting into the balance equations of Eq. (2), we obtain the balance equations

$$0 = -\lambda p_0 + \mu p_1$$

$$0 = -(\lambda + n\mu)p_n + \lambda p_{n-1} + (n+1)\mu p_{n+1} \qquad 1 \leq n < c \qquad (27)$$

$$0 = -(\lambda + c\mu)p_n + \lambda p_{n-1} + c\mu p_{n+1} \qquad n \geq c$$

and solving iteratively leads to the steady-state distribution

$$p_n = \begin{cases} \dfrac{(\lambda/\mu^n}{n!} p_0 & n \leq c \\[3mm] \dfrac{(\lambda/\mu)^n}{c!c^{n-c}} p_0 & n \geq c \end{cases} \qquad (28)$$

where

$$p_0 = \left[\sum_{k=0}^{c-1} \frac{(\lambda/\mu)^k}{k!} + \frac{(\lambda/\mu)^c}{c!\left(1 - \dfrac{\lambda}{c\mu}\right)} \right]^{-1}$$

provided the utilization factor $\rho = \lambda/c\mu < 1$.

The measures of effectiveness for this system are conveniently found by first computing the steady-state queue length from Eq. (5),

$$L_q = \sum_{n=c}^{x} (n - c)p_n$$

which, after considerable algebraic manipulation, reduces to

$$L_q = \left[\frac{(\lambda/\mu)^c \lambda\mu}{(c-1)!(c\mu - \lambda)^2} \right] p_0 \qquad (29)$$

From Little's formula

$$W_q = L_q/\lambda = \left[\frac{\mu(\lambda/\mu)^c}{(c-1)!(c\mu - \lambda)^2} \right] p_0 \qquad (30)$$

and the long run expected time in the system is

$$W = W_q + E(\text{service time}) = W_q + 1/\mu \qquad (31)$$

from which we obtain the expected number in the system

$$L = \lambda W = \left[\frac{(\lambda/\mu)^c \lambda \mu}{(c-1)!(c\mu - \lambda)^2} \right] p_0 + \frac{\lambda}{\mu} \tag{32}$$

M/M/∞—Unlimited Servers. Interarrival times iid exponential with mean $1/\lambda$, infinite number of parallel servers, service time iid exponential with mean $1/\mu$. The birth-death coefficients are

$$\lambda_n = \lambda \qquad n \geq 0$$

$$\mu_n = n\mu \qquad n \geq 1$$

Substituting into Eq. (2) results in

$$p_n = \frac{(\lambda/\mu)^n}{n!} p_0$$

where

$$p_0 = \left[\sum_{n=0}^{\infty} \frac{(\lambda/\mu)^n}{n!} \right]^{-1} \rightarrow e^{-\lambda/\mu}$$

and hence

$$p_n = \frac{(\lambda/\mu)^n e^{-\lambda/\mu}}{n!} \qquad n \geq 0 \tag{33}$$

which is the Poisson distribution. With an infinite number of servers there is no queue; therefore, the long run average number in the system is the average number in service,

$$L = \lambda/\mu \tag{34}$$

and from Little's formula

$$W = 1/\mu \tag{35}$$

M/M/c/K—Finite Capacity. Interarrival times iid exponential with mean $1/\lambda$, c parallel servers, each with service time iid exponential with mean $1/\mu$, a FIFO discipline, and no more than K customers permitted in the system.

Here we have birth-death coefficients

$$\lambda_n = \begin{cases} \lambda & 0 \leq n \leq K \\ 0 & n \geq K \end{cases}$$

$$\mu_n = \begin{cases} n\mu & 0 \leq n < c \\ c\mu & c \leq n \leq K \end{cases}$$

from which, in Eq. (2), we obtain

$$p_n = \begin{cases} \dfrac{(\lambda/\mu)^n}{n!} p_0 & 0 \leq n \leq c \\[3mm] \dfrac{(\lambda/\mu)^n}{c! c^{n-c}} p_0 & c < n \leq K \end{cases} \tag{36}$$

where

$$p_0 = \left[\sum_{n=0}^{c-1} \frac{(\lambda/\mu^n)}{n!} + \frac{(\lambda/\mu)^c}{c!} \left(\frac{1 - \rho^{K-c+1}}{1 - \rho} \right) \right]^{-1}$$

for $\rho = \lambda/c\mu \neq 1$. Note that for the case of $\rho = \lambda/c\mu = 1$

$$\frac{1 - \rho^{K-c+1}}{1 - \rho} \rightarrow (K - c + 1)$$

and hence

$$p_0 = \left[\sum_{n=0}^{c-1} \frac{(\lambda/\mu)^n}{n!} + \frac{(\lambda/\mu)^c}{c!} (K - c + 1) \right]$$

The long run average queue length for $\rho \neq 1$ is given by

$$L_q = \sum_{n=c}^{K} (n - c)p_n = \frac{(\lambda/\mu)^c}{c!} p_0 \sum_{n=c}^{K} (n - c)p^{n-c}$$

which simplifies to

$$L_q = \frac{(\lambda/\mu)^c \rho p_0}{c!(1 - \rho)^2} \{ 1 - [(K - c)(1 - \rho) + 1]\rho^{K-c} \} \tag{37}$$

Since

$$L = L_q + E \text{ [number in service]} = L_q + \sum_{n=0}^{c-1} np_n + c \left(1 - \sum_{0}^{c-1} p_n \right)$$

then

$$L = L_q + c - \sum_{n=0}^{c-1} (c - n)p_n \tag{38}$$

The effective arrival rate is

$$\lambda' = \lambda(1 - p_K) \tag{39}$$

therefore,

$$W_q = L_q/\lambda' = \frac{L_q}{\lambda(1 - p_K)} \tag{40}$$

$$W = L/\lambda' = \frac{L}{\lambda(1 - p_K)} \tag{41}$$

M/M/c—Finite Population. Interarrival times iid exponential with mean $1/\lambda$, c parallel servers with service times iid exponential with mean $1/\mu$, arrivals come from a finite population of size m. The birth-death coefficients are

$$\lambda_n = \begin{cases} (m - n)\lambda & 0 \leq n < m \\ 0 & n \geq m \end{cases}$$

$$\mu_n = \begin{cases} n\mu & 0 \le n < c \\ c\mu & n \ge c \end{cases}$$

Substituting these into Eq. (2), we get the steady-state distribution

$$p_n = \begin{cases} \binom{m}{n}(\lambda/\mu)^n p_0 & 0 \le n < c \\ \binom{m}{n}\dfrac{n!}{c!c^{n-c}}\left(\dfrac{\lambda}{\mu}\right)^n p_0 & c \le n \le m \end{cases} \tag{42}$$

where

$$p_0 = \left[\sum_{n=0}^{c-1} \binom{m}{n}(\lambda/\mu)^n + \sum_{n=c}^{m} \binom{m}{n}\frac{n!}{c!c^{n-c}}(\lambda/\mu)^n \right]^{-1}$$

The long run average number in the system and queue are computed directly from Eqs. (4) and (5), giving

$$L = \sum_{n=0}^{c-1} n\binom{m}{n}\left(\frac{\lambda}{\mu}\right)^n + \frac{1}{c!}\sum_{n=c}^{m} n\binom{m}{n}\frac{n!}{c^{n-c}}\left(\frac{\lambda}{\mu}\right)^n \tag{43}$$

and

$$L_q = L - c + p_0 \sum_{n=0}^{c-1}(c-n)\binom{m}{n}\left(\frac{\lambda}{\mu}\right)^n \tag{44}$$

The effective arrival rate is given by

$$\lambda' = \sum_{n=0}^{m} \lambda p_n$$

or

$$\lambda' = \lambda(m - L) \tag{45}$$

from which it follows from Little's formula that the average time in the system and queue are

$$W = \frac{L}{\lambda(m - L)} \tag{46}$$

and

$$W_q = \frac{L_q}{\lambda(m - L)} \tag{47}$$

Example.   A paper manufacturer produces paper on five machines, each of which breaks down periodically with failure times exponentially distributed with a mean of 50 hours. Each breakdown is serviced by one of two available repair technicians in the order of failure. Repair times are exponentially distributed with a mean of 10 hours.

The parameters are $m = 5$ machines, $c = 2$ repair technicians, $\lambda = 1/50$, and $\mu = 1/10$ machines per hour. In Eq. (42) we have

$$p_0 = \left[\sum_{n=0}^{1}\binom{5}{n}\left(\frac{1}{5}\right)^n + \sum_{n=2}^{5}\binom{5}{n}\frac{n!}{2!2^{n-2}}\left(\frac{1}{5}\right)^n\right]^{-1} = 0.402$$

from which we compute the steady-state probabilities

$$p_n = \begin{cases} \binom{5}{n}(0.2)^n(0.402), & n = 0,1 \\ \binom{5}{n}\dfrac{n!(0.2)^n}{2 \cdot 2^{n-2}}(0.402) & n = 2,\dots,5 \end{cases}$$

The average number of machines down for repair, from Eq. (43), is

$$L = (0.402)\left[\sum_{n=0}^{1} n\binom{5}{n}(0.2)^n + \frac{1}{2!}\sum_{n=2}^{5} n\binom{5}{n}\frac{n!}{2^{n-2}}(0.2)^n\right] = 0.912 \text{ machines}$$

and the number down awaiting service, Eq. (44), is

$$L_q = .912 - 2 + (0.402)\sum_{n=0}^{1}(2-n)\binom{5}{n}(0.2)^n = 0.12 \text{ machines}$$

The effective arrival rate is

$$\lambda' = \frac{1}{50}(5 - 0.912) = 0.082 \text{ machines/hr}$$

and hence the average total downtime for a machine, from Eq. (46), is

$$W = 0.912/0.082 = 11.2 \text{ hr}$$

and, from Eq. (47), the average time spent waiting for repair is

$$W_q = 0.012/0.082 = 1.46 \text{ hr}$$

Queues with Behavioral Influences.   A variety of behavioral characteristics can be incorporated in the general birth-death model by particular choices of (λ_n, μ_n) in Eq. (2). These state-dependent models can be such that service, arrivals, or both are influenced behaviorally by the state of the system. Closed-form results are an exception rather than the rule for state-dependent models. However, once the steady-state distributions are known the performance measures can be computed directly using the results of Eqs. (3) through (7).

   M/M/1—Two-Speed Service.   Interarrival times iid exponential with mean $1/\lambda$, and a single Markov server whose mean rate μ_a shifts to μ_b when J customers are in the system. The birth-death coefficients are

$$\lambda_n = \lambda \qquad n \geq 0$$

$$\mu_n = \begin{cases} \mu_a & 1 \leq n < J \\ \mu_b & n \geq J \end{cases}$$

From Eq. (2), it follows [see Gross and Harris (1985)] that

$$p_n = \begin{cases} (\lambda/\mu_a)^n p_0 & 0 \leq n < J \\ \dfrac{\lambda^n}{\mu_a^{J-1}\mu_b^{n-J+1}} p_0 & n \geq J \end{cases} \qquad (48)$$

with

$$p_0 = \left[\frac{1 - \rho_a^J}{1 - \rho_a} + \frac{\rho_b \rho_a^{J-1}}{1 - \rho_b} \right]^{-1} \qquad \rho_a = \lambda/\mu_a, \ \rho_b = \lambda/\mu_b < 1$$

The steady-state average number of customers in the system is given by

$$L = p_0 \left\{ \frac{\rho_a[1 + (J - 1)\rho_a^J - J\rho_a^{J-1}]}{(1 - \rho_a)^2} + \frac{\rho_b \rho_a^{J-1}[J - (J - 1)\rho_b]}{(1 - \rho_b)^2} \right\} \qquad (49)$$

from which the number in the queue can then be computed from Eq. (5),

$$L_q = L - (1 - p_0) \qquad (50)$$

The average times in the system and queue can then be computed from Little's formula.

M/M/c—Channel Control. Interarrival times iid exponential with mean $1/\lambda$; service is provided by a team of c available parallel servers each with rate μ; after the first server, each additional server up to c is activated when the number in the system reaches each additional multiple of I. The birth-death coefficients are

$$\lambda_n = \lambda \qquad n \geq 0$$

$$\mu_n = \begin{cases} \mu & 1 \leq n \leq I \\ 2\mu & I < n \leq 2I \\ \cdots \\ (c - 1)\mu & (c - 2)I < n \leq (c - 1)I \\ c\mu & n \geq cI \end{cases}$$

Giffin (1978) provides the results for this model in terms of the function

$$K(n, i) = \frac{(\lambda/\mu)^n}{(i!)^i(i + 1)^{n-iI}} \qquad (51)$$

and

$$P_n = \begin{cases} K(n, i)p_0 & 0 \leq n \leq (i + 1)I; \ 0 \leq i \leq c - 1 \\ \dfrac{(\lambda/\mu)^n}{(c!)^I c^{n-cI}} p_0 & n > cI, \end{cases} \qquad (52)$$

with

$$p_0 = \left[\sum_{i=0}^{c-1} \sum_{n=iI}^{(i+1)I} K(n, i) + \frac{(\lambda/\mu)^{cI}}{(c!)^I(1 - \lambda/c\mu)} \right]^{-1}$$

The long run average number of customers in the system is

$$L = \sum_{n=0}^{cI-1} np_n + \frac{(\lambda/\mu)^{cI} p_0}{(c!)^I} \left[\frac{\lambda/c\mu}{(1 - \lambda/c\mu)^2} + \frac{cI}{1 - \lambda/c\mu} \right] \qquad (53)$$

The expected number of customers in service, S, is given by

$$E[S] = \sum_{i+1}^{c-1} \sum_{n=(i-1)I+1}^{iI} ip_n + c \sum_{n=(c-1)I+1}^{cI} p_n + \frac{c(\lambda/\mu)^{cI}}{(c!)^I}\left[\frac{1}{1-\lambda/c\mu-1}\right]p_0 \qquad (54)$$

from which the average number in the queue can be computed,

$$L_q = L - E[S] \qquad (55)$$

The long run average number in the system and queue can be found by Little's formula.

Example. A bank operates with a team of four tellers, each with an average service rate of 0.75 services per minute. A teller not occupied at the counter performs other clerical functions. Interarrival and service times are exponentially distributed, and the mean time between arrivals is 0.5 minute. Management prefers that no more than three persons should have to wait in line before other tellers are assigned.

Here we have $c = 4$ tellers, $I = 3$ customers, $\lambda = 2$ customers/min, $\mu = 0.75$ customers/min, and hence $\lambda/\mu = 2.67$. Thus, from Eq. (51),

$$K(n, i) = \frac{2.67^n}{(i!)^3(i+1)^{n-3i}}$$

and

$$p_0 = \left[\sum_{i=0}^{3} \sum_{n=3i}^{3(i+1)} K(n, i) + \frac{2.67^{12}}{(4!)^3(1-2.67/4)}\right]^{-1}$$

$$= [K(0, 0) + K(1, 0) + K(2, 0) + K(3, 0) + K(3, 1) + \cdots + K(12, 3) + 28.5]^{-1}$$

$$= 0.00245$$

Therefore, in Eq. (52) we have the steady-state probability distribution

$$p_n = \begin{cases} (0.00245)K(n, i) & 0 \le n \le 3(i+1); 0 \le i \le 3 \\ \dfrac{0.00245(2.67)^n}{(4!)^3 4^{n-12}} & n > 12 \end{cases}$$

M/M/c—Discouraged Arrivals/Pressured Service. Interarrival times iid exponential with mean $1/\lambda$, but arriving customers divert as the number in the system increases; c parallel servers each with iid exponentially distributed service time with mean that decreases (or increases) as the number in the system increases (decreases). Let $\alpha > 0$ be a pressure coefficient of a server and $\beta > 0$ be a discouragement coefficient for potentially arriving customers. The birth-death coefficients are

$$\lambda_n = \begin{cases} \lambda & 0 \le n < c \\ \left(\dfrac{c}{n+1}\right)^\beta \lambda & n \ge c \end{cases}$$

$$\mu_n = \begin{cases} n\mu & 1 \le n \le c \\ \left(\dfrac{n}{c}\right)^\alpha c\mu & n > c \end{cases}$$

The steady-state probability distribution is given by

$$
p_n = \begin{cases} \dfrac{(\lambda/\mu)^n}{n!}p_0 & 0 < n \leq c \\[2ex] \dfrac{(\lambda/\mu)^n p_0}{c!(n!/c!)^\gamma c^{(1-\gamma)(n-c)}} & n > c \end{cases}
\tag{56}
$$

where $\gamma = \alpha + \beta$ and

$$
p_0 = \left[\sum_{n=0}^{c-1} \frac{(\lambda/\mu)^n}{n!} + \frac{(\lambda/\mu)^c \displaystyle\sum_{n=c}^{\infty} (\lambda/\mu)^{n-c}}{c!(n!/c!)^\gamma c^{(1-\gamma)(n-c)}} \right]^{-1}
$$

The results for this model were developed by Hillier, Conway, and Maxwell (1964), who also provided tables for p_0, L, and L_q for various values of γ, c, and λ/μ.

PART 3: ADVANCED MARKOV QUEUING MODELS

The models presented in Part 2 allowed transitions to adjacent states only, and the state of the system was completely described by the number of customers in the system. As a result, the balance equations for these systems were relatively simple recursive equations that could be solved readily. In this part of the chapter, the Markov property is maintained at each transition in the queuing process, but the transition is not necessarily to an adjacent state. As a result, the analysis required to determine steady-state probabilities and performance measures is more complicated. Some of the more useful results for non-birth-death Markov queuing models are summarized here. The interested reader is referred to Gross and Harris (1985) for more details concerning their derivation.

$M/M^{(k)}/1$—Bulk Service.   Arrivals to the system form a Poisson process with parameter λ. Service times are iid exponential with mean $1/\mu$ for a single server processing a batch of k customers. If there are $n < k$ in the system when the server is ready, the smaller batches are processed with the same service distribution applying to any smaller batch size $n < k$. Batches are selected on a FIFO basis, and there is no restriction on queue size. We consider a single server here, but the extension to multiple servers is not difficult.

The steady-state balance equations for this system are

$$
0 = \mu P_{n+k} - (\lambda + \mu)P_n + \lambda P_{n-1} \qquad n \geq 1
$$
$$
0 = \mu P_k = \mu P_{k-1} + \cdots + \mu P_1 - \lambda P_0
\tag{57}
$$

The first equation of Eq. (57) indicates that the rate of entry into state n from state $n + k$ is μ, the rate of batch completions. The second equation indicates that if a batch size less than k is being processed, the transition on completion is to state zero. We will use operator methods to analyze these difference equations. A difference operator is such that

$$
P_{n+2} = D^2 P_n
$$

$$
P_{n-1} = D^{-1} P_n
$$

for example. Thus, the first equation of Eq. (57) may be written

$$[\mu D^{(k+1)} - (\lambda + \mu)D + \lambda]P_n = 0 \qquad n \geq 0$$

The resulting operator equation possesses roots

$$r_1, r_2, \ldots, r_{k+1}$$

which can be used to characterize the steady-state system size probabilities as

$$P_n = \sum_{i=1}^{k+1} c_i r_i^n$$

where $c_i = 0$ if $r_i \geq 1$.

Rouche's theorem may be employed to show that only one root (r_0, say) is less than 1; thus,

$$P_n = c r_0^n \qquad n \geq 0, 0 < r_0 < 1 \tag{58}$$

Note that Eq. (58) has the same form as found in the M/M/1 queue analysis with the role of the traffic intensity ρ played by r_0. Hence, since

$$\Sigma P_n = 1$$

then

$$c = 1 - r_0 = P_0$$

and

$$P_n = (1 - r_0)r_0^n \qquad n \geq 0, 0 < r_0 < 1$$

The measures of performance are familiar, as well, such as

$$L = \frac{r_0}{1 - r_0}$$

$$W = \frac{r_0}{\lambda(1 - r_0)}$$

Important to the analysis of this queuing model is the determination of r_0, which can be efficiently determined numerically using a method for finding the roots of polynomials.

M_i/M/1/Infinity/PRP—Preemptive Priority Queue. If the queue discipline selected the next customer for service on the basis of attributes of the customer such that one customer may have priority over another, analysis becomes much more complex. Balance equations may still be constructed, but they are difficult to evaluate because of the increase in dimensionality; the state definition usually includes the number of each customer class in the queue and the classification of the customer in service. Measures of performance of a Markov priority queue are unaffected by priority structure (except waiting time, which becomes a function of priority class) as long as the selection rule is not a function of relative service time. A *preemptive* priority structure implies that a higher-ranking customer will interrupt the service of a lower-ranking customer immediately on arrival. The lower-priority customer ultimately resumes service at the point of interruption or starts over again. With exponential service each alternative is equivalent with respect to measures of performance, but this is not true for general service distributions. For more details on priority queues and their analysis, refer to Jaiswal (1968).

Example. A company is selling time on its research mainframe computer to outside customers with the understanding that internal customers may bump an outside customer's running job at any point (assuming the computer is completely dedicated to one job at a time). Computer runs average 6 hours across all customers and are exponentially distributed. Customers both in and outside the company require computer services according to Poisson process at a rate of $\frac{1}{8}$/hour and $\frac{1}{30}$/hour, respectively. How similar are the response times for the two customer classes (queue plus service times)?

From the data:

$$\mu = \frac{1}{6} \qquad \lambda_1 = \frac{1}{8} \qquad \lambda_2 = \frac{1}{30}$$

Let k = number of priority classes = 2 and

$$a_j = \sum_{i=1}^{j} \frac{\lambda_i}{\mu} \qquad a_0 = 0$$

The expected time in the system for class j programs is

$$W_j = \frac{1/\mu}{(1 - a_{j-1})(1 - a_j)}$$

$$W_1 = \frac{6}{(1 - 0)(1 - \frac{6}{8})} = 24 \text{ hours (1 day)}$$

$$W_2 = \frac{6}{\left(1 - \frac{6}{8}\right)\left(1 - \frac{6}{8} - \frac{6}{30}\right)} = 480 \text{ hours (20 days)}$$

Thus, the company is not very responsive to outside customers at all, given present high usage levels. Note:

$$\rho = \frac{6}{8} + \frac{6}{30} = 0.95$$

$M_i/M/c/Infinity/NPRP—Nonpreemptive Priority Queue.   In most situations where people are the customers and a customer priority structure is in place, customers of lower priority to an arriving customer are not preempted, but allowed to complete their service. Arrivals from each priority class i form a Poisson process with rate λ_i, and service time is iid exponential with common service rate to all priority classes of μ. Let:

$$a_j = \sum_{i=1}^{j} \frac{\lambda_i}{c\mu} \qquad j = 1, \ldots, k$$

where k = number of priority classes.

$$a_o = 0 \qquad \rho = a_k < 1$$

The probability of a customer having to wait for service is

$$P(n > c) = 1 - \sum_{n=0}^{c-1} P_n$$

$$= 1 - P_0 \sum_{n=0}^{c-1} \frac{(\lambda/\mu)^n}{n!}$$

From Eq. (28),

$$P(n \geq c) = \frac{c(\lambda/\mu)^c}{c!(c - \lambda/\mu)}\left[\sum_{n=0}^{c-1} \frac{1}{n!}\left(\frac{\lambda}{\mu}\right)^n + \frac{1}{c!}\left(\frac{\lambda}{\mu}\right)^c \left(\frac{c\mu}{c\mu - \lambda}\right)\right]^{-1}$$

The mean waiting time in the queue for a class j customer is

$$W_{qj} = \frac{P(n \geq c)}{s\mu(1 - a_{j-1})(1 - a_j)} \tag{59}$$

Example.   Consider the example given for the $M_i/M/1/\infty/$PRP model. Under preemption, the priority 2 customer suffered severe delays. Let W_{qj}(PRP) = mean waiting time in the queue for a class j customer under preemption.

$$W_{qj}(\text{PRP}) = W_j - 1/\mu$$

$$W_{q1}(\text{PRP}) = 24 - 6 = 18 \text{ hours}$$

$$W_{q2}(\text{PRP}) = 480 - 6 = 472 \text{ hours}$$

Similarly, from Eq. (59),

$$W_{qj}(\text{NPRP}) = \frac{P(n \geq 1)}{\mu(1 - a_{j-1})(1 - a_j)}$$

$$W_{q1}(\text{NPRP}) = \frac{0.95}{\left(\frac{1}{6}\right)(1 - 0)\left(1 - \frac{6}{8}\right)} = 22.8 \text{ hours}$$

$$W_{q2}(\text{NPRP}) = \frac{0.95}{\left(\frac{1}{6}\right)\left(1 - \frac{6}{8}\right)\left(1 - \frac{6}{30}\right)} = 28.5 \text{ hours}$$

Thus, if outside users are allowed to complete their use of the computer once service begins, there is a dramatic improvement in their response time with only a modest degradation of the response time to the inside user.

PART 4: EMBEDDED MARKOV CHAIN MODELS

In Part 4, the time between transitions in the queuing process no longer needs to be strictly exponentially distributed. As long as either the time between arrivals or the time to complete services is exponential, a Markov chain may be embedded in an otherwise non-Markov process, and the computational tractability benefits of a Markov chain analysis are available to a certain degree. The Chapman-Kolmogorov equations are no longer a basis for the analysis for the resulting stochastic process, however. The following paragraphs will outline how an analysis of queuing systems with embedded Markov chains may be accomplished. For a more detailed discussion, the interested reader is referred to Gross and Harris (1985).

M/G/1—Poisson Input, General Service.   Such systems must be observed at a service completion. At that point the time to the next service completion is a random variable possessing a general distribution and the time to the next arrival possesses an exponential distribution due to the memoryless property of the exponential. Thus, the transition process can be completely specified with a state description that includes only the number in the system, as in previous single-server systems. However, any number of arrivals may occur between service completions, leading to more computationally cumbersome analysis. Let x_n = number of customers in the system at departure point n and A_{n+1} = number of arrivals that occurred between departure point n and departure point $n + 1$. Then

$$x_{n+1} = \begin{cases} x_n - 1 + A_{n+1} & x_n \geq 1 \\ A_{n+1} & x_n = 0 \end{cases}$$

Dropping the subscript of A_n, since the arrival process is not dependent on the service process, and denoting S as the service time variable and $B(t)$ and $b(t)$ as the service time cumulative density function (CDF) and probability density function (PDF), respectively, we have

$$P(A = a) = \int_0^x P(A = a|s = t)dB(t)$$

Since the arrival process is Poisson,

$$P(A = a|s = t) = \frac{e^{-\lambda t}(\lambda t)^a}{a!}$$

and we may describe the transition process for the embedded Markov chain as

$$P\{x_{n+1} = j|x_n = i\} = P\{A = j - i + 1\} \qquad (60)$$

$$= \int_0^x \frac{e^{-\lambda t}(\lambda t)^{j-i+1}}{(j - i + 1)!} dB(t) \qquad \text{if } j \geq i - 1 \text{ and } i \geq 1$$

$$= 0 \qquad \text{if } j < i - 1 \text{ and } i \geq 1$$

The steady-state system size is measured and probabilities determined at departure points when a service is completed. Let:

$$\pi_n = P\{n \text{ in system at a departure point in steady state}\}$$

$$= P_n, \text{ for an M/G/1 system}$$

$$P_{ij} = P\{x_{n+1} = j|x_n = |i|\} \text{ as defined by Eq. (60)}$$

$$m = j - i + 1 = \text{number of arrivals during a service time } s = t$$

$$K_m = \int_0^x \frac{e^{-\lambda t}(\lambda t)^m}{m} dB(t)$$

Then

$$P = [P_{ij}] = \begin{vmatrix} K_0 & K_1 & K_2 & \rightarrow \\ K_0 & K_1 & K_2 & \rightarrow \\ 0 & K_0 & K_1 & \rightarrow \\ 0 & 0 & K_0 & \rightarrow \\ \downarrow & \downarrow & \downarrow & \searrow \end{vmatrix}$$

To determine the steady-state probabilities, we must evaluate

$$\pi P = \pi$$

which may be written

$$\pi_i = \pi_0 K_i + \sum_{j=1}^{i+1} \pi_j K_{i-j+1} \qquad i = 0, 1, 2, \ldots \tag{61}$$

Defining generating functions for π_i and K_i as $\pi(z)$ and $K(z)$, respectively, multiplying each equation of Eq. (61) by z^i, and summing ultimately gives

$$\pi(z) = \frac{\pi_0(1 - z)K(z)}{K(z) - z} \tag{62}$$

with $\pi_0 = 1 - \rho$. Generating function $\pi(z)$ may be inverted for the probabilities in one of the usual ways. If the inversion cannot be accomplished in closed form, a Maclaurin series expansion solution can be tedious, and a numerical solution is sought.

We can determine the expected number in the system by evaluating

$$L = \pi(z)|_{z-1} = \rho + \frac{\rho^2 + \lambda^2 \sigma_s^2}{2(1 - \rho)} \tag{63}$$

where σ_s^2 is the service time distribution variance. Equation (63) is known as the *Pollaczek-Khintchine formula*.

Example. An auto paint shop offers three levels of paint jobs: economy, classic, and supreme. The difference between the various paint jobs is the degree of surface preparation and number of coats applied. As a result, the time required is 1 hour, 3 hours, and 5 hours, respectively. The proportion of customers requesting each level is 0.5, 0.4, and 0.1, respectively. Customers arrive according to a Poisson process at a rate of 0.25/hour. The shop does not wish to have more than one customer waiting more than 25 percent of the time. Is it meeting its goal?

In many realistic settings, data on service times are gathered in such a way that discrete distributions are based on the frequency that service times fall into prespecified intervals. Although the integral in Eq. (60) is usually difficult to evaluate for continuous $b(t)$, discrete b_t reduces Eq. (60) to a summation and greatly simplifies the remaining analysis for p_n. The summation is

$$K_m = \frac{1}{m!} \sum_{t=1}^{T} e^{-\lambda S_t}(\lambda S_t)^m b_t$$

which has generating function $K(z)$, and S_t is the service time associated with frequency class t. Simplifying Eq. (62) by using this $K(z)$ leads to

$$\pi(z) = \frac{(1 - \rho)\left[1 + \left(\sum_{i=1}^{x} \frac{C_i}{C_0 i!} - \frac{C_{i-1}}{C_0(i-1)!}\right)z^i\right]}{1 + \left(\frac{C_1 - C_0}{C_0}\right)z + \sum_{i=2}^{x} \frac{C_i}{C_0 i!}z^i}$$

where

$$C_i = \sum_{t=1}^{T} b_t e^{-\lambda S_t}(\lambda S_t)^i$$

We use the fact that the ratio of two power series is another power series,

$$\frac{1 + \sum_{i=1}^{x} a_i z^i}{1 + \sum_{i=1}^{x} b_i z^i} = \sum_{i=0}^{x} d_i z^i$$

where

$$d_i = \begin{cases} a_i - \sum_{j=1}^{i} b_j d_{i-j} & \text{if } i \geq 1 \\ 1 & \text{if } i = 0 \end{cases}$$

to generate an expression for the steady-state system size probabilities

$$p_n = (1 - \rho)d_n \qquad n \geq 0$$

The auto paint shop has $T = 3$ frequency classes with associated service times $S_t = (1, 3, 5)$, probabilities $b_t = (.5, .4, .1)$, $E(S_t) = 1/\mu = 2.2$ hours, and $\lambda = 0.25/\text{hour}$. Note that

$$a_i = \frac{C_i}{C_0 i!} - \frac{C_{i-1}}{C_0 (i-1)!} \qquad i \geq 1$$

$$b_i = \begin{cases} \dfrac{C_i}{C_0 i!} & i \geq 2 \\ \dfrac{C_1 - 1}{C_0} & i = 1 \end{cases}$$

Evaluating the various constants gives

$$C_i = (.5)e^{-.25(1)}(.25)^i + (.4)e^{-.25(3)}[(.25)(3)]^i + (.1)e^{-.25(5)}[(.25)(5)]^i$$

$$= .3984(.25)^i + .1889(.75)^i + .0287(1.25)^i$$

$$C_0 = .607, \quad C_1 = .2749, \quad C_2 = .1754$$

$$a_1 = \frac{.2749}{(.607)(1)} - \frac{.607}{(.607)(1)} = -.5471$$

$$a_2 = \frac{.1754}{(.607)(2)} - \frac{.2749}{(.607)(1)} = -.3084$$

$$b_1 = \frac{.2749 - 1}{.607} = -1.1946$$

$$b_2 = \frac{.1754}{(.607)(2)} = .1445$$

$$d_0 = 1$$

$$d_1 = -.5471 - (-1.1945)(1) = .6474$$

$$d_2 = -.3084 - (-1.1945)(.6474) - (.1445)(1) = .3204$$

Thus, the probability of more than 1 customer waiting is

$$P(n \geq 3) = \sum_{n = 3}^{x} = 1 - p_0 - p_1 - p_2$$

$$= 1 - (1 - \rho)(d_0 + d_1 + d_2)$$

$$= 1 - \left(1 - \frac{25}{.4545}\right)(1 + .6474 + .3204)$$

$$= .1145$$

which is less than .25; the shop's goal is realized.

M/G/c/c—Loss System.   The steady-state system size distribution is

$$P_n = \frac{(\lambda/\mu)^n/n!}{\displaystyle\sum_{i = 0}^{c}(\lambda/\mu)^i/i!} \qquad 0 \leq n \leq c \qquad (64)$$

which is the truncated Poisson distribution and

$$L = (\lambda/\mu)(1 - P_c)$$

Equation (64) is valid for any general service time distribution, and p_c is often referred to as *Erlang's loss formula* (the probability an arriving customer is turned away and lost to the system). If c is infinite, Eq. (64) reduces to

$$P_n = \frac{e^{-\lambda/\mu}(\lambda/\mu)^n}{n!} \qquad n \geq 0 \qquad (65)$$

associated with what is referred to as the *self-serve* model, since there are as many servers as potential customers.

   Example.   A 1000-bed hospital admits 800 patients each week with arrivals forming a Poisson process. The average stay is for 1 week, with the time normally distributed and a variance of 2. What is the average patient population and what is the likelihood of an inadequate number of beds?

   Note that $\lambda = 800$/week, $1/\mu = 1$ week, $c = 1000$ and is large. Thus, the use of Eq. (65) seems warranted. Stirling's approximation is used for $n!$ (accurate for $n > 20$), so that

$$P_n = \frac{e^{-\lambda/\mu}(\lambda/\mu)^n}{\sqrt{2\pi n}\, n^n e^{-n}} = \frac{e^{n - \lambda/\mu}(\lambda/n\mu)^n}{\sqrt{2\pi n}}$$

A patient is turned away if all beds are occupied, which has probability

$$P_c = \frac{e^{c - \lambda/\mu}(\lambda/c\mu)^c}{\sqrt{2\pi c}} = \frac{e^{1000 - 800}(.8)^{1000}}{\sqrt{2\pi(1000)}}$$

$$= 1.12(10)^{-12} \approx 0$$

that is, all beds being occupied is not very likely, at all! The expected number of occupied beds is equivalent to the expected number of busy servers

$$L = \frac{\lambda}{\mu}(1 - P_c) \approx \frac{\lambda}{\mu} = 800 \text{ occupied beds}$$

Note that the variance of the service distribution plays no role in this analysis.

PART 5: APPROXIMATION METHODS

The more features attributed to a queuing model, the more complicated the analysis becomes. Balance equations become unwieldy in "Part 3: Advanced Markov Chain Models" (for example, the bulk queuing models), and the overall analysis becomes much more difficult as the Markov transition property is relaxed, as it was for arrivals or for services in "Part 4: Embedded Markov Chain Models." For GI/G/c models, very little useful analytical results are available and analysis must be based on numerical methods, simulation, or approximation. It is the last of these approaches that is the subject of this part of the chapter. A number of techniques will be presented with illustrations using more general queuing models. For more details concerning these techniques, the interested reader is referred to Gross and Harris (1985) and Tijms (1986) and to the references given in them.

Asymptotic Expansion-Based Approximations

M/G/1. Except for the case of markovian queuing models, the determination of $P\{W_q > x\}$ for more general queuing models is, typically, a formidable undertaking. However, in calculating $P\{W_q > x\}$, the entire distribution of W_q is not of interest, only its tail. In many practical cases it is possible to fit an approximating exponential distribution to W_q that is most accurate for larger values of x. The parameters for this approximating exponential function are derived through an asymptotic analysis of $P\{W_q > x\}$. It is useful here to think of a customer bringing some number of units of "work" s to the queuing system that must be accomplished by the server before the customer leaves. The server is able to do σ units of work per unit time. The service time of the customer is s/σ. Let

V_n = number of units of work in the system just prior to the arrival of customer n

D_n = the delay in the queue of customer n

$q(x) = \lim_{n \to \infty} P\{V_n > x\} \qquad x \geq 0$

$q(\sigma x) = \lim_{n \to \infty} P\{D_n > x\} \qquad x \geq 0$

$\rho = \lambda E(s)/\sigma = \lambda/\mu$

The function $q(x)$ solves the integro-differential equation:

$$\frac{dq(x)}{dx} = -\frac{\lambda}{\sigma}[1 - B(x)] + \frac{\lambda}{\sigma}q(x) - \frac{\lambda}{\sigma}\int_0^x q(x - y)b(y)\,dy \qquad x > 0 \tag{66}$$

for the case of a M/G/1 queue with $B(x)$ as the customer work requirement CDF. Equation (66) can be shown to be equivalent to

$$q(x) = a(x) + \int_0^x q(x - y)h(y)\,dy \qquad x \geq 0 \tag{67}$$

where

$$a(x) = q(0) - \frac{\lambda}{\sigma}\int_0^x [1 - B(y)]\,dy$$

and

$$h(x) = \frac{\lambda}{\sigma}[1 - B(x)]$$

Equation (67) is in the form of a standard renewal equation except that $h(x)$ is not a proper probability density, since

$$\int_0^x h(x)\,dx = \frac{\lambda}{\sigma}\int_0^x [1 - B(x)]\,dx = \frac{\lambda E(s)}{\sigma} = \rho < 1 \tag{68}$$

However, the "defect" in the renewal equation [Eq. (67)] may be fixed by finding the value of δ such that

$$\frac{\lambda}{\sigma} \int_0^x e^{\delta y} [1 - B(y)] \, dy = 1 \tag{69}$$

Applying the key renewal theorem to Eq. (67) after letting $\hat{q}(x) = e^{\delta x} q(x)$, $\hat{h}(x) = e^{\delta x} h(x)$, and $\hat{a}(x) = e^{\delta x} a(x)$, yields

$$\lim_{x \to x} \hat{q}(x) = \frac{1 - \rho}{\omega \delta} \tag{70}$$

where

$$\omega = \frac{\lambda}{\sigma} \int_0^x y e^{\delta y} [1 - B(y)] \, dy$$

In terms of the original $q(x)$, Eq. (70) becomes

$$q(x) \approx \gamma e^{-\delta x} \qquad \text{for large } x \tag{71}$$

where

$$\gamma = \frac{(1 - \rho)}{\frac{\lambda \delta}{\sigma} \int_0^x y e^{\delta y} [1 - B(y)] \, dy} \tag{72}$$

Even for relatively small values of x [in the range of $E(s)$ to $2E(s)$], Eq. (71) is reasonably accurate, especially when ρ is not too small (say, $\rho \geq .2$). A better estimate for $q(x)$ over all $x \geq 0$ can be derived by letting

$$q(x) = \alpha e^{-\beta x} + \gamma e^{-\delta x} \qquad x \geq 0 \tag{73}$$

and noting that

$$W_q = \int_0^x q(x) \, dx = \frac{\lambda E(s^2)}{2\sigma(1 - \rho)} \tag{74}$$

and

$$q(0) = \rho \tag{75}$$

we get after evaluating Eq. (73)

$$\alpha = \rho - \gamma \tag{76}$$

$$\beta = \frac{\alpha}{\dfrac{\lambda E(s^2)}{2\sigma(1 - \rho)} - \dfrac{\gamma}{\delta}} \tag{77}$$

which is accurate if $\beta > \delta$ and $\rho \geq .2$.

    In summary, to calculate $P\{W_q > x\}$ for the M/G/1 model, Eq. (69) must be solved for δ, typically using numerical quadrature to evaluate the integral as a polynomial and then finding the positive root of that polynomial equation. Numerical integration will, usually, be required to evaluate Eq. (72) for γ. If x is large, Eq. (71) can be evaluated; if not, Eqs. (76) and (77) give α and β to use in Eq. (73) for an estimate of $P\{W_q > x\}$. If we let

$$W_q(x) = 1 - P\{W_q > x\}$$

a related quantity is

$$\varepsilon(p) = \text{the } p\text{th percentile of } W_q(x)$$

$$\approx \frac{1}{\delta} \ln\left(\frac{\gamma}{1-p}\right) \tag{78}$$

for p not close to 0. For a finite work buffer capacity M/G/1/k queue, let

$$P(k) = \text{steady-state probability of turning work away}$$

$$\approx \frac{\sigma\delta}{\lambda} \gamma e^{-\delta k} \qquad \text{for large } k \tag{79}$$

Most of the discussion has focused on the M/G/1 queue to set ideas, but asymptotic results exist for a number of general queuing models. We offer a sampling of those results in the following.

D/G/1—Scheduled Arrivals.   The following results are most useful for $C_s^2 < \frac{1}{2}$. Let

$$P\{W_q > 0\} = \text{probability that an arrival must wait for service}$$

$$= \frac{1 - B(D)}{\int_D^x e^{\delta(t-D)}b(t)\,dt} \tag{80}$$

and

$$W_q = \text{expected waiting time in the queue}$$

$$= \frac{\int_D^x (t-D)b(t)\,dt}{B(D) - 1 + \int_D^x e^{\delta(t-D)}b(t)\,dt} \tag{81}$$

where δ for both Eqs. (80) and (81) uniquely solves

$$e^{-\delta D}\int_0^x e^{\delta y}b(y)\,dy = 1 \tag{82}$$

M/G/1—Impatient Customers.   Customers are willing to wait for service to begin for τ time units. If service has not begun, the customer reneges, leaves the system, and is lost. This represents a good model for perishable-goods inventory analysis. Let the amount of work each customer brings to the system possess a general distribution and the service rate equal σ units of work per unit time. Then

$$\pi(\tau) = \text{proportion of all customers that renege, given a } \tau \text{ waiting threshold} \tag{83}$$

$$\approx \frac{(1-\rho)(\alpha e^{-\beta\sigma\tau} + \gamma e^{-\delta\sigma\tau})}{1 - \rho(\alpha e^{-\beta\sigma\tau} + \gamma e^{-\delta\sigma\tau})}$$

$$\approx (1-\rho)\gamma e^{-\sigma\tau\delta} \tag{84}$$

if $\sigma\tau$ is large, and if $\pi(\tau) < .01$. In addition,

$$W_q(\tau) = \text{mean waiting in the queue, given a } \tau \text{ waiting threshold}$$

$$\approx \frac{\dfrac{\alpha}{\sigma\beta} - \alpha\left(\dfrac{1}{\sigma\beta} + \rho\tau\right)e^{-\beta\sigma\tau} + \dfrac{\gamma}{\sigma\delta} - \gamma\left(\dfrac{1}{\sigma\delta} + \rho\tau\right)e^{-\delta\sigma\tau}}{1 - \rho(\alpha e^{-\beta\sigma\tau} + \gamma e^{-\delta\sigma\tau})} \tag{85}$$

where δ, γ, α, and β are determined by Eqs. (69), (72), (76), and (77), respectively.

M/M/1/k—Finite Work Buffer Capacity.   The arrival process of customers is Poisson with rate γ. Each customer brings some amount of work to be processed by the server, and the amount of work is exponentially distributed. The server processes the work at rate of σ/unit of time. Only k units of work are allowed in the system at one time. An arrival bringing work in excess of buffer capacity is turned away. Let

$\pi(k)$ = proportion of customers turned away because of work buffer being at capacity k

$$(86)$$

$$\approx \gamma e^{-\delta k}$$

for $\rho < 1$ and k such that $\pi(k) < .001$, where

$$\delta = \frac{1 - \rho}{E(s)} \tag{87}$$

$$\gamma = (1 - \rho)e^{-\rho} \tag{88}$$

M/D/1/k—Finite Work Buffer Capacity.   Arrivals bring along a fixed amount of work to be processed, and work in excess of capacity is turned away. Thus, δ is determined by finding the unique solution of

$$e^{\delta E(s)} = 1 + \frac{\delta \sigma}{\lambda} \tag{89}$$

and γ is given by

$$\gamma = \frac{(1 - \rho)}{\delta E(s) - (1 - \rho)} \tag{90}$$

and $\pi(k)$ is determined by using Eq. (86).

M/G/1/k—Finite Work Buffer Capacity.   Let any work brought by an arrival in excess of the work buffer capacity be lost, but the arrival is not turned away, and let

π_k = the steady-state fraction of arrivals that bring work in excess of capacity

$$(91)$$

$$\approx \frac{\frac{\sigma}{\lambda}(\alpha \beta e^{-\beta k} + \gamma \delta e^{-\delta k})}{1 - \alpha e^{-\beta k} - \gamma e^{-\delta k}}$$

V_k = mean number of units of work in the system when a limit of k units or work exists

$$\approx k - \left[\frac{k - \frac{\alpha}{\beta}(1 - e^{-\beta k}) - \frac{\gamma}{\delta}(1 - e^{-\delta k})}{1 - \alpha e^{-\beta k} - \gamma e^{-\delta k}} \right] \tag{92}$$

where δ, γ, α, and β are determined by Eqs. (69), (72), (76), and (77), respectively.

Example.   A facility that processes fresh oranges close to the groves has truckloads of oranges arriving according to a Poisson process at a rate of 85 trucks/day. Packaging is highly standardized and automated with 100 truckloads being processed per day and negligible variation observed. If the oranges cannot be packaged in one day, they are squeezed for juice at less unit profit. If more than 1 percent of the oranges have to be turned into juice, capacity will need to be increased. At 85 percent facility utilization, should there be any concern? At 98 percent utilization?

An M/D/1 model with "impatience" is used to answer these questions, with the reneging feature being the removal of oranges from the packaging option after 1 day in the queue. The data suggest that λ = 85 trucks/day, μ = $1/D$ = 100 trucks/day, and τ = 1 day. Calculating the parameters needed for the asymptotic approximation formulas, we solve for the root of Eq. (69),

$$(85)(e^{\delta(.01)} - 1) - \delta = 0$$

giving $\delta = 31.6687$. Substituting D for $E(s)$ in Eq. (90) gives

$$\gamma = \frac{\left(1 - \dfrac{85}{100}\right)}{(31.67)(.01) - \left(1 - \dfrac{85}{100}\right)} = .9$$

Solving Eqs. (76) and (77) gives

$$\alpha = .85 - .9 = -.05$$

$$\beta = \frac{(-.05)}{\dfrac{(85)(.01)^2}{2(.15)} - \dfrac{.9}{31.67}} = 607.1$$

The steady-state proportion of all oranges that will be squeezed instead of packaged is given by Eq. (83),

$$\pi(\tau) = \pi(1) = \frac{(.15)[(-.05)e^{-590.12} + (.9)e^{-31.67}]}{1 - (.85)[(-.05)e^{-590.12} + (.9)e^{-31.67}]}$$

$$\approx (.15)(.9)e^{-31.67} = 2.38(10)^{-15}$$

or very little chance of it at all. If the plant is operating at 98 percent of capacity, $\gamma = 98$. Then

$$\delta = 4.027 \qquad \gamma = .98667 \qquad \alpha = -.00667 \qquad \beta = 34.129$$

and

$$\pi(1) = .000358 < .01$$

Thus, the plant's capacity need not be increased. The average amount of time a truckload of oranges waits to be processed, when the plant is at 98 percent of capacity is given by Eq. (85),

$$W_q(\tau) = W_q(1) = \frac{\dfrac{-.00667}{34.129} - (-.00667)\left[\dfrac{1}{34.129} + (.98)(1)\right]e^{-34.129}}{1 - (.98)[(-.00667)e^{-34.129} + (.98667)e^{-4.027}]}$$

$$+ \frac{\dfrac{.98667}{4.027} - (.98667)\left[\dfrac{1}{4.027} + (.98)(1)\right]e^{-4.027}}{1 - (.98)[(-.00667)e^{-34.129} + (.98667)e^{-4.027}]}$$

$$= .2332 \text{ day}$$

Two-Moment Approximations. In many practical situations, it is easy to find the mean and variance of the interarrival times or of the service times. This information can be used to form a linear interpolation of a measure of interest by using simpler models to get the measure of interest for a more complex situation. The squared coefficient of variation of the interarrival times, c_a^2, and the service times, c_s^2, are used as weights in the interpolation formula. The two-moment approximation is most useful when

$$0 \leq c_a^2, c_s^2 \leq 2$$

which covers most cases of interest in industrial applications.

The two-moment approximation method can even be used within an asymptotic approximation to ease the computational burden. The parameter evaluation formulas, Eq. (69) for δ and Eq. (72) for γ, may be approximated by

$$\delta \approx \frac{\delta^{(d)}\delta^{(e)}}{(1 - c_s^2)\delta^{(e)} + c_s^2\delta^{(d)}} \tag{93}$$

and

$$\gamma \approx [\gamma^{(e)}]^q[\gamma^{(d)}]^{1-q} \tag{94}$$

where

$$q = \frac{c_s^2\delta}{\delta^{(e)}} \tag{95}$$

$$\gamma^{(e)} = \rho \tag{96}$$

and $\delta^{(e)}$ is given by Eq. (87), $\gamma^{(d)}$ is given by Eq. (90), and $\delta^{(d)}$ solves Eq. (89). Solving Eq. (89) is the only iterative element of this approximation and is solved much easier than the original Eqs. (69) and (72) for general $B(y)$.

A number of results employing a two-moment approximation are now presented for a variety of queuing models.

$GI/G/1$

$$P\{W_q > 0\} = \begin{cases} \rho + \dfrac{(c_a^2 - 1)\rho(1 - \rho)(1 + c_a^2 + \rho c_s^2)}{1 + \rho(c_s^2 - 1) + \rho^2(4c_a^2 + c_s^2)} & \text{if } c_a^2 \leq 1 \\[4mm] \rho + \dfrac{4(c_a^2 - 1)\rho^2(1 - \rho)}{c_a^2 + \rho^2(4c_a^2 + c_s^2)} & \text{if } C_a^2 > 1 \end{cases} \tag{97}$$

$$W_q = \begin{cases} \dfrac{\rho E(s)}{2(1 - \rho)}(c_a^2 + c_s^2)e^{-1(1-\rho)(1-C_a^2)^2/3\rho(C_a^2 + C_s^2)} & \text{if } c_a^2 \leq 1 \\[4mm] \dfrac{\rho E(s)}{2(1 - \rho)}(c_a^2 + c_s^2)e^{-(1-\rho)(C_a^2 - 1)/(C_a^2 + 4C_s^2)} & \text{if } c_a^2 > 1 \end{cases} \tag{98}$$

for c_a^2 not too large and $P\{W_q > 0\}$ not too small.

$M/G/1/k$—*Finite Work Buffer Capacity.* All arrivals are accepted, but work brought by an arrival in excess of buffer capacity is lost.

$$V_q = \text{expected amount of work residing in the buffer} \tag{99}$$

$$= (1 - c_s^2 V_q^{(d)} + c_s^2 V_q^{(e)}$$

where k = maximum amount of work that may reside in the buffer.

$$V_q^{(d)} = k - \left[\sum_{j = 0}^{r} (- 1)^j \frac{(\lambda k/\sigma - \rho j)^j}{j!} e^{\lambda k/\sigma - \rho j} \right]^{-1} \times$$

$$\times \frac{\sigma(1 - \rho)}{\lambda\left[-(r + 1) + \sum_{j=0}^{r} e^{\lambda k/\sigma - \rho j} \sum_{i=0}^{j} (-1)^i \frac{(\lambda k/\sigma - \rho j)^i}{i!} \right]} \tag{100}$$

$$V_q^{(e)} = \frac{\rho E(s)/(1 - \rho) - [E(s)/(1 - \rho) + k]\rho e^{-(1 - \rho k/E(s))}}{1 - \rho e^{-(1 - \rho)k/E(s)}} \tag{101}$$

$k(v)$ = buffer size k such that the proportion of arrivals causing the buffer to overflow is v in the long run

$$= \frac{1}{\sigma} \ln\left[\frac{\sigma \delta \gamma}{\lambda v} \right] \tag{102}$$

an asymptotic approximation using two-moment approximations for δ [Eq. (93)], and γ [Eq. (94)]. Equation (102) is most appropriate when $v \le .01$ and $\rho \le .8$, or $v \le .001$ for any ρ.

   M/G/1/k—Finite Work Buffer Capacity (Revisited).   This model differs from the previous one in that it will not accept any of the work associated with an arrival exceeding the limit, whereas the previous model would accept the portion of the work brought by the arrival that will fit in the buffer.

$p(k)$ = steady-state proportion of all arrivals turned away because of the work capacity limit being met or exceeded by the arrival

$$\approx \gamma e^{-\delta k} \qquad \text{for large } k \tag{103}$$

and, conversely,

$k(v)$ = work capacity limit (or buffer size) that results in a proportion v of arrivals being turned away \hfill (104)

$$\approx \frac{1}{\delta} \ln \frac{\gamma}{v} \qquad \text{for } v \le 10^{-3} \text{ and any } \rho, \text{ or } v \le .01 \text{ and } \rho < .8$$

where two-moment approximations for δ and γ are given by Eqs. (93) and (94), respectively.

$$V_q = \text{the expected amount of work in the buffer having capacity } k \tag{105}$$

$$= (1 - c_s^2)V_q^{(d)} + c_s^2 V_q^{(e)}$$

where

$$V_q^{(d)} = \lambda\left[1 - p^{(d)}(k)\right]\left\{ \frac{D[a(k) - (k - D)p^{(d)}(k)/\sigma]}{1 - p^{(d)}(k)} + \frac{D^2}{2\sigma} \right\}$$

$$a(k) = \frac{\frac{k - D}{\sigma} - \frac{1}{\lambda}}{1 + \rho e^{\lambda(k-D)/\sigma} \sum_{j=0}^{s-1} (-1)^j \frac{(\lambda(k - D)/\sigma - \rho j)^j}{j!} e^{-\rho j}}$$

$$\times \left[-s + e^{-\lambda(k-D)/\sigma} \sum_{j=0}^{s-1} e^{-\rho j} \sum_{i=0}^{s-1} (-1)^i \frac{\lambda(k - D)/(\sigma - \rho j)^i}{i!} \right] \tag{106}$$

where s is an integer and satisfies $(s - 1)E(s) \le (k - D) < sE(s)$ for all ρ, and $p^{(d)}(k)$ is found by evaluating

$$p^{(d)}(k) = \frac{1}{\rho}\left(\rho - 1 + \frac{1}{1 + \rho e^{\lambda(k-D)/\sigma}\sum\limits_{j=0}^{s-1}(-1)^j\frac{[\lambda k/\sigma - \rho(j+1)]^j}{j!}e^{-\rho j}}\right) \qquad (107)$$

where s is an integer and satisfies $sD \le k < (s + 1)D$, and

$$V_q^{(e)} = V_q^{(e)}(0)\rho E(s)\beta_0^{1-\rho}e^{\rho\beta_0}\left\{\frac{k}{(\rho - 1)E(s)}[e^{-\rho} + \gamma(\rho, \rho)\rho^{-\rho}] + \frac{\beta_0^{\rho-1}e^{-\rho\beta_0} - e^{-\rho}}{(\rho - 1)^2}\right.$$

$$\left. - \frac{\gamma(\rho, \rho) - \gamma(\rho, \rho\beta_0)}{(\rho - 1)^2\rho^{\rho-1}} - \frac{\int_{\rho\beta_0}^{\rho}x^{-1}\gamma(\rho, x)\, dx}{(\rho - 1)\rho^{\rho}}\right\} \qquad (108)$$

$$V_q^{(e)}(0) = \frac{1 - \rho}{1 - \rho\beta_0^{1-\rho}e^{\rho(\beta_0 - 1)} - (\rho\beta_0)^{1-\rho}e^{\rho\beta_0}[\gamma(\rho, \rho) - \gamma(\rho, \rho\beta_0)]} \qquad \rho \ne 1 \quad (109)$$

$$= \text{CDF for the amount of work in the buffer evaluated at zero}$$

$$\gamma(\alpha, x) = \text{incomplete gamma function}$$

$$= \int_0^x u^{\alpha-1}e^{-u}\, du \qquad \alpha, x > 0$$

$$\approx \sum_{n=0}^M \frac{(-1)^n x^{\alpha+n}}{(\alpha + n)n!} \qquad M \text{ suitably large}$$

$$\beta_x = e^{-(k-x)/E(s)} \qquad 0 \le x \le k \qquad (110)$$

Example. Housings arrive at a machine center for primary drilling and machining operations according to a Poisson process at a rate of 1.5/hour. The arriving housings are of differing sizes with the amount of work (number of operations) that is to be done on them in proportion to their size. The number of operations per housing is a random variable with mean = 15 and variance = 25. The machine center can perform machining operations at a fairly constant rate of 30/hour. How much space should be provided in the buffer such that the probability that an arriving housing cannot be accommodated is less than 1 percent?

The data suggest that $\lambda = 1.5$ housings/hour, $E(s) = 15$ machining operations/housing, and Var $(s) = 25$. The service rate of the machining center is 30 operations/hour; thus, $\sigma = 30$ operations. The offered load to the center is $\rho = (1.5)(15)/30 = .75$. We use Eq. (104) to answer the question posed and must complete the calculation by using Eq. (96):

$$\gamma^{(e)} = .75$$

and Eq. (87):

$$\delta^{(e)} = \frac{1 - .75}{15} = .01667$$

and solving Eq. (89):

$$\frac{1.5}{30}\left(e^{15\delta^{(d)}} - 1\right) - \delta^{(d)} = 0$$

for

$$\delta^{(d)} = .03668$$

Then use Eq. (90):

$$\gamma^{(d)} = \frac{1 - .75}{(.03668)(15) - (1 - .75)} = .8328$$

Eq. (93):

$$\delta = \frac{(.03668)(.01667)}{(.8889)(.01667) + (.1111)(.03668)} = .03236$$

Eq. (95):

$$q = \frac{(.1111)(.03236)}{.01667} = .2157$$

Eq. (94):

$$\gamma = (.75)^{.2157}(.8328)^{.7843} = .8142$$

and Eq. (104):

$$k(.01) = \frac{1}{.03236} \ln\left[\frac{.8142}{.01}\right] = 135.9 \sim 136$$

Over the long run, if the number of operations associated with an arriving housing is independently sampled from the work size distribution, the expected number of housings that are present when an arriving housing is blocked is $k(v)/E(s) = 136/15 = 9.064$. Since space is to be allocated, the area of the largest housing that the machine center sees is multiplied by $10 > 9.064$ to assure the 1 percent blocking criterion is met.

Parallel Server Approximations.   When c identical parallel servers are added to the complexities already discussed for the single-server case, approximations must be relied on heavily to provide useful results. Analytical results are very limited and for special cases only. Two assumptions are made with respect to the behavior of the queuing process at service completions:

1. At a service completion, if k are left behind in the system, $0 < k < c$, the system acts like an M/G/∞ queue for which there is a server for each customer in the system. For such a system, the time until the next service completion has a distribution given by the steady-state residual life distribution of renewal theory,

$$B_e(t) = \frac{1}{E(S)}\int_0^t [1 - B(x)]\, dx \qquad t \geq 0 \tag{111}$$

2. At a service completion, if $k \geq c$ are left behind, the system acts like a M/G/1 queue with the service rate equal to $c\mu$. Both assumptions are exact if service times are exponentially distributed. And both assumptions will be used in conjunction with asymptotic and two-moment approximation approaches to develop useful results for multiple-server models.

M/G/c.   The analysis required to determine the steady-state system size probabilities at an arbitrary point in time is made much simpler by the two parallel server assumptions listed above. The defining equations for these probabilities, below, allow a recursive solution:

$$P_j = \begin{cases} \dfrac{(c\rho)^j}{j!} P_0 & 0 \le j \le c - 1 \\[2ex] \lambda\alpha_{j-c}P_{c-1} + \lambda \sum\limits_{k=c}^{j} \beta_{j-k}P_k & j \ge c \end{cases} \tag{112}$$

where

$$\rho = \frac{\lambda}{c\mu}$$

$$P_0 = \frac{1}{\sum\limits_{k=0}^{c-1} \dfrac{(c\rho)^k}{k!} + \dfrac{(c\rho)^c}{c!(1-\rho)}} \tag{113}$$

$$\alpha_n = \int_0^x [1 - B_e(t)]^{c-1}[1 - B(t)]e^{-\lambda t} \frac{(\lambda t)^n}{n!} \, dt \qquad n \ge 0 \tag{114}$$

$$\beta_n = \int_0^x [1 - B(ct)]e^{-\lambda t} \frac{(\lambda t)^n}{n!} \, dt \qquad n \ge 0 \tag{115}$$

In general, the integrations in Eqs. (114) and (115) need to be numerically evaluated by using quadrature techniques. It should be noted that P_j for $j = 0, 1, \ldots, c - 1$ are equal to the corresponding P_j for an M/M/c system and, as a result, the probability for a customer waiting for service is

$$P\{W_q > 0\} \approx \frac{\dfrac{(c\rho)^c}{c!(1-\rho)}}{\sum\limits_{k=0}^{c-1} \dfrac{(c\rho)^k}{k!} + \dfrac{(c\rho)^c}{c!(1-\rho)}} \tag{116}$$

which is the well-known Erlang delay probability. Similarly, the expected number in the queue may be related to the corresponding measure of an M/M/c system

$$L_q = \left[(1 - \rho)\gamma \frac{c}{E(s)} + \rho \frac{E(s^2)}{2E(s)^2} \right] L_q^{(e)} \tag{117}$$

where

$L_q^{(e)}$ = expected number in the queue in an M/M/c system

$$= \frac{(c\rho)^c \rho}{(1-\rho)^2 c!} P_0 \tag{118}$$

$$\gamma = \int_0^x [1 - B_e(t)]^c \, dt$$

$$\approx (1 - c_s^2) \frac{E(s)}{c+1} + c_s^2 \frac{E(s)}{c} \qquad \text{if } 0 \le c_s^2 \le 2 \tag{119}$$

and P_0 is given by Eq. (113).

A two-moment approximation for L_q, preferred when $c_s^2 > 1$, is

$$L_q = \frac{1}{2}(1 + c_s^2)\frac{2L_q^{(d)}L_q^{(e)}}{2aL_q^{(d)} + (1 - a)L_q^{(e)}} \tag{120}$$

where

$$L_q^{(d)} = \frac{1}{2}\left[1 + (1 - \rho)(c - 1)\frac{\sqrt{4 + 5c} - 2}{16\rho c}\right]L_q^{(e)} \tag{121}$$

$$a = \begin{cases} 1 & \text{if } c = 1 \\ \dfrac{1}{c - 1}\left[\dfrac{E(s^2)}{\gamma E(s)} - c - 1\right] & \text{if } c > 1 \end{cases} \tag{122}$$

and $L_q^{(e)}$ and γ are given by Eqs. (118) and (119), respectively. Little's formula can be used to find the expected waiting time in the queue, but there may be instances when the expected waiting time in the queue, conditioned on the waiting time being positive, is of interest. A two-moment approximation for this measure is

$$E(W_q|W_q > 0) = \begin{cases} \left(1 - \dfrac{1}{2}\rho - \dfrac{1}{2}\rho^2\right)\gamma + \dfrac{3(\rho - \rho^3)(1 + c_s^2)}{4(1 - \rho)c}E(s) & \text{if } c_s^2 \leq 1 \\ (1 + \rho)\gamma + \dfrac{\rho^2(1 + c_s^2)}{2(1 - \rho)c}E(s) & \text{if } c_s^2 > 1 \end{cases} \tag{123}$$

where γ is given by Eq. (119).

The probability that a customer waits in the queue for a specified amount of time possesses the following asymptotic approximation:

$$P\{W_q > x\} = \frac{(1 - \rho)P\{W_q > 0\}\displaystyle\int_0^x e^{\delta t}[1 - B_e(t)]^{c-1}[1 - B(t)]\,dt}{\rho\delta\displaystyle\int_0^x te^{\delta t}[1 - B(ct)]\,dt}e^{-\delta x} \tag{124}$$

where δ is the positive solution of

$$\lambda\int_0^x e^{\delta y}[1 - B(cy)]\,dy = 1$$

and $P\{W_q > 0\}$ is given by Eq. (116). Equation (124) is most effective when

$$x \geq \frac{E(s)}{\sqrt{c}}$$

and

$$P\{W_q > 0\} \geq .2$$

M/G/c/N—Finite Capacity Queue. The steady-state system size probabilities p_j possess defining Eqs. (112) with the addition of an equation for the capacity state,

$$P_N = \rho P_{c-1} - (1 - \rho)\sum_{k=c}^{N-1} P_k \tag{125}$$

Probabilities may be computed by finding $[P_j/P_0]$, iteratively, using Eqs. (112) and (125) and then finding P_0 by solving

$$\sum_{j=0}^{N} [P_j/P_0] = \frac{1}{P_0}$$

In turn, the probability of waiting in the queue for a time greater than t is found by using

$$P\{W_q > t\} = \frac{1}{1 - P_N} \sum_{j=c}^{N-1} P_j \sum_{k=0}^{j-c} e^{-c\mu t} \frac{(c\mu t)^k}{k!} \qquad t \geq 0 \qquad (126)$$

M/G/c/N/N—Multiple Repairer Model. Each of N machines break down according to an identical Poisson process with rate λ. Each of the c repairers possess an identical service time distribution $B(t)$. By using the simplifying parallel server assumptions, discussed earlier, the steady-state system size (number of machines needing repair) probabilities at an arbitrary point in time are

$$P_k = \begin{cases} \dfrac{N!}{(N-k)k!} [\lambda E(s)]^k P_0 & 0 \leq k \leq c-1 \\[2ex] (N-c+1)\lambda \alpha_{ck} P_{c-1} + \displaystyle\sum_{j=c}^{k} (N-j)\lambda \beta_{jk} P_j & c \leq k \leq N \end{cases} \qquad (127)$$

where

$$\alpha_{ck} = \int_0^x [1 - B_e(t)]^{c-1}[1 - B(t)]\phi_{ck}(t)\, dt \qquad (128)$$

$$\beta_{jk} = \int_0^x [1 - B(ct)]\phi_{jk}(t)\, dt \qquad (129)$$

$$\phi_{jk}(t) = \left(\frac{(N-j)!}{(N-k)!(k-j)!}\right)(1 - e^{-\lambda t})^{k-j} e^{-\lambda t(N-k)} \qquad t > 0,\ c \leq j \leq k \qquad (130)$$

With these probabilities, it is possible to directly calculate the expected number of down machines

$$L_q = \sum_{k=c+1}^{N} (k-c)P_k \qquad (131)$$

and the average repairer utilization

$$U_c = \frac{E(\text{number of busy repairers})}{c}$$

$$= \left[\sum_{k=0}^{N} (N-k)\lambda P_k\right]\frac{E(s)}{c} \qquad (132)$$

GI/G/c. Although attaining tractable results is a formidable challenge, a two-moment approximation for the expected waiting time in the queue provides good results for $0 \leq c_a^2$, $c_s^2 \leq 1$. The interpolation formula is the weighted harmonic mean of the expected waiting time in the queue for an M/M/c system, $W_q(\text{M/M/c})$, an M/D/c system, $W_q(\text{M/D/c})$, and a D/M/c system, $W_q(\text{D/M/c})$, and has the form

$$W_q \approx \frac{C_a^2 + C_s^2}{\dfrac{1 - C_a^2}{W_q(D/M/c)} + \dfrac{1 - C_s^2}{W_q(M/D/c)} + \dfrac{2(C_a^2 + C_s^2 - 1)}{W_q(M/M/c)}} \tag{133}$$

where

$$W_q(M/M/c) = \frac{\dfrac{(c\rho)^c}{c!c\mu(1 - \rho)^2}}{\displaystyle\sum_{k=0}^{c-1} \dfrac{(c\rho)^k}{k!} + \dfrac{(c\rho)^c}{c!(1 - \rho)}} \tag{134}$$

$$W_q(M/D/c) \approx \tfrac{1}{2}[1 + C(c, \rho)]W_q(M/M/c) \tag{135}$$

$$W_q(D/M/c) \approx \tfrac{1}{2}e^{-2(1-\rho)/3\rho}[1 - 4C(c, \rho)]W_q(M/M/c) \tag{136}$$

$$C(c, \rho) = (1 - \rho)(c - 1)\frac{\sqrt{4 + 5c} - 2}{16c\rho} \tag{137}$$

A different two-moment approximation is possible for the expected waiting time in the queue, conditioned on the fact that the arriving customer has a positive waiting time, which is quite similar to Eq. (123) for the M/G/c queue,

$$E(W_q|W_q > 0) = \begin{cases} \left(1 - \dfrac{1}{2}\rho - \dfrac{1}{2}\rho^2\right)\left(\dfrac{1 - C_s^2}{c + 1} + \dfrac{C_s^2}{c}\right)E(s) + \\ \dfrac{(C_a^2 - 1)(1 - \rho) + (3\rho + \rho^3)(1 + C_s^2)}{4(1 - \rho)c}E(s) & \text{if } 0 \leq c_a^2, c_s^2 \leq 1 \\ (1 + \rho)\left(\dfrac{1 - C_s^2}{c + 1} + \dfrac{C_s^2}{c}\right)E(s) + \dfrac{\rho^2(C_a^2 + C_s^2)}{2(1 - \rho)c}E(s) & \text{otherwise} \end{cases} \tag{138}$$

Example. A PC clone manufacturer offers a warranty service that allows buyers to send back their defective PCs for free repairs during the warranty period. Defective units arrive to the service area according to a Poisson process at a rate of 300 per month. Three technicians are on duty at all times and work independently but have a common service time distribution with $E(s) = .00667$ month and Var $(s) = .0000578$. An analysis is required on how responsive the service facility is to customer needs.

The data suggests that $\lambda = 300$, $\rho = \lambda E(s)/c = (300)(.00667)/3 = .667$, $C_s^2 = $ Var $(s)/E(s)^2 = .0000578/.00667 = 1.3$. To estimate the expected number of units in the queue, calculations are required by using in turn

Eq. (113):

$$P_0 = \frac{1}{1 + \dfrac{(3)(.667)}{1} + \dfrac{[(3)(.667)]^2}{2} + \dfrac{[(3)(.667)]^3}{3!(1 - .667)}} = .1111$$

(the proportion of time all technicians are idle),

Eq. (119):

$$\gamma = (1 - 1.3)\left[\frac{1}{150(4)}\right] + 1.3\left[\frac{1}{150(3)}\right] = .0024$$

Eq. (118):

$$L_q^{(e)} = \frac{[(3)(.667)]^3(.667)}{3!(1 - .667)^2}(.1111) = .8889$$

Eq. (122):

$$a = \frac{1}{2}\left[\frac{(.0001)}{(.0024)(.00667)} - 3 - 1\right] = 1.1395$$

Eq. (121):

$$L_q^{(d)} = \frac{1}{2}\left[1 + (1 - .667)(2)\frac{\sqrt{4 + (5)(3)} - 2}{(16)(.667)(3)}\right](.8889) = .4663$$

and, finally, using Eq. (120), we have the desired result

$$L_q = \frac{1}{2}(1 + 1.3)\left[\frac{(2)(.4663)(.8889)}{(2)(1.1395)(.4663) + (1 - 1.1395)(.8889)}\right] = 1.0155 \text{ PCs}$$

Using Eq. (116), we find the probability of an arriving PC waiting for a technician to become free,

$$P\{W_q > 0\} = \frac{[(3)(.667)]^3}{3!(1 - .667)}(.1111) = .4444$$

If a PC must wait for service, the expected length of the wait is given by Eq. (123),

$$E\{W_q|W_q > 0\} = (1 + .667)(.0024) + \frac{(.667)^2(1 + 1.3)(.00667)}{2(1 - .667)(3)}$$

$$= .0074 \text{ months} \approx \tfrac{1}{4} \text{ day}$$

BIBLIOGRAPHY

Bhat, U. N., *Elements of Applied Stochastic Process,* 2d. ed., New York, 1984.

Conway, R. W., and W. L. Maxwell, "A Queuing Model with State Dependent Service Rates," *Journal of Industrial Engineering,* vol. 12, 1961, pp. 132–136.

Giffin, W. C., *Queuing: Basic Theory and Applications,* Grid Publishing, Columbus, Ohio, 1978.

Gross, D., and C. M. Harris, *Fundamentals of Queuing Theory,* 2d. ed., Wiley, New York, 1985.

Haight, F. A., "Queuing with Balking," *Biometrika,* vol. 44, 1957, pp. 360–369.

Hillier, F. S., R. W. Conway, and W. L. Maxwell, "A Multiple Server Queuing Model with State Dependent Service Rate," *Journal of Industrial Engineering,* vol. 15, 1964, pp. 153–157.

Jaiswal, N. K., *Priority Queues,* Academic Press, New York, 1968.

Kendall, N. K., "Stochastic Processes Occurring in the Theory of Queues and Their Analysis by the Method of Imbedded Markov Chains," *Ann. Math. Stat.,* vol. 24, 1953, pp. 338–354.

Tijms, H. C., *Stochastic Modeling and Analysis: A Computational Approach,* Wiley, New York, 1986.

Special Industry Applications

CHAPTER 1
HEALTH SERVICES

Vinod K. Sahney
Corporate Vice President—Planning and Marketing
Henry Ford Health System
Detroit, Michigan
Professor of Industrial and Manufacturing Engineering
Wayne State University
Detroit, Michigan

Swatantra K. Kachhal
Professor and Chair
Department of Industrial and Systems Engineering
The University of Michigan—Dearborn
Dearborn, Michigan

During the past 200 years, hospitals have acted as a hub for the health care delivery system. Most hospitals were founded as charitable organizations to serve the needs of communities. Although there were a number of facilities on record taking care of patients in the late 1600s, the first incorporated hospital in America was Pennsylvania Hospital, located in Philadelphia—in the year 1751. Benjamin Franklin aided in the design of the hospital and served as its first president from 1755 to 1757. The early hospitals were founded to house people during epidemics. In early 1873, there were only 178 hospitals and fewer than 35,000 beds in the United States. The growth of medical technology changed the role of hospitals from public health functions to patient treatment facilities. The introduction of anesthesia in the 1840s allowed more serious operations to be performed. The understanding of infection, the transmission of infection from doctor to patients, and from patients to other patients, in the period 1850 to 1900, further changed the role of hospitals. Mortality rates from infection dropped significantly during this period. The discovery of the x-ray in 1895 further introduced technology into the hospitals. The introduction of technology also changed the role of hospitals from depositories for poor people to community resources, where high-cost equipment could be shared by community physicians. The number of hospitals grew rapidly, and by 1909 there were 4300 hospitals with over 420,000 beds, a major increase from the 1870s level (Sherlock, 1985).

Gradual growth occurred in the hospital industry from 1910 to 1965. A major construction boom occurred with the passage of the Hill-Burton Law, enacted in 1948. This program was designed to provide matching monies to towns without a community hospital. This federal legislation was responsible for most of the small (fewer than 100-bed hospitals) that developed during the 1950s and 1960s. The next major growth period for hospitals occurred with the passage of Medicare legislation in 1965. The legislation provided care for the nation's elderly (over 65 years). A companion legislation introduced the Medicaid program, meant to cover the poor. By 1979, the total number of hospitals reached 7099, of which 377 were federal and 6722 nonfederal (Mahon, 1978).

HOSPITAL OWNERSHIP

Government-owned hospitals account for 50 percent of the total number of hospitals in the country and include federal, state, county, and city hospitals. Federal hospitals include Army, Navy, Air Force, and Veterans Administration hospitals.

The nongovernment hospitals' ownership is divided among church-related groups, community, not-for-profit hospitals, and investor-owned or for-profit hospitals. Not-for-profit community hospitals constitute the bulk of the hospitals in this country. These hospitals are owned and controlled by the community in which they are located via a voluntary board of community leaders. For-profit hospitals originally were owned by physicians and were small, usually fewer than 100 beds. During the 1960s and 1970s, several investor-owned chains were formed. Notable among them are the Hospital Corporation of America (HCA) and Humana, both of which are traded on the stock exchange.

HOSPITAL REGULATIONS

Hospital operations are regulated by numerous state, federal, county, and city agencies, numbering over several hundred. Hospitals are generally licensed by an agency of the state, usually the health department. These regulations primarily cover physical facilities, management, medical staff, safety, drugs and pharmaceuticals, and personnel, as well as medical records. Most states also regulate the building and expansion of health care facilities through a "certificate-of-need" program, commonly operated by the state department of public health. The objective of this regulatory effort is to control the expansion of facilities and services in the public interest.

During the 1960s and 1970s, many health care regulations were passed. The focus of most of these legislations was the control of hospital growth through the formation of health-care planning agencies and professional standards review organizations (PSRO). The PSRO review determines whether services are medically necessary.

PLANNING

After the enactment of the Hill-Burton program in 1948, many voluntary planning agencies developed at the state level. New York state was the first in 1964 to enact a mandatory health-planning law. This law required prior state approval before any hospital construction or expansion could take place. The criterion for expansion was demonstrated need, and a new facility had to be consistent with the state plan for the delivery of health care. After the Medicare program was enacted, a number of states developed certificate-of-need programs. Any hospital planning to expand or build a new facility required approval from the state planning agency (Mahon, 1978).

In 1974, the U.S. Congress enacted Public Law 93-641, labeled the *National Health Planning and Resources Development Act.* This law created organizations at the national, state, and local levels to conduct health services planning. The law required hospitals which participate in Medicare or Medicaid to obtain planning approval for capital expenditures over $100,000. A huge bureaucracy developed at local, state, and national levels to implement this law. Most experts agree that this legislation was a dismal failure, as evidenced by national hospital occupancy levels of 60 percent (Steinwelds, 1981). Most of the states have revised or drastically reduced the scope of this law. Many states have changed the review limit to $1 million. In other states, only the construction of new hospitals or new beds is reviewed.

HOSPITAL REVENUES

In most nongovernment hospitals, the principal sources of revenue are patient services. The past 30 years have seen a dramatic change in health care revenue sources with increasing numbers of people covered by insurance. The largest of these programs is the Medicare program which covers people over the age of 65 and people eligible for social security benefits. A typical hospital receives approximately 35 percent of its revenue from Medicare patients. This program is managed by the federal government through an agency called the Health Care Financing Administration (HCFA). From the enactment of the Medicare legislation in 1967 until 1983, hospitals were paid by Medicare on the basis of cost. To control the rising cost, a variety of methods were used, including cost increase ceilings, but without much success. Finally, in 1983, a revolutionary new means of payment was introduced, known in the industry as the diagnostic related group (DRG) payment system. Under this system, hospital cases are divided into 487 different classes based on diagnosis, age, complications, and the like. For each of these diagnostic groups, a predetermined payment level is specified. If the hospital can deliver care at a cost lower than the payment, the hospital is entitled to keep the difference. On the other hand, if the hospital costs are higher, the hospital incurs a loss. Many of the commercial insurance companies such as Blue Cross and Blue Shield, as well as state Medicaid programs, have adopted this method of payment for the hospitals.

KEY ISSUES AND TRENDS

Between 1965 and 1980, the hospital industry fell under the umbrella of multiple regulatory approaches. Most of these regulatory approaches failed to control the increases in healthcare expenditures (Steinwelds, 1981). More recently, the industry has moved toward a competitive model. During the past decade, a few key trends have emerged as summarized below.

Economics.   Health care expenditure increases have continued at an average rate approaching 10 percent. The rate of expenditure increase reached a high of 15 to 30 percent in 1983 and a low of 7.0 percent in 1988. The length of hospital stays dropped from 1983 to 1988 and has now stabilized. Hospital admission rates also dropped during this period. Many of the procedures done on an inpatient basis are now being performed on an ambulatory care basis. In contrast to the decline in the inpatient setting, outpatient and ambulatory care is growing at a rate of over 15 percent. Most analysts believe that inpatient admissions and length of stays will stabilize, but outpatient and ambulatory care will continue to grow over the next decade.

Financing.   The private health insurance market has been shifting from indemnity insurance to managed health care. Health maintenance organizations (HMOs) grew at a rapid pace during the past 10 years and now account for 30 million people in some 500 different plans. Many of the analysts predict that by 1995, HMOs and Preferred Provider Organizations (PPOs) will account for 70 percent of the private health-care insurance business. Recently, employers have been changing their health care insurance strategy from providing benefits to providing fixed payments toward health care benefits. More coinsurance and deductibles are being introduced.

At the federal level, Medicare has reduced payments to the providers by recalibrating diagnostic related groups (DRG) payments downward, raising deductibles for the elderly, reducing payments for medical education, and limiting price increases to well below the cost increases in the industry. Current trends point to continued federal retrenchment with projected reductions in capital cost reimbursements and medical education costs (Sahney, 1986).

The Medically Indigent.   The number of people who have neither medical insurance nor coverage by federal and/or state programs continues to rise. Latest estimates indicate that some 37 million people are medically indigent. In the past, hospitals have cross-subsidized indigent care. But as the industry has moved to prospective payment and fixed payment systems, hospitals are unable to transfer the costs of indigent care to other payers. As a result, hospital net patient margin has been decreasing and in 1988 was close to 0 percent.

Medical Staff.   During the past 30 years, there has been a dramatic increase in the number of physicians. Since 1950, the number of physicians has grown 150 percent. The number of medical school graduates has increased from approximately 7000 in 1960 to 17,000 in 1985. Physicians in active practice now number over 600,000.

The number of registered nurses (RNs) has grown from 750,000 in 1970 to over 1.2 million now. Even with this growth, hospitals are currently experiencing tremendous shortages in RN staffing especially in such areas as intensive-care units, special-care units, and emergency rooms. The nursing shortage can be attributed to the high percentage of nurses dropping out of the nursing field. In addition, nurses have many other opportunities within health care aside from the inpatient setting, including such areas as ambulatory care, utilization review, quality assurance, home health agencies, and ambulatory surgery. Nurses have found that these positions do not require night shift rotation or weekend coverage. With the baby boom age group approaching middle age, the nursing shortage is projected to continue in critical areas.

Medical Liability.   The health care industry has gone through several medical malpractice crises in the last decade. Nationwide, the cost of medical malpractice has increased to over $3 billion per year. In some years, malpractice insurance costs jumped 30 to 40 percent. The most common claims deal with surgical cases, followed by claims for improper treatment or failure to diagnose. Certain specialties have been hit harder than others; these include obstetrics and gynecology, orthopedics, and neurosurgery.

Industry Structure.   The health care industry has been experiencing two major structural changes in response to the environmental changes facing the industry. These changes are (1) diversification and (2) industry consolidation. Hospital corporations are increasingly defining their mission as *health care* as opposed to *inpatient hospital care.*

Increasingly, hospitals are developing ambulatory care programs which include ambulatory care clinics and ambulatory surgical centers. Other areas of growth include home health care and durable medical equipment.

Two major reasons for the growth of ambulatory care facilities are the growth of HMOs and the need for hospitals to fill beds. Hospitals have developed free-standing ambulatory care centers as a means of penetrating new markets and gaining referral to the inpatient facilities. The growth of HMOs requires that they be geographically accessible to the population, and free-standing primary care centers are a means of providing accessible primary care services.

The other major trend is consolidation within the industry. Increasingly, hospitals are joining together to form stronger economic entities. In Detroit, Cottage Hospital, Kingswood Hospital, Wyandotte Hospital, Henry Ford Hospital, and Health Alliance Plan have joined to form the Henry Ford Health System. Similar consolidations are taking place all across the country. Industry consolidation and diversification will continue and will gain momentum in the coming years.

INDUSTRIAL ENGINEERING IN HEALTH SERVICES

The use of industrial engineering techniques in health services started with the use of *methods improvement,* which was part of the teachings of Frederick W. Taylor, commonly known as the *principles of scientific management* (Taylor, 1911). Frank Gilbreth is considered to be the first person to use methods improvement in a hospital situation by applying his motion study techniques to surgical procedures (Nock, 1913; Gilbreth, 1916). In the 1940s, Lillian Gilbreth urged hospitals to benefit by the use of industrial engineering tools and techniques (Gilbreth, 1945; Gilbreth, 1950). In 1951–52, Lillian Gilbreth, Ruth Kuehn, and Harold Smalley collaborated in an organized effort to apply methods improvement to the entire hospital organization. A result was a 2-week workshop conducted at the University of Connecticut in July 1952 (Smalley, 1982).

Because of the interest shown by a large number of hospital administrators in the techniques of methods improvement, the American Hospital Association (AHA) created a Committee on Methods Improvement in 1952. This committee prepared several papers on methods improvement activities and published them in its interim report in 1954. By the late 1950s, hospitals, the American Hospital Association, and various universities started promoting methods improvement by offering in-service programs at hospitals, workshops around the country, and various courses in education curricula. Industrial engineering courses were offered in hospital administration curricula. Gradually, other industrial engineering techniques were studied and applied to various hospital problems. In 1961, the Hospital Management Systems Society (HMSS) was founded in Atlanta, with Harold Smalley serving as its first executive director. In 1964, the national offices of HMSS moved to the American Hospital Association building in Chicago. In 1987, recognizing the important role played by information systems in health care, the society was renamed the Health Care Information and Management Systems Society. The Institute of Industrial Engineers also recognized the expanding role of industrial engineering techniques and formed a hospital division in 1964. This section changed its name to the Health Services Division in 1977, reflecting the broader scope of the field. In 1988, the Society for Health Systems was formed to replace the Health Services Division.

During the 1970s and 1980s, the use of industrial engineering techniques in health-care systems continued to grow. Many industrial engineers obtained employment in hospitals and health systems, while others worked as consultants in health care. The use of the term *management engineering* to represent industrial engineering as practiced in the health-care field became common. A detailed history of the development of hospitals and hospital management engineering is covered by Smalley in his book (Smalley, 1982).

Some of the industrial engineering techniques which have been used in health care systems are

- Methods improvement and work simplification
- Performance and productivity measurement
- Staffing and scheduling
- Work sampling
- Queuing and simulation modeling
- Optimization
- Personnel management

  - Job analysis
  - Wage incentives
  - Merit rating
  - Suggestion plans

- Variability analysis and control
- Demand forecasting

- Production and inventory control
- Facilities planning
  - Layout
  - Handling systems
  - Warehousing

- Labor and budget control
- Quality analysis, control, and improvement
- Economic analysis
- Project management
- Capacity analysis
- Product management
- Information systems and request for proposal (RFP) process

Applications of some of the techniques listed above will be discussed, using various hospital departments as examples. A brief introduction will be given for each of the hospital departments used in the examples. The emphasis will not be on the techniques but rather on the application of them in health care. Details about the industrial engineering techniques are given in the other chapters of this handbook. A beneficial description of various hospital departments is available in the book *Hospital Department Profiles* (Goldberg and Denoble, 1986).

METHODS IMPROVEMENT AND WORK SIMPLIFICATION

These are the industrial engineering tools and techniques which are used to improve the work methods applied by health care employees in the delivery of health care. The work methods under consideration are those utilized in direct health care as well as in support activities. Management engineers use a number of different charts to document a process or work method. These charts facilitate the critical evaluation of the various steps of the process and lead to the identification and correction of flaws in the process. Some of these charts are

- Flowchart
- Flow process chart
- Flow diagram
- Paperwork simplification chart

Flowcharts are very commonly used to track an existing process for evaluation or to chart a new process to be implemented. They show each step of the process and appear very similar to the flowcharts associated with computer programs; they show the decision points as well as the various possible courses of action.

Application in Central Supply Departments.   Figure 1.1 shows the flowchart for the distribution of items from the central supply department in a hospital (Kelliher and Meling, 1981). A central supply department in a hospital may also be called *central service department, central medical supply department,* or *central processing department.* This department is responsible for supplying almost all of the nondrug medical supplies to the nursing units, surgery, emergency room, laboratory, radiology, and outpatient department. The department personnel are responsible for supplying, retrieving, sterilizing supplies such as needles, syringes, trays, instrument sets, operating packs, and the like. Use of disposable supplies has changed the nature of some of the activities, since these items need only be

sorted and distributed through the central supply department without requiring sterilization. Some of the basic functions performed are instrument management, purchasing, storage, and distribution. One key issue for the success of such a department is the availability
of all the instrument packs and supplies when needed by user departments. Usually, it is
the responsibility of the central supply department to stock various areas on a par-level
basis and to keep appropriate inventories on hand to meet unexpected demands. The flowchart clearly shows the various steps in the distribution, charging, and restocking process,
and critical examination of each step can be used to make improvements.

Flow process charts show operations, inspections, transportation, storages, and delays
associated with a process. A flow diagram is a layout of the department showing the locations where various activities take place. Paperwork simplification charts are beneficial
for charting the uses, functions, and information flow taking place. They focus on the efficient use of paperwork to provide the needed control and communication. For example,
they can be used to chart the processing done on a service memo and its copies following
an outpatient visit.

STAFFING

Staffing dictates the number of employees of a given skill level needed in the department
to meet a given demand for services. Scheduling determines the days, shifts, and hours
each individual is assigned to work.

The determination of the staffing level requires the listing of all the different tasks done
by a certain skill level in the department and the measurement of the work content associated with each task. The total number of hours of work to be performed by skill level is
determined for a given demand level, which then forms the basis for staffing level after
demand variability, desired coverage, scheduling constraints, and the like are factored in.

The measurement of work content associated with each task can be done by using common work measurement techniques such as stopwatch time study, predetermined motion
time systems, work sampling, or a combination of these techniques. These measures of
work content are commonly called *staffing standards* and represent the standard hours of
work associated with a task. These measured staffing standards can be used to determine
the staffing needs for anticipated demand level or to determine staff utilization for an actual demand level. An inherent drawback to the use of measured staffing standards is that
the process is very complex and time-consuming. Moreover, any time the method
changes, the standards must also be changed.

An alternative to the use of measured staffing standards is the employment of one of
the various standard staffing methodologies available in the health care industry. These
methodologies describe the typical tasks which are performed in a particular hospital department and assign a standard time to each of these tasks. Some of these systems include
a large number of different hospital departments. One such system is called the *resource
monitoring system* (RMS) and was developed by the Hospital Association of New York
State. Under these methodologies, the departmental tasks are classified as constant or
variable. Constant tasks are those tasks which are not directly related to the departmental
output, that is, are independent of the level of demand for services. Variable tasks are
those which are directly related to the output of the department. For example, in the elec-
trocardiograph (EKG) department, a constant task element may be the daily cleaning of
the machine while a variable element is the actual performance of a test.

The total department workload in standard hours can be determined by:

$$T = C + V_1 T_1 + V_2 T_2 + \cdots + V_n T_n$$

where T = total department workload in standard hours for a day
   C = total time consumed per day by all the constant tasks in standard hours
   V_i = number of times variable task i is done during the day ($i = 1, 2, \ldots, n$)

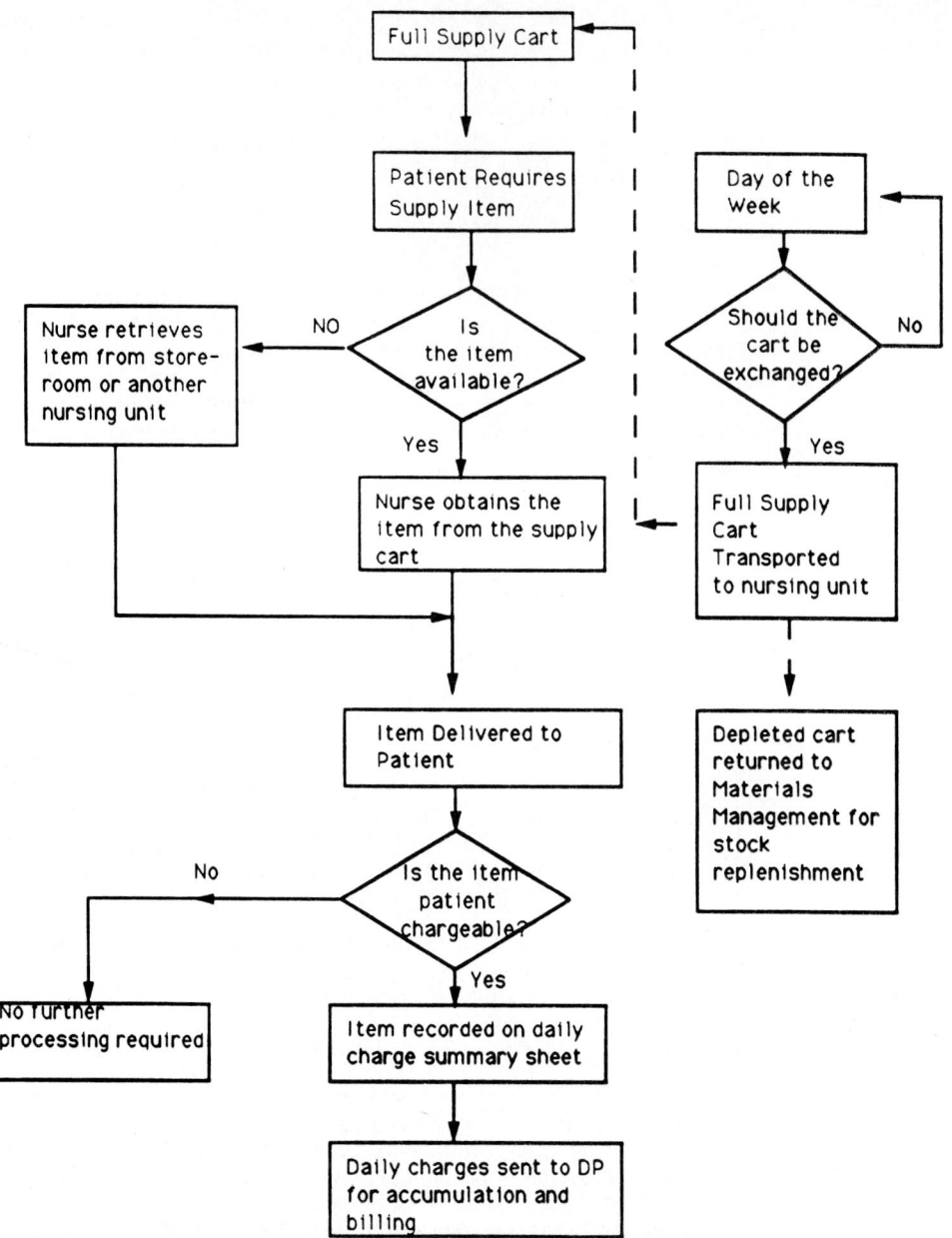

FIGURE 1.1 Flowchart for the distribution of items from the central supply department in a hospital. (*Reprinted with permission from Kelliher, Matthew E., and Robert T. Meling, "Development and Evaluation of a Computer Assisted Supply Distribution System," Proceedings of the 12th Annual Conference of the Health Services Division of the AIIE, New Orleans, 1981.*)

T_i = standard hours associated with the performance of variable task i once (i = 1, 2,..., n)

n = total number of variable tasks

$$\text{Department efficiency} = \frac{T}{\text{actual hours spent}}$$

$$\text{Theoretical FTE needed} = \frac{T}{\text{hours available per FTE per day}}$$

The actual FTE needed will depend on demand variability, coverage issues, scheduling constraints, skill level requirements, and similar factors. One approach is to determine the maximum efficiency level E which could be expected because of these factors. Actual FTE needed would then be given by:

$$\text{Actual FTE needed} = \frac{T}{\text{hours available per FTE per day} \times E}$$

The basic steps needed to apply these methodologies are

1. Document the work load mix for the department for a selected period of 2 to 4 weeks.
2. Check the methods and procedures used in the department against those given in the methodology.
3. Verify the standard times for various tasks as given in the methodology.
4. Obtain standard times for tasks not available in the methodology using work measurement methods such as stopwatch time study.
5. Compute the total work load in standard hours for the study period.
6. Divide the total work load by the actual hours worked to determine efficiency.
7. Compute the efficiency each week to monitor deviations and to determine whether staffing adjustments are needed.

A Case Study in the Pharmacy Department.   Foley and Kubel (1987) applied the workload indicators developed and published by the Canadian Society of Hospital Pharmacists. The pharmacy department in a hospital serves inpatients and outpatients by filling the prescriptions written by the physicians. The major functions performed are the dispensing of medicine, maintenance of medication administration records, maintenance of medication inventory at nursing units and other hospital departments, preparation and provision of intravenous solutions to nursing units, preparation and delivery of unit doses to nursing units, purchase of drugs, and warehousing of drugs. The management engineers are involved in the analysis of overall work flow, staffing, purchasing and inventory analysis, implementation of unit dose systems, charging procedures, paperwork analysis, and evaluation of pharmacy information systems. Figure 1.2 shows the computation of total workload associated with the preparation and delivery of the total parenteral nutrition (TPN) units per day in a hospital pharmacy.

SCHEDULING

Scheduling in health care systems can take two forms: (1) the scheduling of personnel in a department based on the varying demand for services during the various hours of the day while providing equitable scheduling for each individual and (2) the scheduling of work, which may be a possibility in some departments, in order to balance the work load among days of the week and hours of the day, matching the work load with staffing levels.

In personnel scheduling, the staff is assigned to a specific pattern of workdays and off-days. There are basically two types of scheduling patterns—cyclical and noncyclical. In cyclical patterns, each individual's work pattern is repeated after a given number of weeks. Usually, a number of workers share the same cycle. The advantages of the cyclical schedule are the staff members' knowledge of their predictable long-term schedules, and the need for changes only when the work load—that is, the staffing requirements— changes significantly. The disadvantage is the inflexibility of the cyclical schedule to accommodate both demand fluctuations and individual needs of workers.

Ref. no.	Element	Workload indicator	Skill level	Time/ standard	Volume/ day	Total
6.2.3	Calculate TPN orders	TPN orders	RPh	1.68	2	3.36
6.2.5	Prepare TPN solution level	TPN units	RPh	1.21	3	3.63
6.2.6	Prepare clean environment and personnel	Occurrences	RPh	5.34	2	10.68
6.2.7	Obtain and assemble pharmacy items	TPN units	RPh	1.17	3	3.51
6.2.11	Prepare TPN solutions	TPN units	RPh	16.2	3	48.6
6.2.13	Inspect TPN solutions	TPN units	RPh	0.65	3	1.95
6.2.14	Affix labels	TPN units	RPh	0.54	3	1.62
6.2.15	Verify TPN solutions	TPN units	RPh	0.67	3	2.01
6.2.17	Record TPN solutions	TPN orders	RPh	0.64	2	1.28
6.2.18	Cleanup	Occurrences	RPh	3.64	2	7.28
6.2.19	Transfer TPN solutions to delivery area	TPN units	RPh	0.16	3	0.48
6.2.20	Deliver TPN solutions	Ward chart	Tech	3.04	4	12.16
6.4.4	Monitor laboratory orders	Reviews	RPh	1.8	2	3.6

Total min = 100.16
Total TPN units = 3

Total TPN minutes = 100.16
Total TPN units 3   = 33.4 min. per charged TPN solution
= 0.57 hours charged per dose

FIGURE 1.2   Computation of total workload per day associated with TPN units at a hospital pharmacy. (*Reprinted with permission from Foley, Michael F., and Jeffrey E. Kubel, "Comprehensive Pharmacy Department Workload Analysis System—A Case Study,"* Proceedings of the 1988 Annual Healthcare Systems Conference, *vol. 1, 1988.*)

In noncyclical patterns, a new schedule is generated for each scheduling period (usually 2 to 4 weeks) and is based on expected demand and available staff. This approach provides flexibility but is time-consuming and allows little advance planning for workers. Another pattern which can be used allows for four 10-hour days per week or 12-hour shifts or flex-scheduling. Personnel schedules are developed either by using a heuristic, trial-and-error approach or by using some optimization technique. There are several computerized systems available using both types of approaches.

Nurse Scheduling.   The nursing department is responsible for providing direct patient care to all the inpatients and outpatients in conjunction with the physicians. The nursing director is responsible for nursing on medical and surgical inpatient units, specialty units, and, in some cases, in the emergency room, operating room, and ambulatory care. The role of management engineers in a nursing department may involve methods analysis on a nursing unit, quality of care, staffing analysis, scheduling, and computerized nursing information system development. Management engineers are also involved in the development of patient classification systems based on patient acuity. The patient classification system determines the staffing needs relative to the patient census on a nursing unit and the acuity level of these patients.

Warner (1976) developed a computer-aided system for nurse scheduling which maximizes an expression representing the quality of schedule subject to a constraint that minimum coverage be met. This system will be discussed further under "Optimization." Randhawa and Sitompul (1990) provide a detailed review and bibliography of the various nurse-scheduling models which have been developed. They suggest the development of a decision support system which allows the use of a structured model as well as subjective assessments in an interactive mode. This system could provide optimal schedules while allowing the flexibility and speed of heuristic and representational procedures of artificial intelligence.

Work Scheduling.   Some of the areas where work scheduling is possible are elective admission scheduling, case scheduling in operating rooms, appointment scheduling in clinics, and scheduling for radiological procedures and other testing. The most common approach to work scheduling is *peak load analysis,* where the work load and the associated staffing requirements are plotted by the day of the week and hour of the day to identify the peaks and the valleys. Attempts are made to smooth out the peaks and valleys in demand. Another approach is to study the statistical distribution of demand by hour of the day and staff to meet the demand a certain percentage of time (say 90 percent). After the staff schedule is determined, the work schedule basically follows it. Some key issues associated with work scheduling are no-show rates, overbooking, and block scheduling, which must be analyzed to develop appropriate policies. The computerized systems basically allow the computerization of manual appointment books with the capacity to check the availability of resources needed for the task to be performed. These systems can search for available appointments meeting the time restrictions of the patient.

Case Scheduling in the Operating Room.   The Operating Room, or Surgical Department as it is sometimes called, consists of the pre-op room, operating rooms, and recovery room. The surgery may be performed on inpatients who are admitted at least 1 day prior to surgery, outpatients who come in for minor procedures to be done under local anesthetic, or ambulatory surgery patients who have certain procedures done under general anesthesia and are discharged the same day. The role of management engineer in a surgical department may involve methods analysis to reduce delays and improve patient flow, quality, and productivity; staffing analysis; case scheduling analysis; layout development; materials management; and informations system selection or development.

Kachhal and Koch (1989) discuss the functional requirements associated with scheduling a case in the operating room and evaluate the capabilities of the software systems available on the market for meeting these requirements.

In order to carry out the case scheduling function effectively, software should have the capability to:

- Schedule cases a predetermined number of weeks in advance
- Schedule cases for a duration based on the history of the surgeon-specific and procedure-specific times
- Block schedule by surgical speciality
- Restrict scheduling of certain cases in certain rooms
- Check the availability of specific surgeon, anesthesiologist, nurse, and equipment prior to scheduling a case
- Check credential information prior to scheduling a case

There are a number of software programs available which are able to meet most of these requirements. Management engineers are often involved in the development of function requirements in conjunction with the department personnel to be used in a request for proposals to potential vendors.

Another issue in case scheduling is the determination of the optimal sequence in which a given group of cases assigned to a surgical suite must be performed and the assignment of the estimated starting times to these cases. Weiss (1990) discusses a procedure for accomplishing this using an analytical as well as a simulation model while balancing the operating room idle time costs and the surgeon's waiting time costs.

QUEUING AND SIMULATION MODELING

There are a number of situations in health care systems where the customers must wait for service from one or more servers. Some examples of these situations are

- Outpatients waiting for service from a receptionist, nurse, or physician in a clinic
- An emergency room patient awaiting service from a receptionist, nurse, or physician
- A surgeon waiting for the operating room to become available in order to perform the next surgical procedure
- Patients awaiting ancillary services such as blood drawing in the laboratory or x-rays in radiology
- Phone callers waiting to be answered by the receptionists at an appointment scheduling center
- Patients waiting to be moved from one location to another by the patient transporter

The objective is to design the systems with the appropriate number of servers in order to obtain a proper balance between the average customer waiting time and the server idle time. The interarrival times for customers and the service times are random variables. In order to analyze the problems stated above, one needs to first determine the probability distribution of the interarrival times between two consecutive arrivals in addition to the service times. The service discipline and the effect of queue length on arrival rates, if any, also need to be known. In some simple situations, an analytical queuing model can be applied in order to analyze the situations and obtain preliminary results. In more complex situations, simulation modeling is employed to design the system. Simulation models allow you to investigate various assumptions for arrival time and service time distributions, or to compare various designs easily.

Queuing and Simulation Models in the Emergency Room.   An Emergency Room is the central facility in a hospital for the treatment of medical cases which require immediate attention. These cases may include injuries or life-threatening illness such as heart attacks and drug overdoses. Emergency Rooms also treat nonemergency cases after regular clinic hours, such as return visits for change of dressing or suture removal. In an emergency room, a triage nurse determines whether the patient requires immediate medical attention and sets up the sequence in which the attending physicians treat the patients. The treatments are not first-come–first-served generally, but rather follow the priority groups. The Emergency Room also requires additional diagnostic services such as x-ray and laboratory.

One critical issue in the Emergency Room is the proper staffing of nurses and emergency physicians based on the probabilistic demands which may vary by hour of the day and day of the week. Hale (1988) used a multiple-server queuing model with patient arrivals following a Poisson process and exponential distribution for treatment times. From the data collected over a period of time, he was able to verify the validity of assumptions made in the model and to determine the average waiting time and average number of patients waiting for service. Ortiz and Etter (1990) also applied a similar queuing model to the Emergency Room but determined that the assumption of exponential treatment times was not valid for certain types of cases. They found simulation modeling to be a better approach even for simple, less complex systems because of its flexibility and capacity to investigate various alternatives. They used general purpose simulation system (GPSS) to simulate the Emergency Room. Bokhari (1989) used simulation models developed using SLAM II to determine optimal staffing levels.

Simulation models have been extensively used in order to study a variety of departments and activities in health-care systems. Dumas and Hauser (1974) developed a simulation model to study various hospital admission policies. Kachhal, Klutke, and Daniels (1981) used GPSS to simulate outpatient internal medicine clinics in order to investigate various clinic consolidation options. They used another model to determine optimal staffing for audiologists administering hearing tests ordered by ear, nose, and throat (ENT) physicians. Levy, Watford, and Owen (1989) developed a simulation model using SIMAN to assist in the design of a new outpatient service center. Klafehn (1987) used a simulation model to study the patient flow through a radiology department. Meier, Sigal, and Vitale (1985) used a simulation model to plan an ambulatory surgery center. Thomas (1990) used a simulation model to determine staffing needs at a telephone appointment center.

OPTIMIZATION

There are a number of situations in health care systems which require the best possible (optimal) use of resources to meet certain objectives. Many of these problems can be formulated as mathematical models which can then be solved as optimization problems using linear programming, integer programming, branch-and-bound methods, or various nonlinear optimization methods. Some examples of the applications of optimization techniques in health care systems are given below.

Warner (1976) formulates the problem of determining the optimal nurse schedule as a two-phase multiple-choice programming problem and solves it using a modified version of Balintfy's algorithm. In his formulation, the objective is the determination of the schedule which maximizes an expression representing the quality of the schedules, quantifying the degree to which the nurses like the schedules they would work. The solution is determined subject to the constraints that the minimum coverage requirement for each shift and for each skill level be met. In those cases where the minimum coverage cannot be met, the seriousness of coverage violation is minimized. This optimization model is the basis of the ANSOS nurse scheduling software used in the health care industry.

Trivedi (1976) studies the problem of the reallocation of nursing resources at the beginning of each shift. This reallocation is needed because of the daily variation in the demand for nursing time on a unit, resulting from variation in census as well as patient mix. Nursing staff size and mix may also vary because of absenteeism and the 24-hour-a-day, seven-day-a-week coverage requirement. Trivedi develops a *severity index,* which is a measure of the need for an additional staff member by skill level, and obtains regression equations for each of the nursing units relating severity index to staff size, census, and patient classification. The mathematical model has the objectives of minimizing the total severity index for the hospital subject to the constraint of staff availability by skill level, and of keeping the difference in severity indexes between various units less than or equal to a predetermined constant. The model is then solved using a branch-and-bound algorithm.

Calichman (1990) presents a linear programming model to balance the bed demand among various surgical services. The objective is to maximize the total revenue subject to the Operating Room time availability, bed availability, and the demand for services constraints. Fries and Marathe (1981) use mathematical models to determine the optimal variable-sized multiple-block appointment system.

Another use of optimization models has been in the field of epidemiology. Lee and Pierskalla (1988) studied the optimal mass-screening strategies for contagious diseases which have no latent periods using mathematical models. Kaplan (1989) used a mathematical model to study the AIDS epidemic.

QUALITY ANALYSIS, CONTROL, AND IMPROVEMENT

During the late 1980s, there has been a surge in the health care industry toward adoption of the various industrial engineering techniques related to quality in order to improve health care services. Many hospitals have initiated quality improvement programs (QIPs) using the concepts and tools which have already been proved successful in the manufacturing industry. These programs are based on the quality philosophies of Dr. Edward Deming, Dr. J. M. Juran, and Philip Crosby. There are some differences in these philosophies and their respective approaches to quality improvement, but there are more similarities than differences (Sahney, Dutkewych, and Schramm, 1989).

One such quality improvement program based on the Deming philosophy was developed by the Hospital Corporation of America for implementation at its health-care facilities. This program uses a quality improvement strategy called FOCUS-PDCA, an acronym for (Burda, 1988):

- *F*ind opportunity for improvement.
- *O*rganize a team that knows the process.
- *C*larify current knowledge of the process.
- *U*ncover root causes of process variation.
- *S*tart an improvement cycle based on theory.
- *P*lan the process improvement.
- *D*o the improvement.
- *C*heck the results against the theory.
- *A*ct on the process and theory.

Various charts which are helpful in process analysis and improvement are the Pareto chart, cause-effect (or fish bone) diagram, flowcharts, run charts, and control charts. An example of the utilization of the cause-effect diagram to investigate the causes of inappropriate use of Cephalosporin antibiotics is shown in Fig. 1.3.

OTHER INDUSTRIAL ENGINEERING TECHNIQUES

Because of space constraints, the application of other industrial engineering techniques in health services cannot be covered here. References listed in the bibliography include several articles about these applications. Table 1.1 summarizes the various references by the type of technique and the area of application in the health services. The areas of application in health-care functions have been divided into six groupings:

1. Business office
   - Admitting
   - Financial
   - Billing and the like

2. Nursing

3. Ancillary services
   - Physical therapy
   - Laboratory
   - Radiology
   - Pharmacy
   - Occupational therapy and the like

4. Support services
   - Dietary
   - Housekeeping
   - Materials management
   - Medical records
   - Transportation services
   - Laundry
   - Physical facilities and the like

5. Medical services

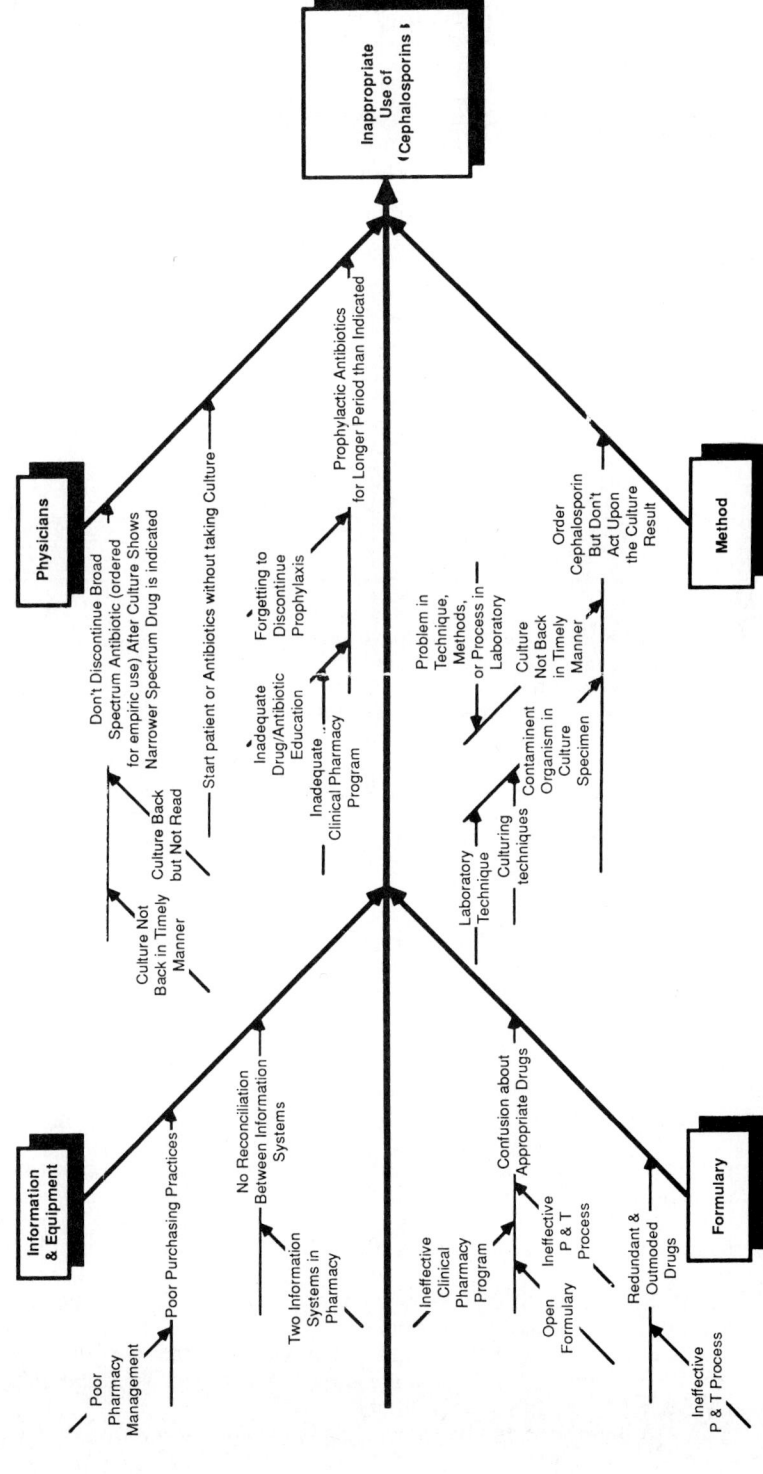

FIGURE 1.3 Cause-effect diagram to investigate the causes of inappropriate use of Cephalosporin antibiotics. (*Reprinted with permission from Nackel, John, and Thomas Collier, "Implementing a Quality Improvement Program," Journal of the Society for Health Systems, vol. 1, no. 1, 1989.*)

TABLE 1.1 Matrix of Reference Numbers for Various Areas of Health Services and Applicable Industrial Engineering Techniques

	Business	Nursing	Ancillary	Support	Medical	General
Methods improvement	1	83	37	3, 32	4, 6, 9, 66	
Staffing and productivity		34, 50, 79, 83, 89, 97	8, 29	17, 52	47, 57, 67, 86, 88	16, 44, 61, 70, 71, 73, 77, 96
Scheduling	27				46	
Queuing and simulation modeling	14		38		4, 26, 33, 43, 49, 60, 88, 94	10
Optimization	7	90, 91, 92			18	35, 41
Personnel management			95	15, 63		40, 76, 81
Production management and inventory				75	25, 42	
Facilities planning				39		
Quality analysis, control, and improvement	1	2		66	11, 66	5, 12, 51, 53, 71, 74, 96
Economic analysis and cost control		34			16, 54, 55, 68	
Information systems and RFP process		50, 89		45	30, 31, 59	13, 56, 72, 78

- Intensive-care unit
- Emergency room
- Operating room
- Outpatient clinics and the like

6. General health care applications

The table may be used to quickly find some representative articles on the applications of a particular industrial engineering technique to a specific type of department in a hospital.

CONCLUSION

Industrial engineers have played a major role during the past 20 years in introducing quality and productivity management tools to the health care industry. Reimbursement systems were one of the major barriers in the introduction of industrial engineering techniques in the hospitals during the 1970s. With the changing reimbursement climate in the '80s, increasing numbers of hospitals are paying attention to issues of quality and productivity. Many of the techniques that were the sole domain of the manufacturing industry are now commonplace in most well-run hospitals. Inventory management, labor staffing, scheduling, methods improvement, and quality management concepts are increasingly being utilized on a daily basis to improve the productivity of the health-care sector.

The Society for Health Systems and the Healthcare Information and Management System Society are the two professional organizations serving the needs of the professionals in this area. Both of these societies offer a joint annual conference in fall or winter. Attendance has grown from 100 in the early '70s to over 1800 in 1990, showing the growth of the field.

As health care costs keep increasing, so will the need for industrial engineers.

BIBLIOGRAPHY

1. Barrett, Pamela, et al., "Quality Begins when the Patient Checks In," *Quest for Quality and Productivity in Health Services 1989 Conference Proceedings,* IIE, pp. 205–208.

2. Batchelor, Gayle, and Roberta Graham, "Quality Management in Nursing," *Journal of the Society for Health Systems,* vol. 1, no. 1, May 1989, pp. 63–88.

3. Beddie, Donald, and Julie Wooden, "Stone Age to Space Age: Automation of the Medical Records Department," *Proceedings of the 1988 Annual Healthcare Systems Conference,* vol. 2, AHA, pp. 31–46.

4. Bokhari, Ijaz, "Reducing Patient Turnaround Time in the Emergency Room," *Proceedings of the 1990 Annual Healthcare Information and Management Systems Conference,* AHA, 1990, pp. 59–70.

5. Burda, D., "Providers Look to Industry for Quality Models," *Modern Healthcare,* July 15, 1988, p. 28.

6. Buttaro, Raymond, "Emergency Department Cost Reduction through Quality and Service Improvements," *Proceedings of the 1990 Annual Healthcare Information and Management Systems Conference,* AHA, 1990, pp. 37–42.

7. Calichman, Murray, "The Use of Linear Programming to Balance Bed Demand," *Proceedings of the 1990 Annual Healthcare Information and Management Systems Conference,* AHA, 1990, pp. 385–390.

8. Carr, Michael, and Ellen Bonk, "Automation of the Physical Therapy Department," *Proceedings of the 1989 Annual Healthcare Systems Conference,* pp. 457–466.

9. Cheatwood, Sharon, "IE's in the Operating Room: Analyzing Surgical Procedures Calls for Special Precautions," *Industrial Engineering,* 1987, vol. 19, no. 7, pp. 34–37.

10. Chizzali, A., and G. Romanin-Jacur, "A General Methodology to Model and Simulate Hospital Departments," *Proceedings of the 1984 Summer Simulation Conference,* Society for Computer Simulation, San Diego, 1984, pp. 948–951.

11. Dahlstrom, Gail, and Joseph Devin, "The Conversion and Expansion of Hospital Quality Assurance to an Ambulatory Care Setting," *Quest for Quality and Productivity in Health Services 1989 Conference Proceedings,* IIE, pp. 75–80.

12. Dasbach, Erik, and David Gustafson, "Impacting Quality in Health Care: The Role of the Health Systems Engineer," *Journal of the Society for Health Systems,* vol. 1, no. 1, May 1989, pp. 75–84.

13. Dowling, Alan, "Health Care Information Systems Architecture of the Near Future," *Journal of the Society for Health Systems,* vol. 1, no. 2, 1989, pp. 77–79.

14. Dumas, M., and N. Hauser, "Hospital Admissions System Simulation for Policy Investigation and Improvement," *Proceedings of the 6th Annual Conference on Modeling and Simulation,* University of Pittsburgh, 1974, pp. 909–920.

15. Dux, Lawrence E., "Gainsharing: Creating Incentives to Improve both Quality and Productivity," *Quest for Quality and Productivity in Health Services 1989 Conference Proceedings,* IIE, pp. 137–143.

16. Eastaugh, S. R., "Improving Hospital Productivity under PPS: Managing Cost Reductions," *Hospital and Health Services Administration,* vol. 30, no. 4, July–August 1985, pp. 97–111.

17. Ennis, James, "Pay for Performance—A Heuristic Approach to Productivity Improvement in a For-Profit Laundry," *Proceedings of the 1989 Annual Healthcare Systems Conference,* pp. 373–384.

18. Fries, B., and V. Marathe, "Determination of Optimal Variable Sized Multiple-Block Appointment Systems," *Operations Research,* 1981, vol. 29, pp. 324–345.

19. "Five Future Areas of Liability Risks Haunt Providers," *Hospitals,* Nov. 20, 1986, pp. 48–53.

20. Foley, Michael F., and Jeffrey E. Kubel, "Comprehensive Pharmacy Department Workload Analysis System—A Case Study," *Proceedings of the 1988 Annual Health Care Systems Conference,* vol. 1, 1988, pp. 9–17.

21. Gilbreth, Frank B., "Motion Study in Surgery," *Canadian Journal of Medical Surgery,* vol. 40, no. 1, July 1916, pp. 22–31.

22. Gilbreth, Lillian M., "Management Engineering and Nursing," *American Journal of Nursing,* vol. 50, no. 12, December 1950, pp. 780–781.

23. Gilbreth, Lillian M., "Time and Motion Study," *Modern Hospital,* vol. 65, no. 3, September 1945, pp. 53–54.

24. Goldberg, Alan J., and Robert A. Denoble (eds.), *Hospital Departmental Profiles,* American Hospital Publishing, 1986.

25. Grimaldi, Paul, and Julie Micheletti, *Diagnostic Related Groups—A Practitioner's Guide,* Pluribus Press, Chicago, 1982.

26. Hale, Jeffrey, "Queuing Theory in the Emergency Department," *Proceedings of the 1988 Annual Health Care Systems Conference,* vol. 1, AHA, pp. 1–7.

27. Hancock, W., et al., "Admission Scheduling and Control Systems," *Cost Control in Hospitals,* Health Administration Press, Ann Arbor, 1976, pp. 150–185.

28. Hanson, Robert, "Observation Methodologies for Work Sampling: Reviewed, Revised and Revisited," *Proceedings of the 1988 Annual Healthcare Systems Conference,* vol. 2, AHA, pp. 1–9.

29. Hutchinson, Randall, and Marvin Christianson, "Productivity Improvements in Rehabilitation Medicine," *Proceedings of the 1989 Annual Healthcare Systems Conference,* AHA, pp. 413–429.

30. Jackovitz, Donald, "Decision Support Systems: An Operating Room Utilization Application," *Proceedings of the 1990 Annual Healthcare Information and Management Systems Conference,* AHA, 1990, pp. 479–489.

31. Kachhal, S. K., and Fran Koch, "Evaluation of the Operating Room Management Information System Packages Available in the Market," *Proceedings of the 1989 Annual Healthcare Systems Conference,* AHA, 1989, pp. 431–441.

32. Kachhal, S. K., and Susan Scott, "An Optical Disk System for Medical Records: A Case Study," *Proceedings of the 1988 Annual Healthcare Systems Conference,* vol. 2, AHA, pp. 47–59.

33. Kachhal, S., G. Klutke, and E. Daniels, "Two Simulation Applications to Outpatient Clinics," *1981 Winter Simulation Conference Proceedings,* IEEE, 1981, pp. 657–665.

34. Kahl, Ken, "Bedside Automation," *Proceedings of the 1990 Annual Healthcare Information and Management Systems Conference,* AHA, 1990, pp. 71–96.

35. Kaplan, Edward, "What are the Risks of Risky Sex?—Modeling the Aids Epidemic," *Operations Research,* 1989, vol. 37, no. 2, pp. 198–209.

36. Kelliher, Matthew E., and Robert T. Meling, "Development and Evaluation of a Computer Assisted Supply Distribution System," *Proceedings of the 12th Annual Conference of the Health Services Division of the AIIE,* New Orleans, 1981, pp. 95–106.

37. Kirtland, Anthony, "Product Performance Improved," *Proceedings of the 1988 Annual Healthcare Systems Conference,* vol. 2, AHA, pp. 133–151.

38. Klafehn, K., "Impact Points in Patient Flows through a Radiology Department Provided through Simulation," *Proceedings of the 1987 Winter Simulation Conference,* IEEE, 1987, pp. 914–918.

39. Kucic, Joseph, and Bruce Hunter, "Estimating Projected Interfacility Traffic Flows at Expanding Medical Center," *Industrial Engineering,* 1988, vol. 20, no. 10, pp. 70–75.

40. Landry, James, "Recognition and Reward," *Productivity and Performance Management in Health Care Institutions,* Americal Publishing, Chicago, 1989, pp. 155–173.

41. Lee, Hau, and William Pierskalla, "Mass Screening Models for Contagious Diseases with No Latent Period," *Operation Research,* 1988, vol. 36, no. 6, pp. 917–928.

42. Levine, Helen, and Elaine Remmlinger, "Product Line Cost Accounting Systems," *Proceedings of the 1989 Annual Healthcare Conference,* pp. 79–89.

43. Levy, Jacqueline, Bevlee Watford, and Valerie Owen, "Simulation Analysis of an Outpatient Service Facility," *Journal of the Society for Health Systems,* vol. 1, no. 2, November 1989, pp. 35–46.

44. Linder, Carl, and Walton Hancock, "Computerized Work Measurement Can Help Hospitals Reduce Costs," *Industrial Engineering,* 1985, vol. 17, no. 3, pp. 70–77.

45. MacArthur, John, and Stephanie Massengill, "Optical Disk Storage and Retrieval for Medical Records Document Management," *Journal of the Society for Health Systems,* 1989, vol. 1, no. 2, pp. 99–109.

46. Magerlein, J., and J. Martin, "Surgical Demand Scheduling: A Review," *Health Services Research,* 1978, vol. 13, pp. 418–433.

47. Mahachek, Arnold, and Marianna Zahurak, "Operating Room Parameters for Utilization Improvement," *Proceedings of the 1988 Annual Health Care Systems Conference,* vol. 1, AHA, pp. 147–157.

48. Mahon's Industry Guides for Accountants and Auditors, Guide 7, *Hospitals,* James J. Mahon (ed.), Warren, Gorham, & Lamont, Boston, 1978.

49. Meier, L., E. Sigal, and F. Vitale, "The Use of a Simulation Model for Planning Ambulatory Surgery," *Proceedings of the 1985 Winter Simulation Conference,* IEEE, 1985, pp. 558–564.

50. Meyer, Diane, *GRASP: A Patient Information and Workload Management System,* MCS, Morgantown, N.C., 1978.

51. Miller, M., and R. Knapp, *Evaluating Quality of Care—Analytic Procedures—Monitoring Techniques,* Aspen Systems, 1979.

52. Nacey, Gene, "Staffing Methodologies for Environmental Service Department," *Proceedings of the 1988 Annual Healthcare Systems Conference,* vol. 2, AHA, pp. 341–345.

53. Nackel, John, and Thomas Callier, "Implementing a Quality Improvement Program," *Journal of the Society for Health Systems,* vol. 1, no. 1, May 1989, pp. 85–100.

54. Nackel, John, G. Kis, and P. J. Fenaroli, *Cost Management for Hospitals,* Aspen Publishers, Rockville, Md., 1987.

55. Nagaprasanna, Bangalore, "Hospital Cost Control Program Saves $1.8 Million per Year," *Industrial Engineering,* 1987, vol. 19, no. 7, pp. 22–37.

56. Napoli, John, "IE's Can Use Computers to Implement Fundamental Changes in Hospitals," *Industrial Engineering,* 1985, vol. 17, no. 3, pp. 64–69.

57. Newcomb, Carol, and S. K. Kachhal, "Successful Management of Resources in an Ambulatory Surgery Center," *Proceedings of the 1988 Annual Healthcare Systems Conference,* vol. 2, AHA, pp. 347–359.

58. Nock, Albert J., "Frank Gilbreth's Great Plan to Introduce Time-Study into Surgery," *American Magazine,* vol. 75, no. 3, March 1913, pp. 48–50.

59. Nolan, Victor, "Automation of Surgeon Preference Cards Using a PC-Based dBase Application," *Proceedings of the 1990 Annual Healthcare Information and Management Systems Conference,* AHA, 1990, pp. 507–512.

60. Ortiz, Alex, and Gary Etter, "Simulation Modeling vs. Queuing Theory," *Proceedings of the 1990 Annual Healthcare Information and Management Systems Conference,* AHA, 1990, pp. 349–357.

61. Page, John, and Mark McDougall, "Staffing for High Productivity," *Productivity and Performance Management in Health Care Institutions,* AHA, 1989, pp. 61–82.

62. Pavlo, Rhoda, and Jay Weinroth, "Using Simulation to Plan for a Major Hospital Expansion," *Proceedings of the 1989 Annual Healthcare Conference,* pp. 111–125.

63. Powasnick, James, and Jim Ekbatani, "An Employee Suggestion Program Called STAR," *Quest for Quality and Productivity in Health Services 1989 Conference Proceedings,* IIE, pp. 144–199.

64. Randhawa, Sabah U., and Darwin Sitompul," Nurse Scheduling: a State-of-the-Art Review," accepted for publication in *The Journal of the Society for Health Systems,* 1990.

65. Resource Monitoring System, Hospital Association of New York State, Albany.

66. Rohe, Duke, and Cheryl Hanks, "Super Unit: The Ideal Process for the Ideal Unit," *Quest for Quality and Productivity in Health Services 1989 Conference Proceedings,* IIE, pp. 64–71.

67. Rowell, Tim, and Bill Schlosser, "A Participative, Manager-Directed Approach to Productivity Improvement," *Proceedings of the 1990 Annual Healthcare Information and Management Systems Conference,* AHA, 1990, pp. 11–22.

68. Rupp, Loralee, "Productivity Costs of Resident Training in Clinical Setting," *Proceedings of the 1990 Annual Healthcare Information and Management Systems Conference,* AHA, 1990, pp. 305–323.

69. Sahney, V. K., D. S. Peters, and S. R. Nelson, "Health Care Delivery System: Current Trends and Prospects for the Future," *Henry Ford Hospital Medical Journal,* vol. 34, no. 4, 1986, pp. 227–232.

70. Sahney, Vinod, and Gail Warden, "The Role of Management in Productivity and Performance Management," *Productivity and Performance Management in Health Care Institutions,* AHA, 1989, pp. 29–44.

71. Sahney, Vinod, and Loralee Rupp, "Quality and Productivity Improvement at Henry Ford Hospital," *Industrial Engineering,* 1988, vol. 20, no. 10, pp. 58–65.

72. Sahney, Vinod, and Swatantra Kachhal, "Managing the Acquisition, Installation and Benefit Re-

alization of Microcomputer Based Systems in Healthcare," *Journal of the Society for Health Systems,* vol. 1, no. 2, 1989, pp. 50–64.

73. Sahney, Vinod, "Managing Variability in Demand: A Strategy for Productivity Improvement in Healthcare," *Health Care Management Review,* vol. 7, no. 2, Spring 1982, pp. 37–41.

74. Sahney, Vinod, Jerry Dutkewych, and William Schramm, "Quality Improvement Process: The Foundation for Excellence in Health Care," *Journal of the Society for Health Systems,* vol. 1, no. 1, May 1989, pp. 17–29.

75. Scavone, Robert, and S. K. Kachhal, "Just-in-Time in Healthcare Delivery Systems," *Proceedings of the 1990 Annual Healthcare Information and Management Systems Conference,* AHA, 1990, pp. 229–238.

76. Schermerhorn, J. R., "Improving Health Care Productivity through High Performance Managerial Development," *Health Care Management Review,* vol. 12, no. 4, Fall 1987, pp. 49–56.

77. Schleichert, Robert, "Flexibility: Tools for Managing Productivity and Quality," *Proceedings of the 1989 Annual Healthcare Conference,* AHA, pp. 45–58.

78. Schmitz, H., *Hospital Information Systems,* Aspen Systems, Rockville, Md., 1979.

79. Shaha, Steven, and Frank Overfelt, "Hospital Wide Multiple Level Acuity Classification Schemes," *Proceedings of the 1988 Annual Health Care Systems Conference,* vol. 1, AHA, pp. 63–72.

80. Sherlock, D. B., *The Emerging Buyers Market: The Traditional Health Services Market,* Solomon Brothers, New York, 1985, pp. 5–6.

81. Shyavitz, L., D. Rosenbloom, and L. Conover, "Financial Incentives for Middle Managers," *Health Care Management Review,* vol. 10, no. 3, Summer 1985, pp. 37–44.

82. Smalley, Harold E., *Hospital Management Engineering,* Prentice-Hall, Englewood Cliffs, N.J., 1982.

83. Smith, Mark, and Maureen Bisognano, "Patient-Centered Process Management—An Alternative to Traditional Nursing Unit Organization," *Proceedings of the 1988 Annual Health Care Systems Conference,* vol. 1, AHA, pp. 47–62.

84. Steinwelds, B., and F. Sloan, "Regulatory Approvals to Hospital Cost Containment: A Synthesis of the Empirical Evidence," in Olsin, M. (ed.), *A New Approach to the Economics of Health Care,* American Enterprise Institute, Washington, D.C., 1981, pp. 274–308.

85. Taylor, Frederick W., *The Principles of Scientific Management,* Harper & Row, 1911.

86. Templin, John, "Conducting the Perfect Post Anesthesia Room Study," *Proceedings of the 1989 Annual Healthcare Systems Conference,* AHA, 1989 pp. 127–146.

87. The Development of a Canadian Hospital Pharmacy Workload Measurement System, Canadian Society of Hospital Pharmacists, Toronto, Ontario.

88. Thomas, Andrew, "Lotus Based Simulation Model for Determining Staffing," *Proceedings of the 1990 Annual Healthcare Information and Management Systems Conference,* AHA, 1990, pp. 261–268.

89. Trimm, J. M., "Using Nursing Decision Support Systems to Improve Quality and Productivity," *Quest for Quality and Productivity in Health Services 1989 Conference Proceedings,* IIE, pp. 87–91.

90. Trivedi, Vandankumar, "Daily Allocation of Nursing Resources," *Cost Control in Hospitals,* Health Administration Press, Ann Arbor, 1976, pp. 202–226.

91. Warner, D. M., and J. Prawda, "A Mathematical Programming Model for Scheduling Personnel in a Hospital," *Management Science,* December 1972, vol. 19, no. 4, pp. 411–422.

92. Warner, D. Michael, "Computer-Aided System for Nurse Scheduling," *Cost Control in Hospitals,* Health Administration Press, Ann Arbor, 1976, pp. 186–201.

93. Weiss, Elliott, "Models for Determining Estimated Start Times and Case Orderings in Hospital Operating Rooms," *IIE Transactions,* 1990, vol. 22, no. 2, pp. 143–150.

94. Weiss, Elliott, and McClain, John, "Administrative Days in Acute Care Facilities: A Queuing-Analytic Approach," *Operations Research,* 1987, vol. 35, no. 1, pp. 35–44.

95. Wells, Mary, "Evaluating an Incentive Plan for Physical Therapists," *Proceedings of the 1989 Annual Healthcare Systems Conference,* pp. 443–449.

96. Werner, John, "Productivity and Quality Management," *Productivity and Performance Management in Health Care Institutions,* AHA, 1989, pp. 83–118.

97. Whaley, Lynne, and Frank Overfelt, "Statistical Validation of Patient Classification Levels," *Proceedings of the 1988 Annual Health Care Systems Conference,* vol. 1, AHA, pp. 73–79.

CHAPTER 2
ELECTRONICS

Arvind Ballakur
Supervisor, Operations Analysis
AT&T Bell Laboratories
Holmdel, New Jersey

The electronics industry consists of consumer electronics (videocassette recorders, audio systems, appliances, and the like), industrial electronics (office automation, medical diagnostic equipment, lasers, and so on), electronic components (such as integrated circuits, capacitors, picture tubes), communications (such as telephones, facsimile), computers (mainframes, personal computers, and others), and related products and services. Electronics deals with the study, application, and control of conduction of electricity in a vacuum, gas, liquid, semiconductor, conductor, or superconductor.[1] Electronic products contains parts, components, subassemblies, and equipment which use the principles of electronics in performing their major functions. These products have made a tremendous impact on our quality of work and life, industrial productivity, employment, international trade, entertainment, and global communications. Electronics technology has revolutionized the way we live and has changed almost everything we do.

Electronics industry is a global business with global markets and global (software and hardware) competitors. In 1989, electronics production totaled $673 billion, and roughly 34 percent being made in the United States, 27 percent in Europe, and another 34 percent in the Far East (Japan, Hong Kong, South Korea, Singapore, and Taiwan).[2] This industry is expected to grow at an average annual rate of 8 percent over the next few years.[2] The electronics industry has gained significant importance in the last decade when compared to other industries. Many countries have gained economic strength and monetary wealth through their competitive strength in their electronics industry. Hence, it is appropriate to examine how industrial engineering applications have contributed in shaping electronics industry processes.

This chapter provides some background on electronics industry processes and then presents specific information on the applications of industrial engineering principles and techniques used in the electronics industry. The main focus of this chapter is on electronics assembly. Electronics fabrication and software development (where industrial engineers are increasingly playing an important role) are dealt with to a limited extent.

ELECTRONICS ASSEMBLY PROCESSES

Electronic products have a wide spectrum of applications ranging from household consumer goods to sophisticated defense systems. *Electronics assembly,* as referred to in this chapter, deals with the assembly of a typical electronic product, which exploits a card-on-board scheme. In such a scheme, various functions are realized on individual printed-circuit boards (PCBs), and integrated into a system by interconnection via a backplane. This section gives an overview of the electronics assembly process. Technical details can be found in Refs. 3 and 4.

A PCB is the board on which electronic components (such as transistors, resistors, and capacitors) are mounted. It has wiring pathways that are printed on the board. A printed-circuit board also has an interconnection device at one of its edges. The system, which could be a cabinet or a collection of shelves, holds the PCBs. It has nesting slots which guide the PCBs onto the backplane of the system via the interconnection device.

To understand the manufacture of PCBs and other electronic products, a basic understanding of electronic principles is necessary. This is important since the manufacture of these assemblies does influence the performance of electrical circuits. For example, the way in which the layout of the wiring or printed circuits is done can affect the capacitive and inductive coupling between circuits and the propagation time through the circuit. This chapter will assume some understanding of electronic principles to explain the electronic assembly processes.

Printed-circuit boards (without components mounted on them) have two primary functions: (1) component support, and (2) circuit interconnection. The component support is provided by a sheet of insulating material and the conductive path for circuit interconnects is provided by etched copper foil. The base material chosen for the PCB will depend on the voltage levels, current levels, and other considerations such as mechanical, environmental, electrical, and chemical requirements. PCBs can be single-sided (copper on one side), double-sided, or multilayer boards (that is, double- and single-sided boards sandwiched together with layers of epoxy-impregnated glass cloth). Multilayer boards are commonly used in applications (such as backplanes) where single-sided and double-sided boards do not suffice to connect the different circuits, and external wiring is complex, incurring high labor costs.

A typical material flow process for PCB assembly is shown in Fig. 2.1. The process starts with components and printed-circuit boards being supplied from a vendor and perhaps provisioned through a storeroom. These components can be attached by either of two technologies: (1) surface mounting and (2) through-hole insertion. Surface-mounting technology (SMT) eliminates leads from components and drastically increases the density of components that can be packed into a certain area of printed-circuit board. This is because leads (used in through-hole technology) comprise a large portion of the average electronic component's volume. On the other hand, through-hole technology uses the leads of components for insertion into holes in the printed-circuit boards. SMTs are increasingly being used in industry because of decreased assembly costs, increased availability of mass assembly techniques, and increased reliability of surface-mounted devices.

For mounting SMT components, the printed-circuit boards are prepared and stencil-printed by cleaning and application of solder paste. The components are then placed by automated equipment (such as pick-and-place robots) and soldered by a technique such as vapor-phase reflow or infrared reflow. The boards are then prepared for through-hole components. These components are inserted by dual in-line (DIP), axial, radial, or variable center distance (VCD) insertion machines, depending on the component package. The board with inserted components is soldered by a technique such as wave soldering.

The printed-circuit board is then given in-circuit and/or functional tests and, when passed, is ready for mounting in a unit or a system. The system may have more than one PCB, other devices, cabling, and the like. Once these are assembled, the system is often subjected to functional testing as a unit in different environments as needed [high temperature, ground vibrations, electromagnetic interference (EMI), electrostatic discharge (ESD), and so on].

The manufacturing process, illustrated in Fig. 2.1, is only a piece of the picture. A successful electronic product design process is driven not only by technological advances but also by perceived market needs of potential customers. These requirements are then con-

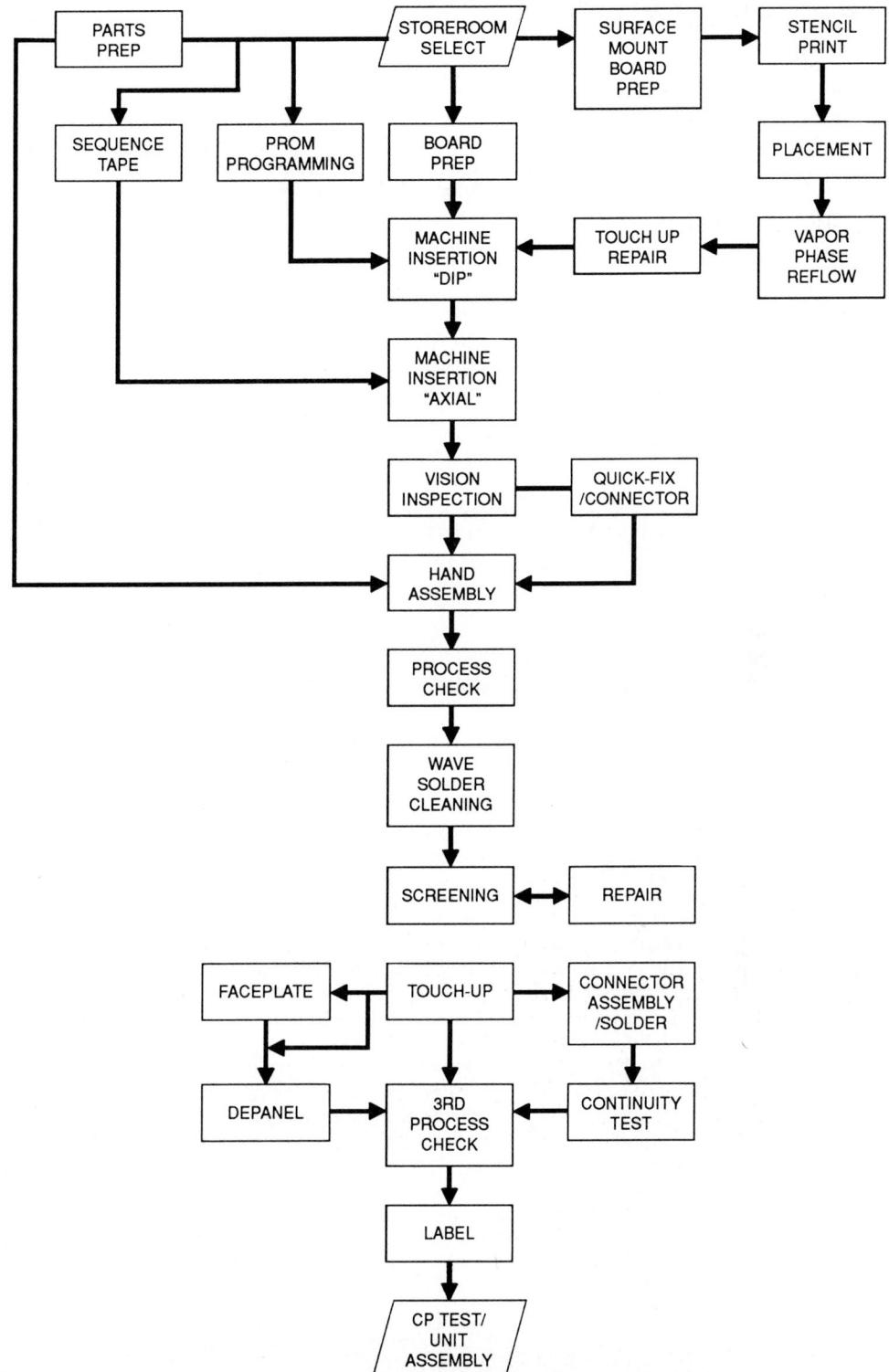

FIGURE 2.1 Material flow process for printed-circuit board assembly. (*Reprinted from Productivity and Quality Improvement in Electronics Assembly. Copyright Institute of Industrial Engineers, 25 Technology Park, Atlanta/Norcross, GA 30092.*)

verted into products (hardware and software) in the product design process. During this phase, it is critical that manufacturing, procurement, and deployment functions be involved early to ensure a smooth transition from design to manufacture and then to distribution. In addition, careful attention should be paid to designing the product right the first time. For this, concurrent engineering and design activities involving circuit design, physical design, model development, software development, and systems testing should also be carried out.

To physically build the products in the required quantity, volume manufacturing operations are required. Procurement processes must ensure that adequate supply and timely delivery of raw materials, components, subassemblies, and the like are guaranteed to meet the manufacturing requirements. To know what to build or what components to order requires good materials management and customer service processes. Different organizations must work together to promise and deliver to the customer in the required quantity at a preestablished schedule.

For delivering the product to the customer, adequate distribution and deployment operations are essential. To sell the product to different customers, the sales force must not only be knowledgeable about the product but also understand the language and needs of the customer. When the product is sold and the payment is collected, the product realization process does not end. Adequate customer support and service (such as maintenance and repair) must be provided throughout the product life cycle to ensure future sales.

MATERIALS MANAGEMENT

The distinct characteristics of the electronics industry in terms of materials management are

1. Short product life cycles
2. High rate of technological advances in products
3. High rate of design changes
4. High number of new product introductions
5. Low product yields
6. High material costs

This is further complicated by global competitive pressures in the industry to continuously reduce product costs, to maintain high quality, and to improve customer service through decreased order-to-installation intervals, and a high degree of after-sales service and support.

Manufacturing requirements planning (MRP) and manufacturing resource planning (MRP II) have been adopted by a number of electronic manufacturers since the early 1970s. The MRP II architecture has been well-deployed by the American Production and Inventory Control Society (APICS) and a number of MRP II systems (IBM COPICS, AMAPS, and so forth) are available commercially. The essential building blocks in most traditional manufacturing resource planning architectures are

- Demand management
- Production planning
- Master scheduling
- Capacity planning and control
- Materials requirement planning
- Material planning
- Order release
- Shop floor control

Supporting high optionality and shortened manufacturing intervals in the electronics manufacturing environment does not require abandoning the planning function. Planning remains a vital necessity in this environment. Component lead times for some materials are such that, if planning a half year to a year in advance did not occur, chronic material shortages would exist. The end-to-end order and manufacturing lead times are also such that without forecast information we would be unable to plan for material and shop capacity. Therefore, the functions performed by the MRP II planning system are critical to successful manufacturing; however, enhancements can be made to improve the functionality to better fit the high optionality environment.

The enhancements to the traditional MRP II environment presented in this chapter include the use of expert systems to address the complexities in the demand management and master scheduling functions; the use of EDI, vendor partnerships, and vendor-to-shop provisioning to address the complexities in the material planning, procurement, and provisioning functions; and the use of daily sequenced production to address the complexities in the shop execution function.

ROLE OF EXPERT SYSTEMS

Knowledge-based expert systems can be used to improve and overcome many of the shortcomings in demand management and master scheduling. One of the advantages of expert systems is that they can tie into existing databases and software systems easily. Another advantage is the flexibility they allow in programming and changing and updating rules. Rules for maintaining families, especially when designs or technologies change, can be coded on the basis of expert knowledge. Algorithms and rules that analyze customer orders and check them against planning bills can be set up using expert systems, making the maintenance of planning bills relatively straightforward. Another application for expert systems is in the analysis of schedule changes. Schedulers can input different schedule changes, and the system then performs capacity and material checks to determine the feasibility of the change and suggests possible alternatives. Expert systems can also be used to determine promise dates for customer orders by checking the availability of each option against the customer-requested date, showing where there may be material or capacity shortages, and suggesting alternative availability dates.

SEQUENCED PRODUCTION

Sequenced production can be used to improve and overcome many of the shop execution complexities which arise in the traditional MRP II environment. Sequencing subassembly production between feeder shops aids in coordinating the arrival of matched sets of subassemblies at the final assembly and system test areas. Sequencing breaks the MRP II planning system's weekly buckets into daily buckets to support the execution of the weekly production plan. Sequencing establishes a series of start-work dates for each product in each feeder shop by using actual manufacturing intervals (rather than the planning intervals, which are typically stated in weeks) and the end-item due date. This results in matched sets of subassemblies arriving in the final assembly and system test area. This decreases the interval and allows the manufacturer to be more responsive to customer order and design changes by delaying the start dates of subassembly production.

Sequencer systems also make possible an electronic link between the design systems and the low-level shop execution systems. The sequenced list of start-work dates can electronically signal a request for a download of machine programs and manufacturing instructions from the design system to the shop execution systems. This helps to reduce the interval by not having to wait for manual requests for program downloads.

ENHANCED MATERIAL PROVISIONING STRATEGIES AND THE ROLE OF EDI

Electronic data interchange (EDI), vendor partnerships, and enhanced material provisioning strategies can be used to improve and overcome many of the complexities which arise in material planning, procurement and provisioning due to the traditional MRP II environment. EDI is a computer-to-computer exchange of standard formatted business transaction information between independent entities. EDI shortens the interval to order material by placing material orders directly on the vendor's system. The sequencer and EDI can be used to improve the material planning and procurement processes by bringing the purchasing organization closer to the shop's execution plan. Typically orders are based on the MRP II planning system. Tying the sequencer output into purchasing and subsequently into the vendor via EDI allows the purchaser to release orders for material which is actually needed by the shop in the short term. This will reduce the cost associated with bringing material in-house for which there no longer is demand.

Allowing the shop to drive short-term material orders requires that the vendor provide very short lead times. This can be accomplished through vendor partnerships. Vendor partnerships bring the vendor closer to the manufacturing environment by making forecast information available to the vendor. Volume, lead time, safety stock, product quality, delivery capabilities, and price are negotiated in advance. Material which is procured in the short term via the sequencer/EDI can then be directly routed to the shop floor on arrival in the factory instead of being stored in the storeroom and pulled out at a later date. This can be accomplished either by a dock-to-shop strategy, where factory employees deliver the material directly from the receiving dock to the shop, or by a vendor-to-shop strategy, where the vendor delivers material directly to the shop floor. This aids in lowering inventory investment and total manufacturing interval.

All the above distinct characteristics presented earlier in this section require that electronics manufacturers must move from traditional weekly bucketed MRP II environments to some combination of MRP II and just-in-time (JIT) environment. The above sections outlined the necessary enhancements to traditional MRP II architecture for electronics industry. References 6 and 7 are good references on this subject. Further, increased attention should be devoted to the design-to-manufacture process to reduce parts proliferation (for the same functionality and/or fit) and control the amount of material obsolescence.[8] This is especially critical in this industry because of the high material costs (and inventory) and increased rate of technological changes experienced in the electronics industry.

QUALITY

Quality management methods in electronics assembly have progressed through several phases: (1) quality inspection, (2) statistical process control, (3) quality improvement, and (4) product and process design for quality.[9,10] Further, with increased global competition, the electronics industry now measures defects in parts per million versus percent defectives measured years ago. Quality metrics are not only being measured now in product assembly, but also in development (for example, software defects, design defects), customer service (customer score cards), field-returned products, and many other parts of the total business process.

Whereas quality inspection involved testing of products at the end of the line, statistical process control measures process parameters periodically using standard deviation as the key parameter. The noted quality guru, Dr. Genichi Taguchi, advocates the use of the quality loss function.[11,12] According to his philosophy, economic loss can be reduced by continually reducing the variability in the process and quantifying the loss associated with this variability in terms of money. The loss function is the loss that occurs when the process is not capable of producing a product meeting the target value for a performance characteristic of the manufacturing process.

Quality improvement techniques,[13] which include fish bone cause-and-effect diagrams and Pareto diagrams, identify root causes of problems and improve the process by eliminating the root cause. On the other hand, product and process design techniques, such as Taguchi's robust design method,[14–15] are continuous improvement techniques which use statistical experimental designs to identify the process variables that affect the mean and variance of the output. These variables are then adjusted to obtain the desired mean and variance.

COST MANAGEMENT

The electronics industry is characterized by short product life cycles, for example, 1 to 2 years; low level of direct labor, typically less than 8 to 10 percent; high level of automation; and high level of overhead. The "traditional" product cost-accounting model used in many electronic firms does not support this manufacturing environment. This traditional model, which was designed decades ago, uses direct labor hours to allocate expenses and does not support managers to do better decision making, particularly in terms of cost cutting or in encouraging designers to focus on nonlabor transactions caused by their designs.

In the past few years, a new model, called *activity-based costing* (ABC),[16–18] has been proposed to promote the right cost-management behavior for the new manufacturing environment, particularly appropriate for the electronics industry. The ABC model treats technology as a direct product cost and allocates expenses of activities on the basis of cost drivers, other than direct labor hours. The basic concept of this model is that product manufacture requires activities, and activities require resources. Therefore, if activities are measured on certain cost drivers, the true cost of a product can be calculated which results in a new, more useful, realistic definition of *product cost.* A detailed application of this model is provided in Ref. 19.

The process of activity analysis for ABC is as follows: Each organization analyzes its expenses to determine the nature of the expenses. Each organization then determines its major activities and allocates its expense dollars to them. All activities identified by each organization and their associated costs need to be reviewed for "non-value-added" content and possible elimination. A cost driver is chosen for every activity. The total expense dollars associated with each activity is divided by the total number of driver occurrences for that activity to arrive at "activity cost rates."

The number of cost drivers chosen should be kept to a minimum, taking into account the product diversity, the volume diversity, and the relative costs of the activities identified. Cost drivers must be selected according to the following criteria: (1) they must *accurately reflect* the cost of the activities they measure, (2) the *cost of measurement* of the cost drivers relative to *accuracy* desired must be low, and (3) cost drivers must encourage management-desired *behavior.* For example, if cycle time of products is too long, then the cost driver chosen for process engineering and operating activities should measure, and aid in the reduction of, cycle time. Examples of the cost drivers chosen in the literature are cycle time, quality defects, number of parts in the bill of material, and machine hours.

The labor and load costing for each end item must be calculated by summing the product of all the activity cost rates and the number of driver occurrences associated with the end item, and dividing the total dollars by the number of units shipped. The unit product cost must then be calculated by adding its direct material cost to its labor and load costs.

ERGONOMICS

Human factors, or ergonomics, is the study of the human being in relationship to the environment in which the individual must operate or interface. In the electronics industry, there are many instances of such human-environment relationships that need to be carefully designed and integrated into a system. This section gives some examples in the electronics industry where industrial engineers are the ergonomic experts. Appropriate references are provided for further details.

In electronics manual assembly operations, an operator usually does repetitive tasks on a routine basis. Some of these tasks affect the operator's long-term safety and health. These must be identified and effective solutions that eliminate or minimize the risk of poor health and safety must be implemented. For example, electronic assembly operators repeatedly use lead-cutting pliers in assembly operations. This method of cutting leads is one of the causes of carpal tunnel syndrome (the inflation of the carpal tunnel at the base of the palm) which results in pain and inability of the operator to effectively use the hand. Therefore, attention has to be given to select an ergonomically correct tool to prevent disability of the hand, particularly in a repetitive task.

Electronics assembly workstation design is another area of concern to electronics manufacturers. Workstation design elements include the material handling interface, test station design, assembly station design, ergonomic seating, and lighting. The effects of poor workstation design include lower productivity and performance, higher absenteeism, and higher worker compensation claims. A systems approach to ergonomically sound design of electronics assembly/test workstations is discussed in Ref. 20.

The widespread use of video display terminals (VDTs) in the electronics industry has produced physical complaints ranging from an assortment of aches and pains and vision problems to wrist ailments (carpal tunnel syndrome, discussed earlier), tendonitis, and other cumulative trauma disorders. Eye fatigue, blurred vision, stiff neck, and back pain are among other discomforts VDT users blame on their machines. Improper workplace designs can contribute to these problems and, therefore, care should be taken to make the VDT workstation more comfortable by positioning the screen so that one can look down, adjusting the screen to avoid glare caused by room lighting and adjusting furniture, such as the chair and stand, to give better support.

Radiation from VDTs is also an issue people may complain about. But it has not been proved that it is a problem. VDTs use the same technology as the television set. Both emit electromagnetic fields, a low-frequency form of radiation. The emission of electromagnetic fields from VDTs are far below applicable health standards. The National Institute for Occupational Safety and Health (NIOSH), part of the U.S. Department of Health and Human Services, keeps track of these data (such as emission from VDTs).

Electrostatic discharge (ESD) is yet another area of concern in electronics assembly. The static discharges in assembly environments can destroy components or degrade them to fail at a later point in time. This results in increased junking expenses, increased cost of repair (both in-house and field returns), and, therefore, reduced profits to electronics manufacturers. To prevent ESD losses, sufficient ESD protection such as static-protective work surfaces, wrist straps, conductive chairs, conductive material-handling devices, and conductive flooring, should be used. ESD protection, and DOD standards are provided in Refs. 20 to 22.

TRADE ASSOCIATIONS

Electronics industry related trade and professional associations are useful in helping the reader locate specific information that may not be included in this chapter. These include the American Electronics Association, Electronics Industries Association, Institute of Electrical and Electronic Engineers, Institute for Interconnecting and Packaging Electronic Circuits, American Production and Inventory Control Society, American Society of Quality Control, Institute of Industrial Engineers, Society of Manufacturing Engineers, and Human Factors Society. A complete listing including addresses and telephone numbers is provided in the bibliography section of Ref. 5. Definitions and terminology of electronic terms can be found in electronic handbooks.[24-25]

REFERENCES

1. *Electronic Market Data Book,* Electronics Industries Association, 1988, Washington, D.C.
2. *IEEE Spectrum,* vol. 24, no. 1, January 1990.
3. Kear, F. W., "Printed Circuit Assembly Manufacturing," Marcel Dekker, New York, 1987.
4. Matisoff, B. S., *Handbook of Electronics Packaging Design and Engineering,* Van Nostrand Reinhold Company, New York, 1982.
5. Edosomwan, J. A., and A. Ballakur, *Productivity and Quality Improvement in Electronics Assembly,* McGraw-Hill, New York, and IIE, Atlanta, 1989.
6. Betts, S. B., A. Ballakur, and J. R. Murray, "Production Planning for Highly Optioned Products," *1990 International Industrial Engineering Conference Proceedings,* San Francisco, May 1990.
7. Chatterjee, A., and A. Ballakur, "Just In Time in Electronics Manufacturing," *Productivity and Quality Improvement in Electronics Assembly,* McGraw-Hill, New York, and IIE, Atlanta, 1989.
8. Ballakur, A., "Assessing the Economic Impact of Parts Proliferation in Electronics Assembly," *1989 International Electronics Assembly Conference Proceedings,* IIE, Atlanta, 1989.
9. Fuchs, E., "Quality: Theory and Practice," *AT&T Technical Journal,* vol. 65, no. 2, 1986, p. 4.
10. Schonberger, R. J., "The Quality Concept: Still Evolving," *National Productivity Review,* Winter 1986, p. 81.
11. Taguchi, G., "Off-Line and On-Line Quality Control Systems," *Proceedings of International Conference on Quality Control,* Tokyo, 1978, pp. B4-1–B4-5.
12. Kackar, R. N., "Off-Line Quality Control, Parameter Design, and the Taguchi Method," *Journal of Quality Technology,* vol. 17, no. 4, 1985, pp. 176–188.
13. Messina, W. S., *Statistical Quality Control for Manufacturing Managers,* Wiley, New York, 1987.
14. Taguchi, G., and D. Clausing, "Robust Quality," *Harvard Business Review,* January–February 1990, pp. 65–75.
15. Box, G. E. P., J. S. Hunter, and W. G. Hunter, *Statistics for Experiments,* Wiley, New York, 1978.
16. Johnson, H. T., and R. S. Kaplan, *Relevance Lost: The Rise and Fall of Management Accounting,* Harvard Business School Press, Boston, 1987.
17. Berliner, C., and J. A. Brimson (eds.), *Cost Management for Today's Advanced Manufacturing: The CAM-I Conceptual Design,* Harvard Business School Press, Boston, 1988.
18. Shank, J. K., and V. Govindarajan, *"Unbundling" the Full Product Line: The Perils of Volume-Based Costing,* Amos Tuck School of Business Administration Report, Dartmouth College, Hanover, N.H., July 1, 1987.
19. Berlant, D., R. Browning, and G. Foster, "How Hewlett-Packard Gets Numbers It Can Trust," *Harvard Business Review,* January–February 1990, p. 178.
20. Pukanic, R. L., and D. L. Morelli, "A Systems Approach to Ergonomically Sound Design of Electronics Assembly/Test Stations," *Industrial Engineering,* July 1985.
21. U.S. Department of Defense, DOD-HDBK-263, May 1980.
22. Testone, A. Q., *Static Electricity in the Electronics Industry,* Testone Enterprises, Lee, Mass., 1985.
23. Pukanic, R. L., "ESD: Understanding the Problems and Methods for Controlling," *1985 Annual International Industrial Engineering Conference Proceedings,* IIE, Atlanta, 1985.
24. ANSI/IEEE, *IEEE Standard Dictionary of Electrical and Electronics Terms,* Standard 100-1984, IEEE and Wiley-Interscience, New York, 1984.
25. Fink, D. G., and D. Christiansen, *Electronic Engineers' Handbook,* 3d ed., McGraw-Hill, New York, 1989.

CHAPTER 3
AEROSPACE AND DEFENSE

Blair H. Schlender
Manager—TQM and Production Engineering
Martin Marietta Corporation, Orlando, Florida

John F. Doran
Manufacturing Consultant
Raytheon Company, Lexington, Massachusetts

The aerospace and defense industry serves two separate markets—commercial and military. These two markets are served with products sufficiently similar that they can be considered as one industry for the purposes of this chapter. On the one hand, is the large industrial complex which produces the military hardware on which the security system of the United States depends. Its customer is the United States government. The second component of the industry is the producers of commercial and private air vehicles. Its customer base consists primarily of large corporations at one end of the spectrum and secondarily of private individuals at the opposite end (general aviation).

The aerospace and defense industry is a significant segment of the total economy of the United States. For fiscal year 1988, the defense budget accounted for $290.4 billion. This represented 6.1 percent of the gross national product (GNP) and 27.3 percent of the federal budget. In addition, for calendar year 1988, sales of aircraft, engines, and parts to entities other than to the United States government totaled $29.4 billion.

Since 1988, the defense budget has trended downward as a percentage of GNP and the federal budget. However, there is another factor at work. The aerospace and defense industry has been a consistent and growing contributor to the United States world trade balance. While the United States trade balance has been negative in recent years, the Aerospace industry alone had a net positive trade balance of $22 billion in 1989. Because of the large percentage of commercial aerospace production that is exported, generally 50 percent of production or more, that favorable effect is likely to continue for some time.

For purposes of this chapter, the aerospace industry produces products that are propelled or deployed above the earth's surface. These include aircraft, long- and short-range missiles, space vehicles, and satellites. All of those categories include both military and commercial products.

The defense industry includes products produced to specifications that have been generated for the military services. The range of products extends from ships, vehicles, firearms, and ammunition to clothing and food.

Military services use many products that are not produced to military specifications. These are not considered to be aerospace and defense industry products. There is an increasing tendency for the military services to procure products to commercial specifications wherever possible. This removes those products from the defense industry category. As a result, costs associated directly with meeting military specifications are eliminated.

INTRODUCTION

The aerospace and defense industry has evolved from a long history and a varied environment. Its history can be traced back to pre-Revolution times when armament manufacture was a one-at-a-time process, such as the crafting of a long rifle. But early in its development, innovation became a recurring characteristic of this industry. An early example was the application of the concept of interchangeable parts manufacture. Eli Whitney applied this previously theoretical approach to the manufacture and assembly of muskets in 1800 in fulfillment of a contract with the United States government.

The growth of the industry over time is logically linked to the war and peace cycles experienced by the United States. During the rise of manufacturing in the nineteenth century, growth in manufacturing was further stimulated by such events as the War of 1812 and the Civil War. Those efforts, however, were small when compared to the spurts experienced during World Wars I and II.

This history of sporadic growth together with the variation of product presents a multiplicity of contrasts. Picture the automated techniques used in the production of small-caliber ammunition on the one hand versus the hand fabrication of a unique space vehicle. This variation of the nature of production has changed with the demands of the times. During major international conflagrations, the United States has served as the "arsenal of the free world," churning out planes, ships, and vehicles in volumes which today, in retrospect, seem unreal. In more recent times the nature of aerospace and defense has been to offset superior volumes of armaments of the competing world political systems with lower volumes of more technically complex weaponry. The development of more complex weapon systems leads to the integration of the specialty products of larger numbers of manufacturers into a single end product.

The growth of the industrial engineering profession, generally considered to have begun with the rise of the concept of scientific management in the late 1800s, occurred in the aerospace and defense industry as well as the rest of corporate America. The usage of the early techniques such as methods analysis, motion studies, and work flow likewise gravitated to usage within the industry. Among the contributions to the tools of industrial engineering, the industry contributed the concept of the learning curve. The learning curve was first presented in print by T. P. Wright of the Wright Patterson Corporation (see Bibliography). The theory was founded on experience and analysis of the results of building aircraft.

Additional tools for the profession resulted from the pressures of the problem solutions of World War II, if not directly from the wartime industrial effort in effect at that time. The multidisciplinary team approach was applied in the development of operations research and ergonomics to solutions required in the usage of the increasingly complex systems and equipment being developed and deployed. The usage of predetermined time standards, value engineering and systems analysis was extensively refined and expanded in industry during that era.

UNIQUE CHARACTERISTICS

Industrial engineers in the aerospace and defense industry utilize all the tools and techniques that their counterparts do with other employers. However, there are unique circumstances in this industry which alter the environment in which the industrial engineer practices.

State of the Art.   Every industry has its state-of-the-art production. However, the nature of this industry is such that the customer requirements drive the introduction of new designs and new technology to a degree far in excess of that common in commercial operations. As new materials and processes are developed, the customer requires rapid transition to production, in some cases before such production is economically feasible from a commercial viewpoint. For some products, continual updating to state-of-the-art materials and processes maintains a production environment of transition from experimental to volume duplication. Products representing this characteristic include radar-absorbing coatings and material, design of deep-space hardware for survivability, noise abatement systems and hardware for underwater craft, and detection systems for oceanic, atmospheric, and space vehicles. These customer requirements are driven by the necessity for national survival with the primary mission being to provide the best product that can be developed for the purpose.

This characteristic of the aerospace and defense industry also manifests itself in the introduction of innovation in the processes, tools, and techniques which, in turn, are adopted by the rest of the manufacturing sector. This leadership role has been demonstrated by such developments as clean rooms applied to large-scale manufacturing and assembly; environmentally severe testing relative to temperature, humidity, vibration, radiation, atmosphere; and application of nondestructive inspection techniques.

Shared Risk/Shared Reward.   The dedication to state-of-the-art products and manufacturing leads to another unique characteristic of aerospace and defense. In pursuing leading-edge technology, government and industry share the costs of development through a variety of means. The government contracts directly for the development phase of a product or process, reviews and reimburses industry for its independent research if it is directly applicable, and develops products and processes directly in its own labs and facilities.

As a result of this up-front sharing of development costs and risks, the profit margins negotiated into contracts are lower than commercial industry standards. This does not imply that risk taking has been removed. There is clearly risk attendant with government contracts, especially with the long-term trend toward fixed-price contracts continuing. Industrial engineers must use all the tools at their disposal, both tried and new, in the assurance of continuing profitability for their employers.

Customer Control.   As stated earlier, most of the aerospace and defense industry sells its products to one customer, the United States government. Because of this dominance and the customer's fiduciary responsibility, customer control of the product and process is far greater than the intervention experienced in most commercial production.

This control extends through the entire life of the program. The impact on the industrial engineer for each phase of the contract is described in the following sections.

Proposal Requirements.   Defense production contracts are usually based on a proposal to a government agency or to another contractor or subcontractor who in turn has made a proposal to a government agency.

For defense items, such as special clothing, that are based on specifications similar to commercial items, a proposal may consist of little more than a dollar quotation plus a repeat of the product specifications and delivery requirements set forth in the request for quotation (RFQ).

For more complicated defense items, the proposal has become a sales tool, whether it is in response to an RFQ or is supplier-originated. This is the type of product with which this chapter is primarily concerned. From an industrial engineering viewpoint, the objective of the proposal is to show the customer that a particular supplier is especially well-qualified to provide the item proposed.

These qualifications can often be shown by descriptions of the production equipment, process capability, and experienced personnel available to produce the item. General statements are frequently supported on major proposals with detailed plant layouts, process flow charts, and process capability studies. Reports of actual production and actual costs are often included where possible.

The main reason for this detail is to provide the customer with information which the customer can use to evaluate the capability of the various potential suppliers. Experience has shown that a low proposal price is no bargain if the supplier is incapable of performing the task required.

For items requiring state-of-the-art production processes, frequently no potential supplier can provide demonstrated capability. In such cases the potential must be based on the capability demonstrated on similar types of processes. Occasionally, the proposal includes descriptions of the capability of specific groups of industrial engineers or even individual industrial engineers.

The customer needs to ensure that the supplier selected possesses the capability required. The supplier uses the proposal to show the customer that it has that capability.

The RFQ may include requirements for documenting the customer's plans to comply with certain military standards or other procedures. The industrial engineering areas frequently specified include:

- Production plans
- Change control plans
- Production control plans
- Work measurement program
- Configuration control plan

Such plans are included in the proposal when they are available. When they must be proposed, a time-phased program is normally involved.

When the customer includes a detailed requirement for a certain area, an implementation schedule for the proposed contract is normally included. For example, a documented work measurement program for a plant may be accompanied by a schedule for installing engineered labor standards. A typical summary chart is shown in Fig. 3.1.

A key element of any proposal is still the proposed cost. A difference between the proposed cost for a defense contract and for a commercial contract is the audit requirement. The cost elements of each proposal are normally subject to audit by the government agency. This is the case even when the contract is for a firm fixed price. Labor costs are normally a significant element of total costs. Frequently, they are the basis for other costs such as support activities and overhead allocations.

Industrial engineers normally are responsible for preparing, as a minimum, those labor costs associated with fabricating, assembling, and testing the product. This is the case whether they are based on estimated actual times, estimated standard times, or engineered standard times. For follow-on contracts, engineered standards may be a contractual requirement.

This need for auditable data significantly changes the task of the industrial engineer responsible for these data from such a role in a commercial operation where the retention of audit capability may be less demanding.

Preproduction Requirements.   Before the actual award of a contract, the customer frequently performs an on-site review of both the contractor's documented procedures and the contractor's actual practice in conformance to those procedures. The contract award is contingent on customer satisfaction with both the documented procedures and observed practice.

This review normally covers the following areas:

- Production plan (Fig. 3.2)
- Process control for each process (Fig. 3.3)
- Method control of each operation of each part (Fig. 3.4)
- Quality control of each process and each material move (Fig. 3.5)
- Material control by contract

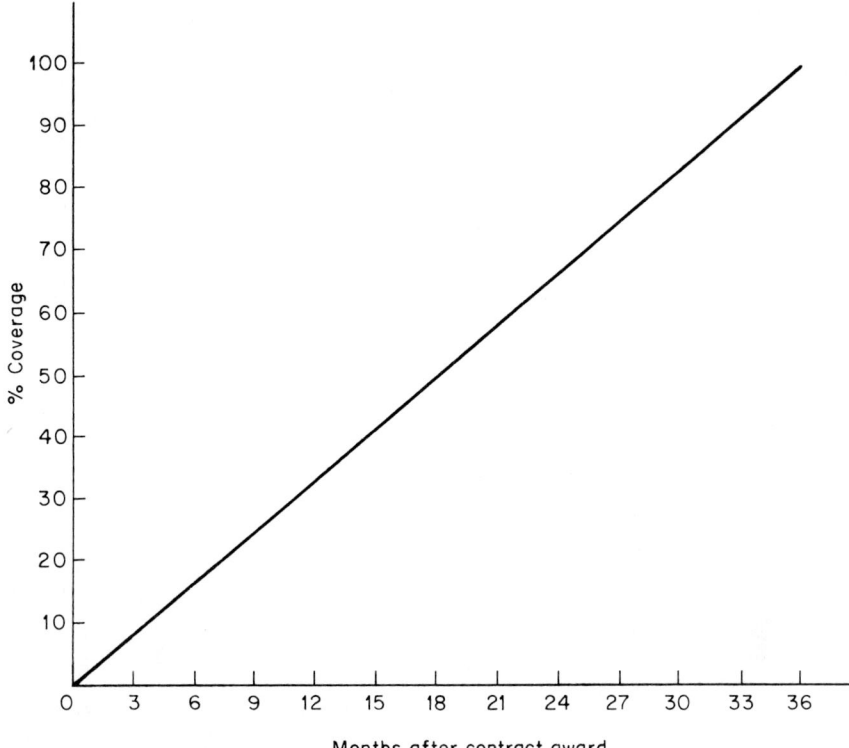

FIGURE 3.1   Labor coverage with engineered standards.

• Traceability of specified components by lot number or serial number when required, showing by serial number the end item in which each is installed (Fig. 3.6)

This on-site review involves what amounts to an audit of actual practice on existing contracts so that the customer is assured that the supplier is actually complying with procedures. This is, of course, the best evidence that the supplier will meet future commitments.

In areas where industrial engineering is involved, the industrial engineer frequently is the leader in presenting the plans and demonstrating actual practice to the customer.

Operating Requirements.   The customer routinely maintains personnel at a contractor's facilities to continually survey the contractor's operations to ensure that procedures and contract requirements are adhered to. Such surveillance is supplemented by teams from off-site to conduct detailed surveys (audits) to further ensure that conformance to procedures continues. These review teams can vary in size and scope from one person for 1 day to as many as 70 persons for a month.

At one extreme, a single customer representative may accompany a single supplier representative on such a task as verifying that the process documentation and labor standards comply with stated policy. A typical result is a single-page document stating the finding and outlining the corrective action, if required. Follow-up reports continue until the corrective action is complete.

At the other extreme, a formal briefing and introduction of the two organization contact points usually starts the review. Detailed discussions and personal observations with brief daily reports, more extensive weekly reports, and exit summaries follow. With teams of

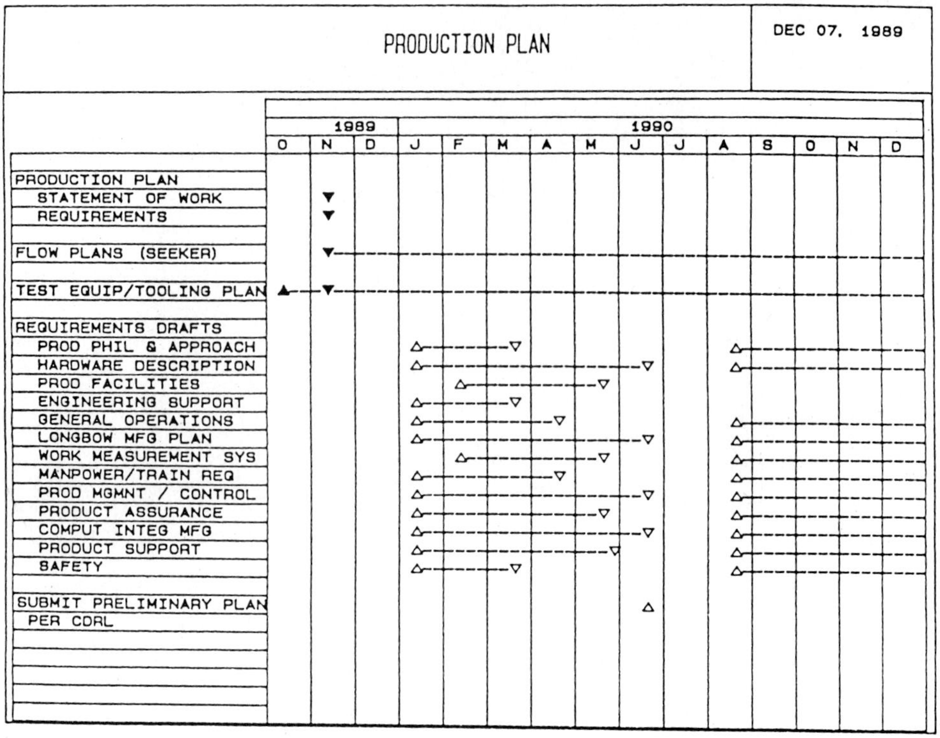

PRODPLAN

FIGURE 3.2 Contents of typical production plan. (*Courtesy of Martin Marietta Electronic Systems, Orlando, Fla.*)

AREA
PWA

SHEET
2 OF

PART NUMBER | PART NAME CIRCUIT CARD ASSY. | EQUIP.

OPER. NO.	WORK CENTER	OP NO & DATE	OPERATION DESCRIPTION	QTY/OL	IEPI	REMARKS
0300	0360-2		MANUALLY ASSEM. TERMINALS		25-152	METHOD A
			PROCESS ILLUSTRATION #4	2		METHOD B
0305	0253		INSPECT PER ITPA 4454			IIS 11456695
0364	0360-2		1. ASSB INDEX PINS TO CONNECTOR	2	25-052	
			2. RIVET CONNECTORS TO GUIDE	1	25-155	
			PROCESS ILLUSTRATION #5 & 6			
0365	0256-2		INSPECT ITP A4596			IIS 11456695
0384	0364		PRE-CLEAN		25-161	
0386	0364		PRE-BAKE		25-254	
0390	0364		SOLDER MASK		25-082	
0394	0364		WAVE SOLDER AND CLEAN		02-003	

(a)

FIGURE 3.3 (*a*) Manual process plan. (*b*) Illustration for manual process plan. (*Courtesy of Raytheon Company, Lexington, Mass.*)

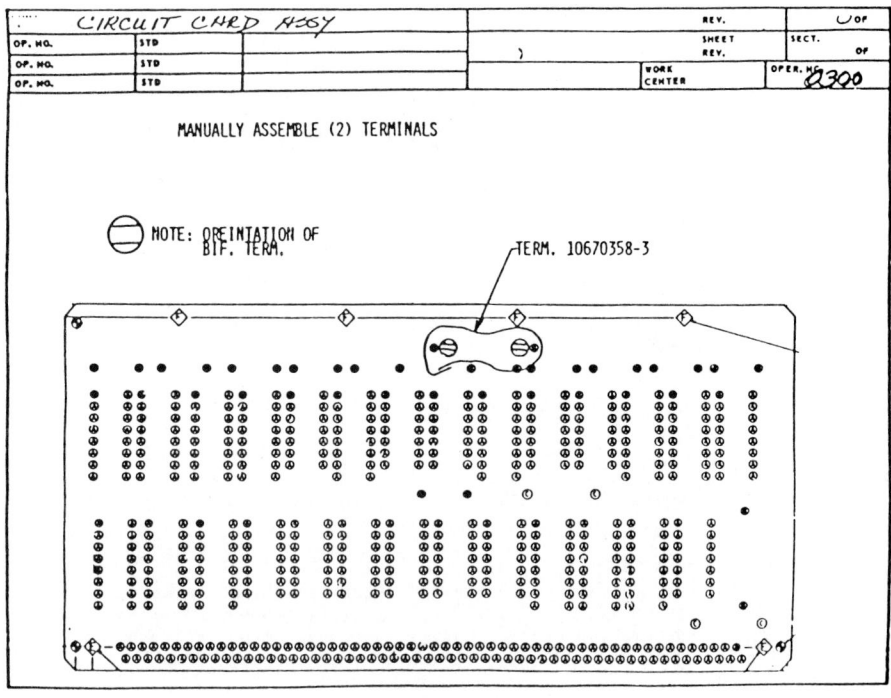

PROCESS ILLUSTRATION # 4

(b)

FIGURE 3.3 *(Contined)* (*b*) Illustration for manual process plan. (*Courtesy of Raytheon Company, Lexington, Mass.*)

this size, considerable time must be spent on communications within and between each organization.

The procedures subject to such surveillance are those for which representative items have been shown under preproduction requirements. One example of requirements that are more demanding than most commercial requirements involves process documentation. If an industrial engineer wishes to change any aspect of processing a part, a document must be prepared that identifies the part, specifies the present and proposed method, the reason for the change, and the cost impact, if any. At least two signatures are required before the document is attached to a process sheet to authorize a temporary change pending a permanent change in process documentation. See Fig. 3.7 for a typical change notification form.

Another example of the surveillance with which the industrial engineer interacts is the monitoring of the product cost profile. The customer is interested in performance of actual costs during the contract versus costs as proposed in the bidding phase. Normally the industrial engineer is charged with variance analysis and reporting to this requirement.

Control of critical parts is frequently required by manufacturing lot and even by serialization of each part. The industrial engineer must provide for recording these data on marriage sheets and the tracing of any necessary replacements.

A contract may include a specifically called-out industrial engineering or management technique which formalizes implementation or reporting tasks beyond that which a company's normal procedure would require. Such tasks might include the application of MIL STD 1567A (a work measurement program) or the templates for the "Transition from Development to Production" (a program management system). Industrial engineers may

OBM7R -AG   MARTIN MARIETTA CORPORATION
                              AEROSPACE GROUP
                             MANUFACTURING PROCESS

                                          MP - 70M32 AP 1 REV K
                                          ISSUED : 08/06/90
                                          REVISED: 08/20/90

DIVISION   - ELECTRONIC SYSTEMS CENTER

THIS DOCUMENT PROVIDES MANUFACTURING DATA TO ACCOMPLISH THE REQUIREMENTS OF:

     SPECIFICATION   REV   NOR   AMD   FINISH1 FINISH2 FINISH3 FINISH4 FINISH5
     DESCRIPTION
     13085036   C   007   000   INFORMATION ONLY
                                        THIS COPY WILL NOT
                                        BE KEPT UP TO DATE

SUBJECT:
     APPLICATION OF BLACK POLYSULFIDE SEALANT

1.
     SCOPE: THIS MP IS WRITTEN TO THE 13085036 SPECIFICATION.

1.1
     THIS PROCESS DESCRIBES THE REQUIREMENTS AND ESTABLISHES THE PROCEDURES
     AND CONTROLS FOR THE USE OF A BLACK, ROOM TEMPERATURE CURING POLYSULFIDE
     SEALANT FOR GENERAL PURPOSE USE IN BONDING, SEALING, OR POTTING VARIOUS
     METALLIC, NON-METALLIC AND OPTICAL COMPONENTS.

     NOTE: IT SHOULD BE UNDERSTOOD THAT ANY ENGINEERING DRAWING EXCEPTIONS
     TO THE PROCEDURES AND/OR REQUIREMENTS OF THIS MP SHALL TAKE
     PRECEDENCE. THEREFORE, THE TERMINOLOGY "UNLESS OTHERWISE SPECIFIED"
     WILL NOT BE REPEATED THROUGHOUT THIS DOCUMENT.

1.2
     APPLICABLE DOCUMENTS:

     THE FOLLOWING DOCUMENTS FORM A PART OF THIS SPECIFICATION TO THE EXTENT
     SPECIFIED HEREIN:

     MP-50046/2   VAPOR DEGREASING OF METAL AND METAL PARTS.

     MP-50047   ABRASIVE CLEANING

HAZARDOUS MATERIAL: POTENTIALLY HAZARDOUS MATERIALS ARE USED IN THE PROCESS.
REFER TO SECTION 8.0 FOR PRESCRIBED SAFETY AND HEALTH GUIDELINES

FIGURE 3.4 Typical manufacturing process to define operation method control. (*Courtesy of Martin Marietta Electronic Systems, Orlando, Fla.*)

be charged with the assessment of cost and system impact to incorporate such requirements, besides being the task leaders of the assignments themselves.

TRENDS

International Competition.   Traditionally, the United States government has retained a self-sufficient aerospace and defense industry on shore. This has been a tenet of the de-

MARTIN MARIETTA ITP INDEX LOADING FORM

FORM D-3201 DEC 81

PROGRAM ______
Q.E. ______ PHONE ______
DATE ______ TIME ______

1	2	3	4	5	6	7	8
R/A/C	SECTION	PLAN NO./ADDENDUM	REVISION	PF	SPEC/DWG – REV	PLAN TITLE	BOOK NO.

(1) ENTER "A" for ADD, "R" for REMOVE, "C" for CHANGE (for CHANGE, complete columns 1, 2, 3, 5, and any other affected columns.)
(2) ENTER Section Number of Index Affected
(3) ENTER Plan Number/Addendum (include Basic Plan No. followed by Addendum Suffix E.G. P103/VL-1)
(4) ENTER Revision Level of Plan Affected (if Addendum, Enter Revision of Addendum ONLY)
(5) ENTER Program Family Code
(6) ENTER SPEC/DWG/REV Level
(7) ENTER Plan Title (use this column to make clarifying statements if necessary)
(8) ENTER Book(s) Number(s) Affected

Attach copy of New/Revised ITP

FIGURE 3.5 Quality control planning form. (*Courtesy of Martin Marietta Electronic Systems, Orlando, Fla.*)

fense preparedness posture. Under this assumption the aerospace and defense industry has been more exempt than most United States industries from competition from foreign companies.

Now that somewhat protected position is changing. The technological lead that the United States once enjoyed in all important technologies has eroded. Today there are technologies where the components can be purchased outside the country to secure the most advanced developments (examples are microchips, optics, radar, high-power microwave devices).

As free trade across borders proliferates and the internationalization of the marketplace continues, specialty manufacturers around the world will continue to take turns leapfrogging existing technology. This assures a continuing worldwide acquisition of material by U.S. industry. Further, an extension of that trend applies also to systems builders and integrators. In an effort to drive down defense costs, total systems from foreign, friendly sources are allowed to compete against domestic producers, thus removing the last vestiges of a protected industry. From 1984 to 1986, military imports of aerospace products alone increased by over 30 percent from $1.1 billion to $1.5 billion. The industrial engineer in productivity improvement plays an increasingly critical part in the economic health of the industry.

Another aspect of competition is resulting in the increasing amount of joint ventures between companies within the United States or the United States and foreign countries. By utilizing the expertise of two or more companies, system development costs are reduced and the combination is more competitive in the world marketplace. Further, if the joint venture participants cross national boundaries, resistance to sales in each of the countries represented is lessened, helping to assure a more competitive position.

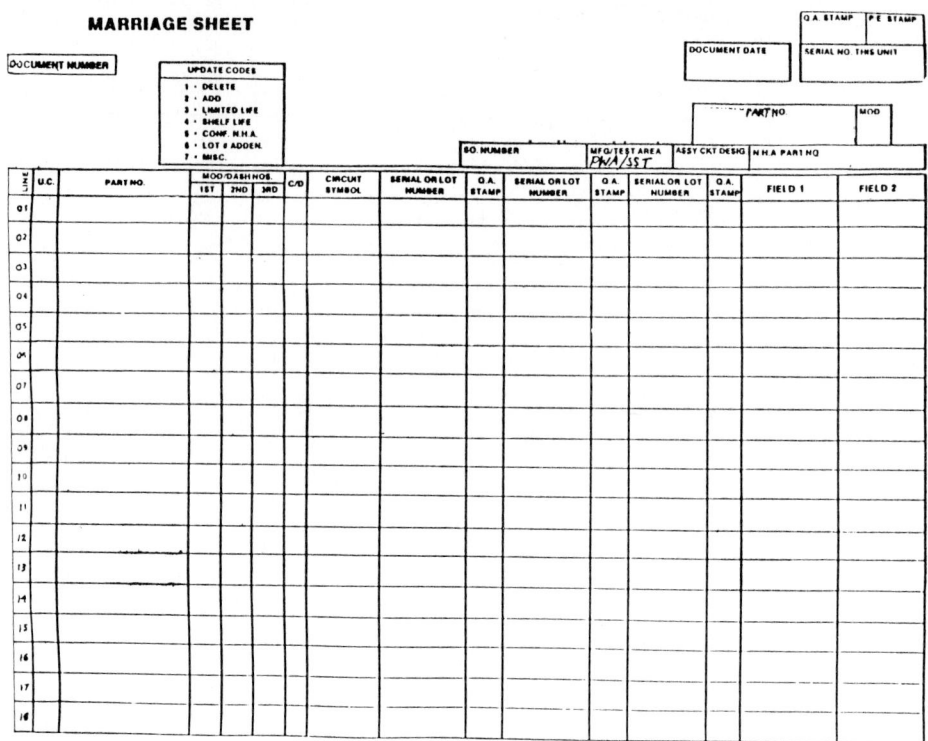

FIGURE 3.6   Form for traceability of serial or lot number. (*Courtesy of Martin Marietta Electronic Systems, Orlando, Fla.*)

International competition also is a pressure to the industry for the incorporation of metrics into product design. Foreign customer specifications and foreign component sources both are drivers for metrication.

Software.   The Department of Defense has pursued the strategy of offsetting superior volumes of conventional weapons by developing weapon systems that are technologically advanced. In development of such systems, computers are integrated to act as a performance multiplier to enhance system sensitivity, rapidity, optimization capability, and survivability. Thus not only is the hardware becoming more complex, but the integration of the controlling software complicates the total system even further.

The implications for the industrial engineer are in the merging of the hardware and software, that is, the assembly process, and the testing of the completed system to confirm a working system. The installation and testing of the software component of the system introduces an entirely new element and further enhances that definition of industrial engineering activity relating to systems integration. In this case it becomes the product itself rather than the process by which the product is built.

But the process also is changing. The planning and control concept of the "paperless" factory has long been talked about—and is finally being realized. Examples already are in existence and, as the systems designers and implementers become more skilled, the quantity and quality of installations (especially smaller ones) will proliferate.

Technology Proof of Concept.   The trend of the aerospace and defense industry to receive a reduced portion of the federal budget has resulted in an increase of a new program type, relatively unknown a few years ago. New concepts now frequently are nurtured through a

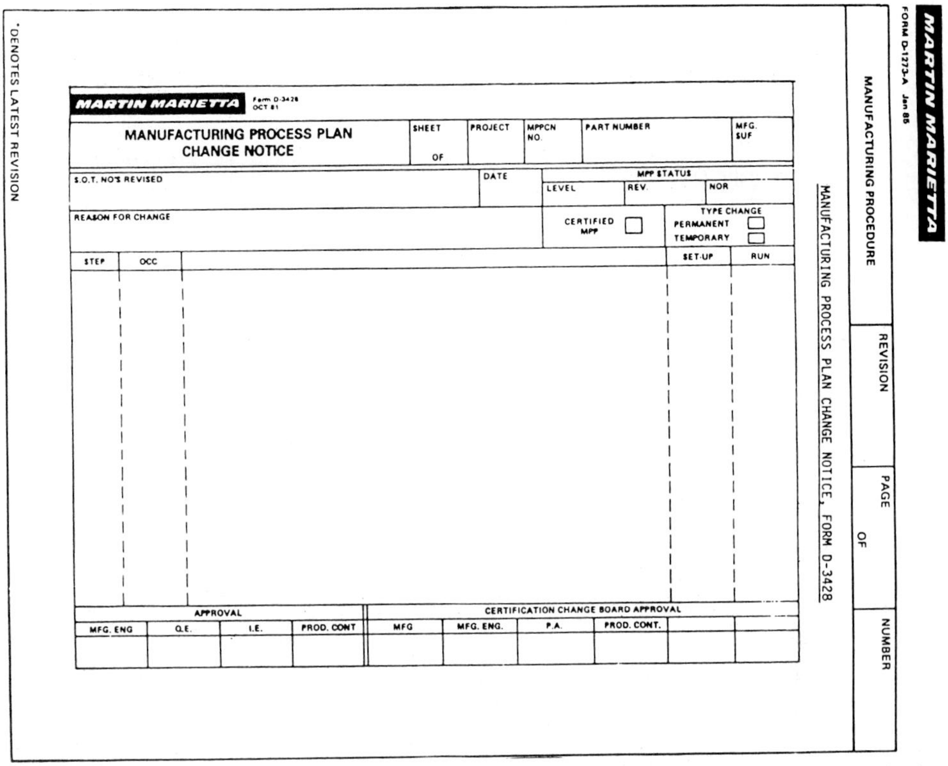
MARTIN MARIETTA
FORM D-1273-A Jan 85
MANUFACTURING PROCEDURE
REVISION
PAGE OF
NUMBER
MANUFACTURING PROCESS PLAN CHANGE NOTICE, FORM D-3428
*DENOTES LATEST REVISION

MARTIN MARIETTA Form D-3428 OCT 81

MANUFACTURING PROCESS PLAN CHANGE NOTICE

SHEET OF	PROJECT	MPPCN NO.	PART NUMBER	MFG. SUF

S.O.T. NO'S REVISED	DATE	MPP STATUS: LEVEL	REV.	NOR

REASON FOR CHANGE	CERTIFIED MPP ☐	TYPE CHANGE: PERMANENT ☐ TEMPORARY ☐

STEP	OCC		SET-UP	RUN

APPROVAL				CERTIFICATION CHANGE BOARD APPROVAL					
MFG. ENG	Q.E.	I.E.	PROD. CONT	MFG	MFG. ENG.	P.A.	PROD. CONT.		

FIGURE 3.7 Part processing change notice. (*Courtesy of Martin Marietta Electronic Systems, Orlando, Fla.*)

proof-of-principle and development phase, but never reach a sustained production phase. As a result, the industrial engineer performs more initial production planning while never reaching the all-up planning stage. It has the same benefits for industrial engineers as for their professional counterparts on such programs in that it incorporates the new technologies involved into industrial engineering practice.

Total Quality Management (TQM). The advent of total quality management will have a long-lasting effect on American industry. The aerospace and defense industry has enthusiastically embraced TQM as has its chief customer, the Department of Defense.

The impact on the function of the industrial engineer is also far-reaching. TQM emphasizes satisfaction of the customer quality requirements and institutionalizes continuous change for improvement, formalizing what has been a sometimes unrecognized factor of continually bettering existing systems. Yet industrial engineers have always been associated with the change process through the systems approach. New techniques shown to cause improvement within an organization should become culturally more acceptable to implement.

The continuous improvement emphasis will exist across the entire organization with a constant emphasis. This implies that a greater percentage of industrial engineering effort will be exerted in the nonproduction floor environment. In the aerospace and defense industry, this represents a large proportion of the population, which has received relatively little industrial engineering concentration. Many of the initiatives which are commonly used under TQM, such as employee involvement team, variants of statistical process control, and process simplification, are being enthusiastically adopted for the white collar environment.

Customer-Supplier Relationship.   Over the past generation the relationship between the customer (U.S. government) and the supplier (aerospace and defense industry) has been characterized by an increasing complexity of regulations. Each supplier sought to comply, and the customer representatives performed reviews and audits to ensure compliance. The inevitable cost associated with the regulations was considered a cost of doing business.

With the trend toward reduction in the defense budget as a percentage of gross national product, both parties in this relationship are reassessing the cost and are attempting to reduce it. A trend is emerging to reduce specific written requirements while increasing the general requirements for professional management. The industrial engineer's role is becoming increasingly one of accomplishing quality production at minimum cost in a professional environment. The customer is fast approaching the point of dealing only with suppliers who voluntarily meet the full spirit of TQM.

BIBLIOGRAPHY

Aerospace Industries Association of America, Inc., *Facts and Figures—Aerospace 89–90,* 1989.

Department of Defense, DOD 4245.7-M, *Transition from Development to Production,* September 1985.

Department of Defense, DOD 5000.51-G, *Total Quality Management Guide,* August 1989.

Department of Defense, MIL-STD-1528A, *Military Standard: Manufacturing Management Program,* September 1986.

Department of Defense, MIL-STD-1567A, *Military Standard: Work Measurement,* January 1987.

Elmaghraby, Salah E., "Operations Research," *Industrial Engineering Handbook,* McGraw-Hill, 1971, pp. 10-115–10-146.

Johnson, Joel L., "World Class Competition," *Aerospace Industries Association Newsletter,* July 1990, pp. 1, 2.

Latham, Jean Lee, "Eli Whitney," *The World Book Encyclopedia,* 1976, vol. W–Z, 1976, pp. 247–248.

Pearson, Richard G., "Human Factors Engineering," *Industrial Engineering Handbook,* McGraw-Hill, 1971, pp. 7-46–7-101.

Wright, T. P., "Factors Affecting the Cost of Airplanes," *Journal of Aeronautical Science,* February 1936, pp. 122–128.

CHAPTER 4
UTILITIES

Joseph H. Redding
Senior Vice President
H. B. Maynard and Company, Inc.
Pittsburgh, Pennsylvania

Lloyd B. Raymond
Principal
H. B. Maynard and Company, Inc.
Pittsburgh, Pennsylvania

The economic climate faced by utilities is important because it is driving programs related to industrial engineering. Utilities have been regulated monopolies, where increased costs were passed on to the consumer through rate increases. But during the 1980s that climate changed. This decade saw the breakup of AT&T and increasing regulatory interest in competition in the previously monopolistic utility field. By 1990, some deregulation had occurred and there were instances where rate increases were refused, or where the utility experienced lower-priced competition. These new developments are causing greater interest in productivity and costs.

In this chapter, utilities cover:

- Electric
- Gas
- Telephone
- Water
- Wastewater

The Role of Industrial Engineering. Utilities have problems which industrial engineers typically solve. They respect the industrial engineering discipline, and have many industrial engineers in their organizations. *But there are virtually no strong industrial engineering organizations in utilities.* As a consequence, there are no typical activities performed by industrial engineers in utilities. Rather, there are programs, initiated by various utility departments, which have an industrial engineering flavor. These areas include:

- *Right sizing:* This human-resources sizing program is typically initiated by a Human Resources department.
- *Work management systems:* These computer systems and (sometimes) expert systems

are typically initiated by a line department, such as Transmission and Distribution or Generation.

- *Incentives:* Some utilities are experimenting with incentives, typically initiated through Human Resources.
- *Supervisory training:* This is typically a Human Resources initiative performed for a line department.

This chapter will focus on these program areas, which are important to utilities and which industrial engineers will find compatible with their training.

RIGHT SIZING

Introduction.   Right sizing, sometimes called *downsizing,* is a type of program which utilities have used to reduce staff, principally administrative staff, as competitive factors and rate pressure have increased.

Software is used to provide summarized data and to highlight opportunities for change in or redeployment of human resources assigned to a function. These outputs include:

- Work activities performed by each work group
- Human and nonlabor input for each activity
- Internal and external customers of the output from each work activity
- Qualitative or quantitative measures of output from key work activities, and their role and value in supporting the company's business objectives
- Measurement of internal customer satisfaction with quality and timeliness of current work outputs
- Data to support reengineering of processes through elimination, combination, or automation to improve quality and timeliness.

A comprehensive right-sizing program that has proven to be very effective involves 10 steps.

1. Conduct Program Introduction.   This task involves an interview with senior management to assure that objectives for the personnel evaluations are clear, and will be met. The composition of the team of middle managers is determined at this time.

Senior management assigns one person to oversee this program from an administrative viewpoint. This project manager can arrange for progress reviews and can help to secure information needed by the team from senior management.

At this time, the program ground rules are set. There are two key considerations:

- Will anyone lose a job over this review?
- Will comments made in interviews be held confidential?

The program schedule is then refined.

2. Train Team.   Heavy team participation requires that the team members have an operational understanding of the techniques which they will be using. Training will include:

Orientation.   The team is first given a thorough description of the process so that the other training makes sense. The three data collection methodologies are discussed:

1. *Direct employee questionnaire.* This details the forms to be used, the type of data to be collected, the role of the task dictionary, separately coded information, and the uses of

the questionnaire data. Employees report directly what they do, the time they spend, and the rework they perform, according to a company dictionary of activities.

2. *Interviews.* These will surface quality and service problems and will identify any cultural factors which may inhibit service and quality. Interviews will cover at least 10 percent of all company employees across all functions and activities.

3. *Process flow charting.* This is done to define the flow of key processes within and across departmental boundaries. These data are a primary source for the activity dictionary.

Interpersonal Skills and Team Building.   This training helps the middle management team provide an overall analysis of personnel requirements. The entire team participates. They will be given the skills to organize, interact effectively, avoid blocks to effective performance, stay on schedule, and represent their efforts clearly to the rest of the organization.

Problem Solving.   A fundamental skill is problem solving. This training provides a process whereby the team can identify and clear away any blocks to problem solving, and will show team members how to generate alternatives, evaluate them and select the best solution, and then plan for implementation, including accountability.

Cost of Quality.   This training describes the cost of quality in terms of: prevention costs; inspection/correction costs; and field failure costs. The team is taught how to identify these three types of costs. They learn how preventive measures provide a cost-effective way to reduce the cost of poor quality and to increase customer satisfaction.

Flow Process Charting.   This training teaches the team to use the basic elements of flow process charting in the context of charting major business processes. Charting identifies flow elements (act, delay, transport, inspect, and store), the flow path through various organizational functions, and the approximate timing involved. The team is trained to handle process steps which occur only part of the time. They are taught how to avoid excessive detail in charting.

3. Map Key Processes.   The team first identifies the processes which are key to the business, that is, strategic processes. These processes include service requests, trouble calls, billing, corporate planning, system planning, construction management, rate case preparation, and commission reporting.

The team then flowcharts each process. The team draws on other personnel as resources to help to describe the appropriate activities. The results of this charting are coded and entered into a computer, and the results will be displayed graphically for the team, and other managers, to use.

During flowcharting, major internal and external customers are identified. Major internal customers are typically the people who receive work from a preceding activity across an organizational boundary, such as when the field receives engineering and customer information prior to performing work. Interviews focus particularly on these customers' perception of service and quality.

Finally, the team validates the process flows which it has charted. This is done through interviews with employees who work the processes, or simply by having the employees mark up the graphic displays of the flows.

4. Validate and Complete Activity Dictionary.   This task is in preparation for the employee questionnaire, described next. An activity dictionary is used. This lists activities involved in the major processes, and other indirect activities, such as preparation of management reports, commission reports, etc. A starter dictionary includes many general business activities. It is expanded, primarily from the major process flowchart information.

5. Collect Data.   Various types of data are collected. Each employee reports his or her typical work on a computer input form. The team interviews at least 10 percent of all employees. They conduct an internal customer survey, to identify internal customer satisfaction with quality and service. And they quantify nonlabor input to the major processes, so that the process result will permit staff budgeting, related to activity levels.

Input Forms.   Each employee is asked to fill out a questionnaire, detailing how he or she spends a typical week. They report their average time against activities in the activity dictionary. They indicate the percent of each activity time spent in rework. This is a primary source of the cost of quality information.

It is essential that all forms be confidential, and that no employee be criticized or reprimanded for anything reported.

Interviews.   The team interviews 10 percent of the total employees. These interviews are structured to assure confidentiality. The middle management team, or persons specially selected, conduct the interviews. In consulting-based programs, the consultant spends a significant amount of time with the interviewers to assure that they are getting unbiased information and that the results will be held in confidence.

Internal Customer Survey.   The team collects information, on the quality and service perceived by employees, related to work performed on major processes, particularly across organizational boundaries.

Nonpersonnel Costs.   The team identifies any nonpersonnel costs associated with major processes and with secondary activities. These costs include computer costs, outside contractor costs, and other costs to permit complete budgeting by using the process results.

6. Code Data.   Additional information is coded onto the computer input form. First is the actual grade level of the employee. Second is the estimated grade level required to perform each reported activity. The organizational structure is also coded into these questionnaires.

7. Process Data.   A computer system is used to analyze employee interview data and to report back to the team the following typical information:

- Activity costs, shown high to low, by the entire organization, by subunits, and by processes
- Fragmentation analysis, by activity and by personnel
- Primary versus secondary activities in each department
- Process cost
- Cost of poor quality
- Compensation versus task complexity
- Activity importance versus resources allocated
- Span of control and levels of management

8. Analyze Data and Develop Measurements.   The questionnaire data are analyzed to ensure that they accurately reflect actual conditions.

Additional data are provided for the basic analysis. The teams are helped to review and analyze the data, and to identify any inhibitions which they may have to addressing the issues which surface. Consulting input is especially helpful both here and in the development of recommendations when an organization first undertakes this type of human-resources review.

9. Develop Recommendations.   The basic approach here is to assure that all problem areas are identified and addressed with recommendations. The consultant provides a double-check on the completeness of the analysis and recommendation, and gets the team over any initial resistance to the task. This task will be much easier the second time it is performed.

A sound recommendation will consider priority, implementation cost, and timing. Consultants will assist the team to make recommendations which senior management can support.

10. Present Recommendations.   The team presents its recommendations to senior management.

WORK MANAGEMENT INFORMATION SYSTEMS

Introduction.   Work management, as related to field operations, is a major area of utility concentration, because so many utility employees are involved in field construction and maintenance activities, and because the result of their work has a major impact on the customer's perception of customer service.

Office operations can also be studied by industrial engineers, but major improvements in the office will more likely come from automation. (Also see Sec. 5, Chap. 7.)

Historically, work management has gone through several phases:

1. *Maintenance management (prior to 1956).* These systems dealt with considerable "firefighting." There was minimal planning, with job times based on estimates or historical job times.

2. *Universal maintenance standards (UMS) or work force management (WFM) systems (1956–1986).* These systems significantly reduced firefighting. The systems were manual, and involved increased planning, with job times based on impartial, engineered standards. WFM systems depended on experienced craft planners. Methods were documented. Limited management information was available. The systems focused on craft or journeyman productivity. (Also see Sec. 5, Chap. 6.)

3. *Work management (WM) (1986 to present).* WM minimizes firefighting. Planning is done with job times based on engineered standards using "best method" multiactivity analysis to optimize crew utilization. Computerized systems and expert systems eliminate the need for special "planner" positions. Complete management information is available from the computer systems. WM focuses on meeting customer requirements with optimum utilization of all resources. (Also see Sec. 10, Chap. 6.)

Maintenance Management.   Initial maintenance management programs focused on managing craft resources in the face of significant firefighting. They depended on two common time-setting techniques: *estimating* and *historical data.*

Estimating-based systems were

- Inconsistent and inaccurate because of variations in personal experience and peer pressure.
- Lacking in credibility and totally dependent on the person making the estimate.
- Limited, with little or no documentation to support the estimate's work content.
- Subject to political pressures where, either intentionally or unintentionally, making the estimate might be worth sacrificing quality, safety, cost, and/or effective utilization of resources.

Programs based on historical data:

- Typically preserved the status quo.
- Did not identify job interferences and delays.
- Gave average job estimates, which ignored job-specific conditions that caused actual job times to be much higher or lower.

- Required work to be performed before the time was established for the next repetition of the job.
- Created times which were comfortable.
- Created peer pressure to have the reported time values "pegged" to match the historical values.
- Typically had little or no work methods documentation to support making improvements in work methods and in justifying equipment, people, dollars, and quality.

Universal Maintenance Standards/Work Force Management.   These programs began about 1956 as universal maintenance standards programs. They used methods time measurement (MTM) data to produce universal standard data (USD). USD tables contained predetermined time values for using various tools and performing short-time-cycle, common, repetitive activities. USD was used to analyze typical maintenance activities, to document work methods, and to generate a work method and activity time, creating UMS bench marks.

These UMS bench marks established the time to perform work and were applied to create work order times by planner analysts (positions created in maintenance departments to help plan, schedule, line up materials, and improve maintenance productivity).

With the use of UMS bench marks and planner analysts, maintenance began moving from experience-based and firefighting styles of management into planned, fact-based management. Knowing the work load led to defining meaningful priority systems, establishing improved frequency of repairs, and formalizing the managing of maintenance. (See Sec. 5, Chap. 6 and Sec. 10, Chap. 6 for further discussions of these techniques.)

Measures were established to assist in documenting the impact of a UMS program and to focus attention on improvement opportunities. Typically the measures consisted of:

- Utilization, which reflected time lost to delays—previously absorbed and not known.
- Performance, a measure of accomplishment as compared to bench marks, that is, actual time minus delay time compared to a work order's estimated time.
- Productivity, the comparison of total time, including delays, calculated as utilization × performance, sometimes called *net effectiveness.*
- Cost per standard hour, base wages divided by productivity to establish a cost of maintenance and provide a means of tracking the impact of productivity improvements in maintenance.
- Coverage, amount of time spent working on jobs covered by UMS bench marks.

UMS/work force management programs were adopted starting in 1960 by a number of utilities including electric, gas, water, and wastewater, and by municipalities concerned with street, park, and golf course maintenance.

UMS, or work force management, programs were sometimes also known as *planning and scheduling* or *planner-analyst programs.* All of these programs had common characteristics:

- Bench marks were developed, using industrial engineering techniques, normally by teams of journeymen.
- Planner analysts applied bench marks to create work order times, and assisted in planning and scheduling.
- A management information system was used to report performance statistics.
- Productivity improved through use of organized approaches to planning, scheduling, and the reduction of emergencies and delays.

The UMS approach produced documentation that assisted maintenance in improving work practices and in training journeymen. As technology changed, the frequency and complexity of maintenance has changed. Documentation has proven to be a valuable tool.

Documentation helps to analyze maintaining versus replacing, to modify designs to simplify maintenance, and to revise the list of parts purchased so as to improve maintainability and life-cycle cost.

These early UMS programs were manually developed and applied and had the above-mentioned advantages over the estimating or historical data approaches. However, because they were manually operated, they were difficult to maintain, supplied limited information, and focused primarily on people, not on all resources.

Work Management.   Today, computer technology, advanced industrial engineering techniques, and management philosophy changes have contributed to the development of *work management.*

Work management shifts the focus from controlling people in work force management programs to resource management. Work management is a data-based, information-structured process that assists all organizational levels in serving their customers while providing good utilization of resources. Resources include materials, equipment, tools, money, time, and people.

Work management systems support comprehensive management processes that assist in supplying management with more objective information. The systems supply management tools, techniques, information and measures that support empowerment with accountability through participation.

Customized, objective management approaches are developed for:

- Establishing the work load of functional areas
- Establishing the expected time for the work
- Planning of resources and coordination
- Scheduling to meet customer's needs
- Assigning work with appropriate instructions and information
- Performing work safely, with the highest quality, and with the best method to optimize resources

A work management system that is designed for the future offers a number of options not available in the days of maintenance management or work force management programs.

Common Characteristics.   Common characteristics of the work management system of today are

- Computerized work measurement techniques are applied by trained, knowledgeable journeymen. They create documented suboperations for identification of a task's work content according to best work methods. The documentation includes a list of proper tools, equipment needed with utilization requirements, a material list, coordination needs, staffing and staff utilization, special instructions, and time. The suboperations form the foundation of the comprehensive estimating system. The time, effort, knowledge, quality, and energy expended in the development of the suboperations establishes the credibility and strength of the foundation of the estimating system. Developed by the journeymen, these suboperations provide the foundation of many benefits gained from the work management process.
- Planner analysts are not required. Engineers, supervisors, foremen, and clerks can be applicators in creating the documented best methods estimates.
- Computer expert system technology is used to create an easy-to-apply estimating process which produces consistent and accurate work orders. The very flexible, custom-designed work orders assist all organizational levels in managing better.
- Management information focuses on information to assist in managing resources.
- On-line, real-time information systems interface with mainframe computer systems on an as-needed basis to provide timely data.

Participation.   A work management program is best founded on participative processes where management is committed to providing its managers with the highest-quality management techniques and processes. These organizations recognize the importance of and are committed to optimizing resources through the use of fact-based, information-structured management. The organizations are continuously training to enhance the benefits of being on the leading edge of technology and management processes.

In a work management system, each work management process is structured. Actions of each organizational level are defined. Structured information is used along with appropriate measures to enhance decision making. Each process interacts to support the next process.

The structured processes include:

1. Work flow
2. Planning
3. Scheduling
4. Assigning
5. Performing work
6. Supervising
7. Clerical/administrative activities
8. Management information
9. Management actions

Work Flow.   All work is tracked. The impact of shifts in work load mix is evaluated in terms of resource requirements. The magnitude of the work load is continuously monitored to *ensure capacity to meet internal and external needs.*

Planning.   Planning is a process that takes place before scheduling. It is a process which establishes:

• Time
• Materials
• Equipment
• Tools
• Coordination
• Engineering, instructions, information
• Work method, job sequence, phases
• Access
• "What if" scenarios
• Dollars
• Completion dates

Other considerations such as weather, vacations, skills, and *special customer needs* are formally reviewed, organized, committed to, and communicated to those appropriate.

Planning.   Planning is also a formal process conducted at *all* organizational levels where the above considerations are reviewed for various windows of time. Windows for ranges from hours to weekly, quarterly, annually, and even multiple years are formally established for evaluating the resources required and/or affected. Compliance to planning is measured by those responsible for the work and for continuous improvement.

Scheduling.   Scheduling is a process of selecting the jobs, projects, and tasks to be worked on from the planned work load. Scheduling details:

• Priorities
• Start and finish times (dates and hours)

- What (scope, limit)
- Who
- How
- Alternatives

Other considerations, such as the use of overtime, contractors, vacations, and last-minute call-ins are reviewed, committed to, and documented. Schedules are distributed to crews and appropriate functions and areas. Schedules are developed for various lengths of time (windows—daily, weekly, monthly, and so on) to *meet internal and external customer needs.*

Compliance to schedule is measured by those performing the work. Continuous improvement is a goal.

Assigning.   Assigning is the process of communicating a request for work to those who will perform it. Assigning takes place after planning and scheduling have been accomplished. It is the process where information is transferred to crews, allowing work to be performed with optimum resource utilization.

Typically a job/project/task work order package is given to a person or crew and may or may not require verbal communication. Assignment packages are available before the scheduled time to assist in maintaining or improving short-range planning by the crews and to provide valuable two-way communications between foremen and crews.

A work order package contents could include:

- Customer needs identification
- Engineering file
- Coordination—operations, dispatcher, customer
- Outage memo
- Material list
- Job phase listing
- Instruction books
- Special instructions
- Safety permits

Anyone, at any time, can obtain the next assignment without waiting for the supervisor.

Performing Work.   Planning, scheduling, and assigning have been done in a way that lets those performing work do it in the safest way, with the highest quality, with maximum productivity, and with optimum utilization of all resources.

Work performance is a process where the primary considerations are

- Safety
- Quality
- The best method

Other considerations, such as training, identification of delays, documentation of test readings, as-found conditions, and engineering files, are monitored. Performance measures and formal site visits are used to evaluate events and to supply data for continuous improvement.

Supervising.   Supervising is interaction to supply support for the optimum utilization of resources in meeting internal and external *customer needs.* Supervising is a process where:

- Subjectivity is replaced with factual data
- Uniform and consistent approaches are used to establish levels of expectations, in time

to perform work, of quality, quantity of work, job content, resource utilization, and actions taken

- Planning is understood, performed, and monitored and includes the training of others
- Scheduling is understood, performed, and monitored and includes the training of others
- Assigning is done in an efficient manner that ensures that job content and decision levels are understood
- Supervisors visit with employees at the work site to monitor safety, observe areas of needed assistance, maintain awareness of technical and physical job changes, monitor quality, monitor utilization of resources, and search for improved work methods
- Effective communication is utilized to provide understanding of company policy, expectations, training, participation in planning, resource utilization, and best methods

Actions speak louder than words, and leadership is demonstrated by supporting, not pushing.

Clerical and Administrative.   These processes can be minimized through existing technology. The objective is to have paperless, on-line, real-time systems. Systems are designed with only two points of input: (1) the creation of the job/project/task work order and (2) the field recording of events. The expert system reduces the preparation of the work order to the level of using a computerized word processor. Hand-held units with or without bar coding can be used to record field events. The expert system and field-reporting units interface to reduce data input requirements.

Data used to support work management are obtained by interfacing with present, mainframe computer systems on an as-needed basis. Information input into work management that is of value in supporting existing computer systems is transferred when required to reduce duplicate entries.

Management Information.   The management information process is structured for each organizational level in detail, and supports decision making. Information needs at the crew level are in detail. A "roll-up" approach is used to generate higher-level information by:

- Job
- Function
- District
- Area
- Company

Management information is grouped to assist with evaluation and decision making. Work load data are available in these categories:

- Backlog (past-due work)
- To be scheduled
- In-progress
- Completed
- Transferred (work done for and by others)

Reports are structured with from/to dates for different windows of information (daily, weekly, monthly, annual). Reports are designed to support resource evaluation in terms of types of work, labor and staffing, equipment, materials, and dates. Reports supporting tracking cover:

- Jobs
- Application approach used to create work orders, including benchmarks, estimates, after-the-fact interferences, times, and opportunities for improvement in future work
- Job variances
- Relative changes in productivity, work load, and customer response

Performance reports include:

- Compliance to planning
- Compliance to scheduling
- Customer response
- Utilization—labor and equipment
- Delays
- Performance
- Productivity
- Cost per standard hour

Management Action.   Management action processes will vary depending on the organizational level and data being evaluated. The basic philosophy is to promote *objective decision making* through the use of fact-based management.

Each management report will support action. A number of reports will be required to evaluate complex relationships. Information is structured in tabular formats as well as in graphic displays. Facts are available on:

- Productivity
- Work load (backlog, in-progress, completed)
- Cost per standard hour (in-house versus contractors)
- Customer service response

Management uses the facts in preparing and evaluating budgets, justifying staffing and equipment, identifying training needs, documenting improvement in productivity and cost reductions, and improving customer service.

Work management improves teamwork and customer service through improving communication processes, using structured processes to move decision making to the lowest appropriate levels with information to support and measure decisions made, and supplying management with information to support management actions regarding budgeting, staffing, justification of tools and equipment, and levels of customer service.

Utilities have used industrial engineering to develop scheduling and productivity standards for field operations.

INCENTIVES

Incentives are just beginning to be used to motivate employees to identify and solve problems. Examples include Orange and Rockland Utilities and Central Vermont Public Service. Utility incentives tend to have relatively modest payouts and incentive pull. Rate commissions would likely veto industrial incentive payouts of 20 to 30 percent for fear of adverse public reaction.

Incentive programs in utilities have the character of gain-sharing programs, but with less simple economic payout formulas than are typical of industrial gain-sharing programs.

An example is Central Vermont Public Service's gain-sharing program, called *CV SharePower,* initiated in 1990.

CV SharePower.   CV SharePower was designed and implemented by a volunteer team of CV employees who were chartered by CV's president and chief executive officer. CV SharePower is described as an involvement program combined with a gain-sharing plan. CV SharePower is built on three principles for organizational effectiveness:

- CV employees must act as a cohesive, single unit.
- Employees can, and are willing to, contribute ideas and suggestions.
- Improving CV improves the future of all employees.

Elements of the Plan.   The elements of CV SharePower are

- Problem-solving process
- Facilitators
- Guidelines for team makeup
- Team member responsibilities
- Team topics (must be business-related)
- Team authority and responsibility
- Team organization
- Meeting structure (special authorization is required to keep a team active beyond 60 days)
- Record keeping and reporting requirements

Problem-Solving Process.   The CV SharePower problem-solving process involves six steps:

1. Define the problem
2. Plan strategy for analyzing the problem (fish bone diagrams are recommended)
3. Collect and analyze information
4. Generate alternative solutions
5. Evaluate and select a solution
6. Plan action steps, accountability, and measurement systems

Payout Criteria.   CV SharePower payout involves several criteria. The first is an affordability criterion. Payouts require that earnings exceed dividends.

Then, employee control is evaluated. Gain-sharing funds are created when actual earnings exceed budgeted earnings before taxes. The CV SharePower steering committee may recommend adjustments for extraordinary items, including unusual weather, tax and regulatory changes, unexpected nuclear plant performance, and gains on the sale of assets.

Gains from increased sales are limited to $500 per employee, unless the gains are directly related to employee efforts. The gain-sharing pool is reduced if capital projects overrun by more than 5 percent. The gain-sharing pool is increased or decreased on the basis of two customer service measurements, as compared with the planned level of those measurements: (1) service reliability (average minutes per outage) and (2) customer favorability as measured by customer surveys. Gains are shared equally by employees and the company. CV SharePower resulted in a payout to CV employees in its first year of operation. Employees increased their participation by looking for efficiencies and suggesting service improvements in their work areas and for the company as a whole.

SUPERVISORY TRAINING

Utilities traditionally promote supervisors up from the ranks. New supervisors are not trained, but left to formulate their own style, or possibly to pattern their behavior after their first supervisor. There is a need for supervisory training which helps supervisors to discover what tools they need to do business, and then gives them the tools. This type of supervisory training can also provide information which can be used to formulate new systems which will have assurance of employee buy-in.

In 1989 AT&T prepared itself for radical change. The chairman declared that his company must escape the passive tendencies of its past. AT&T would no longer sit back silently, satisfied with predictable price increases, while strategic competitive advances and uncontrolled costs eroded its position in the marketplace. The time had come for rapid and aggressive action.

AT&T's Phoenix Works vice president shared the chairman's ambitious outlook. He recognized many of the same problems in his company. Productivity suffered because of poor communication between management and lower-level positions. Employees felt their opinions and suggestions were ignored, and consequently did not feel accountable for business results. Change was fast approaching all areas of AT&T Network Systems. The president of AT&T's Network Cable Systems was already studying costs under the management microscope and expected higher levels of profit performance in the immediate future.

In order to meet these expectations in his own group, the Phoenix Works vice president needed to prepare his people for the upcoming changes. New programs such as just-in-time, total quality management, and total productive maintenance were sure to be introduced as a result of the movement. If his employees were not open to change, however, these programs would fail. He had to find a way to make his people accept the new philosophies with enthusiasm.

The Phoenix Works initiated a program to facilitate these changes. A consulting firm designed the plan to open up both the communication lines within the company and the attitudes toward change among the employees. Employees welcomed other productivity programs more readily, and the Phoenix Works climbed toward world class manufacturing.

The success of AT&T's improvement revolution was significant. Teamwork improved, supervisors acquired a team spirit, and communications and goal-setting skills are continuing to get better. The results? Supervisors improved their management of a number of cost-reduction programs. For example, the maintenance department achieved a 25 percent reduction in maintenance and repair spending—a multimillion dollar annual savings. There was also significant improvement in product quality and equipment availability. A new optimistic attitude toward change is present at the Phoenix Works. Continuous improvements are accelerating their growing success.

How the Training Program Was Designed.   The program focused on the specific needs of the plant. The Phoenix Works did many things right. The vice president wanted to focus only on the skills which needed to be developed.

The program had the support of all staff levels. Plant employees often rejected input from outsiders, but because other AT&T employees conducted the training, everyone accepted it. The program taught new techniques and solutions to real plant problems. Plant employees learned to apply their new skills to everyday situations.

It encouraged senior plant management to support the new supervisory tools. Previous programs foundered when management did not reinforce the new techniques. As a part of the new process, management implemented specific review meetings to assure that the training was being utilized.

The Training Process.   Training followed a nine-step process.

1. Selecting a Training Team.   A team of local supervisors conducts the training of their peers. A typical team includes members from departments which affect productivity improvement.

2. Conducting Confidential Interviews.   The team interviews all management, supervisory, staff, and support levels in the organization. The confidentiality of the interviews encourages candid and honest comments.

3. Assessing Training Needs.   Each organization has its individual strengths and needs. The Phoenix Works team identified the following problems in their program:

- Inadequate planning at all levels
- Poor problem-solving techniques employed by management
- Lack of accountability for results
- Poor measurement systems resulting in a lack of focus on major problems
- Failure to directly relate performance to rewards for supervisors and managers
- Inadequate training at all levels
- Stress on output quantity rather than quality
- Authoritative management style that discouraged teamwork

4. Constructing Training Materials.   The team uses information from the interviews to construct true-to-life case studies that deal with daily problems. They also develop instructor and student manuals, lecture materials, transparencies, and questionnaires from the data.

5. Coaching the Team in Training Techniques.   The team learns how to make effective presentations, how to involve participants in discussions, how to build training enthusiasm, and how to relate the training to real business situations.

6. Conducting the Training.   The questionnaire, interviews, and seminar sessions allow the employees to recognize the need for change in certain areas. A powerful part of the training is a series of real-life case studies that highlight these areas.

7. Presenting the Management Seminar.   Management listens to the problems and "hot" issues that surfaced during the interviews and training sessions.

8. Developing an Implementation Strategy.   Senior management selects the most critical topics and asks for recommendations.

The Phoenix Works management chose these topics:

- How to improve work planning
- How to improve supervisors' understanding of their responsibilities
- How to improve and clarify work communications
- How to institute needed skills training

The Phoenix Works team made a second presentation to senior management, and all its recommendations were accepted.

9. Monitoring and Assisting in Training Implementation.   Management selects the plant departments that would benefit most from reinforced training. The team implements the training with emphasis on topics selected by senior management.

Up until this point, middle management at the Phoenix Works had voiced reservations about the training program. But the hands-on training enabled them to tackle a planning issue with major quality consequences. Middle management became more supportive.

Training Results.   As a follow-up, seminar participants fill out a questionnaire evaluating the overall reaction to the program. At the Phoenix Works, the results were impressive:

Areas of improvement	Percent of trainees seeing positive improvement
Goals are set and results are measured	85
Subordinates are more accountable for measured results	80
Emphasis is shifted from quantity to quality	85
Management style is more participative	67
There is increased teamwork within and among departments	65
Meetings are now more focused on the subject	67
Accountability for results is assigned in meetings	60
More training is provided at all levels	90

When asked their overall opinions of the program's effectiveness, the employees at the Phoenix Works were encouraging. Some of the comments included:

• There is definitely more communication from top management. Results are communicated better. Goals are established.

• Quality is the total business environment, and that message is coming through loud and clear.

• The teams are starting to take action to solve problems sooner and following up to make sure that the problems are truly solved.

The training paved the road for the Phoenix Works' journey toward world-class manufacturing. Cultural changes began with managers and supervisors, and worked their way to all employees. With the support of senior management, it educated the Phoenix Works in the principles of teamwork, communications, goal-setting, and accountability for results. This eased the launching of new programs including just-in-time, total quality management, and total productive maintenance. Business results followed. Costs came down and customer service improved. The Phoenix Works personnel manager pinpointed the secret behind the training's success: "When you change the way people look at their jobs and collectively work toward finding new ways to solve problems, the results can be dramatic." At the Phoenix Works, dramatic results are still taking place.

THE FUTURE OF INDUSTRIAL ENGINEERING IN UTILITIES

The Climate for Change.   Utilities are in a period of flux, when changes will be made. Economically, utilities are experiencing pressure on costs, both from competition and from a reluctance of regulators to grant requests for rate increases.

Utility managements are evolving new styles to deal with change. As with many other U.S. companies, utilities are moving toward participative forms of management based on the idea of empowerment of the work force. Empowerment is described in William C. Byham's *Zapp, the Lightning of Empowerment.*

Right-sizing has flattened organizations, pulling out layers of middle management. Empowerment is a logical conclusion in these flattened, leaner organizations.

While utilities are stressing participation and empowerment, they are not advocating laissez-faire management. As with total quality management practitioners, methods-based standards work still is needed for scheduling and cost management.

The most important factor in change is now employee buy-in. As in other industries, systems were forced on utility employees in the 1960s and 1970s, only to be rejected and fall into disuse. Today, utility top management is especially sensitive to getting buy-in before embarking on any new system or organizational change.

Further, the utility personnel who will willingly work around the clock to repair storm damage are especially critical of any new job requirements which do not help get the job done for the customer. Utilities are focusing strongly on having every new action contribute to improved customer service.

Employee Buy-in.   Recently, the Bonneville Power Administration (BPA) employed a consultant to assist in evaluating the need for a work management program in transmission line and substation maintenance. This evaluation is a model of the correct way to assess a new system. BPA assembled a team of investigators from three organizational levels—union supervisors and the first two levels of management. A questionnaire covering 10 aspects of work management was developed, and the team went to various field locations and interviewed field crews and other supervisors to find out:

1. How effectively are they now able to manage their work?
2. How well could they manage with the right management tools?
3. What was needed to get from where they were to where they wanted to be?

The consultant's role, and an equally valid role for an industrial engineer, was to help the team structure the interview, to train them in interviewing skills and note taking, and to help them debrief after the interview. After much discussion, in which the consultant was a participant, the team drew its own conclusions, and did so in a highly professional manner. The team recommended to BPA management a new and expanded work management capability to help maintain the BPA system. The union has agreed to the proposed new system. The entire process took 7 months. A smaller group may have completed the task in 2 months. However, the recommendations would not have as thoroughly addressed the needs of the field, nor would they have enjoyed as much support as the recommendations of the full committee. In a broader perspective, the 7 months were only a brief period when one considers the strength of the foundation which was built to support later work management system developments, and the fact that such systems tend to operate over a 10-year period before being subject to major revision.

CONCLUSION

Utilities are changing, and will be managing customer service and productivity more closely than ever before. Utility workers, through empowerment, will have a bigger say in what changes are made. We have discussed many of the major types of change programs in this chapter. While there are few formal industrial engineering departments, there are excellent opportunities for an industrial engineer to play an important role in change. Where there is top management commitment and worker level buy-in, an industrial engineer will find it both challenging and rewarding to work on these changes.

REFERENCES

Byham, William, C., *Zapp, The Lightning of Empowerment,* Harmony Books, New York, 1988.
Central Vermont Public Service, *CV SharePower Guidelines,* February 1990.

Dobbs, John H., and Paul A. Elbert, "Performance Improvement in the Utility Industry: One Experience," *Public Utility Fortnightly,* March 6, 1986, pp. 21–24.

Laros, Michael A., "The Changing Dimensions of Utility Human Resources Management," *Public Utilities Fortnightly,* March 18, 1982, pp. 44–51.

Messina, Richard J., and John C. Reece, "Managing Information for Tomorrow," *Public Utilities Fortnightly,* March 6, 1986, pp. 25–31.

Necessary, Curtis, and Donald C. Craft, "Maintenance Performance Indicators," *Maintenance Technology,* January 1991, pp. 31–32.

"On-Board Terminals Boost Productivity," *Electrical World,* January 1990, pp. 49–50.

Ryder, F. H., and D. F. Farr, "Improving the Performance of the Performance Improvement Process at a Power Company," *Industrial Engineering,* July 1991, pp. 51–54.

Tuttle, Thomas C., "Strategic Performance Measurement," *Public Power,* July–August 1991, pp. 16–19.

CHAPTER 5
GOVERNMENT

Marvin E. Mundel
Principal
M. E. Mundel & Associates
Management Consultants
Silver Spring, Maryland

Government organizations (the formal infrastructure that operates the controls on the named sector) may exist at the following levels (the terms may vary from country to country and the list is not exhaustive):

1. Ward or borough
2. City or town
3. County
4. State
5. Country
6. Treaty group
7. International

There are two basic kinds of government organizations: those with objectives which can be stated in physical terms and those with objectives which can be stated only in social terms.

Government Organizations with Physically Stated Objectives. Government organizations which have objectives which can be stated in physical terms exist in fields of endeavor such as

1. Road, bridge, airport, and other infrastructure construction and maintenance activity, including highway grass cutting, cleaning, repairing, and park maintenance
2. Printing, binding, and engraving
3. Production of military supplies and hardware
4. Warehousing and distribution, particularly of, but not limited to, military supplies
5. Waste removal and disposal

Government Organizations with Service Objectives. Government organizations which have objectives that can be stated only in social or service terms are more common, and exist in fields of endeavor such as

1. Educational services
2. Other social and agricultural services
3. Police or military services
4. Loans and grants
5. Procurement
6. Operation of control systems, for example, the Federal Aviation Administration (FAA), the Internal Revenue Service (IRS), and the Patent Office

Effect of Type of Government Organization on Industrial Engineering Techniques. The type of government organization has an important effect on the use of industrial engineering techniques.

Government Organizations with Physically Stated Objectives. As far as problems go, such organizations have many of the same characteristics of the private sector. However, in that the concept of profit does not occur, the usual criteria of success concern productivity and effectiveness: cost, quality, timeliness, and/or social acceptability.

Government Units with Social Objectives. Those whose objectives may be stated only in social terms require additional techniques, resembling those used with white-collar workers, to identify outputs, prior to the use of many industrial engineering techniques.

Effect of Level of Government on Industrial Engineering Techniques. The level of government does not have any special effect on the use of industrial engineering techniques, although it does usually affect the complexity of the problem situations to which the techniques are applied.

SHORT HISTORY OF INDUSTRIAL ENGINEERING IN GOVERNMENT

Industrial engineering in government developed in the manufacturing branch of government (printing and engraving, munitions and arms manufacturing, and other branches with physical outputs) during the period of 1910 to 1917, in parallel with the private sector. Little if any thought was given to applying it in the larger service sector.

Prohibition against Time Study. In 1917, at the urging of organized labor, who felt that time study had been used abusively, Congress added a rider to the appropriations act which stated, essentially:

> None of these monies, nor no part of these monies, shall be used to pay the salary of, or any part of the salary of, any person who makes, or causes to be made, a study of any part of the time expended by a government employee between the time of punching in and the time of punching out.

Later, Congress made the use of time clocks, for punching in and punching out, illegal. To get a feel for the strength of this attitude, it is worth noting that the Treasury Department in Washington, D.C., installed such clocks while Congress was considering the legislation. Out of pique, Congress passed additional legislation forbidding the use of any appropriated monies for repairing time clocks. It is worth noting that the attorney general later ruled that the prohibition was valid only within the bounds of the District of Columbia, as constituted at the time of the legislation. As a result, time clocks were commonly used at other government installations.

The Repeal of the Anti-Time-Study Rider. In 1948, following World War II, Senators Flanders of Vermont and Taft of Illinois argued that the rider was nongermane to the appropriations bill, and in addition, offered for prosecution, for contempt of Congress, various industrial engineers who, during the war, had applied industrial engineering, including time study, to the analysis of various battle operations. The results had been profound: for instance, the time to emplace a battery of field artillery was cut in half; a change in the method of handling antiaircraft ammunition on Navy cruisers had the effect of doubling the firepower of the ships; studies of the methods of target acquisition during kamikaze attacks greatly raised the kill rate; and so on. Nobody in Congress wished to prosecute any of the individuals, and the rider was removed from the appropriations act.

A Presidential Directive on the Employment of Industrial Engineering.   President Truman, in 1949, issued Executive Order 10072, which stated:

> Agency heads are to take steps to assure themselves and the President that operations are being carried out with maximum efficiency and maximum economy.

In the subsequent implementing orders issued by the various departments, bureaus, and agencies, the techniques and procedures to be employed in attaining the goals of maximum efficiency and economy were indicated. However, the scope of the problems involved were usually underestimated. The units in which efficiency was to be measured were seldom stated. Further, economy was thought of only in terms of money, although there are times when economy of time is much more important. Further, in many cases, the required form of implementation became overly specific, substituting form for intent.

Government Training in Industrial Engineering.   In 1950, in an attempt to more fully implement the executive order, the U.S. Army Ordnance Corps appointed a committee of lay engineers, from industry and education, to determine the educational needs of the corps and the means of meeting them. The committee could not find educational resources available and recommended the creation of a corps training unit headquartered at Rock Island Arsenal (but not a part of the arsenal) to meet the educational demands. The school was established in 1951 and was designated the Ordnance Management Engineering Training Agency. Subsequently, it was made into an Army-wide endeavor and renamed the Army Management Engineering Training Agency. In 1988 this was renamed the Army Management Engineering College. The college provides over 100 different short courses and seminars, on both a residential and off-campus basis, rather than a curriculum. The seminars range from 2½ days to 10 weeks in length. Currently the throughput of the college is about 17,000 seminar participants per year. The college has had a profound effect on the sophistication of industrial engineering in the armed services. It is worth noting that the institution frequently has guest enrollees from NATO countries, the State Department, and so forth.

Current Nomenclature for Industrial Engineering in Government.   Currently, in the United States, under the periodic urging of the Office of Personnel Management (formerly the Civil Service Commission) the work of industrial engineering is carried on in the agencies under the titles *management engineering* or *management analysis,* although the title *industrial engineering* can still be found in the manufacturing sector of the government.

Typical Industrial Engineering Techniques in Use in Government.   The following is a partial list of techniques whose use may be found in various government organizations:

1. Process chart—product analysis
2. Process chart—worker analysis
3. Process chart—combined analysis
4. Operation analysis
5. Worker and machine analysis
6. Video analysis

7. Total quality control
8. Just-in-time scheduling
9. Lot size analysis
10. Work measurement for staffing studies
11. Design for effectiveness (value analysis)
12. Plant and office layout
13. Statistical quality control
14. Pareto analysis (for problem identification)
15. Break-even charts
16. Employee participation

The use of the above techniques is explained in other sections of this handbook. As noted, their use, except for the formulation of a criterion of success, is similar to their use in the private sector.

Motivation for Industrial Engineering in Government.   The most desirable form of motivation is to instill the idea that there are limits on expenditures and that the government needs "the most accomplishment for the money," taking timeliness, quality, and customer/client satisfaction into account, as well as economy of expenditure. Unfortunately, most government pay systems give much weight to the number of people supervised and, in addition, most employees consider permanent employment as characteristic of a government job. Hence, there is actually much motivation to increase work forces, retain unnecessary jobs, and avoid improvement. For instance, an industrial engineering study of the U.S. Social Security Administration[1] made from 1978 to 1980 revealed that there were about 23,000 excess employees costing about $1,000,000,000 per year. It took 3 years before sufficient pressure was developed on the agency by the Government Accounting Office, in cooperation with the Bureau of the Budget (as it was then called), to cause the agency to reduce by attrition, over a 6-year period, to the required number of employees.

NATURE OF INDUSTRIAL ENGINEERING TECHNIQUES

Explicit Steps in the Use of the Techniques.   All industrial engineering techniques are specialized forms of problem solving that require the following six steps.[2]

1. Stating the criterion of success
2. Gathering intelligence (new knowledge) with respect to the problem situation
3. Applying knowledge (previously ascertained facts relating to the intelligence) in order to synthesize a solution
4. Testing the solution with respect to the criterion of success, with either a model or paper analysis
5. Applying the proposed solution to the problem situation
6. Standardizing the solution and making it a matter of record and policy

Implicit Steps in the Use of the Techniques.   It should be noted that, inasmuch as most industrial engineering techniques impinge directly or indirectly on people, the above steps

---

[1]This was probably one of the largest work sampling studies ever made, involving over 860,000 observations in a total of 24 locations over a 6-month period.

[2]The steps may be stated in different terms, and divided differently, but the sense will remain the same.

must be carried out in a manner designed to encourage participation rather than resentment and hostility.

DIFFERENCES BETWEEN THE PRIVATE SECTOR AND THE PUBLIC SECTOR AFFECTING THE USE OF INDUSTRIAL ENGINEERING TECHNIQUES

Despite the fact that the same industrial engineering techniques, for the most part, are used in both the private sector and government, the differences in the two sectors create differences in the manner of use, primarily with respect to the stating of the criterion of success.

Characteristics of the Private Sector.   An organization in the private sector, in almost all cases, exists in a competitive environment, has a profit-and-loss statement, and has a balance sheet. The objectives are usually stated in terms of

1. Profit desired
2. Security of the capital
3. Market share

Characteristics of a Government Organization.   A government organization usually has no competition, does not have a profit-and-loss statement, and does not have a balance sheet. The objectives are usually stated as physical or social goals.

INDUSTRIAL ENGINEERING IN GOVERNMENT ORGANIZATIONS WITH PHYSICAL OUTPUTS

Similarities between the Public and Private Sector.   As noted earlier, the major difference between organizations having physical outputs in the private sector and in government are the absence of a profit-and-loss statement and of a balance sheet. Hence, the criterion of success in applying industrial engineering techniques becomes cost, quality, timeliness, or customer satisfaction. It should be noted that these are the same as in the private sector; only profit is missing. However, the priorities may differ.

Differences between the Public and Private Sector.   The absence of a profit-and-loss statement requires different criteria for judging the quality of management of the government organization. A government organization is more properly evaluated with measures of effectiveness and productivity, although, unfortunately, it is not common to find such measures in place. Organizations in the public sector also need a method of budgeting in that one cannot justify additional expenditures as leading to additional profit.

USING MEASURES OF EFFECTIVENESS

Measures of effectiveness are quantifications of the degree of attainment of the organization's objectives. For instance, if an objective exists with respect to timeliness, the measure might be "the percent of products or services produced on time." If an objective exists with respect to quality, then the measure might be "the degree of attainment of the desired quality." If an objective exists with respect to customer satisfaction, then measures of effectiveness may relate to number of valid customer complaints, dollar value of unsatisfactory outputs, percent of unsatisfactory outputs, and so forth, as may be appropriate. One should note that measures of effectiveness cannot be added to produce a meaningful average. There may be tradeoffs among the various objectives.

However, it is worth noting that, in many instances, the objectives of a government organization are not stated in substantive and quantitative terms. Indeed, some government employees will state that such statements of objectives are not possible, although one must wonder how one can make decisions without criteria to guide them. Of course, if the objectives are not stated in substantive and quantitative terms, then measures of effectiveness are not feasible.

USING MEASURES OF PRODUCTIVITY

Productivity measurement in the U.S. federal government has had varied support from successive administrations, although there frequently have been discrepancies between the stated and actual policies. For instance, during the regime of President Reagan the official policy was "Agencies will increase their productivity by 20 percent in ten years." This was later changed to a mandate of a 3 percent increase per year. However, a thorough search of agency documents would have revealed a discrepancy: most of the agencies either did not have productivity measures, or had phony ones. Further, many will claim that productivity measurement is not possible, or that it would cost more than it might save.

Reasons for Avoiding Productivity Measurement.   In some cases the claim that productivity measurement is impossible is an honestly held belief, although usually erroneous. However, the usual reason, not stated, is a desire to avoid the accountability that accompanies proper productivity and effectiveness measurement. An organization with a dormant productivity is one which is not well-managed. The world is changing but that sector is not, but if productivity is not measured this dormancy will not show as readily. Productivity measurement at lower levels of government does exist, but applications are spotty and tend to come and go with changes in administrations. Also, attempts at productivity improvement may take strange forms. For instance, the Pennsylvania legislature passed a law requiring that the word *employee* be spelled in all government publications as *employe* to save keystrokes in document preparation.

METHODS OF BUDGETING IN GOVERNMENT

As noted earlier, one cannot justify additional government expenditures as creating additional profit; some other method of budget justification is needed. Two basically different approaches are used: base budgeting and zero-based budgeting.

Base Budgeting in the U.S. Government.   Base budgeting is the most common and the least appropriate. It consists of taking the previous year's budget, estimating the decreases or increases in the anticipated work load, and adjusting for the anticipated change. It should surprise no one to note that increases are much more common than decreases. This method may be described as a most effective way of increasing annual budgets.

Zero-Based Budgeting in the U.S. Government.   Zero-based budgeting is more of an industrial engineering approach to budgeting. It is a system whereby the needs for each budget period are estimated with an analysis of each program, without regard to previous budgets, starting from zero. There are several versions of zero-based budgets. In the generic form, the needs for each program are analyzed and forecasted, and tradeoffs among programs are made on a political basis. As another approach, President Carter in 1976 attempted to introduce a form of zero-based budgeting which had been developed in the private sector. In this version, programs were "competed" against each other in terms of their merits. As has been noted, the objectives in the private sector relate to profit, stability of capital, and market share, and all of an organization's programs must serve these objectives. Government programs do not have such shared objectives. How is one to compare the desirability among programs of research on a disease, aid to families with dependent children, school lunch programs, clean air and water activities, and military weapons programs, other than via the political process. Hence, despite the fact that zero-based budgeting is still an official program, and although the generic form has replaced President Carter's, the damage from his attempt still shows in the lack of faith in the process.

Responsibility for Budget Preparation.   In national governments, the type of employee responsible for budget preparation varies widely. For instance, in the United States, the Office of Management and Budget has a small number of senior budget officials who supervise a larger group of relatively junior budget examiners who work with agency budget officers in preparing budgets. In contrast, in Australia, the Budget Office is staffed with senior budget officers who come to the positions after serving as agency budget heads. The U.S. theory is that the junior budget examiners owe no allegiance to an agency, but the theory neglects that the only real avenue for promotion of the examiners is through the agency, so that instead of a "silver cord" the examiners have "a golden suction pipe." The Australian theory is that the experienced budget examiners "know where the bodies are buried" and can and will properly squeeze budgets. Other countries have various approaches between these two extremes. At lower levels of government, we may find almost any type of employee, from political appointee to economist, with views ranging from belief in the innate honesty of lower officials, or in the dishonesty of such officials, to a philosophy that the goal of budgeting is to spend the least amount of money possible.

ACTUAL PRODUCTIVITY MEASUREMENT IN GOVERNMENT

As noted earlier, government can be divided readily into two major areas: (1) those producing physical goods, and (2) those producing services. The methods of measuring productivity are different in the two sectors.

Productivity in the Goods-Producing Sector of Government.   The usual algorithm is

$$\text{Productivity} = \frac{\dfrac{\text{aggregated \$ outputs, measured period}}{\text{aggregated \$ inputs, measured period}}}{\dfrac{\text{aggregated \$ outputs, base period}}{\text{aggregated \$ inputs, base period}}}$$

However, this is hardly fit for operational use. The terms need to be further refined. Note that all costs are computed with base-year unit prices so that the productivity equation will reflect changes in operations rather than changes in the external economy.

Let PROD = productivity
m = subscript denoting measured period
b = subscript denoting base period
RIP1 = resource inputs partial, capital
RIP2 = resource inputs partial, direct costs, except material
RIP3 = resource inputs, indirect costs
RIP4 = resource inputs, partial, material costs
AOP1 = aggregated outputs, recovered, partial, from capital costs
AOP2 = aggregated outputs, recovered, partial, from direct costs, except material
AOP3 = aggregated outputs, recovered, partial, from overhead costs
AOP4 = aggregated outputs, recovered, partial, from material costs

Then

$$PROD = \frac{\dfrac{AOP1_m + AOP2_m + AOP3_m + AOP4_m}{RIP1_m + RIP2_m + RIP3_m + RIP4_m}}{\dfrac{AOP1_b + AOP2_b + AOP3_b + AOP4_b}{RIP1_b + RIP2_b + RIP3_b + RIP4_b}}$$

EXAMINATION OF HOW THE TERMS IN THE ALGORITHM ARE COMPUTED

Capital Costs.   The capital costs RIP1 are based on a special form of productivity accounting. RIP1 represents the charge for capital used to produce the goods, but in productivity accounting, in contrast to normal accounting, the equipment depreciation continues indefinitely. For instance, if $100,000 worth of equipment is involved and the depreciation rate is 10 percent, then the charge for RIP1 for each and every year is $10,000. Were we to allow the machines to fully depreciate, we would delude ourselves into thinking that productivity had improved merely because the equipment was obsolete.

Direct Costs.   The direct costs other than materials, RIP2, include wages, power, tooling, and other things consumed in production, but at base-year unit costs, so that changes in the price levels do not skew the productivity computation.

Indirect Costs.   The indirect costs RIP3 include management, material handling, guards, and other wages and central computer depreciation, computed at base-year rates, (except computer depreciation, which is handled like RIP1) for the same reasons as RIP2. Other costs not directly relatable to product should be included in this category.

Material Costs.   The material costs RIP4 represent the direct cost of the material. Many government productivity computations exclude this term for one or more of the following reasons:

1. The materials are furnished by the customer (ordnance)
2. Presumed improvements in productivity, from a cheaper material, may seriously affect quality and hence, should be individually evaluated
3. Materials are such a small part of the expenses that they are easier to ignore
4. Materials are such a large part of the expenses that they would overshadow the other factors if included and they are better considered separately

Capital Recovery.   AOP1, the cost of capital recovered through the use of the equipment to produce outputs, is computed by one of two methods:

1. A transfer price is assigned to each output, taking into account the standard time (during the base year) to process each product on each piece of equipment. These transfer prices of the product (transfer, without profit or management overhead, to another branch of the government) usually also include direct wages, power, and so on, and hence are usually computed as AOP1 + AOP2.
2. The amount of a product processed on each machine in its manufacturing sequence is multiplied by the standard machine time per unit of product during the base year (in hours) and by the hourly depreciation rate, calculated for a standard year, on that machine. The sum of these computations for all products for a period constitutes AOP1.

Direct Cost Recovery.   AOP2, the capital recovered from direct costs, is computed, like AOP1, in one of two ways:

1. The transfer price, used for AOP1, includes direct cost items, at base-year hourly rates, and is included in the AOP1 + AOP2 described under AOP1.
2. The amount of a product processed on each step in its manufacturing sequence is multiplied by the labor standard time per unit of product during the base year (in hours) and by the worker hourly wage during the base year. The sum of these computations for all products for the period is AOP2.

Material Cost Recovery.   AOP4, the capital recovered from material cost, if this term is used, is calculated prior to AOP3, hence its discussion precedes that of AOP3. AOP4 is the total value of all of the quantities of direct materials used in manufacture, each quantity of each material multiplied by its base year unit price. In some cases, this AOP4 is included in the transfer price computed under AOP1. In such cases, a separate computation is not used.

Indirect Cost Recovery.   AOP3, indirect costs, by definition cannot be computed directly. The usual method is to compute AOP3 by the following formula:

$$AOP2_m = \frac{AOP1_m + AOP2_m + AOP4_m}{AOP1_b + AOP2_b + AOP4_b} \times RIP3_b$$

The formula allows increases in overhead costs in direct proportion to increases in capital, direct costs, and material recovery. If the actual indirect costs, $RIP3_m$, are less than the calculated $AOP3_m$, then this also contributes, as it should, to increased productivity.

PERSONAL COMPUTER PROGRAMS

The manual computation of the productivity in an organization of any real size, and with a considerable number of products, is a time-consuming process and subject to many chances for errors. Programs for personal computers are commonly used, and if constructed properly, allow the computation of that productivity in minutes, rather than days, once the data are gathered. Computer programs also allow an easy way to test IF-THEN hypotheses.

Final Productivity Algorithm.   The final equation for computing productivity is

$$Prod = \frac{\dfrac{AOP1_m + AOP2_m + AOP4_m + \dfrac{AOP1_m + AOP2_m + AOP4_m}{AOP1_b + AOP2_b + AOP4_b} \times RIP3_b}{RIP1_m + RIP2_m + RIP3_m + RIP4_m}}{\dfrac{AOP1_b + AOP2_b + AOP4_b + RIP3_b}{RIP1_b + RIP2_b + RIP3_b + RIP4_b}}$$

IF-THEN Hypotheses.   An IF-THEN hypothesis is a question in the form of:

If (AOP1 + AOP2) increases by 10 percent, while other values remain constant, what then would the productivity be?

NOTE: Any change or combination of changes may be tested quickly, to help evaluate the desirability of expending the time and effort to accomplish the changes.

Usual Format of Computer Programs.   For convenience, the suite of programs used usually is divided into six programs, with storage of the computations on disk files. The use of six programs allows part of the computations to be run, instead of being completed all at one sitting. A common set of programs would be

1. RIP1, a program used to compute both $RIP1_b$ and $RIP1_m$.
2. RIP2, a program used to compute both $RIP2_b$ and $RIP2_m$.
3. RIP3, a program used to compute both $RIP3_b$ and $RIP3_m$. NOTE: As noted earlier, RIP4 is usually ignored.
4. AOP1-2, a program used to compute both $AOP1-2_b$ and $AOP1-2_m$.
5. PROD, a program designed to read the disk files of the preceding four programs and compute the productivity of the organization via the algorithm in "Final Productivity Algorithm," above.
6. SUPP, a supplementary program with two uses:
    a. To facilitate correcting errors in the files of any of the first four programs.
    b. To facilitate changing the file values of any of the first four programs in order to run, easily and quickly, any IF-THEN hypothesis.

PRODUCTIVITY MEASUREMENT IN GOVERNMENT SERVICE ACTIVITIES

As noted earlier, government employees can be divided readily into two major areas: (1) those producing physical goods and (2) those producing services, and the methods of measuring productivity are different in the two sectors.

Productivity in the Service-Producing Sector of Government.   In that the outputs, in contrast with the goods-producing sector, cannot be evaluated in terms of monetary value, they are usually evaluated in *earned hours,* the product of the quantity of each output multiplied by its standard time. The usual algorithm is

$$\text{Productivity} = \frac{\dfrac{\text{aggregated earned-hour outputs, measured period}}{\text{aggregated worked-hour inputs, measured period}}}{\dfrac{\text{aggregated earned-hour outputs, base period}}{\text{aggregated worked-hour inputs, base period}}}$$

However, this is hardly fit for operational use. The terms need to be further defined. Note, all earned hours are computed with base year standard times, so that the productivity equation will reflect changes in operations. Further, in that productivity is relative, establishing a base period productivity as less than one (100 percent) is considered an affront to those responsible for productivity and creates unnecessary resistance. Hence, the numerator of the denominator of the algorithm is usually multiplied by a value k such that the value of the total denominator becomes unity, and the standard times are multiplied by the same k, to preserve the equation.

Steps in the Application of Productivity Analysis.   The same steps given for the physical product sector will apply in the service sector of government, but drawings, bills of material, and product lists do not exist in the service sector. Hence, some special techniques must be introduced in gathering intelligence (new knowledge) with respect to the problem situation. These techniques require new concepts for clarifying the knowledge. The first step is to define the objective of the organization in substantive and quantitative terms in order to provide criteria for the identification of the outputs.

STATING THE OBJECTIVES OF A SERVICE ORGANIZATION

As noted earlier, government organizations with service outputs exist in the social service, education, law enforcement, military and similar government organizations. In most cases, the organizations are monopolies. Statements of market share, profit, and so forth, are irrelevant. Other terms are needed. It has been found advantageous, in the interests of clarity, to describe the objective of a service organization under seven headings:

1. *Mode of operation.* Government organizations carry out their functions by providing one or more of the following:
   a. Constructive services. They do things for people that people would otherwise have to do for themselves.
   b. Social constraints. They prevent, or discourage, people from engaging in acts which would infringe on the rights of others.
   c. Making loans. They loan people money, or guarantee loans, for people to purchase houses or to gain an education, or for other purposes that are deemed to serve the public good.
   d. Making grants. They give money to people or organizations in the private sector to carry out activities deemed in the interest of the general public.
   e. Internally consumed activities. Practically all government organizations have groups engaged in activities to facilitate the work of the preceding four objective areas, but whose work does not go outside of the organization. In some instances the attention of a productivity analysis is focused on such a part of an organization.
2. *Mission area.* This term refers to the area of the society, economy, or geography where the effects of the outputs of the organization appear. Examples are
   a. Broad. To enforce the laws of the city.
   b. Narrow. To assure that protein food, of animal origin, in the public market place is free of drug residues from drugs administered in accord with the prescribed regulations.
3. *Intent.* What is the effect sought in the area affected by the outputs? Note, each mission area must have at least one intent. Examples are, for the mission areas given above,
   a.
      1. Deterring violations
      2. Apprehending violators
   b.
      1. Ensuring a percentage of food free of residues from drugs administered in accordance with prescribed regulations
      2. Disseminating knowledge to the trade of the regulations governing the use of animal drugs
4. *Dimension.* In what terms will the desired achievement of each intent be measured? Note, each intent must have at least one dimension. Examples might be, for the dimensions given above:
   a.
      1.   Number of violations
      2.1. Percent of violators of traffic laws apprehended
      2.2.   Percent of violators of criminal laws apprehended, and so on
   b.
      1. Percent
      2. Percent complete and current
5. *Goal.* How much of each intent, measured by each dimension, is sought. Each dimension must have a matching goal, or else it is not a valid dimension. Again, following the same examples used earlier:
   a.
      1.   Zero
      2.1. 100 percent[3]
      2.2. 100 percent[3]
   b.
      1. Zero
      2. 100 percent

---

[3]100 percent may seem like an impossible goal, but it is still realistic even though there will be differences between goal and accomplishment. It is like a person taking a school examination. The goal is seldom, if ever, to write a paper of less than 100 percent, but lower scores are readily accepted.

6. *Limitations.* This is a list of the unique actions not feasible or legal.[4] The numbering of these has no relationship to any specific intent, dimension or goal; they apply in general. For instance, keeping the same examples as have been given:

    a.

        1. People suspected of having lawless tendencies may not be removed from society until they commit an overt act
        2. Criminals do not readily come forward to admit crimes

    b.

        1. Food must be tested to determine if any drug residues exist
        2. Sampling methods have not been developed for determining the knowledge in the trade

7. *Freedoms.* This is a list of types of actions open to choice. For instance, for the same examples:

    a.

        1.1. Public educational programs may be instituted
        1.2. Officers may freely patrol the streets, and so forth

    b.

        1.1. Samples may be taken for testing, at will, from any part of the market chain
        1.2. Suspect food may be ordered held from the market until a reasonable time to complete tests has elapsed

ADDITIONAL NEW TERMS FOR SERVICE ORGANIZATIONS

In defining the subparts of an objective, it is necessary to take words that are synonymous, in common usage, and to give them specific, hierarchical meanings. Using ordinary product terms to describe the service workers' outputs would be very confusing. Certainly, we can speak of the services provided as outputs, but if we start to talk about "models," "parts," and so on, there will be much confusion.

The New Terms. The following new terms have been found useful with service outputs.

    A Work-Unit.[5] Let us define a work-unit as either an amount of work or the results of an amount of work that is convenient to treat as an integer (a "one of," or an "each") when we examine the quantitative aspects of work.

An Order of Work-Units. An order of work-units is a complete list of the outputs of an organization, described at a relatively similar level of aggregation, and in which the items are mutually exclusive and all-inclusive. The word *complete* indicates that the list is not a list of indicators or surrogate measures; it must be a complete list of outputs. A general taxonomy of work-units for any and all organizations is given in Figure 5.1. Although the list is designed to assist with the problems associated with describing the outputs of service workers, it is also applicable to direct outputs.

DIFFERENCES BETWEEN DIRECT OUTPUTS AND THE OUTPUTS OF SERVICE WORKER GROUPS

The following differences are typical.

---

[4]The list should not include items such as: "Limited by budget with respect to staff and money," or "Must conform to law." These are hardly unique; they apply to almost all organizations. Anything contrary would be a unique freedom, such as that of the Social Security Administration, which must and may pay all benefits due to beneficiaries; they have a blank check on the treasury of the United States.

[5]The work-unit terms are taken from *Measuring and Improving Productivity and Effectiveness,* M. E. Mundel, Prentice-Hall, Englewood Cliffs, N.J., 1983, p. 34.

DEFINITIONS OF WORK-UNITS
FROM THE LARGEST TO THE SMALLEST

Numerical designation	Name	Definition
8th-order work-unit	Objective; Results	What is achieved because of the outputs
7th-order work-unit	Gross outputs	A large grouping of products or completed services, sharing some important affinity within the group
6th-order work-unit	Program	A group of outputs or completed services representing part of a 7th-order, but which are a more homogeneous subgroup; they share a greater affinity than the totality of the 7th-order work-unit.
5th-order work-unit	End product	A unit of final output; the units in which a program is quantified; the smallest output which is produced for use outside of the producing organization and which contributes to the objective, without further work being done on that output.
4th-order work-unit	Intermediate product	A part of a unit of final output; the intermediate product may become part of the unit of final output or merely be required to make it feasible to achieve the final output.
3rd-order work-unit	Task	Any part of the activity associated with the performance of a unit of assignment by a person or a group.
2nd-order work-unit	Element	A part of a task which is convenient to separate when analyzing work.
1st-order work-unit	Motion	The performance of a human motion. This is the smallest work-unit usually encountered in the study of work. It is used to facilitate job design or in some forms of work measurement; it never appears in higher level control systems

FIGURE 5.1 A taxonomy of work.

Preplanning. Direct outputs are usually accompanied by one or more drawings describing the output, a parts and material list, and so on. Descriptions of service worker outputs are seldom available.

Visibility. With direct outputs one can see the product; it is a visible, tangible output. One can visit the workplace and see what is being worked on and what is being done to it. With service worker outputs, if one visits the workplace, the real output is frequently not visible. The personnel staff member posting a personnel record may be promoting an employee, but one sees the posting rather than the promotion. Also, it might not be a promotion; it might not be anything but the staff member killing time.

Relationship between the Fourth- and Fifth-Order Work-Units. The direct product usually has a fixed relationship between the fifth-order work-unit and its constituent fourth-order work-units. For instance, we may have a centrifugal pump consisting of a shaft, an impeller, and two halves of a housing, with eight bolt holes and eight bolts. One cannot assemble a pump with two shafts, three halves of a housing, or nine bolts; the relationship between the fifth-order work-unit and the number of fourth-order work-units is fixed! To return to the example of the personnel staff member posting a promotion to a personnel record, all kinds of unnecessary detail could be posted. All sorts of facts could be reviewed prior to the posting. The record could be posted and reposted! The record could be "posted to death" and it would not show in the final output.

Conclusions from the Differences between Direct and Service Worker Work-Units. Because of the differences noted above, most service-worker work-control systems use the fifth-order work-units for planning, control, and productivity measurement.

DELINEATING THE WORK-UNITS FOR SERVICE WORKERS' WORK

The following steps have been found useful.

Preliminary Analysis. We have already noted that, in most cases, one cannot go to the workplace and visually determine what is being done; the nature of the fifth-order work-unit is not usually discernible. The procedure needed is analogous to designing a machine—a process that defines the fifth-order work-units. One needs to state the purpose to be served, divide the allowable machine space into functional areas, create assembly drawings, and from the assembly drawings create the designs of the individual pieces. For service worker's work, one needs to delineate the seventh-order, sixth-order, and then the fifth-order work-units. The delineation of the lists of work-units is usually most conveniently performed by a small group thoroughly familiar with the concept of work-units and with the work under scrutiny. The set of lists, from the eighth-order work-unit down to the lowest order identified, is referred to as a *work-unit structure.*

Guidelines for Separating Seventh-Order Work-Units. The large categories separated should usually clarify the significance of the outputs with respect to one or more of the following:

1. Different mission areas served
2. Radically different methods of producing the outputs
3. Different cost systems, that is, budget money or reimbursed work
4. Different fund accounts

A two-digit number is assigned for each seventh-order work-unit, starting with 01.[6]

Guidelines for Separating Sixth-Order Work-Units. The categories separated should divide the seventh-order work-unit into smaller categories which have a greater affinity among the constituent outputs. It should be noted that the sixth-order work-units are, like the seventh-order, described by a plural term; they are a group of outputs. An additional two-digit code is added to the code from the seventh order, from which the work-units derive, to provide unique identification for each group of outputs. One or more of the following suggestions for the division should be helpful:

1. Independent, separate cause systems which generate work.
2. Useful subaggregations for helping with decisions concerning the mix of benefits to be obtained.
3. Subclasses of the separations initiated at the seventh-order, for the same reasons used for the separation of the seventh-order work-units.

---

[6] The author has not encountered a situation with more seventh-order work-units than the two digits will separate.

4. Outputs which, on the surface, appear alike, but require different amounts of resource support.

5. Subgroups, within a seventh-order work-unit which are more alike, within the subgroup, with respect to mission area served, purpose, and the like than the other outputs in the seventh-order work-unit of which they are a part.

Each sixth-order work-unit is assigned a four-digit number, the first two of which are that of the seventh-order work-unit of which it is a part. The remaining two digits are assigned each time the seventh-order number changes, starting from 01.

Identifying Fifth-Order Work-Units. With an analysis of indirect work, the delineation of a *work-unit structure* usually ends with the listing of the fifth-order work-units. Two additional digits are added to the sixth-order code, from which they derive, to provide a unique identification for each fifth-order work-unit. (*Note:* When the work-units are entered into the PC productivity programs they will be assigned a simple two-digit code number. The six-digit numbers, being described here, keep clear their place in the work-unit structure.) Separation at the fifth-order should serve all of the following criteria, as compared with the separation at the seventh-order and sixth-order, where only one guideline was employed.

1. Identify a normal "each" of actual output, with respect to utility.

2. Identify an "each" in a manner related to the management decision-making process with reference to do or not-do type decisions.

3. Provide a useful basis for determining unit costs of outputs.

4. Provide a work count for that which is contained in each sixth-order work-unit.

5. Provide a work-unit which is convenient for some type of work measurement and some type of work load reporting and forecasting.

6. Identify outputs produced for use outside the organization which directly contribute to the objectives of the organization without further work being done on those outputs.

WORK MEASUREMENT FOR PRODUCTIVITY MEASUREMENT

As noted earlier, service outputs cannot be evaluated in monetary terms and instead are evaluated in terms of *earned hours*. This requires that so-called standard times be set for each kind of output. A choice must be made between the "should-take" times, common in the private sector for direct work, and "did-take" times. Did-take times are values that it took, in the past, to perform the work-units. Such values are preferred for productivity measurement because:

1. They are simpler to establish.

2. Productivity measurement measures improvement over previous performance, and previous times establish a realistic datum.

3. They do not imply any criticism of past performance.

Kinds of Work Measurement Used for Productivity Analysis. One or more methods are chosen, as most convenient, from among the following techniques:

1. *From historical records.* In some cases, data concerning time spent and work count accomplished are available from past records.

2. *Logging.* A record of performance is kept, work-unit by work-unit, over a representa-

8TH-ORDER WORK-UNIT
DIRECTORATE — COMPTROLLER

MODE

Internally consumed service

MISSION AREA

Budget and finances of the Command

PURPOSES (INTENTS AND DIMENSIONS) AND GOALS

NOTE: FOR EASE OF TRACING RELATED ITEMS, EACH PURPOSE (INTENT AND DIMENSION)
AND GOAL ARE GROUPED TOGETHER.
THE ABBREVIATIONS ARE: PI = PURPOSE, INTENT; PD = PURPOSE, DIMENSION; GL = GOAL.

BUDGET SERVICES
1. PI: 1.   To provide a budget matching anticipated command workload
   PD: 1.1 Percent of budget zero base supported
   GL: 1.1 100%
   PD: 1.2 Timeliness of submission
   GL: 1.2 100% within the allowed time
   PD: 1.3 Quality of services
   GL: 1.3 0 percent rejected as unsupported by workload

2. PI: 2.   Execute the budget within all constraints
   PD: 2.1 Amount spent over budget
   GL: 2.1 0
   PD: 2.2 Amount of excesses locally reprogrammable, reprogrammed
   GL: 2.2 100%
   PD: 2.3 Amount of excesses, not locally reprogrammable, reported to
            higher authority
   GL: 2.3 100%

FINANCE AND ACCOUNTING
3. PI: 3.   To account for all funds appropriated
   PD: 3.1 Percent of funds correctly tracked
   GL: 3.1 100%

REVIEW AND ANALYSIS
4. PI: 4.   To meet commanders operations intelligence needs
   PD: 4.1 Percent of Commander's requests met
   GL: 4.1 100%
   PD: 4.2 Timeliness of information furnished
   GL: 4.2 100% on or before requested date
   PD: 4.3 Accurary of information
   GL: 4.4 Zero errors

5. PI: 5.   To carry out special projects, as requested
   PD: 5.1 Percent carried out to requestor's satisfaction
   GL: 5.1 100%
   PD: 5.2 Percent completed on time
   GL: 5.2 100%

ADMINISTRATIVE SERVICES
6. PI: 6.   To provide administrative services for Directorate
   PD: 6.1 Percent of valid complaints re services provided
   GL: 6.1 0%

CAREER COORDINATION
7. PI: 7.   To service career programs
   PD: 7.1 Percent of those enrolled serviced properly and on time
   GL: 7.1 0 valid complaints

8. PI: 8.   EEO program administration in accord with guiding rules and
            regulations
   PD: 8.1 Number of exceptions taken by I.G.
   GL: 8.1 0

MAJOR LIMITATIONS
   1. Cannot control rate of inflow of work
   2. Must work within DoD, Army and OPM rules and regulations
   3. Must conform to Union contract obligations where appropriate
MAJOR FREEDOMS
   1. May establish priorities within major time frames
   2. May travel to field sites as needed
   3. May have direct communication with installations
   4. Good working conditions and telecommunication support
   5. Top management is supportive

FIGURE 5.2   The eighth-order work-unit of the Comptroller Directorate.

DIRECTORATE — COMPTROLLER

7TH-ORDER WORK-UNITS
(Gross groups of outputs)

01 Budget services provided
02 Finance and accounting services provided
03 Review and analysis services provided
04 Administrative services provided (to Directorate)
05 Career coordination services provided

FIGURE 5.3 The seventh-order work-units of the Comptroller Directorate.

COMPTROLLER–DIRECTORATE

6TH-ORDER WORK-UNITS
(Separated by workload generating sources)

BUDGET (NOTE: *The actual acronyms have been replaced.)
  0101 Integrated budget preparation, execution and reprogramming services completed
  0102 AAA* budget preparation, execution and reprogramming services completed
  0103 BBB budget preparation, execution and reprogramming services completed
  0104 CCC budget preparation, execution and reprogramming services completed
  0105 DDD budget preparation, execution and reprogramming services completed
  0106 EEE budget preparation, execution and reprogramming services completed
  0107 FFF budget preparation, execution and reprogramming services completed
  0108 GGG budget preparation, execution and reprogramming services completed
  0109 IIIII budget preparation, execution and reprogramming services completed
  0110 Special projects completed
FINANCE AND ACCOUNTING (NOTE: *The actual acronyms have been replaced.)
  0201 III financial and accounting services provided
  0202 JJJ financial and accounting services provided
  0203 KKK financial and accounting services provided
  0204 LLL financial and accounting services provided
  0205 Special projects completed
REVIEW AND ANALYSIS
  0301 Periodic assessments of Command performance completed
  0302 Periodic assessments of Directorate performance completed
  0303 Periodic assessments of subordinate installation performance completed
  0304 Summary assessments of performance completed
  0305 Procurement assistance provided
  0306 Briefings provided
  0307 Significant events reported
  0308 Productivity improvement assistanace provided
  0309 ADP assistance provided
  0310 Special projects completed.
ADMINISTRATIVE SERVICES
  0401 Requisitions for supply fully processed
  0402 SF 52s prepared and forwarded
  0403 SF 50s fully processed
  0404 Physical equipment inventories reconciled
  0405 Mail and delivery services provided
  0406 Employee services provided
  0407 Records management assistance provided
  0408 Forms management assistance provided
  0409 Library services provided
  0410 Information control and management services provided
  0411 Physical plant alterations services provided
  0412 Training assistance provided
  0413 Review and analysis charts completed
  0414 Special projects completed
CAREER COORDINATION
  0501 Career program managed
  0502 Intern program managed
  0503 Cooperative training and junior fellowship program managed
  0504 Para-professional training program managed
  0505 EEO affirmative action program managed
  0506 Career counseling program provided
  0507 Excecutive development training program managed
  0508 Career appraisal plan meetings planned and conducted
  0509 Career program miscellaneous actions completed
  0510 Special projects completed

FIGURE 5.4 The sixth-order work-units of the Comptroller Directorate.

```
            COMPTROLLER-DIRECTORATE
                 5th-ORDER WORK-UNITS

  FINANCE AND ACCOUNTING SERVICES

  020101.  An III account established
  020102.  An III line item account established
  020103.  An III obligation recorded
  020104.  An III accounts payable recorded
  020105.  An III disbursement recorded
  020106.  An III status of accounts reported
  020107.  An III line item account reviewed and status reported
  020108.  An III earning recorded
  020109.  An III bill issued
  020110.  An III collection received and recorded
  020111.  An III system account action completed

  020201 to 020211; same as above but for JJJ

  020301 to 020311; same as above but for KKK

  020401 to 020411; same as above but for LLL

  020501.  An account quality assurance visit completed
  020502.  A local qual. assurance review completed
  020503.  A vulnerability assessment completed
  020504.  An internal review checklist completed
  020505.  A quality assessment project completed
  020506.  An F & A office monthly operation report completed
  020507.  A propriety decision formalized
  020508.  A customer pricing policy decision formalized
  020509.  A publication (internal to DoD) reviewed and reported on
  020510.  A suggestion system item (re F & A) formally evaluated
  020511.  Special command project completed
  020512.  Special internal project completed
  020513.  Inquiries re U.S. accounts responded to
  020514.  (International) A case closure line processed
  020515.  (International) Inquires re international logistic
           accounts responded to
```

FIGURE 5.5 The fifth-order work-units of Finance and Accounting, Comptroller Directorate.

tive period of time. A separate record is kept of the work count of each work-unit produced.

3. *Fractioned professional estimates.* One or more workers, familiar with the work-unit, list the steps required to perform it and, from their experience, assign a time value to each step.

4. *Work sampling.* See Sec. 4, Chap. 3 in this handbook.

It should be noted that any of the known methods of work measurement may be employed, but the above are the most common. The Department of Defense, under MIL-STD 1567-A, requires the use of engineered time standards.

AN EXAMPLE OF PRODUCTIVITY MEASUREMENT IN A GOVERNMENT SERVICE ACTIVITY

Figure 5.2 gives the details of the objective (eighth-order work-unit) of a major U.S. Army command's comptroller directorate. Figure 5.3 gives the seventh-order work-units; Fig. 5.4 gives the sixth-order work-units, and Fig. 5.5 gives the fifth-order work-units for seventh-order 02, Finance and Accounting Services Provided.

```
COMPTROLLER - FINANCE DIVISION RIA PRODUCTIVITY REPORT
12/30/88  J.JONES  1/2/89

DATA ON WORK-UNITS WITH ZERO WORK COUNTS IN BOTH PERIOD AND CUMULATIVE FILES WILL NOT BE PRINTED
```

WORK-UNIT (* = LE WORK-UNIT)		---- PERIOD ----		-- CUMULATIVE --		PERCENT OF
		W/C	HE	W/C	HE	WORKLOAD
#1 SAPASS ACCOUNT ESTABLISHED		2347	2522.279	2347	2522.279	1.75
#2 SAPASS LINE ITEM ACCOUNT ESTABLISHED		4248	1416.802	4248	1416.802	.95
#3 SAPASS OBLIGATION RECORDED		5649	1884.066	5649	1884.066	1.35
#5 SAPASS DISBURSEMENT RECORDED		13933	3356.139	13933	3356.139	2.35
#6 SAPASS STATUS OF ACCOUNTS REPORTED	*	3	3541.337	3	3541.337	2.45
#7 SAPASS LINE ITEM ACCT REV'D & STAT REPORTED		3948	9070.909	3948	9070.909	6.25
#9 SAPASS BILL ISSUED		3961	2495.375	3961	2495.375	1.75
#10 SAPASS COLLECTION RECEIVED RECORDED		3978	1031.917	3978	1031.917	.75
#11 SAPASS SYST ACC ACT. COMPLETED		2219	575.6219	2219	575.6219	.35
#12 CCSS ACCOUNT ESTABLISHED		18048	4012.937	18048	4012.937	2.75
#13 CCSS LINE ITEM ACCOUNT ESTABLISHED		7960	6194.615	7960	6194.615	4.25
#14 CCSS OBLIGATION RECORDED		555195	10287.21	555195	10287.21	7.15
#15 CCSS ACCOUNTS PAYABLE RECORDED		17460	8411.424	17460	8411.424	5.85
#16 CCSS DISBURSEMENT RECORDED		538	3668.446	538	3668.446	2.55
#17 CCSS STATUS OF ACCOUNTS REPORTED		12630	14743.34	12630	14743.34	10.25
#19 CCSS EARNING RECORDED		200694	7437.319	200694	7437.319	5.15
#20 CCSS BILL ISSUED		744	882.2769	744	882.2769	.65
#21 CCSS COLLECTION RECEIVED RECORDED		1964	2947.668	1964	2947.668	2.05
#22 CCSS SYST ACC ACT. COMPLETED		1319	879.8311	1319	879.8311	.65
#23 AMCOMSYS ACCOUNT ESTABLISHED		169421	9417.606	169421	9417.606	6.55
#24 AMCOMSYS LINE ITEM ACCOUNT ESTABLISHED		23326	2593.245	23326	2593.245	1.75
#25 AMCOMSYS OBLIGATION RECORDED		27046	7015.895	27046	7015.895	4.85
#26 AMCOMSYS ACCOUNTS PAYABLE RECORDED	*	3	3064.678	3	3064.678	2.15
#27 AMCOMSYS DISBURSEMENT RECORDED		44508	8246.888	44508	8246.888	5.75
#29 AMCOMSYS LINE ITEM ACCT REV'D & STAT REPORTED		5221	2611.978	5221	2611.978	1.85
#30 AMCOMSYS EARNING RECORDED		2025	1200.679	2025	1200.679	.85
#31 AMCOMSYS BILL ISSUED		2478	8631.994	2478	8631.994	5.95
#32 AMCOMSYS COLLECTION RECEIVED RECORDED		1122	1101.846	1122	1101.846	.75
#42 PBAS BILL ISSUED		4	204.7825	4	204.7825	.15
#44 AN ACCOUNT QUAL. ASSURANCE VISIT COMPLETED		3	180.6578	3	180.6578	.15
#46 A VULNERABILITY ASSESSMENT COMPLETED		3	67.1491	3	67.1491	.05
#47 AN INTERNAL REVIEW CHECKLIST COMPLETED		109	1559.178	109	1559.178	1.05
#48 QUALITY ASSESSMENT						
...TO BODY REV. & REPORTED ON*		3	3277.181	3	3277.181	2.25
#53 A SUGG. SYST. SUGG. (RE F&A) FORMALLY EVAL'ED		4071	1131.474	4071	1131.474	.75
#54 SPECIAL COMMAND PROJECTS COMPLETED		440	5095.475	440	5095.475	3.55
#55 SPECIAL INTERNAL PROJECTS COMPLETED		1260	1517.525	1260	1517.525	1.05

```
HOURS WORKED THIS PERIOD B=              162966
CUMULATIVE HOURS WORKED   D=             162966
HOURS EARNED THIS PERIOD SA=             144231.7
PERIOD HOURS EARNED INCLUDING SUP RATIO   SC=  162967.4
HOURS EARNED CUMULATIVELY SB=            144231.7
CUMULATIVE HOURS EARNED INCLUDING SUP RATIO  SD=  162967.4


          PRODUCTIVITY THIS PERIOD        100 %
          CUMULATIVE PRODUCTIVITY         100 %
```

FIGURE 5.6   The base period productivity report of Finance and Accounting, Comptroller Directorate (part deleted to compress illustration).

A Productivity Report.   A computer printout of a productivity report for the Finance and Accounting section appears in Fig. 5.6. This report is for the base period wherein the standard times were multiplied by a factor k (built into the program), so that the base period productivity is 100 percent. (Later reports showed a slow but steady increase in productivity.)

Zero-Based Budgeting.   One of the reasons for the slow but steady increase in productivity was the use of a zero-based budget, with the assumption that productivity, as the result of a steady flow of small methods changes, could increase by 3 percent per year. In computing the zero-based budget, forecasts were made of the anticipated work count of each work-unit. A PC program was used to print the work-units and multiply them by the standard time for each work-unit. The resultant total of work time was divided by 1836, the number of hours available per employee per year (yield year). This division produced a staffing value consisting of an integer and a decimal. This was divided by the anticipated productivity, 103 percent, giving a new integer and a decimal. If the new decimal staffing could be handled with an additional increase in productivity of less than 5 percent, it was not granted. If a more than 5 percent increase in productivity was needed to handle the decimal increment to staffing, then the integer was increased by one. The PC program readily accommodated any desired change in the amount of productivity increase before additional staff was added, but 5 percent was used and the computer readily applied the staffing rule. Budgeting, which previously took days, was done in hours, accompanied by a full printout of the data and the results.

BIBLIOGRAPHY

Mundel, Marvin E., *Measuring and Enhancing the Productivity of Service and Government Organizations,* Asian Productivity Organization, Tokyo, 1975.

Mundel, Marvin E., *Motion and Time Study—Improving Productivity,* Prentice-Hall, Englewood Cliffs, N.J., 1983.

Mundel, Marvin E., *Improving Productivity and Effectiveness,* Prentice-Hall, Englewood Cliffs, N.J., and Asian Productivity Organization, Tokyo, 1985.

Mundel, Marvin E., *Measuring Total Productivity of Manufacturing Organizations—Algorithms and PC Programs,* Asian Productivity Organization, Tokyo, and Kraus International Publications, White Plains, N.Y., 1987.

Mundel, Marvin E., *The White-Collar Knowledge Worker, Measuring and Improving Productivity and Effectiveness—Algorithms and PC Programs,* Asian Productivity Organization, Tokyo, and Kraus International Publications, White Plains, N.Y., 1989.

Index